Neuropathology of Drug Addictions and Substance Misuse

Volume 1: Foundations of Understanding, Tobacco, Alcohol, Cannabinoids and Opioids

ELSEVIER *science & technology books*

Companion Web Site:
http://booksite.elsevier.com/9780128002131

Neuropathology of Drug Addictions and Substance Misuse, Volume 1
Victor R. Preedy, Editor

Available Resource:

- Additional Resources and Recommended Reading

ELSEVIER

ACADEMIC PRESS

Neuropathology of Drug Addictions and Substance Misuse

Volume 1: Foundations of Understanding, Tobacco, Alcohol, Cannabinoids and Opioids

Edited by

Victor R. Preedy
King's College London, London, UK

AMSTERDAM • BOSTON • HEIDELBERG • LONDON • NEW YORK • OXFORD • PARIS
SAN DIEGO • SAN FRANCISCO • SINGAPORE • SYDNEY • TOKYO

Academic Press is an imprint of Elsevier

Academic Press is an imprint of Elsevier
125 London Wall, London EC2Y 5AS, UK
525 B Street, Suite 1800, San Diego, CA 92101-4495, USA
50 Hampshire Street, 5th Floor, Cambridge, MA 02139, USA
The Boulevard, Langford Lane, Kidlington, Oxford OX5 1GB, UK

Copyright © 2016 Elsevier Inc. All rights reserved.

No part of this publication may be reproduced or transmitted in any form or by any means, electronic or mechanical, including photocopying, recording, or any information storage and retrieval system, without permission in writing from the publisher. Details on how to seek permission, further information about the Publisher's permissions policies and our arrangements with organizations such as the Copyright Clearance Center and the Copyright Licensing Agency, can be found at our website: www.elsevier.com/permissions.

This book and the individual contributions contained in it are protected under copyright by the Publisher (other than as may be noted herein).

Notices
Knowledge and best practice in this field are constantly changing. As new research and experience broaden our understanding, changes in research methods, professional practices, or medical treatment may become necessary.

Practitioners and researchers must always rely on their own experience and knowledge in evaluating and using any information, methods, compounds, or experiments described herein. In using such information or methods they should be mindful of their own safety and the safety of others, including parties for whom they have a professional responsibility.

To the fullest extent of the law, neither the Publisher nor the authors, contributors, or editors, assume any liability for any injury and/or damage to persons or property as a matter of products liability, negligence or otherwise, or from any use or operation of any methods, products, instructions, or ideas contained in the material herein.

British Library Cataloguing-in-Publication Data
A catalogue record for this book is available from the British Library

Library of Congress Cataloging-in-Publication Data
A catalog record for this book is available from the Library of Congress

ISBN: 978-0-12-800213-1

For information on all Academic Press publications
visit our website at https://www.elsevier.com/

Publisher: Mara Conner
Acquisition Editor: Mara Conner
Editorial Project Manager: Kathy Padilla
Production Project Manager: Julia Haynes
Designer: Matthew Limbert

Typeset by TNQ Books and Journals
www.tnq.co.in

Contents

List of Contributors   xxvii
Preface   xxxvii
Acknowledgments   xxxix

Part I
Setting the Scene: Foundations for Understanding Substance Misuse and Their Effects

1. The Nervous System and Addictions: Essentials for Clinicians

Amelia J. Anderson-Mooney, Jonathan N. Dodd, April Scott and Leila Guller

Introduction	3
A Neuroanatomical Compass	3
General Structure of the Neuroanatomical System	5
Psychoactive Substances and the Brain	7
The Neurobehavioral Facets of Addiction: From Recreational Use to Dependence	8
Applications to Other Addictions and Substance Misuse: New Behavioral Disorders?	11
Societal Impact of Substances	11
Definition of Terms	11
Key Facts	11
Summary Points	12
References	12

2. Pathophysiology-Based Neuromodulation for Addictions: An Overview

Dirk De Ridder, Patrick Manning, Gavin Cape, Sven Vanneste, Berthold Langguth and Paul Glue

Introduction	14
Finding Targets for Neuromodulation Intervention	15
Pathophysiology of Addiction	15
Neuromodulation Techniques and Brain Network Alterations	17
Applications to Other Addictions and Substance Misuse	22
Definition of Terms	22
Key Facts	22
Summary Points	22
References	22

3. Gateway Hypothesis of Addiction and Substance Misuse: The Role of Neurophysiology and Psychological Mechanisms

Hamdy Fouad Moselhy and Mahmoud A. Awara

Introduction	25
Neurophysiological Mechanisms	25
Transition to Addiction and Psychological Mechanisms	28
Personality of the Drug User	29
Conclusion	33
Summary and Key Points	34
References	34

4. Children of Parents with Substance Use Disorder

Iliyan Ivanov, John Leikaif, Juan Pedraza and Jeffrey Newcorn

Introduction	36
Models of Substance Use Disorder Vulnerability	36
From Models to Studies of Neurobiology of Children with Familial Substance Use Disorder	38
Neuroimaging Studies of Children with Familial Substance Use Disorder	40
Future Directions	45
Summary	45
Applications to Other Addictions and Substance Misuse	45
Definition of Terms	46
Key Facts	47
Summary Points	47
References	47

5. Drugs of Abuse and the Internet

Marcelo Dutra Arbo, Rachel Picada Bulcão, Luciana Grazziotin Rossato, Elza Callegari, Lucas Henrique Cendron, Eduardo Rodrigues Cabrera, Marco Antonio de Freitas and Mirna Bainy Leal

Introduction	50
Drugs Sold on the Internet	50
Conclusion	58
Definition of Terms	58
Key Facts on the Internet	58
Summary Points	58
References	58

6. Trauma and Neurological Risks of Addiction

Dessa Bergen-Cico, Sarah Wolf-Stanton, Radmila Filipovic and Jordyn Weisberg

Introduction	61
Overview of Typology of Trauma	62
Neurobiology of Trauma	63
Basic Neurobiology of Drug Use and Addiction	64
Posttraumatic Stress and Cannabinoid Receptors	65
Applications to Other Addictions and Substance Misuse	65
Conclusion	67
Definition of Terms	67
Key Facts	67
Summary Points	68
References	68

7. Emotional Self-Medication and Addiction

Carmen Torres and Mauricio R. Papini

Introduction	71
ESM and Addiction in Humans	72
PSM and ESM in Nonhuman Animals	74
ESM Induced by Reward Loss	75
Future Directions: Bridging ESM and SUDs	77
Applications to Other Addictions and Substance Misuse	79
Definition of Terms	79
Key Facts about Self-Medication	79
Summary Points	79
References	79

8. The Neurobiology of Comorbid Drug Abuse in Schizophrenia and Psychotic Disorders

A. George Awad

Introduction	82
Epidemiological Data	82
Correlates of Comorbid Drug Abuse in Schizophrenia	83
Neurobiology of Comorbid Drug Abuse	84
Applications to Other Addictions and Substance Misuse	86
Definition of Terms	86
Key Facts of the Neurobiology of Comorbid Drug Abuse in Schizophrenia	86
Summary Points	87
References	87

Part II
Tobacco

Section A
General Aspects

9. Nicotine Chemistry, Pharmacology, and Pharmacokinetics

Diana L. McKinney and Andrea R. Vansickel

Introduction	93
Nicotine Chemistry	94
Nicotine Pharmacology	94
Nicotine Pharmacokinetics	97
Summary	98
Afterword	99
Definition of Terms	101
Key Facts: PET Neuroimaging and Nicotine	101
Summary Points	101
References	101

10. Tobacco-Related Mortality among Individuals with Alcohol or Drug Use Disorders

Jodi M. Gatley and Russell C. Callaghan

Introduction	104
Prevalence of Tobacco Use in AODUD Populations	104
Tobacco-Related Diseases	105
A Review of Tobacco-Related Mortality among Individuals with AODUD	107
Best-Practice Guidelines Champion Smoking Cessation Integration into Addictions Treatment	109
Barriers to Implementation of Concurrent Smoking Cessation with AODUD Treatment	109
Staff Knowledge and Attitudes about Concurrent Smoking Cessation during AODUD Treatment	109
Attitudes Toward Smoking Cessation among AODUD Treatment Clients	109
Effectiveness of Treatment for Nicotine Dependence Concurrent with AODUD Treatment	110

Efforts to Increase Access to Nicotine-Dependence Treatment for AODUD Populations	110
Other Policy Interventions to Reduce Tobacco-Related Mortality among Individuals with AODUD	110
Applications to Other Addictions and Substance Misuse	111
Conclusion	111
Definition of Terms	111
Key Facts of Contraband Tobacco Use in the AODUD Population	112
Summary Points	112
References	112

11. Neurological Effects of Nicotine, Tobacco, and Particulate Matter

Bronwyn M. Kivell and Kirsty Danielson

Introduction	115
Neurobiological Effects of Smoking Tobacco	115
The Rewarding Effects of Nicotine	116
Mechanism of Action: Nicotinic Acetylcholine Receptors	117
Addictive Properties of Tobacco Smoke	117
Tobacco Particulate Matter and Extracts	118
Applications to Other Addictions and Substance Misuse	119
Definition of Terms	120
Key Facts	120
Summary Points	120
References	121

Section B
Molecular and Cellular Aspects

12. Rho GTPases and Their Regulators in Addiction: A Focus on the Association of a β2-Chimaerin Polymorphism with Smoking

María J. Caloca, Laura Barrio-Real and Rogelio González-Sarmiento

Introduction	125
Rho GTPases and Nicotine Addiction	125
Chimaerins and Smoking	126
Applications to Other Addictions and Substance Misuse	128
Definition of Terms	130
Key Facts of Smoking Addiction	130
Summary Points	131
References	131

13. Brain Orexin Receptors and Nicotine

Rita Machaalani, Nicholas J. Hunt and Karen A. Waters

Introduction	133
The Orexinergic System	134
The Hypothalamus and Orexin Receptors	134
Nicotine and Nicotinic Acetylcholine Receptors	135
Nicotine and the Orexinergic System	136
Human Studies of Nicotine on Orexin	136
Paradigms of Nicotine Exposure to Replicate Clinical Conditions	136
Nicotine Exposure on Orexin Expression	137
Nicotine, Orexin, Addiction, and Reward	139
Mechanisms Involved in Nicotine-Induced Orexin Expression Changes	140
Conclusion and Future Directions	140
Applications to Other Addictions and Substance Misuse	140
Definition of Terms	141
Key Facts Regarding the Orexinergic System	141
Summary Points	142
References	142

14. Nicotine and Stimulatory Effects on 5-HT DRN Neurons

Salvador Hernández-López, Rene Drucker Colín and Stefan Mihailescu

Introduction	146
Neuronal Nicotinic Acetylcholine Receptors	146
The Dorsal Raphe Nucleus	147
Effects of Nicotine on 5-HT DRN Neuron Firing Rate and 5-HT Release	147
Behavioral Effects of Nicotine Mediated by Increases in Serotonin Release	154
Concluding Remarks	154
Application to Other Addictions and Substance Misuse	155
Definition of Terms	155
Key Facts of Microdialysis	155
Summary Points	155
References	156

15. Critical Role of Cannabinoid CB1 Receptors in Nicotine Reward and Addiction

Ameneh Rezayof and Shiva Hashemizadeh

Introduction	158
Nicotine Reward and Dependence	158
Endocannabinoid System and Signaling	160
Cannabinoid Type 1 Receptors	161
The Role of CB1 Receptors in Brain Reward Processing in Nicotine Administration	161

Effects of CB1 Receptor Agonists and Antagonists on Nicotine Reward	163
CB1 Receptor Ligands and nAChRs	163
Applications to Other Addictions and Substance Misuse	163
Definition of Terms	165
Key Facts	165
Summary Points	165
Acknowledgment	166
References	166

16. Targets of Addictive Nicotine in the Central Nervous System and Interactions with Alcohol

Sodikdjon A. Kodirov

Introduction	168
Effects of Nicotine	169
Concurrent Effects of Nicotine with Alcohol	174
Nicotine	174
Dual Effects of Nicotine	175
Applications to Other Addictions and Substance Misuse	176
Definition of Terms	176
Key Facts of Nicotine and Its Targets	176
Summary Points	176
References	176

17. Cytochrome P450 and Oxidative Stress as Possible Pathways for Alcohol- and Tobacco-Mediated HIV Pathogenesis and NeuroAIDS

Santosh Kumar, P.S.S. Rao, Namita Sinha and Narasimha M. Midde

Introduction	179
Alcohol and NeuroAIDS	180
Tobacco and NeuroAIDS	183
Concluding Remarks	185
Applications to Other Addictions and Substance Misuse	185
Definition of Terms	185
Key Facts of NeuroAIDS	186
Summary Points	186
Acknowledgment	186
References	186

18. Nicotine and Neurokinin Signaling

Mariella De Biasi, Ian McLaughlin and Michelle L. Klima

Introduction	189
Tachykinin Genes and Gene Products	190
Neuropeptide Distribution in the CNS	190
Receptor Distribution in the CNS	191
Neurokinin-Mediated Mechanisms in the CNS	191
Tachykinins and Addiction Mechanisms	192
Neurokinins and Nicotine: Cellular Mechanisms	193
Neurokinins and Nicotine: Regulation of Mood and Affect	194
Neurokinins and the Symptoms of Nicotine Withdrawal	195
Applications to Other Addictions and Substance Misuse	195
Conclusions	196
Definition of Terms	197
Key Facts	197
Summary Points	197
References	197

19. Neurotransmitter Systems and the Nicotine Dependence-Induced Withdrawal Syndrome: Dopamine, Glutamate, GABA, Endogenous Opioids, Endocannabinoids, Noradrenaline, Arginine Vasopressin, Neuropeptide Y, MAO, CREB, and Corticotropin-Releasing Factor

Thakur Gurjeet Singh, Shiwali Sharma and Sonia Dhiman

Introduction	201
Nicotine-Induced Neuroadaptations	201
Neural Basis of Nicotine Withdrawal	202
Neurotransmitters and Mediator Systems Involved in Nicotine Withdrawal	203
Application to Other Addictions and Substance Misuse	206
Definition of Terms	206
Summary Points	206
References	207

20. Genetic Findings on the Relationship between Smoking and the Stress System

Diego L. Rovaris, Nina R. Mota and Claiton H.D. Bau

Introduction	209
The HPA Axis and the Molecular Effects of GCs	209
The Glucocorticoid Receptor-Coding Gene (*NR3C1*)	211
Functional Aspects of *NR3C1* Polymorphisms	212
Effects of *NR3C1* Polymorphisms on Smoking Behavior	213
The MR Coding Gene	213
Functional Aspects of *NR3C2* Polymorphisms	215

Effects of *NR3C2* Polymorphisms on Smoking Behavior	215
Potential for MR/GR Interactions beyond the Statistical Level	215
Perspectives	215
Applications to Other Addictions and Substance Misuse	216
Definition of Terms	216
Key Facts	216
Summary Points	216
References	217

21. Genetic Variants of μ Opioid Receptor and Its Interacting Proteins in Smoking: A Focus on *OPRM1* A118G, *ARRB2*, and *HINT1* 221

Juan Fang and Bei He

Introduction	221
OPRM1 Polymorphisms	223
OPRM1 and Smoking Behaviors	224
Pharmacogenetic Studies of *OPRM1* in Smoking Cessation	227
ARRB2 and *HINT1*	228
Applications to Other Addictions and Substance Misuse	230
Definition of Terms	230
Key Facts of Smoking Related Phenotypes	230
Summary Points	231
References	231

Section C
Structural and Functional Aspects

22. Characterization of Cue-Induced Reinstatement of Nicotine-Seeking Behavior in Smoking Relapse: Use of Animal Models

Xiu Liu

Introduction	237
Rat Model of Smoking Relapse: Response Reinstatement Paradigm	237
Cue-Induced Reinstatement of Nicotine-Seeking Responses	238
Interactions of Nicotine Cue with Stress and Nicotine Priming	240
Pharmacological Substrates Responsible for Cue-Induced Nicotine-Seeking Behavior	241
Applications to Other Addiction and Substance Misuse	242
Definition of Terms	242
Key Facts of Cue-Induced Reinstatement of Nicotine-Seeking Behavior	243
Summary Points	243
References	243

23. Effects of Environmental Enrichment on Nicotine Addiction

Dustin J. Stairs, Megan Kangiser, Tyson Hickle and Charles S. Bockman

Introduction	246
Behavioral Effects of Nicotine in Enriched Animals	247
Neural Effects of Enrichment on Nicotine Effects and Circuits	248
Conclusions and Future Directions	250
Application to Other Addictions and Substance Misuse	251
Definition of Terms	251
Key Facts of Environmental Enrichment	252
Summary Points	252
References	252

24. Smokers with Severe Mental Illness

Maxie Ashton, Benjamin L. Carnell and Cherrie Galletly

Introduction	254
Nicotine Addiction	254
Why Do More People with Mental Illness Smoke?	255
Facilitating Smoking Cessation	256
Conclusion	257
Applications to Other Addictions and Substance Misuse	257
Key Facts of Smoking among Those with Mental Illness	257
Summary Points	258
References	258

25. Nicotine Dependence and Schizophrenia

Aniruddha Basu and Anirban Ray

Introduction	260
Epidemiology	260
Clinical Features	262
Etiological Relationship between Schizophrenia and Nicotine	262
Management	267
Application to Other Addictions and Substance Misuse	269
Definition of Terms	269
Key Facts	269
Summary Points	269
References	269

26. Unraveling the Role of the Amygdala in Nicotine Addiction

Benjamin Becker and René Hurlemann

Introduction	272
The Role of Nicotine in Tobacco Addiction	272
The Amygdala	273
Nicotine Craving	273
Functional Neuroimaging Studies on Cue-Induced Nicotine Craving	273
Functional Neuroimaging Studies on Abstinence-Induced Craving	276
Functional Neuroimaging Studies on Altered Emotion Processing in Smokers	277
Conclusions	277
Applications to Other Addictions and Substance Misuse	278
Definition of Terms	278
Key Facts of Nicotine Addiction	279
Summary Points	279
References	279

27. Nicotine and Cognition: Effects of Nicotine on Attention and Memory Systems in Humans

Anton L. Beer

Introduction	282
Nicotine and Attention	282
Nicotine and Memory	285
Long-Term Effects of Nicotine on Cognition	287
Indirect Effects of Nicotine	287
Applications to Other Addictions and Substance Misuse	287
Definition of Terms	287
Summary Points	288
References	288

28. The Role of Appetitive and Aversive Smoking Cues in Tobacco Use Disorder with a Focus on fMRI

Josiane Bourque, Le-Anh Dinh-Williams and Stéphane Potvin

Introduction	291
Reactivity to Appetitive Smoking Cues—Clinical Relevance	291
Neural Correlates of Smokers to Appetitive Smoking Cues	292
Reactivity to Aversive Smoking Cues—Clinical Relevance	296
Neural Reactivity of Smokers to Aversive Smoking Cues	297

Conclusion	301
Applications to Other Addictions and Substance Misuse	301
Definition of Terms	301
Key Facts of Tobacco Use Disorder	301
Summary Points	302
References	302

29. Different Effects of Cigarette Smoking on Neuropsychological Performance in Psychiatric Disorders

Daniela Caldirola, Giuseppe Iannone, Giuseppina Diaferia and Giampaolo Perna

Introduction	305
Smoking and Cognitive Function in Healthy Subjects	305
Smoking and Cognitive Function in Schizophrenia	307
Smoking and Cognitive Function in Major Depressive Disorder	309
Smoking and Cognitive Function in Bipolar Disorder	311
Smoking and Cognitive Function in Obsessive-Compulsive Disorder	312
Conclusions	312
Applications to Other Addictions and Substance Misuse	313
Definition of Terms	313
Key Facts	313
Summary Points	314
References	314

30. Tobacco Smoke Extract-Produced Behavioral Effects: Locomotor Sensitization and Self-Administration Studies

Katharine Alexandra Brennan and Penelope Truman

Introduction: Tobacco Smoke Extracts in Research	317
Behavioral Tests	318
Conclusions	322
Applications to Other Addictions and Substance Misuse	324
Definition of Terms	324
Key Facts of History of Behavioral Nicotine/Tobacco Addiction Research	324
Summary Points	325
References	325

31. **The Psychological Threat of Mortality and Its Implications for Tobacco and Alcohol Misuse**

    Simon McCabe and Jamie Arndt

Introduction	327
Psychological Defenses against Conscious Mortality Concerns: Implications for Substance Use	330
Applications to Other Addictions and Substance Misuse	333
Definition of Terms	334
Key Facts Concerning the Terror Management Health Model	334
Summary Points	334
References	334

32. **Nicotine Dependence and the Anterior Cingulate-Precuneus Pathway: Using Neuroimaging to Test Addiction Theories**

    Joseph R. DiFranza, Wei Huang and Jean A. King

Physical Dependence	337
Smoking and Brain Structure	338
Learning Theories of Addiction	339
Biological Theories of Addiction	340
Hypothesis Testing	342
What Impact Does This Have for Addiction Theorists?	344
Application to Other Addictions and Substance Misuse	345
Definition of Terms	345
Key Facts of Physical Nicotine Dependence	345
Summary Points	345
References	346

33. **Neural Effects of Nicotine: Peripheral Sensory Systems and Experience-Dependent Neural Sensitization**

    Eugene A. Kiyatkin

Introduction	348
Rapid Neural Activation Induced by Intravenous Nicotine	348
The Role of Peripheral Actions of Nicotine in Mediating Its Acute Neural Effects	349
Experience-Dependent Changes in the Neural, Physiological, and Behavioral Effects of Nicotine	353
The Role of Peripheral Actions of Nicotine in the Development of Nicotine-Induced Neural Sensitization	354
General Discussion	354
Applications to Other Addictions and Substance Misuse	357
Definition of Terms	357
Key Facts	358
Summary Points	358
Acknowledgments	358
References	358

Section D
Methods

34. **Cotinine Urinalysis for Tobacco Use**

    Yatan P.S. Balhara and Siddharth Sarkar

Introduction	363
Pharmacokinetics and Pharmacodynamics	363
Assays for Detection	364
Research and Clinical Applications of Urine Cotinine Detection	366
Sensitivity and Specificity Issues	368
Applications to Other Addictions and Substance Misuse	368
Comparison with Other Biomarkers of Tobacco Use	368
Conclusions	368
Definition of Terms	369
Key Facts	369
Summary Points	369
References	369

Part III
Alcohol

Section A
General Aspects

35. **Ethanol Metabolism and Implications for Disease**

    Roshanna Rajendram, Rajkumar Rajendram and Victor R. Preedy

Introduction	377
The Physical Properties of Ethanol	377
International Variations in Alcohol Consumption	378

Alcohol and the Burden of Disease	378
Absorption and Distribution of Ethanol	379
Metabolism of Ethanol	380
Blood Ethanol Concentration	383
Applications to Other Addictions and Substance Misuse	385
Summary	385
Definition of Terms	385
Key Facts of Acetaldehyde	385
Summary Points	386
References	386

36. Heavy Episodic Drinking or Binge Drinking: A Booming Consumption Pattern

Ana Adan, Irina Benaiges and Diego A. Forero

Introduction	389
Definition of Binge Drinking	389
Consequences on Health and Possible Early Markers	390
Risk and Protection Factors Associated with BD	391
Personality Traits, Circadian Typology, and BD	392
Neuropsychological Impact of BD	393
Future Research on Prevention and/or Treatment on BD	394
Applications to Other Addictions and Substance Misuse	395
Definition of Terms	395
Key Facts of Neuropsychology in Binge Drinkers	395
Summary Points	395
Acknowledgment	395
References	396

Section B
Molecular and Cellular Aspects

37. Alcohol and Endogenous Opioids

Sara Palm and Ingrid Nylander

Neuropeptides as Target System	401
The Endogenous Opioid System	402
The Opioid–Alcohol Link	404
Endogenous Opioids in AUD	406
Applications to Other Addictions and Substance Misuse	408
Definition of Terms	408
Key Facts of Opioids	408
Summary Points	408
References	409

38. Effects of Alcohol on Nicotinic Acetylcholine Receptors and Impact on Addiction

Josephine Tarren, Masroor Shariff, Joan Holgate and Selena E. Bartlett

Introduction	411
Applications to Other Addictions and Substance Misuse	416
Conclusion	416
Definition of Terms	417
Summary Points	417
Acknowledgment	417
References	417

39. Alcohol and Its Impact on Myelin

Consuelo Guerri and María Pascual

Introduction	420
The Myelination Process and Brain Development	421
Composition of Myelin	421
Pathophysiology of the Demyelination and Remyelination Processes in the CNS	423
Alcohol Abuse and White Matter Disturbances	423
Mechanisms of Alcohol-Induced Myelin Alterations	425
Applications to Other Addictions and Substance Misuse	427
Therapeutic Approaches	428
Definition of Terms	428
Key Facts	429
Summary Points	429
Acknowledgments	429
References	429

40. The Effects of Acute and Chronic Ethanol Exposure on GABAergic Neuroactive Steroid Immunohistochemistry: Relationship to Ethanol Drinking

Matthew C. Beattie, Antoniette Maldonado-Devincci, Jason B. Cook and A. Leslie Morrow

Introduction	433
Immunohistochemistry Methodology and Strengths in Neuroactive Steroid Research	434
GABAergic Neuroactive Steroids and Ethanol Interactions in Rats	434
Neuroactive Steroids and Ethanol Drinking in Rats	436
GABAergic Neuroactive Steroids and Ethanol Interactions in Mice	437
Neuroactive Steroids and Ethanol Drinking in Nonhuman Primates	438

Neuroactive Steroids and Ethanol Drinking
  in Humans   439
GABRA2 Genetic Polymorphisms   439
Genetic Polymorphisms and Neuroactive
  Steroids   440
Conclusions   440
Applications to Other Addictions and
  Substance Misuse   440
Definition of Terms   441
Key Facts   441
Summary Points   441
References   442

41. Ethanol and Its Impact on the Brain's Electrical Activity

Rubem Carlos Araújo Guedes, Ranilson de Souza Bezerra and Ricardo Abadie-Guedes

Alcohol Abuse and Its Neurological
  Consequences   445
Influence of Alcohol on the Electrical
  Activity of the Brain   446
CSD as a Tool to Study the Effects of
  Alcohol on the Brain   446
Ethanol Influences CSD Propagation:
  The Role of Antioxidant Molecules   447
Applications to Other Addictions and
  Substance Misuse   449
Definitions and Explanations of Key Terms
  or Words Used in the Chapter   450
Key Facts of Electrophysiology   451
Summary Points   451
Acknowledgments   451
References   451

42. Alcohol and Neuroimmune Interactions

Cynthia J.M. Kane, Susan E. Bergeson and Paul D. Drew

Introduction   454
Microglia   454
Fetal Alcohol Spectrum Disorders   456
Adolescent and College-Age Alcohol
  Consumption   457
Adult Alcohol Consumption   458
Neuroimmune Signaling   459
Anti-inflammatory Therapeutics for
  Treatment of Alcohol Use Disorders   461
Summary   462
Applications to Other Addictions and
  Substance Misuse   462
Definition of Terms   462
Key Facts   462
Summary Points   462
Acknowledgment   463
References   463

43. Role of Glutamate Transport in Alcohol Withdrawal

Osama A. Abulseoud, Christina L. Ruby and Victor Karpyak

Introduction   466
Glutamate Signaling Under Normal
  Physiological Conditions   467
Glutamate Signaling During Alcohol
  Withdrawal   469
Contribution of Genetic Variation in
  Glutamate Signaling to Alcohol
  Withdrawal Risk and Severity   471
Preclinical Evidence for the Use of
  Ceftriaxone to Treat Acute AWS   472
Applications to Other Addictions and
  Substance Misuse   473
Summary and Conclusions   473
Definition of Terms   473
Key Facts on the Tricarboxylic Acid Cycle   473
Summary Points   473
References   474

44. Ethanol, Vitamins, and Brain Dysfunction

Emilio González-Reimers, Camino Fernández-Rodríguez, Emilio González-Arnay and Francisco Santolaria-Fernández

Brain Atrophy Affects Both White Matter
  and Gray Matter   478
Functional Consequences   480
Pathogenesis of Neurotoxicity in Alcoholics   480
Liposoluble Vitamins   480
Hydrosoluble Vitamins   482
Conclusions   484
Applications to Other Addictions and
  Substance Misuse   484
Definition of Terms   485
Key Facts   485
Summary Points   485
References   485

45. Phospholipase D: A Central Switch in the Development of Fetal Alcohol Spectrum Disorder

Ute Burkhardt and Jochen Klein

Fetal Alcohol Spectrum   488
Phospholipases D   489
Regulation of PLDs   490
Enzymatic Reactions of PLDs   491
Brain Cells and PLD   491
PLD Function   492
Effects of Ethanol In Vitro   493
Effects of Ethanol and PLD Deletion In Vivo   494

Applications to Other Addictions and Substance Misuse	496
Conclusion	497
Definition of Terms	497
Key Facts of Phospholipases D	497
Summary Points	497
References	498

46. N-Methyl-D-Aspartate Receptor Subunits and Alcohol

Christina N. Nona and José N. Nobrega

Introduction	500
Ethanol's Actions at the NMDAR: Role in the Subjective and Behavioral Effects of Alcohol	501
NMDAR Subunits and Ethanol	501
Applications to Other Addictions and Substance Misuse	506
Concluding Remarks	506
Definition of Terms	506
Key Facts of Glutamate	507
Summary Points	507
References	507

47. Alcohol Dehydrogenase Alleles and Impact on Neuropathology

Neil C. Dodge, Joseph L. Jacobson and Sandra W. Jacobson

Introduction	510
Alcohol Metabolism	510
ADH Alleles and Alcohol Use Disorders	511
ADH Alleles and Effects on Neuropathology	511
Fetal Alcohol Spectrum Disorders	511
ADH Alleles and Neuropathology in FASD	513
Role of the ADH Alleles in the Fetus	513
Applications to Other Addictions and Substance Misuse	516
Definition of Terms	516
Summary Points	516
References	516

48. ADH Cluster Genes, Genome-Wide Association Studies, and Alcohol Dependence

Byung Lae Park and Hyoung Doo Shin

Introduction	520
Alcohol Metabolism and Major Genetic Factors	520
Genome-Wide Association Study and ADH Gene Cluster	521
Conclusions	527
Applications to Other Addictions and Substance Misuse	527
Definition of Terms	528
Key Facts about Alcohol Dependence	528
Summary	528
References	528

49. Neuropathology of Gene Expression during Alcohol Withdrawal

Harinder Aujla

Introduction	531
Conclusion	538
Applications to Other Addictions and Substance Misuse	538
Definition of Terms	538
Key Facts on the History of Alcohol Consumption	540
Summary Points	540
References	540

50. The Quest of Candidate Gene in Alcohol Dependence: The Dopamine D2 Receptor (*DRD2*) Gene

Sheng-Yu Lee, Yun-Hsuan Chang and Ru-Band Lu

Introduction	543
Molecular Genetics of Alcoholism	543
Subtypes of Alcoholism Identified in Han Chinese	544
Association Studies of the *DRD2* Gene with Alcoholism Subtypes	545
The *DRD2* Gene and Temperament Interaction in Subtyped Alcoholism	547
Summary	548
Definition of Terms	548
Key Facts of Subtyped Alcoholism	549
Summary Points	549
References	549

51. Acetaldehyde: A Reactive Metabolite

Roshanna Rajendram, Rajkumar Rajendram and Victor R. Preedy

Introduction	552
Endogenous Aldehydes	552
Exogenous Acetaldehydes	553
Acetaldehyde Reactivity	554
Protein Adduct Formation	555
Damage to DNA	555

Protein Carbonylation/Oxidation and
  Cytotoxicity   556
Molecular Cytotoxic Mechanisms of
  Aldehydes   556
Acetaldehyde or Ethanol?   557
Human Health Risks   557
Concluding Remarks   559
Applications to Other Addictions and
  Substance Misuse   559
Definition of Terms   559
Key Facts of Brain Metabolism   559
Summary Points   560
References   560

Section C
Structural and Functional Aspects

52. Cortical Morphology in Fetal Alcohol Spectrum Disorders: Insights from Computational Neuroimaging

François De Guio, Ernesta Meintjes, Jean-François Mangin and David Germanaud

Introduction   565
What Is Cortical Morphology?   566
Cortical Morphology in Fetal Alcohol
  Spectrum Disorders   568
Applications to Other Addictions and
  Substance Misuse   571
Definition of Terms   571
Key Facts of Fetal Alcohol Syndrome
  Disorders   571
Summary Points   572
References   572

53. Abnormalities of Cerebellar Structure and Function in Alcoholism and Other Substance Use Disorders

Jessica W. O'Brien and Shirley Y. Hill

Introduction   575
Prevalence of Substance Use Disorders   575
The Cerebellum and Addiction
  Connection   576
Structure and Function of the Cerebellum   576
Development of the Cerebellum   577
Alcohol Use Disorders and the
  Cerebellum   577
Cerebellar Abnormalities in SUD: Cause
  or Consequence?   579
Cerebellar Structure and Function in
  Other Substance Use Disorders   582

Conclusions   583
Applications to Other Addictions and
  Substance Misuse   583
Definition of Terms   583
Key Facts of the Cerebellum   583
Summary Points   583
References   584

54. Brain Volumes in Adolescents with Alcohol Use Disorder

Samanth J. Brooks and Dan J. Stein

Introduction   587
Voxel-Based Morphometry and Related
  Methods of Structural Brain Imaging   587
Hippocampal Volume Is Reduced in
  Adolescents with AUD   588
Prefrontal Cortex Volume Is Differential
  in Adolescents with Alcohol Use
  Disorder   593
Other Brain Regions Showing Differential
  Volumes in Adolescent AUD Compared
  to Controls   594
Moderating and Mediating Factors—Gender,
  Psychiatric Comorbidities, Family History
  of Alcohol Abuse, Childhood Trauma   595
Conclusions   596
Applications to Other Addictions and
  Substance Misuse   596
Definition of Terms   597
Key Facts about Brain Volume Differences
  in Alcohol Use Disorder   597
Summary Points   598
References   598

55. Central Pontine Myelinolysis in Alcoholism: An Osmotic Demyelination Syndrome

Irena Dujmovic

Introduction   600
Pathophysiology   600
Pathology   602
Neuroimaging   603
Clinical Presentation   605
Treatment and Prognosis   606
Applications to Other Addictions and
  Substance Misuse   606
Definition of Terms   607
Key Facts   607
Summary Points   607
References   608

56. The Role of Brain Glucocorticoid Systems in Alcohol Dependence

Annie M. Whitaker, Brittany M. Priddy, Scott Edwards and Leandro F. Vendruscolo

Introduction	610
Alcohol Activates the Hypothalamic–Pituitary–Adrenal Axis	610
Chronic Alcohol Exposure Disrupts HPA Axis Function	611
Role of Brain Glucocorticoid Receptor Activity in the Transition to Alcohol Dependence	611
GR Structure, Chaperone Proteins, and Nuclear Cofactors	612
Induction of Transcriptional Activation by GR	613
Applications to Other Addictions and Substance Misuse	613
Definition of Terms	614
Key Facts Related to Alcohol Misuse	614
Summary Points	614
Acknowledgment	614
References	614

57. Neuropathobiology of Alcohol-Induced Cognitive Deficits

Vinod Tiwari and Kanwaljit Chopra

Introduction	618
Neuropathological Mechanisms Associated with Alcohol-Induced Cognitive Deficits	619
Future Scope and Conclusions	623
Definition of Terms	624
Key Facts on Alcohol-Induced Cognitive Deficit and Its Neuropathobiology	624
Summary Points	624
References	624

58. Neuropathology in Neuropsychiatric Disorders Secondary to Alcohol Misuse

Jagdeo Prasad Rawat, Charles Pinto, Malay Dave and Kirti Yeshwant Tandel

Wernicke Encephalopathy	627
Korsakoff Syndrome	629
Marchiafava–Bignami Disease	630
Alcohol Withdrawal Syndrome	632
Alcohol-Related Dementia	632
Alcoholic Cerebellar Degeneration	633
Other Neuropsychiatric Disorders	633
Application to Other Addictions and Substance Misuse	634
Definition of Terms	634
Key Facts	634
Summary Points	634
References	635

59. Social Isolation and Ethanol Drinking: A Preclinical Model of Addiction Vulnerability in Males, but Not Females

Tracy R. Butler and Jeffrey L. Weiner

Introduction	637
Sex Differences: From Acute Response to Alcohol to Neurobiological Consequences of Alcohol Dependence	637
Epidemiology by Sex: Association with Alcohol Use Disorders	638
Sex Differences: Neuroendocrine Responses in Vulnerable and Addicted Populations	638
Social Isolation as a Model of Addiction Vulnerability in Male Rodents	638
Hormones in Male and Female Rodents Exposed to Early Life Stress	640
Sex Differences in Alcohol Models: Females Drink More!	641
Models of an Addiction-Vulnerable Phenotype in Female Rats	641
Sex Differences in Addiction Vulnerability: Future Directions	641
Applications to Other Addictions and Substance Misuse	643
Definition of Terms	643
Key Facts of Consequences Following Adolescent Social Isolation Model in Male versus Female Rats	644
Summary Points	644
Acknowledgment	644
References	644

Section D
Methods

60. The Ultrarapid Alcohol, Smoking, and Substance Involvement Screening Test (ASSIST-Lite) and Implications for Neuropathology

Robert Ali, Linda Gowing and Jennifer Harland

Introduction	649
Development of the ASSIST	649
ASSIST-Lite Protocol	650
Neuropathology as a Reason for Screening and Intervention	651

Applications to Other Addictions and Substance Misuse	699
Definition of Terms	699
Key Facts of CB1	699
Summary Points	699
References	699

65. Cannabinoid Agonists

Ana Bagues and Carlos Goicoechea

Introduction	702
Endocannabinoids	702
Phytocannabinoids: Δ^9-Tetrahydrocannabinol	707
Synthetic Cannabinoids	708
Cannabinoid Agonists in the Clinic	708
Application to Other Addictions and Substance Misuse	708
Definition of Terms	709
Key Facts	709
Summary Points	710
References	710

66. Molecular Pharmacology of CB1 and CB2 Cannabinoid Receptors

Jahan P. Marcu and Jason B. Schechter

Introduction	713
Introduction to Cannabinoid Receptor Research	713
Overview of Endocannabinoid System Molecular Biology	714
Overview of Cannabinoid Signaling in Cells	717
Summary and Applications to Other Addictions and Substance Misuse	718
Definition of Terms	718
Summary Points	719
References	719

67. Does Agonist Efficacy Alter the In Vivo Effects of Cannabinoids: Girl, Could We Get Much "Higher"?

Roger S. Gifford, Jimit G. Raghav and Torbjörn U.C. Järbe

Introduction	722
Cannabinoid Agonist Classification	722
Herbal Incense: Not Your Parents' Marijuana?	723
Preclinical Studies	724
Marijuana Dependence Studies in Humans	725
Dependence and Withdrawal Studies in Animals	725
Tolerance	725
Drug Discrimination	725
Reinforcement	726
If Not Efficacy, What, Then?	727
Conclusion	728
Applications to Other Addictions and Substance Misuse	728
Definition of Terms	728
Key Facts	728
Summary Points	728
Acknowledgment	729
References	729

68. Cannabis, Cannabinoid Receptors, and Stress-Induced Excitotoxicity

Borja García-Bueno and Javier R. Caso

Introduction	731
Endocannabinoid System and Stress: A Bidirectional Relationship	732
Stress and Cellular Damage/Death by Excitotoxicity	733
Cannabinoid Receptor Activation and Stress-Induced Excitotoxicity	734
Therapeutic Potential in Stress-Related Neuropsychopathologies	735
Definition of Terms	735
Key Facts of Excitotoxicity	735
Summary Points	735
References	737

69. Cannabis, Endocannabinoid CB1 Receptors, and the Neuropathology of Vision

Miguel Dasilva, Kenneth L. Grieve and Casto Rivadulla

Introduction	738
Psychophysical Effects of Cannabinoids on Vision	739
Electrophysiological Effects of Cannabinoids on Vision	741
Conclusions and Future Directions	745
Applications to Other Addictions and Substance Misuse	745
Definition of Terms	745
Key Facts of the Cannabinoid System	745
Summary Points	745
References	746

70. Cannabidiol and 5-HT$_{1A}$ Receptors

Manoela V. Fogaça, Alline C. Campos and Francisco S. Guimarães

Introduction	749
Cannabidiol: A Brief History	749
Multiple Effects of Cannabidiol and Its Therapeutic Potential	749

Conducting a Brief Intervention with an Illicit Drug User	656
Conclusion	657
Key Facts	657
Summary Points	657
References	657

Part IV
Cannabinoids

Section A
General Aspects

61. Overview of Cannabis Use, Misuse, and Addiction

Cristina A.J. Stern, Leandro J. Bertoglio and Reinaldo N. Takahashi

Introduction	665
Acute and Long-Term Effects of Cannabis Use	665
Cannabis Misuse and the Risk of Addiction	666
Translational Research on Cannabis Use in Animal Models	667
Therapies for Cannabis Use and Withdrawal	667
Is There a Potential Therapeutic Value for Medical Cannabis?	668
Policy Changes in Cannabis Regulation	669
Conclusions	669
Applications to Other Addictions and Substance Misuse	669
Definition of Terms	669
Key Facts of the Cannabinoid Tetrad Test	669
Summary Points	670
Acknowledgment	670
References	670

62. An Overview of Major and Minor Phytocannabinoids

Jahan P. Marcu

Introduction	672
Tetrahydrocannabinol Acid, Tetrahydrocannabivarinic Acid, and Cannabigerol Acid	672
Cannabigerol	672
Δ^9-Tetrahydrocannabinol	674
Tetrahydrocannabivarin	675
Cannabidiol	675
Cannabichromene	675
Cannabinol	675
β-Caryophyllene	675
Applications to Other Addictions and Substance Misuse	676
Definition of Terms	676
Summary	676
References	676

63. Treating the Phenomenon of New Psychoactive Substances: Synthetic Cannabinoids and Synthetic Cathinones

Maurizio Coppola, Raffaella Mondola, Francesco Oliva, Rocco Luigi Picci, Daniele Ascheri and Federica Trivelli

Introduction	679
Synthetic Cannabinoids	680
Synthetic Cathinones	683
Applications to Other Addictions and Substance Misuse	684
Definition of Terms	685
Key Facts of Synthetic Cannabinoids and Psychosis	685
Summary Points	685
References	685

Section B
Molecular and Cellular Aspects

64. Role of Lipid Rafts and the Underlying Filamentous-Actin Cytoskeleton in Cannabinoid Receptor 1 Signaling

Dimitra Mangoura, Olga Asimaki, Emmanouella Tsirimonaki and Nikos Sakellaridis

Cannabinoid Receptor 1 Signaling: A Forceful Regulator of Neuronal Development and Function	689
CB1 Signaling to the Major Effector Extracellular Signal-Regulated Kinase	689
CB1 Signaling to ERK Emanates from Lipid Rafts	690
Lipid Rafts, Filamentous-Actin Networks, and Microtubules Regulate CB1 Signaling and Localization	691
Formation of CB1 Signaling Complexes and Transactivation of Tyrosine Kinase Receptors in Lipid Rafts	696
Concluding Remarks	698

Cannabidiol and 5HT$_{1A}$ Receptors	749
How Does Cannabidiol Interact with 5HT$_{1A}$ Receptors?	753
Are All Cannabidiol Effects Mediated by 5-HT$_{1A}$ Receptors?	754
Conclusion	755
Application to Other Addictions and Substance Misuse	756
Definition of Terms	756
Key Facts	756
Summary Points	756
References	757

71. CB1 Receptor-Mediated Signaling Mechanisms in the Deleterious Effects of Spice Abuse

Balapal S. Basavarajappa

Introduction	760
Cannabinoid Receptors	760
Signal Transduction Mechanism of CB1R	762
CB2R Signaling	764
Spice Signaling	764
Applications to Other Addictions and Substance Misuse	766
Definition of Terms	766
Key Facts of Spice Products	767
Summary Points	767
Acknowledgment	767
References	767

Section C
Structural and Functional Aspects

72. Cannabis Use Disorders and Brain Morphology

Valentina Lorenzetti and Janna Cousijn

Introduction	773
Nosology of Cannabis Use Disorder	773
Neurocognitive Aspects of Cannabis Use Disorder	774
Structural Neuroimaging Studies Investigating Cannabis Use Disorder	774
Discussion	781
Applications to Other Addictions and Substance Misuse	782
Concluding Key Facts on the Association between Gray Matter and Cannabis Use Disorder	782
Definition of Terms	783
Key Facts of the Striatum	783
Summary Points	783
References	783

73. Imaging Dopamine Alterations in Cannabis Dependence

Jodi J. Weinstein and Anissa Abi-Dargham

Introduction	786
Imaging Methods Used to Study Dopamine in Humans	787
Effect of Acute Cannabinoid Administration on the Dopamine System	787
Effect of Chronic Cannabis Use on the Dopamine System	788
Conclusion	792
Applications to Other Addictions and Substance Misuse	792
Definition of Terms	792
Key Facts about Imaging the Dopamine System in Humans Using PET	792
Summary Points	793
Acknowledgment	793
References	793

74. Cannabis and the Mesolimbic System

Carla Cannizzaro and Marco Diana

The Dopaminergic Mesolimbic System	795
Cannabinoids in the Reward Circuitry	796
Cannabis Dependence	798
Applications to Other Addictions and Substance Misuse	799
Definition of Terms	801
Key Facts of the "Addictive" Spines in the Nucleus Accumbens	801
Summary Points	801
Acknowledgment	802
References	802

75. Role of the Endocannabinoid System and Major *Cannabis* Constituents in the Reconsolidation and Extinction of Rewarding Drug-Associated Memories

Cristiane R. de Carvalho, Cristina A.J. Stern, Leandro J. Bertoglio and Reinaldo N. Takahashi

Introduction	804
Essential Aspects about Learning and Memory Processing	805
Overview of the Endocannabinoid System	805
Using the Conditioned Place Preference Paradigm to Investigate the Role of the Endocannabinoid System in Memory Reconsolidation and Extinction	807
Role of the Endocannabinoid System in Learning and Memory Processing	807

The Endocannabinoid System and Reconsolidation of Rewarding Drug-Related Memories	808
The Endocannabinoid System and Reconsolidation of Aversive Memories	809
Effects of Δ⁹-Tetrahydrocannabinol and Cannabidiol on Reconsolidation of Drug, Appetitive, and Aversive Memories	809
Concluding Remarks	811
Applications to Other Addictions and Substance Misuse	811
Definition of Terms	811
Key Facts of Memory Reconsolidation	811
Summary Points	812
Acknowledgments	812
References	812

76. Effects of Δ⁹-Tetrahydrocannabinol, Synthetic Cannabinoids, and Fatty Acid Amide Hydrolase Inhibitors on Mood and Serotonin Neurotransmission

Gabriella Gobbi, Nicolas Nuñez, Ryan McLaughin and Francis Bambico

Introduction	815
How to Assess If a Putative Drug Modulates Mood Using Animal Models	816
Δ⁹-Tetrahydrocannabinol and Serotonin Firing Activity	816
The Effect of WIN 55,212-2 on 5-HT Firing and Antidepressant-Like Activity	818
Fatty Acid Amide Hydrolase Inhibition, 5-HT Firing, and Antidepressant-Like Activity	820
Chronic Administration of CB1R Agonists in Adolescence and Adulthood: Implications for 5-HT Firing Activity	821
Conclusion	822
Applications to Other Addictions and Substance Misuse	823
Definition of Terms	823
Key Facts	823
Summary Points	824
References	824

77. CB1 Cannabinoid Receptors and Aggression: Relationship to Cannabis Use

Marta Rodríguez-Arias, José Miñarro, M. Carmen Arenas and María A. Aguilar

Introduction	827
The Endocannabinoid System	828
Cannabis and Aggression in Humans	828
Relation between Aggression and CB1 Receptor in Animal Studies	831
Conclusions	832
Applications to Other Addictions and Substance Misuse	832
Definition of Terms	833
Key Facts of Aggression	833
Summary Points	834
References	834

78. Cannabis and Bipolar Manic Episodes

Jean-Michel Aubry

Substance Use and Bipolar Disorder	836
Childhood Trauma, Cannabis, and Bipolar Disorders	836
Impact of Cannabis Use on Mania Phenomenology	836
Impact of Cannabis Use on Bipolar Disorder Outcome	836
Cannabis Use and Onset Manic Episode	837
Cannabis Use and Psychosis	837
Mechanisms Underlying Cannabis Use and Manic Symptoms	838
Applications to Other Addictions and Substances Misuse	838
Definition of Terms	838
Key Facts of Mania and Bipolar Disorders	838
Summary Points	839
References	839

79. White Matter, Schizophrenia, and Cannabis

Candice E. Crocker and Philip G. Tibbo

Introduction	841
White Matter Changes in Schizophrenia	844
White Matter Changes with Cannabis Use	844
White Matter Changes with Cannabis Use in People with Schizophrenia	847
Applications to Other Addictions and Substance Misuse	848
Definition of Terms	848
Key Facts	848
Summary Points	849
References	849

80. Electroencephalography and Cannabis: From Event-Related Potentials to Oscillations

Patrick D. Skosnik and Jose A. Cortes-Briones

Introduction	851
The Brain Cannabinoid System	851
Electroencephalography	851

Cannabinoids and ERPs	852
Cannabinoids and Neural Oscillations	854
Conclusion and Future Directions	859
Applications to Other Addictions and Substance Misuse	859
Defintion of Terms	860
Key Facts of Cannabis and Electroencephalography	860
Summary Points	860
References	861

Part V
Opioids

Section A
General Aspects

81. Street Level Heroin, an Overview on Its Components and Adulterants

Maryam Akhgari, Afshar Etemadi-Aleagha and Farzaneh Jokar

Introduction	867
An Overview of Opiates	867
Historical Review of Heroin	868
A Focus on the Heroin Production Process	868
Definition of Controlled Substances	868
Controlled Substances Act Schedules	868
Heroin Pharmacology	869
What Does Heroin Look Like?	869
Routes of Administration for Heroin	870
Body Packing of Heroin	870
Street Names and Slang Terms for Heroin	870
Component Analysis of Heroin	870
Medical Consequences of Street Heroin Abuse	873
Concluding Remarks	874
Application to Other Addictions and Substance Misuse	874
Key Facts	874
References	875

82. Opioid Prescription Drug Abuse and Its Relation to Heroin Trends

Dessa Bergen-Cico, Susan Scholl, Nato Ivanashvili and Rachael Cico

Introduction	878
Trends	878
Heroin Availability on the World Market	878
Global Trends in Opioid and Opiate Abuse	879
Opiate and Opioid Overdose Deaths	879
United States	881

Trends in Treatment for Opiates and Opioids	882
Harm Reduction Response to Increasing Opioid and Heroin Abuse	884
Applications to Other Addictions and Substance Misuse	884
Conclusion	885
Definition of Terms	885
Key Facts	885
Summary Points	886
References	886

Section B
Molecular and Cellular Aspects

83. Corticotropin-Releasing Factor Signaling at the Intersection of Pain and Opioid Addiction

M. Adrienne McGinn and Scott Edwards

Introduction	891
The Addicted Phenotype	891
Animal Models of Opioid Intake Escalation and Addiction	892
Chronic Opioid Exposure and Hyperalgesia	892
Opioid Addiction in Relation to Pain Management	892
Shared Neural Mechanisms and Neurocircuitry Underlying Pain and Opioid Addiction	892
Role of Corticotropin-Releasing Factor in Opioid-Induced Hyperalgesia and the Transition to Opioid Dependence	893
Role of Corticotropin-Releasing Factor Signaling in Chronic Pain States	894
Applications to Other Addictions and Substance Misuse	894
Definition of Terms	895
Key Facts Related to Opioid-Induced Hyperalgesia and Addiction	895
Summary Points	895
Acknowledgment	895
References	896

84. The Role of the δ Opioid Receptor Gene, OPRD1, in Addiction

Richard C. Crist and Wade H. Berrettini

Introduction	899
OPRD1 and Human Addiction	899
Rodent Models and Addiction	901
Drugs and δ Opioid Receptor Function	903
Heterodimerization of μ and δ Opioid Receptors	904
Conclusion	905

Applications to Other Addictions and Substance Misuse	905
Definition of Terms	905
Key Facts of Pharmacogenetics of Addiction	905
Summary Points	905
References	906

85. Genome-Wide Association Studies and Human Opioid Sensitivity

Daisuke Nishizawa and Kazutaka Ikeda

Introduction	909
Association Studies on Specific Candidate Genes for Human Opioid Sensitivity	910
Genome-Wide Association Studies of Human Opioid Sensitivity	912
Genome-Wide Association or Linkage Studies of Human Opioid Dependence	915
Applications to Other Addictions and Substance Misuse	917
Definition of Terms	918
Key Facts	918
Summary Points	918
References	919

86. Opiate Receptors and Gender and Relevance to Heroin Addiction

Maja Djurendic-Brenesel and Vladimir Pilija

Introduction	922
Opiate Receptors	923
Classification and Localization of Opiate Receptors	923
Structure of Opiate Receptors and Signal Transduction	925
Tolerance, Dependence, and Addiction	926
Gender Differences in the Pharmacological Effects of Opiates	927
Gender Differences in Heroin Addiction	928
Applications to Other Addictions and Substance Misuse	930
Definition of Terms	930
Key Facts of Opiate Abuse	930
Summary Points	930
References	931

87. Transcriptional Effects of Heroin and Methamphetamine in the Striatum

Ryszard Przewlocki, Michal Korostynski and Marcin Piechota

Introduction	933
Patterns of Gene Expression Regulated by the Acute Administration of Psychostimulants or Opioids	934
Drug-Induced Regulation of Activity-Dependent Genes	934
Activation of Stress-Related Genes by Drugs of Abuse	935
Glucocorticoid Receptor-Dependent Genes Induced by Drugs of Abuse Are Expressed in Astrocytes	936
Gene Expression Regulated by Chronic Administration of Psychostimulants or Opioids	937
Chronic Methamphetamine and Heroin Treatment Activated Genes Enriched in the Nucleus Accumbens	938
Transcriptional Effects of Chronic Methamphetamine and Heroin Treatment in the Striatum Display Similarity	938
Regulation of Glucocorticoid Receptor-Dependent Genes in Response to Chronic Treatment	938
Alteration of Circadian Genes by Chronic Methamphetamine and Heroin Treatment	940
Conclusions	940
Applications to Other Addictions and Substance Misuse	940
Definition of Terms	941
Key Facts of Polydrug Abuse	941
Summary Points	941
Acknowledgments	941
References	941

Section C
Structural and Functional Aspects

88. Neurological Abnormalities in Opiate Addicts

Josef Finsterer and Claudia Stöllberger

Introduction	947
Classification	947
Causes of Neurological Disease in Opiate Addicts	947
Neurological Abnormalities in Opiate Addicts	947
Ischemic Stroke in Opiate Addicts	948
Cerebral Bleeding in Opiate Addicts	949
Epilepsy in Opiate Addicts	949
Trauma	949
Infections	949
Cognitive Impairment	951
Other CNS Abnormalities in Opiate Addicts	951
PNS Disease	952
Neurological Withdrawal Symptoms in Opiate Addicts and in Neonates	952
CNS Imaging in Opiate Addicts	952
Neuropathology in Opiate Addicts	952

Premorbid Neurological Vulnerabilities Fostering the Development of Opiate Addiction	952
Dependency of Neurological Abnormalities on the Mode of Opiate Administration, the Duration of Addiction, and the Opiate Dosage	953
Neurological Complications during General Anesthesia	953
Secondary Disease that Causes Tertiary Neurological Abnormalities	953
Conclusions	953
Applications to Other Addictions and Substance Misuse	953
Definition of Terms	953
Key Facts	954
Summary Points	954
References	954

89. Clinical and Preclinical Molecular Imaging in Chronic Pain—Implications for Analgesic Use and Misuse

Deepak Behera and Nida Ashraf

Introduction	956
Pathophysiology Relevant to Pain Imaging	957
Functional and Molecular Imaging of the Brain	958
Functional and Molecular Imaging of Peripheral Structures	960
Conclusion	962
Applications to Other Addictions and Substance Misuse	962
Definition of Terms	962
Key Facts of Molecular Imaging	963
Summary Points	963
References	963

90. Single-Photon-Emission Computed Tomography Studies with Dopamine and Serotonin Transporters in Opioid Users

Shih-Hsien Lin, Kao Chin Chen and Yen Kuang Yang

Introduction	966
Dopamine	966
Serotonin	968
Discussion	969
Applications to Other Addictions and Substance Misuse	970
Conclusion	970
Definition of Terms	970
Key Facts Related to Opioid Dependency	970
Summary Points	971
References	971

91. Personality Dimensions, Impulsivity, and Heroin

Cuneyt Evren and Muge Bozkurt

Introduction	974
Impulsivity and Heroin Dependence	975
Temperament and Character Dimensions of Personality	976
Relationships of Personality Dimensions with Impulsivity in Heroin Dependents	978
Applications to Other Addictions and Substance Misuse	979
Definition of Terms	979
Key Facts of Impulsivity, Temperament, and Character	980
Summary Points	980
References	980

92. fMRI of Heroin and Ketamine Addiction

Y.L. Jiang, Tony C.H. Chow, Sharon L. Wu, H.C. Tang, Maria S.M. Wai and D.T. Yew

Introduction	983
Application to Types of Addictions	983
Summary Points	987
References	987

93. Heroin-Induced Cerebrovascular Ischemia and Leukoencephalopathy: Role of Mitochondria

Yashar Yousefzadeh Fard, Seyyed A. Hashemi and Megan L. Fitzgerald

Stroke and Heroin: A Chronological Review of Cases	989
Heroin-Induced Leukoencephalopathy: A Chronological Review of Cases	990
Discussion of Possible Pathophysiologies	994
Hypoxia Is a Common Pathway to Induce Leukoencephalopathy	997
Conclusion	1000
Applications to Other Addictions and Substance Misuse	1000
Definition of Terms	1000
Key Facts of Spongiform Leukoencephalopathy	1000
Summary Points	1000
Acknowledgment	1000
References	1000

94. The Hypothalamic–Pituitary–Adrenal Axis and Related Brain Stress-Response Systems and Heroin

Yan Zhou, Hilary Briggs and Mary Jeanne Kreek

Introduction	1003
HPA Axis in Human Heroin Addiction and Rodent Models	1004
Arginine Vasopressin and V1b Receptor Systems	1005
Endogenous Opioid Systems	1006
Orexin and Dynorphin in the Lateral Hypothalamus	1007
Summary	1007
Definition of Terms	1008
Key Facts of Stress Responses in Opiate Addiction	1008
Summary Points	1008
Acknowledgment	1008
References	1009

95. Accelerated Aging in Heroin Abusers: Readdressing a Clinical Anecdote Using Telomerase and Neuroimaging

Gordon L.F. Cheng and Tatia M.C. Lee

Introduction	1012
Neuropathology and Neuropsychology of Heroin Addiction	1013
Biological Indicators of Aging	1014
Heroin, Telomerase, and Neuroimaging: Crossroads between Molecular and Brain System Pathology	1016
A Paradigm Shift in the Understanding of Addiction Pathology	1018
Overall Summary	1019
Definition of Terms	1019
Summary Points	1020
Acknowledgment	1020
References	1020

96. Attention Deficit Hyperactivity Disorder, Substance Use Disorders, and Heroin Addiction

Saad Salman, Jawaria Idrees, Muhammad Anees and Fariha Idrees

Introduction	1023
Applications to Other Addictions and Substance Misuse	1024
Conclusion	1032
Definition of Terms	1032
Key Facts of ADHD and Other Psychiatric Disorders	1033
Summary	1034
References	1034

97. Impaired Cognition Control and Inferior Frontal Cortex Modulation in Heroin Addiction

André Schmidt, Marc Walter and Stefan Borgwardt

Introduction	1037
Behavioral Impairments of Cognitive Control in Heroin Addiction	1038
Inferior Frontal Gyrus, Response Inhibition, and Heroin Addiction	1039
Applications to Other Addictions and Substance Misuse	1042
Conclusions	1043
Suggestions for Further Research	1043
Definition of Terms	1044
Key Facts of Cognitive Control Functions in Heroin Addiction and the Relation to the Inferior Frontal Gyrus	1045
Summary Points	1045
References	1045

Section D
Methods

98. Assay for Opium Alkaloids

Jelena Acevska, Gjoshe Stefkov, Svetlana Kulevanova and Aneta Dimitrovska

Introduction	1051
Types of Biological Material Used for Determination of Opioids	1053
Sample Preparation for Bioanalysis of Opioids	1054
Detection and Quantification of Opioids in Biological Samples	1054
Review of Analytical Methods for Bioassay of Opium Alkaloids and Related Opioids	1056
Applications to Other Addictions and Substance Misuse	1058
Definition of Terms	1058
Key Facts of Bioassay of Opium Alkaloids and Opioid-Related Compounds	1058
Summary Points	1058
References	1059

99. **Features of the Tridimensional Personality Questionnaire Associated with Heroin Users**

Chen-Ying Wu, Wei-Lieh Huang, Yu-Hsuan Lin and Cheryl C.H. Yang

Introduction	1061
The Concepts of the TPQ and the TCI	1062
The TPQ Dimensions and Heroin Use	1062
Categorical Views of Personality in Relation to Heroin Use: the Switch from the TPQ	1063
The Autonomic Basis of Borderline Personality Trait and Heroin Use Disorder	1065
Applications to Other Addictions and Other Substance Misuse	1066
Conclusions	1066
Definition of Terms	1066
Key Facts of Personality Features in Heroin Users	1068
Summary Points	1068
References	1068

Index 1071

List of Contributors

Ricardo Abadie-Guedes Departamento de Fisiologia e Farmacologia, Universidade Federal de Pernambuco, Recife, Pernambuco, Brazil

Anissa Abi-Dargham Departments of Psychiatry and Radiology, Division of Translational Imaging, New York State Psychiatric Institute and Columbia University Medical Center, New York, NY, USA

Osama A. Abulseoud National Institute on Drug Abuse, National Institutes of Health, Baltimore, MD, USA; Department of Psychiatry and Psychology, Mayo Clinic, Rochester, MN, USA

Jelena Acevska Faculty of Pharmacy, Institute of Applied Chemistry and Pharmaceutical Analysis, University Ss. Cyril and Methodius, Skopje, Republic of Macedonia

Ana Adan Department of Psychiatry and Clinical Psychobiology, University of Barcelona, Barcelona, Spain; Institute for Brain, Cognition and Behavior (IR3C), Barcelona, Spain

María A. Aguilar Unidad de Investigación Psicobiología de las Drogodependencias, Departamento de Psicobiología, Facultad de Psicología, Universitat de València, Valencia, Spain

Maryam Akhgari Department of Forensic Toxicology, Legal Medicine Research Center, Legal Medicine Organization, Tehran, Iran

Robert Ali Discipline of Pharmacology, School of Medical Sciences, The University of Adelaide, Adelaide, SA, Australia

Amelia J. Anderson-Mooney Department of Neurology, University of Kentucky College of Medicine, Lexington, KY, USA

Muhammad Anees Department of Neurology and Neurosurgery, Lady Reading Hospital, Post Graduate Medical Institute, Peshawar, Khyber Pakhtunkhwa, Pakistan

M. Carmen Arenas Unidad de Investigación Psicobiología de las Drogodependencias, Departamento de Psicobiología, Facultad de Psicología, Universitat de València, Valencia, Spain

Jamie Arndt Department of Psychology, University of Missouri – Columbia, Columbia, MO, USA

Daniele Ascheri Department of Clinical and Biological Sciences, San Luigi Gonzaga Medical School, University of Turin, Torino, Italy

Nida Ashraf Aga Khan University, Karachi, Pakistan

Maxie Ashton The Adelaide Clinic, Gilberton, SA, Australia

Olga Asimaki Basic Research Center, Biomedical Research Foundation of the Academy of Athens, Athens, Greece

Jean-Michel Aubry Mood Disorders Unit, Division of Psychiatric Specialties, Department of Mental Health and Psychiatry, Geneva University Hospitals, Geneva, Switzerland

Harinder Aujla Department of Psychology, University of Winnipeg, Winnipeg, MB, Canada

A. George Awad Department of Psychiatry and the Institute of Medical Science, University of Toronto, Toronto, Canada

Mahmoud A. Awara Dalhousie University, Halifax, NS, Canada

Ana Bagues A. Farmacología y Nutrición, Departamento de Ciencias Básicas de la Salud. Unidad Asociada de I+D+i al CSIC, Rey Juan Carlos University, Alcorcón, Madrid, Spain

Yatan P.S. Balhara Department of Psychiatry & National Drug Dependence Treatment Centre, WHO Collaborating Centre on Substance Abuse, All India Institute of Medical Sciences, New Delhi, India

Francis Bambico Neurobiological Psychiatry Unit, McGill University, Montreal, QC, Canada; Centre for Addiction and Mental Health, University Toronto, Toronto, ON, USA

Laura Barrio-Real Department of Pharmacology, Perelman School of Medicine, University of Pennsylvania, Philadelphia, PA, USA

Selena E. Bartlett Addiction Neuroscience and Obesity Group, Institute of Health and Biomedical Innovation, School of Clinical Sciences, Faculty of Health, Queensland University of Technology, Translational Research Institute, Woolloongabba, QLD, Australia

Balapal S. Basavarajappa Division of Analytical Psychopharmacology, Nathan Kline Institute for Psychiatric Research, Orangeburg, NY, USA; Department of Psychiatry, College of Physicians and Surgeons, New York State Psychiatric Institute, Columbia University, New York, NY, USA

Aniruddha Basu Department of Psychiatry, National Institute of Mental Health and Neurosciences, Bangalore, India

Claiton H.D. Bau Department of Genetics, Instituto de Biociências, Universidade Federal do Rio Grande do Sul (UFRGS), Porto Alegre, RS, Brazil

Matthew C. Beattie Department of Psychiatry & Pharmacology, Bowles Center for Alcohol Studies, University of North Carolina, Chapel Hill, NC, USA

Benjamin Becker Division of Medical Psychology, Department of Psychiatry and Psychotherapy, University of Bonn, Bonn, Germany

Anton L. Beer Institut für Psychologie, Universität Regensburg, Regensburg, Germany

Deepak Behera Molecular Imaging Program at Stanford, Department of Radiology, Stanford University School of Medicine, Stanford, CA, USA

Irina Benaiges Department of Psychiatry and Clinical Psychobiology, University of Barcelona, Barcelona, Spain

Dessa Bergen-Cico Department of Public Health, Syracuse University, Syracuse, NY, USA

Susan E. Bergeson Department of Pharmacology and Neuroscience, Texas Tech University Health Sciences Center, Lubbock, TX, USA

Wade H. Berrettini Center for Neurobiology and Behavior, Department of Psychiatry, University of Pennsylvania School of Medicine, Philadelphia, PA, USA

Leandro J. Bertoglio Department of Pharmacology, Federal University of Santa Catarina, Florianópolis, Santa Catarina, Brazil

Charles S. Bockman Department of Pharmacology, Creighton University School of Medicine, Omaha, NE, USA

Stefan Borgwardt Department of Psychiatry (UPK), University of Basel, Basel, Switzerland

Josiane Bourque Department of Psychiatry, University of Montreal, Montreal, QC, Canada; GRIP, Centre de recherche CHU Ste-Justine, Montreal, QC, Canada

Muge Bozkurt Research, Treatment and Training Center for Alcohol and Substance Dependence (AMATEM), Bakirkoy Training and Research Hospital for Psychiatry, Neurology and Neurosurgery, Istanbul, Turkey

Katharine Alexandra Brennan School of Psychology, Victoria University of Wellington, Wellington, New Zealand

Hilary Briggs The Laboratory of the Biology of Addictive Diseases, The Rockefeller University, New York, NY, USA

Samanth J. Brooks Department of Psychiatry and Mental Health, Groote Schuur Hospital, University of Cape Town, Cape Town, South Africa

Rachel Picada Bulcão Instituto Médico Legal, Polícia Científica, Curitiba (PR), Brazil

Ute Burkhardt Department of Pharmacology, Goethe University Frankfurt, Frankfurt, Germany

Tracy R. Butler Department of Psychology, University of Dayton, Dayton, OH, USA

Eduardo Rodrigues Cabrera Instituto Médico Legal, Polícia Científica, Curitiba (PR), Brazil

Daniela Caldirola Hermanas Hospitalarias, Villa San Benedetto Menni Hospital, FoRiPsi, Department of Clinical Neurosciences, Albese con Cassano, Italy

Russell C. Callaghan Northern Medical Program, University of Northern British Columbia (UNBC), Prince George, BC, Canada; Human Brain Laboratory, Centre for Addiction and Mental Health (CAMH), Toronto, ON, Canada; Dalla Lana School of Public Health, University of Toronto, Toronto, ON, Canada

Elza Callegari Faculdade de Farmácia, Instituto de Ciências Biológicas, Universidade de Passo Fundo, Passo Fundo (RS), Brazil

María J. Caloca Institute of Molecular Biology and Genetics, Spanish National Research Council (CSIC), University of Valladolid School of Medicine, Valladolid, Spain

Alline C. Campos Department of Pharmacology, Medical School of Ribeirão Preto, University of São Paulo (FMRP-USP), São Paulo, Brazil

Carla Cannizzaro Department of Sciences for Health Promotion and Maternal Care, School of Medicine, University of Palermo, Palermo, Italy

Gavin Cape Department of Psychological Medicine, Dunedin School of Medicine, University of Otago, Dunedin, New Zealand

Benjamin L. Carnell The Adelaide Clinic, Gilberton, SA, Australia

Javier R. Caso Department of Pharmacology, Faculty of Medicine, University Complutense, Madrid, Spain

Lucas Henrique Cendron Faculdade de Farmácia, Instituto de Ciências Biológicas, Universidade de Passo Fundo, Passo Fundo (RS), Brazil

Yun-Hsuan Chang Department of Psychology, College of Medical and Health Science, Asia University, Taichung, Taiwan; Department of Psychiatry, College of Medicine, National Cheng Kung University

Gordon L.F. Cheng Laboratory of Neuropsychology, The University of Hong Kong, Pokfulam, Hong Kong; Laboratory of Cognitive Affective Neuroscience, The University of Hong Kong, Pokfulam, Hong Kong; Institute of Clinical Neuropsychology, The University of Hong Kong, Pokfulam, Hong Kong

Kao Chin Chen Department of Psychiatry, National Cheng Kung University Hospital, College of Medicine, National Cheng Kung University, Tainan, Taiwan

Kanwaljit Chopra Pharmacology Research Laboratory, University Institute of Pharmaceutical Sciences, UGC Centre of Advanced Study Panjab University, Chandigarh, India

Tony C.H. Chow Schools of Biomedical Sciences and Chinese Medicine, The Chinese University of Hong Kong, Hong Kong, China

Rachael Cico Columbia University, New York, NY, USA

Jason B. Cook Department of Psychiatry & Pharmacology, Bowles Center for Alcohol Studies, University of North Carolina, Chapel Hill, NC, USA

Maurizio Coppola Department of Addiction, ASL CN2, Alba, Italy

Jose A. Cortes-Briones Psychiatry Service, VA Connecticut Healthcare System, West Haven, CT, USA; Abraham Ribicoff Research Facilities, Connecticut Mental Health Center, New Haven, CT, USA; Department of Psychiatry, Yale University School of Medicine, New Haven, CT, USA

Janna Cousijn Department of Developmental Psychology, University of Amsterdam, The Netherlands

Richard C. Crist Center for Neurobiology and Behavior, Department of Psychiatry, University of Pennsylvania School of Medicine, Philadelphia, PA, USA

Candice E. Crocker Department of Psychiatry, Dalhousie University, Halifax, NS, Canada; Division of Neurology, Department of Medicine, Dalhousie University, Halifax, NS, Canada

Kirsty Danielson School of Biological Sciences, Centre for Biodiscovery, Victoria University of Wellington, Wellington, New Zealand; Beth Israel Deaconess Medical Centre, Harvard Medical School, Boston, MA, USA

Miguel Dasilva Faculty of Life Sciences, University of Manchester, Manchester, UK

Malay Dave Department of Psychiatry, Jagjivanram Western Railways Hospital, Mumbai, Maharashtra, India

Mariella De Biasi Department of Psychiatry, University of Pennsylvania, Philadelphia, PA, USA; Department of Neuroscience, University of Pennsylvania, Philadelphia, PA, USA; Neuroscience Graduate Group, Perelman School of Medicine, University of Pennsylvania, Philadelphia, PA, USA

Cristiane R. de Carvalho Department of Pharmacology, Federal University of Santa Catarina, Florianópolis, Santa Catarina, Brazil

François De Guio Université Paris Diderot, Sorbonne Paris Cité, UMR-S 1161 INSERM, Paris, France

Dirk De Ridder Section of Neurosurgery, Department of Surgical Sciences, Dunedin School of Medicine, University of Otago, Dunedin, New Zealand

Sonia Dhiman Chitkara College of Pharmacy, Chitkara University, Chandigarh, Punjab, India

Giuseppina Diaferia Hermanas Hospitalarias, Villa San Benedetto Menni Hospital, FoRiPsi, Department of Clinical Neurosciences, Albese con Cassano, Italy

Marco Diana Department of Drug Sciences, School of Pharmacy, University of Sassari, Sassari, Italy

Joseph R. DiFranza Department of Family Medicine and Community Health, University of Massachusetts Medical School, Worcester, MA, USA

Aneta Dimitrovska Faculty of Pharmacy, Institute of Applied Chemistry and Pharmaceutical Analysis, University Ss. Cyril and Methodius, Skopje, Republic of Macedonia

Le-Anh Dinh-Williams Department of Psychiatry, University of Montreal, Montreal, QC, Canada; Centre de recherche de l'Institut universitaire en santé mentale de Montréal, Montreal, QC, Canada

Maja Djurendic-Brenesel Institute of Forensic Medicine, Clinical Center Vojvodina, Novi Sad, Serbia

Jonathan N. Dodd Department of Neurology, Washington University School of Medicine, St. Louis, MO, USA

Neil C. Dodge Department of Psychiatry and Behavioral Neurosciences, Wayne State University School of Medicine, Detroit, MI, USA

Paul D. Drew Department of Neurobiology and Developmental Sciences, University of Arkansas for Medical Sciences, Little Rock, AR, USA

Rene Drucker Colín Departamento de Neurociencias, Instituto de Fisiología Celular, Universidad Nacional Autónoma de México, Mexico DF, Mexico

Irena Dujmovic Department for Immune-Mediated CNS Diseases, Clinic of Neurology, Clinical Centre of Serbia, Belgrade University School of Medicine, Belgrade, Serbia

Marcelo Dutra Arbo Laboratório de Toxicologia, Departamento de Ciências Biológicas, Faculdade de Farmácia, Universidade do Porto, Porto, Portugal

Scott Edwards Department of Physiology, Alcohol and Drug Abuse Center of Excellence, LSU Health Sciences Center, New Orleans, LA, USA

Afshar Etemadi-Aleagha Department of Anesthesiology, Tehran University of Medical Sciences (TUMS), Amir Alam Hospital, Tehran, Iran

Cuneyt Evren Research, Treatment and Training Center for Alcohol and Substance Dependence (AMATEM), Bakirkoy Training and Research Hospital for Psychiatry, Neurology and Neurosurgery, Istanbul, Turkey

Juan Fang Department of Respiratory Medicine, Peking University Third Hospital, Beijing, China

Camino Fernández-Rodríguez Servicio de Medicina Interna, Hospital Universitario de Canarias, Universidad de La Laguna, La Laguna, Tenerife, Canary Islands, Spain

Radmila Filipovic Physiology and Neurobiology Department, University of Connecticut, Storrs, CT, USA

Josef Finsterer Krankenanstalt Rudolfstiftung, Vienna, Austria

Megan L. Fitzgerald Department of Psychiatry, New York State Psychiatry Institute/Columbia University, New York, NY, USA

Manoela V. Fogaça Department of Pharmacology, Medical School of Ribeirão Preto, University of São Paulo (FMRP-USP), São Paulo, Brazil

Diego A. Forero Laboratory of NeuroPsychiatric Genetics, Biomedical Sciences Research Group, School of Medicine, Universidad Antonio Nariño, Bogotá, Colombia

Hamdy Fouad Moselhy College of Medicine and Health Science, United Arab Emirates University, Alain City, UAE

Marco Antonio de Freitas Instituto Médico Legal, Polícia Científica, Curitiba (PR), Brazil

Cherrie Galletly The Adelaide Clinic, Gilberton, SA, Australia

Borja García-Bueno Department of Pharmacology, Faculty of Medicine, University Complutense, Madrid, Spain

Jodi M. Gatley Northern Medical Program, University of Northern British Columbia (UNBC), Prince George, BC, Canada; Human Brain Laboratory, Centre for Addiction and Mental Health (CAMH), Toronto, ON, Canada; Dalla Lana School of Public Health, University of Toronto, Toronto, ON, Canada

David Germanaud AP-HP, Robert-Debré Hospital, DHU PROTECT "Department of Child Neurology", Paris, France; CEA, NeuroSpin Center, UNIACT, Gif-sur-Yvette, France; INSERM, UMR1129, Paris, France

Roger S. Gifford Center for Drug Discovery (CDD), Northeastern University, Boston, MA, USA; Department of Pharmaceutical Sciences, Northeastern University, Boston, MA, USA

Paul Glue Department of Psychological Medicine, Dunedin School of Medicine, University of Otago, Dunedin, New Zealand

Gabriella Gobbi Neurobiological Psychiatry Unit, McGill University, Montreal, QC, Canada

Carlos Goicoechea A. Farmacología y Nutrición, Departamento de Ciencias Básicas de la Salud. Unidad Asociada de I+D+i al CSIC, Rey Juan Carlos University, Alcorcón, Madrid, Spain

Emilio González-Arnay Departamento de Anatomía, Anatomía Patológica e Histología, Hospital Universitario de Canarias, Universidad de La Laguna, La Laguna, Tenerife, Canary Islands, Spain

Emilio González-Reimers Servicio de Medicina Interna, Hospital Universitario de Canarias, Universidad de La Laguna, La Laguna, Tenerife, Canary Islands, Spain

Rogelio González-Sarmiento Molecular Medicine Unit, Department of Medicine, University of Salamanca-IBSAL, Salamanca, Spain; IBMCC, University of Salamanca-CSIC, Salamanca, Spain

Linda Gowing Drug and Alcohol Services of South Australia, Discipline of Pharmacology, Medical School, The University of Adelaide, Adelaide, SA, Australia

Kenneth L. Grieve Faculty of Life Sciences, University of Manchester, Manchester, UK

Rubem Carlos Araújo Guedes Departamento de Nutrição, Universidade Federal de Pernambuco, Recife, Pernambuco, Brazil

Consuelo Guerri Department of Cellular Pathology, Príncipe Felipe Research Centre, Valencia, Spain

Francisco S. Guimarães Department of Pharmacology, Medical School of Ribeirão Preto, University of São Paulo (FMRP-USP), São Paulo, Brazil

Leila Guller Department of Psychology, University of Kentucky, Lexington, KY, USA

Jennifer Harland Discipline of Pharmacology, School of Medical Sciences, The University of Adelaide, Adelaide, SA, Australia

Seyyed A. Hashemi Immunogenetic Research Center, Traditional and Complementary Medicine Research Center, Medical School Mazandaran University of Medical Sciences, Sari, Iran

Shiva Hashemizadeh Department of Animal Biology, School of Biology, College of Science, University of Tehran, Tehran, Iran

Bei He Department of Respiratory Medicine, Peking University Third Hospital, Beijing, China

Salvador Hernández-López Departamento de Fisiología, Facultad de Medicina, Universidad Nacional Autónoma de México, México DF, México

Tyson Hickle Department of Psychology, Creighton University, Omaha, NE, USA

Shirley Y. Hill Department of Psychiatry, University of Pittsburgh School of Medicine, Pittsburgh, PA, USA; Department of Psychology, University of Pittsburgh, Pittsburgh, PA, USA; Department of Human Genetics, Graduate School of Public Health, University of Pittsburgh, Pittsburgh, PA, USA

Joan Holgate Addiction Neuroscience and Obesity Group, Institute of Health and Biomedical Innovation, School of Clinical Sciences, Faculty of Health, Queensland University of Technology, Translational Research Institute, Woolloongabba, QLD, Australia

Wei Huang Center for Comparative NeuroImaging, Department of Psychiatry, University of Massachusetts Medical School, Worcester, MA, USA

Wei-Lieh Huang Department of Psychiatry, National Taiwan University Hospital, Douliou City, Yunlin County, Taiwan; Institute of Brain Science, National Yang-Ming University, Taipei, Taiwan

Nicholas J. Hunt Department of Medicine, University of Sydney, Sydney, NSW, Australia; The BOSCH Institute, University of Sydney, Sydney, NSW, Australia

René Hurlemann Division of Medical Psychology, Department of Psychiatry and Psychotherapy, University of Bonn, Bonn, Germany

Giuseppe Iannone Hermanas Hospitalarias, Villa San Benedetto Menni Hospital, FoRiPsi, Department of Clinical Neurosciences, Albese con Cassano, Italy

Fariha Idrees Department of Chemistry, Islamia College University, Peshawar, Khyber Pakhtunkhwa, Pakistan

Jawaria Idrees Department of Integrated Sciences, Post Graduate Nursing College, Peshawar, Khyber Pakhtunkhwa, Pakistan

Kazutaka Ikeda Addictive Substance Project, Tokyo Metropolitan Institute of Medical Science, Tokyo, Japan

Nato Ivanashvili Department of Public Health, Syracuse University, Tbilisi, Republic of Georgia

Iliyan Ivanov Department of Psychiatry, Icahn School of Medicine at Mount Sinai, New York, NY, USA

Joseph L. Jacobson Department of Psychiatry and Behavioral Neurosciences, Wayne State University School of Medicine, Detroit, MI, USA; Departments of Human Biology and of Psychiatry and Mental Health, Faculty of Health Sciences, University of Cape Town, Detroit, MI, USA

Sandra W. Jacobson Department of Psychiatry and Behavioral Neurosciences, Wayne State University School of Medicine, Detroit, MI, USA; Departments of Human Biology and of Psychiatry and Mental Health, Faculty of Health Sciences, University of Cape Town, Detroit, MI, USA

Torbjörn U.C. Järbe Center for Drug Discovery (CDD), Northeastern University, Boston, MA, USA; Department of Pharmaceutical Sciences, Northeastern University, Boston, MA, USA

Y.L. Jiang Department of Radiology, Kunming Medical University, Yunnan, China

Farzaneh Jokar Department of Forensic Toxicology, Legal Medicine Research Center, Legal Medicine Organization, Tehran, Iran

Cynthia J.M. Kane Department of Neurobiology and Developmental Sciences, University of Arkansas for Medical Sciences, Little Rock, AR, USA

Megan Kangiser Department of Psychology, Creighton University, Omaha, NE, USA

Victor Karpyak Department of Psychiatry and Psychology, Mayo Clinic, Rochester, MN, USA

Jean A. King Center for Comparative NeuroImaging, Department of Psychiatry, University of Massachusetts Medical School, Worcester, MA, USA

Bronwyn M. Kivell School of Biological Sciences, Centre for Biodiscovery, Victoria University of Wellington, Wellington, New Zealand

Eugene A. Kiyatkin Behavioral Neuroscience Branch, National Institute on Drug Abuse – Intramural Research Program, National Institutes of Health, DHHS, Baltimore, MD, USA

Jochen Klein Department of Pharmacology, Goethe University Frankfurt, Frankfurt, Germany

Michelle L. Klima Department of Psychiatry, University of Pennsylvania, Philadelphia, PA, USA

Sodikdjon A. Kodirov Department of General Physiology, Saint Petersburg University, Saint Petersburg, Russia; Pavlov Institute of Physiology, Russian Academy of Sciences, Saint Petersburg, Russia; Neuroscience Institute, Morehouse School of Medicine, Atlanta, GA, USA; Institute of Experimental Medicine, I. P. Pavlov Department of Physiology, Russian Academy of Medical Sciences, Saint Petersburg, Russia

Michal Korostynski Department of Molecular Neuropharmacology, Institute of Pharmacology, Polish Academy of Sciences, Krakow, Poland

Mary Jeanne Kreek The Laboratory of the Biology of Addictive Diseases, The Rockefeller University, New York, NY, USA

Svetlana Kulevanova Faculty of Pharmacy, Institute of Pharmacognosy, University Ss. Cyril and Methodius, Skopje, Republic of Macedonia

Santosh Kumar Pharmaceutical Sciences, College of Pharmacy, University of Tennessee Health Science Center, Memphis, TN, USA

Berthold Langguth Department of Psychiatry and Psychotherapy, University Regensburg, Regensburg, Germany

Mirna Bainy Leal Laboratório de Farmacologia e Toxicologia de Produtos Naturais, Departamento de Farmacologia, Instituto de Ciências Básicas da Saúde, Universidade Federal do Rio Grande do Sul, Porto Alegre (RS), Brazil

Sheng-Yu Lee Department of Psychiatry, Kaohsiung Veteran's Hospital, Kaohsiung, Taiwan; Department of Psychiatry, College of Medicine, National Cheng Kung University

Tatia M.C. Lee Laboratory of Neuropsychology, The University of Hong Kong, Pokfulam, Hong Kong; Laboratory of Cognitive Affective Neuroscience, The University of Hong Kong, Pokfulam, Hong Kong; Institute of Clinical Neuropsychology, The University of Hong Kong, Pokfulam, Hong Kong; The State Key Laboratory of Brain and Cognitive Sciences, The University of Hong Kong, Pokfulam, Hong Kong

John Leikaif Department of Psychiatry, Stanford University, Stanford, CA, USA

Shih-Hsien Lin Department of Psychiatry, National Cheng Kung University Hospital, College of Medicine, National Cheng Kung University, Tainan, Taiwan

Yu-Hsuan Lin Institute of Brain Science, National Yang-Ming University, Taipei, Taiwan

Xiu Liu Department of Pathology, University of Mississippi Medical Center, Jackson, MS, USA

Valentina Lorenzetti Monash Clinical & Imaging Neuroscience, Monash University, Melbourne, VIC, Australia; Melbourne Neuropsychiatry Centre, The University of Melbourne and Melbourne Health, Melbourne, VIC, Australia

Ru-Band Lu Department of Psychiatry, College of Medicine, National Cheng Kung University; Department of Psychiatry, National Cheng Kung University Hospital, Tainan, Taiwan; Institute of Behavioral Medicine, College of Medicine, National Cheng Kung University, Tainan, Taiwan; Addiction Research Center, National Cheng Kung University, Tainan, Taiwan

Rita Machaalani Department of Medicine, University of Sydney, Sydney, NSW, Australia; The BOSCH Institute, University of Sydney, Sydney, NSW, Australia; The Children's Hospital, Westmead, Sydney, NSW, Australia

Antoniette Maldonado-Devincci Department of Psychiatry & Pharmacology, Bowles Center for Alcohol Studies, University of North Carolina, Chapel Hill, NC, USA

Jean-François Mangin CEA, NeuroSpin Center, UNATI, Gif-sur-Yvette, France

Dimitra Mangoura Basic Research Center, Biomedical Research Foundation of the Academy of Athens, Athens, Greece

Patrick Manning Section of Endocrinology, Department of Internal Medicine, Dunedin School of Medicine, University of Otago, Dunedin, New Zealand

Jahan P. Marcu Americans for Safe Access, Washington, DC, USA; Green Standard Diagnostics, Inc., Las Vegas, NV, USA

Simon McCabe Behavioural Science Centre, University of Stirling, Scotland, UK

M. Adrienne McGinn Department of Physiology, Alcohol and Drug Abuse Center of Excellence, LSU Health Sciences Center, New Orleans, LA, USA

Diana L. McKinney RD&E, Altria Client Services, Richmond, VA, USA

Ryan McLaughlin Neurobiological Psychiatry Unit, McGill University, Montreal, QC, Canada; Department of Integrative Physiology and Neuroscience, College of Veterinary Medicine, Washington State University, Pullman, WA, USA

Ian McLaughlin Department of Psychiatry, University of Pennsylvania, Philadelphia, PA, USA; Neuroscience Graduate Group, Perelman School of Medicine, University of Pennsylvania, Philadelphia, PA, USA

Ernesta Meintjes MRC/UCT Medical Imaging Research Unit, Faculty of Health Sciences, University of Cape Town, Cape Town, South Africa

Narasimha M. Midde Pharmaceutical Sciences, College of Pharmacy, University of Tennessee Health Science Center, Memphis, TN, USA

Stefan Mihailescu Departamento de Fisiología, Facultad de Medicina, Universidad Nacional Autónoma de México, México DF, México

José Miñarro Unidad de Investigación Psicobiología de las Drogodependencias, Departamento de Psicobiología, Facultad de Psicología, Universitat de València, Valencia, Spain

Raffaella Mondola Department of Mental Health, ASL CN1, Saluzzo, Italy

A. Leslie Morrow Department of Psychiatry & Pharmacology, Bowles Center for Alcohol Studies, University of North Carolina, Chapel Hill, NC, USA

Nina R. Mota Department of Genetics, Instituto de Biociências, Universidade Federal do Rio Grande do Sul (UFRGS), Porto Alegre, RS, Brazil

Jeffrey Newcorn Department of Psychiatry, Icahn School of Medicine at Mount Sinai, New York, NY, USA

Daisuke Nishizawa Addictive Substance Project, Tokyo Metropolitan Institute of Medical Science, Tokyo, Japan

José N. Nobrega Department of Pharmacology and Toxicology, University of Toronto, Toronto, ON, Canada; Campbell Family Mental Health Research Institute, Behavioural Neurobiology Laboratory, Research Imaging Centre, Centre for Addiction and Mental Health, Toronto, ON, Canada; Department of Psychiatry, University of Toronto, Toronto, ON, Canada; Department of Psychology, University of Toronto, Toronto, ON, Canada

Christina N. Nona Department of Pharmacology and Toxicology, University of Toronto, Toronto, ON, Canada; Campbell Family Mental Health Research Institute, Behavioural Neurobiology Laboratory, Research Imaging Centre, Centre for Addiction and Mental Health, Toronto, ON, Canada

Nicolas Nuñez Neurobiological Psychiatry Unit, McGill University, Montreal, QC, Canada; Hospital Neurospsiquiatrico de Agudos y Cronicos Dr. Alejandro Korn/Universidad, Nacional de La Plata, Buenos Aires, Argentina

Ingrid Nylander Neuropharmacology, Addiction and Behaviour, Department of Pharmaceutical Biosciences, Uppsala University, Uppsala, Sweden

Jessica W. O'Brien Department of Psychiatry, University of Pittsburgh School of Medicine, Pittsburgh, PA, USA; Department of Psychology, University of Pittsburgh, Pittsburgh, PA, USA

Francesco Oliva Department of Clinical and Biological Sciences, San Luigi Gonzaga Medical School, University of Turin, Torino, Italy

Sara Palm Neuropharmacology, Addiction and Behaviour, Department of Pharmaceutical Biosciences, Uppsala University, Uppsala, Sweden

Mauricio R. Papini Department of Psychology, Texas Christian University, Fort Worth, TX, USA

Byung Lae Park Department of Genetic Epidemiology, SNP Genetics, Inc., Seoul, Republic of Korea

María Pascual Department of Cellular Pathology, Príncipe Felipe Research Centre, Valencia, Spain

Juan Pedraza Department of Psychiatry, Icahn School of Medicine at Mount Sinai, New York, NY, USA

Giampaolo Perna Hermanas Hospitalarias, Villa San Benedetto Menni Hospital, FoRiPsi, Department of Clinical Neurosciences, Albese con Cassano, Italy

Rocco Luigi Picci Department of Clinical and Biological Sciences, San Luigi Gonzaga Medical School, University of Turin, Torino, Italy

Marcin Piechota Department of Molecular Neuropharmacology, Institute of Pharmacology, Polish Academy of Sciences, Krakow, Poland

Vladimir Pilija Institute of Forensic Medicine, Clinical Center Vojvodina, Novi Sad, Serbia

Charles Pinto Department of Psychiatry, Jagjivanram Western Railways Hospital, Mumbai, Maharashtra, India

Stéphane Potvin Department of Psychiatry, University of Montreal, Montreal, QC, Canada; Centre de recherche de l'Institut universitaire en santé mentale de Montréal, Montreal, QC, Canada

Jagdeo Prasad Rawat Department of Psychiatry, Jagjivanram Western Railways Hospital, Mumbai, Maharashtra, India

Victor R. Preedy Diabetes and Nutritional Sciences Research Division, Faculty of Life Science and Medicine, King's College London, London, UK

Brittany M. Priddy NIH/NIDA, IRP/INRB, Baltimore, MD, USA

Ryszard Przewlocki Department of Molecular Neuropharmacology, Institute of Pharmacology, Polish Academy of Sciences, Krakow, Poland; Department of Neurobiology and Neuropsychology, Institute of Applied Psychology, Jagiellonian University, Krakow, Poland

Jimit G. Raghav Center for Drug Discovery (CDD), Northeastern University, Boston, MA, USA; Department of Pharmaceutical Sciences, Northeastern University, Boston, MA, USA

Rajkumar Rajendram Department of Anaesthesia and Intensive Care, College of Medicine, King Khalid University Hospital, King Saud University Medical City, Riyadh, Saudi Arabia; Diabetes and Nutritional Sciences Research Division, Faculty of Life Science and Medicine, King's College London, London, UK

Roshanna Rajendram Department of Emergency Medicine, The Royal Woolverhampton NHS Trust, New Cross Hospital, Wolverhampton, UK

P.S.S. Rao Pharmaceutical Sciences, College of Pharmacy, University of Tennessee Health Science Center, Memphis, TN, USA

Anirban Ray Department of Child and Adolescent Psychiatry, National Institute of Mental Health and Neurosciences, Bangalore, India; Department of Psychiatry, Calcutta National Medical College, Calcutta, West Bengal, India

Ameneh Rezayof Department of Animal Biology, School of Biology, College of Science, University of Tehran, Tehran, Iran

Casto Rivadulla NEUROcom, Depto de Medicina, Univ da Coruña, Campus de Oza, A Coruña, Spain; The Institute for Biomedical Research of Coruña (INIBIC), A Coruña, Spain

Marta Rodríguez-Arias Unidad de Investigación Psicobiología de las Drogodependencias, Departamento de Psicobiología, Facultad de Psicología, Universitat de València, Valencia, Spain

Luciana Grazziotin Rossato Faculdade de Farmácia, Instituto de Ciências Biológicas, Universidade de Passo Fundo, Passo Fundo (RS), Brazil

Diego L. Rovaris Department of Genetics, Instituto de Biociências, Universidade Federal do Rio Grande do Sul (UFRGS), Porto Alegre, RS, Brazil

Christina L. Ruby Department of Biology, Indiana University of Pennsylvania, Indiana, PA, USA

Nikos Sakellaridis Pharmacology Department, Medical School, University of Thessaly, Larissa, Greece

Saad Salman Department of Psychiatry and Drug Detoxification Centre, Lady Reading Hospital, Post Graduate Medical Institute, Peshawar, Khyber Pakhtunkhwa, Pakistan

Francisco Santolaria-Fernández Servicio de Medicina Interna, Hospital Universitario de Canarias, Universidad de La Laguna, La Laguna, Tenerife, Canary Islands, Spain

Siddharth Sarkar Department of Psychiatry, Jawaharlal Institute of Postgraduate Medical Education and Research, Pondicherry, India

Jason B. Schechter Cortical Systematics LLC, Tucson, AZ, USA

André Schmidt Department of Psychiatry (UPK), University of Basel, Basel, Switzerland

Susan Scholl Department of Public Health, Syracuse University, Syracuse, NY, USA

April Scott University of Kentucky, Lexington, KY, USA

Masroor Shariff Addiction Neuroscience and Obesity Group, Institute of Health and Biomedical Innovation, School of Clinical Sciences, Faculty of Health, Queensland University of Technology, Translational Research Institute, Woolloongabba, QLD, Australia

Shiwali Sharma Chitkara College of Pharmacy, Chitkara University, Chandigarh, Punjab, India

Hyoung Doo Shin Department of Life Science, Sogang University, Seoul, Republic of Korea

Thakur Gurjeet Singh Chitkara College of Pharmacy, Chitkara University, Chandigarh, Punjab, India

Namita Sinha Pharmaceutical Sciences, College of Pharmacy, University of Tennessee Health Science Center, Memphis, TN, USA

Patrick D. Skosnik Psychiatry Service, VA Connecticut Healthcare System, West Haven, CT, USA; Abraham Ribicoff Research Facilities, Connecticut Mental Health Center, New Haven, CT, USA; Department of Psychiatry, Yale University School of Medicine, New Haven, CT, USA

Ranilson de Souza Bezerra Departamento de Bioquímica, Universidade Federal de Pernambuco, Recife, Pernambuco, Brazil

Dustin J. Stairs Department of Psychology, Creighton University, Omaha, NE, USA

Gjoshe Stefkov Faculty of Pharmacy, Institute of Pharmacognosy, University Ss. Cyril and Methodius, Skopje, Republic of Macedonia

Dan J. Stein Department of Psychiatry and Mental Health, Groote Schuur Hospital, MRC Unit on Anxiety and Stress Disorders, University of Cape Town, Cape Town, South Africa

Cristina A.J. Stern Department of Pharmacology, Federal University of Santa Catarina, Florianópolis, Santa Catarina, Brazil

Claudia Stöllberger Medical Department, Krankenanstalt Rudolfstiftung, Vienna, Austria

Reinaldo N. Takahashi Department of Pharmacology, Federal University of Santa Catarina, Florianópolis, Santa Catarina, Brazil

H.C. Tang Schools of Biomedical Sciences and Chinese Medicine, The Chinese University of Hong Kong, Hong Kong, China

Josephine Tarren Addiction Neuroscience and Obesity Group, Institute of Health and Biomedical Innovation, School of Clinical Sciences, Faculty of Health, Queensland University of Technology, Translational Research Institute, Woolloongabba, QLD, Australia

Philip G. Tibbo Department of Psychiatry, Dalhousie University, Halifax, NS, Canada

Vinod Tiwari Department of Anesthesiology and Critical Care Medicine, Johns Hopkins University, School of Medicine, Baltimore, MD, USA

Carmen Torres Department of Psychology, University of Jaén, Jaén, Spain

Federica Trivelli Department of Clinical and Biological Sciences, San Luigi Gonzaga Medical School, University of Turin, Torino, Italy

Penelope Truman Institute of Environmental Science and Research Ltd, Porirua, New Zealand; The School of Public Health, Massey University, Wellington, New Zealand

Emmanouella Tsirimonaki Basic Research Center, Biomedical Research Foundation of the Academy of Athens, Athens, Greece

Sven Vanneste School of Behavioral and Brain Sciences, The University of Texas at Dallas, TX, USA

Andrea R. Vansickel RD&E, Altria Client Services, Richmond, VA, USA

Leandro F. Vendruscolo NIH/NIDA, IRP/INRB, Baltimore, MD, USA

Maria S.M. Wai Schools of Biomedical Sciences and Chinese Medicine, The Chinese University of Hong Kong, Hong Kong, China

Marc Walter Department of Psychiatry (UPK), University of Basel, Basel, Switzerland

Karen A. Waters Department of Medicine, University of Sydney, Sydney, NSW, Australia; The BOSCH Institute, University of Sydney, Sydney, NSW, Australia; The Children's Hospital, Westmead, Sydney, NSW, Australia

Jeffrey L. Weiner Department of Physiology and Pharmacology, Wake Forest University School of Medicine, Winston–Salem, NC, USA

Jodi J. Weinstein Department of Psychiatry, New York State Psychiatric Institute and Columbia University Medical Center, New York, NY, USA

Jordyn Weisberg Syracuse University, Chicago, IL, USA

Annie M. Whitaker Department of Physiology, Alcohol and Drug Abuse Center of Excellence, LSU Health Sciences Center, New Orleans, LA, USA

Sarah Wolf-Stanton Department of Marriage and Family Therapy, Syracuse University, Syracuse, NY, USA

Chen-Ying Wu Department of Psychiatry, Maimonides Medical Center, Brooklyn, NY, USA

Sharon L. Wu Schools of Biomedical Sciences and Chinese Medicine, The Chinese University of Hong Kong, Hong Kong, China

Cheryl C.H. Yang Institute of Brain Science, National Yang-Ming University, Taipei, Taiwan; Sleep Research Center, National Yang-Ming University, Taipei, Taiwan

Yen Kuang Yang Department of Psychiatry, National Cheng Kung University Hospital, College of Medicine, National Cheng Kung University, Tainan, Taiwan

Kirti Yeshwant Tandel Department of Psychiatry, Jagjivanram Western Railways Hospital, Mumbai, Maharashtra, India

D.T. Yew Schools of Biomedical Sciences and Chinese Medicine, The Chinese University of Hong Kong, Hong Kong, China

Yashar Yousefzadeh Fard Department of Psychiatry, New York State Psychiatry Institute/Columbia University, New York, NY, USA

Yan Zhou The Laboratory of the Biology of Addictive Diseases, The Rockefeller University, New York, NY, USA

Preface

The wellbeing of the individual is highly dependent on maintaining neurophysiological processes in a functional state but also having the ability to adapt to changes in the internal and external milieus. However, adaptive changes may be pathological in some circumstances, with devastating consequences for the individual. Triggers for these neurological abnormalities are varied and may be due to life-stages (e.g., ageing), nutrition (e.g., nutrient deficiency or excess such as iodine and iron, respectively), trauma (e.g., metabolic or physical trauma, such as those due to hypoglycaemia or blunt-instruments) or drugs of addiction and substance misuse (e.g., nicotine, alcohol, caffeine, inhalants and a whole myriad of others). The latter are common and preventable to some extent. For example, in the USA alone there are an estimated 22 million illegal drug users. Sixty million use tobacco and 50 million USA citizens misuse alcohol. Millions are also addicted to, or misuse, caffeine and prescription or over-the counter medications.

As a consequence of addictions and substance misuse, adverse changes occur in affected tissues. These range from molecular and cellular perturbations to structural and functional abnormalities. It is possible that some of the science behind these changes may be applicable to other modes of neurophysiological imbalance. That is, lessons and features in one form of addiction and substance misuse may be transferable to another. Indeed, there are other forms of non-substance addictions such as gambling, gaming, and workaholism which may share common features, mechanisms or outcomes. Understanding commonality provides a platform for studying specific addictions in more depth and allows one to speculate on new modes of understanding, causation, prevention and treatment.

There is some difficulty in describing changes in human tissues since this sort of information is rather limited in scope and analytical depth. Preclinical or non-clinical studies have advanced the detailed understanding of addictions and substance misuse considerably. These range from isolated structures, cells and perfusions to invertebrates, rodents and primates. It is thus essential to have both clinical and preclinical information within the same authoritative textual platform to advance our understanding of addictions and substance misuse.

Understanding neuropathology by itself can be somewhat problematic especially in terms of addictions. This information needs to be placed within its wider context: from procurement of drugs to altered behavior and psychosocial conditions. For some substances there is very little molecular information, whilst for other drugs there is an abundance. The information on behavioral and psychosocial aspects is similarly diverging amongst the different addictions. Thus, any textual information on addictions and substance misuse use requires a scientific continuum of information; with neurological features as a central core.

However, marshalling all the aforementioned information is somewhat difficult due to the wide array of material. To address this, the Editor has compiled *The Neuropathology of Drug Addictions and Substance Misuse*. It has 3 separate Volumes:

Volume 1: Foundations of Understanding, Tobacco, Alcohol, Cannabinoids and Opioids
Volume 2: Stimulants, Club and Dissociative Drugs, Hallucinogens, Steroids, Inhalants and International Aspects
Volume 3: General Processes and Mechanisms, Prescription Medications, Caffeine and Areca, Polydrug Misuse, Emerging Addictions and Non-Drug Addictions

In compiling these volumes we interspersed chapters to aid the holistic understanding of addictions and substance misuse. We have material not only on specific substances but also major sections on the following:

Foundations for Understanding Substance Misuse and their Effects
Emerging Addictions and Drugs of Abuse
International Aspects
Principles of Addictions, Overviews, Detailed Processes and Mechanisms
Dual and Polydrug Abuse
Nondrug Addictions as Comparative Neuropathology

For Volume 1, the main Parts are:

1—[1] Setting the Scene: Foundations for Understanding Substance Misuse and Their Effects
1—[2] Tobacco

1—[3] Alcohol
1—[4] Cannabinoids
1—[5] Opioids

For Volume 2, the main Parts are:

2—[1] Stimulants
2—[2] Club Drugs
2—[3] Dissociative Drugs
2—[4] Hallucinogens
2—[5] Anabolic Steroids, Inhalants and Solvent
2—[6] International Aspects

For Volume 3, the main Parts are:

3—[1] General Aspects: Principles of Addictions, Overviews, Detailed Processes and Mechanisms
3—[2] Prescription Medications
3—[3] Caffeine and Areca (Betal nut)
3—[4] Dual and Polydrug Abuse
3—[5] Emerging Addictions and Drugs of Abuse
3—[6] Nondrug Addictions as Comparative Neuropathology

Each Part is split into different subsections:

General Aspects
Molecular and Cellular Aspects
Structural and Functional Aspects
Methods

It is tempting to focus exclusively on detection, prevention and treatment. However, this will far extend the remit of the book. For example, the analysis of markers in alcoholism itself would merit a single book as would public health prevention or treatment regimes. Instead the book is focused on neuropathology with upstream and downstream causative scenarios, effects and consequences. In the section **General Aspects** basic information is provided to place the substance in context or set the scientific scene. The section **Molecular and Cellular Aspects** provides greater detail. The section **Structural and Functional Aspects** is more broad-based and includes the impact on imaging, psychosocial and behavioral aspects and other wider information. The section **Methods** contains selective techniques for screening and/or analysis. Of course these are generalized divisions and this is recognized by the Editor. Some articles in one section may also be well suited for many other sections. Indeed in a few cases we have located chapters within sections to either complement other chapters, impart a broader example of ideas, coverage or concepts, provide a more in-depth discourse that may be relevant for other drugs and their interactions or provide a greater understanding of substance and polysubstance misuse in general. However, the well-structured and professional index, provided by Elsevier addresses issues in locating information and so relevant material can be quickly located.

Each chapter has the following subheadings:

Applications to Other Addictions and Substance Misuse
Defintion of Terms
Key Facts
Summary Points

These subheadings encompass unique features in the book, which bridge the intellectual divide, so experts in one addiction area may become more knowledgeable in another area. These features are very useful for the novice, student or newly qualified health care professional. Those who wish to gain a wider understanding of addictions and substance misuse will also find these features of benefit.

The subheadings on **Application to Other Addictions and Substance Misuse** is intended to provide either practical, speculative or more broader information. This is particularly useful when applied to those addictions where there is a paucity of scientific material. For example, detailed molecular or functional information gathered from studying one addiction may be applicable to another.

Contributors are authors of either international and national experts, from respected institutions, leaders in the field or trendsetters. Emerging fields of addictions and substance misuse are also incorporated in *Neuropathology of Drug Addictions and Substance Misuse*. This book is essential reading for addiction scientists, health care professionals, research scientists, molecular and cellular biochemists, the medical professions, physicians and other practitioners, as well as those interested in health in general. It is also designed for teachers and lecturers, undergraduates, graduates, post-graduates, professors and libraries.

Victor R. Preedy, Editor

Acknowledgments

The Editor is extremely grateful for the patience, advice, and help of the following, without whom this project would not have been possible:
- April Farr
- Kathy Padilla
- Mica Haley

The entire process from the approval of the original proposal to the submission of manuscripts, copyediting, typesetting, design, and printing is a very long journey. The Editor also wishes to acknowledge the support and help of all those who have made this project possible.

Part I

Setting the Scene: Foundations for Understanding Substance Misuse and Their Effects

Chapter 1

The Nervous System and Addictions: Essentials for Clinicians

Amelia J. Anderson-Mooney[1], Jonathan N. Dodd[2], April Scott[3], Leila Guller[4]

[1]Department of Neurology, University of Kentucky College of Medicine, Lexington, KY, USA; [2]Department of Neurology, Washington University School of Medicine, St. Louis, MO, USA; [3]University of Kentucky, Lexington, KY, USA; [4]Department of Psychology, University of Kentucky, Lexington, KY, USA

Abbreviations

ANS Autonomic nervous system
CG Cingulate gyrus
CN Caudate nucleus
CNS Central nervous system
DA Dopamine
GP Globus pallidus
NA Nucleus accumbens
PNS Peripheral nervous system
PSNS Parasympathetic nervous system
SNS Sympathetic nervous system
VS Ventral striatum
VTA Ventral tegmental area

INTRODUCTION

The study of addictive behavior is a complex pursuit, because it requires understanding of the varied biological, cognitive, affective, and psychosocial factors that contribute to addiction. Neurobiology, however, may be the best foundation on which to build that understanding. In their usual state, neurological operations represent an exacting interplay of interdependent systems cooperating with one another to produce behaviors, thoughts, and emotional states. Addictive behavior alters some of these interactions in critical ways that propagate the drive to seek specific, desirable stimuli, including substances and powerfully gratifying behaviors, such as sexual activity and gambling.

The cortical/mesolimbic system is central to our understanding of addictive behavior. However, addictive behavior can influence all elements of the nervous system either directly or indirectly, so it is helpful to examine mesolimbic operations as a component of the neurological system as a whole. The intricacy of that system can be understandably intimidating, especially when considering that science has not yet fully defined how it works! Despite this, some fundamental concepts exist that can help form a firm foundation for examining addictive behavior.

A NEUROANATOMICAL COMPASS

For some, the most challenging aspect of learning neuroanatomy is conceptualizing the nervous system in three-dimensional space. Diagrams and pictures can certainly be helpful and are included in this chapter to assist you. However, no two-dimensional diagram can represent the relationships among all important structures at once. Thus, neuroanatomists use a system of terms designed to guide you within the brain, much like directions from a global positioning system. These terms may seem opaque on the first review, but they will become quite helpful the more you become comfortable with them.

Some of these terms operate like the cardinal geographical directions: north, south, east, and west. For instance, the frontal lobes are *anterior* or *rostral* to the occipital lobes, which sit *posterior* or *caudal* to the parietal lobes. The *ventral* surface of the brain is toward the underside, and the *dorsal* surface is at the top, like the dorsal fin of a dolphin. The *mesial* temporal lobes are tucked inside the outer surface of the brain, closer to the middle. *Proximal* structures are close to the midline, and *distal* structures are located farther away, or "at a distance." These terms can also be combined to more finely tune directions, like referring to the northeast or southwest. For example, the ventromedial prefrontal cortex directs you to the underside (ventral) of the prefrontal region, toward the middle (medial). Figure 1 provides a general guide to important structures referred to throughout this chapter; visiting these images may be helpful as you orient to these neuroanatomical directions.

Adding complexity, some terms change meaning based on the location within the nervous system. This is based on the process of neural development: As the neural tube (the basic building block of the brain and spinal cord) develops in utero, a bend of about 90° forms at the point that will become the midbrain (Blumenfeld, 2002; Nolte, 2002). Thus, directionality changes at the midbrain for some terms. Recalling the Latin roots for those terms can help provide some consistency; for instance, if you remember that "caudal" is derived from the Latin word for "tail," it becomes clear that the direction refers to posterior regions above the midbrain and inferior regions below the midbrain. Table 1 provides a quick reference for these and other important

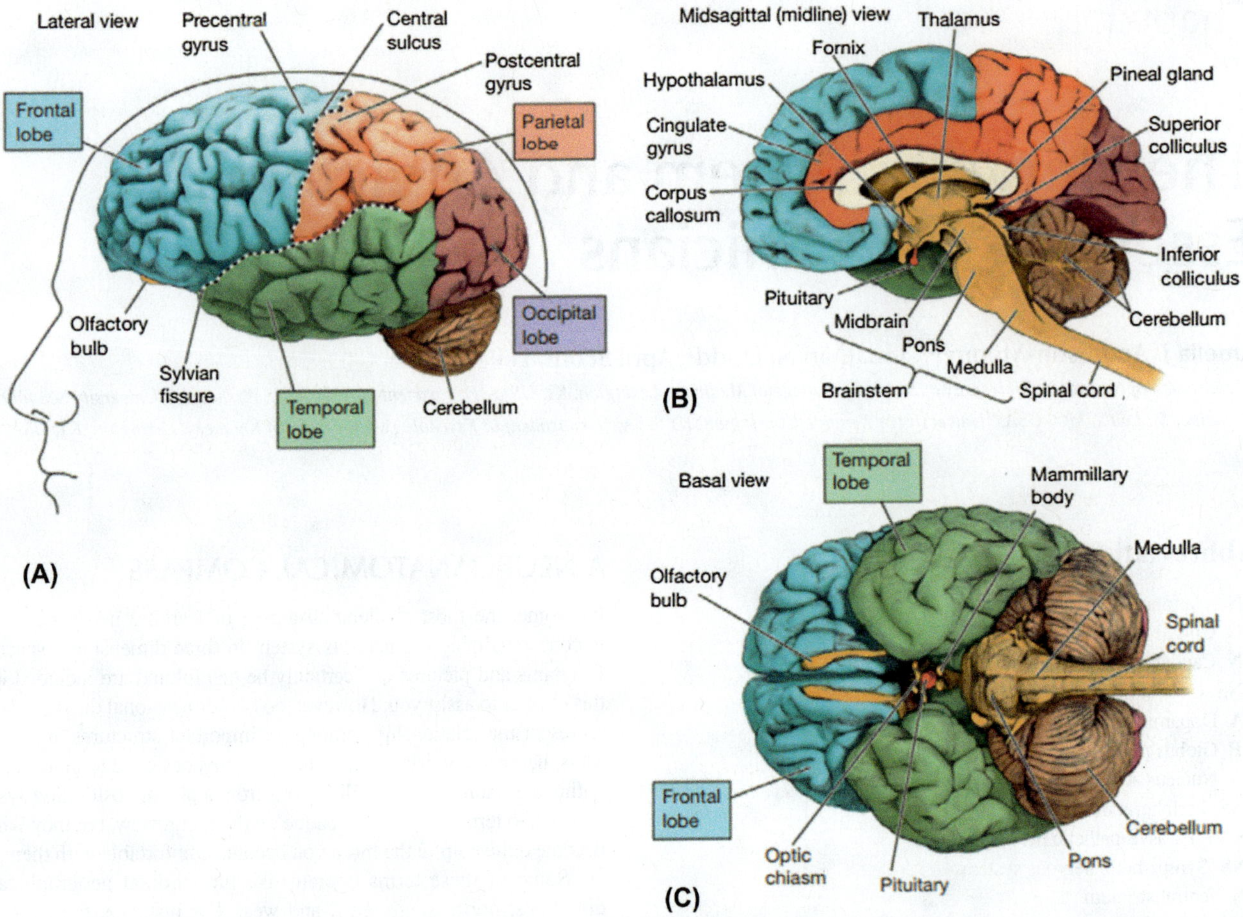

FIGURE 1 **Important cortical and subcortical structures.** (A) Lateral (*side*), (B) midsagittal (*longitudinal midline*), and (C) ventral (*underside*) views of the human brain. Major subdivisions of the cerebral cortex are color coded, and important structures are labeled. *Reprinted from Mair (2012). ©2012, with generous permission from Elsevier.*

TABLE 1 Neuroanatomical Directional Terms			
Term	Translation	Above the Midbrain	Below the Midbrain
Superior		Toward the top	Same
Inferior		Toward the bottom	Same
Anterior		Toward the front	Same
Posterior		Toward the back	Same
Proximal		Close to the midline	Same
Distal		Away from the midline	Same
Mesial/Medial		Toward the middle	Same
Afferent		Arriving, toward a structure	Same
Efferent		Exiting, away from a structure	Same
Ventral	Toward the belly	Toward the underside	Toward the front
Dorsal	Toward the back	Toward the top	Toward the back
Rostral	Toward the beak	Toward the front	Higher/superior
Caudal	Toward the tail	Toward the back	Lower/inferior
Adapted from information provided in Blumenfeld (2002) and Nolte (2002).			

terms. Terms referring to areas above the midbrain are used in this chapter more often, although appropriate terms for structures below the midbrain are provided in Table 1 as well.

There are several excellent online resources for visualizing these three-dimensional relationships, such as Digital Anatomist from the University of Washington's Structural Informatics Group and BrainVoyager by Rainer Goebel. Combining texts like this one with three-dimensional models can be invaluable as you familiarize yourself with neuroanatomical structures.

GENERAL STRUCTURE OF THE NEUROANATOMICAL SYSTEM

In basic, the neuroanatomical system can be divided into the central nervous system (CNS) and the peripheral nervous system (PNS). The CNS is composed of the brain and the spinal cord, and the PNS detects and carries vital information between the CNS and the rest of the body (the periphery). *Afferent* fibers carry information into the CNS, and *efferent* fibers carry information out of the CNS. In other words, afferent fibers "arrive" into the brain and spinal cord with the body's data, and efferent fibers "exit" the brain and spinal cord, returning processed data back to the rest of the body (Blumenfeld, 2002). This division of labor allows seamless, rapid communication between the body and the environment.

Additional classifications provide additional clarity toward how the body interacts with its environment. For instance, the autonomic nervous system (ANS) is further divided into sympathetic (SNS) and parasympathetic nervous systems (PSNS). The SNS functions as an emergency response system, using brain structures to activate bodily responses against environmental threats. Simply put, the SNS uses CNS centers to signal peripheral structures to facilitate either simple reflexive or voluntary, coordinated motor responses to threats via the PNS (Kandel, Schwartz, & Jessell, 2000).

No matter the system, however, the pivotal structure is the key processing center: the brain.

Structure of the Brain

The brain is a highly integrated organ that mediates not only basic bodily functions, but also higher-level decision-making, needs, and desires. In this, the brain also moderates the drives and habits involved in addictive behavior. The brain's varied structures can be grossly divided into four main groups:

1. The processing centers (gray matter),
2. Communicating fibers supporting those processing centers (white matter),
3. Combined structures composed of both gray and white matter, and
4. Cerebrospinal fluid-filled ventricles.

A dense vascular system provides steady, large supplies of oxygen, glucose, and other nutrients vital to maintaining neural function. A basic review of the brain's structures can be found in Table 2.

TABLE 2 Major Structures of the Brain

		Visible Surface	Medial/Interior Areas
Processing centers (groupings of cell bodies)	Cerebral cortex	Frontal lobes Temporal lobes Parietal lobes Occipital lobes	Insular cortex Cingulate gyrus Mesial temporal lobes
	Subcortical structures		Thalamus Hypothalamus Epithalamus Pituitary Caudate Putamen Globus pallidus Amygdala Hippocampus
Communicating fibers (groupings of axons)			Corona radiata Internal capsule Corpus callosum Anterior commissure Fornix
Combined structures (cell bodies + axons)		Pons Cerebellum	Brainstem Midbrain
Ventricles (spaces filled with cerebrospinal fluid)			Lateral ventricles Third ventricle Fourth ventricle Cisterns

Gray Matter in the Brain: Data Processing Centers

Neural data processing is conducted by groupings of cell bodies forming the cerebral cortex and subcortical nuclei. The cerebral cortex is the thin layer of cell bodies covering the surface of the brain. This thin layer, measured in millimeters, naturally appears darker than its supportive tissue, leading to the term "gray matter" (Blumenfeld, 2002; Kolb & Whishaw, 2003). The cortex is divided into two hemispheres by the longitudinal fissure, with four lobes reflected bilaterally. Again, refer to Figure 1 to review key structures. Visual cortices are housed in the occipital lobes, somatosensory cortices in the parietal lobes, auditory cortices in the temporal lobes, and motor cortex in the frontal lobes. The anterior frontal lobes, the *prefrontal cortex*, are fundamental to higher-level cognition and behavior, like problem-solving, attention, decision-making, and impulse control (Nolte, 2002).

The vast majority of the brain, however, is hidden from surface view. The insula, sometimes called the "fifth lobe," is an area of cortex underneath the temporal and posterior frontal lobes. It processes information about gustation, executive functioning, interoception, and other varied functions, the true extent of which has not yet been identified (Blumenfeld, 2002; Verdejo-Garcia, Clark, & Dunn, 2012). Similarly, the mesial temporal lobes are tucked medially inside the lateral surface of the temporal lobes, including the entorhinal, perirhinal, and parahippocampal cortices, in addition to the hippocampus itself. Primary roles for learning and memory have been identified for these areas (Kolb & Whishaw, 2003).

Proceeding medially from the mesial temporal lobes, one encounters the brain's "deep nuclei," collections of cell bodies often termed the *subcortical* structures. These structures include the cingulate gyrus (CG), thalamus, caudate nucleus (CN), amygdala, putamen, and globus pallidus (GP), among others. Some of these structures are strikingly C-shaped, including the CG and CN, located over and underneath the corpus callosum. The deep nuclei play varied roles, some operating in more than one system simultaneously. For instance, the thalamus sits atop the brainstem and midbrain, operating as Grand Central Station for the vast majority of sensory information traveling in the nervous system (Blumenfeld, 2002).

The CN is one of the C-shaped nuclei, with important roles in voluntary movement, language, and behavioral inhibition (Kolb & Whishaw, 2003). The CG is another C-shaped structure layered atop the corpus callosum with important roles in attentional switching, empathy, and decision-making (Nolte, 2002). The amygdala is almond-shaped and located at the tail of the CN, implicated in attaching emotional importance to stimuli. Together, the CN and the putamen are called the *striatum*, because of the striated tissue connecting them laterally. The putamen and the GP, together, are often called the *lentiform nuclei* because they resemble a lentil in coronal sections (Blumenfeld, 2002).

Because of their anatomical and functional relationships, the CG, amygdala, hippocampus and associated mesial temporal cortices, prefrontal cortex (orbitofrontal, specifically), and other associated structures are often termed the *limbic system* (Kolb & Whishaw, 2003). The exact combination of structures incorporated into the limbic system varies depending on the reference source, but the operations are generally agreed to include mediating emotion, reward, and punishment, foreshadowing critical importance in addiction (Blumenfeld, 2002).

White Matter in the Brain: Communication Pathways

Gray matter processing centers communicate with one another and the periphery via collections of axons. Axons are the "tails" of neuronal cell bodies, covered in lipid-rich insulation called *myelin*. High concentrations of fatty compounds in myelin render axons naturally lighter in color than gray matter, leading to the term "white matter" (Kolb & Whishaw, 2003).

We first encounter white matter in the brain after moving through the thin cerebral cortex, moving ventromedially (or subcortically) underneath the surface. Groupings of U-shaped *arcuate* fibers are found here that transfer information between areas of cortex located near one another. Longer-range transmissions travel in *fasciculi*, or *tracts*, and communication between hemispheres is accomplished with dense white matter collections called *commissures* (Blumenfeld, 2002; Schuenke, Schulte, & Schumacher, 2010). These different pathways facilitate progressive integration of data from their primary elements into larger ideas. When those elements have been successfully integrated, some of the longest white matter pathways of all carry vital information back to the body to promote necessary reactions. Although an exhaustive discussion of all white matter pathways is beyond the scope of this chapter (interested readers are referred to Schmahmann & Pandya, 2006 text for more in-depth information), some bear mention.

Moving medially from the cortical surface (or subcortically), we find an impressive fan-like structure called the *corona radiata* (Nolte, 2002), dense axons descending from the cell bodies of the motor cortex. A large band of white matter also forms the *corpus callosum*, a commissure facilitating communication between the two hemispheres of the brain (Kolb & Whishaw, 2003).

Combined Structures and Ventricles

Structures such as the brainstem (including the medulla, pons, and midbrain) and cerebellum are composed of both gray matter and white matter (Blumenfeld, 2002). Many fundamental and complex functions converge in these areas. For instance, the brainstem contains nuclei regulating basic consciousness, breathing, heart rate, and other life-sustaining functions. Injury to these nuclei can easily result in death. The pons is a protuberance on the brainstem's ventral surface also participating in life-sustaining processes, as well as facilitating communication about movement from the cortex to the cerebellum. The cerebellum is situated at the brain's most posterior aspect, involved in coordinating movement as well as some higher-order functions (Kolb & Whishaw, 2003). The midbrain rests atop the brainstem, underneath the thalamus. It has a "mouse ear"-like shape in some cross sections, making it easy to identify. Many complex functions are housed in or travel through the midbrain, including motor control, vision, and hearing (Nolte, 2002).

Also found subcortically is the ventricular system. This system actually represents open spaces filled with cerebrospinal fluid generated by the choroid plexus resting inside. The ventricular system and cerebrospinal fluid are absolutely vital to basic neural maintenance, as the cerebrospinal fluid has been implicated in holding the brain's shape under its own weight, cushioning it against blows to the head, and washing away toxins (Blumenfeld, 2002).

Cellular and Vascular Basis of the Neuroanatomical System

At the cellular level, the brain is composed of multiple cell types. Neurons themselves are the main components of gray matter. Neuronal cell bodies, or *soma*, do the majority of processing, communicating with other cells through projections called *axons* (sending information) and *dendrites* (receiving information). There are many types of neurons subdivided by specific roles in the nervous system, with different structures to complement their dedicated functions. For instance, pyramidal neurons are widely found throughout the nervous system, with multiple roles including supporting motor function in the corticospinal tract and supporting learning and memory in the hippocampus. Although these two functions are both served by pyramidal cells, the cells in those regions have slightly different structures tailored to meet the specific demands made there (Nolte, 2002).

Neurons are accompanied by support cells called *glia* equipped to meet the brain's background operational needs (Kandel et al., 2000). Just as neurons may look and behave differently to achieve varied roles, glia also vary in structure and function. We have already been introduced to one type of glial cell: The myelin sheath insulating axons is composed of glial bodies called oligodendrocytes (in the CNS) or Schwann cells (in the PNS). Refer to Table 3 to learn about some different types of neurons and glia, including examples of locations and functions in the nervous system. Please note that only some types are described; this table does not represent all possible cell types!

As stated, cellular operations are supported by a fantastically dense vascular system composed of arteries delivering freshly oxygenated blood, veins removing deoxygenated blood, and capillaries transferring blood from the arterial system to the venous system. The vessels themselves are a vital part of the *blood–brain barrier* (Blumenfeld, 2002). Elsewhere in the body, endothelial cells building blood vessel walls maintain small gaps between them to allow easy passage of nutrients and other materials. In the brain, endothelial cells in capillaries maintain no such gap, blocking transport except for compounds that are exceptionally small, have special transporters in cell walls, and/or are both water and fat soluble (Kolb & Whishaw, 2003). This barrier prevents many toxic substances from ever reaching the brain.

PSYCHOACTIVE SUBSTANCES AND THE BRAIN

Psychoactive substances, however, thwart this essential defense system. Although some are intended to enhance health, like antidepressants and antiseizure medications, misuse of psychoactive substances can result in harm throughout the neurological system, including neurons themselves, glial support cells, and the blood vessels. For example, substances can damage hippocampal neurons that support learning and memory (Chambers, 2013). Alcohol misuse can sometimes result in Marchiafava-Bignami disease, a disorder that attacks myelin in the corpus callosum (Larner, 2008). Use of stimulants like cocaine, methamphetamines, and ketamine can result in severe vasodilation or vasoconstriction, leading to hemorrhagic or ischemic strokes (Wang, Zheng, Xu, Lam, & Yew, 2013).

Functional Circuitry of Addiction: The Mesocorticolimbic System and Dopamine

Some brain areas are especially vulnerable to psychoactive substances, particularly the limbic (or mesolimbic) system mentioned above. This subcortical system is composed of structures you now know, including the hippocampus, amygdala, CN, and CG. One key mesolimbic region not yet introduced is the *nucleus accumbens* (NA). The NA is a difficult area to highlight, as it is not anatomically separate from other structures. Instead, it represents the area where the putamen fuses with the head of the caudate anteriorly. Thus, the area that includes the NA is referred to as the *ventral striatum* (VS). It can be located in the area labeled VS in Figure 2. The VS, and the NA in particular, is central to the reciprocal neurocircuitry involved in emotions, drives, and the progression of addictive behavior.

Dopamine

This neurocircuitry communicates largely by the propagation of a neurotransmitter called dopamine (DA) from structure to structure. DA is a monoaminergic neurotransmitter produced in an area of the midbrain called the substantia nigra with rich representations in the nearby ventral tegmental area (VTA) of the midbrain (Kandel et al., 2000; Kolb & Whishaw, 2003). Rich white matter projections carry DA away from these deep nuclei, with significant overlap

TABLE 3 Examples of Neurons and Glia in the Nervous System

Type	Name	Location	Function
Neurons	Pyramidal cells	Corticospinal tract Hippocampus Cerebral cortex	Processing and communicating information to regulate movement, cognition, and other important functions
	Purkinje cells	Cerebellum	Processing and communicating information to regulate movement
Glia	Oligodendrocytes/ Schwann cells	Surrounding axons throughout the nervous system	Myelination
	Astrocytes	Throughout the nervous system	Obtaining nutrients Reinforcing blood–brain barrier Maintaining ion concentrations outside neurons

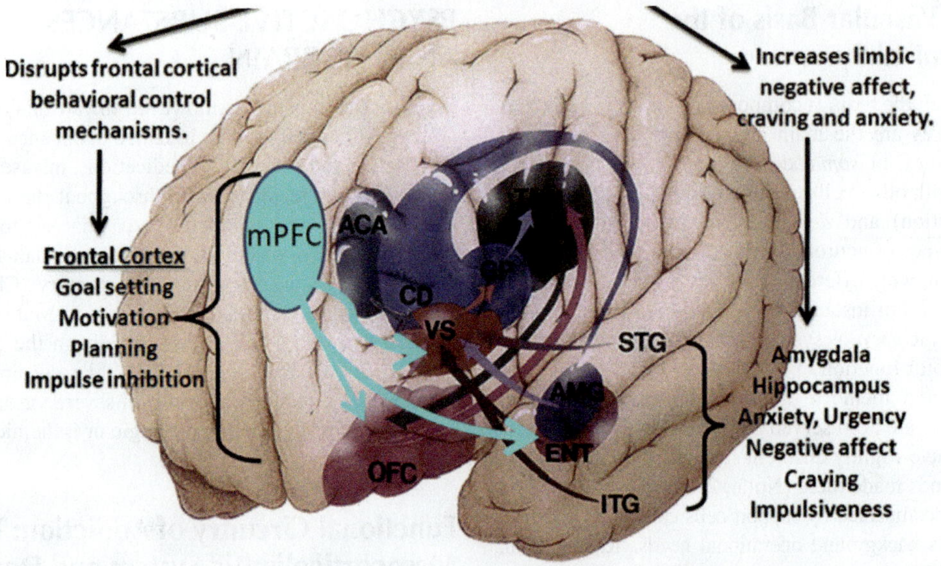

FIGURE 2 Basic schematic of mesolimbic structure and functional operations. mPFC, medial prefrontal cortex; ACA, anterior cingulate; CD, caudate; GP, globus pallidus; VS, ventral striatum; STG, superior temporal gyrus; AMG, amygdala; ENT, entorhinal cortex; ITG, inferior temporal gyrus; OFC, orbitofrontal cortex; Thal, thalamus. *Reprinted from Crews, Zou, and Qin (2011). ©2011, with permission from Elsevier.*

with other functional systems. A more detailed schematic of these functional connections is depicted in Figure 3.

In a more straightforward version of these pathways, DA travels from its source in the substantia nigra by three main pathways, the *nigrostriatal*, *mesolimbic*, and *mesocortical* (Blumenfeld, 2002). The nigrostriatal pathway rises from the substantia nigra to the striatum, contributing heavily to the moderation of movement in the human body. DA deficits in this pathway result in movement disorders, including illnesses like Parkinson disease and Huntington disease. The remaining two pathways, the mesolimbic and the mesocortical, carry DA from its source in the substantia nigra to the VTA. From the VTA, these pathways ascend rostrally to the prefrontal cortex and limbic structures, including the NA, amygdala, and hippocampus (Nolte, 2002). The functional cooperation of these two pathways leads some to term it the *mesocorticolimbic* system, which has heavy importance in the neurobiology of addictive behavior.

The Impact of Substances on the Mesocorticolimbic System: The "Reward Circuit"

These brain structures provide the essential neurobiological foundation for motivation and initiation, impulse control, and the experience of pleasure and reward in the brain. Limbic structures are involved in attaching *hedonic value* (i.e., a sense of pleasure and enjoyment) to stimuli. DA is the key facilitator of those neurological operations, and a vital feature common across all drugs of abuse is enhancement of dopaminergic activity within the mesocorticolimbic system. This leads some to describe this functional system as the "reward circuit." Drug class determines the specific pharmacological mechanism by which DA is increased, either by direct facilitation or by blocking the breakdown or reuptake of DA in the brain. In any case, however, this increase in DA is thought central to the positive reinforcing effects of both naturally rewarding activities and illicit substances (Haber & Knutson, 2009; Taber, Black, Porrino, & Hurley, 2012).

THE NEUROBEHAVIORAL FACETS OF ADDICTION: FROM RECREATIONAL USE TO DEPENDENCE

By this premise, illicit substance use results in a temporary, intensely pleasurable increase in DA in the reward-oriented mesocorticolimbic system. However, when the nature of substance use changes from recreational experimentation to addiction and dependence, the neurobehavioral underpinnings have been shown to change as well. Repeated substance exposure is thought to increase the risk for biological dependence on the substance through chronic alteration of the dopaminergic system.

From Acute Intoxication to Chronic Dependence

Many prominent addiction theories postulate that drug addiction progresses from hedonic voluntary use to more habitual, compulsive use reflected by drug-seeking behavior despite well-understood negative consequences (Taylor, Lewis, & Olive, 2013). Converging lines of evidence reflect this, documenting structural and functional neural alterations and concordant behavioral manifestations. Research has demonstrated that neurochemical activities in response to recreational use and chronic use are indeed different, with a focus on key shifts in the reinforcing nature of substances.

From Positive Reinforcement to Negative Reinforcement

Reinforcement paradigms are thought to shift from positive reinforcement (primary reward) in recreational intoxication to negative reinforcement (avoidance of punishment) in habitual use (Koob, 2013; Koob & Volkow, 2010). This process has been described as a disorder evolving over a three-stage cycle, including

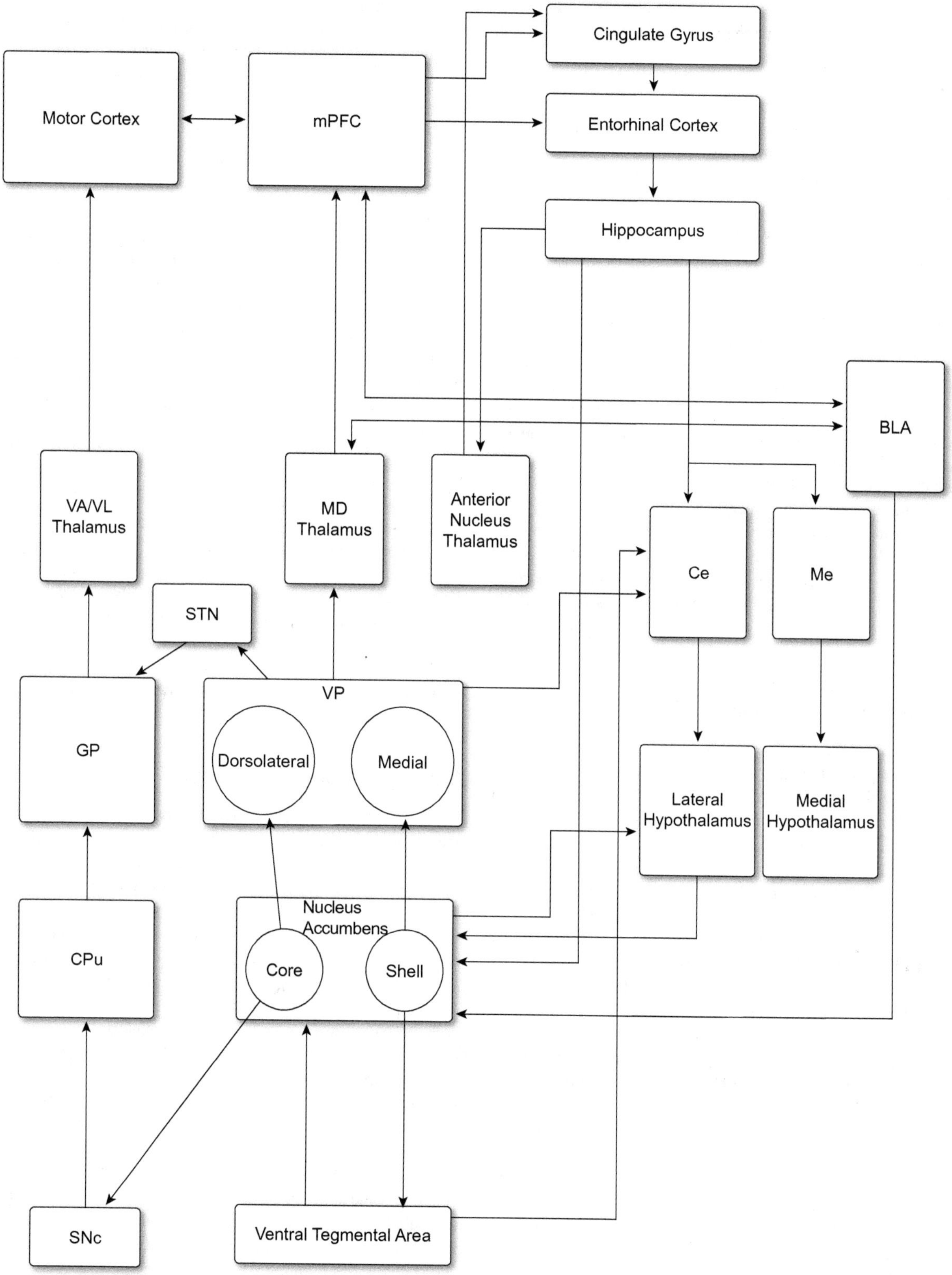

FIGURE 3 Schematic of anatomical and functional mesocorticolimbic connections. BLA, basolateral nucleus of amygdala; Ce, central nucleus of amygdala; CPu, caudate putamen; GP, globus pallidus; MD, mediodorsal nucleus; Me, medial nucleus of amygdala; mPFC, medial prefrontal cortex; SNc, substantia nigra pars compacta; VA/VL, ventral anterior/ventral lateral nuclei; STN, subthalamic nucleus; VP, ventral palladium. *Reprinted from Bari, Niu, Langevin, and Fried (2014). ©2014, with permission from Elsevier.*

(1) binge/intoxication, (2) withdrawal/negative affect, and (3) preoccupation/anticipation, or craving (Koob & Volkow, 2010). In each of these stages, different neuroanatomical substrates are involved, with alterations in normal functioning, emotional processing, and reward circuitry occurring as one progresses from casual to habitual substance use. These changes are thought to account, at least partially, for the "chronically relapsing" nature of substance dependence (Feil & Zangen, 2010; Koob & Volkow, 2010).

Stage 1: Binge/Intoxication

The binge/intoxication stage is influenced by substance pharmacokinetics and administration methods. For instance, rapid administration methods of psychostimulants, such as snorting and injecting cocaine, result in more intense biological responses than slower administration routes, like oral ingestion (Volkow et al., 2001). In addition, different substances also produce vastly different levels of dopaminergic activity. Alcohol produces approximately a 20% increase in extracellular DA in the NA (Doyon et al., 2003; Koob, 2013; Weiss, Markou, Lorang, & Koob, 1992), but cocaine can produce about *200%*.

These increases in dopaminergic activity are related to feelings of pleasure and enjoyment, as discussed. The importance of this affective hedonic state has been highlighted in the risk for chronic relapse in addiction, as the brain reward systems becoming increasingly dysregulated in response to spikes in DA levels (Koob & Le Moal, 1997; Koob & Le Moal, 2001).

Stage 2: Withdrawal/Negative Affect and Changes in Allostasis

Part of that progressively increasing dysregulation is related to the unpleasant physiological and affective symptoms of substance withdrawal. This experience differs considerably depending on the particular substance in question and the duration of use. In-depth discussion of these substance-specific effects is deferred to other excellent writings in this volume, although some illustrations are provided here. For example, the acute phase of heroin withdrawal has been characterized by decreased DA in the NA (Wang et al., 1997), whereas enhanced sensitivity to an inhibitory neurotransmitter called GABA is associated with the acute stages of cocaine withdrawal (Volkow et al., 1998). In any case, once the acute symptoms of withdrawal have subsided, alterations in neurotransmitter activity and functional connectivity can persist, including chronic attenuation of dopaminergic reward pathways as well as decreased expression of DA-specific receptors (Koob & Volkow, 2010). Unfortunately, these pathological decreases in DA may leave an individual in a chronically underactivated, significantly unpleasant anhedonic state.

Such persistent changes have been defined as alterations in neurobiological *allostasis* (Koob & Le Moal, 1997; Koob & Le Moal, 2001; Koob & Volkow, 2010). In the context of substance use, allostatic changes represent downregulation of reward-related neurotransmitter levels, reduced to maintain "normal" reward function when substances are present (Koob & Le Moal, 2001). Over time, maintenance of persistent allostatic changes due to compulsive substance use leads to *allostatic load*, accumulated damage to the neurobiological system caused by chronic, substance-related changes to the homeostatic baseline (Koob & Le Moal, 2001; McEwen & Stellar, 1993). Thus, in essence, the brain is no longer able to regulate its experience of pleasure and enjoyment independently because of chronic substance effects. This change is thought to contribute significantly to the transition from recreational substance use to substance dependence.

Stage 3: Preoccupation/Anticipation (Craving)

Sensibly, then, it follows that individuals in a persistently downregulated state may seek their substance of choice to ameliorate the anhedonia and other unpleasant, undesirable consequences of that state. In habitual use, the reward cues often involved in the binge/intoxication stage and the punishment cues experienced during withdrawal begin to persist even after the acute stages have resolved, leaving the increased sensitivity to conditioned cues associated with the substance in question suggested by allostatic load. Because of this, exposure to such cues (like the neon signs hanging in the liquor store window or illicit drug paraphernalia) is associated with low-level activation of the mesocorticolimbic system in anticipation of receiving the substance. Research has determined that this anticipatory state is powerful, leading to greater neural activation after drug administration than occurs in the absence of those expectations (Volkow et al., 2003).

Activation of the VS has been strongly associated with this anticipatory state (Glimcher, 2011), a phenomenon that has also been connected with the placebo effect in medical treatments. Unfortunately, the impact of life stressors also activates the same neuroanatomical substrates involved in reward processing. This leaves individuals recovering from substance dependence especially vulnerable to chronic relapse as they struggle to cope with even greater deficits of pleasure-related neurotransmission secondary to stress (Duncan et al., 2007; Langleben et al., 2008).

Structure-Specific Changes

It should now be clear that persistent changes in stimulation for the mesolimbic system because of substance use change how that system regulates DA. In fact, these changes have been demonstrated to affect not only dopaminergic systems, but also opioid peptide systems, activity of the hypothalamic–pituitary–adrenal axis, production of corticotropin-releasing factor, and glucocorticoid production (Dackis & O'Brien, 2005; Koob & Le Moal, 1997; Koob & Le Moal, 2001; Leknes & Tracey, 2008). These changes are all related to the experience of pleasure and stress and have the potential to affect the entire body instead of just the brain. This communicates just how globally powerful the impact of substance-related neural changes can be.

Not only does widespread systemic change appear to occur secondary to chronic substance use, change has also been demonstrated to occur in within key structures themselves. Research has demonstrated that specific regions of integral structures, like the striatum and NA, may also response differentially in recreational versus chronic substance use. For instance, the NA can be functionally subdivided into a shell and a core, which have differing roles in addictive behavior. Evidence suggests that DA release in the shell is associated with the initial rewarding effects of substance exposure, whereas the DA activity shifts for the core after prolonged substance use. This had led some researchers to propose that DA activity in the core is more involved with the learned behaviors of drug seeking and craving experienced in substance dependence (Di Chiara et al., 2004; Ito, Robbins, & Everitt, 2004; Taylor et al., 2013). In addition, research on psychostimulants suggests that after chronic use, a shift occurs in the striatum, from increased activity in the VS to the dorsal striatum, perhaps reflecting the progression from voluntary to compulsive use (Everitt et al., 2008).

APPLICATIONS TO OTHER ADDICTIONS AND SUBSTANCE MISUSE: NEW BEHAVIORAL DISORDERS?

These neurological underpinnings of substance addiction have contributed to a great deal of enthusiasm for applying these principles to emerging constructs surrounding compulsive behavioral "addictions," including gambling and hoarding. This young body of research suggests that although there is not an exact correlation between substance-related processes and compulsive behaviors, important behavioral and biological similarities indeed appear to be present.

Pathological Gambling

For instance, pathological gambling has been conceptualized, at least behaviorally, in a cycle-based model similar to those often discussed in substance use. Individuals who gamble compulsively have demonstrated diminished self-control related to gambling, craving states, withdrawal, tolerance, and continued participation despite serious negative consequences (Potenza, 2008). There are biological similarities as well, including increased activation of the VS during anticipation of the gambling stimulus and reduced activation of the prefrontal cortex in response to the gambling stimulus (Bowden-Jones & Clark, 2011; Potenza, 2008). However, the act of gambling has been more strongly associated with autonomic arousal due to increases in a neurotransmitter called noradrenaline rather than DA, although DA is still thought to play a possible role in the development and maintenance of these disorders (Potenza, 2008).

Hoarding

Similarities have also been demonstrated between compulsive hoarding and substance use. The key role of the NA in substance use and dependence has now been clearly demonstrated, and activity in the NA has also been demonstrated in animal models of hoarding. Hoarding rats with chemical lesions introduced into the NA significantly reduced or ceased the hoarding behavior postlesioning. Again highlighting the importance of DA, hoarding behavior returned to prelesion levels when the rats were injected with a precursor of DA called L-dopa, pointing to mesolimbic DA as the main mediator of the behavior (Kelley & Stinus, 1985).

Understanding these biological underpinnings has supported, at least in part, the increasing recognition of such compulsive behaviors as formal disorders and targets for treatment. For instance, compulsive hoarding was recently recognized formally as hoarding disorder in the fifth version of the Diagnostic and Statistical Manual of Mental Disorders (American Psychiatric Association, 2013).

SOCIETAL IMPACT OF SUBSTANCES

In closing, this review clearly highlights how alterations in the brain from substance use can affect emotions, higher cognitive functions, and behavioral patterns. Societally, understanding this effect has never been more salient, as we cope with high mortality rates related to alcohol, tobacco, and illicit substances. Multiple legislations grapple with the question of medicinal and recreational cannabis use, and legislative bodies are also tackling significant issues around opiate analgesics and common over-the-counter decongestants owing to methamphetamine production. A firm understanding of the actual neurological actions of these substances can help inform practical decisions around these issues, as well as supporting efforts around effective treatments and rehabilitation.

DEFINITION OF TERMS

Blood–brain barrier The nervous system's defense against toxins, composed of tight junctions between blood vessel walls and glial astrocytes insulating outer vessel walls.

Central nervous system A subdivision of the neuroanatomical system composed of the brain and spinal cord.

Cerebral cortex A thin layer of cell bodies on the outermost layer of the brain dedicated to processing and integrating environmental information and to higher-level cognitive functioning, including attention, decision-making, and personality.

Dopamine A monoaminergic neurotransmitter with critical roles in the body's responses to psychoactive substances.

Glia Cells in the nervous system primarily performing supporting roles, including obtaining nutrients, clearing waste products, and supporting the brain's defenses against toxins.

Gray matter Collections of neuron cell bodies grouped together for data processing in the brain. The cerebral cortex and deep subcortical nuclei are mainly composed of gray matter.

Mesocorticolimbic system A functional system including the limbic system, the prefrontal cortex, and other associated structures connected by dopaminergic white matter pathways. This system is central in the brain's response to psychoactive substances.

Negative reinforcement The avoidance of unpleasant, noxious consequences that makes a behavior more likely to recur.

Neurons Cells in the nervous system primarily intended for data processing and rapid information transfer.

Peripheral nervous system A subdivision of the neuroanatomical system composed of all nerves outside of the brain and spinal cord that provides an interface between the central nervous system and the environment.

Positive reinforcement The provision of rewarding, desirable consequences that makes a behavior more likely to occur.

Psychoactive substances Substances that can bypass the blood–brain barrier, resulting in both desirable and undesirable changes in thinking, emotion, and behavior.

White matter Collections of myelinated axons departing from cell bodies carrying vital information from one point in the nervous system to another.

KEY FACTS

Key Facts of the CNS

- A working knowledge of neuroanatomy is key to the investigation of addictive behavior.
- General navigation around the nervous system is facilitated by special directional terms, such as rostral/caudal and dorsal/ventral. The meaning of some terms changes at the midbrain to reflect the development of the brain and spinal cord in utero.
- The CNS is composed of the brain and spinal cord, and the PNS is composed of the nerves outside of the brain and spinal

cord. Cooperation between these two systems allows the brain and body to interact smoothly with the environment.
- The nervous system's processing centers are composed of gray matter, or dense collections of neuronal cell bodies.
- Information is transmitted around the nervous system by white matter, or collections of myelinated axons arising from the cell bodies.
- The brain itself requires a large, constant supply of glucose and oxygen to operate properly. This is provided by the dense vascular system, which is also the source of the protective blood–brain barrier.

Key Facts of Neural Structure, Function, and Psychoactive Substances

- Psychoactive substances are compounds that can pass through the blood–brain barrier and cause changes in thoughts, emotions, and behavior because of their effects on the brain.
- Substances have the potential to harm all components of the nervous system, including gray matter, white matter, and vasculature.
- The mesocorticolimbic system is located near the midpoint of the brain and communicates with the prefrontal cortex with DA-mediated white matter pathways.
- The NA is a key site within the mesocorticolimbic pathway. It is found in the area where the putamen and the caudate fuse anteriorly, an area also known as the VS.
- Use of psychoactive substances, particularly drugs of abuse, increases DA in the mesocorticolimbic pathway and the NA, in particular.
- Prolonged use of substances appears to change the brain's fundamental reaction to the substance, changing from a rewarding rush of DA in recreational use to avoidance of unpleasant withdrawal effects in chronic use.
- These changes in reinforcement value are also accompanied by structural changes in the neurobiological response to substances.

SUMMARY POINTS

- The investigation of addictive behavior is greatly facilitated by a strong understanding of the neurobiological system involved in substance use.
- The brain can be divided into many structural and functional systems, although all of those systems cooperate with one another to allow seamless, rapid communication among the brain, the body, and the environment.
- The structure of the brain can be roughly discussed in four parts, including gray matter-centric processing centers, communication fibers made of white matter, combined structures including both gray and white matter, and ventricles filled with cerebrospinal fluid.
- The brain requires a dense vascular system to provide it with constant, large supplies of oxygen and glucose. Special elements of the vascular system form the blood–brain barrier, the brain's primary protection against toxins.
- Psychoactive substances, however, work around this security system, resulting in changes in cognition, behavior, and emotion because of the substance's effects on the brain.
- Certain areas are particularly vulnerable to these effects, especially the mesocorticolimbic system. This system is nested around the midpoint of the brain and communicates with other critical areas, like the prefrontal cortex, by DA-dependent pathways.
- All drugs of abuse increase DA in these pathways, creating a temporary experience of intense enjoyment in response to the substance.
- Prolonged exposure to drugs of abuse is thought to alter DA's operations in the mesocorticolimbic system, leaving individuals prone to the dysphoria, cravings, and compulsive substance use observed in substance addiction.
- Such prolonged exposure is also related to key cognitive and behavioral changes that also propagate substance dependence.
- Although incomplete, our understanding of these processes has supported evolving knowledge around behavioral disorders like compulsive gambling and hoarding.

REFERENCES

American Psychiatric Association. (2013). *Diagnostic and statistical manual for mental disorders* (5th ed.). Washington, DC: Author.

Bari, A., Niu, T., Langevin, J.-P., & Fried, I. (2014). Limbic neuromodulation: implications for addiction, posttraumatic stress disorder, and memory. *Neurosurgery Clinics of North America*, 137–145.

Blumenfeld, H. (2002). *Neuroanatomy through clinical cases*. Sunderland, MA: Sinauer Associates, Inc.

Bowden-Jones, H., & Clark, L. (2011). Pathological gambling: a neurobiological and clinical update. *The British Journal of Psychiatry*, 199(2), 87–89. http://dx.doi.org/10.1192/bjp.bp.110.088146.

Chambers, R. A. (2013). Adult hippocampal neurogenesis in the pathogenesis of addiction and dual diagnosis disorders. *Drug and Alcohol Dependence*, 130(1), 1–12.

Crews, F. T., Zou, J., & Qin, L. (2011). Induction of innate immune genes in brain create the neurobiology of addiction. *Brain, Behavior, and Immunity*, S4–S12.

Dackis, C., & O'Brien, C. (2005). Neurobiology of addiction: treatment and public policy ramifications. *Nature Neuroscience*, 8(11), 1431–1436.

Di Chiara, G., Bassareo, V., Fenu, S., De Luca, M. A., Spina, L., Cadoni, C., … Lecca, D. (2004). Dopamine and drug addiction: the nucleus accumbens shell connection. *Neuropharmacology*, 47, 227–241.

Doyon, W. M., York, J. L., Diaz, L. M., Samson, H. H., Czachowski, C. L., & Gonzales, R. A. (2003). Dopamine activity in the nucleus accumbens during consummatory phases of oral ethanol self-administration. *Alcoholism: Clinical and Experimental Research*, 27(10), 1573–1582.

Duncan, E., Boshoven, W., Harenski, K., Fiallos, A., Tracy, H., Jovanovic, T., … Kilts, C. (2007). An fMRI study of the interaction of stress and cocaine cues on cocaine craving in cocaine–dependent men. *The American Journal on Addictions*, 16(3), 174–182.

Everitt, B. J., Belin, D., Economidou, D., Pelloux, Y., Dalley, J. W., & Robbins, T. W. (2008). Neural mechanisms underlying the vulnerability to develop compulsive drug-seeking habits and addiction. *Philosophical Transactions of the Royal Society B: Biological Sciences*, 363(1507), 3125–3135.

Feil, J., & Zangen, A. (2010). Brain stimulation in the study and treatment of addiction. *Neuroscience & Biobehavioral Reviews*, 34, 559–574.

Glimcher, P. W. (2011). Understanding dopamine and reinforcement learning: the dopamine reward prediction error hypothesis. *Proceedings of the National Academy of Sciences*, 108(Suppl. 3), 15647–15654.

Haber, S. N., & Knutson, B. (2009). The reward circuit: linking primate anatomy and human imaging. *Neuropsychopharmacology, 35*(1), 4–26.

Ito, R., Robbins, T. W., & Everitt, B. J. (2004). Differential control over cocaine-seeking behavior by nucleus accumbens core and shell. *Nature Neuroscience, 7*(4), 389–397.

Kandel, E. R., Schwartz, J. H., & Jessell, T. M. (2000). *Principles of neural science* (4th ed.). New York, NY: McGraw-Hill.

Kelley, A. E., & Stinus, L. (1985). Disappearance of hoarding behavior after 6-hydroxydopamine lesions of the mesolimbic dopamine neurons and its reinstatement with L-dopa. *Behavioral Neuroscience, 99*(3), 531–535.

Kolb, B., & Whishaw, I. Q. (2003). *Fundamentals of human neuropsychology* (5th ed.). New York, NY: Worth Publishers.

Koob, G. F. (2013). Theoretical frameworks and mechanistic aspects of alcohol addiction: alcohol addiction as a reward deficit disorder. *Current Topics in Behavioral Neuroscience, 13*, 3–30.

Koob, G. F., & Le Moal, M. (1997). Drug abuse: hedonic homeostatic dysregulation. *Science, 278*(5335), 52–58.

Koob, G. F., & Le Moal, M. (2001). Drug addiction, dysregulation of reward, and allostasis. *Neuropsychopharmacology, 24*(2), 97–129.

Koob, G. F., & Volkow, N. D. (2010). Neurocircuitry of addiction. *Neuropsychopharmacology, 35*(1), 217–238.

Langleben, D., Ruparel, K., Elman, I., Busch-Winokur, S., Pratiwadi, R., Loughead, J., … Childress, A. R. (2008). Acute effect of methadone maintenance dose on brain fMRI response to heroin-related cues. *The American Journal of Psychiatry, 165*(3), 390–394.

Larner, A. (2008). *Neuropsychological neurology: the neurocognitive impairments of neurological disorders.* New York, NY: Cambridge University Press.

Leknes, S., & Tracey, I. (2008). A common neurobiology for pain and pleasure. *Nature Reviews Neuroscience, 9*(4), 314–320.

Mair, R. G. (2012). Encyclopedia of human behavior (2nd ed.). *The brain*: Vol. 1. (pp. 377–385).

McEwen, B. S., & Stellar, E. (1993). Stress and the individual: mechanisms leading to disease. *Archives of Internal Medicine, 153*(18), 2093–2101.

Nolte, J. (2002). *The human brain: an introduction to its functional anatomy* (5th ed.). St. Louis, MO: Mosby.

Potenza, M. N. (2008). The neurobiology of pathological gambling and drug addiction: an overview and new findings. *Philosophical Transactions of the Royal Society B: Biological Sciences, 363*(1507), 3181–3189.

Schmahmann, J. D., & Pandya, D. N. (2006). *Fiber pathways of the brain.* New York: Oxford University Press.

Schuenke, M., Schulte, E., & Schumacher, U. (2010). *Thieme atlas of anatomy: head and neuroanatomy.* New York, NY: Thieme New York.

Taber, K. H., Black, D. N., Porrino, L. J., & Hurley, R. A. (2012). Neuroanatomy of dopamine: reward and addiction. *The Journal of Neuropsychiatry and Clinical Neurosciences, 24*(1), 1–4.

Taylor, S. B., Lewis, C. R., & Olive, M. F. (2013). The neurocircuitry of illicit psychostimulant addiction: acute and chronic effects in humans. *Substance Abuse & Rehabilitation, 4*, 29–43.

Verdejo-Garcia, A., Clark, L., & Dunn, B. D. (2012). The role of interoception in addiction: a critical review. *Neuroscience & Biobehavioral Reviews, 36*(8), 1857–1869.

Volkow, N. D., Wang, G.-J., Fowler, J. S., Hitzemann, R., Gatley, S. J., Dewey, S. S., & Pappas, N. (1998). Enhanced sensitivity to benzodiazepines in active cocaine-abusing subjects: a PET study. *American Journal of Psychiatry, 155*(2), 200–206.

Volkow, N. D., Wang, G., Fowler, J. S., Logan, J., Gerasimov, M., Maynard, L., … Franceschi, D. (2001). Therapeutic doses of oral methylphenidate significantly increase extracellular dopamine in the human brain. *Journal of Neuroscience, 21*(2), RC121.

Volkow, N. D., Wang, G.-J., Ma, Y., Fowler, J. S., Zhu, W., Maynard, L., … Wong, C. (2003). Expectation enhances the regional brain metabolic and the reinforcing effects of stimulants in cocaine abusers. *The Journal of Neuroscience, 23*(36), 11461–11468.

Wang, C., Zheng, D., Xu, J., Lam, W., & Yew, D. (2013). Brain damages in ketamine addicts as revealed by magnetic resonance imaging. *Frontiers in Neuroanatomy, 7*(23), 1–8.

Wang, G.-J., Volkow, N. D., Fowler, J. S., Logan, J., Abumrad, N. N., Hitzemann, R. J., … Pascani, K. (1997). Dopamine D2 receptor availability in opiate-dependent subjects before and after naloxone-precipitated withdrawal. *Neuropsychopharmacology, 16*(2), 174–182.

Weiss, F., Markou, A., Lorang, M. T., & Koob, G. F. (1992). Basal extracellular dopamine levels in the nucleus accumbens are decreased during cocaine withdrawal after unlimited-access self-administration. *Brain Research, 593*(2), 314–318.

Chapter 2

Pathophysiology-Based Neuromodulation for Addictions: An Overview

Dirk De Ridder[1], Patrick Manning[2], Gavin Cape[3], Sven Vanneste[4], Berthold Langguth[5], Paul Glue[6]

[1]Section of Neurosurgery, Department of Surgical Sciences, Dunedin School of Medicine, University of Otago, Dunedin, New Zealand; [2]Section of Endocrinology, Department of Internal Medicine, Dunedin School of Medicine, University of Otago, Dunedin, New Zealand; [3]Department of Psychological Medicine, Dunedin School of Medicine, University of Otago, Dunedin, New Zealand; [4]School of Behavioral and Brain Sciences, The University of Texas at Dallas, TX, USA; [5]Department of Psychiatry and Psychotherapy, University Regensburg, Regensburg, Germany; [6]Department of Psychological Medicine, Dunedin School of Medicine, University of Otago, Dunedin, New Zealand

Abbreviations

CAS Complex adaptive system
dACC Dorsal anterior cingulate cortex
DBS Deep brain stimulation
DLPFC Dorsolateral prefrontal cortex
EC Effective connectivity
ECT Electroconvulsive therapy
EEG Electroencephalography
FC Functional connectivity
fMRI Functional magnetic resonance imaging
LORETA Low resolution tomography
Nacc Nucleus accumbens
NFB Neurofeedback
NIBS Non-invasive brain stimulation
NIRS Near infrared spectroscopy
OFC Orbitofrontal cortex
PCC Posterior cingulate cortex
pgACC Pregenual anterior cingulate cortex
rACC Rostral anterior cingulate cortex
SMR Sensorimotor rhythm
tACS Transcranial alternating current stimulation
tDCS Transcranial direct current stimulation
tES Transcranial electrical stimulation
TMS Transcranial magnetic stimulation
tRNS Transcranial random noise stimulation

INTRODUCTION

In the past few decades, our knowledge of addiction has changed, with greater emphasis placed on the neurobiological basis of addiction. Based on functional neuroimaging and animal research, brain processes involved in the development of an addiction are progressively being unraveled. However, this information has not yet been translated into effective treatments in clinical practice. Because current treatment success for addiction is poor (Kuhn, Buhrle, Lenartz, & Sturm, 2013), a desperate need for developing effective treatments is essential. For example, with alcohol dependence, the rate of relapse 1 year after standard treatment is reported to range between 40% and 70% (Swift, 1999). In the case of heroin addiction, the current situation is even worse: 90% of treated patients will experience one or multiple relapses. Owing to this failure to attain long-lasting abstinence, substance-related addictions are deemed chronically recurrent diseases with an excessively high rate of therapeutic failure and/or relapse. In addition, existing therapeutic strategies, such as expensive inpatient long-term withdrawal treatment regimens, can often be offered to only a limited number of patients, resulting in many chronic addicts remaining untreated (Kuhn et al., 2013).

Neuromodulation, the targeted alteration of nerve activity through the delivery of neurostimulation to specific sites, could potentially change this, as it may offer a way of normalizing the addiction-related pathological brain activity and connectivity. Neuromodulation is inextricably linked to neuroplasticity. Even though no universally accepted definition exists, neuroplasticity can be operationally defined as the capacity of the nervous system to modify its structural and functional organization, adjusting itself to a changing environment. Therefore neuromodulation can be defined as the induction of neuroplastic changes via local application of electrical, magnetic, acoustic, optic, tactile, or pharmacological stimuli. This is a broader definition than the one used by the International Neuromodulation Society: "Neuromodulation is technology that acts directly upon nerves. It is the alteration—or modulation—of nerve activity by delivering electrical or pharmaceutical agents directly to a target area" (http://www.neuromodulation.com/about-neuromodulation).

The concept of intervening in the brain directly for addiction is not new. Psychosurgery for addiction has a long and turbulent history.

In this review, we will discuss the most important neuromodulation techniques used in addiction, both surgical and nonsurgical. With regard to surgical neuromodulation, two different approaches have been used, namely surgical lesioning and brain stimulation. These two seemingly opposite approaches differ in their reversibility, but exert a similar physiological effect, since the currently used high-frequency electrical neurostimulation creates reversible functional lesions, mimicking irreversible lesioning.

FINDING TARGETS FOR NEUROMODULATION INTERVENTION

Multiple approaches can be used to find targets for neuromodulatory interventions. In the history of functional neurosurgery the identification of treatment targets occurred frequently by serendipity: Initial discovery of promising targets happened by careful observation of patients who showed symptom improvement of a chronic neuropsychiatric disease following the development of a lesion in a specific region due to a tumor, stroke, or trauma.

If serendipity does not help, ideal targets for neuromodulation can be identified based on pathophysiological knowledge. This can be derived from animal studies or functional imaging studies in humans. An alternative approach consists in the identification of the neuronal mechanisms underlying successful pharmacological or cognitive behavioral treatments, e.g., by comparing pre- and post-intervention functional imaging data. Based on a 2013 meta-analysis, successful treatment with either pharmacotherapy or behavioral treatments is related to functional changes in the ventral striatum/nucleus accumbens (Nacc), the inferior frontal gyrus, and the orbitofrontal cortex (OFC) (Konova, Moeller, & Goldstein, 2013). Cognitive behavioral treatments modulated the anterior cingulate cortex, middle frontal gyrus, and precuneus in addition to the above-mentioned areas (Konova et al., 2013). A further approach consists in the analysis of addiction-related networks with network science methods to discover the hubs, i.e., the most critical areas in the network, and choose these as targets (Albert, Jeong, & Barabasi, 2000).

PATHOPHYSIOLOGY OF ADDICTION

The neurobiological substrate of alcohol dependence is progressively being unraveled. It is becoming clear that a complex interaction between genetic setup and environmental factors shape brain mechanisms involved with substance abuse (Koob & Le Moal, 2008), analogous to what is proposed to be the underlying mechanism in many other psychiatric diseases (Caspi & Moffitt, 2006). Based on twin and adoption studies, it is estimated that genetic factors can account for 49% of the total variability in alcohol dependence for men and 64% for women (McGue, Pickens, & Svikis, 1992), analogous to what has been described for other addiction disorders (cocaine, opiates, marijuana, tobacco), with an overall heritability of 40% (Uhl & Grow, 2004). The dopaminergic and opioidergic reward system is the major player in the pathophysiology of alcohol dependence (Koob & Le Moal, 2008). For example, a polymorphism in the dopamine D2 receptor (DRD2-TAQ-1A) is known to reduce the dopamine D2 receptor density by 30%. This reduction has been linked to multiple addictive and compulsive behaviors, and the TAQ-1A polymorphism has also been associated with alcohol dependence (Smith, Watson, Gates, Ball, & Foxcroft, 2008). Mechanistically this genetic variant leads to a selective insensitivity to negative consequences of self-destructive behavior, preventing people with a DRD2 A1 allele from learning through error (Klein et al., 2007). This suggests that when one drinks for relief, but does not learn from the negative consequences of overconsumption, it will lead to greater alcohol consumption and alcohol-related problems.

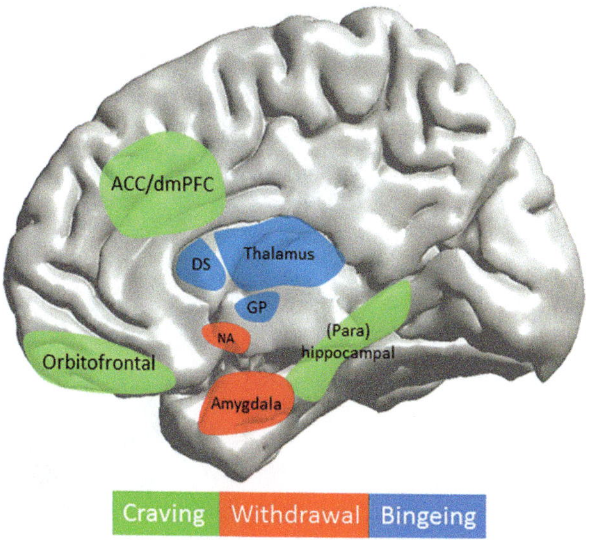

FIGURE 1 Brain areas involved in addiction. Based on (Koob & Volkow, 2010). Note that the PCC, insula, pgACC, frontopolar cortex, and DLPFC are missing. This could be because those areas are not hyperactive, but only more or less densely connected (see Figures 2 and 3). ACC, anterior cingulate cortex; dmPFC, dorsomedial prefrontal cortex; DS, dorsal striatum; GP, globus pallidus; NA, nucleus accumbens.

A pathophysiological model has conceptualized (Koob & Volkow, 2010) addiction as a disorder that involves three stages: (1) bingeing with intoxication, (2) withdrawal because of negative affect, and (3) craving (Figure 1). Animal and human imaging studies have revealed discrete circuits that mediate these three stages of the addiction cycle. The bingeing is related to activity in the ventral tegmental area and ventral striatum (=Nacc). The ventral striatum is involved in learning from trial and error irrespective of the specific nature of provided rewards (Daniel & Pollmann, 2014). It is now generally accepted that addiction is a type of learning, i.e., learned associations between environmental cues that predict drug availability and the rewarding effects of drugs (Feduccia, Chatterjee, & Bartlett, 2012), mediated via the Nacc by mechanisms similar to those involved in natural rewards such as food or sex (Pitchers et al., 2012). Alcohol is used either to feel better (reward drinking) or not to feel bad (relief drinking). Withdrawal and negative affect are related to the extended amygdala complex as well as increased functional connectivity (FC) between the posterior cingulate cortex (PCC)/precuneus, insula, OFC, superior frontal gyrus, and superior and inferior temporal lobe (Huang et al., 2014). Craving is related to a distributed network involving the OFC–dorsal striatum, prefrontal cortex, basolateral amygdala, hippocampus, and insula. However, the rostral (r) to dorsal anterior cingulate cortex (dACC) appear to be the most important regions within the craving network (Kuhn & Gallinat, 2011). Withdrawal-induced tobacco craving correlates highly with activation of the PCC/precuneus, insula, caudate, putamen, dACC, and precentral gyri (Huang et al., 2014). The parahippocampal and hippocampal areas are involved in contextual cues (Aminoff, Gronau, & Bar, 2007) leading to craving (Koob & Le Moal, 2008). In addition, addictive behavior is related to disrupted frontal inhibitory control, which is reflected by alterations in the cingulate gyrus and dorsolateral prefrontal and inferior frontal cortices.

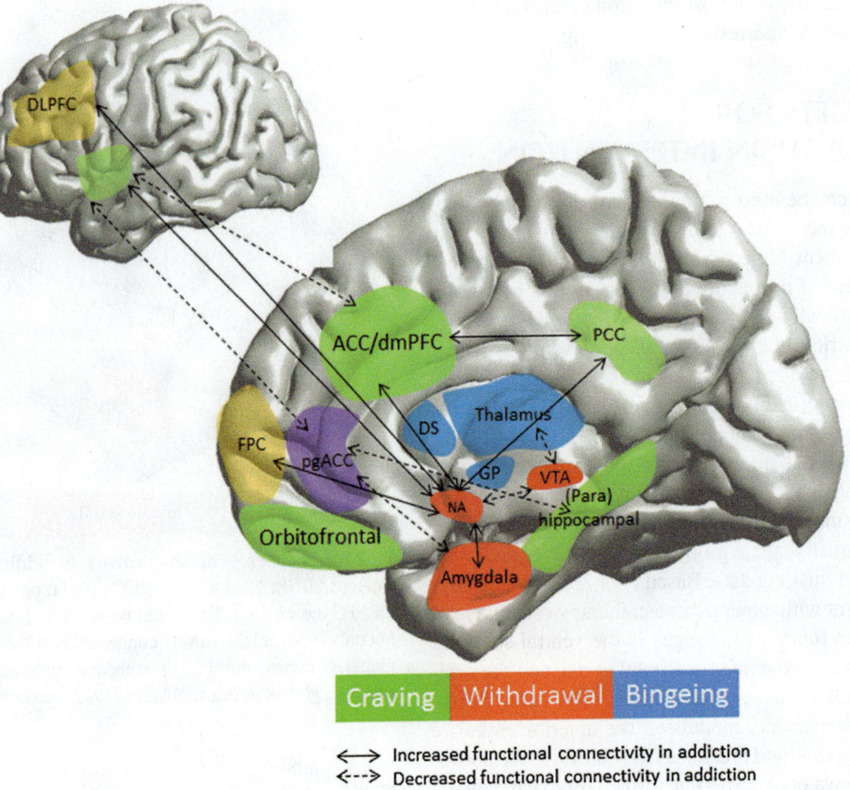

FIGURE 2 **Functional connectivity changes in addiction.** References (Camchong et al., 2014; Gu et al., 2010; Motzkin et al., 2014; Sutherland et al., 2013) can be summarized by increased FC with the Nacc and decreased FC with the pregenual anterior cingulate. Conceptually the Nacc, involved in operant conditioning, is more synchronously active with craving and salience-related areas, suggestive that this could have become the reference state, i.e., allostatic. The substance of abuse has become salient and the reference is reset. Furthermore, the decreased FC with the pgACC suggests there is no more suppression of input, i.e., larger quantities of the drug will be consumed. FPC, frontopolar cortex; VTA, ventral tegmental area.

This model further proposes that the transition to addiction involves neuroplasticity in all of these structures, beginning with changes in the mesolimbic dopamine system in the accumbens and followed by a cascade of neuroadaptations from the ventral striatum to the dorsal striatum and OFC and eventually accompanied by dysregulation of the prefrontal cortex, cingulate gyrus, and extended amygdala.

To account for the fact that craving, withdrawal, and bingeing in addiction are all emergent properties of network activity it is important to look at functional (Figure 2) and effective connectivity changes in addiction (Figure 3).

The development and expression of dependence seem to rely on increased FC between the dACC/insula and the PCC/precuneus (Huang et al., 2014; Lerman et al., 2014), which represents the connection between the salience network (dACC/insula) and the default mode network (PCC/precuneus) (Figure 2). This fits with meta-analytic data indicating that cue-evoked reactivity in alcohol addiction is determined by activation of the rACC and PCC (Schacht, Anton, & Myrick, 2013). In addition, this connection is altered in successful suppression of craving by transcranial magnetic stimulation (TMS) targeting the dACC (De Ridder, Vanneste, Kovacs, Sunaert, & Dom, 2011).

In addition, addiction is characterized by increased FC between the Nacc and areas involved in the salience network (Seeley et al., 2007), i.e., dACC, anterior insula, dorsolateral prefrontal cortex (DLPFC) (Motzkin, Baskin-Sommers, Newman, Kiehl, & Koenigs, 2014), and amygdala (Filbey & Dunlop, 2014) (Figure 2). There is also increased FC between the Nacc and the PCC and frontopolar cortex (Camchong et al., 2014) (Figure 2). On the other hand, there is decreased FC between the Nacc and the ventral tegmental area and between the ventral tegmental area and the thalamus as well as the lentiform nucleus (Gu et al., 2010) (Figure 2). Furthermore, FC is decreased between the anterior insula and the dACC as well as the pregenual anterior cingulate cortex (pgACC) (Sutherland, Carroll, Salmeron, Ross, & Stein, 2013) (Figure 2). The same decrease in FC can be noted between the amygdala and the pgACC. The pgACC furthermore has decreased FC with the insula and hippocampus (Gu et al., 2010) (Figure 2).

FC changes in addiction (Camchong et al., 2014; Gu et al., 2010; Motzkin et al., 2014; Sutherland et al., 2013) can be summarized by increased FC of the Nacc with various brain regions and decreased FC of the pgACC with other regions. Conceptually the Nacc, involved in operant conditioning, is more synchronously active with craving and salience-related areas, reflecting allostatic changes with respect to the urge for the substance of abuse. The decreased FC of the pgACC may reflect impaired suppression of the urge, resulting in the consumption of larger quantities of the drug. Whereas FC reflects only the correlated activity between various brain regions, effective connectivity (EC) indicates directed information flow. EC in addiction is changed as well, with increased information flow from the

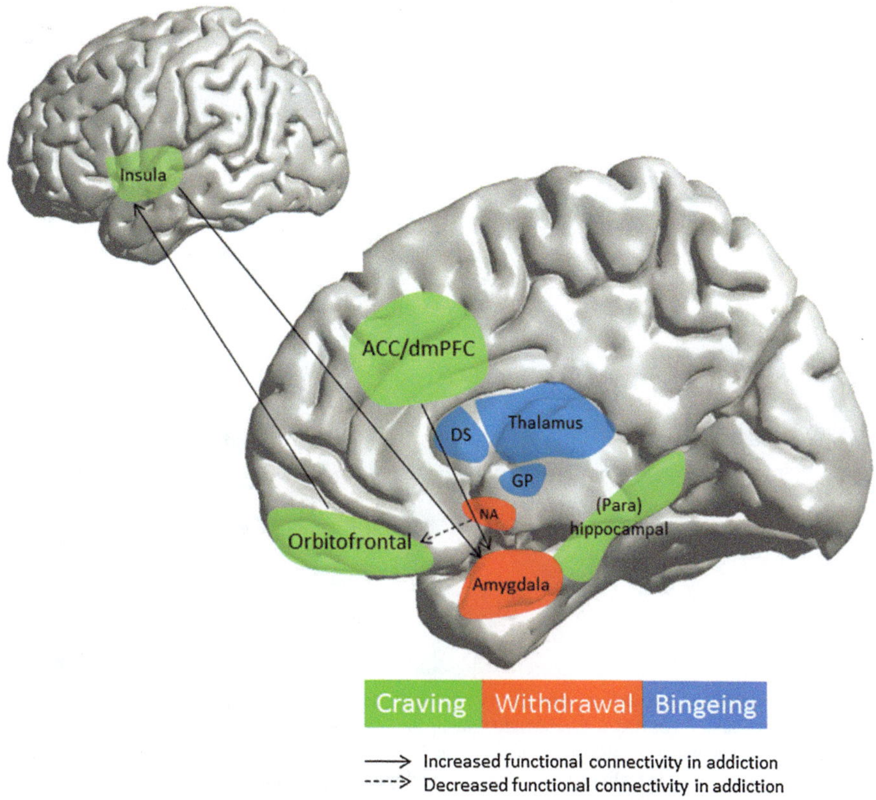

FIGURE 3 **Effective connectivity changes in addiction.** Based on (Filbey & Dunlop, 2014). Conceptually, learning via operant conditioning (Nacc) exerts less hedonic effect (OFC) owing to decreased EC. The OFC exerts more influence on the interoceptive cortex, in other words, the decreased hedonic state (I feel bad) drives the interoceptive state (craving). This drives the amygdala to mobilize an emotional response (urge) under the influence of the dACC, with salience attached to the substance of abuse. The changed EC in this network will, in other words, drive the urge for finding more of the same substance in order to feel less bad.

hippocampus and the anterior cingulate to the Nacc (Filbey & Dunlop, 2014) (Figure 3). There is also decreased information flow from the Nacc to the OFC, suggesting less hedonic activation from the Nacc (Figure 3).

These findings can be integrated in the following model: conceptually, learning via operant conditioning (Nacc) exerts less hedonic effect (OFC) owing to decreased EC. The OFC exerts more influence on the interoceptive cortex (decreased hedonic state; "I feel bad"), which in turn influences the interoceptive state (craving). This in turn drives the amygdala to mobilize an emotional response (urge) under the influence of the dACC, with salience attached to the substance of abuse. The changed EC in this network reflects the urge for finding more of the same substance in order to feel less bad.

From an information theory point of view, craving can be seen as an analog to the requirement of more information to reduce uncertainty (see Figure 4). Uncertainty is defined as a state in which a given representation of the world cannot be adopted as a guide to subsequent behavior, cognition, or emotional processing (Harris, Sheth, & Cohen, 2008). Uncertainty of the environment can be reduced by acquiring more data, by actively looking for more information in the environment (Rushworth & Behrens, 2008) or in memory if not available via the senses (De Ridder, Vanneste, & Freeman, 2014). Uncertainty is encoded by rACC activity (Rushworth & Behrens, 2008) and when more information is required the dACC will be activated, generating craving (Kuhn & Gallinat, 2011; Schacht et al., 2013) and resulting in an urge for

action (Jackson, Parkinson, Kim, Schuermann, & Eickhoff, 2011) in order to get more input. When sufficient input is obtained to reduce uncertainty, further input is not required and is suppressed via the pgACC, as shown both for pain stimuli and for sound (De Ridder, Vanneste, Menovsky, & Langguth, 2012). The same area then encodes a hedonic experience by coactivation of the OFC (Kringelbach, 2005).

Based on the information theory it is assumed that addiction can be conceptualized as an uncertainty disorder, in which the brain tries to get more of the same input, to reduce uncertainty (Figure 4). Through reinforcement learning via the Nacc, the substance of abuse has become salient and the reference is reset (i.e., allostasis), driving the urge for finding more of the same substance in order to feel less bad.

Based on this simplified pathophysiological model, multiple neuromodulation approaches can be considered to normalize the brain circuitry involved in addiction. Neuromodulation can be performed using neurofeedback and noninvasive neurostimulation, which includes transcranial magnetic and electrical stimulation.

NEUROMODULATION TECHNIQUES AND BRAIN NETWORK ALTERATIONS

The link between brain and mind has been a philosophical debate going on for millennia, since Zarathustra conceptualized dualism, the idea that when the body disintegrates the mind lives on. Monism, on the other hand, claims that when the brain dies the

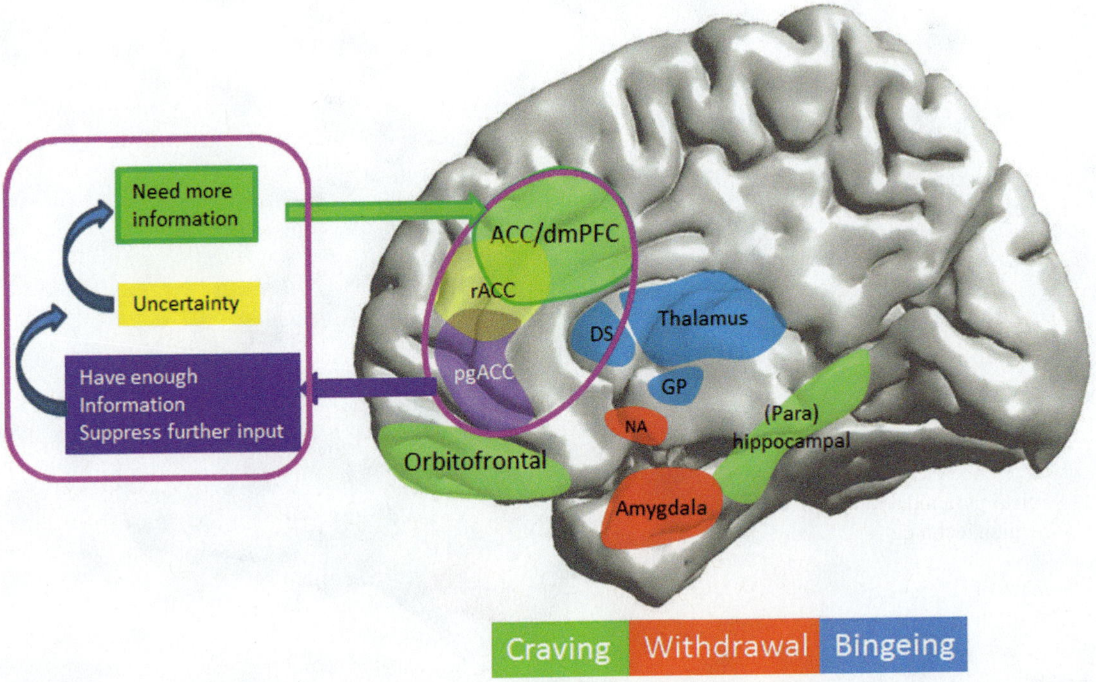

FIGURE 4 **The brain and uncertainty.** From an information theory approach, craving can conceptually be seen as a lack of input/information. The rACC has been linked to encoding uncertainty. The dACC is related to craving and an urge for action signifying that the brain needs more input to resolve uncertainty. The pgACC has been associated with suppression of further input, both nociceptive input and auditory input, suggesting the brain has acquired sufficient input. It can therefore generate a hedonic feeling, another phenomenological expression of pregenual activity.

mind also disappears. Whereas most scientists are or were reductionist monists, more recently emergentist monists suggest that mind is an emergent property of the brain as a complex (adaptive) network. Just like a car is an emergent property of multiple parts put together in a very specific pattern or structure, it is the pattern of a brain network that creates the emergent property, e.g., craving. As such, craving is conceived of as a network property and not a phrenological one-area problem. The importance of this philosophical concept for the application of neuromodulation for the treatment of addiction is that more targets can have similar effects, as it is possible to interfere with the craving or withdrawal network at any of the involved brain areas.

Recent developments of network science enable one to study complex adaptive systems (CAS) and these analyses have been introduced into brain science as well. The underlying idea is that CAS, whether they are the Internet, air-traffic routes, ant societies, human social interactions, the weather, or the economy, can all be described by using similar universal rules (Barabasi, 2009). To be considered as a CAS, only two criteria need to be fulfilled: First, the network structure must have a small-world topology and second, the network must have noise (Amaral, Diaz-Guilera, Moreira, Goldberger, & Lipsitz, 2004). As the brain has a small-world topology and is characterized structurally and functionally by $1/f$ noise, the brain clearly fulfills the requirements of a CAS. Network topology describes how different nodes in a network are connected or linked. It was initially assumed that networks predominantly form randomly, in which each node is connected to another node randomly (Strogatz, 2001). However, this does not permit adaptation, since whenever a stimulus arrives in such a network a random response will arise. On the other side of the spectrum a lattice network or regular network does not permit adaptation either, as every stimulus will always activate a node identical to the ones to which it is connected. More recently, scale-free networks have been described, in which some nodes are more densely connected and are clustered (i.e., they have a shorter path length, turning them into hubs). These scale-free networks have so-called "small-world properties." This suggests that some nodes are clearly more critical with regard to the robustness of the network. The topological scale-free nature of the brain can be identified in its anatomy, electrical and magnetic activity, and blood oxygenation changes (as with functional magnetic resonance imaging (fMRI)) (Buzsaki & Mizuseki, 2014).

Both random and scale-free networks are very robust to random errors, but scale-free networks are more sensitive to attacks on hubs. Translating this to the neuromodulation of addiction, this means that craving or withdrawal in addiction is probably most effectively reduced by targeting the hubs of the craving or withdrawal network. As neuroplasticity means the nervous system can modify its structural and functional organization, neuromodulation can theoretically be used to change both structural and FC, i.e., change the network and thus the emergent problem.

And indeed, neurofeedback (Cannon et al., 2007), TMS (De Ridder, Song, & Vanneste, 2012; De Ridder et al., 2011; Fox, Buckner, White, Greicius, & Pascual-Leone, 2012; Vanneste, Focquaert, Van de Heyning, & De Ridder, 2011), and transcranial direct current stimulation (tDCS) (Hampstead, Brown, & Hartley, 2014; Vanneste et al., 2011) change both functional and effective (=directional) connectivity, thus fulfilling the requirements needed to be able to change emergence.

Neurofeedback

Neurofeedback acts by acquiring brain signals from a patient using electroencephalography (EEG), fMRI, or near-infrared spectroscopy (NIRS) (Birbaumer, Ramos Murguialday, Weber, & Montoya, 2009). The relevant aspects of this signal are extracted and fed back to the participant in real time. In most cases the feedback is visual on a monitor screen. As soon as the signal reaches a predefined target, the participant is rewarded. It is based on the seminal work of Miller (1969), demonstrating that autonomic functions can be modified through operant conditioning. Based on this idea, Sterman and Friar showed that it is possible to use operant conditioning to increase sensorimotor EEG rhythms. This leads to a decrease in seizures in epileptic patients. Successful results were also obtained for attention-deficit hyperactivity disorder by training α and decreasing θ activity.

EEG visualizes the electrical voltage related to synchronous activity of cortical pyramidal neurons. Compared to other functional brain imaging techniques EEG has a high temporal resolution (sensitivity to changes over time) but low spatial resolution. The development of digital EEG has permitted the quantification of activity in different frequency bands and the creation of normative databases, which can aid in the diagnosis of brain-related diseases, such as addiction. Further developments include algorithms for localizing the most likely sources of the electrical signals recorded at the scalp. Neurofeedback can be done based on sensor space (scalp EEG feedback) or source space (source-localized feedback). Furthermore digital quantitative EEG data with high temporal resolution permit a calculation of the relationship between the activity in various locations, also known as FC (lagged phase synchronization), and information transfer between areas, also known as EC, in other words directional FC. EEG neurofeedback links real-time feedback of the signals recorded by the EEG with a target representing the desired brain activity resulting in operant conditioning, i.e., trial-and-error learning. The operant conditioning is usually performed by selecting brain waves that are related to the pathology or by comparing selected brain waves to a normative database (Z score neurofeedback), which aims to down-train activity that is considered pathological.

In an EEG study looking at drug abusers and controls it was shown that drug abusers had higher synchronization levels at low frequencies, mainly in the θ band (4–8 Hz) between frontal and posterior cortical regions. During a counting task, patients showed increased synchronization in the β- (14–35 Hz) and γ- (35–45 Hz) frequency bands, in frontoposterior and interhemispheric temporal regions (Coullaut-Valera et al., 2014). However, the question remains whether that is a cause or consequence of addiction. In chronic opioid users it has been shown that significant negative relations exist between duration (in years) of daily opioid abuse and the number/strength of functional connections in the posterior section of the cortex (Fingelkurts et al., 2006).

Neurofeedback (NFB) for addictive disorders started in 1989 with the presentation of initial results obtained with an α–θ strengthening protocol for alcohol dependence (Peniston & Kulkosky, 1989). In this approach, two slow frequencies are recorded midway back on the skull (PZ), α (8–13 Hz) and θ (4–7 Hz) (Saxby & Peniston, 1995), with eyes closed. The NFB induces a state on the border between sleeping and waking. In a controlled randomized study of 30 patients, 10 were randomized to relaxation training followed by α–θ NFB, one control group followed the 12-step treatment approach, and another group of 10 received traditional treatments. Eight of ten patients in the NFB group became abstinent and seven remained so for 3 years (end of follow-up), while all the patients in the other groups relapsed. Similar results were obtained by adding a β-SMR (sensorimotor rhythm 12–15 Hz) before the α–θ protocol in a larger controlled study of 121 patients with addiction (Scott, Kaiser, Othmer, & Sideroff, 2005). The same protocol was used for opiate addiction in a controlled study of 20 patients, also demonstrating that NFB was effective and superior to conservative medical management (Dehghani-Arani, Rostami, & Nadali, 2013). However, no placebo-controlled NFB study has yet been performed for addiction. Therefore, more research is required, particularly sham-controlled trials, to be able to demonstrate scientifically whether NFB is effective. Based on the published clinical studies and employing (relatively liberal) efficacy criteria adapted by the Association for Applied Psychophysiology and Biofeedback and the International Society for Neurofeedback and Research, α–θ training—either alone for alcoholism or in combination with β training—for stimulant and mixed substance abuse and combined with residential treatment programs is considered to probably be efficacious (Sokhadze, Cannon, & Trudeau, 2008).

Noninvasive Brain Stimulation: TMS and tDCS

As of this writing, there are two noninvasive brain stimulation techniques, TMS and transcranial electrical stimulation (tES). tES can be subdivided into three forms depending on the electrical current delivery: tDCS, transcranial alternating current stimulation (tACS), and transcranial random noise stimulation (tRNS) (Paulus, 2011) (Figure 5).

TMS

TMS is a noninvasive tool provoking a strong impulse of magnetic field that induces an electrical current, which can induce neuronal depolarization at the applied area. This makes it possible to selectively and safely stimulate specific regions of the human brain. Typically TMS is applied with a figure-8 coil (see Figure 5). Positron emission tomography scan studies have demonstrated that TMS not only modulates the directly stimulated cortical area, but also has an effect on remote areas functionally connected to the stimulated area (Hallett, 2000; Kimbrell et al., 2002). If the TMS stimulus is repeated over and over again in trains of stimulation, this is referred to as repetitive TMS (rTMS). A train of rTMS can modulate cortical excitability in a manner that lasts beyond the duration of the rTMS itself, i.e., it can induce a longer lasting modulatory effect, which can be used therapeutically.

It is known that in the motor cortex lower rTMS frequencies (i.e., 1 Hz) can usually suppress cortical excitability, while higher rTMS frequencies (i.e., 5–20 Hz) lead to a transient increase in cortical excitability (Zaghi, Acar, Hultgren, Boggio, & Fregni, 2010). rTMS can also be used to modulate cortical information processing via short-term interference with cortical function, creating a virtual lesion. Furthermore, low-frequency TMS increases FC in the network linked to the stimulated area, whereas high-frequency TMS decreases the FC (Fox et al., 2012).

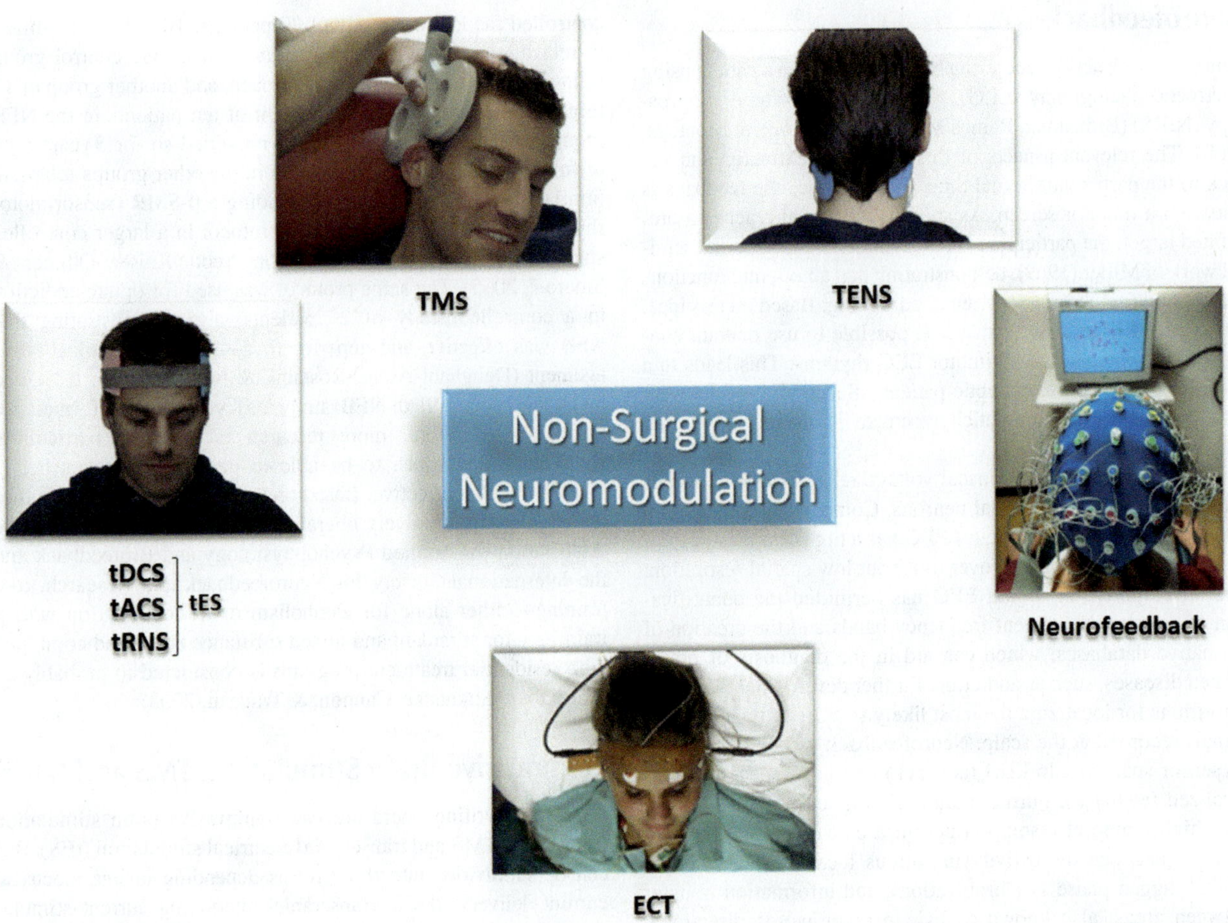

FIGURE 5 **The most commonly used nonsurgical neuromodulation techniques.** Pictures of the most commonly applied nonsurgical neuromodulation techniques are shown.

tDCS

tDCS is a noninvasive method of brain stimulation. When tDCS is applied in humans, a relatively weak constant current (between 0.5 and 2 mA) is passed through the cerebral cortex via scalp electrodes. Depending on the polarity of the stimulation, tDCS can increase or decrease cortical excitability in the brain regions to which it is applied (Miranda, Lomarev, & Hallett, 2006). Currently, tDCS is usually applied through two surface electrodes, one serving as the anode and the other as the cathode. Some of the applied current is shunted through scalp tissue and only a portion of the applied current passes through the brain. Anodal tDCS typically has an excitatory effect on the underlying cerebral cortex by depolarizing neurons, while the opposite occurs under the cathode owing to induced hyperpolarization. This effect of tDCS typically outlasts the stimulation duration by an hour or longer after a single treatment session of about 20 min duration.

Both TMS and tDCS have been used for craving reduction, particularly targeting the DLPFC (Jansen et al., 2013). A 2013 meta-analysis evaluating the effects of noninvasive TMS and tDCS showed a medium effect size favoring active noninvasive neurostimulation over sham stimulation in the reduction of craving (Jansen et al., 2013). No significant differences were found between TMS and tDCS, between the various substances of abuse, between substances of abuse and food, or between left and right DLPFC stimulation. This meta-analysis provides evidence that noninvasive neurostimulation of the DLPFC transiently decreases craving levels in addiction, independent of the substance of abuse (Jansen et al., 2013). TMS has been used for alcohol, cigarette, cocaine, and methamphetamine craving, as has tDCS (Jansen et al., 2013). tDCS is also used for food craving (Goldman et al., 2011; Jansen et al., 2013).

In total, as of this writing, fewer than 20 studies have been published that evaluate the temporary suppression of craving in addictive disorders (Gorelick, Zangen, & George, 2014; Jansen et al., 2013), and only five outpatient clinical trials were phase II studies, with no phase III trials yet conducted (Gorelick et al., 2014). Furthermore all tDCS randomized clinical trial studies consisted of one or two sessions only, and of the rTMS studies only a few involved 10 sessions or more (Jansen et al., 2013). Therefore all results should still be considered experimental, and to date demonstrate only that it is feasible to modulate craving temporarily. Although the available data appear promising, they do not provide sufficient evidence to recommend the use of either TMS or tDCS as a long-term treatment for addiction.

For tobacco addiction, TMS targeting the DLPFC significantly reduced spontaneous or cue-induced nicotine craving in most studies, but changes in craving were not always consistent with changes in smoking. Even if these findings also provide some evidence for the effectiveness of rTMS for smoking cessation, they challenge the use of craving as a surrogate marker for changes in smoking (Gorelick et al., 2014). tDCS seems to exert a beneficial effect on cue-evoked tobacco craving (Fregni et al., 2008).

For substance abuse, two double-blind and sham-controlled outpatient clinical trials have been performed (Gorelick et al., 2014; Politi, Fauci, Santoro, & Smeraldi, 2008). They both found that high-frequency TMS (10 or 20 Hz, respectively) targeted to the left DLPFC significantly decreased spontaneous cocaine craving as well as cocaine use, verified by urine drug testing. Thus, there is promising clinical trial evidence that at least 1 week of high-frequency rTMS targeted to the left DLPFC reduces cocaine craving and use (Gorelick et al., 2014). tDCS has been performed in an attempt to reduce marijuana craving, demonstrating that it can reduce cue-evoked craving (Boggio et al., 2010).

For alcohol addiction, most studies using a figure-8 coil targeting the DLPFC do not yield beneficial results (Gorelick et al., 2014). In contrast, a single outpatient case targeting low-frequency (1 Hz) TMS to the dACC with a double-cone coil (De Ridder et al., 2011) and three outpatient cases using high-frequency (20 Hz) rTMS to the DLPFC bilaterally using the so-called H-coil (Rapinesi et al., 2013) all reported decreased spontaneous alcohol craving after several sessions. The coils used in these studies have in common that they reach deeper brain structures compared to the figure-8 coils, which may explain the better efficacy. This would also fit with the above-mentioned knowledge about the pathophysiology of addiction. tDCS seems to yield better results when using bilateral tDCS delivered to DLPFC (anodal left/cathodal right and anodal right/cathodal left) (Boggio et al., 2008).

Brain Lesioning and Deep Brain Stimulation

Cingulotomies have been successfully used for alcohol dependence (Kanaka & Balasubramaniam, 1978). Even though they seem to have few side effects, apart from a decrease in sustained attention, they have the disadvantage of being irreversible. Furthermore, neuromodulation by implanted electrodes has the theoretical benefit that, depending on the stimulation design, different brain areas can be modulated (Kovacs et al., 2011) and therefore it has the advantage of greater versatility. Nacc lesions have also been performed for alcohol addiction and appear to be a safe method to alleviate alcohol craving, reduce relapse rates, and improve quality of life in patients with alcohol dependence (Wu et al., 2010), but have the same problem of irreversibility. Accumbens ablation in heroin addicts is better studied and carries a long-term relapse rate of 42% and some important, albeit rare, side effects. Therefore the authors concluded that "because ablation is irreversible, nacc surgery for addiction should be performed with caution, and DBS is an ideal alternative" (Li et al., 2013).

Deep brain stimulation (DBS) is a method in which electrodes are surgically introduced into the brain and these electrodes are connected to a programmable generator, which is placed under the skin. This internal pulse generator also contains a battery and delivers electrical pulses through the electrodes to a precisely defined area of the brain. DBS was initially used as a treatment for Parkinson disease in the late twentieth century. From the late 1990s, it was also applied in psychiatry. In 2005 it was first reported that in two patients with Parkinson disease comorbid addictive behavior improved after DBS of the nucleus subthalamicus (Witjas et al., 2005). In the following years more cases were described and also in a patient who received DBS of the Nacc for depression (Kuhn et al., 2009; Voges, Muller, Bogerts, Munte, & Heinze, 2013), and since then, DBS has been further investigated as a possible treatment for addiction, with 15 articles published on DBS in addiction as of this writing. In five of these, addiction was the primary indication for DBS, and in the other 10 articles effects on addictive behavior are described in patients in which DBS was performed for other indications.

A literature review compared the results of DBS of the Nacc and nucleus subthalamicus in both human and animal studies and concluded that the Nacc represents currently the more promising brain region for DBS in addiction (Luigjes et al., 2013). Nacc stimulation has been performed for alcohol dependence in few patients (only six cases published) (Kuhn et al., 2009; Voges et al., 2013), of whom three attained remission.

An improvement in addiction after DBS to the subthalamic nucleus has been described only in patients with Parkinson disease, and even among these patients results are highly variable, ranging from improvement to worsening of addictive behavior. Moreover, a major limitation of these studies is that the use of dopamine agonists or levodopa (which may be an important factor in the emergence or worsening of addictive behavior) is often reduced after treatment with DBS, making it difficult to determine whether the effect on addictive behavior in patients with Parkinson disease is directly related to DBS or rather an indirect effect due to the change in medication (Luigjes et al., 2013).

The addictions for which DBS is currently under investigation include alcohol, cocaine, and heroin addiction (Luigjes et al., 2013). Because clinical studies are lacking, little is known regarding the possible side effects of DBS in addiction. The few adverse events reported to date in subthalamic nucleus DBS include mild apathy in two patients and emotional lability in a single patient (Luigjes et al., 2013). Reported side effects for Nacc DBS include a hypomanic episode for 2 weeks in one patient and incontinence and confusion in the 12 h following surgery in another patient (Luigjes et al., 2013). In summary, as of this writing it is still too early to judge the efficacy and safety of DBS in addiction, as only case reports have been published.

Taken together, neuromodulation offers a unique opportunity to treat addiction, as it can build on the recently acquired pathophysiological knowledge regarding addiction. The initial experimental studies indicated that neuromodulation has the potential to reduce craving, cue reactivity, and substance use. However, most of these studies are proof-of-principle studies. With the exception of EEG NFB, the neuromodulation techniques discussed above as an intervention to treat addiction are relatively new. In particular, research on DBS is still in its infancy. The few studies that were published describe very small groups or individual patients. Regarding TMS/tDCS and EEG NFB there is currently a paucity of sham-controlled clinical trials such that any claims about their effectiveness remain questionable. As a result, there is still very little known about the effects of the different techniques.

Furthermore little is known regarding integrating neuromodulation in a multimodal approach to addiction. It is quite possible that the efficacy of neuromodulation will improve only when it is offered in combination with other interventions. As of this writing only EEG NFB has been studied in combination with behavioral therapy and medication.

Particularly in prolonged and severe addiction it is likely that behavioral support will remain important as a key role in the success of the treatment.

APPLICATIONS TO OTHER ADDICTIONS AND SUBSTANCE MISUSE

A heated debate exists as to whether obesity is a food addiction (Ziauddeen, Farooqi, & Fletcher, 2012). Although it has long been recognized that homeostatic centers in the brain play a pivotal role in body weight regulation, more recently brain areas similar to those involved in drug addiction have been implicated in food consumption. The question arises, however, whether there is indeed a subset of obese people who are food addicted. A quantitative measure of food addiction has been developed, the Yale Food Addiction Scale. Analogous to the fact that there is a fundamental difference between a social consumer of substances and an addicted person, there is most likely a subgroup of obese people who are food addicted, but the majority probably have another pathophysiological problem, which could be interoceptive in nature. Further studies have to be done to evaluate the neural correlates of both the food-addicted obese and the non-food-addicted obese. Lessons learned from addiction can help delineate the food-addicted group, and lessons learned from obesity can help further unravel addiction.

DEFINITION OF TERMS

Neuroplasticity The capacity of the nervous system to modify its structural and functional organization, adjusting itself to a changing environment.

Neuromodulation The induction of directed neuroplastic changes via local application of electrical, magnetic, acoustic, optic, tactile, or pharmacological stimuli.

Emergence A process whereby new properties arise through interactions among smaller or simpler entities that themselves do not exhibit such properties.

Uncertainty A state in which a given representation of the world cannot be adopted as a guide to subsequent behavior, cognition, or emotional processing

KEY FACTS

Key Facts of Addiction

- Addiction, also known as substance dependence, is a prevalent disease, with prevalence rates of 24% for tobacco, 14% for alcohol, and 7.5% for illicit drugs.
- Current treatment success for addiction is poor, with rate of relapse 1 year after standard treatment of between 40% and 70%.

Key Facts of Neuromodulation

- Neuromodulation is inextricably linked to neuroplasticity, which is the capacity of the nervous system to modify its structural and functional organization, adjusting itself to a changing environment. Therefore, neuromodulation can be defined as the induction of neuroplastic changes via local application of electrical, magnetic, acoustic, optic, tactile, or pharmacological stimuli.
- Neuromodulation can be nonsurgical, via TMS; tES consisting of tDCS, tACS, or tRNS; or NFB.
- Neuromodulation can be surgical via implanted electrodes.

Key Facts of Neuromodulation for Addiction

- Based on a better understanding of the neurobiological mechanisms involved in addiction, combined with historical data obtained from psychosurgery through brain lesioning, a theoretical heuristic pathophysiological framework can be developed delineating neuromodulation targets for addiction.
- The various targets used share a common final pathway, modulating the pgACC, which is involved in both processing hedonia and suppressing further input.
- A review of the neuromodulation techniques that have been applied does not permit us to propose any of these techniques to be used in routine clinical practice, but the encouraging preliminary results warrant further studies.

SUMMARY POINTS

- Treatments for addiction are of limited effectivity and characterized by high relapse rates, requiring the development of novel pathophysiology-based treatment approaches.
- One such option is to use neuromodulation.
- Neuromodulation can be defined as the use of techniques to focally induce directed neuroplasticity.
- Both nonsurgical and surgical neuromodulation have been experimentally used for the suppression of craving.
- Noninvasive techniques used include TMS, tES, and NFB.
- Surgical neuromodulation techniques involve lesioning and electrical stimulation via implanted electrodes.
- Most studies using noninvasive stimulation investigated the DLPFC as target.
- The two brain areas most commonly used as targets for lesioning are the anterior cingulate and Nacc.
- For implanted electrodes the Nacc and subthalamic nucleus have been investigated.
- The targets used can be linked to brain circuits involved in craving or withdrawal.
- For noninvasive neuromodulation no long-term studies have been performed and for implanted electrodes only small case series have been reported.
- Thus even if results seem to be promising, they still have to be considered preliminary.

REFERENCES

Albert, R., Jeong, H., & Barabasi, A. L. (2000). Error and attack tolerance of complex networks. *Nature, 406*(6794), 378–382.

Amaral, L. A., Diaz-Guilera, A., Moreira, A. A., Goldberger, A. L., & Lipsitz, L. A. (2004). Emergence of complex dynamics in a simple model of signaling networks. *Proceedings of the National Academy of Sciences of the United States of America, 101*(44), 15551–15555.

Aminoff, E., Gronau, N., & Bar, M. (2007). The parahippocampal cortex mediates spatial and nonspatial associations. *Cerebral Cortex, 17*(7), 1493–1503.

Barabasi, A. L. (2009). Scale-free networks: a decade and beyond. *Science, 325*(5939), 412–413.

Birbaumer, N., Ramos Murguialday, A., Weber, C., & Montoya, P. (2009). Neurofeedback and brain-computer interface clinical applications. *International Review of Neurobiology, 86*, 107–117.

Boggio, P. S., Sultani, N., Fecteau, S., Merabet, L., Mecca, T., Pascual-Leone, A., ... Fregni, F. (2008). Prefrontal cortex modulation using transcranial DC stimulation reduces alcohol craving: a double-blind, sham-controlled study. *Drug and Alcohol Dependence, 92*(1–3), 55–60.

Boggio, P. S., Zaghi, S., Villani, A. B., Fecteau, S., Pascual-Leone, A., & Fregni, F. (2010). Modulation of risk-taking in marijuana users by transcranial direct current stimulation (tDCS) of the dorsolateral prefrontal cortex (DLPFC). *Drug and Alcohol Dependence, 112*(3), 220–225.

Buzsaki, G., & Mizuseki, K. (2014). The log-dynamic brain: how skewed distributions affect network operations. *Nature Reviews Neuroscience, 15*(4), 264–278.

Camchong, J., Macdonald, A. W., 3rd, Mueller, B. A., Nelson, B., Specker, S., Slaymaker, V., & Lim, K. O. (2014). Changes in resting functional connectivity during abstinence in stimulant use disorder: a preliminary comparison of relapsers and abstainers. *Drug and Alcohol Dependence, 139*, 145–151.

Cannon, R., Lubar, J., Congedo, M., Thornton, K., Towler, K., & Hutchens, T. (2007). The effects of neurofeedback training in the cognitive division of the anterior cingulate gyrus. *International Journal of Neuroscience, 117*(3), 337–357.

Caspi, A., & Moffitt, T. E. (2006). Gene-environment interactions in psychiatry: joining forces with neuroscience. *Nature Reviews Neuroscience, 7*(7), 583–590.

Coullaut-Valera, R., Arbaiza, I., Bajo, R., Arrue, R., Lopez, M. E., Coullaut-Valera, J., ... Papo, D. (2014). Drug polyconsumption is associated with increased synchronization of brain electrical-activity at rest and in a counting task. *International Journal of Neural Systems, 24*(1), 1450005.

Daniel, R., & Pollmann, S. (2014). A universal role of the ventral striatum in reward-based learning: evidence from human studies. *Neurobiology of Learning and Memory, 114C*, 90–100.

Dehghani-Arani, F., Rostami, R., & Nadali, H. (2013). Neurofeedback training for opiate addiction: improvement of mental health and craving. *Applied Psychophysiology and Biofeedback, 38*(2), 133–141.

De Ridder, D., Song, J. J., & Vanneste, S. (2012). Frontal cortex TMS for tinnitus. *Brain Stimulation, 6*(3), 355–362.

De Ridder, D., Vanneste, S., & Freeman, W. (2014). The Bayesian brain: phantom percepts resolve sensory uncertainty. *Neuroscience & Biobehavioral Reviews, 44C*, 4–15.

De Ridder, D., Vanneste, S., Kovacs, S., Sunaert, S., & Dom, G. (2011). Transient alcohol craving suppression by rTMS of dorsal anterior cingulate: an fMRI and LORETA EEG study. *Neuroscience Letters, 496*(1), 5–10.

De Ridder, D., Vanneste, S., Menovsky, T., & Langguth, B. (2012). Surgical brain modulation for tinnitus: the past, present and future. *Journal of Neurosurgical Sciences, 56*(4), 323–340. doi:R38122449 [pii].

Feduccia, A. A., Chatterjee, S., & Bartlett, S. E. (2012). Neuronal nicotinic acetylcholine receptors: neuroplastic changes underlying alcohol and nicotine addictions. *Frontiers in Molecular Neuroscience, 5*, 83.

Filbey, F. M., & Dunlop, J. (2014). Differential reward network functional connectivity in cannabis dependent and non-dependent users. *Drug and Alcohol Dependence, 140*, 101–111.

Fingelkurts, A. A., Fingelkurts, A. A., Kivisaari, R., Autti, T., Borisov, S., Puuskari, V., ... Kahkonen, S. (2006). Increased local and decreased remote functional connectivity at EEG alpha and beta frequency bands in opioid-dependent patients. *Psychopharmacology (Berl), 188*(1), 42–52.

Fox, M. D., Buckner, R. L., White, M. P., Greicius, M. D., & Pascual-Leone, A. (2012). Efficacy of transcranial magnetic stimulation targets for depression is related to intrinsic functional connectivity with the subgenual cingulate. *Biological Psychiatry, 72*(7), 595–603.

Fregni, F., Liguori, P., Fecteau, S., Nitsche, M. A., Pascual-Leone, A., & Boggio, P. S. (2008). Cortical stimulation of the prefrontal cortex with transcranial direct current stimulation reduces cue-provoked smoking craving: a randomized, sham-controlled study. *The Journal of Clinical Psychiatry, 69*(1), 32–40.

Goldman, R. L., Borckardt, J. J., Frohman, H. A., O'Neil, P. M., Madan, A., Campbell, L. K., ... George, M. S. (2011). Prefrontal cortex transcranial direct current stimulation (tDCS) temporarily reduces food cravings and increases the self-reported ability to resist food in adults with frequent food craving. *Appetite, 56*(3), 741–746.

Gorelick, D. A., Zangen, A., & George, M. S. (2014). Transcranial magnetic stimulation in the treatment of substance addiction. *Annals of the New York Academy of Sciences, 1327*, 79–93. http://dx.doi.org/10.1111/nyas.12479.

Gu, H., Salmeron, B. J., Ross, T. J., Geng, X., Zhan, W., Stein, E. A. & Yang, Y. (2010). Mesocorticolimbic circuits are impaired in chronic cocaine users as demonstrated by resting-state functional connectivity. *Neuroimage, 53*(2), 593–601.

Hallett, M. (2000). Transcranial magnetic stimulation and the human brain. *Nature, 406*(6792), 147–150.

Hampstead, B. M., Brown, G. S., & Hartley, J. F. (2014). Transcranial direct current stimulation modulates activation and effective connectivity during spatial navigation. *Brain Stimulation, 7*(2), 314–324.

Harris, S., Sheth, S. A., & Cohen, M. S. (2008). Functional neuroimaging of belief, disbelief, and uncertainty. *Annals of Neurology, 63*(2), 141–147.

Huang, W., King, J. A., Ursprung, W. W., Zheng, S., Zhang, N., Kennedy, D. N., ... DiFranza, J. R. (2014). The development and expression of physical nicotine dependence corresponds to structural and functional alterations in the anterior cingulate-precuneus pathway. *Brain and Behavior, 4*(3), 408–417.

Jackson, S. R., Parkinson, A., Kim, S. Y., Schuermann, M., & Eickhoff, S. B. (2011). On the functional anatomy of the urge-for-action. *Cognitive Neuroscience, 2*(3–4), 227–243.

Jansen, J. M., Daams, J. G., Koeter, M. W., Veltman, D. J., van den Brink, W., & Goudriaan, A. E. (2013). Effects of non-invasive neurostimulation on craving: a meta-analysis. *Neuroscience & Biobehavioral Reviews, 37*(10 Pt 2), 2472–2480.

Kanaka, T. S., & Balasubramaniam, V. (1978). Stereotactic cingulumotomy for drug addiction. *Applied Neurophysiology, 41*(1–4), 86–92.

Kimbrell, T. A., Dunn, R. T., George, M. S., Danielson, A. L., Willis, M. W., Repella, J. D., ... Wassermann, E. M. (2002). Left prefrontal-repetitive transcranial magnetic stimulation (rTMS) and regional cerebral glucose metabolism in normal volunteers. *Psychiatry Research, 115*(3), 101–113.

Klein, T. A., Neumann, J., Reuter, M., Hennig, J., von Cramon, D. Y., & Ullsperger, M. (2007). Genetically determined differences in learning from errors. *Science, 318*(5856), 1642–1645.

Konova, A. B., Moeller, S. J., & Goldstein, R. Z. (2013). Common and distinct neural targets of treatment: changing brain function in substance addiction. *Neuroscience & Biobehavioral Reviews, 37*(10 Pt 2), 2806–2817.

Koob, G. F., & Le Moal, M. (2008). Addiction and the brain antireward system. *Annual Review of Psychology, 59*, 29–53.

Koob, G. F., & Volkow, N. D. (2010). Neurocircuitry of addiction. *Neuropsychopharmacology, 35*(1), 217–238.

Kovacs, S., Peeters, R., De Ridder, D., Plazier, M., Menovsky, T., & Sunaert, S. (2011). Central effects of occipital nerve electrical stimulation studied by functional magnetic resonance imaging. *Neuromodulation, 14*(1), 46–55; discussion 56–47.

Kringelbach, M. L. (2005). The human orbitofrontal cortex: linking reward to hedonic experience. *Nature Reviews Neuroscience, 6*(9), 691–702.

Kuhn, J., Buhrle, C. P., Lenartz, D., & Sturm, V. (2013). Deep brain stimulation in addiction due to psychoactive substance use. *Handbook of Clinical Neurology, 116*, 259–269.

Kuhn, J., Lenartz, D., Huff, W., Lee, S. H., Koulousakis, A., Klosterkoetter, J., & Sturm, V. (2009). Remission of alcohol dependency following deep brain stimulation of the nucleus accumbens: valuable therapeutic implications? *BMJ Case Reports, 2009*.

Kuhn, S., & Gallinat, J. (2011). Common biology of craving across legal and illegal drugs - a quantitative meta-analysis of cue-reactivity brain response. *European Journal of Neuroscience, 33*(7), 1318–1326.

Lerman, C., Gu, H., Loughead, J., Ruparel, K., Yang, Y., & Stein, E. A. (2014). Large-scale brain network coupling predicts acute nicotine abstinence effects on craving and cognitive function. *JAMA Psychiatry, 71*(5), 523–530.

Li, N., Wang, J., Wang, X. L., Chang, C. W., Ge, S. N., Gao, L., ... Gao, G. D. (2013). Nucleus accumbens surgery for addiction. *World Neurosurgery, 80*(3–4), S28.e9–S28.e19.

Luigjes, J., de Kwaasteniet, B. P., de Koning, P. P., Oudijn, M. S., van den Munckhof, P., Schuurman, P. R., & Denys, D. (2013). Surgery for psychiatric disorders. *World Neurosurgery, 80*(3–4), S31.e17–S31.e28.

McGue, M., Pickens, R. W., & Svikis, D. S. (1992). Sex and age effects on the inheritance of alcohol problems: a twin study. *Journal of Abnormal Psychology, 101*(1), 3–17.

Miller, N. E. (1969). Learning of visceral and glandular responses. *Science, 163*(866), 434–445.

Miranda, P. C., Lomarev, M., & Hallett, M. (2006). Modeling the current distribution during transcranial direct current stimulation. *Clinical Neurophysiology, 117*(7), 1623–1629.

Motzkin, J. C., Baskin-Sommers, A., Newman, J. P., Kiehl, K. A., & Koenigs, M. (2014). Neural correlates of substance abuse: reduced functional connectivity between areas underlying reward and cognitive control. *Human Brain Mapping, 35*(9), 4282–4292.

Paulus, W. (2011). Transcranial electrical stimulation (tES - tDCS; tRNS, tACS) methods. *Neuropsychological Rehabilitation, 21*(5), 602–617.

Peniston, E. G., & Kulkosky, P. J. (1989). Alpha-theta brainwave training and beta-endorphin levels in alcoholics. *Alcoholism Clinical and Experimental Research, 13*(2), 271–279.

Pitchers, K. K., Schmid, S., Di Sebastiano, A. R., Wang, X., Laviolette, S. R., Lehman, M. N., & Coolen, L. M. (2012). Natural reward experience alters AMPA and NMDA receptor distribution and function in the nucleus accumbens. *PLoS One, 7*(4), e34700.

Politi, E., Fauci, E., Santoro, A., & Smeraldi, E. (2008). Daily sessions of transcranial magnetic stimulation to the left prefrontal cortex gradually reduce cocaine craving. *The American Journal of Addictions, 17*(4), 345–346.

Rapinesi, C., Kotzalidis, G. D., Serata, D., Del Casale, A., Bersani, F. S., Solfanelli, A., ... Girardi, P. (2013). Efficacy of add-on deep transcranial magnetic stimulation in comorbid alcohol dependence and dysthymic disorder: three case reports. *The Primary Care Companion for CNS Disorders, 15*(1).

Rushworth, M. F., & Behrens, T. E. (2008). Choice, uncertainty and value in prefrontal and cingulate cortex. *Nature Neuroscience, 11*(4), 389–397.

Saxby, E., & Peniston, E. G. (1995). Alpha-theta brainwave neurofeedback training: an effective treatment for male and female alcoholics with depressive symptoms. *Journal of Clinical Psychology, 51*(5), 685–693.

Schacht, J. P., Anton, R. F., & Myrick, H. (2013). Functional neuroimaging studies of alcohol cue reactivity: a quantitative meta-analysis and systematic review. *Addiction Biology, 18*(1), 121–133.

Scott, W. C., Kaiser, D., Othmer, S., & Sideroff, S. I. (2005). Effects of an EEG biofeedback protocol on a mixed substance abusing population. *American Journal of Drug and Alcohol Abuse, 31*(3), 455–469.

Seeley, W. W., Menon, V., Schatzberg, A. F., Keller, J., Glover, G. H., Kenna, H., ... Greicius, M. D. (2007). Dissociable intrinsic connectivity networks for salience processing and executive control. *The Journal of Neuroscience, 27*(9), 2349–2356.

Smith, L., Watson, M., Gates, S., Ball, D., & Foxcroft, D. (2008). Meta-analysis of the association of the Taq1A polymorphism with the risk of alcohol dependency: a HuGE gene-disease association review. *American Journal of Epidemiology, 167*(2), 125–138.

Sokhadze, T. M., Cannon, R. L., & Trudeau, D. L. (2008). EEG biofeedback as a treatment for substance use disorders: review, rating of efficacy, and recommendations for further research. *Applied Psychophysiology and Biofeedback, 33*(1), 1–28.

Strogatz, S. H. (2001). Exploring complex networks. *Nature, 410*(6825), 268–276.

Sutherland, M. T., Carroll, A. J., Salmeron, B. J., Ross, T. J., & Stein, E. A. (2013). Insula's functional connectivity with ventromedial prefrontal cortex mediates the impact of trait alexithymia on state tobacco craving. *Psychopharmacology (Berl), 228*(1), 143–155.

Swift, R. M. (1999). Drug therapy for alcohol dependence. *The New England Journal of Medicine, 340*(19), 1482–1490.

Uhl, G. R., & Grow, R. W. (2004). The burden of complex genetics in brain disorders. *Archives of General Psychiatry, 61*(3), 223–229.

Vanneste, S., Focquaert, F., Van de Heyning, P., & De Ridder, D. (2011). Different resting state brain activity and functional connectivity in patients who respond and not respond to bifrontal tDCS for tinnitus suppression. *Experimental Brain Research, 210*(2), 217–227.

Voges, J., Muller, U., Bogerts, B., Munte, T., & Heinze, H. J. (2013). Deep brain stimulation surgery for alcohol addiction. *World Neurosurgery, 80*(3–4), S28.e21–S28.e31.

Witjas, T., Baunez, C., Henry, J. M., Delfini, M., Regis, J., Cherif, A. A., ... Azulay, J. P. (2005). Addiction in Parkinson's disease: impact of subthalamic nucleus deep brain stimulation. *Movement Disorders: Official Journal of the Movement Disorder Society, 20*(8), 1052–1055.

Wu, H. M., Wang, X. L., Chang, C. W., Li, N., Gao, L., Geng, N., ... Gao, G. D. (2010). Preliminary findings in ablating the nucleus accumbens using stereotactic surgery for alleviating psychological dependence on alcohol. *Neuroscience Letters, 473*(2), 77–81.

Zaghi, S., Acar, M., Hultgren, B., Boggio, P. S., & Fregni, F. (2010). Noninvasive brain stimulation with low-intensity electrical currents: putative mechanisms of action for direct and alternating current stimulation. *Neuroscientist, 16*(3), 285–307.

Ziauddeen, H., Farooqi, I. S., & Fletcher, P. C. (2012). Obesity and the brain: how convincing is the addiction model? *Nature Reviews Neuroscience, 13*(4), 279–286.

Chapter 3

Gateway Hypothesis of Addiction and Substance Misuse: The Role of Neurophysiology and Psychological Mechanisms

Hamdy Fouad Moselhy[1], Mahmoud A. Awara[2]
[1]*College of Medicine and Health Science, United Arab Emirates University, Alain City, UAE;* [2]*Dalhousie University, Halifax, NS, Canada*

INTRODUCTION

Although there are many studies that favor the gateway hypothesis, more recent results have challenged the universal applicability of this hypothesis (Bretteville-Jensen, Melberg, & Jones, 2008; Melberg, Jones, & Bretteville-Jensen, 2010). However, understanding the pathways of progression of substance use disorders is important from the standpoint of learning about the evolution of substance use disorders in the community. This is also potentially helpful in predicting specific individuals who are at risk so that they do not progress with further indulgence in other substance misuse (Botvin, Scheier, & Griffin, 2002). While pharmacological differences exist among the various classes of abused drugs, the one-shared factor among all drugs with a strong dependence potential is a rewarding or reinforcing property. Research on the neurobiology of addiction has shown that the reinforcing properties of most misused drugs are mediated by activation of the mesolimbic dopaminergic system, the orbitofrontal cortex, and the extended amygdala (Cunha-Oliveira, Rego, & Oliveira, 2008). Alterations in the intracellular messenger pathways, transcription factors, and immediate early gene expression in these reward circuits are believed to be important for the development of addiction and chronic drug abuse (Koob & Volkow, 2010). There are several lines of evidence suggesting that dopaminergic pathways are implicated in at least some reward circuits, and various drugs may activate or "switch on" the circuits at different points (Pradhan & Dutta, 1977). Several studies consistently show that individual variations (e.g., genetic makeup, personality traits) and environmental factors (e.g., drug availability, peer influences, life events) may explain why young people initiate drug use and move on to additional forms of drug use later in life. In general we can improvise that drug-related behavior is the consequence of a chain reaction between the drugs, the individual, and the society—and the key word to describe this would be "interaction." None of the component factors alone is sufficient to cause drug dependence (figure 1).

NEUROPHYSIOLOGICAL MECHANISMS

Drug addiction has been conceptualized as a disorder that involves elements of both impulsivity and compulsivity that yield a composite addiction cycle composed of three stages: binge/intoxication, withdrawal/negative affect, and preoccupation/anticipation (craving). Animal and human imaging studies have revealed discrete circuits that mediate the three stages of the addiction cycle with key elements of the ventral tegmental area and ventral striatum as a focal point for the binge/intoxication stage, a key role for the extended amygdala in the withdrawal/negative affect stage, and a key role in the preoccupation/anticipation stage for a widely distributed network involving the orbitofrontal cortex–dorsal striatum, prefrontal cortex, basolateral amygdala, hippocampus, and insula, involved in craving, and the cingulate gyrus, dorsolateral prefrontal, and inferior frontal cortices in disrupted inhibitory control. The transition to addiction involves neuroplasticity in all of these structures that may begin with changes in the mesolimbic dopamine system and a cascade of neuroadaptations from the ventral striatum to the dorsal striatum and orbitofrontal cortex and, eventually, dysregulation of the prefrontal cortex, cingulate gyrus, and extended amygdala (Koob & Volkow. 2010). This is why the concept of addiction as a brain disease has been redefined as chronic relapsing brain diseases characterized by the central feature of compulsive drug seeking and use (Leshner, 1997).

The most prominent neurotransmitters and neuropeptides in this process are dopamine, serotonin, glutamate, γ-aminobutyric acid (GABA), and the endorphins (endogenous opioids) (Hayman, 1993). Dopamine is credited with playing the primary role in producing most of the euphoria seen with cocaine, nicotine, alcohol, and

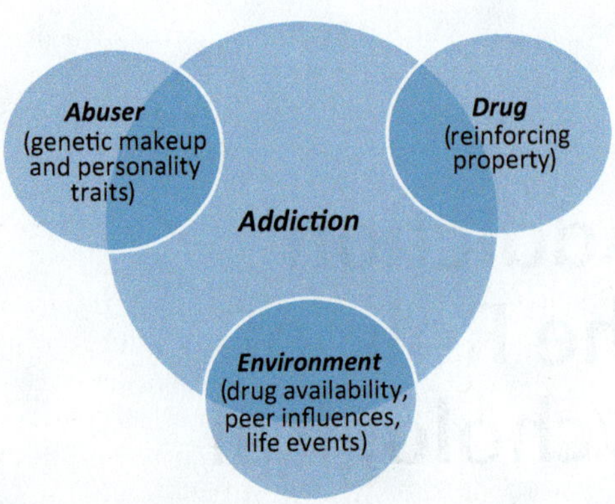

FIGURE 1 Interaction model of addiction.

other stimulants and is known to contribute to the positive reinforcing effects of heroin and other opiates (Kreek & Koob, 1998) (Figure 2).

The Role of Dopamine

Most drugs that produce elevation of mood or euphoria, including nicotine and alcohol, release dopamine into either the nucleus accumbens or the prefrontal cortex of animals, as demonstrated by brain dialysis (Di Chiara & Imperto, 1988). Dopamine release can be either direct (for example, the stimulants, e.g., cocaine and amphetamine, release dopamine) or indirect (opioids switch off the firing of GABA neurons that tonically inhibit dopamine cell firing). Several studies have shown that blockade of either D1 or D2 dopamine receptors attenuates the reinforcing actions of both these

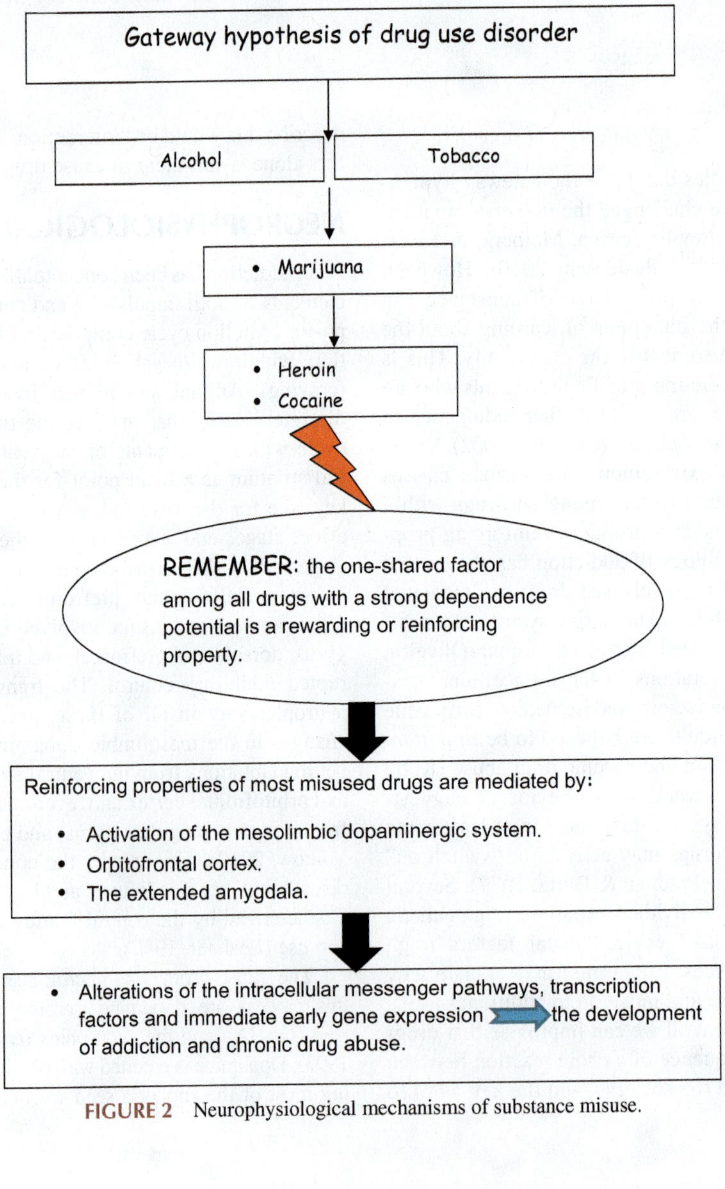

FIGURE 2 Neurophysiological mechanisms of substance misuse.

classes of drugs, which argues for a central mediating role of dopamine receptor activation in the initiation of addiction (Hubner & Moreton, 1991). Of clinical relevance is the suggestion that a genetic polymorphism of the D2 receptor is strongly linked to drug misuse, but this is still controversial (Goldman, 1995).

Homeostatic adaptation to the dopamine-increasing actions of drugs occurs, so that when the drug is stopped, dopamine release is subsequently decreased below normal; this explains the "crash" after stimulant discontinuation (Gawin, 1991) and some aspects of nicotine and alcohol withdrawal. Drugs that block dopamine reuptake (e.g., desipramine and mazindol, which are used to treat cocaine withdrawal) presumably work by increasing dopamine concentrations (Gawin, 1991). Dopamine overactivity probably underlies alcohol-related delirium tremens and the need to treat with dopamine D2 receptor-blocking agents such as haloperidol (Glue & Nutt, 1990). Chronic dysregulation of dopamine function in detoxified alcoholics as revealed by a decreased number of uptake sites in single-photon emission computed tomography (SPECT) studies may explain the new finding that the low-potency neuroleptic tiapride reduces relapse (Shaw et al., 1994).

It is now possible to measure both D1 and D2 dopamine receptors and the dopamine uptake site in human beings with neuroimaging techniques. Cocaine has been shown with positron emission tomography (PET) to bind predominantly to the dopamine-rich areas of the basal ganglia (Pike, 1993). Now that D2 receptors and D1 receptors can be visualized, alterations in function in addicts can be studied (Pike, 1993). Dopamine metabolism can also be monitored in vivo with fluorodopa (Volpicelli, Alterman, Hayashida, & O'Brien, 1993). Similarly, the local metabolic effects of cocaine can be studied with fluorodeoxyglucose uptake. These studies have shown that cocaine globally decreases brain metabolic activity (London et al., 1990). An exciting potential development of PET/SPECT technology is to measure endogenous dopamine release; if cocaine and amphetamine do act by releasing dopamine this should be seen as a displacement of radiolabeled receptor ligand (Nutt, 1996).

Endogenous Agonist Opioids

The brain makes a complex mixture of peptides that act as endogenous transmitters at opioid receptors—the β-endorphins and enkephalins; these are involved in appetite, pain, and response to stress (London et al., 1990). Misused opioids such as heroin act at the same receptors as the natural opioid system. However, they have much higher efficacy than the endogenous transmitter, they "hijack" the natural system by producing a much exaggerated response. Endogenous opioids are thought to be involved in the actions of other misused drugs such as alcohol and stimulants. For example, alcohol may cause dependence because it releases endogenous opioids; this could explain the therapeutic benefit of opioid antagonists such as naltrexone in breaking the dependence cycle (Volpicelli et al., 1993).

There are three types of opioid receptors (μ, δ, and κ) that are distinguished by selective agonists and in some cases antagonists. The μ and δ receptors mediate the euphoric actions of opioids (Koob, 1992), with κ receptors being aversive, and could explain some aspects of opioid actions including dysphoric symptom of withdrawal (Spanagel, Herz, & Shippenberg, 1990). Many misused opioids have activity at all three receptor types so adaptive changes in each may be important in the process of addiction (Nutt, 1996).

In light of the above, it is noticeable that advances in basic and clinical neurosciences and behavioral sciences have revolutionized our understanding of the neurobiology of addiction.

Interactions of different classes of drugs have been proposed to support the physiological mechanism of the gateway hypothesis. Historically cannabinoids were used in combination with opioids for the treatment of various types of pain in humans because of their synergistic interactions in the modulation of noxious stimuli. Cannabinoids produce a variety of pharmacological effects very similar to those elicited by opioids. Direct and indirect interactions with the opioid system have been implicated to explain some cannabinoid effects such as analgesia and attenuation of opioid-withdrawal syndrome. Evidence has been found to support the notion that the rewarding properties of cannabinoids and opioids might be functionally linked. In particular, there is growing evidence from research studies that highlight the important role of the endogenous cannabinoid system in the modulation of the opioid rewarding effect and its addictive impact. It is intuitively known that the use of cannabinoids is related to their positive modulatory effects on brain-rewarding processes along with their ability to positively influence emotional states and remove stress responses to environmental stimuli. Cannabinoids have been tested on a variety of behavioral models of addiction, most of which revealed functional interactions between the endocannabinoid and the opioid systems in the modulation of reciprocal rewarding and addictive effects. Cannabis combines many of the properties of alcohol, tranquilizers, opioids, and hallucinogens; it is anxiolytic, sedative, analgesic, and psychedelic (Ashton, 2001).

Cannabinoids and opioids share many pharmacological properties, including antinociception, hypothermia, sedation, and inhibition of intestinal motility. Chronic administration of both agents produces tolerance to their analgesic and hypothermic effects and leads to the development of physical dependence, but with different intensities nevertheless. In addition to analgesia, endogenous cannabinoids interact with the opioid system in a variety of biological functions, including emesis, intestinal motility, and immune activity as well as modulation of anxiety, stress, emotion, and exploratory behavior. When the caudate-putamen (CP) of rats is treated with repeated administration of the central cannabinoid (CB1) receptor ligand, Δ^9-tetrahydrocannabinol, there are an increase in proenkephalin gene expression and μ-opioid receptor activation of G proteins, a time-related decrease in CB1 receptor gene expression, and a reduction in CB1 receptor activation of G proteins. These findings suggest a possible interaction between the cannabinoid and the opioid systems in a brain area (i.e., CP) (Maldonado & Berrendero, 2010).

A significant amount of research studies suggest that the endogenous cannabinoid and opioid systems play a pivotal modulatory effect on the reward circuitry and participate in addictive propensities of most drug abuse, including that of nicotine. Clinical trials suggest that manipulation of these systems with cannabinoid or opioid antagonists could be of potential therapeutic strategy for treating nicotine addiction (Maldonado & Berrendero, 2010).

TRANSITION TO ADDICTION AND PSYCHOLOGICAL MECHANISMS

There is no doubt that the sociological, economic, and psychological factors leading to the inappropriate use of mood-altering drugs are of paramount importance. Likewise, the use of soft drugs may contribute to further abuse of harder drugs through psychological processes. Several hypotheses have focused on either personality characteristics or environmental circumstances of drug use and its progression. The social environment of an addict, in which drug abuse behavior is accepted and predominant among friends and associates, may be an important factor in inducing the individual to drug use. The environment, settings, and cultural rituals may serve as a secondary reinforcement. Sometimes such social reinforcement maintains drug use behavior in the initial experimental stage until the naïve user begins to appreciate the primary effect of the drug or becomes tolerant to some initial aversive effects of a particular drug. For example, some young people may not like the initial effects of smoking tobacco, or may not experience anything pleasurable in smoking marijuana, or may experience nausea and vomiting with an initial dose of heroin; the social reinforcers may maintain the drug use behavior until these initial adversities are overcome (Abdel Wahab, 2004; Pradhan & Dutta, 1977).

Experimental studies have substantiated the secondary reinforcing effects of environment. Acute drug effects, withdrawal manifestations, and their relief by reinstatement have been shown to be conditioned to environmental stimuli. These phenomena may also play an important role in the relapse of drug addiction. Thus a former narcotic addict may experience cravings for the drug when he or she returns to familiar environmental cues; an alcoholic may have a similar experience when exposed to the sight and smell of alcohol (Jaffe, Cascella, Kumor, & Sherer, 1989). These phenomena and other factors contribute to initiating and maintaining compulsive drug abuse behavior.

In addition to the impact of the environment on the evolution of addiction, social factors play an important role in the development of addiction and the pathways from one drug to another. The most widely accepted sociological theories of addiction fall into three classes: differential association, control, and strain (figure 3, and 4).

Differential Association Theory

This theory postulates that deviance arises when the values of membership in one group come into conflict with the values of more powerful groups who are able to empower their values into law (Sutherland, 1947). People opposing the law-enforced values are defined as deviant because of the values of their countergroups, not because of any unlawful behavior. Presumably people learn to deviate and conform using the same mechanisms, thus, the differential association process has come to incorporate the assumptions of social learning theory. Using the social learning model, Akers, Krohn, Lanza-Kaduce, & Radosevich (1979) demonstrated that differential

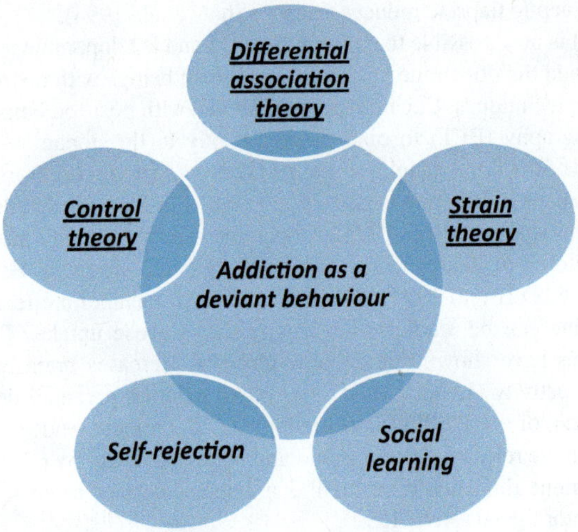

FIGURE 3 Social theories of addiction.

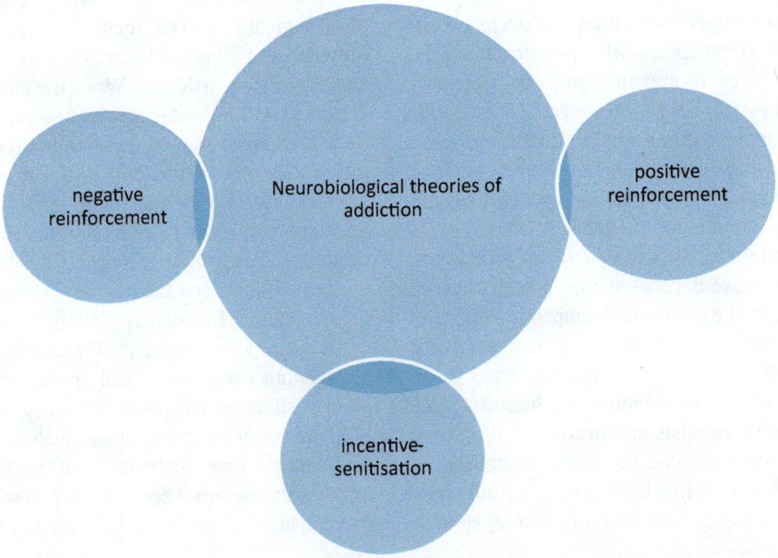

FIGURE 4 The interface between biological and psychosocial theories of addiction.

association variables explain the most variance in both alcohol and drug use among adolescents. In agreement with prior research (Jessor & Jessor, 1977; Kandel, 1978) their strongest predictor of substance use was having friends who use drugs and alcohol.

Control Theory

This theory posits that deviance arises when young people lack sufficient ties to conventional social groups such as families, schools, and churches (Hirschi, 1969). From this point of view, youth deviate not through frustrated desires or actions in accordance with their own reference groups, but because their ties to conventional groups are broken or underdeveloped. While it has become common in the delinquency literature to test both control and differential association theories simultaneously (Conger, 1973; Johnson, 1979; Matsueda, 1982), the limited research on substance use shows little association between the two (Dull, 1984).

Strain Theory

This theory emphasizes entirely different processes compared to differential association or control theory. Its central thesis is that people become prone to deviate when society is unable to satisfy their fundamental needs. A broad version of strain theory views various forms of deviance as mechanisms that allow people to cope with the stresses of everyday life (Dohrenwend & Dohrenwend, 1984; Pearlin, 1981). If strain theory is correct, substance use might be especially prevalent among adolescents who are experiencing the greatest amount of strain. While strain theory has not had much success in explaining delinquent behavior (Hirschi, 1969; Johnson, 1979; Kornhauser, 1978), it is possible that it is a more powerful predictor of adolescent substance use (White, Johnson, & Howrwitz, 1986).

A number of other theories and works have tried to explain various aspects of drug use behaviors including social learning (Akers, Krohn, Lanza-Kaduce, Radosevich, 1979), socialization (Kandel, 1978), social–psychological field theory, and self-rejection (Jessor & Jessor, 1977). However most of these works examine the extent to which a single theoretical perspective fits the data at hand or incorporates elements of several theories into one overarching model (White et al., 1986).

PERSONALITY OF THE DRUG USER

Although there is no consensus that drug users have special personality characteristics, the personality and the drug habit of the abuser may play important roles in initiation and maintenance of compulsive drug abuse. The personality profile of a person may be a factor in drug abuse; individuals with certain types of personality deficits may be more vulnerable to drug abuse. However, most people with such a persona do not become compulsive drug users unless other predisposing factors are present. When such individuals are within a drug misuse-facilitating environment, predisposing factors and availability of psychoactive drugs that alter mood and perception serve as reinforcers for maintaining the drug misusing habit (Pradhan & Dutta, 1977). Drugs are used because they produce alterations in brain function that result in positive changes in mood; this could be an elevation in mood from normal (euphoria) or the reduction of a negative dysphoric mood as in withdrawal. These changes are affected by interactions with neurochemical processes, usually by mimicking or increasing the action of endogenous transmitters (Nutt, 1996). The most widely accepted theories of addiction currently fall into three classes: negative reinforcement models (e.g., drugs are taken to avoid the symptoms of withdrawal) (Dackis & Gold, 1985; Jaffe, 1990; Koob & Bloom, 1988), positive reinforcement models (i.e., drugs reinforce self-administration behavior by producing pleasure) (Stewart, de Wit, & Eikelboom, 1984; Wise, 1988; Wise & Bozarth, 1987), and the incentive-sensitization theory (Robinson & Berridge, 1993). None of these theories are mutually exclusive. Pleasure seeking, escapism from distress and life vicissitudes, and incentive sensitization probably each play some interchangeable role in drug misuse behavior (Table 1).

TABLE 1 Key Points in the Personality Profile of a Vulnerable Person Who May Develop Addiction

1. Clients with **paranoid personality disorder** can be attracted to the dominance drugs (alcohol, cocaine, and amphetamines), because they enhance the need for control that is central to the disorder (Benjamin, 1993). These drugs allow individuals with this personality to feel more powerful in a world that seems dangerous and hostile.
2. Clients with **schizoid personality disorder** may be attracted to psychedelic drugs and become addicted to the state of arousal and satisfaction involved in facilitated fantasy (Milkman & Sunderwirth, 1987).
3. Drugs such as marijuana and LSD may replicate the digressive, tangential quality of thought patterns already present in individuals with **schizotypal personality disorder**; and mere drug use can be enough to precipitate a psychiatric crisis. Psychoeducation is vital with these clients (Ekleberry, 1996).
4. Clients with **antisocial personality disorders**, perhaps due to low neurological arousal, often seek thrills and are likely to be most attracted to stimulants. Their use of alcohol and drugs bothers them only in terms of the pressure they receive from employers, family, or the criminal justice system (Ekleberry, 1996).
5. Clients with **borderline personality disorder** are the best candidates of all those with personality disorders for developing addictive disorders; they will use almost any drug of choice to worst advantage (Richards, 1993).
6. Clients with histrionic **personality disorder** may value drugs and alcohol or compulsive behaviors for social enhancement. Antianxiety drugs are often sought; but stimulants provide them with dramatic mood boosts (Richard, 1993).
7. Clients with **avoidant personality disorder (AvPD)** are vulnerable to substance use that can reduce interpersonal vulnerability or ease social paralysis. Drugs that will make a difference include sedative–hypnotics that calm anxiety and stimulants that provide a sense of strength or reduced vulnerability. Mild hallucinogens facilitate escape into fantasy and distract the AvPD client from the pain of his or her own self-absorption (Stone, 1993).
8. Clients with **dependent personality disorder** may use alcohol and other substance as a passive way to escape from problems (Beck, 1993).

Negative Reinforcement Views of Addiction (Escapism from Distress)

Historically, the aversive consequences of discontinuing drug use (the withdrawal syndrome) have been a central focus of research on addiction, in part because many previous studies were conducted on opiate addiction, which produces clear tolerance and withdrawal symptoms. This research emphasized the action of drugs as negative reinforcers (Dackis & Gold, 1985; Jaffe, 1990; Koob & Bloom, 1988). Negative reinforcers sustained behavior (drug seeking and drug taking) not because of the symptoms they produce, but because of the symptoms they alleviate. According to negative reinforcement theory, the addictive drug misuse behavior is maintained because the aversive symptoms associated with withdrawal are alleviated by the reuse of the drug. Addictive drugs that do not result in overt physical withdrawal symptoms, such as cocaine and amphetamines, are thought to act as negative reinforcers by alleviating a "psychological distress syndrome" produced by the discontinuation of drug use. In addition, previously neutral environmental stimuli associated with withdrawal can themselves elicit withdrawal-like symptoms, by secondary conditioning (Siegel, 1977, 1988). Thus, drugs may alleviate not only "primary" withdrawal symptoms, but also the conditioned withdrawal symptoms induced by exposure to drug-related stimuli. A second negative reinforcement view is that drugs are sometimes used to "self-medicate," relieving preexistent symptoms such as pain, anxiety, or depression or negative symptoms that are not directly related to the drug misuse problem (Robinson & Berridge, 1993).

The traditional focus on withdrawal and tolerance was driven by the assumption that physical dependence is critical for the development and maintenance of addictive behavior. It is now clear, however, that alleviating withdrawal symptoms is not the most important factor in maintaining nonaddictive behavior, although the avoidance of withdrawal may certainly motivate drug-seeking and -taking behavior in some instances. A significant number of research studies on drug addiction behavior have noted that "physical dependence is neither necessary nor sufficient for maintaining the addictive behavior" (Wise, 1988). Furthermore, for rats and monkeys, opiate physical dependence is neither a necessary nor a sufficient condition to act as a reinforcer (Schuster, 1990), and physical dependence is currently viewed as not a direct cause of drug dependency but as one of several factors that contribute to its development (Robinson & Berridge, 1993). A number of critical studies to a negative reinforcement theory of addiction have been published (Stewart et al., 1984; Wise, 1988; Wise & Bozarth, 1987) and the major shortcomings of a negative reinforcement theory in explaining addiction are briefly summarized below (Table 2).

Problems with Negative Reinforcement Views

Both humans and animals will self-administer opioids in the absence of withdrawal symptoms or physical dependence (Ternes, Ehrman, & O'Brien, 1985; Woods & Schuster, 1968). For example, Ternes et al. found that cynomolgus monkeys given the opioid hydromorphone maintained self-administration at doses that produced neither tolerance nor physical dependence, the latter indicated by the absence of any effect of a naloxone challenge.

TABLE 2 Key Points of Negative Reinforcement Theory

1. Drugs are taken to avoid the symptoms of withdrawal of the drug misused.
2. Drugs are sometimes used to "self-medicate," relieving preexistent symptoms such as pain, anxiety, or depression or negative symptoms that not directly related to the drug misuse problem.

Problems with Negative Reinforcement Views

1. Both humans and animals will self-administer opioids in the absence of withdrawal symptoms or physical dependence.
2. There is a lack of correlation between withdrawal distress and drug-seeking behavior.
3. There are many drugs that are being used medically that produce withdrawal syndromes but are not typically self-administered for nonmedical purposes, including certain tricyclic antidepressants.
4. There are numerous reports that the relief of withdrawal is minimally effective in treating addiction.
5. There is a high tendency to relapse even after an extended period of abstinence from drugs, long after overt withdrawal symptoms have subsided.
6. Self-reported craving for some drugs, such as cocaine, is often highest immediately after drug administration, when the drug is producing subjective pleasure (a high) and withdrawal symptoms are at their lowest.
7. Animals will avidly self-administer a variety of drugs directly into brain regions that do not produce withdrawal symptoms.

Maximal periods of drug self-administration often do not temporally coincide with periods of maximal withdrawal distress (Wise & Bozarth, 1987). This lack of correlation between withdrawal distress and drug-seeking behavior is also evident in comparisons made across drug classes. Jaffe (1990) notes that, although the severity of the withdrawal syndrome associated with different drugs varies dramatically, ranging from very subtle physiological signs to life-threatening consequences, "there is little correlation between the visibility of physiological seriousness of withdrawal signs, and their motivational force" in maintaining addictive behavior.

There are many drugs that are being used medically that produce withdrawal syndromes but are not typically self-administered for nonmedical purposes, including certain tricyclic antidepressants (imipramine, amitriptyline), anticholinergics, and κ-opioid agonists (Jaffe, 1990).

There are numerous reports that the relief of withdrawal is minimally effective in treating addiction (Wise & Bozarth, 1987).

There is a high tendency to relapse even after an extended period of abstinence from drugs, long after overt withdrawal symptoms have subsided. This is usually explained in the context of conditioned withdrawal effects, whereby environmental stimuli associated with withdrawal come to elicit withdrawal-like symptoms (Siegle, 1988; Wilker, 1948). There are, however, a number of problems with this explanation. (1) At least a third of opiate addicts deny that they experience conditioned withdrawal symptoms when they are exposed to drug-related stimuli (Childress, McLellan, Ehrman, & O'Brien, 1988). (2) Although many opiate addicts experience conditioned withdrawal symptoms very few cite this as the reason for resuming drug use behavior (McAuliffe, 1982).

There is, in fact, a poor correlation between craving and withdrawal signs (Childress et al., 1988). Even withdrawal-like physiological symptoms induced by drug-associated cues (e.g., temperature, skin resistance, heart rate) are not highly correlated with reports of subjective state (Ehrman, Ternes, O'Brien, & McLellan, 1992). (3) Stewart et al. (1984) argued that attempts to demonstrate a correlation between the conditioned withdrawal symptoms and the increase in probability of drug-taking behavior and relapse in animals have been unsuccessful (Stewart & Wise, 1992).

A number of researchers have noted that self-reported craving for some drugs, such as cocaine, is often highest immediately after drug administration, when the drug is producing subjective pleasure (a high) and withdrawal symptoms are at their lowest (Childress et al., 1988; Ehrman et al., 1992; Foltin & Fischman, 1991; Jaffe, 1990). If drug craving were due to a desire to relieve withdrawal symptoms (Camp & Robinson, 1988), it would be expected that craving would dramatically decrease when the drug alleviated withdrawal and was producing pleasure and not that craving would be sustained even after drug use behavior.

Finally, animals will avidly self-administer a variety of drugs directly into brain regions that do not produce withdrawal symptoms (Wise & Hoffman, 1992). Furthermore, the infusion of drugs into the same brain regions can "prime" or reinstate responding in animals in which drug responding has been extinguished (Stewart et al., 1984). These studies have established that the incentive motivational effects of centrally applied drugs can be dissociated from their negative reinforcing or withdrawal-related effects.

So the alleviation of withdrawal distress may indeed sometimes motivate drug-seeking and drug-taking behavior. However, relief from withdrawal symptoms cannot be the sole cause or even the primary cause of drug craving and compulsive drug-taking behavior.

Positive Reinforcement View of Addiction (Pleasure Seeking)

Owing to the shortcomings of negative reinforcement theories of addiction more recent formulations have focused on the role of drugs as positive reinforcers (Jaffe, 1990; Stewart et al., 1984; Wise, 1988; Wise & Bozarth, 1987). Most drugs that are self-administered by people also act as positive reinforcers for animals. Thus, a positive reinforcement view of addiction hypothesizes that drug self-administration is maintained because of the state the drugs would be inducing rather than to alleviate unpleasant withdrawal symptoms/state (Stewart et al., 1984; Wise, 1988; Wise & Bozarth, 1987). However, stating that addictive drugs are positive reinforcers does not explain addiction behavior. As pointed out by Wise and Bozarth (1987) "identifying a drug as reinforcing goes no further than identifying it as addicting!" because it is the common observation of habitual self-administration that serves as the basis for most definitions of both drug reinforcement and drug addiction. A theory of addiction based on the concept of reinforcement would have to identify actions of drugs that are operationally independent of self-administration habits to offer insight as to why drugs are addictive. That is, positive reinforcement is merely a description of a behavioral effect, not an explanation of the effect (Skinner, 1950). The critical question is, why are some drugs positively reinforcing and why do other

TABLE 3 Key Points of Positive Reinforcement Theory

1. Drugs reinforce self-administration behavior by producing pleasure: drug self-administration is maintained because of the state drugs induce rather than to alleviate unpleasant withdrawal symptoms/state.

Problems with a Positive Reinforcement/Euphoria View of Addiction

1. There is no clear relationship between the ability of individual drugs to produce euphoria and their addictive potential; e.g., nicotine is considered highly addictive, but nicotine does not produce marked euphoria or other strong hedonic states.
2. The magnitudes of negative consequences of continued drug use often far outweigh the magnitude of drug pleasure or the memory of drug pleasure.
3. The negative consequences of continued drug use, including ailing health and loss of family, home, and job, seem enormous compared with the pleasure derived from drugs' euphoric effect.
4. This view does not adequately explain drug craving or relapse elicited by environmental stimuli associated with drug taking.
5. Studies showing that drug self-administration can be maintained in the absence of subjective pleasure.

drugs become more effective reinforcers as addiction develops? It is usually assumed that drugs act as positive reinforcers because they produce pleasure. Thus, Wise and Bozarth (1987) state that the only existing positive reinforcement view of addiction that might qualify as an explanatory theory identifies positive reinforcement with drug euphoria. In this view drugs are addicting (establish compulsive habits) because they produce euphoria or positive affect (McAuliffe & Gordon, 1974). Similarly, Stewart et al. (1984) argued that compulsive drug use is maintained by repetitive motivational states generated by the ability of drugs to produce positive affective states.

There are, however, a number of problems with the hypothesis that the subjective pleasurable (hedonic) effects of drugs are either necessary or sufficient to induce and motivate compulsive drug-seeking and drug-taking behavior (Dews, 1977) (Table 3).

Problems with a Positive Reinforcement/ Euphoria View of Addiction

Although addictive drugs can indeed produce extremely pleasant affective states, it is difficult to believe that this property of drugs alone is sufficient to account for addiction. For one, there is no clear relationship between the ability of individual drugs to produce euphoria and their addictive potential. For example, nicotine is considered highly addictive, but nicotine does not produce marked euphoria or other strong hedonic states (Robinson & Pritchard, 1992). Also, many addictive drugs can actually produce strong dysphoric states, especially with initial use. Second, it could be argued that in addicts the magnitudes of negative consequences of continued drug use often far outweigh the magnitude of drug pleasure or the memory of drug pleasure. In fact, to most people, the negative consequences of continued drug use, including ailing health and loss of family, home, and job, seem enormous compared with the pleasure derived from a drug's euphoric effect. So the drug use is not only irrational behavior, but also shows an apparent disparity between the immediate consequences of the behavior and its strength (Robinson & Berridge, 1993).

A positive reinforcement/euphoria view of addiction does not adequately explain drug craving or relapse elicited by environmental stimuli associated with drug taking. Both Stewart et al. (1984) and Wise and Bozarth (1987) have argued that drug-related stimuli can evoke "drug-like" effects that serve to motivate further drug-seeking and drug-taking behavior. They termed this view a "proponent-process theory" (Stewart & Wise, 1992), in contrast with the "opponent-process" view associated with negative reinforcement models (Koob, Stinus, Le Moal, & Bloom, 1989). Stewart and Wise (1992) argued that it is a drug-like process rather than drug-negating process that whets the appetite and stimulates renewed responding. In this view it is the drug "taste" or the experience of stimuli that through Pavlovian conditioning causes drug-like central effects that motivate drug intake in the experienced subject.

The question remains, however, what exactly is this "drug-like process"? One possibility is that it resembles the positive affective state, the pleasurable state, evoked by the drug itself—that it is equivalent to what has been called a conditioned "high" (Childress et al., 1988). In this case, the drug-associated stimuli may evoke "conditioned pleasure," which reminds the addict of the even greater pleasure of the drug itself, thus motivating the individual to once again obtain the drug (Wise, 1988). However, in laboratory studies self-reports of conditioned highs occur much less frequently than self-reports of either conditioned craving or conditioned withdrawal-like highs (Childress et al., 1988). A second possibility is that relapse in a "recovered" addict is triggered by cues that evoke an explicit memory or representation of past drug experiences where an explicit memory need not be a pleasant one. It recalls past pleasure in a cognitive form, as a semantic proposition or as a conscious image of the act of drug taking and spurs the addict to attempt to regain the remembered experience of pleasure once again.

The most compelling evidence against the idea that drug taking is necessarily motivated by the subjective pleasurable effects of drugs comes from studies showing that drug self-administration can be maintained in the absence of subjective pleasure. Lamb et al. (1991) found that an opiate "postaddict" would work (press a lever) to get an injection of a low dose of morphine, despite the fact that four of five people could not distinguish the subjective effects of the morphine from the placebo. Similar results have been reported by Fischman and Foltin (1992) from laboratory studies of cocaine self-administration behavior in humans. Cocaine users reliably choose a low dose of cocaine (4 mg) over placebo, although this dose produces no self-reported subjective effects or cardiovascular effects. In addition, Fischman and Foltin (1992) reported that within drug-taking sessions, tolerance to the cardiovascular and subjective (euphoric) effects that is produced by higher doses of cocaine is not accompanied by changes in drug-taking behavior; i.e., within a self-administration session a dissociation develops between the subjective effects of cocaine and its self-administration behavior. Similarly, studies in rats on the affective vs reinforcing properties of opiates suggest these actions may be mediated by a separable neural system (Martin, Bechara, & van der Kooy, 1991).

The Incentive-Sensitization Theory of Addiction

The incentive-sensitization theory (Robinson & Berridge, 1993) of addiction posits that addictive behavior is due largely to progressive and persistent neuroadaptations caused by repeated drug use. It is a neuroadaptationist model. It is postulated that these drug-induced changes in the nervous system are manifested both neurochemically and behaviorally by the phenomenon of "sensitization," which refers to a progressive increase in a drug effect with repeated use (Robinson & Becker, 1986).

The neural system that is rendered hypersensitive (sensitized) to activating stimuli is hypothesized to mediate a specific psychological function involved in the process of incentive motivation: namely the attribution of incentive solely to the perception and mental representation of stimuli and its consequences. This makes stimuli and their representations highly salient, attractive, and "desired." It is the activation of this neural system that results in the experience of "wanting" and transforms ordinary stimuli into incentive stimuli (Robinson & Berridge, 1993).

Sensitization of this neural system by drugs results in a pathological enhancement in the reward system and the nervous system attributes the reward effect to the drug-taking behavior. The coactivation of associative learning directs the focus of this neurobehavioral system to specific targets that are associated with drugs and leads to an increasing pathological focus of "incentive salience" on drug-related stimuli. Thus, with repeated drug use the drug-taking behavior and drug-associated stimuli gradually become more attractive. Drug-associated stimuli become more able to control behavior, because the neural system that mediates wanting becomes progressively sensitized. Wanting evolves into obsessive craving and this is manifest behaviorally as compulsive drug seeking and drug taking. Therefore, by this view, drug craving and addictive behavior are due specifically to sensitization of incentive salience (Robinson & Berridge, 1993).

However, "wanting" is not "liking." The neural system responsible for wanting incentives is proposed to be separable from those responsible for liking incentives (i.e., for mediating pleasure) and repeated drug use sensitizes only the neural system responsible for the pleasure seeking theory of addiction, which explicitly assumes that the incentive motivational properties of drugs are due directly to their subjective pleasurable effects; i.e., their ability to produce positive affective states (Robinson & Berridge, 1993).

In addition, the neuroadaptations underlying behavioral sensitization are long lasting and in some cases they may be permanent. It is hypothesized that it is the persistence of sensitization-related neuroadaptations that renders addicts hypersensitive to drugs and to drug-related stimuli, even after years of abstinence. It is, therefore, the permanence of sensitization that is thought to render drug-related stimuli so effective in precipitating relapse, even in detoxified, "recovered" addicts (Robinson & Berridge, 1993).

Finally, it is postulated that the neural substrate for incentive sensitization is the mesotelencephalic dopamine system. Sensitization results in an increase in the responsiveness of the dopamine system to the activating stimuli. Subsequently the activating stimuli produce a greater increase in dopamine neurotransmission in sensitized more than in nonsensitized individuals (Robinson & Berridge, 1993) (Table 4).

TABLE 4 Key Points of the Incentive-Sensitization Theory

1. Addictive behavior is due largely to progressive and persistent neuroadaptations caused by repeated drug use.
2. These drug-induced changes in the nervous system are manifested both neurochemically and behaviorally by the phenomenon of "sensitization," which refers to a progressive increase in a drug effect with repeated use.
3. Sensitization of this neural system results in a pathological enhancement in the reward system.
4. The coactivation of associative learning directs the focus of this neurobehavioral system to specific targets that are associated with drugs and leads to an increasing pathological focus of "incentive salience" on drug-related stimuli.
5. With repeated drug use the drug-taking behavior and drug-associated stimuli gradually become more attractive.
6. Drug-associated stimuli become more able to control behavior, because the neural system that mediates wanting becomes progressively sensitized. Wanting evolves into obsessive craving and this is manifest behaviorally as compulsive drug seeking and drug taking.
7. Sensitization results in an increase in the responsiveness of the dopamine system to the activating stimuli. Subsequently the activating stimuli produce a greater increase in dopamine neurotransmission in sensitized than in nonsensitized individuals.

Problems with the Incentive-Sensitization Theory of Addiction

1. Wanting is not liking. The neural system responsible for wanting incentives is proposed to be separable from those responsible for liking incentives (i.e., for mediating pleasure) and repeated drug use sensitizes only the neural system responsible for the pleasure-seeking theory of addiction, which explicitly assumes that the incentive motivational properties of drugs are due directly to their subjective pleasurable effects.
2. The craving phenomenon has been subdivided into its physiological and cognitive components, suggesting different constructs for addictions with evidence that opiate addicts process craving-related information different from alcoholics.
3. In cocaine misusers, methylphenidate increases brain glucose metabolism in the orbitofrontal or prefrontal cortex in subjects in whom it enhanced craving and mood, respectively, indicating that dopamine enhancement is not sufficient per se to increase metabolism in these frontal regions.

Problems with the Incentive-Sensitization Theory

A review of studies relating craving to relapse suggested that craving is not associated with drug use (Tiffany, 1990). The complexity of the craving phenomenon has been subdivided into its physiological and cognitive components suggesting different constructs for addictions with evidence that opiate addicts process craving-related information different from alcoholics (Weinstein, Feldtkeller, & Malizia, 1998).

Using functional magnetic resonance imaging (fMRI), one study (Breiter, Gollub, & Weisskoff, 1997) reported that craving induced by cocaine was not associated with activity in any distinct brain region, but that it was related to a change in the pattern of brain activity over time; the nucleus accumbens and the right parahippocampal gyrus remain activated after all other brain regions have returned to normal baseline. Cue-induced craving in cocaine addicts was associated with increased activity in the anterior cingulate and the left dorsolateral prefrontal cortex using fMRI (Maas, Lukas, & Kaufman, 1998). In cocaine misusers, craving elicited by recall of previous drug experiences using PET showed increased activity in the temporal insula, a brain region involved in autonomic control, and the orbitofrontal cortex, a brain region involved with expectancy and reinforcing salience of stimuli (Wang, Volkow, & Fowler, 1999). In cocaine misusers, methylphenidate increases brain glucose metabolism in orbitofrontal or prefrontal cortex in subjects in whom it enhanced craving and mood, respectively, indicating that dopamine enhancement is not sufficient per se to increase metabolism in these frontal regions (Volkow, Wang, & Fowler, 1999). This study also showed a predominant association of craving with the right orbitofrontal cortex and striatum, brain regions that were also found to be abnormal in compulsive disorders (Abou-Saleh, 1999), suggesting a laterality of reinforcing and/or conditioned responses. Cue-induced cocaine craving in cocaine misusers is associated with increased regional cerebral blood flow (rCBF) in the limbic region (amygdala and anterior cingulate), compared with a decrease in rCBF in basal ganglia (Childress et al., 1988). This change is specific to these regions since there is no increase in nonlimbic regions and there is an absence of similar limbic activation in non-cocaine-using subjects in response to cocaine external cues (Abou-Saleh, 1999).

The coactivation of the amygdala and anterior cingulate regions during cue-induced craving is consistent with the importance of these regions in affective behavior and in emotional learning. The drop in basal ganglia rCBF during craving may also reflect the influence of these interconnected limbic structures, perhaps representing active inhibition of reward-irrelevant responses (Abou-Saleh, 1999). Childress et al. (1988) concluded that the developing brain signature of cue-induced craving is thus consistent with its clinical phenomenology: the drug user is gripped by a visceral emotional state, experiences a highly focused incentive to act, and is remarkably unencumbered by the memory of negative consequences of drug taking.

CONCLUSION

Although there are several studies in favor of the gateway hypothesis, more recently studies have challenged the universal applicability of the gateway theory (Degenhardt et al., 2009; Kandel, 2002). The multicountry World Mental Health Survey (Degenhardt et al., 2010) suggested that alcohol and tobacco use were precursors to other illicit substances and this is exponentially related. Cannabis use preceded the use of other substances in areas where cannabis use was highly prevalent (Elhamady, Mobasher, Yousef, & Moselhy, 2013), while such a direct association was not found where cannabis use was low. Other studies have, however, found that tobacco or alcohol use alone may not completely predict progression to other illicit substances (Van Leeuwen et al., 2011).

SUMMARY AND KEY POINTS

- This chapter summarizes the interface between neurophysiology and psychological aspects of addiction.
- The impact of neurotransmitters, e.g., dopamine, and endogenous opioids on addictive behavior has been highlighted.
- Psychological reinforcements and incentive sensitization and their impact on maintaining addiction behavior have been illustrated.
- The interpersonal factors in relation to addiction have been elucidated.

REFERENCES

Abdel Wahab, M. (2004). Drug dependence: the magnitude of the problem in Egypt. *Egyptian Journal of Psychiatry, 23*(1), 10–12.

Abou-Saleh, M. T. (1999). Neurobiology of drug use and addiction. *CNS, 2*(2), 19–22.

Akers, R. L., Krohn, M. D., Lanza-Kaduce, L., & Radosevich, M. (1979). Social learning and deviant behaviour: a specific test of a general theory. *American Sociological Review, 44*, 636–655.

Ashton, C. H. (2001). Pharmacology and effects of cannabis: a brief review. *The British Journal of Psychiatry: The Journal of Mental Science, 178*, 101–106.

Beck, A. (1993). *Cognitive therapy of personality disorders*. New York: The Guilford Press.

Benjamin, L. S. (1993). Personality disorders: models for treatment and strategies for treatment development. *Comprehensive Psychiatry, 34*, 87–94.

Botvin, G. J., Scheier, L. M., & Griffin, K. W. (2002). Preventing the onset and developmental progression of adolescent drug use: implications for the gateway hypothesis. In D. B. Kandel (Ed.), *Stages and pathways of drug involvement: Examining the gateway hypothesis* (pp. 115–138). Cambridge, UK: Cambridge University Press.

Breiter, H. C., Gollub, R. L., & Weisskoff, R. M. (1997). Acute effects of cocaine on human brain activity and emotion. *Neuron, 19*, 591–611.

Bretteville-Jensen, A. L., Melberg, H. O., & Jones, A. M. (2008). Sequential patterns of drug use initiation–can we believe in the gateway theory. *The BE Journal of Economic Analysis & Policy, 8*, 1–29.

Camp, D. M., & Robinson, T. E. (1988). Susceptibility to sensitisation. *Behaviour Brain Research, 30*, 55–68.

Childress, A. R., McLellan, A. T., Ehrman, R., & O'Brien, C. P. (1988). Classically conditioned responses in opioid and cocaine dependence: a role in relapse. *NIDA Research Monograph, 84*, 25–43.

Conger, J. J. (1973). *Adolescence and youth: Psychological development in a changing world*. New York: Harper & Row.

Cunha-Oliveira, T., Rego, A. C., & Oliveira, C. R. (2008). Cellular and molecular mechanisms involved in the neurotoxicity of opioid and psychostimulant drugs. *Brain Research Reviews, 58*, 192–208 (Cambridge, England: Cambridge University Press).

Dackis, C. A., & Gold, M. S. (1985). New concepts in cocaine addiction: the dopamine depletion hypothesis. *Neuroscience & Behaviour Review, 9*, 469–477.

Degenhardt, L., Chiu, W. T., Conway, K., Dierker, L., Glantz, M., Kalaydijian, K., ... Kessler, R. C. (2009). Does the gateway matter? Associations between the order of drug initiation and development of drug dependence in the national comorbidity study replication. *Psychological Medicine, 39*, 157–167.

Degenhardt, L., Dierker, L., Chiu, W. T., Medina-Mora, M. E., Neumark, Y., Sampson, N., ... Kessler, R. C. (2010). Evaluating the drug use "gateway" theory using cross-national data: consistency and associations of the order of initiation of drug use among participants in the WHO world Mental Health Surveys. *Drug and Alcohol Dependence, 108*, 84–97.

Dews, P. (1977). Remarks. In T. Thomson, & K. Runna (Eds.), *Predicting dependence liability of stimulant and depressant drugs* (pp. 75–79). Baltimore: University Park Press.

Di Chiara, G., & Imperto, A. (1988). Drugs abused by humans preferentially increase synaptic dopamine concentration in the mesolimbic system of freely moving rats. *Process of Natural Academy of Science, 85*, 5274.

Dohrenwend, B. S., & Dohrenwend, B. P. (1984). *Stressful life events and their contexts*. New Brunswick, NJ: Rutgers University Press.

Dull, R. T. (1984). An empirical examination of the social bond theory of drug use. *The International Journal of the Addictions, 19*, 265–286.

Ehrman, R., Ternes, J., O'Brien, C. P., & McLellan, A. T. (1992). Conditioned tolerance in human opiate addicts. *Psychopharmacology, 108*, 218–224.

Ekleberry, S. (1996). Dual diagnosis: addiction and axis II personality disorders. The Counselor, March/April 7–13. Retrieved on 11/17/06 from http://www.toad.net/~arcturus/dd/ddhome.htm.

Elhamady, M., Mobasher, M., Yousef, S., & Moselhy, H. (2013). Association between the order of drug use initiation and the development of opioid dependence among Egyptian adults: cannabis as a gateway and culture specific drug. *Addictive Disorders and Their Treatment Journal, 12*(2), 91–98.

Fischman, M. W., & Foltin, R. W. (1992). Self-administration of cocaine by humans: a laboratory perspective. In G. R. Bock, & J. Whelan (Eds.), *Cocaine: Scientific and social dimensions. CIBA foundation symposium no 166* (pp. 165–180). Chichester, UK: Wiley.

Foltin, R. W., & Fischman, M. W. (1991). Assessment of a liability of stimulant drugs in humans: a methodological survey. *Drug and Alcohol Dependence, 28*, 3–48.

Gawin, F. H. (1991). Cocaine addiction: psychology and neurophysiology. *Science, 251*, 1580–1586.

Glue, P., & Nutt, D. J. (1990). Overexcitment and disinbition: dynamic neurotransmitter interactions in alcohol withdrawal. *The British Journal of Psychiatry: The Journal of Mental Science, 157*, 491–499.

Goldman, D. (1995). Dopamine transporter, alcoholism and other disease. *Nature Medicine, 1*, 624–625.

Hayman, S. E. (1993). Molecular & cell biology of addiction. *Current Opinion in Neurology and Neorosurgery, 6*, 609–613.

Hirschi, T. (1969). *Causes of delinquency*. Berkeley: University of California Press.

Hubner, C. B., & Moreton, J. E. (1991). Effects of selective D1 and D2 dopamine antagonists on cocaine self-administration in the rat. *Psychopharmacology, 105*, 151–156.

Jaffe, J. H. (1990). Drug addiction and drug abuse. In A. G. Gillman, T. W. Rall, A. S. Nies, & P. Taylor (Eds.), *The pharmacological basis of therapeutics*. New York, USA: Pergamon Press.

Jaffe, J. H., Cascella, N. G., Kumor, K. M., & Sherer, M. A. (1989). Cocaine-induced craving. *Psychopharmacology, 97*, 59–64.

Jessor, R., & Jessor, S. (1977). *Problem behaviour and psychosocial development–A longitudinal study of youth*. New York: Academy Press.

Johnson, R. E. (1979). *Juvenile delinquency and its origins: An integrated approach*. New York: Cambridge University Press.

Kandel, D. B. (1978). *Longitudinal research on drug use: Empirical findings and methodological issues*. New York: Wiley.

Kandel, D. B. (2002). *Stages and pathways of drug involvement: Examining gateway hypothesis* Cambridge, UK; New York: Cambridge University Press.

Koob, G. F. (1992). Drugs of abuse: anatomy, pharmacology and function of reward pathways. *TIPS, 13*, 177–184.

Koob, G. F., & Bloom, F. E. (1988). Cellular and molecular mechanisms of drug dependence. *Science, 242*, 715–723.

Koob, G. F., Stinus, L., Le Moal, M., & Bloom, F. E. (1989). Opponent process theory of motivation: neurobiological evidence from studies of opiate dependence. *Neuroscience Biobehaviour Review, 13*, 135–140.

Koob, G. F., & Volkow, N. D. (2010). Neurocircuitry of addiction. *Neuropsychopharmacology, 35*, 217–238.

Kornhauser, R. R. (1978). *Social sources of delinquency: An appraisal of analytic models*. Chicago: University of Chicago Press.

Kreek, M. J., & Koob, G. F. (1998). Drug dependence: stress and dysregulation of brain reward pathways. *Drug & Alcohol Dependence, 51*, 23–47.

Lamb, R. J., Preston, K. L., Schindler, C., Meisch, R. A., Davis, F., Katz, J. L., ... Goldberg, S. R. (1991). The reinforcing and subjective effects of morphine in post-addicts: a dose-response study. *Journal of Pharmacology & Experimental Therapy, 259*, 1165–1173.

Leshner, A. I. (1997). Addiction is a brain disease and it matters. *Science, 278*, 45–47.

London, E. D., Cascella, N. G., Wong, D. F., Phillips, R. L., Dannals, R. F., Links, J. M., ... Wagner, H . N. Jr. (1990). Cocaine-induced reduction of glucose utilization in human brain. A study using positron emission tomography and [fluorine 18]-fluorodeoxyglucose. *Archive General Psychiatry, 47*(6), 567–574.

Maas, L. C., Lukas, S. E., & Kaufman, M. J. (1998). Functional magnetic resonance imaging of human brain activation during cue-induced cocaine craving. *American Journal of Psychiatry, 155*(1), 124–126.

Maldonado, R., & Berrendero, F. (2010). Endogenous cannabinoid and opioid systems and their role in nicotine addiction. *Current Drug Targets, 11*(4), 440–449.

Martin, G. M., Bechara, A., & van der Kooy, D. (1991). The perception of emotion: parallel neural processing of the effective and discriminative properties of opiates. *Psychobiology, 19*, 147–152.

Matsueda, R. L. (1982). Testing control theory and differential association. *American Sociological Review, 47*(4), 489–504.

McAuliffe, W. E. (1982). A test of Wikler's theory of relapse due to conditioned withdrawal sickness. *The International Journal of the Addictions, 17*, 19–33.

McAuliffe, W. E., & Gordon, R. A. (1974). A test of Lindesmith's theory of addiction: the frequency of euphoria among long-term addicts. *American Journal of Sociology, 79*, 795–840.

Melberg, H. O., Jones, A. M., & Bretteville-Jensen, A. L. (2010). Is cannabis a gateway to hard drugs? *Empirical Economics, 38*(3), 583–603.

Milkman, H., & Sunderwirth, S. (1987). *Craving for ecstasy: the consciousness & chemistry of escape*. Lexington, MA: Lexington Books.

Nutt, D. J. (1996). Addictive brain mechanisms and their treatment implications. *Lancet, 347*, 31–36.

Pearlin, L. (1981). The stress process. *Journal of Health and Social Behaviour, 22*, 337–356.

Pike, V. W. (1993). Positron-emitting radioligands for studies in vivo: probes for human psychopharmacology. *Journal of Psychopharmacology, 7*, 139–158.

Pradhan, S. N., & Dutta, S. N. (1977). *Drug abuse: Clinical and basic aspects*. St Louis, Missouri, USA: Mosby Company, 3–4.

Richards, H. J. (1993). *Therapy of the substance abuse syndromes*. Northvale, NJ: Jason Aronson, Inc.

Robinson, T. E., & Becker, J. B. (1986). Enduring changes in brain and behavior produced by chronic amphetamine administration: a review and evaluation of animal models of amphetamine psychosis. *Brain Research Reviews, 396*, 157–198.

Robinson, T. E., & Berridge, K. C. (1993). The neural basis of drug craving: an incentive-sensitization theory of addiction. *Brain Research. Brain Research Reviews, 18*, 247–291.

Robinson, J. H., & Pritchard, W. S. (1992). The meaning of addiction: reply to west. *Psychopharmacology, 108*, 411–416.

Schuster, C. (1990). Drug-seeking behaviour: implications for theories of drug dependence. In G. Edwards, & M. Lader (Eds.), *The nature of drug dependence* (pp. 171–193). New York: Oxford University Press.

Shaw, G. K., Waller, S., Majumdar, S. K., Alberts, J. L., Latham, C. H., & Dunn, G. (1994). Tiapride in the prevention of relapse in recently detoxified alcoholics. *The British Journal of Psychiatry: The Journal of Mental Science, 165*, 515–523.

Siegel, S. (1977). Learning and psychopharmacology. In M. E. Jarvik (Ed.), *Psychopharmacology in practice of medicine*. New York: Appleton-Century-Crofts.

Siegle, S. (1988). Drug anticipation and drug tolerance. In M. Lader (Ed.), *The psychopharmacology of addiction* (pp. 73–97). New York: Oxford University Press.

Skinner, B. F. (1950). Are theories of learning necessary? *Psychological Review, 57*, 193–216.

Spanagel, R., Herz, A., & Shippenberg, T. S. (1990). The effects of opioid peptide on dopamine release in the nucleus accumbens: an in vivo microdialysis study. *Journal of Neurochemistry, 55*, 1740–1743.

Stewart, J., & Wise, R. A. (1992). Reinstatement of heroin self-administration habits: morphine prompts and naltrexone discourages renewed responding after extinction. *Psychopharmacology, 108*, 79–84.

Stewart, J., de Wit, H., & Eikelboom, R. (1984). Role of unconditioned and conditioned drug effects in self-administration of opiates and stimulants. *Psychological Review, 91*, 251–268.

Stone, M. H. (1993). Long-term outcome in personality disorders. *British Journal of Psychiatry, 162*, 299–313.

Sutherland, E. H. (1947). *Principles of criminology*. Philadelphia: Lippincott.

Ternes, J. W., Ehrman, R. N., & O'Brien, C. P. (1985). Non-dependent monkeys self-administer hydromorphone. *Behavioural Neuroscience, 99*, 583–588.

Tiffany, S. T. (1990). A cognitive model of drug urges and drug-use behavior: role of automatic and non-automatic processes. *Psychological Review, 97*, 147–168.

Van Leeuwen, A. P., Verhulst, F. C., Rejineveld, S. A., Vollebergh, W. A. M., Ormel, J., & Huizink, A. C. (2011). Can the gateway hypothesis, the common liability model and/or, the route of administration model predict initiation of cannabis use during adolescence? A survival analysis-the TRAILS study. *The Journal of Adolescent Health, 48*, 73–78.

Volkow, N. D., Wang, G. J., & Fowler, J. S. (1999). Association of methylphenidate-induced craving with changes in right striato-orbitofrontal metabolism in cocaine abusers: implications in addiction. *American Journal of Psychiatry, 156*, 19–26.

Volpicelli, J. R., Alterman, A. J., Hayashida, M., & O'Brien, C. P. (1993). Naltrexone in the treatment of alcohol dependence. *Archive General Psychiatry, 49*, 876–880.

Wang, G. J., Volkow, N. D., & Fowler, J. S. (1999). Regional brain metabolic activation during craving elicited by recall of previous drug experiences. *Life Science, 64*(9), 775–784.

Weinstein, A., Feldtkeller, B., & Malizia, A. (1998). Integrating the cognitive and physiological aspects of craving. *Journal of Psychopharmacology, 12*(1), 31–38.

White, H. R., Johnson, V., & Howrwitz, A. (1986). An application of three deviance theories to adolescent substance use. *The International Journal of the Addictions, 21*(3), 347–366.

Wilker, A. (1948). Recent progress in research on the neurophysiological basis of morphine addiction. *American Journal of Psychiatry, 105*, 329–338.

Wise, R. A. (1988). The neurobiology of craving: implications for the understanding and treatment of addiction. *Journal of Abnormal Psychology, 97*, 118–132.

Wise, R. A., & Bozarth, M. A. (1987). A psychomotor stimulant theory of addiction. *Psychological Review, 94*, 469–492.

Wise, R. A., & Hoffman, D. C. (1992). Localization of drug reward mechanisms by intracranial injections. *Synapse, 10*, 247–263.

Woods, J. H., & Schuster, C. R. (1968). Reinforcement properties of morphine, cocaine as a function of unit dose. *The International Journal of the Addictions, 3*, 231–237.

Chapter 4

Children of Parents with Substance Use Disorder

Iliyan Ivanov[1], John Leikaif[2], Juan Pedraza[1], Jeffrey Newcorn[1]
[1]Department of Psychiatry, Icahn School of Medicine at Mount Sinai, New York, NY, USA; [2]Department of Psychiatry, Stanford University, Stanford, CA, USA

INTRODUCTION

Variation in the risk for and severity of substance use disorders (SUDs) reflects the interaction of environmental variables with numerous biological and behavioral traits (Tarter et al., 2003). This chapter focuses on the neurobiological basis underlying the heritable transmission of risk for SUDs. However, at the same time, it is important to recognize that parental SUDs can profoundly influence children via both environmental and hereditary effects (Johnson & Leff, 1999). The earliest clear environmental effect would be in utero exposure to drugs of abuse, which can lead to cognitive and motor delays of significant magnitude (Wouldes et al., 2014). Fetal alcohol syndrome is a well-known example (Ungerer, Knezovich, & Ramsay, 2013) associated with loss of white matter integrity and neurocognitive deficits in affected individuals (Spottiswoode et al., 2011). In utero exposure to drugs of abuse is also associated with subsequent behavioral, cognitive, and functional impairments in adolescence (Buckingham-Howes, Berger, Scaletti, & Black, 2013) and beyond.

Postnatal environmental effects of parental substance use have also been demonstrated for children who continue to live in households of substance-using parents. Parental SUD has been shown to increase the risk for child abuse, injury, and medical hospitalizations (Raitasalo, Holmila, Autti-Rämö, Notkola, & Tapanainen, 2014), although the findings are not uniform regarding the increased risk for violence (Johnson & Leff, 1999). Furthermore, peer deviance and neighborhood environment have been shown to have an impact on a child's risk of developing SUD (Kendler, Maes, Sundquist, Ohlsson, & Sundquist, 2014; Kendler, Ohlsson, Sundquist, & Sundquist, 2014; Keyes et al., 2012).

Nevertheless, while environmental effects are important, children of parents with SUD have significantly increased risk for SUD by adolescence above and beyond the effects of environment. The most convincing evidence of biological influences for SUD comes from family and adoption studies, particularly in families with alcoholism. The heritability of alcoholism and other SUDs has been estimated to be in the range of 40–70% (Goldman, Oroszi, & Ducci, 2005). These estimates are in contrast to the much smaller estimates for the contributions of specific genes. Although there are a number of candidate genes that have been studied in relation to SUD, no particular gene or combination of genes has been found to play a decisive role in the development of any particular SUD, raising the question of so-called "missing heritability." However, the strong evidence from family studies supports the notion that heritable biological factors play an important role in the pathophysiology of SUD.

Additional evidence for the neurobiological basis of SUD comes from studies showing that childhood conditions that have been recognized as highly heritable also represent clinical risk factors for the development of drug use early in life. Youth at heritable risk for SUD commonly exhibit impulsivity, reactive aggression, sensation seeking, and excessive risk taking (Tarter et al., 1999). Irritability, negative affect, and a difficult temperament have also been frequently documented in youth at high risk for SUD (Blackson et al., 1999). Furthermore, compared with the general population, high-risk youth more frequently meet criteria for conduct disorder, attention deficit hyperactivity disorder (ADHD), oppositional defiant disorder, anxiety disorders, and depressive disorders (Clark, Parker, & Lynch, 1999). These types of behavioral disturbances have been linked to delayed maturation of the prefrontal cortex (Tarter et al., 1999). Considering the fact that both externalizing and internalizing disorders also show considerable heritability, the elevated comorbidity between SUD and these disorders suggests that there may be shared genetic or epigenetic risk factors. This heritability is now known to be multifactorial as no direct relationship between the clinical syndrome and the biological markers has been documented. There are probably many different mechanisms by which risk for SUD is inherited. Neurobiological mechanisms that could be influenced by both environmental and genetic contributions and that could explain the observed overlap between externalizing, internalizing disorders and SUDs include alterations in the reward and inhibitory control systems. In the next section we discuss possible models for SUD vulnerability.

MODELS OF SUBSTANCE USE DISORDER VULNERABILITY

Several models have been developed to account for and explain the progression from drug experimentation to regular use and addiction in persons with heritable risk. It is worth noting that many people experiment with alcohol and drugs of abuse, but only a subset of individuals go on to develop SUDs. In some cultural settings, experimentation with or use of alcohol or certain other potential drugs of abuse is normative.

Externalizing and Internalizing Pathways

One hypothesis that explains the observed association of certain childhood clinical conditions with risk for alcoholism and, by extension, SUD is that there are parallel "externalizing" and "internalizing" pathways for SUD onset and development (Tessner & Hill, 2010). The *externalizing pathway*, associated clinically with disruptive behavior disorders including ADHD, oppositional defiant disorder, and conduct disorder, would be characterized by altered functions of executive control circuits and behavioral inhibition, in turn linked to abnormalities in the cerebellum, thalamus, and prefrontal cortex. The *internalizing pathway*, clinically characterized by mood problems, negative affect, and hypersensitivity to rewards and stress, would be associated with alterations in the orbitofrontal cortex, the extended amygdala (including connections to the ventral striatum and limbic lobe), the hypothalamus, and the hypothalamic–pituitary axis. These pathway models further suggest that childhood disorders, known to have strong heritable influence, also exhibit alterations in neuronal networks related to reward processing and behavioral control that may be associated with heritability for SUD.

Behavioral Activation and Behavioral Control

Impulsivity is defined as action without planning and with little consideration for possible negative outcomes, and impulsive behaviors have been strongly associated with SUDs (Gullo, Loxton, & Dawe, 2014). However, impulsivity is not a unitary construct. Findings from the personality, behavioral, and neurophysiology literature suggest that impulsivity can be most parsimoniously summarized in a two-factor model. The first is increased "bottom-up" activity involving activation of the amygdala and ventral striatum, which manifests as high motivation to obtain rewarding stimuli. Similar phenotypes have also been termed "sensation seeking" or "hyper-reward sensitivity" in the personality literature and "delay discounting," "choice impulsivity," or "reward-delay impulsivity" in the behavioral literature. The second is decreased "top-down" behavioral control, related to dysfunction of cortical systems, including the ventromedial and dorsolateral prefrontal cortex, the anterior cingulate cortex, and the insula. Corresponding phenotypes have been referred to as "low constraint" or "impulsivity" in the personality literature but also as "motor disinhibition," "poor self-regulation," or "impaired response inhibition" in the behavioral literature. For the purpose of consistency, we will use the terms "behavioral activation" and "behavioral control" for the rest of this chapter and will occasionally use the term "impulsivity" or "reward sensitivity" when referring to specific theories that have originally used these (or similar) terms.

Model of Altered Behavioral Activation

The *Reward Deficiency* and the *Incentive-Salience Hypotheses* are examples of models that place emphasis on neurocircuits related to behavioral activation, reward sensitivity, and processing. The Reward Deficiency Hypothesis proposes that a reduced response of the so-called "reward system" (primarily referring to the nucleus accumbens in the ventral striatum, the ventral tegmental area, and the mesial frontal cortex) to nondrug rewards predisposes some individuals to preferentially seek the rewarding effects of drugs of abuse as opposed to other environmental stimuli (e.g., success in school and sports or the approval of parents and teachers). Support for this model is provided by findings of reduced striatal dopamine receptor density in individuals who also report more pleasurable psychological experiences with the administration of methylphenidate (Volkow et al., 1999). It is plausible that neuroanatomical features of the reward system, such as dopamine receptor density in the striatum, might be influenced by one or more heritable mechanisms, which predispose affected individuals to experiencing effects of drugs as more highly rewarding than persons with no such heritable traits. Although this hypothesis makes a compelling argument for the role of neurobiological factors related to behavioral motivation and reward processing in the development of SUD, it does not adequately consider possible relationships between behavioral activation and behavioral control (Hommer, Bjork, & Gilman, 2011).

The Incentive-Salience Hypothesis in turn suggests that there is not hereditary dysfunction in reward circuits to nondrug cues. Rather, addiction occurs purely through drug-mediated sensitization of reward circuitry and overlearning of drug-associated cues. In its purest form, the Incentive-Salience Hypothesis is most useful as a heuristic, "null" hypothesis, positing that heritable risk is not transmitted through sensitivity to reward at all. Rather, effects on behavioral activation and behavioral control are seen as secondary to the effects of drug use (Hommer et al., 2011). Based on this hypothesis is the model of Incentive-Sensitization described in detail below.

Model of Altered Behavioral Control

The *Impulsivity Hypothesis* refers to a model that implicates decreased behavioral control combined with increased behavioral activation (or "sensitivity to reward") in determining constitutional risk for SUDs (Hommer et al., 2011). Increased sensitivity to the rewarding effects of drugs of abuse leads to earlier onset of drug misuse, which, in the presence of poor behavior control, can predispose some individuals to the development of SUDs. Presumably there is genetic variation in the degree of sensitivity to reward as well as the relative strength of inhibitory control within the population. It is speculated that these hypotheses could be tested via the use of functional neuroimaging because the Impulsivity Hypothesis would predict increased activation in the reward system (specifically in the ventral striatum and mesial frontal cortex) in high-risk individuals, while the Reward Deficiency Hypothesis would predict decreased activation.

Incentive Sensitization

The construct of behavioral activation is theoretically composed of three psychological components: (1) liking, (2) wanting, and (3) learning—and each of these components corresponds to a distinguishable neurobiological network (Berridge & Kringelbach, 2008). The brain areas associated with behavioral activation include the orbitofrontal cortex, anterior cingulate cortex, insular cortex, amygdala, nucleus accumbens/ventral striatum, ventral pallidum, and brainstem sites and their mesolimbic dopaminergic projections. Of these, the orbitofrontal cortex seems to be the most significant nexus for sensory integration, emotional processing, and hedonic experience. The orbitofrontal cortex also has an important role in emotional disorders such as depression, and it is now possible to offer a tentative model of the functional neuroanatomy of the human orbitofrontal cortex in pleasure (Kringelbach, 2005). According to this model, sensory information from the periphery is decoded into stable cortical representations (Berridge & Kringelbach, 2008; Schoenbaum, Roesch, & Stalnaker, 2006) and is then further integrated in the posterior parts of the orbitofrontal cortex. The reward value of any stimulus can be used to influence subsequent decisions and behavior (via lateral aspects of the anterior orbitofrontal cortex and connections to the anterior cingulate cortex), stored for learning/memory (in medial parts of the anterior orbitofrontal cortex), and perhaps made available for subjective hedonic experience (via the midanterior orbitofrontal cortex). Reciprocal information flows between the various regions of the orbitofrontal cortex and other brain regions, including the insular cortex, anterior cingulate cortex, nucleus accumbens, ventral pallidum, and amygdala.

The *Incentive-Sensitization Model* postulates that: (1) addictive drugs share the ability to alter brain organization; (2) these alterations occur in brain systems normally involved in the processing of incentive, motivation, and reward; (3) as a result of substance use these brain reward systems become "sensitized" to drugs and drug-associated stimuli; and (4) such sensitized brain systems mediate reward experience that has been termed incentive salience (e.g., drug "wanting") (Robinson & Berridge, 2001). These processes have been observed in animal models, which suggests that some individuals could be at elevated risk for later SUD if exposed to abusable drugs in early life; on the other hand several longitudinal studies have documented that individuals who received stimulant treatment for childhood ADHD did not show increased rates of adolescent SUD.

The Addiction Cycle

An integrative model for the development of the so-called "addiction cycle" has been proposed by Koob and Volkow (2010). This model distinguishes between mechanisms that are involved during initial drug use and those involved in chronic, problematic use (which corresponds to SUDs in the *Diagnostic and Statistical Manual of Mental Disorders* (DSM)). The model emphasizes the fact that SUDs share certain characteristics with both impulse control disorders and compulsive disorders. The former are defined as disorders characterized by positive reinforcement, gratification, arousal, pleasure, or the release of tension, while the latter are characterized by negative reinforcement mechanisms associated with relief from anxiety and dysphoric states. This model conceptualizes the positive and negative reinforcement phases associated with SUDs as characteristic of different stages of a single addiction cycle, rather than two separate pathways (i.e., the externalizing and internalizing pathways, respectively) leading to SUDs. Further, the model suggests that addiction can be described as following a three-stage cycle: binge/intoxication, withdrawal/negative affect, and preoccupation/anticipation (Figure 1). It is proposed that impulsive use is present initially, but that impulsive behaviors transform to compulsive drug use with repetition over time. This behavioral transition seems to be associated with neuroplastic changes and is therefore a potential target for investigations into neurobiological risk factors. The authors identify five brain regions/circuits that are thought to be involved in the addiction cycle: the mesolimbic dopamine system, the ventral striatum, the ventral striatum/dorsal striatum/thalamus circuit, the dorsolateral frontal cortex/inferior frontal cortex/hippocampus circuits, and the extended amygdala.

The mesolimbic dopamine system and ventral striatum are suggested to play the greatest role in early engagement with drugs of abuse and show plasticity in response to repeated drug administration, which is thought to lead to repeated drug seeking. These changes are believed to be modulated primarily by neurons projecting from mesolimbic dopamine systems to the nucleus accumbens in the ventral striatum, possibly also modulated by glutamate via α-amino-3-hydroxy-5-methyl-4-isoxazolepropionic acid (AMPA) receptors in the nucleus accumbens. It has been hypothesized that either over- or underresponse to rewarding stimuli could increase the risk for SUDs through the early phases of the addiction cycle (either could theoretically lead to a greater difference between the rewarding effects of a drug compared with environmental reinforcers). The ventral striatum/dorsal striatum/thalamus circuit appears to be involved in the transition from early use to habitual drug seeking, as evidence suggests that these regions are central to habit learning, action initiation, and cue/context-dependent craving, as well as arousal and attention modulation (Volkow, Fowler, Wang, & Swanson, 2004; Yin, Ostlund, Knowlton, & Balleine, 2005). As the addiction process progresses, drug-related reinforcers are overvalued relative to other reinforcers. Loss of inhibitory control in response to drug-related cues accompanies these changes, implicating the dorsolateral frontal cortex/inferior frontal cortex/hippocampus circuit. Finally, the extended amygdala is proposed to mediate the negative reinforcement pathways that appear to drive drug-seeking behaviors later in the addiction cycle (Koob & Volkow, 2010). The Addiction Cycle Model and associated neurobiological correlates provide a useful framework with which to understand the nature of risk transmitted from parents with SUD to their children.

FROM MODELS TO STUDIES OF NEUROBIOLOGY OF CHILDREN WITH FAMILIAL SUBSTANCE USE DISORDER

While the models described above tend to emphasize different aspects of the addiction syndrome, there are important similarities across the various models. One overarching construct, which has been discussed in numerous theoretical and data-driven papers, is that SUD risk involves the relationship between behavioral activation (i.e., reward sensitivity, reward-motivated behaviors) and behavioral control (i.e., impulsivity, response inhibition, etc.). It has been well established that individuals with the propensity to engage in impulsive behaviors are more likely to develop problem drug use, have more difficulties managing the negative consequences of such drug use, and have deficits in adjusting their behavior according to associated negative consequences. It therefore seems reasonable to suggest that impaired behavioral control coupled with some type of altered sensitivity to reward represents a diathesis that places an individual at increased risk for developing SUD. The types of impairments associated with sensitivity of the reward system, however, have not been clearly delineated. Competing hypotheses suggest that either diminished or elevated reward sensitivity might be responsible for the development of drug use,

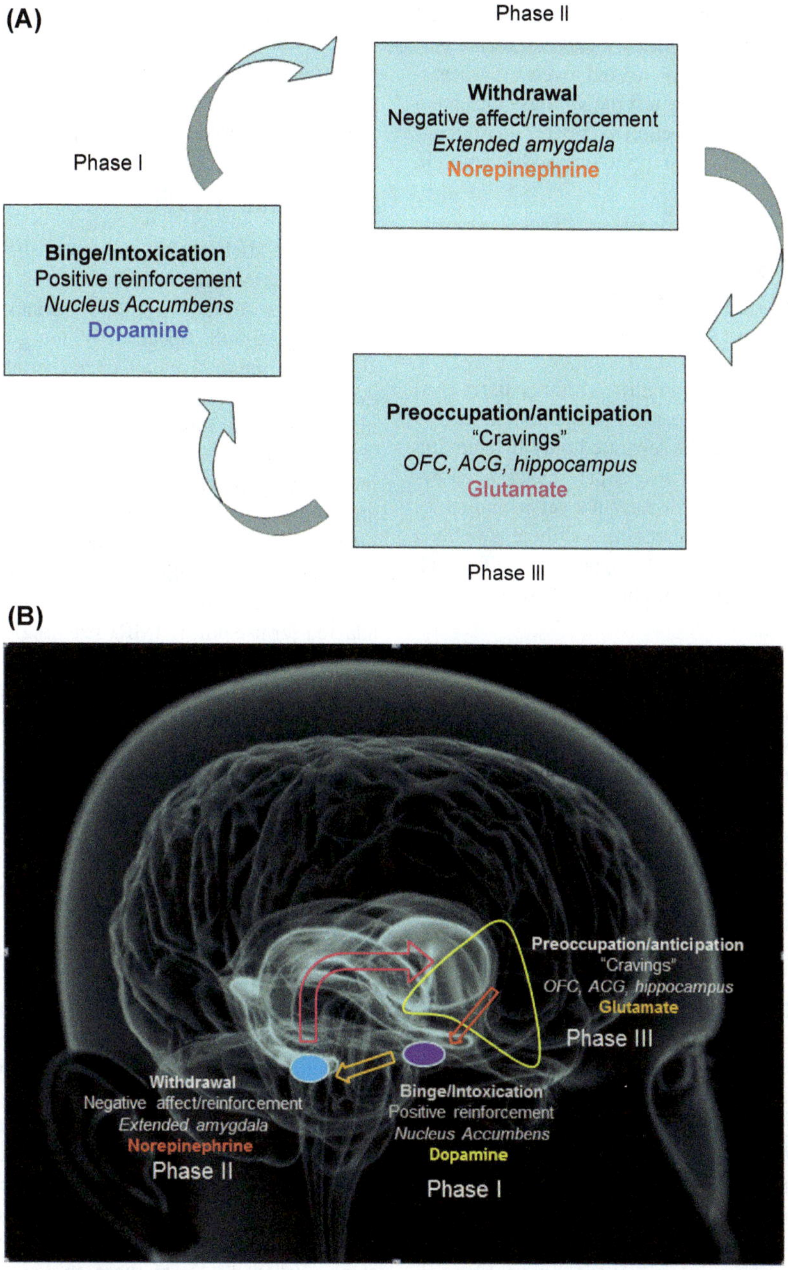

FIGURE 1 (A) A schematic summarizing the addiction cycle is presented. (B) The stages of the addiction cycle.
Phase I, *Binge/Intoxication Stage*. This stage is associated with the positive/reinforcing effects of drugs, which may engage the nucleus accumbens. Neurotransmitters mediating the rewarding effects of drugs of abuse are dopamine and opioid peptides.
Phase II, *Withdrawal Stage*. In this stage, negative affect associated with withdrawal may engage the extended amygdala via norepinephrine-mediated activation.
Phase III, *Preoccupation/Anticipation (Craving) Stage*. This stage involves the processing of contextual information by the hippocampus, as well as executive control and feedback from the orbital frontal cortex (OFC) and anterior cingulate gyrus (ACG) via glutamate-mediated activation.

and it is likely that these different hypotheses describe different pathways to SUD in individuals at constitutional risk. A possible shortcoming of any of these models is that most of the supporting evidence comes from animal studies, cross-sectional studies of humans already exposed to substances of abuse, and clinical observation—none of which are ideal for prospectively delineating predictors of subsequent drug use. With the advancement of neuroimaging techniques that can be safely used in children (e.g., structural and functional magnetic resonance imaging (MRI)), investigators have begun to study youth with family histories of SUD to identify potential neurobiological underpinnings of risk for SUD that precede exposure to substances of abuse. Emerging data suggest that offspring of individuals with SUD demonstrate a variety of abnormalities in brain structure and function in comparison to youth with no heritable risk for SUD. Functional neuroimaging offers the potential to identify and categorize purported differences in neuronal activation related to reward processing and behavioral control and thereby test the various conceptual models described above. In the next section we review neuroimaging studies that have examined brain morphology and function in children of parents with SUD.

NEUROIMAGING STUDIES OF CHILDREN WITH FAMILIAL SUBSTANCE USE DISORDER

One of the key reasons to study children of substance users is to elucidate trajectories leading to SUDs since heritable risk factors and consequences of drug use can be separated only by studying individuals at risk before the onset of drug experimentation. However, studying exclusively children who have not yet been directly exposed to substances of abuse is challenging and there are only a limited number of studies that have used this approach (see Table 1). In the ensuing discussion, we will highlight these studies.

Morphological Studies

Several studies have used structural MRI to investigate morphological brain differences between individuals with a parental history of SUD and controls. A study of 20 alcohol-naïve, male offspring of early onset (prior to age 25) alcohol-dependent individuals (high-risk) and 21 alcohol-naïve, low-risk controls, all ages 8 to 24 (Benegal, Antony, Venkatasubramanian, & Jayakumar, 2007), found that the high-risk participants had lower volumes in the left superior frontal and cingulate cortices, the parahippocampal gyri, bilateral amygdala, thalamus, and cerebellum compared to the low-risk controls. Furthermore, the volume in these areas was negatively correlated with ratings of externalizing behavior, supporting a link between lower volumes in these regions, externalizing behaviors, and inherited SUD risk. In another structural MRI study, 63 non-drug-naïve, high-risk subjects from families with multiple cases of alcoholism were compared with 44 low-risk controls without a family history of alcohol use or Axis I disorders; the authors found decreased bilateral orbitofrontal cortical volume among the high-risk subjects (Hill et al., 2009). Although the study did not exclude subjects who had already used alcohol or drugs of abuse prior to enrollment, and included some subjects in each group who met criteria for SUD, the findings are congruent, if not identical, with those of Benegal et al. In a separate study, the same group reported decreased right amygdala volume in 17 high-risk subjects from families with multiple cases of alcoholism compared with 17 control subjects (Hill et al., 2001). This study similarly did not exclude subjects with early alcohol exposure, and, in fact, several of the high-risk subjects were diagnosed with alcohol or drug dependence, but the main findings were similar to the results of Benegal's group.

In contrast to these findings of decreased cortical, cerebellar, thalamic, and amygdalar volumes, two other studies derived from the same longitudinal cohort of children from families with multiple cases of alcoholism showed increased, rather than decreased, cerebellar and total gray matter volumes in the family history-positive groups (Hill et al., 2007, 2011). As it is difficult to reconcile such divergent results, even as they come from the same sample, these reports also illustrate the idea that the biological correlates of SUD risk are not a uniform constellation of factors and are likely to differ among individuals pending on their unique biological characteristics.

Functional Studies

Functional MRI (fMRI) allows for the assessment of brain activation in association with behavioral performance on tasks that can measure levels of inhibitory control as well as sensitivity to rewarding stimuli. Specifically, the go/no-go task and the stop-signal task measure motor response inhibition, while the Stroop, the Eriksen flanker, and the reversal learning task all measure cognitive inhibition. Reward tasks in general measure an individual's ability to inhibit responses to high-stakes/high-risk stimuli in favor of safer/low-risk trials or preference for larger rewards delivered after a delay period vs immediate but smaller rewards. Examples include the monetary incentive delay (MID) or delay discounting task, the Iowa gambling task, the balloon analog risk task, and the Rogers decision-making task. Most of these tasks have been adapted for use during fMRI scanning.

Studies Using Inhibitory Control Tasks

The utilization of inhibitory control tasks during fMRI aims to assess for purported functional deficits in brain regions known to be involved in behavioral control, including a wide distributed network comprising the frontal and parietal regions (Table 1). One advantage of fMRI is that it is suitable for scanning children, facilitating the investigation of brain activity before exposure to drugs. Reports from studies that compared youth with family history of alcoholism or other SUD who had no prior exposure to substances of abuse have shown decreased activation in the left middle frontal gyrus (Schweinsburg et al., 2004) and frontoparietal regions (Norman et al., 2011) during successful inhibition. The latter report was a longitudinal study that compared baseline scans of subjects who went on to develop heavy alcohol use in a period of 4.2 years vs 17 matched subjects who did not transition to alcohol use. Another longitudinal study of 43 children of alcoholics and 30 controls found blunted activation during inhibition in the right caudate, right middle cingulate, and right middle frontal gyrus among individuals with a family history of SUD (Hardee et al., 2014). As all of these studies showed decreased frontal activation during inhibition tasks among children of substance users or children at high risk, they further complement structural reports of smaller frontal cortical gray matter volumes. Together, these studies suggest that decreased activity in the frontal lobes in response to inhibition tasks, possibly due to structural differences of the frontal cortex, may represent a heritable risk factor for the behavioral deficits seen in children at high risk for substance use. This interpretation is also consistent with models that implicate decreased frontal control in impulsivity. However, it is possible that this finding could be explained by the presence of comorbid disorders characterized by high impulsivity (e.g., ADHD) and, as such, may not be specific for SUD risk.

Studies of children with a family history of SUD and some exposure to alcohol and drugs have produced mixed results. One longitudinal study from a cohort of children of alcoholic parents

Children of Parents with Substance Use Disorder **Chapter** | **4** 41

TABLE 1 Summary of Structural and Functional Studies That Have Examined Children of Parents With SUDs

Author	Method	Population	Results	Conclusions
Structural Studies				
1. Weiland et al. (2014)	fMRI; neuroanatomical volumes were compared between groups and correlated with behavioral measures	106 subjects, ages 18–23, controls (n=64), early risk (ER; n=42); externalizing behaviors were also previously assessed during adolescence	Externalizing behaviors ER>controls, L frontal and L superior frontal cortex volumes ER<controls, while controlling for Fam Hx of alcoholism and current substance use. Total gray matter volumes were negatively associated with substance risk scores. Externalizing behavior scores were negatively associated with both L superior L and L total cortical volumes	Smaller frontal cortical volumes, specifically L superior frontal cortex, may represent an underlying risk factor for SUD
2. Hill et al. (2001)	MRI	34 adolescent subjects; 17 HR with multiplex alcoholism families; 17 controls from families without Axis I psychopathology	R amygdala volume HR<controls, R amygdala volume correlated with visual P300 amplitude	Reduced amygdala volume may represent a risk factor for alcohol dependence
3. Benegal et al. (2007)	MRI	41 male subjects ages 8–24; 20 HR offspring of early onset alcohol-dependent individuals and 21 LR age-matched controls	Volumes of superior frontal, cingulate, and parahippocampal gyri; amygdala; thalamus; and cerebellum HR<controls	HR subjects have smaller gray matter volume in several regions associated with behavioral control
4. Hill et al. (2007)	MRI	33 adolescent subjects; 17 HR with multiplex alcoholism families; 16 age-, gender-, IQ-matched controls	Total cerebellum volume and total gray matter HR>controls. Age-related decreases in total gray volume were documented in controls but not in HR offspring	Offspring from multiplex families for alcohol dependence show susceptibility by having larger cerebellar volume, possibly related to less gray matter pruning with age
5. Hill et al. (2009)	MRI	107 adolescent and young adult subjects (mean age 17); 63 HR individuals from multiplex alcoholism families and 44 LR controls	Bilateral OFC volumes HR<controls, with a greater between-group difference in males; smaller R volumes were associated with 5-HT and BDNF gene variants, and decreased white matter volumes correlated inversely with impulsivity scores	Decreased OFC L volume may mediate risk for alcohol dependence via high impulsivity
6. Hill et al. (2011)	MRI	131 adolescent/young adult subjects; 71 HR subjects from multiplex alcoholism families; 60 LR controls	Total cerebellum and gray matter volumes HR>controls. GABRA2 and BDNF variants were associated with volume	Structural changes in HR associated with genetic variants suggest heritable risk for SUD

Continued

TABLE 1 Summary of Structural and Functional Studies That Have Examined Children of Parents With SUDs—cont'd

Author	Method	Population	Results	Conclusions
Functional Studies				
1. Ivanov et al. (2012)	fMRI; anticipation–conflict–reward task (ACR)	20 drug-naive children with ADHD ages 8–13; separated into LR and HR groups by parental substance abuse	Anticipation: HR>LR activation in L IFG, L anterior insula Reward: HR>LR activation in L OFC and L anterior insula Punishment: HR>LR activation in R anterior insula Conflict: LR>HR activation in ACC	HR children show greater activation with reward components of ACR task in areas related to reward-motivation system HR children show less activation during the cognitive conflict component of the ACR task in the behavioral inhibition system
2. Yau et al. (2012)	fMRI; MID task	40 subjects ages 18–22; 20 children of alcoholics (COA), 20 controls matched for substance use and externalizing behavior	Anticipation (reward and loss): COA<controls for activation in NAcc NAcc response was positively correlated with externalizing risk and lifetime alcohol consumption in COA	Blunted NAcc response may represent a protective or resilience factor
3. Heitzeg et al. (2010)	fMRI; go/no-go task	61 subjects ages 16–22; 41 FH+ for alcohol use disorder divided into 20 with low alcohol problems and 21 with high alcohol problems, 20 FH–	FH–>FH+ deactivation of ventral caudate during successful inhibition FH+>FH– deactivation of medial PFC associated with high alcohol problems	Lower magnitude VS deactivation during inhibition associated with vulnerability in FH+ children Abnormal responding in frontal regions associated with volume of alcohol use
4. Weiland et al. (2013)	fMRI MID task	70 subjects ages 18–22; 49 FH+, 21 FH–	FH–<FH+ for NAcc connectivity with paracentral lobule/precuneus and sensorimotor areas during incentive anticipation In FH+, coupling between L NAcc and supplementary sensorimotor area and R precuneus correlated positively with sensation seeking and drinking volume	FH+>FH– connectivity between NAcc and cortical areas suggests that connectivity with cortex, rather than over- or underactivation, may play a role in SUD risk
5. Bjork et al. (2008)	fMRI MID task	26 subjects ages 12–16; 13 subjects with at least one biological parent with lifetime alcohol dependence and 13 age- and gender-matched controls	No between-group differences for reward anticipation or outcome notification. NAcc activation positively correlated with sensation seeking scores and self-reported excitement to reward across groups	Individual differences in sensation seeking may drive differences in NAcc activation more than parental alcoholism
6. Hardee et al. (2014)	fMRI and go/no-go task imaging at intervals of 1–2 years, with two to four scans performed per subject	Children of alcoholic families (FH+) (n=43) and children of control families (FH–) (n=30) starting at ages 7–12 years	Baseline activation in caudate, middle frontal gyrus, and middle cingulate: alcohol- and drug-naive participants with FH+<FH– FH– show decrease in age-related activation in the R caudate and middle frontal gyrus FH+ show increase in age-related activation in the middle cingulate	Differences in activation in behavioral control circuitry in FH+ vs FH– can be identified in childhood; during adolescence FH+ group shows activation changes inconsistent with normal development. These different activation patterns precede problem drinking and may be a marker for later SUD

7. Schweinsburg et al. (2004)	fMRI go/no-go task	26 subjects ages 12–14; 12 FH+ and 14 controls FH–	Activation in L middle frontal gyrus during inhibition FH+ < FH–	Supports the theory of frontal deficits in relation to behavioral control in FH+ youth
8. McNamee et al. (2008)	fMRI antisaccade task	25 subjects ages 12–19; 17 had parents with SUDs, 8 had no parental SUD history	Composite neurobehavioral inhibition scores negatively correlated with total frontal activation	Low inhibition behavioral profiles in youth at risk is related to frontal hypoactivation
9. Silveri et al. (2011)	fMRI Stroop color naming, word reading, interference	32 subjects ages 8–19; 18 with FH+, 16 controls (FH–)	Activation in ACC, middle frontal gyrus, superior frontal gyrus, and insula FH+ > FH–	Increased recruitment of frontal cortical areas in FH+ during inhibition Stroop task
10. Norman et al. (2011)	fMRI go/no-go task	38 subjects ages 12–14 followed longitudinally and classified as heavy alcohol users (n = 21) and controls (n = 17)	Youth who transitioned to heavy alcohol use showed less activation in 12 frontoparietal regions during response inhibition at baseline	Less activation of frontal areas during inhibition may predict future substance use
11. Mahmood et al. (2013)	fMRI go/no-go task	80 subjects ages 16–19; followed for 18 months and classified as high-frequency users (n = 39) and matched low-frequency users (n = 41)	Lower ventromedial prefrontal activation but greater left angular gyrus activation predicted higher levels of substance abuse and dependence	Different activation patterns possibly related to functional connectivity between prefrontal cortex and left angular gyrus, which might be further complicated by exposure to substances
12. Acheson et al. (2009)	fMRI Iowa gambling task	34 young adult subjects, mean age 23; 15 FH+, 19 FH–	Activation in L dorsal anterior cingulate cortex and L caudate nucleus FH+ > FH–	Youth at risk may need to recruit more cognitive effort during decision-making associated with possible loss
13. Spadoni et al. (2008)	fMRI spatial working memory and vigilance task	72 subjects ages 12–14; an index of family history of alcohol use disorders was calculated for each subject rather than dividing subjects into cases/controls	Greater family history density correlated with less activation in cingulate and medial frontal gyri during simple vigilance condition	Youth with family histories of alcohol use disorders may have more difficulty modulating the default mode network and regulating goal-directed behavior
14. Tapert et al. (2007)	fMRI go/no-go task	33 adolescent subjects; 16 with history of cannabis use after 28 days of monitored abstinence, 17 controls	Cannabis users showed greater activation of R dorsolateral prefrontal, bilateral medial frontal, bilateral inferior, and superior parietal lobules, R occipital gyrus during inhibition, and greater activation to go trials in R prefrontal, insular and parietal cortices	Cannabis use associated with greater activation during both inhibition and go trials in frontoparietal and insular cortex. Increased activation hypothesized to reflect increased processing "effort"

This table summarizes the various structural and functional studies that have evaluated models of SUD predisposition in offspring of substance users. These studies have used functional magnetic resonance imaging (fMRI) and MRI. We found six studies that assessed morphological differences in offspring of drug abusers and 14 studies that assessed functional differences in brain regions in offspring of substance abusers. These studies represent an effort to better understand the relationships between a biological predisposition for SUDs and morphological and functional brain changes in offspring of substance abusers. Unfortunately, most studies have limited power and significant selection bias. These limitations make it difficult to draw definitive conclusions. FH+, family history positive for alcoholism; FH–, family history negative for alcoholism; HR, high risk; LR, low risk; L, left; R, right; 5-HT, serotonin; ACC, anterior cingulate cortex; BDNF, brain-derived neurotrophic factor; Fam Hx, family history; GABRA2, gene that encodes the α2 subunit of the γ-aminobutyric acid A receptor (GABAA); IFG, inferior frontal gyrus; MID, monetary incentive delay task; NAcc, nucleus accumbens; OFC, orbital frontal cortex; PFC, prefrontal cortex; VS, ventral striatum.

examined subjects with a positive family history for SUD with low and high alcohol problems, as well as a negative family history group (Heitzeg, Nigg, Yau, Zucker, & Zubieta, 2010). Participants without family history showed greater deactivation of the ventral caudate during successful inhibition compared to the two family history-positive groups, suggesting that reduced activation in the ventral caudate represents a risk factor for familial SUD. Among the participants with a positive family history, the ones with high alcohol problems had less deactivation of the medial prefrontal cortex during successful inhibition—suggesting that altered activation in the prefrontal cortex is related to alcohol exposure. This interpretation is somewhat at odds with findings from the studies discussed above (Norman et al., 2011; Schweinsburg et al., 2004), which speculate that deficits in inhibitory control due to prefrontal cortex dysfunction may precede exposure to drugs. However, it is worth noting that the patterns of activation changes during inhibitory tasks are not consistent among individuals with family history and active drug use. For instance, adolescents with high-frequency drug use ($n = 39$) performing the go/no-go task showed decreased ventromedial prefrontal activation and greater left angular gyrus activation with inhibition compared to low-frequency users (Mahmood et al., 2013). A different study of 16 adolescents with a history of cannabis use and 17 controls showed greater activation with inhibition in the right dorsolateral prefrontal, bilateral medial frontal, bilateral inferior, and superior parietal lobules among cannabis users (Tapert et al., 2007). Together these results suggest that behavioral control deficits in children with parental SUD and no drug exposure might be linked to a constitutional decrease in frontal activation, but that activation patterns can change as a result of drug use.

Prefrontal activation changes in children at familial risk might also be affected by the use of different inhibitory tasks. This is illustrated by the result from one group that used the Stroop color naming, word reading, and interference tasks to compare children and adolescents (ages 8–19) with and without a family history of alcoholism, all with three or fewer lifetime episodes of alcohol or drug use (Silveri, Rogowska, McCaffrey, & Yurgelun-Todd, 2011). In contrast to prior studies of children without substance exposure, the children of substance users in this study had increased frontal activation in the anterior cingulate cortex, middle frontal gyrus, superior frontal gyrus, and insula during performance of the interference condition of the Stroop task. The authors suggest that this increased activation may actually represent a need for greater neuronal activity to sustain performance comparable to that of the subjects without a family history of SUD. It is possible that unknown task-specific differences or unspecified differences in the study populations may also have contributed to the disparate findings.

Our group has examined youths with no prior exposure to drugs (including in utero and lifetime exposure), also excluding any exposure to stimulant medications in children with ADHD, using a novel task that combines elements of reward incentive and outcomes as well as cognitive inhibition and conflict resolution (e.g., flanker task) (Ivanov et al., 2012). We studied 20 children at risk for SUD, all of whom had ADHD; of those children, 10 also had at least one biological parent with SUD (e.g., cumulative risk). The results indicate that elements of the reward processing system (e.g., anterior insula and the left orbitofrontal cortex) exhibited higher activation with reward cues and outcomes for children with ADHD who also had parental SUD compared to children with ADHD only. In other words, youth with cumulative clinical risk including childhood disruptive behavior disorder and parental SUD had elevated sensitivity to rewarding stimuli. Further, we found lower activation in the anterior cingulate gyrus in relation to the flanker task component among participants with cumulative risk for SUD, probably reflecting deficits in their ability to resolve cognitive conflict. This is in accordance with findings from studies using the go/no-go task and stop-signal task reporting decreased activation in the right prefrontal cortex. These results provide preliminary support for the idea that children with elevated cumulative risk for later SUD may have features of increased reward sensitivity paired with deficits in cognitive control and impairments in right inferior frontal activity.

Additional executive functions that are involved in behavioral control include working memory and affect regulation. Studies that assess these functions in youths with a family history of SUD have documented decreased activation during working-memory tasks in areas of the prefrontal cortex, including the right anterior prefrontal, left dorsolateral prefrontal cortex, right inferior frontal gyrus, and right cingulate cortex (Cservenka, Herting, & Nagel, 2012) and in the cingulate and medial frontal gyri (Spadoni, Norman, Schweinsburg, & Tapert, 2008), among subjects with positive family histories of substance abuse. Conversely, others reported greater activation in the left precuneus and bilateral midoccipital gyri during a face-matching task designed to probe affective regulation in 19 children of substance users, ages 10–14, with current externalizing pathology and fewer than five lifetime instances of alcohol or drug use (Hulvershorn et al., 2013). The authors suggested that increased activation represents a deficit in the regulation of affective processing.

The idea that prefrontal hypoactivation may be a precursor for later SUD is supported by several reports of physiological measures, including electroencephalogram (EEG) markers. It has been suggested that decreased amplitude of the P300 event-related potentials (ERPs) and event-related oscillations on EEG reflect a general decrease in neurocognitive inhibition (Tessner & Hill, 2010), and lower P300 amplitudes have been repeatedly associated with early onset of alcohol use (Porjesz et al., 2005). Moreover, low-amplitude P300 activation is not specific to alcoholism per se, but rather to a behaviorally disinhibited phenotype that predisposes to SUD risk, especially via early onset drinking, as well as conduct problems and externalizing disorders (Iacono & McGue, 2006; Porjesz & Rangaswamy, 2007; Tessner & Hill, 2010). Two studies of children of substance users reported decreased activity across delta, theta, and alpha1 bands during the go/no-go task (Kamarajan et al., 2006) and lower P300 peak amplitude during the visual oddball inhibitory control task (Rangaswamy et al., 2007). In summary, with the caveat that EEG lacks the anatomic resolution of fMRI, these between-group differences suggest that differences in ERP amplitude may be a heritable marker related to decreased inhibitory control and subsequent SUD risk.

Studies with Reward Tasks

Several groups have utilized the MID task to probe the reward system in children of substance abusers with limited early use. As with studies using inhibitory control tasks, the results have been somewhat mixed. One study of 20 children of alcoholic parents (high risk) compared with 20 control subjects (ages 18–22) found blunted ventral striatal response among those high-risk subjects without current problem-drinking behaviors (Yau et al., 2012). This is an intriguing finding and supports the reward deficiency hypothesis because the blunted response was found in high-risk subjects without previous problematic substance use. Nucleus accumbens response was correlated with lifetime alcohol use only within the high-risk group. Another report documented no between-group difference in nucleus accumbens activation in a sample of 49 participants (ages 18–22) with a family history of SUD and a control group of 21 participants without such a history using the MID task (Weiland et al., 2013). However, this study did find greater connectivity between the nucleus accumbens and the cortical areas among the high-risk group. A third MID study of 13 children of alcoholics, ages 12–16, compared with 13 controls found no between-group differences in striatal activation (Bjork, Knutson, & Hommer, 2008). Neither of the studies excluded subjects with early substance use, but the Bjork et al. study did exclude subjects meeting the full criteria for SUD. No differences in the activation of the nucleus accumbens or the ventral striatum were reported by Acheson, Robinson, Glahn, Lovallo, and Fox (2009) utilizing the Iowa gambling task to probe decision-making involving rewards between groups of 15 family history-positive and 19 family history-negative children. However, they documented greater activation of the anterior cingulate and the caudate among those with a family history of substance abuse. As the caudate receives significant dopaminergic input, it is unclear how to interpret these findings with respect to the reward deficiency/reward sensitivity hypotheses. These results may have also been influenced by the effects of substances, since subjects with frank SUDs were excluded, but subjects with previous substance use were not. Although the use of fMRI in unexposed children of substance users remains a promising approach, the evidence at this time is insufficient to draw a firm conclusion as to whether over- or underresponsive reward systems mediate the genetic risk for development of SUDs.

FUTURE DIRECTIONS

One can argue that hypotheses suggesting a simple, two-factor model of the relationship between behavioral activation (i.e., decreased reward sensitivity) and behavioral control (i.e., decreased behavioral control) might be somewhat simplistic. Given the complexity of the syndrome and the multiplicity of clinical risk factors, it is hard to imagine that one single neurobiological model will fit all individuals at heritable risk. An alternative approach might propose that different groups of individuals at risk exhibit different patterns of behavioral deficits, in association with different profiles of brain activity in relation to reward processing. For instance, youth with ADHD may exhibit low responses to rewarding stimuli (i.e., reward deficiency) and particular inhibitory control deficits (e.g., motor inhibition) (Scheres, Milham, Knutson, & Castellanos, 2007), whereas youth with a propensity for sensation/reward seeking might also exhibit deficits in different types of behavioral control (i.e., delay discounting and preference for small immediate vs larger but delayed rewards). Therefore, it is likely that an approach that takes into account the various relationships between behavioral activation and behavioral control in specific high-risk groups might be most useful. As a hypothetical example, youths with greater sensitivity to reward incentives might be more likely to respond to behavioral modification paired with reward contingencies, whereas youth with diminished reward sensitivity might be better treated with pharmacological agents that modulate the activity of the reward/motivation brain systems.

SUMMARY

Variation in the risk for and severity of SUDs is the product of numerous behavioral traits and environmental variables. Children of parents with SUDs have an increased risk of SUDs, with heritability estimates as high as 70%. However, no particular gene or gene combination has been found to be strongly associated with the development of SUD. Several studies have shown that children with disorders involving impulsivity, reactive aggression, sensation seeking, and excessive risk taking are at high risk of developing SUDs. Existing models of SUD vulnerability emphasize different aspects of brain function and behavioral risk factors, all of which may contribute to the addiction syndrome. A key overarching construct is that SUD risk involves both altered motivation (i.e., reward sensitivity, reward-motivated behaviors) and deficient behavioral control (i.e., impulsivity, response inhibition, etc.). Evidence from structural and functional imaging studies provides preliminary support for the hypothesis that altered morphology and activation in brain regions associated with behavioral motivation and behavioral control are associated with familial SUD. For the most part, results have shown decreased frontal activation during inhibition tasks among children of substance users or children at high risk. However, it remains unclear whether over- or underresponsive reward systems mediate the genetic risk for developing SUDs. It is hoped that the many remaining unanswered questions will stimulate further research to identify specific processes that lead to the development of familial SUD (Figure 2).

APPLICATIONS TO OTHER ADDICTIONS AND SUBSTANCE MISUSE

1. Heritability of SUDs is ubiquitous regardless of any particular substance.
2. Studies of children of parents with SUDs have provided preliminary information for the possible neurobiological mechanisms for the development of SUDs.
3. Common features of the various biological models of addiction vulnerability include purported deficits in neuronal systems that process information related to rewarding stimuli and the ability to modulate behavioral responses to rewards.
4. Understanding the possible mechanisms that may lead to the development of the addiction syndrome can provide targets for therapeutic interventions that may influence reward processing and behavioral control and in turn will alter the development of behaviors leading to substance misuse and addiction.

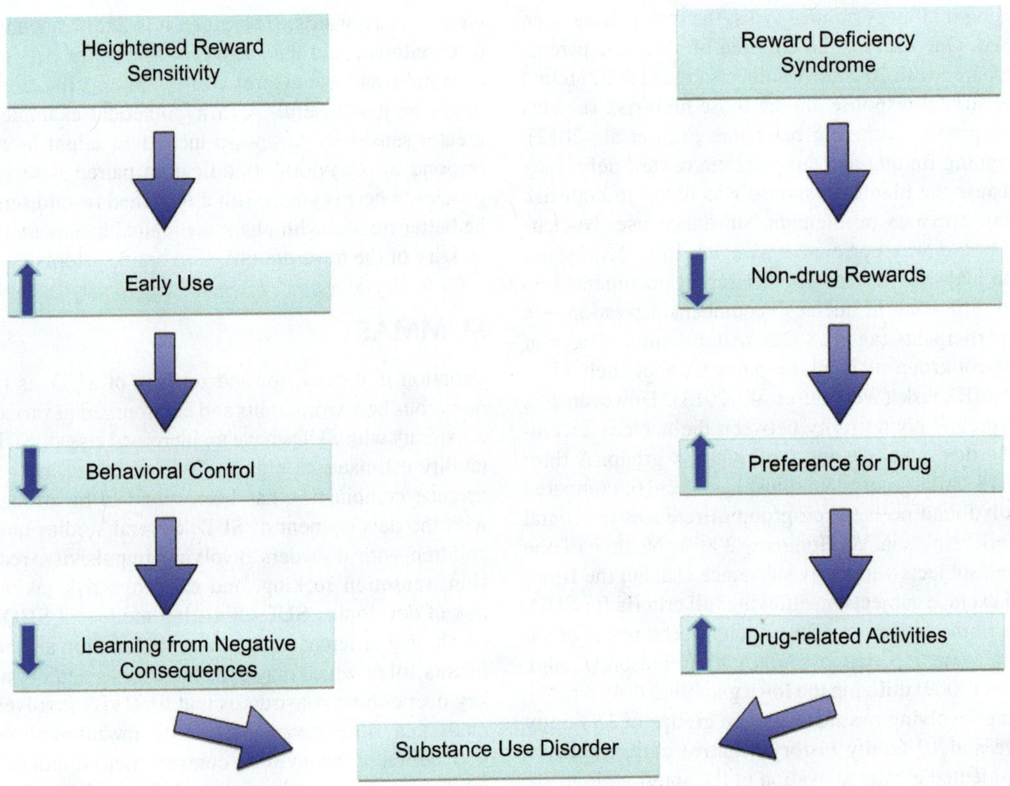

FIGURE 2 Model pathways of SUD development.
- Heightened sensitivity to the rewarding aspects of drug use may lead to experimentation early in life; in this model, heightened sensitivity to the rewarding aspects of drugs is paired with deficits in behavioral inhibition—which in turn produce impulsive behaviors associated with increased incidences of negative consequences. Further, the affected individuals fail to adequately learn from such consequences, which is a prominent feature of SUDs.
- Reward deficiency syndrome is associated with dampened sensitivity to the rewarding aspects of nondrug rewards; alternatively, there is preference for drug experimentation/use that produces highly rewarding experiences, leading to behaviors of persistent seeking/obtaining of drugs, which also is a prominent feature of SUDs.

DEFINITION OF TERMS

Impulsivity This is characterized by action without planning and with little consideration for possible negative outcomes.

Reward sensitivity This hypothesis proposes that a reduced response of the so-called "reward system" (primarily referring to the nucleus accumbens in the ventral striatum, the ventral tegmental area, and the mesial frontal cortex) to nondrug rewards predisposes some individuals to preferentially seek the rewarding effects of drugs of abuse as opposed to other environmental stimuli.

Frontal cortex This is one of the four major lobes of the cortex that make up the anterior portion of the brain, thought to be involved in many executive functions, including inhibitory control.

Anterior cingulate cortex This is a region of the cingulate cortex of the medial aspect of the cortex, which surrounds the corpus callosum and is thought to be involved in autonomic and executive functions as well as anticipation and reward processing.

Ventral striatum This is a region of the basal ganglia that consists of the olfactory tubercle and nucleus accumbens and is thought to be involved in reward processing.

Nucleus accumbens This is a region of the basal forebrain that, with the olfactory tubercle, forms the ventral striatum and has been shown to be involved in reward processing.

Functional magnetic resonance imaging This technique measures changes in blood flow to regions of the brain via the change in magnetization between oxygen-rich and oxygen-poor blood. fMRI allows one to register activation of various brain regions in relation to behavioral or pharmacological stimuli.

Positron emission tomography This is a functional imaging technique that uses computed tomography combined with a positron-emitting radionuclide connected to biologically active molecules, which allows one to register the distribution of particular molecules in the brain.

Internalizing disorder This is A term for emotional or behavioral problems such as anxiety, depression, and eating disorders that are characterized by self-reported symptoms.

Externalizing disorder This is a term for emotional or behavioral problems such as disruptive behavior disorders and conduct problems that are characterized by symptoms that are self-reported and reported by others.

KEY FACTS

- SUDs are defined by the DSM-5 as a pattern of drug use causing clinically significant impairment or distress. The DSM-5 uses the diagnosis of SUD for each substance of abuse, whereas the DSM-IV-TR previously categorized SUDs as either "abuse" or "dependence."
- The children of parents with SUDs have an increased risk of SUDs, with heritability estimates as high as 70%.
- The heritability of SUDs is now known to be multifactorial; however, no specific genes have been identified as biological markers of SUD.
- Behavioral features like impulsivity, excessive risk taking, and sensation seeking are thought to contribute to the development of adolescent SUD.
- Several studies have shown that children that present with high levels of impulsivity, excessive risk taking, and sensation seeking also exhibit dysfunctions in various brain regions including the prefrontal cortex, the anterior cingulate gyrus, and the limbic system.
- Both structural and functional abnormalities have been observed in the prefrontal cortex, the anterior cingulate gyrus, and the limbic system in children of parents with SUDs.
- Alterations in brain structures and functions related to reward processing and behavioral control may present biological vulnerabilities for the development of SUDs in children of substance-using parents.

SUMMARY POINTS

- Several models have been developed to account for and explain the progression from drug experimentation to regular use and addiction in persons with heritable risk. It is worth noting that many people experiment with alcohol and drugs of abuse, but only a subset of individuals go on to develop SUDs. In some cultural settings, experimentation with or use of alcohol or certain other potential drugs of abuse is normative.
- One hypothesis that explains the observed association of certain childhood clinical conditions with risk for alcoholism and, by extension, SUD is that there are parallel "externalizing" and "internalizing" pathways for SUD onset and development.
- SUD risk involves both altered motivation (i.e., reward sensitivity, reward-motivated behaviors) and deficient behavioral control (i.e., impulsivity, response inhibition, etc.).
- Physiological and neuroimaging studies have identified brain regions in individuals with familial SUD that show different structural and functional patterns compared to unaffected controls.
- Structural studies have shown decreased cortical, cerebellar, thalamic, and amygdalar volumes in children of parents with SUDs compared to controls without familial risk (i.e., effects possibly mediated by heritability of SUD).
- Functional neuroimaging studies have shown decreased frontal activation during inhibition tasks among children of parents with SUDs. Existing evidence prevents us from drawing a firm conclusion as to whether over- or underresponsive reward systems mediate the genetic risk for developing SUDs.
- Competing models for SUD vulnerability include a two-factor model of the relationship between behavioral activation (i.e., decreased reward sensitivity) and behavioral control (i.e., decreased behavioral control); a more complex model suggests that different groups of individuals at risk exhibit different patterns of behavioral deficits, with some at-risk individuals showing increased reward/sensation seeking and preference for smaller immediate rewards and others showing decreased reward sensitivity but also compromised ability to withhold behaviors that lead to negative outcomes.

REFERENCES

Acheson, A., Robinson, J. L., Glahn, D. C., Lovallo, W. R., & Fox, P. T. (2009). Differential activation of the anterior cingulate cortex and caudate nucleus during a gambling simulation in persons with a family history of alcoholism: studies from the Oklahoma Family Health Patterns Project. *Drug and Alcohol Dependence*, *100*(1–2), 17–23. http://dx.doi.org/10.1016/j.drugalcdep.2008.08.019.

Benegal, V., Antony, G., Venkatasubramanian, G., & Jayakumar, P. N. (2007). Gray matter volume abnormalities and externalizing symptoms in subjects at high risk for alcohol dependence. *Addiction Biology*, *12*(1), 122–132. http://dx.doi.org/10.1111/j.1369-1600.2006.00043.x.

Berridge, K. C., & Kringelbach, M. L. (2008). Affective neuroscience of pleasure: reward in humans and animals. *Psychopharmacology*, *199*(3), 457–480. http://dx.doi.org/10.1007/s00213-008-1099-6.

Bjork, J. M., Knutson, B., & Hommer, D. W. (2008). Incentive-elicited striatal activation in adolescent children of alcoholics. *Addiction (Abingdon, England)*, *103*(8), 1308–1319. http://dx.doi.org/10.1111/j.1360-0443.2008.02250.x.

Blackson, T. C., Butler, T., Belsky, J., Ammerman, R. T., Shaw, D. S., & Tarter, R. E. (1999). Individual traits and family contexts predict sons' externalizing behavior and preliminary relative risk ratios for conduct disorder and substance use disorder outcomes. *Drug and Alcohol Dependence*, *56*(2), 115–131. Retrieved from http://www.ncbi.nlm.nih.gov/pubmed/10482403.

Buckingham-Howes, S., Berger, S. S., Scaletti, L. A., & Black, M. M. (2013). Systematic review of prenatal cocaine exposure and adolescent development. *Pediatrics*, *131*(6), e1917–e1936. http://dx.doi.org/10.1542/peds.2012-0945.

Clark, D. B., Parker, A. M., & Lynch, K. G. (1999). Psychopathology and substance-related problems during early adolescence: a survival analysis. *Journal of Clinical Child Psychology*, *28*(3), 333–341. http://dx.doi.org/10.1207/S15374424jccp280305.

Cservenka, A., Herting, M. M., & Nagel, B. J. (2012). Atypical frontal lobe activity during verbal working memory in youth with a family history of alcoholism. *Drug and Alcohol Dependence*, *123*(1–3), 98–104. http://dx.doi.org/10.1016/j.drugalcdep.2011.10.021.

Goldman, D., Oroszi, G., & Ducci, F. (2005). The genetics of addictions: uncovering the genes. *Nature Reviews. Genetics*, *6*(7), 521–532. http://dx.doi.org/10.1038/nrg1635.

Gullo, M. J., Loxton, N. J., & Dawe, S. (2014). Impulsivity: four ways five factors are not basic to addiction. *Addictive Behaviors*. http://dx.doi.org/10.1016/j.addbeh.2014.01.002.

Hardee, J. E., Weiland, B. J., Nichols, T. E., Welsh, R. C., Soules, M. E., Steinberg, D. B., ... Heitzeg, M. M. (2014). Development of impulse control circuitry in children of alcoholics. *Biological Psychiatry*. http://dx.doi.org/10.1016/j.biopsych.2014.03.005.

Heitzeg, M. M., Nigg, J. T., Yau, W.-Y. W., Zucker, R. A., & Zubieta, J.-K. (2010). Striatal dysfunction marks preexisting risk and medial prefrontal dysfunction is related to problem drinking in children of alcoholics. *Biological Psychiatry, 68*(3), 287–295. http://dx.doi.org/10.1016/j.biopsych.2010.02.020.

Hill, S. Y., De Bellis, M. D., Keshavan, M. S., Lowers, L., Shen, S., Hall, J., & Pitts, T. (2001). Right amygdala volume in adolescent and young adult offspring from families at high risk for developing alcoholism. *Biological Psychiatry, 49*(11), 894–905. Retrieved from http://www.ncbi.nlm.nih.gov/pubmed/11377407.

Hill, S. Y., Muddasani, S., Prasad, K., Nutche, J., Steinhauer, S. R., Scanlon, J., ... Keshavan, M. (2007). Cerebellar volume in offspring from multiplex alcohol dependence families. *Biological Psychiatry, 61*(1), 41–47. http://dx.doi.org/10.1016/j.biopsych.2006.01.007.

Hill, S. Y., Wang, S., Carter, H., Tessner, K., Holmes, B., McDermott, M., ... Stiffler, S. (2011). Cerebellum volume in high-risk offspring from multiplex alcohol dependence families: association with allelic variation in GABRA2 and BDNF. *Psychiatry Research, 194*(3), 304–313. http://dx.doi.org/10.1016/j.pscychresns.2011.05.006.

Hill, S. Y., Wang, S., Kostelnik, B., Carter, H., Holmes, B., McDermott, M., ... Keshavan, M. S. (2009). Disruption of orbitofrontal cortex laterality in offspring from multiplex alcohol dependence families. *Biological Psychiatry, 65*(2), 129–136. http://dx.doi.org/10.1016/j.biopsych.2008.09.001.

Hommer, D. W., Bjork, J. M., & Gilman, J. M. (2011). Imaging brain response to reward in addictive disorders. *Annals of the New York Academy of Sciences, 1216*, 50–61. http://dx.doi.org/10.1111/j.1749-6632.2010.05898.x.

Hulvershorn, L. A., Finn, P., Hummer, T. A., Leibenluft, E., Ball, B., Gichina, V., & Anand, A. (2013). Cortical activation deficits during facial emotion processing in youth at high risk for the development of substance use disorders. *Drug and Alcohol Dependence, 131*(3), 230–237. http://dx.doi.org/10.1016/j.drugalcdep.2013.05.015.

Iacono, W. G., & McGue, M. (2006). Association between P3 event-related brain potential amplitude and adolescent problem behavior. *Psychophysiology, 43*(5), 465–469. http://dx.doi.org/10.1111/j.1469-8986.2006.00422.x.

Ivanov, I., Liu, X., Clerkin, S., Schulz, K., Friston, K., Newcorn, J. H., & Fan, J. (2012). Effects of motivation on reward and attentional networks: an fMRI study. *Brain and Behavior, 2*(6), 741–753. http://dx.doi.org/10.1002/brb3.80.

Johnson, J. L., & Leff, M. (1999). Children of substance abusers: overview of research findings. *Pediatrics, 103*(5 Pt 2), 1085–1099. Retrieved from http://www.ncbi.nlm.nih.gov/pubmed/10224196.

Kamarajan, C., Porjesz, B., Jones, K., Chorlian, D., Padmanabhapillai, A., Rangaswamy, M., ... Begleiter, H. (2006). Event-related oscillations in offspring of alcoholics: neurocognitive disinhibition as a risk for alcoholism. *Biological Psychiatry, 59*(7), 625–634. http://dx.doi.org/10.1016/j.biopsych.2005.08.017.

Kendler, K. S., Maes, H. H., Sundquist, K., Ohlsson, H., & Sundquist, J. (2014). Genetic and family and community environmental effects on drug abuse in adolescence: a Swedish national twin and sibling study. *The American Journal of Psychiatry, 171*(2), 209–217. http://dx.doi.org/10.1176/appi.ajp.2013.12101300.

Kendler, K. S., Ohlsson, H., Sundquist, K., & Sundquist, J. (2014). Peer deviance, parental divorce, and genetic risk in the prediction of drug abuse in a nationwide Swedish sample: evidence of environment-environment and gene-environment interaction. *JAMA Psychiatry, 71*(4), 439–445. http://dx.doi.org/10.1001/jamapsychiatry.2013.4166.

Keyes, K. M., Schulenberg, J. E., O'Malley, P. M., Johnston, L. D., Bachman, J. G., Li, G., & Hasin, D. (2012). Birth cohort effects on adolescent alcohol use: the influence of social norms from 1976 to 2007. *Archives of General Psychiatry, 69*(12), 1304–1313. http://dx.doi.org/10.1001/archgenpsychiatry.2012.787.

Koob, G. F., & Volkow, N. D. (2010). Neurocircuitry of addiction. *Neuropsychopharmacology: Official Publication of the American College of Neuropsychopharmacology, 35*(1), 217–238. http://dx.doi.org/10.1038/npp.2009.110.

Kringelbach, M. L. (2005). The human orbitofrontal cortex: linking reward to hedonic experience. *Nature Reviews. Neuroscience, 6*(9), 691–702. http://dx.doi.org/10.1038/nrn1747.

Mahmood, O. M., Goldenberg, D., Thayer, R., Migliorini, R., Simmons, A. N., & Tapert, S. F. (2013). Adolescents' fMRI activation to a response inhibition task predicts future substance use. *Addictive Behaviors, 38*(1), 1435–1441. http://dx.doi.org/10.1016/j.addbeh.2012.07.012.

McNamee, R. L., Dunfee, K. L., Luna, B., Clark, D. B., Eddy, W. F., & Tarter, R. E. (2008). Brain activation, response inhibition, and increased risk for substance use disorders. *Alcoholism, Clinical and Experimental Research, 32*(3), 405–413.

Norman, A. L., Pulido, C., Squeglia, L. M., Spadoni, A. D., Paulus, M. P., & Tapert, S. F. (2011). Neural activation during inhibition predicts initiation of substance use in adolescence. *Drug and Alcohol Dependence, 119*(3), 216–223. http://dx.doi.org/10.1016/j.drugalcdep.2011.06.019.

Porjesz, B., & Rangaswamy, M. (2007). Neurophysiological endophenotypes, CNS disinhibition, and risk for alcohol dependence and related disorders. *The Scientific World Journal, 7*, 131–141. http://dx.doi.org/10.1100/tsw.2007.203.

Porjesz, B., Rangaswamy, M., Kamarajan, C., Jones, K. A., Padmanabhapillai, A., & Begleiter, H. (2005). The utility of neurophysiological markers in the study of alcoholism. *Clinical Neurophysiology: Official Journal of the International Federation of Clinical Neurophysiology, 116*(5), 993–1018. http://dx.doi.org/10.1016/j.clinph.2004.12.016.

Raitasalo, K., Holmila, M., Autti-Rämö, I., Notkola, I.-L., & Tapanainen, H. (2014). Hospitalisations and out-of-home placements of children of substance-abusing mothers: a register-based cohort study. *Drug and Alcohol Review*. http://dx.doi.org/10.1111/dar.12121.

Rangaswamy, M., Jones, K. A., Porjesz, B., Chorlian, D. B., Padmanabhapillai, A., Kamarajan, C., ... Begleiter, H. (2007). Delta and theta oscillations as risk markers in adolescent offspring of alcoholics. *International Journal of Psychophysiology: Official Journal of the International Organization of Psychophysiology, 63*(1), 3–15. http://dx.doi.org/10.1016/j.ijpsycho.2006.10.003.

Robinson, T. E., & Berridge, K. C. (2001). Incentive-sensitization and addiction. *Addiction (Abingdon, England), 96*(1), 103–114. http://dx.doi.org/10.1080/09652140020016996.

Scheres, A., Milham, M. P., Knutson, B., & Castellanos, F. X. (2007). Ventral striatal hyporesponsiveness during reward anticipation in attention-deficit/hyperactivity disorder. *Biological Psychiatry, 61*(5), 720–724. http://dx.doi.org/10.1016/j.biopsych.2006.04.042.

Schoenbaum, G., Roesch, M. R., & Stalnaker, T. A. (2006). Orbitofrontal cortex, decision-making and drug addiction. *Trends in Neurosciences, 29*(2), 116–124. http://dx.doi.org/10.1016/j.tins.2005.12.006.

Schweinsburg, A. D., Paulus, M. P., Barlett, V. C., Killeen, L. A., Caldwell, L. C., Pulido, C., ... Tapert, S. F. (2004). An FMRI study of response inhibition in youths with a family history of alcoholism. *Annals of the New York Academy of Sciences, 1021*, 391–394. http://dx.doi.org/10.1196/annals.1308.050.

Silveri, M. M., Rogowska, J., McCaffrey, A., & Yurgelun-Todd, D. A. (2011). Adolescents at risk for alcohol abuse demonstrate altered frontal lobe activation during Stroop performance. *Alcoholism, Clinical and Experimental Research*, *35*(2), 218–228. http://dx.doi.org/10.1111/j.1530-0277.2010.01337.x.

Spadoni, A. D., Norman, A. L., Schweinsburg, A. D., & Tapert, S. F. (2008). Effects of family history of alcohol use disorders on spatial working memory BOLD response in adolescents. *Alcoholism, Clinical and Experimental Research*, *32*(7), 1135–1145. http://dx.doi.org/10.1111/j.1530-0277.2008.00694.x.

Spottiswoode, B. S., Meintjes, E. M., Anderson, A. W., Molteno, C. D., Stanton, M. E., Dodge, N. C., … Jacobson, S. W. (2011). Diffusion tensor imaging of the cerebellum and eyeblink conditioning in fetal alcohol spectrum disorder. *Alcoholism, Clinical and Experimental Research*, *35*(12), 2174–2183. http://dx.doi.org/10.1111/j.1530-0277.2011.01566.x.

Tapert, S. F., Schweinsburg, A. D., Drummond, S. P. A., Paulus, M. P., Brown, S. A., Yang, T. T., & Frank, L. R. (2007). Functional MRI of inhibitory processing in abstinent adolescent marijuana users. *Psychopharmacology*, *194*(2), 173–183. http://dx.doi.org/10.1007/s00213-007-0823-y.

Tarter, R. E., Kirisci, L., Mezzich, A., Cornelius, J. R., Pajer, K., Vanyukov, M., … Clark, D. (2003). Neurobehavioral disinhibition in childhood predicts early age at onset of substance use disorder. *The American Journal of Psychiatry*, *160*(6), 1078–1085. Retrieved from http://www.ncbi.nlm.nih.gov/pubmed/12777265.

Tarter, R., Vanyukov, M., Giancola, P., Dawes, M., Blackson, T., Mezzich, A., & Clark, D. B. (1999). Etiology of early age onset substance use disorder: a maturational perspective. *Development and Psychopathology*, *11*(4), 657–683. Retrieved from http://www.ncbi.nlm.nih.gov/pubmed/10624720.

Tessner, K. D., & Hill, S. Y. (2010). Neural circuitry associated with risk for alcohol use disorders. *Neuropsychology Review*, *20*(1), 1–20. http://dx.doi.org/10.1007/s11065-009-9111-4.

Ungerer, M., Knezovich, J., & Ramsay, M. (2013). In utero alcohol exposure, epigenetic changes, and their consequences. *Alcohol Research: Current Reviews*, *35*(1), 37–46. Retrieved from http://www.pubmedcentral.nih.gov/articlerender.fcgi?artid=3860424&tool=pmcentrez&rendertype=abstract.

Volkow, N. D., Fowler, J. S., Wang, G.-J., & Swanson, J. M. (2004). Dopamine in drug abuse and addiction: results from imaging studies and treatment implications. *Molecular Psychiatry*, *9*(6), 557–569. http://dx.doi.org/10.1038/sj.mp.4001507.

Volkow, N. D., Wang, G. J., Fowler, J. S., Logan, J., Gatley, S. J., Gifford, A., … Pappas, N. (1999). Prediction of reinforcing responses to psychostimulants in humans by brain dopamine D2 receptor levels. *The American Journal of Psychiatry*, *156*(9), 1440–1443. Retrieved from http://www.ncbi.nlm.nih.gov/pubmed/10484959.

Weiland, B. J., Korycinski, S. T., Soules, M., Zubieta, J. K., Zucker, R. A., & Heitzeg, M. M. (2014). Substance abuse risk in emerging adults associated with smaller frontal gray matter volumes and higher externalizing behaviors. *Drug and Alcohol Dependence*, *137*, 68–75.

Weiland, B. J., Welsh, R. C., Yau, W.-Y. W., Zucker, R. A., Zubieta, J.-K., & Heitzeg, M. M. (2013). Accumbens functional connectivity during reward mediates sensation-seeking and alcohol use in high-risk youth. *Drug and Alcohol Dependence*, *128*(1–2), 130–139. http://dx.doi.org/10.1016/j.drugalcdep.2012.08.019.

Wouldes, T. A., Lagasse, L. L., Huestis, M. A., Dellagrotta, S., Dansereau, L. M., & Lester, B. M. (2014). Prenatal methamphetamine exposure and neurodevelopmental outcomes in children from 1 to 3 years. *Neurotoxicology and Teratology*, *42*, 77–84. http://dx.doi.org/10.1016/j.ntt.2014.02.004.

Yau, W.-Y. W., Zubieta, J.-K., Weiland, B. J., Samudra, P. G., Zucker, R. A., & Heitzeg, M. M. (2012). Nucleus accumbens response to incentive stimuli anticipation in children of alcoholics: relationships with precursive behavioral risk and lifetime alcohol use. *The Journal of Neuroscience: The Official Journal of the Society for Neuroscience*, *32*(7), 2544–2551. http://dx.doi.org/10.1523/JNEUROSCI.1390-11.2012.

Yin, H. H., Ostlund, S. B., Knowlton, B. J., & Balleine, B. W. (2005). The role of the dorsomedial striatum in instrumental conditioning. *The European Journal of Neuroscience*, *22*(2), 513–523. http://dx.doi.org/10.1111/j.1460-9568.2005.04218.x.

Chapter 5

Drugs of Abuse and the Internet

Marcelo Dutra Arbo[1], Rachel Picada Bulcão[3], Luciana Grazziotin Rossato[2], Elza Callegari[2], Lucas Henrique Cendron[2], Eduardo Rodrigues Cabrera[3], Marco Antonio de Freitas[3], Mirna Bainy Leal[4]

[1]*Laboratório de Toxicologia, Departamento de Ciências Biológicas, Faculdade de Farmácia, Universidade do Porto, Porto, Portugal;* [2]*Faculdade de Farmácia, Instituto de Ciências Biológicas, Universidade de Passo Fundo, Passo Fundo (RS), Brazil;* [3]*Instituto Médico Legal, Polícia Científica, Curitiba (PR), Brazil;* [4]*Laboratório de Farmacologia e Toxicologia de Produtos Naturais, Departamento de Farmacologia, Instituto de Ciências Básicas da Saúde, Universidade Federal do Rio Grande do Sul, Porto Alegre (RS), Brazil*

Abbreviations

25I-NBOMe 2-(4-Iodo-2,5-dimethoxyphenyl)-*N*-[(2-methoxyphenyl)methyl]ethanamine
5-HT Serotonin
AIDS Acquired immunodeficiency syndrome
BZP *N*-Benzylpiperazine
DA Dopamine
MBDP 1-(3,4-Methylenedioxybenzyl)piperazine
***m*CPP** 1-(3-Chlorophenyl)piperazine
MeOPP 1-(4-Methoxyphenyl)piperazine
MDMA 3,4-Methylenedioxymethamphetamine
MDPV 3,4-Methylenedioxypyrovalerone
NE Norepinephrine
NMDA *N*-Methyl-D-aspartate
***p*FPP** 1-(4-Fluorophenyl)piperazine
TFMPP 1-(3-Trifluoromethylphenyl)piperazine
THC Δ^9-Tetrahydrocannabinol

INTRODUCTION

The easy access to new drugs of abuse through the Internet sharpens the curiosity of youngsters, leading them to believe they are enjoying a low-risk entertainment. The fight against drugs of abuse has in the Internet a great enemy, because it is simple to find, order, and buy these substances. It is a big challenge to all countries, and it results in uncountable losses for both public health and safety, from the economic point of view related to expenses for prevention and treatment to the increase in morbidity and mortality (INCB, 2011a). In addition, addiction reaches all social classes, independent of age, and it strengthens crime and corruption (INCB, 2011b).

Schifano et al. (2006) point out that among 1633 Web sites on the Internet, 10% of them contained psychoactive drugs for sale and 9% contained detailed information on the techniques of synthesis and/or extraction procedures for a series of recreational psychoactive compounds.

In this context, the "legal highs" appear as natural compounds or synthetic drugs presenting the same effects of illicit drugs; however, these compounds are not listed as being controlled by legislation. Users can buy them easily on the Internet without restrictions.

The United Nations Office on Drugs and Crime detected an increase in new psychoactive substances in 2012. At the end of 2009, there were 166, but in 2012 they reached the number of 251 new drugs of abuse. In 2011, it was estimated that between 167 and 315 million youngsters and adults, between 15 and 64 years of age, used illicit substances, corresponding to 3–7% of the population.

Psychoactive substances affect the central nervous system, as well as other systems, and could lead to addiction. Some of them are licit, like nicotine, alcohol, and caffeine, and they are regulated by specialized institutions, but new drugs of abuse are being synthesized and commercialized so fast that it is difficult for countries to control them. Moreover, the adverse effects of the new psychoactive substances are poorly understood and represent a risk to global public health (Rang, Dale, Ritter, Flower, & Henderson, 2011; UNODC, 2013).

DRUGS SOLD ON THE INTERNET

Users can easily buy drugs on the Internet without prescriptions or legal restrictions. Gaps in supervision and control of drugs in this virtual area provide a new market focused on Web sites that sell psychoactive substances on the Internet (Jones, 2010).

This facility represents a major public health challenge to the extent that for some of those substances there are few pharmacological and toxicological studies demonstrating their safety in humans, and sometimes usage information, risks, and effects are obtained from the Internet at informal sites or from the seller of the substance (Coppola & Mondola, 2012a).

In Table 1 are listed some Web sites that commercialize drugs of abuse. The Web sites were accessed between March and July of 2014. The main drugs found, mechanism of action, and effects are summarized in Table 2.

Cocaine

Cocaine is an alkaloid extracted from the leaves of the coca bush (*Erythroxylum coca* Lam. Erythroxylaceae), which is native to the

TABLE 1 Web Sites from Which It Is Possible to Find and Buy Several Types of Drugs of Abuse

Drug	Website
Cannabis sativa	http://portugues.shayanashop.com/Cannabis_Legal-45-hc.aspx http://azarius.pt/lifestyle/hemp_products/hemp_food/cannabis-flavoured-lollipops/ http://www.muito-doido.com/MaconhaLegal.aspx http://www.muito-doido.com/HaxixeLegal.aspx http://www.seeds4free.com/pt/Seeds.aspx http://www.sementemaconha.com/dutch_orange_mix_em_pacote-p790/ http://www.buydutchseeds.com/medical-marijuana-seeds http://www.brasilplayforever.com/t32487-vendo-crack-cocaina-e-maconhaprecos
Methylone	http://buy-methylone-buymethyloneonline.blogspot.com.br/ http://htguimaraes.blogspot.com.br/ http://www.plantfeedshop.com/index.php?option=com_content&view=article&id=49&Itemid=56
Cocaine (crack)	http://www.brasilplayforever.com/t32487-vendo-crack-cocaina-e-maconhaprecos https://twitter.com/juancho_lin/status/297396911688785920 https://twitter.com/juancho_lin/status/297396911688785920
Mushrooms	http://azarius.pt/smartshop/magic-mushrooms/ http://www.naturezadivina.com.br/loja/index.php?cPath=48_31&osCsid=tsuemrpvgnc5bnfn387tjganm7 http://portugues.shayanashop.com/Cogumelos_M%C3%A1gicos-14-hc.aspx
Mephedrone "meow, m-smack, Mcat, drone, charge"	http://www.plantfeedshop.com/index.php?option=com_content&view=article&id=49&Itemid=56
MDPV	http://www.plantfeedshop.com/index.php?option=com_content&view=article&id=49&Itemid=56
Synthetic cannabinoids "spice"	http://portuguese.alibaba.com/promotion/promotion_smoke-spice-promotion-list.html http://www.cannabiscafe.net/foros/search.php?s=43767e9103eaf5158c24a511e1f39928&do=getdaily&contenttype=vBForum_Post http://azarius.pt/smartshop/herb_extracts/spice_diamond/ http://spice-gold-direct.com/ http://spiceworld420.com/
Piperazines "*m*CPP, A2, legal X, legal E, pepe"	http://buybzponline.blogspot.com.br/2013/05/comprar-bzp-bencilpiperazina-online.html
Salvia divinorum	http://azarius.pt/smartshop/salvia_divinorum/salvia_divinorum/ http://portugues.shayanashop.com/Salvia_Divinorum-38-hc.aspx http://salviadf.mx/productos/extracto-de-salvia/ http://www.sementemaconha.com/salvia-l38/ http://www.arenaethnobotanicals.com/organic-salvia-divinorum-c-1.html?ref=17 http://conscioswholesale.com/smartshop/salvia_divinorum/
Ecstasy (MDMA)	http://www.muito-doido.com/Energ%C3%A9ticosLegais.aspx http://www.sementemaconha.com/extasy_legal-l2009/ http://ecstasy2013.blogspot.com.br/2013/01/vendo-lanca-perfumemdlsdecstasymicropon.html http://balasedoces2013.blogspot.com.br/ http://r0xstore.forumeiros.com/t4-comprar-lsd-e-ecstasy-drogas-sinteticas-vendo-bala-vendo-doce-ecstasy http://universoparalello2.forumeiros.com/t1-comprar-lsd-25-cartelas-avatar-bike-100-years-shivas-chapeleiro-avatar-ursinhos-ecstasy-acima-de-10-unidades-orbital-roxa-love-rosa-alien-verde
Heroin "smack, brown sugar"	https://twitter.com/juancho_lin/status/297396911688785920
LSD "candy"	http://lsd25online.blogspot.com.br/ http://vendolsdbalas.blogspot.com.br/ http://ecstasy2013.blogspot.com.br/2013/01/vendo-lanca-perfumemdlsdecstasymicropon.html http://balasedoces2013.blogspot.com.br/ http://blog.clickgratis.com.br/vendosinteticos2014/ http://htguimaraes.blogspot.com.br/ http://r0xstore.forumeiros.com/t4-comprar-lsd-e-ecstasy-drogas-sinteticas-vendo-bala-vendo-doce-ecstasy
25I-NBOMe	http://htguimaraes.blogspot.com.br/ http://universoparalello2.forumeiros.com/t1-comprar-lsd-25-cartelas-avatar-bike-100-years-shivas-chapeleiro-avatar-ursinhos-ecstasy-acima-de-10-unidades-orbital-roxa-love-rosa-alien-verde http://www.muito-doido.com/Psicod%C3%A9licosLegais/LutetiumDoceLegal.aspx
Ketamine	http://ketamine.rdccmaud.org/

TABLE 2 Summary of the Most Common Drugs Sold on the Internet and Their Effects

Chemical Class	Main Drugs	Toxicodynamics	Effects
Piperazines	BZP, TFMPP, mCPP, pFPP, MeOPP	Direct and indirect sympathomimetic activity, release of DA and 5-HT at synaptic cleft and reuptake blockade	Anxiety, agitation, palpitations, vomiting, confusion, insomnia, headache, metabolic acidosis, seizures, and serotonin syndrome
Cathinones	Cathinone (*Catha edulis*), mephedrone, methcathinone, methedrone, methylone, MDPV	Release of NE, DA, and 5-HT at synaptic cleft and reuptake blockade	Sympathomimetic effects, sweating, headache, palpitations, nausea, agitation, tachycardia, confusion, convulsions, psychotic episodes, myocardial infarction, death
Synthetic cannabinoids	JWH-018, JWH-073, JWH-200, JWH-250, JWH-398, CP47-497, HU210, and many others	CB1 and CB2 cannabinoid receptor agonists	Conjunctiva edema, tachycardia, dry mouth, changes in mood and perception, psychotic episodes
Phenylethylamines	MDMA, 2Cs, TMA, NBOMe, and others	Release of NE, DA, and 5-HT at synaptic cleft and reuptake blockade, agonist activity at $5-HT_{2A}$ receptors	Nausea, vomiting, diarrhea, symptoms of panic and confusion
Plants	*Salvia divinorum* (salvinorin A)	κ opioid receptor-selective agonist	Confusion, disorientation, hallucinations, dizziness, tachycardia, palpitations, and sweating

Andes (Brunton, Chabner, & Knollmann, 2012). The base paste is an intermediate in the process of extraction and purification of the drug. From it, it is possible to refine powder products (cocaine hydrochloride), solid products (crack), and paste products (merle).

Cocaine in powder is inhaled or diluted to intravenous injection, while crack and merle are smoked. Crack is a crystalline inodorous white powder with bitter flavor and insoluble in water. Popular in South America, the substance receives various names such as blizzard, Big C, coca, bones, boost, and base. It is normally consumed by inhalation under the form of salt, which can also be dissolved in water and used intravenously (Brunton et al., 2012; Oga, Camargo, & Batistuzzo, 2014; Rang et al., 2011).

Absorbed by oral, nasal, gastrointestinal, rectal, and vaginal routes, as well as by the lung alveoli, after inhalation and by intravenous injection, cocaine leads to plasma concentration peak in 30–60 min. If it is smoked, the effect begins in around 10 s, reaches its maximum in around 5 min, and lasts half an hour. The metabolism is fast and occurs in the liver, leading to the main metabolite, ecgonine methyl ester, which originates from hydrolysis (Oga et al., 2014).

Cocaine is a powerful stimulant of the central nervous system, causing euphoria and lack of appetite. Users report an increase in pleasure, self-confidence, aggressiveness, and sexual desire, but it may also cause paranoia. Blood pressure and body temperature rise, heart beat and breath rhythm accelerate. When the effects cease, the user feels apathetic, depressed, and tired. Regarding crack and merle, all the effects are more intense (Rang et al., 2011).

The toxic effects of cocaine come from the decrease in monoamines reuptake, especially dopamine (DA), in the peripheral and central nervous systems, which explains its euphoric properties. Nonetheless, cocaine blocks the reuptake of norepinephrine (NE) and serotonin (5-HT) and its chronic use produces disturbances in these neurotransmitter systems (Rang et al., 2011). This potentiates the peripheral effects of sympathetic nervous activity and produces a sharp psychomotor effect (Brunton et al., 2012).

Cocaine causes strong dependence, which is characterized by a state of adaptation due to the need for a readjustment of physiological mechanisms to maintain normal function despite the chronic use of the drug (Rang et al., 2011). In this way, the abrupt interruption of the ingestion of cocaine in dependent individuals leads to the abstinence syndrome, whose classical symptoms tend to be opposite those that occur during exposure to the drug, characterized by fatigue, sedation, and depression. The acute withdrawal from cocaine, though not life-threatening, causes an initial "crash" phase lasting several hours that consists of anxiety, depression, drug craving, exhaustion, and hypersomnolence (Oga et al., 2014).

Problems during pregnancy and fetal development can occur in female users of cocaine, as the drug acts in the noradrenergic system and leads to vasoconstriction, thus reducing the arrival of oxygen and nutrients to the placenta. As a consequence, it can cause a sudden delivery or the displacement or rupture of the placenta (Brunton et al., 2012; Rang et al., 2011).

Mushrooms

There are hundreds of hallucinogenic mushrooms, and *Psilocybe cubensis* (Earle) Strophariaceae is the most well known and used by humans since ancient times, owing to the fact that it grows easily. The psychotropic effects are given by the most important active principles, psilocybin (*O*-phosphoryl-4-hydroxy-*N*,*N*-dimethyltryptamine) and psilocin (4-hydroxy-*N*,*N*-dimethyltryptamine). They are indol alkaloids derived from tryptamine with potent agonist activity at 5-HT receptors and were identified in 1958 by the chemist Albert Hofmann (Kirsten & Bernardi, 2010).

They are both classified as prohibited drugs by the 1971 United Nations convention on psychotropic drugs. The main difference between them is that psilocybin, which is the psilocin precursor, is stable in environmental conditions and more stable. The ratio of psilocybin to psilocin varies from species to species, although only psilocin has a psychoactive effect (Oga et al., 2014).

Regarding the administration, oral intake is the chosen method, through mushroom infusion. The effects are characterized as those of lysergic acid diethylamine (LSD); nevertheless, psilocybin is 200 times less potent and has a shorter duration time, which varies from 2 to 6 h (Schwartz & Smith, 1988).

Especially for users with mental disorders, the ingestion of mushrooms can result in "bad trips," which are characterized by self-destructive and suicidal thoughts (Müller, Püschel, & Iwersen-Bergmann, 2013). A bad trip is usually followed by weakness, sadness, depression, and paranoid interpretations, which can last for days, weeks, or months. In these cases, the intoxicated individuals present themselves as agitated, confused, extremely anxious, and disoriented without the capability of concentration or judgment. Moreover, acute psychotic episodes can occur in severe cases, with strong paranoia and total loss of a sense of reality, which can lead to accidents, self-mutilation, or suicide attempts. The combination of mushrooms with other psychoactive drugs, including alcohol, increases the risk of a bad trip (Satora, Goszcz, & Ciszowski, 2005).

Common physiological reactions include muscle relaxation and pupil dilation. In greater doses, mushrooms can cause visual and mental hallucinations, space distortion, increased creative visualization, time-distortion feelings, and so on. These effects occur more often in users with previous experience with mushrooms (Cuomo, Dyment, & Gammino, 1994).

Fatal intoxications due to exposure to mushrooms are rare and, many times, are a result of the combination with other drugs of abuse, mainly alcohol.

Ecstasy

MDMA, chemically 3,4-methylenedioxymethamphetamine or popularly nicknamed as "E" or "ecstasy," was synthesized in Germany in 1912. At the time, Merck wanted to find and patent pathways leading to hemostatic substances, not appetite suppressors, as sometimes reported (Freudenmann, Öxler, & Bernschneider-Reif, 2006). After 1983, MDMA became a popular recreational drug of abuse, especially among students (Seibel & Toscano, 2001).

Ecstasy is a phenylethylamine and presents stimulating and psychedelic properties (Brunton et al., 2012). It is commercialized in tablets or capsules, with various sizes, colors, and aspects. The most common administration route is oral, although the drug can be macerated and inhaled or even used via the anal route (Capela et al., 2009).

Moderate doses of ecstasy taken orally correspond to one or two tablets (75–100 mg), with psychoactive effects observed in about 20–60 min after ingestion and lasting from 2 to 8 h. However, the dose and the potency can vary according to the quality and purity of the product (Brunton et al., 2012).

After use, altered time perception and pleasant sensory experiences, euphoria with an increase in physical and emotional energy, are reported. These experiences render people more affective and emotionally sensitive. Despite that, people report common effects such as rising blood pressure, heart beat, and body temperature; dry mouth; and difficulty sleeping. High-dose effects include visual hallucinations, agitation, hyperthermia, panic, and anxiety crises (Brunton et al., 2012).

The hallucinogenic effects are due to the inhibition of monoamine transporters and chiefly the great affinity to the 5-HT transporters, which leads to an increase in free 5-HT in the brain. Hence, the effects involving 5-HT determine the psychomimetic effects. In their turn, DA and NE are related to the initial euphoria and the later rebound dysphoria (Capela et al., 2009).

Even in small doses, MDMA may cause diseases and sudden death. These effects can be related to the excessive consumption or retention of water by the organism, as MDMA causes inappropriate antidiuretic hormone secretion, most times associated with acute hyperthermic reaction. Heart failure can occur as well in users with undiagnosed cardiac conditions (Rang et al., 2011).

Heroin

Heroin, also called "brown," "H," and "smack," was introduced into the market by Bayer in 1898. Heroin was synthesized by the acetylation of morphine, a natural alkaloid from opium, which is a resin derived from the flowers of the poppy (*Papaver somniferum* L. Papaveraceae) (Oga et al., 2014).

During a long period of time, heroin was administered intravenously, although there has been a change in the preference of its users to intranasal and inhaled administration, owing to the rise of the acquired immunodeficiency syndrome associated with a high incidence in heroin users. Moreover, smoking it avoids frequent superdoses, injuries caused by the use of needles, and viral infections. When injected or smoked, it causes an initial state of euphoria that many people compare to an orgasm (Oga et al., 2014). Under the form of powder, heroin can be adulterated with sugar, baking soda, borax, quinine, lactose, mannitol, talc, and other depressants of the central nervous system such as barbiturates and sedatives (Seibel & Toscano, 2001).

Owing to its high liposolubility, heroin is well absorbed by nasal, lung, and rectal routes and by the mucosas. It leaves the bloodstream rapidly and reaches the central nervous system. The effects can start in a few seconds (inhalation or smoking) or minutes (sniffing or ingestion), usually lasting from 2 to 6 h (Oga et al., 2014; Seibel & Toscano, 2001).

Heroin binds chiefly to μ receptors and, to a lesser extent, δ receptors. Its pure agonist action in the μ receptors explains most of the clinical effects: central analgesia, respiratory depression, difficulty in urinating, euphoria, myosis, abdominal cramps, cough reflex suppression, and, sometimes, low heart beat and hypertension. The person can also turn cold and pale. It shall be noted that the drug confers on its users a state of relaxation and delightful sleep, in addition to reducing pain. These people enter a state of torpor and forget worries and problems (Oga et al., 2014).

Its biological half-life is between 3 and 10 min, after that it is metabolized into 6-monoacetylmorphine and then into morphine in the liver and, like every basic amine, abandons the bloodstream and concentrates in the parenchymatic tissues, like spleen, liver, lung, and kidney. Skeletal muscle tissues contain smaller quantities of the drug, but owing to their high presence in the body, they concentrate most of it. The half-life of morphine lasts from 1.5 to 4 h (Brunton et al., 2012; Oga et al., 2014).

Small amounts of free morphine are found in urine and higher quantities of the conjugated form, which are the two elimination forms of heroin. About 90% of the total administered drug is excreted in the first 24 h by glomerular filtration.

Another elimination form is the stools, and 7–10% of the drug is eliminated in the bile (Silva, 2006).

Chronic use of heroin produces tolerance rapidly. Tolerance reduces first the resistance to pain and the initial euphoria and then the effect in the respiratory system (Silva, 2006). The abstinence syndrome is intense and painful. After some time without the drug, which can be just a few hours, the individual starts to sweat, to tear, and to have a runny nose, as if having a cold. Afterward, the user becomes angry and agitated. Finally, the person has fever, nausea, vomiting, diarrhea, muscle pains, strong abdominal cramps, and even hallucinations. During this period and after that, the user has an uncontrollable desire to use the drug. These symptoms last from 1 to 2 weeks. Craving and the discomfort of abstinence render the dependence on opioids very hard to treat (Oga et al., 2014).

Ketamine

Ketamine (DL-2-(O-chlorophenyl)-2-(methylamine)cyclohexane hydrochloride) is an arylcycloalkylamine that is similar in structure to phencyclidine (PCP), but about four times less potent. Ketamine is used in human and veterinary medicine for hospital purposes, but it is used more in veterinary medicine, because of its psychological effects. Nevertheless, it is also known as a drug of abuse, when used for recreation purposes at parties. Popularly it is known as "K," "Keller," "keta," "new ecstasy," "angel powder," "psychedelic heroin," "Special K," and "Super K" (Silva, 2006).

It can be found in powder, liquid, or tablet forms and can be administered by oral, nasal, or intravenous routes. The effects of the drug vary according to the dose, as low doses produce an experience known as "K Land"—described as "a colorful and wonderful world," whereas high doses are referred as "K hole," when users experience out-of-body experiences or a sensation of fusion with other people or objects (Oga et al., 2014).

Ketamine is very liposoluble and acts quickly in the central nervous system, where it is a noncompetitive antagonist of the N-methyl-D-aspartate (NMDA) receptor, interfering with the action of excitatory amino acids including glutamate and aspartate (Silva, 2006). Nonetheless, it also alters the function of dopaminergic, serotonergic, cholinergic, and opioid receptors as well as that of sodium channels (Kamiyama et al., 2011). The half-life of the drug is about 2 h. Its metabolism occurs in the liver, more precisely by the cytochrome P450 3A4 isoenzyme, which demethylates ketamine and yields norketamine, which is, in turn, hydroxylated (hydroxyketamine) and conjugated with glucuronic acid, which forms hydrosoluble compounds that are excreted in urine (90%) and in the bile with the feces (Oga et al., 2014).

Some of the adverse reactions are visual hallucinations—or flashbacks, which can last for days or weeks after exposure, delirium, lively dreams, fast heart beat, hypertension, psychosis, dizziness, insomnia, nausea, vomiting, bronchial dilatation, and aggressive reactions (Oga et al., 2014).

Lysergic Acid Diethylamine

Lysergic Acid Diethylamine (LSD) is derived from "ergot" alkaloids, which naturally occur in the fungus *Claviceps purpurea*. It is a powerful hallucinogenic and was synthesized in 1938 by the German chemist Albert Hofmann, who discovered its effects by accidently ingesting this substance (Oga et al., 2014; Rang et al., 2011).

It is also called "acid," "L," "tripper," and "sweet" and is presented under various forms: bars, capsules, gelatin strips, microdots, or dried paper leafs. The average dose varies from 50 to 75 μg. It can be absorbed by sublingual and oral routes as well as by inhalation or intravenously (Oga et al., 2014). The distribution of LSD is verified in many organs, mainly kidneys, liver, and lungs, while in blood, adipose tissue, and brain the concentration is comparatively smaller (Oga et al., 2014; Seibel & Toscano, 2001). LSD metabolism occurs through hydroxylation followed by conjugation in liver. Conjugates are excreted in the feces. Regarding tolerance and dependence, studies seem to diverge, but tolerance seems to disappear rapidly after some days of abstinence (Oga et al., 2014).

LSD is a potent hallucinogen with mild sympathomimetic effects and it can cause a total change in the perception of reality in its users, as it reinforces the moment emotion or the invoked ones, as well as causing illusions, visual and aural hallucinations, and great sensorial sensitiveness. The effects last from 8 to 12 h and appear about 30–40 min after consumption. Colors become brighter, and synesthesia, mystical experiences, and flashbacks also occur—the last are the main danger, as the user relives the experience he or she had with the drug without consuming it again. This can happen weeks after the ingestion of the drug and the symptoms are paranoia, changes in perception of time and space, confusion, disorganized thoughts, panic, anxiety, depression, and difficulty in concentration. Tachycardia, hypertension, and a mild muscle relaxation normally occur. Other effects are pupil dilatation, body hair erection, nausea, trembling, and a discrete increase in body temperature (Oga et al., 2014).

LSD interacts with 5-HT receptors in the brain, by acting as an agonist/partial agonist (Rang et al., 2011). Besides, LSD imitates 5-HT in 5-HT$_{1A}$ autoreceptors in the raphe cellular bodies and leads to an acute decrease in the discharge rate of serotonergic neurons (Brunton et al., 2012; Oga et al., 2014; Seibel & Toscano, 2001).

Designer Drugs

These drugs have been synthesized in an attempt to escape the laws that prohibit them from being used legally, and as well they are deceivingly packaged to give the impression that they are "research chemicals," "frankincense," "bath salts," or "plant food." It is worth nothing that on the labels of these products there are warnings like "not for human consumption" or "prohibited for sale to minors" (Musselman & Hampton, 2014).

Designer drugs are always synthetic and their classification is based on the parent compound. The ability of clandestine producers to modify the structures without loss of psychotropic effects is what keeps them in the market of designer drugs. They are synthesized by the manipulation of the chemical structure of another already existing psychoactive drug, usually illicit, originating a new structurally similar product. Moreover, the users can buy them without legal restrictions (Alves, Spaniol, & Linden, 2012; Musselman & Hampton, 2014).

The most popular synthetic drugs nowadays can be divided into synthetic cannabinoids (known as "spice" or "incense"), synthetic cathinones (commercially known as "bath salts" or "plant feeders"), piperazine, tryptamine, phenylethylamine, pipradrol, and fentanyl derivatives (Bulcão et al., 2012).

The clinical effects of most of the new synthetics are described as hallucinogenic, stimulant, or depressant. The easy access associated with the accelerated synthesis of new drugs challenges health care providers in the treatment of users with acute toxicity, owing to the difficulty in identifying the chemical composition and the potency, as those are variable among the packages and drug distributors (Musselman & Hampton, 2014).

Synthetic Cannabinoids

Synthetic cannabinoids represent the most recent advance of the designer drug market. The development of this kind of drug comes from alternative opium esters and synthetic hallucinogens produced from modifications of LSD, PCP, MDMA, methcathinone, and anabolic steroid derivatives, becoming more common since 2005 (Alves et al., 2012).

They appeared for the first time in Europe in 2004, as "spice," and in the United States as "K2" in 2008. Nowadays, the products are commercialized with many different names, in mixtures with herbs, in products globally known as "Aztec gold," "Black Mamba," "Cloud 9," "Mad Hatter," "synthetic marijuana," or "false marijuana," and they are most times packaged in colorful containers aiming to attract teenagers and young adults (Alves et al., 2012; Pourmand, Armstrong, Mazer-Amirshahi, & Shokoohi, 2014). It is a mixture constituted by exotic herbs and aromatic plant extracts commercialized as similar to frankincense for use in aromatherapy, yoga, or meditation. It is possible to identify in the label that the packages contain between 0.4 and 3.0 g of various vegetable species; among them we can emphasize *Pedicularis densiflora* Benth. ex Hook. Orobanchaceae and *Leonotis leonurus* (L.) R. Br. Lamiaceae, because of their psychoactive properties traditionally recognized as substitutes for marijuana (Alves et al., 2012).

The synthetic cannabinoids seem to be three to five times more potent than traditional marijuana and, hence, are more inclined to be associated with sympathomimetic and hallucinogenic effects (Forrester, Kleinschmidt, Schwarz, & Young, 2012). They mimic the effect of Δ^9-tetrahydrocannabinol (THC), the active substance of cannabis, and also have weak inhibition of monoamine oxidase, which means more risk for the serotonergic syndrome (Pourmand et al., 2014; Wells & Ott, 2011).

Synthetic cannabinoids have been associated with acute renal lesions in previously healthy young patients (Kazory & Aiyer, 2013). Other serious complications associated with the use of these drugs are heart arrhythmia and acute myocardial injury (Mir, Obafemi, Young, & Kane, 2011). The analysis of various products called "K2" suggested the presence of a high concentration of chemicals, including vitamin E and clenbuterol. In these mixtures, the heart arrhythmia can be associated with the β-agonist activity of clenbuterol (Wells & Ott, 2011). There is concern regarding the chronic use of synthetic cannabinoids, associated with the chronic use of marijuana, because this association increased the rate of schizophrenia and memory loss (D'Souza, Sewell, & Ranganathan, 2009).

There are many formulations circulating throughout the world at present, but the synthetic cannabinoids contained in each mixture change rapidly with time. From 468 samples of "spice" analyzed by the state of Virginia (USA), in July and August 2011, only 101 contained prohibited compounds, what indicates that "spice" maintains its "legal" drug of abuse status (Johnson, Johnson, & Portier, 2013). Most of the common compounds found in synthetic cannabinoids are JWH-018, JWH-073, JWH-200, and CP47,497 (Figure 1). Although few of them resemble the structure of THC, the original compound, all of them are potent cannabinoid receptor agonists (Pourmand et al., 2014).

The clinical symptoms of exposure to synthetic cannabinoids can be variable, depending on the dose, exposure route, and agent involved. Nevertheless, patients usually have more profound excitatory effects compared to traditional marijuana. Synthetic cannabinoid users report agitation, anxiety, nausea, vomiting, tachycardia, increase in blood pressure, trembling, convulsions, hallucinations, paranoid behavior, and decreased responsiveness (Schneir, Cullen, & Ly, 2011).

There is no specific antidote to synthetic cannabinoid overdose but all support measures are used in patients. The synthetic cannabinoids are currently not identified by routine screening tests, and the development of new products renders the situation even more difficult (Pourmand et al., 2014).

Synthetic Cathinones

Synthetic cathinones are popularly known in the media as "bath salts" (USA) or "plant feeders" (Europe) and are becoming more visible on streets and retail environments, with an increasing presence in emergency rooms and psychiatric units around the world (AAPCC, 2011). They are synthetic derivatives of cathinone, the psychoactive substance found in khat (*Catha edulis* (Vahl) Forssk. ex Endl. Celastraceae). Consumers can buy synthetic cathinones in online stores or convenience outlets. They are sold in packs that contain a white or light brown powder that can be ingested, inhaled, or smoked. Common names include "ivory wave," "vanilla sky," and "white ice" (Valente, Pinho, Bastos, Carvalho, & Carvalho, 2014).

Synthetic cathinones became popular in Europe at the beginning of 2000, and users all over the world started to appreciate this legal alternative to ecstasy and methamphetamine. In the same way as the synthetic cannabinoids, sales on the Internet or at convenience stores, with warnings saying "vegetable food" or "not for human consumption," allowed these products to escape regulation (Pourmand et al., 2014).

Their effects resemble those of ecstasy or cocaine. Some of the desirable effects include euphoria, excitement, and increase in alertness and libido. The important adverse effects include chest pain, sweating, headache, palpitation, hallucinations, secondary trauma, mydriasis, paranoia, confusion, and convulsions (Prosser & Nelson, 2012).

The most common synthetic cathinones are mephedrone, methylone, and 3,4-methylenedioxypyrovalerone (MDPV; Figure 2) (Brunton et al., 2011; López-Arnau et al., 2014). Nevertheless, the increase in abuse was repressed by legislation, as MDPV, mephedrone, and their analogues have been prohibited permanently in many countries (Valente et al., 2014).

Methylone (3,4-methylenedioxi-*N*-methylcathinone), also called "bk-MDMA" or "M1," was synthesized for the first time as an antidepressant, but around 2004, it arose as a recreational drug commercialized online (Bossong, Van Dijk, & Niesink, 2005). It is used via the oral or nasal route (López-Arnau et al., 2014). Human data on methylone abuse obtained from consumer reports indicate that 100–250 mg in each intake is a common oral dose, and users evidence a desire to redose, leading them to ingest large quantities of the drug (López-Arnau et al., 2013). Peripheral blood

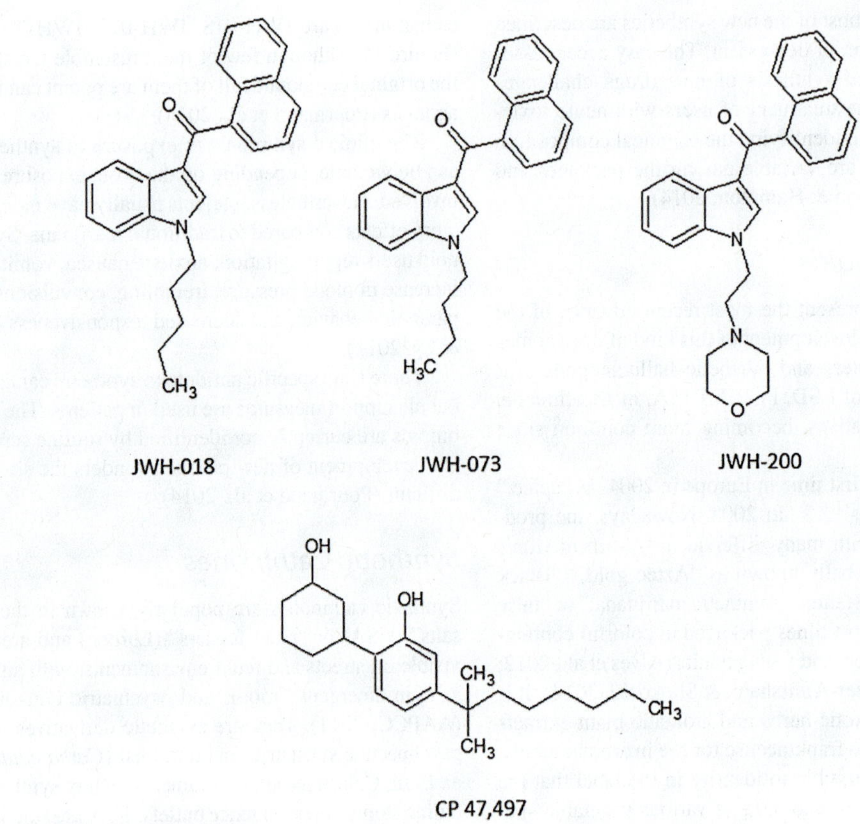

JWH-018 JWH-073 JWH-200

CP 47,497

FIGURE 1 Chemical structures of the most common cannabinoid derivatives.

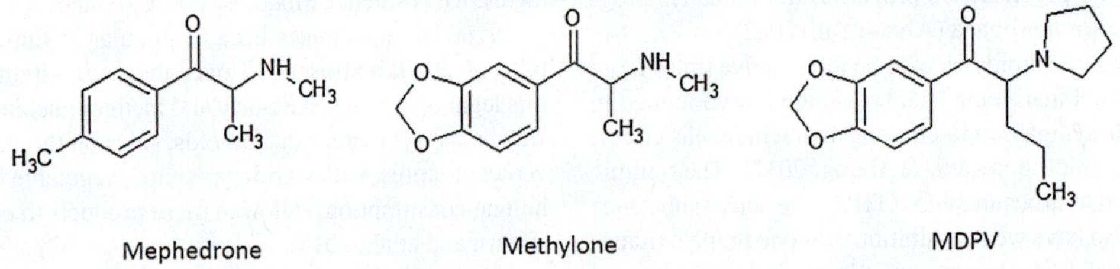

Mephedrone Methylone MDPV

FIGURE 2 Chemical structures of the most common synthetic cathinones.

methylone concentrations greater than 0.5 mg/l may be lethal, and the symptoms of the intoxication are similar to those presented by MDMA (Pearson et al., 2012).

Mephedrone (4-methylmethcathinone), known on streets as "meow–meow," "Mcat," and "4MMC" (Johnson et al., 2013), is more commonly used by the intranasal or oral route, and doses vary from 25 to 75 mg and from 150 to 250 mg, respectively. Regarding the effects, via the intranasal route, they occur in a few minutes, with a peak of efficacy in less than half an hour and a rapid reduction. On the other hand, if the oral route is used, effects begin in 45 min to 2 h, although it provides a 2- to 4-h experience (Sumnall & Wooding, 2010).

MDPV is a pyrrolidine derivative of the synthetic cathinone pyrovalerone (Coppola & Mondola, 2012b). Commercialized as "ocean burst" or "sextacy," it can be administered by oral, sublingual, intramuscular, and intranasal routes. Hence, it is an inhibitor of monoamine absorption that is more lipophilic and more potent that other cathinone derivatives (Coppola & Mondola, 2012b).

Laboratory assays have been developed for cathinones, but they are not readily available in clinical practice. Diagnostics can be based on history and the physical and clinical exams of the patient (Gerona & Wu, 2012). If necessary, the standard support treatment is carried out for synthetic cathinones (Pourmand et al., 2014).

Piperazine Derivatives

Piperazine derivatives are commercialized under the names of "rapture," "charge," "herbal ecstasy," "frenzy," "bliss," "A2," "legal X," and "legal E." They can be divided into two classes, the benzylpiperazines such as *N*-benzylpiperazine (BZP) and its methylenedioxy analogue 1-(3,4-methylenedioxybenzyl)piperazine (MDBP), and the phenylpiperazines such as 1-(3-chlorophenyl)piperazine (*m*CPP), 1-(4-fluorophenyl)piperazine (*p*FPP), 1-(3-trifluoromethylphenyl) piperazine (TFMPP), and 1-(4-methoxyphenyl)piperazine (MeOPP; Figure 3). Most times, they are in the form of capsules,

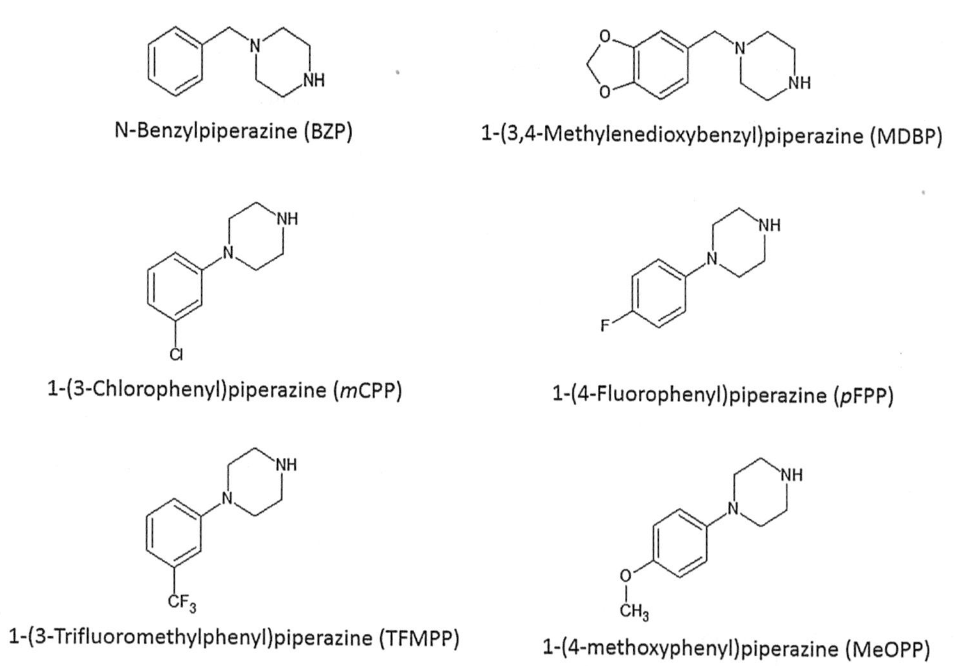

FIGURE 3 Chemical structures of the most common piperazine designer drugs.

tablets, or pills, but also may come as powders or liquids (Arbo, Bastos, & Carmo, 2012).

The most used piperazines are mCPP, BZP, MeOPP, and TFMPP. Also, mixtures of two or more derivatives in the same tablet are common; the best-known combination is BZP:TFMPP (2:1) (Bulcão et al., 2012).

Piperazines are readily absorbed from the gastrointestinal tract. A portion of the absorbed drug is metabolized and excreted in the urine. There is a wide variation in the rates at which piperazines are excreted by different individuals, which adds to the variability of their toxicity (Austin & Monasterio, 2004). The piperazine designer drugs are mainly metabolized in the liver, the phenylpiperazines being more extensively metabolized than the benzylpiperazines, and excreted almost exclusively as metabolites (Maurer, Kraemer, Springer, & Staack, 2004).

Pharmacodynamic effects of piperazine derivatives are related to their action on the neurotransmitters 5-HT, DA, and NE (Nikolova & Danchev, 2008). They raise the extracellular concentrations of neurotransmitters acting both by inhibition of reuptake of the neurotransmitters and by stimulation of the presynaptic receptors (Bankson & Cunningham, 2001). Studies demonstrate that piperazine derivatives and some of their metabolites interact with 5-HT receptors, producing effects similar to those caused by other illicit compounds such as amphetamines and ecstasy (Nikolova & Danchev, 2008). Negative effects of these drugs can be anxiety, confusion, trembling, sensitivity to light and noise, fear of losing control, and panic attacks (Van Veen, Van der Wee, Fiselier, Van Vliet, & Westenberg, 2007).

New Phenylethylamine Derivatives

New drugs keep being developed from substitutions in the base structure of phenylethylamine (phenylethylamine-N-benzyl derivatives). In 2000, the NBOMe substances were synthesized as potent hallucinogens. Among them is 2-(4-iodo-2,5-dimethoxyphenyl)-N-[(2-methoxyphenyl)methyl] ethanamine, best known as "25I-NBOMe," "25I," "N-bomb," or "Cimbi-5" (Stellpflug, Kealey, Hegarty, & Janis, 2014).

They are sold over the Internet as "bath salts," "plant food," "fertilizer," "water cleanser," "vacuum purifier," or "insect repellent" with the notice "not for human consumption" (Zuba, Sekuła, & Buczek, 2013). They are commercialized impregnated in blotting paper or in the form of powder, and low doses should be used (200–1000 µg), because there is a high risk of overdose once they seem to produce sympathomimetic effects more easily (Hill et al., 2013).

The compounds of the NBOMe class are potent agonists of 5-HT$_{2A}$ receptors (Lawn, Barratt, Williams, Horne, & Winstock, 2014). This receptor is associated with complex behaviors such as the working memory, cognitive processes, and affective disturbances such as schizophrenia (Rose, Poklis, & Poklis, 2013).

Salvia

Salvia divinorum Epling and Játiva Lamiaceae is also known as "Eclipse," "Mexican magic mint," "Saint Salvia," "The Shepherd," and "Mary Herb" (Vortherms & Roth, 2006) and is native from Mexico, particularly the mountains of Oaxaca. Its leaves and seeds have been consumed by the Mazatec Indians in rituals since ancient times (Wolowich, Perkins, & Cienki, 2006).

Its use became popular in the 1990s, when it started to be sold on the Internet, and the interest in the plant came to the attention of recreational users in Europe and North America (González, Riba, Bouso, Gómez-Jarabo, & Barbanoj, 2006). *Salvia* and its active compound, salvinorin A, have been used as a legal hallucinogen (Roth et al., 2002). However, the sales are now prohibited in most countries.

Salvia can be chewed using the dried leaves, smoked, or taken by the sublingual route, as well as in a water infusion containing the extract of macerated leaves (Casselman, Nock, Wohlmuth, Weatherby, & Heinrich, 2014). The start of action depends on the form of consumption, on the individual sensitivity of each user, and on the amount of substance used and varies from 5 to 10 min with oral absorption and 30–60 s in the smoked form. The effect has a short duration, between 15 and 30 min (Johnson, MacLean, Reissig, Prisinzano, & Griffths, 2011).

Salvinorin A is an unusual potent agonist of κ opioid receptors, highly selective and not nitrogenated (Grundmann, Phipps, Zadezensky, & Butterweck, 2007). It has no significant affinity for μ opioid receptors (Rothman et al., 2007) or for δ receptors (Ansonoff et al., 2006) and has no action on $5\text{-}HT_{2A}$ receptors (González et al., 2006).

Acute intoxication does not present specific signs. Nonetheless, in some cases, tachycardia and hypertension are reported. When associated with other hallucinogens or alcohol, the risk for cardiovascular toxicity or neuropsychiatric disturbances increases (Vohra, Seefeld, Cantrell, & Clark, 2011).

CONCLUSION

In recent years, the Internet has become a worldwide tool to obtain drugs of abuse. New drugs are arising every day, and the abuse is increasing remarkably worldwide. Emerging drugs of abuse usually are commercialized as being safe or under false advertising as "not for human consumption," making control even more difficult. Nevertheless, they present stimulating and hallucinogenic effects and cause unusual perceptions and experiences after ingestion. Toxicokinetic and toxicodynamic knowledge of the so-called "legal" highs is still scarce but much needed. In addition, users generally experience a combination of drugs of abuse, which also include alcohol and tobacco. Attention should be given to such combinations, since they may produce marked and unexpected toxicity.

DEFINITION OF TERMS

Tolerance when the user needs to increase the dose of the drug to experience the same effects
Dependence compulsive need to use drugs to function normally
Abstinence syndrome symptoms observed by addicts when the exposure to the drug is withdrawn

KEY FACTS ON THE INTERNET

- The Internet became popular worldwide in the household environment at the beginning of the twenty-first century; however, in the 1970s a packet switching network (ARPANET) was already being used to arrange a cannabis sale between students.
- Nowadays, there are legitimate online pharmacies, which sell drugs under prescription and have benefited patients through lower costs, and illegitimate sites that promote unfettered access to both ordinary prescription medications and controlled substances without any direct physician examination or approval.
- An increase has been observed in online sites selling any kind of drugs; this has also been accompanied by an increase in substance abuse; however, there is still limited evidence of the contribution of the Internet to drug misuse.
- The Internet has been responsible for the popularization of the emerging legal highs, which include nonprohibited amphetamine, piperazine, and cathinone derivatives and natural products derived from plants and mushrooms.
- Despite many countries having restricted laws regarding commercialization of the so-called legal highs and the smart shops, the difficulty in controlling the commerce through the Internet makes it a promising environment for drug misuse.

SUMMARY POINTS

- The abuse of drugs is worldwide and especially common among young people, who find in the Internet a simple way to find, order, and buy these substances.
- Over the past several years, legal highs, natural or synthetic drugs created or modified using the molecular structure of a known illegal substance without loss of psychotropic effects, have appeared on the market.
- This chapter summarizes this issue, as well as giving general information about the main drugs of abuse sold by Internet Web sites.
- On the Internet it is possible to find classical drugs of abuse, such as cocaine, heroin, LSD, MDMA, and the emergent legal highs.
- The legal highs rapidly spread throughout the world and include natural compounds, such as salvinorin A, from *S. divinorum*, and synthetic ones (e.g., synthetic cannabinoids and cathinone, piperazine, and phenylethylamine derivatives).

REFERENCES

AAPCC – American Association of Poison Control Centers. (2011). *Synthetic marijuana data: Updated February 8, 2012.* http://www.aapcc.org/alerts/synthetic-marijuana/.

Alves, A. O., Spaniol, B., & Linden, R. (2012). Canabinoides sintéticos: drogas de abuso emergentes. *Revista de Psiquiatria Clínica, 39*, 142–148.

Ansonoff, M. A., Zhang, J., Czyzyk, T., Rothman, R. B., Stewart, J., Xu, H., ... Pintar, J. E. (2006). Antinociceptive and hypothermic effects of Salvinorin A are abolished in a novel strain of k-opioid receptor-1 knockout mice. *Journal of Pharmacology and Experimental Therapeutics, 318*, 641–648.

Arbo, M. D., Bastos, M. L., & Carmo, H. F. (2012). Piperazine compounds as drugs of abuse. *Drug and Alcohol Dependence, 122*, 174–185.

Austin, H., & Monasterio, E. (2004). Acute psychosis following ingestion of 'Rapture'. *Australasian Psychiatry, 12*, 406–408.

Bankson, M., & Cunningham, K. (2001). 3,4-Methylenedioxymethamphetamine (MDMA) as a unique model of serotonin receptor function and serotonin-dopamine interactions. *Journal of Pharmacology and Experimental Therapeutics, 297*, 846–852.

Bossong, M. G., Van Dijk, J. P., & Niesink, R. J. (2005). Methylone and mCPP, two new drugs of abuse? *Addiction Biology, 10*, 321–323.

Brunton, L. L., Chabner, B., & Knollmann, B. C. (2012). *Goodman & Gilman as bases farmacológicas da terapêutica* (12th ed.). Porto Alegre: AMGH.

Bulcão, R., Garcia, S. C., Limberger, R. P., Baierle, M., Arbo, M. D., Chasin, A. A. M., ... Tavares, R. (2012). Designer drugs: aspectos analíticos e biológicos. *Química Nova, 35*, 149–158.

Capela, J. P., Carmo, H., Remião, F., Bastos, M. L., Meisel, A., & Carvalho, F. (2009). Molecular and cellular mechanisms of ecstasy-induced neurotoxicity: an overview. *Molecular Neurobiology, 39*, 210–271.

Casselman, I., Nock, C. J., Wohlmuth, H., Weatherby, R. P., & Heinrich, M. (2014). From local to global-fifty years of research on *Salvia divinorum*. *Journal of Ethnopharmacology, 151*, 768–783.

Coppola, M., & Mondola, R. (2012a). Research chemicals marketed as legal highs: the case of pipradol derivatives. *Toxicology Letters, 212*, 57–60.

Coppola, M., & Mondola, R. (2012b). 3,4-Methylenedioxypyrovalerone (MDPV): chemistry, pharmacology and toxicology of a new designer drug of abuse marketed online. *Toxicology Letters, 208*, 12–15.

Cuomo, M. J., Dyment, P. G., & Gammino, V. M. (1994). Increasing use of "ecstasy" (MDMA) and other hallucinogens on a college campus. *Journal of American College Health, 42*, 271–274.

D'Souza, D. C., Sewell, R. A., & Ranganathan, M. (2009). Cannabis and psychosis/schizophrenia: human studies. *European Archives of Psychiatry and Clinical Neuroscience, 259*, 413–431.

Forrester, M. B., Kleinschmidt, K., Schwarz, E., & Young, A. (2012). Synthetic cannabinoid and marijuana exposures reported to poison centers. *Human & Experimental Toxicology, 31*, 1006–1011.

Freudenmann, R. W., Öxler, F., & Bernschneider-Reif, S. (2006). The origin of MDMA (ecstasy) revisited: the true story reconstructed from the original documents. *Addiction, 101*, 1241–1245.

Gerona, R. R., & Wu, A. H. (2012). Bath salts. *Clinics in Laboratory Medicine, 32*, 415–427.

González, D., Riba, J., Bouso, J. C., Gómez-Jarabo, G., & Barbanoj, M. J. (2006). Pattern of use and subjective effects of Salvia divinorum among recreational users. *Drug and Alcohol Dependence, 85*, 157–162.

Grundmann, O., Phipps, S. M., Zadezensky, I., & Butterweck, V. (2007). *Salvia divinorum* and salvinorin A: an update on pharmacology and analytical methodology. *Planta Medica, 73*, 1039–1346.

Hill, S. L., Doris, T., Gurung, S., Katebe, S., Lomas, A., Dunn, M., … Thomas, S. H. (2013). Severe clinical toxicity associated with analytically confirmed recreational use of 25I – NBOMe: case series. *Clinical Toxicology, 51*, 487–492.

INCB, International Narcotics Control Board. (2011a). *Consecuencias económicas del uso indebido de droga*. http://www.incb.org/documents/Publications/AnnualReports/Thematic_chapters/Spanish/AR_2013_S_Chapter_I.pdf.

INCB, International Narcotics Control Board. (2011b). *Cohesión social, desorganización social y drogas ilegales*. http://www.incb.org/documents/Publications/AnnualReports/Thematic_chapters/Spanish/AR_2011_S_Chapter_I.pdf.

Johnson, L. A., Johnson, R. L., & Portier, R. B. (2013). Current "legal highs". *The Journal of Emergency Medicine, 44*, 1108–1115.

Johnson, M. W., MacLean, K. A., Reissig, C. J., Prisinzano, T. E., & Griffths, R. R. (2011). Human psychopharmacology and dose-effects of salvinorin A, a Kappa opioid agonist hallucinogen present in the plant *Salvia divinorum*. *Drug and Alcohol Dependence, 115*, 150–155.

Jones, A. L. (2010). Legal "highs" available through the internet-implications and solution? *QJM, 103*, 535–536.

Kamiyama, H., Matsumoto, M., Otani, S., Kimura, S. I., Shimamura, K. I., Ishikawa, S., … Togashi, H. (2011). Mechanisms underlying ketamine-induced synaptic depression in rat hippocampus-medial. *Neuroscience, 17*, 159–169.

Kazory, A., & Aiyer, R. (2013). Synthetic marijuana and acute kidney injury: an unforeseen association. *Clinical Kidney Journal, 6*, 330–333.

Kirsten, T. B., & Bernardi, M. M. (2010). Acute toxicity of *Psilocybe cubensis* (Ear.) Sing., Strophariaceae, aqueous extract in mice. *Brazilian Journal of Pharmacognosy, 20*, 397–402.

Lawn, W., Barratt, M., Williams, M., Horne, A., & Winstock, A. (2014). The NBOMe hallucinogenic drug series: patterns of use, characteristics of users and self-reported effects in a large international sample. *Journal of Psychopharmacology, 28*, 780–788.

López-Arnau, R., Martínez-Clemente, J., Abad, S., Pubill, D., Camarasa, J., & Escubedo, E. (2014). Repeated doses of methylone, a new drug of abuse, induce changes in serotonin and dopamine systems in the mouse. *Psychopharmacology, 231* (Article in press).

López-Arnau, R., Martínez-Clemente, J., Carbó, M. L., Pubill, D., Escubedo, E., & Caramasa, J. (2013). An integrated pharmacokinetic and pharmacodynamic study of a new drug of abuse, methylone, a synthetic cathinone sold as "bath salts". *Progress in Neuro-Psychopharmacology & Biological Psychiatry, 45*, 64–72.

Maurer, H. H., Kraemer, T., Springer, D., & Staack, R. F. (2004). Chemistry, pharmacology, toxicology and hepatic metabolism of designer drugs of the amphetamine (ecstasy), piperazine, and pyrrolidinophenone types. *Therapeutic Drug Monitoring, 26*, 127–131.

Mir, A., Obafemi, A., Young, A., & Kane, C. (2011). Myocardial infarction associated with use of the synthetic cannabinoid K2. *Pediatrics, 128*, 1622–1627.

Musselman, M. E., & Hampton, J. P. (2014). "Not for human consumption": a review of emerging designer drugs. *Pharmacotherapy, 34*, 745–757.

Müller, K., Püschel, K., & Iwersen-Bergmann, S. (2013). Suicide under the influence of "magic mushrooms". *Archiv für Kriminologie, 231*, 193–198.

Nikolova, I., & Danchev, N. (2008). Piperazine based substances of abuse: a new party pills on bulgarian drug market. *Biotechnology & Biotechnological Equipment, 22*, 652–655.

Oga, S., Camargo, M. M. A., & Batistuzzo, J. A. O. (2014). *Fundamentos de toxicologia* (4th ed.). São Paulo: Atheneu.

Pearson, J. M., Hargraves, T. L., Hair, K. L. S., Massucci, C. J., Frazee, C. C., Garg, U., & Pietak, B. R. (2012). Three fatal intoxications due to methylone. *Journal of Analytical Toxicology, 36*, 444–451.

Pourmand, A., Armstrong, P., Mazer-Amirshahi, M., & Shokoohi, H. (2014). The evolving high: new designer drugs of abuse. *Human & Experimental Toxicology, 33* (Article in press).

Prosser, J. M., & Nelson, L. S. (2012). The toxicology of bath salts: a review of synthetic cathinones. *Journal of Medical Toxicology, 8*, 33–42.

Rang, H. P., Dale, M. M., Ritter, J. M., Flower, R. J., & Henderson, G. (2011). *Rang & Dale Farmacologia* (7th ed.). Rio de Janeiro: Elsevier.

Rose, S. R., Poklis, J. L., & Poklis, A. (2013). A case of 25I-NBOMe (25-I) intoxication: a new potent 5-HT2A agonist designer drug. *Clinical Toxicology, 51*, 174–177.

Roth, B. L., Baner, K., Westkaemper, R., Siebert, D., Rice, K. C., Steinberg, S., … Rothman, R. B. (2002). Salvinorin A: a potent naturally occurring nonnitrogenous k-opioid selective agonist. *Proceedings in the National Academy of Sciences of the United States of America, 99*, 11934–11939.

Rothman, R. B., Murphy, D. L., Xu, H., Godin, J. A., Dersch, C. M., Partilla, J. S., … Prisinzano, T. E. (2007). Salvinorin A: allosteric interactions at the k-opioid receptor. *Journal of Pharmacology & Experimental Therapeutics, 320*, 801–810.

Satora, L., Goszcz, H., & Ciszowski, K. (2005). Poisonings resulting from the ingestion of magic mushrooms in Kraków. *Przegląd Lekarski, 62*, 394–396.

Schifano, F., Deluca, P., Baldacchino, A., Peltoniemi, T., Scherbaum, N., Torrens, M., ... Ghodse, A. H. (2006). Drugs on the web; the Psychonaut 2002 EU project. *Progress Neuropsychopharmacology & Biological Psychiatry, 30*, 640–646.

Schneir, A. B., Cullen, J., & Ly, B. T. (2011). "Spice" girls: synthetic cannabinoid intoxication. *Journal of Emergence Medicine, 40*, 296–299.

Schwartz, R. H., & Smith, D. (1988). Hallucinogenic mushrooms. *Clinical Pediatrics, 27*, 70–73.

Seibel, S., & Toscano, A., Jr. (2001). *Dependência de drogas*. São Paulo: Atheneu.

Silva, P. (2006). *Farmacologia* (7th ed.). Rio de Janeiro: Guanabara Koogan.

Stellpflug, S. J., Kealey, S. E., Hegarty, C. B., & Janis, G. C. (2014). 2-(4-Iodo-2,5-dimethoxyphenyl)-N-[(2-methoxyphenyl)methyl]ethanamine (25I-NBOMe): clinical case with unique confirmatory testing. *Journal of Medical Toxicology, 10*, 45–50.

Sumnall, H., & Wooding, O. (2010). *Mephedrone: An update on current knowledge*. Liverpool, UK: Centre for Public Health, Liverpool John Moores University.

UNODC, United Nations Office on Drugs and Crime. (2013). *World drug report 2013*. https://www.unodc.org/unodc/secured/wdr/wdr2013/World_Drug_Report_2013.pdf.

Valente, M. J., Pinho, P. G., Bastos, M. L., Carvalho, F., & Carvalho, M. (2014). Khat and synthetic cathinones: a review. *Archives of Toxicology, 88*, 15–45.

Van Veen, J. F., Van der Wee, N. J., Fiselier, J., Van Vliet, I. M., & Westenberg, H. G. (2007). Behavioural effects of rapid intravenous administration of meta-chlorophenylpiperazine (m-CPP) in patients with generalized social anxiety disorder, panic disorder and healthy controls. *European Neuropsychopharmacology, 17*, 637–642.

Vohra, R., Seefeld, A., Cantrell, F. L., & Clark, R. F. (2011). *Salvia divinorum*: exposures reported to a statewide poison control system over 10 years. *Journal of Emergency Medicine, 40*, 643–650.

Vortherms, T. A., & Roth, B. L. (2006). Salvinorin A: from natural product to human therapeutics. *Molecular Interventions, 6*, 257–265.

Wells, D. L., & Ott, C. A. (2011). The "new" marijuana. *Annals of Pharmacotherapy, 45*, 414–417.

Wolowich, W. R., Perkins, A. M., & Cienki, J. J. (2006). Analysis of the psychoactive terpenoid Salvinorin A content in five *Salvia divinorum* herbal products. *Pharmacotherapy, 26*, 1268–1272.

Zuba, D., Sekuła, K., & Buczek, A. (2013). 25C-NBOMe: new potent hallucinogenic substance identified on the drug market. *Forensic Science International, 227*, 7–14.

Chapter 6

Trauma and Neurological Risks of Addiction

Dessa Bergen-Cico[1], Sarah Wolf-Stanton[2], Radmila Filipovic[3], Jordyn Weisberg[4]

[1]Department of Public Health, Syracuse University, Syracuse, NY, USA; [2]Department of Marriage and Family Therapy, Syracuse University, Syracuse, NY, USA; [3]Physiology and Neurobiology Department, University of Connecticut, Storrs, CT, USA; [4]Syracuse University, Chicago, IL, USA

Abbreviations

CRH Corticotropin-releasing hormone
DA Dopamine
eCB Endocannabinoid
GABA γ-Amino butyric acid
HPA Hypothalamic–pituitary–adrenal
NE Norepinephrine
PTSD Posttraumatic stress disorder
SUD Substance use disorders

INTRODUCTION

Traumatic stress can have an impact on individuals in a multitude of ways, including but not limited to neurological alterations and changes in cognitive function. Memories associated with traumatic events generally do not extinguish and they are reinforced because of their emotional salience with subsequent repeated retrieval. From an evolutionary perspective such hardwired memories are conducive to survival; however, for people with posttraumatic stress disorder (PTSD), persistent traumatic emotional memories, also known as *intrusive memories,* can become maladaptive. PTSD symptoms are grouped into three symptomatic categories: (1) hyperarousal, hypervigilance, and increased startle responses; (2) reexperiencing of the traumatic event; and (3) withdrawal or avoidance behaviors and emotional numbing (Tannien & Jaycox, 2008). These three symptomatic categories often trigger one another and do not exist in isolation. The hyperarousal and hypervigilance involve multiple neural systems that foster anxiety, impulsivity, and aggression. Reexperiencing fosters flashbacks and nightmares, whereas the preceding symptoms foster avoidance, withdrawal, and numbing, all of which are distinctive of PTSD symptomology. Among the most common characteristics of PTSD is the persistence of traumatic memories coupled with an inability to extinguish associated fear. Memories associated with traumatic events do not extinguish and are reinforced over time because they have emotional salience and are repeatedly retrieved. This reexperiencing is a trigger that incites the need to numb and avoid the intense feelings that these memories elicit. As such, an individual's response to these symptoms may be to avoid or inhibit the stress-related response, whereby the use of alcohol and other drugs is often adapted as a coping mechanism for emotional numbing and avoidance (Stam, 2007). From an evolutionary perspective such hardwired memories were conducive to survival; however, for people with PTSD persistent traumatic emotional memories can become maladaptive.

The literature states that a significant percentage, possibly as high as 95%, of those who have sought treatment for substance use have experienced trauma (Persons, 2009). People who use opiates are nearly three times more likely to have a history of childhood sexual and/or physical abuse compared to people who do not use opiates, even when accounting and controlling for diagnostic and sociodemographic factors (Heffernan et al., 2000). Alcohol use is also prevalent among people with PTSD (Volpicelli, Balaraman, Hahn, Wallace, & Bux, 1999). People diagnosed with PTSD are more likely to use cannabis than people who do not have PTSD (Cougle, Bonn-Miller, Vujanovic, Zvolensky, & Hawkins, 2011).

Adverse Childhood Experiences

Addiction correlates significantly with adverse childhood experiences, which may result in psychological distress and physiological dysregulation during critical periods of development that present long-standing risks coupled with genetic and neural changes (Davidson et al., 2012; Dube et al., 2003; McGowan et al., 2009). People who have experienced trauma, particularly sexual abuses during childhood, are at greater risk of developing eating disorders, which are categorized as a behavioral addiction (Brewerton, 2007). Accounting for the impact child abuse and trauma can have on drug use can reframe how we view addiction, thus moving us away from a model in which addiction is viewed as being caused primarily by the chemical properties of the substance that is repeatedly used.

War-Related Trauma

Trauma is also demonstrably captured through the ripple effects that each war has had on the evolution of addiction. The intergenerational impact of trauma echoes across societies, affecting civilians, members of the military, veterans, and persecuted groups (refugees, targets of genocide). Over the past century, it is estimated that nearly 200 million human beings, mostly civilians, have died in wars (Smith, 2009). PTSD among those directly involved in battle has become better understood in recent years, and we are beginning to recognize that civilians from war-torn regions also develop PTSD. The connection between mind-altering substances and armed fighting has a long and complicated history. Civilians, veterans, and society at large are at increased risk of substance use and addiction stemming from the impact of war; this is reflected in the surge of drug use among civilians during and following wartime (Bergen-Cico, 2012). Regardless of formal diagnosis, veterans and survivors of war-torn regions have long suffered from the psychological trauma of what they have witnessed, and many spend a lifetime suppressing these memories with alcohol and other drugs. For decades following the Vietnam War, society witnessed growing numbers of homeless and addicted veterans who struggled socially, psychologically, politically, and economically because of unaddressed and underlying symptoms of PTSD. Failure to address postcombat challenges often leads to problematic relationships, depression, anxiety, and addiction.

PTSD is a hallmark of war and civil unrest and has been linked directly to atrocities that both veterans and civilians have witnessed. Similarly, first responders (firefighters, police, paramedics, medical personnel) often experience similar traumatic events involving decimation of the human body and witness of immense suffering. Research has shown that veterans and first responders who have handled human remains have higher levels of PTSD symptoms than their colleagues who have not (Alexander & Klein, 2009; Sutker, Uddo, Brailey, Vasterling, & Errera, 1994). The events that many military and first responders have witnessed have increased their risk for PTSD, suicidal ideation, and substance abuse (Bergen-Cico, 2012). The risks inherent in the use of substances to manage PTSD symptoms extend beyond a drug's toxicity and risk of overdose; when PTSD is coupled with alcoholism the likelihood of suicidal ideation increases 10-fold (Violanti et al., 2007). It is important to note that the impact of trauma extends beyond the persons directly affected by traumatic stress and affects both intra- and extrafamilial systems that may contribute to intergenerational transmission of trauma and associated substance abuse (Dekel & Goldblatt, 2008).

Intergenerational Trauma

Understanding the effects of intergenerational trauma among descendants of slavery, Holocaust and genocide survivors, and First Nations people provides insight into the ways in which unresolved trauma results in collective emotional and psychological injury over an individual's life span and across generations. Native Americans' history of genocide, racism, and refugee resettlement to reservations has resulted in disproportionate rates of alcoholism (five times the national average) and suicide (more than double the national average). Among Canada's First Nations people of the Sioux Lookout Zone, it has been estimated that 80% of the adult population uses prescription drugs illicitly (Timpson & O'Gorman, 2010). Refugees and survivors of war and genocide are susceptible to addiction as a means of escaping historical and recent trauma. These regions are often flooded with drugs because of security vulnerabilities. Case in point: Afghanistan, which produces 90% of the world's illicit heroin supply, has flooded not only Afghanistan, but Iran, Pakistan, Russia, and Europe with heroin, and these regions are now subjected to escalating rates of addiction (Bergen-Cico, 2012).

OVERVIEW OF TYPOLOGY OF TRAUMA

The estimated lifetime prevalence of PTSD is estimated to be 8–9% in Europe and North America (Darves-Bornoz et al., 2008; Van Ameringen, Mancini, Patterson, & Boyle, 2008). The development of posttraumatic stress (PTS)/PTSD encompasses neurological alterations when contrasted to non-PTSD persons, and there are delineations in the typology of trauma (e.g., PTSD, complex PTSD) that may lead to a culmination of symptoms for persons having experiences of trauma. Perceived control of the event and controllability of the response appear to be a significant factor in later development of PTSD. This parallels the notion of maintenance of control and perception of the event as being an overwhelming threat, which greatly influences how the event(s) influences the individual. Regardless of the cause or typology of trauma, it is experienced as feelings and a cascade of bodily sensations.

Posttraumatic Stress

PTSD refers to the psychophysiological responses that may develop in a person following a significant or life-threatening event. PTSD is characterized by hypervigilance, hyperarousal, intrusive memories, panic attacks, and the avoidance of situations that trigger memories and result in anxiety attacks and reliving of traumatic experiences. In this chapter we use the terms PTS and PTSD interchangeably because there is a move away from inclusion of the term "disorder" in PTS stemming from the recognition that the sequela associated with traumatic stress is the natural survival response rather than a disorder.

Complex Trauma

Complex trauma is a type of PTS, which ensues after chronic, repeated, or cumulative types of traumatic experiences (Cloitre et al., 2009; Taylor, Asmundson, & Carleton, 2006). Examples of complex trauma include childhood sexual abuses, family violence, or combat, among a host of other circumstances that involve extensive distress or culminated confusion. Some studies suggest that accumulations of traumatic experiences during developmental phases of life are associated with an adulthood amalgam of symptomology (Cloitre et al., 2009).

Chronic Toxic Stress

PTS and PTSD include the key word "post," indicating the life-threatening traumatic event has ended, such as when a veteran returns home from war or a person has survived a physical or sexual assault. However, people living in chaotic or violent environments, household dysfunction, marginalization, civil and political unrest, and conditions of chronic toxic stress experience these same psychophysiological responses in an effort to cope with their environment (Bergen-Cico, Haygood-El, Jennings-Bey, & Lane, 2014; Jennings-Bey et al., 2015). Since the events and the environment are unending, there is no "post" to their traumatic stress, yet they are likely to develop the same types of traumatic stress responses. With chronic toxic stress individuals are in a persistent state of hypervigilance and physiological arousal.

NEUROBIOLOGY OF TRAUMA

Persons experiencing an acute or chronic traumatic event(s) may respond physiologically, psychologically, and behaviorally in an immediate reaction to the event(s). Whereas the nervous system becomes dysregulated during traumatic events, the mind and body generally reestablish equanimity after the traumatic event has ended and the immediate threat dissipates. However, in some persons, the dysregulation becomes a persistent state with seemingly intractable patterns of physiological and psychological responses to environmental stimuli. When the subsequent physiological and psychological response is maladaptive it becomes very difficult for an individual to regulate his or her physiology, emotions, and thoughts. Moreover, the physiological dysregulation and psychological distress can continue to affect a person for a lifetime of chronic turmoil that may seethe below the surface or be expressed as outward volatile chaos. As such the dysregulation and distress associated with PTSD increase a person's risk of cultivating chemical and/or behavioral addictions in an attempt to regulate the imbalances that ensue post-trauma exposure.

During the time of the trauma, endorphins (endogenous opioids) are released in the body; as a result, during the posttrauma period one may experience a physiological withdrawal (Volpicelli et al., 1999). This endorphin effect is much more prominent in uncontrollable trauma versus a trauma for which a person is able to mobilize sympathetic activation (Van der Kolk, Greenberg, Orr, & Pitman, 1988; Volpicelli et al., 1999). This natural numbing agent (endorphins) has an obvious adaptive component and function. However, the aftermath can conceive an emotional downward spiral that may perpetuate the need to manually seek out a drug or other substance to reestablish the endorphinergic state. The relationship between traumatic events and substance use may begin soon after the event because human exposure to traumatic stress influences endorphins, which may become downregulated following the event. Owing to this refractory period of downregulation, people who experienced trauma could have increased risk for alcohol and drug abuse, because these substances can compensate for posttraumatic endorphin withdrawal by increasing endorphin activity in the brain (Olive, Koenig, Nannini, & Hodge, 2001; Volpicelli et al., 1999). Here we see the early connection between PTSD and substance use as it relates to the psychopharmacological effects of the substances themselves. Thus substances, such as alcohol, cocaine, and amphetamines, may be used to increase endorphin release and manage PTSD symptomology. The endogenous opioid systems within the brain's reward circuit appear to play an important role in drug reinforcement and drug-seeking behavior (Olive et al., 2001). The endogenous opioid system is separate and different from the activation of the mesolimbic (midbrain to nucleus accumbens) dopamine (DA) reward pathway, which is also associated with motivation and reinforcement of drug use and addiction (Nestler, 2005) and which will be not discussed in this chapter.

Among people who are suffering from PTSD, the activity of the limbic system is altered. The amygdala, a mass of gray matter situated in the anteromedial portion of the temporal lobe, is increased while the other parts of the limbic system, such as the hippocampus, may also become abnormal. In addition, there is decreased regulation of fear extinction in the prefrontal cortex (PFC) (Neumeister, 2013). In PTSD there is a dysfunction in connections between the ventromedial PFC and the hippocampus that causes an inability to diminish fear responses to trauma-associated cues or memories (Milad et al., 2009). Subcortical regions of the brain induce the sequela of responses to fear-potentiated stimuli and associated conditioning (LeDoux, 1998), which foster dysregulation of the nervous system endemic to PTSD symptomology. Moreover, persons with a diagnosis of PTSD have altered structural volume in distinct areas of the brain, such as the amygdala, as mentioned previously, which may contribute to addictive patterns and use of substances. During trauma and threat there is an exchange of neuronal activity between the hippocampal region and the amygdala (emotional response center) as part of the paleomammalian complex, which consolidates emotionally latent memories (Akirav, Sandi, & Richter-Levin, 2001). As a result, the hippocampus may become overwhelmed and not function properly, in a way such that the traumatic memories do not become explicitly encoded into autobiographical and chronological form, and therefore a flashback of traumatic memory is perceived as a present event. A variety of stimuli induced by physical or emotional stress activate the hypothalamic–pituitary–adrenal (HPA) axis, which involves secretion of corticotropin-releasing hormone (CRH) from the hypothalamus that stimulates the anterior pituitary gland to secrete adrenocorticotropic hormone and plays a critical role in the allostatic stress response. That pathway is often dysregulated among people with PTSD and as result is associated with abnormal secretion of the hormone cortisol from the adrenal gland (Bergen-Cico, Pigeon, & Possemato, 2014).

It was demonstrated that people who suffered from PTSD caused by chronic familial childhood trauma have a smaller volume of structures that are part of the limbic system, such as the hippocampus, anterior cingulate cortex, and orbitofrontal cortex (OFC), compared to people without a history of trauma (Thomaes et al., 2010). As the OFC plays roles in the integration, regulation, decision-making, punishment, and reward systems, these functions of the OFC might be altered in PTSD and addiction (Inaba & Cohen, 2014; Thomaes et al., 2010). Childhood abuse (physical and psychological) has been linked to specific alterations in the expression of genes for glucocorticoid receptors involved in the regulation of the stress-related neuroendocrine response (Davidson et al., 2012; McGowan et al., 2009); and alterations in prefrontal structures are critical to emotion regulation (Hanson et al., 2010). Such abuse creates an environment of inexorable toxic stress, which results in a cumulative allostatic load (cumulative stress response) that fosters the development of cognitive deficits, anxiety-related disorders, and depression (Shirtcliff, Dahl, & Pollack, 2009; Yehuda et al. 2010), thereby increasing susceptibility to alcohol and other drug abuse (Moffitt et al., 2011). Similarly traumatic stress, such as that caused by war or sexual trauma, can produce similar alterations in cortisol, gene expression, neural networks, and cognitive functions that regulate emotion, attention, and behavior (Almli, Fani, Smith & Ressler, 2014; Bergen-Cico, Pigeon et al., 2014; Dekel &

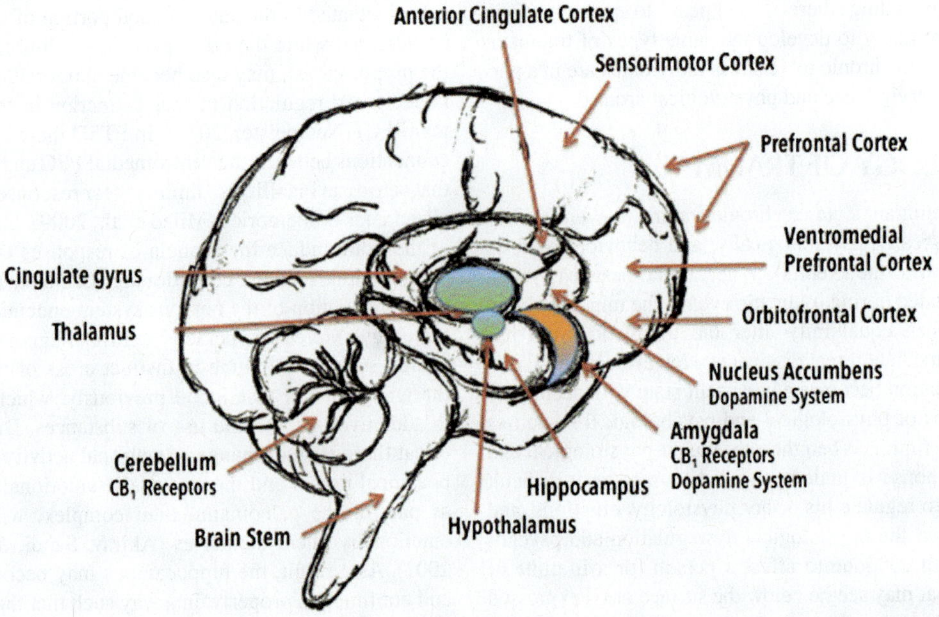

FIGURE 1 **Neuroanatomy of key parts of the brain involved in trauma.** The diagram illustrates key neuroanatomical parts of the brain affected by trauma. Note that the drawing shows the relative positions of the brain's anatomical structures and therefore displays some parts of the brain that would not be simultaneously visible in the sagittal view of the brain shown here. In the midsagittal plane of the brain, where the thalamus and hypothalamus are clearly visible, in the actual brain, the hippocampus and amygdala would be positioned more laterally.

Goldblatt, 2008; Yehuda et al., 2010). As previously mentioned, the endorphin influx and subsequent refractory period, which occur during trauma, leave the person in a withdrawal state from the homeostatic balance of naturally occurring opiates. Other various neurohormonal changes include alterations in catecholamines, glucocorticoids, serotonin, and neurotransmitters that affect memory, including norepinephrine (NE), vasopressin, oxytocin, and endogenous opioids (Van der Kolk, McFarlane, Weisaeth, 1996, p. 220).

BASIC NEUROBIOLOGY OF DRUG USE AND ADDICTION

There is ample research demonstrating the increased risk of substance abuse and addiction among people with PTSD; however, there is less of an appreciation of the physiological reasons for the comorbidity of substance use disorders and PTSD. Addictions involve genetic and acquired alterations in: (1) motivational systems in the brain (OFC, subcallosal cortex), (2) emotional systems in the brain, (3) the ability to self-regulate (anterior cingulate gyrus, PFC), and (4) the engagement of regions of the reward–reinforcement system (amygdala, cortex, OFC, PFC, hippocampus, hypothalamus, and regions in midbrain called the ventral tegmental area) (Inaba & Cohen, 2014; Tschemegg et al., 2013). Alcohol and other drug use provides the user with a chemical means of managing symptoms and extinguishing memories in the short term and produces sedation and relaxation, which facilitate sleep induction thereby providing temporary relief of some PTSD symptoms (Bonn-Miller, Babson, & Vandrey, 2014; Brown & Wolfe, 1994; Jacobsen, Southwick, & Kosten, 2001). Owing to the high prevalence of insomnia among people with PTSD (70–90%), facilitating sleep is desirable and persons with PTSD are more likely to use cannabis, alcohol, and other depressants to facilitate sleep and suppress persistent streams of thought and hypervigilance (Maher, Rego, & Asnis, 2006).

As noted in the section on the neurobiology of trauma, the HPA axis is affected by chronic stress and PTS/PTSD. There are also alterations in the functioning of the HPA axis that result from drug use and addiction and predispose individuals to the vulnerability of addiction. Alcohol and nicotine have distinct forms of interaction with HPA functioning. On one hand chronic heavy use of alcohol and nicotine appear to alter the frontal–limbic interactions and may account for altered HPA response evident in alcoholics and smokers. On the other hand, preexisting alterations in the frontal–limbic interactions with the HPA may increase addiction proneness (Lovallo, 2006). Figure 2 illustrates the key areas of the brain involved in addiction. Note that both Figures 1 and 2 represent the relative positions of the brain structures, because the thalamus and hypothalamus are located deeper within the brain and are not visible from the same plane displayed in the drawing.

The prolonged and repetitive use of a drug produces tissue dependency and neural changes, including changes in the reward and reinforcement centers of the brain. Some of the changes in the brain are substance specific and depend on the extent to which the DA or serotonin pathways are engaged. These changes are secondary to substance abuse and dependency, whereas PTS/PTSD produce neural changes and alterations in neural interactions that make people more vulnerable to substance use and abuse. Table 1 provides an overview of the neuroanatomical, neurobiological, and cognitive aspects that are dually affected by trauma, substance use, and addiction. Although this table is certainly not exhaustive in its coverage of the areas affected by traumatic stress and addiction, it serves to illustrate the common areas that are affected.

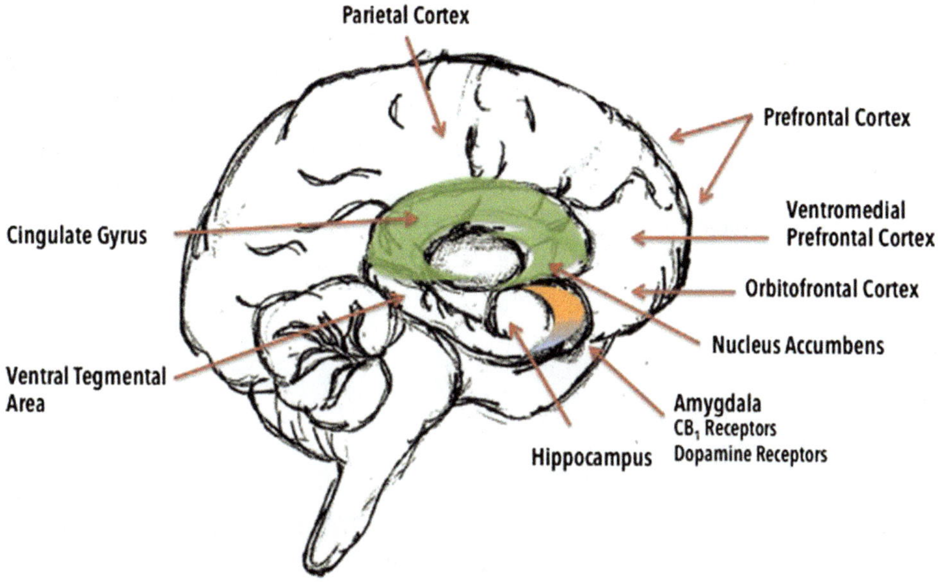

FIGURE 2 **Neuroanatomy of key parts of the brain involved in addiction.** The diagram illustrates key neuroanatomical parts of the brain engaged in addiction. Note that the drawing shows the relative positions of the brain's anatomical structures and therefore shows parts of the brain that would not be simultaneously visible from this plane of view of the brain. With an actual midsagittal plane of the brain the hippocampus and amygdala will not be visible in the same view showing the thalamus and hypothalamus.

POSTTRAUMATIC STRESS AND CANNABINOID RECEPTORS

An endocannabinoid (eCB) signaling system is present within the stress-sensitive cells in the hypothalamus and amygdala, and this points to the significance of the eCB system in the regulation of neuroendocrine and behavioral responses to stress (Passie, Emrich, Karst, Brandt, & Halpern, 2012). Evidence is increasingly accumulating that cannabinoids (CBs) may play a role in fear extinction and have antidepressive effects. In the context of the unique pathophysiology of trauma as it relates to CB receptors we will examine the potential regulatory effects of cannabis for traumatic stress. There are two types of CB receptors in the human brain, CB_1 and CB_2. These receptors bind to anandamide, the body's mediate eCB, and to phytocannabinoids (plant-based CBs, notably Δ^9-tetrahydrocannabinol (THC)) found in cannabis. Impaired CB_1 receptor function leads to chronic anxiety and depressive symptoms associated with PTSD (Neumeister, 2013). Research has found that the eCB that binds to CB_1 receptors is different in people with PTSD compared to people without a history of trauma, indicating that abnormal CB_1 receptor-mediated anandamide signaling is implicated in the etiology of PTSD (Neumeister, 2013; Neumeister et al., 2015). This is important because eCB signaling plays a key role in extinguishing fear associated with traumatic memories. It was determined that anandamide (eCB) and THC facilitate forgetting, and therefore the use of cannabis by people with PTS provides them with some relief from emotional and cognitive symptoms associated with PTSD.

Clinical research has found that THC blocks CB_1 receptors and this helps to suppress recall of fear-based memories and facilitates the extinction of learned fears (Rabinak & Phan, 2014). Therefore, if clinical pharmaceutical delivery of THC modulates the CB_1 system and suppresses the fear-based response and salience of traumatic memories, certainly one can see how nonclinical self-administration of cannabis would also provide individuals with similar relief from anxiety and fear-based symptoms associated with PTSD (Passie et al., 2012). In addition to the use of cannabis recreationally, research shows the potential benefit of medicinal or pharmacological use of cannabis to treat PTSD symptoms (Treza & Campolongo, 2013). However, long-term use of cannabis leads to downregulation of CB_1 receptors, which results in a decrease in the quantity of CB_1 receptors on a cell owing to the persistent presence of the phytocannabinoid THC (a.k.a. delta-9-THC, Δ^9-THC).

Both lifetime and current use of cannabis is statistically significantly greater among people with PTSD compared to people without when adjusting for co-occurring anxiety and mood disorders and trauma type frequency (Cougle et al., 2011). The high prevalence of cannabis use among people with PTSD is quite likely to be due to the benefits they experience. Cannabis appears to dampen the strength or emotional impact of traumatic memories and PTSD symptomology (anxiety, flashbacks), which enables people with PTSD to sleep (Passie et al., 2012). Thus the rewards and benefits of using cannabis to alleviate insomnia and anxiety may reinforce regular cannabis use.

APPLICATIONS TO OTHER ADDICTIONS AND SUBSTANCE MISUSE

The information included in this chapter provides an overview of how traumatic events and chronic toxic stress can affect individuals on the physiological and cognitive level in ways that increase the risk of alcohol and other drug use and behavioral addictions. Adverse childhood experiences, environments of chronic toxic stress, and traumatic events throughout the life cycle can influence

TABLE 1 Neurobiological and Cognitive Areas Affected by Trauma and Substance Use

Feature Mechanisms of Physiological and Psychological Regulation	Biology Affected by Traumatic Stress	Biology Affected by Substance Use and Addiction
Neurotransmitters and Associated Neuroreceptors		
Cannabinoids and eCBs (anandamide)	X	X
Catecholamines	X	X
Epinephrine (adrenaline)	X	X
NE (noradrenaline)	X	X
Endorphins (endogenous opioids)	X	X
DA	X	X
γ-Amino butyric acid (GABA)	X	X
Serotonin	X	X
HPA Axis and Associated Hormones		
Adrenocorticotropic hormone (a.k.a. corticotropin)	X	X
CRH	X	X
Cortisol	X	X
Glucocorticoids	X	X
Neuroanatomical Areas		
Amygdala	X	X
Anterior cingulate cortex	X	X
Cingulate gyrus	X	X
Hippocampus	X	X
Hypothalamus	X	X
OFC	X	X
PFC	X	X
Thalamus	X	X
Ventromedial PFC–hippocampus	X	X
Cognitive Functions		
Cognitive function of attention and concentration	X	X
Cognitive function of memory and learning	X	X
Cognitive processing	X	X
Emotional Function		
Emotional regulation	X	X
Experience of affective states	X	X
Fear extinction	X	X
Impulse control	X	X

Table 1 lists the key neuroanatomical areas and neurobiological factors implicated dually in trauma and addiction. The table also lists the key cognitive and emotional functions that are implicated in both posttraumatic stress and addiction.

people in ways that produce a cascade of neurological changes that prime the human brain to use substances or engage in self-soothing behaviors that can become habitual. Whereas people may use any number of psychoactive or pharmaceutical drugs in an effort to manage PTS symptoms, there is particular evidence of the neurological changes that may increase the propensity for cannabis as evidenced by the increase in CB_1 receptors in the brain. One of the evolutionary reasons for this link appears to stem from the fact that these receptors in the brain bind to anandamide, the body's mediate eCB, and to THC (plant-based phytocannabinoids). Anandamide and THC block the receptors, which helps to suppress recall of fear-based memories, facilitate forgetting, and facilitate the extinction of learned fears. Cannabis is certainly not the only substance that people with PTS may be drawn to using. There is also evidence that suggests that the physiological response to substances and behaviors that increase endorphin release may also help suppress PTS symptoms.

CONCLUSION

Understanding the neurobiological factors that increase risk for addiction can reframe the way we view people with substance abuse problems and broaden our capacity to respond to people who use drugs. Currently screening for trauma and PTSD is not systematically conducted as part of screening and therefore it is not addressed in standard substance use disorder treatment (Najavits, Sullivan, Schmitz, Weiss, & Lee, 2004; Young, Rosen, & Finney, 2005). However, research clearly shows that PTS can predispose people to the risk of developing substance use disorders stemming from trauma survivors' use of alcohol and other drugs to ameliorate PTS symptoms. Therefore the field needs to consider the role of trauma and PTSD in the prevention and treatment of substance abuse disorders.

DEFINITION OF TERMS

Allostatic load The physiological impact of cumulative and chronic stress that results in neurological and neuroendocrinological changes that can negatively affect tissues and organ systems of the individual. This maladaptive "stress response" can occur under conditions of chronic toxic stress as well as posttraumatic stress.

Anandamide The body's endogenous cannabinoids (eCB) that facilitates forgetting.

Complex trauma Complex trauma is a category of posttraumatic stress, which ensues as a result of chronic, repeated, or cumulative types of traumatic experiences.

Cannabinoids The active organic chemical compounds in the plant cannabis (marijuana).

Downregulation A decrease in the number of receptors from overstimulation by a chemical or hormone.

Drugs Drugs are defined in this chapter as psychoactive substances that are capable of being used recreationally with some addictive potential. Drugs include alcohol, pharmaceutical medications, and illicit recreational substances.

Endocannabinoids Endogenous (naturally occurring) neuromodulatory lipids capable of binding to, and functionally activating, the cannabinoid receptors in the brain and body. Two of the most common and well-known endocannabinoids are anandamide (N-arachidonoylethanolamine) and 2-arachidonoylglycerol.

Endorphins The body's natural endogenous opioid neurotransmitters that are produced by the central nervous system and pituitary gland and are released as an inhibitory response to pain and stress.

Phytocannabinoids Plant-based cannabinoids notably found in cannabis (marijuana); the most notable is delta-9-tetrahydrocannabinol (THC), which is also known as delta-9-THC or Δ^9-THC or simply THC.

Posttraumatic stress (PTS) and posttraumatic stress disorder (PTSD) The exposure to traumatic events such as witnessing a death, war, or enduring an injury, whereby the body produces a physiological and behavioral response to traumatic stress through subconscious reexperiences of the event, avoidance of actual or similar events, and increased hyperarousal.

Sequela A secondary negative or abnormal health condition that develops or emanates from another health condition.

Upregulation An increase in cellular response to a molecular stimulus that is caused by an increase in the number or density of receptors on the surface of a cell.

KEY FACTS

Key Facts of Trauma

A trauma is an experience that is considered an overwhelming event, coupled with fear, helplessness and/or horror. There is a tendency during trauma to experience disempowerment and/or lack of control. Additionally the person experiencing this will perceive the event as threatening. This event can be isolated or chronic/cumulative. Trauma can occur both in veteran and in civilian cultures. Trauma can be categorized as "big T" or "little t" traumas and may include but are not limited to combat, rape/sexual assault, neglect, abuse, violence, death, serious injury, oppression, and car accidents. Witnessing a traumatic event is also said to be equally impactful on a person. Persons close to individuals who have experienced trauma or mental health professionals and first responders are at greater risk for developing secondary traumatic exposure through vicarious experiencing of traumatic events through the original victim.

Key Facts of Posttraumatic Stress Disorder (PTSD)

This is a disorder that ensues after exposure to a traumatic event. If and only if an individual has experienced or witnessed a trauma can one be diagnosed with PTSD. There are five criteria according to the *Diagnostic and Statistical Manual of Mental Disorders,* which include: (1) traumatic event, (2) reexperiencing, (3) avoidance, (4) negative cognitions, and (5) hyperarousal. These categories describe the main symptoms that people with PTSD experience. There is an additional subtype of dissociation. PTSD that can also be complex or noncomplex. Complex PTSD is most commonly associated with cumulative forms of trauma coupled with later traumatic exposures. Traumas that lead to complex-type PTSD are prototypically repetitive in nature and experienced many times with longer duration. Examples are long-term child abuse, sexual abuses, domestic violence, or prisoner of war camps.

Key Facts of Posttrauma Response Trajectory

Responses to traumatic events vary greatly. Those who are less susceptible to PTSD and have an increased sense of resilience can integrate the experience and move forward through posttraumatic growth. However, many suffer the grave ramifications of traumatic exposure and begin to respond to a neutral present environment with patterns of response paralleling that of enactment of the original traumatic event with hyperarousal symptoms. Hypervigilance and reexperiencing through intrusive memories lead to subsequent withdrawal and avoidant tendencies, which deprive persons of optimal functioning and engagement in life, regularly.

Key Facts of Neurological and Physiological Alterations

Endorphins play a large role in that these endogenous opioids are released during traumatic exposure, relinquishing any sense of emotional or physical pain. There is a posttrauma refractory period, however, which leaves persons susceptible to depressive mood states. Thus they seek out supplementary substances that will elicit activation of endorphinergic pathways in the brain. This explains addictive tendencies associated with traumatic exposures. Additionally anatomical alterations in the brain and autonomic nervous system ensue after trauma. The HPA axis alters the levels of cortisol both during the exposure and chronically, which can lead to adrenal exhaustion if prolonged, with other negative health outcomes. The amygdala, responsible for emotionally charged or salient responses, increases in activation for those who have experienced trauma. Located near the amygdala, the hippocampus, which is used for encoding new memories, stores traumatic memories in implicit rather than explicit memory and thus these emotionally charged events remain coupled with fear and do not encode with context of time. As the limbic system is activated, the prefrontal regions of the brain are diminished and thus uncoupling fear from associative cues is decreased, explaining chronic high arousal states for those suffering from PTSD.

Key Facts of the Cannabinoid Systems and Traumatic Stress

Traumatic stress affects the endocannabinoid (eCB) signaling system within the stress-sensitive cells in the hypothalamus and amygdala, which points to the significance of the eCB system in the regulation of neuroendocrine and behavioral responses to stress. Cannabis affects the eCB system and may provide some temporary relief of PTSD symptoms. The high prevalence of cannabis use among persons with PTSD is quite probably due to the benefits they experience in relation to dampening of PTSD symptoms.

SUMMARY POINTS

- PTS can predispose people to the risk of developing substance use disorders stemming from the use of alcohol and other drugs to ameliorate PTS symptoms.
- It is important to screen for a history of trauma, so that the impact of traumatic stress can be addressed as a factor in the development of substance abuse disorders.
- Traumatic events may occur at any point in time across the life span and increase the risk of substance use and addiction owing to neurological and psychological risks.
- Endorphins are endogenous opioids that are released during traumatic exposure and suppress emotional or physical pain temporarily. After trauma there is a refractory period with low endorphin levels, which leaves persons susceptible to drug-seeking behavior to activate the endorphinergic pathways in the brain.
- Traumatic stress affects the eCB signaling system within the stress-sensitive cells in the hypothalamus and amygdala, which points to the significance of the eCB system in the regulation of neuroendocrine and behavioral responses to stress.
- Cannabis affects the eCB system and may provide some temporary relief of PTSD symptoms. The high prevalence of cannabis use among persons with PTSD is quite probably due to the benefits they experience in relation to dampening of PTSD symptoms.
- Recurrent negative memories and flashbacks are key symptoms of traumatic stress that foster drug-seeking behavior, because substance use can suppress thoughts and facilitate sleep.
- Anxiety and hypervigilance are also key symptoms of traumatic stress that foster drug-seeking behavior, because substance use can suppress cognitive functions and sympathetic nervous system arousal.

REFERENCES

Akirav, I., Sandi, C., & Richter-Levin, G. (2001). Differential activation of hippocampus and amygdala following spatial learning under stress. *European Journal of Neuroscience, 14*(4), 719–725.

Alexander, D. A., & Klein, S. (2009). First responders after disasters: a review of stress reactions, at-risk, vulnerability, and resilience factors. *Prehospital and Disaster Medicine, 24*(02), 87–94.

Almli, L. M., Fani, N., Smith, A. K., & Ressler, K. J. (2014). Genetic approaches to understanding post-traumatic stress disorder. *International Journal of Neuropsychopharmacology, 17*(2), 355–370.

Bergen-Cico, D. (2012). *War and drugs: The role of military conflict in the development of substance abuse*. New York: Routledge.

Bergen-Cico, D., Haygood-El, A., Jennings-Bey, T., & Lane, S. D. (2014). Street addiction: a proposed theoretical model for understanding the draw of street life and gang activity. *Journal of Addiction Research and Theory, 22*(1), 15–26. http://dx.doi.org/10.3109/16066359.2012.759942.

Bergen-Cico, D., Pigeon, W., & Possemato, K. (2014). Reductions in cortisol associated with a primary care mindfulness-based intervention for veterans with PTSD. *Medical Care, 52*, S25–S31.

Bonn-Miller, M. O., Babson, K. A., & Vandrey, R. (2014). Using cannabis to help you sleep: heightened frequency of medical cannabis use among those with PTSD. *Drug and Alcohol Dependence, 136*, 162–165.

Brewerton, T. D. (2007). Eating disorders, trauma, and comorbidity: focus on PTSD. *Eating Disorders, 15*(4), 285–304.

Brown, P. J., & Wolfe, J. (1994). Substance abuse and post-traumatic stress disorder comorbidity. *Drug and Alcohol Dependence, 35*(1), 51–59.

Cloitre, M., Stolbach, B. C., Herman, J. L., Kolk, B. V. D., Pynoos, R., Wang, J., & Petkova, E. (2009). A developmental approach to complex PTSD: childhood and adult cumulative trauma as predictors of symptom complexity. *Journal of traumatic stress, 22*(5), 399–408.

Cougle, J. R., Bonn-Miller, M. O., Vujanovic, A. A., Zvolensky, M. J., & Hawkins, K. A. (2011). Posttraumatic stress disorder and cannabis use in a nationally representative sample. *Psychology of Addictive Behaviors*, 25(3), 554–558.

Darves-Bornoz, J. M., Alonso, J., de Girolamo, G., Graaf, R. D., Haro, J. M., Kovess–Masfety, V., & Gasquet, I. (2008). Main traumatic events in Europe: PTSD in the European study of the epidemiology of mental disorders survey. *Journal of Traumatic Stress*, 21(5), 455–462.

Davidson, R. J., Dunne, J., Eccles, J. S., Engle, A., Greenberg, M., Jennings, P., ... Vago, D. (2012). Contemplative practices and mental training: prospects for American education. *Child Development Perspectives*, 6(2), 146–153. http://dx.doi.org/10.1111/j.1750-8606.2012.00240.x.

Dekel, R., & Goldblatt, H. (2008). Is there intergenerational transmission of trauma? the case of combat veterans' children. *American Journal of Orthopsychiatry*, 78(3), 281.

Dube, S. R., Felitti, V. J., Dong, M., Chapman, D. P., Giles, W. H., & Anda, R. F. (2003). Childhood abuse, neglect, and household dysfunction and the risk of illicit drug use: the adverse childhood experiences study. *Pediatrics*, 111(3), 564–572.

Hanson, J. L., Chung, M. K., Avants, B. B., Shritcliff, E. A., Gee, J. C., Davidson, R. J., & Pollak, S. D. (2010). Early stress is associated with alterations in the orbitofrontal cortex: a tensor-based morphometry investigation of brain structure and behavioral risk. *Journal of Neuroscience*, 30, 7466–7472.

Heffernan, K., Cloitre, M., Tardiff, K., Marzuk, P. M., Portera, L., & Leon, A. C. (2000). Childhood trauma as a correlate of lifetime opiate use in psychiatric patients. *Addictive Behaviors*, 25(5), 797–803.

Inaba, D., & Cohen, W. (2014). *Uppers, downers, all arounders*. Medford, OR: CNS Publications.

Jacobsen, L. K., Southwick, S. M., & Kosten, T. R. (2001). Substance use disorders in patients with posttraumatic stress disorder: a review of the literature. *American Journal of Psychiatry*, 158(8), 1184–1190.

Jennings-Bey, T., Lane, S., Rubenstein, R., Bergen-Cico, D., Haygood-El, A., Hudoson, H., Sanchez, S., & Fowler, F. (2015). The trauma response team: a community intervention for gang violence. *Journal of Urban Health*, 92(5), 947–954. http://dx.doi.org/10.1007/s11524-015-9978-8.

LeDoux, J. (1998). Fear and the brain: where have we been, and where are we going? *Biological Psychiatry*, 44(12), 1229–1238.

Lovallo, W. R. (2006). Cortisol secretion patterns in addiction and addiction risk. *International Journal of Psychophysiology*, 59(3), 195–202.

Maher, M. J., Rego, S. A., & Asnis, G. M. (2006). Sleep disturbances in patients with post-traumatic stress disorder: epidemiology, impact and approaches to management. *CNS Drugs*, 20(7), 567–590.

McGowan, P. O., Sasaki, A., D'Alessio, A. C., Dymov, S., Labonté, B., Szyf, M., ... Meaney, M. J. (2009). Epigenetic regulation of the glucocorticoid receptor in human brain associates with childhood abuse. *Nature Reviews Neuroscience*, 12, 342–348.

Milad, M. R., Pitman, R. K., Ellis, C. B., Gold, A. L., Shin, L. M., Lasko, N. B., ... Rauch, S. L. (2009). Neurobiological basis of failure to recall extinction memory in posttraumatic stress disorder. *Biological Psychiatry*, 66, 1075–1082.

Moffitt, T. E., Arseneault, L., Belsky, D., Dickson, N., Hancox, R. J., Harrington, H., ... Caspi, A. (2011). A gradient of childhood self-control predicts health, wealth, and public safety. *Proceedings of the National Academy of Sciences*, 108, 2693–2698.

Najavits, L. M., Sullivan, T. P., Schmitz, M., Weiss, R. D., & Lee, C. S. (2004). Treatment utilization by women with PTSD and substance dependence. *American Journal on Addictions*, 13(3), 215–224.

Nestler, E. J. (2005). Is there a common molecular pathway for addiction? *Nature Neuroscience*, 8, 1445–1449.

Neumeister, A. (2013). The endocannabinoid system provides an avenue for evidence-based treatment development for PTSD. *Depression and Anxiety*, 30(2), 93–96.

Neumeister, A., Seidel, J., Ragen, B. J., & Pietrzak, R. H. (2015). Translational evidence for a role of endocannabinoids in the etiology and treatment of posttraumatic stress disorder. *Psychoneuroendocrinology*, 51, 577–584.

Olive, M. F., Koenig, H. N., Nannini, M. A., & Hodge, C. W. (2001). Stimulation of endorphin neurotransmission in the nucleus accumbens by ethanol, cocaine, and amphetamine. *Journal of Neuroscience*, 21(23), RC184 1–5.

Passie, T., Emrich, H. M., Karst, M., Brandt, S. D., & Halpern, J. H. (2012). Mitigation of post–traumatic stress symptoms by Cannabis resin: a review of the clinical and neurobiological evidence. *Drug Testing and Analysis*, 4(7), 649–659.

Persons, A. T. E. (2009). Trauma, posttraumatic stress disorder, and addiction among women. In *Women and addiction: A comprehensive handbook* (p. 242).

Rabinak, C. A., & Phan, K. L. (2014). Cannabinoid modulation of fear extinction brain circuits: a novel target to advance anxiety treatment. *Current Pharmaceutical Design*, 20(13), 2212–2217.

Shirtcliff, E. A., Dahl, R. E., & Pollack, S. D. (2009). Pubertal development: correspondence between hormonal and physical development. *Child Development*, 80, 327–337.

Smith, D. L. (2009). *The most dangerous animal: Human nature and the origins of war*. Macmillan.

Stam, R. (2007). PTSD and stress sensitization: a tale of brain and body: part 1: human studies. *Neuroscience & Biobehavioral Reviews*, 31(4), 530–557.

Sutker, P. B., Uddo, M., Brailey, K., Vasterling, J. J., & Errera, P. (1994). Psychopathology in war-zone deployed and nondeployed operation desert storm troops assigned graves registration duties. *Journal of Abnormal Psychology*, 103(2), 383–390.

Tanielian, T., & Jaycox, L. H. (Eds.). (2008). *Invisible wounds of war: Psychological and cognitive injuries, their consequences, and services to assist recovery*. RAND Corporation, Monograph MG-720. Accessed at: http://www.rand.org/pubs/monographs/2008/RAND_MG720.pdf. Accessed on 01.05.10.

Taylor, S., Asmundson, G. G., & Carleton, R. N. (2006). Simple versus complex PTSD: a cluster analytic investigation. *Journal of Anxiety Disorders*, 20(4), 459–472. http://dx.doi.org/10.1016/j.janxdis.2005.04.003.

Thomaes, K., Dorrepaal, E., Draijer, N., de Ruiter, M. B., van Balkom, A. J., Smit, J. H., & Veltman, D. J. (2010). Reduced anterior cingulate and orbitofrontal volumes in child abuse-related complex PTSD. *The Journal of Clinical Psychiatry*, 71(12), 1636–1644.

Timpson, J., & O'Gorman, K. E. (2010). *Community-based options for addressing prescription drug abuse in remote northwestern Ontario First Nations*. Thunder Bay, ON: Nishnawbe Aski Nation, iv.

Treza, V., & Campolongo, P. (2013). The endocannabinoid system as a possible target to treat PTSD. *Frontiers in Neuroscience*, 7(100), 1–5.

Tschernegg, M., Crone, J. S., Eigenberger, T., Schwartenbeck, P., Fauth-Bühler, M., Lemènager, T., & Kronbichler, M. (2013). Abnormalities of functional brain networks in pathological gambling: a graph-theoretical approach. *Frontiers in Human Neuroscience*, 7(625), 1–10.

Van Ameringen, M., Mancini, C., Patterson, B., & Boyle, M. H. (2008). Post–traumatic stress disorder in Canada. *CNS Neuroscience & Therapeutics*, 14(3), 171–181.

Van der Kolk, B. A., Greenberg, M. S., Orr, S. P., & Pitman, R. K. (1988). Endogenous opioids, stress induced analgesia, and posttraumatic stress disorder. *Psychopharmacology Bulletin, 25*(3), 417–421.

Van der Kolk, B. A., McFarlane, A. C., & Weisaeth, L. (Eds.). (1996). *Traumatic stress: The effects of overwhelming experience on mind, body and society.* New York: Guilford Press.

Violanti, J. M., Cecil, M., Burchfiel, T., Hartley, A., Charles, L. E., & Miller, D. B. (2007). Post-traumatic stress symptoms and cortisol patterns among police officers. *Policing: An International Journal of Police Strategies and Management, 30*(2), 189–202 Print.

Volpicelli, J., Balaraman, G., Hahn, J., Wallace, H., & Bux, D. (1999). The role of uncontrollable trauma in the development of PTSD and alcohol addiction. *Alcohol Research and Health, 23*(4), 256–262.

Yehuda, R., Flory, J. D., Pratchett, L. C., Buxbaum, J., Ising, M., & Holsboer, F. (2010). Putative biological mechanisms for the association between early life adversity and the subsequent development of PTSD. *Psychopharmacology, 212*(3), 405–417.

Young, H. E., Rosen, C. S., & Finney, J. W. (2005). A survey of PTSD screening and referral practices in VA addiction treatment programs. *Journal of Substance Abuse Treatment, 28*(4), 313–319.

Chapter 7

Emotional Self-Medication and Addiction

Carmen Torres[1], Mauricio R. Papini[2]

[1]Department of Psychology, University of Jaén, Jaén, Spain; [2]Department of Psychology, Texas Christian University, Fort Worth, TX, USA

Abbreviations

cE Consummatory appetitive extinction
cSNC Consummatory successive negative contrast
ESM Emotional self-medication
iE Instrumental appetitive extinction
iSNC Instrumental successive negative contrast
PRCE Partial reinforcement contrast effect
PREE Partial reinforcement extinction effect
PSM Physical self-medication
PTSD Posttraumatic stress disorder
RHA Roman high-avoidance rats
RLA Roman low-avoidance rats
SNC Successive negative contrast
SUD Substance use disorder

INTRODUCTION

Humans have traditionally used psychoactive substances with a variety of purposes, including mystical ecstasy, search for pleasure, relief from physical pain, and madness treatment. Many of these practices may have been the result of the observation of self-medication behaviors in nonhuman animals. These behaviors constitute a source of knowledge regarding the therapeutic and psychoactive properties of natural substances (Huffman, 2010). Some of these drugs have been shown to have addictive effects, so it has been proposed that the scientific study of self-medication could shed light on the understanding of substance use disorders (SUDs) (Darke, 2012; Khantzian, 2013).

The *Diagnostic and Statistical Manual for Mental Disorders* (American Psychiatric Association, 2013) provides detailed criteria for the diagnosis of SUDs, including sustained excessive consumption, cravings, tolerance, and withdrawal symptoms. Unlike previous diagnostic criteria (in which abuse and dependence constructs were considered separately), addiction is therein defined as a continuum from mild to severe SUD based on the number of items met of a total of 11. SUDs are viewed as neurobehavioral disorders resulting from persistent dysregulation of neural circuits that mediate motivation, cognition, learning/memory, habits, and stress reactivity (Koob, 2013).

Several approaches have identified the motivations leading a person to consume drugs. Some views state that the positive reinforcing value of abuse drugs constitutes the critical factor that initiates consumption. This value is related to the activity of midbrain dopaminergic neurons that project to the basal forebrain—known as the reward system. Although drugs of abuse differ in their pharmacological profile, most increase dopamine levels within this circuit, particularly in the nucleus accumbens (Koob, Arenns, & LeMoal, 2014). Because the assumption underlying this approach suggests that the onset of addictive behavior relates to the "pleasure" initially induced by these substances, we will characterize this mechanism as one relying on drive induction. Thus, *drive induction* refers to the reinforcing effect of an increase in stimulation in an organism that is not necessarily in a negative internal state. For example (Figure 1), animals learn a new response, whose consequence is a simple change in stimulation (e.g., turning on a light; Levin & Forgays, 1960).

Adaptive theories postulate that environmental influence constitutes a critical determinant of addiction. Within this context, the self-medication hypothesis states that the drug chosen to be consumed depends on the drug's ability to either relieve a psychiatric disorder or reduce occasional or persistent negative emotional states induced by aversive events (Khantzian, 1985, 2013). Experimental and clinical evidence supports the self-medication hypothesis, although some contradictory results invite caution. This chapter focuses on a special case of self-medication, namely, the consumption of psychoactive drugs that modulate negative emotions induced by:

1. Acute and/or chronic stressful experiences in healthy individuals,
2. Psychiatric conditions with negative affect as a prominent symptom, and
3. Distress/dysphoria associated with drug withdrawal.

As shown below, this special case, termed emotional self-medication (ESM) hereafter, underlies some forms of drug use behaviors that contribute to the onset, progression, maintenance, and/or relapse of an SUD. This chapter is organized into four major sections. First, we look at clinical evidence testing the ESM hypothesis. Second, we review physical self-medication (PSM) and ESM behavior in nonhuman animals with two objectives: (1) identify similarities between these two forms of self-medication and (2) analyze the extent to which ESM can influence drug consumption and abuse. Third, we review studies on ESM aiming at bridging the gap between experimental and clinical research. Finally, we identify areas in which future research can have a significant impact on our understanding of the connection between ESM and addiction.

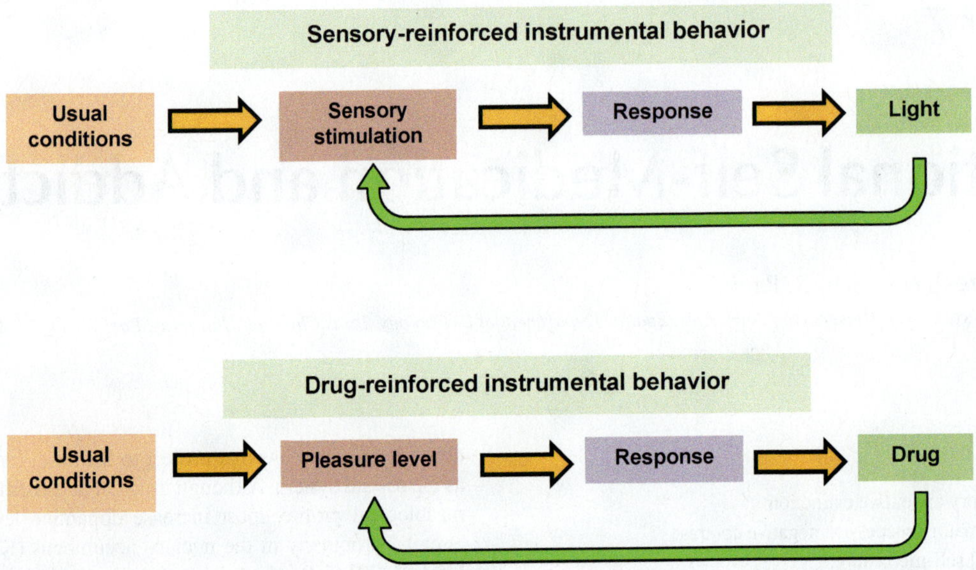

FIGURE 1 Examples of instrumental reinforcement by drive induction. A neutral internal state is increased in intensity by the outcome of an instrumental response (positive reinforcement). Top: in sensory reinforcement, an instrumental response is reinforced by the presentation of a light. Bottom: the initial exposure to a drug increases positive affect.

ESM AND ADDICTION IN HUMANS

Is ESM related to addictive behavior in humans? Addiction has become a public health issue with medical, social, legal, and political implications (Koob et al., 2014). Worldwide interest in this problem notwithstanding, there are few theories explaining addiction, a behavior maintained despite its self-destructive consequences. The self-medication hypothesis was initially conceived as a psychodynamic model of substance dependence in which drug consumption was viewed as a strategy to cope with disordered emotions, self-care, self-esteem, and personal relationships triggering painful and threatening emotions (Khantzian, 1985, 2013; Koob et al., 2014). The self-medication hypothesis was revised from a behavioral perspective according to which self-medication behavior is initiated and maintained by negative reinforcement (Blume, Chmaling, & Marlatt, 2000). This perspective assumes that negative reinforcement occurs when a behavior followed by the removal of an *internal* aversive state increases its future probability (Koob et al., 2014). This notion has been important in the development of psychobiological models of addiction (Koob, 2013), but it conflicts with traditional definitions of reinforcement contingencies (Ferster & Skinner, 1957). Reinforcement contingencies have been defined in terms of the response being followed by either the presentation of an *external* outcome (positive reinforcement) or the removal of an *external* outcome (negative reinforcement). Therefore, reducing an aversive internal state (e.g., hunger, negative emotion, withdrawal symptoms) by procuring an outcome is an instance of positive reinforcement, as illustrated in Figure 2. In these cases, reinforcement may be characterized in terms of drive reduction. *Drive reduction* refers to the strengthening effects of an outcome that compensates for a negative internal state, be it motivational (e.g., thirst, hunger) or emotional (e.g., frustration, drug withdrawal).

The ESM hypothesis of addiction is based on two main assumptions (Darke, 2012). First, the *psychopathology postulate* states that substance use relates to the relief of symptoms of psychological distress. Therefore, the relief of negative affect constitutes the main motivation and provides reinforcement for substance use in distressed individuals (Li, Lu, & Miller, 2013). Consistent with this assumption, patients who suffer from anxiety disorders and cite symptom relief as the main reason for drug use are more likely to develop an SUD (Ruglass et al., 2014). From this perspective, the relationship between alcohol consumption and anxiety disorders would be explained in terms of the tension-reducing effects of alcohol (Conger, 1956). This assumption implies that it is the symptoms of negative affect accompanying a psychiatric condition, rather than the primary psychiatric condition per se, that provides the basis for ESM (Darke, 2012). Moreover, healthy individuals may use ESM as a coping strategy to deal with painful and unbearable feelings; that is, ESM need not be associated with a primary psychiatric disorder (Khantzian, 2013).

The second postulate of the ESM hypothesis, *drug specificity*, suggests that the choice reflects the drug's ability to ameliorate specific distressing symptoms. Thus, psychostimulants would be used by hypomanic patients; anxiolytics and alcohol to cope with anxiety; opiates to reduce anger, rage, or physical pain; energizers to reduce depressed mood; cognitive enhancers for attention deficits; nicotine for the negative symptoms of schizophrenia; and so on (Darke, 2012; Koob et al., 2014; Kumari & Postma, 2005). Therefore, the ESM hypothesis would be supported by showing that patients with similar psychiatric problems displayed similar patterns of substance use (Blume et al., 2000). This assumption also implies that drugs with comparable pharmacological profile would be interchangeably used for similar symptoms.

Tests of the ESM hypothesis have shown mixed results. First, several studies suggest that emotionally painful experiences are associated with the initial use, loss of control, and relapse into drug abuse (Briand & Blendy, 2009). Consistent with this, individuals exposed to either acute or chronic emotional stress are more likely to use drugs. For example, the rates of use of alcohol

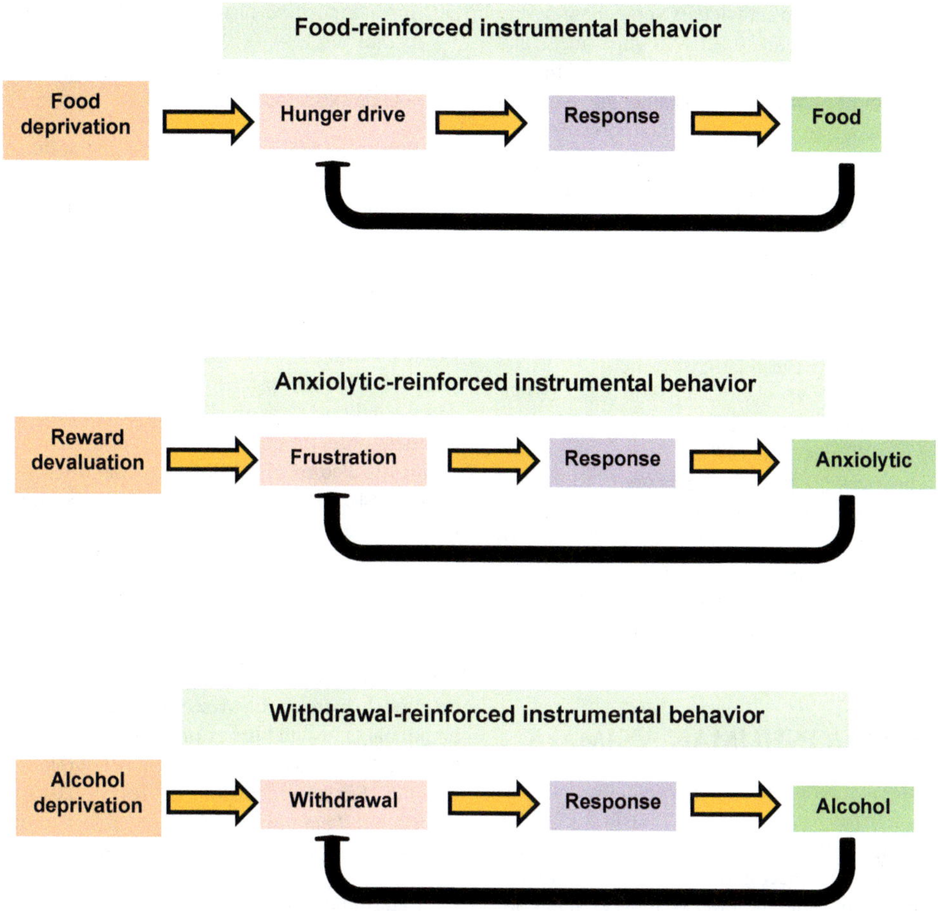

FIGURE 2 **Examples of instrumental reinforcement by drive reduction.** An aversive internal state is reduced by the outcome of an instrumental response (positive reinforcement). Top: conventional food-reinforced instrumental behavior. Middle: emotional self-medication induced by reward loss and access to an anxiolytic. Bottom: relapse of an already established addiction induced by withdrawal symptoms resulting from drug deprivation. In all cases, behavior is reinforced by a reduction in negative emotion.

in war combatants suffering from posttraumatic stress disorder (PTSD) are higher than in veterans without PTSD (Enman, Zhang, & Unterwald, 2014; Mantsch et al., 2014). Similarly, a significant portion of individuals experiencing social phobia consume alcohol to cope with the distress derived from social situations (Carrigan & Randall, 2003). In addition, people exposed to adverse life events (e.g., physical, sexual, or domestic abuse; bullying; natural catastrophes; divorce; death of a loved one; losing a job; family conflicts; poverty; chronic pain; etc.) show higher consumption rates of alcohol, benzodiazepines, and illicit drugs (Duffing, Greiner, Mathias, & Dougherty, 2014; Hassanbeigi, Askari, Hassanbeigi, & Pourmovahed, 2013; Konopka, Pełka-Wysiecka, Grzywacz, & Samochowiec, 2013; Spanagel, Noori, & Heilig, 2014). Therefore, there is evidence of a positive correlation between emotional distress and drug consumption, some with anxiolytic effects (e.g., alcohol and benzodiazepines; Konopka et al., 2013) and some relieving from psychological pain (e.g., opioids; Hassanbeigi et al., 2013; Papini, Fuchs, & Torres, 2015).

Drug addiction is a chronically relapsing disorder characterized by a compulsion to seek and take drugs, loss of control, and the emergence of a negative emotional state (e.g., dysphoria, anxiety, irritability, physical and emotional pain) when the drug is not available (Koob et al., 2014). Additional support for the ESM hypothesis comes from studies showing that relapse in drug abusers is related to the reinforcing effects derived from the relief of drug-withdrawal symptoms. This drive-reduction mechanism has been proposed as critical to transform drug taking from an impulse-control disorder into compulsive behavior, the latter maintained by relief from withdrawal/negative affect (Koob, 2013). Accordingly, it has been shown that emotional distress can trigger drug abuse reinstatement (drug seeking) and relapse (drug taking). Such behaviors are dependent on neuroadaptive processes in brain circuits that counteract the reinforcing effects of drugs (Ahmed, 2012; Mantsch et al., 2014). This view involves ESM, that is, alleviation from emotional discomfort, as a motivational basis for the maintenance and relapse of drug abuse.

Another source of evidence for the ESM hypothesis comes from comorbidity studies. *Comorbidity* refers to individuals with an existing behavioral disorder who are at a higher risk of developing another disorder (Li et al., 2013). The ESM hypothesis implies that SUDs should be correlated with psychiatric conditions involving emotional distress as a core symptom. Accordingly, clinical evidence shows that a high percentage of adults and adolescents meeting diagnostic criteria for anxiety disorders, depression, or both also meet criteria for SUDs (Enman et al., 2014; Menary et al., 2011; Robinson, Sareen, Cix, & Bolton, 2011;

Tomlinson & Brown, 2012). These individuals show more severe symptomatology, health problems, and functional impairment, and relapse more frequently into drug abuse (Ruglass et al., 2014).

By contrast, other studies show that substance use is sometimes associated with the aggravation of psychiatric symptoms, rather than their amelioration (Castaneda, Galanter, & Franco, 1989). In addition, only a low percentage of individuals with anxiety disorders overtly assert to self-medicate by consuming alcohol (Menary et al., 2011), making it difficult to objectively assess the involvement of ESM in alcohol intake. In fact, most studies do not systematically explore whether drug intake actually reduces emotional distress. Finally, the direction of the causal mechanisms underlying the relationship between SUDs and psychiatric disorders is not yet clear (Robinson et al., 2011); psychiatric symptoms could be postmorbid rather than premorbid relative to substance use (Blume et al., 2000).

Clinical research provides insights into the nature of SUDs and identifies major factors in the development of addictive behavior. Animal models may contribute to isolating these factors; to identifying some of them as causally related to, rather than just correlated with, addictive behavior; and to understanding the neurobiological mechanisms underlying addiction. Animal models of self-medication are reviewed in the following section.

PSM AND ESM IN NONHUMAN ANIMALS

Animals have evolved behavioral and physiological mechanisms to treat and control disease that improve health and enhance reproductive fitness (Huffman, 2010). This field of study, referred to as PSM (or zoopharmacognosy), includes behaviors with therapeutic or symptom-relieving effects. PSM covers both prophylactic and therapeutic practices, from behaviors that prevent or reduce the risk of sickness in healthy individuals to those aimed at treating an illness in diseased individuals (Lozano, 1998; Villalba, Miller, Ungar, Landau, & Glendinning, 2014). PSM includes ingestion, absorption, topical application, and proximity to medicinal substances (Clayton & Wolfe, 1993; Huffman, 2010). PSM also includes transgenerational prophylaxis and therapeutic medication directed at offspring and social prophylaxis directed at conspecifics (De Roode, Lefèvre, & Hunter, 2013). Here we consider only ingestive behaviors with therapeutic effects.

The ability of individuals to behaviorally defend themselves against life-threatening diseases provides an adaptive advantage extensively documented across species (Huffman, 2010). Such ability includes species-typical actions patterns as well as responses emerging from individual experience, feedback mechanisms, and learning (Villalba et al., 2014). Species-typical PSM behavior is commonly observed in insects (e.g., flies, ants, butterflies, caterpillars), whereas PSM behavior based on individual and social learning is observed mainly in vertebrates (e.g., snow geese, lambs, goats, sheep, civets, chimpanzees, gorillas, bonobos; De Roode et al., 2013; Huffman, 2010).

Four conditions must be fulfilled for a behavior to be an example of PSM (Clayton & Wolfe, 1993; Huffman, 2010):

- Identify the disease or symptoms being treated.
- Distinguish the use of a therapeutic agent from that of everyday food items.
- Show a positive change in health condition following self-medication.
- Provide evidence of the pharmacological activity of the compounds extracted from therapeutic agents.

Although there is evidence that sick animals are able to select corrective dietary components that are not otherwise consumed in significant quantities, whether these examples meet the requirements listed above remains to be shown (De Roode et al., 2013).

Defensive behavior against parasites constitutes a common form of PSM (Lozano, 1998). Woolly bear caterpillars experimentally infected with parasitoid flies increase their ingestion of alkaloids (Singer, Mace, & Bernays, 2009). Monarch butterflies infected with the protozoan parasite *Ophryocystis elektroscirrha* use milkweed as medication (Lefèvre et al., 2012). Surveys and field observations, as well as controlled studies, indicate that ruminants also exhibit PSM behavior when parasitized (Villalba et al., 2014). Chimpanzees and bonobos consume plant leaves that reduce endoparasite proliferation (Fruth et al., 2014; Masi et al., 2012). The diversity of species exhibiting, at least potentially, PSM suggests the adaptive advantage of identifying, curing, and reducing the negative impact of physical diseases.

Parasitism is a main source of biotic stress faced by many species (Lozano, 1998), but would animals use substances to reduce emotional distress? ESM seems to occur in relation to negative emotional states and drugs of abuse. However, studies must meet four conditions to identify ESM behavior (Huffman, 2010; Manzo, Donaire, Sabariego, Papini, & Torres, 2015). In the following description, "emotional activation" refers to any of a number of aversive states, including anxiety, conflict, depression, and stress:

- Determine that emotional activation is present during or before substance consumption.
- Demonstrate that consumption is selectively directed at substances that reduce emotional activation.
- Show that substance consumption actually reduces emotional activation.
- Provide independent evidence that the consumed substance reduces emotional activation.

Most available evidence involves exposing animals to acute or chronic stress, having simultaneous or subsequent access to drugs such as ethanol (see Becker, Lopez, & Doremus-Fitzwater, 2011; Spanagel et al., 2014). For example, Anisman and Waller (1974) administered inescapable electric shocks to rats with simultaneous access to 10% ethanol. Stressed animals showed an increase in ethanol consumption relative to controls. Additional studies also suggested that the impact of electric shocks on ethanol intake depends on several factors, including control over shock delivery, baseline preference for ethanol versus water, and availability of a safe place with access to ethanol (Becker et al., 2011; Manzo, Gómez, Callejas-Aguilera, Fernández-Teruel, et al., 2014). Neuropathic pain induced by sciatic nerve ligation promotes cannabinoid and opioid self-administration in rats (Ewan & Martin, 2013; Gutiérrez, Crystal, Zvonok, Makriyannis, & Hohmann, 2011). These results can be interpreted in terms of ESM given the negative hedonic state that accompanies physical pain.

Chronic stress (e.g., anxiogenic social stimuli, physical restraint, isolation) may also induce self-administration of ethanol and opioids (Becker et al., 2011). Similarly, the contribution of stress to drug reinstatement was shown using designs in which stressful stimuli reestablish a previously extinguished drug-seeking behavior (Ahmed, 2012).

Whereas the relationship between ESM and drug-taking behavior seems consistent, some results show the complexity of this relationship. The type of stressful experience (e.g., duration, intensity, quality), biological factors (e.g., strain, age, sex), and procedural variables (e.g., dose, simultaneous vs subsequent drug access, free choice vs operant self-administration, unlimited vs time-restricted access) all modulate the relationship (Becker et al., 2011; Spanagel et al., 2014).

It is safe to argue that most of the experimental evidence on ESM reviewed thus far fails to provide evidence concerning one or more of the conditions indicated above. To some extent this is understandable because most of these studies were not designed with the ESM concept in mind, but as a way of illustrating the effects of negative emotion on drug consumption. The ESM hypothesis requires not only that distress leads to drug consumption, but also that drug consumption reduces distress so as to produce a reinforcing effect (Figure 2).

There are three main problems with the available evidence. First, some tests do not provide an assessment of the animal's aversive/negative state (e.g., forced restraint) and of the extent to which this state relates to drug use. Similarly, whether drug intake causes a reduction in emotional distress is rarely assessed, making it difficult to explain the results in terms of ESM. Second, many studies do not consider individual differences in the proneness to taking drugs and in emotional reactivity (Torres & Sabariego, 2014). Such differences depend on genetic factors that contribute, separately or simultaneously, to vulnerability to stress and drug addiction (Vengeliene et al., 2003). Third, there are issues of ecological validity related to the use of certain noxious stimuli (e.g., electric shock) and routes of administration (e.g., intravenous self-administration). Although pain induced by shock and the neurochemical effects of intravenous drug infusion are ecologically valid, with some exceptions (e.g., heroin intravenous administration), these procedures do not mimic typical conditions in human addiction. To tackle some of these limitations, we now turn to studies based on procedures that more closely mimic typical conditions.

ESM INDUCED BY REWARD LOSS

When individuals rank the most stressful daily life events, most of these events are arguably related to reward loss (e.g., divorce, death of a loved one, dismissal from a job, social exclusion, natural catastrophes; Papini, Wood, Daniel, & Norris, 2006). These events trigger behavioral, affective, autonomic, hormonal, and immunological consequences that can negatively affect behavior and health (Papini et al., 2015). Interestingly, the correlation between reward loss and drug consumption is suggested by several clinical studies (Duffing et al., 2014; Egli, Koob, & Edwards, 2012; Hassanbeigi et al., 2013; Konopka et al., 2013; Spanagel et al., 2014). This link could be explained in terms of ESM, that is, the consumption of substances that reduce the negative emotional impact of reward loss.

Reward loss has been systematically studied in the animal laboratory through the devaluation or omission of an expected reward that triggers a negative emotional reaction called frustration, disappointment, anxiety, or, more recently, psychological pain (Papini et al., 2015). These models include consummatory (cSNC) and instrumental successive negative contrast (iSNC), consummatory (cE) and instrumental appetitive extinction (iE), and partial reinforcement extinction effect (PREE) and partial reinforcement contrast effect (PRCE), among others (see Minidictionary of Terms), providing a useful tool to test the notion that reward loss induces emotional distress from a psychobiological perspective (see Papini et al., 2015; Torres & Sabariego, 2014). First, the omission of an appetitive event (or its signal) has consequences similar to those induced by an aversive event, including aggressive behavior, agitation, and a variety of anxiety-like responses (Papini & Dudley, 1997). These behaviors are selectively reduced or abolished by anxiolytics and analgesics, including alcohol, benzodiazepines, barbiturates, opioids, and cannabinoids (Flaherty, 1996; Genn, Tucci, Parikh, & File, 2004; Papini, 2009). In addition, situations involving reward devaluation or omission increase plasma levels of stress hormones (Flaherty, 1996). Lesion studies suggest that pain/fear and frustration are influenced by damage to similar brain areas (e.g., hippocampus, amygdala, prefrontal cortex; Becker, Jarvis, Wagner, & Flaherty, 1984; Flaherty, Coppotelli, Hsu, & Otto, 1998; Lin, Roman, & Reilly, 2009; Ortega, Uhelski, Fuchs, & Papini, 2011). Emotional responses triggered by exposure to aversive stimuli or removal of appetitive stimuli are also partially modulated by common genetic factors. Studies of reward loss in inbred Roman high-avoidance (RHA) and Roman low-avoidance (RLA) rats, initially selected on the basis of their good (RHA) versus poor (RLA) acquisition of a two-way active avoidance response, are especially relevant. As a result of this psychogenetic selection, Roman strains differ in anxiety-inducing situations, with RLA rats exhibiting higher anxiety levels than RHA rats (Torres & Sabariego, 2014). These strains also differ in behavioral traits associated with vulnerability to drug abuse, including novelty seeking (Manzo, Gómez, Callejas-Aguilera, Donaire, et al., 2014), impulsivity (Moreno et al., 2010), and voluntary consumption and preference for alcohol versus water (Manzo et al., 2012), with RHA exhibiting higher vulnerability than RLA rats. RLA rats are also more vulnerable than RHA rats in situations involving reward devaluation and omission: cSNC and iSNC (Gómez, Escarabajal, et al., 2009; Rosas et al., 2007; Torres et al., 2005), iE (Gómez, de la Torre, et al., 2009), PREE (Gómez et al., 2008), and PRCE (Cuenya et al., 2012). Therefore, Roman rats provide a valid animal model to assess the influence of genetic factors on individual reactivity to anxiety-provoking reward-loss events.

Although the evidence described above connects reward loss to emotional distress, the relationship between reward loss and drug consumption has been barely explored in laboratory animals. Kamenetzky, Cuenya, Pedrón, and Mustaca (2009) reported that reward devaluation (cSNC) increased approach to cues previously paired with systemic ethanol administration. The relationship between reward loss and voluntary consumption of ethanol has been more directly studied in a model involving two tasks in tandem (Figure 3). First, RHA and RLA rats were exposed to two appetitive reward omission situations: (1) cE: a downshift from 22% sucrose to water in a consummatory task and (2) iE: a downshift from 12 pellets to nonreward in the goal box of a runway. Both tests were immediately followed by a 2-h alcohol (2%) versus water preference test. Whereas RHA rats prefer ethanol over water under resting environmental conditions (Manzo et al., 2012), when exposed to cE or iE it was

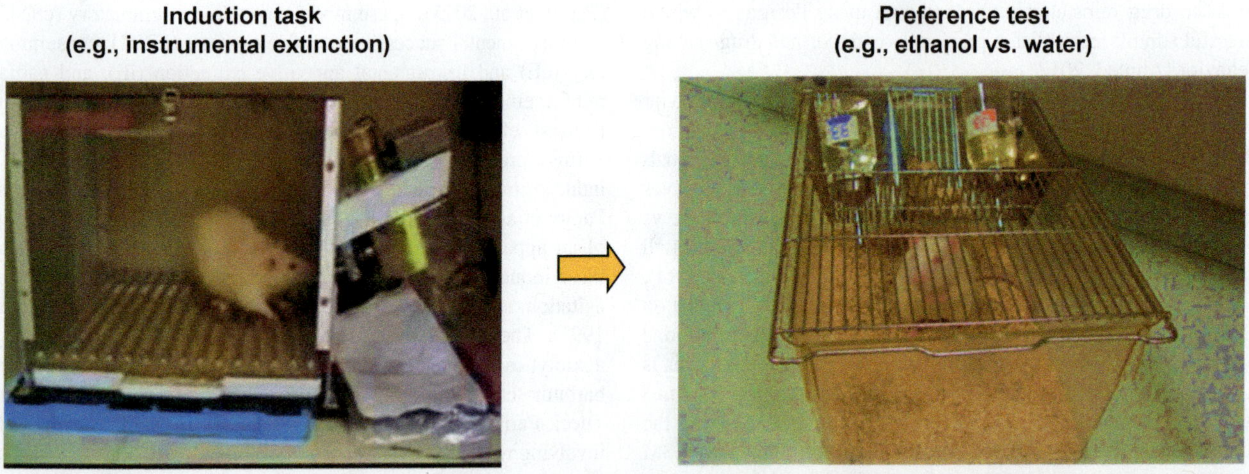

FIGURE 3 Design of emotional self-medication studies. These two tasks are presented everyday to the animals. The first task ("induction task") is designed to produce emotional distress. The second task ("preference test") is designed to measure choice of a substance that reduces negative emotion.

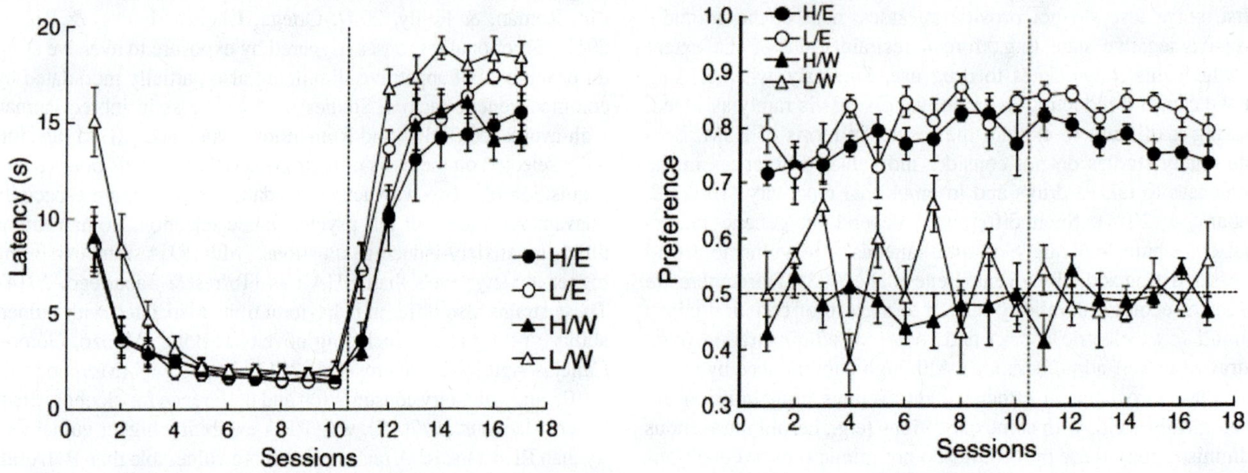

FIGURE 4 Effects of instrumental extinction on postsession ethanol consumption. Left (induction task): mean (±SEM) latency to traverse the runway in groups reinforced with food pellets on sessions 1–10 (acquisition), but not on sessions 11–17 (extinction). There were six trials per session. RLA rats (L) extinguished faster than RHA rats (H), thus showing greater vulnerability to reward loss. Right (preference test): mean (±SEM) preference for ethanol (E) or water (W) during the two-bottle test. Each day, rats received a training session in the runway (induction task) followed by access to ethanol and water in their cage (preference test). RLA rats displayed a greater preference for ethanol after extinction sessions than RHA rats. *Reproduced with permission from Manzo, Gómez, Callejas-Aguilera, Fernández-Teruel, et al. (2014), Elsevier.*

the more anxious RLA rats that showed greater preference for and consumption of ethanol (Manzo, Gómez, Callejas-Aguilera, Fernández-Teruel, et al., 2014). Figure 4 shows the results of an experiment inducing frustration via iE. Notice that all animals preferred ethanol over water; this preference was maintained after extinction sessions in RLA rats, but it was reduced in RHA rats. Moreover, preference for one of the bottles in control groups receiving iE, but only water during the preference test, showed no changes in extinction. Thus, iE did not just increase fluid consumption, but selectively increased consumption of ethanol, although only in RLA rats. These results constitute the first demonstration that reward omission increases voluntary ethanol consumption in emotionally reactive subjects. Because ethanol has anxiolytic effects in situations involving reward loss (Kamenetzky, Mustaca, & Papini, 2008), the authors interpreted the results in terms of the ESM hypothesis.

Second, the increased ethanol consumption observed in the RLA strain after iE was reduced when extinction occurred after partial reinforcement training, as opposed to continuous reinforcement (Manzo, Gómez, et al., 2015). Partial reinforcement attenuates the disrupting effects of extinction and can therefore be conceptualized as a treatment for developing resilience to loss-induced anxiety (Pellegrini, Muzio, Mustaca, & Papini, 2004). Using the same procedure described for Figure 3, although RLA rats showed the conventional PREE during runway training (see Definition of Terms), partially reinforced rats displayed lower ethanol consumption than continuously reinforced rats after extinction sessions. Controls receiving access to water during the preference tests showed no change in preference (Figure 5). These results suggest that ESM is reduced even in individuals genetically vulnerable to anxiety by extensive experience with reward uncertainty.

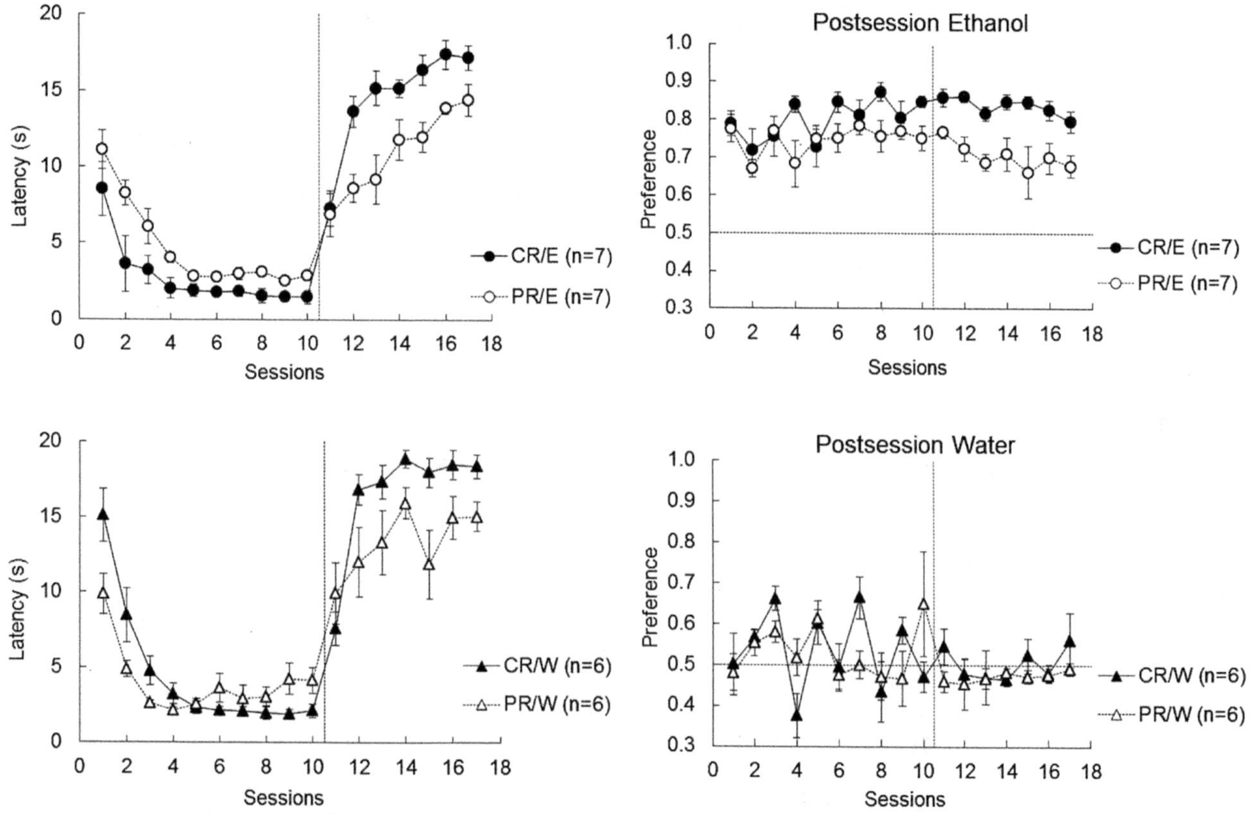

FIGURE 5 **Effects of partial reinforcement (PR) on postsession ethanol consumption.** Left (induction task): mean (±SEM) latency to traverse the runway in RLA groups reinforced with food pellets on sessions 1–10 (acquisition), but not on sessions 11–17 (extinction). There were six trials per session. Two groups received 50% PR in which a random half of the trials ended with 12 food pellets and the rest ended in nonreward. Two groups received continuous reinforcement (CR) in which each trial ended with 12 food pellets. Both pairs of groups showed the PREE (see Minidictionary of Terms). Right (preference test): mean (±SEM) preference for ethanol (E) or water (W) during the two-bottle test. Each day, rats received a training session in the runway (induction task) followed by access to ethanol and water in their cage (preference test). CR rats displayed a greater preference for ethanol after extinction sessions than PR rats. Thus, ESM can be attenuated after extended PR training, even in animals genetically vulnerable to anxiety. *Reproduced with permission from Manzo, Gómez, et al. (2015).*

Third, another study extended these results to nonselected Wistar rats and to a prescription anxiolytic, the benzodiazepine chlordiazepoxide (Manzo, Donaire, et al., 2015). This experiment followed the procedure outlined for Figure 3, except that the induction task was cSNC (see Definition of Terms). During the preference test one bottle contained water, whereas the other contained chlordiazepoxide (1 mg/kg), ethanol (2%), or water for different groups. Chlordiazepoxide is a potentially addictive benzodiazepine anxiolytic used in the treatment of anxiety (Konopka et al., 2013). Rats showed the cSNC effect, that is, a suppression of consummatory behavior after reward downshift relative to unshifted controls. This effect was accompanied by a selective increase in oral consumption of chlordiazepoxide and ethanol (Figure 6). Downshifted animals with access to water and unshifted controls with access to water, chlordiazepoxide, or ethanol exhibited no changes in preference. The results were again explained in terms of ESM, that is, the negative emotional state induced by reward devaluation (cSNC) caused an increase in the consumption of an anxiolytic substance, which, in turn, reduced the negative emotion. Both the cSNC and the ESM effects were transient, suggesting that the rewarding value of the anxiolytic ceased once the animals recovered from the effects of reward devaluation. Such reversibility suggests that animal models of reward loss may provide insights into the connection between negative emotions and drug consumption before the animal develops full-blown addictive behavior. Such drive-reduction model (Figure 2) provides an alternative (probably complementary) to the drive-induction model of Figure 1 for an understanding of the initial stages of an SUD.

FUTURE DIRECTIONS: BRIDGING ESM AND SUDs

Over the past decades, substantial research has centered on modeling human addiction in nonhuman animals. Ahmed (2012) conceptualized such animal research as reverse psychiatry because, whereas clinicians seek to help people suffering from SUDs, preclinical research induces an addiction in a drug-naïve animal. This chapter reviewed the ESM hypothesis of addiction as a theoretical framework that integrates clinical and preclinical studies that have been often poorly connected. We first defined basic concepts, the role of ESM in addiction, the relationship between PSM and ESM in animals, and the ESM

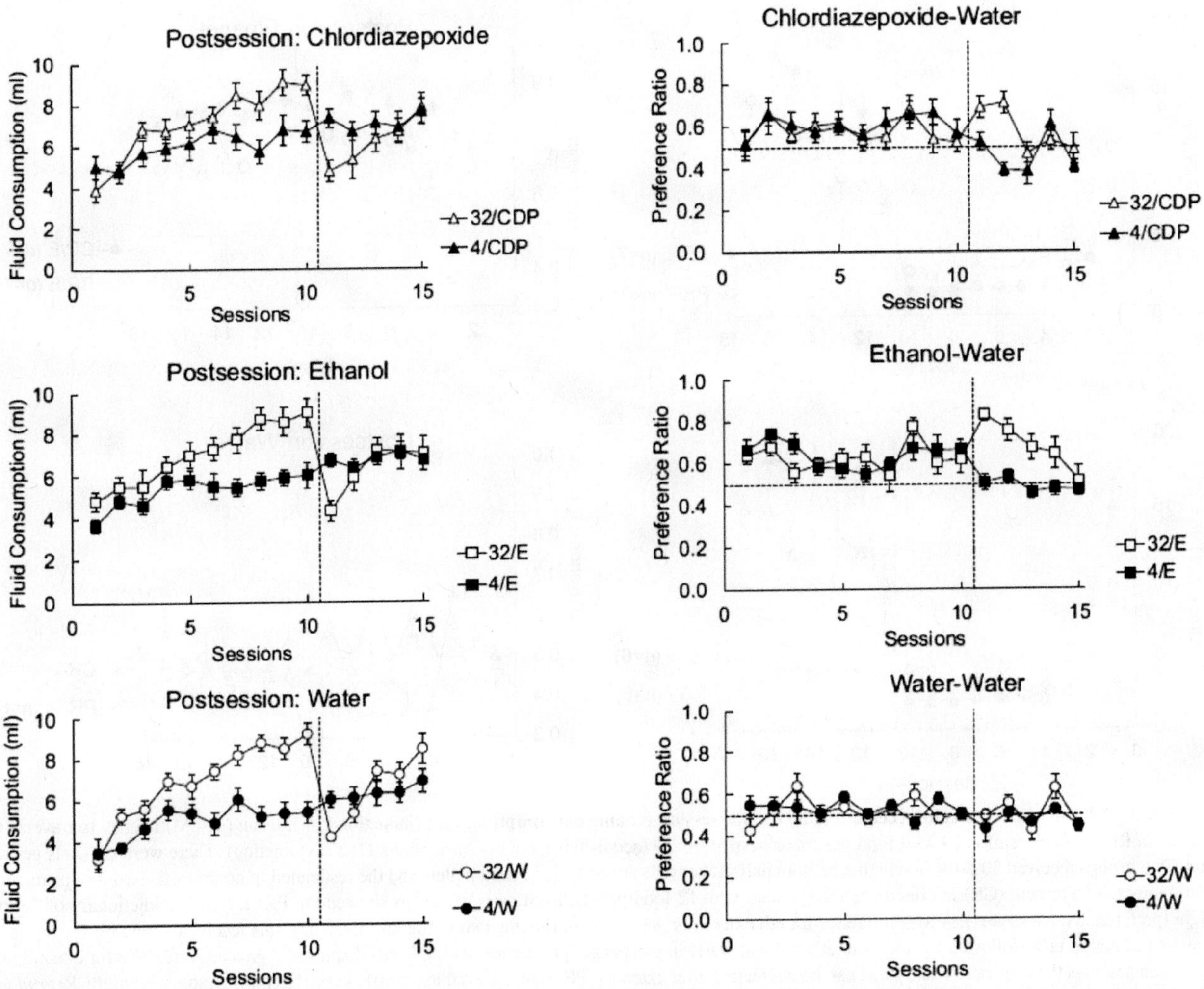

FIGURE 6 **Effects of consummatory successive negative contrast on postsession consumption of chlordiazepoxide and ethanol.** Left (induction task): mean (±SEM) latency to consume either 32% or 4% sucrose in Wistar rats. On sessions 11–15 all animals had access to 4% sucrose. There was a single 5-min session per day. All three pairs of groups showed the cSNC effect (see Minidictionary of Terms). Right (preference test): mean (±SEM) preference for chlordiazepoxide (CDP), ethanol (E), or water (W) during the two-bottle test. Rats displayed greater preference for chlordiazepoxide and ethanol after the initial sessions of reward devaluation. These results provide evidence of ESM tightly correlated with an event involving reward loss. 32: 32-to-4% sucrose downshift. 4: only 4% sucrose. *Reproduced with permission from Manzo, Donaire, et al. (2015), Elsevier.*

hypothesis as being developed in animal models. Although supporting evidence for the ESM hypothesis appears to be broad, several limitations make it difficult to draw firm conclusions about the implications of ESM in the initial and later stages of drug use:

1. There is insufficient *experimental* research on ESM in humans, so most evidence is correlational. For example, comorbidity studies do not determine whether the psychiatric disorder associated with addictive behavior is the cause or the consequence or whether both develop in parallel.
2. It is also unclear whether awareness of the relationship between the consumption of a drug and the reduction of emotional distress is important for the development of ESM. Given that some patients report that symptoms increase in strength, rather than decrease, after drug consumption, it is unclear whether drug use is driven by *drive reduction* or outcome-independent *habitual behavior*.
3. Experimental studies do not always objectively record the emotional state of the organism before and after drug consumption. Therefore, the *reinforcing* properties of the drug are not always empirically supported.

These limitations are even more serious given the restricted number of experimental studies based on the ESM hypothesis. In fact, as far as we know, this is the first review relating PSM and ESM in nonhuman animals to clinical studies. This chapter highlights the need for a set of basic conditions before any given behavior can be considered an example of ESM. Recent studies reviewed in the previous section can shed light on the validity of the ESM hypothesis because (1) they are based on situations involving reward loss, arguably a frequent, yet experimentally neglected, source of emotional distress in humans and also known to be associated with SUDs; (2) they allow for a clear record of behavioral indices of emotional distress; and (3) they relate ESM with individual differences in emotional reactivity and sensitivity to drugs of abuse. This framework allows for the study of vulnerability and resilience to the effects of reward loss on drug consumption, enabling a better understanding of the relationship between ESM and addiction.

APPLICATIONS TO OTHER ADDICTIONS AND SUBSTANCE MISUSE

ESM offers a general framework that can potentially be applied to an understanding of several types of addictive behaviors. Available data from animal models suggest that a variety of distressing situations can trigger corrective consummatory actions, including physical pain (e.g., sciatic nerve ligation), pain-induced fear (inescapable shock), and psychological pain (e.g., reward loss). Such research has also identified voluntary anxiolytic intake as substances preferred and consumed during periods of distress. Theoretically, there are reasons to predict that ESM will be triggered by other sources of emotional distress induced by negative changes (e.g., physical restraint, escape behavior, open spaces) and supported by the consumption of a variety of substances (e.g., opioids, over-the-counter analgesics, serotonergic anxiolytics). ESM may also be supported by nondrug substances, making the framework potentially relevant to obesity, video game addiction, and other activities that can develop properties similar to those of more traditional addictions. Furthermore, the ability to regulate emotional states suggests that the consumption of substances with anxiogenic effects may be selectively inhibited during periods of stress. Finally, the bridge between consumption limited to periods of distress and excessive consumption will require additional research. One possibility is anticipatory ESM, that is, consumption triggered in anticipation of an anxiogenic situation. If sustained, such behavior could become habitual, occurring even in the absence of a negative emotion, or acquire a new goal—reducing withdrawal symptoms generated by the very substance originally used for self-medication. Thus, a behavior that started as a way of reducing negative emotions induced by specific external events (e.g., reward loss) may become co-opted to reduce new negative emotions induced by other events, including drug deprivation.

DEFINITION OF TERMS

Comorbidity Individuals with an existing behavioral disorder are at higher risk of developing another disorder.

Consummatory extinction Following acquisition with a reward, animals are downshifted to no reward and their consummatory behavior is measured (*reward omission*). See also instrumental extinction.

Consummatory successive negative contrast Following acquisition with a large reward, animals are downshifted to a small reward and their consummatory behavior is compared to that of unshifted controls always exposed to the small reward (*reward devaluation*). See also instrumental successive negative contrast.

Drive induction An increase in motivation/emotion that can have positive reinforcing effects on contingent responses.

Drive reduction A decrease in motivation/emotion that can have positive reinforcing effects on contingent responses.

Drug specificity A postulate of ESM stating that the chosen drug reflects its ability to ameliorate distress.

Emotional self-medication Consummatory behavior motivated by hedonically aversive affective states and reinforced by a reduction in those states.

Instrumental extinction Same as consummatory extinction, except that an instrumental behavior is measured (*reward omission*).

Instrumental successive negative contrast Same as consummatory successive negative contrast, except that an instrumental behavior is measured (*reward devaluation*).

Partial reinforcement contrast effect Acquisition of a response under partial reinforcement attenuates the behavioral effects of reward downshift relative to acquisition under continuous reinforcement.

Partial reinforcement extinction effect Acquisition of a response under partial reinforcement results in slower extinction compared to acquisition under continuous reinforcement.

Physical self-medication Consummatory behavior motivated by therapeutic or symptomatic amelioration in organisms suffering from some pathology or disorder.

Psychopathology postulate A postulate of ESM stating that the onset of substance use relates to the relief of psychological distress produced by the substance consumed.

KEY FACTS ABOUT SELF-MEDICATION

- Humans consume drugs to alleviate acute or chronic distress states.
- Comorbidity studies show a positive correlation between SUDs and anxiety disorders.
- Animals self-medicate to treat physical diseases.
- PSM and ESM share behavioral components.
- ESM can underlie some forms of addictive behavior.
- ESM can be induced by reward loss.

SUMMARY POINTS

- SUDs are neurobehavioral conditions resulting from persistent dysregulation of neural circuits mediating reward, learning/memory, and stress.
- The ESM hypothesis states that the drug consumed depends on its ability to relieve a preexistent psychiatric disorder or reduce negative emotions.
- Clinical evidence suggests that stress, psychiatric condition, and drug withdrawal can sustain ESM.
- Animals can behaviorally treat and control physical disease (PSM) and emotional distress (ESM).
- Reward loss triggers a negative emotional reaction that shares commonalities with pain, fear, and anxiety.
- Reward loss supports ESM behavior.
- ESM provides a framework to understand addictions.

REFERENCES

Ahmed, S. H. (2012). The science of making drug-addicted animals. *Neuroscience, 211*, 107–125.

American Psychiatric Association. (2013). *DSM-V. Diagnostic and statistical manual of mental disorders* (5th ed.). Author, Washington, DC.

Anisman, H., & Waller, T. G. (1974). Effects of inescapable shock and shock-produced conflict on self-selection of alcohol in rats. *Pharmacology, Biochemistry and Behavior, 2*, 27–33.

Becker, H. C., Jarvis, M., Wagner, G., & Flaherty, C. F. (1984). Medial and lateral amygdala lesions differentially influence contrast with sucrose solutions. *Physiology and Behavior, 33*, 707–712.

Becker, H. C., Lopez, M. F., & Doremus-Fitzwater, T. L. (2011). Effects of stress on alcohol drinking: a review of animal studies. *Psychopharmacology (Berlin)*, *218*, 131–156.

Blume, A. W., Chmaling, K. B., & Marlatt, G. A. (2000). Revisiting the self-medication hypothesis from a behavioral perspective. *Cognitive and Behavioral Practice*, *7*, 379–384.

Briand, L. A., & Blendy, J. A. (2009). Molecular and genetic substrates linking stress and addiction. *Brain Research*, *1314*, 219–234.

Carrigan, M. H., & Randall, C. L. (2003). Self-medication in social phobia: a review of the alcohol literature. *Addictive Behavior*, *28*, 269–284.

Castaneda, R., Galanter, M., & Franco, H. (1989). Self-medication among addicts with primary psychiatric disorders. *Comprehensive Psychiatry*, *30*, 80–83.

Clayton, D. H., & Wolfe, N. D. (1993). The adaptive significance of self-medication. *Trends in Ecology and Evolution*, *8*, 60–63.

Conger, J. (1956). Reinforcement theory and the dynamics of alcoholism. *Quarterly Journal of Studies on Alcohol*, *17*, 296–305.

Cuenya, L., Sabariego, M., Donaire, R., Fernández-Teruel, A., Tobeña, A., Gómez, M. J., … Torres, C. (2012). The effect of partial reinforcement on instrumental successive negative contrast in inbred Roman high- (RHA-I) and low- (RLA-I) avoidance rats. *Physiology and Behavior*, *105*, 1112–1116.

Darke, S. (2012). Pathways to heroin dependence: time to re-appraise self-medication. *Addictions*, *198*, 659–667.

De Roode, J. C., Lefèvre, T., & Hunter, M. D. (2013). Self-medication in animals. *Science*, *340*, 150–151.

Duffing, T. M., Greiner, S. G., Mathias, C. W., & Dougherty, D. M. (2014). Stress, substance abuse, and addiction. *Current Topics in Behavioral Neurosciences*, *18*, 237–263.

Egli, M., Koob, G. F., & Edwards, S. (2012). Alcohol dependence as a chronic pain disorder. *Neuroscience and Biobehavioral Reviews*, *36*, 2179–2192.

Enman, N. M., Zhang, Y., & Unterwald, E. M. (2014). Connecting the pathology of posttraumatic stress and substance use disorders: monoamines and neuropeptides. *Pharmacology Biochemistry and Behavior*, *117*, 61–69.

Ewan, E. E., & Martin, T. J. (2013). Analgesics as reinforcers with chronic pain: evidence from operant studies. *Neuroscience Letters*, *557*, 60–64.

Ferster, C. B., & Skinner, B. F. (1957). *Schedules of reinforcement*. Englewood Cliffs, NJ: Prentice-Hall.

Flaherty, C. F. (1996). *Incentive relativity*. Cambridge, UK: Cambridge University Press.

Flaherty, C. F., Coppotelli, C., Hsu, D., & Otto, T. (1998). Excitotoxic lesions of the hippocampus disrupt runway but not consummatory contrast. *Behavioural Brain Research*, *93*, 1–9.

Fruth, B., Ikombe, N. B., Matshimba, G. K., Metzger, S., Muganza, D. M., Mundry, R., & Fowler, A. (2014). New evidence for self-medication in bonobos: *Manniophyton fulvum* leaf-and stremtrip-swallowing from LuiKotale, Salonga National Park. *American Journal of Primatology*, *76*, 146–158.

Genn, R. F., Tucci, S., Parikh, S., & File, S. E. (2004). Effects of nicotine and a cannabinoid receptor agonist on negative contrast: distinction between anxiety and disappointment? *Psychopharmacology*, *177*, 93–99.

Gómez, M. J., de la Torre, L., Callejas-Aguilera, J. E., Lerma-Cabrera, J. M., Rosas, J. M., Escarabajal, M. D., … Torres, C. (2008). The partial reinforcement extinction effect (PREE) in female Roman high- (RHA-I) and low-avoidance (RLA-I) rats. *Behavioural Brain Research*, *194*, 187–192.

Gómez, M. J., de la Torre, L., Callejas-Aguilera, J. E., Rosas, J. M., Escarabajal, M. D., Agüero, Á., … Torres, C. (2009). Differences in extinction of an appetitive instrumental response in female inbred Roman high- (RHA-I) and low- (RLA-I) avoidance rats. *Psicológica*, *30*, 181–188.

Gómez, M. J., Escarabajal, M. D., de la Torre, L., Tobeña, A., Fernández-Teruel, A., & Torres, C. (2009). Consummatory successive negative and anticipatory contrast effects in inbred Roman rats. *Physiology and Behavior*, *97*, 374–380.

Gutiérrez, T., Crystal, J., Zvonok, A. M., Makriyannis, A., & Hohmann, A. G. (2011). Self-medication of a cannabinoid CB2 agonist and an animal model of neuropathic pain. *Pain*, *152*, 1976–1987.

Hassanbeigi, A., Askari, J., Hassanbeigi, D., & Pourmovahed, Z. (2013). The relationship between stress and addiction. *Proceedings of the Society for Behavioral Sciences*, *84*, 1333–1340.

Huffman, M. A. (2010). Self-medication: passive prevention and active treatment. In M. D. Breed, & J. Moore (Eds.), *Encyclopedia of animal behavior test* (pp. 125–131). New York: Academic Press.

Kamenetzky, G. V., Cuenya, L., Pedrón, V. T., & Mustaca, A. E. (2009). Frustración y respuestas a contextos asociados al ethanol. *Revista Mexicana de Psicología*, *26*, 193–201.

Kamenetzky, G. V., Mustaca, A. E., & Papini, M. R. (2008). An analysis of the anxiolytic effects of ethanol on consummatory successive negative contrast. *Avances en Psicología Latinoamericana*, *26*, 135–144.

Khantzian, E. J. (1985). The self-medication hypothesis of addictive disorders: focus on heroin and cocaine dependence. *American Journal of Psychiatry*, *142*, 1259–1264.

Khantzian, E. J. (2013). Addiction as a self-regulation disorder and the role of self-medication. *Addiction*, *108*, 668–674.

Konopka, A., Pełka-Wysiecka, J., Grzywacz, A., & Samochowiec, J. (2013). Psychosocial characteristics of benzodiazepine addicts compared to not addicted benzodiazepine users. *Progress in Neuropsychopharmacology and Biological Psychiatry*, *40*, 229–235.

Koob, G. F. (2013). Negative reinforcement in drug addiction: the darkness within. *Current Opinion in Neurobiology*, *23*, 559–563.

Koob, G. F., Arenns, M. A., & LeMoal, M. (2014). *Drugs, addiction, and the brain*. New York: Elsevier.

Kumari, V., & Postma, P. (2005). Nicotine use in schizophrenia: the self-medication hypotheses. *Neuroscience and Biobehavioral Reviews*, *29*, 1021–1034.

Lefèvre, T., Chiang, A., Kelavkar, M., Li, H., Li, J., de Castillejo, C. L. F., … de Roode, J. C. (2012). Behavioural resistance against a protozoan parasite in the monarch butterfly. *Journal of Animal Ecology*, *81*, 70–79.

Levin, H., & Forgays, D. G. (1960). Sensory change as immediate and delayed reinforcement for maze learning. *Journal of Comparative and Physiological Psychology*, *53*, 194–196.

Li, X., Lu, S., & Miller, R. (2013). Self-medication and pleasure seeking as dichotomous motivations underlying behavioral disorders. *Journal of Business Research*, *66*, 1598–1604.

Lin, J. Y., Roman, C., & Reilly, S. (2009). Insular cortex and consummatory successive negative contrast in the rat. *Behavioral Neuroscience*, *123*, 810–814.

Lozano, G. A. (1998). Parasitic stress and self-medication in wild animals. *Advances in the Study of Behavior*, *27*, 291–317.

Mantsch, J. R., Vranjkovic, O., Twining, R. C., Gasser, P. J., McReynolds, J. R., & Blacktop, J. M. (2014). Neurobiological mechanisms that contribute to stress-related cocaine use. *Neuropharmacology*, *76*, 383–394.

Manzo, L., Donaire, R., Sabariego, M., Papini, M. R., & Torres, C. (2015). Anti-anxiety self-medication: oral consumption of chlordiazepoxide and ethanol after reward devaluation. *Behavioural Brain Research*, *278*, 90–97.

Manzo, L., Gómez, M. J., Callejas-Aguilera, J. E., Donaire, R., Sabariego, M., Fernández-Teruel, A., ... Torres, C. (2014). Relationship between ethanol preference and sensation/novelty seeking. *Physiology and Behavior*, *133*, 53–60.

Manzo, L., Gómez, M. J., Callejas-Aguilera, J. E., Fernández-Teruel, A., Papini, M. R., & Torres, C. (2012). Oral ethanol self-administration in inbred Roman high- and low-avoidance rats: gradual versus abrupt ethanol presentation. *Physiology and Behavior*, *108*, 1–5.

Manzo, L., Gómez, M. J., Callejas-Aguilera, J. E., Fernández-Teruel, A., Papini, M. R., & Torres, C. (2014). Anti-anxiety self-medication induced by incentive loss in rats. *Physiology and Behavior*, *123*, 86–92.

Manzo, L., Gómez, M. J., Callejas-Aguilera, J. E., Fernández-Teruel, A., Papini, M. R., & Torres, C. (2015). Partial reinforcement reduces vulnerability to anti-anxiety self-medication during appetitive extinction. *International Journal of Comparative Psychology*, *28*, 1–8.

Masi, S., Gustafsson, E., Jalme, M. S., Narat, V., Todd, A., Bomsel, M.-C., & Krief, S. (2012). Unusual feeding behavior in wild great apes, a window to understand origins of self-medication in humans: role of sociality and physiology of learning process. *Physiology and Behavior*, *105*, 337–349.

Menary, K. R., Kushner, M. G., Maurer, E., & Thuras, P. (2011). The prevalence and clinical implications of self-medication among individuals with anxiety disorders. *Journal of Anxiety Disorders*, *25*, 335–339.

Moreno, M., Cardona, D., Gómez, M. J., Sánchez-Santed, F., Tobeña, A., Fernández-Teruel, A., ... Flores, P. (2010). Impulsivity characterization in the Roman high- and low- avoidance rat strains: behavioral and neurochemical differences. *Neuropsychopharmacology*, *35*, 1198–1208.

Ortega, L. A., Uhelski, M., Fuchs, P. N., & Papini, M. R. (2011). Impairment of recovery from incentive downshift after lesions of the anterior cingulate cortex: emotional or cognitive deficits? *Behavioral Neuroscience*, *125*, 988–995.

Papini, M. R. (2009). Role of opioid receptors in incentive contrast. *International Journal of Comparative Psychology*, *22*, 170–187.

Papini, M. R., & Dudley, R. T. (1997). Consequences of surprising reward omissions. *Review of General Psychology*, *1*, 175–197.

Papini, M. R., Fuchs, P., & Torres, C. (2015). Behavioral neuroscience of psychological pain. *Neuroscience and Biobehavioral Reviews*, *48*, 53–69.

Papini, M. R., Wood, M. D., Daniel, A. M., & Norris, J. N. (2006). Reward loss as psychological pain. *International Journal of Psychology and Psychological Therapy*, *6*, 189–213.

Pellegrini, S., Muzio, R. N., Mustaca, A. E., & Papini, M. R. (2004). Successive negative contrast after partial reinforcement in the consummatory behavior of rats. *Learning and Motivation*, *35*, 303–321.

Robinson, J., Sareen, J., Cix, B. J., & Bolton, J. M. (2011). Role of self-medication in the development of comorbid anxiety and substance use disorders. *Archives of General Psychiatry*, *68*, 800–807.

Rosas, J. M., Callejas-Aguilera, J. E., Escarabajal, M. D., Gómez, M. J., de la Torre, L., Agüero, Á., ... Torres, C. (2007). Successive negative contrast effect in instrumental runway behaviour: a study with Roman high- (RHA) and Roman low- (RLA) avoidance rats. *Behavioural Brain Research*, *185*, 1–8.

Ruglass, L. M., Lopez-Castro, T., Cheref, S., Papini, S., Hien, D. A. (2014). At the crossroads: The intersection of substance use disorders, anxiety disorders, and posttraumatic stress disorder. *Current Psychiatry Reports*, *16*, 505–513.

Singer, M. S., Mace, K. C., & Bernays, E. A. (2009). Self-medication as adaptive plasticity: increased ingestion of plant toxins by parasitized caterpillars. *PLoS One*, *4*, 1–8.

Spanagel, R., Noori, H. R., & Heilig, M. (2014). Stress and alcohol interactions: animal studies and clinical significance. *Trends in Neuroscience*, *37*, 219–227.

Tomlinson, K. L., & Brown, S. A. (2012). Self-medication or social learning? A comparison of models to predict early adolescent drinking. *Addictive Behavior*, *37*, 179–186.

Torres, C., Cándido, A., Escarabajal, M. D., de la Torre, L., Maldonado, A., Tobeña, A., & Fernández-Teruel, A. (2005). Successive negative contrast in one-way avoidance learning in female Roman rats. *Physiology and Behavior*, *85*, 377–382.

Torres, C., & Sabariego, M. (2014). Incentive relativity: gene-environment interactions. *International Journal of Comparative Psychology*, *27*, 446–458.

Vengeliene, V., Siegmund, S., Singer, M. V., Sinclair, J. D., Li, T. K., & Spanagel, R. (2003). A comparative study on alcohol-preferring rat lines: effects of deprivation and stress phases on voluntary alcohol intake. *Alcoholism, Clinical and Experimental Research*, *27*, 1048–1054.

Villalba, J. J., Miller, J., Ungar, E. D., Landau, S. Y., & Glendinning, J. (2014). Ruminant self-medication against gastrointestinal nematodes: evidence, mechanism and origins. *Parasite*, *21*, 1–10.

Chapter 8

The Neurobiology of Comorbid Drug Abuse in Schizophrenia and Psychotic Disorders

A. George Awad
Department of Psychiatry and the Institute of Medical Science, University of Toronto, Toronto, Canada

Abbreviations

GABA γ-aminobutyric acid
CATIE Clinical antipsychotic trial of intervention effectiveness
CAMH Centre for addiction and mental health
CB1 Cannabinoid receptor type 1
CB2 Cannabinoid receptor type 2
THC Δ^9-tetrahydrocannabinol
GROP Genetic risk and outcome in psychosis study
COMT Catechol-O-methyltransferase
SPECT Single-photon emission computerized tomography

INTRODUCTION

Schizophrenia is a serious and disabling psychiatric disorder that is relatively common. It is estimated to affect about 1% of the population. Generally, it pursues a chronic course, with acute psychotic exacerbations that frequently may require hospitalization. Depending on the stage of the illness, the clinical picture includes a broad range of symptoms: positive symptoms such as delusions and hallucinations and a range of disabling deficit and negative symptoms such as social withdrawal, lack of spontaneity, and poverty of speech. It also includes a range of neurocognitive deficits, as well as affective changes that may include blunted affect, depressive symptoms, and also agitation. As a result, the majority of persons with schizophrenia suffer from poor quality of life and an impaired ability to function, particularly in the vocational and social domains.

Despite extensive research, the etiology of schizophrenia remains mostly unclear. It is recognized that schizophrenia and other psychotic disorders are likely to be heterogeneous in nature, with variable etiological and treatment response patterns. Apart from schizophrenia, other psychotic conditions can arise in the context of medical, metabolic, and endocrine dysfunctions, as well as being induced by drugs of abuse. Alterations in dopamine functioning are considered to play a significant role in the neurobiology of schizophrenia. Other neurotransmitters, such as glutamate and γ-aminobutyric acid (GABA) have also been explored, though their role is still not clear.

Although the "dopamine hypothesis" continues to be the most dominant neurobiological hypothesis, other etiological alternatives, such as early neurodevelopmental factors, have also been proposed. In support of the dopamine hypothesis, symptomatic improvement in schizophrenia can be achieved through the use of medications with dopamine-blocking properties. Indeed, the development of antipsychotic medications since the middle of the twentieth century has been based on their ability to block dopamine.

EPIDEMIOLOGICAL DATA

Comorbidity between schizophrenia and drug use disorders has been consistently reported as being high, though estimates among various surveys from different countries may vary as a result of methodological differences and reliance on self-reports, which frequently lack reliability and validity. Similarly, issues related to the accuracy of diagnosis for both schizophrenia and drug use disorders, as well as the evolving changes in diagnostic and classificatory criteria, may also account for some of the variance.

The Epidemiological US Catchment Area Survey, which was one of the first large-scale community-based epidemiological studies, undertaken between the years 1980 and 1984 and including over 20,000 respondents ages 18 years and older, was conducted in three US catchment areas (Regier et al., 1990). According to data from such survey, it was estimated that 47% of persons with schizophrenia, compared with 13.5% of the general population, have or had evidence of drug abuse. Persons with schizophrenia have been estimated to be 4.6 times more likely to have drug abuse problems than persons without psychiatric illness (Regier et al., 1990). This means that almost half of the persons suffering from schizophrenia have a lifetime history of substance abuse disorder.

In the United Kingdom, the National Psychiatric Morbidity Survey (Jenkins et al., 1997) reported a lifetime substance use prevalence rate of 7% for those with schizophrenia and other psychotic disorders, which is rather low compared to US figures. Another London-based survey reported a lifetime prevalence rate of 16% for substance use by patients with schizophrenia (Duke, Pantelis, McPhilips, & Barnes, 2001). A 2006 study conducted in West London, England, which included first-episode schizophrenia, reported 27% with a history of problems with alcohol use. Thirty-five percent of the sample reported current substance use (not including alcohol) and 68% reported lifetime substance use (cannabis and psychostimulants were most commonly used) (Barnes, Mutsata, Hutton, Watt, & Joyce, 2006).

In the US National Institute of Mental Health study Clinical Antipsychotic Trial of Intervention Effectiveness, approximately 60% of a sample of 1460 subjects with a diagnosis of schizophrenia were found to use substances, including 37% with current evidence of substance use disorders (Lieberman et al., 2005).

The high comorbidity of schizophrenia and drug use disorders has significant clinical, economic, and resource use implications. Persons with schizophrenia and comorbid drug abuse tend to have a more complicated clinical course with frequent relapses that may require long hospitalizations. Outcomes are generally poor, including a decline in vocational and psychosocial functioning. Comorbid drug use subjects tend to use more medical and psychiatric resources such as crisis interventions and frequent hospitalizations, as well as indefinite financial and housing support. A study estimated that mental and substance abuse disorders cost, in the province of Ontario, Canada, for the year 2000, about $33.9 billion, with 85% of the total cost related to loss in productivity; $5.1 billion representing direct costs, of which $2.1 billion was for mental disorders; and $3 billion for drug abuse disorders, of which $2 billion was spent on law enforcement (CAMH Media Release 2006). Such a high frequency of comorbidity of drug abuse and schizophrenia clearly exacts a high cost, notwithstanding individual and family suffering.

An important question that arises, then, and continues to be not fully understood, is how to explain the frequent drug comorbidity with schizophrenia. That, by itself, raises a number of other questions: Does drug abuse contribute to the development of schizophrenia? Conversely, does the schizophrenia disorder enhance the vulnerability and predisposition for drug abuse? Alternatively, both schizophrenia and drug abuse disorders may share a common neurobiological origin and relate to each other in a bidirectional mode, which seems to be the likely model, based on the available information so far.

CORRELATES OF COMORBID DRUG ABUSE IN SCHIZOPHRENIA

Demographics, Personality, and Clinical Characteristics

Persons with schizophrenia and comorbid drug abuse are more likely to be males, younger, and less well-educated and to have a positive family history of drug abuse (Barnes et al., 2006; Cantwell, 2003). A number of studies have identified earlier onset of schizophrenia in patients with drug abuse (Mauri et al., 2006; Kovasznay, Fleisher, & Tenenberg-Karent, 1997). Attempts to establish a temporal order for which disorder came first proved inconsistent. One study found that 60% of persons with schizophrenia who were admitted to hospital for the first time had abused drugs before (Silver & Abboud, 1994). A retrospective study by Hambrect and Hafner (1996), identified about one-third of their sample as persons with schizophrenia who had used drugs before the onset of psychosis, another one-third had drug-use problems and onset of psychosis developing at the same time, while the last one-third had the onset of psychosis predating drug abuse by over 1 year. A more recent study (Degenhardt, Hall, & Lynskey, 2003) found no link between the onset of drug abuse and later development of a psychiatric disorder, except, possibly, for early use of cannabis in adolescents, which will be discussed in a later section about cannabis use in schizophrenia.

Similarly, data from a number of studies attempting to document the relationship of symptoms severity and levels of drug abuse proved to be inconsistent (Brunette, Mueser, Xie, & Drake, 1997). Linking drug abuse to specific clinical symptoms has also proved contradictory. While Pencer and Addington (2003) identified a link between drug abuse and positive symptoms in schizophrenia, a 2006 meta-analysis from eight cross-sectional studies confirmed the conclusions of the Pencer and Addington study and, in addition, reported a link between drug abuse and a lower negative symptoms score (Talamo et al., 2006). On the other hand, Scheller-Gilkey, Moynes, Cooper, Kant, and Miller (2004) found no differences in either positive or negative symptoms in persons with or without a drug abuse history.

Overall, then, a review of the extensive literature about demographics, personality, and clinical characteristics has yielded inconsistent and, at times, contradictory results, which is largely due to differences in methodology, study populations, and the rigor of the diagnostic approaches. In general, then, though none of the demographic, personality, or clinical issues have been consistently proven individually as contributing to the development of comorbid drug abuse, collectively they can be construed as just risk factors among the many other risk factors. The only exception is the evidence of the link between early cannabis use in adolescents and the subsequent development of schizophrenia, which seems to be rather stronger, as reviewed in the following section.

Early Cannabis Use and Development of Schizophrenia

Cannabis is one of the most frequently used and abused drugs among the psychiatric population. It owes its pharmacological and clinical effects to its active ingredient, Δ^9-tetrahydrocannabinol, which mimics the action of natural cannabinoids. It binds to cannabinoid receptors CB1 and CB2, which are found in a number of brain regions, including the frontal cortex, striatum, and hippocampus. CB1 has been reported to interact with the dopamine D2 receptors, which coexist in similar brain regions. Such interaction has been related to motor activity, endocrine regulation, cognition, and mood (D'Souza et al., 2005). The molecular biological mechanism of such dopaminergic/cannabinoid interactions is not yet clearly elucidated.

Although, generally, there is little evidence to incriminate drugs of abuse in the later development of schizophrenia or other major psychiatric disorders, the evidence for the link to cannabis use seems to be rather stronger. A number of large cohort studies not only documented significant odds ratios for use of cannabis and later development of schizophrenia (Moore et al., 2007), but also confirmed a dose-dependent effect (Zammit et al., 2002). A study about whether cannabis use predicts the first incidence of mood and anxiety disorder confirmed a strong association, which remained significant after adjustment for strong confounders (Van Laar, van Dorsselaer, Monshouwer, & de Graff, 2007).

More recently, the Genetic Risk and Outcome in Psychosis Study provided evidence that familial liability for psychosis is expressed as differential sensitivity to cannabis (GROP, 2011).

Considering the widespread use of cannabis among the general population and, particularly so among the psychiatric population, the disproportionately lower rate of schizophrenia, one can only infer that the effect of cannabis can be manifested only in individuals who possibly have other vulnerabilities for psychosis and schizophrenia (Caton, Drake, & Hasin, 2005).

Approaching the issue from a different perspective, Caspi et al. (2005) found that those adolescents with the Val158Met polymorphism of the catechol-O-methyltransferase (*COMT* gene) are more vulnerable for the effects of cannabis.

In summary, then, though the evidence of a link between early use of cannabis and later development of psychosis seems to be relatively strong, the issue is still open and requires more confirmatory research.

Genetics and Family Studies

Both schizophrenia and vulnerability for drug abuse disorders are generally believed to include a genetic predisposition. It is not clear whether genetic vulnerability in one disorder can also contribute to increased risk for the other disorder or whether such common genetic factors increase the risk for comorbidity. The answer to this question can be ascertained only by demonstrating a higher frequency than expected of psychiatric problems in relatives of those with drug abuse disorders, and vice versa. A number of studies have reported that psychiatric patients with comorbid drug abuse are likely to have relatives with drug abuse problems in their family background, compared to families of patients without drug abuse problems (Smith, Bark, Wolf, Mamah, & Csernansky, 2008; Tswang et al., 1998). More recently, a few studies reported increased rates of affective disorders in the families of persons with schizophrenia and drug abuse (Tswang, Bar, Harley, & Lyons, 2001). Another question that has not been answered adequately is how much contributions are accounted for by genetic factors; in particular, many genetic studies have not taken into account the possible environment/gene interactions. In the Tswang et al., 1998 study, genetic effects were estimated to contribute only 22% of the variance for cannabis use, which is a rather low contribution, indicating the possible role of many other issues, including environmental factors.

The Self-Medication Hypothesis

In the mid-1980s, Khantzian (1985) introduced the self-medication hypothesis as an explanation for the high drug comorbidity rates in schizophrenia. According to such a hypothesis, persons with schizophrenia take to drug abuse as a direct consequence of dealing with aspects of their illness experience or to alleviate some of the side effects of antipsychotic medications. In a subsequent reformulation of his hypothesis, Khantzian et al. abandoned the notion that the choice of the drug of abuse is specific to certain symptoms (Khantzian, 1997). The self-medication hypothesis quickly proved to be influential, as a result of being clinically intuitive. In support of the self-medication hypothesis, we reported a significant association between neuroleptic-induced dysphoria and comorbid drug abuse, with an odds ratio of 4.08, $\chi^2 = 21.8$, $p > 0.0001$ (Voruganti, Heselgrave, & Awad, 1997). Although it is recognized that association is not causation, the significant odds ratio attracted our attention and allowed us to conclude that neuroleptic dysphoria is likely to be the missing link between schizophrenia and comorbid drug abuse. Other studies followed, confirming our data and contributing to the prominence of such a hypothesis (Dalak, Healy, & Meador-Woodruff, 1998). On the other hand, a rival, but less known, version of the self-medication hypothesis, advanced over a number of years by Duncan (1974, 1975), Duncan and Gold (1983) differed from the Khantzian psychoanalytical model by providing another behaviorally based formulation. The Duncanian model, extensively reviewed by Achalu (2002), makes a clear distinction between drug use and abuse and ascertains that most persons who take to illicit substances do not meet the criteria for substance abuse, let alone for drug dependence. The Duncanian model, then, has been concerned with describing why a minority of those who take illicit drugs nonmedically lose control over their use and become seriously addicted.

Additional criticism of the Khantzian model came from other studies, concluding that the Khantzian model is incapable of explaining the complexity and the full picture of comorbid drug abuse. A number of studies failed to correlate patients' self-reports with alleviation of symptoms (Addington & Duchak, 1997). Additionally, a number of patients with schizophrenia experienced their drug abuse behavior before any early development of psychotic symptoms (Hambrect & Hafner, 1996; Sevy, Robinson, & Solloways, 2001). As a result, the self-medication hypothesis, though it has not completely disappeared, has become less influential owing to a lack of direct and empirical data to confirm its assumptions (Awad & Voruganti, 2005). With the advent of neuroimaging and more sophisticated neurobiological techniques, our findings that correlated low striatal dopamine with neuroleptic dysphoria (Voruganti, Slomka, Zabel, Mattar, & Awad, 2001; Voruganti, Slomka, Zabel, Costa, et al., 2001) and the reports implicating low striatal dopamine function with comorbid drug abuse in schizophrenia have led us to revisit the self-medication hypothesis, as reviewed in the next section.

NEUROBIOLOGY OF COMORBID DRUG ABUSE

The elucidation of the neurobiology of comorbidity in drug abuse in schizophrenia has been in the forefront of basic and clinical research since 1995 and, so far, is the most comprehensive, compared to neurobiological research in other psychiatric disorders such as anxiety, attention deficit, and mood disorders. It is recognized that both schizophrenia and comorbid drug abuse share common neuronal circuitry that involves dopamine and possibly other neurotransmitters that mediate reward and reinforcement behavior (Chambers, Krystal, & Self, 2001; Wise, 2009; Volkow, Wang, Foroler, Tomasi, & Telans, 2011). Researchers examining hedonic homeostatic dysregulation in drug abuse reported that dopamine activity in the nucleus accumbens is implicated in the mechanism of reinforcement for almost all drugs of abuse (Koob & LeMoal, 1997). In an in vivo single-photon emission computerized tomography (SPECT) case study, we documented that cannabis use resulted in an increase in the synaptic dopamine activity in the striatum (Voruganti, Slomka, Zabel, Mattar, &

Awad, 2001). Hyperdopaminergic states, as is the case proposed in schizophrenia, have been reported to disrupt the adaptive mechanism of the dopamine neurons in response to novel rewards. A hyperdopaminergic state resulting from dysfunctional cortical/hippocampal input to the nucleus accumbens has been implicated in the continuous neural representation of reward as a novel stimulus, a mechanism central to initiation and maintenance of addictive behavior (Brady & Sinha, 2005; Chambers et al., 2001). All in all, there is adequate and converging evidence to implicate dopamine in the development of vulnerability to comorbid drug abuse (Volkow, Fowler, Wang, Baler, & Telang, 2009). Moreover, a number of studies have proposed that the vulnerability of patients with schizophrenia to drug abuse may represent a primary symptom of the disease process itself (Chambers et al., 2001). Such a proposal has raised the question of whether it is time to consider comorbid drug abuse as a new indication for antipsychotic drug development (Awad, 2012). A more recent confirmatory study in support of the role of striatal dopamine functioning in the genesis of comorbid abuse in schizophrenia suggested that patients with comorbid schizophrenia and drug abuse show significant generalized blunting of amphetamine-induced dopamine release in all striatal subregions (Thompson et al., 2013). Despite this blunting, dopamine release was significantly associated with the expected amphetamine-induced changes in positive symptoms, as previously observed in schizophrenia alone. These findings suggest, then, that schizophrenia comorbid with drug abuse is associated with abnormal postsynaptic dopamine D2 functioning. Such confirmation of the role of low striatal dopamine has significant clinical implications in terms of the choice of the antipsychotic for the treatment of schizophrenia with comorbid drug abuse (Awad, 2012). The use of strong dopamine D2 blockers has to be avoided, in order not to further compromise striatal dopamine functioning, which can further enhance the vulnerability for comorbid drug abuse. Indeed, a number of reports have raised the question of whether antipsychotic treatment, particularly the old antipsychotics, such as haloperidol, can contribute to drug addiction in schizophrenia (Samaha, 2013; Green, 2005).

Although dopamine has received the more extensive research, other neurotransmitters have also been implicated, though their role continues to be unclear (Thoma & Daum, 2013). Reduced activity in glutamatergic and GABAergic systems has also been proposed (Katon, Sperlagh, & Magloczky, 2000). There is evidence of ethanol-induced interaction between GABAergic and dopaminergic systems in the ventral tegmental area (Pierce & Kumaresan, 2006). Additionally, cannabinoid receptors interact with GABAergic transmission in both the cortical and the subcortical area (Katon et al., 2000).

Another emerging area of new neurobiological exploration has dealt with the role of impulsivity followed by compulsivity as a central theme in the development of comorbid addictive states (Everitt et al., 2008). Such proposition was advanced earlier by Jentsch and Taylor (1999), in their study exploring impulsivity resulting from frontal striatal dysfunction and its role in the development of vulnerability for drug abuse, as well as its implications for the control of behavior by reward-relating stimuli. Such observations have been supported by data from experimental studies on rats demonstrating enhanced receptor impulsivity and cocaine reinforcement in rats, following reductions in the dopamine D2 binding ratio (Dalley, 2007). Another 2010 study about impulsivity-related brain volume deficits in schizophrenia and comorbid addiction reported that comorbid patients compared to nonaddicted persons with schizophrenia showed some significant volume decreases in anterior cingulate, frontal polar, and superior parietal regions (Schiffer et al., 2010). Additionally, an increased nonplanning compulsivity was reported and has been shown to negatively relate to gray matter volumes in the same regions, except for parietal ones. The conclusions of that study confirm the link between severe gray matter volume and functional executive deficits in schizophrenia. The relationship between nonplanning impulsivity and anterior cingulate and frontal polar gray matter volumes points to a specific structure/function relationship that seems to be impaired in schizophrenia with drug comorbidity. Obviously, impulsivity is a neurobiological complex behavior pattern that probably relates to multiple neurotransmitter systems, including serotonin receptors, which are strongly implicated in obsessive–compulsive disorder. Genetic variations may partially underlie complex personality and physiological traits such as impulsivity, risk taking, and stress responsiveness, all implicated in contributing a substantial portion of vulnerability to addictive disease (Kreek, Nielsen, Butelman, & LaForge, 2005). Furthermore, personality and physiological traits, themselves, may differentially affect the various stages of addictions, defined chronologically as initiation of drug use, regular use, addiction/dependence, and potential relapse. In exploring neuromechanisms underlying the vulnerability to develop compulsive drug-seeking habits and addictions, Everitt et al. (2008) hypothesized that drug addiction can be considered as the end point of a series of transactions, from the initial voluntary drug use through to the loss of control over this behavior, such that it becomes habitual and, ultimately, compulsive. As suggested by Everitt et al. (2008), there is evidence that the switch from controlled to compulsive drug-seeking behavior represents a transition at the neural level, from prefrontal cortical to striatal control over drug-seeking and drug-taking behavior, as well as a progression from ventral to more dorsal domains of the striatum, mediated by its serially interconnected dopaminergic circuitry. Furthermore, the same report (Everitt et al., 2008) provides summary evidence showing that impulsivity and spontaneously occurring behavioral tendencies in outbred rats, which are associated with low dopamine D2/3 receptors in the nucleus accumbens, predict both the propensity to enhance cocaine intake and the switch to compulsive drug seeking.

Revisiting the Self-Medication Hypothesis

Revisiting the self-medication hypothesis, in light of the reports implicating the role of low striatal dopamine functioning in comorbid substance abuse, has led us to rethink our early reports about the strong association of neuroleptic dysphoria and comorbid drug abuse. Our early experimental dopamine depletion SPECT study in drug-free patients with schizophrenia demonstrated an inverse relationship between the dopamine D2 binding ratio in the striatum and the development of neuroleptic dysphoria (Voruganti, Slomka, Zabel, Costa, et al., 2001). Observing the patients for the 48 h following dopamine depletion allowed us to identify the cascade of subjective and behavioral changes that followed, as well as their temporal relationship. The altered subjective state was reported as the earliest change experienced by patients, to be followed by affective changes, motor extrapyramidal symptoms, and, finally, cognitive alterations (Voruganti & Awad, 2006). Our original data received confirmation from a number of neuroimaging studies that followed (De Haan, Lavalaye, & Lenszend, 2000; Mizrahi, Mamo, & Pausion, 2009).

With the converging new neurobiological findings implicating low striatal dopamine functioning in the genesis of neuroleptic dysphoria, as well as in the development of comorbid drug abuse (Thompson et al., 2013), we are now able to connect the dots by proposing a neurobiological link between neuroleptic dysphoria and drug abuse, which can be construed as the first likely neurobiological confirmation of the self-medication hypothesis (Awad & Voruganti, 2015). Apart from its theoretical significance, a number of important clinical implications follow. As we propose that the person with schizophrenia who develops neuroleptic dysphoria is likely to be one and the same person who develops vulnerability for drug abuse, it is important to avoid the use of potent dopamine D2-blocking antipsychotics in order not to further impair striatal dopamine functioning (Awad & Voruganti, 2015; Samaha, 2013). Chronic dopamine blockade can lead to postsynaptic upregulation, which, in turn, enhances the reinforcement properties of drugs of abuse. Such information can explain the reported benefits from treatments by clozapine, agonist/antagonists such as aripiprazole, and, possibly, other new atypical antipsychotics (Awad, 2012). Meanwhile, our proposed hypothesis of the neurobiological link between neuroleptic dysphoria and vulnerability for drug abuse can provide an explanation of why not every subject with schizophrenia using drugs ends up being addicted and losing control over their drug use, as initially queried by the Duncan et al. version of the self-medication hypothesis (Duncan, 1974). In essence, the dynamic interaction between the striatal dopamine state and the pharmacological dopamine-blocking properties of the particular antipsychotic determines the outcome and whether the person experiences a dysphoric reaction as well as developing a vulnerability to an addictive state. Since our studies demonstrated that the earliest behavioral change following reduction of striatal dopamine functioning is subjective in nature and appears shortly after the ingestion of the medication, such early subjective dysphoric responses can serve as an early clinical marker for the patient that is likely to develop drug abuse vulnerability (Voruganti & Awad, 2006).

APPLICATIONS TO OTHER ADDICTIONS AND SUBSTANCE MISUSE

The new insights into the neurobiology of comorbid addictive states in schizophrenia have implicated low striatal dopamine development of comorbid addictive behavior. In essence, the group of persons with schizophrenia who develop vulnerability to comorbid drug abuse is neurobiologically different. These findings can explain why not everyone that uses drugs nonmedically ends up getting addicted and losing control of their drug use. These neurobiological insights run in contrast to the recently introduced changes in addictive disorders in the DSM5, which abolish the distinction between use and abuse in favor of symptom-based diagnoses ranging from mild to severe. Such a new classificatory approach not only ignores recent neurobiological findings, but also brings serious implications. The vast majority who occasionally use drugs for recreational purposes would be eligible for a diagnosis of "mild addictive disorder," which may lead to unnecessary treatments or, worse, more legal problems as offenders.

DEFINITION OF TERMS

Clinical symptoms Delusions, hallucinations, disordered thinking, cognitive, and functional deficits.
Clozapine An old antipsychotic with a mixed neurotransmitter profile, less potent as dopamine blocker, used in treatment resistance.
DSM5 Diagnostic and Statistical Manual of Mental Disorders, fifth edition.
Haloperidol Potent dopamine D2 blocker, antipsychotic.
Negative symptoms Social withdrawal, lack of spontaneity, poverty of speech.
Neuroleptic dysphoria Medication-induced alteration in subjective and affective state.
Schizophrenia A chronic psychotic disorder with acute exacerbations.
Self-medication hypothesis Use of a drug to cope with the illness and side effects of medications.
Striatum Subcortical part of the forebrain and a component of the basal ganglia, receives input from cerebral cortex.

KEY FACTS OF THE NEUROBIOLOGY OF COMORBID DRUG ABUSE IN SCHIZOPHRENIA

- Schizophrenia and comorbid drug abuse share neuronal circuitry.
- Both schizophrenia and comorbid drug abuse involve alterations in striatal dopamine functioning.
- Low striatal dopamine functioning is noted in persons with comorbid drug abuse compared to schizophrenia without comorbid drug abuse.
- Low striatal dopamine functioning is also implicated in the development of neuroleptic-induced dysphoria.
- Frequent associations of comorbid drug abuse and neuroleptic dysphoria have been reported.
- Low striatal dopamine functioning in both neuroleptic dysphoria and comorbid drug abuse provides a neurobiological link between these conditions.
- Low striatal dopamine functioning in schizophrenia clarifies why not every person using drugs ends up losing control and becoming addicted.
- Low striatal dopamine in schizophrenia comorbid with drug abuse can explain why patients treated with potent dopamine-blocking antipsychotics, such as haloperidol, continue to abuse drugs.
- In persons with schizophrenia and comorbid drug abuse who require an antipsychotic, the choice has to be one of the new less potent dopamine-blocking medications or medications with fast dissociative pharmacological properties on the dopamine D2 receptor.
- Possible link between neuroleptic dysphoria and vulnerability for development of addictive behavior may serve as an early clinical marker, since dysphoria develops shortly after ingestion of the antipsychotic medication.

SUMMARY POINTS

- High prevalence of comorbid drug abuse in psychiatric disorders has been consistently identified by several large national surveys.
- To explain such a high frequency of drug comorbidity, several factors have been explored including demographics, clinical characteristics, genetics, and family factors.
- Several hypotheses have been advanced to explain the high drug comorbidity; prominent among them has been the self-medication hypothesis in both its Khantzian and its Duncanian versions.
- Support of the self-medication hypothesis was provided by data from clinical studies reporting a strong association between neuroleptic-induced dysphoria and comorbid drug abuse in schizophrenia.
- New insights into the neurobiology of addictive behavior strongly implicate dopamine and probably other lesser explored neurotransmitters as mediating reward and reinforcement behavior.
- Recent evidence points to significant generalized blunting of amphetamine-induced dopamine release in all striatal subregions in patients with schizophrenia and comorbid drug abuse compared to subjects with schizophrenia and no comorbid drug abuse.
- The role of impulsivity and its contribution to risk taking, drug seeking, and the development of addictive behavior has been recently supported by a number of experimental and clinical studies.
- There are significant implications, both etiological and clinical, for the reported low striatal dopamine functioning.

REFERENCES

Achalu, E. D. (2002). The self-medication hypothesis: a review of the two major theories and the research evidence. *SMH Recent Developments on the Self-Medication Hypothesis, 1,* 1–17.

Addington, J., & Duchak, V. (1997). Reasons for substance use in schizophrenia. *Acta Psychiatrica Scandinavica, 96,* 329–333.

Awad, A. G. (2012). Is it time to consider comorbid substance abuse as a new indication for antipsychotic drug development? *Journal of Psychopharmacology, 26,* 953–957.

Awad, A. G., & Voruganti, L. N. P. (2005). Neuroleptic dysphoria, comorbid drug abuse in schizophrenia and the emerging science of subjective tolerability towards a new synthesis. *Journal of Dual Diagnosis, 1,* 83–93.

Awad, A. G., & Voruganti, L. N. P. (2015). Revisiting the 'self-medication' hypothesis in light of the new data linking low striatal dopamine to comorbid addictive behavior. *Therapeutic Advances in Psychopharmocology, 5,* 172–178.

Barnes, T. R., Mutsata, S. H., Hutton, S. B., Watt, H. C., & Joyce, E. M. (2006). Comorbid substance use and age at onset of schizophrenia. *British Journal of Psychiatry, 188,* 237–242.

Brady, K. T., & Sinha, R. (2005). Co-occurring mental and substance use disorder: the neurobiological effects of chronic stress. *The American Journal of Psychiatry, 162,* 1483–1493.

Brunette, M. F., Mueser, K. T., Xie, H., & Drake, R. (1997). Relationship between symptoms of schizophrenia and substance abuse. *The Journal of Nervous and Mental Disease, 185,* 13–20.

Cantwell, R. (2003). Substance use and schizophrenia: effects on symptoms, social functioning and service use. *The British Journal of Psychiatry, 182,* 324–329.

Caspi, A., Moffitt, T. E., Cannon, M., McClay, J., Murray, R., Harrington, H., ... Craig, I. W. (2005). Moderation of the effect of adolescent-onset cannabis use on adult psychosis by a functional polymorphism in the catechol-o-methyltransferase gene: longitudinal evidence of a gene x environment interaction. *Biological Psychiatry, 57,* 1117–1127.

Caton, C. L., Drake, R. E., & Hasin, D. S. (2005). Differences between early-phase primary psychotic disorders with concurrent substance use and substance-induced psychosis. *Archives of General Psychiatry, 62,* 137–145.

Centre for Addiction and Mental Health (CAMH) Media Release. (2006). *The economic costs of mental disorders, alcohol, tobacco and illicit drug abuse in Ontario, 2000.* Released November 14, 2006, Toronto.

Chambers, R. A., Krystal, J. H., & Self, D. W. (2001). A neurobiological basis for substance abuse comorbidity in schizophrenia. *Biological Psychiatry, 50,* 71–83.

Dalak, G. W., Healy, D. J., & Meador-Woodruff, M. (1998). Nicotine dependency in schizophrenia: clinical phenomena and laboratory findings. *The American Journal of Psychiatry, 155,* 1490–1501.

Dalley, J. W. (2007). Nucleus accumbens D2/3 receptors predict trait impulsivity and cocaine reinforcement. *Science, 315,* 1267–1270.

De Haan, L., Lavalaye, J., & Lenszend, D. (2000). Subjective experience and striatal dopamine D2 receptor occupancy in patients with schizophrenia treated by olanzapine and Risperidone. *The American Journal of Psychiatry, 157,* 1019–1020.

Degenhardt, L., Hall, W., & Lynskey, M. (2003). Testing hypothesis about the relationship between cannabis use and psychosis. *Drug and Alcohol Dependence, 71,* 37–48.

D'Souza, D. C., Abi-Saab, W. M., Madonick, S., Forselius-Bielin, K., Doersch, A., & Braley, G. (2005). Delta-9-tetrahydro-cannabinol effects in schizophrenia: implications for cognition, psychosis and addiction. *Biological Psychiatry, 57,* 594–608.

Duke, P. J., Pantelis, C., McPhilips, M. A., & Barnes, T. R. E. (2001). Comorbid non-alcohol substance misuse among people with schizophrenia. *The British Journal of Psychiatry, 179,* 509–513.

Duncan, D. F. (1974). Reinforcement of drug abuse: implications for prevention. *Clinical Toxicology Bulletin, 4,* 69–75.

Duncan, D. F. (1975). The acquisition, maintenance and treatment of polydrug dependence: a public health model. *Journal of Psychoactive Drugs, 7,* 201–213.

Duncan, D. F., & Gold, R. S. (1983). Cultivating drug use: a strategy for the 80s. *Bulletin of the Society Psychologist in Addictive Behaviors, 2,* 143–147.

Everitt, B. J., Belin, D., Economidou, D., Pelloux, X., Dalley, J., & Robins, T. W. (2008). Neural mechanisms underlying the vulnerability to develop compulsive drug-seeking habits and addictions. *Philosophical Transactions of the Royal Society B, 363,* 3125–3135.

Genetic Risk and Outcomes in Psychosis (GROP) Investigators. (2011). Evidence that familial liability for psychosis is expressed as differential sensitivity to cannabis. *Archives of General Psychiatry, 68,* 138–147.

Green, A. I. (2005). Schizophrenia and comorbid substance use disorder: effects of antipsychotics. *Journal of Clinical Psychiatry, 66*(Suppl. 6), 21–26.

Hambrect, M., & Hafner, H. (1996). Substance abuse and the cost of schizophrenia. *Biological Psychiatry, 40,* 1155–1163.

Jenkins, R., Bebbington, P., Brugha, T., Farrell, M., Gill, B., Lewis, G., ... Pettigrew, M. (1997). The national psychiatric morbidity surveys of Great Britain—initial findings from the household survey. *Psychological Medicine, 27*, 775–789.

Jentsch, J. D., & Taylor, J. R. (1999). Impulsivity resulting from franlostriatal dysfunction in drug abuse: implications for the control of behaviour by reward-related stimuli. *Psychopharmacology (Berlin), 146*, 373–390.

Katon, I., Sperlagh, B., & Magloczky, Z. (2000). GABAergic interneurons are the targets of cannabinoid actions as the hippocamus. *Neuroscience, 100*, 797–804.

Khantzian, E. J. (1985). The self-medication hypothesis of addictive disorders: focus on heroin and cocain dependance. *The American Journal of Psychiatry, 142*, 1259–1264.

Khantzian, E. J. (1997). The self-medication hypothesis of substance use disorders: a reconsideration and recent applications. *Harvard Review of Psychiatry, 4*, 231–244.

Koob, G. F., & LeMoal, M. (1997). Drug abuse hedonic homeostatic dysregulation. *Science, 278*, 52–58.

Kovasznay, B., Fleisher, J., & Tenenberg-Karent, M. (1997). Substance use disorder and the early course of illness in schizophrenia and affective psychosis. *Schizophrenia Bulletin, 23*, 195–201.

Kreek, M. J., Nielsen, D. A., Butelman, E. R., & LaForge, K. S. (2005). Genetics influences on impulsivity and vulnerability to drug abuse and addiction. *Nature Neuroscience, 8*, 1450–1457.

Lieberman, J. A., Stroup, S., McEvoy, J., Swartz, M., Rosenbeck, R., Perkins, D., ... Hsiao, J. (2005). Effectiveness of antipsychotic drugs in patients with chronic schizophrenia. *The New England Journal of Medicine, 353*, 1209–1223.

Mauri, M., Volontri, L., De Gaspari, I., Colasanti, A., Brambilla, M., & Cerutti, L. (2006). Substance abuse in first-episode schizophrenia patients: a retrospective study. *Clinical Practice and Epidemiology in Mental Health, 23*, 2–4.

Mizrahi, R., Mamo, D., & Pausion, P. (2009). The relationship between subjective well-being and dopmanine D2 in patients treated with dopamine partial agonists and full antagonist antipsychotics. *The International Journal of Neuropsychopharmacology, 5*, 715–721.

Moore, T. H., Zammit, S., Lingford-Hughes, A., Barnes, T. R., Jones, P. B., Burke, M., & Lewis, G. (2007). Cannabis use and risk of psychotic or affective mental health outcomes: a systematic review. *Lancet, 370*, 319–328.

Pencer, A., & Addington, J. (2003). Substance use and cognition in early psychosis. *J Psychiatry and Neuroscience, 28*, 48–54.

Pierce, R. C., & Kumaresan, V. (2006). The mesolimbic dopamine system: the final common pathway for the reinforcing effect of drugs of abuse? *Neuroscience and Bio Behavioral Reviews, 30*, 215–238.

Regier, D. A., Farmer, M. E., Rae, D. S., Lock, B. Z., Keith, S., Judd, L. L., & Goodwin, F. K. (1990). Comorbidity of mental disorders with alcohol and other drug abuse: results from the Epidemiologic Catchment Area (ECA) study. *JAMA, 264*, 2511–2518.

Samaha, A.-N. (2013). Can antipsychotic treatment contribute to drug addiction in schizophrenia? *Progress in Neuro-psychopharmacology and Biological Psychiatry, 52*, 9–16. http://dx.doi.org/10.1016/jpnpbp.2013.06.008.

Scheller-Gilkey, G., Moynes, K., Cooper, I., Kant, C., & Miller, A. H. (2004). Early life stress and PTSD. Symptoms in patients with comorbid schizophrenia and substance use. *Schizophrenia Research, 48*, 109–123.

Schiffer, B., Muller, B. W., Scherbaum, N., Forsting, M., Wiltfang, J., Leygraf, N., & Gizewski, E. R. (2010). Impulsivity-related brain volume deficits in schizophrenia-addiction comorbidity. *Brain, 133*, 3093–3103.

Sevy, S., Robinson, D. G., & Solloways, S. (2001). Correlates of substance misuse in patients with first-episode schizophrenia and schizoaffective disorder. *Acta Psychiatrica Scandinavica, 104*, 367–374.

Silver, H., & Abboud, E. (1994). Drug abuse in schizophrenia: comparison of patients who began drug abuse before their first admission with those who began abusing drugs after their first admission. *Schizophrenia Research, 13*, 57–63.

Smith, M., Bark, D. M., Wolf, T. J., Mamah, D., & Csernansky, J. (2008). Elevated rates of substance use disorders in non-psychotic siblings of individuals with schizophrenia. *Schizophrenia Research, 106*, 294–299.

Talamo, A., Centorrino, F., Tondo, L., Dimitri, A., Henner, J., & Baldessarini, R. J. (2006). Comorbid substance use in schizophrenia: relations to positive and negative symptoms. *Schizophrenia Research, 86*, 251–255.

Thoma, P., & Daum, I. (2013). Comorbid substance use disorder in schizophrenia: a selective overview of neurobiological and cognitive underpinnings. *Psychiatry and Clinical Neurosciences, 67*, 367–383.

Thompson, J. L., Urban, N., Slifstein, M., Xu, X., Kegeles, L. S., Girgis, R. R., ... Abi-Dargham, A. (2013). Striatal dopamine release in schizophrenia comorbid with substance dependence. *Molecular Psychiatry, 18*, 909–915.

Tswang, M. T., Bar, J. L., Harley, R. M., & Lyons, M. J. (2001). The Harvard twin study of substance abuse. What we have learned. *Harvard Review of Psychiatry, 9*, 267–279.

Tswang, M. T., Lyons, M. J., Meyer, J. M., Doyle, T., Eisen, S. A., Goldberg, J., ... Eaves, L. (1998). Co-occurrence of abuse of different drugs in men: the role of drug-specific and shared vulnerabilities. *Archives of General Psychiatry, 55*, 967–972.

Van Laar, M., van Dorsselaer, Monshouwer, K., & de Graff, R. (2007). Does cannabis use predict the first incidence of mood and anxiety disorders in the adult population? *Addiction, 102*, 1251–1260.

Volkow, N. D., Fowler, J. S., Wang, G. I., Baler, R., & Telang, F. (2009). Imaging dopamine's role in drug abuse and addiction. *Neuropharmacology, 56*(Suppl. 1), 3–8.

Volkow, N. D., Wang, G. K., Foroler, J. S., Tomasi, D., & Telans, F. (2011). Addiction: beyond dopamine circuitry. *PNAS, 108*, 15037–15042.

Voruganti, L. N. P., & Awad, A. G. (2006). Subjective behavioural consequences of striatal dopamine depletion in schizophrenia. Findings from in-vivo SPECT study. *Schizophrenia Research, 88*, 179–186.

Voruganti, L. N. P., Heselgrave, R. J., & Awad, A. G. (1997). Neuroleptic dysphoria may be the missing link between schizophrenia and substance abuse. *The Journal of Nervous and Mental Disease, 185*, 463–465.

Voruganti, L. N. P., Slomka, P., Zabel, P., Costa, G., So, A., Mattar, A., & Awad, A. G. (2001). Subjective effects of AMPT induced dopamine depletion in schizophrenia; correlation between dysphoric response and striatal D2 binding ratios on SPECT neuroimaging. *Neuropsychopharmacology, 25*, 642–650.

Voruganti, L. N. P., Slomka, P., Zabel, P., Mattar, A., & Awad, A. G. (2001). Cannabis induced dopamine release: an in-vivo SPECT study. *Psychiatry Research, 107*, 173–177.

Wise, R. A. (2009). Roles for nigrostriatal—not just mesocorticolimbic—dopamine in reward and addiction. *Trends in Neurosciences, 32*, 517–524.

Zammit, S., Allebeck, P., Andreasson, S., Lundberg, I., & Lewis, G. (2002). Self-reported cannabis use as a risk factor for schizophrenia in Swedish conscripts of 1969: historical cohort study. *BMJ, 325*, 1199.

Part II

Tobacco

Section A

General Aspects

Chapter 9

Nicotine Chemistry, Pharmacology, and Pharmacokinetics

Diana L. McKinney, Andrea R. Vansickel
RD&E, Altria Client Services, Richmond, VA, USA

Abbreviations

ACh Acetylcholine
Ca^{2+} Calcium ions
C_{max} Maximal concentration
CNS Central nervous system
CT Computed tomography
CYP 2A6 Cytochrome P450 2A6 enzyme
e.g. For example
ELISA Enzyme-linked immunosorbent assay
FCTC Framework Convention on Tobacco Control
FDA Food and Drug Administration
FMO3 Flavin-containing monooxygenase 3
FTC Federal Trade Commission
GABA γ-aminobutyric acid
GC Gas chromatography
Hcrt-1 Hypocretin receptor
i.e. that is
K^+ Potassium ions
kg Kilograms
L Liters
LC Liquid chromatography
MAO Monoamine oxidase
mg Milligrams
min Minutes
ml Milliliters
MRI Magnetic resonance imaging
MS Mass spectrometry
Na^+ Sodium ions
nAChR Nicotinic acetylcholine receptor
ng Nanograms
PET Positron emission tomography
pK_a Acid dissociation constant
RIA Radioimmunoassay
ROI Region of interest
$t_{1/2}$ Elimination half-life
T_{max} Time to maximal concentration
UGT Uridine diphosphate-glucuronosyltransferase
UV Ultraviolet
WHO World Health Organization
°C Degrees celsius

INTRODUCTION

This chapter presents a review of the scientific literature and textbook information about nicotine chemistry, pharmacology, and pharmacokinetics. It is intended to be a high-level overview for the reader and not a comprehensive synthesis of the work around nicotine and tobacco. More details about any of the topics presented below can be found in the cited literature.

Nicotiana (tobacco) has long played a role in Native American culture, with uses both recreational and therapeutic (Domino, 1999). Namesake Frenchman Jean Nicot first brought tobacco seeds from the New World back to Paris via Portugal in the 1500s, and from there tobacco use spread worldwide (Domino, 1999). Tobacco leaves were generally smoked or, in more tropical areas of the Americas, extracted with water for use in herbal medicines. Nicotine was first isolated from tobacco extract in Germany in 1828 by Posselt and Reimann (Domino, 1999). Melsens elucidated the chemical structure of nicotine in the mid-1800s, shortly after the discovery of the tetrahedral carbon atom (Domino, 1999). The synthesis of nicotine was first achieved by Pictet and Crepieux in 1893 (Domino, 1999). This early work occurred without the benefit of modern-day analytical techniques or instruments. Today, more than 100 countries worldwide produce tobacco leaf, with different climates and soil types yielding different tobacco varieties (World Health Organization (WHO), 2002). The United States, China, India, Brazil, and Turkey are the world leaders in annual tobacco production (WHO, 2002).

Nicotine is one of many chemical compounds found in tobacco smoke. It is the major naturally occurring alkaloid found in tobacco (>90% of the total alkaloid content) and is considered to be the addictive component of tobacco products (Stratton, Shetty, Wallace, & Bondurant, 2001). *Nicotiana* leaves also contain other alkaloids such as anatabine, anabasine, and nornicotine, albeit at much lower levels. Interestingly, several plant-based food sources such as tomatoes and potatoes contain trace amounts of nicotine (Castro & Monji, 1986; Domino, Hornbach, & Demana, 1993). In addition, nuts, milk, and cereals contain the tobacco alkaloid myosmine (Tyroller, Zwickenpflug, & Richter, 2002). The agricultural industry uses nicotine as an insecticide (Stratton et al., 2001).

The composition of the tobacco blend along with any added flavors or humectants can influence the alkalinity or acidity of the smoke (Stratton et al., 2001). Bright tobacco, the largest component of most US blended tobaccos, is flue-cured and contains a moderate level of nicotine (Stratton et al., 2001). Relative to other tobaccos, air-cured (Burley) tobacco smoke has higher alkalinity and contains higher levels of nicotine (Stratton et al., 2001). Oriental tobaccos have a more complex, aromatic smoke and have lower levels of nicotine (Stratton et al., 2001). Nicotine yields of US cigarettes range from approximately 0.2 to 1.14 mg/cigarette (Federal Trade Commission smoking condition) (National Institutes of Health, 1996).

NICOTINE CHEMISTRY

Nicotine (Figure 1), or 3-(1-methyl-2-pyrrolidinyl) pyridine ($C_{10}H_{14}N_2$, molecular weight 162.23), is a colorless liquid in its pure state and has a boiling point around 246 °C. Nicotine is a weak base with $pK_a = 8.5$ and turns brown via oxidation processes when exposed to light and air. The majority of total nicotine found in tobacco exists in the $S(-)$ form. A racemic $R(+)$ form occurs upon pyrolysis (Armstrong, Wang, & Ercal, 1998). The $S(-)$ form of nicotine readily binds to nicotine receptors, while the $R(+)$ form is a weaker agonist. Ionization of the pyrrolidine nitrogen occurs at pH 7.4 (37 °C), while the pyridine nitrogen remains un-ionized, yielding a compound that exists in two states (protonated and nonprotonated). The nonprotonated, more lipophilic, form of nicotine is easily absorbed through body membranes, whereas the protonated hydrophilic form is not (Domino, 1999; Stratton et al., 2001).

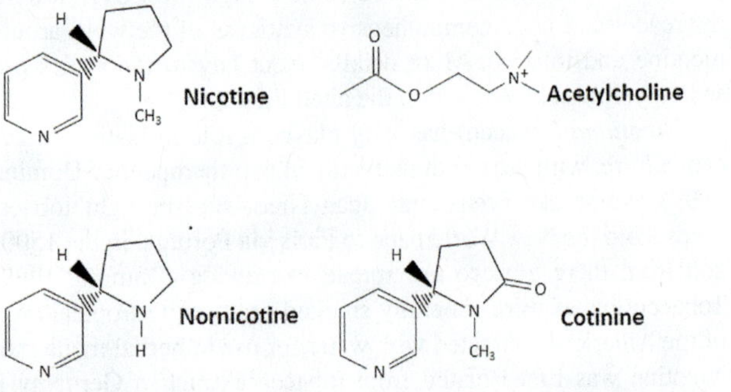

Willits, Swaim, Connelly, and Brice (1950) were the first to describe the ultraviolet spectrophotometric determination of nicotine from tobacco, tobacco extracts, and distillates. Since then, researchers have developed myriad methods for analysis of nicotine. Because of its relative thermal stability and volatility, nicotine is easily characterized in multiple matrices by gas chromatography (GC) methodology. Addition of mass spectrometry (GC–MS) allows quantification of nicotine based on retention time and comparison of the mass spectrum to a standard (nicotine m/z 162 for the molecular ion and m/z 84 for the base peak representing the pyrrolidine unit). Nicotine tartrate salt is often used as a standard since it is more stable and does not discolor (Jacob, Wilson, & Benowitz, 1981). For internal GC standards, investigators have used commercially available or synthesized nicotine analogs and metabolites with deuterium labels. Liquid chromatography–MS and immunoassay techniques (radioimmunoassay, enzyme-linked immunosorbent assay) are also available for detection of nicotine and its metabolites.

NICOTINE PHARMACOLOGY

Structure of Nicotinic Acetylcholine Receptors

Nicotine is structurally similar to the endogenous neurotransmitter acetylcholine (ACh), binding to nicotinic ACh receptors (nAChRs). Although nAChRs are ubiquitously expressed throughout the human body, this chapter will focus on those located in the central nervous system (CNS) and their associated effects. Neuronal nicotinic receptors in the autonomic ganglia and brain exist as pentameric structures (Figure 2), composed of homomeric or heteromeric combinations of α2–α10 and β2–β4 subunits

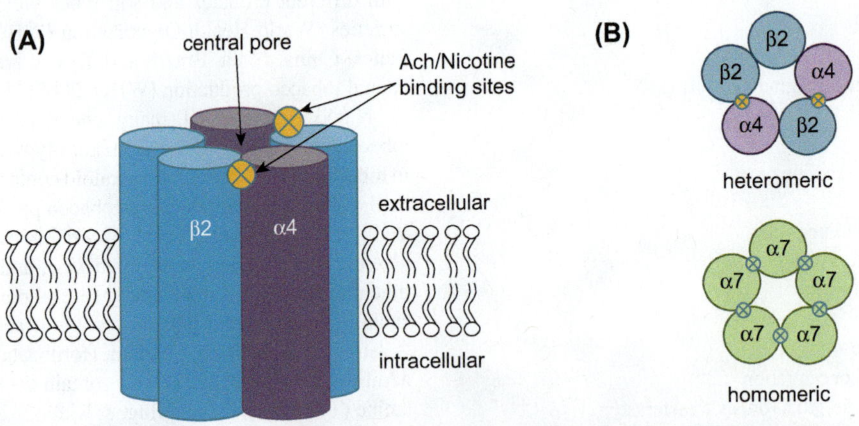

FIGURE 1 Chemical structures of nicotine, acetylcholine, and two metabolites of nicotine. Nicotine is structurally related to the endogenous neurotransmitter acetylcholine. In tobacco, nicotine is primarily the levorotatory (S)-isomer. Nornicotine is one of the most abundant metabolites of nicotine, termed a minor alkaloid. Cotinine is a nicotine metabolite often measured in plasma or urine as indication of nicotine levels after cigarette smoking.

FIGURE 2 A schematic representation of the structure of nAChRs and organization of subunits. (A) The nAChR is a pentameric structure comprising five subunits embedded in the cellular membrane of CNS neurons. The assembled receptor forms around a central pore, allowing ion influx and efflux. Agonists bind at specific sites between subunits. (B) Subunits can be combined to form heteromeric (e.g., α4β2) or homomeric (e.g., α7) receptor subtypes, each with different affinities for cholinergic agonists.

forming a ligand-gated ion channel with a central, water-filled pore (for a thorough review on nAChRs, see Dani & Bertrand, 2007). Each subunit consists of a long extracellular N-terminus, four transmembrane regions with an intracellular loop between the third and the fourth regions, and a short extracellular C-terminus (Dani & Bertrand, 2007). A large number of possible permutations and combinations of α and β subunits are possible but not all yield functional nAChRs.

The most abundant type of nAChR in the brain are the α4β2 receptors, expressed in high numbers in the cortex, striatum, superior colliculus, lateral geniculate nucleus, and cerebellum (Gotti, Zoli, & Clementi, 2006). Homomeric α7 receptors are also prevalent in various brain regions such as cortex, hippocampus, and subcortical limbic regions (Gotti et al., 2006). Complex receptors formed from more than one type of α or β subunit (e.g., α4β2β4) have also been identified using heterologous receptor expression systems (McGehee & Role, 1995). Subunit composition can influence the receptor affinity for agonists such as nicotine. The widespread variety of α and β subunit expression throughout the CNS indicates the potential for functional diversity (Dani, 2001).

Early nAChR isolation studies in the *Torpedo californica* electric organ (Unwin et al., 2002), along with examination of an ACh-binding protein from mollusks (Smit et al., 2001), helped elucidate the crystalline structure of nAChRs and their interactions with agonists. Nicotine binding sites on nAChRs comprise specific amino acid residues found at the junction of two adjacent subunits (e.g., α4 and β2 or two α7). It is common for an individual neuron to express multiple classes of nAChR and for neighboring neurons to express dissimilar nAChRs (Dani, 2001).

Localization and Function of nAChRs within Neurons

The nAChRs in the CNS exist at any moment in one of several conformations: closed (resting), open (active), desensitized, or inactive (a long-lasting desensitized state) (Heidmann & Changeux, 1980). Binding of a brief, relatively high concentration of agonist to the nAChR causes a conformational change from a closed to an open state, allowing ion exchange. Upon agonist removal, the channel closes and the nAChR returns to the resting state. During prolonged agonist exposure and binding, the receptor transforms into a desensitized state and is unavailable to undergo activation (Barik & Wonnacott, 2009). The potential for desensitization is dependent on the subunit composition, affinity of the nAChR for nicotine, and agonist concentration (Barik & Wonnacott, 2009). For instance, the α7 homomeric nAChRs have a lower affinity for ACh than α4β2, but high calcium permeability, and can be rapidly activated and desensitized (Dani, 2001; Dani & Bertand, 2007). Upon binding of ACh or nicotine, the receptor opens to facilitate an influx of sodium (Na^+), potassium (K^+), and calcium (Ca^{2+}) ions, leading to cell depolarization and signal transduction. The intracellular calcium increases via release of Ca^{2+} from internal calcium stores through voltage-gated channels (Barik & Wonnacott, 2009). Thus, nAChR activation influences a range of calcium-dependent neuronal processes and pharmacological effects.

The nAChRs reside in neuronal membranes in dendritic, somal, axonal, presynaptic, and postsynaptic locations (Dani & Bertrand, 2007). The most prevalent and well-characterized nAChR action in the CNS is presynaptic activation, which stimulates neurotransmitter release (ACh, dopamine, norepinephrine, serotonin, glutamate, and γ-aminobutyrate (GABA)) from the associated neuron (Dani, 2001). Many of the presynaptic nAChRs mediating excitatory input are α7 and influence glutamatergic transmission (Dani, 2001; Gotti et al., 2006). However, α7, as well as a variety of other receptor subtypes, is also involved in the release of other neurotransmitters. Preterminal or axonal nAChRs contribute to downstream neurotransmitter release by enhancing voltage-gated calcium release into the neuron as well as stimulating the release of calcium from intracellular stores and increasing conductance of action potentials (Dani & Bertrand, 2007). This mechanism is well observed at GABAergic synapses where calcium influx influences GABA release (Alkondon, Pereira, Barbosa, & Albuquerque, 1997). Nicotinic receptors located postsynaptically (on the soma or dendrite of a neuron) receive a signal from ACh in the synaptic cleft and potentiate fast nicotinic synaptic transmission, but this mechanism is uncommon and less understood (Barik & Wonnacott, 2009). Stimulation of nAChRs results in various effects dependent on the location within the CNS and the conformation of the receptors.

Regional Differences and CNS Effects of Nicotine

Nicotine-induced actions at nAChRs can be either excitatory or inhibitory depending on the site of action (Figure 3). The rewarding or reinforcing effects of nicotine are primarily due to direct binding and activation of presynaptic α7 nAChRs on cholinergic afferents, enhancing glutamate release onto dopaminergic neurons in the midbrain (Barik & Wonnacott, 2009). In the striatum, these dopaminergic nerve terminals containing mainly β2 nAChR subunits are also directly stimulated by nicotine (Dani & Bertrand, 2007). Other inputs from nicotine-induced release of GABA (an inhibitory neurotransmitter) onto dopaminergic neurons modulate the magnitude of the dopamine response (Dani & Bertrand, 2007). Blockade of dopamine release in the nucleus accumbens of the striatum attenuates nicotine self-administration in animal models, providing evidence that this mesostriatal dopamine pathway is important for the rewarding effects of nicotine (Corrigall, 1999).

Products containing nicotine may influence the dopaminergic system in an indirect manner as well. Tobacco smoke contains inhibitors of monoamine oxidase (MAO), the enzyme responsible for the metabolism of dopamine (Fowler, Logan, Wang, & Volkow, 2003). Smokers exhibit lower levels of MAO-A and MAO-B in the brain than nonsmokers (Fowler et al., 2003) and therefore have higher levels of dopamine in the synaptic cleft. Thus, an observed high rate of smoking in patients diagnosed with depression and schizophrenia could in part be due to the resulting enhancement of brain dopamine and the likelihood of these patients to self-medicate (Fowler et al., 2003). MAO inhibition has also been suggested as neuroprotective against Parkinson's disease (Ryan, Ross, Drago, & Loiacono, 2001).

The insular cortex is also implicated as a key brain region in nicotine addiction. In one study, damage to the insula of smokers disrupted smoking behavior and resulted in decreased urge to smoke (Naqvi, Rudrauf, Damasio, & Bechara, 2007). Blockade of hypocretin receptors in the insula of rats significantly reduced

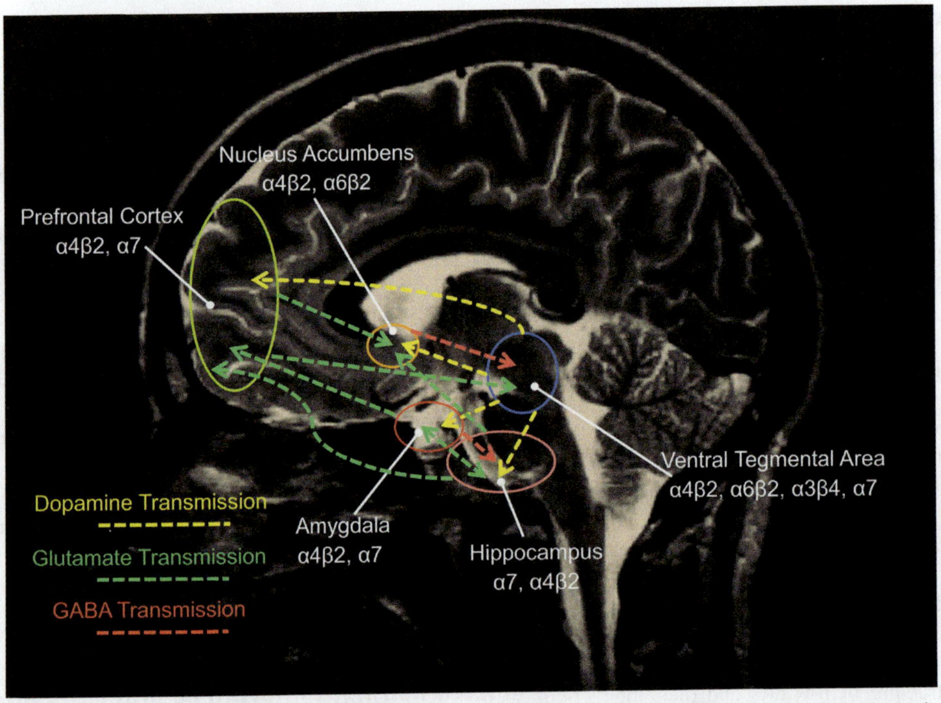

FIGURE 3 **Nicotine's effects on the mesocorticolimbic pathway in the brain.** Nicotine-induced actions at nAChRs in the brain can be either excitatory or inhibitory depending on the site of action. The neurotransmitters dopamine, glutamate, and GABA mediate nicotine's effects through a complex network of projections between brain regions key in reward-seeking behavior, learning and memory, emotion, and daily function. The expression and function of nicotinic receptor subtypes vary across brain regions.

nicotine reinforcement (Hollander, Lu, Cameron, Kamenecka, & Kenny, 2008), implicating this neuropeptide in the maintenance of nicotine-related reward.

In the hippocampus, nicotine enhances gating of relevant sensory stimuli and influences neuronal plasticity via α7 receptors (Dani & Bertrand, 2007). The α4β2 receptors are also present in the hippocampus and contribute to GABA and glutamate release upon stimulation of presynaptic and preterminal nAChRs (Dani & Bertrand, 2007). Studies have shown that schizophrenic patients consistently exhibit a decrease in α7 immunoreactivity in the hippocampus (Leonard et al., 1996). An abnormality in the gene encoding the homomeric α7 nAChR, which modulates auditory sensory gating, may contribute to the link between schizophrenia and smoking (Dalack, Healy, & Meador-Woodruff, 1998); a high percentage (in the range of 70–90%) of schizophrenics smoke cigarettes (Goff, Henderson, & Amico, 1992).

Positive cognitive effects of nicotinic agonists such as memory enhancement and increased concentration are mediated in the hippocampus as well as the frontal and parietal brain cortices (Kumari et al., 2003; Levin, McClernon, & Rezvani, 2006). In some studies, working memory is improved following acute or chronic nicotine administration in animal models (Levin et al., 2006), but not all forms of memory are enhanced by nicotine in this manner. Studies examining the effect of nicotinic agonists in aged rodents show mixed results (e.g., Levin & Torry, 1996), possibly due to age-related decrease in nAChRs (Dani & Bertrand, 2007). Knockout mice in which one or more genes encoding nAChR subunits are mutated or silenced have become extremely useful models for understanding nAChR pharmacology. For example, β2 is implicated in learning and memory based on the results from behavioral testing with β2 knockouts (Picciotto et al., 1995). Alzheimer's disease is associated with a decrease in brain nAChRs, specifically α4 subunits in cortex and hippocampus (Gotti & Clementi, 2004). Nicotinic receptor stimulation as a clinical therapy for Alzheimer's disease shows some early promise, but more work is needed in this area to maximize results (Gotti & Clementi, 2004).

The substantia nigra (a brain region associated with motor function and control) also contains nAChRs and is implicated in the pathogenesis of Parkinson's disease (Grenhoff & Svensson, 1989). Loss of dopaminergic neurons (primarily containing α4β2, α6α4β2, and α7 nAChRs) is a hallmark of this disease (Gotti & Clementi, 2004). However, clinical trials using nicotinic agonists have been limited and relatively ineffective owing to the widespread distribution of nAChRs throughout the dopamine pathway (Gotti & Clementi, 2004).

Desensitization and Upregulation

Long-term exposure to nicotine from tobacco smoke produces neuroadaptation (physiological dependence) and tolerance (Benowitz, 1999). Desensitization of a receptor is an agonist-dependent conformational transition from an active state to an inactive state whereby the receptor is no longer sensitive to agonist challenge. Different nAChRs may desensitize at different rates or to different degrees; this mechanism is not entirely understood (Dani, 2001). In smokers, continued exposure to low nicotine concentrations throughout the day yields a strong, deep desensitization of primarily the β2-type receptors (Dani, 2001). The α7 nAChRs, with their low affinity for nicotine, do not significantly desensitize with the low concentrations of plasma nicotine in smokers (Wooltorton, Pidoplichko, Broide, &

Dani, 2003). Chronic desensitization leads to a dysregulation of nAChR number due to improper recycling of receptors and is believed to play a major role in the development and maintenance of nicotine addiction (Dani, 2001). Prolonged desensitization can lead to eventual upregulation of nAChRs, a possible homeostatic response (Benowitz, 1999). Evidence of increased nAChR number has been shown in the brains of smokers upon autopsy (Benwell, Balfour, & Anderson, 1988). However, these seemingly opposed synaptic neuroadaptations (desensitization and upregulation) are both thought to be involved in nicotine addiction (Barik & Wonnacott, 2009; Picciotto, Addy, Mineur, & Brunzell, 2008).

The sudden removal of agonist stimulation at nAChRs results in deficiencies in neurotransmitter release (Benowitz, Hukkanen, & Jacob, 2009) thereby altering homeostasis. Cessation of smoking in long-term users often manifests in a relatively mild but uncomfortable withdrawal syndrome, at its worst within the first week of abstinence and eventually disappearing over a period of weeks. Restlessness, irritability, depression, anxiety, insomnia, increased appetite, and cognitive deficits are all characteristic of nicotine withdrawal.

NICOTINE PHARMACOKINETICS

Pharmacokinetics refers to the fate of a drug once it enters the body (absorption, distribution, biotransformation/metabolism, and excretion). Physiochemical properties, route of administration, and human physiology have an impact on the pharmacokinetic characteristics of nicotine. Humans self-administer nicotine via several different routes, including inhalation, ingestion, and intranasal, buccal, and transdermal absorption. Nicotine in the form of tobacco is smoked (e.g., cigarettes, cigars, hookah) as well as chewed, sucked, or placed in the mouth (e.g., chewing tobacco, snuff, dissolvable tobacco lozenges). Pharmaceutical nicotine products include gums, dissolvable lozenges, vapor inhalers, skin patches, and nasal sprays.

Absorption

Nicotine is both lipid and water soluble and is therefore readily absorbed into the mucosa, respiratory tissues, and skin. Oral products such as chewing tobacco, nicotine gum, nicotine lozenge, nicotine inhaler, snus, and snuff are buffered to alkaline (higher) pH, which aids absorption through the mucous membranes lining the mouth (Benowitz et al., 2009). Individuals may also swallow some nicotine while using oral tobacco or nicotine-replacement products. Very little nicotine is absorbed in the stomach because of the highly acidic environment (Hukkanen, Jacob, & Benowitz, 2005). Nicotine, though readily absorbed in the small intestine, is subject to a considerable first-pass effect, whereby initial hepatic metabolism greatly reduces nicotine bioavailability (i.e., the fraction of nicotine dose released into the systemic circulation) (Benowitz, Porchet, Sheiner, & Jacob, 1988, 2009). Nicotine bioavailability varies across routes of administration. For example, nicotine bioavailability via smoking is 80–90%, whereas nicotine bioavailability via the nicotine inhaler ranges from 51% to 56% (Benowitz et al., 2009). This discrepancy in bioavailability across products relates to the primary sites of absorption. Nicotine delivered via nasal spray is absorbed primarily through the mucosa of the oral cavity and a large amount is ingested, meaning a considerable portion of the nicotine is metabolized before reaching systemic circulation (Hukkanen et al., 2005).

Some scientists hypothesize that the alkalinity or acidity of tobacco smoke from cigarettes influences the rate and extent of nicotine absorption (Willems, Rambali, Vleeming, Opperhuizen, & van Amsterdam, 2006). This hypothesis hinges on the notion that the presence of nonprotonated, free-base nicotine in tobacco smoke equates to better nicotine transfer into the respiratory epithelium (Willems et al., 2006). Public health officials suggest, based on the Willems hypothesis mentioned above, that the addition of ammonia-forming compounds to tobacco increases nicotine delivery. However, several recent studies provide evidence to the contrary. Armitage and colleagues found that there was no difference in venous nicotine levels after smoking cigarettes with low, medium, and high smoke ammonia levels (Armitage, Dixon, Frost, Mariner, & Sinclair, 2004). Similarly, in examining two commercial cigarettes with a fourfold difference in ammonia content, there were no differences in venous blood levels of nicotine after smoking (van Amsterdam et al., 2011). More recently, a study of arterial blood nicotine levels (the best indicator of transfer from alveoli to blood) after controlled smoking of two test cigarettes with a twofold difference in smoke ammonia levels showed that there was no difference in rate or amount of nicotine absorption (McKinney et al., 2012).

An alternative hypothesis suggests that nicotine protonation in smoke does not influence the rate and extent of nicotine absorption via passive diffusion for several reasons, including the large surface area and buffering capacity of the lung (Seeman & Carchman, 2008). Nicotine in smoke interacts with the alveoli–blood interface in the lung via direct deposition of smoke particles or via evaporation from the particles followed by deposition and passive diffusion across the alveolar membrane (Seeman & Carchman, 2008). Nicotine in its nonprotonated state passively crosses membranes more rapidly than in its protonated state. As nonprotonated nicotine transfers easily into the blood, the equilibrium between protonated and nonprotonated nicotine is quickly reestablished. There is also evidence of active membrane transport of protonated nicotine across the lung–blood barrier (Nair, Chetty, Ho, & Chien, 1997).

Distribution

As noted above, nicotine transports rapidly into arterial and venous circulation once it reaches the small airways and alveoli of the lung. Arterial and venous blood plasma nicotine accumulates quickly after smoking a single cigarette, with maximal venous blood plasma concentration ranging between 15 and 30 ng/ml within 5 min (Hukkanen et al., 2005). Nicotine reaches the arterial circulation within 7–24 s after the first puff inhalation (McKinney et al., 2012). Positron emission tomography (PET) studies demonstrate that brain nicotine levels rise rapidly following inhalation of cigarette smoke, reaching the brain within 5–7 s of the start of an initial puff and reaching maximal brain concentration within 3–5 min (Berridge et al., 2010; Rose et al., 2010). We present some key facts about PET at the end of this chapter.

The average maximal blood nicotine concentration (C_{max}), time to maximal blood nicotine concentration (T_{max}), and overall

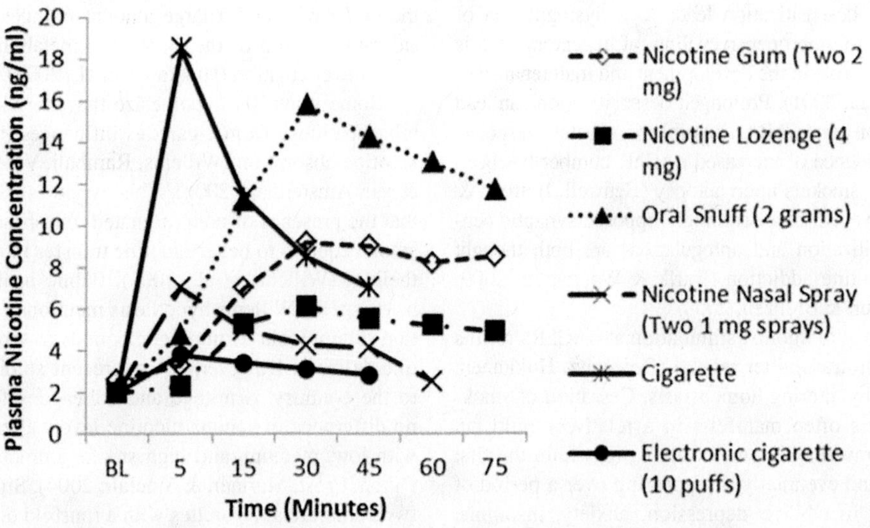

FIGURE 4 **Nicotine pharmacokinetic profiles for various tobacco and nicotine products.** Nicotine pharmacokinetic profiles for nicotine gum (two 2-mg nicotine pieces; *adapted from Benowitz et al. (1988)*), nicotine lozenge (4-mg nicotine piece; *adapted from Kotlyar et al. (2007)*), 2-g "dip" of oral snuff *(adapted from Kotlyar et al.(2007))*, nicotine nasal spray (two 1-mg nicotine sprays; *adapted from Schneider, Lunell, Olmstead, and Fagerström (1996)*), 10 puffs from an electronic cigarette, and one conventional cigarette *(adapted from Vansickel et al. (2010))*.

bioavailability of nicotine vary considerably across tobacco and nicotine replacement therapy products (Figure 4). The nicotine pharmacokinetic profile of tobacco or nicotine products influences the subjective experience following use of that product (e.g., Gray, Breland, Weaver, & Eissenberg, 2008; Kotlyar et al., 2007; Lunell & Curvall, 2011; Vansickel, Cobb, Weaver, & Eissenberg, 2010). When administered via a similar route, subjective ratings of nicotine effects and craving or withdrawal suppression tend to be greater with higher rate and extent of nicotine delivery (Gray et al., 2008; Kotlyar et al., 2007; Lunell & Curvall, 2011; Vansickel et al., 2010). Table 1 shows examples of averages or ranges for C_{max} and T_{max} for a variety of tobacco and nicotine-containing products.

Nicotine distributes through the entire body with a steady-state distribution volume ranging between 2.2 and 3.3 L/kg (Hukkanen et al., 2005; Yildiz, 2004). Nicotine concentrates largely in the lung, liver, kidney, spleen, and brain. It freely crosses the placental barrier and passes into breast milk and cervical fluid (Benowitz et al., 2009). The distribution half-life of nicotine is approximately 9 min (Feyerabend, Ings, & Russell, 1985).

Metabolism

Biotransformation of nicotine occurs primarily in the liver (80%) but some transformation also occurs in the lung and kidneys (Hukkanen et al., 2005). The major metabolite of nicotine is cotinine. See Figure 5 for a schematic representation of the nicotine metabolic pathways. Approximately 80% of nicotine is transformed to cotinine via a two-step process. The first step, facilitated by the cytochrome P450 2A6 enzyme, produces the nicotine-iminium ion via hydroxylation of nicotine to 5-hydroxynicotine. The second step is catalyzed by aldehyde oxidase to produce cotinine (Benowitz et al., 2009; Yildiz, 2004). Roughly 4–7% of nicotine is transformed via N' oxidation to nicotine N'-oxide, facilitated by flavin-containing monooxygenase 3 (Hukkanen et al., 2005). About 3–5% of nicotine is transformed via glucuronidation, catalyzed by uridine diphosphate-glucuronosyltransferase, to nicotine glucuronide (Yildiz, 2004). A small amount of nicotine (0.4–0.8%) is also converted to nornicotine, and that process is thought to be mediated by the cytochrome P450 system (Hukkanen et al., 2005). Several factors can have an impact on nicotine metabolism, including genetic polymorphisms, age, gender, diet, certain health conditions, and use of some medications. Racial and ethnic differences in nicotine metabolism also exist (Hukkanen et al., 2005).

Excretion

The elimination half-life ($t_{1/2}$) of nicotine averages between 100 and 150 min (Hukkanen et al., 2005). Approximately 15% of nicotine is excreted in the urine unchanged. Renal excretion of nicotine is accomplished via glomerular filtration and tubular secretion (Hukkanen et al., 2005). Acidification of the urine increases renal clearance through reduction of free nicotine, thereby limiting reabsorption (Hukkanen et al., 2005). The rate of renal clearance averages approximately 35–90 ml/min (Hukkanen et al., 2005). A small amount of nicotine excretion also occurs in feces and sweat (Hukkanen et al., 2005). Nicotine's metabolites, including nornicotine, nicotine N' oxide, cotinine, and nicotine glucuronide, also appear in urine. Approximately 10–15% of nicotine and its metabolites are excreted in urine as 4-oxo-4-(3-pyridyl)butanoic acid and 4-hydroxy-4-(3-pyridyl)butanoic acid (Hukkanen et al., 2005).

SUMMARY

This chapter reviewed the current body of literature on nicotine chemistry, pharmacology, and pharmacokinetics. Topics covered included the chemical structure of nicotine and several of its

TABLE 1 Maximal Concentration (C_{max}) and Time to Maximal Concentration (T_{max}) for Various Tobacco and Nicotine Products

Product Type	Administration or Dosing Regimen	C_{max} (ng/ml)	T_{max} (min)
Tobacco Products			
Cigarette	One cigarette (~5 min)[a]	15–30	5–8
Cigarillo	10 puffs[b]	5.3	5
Hookah	30–60 min[c,d]	10–12	45
Smokeless tobacco (snuff)	2 g placed between lip and gum for 30 min[e]	16.1	30
Smokeless tobacco (snus)			
Pouched 10.7 mg	Upper lip placement for 60 min[f]	10.8	60
Pouched 14.7 mg	Upper lip placement for 60 min[f]	13.4	60
Electronic cigarettes			
First generation (inexperienced users)	10 puffs (16–18 mg/ml nicotine liquid)[g]	2–3.5	5
First generation (experienced users)	10 puffs (18 mg/ml nicotine liquid)[h,i]	5–7	5–10
New generation (experienced users)	10 puffs (9–24 mg/ml nicotine liquid)[i,j]	6–10	5
Nicotine Replacement Therapies			
Nicotine gum	30 min chewing[a]		
2 mg		6–9	30
4 mg		10–17	30
Nicotine patch	Transdermal, upper arm placement[k]		
21 mg/16 h		18.3	360
25 mg/24 h		16.6	720
Nicotine lozenge	20–30 min in mouth[a]		
2 mg		4.4	60
4 mg		10.8	66
Nicotine nasal spray (1 mg)	Single dose[l]	5–12	11–13

[a]Benowitz et al. (2009).
[b]Blank, Nasim, Hart, and Eissenberg (2011).
[c]Eissenberg and Shihadeh (2009).
[d]Jacob et al. (2011).
[e]Kotlyar et al. (2007).
[f]Digard, Proctor, Kulasekaran, Malmqvist, and Richter (2013).
[g]Vansickel et al. (2010).
[h]Dawkins and Corcoran (2014).
[i]Farsalinos et al. (2014).
[j]Vansickel and Eissenberg (2013).
[k]Devaugh-Geiss, Chen, Kotlyar, Ramsay, and Durcan (2010).
[l]Schneider et al. (1996).

metabolites and its similarity to the endogenous neurotransmitter ACh. Nicotine's interactions with nAChRs throughout the CNS, and its effects, dependent on receptor structure, receptor localization, and function of the cell expressing the receptor, were also discussed. The fate of nicotine upon entering the body (pharmacokinetics) was described, including absorption, distribution, biotransformation, and excretion. The plasma levels of nicotine and the rate of increase after use of various forms of nicotine-containing products were also presented and contrasted.

AFTERWORD

Several countries, including the United States, have implemented comprehensive tobacco control policies and programs, which have led to a substantial decline in cigarette smoking rates (USDHHS, 2014). The WHO Framework Convention on Tobacco Control treaty was enacted in 2005, and the US Food and Drug Administration Family Smoking Prevention and Control Act was signed into law in 2009. Public health officials and regulatory bodies

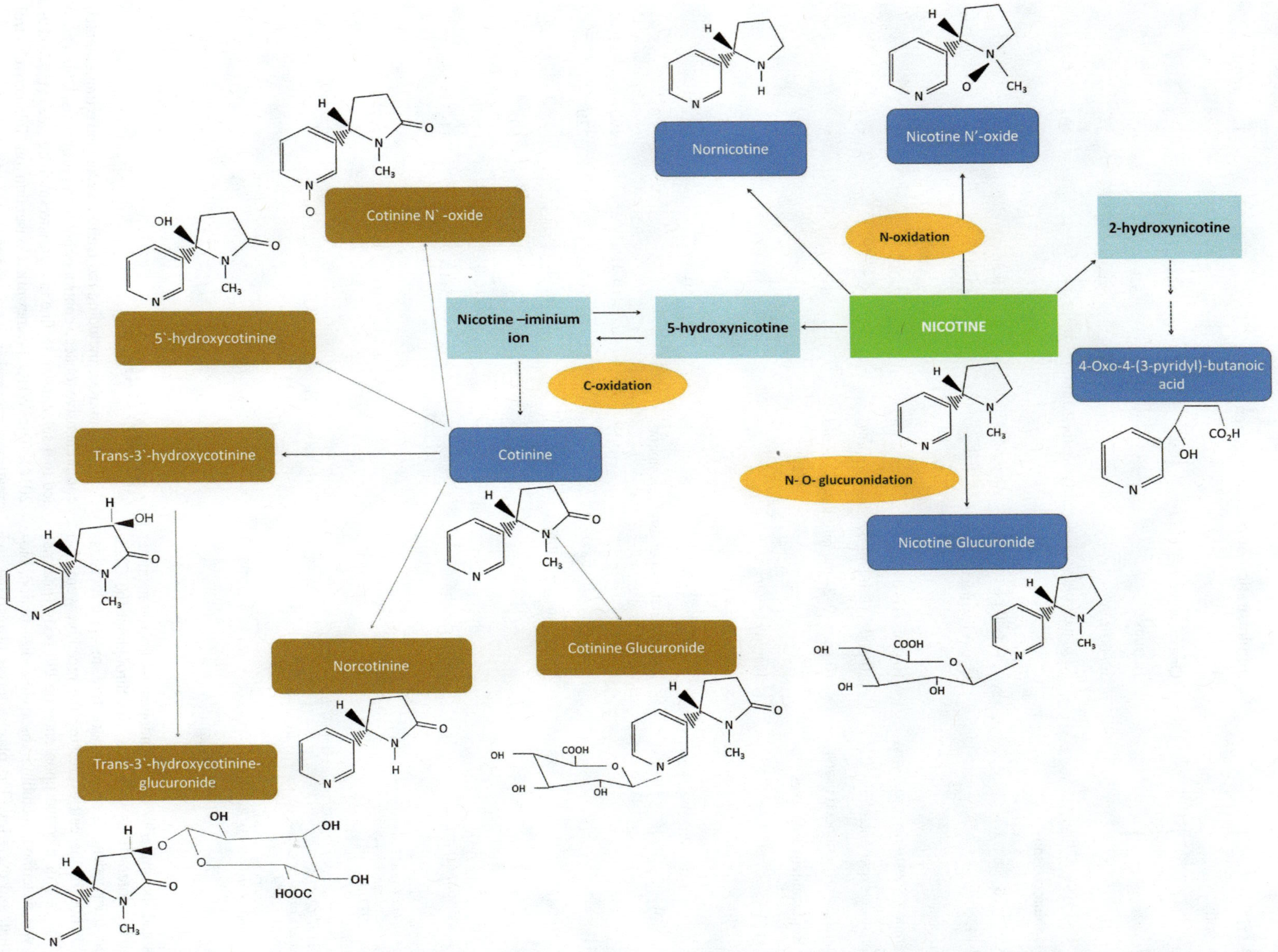

FIGURE 5 Schematic representation of nicotine and cotinine metabolism. Nicotine metabolism, including the major conversion processes of nicotine to its major metabolites, is shown. Dark blue boxes represent nicotine primary metabolites and golden boxes represent cotinine metabolites. *Information was gathered primarily from three sources: Benowitz et al. (2009), Hukannen et al. (2005), and Yildiz (2004).*

require scientific bases for implementing appropriate regulations. As a result, numerous opportunities for further research in the areas of tobacco and nicotine are likely.

DEFINITION OF TERMS

Alkaloid This is a chemical compound that is plant-derived and contains mainly carbon, nitrogen, and oxygen atoms. Alkaloids are also called organic bases.

Bioavailability This is the percentage/fraction of an administered drug dose that reaches circulation.

Biotransformation This is a series of chemical changes in a compound within the body.

Desensitization This is a process by which prolonged drug exposure leads to reduced receptor sensitivity. Desensitization can lead to drug tolerance, in which higher and higher amounts of the drug are required to cause the same stimulation effect.

Distribution half-life This refers to the length of time it takes for a drug to reach 50% distribution in tissues.

Elimination half-life This is also known as terminal half-life. It refers to the length of time it takes for elimination of a drug to half of its original concentration; expressed as $t_{1/2}$.

Homeostasis This is a state of balance or stability that the body continuously maintains by managing compensatory changes in physiological systems.

Neurotransmitter This is a chemical that mediates communication between nerve cells.

Pharmacokinetics This science describes the fate of a drug once it enters the body. Pharmacokinetics consists of four main parts: absorption of the drug into body tissues, distribution of the drug throughout the system, metabolism (or biotransformation) of the drug into other related compounds or breakdown products, and excretion of the drug, either unchanged or metabolized, in waste form.

Pharmacology This is the science of drugs and how they interact with the body.

Volume of distribution This refers to the volume of drug in the body, expressed as a concentration.

KEY FACTS: PET NEUROIMAGING AND NICOTINE

- PET is an imaging technique whereby receptor activity/functionality and distribution of radiolabeled drugs or ligands can be viewed in live humans or animals.
- PET creates a three-dimensional image of total radioactive tissue. Researchers often use magnetic resonance imaging or computed tomography scans of subjects' brains to help map the PET images and identify regions of interest.
- Researchers use PET to characterize the cerebral kinetics of drugs.
- PET neuroimaging has given researchers insights into brain nicotine distribution and rate of increase following smoking through the use of special cigarettes containing radiolabeled racemic [^{11}C]nicotine (Berridge et al., 2010; Rose et al., 2010).
- One benefit of using PET imaging techniques is that radiolabeled drug actions can be viewed in real time.

SUMMARY POINTS

- This chapter describes the basic chemical structure and function of nicotine, the major alkaloid found in tobacco.
- Structurally related to the endogenous neurotransmitter ACh, nicotine exerts a variety of stimulatory and inhibitory actions mediated by nAChRs ubiquitously expressed throughout the body.
- This chapter focuses on the CNS effects of nicotine.
- Presynaptic nAChR activation in the CNS stimulates release of several neurotransmitters, including ACh, dopamine, norepinephrine, serotonin, glutamate, and GABA, from the associated neurons.
- The varying effects of nAChR activation depend on their location within the CNS and the conformation of the receptors.
- This chapter also describes the absorption, distribution, biotransformation, and excretion of nicotine in the human body (pharmacokinetics).
- Nicotine pharmacokinetic profiles vary considerably across tobacco and nicotine products depending on the route of administration.
- Finally, this chapter presents key points describing a unique method for measuring the brain kinetics of nicotine.

REFERENCES

Alkondon, M., Pereira, E. F., Barbosa, C. T., & Albuquerque, E. X. (1997). Neuronal nicotinic acetylcholine receptor activation modulates gamma-aminobutyric acid release from CA1 neurons of rat hippocampal slices. *Journal of Pharmacology and Experimental Therapeutics*, 283, 1396–1411.

van Amsterdam, J., Sleijffers, A., van Spiegel, P., Blom, R., Witte, M., van de Kassteele, J., … Opperhuizen, A. (2011). Effect of ammonia in cigarette tobacco on nicotine absorption in human smokers. *Food and Chemical Toxicology*, 49, 3025–3030.

Armitage, A. K., Dixon, M., Frost, B. E., Mariner, D. C., & Sinclair, N. M. (2004). The effect of tobacco blend additives on the retention of nicotine and solanesol in the human respiratory tract and on subsequent plasma nicotine concentrations during cigarette smoking. *Chemical Research in Toxicology*, 17, 537–544.

Armstrong, D. W., Wang, X., & Ercal, N. (1998). Enantiomeric composition of nicotine in smokeless tobacco, medicinal products and commercial reagents. In *Chirality: Special issue: Proceedings from the Eighth International Symposium on Chiral Discrimination*, 10(7), 587–591.

Barik, J., & Wonnacott, S. (2009). Molecular and cellular mechanisms of action of nicotine in the CNS. *Handbook of Experimental Pharmacology*, 192, 173–207.

Benowitz, N. L. (1999). Nicotine addiction. *Primary Care*, 26(3), 611–631.

Benowitz, N. L., Hukkanen, J., & Jacob, P. J. (2009). Nicotine chemistry, metabolism, kinetics, and biomarkers. *Handbook of Experimental Pharmacology*, 192, 29–60.

Benowitz, N. L., Porchet, H., Sheiner, L., & Jacob, P. (1988). Nicotine absorption and cardiovascular effects with smokeless tobacco use: comparison with cigarettes and nicotine gum. *Clinical Pharmacology and Therapeutics*, 44, 23–28.

Benwell, M. E. M., Balfour, D. J. K., & Anderson, J. M. (1988). Evidence that tobacco smoking increases the density of (−)-[^{3}H] nicotine binding sites in human brain. *Journal of Neurochemistry, 50*, 1243–1247.

Berridge, M. S., Apana, S. M., Nagano, K. K., Berridge, C. E., Leisure, G. P., & Boswell, M. V. (2010). Smoking produces rapid rise of [^{11}C] nicotine in human brain. *Psychopharmacology, 209*(4), 389–394.

Blank, M. D., Nasim, A., Hart, A., & Eissenberg, T. (2011). Acute effects of cigarillo smoking. *Nicotine and Tobacco Research, 13*(9), 874–879.

Castro, A., & Monji, N. (1986). Dietary nicotine and its significance in studies on tobacco smoking. *Biochemistry Archives, 2*, 91–97.

Corrigall, W. A. (1999). Nicotine self-administration in animals as a dependence model. *Nicotine and Tobacco Research, 1*(1), 11–20.

Dalack, G. W., Healy, D. J., & Meador-Woodruff, J. H. (1998). Nicotine dependence in schizophrenia: clinical phenomena and laboratory findings. *American Journal of Psychiatry, 155*, 1490–1501.

Dani, J. A. (2001). Overview of nicotinic receptors and their roles in the central nervous system. *Biological Psychiatry, 49*, 166–174.

Dani, J. A., & Bertrand, D. (2007). Nicotinic acetylcholine receptors and nicotinic cholinergic mechanisms of the central nervous system. *Annual Review of Pharmacology and Toxicology, 49*, 699–729.

Dawkins, L., & Corcoran, O. (2014). Acute electronic cigarette use: nicotine delivery and subjective effects in regular users. *Psychopharmacology, 231*(2), 401–407.

Devaugh-Geiss, A. M., Chen, L. H., Kotlyar, M., Ramsay, L. R., & Durcan, M. J. (2010). Pharmacokinetic comparison of two nicotine transdermal systems, a 21-mg/24-hour patch and a 25-mg/16-hour patch: a randomized, open-label, single-dose, two-way crossover study in adult smokers. *Clinical Therapeutics, 32*(6), 1140–1148.

Digard, H., Proctor, C., Kulasekaran, A., Malmqvist, U., & Richter, A. (2013). Determination of nicotine absorption from multiple tobacco products and nicotine gum. *Nicotine and Tobacco Research, 15*(1), 255–261.

Domino, E. F. (1999). Pharmacological significance of nicotine. In J. W. Gorrod, & P. Jacob (Eds.), *Analytical determination of nicotine and related compounds and their metabolites*. New York: Elsevier.

Domino, E. F., Hornbach, E., & Demana, T. (1993). To the editor: the nicotine content of common vegetables. *New England Journal of Medicine, 369*, 437.

Eissenberg, T., & Shihadeh, A. (2009). Waterpipe tobacco and cigarette smoking: direct comparison of toxicant exposure. *American Journal of Preventive Medicine, 37*(6), 518–523.

Farsalinos, K. E., Spyrou, A., Tsimopoulou, K., Stefopoulos, C., Romagna, F., & Voudris, V. (2014). Nicotine absorption from electronic cigarette use: comparison between first and new-generation devices. *Scientific Reports, 4*(4133), 1–7.

Feyerabend, C., Ings, R. M., & Russell, M. A. (1985). Nicotine pharmacokinetics and its application to intake from smoking. *British Journal of Clinical Pharmacology, 19*(2), 239–247.

Fowler, J. S., Logan, J., Wang, G.-J., & Volkow, N. (2003). Monoamine oxidase and cigarette smoking. *NeuroToxicity, 24*, 75–82.

Goff, D. C., Henderson, D. C., & Amico, E. (1992). Cigarette smoking in schizophrenia: relationship to psychopathology and medication side effects. *American Journal of Psychiatry, 149*(9), 1189–1194.

Gotti, C., & Clementi, F. (2004). Neuronal nicotinic receptors: from structure to pathology. *Progress in Neurobiology, 74*, 363–396.

Gotti, C., Zoli, M., & Clementi, F. (2006). Brain nicotinic acetylcholine receptors: native subtypes and their relevance. *Trends in Pharmacological Sciences, 27*(9), 482–491.

Gray, J. N., Breland, A. B., Weaver, M., & Eissenberg, T. (2008). Potential reduced exposure products (PREPs) for smokeless tobacco users: clinical evaluation methodology. *Nicotine and Tobacco Research, 10*(9), 1441–1448.

Grenhoff, J., & Svensson, T. H. (1989). Pharmacology of nicotine. *British Journal of Addiction, 84*, 477–492.

Heidmann, T., & Changeux, J.-P. (1980). Interaction of a fluorescent agonist with the membrane-bound acetylcholine receptor from *Torpedo marmorata* in the millisecond time range: resolution of an "intermediate" conformational transition and evidence for positive cooperative effects. *Biochemical and Biophysical Research Communications, 97*, 889–896.

Hollander, J. A., Lu, Q., Cameron, M. D., Kamenecka, T. M., & Kenny, P. J. (2008). Insular hypocretin transmission regulates nicotine reward. *Proceedings of the National Academy of Sciences, 105*(49), 19480–19485.

Hukkanen, J., Jacob, P., III, & Benowitz, N. L. (2005). Metabolism and disposition kinetics of nicotine. *Pharmacological Reviews, 57*, 79–115.

Jacob, P., III, Abu Raddaha, A. H., Dempsey, D., Havel, C., Peng, M., Yu, L., & Benowitz, N. L. (2011). Nicotine, carbon monoxide, and carcinogen exposure after a single use of water pipe. *Cancer Epidemiology Biomarkers and Prevention, 20*(11), 2345–2353.

Jacob, P., III, Wilson, M., & Benowitz, N. (1981). Improved gas chromatographic method for determination of nicotine and cotinine in biological fluids. *Journal of Chromatography, 222*, 61–70.

Kotlyar, M., Mendoza-Baumgart, M. I., Zhong-ze, L., Pentel, P. R., Barnett, B. C., Feuer, R. M., … Hatsukami, D. K. (2007). Nicotine pharmacokinetics and subjective effects of three potential reduced exposure products, moist snuff and nicotine lozenge. *Tobacco Control, 16*, 138–142.

Kumari, V., Gray, J. A., ffytche, D. A., Mitterschiffthaler, M. T., Das, M., Zachariah, E., … Sharma, T. (2003). Cognitive effects of nicotine in humans: an fMRI study. *NeuroImage, 19*, 1002–1013.

Leonard, S., Adams, C., Breese, C. R., Adler, L. E., Bickford, P., Byerly, W., … Freedman, R. (1996). Nicotinic receptor function in schizophrenia. *Schizophrenia Bulletin, 22*, 431–445.

Levin, E. D., McClernon, F. J., & Rezvani, A. H. (2006). Nicotinic effects on cognitive function: behavioral characterization, pharmacological specification, and anatomic localization. *Psychopharmacology, 184*, 523–539.

Levin, E. D., & Torry, D. (1996). Acute and chronic nicotine effects on working memory in aged rats. *Psychopharmacology, 123*, 88–97.

Lunell, E., & Curvall, M. (2011). Nicotine delivery and subjective effects of Swedish portion snus compared with 4 mg nicotine polacrilex chewing gum. *Nicotine and Tobacco Research, 13*(7), 573–578.

McGehee, D. S., & Role, L. W. (1995). Physiological diversity of nicotinic acetylcholine receptors expressed by vertebrate neurons. *Annual Review of Physiology, 57*, 521–546.

McKinney, D. L., Gogova, M., Davies, B. D., Ramakrishnan, V., Fisher, K., Carter, W. H., … Barr, W. H. (2012). Evaluation of the effect of ammonia on nicotine pharmacokinetics using rapid arterial sampling. *Nicotine and Tobacco Research, 14*, 586–595.

Nair, M. K., Chetty, D. J., Ho, H., & Chien, Y. W. (1997). Biomembrane permeation of nicotine: mechanistic studies with porcine mucosae and skin. *Journal of Pharmacological Sciences, 86*(2), 257–262.

Naqvi, N. H., Rudrauf, D., Damasio, H., & Bechara, A. (2007). Damage to the insula disrupts addiction to cigarette smoking. *Science, 315*, 531–534.

National Institutes of Health (NIH). (1996). *The FTC cigarette test method for determining tar, nicotine and carbon monoxide yields of U.S. cigarettes. Smoking and tobacco control monograph 7*. Bethesda, MD: National Institutes of Health.

Picciotto, M. R., Addy, N. A., Mineur, Y. S., & Brunzell, D. H. (2008). It is not "either/or": activation and desensitization of nicotinic acetylcholine receptors both contribute to behaviors related to nicotine addiction and mood. *Progress in Neurobiology, 84*, 329–342.

Picciotto, M. R., Zoli, M., Lena, C., Bessis, A., Lallemand, Y., Le Novere, N., ... Changeux, J.-P. (1995). Abnormal avoidance learning in mice lacking functional high-affinity nicotine receptor in the brain. *Nature, 374*, 65–67.

Rose, J. E., Mukhin, A. G., Lokitz, S. J., Turkington, T. G., Herskovic, J., Behm, F. M., ... Garg, P. K. (2010). Kinetics of brain nicotine accumulation in dependent and nondependent smokers assessed with PET and cigarettes containing ^{11}C-nicotine. *Proceedings of the National Academy of Sciences, 107*(11), 5190–5195.

Ryan, R. E., Ross, S. A., Drago, J., & Loiacono, R. E. (2001). Dose-related neuroprotective effects of chronic nicotine in 6-hydroxydopamine treated rats, and loss of neuroprotection in alpha4 nicotinic receptor subunit knockout mice. *British Journal of Pharmacology, 132*, 1650–1656.

Schneider, N. G., Lunell, E., Olmstead, R. E., & Fagerström, K. O. (1996). Clinical pharmacokinetics of nasal nicotine delivery: a review and comparison to other nicotine systems. *Clinical Pharmacokinetics, 31*(1), 65–80.

Seeman, J. I., & Carchman, R. A. (2008). The possible role of ammonia toxicity on the exposure, deposition, retention, and the bioavailability of nicotine during smoking. *Food and Chemical Toxicology, 46*, 1863–1881.

Smit, A. B., Syed, N. I., Schaap, D., van Minnen, J., Klumperman, J., Kits, K. S., ... Geraerts, W. P. (2001). A glia-derived acetylcholine binding protein that modulates synaptic transmission. *Nature, 411*, 261–268.

Stratton, K., Shetty, P., Wallace, R., & Bondurant, S. (Eds.). (2001). *Institute of Medicine: Clearing the smoke: Assessing the science base for tobacco harm reduction*. Washington, DC: National Academy Press (Chapters 4 and 9).

Tyroller, S., Zwickenpflug, W., & Richter, E. (2002). New sources of dietary myosmine uptake from cereals, fruits, vegetables, and milk. *Journal of Agricultural and Food Chemistry, 50*(17), 4909–4915.

Unwin, N., Miyazawa, A., Li, J., & Fujiyoshi, Y. (2002). Activation of the nicotinic acetylcholine receptor involves a switch in conformation of the alpha subunits. *Journal of Molecular Biology, 319*, 1165–1176.

U.S. Department of Health and Human Services (USDHHS). (2014). *The health consequences of smoking-50 years of progress: A report of the surgeon general*. Atlanta, GA: U.S. Department of Health and Human Services, Centers for Disease Control and Prevention, National Center for Chronic Disease Prevention and Health Promotion, Office on Smoking and Health.

Vansickel, A. R., Cobb, C. O., Weaver, M. F., & Eissenberg, T. E. (2010). A clinical laboratory model for evaluating the acute effects of electronic "cigarettes": nicotine delivery profile and cardiovascular and subjective effects. *Cancer Epidemiology Biomarkers and Prevention, 19*(8), 1945–1953.

Vansickel, A. R., & Eissenberg, T. (2013). Electronic cigarettes: effective nicotine delivery after acute administration. *Nicotine and Tobacco Research, 15*(1), 267–270.

Willems, E. W., Rambali, B., Vleeming, W., Opperhuizen, A., & van Amsterdam, J. G. C. (2006). Significance of ammonium compounds on nicotine exposure to cigarette smokers. *Food and Chemical Toxicology, 44*, 678–688.

Willits, C. O., Swaim, M. L., Connelly, J. A., & Brice, B. A. (1950). Spectrophotometric determination of nicotine. *Analytical Chemistry, 22*, 430–433.

Wooltorton, J. R., Pidoplichko, V. I., Broide, R. S., & Dani, J. A. (2003). Differential desensitization and distribution of nicotinic acetylcholine receptor subtypes in midbrain dopamine areas. *Journal of Neuroscience, 23*, 3176–3185.

World Health Organization (WHO). (2002). *The tobacco atlas*. Geneva: Dr. Jusith Mackay & Dr. Michael Eriksen. Retrieved from: http://www.who.int/tobacco/statistics/tobacco_atlas/en/print.html.

Yildiz, D. (2004). Nicotine, its metabolism and an overview of its biological effects. *Toxicon, 43*, 619–632.

Chapter 10

Tobacco-Related Mortality among Individuals with Alcohol or Drug Use Disorders

Jodi M. Gatley[1,2,3], Russell C. Callaghan[1,2,3]

[1]Northern Medical Program, University of Northern British Columbia (UNBC), Prince George, BC, Canada; [2]Human Brain Laboratory, Centre for Addiction and Mental Health (CAMH), Toronto, ON, Canada; [3]Dalla Lana School of Public Health, University of Toronto, Toronto, ON, Canada

Abbreviations

AODUD Alcohol and other drug use disorder
CDC Centers for Disease Control and Prevention
ICD-9 International Classification of Diseases, 9th revision
ICD-10 International Classification of Diseases, 10th revision
SAF Smoking-attributable fraction
SAM Smoking-attributable mortality
SAMMEC Smoking-Attributable Mortality, Morbidity, and Economic Costs

INTRODUCTION

Individuals with alcohol and other drug use disorder (AODUD) not only have an elevated prevalence of tobacco use (ranging from two to four times that of the general population (see Figure 1)), but also tend to smoke more heavily. Tobacco smoking is a well-known, causal factor contributing to a number of tobacco-related mortality conditions, including cardiovascular conditions, respiratory diseases, and malignant neoplasms. While the prominence of tobacco use among individuals with AODUD points toward a potentially increased risk for tobacco-related illnesses, there is a surprising lack of research on patterns of tobacco-related mortality in these groups. Only a few cohort studies have directly examined the impact of smoking on mortality outcomes among persons with AODUD, and most of the relevant scientific literature has not estimated patterns of specific tobacco-related mortality in these groups. It is critically important to describe the long-term impacts of tobacco use among individuals with AODUD on mortality, not only to document the potentially substantial burden of tobacco-related diseases in this population, but also to encourage and support further integration of smoking cessation interventions into alcohol/drug treatment programs. At this time, nicotine dependence is not well addressed in clinical or public health interventions designed for these groups. This chapter aims to: (1) review key evidence describing the risk of tobacco-related mortality among individuals with AODUD, (2) describe the scope of current integration of smoking cessation protocols into addiction treatment, (3) assess the literature describing outcomes of smoking cessation interventions within substance abuse treatment settings, and (4) address the key barriers to fully incorporating smoking cessation into the status quo and strategies for improvement.

PREVALENCE OF TOBACCO USE IN AODUD POPULATIONS

The first step in estimating the full burden of tobacco-related mortality is to assess smoking prevalence among individuals with AODUD. In a systematic review, Guydish et al. (2011) estimated current smoking status among treatment-seeking individuals with alcohol, cocaine, opioids/narcotics, or polydrug use as their primary concern. This large-scale review provided current tobacco smoking prevalence estimates across cohorts as follows: alcohol—74%, cocaine—75%, opioids/narcotics—79%, and alcohol and/or drug—75%. As of this writing, to our understanding, this chapter provides the most up-to-date appraisal of smoking prevalence among persons engaged in substance abuse treatment.

Nonetheless, this review has a number of limitations, most notably the lack of smoking prevalence estimates for treatment-seeking methamphetamine or marijuana users—two groups comprising a substantial segment of treatment admissions. For example, in the Treatment Episode Data Set (TEDS)—a database capturing all publicly funded addiction treatment admissions in the United States—marijuana/hashish as the primary drug problem accounted for 17.5% of all addiction treatment admissions in 2012, while methamphetamine/amphetamines accounted for 7.1%. To supplement the Guydish et al. (2011) review, we conducted a search for all studies assessing smoking prevalence among primary methamphetamine users or primary marijuana users enrolled in formal addiction treatment. In Figure 1, we provide visual estimates for the alcohol, cocaine, and opioid/narcotics cohorts—derived from the Guydish et al. (2011) review—and for the methamphetamine and marijuana cohorts from our informal review.

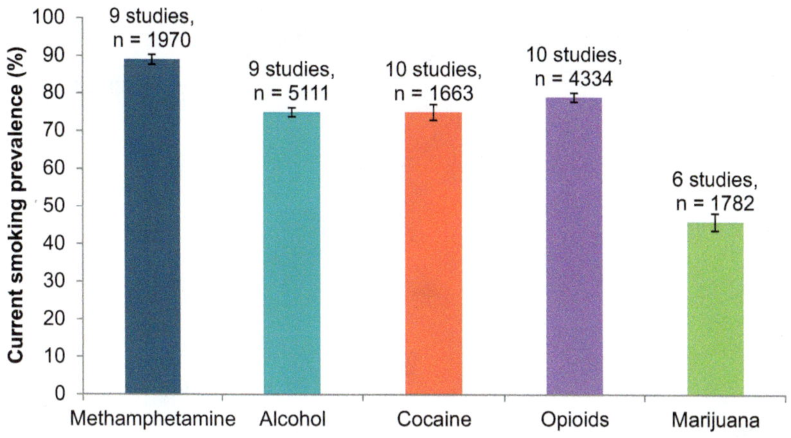

FIGURE 1 Prevalence of current smoking across alcohol or drug use disorder treatment populations (Callaghan, Gatley, Sykes, & Taylor, 2015). The cohort smoking prevalences presented are weighted estimates from studies of substance users assessing treatment services. Citations are available from the authors upon request. *Note: The lines at the end of the bars represent the upper and lower 95% confidence intervals for each estimate.*

Not only do individuals with AODUD have a significantly higher prevalence of smoking, but they also smoke with greater intensity. For example, a national US survey in 1991–1992 showed that individuals with psychiatric disorders and substance abuse disorders accounted for 44% of all cigarettes sold, a disproportionately large value considering they comprised less than one-quarter of the general population (Lasser et al., 2000). Daily smokers in substance abuse treatment also smoke substantially more cigarettes per day than do daily smokers in the general population; for example, a review of individuals in substance abuse treatment found that they smoked on average 24.4 cigarettes per day (Prochaska, Delucchi, & Hall, 2004)—an amount that was approximately 50% greater than the daily quantity consumed by smokers in the US general population in 2005 (Agaku, King, & Dube, 2014). It is not clear why smoking prevalence is so elevated among the AODUD population. There may be shared factors between nicotine addiction and other substance addictions that promote their comorbidity, including genetic predisposition, neurophysiological abnormalities, and neurotransmitter pathways (Grant, Hasin, Chou, Stinson, & Dawson, 2004; Kalman, Morissette, & George, 2005).

TOBACCO-RELATED DISEASES

Tobacco users have poorer overall somatic and mental health, higher medical service utilization and work absenteeism, and higher risk of all-cause mortality than nonsmokers (U.S. Department of Health and Human Services., 2014). There is a large body of evidence linking tobacco use to a number of mortality outcomes, mainly from cardiovascular diseases, respiratory diseases, and malignant neoplasms. The Centers for Disease Control and Prevention (CDC) have a system to estimate the burden of tobacco-related morbidity and mortality in the United States—Smoking-Attributable Mortality, Morbidity, and Economic Costs (SAMMEC) (U.S. Department of Health and Human Services., 2014). SAMMEC lists diseases that are considered to be caused by smoking and allows the calculation of smoking-attributable mortality (SAM). The SAM for a disease is determined using the smoking-attributable fraction (SAF), which is the proportion of the mortality from a particular disease that is attributable to smoking. As of 2014 the SAMMEC list includes 23 diseases and their International Classification of Diseases (ICD), revisions 9 and 10, definitions (see Table 1).

The SAF is defined as follows:

$$SAF = \frac{[(p_0 + p_1(RR_1) + p_2(RR_2)) - 1]}{[p_0 + p_1(RR_1) + p_2(RR_2)]}$$

The proportion of never smokers in the cohort is p_0, p_1 is the proportion of current smokers in the cohort, p_2 is the proportion of former smokers in the cohort, RR_1 is the relative risk of mortality from a condition for current smokers in comparison to never smokers, and RR_2 is the relative risk of mortality from a condition for former smokers in comparison to never smokers.

This equation relies on a number of factors: the estimated prevalence of current smoking in the target population, the estimated prevalence of former smokers in the target population, and the relative risk of developing the tobacco-related condition in the target group vis-à-vis never smokers—all of whom must be at least 35 years of age. Relative risk estimates for each of the CDC's tobacco-related conditions are drawn from the American Cancer Society's Cancer Prevention Study II (CPS-II), an ongoing longitudinal, population-based cohort study of smokers, former smokers, and never smokers in the United States, with 24 years of mortality data currently available since it began in 1982 (American Cancer Society, 2014). The CDC relative risk estimates are based on findings from a 6-year period of the CPS-II (1982–1988). Stratified age-based relative risks are also considered for ischemic (coronary) heart disease and cerebrovascular disease, with different risk ratios for adults ages 35–64 years and those ages 65 years and over, since the risk of mortality decreases steeply at the age of 65 years and onward (U.S. Department of Health and Human Services, 2014).

The SAF approach does raise some potentially imposing obstacles for researchers to estimate smoking-related morbidity and mortality among individuals with AODUD. The SAF approach works for populations at least 35 years of age. The primary reason for this age cutoff is that tobacco-related conditions tend to have long latency periods (usually lasting several decades) between smoking initiation and outcome, and in order not to overestimate tobacco-related burdens, the field has adopted this age standard (U.S. Department of Health and Human Services, 2014). A large proportion of individuals with AODUD, however, tend to be younger than the traditional age cutoff for SAM or morbidity, especially among those with drug use disorders. In the TEDS admissions

TABLE 1 ICD-9 and ICD-10 Codes and Comparability Ratios for 19 Tobacco-Related Mortality Conditions

Disease Category	ICD-10	ICD-9	Comparability Ratio
Malignant Neoplasms			
Lip, oral cavity, pharynx	C00–C14	140–149	0.9600
Esophagus	C15	150	0.9970
Stomach	C16	151	1.0063
Pancreas	C25	157	0.9980
Larynx	C32	161	1.0050
Trachea, lung, bronchus	C33–C34	162	0.9840
Cervix uteri	C53	180	0.9870
Kidney and renal pelvis	C64–C65	189	1.00
Urinary bladder	C67	188	0.9970
Acute myeloid leukemia	C92	205	1.0119
Colon, rectum, and anus	C18–C21	153–154	0.9993
Liver and intrahepatic bile ducts	C22	155	0.9634
Cardiovascular Disease			
Ischemic heart disease	I20–I25	410–414, 429.2	0.9990
Other heart disease	I00–I09, I26–I51	390–398, 415–417, 420–429.1, 429.3–429.9	0.9690
Cerebrovascular disease	I60–I69	430–438	1.0590
Atherosclerosis	I70	440	0.9640
Aortic aneurysm	I71	441	1.0010
Other arterial diseases	I72–I78	442–448	0.8500
Respiratory Diseases			
Pneumonia, influenza	J10–J18	480–487	0.6980
Bronchitis, emphysema	J40–J43	490–492	0.8940
Chronic airway obstruction	J44	496	1.0970
Tuberculosis	A16–A19	010–018	0.8547
Other Diseases			
Diabetes mellitus	E10–E14	250	1.0082

Centers for Disease Control and Prevention (2013).

system, the average ages of admissions presenting with AODUDs to publicly funded addiction treatment in the United States in 2012 were as follows: alcohol—41 years, cocaine—40.4 years, methamphetamine/amphetamines—33 years, marijuana—25 years, and opiates—33.3 years (Center for Behavioral Health Statistics and Quality, Substance Abuse and Mental Health Services Administration, Treatment Episode Data Set (TEDS), 2013). Of the 1.75 million total treatment admissions to publicly funded addiction programs in the United States in 2012, approximately 55% were less than 35 years of age at treatment entry; the average age for persons admitted to treatment with primary methamphetamine/amphetamines, marijuana, or opioid problems was less than 35 years. So, if a researcher wanted to select a treatment sample of individuals with AODUD, it is likely that a substantial proportion of those engaged in addiction treatment would be excluded from studies attempting to assess tobacco-related mortality in samples at least 35 years of age.

Aside from the traditional age cutoff, there are number of other obstacles for using the SAF approach in AODUD groups. The standard approach draws upon relative risk estimates comparing risk trajectories of smokers (or former smokers) versus nonsmokers. At this time, the research field has not yet provided relative risk estimates specifically for AODUD populations, probably because of the research challenges in doing so. The key obstacles in such research are collecting large samples of individuals with AODUD older than 35 years at baseline, ensuring an adequate prevalence of never smokers to serve as the reference group, generating sufficient person-years of follow-up to estimate even the low-incident tobacco-related morbidity and mortality conditions, and preventing loss to follow-up in hard-to-track AODUD groups. Estimating relative risks for tobacco-related conditions in AODUD populations is extremely important, especially as standard, general population, relative risk estimates may not hold. For example, the relative risks of head and neck cancers among individuals with alcohol use disorders are likely to be different from those found in the general population, primarily because of the synergistic relation between alcohol use and smoking vis-à-vis head and neck cancers (Goldstein, Chang, Hashibe, La Vecchia, & Zhang, 2010; Pelucchi, Gallus, Garavello, Bosetti, & La Vecchia, 2008). At this time, very little is known about the potential synergistic effects of tobacco use and other illicit drugs on the development of tobacco-related morbidity and mortality. Future research should aim to close this gap in the literature.

A REVIEW OF TOBACCO-RELATED MORTALITY AMONG INDIVIDUALS WITH AODUD

Given the potential difficulties in estimating tobacco-related morbidity and mortality in AODUD populations, it is, perhaps, not surprising that little is known about the long-term impact of tobacco use on mortality patterns among individuals with AODUD. While a number of studies have estimated mortality for broad classifications of cardiovascular disease, respiratory illnesses, and cancers, or the relation between smoking frequency and overall mortality in AODUD populations (Engstrom, Adamsson, Allebeck, & Rydberg, 1991; Hiroeh et al., 2008; Hser, Anglin, & Powers, 1993; Roerecke & Rehm, 2013; Rossow & Amundsen, 1997), this literature has rarely addressed tobacco-specific morbidity or mortality conditions, as defined by the international community. Usually, this literature provides estimates for broad outcome classifications, such as "neoplasms"—which includes both tobacco-related and non-tobacco-related cancers. This literature does provide some clues to the elevated risk of mortality from cardiovascular conditions, respiratory illnesses, and cancers, but it does not provide adequate information for the calculation of the full burden of tobacco-related illnesses in this population. Only three cohort studies have directly examined the impact of tobacco use on tobacco-related mortality patterns among AODUD treatment samples, either by calculating relative mortality risk in relation to smoking prevalence or by defining a subset of smoking-related diseases to estimate smoking-related mortality. There are several limitations to the reliability of the prior literature, including small sample sizes; a lack of, or changes in, definitions of tobacco-related mortality; primarily male samples; lack of analysis by gender; unknown smoking prevalence of the sample; and no population-level control groups.

Using a 20-year follow-up design, Hser, McCarthy, and Anglin (1994) conducted the first longitudinal study of tobacco-related mortality among drug users—male narcotics offenders mandated to inpatient/outpatient treatment primarily for heroin dependence in California from 1962 to 1964 ($n=405$), with 20 years of follow-up. The researchers found not only that current smokers had a fourfold higher death rate than nonsmokers, but also that 16% of the cumulative mortality was tobacco-related as defined by the CDC (a statistic based on 77 total deaths in the cohort) (Hser et al., 1994).

In the touchstone study in the field, Hurt and his colleagues assessed tobacco-related mortality in a sample of mainly white males ($n=845$; 65% men) undertaking residential addiction treatment (primarily for alcohol-related problems) in Minnesota from 1972 to 1983. After a 22-year follow-up, they identified tobacco-related deaths using CDC guidelines and generated standardized mortality rates in their sample versus the US population. Just over half of participants' cumulative mortality (based on 214 total death records) was due to tobacco-related conditions—a rate surpassing even that from alcohol-related conditions, which accounted for one-third of the deaths (Hurt et al., 1996). The risk of tobacco-related deaths for substance abuse treatment patients over the study period was double the risk in the general population. The most commonly mentioned condition on death certificates (46%) was coronary artery disease, which is tobacco-related.

A study on smoking-related mortality followed large alcohol and other drug-dependent cohorts, approximately 650,000 individuin total, in California from 1990 to 2005 for up to 16 years using inpatient hospital records linked to death records for the following diagnoses (Callaghan et al., 2015): alcohol-, cocaine-, opioid-, marijuana-, and methamphetamine-dependence. Individuals admitted with appendicitis were used as a population proxy comparison group. Mortality was identified as tobacco-related if the primary cause was any the 19 diseases identified by the CDC as being causally linked to tobacco use (see Table 1). Tobacco-related conditions comprised a substantial portion of total deaths in the alcohol and other drug-dependent cohorts, ranging from 36% to 49%. SAM fractions of total mortality for males and females in the cohorts ranged from approximately 20% to 28% and were similar for males and females. Alcohol and drug-dependent cohorts had substantially higher SAM fractions than their demographically matched (e.g., age, sex, and race) population-proxy controls (see Figure 2). Standardized mortality ratios (SMRs), adjusted for age, sex, race, date of index admission, and county of residence, were calculated relative to the general population in 2000. Cohort SMRs suggested that the alcohol and other drug-dependent samples were 2.4–4.3 times more likely to die from any tobacco-linked disease than the controls (see Figure 3). Furthermore, analyses of SMRs by the broad disease categories of cardiovascular diseases, malignant neoplasms, and respiratory diseases demonstrated similarly elevated mortality risks across categories and cohorts (see Figure 4). This research provided one of the first comprehensive assessments of the relative risk of tobacco-related mortality in AODUD populations in comparison to population-level patterns.

This small literature has provided important findings about the large burden of mortality from numerous tobacco-related diseases that AODUD populations face. These individuals are at much greater risk of dying from a tobacco-related disease than someone in the general population. Additionally, results suggest that tobacco smoking may be a more urgent contributing factor to mortality among AODUD populations than the primary substance of abuse.

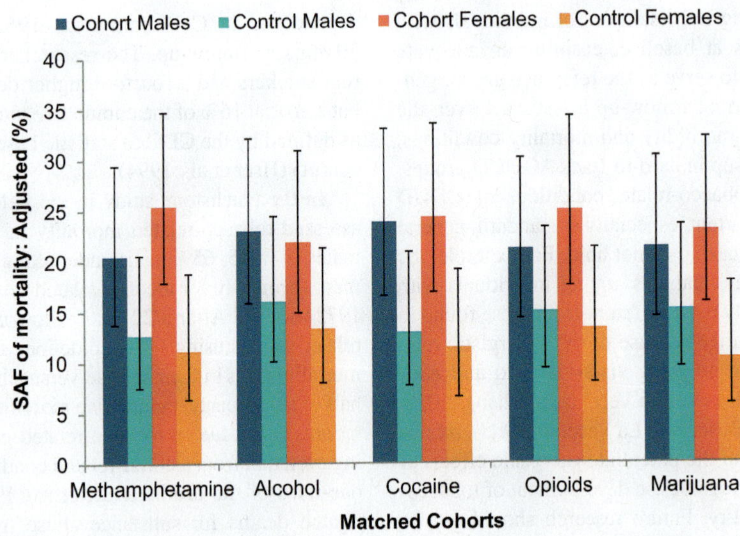

FIGURE 2 SAFs of mortality for appendicitis and drug/alcohol cohort groups matched by age, sex, race, date of index admission, and county of residence.

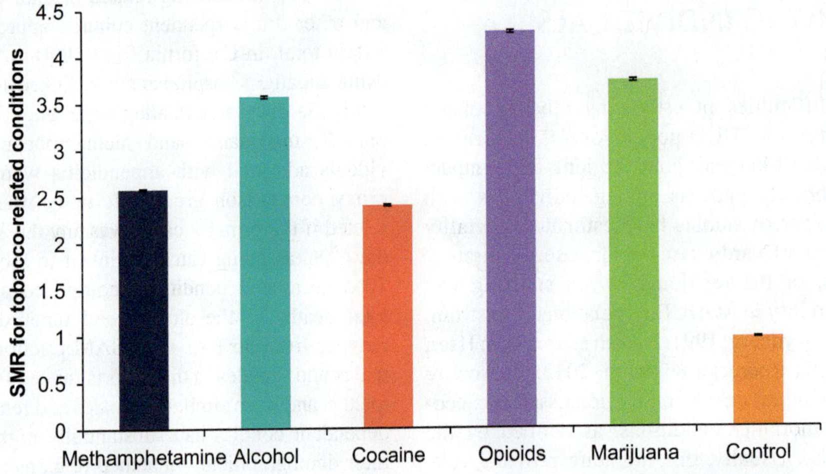

FIGURE 3 **SMRs for tobacco-related conditions across drug/alcohol and appendicitis cohorts**. *Note: The lines at the end of the bars represent the upper and lower 95% confidence intervals for each estimate.*

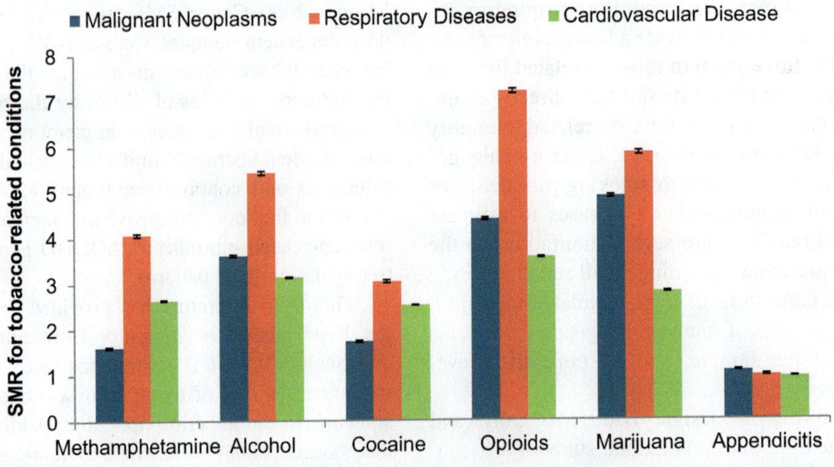

FIGURE 4 **SMRs for tobacco-related conditions across drug/alcohol and appendicitis cohorts, for malignant neoplasms, respiratory diseases, and cardiovascular disease**. *Note: The lines at the end of the bars represent the upper and lower 95% confidence intervals for each estimate.*

BEST-PRACTICE GUIDELINES CHAMPION SMOKING CESSATION INTEGRATION INTO ADDICTIONS TREATMENT

Given the high prevalence of smoking in AODUD populations, a number of national best-practice guidelines have championed the smoking cessation protocols in these groups. For example, the 2011 Canadian Smoking Cessation Clinical Practice Guideline stated that health care providers should do the following for all patients with substance abuse issues: assess them for tobacco addiction, offer counseling and pharmacotherapy, and monitor their addictions to other substances during concurrent treatment (CAN-ADAPTT, 2011). In the United States, the 2008 Clinical Practice Guidelines for Treating Tobacco Use and Dependence recommended that tobacco cessation treatment be offered to all individuals with substance use disorders (Fiore et al., 2008). Similar guidelines in Australia also recommend smoking cessation treatment be offered to AODUD populations (Zwar et al., 2011); however, in the United Kingdom, the most recent national guideline for health professionals does not specifically discuss smoking cessation for individuals with AODUD or identify them as a high-risk group (West, McNeill, & Raw, 2000). Endorsements of concurrent tobacco cessation treatment also suggest delaying it during periods of serious psychiatric symptoms, however, as nicotine withdrawal may have a negative effect on outcomes (Fiore et al., 2008).

BARRIERS TO IMPLEMENTATION OF CONCURRENT SMOKING CESSATION WITH AODUD TREATMENT

Despite the recent calls for increased recognition and treatment of nicotine dependence in AODUD populations, there is still a need for improved integration of accessible smoking cessation services with other addiction treatment services. Although this population has been recognized to have increased tobacco use and to be especially at risk for tobacco-related harms, and concurrent treatment is recommended in national clinical guidelines, few alcohol and drug treatment facilities offer smoking cessation protocols. In the United States, for example, staff surveys of inpatient/outpatient treatment have found that some form of integrated smoking cessation was available at only 17–41% of facilities (Cupertino et al., 2013; Friedmann, Jiang, & Richter, 2008; Knudsen, Studts, & Studts, 2012); and pharmacotherapy was offered at only 14–16% (Cupertino et al., 2013). The situation appears similar in Australia (Walsh, Bowman, Tzelepis, & Lecathelinais, 2005) and Canada (Currie, Nesbitt, Wood, & Lawson, 2003). At a minimum most facilities record smoking status; providing brief advice to quit is also common (Cupertino et al., 2013). However, the treatment services that are demonstrated to be most effective, counseling and pharmacotherapy in combination, are much less available to clients in substance abuse treatment (Cupertino et al., 2013).

STAFF KNOWLEDGE AND ATTITUDES ABOUT CONCURRENT SMOKING CESSATION DURING AODUD TREATMENT

Improving access to tobacco cessation treatment for substance abuse treatment patients is difficult because of the numerous barriers, mainly resource limitations, including low levels of staff training and knowledge, limited staff time/availability, and a need for reimbursement, as well as staff attitudes and their own use of tobacco (Guydish, Passalacqua, Tajima, & Manser, 2007). The barrier to concurrent smoking cessation treatment most often reported is a lack of training and knowledge among providers to assess and treat client smoking (Guydish et al., 2007). In a study of Australian substance abuse treatment programs, 55% of managers and 67% of other staff had never been trained in administering smoking cessation treatment (Walsh et al., 2005). There has also been a trend in normalization of patient cigarette smoking in mental health and addiction treatment settings over the past several decades, due to an exclusive focus on mental health or addiction treatment, apparent unawareness of associated tobacco-related health detriments, and beliefs that clients do not want to quit smoking or, even if motivated, cannot achieve success in this area (Richter, Hunt, Cupertino, Garrett, & Friedmann, 2012). It may also be that tobacco, relative to the primary substance of abuse, is not perceived to cause immediate legal, social, or health problems by treatment providers (Richter et al., 2012). Providers also frequently report views that treating smoking concurrent with another drug will harm outcomes for that drug, that treating other drugs is more important, that smoking provides beneficial stress relief during drug treatment, and that it is not their responsibility (Guydish et al., 2007; Hunt et al., 2014). Prevalent provider smoking of 14–40% is an additional barrier because it influences willingness to treat smoking (Guydish et al., 2007; Hunt, Cupertino, Garrett, Friedmann, & Richter, 2012). Provider smoking was associated with a lower chance of having provided smoking cessation treatment or supporting concurrent treatment compared to nonsmoking providers (Guydish et al., 2007). Furthermore, provider smoking leads to an ambivalent attitude toward smoking treatment since it is perceived that including tobacco treatment policy would place pressure on staff to quit themselves or risk undermining the tobacco treatment of clients and possibly threatening their job security (Richter et al., 2012). There is a need to increase resources and incentives for AODUD treatment facilities to implement or improve smoking cessation initiatives, such as through public health policy mandates and government funding.

ATTITUDES TOWARD SMOKING CESSATION AMONG AODUD TREATMENT CLIENTS

The view that individuals in AODUD treatment do not want to quit smoking is not supported by surveys of these patients; self-reported interest in nicotine dependence treatment is high. Between 58% and 81% of AODUD treatment patients reported a desire to quit smoking, and at AODUD treatment facilities that did not offer smoking cessation treatment (Clarke, Stein, McGarry, & Gogineni, 2001; Clemmey, Brooner, Chutuape, Kidorf, & Stitzer, 1997; DiFranza & Guerrera, 1990; Frosch, Shoptaw, Jarvik, Rawson, & Ling, 1998; Joseph, Lexau, Willenbring, Nugent, & Nelson, 2004; Nahvi, Richter, Li, Modali, & Arnsten, 2006; Richter, Gibson, Ahluwalia, & Schmelzle, 2001), 57–76% of patients stated that they would participate in concurrent smoking cessation treatment if it were available (Clemmey et al., 1997; Frosch et al., 1998; Nahvi et al., 2006; Richter et al., 2001).

The high level of interest in concurrent smoking cessation treatment among those in treatment for AODUD, in relation to the low availability of formal programs, demonstrates that this group is probably underserved in this area.

EFFECTIVENESS OF TREATMENT FOR NICOTINE DEPENDENCE CONCURRENT WITH AODUD TREATMENT

There is strong support that nicotine cessation treatment concurrent with substance abuse treatment, including counseling and nicotine replacement therapy, can improve smoking cessation outcomes, especially in the short term, and that smoking cessation interventions are, at least, not harmful and, at best, beneficial to primary substance outcomes. A review of 19 studies on concurrent smoking cessation treatment in substance abuse treatment/recovery programs evaluated smoking and substance use outcomes immediately posttreatment and at a follow-up of at least 6 months (Prochaska et al., 2004). Recipients of concurrent smoking cessation treatment had a twofold higher chance of posttreatment smoking abstinence than those not receiving it; however, there was no long-term difference in smoking cessation. Additionally, smoking cessation treatment was associated with a long-term decrease in primary substance abuse, which decreased by 25% relative to those not receiving it (Prochaska et al., 2004). Several more recent trials of smoking cessation interventions concurrent with other treatment for AODUD have found good short-term smoking abstinence rates (Winhusen et al., 2013); in the long-term smoking abstinence tends to decrease sharply, although continued effects have been noted (Baca & Yahne, 2009; Reid et al., 2008; Shoptaw et al., 2002; Winhusen et al., 2013). Importantly, in most cases concurrent smoking cessation does not have a negative impact on long-term abstinence from the primary substance of abuse (Hughes & Kalman, 2006; Reid et al., 2008) and may even be associated with better outcomes for primary substance abstinence (Shoptaw et al., 2002; Tsoh, Chi, Mertens, & Weisner, 2011; Winhusen et al., 2013). At least one study has suggested that individuals in treatment for AODUD were more willing to participate in smoking cessation treatment when it was offered concurrent with their other addictions treatment, rather than being delayed until the end of treatment for their primary substance (Kalman et al., 2001). Thus, an expansion of the range and availability of concurrent smoking cessation treatment offered to individuals in treatment with AODUD is a promising and safe method to address their very high smoking prevalence.

EFFORTS TO INCREASE ACCESS TO NICOTINE-DEPENDENCE TREATMENT FOR AODUD POPULATIONS

Some progress has been made to better evaluate the quality of nicotine-dependence treatment services in substance abuse treatment facilities and to improve staff perceptions of, and capability to provide, nicotine-dependence treatment. The Index of Tobacco Treatment Quality (ITTQ), a survey for treatment facility representatives, was developed by Cupertino et al. (2013), using input from experts, association with smoking cessation in the general population, and practical consideration. The ITTQ was intended to quickly assess the completeness of tobacco cessation services specifically for substance abuse treatment facilities and was found to have good test–retest reliability. Another evaluative tool, which aims to measure staff attitudes to smoking cessation treatment, the Tobacco Treatment Commitment Scale, was created by Hunt et al. (2014) using prior research and expert input and incorporating qualitative observations. Additionally, a 12-step model was proposed that would create the infrastructure for smoking cessation services and encourage positive staff participation, e.g., by offering smoking cessation services to staff (Ziedonis, Guydish, Williams, Steinberg, & Foulds, 2006). A collaborative initiative between the Mental Health Services Administration, a federal agency, and the Smoking Cessation Leadership Center at the University of California at San Francisco to improve smoking cessation for mental health and substance abuse populations had promising results. They established Leadership Academies, collaborations of smoking cessation agencies, in seven states between 2010 and 2012; at the 6-month follow-up most (six of seven) states reported increases in quit attempts among substance users and increased importance of smoking cessation to providers of substance abuse/mental health treatment. The report stressed the need for further state and national policy changes, including higher taxation, funding for tobacco control, and tobacco-free policies, to improve outcomes (Santhosh et al., 2014). Thus implementing concurrent smoking cessation treatment across AODUD treatment facilities will be a complex task requiring interventions at the institutional, local, and national levels.

OTHER POLICY INTERVENTIONS TO REDUCE TOBACCO-RELATED MORTALITY AMONG INDIVIDUALS WITH AODUD

Even though this chapter has emphasized the importance of more comprehensive integration of smoking cessation protocols into addiction treatment to reduce the burden of tobacco-related mortality in populations with AODUD, it is important to acknowledge the possible effects of other, standard tobacco-control policies. Since 1965, researchers have demonstrated consistently that taxation strategies designed to increase the price of tobacco products are the most effective means to reduce population-level smoking prevalence (Chaloupka, Yurekli, & Fong, 2012). However, the effects of taxation-based price increases are poorly understood, and may be less beneficial in populations with AODUD (or other psychiatric conditions) (Bader, Boisclair, & Ferrence, 2011)—a segment of the general population spending a relatively much greater proportion of annual income on tobacco products (up to 24% in New York, 2010–2011) (Farrelly, Nonnemaker, & Watson, 2012).

Some studies have suggested that poor, heavily nicotine-dependent smokers frequently turn to contraband (illicit) tobacco products to mitigate population-based price increases on tobacco. For example, while substantial tobacco tax increases reduced smoking among the general population in New York City, researchers found that these taxation strategies did not affect smoking prevalence of individuals in the lowest income strata in the population (which remained double that of the richest income strata)—probably because of the turn to much cheaper tobacco products in a surging black market (Farrelly et al., 2012). Studies of individuals in treatment for AODUD or psychiatric disorders have found that they consumed a larger proportion of contraband cigarettes than the general population (Callaghan, Tavares, & Taylor, 2008). Also, in the presence of a widespread illicit tobacco market, contraband smokers appear to be less likely to make a quit attempt or to quit smoking over time (Gariti et al., 2002; Mecredy, Diemert, Callaghan, & Cohen, 2013). It is also of concern that contraband tobacco use has been associated with other indicators of poor health, including an almost twofold chance of having a disability, as well as increased likelihood of poor mental health, physical pain, high frequency of smoking (greater than 120 cigarettes per week), and beginning smoking at younger than the legal age (Aitken, Fry, Farrell, & Pellegrini, 2009). A number of researchers have acknowledged that while population-level tobacco taxation may reduce general-population smoking prevalence, such taxation strategies place a disproportionately heavy burden on low socioeconomic status groups (Farrelly et al., 2012; Remler, 2004). As a result, these scholars have called for complementary outreach and targeted smoking cessation programs for low socioeconomic groups (and this would include a large proportion of individuals with AODUD) in the face of population-based taxation increases on tobacco products (Farrelly et al., 2012; Remler, 2004).

APPLICATIONS TO OTHER ADDICTIONS AND SUBSTANCE MISUSE

Other populations who are likely to be at a high risk of tobacco-related mortality due to heavy smoking are those with psychiatric disorders (Grant et al., 2004) and nonsubstance addictions such as gambling; these conditions also have a dramatic comorbidity with AODUD (Peters, Schwartz, Wang, O'Grady, & Blanco, 2014). Individuals with psychiatric disorders tend to be heavy smokers and are at increased risk of tobacco dependence (Grant et al., 2004). There is evidence that populations with psychiatric disorders have high levels of smoking and associated mortality. A large cohort study in California from 1990 to 2005 found that individuals ($n \sim 600,000$) diagnosed with schizophrenia, depressive disorders, or bipolar disorder were 1.6–2.5 times more likely to die from a tobacco-related condition than the general population. Mortality risk was elevated for all three broad disease categories (malignant neoplasms, cardiovascular diseases, respiratory diseases) across cohorts, with the exception of neoplasms for the bipolar cohort (Callaghan et al., 2013). Individuals in treatment for psychiatric disorders have similar difficulties in accessing concurrent tobacco cessation treatment, including staff with poor training and unsupportive views (Schroeder & Morris, 2010). Individuals with pathological gambling addiction also have a high prevalence of nicotine dependence (60%) and/or other substance use disorders (58–73%) (Lorains, Cowlishaw, & Thomas, 2011; McGrath & Barrett, 2009). Survey studies of pathological gamblers have also found associations between daily smoking, increased severity of gambling behavior, and psychiatric distress (McGrath & Barrett, 2009). There are no studies on tobacco-related morbidity or mortality among problem gamblers, although the SAM is likely to be elevated (given the relatively high prevalence of tobacco use). The integration of smoking cessation treatment into gambling problem treatment has been tentatively suggested; however, much more research is needed to determine the most appropriate course of treatment for this population (McGrath & Barrett, 2009).

CONCLUSION

There is compelling evidence that the AODUD population suffers from much higher rates of tobacco-attributable mortality than the general population. It is difficult to estimate confidently the actual health burden, as the research literature is still quite small. However, smoking is widespread and of greater intensity among those with AODUD, which in addition to the burden of premature smoking-related mortality, probably also increases their morbidity and negatively affects their quality of life. The beginning of treatment for AODUD is a critical point of clinical contact to offer concurrent treatment for tobacco cessation for these individuals. The AODUD treatment population shows demand for, and can significantly benefit from, simultaneous treatment for nicotine dependence. However, this critical intervention opportunity goes largely unused, owing to impeding factors including staff training and misconceptions, as well as perceived facility resource limitations. Several attempts have been made, as of this writing, to better assess and integrate smoking cessation treatment with treatment for AODUD. There remains an urgent need to improve the availability of smoking cessation treatment for those with AODUD so that their smoking prevalence may be reduced, as well as to study their tobacco-related harms in detail to inform health policy and clinical practice.

DEFINITION OF TERMS

Alcohol and other drug use disorder AODUD and other psychiatric disorders are most commonly defined/diagnosed using two major guidelines for researchers and clinicians: the *Diagnostic and Statistical Manual of Mental Disorders* (DSM), fifth edition, used primarily in the United States (it replaced the DSM, fourth edition (DSM-IV), in May 2013), or the ICD-10, used in most other countries. This chapter uses AODUD to refer either to dependence or to harmful use/abuse of alcohol or other drugs. Comparative reviews have found that the DSM-IV, the version utilized in most US research discussed in the chapter, and ICD-10, have very similar definitions of dependence on a substance, with some differences in defining abuse or harmful use. Overall the two schemes are considered to be generally comparable, especially for diagnosing dependence (Hasin, Hatzenbuehler, Keyes, & Ogburn, 2006).

Cohort study design This is a study in which an initial index sample, or cohort, is tracked over a period of time for outcomes of interest. The statistical soundness of cohort studies may be improved by a larger index sample size and a greater follow-up time.

Concurrent smoking cessation treatment This refers to smoking cessation treatment (counseling and/or pharmacotherapy, e.g., nicotine replacement therapy) provided at the same time as other outpatient/inpatient substance abuse treatment.

Contraband tobacco This is any tobacco product that is not in compliance with applicable laws (e.g., federal and provincial statutes in Canada), including requirements for importation, stamping, marking, manufacturing, distributing, and/or payment of duties and taxes (Royal Canadian Mounted Police, 2008). The Royal Canadian Mounted Police (2008) have proposed that a substantial proportion of the Canadian contraband tobacco market is manufactured on First Nations reserves (in Canada) or Native American communities in northern New York State.

International Classification of Diseases It is a diagnostic tool used as a standard for health research, policy, and clinical practices, established by the World Health Organization (WHO) to facilitate the tracking of morbidity and mortality statistics internationally (WHO member states). ICD coding is entered on many medical records, including death certificates and health records. Most Western countries have adopted the ICD-10, with the exception of the United States, where the ICD-9 is still the primary diagnostic tool, while ICD-10 is used to code mortality records.

Malignant neoplasm This is a general term for cancer.

Population proxy This is a sample selected for control purposes, which is considered to be reasonably representative of the general population in observed and unobserved characteristics.

Standardized mortality ratio This is a ratio of the mortality rate for the population of interest, e.g., individuals with substance abuse disorders, over the mortality rate for the general population, adjusted for demographic variables (such as age, race, and sex).

Smoking-attributable fraction This is the proportion of mortality in a given population, either of the total or caused by a given disease or condition, that can be directly attributed to tobacco use by that population.

Smoking-attributable mortality This is the proportion of mortality from tobacco-related conditions in a population that is considered to be causally related to that population's tobacco use patterns.

Tobacco-related condition This is any disease that has been listed by the CDC as being potentially caused by tobacco use among adults age 35 years and older.

KEY FACTS OF CONTRABAND TOBACCO USE IN THE AODUD POPULATION

- Individuals in treatment settings for alcohol or drug disorders are more likely to use contraband tobacco products, which are not taxed, and may promote greater smoking prevalence owing to their lower cost.
- A study that identified cigarette brands by filter-tip logos on discarded butts in Toronto, Ontario, Canada, found that at an inpatient psychiatric hospital and an outpatient addiction treatment and research facility, 54% and 16% of cigarettes, respectively, were contraband, compared to 6% at a general hospital (Callaghan et al., 2008).
- Another US sample of individuals in inpatient treatment for alcohol or drug use disorders found that 10% reported smoking primarily contraband tobacco, which may have contributed to their poor smoking cessation outcomes (Gariti et al., 2002).
- The higher prevalence of illicit tobacco use by alcohol- and drug-dependent individuals may increase their risk of tobacco-related health harms and make quitting more difficult. A survey of tobacco usage in Ontario found that 30-day abstinence was negatively associated with regular use of contraband cigarettes (Mecredy et al., 2013). Respondents who smoked contraband cigarettes also had higher frequency of smoking at baseline, had more severe perceived addiction, reported more barriers to quitting, and were more likely have had pharmacotherapy treatment for smoking cessation.
- A survey of Australian current smokers found that illicit tobacco use was associated with an almost twofold chance of having a disability, and lifetime illicit tobacco use was associated with poor mental health, increased physical pain, high frequency of smoking (greater than 120 cigarettes per week), and beginning smoking at younger than the legal age (Aitken et al., 2009).

SUMMARY POINTS

- Individuals with AODUD may be especially vulnerable to the serious health harms associated with smoking because of the high smoking prevalence and smoking intensity in this population.
- Mortality rates from all causes are elevated for substance abusers, but there is a lack of research on their tobacco-related mortality.
- New evidence estimating patterns of mortality from tobacco-related diseases among individuals with AODUD suggests that their risk of tobacco-related mortality is up to four times greater than in the general population.
- Access to tobacco cessation treatment for those in treatment with AODUD is limited, despite demonstrated interest in quitting smoking and evidence of the effectiveness and safety of concurrent smoking cessation treatment in alcohol/drug treatment programs.
- Difficulties exist in improving access to nicotine-dependency treatment in AODUD treatment facilities, including staff training and attitudes and institutional resource limitations.

REFERENCES

Agaku, I. T., King, B. A., & Dube, S. R. (2014). Current cigarette smoking among adults — United States, 2005–2012. *Morbidity and Mortality Weekly Report (MMWR)*, *63*(02), 29–34.

Aitken, C. K., Fry, T. R., Farrell, L., & Pellegrini, B. (2009). Smokers of illicit tobacco report significantly worse health than other smokers. *Nicotine & Tobacco Research*, *11*(8), 996–1001.

American Cancer Society. (2014). *Cancer prevention study II (CPS II)*. Retrieved July 30, 2014, from: http://www.cancer.org/research/researchtopreventcancer/currentcancerpreventionstudies/cancer-prevention-study.

Baca, C. T., & Yahne, C. E. (2009). Smoking cessation during substance abuse treatment: what you need to know. *Journal of Substance Abuse Treatment*, *36*(2), 205–219.

Bader, P., Boisclair, D., & Ferrence, R. (2011). Effects of tobacco taxation and pricing on smoking behavior in high risk populations: a knowledge synthesis. *International Journal of Environmental Research and Public Health*, *8*(11), 4118–4139.

Callaghan, R., Gatley, J. M., Sykes, J., & Taylor, L. (2015). The prominence of tobacco-related mortality among individuals with alcohol- or drug-use disorders. (Unpublished).

Callaghan, R. C., Tavares, J., & Taylor, L. (2008). Another example of an illicit cigarette market: a study of psychiatric patients in Toronto, Ontario. *American Journal of Public Health*, *98*(1), 4–5.

Callaghan, R. C., Veldhuizen, S., Jeysingh, T., Orlan, C., Graham, C., Kakouris, G., ... Gatley, J. (2013). Patterns of tobacco-related mortality among individuals diagnosed with schizophrenia, bipolar disorder, or depression. *Journal of Psychiatric Research*, *48*(1), 102–110.

CAN-ADAPTT. (2011). *Canadian smoking cessation clinical practice guideline*. Toronto, ON: Canadian Action Network for the Advancement, Dissemination, and Adoption of Practice-informed Tobacco Treatment, Centre for Addiction and Mental Health.

Center for Behavioral Health Statistics and Quality, Substance Abuse and Mental Health Services Administration, Treatment Episode Data Set (TEDS). (2013). *Table 2.1a. Admissions aged 12 and older, by gender and age at admission according to primary substance of abuse: 2012*. Retrieved July 30, 2014, from: http://www.samhsa.gov/data/2K14/TEDS2012NA/TEDS2012NTbl2.1a.htm.

Centers for Disease Control and Prevention. (2013). *Smoking-attributable mortality, morbidity, and economic costs (SAMMEC)*. Retrieved July 30, 2014, from: https://apps.nccd.cdc.gov/sammec/index.asp.

Chaloupka, F. J., Yurekli, A., & Fong, G. T. (2012). Tobacco taxes as a tobacco control strategy. *Tobacco Control, 21*(2), 172–180.

Clarke, J. G., Stein, M. D., McGarry, K. A., & Gogineni, A. (2001). Interest in smoking cessation among injection drug users. *The American Journal on Addictions, 10*(2), 159–166.

Clemmey, P., Brooner, R., Chutuape, M. A., Kidorf, M., & Stitzer, M. (1997). Smoking habits and attitudes in a methadone maintenance treatment population. *Drug and Alcohol Dependence, 44*(2–3), 123–132.

Cupertino, A. P., Hunt, J. J., Gajewski, B. J., Jiang, Y., Marquis, J., Friedmann, P. D., ... Richter, K. P. (2013). The index of tobacco treatment quality: development of a tool to assess evidence-based treatment in a national sample of drug treatment facilities. *Substance Abuse Treatment, Prevention, and Policy, 8*(1), 13.

Currie, S. R., Nesbitt, K., Wood, C., & Lawson, A. (2003). Survey of smoking cessation services in Canadian addiction programs. *Journal of Substance Abuse Treatment, 24*(1), 59–65.

DiFranza, J. R., & Guerrera, M. P. (1990). Alcoholism and smoking. *Journal of Studies on Alcohol, 51*(2), 130–135.

Engstrom, A., Adamsson, C., Allebeck, P., & Rydberg, U. (1991). Mortality in patients with substance abuse: a follow-up in Stockholm County, 1973–1984. *The International Journal of the Addictions, 26*(1), 91–106.

Farrelly, M. C., Nonnemaker, J. M., & Watson, K. A. (2012). The consequences of high cigarette excise taxes for low-income smokers. *PloS One, 7*(9), e43838.

Fiore, M. C., Jaen, C. R., Baker, T. B., Bailey, W. C., Benowitz, N. L., Curry, S. J., ... Wewers, M. E. (2008). *Treating tobacco use and dependence: 2008 update. Clinical practice guideline*. Rockville, MD: U.S. Department of Health and Human Services. Public Health Service.

Friedmann, P. D., Jiang, L., & Richter, K. P. (2008). Cigarette smoking cessation services in outpatient substance abuse treatment programs in the United States. *Journal of Substance Abuse Treatment, 34*(2), 165–172.

Frosch, D. L., Shoptaw, S., Jarvik, M. E., Rawson, R. A., & Ling, W. (1998). Interest in smoking cessation among methadone maintained outpatients. *Journal of Addictive Diseases, 17*(2), 9–19.

Gariti, P., Alterman, A., Mulvaney, F., Mechanic, K., Dhopesh, V., Yu, E., ... Sacks, D. (2002). Nicotine intervention during detoxification and treatment for other substance use. *The American Journal of Drug and Alcohol Abuse, 28*(4), 671–679.

Goldstein, B. Y., Chang, S. C., Hashibe, M., La Vecchia, C., & Zhang, Z. F. (2010). Alcohol consumption and cancers of the oral cavity and pharynx from 1988 to 2009: an update. *European Journal of Cancer Prevention, 19*(6), 431–465.

Grant, B. F., Hasin, D. S., Chou, S. P., Stinson, F. S., & Dawson, D. A. (2004). Nicotine dependence and psychiatric disorders in the United States: results from the national epidemiologic survey on alcohol and related conditions. *Archives of General Psychiatry, 61*(11), 1107–1115.

Guydish, J., Passalacqua, E., Tajima, B., Chan, M., Chun, J., & Bostrom, A. (2011). Smoking prevalence in addiction treatment: a review. *Nicotine & Tobacco Research, 13*(6), 401–411.

Guydish, J., Passalacqua, E., Tajima, B., & Manser, S. T. (2007). Staff smoking and other barriers to nicotine dependence intervention in addiction treatment settings: a review. *Journal of Psychoactive Drugs, 39*(4), 423–433.

Hasin, D., Hatzenbuehler, M. L., Keyes, K., & Ogburn, E. (2006). Substance use disorders: diagnostic and statistical manual of mental disorders, fourth edition (DSM-IV) and international classification of diseases, tenth edition (ICD-10). *Addiction, 101*(Suppl. 1), 59–75.

Hiroeh, U., Kapur, N., Webb, R., Dunn, G., Mortensen, P. B., & Appleby, L. (2008). Deaths from natural causes in people with mental illness: a cohort study. *Journal of Psychosomatic Research, 64*(3), 275–283.

Hser, Y. I., Anglin, D., & Powers, K. (1993). A 24-year follow-up of California narcotics addicts. *Archives of General Psychiatry, 50*(7), 577–584.

Hser, Y. I., McCarthy, W. J., & Anglin, M. D. (1994). Tobacco use as a distal predictor of mortality among long-term narcotics addicts. *Preventive Medicine, 23*(1), 61–69.

Hughes, J. R., & Kalman, D. (2006). Do smokers with alcohol problems have more difficulty quitting? *Drug and Alcohol Dependence, 82*(2), 91–102.

Hunt, J. J., Cupertino, A. P., Gajewski, B. J., Jiang, Y., Ronzani, T. M., & Richter, K. P. (2014). Staff commitment to providing tobacco dependence in drug treatment: reliability, validity, and results of a national survey. *Psychology of Addictive Behaviors, 28*(2), 389–395.

Hunt, J. J., Cupertino, A. P., Garrett, S., Friedmann, P. D., & Richter, K. P. (2012). How is tobacco treatment provided during drug treatment? *Journal of Substance Abuse Treatment, 42*(1), 4–15.

Hurt, R. D., Offord, K. P., Croghan, I. T., Gomez-Dahl, L., Kottke, T. E., Morse, R. M., & Melton, L. J., 3rd (1996). Mortality following inpatient addictions treatment. Role of tobacco use in a community-based cohort. *The Journal of the American Medical Association, 275*(14), 1097–1103.

Joseph, A., Lexau, B., Willenbring, M., Nugent, S., & Nelson, D. (2004). Factors associated with readiness to stop smoking among patients in treatment for alcohol use disorder. *The American Journal on Addictions, 13*(4), 405–417.

Kalman, D., Hayes, K., Colby, S. M., Eaton, C. A., Rohsenow, D. J., & Monti, P. M. (2001). Concurrent versus delayed smoking cessation treatment for persons in early alcohol recovery. A pilot study. *Journal of Substance Abuse Treatment, 20*(3), 233–238.

Kalman, D., Morissette, S. B., & George, T. P. (2005). Co-morbidity of smoking in patients with psychiatric and substance use disorders. *The American Journal on Addictions, 14*(2), 106–123.

Knudsen, H. K., Studts, C. R., & Studts, J. L. (2012). The implementation of smoking cessation counseling in substance abuse treatment. *The Journal of Behavioral Health Services & Research, 39*(1), 28–41.

Lasser, K., Boyd, J. W., Woolhandler, S., Himmelstein, D. U., McCormick, D., & Bor, D. H. (2000). Smoking and mental illness: a population-based prevalence study. *The Journal of the American Medical Association, 284*(20), 2606–2610.

Lorains, F. K., Cowlishaw, S., & Thomas, S. A. (2011). Prevalence of comorbid disorders in problem and pathological gambling: systematic review and meta-analysis of population surveys. *Addiction, 106*(3), 490–498.

McGrath, D. S., & Barrett, S. P. (2009). The comorbidity of tobacco smoking and gambling: a review of the literature. *Drug and Alcohol Review, 28*(6), 676–681.

Mecredy, G. C., Diemert, L. M., Callaghan, R. C., & Cohen, J. E. (2013). Association between use of contraband tobacco and smoking cessation outcomes: a population-based cohort study. *Canadian Medical Association Journal, 185*(7), E287–E294.

Nahvi, S., Richter, K., Li, X., Modali, L., & Arnsten, J. (2006). Cigarette smoking and interest in quitting in methadone maintenance patients. *Addictive Behaviors, 31*(11), 2127–2134.

Pelucchi, C., Gallus, S., Garavello, W., Bosetti, C., & La Vecchia, C. (2008). Alcohol and tobacco use, and cancer risk for upper aerodigestive tract and liver. *European Journal of Cancer Prevention, 17*(4), 340–344.

Peters, E. N., Schwartz, R. P., Wang, S., O'Grady, K. E., & Blanco, C. (2014). Psychiatric, psychosocial, and physical health correlates of co-occurring cannabis use disorders and nicotine dependence. *Drug and Alcohol Dependence, 134*, 228–234.

Prochaska, J. J., Delucchi, K., & Hall, S. M. (2004). A meta-analysis of smoking cessation interventions with individuals in substance abuse treatment or recovery. *Journal of Consulting and Clinical Psychology, 72*(6), 1144–1156.

Reid, M. S., Fallon, B., Sonne, S., Flammino, F., Nunes, E. V., Jiang, H., … Rotrosen, J. (2008). Smoking cessation treatment in community-based substance abuse rehabilitation programs. *Journal of Substance Abuse Treatment, 35*(1), 68–77.

Remler, D. K. (2004). Poor smokers, poor quitters, and cigarette tax regressivity. *American Journal of Public Health, 94*(2), 225–229.

Richter, K. P., Gibson, C. A., Ahluwalia, J. S., & Schmelzle, K. H. (2001). Tobacco use and quit attempts among methadone maintenance clients. *American Journal of Public Health, 91*(2), 296–299.

Richter, K. P., Hunt, J. J., Cupertino, A. P., Garrett, S., & Friedmann, P. D. (2012). Understanding the drug treatment community's ambivalence towards tobacco use and treatment. *The International Journal on Drug Policy, 23*(3), 220–228.

Roerecke, M., & Rehm, J. (2013). Alcohol use disorders and mortality: a systematic review and meta-analysis. *Addiction, 108*(9), 1562–1578.

Rossow, I., & Amundsen, A. (1997). Alcohol abuse and mortality: a 40-year prospective study of Norwegian conscripts. *Social Science & Medicine, 44*(2), 261–267.

Royal Canadian Mounted Police. (2008). *2008 contraband tobacco enforcement strategy*. Ottawa, ON: Royal Canadian Mounted Police, Customs and Exercise Branch. Retrieved July 30, 2014, from: http://www.rcmp-grc.gc.ca/ce-da/tobacco-tabac-strat-2008-eng.htm.

Santhosh, L., Meriwether, M., Saucedo, C., Reyes, R., Cheng, C., Clark, B., … Schroeder, S. A. (2014). From the sidelines to the frontline: how the substance abuse and mental health services administration embraced smoking cessation. *American Journal of Public Health, 104*(5), 796–802.

Schroeder, S. A., & Morris, C. D. (2010). Confronting a neglected epidemic: tobacco cessation for persons with mental illnesses and substance abuse problems. *Annual Review of Public Health, 31*, 297–314.

Shoptaw, S., Rotheram-Fuller, E., Yang, X., Frosch, D., Nahom, D., Jarvik, M. E., … Ling, W. (2002). Smoking cessation in methadone maintenance. *Addiction, 97*(10), 1317–1328.

Tsoh, J. Y., Chi, F. W., Mertens, J. R., & Weisner, C. M. (2011). Stopping smoking during first year of substance use treatment predicted 9-year alcohol and drug treatment outcomes. *Drug and Alcohol Dependence, 114*(2–3), 110–118.

U.S. Department of Health and Human Services. (2014). *The health consequences of smoking—50 years of progress. A report of the surgeon general*. Atlanta, GA: U.S. Department of Health and Human Services, Centers for Disease Control and Prevention, National Center for Chronic Disease Prevention and Health Promotion, Office on Smoking and Health.

Walsh, R. A., Bowman, J. A., Tzelepis, F., & Lecathelinais, C. (2005). Smoking cessation interventions in Australian drug treatment agencies: a national survey of attitudes and practices. *Drug and Alcohol Review, 24*(3), 235–244.

West, R., McNeill, A. D., & Raw, M. (2000). Smoking cessation guidelines for health professionals: an update. *Thorax, 55*, 987–999.

Winhusen, T. M., Brigham, G. S., Kropp, F., Lindblad, R., Gardin, J. G., 2nd., Penn, P., … Ghitza, U. (2013). A randomized trial of concurrent smoking-cessation and substance use disorder treatment in stimulant-dependent smokers. *The Journal of Clinical Psychiatry, 75*(4), 336–343.

Ziedonis, D. M., Guydish, J., Williams, J., Steinberg, M., & Foulds, J. (2006). Barriers and solutions to addressing tobacco dependence in addiction treatment programs. *Alcohol Research & Health, 29*(3), 228–235.

Zwar, N., Richmond, R., Borland, R., Peters, M., Litt, J., Bell, J., … Ferretter, I. (2011). *Supporting smoking cessation: A guide for health professionals*. Melbourne, Australia: The Royal Australian College of General Practitioners.

Chapter 11

Neurological Effects of Nicotine, Tobacco, and Particulate Matter

Bronwyn M. Kivell[1], Kirsty Danielson[1,2]

[1]School of Biological Sciences, Centre for Biodiscovery, Victoria University of Wellington, Wellington, New Zealand; [2]Beth Israel Deaconess Medical Centre, Harvard Medical School, Boston, MA, USA

Abbreviations

CPA Conditioned place aversion
CPP Conditioned place preference
DA Dopamine
dStr Dorsal striatum
GABA γ-Aminobutyric acid
ICSS Intracranial self-stimulation
MAO Monoamine oxidase
nAcb Nucleus accumbens
nAChR Nicotinic acetylcholine receptor
PFC Prefrontal cortex
VTA Ventral tegmental area

INTRODUCTION

Tobacco and tobacco smoke are made up of more than 9000 chemicals (Rodgman & Perfetti, 2013), many of which are toxic and/or carcinogenic and contribute to numerous diseases including cancer, stroke, heart and vascular disease, and emphysema. Nearly 6 million deaths each year are a result of smoking and secondhand smoke exposure (WHO, 2014), including approximately 90% of lung cancers. Cigarette smoking is the leading cause of preventable death worldwide and if global use of tobacco persists, it has been projected to cause up to 1 billion deaths in the twenty-first century (WHO, 2014).

The tobacco alkaloid nicotine is considered to be the major addictive component found in cigarette smoke (Stolerman & Jarvis, 1995). However, nicotine alone is considered to be a relatively weak reinforcer in self-administration studies in animals (Stolerman & Jarvis, 1995). This is contradictory to the apparently highly addictive nature of tobacco smoking in humans (Kandel, Chen, Warner, Kessler, & Grant, 1997). It is becoming increasing clear that there are more factors contributing to addiction to tobacco smoke than nicotine alone (Brennan, Laugesen, & Truman, 2014). Despite this, the majority of preclinical and cellular studies have used nicotine alone to model addiction to tobacco and it is apparent that owing to the complex nature of the constituents of tobacco smoke (Rodgman & Perfetti, 2013), better preclinical and cellular models need to be used to fully understand tobacco addiction in humans. Current pharmacotherapies used to treat smoking addiction have very low success rates (West et al., 2000). Therefore, gaining a better understanding of the pathways modulated by tobacco smoke may aid in the development of better pharmacotherapies to treat tobacco addiction. Here we outline the neurobiological effects of nicotine, tobacco, and particulate matter and the role they play in addiction to tobacco smoke.

NEUROBIOLOGICAL EFFECTS OF SMOKING TOBACCO

Tobacco is an addictive substance that produces both rewarding and negative effects, a combination of which ultimately leads to tobacco use, relapse, and dependence. The positive reinforcing effects of tobacco are involved in the initiation stage of dependence and include mild euphoria, increased memory, and relaxation, while the negative effects of smoking are associated with cessation or withdrawal and include depressed mood, anxiety, and impaired memory (see Wise, 1996; Wise & Koob, 2014, for review). Like all drugs of abuse, nicotine, cigarettes, and tobacco products affect the mesocorticolimbic reward pathway, the major pathway in the brain for reward and reinforcement for addictive drugs as well as natural rewarding stimuli (Volkow, Fowler, Wang, Swanson, & Telang, 2007). This pathway comprises cell bodies in the ventral tegmental area (VTA) and their projections to the nucleus accumbens (nAcb) and prefrontal cortex (PFC) (Figure 1). Central to the classical neurobiological model of addiction is the release of dopamine (DA) in the nAcb following stimulation of this pathway, which is associated with reward, reinforcement, and the initiation of drug abuse (Volkow et al., 2007). Both nicotine and tobacco smoke induce increases in DA levels in the nAcb through the binding of nicotine to nicotinic acetylcholine receptors (nAChRs) in the VTA. Persistent smoking, like with other drugs of abuse, leads to tolerance and withdrawal after cessation. Both the positive hedonic effects and the avoidance of negative withdrawal effects contribute to tobacco dependence. With repeated use, the positive reinforcing effects become reduced and the mechanisms involving negative reinforcement increase. There is also evidence for neuroadaptive changes occurring with a switch from ventral striatal (nAcb) to dorsal striatal centers linked to a loss of control of drug use. Human smokers also have reduced DA function and reduced DA transporter availability compared to nonsmokers. For a review of neuroadaptive changes in human smokers see Martin-Soelch (2013).

The dopaminergic nigrostriatal pathway consisting of neurons projecting from the substantia nigra to the dorsal striatum (dStr) has also been implicated in smoking addiction, particularly in the hedonic effects of smoking. Related to the mesocorticolimbic reward pathway are the monoaminergic pathways for serotonin and norepinephrine. In addition to DA, serotonin is involved in the rewarding and stimulant effects of nicotine as well as nicotine withdrawal (Seth, Cheeta, Tucci, & File, 2002). Serotonergic neuronal cell bodies reside in the dorsal raphe nuclei and project to the dStr, while norepinephric neurons project from the locus coeruleus to the dStr, hippocampus, and PFC. Similar to DA, nicotine induces the release of serotonin in the dStr (Scholze, Orr-Urtreger, Changeux, McIntosh, & Huck, 2007) and norepinephrine into the PFC, hippocampus, and dStr (Scholze et al., 2007). Release of the monoamines into the synapse results in binding to neurotransmitter receptors and propagation of signaling to neighboring neurons. This is ultimately terminated via their reuptake by the monoamine transporters or degradation by the monoamine oxidases (MAOs). Integration of the monoaminergic pathways as well as the endogenous opioid, glutamate, and γ-aminobutyric acid (GABA) systems produces vast global changes in neurochemistry in response to tobacco smoking.

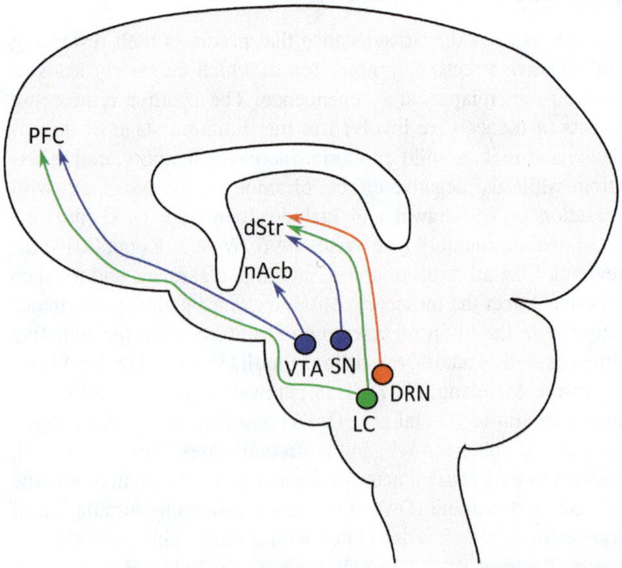

FIGURE 1 Monoaminergic pathways relevant to smoking addiction. Nicotine and cigarette smoking stimulate dopaminergic (blue), serotonergic (red), and norepinephric (green) pathways in the brain. The brain's major reward system, the mesocorticolimbic system, consists of dopaminergic cell bodies in the VTA that project to the dorsal striatum (dStr) and PFC. The nigrostriatal pathway, projecting from the substantia nigra (SN) to the dStr, has also been implicated in behaviors related to smoking addiction. Other monoaminergic pathways of relevance include serotonergic neurons projecting from the dorsal raphe nuclei (DRN) to the dStr and norepinephric neurons projecting from the locus coeruleus (LC) to the PFC.

THE REWARDING EFFECTS OF NICOTINE

Nicotine is self-administered in humans, nonhuman primates, dogs, rats, and mice (reviewed in Le Foll & Goldberg, 2009). Studying nicotine addiction preclinically is complicated as it has both rewarding and aversive properties depending on the dose, timing of administration, mode of administration, and presence of associated cues. Often the same doses of nicotine show both aversive and rewarding effects depending on experimental conditions and the animal model utilized. Some studies have reported a flat dose response with nicotine (Brennan et al., 2013), other studies show nicotine having an inverted U-shaped dose–response curve in that low doses of nicotine have aversive effects (Lynch & Carroll, 1999) and rewarding effects are seen with increased doses of nicotine up to a maximum level, where it becomes aversive again. Aversion is typically seen at short intervals following high doses of nicotine, and positive stimulant effects are typically seen following longer intervals of lower controlled nicotine release (Le Foll & Goldberg, 2009). The gold standard method for studying addiction preclinically is self-administration, as it is considered to be the most direct measure of the reinforcing effects of drugs. In contrast to many psychostimulants, nicotine self-administration in rodents often requires food deprivation, cue association, or preadministration of other abused drugs to maintain nicotine self-administration. Notably, in squirrel monkeys, nicotine self-administration showed high motivational value, although this is less apparent in other primates (Le Foll & Goldberg, 2009).

In addition to studying the rewarding properties of nicotine, the self-administration paradigm has also helped elucidate the pathways and neurotransmitter systems underlying nicotine addiction. For example, the mesocorticolimbic pathway has been shown to be essential for nicotine self-administration, as the nicotinic antagonist dihydro-β-erythroidine, when administered into the VTA but not nAcb of rats, reduced self-administration of nicotine (Corrigall, Coen, & Adamson, 1994). It is believed that the VTA is the site for the initiation of processes responsible for the reinforcing effects of nicotine.

The effects of nicotine have also been studied using conditioned place preference (CPP) paradigms and intracranial self-stimulation (ICSS) models. Animals, like humans, show preference to an environment repeatedly associated with positive effects and this forms the behavioral basis for the preclinical CPP paradigm and, conversely, conditioned place aversion. Nicotine induces CPP in rats over a range of doses from 0.1 to 1.4 mg/kg. Increases in pCREB and Fos levels in the nAcb, VTA, PFC, dStr, amygdala, and hippocampus indicate that induction of CPP is associated with neural activity in these brain regions. Nicotine-induced CPP appears to be nAChR-dependent, as administration of the nAChR antagonist mecamylamine abolished both the preference behavior and the associated increases in neural activity (de la Pena et al., 2014).

ICSS models rely on the animal learning to deliver electrical impulses to the reward pathway in the brain (via an implanted electrode). When ICSS thresholds are lowered, this indicates reward, whereas elevations in the ICSS threshold reflect an anhedonic state. ICSS has been used extensively to model nicotine and tobacco addiction (reviewed by Le Foll & Goldberg, 2009). Nicotine reduces ICSS reward thresholds between 0.5 and 1 mg/kg/day but not at 2 mg/kg (Bozarth, Pudiak, & KuoLee, 1998) and is in agreement with studies utilizing the CPP model.

It is important to note that differences are seen between adult and adolescent animals in all of these preclinical models of nicotine addiction. Adolescence is widely accepted to be a period of greater vulnerability to the development of addictions, and in light of this it is no surprise that differences would be seen between animals of different ages (de la Pena et al., 2014). In a recent study by Ahsan et al. (2014) adolescent rats displayed CPP for lower doses of nicotine (0.2 mg/kg) than adults (0.6 mg/kg) and in self-administration models also showed response to nicotine at 0.03 mg/kg per infusion. Both tests show that adolescent rats are more sensitive to the rewarding and reinforcing effects of nicotine compared to adult rats (Ahsan et al., 2014). There is also evidence for reduced withdrawal and negative effects associated with nicotine in adolescents. Age is a known vulnerability factor for drug use, with early use increasing the probability of dependence. This is significant as the majority of youths have tried smoking tobacco at least once and adolescence is a time at which brain maturation is occurring. There is also growing evidence suggesting that drug use during adolescence may impair the development of the PFC and may lead to cognitive impairments; for a review on the effects of nicotine during adolescence see Goriounova and Mansvelder (2012).

MECHANISM OF ACTION: NICOTINIC ACETYLCHOLINE RECEPTORS

Nicotine exerts its effects on the brain by binding to nAChRs. nAChRs are made up of various combinations of α and β subunits, with each assembly having different properties, including differences in expression level, localization, and nicotine binding affinity (Le Novere, Corringer, & Changeux, 2002; Tuesta, Fowler, & Kenny, 2011). For a review on the role of nAChR subtypes and their expression and function see Tuesta et al. (2011).

In the brain the α4β2 nAChR subunit is the most abundant and is upregulated following exposure to nicotine in the rat brain (Gentry & Lukas, 2002) as well as the brain of human smokers (Breese et al., 1997). Expression of α4 or β2 subunits in the brain is essential for the rewarding effects of nicotine and the α4β2 agonist is actively self-administered by rats (Liu et al., 2003). Considerable evidence supports the fact that the α4β2 nAChRs in the VTA are largely responsible for the stimulatory effects of nicotine, whereas, α4β2 subunits expressed on GABAergic neurons in the VTA are responsible for disrupting endogenous cholinergic signaling.

In humans, variations in genes coding for α5, α3, and β4 nAChRs increase vulnerability to tobacco addiction. Recently Fowler, Lu, Johnson, Marks, and Kenny (2011) showed that mice with a null mutation in the α5 subunit self-administered higher levels of nicotine. The α3 nAChRs play a role in regulating the stimulatory effects of nicotine via DA transmission within the nAcb and striatum, whereas α6 subunits in the striatum are responsible for stimulating DA release. Identification of the precise nAChR subtypes that regulate addiction to tobacco smoke is an important goal for current research to address and may aid in the identification of subtype-selective ligands for the development of more effective cessation pharmacotherapies. The role of particulate matter in the direct or indirect regulation of each nAChR subunit combination and the contribution this may play in smoking addiction remain to be determined.

ADDICTIVE PROPERTIES OF TOBACCO SMOKE

While it is clear that nicotine is addictive, nicotine by itself has been shown to be a relatively poor reinforcer in animal self-administration studies (Stolerman & Jarvis, 1995), in contrast to the highly addictive effects of tobacco smoking in humans (Kandel et al., 1997). In further support of a discrepancy between the effects of nicotine alone in comparison to cigarette smoke is the fact that nicotine replacement therapies (NRTs) have low abuse liability and are only partially effective as cessation treatments (West et al., 2000). Probably most interesting, however, are the relatively few studies performed in human subjects given a choice between different forms of cigarette smoke and nicotine. Smokers will choose to self-administer denicotinized tobacco smoke over intravenous nicotine (Rose, Salley, Behm, Bates, & Westman, 2010), and test subjects reported that denicotinized tobacco smoke decreased cigarette cravings more than a nicotine inhaler (Darredeau & Barrett, 2010). The preference for tobacco smoke over nicotine could be due to several factors, including pharmacologically active compounds in tobacco smoke (Table 1), sensory cues, perception of experience, or, most likely, a combination of these factors. Unfortunately, owing to the aversive nature of cigarette smoke, there is no preclinical self-administration model for cigarette smoke in animals. However, animal models on smoke inhalation have been developed whereby animals passively inhale cigarette smoke in a tightly sealed box. Utilizing this model CPP has been observed (Ypsilantis, Politou, Anagnostopoulos, Kortsaris, & Simopoulos, 2012) and several studies have shown differences between the effects of nicotine and those of tobacco smoke in the rat midbrain.

TABLE 1 Neurologically Active Compounds Found in Tobacco Smoke

Tobacco Smoke Constituent	Description
Nicotine	The major neurologically active compound in cigarette smoke that exerts its effects through binding to the nAChRs.
Acetaldehyde	A major component of cigarette smoke that correlates with tar and carbon monoxide in yield and is produced by the burning of natural tobacco polysaccharides including cellulose.
Minor tobacco alkaloids: Nornicotine Anabasine Cotinine Myosmine	Make up a total of 2–3% of the total alkaloid content in cigarette smoke. They are closely related to nicotine in structure, are all found in tobacco smoke, and have some affinity for nAChRs.
MAO inhibitors: Harman Norharman	Harman and norharman are β-carboline alkaloids that prevent degradation of monoamines by MAOs. This is achieved through inhibition of both MAO-A and MAO-B.

For example, there are decreased levels of tyrosine hydroxylase in the VTA of the smoke-exposed group compared to the nicotine-exposed group (Li et al., 2004). Also, smoke but not nicotine increased DA D1 receptor levels in the rat caudate putamen (Naha, Li, Yang, Park, & Kim, 2009).

Nicotine, as outlined in the previous section, has been used extensively to investigate the cellular and molecular effects of tobacco and to model behavioral responses preclinically (Le Foll & Goldberg, 2009). However, it is becoming increasingly clear that there are differences in the neurobiological effects of tobacco smoke compared to nicotine alone. Furthermore, nonnicotinic components found in tobacco smoke may also contribute to the addictive properties of tobacco and either are rewarding in themselves or potentiate the effects of nicotine (for review see Brennan et al., 2014). This includes acetaldehyde, the minor tobacco alkaloids, and MAO inhibitors. The sections below outline findings on the addictive properties and neurobiological effects of tobacco extract and several individual constituents of tobacco smoke (see Table 2 for a summary of the behavioral effects of individual tobacco smoke constituents).

TOBACCO PARTICULATE MATTER AND EXTRACTS

Owing to the impossibility of a cigarette smoke self-administration regime in animals, researchers have turned to tobacco extracts as a drug model more closely resembling the chemical exposure that human smokers experience. Tobacco particulate matter or tobacco extracts are collected by several different methods, including collection of combustion products on a filter pad, which is subsequently extracted in ethanol (Brennan et al., 2013; Danielson, Truman, & Kivell, 2011; Lewis, Miller, & Lea, 2007), or "bubbling" of tobacco smoke through sterile saline to create an aqueous tobacco solution (Costello et al., 2014) (Figure 2). Tobacco extracts have a different pharmacological profile compared to nicotine alone and have been shown to produce stronger inhibition of serotonergic neurons in the dorsal raphe nucleus (Touiki, Rat, Molimard, Chait, & de Beaurepaire, 2007) and to increase nAChR expression to a greater extent in vitro (Ambrose et al., 2007). More recently, tobacco smoke particulate matter preparations from roll your own cigarettes have been shown to be more addictive than the matched level of nicotine in a rat model of self-administration (Brennan et al., 2013). Similarly, aqueous cigarette smoke extract was more powerful than nicotine for acquisition and maintenance of self-administration in rats in a nose-poke paradigm (Costello et al., 2014). Curiously, mecamylamine attenuated self-administration to similar levels for both nicotine and the smoke extract. This suggests that although cigarette smoke extract may be a more potent

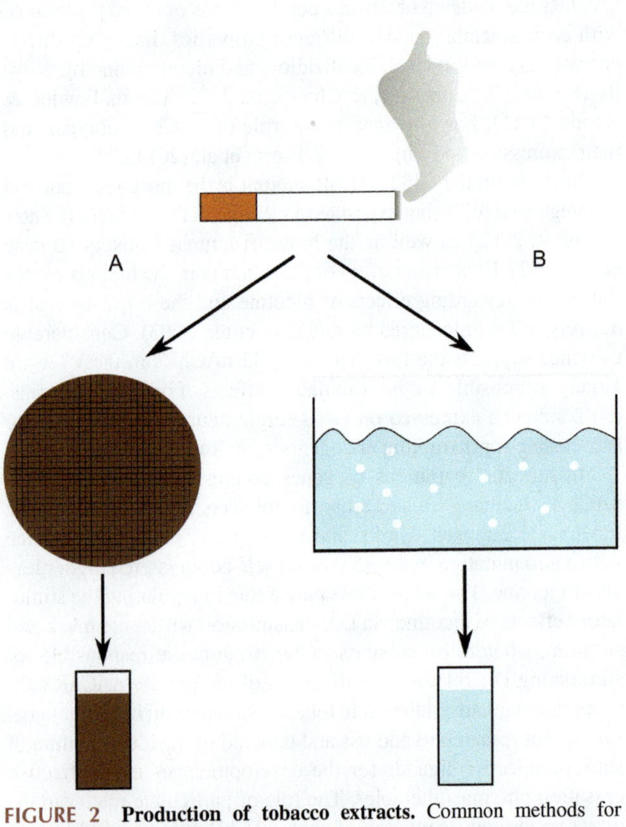

TABLE 2 Summary of the Rewarding Effects of Nonnicotinic Components of Cigarette Smoke

Tobacco Smoke Constituents	Behavioral Effects
Denicotinized tobacco smoke	• Smokers will self-administer in preference to intravenous nicotine • Decreases cigarette cravings more than a nicotine inhaler in human subjects
Whole cigarette smoke	• Induces CPP in rats
Tobacco particulate matter and extracts	• More addictive than matched nicotine doses in rat models of self-administration • Animals self-administering smoke extract are more prone to stress-induced reinstatement than those self-administering nicotine
Acetaldehyde	• Induces CPP in rodents • Is self-administered by rats
Minor tobacco alkaloids: Nornicotine Anabasine Cotinine Anatabine Myosmine	• Addition of minor alkaloids increases nicotine-induced locomotor activity • Nornicotine is self-administered by rats • Anabasine dose-dependently increases (0.02 mg/kg, 25% increase) and decreases (2.0 mg/kg, 50% reduction) nicotine self-administration in rats • Anabasine reduces nicotine withdrawal in mice • Anatabine attenuates nicotine self-administration in rats and rhesus monkeys
MAO inhibitors: Harman Norharman	• Chronic and acute treatment increases nicotine self-administration in rats

FIGURE 2 Production of tobacco extracts. Common methods for obtaining tobacco extracts include (A) collection of combustion products on a filter pad followed by extraction into a solvent and (B) "bubbling" of the combustion products through a sterile saline solution.

reinforcer than nicotine alone, its actions are still dependent upon nAChR activation. In further support of cigarette smoke extract as a more potent addictive agent, the extract self-administering animals were more sensitive to stress-induced reinstatement following administration of yohimbine (Costello et al., 2014). The exact subset of compounds responsible for the effects observed in response to tobacco extract administration is not currently known. However, several compounds have previously been described as pharmacologically active and potentially reinforcing, and they are described below.

Acetaldehyde

Acetaldehyde is a by-product of cigarette smoke formed by the burning of natural sugars. In addition to the presence of natural sugars, flavorings and sugars are often added to tobacco during cigarette manufacture to increase palatability and flavor. High levels of acetaldehyde are believed to increase the rewarding effects of nicotine. Levels of acetaldehyde present in tobacco smoke vary widely and can be as high as 400–1400 μg per cigarette (Hoffman & Evans, 2013). This is relevant because preclinical studies in rodents have shown that acetaldehyde induces CPP (Plescia, Brancato, Marino, & Cannizzaro, 2013) and is self-administered by rats (Myers, Ng, & Singer, 1982; Peana, Muggironi, & Diana, 2010). In CPP studies, adult rats given acetaldehyde at a dose of 10–40 mg/kg (intraperitoneally) showed significant preference for the paired chamber (Quertemont & De Witte, 2001) suggestive of "liking." An inverted U-shaped response was seen in male Long-Evans rats with 2.89 mg/kg infusions, but not with lower doses (Myers et al., 1982). This is supported by studies in rats using a nose-poke operant oral administration regime, with responding to 0.1–3.2% acetaldehyde observed (Peana et al., 2010). Responding on the active nose-poke hole also followed an inverted U-shaped dose–response curve. Neurons in the VTA have also been shown to increase their firing rates following acetaldehyde. For a recent review on the addictive effects of acetaldehyde see Hoffman and Evans (2013) or Plescia et al. (2013).

Minor Tobacco Alkaloids

The minor alkaloids nornicotine, cotinine, anabasine, anatabine, and myosmine are structurally similar to nicotine and together make up 2–3% of the total alkaloid content of tobacco (Clemens, Caille, Stinus, & Cador, 2009). The addition of the collective minor tobacco alkaloids to nicotine has been shown to increase nicotine-induced locomotor activity in rats, with anatabine, cotinine, and myosmine able to individually increase activity (Clemens et al., 2009). Nornicotine is a metabolite of nicotine that has been shown to evoke DA release in the nAcb of the rat in brain slice preparations (Green, Crooks, Bardo, & Dwoskin, 2001). It is also self-administered by rats, suggesting that it has rewarding properties and, like nicotine and acetaldehyde, behavioral effects follow an inverted U-shaped dose–response pattern, with doses between 0.15 and 0.3 mg/kg per infusion supporting self-administration studies. A recent review by Hoffman and Evans (2013) concluded that both acetaldehyde and nornicotine have likely abuse potential. Conversely, there have also been suggestions that the other minor alkaloids, particularly anatabine, have low abuse potential and could potentially be used as alternative smoking cessation therapies. Anabasine is a putative partial agonist of α4β2 nAChRs (Stolerman & Jarvis, 1995) and, when administered prior to nicotine self-administration sessions, has been shown to reduce nicotine self-administration in rats (Caine et al., 2014; Hall et al., 2014). Furthermore, it attenuates nicotine withdrawal in mice (Caine et al., 2014). Similarly, anatabine reduced nicotine self-administration in both rhesus monkeys (Costello et al., 2014) and rats (Caine et al., 2014; Hall et al., 2014). A caveat that should be noted here is that this effect is dose dependent, and in the case of anabasine, the lower dose (0.02 mg/kg) resulted in an increase in self-administration of 25%, contrary to the 50% reduction in self-administration at 2.0 mg/kg. This highlights the need for further studies into the role of minor tobacco alkaloids in reward and their abuse potential.

Monoamine Oxidase Inhibitors

MAO inhibitors contribute to smoking addiction by preventing the degradation of monoamines including DA, serotonin, and norepinephrine. This functions to potentiate the reward signal evoked via DA release in the mesocorticolimbic pathway following nicotine (Lewis et al., 2007). In humans, MAO-A inhibitors have antidepressant properties and human smokers have been shown to have decreased MAO activity in the brain (Fowler et al., 1996). In cigarette smoke the major MAO inhibitors are the β-carbolines harman and norharman, which have known psychoactive effects (for review see Bruijnzeel, 2012). Harman is a reversible inhibitor of MAO-A and norharman is a reversible inhibitor of MAO-A and -B (Herraiz & Chaparro, 2005).

Several studies have found that MAO inhibition increases nicotine self-administration in the rat (Guillem et al., 2005, 2006; Villegier, Lotfipour, McQuown, Belluzzi, & Leslie, 2007). This has been shown with both acute (Villegier et al., 2007) and chronic inhibition (Guillem et al., 2005) and is believed to be mediated predominantly via inhibition of MAO-A rather than MAO-B. Harman and norharman also have antidepressant-like effects in rodents and indeed, some of the effects of tobacco extracts discussed in the previous section are believed to be due to MAO inhibitor activity, given their ability to inhibit MAO-A and -B where nicotine does not (Costello et al., 2014). Studies identifying additional MAO inhibitors present in tobacco smoke are required to further evaluate the differences seen between nicotine and cigarette smoke.

APPLICATIONS TO OTHER ADDICTIONS AND SUBSTANCE MISUSE

There are strong links between the abuse of tobacco, alcohol, cannabinoids, and opioids, as well as other drugs of abuse. It has been estimated from epidemiological studies that between 80% and 90% of alcoholics also smoke tobacco products (Miller & Gold, 1998). It is also accepted that smokers consume more alcohol than nonsmokers (Carmody, Brischetto, Matarazzo, O'Donnell, & Connor, 1985). This has also been modeled in preclinical studies, with ethanol-preferring rats showing increased levels of nicotine self-administration (Le et al., 2006). Genetic variations in nAChR subunits have been shown to modulate alcohol consumption in mice, particularly α7, α4, β2, and α3β2 subunits (Tuesta et al., 2011), although the exact role of other subunits needs to be explored further.

The α7 nAChR subunit has been shown to modulate DA levels following tetrahydrocannabinol, and the selective α7 nAChR antagonist mecamylamine decreases self-administration of the cannabinoid receptor type 1 agonist WIN 55,212-2 (Solinas et al., 2007). It is also interesting that the cannabinoid receptor type 1 agonist rimonabant prevents nicotine CPP in mice (Le Foll & Goldberg, 2004) and also decreases smoking and promotes weight loss in humans (Fagerstrom & Balfour, 2006), although safety concerns have resulted in the removal of rimonabant from human use.

DEFINITION OF TERMS

Conditioned place preference This is a preclinical behavioral model utilizing Pavlovian conditioning to assess the rewarding effects of a drug. The time spent in a drug-paired versus a nonpaired chamber is determined and an increase indicates an association with a positive or rewarding stimulus.

Intracranial self-stimulation This is a preclinical behavioral model used to assess reward and motivation. The subject performs a task to electrically stimulate a brain region within the brain reward pathway resulting in pleasure. Changes in these responses are used to assess reward and motivation.

Mesocorticolimbic pathways These are a group of brain connections making up the brain's endogenous DA reward pathway. This pathway is implicated in reward to natural stimuli and is activated following exposure to drugs of abuse.

Neuroadaptation This describes changes in the brain or central nervous system structure and/or function that occur over time following exposure to an event, chemical, or drug.

Pharmacotherapy This is the use of drugs to treat a medical condition.

Self-administration This is a preclinical behavioral model used to assess the abuse potential and rewarding effects of drugs of abuse and potential therapeutics. Typically this model has an implanted intravenous cannula and the subject performs a task to receive a drug infusion.

Total particulate matter This is a collection of chemicals isolated from cigarette smoke.

KEY FACTS

Key Facts about Tobacco

- In 1964 the US Surgeon General's report established the link between smoking and various disease outcomes.
- Tobacco is one of the most widely abused drugs globally.
- Tobacco use is the single most preventable cause of death worldwide, with approximately 6 million people dying each year as a result of tobacco smoking.
- Smoking tobacco has wider effects than just on the smoker. Secondhand smoke is also responsible for a considerable disease burden.
- Tobacco smoke contains a mixture of over 9000 different chemicals, many of which are known to be toxic or carcinogenic.

Key Facts about the Mesocorticolimbic Brain Reward Pathway

- This is the natural reward pathway in the brain and the main pathway for reward and reinforcement for addictive drugs as well as natural reward.
- It comprises dopaminergic cell bodies in the VTA and their projections to the nAcb and PFC.
- The PFC functions in the planning of complex cognitive decisions and behaviors, and neuroadaptations to this brain region following tobacco use play an important role in addiction behaviors, particularly in adolescence.
- The nAcb is the main location for the release of DA following rewarding stimuli and functions in reinforcement and attention.

Key Facts about Nicotinic Acetylcholine Receptors

- Neurons in the VTA express nAChRs and when activated by nicotine increase DA release in the nAcb.
- nAChRs also modulate the levels of many other neurotransmitters, including acetylcholine, glutamate, serotonin, GABA, and norepinephrine.
- There are 12 α and β subunits, with 9 α and 3 β subunits that combine in different combinations to form five membrane-spanning regions of the functional receptor.
- Each subunit combination has a unique expression pattern in the brain and periphery and they have different pharmacological effects.

SUMMARY POINTS

- This chapter focuses on the effects of chemicals present in tobacco smoke and their role in smoking addiction. With tobacco smoke containing over 9000 chemicals there are many potential candidates for modulating tobacco addiction other than nicotine.
- While it is clear that nicotine is the most significant addictive compound present in tobacco smoke, differences are observed between studies administering nicotine alone and those administering extracts of tobacco smoke, particulate matter, and inhaled smoke. Differences are also observed between species and between the various preclinical models of addiction utilized.
- Multiple nAChR subunit combinations, each with specific localization and expression patterns, also contribute to the complex neurological effects of tobacco smoke.
- Preclinical models using nicotine typically administer low levels in contrast to the highly addictive effects of smoking tobacco in humans.
- NRTs replace nicotine for tobacco smoke to help smokers quit. However, these pharmacotherapies show limited success, suggesting other factors may play a contributing role.
- The leading classes of compounds that may modulate tobacco addiction other than nicotine include acetaldehyde, minor tobacco alkaloids, and MAO inhibitors such as harman and norharman.
- If other factors contribute to smoking addiction, better, more effective pharmacotherapies may be developed in the future to combat the significant global health burden as a result of smoking tobacco.

REFERENCES

Ahsan, H. M., de la Pena, J. B., Botanas, C. J., Kim, H. J., Yu, G. Y., & Cheong, J. H. (2014). Conditioned place preference and self-administration induced by nicotine in adolescent and adult rats. *Biomolecules & Therapeutics (Seoul), 22*(5), 460–466.

Ambrose, V., Miller, J. H., Dickson, S. J., Hampton, S., Truman, P., Lea, R. A., & Fowles, J. (2007). Tobacco particulate matter is more potent than nicotine at upregulating nicotinic receptors on SH-SY5Y cells. *Nicotine & Tobacco Research, 9*(8), 793–799.

Bozarth, M. A., Pudiak, C. M., & KuoLee, R. (1998). Effect of chronic nicotine on brain stimulation reward. II. An escalating dose regimen. *Behavioural Brain Research, 96*(1–2), 189–194.

Breese, C. R., Marks, M. J., Logel, J., Adams, C. E., Sullivan, B., Collins, A. C., & Leonard, S. (1997). Effect of smoking history on [3H]nicotine binding in human postmortem brain. *Journal of Pharmacology and Experimental Therapeutics, 282*(1), 7–13.

Brennan, K. A., Crowther, A., Putt, F., Roper, V., Waterhouse, U., & Truman, P. (2013). Tobacco particulate matter self-administration in rats: differential effects of tobacco type. *Addiction Biology, 20*(2), 227–235.

Brennan, K. A., Laugesen, M., & Truman, P. (2014). Whole tobacco smoke extracts to model tobacco dependence in animals. *Neuroscience and Biobehavioral Reviews, 47C*, 53–69.

Bruijnzeel, A. W. (2012). Tobacco addiction and the dysregulation of brain stress systems. *Neuroscience Biobehavioral Reviews, 36*(5), 1418–1441.

Caine, S. B., Collins, G. T., Thomsen, M., Wright, C., Lanier, R. K., & Mello, N. K. (2014). Nicotine-like behavioral effects of the minor tobacco alkaloids nornicotine, anabasine, and anatabine in male rodents. *Experimental and Clinical Psychopharmacology, 22*(1), 9–22.

Carmody, T. P., Brischetto, C. S., Matarazzo, J. D., O'Donnell, R. P., & Connor, W. E. (1985). Co-occurrent use of cigarettes, alcohol, and coffee in healthy, community-living men and women. *Health Psychology, 4*(4), 323–335.

Clemens, K. J., Caille, S., Stinus, L., & Cador, M. (2009). The addition of five minor tobacco alkaloids increases nicotine-induced hyperactivity, sensitization and intravenous self-administration in rats. *The International Journal of Neuropsychopharmacology, 12*(10), 1355–1366.

Corrigall, W. A., Coen, K. M., & Adamson, K. L. (1994). Self-administered nicotine activates the mesolimbic dopamine system through the ventral tegmental area. *Brain Research, 653*(1–2), 278–284.

Costello, M. R., Reynaga, D. D., Mojica, C. Y., Zaveri, N. T., Belluzzi, J. D., & Leslie, F. M. (2014). Comparison of the reinforcing properties of nicotine and cigarette smoke extract in rats. *Neuropsychopharmacology, 39*(8), 1843–1851.

Danielson, K., Truman, P., & Kivell, B. M. (2011). The effects of nicotine and cigarette smoke on the monoamine transporters. *Synapse, 65*(9), 866–879.

Darredeau, C., & Barrett, S. P. (2010). The role of nicotine content information in smokers' subjective responses to nicotine and placebo inhalers. *Human Psychopharmacology, 25*(7–8), 577–581.

Fagerstrom, K., & Balfour, D. J. (2006). Neuropharmacology and potential efficacy of new treatments for tobacco dependence. *Expert Opinion on Investigational Drugs, 15*(2), 107–116.

Fowler, C. D., Lu, Q., Johnson, P. M., Marks, M. J., & Kenny, P. J. (2011). Habenular alpha5 nicotinic receptor subunit signalling controls nicotine intake. *Nature, 471*(7340), 597–601.

Fowler, J. S., Volkow, N. D., Wang, G. J., Pappas, N., Logan, J., MacGregor, R., ... Cilento, R. (1996). Inhibition of monoamine oxidase B in the brains of smokers. *Nature, 379*(6567), 733–736.

Gentry, C. L., & Lukas, R. J. (2002). Regulation of nicotinic acetylcholine receptor numbers and function by chronic nicotine exposure. *Current Drug Targets CNS and Neurological Disorders, 1*(4), 359–385.

Goriounova, N. A., & Mansvelder, H. D. (2012). Nicotine exposure during adolescence alters the rules for prefrontal cortical synaptic plasticity during adulthood. *Frontiers in Synaptic Neuroscience, 4*, 3.

Green, T. A., Crooks, P. A., Bardo, M. T., & Dwoskin, L. P. (2001). Contributory role for nornicotine in nicotine neuropharmacology: nornicotine-evoked [H-3]dopamine overflow from rat nucleus accumbens slices. *Biochemical Pharmacology, 62*(12), 1597–1603.

Guillem, K., Vouillac, C., Azar, M. R., Parsons, L. H., Koob, G. F., Cador, M., & Stinus, L. (2005). Monoamine oxidase inhibition dramatically increases the motivation to self-administer nicotine in rats. *The Journal of Neuroscience, 25*(38), 8593–8600.

Guillem, K., Vouillac, C., Azar, M. R., Parsons, L. H., Koob, G. F., Cador, M., & Stinus, L. (2006). Monoamine oxidase A rather than monoamine oxidase B inhibition increases nicotine reinforcement in rats. *The European Journal of Neuroscience, 24*(12), 3532–3540.

Hall, B. J., Wells, C., Allenby, C., Lin, M. Y., Hao, I., Marshall, L., & Levin, E. D. (2014). Differential effects of non-nicotine tobacco constituent compounds on nicotine self-administration in rats. *Pharmacology Biochemistry & Behavior, 120*, 103–108.

Herraiz, T., & Chaparro, C. (2005). Human monoamine oxidase is inhibited by tobacco smoke: beta-carboline alkaloids act as potent and reversible inhibitors. *Biochemical and Biophysical Research Communications, 326*(2), 378–386.

Hoffman, A. C., & Evans, S. E. (2013). Abuse potential of non-nicotine tobacco smoke components: acetaldehyde, nornicotine, cotinine, and anabasine. *Nicotine & Tobacco Research, 15*(3), 622–632.

Kandel, D., Chen, K., Warner, L. A., Kessler, R. C., & Grant, B. (1997). Prevalence and demographic correlates of symptoms of last year dependence on alcohol, nicotine, marijuana and cocaine in the U.S. population. *Drug and Alcohol Dependence, 44*(1), 11–29.

Le, A. D., Li, Z., Funk, D., Shram, M., Li, T. K., & Shaham, Y. (2006). Increased vulnerability to nicotine self-administration and relapse in alcohol-naive offspring of rats selectively bred for high alcohol intake. *The Journal of Neuroscience, 26*(6), 1872–1879.

Le Foll, B., & Goldberg, S. R. (2004). Rimonabant, a CB1 antagonist, blocks nicotine-conditioned place preferences. *Neuroreport, 15*(13), 2139–2143.

Le Foll, B., & Goldberg, S. R. (2009). Effects of nicotine in experimental animals and humans: an update on addictive properties. *Handbook of Experimental Pharmacology, 192*, 335–367.

Le Novere, N., Corringer, P. J., & Changeux, J. P. (2002). The diversity of subunit composition in nAChRs: evolutionary origins, physiologic and pharmacologic consequences. *Journal of Neurobiology, 53*(4), 447–456.

Lewis, A., Miller, J. H., & Lea, R. A. (2007). Monoamine oxidase and tobacco dependence. *Neurotoxicology, 28*(1), 182–195.

Li, S. P., Kim, K. Y., Kim, J. H., Kim, J. H., Park, M. S., Bahk, J. Y., & Kim, M. O. (2004). Chronic nicotine and smoking treatment increases dopamine transporter mRNA expression in the rat midbrain. *Neuroscience Letters, 363*(1), 29–32.

Liu, X., Koren, A. O., Yee, S. K., Pechnick, R. N., Poland, R. E., & London, E. D. (2003). Self-administration of 5-iodo-A-85380, a beta2-selective nicotinic receptor ligand, by operantly trained rats. *Neuroreport, 14*(11), 1503–1505.

Lynch, W. J., & Carroll, M. E. (1999). Regulation of intravenously self-administered nicotine in rats. *Experimental and Clinical Psychopharmacology, 7*(3), 198.

Martin-Soelch, C. (2013). Neuroadaptive changes associated with smoking: structural and functional neural changes in nicotine dependence. *Brain Sciences, 3*(1), 159–176.

Miller, N. S., & Gold, M. S. (1998). Comorbid cigarette and alcohol addiction: epidemiology and treatment. *Journal of Addictive Diseases, 17*(1), 55–66.

Myers, W. D., Ng, K. T., & Singer, G. (1982). Intravenous self-administration of acetaldehyde in the rat as a function of schedule, food deprivation and photoperiod. *Pharmacology, Biochemistry, and Behavior, 17*(4), 807–811.

Naha, N., Li, S. P., Yang, B. C., Park, T. J., & Kim, M. O. (2009). Time-dependent exposure of nicotine and smoke modulate ultrasubcellular organelle localization of dopamine D1 and D2 receptors in the rat caudate-putamen. *Synapse, 63*(10), 847–854.

Peana, A. T., Muggironi, G., & Diana, M. (2010). Acetaldehyde-reinforcing effects: a study on oral self-administration behavior. *Frontiers in Psychiatry, 1*, 23.

de la Pena, J. B., Ahsan, H. M., Botanas, C. J., Sohn, A., Yu, G. Y., & Cheong, J. H. (2014). Adolescent nicotine or cigarette smoke exposure changes subsequent response to nicotine conditioned place preference and self-administration. *Behavioural Brain Research, 272*, 156–164.

Plescia, F., Brancato, A., Marino, R. A., & Cannizzaro, C. (2013). Acetaldehyde as a drug of abuse: insight into AM281 administration on operant-conflict paradigm in rats. *Frontiers in Behavioral Neuroscience, 7*, 64.

Quertemont, E., & De Witte, P. (2001). Conditioned stimulus preference after acetaldehyde but not ethanol injections. *Pharmacology, Biochemistry, and Behavior, 68*(3), 449–454.

Rodgman, A., & Perfetti, T. A. (2013). *The chemical components of tobacco and tobacco smoke* (2nd ed.). Boca Raton, Florida: CRC Press.

Rose, J. E., Salley, A., Behm, F. M., Bates, J. E., & Westman, E. C. (2010). Reinforcing effects of nicotine and non-nicotine components of cigarette smoke. *Psychopharmacology (Berl), 210*(1), 1–12.

Scholze, P., Orr-Urtreger, A., Changeux, J. P., McIntosh, J. M., & Huck, S. (2007). Catecholamine outflow from mouse and rat brain slice preparations evoked by nicotinic acetylcholine receptor activation and electrical field stimulation. *British Journal of Pharmacology, 151*(3), 414–422.

Seth, P., Cheeta, S., Tucci, S., & File, S. E. (2002). Nicotinic–serotonergic interactions in brain and behaviour. *Pharmacology, Biochemistry, and Behavior, 71*(4), 795–805.

Solinas, M., Scherma, M., Fattore, L., Stroik, J., Wertheim, C., Tanda, G., & Goldberg, S. R. (2007). Nicotinic alpha7 receptors as a new target for treatment of cannabis abuse. *The Journal of Neuroscience, 27*(21), 5615–5620.

Stolerman, I. P., & Jarvis, M. J. (1995). The scientific case that nicotine is addictive. *Psychopharmacology (Berl), 117*(1), 2–10.

Touiki, K., Rat, P., Molimard, R., Chait, A., & de Beaurepaire, R. (2007). Effects of tobacco and cigarette smoke extracts on serotonergic raphe neurons in the rat. *Neuroreport, 18*(9), 925–929.

Tuesta, L. M., Fowler, C. D., & Kenny, P. J. (2011). Recent advances in understanding nicotinic receptor signaling mechanisms that regulate drug self-administration behavior. *Biochemical Pharmacology, 82*(8), 984–995.

Villegier, A. S., Lotfipour, S., McQuown, S. C., Belluzzi, J. D., & Leslie, F. M. (2007). Tranylcypromine enhancement of nicotine self-administration. *Neuropharmacology, 52*(6), 1415–1425.

Volkow, N. D., Fowler, J. S., Wang, G. J., Swanson, J. M., & Telang, F. (2007). Dopamine in drug abuse and addiction – Results of imaging studies and treatment implications. *Archives of Neurology, 64*(11), 1575–1579.

West, R., Hajek, P., Foulds, J., Nilsson, F., May, S., & Meadows, A. (2000). A comparison of the abuse liability and dependence potential of nicotine patch, gum, spray and inhaler. *Psychopharmacology (Berl), 149*(3), 198–202.

WHO. (May 2014). *Tobacco fact sheet*. World Health Organisation.

Wise, R. A. (1996). Neurobiology of addiction. *Current Opinion in Neurobiology, 6*(2), 243–251.

Wise, R. A., & Koob, G. F. (2014). The development and maintenance of drug addiction. *Neuropsychopharmacology, 39*(2), 254–262.

Ypsilantis, P., Politou, M., Anagnostopoulos, C., Kortsaris, A., & Simopoulos, C. (2012). A rat model of cigarette smoke abuse liability. *Comparative Medicine, 62*(5), 395–399.

Section B

Molecular and Cellular Aspects

Chapter 12

Rho GTPases and Their Regulators in Addiction: A Focus on the Association of a β2-Chimaerin Polymorphism with Smoking

María J. Caloca[1], Laura Barrio-Real[2], Rogelio González-Sarmiento[3,4]

[1]*Institute of Molecular Biology and Genetics, Spanish National Research Council (CSIC), University of Valladolid School of Medicine, Valladolid, Spain;* [2]*Department of Pharmacology, Perelman School of Medicine, University of Pennsylvania, Philadelphia, PA, USA;* [3]*Molecular Medicine Unit, Department of Medicine, University of Salamanca-IBSAL, Salamanca, Spain;* [4]*IBMCC, University of Salamanca-CSIC, Salamanca, Spain*

Abbreviations

BDNF Brain-derived neurotrophic factor
DAG Diacylglycerol
DSM-IV Diagnostic and statistical manual of mental disorders, fourth edition
GAP GTPase-activating protein
GDI Guanine nucleotide dissociation inhibitor
GEF Guanine nucleotide exchange factor
GTPase Guanosine-5′-triphosphatase
GWAS Genome-wide association study
JNK c-Jun-*N*-terminal kinase
LTP Long-term potentiation
mAChR Muscarinic acetylcholine receptor
mGluR Metabotropic glutamate receptor
NMDAR *N*-methyl-D-aspartate receptor
SNP Single-nucleotide polymorphism

INTRODUCTION

Tobacco addiction is a complex behavioral condition that is significantly influenced by genetic factors. Nicotine is the major psychoactive ingredient of tobacco responsible for addiction. Nicotine shares with other drugs of abuse effects on the mesolimbic dopamine system responsible for the rewarding properties of tobacco (Balfour, Wright, Benwell, & Birrell, 2000). However, nicotine is a weak primary reinforcer but has a high addictive liability in comparison with other addictive substances (Chaudhri et al., 2006). The structural changes induced by nicotine in neurons and its effect on synaptic plasticity are mechanisms that contribute to the long-lasting nature of tobacco addiction (Nestler, 2001). The actin cytoskeleton drives these neuroadaptations produced by nicotine, and accordingly, several genes encoding proteins involved in actin cytoskeleton regulation are candidate genes for nicotine dependence (Wang & Li, 2009). Among these genes are members of the Rho family of small GTPases as well as their activators and inhibitors (Chen et al., 2007; Lind et al., 2010).

In this chapter we briefly describe the potential relationship between the biological functions controlled by the Rho GTPases and nicotine addiction, with special focus on the association of the GTPase regulator β2-chimaerin with smoking.

Rho GTPases AND NICOTINE ADDICTION

Role of Rho GTPases in Synaptic Plasticity

Repeated exposure to nicotine induces structural plasticity in various brain regions, including the brain reward circuits. This plasticity involves changes in the density and morphology of dendritic spines, which results in the reorganization of patterns of synaptic connectivity (Robinson & Kolb, 2004). For example, nicotine administered to mice induces long-term potentiation (Tang & Dani, 2009), a long-lasting enhancement of synaptic strength that promotes spine enlargement and new spine formation.

The changes in spine morphology are mainly dependent on the polymerization and arrangement of actin, which makes up the structural core of the spine. These processes are regulated, at least in part, by Rho GTPases (Penzes & Rafalovich, 2012).

The Rho GTPases are a family of 20 proteins that belong to the Ras superfamily of small GTPases. These proteins control fundamental cellular functions such as cell morphology, motility, and cell cycle, through the regulation of actin and the microtubule cytoskeleton, gene expression, and enzymatic activities (Jaffe & Hall, 2005). Rho GTPases are active when they are bound to GTP and inactive when they are bound to GDP. Their ability to rapidly cycle between the active and the inactive state allows them to function as molecular switches. This cycling is tightly regulated by three classes of regulatory proteins that control the nucleotide state of Rho proteins. Activation of Rho GTPases is controlled by guanine nucleotide exchange factors (GEFs), which catalyze the exchange of GDP for GTP, while inhibition is regulated by GTPase activating proteins (GAPs), which enhance the intrinsic GTPase activity of Rho GTPases. An additional level of regulation is exerted by guanine nucleotide dissociation inhibitors, which sequester inactive Rho GTPases in the cytoplasm, thus preventing the GDP/GTP exchange (Jaffe & Hall, 2005).

Of the 20 Rho GTPases encoded in the human genome, RhoA, Rac1, and Cdc42 play major roles in the regulation of dendritic spine dynamics and spine development. These proteins seem to play antagonistic roles; while Rac1 and Cdc42 promote spine formation and growth, RhoA promotes spine retraction and decreases in spine density (Newey, Velamoor, Govek, & Van Aelst, 2005). Among the 83 Rho GEFs and 67 Rho GAPs in the human genome, only a limited number have been proposed as potential regulators of the Rho GTPase functions in dendritic spines. GEFs such as Kalirin, Trio, Tiam1, Lfc, ephexin, and β-Pix act downstream of N-methyl-D-aspartate (NMDA) or ephrin receptors, two receptor systems involved in the plasticity of the postsynaptic neuron, whereas GAP proteins, including BCR, Abr, α- and β-chimaerins, oligophrenin-1, and p190 Rho GAP, impair neurite growth (Kiraly, Eipper-Mains, Mains, & Eipper, 2010; Tolias, Duman, & Um, 2011). Given the strong association between dendritic spine alterations and drug addiction, these proteins are likely to be involved in smoking-related behaviors. Next, we will summarize the research that supports a role for RhoA and the Rac regulator β2-chimaerin in smoking.

Association of RhoA with Smoking

The association of RhoA with smoking was discovered using high-density oligonucleotide analysis in a mouse model of acute nicotine exposure (Chen, et al., 2007). This analysis identified 30 genes encoding proteins with functions relevant to neuronal plasticity that were regulated in the ventral tegmental area, part of the mesocorticolimbic reward circuitry in the brain that is activated by nicotine. One of these genes was *RhoA*, which showed consistent time-dependent changes in expression in mice treated with nicotine. Then, five single-nucleotide polymorphisms (SNPs) for the *RhoA* gene were analyzed in human association studies. Single-marker association tests showed that the rs2878298 polymorphism had significant genotypic association with smoking initiation and nicotine dependence, although allelic association of this marker was significant only for smoking initiation. This association of *RhoA* with smoking was further confirmed in haplotype association analysis. Multimarker analysis (two-, three-, four-, and five-marker haplotype analysis) that included *RhoA* SNPs identified risk and protective haplotypes for smoking initiation but not for nicotine dependence. All multimarker combinations including the rs2878298 polymorphism showed a highly significant association. This study proved the relevance of Rho GTPases in smoking and identified *RhoA* as a gene that influences a specific aspect of smoking behavior.

CHIMAERINS AND SMOKING

Among the regulators of the Rho GTPases, the chimaerins have been associated with smoking behavior. These proteins are a family of Rho GAPs highly expressed in brain, where they have specific roles in the formation of neuronal networks. Next we summarize data that identify various chimaerin isoforms acting downstream of receptors involved in addiction pathways and the specific association of β2-chimaerin with smoking.

Chimaerins and the Nervous System

Chimaerins were first identified in the human brain (Hall et al., 1990). To date, four members of this family of proteins have been described in humans: α1-, α2-, β1-, and β2-chimaerin. All chimaerins have a unique combination of structural domains, with a C1 domain that binds to the lipid second messenger diacylglycerol (DAG) and a catalytic GAP domain with specificity for the Rac GTPase (Caloca, Wang, & Kazanietz, 2003) (Yang & Kazanietz, 2007). α2- and β2-chimaerins also have an N-terminal SH2 domain, most likely involved in heteromolecular interactions with phosphotyrosine proteins (Canagarajah et al., 2004). Therefore, chimaerins can function as intracellular mediators between neuronal receptors coupled to DAG generation and Rac activation to regulate cytoskeletal rearrangements (Figure 1).

α1- and α2-chimaerins are products of the *CHN1* gene, which maps to chromosome 2q31.1. Both isoforms are expressed in postnatal and adult brain and participate in downstream signaling pathways from different receptors. α1-Chimaerin participates in the regulation of dendritic morphology, acting downstream of muscarinic acetylcholine receptors (mAChRs), metabotropic glutamate receptors (mGluRs), and NMDA receptors (NMDARs). Activation of mAChRs and mGluRs induces the translocation of α1-chimaerin to the plasma membrane by binding to the DAG through its C1 domain (Buttery et al., 2006). However, activation of the NMDAR promotes the binding of α1-chimaerin to the receptor through a DAG-independent mechanism that involves direct binding of the GAP domain to the NR2A subunit of the NMDAR (Van de Ven, VanDongen, & VanDongen, 2005).

α2-Chimaerin regulates the dynamics of neurite outgrowth, acting at the level of the semaphorin pathway (Brown et al., 2004; Hall et al., 2001). This chimaerin isoform also displays an essential and specific role by linking EphA4 signaling with the regulation of actin cytoskeleton dynamics during axon pathfinding (Miyake et al., 2008; Wegmeyer et al., 2007).

Whether these functions of α-chimaerins are involved in nicotine addiction is currently unknown. There is, however, evidence for the association of the *CHN1* gene with alcohol dependence combined with conduct disorder and suicide attempts (Dick et al., 2010), which suggests a role for α-chimaerins in addiction-related disorders.

β1- and β2-chimaerins are the products of alternative transcription of the *CHN2* gene, which maps to chromosome 7p15.3 (Ahmed et al., 1990; Leung, How, Manser, & Lim, 1994). Of these two transcripts, only β2-chimaerin has been shown to be expressed in the brain. β2-Chimaerin can be transcriptionally upregulated after exposure to brain-derived neurotrophic factor (BDNF) (Mizuno, Yamashita, & Tohyama, 2004). A study in animal models has demonstrated that β2-chimaerin has a key role in neural circuit formation, controlling axon pruning and synapse elimination in the hippocampus. This activity is mediated by the inhibition of Rac activity by β2-chimaerin upon activation of the neuropilin-2 receptor by semaphorin-3F (Riccomagno et al., 2012).

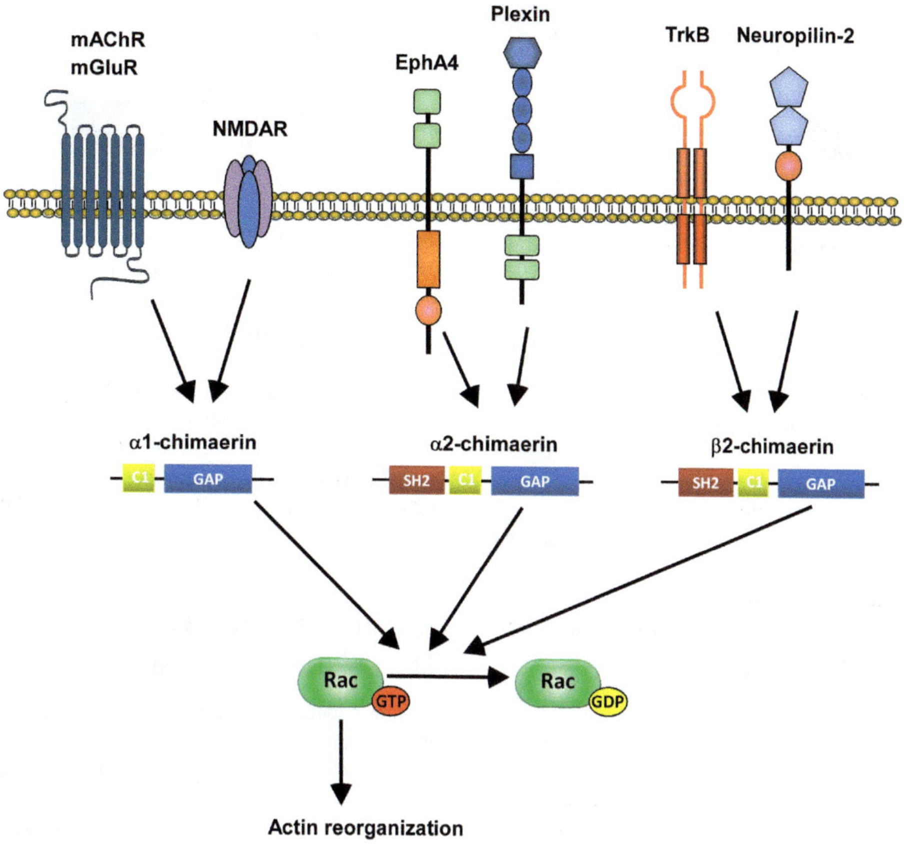

FIGURE 1 Receptors involved in the regulation of various chimaerin isoforms in neurons. Various receptors regulate specific chimaerin isoforms to control Rac activity. Metabotropic glutamate receptors (mGluR), muscarinic acetylcholine receptors (mAChR), and NMDA receptors (NMDAR) can activate α1-chimaerin. α2-Chimaerin participates in semaphorin signaling pathways (through plexin receptors) and ephrin signaling pathways (through Eph receptors). Semaphorins also regulate β2-chimaerin through activation of neuropilin receptors. This chimaerin isoform is also modulated by brain-derived neurotrophic factor, ligand of the TrkB receptor. Activated chimaerins locally inactivate Rac1 and regulate actin cytoskeleton dynamics.

Proper function of these processes is essential since unbalanced synaptic and axonal pruning have been associated with several mental illnesses, including schizophrenia, a psychotic disorder of high comorbidity with smoking (Lewis & Levitt, 2002). In humans, genetic analysis supports the relevance of β2-chimaerin in schizophrenia, since a missense polymorphism (H204R) on the CHN2 gene has been reported to be associated with this mental disorder (Hashimoto et al., 2005). Interestingly, genome-wide association studies of nicotine dependence have identified CHN2 as a candidate gene to contain variants that contribute to smoking addiction (Liu et al., 2006). Further supporting the role for β2-chimaerin in smoking behavior, a study by our group has identified a new SNP on the CHN2 gene that is associated with smoking.

Association of the β2-Chimaerin Gene with Smoking

Considering the biological role of β2-chimaerin and the evidence for the involvement of the β2-chimaerin gene (CHN2) in tobacco addiction, our laboratory searched for the association of individual SNPs on the CHN2 gene with smoking (Barrio-Real, Barrueco, González-Sarmiento, & Caloca, 2013). We analyzed three polymorphisms of potential relevance for β2-chimaerin function: rs3750103 c.611A→G (p.H204R), a missense polymorphism in exon 7 previously associated with schizophrenia; rs12112301 (IVS5+7C→T), a polymorphism within intron 5 that could influence the mRNA splicing; and rs186911567 c.366G→A (p.S122S), a novel synonymous polymorphism in exon 6 first identified in our laboratory (Figure 2). This study was conducted on a sample of 173 Spanish smokers (mean age 60.4 ± 1.4 years, 74.0% males) who fulfilled the criteria for nicotine dependence provided in the *Diagnostic and Statistical Manual of Mental Disorders*, fourth edition (American Psychiatric Association, 2000), and 188 healthy individuals (mean age 45.9 ± 1.4 years, 75.0% males) who had a lifetime history of no more than 100 cigarettes without developing an addiction to tobacco.

In this study, we identified the association of the rs186911567 polymorphism with smoking. While no carriers of the A allele were found in the control group, 4.6% of smokers had the GA or AA genotype. Allele A frequency was statistically different between the two groups as determined by Fisher's exact test ($p=0.003$). No differences were observed in genotypes and allelic frequencies of polymorphisms rs12112301 and rs3750103 among smokers and the control group (Table 1).

The association of the rs186911567 polymorphism showed no correlation with smoking habits, since genotype frequencies were similar between smokers grouped according to daily cigarette consumption or years of smoking (Table 2). Interestingly, when smokers

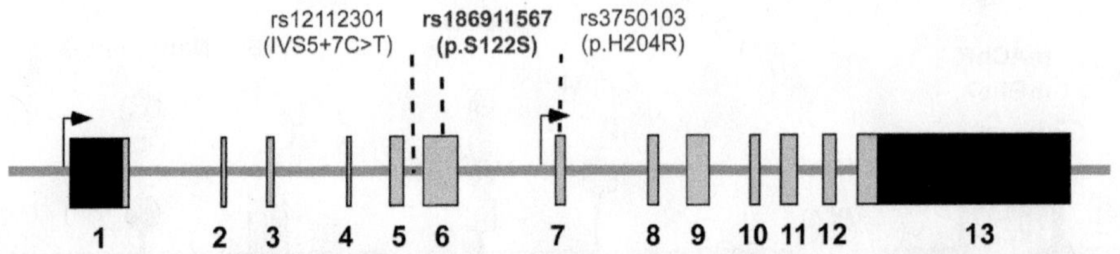

FIGURE 2 Structure of the *CHN2* gene. The *CHN2* gene is located on chromosome 7 in humans and encodes the β1- and β2-chimaerin isoforms that originate from two different transcription start sites (indicated with arrows). The exon structure is shown schematically with exon numbering. The location of the three examined polymorphisms is shown. In bold is indicated the SNP rs186911567, which is associated with smoking.

were grouped according to their age, there was a trend toward an increased frequency of the polymorphism in older smokers.

Many factors could explain the prevalence of the rs186911567 polymorphism in older smokers. Since *CHN2* was identified as a gene harboring allelic variants that distinguish successful versus unsuccessful abstainers in a smoking cessation trial (Uhl et al., 2008), one interpretation is that carriers of the rs186911567 polymorphism have increased the difficulty in quitting smoking, thus maintaining tobacco addiction throughout their lives.

The functional relevance of the synonymous rs186911567 polymorphism is currently unknown, but it is possible that this polymorphism influences smoking behavior by altering the protein function. The G→A substitution at position 366 in exon 6 does not result in an amino acid change. However, it is now widely accepted that synonymous polymorphisms may have an impact at the protein level on structure and function through various effects on RNA conformation and stability (Sauna & Kimchi-Sarfaty, 2011). Bioinformatics analysis predicts that the c.366G→A change can disrupt exonic splicing enhancer motifs of the *CHN2* gene and lead to the loss of a binding site for the splicing factor SC35 (Human Splicing Finder software) (Desmet et al., 2009). The resulting transcript would skip exon 6 of the *CHN2* gene and render a truncated protein lacking functional C1 and GAP domains owing to the generation of a premature stop codon. Inactivation of β2-chimaerin in neurons could result in abnormal levels of active Rac with the subsequent alteration of actin cytoskeleton remodeling, an important process for the nicotine effect on neural plasticity.

Although there are no studies regarding a direct link between nicotine and chimaerin function, several data provide evidence about the signaling pathways that may be involved. Chimaerins may act at the level of glutamatergic transmission through the NMDAR since chimaerin isoforms can inactivate Rac at the synapses by binding to the NR2A receptor subunit (Nesic, Duka, Rusted, & Jackson, 2011; Van de Ven et al., 2005). In addition, BDNF can upregulate β2-chimaerin as part of a mechanism for controlling neurite outgrowth (Mizuno et al., 2004). Since BDNF expression is regulated by nicotine (Kenny, File, & Rattray, 2000; Kim, Kim, Lee, & Kim, 2007) and BDNF levels associate with nicotine dependence (Lang et al., 2007), it is plausible that β2-chimaerin and BDNF signaling pathways may interplay to regulate smoking-related behaviors.

Finally, the high addictive liability of nicotine seems to be mediated by its ability to alter hippocampus-dependent learning (Kenney & Gould, 2008). This effect of nicotine on learning and memory is in part mediated by hippocampal c-Jun-*N*-terminal kinase (JNK) signaling (Kenney, Florian, Portugal, Abel, & Gould, 2009). β2-Chimaerin is a potential candidate to participate in nicotine control over JNK, since the ability of this protein to regulate JNK has been demonstrated in epithelial cells (Notcovich et al., 2010) and data demonstrate the expression of β2-chimaerin in the hippocampus (Riccomagno et al., 2012).

In summary, these studies highlight the fact that Rho GTPase signaling plays an important role in smoking, most likely through the modulation of actin cytoskeleton dynamics during the neuroadaptations induced by nicotine. The identification of SNPs on the genes encoding Rho GTPase proteins and their regulators may help to predict susceptibility to specific smoking behaviors.

APPLICATIONS TO OTHER ADDICTIONS AND SUBSTANCE MISUSE

It is now widely accepted that the structural plasticity of neurons is an important contributor in the development of addiction. Most drugs of abuse induce changes in the morphology of dendrites and dendritic spines in various brain regions, thus sharing a common target, the actin cytoskeleton, which is an essential component of these synaptic changes. However, different drugs can produce different effects in dendritic spines. For example, studies with animal models have demonstrated that chronic treatment with the psychostimulant drug cocaine or amphetamine increases dendritic spine density in the nucleus accumbens and prefrontal cortex, alterations similar to those produced by nicotine. In contrast, chronic morphine treatment markedly decreases spine density in these brain regions, changes that are also observed in dendritic spines in the nucleus accumbens after alcohol exposure (Robinson & Kolb, 2004; Zhou et al., 2007).

The molecular mechanisms that mediate the effects in dendritic spines of different drugs are currently under investigation. Owing to their role in actin cytoskeleton regulation and their expression patterns, Rho GTPases and their regulators are good potential drug targets to drive these drug actions. A few examples support this role for Rho GTPases. For example, cells chronically exposed to ethanol show an increase in the activity of p190 Rho GAP and, consequently, decreased levels of active RhoA (Selva & Egea, 2011). In addition, experiments in *Drosophila* have identified that Rac1 and Rho GAP 18B participate in the regulation of ethanol-induced behaviors (Peru y Colón de Portugal et al., 2012).

Rho GTPases are also targets for cocaine. A single intraperitoneal injection of cocaine in mice can regulate RhoA signaling in the nucleus accumbens, which may contribute to cocaine-induced synaptic plasticity in this site leading to drug addiction (Kim, Shin, Kim, Jeon, & Kim, 2009). Rac1 also has an essential role in cocaine-induced structural plasticity of the nucleus accumbens (Dietz et al., 2012)

Genetic studies have also identified that the signaling pathways involved in the regulation of the actin cytoskeleton are significantly enriched in addiction-related genes (Li, Mao, & Wei,

TABLE 1 Genotype Frequencies of the *CHN2* Polymorphisms, rs12112301 (IVS5+7C→T), rs3750103 c.611A→G (p.H204R), and rs186911567 c.366G→A (p.S122S) in Smokers and Control Subjects

rs12112301 (IVS5+7C→T)

	No. of Patients	Genotype			Grouped Genotypes		χ^2	df	p	HWE Controls	
		CC	CT	TT	CC	CT/TT				χ^2	p
Control subjects	188	151 (0.803)	34 (0.181)	3 (0.016)	0.803	0.197	0.071	1.000	0.790	0.448	0.503
Smokers	173	137 (0.792)	36 (0.208)	0 (0.000)	0.792	0.208					

rs186911567 c.366G→A (p.S122S)

	No. of Patients	Genotype			Grouped Genotypes		p
		GG	GA	AA	GG	GA/AA	
Control subjects	188	188 (1.000)	0 (0.000)	0 (0.000)	1.000	0.000	0.003[a]
Smokers	173	165 (0.954)	7 (0.040)	1 (0.006)	0.954	0.046	

rs3750103 c.611A→G (p.H204R)

	No. of Patients	Genotype			Grouped Genotypes		χ^2	df	p	HWE Controls	
		His/His	His/Arg	Arg/Arg	His	His/Arg and Arg/Arg				χ^2	p
Control subjects	188	164 (0.872)	24 (0.128)	0 (0.000)	0.872	0.128	0.489	1.000	0.485	0.874	0.350
Smokers	173	155 (0.896)	18 (0.104)	0 (0.000)	0.896	0.104					

The three SNPs were genotyped in 361 participants from Castilla y León (Spain). Smokers had a mean age of 60.4 ± 1.4 years (74.0% males) and the control group had a mean age of 45.9 ± 1.4 years (75.0% males). Hardy–Weinberg equilibrium was determined with the χ^2 test. The polymorphism rs186911567 had only the GG genotype in control groups and therefore Hardy–Weinberg equilibrium does not apply. Genotype frequencies of smokers and controls were compared by means of χ^2 test and Fisher's exact test when necessary (expected values below 5). $p < 0.05$ was considered significant for the differences between the genotypes.
[a]Fisher's exact test.
Data are from Barrio-Real et al. (2013) with permission from the publishers.

TABLE 2 Genotype Frequencies of the *CHN2* rs186911567 c.366G→A (p.S122S) Polymorphism in Smokers with Different Ages and Smoking Behaviors

	No. of Patients	Genotype			Grouped Genotypes		p
		GG	GA	AA	GG	GA/AA	
Age (Years)							
<65	80	79 (0.988)	1 (0.012)	0 (0.000)	0.988	0.012	0.070[a]
>65	93	86 (0.925)	6 (0.065)	1 (0.011)	0.925	0.075	
Cigarettes/Day							
Up to 20	97	91 (0.938)	5 (0.052)	1 (0.010)	0.928	0.062	0.468[a]
>20	76	74 (0.974)	2 (0.026)	0 (0.000)	0.974	0.026	
Years Smoking							
Up to 30	78	76 (0.974)	2 (0.026)	0 (0.000)	0.974	0.026	0.297[a]
>30	95	89 (0.937)	5 (0.053)	1 (0.011)	0.937	0.046	

Genotype frequencies for rs186911567 are shown for smokers grouped according to their age, daily cigarette consumption, and years of smoking. Genotype frequencies between smokers' subgroups were compared using Fisher's exact test. The differences did not reach statistical significance ($p > 0.05$). However, there was a trend toward an increased frequency of the polymorphism in older smokers ($p = 0.07$).
[a]*Fisher's exact test*.
Data are from Barrio-Real et al. (2013) with permission from the publishers.

2008). As we described in this chapter, a few studies show the association of genetic variants of the *RhoA* and *CHN2* genes with specific aspects of smoking behavior. It will be interesting in future studies to examine whether the genes encoding Rho GTPases, their regulators, and their effector molecules involved in the regulation of the cytoskeleton harbor allelic variants that contribute to addiction vulnerability to specific drugs.

DEFINITION OF TERMS

Actin cytoskeleton This is a network of filaments made up of polymerized actin that provides mechanical support and determines cell shape.

Dendritic spine Dendritic spines are small actin-rich protrusions that emerge from the surface of neuronal dendrites and form the postsynaptic part of most excitatory synapses.

GTPase activating protein This is a major class of regulators of small GTPases that stimulate the intrinsic GTPase activity of GTP binding. These proteins are essential for the termination of the functions of small G proteins.

Genome-wide association study This is a research technique that compares a broad number of SNPs across the genome in different populations and it is aimed at identifying genetic markers for a particular disease or trait.

Long-term potentiation This is a form of synaptic plasticity that consists in a persistent strengthening of synapses based on recent patterns of activity.

Missense polymorphism This is a SNP in the coding region of a gene that causes a change in the amino acid, thus affecting the protein sequence.

Single-nucleotide polymorphism This is a variation at a single position in a DNA sequence that occurs in more than 1% of the general population.

Synaptic plasticity This is the ability of synapses to change their strength by affecting their morphology, composition, or signal transduction efficiency in response to intrinsic or extrinsic signals.

Synonymous polymorphism This is a SNP in the coding region of a gene that does not affect the protein sequence.

Rho GTPases This is a family of small monomeric G proteins that are activated by exchanging bound GDP for GTP and regulate many aspects of actin cytoskeleton dynamics.

KEY FACTS OF SMOKING ADDICTION

- Smoking addiction is caused by nicotine through multiple molecular mechanisms and is strongly influenced by genetic factors.
- Smoking addiction includes various stages such as smoking initiation, nicotine dependence, cessation, and relapse.
- Genetics makes a substantial contribution to most aspects of smoking behaviors with different heritability estimates for smoking initiation, number of cigarettes smoked, nicotine dependence, and smoking cessation.
- Genes encoding different proteins involved in the control of the actin cytoskeleton harbor genetic variants that influence vulnerability to different smoking behaviors.
- The regulation of the actin cytoskeleton is central to the synaptic plasticity induced by nicotine that underlies the modifications in behavior that occur during the progression from smoking initiation to addiction.

SUMMARY POINTS

- This chapter reviews the role of Rho GTPases in nicotine addiction with a focus on the association of β2-chimaerin with smoking.
- β2-Chimaerin is a negative regulator of the Rac GTPase highly expressed in the brain that links signaling from receptors involved in addiction with Rac activation.
- Rac, together with Rho and Cdc42, is a master regulator of actin cytoskeleton dynamics in the nervous system.
- An important consequence of nicotine exposure is the induction of synaptic plasticity, a process that involves rearrangements of the actin cytoskeleton in dendritic spines and thus is controlled by Rho GTPases.
- According to this function, it has been reported that genetic variants of the *RhoA* and the β2-chimaerin gene (*CHN2*) may influence specific smoking behaviors.
- The SNP rs186911567 in the *CHN2* gene has been reported to be associated with smoking and most likely it influences vulnerability to maintaining nicotine use.

REFERENCES

Ahmed, S., Kozma, R., Monfries, C., Hall, C., Lim, H. H., Smith, P., & Lim, L. (1990). Human brain N-chimaerin cDNA encodes a novel phorbol ester receptor. *Biochemical Journal*, 272(3), 767–773.

American Psychiatric Association. (2000). *Diagnostic and statistical manual of mental disorders*. Text revision (fourth ed.). Washington DC: American Psychiatric Association.

Balfour, D. J. K., Wright, A. E., Benwell, M. E. M., & Birrell, C. E. (2000). The putative role of extra-synaptic mesolimbic dopamine in the neurobiology of nicotine dependence. *Behavioural Brain Research*, 113(1–2), 73–83.

Barrio-Real, L., Barrueco, M., González-Sarmiento, R., & Caloca, M. J. (2013). Association of a novel polymorphism of the beta2-chimaerin gene (*CHN2*) with smoking. *Journal of Investigative Medicine*, 61(7), 1129–1131.

Brown, M., Jacobs, T., Eickholt, B., Ferrari, G., Teo, M., Monfries, C., ... Hall, C. (2004). Alpha2-chimaerin, cyclin-dependent kinase 5/p35, and its target collapsin response mediator protein-2 are essential components in semaphorin 3A-induced growth-cone collapse. *The Journal of Neuroscience*, 24(41), 8994–9004.

Buttery, P., Beg, A. A., Chih, B., Broder, A., Mason, C. A., & Scheiffele, P. (2006). The diacylglycerol-binding protein alpha1-chimaerin regulates dendritic morphology. *Proceedings of the National Academy of Sciences of the United States of America*, 103(6), 1924–1929.

Caloca, M. J., Wang, H., & Kazanietz, M. G. (2003). Characterization of the Rac-GAP (Rac-GTPase-activating protein) activity of beta2-chimaerin, a 'non-protein kinase C' phorbol ester receptor. *Biochemical Journal*, 375(Pt 2), 313–321.

Canagarajah, B., Leskow, F. C., Ho, J. Y., Mischak, H., Saidi, L. F., Kazanietz, M. G., & Hurley, J. H. (2004). Structural mechanism for lipid activation of the Rac-specific GAP, beta2-chimaerin. *Cell*, 119(3), 407–418.

Chaudhri, N., Caggiula, A., Donny, E., Palmatier, M., Liu, X., & Sved, A. (2006). Complex interactions between nicotine and nonpharmacological stimuli reveal multiple roles for nicotine in reinforcement. *Psychopharmacology*, 184(3), 353–366.

Chen, X., Che, Y., Zhang, L., Putman, A. H., Damaj, I., Martin, B. R., ... Miles, M. F. (2007). RhoA, encoding a Rho GTPase, is associated with smoking initiation. *Genes, Brain and Behavior*, 6(8), 689–697.

Desmet, F.-O., Hamroun, D., Lalande, M., Collod-Béroud, G., Claustres, M., & Béroud, C. (2009). Human splicing finder: an online bioinformatics tool to predict splicing signals. *Nucleic Acids Research*, 37(9), e67.

Dick, D. M., Meyers, J., Aliev, F., Nurnberger, J., Kramer, J., Kuperman, S., ... Bierut, L. (2010). Evidence for genes on chromosome 2 contributing to alcohol dependence with conduct disorder and suicide attempts. *American Journal of Medical Genetics Part B: Neuropsychiatric Genetics*, 153B(6), 1179–1188.

Dietz, D. M., Sun, H., Lobo, M. K., Cahill, M. E., Chadwick, B., Gao, V., ... Nestler, E. J. (2012). Rac1 is essential in cocaine-induced structural plasticity of nucleus accumbens neurons. *Nature Neuroscience*, 15(6), 891–896.

Hall, C., Michael, G. J., Cann, N., Ferrari, G., Teo, M., Jacobs, T., ... Lim, L. (2001). Alpha2-chimaerin, a Cdc42/Rac1 regulator, is selectively expressed in the rat embryonic nervous system and is involved in neuritogenesis in N1E-115 neuroblastoma cells. *The Journal of Neuroscience*, 21(14), 5191–5202.

Hall, C., Monfries, C., Smith, P., Lim, H. H., Kozma, R., Ahmed, S., ... Lim, L. (1990). Novel human brain cDNA encoding a 34,000 Mr protein N-chimaerin, related to both the regulatory domain of protein kinase C and BCR, the product of the breakpoint cluster region gene. *Journal of Molecular Biology*, 211(1), 11–16.

Hashimoto, R., Yoshida, M., Kunugi, H., Ozaki, N., Yamanouchi, Y., Iwata, N., ... Kamijima, K. (2005). A missense polymorphism (H204R) of a Rho GTPase-activating protein, the chimaerin 2 gene, is associated with schizophrenia in men. *Schizophrenia Research*, 73(2–3), 383–385.

Jaffe, A. B., & Hall, A. (2005). Rho GTPases: biochemistry and biology. *Annual Review of Cell and Developmental Biology*, 21(1), 247–269.

Kenney, J. W., Florian, C., Portugal, G. S., Abel, T., & Gould, T. J. (2009). Involvement of hippocampal jun-N-terminal kinase pathway in the enhancement of learning and memory by nicotine. *Neuropsychopharmacology*, 35(2), 483–492.

Kenney, J., & Gould, T. (2008). Modulation of hippocampus-dependent learning and synaptic plasticity by nicotine. *Molecular Neurobiology*, 38(1), 101–121.

Kenny, P. J., File, S. E., & Rattray, M. (2000). Acute nicotine decreases, and chronic nicotine increases the expression of brain-derived neurotrophic factor mRNA in rat hippocampus. *Molecular Brain Research*, 85(1–2), 234–238.

Kim, T.-S., Kim, D.-J., Lee, H., & Kim, Y.-K. (2007). Increased plasma brain-derived neurotrophic factor levels in chronic smokers following unaided smoking cessation. *Neuroscience Letters*, 423(1), 53–57.

Kim, W. Y., Shin, S. R., Kim, S., Jeon, S., & Kim, J. H. (2009). Cocaine regulates ezrin–radixin–moesin proteins and RhoA signaling in the nucleus accumbens. *Neuroscience*, 163(2), 501–505.

Kiraly, D. D., Eipper-Mains, J. E., Mains, R. E., & Eipper, B. A. (2010). Synaptic plasticity, a symphony in GEF. *ACS Chemical Neuroscience*, 1(5), 348–365.

Lang, U., Sander, T., Lohoff, F., Hellweg, R., Bajbouj, M., Winterer, G., & Gallinat, J. (2007). Association of the met66 allele of brain-derived neurotrophic factor (BDNF) with smoking. *Psychopharmacology*, 190(4), 433–439.

Leung, T., How, B. E., Manser, E., & Lim, L. (1994). Cerebellar beta 2-chimaerin, a GTPase-activating protein for p21 Ras-related Rac is specifically expressed in granule cells and has a unique N-terminal SH2 domain. *Journal of Biological Chemistry, 269*(17), 12888–12892.

Lewis, D. A., & Levitt, P. (2002). Schizophrenia as a disorder of Neurodevelopment. *Annual Review of Neuroscience, 25*(1), 409–432.

Li, C.-Y., Mao, X., & Wei, L. (2008). Genes and (common) pathways underlying drug addiction. *PLoS Computational Biology, 4*(1), e2. http://dx.doi.org/10.1371/journal.pcbi.0040002.

Lind, P. A., Macgregor, S., Vink, J. M., Pergadia, M. L., Hansell, N. K., de Moor, M. H., ... Madden, P. A. (2010). A genomewide association study of nicotine and alcohol dependence in Australian and Dutch populations. *Twin Research and Human Genetics, 13*(1), 10–29.

Liu, Q.-R., Drgon, T., Johnson, C., Walther, D., Hess, J., & Uhl, G. R. (2006). Addiction molecular genetics: 639,401 SNP whole genome association identifies many "cell adhesion" genes. *American Journal of Medical Genetics Part B: Neuropsychiatric Genetics, 141B*(8), 918–925.

Miyake, N., Chilton, J., Psatha, M., Cheng, L., Andrews, C., Chan, W.-M., ... Engle, E. C. (2008). Human CHN1 mutations hyperactivate α2-chimaerin and cause Duane's retraction syndrome. *Science, 321*(5890), 839–843.

Mizuno, T., Yamashita, T., & Tohyama, M. (2004). Chimaerins act downstream from neurotrophins in overcoming the inhibition of neurite outgrowth produced by myelin-associated glycoprotein. *Journal of Neurochemistry, 91*(2), 395–403.

Nesic, J., Duka, T., Rusted, J., & Jackson, A. (2011). A role for glutamate in subjective response to smoking and its action on inhibitory control. *Psychopharmacology*, 1–14.

Nestler, E. J. (2001). Molecular basis of long-term plasticity underlying addiction. *Nature Reviews: Neuroscience, 2*(2), 119–128. http://dx.doi.org/10.1038/35053570.

Newey, S. E., Velamoor, V., Govek, E.-E., & Van Aelst, L. (2005). Rho GTPases, dendritic structure, and mental retardation. *Journal of Neurobiology, 64*(1), 58–74.

Notcovich, C., Diez, F., Tubio, M. R., Baldi, A., Kazanietz, M. G., Davio, C., & Shayo, C. (2010). Histamine acting on H1 receptor promotes inhibition of proliferation via PLC, RAC, and JNK-dependent pathways. *Experimental Cell Research, 316*(3), 401–411.

Penzes, P., & Rafalovich, I. (2012). Regulation of the actin cytoskeleton in dendritic spines. In M. R. Kreutz, & C. Sala (Eds.), *Synaptic plasticity* (Vol. 970) (pp. 81–95). Vienna: Springer.

Peru y Colón de Portugal, R. L., Acevedo, S. F., Rodan, A. R., Chang, L. Y., Eaton, B. A., & Rothenfluh, A. (2012). Adult neuronal Arf6 controls ethanol-induced behavior with arfaptin downstream of Rac1 and RhoGAP18B. *The Journal of Neuroscience, 32*(49), 17706–17713.

Riccomagno, M. M., Hurtado, A., Wang, H., Macopson, J. G. J., Griner, E. M., Betz, A., ... Kolodkin, A. L. (2012). The RacGAP β2-chimaerin selectively mediates axonal pruning in the hippocampus. *Cell, 149*(7), 1594–1606.

Robinson, T. E., & Kolb, B. (2004). Structural plasticity associated with exposure to drugs of abuse. *Neuropharmacology, 47*(Suppl. 1), 33–46.

Sauna, Z. E., & Kimchi-Sarfaty, C. (2011). Understanding the contribution of synonymous mutations to human disease. *Nature Reviews: Genetics, 12*(10), 683–691. http://dx.doi.org/10.1038/nrg3051.

Selva, J., & Egea, G. (2011). Ethanol increases p190RhoGAP activity, leading to actin cytoskeleton rearrangements. *Journal of Neurochemistry, 119*(6), 1306–1316.

Tang, J., & Dani, J. A. (2009). Dopamine enables in vivo synaptic plasticity associated with the addictive drug nicotine. *Neuron, 63*(5), 673–682.

Tolias, K. F., Duman, J. G., & Um, K. (2011). Control of synapse development and plasticity by Rho GTPase regulatory proteins. *Progress in Neurobiology, 94*(2), 133–148. http://dx.doi.org/10.1016/j.pneurobio.2011.04.011.

Uhl, G. R., Liu, Q.-R., Drgon, T., Johnson, C., Walther, D., Rose, J. E., ... Lerman, C. (2008). Molecular genetics of successful smoking cessation: convergent genome-wide association study results. *Archives of General Psychiatry, 65*(6), 683–693.

Van de Ven, T. J., VanDongen, H. M. A., & VanDongen, A. M. J. (2005). The nonkinase phorbol ester receptor {alpha}1-chimaerin binds the NMDA receptor NR2A subunit and regulates dendritic spine density. *The Journal of Neuroscience, 25*(41), 9488–9496.

Wang, J., & Li, M. D. (2009). Common and unique biological pathways associated with smoking initiation/progression, nicotine dependence, and smoking cessation. *Neuropsychopharmacology, 35*(3), 702–719.

Wegmeyer, H., Egea, J., Rabe, N., Gezelius, H., Filosa, A., Enjin, A., ... Betz, A. (2007). EphA4-dependent axon guidance is mediated by the RacGAP α2-chimaerin. *Neuron, 55*(5), 756–767.

Yang, C., & Kazanietz, M. G. (2007). Chimaerins: GAPs that bridge diacylglycerol signalling and the small G-protein Rac. *Biochemical Journal, 403*(1), 1–12.

Zhou, F. C., Anthony, B., Dunn, K. W., Lindquist, W. B., Xu, Z. C., & Deng, P. (2007). Chronic alcohol drinking alters neuronal dendritic spines in the brain reward center nucleus accumbens. *Brain Research, 1134*, 148–161.

Chapter 13

Brain Orexin Receptors and Nicotine

Rita Machaalani[1,2,3], Nicholas J. Hunt[1,2], Karen A. Waters[1,2,3]
[1]Department of Medicine, University of Sydney, Sydney, NSW, Australia; [2]The BOSCH Institute, University of Sydney, Sydney, NSW, Australia; [3]The Children's Hospital, Westmead, Sydney, NSW, Australia

Abbreviations

ACh Acetylcholine
ARC Hypothalamic arcuate nucleus
cAMP Cyclic adenosine monophosphate
ChAT Choline acetyltransferase
CNS Central nervous system
CSF Cerebral spinal fluid
DMH Dorsal medial hypothalamus
DMN Dorsal medial nucleus
DORA Dual orexin receptor antagonist
GABA γ-aminobutyric acid
GPCR G-protein-coupled receptor
IF Immunoflourescence
IHC Immunohistochemistry
ip Intraperitoneal
ISH In situ hybridization
LHA Lateral hypothalamic area
MHN Medial hypothalamic nucleus
mRNA Messenger ribonucleic acid
NAc Nucleus accumbens
nAChR Nicotinic acetylcholine receptor
NMDA N-methyl-D-aspartate
NPY Neuropeptide Y
Ox Orexin
OxA Orexin A
OxB Orexin B
OX1R Orexin receptor 1
OX2R Orexin receptor 2
P Postnatal day
PCR Polymerase chain reaction
PFA Perifornical area
PLC Phospholipase C
PPO Prepro-orexin
PPT Pedunculopontine tegmental nucleus
PVN Paraventricular nucleus
REM Rapid eye movement
RIA Radioimmunoassay
sc Subcutaneous
SON Supraoptic nucleus
TMN Tuberal mammillary nucleus
VMH Ventral medial hypothalamus
VMN Ventral medial nucleus
VTA Ventral tegmental area
WB Western blot

INTRODUCTION

Since the discovery of orexin (also known as hypocretin) in 1998 (de Lecea et al., 1998; Sakurai et al., 1998) much has been studied regarding the function of the orexinergic system and its role in clinical diseases/disorders, particularly eating behavior and sleep-related disorders. Initially, orexin was identified as having a prominent role in feeding behavior based on its pharmacological activity and its localization to the lateral hypothalamus (also known as the feeding center), with the name "orexin," from the Greek word *orexis,* meaning appetite (Sakurai et al., 1998). Subsequently, orexin neuronal activity was shown to regulate wakefulness and sleep states, based on the finding that sufferers of the sleep disorder narcolepsy had deficient orexin signaling (Crocker et al., 2005; Thannickal et al., 2000). More recently, the role of orexin in addiction and reward has been identified (Aston-Jones, Smith, Moorman, & Richardson, 2009; Georgescu et al., 2003; Sharf, Sarhan, & Dileone, 2010).

Nicotine is the major addictive component and neurotoxic agent of cigarettes/tobacco (Benowitz, 1996; Dani & Heinemann, 1996). Nicotine has been extensively studied in relation to addiction and reward, appetite, sleep, and arousal, with studies showing that nicotine is highly addictive (Benowitz, 1996), increases reward-seeking behaviors (De Biasi & Dani, 2011), decreases appetite (Jo, Talmage, & Role, 2002; Klesges, Meyers, Klesges, & La Vasque, 1989), and decreases sleep quantity and quality (Davila, Hurt, Offord, Harris, & Shepard, 1994; Jaehne, Loessl, Barkai, Riemann, & Hornyak, 2009).

Given the commonalities between nicotine and orexin in regulating these physiological parameters, studies were commenced to understand the role nicotine has on orexin expression and function. This chapter reviews studies aimed at determining how acute and chronic nicotine exposure, both pre- and postnatally through to adulthood, alters expression of the orexinergic system, including prepro-orexin (PPO), orexin A (OxA), orexin B (OxB), and the two orexin receptors 1 and 2 (OX1R, OX2R).

THE OREXINERGIC SYSTEM

The orexinergic system consists of three genes, the precursor PPO gene, which encodes two neuropeptides, OxA and OxB, and a separate gene for each of the receptors OX1R and OX2R (de Lecea et al., 1998; Sakurai et al., 1998). The human PPO gene encodes 131 amino acids and is located on chromosome 17q21 (Sakurai et al., 1999). The human and rat sequences are 83% equivalent, while between rats and mice the polypeptide is 95% identical (Sakurai et al., 1999). The gene consists of two exons and an intervening intron (Figure 1). The PPO sequence including the OxA and OxB sequences is encoded on the second exon. Separating OxA and OxB in the PPO polypeptide sequence is a Lys–Arg basic pair that is thought to be a recognition site for a prohormone convertase (Sakurai et al., 1999). OxA and OxB are produced by the cleavage of PPO at the two sites containing the basic amino acid residues GKR and GRR (Figure 1). Both OxA and OxB contain an amidated C-terminus, with 46% sequence identity to each other, the length of which is 33 and 28 amino acid residues, respectively (Sakurai et al., 1998). OxA has a polyglutamylated N-terminus where the four cysteine residues form two intramolecular disulfide bonds within the chain, while OxB is a linear peptide (Sakurai et al., 1998) (Figure 1).

OX1R and OX2R are class A, G-protein-coupled receptors (GPCRs) of 425 and 444 amino acid sequences, respectively, with 64% identical sequencing (Sakurai et al., 1998). OxA has 10 times higher affinity for OX1R than OxB, while both bind to OX2R with equal affinity (de Lecea et al., 1998; Sakurai et al., 1998). Studies have shown that other neuropeptides do not bind to orexin receptors (Holmqvist, Åkerman, & Kukkonen, 2001), as the orexin receptors have a low (30%) homology with other GPCRs (Kukkonen, 2013). This suggests that the binding electrochemical and stereochemistry of the orexin neuropeptides is specific to the epitopes expressed on orexin receptors. Generally, OX1R signals via G_q-coupled proteins and OX2R signals via G_q or $G_{i/o}$ (Zhu et al., 2003).

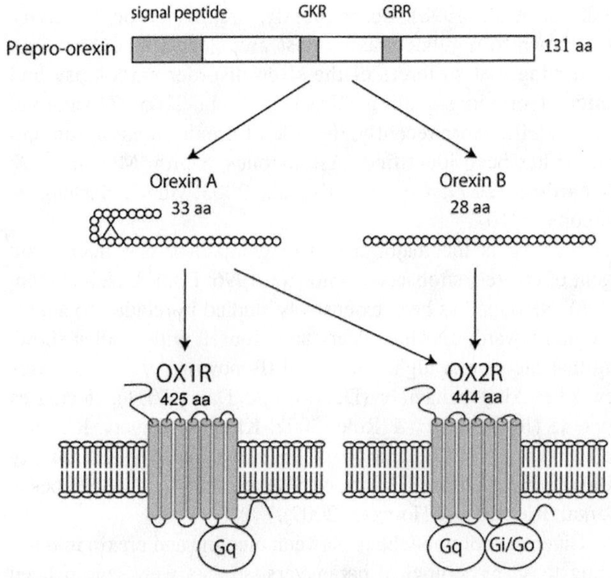

Upon activation of the orexin receptors, the resulting increased intracellular Ca^{2+} levels and receptor-mediated Ca^{2+} influx prolong postsynaptic excitation in the target neurons via leveraging activities of various signal transduction pathways and ion channels (reviewed in Gotter et al., 2012; Kukkonen, 2013). This results in the stimulation of second messengers, such as phospholipase C (PLC) and cyclic adenosine monophosphate (cAMP), and downstream activation of tyrosine kinase receptors involved in cell growth, survival, and death and synaptic plasticity. Hence, this substantial synergistic effect maintains neurotransmitter release postsynaptically, leading to a prolonged response output (reviewed in Gotter et al., 2012; Kukkonen, 2013). The difference in selectivity of OxA and OxB for the receptors also suggests different functionalities of the receptors (reviewed in Mieda et al., 2004), as does the divergence in their distribution in the brain (Marcus et al., 2001).

THE HYPOTHALAMUS AND OREXIN RECEPTORS

Orexin neurons project to most regions in the hypothalamus (Nambu et al., 1999). From the hypothalamus, orexin projections are distributed widely, albeit with variable density, throughout the brain except the cerebellum (Peyron et al., 1998). The number of orexin-containing neurons is relatively small compared to neurotransmitter systems in the brain (Deurveilher, Lo, Murphy, Burns, & Semba, 2006). However, they receive major inputs from glutamatergic, γ-aminobutyric acid (GABA)-ergic, muscarinic, serotonergic, adenosine, and catecholaminergic receptors (Sakurai, 2007) as well as glucose, leptin, and neuropeptide Y (NPY) (Machaalani, Hunt, & Waters, 2013; Tsuneki, Wada, & Sasaoka, 2010). As they also regulate the release of acetylcholine, serotonin, and noradrenaline, the orexin neurons have a prominent role in many physiological functions including arousal, appetite, addiction and reward responses, neurological stress and panic, respiratory and cardiovascular regulation, pain modulation, rapid eye movement (REM) sleep, and non-REM sleep (Aston-Jones et al., 2009; Lu, Sherman, Devor, & Saper, 2006; Luong & Carrive, 2012; Sakurai, 2007; Saper, Scammell, & Lu, 2005). These physiological roles are induced, mediated, and/or facilitated by orexinergic activity within the hypothalamic brain regions that express orexin neurons or orexin receptors.

The hypothalamus is located in the ventralmost portion of the diencephalon, underneath the thalamus, lying along the anterior inferior region of the third ventricle below the hypothalamic sulcus (Swaab, 2003). It can be grossly divided in lateral–medial dimensions into arbitrarily designated preoptic, supraoptic, tuberal, and mammillary regions (Swaab, 2003). The tuberal region contains the majority of orexinergic neurons in humans and rats (Fronczek et al., 2007; Hunt, Rodriguez, Waters, & Machaalani, 2015). The hypothalamus is an integrative center for neural control of various autonomic and neuroendocrine activities since it

FIGURE 1 The production of the orexinergic system. Upon removal of the signal peptide, OxA (33 amino acids) and OxB (28 amino acids) are produced after cleavage of PPO (130 amino acids) at the two sites containing the basic amino acid residues GKR and GRR. OX1R (425 amino acids) is selective for OxA, whereas OX2R (444 amino acids) is nonselective for either OxA or OxB. The two receptors are further coupled to different subclasses of heterotrimeric G proteins to mediate the action of orexin. *Adapted from Sakurai (2007).*

regulates innervation of the pituitary gland. The hypothalamus thus regulates multiple functions including circadian rhythms, temperature regulation, water and food satiety via vasopressin release, energy metabolism, cardiovascular and respiratory sympathetic stress responses, sleep and wakefulness states, and addiction and reward pathways (Swaab, 2003). These functions are often attributed to one particular hypothalamic region over another, as summarized in Figure 2. The hypothalamic regions include the arcuate nucleus (ARC), dorsal medial hypothalamus (DMH) and nucleus (DMN), lateral hypothalamic area (LHA), perifornical area (PFA), paraventricular nucleus (PVN), supraoptic nucleus (SON), ventral medial hypothalamus (VMH) and nucleus (VMN), and tuberal mammillary nucleus (TMN).

OX1R and OX2R are expressed with differing densities in all of these hypothalamic regions (Figure 2). The TMN is consistently reported as being devoid of OX1R expression (Hervieu, Cluderay, Harrison, Roberts, & Leslie, 2001; Hunt, Waters, & Machaalani, 2013; Marcus et al., 2001). As the TMN modulates arousal (Huang et al., 2001; Yamanaka et al., 2002), a lack of OX1R here indicates that orexinergic modulation of arousal must occur via other brain regions, most likely the brain-stem regions of the locus coeruleus, which expresses only OX1R, and the dorsal raphe.

NICOTINE AND NICOTINIC ACETYLCHOLINE RECEPTORS

Nicotine is the major addictive component and neurotoxic agent of cigarettes/tobacco (Benowitz, 1996; Dani & Heinemann, 1996). Unlike many of the compounds present in tobacco smoke, including carbon monoxide, benzene, and acrolein, nicotine has no major environmental sources other than tobacco (US Environmental Protection Agency and Office of Health and Environment, 1992). Nicotine is a tertiary amine consisting of a pyridine and a pyrrolidine ring and can exist in two enantiomer forms: (S)-nicotine and (R)-nicotine (Figure 3). Nicotine's predominant effects are induced by acting on and activating the nicotinic acetylcholine receptor (nAChR). (S)-Nicotine is found in high quantities in tobacco and binds stereoselectively to the nAChR. However, (R)-nicotine is found in small quantities in tobacco smoke owing to its production during the pyrolysis process, is a weak agonist, and tends to be the common form used in pharmacological studies (Benowitz, 1996). Cotinine is the major metabolite of nicotine and has a longer half-life than nicotine. Thus, cotinine is generally the preferred biomarker of exposure to cigarette smoke, with a half-life of 15–20 h compared to nicotine, which has a half-life of 1–2 h (Benowitz, 1996).

The nAChRs belong to the Cys-loop family of ligand-gated ion channels that exist as pentamers of subunits, arranged symmetrically around a central pore (Cooper, Couturier, & Ballivet, 1991). Genes encoding a total of 17 subunits (α1–10, β1–4, δ, γ, and ε) have been identified through gene studies. All subunits are of mammalian origin with the exception of α8, which is avian (Papke et al., 2008). nAChRs are found at skeletal neuromuscular junctions and autonomic ganglia as either homopentamers or heteropentamers. The predominant conformation of these subunits within the central nervous system (CNS) is heteromeric, although α7 and α9 homopentamers exist. The α1, β1, γ, δ, and ε subunits are classified as muscle type (Ke, Eisenhour, Bencherif, & Lukas, 1998), while the remaining are classed as neuronal. In general, neuronal heteromeric nAChRs have higher affinities for nicotine than α7 homomeric nAChRs (Ke et al., 1998). The composition of

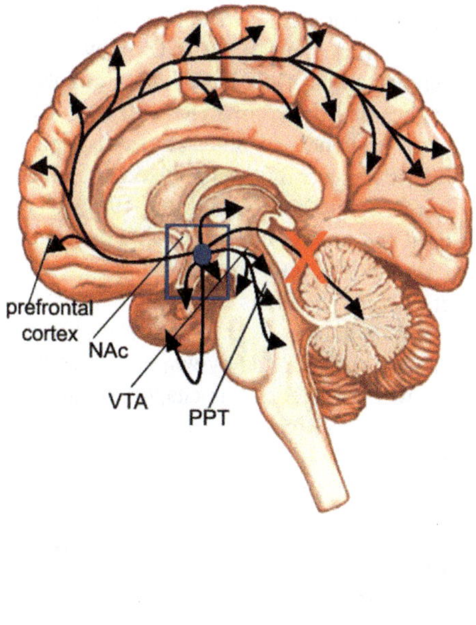

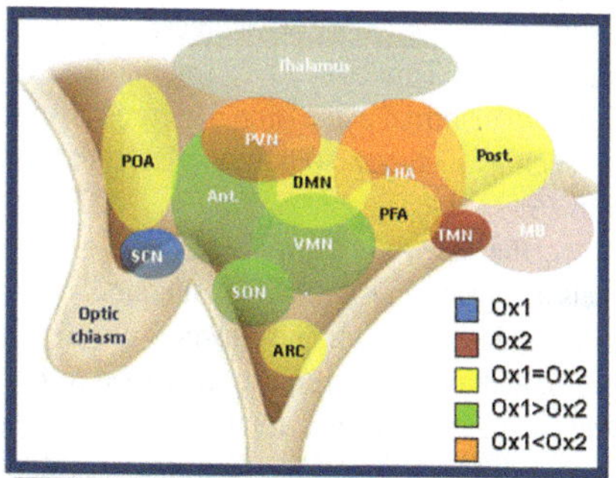

Nuclei	Functional regulation
POA	Sleep/wake, sexual arousal.
SON	Release of vasopressin
SCN	Regulates biological clock
Ant.	Arousal via the release of histamine
PVN	Release of adrenocorticotropin
PFA	Cardiovascular activity and stress
DMH	Regulation of circadian rhythms
LHA	Feeding, locomotion, thermogenesis, addiction/reward seeking
ARC	Feeding
TMN	Arousal via the release of histamine
Post.	Sleep (stabilisation), pain

FIGURE 2 The human hypothalamus and orexin receptor expression in its subregions (inset). The orexinergic neurons of the hypothalamus send projections to all brain regions except for the cerebellum (marked with red X). The hypothalamus is separated into several regions/nuclei, which contain different amounts of OX1R and OX2R, as indicated by the color coding in the inset, and have various functional roles summarized in the table. Regions to which orexin projects and that are important for addiction include the prefrontal cortex, nucleus accumbens (NAc), ventral tegmental area (VTA), and pedunculopontine tegmental nucleus (PPT).

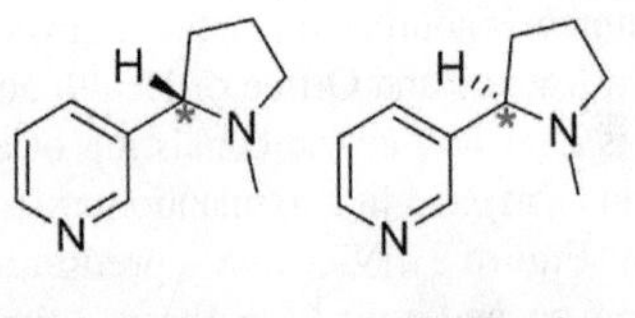

FIGURE 3 The two enantiomer structures of nicotine: (S)- and (R)-nicotine. The red asterisk indicates the chiral carbon on which the hydrogen in (S)-nicotine is pointing toward the reader, but in (R)-nicotine is pointing away from the reader.

subunits in the receptor determines ligand specificity and affinity, cation permeability, and channel kinetics.

In the brain, nAChRs contribute to a wide range of activities and influence a number of physiological functions, and this is via their response to the endogenous neurotransmitter acetylcholine (ACh). ACh binds to ion channels in the target cell membrane and opens them transiently, allowing selected ions to flow through and produce a change in membrane potential (Unwin, 2003). This is important since it regulates "normal" processes such as transmitter release, cell excitability, and neuronal integration, all of which are crucial for network operations and influence physiological functions such as sleep, fatigue, anxiety, the central processing of pain, food intake, and a number of cognitive functions (Gotti, Fornasari, & Clementi, 1997; Lindstrom, 1997; Role & Berg, 1996). However, abnormal (either excessive or deficient) ACh innervation of the nAChRs can lead to various diseases throughout the life span, including frontal lobe epilepsy (Steinlein, 2000), schizophrenia (Freedman, Adams, & Leonard, 2000), and Alzheimer's disease (Wang et al., 2000).

On the other hand, activation of the nAChRs by exogenous nicotine, which occurs when a person actively or passively ingests nicotine, results in unfavorable neuropathologies and functions (reviewed in Slotkin, 1998), including addiction, increased apoptosis (cell death) (Machaalani, Waters, & Tinworth, 2005; Trauth, McCook, Seidler, & Slotkin, 2000), and abnormal expression of neurotransmitters and their receptors, such as glutamate (Fanous, Machaalani, & Waters, 2006), serotonin (Say, Machaalani, & Waters, 2007), and dopamine (Slotkin, 1998), the last thought to be associated with increased endorphin levels and increased addiction to nicotine. Moreover, a direct change in nAChR expression and function occurs. An example is the upregulation and increase in density (Buisson & Bertrand, 2001; Sparks & Pauly, 1999) and expression (Browne, Sharma, Waters, & Machaalani, 2010) of α4β2 and α7 nicotinic receptors; however, this tends to be both dose and time dependent (Ke et al., 1998). More regarding the disparities in results due to experimental paradigms used to study nicotine exposure will be addressed below.

NICOTINE AND THE OREXINERGIC SYSTEM

Why Study Nicotine in Relation to the Orexinergic System?

Studies on the effects of nicotine on the orexinergic system arose for two main reasons:

1. *Tobacco effects on body weight:* Both human and animal studies (reviewed in Jo et al., 2002; Klesges et al., 1989) have shown that cigarette smoking and/or nicotine administration is associated with decreased body weight. The mechanism(s) by which body weight is affected by smoking or nicotine is not well understood, but the two main suggestions are that nicotine suppresses food intake (Bowen, Eury, & Grunberg, 1986; Grunberg, Bowen, & Winders, 1986) or that nicotine increases the metabolic rate (Perkins, Epstein, Stiller, Marks, & Jacob, 1989; Sztalryd, Hamilton, Horwitz, Johnson, & Kraemer, 1996). Regardless, it is clear that nicotine does not do this on its own per se, but rather by acting on other appetite- or metabolism-regulating genes/proteins, including orexin.

2. *Tobacco effects on sleep:* Nicotine consumption (humans) and administration (animal models) result in decreased sleep quantity and quality (reviewed Jaehne et al., 2009). Specifically, nicotine was found to induce symptoms of insomnia, such as increased sleep latency, sleep fragmentation, and decreased slow-wave sleep, with reduced sleep efficiency and increased daytime sleepiness. Nicotine also induces REM sleep suppression and decreases time spent in REM sleep (Davila et al., 1994).

Given that these functions are predominantly regulated by the hypothalamus, and that the orexins are made within the hypothalamus, research commenced to look at the effects of nicotine on orexin expression and function within the brain, with the majority of the studies focusing on the hypothalamus.

HUMAN STUDIES OF NICOTINE ON OREXIN

As of this writing, only two studies have been conducted in humans to determine the effects of nicotine (via cigarette smoking) on the orexinergic system (Aksu et al., 2009; von der Goltz et al., 2010). Both studies undertook radioimmunoassays (RIA) of plasma or serum since, in a clinical setting, the most common orexin marker measured is OxA by RIA. This is performed on cerebral spinal fluid (CSF) or serum/plasma samples. Only one study has also performed RIA for OxB (Ebrahim et al., 2003). The greater number of studies of OxA compared to OxB is due to the higher permeability of the blood–brain barrier to OxA compared to OxB (Kastin & Akerstrom, 1999). The two studies that have been conducted comparing OxA in serum of current ex-smokers (Aksu et al., 2009) and plasma of current smokers (von der Goltz et al., 2010) to adults who never smoked found no change in OxA levels, although a negative correlation between OxA and nicotine craving was reported by von der Goltz et al. (2010). However, as we will see later in this chapter, findings based on serum orexin levels do not necessarily reflect changes in brain expression, since discrepancies between them are a common phenomenon when studying orexins in animal models (Machaalani et al., 2013).

PARADIGMS OF NICOTINE EXPOSURE TO REPLICATE CLINICAL CONDITIONS

Given the limited ability to study the orexin system in humans, with study of orexin receptors possible only on direct brain tissue, hence necessitating autopsy material, animal studies have been developed to provide the much needed data. Prior to commencing studies, the animal model needs to be verified as a good representation of the clinical condition being modeled, each model having a specific target in mind (Cohen & George, 2013). With regard to nicotine exposure, the most common method of verifying the optimal dose of nicotine administration is by measuring either nicotine or cotinine levels in the blood or urine, with measurement of cotinine preferred, given its longer half-life, 16 h compared to 2 h for nicotine (Benowitz & Jacob, 1993). The metabolism, distribution, and excretion rate of nicotine differ with species, hence slight variations in cotinine levels would be anticipated (Bramer & Kallungal, 2003). In general, to mimic maternal (*prenatal*) cigarette smoke exposure, using the rat, doses should be within 2–6 mg/kg/day, resulting in plasma cotinine levels of >100 ng/ml (Lichtensteiger, Ribary, Schlumpf, Odermatt, & Widmer, 1988; Murrin, Ferrer, Zeng, & Haley, 1987). For early *postnatal* exposure, as would be experienced by babies either through breast milk or through environmental passive smoke exposure, doses should be within 2–4 mg/kg/day resulting in plasma and urine cotinine levels >5 and >10 ng/ml, respectively (Luck & Nau, 1985). For childhood and adult cigarette exposure paradigms, doses should be around 6 mg/kg/day to produce levels of cotinine indicative of active smoking, which in serum is >220 ng/ml or 100 nmol/L and in urine is >3516 ng/ml or 1700 nmol/L (SSWPS Handbook, 2014; Vine et al., 1993).

Dose aside, the other consideration when developing the animal model of nicotine exposure is the mode of administration. The common methods to date include orally, by mixing it with drinking water; injections (intraperitoneal or subcutaneous); infusions; minipumps; dermal patches; or placing the animal in a cigarette smoking chamber. Current studies lean toward the use of minipumps and nicotine patches, especially for the prenatal animal models, given that they result in steady states of plasma nicotine levels and overcome the hypoxia–ischemia effects that were seen to occur when using the injection method (Slotkin, 1998). A review addressing the various nicotine animal models and their results is provided by Cohen and George (2013).

NICOTINE EXPOSURE ON OREXIN EXPRESSION

A modest number of studies have examined the effects of nicotine on the expression of the orexinergic system, with the majority consistently linking nicotine to increased expression of orexin and its receptors (summarized Table 1). All of the studies in Table 1 were conducted in the rat species, with the exception of our own study in piglets (Hunt et al., 2013) and the study by Plaza-Zabala, Flores, Maldonado, and Berrendero (2012) performed on mice. Moreover, the majority focused on the male species and adult age. It is only recently that studies have commenced to look at the effects of prenatal nicotine on orexin expression in the offspring (Boychuk et al., 2011; Chang, Karatayev, & Leibowitz, 2013; Morgan, Harrod, Lacy, Stanley, & Fadel, 2013).

The pioneering study in this field was by Kane et al. (2000). In adult male rats, nicotine was delivered by intraperitoneal injections five times per day at a concentration of 2–6 mg/kg/day for a total of 10 days. Studying all five markers of the orexinergic system, they found that all were increased in expression by up to 50% (Table 1). This initial study was closely followed by a binding study for OxA by the same group with results showing that binding capacity (affinity and density) decreased by 60% (Kane, Parker, & Li, 2001). No other study as of this writing has looked at orexin receptor binding after nicotine exposure. However, another method of studying orexin activity is to measure c-fos expression in orexin-containing neurons, an increase in c-fos within orexin neurons indicating an increase in activity. The two studies performing such an investigation (Pasumarthi, Reznikov, & Fadel, 2006; Plaza-Zabala, Martin-Garcia, de Lecea, Maldonado, & Berrendero, 2010) showed that nicotine did indeed increase c-fos expression in orexin neurons of the LHA and PFA, and this was in both adult rats (Pasumarthi et al., 2006) and adult mice (Plaza-Zabala et al., 2010), adding weight to the data by Kane et al. (2000) that nicotine has a stimulatory effect on the orexinergic system. In 2010, three other papers were published studying various paradigms of nicotine exposure on the orexin system (LeSage, Perry, Kotz, Shelley, & Corrigall, 2010; Liu et al., 2010; Pasumarthi & Fadel, 2010). Liu et al. (2010) performed a study similar to that of Kane et al. (2000), examining the whole orexinergic system, but this time, the nicotine exposure paradigm was of cigarette smoke to adult male rats in a chamber, and the orexin receptors were studied in the brain-stem medulla (Table 1). Consistent with Kane et al. (2000), they found an increase in all markers. For Pasumarthi and Fadel (2010) and LeSage et al. (2010), the paradigms used were very different, with Pasumarthi and Fadel (2010) studying an acute (1 day) exposure compared to LeSage et al. (2010), who studied a chronic exposure (19 days). Moreover, Pasumarthi and Fadel (2010) investigated OxA expression focusing only on the hypothalamus, while LeSage et al. (2010) investigated the expression of PPO and the receptors, branching into other brain regions, including the ventral tegmental area, prefrontal cortex, and nucleus accumbens, regions involved in reward and addiction. Pasumarthi and Fadel (2010) showed that a single subcutaneous injection of 2 mg/kg in the adult male rat increased c-fos expression in OxA-positive neurons in the LHA and PFA. However, this increase in activity was dependent on cholinergic neurons from the basal forebrain. Loss of these neurons inhibited the increased c-fos expression in orexin neurons. LeSage et al. (2010) found that although PPO expression was not changed, OX1R was decreased with immediate euthanasia after the final nicotine dose, but with a 5-h window after the last dose, OX1R increased, with no effects on OX2R at either time point. Moreover, these results were restricted to the rostral LHA and ARC of the hypothalamus (LeSage et al., 2010). Following these studies, a third paradigm looked for changes caused by early postnatal nicotine exposure on the expression of OxA/B in the PVN and LHA in adulthood and found no change (Younes-Rapozo et al., 2013). However, a closer look at their data when combining the regions shows that an increase in expression seems to be likely. We, ourselves, were unique in examining the effects of nicotine given in the immediate postnatal period utilizing the piglet as the animal model, and we found that both orexin receptors were increased, but only in three of the eight hypothalamic regions studied (Hunt et al., 2013). Combined, the above studies indicate that the orexinergic system tends to show increased expression immediately after nicotine exposure in both the postnatal and the adult periods, albeit not consistently across all subregions of the hypothalamus.

Prenatal nicotine exposure also seems to affect orexin expression at all stages of development, infancy, childhood, and

TABLE 1 Expression of Orexin and Its Receptors in the Brain of Animal Models of Nicotine Exposure

Nicotine Exposure Paradigm	Animal and Age	Marker (Technique)	Brain Region	Findings	References
Five ip injections/day; 2–6 mg/kg/day; total 10 days	Male adult rat	OxA/OxB protein (RIA) PPO, OX1R, OX2R mRNA (PCR)	Hypothalamus (DMH, LHA, PVN, MHN)	↑PPO (MHN) ↑OxA (DMH) ↑OxB (PVN/DMH) ↑OX1R (MHN) ↑OX2R (MHN)	Kane et al. (2000)
4 mg/kg/day; osmotic minipump for 14 days	Male adult rat	[^{125}I]OxA (competitive binding)	Hypothalamus (anterior, posterior)	↓OxA binding capacity	Kane et al. (2001)
Single ip injection of 0.4 or 2.0 mg/kg	Male adult rat	OxA/c-fos protein (IHC)	Hypothalamus (LHA, PFA)	↑% OxA/c-fos-expressing neurons	Pasumarthi et al. (2006)
Single sc injection of 2.0 mg/kg	Male adult rat	OxA protein (IHC)	Hypothalamus (LHA, PFA)	↑OxA	Pasumarthi and Fadel (2010)
25 mg/kg/day, osmotic minipump for 14 days	Male adult C57BL/6J mouse	OxA/c-fos protein (IF)	Hypothalamus (PFA/DMH/LHA)	↑% OxA/c-fos-expressing neurons	Plaza-Zabala et al. (2012)
0.03 mg/kg/infusions (8–60 infusions/day) for 19 days (~0.24–1.8 mg/kg/day); euthanized 1–11 min after or 5 h after last treatment.	Male adult Long-Evans rat	PPO (in LH only), OX1R and OX2R mRNA (PCR)	Caudal and rostral LHA, ARC, PVN, ^{a}PPT, ^{a}VTA, ^{a}prefrontal cortex, ^{a}NAc	1–11 min: ↓OX1R rostral LHA 5 h: ↑OX1R ARC	LeSage et al. (2010)
Postnatal exposure via lactation from dams receiving 6 mg/kg/day osmotic minipump for 2 weeks; brains collected from pups at P180	Male adult rat	OxA/B protein (IHC)	PVN and LHA	No change	Younes-Rapozo et al. (2013)
2.0 mg/kg/day; osmotic minipump for 14 days	Infant male and female piglet	OX1R, OX2R protein (IHC)	Hypothalamus (DMH, PVN, ARC, VMN, LHA, SONr, TMN)	↑OX1R (DMH, SONr), ↑OX2R (DMH, VMN, TMN)	Hunt et al. (2013)
Cigarette Smoke					
Exposure: 2/h/day for 12 days	Male adult rat	PPO, OX1R, OX2R mRNA (PCR) OxA, OX1R, OX2R protein (IHC, RIA)	Hypothalamus (OxA, PPO), medulla (OX1R and OX2R)	↑PPO, OX1R, OX2R (mRNA) ↑OxA, OX1R, OX2R (protein)	Liu et al. (2010)
Prenatal Nicotine					
Osmotic pump 17.6 mg/kg/day for 21 days, brains collected from pups at P24–28	Male and female infant rat	PPO mRNA (PCR), OxA protein (IHC)	Hypothalamus	↓PPO females. No change OxA	Boychuk et al. (2011)

TABLE 1 Expression of Orexin and Its Receptors in the Brain of Animal Models of Nicotine Exposure—cont'd

Nicotine Exposure Paradigm	Animal and Age	Marker (Technique)	Brain Region	Findings	References
Osmotic pump 1.5 mg/kg/day for 14 days, brains collected from pumps at P15 and P40	Male infant and child rat	PPO mRNA (ISH) and protein (IHC)	Hypothalamus	↑PPO mRNA and protein	Chang et al. (2013)
Injections 3 × 0.05 mg/kg/day for 14 days, brains collected from pups at P130	Male adult rat	OxA protein (IHC)	LHA/PFA and ^aVTA	↑OxA in both regions	Morgan et al., 2013

ARC, hypothalamic arcuate nucleus; DMH, dorsal medial hypothalamus; IHC, immunohistochemistry; IF, immunofluorescence; ISH, in situ hybridization; ip, intraperitoneal; LHA, lateral hypothalamic area; MHN, medial hypothalamic nucleus; NAc, nucleus accumbens shell; OX1R, orexin receptor 1; OX2R, orexin receptor 2; P, postnatal day; PCR, polymerase chain reaction; PFA, perifornical area; PPO, prepro-orexin; PPT, pedunculopontine tegmental nucleus; PVN, paraventricular nucleus; RIA, radioimmunoassay; sc, subcutaneous; SONr, supraoptic nucleus retrochiasmatic part; TMN, tuberal mammillary nucleus; VMN, ventral medial nucleus; VTA, ventral tegmental area.
^aNuclei involved in addiction and reward.

adulthood (Boychuk et al., 2011; Chang et al., 2013; Morgan et al., 2013), yet the results by Boychuk et al. (2011) seem to be contradictory. Boychuk et al. (2011) found that an extremely high dose of nicotine (17.6 mg/kg/day) for 21 days during pregnancy decreased PPO levels in the infant (postnatal day (P) 24–28) female hypothalamus but there was no change for OxA. Chang et al. (2013), on the other hand, found that a low dose of nicotine (1.5 mg kg/day) for 14 days during pregnancy increased PPO levels in the infant (P15) and child (P40) male hypothalamus. Morgan et al. (2013) found that injecting pregnant rats with 0.05 mg/kg/day of nicotine for 14 days increased OxA in the adult (P130) male LHA/PFA and ventral tegmental area (VTA). Combined, the data of these studies add to the common clinical finding that maternal smoking predisposes the offspring to future health problems. For the orexinergic system possible links include increase in obesity (Oken, Levitan, & Gillman, 2008), increase in dysfunctional sleep and arousal (Franco et al., 1999), and increased nicotine dependence (Buka, Shenassa, & Niaura, 2003).

NICOTINE, OREXIN, ADDICTION, AND REWARD

The role of orexin in addiction was first reported in PPO-knockout mice that showed a decrease in withdrawal from morphine (hence increased dependence) (Georgescu et al., 2003). Studies thereafter showed that these knockout mice also had an attenuation of the conditioned reward after morphine (Narita et al., 2006, Sharf et al., 2010) and after nicotine (Plaza-Zabala et al., 2012). The development of OX1R-knockout mice showed that this receptor plays a prominent role in addiction, whereby in its absence, animals showed reduced cocaine and cannabinoid self-administration (Flores, Maldonado, & Berrendero, 2013; Hollander, Pham, Fowler, & Kenny, 2012). Detailed reviews discussing the potential role of orexin in addiction, including nicotine, are provided by Corrigall (2009), Kenny (2011), Merlo Pich and Melotto (2014).

Of the studies summarized in Table 1, the only two that developed their models with the intention of determining the role of nicotine in addiction/dependence and reward were LeSage et al. (2010) and Morgan et al. (2013). The key factor of these two studies was that they determined changes in the orexin system within brain regions highly involved in drug addiction and dependence, including prefrontal cortex, VTA, nucleus accumbens (NAc), and pedunculopontine tegmental nucleus (PPT) (Figure 1). Although LeSage et al. (2010) did not find any change in the expression of OX1R and OX2R in these regions, they did find that both the selective OX1R antagonist SB-334867 and the mixed OXR1/2 antagonist almorexant reduced nicotine self-administration, thus indirectly indicating that the orexin receptors are involved in the addictive nature of nicotine, although evidence for OX1R is greater. The finding with SB-334867 supported the findings of the only other study conducted with a similar aim (Hollander, Lu, Cameron, Kamenecka, & Kenny, 2008). Thus, combined, the data lead to the conclusion that nicotine self-administration (dependence) is regulated by the orexin system and is partly mediated by the activation of OX1R, but it does so broadly (not specific to addiction/reward areas such as the VTA and NAc), leading to changes in several other brain regions.

In contrast to LeSage et al. (2010), Morgan et al. (2013) found that prenatal exposure to nicotine does have a direct effect on the addiction pathway in adulthood as indicated by the increase in OxA expression in the VTA of the adult male rats. Thus, it seems that prenatal nicotine exposure does indeed predispose to future nicotine dependence in the offspring. Further studies are required to determine whether these effects are mediated via OX1R, OX2R, or both, and whether other brain regions are also involved.

MECHANISMS INVOLVED IN NICOTINE-INDUCED OREXIN EXPRESSION CHANGES

The nicotine-induced change in the orexin receptors has been shown to be mediated by the nAChRs, and blocking these receptors reduced the amount of orexin activation (c-fos/OxA expression) (Pasumarthi et al., 2006). Furthermore, expression of nAChRs within other brain regions, such as the basal forebrain and paraventricular thalamus, correlated with increased orexin-induced neural activity after acute doses of nicotine (2.0 mg/kg) (Pasumarthi & Fadel, 2010). Thus, nicotine related changes are closely related to the nAChRs. Other linked systems include the dopaminergic, N-methyl-D-aspartate (NMDA), and GABA systems. Three excellent reviews addressing these various pathways are provided by Baimel, Borgland, and Corrigall (2012), Mahler, Smith, Moorman, Sartor, and Aston-Jones (2012), and Sharf et al. (2010). A schematic of a common pathway is provided in Figure 4. Both acute and chronic nicotine exposures excite orexin neurons (Kane et al., 2000; Pasumarthi et al., 2006). This excitation is dependent on cholinergic neurons in the basal forebrain; loss of these neurons inhibits the excitation of orexin neurons by systemically delivered nicotine (Pasumarthi & Fadel, 2010). Localized delivery of nicotine to the PFA/LHA may excite orexin neurons (Pasumarthi & Fadel, 2010); however, all nicotine delivery pathways in humans are via systemic nicotine ingestion or inhalation. Nicotine promotes activation of cholinergic neurons in the basal forebrain via $\alpha 4\beta 2$, $\alpha 4\alpha 5\beta 2$, and $\alpha 7$ nAChRs (Tuesta, Fowler, & Kenny, 2011) (Figure 4). Cholinergic and glutamatergic basal forebrain neurons project directly to orexin neurons in the hypothalamus, and glutamatergic innervation is substantially more than cholinergic (Henny & Jones, 2006). Pasumarthi and Fadel (2010) observed that if choline acetyltransferase (ChAT) neurons in the basal forebrain were ablated, then there was a reduction, to basal levels, in the c-fos-expressing OxA neuronal staining induced by nicotine. However, ablation also corresponded to reduced glutamatergic and acetylcholinergic staining in the hypothalamus (Pasumarthi & Fadel, 2010). This indicates that orexin neurons receive glutamatergic and cholinergic innervation in response to nicotine binding in the basal forebrain to increase c-fos expression (Pasumarthi & Fadel, 2010) (Figure 4). Future studies are required to determine the localization of muscarinic and nicotinic receptors on orexin neurons to provide detailed information of the precise pathway(s) involved.

CONCLUSION AND FUTURE DIRECTIONS

From this review, the majority of the paradigms involving nicotine exposure result in increased expression of all the markers of the orexinergic system, including PPO, OxA, OxB, and the receptors OX1R and OX2R. However, the majority of studies were undertaken in adult rats. More studies are required to determine with more consistency the changes induced at other developmental stages. It also appears that the different nicotine exposure paradigms could affect reproducibility of the outcomes. The majority of the studies looked at orexin peptide expression, so effects of nicotine on the orexin receptors need to be investigated further as potential therapeutic targets. Already, an increasing number of selective OX1R and OX2R antagonists as well as dual antagonists (referred to as DORA) are being developed, patented, and tested clinically, although their current aim is treatment of sleep disorders (Coleman & Renger, 2010; Equihua, De La Herran-Arita, & Drucker-Colin, 2013; summarized in Table 2). These may also have a role in treating addiction disorders (Gotter et al., 2012). Therapy for metabolic disorders is complicated by the close associations with changes in sleep, so future therapeutic targets in the orexin system are likely to be more successful in the treatment of sleep disorders or addiction and potentially mitigating the effects of noxious insults. A detailed review of the literature on the therapeutic uses of these drugs is provided by Equihua et al. (2013) and Gotter et al. (2012).

APPLICATIONS TO OTHER ADDICTIONS AND SUBSTANCE MISUSE

In this chapter, we focused on the effects of nicotine on the expression of the orexinergic system in the brain. However, as mentioned earlier, the role of orexin in addiction was first reported when looking at morphine (Georgescu et al., 2003; Narita et al., 2006; Sharf et al., 2010), with subsequent studies looking at the role of ethanol (Lawrence, 2009; Morganstren et al., 2010), nicotine (Plaza-Zabala et al., 2012), cocaine, and cannabinoid (Flores et al., 2013; Hollander et al., 2012) effects. We have published a review summarizing the data of these studies (Machaalani et al., 2013). In general, it was found that nicotine, ethanol, and methamphetamines increased OxA and PPO expression in the hypothalamus. Morphine and cocaine, on the other hand, induced no change in OxA or PPO levels, yet their withdrawal leads to an increase (Zhou et al., 2006; reviewed in Machaalani et al., 2013). As with nicotine, the study of the effects on the orexin receptors is limited,

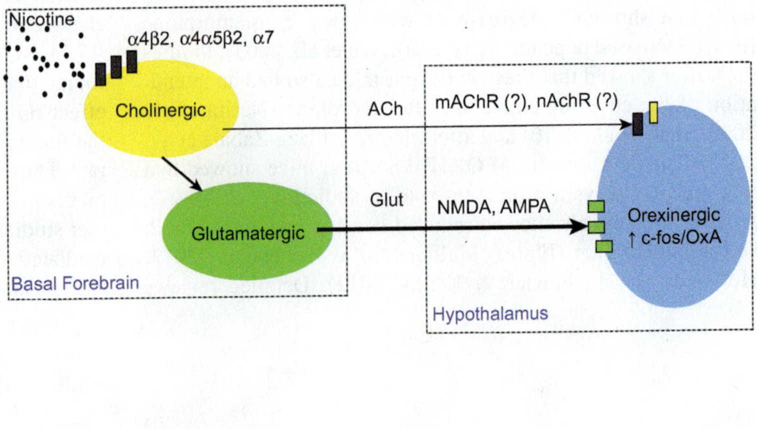

FIGURE 4 **Schematic of a plausible mechanism for increased orexin expression/function after nicotine exposure.** Inhaled or ip-delivered nicotine binds to nAChRs on cholinergic neurons in the basal forebrain. Cholinergic neurons excite neighboring glutamatergic neurons. Orexin neurons in the hypothalamus are innervated by both cholinergic and glutamatergic neurons. Excitation promotes neuronal activity; but for the orexin system, the only data available are via colocalization of OxA and c-fos expression: increased c-fos is indicative of increased activation of orexinergic neurons. mAChR, muscarinic AChR.

with findings showing that ethanol and cocaine increase expression (reviewed in Machaalani et al., 2013).

DEFINITION OF TERMS

c-fos A member of the proto-oncogene protein group that regulates cell proliferation, differentiation, and transformation and may be associated with apoptosis.

Hypothalamus A region of the brain, between the thalamus and the midbrain, that functions as the main control center for the autonomic nervous system by regulating sleep cycles, body temperature, and appetite.

Immunohistochemistry An experimental assay that shows specific antigens in tissues by the use of markers that are either fluorescent dyes or enzymes (such as horseradish peroxidase). A means of measuring the protein expression of a certain gene of interest allowing for cellular localization of the gene viewed by microscopy.

Orexin From the Greek word orexis, meaning appetite, a neuropeptide produced in the hypothalamus.

Polymerase chain reaction A technique for amplifying DNA sequences (as many as 1 billion times) in vitro by separating the DNA into two strands and incubating it with oligonucleotide primers and DNA polymerase.

Radioimmunoassay An experimental assay that measures minute amounts of a substance, such as a hormone or drug, by quantitating the binding, or the inhibition of binding, of a radiolabeled substance to an antibody. A means of measuring protein expression in body fluids or tissue homogenates.

REM sleep A period of sleep characterized by rapid periodic twitching movements of the eye muscles and during which dreaming takes place. Other physiological changes, such as accelerated respiration and heart rate, increased brain activity, and muscle relaxation, also occur. It is the fifth and last stage of sleep that occurs in the sleep cycle, preceded by four stages of non-REM sleep.

Non-REM sleep A period of sleep that occurs in the first four of the five stages of the sleep cycle. It is characterized by decreased metabolic activity, slowed breathing and heart rate, and the absence of dreaming.

Nicotine The chief active constituent of tobacco that is toxic. It acts as a stimulant in small doses, but in larger amounts blocks the action of autonomic nerve and skeletal muscle cells. It is an alkaloid with chemical formula $C_{10}H_{14}N_2$.

Prenatal Period before birth.

Postnatal Period immediately after birth; usually up to 21 days after birth.

Western blot An experimental procedure where proteins in body fluids or tissue homogenates are separated by electrophoresis in gels and transferred (blotted) onto special membranes on which specific complexing with antibodies that are either pre- or post-tagged with a labeled secondary protein are identified.

KEY FACTS REGARDING THE OREXINERGIC SYSTEM

- Orexin is a neuropeptide specifically made in the hypothalamus region of the brain.
- Orexin is synthesized as a 131-amino-acid peptide termed prepro-orexin.
- Prepro-orexin is cleaved into OxA and OxB, of 33 and 29 amino acids, respectively.

TABLE 2 Main Orexin Receptor Antagonists in Clinical Use or Clinical Trials

Receptor Target	Pharmacological Agent	Brain Area of Target	Clinical Application
OX1R antagonist	SB-334,867	DR; LDT/PPT; LC	Withdrawal, addiction, panic disorder, obesity
	SB-408,124		
	SB-674042		
	GSK-1059865		
	ACT-335827		
OX2R antagonist	TCS-OX2-29	DR and TMN	Promotion of sleep
	EMPA		
	JNJ-10397049		
Dual OX1R and OX2R antagonists (DORAs)	Almorexant (Actelion)[a]	DR; LDT/PPT; LC; TMN	Insomnia
	Suvorexant (MK-4305)[b]		
	SB-649,868[c]		
	MK-6096		
	DORA 30		Promotion of sleep

EMPA, (N-ethyl-2-[(6-methoxypyridin-3-yl)-(toluene-2-sulfonyl)-amino]-N-pyridin-3-yl-methylacetamide); DR, dorsal raphe; LDT, laterodorsal tegmental nucleus; LC, Locus coeruleus.
[a]Reached phase III but was discontinued.
[b]Reached phase II and completed.
[c]In phase III and awaiting approval at the time of writing.

- These bind to the two orexin receptors termed OX1R and OX2R.
- Orexin receptors are expressed in most of the CNS except for the cerebellum.
- The expression of orexin and its receptors is generally increased after nicotine exposure.

SUMMARY POINTS

- The effects of cigarette smoking on the *expression* (mRNA and protein) of the orexinergic system have not yet been determined in the human brain, so the evidence is obtained from animal models.
- Animal models show that the expression of orexin and its receptors is generally increased after nicotine exposure.
- In relation to addiction, prenatal nicotine exposure predisposes to future nicotine dependence in the offspring (as indicated by the increase in OxA expression in the VTA).
- There is more evidence for involvement of OX1R in the addiction pathway compared to OX2R.
- The direct mechanism by which nicotine leads to the increase in orexin expression and function is still unclear, although the involvement of the nAChRs within the basal forebrain is implicated, with involvement of the muscarinic AChRs leading to a feedback loop to orexin in the hypothalamus; yet more work is required to determine this.

REFERENCES

Aksu, K., Firat Guven, S., Aksu, F., Ciftci, B., Ulukavak Ciftci, T., Aksaray, S., ... Peker, Y. (2009). Obstructive sleep apnoea, cigarette smoking and plasma orexin-A in a sleep clinic cohort. *The Journal of International Medical Research*, 37, 331–340.

Aston-Jones, G., Smith, R. J., Moorman, D. E., & Richardson, K. A. (2009). Role of lateral hypothalamic orexin neurons in reward processing and addiction. *Neuropharmacology*, 56(Suppl. 1), 112–121.

Baimel, C., Borgland, S. L., & Corrigall, W. (2012). Cocaine and nicotine research illustrates a range of hypocretin mechanisms in addiction. *Vitamins and Hormones*, 89, 291–313.

Benowitz, N. L. (1996). Pharmacology of nicotine: addiction and therapeutics. *Annual Review of Pharmacology and Toxicology*, 36, 597–613.

Benowitz, N. L., & Jacob, P., 3rd. (1993). Nicotine and cotinine elimination pharmacokinetics in smokers and nonsmokers. *Clinical Pharmacology and Therapeutics*, 53, 316–323.

Bowen, D. J., Eury, S. E., & Grunberg, N. E. (1986). Nicotine's effects on female rats' body weight: caloric intake and physical activity. *Pharmacology, Biochemistry, and Behavior*, 25, 1131–1136.

Boychuk, C. R., & Hayward, L. F. (2011). Prenatal nicotine exposure alters postnatal cardiorespiratory integration in young male but not female rats. *Experimental Neurology*, 232, 212–221.

Bramer, S. L., & Kallungal, B. A. (2003). Clinical considerations in study designs that use cotinine as a biomarker. *Biomarkers*, 8, 187–203.

Browne, C. J., Sharma, N., Waters, K. A., & Machaalani, R. (2010). The effects of nicotine on the alpha-7 and beta-2 nicotinic acetycholine receptor subunits in the developing piglet brainstem. *International Journal of Developmental Neuroscience*, 28, 1–7.

Buisson, B., & Bertrand, D. (2001). Chronic exposure to nicotine upregulates the human (alpha)4(beta)2 nicotinic acetylcholine receptor function. *The Journal of Neuroscience*, 21, 1819–1829.

Buka, S. L., Shenassa, E. D., & Niaura, R. (2003). Elevated risk of tobacco dependence among offspring of mothers who smoked during pregnancy: a 30-year prospective study. *The American Journal of Psychiatry*, 160, 1978–1984.

Chang, G. Q., Karatayev, O., & Leibowitz, S. F. (2013). Prenatal exposure to nicotine stimulates neurogenesis of orexigenic peptide-expressing neurons in hypothalamus and amygdala. *Journal of Neuroscience*, 33, 13600–13611.

Cohen, A., & George, O. (2013). Animal models of nicotine exposure: relevance to second-hand smoking, electronic cigarette use, and compulsive smoking. *Frontiers in Psychiatry*, 4, 41.

Coleman, P. J., & Renger, J. J. (2010). Orexin receptor antagonists: a review of promising compounds patented since 2006. *Expert Opinion on Therapeutic Patents*, 20, 307–324.

Cooper, E., Couturier, S., & Ballivet, M. (1991). Pentameric structure and subunit stoichiometry of a neuronal nicotinic acetylcholine receptor. *Nature*, 350, 235–238.

Corrigall, W. A. (2009). Hypocretin mechanisms in nicotine addiction: evidence and speculation. *Psychopharmacology (Berlin)*, 206, 23–37.

Crocker, A., España, R. A., Papadopoulou, M., Saper, C. B., Faraco, J., Sakurai, T., ... Scammell, T. E. (2005). Concomitant loss of dynorphin, NARP, and orexin in narcolepsy. *Neurology*, 65, 1184–1188.

Dani, J. A., & Heinemann, S. (1996). Molecular and cellular aspects of nicotine abuse. *Neuron*, 16, 905–908.

Davila, D. G., Hurt, R. D., Offord, K. P., Harris, C. D., & Shepard, J. W., Jr. (1994). Acute effects of transdermal nicotine on sleep architecture, snoring, and sleep-disordered breathing in nonsmokers. *American Journal of Respiratory and Critical Care Medicine*, 150, 469–474.

De Biasi, M., & Dani, J. A. (2011). Reward, addiction, withdrawal to nicotine. *Annual Review of Neuroscience*, 34, 105–130.

Deurveilher, S., Lo, H., Murphy, J. A., Burns, J., & Semba, K. (2006). Differential c-Fos immunoreactivity in arousal-promoting cell groups following systemic administration of caffeine in rats. *Journal of Comparative Neurology*, 498, 667–689.

Ebrahim, I. O., Semra, Y. K., De Lacy, S., Howard, R. S., Kopelman, M. D., Williams, A., & Sharief, M. K. (2003). CSF hypocretin (Orexin) in neurological and psychiatric conditions. *Journal of Sleep Research*, 12, 83–84.

Equihua, A. C., De La Herran-Arita, A. K., & Drucker-Colin, R. (2013). Orexin receptor antagonists as therapeutic agents for insomnia. *Frontiers in Pharmacology*, 4, 163.

Fanous, A. M., Machaalani, R., & Waters, K. A. (2006). N-methyl-D-aspartate receptor 1 changes in the piglet brainstem after nicotine and/or intermittent hypercapnic-hypoxia. *Neuroscience*, 142, 401–409.

Flores, A., Maldonado, R., & Berrendero, F. (2013). The hypocretin/orexin receptor-1 as a novel target to modulate cannabinoid reward. *Biological Psychiatry*, 75, 499–507.

Franco, P., Groswasser, J., Hassid, S., Lanquart, J. P., Scaillet, S., & Kahn, A. (1999). Prenatal exposure to cigarette smoking is associated with a decrease in arousal in infants. *The Journal of Pediatrics*, 135, 34–38.

Freedman, R., Adams, C. E., & Leonard, S. (2000). The alpha7-nicotinic acetylcholine receptor and the pathology of hippocampal interneurons in schizophrenia. *The Journal of Chemical Neuroanatomy*, 20, 299–306.

Fronczek, R., Overeem, S., Lee, S. Y., Hegeman, I. M., Van Pelt, J., Van Duinen, S. G., ... Swaab, D. F. (2007). Hypocretin (orexin) loss in Parkinson's disease. *Brain*, 130, 1577–1585.

Georgescu, D., Zachariou, V., Barrot, M., Mieda, M., Willie, J. T., Eisch, A. J., ... DiLeone, R. J. (2003). Involvement of the lateral hypothalamic peptide orexin in morphine dependence and withdrawal. *The Journal of Neuroscience*, 23, 3106–3111.

von der Goltz, C., Koopmann, A., Dinter, C., Richter, A., Rockenbach, C., Grosshans, M., ... Kiefer, F. (2010). Orexin and leptin are associated with nicotine craving: a link between smoking, appetite and reward. *Psychoneuroendocrinology, 35*, 570–577.

Gotter, A. L., Roecker, A. J., Hargreaves, R., Coleman, P. J., Winrow, C. J., & Renger, J. J. (2012). Orexin receptors as therapeutic drug targets. *Progress in Brain Research, 198*, 163–188.

Gotti, C., Fornasari, D., & Clementi, F. (1997). Human neuronal nicotinic receptors. *Progress in Neurobiology, 53*, 199–237.

Grunberg, N. E., Bowen, D. J., & Winders, S. E. (1986). Effects of nicotine on body weight and food consumption in female rats. *Psychopharmacology (Berlin), 90*, 101–105.

Henny, P., & Jones, B. E. (2006). Vesicular glutamate (VGlut), GABA (VGAT), and acetylcholine (VACht) transporters in basal forebrain axon terminals innervating the lateral hypothalamus. *The Journal of Comparitive Neurology, 496*, 453–467.

Hervieu, G., Cluderay, J., Harrison, D., Roberts, J., & Leslie, R. (2001). Gene expression and protein distribution of the orexin-1 receptor in the rat brain and spinal cord. *Neuroscience, 103*, 777–797.

Hollander, J. A., Lu, Q., Cameron, M. D., Kamenecka, T. M., & Kenny, P. J. (2008). Insular hypocretin transmission regulates nicotine reward. *Proceedings of the National Academy of Science United States of America, 105*, 19480–19485.

Hollander, J. A., Pham, D., Fowler, C. D., & Kenny, P. J. (2012). Hypocretin-1 receptors regulate the reinforcing and reward-enhancing effects of cocaine: pharmacological and behavioral genetics evidence. *Frontiers in Behavioral Neurosciences, 6*, 47.

Holmqvist, T., Åkerman, K. E., & Kukkonen, J. P. (2001). High specificity of human orexin receptors for orexins over neuropeptide Y and other neuropeptides. *Neuroscience Letters, 305*, 177–180.

Huang, Z.-L., Qu, W.-M., Li, W.-D., Mochizuki, T., Eguchi, N., Watanabe, T., ... Hayaishi, O. (2001). Arousal effect of orexin A depends on activation of the histaminergic system. *Proceedings of the National Academy of Sciences, 98*, 9965–9970.

Hunt, N. J., Rodriguez, M. L., Waters, K. A., & Machaalani, R. (2015). Changes in orexin (hypocretin) neuronal expression with normal aging in the human hypothalamus. *Neurobiology of Aging, 36*, 292–300.

Hunt, N. J., Waters, K. A., & Machaalani, R. (2013). Orexin receptors in the developing piglet hypothalamus, and effects of nicotine and intermittent hypercapnic hypoxia exposures. *Brain Research, 1508*, 73–82.

Jaehne, A., Loessl, B., Barkai, Z., Riemann, D., & Hornyak, M. (2009). Effects of nicotine on sleep during consumption, withdrawal and replacement therapy. *Sleep Medicine Reviews, 13*, 363–377.

Jo, Y. H., Talmage, D. A., & Role, L. W. (2002). Nicotinic receptor-mediated effects on appetite and food intake. *Journal of Neurobiology, 53*, 618–632.

Kane, J., Parker, S., Matta, S., Fu, Y., Sharp, B., & Li, M. (2000). Nicotine up-regulates expression of orexin and its receptors in rat brain. *Endocrinology, 141*, 3623–3629.

Kane, J. K., Parker, S. L., & Li, M. D. (2001). Hypothalamic orexin-A binding sites are downregulated by chronic nicotine treatment in the rat. *Neuroscience Letters, 298*, 1–4.

Kastin, A. J., & Akerstrom, V. (1999). Orexin A but not orexin B rapidly enters brain from blood by simple diffusion. *Journal of Pharmacology and Experimental Therapeutics, 289*, 219–223.

Ke, L., Eisenhour, C. M., Bencherif, M., & Lukas, R. J. (1998). Effects of chronic nicotine treatment on expression of diverse nicotinic acetylcholine receptor subtypes. I. Dose- and time-dependent effects of nicotine treatment. *The Journal of Pharmacology and Experimental Therapeutics, 286*, 825–840.

Kenny, P. J. (2011). Tobacco dependence, the insular cortex and the hypocretin connection. *Pharmacology, Biochemistry and Behavior, 97*, 700–707.

Klesges, R. C., Meyers, A. W., Klesges, L. M., & La Vasque, M. E. (1989). Smoking, body weight, and their effects on smoking behavior: a comprehensive review of the literature. *Psychological Bulletin, 106*, 204–230.

Kukkonen, J. P. (2013). Physiology of the orexinergic/hypocretinergic system: a revisit in 2012. *American Journal of Physiology—Cell Physiology, 304*, C2–C32.

Lawrence, A. J. (2009). Regulation of alcohol-seeking by orexin (hypocretin) neurons. *Brain Research, 1314*, 124–129.

de Lecea, L., Kilduff, T., Peyron, C., Gao, X.-B., Foye, P., Danielson, P. E., ... Sutcliffe, J. G. (1998). The hypocretins: hypothalamus-specific peptides with neuroexcitatory activity. *Proceedings of the National Academy of Sciences, 95*, 322–327.

LeSage, M. G., Perry, J. L., Kotz, C. M., Shelley, D., & Corrigall, W. A. (2010). Nicotine self-administration in the rat: effects of hypocretin antagonists and changes in hypocretin mRNA. *Psychopharmacology (Berlin), 209*, 203–212.

Lichtensteiger, W., Ribary, U., Schlumpf, M., Odermatt, B., & Widmer, H. R. (1988). Prenatal adverse effects of nicotine on the developing brain. *Progress in Brain Research, 73*, 137–157.

Lindstrom, J. (1997). Nicotinic acetylcholine receptors in health and disease. *Molecular Neurobiology, 15*, 193–222.

Liu, Z. B., Song, N. N., Geng, W. Y., Jin, W. Z., Li, L., Cao, Y. X., ... Shen, L. L. (2010). Orexin-A and respiration in a rat model of smoke-induced chronic obstructive pulmonary disease. *Clinical and Experimental Pharmacology and Physiology, 37*, 963–968.

Lu, J., Sherman, D., Devor, M., & Saper, C. B. (2006). A putative flip–flop switch for control of REM sleep. *Nature, 441*, 589–594.

Luck, W., & Nau, H. (1985). Nicotine and cotinine concentrations in serum and urine of infants exposed via passive smoking or milk from smoking mothers. *The Journal of Pediatrics, 107*, 816–820.

Luong, L., & Carrive, P. (2012). Orexin microinjection in the medullary raphe increases heart rate and arterial pressure but does not reduce tail skin blood flow in the awake rat. *Neuroscience, 202*, 209–217.

Machaalani, R., Hunt, N. J., & Waters, K. A. (2013). Effects of changes in energy homeostasis and exposure of noxious insults on the expression of orexin (hypocretin) and its receptors in the brain. *Brain Research, 1526*, 102–122.

Machaalani, R., Waters, K. A., & Tinworth, K. D. (2005). Effects of postnatal nicotine exposure on apoptotic markers in the developing piglet brain. *Neuroscience, 132*, 325–333.

Mahler, S. V., Smith, R. J., Moorman, D. E., Sartor, G. C., & Aston-Jones, G. (2012). Multiple roles for orexin/hypocretin in addiction. *Progress in Brain Research, 198*, 79–121.

Marcus, J. N., Aschkenasi, C. J., Lee, C. E., Chemelli, R. M., Saper, C. B., Yanagisawa, M., & Elmquist, J. K. (2001). Differential expression of orexin receptors 1 and 2 in the rat brain. *Journal of Comparative Neurology, 435*, 6–25.

Merlo Pich, E., & Melotto, S. (2014). Orexin 1 receptor antagonists in compulsive behavior and anxiety: possible therapeutic use. *Frontiers in Neuroscience, 8*, 26.

Mieda, M., Willie, J. T., Hara, J., Sinton, C. M., Sakurai, T., & Yanagisawa, M. (2004). Orexin peptides prevent cataplexy and improve wakefulness in an orexin neuron-ablated model of narcolepsy in mice. *Proceedings of the National Academy of Sciences of the United States of America, 101*, 4649–4654.

Morgan, A. J., Harrod, S. B., Lacy, R. T., Stanley, E. M., & Fadel, J. R. (2013). Intravenous prenatal nicotine exposure increases orexin expression in the lateral hypothalamus and orexin innervation of the ventral tegmental area in adult male rats. *Drug and Alcohol Dependence, 132*, 562–570.

Morganstern, I., Chang, G. Q., Barson, J. R., Ye, Z., Karatayev, O., & Leibowitz, S. F. (2010). Differential effects of acute and chronic ethanol exposure on orexin expression in the perifornical lateral hypothalamus. *Alcoholism Clinical and Experimental Research, 34*, 886–896.

Murrin, L. C., Ferrer, J. R., Zeng, W. Y., & Haley, N. J. (1987). Nicotine administration to rats: methodological considerations. *Life Sciences, 40*, 1699–1708.

Nambu, T., Sakurai, T., Mizukami, K., Hosoya, Y., Yanagisawa, M., & Goto, K. (1999). Distribution of orexin neurons in the adult rat brain. *Brain Research, 827*, 243–260.

Narita, M., Nagumo, Y., Hashimoto, S., Khotib, J., Miyatake, M., Sakurai, T., ... Suzuki, T. (2006). Direct involvement of orexinergic systems in the activation of the mesolimbic dopamine pathway and related behaviors induced by morphine. *The Journal of Neuroscience, 26*, 398–405.

Oken, E., Levitan, E. B., & Gillman, M. W. (2008). Maternal smoking during pregnancy and child overweight: systematic review and meta-analysis. *Internation Journal of Obesity (London), 32*, 201–210.

Papke, R. L., Dwoskin, L. P., Crooks, P. A., Zheng, G., Zhang, Z., McIntosh, J. M., & Stokes, C. (2008). Extending the analysis of nicotinic receptor antagonists with the study of alpha6 nicotinic receptor subunit chimeras. *Neuropharmacology, 54*, 1189–1200.

Pasumarthi, R. K., & Fadel, J. (2010). Stimulation of lateral hypothalamic glutamate and acetylcholine efflux by nicotine: implications for mechanisms of nicotine-induced activation of orexin neurons. *Journal of Neurochemistry, 113*, 1023–1035.

Pasumarthi, R. K., Reznikov, L. R., & Fadel, J. (2006). Activation of orexin neurons by acute nicotine. *European Journal of Pharmacology, 535*, 172–176.

Perkins, K. A., Epstein, L. H., Stiller, R. L., Marks, B. L., & Jacob, R. G. (1989). Acute effects of nicotine on resting metabolic rate in cigarette smokers. *The American Journal of Clinical Nutrition, 50*, 545–550.

Peyron, C., Tighe, D. K., van den Pol, A. N., de Lecea, L., Heller, H. C., Sutcliffe, J. G., & Kilduff, T. S. (1998). Neurons containing hypocretin (orexin) project to multiple neuronal systems. *The Journal of Neuroscience, 18*, 9996–10015.

Plaza-Zabala, A., Flores, A., Maldonado, R., & Berrendero, F. (2012). Hypocretin/orexin signaling in the hypothalamic paraventricular nucleus is essential for the expression of nicotine withdrawal. *Biological Psychiatry, 71*, 214–223.

Plaza-Zabala, A., Martin-Garcia, E., de Lecea, L., Maldonado, R., & Berrendero, F. (2010). Hypocretins regulate the anxiogenic-like effects of nicotine and induce reinstatement of nicotine-seeking behavior. *The Journal of Neuroscience, 30*, 2300–2310.

Role, L. W., & Berg, D. K. (1996). Nicotinic receptors in the development and modulation of CNS synapses. *Neuron, 16*, 1077–1085.

Sakurai, T. (2007). The neural circuit of orexin (hypocretin): maintaining sleep and wakefulness. *Nature Reviews Neuroscience, 8*, 171–181.

Sakurai, T., Amemiya, A., Ishii, M., Matsuzaki, I., Chemelli, R. M., Tanaka, H., ... Yanagisawa, M. (1998). Orexins and orexin receptors: a family of hypothalamic neuropeptides and G protein-coupled receptors that regulate feeding behavior. *Cell, 92*, 573.

Sakurai, T., Moriguchi, T., Furuya, K., Kajiwara, N., Nakamura, T., Yanagisawa, M., & Goto, K. (1999). Structure and function of human prepro-orexin gene. *Journal of Biological Chemistry, 274*, 17771–17776.

Saper, C. B., Scammell, T. E., & Lu, J. (2005). Hypothalamic regulation of sleep and circadian rhythms. *Nature, 437*, 1257–1263.

Say, M., Machaalani, R., & Waters, K. A. (2007). Changes in serotoninergic receptors 1A and 2A in the piglet brainstem after intermittent hypercapnic hypoxia (IHH) and nicotine. *Brain Research, 1152*, 17–26.

Sharf, R., Sarhan, M., & Dileone, R. J. (2010). Role of orexin/hypocretin in dependence and addiction. *Brain Research, 1314*, 130–138.

Slotkin, T. A. (1998). Fetal nicotine or cocaine exposure: which one is worse? *The Journal of Pharmacology Experimental Therapeutics, 285*, 931–945.

Sparks, J. A., & Pauly, J. R. (1999). Effects of continuous oral nicotine administration on brain nicotinic receptors and responsiveness to nicotine in C57Bl/6 mice. *Psychopharmacology (Berlin), 141*, 145–153.

Steinlein, O. K. (2000). Neuronal nicotinic receptors in human epilepsy. *European Journal of Pharmacology, 393*, 243–247.

Swaab, D. F. (2003). The human hypothalamus. In M. J. Aminoff, F. Boller, & D. F. Swaab (Eds.), *Basic and clinical aspects*. Amsterdam: Elsevier.

Sydney South West Pathology Service (SSWPS) Handbook. (2014). http://www.swslhd.nsw.gov.au/sswps/handbook.

Sztalryd, C., Hamilton, J., Horwitz, B. A., Johnson, P., & Kraemer, F. B. (1996). Alterations of lipolysis and lipoprotein lipase in chronically nicotine-treated rats. *The American Journal of Physiology, 270*, E215–E223.

Thannickal, T. C., Moore, R. Y., Nienhuis, R., Ramanathan, L., Gulyani, S., Aldrich, M., ... Siegel, J. M. (2000). Reduced number of hypocretin neurons in human narcolepsy. *Neuron, 27*, 469–474.

Trauth, J. A., McCook, E. C., Seidler, F. J., & Slotkin, T. A. (2000). Modeling adolescent nicotine exposure: effects on cholinergic systems in rat brain regions. *Brain Research, 873*, 18–25.

Tsuneki, H., Wada, T., & Sasaoka, T. (2010). Role of orexin in the regulation of glucose homeostasis. *Acta Physiologica, 198*, 335–348.

Tuesta, L. M., Fowler, C. D., & Kenny, P. J. (2011). Recent advances in understanding nicotinic receptor signaling mechanisms that regulate drug self-administration behavior. *Biochemical Pharmacology, 82*, 984–995.

Unwin, N. (2003). Structure and action of the nicotinic acetylcholine receptor explored by electron microscopy. *FEBS Letters, 555*, 91–95.

Vine, M. F., Hulka, B. S., Margolin, B. H., Truong, Y. K., Hu, P. C., Schramm, M. M., ... Everson, R. B. (1993). Cotinine concentrations in semen, urine, and blood of smokers and nonsmokers. *American Journal of Public Health, 83*, 1335–1338.

Wang, H. Y., Lee, D. H., D'Andrea, M. R., Peterson, P. A., Shank, R. P., & Reitz, A. B. (2000). Beta-Amyloid(1-42) binds to alpha7 nicotinic acetylcholine receptor with high affinity. Implications for Alzheimer's disease pathology. *The Journal of Biological Chemistry, 275*, 5626–5632.

Yamanaka, A., Tsujino, N., Funahashi, H., Honda, K., Guan, J.-L., Wang, Q.-P., ... Sakurai, T. (2002). Orexins activate histaminergic neurons via the orexin 2 receptor. *Biochemical and Biophysical Research Communications, 290*, 1237–1245.

Younes-Rapozo, V., Moura, E. G., Manhaes, A. C., Pinheiro, C. R., Santos-Silva, A. P., de Oliveira, E., … Lisboa, P. C. (2013). Maternal nicotine exposure during lactation alters hypothalamic neuropeptides expression in the adult rat progeny. *Food and Chemical Toxicology, 58*, 158–168.

Zhou, Y., Bendor, J., Hofmann, L., Randesi, M., Ho, A., & Kreek, M. J. (2006). Mu opioid receptor and orexin/hypocretin mRNA levels in the lateral hypothalamus and striatum are enhanced by morphine withdrawal. *Journal of Endocrinology, 191*, 137–145.

Zhu, Y., Miwa, Y., Yamanaka, A., Yada, T., Shibahara, M., Abe, Y., … Goto, K. (2003). Orexin receptor type-1 couples exclusively to pertussis toxin-insensitive G-proteins, while orexin receptor type-2 couples to both pertussis toxin-sensitive and -insensitive G-proteins. *Journal of Pharmacological Sciences, 92*, 259.

Chapter 14

Nicotine and Stimulatory Effects on 5-HT DRN Neurons

Salvador Hernández-López[2], Rene Drucker Colín[1], Stefan Mihailescu[2]

[1]Departamento de Neurociencias, Instituto de Fisiología Celular, Universidad Nacional Autónoma de México, Mexico DF, Mexico; [2]Departamento de Fisiología, Facultad de Medicina, Universidad Nacional Autónoma de México, México DF, México

Abbreviations

8-OH-DPAT 8-Hydroxy-2-dipropylaminotetralin
ACh Acetylcholine
AHP Afterhyperpolarization
AMPA α-Amino-3-hydroxy-5-methyl-4-isoxazolepropionic acid
AP Action potential
CICR Calcium-induced calcium release
CNQX 6-Cyano-7-nitroquinoxaline-2,3-dione
CNS Central nervous system
CPA Cyclopiazonic acid
DHβE Dihydro-β-erythroidine
DMPP Dimethylphenylpiperazinium
DRN Dorsal raphe nucleus
EPSC Excitatory postsynaptic current
IPSCs Inhibitory postsynaptic currents
MLA Methyllycaconitine
Nac Nucleus accumbens
nAChR Nicotinic acetylcholine receptor
NMDA N-Methyl-D-aspartate
PGO Ponto-geniculo-occipital waves
REM Rapid eye movement
RJR-2403 (E)-N-Methyl-4-(3-pyridinyl)-3-buten-1-amine oxalate
SERCA Sarcoplasmic/endoplasmic reticulum calcium-ATPase
TTX Tetrodotoxin
VGCC Voltage-gated calcium channel
VTA Ventrotegmental area

INTRODUCTION

Cigarette smoking is the most important preventable cause of morbidity and mortality worldwide. It represents a major risk factor for chronic obstructive pulmonary disease, lung cancer, ischemic heart disease, and many other pathological conditions (Centers for Disease Control and Prevention, 2008). However, nicotine and synthetic analogs show beneficial effects in the treatment or prevention of many neuropsychiatric disorders. For example, it has been shown that nicotine has antidepressant properties in nonsmoking patients with major depression (Salín-Pascual & Drucker-Colín, 1998). Likewise, smoking produces anxiolytic effects, and this may be one factor that contributes to nicotine dependence (Cheeta, Irvine, Kenny, & File, 2001). Nicotine treatment provides cognitive benefits to patients with Alzheimer disease, schizophrenia, and attention-deficit hyperactivity disorder (Levin, McClernon, & Rezvani, 2006).

The neurons of the dorsal raphe nucleus (DRN) are among the targets of nicotine in the central nervous system (CNS). In the study of Vázquez, Guzmán-Marín, Salín-Pascual, and Drucker-Colín (1996), systemic administration of nicotine in freely moving cats inhibited the ponto-geniculo-occipital (PGO) spikes of rapid eye movement (REM) sleep. Since the activation of DRN neurons has been related to the tonic inhibition of PGO spikes (Jacobs, Henriksen, & Dement, 1972), it was assumed that nicotine increases 5-HT DRN neuron firing rate and 5-HT release. Later studies demonstrated that 5-HT DRN neurons express postsynaptic nicotinic acetylcholine receptors (nAChRs), which, upon stimulation, transiently increase the firing activity of these neurons. It was also shown that noradrenergic and glutamatergic afferents to 5-HT DRN neurons present presynaptic nAChRs that produce long-term increases in the 5-HT DRN neuron firing rate by indirect mechanisms. The following sections describe in detail the interactions between nicotine and 5-HT DRN neurons.

NEURONAL NICOTINIC ACETYLCHOLINE RECEPTORS

Like their muscular analogs, neuronal nAChRs are ligand-gated cation channels with a pentameric structure. As of this writing, 11 types of neuronal nAChR subunits have been discovered: eight of these belong to the "α" family (α2 to α10), whereas the other three types belong to the "β" family (β2 to β4). According to their subunit composition, neuronal nAChRs may be classified as homomeric (containing only α7 or α8 or α9 subunits) or heteromeric, which are formed by assemblies of various types of α and β subunits. Theoretically, more than 1000 types of nAChRs may result from combinations between different types of nAChR subunits. However, experimental studies have demonstrated that in the CNS only two types of nAChR are encountered with high incidence: the $(\alpha 4)_2 (\beta 2)_3$ and the $(\alpha 7)_5$. Neuronal nAChRs exhibit a higher permeability for Ca^{2+} than for Na^+, are located preferentially on axon terminals (presynaptically), and act to increase the release of various neurotransmitters in the brain (Wonnacott, 1997). The electrophysiological and pharmacological properties of nAChRs, as well as their Ca^{2+} permeability, depend on their subunit composition. It is well established that the homomeric α7 nAChR exhibits the highest Ca^{2+} permeability, similar to that of the N-methyl-D-aspartate (NMDA) receptor. Administration of high concentrations of nicotine or long-term exposure to the drug desensitizes nAChRs. Chronic exposure to nicotine produces the upregulation of nAChRs containing the β2 subunit (($α4)_2(β2)_3$ and $(α3)_2(β2)_3$). The high variety of neuronal nAChRs in the brain explains their involvement in the modulation of many nervous processes and behaviors.

THE DORSAL RAPHE NUCLEUS

The DRN belongs to the superior (rostral) group of raphe nuclei together with the median raphe nucleus, nucleus pontis centralis, and caudal linear nucleus. Among the raphe nuclei, DRN contains the largest number of 5-HT neurons (~15,000 cells in the rat, ~24,000 in the cat, and ~235,000 in humans) and provides most of the forebrain serotonergic innervation. It is divided into six regions: rostral, dorsal, ventral, ventrolateral, interfascicular, and caudal, each one with afferents originating in and efferents directed to specific brain areas (Hale & Lowry, 2011). The cellular structure of the DRN is represented by 5-HT neurons (50–70%), γ-aminobutyric acid (GABA)-ergic neurons (20–25%), and small populations of glutamatergic and dopaminergic neurons (Johnson, 1994).

The activity of 5-HT DRN neurons is regulated by a large number of afferents. The most numerous are serotonergic inhibitory, which originate in the raphe nuclei. Other important afferents, also inhibitory, are the GABAergic ones, with numerous origins including the DRN itself. Details concerning the various afferents to 5-HT DRN neurons (origin, cytochemical types, effects on 5-HT DRN neurons) may be obtained from the review articles of Jacobs and Azmitia (1992) and Hornung (2003).

The axons of 5-HT neurons are highly collateralized and innervate more than one terminal area (Köhler & Steinbusch, 1982). This morphologic characteristic allows the simultaneous 5-HT modulation of functionally related neuronal circuits, which, together, generate a particular type of behavior, such as the response to stress (Lowry, 2002). The efferents of the DRN are most frequently ipsilateral (Jacobs, Foote, & Bloom, 1978) and topographically organized (Imai Steindler, & Kitai, 1986). The rostral part of the DRN projects to the caudate putamen, substantia nigra, and all neocortical regions (Vertes, 1991). The 5-HT neurons belonging to caudal regions of the DRN project to other forebrain regions: septum, hippocampus, and entorhinal cortex (Vertes, 1991). The amygdala receives innervation from both the rostral and the interfascicular regions of DRN. It is considered that the dorsal raphe and the raphe magnus nuclei are the main sources of the 5-HT innervation of forebrain stress circuits (Lowry, 2002). Cortical 5-HT afferents originating in the DRN present small, spaced varicosities and form a paracrine modulatory system instead of true chemical synapses (Lambe, Krimer, & Goldman-Rakic, 2000).

The electrical activity of 5-HT DRN neurons obtained with extracellular recordings present a regular, clocklike discharge of biphasic or triphasic action potentials with low frequency (0.5–2.5 Hz) (Vandermaelen & Aghajanian, 1983). With intracellular recordings, 5-HT DRN neurons present wide action potentials (3.34 ± 0.19 ms, $n=17$, measured at the base), with an inflection (shoulder) in the descending phase and long-lasting (150 ± 10.21 ms, $n=17$) and large (10–20 mV) after-hyperpolarizations (AHPs), as well as a low firing rate upon application of depolarizing stimuli (Frías–Dominguez, Garduño, Hernández, Drucker-Colín, & Mihailescu, 2013; Vandermaelen & Aghajanian, 1983). The above-described patterns of electrical activity of 5-HT DRN neurons are used for online identification of 5-HT DRN neurons (Figure 1(A and C)). In comparison, most non-5-HT DRN neurons present action potentials with lower duration (1.77 ± 0.07 ms, $n=7$) and shorter AHPs (108.45 ± 19.65 ms, $n=7$) with smaller amplitude and higher frequency of firing upon application of depolarizing currents (15–20 Hz) (Figure 1(B and D)) (Frías-Dominguez et al., 2013). The most preeminent pharmacological characteristics of 5-HT DRN neurons are their inhibition by agonists of the 5-HT$_{1A}$ receptors (serotonin or 8-hydroxy-2-dipropylaminotetralin (8-OH-DPAT)) and their stimulation by agonists of $α_1$-adrenoreceptors (noradrenaline or phenylephrine) (Vandermaelen & Aghajanian, 1983).

Serotonergic DRN neurons modulate a large number of bodily functions, which depend on sensory, motor, and integrative structures of the brain. 5-HT DRN neurons participate in the regulation of the sleep–wake cycle, cardiovascular and respiratory functions, body temperature, circadian rhythm entrainment, appetite, aggression, sexual behavior, sensorimotor reactivity, pain sensitivity, mood, and learning (Lucki, 1998). Functional suppression or lesions of the 5-HT systems do not produce the disappearance of any of the above-mentioned functions but only the alteration of them. Dysfunctions of the brain serotonergic system generate a wide range of psychiatric disorders such as depression, anxiety, obsessive–compulsive disorder, social phobia, schizophrenia, and nervous anorexia. Other psychiatric disorders produced by alterations of the serotonergic system are the impulse-related ones: aggression, substance abuse, gambling, obsessive control, and attention-deficit disorder (Lucki, 1997).

EFFECTS OF NICOTINE ON 5-HT DRN NEURON FIRING RATE AND 5-HT RELEASE

Microdialysis Studies

Systemic administration of nicotine (0.2 and 8 mg/kg) or RJR-2403 (a selective agonist at α4β2 nAChRs) increased 5-HT release in the frontal and parietal cortices of anesthetized (Ribeiro, Bettiker, Bogdanov, & Wurtman, 1993) and awake rats (Dominiak, Kees, & Grobecker, 1984; Summers & Giacobini, 1995). These results suggest that systemic nicotine increases the firing activity of 5-HT DRN neurons, which leads to an increase in 5-HT release at the projection sites of these neurons. A nicotine-induced release of 5-HT through stimulation of presynaptic nAChRs in the cortex is unlikely, since local cortical administration of low doses of nicotine (250 μM to 2.5 mM) did not produce any effect on 5-HT release (Summers & Giacobini, 1995) and the presence of presynaptic nAChRs on 5-HT axon terminals in the cerebral cortex could not be demonstrated (Schwartz, Lehmann, & Kellar, 1984). In contrast to these data, in the study of Toth, Sershen, Hashim, Vizi, and Lajtha (1992) local administration of high concentrations of nicotine (100 mM) increased

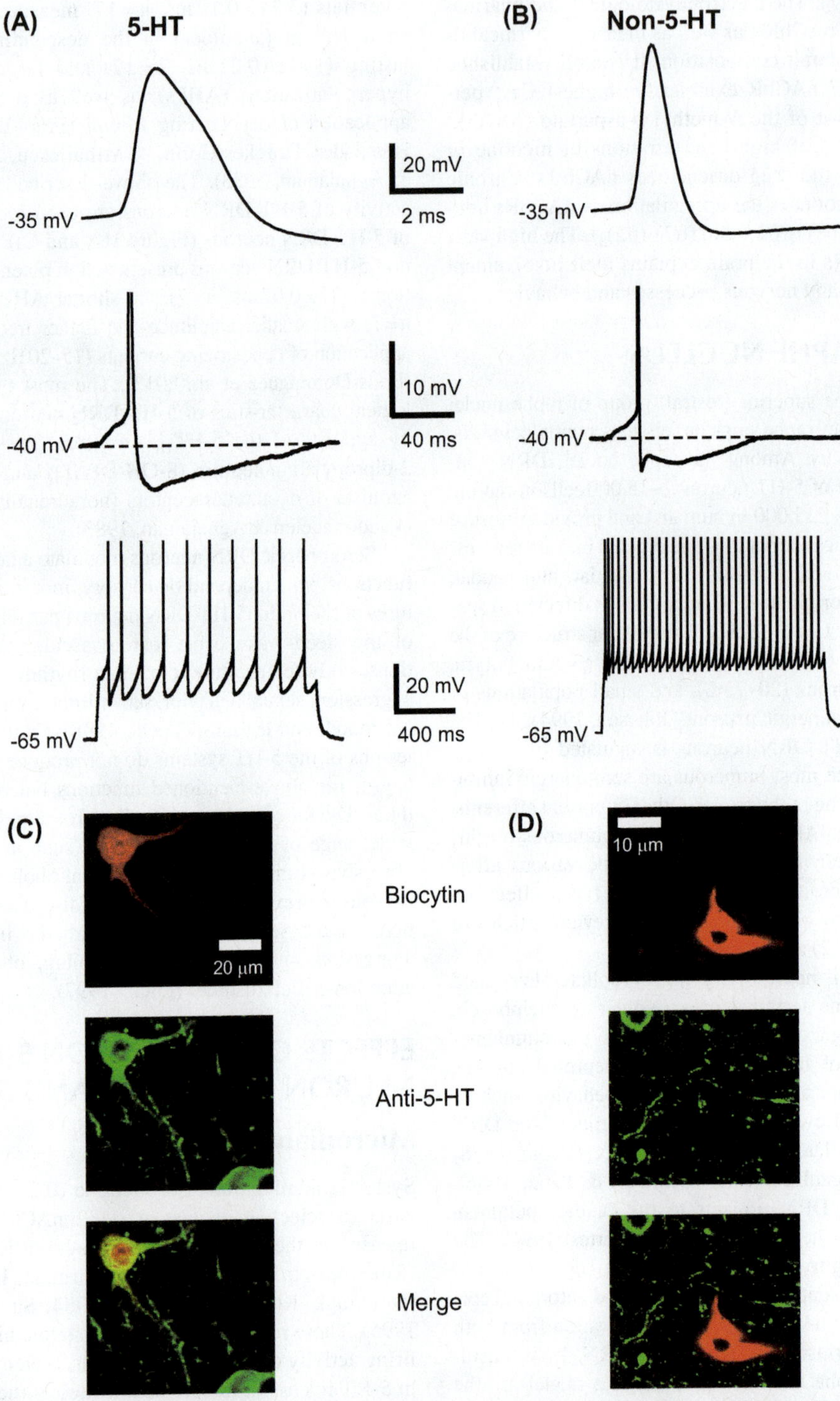

FIGURE 1 Electrophysiological and immunocytochemical characteristics of serotonergic and non-serotonergic dorsal raphe nucleus (DRN) neurons. (A) Electrophysiological characteristics of a serotonergic (5-HT) DRN neuron. Note the long duration of the spike (3.5 ms measured at the base, top trace) and afterhyperpolarization (AHP) (129 ms, middle trace), as well as the low frequency of firing upon application of a depolarizing pulse of current (7 Hz at 175 pA, bottom trace). (B) Electrophysiological characteristics of a non-5-HT DRN neuron. Note the short duration of the action potential (1.6 ms at the base, upper trace) and of the AHP (78 ms, middle trace), as well as the high firing rate upon application of a depolarizing pulse (21 Hz at 175 pA, bottom trace). (C) 5-HT immunoreactivity of the cell whose electrophysiological characteristics are presented in Figure 1(A); upper image presents biocytin staining of the recorded cell, middle image presents positive 5-HT immunoreactivity, bottom image represents the superposition of the upper and middle images, allowing one to identify the recorded cell as 5-HT. (D) 5-HT immunoreactivity of the cell whose electrophysiological characteristics are presented in Figure 1(B); upper image presents biocytin staining of the recorded cell, middle image 5-HT immunoreactivity, bottom image superposition of the upper and middle images allowing one to identify the recorded cell as non-5-HT. *From Frías-Domínguez et al. (2013, Figure 2, p. 14).*

cortical 5-HT release. This effect was, however, mediated by local release of glutamate, since kynurenic acid, an excitatory amino acid antagonist, blocked it.

Systemic administration of nicotine increased extracellular 5-HT levels in the medial temporal cortex, ventrotegmental area (VTA) (Rossi, Singer, Shearman, Sershen, & Lajtha, 2005; Singer et al., 2004), nucleus accumbens (Nac) (Liang et al., 2008), and hypothalamus (Dominiak et al., 1984).

Chronic administration of nicotine in freely moving rats increased 5-HT levels in the hypothalamus, hippocampus, and cerebellum (Takada, Urano, Ihara, & Takada, 1995). The duration of nicotine-induced 5-HT release was larger in awake than in anesthetized animals, which indicates that the nicotine stimulatory effect on 5-HT neurons depends on the previous level of activity of these neurons (Ribeiro et al., 1993).

In Vitro Electrophysiological Experiments

Early studies performed by Mihailescu, Guzmán-Marín, Domínguez Mdel, and Drucker-Colín (2002), Mihailescu, Guzmán-Marín, and Drucker-Colín (2001), and Mihailescu, Palomero-Rivero, Meade-Huerta, Maza-Flores, and Drucker-Colín (1998) in rat midbrain slices using extracellular recordings showed that nicotine increases the firing frequency of 60–80% of 5-HT DRN neurons (Figure 2) and inhibits it in the remaining 20–40%. The stimulatory effect of nicotine on 5-HT DRN neurons was blocked by mecamylamine (a nonselective nAChR antagonist) and was accompanied by an increase in 5-HT release in the DRN. In the study of Li, Rainnie, McCarley, and Greene (1998), nicotine, as well as acetylcholine (ACh) and dimethylphenylpiperazinium (DMPP), depolarized 60% of 5-HT DRN neurons and hyperpolarized 30% of them. The stimulatory effect of nicotine was explained by activation of α7 nAChRs located on DRN noradrenergic terminals, which induced noradrenaline release. In its turn, noradrenaline stimulated $α_1$-adrenoreceptors of 5-HT DRN neurons and thereby produced a G_q protein-dependent closure of an inward-rectifying K^+ channel and membrane depolarization. In the study of Li et al. (1998), the depolarizing effect of nicotine and of other nAChR agonists was blocked by prazosin, an $α_1$-adrenoreceptor antagonist, and by the α7 nAChR antagonist methyllycaconitine (MLA). It is noticeable, however, that in the study of Mihailescu et al. (1998), nicotine's stimulatory effect on 5-HT DRN neurons was not occluded by high noradrenaline concentrations (50 μM); this suggested that a neurotransmitter other than noradrenaline is involved in nicotine-induced increase in firing rate of 5-HT DRN neurons. As discussed later, this other excitatory neurotransmitter is glutamate (Garduño et al., 2012). The hyperpolarizing effect of nicotine observed in a subpopulation of 5-HT DRN neurons was dependent on 5-HT release inside the DRN, which stimulated somatodendritic 5-HT$_{1A}$ autoreceptors. This conclusion is supported by the reduction in nicotine's inhibitory effects by administration of the selective antagonist of 5-HT$_{1A}$ receptors, WAY-100635. The neuronal compartment responsible for nicotine-induced intra-raphe 5-HT release was the somatodendritic one and the nAChR subtype involved in this effect was α4β2 (Frías-Domínguez et al., 2013; Li et al., 1998). Recent studies performed in our laboratory (data in press) indicated that nicotine also increases the GABAergic input to 5-HT DRN neurons, an effect dependent on stimulation of α7 nAChRs located on DRN GABAergic terminals.

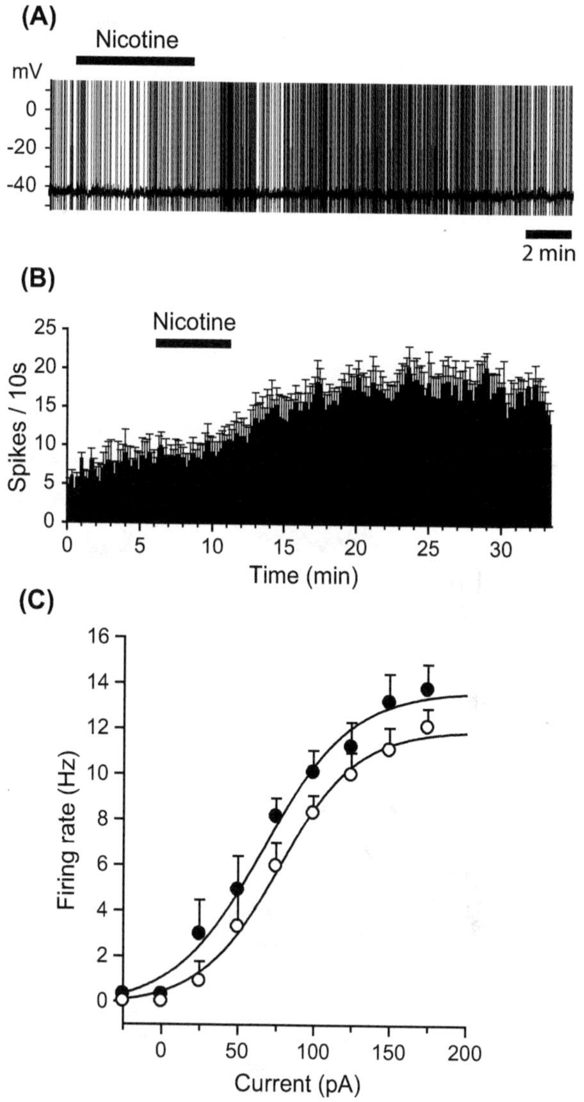

FIGURE 2 **Stimulatory effects of nicotine on identified serotonergic dorsal raphe nucleus (DRN) neurons.** (A) Stimulatory effect of bath application of nicotine (1 μM) on the firing rate of a serotonergic (5-HT) DRN neuron. (B) Frequency histogram showing the increase in firing rate induced by nicotine (1 μM) in 5-HT DRN neurons; data represent means ±SEM ($n=15$). (C) Upward shift of the intensity–frequency curves of 5-HT DRN neurons ($n=7$) after nicotine administration (closed circles). *Modified from Frías-Domínguez et al. (2013, Figure 3, p.15).*

Immunocytochemical studies performed by Bitner et al. (2000) and Bitner and Nikkel (2002) revealed the presence of α7 and α4β2 subunit-containing nAChRs in 5-HT and non-5-HT DRN neurons. The functionality and physiological roles of these receptors were assessed in the study of Galindo-Charles et al. (2008), performed in rat midbrain slices using the whole-cell patch-clamp technique. ACh (1 mM) applied by pressure injection in the presence of atropine (5 μM) and tetrodotoxin (TTX) (1 μM) induced two types of inward currents in identified 5-HT and non-5-HT DRN neurons (Figure 3). One of these currents exhibited the electrophysiological and pharmacological profiles of α7 nAChR-mediated currents: high amplitude, fast rising time, short decay time, and sensitivity to MLA (Figure 3(A)).

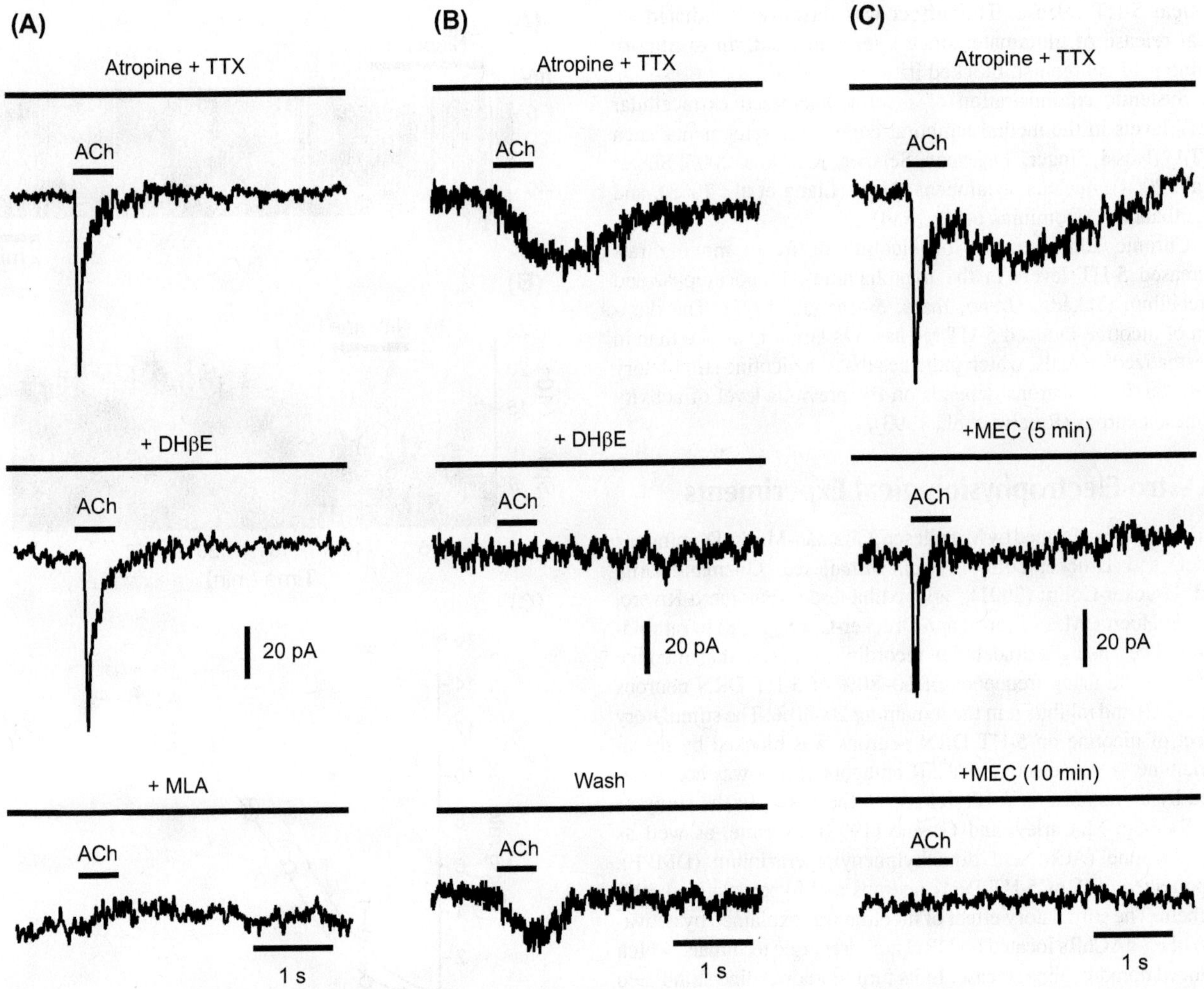

FIGURE 3 Local administration of acetylcholine (ACh) evokes inward currents in dorsal raphe nucleus (DRN) serotonergic neurons. (A) From top to bottom: the recorded cell exhibited a fast current after the "puff" of ACh (1 mM, 500 ms, top trace) was applied on top of the serotonergic (5-HT) neuron. The fast current was insensitive to the α4β2 nAChR-selective antagonist DHβE (100 nM, middle trace). However it was blocked by the α7 nAChR-selective antagonist methyllycaconitine (MLA) (100 nM, bottom trace). (B) This cell showed a slow current after the "puff" of ACh (top trace). This slow current was sensitive to DHβE (middle trace). The antagonist effect was reversible (bottom trace). (C) The "puff" of ACh evoked an inward current exhibiting fast and slow components (top trace). The slow component was completely blocked 5 min after the application of mecamylamine (MEC; 10 μM, middle trace), while the fast component was blocked subsequently, 10 min after the application of the drug (bottom trace). All recordings were performed in the presence of atropine (5 μM) and TTX (1 μM) to avoid muscarinic and indirect actions of ACh, respectively. All three cells shown were 5-HT immunoreactive (not shown). *Modified from Galindo-Charles et al. (2008, Figure 5, p. 610).*

The second inward current had smaller amplitude and slower kinetics and was suppressed by low doses (100 nM) of dihydro-β-erythroidine (DHβE); these characteristics belong to the α4β2 nAChR-mediated current (Figure 3(B)). In several neurons a mixture of α7 and α4β2 nAChR-dependent currents could be detected, which indicates the presence of both types of nAChR in the somatodendritic region of 5-HT and non-5-HT DRN neurons (Figure 3(C)). Under the conditions of our experiments, the expression of postsynaptic homomeric α7 nAChRs by 5-HT and non-5-HT DRN neurons was more reduced than that of α4β2 nAChRs. Chang et al. (2011), using an experimental model similar to that of Galindo-Charles et al. (2008), also reported that ACh, locally applied to 5-HT DRN neurons, induces inward currents dependent on α4β2 and α7 nAChRs. Altogether, the studies of Chang et al. (2011) and Galindo-Charles et al. (2008) indicated that activation of postsynaptic α4β2 and α7 nAChRs contributes to the stimulatory effects of nicotine on 5-HT DRN neurons.

One of the most frequently reported actions of nicotine is the presynaptic release of glutamate (McGehee, Heath, Gelber, Devay, & Role, 1995). DRN neurons receive glutamatergic afferents from the cortex, subcortical nuclei, and glutamatergic DRN interneurons (Soiza-Reilly & Commons, 2011). It was also shown that 5-HT DRN neurons express glutamatergic α-amino-3-hydroxy-5-methyl-4-isoxazolepropionic acid (AMPA)–kainate and NMDA receptors (Gartside, Cole, Williams, McQuade, & Judge, 2007). Taking into account these premises, Garduño et al. (2012) tested if nicotine's stimulatory effect on 5-HT DRN neurons depends on glutamate release inside the DRN. Experiments were performed in visualized 5-HT neurons of rat midbrain slices, using the whole-cell voltage-clamp technique. The effects of nicotine (1 μM) on the frequency and amplitude of spontaneous and miniature excitatory postsynaptic currents (EPSCs) were recorded in the presence of bicuculline (10 μM), a $GABA_A$ receptor blocker.

Nicotine increased the frequency of spontaneous EPSCs by ~88% with respect to the baseline in 16 of 18 identified 5-HT neurons (Figure 4(A, B, D, and E)). Glutamatergic spontaneous EPSC frequency was also increased by administration of eserine, an acetylcholinesterase blocker, which indicates that ACh is tonically released in the DRN (Figure 4(D and E)). Nicotine-induced glutamate release was presynaptic, since nicotine did not change the amplitude of spontaneous EPSCs and TTX did not alter the nicotine-induced increase in frequency of EPSCs (Figure 4(C and E)). Nicotine-induced increase in glutamatergic spontaneous EPSCs

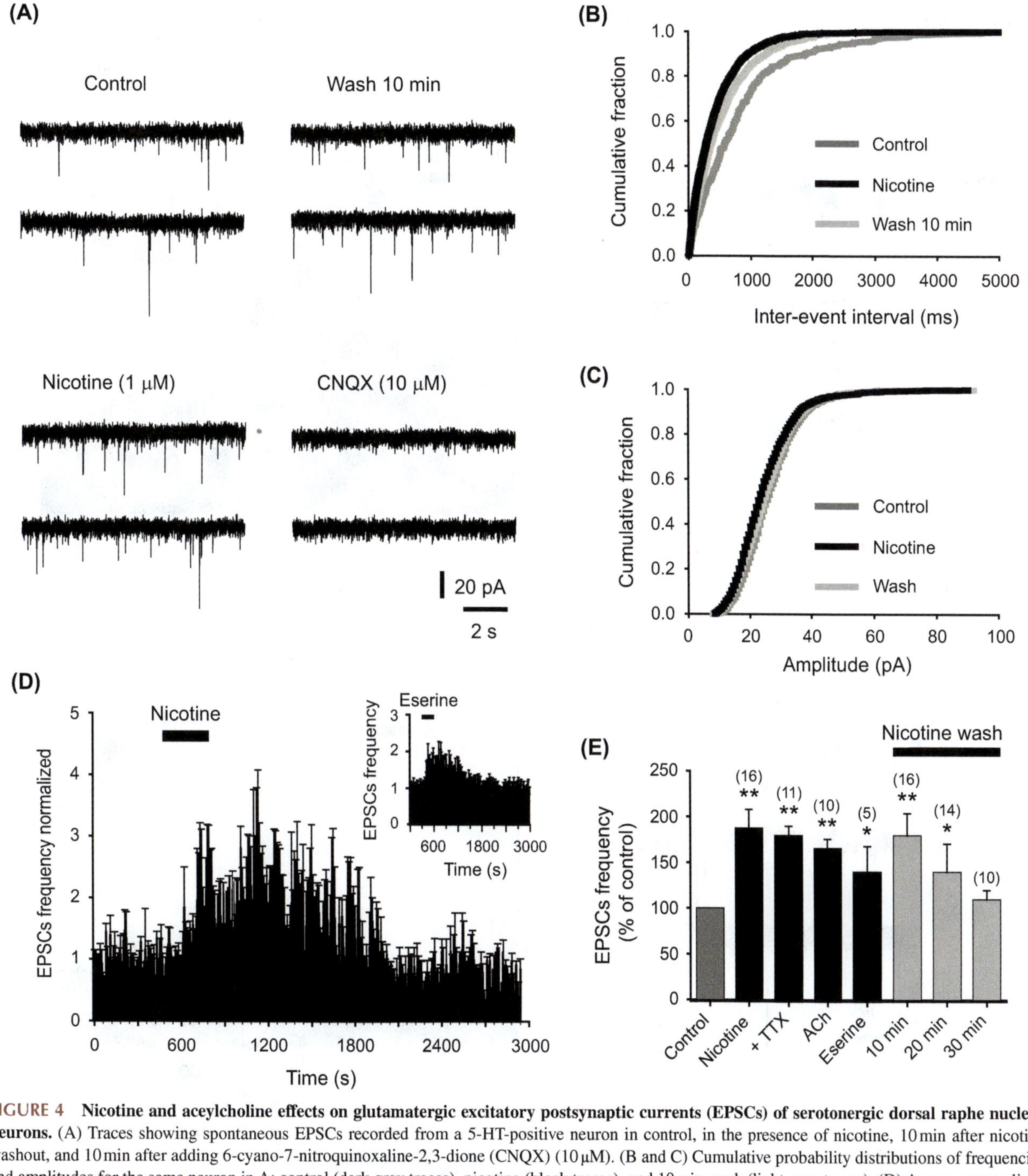

FIGURE 4 **Nicotine and acetylcholine effects on glutamatergic excitatory postsynaptic currents (EPSCs) of serotonergic dorsal raphe nucleus neurons.** (A) Traces showing spontaneous EPSCs recorded from a 5-HT-positive neuron in control, in the presence of nicotine, 10 min after nicotine washout, and 10 min after adding 6-cyano-7-nitroquinoxaline-2,3-dione (CNQX) (10 μM). (B and C) Cumulative probability distributions of frequencies and amplitudes for the same neuron in A: control (dark gray traces), nicotine (black traces), and 10 min wash (light gray traces). (D) Average normalized time–frequency histogram from 16 5-HT neurons tested with nicotine (1 μM). Inset, normalized time–frequency histogram from five 5-HT neurons tested with eserine (10 μM). (E) Summary of the results showing the effect on spontaneous EPSC frequency with nicotine alone and nicotine in the presence of TTX, acetylcholine (ACh), and eserine. All experimental groups were compared against control (before drug application). The last three bars represent the wash time after nicotine alone. All the experiments were done in the presence of bicuculline (10 μM). ACh experiments were performed in the presence of bicuculline and atropine (10 μM) (*$p<0.05$, **$p<0.01$). *From Garduño et al. (2012, Figure 3, p. 15151).*

persisted 10–20 min after drug withdrawal (Figure 4(D and E)), which suggests that nicotine, even after a single administration, induces long-term potentiation of glutamate release in the DRN. This effect was mimicked by the α4β2-selective agonist RJR-2403 and blocked by the α4β2-selective antagonist DHβE (Figure 5(C and D)). In contrast, nicotine-induced increase in spontaneous EPSC frequency was not affected by administration of the selective α7 nAChR blocker MLA (Figure 5(A and B)), while the application of the α7-selective agonist PNU-282987 was without effect. Taken together, these data indicate that nicotine increases glutamate release by stimulating presynaptic α4β2 nAChRs in the DRN. Similar results were obtained by Chang et al. (2011), who reported a stimulating effect of nicotine on the glutamatergic input to 5-HT DRN neurons.

Studies by Dickinson, Kew, and Wonnacott (2008) showed that activation of presynaptic α4β2 nAChRs is not sufficient to increase the intracellular Ca^{2+} concentration to the level required for neurotransmitter release. Instead, the inward

FIGURE 5 β2 subunit-containing nicotinic acetylcholine receptors mediate nicotine's effects on spontaneous glutamatergic excitatory postsynaptic currents. (A) Traces showing spontaneous excitatory postsynaptic currents (EPSCs) recorded from a 5-HT-positive neuron in the presence of methyllycaconitine (MLA) (100 nM, top) plus nicotine (1 μM, middle) and nicotine washout (bottom). (B) Time–frequency histograms from 10 identified 5-HT neurons tested with nicotine (1 μM) in the presence of MLA. The inset shows the normalized spontaneous EPSC frequency. (C) Traces showing spontaneous EPSCs recorded from a 5-HT-positive neuron in the presence of DHβE (100 nM, top), DHβE plus nicotine (1 μM, middle), and nicotine washout (bottom). (D) Time–frequency histogram from 12 identified 5-HT neurons tested with nicotine (1 μM) in the presence of DHβE. The inset shows the normalized spontaneous EPSC frequency. In all the experiments, nicotine was applied after 10 min pretreatment with DHβE or MLA. *From Garduño et al. (2012, Figure 4, p. 15152).*

current induced by presynaptic α4β2 nAChR activation depolarizes axon terminals and thus opens voltage-gated calcium channels (VGCCs); the resulting Ca^{2+} influx produces Ca^{2+} release from the endoplasmic reticulum (calcium-induced calcium release or CICR), which generates neurotransmitter release. In our experiments (Garduño et al., 2012), pretreatment of brain slices with a mixture of L-, N-, and P/Q-selective VGCC blockers suppressed the nicotine-induced increase in glutamatergic spontaneous EPSC frequency (Figure 6(B)). The same effect was observed when the slices were pretreated with ryanodine, a blocker of endoplasmic reticulum calcium channels, or with noncompetitive inhibitors of the sarcoplasmic/endoplasmic reticulum Ca^{2+}-ATPase, thapsigargin and cyclopiazonic acid (CPA) (Figure 6(C and D)).

Interestingly, the nonspecific blocker of VGCCs, cadmium chloride, potentiated the nicotine-induced increase in frequency of spontaneous EPSCs (Figure 6(A and D)), which may be explained by the potentiation of α4β2 nAChRs by cadmium (Hsiao, Dweck, & Luetje, 2001).

Taken together, the studies of Chang et al. (2011) and Garduño et al. (2012) indicate that intra-raphe increase in glutamate release represents an important mechanism for nicotine-dependent long-lasting increase in 5-HT DRN neuron firing rate.

In Vivo Electrophysiological Experiments

Engberg, Erhardt, Sharp, and Hajós (2000) reported that nicotine (50–400 μg/kg, intravenous) induced a transient (1–3 min), short-latency (30 s) inhibition of putative 5-HT DRN neurons in anesthetized rats. This effect was antagonized by mecamylamine and by the specific 5-HT_{1A} receptor antagonist WAY-100635. In the same study, iontophoretic administration of nicotine in the DRN failed

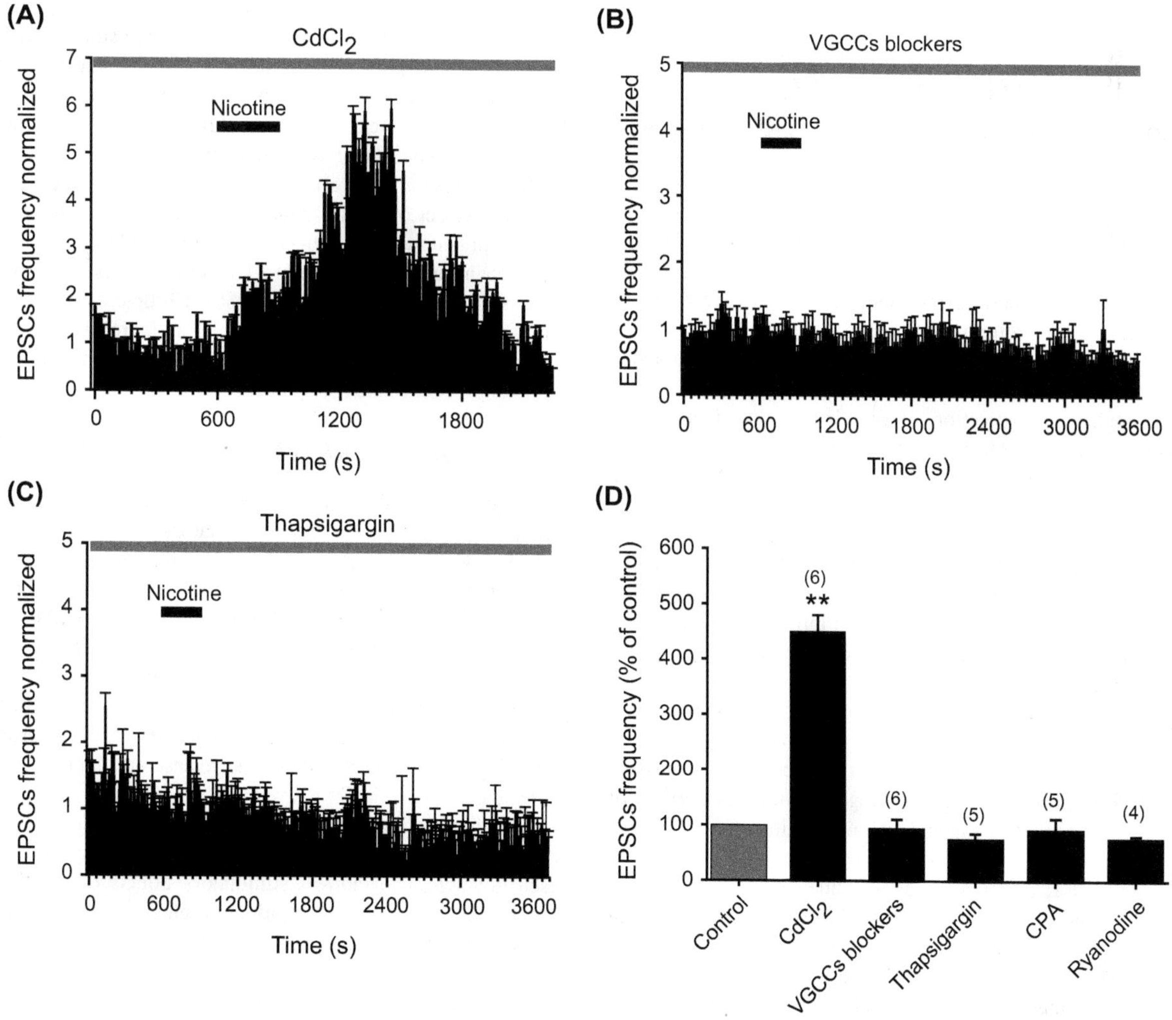

FIGURE 6 **Nicotine's stimulatory effects on spontaneous glutamatergic excitatory postsynaptic currents depend on VGCCs and CICR.** (A) Time–frequency histogram shows the effect of nicotine on the spontaneous EPSC frequency in the presence of $CdCl_2$ (gray bar). (B) Time–frequency histogram shows the lack of effect of nicotine on the spontaneous EPSC frequency in the presence of a mixture containing the Ca^{2+} channel blockers ω-agatoxin-TK, ω-conotoxin-GVIA, and nitrendipine (gray bar). (C) Time–frequency histogram shows the lack of effect of nicotine on the spontaneous EPSC frequency in the presence of the sarcoplasmic/endoplasmic reticulum calcium-ATPase blocker thapsigargin (gray bar). (D) Bar graph shows the effect of nicotine on the spontaneous EPSC frequency in slices pretreated with $CdCl_2$, Ca^{2+} channel blockers, thapsigargin, cyclopiazonic acid (CPA), or ryanodine (**$p<0.01$). Gray bar represents the control (before nicotine application). *From Garduño et al. (2012, Figure 5, p.15153).*

to produce any effect on 5-HT neurons. It is noticeable that the reactivity of 5-HT DRN neurons in vivo depends on the arousal state (Ranade & Mainen, 2009). Therefore, the lack of response of 5-HT DRN neurons to local nicotine administration may be explained by the inhibitory effects of anesthesia on the activity of 5-HT DRN neurons.

In the study of Guzman-Marin et al. (2001), performed in freely moving rats, nicotine (0.1 mg/kg, subcutaneous) did not change significantly the firing rate of putative 5-HT DRN neurons while awake or during slow-wave sleep, but increased it significantly prior to the beginning of REM sleep, when the noradrenergic and histaminergic stimulatory inputs to 5-HT DRN neurons are reduced (Sakai & Crochet, 2000).

It may be concluded that the few in vivo electrophysiological experiments performed to date did not reproduce the results obtained with in vitro experiments owing to the involvement of much more complex systems of regulation of 5-HT DRN neuron activity in vivo.

BEHAVIORAL EFFECTS OF NICOTINE MEDIATED BY INCREASES IN SEROTONIN RELEASE

Nicotine-dependent behaviors such as anxiolysis, increase in locomotor activity, and improvement of memory and cognition may be blocked or enhanced by agonists and antagonists of 5-HT receptors (Müller & Homberg, 2014; Seth et al., 2004 for reviews). This suggests that serotonin plays a permissive role in nicotine's behavioral effects. Several examples of 5-HT-dependent nicotine behavioral effects are presented as follows.

Intra-raphe microinjection of nicotine produced anxiolytic effects, which were blocked by intra-raphe administration of either DHβE, a selective blocker of α4β2 nAChRs, or WAY-100635, a selective 5-HT_{1A} receptor blocker. These experimental facts suggest that nicotine induces intra-DRN 5-HT release by stimulation of α4β2 nAChRs. In its turn, serotonin inhibits the activity of 5-HT neurons by stimulating 5-HT_{1A} autoreceptors (Cheeta et al., 2001). Consequently, 5-HT release decreases in terminal regions of the limbic system, which generates the anxiolytic effect (Sprouse & Aghajanian, 1987).

In rats, nicotine withdrawal produces anxiogenic responses in the social interaction and elevated-plus maze tests (Cheeta, 2001), as well as increasing the startle response (Rasmussen, Kallman, & Helton, 1997). The anxiogenic withdrawal effect of nicotine was reversed by systemic administration of the selective 5-HT_3 receptor antagonist ondansetron (Barnes, Costall, Kellyu, Onaivi, & Naylor, 1990), whereas the enhanced startle response was reversed by the 5-HT_{1A} antagonist WAY-100635 (Rasmussen et al., 1997). Intra-DRN administration of ondansetron also reversed the anxiogenic withdrawal response, which suggests that the DRN is the primary site of action of systemically administered 5-HT_3 receptor antagonists.

Repeated nicotine administration produces behavioral disinhibition on the elevated-plus maze test. This effect was counteracted by 5-HT_{1A} receptor stimulation (Olausson, Akesson, Engel, & Söderpalm, 2001). Two days after nicotine withdrawal an enhancement of the sensitivity of 5-HT_{1A} autoreceptors in the DRN could be observed (Rasmussen & Czachura, 1997).

Under experimental conditions, treatment with 5-HT_{2A} receptor antagonists alleviates depressed mood associated with nicotine withdrawal (Zaniewska, McCreary, Wydra, & Filip, 2010). The behavioral response to a 5-HT_{2A} receptor agonist was increased up to 24 days following nicotine withdrawal, which suggests a sensitization of the 5-HT_{2A} receptor after nicotine withdrawal (Yasuda, Suemaru, Araki, & Gomita, 2002). Accordingly, both chronic and acute treatment with nicotine attenuated the response to 5-HT_{2A} receptor stimulation (Tizabi, Russell, Johnson, & Darmani, 2001).

Acute administration of low doses of nicotine produced small increases in locomotor activity. This effect, dependent on dopamine release in the Nac, is subject to sensitization (it increases in magnitude with repeated administration of nicotine) (Benwell & Balfour, 1992). The acute locomotor effect of nicotine was attenuated by selective stimulation of the 5-HT_{2C} receptor (Grottick, Corrigall, & Higgins, 2001), whereas pharmacological antagonism of this receptor potentiated the locomotor effects of nicotine (Fletcher, Sinyard, & Higgins, 2006). Systemic administration of 8-OH-DPAT potentiated the acute effect of nicotine on locomotor activity, whereas parachlorophenylalanine (a drug that depletes 5-HT stores) decreased the effect of chronic nicotine on locomotor activity (Olausson, Engel, & Söderpalm, 1999).

Nicotine is very probably responsible for the tobacco-induced inhibitory effect on eating. Studies in rats demonstrated that the chronic infusion of nicotine decreases both food intake and body-weight gain and that cessation of nicotine reverses the process (Grunberg, Bowen, & Winders, 1986). Chronic nicotine administration in freely moving rats produced an increase in 5-HT and dopamine levels in the lateral hypothalamus, which resulted in a decrease in food intake (Meguid et al., 2000).

CONCLUDING REMARKS

Electrophysiological studies performed in midbrain slices indicated that nicotine triggers both stimulatory and inhibitory effects on 5-HT DRN neurons. Moreover, each of the two types of nAChR present in the DRN neurons, α4β2 and α7, generates, simultaneously, opposite effects on 5-HT DRN neuron activity. Thus, the α4β2 nAChRs stimulate 5-HT DRN neurons by increasing intra-DRN glutamate release and by generating inward somatodendritic currents and inhibit 5-HT DRN neurons by increasing intra-DRN 5-HT release. The α7 nAChRs stimulate 5-HT DRN neurons by increasing intra-raphe noradrenaline release and inhibit them by increasing presynaptic GABA release.

The direct effects of nicotine, generated by stimulation of somatodendritic receptors, are short lasting owing to nAChR desensitization. The indirect stimulatory effects of nicotine are long lasting owing to potentiation of glutamate release inside the DRN by nicotine. The inhibitory effects of nicotine observed in ~20–30% of 5-HT DRN neurons depend mostly on nicotine-induced intra-DRN 5-HT and GABA release.

In agreement with the in vitro electrophysiological data, microdialysis studies indicated that acute and chronic administration of nicotine increases extracellular levels of serotonin in a large number of brain areas, including the frontal, parietal, and temporal cortices; VTA; Nac; amygdala; DRN; and hypothalamus, but also diminishes 5-HT release in other areas like the dorsal and ventral hippocampus.

Therefore, in brain slices, nicotine triggers stronger stimulatory than inhibitory effects on 5-HT DRN neurons.

The few in vivo electrophysiological studies concerning the effects of nicotine on 5-HT DRN neurons indicated either transient, 5-HT-dependent inhibitory effects or slight stimulatory effects during sleep. The interpretation of these data is difficult because the response of 5-HT DRN neurons to nicotine depends on their previous excitatory state.

Behavioral experiments demonstrated that the serotonergic system plays a permissive role for various nicotine effects such as anxiolysis, depression, and locomotor activity. Moreover, it was found that chronic exposure to nicotine alters the sensitivity of 5-HT receptors: it sensitizes 5-HT$_{1A}$ DRN receptors and desensitizes 5-HT$_{2C}$ receptors.

APPLICATION TO OTHER ADDICTIONS AND SUBSTANCE MISUSE

- Nicotine is not the only addictive drug that uses the serotonergic system to produce behavioral effects.
- As a matter of fact, the majority of the psychotropic drugs (cocaine, amphetamine, methamphetamine, MDMA (ecstasy), heroin, morphine, alcohol) produce adaptive changes in the 5-HT system that result in characteristic behaviors (see Müller & Homberg, 2014 for reviews).
- These adaptive changes consist in alterations of extracellular levels of 5-HT and in increases or decreases in 5-HT receptor sensitivity.
- The drugs of abuse, with the exception of cannabis, increase extracellular levels of 5-HT after acute administration and decrease 5-HT levels after chronic administration.
- The conversion of the controlled use of a drug into addiction to that drug depends on the reduction in extracellular 5-HT levels, the downregulation or inhibition of 5-HT$_{2C}$ receptors, and the sensitization of 5-HT$_{2A}$ receptors. These three conditions are met because of the frequent use of the drug of abuse and/or the existence of genetic predisposing factors (Caspi et al., 2003).

DEFINITION OF TERMS

Rapid eye movement sleep This is also known as paradoxical sleep, a stage of sleep characterized by loss of skeletal muscle tone and rapid, disorganized movements of the eyeballs. During REM sleep the brain electrical activity is desynchronized and metabolism is increased as during the wake state. There are four or five episodes of REM sleep during one night's sleep, with a total duration of 90–120 min (20–25% of total sleep duration). REM sleep is generated by cholinergic tegmental nuclei located in the pons.

Ponto-geniculo-occipital waves These are electrical waves of 300 μV of amplitude and 250 ms of duration (in the cat) that can be recorded in the pons, lateral geniculate nuclei, and occipital cortex, immediately before and during REM sleep episodes. PGO waves could be recorded in cats, humans, and nonhuman primates. An equivalent of PGO waves exists in rats ("P waves," recordable only in the pons).

Tetrodotoxin It is a neurotoxin extracted from the pufferfish, toads, and various species of octopuses. It is a blocker of voltage-gated sodium channels and, consequently, of action potential generation and conduction. TTX is used frequently in neuropharmacology to determine if an event (release of a neurotransmitter e.g.) is dependent on action potential generation.

Excitatory postsynaptic current This is produced by the opening of specific membrane channels, which allows the entry of cations or the exit of anions from the cell. The occurrence of EPSCs depolarizes the cell's membrane and increases the cell's excitability. Various excitatory neurotransmitters (glutamate or ACh) generate EPSCs by coupling with ionotropic channel receptors. Intracellular recording of EPSCs is used in neuropharmacology to determine if a substance increases the excitability of a cell using a specific excitatory path.

Afterhyperpolarization It represents the terminal phase of an action potential, during which the membrane potential is more negative than the resting potential. A prolonged opening of various types of voltage-gated K$^+$ channels produces AHP.

Elevated-plus maze test It is used in rodent experiments to detect the degree of anxiety or antianxiety produced by various agents (neurotransmitters, drugs, hormones, etc.). Rodents present aversion for open spaces and prefer, if there is a choice, the enclosed ones. The setup used for this experimental model consists in two open and two closed arms connected to form a plus (cross) shape, located 40–70 cm above the floor. The animal is placed at the junction of the open and closed arms. During the test, the time spent in the open and closed arms is measured. An increase in the time spent in open arms reflects a reduction in anxiety.

KEY FACTS OF MICRODIALYSIS

- Microdialysis is a technique for continuous measurement of extracellular concentrations of various substances.
- A small tubular probe, endowed with a semipermeable membrane, is introduced into a tissue.
- A dialysis liquid with a composition similar to that of the extracellular fluid is circulated through the dialysis probe at a flow of 1–3 μl/min.
- The substances present in extracellular liquid (neurotransmitters, e.g.) diffuse into the dialysis fluid.
- Posteriorly, the concentration of the substances of interest is measured in the dialysis fluid using specific biochemical techniques.

SUMMARY POINTS

- This chapter focuses on the stimulatory effects of nicotine on serotonergic (5-HT) DRN neurons.
- Clinical and experimental studies suggested that certain behavioral effects of nicotine are mediated by serotonin.
- Most of the forebrain 5-HT innervation originates in 5-HT neurons of the DRN.
- α7 and α4β2 nAChRs were identified in both 5-HT and non-5-HT DRN neurons.
- Stimulation of postsynaptic (somatodendritic) nAChRs induced brief stimulatory effects in 5-HT DRN neurons.
- Stimulation of presynaptic α4β2 and α7 nAChRs, present in DRN glutamatergic and noradrenergic terminals. respectively, induced long-lasting excitatory effects on 5-HT DRN neurons by increasing glutamate and noradrenaline release.

- Intra-raphe glutamate release depends on the opening of VGCCs and on CICR in glutamatergic DRN terminals and is potentiated even after a single nicotine administration.
- Nicotine also produced indirect inhibitory effects on 5-HT DRN neurons by increasing intra-raphe serotonin and GABA release.
- In most 5-HT DRN neurons of brain slices, the stimulatory effects of nicotine are stronger than the inhibitory ones.
- In vivo microdialysis studies indicated that nicotine increases 5-HT release in various forebrain areas and thus support the results obtained in brain slice experiments.
- Behavioral experiments demonstrated that a large number of nicotine behavioral effects are dependent on stimulation or inhibition of various types of 5-HT receptors.

REFERENCES

Barnes, N. M., Costall, B., Kellyu, M. E., Onaivi, E. S., & Naylor, R. J. (1990). Ketotifen and its analogues reduce aversive responding in the rodent. *Pharmacology Biochemistry and Behavior, 37*, 785–793.

Benwell, M. E., & Balfour, D. J. (1992). The effects of acute and repeated nicotine treatment on nucleus accumbens dopamine and locomotor activity. *British Journal of Pharmacology, 105*(4), 849–856.

Bitner, R. S., & Nikkel, A. L. (May 31, 2002). Alpha-7 nicotinic receptor expression by two distinct cell types in the dorsal raphe nucleus and locus coeruleus of rat. *Brain Research, 938*(1–2), 45–54.

Bitner, R. S., Nikkel, A. L., Curzon, P., Donnelly-Roberts, D. L., Puttfarcken, P. S., Namovic, M., ... Decker, M. W. (2000). Reduced nicotinic receptor-mediated antinociception following in vivo antisense knock-down in rat. *Brain Research, 871*(1), 66–74.

Caspi, A., Sugden, K., Moffitt, T. E., Taylor, A., Craig, I. W., Harrington, H., ... Poulton, R. (2003). Influence of life stress on depression: moderation by a polymorphism in the 5-HTT gene. *Science, 301*, 386–389.

Centers for Disease Control and Prevention. (2008). *Smoking-attributable mortality, years of potential life lost, and productivity losses*. Atlanta, GA: U.S. Department of Health and Human Services.

Chang, B., Daniele, C. A., Gallagher, K., Madonia, M., Mitchum, R. D., Barrett, L., ... McGehee, D. S. (2011). Nicotinic excitation of serotonergic projections from dorsal raphe to the nucleus accumbens. *Journal of Neurophysiology, 106*(2), 801–808.

Cheeta, S., Irvine, E. E., Kenny, P. J., & File, S. E. (2001). The dorsal raphe nucleus is a crucial structure mediating nicotine's anxiolytic effects and the development of tolerance and withdrawal responses. *Psychopharmacology (Berlin), 155*(1), 78–85.

Dickinson, J. A., Kew, J. N., & Wonnacott, S. (2008). Presynaptic alpha 7- and beta 2-containing nicotinic acetylcholine receptors modulate excitatory amino acid release from rat prefrontal cortex nerve terminals via distinct cellular mechanisms. *Molecular Pharmacology, 74*(2), 348–359.

Dominiak, P., Kees, F., & Grobecker, H. (1984). Changes in peripheral and central catecholaminergic and serotoninergic neurons of rats after acute and subacute administration of nicotine. *Klinische Wochenschrift, 62*(Suppl. 2), 76–80.

Engberg, G., Erhardt, S., Sharp, T., & Hajós, M. (2000). Nicotine inhibits firing activity of dorsal raphe 5-HT neurones in vivo. *Naunyn-Schmiedeberg's Archives of Pharmacology, 362*(1), 41–45.

Fletcher, P. J., Sinyard, J., & Higgins, G. A. (2006). The effects of the 5-HT(2C) receptor antagonist SB242084 on locomotor activity induced by selective, or mixed, indirect serotonergic and dopaminergic agonists. *Psychopharmacology (Berlin), 187*(4), 515–525.

Frías-Domínguez, C., Garduño, J., Hernández, S., Drucker-Colín, R., & Mihailescu, S. (2013). Flattening plasma corticosterone levels increases the prevalence of serotonergic dorsal raphe neurons inhibitory responses to nicotine in adrenalectomised rats. *Brain Research Bulletin, 98*, 10–22.

Galindo-Charles, L., Hernandez-Lopez, S., Galarraga, E., Tapia, D., Bargas, J., Garduño, J., ... Mihailescu, S. (2008). Serotoninergic dorsal raphe neurons possess functional postsynaptic nicotinic acetylcholine receptors. *Synapse, 62*(8), 601–615.

Garduño, J., Galindo-Charles, L., Jiménez-Rodríguez, J., Galarraga, E., Tapia, D., Mihailescu, S., & Hernandez-Lopez, S. (2012). Presynaptic α4β2 nicotinic acetylcholine receptors increase glutamate release and serotonin neuron excitability in the dorsal raphe nucleus. *Journal of Neuroscience, 32*(43), 15148–15157.

Gartside, S. E., Cole, A. J., Williams, A. P., McQuade, R., & Judge, S. J. (2007). AMPA and NMDA receptor regulation of firing activity in 5-HT neurons of the dorsal and median raphe nuclei. *European Journal of Neuroscience, 25*(10), 3001–3008.

Grottick, A. J., Corrigall, W. A., & Higgins, G. A. (2001). Activation of 5-HT(2C) receptors reduces the locomotor and rewarding effects of nicotine. *Psychopharmacology (Berlin), 157*(3), 292–298.

Grunberg, N. E., Bowen, D. J., & Winders, S. E. (1986). Effects of nicotine on body weight and food consumption in female rats. *Psychopharmacology, 90*, 101–105.

Guzmán-Marín, R., Alam, M. N., Mihailescu, S., Szymusiak, R., McGinty, D., & Drucker-Colín, R. (2001). Subcutaneous administration of nicotine changes dorsal raphe serotonergic neurons discharge rate during REM sleep. *Brain Research, 888*(2), 321–325.

Hale, M. W., & Lowry, C. A. (2011). Functional topography of midbrain and pontine serotonergic systems: implications for synaptic regulation of serotonergic circuits. *Psychopharmacology (Berlin), 213*(2–3), 243–264.

Hornung, J. P. (2003). The human raphe nuclei and the serotonergic system. *Journal of Chemical Neuroanatomy, 26*(4), 331–343.

Hsiao, B., Dweck, D., & Luetje, C. W. (2001). Subunit-dependent modulation of neuronal nicotinic receptors by zinc. *Journal of Neuroscience, 21*(6), 1848–1856.

Imai, H., Steindler, D. A., & Kitai, ST. (1986). The organization of divergent axonal projections from the midbrain raphe nuclei in the rat. *Journal of Comparative Neurology, 243*(3), 363–380.

Jacobs, B. L., & Azmitia, E. C. (1992). Structure and function of the brain serotonin system. *Physiological Reviews, 72*(1), 165–229.

Jacobs, B. L., Foote, S. L., & Bloom, F. E. (1978). Differential projections of neurons within the dorsal raphe nucleus of the rat: a horseradish peroxidase (HRP) study. *Brain Research, 147*(1), 149–153.

Jacobs, B. L., Henriksen, S. J., & Dement, W. C. (1972). Neurochemical bases of the PGO wave. *Brain Research, 48*, 406–411.

Johnson, M. D. (1994). Electrophysiological and histochemical properties of postnatal rat serotonergic neurons in dissociated cell culture. *Neuroscience, 63*(3), 775–787.

Köhler, C., & Steinbusch, H. (1982). Identification of serotonin and non-serotonin-containing neurons of the mid-brain raphe projecting to the entorhinal area and the hippocampal formation. A combined immunohistochemical and fluorescent retrograde tracing study in the rat brain. *Neuroscience, 7*(4), 951–975.

Lambe, E. K., Krimer, L. S., & Goldman-Rakic, P. S. (2000). Differential postnatal development of catecholamine and serotonin inputs to identified neurons in prefrontal cortex of rhesus monkey. *Journal of Neuroscience, 20*(23), 8780–8787.

Levin, E. D., McClernon, F. J., & Rezvani, A. H. (2006). Nicotinic effects on cognitive function: behavioral characterization, pharmacological specification, and anatomic localization. *Psychopharmacology, 184*, 523–539.

Liang, Y., Boules, M., Shaw, A. M., Williams, K., Fredrickson, P., & Richelson, E. (2008). Effect of a novel neurotensin analog, NT69L, on nicotine-induced alterations in monoamine levels in rat brain. *Brain Research, 1231*, 6–15.

Li, X., Rainnie, D. G., McCarley, R. W., & Greene, R. W. (1998). Presynaptic nicotinic receptors facilitate monoaminergic transmission. *Journal of Neuroscience, 18*(5), 1904–1912.

Lowry, C. A. (2002). Functional subsets of serotonergic neurones: implications for control of the hypothalamic-pituitary-adrenal axis. *Journal of Neuroendocrinology, 14*(11), 911–923.

Lucki, I. (1997). The forced swimming test as a model for core and component behavioral effects of antidepressant drugs. *Behavioral Pharmacology, 8*(6–7), 523–532.

Lucki, I. (1998). The spectrum of behaviors influenced by serotonin. *Biological Psychiatry, 44*(3), 151–162.

McGehee, D. S., Heath, M. J., Gelber, S., Devay, P., & Role, L. W. (1995). Nicotine enhancement of fast excitatory synaptic transmission in CNS by presynaptic receptors. *Science, 269*, 1692–1696.

Meguid, M. M., Fetissov, S. O., Varma, M., Sato, T., Zhang, L., Laviano, A., & Rossi-Fanelli, F. (2000). Hypothalamic dopamine and serotonin in the regulation of food intake. *Nutrition, 16*(10), 843–857.

Mihailescu, S., Guzmán-Marín, R., Domínguez Mdel, C., & Drucker-Colín, R. (2002). Mechanisms of nicotine actions on dorsal raphe serotoninergic neurons. *European Journal of Pharmacology, 452*(1), 77–82.

Mihailescu, S., Guzmán-Marín, R., & Drucker-Colín, R. (2001). Nicotine stimulation of dorsal raphe neurons: effects on laterodorsal and pedunculopontine neurons. *European Neuropsychopharmacology, 11*(5), 359–366.

Mihailescu, S., Palomero-Rivero, M., Meade-Huerta, P., Maza-Flores, A., & Drucker-Colín, R. (1998). Effects of nicotine and mecamylamine on rat dorsal raphe neurons. *European Journal of Pharmacology, 360*(1), 31–36.

Müller, C. P., & Homberg, J. R. (2014). The role of serotonin in drug use and addiction. *Behavioural Brain Research, 277*, 146–192. http//dx.do.org/j.bbr.2014.04.007.

Olausson, P., Akesson, P., Engel, J. A., & Söderpalm, B. (2001). Effects of 5-HT$_{1A}$ and 5-HT$_2$ receptor agonists on the behavioral and neurochemical consequences of repeated nicotine treatment. *European Journal of Pharmacology, 420*(1), 45–54.

Olausson, P., Engel, J. A., & Söderpalm, B. (1999). Behavioral sensitization to nicotine is associated with behavioral disinhibition; counteraction by citalopram. *Psychopharmacology (Berlin), 142*(2), 111–119.

Ranade, S. P., & Mainen, Z. F. (2009). Transient firing of dorsal raphe neurons encodes diverse and specific sensory, motor, and reward events. *Journal of Neurophysiology, 102*(5), 3026–3037.

Rasmussen, K., & Czachura, J. F. (1997). Nicotine withdrawal leads to increased sensitivity of serotonergic neurons to the 5-HT$_{1A}$ agonist 8-OH-DPAT. *Psychopharmacology (Berlin), 133*(4), 343–346.

Rasmussen, K., Kallman, M. J., & Helton, D. R. (1997). Serotonin-1A antagonists attenuate the effects of nicotine withdrawal on the auditory startle response. *Synapse, 27*(2), 145–152.

Ribeiro, E. B., Bettiker, R. L., Bogdanov, M., & Wurtman, R. J. (1993). Effects of systemic nicotine on serotonin release in rat brain. *Brain Research, 621*(2), 311–318.

Rossi, S., Singer, S., Shearman, E., Sershen, H., & Lajtha, A. (2005). The effects of cholinergic and dopaminergic antagonists on nicotine-induced cerebral neurotransmitter changes. *Neurochemical Research, 30*(4), 541–558.

Sakai, K., & Crochet, S. (2000). Serotonergic dorsal raphe neurons cease firing by disfacilitation during paradoxical sleep. *NeuroReport, 11*(14), 3237–3241.

Salín-Pascual, R. J., & Drucker-Colín, R. (1998). A novel effect of nicotine on mood and sleep in major depression. *NeuroReport, 9*, 57–60.

Schwartz, R. D., Lehmann, J., & Kellar, K. J. (1984). Presynaptic nicotinic cholinergic receptors labeled by [3H]acetylcholine on catecholamine and serotonin axons in brain. *Journal of Neurochemistry, 42*(5), 1495–1498.

Seth, P., Cheeta, S., Tucci, S., & File, S. E. (2004). Nicotinic–serotonergic interactions in brain and behaviour. *Pharmacology Biochemistry and Behavior, 71*(4), 795–805.

Singer, S., Rossi, S., Verzosa, S., Hashim, A., Lonow, R., Cooper, T., ... Lajtha, A. (2004). Nicotine-induced changes in neurotransmitter levels in brain areas associated with cognitive function. *Neurochemical Research, 29*(9), 1779–1792.

Soiza-Reilly, M., & Commons, K. G. (2011). Glutamatergic drive of the dorsal raphe nucleus. *Journal of Chemical Neuroanatomy, 41*(4), 247–255.

Sprouse, J. S., & Aghajanian, G. K. (1987). Electrophysiological responses of serotoninergic dorsal raphe neurons to 5-HT$_{1A}$ and 5-HT$_{1B}$ agonists. *Synapse, 1*(1), 3–9.

Summers, K. L., & Giacobini, E. (1995). Effects of local and repeated systemic administration of (-)nicotine on extracellular levels of acetylcholine, norepinephrine, dopamine, and serotonin in rat cortex. *Neurochemical Research, 20*(6), 753–759.

Takada, Y., Urano, T., Ihara, H., & Takada, A. (1995). Changes in the central and peripheral serotonergic system in rats exposed to water-immersion restrained stress and nicotine administration. *Neuroscience Research, 23*(3), 305–311.

Tizabi, Y., Russell, L. T., Johnson, M., & Darmani, N. A. (2001). Nicotine attenuates DOI-induced head-twitch response in mice: implications for Tourette syndrome. *Progress in Neuro-Psychopharmacology & Biological Psychiatry, 25*(7), 1445–1457.

Toth, E., Sershen, H., Hashim, A., Vizi, E. S., & Lajtha, A. (1992). Effect of nicotine on extracellular levels of neurotransmitters assessed by microdialysis in various brain regions: role of glutamic acid. *Neurochemical Research, 17*(3), 265–271.

Vandermaelen, C. P., & Aghajanian, G. K. (1983). Electrophysiological and pharmacological characterization of serotonergic dorsal raphe neurons recorded extracellularly and intracellularly in rat brain slices. *Brain Research, 289*(1–2), 109–119.

Vázquez, J., Guzmán-Marín, R., Salín-Pascual, R., & Drucker-Colín, R. (1996). Transdermal nicotine effects on sleep and PGO spikes. *Brain Research, 737*, 317–320.

Vertes, R. P. (1991). A PHA-L analysis of ascending projections of the dorsal raphe nucleus in the rat. *Journal of Comparative Neurology, 313*(4), 643–668.

Wonnacott, S. (1997). Presynaptic nicotinic ACh receptors. *Trends in Neurosciences, 20*(2), 92–98.

Yasuda, K., Suemaru, K., Araki, H., & Gomita, Y. (2002). Effect of nicotine cessation on the central serotonergic systems in mice: involvement of 5-HT(2) receptors. *Naunyn-Schmiedeberg's Archives of Pharmacology, 366*(3), 276–281.

Zaniewska, M., McCreary, A. C., Wydra, K., & Filip, M. (2010). Effects of serotonin (5-HT)2 receptor ligands on depression-like behavior during nicotine withdrawal. *Neuropharmacology, 58*(7), 1140–1146.

Chapter 15

Critical Role of Cannabinoid CB1 Receptors in Nicotine Reward and Addiction

Ameneh Rezayof, Shiva Hashemizadeh

Department of Animal Biology, School of Biology, College of Science, University of Tehran, Tehran, Iran

Abbreviations

2-AG 2-Arachidonylglycerol
AEA Anandamide
CB1 Cannabinoid receptor type 1
CB-R Cannabinoid receptor
CNS Central nervous system
CPP Conditioned place preference
DA Dopamine
DSE Depolarization-induced suppression of excitation
DSI Depolarization-induced suppression of inhibition
eCB Endocannabinoid
ECS Endocannabinoid system
FAAH Fatty acid amide hydrolase
LTP Long-term potentiation
mGluR Metabotropic glutamate receptor
nAChR Nicotinic acetylcholine receptor
NAc Nucleus accumbens
NADA N-arachidonoyl dopamine
NMDA N-methyl-D-aspartate
PFC Prefrontal cortex
THC Δ^9-Tetrahydrocannabinol
VTA Ventral tegmental area

INTRODUCTION

Nicotine is the principal component alkaloid of tobacco that acts on the brain to induce rewarding effects, particularly following a stressful or aversive event. Owing to significant overlapping anatomical distribution, there seems to be a functional interaction between endocannabinoids (eCBs) and the cholinergic nicotinic system at both cellular and neuronal network levels. The overlapping of the eCB and cholinergic nicotine systems contributes to some of the addictive properties of nicotine. Numerous studies have reported changes in the levels of endogenous cannabinoids following nicotine administration, suggesting a regulatory role for nicotine receptors in endogenous cannabinoid release (Pacher & Kunos, 2013). Among researchers, it has long been believed that pharmacologically active cannabinoids show neuromodulatory action on acetylcholine release and turnover in various brain regions involved in nicotine-induced behavioral responses. A significant role of eCB transmission in synaptic plasticity in nicotine reward and addiction may be mediated by the abundance of cannabinoid receptors (CB-Rs) in brain structures. It should be noted that neuroplastic changes in the reward region of the mesocorticolimbic system, including the ventral tegmental area (VTA), nucleus accumbens (NAc), prefrontal cortex (PFC), hippocampus, and amygdala, participate in mediating nicotine addiction (Sidhpura & Parsons, 2011). eCBs mediate retrograde synaptic signaling to inhibit neurotransmitter release by the presynaptic CB-Rs at both excitatory and inhibitory synapses with short- and long-term effects. Evidence suggests that the rewarding effects of nicotine may depend in part on eCB-induced retrograde signaling on synaptic transmission (Murray, Wells, Lyford, & Bevins, 2009). Given the facts that nicotine administration can influence synaptic transmission and repeated or prolonged drug use leads to long-lasting changes in synaptic plasticity, it is theorized that drug-induced synaptic plasticity mediates the development of nicotine dependence (Feduccia, Chatterjee, & Bartlett, 2012). This chapter focuses on recent advances in research and theory regarding the involvement of eCB signaling at central synapses via cannabinoid type 1 (CB1) receptors in nicotine reward and addiction.

NICOTINE REWARD AND DEPENDENCE

Nicotine is responsible for the addictive properties of cigarettes. Nicotine exerts its reward and reinforcement actions by activating nicotinic acetylcholine receptors (nAChRs). In the central nervous system (CNS), nine α-subunits (α2–α10) and three β-subunits (β2–β4) assemble to form hetero- (α- and β-subunits) or homo- (α-subunit only) pentameric ionotropic nAChRs with different properties including channel open time, ion permeability, and rate of desensitization. The binding of nicotine to an nAChR causes a Ca^{2+}, Na^+, or K^+ influx (depending on the nAChR subtype) by rapidly opening the central channel. The entry of cations (positively charged ions) leads to rapid excitatory postsynaptic potential. This is a transient depolarization of a neuron, while calcium flux through homomeric α7 nAChRs, either directly or indirectly, is responsible for synaptic plasticity. Nicotine shows a high affinity for α4β2 nAChRs, which are the predominant nicotinic receptor subtype in the brain. The importance of this subtype in nicotine's reinforcing effects, and physical dependence on the drug, has been widely investigated. For example, nicotine reward could be induced in an α7-knocked-out mouse, but not in a mouse lacking the β2 nAChR subunit using conditioned place preference procedure Walters, Brown, Changeux, Martin, & Damaj, 2006.

Evidence suggests that nicotine abuse stimulates the mesocorticolimbic dopaminergic system, which originates from the VTA and projects to the NAc, the PFC, the hippocampus, and the amygdala (Mark, Shabani, Dobbs, & Hansen, 2011). The VTA nAChRs mediate reward and reinforcement effects of nicotine via stimulation of the mesolimbic dopamine (DA) neurons (Balfour, 2009). Several studies have demonstrated that activation or blockade of nAChRs in the VTA plays a critical role in nicotine's addictive properties (for review see Changeux, 2010). Behavioral studies have shown that the blockade of the VTA nAChRs inhibited systemic nicotine self-administration and nicotine-induced NAc DA release. (Gotti et al., 2010; Kenny & Markou, 2006). Pharmacologically distinct nAChR subtypes expressed in the VTA mediate nicotine-induced increase in DA release (Dani & De Biasi, 2001; Mansvelder, De Rover, McGehee, & Brussaard, 2003). It is important to note that the dopaminergic system may not be the only neurotransmitter responsible for nicotine reward. Other neurotransmitter systems, including cholinergic, glutamatergic, and γ-aminobutyric acid (GABA)-ergic, have been implicated in the formation of stimulus–nicotine associations (see Figure 1 to review the reward pathways). Since nAChRs are widely expressed in somatodendritic, axonal, and postsynaptic sites of various brain regions, nicotine administration may stimulate the release of most neurotransmitters throughout the brain. The VTA dopaminergic, GABAergic, and glutamatergic presynaptic terminals that synapse onto DA neurons express different subtypes of nAChRs. Nicotine increases glutamate release by activation of α7 nAChRs of the VTA glutamatergic terminals and also induces glutamatergic synaptic plasticity in the VTA DA neurons. Therefore, it appears that there is a functional interaction between α7 nAChRs and AMPA/NMDA receptors in the mediation of nicotinic self-administration, nicotine reward, tolerance, and dependence (Jin, Yang, Wang, & Wu, 2011). It is noteworthy that the activity of the VTA dopaminergic neurons also depends on inhibitory GABAergic neurotransmission, which leads to a decrease in the downstream release of DA in the NAc. Much of the current literature on the stimulatory effect of nicotine on the VTA DA neurons pays particular attention to the induction of long-term potentiation of the excitatory glutamatergic input and the depression of GABAergic transmission during nicotine exposure (Mansvelder & McGehee, 2002). On the other hand, recent studies provide evidence linking the eCB system to nicotine reward and dependence directly or indirectly via the regulation of cholinergic, glutamatergic, and GABAergic neurotransmission in the mesocorticolimbic system. Moreover, the eCB system is central for the control of DA-dependent behavior, which has a predominant role in motivational properties of the drug. A considerable amount of literature has been published on the role of the eCB system in modulating the reward and reinforcement effects of drugs, including nicotine (for review see Maldonado & de Fonseca, 2002).

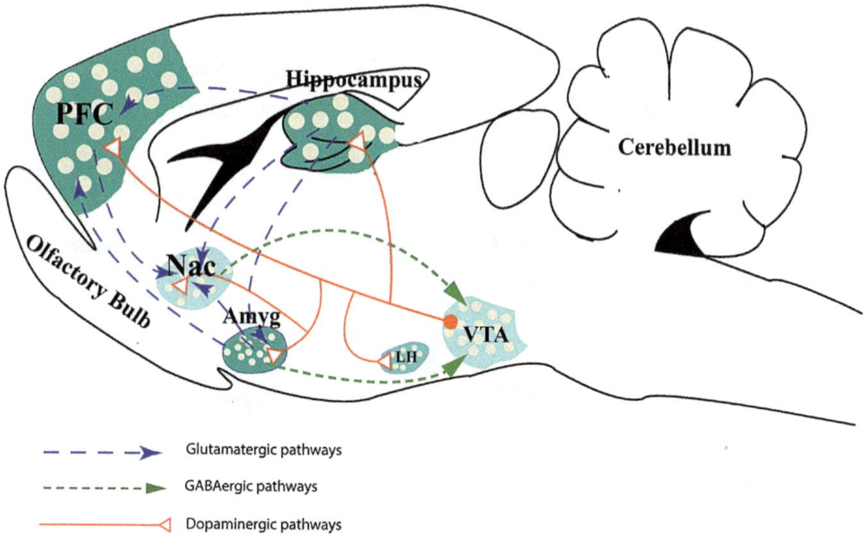

FIGURE 1 A set of simplified schematic diagrams of the excitatory/inhibitory circuits among the VTA, the other limbic structures, and the PFC with the distributions of cannabinoid and nAChRs in these areas. The VTA dopaminergic pathways (red lines) project to the lateral hypothalamus (LH), amygdala (Amyg), NAc, hippocampus, and prefrontal cortex (PFC). The VTA also receives glutamatergic afferents (blue lines) from the PFC and GABAergic (green lines) innervations from the NAc and amygdala. The blue lines represent excitatory afferents to the NAc from the hippocampus, amygdala, and mPFC. The PFC also receives glutamatergic afferents from the hippocampus and amygdala. The diagram shows coexpression of CB1 and nicotine receptors. Color gradient illustrates the changes in CB1 receptor density in the various rat brain regions involved in the nicotine reward. The gradient shows higher CB1 receptor level in the amygdala, hippocampus, and PFC and medium CB1 receptor level in the hypothalamus and also lower CB1 receptor level in the NAc and VTA. The dot pattern represents the expression of nAChRs in these regions. Higher density is represented by larger dots and smaller dots indicate a low density.

ENDOCANNABINOID SYSTEM AND SIGNALING

The eCB system consists of endogenous lipid ligands or eCBs, CB-Rs, and synthetic and metabolizing enzymes (Regehr, Carey, & Best, 2009). In contrast to classical neurotransmitters, which release from presynaptic neurons and stimulate, inhibit, or modulate the postsynaptic neuron, eCBs release from postsynaptic neurons and act as retrograde synaptic messengers of suppressing neurotransmitter release. See Figure 2 for an explanation of the molecular structures of eCBs. It is important to note that eCBs are synthesized in a calcium-dependent manner. Upon strong depolarization of the postsynaptic neurons, Ca^{2+} influx, via voltage-gated Ca^{2+} channels such as the P/Q-type and N-type Ca^{2+} channels, rapidly activates the enzymatic machinery for eCB biosynthesis. Considering that the intracellular Ca^{2+}-concentration is believed to be involved in the synthesis of eCBs, it seems that the activation of G_q-protein-coupled receptors such as group 1 metabotropic glutamate receptors can regulate eCB synthesis. A Ca^{2+}-dependent N-acyltransferase and N-acylphosphatidylethanolamine-hydrolyzing phospholipase D are responsible for synthesizing N-arachidonoylethanolamine (AEA), while phospholipase C and diacylglycerol lipase contribute to the biosynthesis of 2-arachidonylglycerol (2-AG). Evidence suggests that the enzymatic system is responsible for eCB biosynthesis, uptake, and degradation (Pertwee et al., 2010).

Since 1995, a large body of evidence has shown that eCBs are critical regulators of synaptic function. eCBs escape the postsynaptic neuron and bind to adjacent presynaptic CB1 receptors located on the GABAergic and glutamatergic axon terminals, to regulate neurotransmitter release. Research has shown that eCBs via CB1 receptors directly modulate the fine-tuning of dopaminergic neurons of the mesocorticolimbic pathway (Kano, Ohno-Shosaku, Hashimotodani, Uchigashima, & Watanabe, 2009). In recent decades, major progress has been made in our understanding of the origins and functions of eCB signaling as a neuromodulatory neurotransmitter in the CNS (Marsicano & Lutz, 2006) and an immunomodulation factor in peripheral tissue. Central eCBs may serve important functions in neural development (Fride, 2008), pain regulation (Guindon & Hohmann, 2009), memory formation (Heifets & Castillo, 2009), and the expression of emotional behaviors and reward (Lutz, 2009). The diverse functional and regulatory actions of cannabinoid ligands such as AEA and 2-AG can be mediated through the activation of CB-Rs. There are two main categories of CB-R, CB1/CB2 receptors and non-CB1/CB2 receptors, including the vanilloid receptor TRPV1 ion channel and the orphan receptor GPR55, referred to as cannabinoid receptor 3 (CB3) or GPR119 (Sharir & Abood, 2010). Recently new types of nuclear receptors, such as the peroxisome proliferator-activated receptors, have been added to the list of eCB receptors (Pertwee et al., 2010). It is noteworthy that AEA and 2-AG are putative endogenous ligands for all subtypes of CB-R. However, the ability of ligands to target the receptors differ. For example, AEA binds more readily to CB1 than to CB2 receptors, while 2-AG displays slightly greater efficacy than AEA with both CB1 and CB2 receptors (Pertwee, 2008). The human CB1 and CB2 receptors share 44% overall amino acid identity, whereas hGPR55 shows low amino acid identity to CB1 (13.5%) or CB2 (14.4%) receptors. Both CB1 receptors and GPR55 are expressed on neurons, whereas CB2 receptors are primarily found on microglia (Stella, 2010) and in peripheral tissue. It should be noted that other studies have identified other arachidonic acid derivatives such as 2-arachidonyl glyceryl ether (Noladin), O-arachidonoylethanolamine (virodhamine), and N-arachidonoyl dopamine, which are also

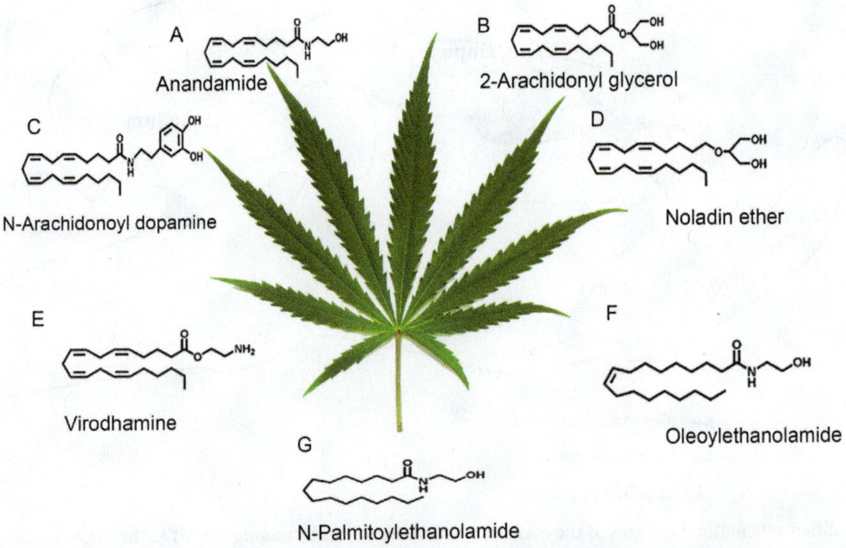

FIGURE 2 Molecular structure of eCBs. (A) Anandamide (N-arachidonoyl ethanolamine; AEA), which is synthesized from N-arachidonoyl phosphatidylethanolamine via multiple signaling pathways, binds to CB1 and CB2 receptors and is degraded by the fatty acid amide hydrolase enzyme. (B) 2-Arachidonyl glycerol (2-AG) is formed from membrane phospholipids through activating phospholipase Cβ/diacylglycerol lipase and acts as a retrograde messenger binding presynaptic CB1 receptors. (C) N-Arachidonoyl dopamine (NADA) is synthesized in dopaminergic terminals where tyrosine hydroxylase is located and binds as an agonist to the CB1 receptor or the transient receptor potential vanilloid type 1 channel. (D) Noladin ether (NE; 2-arachidonyl glyceryl ether), a stable analog of 2-AG, is a putative eCB that mediates appetite and eating behaviors via CB1 receptors. (E) Virodhamine (O-arachidonoyl ethanolamine), which is the ester of arachidonic acid and ethanolamine, acts as an agonist for the CB2 receptor and antagonist for the CB1 receptor. (F) Oleoylethanolamide (OEA; N-oleoyl ethanolamine) is the monounsaturated analog of AEA. It does not activate the CB-Rs. (G) N-Palmitoylethanolamide (PEA), a putative eCB, is an endogenous fatty acid amide that activates the cannabinoid-like G-coupled receptors GPR55 and GPR119.

considered eCBs; however, the physiological functions of these lipids have not been well studied (Piomelli, 2003).

CANNABINOID TYPE 1 RECEPTORS

CB1 receptors, which were first cloned in 1990 (Matsuda, Lolait, Brownstein, Young, & Bonner, 1990), are one of the most abundant metabotropic receptors in the mammalian brain with a highly heterogeneous pattern (see Figure 3 to review CB1 receptor signaling). Most of the biological effects of endo- and exocannabinoids in the CNS are associated with the activation of CB1 receptors. Brain mapping studies reveal ubiquitous distribution of CB1 receptors. Briefly, these receptors can be found in the cortex, the limbic system, the basal ganglia, and the cerebellum (for review see Svizenska, Dubovy, & Sulcova, 2008). At the subcellular level, CB1 receptors are primarily expressed in two major neuronal populations: inhibitory GABAergic interneurons and excitatory glutamatergic principal neurons (Marsicano & Kuner, 2008). Their localization in inhibitory GABAergic interneurons may be 3 to 10 times more than in glutamatergic principal neurons (Kawamura et al., 2006). The high level of expression of CB1 receptors at various presynaptic excitatory and inhibitory neurons implies that they play a prominent role in the release of several neurotransmitters such as glutamate (Figure 3), GABA, acetylcholine, norepinephrine, DA, and serotonin (Kano et al., 2009).

Signal transduction of CB-Rs in the brain can be activated by the principal psychoactive constituent of Cannabis sativa, Δ^9-tetrahydrocannabinol (THC), as well as by synthetic and endogenous ligands that facilitate multiple downstream signals through canonical G-protein-mediated pathways. CB1 receptors have seven transmembrane α-helical domains with an extracellular N-terminus and an intracellular C-terminus, which couples with $G_{i/o}$ proteins for modulating adenylyl cyclase, ion channels, and extracellular signal-regulated kinases (Castillo, Younts, Chávez, & Hashimotodani, 2012). The administration of CB1 receptor agonists has been shown to decrease cAMP production and Ca^{2+} conductance in the striatum (Wade, Tzavara, & Nomikos, 2004), while they increase potassium conductance and the activity of mitogen-activated protein kinases in the hippocampal neurons (Derkinderen et al., 2003). Based on these findings, it has been suggested that endogenous cannabinoid signaling through CB1 receptors induces either short-term or long-term suppression of vesicular neurotransmitter release (Brown, Brenowitz, & Regehr, 2003). Regional distribution of CB1-expressing neurons may be related to key functions of eCBs such as (1) autonomic function, (2) motor coordination, (3) sensation, and (4) cognition.

THE ROLE OF CB1 RECEPTORS IN BRAIN REWARD PROCESSING IN NICOTINE ADMINISTRATION

There is a general agreement among researchers that the VTA and the NAc are key neuroanatomical regions in the brain's rewarding circuitry. The reciprocal projections from the VTA to the NAc and projections from the NAc to other corticolimbic areas, including the PFC, the hippocampus, and the amygdala, mediate the drug-use reward and reinforcement circuitry. Although mRNA and protein levels of CB1 receptors within the VTA and the NAc are low, it is believed that these receptors play a critical role in reward-relevant aspects of cannabinoid exposure. A 2013 study targeting the VTA and the NAc CB1-expressing neurons found that the CB1 receptors of the VTA, rather than the NAc, control nicotine reinforcement and subsequent nicotine-use behavior (Simonnet, Cador, & Caille, 2013).

FIGURE 3 eCB retrograde signaling through presynaptic CB1 receptors in a glutamatergic synapse. Postsynaptic activities of metabotropic glutamate receptors induce a conformational change in the receptors, which activates a G_q protein to form diacylglycerol (DAG) and inositol 1,4,5-trisphosphate (IP3) by phospholipase C (PLC) activation. 2-AG is synthesized from DAG by a DAG lipase, and a PLC- or PLD-dependent pathway may synthesize AEA from N-arachidonoyl phosphatidylethanolamine. eCBs, as a retrograde messenger, stimulate CB1 receptors at the presynaptic axon terminal to inhibit neurotransmitter release by: (1) blocking voltage-sensitive Ca^{2+} and K^+ channels, (2) inhibiting adenylyl cyclase (AC) activity and reducing cAMP levels, and (3) activating protein kinase cascades such as the mitogen-activated protein kinase (MAPK) pathway, glutamate (Glu), metabotropic receptor (M.R.), phosphatidylinositol 4,5-bisphosphate (PIP2), protein kinase C (PKC), and ionotropic receptor (I.R.).

On the other hand, the amygdala regulates both reward-related learning and aversive fear memory. Reciprocal projections between the NAc and the amygdala are believed to be important for the formation of stimulus–reward association. In terms of the amygdala CB1 receptor modulation, this region shows a wide distribution of CB1 receptors (see Figure 1). Social play in adolescent rats results in increased levels of phosphorylated CB1 receptors in the amygdala and the NAc, suggesting eCBs may modulate the rewarding properties of social interactions (Trezza et al., 2012).

Numerous studies have shown that learning and memory processes play a key role in addiction. In view of the facts that the hippocampus expresses high-density levels of the CB1 receptor and the hippocampus also sends a major output to the reward system, it appears that the eCB system may be involved in reward-related learning and addiction. Moreover, it has been suggested that eCBs control consumption of nicotine by affecting the brain's reward system. In light of the research literature that has been presented thus far, it can be inferred that the effects of eCBs depend on the functional balance between the inhibitory and the excitatory afferents in the mesolimbic dopaminergic reward circuitry. Therefore, the specific interactions of the nicotine and eCB systems may affect reward-related behaviors.

Neurobiological studies also show that nicotine affects reward pathways via an increase in DA function in the VTA. As noted earlier, the activation of nAChRs of the VTA dopaminergic neurons increases the firing rate and phasic burst activity of these neurons and subsequently leads to DA release in the PFC and the NAc. A substantial body of research indicates that nicotine reward depends on the modulation of DA neurons in the mesocorticolimbic pathways. Considering that the dopaminergic terminals lack CB1 receptors (Julian et al., 2003; Lupica & Riegel, 2005), it is hypothesized that eCBs may mediate nicotine-induced modulation of dopaminergic neurons via other neurotransmitter systems. Indeed, presynaptic CB1 receptors, which are located on the VTA GABAergic and glutamatergic terminals, can modulate DA neuron firing. As shown in Table 1, the potential of eCB system modulation in treating nicotine addiction has been highlighted in animal and human studies using various experimental models. There is an increasing concern that using cannabis might make a person more vulnerable to nicotine addiction. On the other hand, there are several studies that report that inactivation of CB1 receptors by antagonist administration, or genetic knockout of these receptors, decreased the reinforcing effects of nicotine (Castane et al., 2002). These experimental animal manipulations have also been

TABLE 1 Effects of CB1 Receptor Antagonists on Nicotine-Induced Behaviors

Experimental Model	Cannabinoid	Drug	Main Result	References
Discrimination	CB1 antagonist receptor	SR141716 Rimonabant	No effect in nicotine discrimination (in rats)	Cohen et al. (2002) Le Foll and Goldberg (2004)
Self-administration	CB1 antagonist receptor	SR141716	Suppressed intravenous nicotine, self-administration (in rats)	Cohen et al. (2002)
		AM251	Reduces nicotine intravenous self-administration (in rats)	Shoaib (2008)
		SLV330	Reducing nicotine self-administration (in rats)	De Bruin et al. (2011)
			Intra-VTA infusions of AM 251 dose dependently reduced nicotine intravenous self-administration (in rats)	Simonnet et al. (2013)
			Intra-NAc injection did not alter nicotine behavior (in rats)	
Expression of CPP	CB1 antagonist receptor	SR141716	Rimonabant blocked preferences for Nicotine-paired environment (in rats)	Le Foll and Goldberg (2004) Forget, Hamon, and Thiebot (2005)
		AM251	Intra-BLA administration of AM251 inhibited the acquisition of nicotine-induced place preference (in mice)	Hashemizadeh et al. (2014)
	CB1 knockout (KO)		Failed to display nicotine-induced place preference (in mice)	Merritt et al. (2008) Castane, Berrendero, and Maldonado (2005)
Nicotine withdrawal	CB1 knockout (KO)	SR141716	No change in nicotine withdrawal intensity (in mice)	Castane et al. (2002, 2005) Merritt et al. (2008)
	CB1 antagonist receptor		Failed to precipitate somatic signs in nicotine-dependent (in mice)	Balerio, Aso, Berrendero, Murtra, and Maldonado (2004)
			Reduced responding maintained by nicotine-associated cues (in mice)	Cohen et al. (2005)

shown to inhibit nicotine-induced place preference. On the other hand, chronic exposure to nicotine induces changes in eCB transmission. Gonzalez et al. (2002) reported that chronic administration of nicotine for 1 week altered the tissue content of eCBs in various brain regions, including the limbic forebrain, brainstem, hippocampus, striatum, and cerebral cortex. Moreover, subchronic administration of nicotine increases hippocampal CB1 receptor levels and downregulates the striatal CB1 receptors. These effects are probably similar in both genders (Marco et al., 2007). Nicotine self-administration also changes the brain eCB levels of AEA in the VTA, showing that the changes in eCB signaling may depend on alterations in their nicotine-induced biosynthesis (Buczynski, Polis, & Parsons, 2012). It is believed that nicotine exposure alters CB1 receptor expression and eCB levels in the brain areas associated with drug addiction. It appears that alternation of AEA levels may regulate nicotine receptors. In addition, the increase in AEA by pharmacological inhibition of fatty acid amide hydrolase (FAAH) leads to increased nicotine-induced conditioned place preference (CPP) in mice (Merritt et al., 2008; Muldoon, Lichtman, Parsons, & Damaj, 2013). In contrast to this, it has been reported that the inhibition of FAAH by URB597 injection blocked both CPP and self-administration of nicotine in rats (Forget, Coen, & Le Foll, 2009; Scherma et al., 2008). Also, this treatment blocked DA release in the NAc. Interestingly, the inhibition of the putative AEA membrane transporter by the administration of AM404 reduced nicotine-induced CPP (Scherma et al., 2012). Therefore, it is likely that eCBs modulate synaptic plasticity in the mesocorticolimbic reward system by interacting with nicotine signaling pathways.

EFFECTS OF CB1 RECEPTOR AGONISTS AND ANTAGONISTS ON NICOTINE REWARD

The capability of cannabinoid ligands to modulate excitatory and inhibitory transmission in the CNS is well established. As previously noted, these ligands are released from the postsynaptic neurons and travel backward to the presynaptic receptors inhibiting neurotransmitter release. Electrophysiological studies have also shown that eCBs mediate some forms of short-term retrograde synaptic plasticity, which are known as depolarization-induced suppression of inhibition and depolarization-induced suppression of excitation, in numerous brain regions (for review see Kano et al., 2009). It is important to note that CB1 receptor agonists have biphasic activity on cholinergic neurotransmission. Research has indicated that low doses of CB1 receptor agonists induced a prolonged inhibition of acetylcholine (ACh) efflux in various neuroanatomical sites such as the hippocampus (Nava, Carta, Battasi, & Gessa, 2000), while high doses of the agonists increased ACh release in the hippocampus (Degroot et al., 2006). To support the involvement of CB1 receptors in cholinergic neurotransmission, the investigators showed that the effects of cannabinoid agonists could be reversed by the potent and selective CB1 receptor antagonist SR141716A. On the other hand, coadministration of nicotine and THC (the major active constituent of marijuana) or WIN55,212-2 (a synthetic cannabinoid) increased the rewarding effect of nicotine in a place preference paradigm (Hashemizadeh, Sardari, & Rezayof, 2014) (see Figure 4). In keeping with the idea that CB1 receptors have a critical role in this interaction, Gamaleddin et al. (2012) showed that the administration of WIN55,212-2, a synthetic nonselective CB1/CB2 receptor agonist, increased nicotine-induced self-administration. Moreover, systemic administration of SR141716A (selective antagonist of brain CB-Rs) reduced both nicotine-induced self-administration and CPP in rodents (Cohen, Perrault, Griebel, & Soubrie, 2005; Le Foll & Goldberg, 2004; Shoaib, 2008). The blockade of CB1 receptors by the antagonist administration reduced nicotine-induced DA release in the NAc (Cohen, Perrault, Voltz, Steinberg, & Soubrie, 2002; Merrit et al., 2008). Disruption of eCB signaling in the VTA by microinjection of a selective CB1 antagonist, AM251, blocked the reinforcing properties of nicotine. Moreover, intra-basolateral amygdala (BLA) administration of rimonabant, a potent and selective CB1 receptor antagonist, dose-dependently reduced nicotine-seeking behavior (Kodas, Cohen, Louis, & Griebel, 2007). In agreement with other CB1 antagonists, SLV330 decreased nicotine self-administration and cue-induced behavior (De Bruin et al., 2011). In view of all that has been mentioned thus far, one may infer that the inhibition of CB-Rs may be an effective means of treating nicotine dependence. In a major study, Chen et al. (2008) reported that the CB1 gene has been associated with nicotine dependence. In previous human studies, a clinical trial smoking cessation study was conducted using the administration of a CB1 receptor antagonist. Some studies with rimonabant and tobacco use show promising results for an effective role of rimonabant in antismoking treatment. It is important to note that rimonabant increased rates of smoking cessation in this type of trial (Cahill & Ussher, 2011). In total four clinical trials were completed to directly measure the efficacy of rimonabant in smoking cessation. The four Studies with Rimonabant and Tobacco Use (STRATUS) include STRATUS Europe, STRATUS United States, STRATUS Worldwide, and STRATUS META. These studies are outlined in Table 2.

CB1 RECEPTOR LIGANDS AND nAChRs

An overlapping distribution of CB1 receptors and nAChRs has been reported in several brain structures such as the hippocampus and the amygdala (Picciotto, Caldarone, King, & Zachariou, 2000). This overlapping suggests the possibility of functional interactions between these two receptor mechanisms as depicted in Figure 2. There has been growing interest in these interactions. These studies have found that endogenous cannabinoids and synthetic agonists, such as WIN55,212-2, act directly on nAChRs and inhibit α7 nicotinic receptors. It is noteworthy that cannabinoids specifically bind nAChRs, which are located on pre- and postsynaptic neurons to modulate excitatory and inhibitory synaptic neurotransmission. There is evidence that AEA modulates GABA neurotransmission through an effect on α7 nicotinic receptors. Studies have shown that CB1 receptor antagonists could not reverse some pharmacological effects of AEA. In addition, it has been reported that the increased AEA level boosts nicotine withdrawal, through non-CB1 mechanisms. Regulation mechanism of nAChRs by cannabinoid ligands may represent new therapeutic options for patients who suffer from nicotine abstinence (for review see Oz, Al Kury, Keun-Hang, Mahgoub, & Galadari, 2014). On the other hand, experimental studies have shown that the blockage of α7 nicotinic receptors abolished cannabis-dependent behaviors. Further studies are needed to elucidate the mechanisms of the cannabinoid ligands that modulate the activity of nAChRs in nicotine-seeking behavior.

APPLICATIONS TO OTHER ADDICTIONS AND SUBSTANCE MISUSE

This chapter provides information on the functional interaction between the eCB system and the central nicotinic receptors at both cellular and neuronal network levels in nicotine reward and

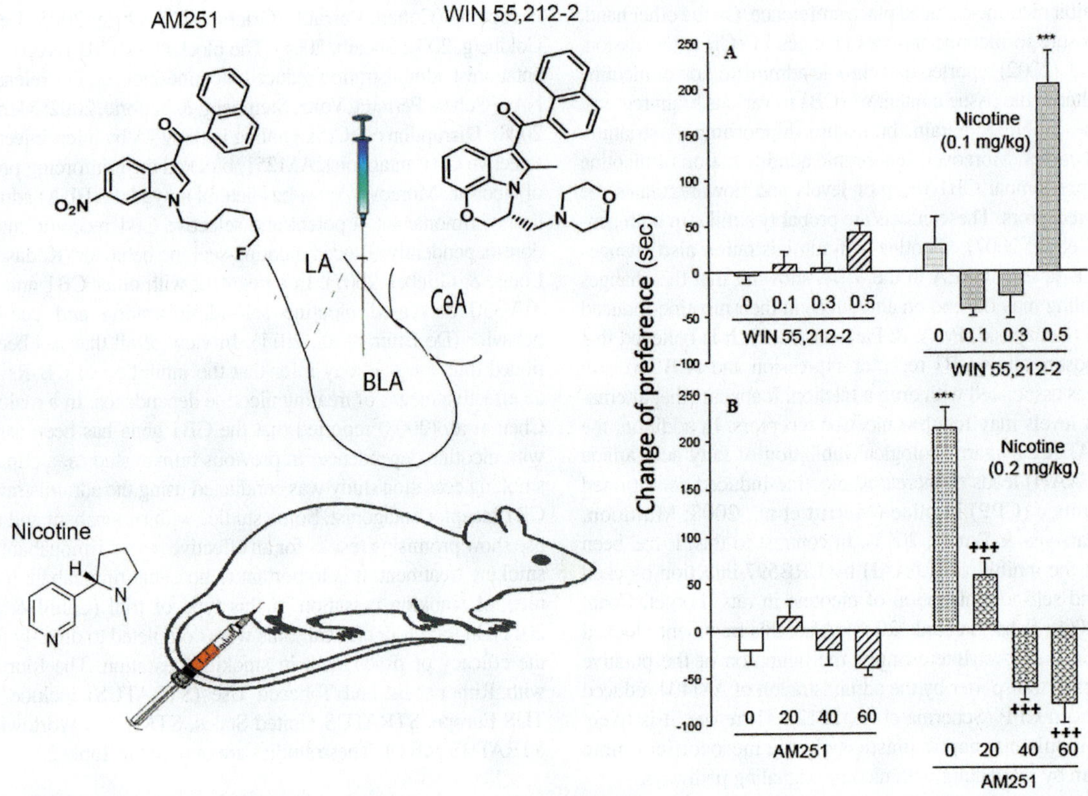

FIGURE 4 Effects of bilateral microinjections of a CB1 receptor agonist (WIN55,212-2) or antagonist (AM251) into the basolateral amygdala (BLA) alone or with nicotine on the acquisition of CPP. In Graph A, the animals received intra-BLA microinjections of WIN55,212-2 (0, 0.1, 0.3, and 0.5 μg/rat) with or without an ineffective dose of nicotine (0.1 mg/kg; subcutaneous (sc)) once daily on a 3-day schedule. In Graph B, the animals received intra-BLA microinjections of AM251 (0, 20, 40, and 60 ng/rat) with or without an effective dose of nicotine (0.2 mg/kg; sc) once daily on a 3-day schedule (conditioning phase). The change of preference was assessed as the difference between the time spent on the day of testing and the time spent on the day of the preconditioning session. The data are expressed as the mean ± SEM of six animals per group. Graph A: ***$p<0.01$, difference from nicotine control group. Graph B: ***$p<0.001$, difference from control group; ++$p<0.01$, +++$p<0.001$ difference from nicotine control group. CeA, central amygdala; LA, lateral amygdala. See Hashemizadeh et al. (2014) for more detail.

TABLE 2 A Series of Studies Designed to Demonstrate the Benefits of Rimonabant for Smoking Cessation

Title of the Study
STRATUS Europe
Comparison of the efficacy and safety of two oral doses of rimonabant, 5 or 20 mg/day, versus placebo, as an aid to smoking cessation
STRATUS United States
Comparison of the efficacy and safety of two oral doses of rimonabant, 5 or 20 mg/day, versus placebo, as an aid to smoking cessation
STRATUS Worldwide
Efficacy and safety of rimonabant as an aid to maintenance of smoking cessation
STRATUS META
Comparison of efficacy and safety of rimonabant 20 mg/day versus placebo in smoking cessation
These were the STRATUS (Studies with Rimonabant and Tobacco Use) trials. Two trials of identical design (STRATUS Europe 2006; STRATUS United States 2006) evaluated rimonabant for smoking cessation, and one trial evaluated rimonabant as an aid to relapse prevention (STRATUS Worldwide 2005).

addiction. A large body of evidence also indicates that there is cross talk among eCBs and other drugs of abuse, such as ethanol (alcohol), opioids, and psychostimulants, in different cognitive and physiological processes associated with addiction. The eCB system via the CB1 receptors plays a critical role in the activating effects of the drugs on the mesocorticolimbic dopaminergic reward system, which is involved in reward, motor control, cognition, and memory formation. It has been shown that acute administration of ethanol inhibits eCB signaling and its chronic administration downregulates CB1 receptors in the brain reward sites of laboratory animals. In contrast, the administration of CB1 receptor agonists increases ethanol-seeking and self-administration during relapse and maintenance conditions. In addition, the involvement of CB1 receptor mechanisms in the rewarding/reinforcing properties of opiates has been reported in several investigations, suggesting a correlation between the eCB and the opioid systems via the activation of the dopaminergic system or facilitation of their signal transduction mechanisms. On the other hand, cannabis is frequently coabused with 3,4-ethylenedioxymethamphetamine (MDMA), or ecstasy, in an effort to relieve MDMA-induced anhedonia, dysphoria, and depression. It should be noted that MDMA and cannabis produce opposite effects on neurotransmitter systems when administered alone or independent of one another. For example, MDMA induces rewarding effects by increasing the release of serotonin and DA, while cannabis inhibits the release of various neurotransmitters such as glutamate and DA via presynaptic CB1 receptor activation. Therefore, further studies would be needed to investigate the complex relationship between MDMA and cannabis in drug dependency. In addition, considering that studies using CB1-knockout mice have revealed the crucial role of CB1 receptors in drug addiction, the potential utility of CB-R antagonists might be considered as a pharmacological tool to treat drug dependence. Taken together, more information on CB1 receptor mechanisms would help the field to establish a greater degree of accuracy on illuminating the drug dependency problem.

DEFINITION OF TERMS

Anandamide An endogenous cannabinoid neurotransmitter, which is synthesized from *N*-arachidonoyl phosphatidylethanolamine by multiple pathways and also named *N*-arachidonoylethanolamine or AEA.

Brain reward system A set of brain regions that are functionally involved in reward and reinforcement effects of psychoactive drugs. These regions include dopaminergic neurons in the VTA, NAc, PFC, hippocampus, and amygdala.

Cannabinoid receptors A class of G-protein-coupled receptors in the brain, which mediate a variety of physiological processes including reward and reinforcement, feeding, pain modulation, learning, and memory.

The cannabinoid receptor type 1 A G-protein-coupled CB-R that possesses seven transmembrane domains connected by six extra- and intracellular loops with an extracellular N-terminal tail and an intracellular C-terminal tail. It is a presynaptic receptor that regulates neurotransmitter release.

Cation A positively charged ion.

Conditioned place preference A standard preclinical behavioral model to measure the rewarding and aversive effects of a drug on the basis of Pavlovian conditioning.

Dopamine A neurotransmitter produced in the basal forebrain and diencephalon that controls movement, emotional response, and ability to experience pleasure. It plays a critical role in addiction.

Drug self-administration in animals An experimental procedure to study addiction under controlled conditions. In this method, a laboratory animal presses a lever for delivering a dose of a drug, such as nicotine or cocaine.

Endocannabinoid system A ubiquitous lipid signaling system that consists of eCBs and related enzymes and their receptors that is essential in a variety of physiological processes and mediates the drug's rewarding effects.

Nicotinic acetylcholine receptors An excitatory ligand-gated ion channel in neuron and skeletal muscle cell membranes, which is permeable to Na^+, Ca^{2+}, and K^+.

Nucleus accumbens A subcortical region, also known as the pleasure center, involved in the reward and reinforcement circuit and control planned movement.

Δ^9-Tetrahydrocannabinol A major psychoactive component in the cannabis plant that exerts its effects only after binding to CB-Rs.

Ventral tegmental area A collection of dopaminergic neurons in the center of the midbrain with important roles in motivation, cognition, and drug addiction.

KEY FACTS

Key Facts of the Brain Reward System

The mesocorticolimbic dopaminergic system is the main reward pathway that detects and interprets rewarding stimuli such as food, sex, and drugs of abuse. This system includes the VTA, which projects dopaminergic afferents to the NAc; the PFC; the hippocampus; and the amygdala. The activation of the dopaminergic pathway through interactions with a variety of neurotransmitter systems leads to drug-seeking and drug-taking behaviors.

Key Facts of Drug Addiction

Addiction is a major public health problem that affects many aspects of an individual's life. A drug addiction is a chronic relapsing disease that is often associated with serious negative consequences. Drugs of abuse produce both rewarding and reinforcing effects on the CNS. Repeated drug use leads to unwanted changes in the brain, including drug-induced dependence, sensitization, and tolerance.

Key Facts of Retrograde Signaling

Several types of messengers such as lipids, gases, peptides, and growth factors act as retrograde signals to change the function of presynaptic neurons. The messengers are released from the somatodendritic domain of a postsynaptic neuron and bind to presynaptic receptors to modify the neurotransmitter release mechanisms of the afferent axon terminal.

Key Facts of Nicotine

Nicotine, as a powerful addictive drug, induces psychological and physical dependence on tobacco. Binding of nicotine to the central nAChRs increases DA levels in the brain reward pathway. Nicotine administration leads to changes in the level of the eCB AEA in various rat brain regions.

SUMMARY POINTS

- Nicotine produces rewarding effects by interacting with nAChRs in the brain reward system.
- The eCB system, via CB1 receptors, modulates the reinforcing/rewarding effects of nicotine.
- eCBs, as retrograde synaptic messengers, are released from postsynaptic neurons to suppress neurotransmitter release.

- Nicotine administration changes the brain eCB levels, including AEA, in the VTA.
- There is a functional interaction between nicotinic and CB-R mechanisms.
- eCBs mediate synaptic plasticity in the reward system by interacting with nicotine signaling pathways.
- There is an overlapping distribution of CB1 receptors and nAChRs in several brain regions.
- Blockage of CB1 receptors, or the receptor knockout, decreases nicotine's reinforcing effects.
- Regulation of nAChRs by eCB mechanisms may open the door to a new therapy for smoking cessation.

ACKNOWLEDGMENT

The authors thank Dr. Dessa Bergen-Cico and Dr. Mahmoud Efatmaneshnik for their assistance in the preparation of the manuscript.

REFERENCES

Balerio, G. N., Aso, E., Berrendero, F., Murtra, P., & Maldonado, R. (2004). Delta9-tetrahydrocannabinol decreases somatic and motivational manifestations of nicotine withdrawal in mice. *European Journal of Neuroscience, 20*, 2737–2748.

Balfour, D. J. (2009). The neuronal pathways mediating the behavioral and addictive properties of nicotine. *Handbook of Experimental Pharmacology, 209*–233.

Brown, S. P., Brenowitz, S. D., & Regehr, W. G. (2003). Brief presynaptic bursts evoke synapse-specific retrograde inhibition mediated by endogenous cannabinoids. *Nature Neuroscience, 6*, 1048–1057.

Buczynski, M. W., Polis, I. Y., & Parsons, L. H. (2012). The volitional nature of nicotine exposure alters anandamide and oleoylethanolamide levels in the ventral tegmental area. *Neuropsychopharmacology, 38*, 574–584.

Cahill, K., & Ussher, M. H. (2011). Cannabinoid type 1 receptor antagonists for smoking cessation. *The Cochrane Database of Systematic Reviews, 3*.

Castane, A., Berrendero, F., & Maldonado, R. (2005). The role of the cannabinoid system in nicotine addiction. *Pharmacology Biochemistry and Behavior, 81*, 381–386.

Castane, A., Valjent, E., Ledent, C., Parmentier, M., Maldonado, R., & Valverde, O. (2002). Lack of CB1 cannabinoid receptors modifies nicotine behavioural responses, but not nicotine abstinence. *Neuropharmacology, 43*, 857–867.

Castillo, P. E., Younts, T. J., Chávez, A. E., & Hashimotodani, Y. (2012). Endocannabinoid signaling and synaptic function. *Neuron, 76*, 70–81.

Changeux, J. P. (2010). Nicotine addiction and nicotinic receptors: lessons from genetically modified mice. *Nature Review: Neuroscience, 11*, 389–401.

Chen, X., Williamson, V. S., An, S.-S., Hettema, J. M., Aggen, S. H., Neale, M. C., & Kendler, K. S. (2008). Cannabinoid receptor 1 gene association with nicotine dependence. *Archives of General Psychiatry, 65*, 816–823.

Cohen, C., Perrault, G., Griebel, G., & Soubrie, P. (2005). Nicotine-associated cues maintain nicotine-seeking behavior in rats several weeks after nicotine withdrawal: reversal by the cannabinoid (CB1) receptor antagonist, rimonabant (SR141716). *Neuropsychopharmacology, 30*, 145–155.

Cohen, C., Perrault, G., Voltz, C., Steinberg, R., & Soubrie, P. (2002). SR141716, a central cannabinoid (CB(1)) receptor antagonist, blocks the motivational and dopamine-releasing effects of nicotine in rats. *Behavioural Pharmacology, 13*, 451–463.

Dani, J. A., & De Biasi, M. (2001). Cellular mechanisms of nicotine addiction. *Pharmacology Biochemistry and Behavior, 70*, 439–446.

De Bruin, N., Lange, J., Kruse, C., Herremans, A., Schoffelmeer, A., Van Drimmelen, M., & De Vries, T. (2011). SLV330, a cannabinoid CB1 receptor antagonist, attenuates ethanol and nicotine seeking and improves inhibitory response control in rats. *Behavioural Brain Research, 217*, 408–415.

Degroot, A., Köfalvi, A., Wade, M. R., Davis, R. J., Rodrigues, R. J., Rebola, N., ... Nomikos, G. G. (2006). CB1 receptor antagonism increases hippocampal acetylcholine release: site and mechanism of action. *Molecular Pharmacology, 70*, 1236–1245.

Derkinderen, P., Valjent, E., Toutant, M., Corvol, J.-C., Enslen, H., Ledent, C., ... Girault, J.-A. (2003). Regulation of extracellular signal-regulated kinase by cannabinoids in hippocampus. *The Journal of Neuroscience, 23*, 2371–2382.

Feduccia, A. A., Chatterjee, S., & Bartlett, S. E. (2012). Neuronal nicotinic acetylcholine receptors: neuroplastic changes underlying alcohol and nicotine addictions. *Frontiers in Molecular Neuroscience, 5*, 83.

Forget, B., Coen, K. M., & Le Foll, B. (2009). Inhibition of fatty acid amide hydrolase reduces reinstatement of nicotine seeking but not break point for nicotine self-administration–comparison with CB(1) receptor blockade. *Psychopharmacology (Berlin), 205*, 613–624.

Forget, B., Hamon, M., & Thiebot, M. H. (2005). Cannabinoid CB1 receptors are involved in motivational effects of nicotine in rats. *Psychopharmacology (Berlin), 181*, 722–734.

Fride, E. (2008). Multiple roles for the endocannabinoid system during the earliest stages of life: pre- and postnatal development. *Journal of Neuroendocrinology, 20*(Suppl. 1), 75–81.

Gamaleddin, I., Wertheim, C., Zhu, A. Z., Coen, K. M., Vemuri, K., Makryannis, A., ... Le Foll, B. (2012). Cannabinoid receptor stimulation increases motivation for nicotine and nicotine seeking. *Addiction Biology, 17*, 47–61.

Gonzalez, S., Cascio, M. G., Fernandez-Ruiz, J., Fezza, F., Di Marzo, V., & Ramos, J. A. (2002). Changes in endocannabinoid contents in the brain of rats chronically exposed to nicotine, ethanol or cocaine. *Brain Research, 954*, 73–81.

Gotti, C., Guiducci, S., Tedesco, V., Corbioli, S., Zanetti, L., Moretti, M., ... Clementi, F. (2010). Nicotinic acetylcholine receptors in the mesolimbic pathway: primary role of ventral tegmental area $\alpha 6\beta 2^*$ receptors in mediating systemic nicotine effects on dopamine release, locomotion, and reinforcement. *The Journal of Neuroscience, 30*, 5311–5325.

Guindon, J., & Hohmann, A. G. (2009). The endocannabinoid system and pain. *CNS & Neurological Disorders Drug Targets, 8*, 403–421.

Hashemizadeh, S., Sardari, M., & Rezayof, A. (2014). Basolateral amygdala CB1 cannabinoid receptors mediate nicotine-induced place preference. *Progress in Neuropsychopharmacology & Biological Psychiatry, 51*, 65–71.

Heifets, B. D., & Castillo, P. E. (2009). Endocannabinoid signaling and long-term synaptic plasticity. *Annual Review of Physiology, 71*, 283–306.

Jin, Y., Yang, K., Wang, H., & Wu, J. (2011). Exposure of nicotine to ventral tegmental area slices induces glutamatergic synaptic plasticity on dopamine neurons. *Synapse, 65*, 332–338.

Julian, M., Martin, A., Cuellar, B., Rodriguez De Fonseca, F., Navarro, M., Moratalla, R., & Garcia-Segura, L. M. (2003). Neuroanatomical relationship between type 1 cannabinoid receptors and dopaminergic systems in the rat basal ganglia. *Neuroscience, 119*, 309–318.

Kano, M., Ohno-Shosaku, T., Hashimotodani, Y., Uchigashima, M., & Watanabe, M. (2009). Endocannabinoid-mediated control of synaptic transmission. *Physiological Reviews, 89*, 309–380.

Kawamura, Y., Fukaya, M., Maejima, T., Yoshida, T., Miura, E., Watanabe, M., ... Kano, M. (2006). The CB1 cannabinoid receptor is the major cannabinoid receptor at excitatory presynaptic sites in the hippocampus and cerebellum. *The Journal of Neuroscience, 26*, 2991–3001.

Kenny, P. J., & Markou, A. (2006). Nicotine self-administration acutely activates brain reward systems and induces a long-lasting increase in reward sensitivity. *Neuropsychopharmacology, 31*, 1203–1211.

Kodas, E., Cohen, C., Louis, C., & Griebel, G. (2007). Cortico-limbic circuitry for conditioned nicotine-seeking behavior in rats involves endocannabinoid signaling. *Psychopharmacology (Berlin), 194*, 161–171.

Le Foll, B., & Goldberg, S. R. (2004). Rimonabant, a CB1 antagonist, blocks nicotine-conditioned place preferences. *Neuroreport, 15*, 2139–2143.

Lupica, C. R., & Riegel, A. C. (2005). Endocannabinoid release from midbrain dopamine neurons: a potential substrate for cannabinoid receptor antagonist treatment of addiction. *Neuropharmacology, 48*, 1105–1116.

Lutz, B. (2009). Endocannabinoid signals in the control of emotion. *Current Opinion in Pharmacology, 9*, 46–52.

Maldonado, R., & de Fonseca, F. R. (2002). Cannabinoid addiction: behavioral models and neural correlates. *The Journal of Neuroscience, 22*, 3326–3331.

Mansvelder, H. D., De Rover, M., McGehee, D. S., & Brussaard, A. B. (2003). Cholinergic modulation of dopaminergic reward areas: upstream and downstream targets of nicotine addiction. *European Journal of Pharmacology, 480*, 117–123.

Mansvelder, H. D., & McGehee, D. S. (2002). Cellular and synaptic mechanisms of nicotine addiction. *Journal of Neurobiology, 53*, 606–617.

Marco, E. M., Granstrem, O., Moreno, E., Llorente, R., Adriani, W., Laviola, G., ... Viveros, M. P. (2007). Subchronic nicotine exposure in adolescence induces long-term effects on hippocampal and striatal cannabinoid-CB1 and mu-opioid receptors in rats. *European Journal of Pharmacology, 557*, 37–43.

Mark, G. P., Shabani, S., Dobbs, L. K., & Hansen, S. T. (2011). Cholinergic modulation of mesolimbic dopamine function and reward. *Physiology & Behavior, 104*, 76–81.

Marsicano, G., & Kuner, R. (2008). Anatomical distribution of receptors, ligands and enzymes in the brain and in the spinal cord: circuitries and neurochemistry. In *Cannabinoids and the brain* (pp. 161–201). Springer.

Marsicano, G., & Lutz, B. (2006). Neuromodulatory functions of the endocannabinoid system. *Journal of Endocrinological Investigation, 29*, 27–46.

Matsuda, L. A., Lolait, S. J., Brownstein, M. J., Young, A. C., & Bonner, T. I. (1990). Structure of a cannabinoid receptor and functional expression of the cloned cDNA. *Nature, 346*, 561–564.

Merritt, L. L., Martin, B. R., Walters, C., Lichtman, A. H., & Damaj, M. I. (2008). The endogenous cannabinoid system modulates nicotine reward and dependence. *The Journal of Pharmacology and Experimental Therapeutics, 326*, 483–492.

Muldoon, P. P., Lichtman, A. H., Parsons, L. H., & Damaj, M. I. (2013). The role of fatty acid amide hydrolase inhibition in nicotine reward and dependence. *Life Sciences, 92*, 458–462.

Murray, J. E., Wells, N. R., Lyford, G. D., & Bevins, R. A. (2009). Investigation of endocannabinoid modulation of conditioned responding evoked by a nicotine CS and the Pavlovian stimulus effects of CP 55,940 in adult male rats. *Psychopharmacology (Berlin), 205*, 655–665.

Nava, F., Carta, G., Battasi, A., & Gessa, G. (2000). D_2 dopamine receptors enable Δ^9-tetrahydrocannabinol induced memory impairment and reduction of hippocampal extracellular acetylcholine concentration. *British Journal of Pharmacology, 130*, 1201–1210.

Oz, M., Al Kury, L., Keun-Hang, S. Y., Mahgoub, M., & Galadari, S. (2014). Cellular approaches to the interaction between cannabinoid receptor ligands and nicotinic acetylcholine receptors. *European Journal of Pharmacology, 731*, 100–105.

Pacher, P., & Kunos, G. (2013). Modulating the endocannabinoid system in human health and disease–successes and failures. *The FEBS Journal, 280*, 1918–1943.

Pertwee, R. G. (2008). CB1 and CB2 receptor pharmacology. In *Cannabinoids and the brain* (pp. 91–99). Springer.

Pertwee, R. G., Howlett, A. C., Abood, M. E., Alexander, S. P., Di Marzo, V., Elphick, M. R., ... Ross, R. A. (2010). International Union of Basic and Clinical Pharmacology. LXXIX. Cannabinoid receptors and their ligands: beyond CB(1) and CB(2). *Pharmacological Reviews, 62*, 588–631.

Picciotto, M. R., Caldarone, B. J., King, S. L., & Zachariou, V. (2000). Nicotinic receptors in the brain: links between molecular biology and behavior. *Neuropsychopharmacology, 22*, 451–465.

Piomelli, D. (2003). The molecular logic of endocannabinoid signalling. *Nature Review: Neuroscience, 4*, 873–884.

Regehr, W. G., Carey, M. R., & Best, A. R. (2009). Activity-dependent regulation of synapses by retrograde messengers. *Neuron, 63*, 154–170.

Scherma, M., Justinova, Z., Zanettini, C., Panlilio, L. V., Mascia, P., Fadda, P., ... Goldberg, S. R. (2012). The anandamide transport inhibitor AM404 reduces the rewarding effects of nicotine and nicotine-induced dopamine elevations in the nucleus accumbens shell in rats. *British Journal of Pharmacology, 165*, 2539–2548.

Scherma, M., Panlilio, L. V., Fadda, P., Fattore, L., Gamaleddin, I., Le Foll, B., ... Goldberg, S. R. (2008). Inhibition of anandamide hydrolysis by cyclohexyl carbamic acid 3'-carbamoyl-3-yl ester (URB597) reverses abuse-related behavioral and neurochemical effects of nicotine in rats. *The Journal of Pharmacology and Experimental Therapeutics, 327*, 482–490.

Sharir, H., & Abood, M. E. (2010). Pharmacological characterization of GPR55, a putative cannabinoid receptor. *Pharmacology & Therapeutics, 126*, 301–313.

Shoaib, M. (2008). The cannabinoid antagonist AM251 attenuates nicotine self-administration and nicotine-seeking behaviour in rats. *Neuropharmacology, 54*, 438–444.

Sidhpura, N., & Parsons, L. H. (2011). Endocannabinoid-mediated synaptic plasticity and addiction-related behavior. *Neuropharmacology, 61*, 1070–1087.

Simonnet, A., Cador, M., & Caille, S. (2013). Nicotine reinforcement is reduced by cannabinoid CB1 receptor blockade in the ventral tegmental area. *Addiction Biology, 18*, 930–936.

Stella, N. (2010). Cannabinoid and cannabinoid-like receptors in microglia, astrocytes, and astrocytomas. *Glia, 58*, 1017–1030.

Svizenska, I., Dubovy, P., & Sulcova, A. (2008). Cannabinoid receptors 1 and 2 (CB1 and CB2), their distribution, ligands and functional involvement in nervous system structures–a short review. *Pharmacology Biochemistry and Behavior, 90*, 501–511.

Trezza, V., Damsteegt, R., Manduca, A., Petrosino, S., Van Kerkhof, L. W., Pasterkamp, R. J., ... Di Marzo, V. (2012). Endocannabinoids in amygdala and nucleus accumbens mediate social play reward in adolescent rats. *The Journal of Neuroscience, 32*, 14899–14908.

Wade, M. R., Tzavara, E. T., & Nomikos, G. G. (2004). Cannabinoids reduce cAMP levels in the striatum of freely moving rats: an in vivo microdialysis study. *Brain Research, 1005*, 117–123.

Walters, C. L., Brown, S., Changeux, J. P., Martin, B., & Damaj, M. I. (2006). The β2 but not α7 subunit of the nicotinic acetylcholine receptor is required for nicotine-conditioned place preference in mice. *Psychopharmacology (Berlin), 184*, 339–344.

Chapter 16

Targets of Addictive Nicotine in the Central Nervous System and Interactions with Alcohol

Sodikdjon A. Kodirov

Department of General Physiology, Saint Petersburg University, Saint Petersburg, Russia; Pavlov Institute of Physiology, Russian Academy of Sciences, Saint Petersburg, Russia; Neuroscience Institute, Morehouse School of Medicine, Atlanta, GA, USA; Institute of Experimental Medicine, I. P. Pavlov Department of Physiology, Russian Academy of Medical Sciences, Saint Petersburg, Russia

Abbreviations

ACh Acetylcholine
AP Action potential
ADP After-depolarizing potential
BTx Bungarotoxin
CA1 Cornu ammonis 1 (or 2, 3, 4) area of the hippocampus
CDI Chlorisondamine diiodide
CHRNA Gene responsible for cholinergic receptor of nicotinic α subunit types
CHRNB Gene encoding cholinergic receptor of nicotinic β subunit types
DHβE Dihydro-β-erythroidine
DMAB 3-(4)-Dimethylaminobenzylidine anabaseine
DMPP 1,1-Dimethyl-4-phenylpiperazinium
d-TC D-tubocurarine
GABA γ-Aminobutyric acid
HCN channel Hyperpolarization-activated cyclic nucleotide gated nonselective cation channel
I_f Funny current
I_q Queer current
LA Lateral amygdala
LS Lateral septum
LTP Long-term potentiation
MEC Mecamylamine
MLA Methyllycaconitine citrate
MS Medial septum
nAChR Nicotinic acetylcholine receptor
PPD Paired-pulse depression
PPF Paired-pulse facilitation
RAP Rebound action potential
RTP Rebound tail potential
α7* Heteromeric receptor containing (*) α7 or other given subunits

---

"…Dr. McEvoy remembered he had a bottle of nicotine in his dispensary, which he had been using in a case of puerperal tetanus."

Haughton (1872).

INTRODUCTION

Nicotine is one of the oldest known compounds. In earlier times some doctors prescribed nicotine even in cases of strychnine poisoning and advised the smoking of tobacco (Haughton, 1872). Its effects have been shown in both vertebrates and lower order animals, including snails (Dougherty & Lester, 2001).

Nicotine is a ligand for the nicotinic acetylcholine receptor (nAChR), and its α(2–7) and β(2–4) subunits are present in one or another region of the mammalian brain (Porter et al., 1999). The number of neuronal-type subunits is more, and α8–10 are also described. Earlier α-BTx binding assays (Segal, Dudai, & Amsterdam, 1978) show differential activity in distinct regions of the rat brain in toto: hypothalamus > hippocampus > cortex > medulla > septum > thalamus > caudate > cerebellum. In the seventh lumbar (L7) ventral root, nicotine, for a longer period, but also acetylcholine (ACh), transiently inhibited the amplitude of spikes, which were evoked monosynaptically by stimulation of L7 dorsal root volley at 1 Hz (Curtis, Eccles, & Eccles, 1957). "The animal (cat)" was anesthetized and its spinal cord was transected and, if required, segments L2 to the third sacral were crushed.

Embryonic chicken ciliary ganglion neurons in culture respond with inward currents to 20 µM nicotine (Neff, Conroy, Schoellerman, & Berg, 2009). Knockdown of PSD-95 (postsynaptic density protein 95) and SAP-102 (synapse-associated protein 102), but not SAP-97, with corresponding RNAi reduced the amplitude of currents without changing the sensitization and desensitization profiles. This procedure also decreased the frequency of miniature excitatory postsynaptic currents (mEPSCs) and significantly prevented paired-pulse depression (PPD). Therefore, the role of "PSD-95 family members at nicotinic synapses" was concluded. However, there is no endogenous nicotine or nicotinic neurotransmission, so there are no dedicated synapses, and the responsible ones are cholinergic.

One of nicotine's targets is DARPP-32 (dopamine- and cAMP-regulated phosphoprotein with molecular mass of 32 kDa). DARPP-32 is also a target of all major substances of abuse: caffeine, ethanol, lysergic acid diethylamide, phencyclidine, amphetamine, and cocaine (Svenningsson, Nairn, & Greengard, 2005). In the cornu ammonis 3 (CA3) region of the hippocampus 1 μM nicotine enables θ oscillations (Lu and Henderson, 2010) and increases time dependently the peak frequency (Hz), area (uV2), and power (uV2/Hz). The area increases gradually and reaches a plateau in ~1 h, and these effects are not observed in the concurrent presence of dihydro-β-erythroidine (DHβE), mecamylamine (MEC), and methyllycaconitine citrate (MLA). These effects were mediated mainly by DHβE. One micromolar PNU-282987 also had similar effects, but not epibatidine. Nicotine-induced θ rhythm was blocked by 2,3-dihydroxy-6-nitro-7-sulfamoylbenzo[f]quinoxaline-2,3-dione (NBQX) and bicuculline.

Nicotine's effects cannot be attributed only to its multiple receptors, and no single brain area is a definitive target. Smoking increases nicotine levels in many regions, and every nucleus is connected with another in first- or second-order fashion (Mouly & Di Scala, 2006). Nicotine at 10 μM–1 mM inhibits Cav, I_{Kr}, I_{Ks}, and I_{K1} channels in cardiomyocytes (Satoh, 2002), and similar effects may take place in the brain as well. The earliest studies focused on nicotine effects per se, and more recent ones are starting to define a dose that mimics the levels achieved after smoking tobacco in plasma, blood, and tissues, including those of the brain–the czar of the body in all animals.

EFFECTS OF NICOTINE

In the brain, the effects of nicotine cannot be described by a simple and single target, and they may involve not only key neurotransmitters, but even calcium/calmodulin-dependent protein kinase II (Welsby, Rowan, & Anwyl, 2009) and angiotensin II (Wayner, Armstrong, & Phelix, 1996). Moreover, "the nicotinic acetylcholine receptor belongs to a superfamily that also contains receptors for the neurotransmitters serotonin, GABA, glycine, and glutamate" (Dougherty & Lester, 2001).

Many individuals smoke ~20 cigarettes per day (Carter et al., 2009). Inhaling a puff of tobacco leads to a rapid increase in nicotine levels in the whole brain with a latency of only 10 s (Berridge et al., 2010). The peak is reached within 60 s and effects are similar in the cerebellum, nucleus accumbens, amygdala, and cortex, but slightly higher in the thalamus. An uptake curve by the whole brain in all subjects had similar tendencies, albeit different magnitudes. Before smoking, the nicotine plasma levels in veins are higher compared to those in arteries (Henningfield, Stapleton, Benowitz, Grayson, & London, 1993), but after smoking of only one cigarette this is reversed. The blood plasma level of nicotine in arteries increases from ~5 to 50 ng/ml within 5 min, while the venous level increases from ~7 to 23 ng/ml. After 10 min these values decrease to ~30 and 20 ng/ml, respectively.

Also, after nasal spray of nicotine the peak arterial plasma levels (~9 vs 4 ng/ml) are higher compared to venous (Gourlay & Benowitz, 1997). Similar tendencies persist during smoking and intravenous injection, and effects decline only upon termination of nicotine. During 10 min of smoking multiple peaks are observed. This is different compared with nasal spray, since the peak is reached later and it declines spontaneously within ~10 min. Effects were accompanied by an increase in heart rate in all three groups. Nasal nicotine sprayed for 30 min increased cotinine for at least 5.5 h, and venous/arterial plasma levels were not far apart compared to nicotine, although they were higher in the arteries. Levels of stress hormones, epinephrine and norepinephrine (Kodirov, 2012), were also higher in the arteries, and nicotine increased only epinephrine (Gourlay & Benowitz, 1997). Self-reported arousal in response to either cigarette or neutral pictures was equally higher in a control group compared to abstainers (Carter et al., 2009), while craving (self-report as well) levels were similar. However, no correlation was found between arousal and craving.

Both an acute and a chronic exposure to nicotine facilitates the induction of long-term potentiation (LTP) in the CA1 area of the hippocampus in vitro (Fujii, Ji, Morita, & Sumikawa, 1999). Interestingly, 1 μM acute nicotine caused additive facilitation when slices were obtained from animals that were subjected to chronic exposure. This hints at distinct pathways involved during acute and chronic procedures. However, note that acute 5 μM nicotine did not influence its chronic effects on LTP in the dentate gyrus (Welsby et al., 2009). An acute treatment with nicotine also enhances learning and hippocampus-dependent memory via β2 nAChRs in vivo (Kenney, Florian, Portugal, Abel, & Gould, 2009). This enhancement is accompanied by the transcriptional upregulation of Jun-N-terminal kinase 1 (JNK1) in the hippocampus of wild-type (WT) mice. Neither learning/memory enhancements nor increases in JNK1 were observed in β2-subunit nAChR-knockout mice. If currents induced by activation of α7* are normalized correctly (since 1000%, but not 100%, is presented as stated), nicotine is much more potent than ACh, or 1,1-dimethyl-4-phenylpiperazinium (DMPP) and cytosine (Seguela, Wadiche, Dineley-Miller, Dani, & Patrick, 1993). Note that effects were normalized to 10 μM nicotine, but up to 100 μM was used. One-third of nicotinic currents were blocked by niflumic and flufenamic acids, Ca^{2+}-activated Cl^- channel blockers, although oocytes were injected only with α7. When extracellular NaCl was increased from 115 to 350 mM, nicotine induced small inward followed by bigger outward currents. Acids blocked selectively outward components paralleled by an increase in inward currents as they were masked. Neurotoxin I of *Naja oxiana* inhibits ACh-induced inward currents via α7 subunits by ~50% at 30 nM and almost completely at 100 nM (Lyukmanova et al., 2007).

Nicotine increases the frequency of recurrent spikes in cholinergic medial habenular (MHb) neurons (Gorlich et al., 2013). The authors observed complex responses to nicotine in transgenic α3β4α5 cluster (Tabac) mice, which comprise an initial increase in firing rate and sudden cessation of spikes. There is a possibility that the inhibition developed despite the presence of nicotine, but was reversible upon washout. Note that the neuron resumed firing with a rebound increase in rate. The conclusion that "nicotine-dependent increase in the number of spontaneous APs in the MHb is therefore mediated by nAChRs and is not caused by effects of nicotine on

HCN, L-type calcium, or BK channels" is derived from different experiments and warrants the coapplication of nicotine with either antagonist. Although 1 µM nicotine similarly increased whole-cell currents (mediated by A1 subfamily of transient receptor potential cation channels (TRPA1)) in both inward and outward directions, the single-channel recordings revealed different outcomes (Talavera et al., 2009). Inward and outward conductances were ~110 pS; nicotine, despite the higher concentration (1 mM), did not affect the former, but decreased the latter to ~80 pS. Thus, the value of traces and the notion that "nicotine activates TRPA1 in cell-free inside-out patches" are not reflected in the bar graphs.

The diversity of endogenous genes encoding α (*CHRNA3, 4, 5, 7*) and β (*CHRNB4*) subunits are also expressed in PC12 rat adrenal medulla pheochromocytoma cells (Virginio, Giacometti, Aldegheri, Rimland, & Terstappen, 2002). Choline dose dependently activates inward currents in PC12 cells, revealing functional receptors. These currents have two components at 3 mM choline, and 1 nM MLA blocks a rapid one. The EC_{50} of choline (2 mM) in PC12 cells was higher than that of ACh (389 µM), and the lowest was that of nicotine (52 µM), which is similar to that of cytisine (51 µM). Agonists of the nAChR enhance the binding sites for nicotine in numbers (Wonnacott, 1990).

Amygdala

Chronic exposure to nicotine facilitates the induction of LTP at synapses between the cortical, but not the thalamic, inputs and the lateral amygdala (LA) neurons (Huang, Kandel, & Levine, 2008). The presence of α7 in the pyramidal neurons of the basolateral amygdala (BLA) has been shown indirectly by application of 100 µM MLA (Ashenafi et al., 2005). It decreased the amplitude of evoked excitatory postsynaptic potentials (eEPSPs) and evoked inhibitory postsynaptic potentials (eIPSPs) and increased paired-pulse facilitation (PPF) at 60-ms intervals, but the scaling in the article's figure might be incorrect. In the presence of MLA or alone, 800 nM β-amyloid$_{25-35}$ decreases the amplitude of EPSPs. However, the aim of concurrent experiments is not clear, since MLA inhibited EPSPs, which was compensated for by increasing the stimulus intensity before the β-amyloid$_{25-35}$ application.

Pulses of 2 mM ACh trigger inward currents and action potentials (APs) in LA and BLA neurons (Klein & Yakel, 2006). Recovery from desensitization occurs after 30 s. Magnitude of currents was voltage dependent and was higher during hyperpolarization and completely inhibited by 10 nM MLA. Nicotine (1 µM) transiently increased the frequency of spontaneous excitatory postsynaptic currents (sEPSCs) in olfactory bulb (OB) explants and amygdalar neurons coculture (Barazangi & Role, 2001). Similar effects perhaps took place also with regard to sEPSC amplitude (not addressed), since nicotine increases eEPSC amplitude and decreases failures. Neurotransmission was thought to have been established between OB tissues and amygdalar neurons, but amygdalar neurons are also connected among themselves, and nicotine increases spontaneous inhibitory postsynaptic current (sIPSC).

Cerebellum

In humans and mice, 2–20 µM nicotine locally (independent of inputs) induces a gamma rhythm and very fast oscillations (VFO) in the Purkinje cell layer, which are antagonized by D-tubocurarine (d-TC), but not chlorisondamine diiodide, DHβE, or MLA (Middleton et al., 2008). Nicotine decreases AP frequencies at resting membrane potential (RMP), and now instead of spikes (exhibiting significant amplitude variations), spikelets are observed, which compensate for the initial frequency. Similar spikelets in the presence of nicotine are also seen in basket and stellate cells.

Gamma oscillation at ~40 Hz is caused by nicotine alone, while VFO (~150 Hz) is seen when nicotine is combined with 2 µM gabazine (to block γ-aminobutyric acid (GABA) receptors). The peak frequency of VFOs is moderately decreased by 10 µM β-pompilidotoxin, suggesting the involvement of Nav1 (Traub, Middleton, Knopfel, & Whittington, 2008). Since nicotine modulates hyperpolarization-activated cyclic nucleotide gated nonselective cation (HCN) channels that contribute to resonance frequency (Griguoli et al., 2010; Kodirov, Wehrmeister, & Colom, 2014), it would be meaningful to test the effects of ZD7288 on both gamma rhythm and VFO.

Cortex

Several mutations are associated with autosomal dominant nocturnal frontal lobe epilepsy syndrome and 1 µM nicotine increases the amplitude and frequency of sIPSC in layer 2/3 (LII/III) pyramidal neurons of the frontal cortex of heterozygote *Chrna4*[S252F/wt] mice, while minor effects are observed in WT (Klaassen et al., 2006). Nicotine effects on mIPSCs persisted in *Chrna4*[S252F/wt], and magnitudes are much higher than stated in the text, ~20 and 80 vs 17 and 42% for amplitude and frequency, respectively. It is also not clear what the statement "…combined effect of nicotine on mIPSC frequency and amplitude" refers to in that article.

Another study also describes the absence of nicotine effects on mIPSC at 1 µM (Yamamoto et al., 2007). However, there might be several shortcomings in the recordings and presentations. Frequency (~10 Hz) and amplitude (~100 pA) of mIPSCs are high, and a trace without tetrodotoxin (TTX) could substantiate adequate comparisons. Nicotine was concluded to increase mIPSC, when midazolam (MDZ) was applied ~5 min prior, but positive effects of this benzodiazepine started to develop right before the application of nicotine. Similarly, diazepam's effects on frequency were analyzed and considered absent within ~5 min, but immediately after, the frequency and amplitude of mIPSC were increased. Prevention of MDZ effects by 100 nM MLA and 1 µM calphostin C was also not convincing as judged by representative traces. Nicotine effects were considered presynaptic and GABAergic, since the frequency of mIPSC was affected, but not mEPSC. Insertion of α7 by MDZ was concluded to contribute to nicotine effects.

In LVI of the prefrontal cortex nicotine increases the holding current and baseline noise in a concentration-dependent manner (Bailey, De Biasi, Fletcher, & Lambe, 2010). These effects were absent in α5$^{-/-}$ mice, but nicotine increased EPSC amplitudes, which, however, was not discussed. The transgenic model is an adequate tool, but concurrent application of nicotine with DHβE as used during ACh experiments would be insightful. The relationship between α5 and α4β2* (and its antagonist) is not clear.

Interneurons of LII/III and pyramidal cells of LVI express α5 and similarly respond to 1 mM ACh (Poorthuis, Bloem, Verhoog, & Mansvelder, 2013). Interestingly, the effects persisted in α5-null knockout mice. Nevertheless, the authors discovered differences in α5-null animals, namely, priming with 300 nM nicotine desensitized β2* nAChRs to ACh in interneurons, similar to WT. However, WT

TABLE 1 Experimentally Defined EC$_{50}$ of Nicotine Differs Greatly across Regions or Preparations

Brain Region	Cell Type	Tested Range (µM)	EC$_{50}$ (µM)	Source
Hippocampus	Mossy fiber synaptosomes	0.1–50	3	Bancila et al. (2009)
CA1	Pyramidal layer	0.1–10	–	Fujii et al. (1999)
	Interneurons	0.01–10	0.062	Griguoli et al. (2010)
Hypothalamus	E18	10	–	Belousov et al. (2001)
Lateral septum	LSD–GABA	1–3	0.92	Kodirov et al. (2014)
LDTN	Cholinergic	1–10	–	Ishibashi et al. (2009)
Cell culture	CHO–TRPA1	0.1–300	~10	Talavera et al. (2009)
	GH4C1–α7	–	41	Virginio et al. (2002)
	PC12–α7	–	52	

LDTN, laterodorsal tegmental nucleus; LSD, lateral septal nucleus, dorsal part.

pyramidal cells were only partially desensitized by nicotine after 10 min in contrast to α5-null cells. The authors attributed the effects of nicotine to expression of α5 in specific layers, although different cell types were studied in two distinct laminas.

Hippocampus

Acute 1 µM nicotine transforms the short-term potentiation (STP) of both field EPSPs and population spikes (PS) into LTP in the CA1 (Fujii et al., 1999). LTP is prevented by 3 µM MEC and similar results were obtained also during chronic treatments. Relatively weak stimulation of Schaffer collaterals (SC) (15 times at 100 Hz) induces only STP at 100 nM nicotine or its absence. No pharmacological LTP was observed for EPSP, but PS amplitudes were depressed.

In the stratum oriens interneurons of the CA1 (Griguoli et al., 2010) nicotine inhibits HCN channels by direct binding. In contrast it has dual effects on lateral septum (LS) neurons (Kodirov et al., 2014). Nevertheless, one neuron remained unaffected by nicotine (Griguoli et al., 2010). Cells were defined as oriens–lacunosum moleculare interneurons, since their dendrites branch up to stratum-lacunosum moleculare (SL–M). Nicotine and ZD7288 reduced the frequency from an almost identical 4 to ~2 Hz at −90 but not −70 mV. Although the EC$_{50}$ was estimated to be 62 nM, nicotine did not completely block I_h at 1 µM, as also judged by the magnitude of remaining tail currents (Kodirov et al., 2014).

Cells (not clear whether interneurons or pyramidal) of a P0–3 rat hippocampus responded to 500 µM nicotine with similarly sensitizing and desensitizing inward currents, which was followed by a [Ca^{2+}]$_o$-dependent transient increase in mEPSC frequency and amplitude (Radcliffe & Dani, 1998). However, based on cumulative probability, it was concluded that the amplitude was insensitive. Normalization of the frequency is also not clear, since all data points aligned around ~1 Hz, while there was one at ~10 Hz. Nicotine-induced current was completely and reversibly blocked by 5 nM MLA, which in representative traces also blocked mEPSC. Nicotine enhanced the eEPSC leading to PPD (100-ms interval) compared to mixed PPF and PPD under control conditions.

In hippocampal mossy fiber synaptosomes (which possess α7) 20 µM nicotine triggers glutamate release (Bancila, Cordeiro, Bloc, & Dunant, 2009). Its amplitude and rising kinetics were similar to those evoked by 5 pM exogenous glutamate, but decayed more slowly (~3 vs 5 min). A similar concentration of nicotine or 200 µM ACh did not affect the membrane potential (MP) and H$^+$ gradient, which is sensitive to the ionophore CCCP. Nicotine was more potent, since it released ~15 pM/mg glutamate, similar to 30 mM KCl and 200 µM ACh. No glutamate was released by 5 mM Ca^{2+}. The concentration used in these experiments was higher than the EC$_{50}$ and most probably was saturated (Table 1).

Nicotine at only 5 µM concentration facilitated the LTP of dentate gyrus (DG) EPSP in vitro (Welsby et al., 2009), but had no effects in vivo when rats were injected subcutaneously with up to 0.4 mg/kg (Wayner et al., 1996). Responses were referred to as population EPSPs (no traces were shown) of granular cells (GCs) and evoked at an intensity that did not trigger an established PS. However, in the presence of nicotine, responses of GCs to four tetani were robust compared to control and gradually increased after each tetanus. In general, comparing and finding mild effects on LTP is difficult in brain slices, and this is evidenced also in this article in vivo, since magnitudes and patterns of GC responses differ significantly (~35 vs 45% of baseline) when just two control groups were subjected separately to the LTP paradigm.

Hypothalamus

In cultured embryonic (E18) hypothalamic neurons nAChRs are present (Belousov, O'Hara, & Denisova, 2001). However, the proportion of responsive neurons is very low (2 of 10), which may be related to long-term culture—up to 21 days in vitro (DIV). To the authors' surprise, identical concentrations of 6-cyano-7-nitroquinoxaline-2,3-dione (CNQX) (10 µM) and D-2-amino-5-phosphonovalerate (D-APV) (100 µM) did not block sEPSP when treated chronically up to 17 DIV, unlike the complete inhibition seen when applied acutely. In some cells 10 µM nicotine increased the intracellular calcium concentration. Under control conditions, both glutamatergic and GABAergic neurotransmissions were completely abolished by 1 µM TTX, not revealing the adequacy of the culture approach, since the AP-independent events (mEPSP and mIPSP) were missing.

Raphe Nucleus

Serotonergic neurons of dorsal and median raphe and their terminals (in hippocampus and septum) express α7 (Aznar, Kostova, Christiansen, & Knudsen, 2005). ACh either depolarizes or hyperpolarizes dorsal raphe neurons at identical concentrations (30 μM). Nicotine (20 μM) and DMPP also depolarize MP in these cells by agonizing the nAChR (Li, Rainnie, McCarley, & Greene, 1998). During either effect the input resistance (IR) decreases.

Septum

Within this structure there are two major nuclei defined as LS and medial septum (MS). LS and its nAChR are involved in anxiety (Ouagazzal, Kenny, & File, 1999), since social interactions of rats and amount of time spent in the open arms of an elevated maze decrease after injection of nicotine into the LS. However, MEC increased the social interaction at 15 ng, whereas increasing the concentration to 30 and 50 ng had no effect. In contrast, nicotine at 1 and 4 μM similarly decreased those two parameters.

Nicotine and DMPP reversibly hyperpolarize MP of LS neurons and block APs (Sorenson & Gallagher, 1996). Effects of DMPP were abolished in the presence of 500 μM, but not 100 μM, chlorisondamine. DMPP's effect is [Ca]$_{in}$ dependent, since chelating neurons with BAPTA abolished responses, as did GTPγS treatment. The frequency of APs was increased by 100 nM trimethaphan camsylate (TMP) with no apparent effects on RMP, but at 10 μM it hyperpolarized the cells and blocked APs. TMP, to some extent, also promoted the bursting activity, which is abolished by 50 μM MEC.

AP waveforms of LS neurons are similar, when recorded from a small area (Figure 1). Despite these similarities, two subpopulations are distinguished by the absence and presence/contribution of HCN channels to MP (Figure 2). During sudden hyperpolarizations HCN channels are activated; however, maximal activation is reached gradually resulting in sag—slow depolarization in MP at a hyperpolarized state. In neurons, the termination of hyperpolarization results in rebound tail potential (RTP) that is equivalent to a tail current. When the amplitude of RTP is sufficient to reach the threshold, then rebound action potentials are triggered. Nicotine either inhibits or facilitates the excitability of the subset of neurons via modulation of HCN channels (Kodirov et al., 2014). However, other interpretations are plausible, since it increased (Figure 3) the IR, which could be due to effects on additional channels or desensitization of nAChR. This could alter the AP threshold, firing rate, and shunting of sag currents. The sag amplitude was increased by 3 μM nicotine despite the concurrent 10 μM MEC (Figure 3). Nicotine increased the number of spikelets and AP frequency with a slight reversibility (Figure 4). In some cells only after-depolarizing potentials are observed, instead of double spikes, which are triggered after application of nicotine.

Nicotine does not influence eEPSC in MS, but it enhances the inhibitory effects of β-amyloid (Santos-Torres et al., 2007). EPSC amplitude is further decreased by d-TC in the presence of β-amyloid(25–35). In MS the sag (HCN mediated)-expressing neurons respond to ACh with slowly activating and desensitizing inward currents compared to other cell types (Thinschmidt, Frazier, King, Meyer, & Papke, 2005). About 50% of currents are blocked by 50 nM MLA, and complete inhibition is achieved by concurrent 5 μM MEC.

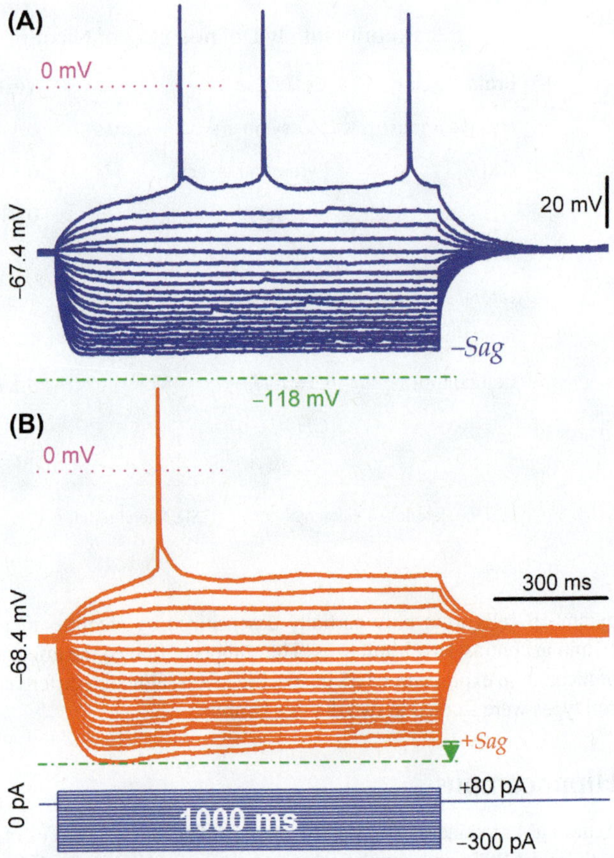

FIGURE 1 Contribution of HCN channels to MP. (A, B) The absence (−Sag) and presence of I_h (+Sag) in neurons of the LS. Note that RMP, latency, threshold, rheobase, overshoot, amplitude of APs, AHP, and rebound responses of MP are similar, albeit there is higher excitability in the absence of sag.

Striatum

Local stimulation of striatum always evokes polysynaptic IPSCs in cholinergic interneurons with intensity-dependent latency and was completely blocked by 500 nM nicotine (Sullivan, Chen, & Morikawa, 2008). Inhibition occurred fast and suddenly after ~2 min. Although nicotine was applied for only 5 min, the inhibition lasted longer than ~30 min, and only thereafter did it gradually reverse. Despite the fact that nicotine, 100 nM DHβE, 10 μM MEC, and 50 μM hexamethonium similarly and completely blocked the evoked IPSCs, the β2 was defined as a target. The latency of IPSC differs under control conditions, before nicotine (~10 ms) and after application of antagonists (~1 ms), so comparison is not adequate. The first peak in the complex polysynaptic response was considered as monosynaptic, which was absent during the nicotine experiment and unaffected by DHβE, MEC, and hexamethonium. Properties of striatal interneurons resemble those of tonically active neurons, and similar spiking (albeit distinct neurotransmitters) is also observed in invertebrates (Kodirov, 2011). ACh effects and acetylcholinesterase activity are shown in identified neurons.

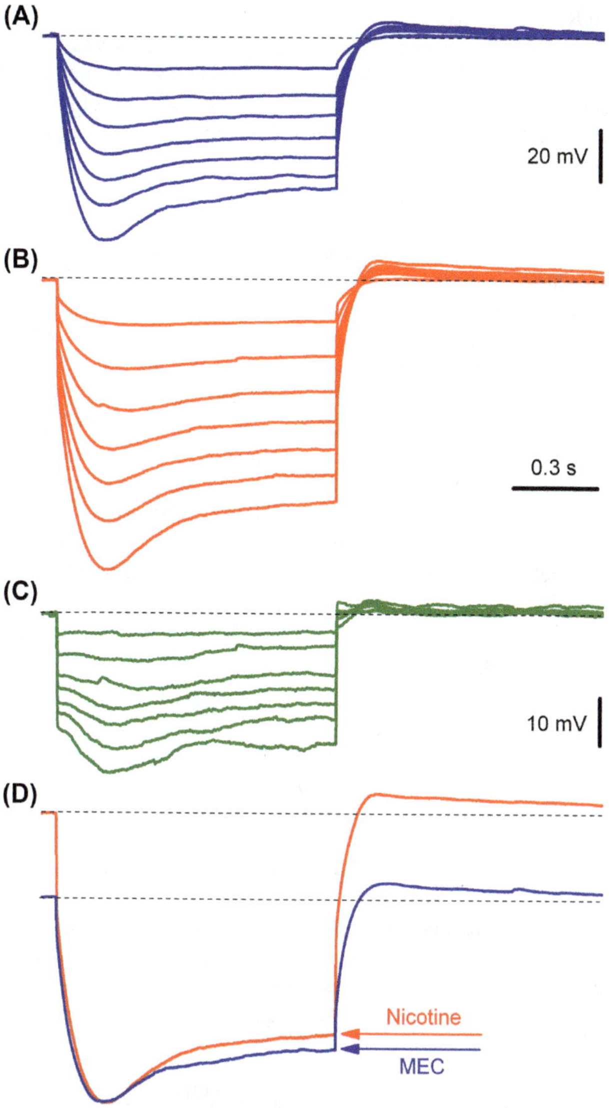

Nicotine Effects on Receptors and Channels **Chapter | 16** 173

FIGURE 2 Modulation of sag by nicotine in the presence of a nonselective inhibitor of nAChR. (A, B) The MP response after application of 10 μM MEC and nicotine in the same neuron, respectively. (C) The subtracted traces show an increase in sag; note linear response at moderate current injections indicative of IR changes. (D) The superimposed traces reveal a significant increase in sag amplitude by nicotine. The time scales for A–D are identical, as is amplitude for A, B, and D.

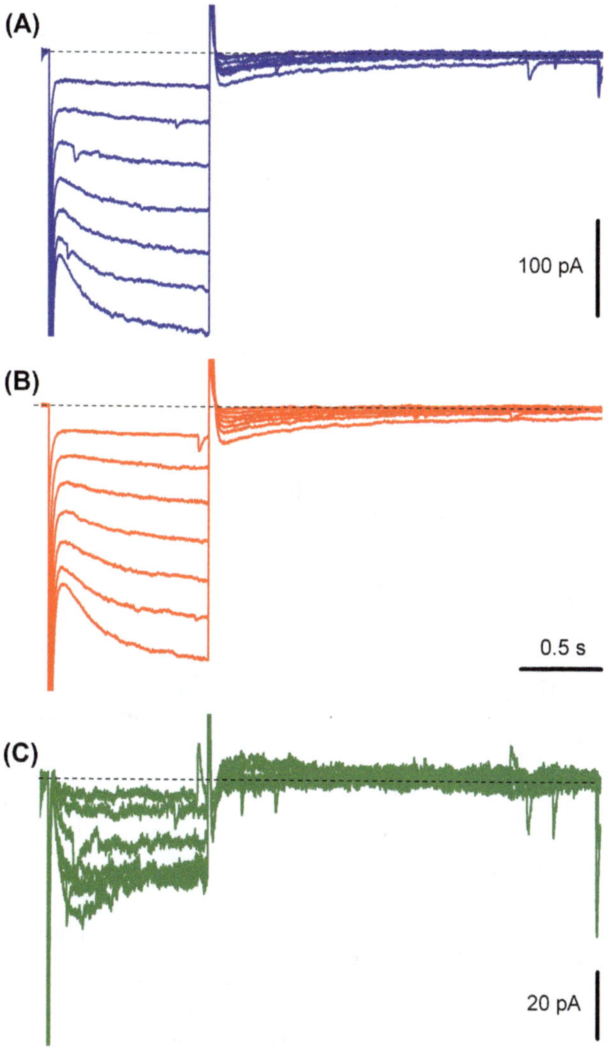

Brainstem

Ventral Tegmental Area

The cholinergic terminals into the neurons of the ventral tegmental area (VTA) originate from the pedunculopontine tegmentum and the laterodorsal pontine tegmentum. Some of the VTA dopaminergic neurons respond to 500 nM nicotine rapidly with increased sEPSC (Pidoplichko et al., 2004). The amplitude of the sEPSCs remained unaffected, but that of evoked ones was increased by 1 μM nicotine. Effects were significantly reversible even after up to ~20 min treatment with nicotine. Nicotine at 500 nM had dual effects, an initial transient increase followed by a permanent decrease in frequency of sIPSCs despite the washout. MLA (10 nM) decreased the amplitude

FIGURE 3 Modulation of I_h by nicotine. (A) I_h waveforms in the presence of 10 μM MEC (protocol in Figure 5C). (B) Application of 3 μM nicotine for 10 min decreased the amplitude of I_h. (C) Subtraction revealing the small nicotine-sensitive currents. Time scales for A–C and amplitude scale for A and B are identical.

of eEPSCs (3 of 10), but not eIPSCs, which was inhibited by 1 μM DHβE (4 of 10). Why fewer numbers of neurons responded to these agonists and antagonists needs to be defined. ACh-induced currents were partially diminished by 20 nM, but completely by 500 nM nicotine in WT. Neurons from β2–null mice do not respond to nicotine, and MLA prevents ACh-evoked currents.

Chronic nicotine decreased the cumulative probability of spike numbers within the bursts, but the effects were not statistically significant (Besson et al., 2007). The absence of effects was considered a result of the continuous presence of nicotine. In WT mice changes in spike properties are expected upon withdrawal.

Laterodorsal Tegmental Nucleus

A short 40-ms application of 10 μM nicotine evokes burst of APs (Ishibashi, Leonard, & Kohlmeier, 2009) in cholinergic neurons of the laterodorsal tegmental nucleus. When spike generation was

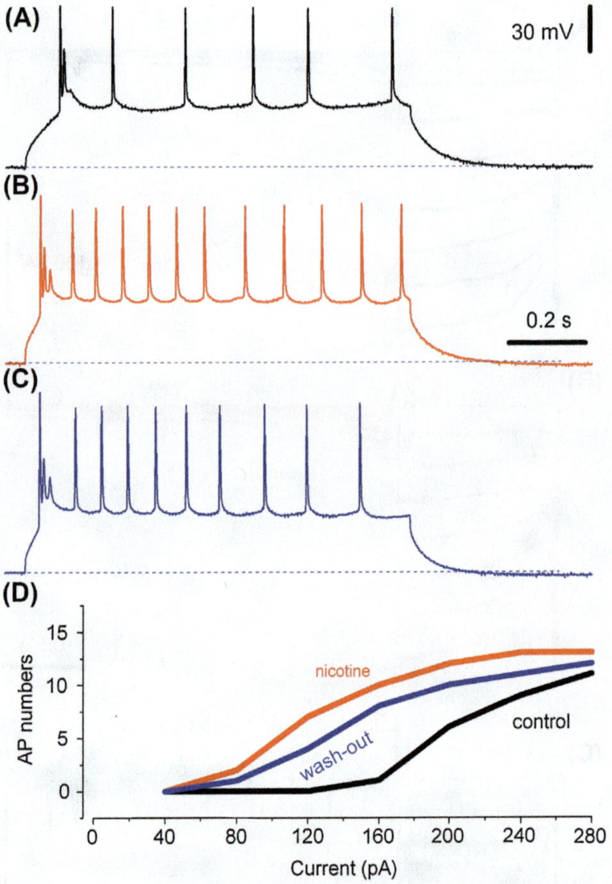

FIGURE 4 Modulation of excitability by nicotine. (A) Evoked APs by 200 pA current application for 1 s under control conditions. Note a double spiking at the start. (B) Decreased latency to the first AP, addition of spikelet to double spikes, and increased AP rate after application of 3 μM nicotine for 17 min. (C) Washout of nicotine for 35 min. (D) Response of same neuron to increasing current magnitudes revealing significant reversibility of nicotine effects. Time and amplitude scales are identical.

prevented by TTX, nicotine depolarized MP by ~20 mV. Priming with a 10 mM (not 10 μM as stated) nicotine puff subsequently increased the number of evoked APs by stepwise depolarization. Nicotine, DMPP, and ACh evoke inward currents in the presence of GABAergic and glutamatergic inhibitors. ACh-induced currents were inhibited by MLA, MEC, and DHβE by up to 40%, and their concurrent application led to ~95% blockade. The conclusion "...failure of TTX, in abolishment of the nicotine-induced changes in frequency of EPSPs indicate that nicotine is acting on presynaptic, glutamatergic terminals" warrants some clarifications. 1) Not EPSP, but EPSC was recorded. 2) Properties of sEPSC remained unaffected by TTX and thus mEPSC was absent. 3) Nicotine similarly increased the amplitude and frequency of both sEPSC and mEPSC. Therefore, changes are targeted equally to both presynaptic terminals and postsynaptic soma.

Spinal Cord

Nicotine and mustard oil increase TRPA1 currents in trigeminal neurons of mice (Talavera et al., 2009). Currents are prevented by 5 mM MEC, but not hexamethonium. Evoked synaptic currents in Renshaw cells (interneurons) caused by antidromic stimulation in spinal cord slices are decreased by 10 nM MLA and blocked by subsequent 3 μM DHβE (Lamotte d'Incamps & Ascher, 2008). Identical stimulation evoked a complex AP (hallmark of Renshaw cell) and D-APV blocked two later components, leaving in isolation single/simple spikes, which were blocked by concurrent MLA, DHβE, and NBQX. Ideally these effects will not lead to any conclusion, since it could be effects of an AMPA antagonist alone. Note an EPSP-like component in the presence of all four agents. It is also not clear why 2 μM NBQX does or does not completely block or leaves "residual currents." Currents sensitive to MLA and DHβE are labeled vice versa.

About 10% of dorsal root ganglion neurons are excited by 500 μM nicotine, while 50 μM ATP excites ~90% (Dussor, Zylka, Anderson, & McCleskey, 2008). When isolated nociceptive neurons do not express the *Mrgprd* (Mas-related G-protein-coupled receptor) gene, these values are ~15% and 30%, respectively. The current waveform in response to ATP is similar to that of homomeric pyrinergic P2X3 receptors. Whether *Mrgprd* is present or not, cells fire similar APs with typical repolarization phase and after-hyperpolarizing potential (AHP). These cells possess Nav1 and Cav1 channels and may exhibit recurrent activity (Kodirov, Plakhova, Krylov, unpublished observation).

Nucleus Hypoglossus

Motoneurons of the hypoglossal nucleus in a neonatal rat brainstem respond to 10 μM nicotine bath application with increased holding current that gradually transits into oscillatory (~10 Hz) inward currents (Lamanauskas & Nistri, 2008). Concurrent 10 μM riluzole completely blocks θ oscillations in a subgroup of neurons, while in others it decreased frequency (10 vs 3 Hz) and amplitude (87 vs 40 pA). Oscillations by nicotine started after ~5 min and lasted up to ~10 min, and the amplitude gradually increased, while frequency reached stable values relatively fast.

CONCURRENT EFFECTS OF NICOTINE WITH ALCOHOL

NICOTINE

Nicotine and its receptors also interact with other compounds. One of the first substances of interest and relevance to people is alcohol, since its concurrent consumption with nicotine occurs frequently both socially and in isolation. It is not logical to expect that a decrease in the brain's performance during intoxication with ethanol will be reversed by smoking tobacco, but concurrent ethanol and nicotine in rats did not decrease the working memory, as does ethanol alone (Tracy, Wayner, & Armstrong, 1999). Ethanol influences nAChRs in postsynaptic membranes of electroplaques of the fish *Torpedo nobiliana* (Forman et al., 1989). Affinity of ACh is increased by 1 M ethanol from 79 to 5 μM, and carbamycholine-induced lslow desensitization of nAChRs is accelerated. This perhaps was one of the initiating points for concurrent nicotine and ethanol studies in mammals.

Concurrent nicotine and alcohol effects are complex when targeted to brain tissue and depend on the region. One cannot analyze first nicotine and then the sequential application of ethanol in vivo, since its effects do not decline quickly. In DG, alcohol alone decreases, nicotine alone increases, but both further increase the number of c-Fos-positive neurons (Bachtell & Ryabinin, 2001). In LA, alcohol and nicotine alone increased c-Fos-positive neurons; however, an increase during concurrent application was less pronounced than alcohol alone and was more than nicotine alone. Therefore, it is hard to conclude that they may have additive effects when applied simultaneously, or offset, or antagonize each other.

It has been suggested the α3* nAChRs in VTA dopaminergic neurons may be a target of acute ethanol (Kamens et al., 2009). Priming with nicotine for 10 days increases both ethanol preference and ethanol intake in rats. Although these effects gradually declined, they remained increased up to 25 days after nicotine withdrawal (Blomqvist, Ericson, Johnson, Engel, & Soderpalm, 1996). Concurrent nicotine does not influence the withdrawal symptoms of ethanol (Penland, Hoplight, Obernier, & Crews, 2001). There was no impact of nicotine on alcohol blood levels in two groups of rats at day 4, but a decrease was observed at day 3. Ethanol damaged the cortex and OB, and nicotine only the latter. However, interestingly, nicotine reduced the neurodegeneration of OB by ethanol, while in the cortex it led to further damage. Similar to other classical receptors (Kodirov et al., 2010), the nAChR is sensitive to endocannabinoids (Butt, Alptekin, Shippenberg, & Oz, 2008). ACh-induced $^{86}Rb^+$ efflux in thalamic synaptosomes was decreased by 3 μM arachidonylethanolamide (AEA), also known as anandamide. Note that the effect of ACh is concentration dependent and saturates at 1 mM, but in the presence of AEA it saturates at 100 μM. Nevertheless, cannabinoid effects on nAChR or the coexistence of both receptor families in neurons could not be advantageous and straightforwardly promising, since "there is anecdotal evidence for the beneficial effects of marijuana use in managing some forms of epilepsy" (Mann & Mody, 2008).

DUAL EFFECTS OF NICOTINE

We have demonstrated and directly stated the dual effects of nicotine in regard to the excitability of LS neurons (Kodirov et al., 2014). This is not surprising taking into account the complexity of the mammalian brain and diversity of neurons, of which no region or subtype of cells could be considered a minority when compared directly. The effects of nicotine were complex when details were further discriminated (Figure 5). As the reader can observe, the complex effects of nicotine should be borne in mind, and a multiplicity of actions of known compounds is not a new idea (Forman, Righi, & Miller, 1989).

Nicotine exerts dual effects depending on neuronal types not only within the hippocampus, but within each subtype (Segal, 1978). It excited only two of 70 pyramidal (bursting) cells, while it inhibited 34. In nonpyramidal neurons, nicotine inhibited 14 of 18 cells and excited one. The inhibition is prevented by d-TC in some neurons. Note that ACh only excited neurons, while nicotine prevailingly inhibited them. In both cell types a portion of neurons were nonresponsive to either drug.

Nicotine has also dual concentration-dependent effects in the neostriatum. In the presence of 10 μM nomifensine (a dopamine uptake inhibitor), it decreased the phosphorylation of DARPP–32 at Thr 34 at 1 μM, transiently (up to 1 min) increased

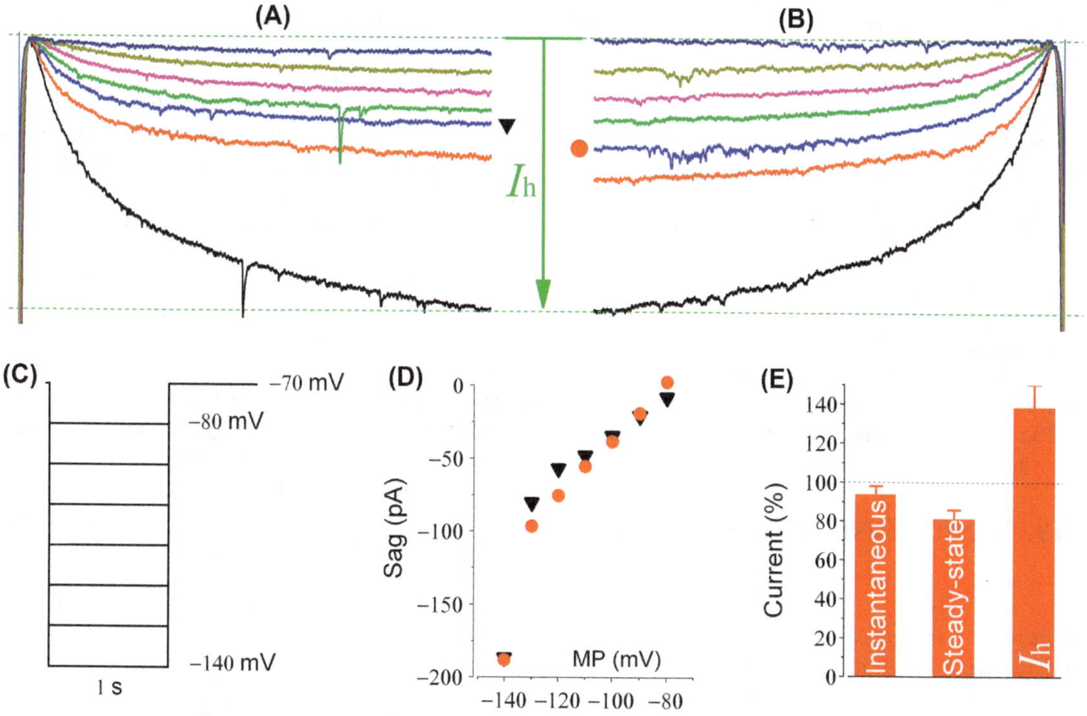

FIGURE 5 **Dual effects of nicotine on HCN channels.** (A) I_h was recorded at holding potential of −70 mV, similar to Figure 4, using an identical protocol (C). Currents are shown after the subtraction of the instantaneous component and enabling a complete isolation of gradually increasing I_h. (B) After 15 min application of 3 μM nicotine. For clarity the x axis is inverted in (B), and traces under both conditions are color matched based on potentials ranging from −140 to −80 mV. (D) The amplitudes of I_h–sag currents before and after application of nicotine, respectively. Note a decrease, an increase, or no effect at tested potentials. (E) Normalized instantaneous, steady-state, and sag currents at −120 mV (n = 6 of 11). Identical time and amplitude scales are indicated.

it at 10/100 μM, and had no effects at 100 nM (Hamada, Higashi, Nairn, Greengard, & Nishi, 2004). Inhibition of phosphorylation was dopaminergic receptor type 2 (D2R) dependent, since raclopride not only prevented nicotine effects, but now 1 μM nicotine transiently increased the phosphorylation of DARPP–32. MEC and DHβE prevented the effects of 1 μM nicotine, while α–BTx did not. Note that all these nAChR antagonists slightly increased phosphorylation. The Dopaminergic receptor type 1 (D1R) antagonist SCH23390 did not prevent the inhibition of phosphorylation by 1 μM nicotine, but blocked the transient increase by 100 μM.

APPLICATIONS TO OTHER ADDICTIONS AND SUBSTANCE MISUSE

- Nicotine often has psychologically calming effects during smoking, but some individuals will become anxious.
- Many individuals will smoke after they drink or alternate heavily during socializing.
- Increasing numbers of marijuana users are also commonly tobacco smokers.
- Nicotine use in youth will not be declining because of new popular ways of delivery including the ingenious discovery of electronic cigarettes.

DEFINITION OF TERMS

Cotinine This is a tobacco constituent.
Desensitization This is the slowly decaying phase of activated currents despite the continued presence or repeated exposure to a ligand.
Excitability This is the ability of neuronal and muscle cells to trigger APs.
Long-term depression This is a decrease in synaptic strength and perhaps a counterpart to LTP.
Long-term potentiation This is memory formation at the cellular level when synapses undergo facilitated plasticity in response to a Pavlovian paradigm.
Miniature events This is neurotransmitter release in the absence of excitability.
Pacemaker channels These are channels responsible for recurrent depolarization leading to rhythmic APs.
Pavlovian classical conditioning This is an associative pairing paradigm that increases synaptic strength within circuits.
Recurrent spikes These are spontaneous APs with a stable rate.
Sensitization This is the fast rising phase of activated currents in response to a ligand.

KEY FACTS OF NICOTINE AND ITS TARGETS

- About 150 years ago liquid nicotine was used to remedy puerperal tetanus in Europe.
- Although there is no endogenous nicotine in the human body, and these creatures initially had no knowledge of tobacco, multiple nAChRs are readily expressed across the brain.
- Nicotine modulates channels with nontraditional rectification accordingly termed as I_f or I_q (earlier as ik_2 and I_{AB}), funny or queer, respectively.
- I_f and I_q are activated by hyperpolarization and their names evolved multiple times.

- Why does the ligand nicotine have dedicated receptors, but addictive ethanol does not?
- Even a brief cessation in heavy smokers leads to craving and physiological arousal after exposure to a cue—images of cigarettes.
- This associative process can be considered as classical Pavlovian conditioning and subsequently as a learning event.

SUMMARY POINTS

- Targeted and differential nicotine effects on certain subunits of nAChRs are of primary importance.
- Indirect (not as a ligand) nicotinic effects are important to study, since the nAChR desensitizes quickly.
- Nicotine directly modulates HCN channels in the septum and hippocampus and presumably in other regions.
- Dual effects of nicotine, similar to those in a subgroup of LS neurons in regard to HCN channels, might be an universal phenomenon in the brain.
- Concurrent nicotine with other drugs could be used to study and unravel whether they may counterbalance one another's effects.

REFERENCES

Ashenafi, S., Fuente, A., Criado, J. M., Riolobos, A. S., Heredia, M., & Yajeya, J. (2005). β–Amyloid peptide$_{25-35}$ depresses excitatory synaptic transmission in the rat basolateral amygdala "in vitro". *Neurobiology of Aging, 26*(4), 419–428.

Aznar, S., Kostova, V., Christiansen, S. H., & Knudsen, G. M. (2005). α7 nicotinic receptor subunit is present on serotonin neurons projecting to hippocampus and septum. *Synapse, 55*(3), 196–200.

Bachtell, R. K., & Ryabinin, A. E. (2001). Interactive effects of nicotine and alcohol co-administration on expression of inducible transcription factors in mouse brain. *Neuroscience, 103*(4), 941–954.

Bailey, C. D. C., De Biasi, M., Fletcher, P. J., & Lambe, E. K. (2010). The nicotinic acetylcholine receptor α5 subunit plays a key role in attention circuitry and accuracy. *Journal of Neuroscience, 30*(27), 9241–9252.

Bancila, V., Cordeiro, J. M., Bloc, A., & Dunant, Y. (2009). Nicotine-induced and depolarisation-induced glutamate release from hippocampus mossy fibre synaptosomes: two distinct mechanisms. *Journal of Neurochemistry, 110*(2), 570–580.

Barazangi, N., & Role, L. W. (2001). Nicotine-induced enhancement of glutamatergic and GABAergic synaptic transmission in the mouse amygdala. *Journal of Neurophysiology, 86*(1), 463–474.

Belousov, A. B., O'Hara, B. F., & Denisova, J. V. (2001). Acetylcholine becomes the major excitatory neurotransmitter in the hypothalamus in vitro in the absence of glutamate excitation. *Journal of Neuroscience, 21*(6), 2015–2027.

Berridge, M. S., Apana, S. M., Nagano, K. K., Berridge, C. E., Leisure, G. P., & Boswell, M. V. (2010). Smoking produces rapid rise of [^{11}C] nicotine in human brain. *Psychopharmacology (Berlin), 209*(4), 383–394.

Besson, M., Granon, S., Mameli-Engvall, M., Cloez-Tayarani, I., Maubourguet, N., Cormier, A., ... Faure, P. (2007). Long-term effects of chronic nicotine exposure on brain nicotinic receptors. *Proceedings of the National Academy of Sciences of the United States of America, 104*(19), 8155–8160.

Blomqvist, O., Ericson, M., Johnson, D. H., Engel, J. A., & Soderpalm, B. (1996). Voluntary ethanol intake in the rat: effects of nicotinic acetylcholine receptor blockade or subchronic nicotine treatment. *European Journal of Pharmacology, 314*(3), 257–267.

Butt, C., Alptekin, A., Shippenberg, T., & Oz, M. (2008). Endogenous cannabinoid anandamide inhibits nicotinic acetylcholine receptor function in mouse thalamic synaptosomes. *Journal of Neurochemistry, 105*(4), 1235–1243.

Carter, B. L., Lam, C. Y., Robinson, J. D., Paris, M. M., Waters, A. J., Wetter, D. W., & Cinciripini, P. M. (2009). Generalized craving, self-report of arousal, and cue reactivity after brief abstinence. *Nicotine & Tobacco Research, 11*(7), 823–826.

Curtis, D. R., Eccles, J. C., & Eccles, R. M. (1957). Pharmacological studies on spinal reflexes. *Journal of Physiology, 136*(2), 420–434.

Dougherty, D. A., & Lester, H. A. (2001). Neurobiology. Snails, synapses and smokers. *Nature, 411*(6835), 252–255.

Dussor, G., Zylka, M. J., Anderson, D. J., & McCleskey, E. W. (2008). Cutaneous sensory neurons expressing the Mrgprd receptor sense extracellular ATP and are putative nociceptors. *Journal of Neurophysiology, 99*(4), 1581–1589.

Forman, S. A., Righi, D. L., & Miller, K. W. (1989). Ethanol increases agonist affinity for nicotinic receptors from *Torpedo*. *Biochimica et Biophysica Acta, 987*(1), 95–103.

Fujii, S., Ji, Z., Morita, N., & Sumikawa, K. (1999). Acute and chronic nicotine exposure differentially facilitate the induction of LTP. *Brain Research, 846*(1), 137–143.

Gorlich, A., Antolin-Fontes, B., Ables, J. L., Frahm, S., Slimak, M. A., Dougherty, J. D., & Ibanez-Tallon, I. (2013). Reexposure to nicotine during withdrawal increases the pacemaking activity of cholinergic habenular neurons. *Proceedings of the National Academy of Sciences of the United States of America, 110*(42), 17077–17082.

Gourlay, S. G., & Benowitz, N. L. (1997). Arteriovenous differences in plasma concentration of nicotine and catecholamines and related cardiovascular effects after smoking, nicotine nasal spray, and intravenous nicotine. *Clinical Pharmacology & Therapeutics, 62*(4), 453–463.

Griguoli, M., Maul, A., Nguyen, C., Giorgetti, A., Carloni, P., & Cherubini, E. (2010). Nicotine blocks the hyperpolarization-activated current I_h and severely impairs the oscillatory behavior of oriens-lacunosum moleculare interneurons. *Journal of Neuroscience, 30*(32), 10773–10783.

Hamada, M., Higashi, H., Nairn, A. C., Greengard, P., & Nishi, A. (2004). Differential regulation of dopamine D1 and D2 signaling by nicotine in neostriatal neurons. *Journal of Neurochemistry, 90*(5), 1094–1103.

Haughton, S. (1872). Case of strychnia poisoning, successfully treated by nicotine. *British Medical Journal*, 660–661.

Henningfield, J. E., Stapleton, J. M., Benowitz, N. L., Grayson, R. F., & London, E. D. (1993). Higher levels of nicotine in arterial than in venous blood after cigarette smoking. *Drug and Alcohol Dependence, 33*(1), 23–29.

Huang, Y. Y., Kandel, E. R., & Levine, A. (2008). Chronic nicotine exposure induces a long-lasting and pathway-specific facilitation of LTP in the amygdala. *Learning & Memory, 15*(8), 603–610.

Ishibashi, M., Leonard, C. S., & Kohlmeier, K. A. (2009). Nicotinic activation of laterodorsal tegmental neurons: implications for addiction to nicotine. *Neuropsychopharmacology, 34*(12), 2529–2547.

Kamens, H. M., McKinnon, C. S., Li, N., Helms, M. L., Belknap, J. K., & Phillips, T. J. (2009). The α3 subunit of the nicotinic acetylcholine receptor is a candidate gene for ethanol stimulation. *Genes Brain Behavior, 8*(6), 600–609.

Kenney, J. W., Florian, C., Portugal, G. S., Abel, T., & Gould, T. J. (2009). Involvement of hippocampal jun-N terminal kinase pathway in the enhancement of learning and memory by nicotine. *Neuropsychopharmacology, 35*(2), 483–492.

Klaassen, A., Glykys, J., Maguire, J., Labarca, C., Mody, I., & Boulter, J. (2006). Seizures and enhanced cortical GABAergic inhibition in two mouse models of human autosomal dominant nocturnal frontal lobe epilepsy. *Proceedings of the National Academy of Sciences of the United States of America, 103*(50), 19152–19157.

Klein, R. C., & Yakel, J. L. (2006). Functional somato-dendritic α7-containing nicotinic acetylcholine receptors in the rat basolateral amygdala complex. *Journal of Physiology, 576*(Pt 3), 865–872.

Kodirov, S. A. (2011). The neuronal control of cardiac functions in Molluscs. *Comparative Biochemistry and Physiology, Part A: Molecular & Integrative Physiology, 160*(2), 102–116.

Kodirov, S. A. (2012). The role of norepinephrine in amygdala dependent fear learning and memory. In B. Ferry (Ed.), *The amygdala – A discrete multitasking manager* (pp. 121–140).

Kodirov, S. A., Jasiewicz, J., Amirmahani, P., Psyrakis, D., Bonni, K., Wehrmeister, M., & Lutz, B. (2010). Endogenous cannabinoids trigger the depolarization-induced suppression of excitation in the lateral amygdala. *Learning & Memory, 17*(1), 832–838.

Kodirov, S. A., Wehrmeister, M., & Colom, L. V. (2014). Modulation of HCN channels in lateral septum by nicotine. *Neuropharmacology, 81*, 274–282.

Lamanauskas, N., & Nistri, A. (2008). Riluzole blocks persistent Na^+ and Ca^{2+} currents and modulates release of glutamate via presynaptic NMDA receptors on neonatal rat hypoglossal motoneurons in vitro. *European Journal of Neuroscience, 27*(10), 2501–2514.

Lamotte d'Incamps, B., & Ascher, P. (2008). Four excitatory postsynaptic ionotropic receptors coactivated at the motoneuron-Renshaw cell synapse. *Journal of Neuroscience, 28*(52), 14121–14131.

Li, X., Rainnie, D. G., McCarley, R. W., & Greene, R. W. (1998). Presynaptic nicotinic receptors facilitate monoaminergic transmission. *Journal of Neuroscience, 18*(5), 1904–1912.

Lu, C. B., & Henderson, Z. (2010). Nicotine induction of theta frequency oscillations in rodent hippocampus in vitro. *Neuroscience, 166*(1), 84–93.

Lyukmanova, E. N., Shenkarev, Z. O., Schulga, A. A., Ermolyuk, Y. S., Mordvintsev, D. Y., Utkin, Y. N., … Kirpichnikov, M. P. (2007). Bacterial expression, NMR, and electrophysiology analysis of chimeric short/long-chain α-neurotoxins acting on neuronal nicotinic receptors. *Journal of Biological Chemistry, 282*(34), 24784–24791.

Mann, E. O., & Mody, I. (2008). The multifaceted role of inhibition in epilepsy: seizure-genesis through excessive GABAergic inhibition in autosomal dominant nocturnal frontal lobe epilepsy. *Current Opinion in Neurology, 21*(2), 155–160.

Middleton, S. J., Racca, C., Cunningham, M. O., Traub, R. D., Monyer, H., Knopfel, T., … Whittington, M. A. (2008). High-frequency network oscillations in cerebellar cortex. *Neuron, 58*(5), 763–774.

Mouly, A. M., & Di Scala, G. (2006). Entorhinal cortex stimulation modulates amygdala and piriform cortex responses to olfactory bulb inputs in the rat. *Neuroscience, 137*(4), 1131–1141.

Neff, R. A., 3rd, Conroy, W. G., Schoellerman, J. D., & Berg, D. K. (2009). Synchronous and asynchronous transmitter release at nicotinic synapses are differentially regulated by postsynaptic PSD-95 proteins. *Journal of Neuroscience, 29*(50), 15770–15779.

Ouagazzal, A. M., Kenny, P. J., & File, S. E. (1999). Stimulation of nicotinic receptors in the lateral septal nucleus increases anxiety. *European Journal of Neuroscience, 11*(11), 3957–3962.

Penland, S., Hoplight, B., Obernier, J., & Crews, F. T. (2001). Effects of nicotine on ethanol dependence and brain damage. *Alcohol, 24*(1), 45–54.

Pidoplichko, V. I., Noguchi, J., Areola, O. O., Liang, Y., Peterson, J., Zhang, T., & Dani, J. A. (2004). Nicotinic cholinergic synaptic mechanisms in the ventral tegmental area contribute to nicotine addiction. *Learning & Memory, 11*(1), 60–69.

Poorthuis, R. B., Bloem, B., Verhoog, M. B., & Mansvelder, H. D. (2013). Layer-specific interference with cholinergic signaling in the prefrontal cortex by smoking concentrations of nicotine. *Journal of Neuroscience, 33*(11), 4843–4853.

Porter, J. T., Cauli, B., Tsuzuki, K., Lambolez, B., Rossier, J., & Audinat, E. (1999). Selective excitation of subtypes of neocortical interneurons by nicotinic receptors. *Journal of Neuroscience, 19*(13), 5228–5235.

Radcliffe, K. A., & Dani, J. A. (1998). Nicotinic stimulation produces multiple forms of increased glutamatergic synaptic transmission. *Journal of Neuroscience, 18*(18), 7075–7083.

Santos-Torres, J., Fuente, A., Criado, J. M., Riolobos, A. S., Heredia, M., & Yajeya, J. (2007). Glutamatergic synaptic depression by synthetic amyloid β-peptide in the medial septum. *Journal of Neuroscience Research, 85*(3), 634–648.

Satoh, H. (2002). Modulation by nicotine of the ionic currents in guinea pig ventricular cardiomyocytes. Relatively higher sensitivity to I_{Kr} and I_{K1}. *Vascular Pharmacology, 39*(1–2), 55–61.

Segal, M. (1978). The acetylcholine receptor in the rat hippocampus; nicotinic, muscarinic or both? *Neuropharmacology, 17*(8), 619–623.

Segal, M., Dudai, Y., & Amsterdam, A. (1978). Distribution of an α-bungarotoxin-binding cholinergic nicotinic receptor in rat brain. *Brain Research, 148*(1), 105–119.

Seguela, P., Wadiche, J., Dineley-Miller, K., Dani, J. A., & Patrick, J. W. (1993). Molecular cloning, functional properties, and distribution of rat brain α7: a nicotinic cation channel highly permeable to calcium. *Journal of Neuroscience, 13*(2), 596–604.

Sorenson, E. M., & Gallagher, J. P. (1996). The membrane hyperpolarization of rat dorsolateral septal nucleus neurons is mediated by a novel nicotinic receptor. *Journal of Pharmacology and Experimental Therapeutics, 277*(3), 1733–1743.

Sullivan, M. A., Chen, H., & Morikawa, H. (2008). Recurrent inhibitory network among striatal cholinergic interneurons. *Journal of Neuroscience, 28*(35), 8682–8690.

Svenningsson, P., Nairn, A. C., & Greengard, P. (2005). DARPP-32 mediates the actions of multiple drugs of abuse. *AAPS Journal, 7*(2), E353–E360.

Talavera, K., Gees, M., Karashima, Y., Meseguer, V. M., Vanoirbeek, J. A., Damann, N., ... Vennekens, R. (2009). Nicotine activates the chemosensory cation channel TRPA1. *Nature Neuroscience, 12*(10), 1293–1299.

Thinschmidt, J. S., Frazier, C. J., King, M. A., Meyer, E. M., & Papke, R. L. (2005). Medial septal/diagonal band cells express multiple functional nicotinic receptor subtypes that are correlated with firing frequency. *Neuroscience Letters, 389*(3), 163–168.

Tracy, H. A., Jr., Wayner, M. J., & Armstrong, D. L. (1999). Nicotine blocks ethanol and diazepam impairment of air righting and ethanol impairment of maze performance. *Alcohol, 18*(2–3), 123–130.

Traub, R. D., Middleton, S. J., Knopfel, T., & Whittington, M. A. (2008). Model of very fast (>75 Hz) network oscillations generated by electrical coupling between the proximal axons of cerebellar Purkinje cells. *European Journal of Neuroscience, 28*(8), 1603–1616.

Virginio, C., Giacometti, A., Aldegheri, L., Rimland, J. M., & Terstappen, G. C. (2002). Pharmacological properties of rat α7 nicotinic receptors expressed in native and recombinant cell systems. *European Journal of Pharmacology, 445*(3), 153–161.

Wayner, M. J., Armstrong, D. L., & Phelix, C. F. (1996). Nicotine blocks angiotensin II inhibition of LTP in the dentate gyrus. *Peptides, 17*(7), 1127–1133.

Welsby, P. J., Rowan, M. J., & Anwyl, R. (2009). Intracellular mechanisms underlying the nicotinic enhancement of LTP in the rat dentate gyrus. *European Journal of Neuroscience, 29*(1), 65–75.

Wonnacott, S. (1990). The paradox of nicotinic acetylcholine receptor upregulation by nicotine. *Trends in Pharmacological Sciences, 11*(6), 216–219.

Yamamoto, S., Yamada, J., Ueno, S., Kubota, H., Furukawa, T., & Fukuda, A. (2007). Insertion of α7 nicotinic receptors at neocortical layer V GABAergic synapses is induced by a benzodiazepine, midazolam. *Cerebral Cortex, 17*(3), 653–660.

Chapter 17

Cytochrome P450 and Oxidative Stress as Possible Pathways for Alcohol- and Tobacco-Mediated HIV Pathogenesis and NeuroAIDS

Santosh Kumar*, P.S.S. Rao*, Namita Sinha, Narasimha M. Midde
Pharmaceutical Sciences, College of Pharmacy, University of Tennessee Health Science Center, Memphis, TN, USA

Abbreviations

AIDS Acquired immune deficiency syndrome
ART Antiretroviral therapy
BBB Blood–brain barrier
CNS Central nervous system
CYP Cytochrome P450
HAD HIV-associated dementia
HAND HIV-associated neurocognitive disorders
HIV Human immunodeficiency virus
MLC Myosin light chain
NNK Nicotine-derived nitrosamine ketone
NNRTI Nonnucleoside reverse transcriptase inhibitor
NRTI Nucleoside reverse transcriptase inhibitor
PI Protease inhibitor
ROS Reactive oxygen species

INTRODUCTION

Human immunodeficiency virus (HIV) infection has become a global pandemic since the beginning of its outbreak in 1980s. The World Health Organization estimates that approximately 34 million people worldwide are infected with HIV. The majority of HIV-infected cases are from sub-Saharan Africa, accounting for 67% of the total HIV-infected population. A report from UNAIDS, the United Nations Acquired Immune Deficiency Syndrome program, reveals that HIV infection is highly prevalent among pregnant women in Africa, which rapidly increases the rate of infection in newly born babies through mother-to-child transmission. However, the majority of the African population does not have access to effective treatment to circumvent infection among the pediatric population. The introduction of antiretroviral therapy (ART) in 1990s and its rapid progress in the twenty-first century has substantially decreased the mortality rates among individuals living with HIV/AIDS, especially in the Western countries. In addition, the HIV treatment strategy has significantly increased the life span of HIV-infected patients, transforming AIDS from a deadly disease of the twentieth century to a manageable chronic disease in the twenty-first century.

Since HIV-positive patients now live longer, HIV-infected macrophages (secondary target of HIV and major viral reservoir) persistently infiltrate into the central nervous system (CNS) and infect microglia and astrocytes (Williams et al., 2014). HIV-infected brain cells release toxic viral proteins and proinflammatory cytokines, as well as producing oxidative stress resulting in neuroAIDS/HIV-associated neurocognitive dysfunction (HAND) (Peruzzi et al., 2005). In addition to causing slowness of physical movements, neuroAIDS/HAND leads to reduced concentration and memory, which significantly compromises the lifestyle of HIV patients and increases social and financial burdens. Furthermore, the increased prevalence of neuroAIDS makes HIV-infected individuals vulnerable to other neuropsychological and neurobehavioral impairments (Mothobi & Brew, 2012).

Illicit drug use is known to increase the risk of HIV transmission through sharing of the infected needles while administering these drugs. Use of these drugs also drives individuals toward unsafe and inappropriate sexual practices, leading to increased chances of propagating HIV infection (Ross & Williams, 2001). The National Survey on Drug Use and Health data collected from 2005 to 2009 revealed that approximately 20% of individuals with HIV/AIDS have used an intravenous illicit drug in their lifetime. In addition, nearly two-thirds of the HIV-infected population have acknowledged using a nonintravenous illicit drug. Moreover, approximately 24% of HIV-infected individuals were in need of treatment for alcoholism or other substance abuse. Importantly, neuropsychological and neurobehavioral impairments in HIV patients lead to an increase in the prevalence of drug abuse behavior. In turn, addiction to these substances is known to cause severe

---
*Equal contribution.

neurological and psychological impairments, which will further exacerbate the neurological impairments caused by HIV infection and neuroAIDS.

The interaction of drugs of abuse with HIV is very complex and occurs at many levels, e.g., those of biology, behavior, and personality. The impact of HIV pathogenesis, substance of abuse, and AIDS/neuroAIDS development and progression has been discussed in the 2011 edition of *The Neurology of AIDS* by Gendelman et al. published by Oxford University Press. This book also covers perspectives of HIV-infected patients who suffer from neuroAIDS. Their perspectives are intriguing and serve as an eye-opener for researchers in the field of substance of abuse and neuroAIDS. Further, the book highlights several unresolved issues that need to be addressed to rescue/decrease the development of neuroAIDS, especially in populations who regularly abuse drugs. Thus, it is imperative to examine pathways that are responsible for neuroAIDS among HIV-infected drug abusers (Figure 1) and find novel interventions for these individuals. Therefore, this chapter covers the impact of the two most common drugs of abuse, alcohol and tobacco, on HIV pathogenesis and neuroAIDS. Furthermore, this chapter suggests possible pathways mediated through cytochrome P450 (CYP) and oxidative stress that may be involved in alcohol- and tobacco-mediated HIV pathogenesis and neuroAIDS.

ALCOHOL AND NeuroAIDS

Prevalence of Alcohol Use in the HIV Population and Its Impact on Adherence to HIV Medication

Alcohol consumption is very common among HIV-infected individuals (Galvan et al., 2002). Mild to moderate drinking (7–14 standard drinks/week for men and 4–7 standard drinks/week for women) is approximately 2.5 times more prevalent among the HIV-infected population (50–60%) than the general population (20–25%). Similarly, heavy alcohol drinking behavior among HIV-infected individuals is almost twice that of the general population. Needless to say, alcohol drinking, especially heavy drinking, affects adherence to medication among HIV-infected patients. Patients with an alcohol drinking problem are likely to take medicine off schedule or skip dosing intentionally as a result of impaired psychological behavior (Kalichman et al., 2013). Physicians and health care professionals recommend that HIV-infected patients, especially while on ART treatment, abstain from alcohol. However, alcohol users, especially heavy drinkers, do not comply with their recommendations (Hendershot, Stoner, Pantalone, & Simoni, 2009). Current practice guidelines for ART do not exclude patients who regularly consume mild to moderate or even heavy amounts of alcohol. Since patients' compliance on abstaining from alcohol is very low and there are no strict guidelines for alcohol abstention, there is an urgent need to study the impact of alcohol use in HIV-infected individuals to help tailor ART regimens or to develop novel drugs for HIV-infected alcohol drinkers.

Impact of Alcohol Use on HIV Pathogenesis and NeuroAIDS

Heavy as well as mild to moderate alcohol consumption among HIV patients who are on HIV medication is associated with increased viral burden (Miguez, Shor-Posner, Morales, Rodriguez, & Burbano, 2003). Several reports of in vitro and in vivo studies have suggested that alcohol increases HIV replication leading to enhanced immune suppression and rapid progression to AIDS (Haorah et al., 2004; Kumar et al., 2005). Similarly, increased replication of simian immunodeficiency virus in macaques has been reported following chronic ethanol consumption (Kumar et al., 2005). In addition, patients consuming heavy amounts of alcohol while on ART medication are four times less likely to achieve a

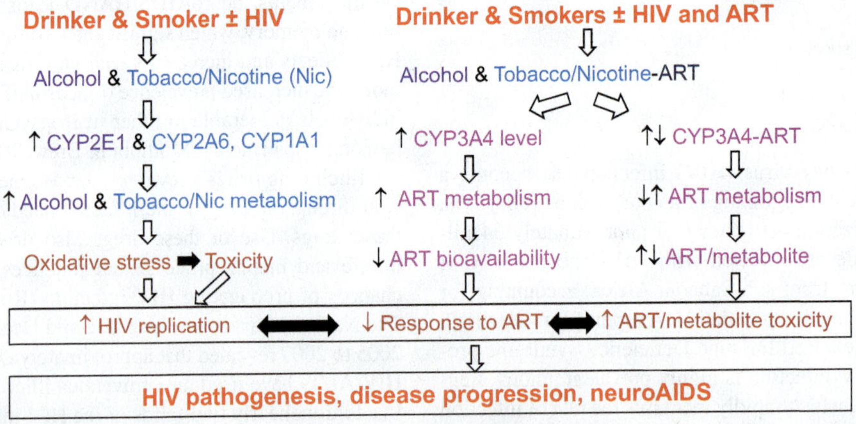

FIGURE 1 Proposed cytochrome P450 (CYP) and oxidative stress pathways for alcohol- and tobacco-mediated HIV progression and neuroAIDS. Alcohol drinking and tobacco use by ART-naïve HIV-positive individuals increase the induction of CYP enzymes, which in turn accelerate the metabolism of these abused drugs, leading to the generation of secondary metabolites and reactive oxygen species. The accumulation of these compounds in the system causes cellular toxicity and eventually enhances HIV replication. Likewise, HIV-positive patients who are on the antiretroviral therapy (ART) regimen and concomitantly use alcohol and/or tobacco are prone to having suboptimal drug concentrations in the system and reduced therapeutic effects on the HIV-infected cells. This is due to the interaction of the abused as well as the antiretroviral drugs with CYP enzymes. These interactions either induce or inhibit the CYP enzymes that can alter the availability of the ART drugs, resulting in virological failure by poorly responding to the medication and/or adverse effects due to metabolite toxicity. Overall, additive or synergistic effects of alcohol, tobacco, and ART drugs mediated through CYP and oxidative stress routes increase the severity and progression of the HIV infection and exacerbate the neurocognitive deficits in the HIV-positive population.

positive virological response. Moreover, heavy alcohol drinkers on ART are two times more likely to have their CD4 counts below 500/mm³ than nondrinkers. Furthermore, chronic alcohol consumption during ART is associated with increased neuronal toxicity (Ferrari & Levine, 2010). Similarly, long-term moderate or heavy alcohol consumption has been shown to exacerbate HIV pathogenesis, particularly in patients with neuroAIDS, resulting in decreased survival of HIV-infected patients (Persidsky et al., 2011).

Monocytes/macrophages are one of the major cellular targets of HIV and function as an important viral reservoir (Igarashi et al., 2001). Compared to activated CD4 T lymphocytes, macrophages are more resistant to the cytopathic effects of HIV and survive HIV infection for extended periods of time (Kedzierska & Crowe, 2002). Furthermore, infiltration of macrophages into the brain results in spreading of the virus to resident glia, including perivascular macrophages, microglia, and astrocytes (Williams et al., 2014). Migration of HIV-infected monocytes through the blood–brain barrier (BBB) is thought to be one of the major mechanisms responsible for the neuropathogenic effects of HIV, such as HIV-associated dementia (HAD) and HIV encephalitis (HIVE). Since alcohol is known to disrupt the BBB and increase infiltration of HIV-infected monocytes/macrophages into the CNS, this is one of the mechanisms by which alcohol can further exacerbate neuroinflammation and neuropsychological impairment (Pfefferbaum et al., 2012).

Studies have shown that the antiviral activity of protease inhibitors (PIs), which are a major part of ART regimes, in chronically infected macrophages is approximately 2- to 10-fold lower than in chronically infected lymphocytes (Perno et al., 2004). However, the antiviral activity of nonnucleoside reverse transcriptase inhibitors (NNRTIs) and nucleoside reverse transcriptase inhibitors (NRTIs) is similar in both cell types. Therefore, strategies that target HIV in chronically infected macrophages need to be developed to treat HIV-infected individuals (Aquaro et al., 2006). Based on findings described in the next section Possible Involvement of Cytochrome P450 and Oxidative Stress Pathways in Alcohol-Mediated HIV Pathogenesis and NeuroAIDS, it is likely that alcohol would further decrease the efficacy of PIs in chronically infected macrophages and lead to suboptimal treatment outcomes in alcohol-consuming HIV-infected individuals. Thus, determination of the pathway(s) involved in alcohol–antiretroviral interactions, especially in HIV-infected macrophages, is important for the development of novel therapeutics and to optimize therapeutic regimens.

Astrocytes are the most abundant glial cells in the brain and are important for protecting the integrity and nourishment of neurons. They play a major role in the pathology of neuroAIDS (Li, Bentsman, Potash, & Volsky, 2007). Astrocytes are also known to be infected by HIV, though to a lesser extent compared to microglia (Wang et al., 2008). In addition, viral proteins and proinflammatory cytokines released by HIV-infected microglia are known to damage astrocytes (Li et al., 2007), which may ultimately lead to neuroAIDS. It has been shown that the HIV protein gp120 induces proinflammatory cytokines such as interleukin-6 in astrocytes (Shah et al., 2011). Furthermore, recent reports show that other HIV proteins such as Nef, Vpr, and Tat also induce CCL5 in astrocytes (Gangwani, Noel, Shah, Rivera-Amill, & Kumar, 2013; Liu et al., 2014; Nookala, Shah, Noel, & Kumar, 2013). CCL5, or RANTES, is a β-chemokine and elevated levels of CCL5 have been observed in neuroinflammatory conditions, including HAD. Tat is one of the two essential regulatory proteins expressed by HIV, while Vpr and Nef are HIV accessory proteins with prominent roles in the HIV life cycle. These neurotoxic proteins are secreted from HIV-infected microglia, peripheral macrophages, and astrocytes in the brain and may ultimately lead to neuroAIDS (Peruzzi et al., 2005). Since alcohol is also known to increase some of these cytokines/chemokines (Valles, Blanco, Pascual, & Guerri, 2004), it is expected that alcohol and viral proteins together will enhance inflammatory cytokines/chemokines in an additive/synergistic manner. Thus, astrocyte-mediated inflammation in HIV-infected alcohol users may play a significant role in neuroAIDS.

Possible Involvement of Cytochrome P450 and Oxidative Stress Pathways in Alcohol-Mediated HIV Pathogenesis and NeuroAIDS

An increased viral load and decreased CD4 counts in HIV-infected drinkers who are not on ART are likely to be a direct effect of oxidative stress mediated through CYP2E1, which has been previously linked to viral replication (Figure 2). Similarly, a decrease in the response to ART drugs leading to increased HIV replication in HIV-infected drinkers is likely to be through CYP3A4-mediated metabolism of ART regimens, such as NNRTIs and PIs (Figure 1). The following sections provide evidence for these hypotheses.

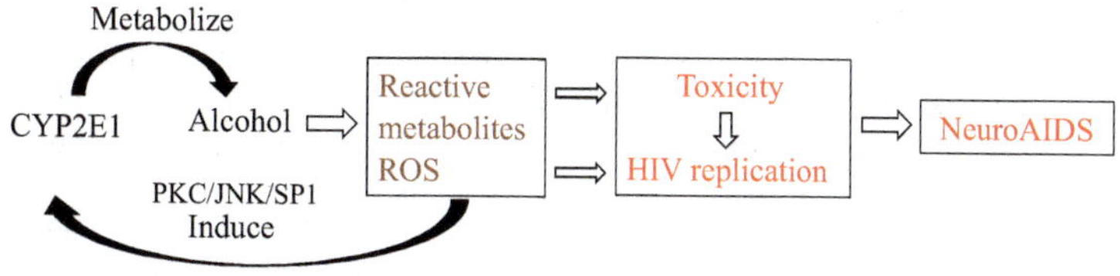

FIGURE 2 **Role of CYP2E1 in alcohol-mediated oxidative stress leading to cellular toxicity and HIV replication and ultimately neuroAIDS.** Metabolism of alcohol by CYP2E1 generates reactive oxygen species and the reactive metabolite acetaldehyde, which can damage cell membranes through lipid peroxidation and cellular organelles through DNA oxidation. This oxidative stress in turn can also induce the expression of CYP2E1 through the protein kinase C/c-Jun N-terminal kinase/specificity protein 1 (PKC/JNK/SP1) pathway, at least in astrocytes and monocytes. In addition to cell toxicity, oxidative stress directly contributes to HIV viral replication. Increased viral replication and oxidative stress-mediated cell death directly or indirectly facilitates HIV disease progression and the pathogenesis of neuroAIDS.

CYP2E1 in Alcohol-Mediated Oxidative Stress in Monocytes/Macrophages, Astrocytes, and Neurons

We have shown the expression and possible role of CYP2E1 in monocytes (Jin et al., 2011). Although its total amount is much lower in monocytes/macrophages than in the liver, the relative amount of CYP2E1 (~3%) to the total level of CYP isoforms is comparable in the U937 monocytic cell line and the liver. Thus, CYP2E1 appears to play an important role in alcohol metabolism and resultant oxidative stress in monocytes/macrophages that may be responsible for exacerbating HIV pathogenesis (Jin, Kumar, & Kumar, 2012). An increase in the expression of CYP2E1, production of reactive oxygen species (ROS), and induction of the antioxidant enzymes superoxide dismutase 1 and catalase by acute alcohol treatment in these cells further strengthened our hypothesis (Jin et al., 2011). Oxidative stress generated through ROS, as well as the imbalance in the levels of glutathione, has been shown to be associated with increased HIV replication in monocytes/macrophages (Mialocq et al., 2001). Together, these findings further suggest that CYP2E1 plays a key role in alcohol-mediated oxidative stress, and this may be a major mechanism responsible for the observed increase in HIV replication in HIV-positive individuals who consume alcohol.

Recent reports have documented the expression of several CYP enzymes, including CYP2E1, in astrocytes (Malaplate-Armand, Leininger-Muller, & Batt, 2004). However, the relative expression level of CYP2E1 (<1%) compared to the total CYP enzymes in the SVGA astrocytic cell line is much lower than its relative expression in the liver (4% of the total amount of CYP isoforms expressed) (Ande et al., 2012). Needless to say, the total levels of CYP enzymes in astrocytes are much lower than in the liver. Although a minor pathway, the presence of CYP2E1 in astrocytes suggests that CYP2E1-mediated alcohol metabolism and the resultant oxidative stress could also be involved in exacerbating the effects of HIV-1 on astrocytes. This hypothesis is supported by our finding that CYP2E1-mediated metabolism of alcohol produces oxidative stress, which causes DNA damage and cell toxicity in astrocytes (Jin, Ande, Kumar, & Kumar, 2013).

Findings from the Haorah group suggest that the metabolism of ethanol followed by the formation of ROS in primary human neurons is mediated mainly through CYP2E1 (Haorah et al., 2008). They have also shown an association between a marked increase in a lipid peroxidation product (4-hydroxynonenal) and enhanced ROS generation with decreased neuronal viability and diminished expression of neurofilament protein, a neuronal marker (Haorah et al., 2008). Furthermore, an in vivo study with rats and an in vitro study with human neuroblastoma IMR-32 cells have shown that exposure to alcohol also induces CYP2E1, resulting in increased ROS generation (Haorah, Knipe, Leibhart, Ghorpade, & Persidsky, 2005; Haorah et al., 2008; Howard, Miksys, Hoffmann, Mash, & Tyndale, 2003). This phenomenon is consistent with alcohol-mediated activation of myosin light chain (MLC) kinase, phosphorylation of MLC and tight-junction proteins, decreased BBB integrity, and enhanced monocyte migration across the BBB. Thus, oxidative stress through the CYP2E1 pathway in brain microvascular endothelial cells may cause BBB breakdown in alcohol abusers and serve as an aggravating factor in neuroinflammatory disorders. Although the role of CYP2E1 is well documented in alcohol-mediated BBB and neuronal damage, there is nothing known about the role of CYP2E1 in HIV-infected individuals who consume alcohol. Since both HIV and alcohol are known to damage the BBB, it is likely that migration of HIV-infected monocytes/macrophages to the CNS is increased in HIV-infected alcohol users, which further leads to neuronal damage in an additive/synergistic manner. Therefore, it is imperative to study the CNS effects mediated by CYP2E1 in the context of HIV-1 infection/pathogenesis.

CYP3A4 in Monocytes/Macrophages, Astrocytes, and Neurons and the Effect of Alcohol on CYP3A4–ART Interactions

CYP3A4 is responsible for the metabolism of all the NNRTIs and PIs that are an integral part of ART regimens. In addition, NNRTIs and PIs are known to induce CYP3A4 protein levels as well as inhibiting CYP3A4 enzyme activity, suggesting drug–drug interactions (Walubo, 2007). Significant variability in the activity of hepatic CYP3A4 has been documented in HIV patients compared to control subjects, suggesting a wide range of inherent drug metabolism in patients (Slain, Pakyz, Israel, Monroe, & Polk, 2000). Since PIs must act in HIV-infected macrophages as well as in T cells, their optimal concentrations in these cells are critical. However, the EC_{50} of PIs is higher in chronically infected macrophages than in T lymphocytes (Aquaro et al., 2006), suggesting that the level of PIs is lower in macrophages than in T cells. This could be a result of a relatively higher level of CYP3A4 in macrophages compared to lymphocytes. Consistent with this, we have shown that CYP3A4 is significantly expressed in U937 monocytic cell lines as well as in primary monocytes/macrophages (Jin et al., 2011). In addition, we have shown that alcohol increases the expression of CYP3A4 in U937 cells (Jin et al., 2011), which is expected to decrease the bioavailability of PIs in monocytes/macrophages, leading to further decrease in antiviral activity in these cells.

CYP3A4 is also expressed in endothelial cells in the BBB and neurons. In general, the levels of CYP3A4 in neurons are very low; however, local concentrations of CYP3A4 in certain regions of the brain could reach as high as the liver CYP3A4 concentrations (Ferguson & Tyndale, 2011). Thus, brain CYP3A4 plays an important role in drug metabolism in the CNS. A recent study also suggests its involvement in cytoprotective mechanisms, because CYP3A4 expression inversely correlates with DAPI nuclear condensation and positively correlates with the amount of carbamazepine metabolism. Accumulation of carbamazepine is known to cause cell toxicity. Although the presence of CYP3A4 is documented in monocytes/macrophages, astrocytes, the BBB, and neurons, there is very little known in terms of the metabolism of PIs or NNRTIs in the context of both HIV infection and alcohol in these cells.

Alcohol can increase the transport of ART drugs through a leaky BBB. Although increased CNS bioavailability of ART drugs will help to decrease the viral load in the brain, increased accumulation of ART drugs in the brain may also cause neurotoxicity. Since CYP3A4 is widely expressed in the CNS (Ferguson & Tyndale, 2011), CYP3A4 may play an important role in alcohol–ART interactions in the CNS. In fact, chronic ethanol consumption is known to induce CYP3A4 in the liver, leading to increased drug metabolism and subsequently decreased drug efficacy and increased drug toxicity (Niemela et al., 1998). This is consistent with the report that HIV-infected chronic drinkers show a significant effect on the therapeutic steady-state plasma drug concentrations of HIV medications including stavudine, lamivudine, and nevirapine (Bbosa et al., 2014).

CYP3A4 induction by alcohol and the impact of CYP3A4 on drug metabolism and toxicity in the liver are well known, and the same phenomenon is expected in monocytes, astrocytes, and neurons. However, CYP3A4–alcohol physical interactions and their impact on drug binding, inhibition, or metabolism have not been investigated. Toward this end, we have shown that alcohol binds to the CYP3A4 active site at physiological concentrations in a type I manner, which is characterized by a noncovalent binding of alcohol with the heme-Fe of CYP3A4 (Kumar, Earla, Jin, Mitra, & Kumar, 2010). Furthermore, we showed that physiological concentrations of ethanol alter the binding and inhibition characteristics of the PIs in a differential manner based on the binding characteristics of the PIs (Kumar & Kumar, 2011). For example, alcohol decreased the maximum spectral change (δA_{max}) with the type I PIs lopinavir and saquinavir, but it did not alter the δA_{max} with other PIs. Similarly, alcohol decreased the IC_{50} of type II PIs, indinavir and ritonavir, and markedly increased the IC_{50} of amprenavir and darunavir. This study suggests a differential effect of ethanol on PIs binding with CYP3A4, which may ultimately alter the metabolism and thus the bioavailability of these drugs in HIV-infected patients. An altered bioavailability of these PIs is likely to alter the efficacy of the PI, as well as increasing PI- or PI metabolite-mediated toxicity in HIV-infected alcohol users (Kumar et al., 2012). Therefore, it is imperative to further study these interactions, so that ART regimens containing appropriate PIs can be tailored for HIV-infected alcohol users.

TOBACCO AND NeuroAIDS

Prevalence of Tobacco Smoking among the HIV Population and Its Impact on Adherence to HIV Medication

Tobacco smoking is very common among HIV-infected individuals, with approximately 50% of HIV patients identified as smokers/nicotine dependent in various parts of the world (Helleberg et al., 2013; Luo et al., 2014). In the United States, approximately 40% of HIV-infected individuals reported smoking and/or nicotine dependence compared to approximately 15% of the uninfected population (Lifson et al., 2010). In addition, the prevalence of smoking in general, as well as in the HIV-infected population, is higher among males compared to females across the world. Smoking/nicotine dependence has been identified as a critical independent predictor of nonadherence to ART among HIV-infected patients (King et al., 2012). More specifically, a study has demonstrated that smoking among female HIV patients is associated with relatively low adherence to ART (Feldman et al., 2006). Compared to nonsmokers, smoking associated with increased levels of depression is likely to be the reason for nonadherence to ART in HIV-infected smokers (Webb, Vanable, Carey, & Blair, 2009).

Impact of Tobacco Smoking on HIV Pathogenesis and NeuroAIDS

The mortality rates among HIV/AIDS patients who smoke are two times higher than among HIV/AIDS patients who do not smoke (Helleberg et al., 2013). In part, increased death rate among HIV-infected smokers is associated with an increase in the prevalence of other viral infections. For example, human papillomavirus infection leading to the risk of cervical cancer has been reported in HIV-seropositive women. Similarly, the incidence of emphysema has been shown to occur earlier, and at an elevated level, among HIV-infected smokers compared to uninfected smokers. The literature suggests that smoking is associated with a decline in CD4 cell counts in the HIV-infected population, even though the CD4 level is higher in uninfected smokers than in nonsmokers (Marshall, McCormack, & Kirk, 2009). However, the role of smoking in the progression to AIDS remains unclear. A 2014 report has demonstrated that neuronal exposure to nicotine significantly increases HIV infectivity (Atluri et al., 2014). Similarly, nicotine treatment has been shown to increase viral replication in microglial cells (Rock et al., 2008). In addition, a multisite longitudinal study for up to 7.9 years of 924 women has shown that tobacco smoking negates the effect of ART medication in relation to CD4 recovery (Feldman et al., 2006). However, the exact mechanism of tobacco-mediated toxicity, increased HIV replication, and decreased efficacy of ART is largely unknown.

There are limited studies with regard to the effects of smoking on cognitive function and neurological disorders in HIV-infected individuals (Wojna et al., 2007). Of relevance, chronic nicotine exposure has been reported to disrupt the BBB integrity by modulating tight-junction proteins (Manda, Mittapalli, Geldenhuys, & Lockman, 2010). Since BBB integrity is critical to prevent the development of neuroAIDS/HAND, smoking could enhance the development and progression of this disorder through nicotine-mediated increased BBB permeability. Hence it is vital to understand the interactions of cigarette smoking and ART with the BBB, which may affect the pharmacokinetic profile of these drugs, leading to decreased efficacy and increased toxicity in the brain. The exposure of these agents to the CNS is likely to exacerbate neuroAIDS and neurodegenerative diseases (Manda et al., 2010). However, another report from a cross-sectional study conducted in 56 women suggests no correlation between smoking and cognitive dysfunction in the HIV-infected population (Wojna et al., 2007). Collectively, the available evidence suggests that there is an increase in smoking-mediated HIV pathogenesis and therefore it is imperative to examine the underlying mechanism to develop novel interventions/drug-dose adjustments to treat HIV-infected smokers effectively.

There is very limited information on the effects of smoking on neuroAIDS/HAND. A 2011 report shows the role of viral proteins in nicotine-induced behavioral sensitization in rats (Midde, Gomez, Harrod, & Zhu, 2011). This sensitization is associated with an altered cAMP and kinase signaling pathway in the mesocorticolimbic system. In addition, in vitro and in vivo studies have revealed that nitrogen-derived nitrosamine ketone (NNK), a tobacco constituent, is known to cause neuroinflammation in the brain, perhaps by activating microglia and astrocytes to induce proinflammatory cytokines/chemokines (Ghosh, Mishra, Das,

Kaushik, & Basu, 2009). In addition, NNK treatment increases phosphorylation of inflammatory signaling molecules such as NF-κB, ERKs, JNK, and p38 MAPK in the brains of BALB/c mice (Ghosh et al., 2009). Since HIV proteins such as gp120, Tat, Vpr, and Nef induce proinflammatory cytokines/chemokines in astrocytes (Gangwani et al., 2013; Liu et al., 2014; Nookala et al., 2013; Shah et al., 2011), NNK-mediated neuroinflammation could further aggravate HAND.

Possible Involvement of Cytochrome P450 and Oxidative Stress Pathways in Smoking-Mediated HIV Pathogenesis and NeuroAIDS

Nicotine, the major constituent of tobacco smoke, is primarily metabolized by CYP2A6 (liver) and CYP2A13 (lungs) to cotinine and other minor metabolites and produces oxidative stress as well as reactive metabolites such as NNK (Benowitz, Hukkanen, & Jacob, 2009). Similarly, CYP1A1-mediated metabolism of polycyclic aromatic hydrocarbons (PAHs) is known to produce oxidative stress and reactive metabolites leading to toxicity and carcinogenicity (Nebert, Dalton, Okey, & Gonzalez, 2004). PAH metabolite-derived DNA adducts have been shown to stimulate HIV replication (Yao, Hoffer, Chang, & Puga, 1995). Feldman et al. has also found a direct association of CYP1A1 with a reduced responsiveness to ART in female smokers with HIV infection, thereby implicating a CYP pathway for the ineffectiveness of ART in smokers (Feldman et al., 2009). However, the mechanism by which smoking/nicotine may reduce efficacy of ART is unknown.

Increased viral loads and decreased CD4 counts in HIV-infected smokers are likely to be a direct effect of oxidative stress through the CYP1A1 and CYP2A6 pathways, which may be linked to viral replication (Figure 3). Similarly, a decrease in the response to ART leading to increased HIV replication in HIV-infected smokers is likely to be mediated through CYP3A4-mediated enhanced metabolism of ART medications. The following sections provide evidence for these hypotheses.

Tobacco/Nicotine-Mediated Oxidative Stress via CYP Pathways in Monocytes/Macrophages, Astrocytes, and Neurons

The major nicotine-metabolizing CYP2A6 enzyme is expressed at higher levels than other CYP enzymes in the U937 monocytic cell line (Jin et al., 2012). A relatively high abundance of CYP2A6 in U937 cells is intriguing, because it suggests an important role for CYP2A6 in these cells. Further investigation in U937 cells shows that nicotine is metabolized to cotinine and NNK by CYP2A6 and generates ROS (Jin et al., 2012). This study also shows the presence in U937 cells of CYP1A1, which is known to metabolize PAHs and produce oxidative stress in the lung and liver cells. However, its role in monocytes/macrophages is yet to be determined. More recently, our study with human cohorts shows an increase in nicotine metabolism in HIV-infected smokers compared with uninfected smokers (Earla, Ande, McArthur, Kumar, & Kumar, 2014). Furthermore, we have demonstrated an increase in oxidative stress and viral replication in HIV-infected smokers compared with uninfected smokers or HIV-infected nonsmokers (unpublished observations). Together, as proposed in a 2013 review (Ande, McArthur, Kumar, & Kumar, 2013), these findings suggest a role for CYP2A6-mediated nicotine metabolism in HIV replication through an oxidative stress pathway. Therefore, further investigations using HIV-infected macrophages (in vitro), HIV-infected humanized mice (in vivo), and human cohorts (ex vivo) are under way.

Investigations also show significant expression of CYP1A1 and CYP2A6 in SVGA astrocytes (Ande et al., 2012). However, their levels in astrocytic cells are much lower than in monocytic cells. In addition, this study demonstrates that, unlike monocytes, CYP1A1 and CYP2A6 are inducible by nicotine in astrocytes (Ande et al., 2012), perhaps through oxidative stress mediated by the PKC/JNK/SP1 pathway, as observed previously (Jin et al., 2013). Furthermore, similar to monocytes, this study demonstrates the role of CYP2A6 in nicotine metabolism and generation of ROS in astrocytes. Taken together, this study suggests a possible role for CYP2A6-mediated nicotine metabolism and oxidative stress in the brain of HIV-infected patients that may ultimately lead to neuroAIDS. Another nicotine-metabolizing enzyme, CYP2B6, is expressed in both astrocytes and neurons (Miksys, Lerman, Shields, Mash, & Tyndale, 2003). An earlier report has shown that rat CYP2B1 (homologous to human CYP2B6) is induced by nicotine in the brain (Khokhar, Miksys, & Tyndale, 2010). Similarly, the level of CYP2B6 is increased in the brains of smokers compared to nonsmokers, specifically in cerebellar Purkinje cells and hippocampal pyramidal neurons (Miksys et al., 2003). Consistent with these observations, a study in African green monkeys

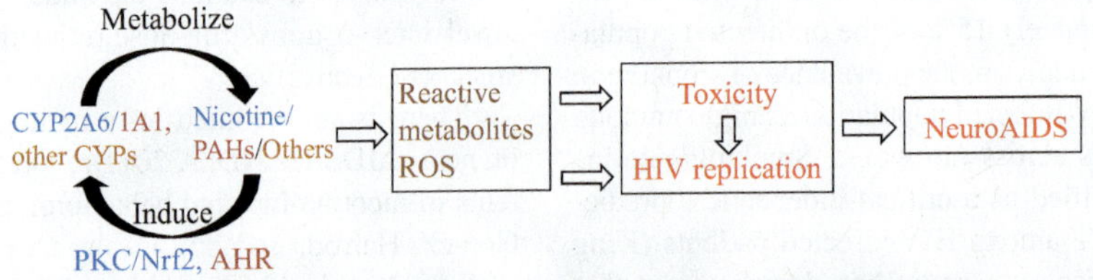

FIGURE 3 **Role of CYP1A1, CYP2A6, and other CYPs in PAH-, nicotine-, and other tobacco constituent-mediated oxidative stress leading to cellular toxicity and HIV replication and ultimately neuroAIDS.** In monocytes and astrocytes of the brain, CYP1A1, CYP2A6, and other CYPs metabolize nicotine, PAHs, and other compounds of tobacco smoke, leading to the generation of toxic metabolites and ROS. The overproduction of these reactive molecules may activate the PKC/Nrf2 pathway (protein kinase C-mediated phosphorylation and translocation of nuclear erythroid 2-related factor 2 to the nucleus), which in turn induces the respective CYPs. This continuous catalytic process triggers the accumulation of more oxidants, resulting in oxidative stress and cellular toxicity, leading to enhanced HIV replication in macrophages and microglia. Increased viral load due to active viral replication concurrent with degeneration of the cells in the brain due to oxidative stress plays an essential role in the neuropathogenesis of HIV infection.

shows the induction of CYP2B6 in the brain following chronic nicotine treatment (Lee, Miksys, Palmour, & Tyndale, 2006). An increased level of CYP2B6 in smokers could lead to an altered metabolism of CYP2B6 substrates. For example, CYP2B6 can metabolize nicotine and activate NNK, leading to increased oxidative stress-mediated damage to the brain (Dicke, Skrlin, & Murphy, 2005). In addition, CYP2B6 metabolizes an important ART drug, efavirenz, which may cause tobacco–ART interactions. Overall, further studies are needed to examine the role of CYP enzymes in tobacco/nicotine-mediated neuropathogenesis and tobacco–ART interactions in neurons.

CYP3A4 in Monocytes/Macrophages, Astrocytes, and Neurons and the Effect of Tobacco/Nicotine on CYP3A4

CYP3A4 is mainly involved in the metabolism of ART drugs, including NNRTIs, PIs, and integrase inhibitors. Altered levels of CYP3A4 in the HIV model systems mediated by tobacco/nicotine are expected to affect the response to ART drugs. As described in the previous section, CYP3A4 is expressed in monocytes, astrocytes, and neurons. However, its role in tobacco/nicotine and HIV has not been elucidated. In this regard, our unpublished findings show a significant induction of CYP3A4 in uninfected smokers, HIV-infected nonsmokers, and HIV-infected smokers compared to uninfected nonsmokers. These results predict that HIV-infected smokers would have increased metabolism of ART drugs, especially NNRTIs, PIs, and integrase inhibitors. This would lead to a decreased response to these ART drugs and increased viral replication in HIV-infected smokers. Therefore, the contribution of the CYP3A4 pathway to increased HIV pathogenesis, via decreased efficacy of ART drugs and increased HIV replication, warrants further investigations.

CONCLUDING REMARKS

In the twenty-first century the advancements in the field of neuroAIDS/HAND research have clearly demonstrated interactions between HIV, ART drugs, and substances of abuse, especially alcohol and tobacco/nicotine. However, the underlying mechanisms for these interactions are poorly understood. Alcohol users and smokers are strongly recommended to abstain from drinking and smoking after they are diagnosed with HIV infection, especially when they are on ART medication. However, most HIV-infected individuals do not follow these recommendations, especially if they are moderate to heavy drinkers/smokers. In addition, these individuals have poor adherence to medication. Thus, these individuals are at higher risk of developing AIDS and neuroAIDS, resulting in early death. Therefore, there is an urgent need to investigate the underlying mechanisms of alcohol- and smoking-mediated HIV pathogenesis and development of neuroAIDS. This will help potentially to develop novel drugs and to adjust drug doses and drug regimens for HIV-infected drinkers and smokers. Research on the possible involvement of CYP and oxidative stress pathways in alcohol- and smoking-mediated HIV pathogenesis and neuroAIDS is in the positive direction toward treating HIV-infected drinkers and smokers effectively. The CYP-mediated oxidative stress pathway has been well documented in alcohol-mediated liver damage, as well as smoking-mediated lung cancer. In fact, several selective inhibitors of CYPs and specific antioxidants are being considered as potential drug candidates to target alcohol-mediated liver damage and smoking-mediated lung cancer. In the near future we expect a greater advancement in the field of CYP and oxidative stress pathways in alcohol- and smoking-mediated HIV pathogenesis and better treatment strategies for HIV-infected drinkers and smokers.

APPLICATIONS TO OTHER ADDICTIONS AND SUBSTANCE MISUSE

In addition to alcohol and smoking, other drugs of abuse, such as cocaine, methamphetamine, opioids, and marijuana, are also highly prevalent among HIV-infected individuals (*The Neurology of AIDS*, by Gendelman et al., published by Oxford University Press, 2011). These substances are also known to modulate CYPs and increase oxidative stress. For example, cocaine, methamphetamine, and opioids have been shown to significantly increase the expression of various CYP isozymes, CYP3A4 in particular, in in vitro studies. CYP enzymes upon induction can metabolize these substances and produce ROS and reactive metabolites, ultimately leading to oxidative stress. Importantly, since induction of CYP enzymes is also known to alter neuronal homeostasis toward proinflammatory/oxidative stress conditions, these observations can be expected to apply to most substances. Drugs of abuse, therefore, may lead to enhanced neuronal death and hasten the progression of neuroAIDS as observed with alcohol and tobacco/smoking.

DEFINITION OF TERMS

Alcoholism This is a chronic and often progressive disease that includes problems controlling the amount of alcohol being consumed, compulsion to drink more to get the same effect (physical dependence), or having withdrawal symptoms when alcohol intake is rapidly decreased or stopped.

Antiretroviral therapy This is a combination of antiretroviral drugs (typically one NRTI, one NNRTI, and two PIs/integrase inhibitor) used for the treatment of HIV infection.

Cytochrome P450 isozymes These are a family of heme-containing enzymes that are responsible for metabolizing xenobiotics such as the majority of marketed drugs, substances of abuse, industrial contaminants, and food and herbal supplements, as well as endogenous compounds such as steroids, vitamins, and fatty acids. They are ubiquitously present in almost all organisms and in many tissue/organ systems.

Heavy alcohol drinker This is an individual who drinks more than 14 (men) or 7 drinks (women) per week. One drink is equivalent to 14 g of pure alcohol.

HIV-associated neurocognitive dysfunction This is an infection of the CNS by HIV leading to impairment of cognition.

HIV pathogenesis This is the replication of HIV and production of new viruses following entry into host (human body) cells (lymphocytes and monocytes).

Humanized mouse This is an experimental model in which a mouse gene is replaced by a portion of or a complete human gene.

NeuroAIDS This refers to neurological conditions associated with HIV infection leading to neurocognitive disorders as well as neurological and psychological impairments.

Nonadherence This describes the lack of administration of prescribed drugs at the recommended dosage and schedule. In general, compliance below 95% is considered suboptimal or nonadherence.

Oxidative stress This is an imbalance between the antioxidant defense mechanism and the production of reactive oxygen species. Increased oxidative stress often leads to cellular toxicity.

KEY FACTS OF NeuroAIDS

- NeuroAIDS is defined as HIV infection of the brain leading to neurocognitive disorders, which are also associated with neurological and psychological impairments.
- NeuroAIDS is also defined by HAND, in which HIV patients show lack of attention and decreased motor activity such as eye–hand coordination.
- The severe forms of HAND are HAD and HIVE.
- NeuroAIDS occurs when HIV-infected macrophages enter into the brain and infect microglia, perivascular macrophages, and astrocytes.
- The HIV-infected brain cells secrete viral proteins and inflammatory cytokines and produce oxidative stress, which together damage neurons eventually leading to neuroAIDS/HAND.
- It is very difficult to prevent and/or treat neuroAIDS because most ART regimens do not cross the BBB.
- The use of various substances of abuse including alcohol and smoking further exacerbates neuroAIDS.

SUMMARY POINTS

- The prevalence of drug abuse, especially alcohol drinking and tobacco smoking, is significantly higher among HIV patients compared to the general population.
- Drug use among HIV patients is associated with exacerbation of neurological deficits and pathophysiology of HIV and neuroAIDS.
- Alcohol-mediated induction of CYP2E1 in macrophages, astrocytes, and neurons is likely to enhance oxidative stress in these cells and cause neuronal damage in HIV patients.
- Nicotine/tobacco-mediated induction of CYP1A1 and CYP2A6 in macrophages, astrocytes, and neurons is likely to enhance oxidative stress in these cells and cause neuronal damage in HIV patients.
- Alcohol- and tobacco/nicotine-mediated induction of CYP3A4 in the liver and macrophages is likely to enhance the metabolism of antiretroviral drugs, leading to a decrease in response to HIV medication.
- A better understanding of the role of CYP and oxidative stress pathways in HIV-infected alcohol and tobacco users would potentially help find novel therapeutics and drug dose/regimen adjustments to treat these patients effectively.

ACKNOWLEDGMENT

The authors acknowledge financial support from National Institutes of Health Grant AA022063.

REFERENCES

Ande, A., Earla, R., Jin, M., Silverstein, P. S., Mitra, A. K., Kumar, A., & Kumar, S. (2012). An LC-MS/MS method for concurrent determination of nicotine metabolites and the role of CYP2A6 in nicotine metabolite-mediated oxidative stress in SVGA astrocytes. *Drug and Alcohol Dependence*, 125(1–2), 49–59.

Ande, A., McArthur, C., Kumar, A., & Kumar, S. (2013). Tobacco smoking effect on HIV-1 pathogenesis: role of cytochrome P450 isozymes. *Expert Opinion on Drug Metabolism & Toxicology*, 9(11), 1453–1464.

Aquaro, S., Svicher, V., Schols, D., Pollicita, M., Antinori, A., Balzarini, J., & Perno, C. F. (2006). Mechanisms underlying activity of antiretroviral drugs in HIV-1-infected macrophages: new therapeutic strategies. *Journal of Leukocyte Biology*, 80(5), 1103–1110.

Atluri, V. S., Pilakka-Kanthikeel, S., Samikkannu, T., Sagar, V., Kurapati, K. R., Saxena, S. K., ... Nair, M. P. (2014). Vorinostat positively regulates synaptic plasticity genes expression and spine density in HIV infected neurons: role of nicotine in progression of HIV-associated neurocognitive disorder. *Molecular Brain*, 7, 37.

Bbosa, G. S., Kyegombe, D. B., Anokbonggo, W. W., Ogwal-Okeng, J., Musoke, D., Odda, J., ... Ntale, M. (2014). Chronic ethanol use in alcoholic beverages by HIV-infected patients affects the therapeutic window of stavudine, lamivudine and nevirapine during the 9-month follow-up period: using chronic alcohol-use biomarkers. *Journal of Basic and Clinical Physiology Pharmacology*, 1–12.

Benowitz, N. L., Hukkanen, J., & Jacob, P., 3rd. (2009). Nicotine chemistry, metabolism, kinetics and biomarkers. *Handbook of Experimental Pharmacology*, 192, 29–60.

Dicke, K. E., Skrlin, S. M., & Murphy, S. E. (2005). Nicotine and 4-(methylnitrosamino)-1-(3-pyridyl)-butanone metabolism by cytochrome p450 2B6. *Drug Metabolism and Disposition*, 33(12), 1760–1764.

Earla, R., Ande, A., McArthur, C., Kumar, A., & Kumar, S. (2014). Enhanced nicotine metabolism in HIV-1-positive smokers compared with HIV-negative smokers: simultaneous determination of nicotine and its four metabolites in their plasma using a simple and sensitive electrospray ionization liquid chromatography-tandem mass spectrometry technique. *Drug Metababolism and Disposition*, 42(2), 282–293.

Feldman, D. N., Feldman, J. G., Greenblatt, R., Anastos, K., Pearce, L., Cohen, M., ... Burk, R. (2009). Cyp1a1 genotype modifies the impact of smoking on effectiveness of HAART among women. *AIDS Education and Prevention*, 21(3), 81–93.

Feldman, J. G., Minkoff, H., Schneider, M. F., Gange, S. J., Cohen, M., Watts, H., ... Anastos, K. (2006). Association of cigarette smoking with HIV prognosis among women in the HAART era: a report from the Women's Interagency HIV study. *American Journal of Public Health*, 96(6), 1060–1065.

Ferguson, C. S., & Tyndale, R. F. (2011). Cytochrome P450 enzymes in the brain: emerging evidence of biological significance. *Trends in Pharmacological Sciences*, 32(12), 708–714.

Ferrari, L. F., & Levine, J. D. (2010). Alcohol consumption enhances antiretroviral painful peripheral neuropathy by mitochondrial mechanisms. *European Journal of Neuroscience*, 32(5), 811–818.

Galvan, F. H., Bing, E. G., Fleishman, J. A., London, A. S., Caetano, R., Burnam, M. A., ... Shapiro, M. (2002). The prevalence of alcohol consumption and heavy drinking among people with HIV in the United States: results from the HIV cost and services utilization study. *Journal of Studies on Alcohol*, 63(2), 179–186.

Gangwani, M. R., Noel, R. J., Shah, A., Rivera-Amill, V., & Kumar, A. (2013). Human immunodeficiency virus type 1 viral protein R (Vpr) induces CCL5 expression in astrocytes via PI3K and MAPK signaling pathways. *Journal of Neuroinflammation, 10*.

Ghosh, D., Mishra, M. K., Das, S., Kaushik, D. K., & Basu, A. (2009). Tobacco carcinogen induces microglial activation and subsequent neuronal damage. *Journal of Neurochemistry, 110*(3), 1070–1081.

Haorah, J., Heilman, D., Diekmann, C., Osna, N., Donohue, T. M., Jr., Ghorpade, A., & Persidsky, Y. (2004). Alcohol and HIV decrease proteasome and immunoproteasome function in macrophages: implications for impaired immune function during disease. *Cellular Immunology, 229*(2), 139–148.

Haorah, J., Knipe, B., Leibhart, J., Ghorpade, A., & Persidsky, Y. (2005). Alcohol-induced oxidative stress in brain endothelial cells causes blood–brain barrier dysfunction. *Journal of Leukocyte Biology, 78*(6), 1223–1232.

Haorah, J., Ramirez, S. H., Floreani, N., Gorantla, S., Morsey, B., & Persidsky, Y. (2008). Mechanism of alcohol-induced oxidative stress and neuronal injury. *Free Radical Biology and Medicine, 45*(11), 1542–1550.

Helleberg, M., Afzal, S., Kronborg, G., Larsen, C. S., Pedersen, G., Pedersen, C., ... Obel, N. (2013). Mortality attributable to smoking among HIV-1-infected individuals: a nationwide, population-based cohort study. *Clinical Infectious Diseases, 56*(5), 727–734.

Hendershot, C. S., Stoner, S. A., Pantalone, D. W., & Simoni, J. M. (2009). Alcohol use and antiretroviral adherence: review and meta-analysis. *Journal of Acquired Immune Deficiency Syndromes, 52*(2), 180–202.

Howard, L. A., Miksys, S., Hoffmann, E., Mash, D., & Tyndale, R. F. (2003). Brain CYP2E1 is induced by nicotine and ethanol in rat and is higher in smokers and alcoholics. *British Journal of Pharmacology, 138*(7), 1376–1386.

Igarashi, T., Brown, C. R., Endo, Y., Buckler-White, A., Plishka, R., Bischofberger, N., ... Martin, M. A. (2001). Macrophage are the principal reservoir and sustain high virus loads in rhesus macaques after the depletion of CD4+ T cells by a highly pathogenic simian immunodeficiency virus/HIV type 1 chimera (SHIV): implications for HIV-1 infections of humans. *Proceedings of the National Academy of Sciences of the United States of America, 98*(2), 658–663.

Jin, M., Ande, A., Kumar, A., & Kumar, S. (2013). Regulation of cytochrome P450 2e1 expression by ethanol: role of oxidative stress-mediated pkc/jnk/sp1 pathway. *Cell Death & Disease, 4*.

Jin, M., Arya, P., Patel, K., Singh, B., Silverstein, P. S., Bhat, H. K., ... Kumar, S. (2011). Effect of alcohol on drug efflux protein and drug metabolic enzymes in U937 macrophages. *Alcoholism-Clinical and Experimental Research, 35*(1), 132–139.

Jin, M., Earla, R., Shah, A., Earla, R. L., Gupte, R., Mitra, A. K., ... Kumar, S. (2012). A LC-MS/MS method for concurrent determination of nicotine metabolites and role of CYP2A6 in nicotine metabolism in U937 macrophages: implications in oxidative stress in HIV+smokers. *Journal of Neuroimmune Pharmacology, 7*(1), 289–299.

Jin, M., Kumar, A., & Kumar, S. (2012). Ethanol-mediated regulation of cytochrome P450 2A6 expression in monocytes: role of oxidative stress-mediated PKC/MEK/Nrf2 pathway. *PLoS One, 7*(4).

Kalichman, S. C., Grebler, T., Amaral, C. M., McNerey, M., White, D., Kalichman, M. O., ... Eaton, L. (2013). Intentional non-adherence to medications among HIV positive alcohol drinkers: prospective study of interactive toxicity beliefs. *Journal of General Internal Medicine, 28*(3), 399–405.

Kedzierska, K., & Crowe, S. M. (2002). The role of monocytes and macrophages in the pathogenesis of HIV-1 infection. *Current Medicinal Chemistry, 9*(21), 1893–1903.

Khokhar, J. Y., Miksys, S. L., & Tyndale, R. F. (2010). Rat brain CYP2B induction by nicotine is persistent and does not involve nicotinic acetylcholine receptors. *Brain Research, 1348*, 1–9.

King, R. M., Vidrine, D. J., Danysh, H. E., Fletcher, F. E., McCurdy, S., Arduino, R. C., & Gritz, E. R. (2012). Factors associated with nonadherence to antiretroviral therapy in HIV-positive smokers. *AIDS Patient Care and STDs, 26*(8), 479–485.

Kumar, S., Earla, R., Jin, M., Mitra, A. K., & Kumar, A. (2010). Effect of ethanol on spectral binding, inhibition, and activity of CYP3A4 with an antiretroviral drug nelfinavir. *Biochemical and Biophysical Research Communications, 402*(1), 163–167.

Kumar, S., Jin, M. Y., Ande, A., Sinha, N., Silverstein, P. S., & Kumar, A. (2012). Alcohol consumption effect on antiretroviral therapy and HIV-1 pathogenesis: role of cytochrome P450 isozymes. *Expert Opinion on Drug Metabolism & Toxicology, 8*(11), 1363–1375.

Kumar, S., & Kumar, A. (2011). Differential effects of ethanol on spectral binding and inhibition of cytochrome P450 3A4 with eight protease inhibitors antiretroviral drugs. *Alcoholism-Clinical and Experimental Research, 35*(12), 2121–2127.

Kumar, R., Perez-Casanova, A. E., Tirado, G., Noel, R. J., Torres, C., Rodriguez, I., ... Kumar, A. (2005). Increased viral replication in simian immunodeficiency virus/simian-HIV-infected macaques with self-administering model of chronic alcohol consumption. *Journal of Acquired Immune Deficiency Syndromes, 39*(4), 386–390.

Lee, A. M., Miksys, S., Palmour, R., & Tyndale, R. F. (2006). CYP2B6 is expressed in African Green monkey brain and is induced by chronic nicotine treatment. *Neuropharmacology, 50*(4), 441–450.

Li, J. L., Bentsman, G., Potash, M. J., & Volsky, D. J. (2007). Human immunodeficiency virus type 1 efficiently binds to human fetal astrocytes and induces neuroinflammatory responses independent of infection. *BMC Neuroscience, 8*.

Lifson, A. R., Neuhaus, J., Arribas, J. R., van den Berg-Wolf, M., Labriola, A. M., Read, T. R. H., & Grp, I. S. S. (2010). Smoking-related health risks among persons with HIV in the strategies for management of antiretroviral therapy clinical trial. *American Journal of Public Health, 100*(10), 1896–1903.

Liu, X., Shah, A., Gangwani, M. R., Silverstein, P. S., Fu, M. G., & Kumar, A. (2014). HIV-1 Nef induces CCL5 production in astrocytes through p38-MAPK and PI3K/Akt pathway and utilizes NF-kB, CEBP and AP-1 transcription factors. *Scientific Reports, 4*.

Luo, X. F., Duan, S., Duan, Q. X., Pu, Y. C., Yang, Y. C., Ding, Y. Y., ... He, N. (2014). Tobacco use among HIV-infected individuals in a rural community in Yunnan Province, China. *Drug and Alcohol Dependence, 134*, 144–150.

Malaplate-Armand, C., Leininger-Muller, B., & Batt, A. M. (2004). Astrocytic cytochromes p450: an enzyme subfamily critical for brain metabolism and neuroprotection. *Revue Neurologique, 160*(6–7), 651–658.

Manda, V. K., Mittapalli, R. K., Geldenhuys, W. J., & Lockman, P. R. (2010). Chronic exposure to nicotine and saquinavir decreases endothelial Notch-4 expression and disrupts blood–brain barrier integrity. *Journal of Neurochemistry, 115*(2), 515–525.

Marshall, M. M., McCormack, M. C., & Kirk, G. D. (2009). Effect of cigarette smoking on HIV acquisition, progression, and mortality. *AIDS Education and Prevention, 21*(Suppl. 3), 28–39.

Mialocq, P., Oiry, J., Puy, J. Y., Rimaniol, A. C., Imbach, J. L., Dormont, D., & Clayette, P. (2001). Oxidative metabolism of HIV-infected macrophages: the role of glutathione and a pharmacologic approach. *Pathologie Biologie (Paris), 49*(7), 567–571.

Midde, N. M., Gomez, A. M., Harrod, S. B., & Zhu, J. (2011). Genetically expressed HIV-1 viral proteins attenuate nicotine-induced behavioral sensitization and alter mesocorticolimbic ERK and CREB signaling in rats. *Pharmacology Biochemistry and Behavior, 98*(4), 587–597.

Miguez, M. J., Shor-Posner, G., Morales, G., Rodriguez, A., & Burbano, X. (2003). HIV treatment in drug abusers: impact of alcohol use. *Addiction Biology, 8*(1), 33–37.

Miksys, S., Lerman, C., Shields, P. G., Mash, D. C., & Tyndale, R. T. (2003). Smoking, alcoholism and genetic polymorphisms alter CYP2B6 levels in human brain. *Neuropharmacology, 45*(1), 122–132.

Mothobi, N. Z., & Brew, B. J. (2012). Neurocognitive dysfunction in the highly active antiretroviral therapy era. *Current Opinion in Infectious Diseases, 25*(1), 4–9.

Nebert, D. W., Dalton, T. P., Okey, A. B., & Gonzalez, F. J. (2004). Role of aryl hydrocarbon receptor-mediated induction of the CYP1 enzymes in environmental toxicity and cancer. *Journal of Biological Chemistry, 279*(23), 23847–23850.

Niemela, O., Parkkila, S., Pasanen, M., Iimuro, Y., Bradford, B., & Thurman, R. G. (1998). Early alcoholic liver injury: formation of protein adducts with acetaldehyde and lipid peroxidation products, and expression of CYP2E1 and CYP3A. *Alcoholism-Clinical and Experimental Research, 22*(9), 2118–2124.

Nookala, A. R., Shah, A., Noel, R. J., & Kumar, A. (2013). HIV-1 Tat-mediated induction of CCL5 in astrocytes involves NF-kappaB, AP-1, C/EBPalpha and C/EBPgamma transcription factors and JAK, PI3K/Akt and p38 MAPK signaling pathways. *PLoS One, 8*(11), e78855.

Perno, C. F., Balestra, E., Francesconi, M., Abdelahad, D., Calio, R., Balzarini, J., & Aquaro, S. (2004). Antiviral profile of HIV inhibitors in macrophages: Implications for therapy. *Current Topics in Medicinal Chemistry, 4*(9), 1009–1015.

Persidsky, Y., Ho, W. Z., Ramirez, S. H., Potula, R., Abood, M. E., Unterwald, E., & Tuma, R. (2011). HIV-1 infection and alcohol abuse: neurocognitive impairment, mechanisms of neurodegeneration and therapeutic interventions. *Brain Behavior and Immunity, 25*, S61–S70.

Peruzzi, F., Bergonzini, V., Aprea, S., Reiss, K., Sawaya, B. E., Rappaport, J., ... Khalili, K. (2005). Cross talk between growth factors and viral and cellular factors alters neuronal signaling pathways: implication for HIV-associated dementia. *Brain Research Reviews, 50*(1), 114–125.

Pfefferbaum, A., Rosenbloom, M. J., Sassoon, S. A., Kemper, C. A., Deresinski, S., Rohlfing, T., & Sullivan, E. V. (2012). Regional brain structural dysmorphology in human immunodeficiency virus infection: effects of acquired immune deficiency syndrome, alcoholism, and age. *Biological Psychiatry, 72*(5), 361–370.

Rock, R. B., Gekker, G., Aravalli, R. N., Hu, S. X., Sheng, W. S., & Peterson, P. K. (2008). Potentiation of HIV-1 expression in microglial cells by nicotine: involvement of transforming growth factor-beta 1. *Journal of Neuroimmune Pharmacology, 3*(3), 143–149.

Ross, M. W., & Williams, M. L. (2001). Sexual behavior and illicit drug use. *Annual Review of Sex Research, 12*, 290–310.

Shah, A., Verma, A. S., Patel, K. H., Noel, R., Rivera-Amill, V., Silverstein, P. S., ... Kumar, A. (2011). HIV-1 gp120 induces expression of IL-6 through a nuclear factor-kappa B-dependent mechanism: suppression by gp120 specific small interfering RNA. *PLoS One, 6*(6).

Slain, D., Pakyz, A., Israel, D. S., Monroe, S., & Polk, R. E. (2000). Variability in activity of hepatic CYP3A4 in patients infected with HIV. *Pharmacotherapy, 20*(8), 898–907.

Valles, S. L., Blanco, A. M., Pascual, M., & Guerri, C. (2004). Chronic ethanol treatment enhances inflammatory mediators and cell death in the brain and in astrocytes. *Brain Pathology, 14*(4), 365–371.

Walubo, A. (2007). The role of cytochrome P450 in antiretroviral drug interactions. *Expert Opinion on Drug Metabolism and Toxicology, 3*(4), 583–598.

Wang, T., Gong, N., Liu, J. U., Kadiu, I., Kraft-Terry, S. D., Schlautman, J. D., ... Gendelman, H. E. (2008). HIV-1-infected astrocytes and the microglial proteome. *Journal of Neuroimmune Pharmacology, 3*(3), 173–186.

Webb, M. S., Vanable, P. A., Carey, M. P., & Blair, D. C. (2009). Medication adherence in HIV-infected smokers: the mediating role of depressive symptoms. *AIDS Education and Prevention, 21*(3), 94–105.

Williams, D. W., Veenstra, M., Gaskill, P. J., Morgello, S., Calderon, T. M., & Berman, J. W. (2014). Monocytes mediate HIV neuropathogenesis: mechanisms that contribute to HIV associated neurocognitive disorders. *Current HIV Research, 12*(2), 85–96.

Wojna, V., Robles, L., Skolasky, R. L., Mayo, R., Selnes, O., de la Torre, T., ... Lasalde-Dominicci, J. (2007). Associations of cigarette smoking with viral immune and cognitive function in human immunodeficiency virus-seropositive women. *Journal of Neurovirology, 13*(6), 561–568.

Yao, Y., Hoffer, A., Chang, C. Y., & Puga, A. (1995). Dioxin activates HIV-1 gene-expression by an oxidative stress pathway requiring a functional cytochrome-p450 Cyp1a1 enzyme. *Environmental Health Perspectives, 103*(4), 366–371.

Chapter 18

Nicotine and Neurokinin Signaling

Mariella De Biasi[1,2,3], Ian McLaughlin[1,3], Michelle L. Klima[1]

[1]Department of Psychiatry, University of Pennsylvania, Philadelphia, PA, USA; [2]Department of Neuroscience, University of Pennsylvania, Philadelphia, PA, USA; [3]Neuroscience Graduate Group, Perelman School of Medicine, University of Pennsylvania, Philadelphia, PA, USA

Abbreviations

5-HT 5-hydroxytryptamine, serotonin
ACh Acetylcholine
AM Amygdala
AMPA α-amino-3-hydroxy-5-methyl-4-isoxazolepropionic acid
BF Basal forebrain
BNST Bed nuclei stria terminalis
CNQX 6-cyano-7-nitroquinoxaline-2,3-dione
CNS Central nervous system
CP Caudate putamen
CPP Conditioned place preference
DA Dopamine
dMHb Dorsal medial habenula
DR Dorsal raphe
EKA Endokinin A
EKB Endokinin B
EKC Endokinin C
EKD Endokinin D
EPM Elevated plus maze
EPSC Excitatory post synaptic current
FC Frontal cortex
GABA γ-aminobutyric acid
GnRH Gonadotropin-releasing hormone
HC Hippocampus
HK-1 Hemokinin-1
HTH Hypothalamus
ICSS Intracranial self-stimulation paradigm
ICV Intracerebroventricular
IPN Interpeduncular nucleus
LC Locus coeruleus
MHb Medial habenula
MOR μ-opioid receptor
mRNA Messenger ribonucleic acid
NAcc Nucleus accumbens
nAChR Nicotinic acetylcholine receptors
NK1R Neurokinin 1 receptor
NK2R Neurokinin 2 receptor
NK3R Neurokinin 3 receptor
NKA Neurokinin A
NKB Neurokinin B
NMDA N-methyl-D-aspartate
OB Olfactory bulb
PAG Periaqueductal gray
PBN Parabrachial nuclei
PNS Peripheral nervous system
PPT Preprotachykinin
PVN Paraventricular nucleus
SC Superior colliculus
SEP Septum
SN Substantia nigra
SP Substance P
Tac1 Mouse tachykinin 1 gene
TAC1 Tachykinin 1 gene
Tac2 Mouse tachykinin 2 gene
TAC2 Tachykinin 2 gene
TAC3 Tachykinin 3 gene
TAC4 Tachykinin 4 gene
TH Thalamus
vMHb Ventral medial habenula
VTA Ventral tegmental area
ZI Zona incerta

INTRODUCTION

Neurokinins belong to the tachykinin family of proteins, which comprises a series of structurally related neuropeptides found in many species, from amphibians to mammals. Tachykinins are expressed throughout the nervous and immune systems and participate in a variety of physiological processes, including inflammation, nociception, smooth muscle contractility, epithelial secretion, and proliferation. They also contribute to various disease processes, such as acute and chronic inflammation and pain, fibrosis, bladder and intestinal disorders, infection, and cancer. In addition, pharmacological and biochemical evidence implicates the tachykinin signaling system in the mechanisms underlying aversive affective states, as well as several pathological brain states. Animal models have shown long-lasting effects of stress on neurokinin receptor regulation, and pharmacological experiments have demonstrated an ability of neurokinin receptor agonists and antagonists to modulate anxiety levels. Coupled with the distribution of peptides and receptors in the central nervous system (CNS), preclinical and clinical experiments are identifying the tachykinin system as a clear regulator of aversive affect and addiction.

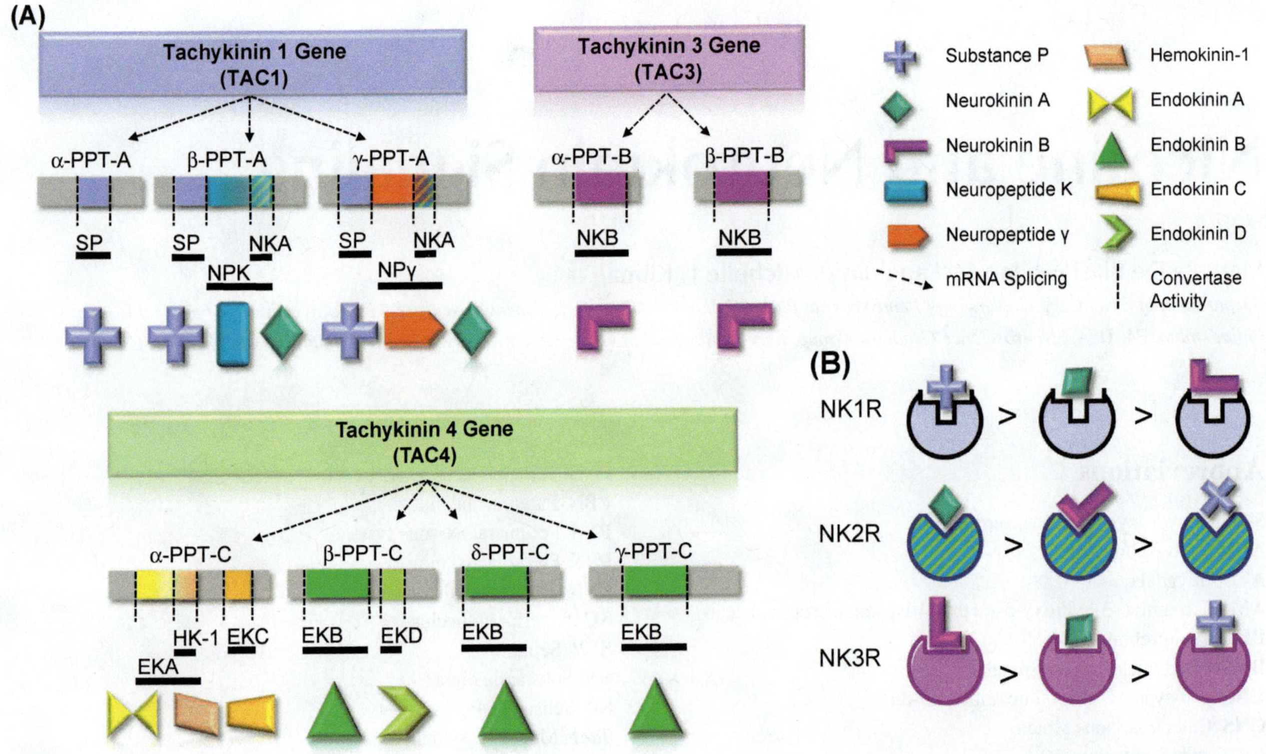

FIGURE 1 Coding and processing of tachykinin peptides. (A) Neurokinins are the product of three genes: tachykinin 1 (*TAC1*), *TAC3*, and *TAC4*. The mRNA from these genes is spliced to form precursor proteins, preprotachykinin (PPT) A, B, and C, which are further cleaved by convertases to give the final peptide products. Substance P, neurokinin A, and neurokinin B are the major neurokinins produced. (B) Neurokinins bind to three G-protein-coupled receptors called neurokinin 1 receptor (NK1R), NK2R, and NK3R. Although all neurokinins can bind to these receptors, each neurokinin has a preferential affinity for one NKR.

TACHYKININ GENES AND GENE PRODUCTS

In humans, there are three tachykinin genes: *TAC1* and *TAC3*, which are equivalent to *Tac1* and *Tac2* in the mouse, respectively, and *TAC4* (Figure 1). These genes encode precursor proteins, called preprotachykinins, which undergo posttranslational proteolytic cleavage to yield the active peptides (Steinhoff, von Mentzer, Geppetti, Pothoulakis, & Bunnett, 2014). *TAC1* encodes neurokinin A (NKA; also known as neurokinin α and neuromedin L), neuropeptide γ, neuropeptide K, and substance P (SP). Three splice variants of *TAC1* are known and produce different sets of peptides. All three splice forms of *TAC1* produce SP, but only the β and γ forms produce the other three peptides. *TAC2* was initially assigned to the gene encoding the NKA precursor, but was subsequently found to be identical to *TAC1*. *TAC3* encodes neurokinin B (NKB; also known as neurokinin β and neuromedin K). Finally, *TAC4* encodes hemokinin-1 and its N-terminal extended forms, endokinin A (EKA) and EKB (Figure 1). EKA and EKB have biological actions similar to those of SP and can interact with neurokinin receptors (NKRs), while EKC and EKD lack the tachykinin sequence, have minimal tachykinin-like actions, and show negligible affinity for the NKRs (Steinhoff et al., 2014).

NEUROPEPTIDE DISTRIBUTION IN THE CNS

Neurokinins and their receptors are found throughout the brain, where they have both overlapping and distinct patterns of expression. High levels of SP immunoreactivity are found in many forebrain, midbrain, and brainstem areas. Those areas include the cingulate cortex, caudate putamen, nucleus accumbens, septum, hippocampus, amygdala, various hypothalamic nuclei, periaqueductal gray, dorsal raphe nucleus, locus coeruleus, various parabrachial nuclei, and the nucleus of the tractus solitarius (Hökfelt, Kuteeva, Stanic, & Ljungdahl, 2004). In general, the distribution of SP in the human brain is quite similar to that in the rat brain, with particularly dense distributions of immunoreactive fibers or SP-positive neurons in cortex, hypothalamus, hippocampus, substantia nigra, and brain stem (Hökfelt et al., 2004).

The immunoreactivity for NKA is highly colocalized with that of SP owing to the shared precursor gene (Figures 1 and 2). However, despite being colocalized, NKA and SP may be expressed at different levels in a given tissue. For example, SP is more abundant than NKA in striatum and substantia nigra, while NKA levels are higher in the hippocampus (Arai & Emson, 1986).

The distribution of NKB-positive neurons and fibers is, in general, distinct from that of other neuropeptides. NKB is highly expressed in the olfactory bulb, certain hypothalamic nuclei, the basal forebrain, and the habenulointerpeduncular tract. The bed nucleus of the stria terminalis presents some of the highest immunoreactivities and mRNA levels for both NKB and SP (Warden & Young, 1988). The concentration of NKB is also high in the dorsal and ventral portions of the medial habenula (MHb) (Marksteiner, Sperk, & Krause, 1992). This is in contrast with the distribution of SP immunoreactivity and mRNA, which are mainly present in the dorsal part of the MHb. MHb neurons send projections to the

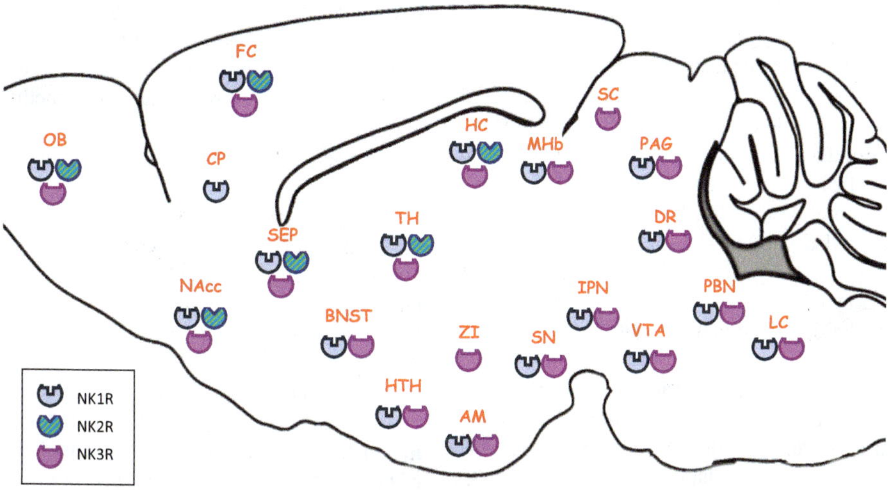

FIGURE 2 **NK1R, NK2R, and NK3R are located throughout the CNS.** The receptors are predominantly found on limbic structures and often overlap with serotonergic and noradrenergic pathways, implicating neurokinins in affective behavior and mood regulation. OB, olfactory bulb; FC, frontal cortex; CP, caudate putamen; SEP, septum; NAcc, nucleus accumbens; BNST, bed nuclei of stria terminalis; HTH, hypothalamus; AM, amygdala; TH, thalamus; ZI, zona incerta; HC, hippocampus; MHb, medial habenula; SC, superior colliculus; PAG, periaqueductal gray; DR, dorsal raphe; PBN, parabrachial nuclei; LC, locus coeruleus; VTA, ventral tegmental area; IPN, interpeduncular nucleus; SN, substantia nigra.

interpeduncular nucleus (IPN) that form an extremely dense triangular network of NKB-immunoreactive fibers, visible especially in the ventrolateral portion of the rostral IPN. At a more caudal level, this dense immunoreactive plexus separates into the intermediate and lateral nuclei. No NKB mRNA is observed within IPN neurons (Marksteiner et al., 1992). The IPN also receives SP-rich projections from the MHb (Cuello & Kanazawa, 1978). The striatum and substantia nigra pars reticulata contain very few NKB-positive neurons, in contrast with the prominent presence of SP-positive neurons (Burgunder & Young, 1989). Conversely, the hippocampal formation expresses very low levels of SP but contains rather high concentrations of NKB mRNA and immunoreactivity (Ribeiro-da-Silva & Hokfelt, 2000).

RECEPTOR DISTRIBUTION IN THE CNS

NK1, NK2, and NK3 are the three NKR subtypes found in mammals. SP binds preferentially to NK1R, NKA to NK2R, and NKB to NK3R. Despite binding with high affinity to one receptor subtype, each tachykinin can activate all three receptors, a feature that may contribute to pathophysiological conditions characterized by increased tachykinin release (Ebner, Sartori, & Singewald, 2009).

NK1Rs are widely distributed in the CNS, the peripheral nervous system (PNS), and various peripheral organs. They are highly expressed in brain areas known to regulate emotions and the response to stress, such as the amygdala, hypothalamus, hippocampus, frontal cortex, raphe, and locus coeruleus. In those areas, SP and NK1Rs are colocalized. However, in regions such as the substantia nigra and the lateral interpeduncular nuclei, the intense SP staining does not always match the signal for NK1Rs (Mantyh, 2002). In those areas, SP may bind to closely related NK2Rs or NK3Rs (Saria, 1999).

NK2Rs have a limited distribution in the CNS and are most abundant in peripheral tissues. In the CNS, NK2Rs are found in some cortical areas, the hippocampus, the nucleus accumbens, parts of the thalamus, and the lateral septum (Hökfelt et al., 2004; Saffroy, Torrens, Glowinski, & Beaujouan, 2003). As discussed for NK1Rs, the presence of NK2Rs in several limbic structures is consistent with a role for NKA in the processing of emotions (Hökfelt et al., 2004).

NK3Rs are mostly found in the CNS. They are expressed in the cerebral cortex, the zona incerta, the MHb, the amygdala, the superior colliculus, and the IPN, as well as the septum, dorsal raphe, periaqueductal gray, and locus coeruleus (Ebner et al., 2009). Despite considerable anatomical overlap of NK1Rs and NK3Rs, there are marked differences in their distribution patterns in certain brain areas. For example, strong NK1R signals are found in the medial and cortical portions of the amygdala, while NK3R expression is mainly localized to basomedial and basolateral nuclei. Distinct patterns of expression for the two receptors are also found in septum, hypothalamus, and prefrontal cortex (Ebner et al., 2009; Hökfelt et al., 2004; Saffroy et al., 2003).

NEUROKININ-MEDIATED MECHANISMS IN THE CNS

Tachykinins are found in the axons of many neuronal populations, including those that release glutamate, γ-aminobutyric acid (GABA), serotonin, and acetylcholine (Commons, 2010). In addition to axons, neurokinin-containing large dense-core vesicles are found in somata, dendrites, and neuronal varicosities. Neuropeptides are packaged in large dense-core vesicles that are distinct from the small synaptic vesicles containing the coexisting neurotransmitter (Keegan, Woodruff, & Pinnock, 1992; Mantyh, 2002). The neuropeptides present in various neuronal classes are coreleased with classical neurotransmitters or neuromodulators, especially upon strong activation of the parent cell. The mechanisms that regulate neurokinin release are different from those that regulate fast synaptic transmission, as neurokinin-mediated responses have a slower time course of activation compared to those triggered by ionotropic neurotransmitters. Owing to these characteristics, neurokinins provide a spatially and temporally defined refinement of neuronal circuits.

Dopamine

SP is prominently expressed in striatal structures and has the ability to increase extracellular dopamine (DA) levels (Ribeiro-da-Silva & Hokfelt, 2000). In particular, there seems to be a positive feedback loop involving the striatum and the mesencephalon in which the DA released from ventral tegmental area (VTA) projections can induce the expression of TAC1 mRNA coding for SP in the striatum (Campbell & Walker, 2001). Conversely, the SP released from striatal projections onto the VTA can increase DA release from midbrain neurons (Kovács, Steinmann, Magistretti, Halfon, & Cardinaux, 2006). Given that DA release is associated with reward and reward prediction, such a positive feedback loop is likely to play an important role in motivation and addiction. In addition to NK1Rs, the VTA and substantia nigra pars compacta express NK3Rs, and application of the selective NK3R agonist senktide to those two areas can also enhance the firing of DA neurons (Keegan et al., 1992) and trigger DA release in projection sites (Marco et al., 1998).

Serotonin

A number of studies suggest that neurokinins may induce serotonin (5-HT)-mediated behaviors by enhancing the release of endogenous 5-HT through the increase of 5-HT cell firing in the raphe nuclei (Gradin, Qadri, Nomikos, Hillegaart, & Svensson, 1992). Electrophysiology studies in brain slice preparations showed that both NK1 and NK3 receptor agonists, rather than acting directly on 5-HT neurons, activate them indirectly, by increasing excitatory postsynaptic currents (EPSCs) (Liu, Ding, & Aghajanian, 2002). The EPSCs are blocked by the α-amino-3-hydroxy-5-methyl-4-isoxazolepropionic acid (AMPA)/kainate glutamate receptor antagonist 6-cyano-7-nitroquinoxaline-2,3-dione and can be prevented by tetrodotoxin, suggesting that neurokinins enhance local glutamatergic neuronal afferents (Liu et al., 2002).

Norepinephrine

Locus coeruleus (LC) neurons are innervated by SP-containing fibers (Baker et al., 1991), express NK1Rs (Chen, Wei, Liu, & Rao, 2000), and are activated by SP (Cheeseman, Pinnock, & Henderson, 1983). Acute exposure to SP increases LC firing rates and chronic, but not acute, treatment with an SP receptor antagonist changes firing properties from tonic to phasic burst firing (Maubach et al., 2002). This effect resembles that of the conventional antidepressant imipramine. Phasic burst firing activity in the LC occurs during focused attention and orienting behavior, while tonic firing is increased during acute stress (Aston-Jones, Rajkowski, & Cohen, 1999). Based on this evidence, it has been hypothesized that SP might participate in the processing of stressful experiences and that the use of NK1R antagonists might facilitate adaptations to stress (Maubach et al., 2002).

Acetylcholine

In addition to modulating dopaminergic activity, tachykinins exert a profound influence on cholinergic activity. This effect has been documented in both the striatum and the septum (Schable, Huston, & Silva, 2012). In the striatum, tachykinins (SP and NKA) and GABA have opposite effects on the regulation of acetylcholine (ACh) release from cholinergic interneurons (Blanchet et al., 2000). While GABA has a facilitatory effect on ACh release, tachykinins inhibit it, especially under maximal N-methyl-D-aspartate receptor activation.

In the medial septum, which sends cholinergic projections to the hippocampus and the amygdala, tachykinins increase ACh release. The medial septum has among the highest levels of NK2Rs (Saffroy et al., 2003) and expresses both NK1Rs and NK3Rs, which are located on the cholinergic population of septal neurons (Schable et al., 2011). Injection of NK1R, NK2R, or NK3R agonists into the medial septum increases ACh levels in the hippocampus, and NK2Rs have specifically been implicated in stress-induced hippocampal ACh release (Schable et al., 2012). It has been proposed that these tachykinin-mediated alterations in septal cholinergic activity might underlie the reported effects of neurokinins on anxiety and learning processes (Schable et al., 2012). It should also be noted that intraseptal injection of NKA and NKB increases ACh release in the amygdala but not in the frontal cortex (Schable et al., 2012).

The MHb is another brain area where tachykinins modulate cholinergic activity. MHb neurons are glutamatergic, but the nucleus can subdivided into two principal subnuclei based on the use of ACh as a cotransmitter in the ventral MHb (vMHb) and the expression of Substance P (SP) in the dorsal MHb (Dao, Salas, & De Biasi, 2015). MHb neurons respond to NK1R and NK3R stimulation with a rapid and concentration-dependent increase in firing rates, but no effect is seen following NK2R activation (Norris, Boden, & Woodruff, 1993).

TACHYKININS AND ADDICTION MECHANISMS

Tachykinins and their receptors are expressed in areas that are prominently involved in addictive processes such as the nucleus accumbens and the habenula (Luthi & Luscher, 2014). Drug abuse often occurs in subjects experiencing mood- and anxiety-related symptoms, and stress is a major cause of drug seeking and relapse. Therefore, when considering the regulation of behaviors that lie at the intersection of stress and addiction, the ability of neurokinins to modulate neurotransmitter function is particularly relevant. For example, norepinephrine mediates stress-induced reinstatement and escalates self-administration of multiple addictive drugs (Erb et al., 2000; Gilpin & Koob, 2010; Leri, Flores, Rodaros, & Stewart, 2002; Mantsch et al., 2010; Vranjkovic, Hang, Baker, & Mantsch, 2012). NK1R ligands, by modulating the firing patterns of the LC, can affect norepinephrine release. Dopaminergic activity, while best known for its role in reward-related mechanisms, is also increased during exposure to some stressors. Interestingly, NK1R antagonists can prevent DA release in the prefrontal cortex induced by immobilization stress (Hutson, Patel, Jay, & Barton, 2004).

Neurokinins might be directly involved in the mechanisms of drug abuse, as systemic and intracranial administration of SP (Hasenohrl et al., 2000) and the NK3R-selective agonist aminosenktide is sufficient to produce conditioned place preference (CPP) (Ciccocioppo et al., 1998). The effects of SP are probably mediated by its active C-terminal fragment, which exhibits greater affinity for NK3Rs than for NK1Rs (Hasenohrl, Gerhardt, & Huston, 1992), suggesting that NK3Rs contribute to the reinforcing effects of SP.

NEUROKININS AND NICOTINE: CELLULAR MECHANISMS

Most of what is known about the interactions between neurokinins and the nicotine contained in tobacco comes from studies conducted on peripheral organs. Those studies established that SP and NKA, expressed in sensory neurons and immune-system cells such as macrophages and dendritic cells, participate in the mechanisms of bronchoconstriction and inflammation produced by smoking (Joos, De Swert, & Pauwels, 2001). Initially, smoke activates bronchopulmonary C-fiber afferents, at least in part by binding to nicotinic acetylcholine receptors (nAChRs) expressed in nearby epithelial cells (Lee et al., 2007). This produces a vagal reflex that leads to the release of neurokinins (Hong, Rodger, & Lee, 1995). Neurokinins, in turn, facilitate ACh release from postganglionic parasympathetic neurons, thereby increasing the contractility of airway smooth muscle cells. SP and NK1Rs also promote the accumulation of macrophages and dendritic cells in the airways following exposure to cigarette smoke. Through this mechanism, they contribute to the development of smoking-induced emphysema and chronic obstructive pulmonary disease (Joos et al., 2001).

In the CNS, nicotine acts through multiple intricate mechanisms to facilitate the release of DA and several other neurotransmitters (Wang et al., 2014). In the VTA and substantia nigra, chronic nicotine treatment causes a decrease in SP immunoreactivity, consistent with enhanced SP release and consequent tissue depletion of neuropeptide (Alburges, Frankel, Hoonakker, & Hanson, 2009). These nicotine-mediated changes in SP immunoreactivity can be blocked by the nonselective nAChR antagonist mecamylamine, as well as a DA D1 or D2 receptor antagonist. This observation, considered together with the fact that SP injection into the VTA increases neuronal firing and DA turnover (Kelley & Delfs, 1991), suggests that part of the effects of nicotine on mesolimbic function could be derived from an interaction with the neurokinin system.

The MHb is another area of interaction between the cholinergic and the tachykinin systems that is involved in nicotine addiction. All MHb neurons are cholinoceptive and express high levels of nAChRs (Dao et al., 2015). In addition, the vMHb contains neurons that synthesize and release ACh onto the IPN and other projection sites. Our lab examined the effects of nicotine on the cholinoceptive and cholinergic neurons located in the vMHb (Figure 3A) and showed that activation of nAChRs by nicotine

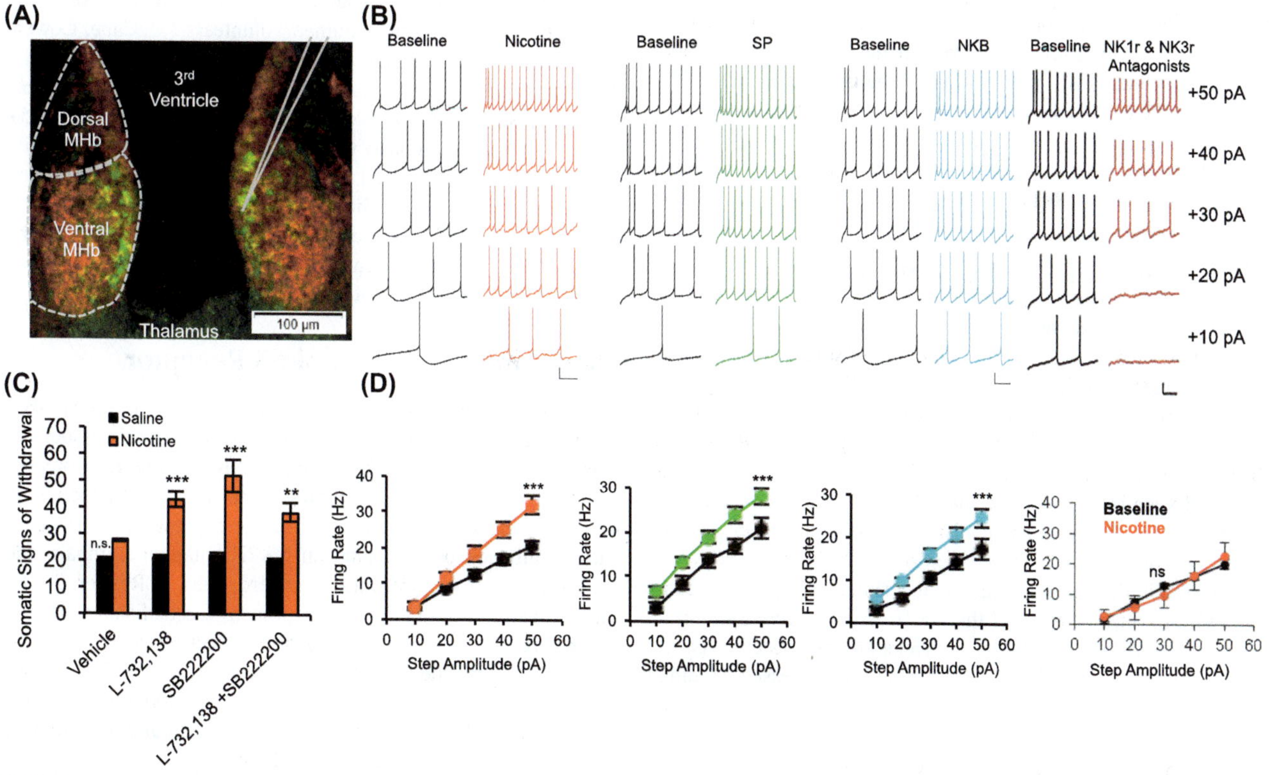

FIGURE 3 SP and NKB increase MHb neuronal firing rates similar to the effect of nicotine, and NKR antagonists inhibit nicotine-induced facilitation of excitability. (A) A brain slice from a mouse expressing enhanced yellow fluorescent protein (green) driven by the choline acetyltransferase (ChAT) promoter, stained with antibodies targeting the NK1R (red), reveals a dense presence of both ChAT and the NK1R in the ventral subregion of the MHb. Lines indicate the placement of recording electrodes. (B) Cholinoceptive neurons located in the vMHb held at −70 mV in current-clamp were stimulated with 2-s depolarizing current steps of increasing amplitudes (10–50 pA) to evaluate action potential behavior. Action potential frequency elicited by each current step increased relative to baseline, suggesting increased intrinsic excitability. Bath application of nicotine (1 μM), as well as both SP and NKB, augmented the increase in firing frequencies produced by cell depolarization. This potentiation was abolished by exposure to antagonists targeting NK1R (L-732,138, 10 μM) and NK3R (SB222200, __uM). Relative frequencies are quantified and graphed in (D). (C) Infusion of NK1R and NK3R antagonists into the MHb of mice chronically treated with nicotine precipitated somatic signs of withdrawal. Following 2 weeks of 8.4 mg/kg/day delivery of nicotine via an osmotic minipump, infusions of L-732,138 (NK1R antagonist), SB222200 (NK3R antagonist), or both simultaneously increased the number of somatic signs over the course of a 20-min examination period relative to vehicle-microinjected controls. *Figure partially adapted from Dao et al. (2014).*

enhances the intrinsic excitability of MHb neurons in a fashion similar to that produced by tachykinins (Figure 3B) (Dao, Perez, Teng, Dani, & De Biasi, 2014). We found that enhancement of excitability by nicotine does not require postsynaptic nAChR function nor ionotropic glutamatergic or GABAergic transmission. Rather, nicotine's enhancement of MHb excitability is abolished by blockade of either NK1Rs or NK3Rs. Thus, under basal conditions, SP and NKB release onto MHb neurons modulates their intrinsic excitability, but this effect can be further amplified by local release of ACh or by nicotine. Our work suggests that nAChRs containing the α5 subunit modulate MHb excitability, largely by facilitating NKB release onto MHb neurons, with additional facilitation of SP release. Because of the anatomical expression, we suspect that the source of NKB release onto the cholinoceptive/cholinergic MHb originates from local neurons, in addition to neurons throughout the MHb. Conversely, the source of possible SP release onto the vMHb might be from the dorsal one-third of the MHb (Dao et al., 2015).

NEUROKININS AND NICOTINE: REGULATION OF MOOD AND AFFECT

Owing to their expression in the limbic circuits that control stress and anxiety, tachykinins participate in the regulation of mood and affect (Ebner et al., 2009). Most of the information available focuses on the roles played by NK1R and its preferred ligand, SP, although more recent animal studies have revealed that NK3R and its cognate ligand, NKB, can also contribute to aversive emotional states. NKA and NK2Rs appear to play a modulatory role as well.

Substance P/Neurokinin 1 Receptor

SP infusion induces anxiety-like behavior in rodents that is NK1R-dependent (Ebner et al., 2009). Early studies with guinea pigs, characterizing the effects of systemic SP or NK1R agonist injections, reported behaviors associated with anxiety, such as reduced food intake (Hasenohrl, Schwarting, Gerhardt, Privou, & Huston, 1994), grooming behaviors (Takahasi et al., 1987), and increased vocalizations. These manifestations were abolished by pretreatment with NK1R antagonists (Kramer et al., 1998). Acute administration of an NK1R antagonist to rats is also anxiolytic when tested in the social interaction test (File, 1997). In addition, mice with genetic deletion of the NK1R exhibit decreased anxiety in behavioral paradigms that provoke anxious states, such as the elevated-plus maze (EPM) and novelty-suppressed feeding test, and produce fewer ultrasonic vocalizations during maternal separation (Santarelli et al., 2001). Additionally, transgenic mice that lack *Tac1*, and therefore do not express SP or NKA, exhibit significantly lower levels of anxiety relative to wild-type controls in several behavioral paradigms (Bilkei-Gorzo, Racz, Michel, & Zimmer, 2002).

Although systemic injection of SP is typically anxiogenic, and NK1R antagonism is anxiolytic, there is evidence for variable regulation of anxiety by the SP/NK1R system in different anatomical regions. For example, selective ablation of NK1R-expressing neurons within the amygdala generates increased anxiety-associated behavior in the EPM (Gadd, Murtra, De Felipe, & Hunt, 2003). Immobilization stress and the EPM test elevate SP levels in the medial subnucleus of the amygdala, an effect that can be blocked by microinfusion of NK1R antagonists into the nucleus (Ebner et al., 2009). Additionally, direct infusion of SP into the central or medial—but not basolateral—subnuclei of the amygdala is anxiogenic in the EPM, suggesting distinct contributions of amygdalar subnuclei to the behavior (Bassi, de Carvalho, & Brandao, 2014).

In addition to the amygdala, other studies have shown that the levels of SP in the lateral septum increase in response to the forced swim test—with NK1R antagonists enhancing intraseptal serotonergic transmission, which increases active stress-coping strategies (Ebner et al., 2009). The LC, paraventricular nucleus (PVN) of the hypothalamus, and lateral septum may also contribute to the behavioral effects of SP, as NK1R antagonists counteract stress-induced elevations of c-Fos expression in those three brain areas (Vendruscolo, Takahashi, Bruske, & Ramos, 2003). Further, speaking to the complex effects of neurokinins in the CNS, while some anxiety-provoking behavioral tests elevate SP in anatomical structures associated with anxiety and stress, a 20-min foot-shock paradigm reduced SP levels in the VTA and IPN (Lisoprawski, Blanc, & Glowinski, 1981).

Several studies in humans have validated the preclinical work described above. Functional magnetic resonance imaging highlighted altered signaling in limbic structures following NK1R blockade in healthy human volunteers (McCabe, Cowen, & Harmer, 2009), and the administration of a different NK1R antagonist for 6 weeks in patients diagnosed with social phobia resulted in significant symptom alleviation (Furmark et al., 2005). An early NK1R antagonist yielded significant anxiolytic activity in a population of depressed patients exhibiting high levels of anxiety (Kramer et al., 1998). Since this study, additional human studies have not yielded consistently encouraging outcomes (McCabe et al., 2009), possibly because of pharmacokinetic and genetic variables that were not considered in the experimental designs.

Neurokinin B/Neurokinin 3 Receptor

More recent efforts have revealed signaling contributions by NKB and its preferred receptor, NK3R, in areas of the brain associated with anxiety and stress. For example, NKB is highly expressed in the amygdala. Fear conditioning, a paradigm that depends on intact amygdalar function, increases NKB expression significantly and the infusion of osanetant, an NK3R antagonist, impairs the consolidation of fear memories (Andero, Dias, & Ressler, 2014). Interestingly, a loss-of-function approach to establishing the role of NKB/NK3R signaling in modulating affect revealed a nuanced story. Absence of the NK3R impairs both the acquisition of conditioned avoidance and performance in the Morris water maze (Siuciak et al., 2007), further suggesting a particular involvement of NKB/NK3R in aversive emotional memory.

Other peptides might also contribute to the effects of SP on mood and affect. Restraint stress results in altered trafficking of NK3Rs in vasopressin-expressing cells of the PVN (Miklos, Flynn, & Lessard, 2014). Vasopressin and the PVN are part of a signaling system that regulates social behavior and stress, the release of adrenocorticotropic hormone, and the hypothalamic–pituitary–adrenal axis (Stevenson & Caldwell, 2012). In addition, an NK3R agonist could inhibit gonadotropin-releasing hormone (GnRH) secretion in mice (Navarro et al., 2009), pointing to another possible mechanism for the interaction between neurokinins and anxiety/stress via GnRH secretion (Umathe, Bhutada, Jain, Dixit, & Wanjari, 2008). Indeed, the administration of an NK3R antagonist resulted in anxiolysis (Salome, Stemmelin, Cohen, & Griebel, 2006), performing similar to diazepam and buspirone—both of which are used to treat anxiety. Contrary to the lack of effect upon systemic injection of an NK3R agonist, anxiolysis in the EPM was observed following intracerebroventricular (ICV) infusion in the mouse. Pretreatment with naloxone enhanced this effect, while naloxone alone did not alter behavior, suggesting a possible interaction between NK3R and the opioid signaling systems (Ribeiro & De Lima, 1998). Taken together, these data suggest that NKB/NK3R also participates in anxiety-associated signaling.

Neurokinin A/Neurokinin 2 Receptor

Perhaps because of the restricted expression profile of NKA/NK2 in the CNS, relatively few studies have evaluated the role of NK2R activation in anxiety and stress. However, the administration of NK2R antagonists has proven to be consistently anxiolytic. Saredutant, an NK2R antagonist, was anxiolytic in gerbils undergoing the social interaction test (Salome et al., 2006), mice tested in the light–dark box paradigm, and primates evaluated by the human intruder response test (Walsh, Stratton, Harvey, Beresford, & Hagan, 1995). Additionally, NK2R antagonists reduce defensive biting and escape attempts in mice (Griebel, Perrault, & Soubrie, 2001) and reduce anxiety in the EPM in both mice (Teixeira et al., 1996) and rats (Griebel et al., 2001). NK2R agonists are anxiogenic in the EPM (Teixeira et al., 1996) and are capable of inhibiting the anxiolytic effects of diazepam (Ribeiro & De Lima, 2002). An investigation of the long-term consequences of foot-shock treatment in rats revealed significantly changed *Tac1* mRNA in a variety of anatomical structures associated with the regulation of emotional states and autoradiography studies revealed reduced NK2R binding in the amygdala 2 weeks after treatment. While the mechanisms of this receptor downregulation are unclear, the authors posit that it may result from an adaptive response to limit excessive SP release (de Lange, Wiegant, & Stam, 2008).

NEUROKININS AND THE SYMPTOMS OF NICOTINE WITHDRAWAL

Although reward significantly contributes to the addictive properties of nicotine, smoking is often motivated by the urge to alleviate the affective, cognitive, and somatic symptoms of withdrawal (McLaughlin, Dani, & De Biasi, 2015). Affective and cognitive signs of withdrawal are produced by CNS mechanisms. Anxiety is a prominent affective symptom, and it acts as a potent negative reinforcer that promotes smoking. Somatic signs reflect mainly peripheral, "bodily" mechanisms such as decreased heart rate and constipation. However, in addition to a peripheral nAChR component, somatic signs have a central component that is thought to translate to a dysphoric state of heightened irritability. These withdrawal symptoms are effectively recapitulated in animal models; upon nicotine withdrawal, both rats and mice exhibit stereotypies, including excessive grooming, chewing, tremors, "wet-dog shakes," yawns, and teeth chattering—as well as behaviors that have been associated with anxiety, anhedonia, depression, dysphoria, and hyperalgesia (McLaughlin et al., 2015). Despite the abundance of data implicating neurokinin signaling in anxiety and stress, to our knowledge, no studies have been conducted so far to precisely characterize the role of neurokinins and their receptors in the anxiety-related symptoms of nicotine withdrawal. In addition to the areas in which tachykinin expression overlaps with anatomical structures that synthesize and release 5-HT, norepinephrine, and DA, the MHb and IPN are potential sites for interactions between signaling systems. This hypothesis is derived from the high levels of SP/NK1R, NKB/NK3R, and nAChRs in the MHb and the fact that the MHb–IPN circuit plays a prominent role in anxiety in both humans and rodents (Dao et al., 2015). In addition, the interplay between cholinergic and neurokinin signaling within the MHb may represent a critical anatomical region wherein neuroadaptations occur, resulting in the signaling that causes anxiety exhibited during withdrawal from the chronic use of nicotine and other drugs (Dao et al., 2015).

The MHb–IPN circuit has an established role in the somatic symptoms of nicotine, ethanol, and, probably, morphine abstinence (Dao et al., 2015; Muldoon et al., 2014; Perez, Quijano-Carde, & De Biasi, 2015). Interestingly, intraperitoneal and ICV infusions of SP in rats elicit physiological and behavioral responses such as increased heart rates, wet-dog shakes, and increased grooming (Tschope, Jost, Unger, & Culman, 1995). Those symptoms are observed during stress as well as nicotine withdrawal (McLaughlin et al., 2015). To that end, we were able to show that administration of NK1R and NK3R antagonists in mice chronically treated with nicotine can precipitate somatic symptoms of withdrawal (Figure 3C)—supporting a model wherein nicotine directly modulates neurokinin signaling in the MHb to promote intrinsic neuronal excitability (Dao et al., 2015).

APPLICATIONS TO OTHER ADDICTIONS AND SUBSTANCE MISUSE

Several groups have examined the involvement of tachykinins in drug abuse. Most studies addressed the role of SP and NK1Rs and have shown their involvement in the mechanisms of dependence from opioids, cocaine, and alcohol. The following sections describe the involvement of neurokinin signaling in addictions to other substances.

Opiates

SP and NK1Rs modulate the rewarding effects of opiates (Commons, 2010). NK1R antagonism suppresses heroin self-administration in rats (Barbier et al., 2013). In addition, NK1R-null mice display morphine reward deficits in the CPP paradigm (Gadd et al., 2003; Murtra, Sheasby, Hunt, & De Felipe, 2000) and self-administer less morphine than their control littermates, without any change in natural reward (Ripley, Gadd, De Felipe, Hunt, & Stephens, 2002). Morphine's ability to potentiate reward is also attenuated by NK1R antagonism in the intracranial self-stimulation paradigm (Robinson et al., 2012). Based on lesion studies, the amygdala, rather than the nucleus accumbens, is where the presence of NK1Rs affects morphine reward (Gadd et al., 2003). Lack of NK1Rs also attenuates morphine-induced locomotor activation and psychomotor sensitization (Murtra et al., 2000; Ripley et al., 2002). There are also preclinical data indicating that NK1R blockade might attenuate the symptoms of opioid withdrawal

(Maldonado, Girdlestone, & Roques, 1993) and that administration of SP or its N-terminal peptide can prevent the development of morphine-withdrawal symptoms (Kreeger & Larson, 1996). A small study conducted in methadone-maintained patients showed that the US FDA-approved NK1R antagonist aprepitant might have utility for alleviating opioid withdrawal and craving (Jones et al., 2013).

In addition to the influence of SP on the brain circuits underlying morphine-related behaviors, NK1Rs and μ-opioid receptors (MORs) may interact at the cellular level. Specifically, SP prevents morphine-induced internalization and acute desensitization of MORs (Yu, Arttamangkul, Evans, Williams, & von Zastrow, 2009). This phenomenon was observed in both amygdala and LC neurons, which can coexpress NK1Rs and MORs. Finally, chronic morphine administration results in NK1R upregulation (Wan et al., 2006). Overall, these data point to a rich interaction between the opioid and the tachykinin systems and suggest that NK1R blockade could be one approach to reducing opiate abuse by limiting their rewarding and reinforcing properties, as well as mitigating withdrawal symptoms.

Cocaine

The effects of SP on cocaine-induced behaviors differ from those observed for morphine. When cocaine-induced locomotion and DA release in the dorsal striatum were examined, it was found that NK1R blockade attenuated both responses (Loonam, Noailles, Yu, Zhu, & Angulo, 2003). However, lack of NK1Rs did not affect locomotor sensitization to cocaine in mice (Ripley et al., 2002), nor did it affect cocaine CPP (Murtra et al., 2000). Furthermore, in contrast with what was found for morphine, targeted lesions of the amygdala did not affect CPP (Gadd et al., 2003). Cocaine self-administration rates were not affected by NK1R deletion in mice (Ripley et al., 2002) or NK1R antagonism in rats (Schank et al., 2014). Interestingly, NK1R blockade could prevent stress-induced reinstatement but had no effect on cocaine seeking triggered by a cocaine priming injection (Placenza, Vaccarino, Fletcher, & Erb, 2005).

Other studies showed that NK3Rs are involved in cocaine-induced hyperlocomotion and DA release within the nucleus accumbens, but not in cocaine-induced CPP (Jocham, Lauber, Müller, Huston, & De Souza Silva, 2007). In summary, although neurokinins might be involved in some aspects of cocaine dependence, there is no evidence that they influence its rewarding properties.

Alcohol

Preclinical and clinical data, which are discussed in detail elsewhere in this volume, indicate that neurokinins are involved in the mechanisms of alcohol abuse. Briefly, similar to morphine, NK1R-null mice do not display alcohol CPP (Thorsell, Schank, Singley, Hunt, & Heilig, 2010). They also consume less alcohol in the two-bottle choice drinking paradigm (George et al., 2008), a phenomenon also observed when using both NK1R antagonists (Thorsell et al., 2010) and NK1R knockdown (Baek et al., 2010). In addition, NK1R-null mice do not display increased alcohol consumption after repeated cycles of deprivation, suggesting that SP influences the neuroadaptations that occur during repeated withdrawal and lead to drinking escalation (Thorsell et al., 2010).

Interestingly, blockade of NK1R does not affect baseline operant alcohol self-administration in the rat, but, similar to what was found for cocaine, it suppresses stress-induced reinstatement of alcohol seeking (Schank et al., 2014).

Information on the effects of NK1R blockade is also available for humans. A clinical trial of the NK1R antagonist LY686017 in detoxified alcoholic inpatients demonstrated suppression of spontaneous alcohol cravings and improved overall well-being (George et al., 2008). Subsequently, a case–control association study conducted on Caucasian subjects concluded that polymorphisms of the NK1R are significantly associated with the development of alcohol dependence (Seneviratne et al., 2009). Based on this evidence, NK1R antagonists may be a treatment strategy for some subpopulations of alcohol abusers, such as those with certain genetic variants of the *TACR1* gene.

CONCLUSIONS

Tachykinins and their receptors are expressed in brain areas and circuits that control stress and anxiety. These areas also participate in the mechanisms of drug abuse, and there is evidence for the involvement of tachykinins in the abuse of opiates, cocaine, and alcohol. For opiates, NK1Rs affect baseline reward and reinforcement, as well as escalated self-administration rates. For cocaine and alcohol, NK1Rs mediate stress-induced reinstatement of drug seeking and escalated drug self-administration, but do not affect baseline self-administration. Very little is known regarding the role of tachykinins in the manifestations of withdrawal, except for reports that link SP to some of the symptoms of opiate abstinence.

Surprisingly, despite a clear role of tachykinins in many peripheral effects of nicotine, there have been no studies on their roles in nicotine reward and seeking. As for the mechanisms of withdrawal, our lab was the first to report a link between SP and NKB and the manifestations of physical dependence on nicotine, highlighting the MHb–IPN circuit as a major site for those interactions. Therefore, the field is ripe for further investigations into mechanisms and the potential application of NK1R antagonists to facilitate smoking cessation. Even less is known about the role of NK2R and NK3R in drug addiction. However, given the growing availability of NKR subtype-specific antagonists, genetically engineered preclinical models, and many additional molecular and genetic tools, future studies should be able to address the specific role of NKA and NKB in nicotine addiction.

As for the potential use in the clinic, NK1R antagonists are generally well tolerated, warranting further investigation into their effectiveness for the treatment of nicotine and other drug abuse disorders. The clinical trials examining the effects of NK1R antagonists on alcohol abuse have yielded mixed results, possibly due, among other factors, to relatively small sample sizes and variable drug pharmacokinetics. As pharmacogenetic analyses become cheaper and more easily incorporated into clinical trials, investigators will be able to determine the impact of genetic variations in neurokinin genes on both the risk and the response to treatment in addicted subjects. The use of empirically derived subphenotypes (e.g., comorbid anxiety) will be particularly helpful in identifying the subjects that would benefit most from the addition of neurokinin antagonists to their treatment strategies.

DEFINITION OF TERMS

Cholinergic neuron This is a neuron that possesses the intracellular machinery for the synthesis and release of ACh.

Cholinoceptive neuron This is a neuron that expresses receptors for ACh.

Electrophysiology This is a technique that allows measurement of the electrical activity of neurons.

Feedback loop This is a mechanism through which a neuronal system can self-adjust to optimize function and maintain equilibrium.

Immunoreactivity This is the measure of the signal for a particular protein that can be detected in the brain by antibodies that bind with high specificity to that protein.

Intracranial self-stimulation This is an experimental paradigm in which animals self-stimulate certain brain regions through electrodes surgically implanted into the brain. The stimulation activates the brain reward system.

Mecamylamine This is a nicotinic acetylcholine or nAChR receptor antagonist used to precipitate nicotine withdrawal in mice chronically exposed to nicotine.

Pharmacogenetic analyses These are studies that examine the response to a certain drug based on the genetic makeup of an individual.

Somatic signs of withdrawal These are the physical symptoms associated with nicotine abstinence. The nicotine abstinence syndrome comprises affective, cognitive, and somatic signs.

Ultrasonic vocalizations These are calls of distress inaudible to humans. They are thought to reflect negative affective states.

KEY FACTS

Tachykinin signaling is involved in anxiety, stress, and addiction.

- Tachykinin signaling plays a modulatory role in the signaling of DA, 5-HT, norepinephrine, and ACh—all of which are known to play roles in the pathophysiology of mood-related disorders and addiction.
- Systemic injection of SP is generally anxiogenic, while antagonists targeting SP's cognate receptor are anxiolytic.
- NK2R antagonists have proven to be anxiolytic in preclinical models, and stress induces long-lasting alterations of tachykinin gene product levels.
- Pharmacological manipulations of the NK3R result in modified anxiety-associated behavior in animal models, depending upon the particular CNS structure into which they are infused.
- The MHb–IPN circuit, which is critically involved in anxiety and symptoms of withdrawal from nicotine and other drugs, hosts considerable tachykinin signaling. Pharmacological manipulations of NK1R and NK3R-dependent signaling in this circuit are sufficient to induce withdrawal symptoms in mice chronically treated with nicotine.
- Many questions remain unanswered concerning the functional roles played by each receptor type, as well as the roles of other tachykinin signaling peptides.
- Clinical applications of NKR-targeting drugs have not consistently proven superior to the standards of care for mood-related disorders or addiction.
- More nuanced approaches to clinical trials, such as genomic screening, combinatorial pharmacological strategies, and increased receptor specificity for more targeted action, may yield more positive results for the treatments of mood-related disorders and addiction.

SUMMARY POINTS

- SP, NKA, and NKB are all members of the tachykinin family of signaling peptides and are expressed and released in areas of the brain associated with anxiety, stress, and drug addiction.
- NK1R and NK3R are predominantly expressed in the CNS, while NK2R is most densely expressed in the PNS.
- NK1Rs bind SP, NK2Rs bind NKA, and NK3Rs bind NKB with the greatest affinities—but each of the peptides is capable of binding all of the receptors.
- While there is substantial overlap in the anatomical distribution of expression and release, each peptide and its receptor also occupy distinct anatomical regions.
- The receptors for all three neurokinins are G-protein coupled and have been shown to modulate DA, 5-HT, and ACh signaling.
- Tachykinin signaling has been implicated in the effects of nicotine and other drugs, as well as the neuroadaptations that occur as a result of chronic use.
- Pharmacological manipulation of NKRs has been shown to alter anxiety-, depression-, and stress-associated behavior in preclinical and clinical studies.

REFERENCES

Alburges, M. E., Frankel, P. S., Hoonakker, A. J., & Hanson, G. R. (2009). Responses of limbic and extrapyramidal substance P systems to nicotine treatment. *Psychopharmacology, 201*(4), 517–527.

Andero, R., Dias, B. G., & Ressler, K. J. (2014). A role for Tac2, NkB, and Nk3 receptor in normal and dysregulated fear memory consolidation. *Neuron, 83*(2), 444–454.

Arai, H., & Emson, P. C. (1986). Regional distribution of neuropeptide K and other tachykinins (neurokinin A, neurokinin B and substance P) in rat central nervous system. *Brain Research, 399*(2), 240–249.

Aston-Jones, G., Rajkowski, J., & Cohen, J. (1999). Role of locus coeruleus in attention and behavioral flexibility. *Biological Psychiatry, 46*(9), 1309–1320.

Baek, M. N., Jung, K. H., Halder, D., Choi, M. R., Lee, B. H., Lee, B. C., … Chai, Y. G. (2010). Artificial microRNA-based neurokinin-1 receptor gene silencing reduces alcohol consumption in mice. *Neuroscience Letters, 475*(3), 124–128.

Baker, K. G., Halliday, G. M., Hornung, J. P., Geffen, L. B., Cotton, R. G., & Tork, I. (1991). Distribution, morphology and number of monoamine-synthesizing and substance P-containing neurons in the human dorsal raphe nucleus. *Neuroscience, 42*(3), 757–775.

Barbier, E., Vendruscolo, L. F., Schlosburg, J. E., Edwards, S., Juergens, N., Park, P. E., … Heilig, M. (2013). The NK1 receptor antagonist L822429 reduces heroin reinforcement. *Neuropsychopharmacology, 38*(6), 976–984.

Bassi, G. S., de Carvalho, M. C., & Brandao, M. L. (2014). Effects of substance P and Sar-Met-SP, a NK1 agonist, in distinct amygdaloid nuclei on anxiety-like behavior in rats. *Neuroscience Letters, 569*, 121–125.

Bilkei-Gorzo, A., Racz, I., Michel, K., & Zimmer, A. (2002). Diminished anxiety- and depression-related behaviors in mice with selective deletion of the Tac1 gene. *The Journal of Neuroscience: The Official Journal of the Society for Neuroscience, 22*(22), 10046–10052.

Blanchet, F., Gauchy, C., Pérez, S., Soubrié, P., Glowinski, J., & Kemel, M.-L. (2000). Control by GABA and tachykinins of the evoked release of acetylcholine in striatal compartments under different modalities of NMDA receptor stimulation. *Brain Research, 853*(1), 142–150.

Burgunder, J. M., & Young, W. S., 3rd (1989). Distribution, projection and dopaminergic regulation of the neurokinin B mRNA-containing neurons of the rat caudate-putamen. *Neuroscience, 32*(2), 323–335.

Campbell, B. M., & Walker, P. D. (2001). NMDA receptor antagonism modifies the synergistic regulation of striatal tachykinin gene expression induced by dopamine D(1) and serotonin(2) receptor stimulation following neonatal dopamine depletion. *Brain Research. Molecular Brain Research, 93*(1), 90–94.

Cheeseman, H. J., Pinnock, R. D., & Henderson, G. (1983). Substance P excitation of rat locus coeruleus neurones. *European Journal of Pharmacology, 94*(1–2), 93–99.

Chen, L. W., Wei, L. C., Liu, H. L., & Rao, Z. R. (2000). Noradrenergic neurons expressing substance P receptor (NK1) in the locuEuropean Journal of Pharmacologys coeruleus complex: a double immunofluorescence study in the rat. *Brain Research, 873*(1), 155–159.

Ciccocioppo, R., Panocka, I., Polidori, C., Froldi, R., Angeletti, S., & Massi, M. (1998). Mechanism of action for reduction of ethanol intake in rats by the tachykinin NK-3 receptor agonist aminosenktide. *Pharmacology Biochemistry and Behavior, 61*(4), 459–464.

Commons, K. G. (2010). Neuronal pathways linking substance P to drug addiction and stress. *Brain Research, 1314*, 175–182.

Cuello, A. C., & Kanazawa, I. (1978). The distribution of substance P immunoreactive fibers in the rat central nervous system. *The Journal of Comparative Neurology, 178*(1), 129–156.

Dao, D. Q., Perez, E. E., Teng, Y., Dani, J. A., & De Biasi, M. (2014). Nicotine enhances excitability of medial habenular neurons via facilitation of neurokinin signaling. *The Journal of Neuroscience: The Official Journal of the Society for Neuroscience, 34*(12), 4273–4284.

Dao, D. Q., Salas, R., & De Biasi, M. (2015). Nicotinic acetylcholine receptors along the habenulo-, interpeduncular pathway: roles in nicotine withdrawal and other aversive aspects. In R. A. J. Lester (Ed.), *Nicotinic receptors* (Vol. 26). New York: Springer Science & Business Media.

Ebner, K., Sartori, S. B., & Singewald, N. (2009). Tachykinin receptors as therapeutic targets in stress-related disorders. *Current Pharmaceutical Design, 15*(14), 1647–1674.

Erb, S., Hitchcott, P. K., Rajabi, H., Mueller, D., Shaham, Y., & Stewart, J. (2000). Alpha-2 adrenergic receptor agonists block stress-induced reinstatement of cocaine seeking. *Neuropsychopharmacology, 23*(2), 138–150.

File, S. E. (1997). Anxiolytic action of a neurokinin1 receptor antagonist in the social interaction test. *Pharmacology Biochemistry and Behavior, 58*(3), 747–752.

Furmark, T., Appel, L., Michelgard, A., Wahlstedt, K., Ahs, F., Zancan, S., ... Fredrikson, M. (2005). Cerebral blood flow changes after treatment of social phobia with the neurokinin-1 antagonist GR205171, citalopram, or placebo. *Biological Psychiatry, 58*(2), 132–142.

Gadd, C. A., Murtra, P., De Felipe, C., & Hunt, S. P. (2003). Neurokinin-1 receptor-expressing neurons in the amygdala modulate morphine reward and anxiety behaviors in the mouse. *The Journal of Neuroscience: The Official Journal of the Society for Neuroscience, 23*(23), 8271–8280.

George, D. T., Gilman, J., Hersh, J., Thorsell, A., Herion, D., Geyer, C., ... Heilig, M. (2008). Neurokinin 1 receptor antagonism as a possible therapy for alcoholism. *Science, 319*(5869), 1536–1539.

Gilpin, N. W., & Koob, G. F. (2010). Effects of β-adrenoceptor antagonists on alcohol drinking by alcohol-dependent rats. *Psychopharmacology (Berl), 212*(3), 431–439.

Gradin, K., Qadri, F., Nomikos, G. G., Hillegaart, V., & Svensson, T. H. (1992). Substance P injection into the dorsal raphe increases blood pressure and serotonin release in hippocampus of conscious rats. *European Journal of Pharmacology, 218*(2–3), 363–367.

Griebel, G., Perrault, G., & Soubrie, P. (2001). Effects of SR48968, a selective non-peptide NK2 receptor antagonist on emotional processes in rodents. *Psychopharmacology (Berl), 158*(3), 241–251.

Hasenohrl, R. U., Gerhardt, P., & Huston, J. P. (1992). Positively reinforcing effects of the neurokinin substance P in the basal forebrain: mediation by its C-terminal sequence. *Experimental Neurology, 115*(2), 282–291.

Hasenohrl, R. U., Schwarting, R. K., Gerhardt, P., Privou, C., & Huston, J. P. (1994). Comparison of neurokinin substance P with morphine in effects on food-reinforced operant behavior and feeding. *Physiology & Behavior, 55*(3), 541–546.

Hasenohrl, R. U., Souza-Silva, M. A., Nikolaus, S., Tomaz, C., Brandao, M. L., Schwarting, R. K., & Huston, J. P. (2000). Substance P and its role in neural mechanisms governing learning, anxiety and functional recovery. *Neuropeptides, 34*(5), 272–280.

Höckfelt, T., Kuteeva, E., Stanic, D., Ljungdahl, Å. (2004). The histochemistry of tachykinin system in the brain. In P. Holzer (Ed.), *Tachykinins. Handbook of Experimental Pharmacology* (pp. 63–120). Springer-Verlag, Berlin: Heidelberg.

Hong, J. L., Rodger, I. W., & Lee, L. Y. (1995). Cigarette smoke-induced bronchoconstriction: cholinergic mechanisms, tachykinins, and cyclooxygenase products. *Journal of Applied Physiology, 78*, 2260–2266.

Hutson, P. H., Patel, S., Jay, M. T., & Barton, C. L. (2004). Stress-induced increase of cortical dopamine metabolism: attenuation by a tachykinin NK1 receptor antagonist. *European Journal of Pharmacology, 484*(1), 57–64.

Jocham, G., Lauber, A. C., Müller, C. P., Huston, J. P., & De Souza Silva, M. A. (2007). Neurokinin 3 receptor activation potentiates the psychomotor and nucleus accumbens dopamine response to cocaine, but not its place conditioning effects. *European Journal of Neuroscience, 25*(8), 2457–2472.

Jones, J. D., Speer, T., Comer, S. D., Ross, S., Rotrosen, J., & Reid, M. S. (2013). Opioid-like effects of the neurokinin 1 antagonist aprepitant in patients maintained on and briefly withdrawn from methadone. *American Journal of Drug and Alcohol Abuse, 39*(2), 86–91.

Joos, G. F., De Swert, K. O., & Pauwels, R. A. (2001). Airway inflammation and tachykinins: prospects for the development of tachykinin receptor antagonists. *European Journal of Pharmacology, 429*(1–3), 239–250.

Keegan, K. D., Woodruff, G. N., & Pinnock, R. D. (1992). The selective NK3 receptor agonist senktide excites a subpopulation of dopamine-sensitive neurones in the rat substantia nigra pars compacta in vitro. *British Journal of Pharmacology, 105*(1), 3–5.

Kelley, A. E., & Delfs, J. M. (1991). Dopamine and conditioned reinforcement. II. Contrasting effects of amphetamine microinjection into the nucleus accumbens with peptide microinjection into the ventral tegmental area. *Psychopharmacology (Berl), 103*(2), 197–203.

Kovács, K. A., Steinmann, M., Magistretti, P. J., Halfon, O., & Cardinaux, J.-R. (2006). C/EBPβ couples dopamine signalling to substance P precursor gene expression in striatal neurones. *Journal of Neurochemistry, 98*(5), 1390–1399.

Kramer, M. S., Cutler, N., Feighner, J., Shrivastava, R., Carman, J., Sramek, J. J., ... Rupniak, N. M. (1998). Distinct mechanism for antidepressant activity by blockade of central substance P receptors. *Science, 281*(5383), 1640–1645.

Kreeger, J. S., & Larson, A. A. (1996). The substance P amino-terminal metabolite substance P(1-7), administered peripherally, prevents the development of acute morphine tolerance and attenuates the expression of withdrawal in mice. *Journal of Pharmacology and Experimental Therapeutics, 279*(2), 662–667.

de Lange, R. P., Wiegant, V. M., & Stam, R. (2008). Altered neuropeptide Y and neurokinin messenger RNA expression and receptor binding in stress-sensitised rats. *Brain Research, 1212*, 35–47.

Lee, L. Y., Burki, N. K., Gerhardstein, D. C., Gu, Q., Kou, Y. R., & Xu, J. (2007). Airway irritation and cough evoked by inhaled cigarette smoke: role of neuronal nicotinic acetylcholine receptors. *Pulmonary Pharmacology & Therapeutics, 20*(4), 355–364.

Leri, F., Flores, J., Rodaros, D., & Stewart, J. (2002). Blockade of stress-induced but not cocaine-induced reinstatement by infusion of noradrenergic antagonists into the bed nucleus of the stria terminalis or the central nucleus of the amygdala. *The Journal of Neuroscience: The Official Journal of the Society for Neuroscience, 22*(13), 5713–5718.

Lisoprawski, A., Blanc, G., & Glowinski, J. (1981). Activation by stress of the habenulo-interpeduncular substance P neurons in the rat. *Neuroscience Letters, 25*(1), 47–51.

Liu, R., Ding, Y., & Aghajanian, G. K. (2002). Neurokinins activate local glutamatergic inputs to serotonergic neurons of the dorsal raphe nucleus. *Neuropsychopharmacology, 27*(3), 329–340.

Loonam, T. M., Noailles, P. A., Yu, J., Zhu, J. P., & Angulo, J. A. (2003). Substance P and cholecystokinin regulate neurochemical responses to cocaine and methamphetamine in the striatum. *Life Sciences, 73*(6), 727–739.

Luthi, A., & Luscher, C. (2014). Pathological circuit function underlying addiction and anxiety disorders. *Nature Neuroscience, 17*(12), 1635–1643.

Maldonado, R., Girdlestone, D., & Roques, B. P. (1993). RP 67580, a selective antagonist of neurokinin-1 receptors, modifies some of the naloxone-precipitated morphine withdrawal signs in rats. *Neuroscience Letters, 156*(1–2), 135–140.

Mantsch, J. R., Weyer, A., Vranjkovic, O., Beyer, C. E., Baker, D. A., & Caretta, H. (2010). Involvement of noradrenergic neurotransmission in the stress- but not cocaine-induced reinstatement of extinguished cocaine-induced conditioned place preference in mice: role for β-2 adrenergic receptors. *Neuropsychopharmacology, 35*(11), 2165–2178.

Mantyh, P. W. (2002). Neurobiology of substance P and the NK1 receptor. *Journal of Clinical Psychiatry, 63*(Suppl. 11), 6–10.

Marco, N., Thirion, A., Mons, G., Bougault, I., Le Fur, G., Soubrie, P., & Steinberg, R. (1998). Activation of dopaminergic and cholinergic neurotransmission by tachykinin NK3 receptor stimulation: an in vivo microdialysis approach in guinea pig. *Neuropeptides, 32*(5), 481–488.

Marksteiner, J., Sperk, G., & Krause, J. E. (1992). Distribution of neurons expressing neurokinin B in the rat brain: Immunohistochemistry and in situ hybridization. *The Journal of Comparative Neurology, 317*(4), 341–356.

Maubach, K. A., Martin, K., Chicchi, G., Harrison, T., Wheeldon, A., Swain, C. J., … Seabrook, G. R. (2002). Chronic substance P (NK1) receptor antagonist and conventional antidepressant treatment increases burst firing of monoamine neurones in the locus coeruleus. *Neuroscience, 109*(3), 609–617.

McCabe, C., Cowen, P. J., & Harmer, C. J. (2009). NK1 receptor antagonism and the neural processing of emotional information in healthy volunteers. *International Journal of Neuropsychopharmacology, 12*(9), 1261–1274.

McLaughlin, I., Dani, J. A., & De Biasi, M. (2015). Nicotine withdrawal. *Current Topics in Behavioural Neurosciences, 24*, 99–123.

Miklos, Z., Flynn, F. W., & Lessard, A. (2014). Stress-induced dendritic internalization and nuclear translocation of the neurokinin-3 (NK3) receptor in vasopressinergic profiles of the rat paraventricular nucleus of the hypothalamus. *Brain Research, 1590*, 31–44.

Muldoon, P. P., Jackson, K. J., Perez, E., Harenza, J. L., Molas, S., Rais, B., … Damaj, M. I. (2014). The α3β4* nicotinic ACh receptor subtype mediates physical dependence to morphine: mouse and human studies. *British Journal of Pharmacology, 171*(16), 3845–3857.

Murtra, P., Sheasby, A. M., Hunt, S. P., & De Felipe, C. (2000). Rewarding effects of opiates are absent in mice lacking the receptor for substance P. *Nature, 405*(6783), 180–183.

Navarro, V. M., Gottsch, M. L., Chavkin, C., Okamura, H., Clifton, D. K., & Steiner, R. A. (2009). Regulation of gonadotropin-releasing hormone secretion by kisspeptin/dynorphin/neurokinin B neurons in the arcuate nucleus of the mouse. *The Journal of Neuroscience: The Official Journal of the Society for Neuroscience, 29*(38), 11859–11866.

Norris, S. K., Boden, P. R., & Woodruff, G. N. (1993). Agonists selective for tachykinin NK1 and NK3 receptors excite subpopulations of neurons in the rat medial habenula nucleus in vitro. *European Journal of Pharmacology, 234*(2–3), 223–228.

Perez, E., Quijano-Carde, N., & De Biasi, M. (2015). Nicotinic mechanisms modulate ethanol withdrawal and modify time course and symptoms severity of simultaneous withdrawal from alcohol and nicotine. *Neuropsychopharmacology, 40*(10), 2327–2336.

Placenza, F. M., Vaccarino, F. J., Fletcher, P. J., & Erb, S. (2005). Activation of central neurokinin-1 receptors induces reinstatement of cocaine-seeking behavior. *Neuroscience Letters, 390*(1), 42–47.

Ribeiro, R. L., & De Lima, T. C. (2002). Participation of GABAA receptors in the modulation of experimental anxiety by tachykinin agonists and antagonists in mice. *Progress in Neuro-Psychopharmacology & Biological Psychiatry, 26*(5), 861–869.

Ribeiro, S. J., & De Lima, T. C. (1998). Naloxone-induced changes in tachykinin NK3 receptor modulation of experimental anxiety in mice. *Neuroscience Letters, 258*(3), 155–158.

Ribeiro-da-Silva, A., & Hokfelt, T. (2000). Neuroanatomical localisation of substance P in the CNS and sensory neurons. *Neuropeptides, 34*(5), 256–271.

Ripley, T. L., Gadd, C. A., De Felipe, C., Hunt, S. P., & Stephens, D. N. (2002). Lack of self-administration and behavioural sensitisation to morphine, but not cocaine, in mice lacking NK1 receptors. *Neuropharmacology, 43*(8), 1258–1268.

Robinson, J. E., Fish, E. W., Krouse, M. C., Thorsell, A., Heilig, M., & Malanga, C. J. (2012). Potentiation of brain stimulation reward by morphine: effects of neurokinin-1 receptor antagonism. *Psychopharmacology (Berl), 220*(1), 215–224.

Saffroy, M., Torrens, Y., Glowinski, J., & Beaujouan, J. C. (2003). Autoradiographic distribution of tachykinin NK2 binding sites in the rat brain: comparison with NK1 and NK3 binding sites. *Neuroscience, 116*(3), 761–773.

Salome, N., Stemmelin, J., Cohen, C., & Griebel, G. (2006). Selective blockade of NK2 or NK3 receptors produces anxiolytic- and antidepressant-like effects in gerbils. *Pharmacology Biochemistry and Behavior, 83*(4), 533–539.

Santarelli, L., Gobbi, G., Debs, P. C., Sibille, E. T., Blier, P., Hen, R., & Heath, M. J. (2001). Genetic and pharmacological disruption of neurokinin 1 receptor function decreases anxiety-related behaviors and increases serotonergic function. *Proceedings of the National Academy of Sciences of the United States of America, 98*(4), 1912–1917.

Saria, A. (1999). The tachykinin NK1 receptor in the brain: pharmacology and putative functions. *European Journal of Pharmacology, 375*(1–3), 51–60.

Schable, S., Huston, J. P., & Silva, M. A. (2012). Neurokinin2-R in medial septum regulate hippocampal and amygdalar ACh release induced by intraseptal application of neurokinins A and B. *Hippocampus*, *22*(5), 1058–1067.

Schable, S., Topic, B., Buddenberg, T., Petri, D., Huston, J. P., & de Souza Silva, M. A. (2011). Neurokinin3-R agonism in aged rats has anxiolytic-, antidepressant-, and promnestic-like effects and stimulates ACh release in frontal cortex, amygdala and hippocampus. *European Neuropsychopharmacology*, *21*(6), 484–494.

Schank, J. R., King, C. E., Sun, H., Cheng, K., Rice, K. C., Heilig, M., … Schroeder, J. P. (2014). The role of the neurokinin-1 receptor in stress-induced reinstatement of alcohol and cocaine seeking. *Neuropsychopharmacology*, *39*(5), 1093–1101.

Seneviratne, C., Ait-Daoud, N., Ma, J. Z., Chen, G., Johnson, B. A., & Li, M. D. (2009). Susceptibility locus in Neurokinin-1 receptor gene associated with alcohol dependence. *Neuropsychopharmacology*, *34*(11), 2442–2449.

Siuciak, J. A., McCarthy, S. A., Martin, A. N., Chapin, D. S., Stock, J., Nadeau, D. M., … McLean, S. (2007). Disruption of the neurokinin-3 receptor (NK3) in mice leads to cognitive deficits. *Psychopharmacology (Berl)*, *194*(2), 185–195.

Steinhoff, M. S., von Mentzer, B., Geppetti, P., Pothoulakis, C., & Bunnett, N. W. (2014). Tachykinins and their receptors: contributions to physiological control and the mechanisms of disease. *Physiological Reviews*, *94*(1), 265–301.

Stevenson, E. L., & Caldwell, H. K. (2012). The vasopressin 1b receptor and the neural regulation of social behavior. *Hormones and Behavior*, *61*(3), 277–282.

Takahasi, K., Sakurada, T., Sakurada, S., Kuwahara, H., Yonezawa, A., Ando, R., & Kisara, K. (1987). Behavioural characterization of substance P-induced nociceptive response in mice. *Neuropharmacology*, *26*(9), 1289–1293.

Teixeira, R. M., Santos, A. R., Ribeiro, S. J., Calixto, J. B., Rae, G. A., & De Lima, T. C. (1996). Effects of central administration of tachykinin receptor agonists and antagonists on plus-maze behavior in mice. *European Journal of Pharmacology*, *311*(1), 7–14.

Thorsell, A., Schank, J. R., Singley, E., Hunt, S. P., & Heilig, M. (2010). Neurokinin-1 receptors (NK1R:s), alcohol consumption, and alcohol reward in mice. *Psychopharmacology (Berl)*, *209*(1), 103–111.

Tschope, C., Jost, N., Unger, T., & Culman, J. (1995). Central cardiovascular and behavioral effects of carboxy- and amino-terminal fragments of substance P in conscious rats. *Brain Research*, *690*(1), 15–24.

Umathe, S. N., Bhutada, P. S., Jain, N. S., Dixit, P. V., & Wanjari, M. M. (2008). Effects of central administration of gonadotropin-releasing hormone agonists and antagonist on elevated plus-maze and social interaction behavior in rats. *Behavioural Pharmacology*, *19*(4), 308–316.

Vendruscolo, L. F., Takahashi, R. N., Bruske, G. R., & Ramos, A. (2003). Evaluation of the anxiolytic-like effect of NKP608, a NK1-receptor antagonist, in two rat strains that differ in anxiety-related behaviors. *Psychopharmacology (Berl)*, *170*(3), 287–293.

Vranjkovic, O., Hang, S., Baker, D. A., & Mantsch, J. R. (2012). β-adrenergic receptor mediation of stress-induced reinstatement of extinguished cocaine-induced conditioned place preference in mice: roles for $β_1$ and $β_2$ adrenergic receptors. *Journal of Pharmacology and Experimental Therapeutics*, *342*(2), 541–551.

Walsh, D. M., Stratton, S. C., Harvey, F. J., Beresford, I. J., & Hagan, R. M. (1995). The anxiolytic-like activity of GR159897, a non-peptide NK2 receptor antagonist, in rodent and primate models of anxiety. *Psychopharmacology (Berl)*, *121*(2), 186–191.

Wan, Q., Douglas, S. D., Wang, X., Kolson, D. L., O'Donnell, L. A., & Ho, W. Z. (2006). Morphine upregulates functional expression of neurokinin-1 receptor in neurons. *Journal of Neuroscience Research*, *84*(7), 1588–1596.

Wang, L., Shang, S., Kang, X., Teng, S., Zhu, F., Liu, B., … Zhou, Z. (2014). Modulation of dopamine release in the striatum by physiologically relevant levels of nicotine. *Nature Communications*, *21*(5), 3925.

Warden, M. K., & Young, W. S., 3rd (1988). Distribution of cells containing mRNAs encoding substance P and neurokinin B in the rat central nervous system. *The Journal of Comparative Neurology*, *272*(1), 90–113.

Yu, Y. J., Arttamangkul, S., Evans, C. J., Williams, J. T., & von Zastrow, M. (2009). Neurokinin 1 receptors regulate morphine-induced endocytosis and desensitization of mu-opioid receptors in CNS neurons. *The Journal of Neuroscience: The Official Journal of the Society for Neuroscience*, *29*(1), 222–233.

Chapter 19

Neurotransmitter Systems and the Nicotine Dependence-Induced Withdrawal Syndrome: Dopamine, Glutamate, GABA, Endogenous Opioids, Endocannabinoids, Noradrenaline, Arginine Vasopressin, Neuropeptide Y, MAO, CREB, and Corticotropin-Releasing Factor

Thakur Gurjeet Singh, Shiwali Sharma, Sonia Dhiman
Chitkara College of Pharmacy, Chitkara University, Chandigarh, Punjab, India

Abbreviations

5-HT 5-Hydroxytryptamine
AMPA α-Amino-3-hydroxy-5-methyl-4-isoxazolepropionic acid
AMYG Amygdala
AVP Arginine vasopressin
cAMP Cyclic adenosine monophosphate 3,5-monophosphate
CREB Cyclic adenosine monophosphate response-element binding protein
CRF Corticotropin-releasing factor
DA Dopamine
GABA γ-Aminobutyric acid
GRK3 G-Protein receptor kinase-3
KOR κ Opioid receptor
MAO Monoamine oxidase
nAChR Nicotinic acetylcholine receptor
NMDA N-Methyl-D-aspartate receptor
NPY Neuropeptide Y
NTS Nucleus tractus solitarius
PFC Prefrontal cortex
PVN Hypothalamic paraventricular nucleus
VTA Ventral tegmental area

INTRODUCTION

Nicotine is one of the main psychoactive components in tobacco smoke and contributes to high morbidity and mortality throughout the world (Murray & Short, 1997). Nicotine dependence is more prevalent compared to other drugs of abuse. Quit rates (or *smoking cessation* rates) stay low despite the availability of several pharmacological and nonpharmacological treatments for either cessation or attenuation of withdrawal symptoms (Haas, Munoz, Humfleet, Reus, & Hall, 2004; Zaparoli & Galduróz, 2012). Nicotine is an alkaloid that acts as an agonist for nicotinic acetylcholine receptors (nAChRs) within the brain and peripheral nervous system. Neuronal nAChRs are ligand-gated ion channels comprising five membrane-spanning subunits that combine to form a functional receptor and include nine isoforms of the neuronal α-subunit (α2–α10) and three isoforms of the neuronal β-subunit (β2–β4) (Changeux & Taly, 2008). These subunits combine with a stoichiometry of two α- and three β-subunits or five β7-subunits to form nAChRs with distinct pharmacologic and kinetic properties. Acetylcholine is the endogenous neurotransmitter that binds and activates nAChRs. Numerous nAChRs are situated on somatodendritic, axonal, postsynaptic sites and presynaptic terminals and modulate the release of various neurotransmitters throughout the brain during chronic exposure to nicotine (McGehee & Role, 1995; Wonnacott, 1977). Therefore, various neurotransmitter systems are involved in the rewarding and reinforcing effects of nicotine and the adaptations that occur in response to chronic nicotine exposure that give rise to dependence and withdrawal symptoms.

NICOTINE-INDUCED NEUROADAPTATIONS

Nicotine addiction involves long-lasting malfunctioning adaptive changes including development of disruptive nicotine stimulus associations. Nicotine-induced neuroplasticity underlies the development

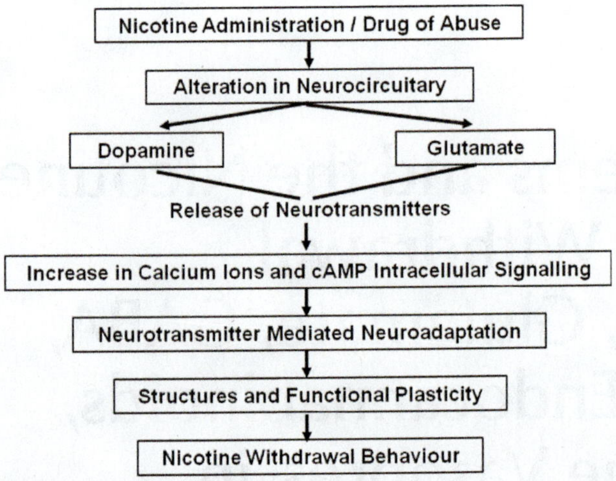

FIGURE 1 Nicotine-induced neuroadaptations. Tubular flow of events showing effect of nicotine on neurocircuitry-mediated flow of neurotransmitter-induced neuroadaptations.

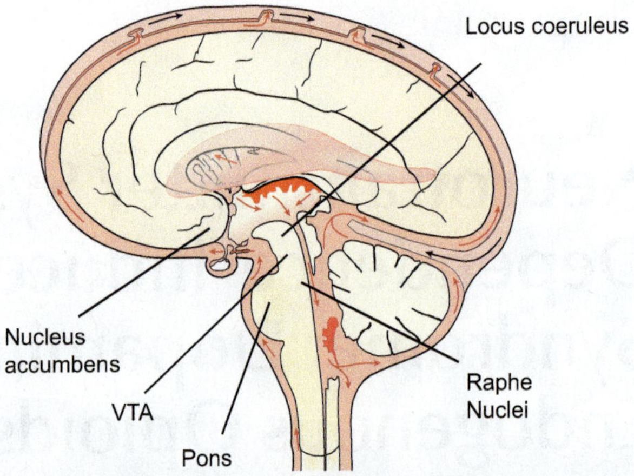

FIGURE 2 Brain parts involved in nicotine withdrawal syndrome. The flow of impulse due to chronic administration of nicotine interconnects various parts of the brain that are involved in nicotine dependence.

of tobacco addiction (in regions such as the hippocampus) and the enhancement of cognitive capabilities (Molas et al., 2014). Nicotine dependence is accompanied by neuroadaptive changes that occur in the circuits underlying emotion and motivation (Koob & Volkow, 2010). Preclinical studies of nicotine dependence suggest important neurotransmitter interactions between various transmitters such as glutamate, γ-aminobutyric acid (GABA), acetylcholine, and dopamine in the ventral tegmental area, the central nucleus of the amygdala, and the prefrontal cortex (Markou, 2008). The upregulation of neuronal nAChRs is a significant adaptive change that arises as a consequence of continuous exposure to nicotine and leads to the addictive properties of nicotine (Wonnacott, 1990). Nicotine desensitizes nAChRs and this leads to an increase in the upregulation of receptors for maintaining circuit-level homeostasis (Fenster, Whitworth, Sheffield, Quick, & Lester, 1999). This upregulation of nAChRs as a consequence of chronic exposure to nicotine differs among receptor subtypes. Alterations in brain regions for the same subtype occur via one or more regulatory processes such as receptor assembly, trafficking, and/or degradation (Gentry & Lukas, 2002; Lester et al., 2009). Isomerization of surface nAChRs to high-affinity nicotinic sites is due to binding with nicotine as the agonist for its active sites (Vallejo, Buisson, Bertrand, & Green, 2005). Repeated nicotine exposure produces heterologous neuroadaptation, which alters the cholinergic inputs and modulates the release of various neurotransmitters. Nicotine increases α-amino-3-hydroxy-5-methyl-4-isoxazolepropionic acid (AMPA)/N-methyl-D-aspartate (NMDA) current ratios in dopamine neurons (Gao et al., 2010; Placzek & Dani, 2009) and other locations involved in drug-associated memory (Dani & De Biasi, 2001; Kauer & Malenka, 2007). In the nucleus accumbens, administration of nicotine leads to an increase in high-affinity dopamine (DA) D2 receptors (Novak, Seeman, & Le Foll, 2010), which leads to an increase in G-protein-coupled DA D2 receptors. This results in DA supersensitivity in nicotine-treated animals (Balfour, 2009). Nicotine self-administration also decreases expression of the cystine–glutamate exchanger, in the nucleus accumbens and ventral tegmental area (VTA), and decreases the glial glutamate transporter, in the nucleus accumbens (Knackstedt et al., 2009).

Nicotine inhibits proteasomal function by altering the scaffolding proteins at the synaptic junction (Rezvani et al., 2009) and also alters the functioning of the endogenous opioid system that regulates both negative and positive motivational and affective states (Hadjiconstantinou & Neff, 2011; Trigo, Martin-Garcia, Berrendero, Robledo, & Maldonado, 2010). Opioid peptides like dynorphins and enkephalins affect DA function in the VTA and the striatum (Di Chiara & Imperato, 1988). DA controls the formation of striatal dynorphin and enkephalin by affecting their transcription (Angulo & McEwen, 1994). Nicotine has an impact on these system- and cellular-level interactions by altering the synthesis and release of opioid peptides in a time- and peptide-specific manner.

Agonist-based nAChR activation (i.e., by nicotine) affects the release of various neurotransmitters (Kenny, 2011). In the chronic time scale, these nicotine-induced changes in neurotransmission are responsible for system adaptations (Figure 1).

NEURAL BASIS OF NICOTINE WITHDRAWAL

Sudden stoppage of nicotine (i.e., smoking cessation or quitting) alters the neurochemistry of nAChRs in the brain of addicted subjects and triggers the affective and somatic signs of withdrawal. In the withdrawal syndrome induced by removal of nicotine, there are decreases in the activity of the mesolimbic dopaminergic system, characterized by both reduced DA release and an increase in DA reuptake (Duchemin, Zhang, Neff, & Hadjiconstantinou, 2009; Rahman, Zhang, & Corrigall, 2004). Nicotine-induced withdrawal symptoms are due to deficits in DA transmission in the nucleus accumbens and increased upregulation of DA transporters in the prefrontal cortex (PFC) (Figure 2; Carboni, Bortone, Giua, & Di Chiara, 2000; Hadjiconstantinou, Duchemin, Zhang, & Neff, 2011). This increase in DA levels in the PFC is triggered by stressful and aversive stimuli that contribute anxiety-like behaviors in nicotine-dependent subjects. The effect of this is to attenuate aversive stimuli. Smoking and administration of drugs of abuse impart a calming effect and act as potent negative reinforcers in nicotine addicts. This in turn leads to increased craving and relapse (Brown, Gonzalez, Whishaw, & Kolb, 2001; Kawasaki et al., 2011; Morissette, Tull, Gulliver, Kamholz, & Zimering, 2007). The extended amygdala and the hypothalamic–pituitary–adrenal axis play important roles in the negative affective states linked with nicotine-induced withdrawal syndrome (Koob & Volkow, 2010). There are increased levels of the stress hormones corticosterone and corticotropin-releasing factor (CRF) during nicotine-induced withdrawal symptoms. A study by George et al. (2007) showed that levels of CRF were increased by 500 times in the central nucleus of the amygdala (CeA) after nicotine withdrawal and can also be precipitated with the nonselective nAChR antagonist mecamylamine. Intracerebral injection of a CRF1 receptor antagonist into the CeA precipitates anxiety-like behavior in nicotine-dependent rodents (George et al., 2007). Nicotine-induced neuroadaptations occur at various brain sites, such as the amygdala, nucleus of the stria terminalis, VTA, nucleus accumbens, and interpeduncular nucleus, and affect the release of several neurotransmitter and neuropeptide systems that disrupt nicotine withdrawal syndrome (George et al., 2007; Koob & Volkow, 2010). Opioid peptides and the serotonergic and noradrenergic systems are known to modulate various mechanisms of nicotine withdrawal (Hadjiconstantinou & Neff, 2011; Semenova & Markou, 2010).

NEUROTRANSMITTERS AND MEDIATOR SYSTEMS INVOLVED IN NICOTINE WITHDRAWAL

Acute and continuous exposure to nicotine or drugs of abuse increases the release of the neurotransmitter DA from dopaminergic neurons originating in the VTA, nucleus accumbens, PFC, and hippocampus. These changes in brain neurochemistry are among the most important causes of nicotine addiction (Balfour, 2009). Binding of nicotine with nAChRs leads to activation of the VTA that results in DA release in the nucleus accumbens (Figure 3). Continuous nicotine exposure results in upregulation of nAChRs at different rates. Such nAChRs become desensitized and provoke glutamate-mediated excitation that increases the firing frequency of dopaminergic neurons. This releases DA and diminishes the GABA-mediated inhibitory tone, which enhances the responsiveness to nicotine (Benowitz, 2010).

Below we describe various neurotransmitters in relation to nicotine withdrawal:

Dopamine

DA is an excitatory neurotransmitter that is classified into two types of receptors D1 and D2, based on their pharmacological action. They are further subdivided into their subtypes as follows: D1 into D1 and D5 and D2 into D2, D3, and D4. They are mostly present in the central nervous system. Both D1 and D2 receptors act opposite to each other with respect to their activation. Activation of the D1 family receptors increases cyclic adenosine monophosphate 3,5-monophosphates (cAMP) by the enzyme adenylate cyclase through stimulatory G_s protein of the G-protein-coupled receptors. The activation of D2 receptors decreases cAMP through a G-inhibitory-protein pathway. Increases in intracellular cAMP lead to the activation of protein kinase and G-protein receptor kinase-3 that phosphorylates various receptors and channels of DA, which leads to the activation of transcription factors like cyclic adenosine monophosphate response element-binding protein (CREB). Nicotine's addictive properties are due to modulation of these intracellular transduction pathways and alterations in the expression of gene products upon continuous activation of these intracellular pathways

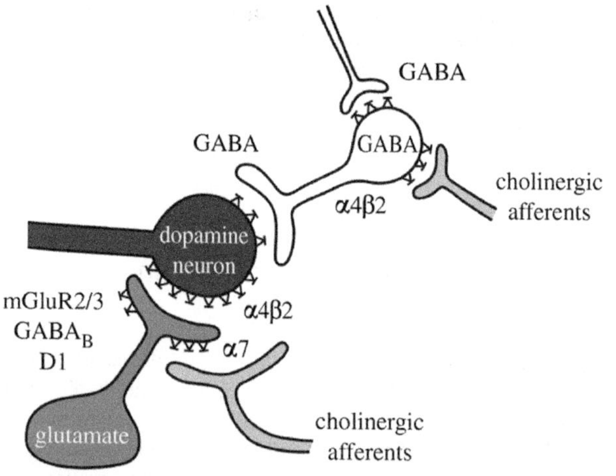

FIGURE 3 Role of various neurotransmitters in nicotine withdrawal. Pictorial presentation of nicotine–acetylcholine–glutamate–GABA–DA interactions involved in mediating the effects of nicotine withdrawal. *Adapted from Mansvelder and McGehee (2002) (License Number: 3546850493873 with permission from publisher).*

(Reddy, 2013). D3 receptors play a significant role in nicotine withdrawal and addiction. Inhibition of DA transmission in the reward circuit of the brain acts as antinicotine therapy. The partial agonist of DA D3 receptors BP897 dose-dependently reduces the nicotine dependence-induced withdrawal symptoms in rodents. SB-277011A, a selective D3 receptor antagonist, has shown its efficacy in abolishing nicotine-seeking behavior in dependent animals. Nicotine enhances DA clearance in rat nucleus accumbens, suggesting that nicotine regulates DA concentration by the DA transporter (DAT). The nicotinic receptor antagonist DHβE decreases the level of DA by modulating the functioning of DAT (Xi, Spiller, & Gardner, 2009).

Dopamine and Glutamate

Nicotine-based activation of NMDA receptors in the VTA increases release of DA in the nucleus accumbens. Acute exposure of nicotine activates nAChRs located presynaptically in the glutamatergic pathway, leading to increased glutamate release that enhances the firing of these neurons (Watkins, Koob, & Markou, 2000). Studies have shown that nicotine elevates glutamate levels in the cerebral cortex, dorsal striatum, hippocampus, hypothalamus, locus coeruleus, and cerebellum, which result in increased excitatory postsynaptic firing and neural activity mediated via the glutamate receptor. These events lead to long-lasting synaptic plasticity in numerous brain parts, which promote nicotine addiction and neuroadaptive changes in the expression of various proteins related to glutamate neurotransmission. Chronic nicotine self-administration in rats also increases glutamate receptor 2/3 expression in the VTA and upregulation of the glutamate transporter EAAT2 in the nucleus accumbens (Figure 3).

The adaptive changes in glutamate transmission produced by the continuous use of nicotine for a prolonged period of time may provide a neural basis for the enhanced susceptibility to nicotine addiction during puberty in both humans and animals. Thus, nicotine produces neuroadaptations in glutamatergic transmission (Gass & Olive, 2008). DA and glutamate are extensively dispersed in the cortex, limbic system, and basal ganglia and are responsible for nicotine dependence of behavior motivation, learning, and memory. Synchronized behavior of DA and glutamate is prominent in nicotine dependence, owing to a mechanism related to their DA D1, glutamate, NMDA, and AMPA receptors. These receptors are further responsible for adaptive changes in gene expression and results in synaptic plasticity (Reddy, 2013).

γ-Aminobutyric Acid and Dopamine Interaction

GABAergic afferent neurons are intricately connected with the dopaminergic neurons in the VTA and mediate inhibition of mesolimbic DA release and limit both reward and reinforcement activities (Watkins et al., 2000). The neurotransmitter GABA is released in the VTA by neurons that originate from the pedunculopontine tegmental nucleus, ventral pallidum, and nucleus accumbens, as well as from interneurons located in the VTA. Endogenous GABA acts via ionotropic $GABA_A$ and $GABA_C$ receptors and metabotropic $GABA_B$ receptors. Acute nicotine exposure in rodents increases GABA release by activating the excitatory α4β2-subunit of nAChRs that are present on GABAergic neurons in the VTA. First, nicotine-mediated GABA release inhibits the rewarding effects of nicotine and later on results in desensitization of the α4β2-containing subunit of nAChRs on GABAergic receptors. This desensitization inhibits nicotine-induced GABA release, leading to decreased inhibition of dopaminergic neurons in the VTA and increased DA level in the nucleus accumbens. This in turn facilitates the reinforcing effects of nicotine. Activation of GABAergic neurotransmission decreases both the reinforcing effects of nicotine and the reinstatement of cue-induced nicotine-seeking behavior in rats and prevents relapse to tobacco smoking (D'Souza & Markou, 2011).

Effects of Nicotine on Dopaminergic Tracts

The Nigrostriatal Dopamine System: Chronic nicotine exposure activates the nigrostriatal dopaminergic system, which is responsible for movement disorders related to nicotine dependence (Ziedonis & George, 1997).

Mesolimbic Structures: Administration of nicotine leads to activation of the nucleus accumbens and VTA, which causes physiological alterations in emotional expression, including positive symptoms of psychosis, drug reinforcement, and reward (Ziedonis & George, 1997).

The Mesocortical System: Repeated experience of nicotine activates ventral tegmental projections to the PFC area by modulating release of DA. In schizophrenia, deficiency of DA in the PFC leads to negative symptoms such as social withdrawal and apathy. Continuous administration of nicotine increases release of DA, so smoking may help to correct DA deficiency in the PFC of schizophrenia patients and thus relieve negative symptoms. So, nicotine positively affects schizophrenia (Ziedonis & George, 1997).

Serotonin (5-Hydroxytryptamine) and Nicotine Addiction

Projections of serotonergic neurons in the dorsal raphe nuclei run to the dopaminergic area of the midbrain, mesocorticolimbic system, and hippocampus. Serotonin (also called 5-hydroxytryptamine (5-HT)) is an important mediator in nicotine withdrawal. Serotonin receptors have been classified into three subclasses: $5-HT_1$, $5-HT_2$, and $5-HT_3$. The most abundant form of 5-HT receptor in the VTA is $5-HT_{1B}$ (Reddy, 2013). Mannucci, Pieratti, Firenzuoli, Caputi, and Calapai (2007) proved that chronic administration of nicotine leads to an increase in the density of $5-HT_{1A}$ receptors in the brain and reduces the effect of stress by acting on 5-HT neurons within the hippocampus. The serotonergic system enhances the positive reinforcing effects of nicotine, but it has also been shown to act on the negative reinforcing effects of nicotine withdrawal (Mannucci et al., 2007). Release of 5-HT is regulated by the autoreceptors present in the nerve endings and heteroreceptors located on dopaminergic, glutamatergic, GABAergic, or cholinergic neurons. Together they play prominent roles in nicotine addiction (Reddy, 2013).

The Endogenous Opioid System

The endogenous opioid system plays an important role in nicotine-induced withdrawal symptoms. The opioid peptide system includes β-endorphins, met- and leu-enkephalins, dynorphins, and nociceptin/orphanin FQ. Various studies have shown that nicotine in tobacco smokers releases brain opioid peptides by modulating the μ opioid receptors. The μ opioid receptors are responsible for increased nicotine binding in the amygdala, thalamus, and VTA, which are involved in the anticipation of reward in nicotine dependence. Further, findings show that nicotine enhances the level of endogenous opioids, peptides derived from preproenkephalin, which is involved in the antinociceptive (by which the sensitivity to painful stimuli is decreased) effects of nicotine dependence (Xue & Domino, 2008). The κ opioid receptor (KOR) plays a role in mediating the withdrawal aspects of nicotine dependence. Long-term exposure to nicotine is associated with alterations in the functioning of the endogenous opioid system and the hypothalamic–pituitary–adrenal axis that contribute to nicotine dependence. KOR activation elicits negative affect via modulation of mesolimbic DA transmission in the nucleus accumbens that modulates nicotine withdrawal. Chronic nicotine use dysregulates the KOR system in a manner that modulates negative reinforcement processes that contribute to nicotine-dependence-induced withdrawal syndrome (Jackson, Carroll, Negus, & Damaj, 2010).

Endocannabinoids

The endocannabinoid system modulates the addictive properties of nicotine (Cippitelli et al., 2011). This system is responsible for the rewarding effects of cannabinoids, nicotine, alcohol, and opioids, through the release of endocannabinoids in the VTA of the mesolimbic pathway of the brain. Cannabinoid receptors and nAChRs are widely distributed in the hippocampus and the amygdala, parts of the brain that are responsible for nicotine-seeking behavior. Cannabinoid receptor activation has been shown to modulate nAChRs, which promotes release of acetylcholine into various brain parts. Recent clinical trials have suggested that the cannabinoid receptor type 1 (CB_1) antagonist rimonabant can be used for smoking cessation. Thus, CB_1 antagonists could represent a new generation of compounds to treat nicotine addiction (Maldonado, Valverde, & Berrendero, 2006). Chronic nicotine administration increases levels of arachidonoylethanolamide (an endogenous cannabinoid neurotransmitter) in the limbic forebrain and brainstem but decreases levels in the hippocampus, striatum, and cerebral cortex. Rimonabant, a CB_1 antagonist, decreases nicotine self-administration and conditioned place preference in rats, suggesting that endocannabinoid signaling is involved in nicotine reinforcement (Merritt, Martin, Walters, Lichtman, & Damaj, 2008).

Noradrenaline

Nicotine dependence-induced withdrawal syndrome is characterized by depression-like behavior that may be mediated by dysregulation of norepinephrine transmission (Paterson, Semenova, & Markou, 2008). Nicotine-mediated activation of nAChRs in postganglionic sympathetic nerve endings enhances the level of noradrenaline in various tissues (such as the heart), indicating exocytotic calcium-independent release of various neurotransmitters (Richardt et al., 1994). Noradrenergic neuronal pathways from the brainstem nucleus tractus solitarius (NTS) to the hypothalamic paraventricular nucleus (PVN) and amygdala are involved in nicotine-related stress responses and craving. These effects are mediated entirely through the impact of nicotine on glutamate afferents in the NTS and also NMDA receptors that, in part, stimulate nitric oxide production, which in turn results in activation of noradrenergic neurons (Zhao, Chen, & Sharp, 2007). Noradrenergic cell bodies in the nucleus accumbens and VTA project afferent neurons to the A_1 and A_2 areas of the brainstem as well as the locus coeruleus. Noradrenaline acts through various excitatory mechanisms, for example, directly stimulating the firing of dopaminergic neurons and indirectly potentiating the activity of noradrenergic neurons projecting to the PFC and the nucleus accumbens. Overall these effects contribute to the phenomenon of nicotine dependence (Reddy, 2013).

Arginine Vasopressin

Nicotine from cigarette smoke produces a rise in blood pressure. This is associated with an increased level of catecholamines and other vasoactive hormones, including arginine vasopressin, in the plasma (Marano, Ramirez, Mori, & Ferrari, 1999). Nicotine-mediated changes in glutamate and arginine vasopressin release in the nucleus accumbens modulates neuropharmacological interactions during the initial stages of chronic nicotine withdrawal. Nicotine induces stress because of neuroendocrine responses, such as the release of arginine vasopressin from the PVN and activation of the hypothalamic–pituitary–adrenal axis (Suzuki et al., 2009).

Neuropeptide Y

The orexigenic neurotransmitter neuropeptide Y (NPY) attenuates somatic withdrawal signs linked with nicotine dependence. NPY has been reported to decrease neuronal excitability by decreasing synaptic transmission of glutamate and increasing inhibitory synaptic transmission of GABA in brainstem areas such as the locus coeruleus (Bacci, Huguenard, & Prince, 2002). Therefore, stimulation of Y1 receptors (one of the family of NPY receptors) attenuates the somatic signs associated with nicotine withdrawal (Rylkova et al., 2008).

Monoamine Oxidase A and B

Cigarette smoking causes a distinct decrease in the levels of monoamine oxidase (MAO). MAO is responsible for oxidation of monoamines and breakdown of DA. Smokers may exhibit higher levels of DA in the nucleus accumbens, which is known to be important in nicotine reinforcement. Increased levels of DA may in turn increase nicotine withdrawal syndrome in addicts. Chronic nicotine exposure enhances the brain DA level, leading to nicotine dependence indirectly (Jain, 2003). MAO inhibitors contained in cigarette smoke are persuasive modulators of the rewarding and withdrawal effects of nicotine. Pretreatment with MAO inhibitors induces a long-lasting conditioned placed aversion associated with nicotine withdrawal syndrome (Guillem, Vouillac, Koob, Cador, & Stinus, 2008).

Cyclic Adenosine Monophosphate Response Element-Binding Protein

CREB is not a neurotransmitter but a cellular transcription factor, which binds to DNA sequences called cAMP response elements, thereby increasing or decreasing the transcription of downstream genes (Bourtchuladze et al., 1994). It is important to mention that CREB is interlinked with the phenomena of addiction and withdrawal in relation to nicotine. Chronic use of nicotine upregulates the μ opioid receptors and alters the transcription factor CREB, which leads to nicotine dependence and reinforcing behavior like craving and relapse (Dani & Harris, 2005). CREB acts through calcium and the protein kinase A pathway in the corticostriatal region (a component of the central nervous system that plays a major role in learning). This process is mediated by glutamate and DA signals. Many genes that are responsible for the generation and phosphorylation of CREB are dependent on NMDA and DA D1 receptors. Drugs that antagonize the actions of NMDA and/or DA D1 also inhibit the generation of CREB. Thus, theoretically CREB is a putative therapeutic agent in nicotine addiction (Reddy, 2013).

Corticotropin-Releasing Factor

Various studies have confirmed the role of CRF-like peptides in the negative affective state associated with acute and protracted withdrawal related with nicotine use (George et al., 2007). Increased CRF and, concomitantly, norepinephrine transmission play a role in the stress-induced reinstatement of drug-seeking behavior. Thus, the activation of brain stress systems mediates stress-induced relapse to smoking. George et al. (2007) have reported that precipitation of nicotine withdrawal in rats is principally modulated by an increase in the activation of the CRF–CRF1 receptor system in the CeA. Furthermore, the CRF–CRF1 system is likewise recognized to mediate the biology and conduct linked to negative reinforcement (i.e., setting off the unpleasantness of cessation) and relapse (i.e., a return to self-administration) also usually attributed to nicotine addiction (see Figure 4).

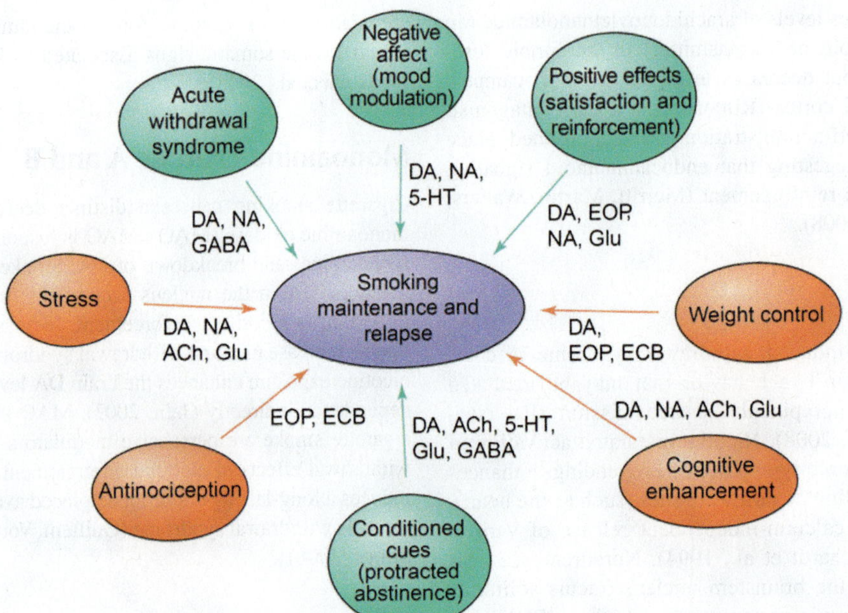

FIGURE 4 **Various factors affecting smoking maintenance and relapse.** Various factors that mediate smoking maintenance and relapse in nicotine dependence are shown. The green circles represent primary contributors to smoking maintenance and relapse, whereas the red circles represent secondary contributors to these processes. ACh, acetylcholine (nicotinic ACh receptor); DA, dopamine; ECB, endocannabinoid (CB1 receptor); EOP, endogenous opioid peptide; Glu, glutamate; 5-HT, 5-hydroxytryptamine (also called serotonin); NA, noradrenaline; GABA, γ-aminobutyric acid. *Data from George and O'Malley (2004) with permission from publisher Elsevier under License Number 3638071207273.*

APPLICATION TO OTHER ADDICTIONS AND SUBSTANCE MISUSE

Tobacco use is the leading preventable cause of disease, disability, and death all over the world. Similar to other addictive drugs or substances like alcohol, morphine, cocaine, and heroin, nicotine exposure increases levels of the neurotransmitter DA, which affects the brain pathways that control stimulus–reward learning processes. Various preclinical and clinical studies have indicated that a strong association exists between cigarette smoking and the use of alcohol and other addictive drugs. The frequency of alcohol and illicit drug use is greater among nicotine addicts than nonsmokers. Ethanol interacts with nAChRs in the mesocorticolimbic dopaminergic reward circuitry to affect brain reward systems. Like nicotine, ethanol activates dopaminergic neurons of the VTA, which project to the nucleus accumbens, which provokes the release of DA and other excitatory neurotransmitters that play important roles in their dependence. Nicotine acts as a strong conditioned cue for alcohol and illicit substance use, owing to its direct and indirect activation of various transduction systems that mediate the release of various neurotransmitters like DA, GABA, glutamate, and epinephrine that also play prominent roles in substance dependence and neuroadaptation of their respective receptors. Thus, attempts to treat nicotine dependence in psychiatric patients must involve treatment of the nonnicotine substance abuse disorder- or dependence-induced withdrawal syndrome. Owing to the serious consequences of smoking and other substance addictions, the development of effective therapies based on these common neurochemical systems helps us to attenuate the adverse effects of the various abuse substance addictions.

DEFINITION OF TERMS

Extended amygdala The CeA, together with the bed nucleus of the stria terminalis and the posterior shell of the nucleus accumbens, is known as the "extended amygdala."

Neuroadaptation This refers to the process whereby the sensory system of the body compensates for the presence of a constant stimulus in the form of a chemical or drug in the body so that it can continue to function normally.

Neuropeptide These are small protein-like molecules used by neurons to communicate with one another, for example, NPY.

Corticotropin-releasing factor: This is a neuropeptide present in the paraventricular nucleusPVN of the hypothalamus that regulates the release of corticotropin (ACTH) from the pituitary gland during stress conditions.

***N*-Methyl-D-aspartate** This is an amino acid that acts as a specific agonist for the NMDA receptor.

SUMMARY POINTS

- Nicotine is the main constituent of tobacco smoke that plays a prominent role in maintaining the smoking habit and develops addiction.
- Nicotine serves as a major reinforcer in both humans and animals. It produces a complex behavioral phenomenon comprising effects on several neural systems.
- When a nicotine addict tries to quit, he or she experiences nicotine dependence-induced withdrawal symptoms like anxiety, irritability, attention difficulties, sleep disturbances, increased appetite, and cravings for tobacco.

- The adaptive changes in nAChRs produced by chronic exposure to nicotine play a crucial role in the neuroadaptations and DA transmission in the nucleus accumbens that sustain the reinforcing effects of smoking.
- Various neurotransmitters and transmitter-related processes participate in the addictive and withdrawal effects of nicotine, including DA, glutamate, GABA, endogenous opioids, endocannabinoids, noradrenaline, arginine vasopressin, NPY, MAO, CREB, and CRF. Various preclinical and clinical studies have indicated that nAChRs have an important role in the addictive properties of ethanol, as well as in nicotine and alcohol codependence.
- The development of drugs that act on DA, glutamate, GABA, endogenous opioids, endocannabinoids, noradrenaline, arginine vasopressin, NPY, MAO, CREB, and CRF and other pathways that determine nicotine addiction or withdrawal effects could enhance the effectiveness of the current methods used for smoking cessation.

REFERENCES

Angulo, J. A., & McEwen, B. S. (1994). Molecular aspects of neuropeptide regulation and function in the corpus striatum and nucleus accumbens. *Brain Research. Brain Research Reviews*, *19*(1), 1–28.

Bacci, A., Huguenard, J. R., & Prince, D. A. (2002). Differential modulation of synaptic transmission by neuropeptide Y in rat neocortical neurons. *Proceedings of the National Academy of Sciences of the United States of America*, *99*, 17125–17130.

Balfour, D. J. (2009). The neuronal pathways mediating the behavioral and addictive properties of nicotine. *Handbook of Experimental Pharmacology*, *192*, 209–233.

Benowitz, N. L. (2010). Mechanisms of disease. *The New England Journal of Medicine*, *362*(24), 2297–2303.

Bourtchuladze, R., Frenguelli, B., Blendy, J., Cioffi, D., Schutz, G., & Silva, A. J. (1994). Deficient long-term memory in mice with a targeted mutation of the cAMP-responsive element-binding protein. *Cell*, *79*(1), 59–68.

Brown, R. W., Gonzalez, C. L., Whishaw, I. Q., & Kolb, B. (2001). Nicotine improvement of Morris water task performance after fimbria-fornix lesion is blocked by mecamylamine. *Behavioural Brain Research*, *119*(2), 185–192.

Carboni, E., Bortone, L., Giua, C., & Di Chiara, G. (2000). Dissociation of physical abstinence signs from changes in extracellular dopamine in the nucleus accumbens and in the prefrontal cortex of nicotine dependent rats. *Drug and Alcohol Dependence*, *58*(1–2), 93–102.

Changeux, J. P., & Taly, A. (2008). Nicotinic receptors, allosteric proteins and medicine. *Trends in Molecular Medicine*, *14*, 93–102.

Cippitelli, A., Astarita, G., Duranti, A., Caprioli, G., Ubaldi, M., Stopponi, S., ... Ciccocioppo, R. (2011). Endocannabinoid regulation of acute and protracted nicotine withdrawal: effect of FAAH inhibition. *PLoS One*, *6*(11), e28142.

Dani, J. A., & De Biasi, M. (2001). Cellular mechanisms of nicotine addiction. *Pharmacology, Biochemistry, and Behavior*, *70*(4), 439–446.

Dani, J. A., & Harris, R. A. (2005). Nicotine addiction and comorbidity with alcohol abuse and mental illness. *Nature Neuroscience*, *8*(11), 1465–1470.

Di Chiara, G., & Imperato, A. (1988). Opposite effects of mu and kappa opiate agonists on dopamine release in the nucleus accumbens and in the dorsal caudate of freely moving rats. *The Journal of Pharmacology and Experimental Therapeutics*, *244*(3), 1067–1080.

D'Souza, M. S., & Markou, A. (2011). Neuronal mechanisms underlying development of nicotine dependence: implications for novel smoking-cessation treatments. *Addiction Science & Clinical Practice*, *6*(1), 4–16.

Duchemin, A. M., Zhang, H., Neff, N. H., & Hadjiconstantinou, M. (2009). Increased expression of VMAT2 in dopaminergic neurons during nicotine withdrawal. *Neuroscience Letters*, *467*(2), 182–186.

Fenster, C. P., Whitworth, T. L., Sheffield, E. B., Quick, M. W., & Lester, R. A. (1999). Up-regulation of surface alpha4beta2 nicotinic receptors is initiated by receptor desensitization after chronic exposure to nicotine. *The Journal of Neuroscience*, *19*(12), 4804–4814.

Gao, M., Jin, Y., Yang, K., Zhang, D., Lukas, R. J., & Wu, J. (2010). Mechanisms involved in systemic nicotine-induced glutamatergic synaptic plasticity on dopamine neurons in the ventral tegmental area. *The Journal of Neuroscience*, *30*(41), 13814–13825.

Gass, J. T., & Olive, M. F. (2008). Glutamatergic substrates of drug addiction and alcoholism. *Biochemical Pharmacology*, *75*(1), 218–265.

Gentry, C. L., & Lukas, R. J. (2002). Regulation of nicotinic acetylcholine receptor numbers and function by chronic nicotine exposure. *Current Drug Targets CNS & Neurological Disorders*, *1*(4), 359–385.

George, O., Ghozland, S., Azar, M. R., Cottone, P., Zorrilla, E. P., Parsons, L. H., ... Koob, G. F. (2007). CRF-CRF1 system activation mediates withdrawal-induced increases in nicotine self-administration in nicotine-dependent rats. *Proceedings of the National Academy of Sciences of the United States of America*, *104*(43), 17198–17203.

George, T. P., & O'Malley, S. S. (2004). Current pharmacological treatments for nicotine dependence. *Trends in Pharmacological Sciences*, *25*(1), 42–48.

Guillem, K., Vouillac, C., Koob, G. F., Cador, M., & Stinus, L. (2008). Monoamine oxidase inhibition dramatically prolongs the duration of nicotine withdrawal-induced place aversion. *Biological Psychiatry*, *63*(2), 158–163.

Haas, A. L., Munoz, R. F., Humfleet, G. L., Reus, V. I., & Hall, S. M. (2004). Influences of mood, depression history, and treatment modality on outcomes in smoking cessation. *Journal of Consulting and Clinical Psychology*, *72*, 563–570.

Hadjiconstantinou, M., Duchemin, A. M., Zhang, H., & Neff, N. H. (2011). Enhanced dopamine transporter function in striatum during nicotine withdrawal. *Synapse*, *65*(2), 91–98.

Hadjiconstantinou, M., & Neff, N. H. (2011). Nicotine and endogenous opioids: neurochemical and pharmacological evidence. *Neuropharmacology*, *60*(7–8), 1209–1220.

Jackson, K. J., Carroll, F. I., Negus, S. S., & Damaj, M. I. (2010). Effect of the selective kappa-opioid receptor antagonist JDTic on nicotine antinociception, reward, and withdrawal in the mouse. *Psychopharmacology*, *210*(2), 285–294.

Jain, A. (2003). Treating nicotine addiction. *BMJ*, *327*(7428), 1394–1395.

Kauer, J. A., & Malenka, R. C. (2007). Synaptic plasticity and addiction. *Nature Reviews Neuroscience*, *8*(11), 844–858.

Kawasaki, Y., Araki, H., Suemaru, K., Kitamura, Y., Gomita, Y., & Sendo, T. (2011). Involvement of dopaminergic receptor signaling in the effects of glutamatergic receptor antagonists on conditioned place aversion induced by naloxone in single-dose morphine-treated rats. *Journal of Pharmacological Sciences*, *117*(1), 27–33.

Kenny, P. J. (2011). Tobacco dependence, the insular cortex and the hypocretin connection. *Pharmacology, Biochemistry and Behavior, 97*(4), 700–707.

Knackstedt, L. A., LaRowe, S., Mardikian, P., Malcolm, R., Upadhyaya, H., Hedden, S., ... Kalivas, P. W. (2009). The role of cystine–glutamate exchange in nicotine dependence in rats and humans. *Biological Psychiatry, 65*(10), 841–845.

Koob, G. F., & Volkow, N. D. (2010). Neurocircuitry of addiction. *Neuropsychopharmacology, 35*(1), 217–238.

Lester, H. A., Xiao, C., Srinivasan, R., Son, C. D., Miwa, J., Pantoja, R., ... Wang, J. C. (2009). Nicotine is a selective pharmacological chaperone of acetylcholine receptor number and stoichiometry. Implications for drug discovery. *The AAPS Journal, 11*(1), 167–177.

Maldonado, R., Valverde, O., & Berrendero, F. (2006). Involvement of the endocannabinoid system in drug addiction. *Trends in Neurosciences, 29*(4), 225–232.

Mannucci, C., Pieratti, A., Firenzuoli, F., Caputi, A. P., & Calapai, G. (2007). Serotonin mediates beneficial effects of *Hypericum perforatum* on nicotine withdrawal signs. *Phytomedicine, 14*(10), 645–651.

Mansvelder, H. D., & McGehee, D. S. (2002). Cellular and synaptic mechanisms of nicotine addiction. *Journal of Neurobiology, 53*(4), 606–617.

Marano, G., Ramirez, A., Mori, I., & Ferrari, A. U. (1999). Sympathectomy inhibits the vasoactive effects of nicotine in conscious rats. *Cardiovascular Research, 42*(1), 201–205.

Markou, A. (2008). Review. Neurobiology of nicotine dependence. *Philosophical Transactions of the Royal Society of London. Series B, Biological Sciences, 363*(1507), 3159–3168.

McGehee, D. S., & Role, L. W. (1995). Physiological diversity of nicotinic acetylcholine receptors expressed by vertebrate neurons. *Annual Review of Physiology, 57*, 521–546.

Merritt, L. L., Martin, B. R., Walters, C., Lichtman, A. H., & Damaj, M. I. (2008). The endogenous cannabinoid system modulates nicotine reward and dependence. *The Journal of Pharmacology and Experimental Therapeutics, 326*(2), 483–492.

Molas, S., Gener, T., Güell, J., Martín, M., Ballesteros-Yáñez, I., Sanchez-Vives, M. V., & Dierssen, M. (November 11, 2014). Hippocampal changes produced by overexpression of the human CHRNA5/A3/B4 gene cluster may underlie cognitive deficits rescued by nicotine in transgenic mice. *Acta Neuropathologica Communications, 2*(1), 147.

Morissette, S. B., Tull, M. T., Gulliver, S. B., Kamholz, B. W., & Zimering, R. T. (2007). Anxiety, anxiety disorders, tobacco use, and nicotine: a critical review of interrelationships. *Psychological Bulletin, 133*(2), 245–272.

Murray, D. M., & Short, B. J. (1997). Intra class correlation among measures related to tobacco use by adolescents: estimates, correlates, and applications in intervention studies. *Addictive Behaviors, 22*(1), 1–12.

Novak, G., Seeman, P., & Le Foll, B. (2010). Exposure to nicotine produces an increase in dopamine D2(High) receptors: a possible mechanism for dopamine hypersensitivity. *The International Journal of Neuroscience, 120*(11), 691–697.

Paterson, N. E., Semenova, S., & Markou, A. (2008). The effects of chronic versus acute desipramine on nicotine withdrawal and nicotine self-administration in the rat. *Psychopharmacology, 198*(3), 351–362.

Placzek, A. N., & Dani, J. A. (2009). Synaptic plasticity within midbrain dopamine centers contributes to nicotine addiction. *Nebraska Symposium on Motivation, 55*, 5–15.

Rahman, S., Zhang, J., & Corrigall, W. A. (2004). Local perfusion of nicotine differentially modulates somatodendritic dopamine release in the rat ventral tegmental area after nicotine preexposure. *Neurochemical Research, 29*(9), 1687–1693.

Reddy, D. S. (2013). Current pharmacotherapy of attention deficit hyperactivity disorder. *Drugs of Today (Barcelona), 49*(10), 647–665.

Rezvani, K., Teng, Y., Pan, Y., Dani, J. A., Lindstrom, J., Garcia Gras, E. A., ... De Biasi, M. (2009). UBXD4, a UBX-containing protein, regulates the cell surface number and stability of alpha3-containing nicotinic acetylcholine receptors. *The Journal of Neuroscience, 29*(21), 6883–6896.

Richardt, G., Brenn, T., Seyfarth, M., Haass, M., Schömig, E., & Schömig, A. (1994). Dual effect of nicotine on cardiac noradrenaline release during metabolic blockade. *Basic Research in Cardiology, 89*(6), 524–534.

Rylkova, D., Boissoneault, J., Isaac, S., Prado, M., Shah, H. P., & Bruijnzeel, A. W. (2008). Effects of NPY and the specific Y1 receptor agonist [D-His(26)]-NPY on the deficit in brain reward function and somatic signs associated with nicotine withdrawal in rats. *Neuropeptides, 42*(3), 215–227.

Semenova, S., & Markou, A. (2010). The alpha2 adrenergic receptor antagonist idazoxan, but not the serotonin-2A receptor antagonist M100907, partially attenuated reward deficits associated with nicotine, but not amphetamine, withdrawal in rats. *European Neuropsychopharmacology, 20*(10), 731–746.

Suzuki, H., Kawasaki, M., Ohnishi, H., Otsubo, H., Ohbuchi, T., Katoh, A., ... Ueta, Y. (2009). Exaggerated response of a vasopressin-enhanced green fluorescent protein transgene to nociceptive stimulation in the rat. *The Journal of Neuroscience, 29*(42), 13182–13189.

Trigo, J. M., Martin-Garcia, E., Berrendero, F., Robledo, P., & Maldonado, R. (2010). The endogenous opioid system: a common substrate in drug addiction. *Drug and Alcohol Dependence, 108*(3), 183–194.

Vallejo, Y. F., Buisson, B., Bertrand, D., & Green, W. N. (2005). Chronic nicotine exposure upregulates nicotinic receptors by a novel mechanism. *The Journal of Neuroscience, 25*(23), 5563–5572.

Watkins, S. S., Koob, G. F., & Markou, A. (2000). Neural mechanisms underlying nicotine addiction: acute positive reinforcement and withdrawal. *Nicotine & Tobacco Research, 2*(1), 19–37.

Wonnacott, S. (1977). Presynaptic nicotinic ACh receptors. *Trends in Neurosciences, 20*, 92–98.

Wonnacott, S. (1990). The paradox of nicotinic acetylcholine receptor upregulation by nicotine. *Trends in Pharmacological Sciences, 11*(6), 216–219.

Xi, Z. X., Spiller, K., & Gardner, E. L. (2009). Mechanism-based medication development for the treatment of nicotine dependence. *Acta Pharmacologica Sinica, 30*(6), 723–739.

Xue, Y., & Domino, E. F. (2008). Tobacco/nicotine and endogenous brain opioids. *Progress in Neuropsychopharmacology and Biological Psychiatry, 32*(5), 1131–1138.

Zaparoli, J. X., & Galduróz, J. C. (2012). Treatment for tobacco smoking: a new alternative? *Medical Hypotheses, 79*(6), 867–868.

Zhao, R., Chen, H., & Sharp, B. M. (2007). Nicotine-induced norepinephrine release in hypothalamic paraventricular nucleus and amygdala is mediated by *N*-methyl-D-aspartate receptors and nitric oxide in the nucleus tractus solitarius. *The Journal of Pharmacology and Experimental Therapeutics, 320*(2), 837–844.

Ziedonis, D. M., & George, T. P. (1997). Schizophrenia and nicotine use: report of a pilot smoking cessation program and review of neurobiological and clinical issues. *Schizophrenia Bulletin, 3*(2), 247–254.

Chapter 20

Genetic Findings on the Relationship between Smoking and the Stress System

Diego L. Rovaris, Nina R. Mota, Claiton H.D. Bau

Department of Genetics, Instituto de Biociências, Universidade Federal do Rio Grande do Sul (UFRGS), Porto Alegre, RS, Brazil

Abbreviations

ACTH Adrenocorticotropic hormone
AF1 Activation function 1
COPD Chronic obstructive pulmonary disease
CRH Corticotropin-releasing hormone
CRHR1 CRH receptor type 1
DBD DNA-binding domain
FKBP4 FK506 binding protein 4
FKBP5 FK506 binding protein 5
FTND Fagerström Test for Nicotine Dependence
GC Glucocorticoid
GR Glucocorticoid receptor
GRE Glucocorticoid-response element
GWAS Genome-wide association studies
HPA Hypothalamic–pituitary–adrenal axis
LBD Ligand-binding domain
mRNA Messenger RNA
MR Mineralocorticoid receptor
nAChRs Nicotinic acetylcholine receptors
NPC Nuclear pore complex
NR3C1 Nuclear receptor subfamily 3, group C, member 1
NR3C2 Nuclear receptor subfamily 3, group C, member 2
NTD *N*-terminal transactivation domain
POMC Pro-opiomelanocortin
PPI Protein–protein interaction
SNP Single-nucleotide polymorphism
TSST Trier social stress test

INTRODUCTION

There are several studies showing that the dysregulation of the hypothalamic–pituitary–adrenal (HPA) axis plays a fundamental role in nicotine dependence and that stress has an important effect on several aspects of smoking behavior, including initiation, maintenance, and relapse (Bruijnzeel, 2012). Smokers present higher cortisol levels (Mendelson, Goletiani, Sholar, Siegel, & Mello, 2008), and nicotine, the most important compound of tobacco cigarettes, activates the HPA axis in a dose-dependent manner (Lutfy et al., 2012). Additionally, it has been demonstrated that glucocorticoids (GCs) increase dopamine release in mesocorticolimbic circuits involved in the reward and reinforcement effects of drugs of abuse (Wand et al., 2007) and that the corticotropin-releasing hormone (CRH) may also operate independent of the HPA axis, activating the sympathetic nervous system (Sutton, Koob, Le Moal, Rivier, & Vale, 1982).

As a general scenario, the literature suggests a hyperreactivity of the HPA axis related to smoking, which affects the reward and reinforcement effects of nicotine on the brain. Conversely, abstinence states may result in hyporeactivity of the HPA axis, with a consequent decrease in the levels of circulating GCs (Bruijnzeel, 2012). The reduction of cortisol levels is associated with increased relapse rates (al'Absi, Hatsukami, Davis, & Wittmers, 2004) and this may be related to the modulating effects of GCs on nicotinic acetylcholine receptors (nAChRs). Animal model findings have suggested that, similar to nicotine, corticosterone desensitizes the nAChRs (Robinson, Grun, Pauly, & Collins, 1996), and this could add to the explanation of withdrawal symptoms, considering the sudden reduction in cortisol levels in abstainers.

In this chapter, we will briefly describe the functioning of the HPA axis and the possible implications of the variability of stress-related genes in smoking behaviors. In particular, we will characterize on a molecular level the structure and mode of action of GC receptors (GRs) and mineralocorticoid receptors (MRs), which are considered the maestros that orchestrate the events of response and adaptation to stressors. Polymorphisms in the genes encoding these receptors have been implicated in various stress response patterns and have been associated with several psychiatric disorders, including substance use disorders (SUDs). Here we will discuss association findings from candidate gene and genome-wide association studies (GWAS) involving GR- and MR-coding genes and smoking, suggesting new research approaches.

THE HPA AXIS AND THE MOLECULAR EFFECTS OF GCs

The HPA axis is the main stress response system and it is involved in several relevant biological processes, such as the modulation of normal and pathologic behavior (Sapolsky, Romero, & Munck, 2000). The most important mediators of

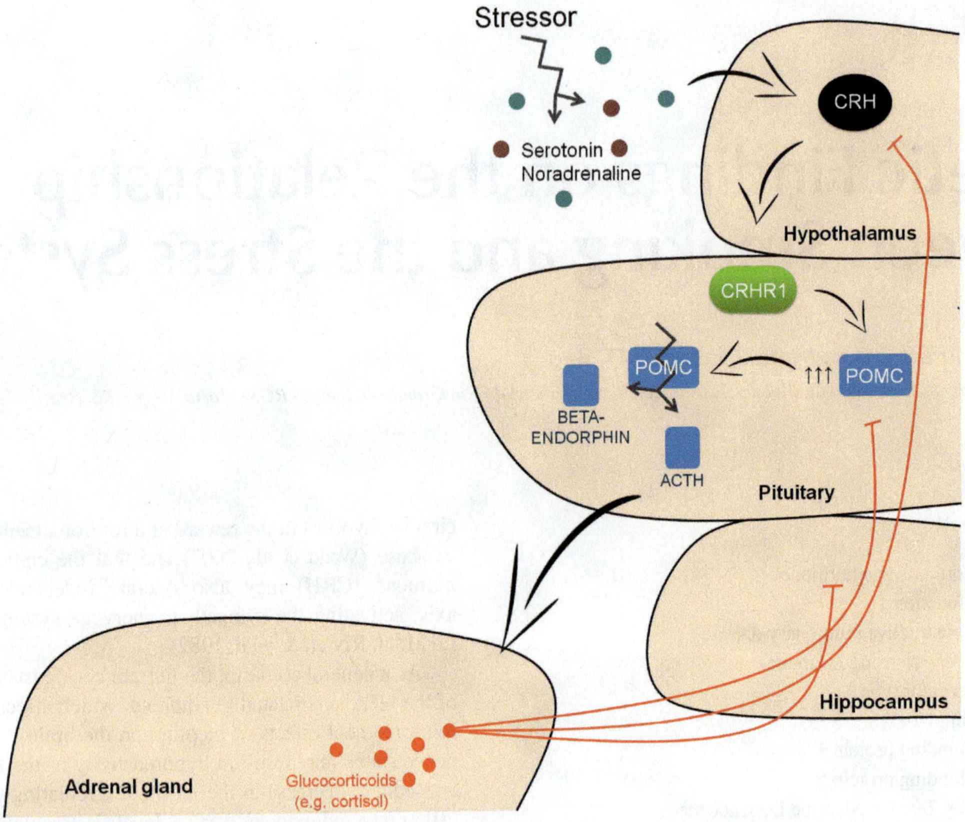

FIGURE 1 HPA axis functioning in humans. When the human body is subjected to a threat, be it of physiological (e.g., drug withdrawal symptoms), psychological (e.g., death of a family member and loss of relationship), or pharmacological origin (e.g., drugs, including nicotine and cocaine), the activation of the HPA axis occurs. The release of serotonin and noradrenaline stimulates the production of CRH by the hypothalamus. In the anterior pituitary, CRH activates the CRH receptor type 1 (CRHR1), resulting in increased expression of pro-opiomelanocortin (POMC). The POMC is cleaved, generating β-endorphin and ACTH. Acting as an endocrine hormone, ACTH stimulates the production of GCs (e.g., cortisol) in the adrenal cortex. Increased levels of cortisol result in negative feedback on the hypothalamus, anterior pituitary, and hippocampus through the activation of GRs and MRs.

this neuroendocrine circuit are CRH, adrenocorticotropic hormone (ACTH), and cortisol (Figure 1). Following a challenge or a threat, there is activation of the HPA axis by noradrenaline and serotonin and, consequently, an increase in GC levels (cortisol in humans and corticosterone in rodents). In response to a stressor agent (nicotine, e.g.), the hypothalamus secretes CRH, which in the anterior pituitary stimulates the expression of the gene encoding pro-opiomelanocortin (POMC). POMC is then processed, resulting in ACTH and other peptides, including β-endorphin (Zhou, Proudnikov, Yuferov, & Kreek, 2010). Subsequently, ACTH is released into the bloodstream and, in the adrenal cortex, it stimulates the enzymes that are involved in producing GCs (e.g., cortisol).

GCs exert their effects through their binding to GRs and MRs, which are ligand-dependent transcription factors (Derijk, 2009). The GR (Nr3c1) and the MR (Nr3c2), along with the progesterone and the androgen receptors (Nr3c3 and Nr3c4, respectively), are part of nuclear receptor subfamily 3 (Kassahn, Ragan, & Funder, 2011). These receptors are structurally similar and present three functionally distinct regions: an N-terminal transcription activation domain (NTD), a central zinc finger DNA-binding domain (DBD), and a C-terminal ligand-binding domain (LBD) (Oakley & Cidlowski, 2013). The interactions between GCs and the GRs and MRs result in conformational changes that allow these receptors to interact with specific DNA sequences and, consequently, modulate the expression of several genes (Schoneveld, Gaemers, & Lamers, 2004; Figure 2(A)).

GRs and MRs are ubiquitously distributed in the brain and regulate a series of neurophysiologic processes related to stress reactions. Among the cerebral effects resulting from the activation of GRs and MRs are the facilitation of alertness, arousal, vigilance, attention, and aggression, in addition to the maintenance of cognitive functions and the activation of negative feedback loops (Joels, Karst, DeRijk, & de Kloet, 2008). The HPA axis presents an inverted U-shaped dose–response curve (Figure 3), in which homeostasis is maintained when the system activity (GC production) is optimal (Chrousos, 2009). However, inappropriate responses can result in inefficient adaptation states (increased and insufficient levels of GCs), which may increase the risk or moderate the course of stress-related diseases, including smoking.

Traditionally, the GR and MR were known as cytosolic receptors that, when activated, would translocate to the nucleus; however, in the past few years, several pieces of evidence have emerged indicating that these receptors could also be located on the membrane (as G-protein-coupled receptors), mediating fast and nongenomic actions of GCs (Prager & Johnson, 2009; Figure 2(B)). In this sense, the cytosolic MR seems to be involved in the maintenance of cellular excitability and also with the stress response initiation, while the membrane MR is thought to amplify such signals (Joels, et al., 2008). Meanwhile, the membrane and

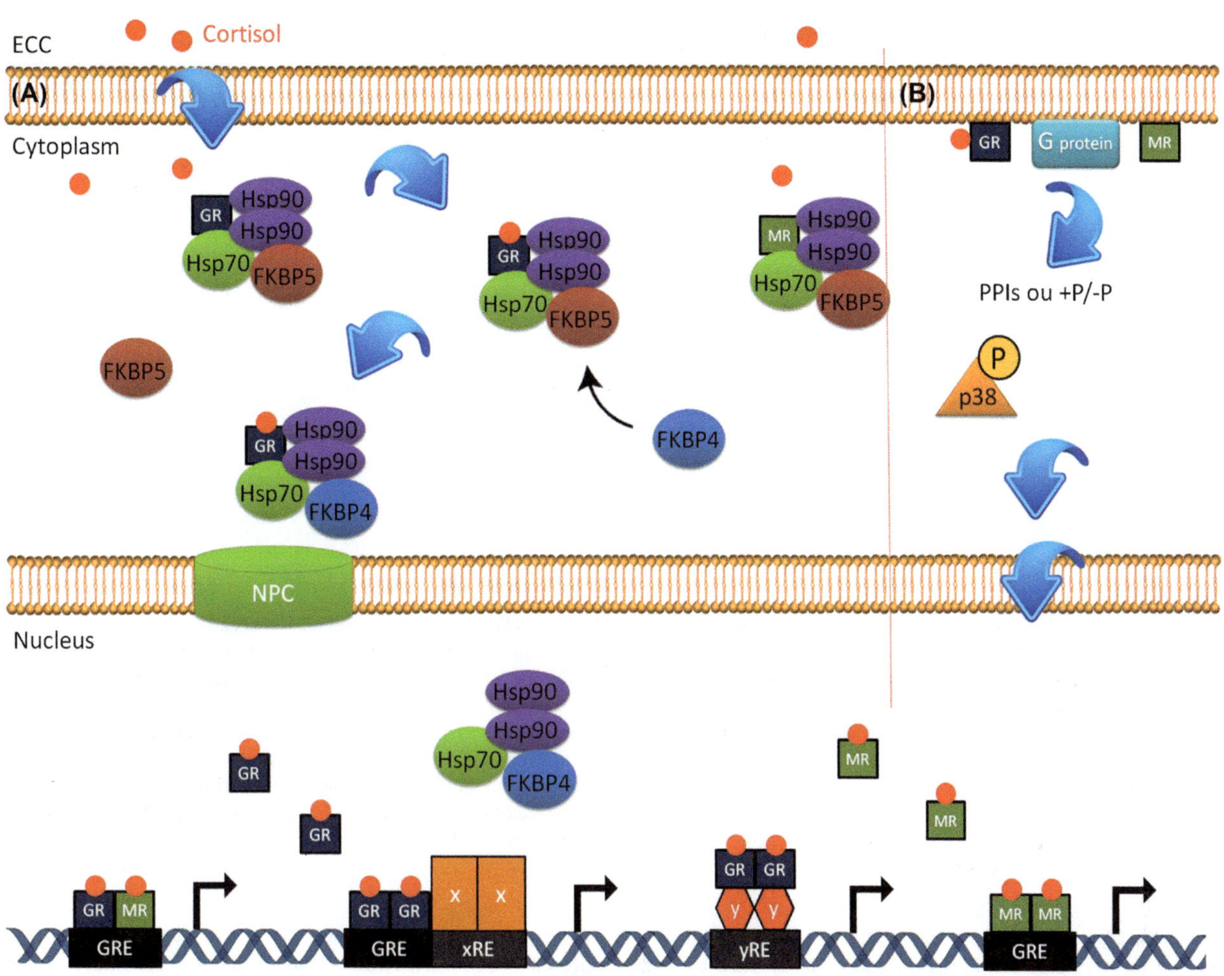

FIGURE 2 **Mechanisms of GR and MR actions.** Cortisol (red dots) spreads from the extracellular compartment (ECC) to the intracellular compartment (cytoplasm) and it interacts with the cytosolic GR or MR. (A) The binding to cortisol promotes conformational changes in the receptors allowing the exchange of FKBP4 for FKBP5 and subsequent nuclear transfer through the nuclear pore complex (NPC). In the nucleus, the chaperone proteins (Hsp70, Hsp90, and FKBP4) dissociate and the activated GRs and MRs interact with glucocorticoid response elements (GREs). Through the interaction with other transcription factors (shown as X or Y), these receptors may also modulate different responsive elements (shown as xRE or yRE). GRs and MRs act as homodimers or heterodimers. (B) In addition to this classical pathway controlling gene expression, there is another one comprising rapid and nongenomic effects, which involves phosphorylation (+P), dephosphorylation (−P), and protein–protein interactions (PPIs). This pathway indirectly modulates gene expression. P38 = p38 MAP kinase; TF, transcription factor.

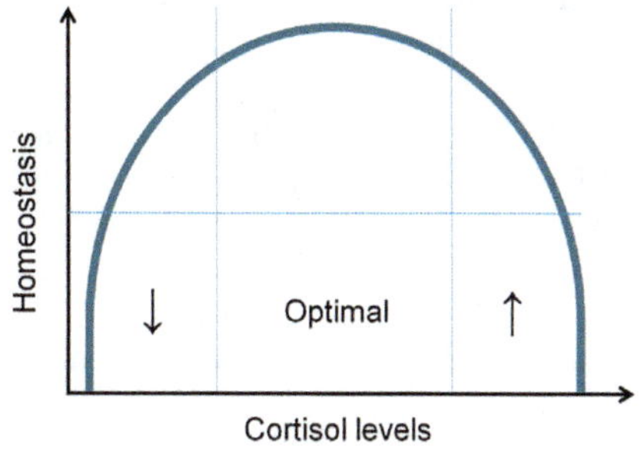

FIGURE 3 **Inverted U-shaped dose–response curve.** Inappropriate cortisol exposures (deficiency or excess) may be a consequence of inefficient adaptive responses, which are associated with susceptibility to stress-related diseases, including SUDs.

cytosolic GRs are responsible for the recovery and consolidation of the response to stress, allowing the storage of this information in memory (Oitzl, Champagne, van der Veen, & de Kloet, 2010). Such temporally coordinated action mechanism explains why MRs are always occupied (either while activated or not) by GCs, even at low levels of these hormones, whereas GRs are occupied only after GCs reach plasma peak (Martinerie et al., 2013).

THE GLUCOCORTICOID RECEPTOR-CODING GENE (*NR3C1*)

The GR is encoded by the *NR3C1* gene, which is located on chromosome region 5q31–q32 and can be considered highly complex, as its organization differs from that of a classical gene. The first exon of this gene is not translated and presents nine different promoters, leading to the so-called nine alternative untranslated first exons (1A, 1I, 1D, 1J, 1E, 1B, 1F, 1C, and 1H) (Turner et al., 2010; Figure 4(A)). The coding region of *NR3C1* is composed by eight other exons and alternative splicing processes, influenced by the different promoters, result in the expression of five main GR isoforms: GRα, GRβ, GRγ, GR-A, and GR-P. These isoforms differ in their *C*-terminal extremities and display different functions

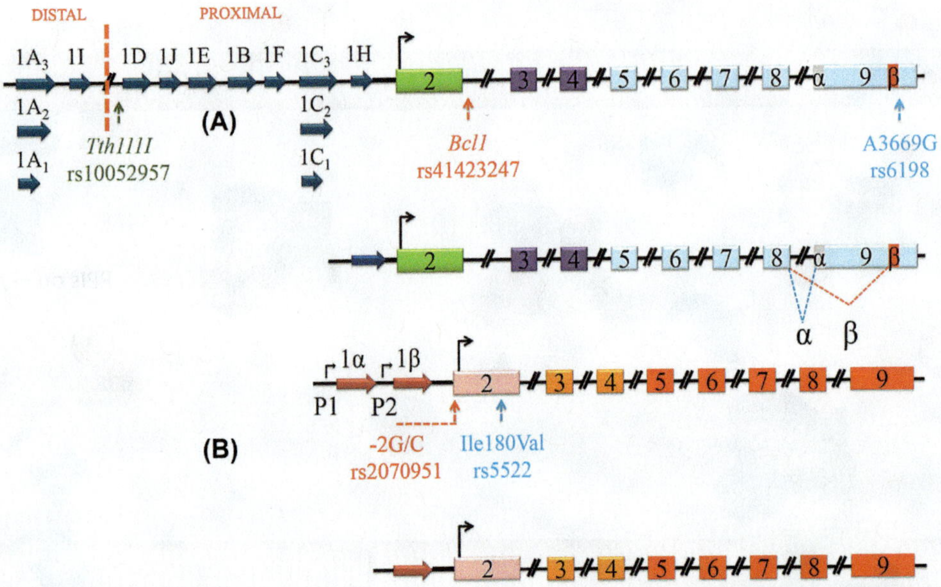

FIGURE 4 Schematic representation of the structure of the GR and the MR. The NTD (exon 2), the central DBD (exons 3 and 4), and the C-terminal LBD (exons 5–9) are indicated in different colors. (A) The *NR3C1* gene has nine alternative first exons that are transcribed but not translated. Different mRNAs are generated from the selection of the first alternative exon. The promoter region is formed by a distal portion (control of exons 1A and 1I) and a proximal portion (control of exons 1L, 1M, 1E, 1B, 1F, 1C, and 1H). The selection of the first exon appears to influence the isoform that will be expressed. For example, the use of promoter 1C increases the expression of GRα. The most common isoforms are α and β. The first is generated by an alternative splicing between the end of exon 8 and exon 9, while GRβ is formed by joining the end of exon 8 with a splicing site located downstream of exon 9. Exon 2 has eight alternative translation initiation codons, which enables the formation of different variants of GR (not shown) with shorter N-terminal ends. The rs6198, rs10052967, and rs414123247 SNPs are indicated. (B) The *NR3C2* gene has two first untranslated exons (1α and 1β) that are alternatively transcribed. The promoter region is formed by two sequences called P1 and P2, which are located upstream of exons 1α and 1β, respectively. Multiple isoforms of MR are also formed from alternative splicing and alternative translation initiation. The rs2070951 and rs5522 SNPs are indicated.

(Kino, Su, & Chrousos, 2009). Moreover, additional variants can be formed according to the initiation codon used, since exon 2 has eight ATG initiation codons (on residues 1, 27, 86, 90, 98, 316, 330, and 336), which result in eight variants of GRα (and possibly of other isoforms) with progressively shorter N-terminal extremities (GRα-A, GRα-B, GRα-C1, GRα-C2, GRα-C3, GRα-D1, GRα-D2, and GRα-D3) (Oakley & Cidlowski, 2013).

The GR is expressed in virtually all tissues and, in the nervous system, it can be traced to almost every area of the brain, although high GR levels are found only on the hippocampus and the amygdala (Sapolsky, et al., 2000). GRα is a cytosolic functional isoform and it is involved in the transactivation and transrepression of multiple genes regulated by GCs (Saif et al., 2014). GRβ is located in the cytoplasm and in the nucleus and does not bind to the GCs as does the classic GRα. When these two isoforms are coexpressed, GRβ acts as a dominant-negative inhibitor of GRα by avoiding the binding of this isoform to glucocorticoid response elements, competing for transcriptional coregulators and/or forming inactive GRα/GRβ dimers (Oakley & Cidlowski, 2013). In this sense, high levels of GRβ seem to confer resistance to GCs in the tissues in which it is expressed, while low GRβ levels may be related to increased sensitivity to these hormones (Lewis-Tuffin & Cidlowski, 2006).

High-throughput gene expression findings have suggested that GRβ might also act directly as a transcriptional factor and that such mechanism is independent of its activity as a dominant-negative inhibitor of GRα (Kino et al., 2009; Lewis-Tuffin, Jewell, Bienstock, Collins, & Cidlowski, 2007). Corticosteroid receptors have two activation function domains (AF1 and AF2), which are capable of interactions with other proteins, integrating signals from different pathways. As a consequence of GR's alternative splicing, the AF2 region is removed from the RNA that originates the GRβ isoform. Nevertheless, GRβ preserves the AF1 domain intact and this specific feature permits several interactions with transcriptional machinery, enabling gene expression modulation through GRβ (Kino, Su, et al., 2009). Corroborating this idea, transcriptome assays showed that mifepristone, a GC antagonist, is able to bind to GRβ and abolish the gene expression changes induced by this isoform, similar to GRα (Lewis-Tuffin, et al., 2007).

With respect to the remaining isoforms, scarce information can be found in the literature. It is known that GRγ binds to GCs but it has a limited ability in stimulating the target genes of these hormones (Meijsing et al., 2009). It has been shown that the GR-A and GR-P isoforms, along with GRβ and the different GRα variants, can be found on the human placenta and that there is a distinct expression pattern according to gender. These data suggest that nonclassical isoforms of GR mediate differences in the sensitivity to cortisol between male and female fetuses during the intrauterine period (Saif, et al., 2014).

FUNCTIONAL ASPECTS OF *NR3C1* POLYMORPHISMS

The *NR3C1* gene is quite polymorphic and some of its SNPs (single-nucleotide polymorphisms) have been investigated with respect to their effects on HPA axis responsiveness, especially rs10052957 (*Tht111I*), rs41423247 (*Bcl1*), and rs6198 (A3669G). The rs10052957 SNP is characterized by a cytosine-to-thymine change in the intron between untranslated exons $1A_{1-3}$ and 1D (Figure 4(A)) and it has been associated with higher total and evening cortisol levels (Rosmond et al., 2000). The rs41423247 SNP is a cytosine-to-guanine substitution 646 nucleotides downstream of exon 2 (Figure 4(A)) that has been associated with response differences during the Trier Social Stress Test (TSST) (Kumsta et al., 2007; Wust et al., 2004) and with variations on corticosteroid sensitivity (Panarelli et al., 1998; van Rossum & Lamberts, 2004; Stevens et al., 2004). It also has been shown that both SNPs modulate the differences in the expression of GR isoforms on the dorsolateral prefrontal cortex between individuals with and without schizophrenia and bipolar disorder (Sinclair, Fullerton, Webster, & Shannon Weickert, 2012).

The rs6198 SNP consists in an adenine-to-guanine substitution at the 3669 nucleotide position (Figure 4(A)). This change occurs on the first A (adenine) of an ATTTA sequence, which is responsible for a GRβ mRNA destabilization. The presence of the G (guanine) allele of rs6198 in this motif results in a better stability and, consequently, in an increased GRβ expression (Derijk et al., 2001). Clinically, G carriers present an exacerbated response to cortisol during the TSST or after the dexamethasone suppression test, possibly owing to a less efficient negative feedback (Kumsta et al., 2007, 2009).

EFFECTS OF *NR3C1* POLYMORPHISMS ON SMOKING BEHAVIOR

As of this writing, there are three candidate gene association studies investigating the role of *NR3C1* SNPs in smoking, and two of the SNPs mentioned above (rs41423247 and rs6198) have been implicated in such behavior (Rogausch, Kochen, Meineke, & Hennig, 2007; Rovaris, Mota, de Azeredo et al., 2013; Siiskonen et al., 2009). These studies based their hypotheses on results from clinical and preclinical findings showing that tobacco stimulates the release of GCs (Badrick, Kirschbaum, & Kumari, 2007; Lutfy, et al., 2012; Mendelson, et al., 2008) and that such release can modify the reward and reinforcement effects of nicotine by activating GRs on dopaminergic neurons (de Jong & de Kloet, 2004; Rouge-Pont, Deroche, Le Moal, & Piazza, 1998).

The first study tested the association between the *NR3C1* rs41423247 SNP and smoking status and severity (Rogausch, et al., 2007). They analyzed smoking behavior data from 327 Caucasian patients with asthma or chronic obstructive pulmonary disease, which were classified as "smokers" or "nonsmokers." Subjects from the first group represented 23.2% of the total sample and were further classified as "light" and "heavy" smokers according to their scores on the Fagerström Test for Nicotine Dependence (FTND). The rs41423247 G allele was significantly associated with an increased probability of being a smoker, since more than 25% of the individuals carrying the G allele were smokers, compared to only 17.3% of the CC individuals. Additionally, G carriers had significantly higher daily cigarette consumption than C homozygotes.

The influence of the rs41423247 SNP on smoking was also investigated in a large population cohort sample, in which 6358 individuals were evaluated three times after baseline during a follow-up period of 12 years (mean = 6.2 years) (Siiskonen, et al., 2009). The main outcome was a polytomous variable: current smokers, ex-smokers, and nonsmokers. Additionally, the role of the rs41423247 SNP on smoking severity was assessed in the current smokers group and survival analyses tested its effect on smoking cessation during follow-up. At baseline, 22.8% of individuals were current smokers, 41.8% were ex-smokers, and 35.4% were nonsmokers. No significant difference in smoking status or severity was observed between the two genotype groups (CC vs G carriers). During follow-up, 407 individuals had stopped smoking; however, no influence of the rs41423247 SNP was detected on smoking cessation.

The effects of the rs6198 SNP on smoking behavior were assessed in a population sample of 627 individuals, for which the main outcome was smoking status (lifetime smokers vs nonsmokers) (Rovaris, Mota, de Azeredo, et al., 2013). Additionally, analysis of smoking severity was conducted within the lifetime smokers group. No main effect of the rs6198 SNP on smoking status was observed. However, when a combined analysis considered its effect together with an SNP in the gene that encodes the MR (*NR3C2*), the presence of the rs6198 G allele was associated with either risk or protection depending on the *NR3C2* alleles present, suggesting a gene–gene interaction effect. Additionally, the presence of the rs6198 G allele in lifetime smokers was significantly associated with a higher number of cigarettes smoked per day and to greater severity according to the FTND scores.

To the best of our knowledge, no other *NR3C1* SNP has been associated with smoking by candidate gene studies nor has any significant hit been detected near this gene by GWAS on smoking (Table 1).

THE MR CODING GENE

As mentioned above, the MR is encoded by the *NR3C2* gene (chromosome 4q31.1), which can be expressed as distinct isoforms (Pascual-Le Tallec & Lombes, 2005; Viengchareun et al., 2007; Figure 4(B)). MRα and MRβ are the products of splicing of the first two alternative untranslated exons, 1α and 1β, respectively, and they are expressed in several tissues (Zennaro, Farman, Bonvalet, & Lombes, 1997). Moreover, alternative splicing events involving the 3′ region, as well as alternative translation initiation sites, increase the diversity of MR variants encoded by the *NR3C2* gene (Pascual-Le Tallec & Lombes, 2005). At least two of these variants, MRA and MRB, with different transactivation abilities (MRA > MRB), are produced by the use of different alternative translation initiation sites (Pascual-Le Tallec, Demange, & Lombes, 2004).

The transcription of *NR3C2* is regulated by two promoters, known as P1 and P2, which are located upstream of the two first alternative exons (Listwak, Gold, & Whitfield, 1996; Figure 4(B)). Results from in vivo experiments showed a tissue- and a temporal-specific pattern of use of these promoters. While P1 is active in aldosterone target organs, such as brain, kidneys, colon, heart, lungs, adipose tissue, and liver, activation of P2 is restricted to some developmental stages (Le Menuet et al., 2000).

In the brain specifically, the MR is expressed in regulatory structures involved in the maintenance of the electrolyte balance, such as the organum vasculosum of the lamina terminalis, amygdala, and other periventricular regions (Funder, Pearce, Smith, & Smith, 1988).

TABLE 1 Results Involving NR3C1 and NR3C2 Genes from GWAS and Meta-analysis of GWAS. The Outcomes Were Any Smoking Phenotype

Article	Initial Sample	Replication Sample	P-Value ≤5×10⁻⁸	P-Value between 5×10^{-8} and 5×10^{-5}	Gene	SNP	PubMed PMID
Loukola et al. (2014)	1114[a]	5294[a]	No	No	—		23752247
McGue et al. (2013)	7188	—	No	No	—		23942779
Kumasaka et al. (2012)	11,696	5462	No	No	—		23049750
David et al. (2012)	>32,000	—	No	No	—		22832964
Rice et al. (2012)	3365	835	No	No	—		22524403
Wang et al. (2012)	2420	3304	No	No	—		22377092
Yoon et al. (2012)	8442	1366	No	No	—		22006218
Liu et al. (2010)	41,150	120,516	No	No	—		20418889
TAG Consortium (2010)	>74,000	>68,000	No	No	—		20418890
Thorgeirsson et al. (2010)	31,266	54,731	No	No	—		20418888
Uhl et al. (2010a)	369	—	No	No	—		20811658
Uhl et al. (2010b)	324	—	No	No	—		20235792
Vink et al. (2009)	3497	6215	No	No	—		19268276
Caporaso et al. (2009)	4342	—	No	No	—		19247474
Liu et al. (2009)	840	8874	No	No	—		19188921
Drgon et al. (2009)	480	—	No	No	—		19009022
Uhl et al. (2008)	550	—	No	No	—		18519826
Thorgeirsson et al. (2008)	10,995	4848	No	No	—		18385739
Berrettini et al. (2008)	7481	~2000	No	Yes	NR3C2	rs5522, rs5525	18227835
Uhl et al. (2007)	454	—	No	No	—		17407593
Bierut et al. (2007)	948	981	No	No	—		17158188

[a]Twins.
From GWAS catalog, http://www.genome.gov/gwastudies/.

Furthermore, high MR levels are found in the hippocampus and other limbic regions, where this receptor binds to GCs as well as synthetic corticosteroids and progesterone (Baker, Funder, & Kattoula, 2013; Le Menuet & Lombes, 2014).

FUNCTIONAL ASPECTS OF *NR3C2* POLYMORPHISMS

The *NR3C2* is also a highly polymorphic gene; however, only a few of its SNPs have been analyzed in relation to their effects on HPA axis reactivity. Among these, particularly notable are rs2070951 (−2G/C) and rs5522 (I180V) (Figure 4(B)). The rs2070951 SNP, located on exon 2, is characterized by a change from a guanine to a cytosine two nucleotides before the first ATG start codon, whereas the rs5522 SNP, also located on exon 2, is a thymine-to-cytosine variation at codon 180, resulting in a change of the amino acid isoleucine (Ile) to a valine (Val) (Derijk, 2009).

Results from in vitro assays have shown that these two SNPs affect the MR transactivation activity when cortisol or dexamethasone is used as ligand (Arai et al., 2003; DeRijk et al., 2006; van Leeuwen et al., 2011). Moreover, these SNPs have been associated with plasmatic and salivary cortisol level changes and with variation in HPA axis suppression by dexamethasone (DeRijk, et al., 2006; Kuningas et al., 2007; van Leeuwen et al., 2010). In addition, differences have been found in responses to the TSST and the Trier Inventory for the Assessment of Chronic Stress (van Leeuwen, et al., 2011); however, these results could not be replicated by others (Bouma et al., 2011).

EFFECTS OF *NR3C2* POLYMORPHISMS ON SMOKING BEHAVIOR

Regarding GWAS, the only two nominal associations of SNPs in corticosteroid receptor genes and smoking behavior come from the *NR3C2* gene. A GWAS focusing on heavy smoking, with a total of about 9500 individuals of European descent, in which subjects who smoked 25 or more cigarettes per day were considered "cases," was conducted (Berrettini et al., 2008). In the pooled analysis the rs5522 Val allele and the rs5525 A allele, which is another SNP located in exon 2, were both nominally associated with heavy smoking (P=1.52E-5 and 3.78E-5, respectively).

A subsequent candidate gene study evaluated, in a Brazilian sample of European descent, the role of the rs5522 polymorphism and 20 other SNPs in genes previously associated with smoking behaviors (dos Santos et al., 2012). This study included 168 lifetime smokers and 363 controls who had never smoked or who had smoked ≤100 cigarettes in their lifetime. No effect of the 5522 Val allele on smoking status was found in the case–control analysis.

As mentioned before, a study evaluated the effects of functional SNPs in the *NR3C1* and *NR3C2* genes on smoking susceptibility and smoking severity (Rovaris, Mota, de Azeredo, et al., 2013). This study also investigated if epistatic interactions between *NR3C1* rs6198 and *NR3C2* rs5522 could play a role in addiction. In the case–control analysis, an interaction effect between the *NR3C1* rs6198 and the *NR3C2* rs5522 SNPs was observed on susceptibility to smoking, but not on smoking severity. The presence of the rs6198 G allele in rs5522 Val carriers was significantly associated with a protective effect against smoking, while in Val noncarriers a risk effect was observed. Afterward, the same interaction pattern was replicated in smoked cocaine addiction, highlighting the need for functional studies (Rovaris et al., 2015).

POTENTIAL FOR MR/GR INTERACTIONS BEYOND THE STATISTICAL LEVEL

The interdependent actions of GRs and MRs reinforce the potential for gene–gene interactions. The binding of GCs to MRs and GRs results in conformational changes that culminate in receptor dimerization on DNA (Figure 2(A)). Such processes may result in MR/MR, GR/MR, or GR/GR dimers with different transactivation abilities (Liu, Wang, Sauter, & Pearce, 1995; Savory et al., 2001; Trapp, Rupprecht, Castren, Reul, & Holsboer, 1994). Additionally, the efficiency of the heterodimerization process can differ according to the presence or absence of the GRβ isoform, which acts as an inhibitor of the classical MR and GR isoforms (Bamberger, Bamberger, Wald, Chrousos, & Schulte, 1997; Oakley & Cidlowski, 2013).

An MR/GR balance hypothesis has been proposed, since these receptors necessarily need to act together in regulating different steps of the stress response (Joels, et al., 2008; Oitzl, et al., 2010). In this way, inadequate operation of either receptor could compromise stress reactions, increasing the risk of developing stress-related disorders, including smoking. A study was designed to explore if the impact of MR levels on the HPA axis or cognitive functions of mice is influenced by the levels of GR and vice versa (Harris, Holmes, de Kloet, Chapman, & Seckl, 2013). Their results showed MR/GR interactions controlling the HPA-axis activity under stress, but not under basal conditions, as well as main and interactive effects modulating a range of cognitive functions.

PERSPECTIVES

Several genes have been significantly (or nominally) associated with smoking and other addictions in GWAS. However, replication of such findings did not succeed in many subsequent studies (Ho et al., 2010), making the understanding and application of these results difficult. In addition, the effect of all variants already associated with multifactorial traits, either smoking or any other, explains only a small fraction of their heritabilities (Lander, 2011). Thus, there is a remaining undiscovered genetic contribution, which has been called "missing heritability." Several explanations for the phenomenon of missing heritability have been proposed in the literature, such as the existence of multiple interactions among genes, as well as the possibility of gene–environment interactions (Zuk, Hechter, Sunyaev, & Lander, 2012).

Although the effect of genetic interactions in complex phenotypes is usually mentioned as a possibility, it has been neglected in practice (Rovaris, Mota, Callegari-Jacques, & Bau, 2013). The majority of GWAS and candidate gene studies carried out to date used single-locus approaches, making epistatic effects undetectable (Cordell, 2009). In this way, the study of the impact of gene–gene interactions on smoking behaviors could help us understand why results of preclinical and clinical studies are so distant from GWAS findings. Several studies have shown that dysregulation of the stress system is crucial in all aspects of smoking (Bruijnzeel, 2012), while there is no GWAS showing genome-wide significant signals for genes of the HPA axis (Table 1). It is possible that the additive model is not capable of detecting significant effects of SNPs in genes coding for proteins that have a high probability of interacting, as in the case of GR and MR.

Therefore, limitations of the additive model in explaining missing heritability, including the fact that epistasis can be a major determinant of the additive genetic variance (Monnahan & Kelly, 2015), indicate that the study of gene–gene interactions may be a relevant approach for the genetic analyses of phenotypes related to drug dependence. One of the challenges would be to develop models showing when addictive, epistatic, and other complex mechanisms could be considered in the molecular pathways of addiction to each substance of abuse. Also, different sources of biological and biochemical data could be integrated and used to help optimize the analyses.

Additionally, new strategies in the search of relevant SNPs should be applied. This step forward was reached in a genome-wide methylation study in postmortem brains investigating epigenetic associations with suicidal behaviors (Guintivano et al., 2014). The results demonstrated a relevant methylation site located right on top of the rs7208505 SNP. Interestingly, this SNP is located in the gene encoding SKA2, a protein involved in chaperoning GRs from the cytoplasm to the nucleus, which may have altered expression when hypermethylated. The thymine allele of this SNP eliminates the cytosine–guanine (CpG) dinucleotide, making the methylation on this site impossible, and significantly reduces the risk for suicidal behavior and progression from suicidal ideation to suicide attempt. So, could events like that occurring in other genes related to the stress systems, such as *NR3C1* and *NR3C2*, affect SUD susceptibility? Since stress-related genes are rich in CpG islands and highly polymorphic, this may be another perspective for the study of smoking behavior, as well as for other psychiatric disorders.

APPLICATIONS TO OTHER ADDICTIONS AND SUBSTANCE MISUSE

Since the dysregulation of the stress system is involved in several aspects of addiction to alcohol, nicotine, cocaine, and other drugs of abuse (Richards et al., 2011; Sinha, 2008), genetic investigations should give special emphasis to this system. Additionally, more attention should be given to analyses involving severity and response to treatment of each SUD. It is reasonable to think that variations in stress-related genes may have an impact on the clinical heterogeneity of addictive behaviors.

DEFINITION OF TERMS

Alternative splicing In this process, particular introns of a gene may be included as exons in the processed mRNA and vice versa. By this process, a single gene encodes two or more distinct protein isoforms.

Candidate gene study This approach to association studies focuses on genes selected considering prior evidence suggesting possible involvement in the biological underpinnings of the phenotype of interest. Genes in regions appointed by genome scans are also called candidate genes.

Dominant-negative inhibitor A mutated protein that interferes in the functionality of the corresponding normal protein.

Gene–gene interactions Interactions between genetic loci. Often "gene–gene interaction" is used as a synonym of "epistasis."

Genome-wide association study An association study that examines hundreds of thousands of common genetic variants, especially SNPs, throughout the whole genome.

Heritability The proportion of observed variation of a phenotype in a population that is due to genetic differences.

Homeostasis In an organism, this concept means to maintain the internal environment in a stable condition.

Stressor agent A psychological, physiological, or pharmacological "event" that disturbs the homeostasis.

Transactivation A process in which the expression of multiple genes can be increased by transcription factors.

KEY FACTS

Key Facts of Smoking Behaviors

- Smoking behaviors are influenced by genetic and environmental factors.
- Smoking traits have heritability estimates ranging from 30% to 80%.
- Dysregulation of the stress system has an important role in smoking behaviors.
- Some loci influencing smoking behaviors have been revealed by GWAS.
- Loci revealed by GWAS explain a small proportion of the variance of smoking traits.
- Gene–gene interactions may be involved in the missing heritability of smoking behaviors.

Key Facts of the Stress System

- Homeostasis is disturbed when an organism faces a stressor.
- The HPA axis is activated by stressful situations.
- Nicotine is a pharmacological stressor that activates the HPA axis.
- Cortisol release is a consequence of the activation of the HPA axis.
- GRs and MRs are the "maestros" that orchestrate stress reactions.
- MRs and GRs interact directly and indirectly.

SUMMARY POINTS

- This chapter focuses on the functioning of the HPA axis and the possible implications of related genetic variation in smoking.
- The HPA axis seems to have a putative role in various aspects of smoking behaviors.
- The GRs and MRs are expressed in several brain areas and play a key role in negative feedback of the HPA axis.
- An MR/GR balance hypothesis suggests that an imbalance between these receptors may result in dysfunctional stress responses, increasing susceptibility to stress-related diseases, including smoking.

- Functional polymorphisms in the *NR3C1* and *NR3C2* genes (encoding the GR and MR, respectively) become candidates for association studies in smoking.
- Association findings, from candidate gene studies and GWAS, involving the *NR3C1* and *NR3C2* genes and smoking are discussed in this chapter.

REFERENCES

al'Absi, M., Hatsukami, D., Davis, G. L., & Wittmers, L. E. (2004). Prospective examination of effects of smoking abstinence on cortisol and withdrawal symptoms as predictors of early smoking relapse. *Drug Alcohol Dependence, 73*(3), 267–278. http://dx.doi.org/10.1016/j.drugalcdep.2003.10.014. pii:S0376871603003089.

Arai, K., Nakagomi, Y., Iketani, M., Shimura, Y., Amemiya, S., Ohyama, K., & Shibasaki, T. (2003). Functional polymorphisms in the mineralocorticoid receptor and amirolide-sensitive sodium channel genes in a patient with sporadic pseudohypoaldosteronism. *Human Genetics, 112*(1), 91–97. http://dx.doi.org/10.1007/s00439-002-0855-7.

Badrick, E., Kirschbaum, C., & Kumari, M. (2007). The relationship between smoking status and cortisol secretion. *The Journal of Clinical Endocrinology and Metabolism, 92*(3), 819–824. http://dx.doi.org/10.1210/jc.2006-2155.

Baker, M. E., Funder, J. W., & Kattoula, S. R. (2013). Evolution of hormone selectivity in glucocorticoid and mineralocorticoid receptors. *The Journal of Steroid Biochemistry and Molecular Biology, 137*, 57–70. http://dx.doi.org/10.1016/j.jsbmb.2013.07.009. pii:S0960-0760(13)00138-6.

Bamberger, C. M., Bamberger, A. M., Wald, M., Chrousos, G. P., & Schulte, H. M. (1997). Inhibition of mineralocorticoid activity by the beta-isoform of the human glucocorticoid receptor. *The Journal of Steroid Biochemistry and Molecular Biology, 60*(1–2), 43–50.

Berrettini, W., Yuan, X., Tozzi, F., Song, K., Francks, C., Chilcoat, H., ... Mooser, V. (2008). Alpha-5/alpha-3 nicotinic receptor subunit alleles increase risk for heavy smoking. *Molecular Psychiatry, 13*(4), 368–373. http://dx.doi.org/10.1038/sj.mp.40021544002154.

Bierut, L. J., Madden, P. A., Breslau, N., Johnson, E. O., Hatsukami, D., Pomerleau, O. F., ... Ballinger, D. G. (2007). Novel genes identified in a high-density genome wide association study for nicotine dependence. *Human Molecular Genetics, 16*(1), 24–35. http://dx.doi.org/10.1093/hmg/ddl441. ddl441 [pii].

Bouma, E. M., Riese, H., Nolte, I. M., Oosterom, E., Verhulst, F. C., Ormel, J., & Oldehinkel, A. J. (2011). No associations between single nucleotide polymorphisms in corticoid receptor genes and heart rate and cortisol responses to a standardized social stress test in adolescents: the TRAILS study. *Behavior Genetics, 41*(2), 253–261. http://dx.doi.org/10.1007/s10519-010-9385-6.

Bruijnzeel, A. W. (2012). Tobacco addiction and the dysregulation of brain stress systems. *Neuroscience and Biobehavioral Reviews, 36*(5), 1418–1441. http://dx.doi.org/10.1016/j.neubiorev.2012.02.015. pii:S0149-7634(12)00041-3.

Caporaso, N., Gu, F., Chatterjee, N., Sheng-Chih, J., Yu, K., Yeager, M., ... Bergen, A. W. (2009). Genome-wide and candidate gene association study of cigarette smoking behaviors. *PLoS One, 4*(2), e4653. http://dx.doi.org/10.1371/journal.pone.0004653.

Chrousos, G. P. (2009). Stress and disorders of the stress system. *Nature Reviews Endocrinology, 5*(7), 374–381. http://dx.doi.org/10.1038/nrendo.2009.106.

Cordell, H. J. (2009). Detecting gene-gene interactions that underlie human diseases. *Nature Reviews Genetics, 10*(6), 392–404. http://dx.doi.org/10.1038/nrg2579.

David, S. P., Hamidovic, A., Chen, G. K., Bergen, A. W., Wessel, J., Kasberger, J. L., ... Furberg, H. (2012). Genome-wide meta-analyses of smoking behaviors in African Americans. *Translational Psychiatry, 2*, e119. http://dx.doi.org/10.1038/tp.2012.41. tp201241 [pii].

Derijk, R. H. (2009). Single nucleotide polymorphisms related to HPA axis reactivity. *Neuroimmunomodulation, 16*(5), 340–352. http://dx.doi.org/10.1159/000216192000216192.

Derijk, R. H., Schaaf, M. J., Turner, G., Datson, N. A., Vreugdenhil, E., Cidlowski, J., ... Detera-Wadleigh, S. D. (2001). A human glucocorticoid receptor gene variant that increases the stability of the glucocorticoid receptor beta-isoform mRNA is associated with rheumatoid arthritis. *The Journal of Rheumatology, 28*(11), 2383–2388.

DeRijk, R. H., Wust, S., Meijer, O. C., Zennaro, M. C., Federenko, I. S., Hellhammer, D. H., ... de Kloet, E. R. (2006). A common polymorphism in the mineralocorticoid receptor modulates stress responsiveness. *The Journal of Clinical Endocrinology and Metabolism, 91*(12), 5083–5089. http://dx.doi.org/10.1210/jc.2006-0915.

Drgon, T., Montoya, I., Johnson, C., Liu, Q. R., Walther, D., Hamer, D., & Uhl, G. R. (2009). Genome-wide association for nicotine dependence and smoking cessation success in NIH research volunteers. *Molecular Medicine, 15*(1–2), 21–27. http://dx.doi.org/10.2119/molmed.2008.00096.

Funder, J. W., Pearce, P. T., Smith, R., & Smith, A. I. (1988). Mineralocorticoid action: target tissue specificity is enzyme, not receptor, mediated. *Science, 242*(4878), 583–585.

Guintivano, J., Brown, T., Newcomer, A., Jones, M., Cox, O., Maher, B. S., ... Kaminsky, Z. A. (2014). Identification and replication of a combined epigenetic and genetic biomarker predicting suicide and suicidal behaviors. *The American Journal of Psychiatry, 171*(12), 1287–1296. http://dx.doi.org/10.1176/appi.ajp.2014.140100081892819.

Harris, A. P., Holmes, M. C., de Kloet, E. R., Chapman, K. E., & Seckl, J. R. (2013). Mineralocorticoid and glucocorticoid receptor balance in control of HPA axis and behaviour. *Psychoneuroendocrinology, 38*(5), 648–658. http://dx.doi.org/10.1016/j.psyneuen.2012.08.007. pii:S0306-4530(12)00298-3.

Ho, M. K., Goldman, D., Heinz, A., Kaprio, J., Kreek, M. J., Li, M. D., ... Tyndale, R. F. (2010). Breaking barriers in the genomics and pharmacogenetics of drug addiction. *Clinical Pharmacology and Therapeutics, 88*(6), 779–791. http://dx.doi.org/10.1038/clpt.2010.175.

Joels, M., Karst, H., DeRijk, R., & de Kloet, E. R. (2008). The coming out of the brain mineralocorticoid receptor. *Trends in Neuroscience, 31*(1), 1–7. http://dx.doi.org/10.1016/j.tins.2007.10.005. pii:S0166-2236(07)00296-2.

de Jong, I. E., & de Kloet, E. R. (2004). Glucocorticoids and vulnerability to psychostimulant drugs: toward substrate and mechanism. *Annals of the New York Academy Sciences, 1018*, 192–198. http://dx.doi.org/10.1196/annals.1296.022.

Kassahn, K. S., Ragan, M. A., & Funder, J. W. (2011). Mineralocorticoid receptors: evolutionary and pathophysiological considerations. *Endocrinology, 152*(5), 1883–1890. http://dx.doi.org/10.1210/en.2010-1444.

Kino, T., Manoli, I., Kelkar, S., Wang, Y., Su, Y. A., & Chrousos, G. P. (2009). Glucocorticoid receptor (GR) beta has intrinsic, GRalpha-independent transcriptional activity. *Biochemical and Biophysical Research Communications, 381*(4), 671–675. http://dx.doi.org/10.1016/j.bbrc.2009.02.110. pii:S0006-291X(09)00394-5.

Kino, T., Su, Y. A., & Chrousos, G. P. (2009). Human glucocorticoid receptor isoform beta: recent understanding of its potential implications in physiology and pathophysiology. *Cellullar and Molecular Life Sciences*, *66*(21), 3435–3448. http://dx.doi.org/10.1007/s00018-009-0098-z.

Kumasaka, N., Aoki, M., Okada, Y., Takahashi, A., Ozaki, K., Mushiroda, T., … Kubo, M. (2012). Haplotypes with copy number and single nucleotide polymorphisms in CYP2A6 locus are associated with smoking quantity in a Japanese population. *PLoS One*, *7*(9), e44507. http://dx.doi.org/10.1371/journal.pone.0044507PONE-D-12-17352 [pii].

Kumsta, R., Entringer, S., Koper, J. W., van Rossum, E. F., Hellhammer, D. H., & Wust, S. (2007). Sex specific associations between common glucocorticoid receptor gene variants and hypothalamus-pituitary-adrenal axis responses to psychosocial stress. *Biological Psychiatry*, *62*(8), 863–869. http://dx.doi.org/10.1016/j.biopsych.2007.04.013. pii:S0006-3223(07)00331-9.

Kumsta, R., Moser, D., Streit, F., Koper, J. W., Meyer, J., & Wust, S. (2009). Characterization of a glucocorticoid receptor gene (GR, NR3C1) promoter polymorphism reveals functionality and extends a haplotype with putative clinical relevance. *American Journal of Medical Genetics: Part B, Neuropsychiatric Genetics*, *150B*(4), 476–482. http://dx.doi.org/10.1002/ajmg.b.30837.

Kuningas, M., de Rijk, R. H., Westendorp, R. G., Jolles, J., Slagboom, P. E., & van Heemst, D. (2007). Mental performance in old age dependent on cortisol and genetic variance in the mineralocorticoid and glucocorticoid receptors. *Neuropsychopharmacology*, *32*(6), 1295–1301. http://dx.doi.org/10.1038/sj.npp.1301260.

Lander, E. S. (2011). Initial impact of the sequencing of the human genome. *Nature*, *470*(7333), 187–197. http://dx.doi.org/10.1038/nature09792.

Le Menuet, D., & Lombes, M. (2014). The neuronal mineralocorticoid receptor: from cell survival to neurogenesis. *Steroids*, *91*. http://dx.doi.org/10.1016/j.steroids.2014.05.018 pii:S0039-128X(14)00131-7.

Le Menuet, D., Viengchareun, S., Penfornis, P., Walker, F., Zennaro, M. C., & Lombes, M. (2000). Targeted oncogenesis reveals a distinct tissue-specific utilization of alternative promoters of the human mineralocorticoid receptor gene in transgenic mice. *The Journal of Biological Chemistry*, *275*(11), 7878–7886.

van Leeuwen, N., Bellingrath, S., de Kloet, E. R., Zitman, F. G., DeRijk, R. H., Kudielka, B. M., & Wust, S. (2011). Human mineralocorticoid receptor (MR) gene haplotypes modulate MR expression and transactivation: implication for the stress response. *Psychoneuroendocrinology*, *36*(5), 699–709. http://dx.doi.org/10.1016/j.psyneuen.2010.10.003. pii:S0306-4530(10)00262-3.

van Leeuwen, N., Kumsta, R., Entringer, S., de Kloet, E. R., Zitman, F. G., DeRijk, R. H., & Wust, S. (2010). Functional mineralocorticoid receptor (MR) gene variation influences the cortisol awakening response after dexamethasone. *Psychoneuroendocrinology*, *35*(3), 339–349. http://dx.doi.org/10.1016/j.psyneuen.2009.07.006. pii:S0306-4530(09)00219-4.

Lewis-Tuffin, L. J., & Cidlowski, J. A. (2006). The physiology of human glucocorticoid receptor beta (hGRbeta) and glucocorticoid resistance. *Annals of the New York Academy Sciences*, *1069*, 1–9. http://dx.doi.org/10.1196/annals.1351.001.

Lewis-Tuffin, L. J., Jewell, C. M., Bienstock, R. J., Collins, J. B., & Cidlowski, J. A. (2007). Human glucocorticoid receptor beta binds RU-486 and is transcriptionally active. *Molecular and Cellular Biology*, *27*(6), 2266–2282. http://dx.doi.org/10.1128/MCB.01439-06.

Listwak, S. J., Gold, P. W., & Whitfield, H. J., Jr. (1996). The human mineralocorticoid receptor gene promoter: its structure and expression. *The Journal of Steroid Biochemistry and Molecular Biology*, *58*(5–6), 495–506.

Liu, Y. Z., Pei, Y. F., Guo, Y. F., Wang, L., Liu, X. G., Yan, H., … Deng, H. W. (2009). Genome-wide association analyses suggested a novel mechanism for smoking behavior regulated by IL15. *Molecular Psychiatry*, *14*(7), 668–680. http://dx.doi.org/10.1038/mp.2009.3. mp20093 [pii].

Liu, J. Z., Tozzi, F., Waterworth, D. M., Pillai, S. G., Muglia, P., Middleton, L., … Marchini, J. (2010). Meta-analysis and imputation refines the association of 15q25 with smoking quantity. *Nature Genetics*, *42*(5), 436–440. http://dx.doi.org/10.1038/ng.572. ng.572 [pii].

Liu, W., Wang, J., Sauter, N. K., & Pearce, D. (1995). Steroid receptor heterodimerization demonstrated in vitro and in vivo. *Proceedings of the National Academy of Sciences USA*, *92*(26), 12480–12484.

Loukola, A., Wedenoja, J., Keskitalo-Vuokko, K., Broms, U., Korhonen, T., Ripatti, S., … Kaprio, J. (2014). Genome-wide association study on detailed profiles of smoking behavior and nicotine dependence in a twin sample. *Molecular Psychiatry*, *19*(5), 615–624. http://dx.doi.org/10.1038/mp.2013.72. mp201372 [pii].

Lutfy, K., Aimiuwu, O., Mangubat, M., Shin, C. S., Nerio, N., Gomez, R., … Friedman, T. C. (2012). Nicotine stimulates secretion of corticosterone via both CRH and AVP receptors. *Journal of Neurochemistry*, *120*(6), 1108–1116. http://dx.doi.org/10.1111/j.1471-4159.2011.07633.x.

Martinerie, L., Munier, M., Le Menuet, D., Meduri, G., Viengchareun, S., & Lombes, M. (2013). The mineralocorticoid signaling pathway throughout development: expression, regulation and pathophysiological implications. *Biochimie*, *95*(2), 148–157. http://dx.doi.org/10.1016/j.biochi.2012.09.030. pii:S0300-9084(12)00390-2.

McGue, M., Zhang, Y., Miller, M. B., Basu, S., Vrieze, S., Hicks, B., … Iacono, W. G. (2013). A genome-wide association study of behavioral disinhibition. *Behavior Genetics*, *43*(5), 363–373. http://dx.doi.org/10.1007/s10519-013.

Meijsing, S. H., Pufall, M. A., So, A. Y., Bates, D. L., Chen, L., & Yamamoto, K. R. (2009). DNA binding site sequence directs glucocorticoid receptor structure and activity. *Science*, *324*(5925), 407–410. http://dx.doi.org/10.1126/science.1164265.

Mendelson, J. H., Goletiani, N., Sholar, M. B., Siegel, A. J., & Mello, N. K. (2008). Effects of smoking successive low- and high-nicotine cigarettes on hypothalamic-pituitary-adrenal axis hormones and mood in men. *Neuropsychopharmacology*, *33*(4), 749–760. http://dx.doi.org/10.1038/sj.npp.1301455.

Monnahan, P. J., & Kelly, J. K. (2015). Epistasis is a major determinant of the additive genetic variance in mimulus guttatus. *PLoS Genetics*, *11*(5), e1005201. http://dx.doi.org/10.1371/journal.pgen.1005201. PGENETICS-D-15-00240 [pii].

Oakley, R. H., & Cidlowski, J. A. (2013). The biology of the glucocorticoid receptor: new signaling mechanisms in health and disease. *The Journal of Allergy and Clinical Immunology*, *132*(5), 1033–1044. http://dx.doi.org/10.1016/j.jaci.2013.09.007. pii:S0091-6749(13)01388-2.

Oitzl, M. S., Champagne, D. L., van der Veen, R., & de Kloet, E. R. (2010). Brain development under stress: hypotheses of glucocorticoid actions revisited. *Neuroscience and Biobehavioral Reviews*, *34*(6), 853–866. http://dx.doi.org/10.1016/j.neubiorev.2009.07.006. pii:S0149-7634(09)00103-1.

Panarelli, M., Holloway, C. D., Fraser, R., Connell, J. M., Ingram, M. C., Anderson, N. H., & Kenyon, C. J. (1998). Glucocorticoid receptor polymorphism, skin vasoconstriction, and other metabolic intermediate phenotypes in normal human subjects. *The Journal of Clinical Endocrinology and Metabolism*, *83*(6), 1846–1852. http://dx.doi.org/10.1210/jcem.83.6.4828.

Pascual-Le Tallec, L., Demange, C., & Lombes, M. (2004). Human mineralocorticoid receptor A and B protein forms produced by alternative translation sites display different transcriptional activities. *European Journal of Endocrinology, 150*(4), 585–590.

Pascual-Le Tallec, L., & Lombes, M. (2005). The mineralocorticoid receptor: a journey exploring its diversity and specificity of action. *Molecular Endocrinology, 19*(9), 2211–2221. http://dx.doi.org/10.1210/me.2005-0089.

Prager, E. M., & Johnson, L. R. (2009). Stress at the synapse: signal transduction mechanisms of adrenal steroids at neuronal membranes. *Science Signaling, 2*(86), re5. http://dx.doi.org/10.1126/scisignal.286re5.

Rice, J. P., Hartz, S. M., Agrawal, A., Almasy, L., Bennett, S., Breslau, N., ... Bierut, L. J. (2012). CHRNB3 is more strongly associated with Fagerstrom test for cigarette dependence-based nicotine dependence than cigarettes per day: phenotype definition changes genome-wide association studies results. *Addiction, 107*(11), 2019–2028. http://dx.doi.org/10.1111/j.1360-0443.2012.03922.x.

Richards, J. M., Stipelman, B. A., Bornovalova, M. A., Daughters, S. B., Sinha, R., & Lejuez, C. W. (2011). Biological mechanisms underlying the relationship between stress and smoking: state of the science and directions for future work. *Biological Psychology, 88*(1), 1–12. http://dx.doi.org/10.1016/j.biopsycho.2011.06.009. pii:S0301-0511(11)00153-0.

Robinson, S. F., Grun, E. U., Pauly, J. R., & Collins, A. C. (1996). Changes in sensitivity to nicotine and brain nicotinic receptors following chronic nicotine and corticosterone treatments in mice. *Pharmacology, Biochemistry, and Behavior, 54*(3), 587–593. pii:0091-3057(95)02281-3.

Rogausch, A., Kochen, M. M., Meineke, C., & Hennig, J. (2007). Association between the BclI glucocorticoid receptor polymorphism and smoking in a sample of patients with obstructive airway disease. *Addiction Biology, 12*(1), 93–99. http://dx.doi.org/10.1111/j.1369-1600.2006.00045.x.

Rosmond, R., Chagnon, Y. C., Chagnon, M., Perusse, L., Bouchard, C., & Bjorntorp, P. (2000). A polymorphism of the 5'-flanking region of the glucocorticoid receptor gene locus is associated with basal cortisol secretion in men. *Metabolism, 49*(9), 1197–1199. http://dx.doi.org/10.1053/meta.2000.7712. pii:S0026-0495(00)90112-4.

van Rossum, E. F., & Lamberts, S. W. (2004). Polymorphisms in the glucocorticoid receptor gene and their associations with metabolic parameters and body composition. *Recent Progress in Hormone Research, 59*, 333–357.

Rouge-Pont, F., Deroche, V., Le Moal, M., & Piazza, P. V. (1998). Individual differences in stress-induced dopamine release in the nucleus accumbens are influenced by corticosterone. *The European Journal of Neuroscience, 10*(12), 3903–3907.

Rovaris, D. L., Mota, N. R., Bertuzzi, G. P., Aroche, A. P., Callegari-Jacques, S. M., Guimaraes, L. S., ... Grassi-Oliveira, R. (2015). Corticosteroid receptor genes and childhood neglect influence susceptibility to crack/cocaine addiction and response to detoxification treatment. *Journal of Psychiatric Research, 68*, 83–90. http://dx.doi.org/10.1016/j.jpsychires.2015.06.008. S0022-3956(15)00179-X [pii].

Rovaris, D. L., Mota, N. R., Callegari-Jacques, S. M., & Bau, C. H. (2013). Approaching "phantom heritability" in psychiatry by hypothesis-driven gene-gene interactions. *Frontiers in Human Neuroscience, 7*, 210. http://dx.doi.org/10.3389/fnhum.2013.00210.

Rovaris, D. L., Mota, N. R., de Azeredo, L. A., Cupertino, R. B., Bertuzzi, G. P., Polina, E. R., ... Bau, C. H. (2013). MR and GR functional SNPs may modulate tobacco smoking susceptibility. *Journal of Neural Transmission, 120*(10), 1499–1505. http://dx.doi.org/10.1007/s00702-013-1012-2.

Saif, Z., Hodyl, N. A., Hobbs, E., Tuck, A. R., Butler, M. S., Osei-Kumah, A., & Clifton, V. L. (2014). The human placenta expresses multiple glucocorticoid receptor isoforms that are altered by fetal sex, growth restriction and maternal asthma. *Placenta, 35*(4), 260–268. http://dx.doi.org/10.1016/j.placenta.2014.01.012. pii:S0143-4004(14)00033-2.

dos Santos, V. A., Chatkin, J. M., Bau, C. H., Paixao-Cortes, V. R., Sun, Y., Zamel, N., & Siminovitch, K. (2012). Glutamate and synaptic plasticity systems and smoking behavior: results from a genetic association study. *PLoS One, 7*(6), e38666. http://dx.doi.org/10.1371/journal.pone.0038666. pii:PONE-D-11-13216.

Sapolsky, R. M., Romero, L. M., & Munck, A. U. (2000). How do glucocorticoids influence stress responses? Integrating permissive, suppressive, stimulatory, and preparative actions. *Endocrine Reviews, 21*(1), 55–89. http://dx.doi.org/10.1210/edrv.21.1.0389.

Savory, J. G., Prefontaine, G. G., Lamprecht, C., Liao, M., Walther, R. F., Lefebvre, Y. A., & Hache, R. J. (2001). Glucocorticoid receptor homodimers and glucocorticoid-mineralocorticoid receptor heterodimers form in the cytoplasm through alternative dimerization interfaces. *Molecular and Cellular Biology, 21*(3), 781–793. http://dx.doi.org/10.1128/MCB.21.3.781-793.2001.

Schoneveld, O. J., Gaemers, I. C., & Lamers, W. H. (2004). Mechanisms of glucocorticoid signalling. *Biochimica et Biophysica Acta, 1680*(2), 114–128. http://dx.doi.org/10.1016/j.bbaexp.2004.09.004. pii:S0167-4781(04)00179-4.

Siiskonen, S. J., Visser, L. E., Tiemeier, H., Hofman, A., Lamberts, S. W., Uiterlinden, A. G., & Stricker, B. H. (2009). BclI glucocorticoid receptor polymorphism and smoking in the general population. *Addiction Biology, 14*(3), 349–355. http://dx.doi.org/10.1111/j.1369-1600.2009.00154.x.

Sinclair, D., Fullerton, J. M., Webster, M. J., & Shannon Weickert, C. (2012). Glucocorticoid receptor 1B and 1C mRNA transcript alterations in schizophrenia and bipolar disorder, and their possible regulation by GR gene variants. *PLoS One, 7*(3), e31720. http://dx.doi.org/10.1371/journal.pone.0031720. pii:PONE-D-11-22200.

Sinha, R. (2008). Chronic stress, drug use, and vulnerability to addiction. *Annals of the New York Academy Sciences, 1141*, 105–130. http://dx.doi.org/10.1196/annals.1441.030.

Stevens, A., Ray, D. W., Zeggini, E., John, S., Richards, H. L., Griffiths, C. E., & Donn, R. (2004). Glucocorticoid sensitivity is determined by a specific glucocorticoid receptor haplotype. *The Journal of Clinical Endocrinology and Metabolism, 89*(2), 892–897. http://dx.doi.org/10.1210/jc.2003-031235.

Sutton, R. E., Koob, G. F., Le Moal, M., Rivier, J., & Vale, W. (1982). Corticotropin releasing factor produces behavioural activation in rats. *Nature, 297*(5864), 331–333.

TAG Consortium. (2010). Genome-wide meta-analyses identify multiple loci associated with smoking behavior. *Nature Genetics, 42*(5), 441–447. http://dx.doi.org/10.1038/ng.571. ng.571 [pii].

Thorgeirsson, T. E., Geller, F., Sulem, P., Rafnar, T., Wiste, A., Magnusson, K. P., ... Stefansson, K. (2008). A variant associated with nicotine dependence, lung cancer and peripheral arterial disease. *Nature, 452*(7187), 638–642. http://dx.doi.org/10.1038/nature06846. nature06846 [pii].

Thorgeirsson, T. E., Gudbjartsson, D. F., Surakka, I., Vink, J. M., Amin, N., Geller, F., ... Stefansson, K. (2010). Sequence variants at CHRNB3-CHRNA6 and CYP2A6 affect smoking behavior. *Nature Genetics, 42*(5), 448–453. http://dx.doi.org/10.1038/ng.573. ng.573 [pii].

Trapp, T., Rupprecht, R., Castren, M., Reul, J. M., & Holsboer, F. (1994). Heterodimerization between mineralocorticoid and glucocorticoid receptor: a new principle of glucocorticoid action in the CNS. *Neuron*, *13*(6), 1457–1462. pii:0896-6273(94)90431-6.

Turner, J. D., Alt, S. R., Cao, L., Vernocchi, S., Trifonova, S., Battello, N., & Muller, C. P. (2010). Transcriptional control of the glucocorticoid receptor: CpG islands, epigenetics and more. *Biochemical Pharmacology*, *80*(12), 1860–1868. http://dx.doi.org/10.1016/j.bcp.2010.06.037. pii:S0006-2952(10)00471-5.

Uhl, G. R., Drgon, T., Johnson, C., Ramoni, M. F., Behm, F. M., & Rose, J. E. (2010a). Genome-wide association for smoking cessation success in a trial of precessation nicotine replacement. *Molecular Medicine*, *16*(11–12), 513–526. http://dx.doi.org/10.2119/molmed.2010.00052. molmed.2010.00052 [pii].

Uhl, G. R., Drgon, T., Johnson, C., Walther, D., David, S. P., Aveyard, P., ... Munafo, M. R. (2010b). Genome-wide association for smoking cessation success: participants in the patch in practice trial of nicotine replacement. *Pharmacogenomics*, *11*(3), 357–367. http://dx.doi.org/10.2217/pgs.09.156.

Uhl, G. R., Liu, Q. R., Drgon, T., Johnson, C., Walther, D., & Rose, J. E. (2007). Molecular genetics of nicotine dependence and abstinence: whole genome association using 520,000 SNPs. *BMC Genetics*, *8*, 10. http://dx.doi.org/10.1186/1471-2156-8-10. 1471-2156-8-10 [pii].

Uhl, G. R., Liu, Q. R., Drgon, T., Johnson, C., Walther, D., Rose, J. E., ... Lerman, C. (2008). Molecular genetics of successful smoking cessation: convergent genome-wide association study results. *Archives of General Psychiatry*, *65*(6), 683–693. http://dx.doi.org/10.1001/archpsyc.65.6.683. 65/6/683 [pii].

Viengchareun, S., Le Menuet, D., Martinerie, L., Munier, M., Pascual-Le Tallec, L., & Lombes, M. (2007). The mineralocorticoid receptor: insights into its molecular and (patho)physiological biology. *Nuclear Receptor Signaling*, *5*, e012. http://dx.doi.org/10.1621/nrs.05012.

Vink, J. M., Smit, A. B., de Geus, E. J., Sullivan, P., Willemsen, G., Hottenga, J. J., ... Boomsma, D. I. (2009). Genome-wide association study of smoking initiation and current smoking. *American Journal of Human Genetics*, *84*(3), 367–379. http://dx.doi.org/10.1016/j.ajhg.2009.02.001. S0002-9297(09)00062-7 [pii].

Wand, G. S., Oswald, L. M., McCaul, M. E., Wong, D. F., Johnson, E., Zhou, Y., ... Kumar, A. (2007). Association of amphetamine-induced striatal dopamine release and cortisol responses to psychological stress. *Neuropsychopharmacology*, *32*(11), 2310–2320. http://dx.doi.org/10.1038/sj.npp.1301373.

Wang, K. S., Liu, X., Zhang, Q., & Zeng, M. (2012). ANAPC1 and SLCO3A1 are associated with nicotine dependence: meta-analysis of genome-wide association studies. *Drug and Alcohol Dependence*, *124*(3), 325–332. http://dx.doi.org/10.1016/j.drugalcdep.2012.02.003. S0376-8716(12)00045-2 [pii].

Wust, S., Van Rossum, E. F., Federenko, I. S., Koper, J. W., Kumsta, R., & Hellhammer, D. H. (2004). Common polymorphisms in the glucocorticoid receptor gene are associated with adrenocortical responses to psychosocial stress. *The Journal of Clinical Endocrinology and Metabolism*, *89*(2), 565–573. http://dx.doi.org/10.1210/jc.2003-031148.

Yoon, D., Kim, Y. J., Cui, W. Y., Van der Vaart, A., Cho, Y. S., Lee, J. Y., ... Park, T. (2012). Large-scale genome-wide association study of Asian population reveals genetic factors in FRMD4A and other loci influencing smoking initiation and nicotine dependence. *Human Genetics*, *131*(6), 1009–1021. http://dx.doi.org/10.1007/s00439-011-1102-x.

Zennaro, M. C., Farman, N., Bonvalet, J. P., & Lombes, M. (1997). Tissue-specific expression of alpha and beta messenger ribonucleic acid isoforms of the human mineralocorticoid receptor in normal and pathological states. *The Journal of Clinical Endocrinology and Metabolism*, *82*(5), 1345–1352. http://dx.doi.org/10.1210/jcem.82.5.3933.

Zhou, Y., Proudnikov, D., Yuferov, V., & Kreek, M. J. (2010). Drug-induced and genetic alterations in stress-responsive systems: implications for specific addictive diseases. *Brain Research*, *1314*, 235–252. http://dx.doi.org/10.1016/j.brainres.2009.11.015. pii:S0006-8993(09)02407-X.

Zuk, O., Hechter, E., Sunyaev, S. R., & Lander, E. S. (2012). The mystery of missing heritability: genetic interactions create phantom heritability. *Proceedings of the National Academy of Sciences USA*, *109*(4), 1193–1198. http://dx.doi.org/10.1073/pnas.1119675109.

Chapter 21

Genetic Variants of μ Opioid Receptor and Its Interacting Proteins in Smoking: A Focus on *OPRM1* A118G, *ARRB2*, and *HINT1*

Juan Fang, Bei He

Department of Respiratory Medicine, Peking University Third Hospital, Beijing, China

Abbreviations

CPA Conditioned place aversion
EOT End of treatment
FTND Fagerström Test for Nicotine Dependence
GPCR G-protein-coupled receptor
GWAS Genome-wide association study
HSI Heaviness of Smoking Index
LD Linkage disequilibrium
MAF Minor allele frequency
mPKCI-1 Protein kinase C-interacting protein
ND Nicotine dependence
NRT Nicotine replacement therapy
NS Nasal spray
PET Positron emission tomography
SI Smoking initiation
TN Transdermal nicotine

INTRODUCTION

The endogenous opioid system has been demonstrated to play an important role in brain reward response and regulation of the behavioral and neurochemical effects of multiple drugs of abuse. The μ opioid receptor (MOR) has received the most focus in genetic and pharmacological studies and plays an important role in smoking behaviors. β-Endorphin binds with a higher affinity to MOR than to the δ (DOR) or κ opioid receptor (KOR) and is known as a main endogenous ligand of MOR (Roth-Deri, Green-Sadan, & Yadid, 2008). After nicotine exposure, endogenous opioids, especially β-endorphins, are released and activate the MORs in the ventral tegmental area. The binding of β-endorphins to MOR is essential for nicotine-induced dopamine release in the nucleus accumbens, which gives feelings of reward and reinforcement (Balfour, 2004; Bond et al., 1998).

Animal model and pharmacological studies have indicated that the endogenous opioid system, particularly MOR and its ligand β-endorphin, is an important neurobiological mechanism for nicotine addiction (Tables 1 and 2). Nicotine produced a rewarding response in wild-type mice, but failed in MOR-knockout mice, demonstrating that MORs are crucial to the property of nicotine to induce a rewarding response (Berrendero, Kieffer, & Maldonado, 2002). By selectively blocking the MOR, DOR, and KOR in rat models of nicotine self-administration, activation of the MOR, but not the DOR and KOR, is found to be required for the reinforcement of nicotine (Liu & Jernigan, 2011). In animal models, nicotine-induced rewarding effects were attenuated in β-endorphin-knockout mice compared to wild-type mice (Trigo, Zimmer, & Maldonado, 2009). In a human behavioral pharmacology study, the opioid receptor antagonist naltrexone significantly reduced the relative reinforcing value of nicotine as measured by the number of nicotine cigarette choices (nicotine vs denicotinized cigarettes) compared to placebo, which supports the involvement of the opioid receptor in the reinforcement effects of tobacco smoking (Rukstalis et al., 2005). These studies indicate that the MOR and its ligand β-endorphin are involved in rewarding and reinforcing effects of nicotine and mediate the development of nicotine dependence.

The endogenous opioid system participates in nicotine-induced behavioral responses, including the opposing effects of anxiolytic- and anxiogenic-like behaviors, behavioral sensitization, and antinociceptive response, in relation to rewarding properties and physical dependence on nicotine. The MOR antagonist abolished nicotine-induced anxiolytic-like effects, but not the DOR or the KOR antagonist, suggesting that MORs are involved in the anxiolytic effects of nicotine and facilitate the development of nicotine tolerance (Balerio, Aso, & Maldonado, 2005). Compared to wild-type mice, β-endorphin-knockout mice exhibited hypoalgesia and hyperlocomotion, and nicotine-induced anxiogenic response was attenuated (Trigo et al., 2009). In addition, repeated administration of nicotine produces behavioral sensitization. In MOR-knockout mice, nicotine challenge failed to induce behavioral sensitization (Yoo, Lee, Loh, Ho, & Jang, 2004), suggesting the role of the MOR

TABLE 1 Behavioral Effects of Nicotine in μ Opioid Receptor- or β-Endorphin-Knockout Mice

Knockout Mice	Behavioral Test	Nicotine-Related Phenotype	Effect	Study
μ Opioid receptor	CPP	Nicotine reward	Attenuated	Berrendero et al. (2002)
	Mecamylamine-precipitated nicotine withdrawal	Physical dependence	Attenuated	Berrendero et al. (2002)
	Locomotor activity	Behavioral sensitization	Attenuated	Yoo et al. (2004)
	The tail-immersion test	Nicotine tolerance	Faster developed	Galeote et al. (2006)
β-Endorphin	CPP	Nicotine reward	Attenuated	Trigo et al. (2009)
	Mecamylamine-precipitated nicotine withdrawal	Physical dependence	Unchanged	Trigo et al. (2009)

The table shows the MOR-knockout and β-endorphin-knockout mouse lines, addressing nicotine-induced responses by various behavioral tests. Attenuated: knockout mice show lower response compared to wild-type mice; Faster developed: knockout mice show faster development of tolerance to nicotine compared to wild-type mice; Unchanged: no differences between knockout and wild-type mice. CPP, conditioned place preference.

TABLE 2 Effects of Pharmacological Interventions in the Opioid System on Nicotine-Induced Behavioral Responses

Drug	Dose and Route	Animal	Behavioral Test	Nicotine-Induced Response	Effect of Drug	Study
Naloxone	3 mg/kg, sc	Mice	CPA	Nicotine withdrawal	Manifestation	Balerio et al. (2004)
β-Funaltrexamine	5 mg/kg, ip	Mice	Elevated-plus maze	Nicotine-induced anxiolytic-like effect	Reduced	Balerio et al. (2005)
Nor-binaltorphimine	2.5 mg/kg, ip	Mice	Elevated-plus maze	Nicotine-induced anxiolytic-like effect	Unmodified	Balerio et al. (2005)
Naltrindole	2.5 mg/kg, ip	Mice	Elevated-plus maze	Nicotine-induced anxiolytic-like effect	Unmodified	Balerio et al. (2005)
Naloxonazine	5, 15 mg/kg, ip	Rats	Self-administration	Nicotine rewarding effect	Reduced	Liu and Jernigan (2011)
Naltrindole	0.5, 5 mg/kg, ip	Rats	Self-administration	Nicotine rewarding effect	Unmodified	Liu and Jernigan (2011)
GNTI	0.25, 1 mg/kg, ip	Rats	Self-administration	Nicotine rewarding effect	Unmodified	Liu and Jernigan (2011)

The table shows the pharmacological studies in animals designed to evaluate the effects of opioid receptor antagonists on nicotine-induced behavioral responses. β-Funaltrexamine and naloxonazine: selective MOR antagonists; naltrexone: nonselective opioid receptor antagonist; naltrindole: selective DOR antagonist; nor-binaltorphimine and GNTI: selective KOR antagonists. CPA, conditioned place aversion; GNTI, 5′-guanidinonaltrindole di (trifluoroacetate) salt hydrate.

in nicotine-induced behavioral sensitization, which may regulate the development of nicotine addiction. Chronic nicotine treatment in mice resulted in tolerance to antinociceptive responses of nicotine. In the nicotine-tolerant mice, autoradiography showed that the density of MORs in the spinal cord was not modified, but the functional activity of these receptors was significantly increased as a result of chronic nicotine exposure (Galeote, Kieffer, Maldonado, & Berrendo, 2006). The nicotine-induced tolerance developed faster in MOR-knockout mice compared to wild-type mice. This study suggests that increased activity of MOR may be an adaptive mechanism to counteract the establishment of nicotine tolerance.

In addition to its role in rewarding effects and behavioral responses of nicotine, the endogenous opioid system is involved in nicotine withdrawal. In rodents, an opioid antagonist, naloxone, induced conditioned place aversion in mice with chronic nicotine treatment and precipitated somatic manifestations of withdrawal, reflecting the motivational manifestations of nicotine withdrawal (Balerio, Aso, Berrendero, Murtra, & Maldonado, 2004). Nicotine withdrawal induced by mecamylamine, a nonselective and noncompetitive antagonist of nicotinic acetylcholine receptors, was significantly attenuated in MOR-knockout mice compared with wild-type mice (Berrendero et al., 2002). However, in β-endorphin-knockout mice, nicotine induced a physical dependence similar to that in wild-type mice, and no differences were found in the somatic signs of mecamylamine-precipitated nicotine withdrawal (Trigo et al., 2009). The MOR appears to be involved in nicotine withdrawal through a neurochemical mechanism other than the activation of β-endorphins.

OPRM1 POLYMORPHISMS

The MOR gene *OPRM1*, located at chromosome 6q24–q25, plays an important role in smoking-related behaviors manifested by multiple genetic studies. By looking for the concordance between gene regions implicated in affected sibling pairs and epistatic linkage analyses with genes suggested from microarray studies of experimental nicotine exposure and the literature, five sets of genes, including *OPRM1*, were identified as logical candidates for searching genetic variation related to nicotine dependence (Sullivan et al., 2004). One genome-wide association study (GWAS) on smoking phenotypes reported the association of *OPRM1* with smoking behavior, with involvement in smoking initiation (Vink et al., 2009). Using large-scale sequencing, eight single-nucleotide polymorphisms (SNPs) have been identified in the human MOR gene, which affect the amino acid sequence (Hoehe et al., 2000; LaForge, Yuferov, & Kreek, 2000) (Table 3).

The most common SNP in *OPRM1* is the A118G variant (rs1799971), located in the coding region of the first exon of *OPRM1*, with an adenine-to-guanine transition at nucleotide position 118 of the coding sequence of the gene. Nucleotide 118 is the first base in codon 40 of the human MOR, and the missense SNP A118G variant, also called the Asn40Asp variant, causes the substitution of an aspartate (Asp) for an asparagine (Asn) at position 40 in the amino acid sequence, a putative N-glycosylation site. However, a multitude of genetic studies since 2005 have made no consistent conclusions regarding the functional characterization of this genetic variant related to smoking-related behavior.

Pros for the Functional *OPRM1* A118G Variant

Although the GWAS on smoking phenotypes suggests the involvement of *OPRM1* in smoking initiation, SNP rs1799971 was not found associated with this phenotype (Vink et al., 2009). The early candidate gene studies recruiting both former heroin addicts in methadone maintenance treatment and normal control subjects indicated that allele distributions were significantly different among ethnic groups and the A118G variant was present in a significantly higher proportion of non-opioid-dependent subjects compared with the opioid-dependent subjects in the Hispanic ethnic group (Bond et al., 1998). In further in vitro experiments, by transfection of the A118G SNP plasmid into AV-12 cells and *Xenopus* oocytes, the MOR encoded by the G allele had increased affinity for β-endorphins, three times more than the receptor with the most common AA genotype (Bond et al., 1998). In vitro studies performed in a variety of cellular expression systems under different experimental conditions have found inconsistent results in terms of the functional variant in the *OPRM1* A118G polymorphism. Several studies indicate that the *OPRM1* A118G SNP influences the expression level of the MOR. By measuring the opioid receptor expression in Chinese hamster ovary cells transfected with the human coding region of *OPRM1*, the *OPRM1* G allele was found to yield lower MOR mRNA and protein levels (Zhang, Wang, Johnson, Papp, & Sadée, 2005). In human embryonic kidney (HEK293) cells stably expressing human MOR, either the wild-type or the Asn40Asp variant, the expression of MOR with the

TABLE 3 Single-Nucleotide Polymorphisms Resulting in Amino Acid Substitutions in the *OPRM1* Gene

dbSNP ID	Nucleotide Substitution	Location in Gene	Amino Acid Substitution	Main Function	Global MAF
rs35174096	C12G	Exon I	Ser4Arg (S4R)	N-terminal	N/A
rs1799972	C17T	Exon I	Ala6Val (A6V)	N-terminal	T = 0.0713
ra1799971	A118G	Exon I	Asn40Asp (N40D)	N-terminal	G = 0.2234
rs17174794	C440G	Exon II	Ser147Cys (S147C)	Transmembrane domain 3	G = 0.0018
rs17174801	A454G	Exon II	Asn152Asp (N152D)	Transmembrane domain 3	G = 0.0024
rs1799974	G779A	Exon III	Arg260His (R260H)	Cytoplasmic loop 3	A = 0.0006
rs376950705	G794A	Exon III	Arg265His (R265H)	Cytoplasmic loop 3	N/A
rs200811844	T802C	Exon III	Ser268Pro (S268P)	Cytoplasmic loop 3	C = 0.0010

The table shows the SNPs in the *OPRM1* gene that result in amino acid substitutions in the predicted primary structure of the receptor. Global MAF indicates the minor allele frequency for each SNP provided in the NCBI SNP database (Build 142). SNP, single-nucleotide polymorphism; N/A, not available.

G allele was reduced compared to the AA wild type (Beyer, Koch, Schroder, Schulz, & Hollt, 2004). Altogether, the enhancement in receptor affinity for β-endorphin might be offset by a reduction in the number of receptors expressed. In HEK293 cells and a Syrian hamster adenovirus-12-induced tumor cell line (AV-12) either stably or transiently expressing the MOR prototype and A118G variant, Kroslak et al. (2007) found lower cell-surface receptor binding site availability (measured by (D-Ala2, N-Me-Phe4, Gly5-ol)-enkephalin, DAMGO) in cell lines with the G allele compared to the AA genotype, but no differences in agonist-mediated cAMP signaling for DAMGO or β-endorphin. Mague et al. (2009) generated a knock-in mouse model possessing the Oprm1 gene with the equivalent of the human OPRM1 gene, based on the high homology between mouse and human sequences at the nucleotide and amino acid levels and similar gene expression levels. The results showed that the expression levels of mRNA and protein levels in the brain reward area decreased in the mice with the GG genotype compared to those with the AA genotype, but no evidence suggesting altered affinity to MOR agonists. Additionally, the GG mice showed deficits in the locomotor sensitization and antinociceptive effect of acute morphine administration, and the altered behavioral responses were most evident in female mice.

Cons for the Functional OPRM1 A118G Variant

Other studies suggest that the OPRM1 Asn40Asp variant has no effect on the expression levels of MOR mRNA and protein or its functional properties. In monkey kidney (COS) cells with transient expression of the wild-type and the Asn40Asp variant receptors, the expression levels of MOR are similar. Binding affinities for the main endogenous opioid peptides including β-endorphin were not significantly changed in the Asn40Asp variant receptors compared to wild type (Befort et al., 2001). In agreement with the findings in COS cells, another study showed that HEK293 cells stably expressing the Asn40Asp variant receptor and wild-type receptor presented similar binding affinities and internalization of β-endorphin and similar desensitization time courses and resensitization rates after prolonged treatment with β-endorphin (Beyer et al., 2004). These studies indicate that the Asn40Asp variant may not affect the conformation of the extracellular N-terminal domain of the MOR.

The Functional OPRM1 A118G Variant in Humans

Human brain imaging studies have provided evidence for a functional role of the OPRM1 A118G polymorphism in smoking. Using [^{11}C]carfentanil positron emission tomography (PET) comparing smokers and nonsmokers balanced for the OPRM1 A118G variant, Ray et al. (2011) showed that smokers with the AA genotype exhibited significantly higher levels of MOR binding potential (or receptor availability) than smokers carrying the G allele in the bilateral amygdala, left thalamus, and left anterior cingulate cortex. Among the smokers carrying the G allele, the extent of nicotine reward was associated with MOR binding potential in regions of the mesolimbic rewarding system. In line with the previous in vitro experiments, this study suggested that the MORs encoded by the AA genotype have relatively lower affinity compared to the receptors with the G allele. Another PET brain imaging study using [^{11}C]raclopride indicated that healthy males with the G allele had greater dopamine release after tobacco smoking than those with the AA genotype in the right caudate and right ventral pallidum (Domino et al., 2012). Chen et al. (2013) assessed the association of the OPRM1 genetic polymorphisms with the severity of cigarette smoking by directly measuring the plasma concentration of cotinine in a Taiwanese methadone maintenance treatment (MMT) cohort. They found that the G-allele carriers of SNP rs1799971 (A118G) had a lower plasma cotinine concentration than the A-allele carriers, suggesting that the G allele had a protective effect against cigarette smoking in patients under MMT.

As previously suggested, it may be oversimplifying to consider the OPRM1 A118G SNP to be either a gain- or a loss-of-function variant, and genetic effects may vary depending on the specific pharmacological challenge in the specific context of the individual.

OPRM1 AND SMOKING BEHAVIORS

A number of genetic studies have explored the possible association of the OPRM1 gene, particularly the A118G variant (rs1799971), with various smoking-related phenotypes, including smoking initiation (SI), initial sensitivity, nicotine dependence (ND), heaviness of smoking, and nicotine withdrawal (Table 4).

The OPRM1 A118G Polymorphism and Smoking Initiation

Genetic factors may alter the risk of ND by influencing the initial sensitivity to acute nicotine exposure. However, the findings about the OPRM1 gene vary depending on the methodology and subject sources. One study exploring the association of functional candidate gene polymorphisms with initial sensitivity to nicotine in young adult nonsmokers of European ancestry showed no significant association of the OPRM1 A118G SNP with initial sensitivity to nicotine (Perkins, Lerman, Coddington et al., 2008). Zhang, Kendler, and Chen (2006) reported a case–control study examining the influence of the OPRM1 gene on SI and ND by recruiting never-smokers and regular smokers and using a haplotype-tagging approach to select 11 SNPs within the OPRM1 gene, including the A118G SNP. The results found that three SNPs, rs2075572, rs10485057, and rs10485058, were associated with SI and one SNP, rs10485057, was associated with ND. There was no evidence for the association of rs1799971 itself with smoking phenotype, but rs1799971 was in high linkage disequilibrium with the core markers rs9479757–rs2075572–rs10485057, which was significantly associated with SI. In addition, several longitudinal studies regarding the OPRM1 gene and SI, initial sensitivity to nicotine, and smoking development in adolescents have been conducted. One study examined the effects of the A118G polymorphism on the initial values and the development of alcohol use and smoking over a 4-year period (Kleinjan, Poelen, Engels, & Verhagen, 2013). In addition to alcohol use as a risk for the development of smoking behavior, a sex specificity of the OPRM1 gene was

TABLE 4 Studies on the Association of OPRM1 Variants with Smoking Behaviors

Study	Subjects	Sample Size, n (Age Range/Mean ± SD)	Study Design	Genotype, n (%)	Phenotype	Result
Perkins, Lerman, Coddington et al. (2008)	Young adult nonsmokers (European ancestry)	101 (ages 21–39 years)	Experimental study: nicotine nasal spray 0, 5, and 10 µg/kg doses; nicotine vs placebo spray choice	AA: male 30 (86%), female 47 (73%); AG/GG: male 5 (14%), female 17 (27%)	Initial sensitivity to nicotine, nicotine reinforcement	No significant association of A118G with initial sensitivity to nicotine and nicotine reinforcement
Zhang et al. (2006)	Nonsmokers and regular smokers (low ND and high ND) (European ancestry)	688	Case–control study: smokers (444) vs nonsmokers (244); low ND (215) vs high ND (229)	Nonsmokers: GG 5, AG 46, AA 187; Low ND: GG 3, AG 49, AA 161; High ND: GG 7, AG 41, AA 173	SI and ND	No association of A118G with SI or ND; A118G in LD with the haplotype associated with SI
Kleinjan et al. (2013)	Dutch adolescents (Caucasian descent)	311 (ages 13–15 years)	Longitudinal four-wave design	AA 78.5%, AG 20.3%, GG 1.0%	Development of smoking behavior	Male A-allele carriers and females G-allele carriers showed a faster development of smoking behavior
Wilkinson et al. (2012)	Youth never-smokers (Mexican origin)	1118 (ages 11–13 years, 52.2% female)	Longitudinal study (3 years)	SNP rs9322451 variant 26.5%	Smoking experimentation	SNP rs9322451 in OPRM1 associated with new experimentation; No evidence for rs1977791
Schuck et al. (2014)	Dutch adolescents (never-smokers)	171 (mean age 13.9 years, 50.9% female)	Cross-sectional study	AA 79.5%, AG/GG 20.1%	Initial sensitivity to nicotine	The G variant carriers more likely to report liking of initial smoking
Ray et al. (2006)	Smokers (European ancestry)	60 (ages 18–70, mean age 43.2 years, 63% female)	Within-subject, double-blind human laboratory study with 4-day follow-up with treatment of naltrexone 50mg or placebo and cigarettes (nicotine vs denicotinized) choice paradigm	AA 30, AG 26, GG 4	Nicotine reinforcement	The G allele associated with a reduced reinforcing value of nicotine in females, not in males
Perkins, Lerman, Grottenthaler et al. (2008)	Smokers of European ancestry	72 (age 28.1 ± 1.3 years)	2 × 2 balanced-placebo design: actual and expected nicotine dose in cigarettes, negative vs positive mood induction	AA: male 25 (73.5%), female 26 (70.3%); AG/GG: male 9 (26.5%), female 11 (29.7%)	Smoking reward and reinforcement	AA genotype experienced higher smoking reward during negative vs positive mood induction; the mood condition made no difference in AG/GG genotypes

Continued

TABLE 4 Studies on the Association of OPRM1 Variants with Smoking Behaviors—cont'd

Study	Subjects	Sample Size, n (Age Range/Mean ± SD)	Study Design	Genotype, n (%)	Phenotype	Result
Daher et al. (2013)	Smokers of Brazilian descent (whites 49%, browns 35.5%, blacks 15%, Amerindian 0.5%)	200 (ages 18–92, mean age 50.8 ± 15.6 years, 39.5% female)	Cross-sectional study	AA 146 (73%), AG 44 (22%), GG 10 (5%)	Heaviness of smoking (pack-years)	The G-allele carriers had higher tobacco exposure than those with the AA genotype
Hirasawa-Fujita et al. (2014)	Smokers and nonsmokers with schizophrenia and bipolar disorder	177 schizophrenia and 113 bipolar disorder	Cross-sectional study	N/A	Smoking status	Subjects with the G allele smoked more cigarettes per day than the AA genotype carriers in schizophrenia group
Verde et al. (2011)	Smokers and never-smokers (Caucasian, Spanish descent)	126 smokers (62 female; mean age 54 ± 14 years) and 80 never-smokers (43 female; mean age 42 ± 11 years)	Case-control study: smokers (126) vs nonsmokers (80)	Nonsmokers: AA 74.0%, AG 24.6%, GG 1.4% Smokers: AA 67.5%, AG 24.7%, GG 7.8%	Smoking consumption (cigarettes per day and pack-years smoked), ND (FTND)	No significant association of A118G with smoking consumption and ND
Fang et al. (2014)	Male current and ex-smokers (Chinese)	284 (mean age 62.6 ± 13.4 years)	Cross-sectional study	Current smokers: AA 64 (46.7%), AG 62 (45.3%), GG 11 (8%); Ex-smokers: AA 72 (49.3%), AG 58 (39.7%), GG 16 (11.0%)	Smoking status, daily cigarette consumption, ND	No association of A118G with smoking status, daily cigarette consumption, and ND
Bergen et al. (2009)	Smokers (white)	828 (mean age 45.5 ± 11.5 years, 50% female)	Candidate gene study	MAF: 13%	ND (FTND)	No association of A118G with ND
Hardin et al. (2009)	Current and former smokers (white, nonwhite/mixed/other)	419 (ages 11–15 years, 51.3% female)	A longitudinal family-based association study	AA 77%, AG/GG 23%	Relapse (analyzed by nicotine withdrawal sensitivity scores)	Eight SNPs in OPRM1 but not rs1799971 associated with nicotine withdrawal sensitivity scores
Wang et al. (2008)	Smokers (European ancestry)	13 (mean age 38.3 ± 2.9, 7 female)	Within-subject, brain imaging study (abstinence vs satiety)	AA 76.9%, AG/GG 23.1%	ND (rCBF changes in abstinence vs satiety)	Smokers with the AA genotype showed significant rCBF increases in abstinence vs satiety

The table lists the genetic studies in humans on the genetic variants in OPRM1, particularly SNP A118G (rs1799971), in relation to smoking behaviors. FTND, Fagerström Test for Nicotine Dependence; LD, linkage disequilibrium; MAF, minor allele frequency; N/A, not available; ND, nicotine dependence; rCBF, regional cerebral blood flow; SD, standard deviation; SI, smoking initiation; SNP, single-nucleotide polymorphism.

found for smoking development: male A-allele carriers showed a faster development of smoking behavior, whereas in females, the G allele led to a faster development of smoking. These results imply that the *OPRM1* gene may have a sex differential effect on smoking behavior, and the underlying biological mechanism needs to be explored. Another study collected prospective data on cigarette experimentation (defined as having tried cigarettes, but not smoking on a monthly basis) among youth of Mexican origin over 3 years and evaluated the role of both genetic and psychosocial factors associated with the transition from never smoking to cigarette experimentation (Wilkinson et al., 2012). The study found that in committed never-smokers the *OPRM1* SNP rs9322451 was associated with smoking experimentation. Schuck, Otten, Engels, and Kleinjan (2014) carried out a cross-sectional survey in Dutch adolescents who had never inhaled on a cigarette and analyzed their self-reported responses to initial smoking. The results revealed that adolescents carrying the G allele were more likely to report liking of cigarettes independent of gender. Although the findings are not identical, three of the above-mentioned studies indicate that the *OPRM1* A118G variant is associated with SI and may be viewed as an early risk of ND.

The *OPRM1* A118G Polymorphism and Nicotine Reward

In addition to SI, the role of the *OPRM1* A118G variant in nicotine rewarding and reinforcing effects has been explored by pharmacological and human laboratory studies. A within-subject, double-blind study with administration of naltrexone versus placebo showed that the G allele was associated with a reduced reinforcement of nicotine in female smokers but not in male smokers; no effect of naltrexone on nicotine reinforcement was found among the genotypes or gender subgroups (Ray et al., 2006). A study was carried out to explore the effects of mood manipulation and actual or expected nicotine dose in cigarettes on genetic associations with smoking reward (measured by the Rose Sensory Questionnaire, rated on a 0 to 100 visual analog scale) and reinforcement (measured by the latency to first puff and the total number of puffs taken). Smokers of European ancestry were randomized to one of four groups in a 2×2 balanced-placebo design, corresponding to manipulation of actual (0.6 mg vs 0.05 mg) and expected (told nicotine, told denicotinized) nicotine dose in cigarettes during each of two sessions (negative vs positive mood induction). The results showed that the *OPRM1* AA genotype experienced higher smoking reward during negative versus positive mood induction, whereas the mood condition made no difference in terms of smoking reward in those with the minor G allele (Perkins, Lerman, Grottenthaler et al., 2008). This study implies that the opioid genes interacting with mood condition may influence smoking reward and reinforcement, in which other neurobiological pathways may be involved.

The *OPRM1* A118G Polymorphism and Heaviness of Smoking

Recent studies have indicated the association of the *OPRM1* A118G polymorphism with the quantity of smoking. The study evaluating the A118G variant related to smoking status among surgical patients of Brazilian descent found that patients with the G allele reported higher tobacco exposure measured in pack-years than those who were homozygous for the A allele (Daher, Costa, & Neves, 2013). Among patients with schizophrenia, the *OPRM1* G-allele carriers smoked more cigarettes per day than the AA-genotype carriers (Hirasawa-Fujita, Bly, Ellingrod, Dalack, & Domino, 2014). These two studies consistently suggest that the G-allele carriers tend to have higher tobacco consumption. However, the candidate gene study in a cohort of Spanish smokers and matched never-smokers showed no association of the *OPRM1* A118G variant with smoking consumption and ND (Verde et al., 2011). Our study also did not show evidence for the association of the A118G variant with smoking behaviors, including smoking status and daily cigarette consumption, and ND in a sample of Chinese men (Fang, Wang, & He, 2014). The discrepancy among these studies may be partially due to ethnic differences, and replication studies with independent samples are needed.

The *OPRM1* A118G Polymorphism and Nicotine Dependence and Withdrawal

Additionally, the involvement of the *OPRM1* gene in ND and withdrawal has also been demonstrated in human and animal studies. In the study of Zhang et al. (2006), one SNP (rs10485057) in the *OPRM1* gene was found to be related with smoking dependence, but no evidence supported the association of SNP rs1799971 and ND. Utilizing a cohort of treatment-seeking white cigarette smokers, Bergen et al. (2009) examined 1123 SNPs at 55 autosomal candidate genes associated with ND measured by the Fagerström Test for Nicotine Dependence (FTND). No association of the *OPRM1* SNPs with ND was found. By using a Rasch-modeled nicotine withdrawal sensitivity score in a multiplex smoking pedigree sample, eight SNPs in the *OPRM1* gene, but not SNP rs1799971, were found significantly associated with the nicotine withdrawal sensitivity score, suggestive of an increased risk of relapse (defined as reengaging in smoking after quitting for 1 week or longer) (Hardin et al., 2009). A brain imaging study investigated whether the A118G variant was associated with regional cerebral blood flow (rCBF) changes induced by nicotine abstinence (Wang et al., 2008). Among smokers with the *OPRM1* AA genotype, the CBF in regions previously related to cigarette cravings significantly increased in abstinence versus satiety, which was greater than in those with the AG and GG genotypes. This result provides evidence for increased nicotine reward and reinforcement in smokers with the AA genotype and is consistent with the human laboratory studies (Perkins, Lerman, Grottenthaler et al., 2008; Ray et al., 2006). Although the existing gene association studies yielded inconsistent results, they have enabled some theoretical biological views and provided insights into the biological mechanistic basis of the *OPRM1* A118G gene.

PHARMACOGENETIC STUDIES OF *OPRM1* IN SMOKING CESSATION

Despite the progress in the pharmacological treatment of ND, the available treatments are limited in efficacy and benefit only a minority of smokers. The individual differences in response to pharmacotherapies for smoking cessation have prompted the investigation of genetic factors. Based on the previous data of functional genetic variants in the *OPRM1* gene, multiple pharmacogenetic studies have been designed to explore the influence of the *OPRM1* A118G variant on smoking cessation. Pharmacological treatments for smoking cessation have been applied including nicotine replacement therapy (NRT) such as the transdermal nicotine (TN) patch and nicotine nasal spray (NS) and bupropion (Table 5).

The earliest study in smokers of European ancestry examined whether the Asn40Asp (A118G) SNP in *OPRM1* predicted the comparative efficacy of TN and nicotine NS for an 8-week treatment period (Lerman et al., 2004). The results showed that smokers carrying the G allele were more likely to be abstinent at the end of the NRT phase than those with the AA genotype, and they experienced less mood disturbance and short-term weight gain. This effect of the A118G SNP was significant only in the TN group, particularly during the first 4 weeks with the 21-mg dose. The smokers with the G allele appeared to gain more reward from nicotine through the TN treatment and decreased aversive effect of abstinence, which facilitated smoking cessation. However, the subsequent studies failed to replicate these findings. Another randomized controlled trial investigated the association of the *OPRM1* genotype with long-term smoking cessation by using TN patch treatment (for 12 weeks) or placebo with follow-up over an 8-year period (Munafò, Elliot, Murphy, Walton, & Johnstone, 2007). The results indicate that the *OPRM1* genotype may moderate the effect of the TN patch compared to placebo during active treatment, with a significant benefit of active NRT compared to placebo in the *OPRM1* AA-genotype group but not in the AG- and GG-genotype groups. This effect was observed only for abstinence up to the end of treatment and not at subsequent time points during the 8-year follow-up, indicating the pharmacogenetic effect of the *OPRM1* genotype. A 2013 study tested a tailored NRT and the effects of *OPRM1* genotype and the severity of ND on smoking cessation at 4- and 26-week follow-up, in which study the participants were given an oral NRT with dose tailored according to the *OPRM1* genotype (those with AA genotype were prescribed a lower dose of NRT and those with one or more copies of the G allele were prescribed a higher dose of NRT) or FTND score (Munafò, Johnstone, Aveyard, & Marteau, 2013). The results showed no effect of *OPRM1* genotype on smoking cessation and no difference in the relationship between NRT dose and smoking cessation between those with the G allele and those with the AA genotype. Conti et al. (2008) performed a systems-based candidate gene study of 1295 SNPs in 58 genes within the neuronal nicotinic receptor and dopamine systems as well as the opioid system to investigate the role of these genes in smoking cessation in a bupropion placebo-controlled randomized clinical trial. They did not find any evidence for the effects of the *OPRM1* SNPs on smoking cessation.

Above all, although *OPRM1* remains a plausible candidate gene for smoking-related behaviors and nicotine addiction, the existing pharmacogenetic studies regarding the effects of *OPRM1* on smoking cessation have not reached a consensus, and explicit genetic replication studies are warranted. Furthermore, given the wide genetic background in relation to the multiple neurobiological mechanisms of nicotine addiction, the contribution of the *OPRM1* gene may be limited to specific aspects of smoking-related behaviors and vary depending on the pharmacotherapy.

ARRB2 AND *HINT1*

The MOR interacts directly with multiple proteins, known as MOR-interacting proteins. Of those, β-arrestin 2 and protein kinase C-interacting protein (mPKCI-1) are of most interest and considered to influence the individual susceptibility to drug dependence.

ARRB2 in Rodent Models

The MOR is a member of the G-protein-coupled receptor (GPCR) family. GPCRs are known to undergo desensitization through phosphorylation of the receptor. By binding to phosphorylated MOR, β-arrestin 2 is identified as an important regulator of signal transduction and plays a role in opioid reward. *ARRB2* is the gene encoding β-arrestin 2. The β-arrestin 2 gene knockout (*ARRB2*-null mutant) mice display remarkable potentiation and prolongation of the antinociceptive effects of morphine compared to wild-type littermates, suggesting the enhancement of GPCR signaling and receptor internalization (Bohn et al., 1999). In addition, morphine induced more dopamine release in striatum in β-arrestin 2-knockout mice, and these mice showed greater rewarding response and increased tolerance to morphine than wild-type mice (Bohn et al., 2003). On account of the role of β-arrestin 2 in mediating the behavioral effects of addictive drugs, Correll et al. (2009) explored nicotine sensitization in β-arrestin 2-knockout mice and found that the locomotor activity was significantly reduced in β-arrestin 2-knockout mice compared to wild-type mice. The study implies the involvement of β-arrestin 2 in nicotine-induced behavioral responses.

HINT1 In Vitro and in Rodent Models

In contrast to β-arrestin 2, mPKCI-1 (known as histidine triad nucleotide-binding protein 1 in humans), encoded by the gene *HINT1*, inhibits PKC-mediated MOR phosphorylation and attenuates receptor desensitization. *HINT1* also appears to have a role in modulating the effects of drugs abuse. It has been shown that mPKCI-1-knockout mice developed tolerance to the analgesic effect of morphine much more quickly than wild-type mice (Guang, Wang, Su, Weinstein, & Wang, 2004). Human postmortem brain expression studies showed a significant increase in *HINT1* at the protein level in the nucleus accumbens after chronic nicotine exposure (Jackson et al., 2011). This is consistent with the findings in mouse models, and nicotine-induced increase in *HINT1* expression could be inhibited by the nicotinic-receptor antagonist mecamylamine or after cessation of nicotine treatment (Jackson et al., 2011). In addition, the findings from male and female *HINT1*-wild-type and -knockout mice revealed the role of *HINT1* in nicotine-mediated behaviors such as anxiogenic effect, anxiolytic effect, antinociception, and locomotor activity and suggested that the *HINT1* gene may have differential effects in males and females (Jackson, Wang, Barbier, Chen, & Damaj, 2012). Furthermore, nicotine reward and withdrawal have been evaluated in *HINT1*-knockout mice compared to wild-type mice

TABLE 5 Pharmacogenetic Studies of the *OPRM1* Variants in Smoking Cessation

Trial	Design	Number of Participants	Ethnicity	Intervention	Follow-Up Time Point of Abstinence	Quit Rate	Genotype, n (%)	Outcome
Lerman et al. (2004)	An open-label randomized clinical trial (TN vs NS)	320 (current smokers)	European ancestry	TN: 21 mg × 4 weeks, 14 mg × 2 weeks, 7 mg × 2 weeks; NS: 1.0 mg dose, 8–40 times per day × 4 weeks, taper by one-third × 2 weeks, taper another third × 2 weeks	EOT and 6-month follow-up	EOT: TN (AA 33.3%, AG/GG 52.4%) vs NS (AA 26.8%, AG/GG 30%); 6-month follow-up: TN (AA 16.2%, AG/GG 31%) vs NS (AA 15%, AG/GG 12.5%)	GG 5 (1.6), AG 77 (24.0), AA 238 (74.4)	The G-allele carriers were significantly more likely to be abstinent than those homozygous for the A allele at EOT
Munafò et al. (2007)	A double-blind randomized placebo-controlled trial	710 (current and ex-smokers, 59% female)	European ancestry	TN 21 mg × 4 weeks, 14 mg × 4 weeks, 7 mg × 4 weeks; placebo × 12 weeks	12-week, 26-week, 1-year, and 8-year follow-up	N/A	AA (78), AG (21), GG (1)	Male subjects in the AA group were more likely to be abstinent than those in the AG/GG group, only for abstinence up to the EOT (12-week follow-up) and not subsequently
Munafò et al. (2013)	An open-label, parallel-groups randomized trial	598 (53.3% female)	European ancestry and non-European ancestry (9%)	All participants were prescribed an NRT patch with dose based on the number of cigarettes per day. The genotype arm: AA 6 mg, AG/GG 12 mg; the phenotype arm: FTND <8, 6 mg; FTND ≥8, 12 mg (oral NRT treatment for 4 weeks)	4- and 26-week follow-up	N/A	AA (79.4), AG/GG (20.6)	No effect of the *OPRM1* genotypes on smoking cessation; no influence on response to high- vs low-dose NRT
Conti et al. (2008)	A systems-based candidate gene study in a randomized placebo clinical trial	417 (54% female)	European ancestry	Bupropion 150 mg/day × 3 days, followed by 300 mg/day up to 10 weeks; placebo × 10 weeks	EOT and 6-month follow-up	EOT: 32.3% vs 21.5%; follow-up: 25.8% vs 17.4% (bupropion vs placebo)	N/A	No evidence for the effect of *OPRM1* SNPs on smoking cessation

The table lists the pharmacogenetic studies of the *OPRM1* variants, particularly the A118G (rs1799971) polymorphism, in smoking cessation with NRT (TN and NS) and bupropion. EOT, end of treatment; N/A, not available; NRT, nicotine replacement therapy; NS, nicotine nasal spray; TN, transdermal nicotine.

(Jackson, Wang, Barbier, Damaj, & Chen, 2013). The findings showed that the *HINT1*-knockout mice failed to develop a significant nicotine preference, which might be associated with dysregulation of postsynaptic dopamine transmission (Barbier et al., 2007). The physical withdrawal signs were attenuated in *HINT1*-knockout mice (Jackson et al., 2013). Altogether, the brain expression studies and mouse behavioral studies implicate a role for *HINT1* in nicotine reward and withdrawal.

ARRB2 and *HINT1* in Relation to Smoking Behaviors

In addition to the animal experiments, human genetic studies have been reported in terms of the association of *ARRB2* and *HINT1* with smoking-related behaviors and pharmacological treatment for smoking cessation. By using two population samples, Jackson et al. (2011) found two SNPs, rs3864283 and rs2526303, in the *HINT1* gene that were significantly associated with ND (measured by the FTND score and number of cigarettes smoked per day). Smokers with variants in rs3864283 had a higher level of *HINT1* mRNA than nonsmokers. The results suggest that genetic variants in *HINT1* inducing higher mRNA expression are associated with increased risk of ND. A family-based association study showed that SNPs in both β-arrestins 1 and 2 were associated with ND in European American smokers, but not in African American smokers (Sun, Ma, Payne, & Li, 2008). Specifically, SNPs rs472112 in *ARRB1* and rs4790694 in *ARRB2* were significantly associated with the Heaviness of Smoking Index (HSI) and FTND score, and the association of rs4790694 in *ARRB2* remained significant after correction for multiple testing. Haplotype analysis revealed that the haplotypes C-G-C-G-G-T in *ARRB1* (formed by SNPs rs528833, rs1320709, rs480174, rs5786130, rs611908, and rs472112) and C-C-A-T in *ARRB2* (formed by SNPs rs3786047, rs4522461, rs1045280, and rs4790694) were positively associated with HSI and FTND (Sun et al., 2008). By single-locus and haplotype-based association analyses, we examined the associations of SNPs in *ARRB2* and *HINT1* with smoking behaviors in Chinese men and found that *HINT1* rs3852209 was significantly associated with smoking status (Fang et al., 2014). Although the findings from the above-mentioned genetic association studies are not concordant because of the different methodologies, these studies provide evidence for the important role of *HINT1* in smoking phenotypes. A few pharmacogenetic studies of smoking cessation with regard to *ARRB2* and *HINT1* have been reported. Ray et al. (2007) reported an open-label randomized trial with TN versus NS treatment for 8 weeks and explored the role of SNPs in *ARRB2* and *HINT1* and their interactions with *OPRM1*. As previously reported (Lerman et al., 2004), the study provides evidence for the role of *OPRM1*: smokers with the G allele had a higher quit rate than smokers with the AA genotype at end of treatment and 6-month follow-up. Additionally, the *HINT1* SNP rs3852209 was associated with abstinence rate at 6-month follow-up and smokers with the T allele had significantly higher abstinence rates than smokers with the CC genotype. An interaction between *OPRM1* and *ARRB2* rs3786047 was indicated in that smokers with the *OPRM1* G allele maintained higher abstinence rates up to 6-month follow-up only among those with the GG genotype in *ARRB2* rs3786047.

Altogether, the preliminary studies imply that genetic variants in *ARRB2* and *HINT1* may be involved in smoking-related behaviors and pharmacotherapies for smoking cessation, but additional validation of the present findings in future clinical trials is warranted.

APPLICATIONS TO OTHER ADDICTIONS AND SUBSTANCE MISUSE

The MOR, encoded by *OPRM1*, plays a pivotal role in the rewarding effects of many drugs of abuse, including opioids, nicotine, and alcohol. In addition to nicotine, genetic variants in *OPRM1*, particularly the A118G polymorphism, have been frequently studied in relation to heroin and alcohol dependence. Heroin is the most common illicit opioid drug and is deacetylated to morphine in the brain, directly targeting the MOR. However, multiple studies on the associations between *OPRM1* polymorphisms and opioid dependence have yielded inconsistent results and suggest that ethnicity differences may modulate the association between genetic variants of *OPRM1* and susceptibility to opioid dependence. In addition, the association between the A118G polymorphism and alcohol dependence may be explained by different physiological responses to alcohol based on the A118G genotypes. Subjects with the G allele have a stronger rewarding response to drink and are susceptible to alcohol dependence. Owing to shared neurobiological mechanisms and concordance in courses of drugs of abuse, ND often co-occurs with other substances abuse. Genetic studies focusing on the co-occurrence of alcohol use and smoking have been reported.

DEFINITION OF TERMS

Conditioned place aversion This refers to Pavlovian conditioning based on the capacity of the animal to associate the drug effect with the context. If the drug is aversive, mice avoid the drug-paired box.

Pack-years This is the number of years of smoking multiplied by the number of packs of cigarettes smoked per day.

Pharmacogenetic clinical trials These trials are used to evaluate the role of genetic variation in drug pharmacokinetics or pharmacodynamics in determining treatment efficacy within patient populations and provide a method to tailor drug treatments to individuals' genetic background to improve overall patient outcomes.

Self-administration By operant training, the animal works to obtain the drug and learns an action–outcome association. This model enables investigation of the rewarding effects of the drug and motivational aspects of drug intake.

The elevated-plus maze test The apparatus consists of open arms and closed arms, crossed in the middle perpendicular to each other, and a center area. Mice are given access to all of the arms and are allowed to move freely among them. The number of entries into the open arms and the time spent in the open arms are used as indices of open-space-induced anxiety in mice.

KEY FACTS OF SMOKING RELATED PHENOTYPES

- Smoking behavior occurring along a trajectory begins with experimentation/initial sensitivity and moves on to initiation, regular smoking, dependence, and then cessation with success or relapse (Figure 1).
- The rigorous defined phenotypes of smoking behavior based on the smoking trajectory have the potential to reduce the bias of classification and lack of specificity within broader existing phenotypes.
- It is noteworthy that nicotine is not the only substance involved in the development of tobacco dependence.

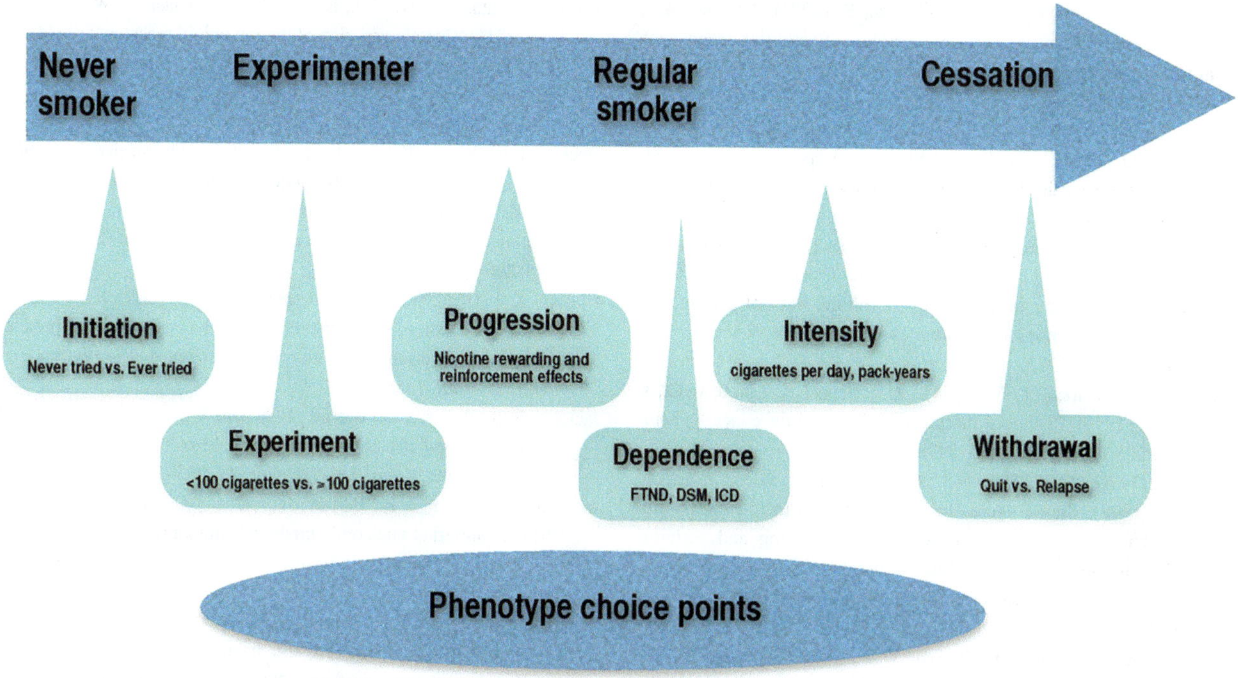

FIGURE 1 **Phenotypes along the smoking trajectory.** Smoking behaviors develop along a trajectory. Various stages along the smoking trajectory represent potential phenotype choice points at which a specific phenotype can be defined based on smoking behavior. DSM, Diagnostic and Statistical Manual; FTND, Fagerström Test for Nicotsine Dependence; ICD, International Classification of Diseases.

- Therefore, the term "tobacco dependence" may be preferred in clinical research rather than nicotine dependence.
- The existing studies on the *OPRM1* A118G SNP in relation to smoking behaviors, yielding inconsistent results, are partially due to the heterogeneity in the definition of smoking phenotypes.
- Refinement of smoking phenotypes may facilitate the exact replication studies and lead to more convictive evidence for the association of genetic variables and smoking behavior.

SUMMARY POINTS

- This chapter focuses on the MOR, a GPCR, which plays an important role in smoking behavior.
- Activation of the MOR by β-endorphins released after nicotine exposure is crucial to nicotine-induced dopamine release in the nucleus accumbens.
- MOR is involved in nicotine-induced behavioral responses, nicotine tolerance, ND, and withdrawal.
- The most common SNP in *OPRM1*, the gene encoding MOR, is the A118G (Asn40Asp) variant (rs1799971), and the discrepancy exists in the biological mechanism.
- The increased affinity of MOR with the G allele for β-endorphins appears to compensate for lower expression of MOR proteins.
- Gene association studies yielded inconsistent results, but provided insights into the biological mechanistic basis of the *OPRM1* A118G variant in smoking behaviors.
- It is noteworthy that sex differences may influence the effects of the *OPRM1* A118G variant on nicotine reinforcement and development of smoking behavior.
- The functional genes *ARRB2* and *HINT1* interacting with MOR are involved in nicotine addiction.
- A few gene association and pharmacogenetic studies on *ARRB2* and *HINT1* have been reported, and replication studies are warranted.

REFERENCES

Balerio, G. N., Aso, E., Berrendero, F., Murtra, P., & Maldonado, R. (2004). Delta 9 tetrahydrocannabinol decreases somatic and motivational manifestations of nicotine withdrawal in mice. *The European Journal of Neuroscience, 20,* 2737–2748.

Balerio, G. N., Aso, E., & Maldonado, R. (2005). Involvement of the opioid system in the effects induced by nicotine on anxiety-like behaviour in mice. *Psychopharmacology, 181,* 260–269.

Balfour, D. J. (2004). The neurobiology of tobacco dependence: a preclinical perspective on the role of the dopamine projections to the nucleus accumbens. *Nicotine & Tobacco Research, 6*(6), 899–912.

Barbier, E., Zapata, A., Oh, E., Liu, Q., Zhu, F., Undie, A., ... Wang, J. B. (2007). Supersensitivity to amphetamine in protein kinase-C interacting protein/*HINT1* knockout mice. *Neuropsychopharmacology, 32*(8), 1774–1782.

Befort, K., Filliol, D., Decaillot, F. M., Gaveriaux-Ruff, C., Hoehe, M. R., & Kieffer, B. L. (2001). A single nucleotide polymorphic mutation in the human mu-opioid receptor severely impairs receptor signaling. *The Journal of Biological Chemistry, 276*(5), 3130–3137.

Bergen, A. W., Conti, D. V., Van Den Berg, D., Lee, W., Liu, J., Li, D., ... Swan, G. E. (2009). Dopamine genes and nicotine dependence in treatment-seeking and community smokers. *Neuropsychopharmacology, 34*(10), 2252–2264.

Berrendero, F., Kieffer, B. L., & Maldonado, R. (2002). Attenuation of nicotine-induced antinociception, rewarding effects, and dependence in mu-opioid receptor knock-out mice. *The Journal of Neuroscience, 22*(24), 10935–10940.

Beyer, A., Koch, T., Schroder, H., Schulz, S., & Hollt, V. (2004). Effect of the A118G polymorphism on binding affinity, potency and agonist-mediated endocytosis, desensitization, and resensitization of the human mu-opioid receptor. *Journal of Neurochemistry, 89*(3), 553–560.

Bohn, L. M., Gainetdinov, R. R., Sotnikova, T. D., Medvedev, I. O., Lefkowitz, R. J., Dykstra, L. A., & Caron, M. G. (2003). Enhanced rewarding properties of morphine, but not cocaine, in beta(arrestin)-2 knock-out mice. *The Journal of Neuroscience, 23*(32), 10265–10273.

Bohn, L. M., Lefkowitz, R. J., Gainetdinov, R. R., Peppel, K., Caron, M. G., & Lin, F. T. (1999). Enhanced morphine analgesia in mice lacking beta-arrestin 2. *Science, 286*, 2495–2498.

Bond, C., LaForge, K. S., Tian, M., Melia, D., Zhang, S., Borg, L., ... Yu, L. (1998). Single-nucleotide polymorphism in the human mu opioid receptor gene alters beta-endorphin binding and activity: possible implications for opiate addiction. *Proceedings of the National Academy of Science of the United States of America, 95*(16), 9608–9613.

Chen, Y. T., Tsou, H. H., Kuo, H. W., Fang, C. P., Wang, S. C., Ho, I. K., ... Liu, Y. L. (2013). *OPRM1* genetic polymorphisms are associated with the plasma nicotine metabolite cotinine concentration in methadone maintenance patients: a cross sectional study. *Journal of Human Genetics, 58*(2), 84–90. http://dx.doi.org/10.1038/jhg.2012.139.

Conti, D. V., Lee, W., Li, D., Liu, J., Berg, D. V. D., Thomas, P. D., ... Lerman, C. (2008). Nicotinic acetylcholine receptor beta2 subunit gene implicated in a systems-based candidate gene study of smoking cessation. *Human Molecular Genetics, 17*(18), 2834–2848.

Correll, J. A., Noel, D. M., Sheppard, A. B., Thompson, K. N., Li, Y., Yin, D., & Brown, R. W. (2009). Nicotine sensitization and analysis of brain-derived neurotrophic factor in adolescent beta-arrestin-2 knockout mice. *Synapse, 63*(6), 510–519.

Daher, M., Costa, F. M., & Neves, F. A. (2013). Genotyping the mu-opioid receptor A118G polymorphism using the real-time amplification refractory mutation system: allele frequency distribution among Brazilians. *Pain Practice: The Official Journal of the World Institute of Pain, 13*(8), 614–620.

Domino, E. F., Evans, C. L., Ni, L., Guthrie, S. K., Koeppe, R. A., & Zubieta, J. K. (2012). Tobacco smoking produces greater striatal dopamine release in G-allele carriers with mu opioid receptor A118G polymorphism. *Progress in Neuropsychopharmacology & Biological Psychiatry, 38*(2), 236–240.

Fang, J., Wang, X. H., & He, B. (2014). Association between common genetic variants in the opioid pathway and smoking behaviors in Chinese men. *Behavioral and Brain Functions, 10*, 2.

Galeote, L., Kieffer, B. L., Maldonado, R., & Berrendo, F. (2006). Mu-opioid receptors are involved in the tolerance to nicotine antinociception. *Journal of Neurochemistry, 97*, 416–423.

Guang, W., Wang, H., Su, T., Weinstein, I. B., & Wang, J. B. (2004). Role of mPKCI, a novel mu-opioid receptor interactive protein, in receptor desensitization, phosphorylation, and morphine-induced analgesia. *Molecular Pharmacology, 66*, 1285–1292.

Hardin, J., He, Y., Javitz, H. S., Wessel, J., Krasnow, R. E., Tildesley, E., ... Bergen, A. W. (2009). Nicotine withdrawal sensitivity, linkage to chr6q26, and association of *OPRM1* SNPs in the SMOking in FAMilies (SMOFAM) sample. *Cancer Epidemiology, Biomarkers & Prevention, 18*(12), 3399–3406.

Hirasawa-Fujita, M., Bly, M. J., Ellingrod, V. L., Dalack, G. W., & Domino, E. F. (2014). Genetic variation of the mu opioid receptor (*OPRM1*) and dopamine D2 receptor (*DRD2*) is related to smoking differences in patients with schizophrenia but not bipolar disorder. *Clinical Schizophrenia & Related Psychoses, 20*, 1–27.

Hoehe, M. R., Köpke, K., Wendel, B., Rohde, K., Flachmeier, C., Kidd, K. K., ... Church, G. M. (2000). Sequence variability and candidate gene analysis in complex disease: association of mu opioid receptor gene variation with substance dependence. *Human Molecular Genetics, 9*(19), 2895–2908.

Jackson, K. J., Chen, Q., Chen, J., Aggen, S. H., Kendler, K. S., & Chen, X. (2011). Association of the histidine-triad nucleotide-binding protein-1 (*HINT1*) gene variants with nicotine dependence. *The Pharmacogenomics Journal, 11*(4), 251–257.

Jackson, K. J., Wang, J. B., Barbier, E., Chen, X., & Damaj, M. I. (2012). Acute behavioral effects of nicotine in male and female *HINT1* knock-out mice. *Genes, Brain, and Behavior, 11*, 993–1000.

Jackson, K. J., Wang, J. B., Barbier, E., Damaj, M. I., & Chen, X. (2013). The histidine triad nucleotide binding 1 protein is involved in nicotine reward and physical nicotine withdrawal in mice. *Neuroscience Letters, 550*, 129–133.

Kleinjan, M., Poelen, E. A., Engels, R. C., & Verhagen, M. (2013). Dual growth of adolescent smoking and drinking: evidence for an interaction between the mu-opioid receptor (*OPRM1*) A118G polymorphism and sex. *Addiction Biology, 18*(6), 1003–1012.

Kroslak, T., Laforge, K. S., Gianotti, R. J., Ho, A., Nielsen, D. A., & Kreek, M. J. (2007). The single nucleotide polymorphism A118G alters functional properties of the human mu opioid receptor. *Journal of Neurochemistry, 103*(1), 77–87.

LaForge, K. S., Yuferov, V., & Kreek, M. J. (2000). Opioid receptor and peptide gene polymorphisms: potential implications for addictions. *European Journal of Pharmacology, 410*(2–3), 249–268.

Lerman, C., Wileyto, E. P., Patterson, F., Rukstalis, M., Audrain-McGovern, J., Restine, S., ... Berrettini, W. H. (2004). The functional mu opioid receptor (*OPRM1*) Asn40Asp variant predicts short-term response to nicotine replacement therapy in a clinical trial. *The Pharmacogenomics Journal, 4*(3), 184–192.

Liu, X., & Jernigan, C. (2011). Activation of the opioid μ1, but not δ or κ, receptors is required for nicotine reinforcement in a rat model of drug self-administration. *Progress in Neuropsychopharmacology & Biological Psychiatry, 35*(1), 146–153.

Mague, S. D., Isiegas, C., Huang, P., Liu-Chen, L. Y., Lerman, C., & Blendy, J. A. (2009). Mouse model of *OPRM1* (A118G) polymorphism has sex-specific effects on drug-mediated behavior. *Proceedings of the National Academy of Science of the United States of America, 106*(26), 10847–10852.

Munafò, M. R., Elliot, K. M., Murphy, M. F., Walton, R. T., & Johnstone, E. C. (2007). Association of the mu-opioid receptor gene with smoking cessation. *The Pharmacogenomics Journal, 7*(5), 353–361.

Munafò, M. R., Johnstone, E. C., Aveyard, P., & Marteau, T. (2013). Lack of association of *OPRM1* genotype and smoking cessation. *Nicotine & Tobacco Research, 15*(3), 739–744.

Perkins, K. A., Lerman, C., Coddington, S., Jetton, C., Karelitz, J. L., Wilson, A., ... Benowitz, N. L. (2008). Gene and gene by sex associations with initial sensitivity to nicotine in nonsmokers. *Behavioural Pharmacology, 19*(5–6), 630–640.

Perkins, K. A., Lerman, C., Grottenthaler, A., Ciccocioppo, M. M., Milanak, M., Conklin, C. A., ... Benowitz, N. L. (2008). Dopamine and opioid gene variants are associated with increased smoking reward and reinforcement owing to negative mood. *Behavioural Pharmacology, 19*(5–6), 641–649.

Ray, R., Jepson, C., Patterson, F., Strasser, A., Rukstalis, M., Perkins, K., … Lerman, C. (2006). Association of *OPRM1* A118G variant with the relative reinforcing value of nicotine. *Psychopharmacology (Berlin), 188*(3), 355–363.

Ray, R., Jepson, C., Wileyto, E. P., Dahl, J. P., Patterson, F., Rukstalis, M., … Lerman, C. (2007). Genetic variation in mu-opioid-receptor-interacting proteins and smoking cessation in a nicotine replacement therapy trial. *Nicotine & Tobacco Research, 9*, 1237–1241.

Ray, R., Ruparel, K., Newberg, A., Wileytoa, E. P., Lougheadb, J. W., Divgi, C., … Lerman, C. (2011). Human mu opioid receptor (*OPRM1* A118G) polymorphism is associated with brain mu-opioid receptor binding potential in smokers. *Proceedings of the National Academy of Science of the United States of America, 108*(22), 9268–9273.

Roth-Deri, I., Green-Sadan, T., & Yadid, G. (2008). β-Endorphin and drug-induced reward and reinforcement. *Progress in Neurobiology, 86*(1), 1–21.

Rukstalis, M., Jepson, C., Strasser, A., Lynch, K. G., Perkins, K., Patterson, F., & Lerman, C. (2005). Naltrexone reduces the relative reinforcing value of nicotine in a cigarette smoking choice paradigm. *Psychopharmacology, 180*, 41–48.

Schuck, K., Otten, R., Engels, R. C., & Kleinjan, M. (2014). Initial responses to the first dose of nicotine in novel smokers: the role of exposure to environmental smoking and genetic predisposition. *Psychology & Health, 29*(6), 698–716.

Sullivan, P. F., Neale, B. M., Van Den Oord, E., Miles, M. F., Neale, M. C., Bulik, C. M., … Kendler, K. S. (2004). Candidate genes for nicotine dependence via linkage, epistasis, and bioinformatics. *American Journal of Medical Genetics, 126B*, 23–36.

Sun, D., Ma, J. Z., Payne, T. J., & Li, M. D. (2008). Beta-arrestins 1 and 2 are associated with nicotine dependence in European American smokers. *Molecular Psychiatry, 13*, 398–406.

Trigo, J. M., Zimmer, A., & Maldonado, R. (2009). Nicotine anxiogenic and rewarding effects are decreased in mice lacking beta-endorphin. *Neuropharmacology, 56*(8), 1147–1153.

Verde, Z., Santiago, C., Rodríguez González-Moro, J. M., de Lucas Ramos, P., López Martín, S., Bandrés, F., … Gómez-Gallego, F. (2011). 'Smoking genes': a genetic association study. *PLoS One, 6*(10), e26668. http://dx.doi.org/10.1371/journal.pone.0026668.

Vink, J. M., Smit, A. B., de Geus, E. J., Sullivan, P., Willemsen, G., Hottenga, J., … Boomsma, D. I. (2009). Genome-wide association study of smoking initiation and current smoking. *American Journal of Human Genetics, 84*(3), 367–379.

Wang, Z., Ray, R., Faith, M., Tang, K., Wileyto, E. P., Detre, J. A., & Lerman, C. (2008). Nicotine abstinence-induced cerebral blood flow changes by genotype. *Neuroscience Letters, 438*(3), 275–280.

Wilkinson, A. V., Bondy, M. L., Wu, X., Wang, J., Dong, Q., D'Amelio, A. M., Jr., … Spitz, M. R. (2012). Cigarette experimentation in Mexican origin youth: psychosocial and genetic determinants. *Cancer Epidemiology, Biomarkers & Prevention : A Publication of the American Association for Cancer Research, Cosponsored by the American Society of Preventive Oncology, 21*(1), 228–238.

Yoo, J. H., Lee, S. Y., Loh, H. H., Ho, I. K., & Jang, C. G. (2004). Altered emotional behaviors and the expression of 5-HT1A and M1 muscarinic receptors in micro-opioid receptor knockout mice. *Synapse, 54*, 72–82.

Zhang, L., Kendler, K. S., & Chen, X. (2006). The mu-opioid receptor gene and smoking initiation and nicotine dependence. *Behavioral and Brain Functions, 2*, 28.

Zhang, Y., Wang, D., Johnson, A. D., Papp, A. C., & Sadée, W. (2005). Allelic expression imbalance of human mu opioid receptor (*OPRM1*) Caused by variant A118G. *The Journal of Biological Chemistry, 280*, 32618–32624.

Section C

Structural and Functional Aspects

Chapter 22

Characterization of Cue-Induced Reinstatement of Nicotine-Seeking Behavior in Smoking Relapse: Use of Animal Models

Xiu Liu
Department of Pathology, University of Mississippi Medical Center, Jackson, MS, USA

Abbreviations

DHβE Dihydro-β-erythroidine
FR Fixed-ratio
GABA γ-Aminobutyric acid
mGluR Metabotropic glutamate receptor
nAChR Nicotinic acetylcholine receptor
NMDAR *N*-Methyl-D-aspartate receptor

INTRODUCTION

Tobacco smoking and nicotine addiction, similar to addiction to other drugs of abuse, is a chronic relapsing disorder. Most smokers (up to 95–97%) who try to quit smoking without medical aids relapse each year. There are several pharmacological treatments for smoking cessation available on the market, i.e., nicotine replacement, bupropion, and varenicline. These medications have been approved by the US Food and Drug Administration. However, the long-term abstinence rates, even when individuals are on these medications, still remain unsatisfactorily low. The 1-year abstinence rates are ≤16.1% for bupropion, ≤20.3% for nicotine replacement, and ≤26.1% for varenicline. The high recidivism of smoking in abstinent smokers presents a formidable challenge for the long-term success of smoking cessation treatment.

One significant factor thought to be important in relapse to drug taking, including smoking, is exposure to environmental stimuli previously associated with drug intake. Conceivably, smoking behavior involves more frequent pairings between environmental stimuli and nicotine intake (cigarette puffs, approximately 70,000 times each year) than any other drug-taking behavior. As such, cigarette smoking may be particularly effective in establishing the incentive properties of nicotine-associated environmental stimuli (cues), such as the smell and taste of cigarettes or contexts within which smoking occurs (Caggiula et al., 2001; Goldberg, Spealman, & Goldberg, 1981). Clinical studies have demonstrated that smoking cues produce physiological responses, enhance the desire to smoke, and increase the rate, intensity, and time of smoking. Smoking denicotinized cigarettes (i.e., cue alone) produces an equal amount of smoke intake and similar levels of satisfaction compared to nicotine-containing cigarettes (i.e., cue plus nicotine) (Butschky, Bailey, Henningfield, & Pickworth, 1995; Gross, Lee, & Stitzer, 1997; Rose, Behm, Westman, & Johnson, 2000).

Since about 2005, animal research from our own and other laboratories has demonstrated the conditioned behavior motivational effects of nicotine-associated cues using the response reinstatement model of relapse. Moreover, these preclinical studies have highlighted an array of biological signaling pathways that are responsible for the mediation of the cue-induced reinstatement of nicotine-seeking behavior and provide insights into the mechanisms that underlie the conditioned incentive properties of nicotine cues.

This chapter presents a brief overview on the experimental procedures used to test reinstatement of nicotine-seeking behavior and characteristics of cue-induced nicotine seeking as well as interactions of cue exposure with stress and nicotine priming, the other two major risk factors of smoking relapse. The neuropharmacological substrates responsible for the cue-induced relapse to nicotine seeking are also summarized.

RAT MODEL OF SMOKING RELAPSE: RESPONSE REINSTATEMENT PARADIGM

Male Sprague-Dawley rats had ad libitum access to water but limited laboratory chow availability (20 g per day). The animals remained on a reversed 12/12-h light/dark cycle with all experimental sessions being conducted during the dark phase. To facilitate learning of operant responding for nicotine self-administration, the rats were trained to press a lever for food reinforcement in standard operant conditioning chambers. The rats were implanted with an indwelling intravenous catheter under isoflurane anesthesia.

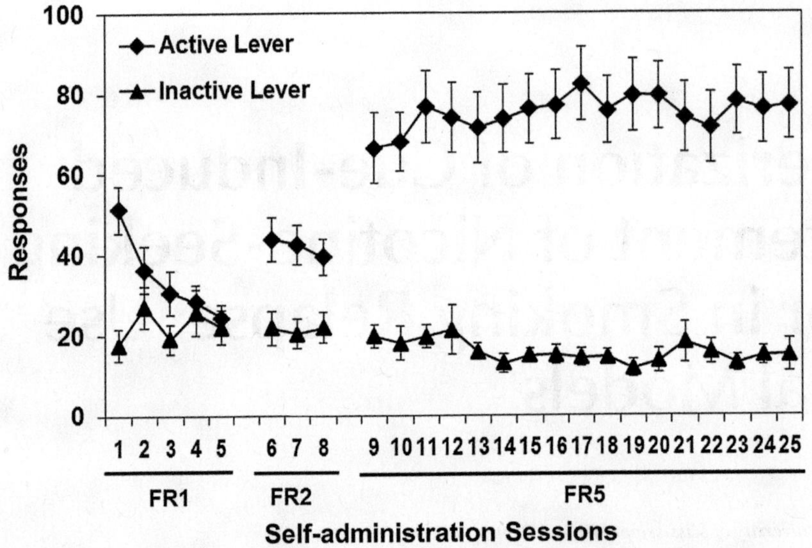

FIGURE 1 A representative behavior profile of intravenous nicotine self-administration in rats. Responses on the active lever were reinforced by nicotine infusion and its associated cue presentation, whereas responses on the inactive lever had no programmed consequences. The number of responses is presented as the mean ± SEM.

The daily 1-h nicotine self-administration sessions were initiated by introduction of the two levers with illumination of the red chamber light. Once a fixed-ratio (FR) schedule of reinforcement on the active lever was met, an infusion of nicotine was dispensed by the drug-delivery system in a volume of 0.1 ml in approximately 1 s, depending on the animal's body weight. Each nicotine infusion was signaled by or associated with presentation of an auditory/visual stimulus consisting of a 5-s tone and illumination of a light above the active lever for 20 s. As such, the stimulus acquired a conditioned incentive value and became a nicotine cue. Following nicotine infusions, there was a 20-s time-out period during which responses were recorded but not reinforced. Responses on the inactive lever had no consequence. An FR1 schedule was used for days 1–5, an FR2 for days 6–8, and an FR5 for the remainder of the experiments. The acquisition and development of a stable level of nicotine self-administration are shown in Figure 1.

After completion of the self-administration and conditioning phase, the nicotine-reinforced responses were extinguished by withholding nicotine and its associated cue. Specifically, the daily 1-h extinction sessions began with introduction of the two levers and illumination of the red chamber light and responses on the active lever resulted in the delivery of saline rather than nicotine and no cue presented. The FR5 schedule and the 20-s time-out period were still in effect for saline infusions. As shown in Figure 2, the criterion for extinction was that for 3 consecutive days the number of responses per session decreased to less than 20% of the number of responses per session that occurred during the last 3 days of the nicotine self-administration and conditioning phase.

The reinstatement test sessions were performed once the rats reached the extinction criterion. As happened in the self-administration/conditioning and extinction phases, the test sessions started with introduction of the two levers and illumination of the red chamber light. During the test sessions, responses on the active lever resulted in presentation of the cue and saline infusion on the FR5 schedule with the 20-s time-out period. As such, there was

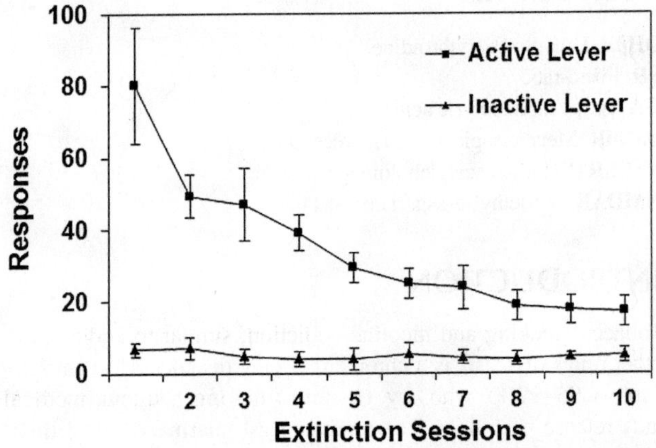

FIGURE 2 A representative behavior profile of extinguishing the previously nicotine-reinforced lever-pressing responses in rats. Responses on the active lever resulted in saline (instead of nicotine) infusion and no nicotine cue presented. Responses on the inactive lever had no programmed consequences. The number of responses is presented as the mean ± SEM.

still no availability of nicotine. Responses on the inactive lever had no consequence. The test sessions lasted 1 h.

CUE-INDUCED REINSTATEMENT OF NICOTINE-SEEKING RESPONSES

In the reinstatement test sessions, response-contingent re-presentation of the cue significantly reinstated the extinguished responding on the active lever previously reinforced by nicotine (Figure 3). Since responses on the inactive lever remained at the extinction level, the response-reinstating effect of the cue was not attributable to any nonspecific arousal and/or locomotor activation as a result of reexposure to the cue. Thus, the reinstatement of responses on the active lever after extinction was selectively

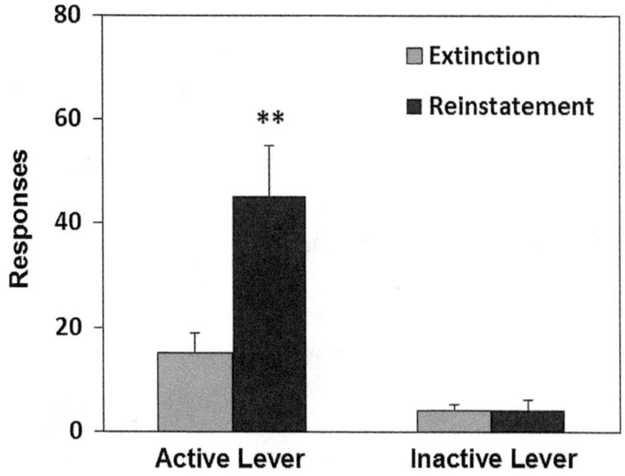

FIGURE 3 Cue-induced reinstatement of nicotine-seeking responses in rats. In the reinstatement test session performed 1 day after extinction, responses on the active lever resulted in re-presentations of the cue but without availability of nicotine. Responses on the inactive lever had no programmed consequences. The number of responses is presented as the mean ± SEM. **$p < 0.01$, significantly different from extinction.

controlled by the response-contingent re-presentation of the cue that had been associated with delivery and reinforcing actions of nicotine during self-administration and conditioning training. This finding is consistent with studies generated from other laboratories (Cohen, Perrault, Griebel, & Soubrie, 2005; LeSage, Burroughs, Dufek, Keyler, & Pentel, 2004; Paterson, Froestl, & Markou, 2005). These findings were supportive of an earlier report showing that exposing rats to the self-administration training chambers after 21 drug-free days, during which time the rats remained in home cages, resulted in recovery of lever-pressing (Shaham, Adamson, Grocki, & Corrigall, 1997).

Because rats were first trained to learn lever-pressing responses by using a food reward, it is speculated whether lever experience with food reinforcement during prior food training produces a response bias, which is then nonspecifically sustained by nicotine and can account for the subsequent nicotine self-administration and reinstatement of nicotine-seeking behavior induced by the cue. To address this issue, we used an active lever switch procedure in which the active and inactive levers were switched after completion of food training (Liu et al., 2006). That is, during nicotine self-administration sessions the previous food-reinforced, active lever was made inactive, while the inactive lever in the food-training phase was made active, at which responses were reinforced by nicotine infusion. The results showed similar behavioral profiles of nicotine self-administration and cue-induced reinstatement of nicotine-seeking responses under the two active-lever-assignment conditions. This demonstrates development of a stable level of nicotine self-administration and establishment of conditioned incentive properties of a nicotine cue via its repeated association with nicotine infusions regardless of whether the prior lever experience had been rewarded by delivery of food pellets. Therefore, with the ease of establishing stable nicotine self-administration, the prior food-training procedure seems to be of the choice of levers.

The testing procedure involves a sensory stimulus that was omitted during extinction after the self-administration/conditioning training but re-presented in the reinstatement tests. In light of the fact that animal studies have shown that some sensory stimuli have intrinsic reinforcing properties and thereby support moderate levels of operant responding such as lever-pressing (Cohen et al., 2005; Donny et al., 2003; Liu, Palmatier, Caggiula, Donny, & Sved, 2007), it is reasonable to speculate that the increased responses with response-contingent re-presentation of the cue in the reinstatement test sessions might be attributable to the possible reinforcing value of the cue stimulus regardless of its association with nicotine; this might then be misinterpreted as response reinstatement induced by the conditioned incentive of the stimulus. To address this issue, in another study (Liu, Caggiula, Palmatier, Donny, & Sved, 2008) we included a control group that received procedures exactly the same as other experimental groups, except that saline rather than nicotine was available during the self-administration and conditioning training phase. Since the stimulus, albeit the same as in other groups, was never associated with nicotine delivery and the subjective actions of nicotine, it did not acquire a conditioned reinforcement value. Responses on the active lever in this control group remained low and constant across the self-administration/conditioning, extinction, and reinstatement test phases. This finding indicates that the specific stimulus used in the procedure does not have intrinsic reinforcing value. Particularly convincing is the observation that removal of the stimulus during extinction in this saline control group did not influence the level of responding. Therefore, the recovery of extinguished responses of the nicotine groups during the reinstatement test sessions resulted from the conditioned incentive value of the stimulus due to its prior repeated association with nicotine infusion and pharmacological actions.

The conditioned incentive value of the nicotine cue was evidenced to be long-lasting. In a study (Liu et al., 2008), the effect of nicotine cue was tested in second and third tests, which were conducted 15 and 30 days after the first test, i.e., 26 and 41 days after completion of nicotine self-administration and conditioning training. Response-contingent re-presentation of the cue still effectively elicited reinstatement of nicotine-seeking responses (Figure 4). Although it was reported that intervening days after extinction would give rise to spontaneous recovery of responses for nicotine (Shaham et al., 1997), increased lever responses during the repeated reinstatement tests were not readily attributable to this phenomenon because responses were reextinguished immediately before each of these tests. Therefore, reinstatement was elicited specifically by re-presentation of the nicotine-conditioned cue. The persistence of cue-induced reinstatement of nicotine-seeking behavior is consistent with a previous report (Cohen et al., 2005) showing that a similar compound auditory/visual cue associated with nicotine self-administration was found to reinstate nicotine-seeking responses after 25 daily extinction sessions. This finding also stays in line with observations showing a long-lasting motivational effect of cues previously associated with other drugs of abuse such as cocaine, heroin, and alcohol (e.g., Di Ciano & Everitt, 2002; Grimm, Hope, Wise, & Shaham, 2001; Weiss et al., 2001).

In light of the fact that the reinstatement tests dissociate the motivation to engage in nicotine-seeking behavior from the direct reinforcing properties of nicotine, the response-reinstatement paradigm may be particularly useful for understanding the factors involved in smoking relapse and in the neurobiological mechanisms involved in nicotine-seeking behavior associated with exposure to environmental cues related to tobacco smoking.

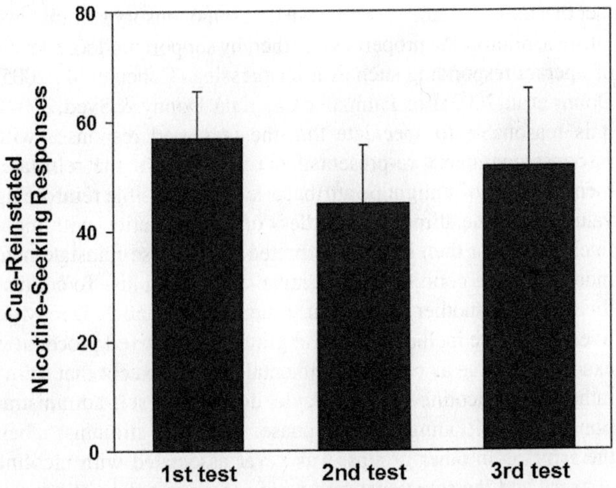

FIGURE 4 Persistence of cue-induced nicotine-seeking behavior across repeated reinstatement tests in rats. The first reinstatement test was conducted the next day after 10 daily extinction sessions. The second test occurred after an intervening 15 days and reextinction, i.e., 25 days after self-administration and conditioning. The third test was performed 15 days after the second test, i.e., 40 days after self-administration and conditioning. The number of responses is presented as the mean ± SEM.

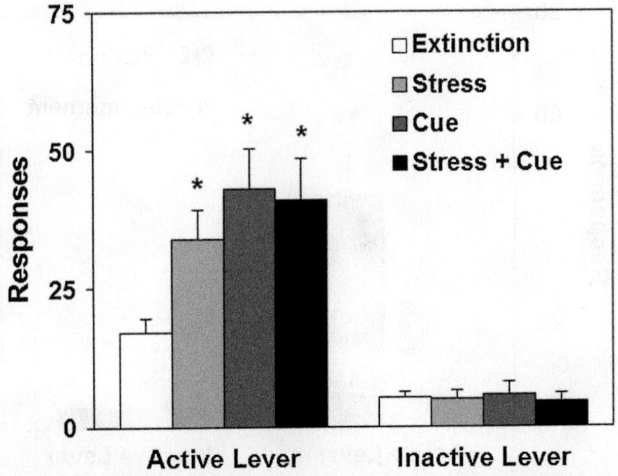

FIGURE 5 Interaction of nicotine cue with stress challenge in reinstating nicotine-seeking behavior in rats. To test the effect of stress, rats received a presession intraperitoneal administration of yohimbine (2 mg/kg), and during the session there was no cue presentation. The effect of cue exposure was tested as described for Figure 3. To test the effect of the combination of stress with cue exposure, the test session was performed with presession yohimbine injection and with response-contingent re-presentation of the cue during the session. The number of responses is presented as the mean ± SEM. *$p < 0.05$, significantly different from extinction. *Reproduced from Liu (2010), with permission of Nova Science Publishers, Inc.*

INTERACTIONS OF NICOTINE CUE WITH STRESS AND NICOTINE PRIMING

In animal models testing the effects of stress exposure on drug-seeking behavior, the pharmacological stress produced by administration of yohimbine has increasingly been employed. Yohimbine, via blockade of the presynaptic α2 adrenergic receptors, increases the activity of noradrenergic systems, including neural structures implicated in stress response, and thereby produces anxiety- and stress-like states in humans and laboratory animals (e.g., Bremner, Krystal, Southwick, & Charney, 1996). The long half-life (7–8 h) of yohimbine has ensured a prolonged stress state throughout the whole reinstatement test session. That could guarantee the overlap between stress challenge and exposure to nicotine cue during the reinstatement test sessions. As shown in Figure 5, after lever-responding was extinguished, an administration of yohimbine prior to the reinstatement test session effectively reinstated nicotine-seeking responses. The unchanged responses on the inactive lever indicate a specific behavioral motivation effect of this stressor on nicotine-seeking responses. This finding is consistent with two previous reports that employed foot shock as a stressor (Buczek, Le, Stewart, & Shaham, 1999; Zislis, Desai, Prado, Shah, & Bruijnzeel, 2007). However, the combination of the presession administration of yohimbine and the response-contingent re-presentations of nicotine cue during the test session did not produce an enhanced motivational effect. The magnitude of reinstatement of nicotine-seeking responses under the stress and cue combination condition was similar to that elicited by nicotine cue alone (Figure 5). This observation is in contrast to previous reports obtained with other drugs of abuse. For instance, we demonstrated that exposure to foot-shock stress and an ethanol cue interacted to produce an additive effect in reinstating ethanol-seeking responses (Liu & Weiss, 2002). Similarly, it has been documented that the stress produced by presession administration of yohimbine, as used in our nicotine study presented above, and a cocaine cue produces an additive effect in reinstating cocaine-seeking response (Feltenstein & See, 2006). At present, we are not able to provide a full explanation for the differences in the interactive nature of stress and cue across nicotine and other drugs of abuse although it may reside in the different categories of the drugs of abuse tested across these experiments. In this respect, it is necessary for future work to elucidate the biological mechanisms underlying this discrepancy between nicotine and ethanol or cocaine.

In human addicts, a lapse or slip to drug use is a good predictor of a full-blown relapse to drug abuse including cigarette smoking (e.g., Marlatt, Curry, & Gordon, 1988). As shown in Figure 6, a subcutaneous administration of nicotine at 0.25 mg/kg prior to the reinstatement test session effectively reinstated extinguished nicotine-seeking responses, while responses on the inactive lever remained unchanged, indicating a specific behavioral motivation effect of nicotine priming. This finding is consistent with several previous studies reporting the response-reinstating effect of nicotine priming in the response-reinstatement paradigm in rats (e.g., Chiamulera, Borgo, Falchetto, Valerio, & Tessari, 1996). In these animal studies, the effective doses of nicotine to prime nicotine-seeking responses varied considerably. It has been argued that in the priming tests nicotine acts as an occasion setter or discriminative stimulus as long as the injected doses are above a threshold that produces pharmacological actions. This notion could be supported from our observation showing that self-administered nicotine across a dose range of 0.015–0.06 mg/kg/infusion could endow its associated stimulus with a similar strength compared to the conditioned incentive value in rats (Liu et al., 2008). However, nicotine priming injection, like the yohimbine stress challenge described above, did not interact with nicotine cue presentation

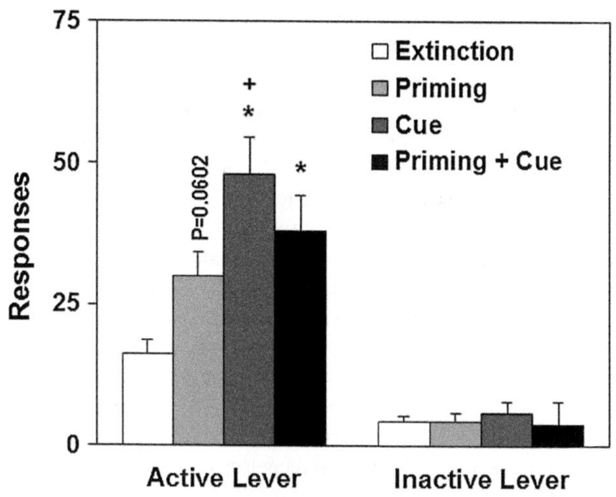

FIGURE 6 Interaction of nicotine cue with nicotine priming in reinstating nicotine-seeking behavior in rats. To test the effect of nicotine priming, rats received a presession subcutaneous administration of nicotine (0.25 mg/kg) and during the session there was no cue presentation. The effect of cue exposure was tested as described for Figure 3. To test the effect of the combination of nicotine priming with cue exposure, the session was performed with presession nicotine priming injection and with response-contingent re-presentation of the cue during the session. The number of responses is presented as the mean ± SEM. *$p < 0.05$, significantly different from extinction; +$p < 0.05$, significantly different from priming. *Reproduced from Liu (2010), with permission of Nova Science Publishers, Inc.*

to produce any additive effect. The magnitude of reinstatement of nicotine-seeking responses under combined exposure to the presession nicotine priming and in-session response-contingent re-presentations of nicotine cue was similar to that elicited by nicotine cue alone (Figure 6).

In summary, these animal studies provide experimental evidence to support the notion that reexposure to smoking-related environmental cues plays a critical role in the high recidivism rates of smoking in abstinent smokers. In fact, compared to other drugs of abuse, environmental stimuli associated with nicotine intake in smokers play a more important role in maintaining nicotine self-administration since tobacco smoking is particularly effective at establishing the incentive properties of accompanying environmental stimuli because smoking behavior involves more frequent pairings between environmental cues and nicotine intake (cigarette puffs, approximately 70,000 times each year) than any other drug-taking behavior. Conceivably, among the three major risk factors for drug relapse, exposure to smoking cues is expected to produce the most robust behavioral motivation effect. Therefore, our results may lend support for the continued effort on cue management as a strategy for the treatment and prevention of smoking relapse.

PHARMACOLOGICAL SUBSTRATES RESPONSIBLE FOR CUE-INDUCED NICOTINE-SEEKING BEHAVIOR

Nicotine exerts its reinforcing actions by activating nicotinic acetylcholine receptors (nAChRs). Among many nAChR subtypes, heteromeric α4β2- and homomeric α7-containing receptors are the most abundant and widespread, comprising more than 90% of the nAChRs in the brain (Albuquerque, Pereira, Alkondon, & Rogers, 2009; Zoli, Lena, Picciotto, & Changeux, 1998). Accumulating studies have established a pivotal role for α4β2 nAChRs in the mediation of the primary reinforcing actions of nicotine (e.g., Mineur & Picciotto, 2008; O'Connor, Parker, Rollema, & Mead, 2010; Watkins, Epping-Jordan, Koob, & Markou, 1999). In contrast, research on α7 nAChRs has been inconclusive, since the majority of the published work has demonstrated that manipulation of α7 nAChRs does not alter the reinforcing actions of nicotine (e.g., Grottick et al., 2000; Pons et al., 2008; Walters, Brown, Changeux, Martin, & Damaj, 2006), with only a few other studies reporting the involvement of α7 nAChRs in nicotine reinforcement (Besson et al., 2012; Brunzell & McIntosh, 2012; Markou & Paterson, 2001). Using the response-reinstatement paradigm, we have demonstrated that mecamylamine, a nonselective nAChR antagonist, effectively reversed the cue-induced reinstatement of nicotine-seeking behavior (Liu, Caggiula, et al., 2007). That has extended the role of nicotinic neurotransmission in the mediation of the primary reinforcing actions of nicotine to the conditioned (secondary) motivational properties of nicotine-associated cues. Using receptor subtype-specific antagonists, we have further demonstrated that α4β2 and α7 subtypes of the nAChRs may play differential roles in nicotine-induced reinforcement and the conditioned reinforcement induced by nicotine cues. Specifically, the α4β2-nAChR-selective antagonist dihydro-β-erythroidine (DHβE) significantly suppressed nicotine self-administration but did not interfere with the conditioned reinforcement induced by nicotine cues, whereas the α7-nAChR-selective antagonist methyllycaconitine effectively attenuated the cue-induced reinstatement of nicotine-seeking responses (Liu, 2014). Interestingly, Li, Li, Pei, Le, and Liu (2012) found that interruption of a complex formation between α7 nAChRs and the ionotropic glutamate *N*-methyl-D-aspartate receptors (NMDARs) blocked cue-induced reinstatement of nicotine-seeking responses.

The finding that DHβE did not change conditioned incentive motivation by nicotine cues was consistent with a study showing that the α4β2 nAChR partial agonist varenicline suppressed nicotine self-administration and the reinstatement of nicotine seeking induced by nicotine priming and the combination of nicotine and its cue, but did not affect reinstatement induced by the nicotine cue alone (O'Connor et al., 2010). The dissociation in the neuropharmacological substrates responsible for primary and secondary (conditioned) reinforcement of nicotine was also revealed at the opioidergic neurotransmission level. Our previous study showed that nonselective blockade of opioid receptors by naltrexone attenuated the cue-induced reinstatement of nicotine seeking but had no effect on nicotine self-administration (Liu et al., 2009). A similar dissociation was found with other drugs of abuse and signaling pathways. For example, a gaseous neurotransmitter nitric oxide synthase inhibitor (N^G-nitro-L-arginine methyl ester) attenuated the cue-induced reinstatement of ethanol seeking but failed to alter the self-administration of ethanol, the primary reinforcing action (Liu & Weiss, 2004). Similarly, antagonism of orphan σ1 receptors attenuated the conditioned incentive effects of a cocaine cue but did not interfere with cocaine primary reinforcement (Martin-Fardon, Maurice, Aujla, Bowen, & Weiss, 2007). Even in cases in which one drug produced effects on both conditioned and primary reinforcement, the sensitivity of the effect was different. For example, responding motivated by stimuli conditioned to cocaine was more sensitive to glutamate antagonists than behavior maintained by cocaine itself (Baptista, Martin-Fardon, & Weiss, 2004; Newman & Beardsley, 2006). Therefore, the conditioned incentive properties of nicotine cues and primary reinforcing actions of nicotine may be mediated by different neurobiological substrates.

Our research demonstrated that antagonism of dopamine D1 or D2 receptors by the selective antagonist SCH 23390 or eticlopride effectively attenuated lever-responding to the presentation of the nicotine cue (Liu et al., 2010). Although dopamine antagonists are in general prone to producing a suppressant effect on locomotor activity, especially the cataleptogenic properties of D2 blockade, their nonspecific suppression of operant behavior was ruled out since SCH 23390 and eticlopride (except at the highest dose, 30 µg/kg) failed to alter food self-administration behavior. Therefore, both SCH 23390 and eticlopride specifically attenuated cue-induced reinstatement of nicotine-seeking behavior, indicating that dopaminergic neurotransmission via D1 and D2 receptors is required for expression of the conditioned incentive properties of nicotine cues. These findings suggest that manipulation of dopaminergic D1 and/or D2 receptor activation may prove to be a potential target for the development of pharmacotherapy for smoking relapse prevention, which lends support to the continued clinical effort to test the effectiveness of antipsychotics for prevention of environmental cue-triggered smoking relapse. For instance, haloperidol has been found to reduce smoking of denicotinized cigarettes, which indicates a decreased reaction to nicotine cue exposure (Brauer, Cramblett, Paxton, & Rose, 2001), and a D2 receptor antagonist, olanzapine, attenuates cue-induced craving for cigarette smoking (Hutchison et al., 2004; Rohsenow et al., 2008). From a clinical perspective, however, a caveat should be noted that careful selection of effective doses of the D2 antagonists seems to be critical because these agents are prone to producing extrapyramidal side effects at higher doses.

Another line of our studies has demonstrated that pretreatment with an opioid receptor antagonist, naltrexone, prior to the reinstatement tests significantly attenuated the cue-induced reinstatement of nicotine-seeking responses (Liu et al., 2009). In addition, in the extinction tests performed in separate groups of rats, naltrexone significantly suppressed the cue-maintained responses because responses on the active lever resulted in only the cue presentation and not nicotine infusion (saline substitution). These results indicate that activation of opioid receptors may play a role in the mediation of conditioned incentive properties of nicotine cues, which underlies the cue-induced reinstatement of nicotine seeking in animals and relapse of smoking behavior in humans. This conclusion is supported by an observation showing that naloxone, another opioid receptor antagonist, blocked expression of nicotine-induced conditioned place preference (Walters, Cleck, Kuo, & Blendy, 2005). It is also in line with clinical studies showing that naltrexone decreased smoking cue-induced urge to smoke (Hutchison et al., 1999; King & Meyer, 2000; Lee et al., 2005). An increasing number of animal studies have shown that blockade of opioid neurotransmission attenuates drug cue-induced reinstatement of operant responding for previously self-administered drugs of abuse, including opiates (e.g., Shaham & Stewart, 1996), alcohol (e.g., Liu & Weiss, 2002), methamphetamine (Anggadiredja, Sakimura, Hiranita, & Yamamoto, 2004), and cocaine (Burattini, Burbassi, Aicardi, & Cervo, 2008). Together, it is suggested that activation of the opioid receptors may to some extent play a general role in the expression of the conditioned incentive properties of environmental stimuli previously associated with drug taking and subjective effects of the drugs. Therefore, it is proposed that naltrexone might have a broad clinical potential for prevention of relapse to drug use, including cigarette smoking, which is associated with exposure to environmental drug cues.

Since 2010, an increasing number of animal studies from other laboratories have recruited other neuropharmacological substrates responsible for expression of the behavioral motivational effect of nicotine cues. For example, the cue-induced reinstatement of nicotine-seeking behavior has been found to be sensitive to antagonists selective for D3 receptors (Khaled et al., 2010), noradrenergic α1 (Forget et al., 2010) and β receptors (Chiamulera, Tedesco, Zangrandi, Giuliano, & Fumagalli, 2010), cannabinoid CB1 receptors (Cohen et al., 2005; De Vries, de Vries, Janssen, & Schoffelmeer, 2005; Shoaib, 2008), metabotropic glutamate receptor (mGluR) 1 (Dravolina et al., 2007) and mGluR5 (Bespalov et al., 2005), ionotropic glutamate NMDARs (Pechnick et al., 2011), and T-type Ca^{2+} channels (Uslaner et al., 2010). Moreover, the behavioral motivational effect of nicotine cues was also attenuated by an mGluR2/3 agonist (Liechti, Lhuillier, Kaupmann, & Markou, 2007), $GABA_B$ receptor agonist (Paterson et al., 2005), and α-type peroxisome proliferator-activated receptor agonist (Panlilio et al., 2012). Taken together, these studies highlight an array of biological signaling pathways that are responsible for the mediation of the cue-induced reinstatement of nicotine-seeking behavior and provide insights into the mechanisms that underlie the conditioned incentive properties of nicotine cues.

APPLICATIONS TO OTHER ADDICTION AND SUBSTANCE MISUSE

The response reinstatement of drug relapse was first described in the 1980s by Stewart (1984). This paradigm has a profound utility in investigating relapse behavior across all classes of drugs of abuse. It has been increasingly employed to examine a similar reinstating behavior for natural rewards such as food and water. Being a preclinical model of relapse to drug-seeking behavior, this paradigm shows excellent face and predictive validities, although it is still far away from fully capturing all aspects of relapse to drug intake in abstinent addicts owing to some inherent shortcomings.

DEFINITION OF TERMS

Receptor antagonist This describes a ligand that, when binding to a neurotransmitter receptor, attenuates or completely blocks the neurotransmitter-mediated response, while on its own does not provoke a biological response.

Response reinstatement This is an animal model of drug relapse. The paradigm involves three phases: (1) Laboratory animals are trained to lever-press or nose-poke for drug self-administration. (2) The drug-reinforced response is extinguished by withholding delivery of the drug and its associated cue as well. (3) The reinstatement of drug-seeking response induced by response-contingent re-presentation of the cue and/or drug priming or stressors is tested.

Cue-induced reinstatement Laboratory animals are first trained to self-administer a drug. To establish a conditioned cue, a discrete stimulus (e.g., tone, light) is temporally associated with each drug delivery. Then, lever-responding is extinguished in the absence of the drug and the cue. During reinstatement tests, response-contingent reexposure to the discrete cue reinstates lever-responding.

Extinction This is a reduction or ceasing of operant behavior that has been previously reinforced by a reinforcer (e.g., drug of abuse or natural reward) due to the reinforcing consequences no longer being available.

Priming-induced reinstatement Laboratory animals are first trained to self-administer a drug and usually each drug delivery is signaled by a discrete cue. Then, lever-responding is extinguished by withholding the drug with or without omission of the cue. During reinstatement tests, noncontingent priming injections of the previously self-administered drug (or in some cases other agents) reinstates lever-responding in the presence or absence of the discrete cue.

Self-administration In animal studies, a subject, by emitting operant responses (e.g., lever-pressing or nose-poking), administers reinforcing substances such as drugs of abuse or natural rewards to itself. Delivery of the reinforcer is usually signaled by the presentation of a discrete stimulus (e.g., tone, light).

Stress-induced reinstatement Laboratory animals are first trained to self-administer a drug and usually each drug delivery is signaled by a discrete cue. Then, lever-responding is extinguished by withholding the drug with or without omission of the cue. During reinstatement tests, presession exposure to certain stressors (typically intermittent foot shock or yohimbine injection) reinstates lever-responding in the presence or absence of the discrete cue.

KEY FACTS OF CUE-INDUCED REINSTATEMENT OF NICOTINE-SEEKING BEHAVIOR

- Exposure to environmental cues is a significant risk factor for smoking relapse in abstinent smokers, contributing to the high recidivism of tobacco smoking.
- Reinstatement of nicotine-seeking behavior is an animal model of smoking relapse.
- The response reinstatement paradigm is validated for examining behavioral effects of cue exposure as well as stress challenge and drug priming.
- Exposure to cues produces the most robust response-reinstating effect, relative to stress challenge or nicotine priming.
- Cue-induced reinstatement of nicotine-seeking behavior is a long-lasting phenomenon.
- The cue effect involves an array of pharmacological substrates.

SUMMARY POINTS

- This chapter presents an overview of cue-induced nicotine relapse in an animal model.
- The animal model has shown good validity for simulating human smoking and relapse after abstinence.
- The model involves three experimental phases: self-administration and conditioning, extinction, and reinstatement test.
- Rats learn to intravenously self-administer nicotine by a pressing a lever.
- A discretely conditioned cue can be established via its repeated association with nicotine infusions.
- Response-contingent re-presentation of the cue after extinction effectively reinstates nicotine-seeking responses.
- The conditioned incentive effect of nicotine cues is the most robust relative to stress challenge or nicotine priming.
- The behavioral motivation effect of nicotine cues is a long-lasting phenomenon.
- Several pharmacological substrates have been found to underlie the conditioned incentive properties of nicotine cues.

REFERENCES

Albuquerque, E. X., Pereira, E. F., Alkondon, M., & Rogers, S. W. (2009). Mammalian nicotinic acetylcholine receptors: from structure to function. *Physiological Reviews, 89*, 73–120.

Anggadiredja, K., Sakimura, K., Hiranita, T., & Yamamoto, T. (2004). Naltrexone attenuates cue- but not drug-induced methamphetamine seeking: a possible mechanism for the dissociation of primary and secondary reward. *Brain Research, 1021*, 272–276.

Baptista, M. A., Martin-Fardon, R., & Weiss, F. (2004). Preferential effects of the metabotropic glutamate 2/3 receptor agonist LY379268 on conditioned reinstatement versus primary reinforcement: comparison between cocaine and a potent conventional reinforcer. *Journal of Neuroscience, 24*, 4723–4727.

Bespalov, A. Y., Dravolina, O. A., Sukhanov, I., Zakharova, E., Blokhina, E., Zvartau, E., ... Markou, A. (2005). Metabotropic glutamate receptor (mGluR5) antagonist MPEP attenuated cue- and schedule-induced reinstatement of nicotine self-administration behavior in rats. *Neuropharmacology, 49*(Suppl. 1), 167–178.

Besson, M., David, V., Baudonnat, M., Cazala, P., Guilloux, J. P., Reperant, C., ... Granon, S. (2012). Alpha7-nicotinic receptors modulate nicotine-induced reinforcement and extracellular dopamine outflow in the mesolimbic system in mice. *Psychopharmacology (Berl), 220*, 1–14.

Brauer, L. H., Cramblett, M. J., Paxton, D. A., & Rose, J. E. (2001). Haloperidol reduces smoking of both nicotine-containing and denicotinized cigarettes. *Psychopharmacology (Berl), 159*, 31–37.

Bremner, J. D., Krystal, J. H., Southwick, S. M., & Charney, D. S. (1996). Noradrenergic mechanisms in stress and anxiety: II. Clinical studies. *Synapse, 23*, 39–51.

Brunzell, D. H., & McIntosh, J. M. (2012). Alpha7 nicotinic acetylcholine receptors modulate motivation to self-administer nicotine: implications for smoking and schizophrenia. *Neuropsychopharmacology, 37*, 1134–1143.

Buczek, Y., Le, A. D., Stewart, J., & Shaham, Y. (1999). Stress reinstates nicotine seeking but not sucrose solution seeking in rats. *Psychopharmacology (Berl), 144*, 183–188.

Burattini, C., Burbassi, S., Aicardi, G., & Cervo, L. (2008). Effects of naltrexone on cocaine- and sucrose-seeking behaviour in response to associated stimuli in rats. *International Journal of Neuropsychopharmacology, 11*, 103–109.

Butschky, M. F., Bailey, D., Henningfield, J. E., & Pickworth, W. B. (1995). Smoking without nicotine delivery decreases withdrawal in 12-hour abstinent smokers. *Pharmacology Biochemistry and Behavior, 50*, 91–96.

Caggiula, A. R., Donny, E. C., White, A. R., Chaudhri, N., Booth, S., Gharib, M. A., ... Sved, A. F. (2001). Cue dependency of nicotine self-administration and smoking. *Pharmacology Biochemistry and Behavior, 70*, 515–530.

Chiamulera, C., Borgo, C., Falchetto, S., Valerio, E., & Tessari, M. (1996). Nicotine reinstatement of nicotine self-administration after long-term extinction. *Psychopharmacology (Berl), 127*, 102–107.

Chiamulera, C., Tedesco, V., Zangrandi, L., Giuliano, C., & Fumagalli, G. (2010). Propranolol transiently inhibits reinstatement of nicotine-seeking behaviour in rats. *Journal of Psychopharmacology, 24*, 389–395.

Cohen, C., Perrault, G., Griebel, G., & Soubrie, P. (2005). Nicotine-associated cues maintain nicotine-seeking behavior in rats several weeks after nicotine withdrawal: reversal by the cannabinoid (CB1) receptor antagonist, rimonabant (SR141716). *Neuropsychopharmacology, 30*, 145–155.

De Vries, T. J., de Vries, W., Janssen, M. C., & Schoffelmeer, A. N. (2005). Suppression of conditioned nicotine and sucrose seeking by the cannabinoid-1 receptor antagonist SR141716A. *Behavioural Brain Research, 161*, 164–168.

Di Ciano, P., & Everitt, B. J. (2002). Reinstatement and spontaneous recovery of cocaine-seeking following extinction and different durations of withdrawal. *Behavioural Pharmacology, 13*, 397–405.

Donny, E. C., Chaudhri, N., Caggiula, A. R., Evans-Martin, F. F., Booth, S., Gharib, M. A., ... Sved, A. F. (2003). Operant responding for a visual reinforcer in rats is enhanced by noncontingent nicotine: implications for nicotine self-administration and reinforcement. *Psychopharmacology (Berl), 169*, 68–76.

Dravolina, O. A., Zakharova, E. S., Shekunova, E. V., Zvartau, E. E., Danysz, W., & Bespalov, A. Y. (2007). mGlu1 receptor blockade attenuates cue- and nicotine-induced reinstatement of extinguished nicotine self-administration behavior in rats. *Neuropharmacology, 52*, 263–269.

Feltenstein, M. W., & See, R. E. (2006). Potentiation of cue-induced reinstatement of cocaine-seeking in rats by the anxiogenic drug yohimbine. *Behavioural Brain Research, 174*, 1–8.

Forget, B., Wertheim, C., Mascia, P., Pushparaj, A., Goldberg, S. R., & Le Foll, B. (2010). Noradrenergic alpha1 receptors as a novel target for the treatment of nicotine addiction. *Neuropsychopharmacology, 35*, 1751–1760.

Goldberg, S. R., Spealman, R. D., & Goldberg, D. M. (1981). Persistent behavior at high rates maintained by intravenous self-administration of nicotine. *Science, 214*, 573–575.

Grimm, J. W., Hope, B. T., Wise, R. A., & Shaham, Y. (2001). Neuroadaptation. Incubation of cocaine craving after withdrawal. *Nature, 412*, 141–142.

Gross, J., Lee, J., & Stitzer, M. L. (1997). Nicotine-containing versus denicotinized cigarettes: effects on craving and withdrawal. *Pharmacology Biochemistry and Behavior, 57*, 159–165.

Grottick, A. J., Trube, G., Corrigall, W. A., Huwyler, J., Malherbe, P., Wyler, R., & Higgins, G. A. (2000). Evidence that nicotinic alpha(7) receptors are not involved in the hyperlocomotor and rewarding effects of nicotine. *Journal of Pharmacology and Experimental Therapeutics, 294*, 1112–1119.

Hutchison, K. E., Monti, P. M., Rohsenow, D. J., Swift, R. M., Colby, S. M., Gnys, M., ... Sirota, A. D. (1999). Effects of naltrexone with nicotine replacement on smoking cue reactivity: preliminary results. *Psychopharmacology (Berl), 142*, 139–143.

Hutchison, K. E., Rutter, M. C., Niaura, R., Swift, R. M., Pickworth, W. B., & Sobik, L. (2004). Olanzapine attenuates cue-elicited craving for tobacco. *Psychopharmacology (Berl), 175*, 407–413.

Khaled, M. A., Farid Araki, K., Li, B., Coen, K. M., Marinelli, P. W., Varga, J., ... Le Foll, B. (2010). The selective dopamine D3 receptor antagonist SB 277011-A, but not the partial agonist BP 897, blocks cue-induced reinstatement of nicotine-seeking. *International Journal of Neuropsychopharmacology, 13*, 181–190.

King, A. C., & Meyer, P. J. (2000). Naltrexone alteration of acute smoking response in nicotine-dependent subjects. *Pharmacology Biochemistry and Behavior, 66*, 563–572.

Lee, Y. S., Joe, K. H., Sohn, I. K., Na, C., Kee, B. S., & Chae, S. L. (2005). Changes of smoking behavior and serum adrenocorticotropic hormone, cortisol, prolactin, and endogenous opioids levels in nicotine dependence after naltrexone treatment. *Progress in Neuro-Psychopharmacology & Biological Psychiatry, 29*, 639–647.

LeSage, M. G., Burroughs, D., Dufek, M., Keyler, D. E., & Pentel, P. R. (2004). Reinstatement of nicotine self-administration in rats by presentation of nicotine-paired stimuli, but not nicotine priming. *Pharmacology Biochemistry and Behavior, 79*, 507–513.

Li, S., Li, Z., Pei, L., Le, A. D., & Liu, F. (2012). The alpha7nACh-NMDA receptor complex is involved in cue-induced reinstatement of nicotine seeking. *Journal of Experimental Medicine, 209*, 2141–2147.

Liechti, M. E., Lhuillier, L., Kaupmann, K., & Markou, A. (2007). Metabotropic glutamate 2/3 receptors in the ventral tegmental area and the nucleus accumbens shell are involved in behaviors relating to nicotine dependence. *Journal of Neuroscience, 27*, 9077–9085.

Liu, X. (2010). Contribution of drug cue, priming, and stress to reinstatement of nicotine-seeking behavior in a rat model of relapse. In J. Egger, & M. Kalb (Eds.), *Smoking relapse: Causes, prevention and recovery* (pp. 143–163). New York: Nova Science Publisher. [ISBN: 978-1-60876-580-5].

Liu, X. (2014). Effects of blockade of alpha4beta2 and alpha7 nicotinic acetylcholine receptors on cue-induced reinstatement of nicotine-seeking behaviour in rats. *International Journal of Neuropsychopharmacology, 17*, 105–116.

Liu, X., Caggiula, A. R., Palmatier, M. I., Donny, E. C., & Sved, A. F. (2008). Cue-induced reinstatement of nicotine-seeking behavior in rats: effect of bupropion, persistence over repeated tests, and its dependence on training dose. *Psychopharmacology (Berl), 196*, 365–375.

Liu, X., Caggiula, A. R., Yee, S. K., Nobuta, H., Poland, R. E., & Pechnick, R. N. (2006). Reinstatement of nicotine-seeking behavior by drug-associated stimuli after extinction in rats. *Psychopharmacology (Berl), 184*, 417–425.

Liu, X., Caggiula, A. R., Yee, S. K., Nobuta, H., Sved, A. F., Pechnick, R. N., & Poland, R. E. (2007). Mecamylamine attenuates cue-induced reinstatement of nicotine-seeking behavior in rats. *Neuropsychopharmacology, 32*, 710–718.

Liu, X., Jernigen, C., Gharib, M., Booth, S., Caggiula, A. R., & Sved, A. F. (2010). Effects of dopamine antagonists on drug cue-induced reinstatement of nicotine-seeking behavior in rats. *Behavioural Pharmacology, 21*, 153–160.

Liu, X., Palmatier, M. I., Caggiula, A. R., Donny, E. C., & Sved, A. F. (2007). Reinforcement enhancing effect of nicotine and its attenuation by nicotinic antagonists in rats. *Psychopharmacology (Berl), 194*, 463–473.

Liu, X., Palmatier, M. I., Caggiula, A. R., Sved, A. F., Donny, E. C., Gharib, M., & Booth, S. (2009). Naltrexone attenuation of conditioned but not primary reinforcement of nicotine in rats. *Psychopharmacology (Berl), 202*, 589–598.

Liu, X., & Weiss, F. (2002). Additive effect of stress and drug cues on reinstatement of ethanol seeking: exacerbation by history of dependence and role of concurrent activation of corticotropin-releasing factor and opioid mechanisms. *Journal of Neuroscience, 22*, 7856–7861.

Liu, X., & Weiss, F. (2004). Nitric oxide synthesis inhibition attenuates conditioned reinstatement of ethanol-seeking, but not the primary reinforcing effects of ethanol. *Alcoholism: Clinical and Experimental Research, 28*, 1194–1199.

Markou, A., & Paterson, N. E. (2001). The nicotinic antagonist methyllycaconitine has differential effects on nicotine self-administration and nicotine withdrawal in the rat. *Nicotine & Tobacco Research, 3*, 361–373.

Marlatt, G. A., Curry, S., & Gordon, J. R. (1988). A longitudinal analysis of unaided smoking cessation. *Journal of Consulting and Clinical Psychology, 56*, 715–720.

Martin-Fardon, R., Maurice, T., Aujla, H., Bowen, W. D., & Weiss, F. (2007). Differential effects of sigma1 receptor blockade on self-administration and conditioned reinstatement motivated by cocaine vs natural reward. *Neuropsychopharmacology, 32*, 1967–1973.

Mineur, Y. S., & Picciotto, M. R. (2008). Genetics of nicotinic acetylcholine receptors: relevance to nicotine addiction. *Biochemical Pharmacology, 75*, 323–333.

Newman, J. L., & Beardsley, P. M. (2006). Effects of memantine, haloperidol, and cocaine on primary and conditioned reinforcement associated with cocaine in rhesus monkeys. *Psychopharmacology (Berl), 185*, 142–149.

O'Connor, E. C., Parker, D., Rollema, H., & Mead, A. N. (2010). The alpha4beta2 nicotinic acetylcholine-receptor partial agonist varenicline inhibits both nicotine self-administration following repeated dosing and reinstatement of nicotine seeking in rats. *Psychopharmacology (Berl), 208*, 365–376.

Panlilio, L. V., Justinova, Z., Mascia, P., Pistis, M., Luchicchi, A., Lecca, S., ... Goldberg, S. R. (2012). Novel use of a lipid-lowering fibrate medication to prevent nicotine reward and relapse: preclinical findings. *Neuropsychopharmacology, 37*, 1838–1847.

Paterson, N. E., Froestl, W., & Markou, A. (2005). Repeated administration of the GABAB receptor agonist CGP44532 decreased nicotine self-administration, and acute administration decreased cue-induced reinstatement of nicotine-seeking in rats. *Neuropsychopharmacology, 30*, 119–128.

Pechnick, R. N., Manalo, C. M., Lacayo, L. M., Vit, J. P., Bholat, Y., Spivak, I., ... Farrokhi, C. (2011). Acamprosate attenuates cue-induced reinstatement of nicotine-seeking behavior in rats. *Behavioural Pharmacology, 22*, 222–227.

Pons, S., Fattore, L., Cossu, G., Tolu, S., Porcu, E., McIntosh, J. M., ... Fratta, W. (2008). Crucial role of alpha4 and alpha6 nicotinic acetylcholine receptor subunits from ventral tegmental area in systemic nicotine self-administration. *Journal of Neuroscience, 28*, 12318–12327.

Rohsenow, D. J., Tidey, J. W., Miranda, R., McGeary, J. E., Swift, R. M., Hutchison, K. E., ... Monti, P. M. (2008). Olanzapine reduces urge to smoke and nicotine withdrawal symptoms in community smokers. *Experimental and Clinical Psychopharmacology, 16*, 215–222.

Rose, J. E., Behm, F. M., Westman, E. C., & Johnson, M. (2000). Dissociating nicotine and nonnicotine components of cigarette smoking. *Pharmacology Biochemistry and Behavior, 67*, 71–81.

Shaham, Y., Adamson, L. K., Grocki, S., & Corrigall, W. A. (1997). Reinstatement and spontaneous recovery of nicotine seeking in rats. *Psychopharmacology (Berl), 130*, 396–403.

Shaham, Y., & Stewart, J. (1996). Effects of opioid and dopamine receptor antagonists on relapse induced by stress and re-exposure to heroin in rats. *Psychopharmacology (Berl), 125*, 385–391.

Shoaib, M. (2008). The cannabinoid antagonist AM251 attenuates nicotine self-administration and nicotine-seeking behaviour in rats. *Neuropharmacology, 54*, 438–444.

Stewart, J. (1984). Reinstatement of heroin and cocaine self-administration behavior in the rat by intracerebral application of morphine in the ventral tegmental area. *Pharmacology Biochemistry and Behavior, 20*, 917–923.

Uslaner, J. M., Vardigan, J. D., Drott, J. M., Uebele, V. N., Renger, J. J., Lee, A., ... Hutson, P. H. (2010). T-type calcium channel antagonism decreases motivation for nicotine and blocks nicotine- and cue-induced reinstatement for a response previously reinforced with nicotine. *Biological Psychiatry, 68*, 712–718.

Walters, C. L., Brown, S., Changeux, J. P., Martin, B., & Damaj, M. I. (2006). The beta2 but not alpha7 subunit of the nicotinic acetylcholine receptor is required for nicotine-conditioned place preference in mice. *Psychopharmacology (Berl), 184*, 339–344.

Walters, C. L., Cleck, J. N., Kuo, Y. C., & Blendy, J. A. (2005). Mu-opioid receptor and CREB activation are required for nicotine reward. *Neuron, 46*, 933–943.

Watkins, S. S., Epping-Jordan, M. P., Koob, G. F., & Markou, A. (1999). Blockade of nicotine self-administration with nicotinic antagonists in rats. *Pharmacology Biochemistry and Behavior, 62*, 743–751.

Weiss, F., Martin-Fardon, R., Ciccocioppo, R., Kerr, T. M., Smith, D. L., & Ben-Shahar, O. (2001). Enduring resistance to extinction of cocaine-seeking behavior induced by drug-related cues. *Neuropsychopharmacology, 25*, 361–372.

Zislis, G., Desai, T. V., Prado, M., Shah, H. P., & Bruijnzeel, A. W. (2007). Effects of the CRF receptor antagonist D-Phe CRF(12-41) and the alpha2-adrenergic receptor agonist clonidine on stress-induced reinstatement of nicotine-seeking behavior in rats. *Neuropharmacology, 53*, 958–966.

Zoli, M., Lena, C., Picciotto, M. R., & Changeux, J. P. (1998). Identification of four classes of brain nicotinic receptors using beta2 mutant mice. *Journal of Neuroscience, 18*, 4461–4472.

Chapter 23

Effects of Environmental Enrichment on Nicotine Addiction

Dustin J. Stairs[1], Megan Kangiser[1], Tyson Hickle[1], Charles S. Bockman[2]

[1]Department of Psychology, Creighton University, Omaha, NE, USA; [2]Department of Pharmacology, Creighton University School of Medicine, Omaha, NE, USA

Abbreviations

ACH Acetylcholine
CPP Condition place preference
DA Dopamine
DAT Dopamine transporter protein
DHβE Dihydro-β-erythroidine hydrobromide
EC Enriched condition
GLU Glutamate
IC Isolated condition
mPFC Medial prefrontal cortex
NAc Nucleus accumbens
nAChR Nicotinic acetylcholine receptor
VTA Ventral tegmental area

INTRODUCTION

Despite the known negative health consequences of tobacco use, there are still approximately 42 million Americans who are current smokers (Agaku, King, Dube, Centers for Disease Control, & Prevention, 2014). The negative health consequences of tobacco use are an even more pressing issue when one considers the growing rates of tobacco use in various developing countries. One study found that there are approximately 3 billion smokers across 16 different countries, and many of these countries show very low quit rates (Giovino et al., 2012). These findings indicate the importance of developing not only effective cessation programs, but also smoking initiation prevention strategies.

A better understanding of the various risk factors and individual differences that are associated with increased likelihood of tobacco addiction could result in improvements in current smoking prevention strategies. Preclinical animal models are useful models for drug abuse researchers to study the various individual difference factors in drug abuse vulnerability (Carroll, Anker, & Perry, 2009). A number of animal models have been developed to study individual differences such as sex, impulsivity, and novelty reactivity, including differential exposure to environmental novelty or enrichment.

In an attempt to better understand the role of environmental novelty in vulnerability to drug abuse, researchers have adapted a rodent environmental enrichment paradigm that was previously used to investigate the effects of enrichment on learning (Renner & Rosenzweig, 1987). With the use of this rodent model, researchers have found that environmental enrichment exposure during adolescence appears to have a protective effect against drug abuse vulnerability in adulthood (Solinas, Thiriet, Chauvet, & Jaber, 2010; Stairs & Bardo, 2009). There are numerous environmental enrichment paradigms used in different laboratories; these paradigms can vary on a number of dimensions, which can make it difficult to compare results across laboratories (see Simpson & Kelly, 2011 for a review). The environmental enrichment paradigm that our laboratory uses most closely resembles that used by the Bardo and colleagues' laboratory. In this environmental enrichment paradigm, male Sprague–Dawley rats are raised from approximately 21 to 51 days of age, typically in one of two environmental conditions: enriched or impoverished (Bardo et al., 1995; Stairs & Bardo, 2009). In these experiments, typical enriched-condition (EC) rats are housed in a large cage with a number of social cohorts and novel objects, which are reconfigured daily, resulting in daily novel physical enrichment as well as novelty from the social interaction with their cohorts. In the impoverished condition (IC), rats are housed individually in hanging stainless-steel cages, without novel objects or social cohorts. A third social condition (SC), in which rats are housed with social cohorts without the novel objects, is also often used. These three conditions allow the researcher to determine the influence of social interaction and novel environmental stimuli on various drug effects. While various laboratories have been using rodent environmental enrichment paradigms to study how differential exposure to environments can alter both the neurochemistry and the neurobehavioral effects of abused drugs such as cocaine (Smith et al., 2009; Solinas, Thiriet, El Rawas, Lardeux, & Jaber, 2009), amphetamines (Bowling & Bardo, 1994; Green, Gehrke, & Bardo, 2002), and opiates (Smith et al., 2005), far less research has been done investigating the effects of enrichment on the more commonly abused psychostimulant nicotine. Given this, our laboratory has been investigating how and if environmental enrichment can alter the neural and behavioral effects of nicotine. The aim of this chapter is to summarize the research investigating the effects of environmental enrichment on the behavioral and neural effects of nicotine. In this chapter we will also integrate the limited nicotine data with the larger existing data of how environmental enrichment alters sensitivity to other abused psychostimulants such as cocaine and amphetamines.

BEHAVIORAL EFFECTS OF NICOTINE IN ENRICHED ANIMALS

One of the first studies looking at the effects of environmental enrichment on the behavioral effects of nicotine was conducted by Green, Cain, Thompson, and Bardo (2003). In this study, the authors examined both the acute and the repeated effects of 0.2 and 0.8 mg/kg (freebase) injections of nicotine on locomotor behavior. The authors found that IC rats showed a greater increase in the locomotor response to the 0.2 mg/kg acute nicotine dose compared to their EC counterparts, while there was no difference between EC and IC animals as to the acute effects of the 0.8 mg/kg dose. Following eight repeated injections of nicotine, EC rats were less sensitive than IC rats to the stimulatory effects of nicotine (Green et al., 2003). A study by Coolon and Cain (2009) expanded on the effects of environmental enrichment on nicotine locomotor behavior by investigating the effects of enrichment on nicotine-induced conditioned hyperactivity. Here the authors repeatedly injected EC, IC, and SC rats with 0.4 mg/kg (freebase) nicotine and then tested for the appearance of Pavlovian hyperactivity response. The authors also tested whether the nicotinic acetylcholine receptor (nAChR) antagonist mecamylamine blocked the nicotine-induced conditioned hyperlocomotor effect. The authors also found that following repeated pretreatments of 0.4 mg/kg nicotine, EC rats exhibited less nicotine-induced locomotor sensitization and less conditioned hyperactivity than IC rats. Finally the authors found that pretreatments of mecamylamine blocked the nicotine-induced conditioned hyperactivity in EC and SC rats but not IC rats. This last finding suggests a potential neural mechanism for how environmental enrichment may alter sensitivity to nicotine, which we will expand upon later in the chapter.

While it appears from the previous two studies that environmental enrichment seems to decrease the sensitivity to the locomotor effects of repeated nicotine exposure, a 2012 study found that, following repeated pretreatments with 0.35 mg/kg (freebase) nicotine, EC rats appeared to exhibit greater nicotine-induced locomotor sensitization compared to their IC counterparts (Gomez, Midde, Mactutus, Booze, & Zhu, 2012). While at first glance this seems incongruent with the previous results, the authors indicate that the EC rats appeared to be more sensitive to nicotine locomotor sensitization only when the data were presented as a percentage change from saline controls, but IC rats were actually more sensitive when raw locomotor data were examined. This study highlights a potential difficulty when using the environmental enrichment paradigm; that is, typically EC and IC rats have differences in the baseline or control rates of behavior, which may confound interpretation of the results. When using the enrichment paradigm, it is important to consider the appropriate method of data presentation when attempting to depict potential environmentally induced differences in the effects of nicotine. The importance of baseline differences between EC and IC rats was also highlighted in a 2009 paper analyzing the effects of enrichment on cocaine sensitivity (Smith et al., 2009).

The ability of environmental enrichment to decrease nicotine-induced locomotor sensitization was expanded upon by a 2013 study in our laboratory examining the ability of environmental enrichment to block the capacity of adolescent nicotine exposure to induce either nicotine sensitization or cross-sensitization to D-amphetamine in adulthood (Adams, Klug, Quast, & Stairs, 2013). In this study, we injected both EC and IC male rats with either saline or 0.4 mg/kg (freebase) nicotine once daily for 7 days from postnatal day 28 to 34. Following a 35-day washout period, both EC and IC animals were challenged with saline, nicotine (0.2 or 0.4 mg/kg), or D-amphetamine (0.5 or 1.0 mg/kg). Once more, environmental enrichment appears to be protective since it blocked the ability of adolescent nicotine exposure to induce nicotine sensitization at either nicotine dose tested. IC rats treated with nicotine in adolescence showed nicotine sensitization following the 0.4 mg/kg nicotine dose. The authors also found that enrichment blocked nicotine-induced cross-sensitization to the 0.5 mg/kg dose of D-amphetamine, while IC rats exposed to nicotine in adolescence displayed cross-sensitization to the 0.5 mg/kg dose of amphetamine in adulthood. However, EC rats did exhibit nicotine-induced cross-sensitization to the 1.0 mg/kg dose of D-amphetamine (see Figure 1). These findings indicate that environmental enrichment may actually counteract the risk factor of adolescent nicotine exposure to increase vulnerability to stimulant addiction in adulthood.

While there is some published work on the effects of environmental enrichment on nicotine-induced locomotor behavior, there is less published research on the effects of environmental enrichment on other behavioral assays, such as the rodent drug-discrimination procedure and the conditioned place preference procedure. Given the lack of published studies, our laboratory conducted a study to determine whether environmental enrichment alters sensitivity to the discriminative stimulus effects of

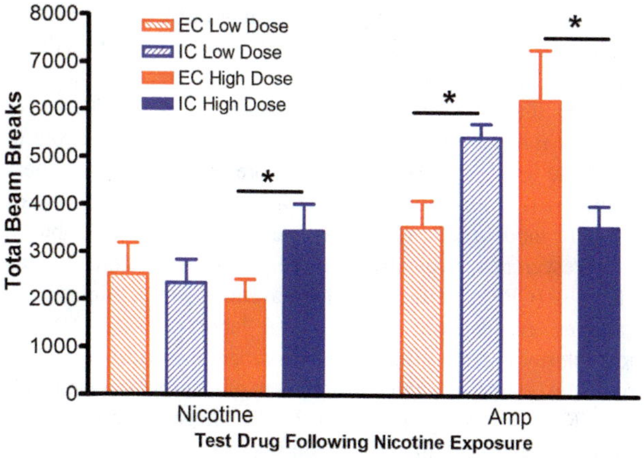

FIGURE 1 **Locomotor sensitization and cross-sensitization in EC and IC rats following adolescent nicotine exposure.** The mean (±SEM) total beam breaks during 45-min locomotor sessions for EC and IC rats following injection of 0.5 (low dose) or 1.0 (high dose) mg/kg amphetamine or 0.2 (low dose) or 0.4 (high dose) mg/kg nicotine ($n=7$ or 8/group). All animals were pretreated with nicotine (0.4 mg/kg) during adolescence. *Significant difference between EC and IC groups ($p \leq 0.05$). These data highlight the protectant effects of environmental enrichment on locomotor sensitization. Adapted from Adams et al. (2013), with permission from Elsevier.

nicotine. We also attempted to characterize the receptors involved in the enrichment-induced differences by investigating the ability of either nicotinic or dopaminergic antagonists to differentially block the discriminative stimulus effects (Stairs, Bockman, Fosdick, Mittelstet, & Schwarzkopf, 2009). In this study, the animals were trained on a two-lever-operant drug-discrimination task to discriminate between saline and a training dose of 0.3 mg/kg (freebase) nicotine. Following acquisition of the discrimination (80% appropriate responding), a nicotine generalization curve was determined (0–0.3 mg/kg). Following completion of the nicotine generalization curve, pretreatments with either the nicotinic antagonist mecamylamine (0.0625–1.0 mg/kg) or the dopamine D_2/D_3 antagonist eticlopride (0.01–0.3 mg/kg) were administered prior to injections of the nicotine training dose. We found that both EC and IC rats acquired nicotine drug discrimination at the 0.3 mg/kg dose and did so in the same number of sessions. Results from the nicotine generalization curves found that EC rats emitted less nicotine-lever responding when tested with the two lowest doses of nicotine. IC rats displayed significantly more nicotine-lever responding at those same two doses, never dropping below 50% of their responding on the nicotine lever. Mecamylamine blocked the discriminative stimulus effects of the training dose of nicotine at lower mecamylamine doses in EC rats than in IC rats. The D_2/D_3 antagonist eticlopride dose-dependently blocked the discriminative stimulus effects of nicotine equally in EC and IC rats. The results of this study indicate that environmental enrichment decreases the sensitivity to the discriminative stimulus effects of nicotine. This outcome is also similar to enrichment effects on the discriminative stimulus properties of amphetamine and cocaine (Fowler et al., 1993). The ability of the antagonists to block the discriminative stimulus effects indicates that both the nicotinic receptor and the D_2/D_3 dopamine receptor mediate the discriminative effects of nicotine. Given the differences in the antagonistic effects of mecamylamine, the differential sensitivity to the discriminative stimulus effects of nicotine may be due to differences in either the number or the function of the nAChRs. Further studies need to be conducted to determine how the nAChRs differ between EC and IC rats.

While the drug discrimination paradigm is considered to model the interoceptive cues of a drug, which can play a role in drug abuse and particularly relapse, more frequently researchers are interested in modeling the rewarding or reinforcing effects of a drug. To model the rewarding and reinforcing effects of abused drugs, researchers typically use the rodent conditioned place preference (CPP) and the rodent intravenous self-administration paradigms, respectively. In 2014, our laboratory completed a study determining the effects of environmental enrichment on the acquisition, extinction, and reinstatement of nicotine-induced CPP (Ewin, Kangiser, & Stairs, 2015). In this study, EC and IC rats were placed into one of three nicotine-conditioning groups: 0.4, 0.6, or 0.8 mg/kg (freebase). Using a three-chamber CPP design, animals first had a 15-min pretest with access to all three chambers to determine their initial side preference. On the next day, the animals started daily conditioning trials in which they received an injection with either nicotine or saline. Following injections, animals were immediately confined to one of the pairing chambers of the CPP apparatus for a 25-min conditioning trial. Conditioning trials were repeated across 8 days, with four alternating saline and nicotine conditioning sessions. Following conditioning, the animals were tested for a CPP response during a 15-min test session in which the animals again had access to all three chambers. Following the test for acquisitions animals had five extinction sessions, which were identical to the test session. After the last day of extinction, the animals were tested for reinstatement of the CPP response by being injected with their training dose of nicotine or saline and then given a 15-min test session. Using this procedure, we found that only EC rats displayed a significant CPP response with all three doses of nicotine tested compared to saline controls. IC rats conditioned with 0.8 mg/kg dose of nicotine showed only a marginally significant ($p=0.07$) CPP response compared to their saline controls. Comparing the CPP response between EC and IC animals, EC rats showed a greater nicotine-induced CPP response when conditioned with the 0.4 and 0.8 mg/kg dose of nicotine compared to their IC counterparts. Extinction of the CPP response occurred in all three doses for only EC rats after 5 days or less without drug exposure. IC rats failed to show extinction at any nicotine dose by the fifth day of extinction. The nicotine doses tested for reinstatement did not significantly increase time spent in the drug-paired side above the levels seen at the last day of extinction in either EC or IC rats (Ewin et al., 2015). These results indicate that environmental enrichment may actually increase the rewarding effect of nicotine but also increases the animals' sensitivity to extinction of a nicotine CPP response. While the CPP results with enriched animals showing an increased sensitivity to nicotine are opposite the decreased sensitivity to nicotine seen in the locomotor (Adams et al., 2013; Coolon & Cain, 2009; Green et al., 2003) and drug-discrimination (Stairs et al., 2009) studies, the findings are consistent with enrichment effects on amphetamine CPP (Bowling & Bardo, 1994). The reason for this discrepancy between the CPP procedure and the locomotor and drug-discrimination procedures is still unclear; future studies will be needed to determine why this discrepancy exists.

Although there are at least a few studies that have investigated the effects of environmental enrichment on nicotine-induced locomotor behavior, CPP, and drug discrimination, no studies to date have investigated the effect of environmental enrichment on nicotine self-administration. The lack of data on the effects of enrichment on the reinforcing effects of nicotine is an important gap in the literature on environmental enrichment that needs to be filled. The most likely reason this has yet to be done is the difficulty that comes from maintaining catheters in group-housed EC rats (Stairs & Bardo, 2009), but also the difficulty in maintaining responding with intravenous nicotine because of its weak primary reinforcing properties (Donny et al., 2000). However, with the more recent understanding of the dual reinforcing properties of intravenous nicotine (Caggiula et al., 2009) and continued improvement in our lab in maintaining catheters in EC rats, our laboratory most likely will be conducting an environmental enrichment nicotine self-administration study in the near future.

NEURAL EFFECTS OF ENRICHMENT ON NICOTINE EFFECTS AND CIRCUITS

There are only a handful of studies that have directly studied the neurochemical effects of nicotine in EC and IC rats. The majority of our understanding of the underlying neural mechanisms to explain the behavioral effects of environmental enrichment comes from behavioral studies in which nicotinic receptor antagonist pretreatments were administered. We also can make inferences from previous studies investigating the neural mechanisms of nicotine self-administration and drug discrimination as well as the effects of enrichment on the neural mechanisms of cocaine or D-amphetamine.

One study examined the neural response of EC and IC rats to an acute dose of nicotine (Zhu, Bardo, Green, Wedlund, & Dwoskin, 2007). Using in vivo voltammetry to assess dopamine transporter (DAT) function in rat striatum and medial prefrontal cortex (mPFC) following an acute injection with 0.4 mg/kg (freebase) nicotine, the authors found that in the mPFC EC rats had an increase in dopamine clearance relative to saline controls, whereas IC rats did not (Zhu et al., 2007). Enrichment differences were not seen between EC and IC in dopamine clearance in the striatum following an acute nicotine injection. The ability of environmental enrichment to alter DAT function in response to nicotine specifically in the mPFC highlights the growing importance of mPFC DAT function in enrichment-induced differences, given that previous research shows that EC rats have differential DAT protein levels in the mPFC at baseline compared to IC rats (Zhu, Apparsundaram, Bardo, & Dwoskin, 2005). Finally, the enrichment-induced changes in the mPFC may take place through changes in the phosphorylation of dopamine- and cAMP-regulated phosphoprotein-32 and cAMP-response element binding protein, as suggested by the 2012 finding that EC rats show greater phosphorylation of these proteins following repeated nicotine pretreatments compared to IC rats (Gomez et al., 2012). The ability of environmental enrichment to alter DAT function in the mPFC, we believe, may be the neural mechanism that explains why EC rats did not display nicotine cross-sensitization to D-amphetamine (Adams et al., 2013).

The protective effects of enrichment on nicotine sensitization as well as nicotine-conditioned hyperlocomotor response most likely result from changes in nAChRs. Coolon and Cain (2009) found that the nAChR antagonist mecamylamine blocked nicotine-conditioned hyperlocomotor activity only in EC and SC rats and not in IC rats. This effect indicated that nAChRs play a role in nicotine-conditioned hyperlocomotor activity and that nAChRs are somehow differentially sensitive between EC and IC rats. Previous data indicate that nicotine exposure results in long-lasting upregulation of nAChRs in midbrain areas (Trauth, Seidler, McCook, & Slotkin, 1999), which are believed to be responsible for the sensitized locomotor response to nicotine (Ksir, Hakan, & Kellar, 1987). Our lab has unpublished autoradiography data using ^{125}I-labeled epibatidine binding, which indicate that EC rats have a lower density of nAChRs in the ventral tegmental area (VTA) compared to their IC counterparts in the absence of any nicotine exposure. Given these data, perhaps EC rats do not show nicotine sensitization (Adams et al., 2013; Coolon & Cain, 2009; Green et al., 2003) because environmental enrichment alters the expression of nAChRs in the VTA, decreasing the amount of dopamine release in the nucleus accumbens (NAc), which has been implicated in nicotine-induced locomotor sensitization (Birrell & Balfour, 1998).

The enrichment-induced changes in the nAChRs in mesolimbic structures also most likely play a role in the differences seen in EC and IC rats in nicotine drug discrimination. Mesocorticolimbic structures have been shown to be involved in the discriminative stimulus of nicotine. For instance, in animals trained to discriminate systemic injections of nicotine from saline, local injections into the mPFC resulted in complete substitution for systemic nicotine injections (Miyata, Ando, & Yanagita, 1999), while local injections into the VTA result in only partial substitution for systemic nicotine (Miyata, Ando, & Yanagita, 2002). The involvement of the NAc in the discriminative effects of nicotine is somewhat mixed, in that Shoaib and Stolerman (1996) found that local injections of nicotine into the NAc resulted in no generalization to systemic injections, whereas another study by Miyata et al. (2002) found full generalization to systemic injections of local injections into the NAc. The mixed results dealing with the NAc may be an effect of the local injection dose of nicotine, since Miyata et al. (2002) used a much higher dose of nicotine compared to Shoaib and Stolerman (1996).

The neural nAChR subtypes that mediate the discriminative stimulus effects of nicotine have also been investigated. Using an agonist substitution method, Smith et al. (2007) found that the $\alpha 4\beta 2$ nAChR plays a crucial role in the discriminative stimulus effects of nicotine, while the homopentamer $\alpha 7$ receptor and $\alpha 3\beta 4$ receptors do not. The specific involvement of the $\alpha 4\beta 2$ receptor was also implicated when $\beta 2$-knockout mice failed to acquire nicotine drug discrimination (Shoaib et al., 2002), while mice lacking the $\alpha 7$ subunit acquired nicotine drug discrimination similar to levels found in wild-type animals. The involvement of the $\alpha 4\beta 2$ receptor in nicotine drug discrimination parallels the importance of this subunit in the reinforcing properties of nicotine (Picciotto et al., 1998; Watkins, Epping-Jordan, Koob, & Markou, 1999).

Given the behavioral differences we see between EC and IC rats in nicotine drug discrimination, and the neural substrates and receptor subtypes involved, our laboratory has been using autoradiography to determine the distribution of nicotinic receptors throughout the brain and in key mesolimbic structures involved in the discriminative stimulus and reinforcing effects of nicotine. As stated earlier, we have data from using ^{125}I-labeled epibatidine binding, which indicate that EC rats have a lower density of $\alpha 4\beta 2$ subtype nAChR in the VTA compared to their IC counterparts. We have not found any significant differences in the $\alpha 4\beta 2$ subtype in either the NAc or the mPFC. We have also characterized the distribution and expression of the $\alpha 7$ nicotinic receptor in EC and IC brains using ^{125}I-labeled bungarotoxin binding; again, in the mesolimbic structures we have tested so far, we have found no differences between EC and IC rats in $\alpha 7$ receptor densities (Stairs, Quast, & Bockman, 2013). Given both our nicotine drug-discrimination and our quantitative autoradiography data, we believe that environmental enrichment may alter the density of $\alpha 4\beta 2$ receptors in the VTA, which decreases EC rats' sensitivity to the discriminative stimulus effects of nicotine.

While we have found structural differences in EC and IC rats in terms of nicotinic receptor densities, to our knowledge there are no data from investigating whether environmental enrichment alters the functioning of the nAChRs. It would be of interest to know, using biotinylation and immunoblotting techniques, whether the changes we see in nAChR densities are due to changes in receptor densities at the cell surface. It would also be informative to use an ex vivo assay to determine if the nAChRs in the VTA in EC and IC rats differentially respond to nicotine. The only data studying EC and IC rats' neurochemical response to nicotine have been collected by Zhu, Bardo, and Dwoskin (2013) and Zhu et al. (2007). In these studies, the authors used an in vivo voltammetry assay to investigate DAT function or DA clearance following nicotine exposure in various mesolimbic structures. Zhu et al. (2007) found that with nicotine-treated EC rats there was an increase in DA clearance in the mPFC compared to saline-treated EC rats, but no effect of nicotine in the mPFC in IC rats. Also, no effects of nicotine in the striatum were seen between EC and IC rats. These results indicate that enrichment enhanced the nicotine-induced increase in DAT function in mPFC. In a later study, Zhu et al. (2013), examining DA clearance in the NAc core and shell, found that acute nicotine administration increased NAc shell DA clearance in EC rats, but not in IC rats. However, in the NAc core, nicotine increased DA clearance in IC rats, but not in EC rats.

While DAT function in the mPFC and NAc has important implications for the differential behavioral responses to nicotine in EC and IC rats, data on nAChR function in mesolimbic structures in EC and IC rats can provide a clearer understanding of how environmental enrichment may alter sensitivity to the reinforcing properties of nicotine. While no studies as of this writing have looked at EC and IC nicotine self-administration, the neural mechanism of nicotine self-administration has been known for some time. The pathway involves the same mesolimbic structures we just discussed that show differences between EC and IC rats. For instance, lesions to the mesolimbic dopamine pathway produced by infusion of 6-hydroxydopamine into the NAc result in decreases in nicotine self-administration (Corrigall, Franklin, Coen, & Clarke, 1992). These lesions resulted in a decrease in the amount of dopamine in the terminal field of the mesolimbic dopamine pathway in the mPFC (Corrigall et al., 1992). The mechanism by which nicotine increases levels of extracellular dopamine in the NAc was clarified when it was found that infusions of the α4β2 nicotinic antagonist dihydro-β-erythroidine hydrobromide (DHβE) into the VTA produced a significant decrease in nicotine self-administration (Corrigall, Coen, & Adamson, 1994). The decrease in responding following the delivery of the antagonist was specific to nicotine self-administration in that infusions of DHβE into the VTA had no effect on cocaine self-administration or food-maintained responding (Corrigall et al., 1994). Also, given the importance of the β2 subunit of the nAChR in nicotine self-administration (Picciotto et al., 1998), the differences our lab has found in the density of α4β2 receptors in the VTA, presumably on the cell body of dopamine neurons, could be a potential mechanism for environmental enrichment to alter the reinforcing properties of nicotine.

CONCLUSIONS AND FUTURE DIRECTIONS

In general, the effects of environmental enrichment on the behavioral and neural effects of nicotine are consistent with the effects of enrichment on other drugs of abuse like cocaine, amphetamines, and opiates (Stairs & Bardo, 2009). Environmental enrichment decreases the sensitivity to both acute and repeated nicotine-induced locomotor stimulatory effects (Adams et al., 2013; Coolon & Cain, 2009; Green et al., 2003). We also believe that differential effects of enrichment on nicotine-induced locomotor behavior may be mediated through the α4β2 nicotinic receptor densities in the NAc. We believe the α4β2 nicotinic receptors in the NAc are involved since we have preliminary data showing a stronger relationship between nicotine-induced locomotor behavior and α4β2 nicotinic receptor densities in the shell and core of the NAc in EC rats, compared to IC rats (see Figures 2 and 3). Also, environmental enrichment decreases the sensitivity to the discriminative stimulus properties of nicotine, similar to those seen with amphetamine (Fowler et al., 1993; Stairs et al., 2009). We also find that environmental enrichment appears to increase the sensitivity to nicotine CPP (Ewin et al., 2015), similar to the effects seen with amphetamine (Bowling & Bardo, 1994). The ability of environmental enrichment to alter the neural effects of nicotine appears to be mediated through changes in DAT function in the mPFC and NAc (Zhu et al., 2007, 2013). The previous research combined with data from our laboratory indicates that enrichment alters the density of nAChRs in the VTA leading to a

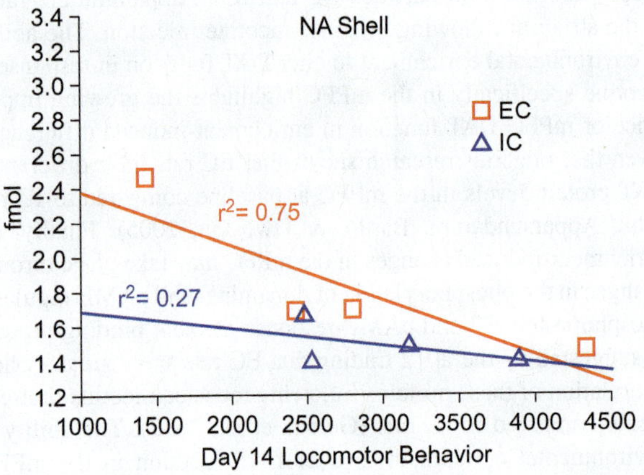

FIGURE 2 Association between nicotine-induced locomotor behavior and nAChR density in the NAc shell. Unpublished work illustrating the relationship between the levels of nicotine-induced locomotor behavior on the 14th day of nicotine exposure of the animals and the corresponding density of nAChRs (quantified in femtomoles) in the shell of the NAc in EC and IC rats ($n=4$/group). The regression line is the line that best fits the data for either EC (red line) or IC (blue line). The r^2 value is the amount of variance in the data accounted for by the regression line.

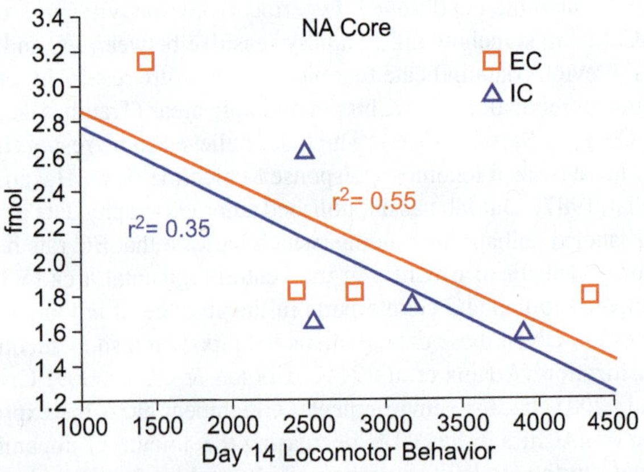

FIGURE 3 Association between nicotine-induced locomotor behavior and nAChR density in the NAc core. Unpublished work illustrating the relationship between the levels of nicotine-induced locomotor behavior on the 14th day of nicotine exposure of the animals and the corresponding density of nAChRs (quantified in femtomoles) in the core of the NAc in EC and IC rats ($n=4$/group). The regression line is the line that best fits the data for either EC (red line) or IC (blue line). The r^2 value is the amount of variance in the data accounted for by the regression line.

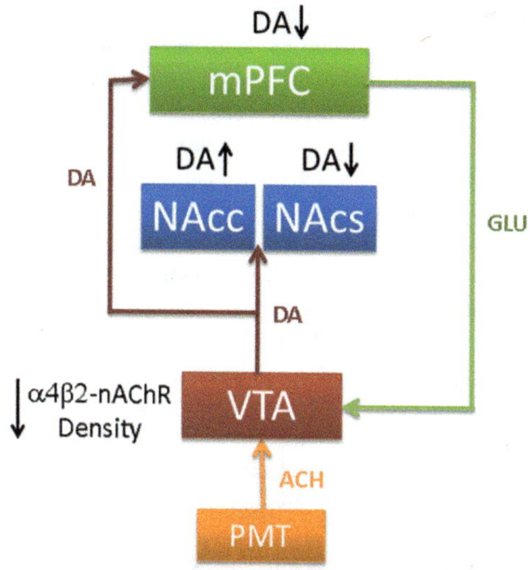

FIGURE 4 Model of the enrichment effect on the neural response to acute nicotine. Model of the enrichment effect on the neural response to acute nicotine in the mesocorticolimbic dopamine system. Environmental enrichment downregulates (↓) α4β2-nAChR density in the VTA. In response to acute administration of nicotine, extracellular dopamine levels are elevated (DA↑) in the core of the NAc (NAcc) and reduced (DA↓) in the shell of the NAc (NAcs) and mPFC in EC rats relative to their IC counterparts. ACH, acetylcholine; GLU, glutamate; PMT, pontomesencephalic tegmental.

decrease in the dopamine response in the NAc shell and mPFC, which could explain the differential behavioral responses to nicotine in enriched animals (see Figure 4).

While a fair amount of research has been conducted investigating the effects of environmental enrichment on various nicotine effects, it is clear in writing this chapter that much more research is needed to fully understand the effects of environmental enrichment on both the behavioral and the neural effects of nicotine. Given the lack of published data, our research laboratory will continue to investigate the effects of environmental enrichment on the addiction-related effects of nicotine. In particular, it will be of interest to see if nAChRs in EC and IC rats differ in an ex vivo assay.

APPLICATION TO OTHER ADDICTIONS AND SUBSTANCE MISUSE

The rodent environmental enrichment model has been shown to alter the behavioral effects of other drugs, including cocaine, opiates, and amphetamines (for a review see Solinas et al., 2010; Stairs & Bardo, 2009). While the effects of enrichment have been investigated on psychostimulants and opiates, no research to the authors' knowledge has explored the effects of enrichment on cannabinoid drugs. This line of research could have relevance in understanding individual differences in sensitivity to marijuana. Just like there is a limited amount of research on the effects of environmental enrichment on the commonly abused drug nicotine, little to no research has been conducted on whether enrichment can alter cannabinoid receptors and the role they play in the abuse-related potential of marijuana. This could be a growing importance as marijuana becomes legal for recreational use in the United States and use may increase.

From a more applied angle, the rodent environmental enrichment paradigm can be seen as a translational model for studying how novelty exposure can alter the sensitivity or vulnerability to drugs of abuse. This has implication for epidemiological research, which shows that individuals with high sensation- or novelty-seeking personality traits report using more drugs than do low sensation- or novelty-seeking individuals (Wills, Vaccaro, & McNamara, 1994; Wills, Windle, & Cleary, 1998). A better understanding of how novelty in the environment may alter neurochemistry or anatomy leading to an increased sensitivity to drugs of abuse has the potential to lead to better treatment outcomes as well as better tailoring of prevention programs to vulnerable populations.

DEFINITION OF TERMS

Locomotor sensitization This is an animal model of neural changes that take place during the development of addiction, which consists of an increase in locomotor activity following successive injections of the drug and an increased sensitivity to a challenge dose following a washout period.

Drug-conditioned place preference This is a Pavlovian animal model that tests the rewarding effects of a drug by differentially pairing two distinct sets of contextual cues with drug injections. Conditioning involves an animal receiving repeated injections of the drug (unconditioned stimulus) in one context (the conditioned stimulus). Intermixed are trials with saline injections paired with the other context (no unconditioned stimulus). Following conditioning is a choice test in which the animal receives unrestricted access to both contexts in the absence of the unconditioned stimulus. An increase in time spent in the drug-paired context is taken as evidence that the drug has rewarding properties.

Drug self-administration This is an animal model used to study the reinforcing effects of a drug. Typically self-administration is a form of operant conditioning in which the reinforcer maintaining behavior is a drug, and the drug is typically administered through an intravenous catheter or through oral consumption.

Drug-primed reinstatement This is an animal model of relapse in which an experimenter administers a drug-priming injection to an animal that has gone through extinction and is in a state of abstinence. Following the drug prime a drug-seeking behavior will be reinstated in the animal. In an operant drug self-administration paradigm this is typically defined as responses made on the previously active drug lever, and in conditioned place preference it is an increase in the time spent in the context previously paired with drug injections.

Ventral tegmental area This is a midbrain structure richly innervated with dopamine neurons. The VTA is part of the mesolimbic dopamine pathway and sends projections to various structures including the NAc and the mPFC.

Nucleus accumbens This is one of the terminal structures of the mesolimbic dopamine pathway in the midbrain. The nucleus receives dopaminergic inputs from the VTA that play an important role in the reinforcing properties of many abused drugs.

Medial prefrontal cortex This is a terminal structure of the mesolimbic dopamine pathway in the cortex. The cortex receives dopaminergic inputs from the VTA and sends glutamatergic feedback to the VTA.

Dopamine This is a catecholamine neurotransmitter found within the central nervous system. Dopamine has numerous functions within the central nervous system, but appears to play a central role in the reinforcing effects of various drugs of abuse.

Nicotinic acetylcholine receptor Neuronal nAChRs are ligand-gated ion channels existing as pentamers of α and β subunits. Multiple nAChR subtypes exist as combinations of 11 neuronal subunits (α2–α7, α9, α10, and β2–β4). Subtypes of nAChRs are referred to by their known subunit composition, e.g., α4β2, α3β4.

KEY FACTS OF ENVIRONMENTAL ENRICHMENT

- Environmental enrichment is an animal model that allows researchers to determine the effects of exposure to both physical and social novelty on drug abuse vulnerability.
- Environmental enrichment has been shown to result in an increased sensitivity to the acute locomotor stimulatory effects of amphetamine compared to isolated animals, although after repeated exposure to amphetamine and other stimulants, EC rats show less locomotor sensitization compared to IC rats.
- When looking at the discriminative stimulus effects of a drug, EC rats appear to be less sensitive to the discriminative stimulus effect of both nicotine and amphetamine compared to IC rats.
- EC rats also are seen to self-administer less cocaine and D-amphetamine than IC rats typically when the unit dose of the drug is low. Once higher doses of drug are tested EC and IC rats self-administer similar amounts.
- In contrast to the effects of enrichment seen on self-administration, EC rats appear to be more sensitive to cocaine, amphetamine, nicotine, and opiate CPP.

SUMMARY POINTS

- Environmental enrichment decreases the sensitivity to both acute and repeated nicotine-induced locomotor stimulatory effects (Adams et al., 2013; Coolon & Cain, 2009; Green et al., 2003).
- Environmental enrichment decreases the sensitivity to the discriminative stimulus properties of nicotine, similar to that seen with amphetamine.
- Environmental enrichment appears to increase the sensitivity to nicotine CPP, as has been found previously with amphetamine (Ewin et al., 2015).
- Environmental enrichment alters the neural effects of nicotine, which appears to be mediated through changes in DAT function in the mPFC and NAc.
- Enrichment may alter the nAChRs in key mesolimbic structures involved with abuse (Stairs et al., 2013).

REFERENCES

Adams, E., Klug, J., Quast, M., & Stairs, D. J. (2013). Effects of environmental enrichment on nicotine-induced sensitization and cross-sensitization to D-amphetamine in rats. *Drug and Alcohol Dependence*, *129*(3), 247–253. http://dx.doi.org/10.1016/j.drugalcdep.2013.02.019.

Agaku, I. T., King, B. A., Dube, S. R., & Centers for Disease Control, & Prevention (2014). Current cigarette smoking among adults - United States, 2005–2012. *MMWR Morbidity and Mortality Weekly Report*, *63*(2), 29–34.

Bardo, M. T., Bowling, S. L., Rowlett, J. K., Manderscheid, P., Buxton, S. T., & Dwoskin, L. P. (1995). Environmental enrichment attenuates locomotor sensitization, but not in vitro dopamine release, induced by amphetamine. *Pharmacology Biochemistry and Behavior*, *51*(2–3), 397–405.

Birrell, C. E., & Balfour, D. J. (1998). The influence of nicotine pretreatment on mesoaccumbens dopamine overflow and locomotor responses to D-amphetamine. *Psychopharmacology (Berlin)*, *140*(2), 142–149.

Bowling, S. L., & Bardo, M. T. (1994). Locomotor and rewarding effects of amphetamine in enriched, social, and isolate reared rats. *Pharmacology Biochemistry and Behavior*, *48*(2), 459–464.

Caggiula, A. R., Donny, E. C., Palmatier, M. I., Liu, X., Chaudhri, N., & Sved, A. F. (2009). The role of nicotine in smoking: a dual-reinforcement model. *Nebraska Symposium on Motivation*, *55*, 91–109.

Carroll, M. E., Anker, J. J., & Perry, J. L. (2009). Modeling risk factors for nicotine and other drug abuse in the preclinical laboratory. *Drug and Alcohol Dependence*, *104*(Suppl. 1), S70–S78.

Coolon, R. A., & Cain, M. E. (2009). Effects of mecamylamine on nicotine-induced conditioned hyperactivity and sensitization in differentially reared rats. *Pharmacology Biochemistry and Behavior*, *93*(1), 59–66.

Corrigall, W. A., Coen, K. M., & Adamson, K. L. (1994). Self-administered nicotine activates the mesolimbic dopamine system through the ventral tegmental area. *Brain Research*, *653*(1–2), 278–284.

Corrigall, W. A., Franklin, K. B., Coen, K. M., & Clarke, P. B. (1992). The mesolimbic dopaminergic system is implicated in the reinforcing effects of nicotine. *Psychopharmacology (Berlin)*, *107*(2–3), 285–289.

Donny, E. C., Caggiula, A. R., Rose, C., Jacobs, K. S., Miekle, M. M., & Sved, A. F. (2000). Differential effects of response-contingent and response-independent nicotine in rats. *European Journal of Pharmacology*, *402*, 231–240.

Ewin, S. E., Kangiser, M. M., & Stairs, D. J. (2015). The effects of environmental enrichment on nicotine condition place preference in male rats. *Experimental Clinical and Psychopharmacololgy*, *23*(5), 387–394. http://dx.doi.org/10.1037/pha0000024.

Fowler, S. C., Johnson, J. S., Kallman, M. J., Liou, J. R., Wilson, M. C., & Hikal, A. H. (1993). In a drug discrimination procedure isolation-reared rats generalize to lower doses of cocaine and amphetamine than rats reared in an enriched environment. *Psychopharmacology (Berlin)*, *110*(1–2), 115–118.

Giovino, G. A., Mirza, S. A., Samet, J. M., Gupta, P. C., Jarvis, M. J., Bhala, N., ... Group, G. C. (2012). Tobacco use in 3 billion individuals from 16 countries: an analysis of nationally representative cross-sectional household surveys. *Lancet*, *380*(9842), 668–679. http://dx.doi.org/10.1016/S0140-6736(12)61085-X.

Gomez, A. M., Midde, N. M., Mactutus, C. F., Booze, R. M., & Zhu, J. (2012). Environmental enrichment alters nicotine-mediated locomotor sensitization and phosphorylation of DARPP-32 and CREB in rat prefrontal cortex. *PLoS One*, *7*(8), e44149. http://dx.doi.org/10.1371/journal.pone.0044149. pii:PONE-D-12-08447.

Green, T. A., Cain, M. E., Thompson, M., & Bardo, M. T. (2003). Environmental enrichment decreases nicotine-induced hyperactivity in rats. *Psychopharmacology (Berlin)*, *170*(3), 235–241.

Green, T. A., Gehrke, B. J., & Bardo, M. T. (2002). Environmental enrichment decreases intravenous amphetamine self-administration in rats: dose-response functions for fixed- and progressive-ratio schedules. *Psychopharmacology (Berlin)*, *162*(4), 373–378.

Ksir, C., Hakan, R. L., & Kellar, K. J. (1987). Chronic nicotine and locomotor activity: influences of exposure dose and test dose. *Psychopharmacology (Berlin)*, *92*(1), 25–29.

Miyata, H., Ando, K., & Yanagita, T. (1999). Medial prefrontal cortex is involved in the discriminative stimulus effects of nicotine in rats. *Psychopharmacology (Berlin)*, *145*(2), 234–236.

Miyata, H., Ando, K., & Yanagita, T. (2002). Brain regions mediating the discriminative stimulus effects of nicotine in rats. *Annals of the New York Academy of Science, 965*, 354–363.

Picciotto, M. R., Zoli, M., Rimondini, R., Lena, C., Marubio, L. M., Pichll, E. M., ... Chanqeux, J. (1998). Acetylcholine receptors containing β2 subunit are involved in the reinforcing properties of nicotine. *Nature, 391*, 173–177.

Renner, M. J., & Rosenzweig, M. R. (1987). *Enriched and impoverished environments: effects on brain and behavior*. New York: Springer-Verlag.

Shoaib, M., Gommans, J., Morley, A., Stolerman, I. P., Grailhe, R., & Changeux, J. P. (2002). The role of nicotinic receptor β-2 subunits in nicotine discrimination and conditioned taste aversion. *Neuropharmacology, 42*(4), 530–539.

Shoaib, M., & Stolerman, I. P. (1996). Brain sites mediating the discriminative stimulus effects of nicotine in rats. *Behavioural Brain Research, 78*(2), 183–188.

Simpson, J., & Kelly, J. P. (2011). The impact of environmental enrichment in laboratory rats–behavioural and neurochemical aspects. *Behavioural Brain Research, 222*(1), 246–264. http://dx.doi.org/10.1016/j.bbr.2011.04.002. pii:S0166-4328(11)00289-0.

Smith, M. A., Chisholm, K. A., Bryant, P. A., Greene, J. L., McClean, J. M., Stoops, W. W., & Yancey, D. L. (2005). Social and environmental influences on opioid sensitivity in rats: importance of an opioid's relative efficacy at the mu-receptor. *Psychopharmacology (Berlin), 181*, 27–37.

Smith, M. A., Iordanou, J. C., Cohen, M. B., Cole, K. T., Gergans, S. R., Lyle, M. A., & Schmidt, K. T. (2009). Effects of environmental enrichment on sensitivity to cocaine in female rats: importance of control rates of behavior. *Behavioural Pharmacology, 20*(4), 312–321. http://dx.doi.org/10.1097/FBP.0b013e32832ec568.

Smith, J. W., Mogg, A., Tafi, E., Peacey, E., Pullar, I. A., Szekeres, P., & Tricklebank, M. (2007). Ligands selective for α4β2 but not α3β4 or α7 nicotinic receptors generalise to the nicotine discriminative stimulus in the rat. *Psychopharmacology (Berlin), 190*(2), 157–170.

Solinas, M., Thiriet, N., Chauvet, C., & Jaber, M. (2010). Prevention and treatment of drug addiction by environmental enrichment. *Progress in Neurobiology, 92*(4), 572–592. http://dx.doi.org/10.1016/j.pneurobio.2010.08.002. pii:S0301-0082(10)00145-0.

Solinas, M., Thiriet, N., El Rawas, R., Lardeux, V., & Jaber, M. (2009). Environmental enrichment during early stages of life reduces the behavioral, neurochemical, and molecular effects of cocaine. *Neuropsychopharmacology, 34*(5), 1102–1111. http://dx.doi.org/10.1038/npp.2008.51.

Stairs, D. J., & Bardo, M. T. (2009). Neurobehavioral effects of environmental enrichment and drug abuse vulnerability. *Pharmacology Biochemistry and Behavior, 92*(3), 377–382. http://dx.doi.org/10.1016/j.pbb.2009.01.016.

Stairs, D. J., Bockman, C. S., Fosdick, J., Mittelstet, B., & Schwarzkopf, L. (2009). Enrichment-induced differences in nicotine drug discrimination in rats is nicotinic receptor mediated. In *Paper presented at the college on problems on drug dependence, Reno, NV*.

Stairs, D. J., Quast, M., & Bockman, C. S. (2013). Relationship between nAChR subtypes and nicotine CPP in differentially-reared rats. In *Paper presented at the American Psychological Association, Honolulu, HI*.

Trauth, J. A., Seidler, F. J., McCook, E. C., & Slotkin, T. A. (1999). Adolescent nicotine exposure causes persistent upregulation of nicotinic cholinergic receptors in rat brain regions. *Brain Research, 851*(1–2), 9–19. pii:S0006-8993(99)01994-0.

Watkins, S. S., Epping-Jordan, M. P., Koob, G. F., & Markou, A. (1999). Blockade of nicotine self-administration with nicotinic antagonists in rats. *Pharmacology Biochemistry and Behavior, 62*(4), 743–751.

Wills, T. A., Vaccaro, D., & McNamara, G. (1994). Novelty seeking, risk taking, and related constructs as predictors of adolescent substance use: an application of Cloninger's theory. *Journal of Substance Abuse, 6*(1), 1–20.

Wills, T. A., Windle, M., & Cleary, S. D. (1998). Temperament and novelty seeking in adolescent substance use: convergence of dimensions of temperament with constructs from Cloninger's theory. *The Journal of Personality and Social Psychology, 74*(2), 387–406.

Zhu, J., Apparsundaram, S., Bardo, M. T., & Dwoskin, L. P. (2005). Environmental enrichment decreases cell surface expression of the dopamine transporter in rat medial prefrontal cortex. *Journal of Neurochemistry, 93*(6), 1434–1443.

Zhu, J., Bardo, M. T., & Dwoskin, L. P. (2013). Distinct effects of enriched environment on dopamine clearance in nucleus accumbens shell and core following systemic nicotine administration. *Synapse, 67*(2), 57–67. http://dx.doi.org/10.1002/syn.21615.

Zhu, J., Bardo, M. T., Green, T. A., Wedlund, P. J., & Dwoskin, L. P. (2007). Nicotine increases dopamine clearance in medial prefrontal cortex in rats raised in an enriched environment. *Journal of Neurochemistry, 103*(6), 2575–2588.

Chapter 24

Smokers with Severe Mental Illness

Maxie Ashton, Benjamin L. Carnell, Cherrie Galletly
The Adelaide Clinic, Gilberton, SA, Australia

INTRODUCTION

Tobacco smoking is the leading behavioral risk factor responsible for disease and premature death. It is linked to nearly 6 million deaths worldwide each year and causes many chronic diseases, including cardiovascular and respiratory diseases (WHO, 2013). Many smokers spend years suffering with related illness and disability caused by their tobacco use.

The World Health Organization's Framework Convention on Tobacco Control has committed governments to implementing strategies to reduce rates of tobacco use. These strategies have included banning tobacco advertising, providing clear warnings, smoke-free policies, higher prices on tobacco, and the provision of smoking cessation services (WHO, 2013). These approaches have been effective in reducing the prevalence of tobacco use. For example, in Australia, the percentage of adults smoking daily has reduced from 24% in 1991 to 15% in 2010 (AIHW, 2011). However, the rates continue to remain high among some specific groups, including people with mental illness. People living with mental illness are now a significant and increasing percentage of the remaining smokers (Australian Bureau of Statistics, 2008).

The 2007 National Survey of Mental Health and Wellbeing (Australian Bureau of Statistics) found that 3.6 million people in Australia identified themselves as current smokers, and of these 32% had experienced mental illness in the previous 12 months. The survey found that those living with mental illness smoke at nearly double the rate of those not living with mental illness. An Australian study (Cooper et al., 2012) of people living with a psychotic disorder in 2010 found that 66.6% smoked tobacco, 71.1% of men and 58.8% of women, and participants smoked an average of 21 cigarettes per day.

Tobacco-related deaths are higher in Australia than the deaths caused by a combination of breast cancer, infections and parasitic diseases, suicide, other drug dependence, falls, road accidents, alcohol, poisoning, and homicide and violence (Begg et al., 2007). The significant percentage of smokers living with mental illness suggests that many of those dying or suffering with disease and disability caused by tobacco are people with mental illness.

People with severe mental illness generally die earlier than the general population, with life expectancy reduced by 18.7 and 16.3 years for men and women with schizophrenia, respectively (Laursen, 2011). Those living with schizophrenia are twice as likely to die of cardiovascular problems and face three times the risk of respiratory disease (Mauer, 2006). This increased rate of illness and premature death is primarily due to risk factors such as smoking, obesity, substance abuse, and inadequate health care (Mauer, 2006). A comprehensive study of adults diagnosed with schizophrenia, bipolar disorder, or depression, with a follow-up period of up to 16 years, found significantly increased tobacco-related mortality (Callaghan et al., 2014).

In addition to the physical health effects, smoking can have a serious impact on the social well-being of smokers. Many community venues and workplaces are now smoke-free, including many outdoor areas, requiring smokers to go elsewhere to smoke, further isolating an already marginalized group. Smoking and its associated behaviors can also be a major source of conflict and difficulty within families, shared accommodation, and social settings.

Heavy smokers, especially those living on a low income or disability pension, can have very little money left for food, recreation, public transport, clothing, and health care. Some smokers with mental illness frequently run out of money and as a result experience nicotine withdrawal and stress, and have more difficulty managing their mental health and everyday tasks. Some resort to picking up butts, begging, and getting into debt (Lawn, Pols, & Barber, 2002). Data from the second national Australian survey of psychosis (Morgan et al., 2011) showed that nearly 90% of participants who were smokers lived on welfare benefits, with 82.2% receiving the disability support pension. The smokers in this survey were on average smoking over 20 cigarettes per day, and a higher proportion of them experienced financial difficulty compared to nonsmokers. These smokers were significantly more likely to report difficulties heating or cooling their home or room, going without meals, and having to seek financial assistance from family or friends (Morgan et al., 2011). In addition to the personal costs of smoking, a study of the economic costs in the United Kingdom in 2009–2010 found about £719 million was spent on directly treating diseases caused by smoking in people with mental disorders (Wu, Szatkowski, Britton, & Parrott, 2014).

It has been (conservatively) estimated that smoking causes at least 4500 fires in Australia each year and is responsible for nearly 25% of all fire-related deaths (Scollo & Winstanley, 2012). People with mental illness who live in group homes, supported residential settings, and boarding houses can be especially at risk.

NICOTINE ADDICTION

Contributing factors to tobacco use, as with other forms of drug use, include the pharmacology of the substance as well as both personal and environmental factors (Benowitz, 2010).

Neuropathology

Of the thousands of chemicals found in tobacco smoke, nicotine is the major component responsible for creating the physical addiction to tobacco. When a person inhales tobacco smoke, nicotine is rapidly absorbed through the lungs and takes only seconds to reach the bloodstream and the brain. In the brain, nicotine binds to nicotine receptors, including the α4β2 nicotinic acetylcholine receptor. The activation of these receptors triggers the release of neurotransmitters, including acetylcholine, dopamine, and serotonin. The release of these neurotransmitters (in particular dopamine) creates the pleasurable effects associated with smoking, consequently reinforcing the act of smoking (Benowitz, 2010).

After repeated exposure to nicotine, smokers develop tolerance, resulting in a physical dependence. Symptoms of withdrawal may occur and can include insomnia, depressed mood, irritability, frustration, anger, anxiety, and difficulty concentrating. These symptoms can cause significant discomfort and stress and consequently smokers often smoke to relieve these symptoms.

The neuropharmacological factors motivating smokers to smoke are therefore a combination of the release of the "pleasurable" neurotransmitters and the desire to avoid the discomfort of nicotine withdrawal (Benowitz, 2010).

Environmental Factors

In addition to the neuropharmacological factors of nicotine, tobacco smoking is also influenced by other personal and environmental factors, including genetics, age, gender, life experiences, beliefs, cultural background, psychological traits, access to tobacco, and acceptance by family, friends, and others.

The US Surgeon General identified a number of the factors that put a person at greater risk of initiation and ongoing use of tobacco. These include low socioeconomic status, lack of support, low levels of academic achievement, lack of skills required to resist influences encouraging tobacco use, belief that tobacco use is functional, low self-esteem, lack of self-efficacy, and tobacco use or approval by family members, peers, and significant others (Ragg & Ahmed, 2008; US Department of Health and Human Services, 1994).

WHY DO MORE PEOPLE WITH MENTAL ILLNESS SMOKE?

Neuropathology

Smokers with mental illness, in the same way as other smokers, smoke to experience the "pleasurable" effects triggered by the release of neurotransmitters and to avoid the discomfort of nicotine withdrawal.

In addition, some people with mental illness find that the symptoms of nicotine withdrawal feel similar to the symptoms of their illness. As a result, these people believe they smoke to feel mentally better, or to "self-medicate." Symptoms of withdrawal (including depression, insomnia, irritability, anxiety, poor concentration, and restlessness) can cause discomfort that may be misinterpreted by the person, their health worker, and others as related to their illness and not recognized as nicotine withdrawal (Lawn & Pols, 2003).

Early studies suggested that biological and genetic factors specific to people with mental illness are linked to the initiation and ongoing use of nicotine (Leonard et al., 2001). Early studies also suggested that smoking cessation can precipitate depressive symptoms (Glassman, 1993; Laje, Berman, & Glassman, 2001) and that nicotine can relieve symptoms of both depression and anxiety (Picciotto, Brunzell, & Caldarone, 2002; Pomerleau & Pomerleau, 1984). However, more recent studies suggest the "relief of symptoms" experienced is more likely to be related to relief from nicotine withdrawal and the pleasurable effects caused by the release of neurotransmitters. Moreover, a systematic review and meta-analysis of observational studies found smoking cessation to be associated with reduced depression, anxiety, and stress as well as improved positive mood and quality of life compared with continuing to smoke. This study found that the positive impact of smoking cessation was equal to or larger than the impact of antidepressant treatment for mood and anxiety disorders (Taylor et al., 2014).

Environmental and Social Factors

On average, people with mental illness are more likely to be affected by the individual and environmental risk factors for smoking identified by the US Surgeon General. People with mental illness are more likely to be unemployed, be on a low income, have interrupted education, and have lowered self-esteem. They are more likely to be socially disadvantaged and/or isolated, to have more limited recreation options, and to spend more time with other smokers. These factors all contribute to higher rates of smoking, lower cessation rates, and increased relapse (US Department of Health and Human Services, 1994). In addition, tobacco use has historically been a part of the culture of mental health services and has often been overlooked, or "accepted," by family, peers, and health workers (Lawn et al., 2002).

Although the rate of smoking by people with mental illness is very high and the impact on their health and quality of life is serious, very few people with mental illness are asked about their tobacco use. Furthermore, very few are encouraged to consider quitting tobacco or provided with smoking cessation information and support that are tailored to their needs (Lawn et al., 2002; Williams & Foulds, 2007).

There are a number of common beliefs that are untrue and yet persist within the health services and prevent people with mental illness from receiving appropriate support (Prochaska, 2011). There is an ongoing belief that tobacco use is an effective form of self-medication for people with mental illness (Prochaska, 2011). Documents from within the tobacco industry have shown that they have deliberately fostered this belief by funding research on the self-medication theory, publishing in the lay press and on the internet, and marketing directly to people with mental illness (Prochaska, Hall, & Bero, 2008). Although nicotine temporarily enhances concentration and attention, regardless of the smoker's mental health status, there is no evidence that in the longer term smoking helps to improve symptoms of mental illness or that mental illness deteriorates after quitting (Prochaska, 2011). On the contrary, there is growing evidence that smoking cessation is associated with improvements in mental health. A longitudinal study with data from 4800 daily smokers found that cessation was associated with reduced risk of mood, anxiety, and alcohol dependence at the 3-year follow-up point (Cavazos-Rehg et al., 2014).

Another common belief is that people with mental illness are not interested in addressing their tobacco use. This belief persists despite the Quitline reporting that between 20% and 33% of the callers seeking help report that they have a mental illness (Morris, Tedeschi, Waxmonsky, May, & Giese, 2009). Of the 66.6% of people in the Australian study of people with a psychotic disorder who were smokers, 70% had tried to stop smoking. Nearly a third of these had made a quit attempt in the previous 12 months (Cooper et al., 2012).

It is also commonly thought that people with mental illness cannot stop smoking. Several studies have shown that this is untrue. The Tobacco and Mental Illness Project in South Australia provided group programs primarily for people living with severe and disabling mental illness and reported a cessation rate of 19% at 12 months (Ashton, Rigby, & Galletly, 2013a). Another Australian study (Baker et al., 2006) provided an eight-session individual intervention for smokers with psychotic disorders and 30% of those who completed all eight sessions achieved cessation at 3 months. A follow-up study found that smokers with a psychotic disorder are able to maintain long-term smoking cessation or reduction and that mental health symptomatology and functioning had improved at the 4-year follow-up point (Baker, Richmond, Lewin, & Kay-Lambkin, 2010). A 2013 review of the literature on smoking cessation interventions for people with schizophrenia found that both pharmacologic and psychosocial smoking cessation treatments were useful in helping people with schizophrenia reduce and quit smoking in the short term (Bennett, Wilson, Genderson, & Saperstein, 2013).

There is a belief that population-wide approaches such as media campaigns, increasing the price of tobacco products, and smoke-free legislation will be as effective for smokers with mental illness as they have been across the general population. There is very little evidence to show that these approaches have helped people with mental illness to stop smoking, and in some cases there is evidence they have added to their difficulties (Ashton, Rigby, & Galletly, 2013b; Bader, Boisclair, & Ferrence, 2011; Greaves et al., 2006; Thornton, Baker, Johnson, & Kay-Lambkin, 2011).

FACILITATING SMOKING CESSATION

Given the high rate of tobacco use by people with mental illness and the very serious impact on their health and quality of life, it is critical that tobacco control services and mental health services work together to overcome the barriers and look to developing and implementing a comprehensive range of effective strategies.

Smoking cessation guidelines for health professionals have identified a number of important strategies, and there is good evidence these strategies are effective at increasing the rate of smoking cessation across the general population (US Department of Health and Human Services, 2008; Zwar et al., 2011). However, there is less clear evidence these approaches alone are effective for people living with mental illness. Much more research focused on cessation approaches for this population is urgently needed.

There is growing evidence that smokers with mental illness can quit tobacco if they are provided with information and support that are similar to those recommended for the general community but adapted and tailored specifically to meet more accurately their needs (Ashton et al., 2013a; Baker et al., 2010; Bennett et al., 2013).

The Role of Health Professionals

Smoking cessation advice and support from health professionals are very important in prompting smokers to think about their tobacco use, motivating them to reduce and stop smoking, and providing information and support.

The US Department of Health and Human Services (2008) recommended that all health professionals systematically identify smokers, assess their smoking status, and offer them advice and cessation treatments. There is evidence that if these are provided by health professionals, the rate of cessation is increased (US Department of Health and Human Services, 2008).

It is clear from several studies that many smokers living with mental illness are concerned about their tobacco use and many have made quit attempts (Cooper et al., 2012; Ashton, Rigby, & Galletly, 2013c). Asking all people with mental illness about tobacco use as a routine component of health care will ensure they are aware that the health worker and service are concerned about tobacco use and are willing to provide information and support that will enable them to seek help more easily.

Pharmacotherapy

Nicotine replacement therapy (NRT), varenicline, and bupropion have all been shown to be effective in increasing cessation rates, especially when provided in combination with counseling and support (Zwar et al., 2011).

A review of the literature by Bennett et al. (2013) on smoking cessation interventions for people with schizophrenia found that the use of NRT, varenicline, and bupropion resulted in increased cessation rates.

Counseling and Psychosocial Support

The review by Bennett et al. (2013) also found that psychosocial interventions can help people with schizophrenia to change their smoking behavior. These programs were mostly delivered in group settings and involved psychoeducation, relapse prevention strategies, cognitive behavioral therapy, and skill-building approaches.

There is other evidence that individual, group, and telephone counseling increases quit rates among people living with mental illness (Ashton et al., 2013a; Baker et al., 2006; Morris et al., 2009).

People with mental illness often face additional challenges in their everyday life and these can have an impact on their efforts to tackle tobacco use. Psychosocial interventions need to be adjusted to help participants meet these additional challenges.

Smokers with mental illness are likely to need specific information and support. Some find that the symptoms of nicotine withdrawal resemble and feel like some of the symptoms of their mental illness. While smoking relieves nicotine withdrawal for a short time it does not help with the symptoms of mental illness. Such individuals may need information and reassurance about this link.

Some people experience underlying symptoms of mental illness every day, which they have to cope with, and nicotine withdrawal can make these symptoms more difficult to manage. They may require specific information and support to deal with symptoms without smoking.

Stress is commonly linked with smoking and often smokers say they smoke more when they feel stressed, or they smoke to relieve stress. This effect is related primarily to the release of neurotransmitters resulting in a pleasurable feeling and the relief of the stress associated with nicotine withdrawal. Smoking also provides the smoker with a distraction or an opportunity to "take a break." These factors can cause the smoker to associate smoking with the relief of stress, and they can begin to rely on the cigarette when they feel stressed rather than develop and use healthier coping mechanisms. Some people with mental illness have high levels of stress caused by their illness, lifestyle, lack of social support, and other environmental factors. Managing stress without smoking is an important step in tackling tobacco use and hence needs to be included in smoking cessation programs.

Some people with mental illness have significant disabilities associated with their mental illness and the side effects of medications. As a result, many of these individuals have difficulties managing everyday tasks. For example, some find it difficult to concentrate and take in information, others have trouble organizing their time and daily needs. They may have poor coping skills, low self-esteem, and more difficulty managing practical tasks. Smoking cessation programs for people with mental illness need to take into account the difficulties that individuals may be experiencing and the programs must be adapted accordingly to address these needs.

Some people with mental illness do not have the support of significant others when attempting to quit smoking. Furthermore, some health workers and caregivers lack the skills necessary to support people to address tobacco. Again, adapting cessation programs to allow for these factors can be a significant advantage.

Further Considerations

Some mental health settings are still places where people smoke together and these settings can make it more difficult for those who want to stop smoking. Smoking cessation programs need to be aware of the environmental factors that may make it more difficult for participants. Providing strategies to assist in coping with these factors would be beneficial. Programs may also need to work with other service providers to encourage them to provide support and supportive environments.

The tar in cigarette smoke (not the nicotine) induces certain liver enzymes and this causes the increased metabolism of some medications. As a result, the smoker may need to take a higher dose of medication. Conversely, if the smoker stops smoking, he or she may need to reduce his or her medication. Smoking cessation can result in a reduction of the liver enzymes and an increased blood serum level of the medication. The increased blood level may result in increased side effects. In a small number of cases a serious event has been reported (Cole, Trigoboff, Demler, & Opler, 2010). It is important that smoking cessation programs encourage all participants who are taking medications to work with their medical providers to monitor any changes that may be required. Some psychiatric medications are metabolized by these enzymes and while the risks of continued smoking far outweigh the risks of stopping medication, the effect on medications has concerned some health workers and been a barrier to people with mental illness getting help to stop smoking. It is important that mental health workers are aware of the interactions between some medications and smoking cessation and that they monitor these appropriately.

Many smokers have identified boredom as a trigger for smoking and a cause of relapse. Social isolation, unemployment, and relative poverty are more common among people with severe mental illness, with many having few activities during their week and living on a disability support pension. As a result, some say that they smoke to fill their time. Dealing with boredom and encouraging individuals to link in with social supports and activities may be important for smokers when attempting to quit.

Some smokers are concerned about putting on weight when they stop smoking. People with mental illness may be especially concerned as some psychiatric medications are also linked with weight gain. They may need specific information about exercise, nutrition, and maintaining a healthy weight. Working with other support providers may be particularly useful in this regard.

CONCLUSION

Linked to a number of diseases, tobacco smoking is a leading behavioral risk factor for many diseases and disability. Targeted strategies to reduce the incidence of smoking have been effective, reducing the number of people smoking significantly. The rates of smoking remain high, however, among certain groups, one of which is those with mental illness. This presents a number of issues. Commonly, smoking is seen as a method to reduce the burden of mental illness. At the same time, however, the risk for other health concerns grows. Further to this, those who smoke and have a mental illness often face other stressors and restrictions, which as a result reduces their quality of life significantly.

As it currently stands, the model of care provided to assist smokers with mental illness to quit falls short in many areas. Many factors are not considered. For example, factors such as financial situation, self-medication, other life stressors, social environment, available support, and the implications for medication and mental illness are often not considered. A holistic approach that addresses the neuropathology, the psychological, the social, and the environmental factors is needed if interventions are to be effective at reducing the prevalence of tobacco use among those with mental illness. Encouraging people with mental illness to quit smoking tobacco is a very important clinical and public health issue, and health professionals need to work together to ensure that interventions are targeted specifically for this population.

APPLICATIONS TO OTHER ADDICTIONS AND SUBSTANCE MISUSE

The provision of tailored holistic support for people living with mental illness for other lifestyle risk factors, such as drug and alcohol use, obesity, and lack of exercise, may be effective.

Tailored smoking cessation programs that include both behavioral support and pharmacology are likely to be effective for other high-risk groups such as indigenous smokers, prisoners, and smokers from low socioeconomic communities.

KEY FACTS OF SMOKING AMONG THOSE WITH MENTAL ILLNESS

- In Australia, effective cessation campaigns have contributed to the percentage of adults smoking daily in 1991 dropping by 9% to 15% in 2010.
- The rate of smoking among those with mental illness, however, has remained steady, suggesting such campaigns have not been as effective for this population.

- Research from Australia suggests that approximately two-thirds of those with a psychotic disorder also smoke.
- More than two-thirds of these adults had tried to stop smoking.
- There are a range of factors specific to the population of smokers with mental illness that have an impact on cessation efforts.

SUMMARY POINTS

- The use of tobacco is a leading behavioral risk factor for disease and premature death.
- The rate of tobacco smoking by people with mental illness is very high and they are a large percentage of the remaining smokers.
- Smokers with mental illness are often concerned about their tobacco use and want to stop smoking.
- There are a range of factors that are relevant to smokers with mental illness that are not considered by general population-wide approaches to smoking cessation.
- When smokers with mental illness are provided with effective cessation support, many can stop smoking.

REFERENCES

AIHW. (2011). *2010 national drug strategy household survey report* Drug statistics series no. 25. Cat. no. PHE 145. Canberra: AIHW.

Ashton, M., Rigby, A., & Galletly, C. (2013a). Evaluation of a community-based smoking cessation programme for people with severe mental illness. *Tobacco Control, 24* tobaccocontrol-2013.

Ashton, M., Rigby, A., & Galletly, C. (2013b). Do population-wide tobacco control approaches help smokers with mental illness? *The Australian and New Zealand Journal of Psychiatry, 48*(2), 121–123.

Ashton, M., Rigby, A., & Galletly, C. (2013c). What do 1000 people with mental illness say about their tobacco use? *The Australian and New Zealand Journal of Psychiatry, 47*(7), 631–636.

Australian Bureau of Statistics. (2008). *National survey of mental health and wellbeing: summary of results, 2007*. ABS Cat 4326.0. Canberra: ABS.

Bader, P., Boisclair, D., & Ferrence, R. (2011). Effects of tobacco taxation and pricing on smoking behavior in high risk populations: a knowledge synthesis. *International Journal of Environmental Research and Public Health, 8*(11), 4118–4139.

Baker, A., Richmond, R., Haile, M., Lewin, T., Carr, V., Taylor, R., ... Wilhelm, K. (2006). A randomized controlled trial of a smoking cessation intervention among people with a psychotic disorder. *The American Journal of Psychiatry, 163*(11), 1934–1942.

Baker, A., Richmond, R., Lewin, T. J., & Kay-Lambkin, F. (2010). Cigarette smoking and psychosis: naturalistic follow up 4 years after an intervention trial. *The Australian and New Zealand Journal of Psychiatry, 44*(4), 342–350.

Begg, S., Vos, T., Barker, B., Stevenson, C., Stanley, L., & Lopez, A. (2007). *Burden of disease and injury in Australia, 2003*. AIHW.

Bennett, M. E., Wilson, A. L., Genderson, M., & Saperstein, A. M. (2013). Smoking cessation in people with schizophrenia. *Current Drug Abuse Reviews, 6*(3), 180–190.

Benowitz, N. L. (2010). Nicotine addiction. *The New England Journal of Medicine, 362*, 2295–2303.

Callaghan, R. C., Veldhuizen, S., Jeysingh, T., Orlan, C., Graham, C., Kakouris, G., ... Gatley, J. (2014). Patterns of tobacco-related mortality among individuals diagnosed with schizophrenia, bipolar disorder, or depression. *Journal of Psychiatric Research, 48*, 102–110.

Cavazos-Rehg, P. A., Breslau, N., Hatsukami, D., Krauss, M. J., Spitznagel, E. L., Grucza, R. A., ... Bierut, L. J. (2014). Smoking cessation is associated with lower rates of mood/anxiety and alcohol use disorders. *Psychological Medicine, 44*, 1–13.

Cole, M. L., Trigoboff, E., Demler, T. L., & Opler, L. A. (2010). Impact of smoking cessation on psychiatric inpatients treated with clozapine or olanzapine. *Journal of Psychiatric Practice, 16*(2), 75–81.

Cooper, J., Mancuso, S. G., Borland, R., Slade, T., Galletly, C., & Castle, D. (2012). Tobacco smoking among people living with a psychotic illness: the second Australian survey of psychosis. *The Australian and New Zealand Journal of Psychiatry, 46*(9), 851–863.

Glassman, A. H. (1993). Cigarette smoking: implications for psychiatric illness. *The American Journal of Psychiatry, 150*, 546–553.

Greaves, L., Johnson, J., Bottorff, J., Kirkland, S., Jategaonkar, N., McGowan, M., ... Battersby, L. (2006). What are the effects of tobacco policies on vulnerable populations? A better practices review. *Canadian Journal of Public Health/Revue Canadienne de Sante'e Publique, 97*, 310–315.

Laje, R. P., Berman, J. A., & Glassman, A. H. (2001). Depression and nicotine: preclinical and clinical evidence for common mechanisms. *Current Psychiatry Reports, 3*(6), 470–474.

Laursen, T. M. (2011). Life expectancy among persons with schizophrenia or bipolar affective disorder. *Schizophrenia Research, 131*, 101–104.

Lawn, S., & Pols, R. (2003). Nicotine withdrawal: pathway to aggression and assault in the locked psychiatric ward? *Australasian Psychiatry, 11*, 199–203.

Lawn, S., Pols, R., & Barber, J. (2002). Smoking and quitting: a qualitative study with community-living psychiatric clients. *Social Science & Medicine, 54*, 93–104.

Leonard, S., Adler, L. E., Benhammou, K., Berger, R., Breese, C. R., Drebing, C., ... Freedman, R. (2001). Smoking and mental illness. *Pharmacology, Biochemistry and Behavior, 70*(4), 561–570.

Mauer, B. (2006). Morbidity and mortality in people with serious mental illnesses. In J. Parks, D. Svendsen, P. Singer, & M. E. Foti (Eds.), *Technical report 13*. Alexandria, Virginia: National Association of State Mental Health Program Directors Council.

Morgan, V. A., Waterreus, A., Jablensky, A., Mackinnon, A., McGrath, J. J., Carr, V., ... Saw, S. (2011). *People living with psychotic illness 2010. Report on the second Australian national survey*. Department of Health and Ageing.

Morris, C. D., Tedeschi, G. J., Waxmonsky, J. A., May, M., & Giese, A. A. (2009). Tobacco quitlines and persons with mental illnesses: perspective, practice, and direction. *Journal of the American Psychiatric Nurses Association, 15*(1), 32–40.

Picciotto, M. R., Brunzell, D. H., & Caldarone, B. J. (2002). Effect of nicotine and nicotinic receptors on anxiety and depression. *Neuroreport, 13*(9), 1097–1106.

Pomerleau, O. F., & Pomerleau, C. S. (1984). Neuroregulators and the reinforcement of smoking: towards a biobehavioral explanation. *Neuroscience & Biobehavioral Reviews, 8*(4), 503–513.

Prochaska, J. J. (2011). Smoking and mental illness—breaking the link. *The New England Journal of Medicine, 365*(3), 196–198.

Prochaska, J. J., Hall, S. M., & Bero, L. A. (2008). Tobacco use among individuals with schizophrenia: what role has the tobacco industry played? *Schizophrenia Bulletin, 34*(3), 555–567.

Ragg, M., & Ahmed, T. (2008). *Smoke and mirrors: a review of the literature on smoking and mental illness*. Cancer Council NSW.

Scollo, M. M., & Winstanley, M. H. (2012). *Tobacco in Australia: facts and issues* (4th ed.). Melbourne: Cancer Council Victoria. Available from www.tobaccoinaustralia.org.au.

Taylor, G., McNeill, A., Girling, A., Farley, A., Lindson-Hawley, N., & Aveyard, P. (2014). Change in mental health after smoking cessation: systematic review and meta-analysis. *BMJ: British Medical Journal, 348*, g1151.

Thornton, L. K., Baker, A. L., Johnson, M. P., & Kay-Lambkin, F. J. (2011). Perceptions of anti-smoking public health campaigns among people with psychotic disorders. *Mental Health and Substance Use, 4*(2), 105–115.

US Department of Health and Human Services. (1994). *Preventing tobacco use among youth and young adults: a report of the surgeon general*. Atlanta, GA: US Department of Health and Human Services, Centers for Disease Control and Prevention, National Center for Chronic Disease Prevention and Health Promotion, Office on Smoking and Health.

US Department of Health and Human Services. (2008). *Treating tobacco use and dependence. Update, clinical practice guideline*. Rockville, MD: US Department of Health and Human Service Public Health Service.

Williams, J., & Foulds, J. (2007). Successful tobacco dependence treatment in schizophrenia. *The American Journal of Psychiatry, 164*(2), 222–227.

World Health Organization. (2013). *WHO report on the global tobacco epidemic, 2013: enforcing bans on tobacco advertising, promotion and sponsorship*. World Health Organization.

Wu, Q., Szatkowski, L., Britton, J., & Parrott, S. (2014). Economic cost of smoking in people with mental disorders in the UK. *Tobacco Control, 24* tobaccocontrol-2014.

Zwar, N., Richmond, R., Borland, R., Peters, M., Litt, J., Bell, J., ... Ferretter, I. (2011). *Supporting smoking cessation: a guide for health professionals*. Melbourne: The Royal Australian College of General Practitioners.

Chapter 25

Nicotine Dependence and Schizophrenia

Aniruddha Basu[1], Anirban Ray[2,3]

[1]Department of Psychiatry, National Institute of Mental Health and Neurosciences, Bangalore, India; [2]Department of Child and Adolescent Psychiatry, National Institute of Mental Health and Neurosciences, Bangalore, India; [3]Department of Psychiatry, Calcutta National Medical College, Calcutta, West Bengal, India

Abbreviations

fMRI Functional magnetic resonance imaging
FTND Fagerström test for nicotine dependence
GABA γ-Amino butyric acid
nAChR Nicotinic acetylcholine receptor
NRT Nicotine replacement therapy
PPI Prepulse inhibition
rTMS Repetitive transcranial magnetic stimulation
SMH Self-medication hypothesis
SPEM Smooth pursuit eye movements
TNP Transdermal nicotine patch

INTRODUCTION

Uses tobacco products constantly, burns fingers on stubs, requires supervision to avoid overuse.

Description of a schizophrenia patient with "severe" smoking as per the Elgin Behavioral Rating Scale (Luchins, Goldman, Lieb, & Hanrahan, 1992).

Tobacco is one of the most commonly used psychoactive substances—as per the World Health Organization it alone causes about 6 million deaths yearly. So it is also the most common substance abused in persons with schizophrenia (Hughes, Hatsukami, Mitchell, & Dahlgren, 1986). This close association between schizophrenia and tobacco has been recognized from the era of Eugene Bleuler. Here we first study the epidemiology and clinical features before going to the more relevant question of why there is more tobacco use in schizophrenia, and finally we embark upon its management. Differences in these aspects between a normal smoker and one with schizophrenia are illustrated in Table 1.

Tobacco is taken in various forms—smoking cigarettes is very common in Western industrialized countries, whereas the smokeless variety is widely used in south Asia. The active agent from tobacco is nicotine and it is strongly addictive. While smoking, inhaled nicotine leads to acute central nervous system effects by activating various nicotinic acetylcholine receptors (nAChRs)—ligand-gated ion channels made up of homo- or heteropentameric combinations of 12 subunits (α2–α10 and β2–β4), encoded by the *CHRNA2–10* and *CHRNB2–4* genes, respectively (Leonard & Bertrand, 2001). There is both rapid and phasic stimulation of dopaminergic neurons in the nucleus accumbens of the mesocorticolimbic pathways leading to the development of positive reinforcing effects. Repeated administration leads to receptor desensitization and upregulation, whereas abstinence leads to resensitization and withdrawal. There is also simultaneous stimulation of the glutamatergic and γ-aminobutyric acid (GABA)-ergic pathways (Koob & Moal, 2005). This neuroadaptation is common in all cases of nicotine dependence with some additional features in schizophrenia.

The two most commonly expressed nAChRs in the brain are high-affinity α4β2 nAChRs and low-affinity α7 nAChRs located at high concentration in the hippocampus, thalamus, and frontal, cingulate, and occipital cortices (Leonard & Bertrand, 2001) (see Figure 1). In a person with schizophrenia, nicotine binds with α4β2 receptors, which are upregulated, although less than usual (only 50%). Similarly, surface expression of low-affinity α7 receptors is also less, unlike normal individuals, as proved by different postmortem studies (Freedman, Hall, Adler, & Leonard, 1995). The α4β2 receptor is more important in the initiation of smoking and with minimal nicotine exposure it gets saturated. However, the action on α7 receptors is different, and they have a pathophysiologic role in schizophrenia (Adler et al., 1998). Owing to the explicit drive to saturate the low-affinity α7 receptors, which are responsible for persistent dopamine release, smokers with schizophrenia continue to smoke vigorously. It has been shown that posttranscriptional mechanisms, aberrant interactions between neuronal calcium sensor protein and nicotinic receptors, and altered *trans*-Golgi membrane trafficking lead to changes in overall neurotransmitter release and thereby constitute a core feature of smoking in schizophrenia (Winterer, 2010).

EPIDEMIOLOGY

Up to 90% of patients with schizophrenia have tobacco dependence and this relationship persists even after controlling for confounders like age, gender, socioeconomic status, marital status, and concurrent alcohol use (Hughes et al., 1986). It is associated with both an increased "current smoking" and an increased "ever smoking" by a weighted average odds ratio of 5.9 (95% CI of 4.9–5.7) and 3.1 (95% CI of 2.4–3.8), respectively

TABLE 1 Comparison of Smoking-related Parameters between Normal Smokers and Smokers with Schizophrenia

Items		Normal Smoker	Smoker with Schizophrenia
Epidemiology	Prevalence	10–20%	80–90%
	Gender	M = F	M > F
	Socioeconomic status	Higher, more employment	Lower, more unemployment
	Geographical location	More in rural, traditional, developing societies	More in modern urban developed societies
Pathophysiology	Genetic	Increased expression and upregulation of α7 and α4β2 receptor genes	Decreased expression and downregulation of α7 and α4β2 receptor genes
	Neurocognitive (sensory gating, working memory, executive functions)	Minimal deficits	Deficits present that improve with acute nicotine use even after prolonged abstinence
	Neuroimaging	Decreased anterior cingulate and parahippocampal gray matter volume	Deficit in functional connectivity between dorsal anterior cingulate cortex and limbic region
	Psychosocial	Fewer temperamental issues and social cognition deficits	Disinhibition, impulsivity, deficit in social cognition, and poor coping strategies
Clinical	Related to psychopathology	Lower rates of depression, lower suicide rates, and better social adjustment	More negative symptoms, higher depression and suicidality rates, poor social adjustment
	Related to smoking topography	Decreased craving and nicotine extraction from lower number of cigarettes	More craving, higher nicotine extraction, deeper puffs, shorter interpuff timings
Management	Treatment related	Better motivation and quit rates	Poorly motivated and poor appreciation of health risks, more hospitalization, higher neuroleptic dose, fewer extrapyramidal side effects, and poor quit rates
	Pharmacological	Nicotine transdermal patch more efficacious	Transdermal nicotine patch much less efficacious
		Varenicline may improve some cognitive deficits	Varenicline use recommended but with caution as per guidelines
	Nonpharmacological	Evidence for efficacy of a wide gamut of interventions, from simple advice by physician to group cognitive behavioral interventions or telephone counseling	Limited evidence for a small number of interventions—prominent among them is contingency management

This table illustrates the significant differences between smoking in normal individuals and smoking in schizophrenia in terms of epidemiology, etiology, clinical features, and management.

(see Table 2). This general trend of association is similar in all cultures irrespective of the general prevalence of smoking (de Leon & Diaz, 2005). Not only is the association between schizophrenia and smoking more than in the general population but also more than among those with other severe mental illnesses (McClave, McKnight-Eily, Davis, & Dube, 2010). In cases of heavy smoking the evidence is less robust—partly because of the heterogeneity in the studies and also the non-uniformity in the definition of "heavy smoking." High tobacco intake is a risk factor for the development of noncommunicable illnesses leading to an about 20% decrease in life expectancy (Goff et al., 2005). This causes a huge drain, not only on their health but also on their finances (Steinberg, Williams, & Ziedonis, 2004). The picture is a little different in developing countries, where there is preliminary evidence suggesting that the rate of smoking among patients with schizophrenia is less (one-third in the developing, but at least two-thirds in the developed world) (Chandra et al., 2005). Different social and family situations may be responsible—for example, in India, because of strong family support, there may be restrictions on tobacco use (Srinivasan & Thara, 2002). Whatever the prevalence of tobacco in schizophrenia is, it contributes significantly to morbidity and mortality—so tobacco cessation services in people with schizophrenia should deserve special attention.

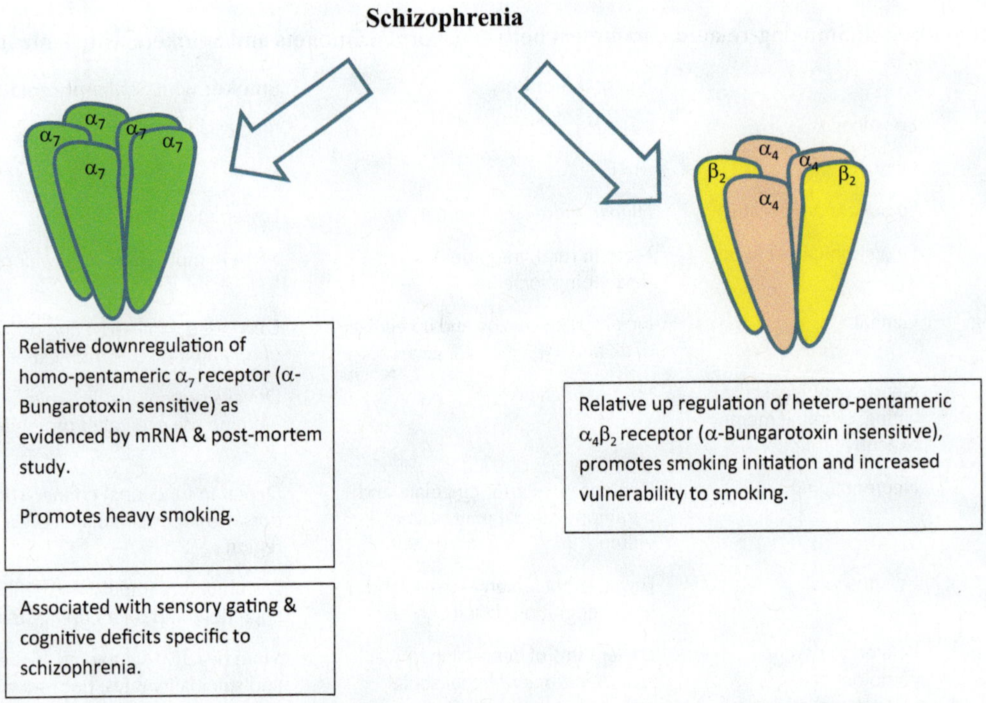

Schizophrenia

Relative downregulation of homo-pentameric α_7 receptor (α-Bungarotoxin sensitive) as evidenced by mRNA & post-mortem study.

Promotes heavy smoking.

Associated with sensory gating & cognitive deficits specific to schizophrenia.

Relative up regulation of hetero-pentameric $\alpha_4\beta_2$ receptor (α-Bungarotoxin insensitive), promotes smoking initiation and increased vulnerability to smoking.

FIGURE 1 **Nicotinic receptor subtypes mainly implicated in schizophrenia.** In schizophrenia, upregulation of $\alpha4\beta2$ leads to early initiation, and downregulation of $\alpha7$ leads to increased persistence and severity of smoking behavior.

CLINICAL FEATURES

The huge burden of tobacco use emphasizes the need for detailed assessment regarding both smoking and smokeless forms in all patients with schizophrenia. Apart from detailed history, patients should be examined for stigmata of tobacco use like nicotine stains in oral mucosa, teeth, and fingers. Breath carbon monoxide measures and biochemical monitoring for nicotine or cotinine levels may be used to corroborate clinical findings. Tobacco use can be assessed with the help of the Fagerström Test of Nicotine Dependence (FTND) for smokers and modified for smokeless tobacco (Steinberg, Williams, Steinberg, Krejci, & Ziedonis, 2005). Although the FTND is a very standard method of assessment, there are certain limitations in severely disabled patients with negative symptoms (Steinberg et al., 2005). Also such patients have very poor appreciation of the health risks, leading to poor motivation for tobacco cessation (Kelly et al., 2012).

The typical clinical profile of patients with nicotine use and schizophrenia is young onset, male, higher extent of negative and cognitive symptoms, more depressive symptoms, high suicidality, lower response to commonly prescribed drugs, frequent relapses and hospitalization, higher dosage requirement of neuroleptic drugs, lesser parkinsonian side effects, and poorer long-term prognosis (Goff, Henderson, & Amico, 1992). However, this generalization is far from being conclusive. Studies correlating the nature of psychotic symptoms (both positive and negative) and smoking have been equivocal (Aguilar, Gurpegui, Diaz, & de Leon, 2005). The severity of dependence may act as a mediator in this relationship—those with severe dependence successfully overcome negative symptoms by increasing their level of nicotine dependence—but which maladaptively leads to increased positive symptoms, treatment with greater antipsychotic dose, poorer social adjustment, and increased morbidity (Krishnadas, Jauhar, Telfer, Shivashankar, & McCreadie, 2012). From a neurobiological perspective, increased release of dopamine in the prefrontal cortex may improve negative symptoms; but release of dopamine in the striatum may give rise to positive symptoms (Brody et al., 2004). The relationship between smoking, negative symptoms, and cognitive symptoms is also complex and not simply related in an inverse manner. After smoking, it has been seen that extrapyramidal symptoms are normalized to a large extent (Goff et al., 1992). The reason may be that nicotine releases dopamine in the nigrostriatal pathway and interacts with hepatic enzymes pharmacologically to induce them, thereby increasing the clearance of antipsychotics. This may cause decreased efficacy of some antipsychotics like clozapine and haloperidol (McEvoy, Freudenreich, Levin, & Rose, 1995). The relationship between tardive dyskinesia and smoking is still inconclusive, the discrepancy being due to methodological differences between studies and also the complex interaction between nicotine, antipsychotic drug level, and nicotinic receptors (Zhang et al., 2011). Smoking topography studies indicate that persons with schizophrenia smoke more intensely, take significantly more puffs, have a shorter interpuff interval, and have larger total cigarette puff volumes compared to matched total healthy control smokers. They also have high craving and withdrawal with the same level of dependence (Tidey, Rohsenow, Kaplan, & Swift, 2005). Differences between normal smokers and schizophrenic smokers are illustrated in Table 2.

ETIOLOGICAL RELATIONSHIP BETWEEN SCHIZOPHRENIA AND NICOTINE

Although some preliminary evidence had shown that schizophrenic patients smoke cigarettes because of significant unemployment or lack of ability to engage themselves in fruitful work,

TABLE 2 Odds Ratios Comparing Frequencies of Smoking-Related Prevalences among Patients with Schizophrenia versus General Population and Patients with Other Mental Illnesses (de Leon & Diaz, 2005)

Items	Comparison Group	Stratification for Gender and Other Variables (After Controlling for Confounders)	Odds Ratio (Confidence Interval)[a]
Current smoking	Whole population		5.3 (4.9–5.7)
		Male	7.2 (6.1–8.3)
		Female	3.3 (3.0–3.6)
	Other severe mental illness		1.9 (1.7–2.1)
		Male	2.3 (2.0–2.7)
		Female	1.8 (1.5–2.3)
		Institutionalized	2.2 (1.6–3.1)
		Other substance abuse	2.7 (1.6–4.4)
		Mental retardation	3.5 (1.1–11.4)
Heavy smoking	Whole population		3.0 (2.3–4.1)[b]
		Male	2.0 (1.5–2.8)[b]
		Female	2.0 (1.1–2.9)[b]
	Other severe mental illness		1.2 (0.8–1.7)
		Male	1.5 (1.05–2.2)
		Female	0.94 (0.45–2.0)
Ever smoking	Whole population		3.1 (2.4–3.8)
		Male	7.3 (1.04–13.6)
		Female	2.8 (1.2–4.4)
	Other mental illness		2.0 (1.6–2.4)
		Male	2.0 (1.5–2.7)
		Female	0.92 (0.44–1.9)
Tobacco cessation	Whole population		0.19 (0.14–0.24)
		Male	0.10 (0.06–0.14)
		Female	0.46 (0.23–0.69)
	Other mental illness		0.55 (0.33–0.90)
		Male	0.34 (0.22–0.54)
		Female	2.7 (0.8–9.0)

Prevalences of smoking-related status such as current smoking, ever smoking, heavy smoking, and smoking cessation are compared between patients with schizophrenia and the general population or those with other mental illnesses after adjustment for gender.
[a]Weighted average of all odds ratios, using schizophrenia sample sizes as weights.
[b]In cases in which the weighted average was not available, we used the odds ratio of individual studies with highest sample size.

such an explanation has been largely shelved because of lack of evidence. The theories used to explain this comorbidity with more empirical evidence are the following (Lybrand & Caroff, 2009), as shown in Figure 2:

- Self-medication hypothesis
- Affect dysregulation model
- Addiction vulnerability hypothesis

Self-Medication Hypothesis

The nomenclature of this hypothesis is self-explanatory and can be traced back to the era of Hippocrates, when he said that wine drunk with an equal volume of water puts away terror and anxiety. However, it was formally propounded in 1985 by eminent psychiatrist Khantzian that patients suffering from psychiatric illnesses take substances to relieve themselves

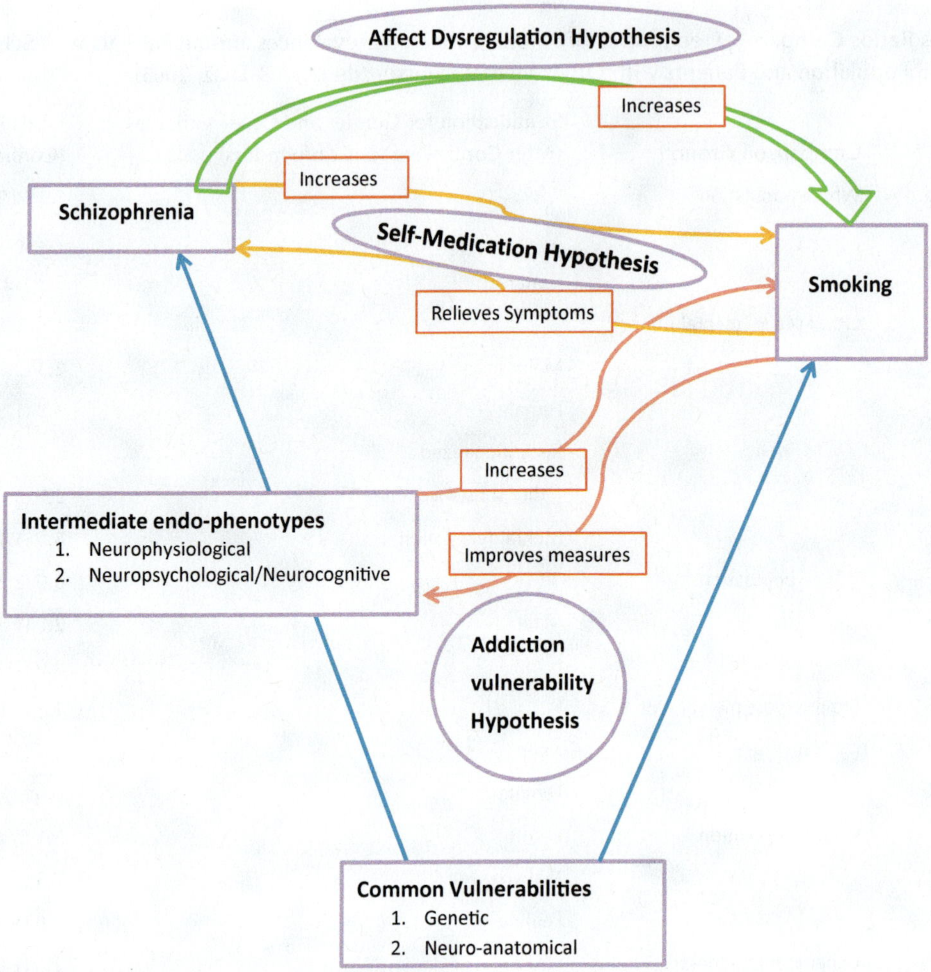

FIGURE 2 Overview of theoretical models of smoking in schizophrenia. Various theoretical models complement each other to explain the increased smoking in schizophrenia.

from the psychopathology or psychological suffering—known as the "causation postulate" (Khantzian, 1985). He also said that the choice of substances is predicted by a particular set of psychopathology symptoms—the "specificity postulate." Implicit in the self-medication hypothesis (SMH) as a corollary is the "treatment postulate"—treating the underlying psychiatric illness will resolve the addictions. The causation postulate of Khantzian was used as an alternative to the commonly used notion that substance precipitates psychosis. For cases in which substance use preceded obvious psychiatric symptoms he argued with empirical evidence that most of the schizophrenic patients have a long prodromal period characterized by social maladjustment and intellectual deterioration. During this period they self-medicate themselves to get rid of the pain, affect, rage, and suffering. So later he revised the "self-medication" to the "self-regulation" hypothesis (Khantzian, 1997). SMH was supported by biological studies that showed that the use of nicotine in schizophrenic patients relieved much of the dysfunction due to negative, cognitive, and mood symptoms (Kumari & Postma, 2005). Increased dopamine secretion in the prefrontal cortex may be responsible. Other biological and neurocognitive markers of schizophrenia may also be normalized, as illustrated in Figure 3. Despite such empirical support, his theory faced widespread criticism.

Although Khantzian had cited empirical research for his causation postulate, those studies were either narrative or cross-sectional in nature. Long-term prospective studies like the Israeli conscript study on 4000 ever-smoker adolescents followed up longitudinally for 4–16 years had higher prevalence of smoking in future schizophrenics—thereby questioning the causation postulate (Weiser et al., 2004). Another Swedish conscript study did not replicate the findings (Zammit et al., 2003). Schuckit and Smith (1996), in a follow-up study involving sons of alcoholics for over 8 years, established that most of the depression and anxiety associated with substance use are related to its acute or chronic effects and clear with significant abstinence. In response to this criticism, Khantzian (1997) asserted that the sample of children of alcoholics suffered from selection bias and that the authors had not considered the subsyndromal difficulties in affect regulation, anger, painful affect, and other developmental suffering that had prompted the subjects to indulge in substance use. Evidence also accumulated against the specificity postulate—in the real world patients with anger and aggression do not always choose opioids and depressives

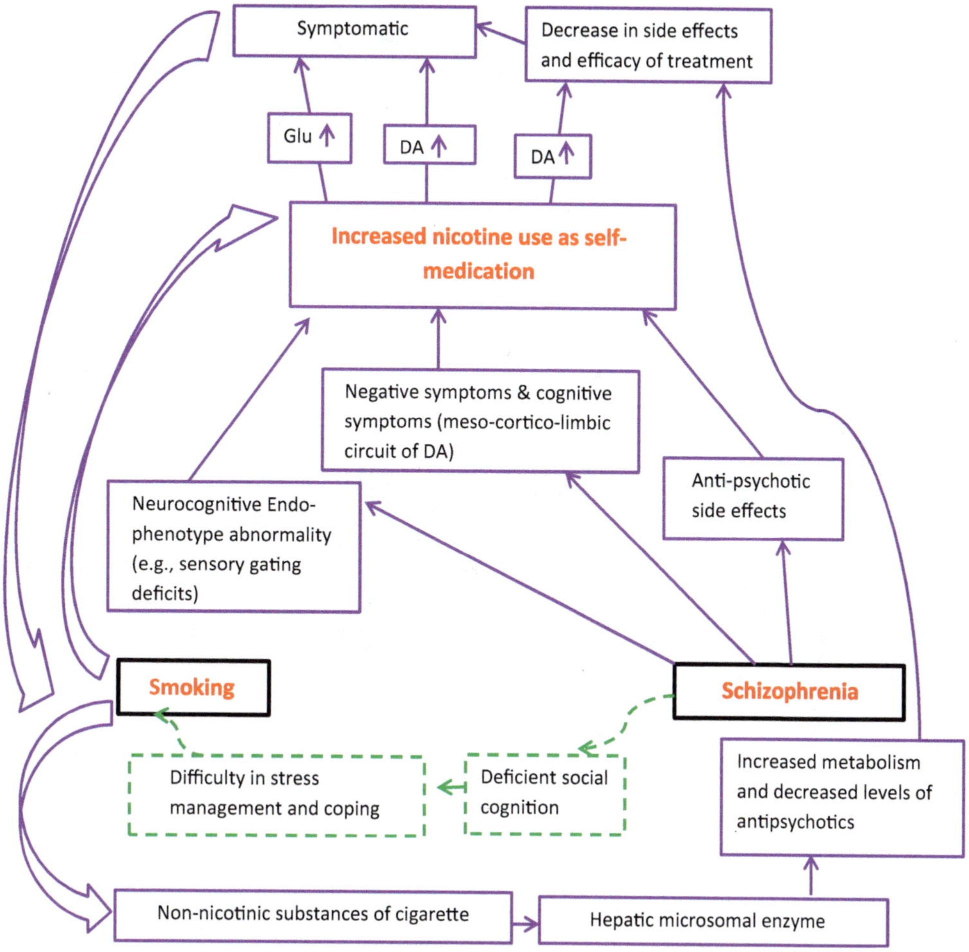

FIGURE 3 Self-medication and affect dysregulation hypotheses in schizophrenia. Violet solid lines (———) denote SMH and green dashed lines (---) denote affect dysregulation hypothesis of higher tobacco dependence among patients with schizophrenia. DA, dopamine; Glu, glutamate.

and instead use stimulants (Aharonovich, Nguyen, & Nunes, 2001). As far as the treatment postulate is concerned, Khantzian faced the strongest criticism. It is well known that treatment of the underlying mood and psychotic symptoms does not treat the substance use disorder. In fact, patients with psychiatric illness also face the usual need for recreation and social interaction and have personality traits that make them more vulnerable to substance use, and for this reason both problems need to be addressed simultaneously. As per Lembke (2012), the SMH has done major harm to the treatment philosophy of schizophrenia and co-occurring substance use disorder, as the patient may consider his or her addiction to be innocuous and indulge in minimization and denial. Hence, the SMH, though useful for many patients, may not be the only explanation for this comorbidity and we need other explanatory models.

Affect Dysregulation Theory

According to this model proposed by Blanchard, Brown, Horan, and Sherwood (2000), temperamental factors like impulsivity/disinhibition, neuroticism/negative affect, and novelty seeking are predictors of increased substance use in schizophrenia. Coping strategies and other problem-solving skills might be the mediators between stress and the development of substance use. Owing to a deficit in social cognition, patients with schizophrenia are more prone to increased stress and substance use, as shown in Figure 3 (Blanchard, Mueser, & Bellack, 1998). Hence, this model points out significant differences between self-regulation or self-medication and affect regulation. Whereas the former deals with the emotional, cognitive, and behavioral disturbances in the acute situation, the latter deals with the same characteristics as part of the pervasive temperamental pattern.

Addiction Vulnerability Hypothesis

The observation of age of initiation of smoking being years before the onset of schizophrenia led to a search for alternatives to the SMH (de Leon & Diaz, 2005). A discordant twin study, with 24 pairs collected from a large twin male sample, found that unaffected co-twins had a frequency of ever daily smoking (88%) similar to that in male schizophrenia probands (83%) and higher than that in male twin controls (66%) (Lyons et al., 2002). This led to the "addiction vulnerability" or "shared neurobiology" hypothesis wherein the genetic, neurophysiological, cognitive, and other neurobiological predisposing factors are found to be responsible for increased nicotine use in schizophrenia, as shown in Figure 4.

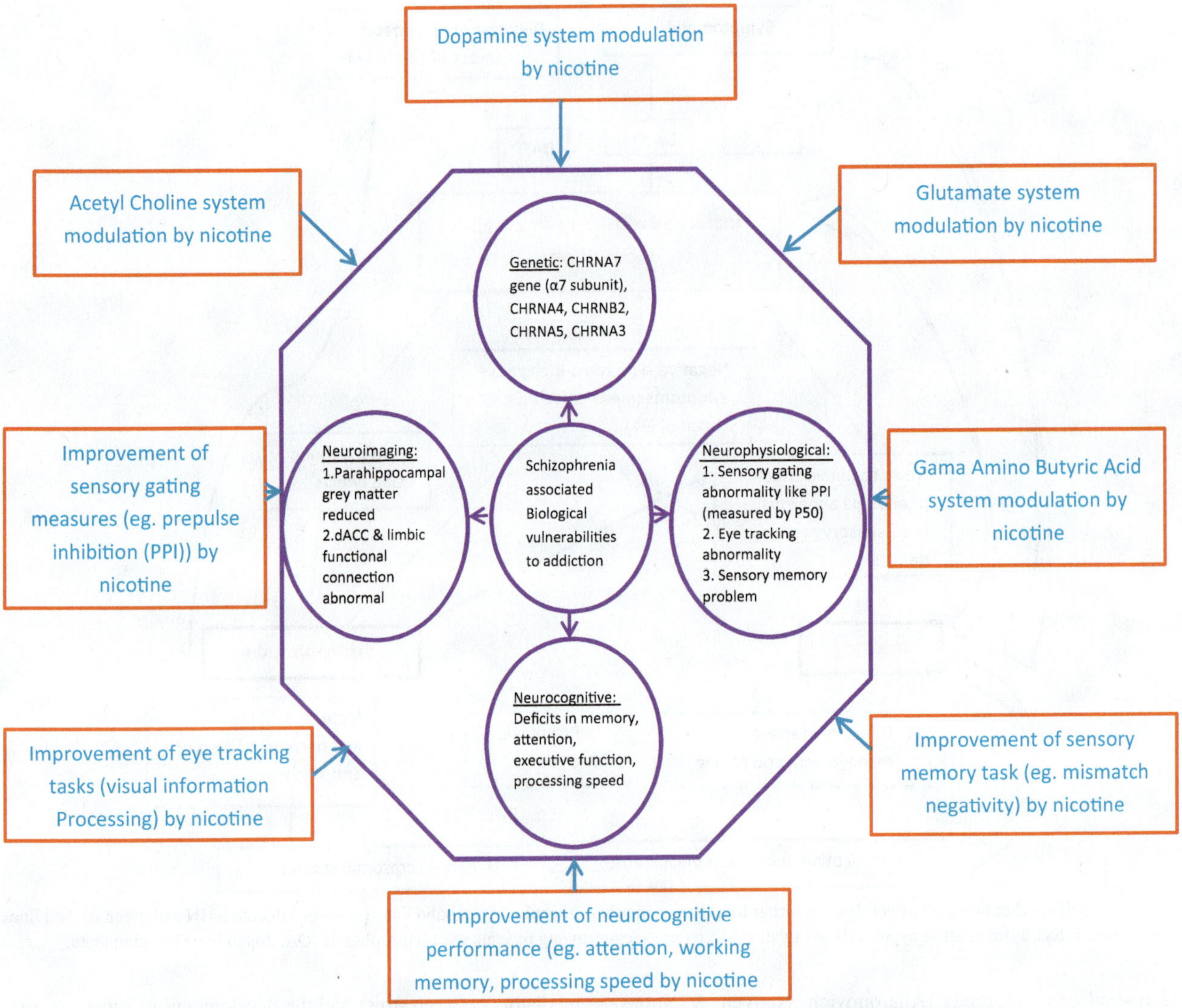

FIGURE 4 Outline of pathophysiology of smoking in schizophrenia—an insight into the addiction vulnerability hypothesis. Common biological vulnerabilities associated with schizophrenia and their interactions with nicotine explain the high comorbidity. *CHRNA7*, genes encoding cholinergic receptor (nicotinic) α7 (likewise A3, A4, A5, B2: similar receptors α3, α4, α5, β2); dACC: dorsal anterior cingulate cortex.

Genetic Studies

Of the α and β genes constituting the nicotinic receptors, particularly α7 is thought to be a common gene implicated in both the processes of cognitive dysfunction of schizophrenia, like prepulse inhibition (PPI), SPEM (smooth pursuit eye movements), and P50 deficits and nicotine use. From linkage studies and candidate gene research we got an early indication that the α7 receptor subunit located at the 15q13–q14 region is associated with schizophrenia and smoking. However, small sample sizes in underpowered replication studies led to inconsistency in the results (Leonard et al., 1996). A meta-analysis of genome-wide association studies in the general population did not find such a result, though it found an association between the related genes *CHRNA3–CHRNA5–CHRNB4* at the 15q25 locus, encoding the α3, α5, and β4 subunits, and smoking (Tobacco and Genetics Consortium, 2010). In schizophrenia patients of both Caucasian and African American origins, a functional smoking-related nAChR gene (*CHRNA5*) polymorphism, Asp389Asn, is associated with "excessive smoking" but not with smoking status only (Hong et al., 2011). Because of differences in ethnicity, allele frequency, and other methodological issues, the results are inconsistent and need replication in other populations. Not only genes related to nicotinic receptors and dopaminergic or monoaminergic pathways, but also the opioid and neuregulin receptor genes are implicated in smokers with schizophrenia (Loukola et al., 2014). These genes, as well as gene × environment interactions in the form of epigenetics, may predict the underlying vulnerability of smoking and schizophrenia. Hence, though the specific genetic associations are not confirmed, there is ample evidence to believe that smoking and schizophrenia are genetically linked.

Neuroimaging Studies

Another approach to understanding the comorbidity between schizophrenia and smoking is to study the related brain connectivity. Early evidence in this regard was obtained from animal studies (Chambers, Sentir, Conroy, Truitt, & Shekhar, 2010). In humans, however, we have to rely on functional magnetic resonance imaging (fMRI) studies, as structural imaging studies in this domain are sparse. Decreased volume in the medial temporal lobe, parahippocampal gray matter, and amygdala is a common finding in schizophrenia. In tobacco users, decreased anterior cingulate and parahippocampal gray matter volume has been found. So, it has been hypothesized and thereafter shown with resting state fMRI that there is a functional connectivity between the dorsal anterior cingulate cortex and the limbic region that is inherently abnormal in schizophrenia, is present in their first-degree relatives, and overlaps with a nicotine-addiction-related circuit (Moran, Sampath, Kochunov, & Hong, 2013). This is also the basis of aberrant salience attribution in schizophrenia leading to increased propensity to developing incentive salience to drug-related cues and thereby nicotine dependence. In diffusion tensor imaging studies, the finding that there is a common and additive effect of schizophrenia and smoking on white matter fibers within the anterior thalamic radiation/anterior limb of the internal capsule indicates an underlying neurobiological link (Zhang, Stein, & Hong, 2010).

Neurophysiological Studies

"Sensory gating" is said to be the central event in explaining the underlying pathophysiology of delusions and hallucinations in schizophrenia. It is measured by PPI in which the prepulse is thought to evoke inhibitory mechanisms that limit response to further stimulation until the stimulation is fully processed. It is assessed by P50, which is an electroencephalographic event-related potential wave elicited 50 ms after exposure to an auditory stimulus. Also the eyeblink response measured by electromyography of the orbicularis oculi muscle has frequently been used to measure PPI, which has strong genetic and biological correlates. Another importance of sensory gating is that it is also deficient in first-degree relatives who have not currently developed the psychopathology. Now, nicotine influences sensory gating in two possible mechanisms (Adler, Hoffer, Wiser, & Freedman, 1993). It has been seen that nicotine challenge acutely improves P50 suppression—though this is dependent on diagnosis, genotype, and baseline gating levels. Nicotine stimulates the release of glutamate following activation of α7 receptors, which facilitate the release of GABA from interneurons, leading to inhibition of the hippocampal CA3 and CA4 pyramidal neurons, the net result being suppression of the subsequent response to the next stimulus (Adler et al., 1998; Kumari & Postma, 2005). The effect is short-lived because nicotinic receptors desensitize rapidly. In chronic smokers with schizophrenia there is also improvement, but the mechanism appears to be through inhibition of monoamine oxidase, which affects dopamine metabolism (Song et al., 2014).

SPEM and antisaccadic tasks are two types of eye movements that have received attention in schizophrenia research. Both experimental and clinical studies have shown that nicotine improves them—in fact, among currently known chemicals, nicotine is the only drug that improves SPEM (Tanabe, Tregellas, Martin, & Freedman, 2006). One of the possible mechanisms might be activation of the nAChRs (α7) on inhibitory interneurons, particularly those in the hippocampus and anterior cingulate gyrus (Avila, Sherr, Hong, Myers, & Thaker, 2003). Mismatch negativity, which is a measure of auditory sensory memory, is similarly deficient in schizophrenia and improves with acute nicotine administration.

Neuropsychological Studies

In schizophrenia, deficits particularly in the domains of attention, verbal memory, working memory, executive functions, and cognitive processing speed are seen (Mackowick et al., 2014). With acute nicotine challenge in abstinent or never-smokers, significant improvements are observed, implying that nicotine really improves cognition in schizophrenia and it is not merely restoration of deficits caused by nicotine deprivation (Harris et al., 2004). However, given the overall harm of tobacco use, this should never be a basis for discouraging patients from its cessation or reduction. More objective evidence from fMRI studies shows that the cognitive effects may involve brain regions such as anterior cingulate and thalamus. In cases of more complex cognitive functions like sustained attention or working memory, normalization or enhancement of brain function is not always observed. Hence, the effect of acute nicotine administration in patients with schizophrenia may depend upon the task involved. Studies related to the chronic effect of nicotine on cognitive functions are still inconclusive and there is a dearth of studies dealing with this issue (Friedman et al., 2008; Thoma & Daum, 2013).

Improvement of cognitive functions in schizophrenia by nicotine may point toward the SMH. However, from the foregoing discussion we understand that in schizophrenia a biological addiction-vulnerable state may be produced through intermediate endophenotypes, namely PPI deficit, dysregulated mesolimbic and mesocortical neurocircuitry, and cognitive deficits like inattention and impaired working memory, which improves with nicotine (Thoma & Daum, 2013). These, along with social, cultural, and environmental risk factors, help to initiate and maintain tobacco addiction in schizophrenia.

MANAGEMENT

Smoking increases medical morbidity and mortality in schizophrenia. So, currently all standard treatment guidelines have made it mandatory for all persons with schizophrenia to be treated for tobacco dependence instead of using it as a reward in contingency management. Owing to the very poor tobacco cessation rates, high risks of relapse, and poor 6-month abstinence rates, treating them is a challenge (de Leon & Diaz, 2005). The current consensus is to approach these patients in the integrated model involving pharmacotherapy, psychotherapy, and various psychosocial interventions by a multispecialty team.

Pharmacological

Among pharmacological treatments, nicotine replacement therapy (NRT), bupropion, and varenicline have been recommended by various guidelines.

Nicotine Replacement Therapy

The various options available are nicotine gums, patches, lozenges, spray, and inhalers. Among them the most studied in schizophrenia patients is the transdermal nicotine patch (TNP), which is one of the efficacious modes of treatment of smokers in the general population. But among patients with schizophrenia, the evidence is less convincing, with one Cochrane review not able to find enough good-quality studies in this field (Punnoose & Belgamwar, 2006). Another Cochrane review even opined that at present there is no evidence supporting the use of nicotine replacement in schizophrenia (Tsoi, Porwal, & Webster, 2013). Several reasons for decreased efficacy have been proposed, like less availability of low-affinity α7 receptors, which are reduced in both number and function in schizophrenia. Hence, to satisfy the high cue-induced craving there should be high plasma nicotine levels, which can be achieved only with rapid modes of nicotine delivery, like nasal spray, compared to TNP (Williams et al., 2008). However, in acute agitated psychotic patients in the emergency ward, application of TNP significantly decreased instances of violence and aggression (Allen et al., 2011). In patients who have assumed abstinence, preliminary studies have shown that TNP along with group sessions can be used for relapse prevention for durations of up to 6 months (Dale Horst, Klein, Williams, & Werder, 2005). Although there has been some limitation in the evidence base of NRT as a singular therapy, its efficacy as an adjunct to psychosocial interventions and other forms of pharmacotherapy like bupropion is well established (Tsoi et al., 2013). NRTs are in general well tolerated with mostly local side effects, like nasal irritation for nasal spray.

Bupropion

This is an atypical antidepressant with both dopaminergic and adrenergic actions. It acts as a noncompetitive antagonist at nAChRs. Bupropion is metabolized by the cytochrome P450 system in the liver and may interact with commonly used antipsychotic medications. The bupropion sustained-release preparation is generally started at the dosage of 150 mg per day and increased to 300 mg within a week of initiation. The maximum dosage is 450 mg per day. However, at higher dosage there is an increased risk of seizures, which may be a contraindication for the use of bupropion. Since the quit date is generally 2 weeks after the onset of medication, patients are actively advised to use NRT in the first few weeks. Initially, bupropion use was discouraged in schizophrenia because bupropion was considered to exacerbate psychotic symptoms. One of the guidelines even opined against bupropion. However, in a 2010 meta-analysis it was seen that smokers with schizophrenia who used bupropion had two and a half times higher rates of abstinence and significant smoking reduction rates at the end of 6 months compared to placebo (Tsoi, Porwal, & Webster, 2010). Further long-term studies with sustained-release bupropion used up to 1 year are also reported to be safe, efficacious, and well tolerated, although randomized controlled trials are not available. Better results were shown after combining bupropion with nicotine patch—however, head-to-head comparisons between bupropion and NRT have not been done (Tsoi et al., 2013). As far as mental status is considered, there has not been any deterioration—nor any reported seizures. Dry mouth, jitteriness, lack of concentration, light headedness, and muscle stiffness were other reported side effects, though overall it was well tolerated.

Varenicline

This is a partial α4β2 and a full α7 nAChR agonist. As a partial agonist it causes some dopamine release in the nucleus accumbens and prefrontal cortex and thereby causes less nicotine withdrawal and craving, at the same time blocking the effects of nicotine in tobacco smoke. There is a much smaller chance of varenicline interacting with other metabolizing enzymes, as it is not metabolized by the cytochrome system and 92% of the varenicline is excreted unchanged in the urine. Apart from nausea, sleep disturbances, and dizziness, which are common side effects, there were also reported suicidality, hostility, and aggression for which the US Federal Drug Administration has issued a black box warning. Also there has been an emergence of cardiovascular adverse events. After 12 weeks of treatment with varenicline it was seen that participants who took varenicline were around five times as likely to quit at the end of 6 months (Williams et al., 2012). For nonabstinent smokers on varenicline, though, there was a significant decrease in smoking at 12 weeks—yet it did not persist up to 6 months. The patients tolerated it well, with no exacerbation of psychiatric symptoms and rather an improvement in cognitive symptoms. Although a Cochrane review illustrates the efficacy of varenicline in smokers with schizophrenia, a meta-analysis is contradictory, leaving it to be settled in future studies (Kishi & Iwata, 2014; Tsoi et al., 2013).

Others

These are the core medications considered for the treatment of psychotic symptoms. The first generation of antipsychotics, particularly haloperidol, is known to exacerbate smoking, for which the reason may be to relieve oneself of the side effects of antipsychotics, like akathisia and psychopathology, particularly depressive symptoms. Among atypical antipsychotics, clozapine has been found to reduce smoking. Medications like nortriptyline, selegiline, and clonidine, which have been very extensively studied in normal smokers, have not been studied in schizophrenia. On the other hand, naltrexone, galantamine, atomoxetine, and topiramate have been studied but not found to be useful in this population (Tsoi et al., 2013). Some somatic therapies may be helpful, like high-frequency (20 Hz) repetitive transcranial magnetic stimulation as an adjunct to NRT and group therapy (Wing, Bacher, Wu, Daskalakis, & George, 2012). E-cigarettes, which are battery-powered nicotine delivery devices, have been found to help unmotivated patients, though as of this writing their use is restricted by the recent caution issued by the World Health Organization (Caponnetto, Auditore, Russo, Cappello, & Polosa, 2013). A harm-reduction approach may be tried by decreasing the nicotine content in cigarettes (Tidey, Rohsenow, Kaplan, Swift, & Ahnallen, 2013).

Nonpharmacological

All smokers with schizophrenia should be offered some psychosocial intervention, ranging from single-session motivational intervention to dyadic psychoeducation, cognitive behavior therapy, and contingency management, in both individual and group settings (Williams & Foulds, 2007), along with medications. If cessation is achieved, relapse prevention becomes the goal, but if complete abstinence is not attained, smoking reduction should be tried. We need to personalize our treatment approaches so that they can be modified as per the psychopathology, motivation, and overall clinical condition.

APPLICATION TO OTHER ADDICTIONS AND SUBSTANCE MISUSE

Similar to the trajectory of tobacco dependence in schizophrenia, other substances also have a difficult course and poor cessation rates (Lybrand & Caroff, 2009). As the addiction vulnerability and self-medication hypotheses have tried to explain smoking in schizophrenia, they may also try to explain other substance use. Whatever theoretical model may be followed, we need an in-depth understanding of the common underlying neurobiological factors. So, the need arises for targeting the biological basis of the intermediate cognitive endophenotypes like the α7 nAChRs, which may play a central role in the pathophysiology, dysfunctional reward circuitry, cognitive impairments, and resulting substance use in schizophrenia (Pohanka, 2012). Hence, though still nascent, this field of nicotine research in schizophrenia can be translated to the management of other substance use in schizophrenia ranging from alcohol to cocaine.

DEFINITION OF TERMS

Smoking topography studies These are studies concerning the typical style of smoking of individuals, including puffs per cigarette, frequency, volume, etc.

Linkage studies These are studies that determine the genes that are closely situated on a particular chromosome and tend to be inherited together—thereby said to be "linked" to one another.

Genome-wide association studies This is a population genetics approach in which whole-genome scans for more than 1 million common marker single-nucleotide polymorphisms are undertaken to find disease association.

Prepulse inhibition When a comparatively low-intensity stimulus is presented just before a high-intensity stimulus, then the response or reaction to the high-intensity stimulus is dampened. This phenomenon is deficient in disorders like schizophrenia.

P50 This is an event-related potential expressed as a positive wave in an electroencephalogram after a 50-ms presentation of a stimulus in the form of an audible click. It can act as a measure of sensory gating.

Noradrenaline dopamine reuptake inhibitor This is a type of antidepressant, like bupropion, which increases both dopamine and noradrenaline in synaptic clefts, secondary to inhibition of reuptake of both these neurotransmitters.

Nicotinic receptor partial agonist These are substrates of nAChRs that can partially stimulate the receptors but prevent other full agonists like nicotine from binding to them.

Contingency management This is a technique of behavior therapy in which desirable behaviors are reinforced positively and undesirable behaviors are not rewarded or are punished (rarely), so that the former increase in frequency.

Antisaccade task This is an eye-tracking task in which the patient is directed to suppress the impulse to look toward an emergent light dot on a computer monitor and look toward the mirror image position.

KEY FACTS

α7 Nicotinic Acetylcholine Receptors: (Pohanka, 2012)

- This is a protein encoded by the gene *CHRNA7*, at 15q14, sharing a locus with juvenile myoclonic epilepsy and schizophrenia.
- It is a homopentameric membrane-bound ligand-gated ion channel with three transmembrane loops and one cytoplasmic loop.
- Its natural agonists are acetylcholine and nicotine, whereas synthetic agonists are clozapine (antipsychotic), CNI 1493, DMXB-A, and ABT-107 (under trial), which improve cognition in schizophrenia and dementia.
- Its natural antagonists are α-conotoxin and α-bungarotoxin, whereas synthetic antagonists are memantine (an NMDA-receptor antagonist—medication for dementia) and HI-6 (asoxime; an antidote for organophosphorus poisoning).
- Peripherally it is associated with the macrophage-mediated inflammatory response.

SUMMARY POINTS

- Prevalence of tobacco use is high in patients with schizophrenia.
- Owing to upregulation of high-affinity α4β2 nAChRs, persons with schizophrenia are more readily addicted to tobacco, and owing to downregulation of low-affinity α7 nAChRs, they smoke more heavily.
- Negative symptoms and antipsychotic-induced extrapyramidal side effects improve with smoking-induced dopamine surge; thus schizophrenia patients are said to self-medicate with tobacco (SMH).
- Some biological vulnerabilities in schizophrenia (addiction vulnerability hypothesis), like genetic (α7 nAChR, α3, β4, α5 nAChRs, neuregulin, etc.), neurological (abnormal connection between anterior cingulate cortex and limbic system), neurophysiologic (sensory gating abnormality, SPEM, antisaccade task abnormality), and neurocognitive deficits (deterioration of cognitive ability, e.g., inattention) predispose to nicotine addiction.
- Treatment of smokers with schizophrenia may include the following:
  - Nonpharmacological therapy: motivation enhancement, contingency management, and cognitive behavior therapy.
  - Pharmacological therapy:
    - NRT: patch, gum, or capsules
    - Bupropion: noradrenaline and dopamine reuptake inhibitor
    - Varenicline: α4β2 partial agonist and α7 full agonist.

Overall, it is a challenge to treat smoking in schizophrenia, with very low cessation rates.

REFERENCES

Adler, L. E., Hoffer, L. D., Wiser, A., & Freedman, R. (1993). Normalization of auditory physiology by cigarette smoking in schizophrenic patients. *The American Journal of Psychiatry*, 150(12), 1856–1861.

Adler, L. E., Olincy, A., Waldo, M., Harris, J. G., Griffith, J., Stevens, K., ... Freedman, R. (1998). Schizophrenia, sensory gating, and nicotinic receptors. *Schizophrenia Bulletin*, 24(2), 189–202.

Aguilar, M. C., Gurpegui, M., Diaz, F. J., & de Leon, J. (2005). Nicotine dependence and symptoms in schizophrenia: naturalistic study of complex interactions. *The British Journal of Psychiatry: The Journal of Mental Science*, 186, 215–221. http://dx.doi.org/10.1192/bjp.186.3.215.

Aharonovich, E., Nguyen, H. T., & Nunes, E. V. (2001). Anger and depressive states among treatment-seeking drug abusers: testing the psychopharmacological specificity hypothesis. *The American Journal on Addictions/American Academy of Psychiatrists in Alcoholism and Addictions*, 10(4), 327–334.

Allen, M. H., Debanné, M., Lazignac, C., Adam, E., Dickinson, L. M., & Damsa, C. (2011). Effect of nicotine replacement therapy on agitation in smokers with schizophrenia: a double-blind, randomized, placebo-controlled study. *The American Journal of Psychiatry, 168*(4), 395–399. http://dx.doi.org/10.1176/appi.ajp.2010.10040569.

Avila, M. T., Sherr, J. D., Hong, E., Myers, C. S., & Thaker, G. K. (2003). Effects of nicotine on leading saccades during smooth pursuit eye movements in smokers and nonsmokers with schizophrenia. *Neuropsychopharmacology: Official Publication of the American College of Neuropsychopharmacology, 28*(12), 2184–2191. http://dx.doi.org/10.1038/sj.npp.1300265.

Blanchard, J. J., Brown, S. A., Horan, W. P., & Sherwood, A. R. (2000). Substance use disorders in schizophrenia: review, integration, and a proposed model. *Clinical Psychology Review, 20*(2), 207–234.

Blanchard, J. J., Mueser, K. T., & Bellack, A. S. (1998). Anhedonia, positive and negative affect, and social functioning in schizophrenia. *Schizophrenia Bulletin, 24*(3), 413–424.

Brody, A. L., Olmstead, R. E., London, E. D., Farahi, J., Meyer, J. H., Grossman, P., ... Mandelkern, M. A. (2004). Smoking-induced ventral striatum dopamine release. *The American Journal of Psychiatry, 161*(7), 1211–1218.

Caponnetto, P., Auditore, R., Russo, C., Cappello, G. C., & Polosa, R. (2013). Impact of an electronic cigarette on smoking reduction and cessation in schizophrenic smokers: a prospective 12-month pilot study. *International Journal of Environmental Research and Public Health, 10*(2), 446–461. http://dx.doi.org/10.3390/ijerph10020446.

Chambers, R. A., Sentir, A. M., Conroy, S. K., Truitt, W. A., & Shekhar, A. (2010). Cortical-striatal integration of cocaine history and prefrontal dysfunction in animal modeling of dual diagnosis. *Biological Psychiatry, 67*(8), 788–792. http://dx.doi.org/10.1016/j.biopsych.2009.09.006.

Chandra, P. S., Carey, M. P., Carey, K. B., Jairam, K. R., Girish, N. S., & Rudresh, H. P. (2005). Prevalence and correlates of tobacco use and nicotine dependence among psychiatric patients in India. *Addictive Behaviors, 30*(7), 1290–1299. http://dx.doi.org/10.1016/j.addbeh.2005.01.002.

Dale Horst, W., Klein, M. W., Williams, D., & Werder, S. F. (2005). Extended use of nicotine replacement therapy to maintain smoking cessation in persons with schizophrenia. *Neuropsychiatric Disease and Treatment, 1*(4), 349–355.

Freedman, R., Hall, M., Adler, L. E., & Leonard, S. (1995). Evidence in postmortem brain tissue for decreased numbers of hippocampal nicotinic receptors in schizophrenia. *Biological Psychiatry, 38*(1), 22–33. http://dx.doi.org/10.1016/0006-3223(94)00252-X.

Friedman, L., Turner, J. A., Stern, H., Mathalon, D. H., Trondsen, L. C., & Potkin, S. G. (2008). Chronic smoking and the BOLD response to a visual activation task and a breath hold task in patients with schizophrenia and healthy controls. *NeuroImage, 40*(3), 1181–1194. http://dx.doi.org/10.1016/j.neuroimage.2007.12.040.

Goff, D. C., Cather, C., Evins, A. E., Henderson, D. C., Freudenreich, O., Copeland, P. M., ... Sacks, F. M. (2005). Medical morbidity and mortality in schizophrenia: guidelines for psychiatrists. *The Journal of Clinical Psychiatry, 66*(2), 183–194; quiz 147, 273–274.

Goff, D. C., Henderson, D. C., & Amico, E. (1992). Cigarette smoking in schizophrenia: relationship to psychopathology and medication side effects. *The American Journal of Psychiatry, 149*(9), 1189–1194.

Harris, J. G., Kongs, S., Allensworth, D., Martin, L., Tregellas, J., Sullivan, B., ... Freedman, R. (2004). Effects of nicotine on cognitive deficits in schizophrenia. *Neuropsychopharmacology: Official Publication of the American College of Neuropsychopharmacology, 29*(7), 1378–1385. http://dx.doi.org/10.1038/sj.npp.1300450.

Hong, L. E., Yang, X., Wonodi, I., Hodgkinson, C. A., Goldman, D., Stine, O. C., ... Thaker, G. K. (2011). A CHRNA5 allele related to nicotine addiction and schizophrenia. *Genes, Brain, and Behavior, 10*(5), 530–535. http://dx.doi.org/10.1111/j.1601-183X.2011.00689.x.

Hughes, J. R., Hatsukami, D. K., Mitchell, J. E., & Dahlgren, L. A. (1986). Prevalence of smoking among psychiatric outpatients. *The American Journal of Psychiatry, 143*(8), 993–997.

Kelly, D. L., Raley, H. G., Lo, S., Wright, K., Liu, F., McMahon, R. P., ... Heishman, S. J. (2012). Perception of smoking risks and motivation to quit among nontreatment-seeking smokers with and without schizophrenia. *Schizophrenia Bulletin, 38*(3), 543–551. http://dx.doi.org/10.1093/schbul/sbq124.

Khantzian, E. J. (1985). The self-medication hypothesis of addictive disorders: focus on heroin and cocaine dependence. *The American Journal of Psychiatry, 142*(11), 1259–1264.

Khantzian, E. J. (1997). The self-medication hypothesis of substance use disorders: a reconsideration and recent applications. *Harvard Review of Psychiatry, 4*(5), 231–244. http://dx.doi.org/10.3109/10673229709030550.

Kishi, T., & Iwata, N. (2014). Varenicline for smoking cessation in people with schizophrenia: systematic review and meta-analysis. *European Archives of Psychiatry and Clinical Neuroscience, 265*(3), 259–268. http://dx.doi.org/10.1007/s00406-014-0551-3.

Koob, G. F., & Moal, M. L. (2005). *Neurobiology of addiction*. San Diego, CA: Academic Press.

Krishnadas, R., Jauhar, S., Telfer, S., Shivashankar, S., & McCreadie, R. G. (2012). Nicotine dependence and illness severity in schizophrenia. *The British Journal of Psychiatry: The Journal of Mental Science, 201*(4), 306–312. http://dx.doi.org/10.1192/bjp.bp.111.107953.

Kumari, V., & Postma, P. (2005). Nicotine use in schizophrenia: the self medication hypotheses. *Neuroscience and Biobehavioral Reviews, 29*(6), 1021–1034. http://dx.doi.org/10.1016/j.neubiorev.2005.02.006.

Lembke, A. (2012). Time to abandon the self-medication hypothesis in patients with psychiatric disorders. *The American Journal of Drug and Alcohol Abuse, 38*(6), 524–529. http://dx.doi.org/10.3109/00952990.2012.694532.

de Leon, J., & Diaz, F. J. (2005). A meta-analysis of worldwide studies demonstrates an association between schizophrenia and tobacco smoking behaviors. *Schizophrenia Research, 76*(2–3), 135–157. http://dx.doi.org/10.1016/j.schres.2005.02.010.

Leonard, S., Adams, C., Breese, C. R., Adler, L. E., Bickford, P., Byerley, W., ... Freedman, R. (1996). Nicotinic receptor function in schizophrenia. *Schizophrenia Bulletin, 22*(3), 431–445.

Leonard, S., & Bertrand, D. (2001). Neuronal nicotinic receptors: from structure to function. *Nicotine & Tobacco Research: Official Journal of the Society for Research on Nicotine and Tobacco, 3*(3), 203–223. http://dx.doi.org/10.1080/14622200110050213.

Loukola, A., Wedenoja, J., Keskitalo-Vuokko, K., Broms, U., Korhonen, T., Ripatti, S., ... Kaprio, J. (2014). Genome-wide association study on detailed profiles of smoking behavior and nicotine dependence in a twin sample. *Molecular Psychiatry, 19*(5), 615–624. http://dx.doi.org/10.1038/mp.2013.72.

Luchins, D. J., Goldman, M. B., Lieb, M., & Hanrahan, P. (1992). Repetitive behaviors in chronically institutionalized schizophrenic patients. *Schizophrenia Research, 8*(2), 119–123.

Lybrand, J., & Caroff, S. (2009). Management of schizophrenia with substance use disorders. *The Psychiatric Clinics of North America, 32*(4), 821–833. http://dx.doi.org/10.1016/j.psc.2009.09.002.

Lyons, M. J., Bar, J. L., Kremen, W. S., Toomey, R., Eisen, S. A., Goldberg, J., ... Tsuang, M. (2002). Nicotine and familial vulnerability to schizophrenia: a discordant twin study. *Journal of Abnormal Psychology, 111*(4), 687–693.

Mackowick, K. M., Barr, M. S., Wing, V. C., Rabin, R. A., Ouellet-Plamondon, C., & George, T. P. (2014). Neurocognitive endophenotypes in schizophrenia: modulation by nicotinic receptor systems. *Progress in Neuro-Psychopharmacology & Biological Psychiatry*, 52, 79–85. http://dx.doi.org/10.1016/j.pnpbp.2013.07.010.

McClave, A. K., McKnight-Eily, L. R., Davis, S. P., & Dube, S. R. (2010). Smoking characteristics of adults with selected lifetime mental illnesses: results from the 2007 National Health Interview Survey. *American Journal of Public Health*, 100(12), 2464–2472. http://dx.doi.org/10.2105/AJPH.2009.188136.

McEvoy, J. P., Freudenreich, O., Levin, E. D., & Rose, J. E. (1995). Haloperidol increases smoking in patients with schizophrenia. *Psychopharmacology*, 119(1), 124–126.

Moran, L. V., Sampath, H., Kochunov, P., & Hong, L. E. (2013). Brain circuits that link schizophrenia to high risk of cigarette smoking. *Schizophrenia Bulletin*, 39(6), 1373–1381. http://dx.doi.org/10.1093/schbul/sbs149.

Pohanka, M. (2012). Alpha7 nicotinic acetylcholine receptor is a target in pharmacology and toxicology. *International Journal of Molecular Sciences*, 13(2), 2219–2238. http://dx.doi.org/10.3390/ijms13022219.

Punnoose, S., & Belgamwar, M. R. (2006). Nicotine for schizophrenia. *The Cochrane Database of Systematic Reviews*, (1), CD004838. http://dx.doi.org/10.1002/14651858.CD004838.pub2.

Schuckit, M. A., & Smith, T. L. (1996). An 8-year follow-up of 450 sons of alcoholic and control subjects. *Archives of General Psychiatry*, 53(3), 202–210.

Song, L., Chen, X., Chen, M., Tang, Y., Wang, J., Zhang, M., … Chen, C. (2014). Differences in P50 and prepulse inhibition of the startle reflex between male smokers and non-smokers with first episode schizophrenia without medical treatment. *Chinese Medical Journal*, 127(9), 1651–1655.

Srinivasan, T. N., & Thara, R. (2002). Smoking in schizophrenia – all is not biological. *Schizophrenia Research*, 56(1–2), 67–74.

Steinberg, M. L., Williams, J. M., Steinberg, H. R., Krejci, J. A., & Ziedonis, D. M. (2005). Applicability of the Fagerström test for nicotine dependence in smokers with schizophrenia. *Addictive Behaviors*, 30(1), 49–59. http://dx.doi.org/10.1016/j.addbeh.2004.04.012.

Steinberg, M. L., Williams, J. M., & Ziedonis, D. M. (2004). Financial implications of cigarette smoking among individuals with schizophrenia. *Tobacco Control*, 13(2), 206.

Tanabe, J., Tregellas, J. R., Martin, L. F., & Freedman, R. (2006). Effects of nicotine on hippocampal and cingulate activity during smooth pursuit eye movement in schizophrenia. *Biological Psychiatry*, 59(8), 754–761. http://dx.doi.org/10.1016/j.biopsych.2005.09.004.

Thoma, P., & Daum, I. (2013). Comorbid substance use disorder in schizophrenia: a selective overview of neurobiological and cognitive underpinnings. *Psychiatry and Clinical Neurosciences*, 67(6), 367–383. http://dx.doi.org/10.1111/pcn.12072.

Tidey, J. W., Rohsenow, D. J., Kaplan, G. B., & Swift, R. M. (2005). Cigarette smoking topography in smokers with schizophrenia and matched nonpsychiatric controls. *Drug and Alcohol Dependence*, 80(2), 259–265. http://dx.doi.org/10.1016/j.drugalcdep.2005.04.002.

Tidey, J. W., Rohsenow, D. J., Kaplan, G. B., Swift, R. M., & Ahnallen, C. G. (2013). Separate and combined effects of very low nicotine cigarettes and nicotine replacement in smokers with schizophrenia and controls. *Nicotine & Tobacco Research: Official Journal of the Society for Research on Nicotine and Tobacco*, 15(1), 121–129. http://dx.doi.org/10.1093/ntr/nts098.

Tobacco and Genetics Consortium. (2010). Genome-wide meta-analyses identify multiple loci associated with smoking behavior. *Nature Genetics*, 42(5), 441–447. http://dx.doi.org/10.1038/ng.571.

Tsoi, D. T., Porwal, M., & Webster, A. C. (2010). Efficacy and safety of bupropion for smoking cessation and reduction in schizophrenia: systematic review and meta-analysis. *The British Journal of Psychiatry: The Journal of Mental Science*, 196(5), 346–353. http://dx.doi.org/10.1192/bjp.bp.109.066019.

Tsoi, D. T., Porwal, M., & Webster, A. C. (2013). Interventions for smoking cessation and reduction in individuals with schizophrenia. *The Cochrane Database of Systematic Reviews*, 2, CD007253. http://dx.doi.org/10.1002/14651858.CD007253.pub3.

Weiser, M., Reichenberg, A., Grotto, I., Yasvitzky, R., Rabinowitz, J., Lubin, G., … Davidson, M. (2004). Higher rates of cigarette smoking in male adolescents before the onset of schizophrenia: a historical-prospective cohort study. *The American Journal of Psychiatry*, 161(7), 1219–1223.

Williams, J. M., Anthenelli, R. M., Morris, C. D., Treadow, J., Thompson, J. R., Yunis, C., & George, T. P. (2012). A randomized, double-blind, placebo-controlled study evaluating the safety and efficacy of varenicline for smoking cessation in patients with schizophrenia or schizoaffective disorder. *The Journal of Clinical Psychiatry*, 73(5), 654–660. http://dx.doi.org/10.4088/JCP.11m07522.

Williams, J. M., & Foulds, J. (2007). Successful tobacco dependence treatment in schizophrenia. *The American Journal of Psychiatry*, 164(2), 222–227. http://dx.doi.org/10.1176/ajp.2007.164.2.222 quiz 373.

Williams, J. M., Gandhi, K. K., Karavidas, M. K., Steinberg, M. L., Lu, S.-E., & Foulds, J. (2008). Open-label study of craving in smokers with schizophrenia using nicotine nasal spray compared to nicotine patch. *Journal of Dual Diagnosis*, 4(4), 355–376. http://dx.doi.org/10.1080/15504260802085919.

Wing, V. C., Bacher, I., Wu, B. S., Daskalakis, Z. J., & George, T. P. (2012). High frequency repetitive transcranial magnetic stimulation reduces tobacco craving in schizophrenia. *Schizophrenia Research*, 139(1–3), 264–266. http://dx.doi.org/10.1016/j.schres.2012.03.006.

Winterer, G. (2010). Why do patients with schizophrenia smoke? *Current Opinion in Psychiatry*, 23(2), 112–119. http://dx.doi.org/10.1097/YCO.0b013e3283366643.

Zammit, S., Allebeck, P., Dalman, C., Lundberg, I., Hemmingsson, T., & Lewis, G. (2003). Investigating the association between cigarette smoking and schizophrenia in a cohort study. *The American Journal of Psychiatry*, 160(12), 2216–2221.

Zhang, X., Stein, E. A., & Hong, L. E. (2010). Smoking and schizophrenia independently and additively reduce white matter integrity between striatum and frontal cortex. *Biological Psychiatry*, 68(7), 674–677. http://dx.doi.org/10.1016/j.biopsych.2010.06.018.

Zhang, X. Y., Yu, Y. Q., Sun, S., Zhang, X., Li, W., Xiu, M. H., … Kosten, T. R. (2011). Smoking and tardive dyskinesia in male patients with chronic schizophrenia. *Progress in Neuro-Psychopharmacology & Biological Psychiatry*, 35(7), 1765–1769. http://dx.doi.org/10.1016/j.pnpbp.2011.06.006.

Chapter 26

Unraveling the Role of the Amygdala in Nicotine Addiction

Benjamin Becker, René Hurlemann
Division of Medical Psychology, Department of Psychiatry and Psychotherapy, University of Bonn, Bonn, Germany

Abbreviations

ASL Arterial spin labeling
CBF Cerebral blood flow
DA Dopamine
fMRI Functional magnetic resonance imaging
nAChR Nicotinic acetylcholine receptor
ROI Region of interest
VTA Ventral tegmental area

INTRODUCTION

Drug addiction is a chronic, relapsing disorder of the brain characterized by escalating drug use, loss of control over drug intake, and continued use in the face of negative consequences. Nicotine addiction in its most prevalent form of smoking tobacco has less severe binge/intoxication and somatic withdrawal stages compared to other drug addictions, such as those of alcohol or psychostimulants (Koob & Volkow, 2010). However, despite being a legal drug, tobacco was rated as being one of the top 10 most harmful drugs by both experts (Nutt, King, Saulsbury, & Blakemore, 2007) and users (Morgan, Muetzelfeldt, Muetzelfeldt, Nutt, & Curran, 2010). In terms of prevalence and clinically significant long-term consequences, tobacco use, at least in its most prevalent form of smoking cigarettes, exceeds most other drugs of abuse.

Approximately 1.3 billion people currently use tobacco worldwide, most commonly in the form of smoking cigarettes. Partly due to strategic marketing by tobacco companies, the number of cigarette smokers is still growing, particularly in low- and middle-income countries. It has been estimated that by the year 2025 the number of people who smoke will have reached 1.6 billion worldwide (Royal College of Physicians, 2007). Smoking is responsible for one in every five deaths in the United States, and the chronic diseases associated with tobacco smoking, such as chronic obstructive pulmonary disease and coronary heart disease, are the leading causes of premature death and disability (United States Department of Health and Human Services, 2010). Indeed, the risk that a chronic smoker will die prematurely from a smoking-related medical complication is estimated up to 50% (Benowitz, 2010). In addition to health risks for users, tobacco smoking poses a heavy economic burden on society as a whole. This includes indirect costs related to workday losses due to morbidity as well as direct costs resulting from inpatient and outpatient care of smokers suffering from the consequences of prolonged tobacco use. The amount of health care costs and loss of productivity is estimated up to €98 billion for EU countries and $193 billion for the United States (United States Department of Health and Human Services, 2010; World Health Organization, 2008), which is approximately between 1.01% and 1.39% of the region's gross domestic product in 2000. Nicotine addiction is difficult to treat, and of those who try to quit smoking each year, only 3–5% are able to sustain nicotine craving without the use of pharmacological nicotine replacement therapies, and no more than 30% are successful with them (Dome, Lazary, Kalapos, & Rihmer, 2010). Meta-analytic retrospective data suggest that only 3–7% of smokers who try to quit remain abstinent for more than 6 months and that 2–15% of smokers relapse after the first year of abstinence (Hughes, Peters, & Naud, 2008).

THE ROLE OF NICOTINE IN TOBACCO ADDICTION

Nicotine is the main psychoactive ingredient in tobacco and is considered a major cause of tobacco addiction (Stolerman & Jarvis, 1995). The addictive potential of nicotine is derived from its stimulatory effects on neuronal nicotinic acetylcholine receptors (nAChRs) in the central nervous system, which facilitate the release of dopamine (DA) in the mesocorticolimbic circuit (Clarke, Fu, Jakubovic, & Fibiger, 1988). The mesocorticolimbic circuit has been extensively implicated in the rewarding properties of natural reinforcers, such as food and sex, as well as addictive drugs (Feltenstein & See, 2008), and comprises dopaminergic projections ascending from the ventral tegmental area (VTA) to the limbic structures, including the nucleus accumbens, hippocampus, amygdala, and prefrontal regions, particularly the prefrontal cortex. Nicotine activates nAChRs located on dopaminergic neurons in the VTA, causing the release of DA in the shell of the nucleus accumbens, a mechanism that is thought to critically underlie the primary rewarding effects of most, if not all, addictive substances (Volkow, Fowler, & Wang, 2003). Chronic exposure to nicotine initiates long-term neuroadaptive changes in the reward system that lead to continued smoking and the development of nicotine tolerance and dependence. Upon cessation of smoking, nicotine-dependent smokers experience somatic and affective withdrawal symptoms such as anger, anxiety, depressed mood, irritability, mild cognitive deficits, and physiological symptoms (Shiffman, West, & Gilbert, 2004). Withdrawal symptoms typically begin within 48 h after smoking cessation, peak within the first week, and last for 2–4 weeks (Hughes, 2007). These distressing withdrawal symptoms can be avoided with the continued intake of nicotine, which maintains the neural homeostasis. Pharmacologically, nicotine addiction can be viewed as an interplay between mechanisms of positive (e.g., enhancement of mood) and negative (e.g., to avoid distressing withdrawal symptoms) reinforcement (Benowitz, 2010). Thus, there is convincing evidence that nicotine exhibits the principal psychopharmacological properties of an addictive substance, namely acute reinforcing and rewarding effects during initial stages of use and withdrawal symptoms when consumption is stopped after a period of chronic use.

Although nicotine is clearly the major factor driving smoking, it has been argued that the reinforcing effects of nicotine per se would not be adequate to explain the high addictive potential of tobacco smoking (Balfour, 2004; Caggiula et al., 2001). Compared with other addictive psychostimulant drugs, such as amphetamine and cocaine, the reinforcing effects of nicotine are relatively subtle and would not readily predict the low success rates of smoking cessation attempts (Caggiula et al., 2001). Additionally, when considered in isolation, the reinforcing effects of nicotine would not sufficiently explain why environmental stimuli frequently associated with smoking can induce craving and relapse even after years of abstinence (Conklin & Tiffany, 2002). In addition to the direct psychopharmacological mechanisms of nicotine, learning processes such as Pavlovian and instrumental learning may contribute to the development and maintenance of nicotine addiction. With repeated exposure, environmental contexts and stimuli associated with smoking become associated with the reinforcing effects of nicotine and gain excessive salience (Davis & Gould, 2008). These drug-associated memories have strong emotional and motivational properties and can elicit nicotine craving and nicotine intake even after long periods of abstinence (Robbins, Ersche, & Everitt, 2008).

THE AMYGDALA

The amygdala plays a pivotal role in emotion processing, including valence detection and emotional learning (LeDoux, 2000). Initial studies emphasized the role of the amygdala in the processing of negative emotions, particularly Pavlovian fear conditioning and the evaluation of threat (LeDoux, 2000). In addition, accumulating evidence from studies in laboratory animals has associated the amygdala with positive emotions, particularly appetitive processing, positive reinforcement (Baxter & Murray, 2002), and reward expectancy (Holland & Gallagher, 2004). Anatomically, the amygdala is centered in a widely interconnected neural circuit encompassing regions involved in learning, such as the hippocampus, and motivation, particularly the striatum and prefrontal cortex, and thus is ideally located to combine the detection of relevant stimuli in the environment with learned associations and translate them into either appetitively or aversively motivated behavior (Everitt, Cardinal, Parkinson, & Robbins, 2003). In congruence with the well-recognized role of the amygdala in negative reinforcement and Pavlovian conditioning, animal models of addiction propose an important role of the amygdala in the negative emotional state during withdrawal and drug-associated memories that induce craving and promote relapse (George & Koob, 2010; Robbins et al., 2008). Given the proposed crucial importance of craving and drug-related memories in maintaining smoking behavior and relapse even after prolonged periods of abstinence, a contribution of the amygdala to nicotine addiction seems likely. To explore the role of the amygdala in nicotine addiction, we here summarize evidence from neuroimaging studies using functional magnetic resonance imaging (fMRI) in smokers (for a more detailed overview on the studies, refer to Mihov & Hurlemann, 2012).

NICOTINE CRAVING

Drug craving and subsequent relapse to drug use represent hallmarks of addiction, including nicotine addiction. Contingent on the presence or absence of smoking-related cues, two forms of nicotine craving can be differentiated. Cue-induced (phasic) nicotine craving is thought to result from conditioning processes in which stimuli associated with cigarette smoking trigger craving and subsequent drug-seeking behavior (Benowitz, 2010; Robinson & Berridge, 2001). In contrast, abstinence-induced (tonic) craving is thought to develop in the absence of smoking-related cues (Tiffany, Cox, & Elash, 2000), is more sensitive to the delivery of nicotine (Jarvik et al., 2000), and consequently develops quickly after smoking cessation.

FUNCTIONAL NEUROIMAGING STUDIES ON CUE-INDUCED NICOTINE CRAVING

It is well known that in nicotine-dependent individuals the sight or smell of a burning cigarette can induce strong nicotine craving (Carter & Tiffany, 1999). Because of its obvious relevance for maintaining smoking behavior and relapse after smoking cessation, several studies have implemented functional neuroimaging experiments, mostly using fMRI, to examine the neural substrates of smoking-cue-elicited cravings in smokers. During these experiments, nicotine-dependent smokers and nonsmoking controls are usually shown pictures with smoking-related content (e.g., people smoking or holding a cigarette) and neutral control pictures that lack smoking-related items while their brain activation is measured. The individual neural networks engaged in cue reactivity are then assessed by subtracting the neural activity during the presentation of neutral stimuli from the neural activity during presentation of smoking-related stimuli. Often, an additional between-group comparison is conducted to compare neural cue reactivity between nicotine-addicted individuals and matched nonusing control subjects (Jasinska, Stein, Kaiser, Naumer, & Yalachkov, 2014). Given the multisensory nature of real-life stimuli associated with smoking (e.g., smelling a burning cigarette, lighting a cigarette) studies have begun using multisensory cues to increase the ecological validity of laboratory experiments (Yalachkov, Kaiser, & Naumer, 2012). For example, in some experiments, participants are presented with smoking-related pictures while simultaneously holding a cigarette to mimic the visuohaptic perception of cigarette smoking (Yalachkov et al., 2012). To further disentangle the specific contributions of the identified brain regions, most studies collect additional variables of relevance, such as self-reported measures of nicotine craving, withdrawal, and severity of nicotine dependence, in addition to collecting task-related responses. The severity of nicotine dependence is often assessed by the Fagerström Test of Nicotine Dependence (FTND) (Table 1), which assesses relevant addiction criteria for nicotine dependence (Fagerström, 1978; Heatherton, Kozlowski, Frecker, & Fagerström, 1991). Several of the test's criteria have been shown to be reliable indicators of the severity of nicotine addiction, particularly early morning smoking, smoking more than 10 cigarettes per day, and repeated, yet unsuccessful, attempts to quit smoking in the past.

Despite a wealth of studies examining smoking-cue-induced craving in smokers, findings regarding the involvement of specific brain regions, including the amygdala, are characterized by a certain degree of inconsistency. These inconsistencies between studies have been attributed to the different statistical analysis methods used (e.g., whole-brain versus regionally restricted analysis of regions of a priori interest), choice of drug cues (e.g., visual versus haptic cues), length of abstinence from nicotine, ambivalence about smoking, motivation to quit smoking, and perceived opportunity to smoke after the experimental assessment (Jasinska et al., 2014; McClernon, Kozink, Lutz, & Rose, 2009; Mihov & Hurlemann, 2012; Wilson, Sayette, & Fiez, 2012; Yalachkov et al., 2012).

Nevertheless, findings from two meta-analyses point to the amygdala as an important node in the neural network constituting the neurobiological basis of cue-induced craving in nicotine addiction (Kuhn & Gallinat, 2011; Mihov & Hurlemann, 2012). A quantitative meta-analysis from Kuhn and Gallinat (2011) analyzed the specific and overlapping networks engaged in nicotine, alcohol, and cocaine addiction. Specifically addressing the networks engaged in smoking-cue reactivity in nicotine addiction, the authors included 13 cue-reactivity studies of smokers with a total of 251 participants. Nicotine-dependent smokers consistently showed greater activity in response to smoking cues compared to neutral cues in the ventral striatum, anterior cingulate, and amygdala. Remarkably, in a separate meta-analysis of alcohol and cocaine, the amygdala was also consistently associated with cue reactivity (Figure 1). Further support for an engagement of the amygdala in cue reactivity comes from a qualitative meta-analysis that addressed altered amygdala functioning in smokers (Mihov & Hurlemann, 2012). Ten of 20 neuroimaging studies of smokers that were included in the qualitative meta-analysis reported altered amygdala reactivity in response to smoking-related stimuli. Interestingly, the authors found that the statistical analysis methods employed by the studies had a dramatic effect on the findings regarding the amygdala. Ten of 12 studies that conducted region-of-interest (ROI) analyses that focused on a priori defined brain regions to increase the regional sensitivity of the fMRI analysis reported altered amygdala reactivity to smoking-associated stimuli; however, none of the eight studies that conducted whole-brain analyses reported the same (Table 2). The lack of amygdala reactivity in the whole-brain analyses and the need to use ROI analysis with a higher regional sensitivity to observe amygdala

TABLE 1 The Fagerström Test of Nicotine Dependence

Item	Choice	Score
How soon after you wake up do smoke your first cigarette?	Within 5 min	3
	Within 6–30 min	2
	Within 31–60 min	1
	After 60 min	0
Do you find it difficult to refrain from smoking in places where it is forbidden, e.g., in church, at the library, in the cinema, etc.?	Yes	1
	No	0
Which cigarette would you hate most to give up?	The first on in the morning	1
	All others	0
How many cigarettes per day do you smoke?	10 or fewer	0
	11–20	1
	21–30	2
	31 or more	3
Do you smoke more frequently the first hours after waking than during the rest of the day?	Yes	1
	No	0
Do you smoke if you are so ill that you are in bed most of the day?	Yes	1
	No	0

Items and multiple-choice answers of the Fagerström Test of Nicotine Dependence. The test assesses the degree of nicotine-addiction severity in smokers. Based on the overall score ranging from 0 to 10, the following degrees of nicotine dependence are often categorized: 0, not dependent; 1–2, slightly dependent; 3–5, moderately dependent; 6–8, highly dependent; 9–10, very dependent.
Adapted with friendly permission from Karl Olov Fagerström.

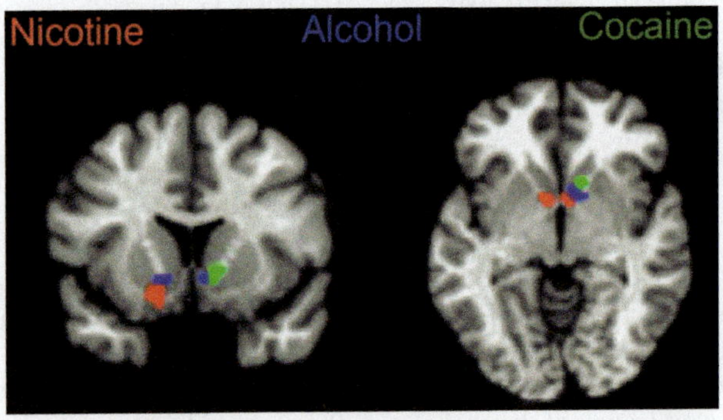

FIGURE 1 Conjunction analysis of nicotine, alcohol, and cocaine cue reactivity. Findings from a conjunction analysis of cue reactivity in nicotine, alcohol, and cocaine addiction showing overlapping cue reactivity for the addictions. The figure shows convergent regions involved in nicotine (red), alcohol (blue), and cocaine (green) cue reactivity. The analysis revealed a direct overlap within the left and right ventral striatum. In addition, the amygdala (found in nicotine, alcohol, and cocaine cue reactivity) and the anterior cingulate cortex (found in nicotine and cocaine cue reactivity) revealed convergent findings. The amygdala revealed convergence with alcohol in the right hemisphere and nicotine and cocaine in the left hemisphere. The meta-analytic findings suggest that the ventral striatum, anterior cingulate cortex, and amygdala are convergently involved in cue reactivity for nicotine, alcohol, and cocaine. *Used with permission from Kuhn and Gallinat (2011), adapted from Figure 2, p. 1323, John Wiley and Sons, 2011.*

TABLE 2 Studies Employing ROI Analysis to Examine Amygdala Reactivity in Smokers

Authors	Samples	Abstinence	Task/Stimuli	Analysis	Amygdala Reactivity
Artiges et al. (2009)	13 smokers 13 nonsmokers	2 h abstinence	Emotion recognition task preceded by neutral or smoking-related images	ROI	Smoking-related cues decreased right amygdala reactivity in smokers
Due, Huettel, Hall, and Rubin (2002)	11 smokers 6 nonsmokers	10 h abstinence	Smoking-related, neutral, and target images	ROI	Smoking cues elicited increased amygdala reactivity in smokers but not in nonsmokers
Franklin et al. (2007)	21 smokers	Satiety	In vivo exposure to smoking-related vs neutral stimuli	ROI	Amygdala perfusion increased for smoking cues
Janes et al. (2010)	21 smokers 9 slip/12 abstinence	Satiety	Smoking-related, neutral, and target images	ROI	Future slippers show greater cue reactivity in the amygdala than future abstainers
Kober et al. (2010)	21 smokers	2 h abstinence	Craving-regulation; smoking-related, neutral images	ROI	Cognitive downregulation of craving was associated with decreased amygdala reactivity
McBride, Barrett, Kelly, Aw, and Dagher (2006)	19 smokers	12 h abstinence versus satiety	Video clips with smoking-related versus neutral content	ROI	No effect
McClernon et al. (2005)	13 smokers	12 h abstinence versus satiety	Smoking-related, neutral, and target images	ROI	No effect
Mc Clernon et al. (2007)	16 smokers	2 h abstinence	Smoking-related, neutral, and target images	ROI	Amygdala smoking-cue reactivity decreased during treatment
Nestor, McCabe, Jones, Clancy, and Garavan (2011)	10 smokers 10 ex-smokers 13 nonsmokers	Satiety	Attentional bias task with smoking-related or neutral images	ROI	Smoking cues elicited increased amygdala reactivity in smokers but not in nonsmokers
Okuyemi et al. (2006)	17 smokers 17 nonsmokers	12 h abstinence	Smoking-related versus neutral images	ROI	Stronger activity to smoking cue in right amygdala
					Stronger in African American than in Caucasian smokers
Rubinstein et al. (2011)	12 smokers 12 nonsmokers	Satiety	Smoking-related versus neutral images	ROI	Increased left-amygdala smoking-cue reactivity in smokers
Stippekohl et al. (2010)	20 deprived smokers 19 satiated smokers 17 controls	12 h abstinence	Video stimuli in an event-related design: arousal ratings and craving ratings	ROI	"Last puff" video evoked amygdala activation bilaterally in nondeprived smokers and in the right amygdala of deprived smokers

Overview of functional neuroimaging studies in smokers that used ROI analysis to examine amygdala reactivity to smoking-related stimuli.
Used with permission from Mihov and Hurlemann (2012), adapted from Table 1, p. 1722, Elsevier, 2012.

smoking-cue reactivity might be attributed to the small signal changes that are usually observed in the amygdala in response to activating stimuli (Breiter et al., 1996). This would also explain the lack of amygdala results in a further quantitative meta-analysis on smoking-cue reactivity that, to account for potential biasing effects in a whole-brain meta-analysis, excluded studies using ROI analysis (Engelmann et al., 2012).

In addition to these meta-analytic findings, some interesting results from individual studies might help to shed some light on the specific contribution of the amygdala in smoking-cue reactivity.

In some cases, an association between the cue-induced activity in the amygdala and the severity of subjective craving in response to smoking cues was observed (Franklin et al., 2011; Goudriaan, de Ruiter, van den Brink, Oosterlaan, & Veltman, 2010), suggesting an association between amygdala cue reactivity and the subjectively experienced urge to smoke. Notably, these studies examined heavy smokers who reported strong to moderate craving in response to the smoking cues, suggesting that this association might become evident only with a certain degree of addiction severity and craving. However, the association between severity of nicotine addiction, subjective craving, and amygdala cue reactivity seems more complex. Vollstadt-Klein et al. (2011) specifically addressed whether the severity of nicotine dependence influences smoking-cue reactivity in a group of 22 smokers. The study used tobacco advertisements as smoking-related cues that appear in everyday life and therefore have a particularly high ecological validity. The study found a negative correlation between the severity of dependence as measured with the FTND and cue-induced activity in limbic regions including the amygdala, indicating that less dependent smokers show higher amygdala reactivity. Furthermore, findings from one study suggest that amygdala cue reactivity in smokers might have a high predictive value for the success of smoking cessation attempts. Janes et al. (2010) used a prospective design to examine whether smoking-cue reactivity would predict relapse susceptibility in a group of 22 smokers. Smokers who subsequently relapsed showed increased prequitting cue reactivity in emotion processing regions, including the amygdala, suggesting that these regions are strongly implicated in relapse vulnerability. Interestingly, findings from a 2013 study suggest gender differences in cue-induced amygdala reactivity in nicotine addiction (Wetherill et al., 2013). In this study, male and female smokers underwent a smoking-cue-reactivity paradigm during arterial spin labeling (ASL) perfusion MRI. A direct comparison of brain responses to smoking cues with those to neutral stimuli between male and female smokers revealed greater smoking-cue-induced activity in the bilateral amygdala of male participants. The authors suggested that this might be explained by sex differences in amygdala-associated emotional memory formation and variations in circulating gonadal hormones due to variations in menstrual cycle phase.

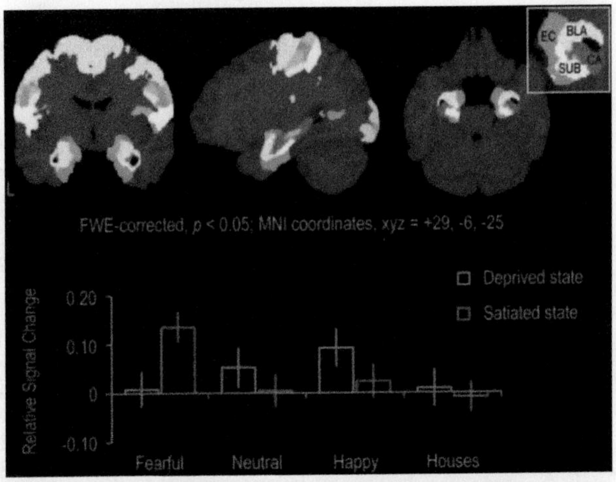

FUNCTIONAL NEUROIMAGING STUDIES ON ABSTINENCE-INDUCED CRAVING

In contrast to cue-induced craving, which has been extensively studied since about 2010, few studies have addressed abstinence-induced craving in smokers. This is surprising because initial behavioral studies yielded a strong predictive value of unprovoked craving during early abstinence after smoking cessation and subsequent relapse (Killen & Fortmann, 1997; Shiffman et al., 1997). In contrast to studies on cue-induced craving that use explicit smoking-related cues, studies on abstinence-induced craving assess unprovoked craving as a function of nicotine satiety versus nicotine deprivation. Usually, these studies measure brain activity in smokers on two separate occasions. Before one occasion, participants are allowed to smoke as usual (nicotine satiety), whereas on the other occasion they have to abstain from smoking for 12–24 h (nicotine deprivation). One study employed ASL perfusion MRI to measure cerebral blood flow (CBF) during the resting state in smokers during smoking satiety and after 12 h of smoking deprivation (Wang et al., 2007). The abstinence state was associated with increased CBF in the anterior cingulate and the medial orbitofrontal cortex. Subjectively experienced levels of craving after 12 h of smoking deprivation were accompanied by increased abstinence-induced CBF in striatal and limbic regions, including the bilateral amygdala. These findings suggest an association between increasing levels of abstinence-induced craving and increasing amygdala activity in the absence of smoking-related stimuli. This association is particularly important, because of those smokers who try to quit

FIGURE 2 Fear-specific amygdala dysfunction in abstinent smokers. Compared to the satiated state, smokers in the nicotine-deprived state following 12 h of nicotine abstinence showed lowered right amygdala reactivity to fearful facial stimuli. In contrast, amygdala reactivity for neutral and happy faces, as well as for nonsocial stimuli (houses), was not altered following 12 h of abstinence. The fear-selective decline was located in the basolateral subregion of the amygdala (BLA). The BLA plays an important role in the detection and validation of fear signals. Abbreviations: BLA, basolateral amygdala; CA, cornu ammonis; EC, entorhinal cortex; FWE, family-wise error; MNI, Montreal Neurological Institute; SUB, subiculum. *Used with permission from Onur et al. (2012), adapted from Figure 3, p. 1413, John Wiley and Sons, 2012.*

without nicotine replacement therapy, 50–75% relapse within the first week of a quit attempt (Garvey, Bliss, Hitchcock, Heinold, & Rosner, 1992; Hughes, Keely, & Naud, 2004). Another study addressed whether abstinence-induced craving in smokers interferes with emotion processing in the amygdala (Onur et al., 2012). This study used emotional stimuli of positive, negative, and neutral valence (happy, fearful, and neutral facial pictures) to address the functional integrity of the amygdala in smokers under conditions of smoking satiety and after 12 h of smoking deprivation. Relative to the satiated state, smokers exhibited attenuated amygdala reactivity to fearful faces after 12 h of smoking deprivation, but not happy or neutral faces, suggesting that abstinence-induced craving interrupts normal emotion processing, particularly fear perception, in the amygdala (Figure 2). Importantly, reduced amygdala fear reactivity during smoking abstinence varied as a function of the severity of nicotine addiction as assessed with the FTND: smokers with the highest addiction severity showed the lowest amygdala reactivity after smoking deprivation (Figure 3).

Finally, two studies examined smokers under both smoking satiety and 24 h of smoking abstinence using smoking cue-reactivity paradigms and might therefore reveal insights into whether the amygdala is differentially involved in both forms of nicotine craving (McClernon, Hiott, Huettel, & Rose, 2005; McClernon et al., 2009). The studies, however, had conflicting results: one found no effect of abstinence on neural reactivity to smoking cues (McClernon et al., 2005), and the other found increased smoking-cue reactivity in parietal, striatal, and frontal regions following abstinence (McClernon et al., 2009). Both studies, however, did not find effects of abstinence on the amygdala's smoking-cue reactivity, suggesting that conditioned smoking cues and abstinence have differential effects on the amygdala.

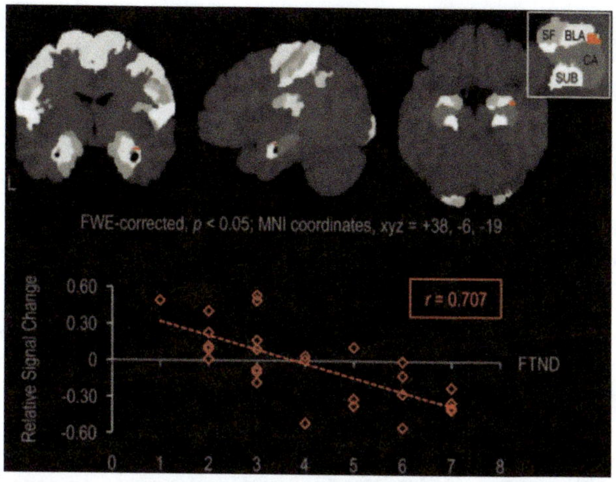

FIGURE 3 Association between higher addiction severity and decreased amygdala fear reactivity. Reduced amygdala fear reactivity during smoking abstinence was associated with the severity of nicotine addiction as assessed with the FTND. Smokers with the highest addiction severity showed the lowest amygdala reactivity to fear signals (fearful faces) following 12 h of nicotine deprivation. The effect was located in the basolateral subregion of the amygdala (BLA). The BLA is involved in the detection and validation of fear signals. Abbreviations: BLA, basolateral amygdala; CA, cornu ammonis; FTND, Fagerström Test of Nicotine Dependence; FWE, family-wise error; MNI, Montreal Neurological Institute; SF, superficial amygdala; SUB, subiculum. *Used with permission from Onur et al. (2012), adapted from Figure 3, p. 1413, John Wiley and Sons, 2012.*

FUNCTIONAL NEUROIMAGING STUDIES ON ALTERED EMOTION PROCESSING IN SMOKERS

In addition to numerous studies that assessed the neural basis of nicotine craving in smokers, a handful of studies have begun to explore associations between nicotine addiction and altered emotion processing. An initial study compared neural activity in nicotine-dependent smokers and matched never-smoking controls while they were shown unpleasant, pleasant, and neutral pictorial scenes from a validated emotional picture system (Kobiella et al., 2010). Relative to never-smoking controls, smokers showed reduced amygdala reactivity to unpleasant stimuli and not to pleasant or neutral stimuli. Another study used a face perception task to specifically address whether abstinence-induced craving interferes with emotion processing in the amygdala (Onur et al., 2012). In this study, smokers underwent fMRI on two separate occasions under conditions of satiety and after 12 h of smoking abstinence. Compared to the satiated state, smokers in the deprived state showed reduced amygdala reactivity to fearful face stimuli, yet not to happy or neutral facial stimuli, suggesting a fear-specific amygdala dysfunction during abstinence. The amygdala has long been involved in threat perception, and amygdala reactivity to threatening stimuli, including fearful faces, has been directly associated with threat sensitivity (Etkin et al., 2004). This fear-specific amygdala functioning thus might reflect reduced threat perception in smokers, which might constitute a mechanism that, although not directly addictive, maintains smoking behavior and promotes relapse during early abstinence.

The study also incorporated a group of never-smokers as a reference group. Interestingly, amygdala functioning in satiated smokers did not differ from that in never-smokers, suggesting that decreased amygdala reactivity to fearful stimuli in smokers normalizes with nicotine ingestion. This effect of nicotine on amygdala functioning would be in line with the observation that nicotine application in never-smokers specifically increased amygdala reactivity to unpleasant stimuli (Kobiella et al., 2011). Further, a prospective study in smokers suggesting an association between reduced amygdala functioning and relapse vulnerability (Jasinska et al., 2012) is in line with the proposed relevance of decreased amygdala reactivity to potentially threatening stimuli for the maintenance of smoking behavior. This study examined neural activity in smokers in response to smoking-cessation messages before starting a Web-based smoking intervention and showed that reduced preintervention amygdala reactivity was associated with quitting the cessation program during the 4-month follow-up.

CONCLUSIONS

In line with the animal literature emphasizing an important role of the amygdala in maintaining drug addiction through its engagement with drug-associated memories and craving, findings from the available human neuroimaging literature on smokers suggest that the amygdala contributes to the high addictive potential of smoking. Specifically, the available human neuroimaging literature suggests that the amygdala plays a crucial role in three mechanisms that promote the maintenance of smoking behavior and relapse during abstinence: (1) smoking-cue-induced craving, (2) abstinence-induced craving, and (3) deficient threat perception.

First of all, studies examining amygdala reactivity to smoking-related cues have consistently reported increased amygdala reactivity to smoking-related cues in smokers. This is in line with the pivotal role of the amygdala in salience processing and emotional learning. Repeated ingestion of nicotine via cigarette smoking may promote the association of reinforcing nicotine effects and smoking-related cues that underlie incentive salience sensitization (Robinson & Berridge, 2001) and drug-related memories (Everitt et al., 2003). After repeated pairings of nicotine self-administration and smoking-related cues, the formerly neutral smoking-related cues themselves gain excessive salience and the property to induce craving and smoking behavior. Studies in laboratory animals and humans have provided compelling evidence that drug-related memories contribute to the persistence of drug use and relapse when addicts or previously addicted animals are reexposed to drug cues (Robbins et al., 2008).

Second, initial studies have begun to elucidate the contribution of the amygdala to abstinence-induced craving that develops independent of the presence of smoking-related cues. The observation that amygdala activity during rest increases with subjective craving suggests that the amygdala is critically involved in the aversive emotional states that smokers experience after cessation of use. The emergence of a negative emotional state when access to the drug is prevented has been conceptualized as a key element of drug addiction. In line with the proposed role of the amygdala in the aversive emotional states that smokers experience during smoking deprivation, animal models of addiction propose a key role for the amygdala in the withdrawal and negative affect stage of the addiction cycle (Koob & Volkow, 2010).

Third, initial studies that assessed emotional processing in smokers revealed evidence of fear-specific dysfunctions of the amygdala, indicating deficits in fear recognition and threat perception in smokers that might present either a predisposition or a consequence of smoking. Although the finding of dysfunctional fear processing in the amygdala in smokers needs to be replicated, the initial findings point to a mechanism that, although not addictive per se, might maintain smoking behavior and contribute to high relapse rates in smokers, particularly during early abstinence. There is a strong consensus that the amygdala extracts biological significance, particularly threat level, from the environment and initiates affective and motivational responses to the stimulus (Adolphs, 2013; LeDoux, 2000). In terms of threat perception, the amygdala's well-known role in emotional memory interacts with a rather principal role in the online detection of potentially threatening stimuli, indicating danger (McGaugh, 2004). Fear appeals are frequently used by public health awareness campaigns in the form of warning labels or showing the severe health consequences of smoking, such as cancer, and have been shown to have little effect on smoking behavior. To this end, some researchers have advocated that disrupted threat perception in the amygdala and consequently reduced avoidance of potential harmful behavior in smokers might promote the maintenance of smoking behavior and contribute to high relapse rates in nicotine addiction (Mihov & Hurlemann, 2012; Onur et al., 2012). In support of the hypothesized contribution of the amygdala, lesions of the amygdala in humans have been associated with impairments in the recognition of stimuli that might signal threat (Becker et al., 2012), deficient fear conditioning (LaBar, LeDoux, Spencer, & Phelps, 1995), lack of loss avoidance (De Martino, Camerer, & Adolphs, 2010), and even the lack of subjective feelings of fear (Feinstein, Adolphs, Damasio, & Tranel, 2011). From an evolutionary perspective, all of these amygdala-associated functions aim at protecting the organism from potential harm. Thus, the compromised threat perception in smokers could undermine the effectiveness of fear-based public health awareness campaigns, such as warning labels on cigarette packaging or alarming advertisements depicting the fatal physiological consequences of cigarette smoking. Disrupted threat perception could thus diminish the effectiveness of these threatening messages for smoking cessation and increase the risk of relapse into smoking behavior.

APPLICATIONS TO OTHER ADDICTIONS AND SUBSTANCE MISUSE

Research on nicotine addiction emphasizes the contribution of nondrug factors to the development and maintenance of addictive behavior. Based on the mesocorticolimbic DA hypothesis, addiction research in humans has predominantly focused on the role of the striatum. Studies in laboratory animals and human nicotine-addicted smokers stress the contribution of learning processes, many of which are mediated by amygdala-centered circuits. Previous quantitative and qualitative studies have consistently associated the amygdala with increased reactivity to drug cues in nicotine (Kuhn & Gallinat, 2011; Mihov & Hurlemann, 2012) and other addictions (Chase, Eickhoff, Laird, & Hogarth, 2011; Wilson,

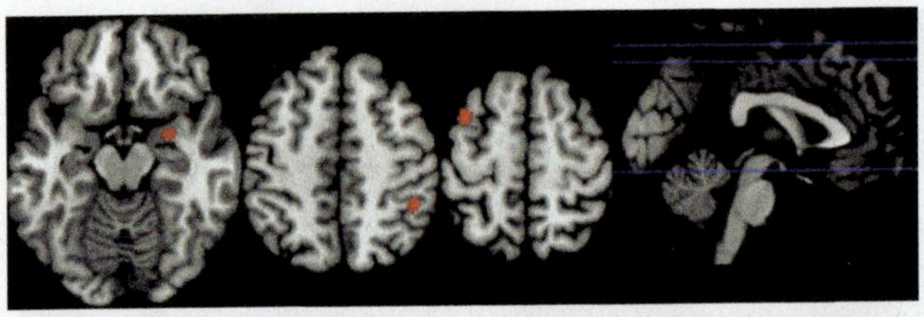

FIGURE 4 Meta-analytic data on neural activation that accompanies increases in craving. Findings from a meta-analysis on studies reporting positive associations between neural activity and craving in substance users are shown. Convergent activation was found in the amygdala and parietal and middle frontal cortices, suggesting that higher craving in substance abusers is accompanied by increased activity in these regions. *Used with permission from Chase et al. (2011), Figure 3, p. 789, Elsevier, 2011.*

Sayette, & Fiez, 2004). Notably, meta-analytic data on neural activity that accompanies increases in craving converged on the amygdala (Chase et al., 2011) (Figure 4). Behavioral studies of nicotine addiction have provided strong evidence for an association between unprovoked abstinence-induced craving and subsequent relapse. Together with findings from initial neuroimaging studies that associated abstinence-induced changes in amygdala functioning with the severity of nicotine dependence and subjective craving, these findings might suggest that altered amygdala functioning during early abstinence could be a reliable predictor of relapse vulnerability. Given that several factors, such as individual drug use history and ambivalence about drug use, might lead to large interindividual variance in neural cue reactivity (Jasinska et al., 2014), unprovoked alterations in amygdala functioning during early abstinence might represent more reliable predictors of treatment outcome and relapse. Finally, findings from neuroimaging studies of smokers suggest that threat-perception deficits in the amygdala, either as a predisposition or as a consequence of nicotine use, might promote the maintenance of nicotine behavior. This mechanism might also undermine the efficiency of treatment approaches and health campaigns for other drugs of abuse.

DEFINITION OF TERMS

Abstinence-induced craving This develops in the absence of drug cues and is more sensitive to the delivery of the drug.

Arterial spin labeling magnetic resonance imaging This is a neuroimaging technique that labels arterial blood before it enters the brain to examine changes in CBF.

Craving This describes the overwhelming desire or urge to take drugs that is often reported by drug-addicted individuals and often leads to continued drug use or relapse after periods of abstinence.

Cue-induced craving This corresponds to craving elicited by a stimulus that has been previously associated with the drug.

Cue-reactivity paradigms These studies present drug-related stimuli (mostly pictures with drug-related content) and neutral stimuli to drug addicts; neural networks engaged in drug-cue reactivity are assessed by subtracting neural activity during neutral stimuli from activity during drug-related stimuli.

Fear This is a negative emotion elicited by the perception of a potential harmful threat, has strong motivational character, and usually induces a behavioral response to the threatening stimuli.

Functional magnetic resonance imaging This is neuroimaging technique that noninvasively measures brain activity by means of magnetic resonance imaging; it measures changes in blood flow and blood oxygenation related to brain activity.

Nicotinic acetylcholine receptors These receptors form ligand-gated ion channels on neurons that can be activated by the endogenous transmitter acetylcholine but also by nicotine.

Reward system This refers to a dopaminergic circuitry in the brain that mediates the reinforcing effects of natural rewards (e.g., food and sex) as well as the reinforcing effects of addictive drugs.

Valence This refers to the intrinsic emotional attractiveness or aversiveness of an event or stimulus.

KEY FACTS OF NICOTINE ADDICTION

- Tobacco use is the leading cause of preventable illness and death in the world today.
- Tobacco use causes 1 in 10 deaths among adults worldwide.
- The risk that a chronic smoker will die prematurely from a medical complication of smoking is estimated to be up to 50%.
- Only 3–5% of smokers who try to quit without treatment achieve prolonged abstinence for 6–12 months.
- Most smokers relapse during the first week of a quit attempt.
- Nicotine, the main psychoactive ingredient in tobacco, drives smoking.
- Like other addictive drugs, nicotine increases DA in the reward centers of the brain.
- Nicotine-dependent smokers experience withdrawal symptoms when they stop smoking.
- Withdrawal symptoms typically begin within 48 h after smoking cessation.
- Withdrawal symptoms include somatic and affective symptoms, such as anger, anxiety, depressed mood, irritability, mild cognitive deficits, and physiological symptoms.
- Tobacco addiction is a complex process; in addition to the reinforcing effects of nicotine, learning and memory processes contribute to the highly addictive nature of smoking.

SUMMARY POINTS

- This chapter focuses on the role of the amygdala in nicotine addiction.
- The reinforcing effects of nicotine are a major factor that drives smoking.
- Learning and memory processes mediated by amygdala-centered circuits may additionally contribute to the highly addictive nature of smoking.
- Animal studies suggest an important role for the amygdala in drug memories and the negative affective state during withdrawal.
- The available human neuroimaging literature suggests that the amygdala may contribute to the maintenance of smoking behavior and relapse.
- Three potential amygdala-centered mechanisms have been identified, namely, increased neural reactivity to smoking-associated cues, aversive state during initial abstinence, and reduced threat perception.

REFERENCES

Adolphs, R. (2013). The biology of fear. *Current Biology, 23*(2), R79–R93.

Artiges, E., Ricalens, E., Berthoz, S., Krebs, M. O., Penttila, J., Trichard, C., & Martinot, J. L. (2009). Exposure to smoking cues during an emotion recognition task can modulate limbic fMRI activation in cigarette smokers. *Addiction Biology, 14*(4), 469–477.

Balfour, D. J. (2004). The neurobiology of tobacco dependence: a preclinical perspective on the role of the dopamine projections to the nucleus accumbens [corrected]. *Nicotine & Tobacco Research, 6*(6), 899–912.

Baxter, M. G., & Murray, E. A. (2002). The amygdala and reward. *Nature Reviews Neuroscience, 3*(7), 563–573.

Becker, B., Mihov, Y., Scheele, D., Kendrick, K. M., Feinstein, J. S., Matusch, A., ... Hurlemann, R. (2012). Fear processing and social networking in the absence of a functional amygdala. *Biological Psychiatry, 72*(1), 70–77.

Benowitz, N. L. (2010). Nicotine addiction. *The New England Journal of Medicine, 362*(24), 2295–2303.

Breiter, H. C., Etcoff, N. L., Whalen, P. J., Kennedy, W. A., Rauch, S. L., Buckner, R. L., ... Rosen, B. R. (1996). Response and habituation of the human amygdala during visual processing of facial expression. *Neuron, 17*(5), 875–887.

Caggiula, A. R., Donny, E. C., White, A. R., Chaudhri, N., Booth, S., Gharib, M. A., ... Sved, A. F. (2001). Cue dependency of nicotine self-administration and smoking. *Pharmacology Biochemistry and Behavior, 70*(4), 515–530.

Carter, B. L., & Tiffany, S. T. (1999). Meta-analysis of cue-reactivity in addiction research. *Addiction, 94*(3), 327–340.

Chase, H. W., Eickhoff, S. B., Laird, A. R., & Hogarth, L. (2011). The neural basis of drug stimulus processing and craving: an activation likelihood estimation meta-analysis. *Biological Psychiatry, 70*(8), 785–793.

Clarke, P. B., Fu, D. S., Jakubovic, A., & Fibiger, H. C. (1988). Evidence that mesolimbic dopaminergic activation underlies the locomotor stimulant action of nicotine in rats. *Journal of Pharmacology and Experimental Therapeutics, 246*(2), 701–708.

Conklin, C. A., & Tiffany, S. T. (2002). Applying extinction research and theory to cue-exposure addiction treatments. *Addiction, 97*(2), 155–167.

Davis, J. A., & Gould, T. J. (2008). Associative learning, the hippocampus, and nicotine addiction. *Current Drug Abuse Reviews, 1*(1), 9–19.

De Martino, B., Camerer, C. F., & Adolphs, R. (2010). Amygdala damage eliminates monetary loss aversion. *Proceedings of the National Academy of Sciences of the United States of America, 107*(8), 3788–3792.

Dome, P., Lazary, J., Kalapos, M. P., & Rihmer, Z. (2010). Smoking, nicotine and neuropsychiatric disorders. *Neuroscience & Biobehavioral Reviews, 34*(3), 295–342.

Due, D. L., Huettel, S. A., Hall, W. G., & Rubin, D. C. (2002). Activation in mesolimbic and visuospatial neural circuits elicited by smoking cues: evidence from functional magnetic resonance imaging. *American Journal of Psychiatry, 159*(6), 954–960.

Engelmann, J. M., Versace, F., Robinson, J. D., Minnix, J. A., Lam, C. Y., Cui, Y., ... Cinciripini, P. M. (2012). Neural substrates of smoking cue reactivity: a meta-analysis of fMRI studies. *NeuroImage, 60*(1), 252–262.

Etkin, A., Klemenhagen, K. C., Dudman, J. T., Rogan, M. T., Hen, R., Kandel, E. R., & Hirsch, J. (2004). Individual differences in trait anxiety predict the response of the basolateral amygdala to unconsciously processed fearful faces. *Neuron, 44*(6), 1043–1055.

Everitt, B. J., Cardinal, R. N., Parkinson, J. A., & Robbins, T. W. (2003). Appetitive behavior: impact of amygdala-dependent mechanisms of emotional learning. *Annals of the New York Academy of Sciences*, *985*, 233–250.

Fagerström, K. O. (1978). Measuring degree of physical dependence to tobacco smoking with reference to individualization of treatment. *Addictive Behaviors*, *3*(3–4), 235–241.

Feinstein, J. S., Adolphs, R., Damasio, A., & Tranel, D. (2011). The human amygdala and the induction and experience of fear. *Current Biology*, *21*(1), 34–38.

Feltenstein, M. W., & See, R. E. (2008). The neurocircuitry of addiction: an overview. *British Journal of Pharmacology*, *154*(2), 261–274.

Franklin, T. R., Wang, Z., Li, Y., Suh, J. J., Goldman, M., Lohoff, F. W., ... Childress, A. R. (2011). Dopamine transporter genotype modulation of neural responses to smoking cues: confirmation in a new cohort. *Addiction Biology*, *16*(2), 308–322.

Franklin, T. R., Wang, Z., Wang, J., Sciortino, N., Harper, D., Li, Y., ... Childress, A. R. (2007). Limbic activation to cigarette smoking cues independent of nicotine withdrawal: a perfusion fMRI study. *Neuropsychopharmacology*, *32*(11), 2301–2309.

Garvey, A. J., Bliss, R. E., Hitchcock, J. L., Heinold, J. W., & Rosner, B. (1992). Predictors of smoking relapse among self-quitters: a report from the normative aging Study. *Addictive Behaviors*, *17*(4), 367–377.

George, O., & Koob, G. F. (2010). Individual differences in prefrontal cortex function and the transition from drug use to drug dependence. *Neuroscience & Biobehavioral Reviews*, *35*(2), 232–247.

Goudriaan, A. E., de Ruiter, M. B., van den Brink, W., Oosterlaan, J., & Veltman, D. J. (2010). Brain activation patterns associated with cue reactivity and craving in abstinent problem gamblers, heavy smokers and healthy controls: an fMRI study. *Addiction Biology*, *15*(4), 491–503.

Heatherton, T. F., Kozlowski, L. T., Frecker, R. C., & Fagerström, K. O. (1991). The Fagerström test for nicotine dependence: a revision of the Fagerström tolerance questionnaire. *British Journal of Addiction*, *86*(9), 1119–1127.

Holland, P. C., & Gallagher, M. (2004). Amygdala-frontal interactions and reward expectancy. *Current Opinion in Neurobiology*, *14*(2), 148–155.

Hughes, J. R. (2007). Effects of abstinence from tobacco: valid symptoms and time course. *Nicotine & Tobacco Research*, *9*(3), 315–327.

Hughes, J. R., Keely, J., & Naud, S. (2004). Shape of the relapse curve and long-term abstinence among untreated smokers. *Addiction*, *99*(1), 29–38.

Hughes, J. R., Peters, E. N., & Naud, S. (2008). Relapse to smoking after 1 year of abstinence: a meta-analysis. *Addictive Behaviors*, *33*(12), 1516–1520.

Janes, A. C., Pizzagalli, D. A., Richardt, S., deB Frederick, B., Chuzi, S., Pachas, G., ... Kaufman, M. J. (2010). Brain reactivity to smoking cues prior to smoking cessation predicts ability to maintain tobacco abstinence. *Biological Psychiatry*, *67*(8), 722–729.

Jarvik, M. E., Madsen, D. C., Olmstead, R. E., Iwamoto-Schaap, P. N., Elins, J. L., & Benowitz, N. L. (2000). Nicotine blood levels and subjective craving for cigarettes. *Pharmacology Biochemistry and Behavior*, *66*(3), 553–558.

Jasinska, A. J., Chua, H. F., Ho, S. S., Polk, T. A., Rozek, L. S., & Strecher, V. J. (2012). Amygdala response to smoking-cessation messages mediates the effects of serotonin transporter gene variation on quitting. *NeuroImage*, *60*(1), 766–773.

Jasinska, A. J., Stein, E. A., Kaiser, J., Naumer, M. J., & Yalachkov, Y. (2014). Factors modulating neural reactivity to drug cues in addiction: a survey of human neuroimaging studies. *Neuroscience & Biobehavioral Reviews*, *38*, 1–16.

Killen, J. D., & Fortmann, S. P. (1997). Craving is associated with smoking relapse: findings from three prospective studies. *Experimental and Clinical Psychopharmacology*, *5*(2), 137–142.

Kober, H., Mende-Siedlecki, P., Kross, E. F., Weber, J., Mischel, W., Hart, C. L., & Ochsner, K. N. (2010). Prefrontal-striatal pathway underlies cognitive regulation of craving. *Proceedings of the National Academy of Sciences of the United States of America*, *107*(33), 14811–14816.

Kobiella, A., Ulshofer, D. E., Vollmert, C., Vollstadt-Klein, S., Buhler, M., Esslinger, C., ... Smolka, M. N. (2011). Nicotine increases neural response to unpleasant stimuli and anxiety in non-smokers. *Addiction Biology*, *16*(2), 285–295.

Kobiella, A., Vollstadt-Klein, S., Buhler, M., Graf, C., Buchholz, H. G., Bernow, N., ... Smolka, M. N. (2010). Human dopamine receptor D2/D3 availability predicts amygdala reactivity to unpleasant stimuli. *Human Brain Mapping*, *31*(5), 716–726.

Koob, G. F., & Volkow, N. D. (2010). Neurocircuitry of addiction. *Neuropsychopharmacology*, *35*(1), 217–238.

Kuhn, S., & Gallinat, J. (2011). Common biology of craving across legal and illegal drugs – a quantitative meta-analysis of cue-reactivity brain response. *European Journal of Neuroscience*, *33*(7), 1318–1326.

LaBar, K. S., LeDoux, J. E., Spencer, D. D., & Phelps, E. A. (1995). Impaired fear conditioning following unilateral temporal lobectomy in humans. *Journal of Neuroscience*, *15*(10), 6846–6855.

LeDoux, J. E. (2000). Emotion circuits in the brain. *Annual Review of Neuroscience*, *23*, 155–184.

McBride, D., Barrett, S. P., Kelly, J. T., Aw, A., & Dagher, A. (2006). Effects of expectancy and abstinence on the neural response to smoking cues in cigarette smokers: an fMRI study. *Neuropsychopharmacology*, *31*(12), 2728–2738.

McClernon, F. J., Hiott, F. B., Huettel, S. A., & Rose, J. E. (2005). Abstinence-induced changes in self-report craving correlate with event-related FMRI responses to smoking cues. *Neuropsychopharmacology*, *30*(10), 1940–1947.

McClernon, F. J., Hiott, F. B., Liu, J., Salley, A. N., Behm, F. M., & Rose, J. E. (2007). Selectively reduced responses to smoking cues in amygdala following extinction-based smoking cessation: results of a preliminary functional magnetic resonance imaging study. *Addiction Biology*, *12*(3–4), 503–512.

McClernon, F. J., Kozink, R. V., Lutz, A. M., & Rose, J. E. (2009). 24-h smoking abstinence potentiates fMRI-BOLD activation to smoking cues in cerebral cortex and dorsal striatum. *Psychopharmacology (Berl)*, *204*(1), 25–35.

McGaugh, J. L. (2004). The amygdala modulates the consolidation of memories of emotionally arousing experiences. *Annual Review of Neuroscience*, *27*, 1–28.

Mihov, Y., & Hurlemann, R. (2012). Altered amygdala function in nicotine addiction: insights from human neuroimaging studies. *Neuropsychologia*, *50*(8), 1719–1729.

Morgan, C. J., Muetzelfeldt, L., Muetzelfeldt, M., Nutt, D. J., & Curran, H. V. (2010). Harms associated with psychoactive substances: findings of the UK National Drug Survey. *Journal of Psychopharmacology*, *24*(2), 147–153.

Nestor, L., McCabe, E., Jones, J., Clancy, L., & Garavan, H. (2011). Differences in "bottom-up" and "top-down" neural activity in current and former cigarette smokers: evidence for neural substrates which may promote nicotine abstinence through increased cognitive control. *NeuroImage*, *56*(4), 2258–2275.

Nutt, D., King, L. A., Saulsbury, W., & Blakemore, C. (2007). Development of a rational scale to assess the harm of drugs of potential misuse. *Lancet, 369*(9566), 1047–1053.

Okuyemi, K. S., Powell, J. N., Savage, C. R., Hall, S. B., Nollen, N., Holsen, L. M., … Ahluwalia, J. S. (2006). Enhanced cue-elicited brain activation in African American compared with Caucasian smokers: an fMRI study. *Addiction Biology, 11*(1), 97–106.

Onur, O. A., Patin, A., Mihov, Y., Buecher, B., Stoffel-Wagner, B., Schlaepfer, T. E., … Hurlemann, R. (2012). Overnight deprivation from smoking disrupts amygdala responses to fear. *Human Brain Mapping, 33*(6), 1407–1416.

Robbins, T. W., Ersche, K. D., & Everitt, B. J. (2008). Drug addiction and the memory systems of the brain. *Annals of the New York Academy of Sciences, 1141*, 1–21.

Robinson, T. E., & Berridge, K. C. (2001). Incentive-sensitization and addiction. *Addiction, 96*(1), 103–114.

Royal College of Physicians. (2007). *Harm reduction in nicotine addiction: Helping people who can't quit: A report by the Tobacco Advisory Group of the Royal College of Physicians*. London: Royal College of Physicians.

Rubinstein, M. L., Luks, T. L., Moscicki, A. B., Dryden, W., Rait, M. A., & Simpson, G. V. (2011). Smoking-related cue-induced brain activation in adolescent light smokers. *Journal of Adolescent Health, 48*(1), 7–12.

Shiffman, S., Engberg, J. B., Paty, J. A., Perz, W. G., Gnys, M., Kassel, J. D., & Hickcox, M. (1997). A day at a time: predicting smoking lapse from daily urge. *Journal of Abnormal Psychology, 106*(1), 104–116.

Shiffman, S., West, R., & Gilbert, D. (2004). Recommendation for the assessment of tobacco craving and withdrawal in smoking cessation trials. *Nicotine & Tobacco Research, 6*(4), 599–614.

Stippekohl, B., Winkler, M., Mucha, R. F., Pauli, P., Walter, B., Vaitl, D., & Stark, R. (2010). Neural responses to BEGIN- and END-stimuli of the smoking ritual in nonsmokers, nondeprived smokers, and deprived smokers. *Neuropsychopharmacology, 35*(5), 1209–1225.

Stolerman, I. P., & Jarvis, M. J. (1995). The scientific case that nicotine is addictive. *Psychopharmacology (Berl), 117*(1) 2–10; discussion 14–20.

Tiffany, S. T., Cox, L. S., & Elash, C. A. (2000). Effects of transdermal nicotine patches on abstinence-induced and cue-elicited craving in cigarette smokers. *Journal of Consulting and Clinical Psychology, 68*(2), 233–240.

United States Department of Health and Human Services. (2010). *How tobacco smoke causes disease: The biology and behavioral basis for smoking-attributable disease: A report of the surgeon general*. Rockville, MD: United States Department of Health and Human Services.

Volkow, N. D., Fowler, J. S., & Wang, G. J. (2003). The addicted human brain: insights from imaging studies. *Journal of Clinical Investigation, 111*(10), 1444–1451.

Vollstadt-Klein, S., Kobiella, A., Buhler, M., Graf, C., Fehr, C., Mann, K., & Smolka, M. N. (2011). Severity of dependence modulates smokers' neuronal cue reactivity and cigarette craving elicited by tobacco advertisement. *Addiction Biology, 16*(1), 166–175.

Wang, Z., Faith, M., Patterson, F., Tang, K., Kerrin, K., Wileyto, E. P., … Lerman, C. (2007). Neural substrates of abstinence-induced cigarette cravings in chronic smokers. *Journal of Neuroscience, 27*(51), 14035–14040.

Wetherill, R. R., Young, K. A., Jagannathan, K., Shin, J., O'Brien, C. P., Childress, A. R., & Franklin, T. R. (2013). The impact of sex on brain responses to smoking cues: a perfusion fMRI study. *Biology of Sex Differences, 4*(1), 9.

Wilson, S. J., Sayette, M. A., & Fiez, J. A. (2004). Prefrontal responses to drug cues: a neurocognitive analysis. *Nature Neuroscience, 7*(3), 211–214.

Wilson, S. J., Sayette, M. A., & Fiez, J. A. (2012). Quitting-unmotivated and quitting-motivated cigarette smokers exhibit different patterns of cue-elicited brain activation when anticipating an opportunity to smoke. *Journal of Abnormal Psychology, 121*(1), 198–211.

World Health Organization. (2008). *Report on the Global Tobacco Epidemic: The MPOWER package*. Geneva: World Health Organization.

Yalachkov, Y., Kaiser, J., & Naumer, M. J. (2012). Functional neuroimaging studies in addiction: multisensory drug stimuli and neural cue reactivity. *Neuroscience & Biobehavioral Reviews, 36*(2), 825–835.

Chapter 27

Nicotine and Cognition: Effects of Nicotine on Attention and Memory Systems in Humans

Anton L. Beer

Institut für Psychologie, Universität Regensburg, Regensburg, Germany

Abbreviations

ACh Acetylcholine
AChE Acetylcholinesterase
AChR ACh receptor
ARAS Ascending reticular activating system
CPT Continuous performance test
EEG Electroencephalography
ERP Event-related potential
FFT Fast Fourier transformation
fMRI Functional magnetic resonance imaging
mAChR Muscarinic ACh receptor
N1, N2 Negative ERP components
nAChR Nicotinic AChR
P1, P2 Positive ERP components
P300 Positive EEG deflection with onset latency around 300 ms post stimulus onset
REM Rapid eye movements
RVIP Rapid visual information processing
SWS Slow-wave sleep

INTRODUCTION

Nicotine is a substance that readily crosses the blood–brain barrier and has a substantial effect on brain activity. As stated in previous chapters of this book, its major neural consequence is to modulate the efficacy of the neurotransmitter acetylcholine (ACh). In particular, nicotine acts as an agonist for acetylcholine receptors (AChR). There are two classes of AChR: nicotinic (nAChR) and muscarinic (mAChR). Nicotine primarily affects nicotinic ACh receptors.

The cholinergic system—that is, the neural circuits that operate with the neurotransmitter ACh—is strongly involved in a number of cognitive processes. It regulates sensory and motor processing, modulates attention, and is involved in learning and memory. As nicotine is an agonist of AChR, its major consequence is to upregulate cholinergic signaling. As of this, it may be considered a cognitive enhancer (but see below for reverse effects). Nicotine strongly affects cognitive processes that involve the cholinergic system. Therefore, this chapter will primarily review the relatively large number of studies showing the cognitive effects of nicotine on attention, learning, and memory, although some studies also showed nicotine-induced performance improvements in other cognitive tasks, such as reasoning (e.g., Foulds et al., 1996).

The cholinergic system is affected by a number of substances that act in a similar or antagonistic manner to nicotine. To understand the cognitive effects of nicotine, research on these other substances is occasionally discussed in this chapter. In particular, mecamylamine is an antagonist of nicotinic AChR and may counteract the effects of nicotine. Scopolamine is an antagonist of muscarinic AChR. Moreover, a number of substances affect the molecular processes of neurotransmitter reuptake from the synaptic cleft, which is primarily accomplished by the enzyme acetylcholinesterase (AChE). Donepezil and physostigmine inhibit this AChE process. Although all of these substances are cholinergic modulators, important differences should be considered when comparing their cognitive effects with those of nicotine. These include the molecular target process (nAChR, mAChR, and AChE), its major modulating effect (agonistic vs. antagonistic), and its half-life (i.e., the time until the substance is metabolized in the body). Whereas nicotine is metabolized in the liver with a half-life of about 2–3 h (Benowitz, Hukkanen, & Jacob, 2009), the AChE inhibitor donepezil has a half-life of about 80 h (Rogers et al., 1998). Due to the relatively brief half-life of nicotine, its cognitive consequences may be considered short term. Nevertheless, as nicotine modulates learning and brain plasticity, it may also exert long-term effects on cognition. Moreover, habitual consumption of nicotine may have reverse effects on cognitive processing due to the desensitization of ACh receptors.

NICOTINE AND ATTENTION

Attention is a complex cognitive process that includes mechanisms of arousal and vigilance, mechanisms for selecting sensory signals and actions, and mechanisms for allocating cognitive resources to multiple tasks. These different mechanisms of attention are ascribed to partially separate brain circuits for alerting, orienting, and executive control (Himelstein, Newcorn, & Halperin, 2000;

Raz & Buhle, 2006). The alerting system involves subcortical brain structures, such as the locus coeruleus and the ascending reticular activating system (ARAS). These subcortical nuclei control the degree of neural processing by projecting to the thalamus (a major relay station for sensory signals) and the cortex (Brown, Basheer, McKenna, Strecker, & McCarley, 2012). The orienting system of attention is primarily mediated by a posterior brain network that includes the pulvinar, the superior colliculus, and the parietal cortex. Orienting requires at least three partially distinct processes (Posner & Rafal, 1987): disengagement from previously attended items or events, shifting attention, and engaging attention to a new item or event. The executive system of attention is primarily mediated by a frontal brain network that includes the anterior cingulate cortex and the lateral prefrontal cortex. Nicotine modulates processing of all three major attentional networks, but it exerts distinct effects on subcomponents of these systems. The effects of nicotine on attention were extensively studied in animals (see Poorthuis & Mansvelder, 2013 for a recent review). This chapter will focus on the role of nicotine in humans as examined by behavioral measures, electroencephalography (EEG), and functional magnetic resonance imaging (fMRI).

Many studies showed that nicotine enhanced performance in a variety of tasks that involve attention. For instance, nicotine leads to more accurate and faster responses in the rapid visual information processing task (RVIP) in smokers (Juliano, Fucito, & Harrell, 2011; Lawrence, Ross, & Stein, 2002; Parrott & Craig, 1992) and nonsmokers (Foulds et al., 1996; Wesnes & Warburton, 1984). In the RVIP task, observers are asked to detect a sequence of three sequential digits (e.g., 2-4-6) in a stream of rapidly presented digits (e.g., 100 per minute). Outcome measures (high number of correct detections, fast response times) indicate a high degree of alertness and vigilance. People that received nicotine responded faster and detected more target sequences in the RVIP task than a placebo control group. On the other hand, scopolamine—a (muscarinic) antagonist of the cholinergic system—significantly reduced performance, suggesting a critical role of the cholinergic system for the RVIP task (Wesnes & Warburton, 1984). Similarly, nonsmokers with a transdermal nicotine patch made fewer errors in the continuous performance test (CPT, a task comparable to the RVIP) than the placebo group (Levin et al., 1998). The performance-enhancing effect of nicotine seems to be dose dependent. A monotonic dose–response function was observed between nicotine (up to 4 mg nicotine gum) and performance in the RVIP task (Foulds et al., 1996; Parrott & Winder, 1989).

The behavioral effects of nicotine are also evident in EEG signals. EEG measures the electrical potentials generated by the brain at the scalp surface. These potentials fluctuate across time and the frequency of fluctuations indicates distinct cognitive states. For instance, frequencies in the range of 12–30 Hz (called beta waves) reflect intense cognitive processing. EEG frequencies below 8 Hz (called delta and theta waves) usually emerge during sleep (Brown et al., 2012). Frequencies in the range from 8 to 12 Hz reflect states of drowsiness or being relaxed. Several studies have shown that nicotine substantially reduces alpha EEG waves in smokers (Golding, 1988) and nonsmokers (Beer, Vartak, & Greenlee, 2013), suggesting that nicotine reduces drowsiness and enhances alertness.

Figure 1 shows the EEG power spectrum of nonsmokers during consumption of chewing tobacco (containing nicotine) or a control substance (without nicotine) while performing a simple visual detection task. EEG waves in the alpha frequency range (8–12 Hz) were substantially reduced in the nicotine group compared to the control group. Moreover, small but significant reductions in theta (but not delta) waves and increased activity of beta waves as a consequence of nicotine consumption were observed (Golding, 1988). Furthermore, subcutaneously injected nicotine was shown to shift the mean alpha frequency of the EEG towards the beta spectrum in nonsmokers by about 2 Hz (Foulds et al., 1994). Some fMRI studies explored the brain areas affected by nicotine. The method of fMRI reveals the hemodynamic response associated with neural processing. Consistent with the view that nicotine enhances alertness aspects of attention, a transdermal nicotine patch in (nonabstinent) smokers enhanced fMRI activity in the parietal cortex, thalamus, caudate, and the occipital cortex during an RVIP task (Lawrence et al., 2002).

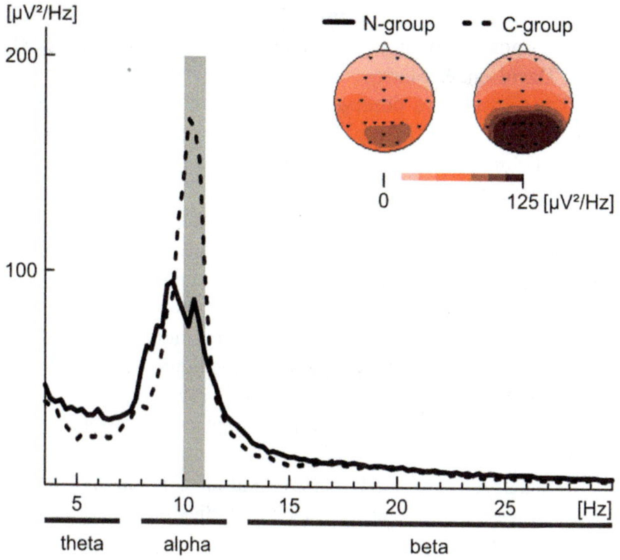

Nicotine also seems to promote the orienting system of attention. This system of attention may be examined by a spatial cuing task. In this task, observers are asked to respond to a visual target stimulus that is presented in the peripheral visual field. Prior to the target stimulus, a cue is presented (e.g., left or right), which is either valid (predicting the location of the subsequent target), neutral, or invalid (predicting a wrong location). Typically, observers benefit from the cue: They respond faster to targets following valid cues than to targets following neutral cues. On the other hand, invalid cues are associated with processing costs: Responses to targets are slower in trials with invalid cues than in trials with neutral cues. The response time difference—also known as validity effect—reflects attentional processes of the orienting system.

FIGURE 1 **Modulation of EEG frequencies by nicotine.** The plot shows the EEG power density per frequency at an occipital-parietal electrode cluster of two groups of nonsmokers while consuming oral nicotine (N-group) or a placebo substance (C-group). Participants performed a simple visual detection task. Alpha power in the frequency range from 10 to 11 Hz (gray bar) was substantially reduced in the nicotine (N) group as compared to the control (C) group. The maximum alpha power was observed at occipital-parietal scalp sites (see topographic maps) in both groups. *Adapted and modified from Beer et al. (2013) with permission from the publishers.*

Smokers responded faster in invalid trials of this task following cigarette smoking (with no difference for valid or neutral trials) (Witte, Davidson, & Marrocco, 1997). Enhanced performance for invalid trials was also observed in casual smokers immediately following cigarette smoking but not when tested one day after smoking (Murphy & Klein, 1998). As performance improvements were only observed for invalid trials but not for neutral or valid trials, the authors concluded that nicotine primarily facilitated the disengagement (but not the engagement) process of the orienting system in attention. Similar modulations of the validity effect were observed in nonsmokers consuming a nicotine gum (Thiel & Fink, 2008; Vossel, Thiel, & Fink, 2008). As the amount of absorbed nicotine correlated with the strength of the validity effect in smokers (Shirtcliff & Marrocco, 2003), a linear relationship between nicotine and its effect on the disengagement process of attention may be assumed.

Correlates of enhanced selective stimulus processing were also found in EEG potentials. Several studies have shown that nicotine substantially modulates the latency and amplitude of the P300 wave (Beer et al., 2013; Gilbert et al., 2007; Houlihan, Pritchard, & Robinson, 1996). The P300 wave is an event-related EEG potential (ERP). Visual events (e.g., a light flash) automatically elicit a sequence of characteristic positive (e.g., P1, P2) and negative (e.g., N1, N2) ERPs (see Figure 2(A)). Although elicited automatically, they are modulated by cognitive states. In particular, a relevant event as compared to an irrelevant event generates a positive ERP deflection with an onset of about 300 ms and a maximum at around 600 ms after stimulus onset. This positive deflection (known as the P300 wave) reflects neural processes associated with the relevancy of a sensory event (Coles & Rugg, 1995). Figure 2(B) shows the P300 wave of nonsmokers who were asked to respond to a visual target event (task relevant) while ignoring other visual events (task irrelevant). While performing this task, one group was consuming chewing tobacco (with nicotine), whereas the other group consumed a tobacco surrogate (without nicotine). Relevant items elicited a more positive ERP than irrelevant items, known as the P300 wave. This P300 wave was observed in both groups. However, its peak latency emerged substantially earlier in the nicotine group compared to the control group, suggesting that nicotine enhanced attentional processes of stimulus selection for visual events.

Another study examined the effect of nicotine on the P300 wave in smokers while performing a visual or an auditory task (Houlihan et al., 1996). This study also found a reduced latency (and an enhanced amplitude) of the P300 wave during nicotine consumption. Moreover, the reduced latency of the P300 wave was associated with faster response times, suggesting measurable behavioral consequences of the EEG waves. Interestingly, P300 modulation was only observed for the visual but not for the auditory task. The authors attributed this difference between vision and audition to the relatively large prevalence of cholinergic neurons in the visual cortex as compared to the auditory cortex.

Gilbert et al. (2007) examined the P3b wave (a subcomponent of the P300) when the relevant target stimulus was preceded by an aversive distractor or a neutral stimulus. They found that nicotine-deprived smokers showed an enhanced P3b wave when tested with a nicotine patch compared to a placebo patch. However, this P3b enhancement was only observed when the target stimulus was preceded by an aversive distractor but was absent when the target was preceded by the neutral stimulus. This finding suggests that nicotine facilitated disengagement processes from irrelevant (aversive) distractors. fMRI studies showed in nonsmokers that the reduced validity effect following consumption of a nicotine gum was associated with reduced hemodynamic responses in the frontal and parietal cortex (Thiel & Fink, 2008; Vossel et al., 2008).

Nicotine also promotes the executive system of attention and response selection. Executive aspects of attention may be tested,

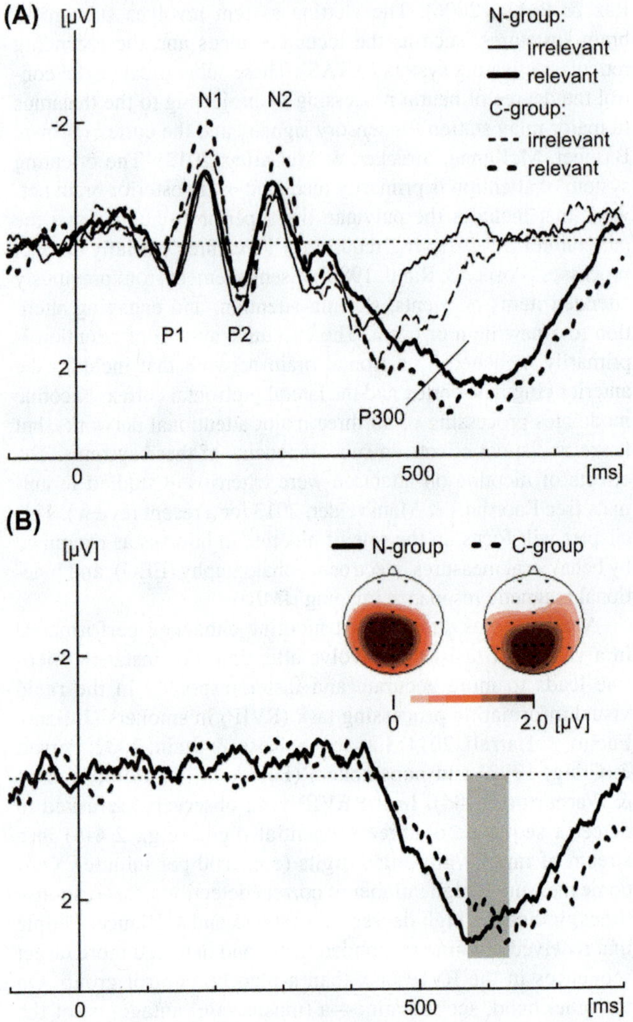

FIGURE 2 **Modulation of the P300 EEG wave by nicotine.** Visual event-related EEG potentials (ERPs) of two groups of nonsmokers while consuming oral nicotine (N-group) or a placebo substance (C-group). Participants performed a visual oddball task on pictures presented in the peripheral visual field. They had to detect a rare target picture (relevant) within a stream of irrelevant pictures. ERPs at a parietal electrode cluster are shown. (A) Pictures elicited a visual ERP consisting of a sequence of P1, N1, P2, and N2 components. Relevant pictures compared to irrelevant pictures elicited more positive ERPs at later latencies (P300 difference wave) in both groups. (B) The difference ERPs (relevant minus irrelevant) depict the P300 wave separate for each group. As highlighted by the gray bar, the latency of the P300 peak was substantially reduced in the nicotine group (570 ms) compared to the control group (630 ms). No difference was found in the peak amplitude of the P300 wave or its scalp topography. *Adapted and modified from Beer et al. (2013) with permission from the publishers.*

TABLE 1 Key Facts about Nicotine and Attention

- Attention involves at least three partially distinct cognitive systems: (1) The alerting system regulates the responsiveness of the brain to internal and environmental events. (2) The orienting system selects relevant events (engagement) and ignores irrelevant events (disengagement). (3) The executive system resolves conflicts between multiple actions.

- Nicotine enhances all three major attention systems. However, nicotine exerts selective effects on subcomponents of these systems.

- Nicotine enhances the alerting system of attention. Nicotine enhances and maintains reactivity to environmental events. Nicotine enhances performance in the rapid visual information processing task.

- Nicotine enhances subcomponents of the orienting system of attention. Nicotine facilitates the disengagement of attention from irrelevant events. Nicotine has no or only moderate effects on the engagement processes.

- Nicotine enhances the executive system of attention and facilitates response preparation and execution.

for instance, by a color-word Stroop task. In this task, observers are asked to name the color of words. The words themselves denote a color and are presented with a congruent (e.g., the word *red* colored red) or incongruent color (e.g., the word *red* colored green). Typically, incongruent items produce a response conflict and result in slower response times. Nonsmokers consuming a nicotine gum responded faster in incongruent trials of this task than a placebo group (Provost & Woodward, 1991). This finding is in line with a number of other studies suggesting that nicotine not only facilitates sensory processing but also motor preparation and execution (Pritchard, Sokhadze, & Houlihan, 2004). For instance, Sherwood and colleagues (Sherwood, 1995; Sherwood, Kerr, & Hindmarch, 1992) found that nicotine administered to tobacco-deprived smokers not only reduced flicker fusion thresholds but also motor reaction times.

In summary, the enhancing effects of nicotine are observed on all major aspects of attention: alertness, orienting, and executive (see Key Facts in Table 1). However, it may affect specific subcomponents of these attentional networks. For instance, nicotine seems to facilitate disengagement processes of orienting, but has no or only moderate effects on maintaining attention to sensory events (engagement processes).

NICOTINE AND MEMORY

Learning and memory may be categorized into two main systems that are partially distinct (Squire, 2004): a declarative and a nondeclarative system. Declarative learning refers to learning of semantic (e.g., facts) and episodic (e.g., experiences) content. It refers to material that may be verbalized (declared) and is also known as explicit learning. Nondeclarative learning refers to learning of perceptual and motor skills or emotional responses. Its content is difficult to verbalize (declare). It is also known as implicit learning. Declarative and nondeclarative learning are mediated by different brain circuits. Whereas declarative learning strongly relies on the hippocampus and the medial temporal lobe, nondeclarative learning is hippocampus independent but instead depends on other brain structures that are relevant for the acquired skill (e.g., motor cortex, basal ganglia, and cerebellum for motor skills; sensory cortices for perceptual skills; amygdala for emotional responses).

Learning and memory involves at least three major processing steps (Stickgold & Walker, 2005): encoding (or memory formation), consolidation, and retrieval (or recall). The material to be learned must be perceived and adequately encoded. However, not everything that is perceived and encoded is maintained for subsequent retrieval. Memory traces may decay or may interfere with old or new material. Memory traces are stabilized by a process called consolidation. The neurotransmitter ACh and sleep seems to play a crucial role in memory consolidation. Sleep consists of several qualitatively distinct phases (Brown et al., 2012). Slow-wave-sleep (SWS) phases are characterized by relatively slow EEG waves (e.g., delta and theta waves). Other stages are characterized by rapid eye movements (REM). During SWS, the neurotransmitter ACh is downregulated in the hippocampus, whereas during REM sleep ACh concentrations are at a normal level (Kametani & Kawamura, 1990; Marrosu et al., 1995). Several studies suggest that SWS sleep promotes consolidation in declarative learning (Rasch, Büchel, Gais, & Born, 2007). On the other hand, REM sleep promotes nondeclarative memory consolidation (Karni, Tanne, Rubenstein, Askenasy, & Sagi, 1994). Finally, acquired knowledge and skills must be retrieved. Retrieval, however, is not a unidirectional process, but may instead modify memory traces (e.g., by retrieval-induced forgetting).

Several studies demonstrated that nicotine enhances memory. In particular, nicotine facilitates short-term or working memory. Nicotine administered to nonsmoking men improved their accuracy in an n-back task (Kumari et al., 2003). In this task, a stream of items (e.g., letters) is presented and participants are asked to press a button whenever an item was shown one (1-back) or two trials (2-back) before. Accordingly, it is a task that tests for short-term or working memory capabilities. On the other hand, nicotine deprivation reduced working memory performance. Schizophrenic smokers who were nicotine abstinent made more errors in a visuospatial working memory task than nonabstinent patients (Ghiasi, Farhang, Farnam, & Safikhanlou, 2013). Improved verbal working memory performance following nicotine administration was observed in elderly nonsmokers (Min, Moon, Ko, & Shin, 2001). Prospective memory refers to the ability to remember an intended action after a delay. Several studies demonstrated enhanced prospective memory in abstinent smokers and nonsmokers who received nicotine as compared to control groups (Dawkins, Turner, & Crowe, 2013; Jansari, Froggatt, Edginton, & Dawkins, 2013; Rusted, Trawley, Heath, Kettle, & Walker, 2005).

Nicotine facilitates declarative learning (Froeliger, Gilbert, & McClernon, 2009; Jubelt et al., 2008; Warburton, Skinner, & Martin, 2001). For instance, Froeliger and colleagues asked smokers and nonsmokers to perform a novelty detection task while wearing a nicotine patch or a placebo patch. Observers had to detect novel pictures or words among a stream of standard items. The next day, their recognition performance for previously presented items was tested. For both smokers and nonsmokers, the groups wearing the nicotine patch scored higher on the recognition test than the control groups. Only a limited number of studies examined the role of nicotine in nondeclarative learning of humans. However, based on research with other cholinergic substances or animals, it may be inferred that nicotine also facilitates nondeclarative learning. For instance, nonsmokers receiving donepezil (AChE inhibitor) prior to a perceptual learning task on motion discrimination performed subsequently superior in this task than the control group (Rokem & Silver, 2010). Animal research showed that nicotine facilitated conditioning of emotional responses (Kenney & Gould, 2008).

During most of these previous studies, nicotine (or a comparable cholinergic modulator) was delivered before or during the learning task. Hence, improved memory performance likely resulted from enhanced encoding (e.g., by enhanced attention) rather than reflecting enhanced retention (consolidation) processes (Warburton, Rusted, & Fowler, 1992). A limited number of studies explicitly investigated the role of nicotine on memory consolidation. These studies suggest a more complex relationship between nicotine and memory.

Our group examined how nicotine affects perceptual learning, which is a special type of nondeclarative learning (Beer et al., 2013). Here, tobacco was administered during a 1 h period after training with a texture discrimination task. Performance in this task was tested 1 day after training at a time when nicotine was fully metabolized. The group that received posttraining nicotine showed larger learning effects (performance difference between test and training session) for this perceptual task than the control group. This finding suggests that nicotine facilitated consolidation processes in nondeclarative learning. Most likely, nicotine facilitated molecular processes of synaptic long-term potentiation that are enhanced by cholinergic agonists (Bröcher, Artola, & Singer, 1992). By contrast, in declarative learning, the consequences of cholinergic agonists such as nicotine on memory consolidation are more entangled. Nicotine administered directly after a word list learning task improved subsequent free recall of these words (Rusted & Warburton, 1992). Similar enhancing effects of postlearning nicotine administration were observed 1 week after learning with a paired-associative learning task (Colrain, Mangan, Pellett, & Bates, 1992). However, low doses of postacquisition nicotine resulted in higher performance than higher doses. Other studies demonstrated enhanced declarative memory consolidation by blocking nicotinic and muscarinic AChR with mecamylamine and scopolamine after a verbal learning task (Rasch, Born, & Gais, 2006). Similarly, postlearning administration of physostigmine—an AChE inhibitor—impaired consolidation of declarative memories for word pairs (Gais & Born, 2004).

This line of research suggests that cholinergic enhancers, such as nicotine, may be detrimental for memory consolidation in declarative learning. It was proposed that hippocampus-driven consolidation in declarative learning requires the suppression of sensory input in order to enable excitatory cortical feedback loops that promote memory consolidation (Hasselmo & McGaughy, 2004). As nicotine enhances sensory signals, it may be detrimental for this type of consolidation processes. Although further research is needed to elucidate the role of nicotine on memory consolidation, the present set of findings suggest a relatively complex pattern: For nondeclarative learning, nicotine seems to facilitate consolidation processes. For declarative learning, low doses of nicotine seem to facilitate memory consolidation, whereas high doses of nicotine may be detrimental, presumably by preventing cortical feedback loops.

TABLE 2 Key Facts about Nicotine and Memory

- Learning and memory relies on at least two relatively distinct systems: The declarative memory system stores knowledge that can be "declared," such as facts and experiences. It is mediated by the hippocampus and the medial temporal lobe. The nondeclarative memory system stores sensory and motor skills and emotional responses. It does not require the hippocampus and the medial temporal lobe.
- Learning is a multistep process that includes encoding, consolidation, and retrieval.
- Nicotine facilitates encoding processes for declarative and nondeclarative learning.
- Nicotine facilitates consolidation processes in nondeclarative learning. Cholinergic enhancers, such as nicotine, seem to have antagonistic effects for consolidation in declarative memory systems: Low doses seem to promote consolidation, whereas high doses seem to inhibit consolidation.
- Nicotine seems to facilitate retrieval processes in declarative memory systems. However, further research is needed in order to incorporate controversial findings.

Nicotine facilitates learning when administered before or during learning. Moreover, it seems to have complex and partially antagonistic effects on consolidation processes. Currently, only a limited number of studies investigated its effect on retrieval processes. Among these studies, retrieval of lexical memory was tested in a lexical decision task (Hale, Gentry, & Meliska, 1999). Abstinent smokers having a nicotine cigarette were faster in discriminating real words from nonwords than those smoking a placebo cigarette, suggesting superior retrieval of lexical memory. Retrieval-induced forgetting is a phenomenon in episodic list learning, by which retrieval of a previously learned item impairs retention of this item for subsequent tests. Nonsmokers who received nasal nicotine showed increased retrieval-induced forgetting compared to a placebo group (Rusted & Alvares, 2008). An fMRI study with elderly people showed that cholinergic stimulation by the AChE inhibitor physostigmine enhanced brain activity associated with encoding processes but reduced brain activity associated with retrieval processes in a spatial source memory task of declarative learning (Kukolja, Thiel, & Fink, 2009). This study suggests that cholinergic enhancers (e.g., nicotine) may impair retrieval performance. However, it must be considered that the AChE inhibitor in this study was administered already during the learning stage. Therefore, it is possible that impaired test performance in this study reflected reduced consolidation rather than a retrieval deficit.

In summary, nicotine facilitates working memory and the encoding processes in declarative and nondeclarative learning (see Key Facts in Table 2). Nicotinic effects on memory consolidation seem to be contingent on the type of learning: Nicotine facilitates consolidation in nondeclarative learning. For declarative learning, only low doses of nicotine promote consolidation, whereas high doses of nicotine likely inhibit consolidation. Nicotine also seems to facilitate memory retrieval. However, further research is needed in order to understand controversial findings.

LONG-TERM EFFECTS OF NICOTINE ON COGNITION

Most research on the effects of nicotine on cognition examined its short-term effects. These studies primarily found an enhancing effect of nicotine on cognition. However, chronic exposure to nicotine for a period of more than 4 weeks leads to neural processes of desensitization (Pietila, Lahde, Attila, Ahtee, & Nordberg, 1998). Abstinent smokers demonstrate inferior cognitive performance compared to nonsmoking controls on some cognitive tasks (De Biasi & Dani, 2011; Foulds et al., 1996). However, the neural processes of desensitization seem to be reversible (Pietila et al., 1998).

Some studies have examined the effects of long-term nicotine consumption on cognition. These studies showed mixed results, in part due to the poor control (e.g., only a few subjects examined). A well-controlled multicenter study showed subtle deficits in visual attention and cognitive impulsivity but not in verbal memory, verbal fluency, or Stroop tasks of 1002 long-term nonabstinent smokers compared to 1161 nonsmokers (Wagner et al., 2013). However, the observed deficits did not correlate with lifetime tobacco consumption, suggesting that they reflect a priori deficits of smokers compared to nonsmokers.

INDIRECT EFFECTS OF NICOTINE

Although nicotine affects cognition primarily via the cholinergic system, it is noteworthy that nicotine also has a substantial impact on other neurotransmitter systems. For instance, nicotine modulates serotonergic neurons in the dorsal raphe nucleus (Hernandez-Lopez, Garduno, & Mihailescu, 2013). Moreover, nicotine stimulates dopaminergic neurons in the midbrain. The dopaminergic effects of nicotine have been associated with the reward system that initiates nicotine addiction (De Biasi & Dani, 2011; Doyon, Thomas, Ostroumov, Dong, & Dani, 2013).

Nicotine is an agonist of the cholinergic system. It is metabolized in the liver. Its main metabolite is cotinine (Benowitz et al., 2009). Contrary to previous beliefs that cotinine is not psychoactive, recent research suggests that it may have cognitive effects distinct from nicotine (Grizzell & Echeverria, 2015).

APPLICATIONS TO OTHER ADDICTIONS AND SUBSTANCE MISUSE

The primary mechanism of nicotine addiction seems to be based on the dopaminergic reward system (De Biasi & Dani, 2011; Doyon et al., 2013). However, the enhancing effect of nicotine on attention likely provides an additional reward signal. Nicotine facilitates learning. In particular, consolidation of nondeclarative or implicit learning is enhanced. This type of learning includes sensory or motor skills and emotional responses. Nondeclarative learning might contribute to the environmental and social setting that maintains tobacco addiction.

Tobacco consumption is often accompanied by the misuse of other substances (e.g., alcohol). The enhancing effect of nicotine may mask or compensate for cognitive deficits resulting from these substances. For instance, neurocognitive effects of nicotine were compared between smokers with and without alcohol addiction (Nixon, Lawton-Craddock, Tivis, & Ceballos, 2007). As expected, alcohol-addicted people performed worse in the cognitive tasks than other people. However, alcohol-addicted people showed larger performance gains due to nicotine than the control group. Other studies found that the compensatory effect of nicotine for the cognitive deficits due to alcohol is additive rather than interactive (Greenstein et al., 2010). Similar compensatory effects may be expected for the cognitive deficits associated with cocaine addiction (Sofuoglu, 2010). As cognitive deficits are associated with poor treatment retention and outcome, supplementing treatment by cholinergic enhancers (e.g., donepezil) may improve treatment approaches.

DEFINITION OF TERMS

Acetylcholine (ACh) ACh is a neurotransmitter. It is released from the synapses into the postsynaptic cleft (gap between nerve cells) and binds on nicotinic or muscarinic receptors of the postsynaptic membrane. Acetylcholine is recycled by the enzyme acetylcholinesterase.

Ascending reticular activating system (ARAS) The ARAS is a brain circuit that connects nuclei of the reticular formation in the brain stem, the thalamus, and the frontal cortex. The thalamus is a midbrain structure that relays sensory information to the cortex. The ARAS regulates sleep–wake mechanisms and controls arousal (responsiveness to environmental events) and vigilance (sustained responsiveness).

Continuous performance test (CPT) The CPT is an established neuropsychological test for attention. The task requires observers to detect a number (e.g., 1) among a stream of other numbers. It measures alertness aspects (arousal, vigilance), as well as orienting aspects of attention.

Declarative learning Declarative learning (also known as explicit learning) refers to the learning of semantic (e.g., facts) and episodic (e.g., experiences) content. It refers to content that may be verbalized (declared). The hippocampus and the medial temporal lobe are crucial brain structures for this type of learning.

Electroencephalography (EEG) EEG is a noninvasive method of measuring the electrical potentials generated by the brain by electrodes attached to the scalp surface. These potentials fluctuate across time. The frequencies of fluctuations reflect different cognitive states. Five major frequency bands are distinguished: alpha (8–12 Hz), beta (12–30 Hz), delta (2–4 Hz), theta (4–8 Hz), gamma (30–60 Hz). In addition to the ongoing EEG signal, event-related EEG waves may be observed. These event-related potentials are elicited by a sensory or motor event and modulated by a mental state (e.g., P300 wave).

Functional magnetic resonance imaging (fMRI) Magnetic resonance imaging (MRI) is a noninvasive method for high-resolution (at a millimeter scale) three-dimensional imaging of brain tissue. Functional MRI is sensitive to the difference in magnetic properties of oxygenated and deoxygenated hemoglobin (blood cells). Hemodynamic responses to environmental or cognitive events may be revealed by continuous fMRI measurements. As neural activity is correlated with hemodynamic responses, fMRI provides an indirect measure of neural activity.

Lexical decision task In this task, people are sequentially presented with a string of letters, which composes either a real word (e.g., *cake*) or a nonword (e.g., *heark*). Nonwords are usually pronounceable pseudowords. Response times reflect the ease of word retrieval from a lexical memory.

N-back task During this memory task, participants perceive a stream of items (e.g., letters) and are asked to press a button whenever an item was presented one (1-back) or two trials (2-back) before. This task measures aspects of working memory.

Nondeclarative learning Nondeclarative learning (also known as *implicit learning*) refers to learning of perceptual and motor skills or emotional responses. It is mediated by brain structures that are relevant for the acquired skill (e.g., sensory or motor areas). It does not depend on the hippocampus.

Paired-associative learning task The paired-associative learning task is a declarative learning task. During a learning session, participants are asked to memorize pairs of words (e.g. *tree–ball*). During a test session, participants are asked to complete the pair while only one item is presented (e.g., *tree–?*).

Rapid visual information processing task (RVIP) The RVIP task asks observers to detect a sequence of three subsequent digits (e.g., 2-4-6) in a stream of rapidly presented digits (e.g., 100 per minute). It measures alertness aspects (arousal, vigilance) of attention.

Spatial cuing task of attention The spatial cuing task also known as the Posner cuing paradigm measures orienting mechanisms of selective spatial attention. In this task, observers are asked to respond to a visual target stimulus that is presented in the peripheral visual field. Prior to the target stimulus observers receive a cue about its location (e.g., left or right). This cue may be valid, neutral, or invalid. Typically, observers respond fastest in trials with valid cues and slowest in trials with invalid cues.

Stroop task In the Stroop task observers are asked to name the color of words. The words themselves denote a color and are presented in congruent (e.g., the word *red* colored red) or incongruent colors (e.g., the word *red* colored green). Typically, incongruent items produce a response conflict and result in slower response times. Performance in the Stroop task reflects executive aspects of attention.

SUMMARY POINTS

- This chapter focuses on the effects of nicotine on cognitive systems of attention and memory.
- Nicotine enhances all major attention systems—the alerting system, the orienting system, and the executive system. The enhancing effects were demonstrated in a number of attention tasks, including the RVIP, the CPT, the spatial cuing task, and the Stroop task. The enhancing effects of nicotine were demonstrated with EEG (alpha activity, P300) and fMRI.
- However, nicotine has selective effects on subcomponents of these attention systems. In particular, it enhances disengagement processes of the orienting system, but seems to have no effect on engagement processes.
- Nicotine enhances learning of both declarative and nondeclarative memory systems. The enhancing effects of nicotine were demonstrated in a number of memory tasks, including working memory tasks, the paired-associative learning task, perceptual learning tasks, and the lexical decision task.
- However, different steps of the learning process are selectively affected by nicotine. Whereas nicotine enhances encoding irrespective of the memory system, it shows antagonistic effects on the consolidation processes. For nondeclarative memory systems, nicotine facilitates consolidation. For declarative memory systems, only low doses of nicotine seem to facilitate consolidation, whereas high doses of cholinergic enhancers seem to inhibit memory consolidation.
- Habitual nicotine consumption may impair cognitive performance during nicotine abstinence due to desensitization of the cholinergic system. Desensitization is reversible.

REFERENCES

Beer, A. L., Vartak, D., & Greenlee, M. W. (2013). Nicotine facilitates memory consolidation in perceptual learning. *Neuropharmacology, 64*(1), 443–451.

Benowitz, N. L., Hukkanen, J., & Jacob, P., 3rd (2009). Nicotine chemistry, metabolism, kinetics and biomarkers. *Handbook of Experimental Pharmacology, 192*, 29–60.

Bröcher, S., Artola, A., & Singer, W. (1992). Agonists of cholinergic and noradrenergic receptors facilitate synergistically the induction of long-term potentiation in slices of rat visual cortex. *Brain Res, 573*(1), 27–36.

Brown, R. E., Basheer, R., McKenna, J. T., Strecker, R. E., & McCarley, R. W. (2012). Control of sleep and wakefulness. *Physiological Review, 92*(3), 1087–1187.

Coles, M. G. H., & Rugg, M. D. (1995). Event-related brain potentials: an introduction. In M. D. Rugg, & M. G. H. Coles (Eds.), *Electrophysiology of mind: Event-related brain potentials and cognition* (pp. 1–26). Oxford: Oxford University Press.

Colrain, I. M., Mangan, G. L., Pellett, O. L., & Bates, T. C. (1992). Effects of post-learning smoking on memory consolidation. *Psychopharmacology (Berlin), 108*(4), 448–451.

Dawkins, L., Turner, J., & Crowe, E. (2013). Nicotine derived from the electronic cigarette improves time-based prospective memory in abstinent smokers. *Psychopharmacology (Berlin), 227*(3), 377–384.

De Biasi, M., & Dani, J. A. (2011). Reward, addiction, withdrawal to nicotine. *Annual Review of Neuroscience, 34*, 105–130.

Doyon, W. M., Thomas, A. M., Ostroumov, A., Dong, Y., & Dani, J. A. (2013). Potential substrates for nicotine and alcohol interactions: a focus on the mesocorticolimbic dopamine system. *Biochemical Pharmacology, 86*(8), 1181–1193.

Foulds, J., McSorley, K., Sneddon, J., Feyerabend, C., Jarvis, M. J., & Russell, M. A. (1994). Effect of subcutaneous nicotine injections of EEG alpha frequency in non-smokers: a placebo-controlled pilot study. *Psychopharmacology (Berlin), 115*(1–2), 163–166.

Foulds, J., Stapleton, J., Swettenham, J., Bell, N., McSorley, K., & Russell, M. A. (1996). Cognitive performance effects of subcutaneous nicotine in smokers and never-smokers. *Psychopharmacology (Berlin), 127*(1), 31–38.

Froeliger, B., Gilbert, D. G., & McClernon, F. J. (2009). Effects of nicotine on novelty detection and memory recognition performance: double-blind, placebo-controlled studies of smokers and nonsmokers. *Psychopharmacology (Berlin), 205*(4), 625–633.

Gais, S., & Born, J. (2004). Low acetylcholine during slow-wave sleep is critical for declarative memory consolidation. *Proceedings of the National Academy of Sciences of the United States of America, 101*(7), 2140–2144.

Ghiasi, F., Farhang, S., Farnam, A., & Safikhanlou, S. (2013). The short term effect of nicotine abstinence on visuospatial working memory in smoking patients with schizophrenia. *Nordic Journal of Psychiatry, 67*(2), 104–108.

Gilbert, D. G., Sugai, C., Zuo, Y., Rabinovich, N. E., McClernon, F. J., & Froeliger, B. (2007). Brain indices of nicotine's effects on attentional bias to smoking and emotional pictures and to task-relevant targets. *Nicotine and Tobacco Research, 9*(3), 351–363.

Golding, J. F. (1988). Effects of cigarette smoking on resting EEG, visual evoked potentials and photic driving. *Pharmacology, Biochemistry and Behavior*, *29*(1), 23–32.

Greenstein, J. E., Kassel, J. D., Wardle, M. C., Veilleux, J. C., Evatt, D. P., Heinz, A. J., ... Yates, M. C. (2010). The separate and combined effects of nicotine and alcohol on working memory capacity in non-abstinent smokers. *Experimental and Clinical Psychopharmacology*, *18*(2), 120–128.

Grizzell, J. A., & Echeverria, V. (2015). New insights into the mechanisms of action of cotinine and its distinctive effects from nicotine. *Neurochemical Research*, *40*(10), 2032–2046.

Hale, C. R., Gentry, M. V., & Meliska, C. J. (1999). Effects of cigarette smoking on lexical decision-making. *Psychological Reports*, *84*(1), 117–120.

Hasselmo, M. E., & McGaughy, J. (2004). High acetylcholine levels set circuit dynamics for attention and encoding and low acetylcholine levels set dynamics for consolidation. *Progress in Brain Research*, *145*, 207–231.

Hernandez-Lopez, S., Garduno, J., & Mihailescu, S. (2013). Nicotinic modulation of serotonergic activity in the dorsal raphe nucleus. *Reviews in Neurosciences*, *24*(5), 455–469.

Himelstein, J., Newcorn, J. H., & Halperin, J. M. (2000). The neurobiology of attention-deficit hyperactivity disorder. *Frontiers in Bioscience*, *5*, D461–D478.

Houlihan, M. E., Pritchard, W. S., & Robinson, J. H. (1996). Faster P300 latency after smoking in visual but not auditory oddball tasks. *Psychopharmacology (Berlin)*, *123*(3), 231–238.

Jansari, A. S., Froggatt, D., Edginton, T., & Dawkins, L. (2013). Investigating the impact of nicotine on executive functions using a novel virtual reality assessment. *Addiction*, *108*(5), 977–984.

Jubelt, L. E., Barr, R. S., Goff, D. C., Logvinenko, T., Weiss, A. P., & Evins, A. E. (2008). Effects of transdermal nicotine on episodic memory in non-smokers with and without schizophrenia. *Psychopharmacology (Berlin)*, *199*(1), 89–98.

Juliano, L. M., Fucito, L. M., & Harrell, P. T. (2011). The influence of nicotine dose and nicotine dose expectancy on the cognitive and subjective effects of cigarette smoking. *Experimental and Clinical Psychopharmacology*, *19*(2), 105–115.

Kametani, H., & Kawamura, H. (1990). Alterations in acetylcholine release in the rat hippocampus during sleep-wakefulness detected by intracerebral dialysis. *Life Sciences*, *47*(5), 421–426.

Karni, A., Tanne, D., Rubenstein, B. S., Askenasy, J. J., & Sagi, D. (1994). Dependence on REM sleep of overnight improvement of a perceptual skill. *Science*, *265*(5172), 679–682.

Kenney, J. W., & Gould, T. J. (2008). Nicotine enhances context learning but not context-shock associative learning. *Behavioral Neuroscience*, *122*(5), 1158–1165.

Kukolja, J., Thiel, C. M., & Fink, G. R. (2009). Cholinergic stimulation enhances neural activity associated with encoding but reduces neural activity associated with retrieval in humans. *The Journal of Neuroscience*, *29*(25), 8119–8128.

Kumari, V., Gray, J. A., ffytche, D. H., Mitterschiffthaler, M. T., Das, M., Zachariah, E., ... Sharma, T. (2003). Cognitive effects of nicotine in humans: an fMRI study. *Neuroimage*, *19*(3), 1002–1013.

Lawrence, N. S., Ross, T. J., & Stein, E. A. (2002). Cognitive mechanisms of nicotine on visual attention. *Neuron*, *36*(3), 539–548.

Levin, E. D., Conners, C. K., Silva, D., Hinton, S. C., Meck, W. H., March, J., & Rose, J. E. (1998). Transdermal nicotine effects on attention. *Psychopharmacology (Berlin)*, *140*(2), 135–141.

Marrosu, F., Portas, C., Mascia, M. S., Casu, M. A., Fa, M., Giagheddu, M., ... Gessa, G. L. (1995). Microdialysis measurement of cortical and hippocampal acetylcholine release during sleep-wake cycle in freely moving cats. *Brain Research*, *671*(2), 329–332.

Min, S. K., Moon, I. W., Ko, R. W., & Shin, H. S. (2001). Effects of transdermal nicotine on attention and memory in healthy elderly non-smokers. *Psychopharmacology (Berlin)*, *159*(1), 83–88.

Murphy, F. C., & Klein, R. M. (1998). The effects of nicotine on spatial and non-spatial expectancies in a covert orienting task. *Neuropsychologia*, *36*(11), 1103–1114.

Nixon, S. J., Lawton-Craddock, A., Tivis, R., & Ceballos, N. (2007). Nicotine's effects on attentional efficiency in alcoholics. *Alcoholism, Clinical and Experimental Research*, *31*(12), 2083–2091.

Parrott, A. C., & Craig, D. (1992). Cigarette smoking and nicotine gum (0, 2 and 4 mg): effects upon four visual attention tasks. *Neuropsychobiology*, *25*(1), 34–43.

Parrott, A. C., & Winder, G. (1989). Nicotine chewing gum (2 mg, 4 mg) and cigarette smoking: comparative effects upon vigilance and heart rate. *Psychopharmacology (Berlin)*, *97*(2), 257–261.

Pietila, K., Lahde, T., Attila, M., Ahtee, L., & Nordberg, A. (1998). Regulation of nicotinic receptors in the brain of mice withdrawn from chronic oral nicotine treatment. *Naunyn-Schmiedeberg's Archives of Pharmacology*, *357*(2), 176–182.

Poorthuis, R. B., & Mansvelder, H. D. (2013). Nicotinic acetylcholine receptors controlling attention: behavior, circuits and sensitivity to disruption by nicotine. *Biochemical Pharmacology*, *86*(8), 1089–1098.

Posner, M. I., & Rafal, R. D. (1987). Cognitive theories of attention and the rehabilitation of attentional deficits. In M. J. Meier, A. L. Benton, & L. Diller (Eds.), *Neuropsychological rehabilitation* (pp. 182–201). Edinburgh: Churchill Livingstone.

Pritchard, W., Sokhadze, E., & Houlihan, M. (2004). Effects of nicotine and smoking on event-related potentials: a review. *Nicotine and Tobacco Research*, *6*(6), 961–984.

Provost, S. C., & Woodward, R. (1991). Effects of nicotine gum on repeated administration of the stroop test. *Psychopharmacology (Berlin)*, *104*(4), 536–540.

Rasch, B. H., Born, J., & Gais, S. (2006). Combined blockade of cholinergic receptors shifts the brain from stimulus encoding to memory consolidation. *Journal of Cognitive Neuroscience*, *18*(5), 793–802.

Rasch, B., Büchel, C., Gais, S., & Born, J. (2007). Odor cues during slow-wave sleep prompt declarative memory consolidation. *Science*, *315*(5817), 1426–1429.

Raz, A., & Buhle, J. (2006). Typologies of attentional networks. *Nature Reviews Neuroscience*, *7*(5), 367–379.

Rogers, S. L., Cooper, N. M., Sukovaty, R., Pederson, J. E., Lee, J. N., & Friedhoff, L. T. (1998). Pharmacokinetic and pharmacodynamic profile of donepezil HCl following multiple oral doses. *British Journal of Clinical Pharmacology*, *46*(Suppl. 1), 7–12.

Rokem, A., & Silver, M. A. (2010). Cholinergic enhancement augments magnitude and specificity of visual perceptual learning in healthy humans. *Current Biology*, *20*(19), 1723–1728.

Rusted, J. M., & Alvares, T. (2008). Nicotine effects on retrieval-induced forgetting are not attributable to changes in arousal. *Psychopharmacology (Berlin)*, *196*(1), 83–92.

Rusted, J. M., Trawley, S., Heath, J., Kettle, G., & Walker, H. (2005). Nicotine improves memory for delayed intentions. *Psychopharmacology (Berlin)*, *182*(3), 355–365.

Rusted, J. M., & Warburton, D. M. (1992). Facilitation of memory by post-trial administration of nicotine: evidence for an attentional explanation. *Psychopharmacology (Berlin)*, *108*(4), 452–455.

Sherwood, N. (1995). Effects of cigarette smoking on performance in a simulated driving task. *Neuropsychobiology, 32*(3), 161–165.

Sherwood, N., Kerr, J. S., & Hindmarch, I. (1992). Psychomotor performance in smokers following single and repeated doses of nicotine gum. *Psychopharmacology (Berlin), 108*(4), 432–436.

Shirtcliff, E. A., & Marrocco, R. T. (2003). Salivary cotinine levels in human tobacco smokers predict the attentional validity effect size during smoking abstinence. *Psychopharmacology (Berlin), 166*(1), 11–18.

Sofuoglu, M. (2010). Cognitive enhancement as a pharmacotherapy target for stimulant addiction. *Addiction, 105*(1), 38–48.

Squire, L. R. (2004). Memory systems of the brain: a brief history and current perspective. *Neurobiology of Learning and Memory, 82*(3), 171–177.

Stickgold, R., & Walker, M. P. (2005). Memory consolidation and reconsolidation: what is the role of sleep? *Trends in Neurosciences, 28*(8), 408–415.

Thiel, C. M., & Fink, G. R. (2008). Effects of the cholinergic agonist nicotine on reorienting of visual spatial attention and top-down attentional control. *Neuroscience, 152*(2), 381–390.

Vossel, S., Thiel, C. M., & Fink, G. R. (2008). Behavioral and neural effects of nicotine on visuospatial attentional reorienting in non-smoking subjects. *Neuropsychopharmacology, 33*(4), 731–738.

Wagner, M., Schulze-Rauschenbach, S., Petrovsky, N., Brinkmeyer, J., von der Goltz, C., Grunder, G., ... Winterer, G. (2013). Neurocognitive impairments in non-deprived smokers–results from a population-based multi-center study on smoking-related behavior. *Addiction Biology, 18*(4), 752–761.

Warburton, D. M., Rusted, J. M., & Fowler, J. (1992). A comparison of the attentional and consolidation hypotheses for the facilitation of memory by nicotine. *Psychopharmacology (Berlin), 108*(4), 443–447.

Warburton, D. M., Skinner, A., & Martin, C. D. (2001). Improved incidental memory with nicotine after semantic processing, but not after phonological processing. *Psychopharmacology (Berlin), 153*(2), 258–263.

Wesnes, K., & Warburton, D. M. (1984). Effects of scopolamine and nicotine on human rapid information processing performance. *Psychopharmacology (Berlin), 82*(3), 147–150.

Witte, E. A., Davidson, M. C., & Marrocco, R. T. (1997). Effects of altering brain cholinergic activity on covert orienting of attention: comparison of monkey and human performance. *Psychopharmacology (Berlin), 132*(4), 324–334.

Chapter 28

The Role of Appetitive and Aversive Smoking Cues in Tobacco Use Disorder with a Focus on fMRI

Josiane Bourque[1,2], Le-Anh Dinh-Williams[1,3], Stéphane Potvin[1,3]

[1]*Department of Psychiatry, University of Montreal, Montreal, QC, Canada;* [2]*GRIP, Centre de recherche CHU Ste-Justine, Montreal, QC, Canada;*
[3]*Centre de recherche de l'Institut universitaire en santé mentale de Montréal, Montreal, QC, Canada*

Abbreviations

ACC Anterior cingulate cortex
dlPFC Dorsolateral prefrontal cortex
dmPFC Dorsomedial prefrontal cortex
DSM-5 Diagnostic and statistical manual of mental disorders, fifth edition
fMRI Functional magnetic resonance imaging
mPFC Medial prefrontal cortex
PCC Posterior cingulate cortex
PFC Prefrontal cortex
SUD Substance use disorder

INTRODUCTION

It is estimated that there are over a billion smokers worldwide, with another billion projected for 2030. Smoking harms nearly every organ in the body and is associated with significant disease and mortality. Among chronic cigarette smokers, it is estimated that about 50% will eventually be killed by tobacco-related diseases. Despite growing awareness of the deleterious effects of smoking on health, smokers continue to smoke and have significant difficulty quitting. More than 70% of smokers report a desire to quit, but only 5–17% of quit attempts are successful without proper aid (Hughes, Peters, & Naud, 2008). For reasons such as these, tobacco is considered one of the most addictive drugs worldwide, and a significant amount of research has been invested in understanding the psychological and biological processes underlying its addictive nature. In this chapter, we discuss how the neurobiological response of smokers to smoking cues, both appetitive (e.g., someone smoking) and aversive (e.g., antismoking campaigns), act as an important mechanism underlying the maintenance of smoking behavior and the inability to quit despite its potentially deleterious effects.

REACTIVITY TO APPETITIVE SMOKING CUES—CLINICAL RELEVANCE

For many, the most important concerns about quitting are the persistent urges, thoughts, or desires to smoke a cigarette that fluctuate through the day. As the abstinence period lengthens, these cravings tend to decrease in strength and frequency, although a majority of ex-smokers still report more or less intense urges at 6 months after quitting (Ussher, Beard, Abikoye, Hajek, & West, 2013). More importantly, empirical studies have underlined that these feelings of craving, considered a core feature of tobacco use disorder in the DSM-5, are one of the most consistent triggers of smoking and predictors of relapse in cigarette smokers (Killen & Fortmann, 1997). Consequently, one of the main focuses of cessation treatments and therapy is helping smokers reduce their urges.

Craving is often depicted as a subjective experience of wanting to use a drug, therefore orienting oneself toward drug-taking behaviors. This concept was originally derived from classical conditioning models, which stipulate that urges are situation specific (meaning that they can be triggered by stimuli previously associated with drug use) and persistent (in the sense that they can be reinstated years after abstinence; Robinson & Berridge, 1993). Tiffany and Conklin (2000) have added to this model by positing that craving is a multifaceted construct implicating cognitive capacities, notably memory recollection of past drug taking, expectancy of subsequent use, interpretation of this subjective experience, and other nonautomatic processes. According to the authors, drug-taking behaviors may occur in the absence of craving because, in the transition phase to dependence, these conducts become automatized processes. The role of craving would therefore consist of a variety of cognitive efforts to either aid or prevent the execution of automatized sequences of drug use.

Craving may be further divided into tonic (background levels) and phasic desires for a drug; the former reflects the slowly changing state induced by abstinence, whereas the latter is a fast onset peak in response to appetitive drug-related cues. Both tonic and phasic craving have been associated with smoking relapse risk (Herd, Borland, & Hyland, 2009; Killen & Fortmann, 1997). For example, background feelings of craving at day 1 postquitting were found to be a significant predictor of subsequent cigarette smoking lapse (Sweitzer, Denlinger, & Donny, 2013). In experimental settings, acute craving can easily be triggered by appetitive drug cues or stress. In both abstinent and nonabstinent smokers, exposure to appetitive smoking-related stimuli (e.g., pictures, videos, a lit cigarette) can provoke intense self-reported cravings (Bedi et al., 2011), even following a decline in tonic craving. Similarly, a variety of laboratory stressors have been shown to induce intense cigarette craving reactions in smokers (Erblich, Boyarsky, Spring, Niaura, & Bovbjerg, 2003). Accordingly, these experimental tools offer a considerable opportunity to improve our understanding of the mechanisms of relapse and the inability to quit.

NEURAL CORRELATES OF SMOKERS TO APPETITIVE SMOKING CUES

In the last decade, substantial efforts have been made to identify and grasp the neurobiological network of cigarette cravings. The majority of these studies were conducted with functional magnetic resonance imaging (fMRI) using either pictures or videos (85% of studies), and in some cases, using virtual reality to depict cigarette smoking situations that evoke a craving response. This appetitive smoking-related material has been shown to consistently elicit, relative to neutral stimuli, a significant brain response distributed throughout the cortex as well as subcortical regions in both abstinent and nonabstinent smokers (see Table 1). A meta-analysis by Engelmann et al. (2012) reported that the significant clusters of activations during the viewing of appetitive cigarette-related cues were predominantly located within the extended visual cortex, followed by the cingulate gyrus (posterior and anterior), the superior and middle temporal gyri, and finally, the prefrontal cortex (medial and dorsolateral). Relatively small clusters were also reported in the insula and the dorsal striatum. Importantly, when looking at each individual study assessing the neural correlates of cigarette craving, one finds that several factors contribute to the heterogeneity of findings between studies, notably abstinence and smoking expectancy levels, tobacco dependence severity, and sex differences. Nevertheless, all together, the widespread cerebral reactivity to drug cues validates the idea that craving is a multidimensional construct implicating a variety of cognitive processes. Based on Engelmann et al.'s findings, it appears clear that the brain craving response is not solely confined to the reward circuitry as classic theories of drug addiction postulate (Robinson & Berridge, 1993); other regions need to be recruited in order to fully experience the complex phenomenon of craving.

The most robust finding of the meta-analysis is activity of the visual cortex that extends to the cuneus, lingual and fusiform gyri, as well as some parietal and temporal areas. This result parallels emotional processing imaging studies in which highly arousing or appetitive stimuli, relative to neutral material, consistently elicit more activation in these areas (Bradley et al., 2003). This represents an attentional bias coupled with enhanced perceptual processing of appetitive smoking-related cues in smokers, thus reflecting an increased sensitivity to the value of these drug cues.

The second most significantly activated cluster during the processing of appetitive smoking cues is the precuneus (located in the parietal cortex), along with the adjacent posterior cingulate cortex (PCC). Interestingly, these structures are consistently reported in studies looking at reactivity to other drug-related cues (e.g., alcohol, cigarette, cocaine) (Chase, Eickhoff, Laird, & Hogarth, 2011), but their specific role in craving is still understudied. With its direct connections to the prefrontal cortex, some authors propose that the complex precuneus/PCC relays smoking cue information from the visual cortex to the prefrontal and subcortical areas implicated in goal-directed behaviors and decision making (e.g., striatum, prefrontal cortex) (Engelmann et al., 2012). Considering the emerging role of the precuneus/PCC in self-referential processes and autobiographic memory retrieval (Cavanna & Trimble, 2006; Torta & Cauda, 2011), others suggest a more central role of this region in the "psychological experience of craving." Indeed, during the passive viewing of cigarette-related pictures, smokers may recall pleasant memories of previous smoking experiences, which in turn elicit cravings and provoke high motivation for drug seeking (Muller, 2013).

Another region that presented significant activation and is known to contribute to the vast and heterogeneous response to appetitive drug-related cues is the prefrontal cortex. Numerous neuroimaging findings demonstrate that both the dorsolateral (dlPFC) and dorsomedial (dmPFC) prefrontal cortex are implicated in sustained attention, regulatory processes, and decision making (Li & Sinha, 2008). More specifically, in fMRI studies of drug craving, these prefrontal structures were shown to orient behaviors (e.g., toward either drug-seeking or drug-avoiding conducts) following an integrative processing of multiple sources of information (McBride, Barrett, Kelly, Aw, & Dagher, 2006). Also part of this large prefrontal network, the anterior cingulate cortex (ACC), a midline structure, is considered a key region in drug craving, as outlined by Engelmann et al. (2012). The dorsal part of the ACC is consistently recruited during cognitively demanding tasks that involve cognitive control and conflict monitoring (Bush, Luu, & Posner, 2000). Together, these findings suggest that dlPFC, dmPFC, and ACC activity during the viewing of appetitive cigarettes cues could refer to decision-making processes and preparing for the act of smoking. Alternatively, these activations could point to implicit cognitive control processes aimed at downregulating drug urges in a context where it is not possible to smoke (e.g., in an fMRI scanner during an experiment).

Lastly, the meta-analysis has reported a small cluster of activity within the insula and the dorsal striatum. Of note, the role of the insula in drug dependence has received an increasing amount of attention. It is thought to play a central role in addiction; different case studies have shown that damage to the insula can result in subsequent loss of craving and smoking cessation (Naqvi, Rudrauf, Damasio, & Bechara, 2007). The involvement of the insula in the craving network may relate to the subjective emotional experience, the representation and awareness of somatic and physiological response to drug-cue stimuli that could contribute to decision-making processes of drug taking (Garavan, 2010). Interestingly, based on Engelmann's results, it is the dorsal part of the striatum, not the ventral part, that presents significant activity during exposure to appetitive cigarette-related cues. Indeed, although various studies of reactivity to cigarette material have observed small clusters of activity in the ventral striatum, the meta-analysis

TABLE 1 Summary of Methodological Details and Results of Functional Imaging Studies of Tobacco Cue Reactivity

Authors	Smokers (N)	Nonsmokers (N)	Men (N)	Cigarettes/Day	fMRI Analysis	Abstinence	Significant Activity for the Contrast: Smoking-Related Cues Minus Neutral Cues
Brody et al. (2007)[a]	42	0	30	23.3 (SD: 8.2)	Exploratory	Ad libitum	Retrosplenial area, cuneus, precuneus, lingual gyrus, supra-marginal gyrus, lateral occipital gyrus, and angular gyrus.
David et al. (2007)[a]	8	0	0	17.9 (SD: 7.6)	Exploratory	• Ad libitum • Overnight (12 h)	• Posterior fusiform gyrus and inferior temporal gyrus. • Posterior fusiform gyrus, inferior temporal gyrus and ventral striatum (as a region of interest).
David et al. (2005)[a]	9	11	8	18.3 (SD: 8.7)	Exploratory	Overnight	**Smokers:** Anterior cingulate cortex, orbitofrontal cortex, superior frontal gyrus, fusiform and lingual gyri, as well as ventral striatum (as a region of interest). **Smokers > nonsmokers:** No significant differences.
Due, Huettel, Hall, and Rubin (2002)[a]	11	6	11	23.5 (SD: 7.7)	Region of interest: Ventral tegmental area, nucleus accumbens, amygdala, hippocampus, thalamus, prefrontal cortex, visual areas and anterior cingulate cortex.	Overnight (10 h)	**Smokers > nonsmokers:** Fusiform gyrus, intraparietal sulcus, hippocampus, thalamus, ventral tegmental area, amygdala, and inferior and middle frontal gyri.
Franklin et al. (2007)	21	0	9	19.6 (SD: 1.7)	Region of interest: Ventral tegmental area, ventral striatum, amygdala, cingulate and prefrontal cortices, hippocampus, thalamus, insula and fusiform gyrus.	Ad libitum	Ventral striatum, amygdala, hippocampus, thalamus, orbitofrontal cortex and insula.
Franklin et al. (2011)[a]	16	0	16	17.5 (SD: 1.6)	Exploratory	Ad libitum	Ventral striatum, orbitofrontal cortex, posterior cingulate cortex, insula and superior temporal gyrus.

Continued

TABLE 1 Summary of Methodological Details and Results of Functional Imaging Studies of Tobacco Cue Reactivity—cont'd

Authors	Smokers (N)	Nonsmokers (N)	Men (N)	Cigarettes/Day	fMRI Analysis	Abstinence	Significant Activity for the Contrast: Smoking-Related Cues Minus Neutral Cues
Goudriaan et al. (2010)	18	17	35	17.2 (SD: 3.8)	Exploratory	Overnight (16–18 h)	**Smokers with severe tobacco dependence > nonsmokers:** Ventromedial and ventrolateral prefontal cortices, anterior cingulate cortex, superior frontal gyrus, and the parietal cortex.
Lee, Lim, Wiederhold, and Graham (2005)[a]	8	0	8	15.3 (SD: 5.0)	Exploratory	7 h abstinence	Prefrontal and orbitofrontal cortices, anterior cingulate cortex, supplementary motor area, parahippocampal gyrus, inferior temporal gyrus, precuneus, lingual gyrus, superior temporal gyrus and inferior occipital gyrus.
McBride et al. (2006)[a]	19	0	10	22 (SD: 6)	**Region of interest:** Dorsolateral prefrontal cortex, anterior cingulate cortex, orbitofrontal cortex, insula and the limbic system.	• Ad libitum • 12 h abstinence	• Anterior and posterior cingulate cortices, precuneus, ventral pallidum and middle temporal gyrus. • No significant brain response.
McClernon, Kozink, and Rose (2008)[a]	30	0	7	23.6 (SD: 1.6)	Exploratory	Smoked 1–2 h before imagery session	Anterior and posterior cingulate cortices, superior frontal gyrus, cerebellum, precuneus, middle occipital gyrus, superior temporal gyrus, precentral gyrus and lingual gyrus.
McClernon, Kozink, Lutz, and Rose (2009)[a]	18	0	7	17.8 (SD: 2.8)	Exploratory	• 24 h abstinence • Ad libitum	• Superior parietal lobule, precuneus, posterior cingulate cortex, striatum, superior frontal gyrus, occipital cortex, thalamus, cerebellum and the precentral and postcentral gyri. • No Significant differences.
Yalachkov et al. (2009)	15	15	12	Heavy and light smokers	Exploratory	Ad libitum	**Smokers > nonsmokers:** Dorsal and ventral striatum, insula, dorsolateral prefrontal cortex, parahippocampal gyrus, medial frontal gyrus, cerebellum, superior parietal cortex and premotor cortex.
Wilson, Sayette, Delgado, and Fiez (2005)	22	0	22	21.6 (SD: 2.7)	**Region of interest** (no further information)	8 h abstinence	Anterior and posterior cingulate cortices, middle temporal gyrus, inferior parietal cortex, superior and middle occipital gyri, cuneus and fusiform gyrus.

Study							Results
Wang et al. (2007)	14	0	6	16.9 (SD: 5.6)	Exploratory	Ad libitum; Overnight (12–14 h)	**Abstinence > ad libitum:** Orbitofrontal cortex and anterior cingulate cortex.
Vollstädt-Klein et al. (2011)	22	21	43	24.6 (SD: 7.2)	Exploratory	Ad libitum	**Smokers:** Inferior occipital and fusiform gyri, cuneus, superior frontal gyrus, middle frontal gyrus, inferior frontal gyrus, precentral gyrus, inferior parietal lobule and cingulate cortex. **Smokers > nonsmokers:** No significant differences.
Hartwell et al. (2011)[a]	32	0	14	17.7 (SD: 6.9)	Exploratory	2 h abstinence	**Smokers:** Anterior cingulate cortex, medial prefrontal cortex, orbitofrontal cortex, posterior cingulate cortex, precuneus and middle cingulate cortex.
Janes et al. (2009)[a]	21	0	0	≥10	Exploratory	Ad libitum	**Smokers who relapsed > smokers still abstinent:** Insula, anterior and posterior cingulate cortices, amygdala, premotor and primary motor cortex, inferior parietal cortex, parahippocampal gyrus, thalamus and putamen.
Rubinstein et al. (2011)[a]	12	12	14	3.6 (SD: 1.3)	Exploratory	Ad libitum	**Smokers:** Anterior and posterior cingulate cortices, orbitofrontal cortex, hippocampus, amygdala, parahippocampal gyrus, middle and inferior temporal gyri as well as middle occipital cortex. **Smokers > nonsmokers:** No significant differences.
Bourque et al. (2013)	31	0	15	19.3 (SD: 5.7)	Exploratory	Ad libitum	**Smokers:** Anterior cingulate cortex, medial and lateral superior frontal gyri as well as posterior cingulate cortex.
Kang et al. (2012)	25	0	25	17.6 (SD: 0.8)	Exploratory	36 h abstinence	**Smokers:** Dorsolateral prefrontal cortex, anterior cingulate gyrus, superior temporal gyrus, premotor cortex, insula, supplementary motor area, thalamus, primary motor cortex, posterior cingulate gyrus, superior parietal gyrus and superior occipital gyrus.

SD, standard deviation; N, number of participants; h, hour.
[a]Studies compiled in the meta-analysis by Engelmann et al. (2012). This table lists studies examining a craving response in addiction and the structures activated when users view drug-related stimuli.

by Engelmann et al. (2012) did not report any significant findings in this specific region. This negative result does not conform to classic theories of addiction, which suggest that the hedonic and reinforcing effects of drug use and drug cues are processed in this core region of the reward system. The dorsal striatum, which receives innervations from the substantia nigra, is implicated in motor and cognitive control and the modulation of stimulus-response learning (Hyman, Malenka, & Nestler, 2006). According to Everitt and Robbins (2013), the shift from voluntary use to compulsive drug taking is a result of the predominant action of the dorsal striatum, the center of automatic behaviors, combined with a decreased in prefrontal cortex (PFC) control processes, rather than activity of the ventral striatum. Therefore, it is suggested that cue-induced craving in chronic smokers (>15 cigarettes/day) may be best characterized by automatic and compulsive drug-seeking behaviors (e.g., dorsal striatum) rather than simply the triggering of hedonic processes.

In addition to the vast neuroimaging literature demonstrating the increased sensitivity to appetitive smoking-related material in smokers, a few authors have further investigated the specific neural correlates of appetitive drug-related processing by comparing it to the processing of other natural appetitive stimuli. As it was postulated by Volkow, Fowler, and Wang (2004), drug addicts present a significant decrease in incentive salience toward natural rewards compared to drug rewards, thus contributing to explanation of the heightened motivation and maintenance of drug use. For instance, in a context of nicotine dependence, it was shown that chronic cigarette smokers were more reactive to cigarette cues relative to erotic pictures; the opposite pattern was observed in nonsmoker controls. This difference in brain reactivity between smokers and nonsmokers was found in the middle frontal gyrus (Brodmann 6/8) (Augustus Diggs, Froeliger, Carlson, & Gilbert, 2013), a region located near the border of the supplementary motor area thought to be implicated in the planning of actions and expectancy. This hyperactivity in smokers suggests a greater orientation toward the appetitive value of drug use than of natural rewards. Moreover, Versace et al. (2014) found that smokers presenting hypoactivity in the dorsal striatum while viewing pleasant (i.e., erotic and romantic) pictures compared to smoking cues were significantly more likely than those who presented hyperactivity to have relapsed 6 months after their quit attempt. The authors proposed that it is more difficult to quit for smokers who present this pattern of hypoactivity to pleasant stimuli because of their reinforced drug habit and biased sensitivity toward the appetitive value of smoking.

Although significant cue effects were reported in the brain reactivity to appetitive material (relative to neutral or natural reward stimuli), specific individual traits and states appear to modulate the brain's response to appetitive smoking cues. For instance, some authors have shown that one's intention to quit can impact prefrontal and limbic activity toward cigarette pictures (Wilson, Sayette, & Fiez, 2004). Also, it was shown that smokers who are discontent with their smoking behavior, relative to those in total acceptance of tobacco use, were more reactive (e.g., orbitofrontal and limbic activations) to appetitive cigarette cues (Stippekohl et al., 2012). Bourque, Mendrek, Dinh-Williams, and Potvin (2013) tested whether personality traits related to substance misuse (e.g., impulsivity) influence the craving response of smokers and found that increasingly poor levels of self-control and self-reflection were associated with enhanced feelings of craving.

Interestingly, this relation was mediated by decreased activity in the PCC while viewing appetitive cigarette cues. Besides its role in self-referential processing, the PCC may be implicated in resisting cigarette cravings. Brody et al. (2007) found that the PCC was specifically engaged while participants were actively trying to suppress their urges, but not when they allowed themselves to crave. It is possible then that those with poor self-control abilities experience greater difficulty in controlling their urges and exhibit a greater craving response to drug cues, and that this is partially mediated by lower activation of the PCC.

In summary, these results highlight the complex and vastly distributed neuronal network underlying smokers' craving response. This brain reactivity to appetitive smoking-related cues relative to neutral and other appetitive material stresses the importance of an attentional bias (e.g., extended visual system), self-referential processing (e.g., precuneus/PCC), planning/regulatory processes (e.g., ACC and dlPFC), emotional responding (e.g., insula), as well as the triggering of automatic conducts (e.g., dorsal striatum) in the maintenance of smoking behavior.

REACTIVITY TO AVERSIVE SMOKING CUES—CLINICAL RELEVANCE

In an effort to decrease smoking rates and cravings, a significant amount of governmental effort and investment has been focused on antismoking advertisements, depicting the negative value of smoking. This smoking-cessation strategy typically involves the use of graphic images with text or televised antismoking campaigns that act as aversive smoking-related cues; they are designed to evoke a negative emotional response and remind users of the negative value of smoking. These health warnings tend to illustrate the serious health effects of smoking on quality of life and the risk of mortality, but also its negative effects on physical appearance, its social acceptability, the perils and costs of addiction, as well as the effects of secondhand smoke.

The use of these messages is based on the premise that viewing how "bad" smoking is will motivate viewers to change their health behavior and will help curb their desire to smoke. Indeed, a few studies suggest that there are circumstances in which these health warnings may be persuasive. Messages that use emotive testimonials depicting loss of family and other smoking-related hardships have been shown to reinforce smokers' intentions to quit (Durkin & Wakefield, 2010). Televised messages that use graphic images to depict the negative health consequences of smoking have been shown to help prevent commencement of tobacco use in preadolescence or early adolescence (Wakefield & Chaloupka, 2000). Similarly, viewing antismoking campaigns has been shown to decrease cravings for a cigarette in smokers and ex-smokers (Kothe & Mullan, 2011). A number of studies have reported that viewing antismoking messages is perceived as an effective way of motivating the desire to quit (Hammond, 2011).

Findings regarding the persuasive impact of antismoking campaigns, however, are far from consistent. For instance, Siegel and Biener (2000) found that exposure to antismoking advertising decreased the likelihood of progression to established smoking for younger adolescents (12–13 years), but it had no effect on older adolescents (aged 14–15). A large study of Australian youths (Mazanov & Byrne, 2007) found that smokers' reported intentions to quit or to continue smoking were unrelated to their subsequent behavior. A number of researchers have even highlighted that exposing users to reasons to quit may do "more harm than good" (Erceg-Hurn & Steed, 2011). Exposure of smokers to antismoking campaigns has been found to paradoxically result in increased anxiety and cravings, rather than decreases in the desire to smoke (Gilbert, 2005; Loeber et al., 2011). Robinson and Killen (1997) found that greater knowledge of cigarette warning labels at baseline was not associated with decreases in tobacco consumption over a 3-month period, but rather increases in smoking behavior.

In summary, the efficiency of graphic antismoking messages in curbing smoking behavior is questionable, particularly in the context of tobacco dependence (Ruiter & Kok, 2005). Advertisements eliciting strong emotional responses are perceived as effective in reducing the desire to smoke in nonsmokers and smokers that are planning to quit, but not in current smokers (Siegel & Biener, 2000). Similarly, greater cigarette consumption (20 or more vs 10 or less per day) and lower motivation to quit are associated with decreases in the perception of effectiveness of these ads (Davis, Nonnemaker, Farrelly, & Niederdeppe, 2011). Studies have also found that current smokers react to antismoking information in a manner that helps to minimize its impact. Smokers have been shown to downplay the likelihood that they will suffer from the health problems depicted in the warnings (Harris, Mayle, Mabbott, & Napper, 2007) or mentally "switch off" when viewing antismoking information (Leshner, Bolls, & Thomas, 2009). Together, these findings highlight that current smokers are not readily persuaded by antismoking information.

Arguably, while it is intuitive that being exposed to the negative consequences of smoking should motivate smoking cessation, this health promotion strategy ignores one of the core features of substance use disorders (SUDs). Addicts are marked by a decreased sensitivity to the negative consequences of their drug consumption. Decades of research on psychosocial interventions have shown that awareness and experience with the harmful consequences of drug use on social, occupational, medical, and psychological health are not sufficient to alter the behavior of an addict (Chamberlain et al., 2013). In a traditional learning paradigm, when the consequences of an act are negative, the behavior tends not to be repeated. An addict, however, typically continues to use drugs despite negative drug-related consequences, such as job loss, rejection from family and friends, and ill health. This inability to discontinue consumption despite negative consequences is recognized in the DSM-5 as a core feature of SUDs, including tobacco, and is regarded as an important mechanism that helps maintain addictive behavior. Indeed, according to Campbell (2003), substance dependence should be viewed not solely as a sensitivity to the appetitive value of drugs, but as a disorder of faulty volition characterized by the inability to use knowledge or experience with the negative consequences of drug consumption to promote healthy decision making and self-control. Campbell states that treatment will continue to achieve limited success until the mechanisms underlying this clinical feature are properly understood. In all, the inability for negative consequences to hinder drug-seeking behavior is an important process by which smoking behavior is maintained in tobacco use disorder; however, little is known on the psychological and biological mechanisms underlying this disregard for negative consequences.

NEURAL REACTIVITY OF SMOKERS TO AVERSIVE SMOKING CUES

In combination with increased sensitivity to appetitive drug cues, drug-addicted individuals, including tobacco users, show a generalized pattern of reduced behavioral reactivity to the negative consequences of their behavior (Bechara & Damasio, 2002) associated to a specific pattern of neurobiological activity. This brain activation pattern may underlie the inability of negative consequences to significantly influence the behavior of drug users.

For instance, in a study by Hester, Bell, Foxe, and Garavan (2013), abstinent cocaine-dependent subjects and control participants performed a response inhibition task that measures their capacity to withhold a prepotent response to act when required to inhibit this response; missed trials resulted in monetary fines. Cocaine-dependent subjects were significantly less sensitive to punishment, such that their capacity for inhibitory control and adapting their performance was not modulated following error, despite knowing it would result in monetary loss. In addition, they showed a blunted neuronal response to failed control attempts (errors), which was associated to this behavioral insensitivity to punishment. Similar findings have been found for other addictions (e.g., cannabis, tobacco) in the context of decreased brain reactivity to error, as well as punishment following error (Hester, Nestor, & Garavan, 2009). The evidence thus far demonstrates that addiction is marked by a generalized pattern of decreased reactivity to negative feedback and consequences, and that this response is associated with a decreased neurobiological response specific to these tasks, compared to nonusers. These studies suggest that decreased activity in regions important for promoting a negative emotional response (i.e., medial prefrontal cortex, insula, anterior cingulate cortex, dorsal striatum) and the processing of self-relevant material (i.e., medial prefrontal cortex) may underlie this decreased reactivity to the negative consequences of behavior in addiction (Gowin, Mackey, & Paulus, 2013). A few studies have examined this response in the context of tobacco dependence more specifically and report similar results. de Ruiter, Oosterlaan, Veltman, van den Brink, and Goudriaan (2012) found failure to activate the insula in a group of chronic smokers compared to controls when processing monetary loss during a gambling task.

Another approach to investigating this facet of tobacco dependence has been to measure the brain reactivity of smokers when viewing aversive smoking-related stimuli, such as text, graphic images, and televised antismoking campaigns that depict the negative consequences of using. Interestingly, these studies demonstrate that increased activity in structures key to the generation of a negative emotional response (i.e., mPFC, insula, ACC, dorsal striatum, lateral prefrontal cortex, amygdala) and self-relevant processing (i.e., mPFC, precuneus) promote an increased sensitivity to the negative consequences of drug use. Increased activity in these regions is associated to greater message persuasiveness, memory for its content, and can predict behavior change (Dinh-Williams, Mendrek, Bourque, & Potvin, 2014; Falk, Berkman, Whalen, & Lieberman, 2011; Jasinska et al., 2012; Ramsay, Yzer, Luciana, Vohs, & MacDonald, 2013; Wang et al., 2013). For instance, greater activity in the medial prefrontal cortex, as well as precuneus, while processing antismoking information has been shown to predict declines in smoking behavior at 1-month follow-up (Falk et al., 2011). Thus, because tobacco dependence is characterized by reduced sensitivity toward the harmful effects of drug use, it is possible that decreased activity in these same regions while viewing aversive smoking-related stimuli underlies this important clinical feature. On the other hand, increased activity reflects an increased sensitivity to aversive depictions of smoking and promotes its impact on behavior.

Interested in this topic, our team investigated whether there were biases in how tobacco-dependent participants processed aversive images depending on whether they are smoking-related (e.g., lung cancer) or non-smoking-related (e.g., old woman on her death bed; Dinh-Williams et al., 2014). Our findings suggest that chronic smokers exhibit an aversive response when processing reminders of the negative value of smoking. However, despite being matched for arousal and valence by nonsmokers, aversive smoking-related stimuli was rated as less arousing and negative compared to aversive non-smoking-related images, and this was associated with lower activity in regions important for a negative emotional response (see Tables 2 and 3; Figure 1). This activation pattern is consistent with

TABLE 2 Cerebral Activations of Chronic Smokers While Viewing of Aversive Smoking-Related Images (Relative to Neutral Images)

Brain Region	R/L	BA	MNI Coordinates			Z-Score	Voxels
			x	y	z		
Inferior occipital/middle temporal	L	19	−46	−70	−14	7.76	2005
Fusiform/inferior occipital	R	37	46	−60	−18	6.61	1063
Supramarginal	R	2	63	−24	38	5.73	65
Middle occipital/superior occipital/inferior parietal	R	39	32	−70	32	4.74	193
Amygdala	R	21	35	0	−28	4.63	94
Inferior frontal triangular part	L	46	−52	35	14	4.47	74
Inferior orbitofrontal	L	11	−38	35	−14	4.03	66
Medial superior frontal/superior frontal	L	10	−4	60	32	4.01	51

R, right; L, left; BA, brodmann area. This table lists all the brain structures that are significantly more activated while viewing aversive smoking-related images than neutral.
Reprinted from Dinh-Williams et al. (2014), p. 70. Copyright 2014 by Elsevier. Adapted with permission.

TABLE 3 Cerebral Activation Differences between the Processing of Aversive Nonsmoking-Related and Aversive Smoking-Related Images in Chronic Smokers

Paired *t*-Tests	Brain Region	R/L	BA	MNI Coordinates			Z-Score	Voxels
				x	y	z		
Aversive smoking—Aversive IAPS	Inferior occipital/inferior temporal	R/L	37	−42	−66	−10	4.55	131
	Inferior parietal	R/L	40	56	−60	42	3.73	42
Aversive IAPS—Aversive smoking	Calcarine/cuneus	R/L	18	−4	−94	14	7.45	3475
	Cerebellum	L	–	−10	−42	−56	4.98	90
	Parahippocampal	R	35	24	−42	−10	4.23	24
	Precuneus	R/L	7	7	−52	56	3.83	110
	Inferior frontal *opercular part*	R	13	42	18	10	3.49	43
	Inferior frontal *triangular part*	R	44	46	21	7	3.45	54
	Precentral	R	6	42	−4	46	3.41	29
	Insula	R	13	38	32	4	3.23	43

R, right; L, left; BA, brodmann area. This table lists all the brain structures that are significantly more activated while viewing aversive nonsmoking-related images (e.g., a person holding a gun) compared to when aversive and related to smoking (e.g., a skeleton holding a cigarette) and inversely, while viewing aversive smoking-related relative to aversive nonsmoking-related images.
Reprinted from Dinh-Williams et al. (2014), p. 71. Copyright 2014 by Elsevier. Adapted with permission.

previous research on negative consequences in addiction (e.g., monetary loss) and further suggests the importance of this network of regions (e.g., insula, ACC, parahippocampal gyrus, dorsal striatum) as a potential mechanism underlying this clinical feature in addiction. Tobacco dependence, and addiction more generally, is marked by decreased activity in brain regions important for an aversive response when processing negative consequences, whether drug- or non-drug-related. This decreased aversive reactivity may play a role in explaining why the negative consequences of one's behavior are not sufficient to curb drug-seeking behavior.

In addition, because addiction is characterized by an increased sensitivity to appetitive drug cues and decreased sensitivity to aversive drug cues, we were interested in examining whether there was a neurophysiological response that can explain this behavioral disposition. Both aversive and appetitive smoking-related images triggered significant activations in the medial prefrontal cortex (mPFC); however the latter, along with the posterior cingulate/precuneus, were significantly more activated during the processing of appetitive smoking cues (see Table 4; Figure 2). Activation of these cortical midline structures is consistently observed when processing material that is relevant to the self and during self-referential thought (Lieberman, 2010). This self-processing mechanism has been implicated as an important factor in promoting motivation and goal-oriented behavior (Walter, Abler, Ciaramidaro, & Erk, 2005). Greater midline cortical activity for appetitive compared to aversive smoking-related cues may then promote a stronger driving response toward smoking cues and weaker driving response when exposed to the detrimental effects of smoking, which ultimately promotes greater orientation toward drug seeking compared to quitting.

Together, these studies highlight a neurophysiological bias in addiction when processing negative consequences, whether drug- or non-drug-related, which may hinder their impact on behavior. Most importantly, decreased activity in structures key to an aversive response (i.e., mPFC, amygdala, insula, ACC, dorsal striatum) and self-relevant processing (i.e., mPFC, posterior cingulate/precuneus) while being exposed to the negative consequences of

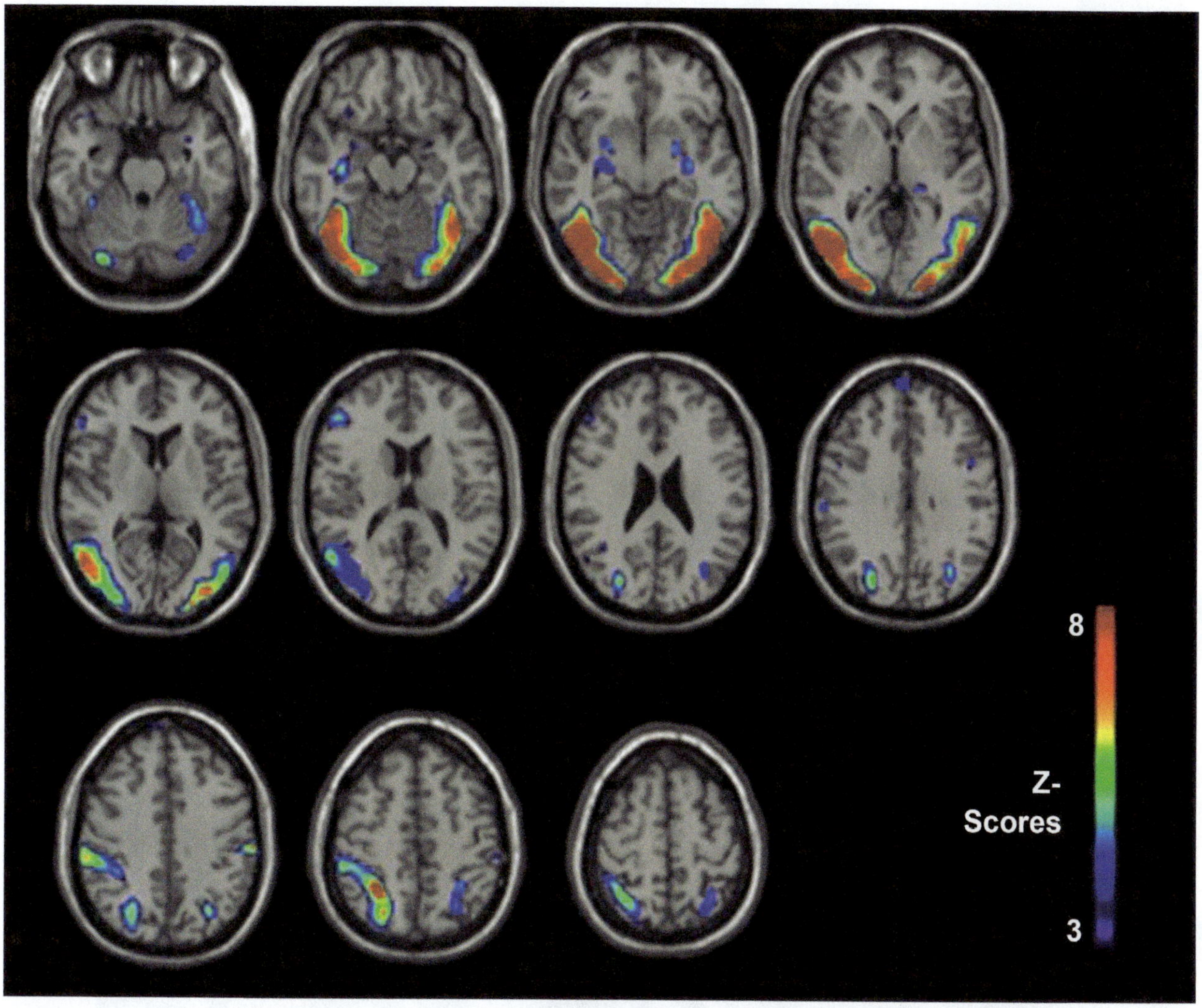

FIGURE 1 **Increased cerebral activity in chronic smokers while processing aversive non-smoking-related images, compared to aversive smoking-related images.** All colored regions were significantly more activated when viewing aversive images that are not related to smoking versus when related to smoking. Transition from purple to red illustrates increases in strength of activity. *Reprinted from Dinh-Williams et al. (2014), p. 69. Copyright 2014 by Elsevier. Adapted with permission.*

TABLE 4 Cerebral Activations Differences between the Processing of Appetitive and Aversive Smoking-Related Images

Paired t-Tests	Brain Region	R/L	BA	MNI Coordinates			Z-Score	Voxels
				x	y	z		
Appetitive – aversive	Precuneus	R	7	14	−63	32	3.87	26
	Hippocampus/posterior cingulate/calcarine	L	–	−21	−42	10	3.84	48
	Inferior parietal	R	40	56	−60	42	3.73	42
	Superior frontal/middle orbitofrontal	L	10	−24	60	0	3.58	63
	Superior frontal	R	10	24	60	7	3.53	28
Aversive – appetitive	Inferior temporal/inferior occipital/calcarine	L	19	−49	−63	−10	6.40	1866
	Inferior temporal/inferior occipital/calcarine	R	37	49	−56	−14	5.93	1481
	Cerebellum	L	–	−21	−66	−52	4.31	36
	Precentral	R	9	46	4	32	3.67	36
	Triangular inferior frontal	L	46	−52	32	21	3.35	30
	Precentral	L	6	−42	−4	42	3.20	26

R, right; L, left; BA, brodmann area. This table lists all the brain structures that are significantly more activated while viewing appetitive versus aversive smoking-related images and inversely, while viewing aversive versus appetitive smoking-related images.
Reprinted from Dinh-Williams et al. (2014), p. 70. Copyright 2014 by Elsevier. Adapted with permission.

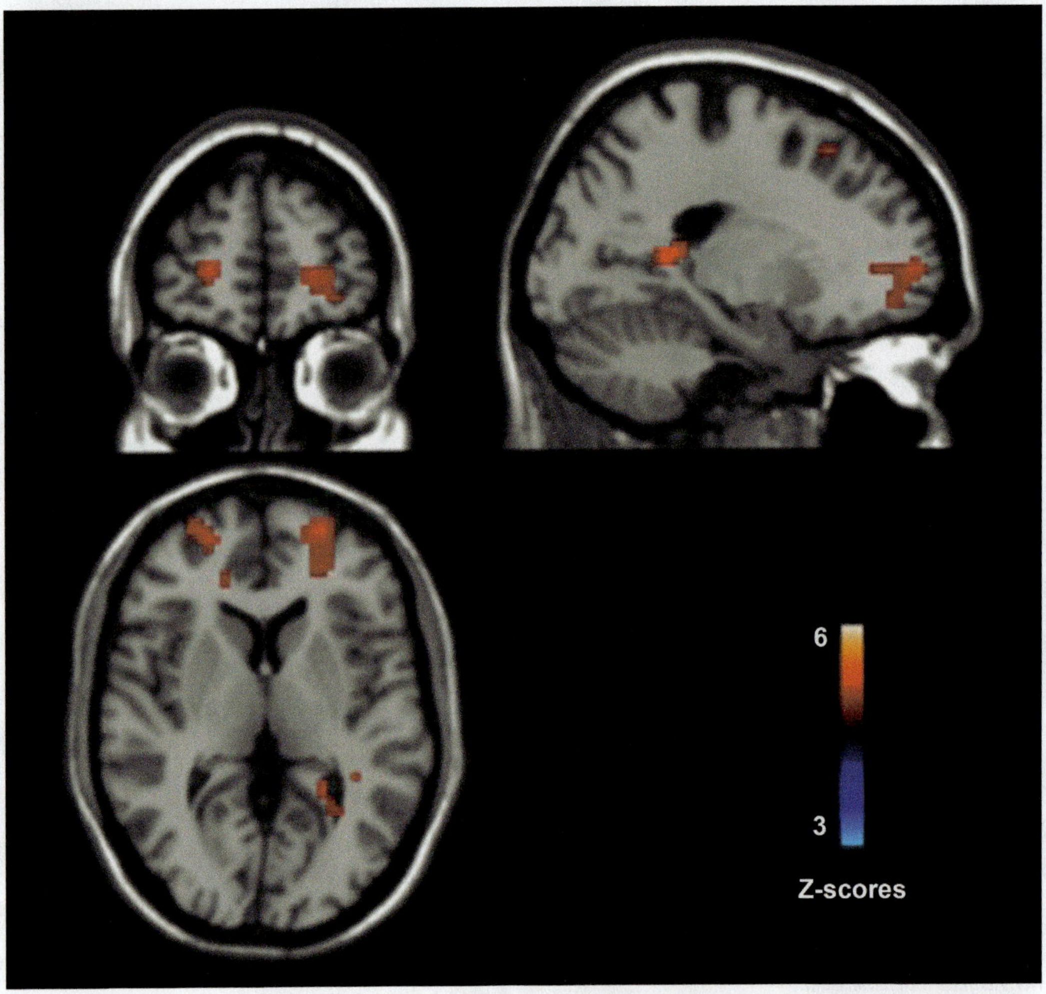

FIGURE 2 Increased cerebral activity in chronic smokers while processing appetitive smoking-related images, compared to aversive smoking-related images. All colored regions were significantly more activated when viewing appetitive images that are related to smoking versus when aversive and related to smoking. All colored regions were significantly activated.

smoking may underlie its inability to influence smoking behavior. In other words, the neurophysiological response of smokers suggest that tobacco use disorder is characterized by a decreased aversive response and a low degree of self-relevant processing when confronted with the negative consequences of smoking. Ultimately, this reactivity helps maintain the habit of smoking despite knowledge of its negative effects.

CONCLUSION

In all, tobacco use disorder is characterized by responses to appetitive and aversive smoking-related cues that help maintain the addictive behavior and significantly impede smoking cessation efforts. Appetitive smoking-related stimuli have the potential to trigger acute feelings of craving in regular smokers and those attempting abstinence. The brain activation pattern suggests the importance of an attentional bias (e.g., extended visual system), self-referential processing (e.g., precuneus/PCC), planning/regulatory processes (e.g., ACC and dlPFC), emotional responding (e.g., insula), as well as the triggering of automatic conducts (e.g., dorsal striatum) in an appetitive response, promoting the urge to consume and inability to resist. In combination with this reactivity to appetitive cues, drug-addicted individuals, including tobacco users, show a generalized pattern of reduced reactivity to the negative consequences of their behavior, such as those outlined in antismoking campaigns. This blunted response is thought to promote the inability of negative consequences to significantly influence the behavior of drug users. Neuroimaging findings propose that this clinical feature is partly explained by hypoactivation of both the aversive system (e.g., ACC, parahippocampal gyrus, insula) and self-relevant processing (e.g., precuneus/PCC, mPFC) to these events. In all, there is a bias in how appetitive and aversive smoking cues are processed by chronic smokers that promotes the maintenance of cigarette smoking and reduces the effectiveness of cessation efforts. Treatment and intervention may benefit by uncovering how to promote decreased activation of the appetitive system in response to appetitive smoking cues, and increased activity in the aversive system while viewing the negative consequences of smoking.

APPLICATIONS TO OTHER ADDICTIONS AND SUBSTANCE MISUSE

This pattern of reactivity in drug users to appetitive stimuli, associated with their respective drug of abuse, has been found across a variety of other substances, such as cocaine (Potenza et al., 2012), alcohol (Filbey et al., 2008), cannabis (Charboneau et al., 2013), methamphetamines (Yin et al., 2012), and heroin (Tabatabaei-Jafari et al., 2014). Similarly, the systems responsible for a cue-elicited appetitive response are activated by current or abstinent drug users when exposed to drug-related stimuli (Jasinska, Stein, Kaiser, Naumer, & Yalachkov, 2014). Furthermore, the responses in these systems are correlated with the severity of dependence and the degree of automaticity of the behavioral responses toward drug cues (Smolka et al., 2006). This line of research highlights the importance of appetitive drug cue reactivity in all forms of addiction.

As for aversive drug cues, only one study to date has investigated this topic for other substances. Similarly to the studies on aversive smoking-related stimuli, this study highlighted that medial and subcortical brain structures (e.g., parahippocampal gyrus, precuneus) are recruited by more persuasive antidrug messages (Ramsay et al., 2013). In addition, they highlighted that increased functional connectivity between the left PFC and subcortical structures (e.g., amygdala, insula) are associated with the processing of more powerful antidrug messages. Together, these findings suggest the recruitment of a similar aversive network for promoting the impact of antidrug information. More research is needed on how current drug users process aversive drug-related stimuli relevant to their specific consumption (e.g., how cocaine users process the negative health consequences of cocaine use); the latter study investigated the reactivity of young adolescents without any substance use issues. However, neuroimaging findings thus far in addiction, punishment, and antidrug processing suggest the importance of aversive and self-relevant processing, mediated by medial and subcortical reactivity, in promoting the impact of negative consequences on drug behavior.

DEFINITION OF TERMS

Appetitive drug (smoking) cues Stimuli associated with drug use that trigger a craving response.
Aversive drug (smoking) cues Stimuli depicting the negative value of drug use (e.g., serious health effects of smoking on quality of life and the risk of mortality, physical appearance) that triggers a negative emotional response.
Classical conditioning An associative learning process that occur when two stimuli (i.e., neutral and unconditioned stimuli) have been repeatedly paired. Through this process, the response that is normally elicited by the unconditioned stimulus can now be elicited by the newly conditioned stimulus (previously, the neutral cue).
Cue-induced craving or phasic craving A fast-onset acute urge in response to appetitive drug-related cues, which declines minutes after the removal of cues.
Decreased sensitivity The ability for stimuli to impact affective, cognitive, and behavioral processes.
Functional magnetic resonance imaging This technology allows us to measure brain activity according to the associated changes in regional blood flow that occur during mental processes.
Incentive salience A motivation ("wanting") for a highly desirable and attractive goal that has been associated with reward.
Psychological and biological mechanisms Psychological or biological processes by which something is brought about.
Self-referential processes The overall concept of self-reference suggests that people interpret incoming information in relation to themselves, using their self-concept as a background for new information.
Self-relevant material Stimuli encoded as related to the self.
Tonic craving A slowly changing urge state induced by abstinence which tends to decline as the abstinence period lengthens to weeks or months.

KEY FACTS OF TOBACCO USE DISORDER

- In North America, the prevalence of cigarette smokers has consistently declined since the 1970s; however, since 2005, this prevalence appears to stagnate at approximately 20% of the general population.
- Craving is now considered a diagnostic criterion of tobacco use disorder in the DSM-5.

- Tobacco SUD is also characterized by continued use despite knowledge or experience with the negative consequences of smoking.
- Increased activity in prefrontal (dlPFC, mPFC) and medial/subcortical structures (precuneus/PCC, ACC, insula, dorsal striatum) when exposed to appetitive smoking cues underlies a craving response.
- Decreased activity in similar regions (mPFC, precuneus/PCC, ACC, insula, dorsal striatum, parahippocampus) underlies the inability of negative consequences to have an impact on smoking behavior.

SUMMARY POINTS

This chapter focused on the two important clinical features of tobacco dependence that help maintain its addictive nature.

- The craving response when one is exposed to explicit appetitive smoking cues or implicit cues such as stress has been related to cessation outcomes.
    - Craving is a multidimensional construct involving a variety of cognitive and other nonautomatic processes that are not restricted to the reward pathway.
    - The brain craving response is characterized by attentional, affective, self-relevant, regulatory and automatic processes which are mediated by the extended visual system, insula, dmPFC, precuneus, dlPFC, ACC, and dorsal striatum.
    - This brain response is largely influenced by abstinence levels, smoking expectancy and individual traits such as a desire to quit, tobacco dependence severity and impulsivity.
- The inability for knowledge or experience with the negative consequences of smoking to greatly impact smoking behavior.
    - Little evidence suggests that exposure to the negative consequences of smoking (e.g., antismoking campaigns) modifies actual smoking behavior in current smokers.
    - A potential mechanism underlying this clinical feature is that addicts are insensitive to negative consequences in general.
    - Hypoactivation in the mPFC, precuneus, ACC, insula, amygdala, dorsal striatum, and parahippocampal gyrus compared to controls may underlie this hyposensitivity to negative drug- or non-drug-related consequences.

REFERENCES

Augustus Diggs, H., Froeliger, B., Carlson, J. M., & Gilbert, D. G. (2013). Smoker-nonsmoker differences in neural response to smoking-related and affective cues: an fMRI investigation. *Psychiatry Research*, 211(1), 85–87. http://dx.doi.org/10.1016/j.pscychresns.2012.06.009.

Bechara, A., & Damasio, H. (2002). Decision-making and addiction (part I): impaired activation of somatic states in substance dependent individuals when pondering decisions with negative future consequences. *Neuropsychologia*, 40(10), 1675–1689.

Bedi, G., Preston, K. L., Epstein, D. H., Heishman, S. J., Marrone, G. F., Shaham, Y., & de Wit, H. (2011). Incubation of cue-induced cigarette craving during abstinence in human smokers. *Biological Psychiatry*, 69(7), 708–711. http://dx.doi.org/10.1016/j.biopsych.2010.07.014.

Bourque, J., Mendrek, A., Dinh-Williams, L., & Potvin, S. (2013). Neural circuitry of impulsivity in a cigarette craving paradigm. *Frontiers in Psychiatry*, 4, 67. http://dx.doi.org/10.3389/fpsyt.2013.00067.

Bradley, M. M., Sabatinelli, D., Lang, P. J., Fitzsimmons, J. R., King, W., & Desai, P. (2003). Activation of the visual cortex in motivated attention. *Behavioral Neuroscience*, 117(2), 369–380.

Brody, A. L., Mandelkern, M. A., Olmstead, R. E., Jou, J., Tiongson, E., Allen, V., ... Cohen, M. S. (2007). Neural substrates of resisting craving during cigarette cue exposure. *Biological Psychiatry*, 62(6), 642–651. http://dx.doi.org/10.1016/j.biopsych.2006.10.026.

Bush, G., Luu, P., & Posner, M. I. (2000). Cognitive and emotional influences in anterior cingulate cortex. *Trends in Cognitive Sciences*, 4(6), 215–222.

Campbell, W. G. (2003). Addiction: a disease of volition caused by a cognitive impairment. *Canadian Journal of Psychiatry*, 48(10), 669–674.

Cavanna, A. E., & Trimble, M. R. (2006). The precuneus: a review of its functional anatomy and behavioural correlates. *Brain*, 129(Pt 3), 564–583. http://dx.doi.org/10.1093/brain/awl004.

Chamberlain, C., O'Mara-Eves, A., Oliver, S., Caird, J. R., Perlen, S. M., Eades, S. J., & Thomas, J. (2013). Psychosocial interventions for supporting women to stop smoking in pregnancy. *The Cochrane Database of Systematic Reviews*, 10, CD001055.

Charboneau, E. J., Dietrich, M. S., Park, S., Cao, A., Watkins, T. J., Blackford, J. U., ... Cowan, R. L. (2013). Cannabis cue-induced brain activation correlates with drug craving in limbic and visual salience regions: preliminary results. *Psychiatry Research*, 214(2), 122–131. http://dx.doi.org/10.1016/j.pscychresns.2013.06.005.

Chase, H. W., Eickhoff, S. B., Laird, A. R., & Hogarth, L. (2011). The neural basis of drug stimulus processing and craving: an activation likelihood estimation meta-analysis. *Biological Psychiatry*, 70(8), 785–793. http://dx.doi.org/10.1016/j.biopsych.2011.05.025.

David, S. P., Munafo, M. R., Johansen-Berg, H., Smith, S. M., Rogers, R. D., Matthews, P. M., & Walton, R. T. (2005). Ventral striatum/nucleus accumbens activation to smoking-related pictorial cues in smokers and nonsmokers: a functional magnetic resonance imaging study. *Biological psychiatry*, 58(6), 488–494.

David, S. P., Munafo, M. R., Johansen-Berg, H., Mackillop, J., Sweet, L. H., Cohen, R. A., ... Walton, R. T. (2007). Effects of Acute Nicotine Abstinence on Cue-elicited Ventral Striatum/Nucleus Accumbens Activation in Female Cigarette Smokers: A Functional Magnetic Resonance Imaging Study. *Brain imaging and behavior*, 1(3–4), 43–57.

Davis, K. C., Nonnemaker, J. M., Farrelly, M. C., & Niederdeppe, J. (2011). Exploring differences in smokers' perceptions of the effectiveness of cessation media messages. *Tobacco Control*, 20, 26–33.

Dinh-Williams, L., Mendrek, A., Bourque, J., & Potvin, S. (2014). Where there's smoke, there's fire: the brain reactivity of chronic smokers when exposed to the negative value of smoking. *Progress in Neuropsychopharmacology & Biological Psychiatry*, 50, 66–73. http://dx.doi.org/10.1016/j.pnpbp.2013.12.009.

Due, D. L., Huettel, S. A., Hall, W. G., & Rubin, D. C. (2002). Activation in mesolimbic and visuospatial neural circuits elicited by smoking cues: evidence from functional magnetic resonance imaging. *The American journal of psychiatry*, 159(6), 954–960.

Durkin, S., & Wakefield, M. (2010). Comparative responses to radio and television anti-smoking advertisements to encourage smoking cessation. *Health Promotion International*, 25(1), 5–13. http://dx.doi.org/10.1093/heapro/dap044.

Engelmann, J. M., Versace, F., Robinson, J. D., Minnix, J. A., Lam, C. Y., Cui, Y., ... Cinciripini, P. M. (2012). Neural substrates of smoking cue reactivity: a meta-analysis of fMRI studies. *NeuroImage*, *60*(1), 252–262. http://dx.doi.org/10.1016/j.neuroimage.2011.12.024.

Erblich, J., Boyarsky, Y., Spring, B., Niaura, R., & Bovbjerg, D. H. (2003). A family history of smoking predicts heightened levels of stress-induced cigarette craving. *Addiction*, *98*(5), 657–664.

Erceg-Hurn, D. M., & Steed, L. G. (2011). Does exposure to cigarette health warnings elicit psychological reactance in smokers. *Journal of Applied Social Psychology*, *41*(1), 219–237.

Everitt, B. J., & Robbins, T. W. (2013). From the ventral to the dorsal striatum: devolving views of their roles in drug addiction. *Neuroscience and Biobehavioral Reviews*, *37*(9 Pt A), 1946–1954. http://dx.doi.org/10.1016/j.neubiorev.2013.02.010.

Falk, E. B., Berkman, E. T., Whalen, D., & Lieberman, M. D. (2011). Neural activity during health messaging predicts reductions in smoking above and beyond self-report. *Health Psychology*, *30*(2), 177–185. http://dx.doi.org/10.1037/a0022259.

Filbey, F. M., Claus, E., Audette, A. R., Niculescu, M., Banich, M. T., Tanabe, J., ... Hutchison, K. E. (2008). Exposure to the taste of alcohol elicits activation of the mesocorticolimbic neurocircuitry. *Neuropsychopharmacology*, *33*(6), 1391–1401. http://dx.doi.org/10.1038/sj.npp.1301513.

Franklin, T. R., Wang, Z., Wang, J., Sciortino, N., Harper, D., Li, Y., ... Childress, A. R. (2007). Limbic activation to cigarette smoking cues independent of nicotine withdrawal: a perfusion fMRI study. *Neuropsychopharmacology : official publication of the American College of Neuropsychopharmacology*, *32*(11), 2301–2309.

Franklin, T., Wang, Z., Suh, J. J., Hazan, R., Cruz, J., Li, Y., ... Childress, A. R. (2011). Effects of varenicline on smoking cue-triggered neural and craving responses. *Archives of general psychiatry*, *68*(5), 516–526.

Garavan, H. (2010). Insula and drug cravings. *Brain Structure & Function*, *214*(5–6), 593–601. http://dx.doi.org/10.1007/s00429-010-0259-8.

Gilbert, E. (2005). Contextualising the medical risks of cigarette smoking: Australian young women's perceptions of anti-smoking campaigns. *Health, Risk & Society*, *7*, 227–245.

Goudriaan, A. E., de Ruiter, M. B., van den Brink, W., Oosterlaan, J., & Veltman, D. J. (2010). Brain activation patterns associated with cue reactivity and craving in abstinent problem gamblers, heavy smokers and healthy controls: an fMRI study. *Addiction biology*, *15*(4), 491–503.

Gowin, J. L., Mackey, S., & Paulus, M. P. (2013). Altered risk-related processing in substance users: imbalance of pain and gain. *Drug and Alcohol Dependence*, *132*(1–2), 13–21. http://dx.doi.org/10.1016/j.drugalcdep.2013.03.019.

Hammond, D. (2011). Health warning messages on tobacco products: a review. *Tobacco Control*, *20*(5), 327–337. http://dx.doi.org/10.1136/tc.2010.037630.

Harris, P. R., Mayle, K., Mabbott, L., & Napper, L. (2007). Self-affirmation reduces smokers' defensiveness to graphic, on-pack cigarette warning labels. *Health Psychology*, *26*, 437–446.

Hartwell, K. J., Johnson, K. A., Li, X., Myrick, H., LeMatty, T., George, M. S., & Brady, K. T. (2011). Neural correlates of craving and resisting craving for tobacco in nicotine dependent smokers. *Addiction biology*, *16*(4), 654–666.

Herd, N., Borland, R., & Hyland, A. (2009). Predictors of smoking relapse by duration of abstinence: findings from the International Tobacco Control (ITC) four country survey. *Addiction*, *104*(12), 2088–2099. http://dx.doi.org/10.1111/j.1360-0443.2009.02732.x.

Hester, R., Bell, R. P., Foxe, J. J., & Garavan, H. (2013). The influence of monetary punishment on cognitive control in abstinent cocaine-users. *Drug and Alcohol Dependence*, *133*(1). http://dx.doi.org/10.1016/j.drugalcdep.2013.05.027.

Hester, R., Nestor, L., & Garavan, H. (2009). Impaired error awareness and anterior cingulate cortex hypoactivity in chronic cannabis users. *Neuropsychopharmacology*, *34*(11), 2450–2458. http://dx.doi.org/10.1038/npp.2009.67.

Hughes, J. R., Peters, E. N., & Naud, S. (2008). Relapse to smoking after 1 year of abstinence: a meta-analysis. *Addiction Behaviors*, *33*(12), 1516–1520. http://dx.doi.org/10.1016/j.addbeh.2008.05.012.

Hyman, S. E., Malenka, R. C., & Nestler, E. J. (2006). Neural mechanisms of addiction: the role of reward-related learning and memory. *Annual Review of Neuroscience*, *29*, 565–598. http://dx.doi.org/10.1146/annurev.neuro.29.051605.113009.

Janes, A. C., Frederick, B., Richardt, S., Burbridge, C., Merlo-Pich, E., Renshaw, P. F., ... Kaufman, M. J. (2009). Brain fMRI reactivity to smoking-related images before and during extended smoking abstinence. *Experimental and clinical psychopharmacology*, *17*(6), 365–373.

Jasinska, A. J., Chua, H. G., Ho, S. S., Polk, T. A., Rozek, L. S., & Strecher, V. J. (2012). Amygdala response to smoking-cessation messages mediates the effects of serotonin transporter gene variation on quitting. *NeuroImage*, *60*, 766–773.

Jasinska, A. J., Stein, E. A., Kaiser, J., Naumer, M. J., & Yalachkov, Y. (2014). Factors modulating neural reactivity to drug cues in addiction: a survey of human neuroimaging studies. *Neuroscience and Biobehavioral Reviews*, *38*, 1–16. http://dx.doi.org/10.1016/j.neubiorev.2013.10.013.

Kang, O. S., Chang, D. S., Jahng, G. H., Kim, S. Y., Kim, H., Kim, J. W., ... Chae, Y. (2012). Individual differences in smoking-related cue reactivity in smokers: an eye-tracking and fMRI study. *Progress in neuro-psychopharmacology & biological psychiatry*, *38*(2), 285–293.

Killen, J. D., & Fortmann, S. P. (1997). Craving is associated with smoking relapse: findings from three prospective studies. *Experimental and Clinical Psychopharmacology*, *5*(2), 137–142.

Kothe, E., & Mullan, B. (2011). Smokers and ex-smokers reaction to anti-smoking advertising: a mixed methods approach. *Orbit*, *2*(1), 29–37.

Lee, J. H., Lim, Y., Wiederhold, B. K., & Graham, S. J. (2005). A functional magnetic resonance imaging (FMRI) study of cue-induced smoking craving in virtual environments. *Applied psychophysiology and biofeedback*, *30*(3), 195–204.

Leshner, G., Bolls, P., & Thomas, E. (2009). Scare 'em or disgust 'em: the effects of graphic health promotion messages. *Health Communication*, *24*, 447–458.

Li, C. S., & Sinha, R. (2008). Inhibitory control and emotional stress regulation: neuroimaging evidence for frontal-limbic dysfunction in psycho-stimulant addiction. *Neuroscience and Biobehavioral Reviews*, *32*(3), 581–597. http://dx.doi.org/10.1016/j.neubiorev.2007.10.003.

Lieberman, M. (2010). *Social cognitive neuroscience* (5th ed.). New York: McGraw-Hill.

Loeber, S., Vollstadt-Klein, S., Wilden, S., Schneider, S., Rockenbach, C., Dinter, C., ... Kiefer, F. (2011). The effect of pictorial warnings on cigarette packages on attentional bias of smokers. *Pharmacology, Biochemistry and Behavior*, *98*, 292–298.

Mazanov, J., & Byrne, D. G. (2007). "Do you intend to smoke?" A test of the assumed psychological equivalence in adolescent smoker and non-smoker intention to change smoking behaviour. *Australian Journal of Psychology*, *59*, 34–42.

McBride, D., Barrett, S. P., Kelly, J. T., Aw, A., & Dagher, A. (2006). Effects of expectancy and abstinence on the neural response to smoking cues in cigarette smokers: an fMRI study. *Neuropsychopharmacology, 31*(12), 2728–2738. http://dx.doi.org/10.1038/sj.npp.1301075.

McClernon, F. J., Kozink, R. V., & Rose, J. E. (2008). Individual differences in nicotine dependence, withdrawal symptoms, and sex predict transient fMRI-BOLD responses to smoking cues. *Neuropsychopharmacology : official publication of the American College of Neuropsychopharmacology, 33*(9), 2148–2157.

McClernon, F. J., Kozink, R. V., Lutz, A. M., & Rose, J. E. (2009). 24-h smoking abstinence potentiates fMRI-BOLD activation to smoking cues in cerebral cortex and dorsal striatum. *Psychopharmacology, 204*(1), 25–35.

Muller, C. P. (2013). Episodic memories and their relevance for psychoactive drug use and addiction. *Frontiers in Behavioral Neuroscience, 7*, 34. http://dx.doi.org/10.3389/fnbeh.2013.00034.

Naqvi, N. H., Rudrauf, D., Damasio, H., & Bechara, A. (2007). Damage to the insula disrupts addiction to cigarette smoking. *Science, 315*(5811), 531–534. http://dx.doi.org/10.1126/science.1135926.

Potenza, M. N., Hong, K. I., Lacadie, C. M., Fulbright, R. K., Tuit, K. L., & Sinha, R. (2012). Neural correlates of stress-induced and cue-induced drug craving: influences of sex and cocaine dependence. *The American Journal of Psychiatry, 169*(4), 406–414. http://dx.doi.org/10.1176/appi.ajp.2011.11020289.

Ramsay, I. S., Yzer, M. C., Luciana, M., Vohs, K. D., & MacDonald, A. W., 3rd (2013). Affective and executive network processing associated with persuasive antidrug messages. *Journal of Cognitive Neuroscience, 25*(7), 1136–1147. http://dx.doi.org/10.1162/jocn_a_00391.

Robinson, T. E., & Berridge, K. C. (1993). The neural basis of drug craving: an incentive-sensitization theory of addiction. *Brain Research: Brain Research Reviews, 18*(3), 247–291.

Robinson, T. N., & Killen, J. D. (1997). Do cigarette warning labels reduce smoking? *Archives of Pediatrics and Adolescent Medicine, 151*, 267–272.

Rubinstein, M. L., Luks, T. L., Moscicki, A. B., Dryden, W., Rait, M. A., & Simpson, G. V. (2011). Smoking-related cue-induced brain activation in adolescent light smokers. *The Journal of adolescent health: official publication of the Society for Adolescent Medicine, 48*(1), 7–12.

Ruiter, R. A. C., & Kok, G. (2005). Saying is not (always) doing: cigarette warnings labels are useless. *European Journal of Public Health, 15*, 329–330.

de Ruiter, M. B., Oosterlaan, J., Veltman, D. J., van den Brink, W., & Goudriaan, A. E. (2012). Similar hyporesponsiveness of the dorsomedial prefrontal cortex in problem gamblers and heavy smokers during an inhibitory control task. *Drug and Alcohol Dependence, 121*(1–2), 81–89. http://dx.doi.org/10.1016/j.drugalcdep.2011.08.010.

Siegel, M., & Biener, L. (2000). The impact of an antismoking media campaign on progression to established smoking: results of a longitudinal youth study. *American Journal of Public Health, 90*(3), 380–386.

Smolka, M. N., Buhler, M., Klein, S., Zimmermann, U., Mann, K., Heinz, A., & Braus, D. F. (2006). Severity of nicotine dependence modulates cue-induced brain activity in regions involved in motor preparation and imagery. *Psychopharmacology (Berlin), 184*(3–4), 577–588. http://dx.doi.org/10.1007/s00213-005-0080-x.

Stippekohl, B., Winkler, M. H., Walter, B., Kagerer, S., Mucha, R. F., Pauli, P., ... Stark, R. (2012). Neural responses to smoking stimuli are influenced by smokers' attitudes towards their own smoking behaviour. *PLoS One, 7*(11), e46782. http://dx.doi.org/10.1371/journal.pone.0046782.

Sweitzer, M. M., Denlinger, R. L., & Donny, E. C. (2013). Dependence and withdrawal-induced craving predict abstinence in an incentive-based model of smoking relapse. *Nicotine & Tobacco Research, 15*(1), 36–43. http://dx.doi.org/10.1093/ntr/nts080.

Tabatabaei-Jafari, H., Ekhtiari, H., Ganjgahi, H., Hassani-Abharian, P., Oghabian, M. A., Moradi, A., ... Zarei, M. (2014). Patterns of brain activation during craving in heroin dependents successfully treated by methadone maintenance and abstinence-based treatments. *Journal of Addiction Medicine, 8*(2), 123–129. http://dx.doi.org/10.1097/adm.0000000000000022.

Tiffany, S. T., & Conklin, C. A. (2000). A cognitive processing model of alcohol craving and compulsive alcohol use. *Addiction, 95*(Suppl. 2), S145–S153.

Torta, D. M., & Cauda, F. (2011). Different functions in the cingulate cortex, a meta-analytic connectivity modeling study. *NeuroImage, 56*(4), 2157–2172. http://dx.doi.org/10.1016/j.neuroimage.2011.03.066.

Ussher, M., Beard, E., Abikoye, G., Hajek, P., & West, R. (2013). Urge to smoke over 52 weeks of abstinence. *Psychopharmacology (Berlin), 226*(1), 83–89. http://dx.doi.org/10.1007/s00213-012-2886-7.

Versace, F., Engelmann, J. M., Robinson, J. D., Jackson, E. F., Green, C. E., Lam, C. Y., ... Cinciripini, P. M. (2014). Prequit fMRI responses to pleasant cues and cigarette-related cues predict smoking cessation outcome. *Nicotine & Tobacco Research, 16*(6), 697–708. http://dx.doi.org/10.1093/ntr/ntt214.

Volkow, N. D., Fowler, J. S., & Wang, G. J. (2004). The addicted human brain viewed in the light of imaging studies: brain circuits and treatment strategies. *Neuropharmacology, 47*(Suppl. 1), 3–13. http://dx.doi.org/10.1016/j.neuropharm.2004.07.019.

Vollstadt-Klein, S., Kobiella, A., Buhler, M., Graf, C., Fehr, C., Mann, K., & Smolka, M. N. (2011). Severity of dependence modulates smokers' neuronal cue reactivity and cigarette craving elicited by tobacco advertisement. *Addiction biology, 16*(1), 166–175.

Wakefield, M., & Chaloupka, F. (2000). Effectiveness of comprehensive tobacco control programs in reducing teenage smoking in the USA. *Tobacco Control, 9*, 177–186.

Walter, H., Abler, B., Ciaramidaro, A., & Erk, S. (2005). Motivating forces of human actions. Neuroimaging reward and social interaction. *Brain Research Bulletin, 67*(5), 368–381. http://dx.doi.org/10.1016/j.brainresbull.2005.06.016.

Wang, Z., Faith, M., Patterson, F., Tang, K., Kerrin, K., Wileyto, E. P., ... Lerman, C. (2007). Neural substrates of abstinence-induced cigarette cravings in chronic smokers. *The Journal of neuroscience : the official journal of the Society for Neuroscience, 27*(51), 14035–14040.

Wang, A. L., Ruparel, K., Loughead, J. W., Strasser, A. A., Blady, S. J., Lynch, K. G., ... Langleben, D. D. (2013). Content matters: neuroimaging investigation of brain and behavioral impact of televised anti-tobacco public service announcements. *The Journal of Neuroscience, 33*(17), 7420–7427. http://dx.doi.org/10.1523/jneurosci.3840-12.2013.

Wilson, S. J., Sayette, M. A., & Fiez, J. A. (2004). Prefrontal responses to drug cues: a neurocognitive analysis. *Nature Neuroscience, 7*(3), 211–214. http://dx.doi.org/10.1038/nn1200.

Wilson, S. J., Sayette, M. A., Delgado, M. R., & Fiez, J. A. (2005). Instructed smoking expectancy modulates cue-elicited neural activity: a preliminary study. *Nicotine & Tobacco Research, 7*(4), 637–645. http://dx.doi.org/10.1080/14622200500185520.

Yalachkov, Y., Kaiser, J., & Naumer, M. J. (2009). Brain regions related to tool use and action knowledge reflect nicotine dependence. *The Journal of neuroscience : the official journal of the Society for Neuroscience, 29*(15), 4922–4929.

Yin, J. J., Ma, S. H., Xu, K., Wang, Z. X., Le, H. B., Huang, J. Z., ... Cai, Z. L. (2012). Functional magnetic resonance imaging of methamphetamine craving. *Clinical Imaging, 36*(6), 695–701. http://dx.doi.org/10.1016/j.clinimag.2012.02.006.

Chapter 29

Different Effects of Cigarette Smoking on Neuropsychological Performance in Psychiatric Disorders

Daniela Caldirola, Giuseppe Iannone, Giuseppina Diaferia, Giampaolo Perna

Hermanas Hospitalarias, Villa San Benedetto Menni Hospital, FoRiPsi, Department of Clinical Neurosciences, Albese con Cassano, Italy

Abbreviations

ACh Acetylcholine
BD Bipolar disorder
MDD Major depressive disorder
nAChRs Brain nicotinic acetylcholine receptors
OCD Obsessive-compulsive disorder
PFC Prefrontal cortex
SZ Schizophrenia
VTA Ventral tegmental area

INTRODUCTION

People with psychiatric disorders consume more than 44% of all the cigarettes smoked in the United States (Ziedonis et al., 2008). This number is incredibly high considering that people with mental illness comprise only 7.1% of the US population. Smoking rates are higher in the psychiatric population (40–88%) than in the general population (25%), and generally smokers with psychiatric disorders present more severe nicotine dependence and withdrawal symptoms (e.g., cigarettes craving, irritability, nervousness) than smokers without psychiatric disorders (Breslau, Kilbey, & Andreski, 1992). However, the proportion of smokers is not homogeneous among different psychiatric disorders. Exceptionally high rates of smokers have been found in patients suffering from schizophrenia (58–88%), bipolar disorder (BD; 40–70%), major depressive disorder (MDD; 30–60%), and anxiety disorders (40–50%). On the contrary, lower rates of smokers both than in the other psychiatric disorders, as well as in the general population (25%), have been found in subjects with obsessive-compulsive disorder (OCD; about 13%) (Abramovitch, Pizzagalli, Geller, Reuman, & Wilhelm, 2014; Dome, Lazary, Kalapos, & Rihmer, 2010; Sacco et al., 2005) (Figure 1). These findings suggest that subjects with psychiatric disorders exhibit peculiar features that might foster smoking habits to a greater extent than in the general population. Also, specific neurobiological substrates and/or behavioral features may underlie the heterogeneity of smoking frequencies among different psychiatric disorders. However, the precise nature of these features remains unclear to date. Socioeconomic and demographic factors, common environmental/genetic risk factors, and the use of smoking to alleviate aversive symptoms or medications side effects ("self-medication hypothesis") might explain the association between smoking and most psychiatric disorders (Barr, Procyshyn, Hui, Johnson, & Honer, 2008). Because impairment across several cognitive domains has been firmly established both in schizophrenia and, to a lesser extent, in MDD and BD (e.g., Hammar & Ardal, 2009), it has been hypothesized that nicotine might exert a procognitive effect and therefore fostering smoking habits. On the contrary, very preliminary findings suggest that nicotine may exert detrimental cognitive effects in subjects with OCD, which may discourage them from smoking. In this chapter, we will specifically delve into this topic. We will explore whether the different effects of nicotine on cognition may explain the different smoking rates observed among subjects suffering from schizophrenia, MDD, BD, or OCD and in healthy subjects.

SMOKING AND COGNITIVE FUNCTION IN HEALTHY SUBJECTS

Nicotine is the dominant component of tobacco smoke. It seems to influence human cognition by stimulating the brain nicotinic acetylcholine receptors (nAChRs), which are activated by the neurotransmitter acetylcholine (ACh) and modulate the cholinergic signaling. The cholinergic system influences several cognitive functions, including learning, memory, attention, and executive function (Table 1) by projecting from the forebrain to areas that are critically involved in cognitive activities, such as the hippocampus, amygdala, ventral tegmental area (VTA), prefrontal cortex (PFC), thalamus-PFC connections, parietal and cingulated cortex (Briton, 2000; Collins, Luo, Selvaag, & Marks, 1994; Mansvelder, van Aerde, Couey, & Brussaard, 2006). Cholinergic signaling plays an important role in brain maturation and synaptic plasticity of the hippocampus and PFC (Poorthuis, Goriounova, Couey, & Mansvelder, 2009). Synaptic/nonsynaptic cholinergic transmission might modulate the neuronal networks involved in cognition and produce both inhibitory and excitatory effects, depending on which nAChRs activates or desensitizes (Mansvelder et al., 2006).

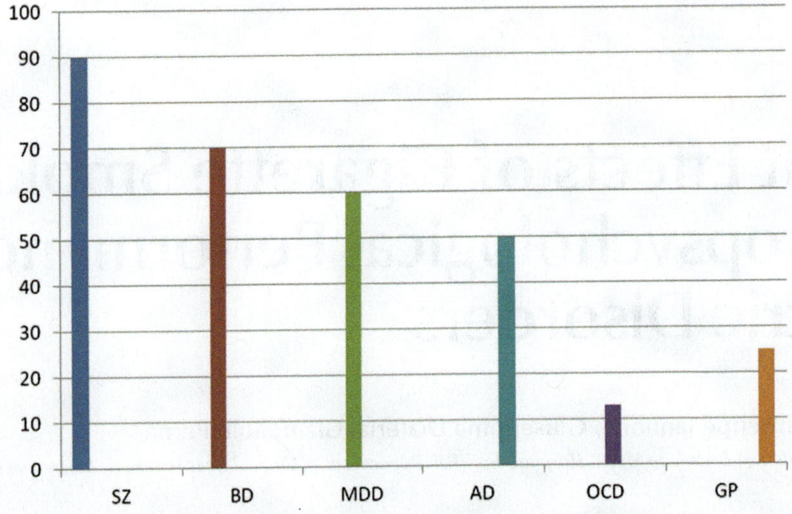

FIGURE 1 Prevalence of nicotine users among different mental illness and in the general population. The prevalence of smokers is not homogeneous among different psychiatric conditions. Moreover, while individuals with SZ, BD, AD, and MDD tend to smoke more than the general population, individuals suffering from OCD seem to smoke less than the general population. SZ, schizophrenia; BD, bipolar disorder; MDD, major depressive disorder; OCD, obsessive-compulsive disorder; AD, anxiety disorders; GP, general population.

TABLE 1 Description of Some Major Cognitive Functions	
Cognitive Function	**Description**
Perception	The cognitive process of organizing, identifying, and interpreting sensory stimuli (touch, smell, hearing, etc.) in order to represent the environment.
Attention	The cognitive process of selectively concentrating on one aspect of the external (e.g., smells, sounds, images) or internal (e.g., thoughts) environment while ignoring other things.
Memory	The cognitive process of encoding, storing, and retrieving information. Usually, it is divided according to how long information has to be remembered for. Sensory memory records information for a few milliseconds. Short-term memory (or working memory) retains information up to minutes and has a limited capacity of storing information. Long-term memory can store information for longer periods of time.
Executive function	A set of mental processes that helps to perform activities such as planning, organizing, strategizing, paying attention to and remembering details, and managing time and space.
Motor function	The ability to mobilize our muscles and bodies in order to perform a goal-oriented task.
Language	Skills allowing us to translate sounds into words and generate verbal output by using systems of complex communication.
Visual and spatial processing	The ability of processing incoming visual stimuli. Understanding spatial relationship between objects and visualizing images and scenarios.

Perception, attention, memory, executive function, motor function, language, and visuo-spatial processing are important cognitive functions that are influenced by nicotine.

Nicotine binds to the nAChRs in the brain and facilitates the release of several neurotransmitters, such as dopamine, serotonin, noradrenaline, GABA, and glutamate, which are all involved in modulating cognitive processes (Di Matteo, Pierucci, Di Giovanni, Benigno, & Esposito, 2007) (Table 2).

Experimental disruption of the cholinergic system resulted in memory and attention impairment both in humans and animals; conversely, pharmacological enhancement of cholinergic neurotransmission yielded the opposite effect (D'Souza & Markou, 2012). Finally, nAChR agonists, especially those acting at $\alpha 4\beta 2$ and $\alpha 7$ nAChRs, improved attentional and working memory deficits in preclinical animal models and in clinical studies (e.g., Howe et al., 2010).

Most studies assessed the impact of acute nicotine administration on the cognitive performance of healthy subjects, while its chronic effects have been investigated to a lesser extent. Acute administration of nicotine enhanced vigilance, selective attention, verbal and working memory, and executive function both in healthy nonsmokers and nondeprived smokers (Dumatar & Chauhan, 2011). A meta-analysis found that acute nicotine administration (via medications, patches, nasal spray, inhalers, or subcutaneous injections)

TABLE 2 Neurotransmitters and Cognitive Function

Neurotransmitter	Cognitive Function
Acetylcholine	Motor control, learning, memory, arousal
Serotonin	Emotion, impulses, thinking, learning, memory
Dopamine	Motivation, motor control, thinking, planning, problem solving, learning, memory, alertness
Noradrenaline	Arousal, vigilance, memory, learning, attention
GABA	Cognition, emotion, memory
Glutamate	Cognition, memory, learning

Cognitive functions are modulated by several neurotransmitters that are involved both in smoking addiction and psychiatric disorders.

or cigarette smoking ameliorated fine motor, alerting attention-accuracy and response time, orienting attention-reaction time, short-term episodic memory-accuracy, and working memory-reaction time performance in healthy nonsmokers or smokers who were not tobacco-deprived (Heishman, Kleykamp, & Singleton, 2010).

On the contrary, chronic cigarette smoking may negatively affect cognitive performance in healthy subjects, even though results are mixed. Adolescent smoking has been strongly associated with cognitive impairment in later life (Wiers et al., 2007). Functional MRI studies showed decreased PFC activation during working memory and attention tasks in adolescent smokers, less efficiency, and altered functional coordination than adolescent nonsmokers. Smoking duration was associated with diminished PFC activity, suggesting that nicotine exerts long-lasting effects on the PFC functioning (Musso et al., 2007). In a cross-sectional study of young adults, smokers showed significant cognitive impairment on sustained attention, spatial working memory, executive planning, and decision making, when compared to nonsmokers matched for age, education, income, and gender (Chamberlain, Odlaug, Schreiber, & Grant, 2012). Conversely, in another study, cognitive differences between young smokers and nonsmokers did not emerge (Paul et al., 2006).

Chronic smokers (50–70 years of age) exhibited impaired language, processing speed, executive functioning, verbal memory and learning, visual memory, and visual construction performance (Shih et al., 2006). A 5-year follow-up study on elderly healthy subjects found that current smokers performed worse in verbal memory, working memory, verbal fluency, inductive reasoning, and prospective memory tests when compared to the nonsmokers; the higher the number of packs of cigarettes smoked per year, the higher the differences between smokers and nonsmokers. Former smokers showed better cognitive performance than current smokers, even after controlling for number of packs-year (Mons, Schötter, Müller, Kliegel, & Brenner, 2013). These results suggested that smokers may be at increased risk for cognitive impairment in later life, with smoking duration and intensity increasing such risk but smoking cessation decreasing it. These results are in line with previous studies suggesting poorer cognitive performance in older healthy smokers than nonsmokers, even after controlling for comorbid medical conditions and other confounding factors (e.g., Paul et al., 2006). However other investigators failed to replicate these results (e.g., Chen et al., 2003). Finally, current or lifetime amount of cigarette smoked was not correlated to cognitive deficits: nondeprived smokers exhibited significant deficits in visual attention and cognitive impulsivity tasks than nonsmokers, whereas verbal episodic memory, fluency, and working memory did not differ between groups, even after controlling for age, gender, and education (Wagner et al., 2013).

The discrepancy of results among different studies may be ascribed to several methodological factors, including sample size differences, sensitivity of the cognitive test batteries, and, in some studies, lack of control for confounding factors, such as physical health, the cognitive effects of smoking deprivation and/or cognitive enhancement following withdrawal relief, and lifetime smoking duration. Finally, differences in the neuroanatomy or genetic polymorphisms among subjects might explain the variability observed in some cognitive outcomes. In fact, individual differences in neuronal location and expression of different nAChRs may exist. Finally, dopamine D2 receptor genetic variation may influence the effects of nicotine on working memory (Heishman et al., 2010). With this in mind, overall nicotine seems to exert acute procognitive effects in healthy subjects, whereas chronic smoking may be associated with no improvement or even cognitive deficit.

The determinants of cognitive impairment in chronic smokers remain unclear. Chronic smoking may result in long-term modulation of receptors density and/or neurotransmitters levels and may be associated with the release of toxic substances, such as cytokines, interleukins, and nitric oxide which negatively affect cognitive function (Weruaga, Balkan, Koylu, Pogun, & Alonso, 2002). Reduced volume of the prefrontal gray matter, anterior cingulate cortex, and cerebellum has been observed in smokers when compared to nonsmokers, suggesting that alterations in the cortical volume may underlie smoking-related cognitive changes (Brody et al., 2004). Finally, age may influence the effects of nicotine exposure on cognition. Adolescence is a period during which the brain is still developing; therefore, nicotine may exert particular long-lasting unfavorable actions. Accordingly, nicotine induced long-term structural and functional modifications in the PFC of adolescent animals that have been associated with impaired attention and cognitive control during adulthood (Counotte et al., 2009).

SMOKING AND COGNITIVE FUNCTION IN SCHIZOPHRENIA

Schizophrenia (SZ) is a chronic and severe psychiatric disorder with a lifetime prevalence of 0.3–0.7% in the general population.

It is characterized by positive symptoms (delusions, hallucinations, disorganized speech, disorganized/catatonic behavior) and negative symptoms (social withdrawal, avolition, diminished emotional expression) (American Psychiatric Association, 2013). Cognitive dysfunction across several domains might be a core feature of this disorder, which significantly impairs functional outcome in these subjects (Aas et al., 2014). Meta-analytic results suggested that subjects with SZ exhibited impaired memory, language, executive function (planning, flexibility, judgment), attention, learning, and reasoning performance when compared to healthy subjects (Fioravanti, Carlone, Vitale, Cinti, & Clare, 2005). Subjects with SZ also showed deficits in elementary sensory processes, such as auditory memory impairment and deficient P50 auditory sensory gating, which consists of the inability to filter out distracting stimuli and focus attention (Dulude, Labelle, & Knott, 2010). These sensory deficits may further impair cognitive function by altering subsequent information processing within higher cortical centers.

Poorer neuropsychological performance has been associated with more severe negative symptoms in patients with a first episode of psychosis (Zanelli et al., 2010). Some studies indicated that cognitive symptoms are often present before SZ onset and persist even after remission of clinical symptoms (Green, Kern, & Heaton, 2004). Deregulation of the cholinergic system may be involved in the pathophysiology of SZ and cognitive deficits. This might explain the exceptionally high cigarette consumption in this population. Postmortem and brain imaging studies showed elevated levels of acetylcholine in several brain areas that are thought to compensate for nAChRs dysfunction in schizophrenia. Indeed, decreased $\alpha 4\beta 2$ and $\alpha 7$ nAChRs expression was found in the hippocampus, thalamus, PFC, and striatum of subjects with SZ compared with controls. SZ-associated polymorphisms in the neuregulin-1 gene have been related to decreased $\alpha 7$ nAChR binding and decreased $\alpha 7$ nAChR mRNA levels in the PFC. Deficient P50 auditory sensory gating has been linked to a polymorphism in the $\alpha 7$ nAChR gene, at chromosome 15q14 (D'Souza & Markou, 2012).

Several studies investigated the impact of nicotine/cigarette smoking on cognitive functions in SZ. A cross-sectional study of outpatients and healthy controls showed that never-smokers with SZ performed worst on measures of sustained attention, processing speed, and response inhibition when compared to nondeprived current smokers or former smokers with SZ, whereas cigarette smoking did not alter cognition in healthy controls. These findings suggest that smoking status and history differentially affect neuropsychological outcomes in SZ compared to nonpsychiatric controls and that never-smokers exhibit even greater cognitive impairment (Wing, Bacher, Sacco, & George, 2011). Similarly, smokers with SZ showed better verbal memory, attention, and working memory performance than nonsmokers, also during first episodes of psychosis (Morisano, Wing, Sacco, Arenovich, & George, 2013).

Several crossover placebo-controlled studies reported improved cognitive performance in patients with SZ after cigarette smoking or nicotine administration (patch, nasal spray) or found better cognitive functioning in smoking subjects with SZ compared to those who stopped smoking. Nicotine administration enhanced attention, visuospatial working memory (VSWM), and verbal memory in chronic smoker patients with schizophrenia who had been abstinent overnight (Smith, Singh, Infante, Khandat, & Kloos, 2002; Smith et al., 2006). Sacco and coworkers showed that overnight nicotine abstinence selectively impaired VSWM in smokers with SZ, whereas this effect was not found in healthy control smokers. Smoking reinstatement enhanced VSWM and attention only in smokers with SZ; this effect was blocked by the cholinergic antagonist mecamylamine (Sacco et al., 2005). In line with this, George and coworkers reported that both smokers with SZ and healthy control smokers exhibited impaired VSWM performance compared to nonsmoker healthy controls. Smokers with SZ who quit smoking had higher VSWM deficits, whereas quitter controls showed improved VSWM performance (George et al., 2002). These findings suggest that smoking exerts some beneficial effects on VSWM performance in SZ, but not control, smokers. VSWM is known to be partly dependent on the prefrontal cortical dopamine (DA) system. The hypothesized cortical DA hypofunction in SZ may explain the VSWM deficits observed in this disorder. Because nicotine stimulates central DA release and its metabolism in both subcortical and cortical areas, the peculiar effects of nicotine on VSWM in these subjects may be related to a partial restoring of cortical DA levels (Figure 2). Conversely, cigarette smoking may augment normal cortical DA in control smokers and negatively affect VSWM performance. In nonsmokers with SZ, transdermal nicotine administration significantly improved episodic memory, attention, and inhibition of impulsive responses compared to nonsmoker healthy controls. Both the striatum and PFC are involved in response inhibition, which is modulated by several systems, including the dopaminergic one (Barr, Culhane, et al., 2008). Thus, nicotine may ameliorate deficits in response inhibition in subjects with SZ more than in controls by augmenting deficient nAChR function, thereby increasing DA activity in brain areas relevant to this process.

Some studies showed nicotine-induced improvement also in elementary sensory processes, which are modulated by the cholinergic system and nAChRs. Acute administration of nicotine ameliorated auditory sensory memory deficits in smokers with SZ compared to healthy control smokers, independently from smoking withdrawal and/or nicotine-induced withdrawal relief (Dulude et al., 2010). Both cigarette smoking and nicotine gums improved sensory gating in subjects with SZ (Ziedonis et al., 2008).

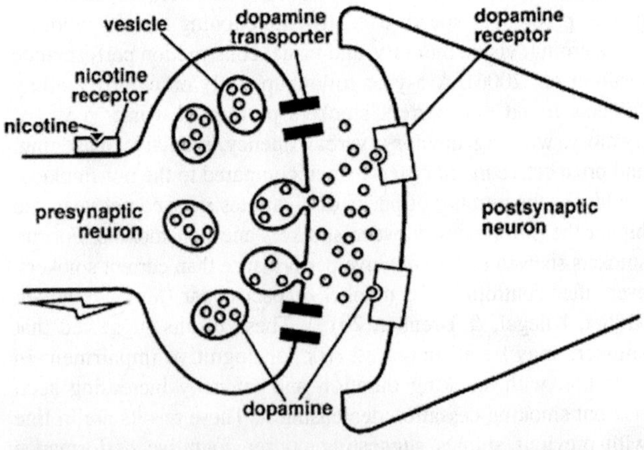

FIGURE 2 **Nicotine increases dopamine release in a synapsis.** Nicotine imitates the action of acetylcholine and binds to a particular type of acetylcholine receptor, known as the nicotinic receptor, where it excites the neuron causing more dopamine release. However, the acute effects of nicotine subsides within minutes; therefore, frequent smoking is required to sustain the pleasurable effects of nicotine and to prevent withdrawal symptoms.

On the contrary, fewer studies failed to find positive effects of nicotine on sustained visual attention performance in smokers with SZ compared with healthy smoking controls and on learning and memory, language, and visuospatial/constructional abilities in both smoker and nonsmoker patients with SZ (Hong et al., 2011). In a longitudinal study, smoking patients with a first-episode SZ and under antipsychotic treatment showed better attention and working memory performance than the nonsmoking patients at baseline. Conversely, after 12-month treatment, performance was similar in both groups: the nonsmokers showed significant cognitive improvement, whereas the smokers lost their superior baseline performance. These results suggest that smoking might improve attention and working memory to the same extent as (atypical) antipsychotics, and that smoking might represent an attempt to ameliorate cognitive dysfunction (Segarra et al., 2011). Finally, a study found worse problem-solving performance in smoking treatment-resistant patients with SZ (TRS) than in nonsmoking TRS patients.

Overall, the available data suggest that cigarette smoking has procognitive effects in SZ, even though different sample sizes, nicotine exposure, design of the studies, cognitive assessment procedures, nicotine withdrawal, and type of nicotine administration may influence nAChRs speed desensitization and yield heterogeneous results. In addition, it is known that antipsychotic medications can also influence cognitive functions (Ziedonis et al., 2008). Moreover, given that SZ is a heterogeneous disorder (Harris et al., 2004), smoker and nonsmoker patients may have distinct neurobiological profiles that might influence both cognitive response to acute and chronic nicotine use. Finally, nicotine may influence different cognitive domains across different subgroups of patients with schizophrenia.

Even though more research on this topic is needed, the self-medication hypothesis is supported by the assumption that nicotine exerts procognitive effects in SZ, which might explain the exceptionally higher rates of smokers, cigarette consumption, and vulnerability to nicotine dependence or to experience more severe withdrawal symptoms in subjects with SZ.

Nicotine may exert its procognitive effects by boosting deficient cholinergic activity, thereby increasing DA release in the hypofunctioning mesocortical projections and restoring glutamatergic dysfunctions in several brain areas implicated in cognitive functioning (Barr, Procyshyn, et al., 2008; Picciotto, Caldarone, King, & Zachariou, 2000; Smith et al., 2002). Other explanations may also be involved in the association between SZ and smoking, including increased vulnerability to addictive behavior (Chambers, Krystal, & Self, 2001), psychosocial factors, nicotine-associated disease, or medication side effects mitigation (Ziedonis et al., 2008). However, the procognitive effects of nicotine in SZ may help to explore new therapeutic approaches. Indeed, in preclinical animal models, administration of nAChR subtype-selective agonists resulted in beneficial cognitive effects on attention and working memory performance. The $\alpha 4\beta 2$ and $\alpha 7$ nAChR are currently being investigated as possible therapy targets for SZ-associated cognitive deficits (D'Souza & Markou, 2012). Cognitive remediation interventions significantly and permanently improved cognitive functioning in patients with schizophrenia (Bowie, McGurk, Mausbach, Patterson, & Harvey, 2012). Both pharmacological and/or nonpharmacological cognitive enhancement may help to decrease smoking morbidity by alleviating the cognitive deficits associated with cigarette consumption and dependence in this disorder and might eventually help patients to quit smoking.

SMOKING AND COGNITIVE FUNCTION IN MAJOR DEPRESSIVE DISORDER

Major depressive disorder (MDD) is a psychiatric disorder characterized by periods of depressed mood and/or loss of interest or pleasure in almost all the activities, and it is associated with several symptoms, including weight loss/gain, insomnia/hypersomnia, psychomotor agitation or retardation, fatigue, diminished ability to think or concentrate, recurrent thoughts of death, and suicidal behaviors. MDD has a 12-month prevalence of about 7% in the United States and depressive episodes may be recurrent (American Psychiatric Association, 2013).

Cognitive impairment across several cognitive domains, including executive function, verbal/visual/episodic memory, attention, verbal fluency, learning, and information processing speed is prominent in MDD and contributes to poorer social, occupational, and educational functioning (McDermott & Ebmeier, 2009). Cognitive deficits differ across subjects with MDD. This discrepancy may be explained by both methodological factors, such as low statistical power or use of different cognitive test batteries, different samples, symptom severity, number of previous depressive episodes, use of psychotropic medication, age, and education. A meta-analysis reported impaired executive function in subjects with MDD when compared to healthy controls, including inhibition processes, set-shifting, categorization and cognitive flexibility, verbal working memory, and verbal fluency tasks deficits, even after controlling for age and education. Depression severity has been associated with greater impairment in many executive domains, but other cognitive areas seem not be influenced by symptom severity and the deficits persisted even in the remission phase. Meta-analytic results showed a correlation between depression severity and poorer memory, some executive domains, and processing speed performance, but they did not find associations with semantic or visuospatial memory, attention, and several executive domains such as switching, verbal fluency, and cognitive flexibility. Even though cognitive deficits may be proportional to number of depressive episodes, they are common also among patients with a first episode of depression (Lee, Hermens, Porter, & Redoblado-Hodge, 2012).

Finally, longitudinal reports showed that impairment may persist even after remission, especially in the attentive and executive functions. Hence, these cognitive deficits might be considered possible trait markers of MDD. Depressive symptoms depend on complex changes in a variety of neurobiological systems in multiple brain areas. Cholinergic dysfunctions may account for the development of cognitive symptoms in MDD, while deficits in the serotonergic, dopaminergic, and noradrenergic systems might be involved in depression-associated anxiety, loss of pleasure, and energy. Structural and functional abnormalities in several brain areas that promote cognitive functions, including the hippocampus, amygdala, prefrontal cortex, and basal ganglia, have been found in subjects with MDD. These abnormalities have been partly attributed to alterations in the central cholinergic system, which may result in impaired neurogenesis and/or in defective functions of several neurotransmitter systems. This, in turn, would lead to the cognitive deficits observed in MDD (Dagyte, Den Boer, & Trentani, 2011; Mineur & Picciotto, 2010). Accordingly, mice lacking the β4 subunit of the nAChRs exhibited deficits in both hippocampus- and amygdala-dependent memory functions and depression-like behaviors (Semenova, Contet, Roberts, & Markou, 2012). The connection between the cholinergic system and cognitive deficits suggests that smoking may exert procognitive effects in MDD, similarly to what was previously found in SZ. Smoking as "cognitive self-medication" may be why about 30% of individuals with current depression and 60% of patients with a lifetime history of depression are current daily smokers (Dome et al., 2010), as well as the higher rates of lifetime depression in smoking compared to nonsmoking populations (Breslau et al., 1992). So far, this topic has been investigated to a very limited extent, and unfortunately only preliminary data are available.

Morisano et al. (2013) compared 6 smokers and 10 nonsmokers with MDD on a set of cognitive performance, including sustained attention, response inhibition, psychomotor speed, verbal learning/memory, and some domains of the executive function (working memory, cognitive set-shifting/flexibility). No differences between smokers and nonsmokers emerged, but the small sample size may have masked potential differences between the two groups (and led to a type II error) (Morisano et al., 2013). Different results have been found by our research group. We investigated the effects of cigarette smoking on the cognitive performance of 100 depressed patients with MDD ($n=61$) or bipolar disorder (BD) ($n=39$), who were hospitalized for a 4-week psychiatric rehabilitation program (Caldirola et al., 2013). Forty-five patients were active regular smokers and 55 were nonsmokers (never smokers). At the beginning and at the end of the hospitalization, patients were administered a comprehensive cognitive test battery evaluating short- and long-term verbal memory, attention, visual-constructive ability, long-term visual-constructive memory, language fluency, working memory, and the ability to understand and process semantic information. Smoking status was assessed by personal interviews. Investigators were blind to the results of the cognitive tests and to the smoking status of the patients. No restrictions on smoking were imposed during hospitalization. At the beginning of the hospitalization, smokers performed better on verbal memory, language fluency, and working memory tasks than nonsmokers (Table 3). At the end of the hospitalization, all the patients significantly improved in most cognitive domains, although smokers performed better on verbal memory, language fluency, and working memory tasks than nonsmokers. Conversely, we did not find differences in depressive symptoms severity. Our study presents several limitations. First, the relatively small sample size did not allow us to compare smokers and nonsmokers with MDD or BD separately. Second, we did not include subjects in different phases of the disorders or with different smoking patterns. However, our preliminary findings suggest that cognitive enhancement, mainly in the verbal and working memory domains, may be associated with nicotine use in depressed patients with MDD or BD and that smoking may represent a form of "cognitive self-medication" in these individuals.

Several other mechanisms have been proposed to explain the association between smoking and depression, such as the effects of nicotine in elevating mood (Cosci, Griez, Nardi, & Schruers, 2014), the potential effect of nicotine in increasing risk of developing depression in vulnerable individuals by causing brain neurobiological changes (Luger, Suls, & Vander Weg, 2014), and common genetic/environmental risk factors (Tsuang, Francis, Minor, Thomas, & Stone, 2012). So far, these contrasting explanations are still being debated. However, the involvement of the cholinergic system in depression, its potential role in depression-associated cognitive impairment, and the preliminary findings of procognitive effects of nicotine in MDD need further exploration. Subjects with MDD reported less severe cognitive impairment than those with SZ (Morisano et al., 2013), and this might partly explain the higher rate of smokers in SZ than in MDD. Nonetheless, the prevalence of smokers among subjects with MDD is considerably higher than in the general population and cognitive deficits in MDD influence their daily-life disability. Should the "cognitive self-medication" hypothesis in MDD be confirmed, cognitive remediation trainings (Lee et al., 2012)

TABLE 3 Results from the Study by Caldirola et al. (2013) on the Effects of Smoking on Neuropsychological Performance of Patients with MDD and BD

Performance	Smokers ($n=45$)	Nonsmokers ($n=55$)	ANOVA	p
Long and short-term verbal memory	11.70 (4.88)	9.40 (3.31)	$F=6.99$	<0.01*
Attention	38.67 (9.78)	37.22 (9.21)	$F=0.30$	0.58
Visuo-constructive ability	26.59 (8.79)	25.19 (9.42)	$F=0.18$	0.67
Long-term visuo-constructive memory	10.51 (6.80)	10.11 (5.01)	$F=0.02$	0.88
Language fluency	31.07 (10.44)	25.69 (8.22)	$F=7.91$	<0.01*
Language fluency	35.71 (8.64)	30.66 (7.11)	$F=10.78$	<0.01*
Semantic information processing	29.41 (3.36)	28.26 (4.08)	$F=2.05$	0.15

Data are presented as means (standard deviations).
*Statistical significance $p<0.01$.
ANOVA, analysis of variance; MDD, major depressive disorder; BD, bipolar disorder.
The higher the score, the better the performance.
At the beginning of the hospitalization, smokers with MDD or BD showed significantly better performance in verbal memory, language fluency, and working memory (all $p<0.01$) than nonsmokers with MDD or BD. No interaction between smoking and diagnosis was found.

and/or pharmacological modulation of the cholinergic system may help subjects with MDD to improve cognitive performance, which might perpetuate a smoking habit and, ultimately, to quit smoking.

SMOKING AND COGNITIVE FUNCTION IN BIPOLAR DISORDER

Bipolar disorder (BD) is a psychiatric disorder characterized by distinct periods of abnormal elevated, expansive, and irritable mood and increased energy and goal-directed activity, lasting at least 1 week (manic episode, type BD I) or at least four consecutive days (hypomanic episode, type BD II). Other symptoms include a decreased need for sleep, higher distractibility, and flight of ideas. The manic/hypomanic episodes alternate to depressive episodes. The 12-month BD prevalence is about 1% in the United States (American Psychiatric Association, 2013). Cognitive impairment in BD has been under investigation during the last decade. Recent meta-analyses reported significant cognitive impairment in euthymic patients when compared to healthy controls across several domains, including working memory, executive control, set-shifting, language fluency, immediate/delayed verbal recall, verbal learning, and sustained attention, which may be implicated in the poor occupational, social, and educational outcomes often found in this population (Kurtz & Gerraty, 2009). Greater cognitive dysfunction has been observed during manic/depressive phases, as well as in elderly patients with a chronic course of the pathology and in patients with a worse course of illness, higher number of episodes, hospitalizations, and duration of illness (Robinson et al., 2006). Preliminary findings suggested that subjects with BD type I, BD type II, or MDD exhibited similar cognitive impairment during the depressive episodes, but subjects with BD type I exhibited worse performance in specific cognitive domains when compared to subjects with BD II and MDD suggesting that individuals with BD I may be at higher risk for cognitive abnormalities (Xu et al., 2012). Cognitive impairment may precede BD onset and some cognitive deficits persist in the euthymic phase, especially in the executive function and verbal memory domains (Robinson et al., 2006). First-degree relatives of patients with BD with no mental illness also performed worse in these cognitive domains than controls. It has been hypothesized that deficits in the executive function and verbal memory may be trait vulnerability factors and candidate endophenotypes for BD (Arts, Jabben, Krabbendam, & van Os, 2008).

Structural and functional abnormalities in the frontal cortex, temporal lobes, basal ganglia, amygdala, and hippocampus have been associated with BD. Aberrant cholinergic modulation in these areas seems to influence both manic and depressive symptoms, as well as the circadian rhythms (Bymaster & Felder, 2002). Altered regulation of $\alpha 7$ nAChR expression in the PFC and hippocampus of subjects with BD, as well as polymorphisms of the neuregulin-1 gene, are involved in BD and in abnormal functioning of the nAChRs in the PFC (Arts et al., 2008). Because the PFC and the other brain areas implicated in the pathophysiology of BD are also involved in modulating cognition, cholinergic dysfunctions may be partly responsible for the neuropsychological deficits observed in BD. So far, the potential impact of smoking on cognitive function in BD has received very little attention, despite that the mechanism may justify the higher rates of smokers in BD (30–60%, which is two to three times higher than in the general population).

Law and coworkers compared euthymic BD (type I/II) current smokers ($n=16$) with nonsmokers ($n=27$), both under pharmacological treatment, on psychomotor speed, attention, memory, learning, and executive function. Although no difference in cognitive performance between smokers and nonsmokers was found, smokers reported higher subjective perception of cognitive failure than nonsmokers, suggesting that expectancies about potential procognitive effects of nicotine might motivate subjects with BD to smoke, independently from the real beneficial effects of nicotine consumption (Law et al., 2009). Similarly, 10 smokers and 6 nonsmokers with BD did not differ on sustained attention, response inhibition, psychomotor speed, verbal learning/memory, working memory, and cognitive set-shifting/flexibility tasks (Morisano et al., 2013). Conversely, a study by our group (described in detail in the previous paragraph) provided preliminary evidence of better cognitive performance in smokers with BD (type I/II) in the depressive phase than nonsmokers, whereas smokers and nonsmokers with BD did not differ on depressive symptom severity. These results suggest that patients with BD might smoke to improve cognitive performance and not to ameliorate depressive symptoms (Caldirola et al., 2013). In fact, smoking has been associated with a more severe clinical course of BD (Ostacher et al., 2006). These findings seem to contrast with the fact that the prevalence of smokers is higher in BD than in the general population. Our findings may explain this discrepancy and suggest that procognitive effects of smoking may encourage subjects with BD to smoke despite that smoking has unfavorable effects on the clinical symptoms. However, future studies are needed to validate this hypothesis.

Some reasons may explain the differences between our results and those obtained from the two abovementioned studies. In the study by Morisano and coworkers, the sample size was very small; thus, the presence of type II error cannot be excluded. The study of Law and coworkers arose from a post-hoc analysis of data collected during a separate study with different aims and included patients during an euthymic phase. Moreover, smoking status was self-reported, other cognitive tasks were performed, and subjects were asked not to smoke before testing. Thus, the results are hardly comparable. Several other factors might explain the association between smoking and BD and are currently under investigation. Subjects with BD might smoke to counteract some of the side effects associated with medication use, especially antipsychotics and antiepileptics, which may prompt individuals with BD to smoke by altering the nicotine metabolism (Heffner, Strawn, DelBello, Strakowski, & Anthenelli, 2011). Smoking may increase the risk of developing BD over time by altering nAChRs functioning (Mineur & Picciotto, 2009). Smoking and BD may share common environmental/genetic risk factors, such as genes encoding COMT, a dopamine/serotonin transporter (McEachin et al., 2010). Finally, lower GABA levels in BD have been associated with excessive engagement in impulsive behavior, including higher cigarette consumption (Benes, 2012).

Much more work is clearly required to clarify the mechanisms underlying the higher prevalence of smokers in BD. Studies of a larger sample, possibly longitudinal and including patients in different phases of the disorder, are needed. Research about the effects of nicotine on the cognitive functions in patients with BD is at its infancy, but it is an area worthy of being extended because it may improve therapeutic strategies for both smoking cessation and amelioration of daily-life functioning in this population.

SMOKING AND COGNITIVE FUNCTION IN OBSESSIVE-COMPULSIVE DISORDER

Obsessive-compulsive disorder (OCD) is characterized by the presence of recurrent and persistent thoughts, urges, and images experienced as intrusive and that cause distress (obsessions), and/or repetitive and time-consuming behaviors performed to alleviate the distress associated with the obsessions (compulsions). The 12-months prevalence of OCD is 1.2–1.8% (American Psychiatric Association, 2013). The prevalence of smokers in subjects with OCD is about 13%, which is significantly lower when compared to both healthy individuals and subjects with MDD, SZ, or BD. In OCD, the reasons for this intriguing exception are unclear. In a rat model of OCD, nicotine administration attenuated compulsive checking (Tizabi et al., 2002). Accordingly, preliminary findings in humans showed that nicotine patches significantly reduced compulsive behaviors in nonsmoker drug-free OCD patients compared with placebo patches (Salin-Pascual & Basanez-Villa, 2003).

Despite these potential benefits, the lower prevalence of smokers suggests that other mechanisms may discourage these individuals from smoking. Cognitive impairment in the executive function, memory, visuospatial abilities, attention, and processing speed has been found in OCD subjects (Abramovitch, Abramowitz, & Mittelman, 2013); especially in executive function, it has been associated with hyperactivity in several brain areas, such as the frontal-striatal circuit and the parietal cortex, which, in turn, may be related to imbalanced monoaminergic neurotransmission (Millet et al., 2013). Because nicotine enhances cortical activity, brain hyperactivity might worsen cognitive function and discourage subjects with OCD from smoking.

To the best of our knowledge, only our group investigated this assumption in OCD (unpublished study). We compared the executive function of 20 current smokers and 20 nonsmokers (i.e., never smokers) with OCD. Participants were consecutively recruited from individuals who were hospitalized for a 4-week psychiatric rehabilitation program. All the subjects were taking medications recommended for OCD. Smokers and nonsmokers did not differ on medication distribution, sociodemographic, and clinical variables. Executive function was assessed by a comprehensive computerized battery (CANTAB) and smoking was assessed by a standardized interview. We applied the same procedure to a sample of 10 smoker and 10 nonsmoker healthy controls, who did not differ on age, gender, and education from the OCD patients. No smoking restriction was imposed on the participants. The inpatients with OCD exhibited impaired executive function across most subdomains compared with controls. OCD smokers performed worst in the reaction time during the decision-making task and in the number of directional errors committed in the motor inhibition task than the nonsmoker OCD patients. Duration of smoking was associated with poorer performance in tasks evaluating planning time, visuospatial short-term memory, visuospatial, and verbal working memory. Conversely, smoking and nonsmoking controls did not differ on executive function performance. In the control smokers, duration of smoking was associated with poorer verbal working memory and longer decision-making time, but with better processing speed, visuo-motor coordination, and set-shifting performance. The main limitation of this study is the risk of type II errors given the small sample sizes; thus, our results should be considered preliminary and confirmed by larger studies.

Our results are in contrast to those showing a procognitive effect of nicotine in SZ, MDD, or BD, and they suggest that nicotine may exert specific cognitive effects in OCD. Overall, our findings suggest that smoking may negatively influence some aspects of executive function in OCD, even to a greater extent than in healthy subjects. Such cognitive effects might be explained by the peculiar neurobiological features characterizing patients with OCD. Hyperactivation of the frontal-striatal circuit and the parietal regions has been implicated both in pathological behavior and cognitive impairment in subjects with OCD and it has been related to dopaminergic hyperactivity in these brain areas, which might arise from central serotonergic-dopaminergic imbalance (Logue & Gould, 2013). Nicotine has stimulating effects on the brain function by acting directly on the cholinergic signaling and mediates the release of several neurotransmitters, including dopamine, serotonin, norepinephrine, and glutamate (Jasinska, Zorick, Brody, & Stein, 2014). Smoking would increase activity in the brain areas that are usually hyperactive, which may result in lack of benefits or even in executive function deterioration in smokers with OCD. On the contrary, in subjects with SZ, nicotine would enhance the reduced frontal lobes activity and result in cognitive functioning improvement. In conclusion, the specific negative cognitive effects observed in subjects with OCD, but also fear of contamination, fear of "uncleanliness" related to smoking, (Tizabi et al., 2002), or genetic factors influencing liability to both OCD and lack of positive reinforcement from nicotine (Bejerot & Humble, 1999), might explain the lower prevalence of smokers in this population. Our findings suggest caution in the use of nicotine/nicotinic agonists as therapeutic agents in this population. However, future research is needed before drawing more reliable conclusions about the effects of smoking in OCD.

CONCLUSIONS

Smoking contributes to the high rate of medical morbidity and mortality in the psychiatric population, therefore promoting smoking prevention and cessation as a crucial requirement for mental health professionals. Understanding some of the possible reasons that might justify comorbidity between smoking and psychiatric disorders is decisive. In this chapter, we suggested that smoking may exert beneficial effects on the cognitive function of subjects with SZ, MDD, or BD, whereas it may have detrimental cognitive effects in healthy subjects and, to a greater extent, in subjects with OCD. The distinct neurobiological patterns underlying the cognitive function of these populations may explain the different impact of smoking on cognition. Although multifactorial aspects are involved in smoking initiation and nicotine dependence, the beneficial cognitive effects of nicotine may prompt smoking initiation and maintenance of tobacco dependence in some psychiatric populations. Pharmacological interventions acting on the cholinergic system, in combination with cognitive remediation trainings, may help to improve cognitive deficits and promote smoking cessation in individuals with mental disorders.

APPLICATIONS TO OTHER ADDICTIONS AND SUBSTANCE MISUSE

- Like nicotine, cannabis can induce memory performance alterations by decreasing glutamate and acetylcholine release within the hippocampus, and by enhancing dopaminergic neurons activity in the prefrontal cortex of both animals and humans.
- Alcohol and tobacco abuse often co-occur, suggesting that these two substances may share some common neurological mechanisms. For instance, both nicotine and ethanol act on the nAChRs, which mediate the effects of nicotine as well as some of the effects of alcohol and ultimately cause dopamine release, which might underlie both addictions.
- Like nicotine, cocaine can increase extracellular dopamine (DA) levels in the nucleus accumbens (NAc). In rats and human subjects, pre-exposure to nicotine increased self-administration or craving of cocaine. Indeed, activation of nAChRs seems to potentiate cocaine reinforcement, whereas inactivation of nAChRs decreases cocaine reinforcement. Therefore, nicotinic antagonist treatments might be of use in decreasing the rewarding efficacy of cocaine.
- Acetylcholine activates the alpha7 nicotinic receptors in the prefrontal cortex, which enhances complex mental performance. Like nicotine, caffeine increases acetylcholine release and therefore may foster important neuropsychological functions such as attention. Finally, like nicotine (and like most drugs), caffeine increases dopamine release in the brain's pleasure centers.
- Mounting evidence suggests the involvement of nAChRs in influencing cognitive dysfunction across several diseases, such as schizophrenia, Alzheimer disease, and Parkinson disease. Exploring new medications that stimulate the nicotinic receptors may be promising for treating neuropsychological impairment.

DEFINITION OF TERMS

Schizophrenia A psychiatric disorder characterized by positive (delusions, hallucinations, disorganized speech, disorganized/catatonic behavior) and negative (social withdrawal, avolition, diminished emotional expression) symptoms.

Major depressive disorder A psychiatric disorder characterized by periods of depressed mood and/or loss of interest or pleasure in almost all the activities, associated with several symptoms, including weight loss/gain, insomnia/hypersomnia, psychomotor agitation or retardation, fatigue, diminished ability to think or concentrate, recurrent thoughts of death and suicidal behaviors.

Bipolar disorder A psychiatric disorder characterized by alternate periods of abnormally and persistently elevated expansive/irritable mood and increased goal-directed activity/energy, lasting at least 1 week (manic episode, type BD (I) or at least four consecutive days (hypomanic episode, type BD (II), associated with other symptoms such as decreased need for sleep, distractibility, flight of ideas.

Obsessive-compulsive disorder A psychiatric disorder characterized by the presence of recurrent and persistent thoughts, urges, and/or images experienced as intrusive and that cause distress (obsessions), and/or repetitive and time-consuming behaviors performed to alleviate the distress associated with the obsessions (compulsions).

Acetylcholine The most common neurotransmitter, which acts in the central nervous system and autonomic nervous system, where it can exert both inhibitory and excitatory effects.

Cognition A set of mental activities that helps associated with knowledge, which includes a set of abilities such as memory, attention, reasoning, decision making, language, computation, judgment, evaluation, and problem solving.

Forebrain Also called prosencephalon, the forebrain is the largest and most frontal area of the brain. It includes the cerebrum, thalamus, hypothalamus, and the limbic system and is involved in sensory integration, control of voluntary movement, and higher intellectual functions, such as speech and abstract thought.

Prefrontal cortex A brain region located in the rostral part of the brain implicated in the execution of complex cognitive functions, abstract thinking, personality expression, decision making, and moderating social behavior.

Hippocampus A limbic structure (one per hemisphere) located in the temporal lobes, which is involved in memory forming, organizing, and storing, emotional responses, navigation, and spatial orientation.

Amygdala An almond-shaped limbic structure (one per hemisphere) located in the temporal lobes, which is involved in memory processing, decision making, and emotional reactions (e.g., fear, aggression, and pleasure).

Receptors Molecules usually located inside or on the surface of a cell that selectively receive chemical signals and bind a specific substance from outside the cell (e.g., neurotransmitters, hormones, etc.).

Neurotransmitters The brain chemical messengers that transmit information across synapses from one neuron to another. Serotonin, dopamine, acetylcholine, GABA, glutamate, epinephrine, and norepinephrine are among the most important neurotransmitters.

P50 This electrical wave occurs in the brain approximately 50 ms after the presentation of an auditory stimulus. It is used to measure sensory gating (i.e., the process of filtering out unnecessary stimuli). When two auditory stimuli are presented 8–10 s apart, the brain produces a smaller P50 wave to the second sound. This diminution in the P50 wave to the second click is known as P50 sensory gating.

KEY FACTS

Key Facts of Smoking

- Smoking is very addictive and causes cancer, cardiovascular problems, lung diseases, diabetes, and other serious diseases.
- Almost 90% of lung cancer deaths are caused by cigarette smoking.
- Cigarette smoke contains 69 chemicals that cause cancer.
- Despite significant declines during the last decades, about 650 million people are current smokers.
- Smoking kills more than 5 million people each year.

Key Facts of Schizophrenia

- A combination of genetic and environmental factors are likely to cause schizophrenia.
- Schizophrenia onset is usually between 15 and 35 years of age.
- The chances of developing schizophrenia are 1 in 100 if none of the parents suffers from schizophrenia, 1 in 10 if one parent

suffers from schizophrenia, 1 in 8 if a heterozygote twin suffers from schizophrenia, and 1 in 2 if a homozygote twin suffers from schizophrenia.
- Treatment for schizophrenia includes pharmacotherapy (e.g., antipsychotic medications) and psychological therapy (e.g., cognitive behavioral therapy and family therapy).
- Having smoked cannabis more than 50 times during early teens has been related to a 6 times higher chance of developing schizophrenia over time.
- Approximately 20% of people suffering from schizophrenia get better within 5 years of the first symptoms, 60% will get better but relapse will occur, and 20% will experience serious symptoms for longer periods of time.

SUMMARY POINTS

- This chapter explores both short- and long-term effects of nicotine on the neuropsychological functions of individuals suffering from schizophrenia, major depressive disorder, bipolar disorder, obsessive-compulsive disorder, and healthy subjects.
- People with mental disorders smoke significantly more than the general population, even though the proportion of smokers is heterogeneous among different psychiatric disorders.
- People suffering from schizophrenia, bipolar disorder, and major depressive disorder smoke more than the general population, whereas people suffering from obsessive-compulsive disorder smoke less than the general population.
- Such variation might be explained by the peculiar effects smoking would exert on the neuropsychological performance across the different psychiatric populations.
- Cognitive impairment is a common feature of mental disorders.
- Nicotine influences cognitive performance, such as learning, memory, attention, and executive function.
- In healthy subjects, acute administration of nicotine ameliorates cognitive functioning, whereas chronic administration negatively affects cognitive performance.
- In patients with schizophrenia, major depressive disorder and bipolar disorder nicotine exerts procognitive effects, probably fostering cigarette consumption in these populations.
- In patients with obsessive-compulsive disorder, nicotine seems to negatively influence some cognitive functions, and perhaps this would discourage these individuals from smoking.
- Pharmacological interventions acting on the cholinergic system and cognitive remediation training may contribute both to improve cognitive deficits and to foster smoking cessation in these populations.

REFERENCES

Aas, M., Dazzan, P., Mondelli, V., Melle, I., Murray, R. M., & Pariante, C. M. (2014). A systematic review of cognitive function in first-episode psychosis, including a discussion on childhood trauma, stress, and inflammation. *Frontiers in Psychiatry*, *4*, 1–13.

Abramovitch, A., Abramowitz, J. S., & Mittelman, A. (2013). The neuropsychology of adult obsessive-compulsive disorder: a meta-analysis. *Clinical Psychology Review*, *33*, 1163–1171.

Abramovitch, A., Pizzagalli, D. A., Geller, D. A., Reuman, L., & Wilhelm, S. (2014). Cigarette smoking in obsessive-compulsive disorder and unaffected parents of OCD patients. *European Psychiatry*, *30*(1). http://dx.doi.org/10.1016/j.eurpsy.2013.12.003.

American Psychiatric Association. (2013). *Diagnostic and statistical manual of mental disorders* (5th ed.). Arlington, VA: American Psychiatric Publishing.

Arts, B., Jabben, N., Krabbendam, L., & van Os, J. (2008). Meta-analyses of cognitive functioning in euthymic bipolar patients and their first-degree relatives. *Psychological Medicine*, *38*, 771–785.

Barr, A. M., Procyshyn, R. M., Hui, P., Johnson, J. L., & Honer, W. G. (2008). Self-reported motivation to smoke in schizophrenia is related to antipsychotic drug treatment. *Schizophrenia Research*, *100*, 252–260.

Barr, R. S., Culhane, M. A., Jubelt, L. E., Mufti, R. S., Dyer, M. A., Weiss, A. P., … Evins, A. E. (2008). The effects of transdermal nicotine on cognition in nonsmokers with schizophrenia and nonpsychiatric controls. *Neuropsychopharmacology*, *33*, 480–490.

Bejerot, S., & Humble, M. (1999). Low prevalence of smoking among patients with obsessive-compulsive disorder. *Comprehensive Psychiatry*, *40*, 268–272.

Benes, F. M. (2012). Nicotinic receptors and functional regulation of GABA cell microcircuitry in bipolar disorder and schizophrenia. *Handbook of Experimental Pharmacology*, 401–417.

Bowie, C. R., McGurk, S. R., Mausbach, B., Patterson, T. L., & Harvey, P. D. (2012). Combined cognitive remediation and functional skills training for schizophrenia: effects on cognition, functional competence, and real-world behavior. *The American Journal of Psychiatry*, *169*, 710–718.

Breslau, N., Kilbey, M. M., & Andreski, P. (1992). Nicotine withdrawal symptoms and psychiatric disorders: findings from an epidemiologic study of young adults. *The American Journal of Psychiatry*, *149*, 464–469.

Briton, J. (2000). *Nicotine addiction in Britain*. London: Pitman Medical Publishing Co.

Brody, A. L., Mandelkern, M. A., Jarvik, M. E., Lee, G. S., Smith, E. C., Huang, J. C., … London, E. D. (2004). Differences between smokers and nonsmokers in regional gray matter volumes and densities. *Biological Psychiatry*, *55*, 77–84.

Bymaster, F. P., & Felder, C. C. (2002). Role of the cholinergic muscarinic system in bipolar disorder and related mechanism of action of antipsychotic agents. *Molecular Psychiatry*, *7*.

Caldirola, D., Dacco, S., Grassi, M., Citterio, A., Menotti, R., Cavedini, P., … Perna, G. (2013). Effects of cigarette smoking on neuropsychological performance in mood disorders: a comparison between smoking and nonsmoking inpatients. *The Journal of Clinical Psychiatry*, *74*, 130–136.

Chamberlain, S. R., Odlaug, B. L., Schreiber, L. R., & Grant, J. E. (2012). Association between tobacco smoking and cognitive functioning in young adults. *American Journal on Addictions*, *21*, 114–119.

Chambers, R. A., Krystal, J. H., & Self, D. W. (2001). A neurobiological basis for substance abuse comorbidity in schizophrenia. *Biological Psychiatry*, *50*, 71–83.

Chen, W. T., Wang, P. N., Wang, S. J., Fuh, J. L., Lin, K. N., & Liu, H. C. (2003). Smoking and cognitive performance in the community elderly: a longitudinal study. *Journal of Geriatric Psychiatry and Neurology*, *16*, 18–22.

Collins, A. C., Luo, Y., Selvaag, S., & Marks, M. J. (1994). Sensitivity to nicotine and brain nicotinic receptors are altered by chronic nicotine and mecamylamine infusion. *The Journal of Pharmacology and Experimental Therapeutics*, *271*, 125–133.

Cosci, F., Griez, E. J., Nardi, A. E., & Schruers, K. R. (2014). Nicotine effects on human affective functions. A systematic review of the literature on a controversial issue. *CNS Neurol Disord Drug Targets, 13*(6), 981–991 submitted for publication.

Counotte, D. S., Spijker, S., Van de Burgwal, L. H., Hogenboom, F., Schoffelmeer, A. N., De Vries, T. J., ... Pattij, T. (2009). Long-lasting cognitive deficits resulting from adolescent nicotine exposure in rats. *Neuropsychopharmacology, 34,* 299–306.

Dagyte, G., Den Boer, J. A., & Trentani, A. (2011). The cholinergic system and depression. *Behavioural Brain Research, 221,* 574–582.

Di Matteo, V., Pierucci, M., Di Giovanni, G., Benigno, A., & Esposito, E. (2007). The neurobiological bases for the pharmacotherapy of nicotine addiction. *Current Pharmaceutical Design, 13,* 1269–1284.

Dome, P., Lazary, J., Kalapos, M. P., & Rihmer, Z. (2010). Smoking, nicotine and neuropsychiatric disorders. *Neuroscience Biobehavioral Reviews, 34,* 295–342.

D'Souza, M. S., & Markou, A. (2012). Schizophrenia and tobacco smoking comorbidity: nAChR agonists in the treatment of schizophrenia-associated cognitive deficits. *Neuropharmacology, 62,* 1564–1573.

Dulude, L., Labelle, A., & Knott, V. J. (2010). Acute nicotine alteration of sensory memory impairment in smokers with schizophrenia. *Journal of Clinical Psychopharmacology, 30,* 541–548.

Dumatar, C., & Chauhan, J. (2011). A study of the effect of cigarette smoking on cognitive parameters in human volunteers. *National Journal of Integrated Research in Medicine, 2,* 71–76.

Fioravanti, M., Carlone, O., Vitale, B., Cinti, M. E., & Clare, L. (2005). A meta-analysis of cognitive deficits in adults with a diagnosis of schizophrenia. *Neuropsychology Review, 15,* 73–95.

George, T. P., Vessicchio, J. C., Termine, A., Sahady, D. M., Head, C. A., Pepper, W. T., ... Wexler, B. E. (2002). Effects of smoking abstinence on visuospatial working memory function in schizophrenia. *Neuropsychopharmacology, 26,* 75–85.

Green, M. F., Kern, R. S., & Heaton, R. K. (2004). Longitudinal studies of cognition and functional outcome in schizophrenia: implications for MATRICS. *Schizophrenia Research, 72,* 41–51.

Hammar, A., & Ardal, G. (2009). Cognitive functioning in major depression – a summary. *Frontiers in Human Neuroscience, 3,* 1–7.

Harris, J. G., Kongs, S., Allensworth, D., Martin, L., Tregellas, J., Sullivan, B., ... Freedman, R. (2004). Effects of nicotine on cognitive deficits in schizophrenia. *Neuropsychopharmacology, 29,* 1378–1385.

Heffner, J. L., Strawn, J. R., DelBello, M. P., Strakowski, S. M., & Anthenelli, R. M. (2011). The co-occurrence of cigarette smoking and bipolar disorder: phenomenology and treatment considerations. *Bipolar Disorders, 13,* 439–453.

Heishman, S. J., Kleykamp, B. A., & Singleton, E. G. (2010). Meta-analysis of the acute effects of nicotine and smoking on human performance. *Psychopharmacology (Berl), 210,* 453–469.

Hong, L. E., Schroeder, M., Ross, T. J., Buchholz, B., Salmeron, B. J., Wonodi, I., ... Stein, E. A. (2011). Nicotine enhances but does not normalize visual sustained attention and the associated brain network in schizophrenia. *Schizophrenia Bulletin, 37,* 416–425.

Howe, W. M., Ji, J., Parikh, V., Williams, S., Mocaer, E., Trocme-Thibierge, C., & Sarter, M. (2010). Enhancement of attentional performance by selective stimulation of alpha4beta2(*) nAChRs: underlying cholinergic mechanisms. *Neuropsychopharmacology, 35,* 1391–1401.

Jasinska, A. J., Zorick, T., Brody, A. L., & Stein, E. A. (2014). Dual role of nicotine in addiction and cognition: a review of neuroimaging studies in humans. *Neuropharmacology, 84,* 111–122.

Kurtz, M. M., & Gerraty, R. T. (2009). A meta-analytic investigation of neurocognitive deficits in bipolar illness: profile and effects of clinical state. *Neuropsychology, 23,* 551–562.

Law, C. W., Soczynska, J. K., Woldeyohannes, H. O., Miranda, A., Brooks, J. O., 3rd, & McIntyre, R. S. (2009). Relation between cigarette smoking and cognitive function in euthymic individuals with bipolar disorder. *Pharmacology Biochemistry & Behavior, 92,* 12–16.

Lee, R. S. C., Hermens, D. F., Porter, M. A., & Redoblado-Hodge, M. A. (2012). A meta-analysis of cognitive deficits in first-episode major depressive disorder. *Journal of Affective Disorders, 140,* 113–124.

Logue, S. F., & Gould, T. J. (2013). The neural and genetic basis of executive function: attention, cognitive flexibility, and response inhibition. *Pharmacology Biochemistry & Behavior, 123C,* 45–54.

Luger, T. M., Suls, J., & Vander Weg, M. W. (2014). How robust is the association between smoking and depression in adults? A meta-analysis using linear mixed-effects models. *Addictive Behaviors, 39,* 1418–1429.

Mansvelder, H. D., van Aerde, K. I., Couey, J. J., & Brussaard, A. B. (2006). Nicotinic modulation of neuronal networks: from receptors to cognition. *Psychopharmacology (Berl), 184,* 292–305.

McDermott, L. M., & Ebmeier, K. P. (2009). A meta-analysis of depression severity and cognitive function. *Journal of Affective Disorders, 119,* 1–8.

McEachin, R. C., Saccone, N. L., Saccone, S. F., Kleyman-Smith, Y. D., Kar, T., Kare, R. K., ... McInnis, M. G. (2010). Modeling complex genetic and environmental influences on comorbid bipolar disorder with tobacco use disorder. *BMC Medical Genetics, 11,* 1–32.

Millet, B., Dondaine, T., Reymann, J. M., Bourguignon, A., Naudet, F., Jaafari, N., ... Le Jeune, F. (2013). Obsessive-compulsive disorder networks: positron emission tomography and neuropsychology provide new insights. *PLoS One, 8,* 532–541.

Mineur, Y. S., & Picciotto, M. R. (2009). Biological basis for the co-morbidity between smoking and mood disorders. *Journal of Dual Diagnosis, 5,* 122–130.

Mineur, Y. S., & Picciotto, M. R. (2010). Nicotine receptors and depression: revisiting and revising the cholinergic hypothesis. *Trends in Pharmacological Sciences, 31,* 580–586.

Mons, U., Schötter, B., Müller, H., Kliegel, M., & Brenner, H. (2013). History of lifetime smoking, smoking cessation and cognitive function in the elderly population. *Neuroepidemiology, 28,* 823–831.

Morisano, D., Wing, V. C., Sacco, K. A., Arenovich, T., & George, T. P. (2013). Effects of tobacco smoking on neuropsychological function in schizophrenia in comparison to other psychiatric disorders and non-psychiatric controls. *American Journal on Addictions, 22,* 46–53.

Musso, F., Bettermann, F., Vucurevic, G., Stoeter, P., Konrad, A., & Winterer, G. (2007). Smoking impacts on prefrontal attentional network function in young adult brains. *Psychopharmacology (Berl), 191,* 159–169.

Ostacher, M. J., Nierenberg, A. A., Perlis, R. H., Eidelman, P., Borrelli, D. J., Tran, T. B., ... Sachs, G. S. (2006). The relationship between smoking and suicidal behavior, comorbidity, and course of illness in bipolar disorder. *The Journal of Clinical Psychiatry, 67,* 1907–1911.

Paul, R. H., Brickman, A. M., Cohen, R. A., Williams, L. M., Niaura, R., Pogun, S., ... Gordon, E. (2006). Cognitive status of young and older cigarette smokers: data from the international brain database. *Journal of Clinical Neuroscience, 13,* 457–465.

Picciotto, M. R., Caldarone, B. J., King, S. L., & Zachariou, V. (2000). Nicotinic receptors in the brain. Links between molecular biology and behavior. *Neuropsychopharmacology, 22,* 451–465.

Poorthuis, R. B., Goriounova, N. A., Couey, J. J., & Mansvelder, H. D. (2009). Nicotinic actions on neuronal networks for cognition: general principles and long-term consequences. *Biochemical Pharmacology, 78*, 668–676.

Robinson, L. J., Thompson, J. M., Gallagher, P., Goswami, U., Young, A. H., Ferrier, I. N., & Moore, P. B. (2006). A meta-analysis of cognitive deficits in euthymic patients with bipolar disorder. *Journal of Affective Disorders, 93*, 105–115.

Sacco, K. A., Termine, A., Seyal, A., Dudas, M. M., Vessicchio, J. C., Krishnan-Sarin, S., … George, T. P. (2005). Effects of cigarette smoking on spatial working memory and attentional deficits in schizophrenia: involvement of nicotinic receptor mechanisms. *Archives of General Psychiatry, 62*, 649–659.

Salin-Pascual, R. J., & Basanez-Villa, E. (2003). Changes in compulsion and anxiety symptoms with nicotine transdermal patches in non-smoking obsessive-compulsive disorder patients. *Revista de investigación clínica, 55*, 650–654.

Segarra, R., Zabala, A., Eguiluz, J. I., Ojeda, N., Elizagarate, E., Sanchez, P., … Gutierrez, M. (2011). Cognitive performance and smoking in first-episode psychosis: the self-medication hypothesis. *European Archives of Psychiatry and Clinical Neuroscience, 261*, 241–250.

Semenova, S., Contet, C., Roberts, A. J., & Markou, A. (2012). Mice lacking the beta4 subunit of the nicotinic acetylcholine receptor show memory deficits, altered anxiety- and depression-like behavior, and diminished nicotine-induced analgesia. *Nicotine & Tobacco Research, 14*, 1346–1355.

Shih, R. A., Glass, T. A., Bandeen-Roche, K., Carlson, M. C., Bolla, K. I., Todd, A. C., & Schwartz, B. S. (2006). Environmental lead exposure and cognitive function in community-dwelling older adults. *Neurology, 67*, 1556–1562.

Smith, R. C., Singh, A., Infante, M., Khandat, A., & Kloos, A. (2002). Effects of cigarette smoking and nicotine nasal spray on psychiatric symptoms and cognition in schizophrenia. *Neuropsychopharmacology, 27*, 479–497.

Smith, R. C., Warner-Cohen, J., Matute, M., Butler, E., Kelly, E., Vaidhyanathaswamy, S., & Khan, A. (2006). Effects of nicotine nasal spray on cognitive function in schizophrenia. *Neuropsychopharmacology, 31*, 637–643.

Tizabi, Y., Louis, V. A., Taylor, C. T., Waxman, D., Culver, K. E., & Szechtman, H. (2002). Effect of nicotine on quinpirole-induced checking behavior in rats: implications for obsessive-compulsive disorder. *Biological Psychiatry, 51*, 164–171.

Tsuang, M. T., Francis, T., Minor, K., Thomas, A., & Stone, W. S. (2012). Genetics of smoking and depression. *Human Genetics, 131*, 905–915.

Wagner, M., Schulze-Rauschenbach, S., Petrovsky, N., Brinkmeyer, J., von der Goltz, C., Grunder, G., … Winterer, G. (2013). Neurocognitive impairments in non-deprived smokers – results from a population-based multi-center study on smoking-related behavior. *Addiction Biology, 18*, 752–761.

Weruaga, E., Balkan, B., Koylu, E. O., Pogun, S., & Alonso, J. R. (2002). Effects of chronic nicotine administration on nitric oxide synthase expression and activity in rat brain. *Journal of Neuroscience Research, 67*, 689–697.

Wiers, R. W., Bartholow, B. D., van den Wildenberg, E., Thush, C., Engels, R. C., Sher, K. J., … Stacy, A. W. (2007). Automatic and controlled processes and the development of addictive behaviors in adolescents: a review and a model. *Pharmacology Biochemistry & Behavior, 86*, 263–283.

Wing, V. C., Bacher, I., Sacco, K. A., & George, T. P. (2011). Neuropsychological performance in patients with schizophrenia and controls as a function of cigarette smoking status. *Psychiatry Research, 188*, 320–326.

Xu, G., Lin, K., Rao, D., Dang, Y., Ouyang, H., Guo, Y., … Chen, J. (2012). Neuropsychological performance in bipolar I, bipolar II and unipolar depression patients: a longitudinal, naturalistic study. *Journal of Affective Disorders, 136*, 328–339.

Zanelli, J., Reichenberg, A., Morgan, K., Fearon, P., Kravariti, E., Dazzan, P., … Murray, R. M. (2010). Specific and generalized neuropsychological deficits: a comparison of patients with various first-episode psychosis presentations. *The American Journal of Psychiatry, 167*, 78–85.

Ziedonis, D., Hitsman, B., Beckham, J. C., Zvolensky, M., Adler, L. E., Audrain-McGovern, J., … Riley, W. T. (2008). Tobacco use and cessation in psychiatric disorders: National Institute of Mental Health report. *Nicotine & Tobacco Research, 10*, 1691–1715.

Chapter 30

Tobacco Smoke Extract-Produced Behavioral Effects: Locomotor Sensitization and Self-Administration Studies

Katharine Alexandra Brennan[1], Penelope Truman[2,3]
[1]School of Psychology, Victoria University of Wellington, Wellington, New Zealand; [2]Institute of Environmental Science and Research Ltd, Porirua, New Zealand; [3]The School of Public Health, Massey University, Wellington, New Zealand

Abbreviations

FR Fixed ratio schedule
MAO Monoamine oxidase enzyme
nAChRs Nicotinic acetylcholine receptors
PR Progressive ratio schedule
TPM Tobacco particulate matter

INTRODUCTION: TOBACCO SMOKE EXTRACTS IN RESEARCH

Early research established that nicotine plays a central role in the addictive effects of tobacco smoking (Benowitz, 1988; Dani & Balfour, 2011). Consequently, behavioral pharmacologists have studied addiction-related behavioral effects produced by nicotine and associated brain changes to elucidate what neurochemical systems/mechanisms are involved in tobacco addiction. Yet, despite a critical role for nicotine in the addictive effects, there is increasing recognition that tobacco is a highly complex mixture of more than 9000 constituents (Rodgman & Perfetti, 2013; Stedman, 1968). Thus, there is presently an endeavor to include whole smoke extracts in addition to pure nicotine to better understand the pharmacology of tobacco addiction.

The use of whole tobacco extracts in animal research involves two different routes of administration: the inhalative route in specialized smoking chambers (Harris, Mattson, Lesage, Keyler, & Pentel, 2010; Small et al., 2010) or systemically administered aqueous tobacco/smoke extracts (Ambrose et al., 2007; Brennan, Crowther, et al., 2013; Brennan, Laugesen, & Truman, 2014; Brennan, Putt, Roper, Waterhouse, & Truman, 2013; Brennan, Putt, & Truman, 2013; Costello et al., 2014; Danielson, Putt, Truman, & Kivell, 2014; Harris, Stepanov, Pentel, & Lesage, 2012; Lewis, Truman, Hosking, & Miller, 2012; Touiki, Rat, Molimard, Chait, & de Beaurepaire, 2007).

Inhalative smoke exposure in animals has been frequently used to model tobacco dependence (Harris et al., 2010; Small et al., 2010). This model best mimics the route of smoke administration used by human smokers, but the main drawback is that animals are involuntarily exposed to smoke. This lack of control over drug intake/exposure has been shown to affect drug-produced neurological changes (Dworkin, Co, & Smith, 1995; Jacobs, Smit, de Vries, & Schoffelmeer, 2003; Stefanski, Ladenheim, Lee, Cadet, & Goldberg, 1999). Additionally, smoke exposure is a considerable stressor, where physical restraint, burning eyes, or respiratory difficulties have been reported (Harris et al., 2010). Because stress can lead to widespread alterations in brain-reward pathways (Abercrombie, Keefe, DiFrischia, & Zigmond, 1989; Carlson, Fitzgerald, Keller, & Glick, 1991; Imperato, Angelucci, Casolini, Zocchi, & Puglisi-Allegra, 1992; Pawlak et al., 2000), forced smoke exposure might produce behavioral effects and/or neurological changes that do not relate specifically to tobacco addiction.

The alternative approach is to use aqueous tobacco/smoke extract solutions that can be systemically administered. Specifically, intravenous administration allows rapid delivery of nicotine and constituents to the brain, while also affording greater precision and control of the quantities of tobacco constituents being delivered compared to inhalative methods (Danielson et al., 2014; Touiki et al., 2007). The most important advantage of this model is that it is possible to conduct low-stress experiments, where the animal has control over drug intake. The limitations are that constituents present in solutions might not accurately reflect smoke composition and that the route of administration is dissimilar to that of human smokers. Furthermore, systemic administration means that the tobacco constituents undergo first-pass metabolism in the liver and constituents might not reach the brain as rapidly. Despite these disadvantages, it has been reported that the pharmacokinetic parameters of nicotine delivered via smoke inhalation versus intravenous infusion in rats were similar (Rotenberg & Adir, 1983; Rotenberg, Miller, & Adir, 1980). Thus, although it does not exactly model the human route of administration, intravenously administered smoke extracts provide a viable option for studying tobacco dependence in laboratory animals.

Cellular and molecular studies show a myriad of physiological changes that occur in the brain as a consequence of tobacco exposure and/or the development of tobacco addiction (Danielson et al., 2014). Emerging research has also demonstrated that inhalative/aqueous tobacco smoke exposure produces distinct neurological changes compared to those of nicotine alone (for a review, see Brennan et al., 2014). Behavioral pharmacologists have sought to determine whether some of these observed changes are functionally or clinically relevant to addiction. To these ends, animal behavioral tests have been selected that have some validity to an aspect of the addiction cycle, such as initiation of drug-taking, long-term maintenance of drug use, and craving/relapse after abstinence. Two frequently used tests are locomotor sensitization and intravenous self-administration.

BEHAVIORAL TESTS
Locomotor Sensitization

Repeated nicotine administration has reliably produced locomotor sensitization in laboratory animals that is dependent on dose/treatment schedule (Brennan et al., 2013; Clarke & Kumar, 1983; DiFranza & Wellman, 2007; Villegier, Blanc, Glowinski, & Tassin, 2003). Nicotine-produced locomotor sensitization tends to be relatively weak and transient with an intermittent dosing regimen, contrasting with the greater and more persistent effects of other psychostimulant drugs, such as amphetamines (Ball, Klein, Plocinski, & Slack, 2011; Brennan et al., 2013; Villegier et al., 2003).

The effects of inhalative smoke exposure on locomotor sensitization have been compared to those of systemically administered nicotine (Bruijnzeel, Rodrick, Singh, Derendorf, & Bauzo, 2011; Harris et al., 2010). Rats were exposed to tobacco smoke for short (Harris et al., 2010) or longer sessions (Bruijnzeel et al., 2011) each day for the same time period. Harris et al. (2010) administered 0.3 mg/kg nicotine for 5 days after the smoke exposure to determine whether this treatment affected the development of nicotine-produced sensitization. The results showed that smoke exposure had no effect on initial response to the first nicotine injection, nor on the development of sensitization to nicotine. In contrast, Bruijnzeel et al. (2011) found that inhalative smoke exposure produced a small sensitized locomotor response to smoke or nicotine challenge on the last treatment day and following a withdrawal period. The disparate results between laboratories are likely attributable to different smoke exposure protocols, withdrawal periods, and challenge nicotine doses. Bruijnzeel et al. (2011) showed that intense smoke exposure can produce sensitization to smoke exposure, but generally, the effects of smoke were less potent than equivalent doses of systemically administered nicotine. In conclusion, it was unclear whether the nonnicotinic components in tobacco smoke were inhibitory to the development of sensitization or whether the different modes of administration (i.e., inhalation versus injection) might explain the more potent effects of nicotine injection.

The ability for aqueous tobacco extracts to produce sensitization has been compared to the effects of nicotine alone (Brennan et al., 2013; Harris et al., 2012). The key advantage of these studies over the aforementioned was that the tobacco extracts and nicotine were delivered via the same mode of administration. Harris et al. (2012) administered the nicotine/extract (0.4 mg/kg nicotine) for two consecutive weeks. The challenge test was then conducted after a 10-day withdrawal, where the results showed that the sensitized response was still present for both groups and was identical. Brennan et al. (2013) conducted similar, but more extensive testing, where two doses of nicotine/extract were examined (0.2 and 0.4 mg/kg), as was persistence of the sensitized response (4- and 15-day withdrawal periods). The results revealed that identical levels of sensitization to 0.4 mg/kg nicotine and extract was evident at the 4-day withdrawal point, where the lowest 0.2 mg/kg dose of nicotine/extract failed to produce reliable sensitization (Figure 1). Furthermore, the response for both groups was gone after 15 days, showing that sensitization is transient following a less intense and intermittent dosing regimen. This study utilized a different method of extract preparation and treatment regimen to that of Harris et al. (2012), yet yielded similar results. Together, these findings suggest that the nonnicotinic tobacco constituents might not alter the sensitizing effects of nicotine (Table 1).

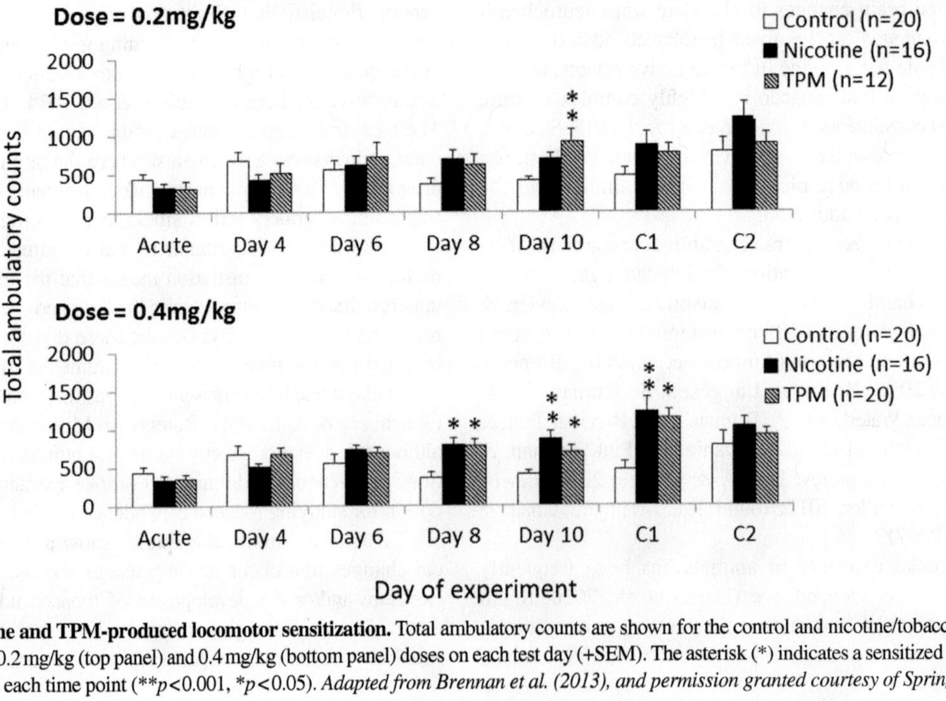

FIGURE 1 **Nicotine and TPM-produced locomotor sensitization.** Total ambulatory counts are shown for the control and nicotine/tobacco particulate matter (TPM) groups at the 0.2 mg/kg (top panel) and 0.4 mg/kg (bottom panel) doses on each test day (+SEM). The asterisk (*) indicates a sensitized locomotor response relative to controls at each time point (**$p<0.001$, *$p<0.05$). *Adapted from Brennan et al. (2013), and permission granted courtesy of Springer publishing.*

TABLE 1 The Effects of Nicotine versus TPM on Locomotor Sensitization

Author	Extract Type	Treatment Regimen	Results	Conclusions
Harris et al. (2010)	Inhalative 2R4F research cigarettes containing 13 mg nicotine per cigarette. Smoking machine delivered nose-only-exposure at 4 puffs/min.	Experiment 1: Male Holtzman Sprague-Dawley rats (275–325 g) were administered 0.1 mg/kg nicotine (SC) or 45 min smoke exposure over 14 days with activity measured daily (days 1–14). Experiment 2: On days 17–21 all groups received 0.3 mg/kg nicotine (SC), activity measured daily.	Experiment 1: Only the 0.1 mg/kg nicotine group developed sensitization by days 11–14. Experiment 2: All groups developed a sensitized response to 0.3 mg/kg nicotine, but previous exposure to 0.1 mg/kg nicotine enhanced this response.	Smoke exposure did not induce locomotor sensitization nor affect the development of sensitization to repeated nicotine (0.3 mg/kg) treatment.
Bruijnzeel et al. (2011)	Inhalative 3R4F research cigarettes. Smoking machine delivered whole-body-exposure at 1 puff/min.	Experiment 1: Male Wistar rats (250–300 g) were exposed to air/smoke for 2 h/day for 14 days with activity recorded on days 1, 7, and 14. A challenge test followed a 3-week withdrawal period, comprising re-exposure to smoke. Experiment 2 and 3: Rats were exposed to smoke/air for 2 h/day for 13 days. On day 14, rats received nicotine (0.0, 0.04, or 0.4 mg/kg) and activity was measured.	Experiment 1: On day 7, tobacco exposed rats had higher activity during 15–30 min period. This group was also more active on day 14 during the 30–60 min period. After withdrawal, the tobacco group was still more active during the first 45 min period. Experiment 2 and 3: The smoke exposed group exhibited an enhanced sensitized response upon administration of either 0.4 mg/kg or 0.04 mg/kg nicotine.	Repeated exposure to smoke produced a sensitized locomotor response to subsequent smoke or nicotine (0.04 and 0.4 mg/kg) exposure.
Harris et al. (2012)	Aqueous Unburnt tobacco extract was prepared by soaking in saline/artificial saliva vehicle solution.	Male Holtzman Sprague-Dawley rats (275–325 g) received treatment for 2 consecutive weeks with 0.4 mg/kg nicotine or extract with matched nicotine content and activity measured daily. There was a 10-day withdrawal period followed by a challenge test where the same extract/nicotine doses were readministered.	Extract and nicotine alone produced comparable levels of sensitization. Activity in these groups was higher than saline controls from days 4–10. Sensitization was still present in both groups after the withdrawal period.	The nicotine and extract groups (0.4 mg/kg nicotine) produced identical levels of locomotor sensitization.
Brennan et al. (2013)	Aqueous Burnt cigarette smoke (Holiday brand) and tobacco (Drum brand) extracts were prepared by collecting tobacco particulate matter on filter papers, and soaking in saline/ethanol.	Male Sprague-Dawley rats (230–260 g) were treated every second day over 10 consecutive days with 0.2 or 0.4 mg/kg nicotine/extract. Activity was recorded daily and at the challenge tests after 4- and 15-day withdrawals.	The lowest 0.2 mg/kg dose produced minimal sensitization; however, the extract group had higher activity levels on day 10. The 0.4 mg/kg dose produced comparable sensitization in both groups that was gone after a 15-day withdrawal.	There was only a small enhancement of sensitization produced by the extract at the 0.2 mg/kg dose. The 0.4 mg/kg nicotine and extract groups produced identical levels of locomotor sensitization.

Self-Administration

Self-administration tests arguably have the best predictive validity to drug dependence. Nicotine is self-administered by laboratory animals, but it has a different profile to most prototypical drugs of abuse. There are relatively strict parameters to optimize acquisition and maintenance of nicotine self-administration, such as rat strain, infusion times, and nicotine dose (Brower, Fu, Matta, & Sharp, 2002; Sorge & Clarke, 2009). There are also several clues indicating that nicotine is weakly reinforcing when compared to more addictive psychostimulant drugs, such as cocaine. For example, motivation to receive nicotine infusions is much lower than for cocaine (Risner & Goldberg, 1983), and cocaine will be selected if a choice between drugs is offered (Manzardo, Stein, & Belluzzi, 2002).

Nicotine self-administration is also different than many other drugs because it might not be primarily maintained by pharmacologically produced reinforcement. Firstly, the role of environmental cues in nicotine self-administration appears to be stronger than that for other drugs (Chaudhri et al., 2007, 2006; Sorge, Pierre, & Clarke, 2009). After drug abstinence, reintroduction of an environmental cue that was previously always delivered with nicotine was more effective than a nicotine priming injection at triggering drug-seeking behavior (Feltenstein, Ghee, & See, 2012; LeSage, Burroughs, Dufek, Keyler, & Pentel, 2004). This contrasts to other psychostimulant drugs, where a priming injection usually produces pronounced drug-seeking (Schenk, Hely, Gittings, Lake, & Daniela, 2008; Schenk & Partridge, 1999). Secondly, the dose–response curve for nicotine self-administration can be flat (Brennan, Crowther, et al., 2013; Brennan, Putt, Roper, et al., 2013; Corrigall & Coen, 1989; Donny et al., 1998; Harris, Pentel, & LeSage, 2009; Manzardo et al., 2002), which reflects little compensatory change in responding when doses are altered. In contrast, cocaine/amphetamine produces the typical inverted U-shaped dose–response curves showing regulation of intake and compensation in responding (Lau & Sun, 2002; Yokel & Piekens, 1974).

Self-administration of inhalative smoke by laboratory animals would be logistically difficult to set up. Thus, aqueous tobacco smoke extracts have been used to establish intravenous self-administration in rats compared to nicotine alone. Two groups have successfully conducted such experiments and published their results.

Brennan, Putt, Roper, et al. (2013) first sought to compare the underlying receptor mechanisms and reinforcing capacity of cigarette smoke extract versus nicotine. Extract/nicotine self-administration was first established on a fixed ratio (FR) schedule and then several antagonist drugs (SCH23390, mecamylamine, and ketanserin) were administered prior to the self-administration sessions to determine which receptors could be important in the maintenance of drug-taking behavior. The results showed that initiation of self-administration and dose–response curves of pure nicotine (7.5, 15, 30, and 60 µg/kg/infusion) were identical to those of extract that contained matched nicotine doses. SCH23390 pretreatment comparably attenuated both nicotine and extract self-administration across all doses, indicating that dopaminergic mechanisms were important for both. In contrast, extract self-administering animals were more resilient to the inhibitory effects on responding of mecamylamine and ketanserin. These results indicate that nicotine is the primary driver behind extract self-administration, but that extract self-administration involves different neurotransmitter/receptor systems compared to those of nicotine alone (Figures 2 and 3).

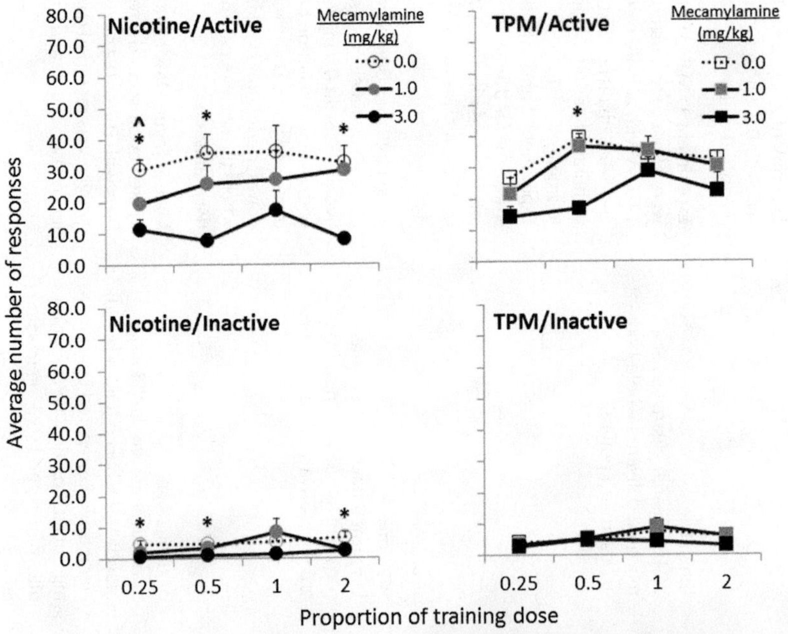

FIGURE 2 Nicotine and TPM-produced self-administration responses following mecamylamine pretreatment. Average number of lever responses across four tobacco particulate matter (TPM)/nicotine doses (0.25, 0.5, 1.0, and 2.0) following mecamylamine pretreatment (0.0, 1.0, or 3.0 mg/kg) (+SEM). The top panel shows active lever responses exhibited by the nicotine ($n=6$) and TPM ($n=8$) groups, whereas the bottom panel shows inactive lever responses exhibited by the nicotine and TPM groups. The accent (^) indicates a significant difference between the 1.0 mg/kg mecamylamine-treated and baseline conditions ($p<0.05$). The asterisk (*) indicates a significant difference between the 3.0 mg/kg mecamylamine-treated and baseline conditions ($p<0.05$). Adapted from Brennan, Putt, Roper, et al. (2013), and permission granted courtesy of Bentham Science publishing.

The second Brennan, Crowther, et al. (2013) study compared self-administration of two different extract types, as these were known to vary in levels of nonnicotine constituents and had different pharmacological properties (Lewis et al., 2012). The reinforcing efficacies of nicotine alone versus two extract types were compared: cigarette and roll-your-own extracts. The results revealed that, as previously reported, the nicotine and cigarette extract were very similar across all schedules. However, the roll-your-own extract group exhibited significantly faster initiation of self-administration and increased motivation to obtain infusions on the progressive ratio (PR) schedule. The roll-your-own extract group also responded significantly more for the 15 and 30 µg/kg/infusion doses, which was in contrast to the very flat dose–response curves for the nicotine and cigarette extract groups. These findings suggest that the nonnicotinic agents could significantly affect the reinforcing properties in some tobacco preparations (Figure 4).

Costello et al. (2014) used different extract preparation methods and experimental parameters, but they also sought to compare the reinforcing efficacy of nicotine to that of extract with matched nicotine doses. Additionally, the effects of several acetylcholine nicotinic receptor (nAChR) antagonists were tested on responding. The results revealed that extract self-administration was acquired at a very low nicotine dose, whereas the same dose of pure nicotine failed to support the acquisition. Changing the doses also yielded significant differences in responding for extract but not for nicotine. These findings were comparable to the Brennan, Crowther, et al. (2013) results, suggesting fairly robust effects.

The PR testing did not reveal any differences in breakpoints between extract and nicotine. Costello et al. (2014) concluded that extract was no more reinforcing than nicotine, but rather was more potent at producing behavioral effects. This was contrary to the Brennan, Crowther, et al. (2013) findings with the PR schedule, where it was observed that the roll-your-own tobacco extracts were more reinforcing than matched doses of nicotine or cigarette extracts. The main differences between these studies were that Brennan, Crowther, et al. (2013) tested a range of doses in PR, whereas Costello et al. (2014) only tested a single dose. Furthermore, Brennan, Crowther, et al. (2013) tested two different tobacco extract types, where only one was significantly more reinforcing. Costello et al. (2014) might have also observed differences in breakpoints with a range of doses and different tobacco types.

Costello et al. (2014) also reported that the extract group was slightly more resistant to extinction, and that stress was significantly better at reinstating drug-seeking behavior in the extract group relative to the nicotine group. These findings suggest that the extract could have been more difficult to "give up" (slower extinction in the extract group) and that the extract group were more susceptible to stress-induced relapse. The reinstatement results could not be considered alone, but the greater stress-produced relapse to extract response fits with the idea that extract could be more reinforcing than the nicotine.

Costello et al. (2014) tested the effects of three nAChR ligands on responding for a single dose of nicotine/extract. Mecamylamine and varenicline both similarly reduced nicotine and extract self-administration, so the authors proposed that extract self-administration was primarily driven by nicotinic mechanisms. In contrast, Brennan, Putt, Roper, et al. (2013) reported that mecamylamine was less effective at reducing responding for extract than nicotine at one dose.

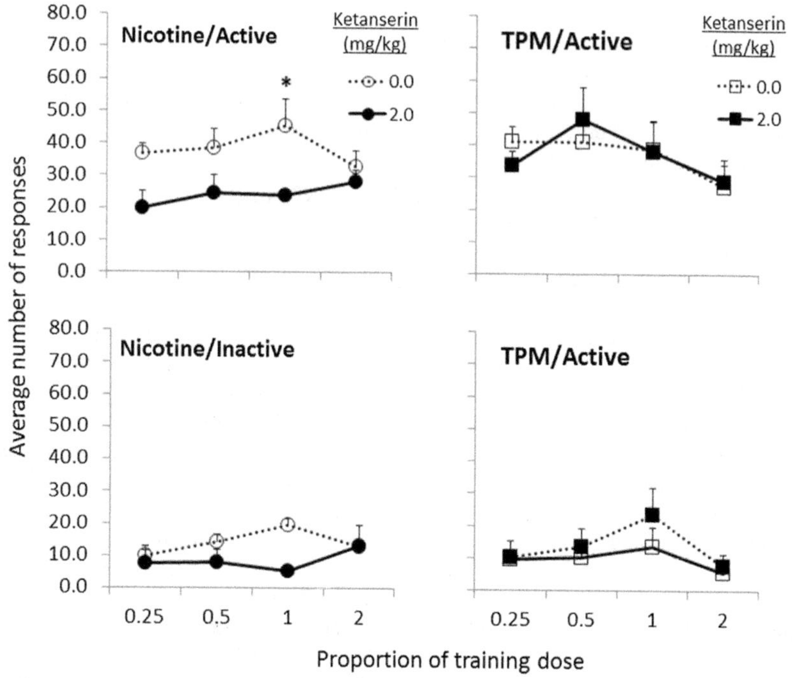

FIGURE 3 **Nicotine and TPM-produced self-administration responses following ketanserin pretreatment.** Average number of lever responses across four tobacco particulate matter (TPM)/nicotine doses (0.25, 0.5, 1.0 and 2.0) following ketanserin pretreatment (0.0 or 2.0 mg/kg). The top panel shows active lever responses exhibited by the nicotine ($n=5$) and TPM ($n=5$) groups, whereas the bottom panel shows inactive lever responses exhibited by the nicotine and TPM groups. The asterisk (*) indicates a significant difference between the ketanserin-treated and baseline conditions ($p<0.01$). *Adapted from Brennan, Putt, Roper, et al. (2013), and permission granted courtesy of Bentham Science publishing.*

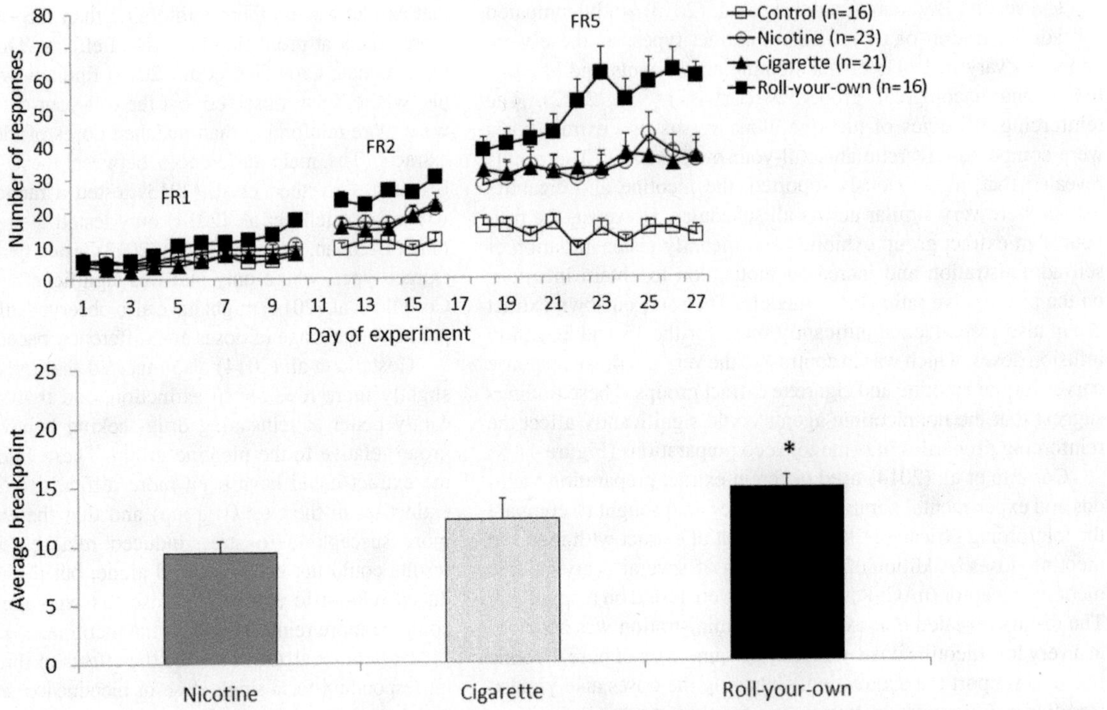

FIGURE 4 Nicotine, cigarette, and roll-your-own TPM-produced self-administration. The top panel shows self-administration on a fixed ratio (FR) schedule. Data points represent the average number of total responses produced by the nicotine, cigarette tobacco particulate matter (TPM) and roll-your-own TPM groups (FR1, Days 1–10, FR2 Days 12–16, FR5 Days 18–27) on the active lever (+SEM). The bottom panel shows self-administration on a progressive ratio schedule. Columns represent the average breakpoint collapsed across doses for each treatment group (+SEM). The asterisk (*) indicates a significant difference from the nicotine group ($p<0.05$). *Adapted from Brennan, Crowther, et al. (2013), and permission granted courtesy of Wiley Publishing.*

These disparate results between laboratories could be explained in that Brennan, Putt, Roper, et al. (2013) tested the effects of several doses of mecamylamine against four nicotine/extract doses. This increases the chances of observing small differences, whereas Costello et al. (2014) only tested a single nicotine/extract dose.

The last antagonist tested by Costello et al. (2014), AT-1001, was effective at reducing nicotine self-administration but responding for extract was resilient to these effects. This supports the Brennan, Putt, Roper, et al. (2013) conclusions that extract self-administration might be sustained by differential neurochemical systems. Logically, it follows that a whole mixture of chemicals (tobacco) would recruit numerous systems, whereas the effects of pure nicotine would be more limited in comparison. This might explain why extract self-administration was generally more resilient to the effects of antagonists across both studies: when specific receptors are blocked, activation of other receptors could mask these effects.

Costello et al. (2014) also sought to determine whether extract could produce differential radioligand binding to nAChRs and monoamine oxidase (MAO) inhibition between extract and nicotine groups. The radioligand binding revealed no differences, but the extract brain preparations exhibited considerable MAO inhibition compared to the nicotine groups. MAO inhibition in the tobacco-exposed group would be attributable to numerous identified (and still unidentified) MAO inhibitor substances known to be present in tobacco (Lewis, Miller, & Lea, 2007), whereas nicotine does not exhibit this property (Fowler et al., 1998). Interestingly, the tobacco extract-produced MAO inhibition could also relate to the differential behavioral effects observed, because MAO affects neurochemical systems involved in reward and reinforcement (content adapted from Brennan et al. (2014), with permission granted courtesy of Elsevier publishing) (Table 2).

CONCLUSIONS

Intense inhalative smoke exposure produced locomotor sensitization in rats, but the behavioral response was weaker than equivalent doses of nicotine. These results suggest that either the nonnicotinic components in tobacco were inhibitory to locomotor sensitization and/or that the inhalative route of administration produced less potent effects than systemic injection. The aqueous extract studies were more controlled as they used the same route of administration and matched doses for extract and nicotine. These results revealed no differences between the sensitizing effects of nicotine versus extract. Thus, the conclusions were that the nonnicotinic components in tobacco do not seem to produce effects on reward-related behavior; however, these results were not being compared across behavioral paradigms.

The most recent self-administration studies have shown that there are behavioral differences between extract and nicotine. Thus, self-administration tests are more sensitive and better suited to detect differences in reinforcing efficacy than locomotor sensitization. It is unclear which nonnicotinic tobacco constituents are responsible for these behavioral differences, so future research will attempt to isolate and identify these compounds. This information would enhance

TABLE 2 The Effects of Nicotine versus TPM on Self-Administration

Author	Extract Type	Treatment Regimen	Results	Conclusions
Brennan, Putt, Roper, et al. (2013)	Aqueous Burnt cigarette smoke (Holiday brand) extract was prepared by collecting tobacco particulate matter on filter papers, and soaking in saline/ethanol.	Experiment 1: Male Sprague-Dawley rats (300–330g) had access to 30 μg/kg/infusion nicotine or extract with matched nicotine dose. Acquisition occurred on a fixed ratio schedule for 25 days, followed by dose response testing (7.5, 15, 30, and 60 μg/kg/infusion nicotine). Experiment 2: Effects of antagonist pretreatment were tested at each nicotine/extract dose. Two separate groups received mecamylamine (0.0, 1.0, or 3.0mg/kg) or SCH23390 (0.0 or 0.02mg/kg) and ketanserin (0.0 or 2.0mg/kg).	Experiment 1: Acquisition of self-administration was identical for nicotine and extract groups. Responding was unaffected by dose changes in both groups. Experiment 2: Responding for nicotine was decreased by mecamylamine, SCH23390, and ketanserin. In contrast, extract groups were more resilient to the effects of mecamylamine and ketanserin.	Nicotine is the primary driver for the reinforcing effects of cigarettes. Nonnicotinic agents in the extract likely recruit different neurochemical systems that contribute to reinforcement processes.
Brennan, Crowther, et al. (2013)	Aqueous Burnt cigarette smoke (Holiday brand) and tobacco (Drum brand) extracts were prepared by collecting tobacco particulate matter on filter papers, and soaking in saline/ethanol.	Experiment 1: Male Sprague-Dawley rats (300–330g) had access to 30 μg/kg/infusion nicotine, cigarette extract or tobacco extract with matched nicotine doses. Acquisition occurred on a fixed ratio schedule for 25 days, followed by dose response testing (7.5, 15, 30, and 60 μg/kg/infusion nicotine). Experiment 2: A subset of rats that completed all the experiment 1 tests underwent progressive ratio testing at each of the 4 doses specified above.	Experiment 1: There were no differences between the nicotine and cigarette extract groups. However, the tobacco extract yielded faster acquisition, higher numbers of responses during the fixed ratio trials, and adjusted responding to dose changes. Phase 2: The tobacco extract group were significantly more motivated to receive infusions of 15 and 30 μg/kg/infusion extract than the other groups.	Nonnicotinic components could significantly contribute to the reinforcing properties of tobacco. Implications are that (Drum) tobacco could be more reinforcing or have higher addictive potential than (Holiday) cigarettes.
Costello et al. (2014)	Aqueous Burnt cigarette smoke (Camel and RJ Reynolds brands) was bubbled through saline.	Experiment 1: Male Sprague–Dawley rats (300–325g) had access to 3.75 μg/kg/infusion nicotine or extract with matched nicotine dose for the acquisition study. Separate groups underwent dose response (0, 3.75, 7.5, and 15 μg/kg/infusion nicotine), progressive ratio (15 μg/kg/infusion) and extinction/reinstatement testing. Experiment 2: Effects of antagonist pretreatment on responding for 15 μg/kg/infusion nicotine/extract were tested; including mecamylamine (0, 0.5, 1.0, and 2.0mg/kg), AT-1001 (0, 0.75, 1.5, and 3mg/kg) or varenicline (0.3, 1, and 3mg/kg). Experiment 3: Brain monoamine oxidase A and B activities and nAChR ligand binding autoradiography were assessed.	Experiment 1: The 3.75 μg/kg/infusion dose did not support nicotine self-administration, whereas extract self-administration was acquired at this nicotine dose. Responding for extract was higher at the 7.5 μg/kg/infusion nicotine than that for pure nicotine. There were no differences between groups in progressive ratio responding. The group that self-administered extract exhibited significantly higher responding to the stress reinstatement condition than the nicotine group. Experiment 2: Varenicline and mecamylamine similarly decreased responding for both extract and nicotine; however AT-1001 was less effective at inhibiting extract self-administration. Experiment 3: Only extract produced monoamine oxidase A and B inhibition. There were no group differences for inhibition of radioligand binding to nAChRs.	Extract was no more reinforcing than nicotine, but rather was more potent. Nonnicotinic agents in the extract likely recruit different neurochemical systems that contribute to reinforcement processes. Extract produces differential non-nAChR-related neuroadaptations to nicotine.

our understanding of the pharmacology of tobacco addiction, potentially improve smoking cessation treatments, and allow more effective monitoring and regulation of different tobacco products.

APPLICATIONS TO OTHER ADDICTIONS AND SUBSTANCE MISUSE

- The nAChRs modulate mesoaccumbens dopamine pathways and thus have a likely role in controlling the consumption of addictive drugs other than nicotine, such as methamphetamine, cocaine, alcohol, opiates, and cannabinoids (Hiranita, Nawata, Sakimura, & Yamamoto, 2008; Tuesta, Fowler, & Kenny, 2011).
- Cannabis is a mixture of many compounds (e.g., tobacco), where tetrahydrocannabinol (THC) has been identified as responsible for a large degree of the psychoactive effects (Adams & Martin, 1996). Experiments to elucidate whether the other non-THC compounds have effects on the brain or behavior could use a similar methodology as described in this chapter, where whole cannabis smoke extracts could be compared in behavioral/neurological tests to pure THC.
- The beta-carbolines (harman/norharman) are suspected of having a role in tobacco dependence and are also found in alcohol (Herraiz & Chaparro, 2005). Thus, establishing the role of these compounds in tobacco addiction could also contribute to a better understanding of the pharmacology of alcohol addiction (Spijkerman, van den Eijnden, van de Mheen, Bongers, & Fekkes, 2002).

DEFINITION OF TERMS

AT-1001 AT-1001 is a α3β4 acetylcholine nicotinic receptor antagonist.

Acquisition This term is used to describe initial learning of the self-administration operant task. There are usually two levers or nose-poke holes in the operant chamber, where activation of only one active lever/hole results in a drug infusion and activation of the inactive lever/hole has no programmed response. The criterion for self-administration acquisition is typically that responses on the active lever/hole are significantly greater than on the inactive.

Behavioral pharmacology The study of animal behavior is used as an index of brain function following the administration of psychoactive drugs.

Breakpoint On a progressive ratio schedule, when an animal is no longer willing to undertake further responses to receive another drug infusion, the breakpoint has been reached (point where the animal "gives up"). The more addictive and reinforcing drugs have higher breakpoints.

Extinction Extinction involves removing access to the drug that has been self-administered and often the accompanying environmental cues (e.g., cue lights, noises) from the operant chamber. When responding for the drug (lever presses/nose pokes) ceases, this is termed extinction. Extinction testing can show how long drug-seeking behavior persists in the absence of the drug, where the analogy to humans is how difficult the drug is to give up.

Fixed ratio schedule This schedule is used in self-administration experiments where a drug infusion is delivered after a set number of responses. This schedule is usually used to assess the initiation of drug-taking behavior.

Ketanserin This is a serotonin 2A/2C receptor antagonist.

Locomotor sensitization Repeated administration of psychostimulant drugs produces an enhanced or sensitized locomotor response to subsequent drug challenges in animals. This sensitized response is thought to reflect neurological changes that occur in the early stages in the development of drug addiction.

Mecamylamine This is a noncompetitive and nonselective acetylcholine nicotinic receptor antagonist.

Monoamine oxidase enzymes These mitochondrial enzymes are found in the brain, liver, gastrointestinal tract, placenta, and blood platelets. These enzymes catalyze the oxidation of monoamines and play a vital role in neurotransmission as they inactivate monoamine neurotransmitters.

Priming Priming involves reintroducing one of the stimuli associated with self-administration, such as the drug or a cue light. Priming could be conducted during extinction to determine whether reinstatement to drug-seeking occurs (i.e., administer an injection of the test drug or restore the test light normally associated with drug delivery).

Progressive ratio schedule This schedule is used in self-administration experiments to determine the degree of addiction or motivation to receive drug infusions, as the animal must work harder to receive subsequent drug infusions (Richardson & Roberts, 1996). Generally, after each infusion earned, the number of responses required to obtain the next increases exponentially.

Reinstatement This self-administration test follows a period of extinction training, where the drug/cues have been removed and drug-seeking has ceased. A reinstatement test can involve reintroducing the drug, cues, or both to determine whether there is a return to previous drug-taking behavior. The analogy to humans is the tendency to relapse to drug-seeking.

SCH23390 This is an antagonist at the dopamine D1-like receptors.

Self-administration Intravenous drug self-administration is a model with high predictive validity to human drug abuse, as animals control their own drug intake. Animals are implanted with intravenous catheters, and then delivery of the test drug occurs following a lever-press/nose-poke operant response. The drug delivery (if it is a reinforcing drug) will positively reinforce the operant behavior; thus, reward is inferred if frequency of responding increases.

Varenicline This is apartial agonist at nAChRs.

KEY FACTS OF HISTORY OF BEHAVIORAL NICOTINE/TOBACCO ADDICTION RESEARCH

- Nicotine was first identified in tobacco in 1828 by Posselt and Reimann.
- The first suggestion that tobacco had addictive potential was in 1849, when Trall denounced the medical use of tobacco and presented a case of tobacco addiction (Maisto, Galizio, & Connors, 2011).
- The first intravenous self-administration paper in rats was published where a large number of drugs were screened for reinforcing potential, including nicotine (Collins, Weeks, Cooper, Good, & Russell, 1984).
- In 1988, the US Surgeon General reported that nicotine was physically addicting and harmful to health; this was the first official public health alarm.
- The first nicotine sensitization experiments were conducted (Clarke & Kumar, 1983; Morrison & Stephenson, 1972).

- The first nicotine self-administration experiments were published, where these findings were very controversial as it was not generally accepted that nicotine was an addictive drug (Corrigall & Coen, 1989).
- The first whole smoke extract locomotor sensitization studies that compared the effects to nicotine were conducted (Brennan et al., 2013; Harris et al., 2010)
- The first whole tobacco smoke extract self-administration studies were published (Brennan, Crowther, et al., 2013; Brennan, Putt, Roper, et al., 2013; Costello et al., 2014).

SUMMARY POINTS

- Nicotine plays a central role in the addictive effects of tobacco smoking and has been used for decades to model tobacco addiction in animals.
- Because tobacco contains many other constituents, whole smoke extracts are now being used in addiction studies as a complement to pure nicotine.
- Tobacco smoke extract exposure produces distinct neurological changes compared to those of nicotine alone. Thus, behavioral tests have been used to determine whether these changes could be functionally or clinically relevant to addiction.
- Behavioral tests have been selected that have some validity to an aspect of the addiction cycle; two frequently used tests are locomotor sensitization and intravenous self-administration.
- Locomotor sensitization tests have shown no differences between extract and nicotine, indicating that the nonnicotinic constituents do not appear to have a role in producing sensitized responses.
- In contrast, self-administration tests revealed differential effects between nicotine and extracts showing that the nonnicotinic constituents could have a previously unrecognized role in the addictive effects of tobacco.
- The general conclusions from the self-administration studies were that extract self-administration was mediated via differential neurochemical systems compared to those of nicotine, and that tobacco could be more reinforcing and/or more difficult to give up compared to pure nicotine.

REFERENCES

Abercrombie, E. D., Keefe, K. A., DiFrischia, D. S., & Zigmond, M. J. (1989). Differential effect of stress on in vivo dopamine release in striatum, nucleus accumbens, and medial frontal cortex. *Journal of Neurochemistry, 52*(5), 1655–1658.

Adams, I. B., & Martin, B. R. (1996). Cannabis: pharmacology and toxicology in animals and humans. *Addiction, 91*(11), 1585–1614.

Ambrose, V., Miller, J. H., Dickson, S. J., Hampton, S., Truman, P., Lea, R. A., & Fowles, J. (2007). Tobacco particulate matter is more potent than nicotine at upregulating nicotinic receptors on SH-SY5Y cells. *Nicotine and Tobacco Research, 9*(8), 793–799.

Ball, K. T., Klein, J. E., Plocinski, J. A., & Slack, R. (2011). Behavioral sensitization to 3,4-methylenedioxymethamphetamine is long-lasting and modulated by the context of drug administration. *Behavioral Pharmacology, 22*(8), 847–850.

Benowitz, N. L. (1988). Drug therapy. Pharmacologic aspects of cigarette smoking and nicotine addition. *The New England Journal of Medicine, 319*(20), 1318–1330.

Brennan, K. A., Crowther, A., Putt, F., Roper, V., Waterhouse, U., & Truman, P. (2013). Tobacco particulate matter self-administration in rats: differential effects of tobacco type. *Addiction Biology, 20*. http://dx.doi.org/10.1111/adb.12099.

Brennan, K. A., Laugesen, M., & Truman, P. (2014). Whole tobacco smoke extracts to model tobacco dependence in animals. *Neuroscience and Biobehavioral Reviews, 47C*, 53–69.

Brennan, K. A., Putt, F., Roper, V., Waterhouse, U., & Truman, P. (2013). Nicotine and tobacco particulate self-administration: effects of mecamylamine, SCH23390 and ketanserin pretreatment. *Current Psychopharmacology, 2*(3), 229–240.

Brennan, K. A., Putt, F., & Truman, P. (2013). Nicotine-, tobacco particulate matter- and methamphetamine-produced locomotor sensitization in rats. *Psychopharmacology (Berlin), 228*(4), 659–672.

Brower, V. G., Fu, Y., Matta, S. G., & Sharp, B. M. (2002). Rat strain differences in nicotine self-administration using an unlimited access paradigm. *Brain Research, 930*(1–2), 12–20.

Bruijnzeel, A. W., Rodrick, G., Singh, R. P., Derendorf, H., & Bauzo, R. M. (2011). Repeated pre-exposure to tobacco smoke potentiates subsequent locomotor responses to nicotine and tobacco smoke but not amphetamine in adult rats. *Pharmacology, Biochemistry, and Behavior, 100*(1), 109–118.

Carlson, J. N., Fitzgerald, L. W., Keller, R. W., Jr., & Glick, S. D. (1991). Side and region dependent changes in dopamine activation with various durations of restraint stress. *Brain Research, 550*(2), 313–318.

Chaudhri, N., Caggiula, A. R., Donny, E. C., Booth, S., Gharib, M., Craven, L., ... Sved, A. F. (2007). Self-administered and noncontingent nicotine enhance reinforced operant responding in rats: impact of nicotine dose and reinforcement schedule. *Psychopharmacology (Berlin), 190*(3), 353–362.

Chaudhri, N., Caggiula, A. R., Donny, E. C., Palmatier, M. I., Liu, X., & Sved, A. F. (2006). Complex interactions between nicotine and non-pharmacological stimuli reveal multiple roles for nicotine in reinforcement. *Psychopharmacology (Berlin), 184*(3–4), 353–366.

Clarke, P. B., & Kumar, R. (1983). The effects of nicotine on locomotor activity in non-tolerant and tolerant rats. *British Journal of Pharmacology, 78*(2), 329–337.

Collins, R. J., Weeks, J. R., Cooper, M. M., Good, P. I., & Russell, R. R. (1984). Prediction of abuse liability of drugs using IV self-administration by rats. *Psychopharmacology (Berlin), 82*(1–2), 6–13.

Corrigall, W. A., & Coen, K. M. (1989). Nicotine maintains robust self-administration in rats on a limited-access schedule. *Psychopharmacology (Berlin), 99*(4), 473–478.

Costello, M. R., Reynaga, D. D., Mojica, C. Y., Zaveri, N. T., Belluzzi, J. D., & Leslie, F. M. (2014). Comparison of the reinforcing properties of nicotine and cigarette smoke extract in rats. *Neuropsychopharmacology, 39*(8), 1843–1851.

Dani, J. A., & Balfour, D. J. (2011). Historical and current perspective on tobacco use and nicotine addiction. *Trends in Neurosciences, 34*(7), 383–392.

Danielson, K., Putt, F., Truman, P., & Kivell, B. M. (2014). The effects of nicotine and tobacco particulate matter on dopamine uptake in the rat brain. *Synapse, 68*(2), 45–60.

DiFranza, J. R., & Wellman, R. J. (2007). Sensitization to nicotine: how the animal literature might inform future human research. *Nicotine and Tobacco Research, 9*(1), 9–20.

Donny, E. C., Caggiula, A. R., Mielke, M. M., Jacobs, K. S., Rose, C., & Sved, A. F. (1998). Acquisition of nicotine self-administration in rats: the effects of dose, feeding schedule, and drug contingency. *Psychopharmacology (Berlin), 136*(1), 83–90.

Dworkin, S. I., Co, C., & Smith, J. E. (1995). Rat brain neurotransmitter turnover rates altered during withdrawal from chronic cocaine administration. *Brain Research, 682*(1–2), 116–126.

Feltenstein, M. W., Ghee, S. M., & See, R. E. (2012). Nicotine self-administration and reinstatement of nicotine-seeking in male and female rats. *Drug and Alcohol Dependence, 121*(3), 240–246.

Fowler, J. S., Volkow, N. D., Logan, J., Pappas, N., King, P., MacGregor, R., ... Gatley, S. J. (1998). An acute dose of nicotine does not inhibit MAO B in baboon brain in vivo. *Life Sciences, 63*(2), PL19–PL23.

Harris, A. C., Mattson, C., Lesage, M. G., Keyler, D. E., & Pentel, P. R. (2010). Comparison of the behavioral effects of cigarette smoke and pure nicotine in rats. *Pharmacology, Biochemistry, and Behavior, 96*(2), 217–227.

Harris, A. C., Pentel, P. R., & LeSage, M. G. (2009). Correlates of individual differences in compensatory nicotine self-administration in rats following a decrease in nicotine unit dose. *Psychopharmacology (Berlin), 205*(4), 599–611.

Harris, A. C., Stepanov, I., Pentel, P. R., & Lesage, M. G. (2012). Delivery of nicotine in an extract of a smokeless tobacco product reduces its reinforcement-attenuating and discriminative stimulus effects in rats. *Psychopharmacology (Berlin), 220*(3), 565–576.

Herraiz, T., & Chaparro, C. (2005). Human monoamine oxidase is inhibited by tobacco smoke: beta-carboline alkaloids act as potent and reversible inhibitors. *Biochemical and Biophysical Research Communications, 326*(2), 378–386.

Hiranita, T., Nawata, Y., Sakimura, K., & Yamamoto, T. (2008). Methamphetamine-seeking behavior is due to inhibition of nicotinic cholinergic transmission by activation of cannabinoid CB1 receptors. *Neuropharmacology, 55*(8), 1300–1306.

Imperato, A., Angelucci, L., Casolini, P., Zocchi, A., & Puglisi-Allegra, S. (1992). Repeated stressful experiences differently affect limbic dopamine release during and following stress. *Brain Research, 577*(2), 194–199.

Jacobs, E. H., Smit, A. B., de Vries, T. J., & Schoffelmeer, A. N. (2003). Neuroadaptive effects of active versus passive drug administration in addiction research. *Trends in Pharmacological Sciences, 24*(11), 566–573.

Lau, C. E., & Sun, L. (2002). The pharmacokinetic determinants of the frequency and pattern of intravenous cocaine self-administration in rats by pharmacokinetic modeling. *Drug Metabolism and Disposition, 30*(3), 254–261.

LeSage, M. G., Burroughs, D., Dufek, M., Keyler, D. E., & Pentel, P. R. (2004). Reinstatement of nicotine self-administration in rats by presentation of nicotine-paired stimuli, but not nicotine priming. *Pharmacology, Biochemistry, and Behavior, 79*(3), 507–513.

Lewis, A., Miller, J. H., & Lea, R. A. (2007). Monoamine oxidase and tobacco dependence. *Neurotoxicology, 28*(1), 182–195.

Lewis, A. J., Truman, P., Hosking, M. R., & Miller, J. H. (2012). Monoamine oxidase inhibitory activity in tobacco smoke varies with tobacco type. *Tobacco Control, 21*(1), 39–43.

Maisto, S. A., Galizio, M., & Connors, G. J. (2011). *Drug use and abuse* (6th ed.). Belmont, CA: Wadsworth: Cengage Learning.

Manzardo, A. M., Stein, L., & Belluzzi, J. D. (2002). Rats prefer cocaine over nicotine in a two-lever self-administration choice test. *Brain Research, 924*(1), 10–19.

Morrison, C. F., & Stephenson, J. A. (1972). The occurrence of tolerance to a central depressant effect of nicotine. *British Journal of Pharmacology, 46*(1), 151–156.

Pawlak, R., Takada, Y., Takahashi, H., Urano, T., Ihara, H., Nagai, N., & Takada, A. (2000). Differential effects of nicotine against stress-induced changes in dopaminergic system in rat striatum and hippocampus. *European Journal of Pharmacology, 387*(2), 171–177.

Richardson, N. R., & Roberts, D. C. (1996). Progressive ratio schedules in drug self-administration studies in rats: a method to evaluate reinforcing efficacy. *Journal of Neuroscience Methods, 66*(1), 1–11.

Risner, M. E., & Goldberg, S. R. (1983). A comparison of nicotine and cocaine self-administration in the dog: fixed-ratio and progressive-ratio schedules of intravenous drug infusion. *Journal of Pharmacology and Experimental Therapeutics, 224*(2), 319–326.

Rodgman, A., & Perfetti, T. A. (2013). *The chemical components of tobacco and tobacco smoke* (2nd ed.). Boca Raton, FL: CRC Press: Taylor & Francis Group.

Rotenberg, K. S., & Adir, J. (1983). Pharmacokinetics of nicotine in rats after multiple-cigarette smoke exposure. *Toxicology and Applied Pharmacology, 69*(1), 1–11.

Rotenberg, K. S., Miller, R. P., & Adir, J. (1980). Pharmacokinetics of nicotine in rats after single-cigarette smoke inhalation. *Journal of Pharmaceutical Sciences, 69*(9), 1087–1090.

Schenk, S., Hely, L., Gittings, D., Lake, B., & Daniela, E. (2008). Effects of priming injections of MDMA and cocaine on reinstatement of MDMA- and cocaine-seeking in rats. *Drug and Alcohol Dependence, 96*(3), 249–255.

Schenk, S., & Partridge, B. (1999). Cocaine-seeking produced by experimenter-administered drug injections: dose-effect relationships in rats. *Psychopharmacology (Berlin), 147*(3), 285–290.

Small, E., Shah, H. P., Davenport, J. J., Geier, J. E., Yavarovich, K. R., Yamada, H., ... Bruijnzeel, A. W. (2010). Tobacco smoke exposure induces nicotine dependence in rats. *Psychopharmacology (Berlin), 208*(1), 143–158.

Sorge, R. E., & Clarke, P. B. (2009). Rats self-administer intravenous nicotine delivered in a novel smoking-relevant procedure: effects of dopamine antagonists. *The Journal of Pharmacology and Experimental Therapeutics, 330*(2), 633–640.

Sorge, R. E., Pierre, V. J., & Clarke, P. B. (2009). Facilitation of intravenous nicotine self-administration in rats by a motivationally neutral sensory stimulus. *Psychopharmacology (Berlin), 207*(2), 191–200.

Spijkerman, R., van den Eijnden, R., van de Mheen, D., Bongers, I., & Fekkes, D. (2002). The impact of smoking and drinking on plasma levels of norharman. *European Neuropsychopharmacology, 12*(1), 61–71.

Stedman, R. L. (1968). The chemical composition of tobacco and tobacco smoke. *Chemical Review, 68*(2), 153–207.

Stefanski, R., Ladenheim, B., Lee, S. H., Cadet, J. L., & Goldberg, S. R. (1999). Neuroadaptations in the dopaminergic system after active self-administration but not after passive administration of methamphetamine. *European Journal of Pharmacology, 371*(2–3), 123–135.

Touiki, K., Rat, P., Molimard, R., Chait, A., & de Beaurepaire, R. (2007). Effects of tobacco and cigarette smoke extracts on serotonergic raphe neurons in the rat. *Neuroreport, 18*(9), 925–929.

Tuesta, L. M., Fowler, C. D., & Kenny, P. J. (2011). Recent advances in understanding nicotinic receptor signaling mechanisms that regulate drug self-administration behavior. *Biochemical Pharmacology, 82*(8), 984–995.

Villegier, A. S., Blanc, G., Glowinski, J., & Tassin, J. P. (2003). Transient behavioral sensitization to nicotine becomes long-lasting with monoamine oxidases inhibitors. *Pharmacology, Biochemistry, and Behavior, 76*(2), 267–274.

Yokel, R. A., & Piekens, R. (1974). Drug level of d- and l-amphetamine during intravenous self-administration. *Psychopharmacologia, 34*(3), 255–264.

Chapter 31

The Psychological Threat of Mortality and Its Implications for Tobacco and Alcohol Misuse

Simon McCabe[1], Jamie Arndt[2]

[1]*Behavioural Science Centre, University of Stirling, Scotland, UK;* [2]*Department of Psychology, University of Missouri – Columbia, Columbia, MO, USA*

Abbreviations

PSA Public service announcement
PTSD Posttraumatic stress disorder
TMHM Terror management health model
TMT Terror management theory

INTRODUCTION

William James (1890) referred to the "skull beneath the skin" to capture the ever-present reality of inevitable mortality with which people must contend. It is an awareness that is difficult to escape. Reminders of mortality permeate much of our social world, whether through news media and health warnings, our own experience of suddenly extreme turbulence on an airplane, or perhaps something as mundane as driving past roadkill. How might the lurking backdrop of our mortal fate and its associated reminders—sometimes pronounced and explicit, sometimes inconspicuous and subtle—relate to substance use, abuse, and addiction?

Consider the atrocity that was the terrorist attacks on the United States on September 11, 2001. Vlahov et al. (2002) randomly telephoned people 5–8 weeks after the attack and found a 28.8% increase in common drug use, with 9.9% of the sample increasing cigarette smoking, 24.6% increasing alcohol consumption, and 3.2% increasing marijuana use. What about in the months that followed? Of 1570 adults surveyed, 9.9% still reported increased cigarette smoking 6–9 months later, 17.5% still reported increased alcohol use, and 2.7% still reported increased marijuana use compared to the month before September 11, 2001 (Vlahov, Galea, Ahern, Resnick, & Kilpatrick, 2004). Furthermore, 18 months after the attacks, elevated levels of substance use remained. Students who had some but minimal exposure were found (controlling for depression and posttraumatic stress disorder (PTSD)) to have a fivefold increase in substance use, whereas those with more intense exposures had a nearly 19-fold increase (Chemtob, Nomura, Josephson, Adams, & Sederer, 2009).

There are, of course, likely to be various factors that contributed to these alarming increases in substance use (e.g., the development of psychological disorders and maladaptive stress coping strategies). However, given the propensity for the 9/11 attacks to elicit thoughts of mortality (e.g., Pyszczynski, Solomon, & Greenberg, 2003), these studies raise the possibility that the existential threat of mortality might have important implications for substance use, abuse, and addiction that warrant further scrutiny. In this chapter, the terror management health model (TMHM; Goldenberg & Arndt, 2008) is used to consider these implications. Broadly stated, the model offers a framework for understanding the health consequences of people's efforts to psychologically manage conscious and nonconscious mortality concerns. The chapter begins with a brief review of the model's theoretical foundation before turning to how the forms of psychological defense that the model describes may facilitate understanding of substance use.

Before proceeding, note that the present focus on generalized substance-related implications of psychological defenses against everyday mortality awareness is not intended to downplay the importance of substance use related to bereavement, nor the importance of unique factors influencing the use and abuse of particular substances, such as alcohol and marijuana.

Conceptual Foundation: Terror Management Theory

The TMHM stems from terror management theory (TMT; Greenberg, Pyszczynski, & Solomon, 1986), which is based largely on the work of cultural anthropologist Ernest Becker (1973). Becker took a multidisciplinary approach, synthesizing a long tradition of scholarly work to describe how the potentially overwhelming anxiety (terror) evoked by peoples' knowledge of mortality—and the efforts to manage such anxiety—have implications across a broad spectrum of human activities. TMT translated these ideas into a formal social psychological theory, allowing Becker's ideas to be examined under the lens of empirical scrutiny (see e.g., Burke, Martens, & Faucher, 2010; Greenberg, Solomon, & Arndt, 2008; for conceptual and meta-analytic reviews, respectively).

The basic premise of TMT is that people evolved to manage the existential threat of mortality awareness by investing in a culturally derived system of beliefs (i.e., a cultural worldview) that imbues their life with a sense of meaning, order, and permanence. Furthermore, the cultural belief system provides prescriptions for valued behavior that allow the individual to obtain a sense of significance (i.e., self-esteem) within the cultural meaning system (an overview of TMT is provided in Figure 1). In this light, one of the broader hypotheses derived from the theory is that awareness of mortality motivates people to identify with their cultural worldview and to live up to its values.

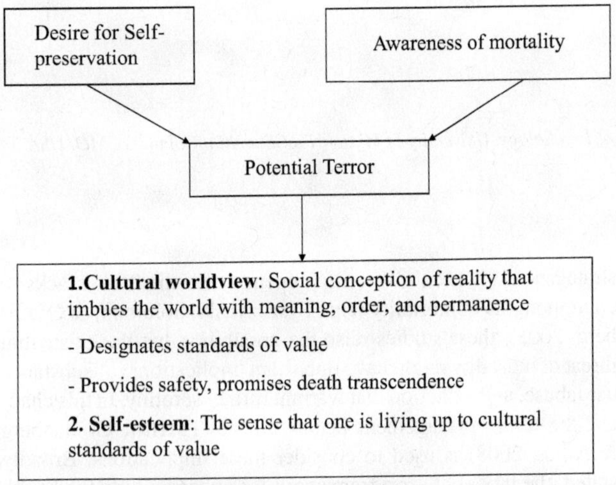

To examine the role of mortality awareness in human social behavior, researchers often either manipulate the salience of mortality (i.e., mortality salience) or measure the cognitive accessibility of death-related thought (i.e., how much people may be thinking about death even if not aware of it; see Hayes, Schimel, Arndt, & Faucher, 2010 for a review). Methods for inducing mortality salience include, for example, answering open-ended questions about death, subliminal primes, proximity to a funeral home, viewing gory images of mortality, health warnings, news footage, and temporal proximity to 9/11. To date, hundreds of studies utilizing a range of samples (e.g., college students, the elderly, judges, soldiers) from at least 15 countries (see e.g., Burke et al., 2010) have demonstrated a role of mortality concerns in a variety of domains of human social behavior (e.g., prejudice, aggression, close relationships, political and legal judgment, consumer decisions). Furthermore, a range of negative and threatening control topics have been used in these studies, suggesting that outcomes are the result of the unique influence of mortality-related cognitions, and not threat or negative cognitions broadly.

FIGURE 1 **Graphic overview of terror management theory.** The conflicting desires for self-preservation and awareness mortality engender the potential for anxiety, which is managed by investing in a cultural worldview and maintaining self-esteem.

Terror Management Health Model

Goldenberg and Arndt (2008) proposed the TMHM (see Figure 2) to explore how terror management processes may operate in health contexts. The model begins with the hypothesis that health conditions and warnings have the capacity to activate death-related concerns to varying degrees. For example, and with particular relevance to the present focus on substance use, presenting people with communications about cigarette smoking or drunk driving increases the accessibility of death-related thought (e.g., Hansen, Winzeler, & Topolinski, 2010; Jessop & Wade, 2008). The model

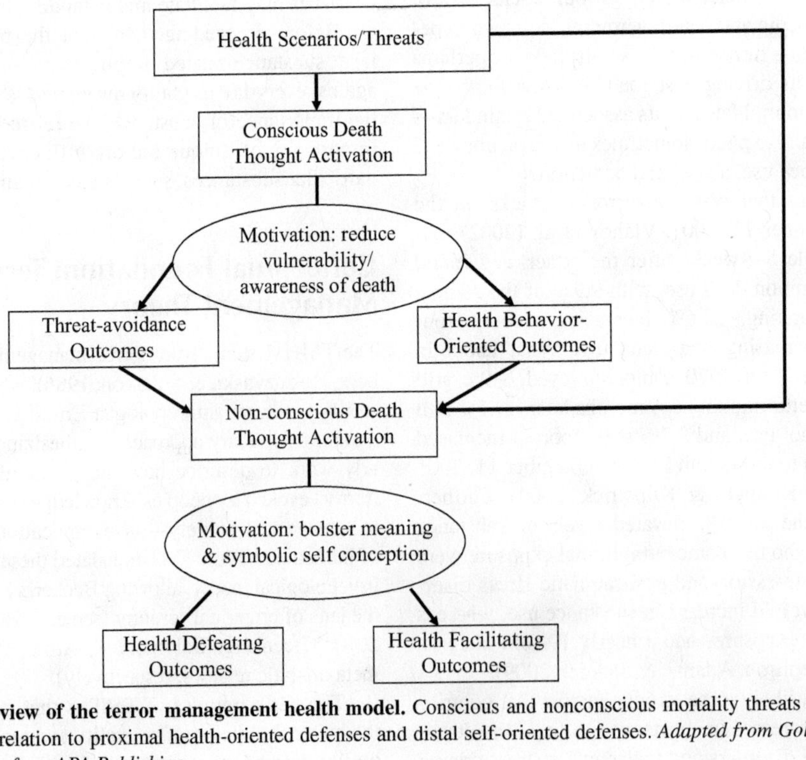

FIGURE 2 **Graphic overview of the terror management health model.** Conscious and nonconscious mortality threats as the result of health scenarios/threats are shown in relation to proximal health-oriented defenses and distal self-oriented defenses. *Adapted from Goldenberg and Arndt (2008). Reproduced with permission from APA Publishing.*

then integrates traditional insights from TMT about how people manage conscious and nonconscious death-related cognitions (Pyszczynski, Greenberg, & Solomon, 1999) with the recognition that health decisions can be influenced by concerns central (or *proximal*; e.g., vulnerability perceptions; Rothman & Schwarz, 1998) and more tangential (or *distal*; e.g., esteem and presentational motivation; Leary, Tchividijian, & Kraxberger, 1994) to the health context.

The guiding idea is that when mortality concerns are conscious, health decisions are largely driven by the proximal motivational goals of reducing perceived vulnerability to a health threat and removing death-related thought from focal attention. These efforts can either take a productive route of engaging in healthy behavior or less adaptive responses of trying to avoid, or deny, the threat. In contrast, when mortality concerns are active but outside of focal attention, health-relevant decisions are guided more by the distal motivational goals of bolstering self-esteem and maintaining one's symbolic conception of self. These efforts can also proceed along adaptive or maladaptive health trajectories, but here motivations are focused on the value and meaning of the self. In this way, the model informs how concerns about mortality can have multifaceted implications for health decisions.

The majority of TMHM research has focused on non-substance-related health domains (e.g., cancer cognitions, sun-tanning, fitness, nutrition, screening decisions). One study illustrating the basic distinction the model hypothesizes comes from McCabe, Vail, Arndt, and Goldenberg (2014). This study examined the efficacy of endorsements for an ostensible new brand of bottled water given by either a medical doctor or a popular-culture celebrity. Which endorsement might be more effective? The TMHM suggests that the answer depends on whether thoughts of death are in focal awareness or active but outside of focal awareness. Accordingly, when participants were reminded of death and immediately after received a water bottle featuring an endorsement by a medical doctor, they consumed more water than participants in the other conditions. Indeed, the cultural celebrity had no effect in this context. However, when thoughts of death were elicited but allowed to fade from focal attention (via the completion of a delay task; see Hayes et al., 2010; Pyszczynski et al., 1999), participants drank more water when it was accompanied with an endorsement by the cultural celebrity. In this context, the medical endorsement had little effect.

These findings illustrate the basic distinction that has been observed in a number of health-oriented domains. Conscious thoughts of death elicit motivations geared toward maintaining perceptions of health (suggested here by the greater efficacy of the medical doctor's endorsement). In contrast, active but nonconscious thoughts of death elicit motivations geared toward maintaining cultural conceptions of value (suggested here by the greater efficacy of the celebrity endorsement). Further evidence for these two distinct forms of psychological defense comes from a number of lines of research, including studies suggesting distinct neural correlates (Klackl, Jonas, & Kronbichler, 2013; Quirin et al., 2011). These forms of psychological defense, in turn, provide a useful heuristic for understanding how awareness of mortality impacts substance use behavior; moreover, they can be a generative platform for future research directions (an overview of these heuristics in relation to substance use is provided in Figure 3). The remainder of this chapter is devoted to considering extant research and elucidating that potential.

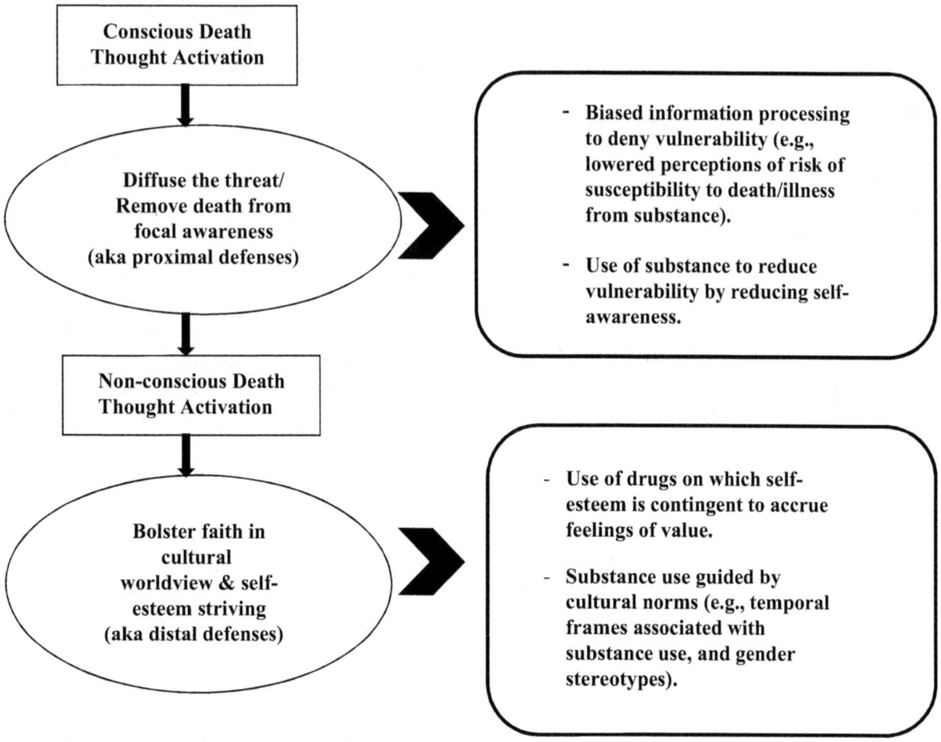

FIGURE 3 Proximal and distal defenses: Implications for alcohol and tobacco use. Implications are reviewed for proximal and distal defenses for substance use.

PSYCHOLOGICAL DEFENSES AGAINST CONSCIOUS MORTALITY CONCERNS: IMPLICATIONS FOR SUBSTANCE USE

The propensity for people to respond to conscious thoughts of mortality with a health orientation that can limit perceptions of vulnerability suggests a number of ways that substance use may be affected by conscious thoughts of death. To the extent that conscious thoughts of mortality engage productive efforts to limit vulnerability to death by improving one's health, dangerous or health threatening use of substances might be reduced. This is similar to the logic underlying appeals that elicit fear over negative health outcomes, as well as efficacy in avoiding the cause (or risk) of that fear (Leventhal, 1971). The hypothesis that conscious thoughts of death will reduce substance use has not been examined directly, but research showing that such cognition can increase productive health intentions (e.g., to exercise more, use sunscreen: Arndt, Schimel, & Goldenberg, 2003; Routledge, Arndt, & Goldenberg, 2004), particularly when perceived efficacy and other such perceptions are high (e.g., Cooper, Goldenberg, & Arndt, 2010), is consistent with this possibility.

Unfortunately, reminders of mortality have also been found to engage biased information processes that limit not so much actual vulnerability, but perceptions thereof through denial or avoidance based strategies. For example, immediately after being reminded of death (vs a control topic of watching television), participants biased their self-assessment of emotionality to deny vulnerability to a short life expectancy (Greenberg, Arndt, Simon, Pyszczynski, & Solomon, 2000). Adopting a threat avoidance orientation that entails a reduction in perceptions of risk may in turn contribute to increased substance use.

Risk perception is an established predictor of substance use and abuse, as lower substance use risk perception, for example, is associated with higher consumption of both legal and illicit substances (e.g., Bejarano et al., 2011; Johnston, 2003). Thus, to the extent that proximal defenses against conscious mortality threats can lead to a denial of vulnerability, at least for some individuals, substance use may be perceived as less risky when mortality concerns are in focal awareness. Although limited, there is some evidence consistent with this theorizing. When confronted with health messages that accentuated the fatal consequences of smoking and increased the accessibility of death-related thought, relative to messages that had less impact on death thought accessibility and instead accentuated social exclusion consequences, smokers rated their health risks as lower (Martin & Kamins, 2010).

Such reactions might then influence actual tobacco consumption. Arndt et al. (2013) examined how conscious thoughts of mortality affect smoking behavior among light habitual smokers depending on their level of craving. High smoking craving is associated with discounting future health risks and a greater focus on more positive (e.g., affective rewards) aspects of smoking (e.g., Sayette, Loewenstein, Kirchner, & Travis, 2005). Accordingly, it was hypothesized that if reminders of death further attenuate risk perceptions, participants high in smoking craving who are reminded of mortality would smoke with greater intensity as measured by the topography of their inhalations (e.g., puff duration). The results of two studies were consistent with this logic, indicating that participants who were high (vs low) in tobacco craving, after writing about death (vs another aversive topic or a more neutral control topic), smoked with greater intensity (see Figure 4).

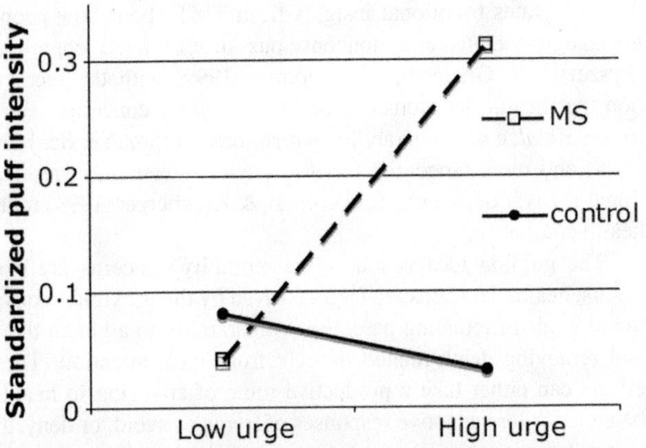

FIGURE 4 Smoking topography (puff intensity) as a function of reminders of mortality (vs control) and craving intensity. Smoking topography (puff intensity) is shown for the interaction between mortality reminder (vs failing an examination) and craving intensity. *Reproduced with permission from APA Publishing.*

Of course, people have a number of biases when it comes to assessing their risk level for negative health consequences, and this is especially true of smokers (e.g., Weinstein, Marcus, & Moser, 2005). One such bias is the underaccumulation bias, wherein people fail to appreciate how risk accumulates, often exponentially (Linville, Fischer, & Fischhoff, 1993). Klein, Koblitz, Kaufman, Vail, and Arndt (2014) thus examined whether reminders of mortality influence the way smokers perceive their death from lung cancer risk over the lifespan. Young adult smokers were reminded of mortality or a control topic about failing an examination and then estimated their 10-year relative lung cancer risk at age 35, 45, and 55 years. As illustrated by Thun et al. (2013), relative lung cancer risk accumulates dramatically during this age span. However, although smokers acknowledged they were at greater risk than nonsmokers at all three time points—consistent with work showing that smokers appreciate their greater cancer risk compared with nonsmokers (e.g., Weinstein et al., 2005)—reminders of mortality decreased these risk estimates and strengthened the underaccumulation bias. This pattern was especially pronounced for smokers with a strong smoker identity. Taken together, extant research suggests that thoughts of mortality, while certainly having the potential to motivate more responsible substance use or cessation intentions, also carry the potential to ironically increase risky substance use through biased processing that lowers perceived risk.

Another potential way that mortality concerns may ironically increase substance use is by helping to attenuate awareness of vulnerability by reducing focus on the self. Noted film writer and director John Cassavetes, in his film *Shadows* (1959), includes the line "man is conscious of his existence and therefore conscious of his non-existence; ergo, he has anxiety." This quote captures the idea that self-awareness can exacerbate conscious mortality concerns because self-awareness makes one more aware of one's existence (Pyszczynski, Greenberg, Solomon, & Hamilton, 1990). Indeed, studies find that not only do reminders of mortality motivate people to avoid focusing attention on the self (Arndt, Greenberg, Simon, Pyszczynski, & Solomon, 1998), but leading people to focus attention on the self also increases the accessibility of death-related thought (Silvia, 2001). The relevance of this becomes apparent when considering that a range of substances have been associated with "fleeing the self," or numbing self-awareness to dull anxiety and stress (e.g., Sayette, 1999). Indeed, metaphors associated with various drugs hint at this effect, such as "getting out of my head" (Lakoff & Johnson, 1980). More substantially, Hull (1981) proposed that alcohol can serve to decrease self-awareness by interfering with encoding processes fundamental to a state of self-awareness— decreasing sensitivity to self-relevant cues and self-evaluative feedback. Alcohol use is therefore argued to provide a form of psychological relief when the self is threatened. The model is supported by evidence that, for example, alcohol (vs placebo) consumption results in participants giving fewer self-focused statements when giving a short speech after imbibing (Hull, Levenson, Young, & Sher, 1983), and that when the self is threatened (e.g., through negative performance feedback), those high in self-awareness consume more alcohol (Hull & Young, 1983). Integrating Hull's model with the TMHM thus suggests that substance use may offer one (ironic) strategy of placating mortality concerns (Ein-Dor et al., 2014). Of course, at this point the connection between mortality concerns, self-awareness, and substance use awaits further research.

Psychological Defenses against Nonconscious Mortality Concerns: Implications for Substance Use

While conscious thoughts of death engage proximal health-oriented defenses, nonconscious death thoughts result in psychological responses that augment the individuals' sense of personal value and overall sense of culturally infused meaning. The elicitation of nonconscious death thoughts, and subsequent engagement of these psychological defenses, has been achieved experimentally with a number of techniques, including explicit reminders of death from which participants are then distracted, or presenting participants with reminders of mortality that are subtle or subliminal. This next section considers the implications of these forms of psychological defense for substance use.

Nonconscious Mortality Concerns, Self-esteem Striving, and Substance Use

Decades of research attest to the ubiquity and central importance of self-esteem to people (Sedikides, Gaertner, & Toguchi, 2003). The domains people can draw self-esteem from, or contingencies of self-worth, are numerous (Crocker & Wolfe, 2001). Of relevance, some individuals derive their self-esteem either directly or indirectly from behaviors that implicate risky substance use. For example, adolescents and young adults particularly may derive self-esteem from smoking or drinking behavior, as such reinforcing a desirable self-image (McCool, Cameron, & Petrie, 2004; Nichter et al., 2006). Indeed, college students who base their self-worth on their appearance tend to engage in heavier drinking, presumably at least in part because it is seen as fostering a popular image (Luhtanen & Crocker, 2005).

An extensive literature supports the terror management hypothesis that nonconscious thoughts of death motivate striving to obtain self-esteem in culturally prescribed ways (Pyszczynski, Greenberg, Solomon, Arndt, & Schimel, 2004). Given that some individuals develop substance-based contingencies of self-esteem, positive associations and consumption behaviors associated with the substance may be exacerbated in an effort to quell nonconscious death concerns. Research has considered this theorizing with regard to two legal and common substances, tobacco and alcohol.

Studies have examined the link between nonconscious mortality concerns, smoking-based self-esteem, and smoking attitudes and behaviors. Hansen et al. (2010) aimed to inform the efficacy of warning labels on cigarette packets that featured mortality reminders. The study had participants complete a measure of smoker-based self-esteem (e.g., "Smoking allows me to feel worthy") and found that mortality salience, as delivered via the warning message on a cigarette packet (vs a control topic of a warning that did not implicate mortality), resulted in more positive attitudes toward smoking for those high (vs low) in smoker self-esteem.

Notably, while the Hansen et al. (2010) study suggests that mortality reminders can increase smoking intentions when the behavior is a basis of self-worth, the overall conceptual analysis also suggests that interventions might capitalize on these motivations to increase cessation intentions. Preliminary research on this possibility is encouraging. In one study from Arndt et al., (2009), researchers first assessed the extent to which young adults smoked for image and esteem-oriented reasons. Participants then either answered questions about mortality (vs failing), followed by a delay task (to allow mortality concerns to fade from focal awareness). Participants then watched a short commercial that highlighted the negative social consequences of smoking (e.g., that it smells bad, lowers dating prospects). Participants then evaluated the commercial (e.g., rated its persuasive impact) and reported their intentions to quit smoking. When participants were reminded of mortality (vs failing), the more they smoked for image and esteem oriented reasons, the more they endorsed quitting after viewing the commercial targeting image consequences of smoking (see Figure 5).

The interactive effect of deriving self-esteem from smoking and antismoking appeals featuring either fatal health risks or social exclusion risks was also explored in the aforementioned study by Martin and Kamins (2010). Smokers who derived self-esteem from smoking reported greater short terms intentions to quit when exposed to an antismoking message that focused on social exclusion consequences rather than health consequences. Collectively, these studies highlight the important role that contingencies of self-esteem have for the efficacy of smoking deterrents.

The role of nonconscious mortality concerns and self-esteem contingencies have also been implicated in alcohol use, binge drinking attitudes, and deterrent efficacy. Jessop and Wade (2008) explored responses to information concerning the health consequences of binge drinking, how this may have different implications for binge drinking when that information is presented with a mortality threat, and how varying degrees of alcohol-related self-esteem affect such responses. Participants first completed measures of the frequency of binge drinking and also the extent to which they derive self-esteem from binge drinking. They were

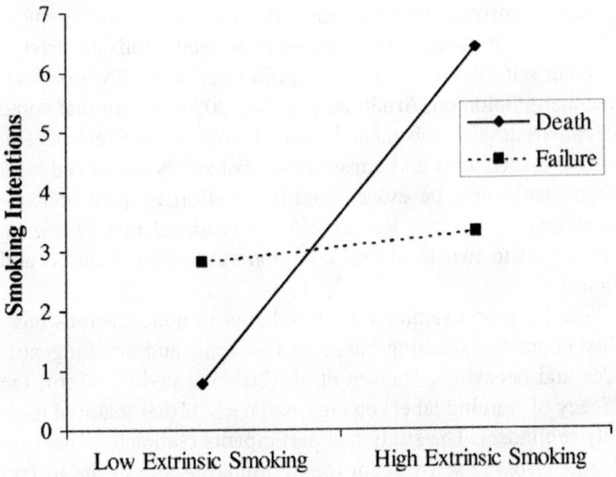

FIGURE 5 Smoking intentions as a function of reminders of mortality (vs control) and extrinsic smoker self-esteem. Smoking intentions are shown for the interaction between mortality reminder (vs examination failure), and extrinsic self-esteem. *Reproduced with permission from APA Publishing.*

then exposed to one of three conditions communicating information about alcohol. One condition featured warnings about binge drinking including a reminder of mortality ("Binge drinking significantly increases your risk of getting liver, mouth and esophagus cancer, all of which can be fatal"), a second featured information on binge drinking without the mortality reminder ("Binge drinking can make you look foolish in front of the people you care about"), and the final condition simply presented neutral information on alcohol ("Each molecule of alcohol is less than a billionth of a meter long"). Relative to communications that did not feature reminders of mortality, the risk communication about binge drinking that also primed thoughts of mortality increased willingness to binge drink for those participants who were labeled as binge drinkers, and, interestingly for non-binge drinkers who perceived binge drinking as benefitting (vs not) self-esteem. Wong and Dunn (2013) replicated and extended these findings. In this research, college students who had high (vs low) binge drinking contingent self-esteem and were exposed to an anti-binge drinking public service announcement (PSA) featuring mortality primes (vs not) reported lower intentions to not binge drink, more negative evaluations of the PSA, greater message avoidance, and more positive binge drinking attitudes.

Similar findings are obtained when examining the efficacy of messages designed to deter drunk driving. Shehryar and Hunt (2005) hypothesized that for those who derive a sense of self-value from drinking, the efficacy of anti-drunk driving appeals featuring mortality reminders would be blunted as alcohol consumption serves as a way to attain self-esteem. To probe this hypothesis, participants completed a measure of identity investment in drinking (e.g., "I like being known as a person who drinks alcohol") and were shown an anti-drunk driving message containing a fear appeal relating to fear of arrest, serious injury, or mortality. Participants whose sense of identity involved a high commitment to drinking (vs low on this measure), when exposed to an anti-drunk driving message that also primed mortality, evaluated drunk driving as more acceptable relative to the same message that replaced mortality threat with the threat of arrest or serious injury.

Nonconscious Mortality Concerns, Cultural Values, and Substance Use

In addition to self-esteem striving, traditional TMT research has focused on awareness of mortality as a key motivational catalyst in peoples' identification with cultural norms and beliefs (Greenberg et al., 2008). Adherence to cultural values is argued to provide existential security because it provides a sense that one is part of something bigger and longer lasting than the self. This suggests that when thoughts of death are active but outside of focal awareness, substance use and related attitudes may vary as a function of prevailing or situationally salient cultural norms and values.

One widespread norm across many cultures is of course the consumption of alcohol, particularly on college campuses. Thus, one possibility is that subtle reminders of mortality might increase interest in alcohol consumption. Ein-Dor et al. (2014) examined this potential in a clever field experiment on a college campus in Israel. Research assistants handed out fliers to people passing by that served either to elicit a subtle reminder of mortality or back pain. In the mortality condition, the flier included the line "Are you concerned about death? We can help! Call us and we can ease your suffering both physically and spiritually"; the back pain flyer had a parallel line. Fifteen meters away, another research assistant was offering an alcoholic or nonalcoholic beverage. Those participants in the subtle (nonconscious) mortality reminder condition were more likely to consume an alcoholic beverage in comparison to those in the back pain condition. This study is important in suggesting that subtle reminders of mortality in an ecologically valid setting can elicit actual alcohol consumption. Of course, while these findings are consistent with subtle reminders of mortality increasing culturally normative alcohol consumption, there are other possible explanations (e.g., distress regulation). We return to this issue later in the chapter.

Cultural meanings also have implications for substance use when considering gender and the cultural value of risk-taking. Many western cultures include stereotypes of males as more likely to take risks, and some have argued males are socialized to be more risk-oriented than females (Kelling, Zerkes, & Myerowitz, 1976). This gender-based difference in risk-taking might have repercussions for substance use when people are reminded of mortality. Indeed, Hirschberger, Florian, Mikulincer, Goldenberg, and Pyszczynski (2002) demonstrated that after being reminded of, and then distracted from, mortality (vs a control topic of visiting a shopping mall), men (but not women) indicated that they were more willing to try different psychoactive drugs such as hashish, heroin, and LSD. This highlights that efforts to manage nonconscious mortality concerns by living up to cultural values pertaining to risk can have negative implications for attitudes toward a range of substances, especially for men for whom risk is culturally valued.

Culture does more than communicate who should engage in particular behaviors. It also communicates when they should do so. The weekend effect is one such temporally demarcated cultural norm, and refers to the tendency for people (particularly college students) to engage in higher levels of alcohol consumption on Fridays and Saturdays, in comparison to other days (e.g., Maggs, Williams, & Lee, 2011). Therefore, if the weekend is associated with the cultural norm of alcohol consumption and mortality reminders spur

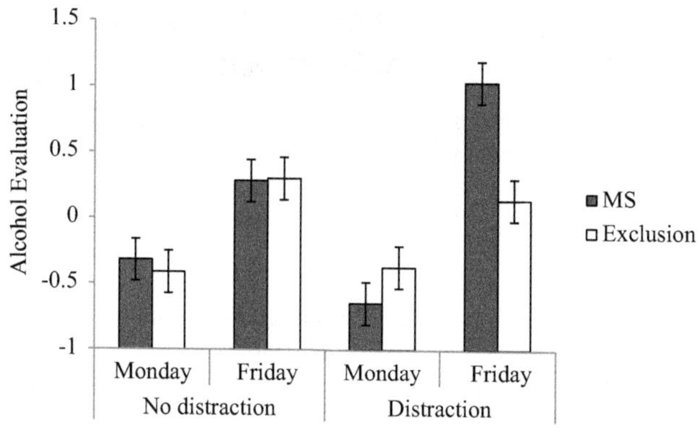

FIGURE 6 **Attitudes toward alcohol as function of reminders of mortality (vs control) and imagined weekday.** Evaluations of alcohol are shown for the interaction between mortality reminder (vs exclusion) and imagined weekday (Monday vs Friday). *Data are from McCabe et al. (2016).*

adherence to cultural norms, then reminders of mortality should exacerbate the tendency to have more positive evaluations of alcohol on a Friday in comparison to a Monday (when it is not the cultural norm to drink alcohol). McCabe, Arndt, Bartholow, and Engelhardt (2016) conducted two studies to examine this hypothesis. Participants were reminded of mortality (vs a control topic of social exclusion), given a delay task (to allow death thoughts to migrate away from consciousness), and then asked to indicate their attitudes and intentions about drinking alcohol on either a Monday or a Friday. As shown in Figure 6, after mortality reminders, students evaluated alcohol more positively on a Friday, when it is the cultural norm to drink, but not Monday, when it is not the cultural norm. These effects were replicated when the study was run on an actual Monday or Friday (as opposed to participants being asked to imagine these days) and participants were asked about their willingness to drink that evening. This suggests that adherence to cultural values may have implications for substance use contingent on temporal demarcations, and the cultural meanings regarding the substance that are yoked to those discrete temporal units.

In addition to culture prescribing temporal periods when substance use might confer value, certain culturally valued individuals (e.g., celebrities) also hold the capacity to steer what is valued in ways that can affect substance use. For example, the more people watch movies portraying characters smoking or drinking, the more they report positive attitudes toward the substances (Dal Cin et al., 2009), and even motivations to initiate the use of such substances (Dalton et al., 2003). Would the psychological threat of mortality increase adherence to the values that are touted by such cultural figures in ways that can reduce substance use?

Recall the illustrative study on celebrity endorsement by McCabe et al. (2014) that was presented earlier in the chapter. McCabe et al. extended this work to examine the interactive effect of mortality reminders and celebrity endorsement on both alcohol- and smoking-related attitudes. In one study, participants were reminded of (and then distracted from) mortality or uncertainty. They then read an ostensible appeal for responsible drinking by a medical doctor, a popular celebrity, or a celebrity whose fame was waning. Mortality reminders increased intentions to drink responsibly (as opposed to excessive drinking) when a "drink responsibly" public service campaign was endorsed by a popular and successful celebrity, but not when endorsed by an unpopular celebrity or a noncelebrity (e.g., medical doctor). An additional study found that when a celebrity endorsed a nonsmoking campaign (vs influenza vaccination), participants reminded of mortality (vs failure) subsequently indicated that it was more important to their sense of identity to remain a nonsmoker and increased the negativity with which they viewed smokers. These findings speak to the power that certain culturally valued individuals (i.e., celebrities) hold and how it may be harnessed to aid efforts to deter substance use.

Summary of a Terror Management Perspective on Substance Use

When adopting a rationalist perspective to thinking about how awareness of mortality might impact substance use, it is tempting to conclude that cognizance of our health vulnerabilities should encourage safer, more responsible behavior. Yet, although research exploring how existential motivations influence substance use is still in its infancy, one overarching contribution of the accumulating research is that, while this certainly can be the case, it is not necessarily, or perhaps even often, so.

When people are explicitly aware of mortality, ensuing psychological responses may feature a reduction in risk perceptions, a desire to escape self-awareness, and other denial or avoidance strategies that can ironically contribute to increased substance use. In comparison, when thoughts of mortality are active but outside of people's focal awareness, motivations for the existential comforts of esteem, social connection, and cultural meaning can again ironically dispose people to be more inclined, at least under certain circumstance, to increase substance use behavior. Yet there is also encouraging potential, under certain circumstances, for the respective motivations engaged by mortality reminders to help reduce substance use.

APPLICATIONS TO OTHER ADDICTIONS AND SUBSTANCE MISUSE

Perhaps the most notable implication of this research, then, is the advice for caution in how health communications that feature reminders of death should be used when the goal is to discourage substance use behavior. While research highlights potential pairings that are promising (e.g., pairing medical and efficacy-oriented information with explicit references to fatal risks, but pairing esteem-oriented information with more subtle references to fatalities), more research is needed. In addition, the extant literature features limitations and gaps that can be informed by further research. For example, research has yet to really consider the potential for people to perceive substances to assist in the management of distress, fear, and other forms of negative affect that might emanate from awareness of mortality. Part of the reason for this neglect might be that TMT has traditionally not found that conventional manipulations of mortality salience increase consciously experienced affect (Greenberg et al., 2008). Further, although studies suggest that conventional inductions of mortality salience can have some impact on participants' reports of fear, these reports do not covary with the cultural identifications that people use as psychological defense (e.g., Lambert et al., 2014). Nonetheless, there is evidence that people are motivated to circumvent the potential for anxiety that may be signaled by subtle reminders of mortality (Greenberg et al., 2003), and clearly there is a long history of research on how substances can be used as a distress regulation strategy (e.g., Holahan, Moos, Holahan, Cronkite, & Randall, 2001). Thus, the intersection between existential insecurity, distress regulation, and substance use could be an important direction for future research.

Another limitation of the extant research is a preponderant (although not exclusive) focus on reports of attitudes and intentions rather than substance use behavior, as well as a dearth of research into the role of existential concerns in longer term addiction. Indeed, the research reviewed here has focused on how awareness of mortality affects single shot reports or behaviors. Research is needed to understand how (both conscious and nonconscious) mortality concerns may operate in the context of sustained addiction. Considering the interplay of TMHM's proximal and distal defenses may afford a novel perspective on how mortality concerns and subsequent defenses (or lack thereof) can serve as part of the foundation for addiction.

One possibility is that individuals who have dispositionally low self-esteem, or experience a threat to their self-esteem, may experience more conscious mortality thoughts (see Hayes, Schimel, & Williams, 2008; Routledge et al., 2010). This might then result in nonconscious mortality thoughts being "kicked up" to a conscious level and engage risk perception biases and self-focus avoidance strategies that encourage more problematic substance use to mitigate the excessive and frequent mortality thoughts. As prolonged substance use may also be associated with lower levels of self-esteem (e.g., Zimmerman, Copeland, Shope, & Dielman, 1997), this may then further degrade the psychological resources that an individual can use to manage underlying existential concerns, which in turn would contribute to a problematic cycle of increasing and continued substance use.

Notably, this account is speculative and is articulated as just one possible direction for future research. Understanding the role of existential concerns and the psychological management thereof in substance use and addiction is still at an early stage. Yet, given the scope of the problems with which society must contend, such a perspective may be positioned to offer novel insights.

DEFINITION OF TERMS

Cultural worldview A socially shared structured system of values, beliefs, scripts, and meanings.

Death thought accessibility Mortality-related cognition that is active but outside of conscious attention; can be measured in a number of ways (e.g., word-stem completion tasks, lexical decision tasks).

Distal defense Psychological responses to mortality concerns that are outside of focal awareness, involving identification with cultural beliefs and striving for self-esteem.

Mortality salience Thoughts of mortality that are brought to mind, generally using experimental manipulations that prime, or activate, thoughts of death.

Proximal defense Psychological responses to mortality concerns in active attention, involving decreasing perceptions of vulnerability and removing such cognition from awareness.

Self-esteem An individuals' sense of their overall worth or value.

KEY FACTS CONCERNING THE TERROR MANAGEMENT HEALTH MODEL

- The model provides an existential-social-psychological framework for understanding the impact of mortality concerns on health attitudes and behaviors.
- The model was developed by Goldenberg and Arndt (2008).
- The model is an extension of terror management theory (Greenberg et al., 1986).
- To date, hundreds of terror management studies have been conducted in at least 15 countries (Burke et al., 2010).
- The model has been used to examine a range of substance use attitudes and behaviors, but has to date focused predominantly on alcohol and tobacco.

SUMMARY POINTS

- This chapter reviews the relationship between how people psychologically manage the awareness of mortality and substance use through the lens of the terror management health model.
- Mortality reminders are ubiquitous and can be explicit and conscious, or subtle and nonconscious.
- When thoughts of mortality are conscious, proximal defenses are engaged that reduce perceived vulnerability to a health threat and remove death-related thought from focal attention. These efforts can either take a productive route of engaging in healthy behavior or less adaptive responses of trying to avoid, or deny, the threat.
- When thoughts of mortality are active but not conscious, distal defenses are engaged to bolster self-esteem and maintain one's symbolic conception of self. These efforts can also proceed along adaptive or maladaptive health trajectories.
- Studies have found that reminders of mortality, and the consequent motivations, can both increase and decrease substance-related attitudes and behaviors, particularly in the areas of alcohol and tobacco use.
- Studies converge to suggest caution is warranted in how messages intended to deter substance use employ reminders of death.
- Research on the role of existential fears of mortality in substance use and addiction is in its infancy, but it is sufficiently promising to warrant further research.

REFERENCES

Arndt, J., Cox, C. R., Goldenberg, J. L., Vess, M., Routledge, C., Cooper, D. P., & Cohen, F. (2009). Blowing in the (social) wind: implications of extrinsic esteem contingencies for terror management and health. *Journal of Personality and Social Psychology*, 96, 1191–1205.

Arndt, J., Greenberg, J., Simon, L., Pyszczynski, T., & Solomon, S. (1998). Terror management and self-awareness: evidence that mortality salience provokes avoidance of the self-focused state. *Personality and Social Psychology Bulletin*, 24, 1216–1227.

Arndt, J., Schimel, J., & Goldenberg, J. L. (2003). Death can be good for your health: fitness intentions as a proximal and distal defense against mortality salience. *Journal of Applied Social Psychology*, 33, 1726–1746.

Arndt, J., Vail, K. E., III, Cox, C. R., Goldenberg, J. L., Piasecki, T. M., & Gibbons, F. X. (2013). The interactive effect of mortality reminders and tobacco craving on smoking topography. *Health Psychology*, 32, 525–532.

Becker, E. (1973). *The denial of death*. New York, NY: Free Press.

Bejarano, B., Ahumada, G., Sánchez, G., Cadenas, N., de Marco, M., Hynes, M., & Cumsille, F. (2011). Perception of risk and drug use: an exploratory analysis of explanatory factors in six Latin American countries. *The Journal of International Drug, Alcohol and Tobacco Research, 1*(1), 9–17.

Burke, B. L., Martens, A., & Faucher, E. H. (2010). Two decades of terror management theory: a meta-analysis of mortality salience research. *Personality and Social Psychology Review, 14*, 155–195.

Chemtob, C. M., Nomura, Y., Josephson, L., Adams, R. E., & Sederer, L. (2009). Substance use and functional impairment among adolescents directly exposed to the 2001 World Trade Center attacks. *Disasters, 33*, 337–352.

Cooper, D. P., Goldenberg, J. L., & Arndt, J. (2010). Examining the terror management health model: the interactive effect of conscious death thought and health-coping variables on decisions in potentially fatal health domains. *Personality and Social Psychology Bulletin, 36*, 937–946.

Crocker, J., & Wolfe, C. T. (2001). Contingencies of self-worth. *Psychological Review, 108*, 593–623.

Dal Cin, S., Worth, K. A., Gerrard, M., Gibbons, F. X., Stoolmiller, M., Wills, T. A., & Sargent, J. D. (2009). Watching and drinking: expectancies, prototypes, and friends' alcohol use mediate the effect of exposure to alcohol use in movies on adolescent drinking. *Health Psychology, 28*, 473–483.

Dalton, M. A., Sargent, J. D., Beach, M. L., Titus-Ernstoff, L., Gibson, J. J., Ahrens, M. B., & Heatherton, T. F. (2003). Effect of viewing smoking in movies on adolescent smoking initiation: a cohort study. *The Lancet, 362*, 281–285.

Ein-Dor, T., Hirschberger, G., Perry, A., Levin, N., Cohen, R., Horesh, H., & Rothschild, E. (2014). Implicit death primes increase alcohol consumption. *Health Psychology, 33*, 748–751.

Goldenberg, J. L., & Arndt, J. (2008). The implications of death for health: a terror management health model for behavioral health promotion. *Psychological Review, 115*, 1032–1053.

Greenberg, J., Arndt, J., Simon, L., Pyszczynski, T., & Solomon, S. (2000). Proximal and distal defenses in response to reminders of one's mortality: evidence of a temporal sequence. *Personality and Social Psychology Bulletin, 26*, 91–99.

Greenberg, J., Martens, A., Jonas, E., Eisenstadt, D., Pyszczynski, T., & Solomon, S. (2003). Psychological defense in anticipation of anxiety: eliminating the potential for anxiety eliminates the effects of mortality salience on worldview defense. *Psychological Science, 14*, 516–519.

Greenberg, J., Pyszczynski, T., & Solomon, S. (1986). The causes and consequences of a need for self-esteem: a terror management theory. In R. F. Baumeister (Ed.), *Public self and private self* (pp. 189–212). New York: Springer-Verlag.

Greenberg, J., Solomon, S., & Arndt, J. (2008). A uniquely human motivation: terror management. In J. Shah, & W. Gardner (Eds.), *Handbook of motivation science* (pp. 113–134). New York: Guilford.

Hansen, J., Winzeler, S., & Topolinski, S. (2010). When the death makes you smoke: a terror management perspective on the effectiveness of cigarette on-pack warnings. *Journal of Experimental Social Psychology, 46*, 226–228.

Hayes, J., Schimel, J., Arndt, J., & Faucher, E. H. (2010). A theoretical and empirical review of the death-thought accessibility concept in terror management research. *Psychological Bulletin, 136*, 699–739.

Hayes, J., Schimel, J., & Williams, T. J. (2008). Evidence for the death thought accessibility hypothesis II: threatening self-esteem increases the accessibility of death thoughts. *Journal of Experimental Social Psychology, 44*, 600–613.

Hirschberger, G., Florian, V., Mikulincer, M., Goldenberg, J. L., & Pyszczynski, T. (2002). Gender differences in the willingness to engage in risky behavior: a terror management perspective. *Death Studies, 26*, 117–141.

Holahan, C. J., Moos, R. H., Holahan, C. K., Cronkite, R. C., & Randall, P. K. (2001). Drinking to cope, emotional distress and alcohol use and abuse: a ten-year model. *Journal of Studies on Alcohol and Drugs, 62*, 190–198.

Hull, J. G. (1981). A self-awareness model of the causes and effects of alcohol consumption. *Journal of Abnormal Psychology, 90*, 586–600.

Hull, J. G., Levenson, R. W., Young, R. D., & Sher, K. J. (1983). Self-awareness-reducing effects of alcohol consumption. *Journal of Personality and Social Psychology, 44*, 461–473.

Hull, J. G., & Young, R. D. (1983). Self-consciousness, self-esteem, and success–failure as determinants of alcohol consumption in male social drinkers. *Journal of Personality and Social Psychology, 44*, 1097–1109.

James, W. (1890). *The principles of psychology* (Vol. 1). Mineola, NY: Dover Publications.

Jessop, D. C., & Wade, J. (2008). Fear appeals and binge drinking: a terror management theory perspective. *British Journal of Health Psychology, 13*, 773–788.

Johnston, L. D. (2003). Alcohol and illicit drugs: the role of risk perceptions. In D. Romer (Ed.), *Reducing adolescent risk: Toward an integrated approach* (pp. 56–74). Thousand Oaks, CA: Sage.

Kelling, G. W., Zerkes, R., & Myerowitz, D. (1976). Risk as value: a switch of set hypothesis. *Psychological Reports, 38*, 655–658.

Klackl, J., Jonas, E., & Kronbichler, M. (2013). Existential neuroscience: neurophysiological correlates of proximal defenses against death-related thoughts. *Social Cognitive and Affective Neuroscience, 8*, 333–340.

Klein, W. M. P., Koblitz, A. R., Kaufman, A. E., Vail, K. E., & Arndt, J. (2014). Ironic effects of mortality salience on relative risk perceptions in smokers. Unpublished manuscript.

Lakoff, G., & Johnson, M. (1980). The metaphorical structure of the human conceptual system. *Cognitive Science, 4*, 195–208.

Lambert, A. J., Eadeh, F. R., Peak, S. A., Scherer, L. D., Schott, J. P., & Slochower, J. M. (2014). Toward a greater understanding of the emotional dynamics of the mortality salience manipulation: revisiting the "affect-free" claim of terror management research. *Journal of Personality and Social Psychology, 106*, 655–678.

Leary, M. R., Tchividijian, L. R., & Kraxberger, B. E. (1994). Self-presentation can be hazardous to your health: impression management and health risk. *Health Psychology, 13*, 461–470.

Leventhal, H. (1971). Fear appeals and persuasion: the differentiation of a motivational construct. *American Journal of Public Health, 61*, 1208–1224.

Linville, P. W., Fischer, G. W., & Fischhoff, B. (1993). AIDS risk perceptions and decision biases. In J. B. Pryor, & G. D. Reeder (Eds.), *The social psychology of HIV infection* (pp. 5–38). Hillsdale, NJ: Erlbaum.

Luhtanen, R. K., & Crocker, J. (2005). Alcohol use in college students: effects of level of self-esteem, narcissism, and contingencies of self-worth. *Psychology of Addictive Behaviors, 19*, 99–103.

Maggs, J. L., Williams, L. R., & Lee, C. M. (2011). Ups and downs of alcohol use among first-year college students: number of drinks, heavy drinking, and stumble and pass out drinking days. *Addictive Behaviors, 36*, 197–202.

Martin, I. M., & Kamins, M. A. (2010). An application of terror management theory in the design of social and health-related anti-smoking appeals. *Journal of Consumer Behaviour, 9*(3), 172–190.

McCabe, S., Arndt, J., Bartholow, B., & Engelhardt, C. (2016). *Mortality salience and the weekend effect: Thoughts of death enhance alcohol-related attitudes on friday but not monday.* in preparation. University of Missouri – Columbia.

McCabe, S., Vail, K. E., III, Arndt, J., & Goldenberg, J. L. (2014). Hails from the crypt: a terror management health model investigation of health and celebrity endorsements. *Personality and Social Psychology Bulletin, 40*, 289–300.

McCool, J. P., Cameron, L., & Petrie, K. (2004). Stereotyping the smoker: adolescents' appraisals of smokers in film. *Tobacco Control, 13*, 308–314.

McEndree, M., Papatakis, N., & Cassavetes, J. (1959). *Shadows* (Motion picture). United States: Lion International.

Nichter, M., Nichter, M., Lloyd-Richardson, E. E., Flaherty, B., Carkaglu, A., & Taylor, N. (2006). Gendered dimensions of smoking among college students. *Journal of Adolescent Research, 21*, 215–243.

Pyszczynski, T., Greenberg, J., & Solomon, S. (1999). A 'dual-process model of defense against conscious and unconscious death-related thoughts: an extension of terror management theory. *Psychological Review, 106*, 835–845.

Pyszczynski, T., Greenberg, J., Solomon, S., Arndt, J., & Schimel, J. (2004). Why do people need self-esteem? A theoretical and empirical review. *Psychological Bulletin, 130*, 435–468.

Pyszczynski, T., Greenberg, J., Solomon, S., & Hamilton, J. (1990). A terror management analysis of self-awareness and anxiety: the hierarchy of terror. *Anxiety Research, 2*, 177–195.

Pyszczynski, T., Solomon, S., & Greenberg, J. (2003). *In the wake of 9/11: The psychology of terror.* Washington, DC: American Psychological Association.

Quirin, M., Loktyushin, A., Arndt, J., Küstermann, E., Lo, Y. Y., Kuhl, J., & Eggert, L. (2011). Existential neuroscience: a functional magnetic resonance imaging investigation of neural responses to reminders of one's mortality. *Social Cognitive and Affective Neuroscience, 7*, 193–198.

Rothman, A., & Schwarz, N. (1998). Constructing perceptions of vulnerability: personal relevance and the use of experiential information in health judgments. *Personality and Social Psychology Bulletin, 10*, 1053–1064.

Routledge, C., Arndt, J., & Goldenberg, J. L. (2004). A time to tan: proximal and distal effects of mortality salience on sun exposure intentions. *Personality and Social Psychology Bulletin, 30*, 1347–1358.

Routledge, C., Ostafin, B., Juhl, J., Sedikides, C., Cathey, C., & Liao, J. (2010). Adjusting to death: the effects of mortality salience and self-esteem on psychological well-being, growth motivation, and maladaptive behavior. *Journal of Personality and Social Psychology, 99*, 897–916.

Sayette, M. A. (1999). Does drinking reduce stress? *Alcohol Research and Health, 23*, 250–255.

Sayette, M. A., Loewenstein, G., Kirchner, T. R., & Travis, T. (2005). Effects of smoking urge on temporal cognition. *Psychology of Addictive Behaviors, 19*, 88–93.

Sedikides, C., Gaertner, L., & Toguchi, Y. (2003). Pancultural self-enhancement. *Journal of Personality and Social Psychology, 84*, 60–79.

Shehryar, O., & Hunt, D. M. (2005). A terror management perspective on the persuasiveness of fear appeals. *Journal of Consumer Psychology, 15*(4), 275–287.

Silvia, P. J. (2001). Nothing or the opposite: intersecting terror management and objective self-awareness. *European Journal of Personality, 15*(1), 73–82.

Thun, M. J., Carter, B. D., Feskanich, D., Freedman, N. D., Prentice, R., Lopez, A. D., ... Gapstur, S. M. (2013). 50-Year trends in smoking-related mortality in the United States. *New England Journal of Medicine, 368*, 351–364.

Vlahov, D., Galea, S., Ahern, J., Resnick, H., & Kilpatrick, D. (2004). Sustained increased consumption of cigarettes, alcohol, and marijuana among Manhattan residents after September 11, 2001. *American Journal of Public Health, 94*, 253–254.

Vlahov, D., Galea, S., Resnick, H., Ahern, J., Boscarino, J. A., Bucuvalas, M., & Kilpatrick, D. (2002). Increased use of cigarettes, alcohol, and marijuana among Manhattan, New York, residents after the September 11th terrorist attacks. *American Journal of Epidemiology, 155*, 988–996.

Weinstein, N. D., Marcus, S. E., & Moser, R. P. (2005). Smokers' unrealistic optimism about their risk. *Tobacco Control, 14*, 55–59.

Wong, N. C., & Dunn, S. S. (2013). Binge drinking and TMT: evaluating responses to anti-binge drinking PSAs from a terror management theory perspective. *Studies in Media and Communication, 1*, 81–95.

Zimmerman, M. A., Copeland, L. A., Shope, J. T., & Dielman, T. E. (1997). A longitudinal study of self-esteem: implications for adolescent development. *Journal of youth and Adolescence, 26*, 117–141.

Chapter 32

Nicotine Dependence and the Anterior Cingulate-Precuneus Pathway: Using Neuroimaging to Test Addiction Theories

Joseph R. DiFranza[1], Wei Huang[2], Jean A. King[2]

[1]Department of Family Medicine and Community Health, University of Massachusetts Medical School, Worcester, MA, USA; [2]Center for Comparative NeuroImaging, Department of Psychiatry, University of Massachusetts Medical School, Worcester, MA, USA

Abbreviations

ACb Anterior cingulate bundle
ACC Anterior cingulate cortex
BOLD Blood oxygen level derived
DMN Default mode network
FA Fractional anisotropy
FTND Fagerström Test for Nicotine Dependence
fMRI Functional magnetic resonance imaging
HONC Hooked on Nicotine Checklist
rsFC Resting-state functional connectivity

For generations, the prevailing paradigm has described addiction as a maladaptive behavior acquired through Pavlovian conditioning (American Psychiatric Association, 1980; Berridge, Robinson, & Aldridge, 2009). Representative of this viewpoint is the incentive-sensitization theory that addiction results from drug cues acquiring excessive salience through repeated pairings with reward (Robinson & Berridge, 1993b). In line with the popularity of learning models of addiction, about 90% of nicotine addiction neuroimaging research has focused on cue reactivity, while very little addresses biological models. Because cue-reactivity research has been reviewed elsewhere (Jasinska, Stein, Kaiser, Naumer, & Yalachkov, 2014), this chapter will initially focus on nicotine addiction from a neurobiological perspective.

Biological models, such as the opponent-process theory, also have a long history (Koob, Caine, Parsons, Markou, & Weiss, 1997). A modification of the opponent-process model is the sensitization-homeostasis theory (DiFranza & Wellman, 2005). This theory postulates that homeostatic neural adaptations in a so-called *craving generation system* produce drug cravings whenever the effect of the drug wears off. In support of biological dependence theories, clinical research establishes that nicotine addiction presents in a stereotypical clinical progression (DiFranza, Ursprung, & Biller, 2012; DiFranza, Ursprung, & Carlson, 2010; DiFranza, Wellman, Mermelstein, et al., 2011). This chapter reviews nicotine neuroimaging research as it relates to biological theories, followed by a discussion that unifies learning and biological models.

PHYSICAL DEPENDENCE

We begin by defining physical dependence as the latent neurophysiological condition that is manifested clinically when drug abstinence triggers withdrawal symptoms. Physical dependence develops through a stereotypical sequence of stages in all smokers (DiFranza, Ursprung, et al., 2012). In this way, physical dependence resembles many other biological disorders in which the pathophysiology causes disease symptoms to present in a set order.

As physical dependence first develops, prolonged abstinence triggers only a mild desire to smoke, termed *wanting*. As dependence progresses, abstinence provokes a stronger desire that intrudes upon the smoker's thoughts, termed *craving*. At the most advanced stage, abstinence triggers an urgent and unrelenting desire to smoke, termed *needing* (DiFranza, Sweet, Savageau, & Ursprung, 2011; DiFranza et al., 2010; DiFranza, Wellman, & Savageau, 2011). Physical dependence always develops through the same sequence (Table 1; DiFranza, Ursprung, et al., 2012; DiFranza et al., 2010; DiFranza, Wellman, Mermelstein, et al., 2011; DiFranza, Wellman, & Savageau, 2011):

Stage 0: no abstinence-induced urges to smoke
Stage 1: wanting
Stage 2: craving
Stage 3: needing

We validated an instrument that assesses the Stage of Physical Dependence (also known as the Levels of Physical Dependence, Table 1; DiFranza, Sweet, et al., 2011; DiFranza, Wellman, et al., 2012). This instrument assigns individuals to a stage based on their most advanced symptom. We reasoned that as the Stages of Physical Dependence identifies homogenous populations along a pathophysiologic progression, it might be possible to identify

TABLE 1 The Stages of Physical Dependence

		This Item Describes Me...			
Wanting	If I go too long without smoking, the first thing I will notice is a mild desire to smoke that I can ignore.	Not at all	A little	Pretty well	A lot
Craving	If I go too long without smoking, the desire for a cigarette becomes so strong that it is hard to ignore and it interrupts my thinking.	Not at all	A little	Pretty well	A lot
Needing	If I go too long without smoking, I just can't function right, and I know I will have to smoke just to feel normal again.	Not at all	A little	Pretty well	A lot

brain areas involved with physical dependence by correlating the Stages of Physical Dependence with measures of brain structure.

SMOKING AND BRAIN STRUCTURE

Efforts to identify the neural correlates of addiction have identified numerous structural differences between the brains of smokers and nonsmokers (Akkus et al., 2013; Brody, Mandelkern, Jarvik, et al., 2004; Gallinat et al., 2006; Hudkins, O'Neill, Tobias, Bartzokis, & London, 2012; McClernon et al., 2010; Paul et al., 2008; Zhang, Salmeron, Ross, Gu, et al., 2011). Smokers had decreased gray matter volume or density in the anterior and posterior cingulum, insula, precuneus, parahippocampal gyrus, thalamus, and prefrontal and orbitofrontal cortex (Almeida et al., 2008; Fritz et al., 2014; Gallinat et al., 2006). Fractional anisotropy (FA) measures microstructural order in white matter, with high values indicating more uniformly ordered structure. Smokers tend to have higher FA than nonsmokers (Hudkins et al., 2012; Jacobsen et al., 2007; Liao et al., 2011; Paul et al., 2008). This may reflect an increase in FA soon after smoking onset, followed by a progressive decline in FA with increasing pack-years of smoking (Hudkins et al., 2012; Jacobsen et al., 2007). FA has correlated negatively with dependence measures in some studies (Huang et al., 2013; Hudkins et al., 2012; Zhang, Salmeron, Ross, Geng, et al., 2011) but not others (Lin, Wu, Zhu, & Lei, 2013; Paul et al., 2008; Zhang, Stein, & Hong, 2010).

In cross-sectional studies, it is possible that findings reflect preexisting conditions in smokers that make them more likely to adopt smoking or develop dependence. A strong argument against this interpretation is the observation that FA correlates *directly* with pack-years in adolescents, but *inversely* with pack-years among adult smokers. Static preexisting differences should not correlate with a dynamic measure such as pack-years of smoking.

Another interpretation is that structural changes in the brains of smokers reflect nonspecific damage caused by years of smoking (Gallinat et al., 2007; Liao et al., 2011; Lin et al., 2013). Such an interpretation would be consistent with the observed correlation of decreasing FA with increasing pack-years of smoking in adults (Fritz et al., 2014; Gallinat et al., 2006), but it would not explain why FA increases with pack-years of exposure in novice smokers. Nonspecific neural toxicity would not explain why structural measures correlate with measures of dependence in some studies.

Studies have not been consistent in identifying the same brain regions as being affected by smoking. Some of these discrepancies can be attributed to differences in measures (FA, volume, neural density), populations (adolescent, young adult, geriatric), and exposures (light and heavy smoking). It is likely that some changes reflect nonspecific damage (Almeida et al., 2008), while others reflect neuroplasticity linked to dependence (Huang et al., 2013) or neuroplasticity unrelated to dependence. For example, some forms of nicotine tolerance do not correlate with indicators of nicotine dependence (Perkins, 2002) but are probably attributable to some form of neuroplasticity.

The challenge is to determine which structural changes are related to addiction. The observation that physical dependence develops through identical stages in all smokers suggests that the neural alterations that underlie physical dependence also develop in sequence. We therefore conducted a study to correlate the Stages of Physical Dependence with measures of brain structure.

We recruited 11 young adult smokers and 10 nonsmokers (Huang et al., 2013). Deviating from the standard protocol used in prior neuroimaging studies, we included light smokers because we wanted our sample to represent all stages of physical dependence. We limited our sample to younger individuals to minimize damage from long-term smoking (Almeida et al., 2008). In addition to the level of physical dependence, we included two other measures of nicotine dependence: the Hooked on Nicotine Checklist (HONC) and the Fagerström Test for Nicotine Dependence (FTND; DiFranza, Sweet, et al., 2011).

Compared to nonsmokers, smokers demonstrated higher FA in the right anterior cingulate bundle (ACb) and a strong trend toward higher FA in the left ACb ($p=0.06$). The anterior cingulate cortex (ACC) responds to smoking cues and activates during craving for nicotine (Jasinska et al., 2014).

Among smokers, the level of physical dependence correlated negatively with FA in the left ACb ($r=-0.68$, $p=0.02$). HONC scores also correlated negatively with FA in the left ACb ($r=-0.65$, $p=0.03$), but FTND scores did not correlate significantly with FA in the left ACb ($r=-0.49$, $p=0.12$) (Figure 1). The correlation of the Level of Physical Dependence instrument with a physical feature of the brain provides validation for this measure.

Having identified the left ACb as a region where FA correlated strongly with the level of physical dependence, we examined the density of neural streamlines between the ACb and other brain structures. Compared to nonsmokers, smokers demonstrated a lower density of neural streamlines between the ACb and the orbital frontal area, the middle cingulum, and the posterior cingulum.

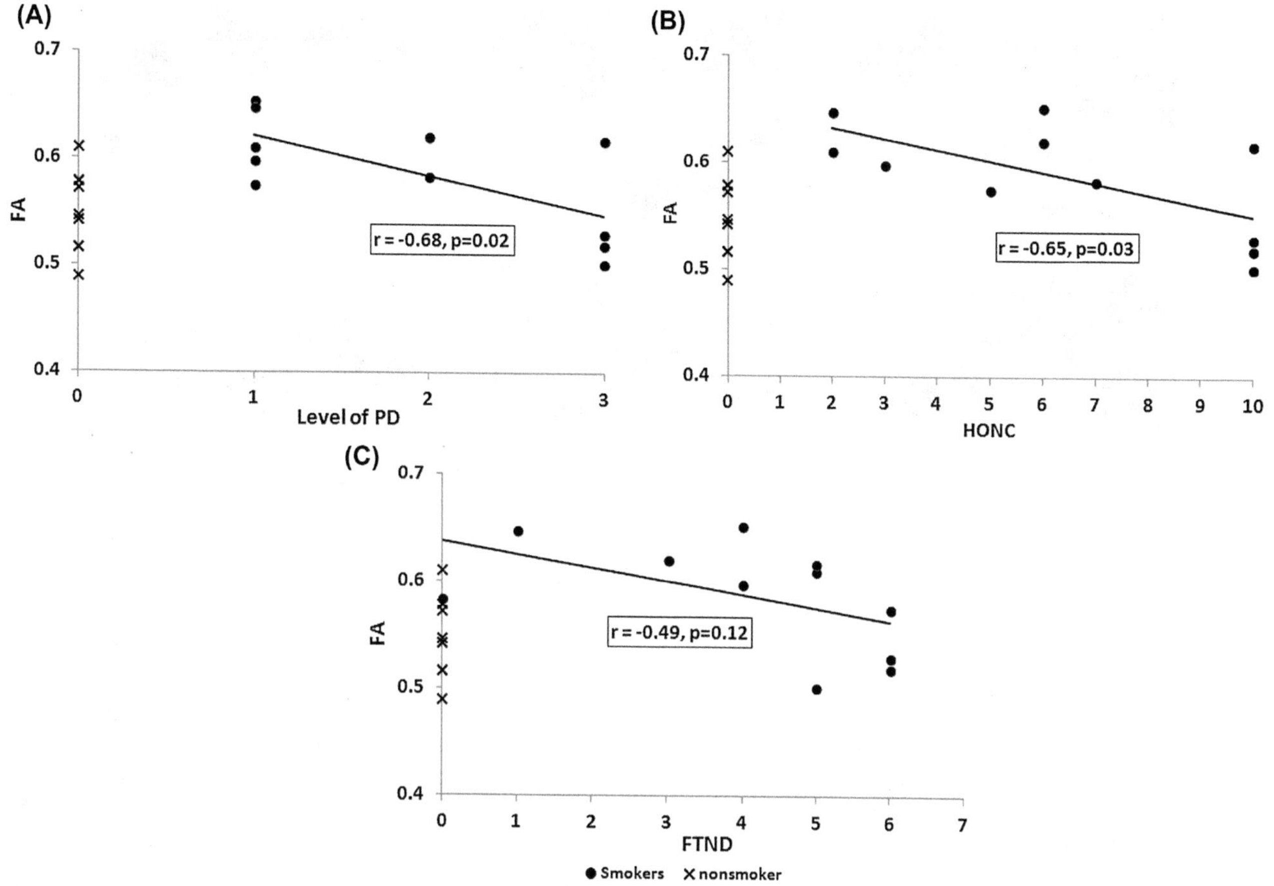

FIGURE 1 Linear correlation between smokers' fractional anisotropy (FA) and three measures of nicotine dependence. X's are FA for nonsmokers (not included in the correlation analysis). The lines indicate the linear relationship between each measure of addiction and FA in the left ACb. (A) Level of Physical Dependence ($r=-0.68$, $p=0.02$); (B) Hooked on Nicotine Checklist (HONC) ($r=-0.65$, $p=0.03$); (C) Fagerström Test for Nicotine Dependence (FTND) ($r=-0.49$, $p=0.12$). *Adapted from Huang et al. (2013).*

Among smokers, increasing physical dependence was associated with *increased* neural connections between the left ACb and the precuneus and *decreased* neural connections between the left ACb and the frontal cortex (Figure 2). These data support the idea that the development of physical dependence is associated with significant remodeling of neural circuits. These findings were replicated using the HONC and the FTND, which also measure physical dependence but not as a sequence (DiFranza, Wellman, Savageau, Beccia, & Ursprung, 2013).

Dozens of studies have associated activation of the ACC with craving and reactivity to cues for a variety of drugs (Engelmann et al., 2012; Jasinska et al., 2014). The precuneus also shows reliable reactivity to smoking cues in functional magnetic resonance imaging (fMRI) studies (Engelmann et al., 2012; Hartwell et al., 2011). The frontal cortex is associated with top-down control over urges (Heatherton & Wagner, 2011). It is notable that advancing levels of physical dependence are associated with increasing neural connections between brain areas that are associated with drug craving and cue reactivity (ACC and precuneus), and decreasing neural connectivity between the ACC and the frontal lobe that may be involved in resisting urges. These changes may be related to increasing salience of smoking cues and decreasing ability to resist craving, respectively. It is interesting to note that decreased FA and decreased gray matter volume in the frontal cortex have been observed with tobacco, alcohol, cocaine, methamphetamine, and opiate dependence/abuse (see Lin et al., 2013 for a review). It was possible to identify the correlates of physical dependence only because we included individuals who were at very early stages of the addiction process. Research is needed to determine if addiction to other substances progresses through physiologically based stages.

LEARNING THEORIES OF ADDICTION

At this point, it may be helpful to compare learning and biological theories of addiction in relation to hypotheses that are testable with fMRI. For more than half a century, the prevailing paradigm has been the idea that addiction is a learned maladaptive behavior with addiction developing through conditioning involving positive and/or negative reinforcement (Robinson & Berridge, 1993a). Positive reinforcement would stem from the intrinsically rewarding euphoric effects of the drug mediated by dopamine acting on the brain's pleasure centers. Negative reinforcement would stem from the relief of withdrawal symptoms. Through repeated pairings with both positive and negative reinforcement, drug cues take on a powerful salience. Cues (both internal and external) become the main driver of addiction, triggering craving, making quitting

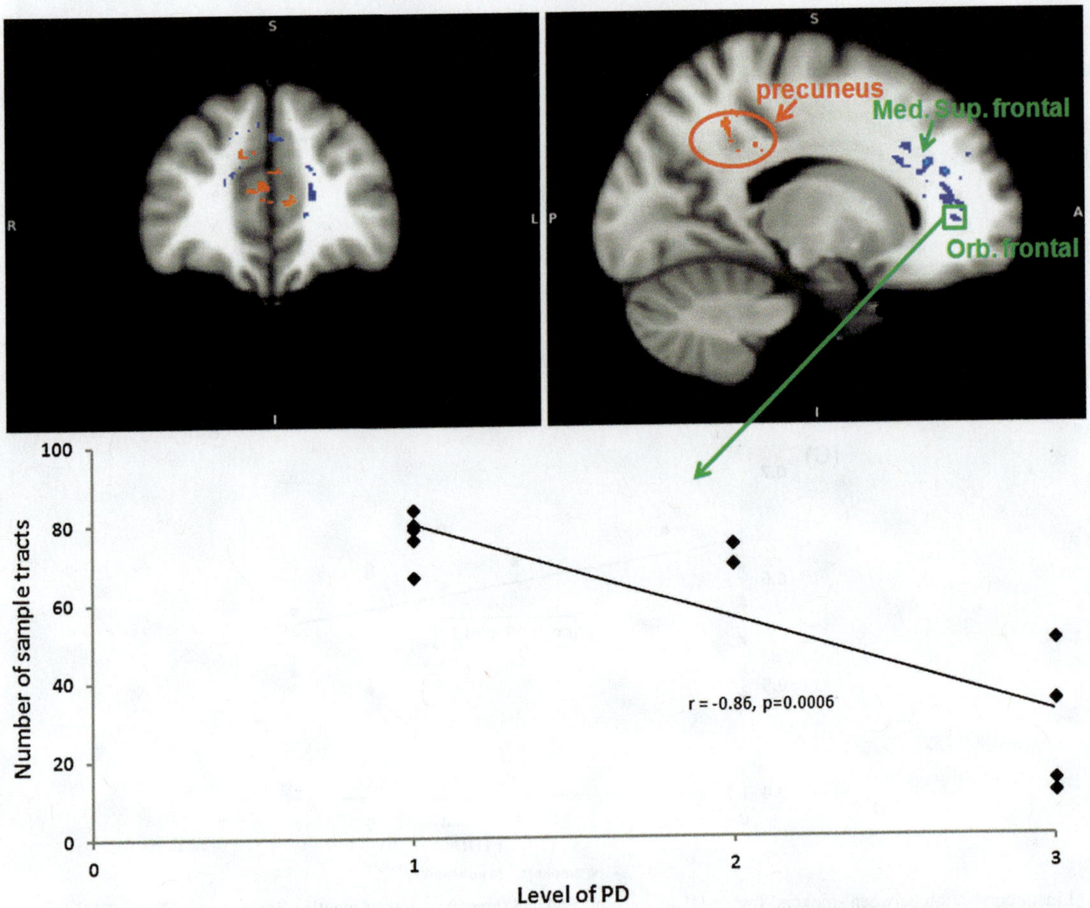

FIGURE 2 Correlation map of smokers' number of white matter tracts with their level of physical dependence (PD). Using the ACb as a seed, the numbers of sample white matter tracts to other brain regions were measured in smokers and correlated with their level of PD. Red indicates areas of positive correlation and blue indicates areas of negative correlation. In the left hemisphere, white matter tracts approaching the superior frontal and medial superior frontal areas showed negative correlation with level of PD ($p<0.03$), whereas tracts approaching the precuneus showed positive correlation ($r=0.75$, $p<0.05$). Fewer correlations were seen in the right hemisphere. The graph shows the correlation between the level of PD and the sample tracts between the left ACb and the superior frontal area indicated by the green square. Level of PD showed a strong negative correlation with sample tracts ($r=-0.86$, $p=0.0006$). *Adapted from Huang et al. (2014).*

difficult, and contributing to relapse. Under this conditioned learning model, physical dependence is not essential, as addiction is defined as a maladaptive behavior (American Psychiatric Association, 2013). Neural remodeling may be involved in the process that gives cues excessive salience.

Testable hypotheses related to conditioning theories are that the salience and valence of smoking cues are products of past associations with reward. Due to their acquired excessive salience, brain reactivity to smoking cues will be greater than reactivity to neutral cues. In the absence of cues, the addicted brain should function normally.

BIOLOGICAL THEORIES OF ADDICTION

Biological theories, such as opponent-process or homeostasis models, hold that drugs induce neuroplasticity that renders the brain unable to function normally in the absence of the drug. These models hold that drugs disrupt brain homeostasis through excessive stimulation or inhibition. Neuroplastic changes restore normal brain function by opposing the action of the drug. However, when the drug effect wears off, the same neuroplastic changes disturb homeostasis in a direction opposite that induced by the drug. Two theories address the question as to how a shift in brain activity in a direction opposite that of the drug causes addiction: hedonic dysregulation theories and sensitization-homeostasis theory.

Hedonic Dysregulation

Hedonic dysregulation theories stipulate that drugs provide excessive stimulation to brain areas involved in the experience of hedonic pleasure (Koob & Le Moal, 1997). Neuroplastic changes that oppose this action depress hedonic status when the drug wears off. The individual is then compelled to use the drug in order to relieve the resulting dysphoria. Similar to the learning models, the hedonic model assumes that addiction liability lies in a drug's ability to stimulate hedonic pleasure centers. Under hedonic dysregulation theories, cue conditioning would develop through pairing of drug delivery with the onset of a positive hedonic state or the relief of a negative hedonic state, but cues would not play a central role because behavior would be driven by the need to regulate hedonic

TABLE 2 Comparison of Addiction Theories

	Incentive-Sensitization Theory	Hedonic Dysregulation	Sensitization-Homeostasis Theory
Nicotine's primary action	Stimulates pleasure	Stimulates pleasure	Inhibits craving
Site of action	Pleasure center	Pleasure center	Craving generation system
Mechanism	Learned behavior	Homeostatic mechanisms related to emotional regulation	Homeostatic mechanisms related to craving and satiety
Mode	Conditioning	Neuroplasticity	Neuroplasticity
Driver of addiction	Cue-exposure	Dysphoria generated by withdrawal-induced homeostatic imbalance	Craving generated by withdrawal-induced homeostatic imbalance
Cue salience	Acquired through positive and negative reward conditioning.	Acquired through positive and negative reward conditioning.	Cue salience derives primarily from brain's need for nicotine to maintain physiologic homeostasis.
Cue reactivity	Cue strength and valence (positive or negative value) derives from past pairings with reward.		Cue strength and valence reflects the brain's momentary physiologic status.

status (Koob et al., 1997). Considering nicotine to be a stimulant, during nicotine deprivation, brain activity would be depressed in relation to that of nonaddicted controls. Also, brain activity should be greater after drug delivery than during withdrawal.

Sensitization-Homeostasis Theory

Both the learning and hedonic dysregulation theories assume that nicotine is a stimulant (Table 2). The sensitization-homeostasis theory (DiFranza & Wellman, 2005) postulates that the primary addiction-related action of nicotine is inhibitory, which is consistent with the fact that most smokers say they smoke for relaxation, not for stimulation. Many smokers smoke before going to bed and when they awaken at night, which would be odd if nicotine is a stimulant (Scharf, Dunbar, & Shiffman, 2008).

While the learning and hedonic dysregulation theories both postulate that the primary site of action of nicotine is a pleasure center, the sensitization-homeostasis theory postulates that nicotine's primary site of action is a neural circuit that normally regulates craving/satiety in regard to physiologic needs (hunger, thirst, salt appetite, etc.). While the other theories postulate that nicotine *stimulates* pleasure, the sensitization-homeostasis theory postulates that nicotine *inhibits* craving. Under this theory, pleasure might provide a motive for use, but it is not the reason why nicotine is addictive.

Under the sensitization-homeostasis theory, there is a *craving generation system* that generates craving whenever a particular behavior is needed to maintain physiologic homeostasis. This system would be responsible for generating craving in the form of hunger, thirst, sexual desire, salt craving, etc. in response to physiologic needs. According to the sensitization-homeostasis theory, nicotine inhibits the *craving generation system*. Neuroplastic changes restore homeostasis to the *craving generation system*, but whenever the use of nicotine is curtailed, homeostasis is disrupted and nicotine is required to restore it. Support for the idea that nicotine is required to restore normal brain function comes from hundreds of studies in which smokers perform better when they receive nicotine than when they are deprived of it. In this way, nicotine is different from most other addictive drugs, which typically impair performance on cognitive and motor tasks.

When nicotine is needed to restore physiologic homeostasis, the *craving generation system* generates a hunger for nicotine in the form of wanting, craving, or needing in a fashion similar to how the brain uses hunger, thirst, libido, sleepiness, and other subjective sensations to motivate the behavior that is needed to restore physiologic homeostasis. This is consistent with case histories in which smokers describe their brains being "hungry" for nicotine (DiFranza, Wellman, Mermelstein, et al., 2011).

Under the sensitization-homeostasis theory, nicotine becomes a physiologic requirement for normal brain function, and it is the brain's ability to regulate behavior in order to maintain homeostasis that is the primary determinant of addiction, not learning or pleasure. In this way, the sensitization-homeostasis theory is a basic physiologic model. Just as food cues take on increased salience when the brain is hungry, smoking cues take on increased salience when the brain is hungry for nicotine. Food cues can stimulate hunger; smoking cues can stimulate a desire to smoke. Food cues become aversive under conditions of satiation ("I'm so full, I don't even want to look at food"); cues for nicotine may be aversive under conditions of nicotine satiation (Gloria et al., 2009). Under the sensitization-homeostasis theory, addiction is not caused by cues acquiring excessive salience; rather, cues acquire salience when nicotine becomes a requirement for the maintenance of brain homeostasis. Smoking cues are salient because the brain needs nicotine and is "looking" for it. That is not to say that cues cannot also acquire salience through conditioning.

Several experiments provide scientific plausibility for the sensitization-homeostasis model. Researchers demonstrated that although rats dislike intensely salty solutions when they are in a state of homeostasis in regard to salt balance, when put in a physiologic state of salt depletion they demonstrate "wanting" for the solution that they dislike (Berridge et al., 2009; Tindell, Smith, Berridge, & Aldridge, 2009; Tindell, Smith, Pecina, Berridge, & Aldridge, 2006). This indicates that the salience of the super-salty solution is not governed by its hedonic value or by conditioning, but rather by the physiologic requirements of the brain at the moment. When the brain is put into physiologic imbalance, it will generate "wanting" for the required substance, even if that substance is distasteful. This is consistent with common reports that smokers can experience intense cravings for nicotine even though they have come to hate the act of smoking.

The sensitization-homeostasis theory indicates that the critical action of nicotine is not on pleasure centers but on a *craving generation system*. An elegant experiment demonstrated that different areas of the ventral striatum are associated with "liking" a drug and "wanting" a drug (Berridge et al., 2009). This supports the idea that somewhat different areas are involved with hedonic reward and the generation of craving.

Specific testable hypotheses under the sensitization-homeostasis theory include the following.

- During withdrawal from nicotine, activity in brain areas involved with craving should increase and should be stronger in abstinent smokers than in nonsmoking controls.
- The spontaneous activations that occur during abstinence should correlate with the intensity of craving for nicotine, indicating that craving generation circuits activate during withdrawal.
- As the theory postulates that both cues and withdrawal stimulate the craving generation system, withdrawal-induced craving and cue-induced craving should be associated with activation in some of the same brain regions.
- Smoking cues should produce greater reactivity under conditions of nicotine deprivation than satiation.

HYPOTHESIS TESTING

Cue-reactivity studies involve a comparison of brain activity during exposure to a smoking cue versus neutral cues such as office supplies. Numerous studies have used blood oxygen level derived (BOLD) contrast to assess cue reactivity for a variety of drugs (Jasinska et al., 2014).

In contrast, the study of the effects of nicotine deprivation does not lend itself to an event-related BOLD analysis because the symptoms of nicotine withdrawal appear too gradually to study in real time and because it would be impractical to study the same subjects dozens of times. There are very few non-task-related neuroimaging studies of brain function in smokers during nicotine deprivation (Cole et al., 2010; Ding & Lee, 2013; Huang et al., 2014; Wang et al., 2007).

Resting-state functional connectivity (rsFC) is an imaging technique that measures the degree to which activity in different brain areas is correlated. Brain activity is measured with the subject at rest; there are no tasks or cue exposures. rsFC does not indicate absolute values of brain activity but rather coordination of activity, which is interpreted to indicate a functional connectivity between structures. Arterial spin labeling is another MRI method that can be used for non-event-related studies.

Using the same sample of smokers and nonsmokers as described above for our analysis of brain structure, we used rsFC to evaluate brain activity (Huang et al., 2014). As our structural analysis demonstrated that changes in FA and neural streamline density in the ACb correlated with the level of physical dependence, we examined rsFC between the dorsal ACC and the rest of the brain. As the sensitization-homeostasis theory predicts that brain activity increases during nicotine deprivation, we compared rsFC between nonsmokers and smokers after an 11-h abstinence.

Compared to nonsmokers, abstinent smokers showed stronger rsFC between the dorsal ACC and the precuneus, caudate, putamen, frontal cortex, temporal cortex, and inferior parietal lobe (Figure 3). Smokers also showed enhanced connectivity between the default mode network (DMN) and areas of the dorsal ACC, caudate, putamen, middle frontal area, precentral gyrus, and the medial frontal gyrus. These data are the first to show that brain activity in smokers during withdrawal exceeds that of nonsmokers, supporting the prediction that withdrawal from nicotine *increases* brain activity.

The smokers were then reevaluated after they had smoked to satiety. Compared to the satiated condition, deprived smokers had increased rsFC between the dorsal ACC and the precuneus, insula, orbital frontal gyrus, superior frontal gyrus, posterior cingulate cortex, superior temporal lobe, and the inferior temporal lobe. Enhanced rsFC was seen between the DMN and areas of the dorsal ACC, precuneus, medial orbital frontal area, insula, superior medial frontal area, middle temporal gyrus, and superior frontal area. These data provided further support for the idea that neuroplastic changes associated with the development of physical dependence trigger activation of neural circuits under conditions of nicotine deprivation. Increased brain activity during nicotine withdrawal was also observed in both prior studies that looked at this (Ding & Lee, 2013; Wang et al., 2007).

Nicotine withdrawal triggers withdrawal-induced craving along with a variety of other symptoms such as difficulty concentrating, anxiety, restlessness, irritability, and increased appetite. As smokers progress through the levels of physical dependence, the intensity of the withdrawal-induced craving they experience increases. We next performed an analysis to determine which changes in rsFC correlated with the change in the intensity of withdrawal-induced craving between the deprived and satiated conditions. The intensity of withdrawal-induced craving correlated with the strength of rsFC between the dorsal ACC and the precuneus, insula, caudate, putamen, middle cingulate gyrus, and precentral gyrus (Huang et al., 2014).

To summarize, the advancement through the levels or stages of physical dependence is associated with decreasing FA in the ACb, increasing number of neural tracts between the ACb and the precuneus, and decreasing number of neural tracts between the ACb and the frontal cortex. During withdrawal from nicotine, rsFC increases between the ACC and other brain areas compared to the satiated state, and in comparison to nonsmoking controls. In several ACC circuits, rsFC correlates with the intensity of withdrawal-induced craving. A key circuit is the ACC-precuneus pathway, which shows increased numbers of neural connections with advancing physical dependence and increased rsFC during withdrawal, which correlates with the intensity of withdrawal-induced craving. Ding and Lee demonstrated that there was no detectable activity in the precuneus-ACC pathway after smoking,

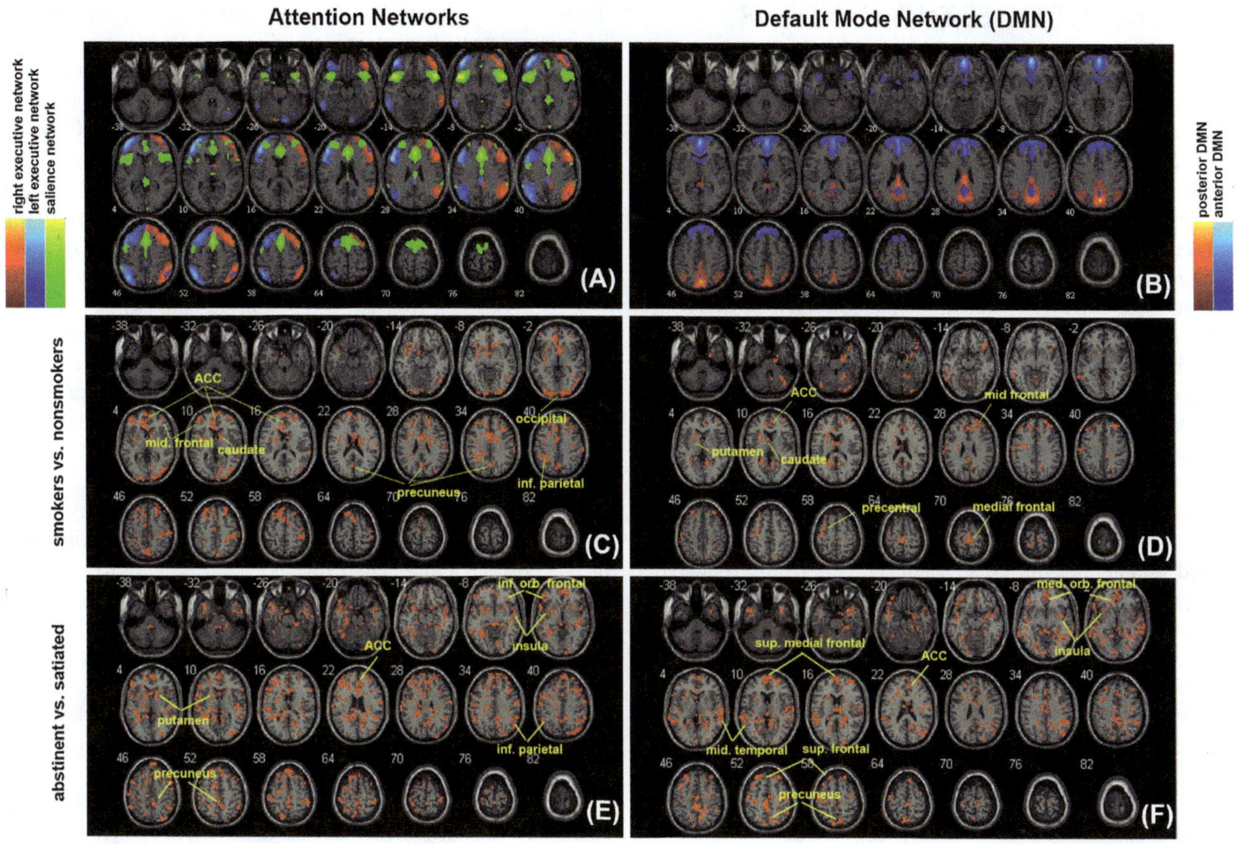

FIGURE 3 Functional magnetic resonance imaging was used to compare brain activity in nonsmokers and in smokers under conditions of nicotine satiety and after 11 h of abstinence. Results from independent component analysis (ICA), particularly in the attention networks and the default mode network (DMN). (A) Components that formed the attention networks, including the left executive network, the right executive network, and the salience network. (B) Components that formed the DMN, including the posterior DMN and the anterior DMN. (C) Difference in the attention networks between nonsmokers and smokers during abstinence, shown in binary mask. Compared to nonsmokers, smokers in the abstinent condition showed enhanced connectivity in their attention networks to the areas of the anterior cingulate cortex (ACC), precuneus, caudate, occipital lobes, inferior parietal lobe, and the middle frontal area. (D) Difference within the DMN between nonsmokers and smokers during abstinence, shown in binary mask. Compared to nonsmokers, smokers in the abstinent condition showed enhanced connectivity in their DMN to areas of the ACC, caudate, putamen, middle frontal area, precentral gyrus, and the medial frontal gyrus. (E) Difference in smokers' attention networks between abstinent (withdrawal) and satiated conditions. When compared to the satiated condition, smokers in the abstinent condition had enhanced connectivity in their attention networks to areas of the ACC, precuneus, putamen, insula, inferior orbital frontal area, and the inferior parietal lobe. (F) Difference in smokers' DMN between abstinent (withdrawal) and satiated conditions. When compared to the satiated condition, DMN of smokers in the abstinent condition had enhanced connectivity to areas of the ACC, precuneus, medial orbital frontal area, insula, superior medial frontal area, middle temporal gyrus, and superior frontal area. Red binary masks in C, D, E, and F indicate differences at a significance level of $p < 0.05$.

but during abstinence there was bidirectional activations between the precuneus and dorsal ACC (Ding & Lee, 2013). These studies support the theory that, during abstinence, the neuroplastic changes that are responsible for physical dependence trigger activation of brain circuits that support withdrawal-induced craving (DiFranza, Huang, et al., 2012).

The ACC and precuneus are both major components of the DMN (Ding & Lee, 2013). Two prior studies suggest that nicotine suppresses activity in the DMN, while nicotine withdrawal appears to activate it (Cole et al., 2010; Sutherland, McHugh, Pariyadath, & Stein, 2012). Our study is consistent with the only other study of withdrawal-induced craving, which found craving to be correlated with increased rsFC between the precuneus and the DMN (Cole et al., 2010). Another study also found increased rsFC in the abstinent state in circuits connecting the ACC, precuneus, and insula; however, they did not measure withdrawal-induced craving (Ding & Lee, 2013). Consequently, three studies show that nicotine withdrawal is associated with increased rsFC in the precuneus, and in two studies, rsFC in precuneus circuits correlated with the severity of withdrawal-induced craving.

Ding and Lee identified a network of circuits that are active during both states of nicotine deprivation and satiation (Ding & Lee, 2013). However, the intensity of connections between individual structures and sometimes the direction of signaling between structures changes between the deprived and satiated states. This suggests that rather than there being one circuit for craving and another for satiation, there is a network wherein activity toggles to generate both craving and satisfaction. The sensitization-homeostasis theory proposed two brain systems, one for craving generation and another for craving inhibition, but these data hint that these functions may be combined in a single network.

WHAT IMPACT DOES THIS HAVE FOR ADDICTION THEORISTS?

As there are few imaging studies of nicotine withdrawal, it would be premature to claim that the major premises of the sensitization-homeostasis theory have been proven, but the data are certainly supportive. Progression through the levels of physical dependence correlates with neuroplastic changes in specific brain pathways. Brain activity increases spontaneously during nicotine withdrawal, and in several circuits activity correlates with the intensity of withdrawal-induced craving. The observed increases in rsFC and brain metabolism (Wang et al., 2007) during nicotine deprivation weigh against the ideas that nicotine is a stimulant and withdrawal results in a depressed state. Increased brain activity triggered by withdrawal is consistent with the clinical picture of nicotine withdrawal which includes anxiety, nervousness, restlessness, increased appetite, irritability, and hand tremor. Some of these withdrawal symptoms correlate with brain activity (Cole et al., 2010).

Consistent with learning theories, smoking-related cues have triggered stronger reactions than neutral stimuli in dozens of studies (Engelmann et al., 2012; Jasinska et al., 2014). However, smoking cues trigger brain activations that are only comparable to those produced by food cues, and less than those produced by money and erotic imagery (Claus, Blaine, Filbey, Mayer, & Hutchison, 2013; Gray et al., 2014; Versace et al., 2011). As most people do not lose control when exposed to money cues or erotic imagery, it seems unlikely that reactions to smoking cues would be sufficient to establish and maintain addiction by themselves.

There is substantial overlap in the brain areas that are activated by smoking and food cues (Claus et al., 2013). This is interesting in relation to the sensitization-homeostasis theory's proposal that the brain circuits involved with nicotine craving are the same as those involved with the maintenance of normal homeostasis.

Two areas that consistently activate in response to smoking cues are the precuneus and the ACC (Engelmann et al., 2012). As we have noted, the number of neural tracts connecting these structures correlates with the level of physical dependence. Activity in this circuit correlates with the intensity of withdrawal-induced craving (Cole et al., 2010), as well as with smoking cue-reactivity and cue-induced craving (Hanlon et al., 2013; Hartwell et al., 2011). While it is clear that smoking cues have increased salience in human smokers, it is not clear whether cues obtain their salience through conditioning as they do in experimental animals, or from the fact that in physically addicted smokers the brain has a physiologic requirement for nicotine to function normally. In clinical studies, smokers' reports of cue-induced urges to smoke parallel the development of physical dependence (DiFranza, Wellman, Ursprung, & Sabiston, 2009). The observations that physical dependence and cue reactivity involve some of the same circuits supports the idea that smoking cues attain some of their salience through physical dependence.

Consistent with the idea that smoking cues derive their valence and salience from the physiologic state of the brain, Lim et al. found activation in the ACC when smoking cues were presented during abstinence, but not after smoking (Lim et al., 2005). In other studies, activation was not seen in the ACC when smoking cues were presented soon after subjects had smoked (Brody et al., 2006; Franklin et al., 2007). Attenuation of cue-induced craving and ACC activation has been observed in smokers treated with the smoking cessation drug bupropion (Brody, Mandelkern, Lee, et al., 2004). These studies suggest that smoking cues lose salience when the brain is replete with nicotine.

A unique study suggests that a cue for impending nicotine delivery can switch valence depending on whether the subject is in a condition of satiation or deprivation (Gloria et al., 2009). The investigators used classic conditioning to associate a signal with the imminent delivery of nicotine. Across multiple brain regions, in abstinent smokers the signal that predicted nicotine delivery triggered greater activation than the signal predicting a dose of saline. However, response to the nicotine signal was markedly reduced when subjects had smoked to satiation. In some areas, including the ACC, the nicotine signal triggered less activation than the saline signal when smokers were satiated. In other words, reaction to the nicotine cue in the ACC was suppressed during satiated conditions. This is consistent with the sensitization-homeostasis model, which indicates that activity in neural circuits that are involved with the generation of craving is actively suppressed by nicotine.

This experiment suggests that when the brain is in physiologic balance in regard to nicotine, additional doses would be disruptive of homeostasis and a signal for another dose may have a negative valence. This suggests that the valence of smoking cues is determined primarily by the physiologic status of the brain (in relation to nicotine), rather than past positive or negative reinforcement. Also supporting this line of reasoning are studies described above showing that super-salty solutions are desired when there is a physiologic sodium deficit but are aversive during a state of sodium balance (Tindell et al., 2009).

As we have discussed, physical dependence develops through stages of wanting, craving, and needing. These symptoms appear whenever an individual goes too long without nicotine. When physical dependence first develops, most smokers can go for several days or weeks without using tobacco before they experience wanting. This is termed the latency to withdrawal (DiFranza & Ursprung, 2008; Fernando, Wellman, & DiFranza, 2006; Ursprung, Morello, Gershenson, & DiFranza, 2010). Over time, the latency to withdrawal shortens progressively, meaning that the smoker must smoke cigarettes at more frequent intervals to relieve withdrawal symptoms. The shortening of the latency to withdrawal is the primary factor determining the trajectory of tobacco use (DiFranza, Wellman, Mermelstein, et al., 2011).

Is it possible to reconcile the incentive-sensitization and sensitization-homeostasis theories? Both theories accommodate the facts that nicotine can deliver pleasure, that individuals become physically dependent on nicotine, that smoking cues take on greater salience to smokers, and that smoking cues can trigger craving, make quitting more difficult, and contribute to relapse. Both theories postulate that sensitized responses to nicotine lead to neuroplastic changes in the brain that make nicotine use a priority. Both theories allow that conditioning through positive and negative reinforcement contributes to the salience of smoking cues. Both theories are compatible with the idea that smoking behavior is driven to some degree by withdrawal-induced craving and cue-induced craving. The central difference between the theories is that the incentive-sensitization theory sees nicotine addiction as a behavior acquired through conditioned learning, while the sensitization-homeostasis theory sees nicotine addiction as a manifestation of an acquired physiologic need for nicotine to maintain homeostasis in the brain.

Neuroimaging studies support the idea that smoking cues obtain salience similar to those of food cues, but do not support the idea that smoking cues take on salience in excess of those for money or sex, nor are there any case histories describing individuals who feel so controlled by smoking cues that it made them addicted to smoking. All available case histories implicate physical dependence (DiFranza et al., 2010). Yet, there is no denying that cues affect smokers and that smokers develop a psychological reliance on smoking.

The first author was curious to determine if there are smokers who have only physical dependence, only psychological reliance on smoking, or only responsivity to smoking cues. If so, this would suggest three different mechanisms leading to addiction. A measure was developed and validated with subscales that assess physical dependence, cue-induced craving, and psychological reliance on tobacco (DiFranza et al., 2009). As it turns out, these three domains are highly correlated. One does not exist in the absence of the others. This suggests a common mechanism leading to physical dependence, cue reactivity, and psychological reliance. As individuals progress through the stages of physical dependence, they become more reactive to smoking cues and develop a greater psychological reliance on smoking. It is clear from a clinical perspective that physical dependence and the shortening of the latency to withdrawal are determinative of the trajectory of nicotine addiction. What is not clear is whether cue reactivity and psychological reliance on tobacco also develop through simple physiologic processes or whether these develop through learning mechanisms. It is possible that cue reactivity develops in part as a reflection of physiologic necessity and in part through conditioning.

APPLICATION TO OTHER ADDICTIONS AND SUBSTANCE MISUSE

Although nicotine is as addictive as any other drug, there are important differences between nicotine addiction and addiction to other drugs. Nicotine suppresses the craving for nicotine. It is not associated with binging, which distinguishes this drug from alcohol, cocaine, and methamphetamine. It is possible that nicotine addiction develops through mechanisms that are different from those involved with binge drugs.

Nicotine does not get you high. In addicted individuals, nicotine does not cause intoxication; it restores normal brain function. Use patterns for nicotine most closely resemble those for opiates in that users may wait between doses until withdrawal symptoms necessitate use. The trajectory of nicotine addiction was only understood when it was recognized that the latency to withdrawal can be very long but shortens over time. It should be considered that the latency to withdrawal in opiate addiction may also be several weeks at the onset of addiction. Likewise, it would be interesting to study the latency to withdrawal with alcohol using more sensitive indicators of alcohol withdrawal. At the onset of physical nicotine dependence, the only symptom of withdrawal is a transient mild desire to smoke. Efforts should be made to identify the earliest symptoms of physical dependence for other drugs.

Researchers should direct their attention to whether addiction to other drugs also involves characteristic physiologically based stages. The emerging symptoms of physical nicotine dependence could only be recognized by setting aside conventional definitions of addiction. Researchers should be aware that it is likely that the first symptoms of other drug addictions likewise fly under the radar of traditional behavior-based definitions of addiction.

DEFINITION OF TERMS

Craving A strong desire to use tobacco that intrudes upon a person's thoughts.
Cue-induced craving A desire to use tobacco that is triggered by exposure to smoking cues.
Hedonic dysregulation theory An addiction theory that posits that addiction is caused by neuroplasticity that counters the mood-enhancing properties of addictive drugs.
Incentive-sensitization theory An addiction theory that posits that addiction develops as a result of Pavlovian conditioning resulting from the pairing of drug cues with reward.
Latency to withdrawal The time elapsed between the last use of tobacco and the onset of withdrawal symptoms such as wanting, craving, and needing.
Level of physical dependence Synonymous with stages of physical dependence.
Needing An urgent and unrelenting need to use tobacco to feel normal and restore normal functioning.
Physical dependence A latent physiologic condition that is manifested clinically when abstinence from the drug triggers withdrawal symptoms.
Sensitization-homeostasis theory An addiction theory that posits that neuroplastic changes that develop to counter the effects of nicotine activate the brain and trigger craving whenever the drug wears off.
Stages of physical dependence Clinical stages in the development of physical dependence on nicotine. The stages are wanting, craving, and needing.
Wanting A mild desire to use tobacco that is transient and easily ignored.
Withdrawal-induced craving A desire to use tobacco that is triggered by abstinence from nicotine.

KEY FACTS OF PHYSICAL NICOTINE DEPENDENCE

- Physical nicotine dependence has been demonstrated with the first dose of nicotine in experimental animals.
- In humans, physical nicotine dependence is sometimes reported after the first cigarette but develops in the majority of smokers by the time they have smoked a pack of cigarettes.
- Once physical dependence develops, withdrawal symptoms always appear if the smoker waits too long before smoking.
- The time elapsed between finishing a cigarette and experiencing withdrawal symptoms is the latency to withdrawal. The latency to withdrawal may be as long as several weeks in novice smokers but shortens progressively with repeated use.

SUMMARY POINTS

- Physical addiction to nicotine develops through a set sequence of clinical stages: wanting, craving, and needing.
- The progression of physical dependence corresponds to increases in the number of neural tracts between the anterior cingulate cortex and the precuneus, and at the same time a marked decrease in tracts connecting the anterior cingulate cortex and the frontal lobe.
- The anterior cingulate and precuneus activate in relation to cue-induced craving and withdrawal-induced craving, suggesting a common pathway.

- Activity in the anterior cingulate-precuneus circuit increases spontaneously during nicotine withdrawal and correlates with the severity of withdrawal-induced craving, supporting a key prediction of the sensitization-homeostasis model.
- Neuroimaging research indicates that both physical dependence and cue reactivity play a role in nicotine addiction.

REFERENCES

Akkus, F., Ametamey, S. M., Treyer, V., Burger, C., Johayem, A., Umbricht, D., ... Hasler, G. (2013). Marked global reduction in mGluR5 receptor binding in smokers and ex-smokers determined by [11C]ABP688 positron emission tomography. *Proceedings of the National Academy of Sciences of the United States of America, 110*(2), 737–742. http://dx.doi.org/10.1073/pnas.1210984110210984110.

Almeida, O. P., Garrido, G. J., Lautenschlager, N. T., Hulse, G. K., Jamrozik, K., & Flicker, L. (2008). Smoking is associated with reduced cortical regional gray matter density in brain regions associated with incipient Alzheimer disease. *American Journal of Geriatric Psychiatry, 16*(1), 92–98. http://dx.doi.org/10.1097/JGP.0b013e318157cad2.

American Psychiatric Association. (1980). *Diagnostic and statistical manual of mental disorders: DSM-3* (3rd ed.). Washington, DC: American Psychiatric Association.

American Psychiatric Association. (2013). *Diagnostic and statistical manual of mental disorders: DSM-5*. Washington, DC: American Psychiatric Association.

Berridge, K. C., Robinson, T. E., & Aldridge, J. W. (2009). Dissecting components of reward: 'liking', 'wanting', and 'learning'. *Current Opinion in Pharmacology, 9*(1), 65–73. http://dx.doi.org/10.1016/j.coph.2008.12.014.

Brody, A. L., Mandelkern, M. A., Jarvik, M. E., Lee, G. S., Smith, E. C., Huang, J. C., ... London, E. D. (2004). Differences between smokers and nonsmokers in regional gray matter volumes and densities. *Biological Psychiatry, 55*(1), 77–84.

Brody, A. L., Mandelkern, M. A., Lee, G., Smith, E., Sadeghi, M., Saxena, S., ... London, E. D. (2004). Attenuation of cue-induced cigarette craving and anterior cingulate cortex activation in bupropion-treated smokers: a preliminary study. [erratum appears in *Psychiatry Research*, December 15, 2004;*132*(2),183–184]. *Psychiatry Research, 130*(3), 269–281.

Brody, A., Mandelkern, M., Olmstead, R., Jou, J., Tiongson, E., Allen, V., ... Cohen, M. (2006). Neural substrates of resisting the urge to smoke. In *Paper presented at the society for research on nicotine and tobacco, Orlando, Florida*.

Claus, E. D., Blaine, S. K., Filbey, F. M., Mayer, A. R., & Hutchison, K. E. (2013). Association between nicotine dependence severity, BOLD response to smoking cues, and functional connectivity. *Neuropsychopharmacology, 38*(12), 2363–2372. http://dx.doi.org/10.1038/npp.2013.134.

Cole, D. M., Beckmann, C. F., Long, C. J., Matthews, P. M., Durcan, M. J., & Beaver, J. D. (2010). Nicotine replacement in abstinent smokers improves cognitive withdrawal symptoms with modulation of resting brain network dynamics. *Neuroimage, 52*(2), 590–599. http://dx.doi.org/10.1016/j.neuroimage.2010.04.251pii:S1053-8119(10)00663-4.

DiFranza, J., & Ursprung, W. (2008). The latency to the onset of nicotine withdrawal: a test of the Sensitization-Homeostasis theory. *Addictive Behaviors, 33*, 1148–1153.

DiFranza, J. R., & Wellman, R. J. (2005). A sensitization-homeostasis model of nicotine craving, withdrawal, and tolerance: integrating the clinical and basic science literature. *Nicotine & Tobacco Research, 7*(1), 9–26. http://dx.doi.org/10.1080/14622200412331328538.

DiFranza, J., Wellman, R., Ursprung, S., & Sabiston, C. (2009). The autonomy over smoking scale. *Psychology of Addictive Behaviors, 23*, 656–665. http://dx.doi.org/10.1037/a0017439.

DiFranza, J., Ursprung, W., & Carlson, A. (2010). New insights into the compulsion to use tobacco from a case series. *Journal of Adolescence, 33*, 209–214. http://dx.doi.org/10.1016/j.adolescence.2009.03.009.

DiFranza, J., Sweet, M., Savageau, J., & Ursprung, W. (2011). An evaluation of a clinical approach to staging tobacco addiction. *Journal of Pediatrics, 159*, 999–1003 e1.

DiFranza, J., Wellman, R., Mermelstein, R., Pbert, L., Klein, J., Sargent, J., ... Winickoff, J. (2011). The natural history and diagnosis of nicotine addiction. *Current Pediatric Reviews, 7*, 88–96.

DiFranza, J., Wellman, R., & Savageau, J. (2011). Does progression through the stages of physical addiction indicate increasing overall addiction to tobacco? *Psychopharmacology, 219*. http://dx.doi.org/10.1007/s00213-011-2411-4.

DiFranza, J., Huang, W., & King, J. (2012). Neuroadaptation in nicotine addiction: update on the Sensitization-Homeostasis model. *Brain Sciences, 2*(4), 523–552. http://dx.doi.org/10.3390/brainsci2040523.

DiFranza, J., Ursprung, W., & Biller, L. (2012). The developmental sequence of tobacco withdrawal symptoms of wanting, craving and needing. *Pharmacology Biochemistry and Behavior, 100*, 494–497.

DiFranza, J., Wellman, R., & Savageau, J. (2012). Does progression through the stages of physical addiction indicate increasing overall addiction to tobacco? *Psychopharmacology, 219*, 815–822. http://dx.doi.org/10.1007/s00213-011-2411-4.

DiFranza, J., Wellman, R., Savageau, J., Beccia, A., & Ursprung, W. (2013). What aspect of dependence does the Fagerstrom test for nicotine dependence measure? *ISRN Addiction*, 8 Article ID 906276.

Ding, X., & Lee, S. W. (2013). Changes of functional and effective connectivity in smoking replenishment on deprived heavy smokers: a resting-state fMRI study. *PLoS One, 8*(3), e59331. http://dx.doi.org/10.1371/journal.pone.0059331.

Engelmann, J. M., Versace, F., Robinson, J. D., Minnix, J. A., Lam, C. Y., Cui, Y., ... Cinciripini, P. M. (2012). Neural substrates of smoking cue reactivity: a meta-analysis of fMRI studies. *Neuroimage, 60*(1), 252–262. http://dx.doi.org/10.1016/j.neuroimage.2011.12.024 pii:S1053-8119(11)01426-1.

Fernando, W., Wellman, R., & DiFranza, J. (2006). The relationship between level of cigarette consumption and latency to the onset of retrospectively reported withdrawal symptoms. *Psychopharmacology, 188*, 335–342. http://dx.doi.org/10.1007/s00213-006-0497-x.

Franklin, T. R., Wang, Z., Wang, J., Sciortino, N., Harper, D., Li, Y., ... Childress, A. R. (2007). Limbic activation to cigarette smoking cues independent of nicotine withdrawal: a perfusion fMRI study. *Neuropsychopharmacology, 32*(11), 2301–2309. http://dx.doi.org/10.1038/sj.npp.1301371.

Fritz, H. C., Wittfeld, K., Schmidt, C. O., Domin, M., Grabe, H. J., Hegenscheid, K., ... Lotze, M. (2014). Current smoking and reduced gray matter volume-a voxel-based morphometry study. *Neuropsychopharmacology, 39*(11), 2594–2600. http://dx.doi.org/10.1038/npp.2014.112.

Gallinat, J., Meisenzahl, E., Jacobsen, L. K., Kalus, P., Bierbrauer, J., Kienast, T., ... Staedtgen, M. (2006). Smoking and structural brain deficits: a volumetric MR investigation. *European Journal of Neuroscience, 24*(6), 1744–1750.

Gallinat, J., Lang, U. E., Jacobsen, L. K., Bajbouj, M., Kalus, P., von Haebler, D., ... Schubert, F. (2007). Abnormal hippocampal neurochemistry in smokers: evidence from proton magnetic resonance spectroscopy at 3 T. *Journal of Clinical Psychopharmacology, 27*(1), 80–84.

Gloria, R., Angelos, L., Schaefer, H. S., Davis, J. M., Majeskie, M., Richmond, B. S., ... Baker, T. B. (2009). An fMRI investigation of the impact of withdrawal on regional brain activity during nicotine anticipation. *Psychophysiology*, *46*(4), 681–693. http://dx.doi.org/10.1111/j.1469-8986.2009.00823.x.

Gray, J. C., Amlung, M. T., Acker, J., Sweet, L. H., Brown, C. L., & MacKillop, J. (2014). Clarifying the neural basis for incentive salience of tobacco cues in smokers. *Psychiatry Research*, *223*(3), 218–225. http://dx.doi.org/10.1016/j.pscychresns.2014.06.003.

Hanlon, C. A., Hartwell, K. J., Canterberry, M., Li, X., Owens, M., Lematty, T., ... George, M. S. (2013). Reduction of cue-induced craving through realtime neurofeedback in nicotine users: the role of region of interest selection and multiple visits. *Psychiatry Research*, *213*(1), 79–81. http://dx.doi.org/10.1016/j.pscychresns.2013.03.003.

Hartwell, K. J., Johnson, K. A., Li, X., Myrick, H., LeMatty, T., George, M. S., & Brady, K. T. (2011). Neural correlates of craving and resisting craving for tobacco in nicotine dependent smokers. *Addiction Biology*, *16*(4), 654–666. http://dx.doi.org/10.1111/j.1369-1600.2011.00340.x.

Heatherton, T. F., & Wagner, D. D. (2011). Cognitive neuroscience of self-regulation failure. *Trends in Cognitive Sciences*, *15*(3), 132–139. http://dx.doi.org/10.1016/j.tics.2010.12.005 pii:S1364-6613(10)00269-X.

Huang, W., DiFranza, J. R., Kennedy, D. N., Zhang, N., Ziedonis, D., Ursprung, S., & King, J. A. (2013). Progressive levels of physical dependence to tobacco coincide with changes in the anterior cingulum bundle microstructure. *PLoS One*, *8*(7), e67837. http://dx.doi.org/10.1371/journal.pone.0067837 pii:PONE-D-13-12855.

Huang, W., King, J., Ursprung, W., Zheng, S., Zhang, N., Kennedy, D., ... DiFranza, J. (2014). The development and expression of physical nicotine dependence corresponds to structural and functional alterations in the anterior cingulate-precuneus pathway. *Brain and Behavior, Open Access*, 1–10. http://dx.doi.org/10.1002/brb3.227.

Hudkins, M., O'Neill, J., Tobias, M., Bartzokis, G., & London, E. (2012). Cigarette smoking and white matter microstructure. *Psychopharmacology*, *221*. http://dx.doi.org/10.1007/s00213-011-2621-9.

Jacobsen, L. K., Picciotto, M. R., Heath, C. J., Frost, S. J., Tsou, K. A., Dwan, R. A., ... Mencl, W. E. (2007). Prenatal and adolescent exposure to tobacco smoke modulates the development of white matter microstructure. *Journal of Neuroscience*, *27*(49), 13491–13498. http://dx.doi.org/10.1523/jneurosci.2402-07.2007.

Jasinska, A. J., Stein, E. A., Kaiser, J., Naumer, M. J., & Yalachkov, Y. (2014). Factors modulating neural reactivity to drug cues in addiction: a survey of human neuroimaging studies. *Neuroscience & Biobehavioral Reviews*, *38*, 1–16. http://dx.doi.org/10.1016/j.neubiorev.2013.10.013.

Koob, G. F., & Le Moal, M. (1997). Drug abuse: hedonic homeostatic dysregulation. *Science*, *278*(5335), 52–58.

Koob, G. F., Caine, S. B., Parsons, L., Markou, A., & Weiss, F. (1997). Opponent process model and psychostimulant addiction. *Pharmacology, Biochemistry and Behavior*, *57*(3), 513–521. pii:S0091305796004388.

Liao, Y., Tang, J., Deng, Q., Deng, Y., Luo, T., Wang, X., ... Hao, W. (2011). Bilateral fronto-parietal integrity in young chronic cigarette smokers: a diffusion tensor imaging study. *PLoS One*, *6*(11), e26460. http://dx.doi.org/10.1371/journal.pone.0026460.

Lim, H. K., Pae, C. U., Joo, R. A., Yoo, S. S., Choi, B. G., Kim, D. J., ... Lee, C. U. (2005). fMRI investigation on cue-induced smoking craving. *Journal of Psychiatric Research*, *39*, 333–335.

Lin, F., Wu, G., Zhu, L., & Lei, H. (2013). Heavy smokers show abnormal microstructural integrity in the anterior corpus callosum: a diffusion tensor imaging study with tract-based spatial statistics. *Drug and Alcohol Dependence*, *129*(1–2), 82–87. http://dx.doi.org/10.1016/j.drugalcdep.2012.09.013.

McClernon, F. J., Froeliger, B., Kozink, R. V., Rose, J. E., Behm, F. M., & Salley, A. N. (2010). Hippocampal and striatal gray matter volume are associated with a smoking cessation treatment outcome: results of an exploratory voxel-based morphometric analysis. *Psychopharmacology*, *210*(4), 577–583. http://dx.doi.org/10.1007/s00213-010-1862-3.

Paul, R. H., Grieve, S. M., Niaura, R., David, S. P., Laidlaw, D. H., Cohen, R., ... Gordon, E. (2008). Chronic cigarette smoking and the microstructural integrity of white matter in healthy adults: a diffusion tensor imaging study. *Nicotine & Tobacco Research*, *10*(1), 137–147. http://dx.doi.org/10.1080/14622200701767829 pii:789470830.

Perkins, K. A. (2002). Chronic tolerance to nicotine in humans and its relationship to tobacco dependence. *Nicotine & Tobacco Research*, *4*, 405–422.

Robinson, T. E., & Berridge, K. C. (1993a). The neural basis of drug craving: an incentive-sensitization theory of addiction. *Brain Research Review*, *18*, 231–241.

Robinson, T. E., & Berridge, K. C. (1993b). The neural basis of drug craving: an incentive-sensitization theory of addiction. *Brain Research. Brain Research Reviews*, *18*(3), 247–291.

Scharf, D. M., Dunbar, M. S., & Shiffman, S. (2008). Smoking during the night: prevalence and smoker characteristics. *Nicotine & Tobacco Research*, *10*(1), 167–178. http://dx.doi.org/10.1080/14622200701767787 pii:789470919.

Sutherland, M. T., McHugh, M. J., Pariyadath, V., & Stein, E. A. (2012). Resting state functional connectivity in addiction: lessons learned and a road ahead. *Neuroimage*, *62*, 2281–2295. http://dx.doi.org/10.1016/j.neuroimage.2012.01.117 pii:S1053-8119(12)00135-8.

Tindell, A. J., Smith, K. S., Berridge, K. C., & Aldridge, J. W. (2009). Dynamic computation of incentive salience: "wanting" what was never "liked". *Journal of Neuroscience*, *29*(39), 12220–12228. http://dx.doi.org/10.1523/jneurosci.2499-09.2009.

Tindell, A. J., Smith, K. S., Pecina, S., Berridge, K. C., & Aldridge, J. W. (2006). Ventral pallidum firing codes hedonic reward: when a bad taste turns good. *Journal of Neurophysiology*, *96*(5), 2399–2409. http://dx.doi.org/10.1152/jn.00576.2006.

Ursprung, S., Morello, P., Gershenson, B., & DiFranza, J. (2010). Development of a measure of the latency to needing a cigarette. *Journal of Adolescent Health*, *48*, 338–343. http://www.journals.elsevierhealth.com/periodicals/jah.

Versace, F., Minnix, J. A., Robinson, J. D., Lam, C. Y., Brown, V. L., & Cinciripini, P. M. (2011). Brain reactivity to emotional, neutral and cigarette-related stimuli in smokers. *Addiction Biology*, *16*(2), 296–307. http://dx.doi.org/10.1111/j.1369-1600.2010.00273.x.

Wang, Z., Faith, M., Patterson, F., Tang, K., Kerrin, K., Wileyto, E. P., ... Lerman, C. (2007). Neural substrates of abstinence-induced cigarette cravings in chronic smokers. *Journal of Neuroscience*, *27*(51), 14035–14040. http://dx.doi.org/10.1523/jneurosci.2966-07.2007.

Zhang, X., Salmeron, B. J., Ross, T. J., Geng, X., Yang, Y., & Stein, E. A. (2011). Factors underlying prefrontal and insula structural alterations in smokers. *Neuroimage*, *54*(1), 42–48. http://dx.doi.org/10.1016/j.neuroimage.2010.08.008 pii:S1053-8119(10)01075-X.

Zhang, X. C., Salmeron, B. J., Ross, T. J., Gu, H., Geng, X. J., Yang, Y. H., & Stein, E. A. (2011). Anatomical differences and network characteristics underlying smoking cue reactivity. *Neuroimage*, *54*(1), 131–141. http://dx.doi.org/10.1016/j.neuroimage.2010.07.063.

Zhang, X., Stein, E. A., & Hong, L. E. (2010). Smoking and schizophrenia independently and additively reduce white matter integrity between striatum and frontal cortex. *Biological Psychiatry*, *68*(7), 674–677. http://dx.doi.org/10.1016/j.biopsych.2010.06.018 pii:S0006-3223(10)00647-5.

Chapter 33

Neural Effects of Nicotine: Peripheral Sensory Systems and Experience-Dependent Neural Sensitization

Eugene A. Kiyatkin
Behavioral Neuroscience Branch, National Institute on Drug Abuse – Intramural Research Program, National Institutes of Health, DHHS, Baltimore, MD, USA

Abbreviations

CNS Central nervous system
EEG Electroencephalic
EMG Electromyographic
iv Intravenous
NAcc Nucleus accumbens
Nicotine$_{PM}$ Nicotine pyrrolidine methiodide
VTA Ventral tegmental area

INTRODUCTION

It is well established that nicotine's action on centrally located nicotinic acetylcholine receptors is essential for the reinforcing properties of this drug and the development of nicotine dependence (Picciotto & Corrigall, 2002). However, before reaching the brain and interacting with these receptors, nicotine transiently activates nicotinic receptors on the afferents of sensory nerves at the sites of its entry (i.e., lung alveoli, nasal and oral cavities) and within the circulatory system (Anand, 1996; Ginzel, 1975; Jonsson et al., 2002; Liu & Simon, 1996; Steen & Reeh, 1993; Walker, Kendal-Reed, Keiger, Bencherif, & Silver, 1996). Although it is known that this peripheral action of nicotine is responsible for mediating its primary sensory and cardiovascular effects (Anand, 1996; Comroe, 1960; Ginzel, 1975), it is usually neglected in considerations of the drug's reinforcing properties and experience-dependent changes in its behavioral and physiological effects that occur as a result of repeated nicotine administration (see, however, Bevins et al., 2012; Engberg & Hajos, 1994). This work is a review of my group's behavioral, electrophysiological, and physiological studies (Lenoir & Kiyatkin, 2011; Lenoir, Tang, Woods, & Kiyatkin, 2013; Tang & Kiyatkin, 2010) that demonstrate the critical role of nicotine's peripheral actions in mediating its neural effects following acute and chronic exposure. Based on these data, we propose that the peripheral actions of nicotine, by creating a rapid neural signal and interacting with subsequent direct drug actions in the brain, play an important role in the development of nicotine-selective neural sensitization—a basic phenomenon underlying experience-dependent changes in its behavioral, physiological, and psychoemotional effects.

RAPID NEURAL ACTIVATION INDUCED BY INTRAVENOUS NICOTINE

To assess the nicotine-induced changes in neural activity and their underlying mechanisms, we used *electroencephalography* (EEG) with subsequent high-speed analysis of its power and frequency characteristics. To assess structural specificity of the neural response, electrical activity was simultaneously recorded from the cortex and the ventral tegmental area (VTA), a critical structure of the motivational-reinforcement circuit (Wise, 2004). The great advantage of EEG is its dynamic nature, which allows for high-speed resolution analysis critical for evaluating the latency and time-course of rapid neural responses. In addition to EEG, we also recorded neck *electromyographic* (EMG) activity, a centrally mediated physiological parameter that reflects tonic and phasic changes in motor output. In contrast to locomotion that is insensitive during certain behavioral responses (i.e., fear-related freezing, intense stereotypy), EMG activity is highly sensitive to even very weak sensory stimuli, such as auditory signals or stress- and cue-free intravenous (iv) saline injection (Kiyatkin & Smirnov, 2010). EEG and EMG signals, moreover, could be monitored for extended periods of time both within one daily session and during repeated sessions, thus allowing the assessment of basal activity state and event-related neural and motor responses. While the total power of the EEG signal is the primary index of electrical synchronization/desynchronization (Buzsaki, 2006; Steriage & McCarley, 2005), we also conducted EEG spectral analysis in order to reveal dynamic, time-dependent alterations in specific activity bands (Lenoir & Kiyatkin, 2011).

A three-point thermorecording paradigm (Kiyatkin, 2010) was used to assess the physiological effects of nicotine and changes following repeated drug exposure. By using miniature thermocouple sensors, temperatures were recorded simultaneously from the temporal muscle and facial skin as well as from the nucleus accumbens (NAcc), a critical structure of the brain motivation-reinforcement circuit (Di Chiara, 2000; Mogenson, Jones, & Yim, 1980; Wise, 2004). The brain site-muscle temperature differences detected the effect of nicotine on intrabrain heat production due to metabolic brain activation, while the skin–muscle difference indicated the change in the state of peripheral blood vessels due to vasoconstriction or vasodilation. Similar to the electrophysiological studies, temperature recordings were conducted in freely moving rats during repeated sessions, thus allowing for an examination of experience-dependent changes in temperature effects of nicotine and related drugs.

In addition to collecting electrophysiological and temperature data, we monitored locomotion as an index of behavioral activation. It is well known that nicotine affects locomotion and its repeated administration results in the progressive enhancement or sensitization of locomotor responses (Benwell & Balfour, 1992; Clarke & Kumar, 1983; Domino, 2001; Mao & McGehee, 2010; Vezina, McGehee, & Green, 2007). Thus, this simple measure allowed us to correlate changes in neural and physiological parameters with animal behaviors elicited by acute and repeated nicotine exposure.

When recorded in freely moving rats extensively habituated to the recording environment, electrical activity in both the cortex and VTA typically shows high-magnitude, low-frequency fluctuations (synchronized activity) transiently interrupted by episodes of low-magnitude, high-frequency activity that occurs either spontaneously or in response to sensory stimuli. While synchronized activity is a typical electrophysiological feature of slow-wave sleep, the so-called desynchronized activity is an index of neural activation (Buzsaki, 2006; Steriage & McCarley, 2005). As shown in Figure 1, iv nicotine injection at low doses (10 and 30 μg/kg) results in a very rapid and almost synchronous change in cortical and VTA EEG signals, closely followed by the appearance of strong EMG activation. During quantitative analyses, we found that EEG total power rapidly decreased after the onset of the injection, reaching significance in the middle of a 15-s injection. Importantly, this rapid effect of nicotine was very similar in both the cortex and VTA and for two low doses (10 and 30 μg/kg), although its duration was clearly shorter for the smaller dose (A and B). The EMG signal showed similar dynamics; however, overall changes were quantitatively larger (C). The total power of the EMG signal began to increase during the nicotine injection, peaked immediately after its end, and slowly decreased thereafter. Importantly, although the initial component of nicotine-induced EEG desynchronization is much stronger and longer quantitatively, it does not differ from the transient activation response that occurs after iv saline injection, a weak visceral sensory stimulus. Saline injection also induced a very small EMG response, which was incomparably weaker than the response induced by nicotine.

Nicotine injected at a dose of 30 μg/kg in drug-naive rats induced relatively weak locomotor activation, weak decreases in NAcc and muscle temperatures, and strongly decreased skin temperature, suggesting peripheral vasoconstriction (Figure 2(A1–C1)). Nicotine also induced a weak but significant increase in the NAcc-muscle differential, suggesting intrabrain heat production due to metabolic brain activation. However, the effects of nicotine rapidly changed during the session, as evidenced by the appearance of increases in NAcc and muscle temperatures. However, the mean of three injections during the first treatment session displayed no significant changes in NAcc and muscle temperatures and revealed persistence of skin vasoconstriction and brain activation (Figure 2(A2–C2)).

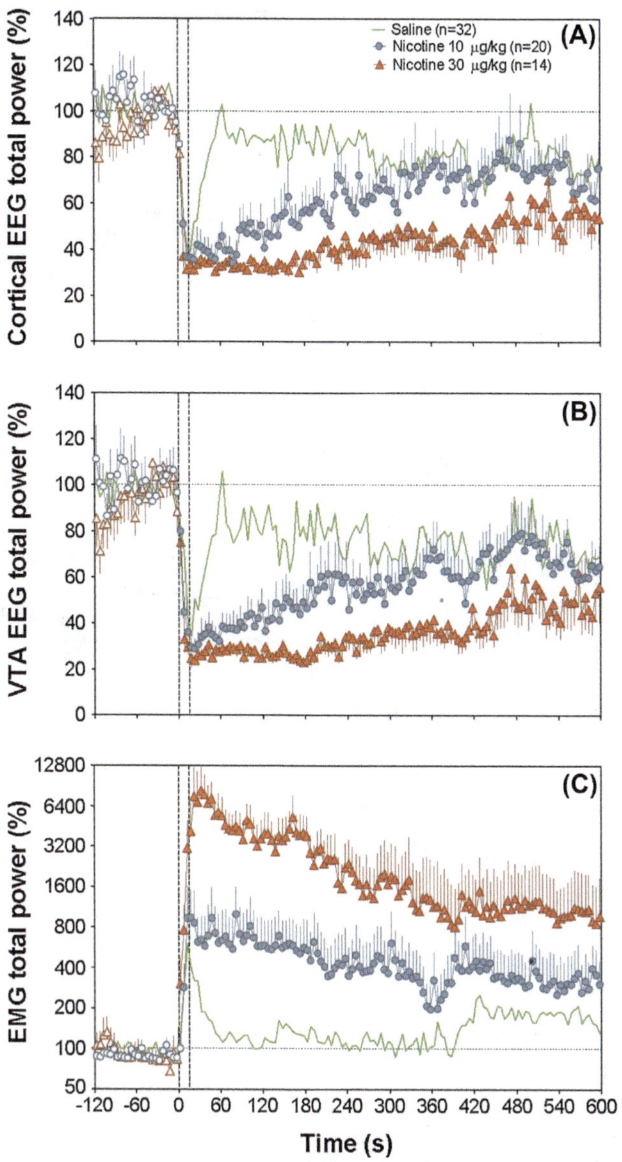

FIGURE 1 Intravenous nicotine at low, self-administering doses induces rapid neural and muscular activation. Changes in cortical (A) and VTA (B) EEG total power as well as neck EMG total power (C) induced by iv nicotine (10 and 30 μg/kg) and saline in drug-naive freely moving rats. *Original data were presented in Lenoir and Kiyatkin (2011), where all details of the protocol and statistical analyses could be found.*

THE ROLE OF PERIPHERAL ACTIONS OF NICOTINE IN MEDIATING ITS ACUTE NEURAL EFFECTS

To evaluate the contribution of the peripheral and central actions of nicotine in mediating its neural effects, we used two pharmacological

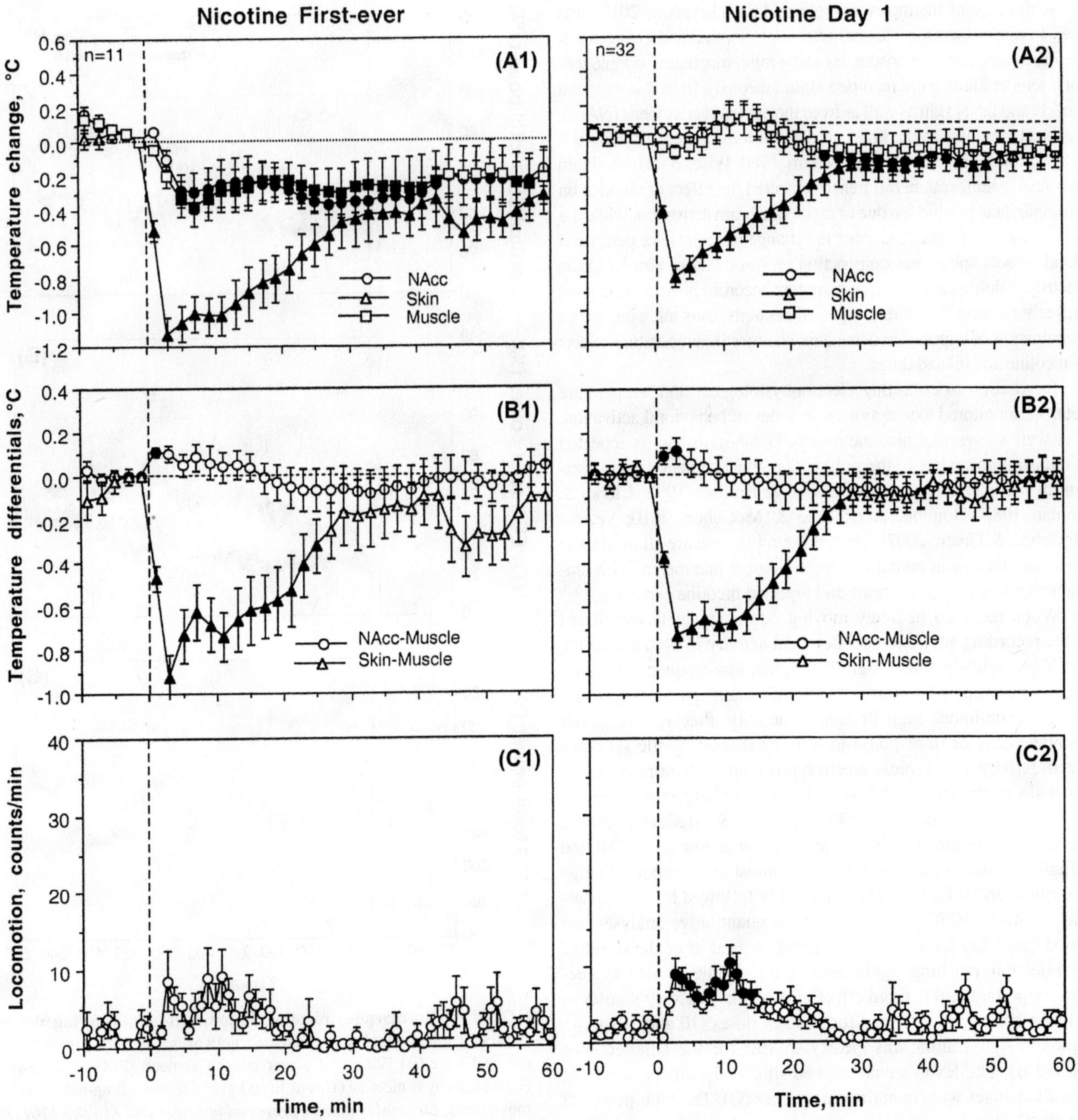

FIGURE 2 Intravenous nicotine induces weak temperature and locomotor effects in drug-naive rats, but these effects are changed following repeated nicotine administration. The left panel shows mean (±SEM) changes in temperature in the NAcc, temporal muscle, and skin (A), mean changes in NAcc-muscle and skin-muscle differentials (B), and locomotor responses (C) induced by iv nicotine (30 μg/kg) after the first-ever injection in drug-naive rats. The right panel shows the same parameters as a mean for three injections during the first treatment session. Filled symbols indicate values significantly different from baseline. *Original data were presented in Tang and Kiyatkin (2010).*

tools: hexamethonium bromide and nicotine pyrrolidine methiodide (nicotine$_{PM}$), which both are highly charged and unable to cross the blood–brain barrier (BBB) (Aceto, Awaya, Martin, & May, 1983; Barlow & Dobson, 1955; Ginzel, 1973, 1975; Oldendorf, Stoller, & Harris, 1993; Wasserman, 1972; Woods, Moyer, & Jackson, 2008). By using hexamethonium pretreatment, we were able to examine how the blockade of peripheral nicotinic receptors affects acute neural effects of iv nicotine. In contrast, nicotine$_{PM}$ selectively activates the peripheral pool of nicotinic receptors, thus allowing for the examination of their impact in mediating the neural and behavioral effects of nicotine. By using ultrasensitive mass spectrometry, we confirmed that nicotine ions cannot be detected in brain tissue after iv administration of nicotine$_{PM}$ at 0.1 mg/kg dose (Lenoir et al., 2013), but direct data on the affinity of nicotine$_{PM}$ with respect to different subtypes of nicotinic and other receptor remains unknown. Nonetheless, a basic similarity in the receptor affinity of both nicotine analogs could be suggested because intraventricular administration of both drugs induces similar antinociceptive effects, which are

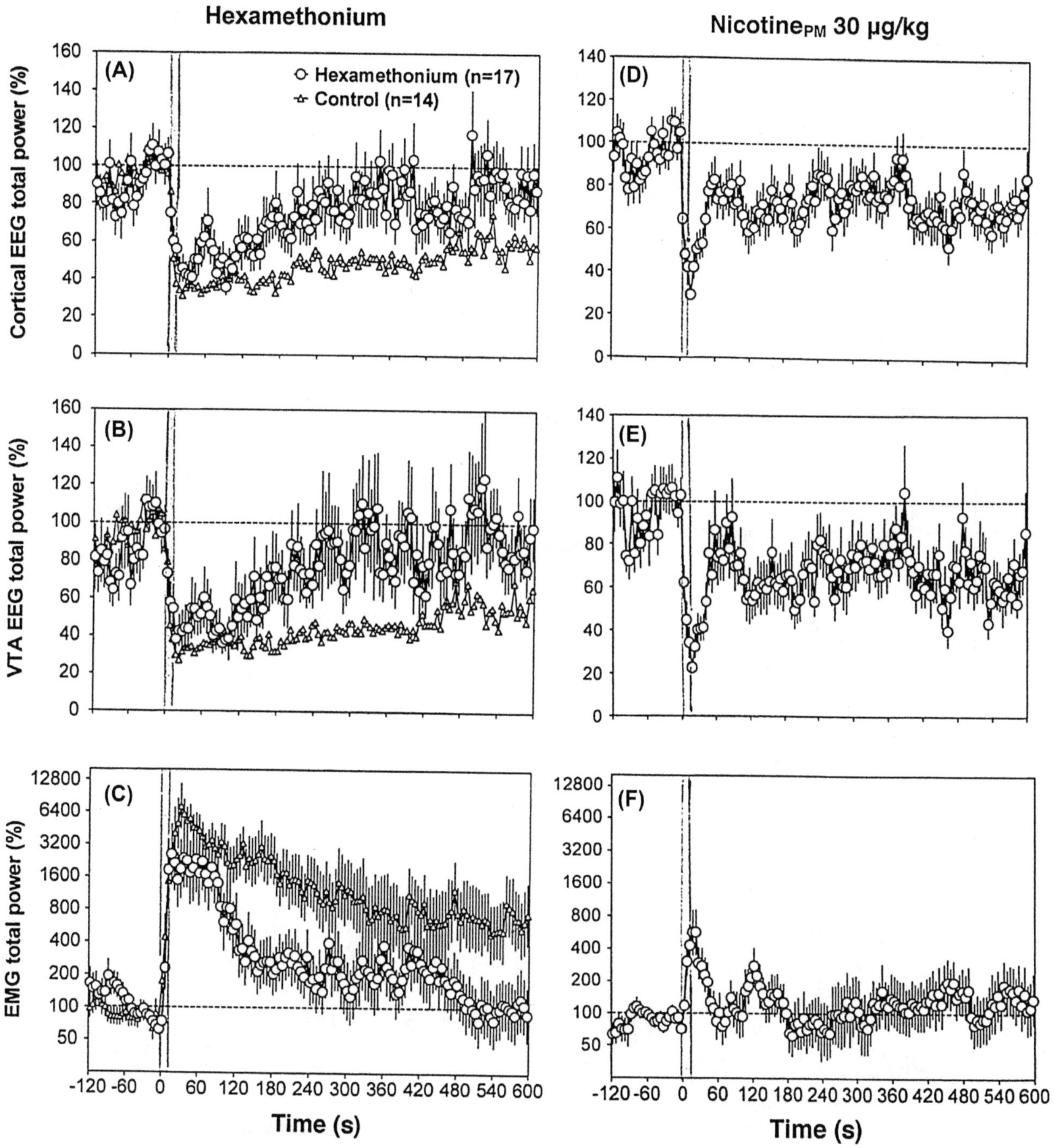

FIGURE 3 **Blockade of peripheral nicotinic receptors by hexamethonium attenuates nicotine-induced neural and motor activation (A–C) and intravenous nicotine$_{PM}$, a peripherally acting nicotine analog, induces weak, transient neuronal and motor activation (D–F).** Graphs show mean (±SEM) changes in cortical (A and D) and VTA (B and E) EEG total power as well as neck EMG total power (C and F) induced by either iv nicotine (30 μg/kg) after pretreatment with hexamethonium (A–C) or by nicotine$_{PM}$ (D–F; 30 μg/kg) in drug-naive freely moving rats. Duration of injection in each graph is shown by two vertical lines at 0 s. *Original data were presented in Lenoir and Kiyatkin (2011).*

greatly attenuated by mecamylamine, a nonselective antagonist of nicotinic receptors (Aceto et al., 1983). Finally, nicotine$_{PM}$ could serve another important purpose when used in nicotine-experienced individuals. If the peripheral actions of nicotine are critical for the development of neural sensitization, activation of peripheral nicotinic receptors in nicotine-experienced individuals should induce a conditioned neural or physiological response distinct from the initial effect of nicotine$_{PM}$ before nicotine treatment.

As shown in Figure 3, hexamethonium pretreatment (5 mg/kg iv, 8 min preceding nicotine injection) significantly attenuated nicotine-induced EEG desynchronization in both the cortex and VTA (A–B). While nicotine administration after hexamethonium pretreatment continued to rapidly decrease EEG total powers in both the cortex and VTA, all effects were significantly weaker than those observed before drug pretreatment and similar to those induced by saline injections. Hexamethonium pretreatment also

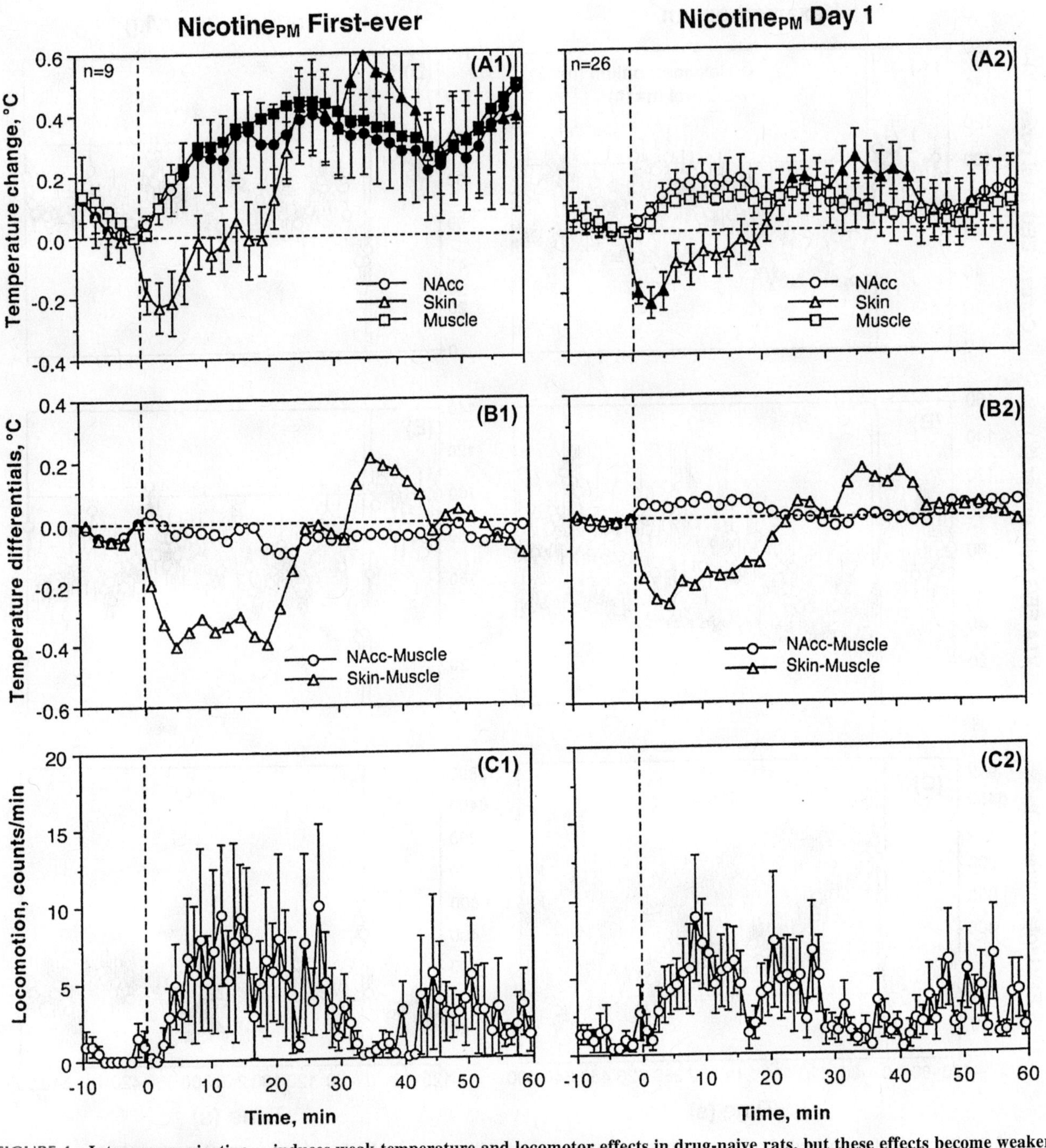

FIGURE 4 Intravenous nicotine$_{PM}$ induces weak temperature and locomotor effects in drug-naive rats, but these effects become weaker following repeated drug administration. The left panel shows mean (±SEM) changes in temperature in the NAcc, temporal muscle and skin (A), mean changes in NAcc-muscle and skin-muscle differentials (B), and locomotor responses (C) induced by nicotine$_{PM}$ (30 μg/kg) after the first-ever injection in drug-naive rats. The right panel shows the same parameters as a mean for three injections during the first treatment session. Filled symbols indicate values significantly different from baseline. *Original data were presented in Tang and Kiyatkin (2010).*

drastically attenuated nicotine-induced EMG activation, but this effect was clearly larger than that induced by saline (C).

While nicotine$_{PM}$ induced rapid EEG desynchronization and EMG activation, its effects were drastically shorter than those of nicotine and only slightly larger than those of saline (Figure 3).

The effects of nicotine$_{PM}$ on NAcc, muscle and skin temperatures, and locomotion differed from those of regular, BBB-permeable nicotine (Figure 4). Following the first-ever injection of this drug, both NAcc and muscle temperature moderately increased, skin temperature showed a biphasic change, and locomotion slightly increased for 20–30 min postinjection (Figure 4(A1–C1)). However, these effects became weaker with subsequent injections. As a mean for three injections during the first treatment session, both NAcc and muscle temperatures did not change significantly (Figure 4(A2–C2)). In contrast to nicotine, nicotine$_{PM}$ did not induce changes in the NAcc-muscle differential, but resulted in much weaker decreases in skin temperature. As such, the central actions of nicotine appear to be vitally important to nicotine's ability to induce metabolic brain activation and peripheral vasoconstriction.

EXPERIENCE-DEPENDENT CHANGES IN THE NEURAL, PHYSIOLOGICAL, AND BEHAVIORAL EFFECTS OF NICOTINE

Similar to other drugs of abuse, the locomotor effects of nicotine are enhanced or sensitized following repeated intermittent drug injections (Benwell & Balfour, 1992; Clarke & Kumar, 1983; Domino, 2001; Mao & McGehee, 2010; Vezina et al., 2007). Consistent with these data, we found that the locomotor effects of iv nicotine (30 μg/kg) are significantly increased in nicotine-experienced rats (six injections of nicotine during two treatment days with a challenge test after one drug-free day; Figure 5(C)).

While the overall magnitude of the sensitized response increased twofold, locomotion occurred with shorter latencies and peaked at earlier times after drug administration. We also found that nicotine experience dramatically affects the resulting temperature responses, with a significant, robust enhancement of both brain and muscle temperature elevations and much stronger decreases in skin temperatures (Figure 5(A)). In addition, an increase in the NAcc-muscle differential, an index reflecting metabolic brain activation (Kiyatkin, 2010), becomes more pronounced, as does a decrease in skin-muscle differential, thus reflecting the stronger vasoconstrictive effects of nicotine (C). Therefore, not only the locomotor but also the physiological effects of nicotine become enhanced or sensitized as a result of previous drug experience.

In contrast to the physiological effects, nicotine-induced EEG desynchronization and EMG activation remained equally strong but more prolonged both in the cortex and VTA in nicotine-experienced animals (Figure 6(A)). A weak enhancement was found only for

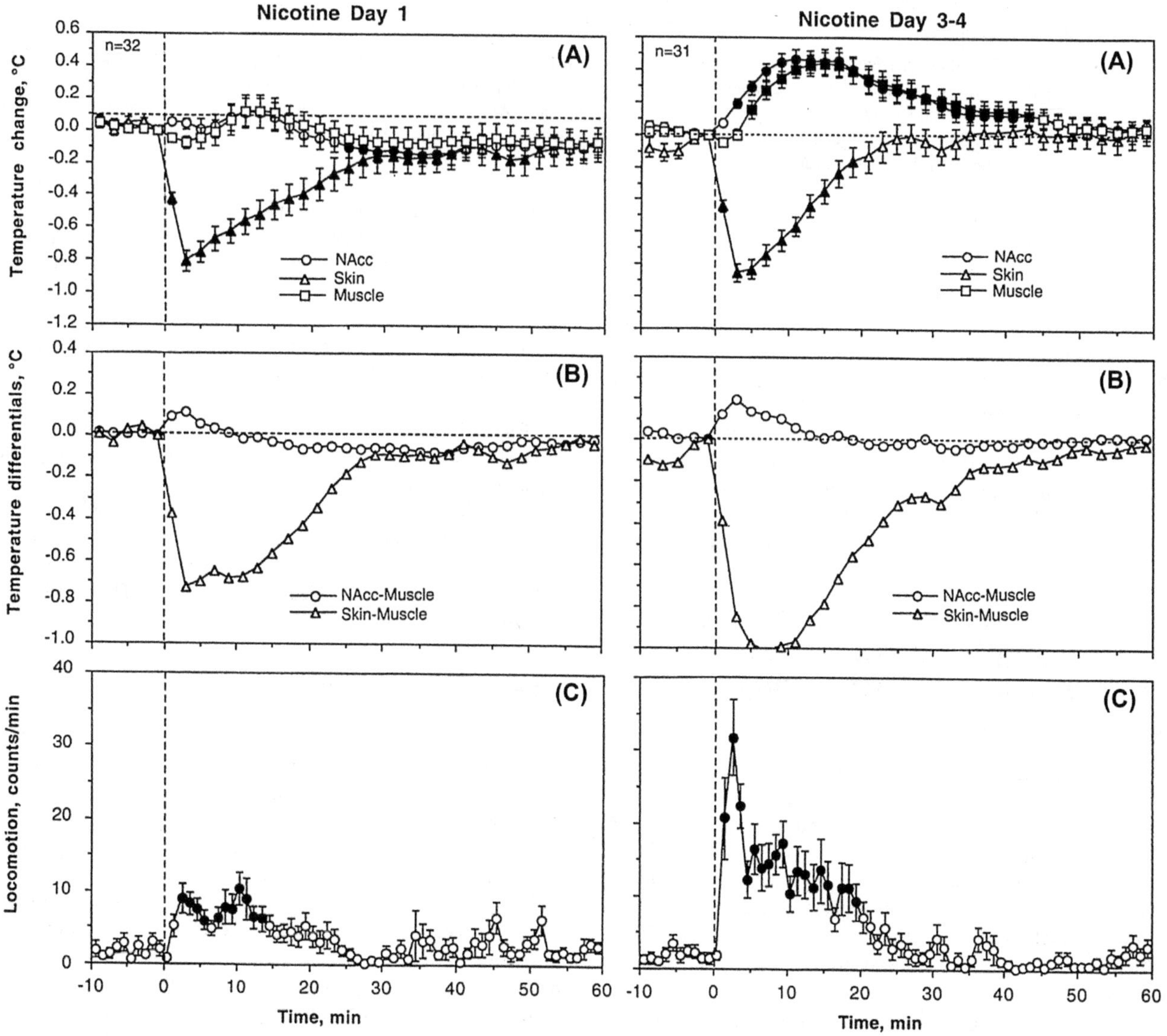

FIGURE 5 Temperature and locomotor effects of nicotine show dramatic enhancement (sensitization) after a relatively short drug experience. Changes in NAcc, muscle, and skin temperatures (A), NAcc-muscle and skin-muscle temperature differentials (B), and locomotion (C) induced by iv nicotine (30 μg/kg) in drug-naive conditions (day 1; left panel) and after previous nicotine exposure (6 injections during 2 days; right panel). Filled symbols indicate values significantly different from baseline. *Original data were presented in Lenoir et al. (2013).*

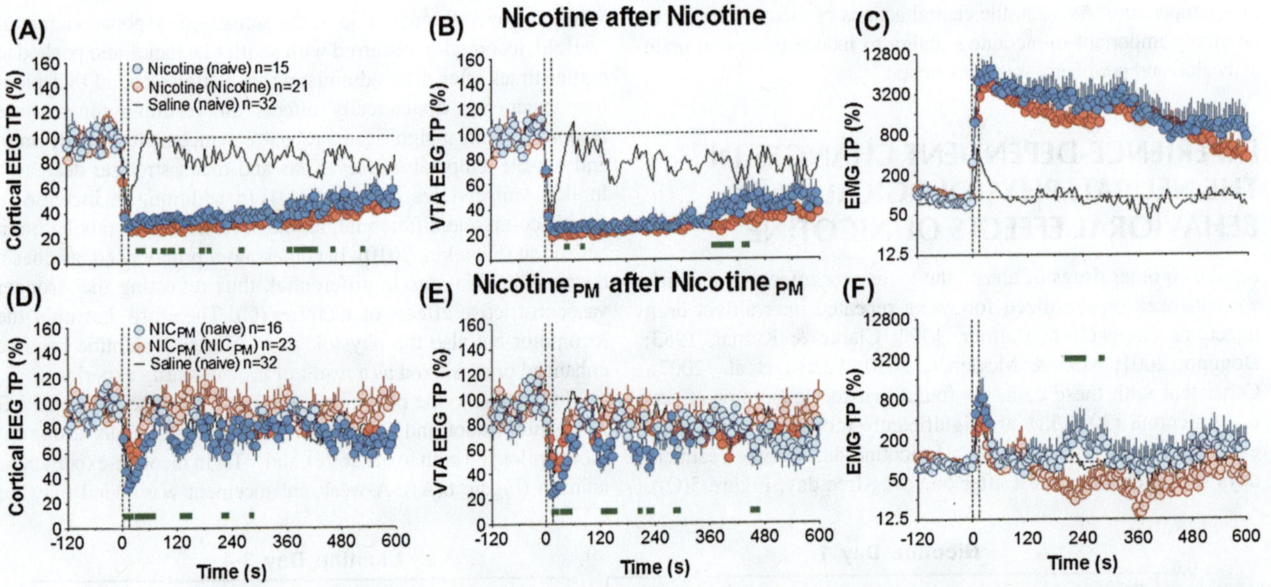

FIGURE 6 Electrophysiological effects of nicotine remain highly stable after drug experience, with a tendency to become more prolonged, but the same effects of nicotine$_{PM}$ habituate following repeated drug exposure. The left panel shows changes in total power of cortical EEG (A), VTA EEG (B), and neck EMG (C) induced by iv nicotine (30 µg/kg) in drug-naive and experienced rats. The right column shows changes in the same parameters for nicotine$_{PM}$ (30 µg/kg). Significant differences in the effects of each drug between two conditions (naive vs experienced) are shown as bold green lines. Original data were presented in Lenoir et al. (2013).

the immediate effects of nicotine in the cortex (A). In contrast to the stability of the electrophysiological effects of nicotine, both the EEG and EMG effects induced by nicotine$_{PM}$ became weaker or habituated following previous experience with this drug (Figure 6(D–F)). This effect was evident in each structure in terms of both the immediate and long-term effects of the drug. Therefore, the acute neural effects of nicotine remain relatively stable following repeated experience, but the effects of selective activation of peripheral nicotinic receptors rapidly habituate following repeated use of nicotine$_{PM}$. A similar weakening was also found in motor activity and temperature responses (Figure 8). In this case, both the temperature and motor effects of nicotine$_{PM}$ after 2 days of previous treatment were almost indistinguishable from those of saline.

THE ROLE OF PERIPHERAL ACTIONS OF NICOTINE IN THE DEVELOPMENT OF NICOTINE-INDUCED NEURAL SENSITIZATION

If the peripheral actions of nicotine are important for the development of nicotine-induced neural and behavioral sensitization, the effects of selective activation of peripheral nicotinic receptors by nicotine$_{PM}$ should be changed in animals with previous nicotine experience. Similar to the appearance of a conditioned response as a result of a sensory stimulus repeatedly paired with natural reinforcers such as food, nicotine$_{PM}$ used in nicotine-experienced subjects could induce a conditioned "nicotine-like" response in the absence of nicotine penetration into the brain.

As shown in Figure 7(A), nicotine$_{PM}$ used in nicotine-experienced rats induced much stronger increases in NAcc and muscle temperatures as well as stronger decreases in skin temperatures than in drug-naive rats. Because the effects of nicotine$_{PM}$ underwent habituation following its repeated exposure, even larger differences were found between the effects of nicotine$_{PM}$ in rats repeatedly exposed to either nicotine or nicotine$_{PM}$. These strong changes in physiological effects are in contrast with nicotine$_{PM}$-induced locomotion, which did not change significantly after nicotine experience (Figure 7(C)). Therefore, a clear conditioned activation was evident in this case at the physiological level, but it was absent at the level of locomotion.

Rapid development of conditioned neural activation was also evident in our electrophysiological data. After a relatively brief nicotine exposure (2 days, six injections), nicotine$_{PM}$ induced a strong, prolonged EEG desynchronization (Figure 8(A) and (B)) and powerfully increased EMG activity (C), greatly exceeding the responses to this drug in both drug-naive and experienced (with repeated nicotine$_{PM}$ injections) conditions. The effects were significant and strong for each parameter.

GENERAL DISCUSSION

Nicotine after iv injection is rapidly distributed by the circulating blood within the body, affecting multiple nicotinic acetylcholine receptors widely expressed in the peripheral and central nervous system (CNS). In this case, local concentrations of nicotine fall rapidly from very large values at the site of administration to much lower levels in distal body locations, including the brain (Berridge et al., 2010; Rose et al., 2010). Although nicotine's action on centrally located nicotinic receptors is essential in mediating its reinforcing properties and its ability to induce nicotine dependence (Picciotto & Corrigall, 2002), the studies reviewed here suggest that the rapid, transient action of nicotine on the afferents of sensory nerves at the sites of administration plays an important role in mediating acute neural effects of this drug and their consistent changes following drug experience.

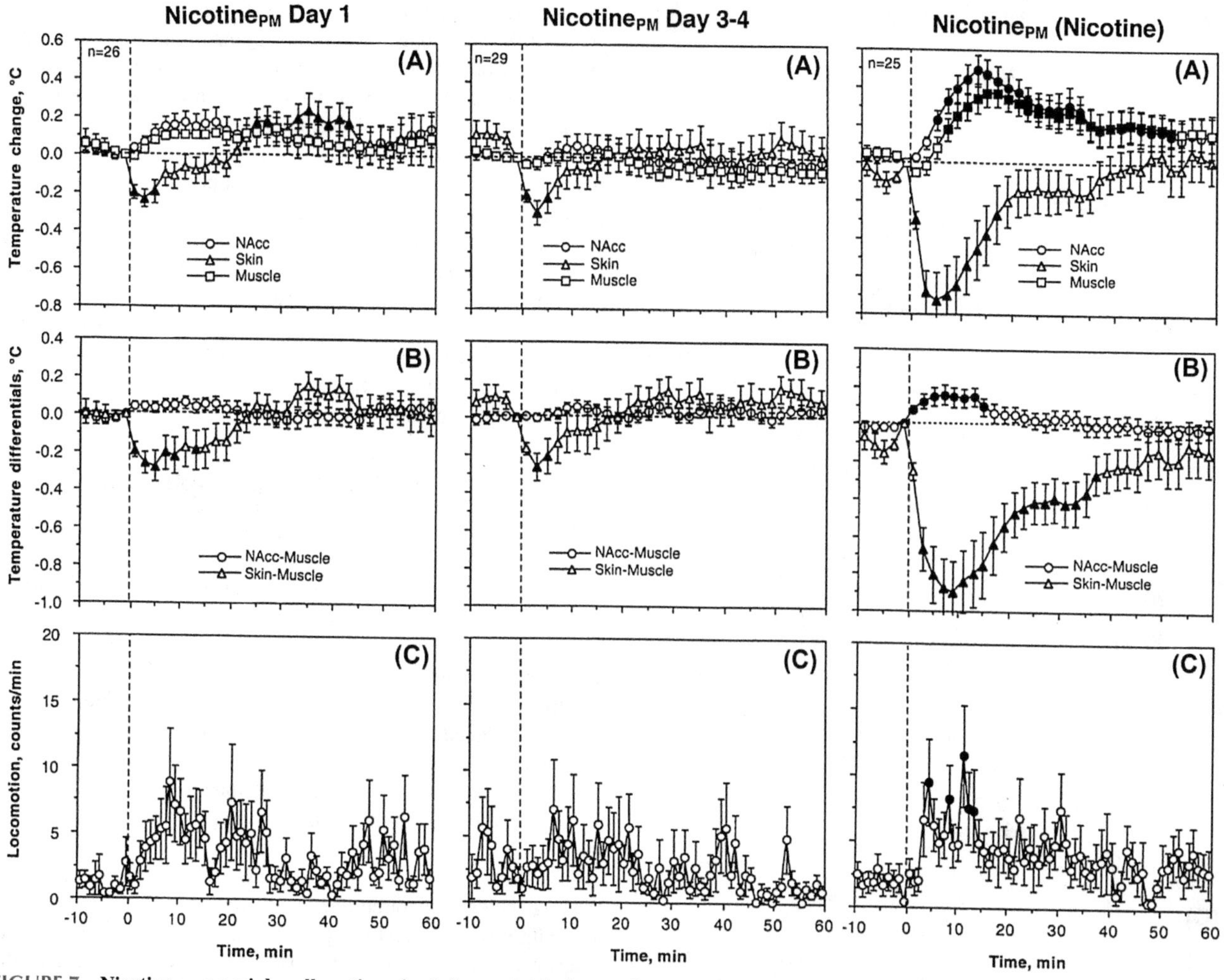

FIGURE 7 Nicotine$_{PM}$, a peripherally acting nicotinic agonist, induces robust temperature effects and weak increase in locomotion after previous nicotine experience. Changes in NAcc, muscle and skin temperatures (A), NAcc-muscle and skin-muscle temperature differentials (B), and locomotion (C) induced by nicotine$_{PM}$ (30 μg/kg) in rats during the first treatment session (day 1; left panel), after 2 days of previous treatment (middle panel) and after nicotine experience (6 injections during 2 days; right panel). *Original data were presented in Lenoir et al. (2013).*

Peripheral and Central Contributions to Nicotine's Neural Effects

Nicotine$_{PM}$, a highly charged nicotine analog that does not enter the brain after systemic administration, is an important tool to dissociate the peripheral actions of nicotine from its central actions. When used in drug-naive conditions, nicotine$_{PM}$ mimicked the electrophysiological effects of weak natural somatosensory stimuli, inducing transient EEG desynchronization and brief EMG activation. Consistent with the known weakening (i.e., habituation) of neural effects induced by simple somatosensory stimuli (Hendry, Hsiao, & Brown, 1999; Sandler & Tsitolovsky, 2008; Stancak, 2006), the electrophysiological effects of nicotine$_{PM}$ habituated following repeated drug exposure. Decreases in neural activation induced by nicotine$_{PM}$ paralleled EMG changes, suggesting that habituation also extends to the peripheral motor output. Therefore, the arousing effects of nicotine$_{PM}$ virtually fully disappeared (i.e., indistinguishable from the effects of iv saline) after minimal experience with this drug.

The central actions of nicotine are likely responsible for its strengthened electrophysiological effects as well as the resistance to habituation following repeated drug exposure as compared to nicotine$_{PM}$. In both drug-naive and drug-experienced conditions, nicotine induced equally strong EEG desynchronization and EMG activation, which became only more prolonged in nicotine-experienced rats. The nicotine-induced EMG activation also remained equally strong in both groups, suggesting the lack of experience-dependent habituation in drug-induced motor output. This finding suggests the differences between EMG, which reflects tonic and phasic changes in muscular electrical activity, and locomotion, which clearly increases following repeated nicotine exposure.

Conditioned Neural Activation Induced by Stimulation of Peripheral Nicotinic Receptors

While the neural and motor effects of nicotine$_{PM}$ rapidly habituated to the levels produced by saline, this drug induced powerful EEG desynchronization in nicotine-experienced animals.

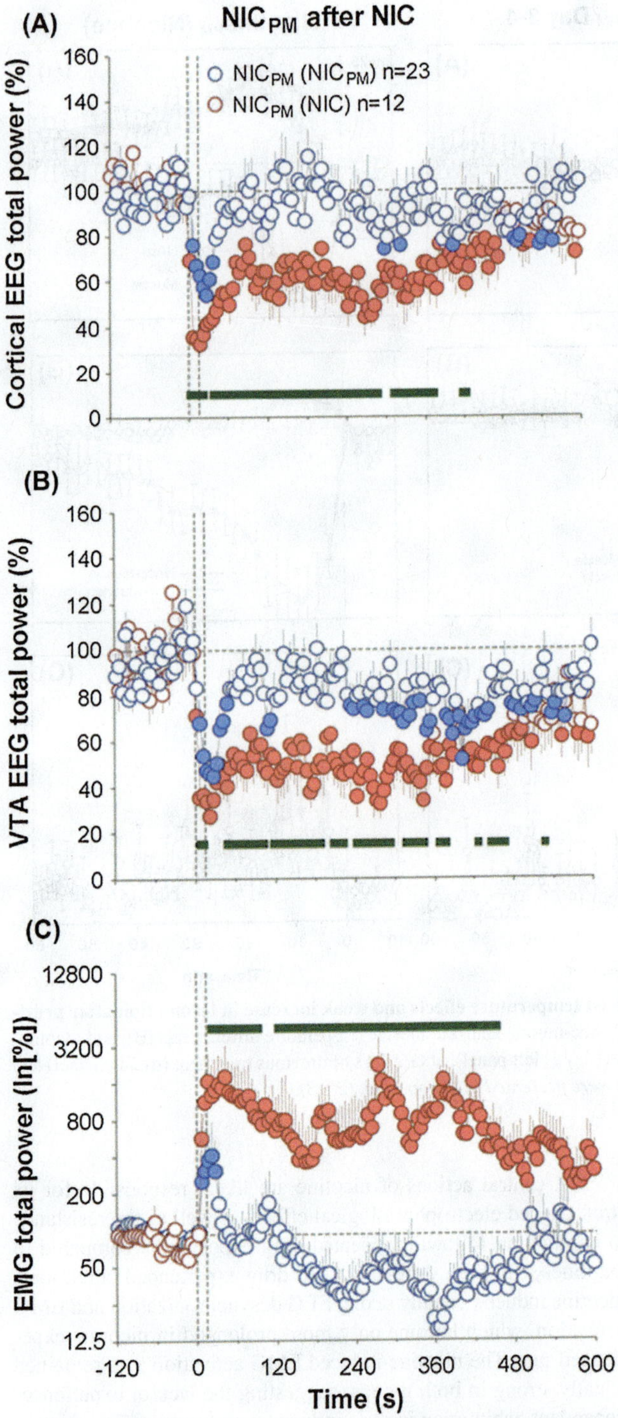

FIGURE 8 Nicotine$_{PM}$, a peripherally acting nicotinic agonist, induces robust and prolonged EEG desynchronization and EMG activation after previous nicotine experience. Changes in total power of cortical EEG (A), VTA EEG (B), and neck EMG (C) induced by nicotine$_{PM}$ (30 μg/kg) in rats that received previous treatment with either nicotine or nicotine$_{PM}$. Green lines in A–C show the time intervals with significant between-group differences. *Original data were presented in Lenoir et al. (2013).*

This effect could be viewed as a manifestation of conditioned neural activation triggered by selective activation of the peripheral pool of nicotinic receptors. This effect was equally strong in the cortex and VTA, mimicking the effect induced by nicotine itself. This change was coupled with strong EMG activation, suggesting that conditioning also occurs at the level of centrally regulated motor output. Robust conditioned activation was also found at the level of physiological parameters. In contrast to the very weak activating effects of nicotine$_{PM}$ in drug-naive conditions, the same drug in the same dose induced strong increases in both brain and muscle temperatures as well as powerful and prolonged peripheral vasoconstriction—novel effects observed neither before nicotine exposure nor after repeated pretreatment with nicotine$_{PM}$. By these parameters, the effects of nicotine$_{PM}$ after previous nicotine exposure were stronger than the initial effects of nicotine itself and similar to the effects induced by nicotine after previous exposure. However, powerful potentiation of electrophysiological and physiological effects of nicotine$_{PM}$ was not accompanied by changes in locomotion, which remained the same regardless of pretreatment.

Conclusions and Functional Implications

Although associative learning is viewed as an important contributor to drug-induced behavioral sensitization (Di Chiara, 2000; Kalivas & Stewart, 1991; Robinson & Berridge, 1993), it is usually considered as the result of the interaction between environmental stimuli that precede drug intake and the drug per se. However, this stimulus–drug interaction could be more complex in the case of nicotine, which by itself has two different, time-shifted actions that are mediated via the drug's interaction with peripherally and centrally located nicotinic receptors. The data presented in this review suggest the critical role of peripheral actions of nicotine in mediating its rapid neural and physiological effects. In addition to the pharmacological evidence provided by using peripherally acting nicotinic agonists and antagonists, the very short, second-scale latencies of nicotine-induced neural activation are inconsistent with possible direct drug actions in the brain. A certain time is always needed for the drug to reach brain vessels from the site of administration, cross the blood–brain barrier, and diffuse passively to appropriate receptor sites. In contrast, nicotine rapidly but transiently activates nicotinic receptors abundantly expressed on the afferents of sensory nerves at the sites of its entry (e.g., lung alveoli, nasal and oral cavities) and within the circulatory system (Anand, 1996; Ginzel, 1975; Jonsson et al., 2002; Juan, 1982; Liu & Simon, 1996; Steen & Reeh, 1993; Walker et al., 1996) and induces an excitatory neural signal that rapidly reaches the brain via visceral somatosensory pathways. While this study clarifies the nature of this first, rapid, and transient action of nicotine, it is still unclear at the mechanistic level how the neural effects triggered by this peripheral action are affected by direct actions of nicotine on brain neurons. This direct central action appears with much longer latencies and is maintained for a longer time after systemic drug administration at human-relevant doses. This direct central action appears to be critical for fixation in memory of traces associated with the initial, sensory effects of nicotine and quite possibly plays a key role in nicotine reinforcement (Huston & Oitzl, 1989).

While it is well established that the direct action of nicotine on brain neurons is essential for the development of nicotine-induced neural sensitization, the present studies provide clear evidence that peripheral actions of nicotine are also critically involved in this process. The coexistence of two initially independent pharmacological actions that are mediated via peripheral and central nicotinic acetylcholine receptors allows for their interaction within the same neural substrates based on principles of Pavlovian conditioning. Due to this interaction, the initial sensory effects of nicotine acquire new properties of interoceptive cues, thus affecting the behavioral and physiological effects of this drug after repeated exposure. Therefore, the development of conditioned association could occur not only for external (exteroceptive) stimuli and the reinforcer (drug), but also to the drug itself (pharmacological or within-drug conditioning), which shares the dual properties of a sensory stimulus and reinforcer. Therefore, the effects of the drug in experienced individuals are always different from those seen in drug-naive individuals and they always represent "conditioned drug effects," reflecting a joint contribution of pharmacological and learning variables (Stewart, 1992).

The ability of nicotine to serve as its own sensory signal could explain its high addictive potential in humans and the rapid transformation of experimental smoking into a highly compulsive habit. Similar to the known shift of neural activation from a primary reinforcer to its sensory predictors (Schultz, 1998), the initially unpleasant or aversive sensory stimulation associated with nicotine consumption (smoking per se) becomes in drug-experienced individuals the source of powerful neural activation and the most desired aspect of smoking. Consistent with this idea, smoking becomes much less rewarding and satisfying when the peripheral actions of nicotine on the sensory afferents of the upper respiratory tract are blocked by local anesthetics (Rose, Tashkin, Ertle, Zinser, & Lafer, 1985). On the other hand, interoceptive actions of nicotine provide only a part of sensory experience of smoking, explaining why habitual smokers often perceive similar immediate subjective and autonomic responses during smoking of nicotine-free cigarettes (Butschky, Bailey, Henningfield, & Pickworth, 1995; Westman, Behm, & Rose, 1996). Obviously, these responses are triggered by multiple nonnicotine sensory stimuli associated with smoking, which in smokers become powerful conditioned stimuli. This basic mechanism of pharmacological conditioning could explain the persistent nature of nicotine addiction in humans and limited success in its pharmacological correction. On the other hand, it suggests that attempts to reduce nicotine addiction could be more effective if they also target the immediate conditioned effects of the drug rather than focusing exclusively on decreasing its unconditioned rewarding effects in the brain.

APPLICATIONS TO OTHER ADDICTIONS AND SUBSTANCE MISUSE

Although this chapter mainly focuses on the role of nicotine's peripheral actions in triggering its rapid neural and physiological effects as well as their changes that occur as the result of repeated drug use, a similar logic could be applied for other drugs of abuse, particularly cocaine. When injected intravenously, cocaine interacts with multiple ionic channels expressed on the afferents of sensory nerves that densely innervate blood vessels. Due to this direct action, cocaine injection results in an excitatory signal that rapidly reaches the CNS via somatosensory pathways. This peripheral action could explain the rapidity of various neural effects of this drug that occur within the injection timeframe and well before the drug is able to reach the brain, cross the blood–brain barrier, diffuse to its receptive substrates, and affect central neurons (Kiyatkin, Kiyatkin, & Rebec, 2000; Kiyatkin & Smirnov, 2010; Wakabayashi & Kiyatkin, 2014).

Although cocaine-methiodide, a peripherally acting cocaine analog, induces similarly rapid neural activation (Kiyatkin & Smirnov, 2010) and rapid glutamate release (Wakabayashi & Kiyatkin, 2014), confirming the role of peripheral neural substrates in triggering the initial effects of cocaine, this action is difficult to verify due to the lack of pharmacological tools for its blockade. In addition to cocaine, this logic could be also applied to opioid drugs, which also interact with peripherally located opioid receptors. Therefore, it appears that neural signals that are generated in the peripheral nervous system by different drugs of abuse and rapidly transmitted to the CNS could be critical in mediating the neural effects of these drugs. These sensory signals elicited by addictive drugs may not only play an important role in experience-dependent changes in neural, physiological, and human psychoemotional effects following repeated drug use, but they also may be tightly involved in the development of drug dependence.

DEFINITION OF TERMS

Nicotine This is a highly selective agonist of nicotinic acetylcholine receptors and the active ingredient in tobacco products responsible for the development of smoking. Nicotine interacts with *nicotinic receptors,* which are expressed on both multiple central neurons and in the peripheral nervous system, particularly on afferents of sensory nerves densely innervating blood vessels.

Neural activation Neural activation generally refers to the excitation of central neurons. Experimentally, neural activation manifests as an increase in impulse activity of central neurons, although many neurons are inhibited when other cells are activated. On a more global level, neural activation manifests as *electroencephalic (EEG) desynchronization,* a specific change in the amplitude and frequency characteristics of the electrical signal directly recorded from the brain. Total power is common measure of EEG signal, where its drop indicates neuronal activation. EEG desynchronization is usually accompanied by *electromyographic (EMG) activation,* a change in amplitude and frequency of electrical activity recorded from the muscle; the increase in this parameter reflects tonic and phasic changes in motor output. Neural activation could manifest at the metabolic level (*metabolic brain activation*) and be assessed as intrabrain heat production as a consequence of increased metabolic activity of brain cells.

Blood–brain barrier (BBB) The BBB separates the brain environment from the periphery and any changes in its permeability have dramatic effects on brain functions. Some drugs (i.e., nicotine) easily penetrate the BBB, thus acting both in the brain and periphery. Other drugs have either weak BBB permeability or cannot enter the brain at all. Two drugs used in this study (*hexamethonium* and *nicotine pyrrolidine methiodide or nicotine$_{PM}$*) are highly charged molecules that cannot cross the BBB, thus serving as tools to selectively block or activate the peripheral nicotinic receptors.

Sensitization Sensitization usually refers to a progressive increase in behavioral or physiological responses to a specific environmental stimulus or a specific drug following its repeated (intermittent) use. Usually, in drug research this term refers to locomotor activity (locomotor sensitization), but physiological responses could also be increased in magnitude and duration following repeated drug use. *Tolerance (habituation)* is typically an opposite change—that is, a progressive decrease in locomotor or physiological responses to a specific stimulus or a drug following its repeated use.

KEY FACTS

Key Facts on Peripheral Sensory Systems and Drugs

- Sensory systems provide the CNS with information about changes in the external and internal environment.
- While direct actions of drugs on central neurons are essential for the drugs' reinforcing properties and addictive potential, drugs also interact with ionic channels and/or receptors located on the afferents of sensory nerves at the sites of administration (i.e., blood vessels, lung alveoli).
- The resulting neural signals rapidly reach the brain via spinal and extraspinal somatosensory pathways, inducing generalized neural activation and multiple physiological effects.

Key Facts on Electroencephalography and Electromyography

- EEG and EMG are two electrophysiological techniques widely used in clinical practice. They are based on the direct recording of sum of electrical signals generated by a large number of neural and muscular cells.
- While EMG signal provides information on tonic and phasic changes in local muscular activity, the amplitude and frequency characteristics of the EEG signal are used to characterize the activity state of the CNS and, with more local recording, the general pattern of neural activity within more local brain areas.
- Although EEG and EMG are often viewed as outdated techniques, they can be used in chronic animals, thus providing valuable information on changes in neural and muscular responses during long-term multisession experiments.

Key Facts on Neural Habituation and Sensitization

- An organism's initial exposure to a novel environmental stimulus elicits a neural and physiological arousal response. If the stimulus is not dangerous, its repeated exposure results in a weakening (habituation) of this arousal response. However, responses to certain stimuli may increase (or sensitize) or may be resistant to habituation despite repeated exposure.
- Sensitization is typical of most addictive drugs, usually manifesting as an increased drug-induced motor response in drug-experienced animals (locomotor sensitization). However, sensitization could also manifest as the enhanced physiological and neural response to the drug.
- The idea of sensitization could be also applied to psychoemotional effects in humans, manifesting in a much stronger, sharper subjective response in drug-experienced individuals as compared to first-time users.

SUMMARY POINTS

- This chapter summarizes the results of my group's recent experiments examining the role of nicotine's actions in the peripheral nervous system in regard to the development of neural sensitization.
- By using EEG and EMG recordings in freely moving rats, we showed that nicotine—at a low, behaviorally active intravenous dose—induces rapid neural and motor activation. These rapid changes in electrophysiological parameters were associated with relatively weak changes in brain and body temperatures and weak increases in locomotion.
- By using peripherally acting agonists and antagonists of nicotine receptors, we showed that rapid neural effects of nicotine are mediated via its direct action in the periphery, involving afferents of sensory nerves that densely innervate blood vessels.
- Both the physiological and locomotor effects of nicotine are rapidly increased or sensitized following repeated nicotine injections, while its neural effects persisted with a tendency to become more prolonged. These changes contrasted with those seen with the peripherally acting nicotine analog, nicotine pyrrolidine methiodide; the effects of this drug progressively decreased (habituated) following repeated use. However, when this peripherally acting drug was used in nicotine-experienced rats, it induced powerful neural and motor activation and robust increases in brain and body temperatures.
- Therefore, the peripheral action of nicotine appears to be important in mediating the acute neural and physiological effects of nicotine. This rapid but transient action that creates a sensory signal to the CNS is followed by slower, more prolonged direct drug action in the brain, resulting in their interaction, contributing to the development of Pavlovian conditioned association.
- This within-drug conditioning mechanism could explain experience-dependent physiological, behavioral, and human psychoemotional effects of nicotine, which in drug-experienced individuals typically represent a combination of pharmacological and learning factors.

ACKNOWLEDGMENTS

This study was supported by the Intramural Research Program of NIDA-IRP. The author greatly appreciates the editorial assistance of Dr. Ken Wakabayashi and Suelynn Ren.

Experimental data presented and discussed in this work were obtained in collaboration with Dr. Magalie Lenoir and Jeremy Tang, who are coauthors on the original papers cited in this review.

REFERENCES

Aceto, M. D., Awaya, H., Martin, B. R., & May, E. L. (1983). Antinociceptive action of nicotine and its methiodide derivatives in mice and rats. *British Journal of Pharmacology, 79*, 869–876.

Anand, A. (1996). Role of aortic chemoreceptors in the hypertensive response to cigarette smoke. *Respiratory Physiology, 106*, 231–238.

Barlow, R. B., & Dobson, N. A. (1955). Nicotine monomethiodide. *Journal of Pharmacy and Pharmacology, 7*, 27–34.

Benwell, M. E., & Balfour, D. J. (1992). The effects of acute and repeated nicotine treatment on nucleus accumbens dopamine and locomotor activity. *British Journal of Pharmacology, 105*, 849–856.

Berridge, M. S., Apana, S. M., Nagano, K. K., Berridge, C. E., Leisure, G. P., & Boswell, M. V. (2010). Smoking produces rapid rise of [^{11}C]nicotine in human brain. *Psychopharmacology (Berlin), 209*, 383–394.

Bevins, R. A., Barrett, S. T., Polewan, R. J., Pittenger, S. T., Swalve, N., & Charntikov, S. (2012). Disentangling the nature of the nicotine stimulus. *Behavioral Processes, 90*, 28–33.

Butschky, M. F., Bailey, D., Henningfield, J. E., & Pickworth, W. B. (1995). Smoking without nicotine delivery decreases withdrawal in 12-hour abstinent smokers. *Pharmacology, Biochemistry & Behavior, 50*, 91–96.

Buzsaki, G. (2006). *Rhythms of the brain*. Oxford: Oxford University Press.

Clarke, P. B., & Kumar, R. (1983). The effects of nicotine on locomotor activity in non-tolerant and tolerant rats. *British Journal of Pharmacology, 78*, 329–337.

Comroe, J. H. (1960). The pharmacological actions of nicotine. *Annals of New York Academy of Sciences, 27*, 48–51.

Di Chiara, G. (2000). Role of dopamine in the behavioral actions of nicotine related to addiction. *European Journal of Pharmacology, 393*, 295–314.

Domino, E. F. (2001). Nicotine induced behavioral locomotor sensitization. *Progress in Neuropsychopharmacology & Biological Psychiatry, 25*, 59–71.

Engberg, G., & Hajos, M. (1994). Nicotine-induced activation of locus coeruleus neurons-an analysis of peripheral versus central induction. *Naunyn-Schmiedeberg's Archive of Pharmacology, 349*, 443–446.

Ginzel, K. H. (1973). Muscle relaxation by drugs which stimulate sensory nerve endings. II. The effects of nicotinic agents. *Neuropharmacology, 13*, 149–164.

Ginzel, K. H. (1975). The importance of sensory nerve endings as sites of drug action. *Naunyn-Schmiedeberg's Archive of Pharmacology, 288*, 29–56.

Hendry, S. C., Hsiao, S. S., & Brown, M. C. (1999). Fundamentals of sensory systems. In M. J. Zigmond, F. E. Bloom, S. C. Landis, J. L. Roberts, & L. R. Squire (Eds.), *Fundamental neuroscience* (pp. 657–670). San Diego: Academic Press.

Huston, J. P., & Oitzl, M. S. (1989). The relationships between reinforcement and memory: parallels in the rewarding and mnemonic effects of the neuropeptide substance P. *Neuroscience & Biobehavioral Reviews, 13*, 171–180.

Jonsson, M., Kim, C., Yamamoto, Y., Runold, M., Lindahl, S. G., & Eriksson, L. I. (2002). Atracurium and vecuronium block nicotine-induced carotid body chemoreceptor responses. *Acta Anaesthesiology Scandinavica, 46*, 488–494.

Juan, H. (1982). Nicotinic nociceptors on perivascular sensory nerve endings. *Pain, 12*, 259–264.

Kalivas, P. W., & Stewart, J. (1991). Dopamine transmission in the initiation and expression of drug- and stress-induced sensitization of motor activity. *Brain Research Reviews, 16*, 223–244.

Kiyatkin, E. A. (2010). Brain temperature homeostasis: physiological fluctuations and pathological shifts. *Frontiers in Bioscience, 15*, 73–92.

Kiyatkin, E. A., Kiyatkin, D. E., & Rebec, G. V. (2000). Phasic inhibition of dopamine uptake by intravenous cocaine in freely moving rats. *Neuroscience, 98*, 729–741.

Kiyatkin, E. A., & Smirnov, M. S. (2010). Rapid EEG desynchronization and EMG activation induced by intravenous cocaine in freely moving rats: a peripheral, nondopamine neural triggering. *American Journal of Physiology, 298*, R285–R300.

Lenoir, M., & Kiyatkin, E. A. (2011). Critical role of peripheral actions of intravenous nicotine in mediating its central effects. *Neuropsychopharmacology, 36*, 2125–2138.

Lenoir, M., Tang, J. S., Woods, A. S., & Kiyatkin, E. A. (2013). Rapid sensitization of physiological, neuronal, and locomotor effects of nicotine: critical role of peripheral drug actions. *Journal of Neuroscience, 33*, 9937–9949.

Liu, L., & Simon, S. A. (1996). Capsaicin and nicotine both activate a subset of rat trigeminal ganglion neurons. *American Journal of Physiology, 270*, C1807–C1814.

Mao, D., & McGehee, D. S. (2010). Nicotine and behavioral sensitization. *Journal of Molecular Neuroscience, 40*, 154–163.

Mogenson, G. J., Jones, D. L., & Yim, C. Y. (1980). From motivation to action: functional interface between the limbic system and the motor system. *Progress in Neurobiology, 14*, 69–97.

Oldendorf, W. H., Stoller, B. E., & Harris, F. L. (1993). Blood–brain barrier penetration abolished by N-methyl quaternization of nicotine. *Proceedings of the National Academy of Sciences of the United States of America, 90*, 307–311.

Picciotto, M. R., & Corrigall, W. A. (2002). Neuronal systems underlying behaviors related to nicotine addiction: neural circuits and molecular genetics. *Journal of Neuroscience, 22*, 3338–3341.

Robinson, T. E., & Berridge, K. C. (1993). The neural basis of drug craving: an incentive-sensitization theory of addiction. *Brain Research Reviews, 18*, 247–291.

Rose, J. E., Mukhin, A. G., Lokitz, S. J., Turkington, T. G., Herskovic, J., Behm, F. M., … Garg, P. K. (2010). Kinetics of brain nicotine accumulation in dependent and nondependent smokers assessed with PET and cigarettes containing ^{11}C-nicotineotine. *Proceedings of the National Academy of Sciences of the United States of America, 107*, 5190–5195.

Rose, J. E., Tashkin, D. P., Ertle, A., Zinser, M. C., & Lafer, R. (1985). Sensory blockade of smoking satisfaction. *Pharmacology, Biochemistry & Behavior, 23*, 289–293.

Sandler, U., & Tsitolovsky, L. (2008). *Neural cell behavior and fuzzy logic*. New York: Springer.

Schultz, W. (1998). Predictive reward signal of dopamine neurons. *Journal of Neurophysiology, 80*, 1–27.

Stancak, A. (2006). Cortical oscillatory changes occurring during somatosensory and thermal stimulation. *Progress in Brain Research, 159*, 237–252.

Steen, K. H., & Reeh, P. W. (1993). Actions of cholinergic agonists and antagonists on sensory nerve endings in rat skin, in vitro. *Journal of Neurophysiology, 70*, 397–405.

Steriage, M., & McCarley, R. W. (2005). *Brain control of wakefulness and sleep*. New York: Kluwer Academic/Plenum Publishers.

Stewart, J. (1992). Neurobiology of conditioning to drugs of abuse. *Annals of New York Academy of Sciences, 654*, 335–346.

Tang, J. S., & Kiyatkin, E. A. (2010). Fluctuations in central and peripheral temperatures induced by intravenous nicotine: central and peripheral contributions. *Brain Research, 1383*, 141–153.

Vezina, P., McGehee, D. S., & Green, W. N. (2007). Exposure to nicotine and sensitization of nicotine-induced behaviors. *Progress in Neuropsychopharmacology & Biological Psychiatry, 31*, 1625–1638.

Wakabayashi, K. T., & Kiyatkin, E. A. (2014). Critical role of peripheral drug actions in experience-dependent changes in nucleus accumbens glutamate release induced by intravenous cocaine. *Journal of Neurochemistry, 128*, 672–685.

Walker, J. C., Kendal-Reed, M., Keiger, C. J., Bencherif, M., & Silver, W. L. (1996). Olfactory and trigeminal responses to nicotine. *Drug Development Research, 38*, 160–168.

Wassermann, O. (1972). Studies on the pharmacokinetics of bis-quaternary ammonium compounds. 3. Autoradiographic studies on the substituent-dependent distribution pattern of 3H-hexamethonium derivatives in mice. *Naunyn Schmiedeberg's Archive of Pharmacology, 275*, 251–261.

Westman, E. C., Behm, F. M., & Rose, J. E. (1996). Dissociating the nicotine and airway sensory effects of smoking. *Pharmacology, Biochemistry & Behavior, 53*, 309–315.

Wise, R. A. (2004). Dopamine, learning and motivation. *Nature Reviews in Neuroscience, 5*, 483–494.

Woods, A. S., Moyer, S. C., & Jackson, S. N. (2008). Amazing stability of phosphate-quaternary amine interactions. *Journal of Proteomic Research, 7*, 3423–3427.

Section D

Methods

Chapter 34

Cotinine Urinalysis for Tobacco Use

Yatan P.S. Balhara[1], Siddharth Sarkar[2]

[1]Department of Psychiatry & National Drug Dependence Treatment Centre, WHO Collaborating Centre on Substance Abuse, All India Institute of Medical Sciences, New Delhi, India; [2]Department of Psychiatry, Jawaharlal Institute of Postgraduate Medical Education and Research, Pondicherry, India

Abbreviations

CYP Cytochrome P450
GCMS Gas chromatography–mass spectrometry
HPLC High-performance liquid chromatography
SHS Secondhand smoke

INTRODUCTION

Biomarkers for tobacco consumption are important in the field of biological and clinical research, as well as clinical practice. They help to objectively ascertain exposure to nicotine and other tobacco-related products due to personal consumption or environmental exposure. The different biomarkers for assessment of exposure to tobacco products include compounds such as nicotine, cotinine, carbon monoxide, thiocyanate, and others. These compounds can be assessed using different biological samples such as urine (cotinine, thiocyanate), plasma (cotinine, nicotine), saliva (cotinine), exhaled air (carbon monoxide), and hair (cotinine). Each of the biological samples has distinct advantages and disadvantages in the measurement of tobacco exposure. Among all the methods, urinary cotinine analysis remains a quite commonly used method of assessment.

Cotinine is chemically (5S)-1-methyl-5-(3-pyridyl) pyrrolidin-2-one and is a metabolite of nicotine. This molecule has been utilized as a marker of exposure to nicotine among those who smoke or consume tobacco-related substances, as well those nonusers who are exposed to the secondhand smoke of tobacco products. The measurement of cotinine rather than nicotine has certain advantages and is preferred over the assessment of nicotine. Firstly, cotinine has a longer half-life than nicotine, and hence can assess exposure to tobacco over a longer duration of time. Secondly, assessment of nicotine can be done through a plasma sample; cotinine can be measured in a urine sample, which is a much less invasive way of sample collection. Hence, the analysis of cotinine has gained widespread use for the objective assessment of nicotine and tobacco exposure.

Cotinine levels might not only give an estimate of whether a person smokes or not, but also provide a quantitative estimate of the degree of smoking. Cotinine often has been used in conjunction with the behavioral measures to estimate the amount of tobacco products consumed on a daily basis. Detection of cotinine levels may be particularly helpful in situations where the respondent is not likely to provide an accurate account. Quantitative estimates of cotinine also are helpful to assess the degree of exposure and the risk of harm to those who do not take tobacco products. Hence, cotinine urinalysis may provide an objective measure of exposure to nicotine.

This chapter deals with cotinine measurement, providing a broad overview of cotinine urinalysis. To begin with, the metabolism of nicotine and cotinine are discussed. This is followed by elaboration of the assays used for the detection of cotinine. Thereafter, the chapter covers the research and clinical application of cotinine detection. This is followed by evaluation of the sensitivity and specificity issues relating to the use of cotinine urinalysis.

PHARMACOKINETICS AND PHARMACODYNAMICS

Cotinine is derived from the metabolism of nicotine. About 70 to 80 percent of the nicotine consumed by a person is converted to cotinine (Hukkanen, Jacob, & Benowitz, 2005). Nicotine is also converted by liver enzymes to other compounds, such as nicotine glucuronide, nicotine N′-oxide, nornicotine, and 2′-hydroxynicotine. The conversion from nicotine to cotinine (the major metabolite) involves two steps. The first step is the conversion of nicotine to nicotine-$\Delta 1'$ (5′)-iminium ion, which is mediated by CYP2A6 isoenzyme. The second step is the conversion of nicotine-$\Delta 1'$ (5′)-iminium ion to cotinine, catalyzed by aldehyde oxidase enzyme in the cytoplasm. The half-life of nicotine has been estimated to be between 100 and 150 min.

Cotinine undergoes further metabolism (Figure 1). Only 10 to 15 percent of the cotinine is excreted as unchanged cotinine in the urine. The remaining part of cotinine is metabolized into *trans*-3′-hydroxycotinine, 5′-hydroxycotinine, cotinine N-oxide, cotinine methonium ion, cotinine glucuronide, and norcotinine. Some of these compounds are further metabolized. Conversion to *trans*-3′-hydroxycotinine is the major pathway of cotinine metabolism and is carried out by CYP2A6 isoenzyme. The major portion of 3′-hydroxycotinine is excreted unchanged through the kidneys, while some amount is excreted as a glucuronide conjugate. Hence, 3′-hydroxycotinine is the major metabolite of nicotine that is excreted in the urine. The half-life of cotinine has been estimated to range from 770 to over 1100 min (Benowitz, Hukkanen, & Jacob, 2009).

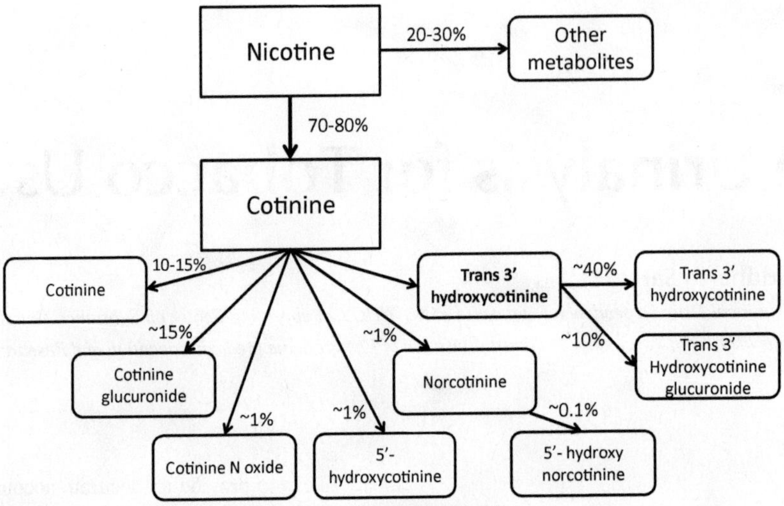

FIGURE 1 Metabolites of cotinine.

CYP2A6 is the major isoenzyme involved in the metabolism of nicotine and cotinine. Hence, the rates of metabolism of cotinine can be influenced by the activity of this isoenzyme. Variants of CYP2A6 isoenzyme have been associated with lower clearance of nicotine and cotinine from the body (Benowitz, Dempsey, Tyndale, St Helen, & Jacob, 2013). The clearance of cotinine has been estimated to be around 50 ml/min, with majority of it attributed to nonrenal hepatic clearance. The volume of distribution has been estimated to be 0.69–0.93 l/kg. Both clearance and volume of distribution of cotinine are significantly lower than that of nicotine (Hukkanen et al., 2005).

Ethnic Differences

Ethnic differences have been noted to be present in the metabolism of cotinine. The clearance of cotinine and the proportion of nicotine converted to cotinine has been found to be higher in whites than African Americans (Benowitz et al., 1999). This has been ascribed to the slower oxidative metabolism of nicotine among African Americans than whites. Similar lower clearance of nicotine through the cotinine pathway has been observed for Chinese Americans (Benowitz, Pérez-Stable, Herrera, & Jacob, 2002), but not Latinos (Caraballo et al., 1998). These ethnic differences in cotinine metabolism are important when interpreting the results of analysis of biological samples.

Gender Differences

Gender differences have been suggested to exist for the metabolism of cotinine. Women show a smaller volume of distribution and a shorter half-life for cotinine as compared to men (Benowitz et al., 1999). Similarly, the literature suggests that the metabolism of cotinine is faster in women than men, and cotinine is cleared more rapidly from the body (Benowitz, Lessov-Schlaggar, Swan, & Jacob, 2006). The use of oral contraceptives increases the rate of metabolism of cotinine even further. Such gender differences in cotinine metabolism are not observed for perimenopausal and menopausal women (Benowitz et al., 2006).

ASSAYS FOR DETECTION

Various methods are available for the detection of cotinine in the urine. Initially, the detection of cotinine in urine samples depended upon the use of high-performance liquid chromatography (HPLC). Briefly, HPLC involves a procedure for separating the components in a mixture and quantifying the amount of each component. In this procedure, a pressurized liquid solvent that contains the index sample is passed through an adsorbent column. Each component of the mixture flows at a different rate through the adsorbent; hence, the components can be separated as they flow out of the column. The principles of this method are graphically represented in Figure 2. This method for the determination of cotinine in urine samples was introduced in the late 1970s (Watson, 1977). HPLC has been refined over time with introduction of techniques such as reverse-phased HPLC, tandem spectroscopy, and electrospray ionization (Apinan, Choemung, & Na-Bangchang, 2010; Baumann, Regenthal, Burgos-Guerrero, Hegerl, & Preiss, 2010; Massadeh, Gharaibeh, & Omari, 2009). HPLC has been found to work well for cotinine in the urine in the range of 1–5000 mcg/l. HPLC can be used to detect cotinine as well as its metabolites, such as glucuronide and hydroxy forms (Piller, Gilch, Scherer, & Scherer, 2014; Rangiah, Hwang, Mesaros, Vachani, & Blair, 2011).

Gas chromatography with mass spectrometry (GCMS) has been introduced subsequently as a means of detecting cotinine in the urine samples. Briefly, it involves a gas base medium as a solvent for eluting out different components of a mixture. The carrier gas is often nitrogen or helium. The run-off is then analyzed using a mass spectrometer, which ionizes the chemical compounds to generate charged molecules and then assesses the charge-to-mass ratio. The procedure of GCMS has been diagrammatically represented in Figure 3. GCMS has been used for effectively detecting cotinine levels in the urine (Chiadmi & Schlatter, 2014; Iwai et al., 2013). Using refined techniques, the limit of detection for cotinine has been reported to be as low as 0.2 mcg/l. Studies have suggested a high degree of correlation between GCMS and HPLC (Massadeh et al., 2009), with a correlation coefficient exceeding 0.99.

Immunoassay is the third type of procedure used for the detection of cotinine in urine samples. In this procedure, antibodies are produced in sheep against cotinine conjugated to a hapten.

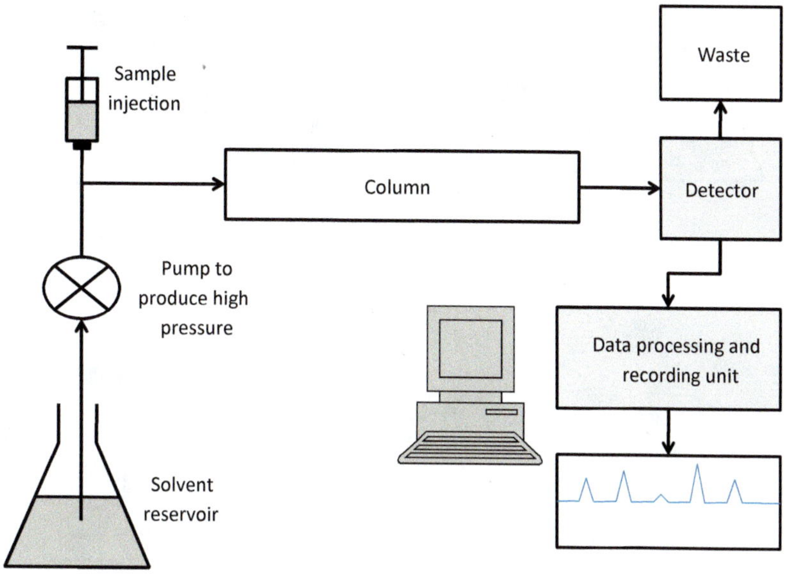

FIGURE 2 Schematic diagram of high-performance liquid chromatography.

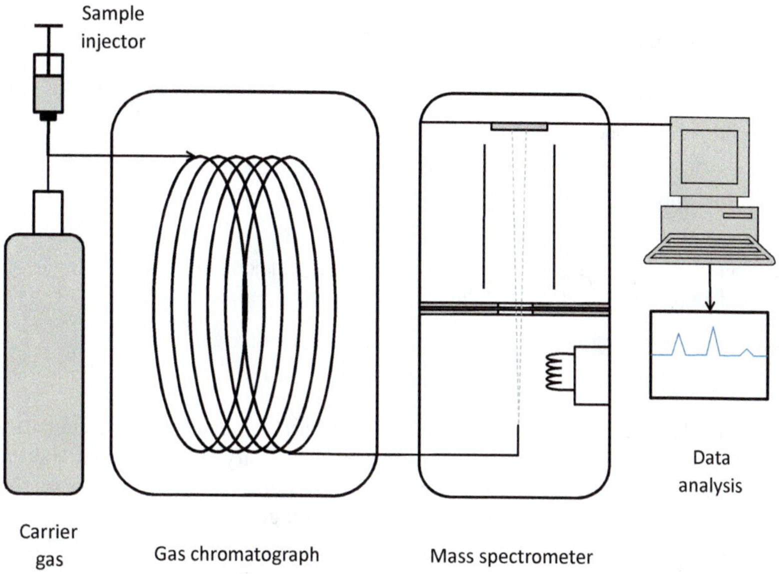

FIGURE 3 Schematic diagram of gas chromatography–mass spectrometry.

These antibodies are recovered and are coupled with labeled cotinine conjugate in a cuvette. When urinary cotinine is introduced, some of the coupled conjugate is eluted out and replaced by cotinine in the urine. This is subsequently used to measure and estimate the presence of cotinine. The principles of immunoassay techniques are represented in Figure 4. A semiquantitative measurement of urine using immunoassay test strips has also been developed for immediate assessment of smoking status (Acosta, Buchhalter, Breland, Hamilton, & Eissenberg, 2004). The method shows good sensitivity but comparatively lower specificity. Such strips have been compared to chromatographic techniques and have shown similar results in detecting tobacco usage (Gaalema, Higgins, Bradstreet, Heil, & Bernstein, 2011; Schepis et al., 2008). The advantages and disadvantages of each method are represented in Table 1.

Different methods of estimating cotinine levels would be useful for different situations. In field settings or when results are required as fast as possible, use of immunoassay strips would be more prudent. When high precision of quantitative analysis is required, HPLC or GCMS may be the preferred choices. GCMS requires more sophisticated equipment than HPLC, but results are probably more accurate. While using quantitative techniques for assessment of cotinine, one should attempt to minimize the chances of evaporation from the urine sample, which might result in a higher concentration of cotinine being reported. Processing of the sample as soon as possible after sample collection might reduce lack of precision due to evaporation. It has been seen that adequately frozen samples of urine ($-20\,°C$) can give accurate results of cotinine levels, even after a long time of storage (Riboli, Haley, De Waard, & Saracci, 1995).

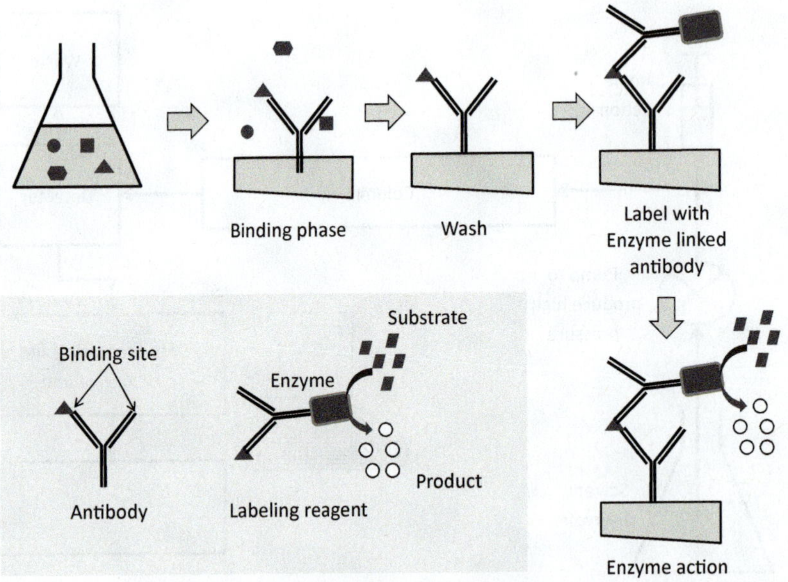

FIGURE 4 Schematic diagram of immunoassay techniques.

TABLE 1 Overview of Different Assay Methods for Cotinine

	HPLC	GCMS	Immunoassay
Technique	Liquid chromatography under high pressure	Gas chromatography coupled with mass spectrometry	Using antibodies against cotinine for estimation
Advantages	Good for quantitative assessment	Good for quantitative assessment	Quick and requires less sophisticated on-site equipment
Limitations	Need for sophisticated equipment, time required for processing	Need for sophisticated equipment, time required for processing	Semiquantitative

HPLC, high-performance liquid chromatography; GCMS, gas chromatography–mass spectrometry.

Urine cotinine assay techniques have been compared to samples from other parts of the body. Urinary cotinine using HPLC-mass spectroscopy has been compared to saliva cotinine levels using dipstick immunoassay (Montalto & Wells, 2007). Urinary cotinine levels have been compared to saliva and hair cotinine levels using HPLC (Shin, Kim, Shin, & Jee, 2002; Toraño & van Kan, 2003).

RESEARCH AND CLINICAL APPLICATIONS OF URINE COTININE DETECTION

There can be many applications of urine cotinine detection. These are summarized in Table 2 and are discussed in the following sections.

Confirmation of Tobacco Abstinence in the Treatment Setting

Self-reporting may not be a robust marker of current smoking status, as patients may knowingly or unknowingly give wrong information (Balhara, Jain, Sundar, & Sagar, 2011, 2012). Urine cotinine levels provide an objective measure for assessing the use of tobacco products in the recent past. Apart from urinary cotinine levels, salivary cotinine levels and breath carbon monoxide levels are other objective markers of recent tobacco usage. Studies have suggested that self-reported tobacco usage did not significantly correlate with urinary cotinine levels for general psychiatry outpatients, patients with somatoform disorder, and patients with bipolar disorder (Balhara et al., 2011, 2012).

Cotinine levels as proxy markers of smoking cessation have been utilized in patients with specific disorders, such as chronic obstructive pulmonary disease (Anthonisen et al., 1994; Hilberink, Jacobs, Bottema, de Vries, & Grol, 2005) and diabetes mellitus (Ardron, MacFarlane, Robinson, van Heyningen, & Calverley, 1988). This has been utilized to confirm the abstinence from smoking, as continuation of smoking in these conditions represents a health hazard. Specific assays have been validated for transplant patients, including those with heart-lung transplants, to assess the smoking status (Chadwick & Keevil, 2007).

There may be discrepancy between the different methods of assessment of nicotine abstinence. A combination of self-report with carbon monoxide and urinary cotinine may give a better picture of abstinence status (Gariti, Alterman, Ehrman, Mulvaney, & O'Brien, 2002). Comparative studies have suggested that urinary

TABLE 2 Clinical and Research Applications of Urine Cotinine Detection
Applications
To confirm tobacco abstinence in treatment setting
As an outcome measure in smoking cessation trials
Determining secondhand smoke exposure among nonsmokers: general public, children of smokers, workplace
Giving objective feedback to parents about the harms caused to children due to the habit of smoking
Determining risk of exposure to fetuses in pregnant women
Ascertaining efficacy of smoking prevention strategies among adolescents

cotinine levels may be better markers than breath carbon monoxide for detecting recent smoking (Fritz et al., 2010).

Patients may inaccurately report their smoking status for many different reasons. Firstly, such concealment may stem out of embarrassment experienced by the patient in disclosing an inability to quit tobacco in front of the physician or family members. Secondly, it may occur due to inability of the patient to recollect accurately their last usage of the tobacco product. Thirdly, there may be incentive for reporting current smoking status as abstinent, especially in de-addiction programs. Other reasons that could contribute to a discrepancy between patient report and urinalysis report could be the mislabeling of urinary samples, improper storage, and inadequate calibration of the assay.

An Outcome Measure in Smoking Cessation Trials

Urinary cotinine levels have been used as a proxy marker of smoking cessation in treatment studies. Many randomized controlled trials have found urinary cotinine levels to be reliable markers of current smoking status. The use of cotinine urinalysis stems from the requirement of an objective biological marker for documenting the cessation of smoking or other tobacco usage. Biological markers are often used to complement self-reports in smoking cessation trials. Urinary cotinine levels coupled with breath carbon monoxide and salivary cotinine have been used as more comprehensive outcome measures.

Differences in urinary cotinine levels have been observed at the outcomes of trials using nicotine gum and inhalers (Kralikova, Kozak, Rasmussen, Gustavsson, & Le Houezec, 2009), varenicline (McClure, Vandrey, Johnson, & Stitzer, 2013), and nortriptyline (Prochazka, Kick, Steinbrunn, Miyoshi, & Fryer, 2004) between active and placebo groups. Similar results also have been achieved for counseling and psychotherapeutic interventions (Dent, Harris, & Noonan, 2009), and repetitive Transcranial Magnetic Stimulation (rTMS) (Amiaz, Levy, Vainiger, Grunhaus, & Zangen, 2009) where the levels of urinary cotinine have been reported to be lower than waitlist or other controls.

Studies in the pediatric and adolescent age groups have also successfully utilized urinary cotinine levels, as outcome measures in the home setting (Yilmaz, Caylan, & Karacan, 2013) and school setting (Krishnan-Sarin et al., 2013). Trials have also been conducted in prison settings with urine cotinine as marker of efficacy of motivational interview-based methods for smoking cessation (Clarke et al., 2013).

Secondhand Smoke Exposure

Smoking not only represents a health hazard to the person who consumes the tobacco product, but also to others in the vicinity through exposure to secondhand smoke (SHS). Exposure to side-stream smoke or SHS has been associated with adverse health consequences. This is more common through exposure in the closed surroundings of a house. A nationally representative cross-sectional survey from Canada suggests that individuals of all age groups are affected by SHS (Centers for Disease Control and Prevention (CDC), 2010; Wong, Malaison, Hammond, & Leatherdale, 2013), but children and adolescents are particularly affected.

Children have been found to be quite vulnerable to SHS. Indoor parental smoking is associated with greater levels of urinary cotinine in children than outdoor smoking (Kehl, Thyrian, Lüdemann, Nauck, & John, 2010). Maternal smoking determined through urinary cotinine levels has been associated with wheezing symptoms in children (Schvartsman, Farhat, Schvartsman, & Saldiva, 2013). Greater time spent inside the house when smoking is permitted is associated with greater urinary cotinine levels of children with asthma (Tung et al., 2013). Apart from parental smoking behavior, dwelling space and social and educational status of the parents also determines the urinary cotinine levels in children (Scherer et al., 2004). SHS may have adverse health outcomes not only in children but also in adults who may be predisposed to bronchitis when exposed to SHS confirmed through urinary cotinine (Wu et al., 2010).

The workplace is another situation that may expose nonsmokers to tobacco smoke and lead to an increase in urinary cotinine levels. A study among nonsmoking workers suggested that SHS exposure in the workplace confirmed through urinary cotinine levels was associated poorer lung function (Lai et al., 2011). Those working in smoking designated areas and nightclubs had higher levels of urinary cotinine (Lazcano-Ponce et al., 2007; Pacheco et al., 2012). A decrement in urinary cotinine levels over a period of time have been found in nonsmoking employees of bars and restaurants with implementation of antismoking laws (Caman, Erguder, Ozcebe, & Bilir, 2013). Similar assessments have found SHS exposure to be high amongst flight attendants (Lindgren, Willers, Skarping, & Norbäck, 1999).

A report of the urinary cotinine levels as a feedback to parents has been studied as a measure to reduce SHS exposure among asthmatic children (Wilson, Farber, Knowles, & Lavori, 2011). Although the trial results were negative, it suggested a unique way to give feedback to parents about their smoking habits so that the frequency of asthma attacks may decrease.

Risk of Harm to Fetus in Pregnant Women

Studies in pregnant women (Panaretto et al., 2009) suggest that urinary cotinine levels correlate significantly with the severity of nicotine dependence scores among smokers. Discrepancies between self-reported smoking and cotinine levels have been found among pregnant women (Britton, Brinthaupt, Stehle, & James, 2004). About one-third of women who denied smoking had cotinine levels above biochemical cut-off, suggesting that reporting of nonsmoking status may not be accurate among pregnant women. It has been seen that higher cotinine levels during pregnancy are related to lower weight of the fetus at delivery (Wang, Tager, Van Vunakis, Speizer, & Hanrahan, 1997).

Studies have also looked at exposure to SHS among pregnant women. It has been seen that having smoked previously, low educational level, and being primiparous were independent predictors of SHS among such women (Aurrekoetxea et al., 2014). Exposure to SHS in cars was associated with about twice the increase in cotinine levels in pregnant women than exposure to SHS at home (Vardavas et al., 2013). A spouse smoking inside the home has also been associated with pregnant women having high urinary cotinine levels (Paek et al., 2009).

Adolescents

Adolescents represent a vulnerable age group where experimentation with cigarettes might proceed on to regular and dependent use of tobacco. Hence, many smoking prevention strategies focus on the adolescent age group. Urinalysis of cotinine levels in this age group may help to discern those who are starting to use cigarettes and other tobacco products. The literature of adolescent reporting of smoking status assessed using urine cotinine level analysis has found discordant results. There are studies to suggest that among the adolescent schoolgoing population, urinary cotinine levels may correlate well with the self-report, with high sensitivity and specificity values (Park & Kim, 2009). On the other hand, studies report that urine cotinine analysis do not correlate well with the reported tobacco product usage. A study from Brazil suggests that adolescents underestimated tobacco consumption and self-reported smoking had low agreement with cotinine concentration (Malcon, Menezes, Assunção, Neutzling, & Hallal, 2008). Nonetheless, urinary cotinine levels can help identify a targeted population and implement measures for smoking prevention.

SENSITIVITY AND SPECIFICITY ISSUES

The main use of urine cotinine analysis is for the assessment of current use of tobacco products. However, the levels of cotinine in the body that should be considered abnormal may be a matter of contention. The doubt may arise for two reasons. Firstly, the cotinine levels may rise due to the consumption of a tobacco product by itself, or due to exposure of nicotine through SHS provided through air pollutants at home or general environment. Secondly, a person may be in the process of quitting tobacco products and cotinine levels may take time to come to lower normative levels after the cessation of the use of tobacco products. Hence, the cutoff levels that should be utilized as a marker of tobacco product usage may be a matter of debate. Each cutoff value may have a related sensitivity and specificity. Therein, a receiver-operator curve can be constructed to find the optimal values for cutoff that ascertain recent tobacco usage (Balhara & Jain, 2013).

Different studies have suggested different cutoffs for cotinine levels in the urine. The cutoff levels of cotinine taken in published studies have usually ranged from 20 ng/ml (Anthonisen et al., 1994) to as high as 500 ng/ml (Holl, Grabert, Heinze, & Debatin, 1998).

Some studies have used a cutoff as low as 3 ng/ml in determining smoking status (Hawkins et al., 2014). The optimal cutoff of smoking status has not been settled well and should take into consideration the ethnicity, gender, age, assay technique, sample collection policy, and other relevant variables. Exposure to SHS has an influence on the optimal cutoff for urinary cotinine values. Cutoffs may need to be up to two times for those who have a higher degree of SHS exposure (Aurrekoetxea et al., 2013). Attempts have been made to identify the degree of dependence by correlating with urinary cotinine levels. Urinary cotinine levels have been found to be associated with nicotine dependence in adults (Jung et al., 2012) as well as adolescents (Carpenter, Baker, Gray, & Upadhyaya, 2010).

APPLICATIONS TO OTHER ADDICTIONS AND SUBSTANCE MISUSE

Cotinine levels in the urine may not have a direct influence on other addictions and substances of misuse. However, it should be considered that individuals with nicotine dependence may have a dependence on other substances as well. There is a high rate of comorbidity of nicotine dependence in individuals with dependence on other drugs of abuse (Compton, Thomas, Stinson, & Grant, 2007). Hence, the presence of nicotine dependence or high levels of urinary cotinine should alert for the presence of additional substance use disorders. It has been observed that benzodiazepines may cause false-positive results for urine screening for cotinine in individuals with a high degree of nicotine dependence (Haj Mouhamed et al., 2012).

COMPARISON WITH OTHER BIOMARKERS OF TOBACCO USE

The advantages of urinary cotinine vis-à-vis other methods of assessment of tobacco exposure need to be examined. Urinary cotinine is a relatively noninvasive method as compared to the plasma assessment of nicotine and cotinine. It also would be considered less invasive than the assessment of cotinine in the hair. However, urinary cotinine requires collection of a sample of voided urine, which may be inconvenient in certain circumstances. Therein, the use of breath carbon monoxide levels and estimation of salivary cotinine may be more acceptable for assessing exposure to nicotine products.

Also, the accurate estimation of urinary cotinine requires sophisticated on-site equipment and proper calibration. Urinary cotinine levels have primarily been used to assess exposure to smoked forms of tobacco. However, some evidence has also been gathered among smokeless tobacco users (Balhara & Jain, 2013; Cok & Ozturk, 2000; Hecht et al., 2007). Studies suggest that urinary cotinine levels may be higher in users of smokeless tobacco products as compared to users of smoked forms (Cok & Ozturk, 2000). It has been suggested that urinary cotinine levels may also be used along with other alkaloid markers in the determination of current usage status of tobacco products.

CONCLUSIONS

Urinary cotinine levels can be utilized as a means of detecting exposure to nicotine due to self-consumption or exposure to secondhand smoke. The use of urinary cotinine levels is relatively noninvasive and a reliable measure of detecting tobacco usage. It can be utilized for both smoked and smokeless forms of tobacco. Various techniques of detection of cotinine in the urine sample are available, including chromatographic methods, mass spectrometry, and enzyme immunoassay. Urinary cotinine can be used either alone or in conjunction with other markers of nicotine exposure, such as breath carbon monoxide and thiocyanate. Although the estimation of urinary cotinine levels has been fairly widely accepted, issues may be encountered in the collection, processing, and storage of the sample.

Urinary cotinine levels have been used fairly regularly in the literature on the efficacy of smoking cessation methods. Further research needs to focus upon refining the immunoassay methods of detecting cotinine levels at the point of collection of the urine sample. Further efforts are also required to deliberate consensus about the optimal cutoff values for defining smoking status and assessing environmental smoke exposure.

DEFINITION OF TERMS

GCMS Gas chromatography is a procedure that uses inert gas as a carrier to separate different components from a mixture, and is combined with mass spectrometry to assess the nature of the chemical compound separated.

HPLC High-performance liquid chromatography is a technique that uses liquid at a high pressure to run through a column to separate components of a mixture.

Secondhand smoke Secondhand smoke is the passive smoke inhaled by a nonsmoker due to smoke released to the environment by a cigarette smoker.

Sensitivity Sensitivity is the rate of identification of true positives out of all the positives identified through a screening test.

Specificity Specificity is the rate of identification of true negatives out of all the negatives identified through a screening test.

KEY FACTS

- About 70 to 80% of the nicotine that enters the body is converted to cotinine.
- CYP2A6 isoenzyme is the major enzyme that influences cotinine metabolism.
- Ethnic and gender differences have been noted in the metabolism of cotinine.
- HPLC and GCMS give better quantitative estimates than immunoassay methods.

SUMMARY POINTS

- Cotinine is the most important metabolite of nicotine.
- Urinary cotinine level has been used as marker of nicotine exposure due to its longer half-life compared to nicotine.
- Estimation of cotinine in the urine is done through chromatographic or immunoassay-based methods.
- Urinary cotinine levels have many clinical applications, such as outcome measures in trials and clinical assessment of abstinence.
- Urinary cotinine levels also have public health applications, such as estimating secondhand smoke exposure.

REFERENCES

Acosta, M., Buchhalter, A., Breland, A., Hamilton, D., & Eissenberg, T. (2004). Urine cotinine as an index of smoking status in smokers during 96-hr abstinence: comparison between gas chromatography/mass spectrometry and immunoassay test strips. *Nicotine & Tobacco Research: Official Journal of the Society for Research on Nicotine and Tobacco*, 6(4), 615–620. http://dx.doi.org/10.1080/14622200410001727867.

Amiaz, R., Levy, D., Vainiger, D., Grunhaus, L., & Zangen, A. (2009). Repeated high-frequency transcranial magnetic stimulation over the dorsolateral prefrontal cortex reduces cigarette craving and consumption. *Addiction (Abingdon, England)*, 104(4), 653–660. http://dx.doi.org/10.1111/j.1360-0443.2008.02448.x.

Anthonisen, N. R., Connett, J. E., Kiley, J. P., Altose, M. D., Bailey, W. C., Buist, A. S., ... O'Hara, P. (1994). Effects of smoking intervention and the use of an inhaled anticholinergic bronchodilator on the rate of decline of FEV1. The Lung Health Study. *JAMA: The Journal of the American Medical Association*, 272(19), 1497–1505.

Apinan, R., Choemung, A., & Na-Bangchang, K. (2010). A sensitive HPLC-ESI-MS-MS method for the determination of cotinine in urine. *Journal of Chromatographic Science*, 48(6), 460–465.

Ardron, M., MacFarlane, I. A., Robinson, C., van Heyningen, C., & Calverley, P. M. (1988). Anti-smoking advice for young diabetic smokers: is it a waste of breath? *Diabetic Medicine: A Journal of the British Diabetic Association*, 5(7), 667–670.

Aurrekoetxea, J. J., Murcia, M., Rebagliato, M., Fernández-Somoano, A., Castilla, A. M., Guxens, M., ... Santa-Marina, L. (2014). Factors associated with second-hand smoke exposure in non-smoking pregnant women in Spain: self-reported exposure and urinary cotinine levels. *The Science of the Total Environment*, 470–471, 1189–1196. http://dx.doi.org/10.1016/j.scitotenv.2013.10.110.

Aurrekoetxea, J. J., Murcia, M., Rebagliato, M., López, M. J., Castilla, A. M., Santa-Marina, L., ... Ballester, F. (2013). Determinants of self-reported smoking and misclassification during pregnancy, and analysis of optimal cut-off points for urinary cotinine: a cross-sectional study. *BMJ Open*, 3(1). http://dx.doi.org/10.1136/bmjopen-2012-002034.

Balhara, Y. P. S., & Jain, R. (2013). A receiver operated curve-based evaluation of change in sensitivity and specificity of cotinine urinalysis for detecting active tobacco use. *Journal of Cancer Research and Therapeutics*, 9(1), 84. http://dx.doi.org/10.4103/0973-1482.110384.

Balhara, Y. P. S., Jain, R., Sundar, S. A., & Sagar, R. (2011). A comparative study of reliability of self report of tobacco use among patients with bipolar and somatoform disorders. *Journal of Pharmacology & Pharmacotherapeutics*, 2(3), 174–178. http://dx.doi.org/10.4103/0976-500X.83282.

Balhara, Y. P. S., Jain, R., Sundar, A. S., & Sagar, R. (2012). Use of cotinine urinalysis to verify self-reported tobacco use among male psychiatric out-patients. *Lung India: Official Organ of Indian Chest Society*, 29(3), 217–220. http://dx.doi.org/10.4103/0970-2113.99102.

Baumann, F., Regenthal, R., Burgos-Guerrero, I. L., Hegerl, U., & Preiss, R. (2010). Determination of nicotine and cotinine in human serum by means of LC/MS. *Journal of Chromatography B: Analytical Technologies in the Biomedical and Life Sciences*, 878(1), 107–111. http://dx.doi.org/10.1016/j.jchromb.2009.11.032.

Benowitz, N. L., Dempsey, D., Tyndale, R. F., St Helen, G., & Jacob, P., 3rd (2013). Dose-independent kinetics with low level exposure to nicotine and cotinine. *British Journal of Clinical Pharmacology*, 75(1), 277–279. http://dx.doi.org/10.1111/j.1365-2125.2012.04327.x.

Benowitz, N. L., Hukkanen, J., & Jacob, P. (2009). Nicotine chemistry, metabolism, kinetics and biomarkers. *Handbook of Experimental Pharmacology, 192*, 29–60. http://dx.doi.org/10.1007/978-3-540-69248-5_2.

Benowitz, N. L., Lessov-Schlaggar, C. N., Swan, G. E., & Jacob, P., 3rd (2006). Female sex and oral contraceptive use accelerate nicotine metabolism. *Clinical Pharmacology and Therapeutics, 79*(5), 480–488. http://dx.doi.org/10.1016/j.clpt.2006.01.008.

Benowitz, N. L., Perez-Stable, E. J., Fong, I., Modin, G., Herrera, B., & Jacob, P., 3rd (1999). Ethnic differences in N-glucuronidation of nicotine and cotinine. *The Journal of Pharmacology and Experimental Therapeutics, 291*(3), 1196–1203.

Benowitz, N. L., Pérez-Stable, E. J., Herrera, B., & Jacob, P., 3rd (2002). Slower metabolism and reduced intake of nicotine from cigarette smoking in Chinese–Americans. *Journal of the National Cancer Institute, 94*(2), 108–115.

Britton, G. R. A., Brinthaupt, J., Stehle, J. M., & James, G. D. (2004). Comparison of self-reported smoking and urinary cotinine levels in a rural pregnant population. *Journal of Obstetric, Gynecologic, and Neonatal Nursing: JOGNN/NAACOG, 33*(3), 306–311.

Caman, O. K., Erguder, B. I., Ozcebe, H., & Bilir, N. (2013). Urinary cotinine and breath carbon monoxide levels among bar and restaurant employees in Ankara. *Nicotine & Tobacco Research: Official Journal of the Society for Research on Nicotine and Tobacco, 15*(8), 1446–1452. http://dx.doi.org/10.1093/ntr/nts345.

Caraballo, R. S., Giovino, G. A., Pechacek, T. F., Mowery, P. D., Richter, P. A., Strauss, W. J., … Maurer, K. R. (1998). Racial and ethnic differences in serum cotinine levels of cigarette smokers: third National Health and Nutrition Examination Survey, 1988–1991. *JAMA: The Journal of the American Medical Association, 280*(2), 135–139.

Carpenter, M. J., Baker, N. L., Gray, K. M., & Upadhyaya, H. P. (2010). Assessment of nicotine dependence among adolescent and young adult smokers: a comparison of measures. *Addictive Behaviors, 35*(11), 977–982. http://dx.doi.org/10.1016/j.addbeh.2010.06.013.

Centers for Disease Control and Prevention (CDC). (2010). Vital signs: nonsmokers' exposure to secondhand smoke – United States, 1999–2008. *MMWR: Morbidity and Mortality Weekly Report, 59*(35), 1141–1146.

Chadwick, C. A., & Keevil, B. (2007). Measurement of cotinine in urine by liquid chromatography tandem mass spectrometry. *Annals of Clinical Biochemistry, 44*(Pt 5), 455–462. http://dx.doi.org/10.1258/000456307781645996.

Chiadmi, F., & Schlatter, J. (2014). Simultaneous determination of cotinine and trans-3-hydroxycotinine in urine by automated solid-phase extraction using gas chromatography-mass spectrometry. *Biomedical Chromatography: BMC, 28*(4), 453–458. http://dx.doi.org/10.1002/bmc.3159.

Clarke, J. G., Stein, L. A. R., Martin, R. A., Martin, S. A., Parker, D., Lopes, C. E., … Bock, B. (2013). Forced smoking abstinence: not enough for smoking cessation. *JAMA Internal Medicine, 173*(9), 789–794. http://dx.doi.org/10.1001/jamainternmed.2013.197.

Cok, I., & Oztürk, R. (2000). Urinary cotinine levels of smokeless tobacco (Maraş powder) users. *Human & Experimental Toxicology, 19*(11), 650–655.

Compton, W. M., Thomas, Y. F., Stinson, F. S., & Grant, B. F. (2007). Prevalence, correlates, disability, and comorbidity of DSM-IV drug abuse and dependence in the United States: results from the national epidemiologic survey on alcohol and related conditions. *Archives of General Psychiatry, 64*(5), 566–576. http://dx.doi.org/10.1001/archpsyc.64.5.566.

Dent, L. A., Harris, K. J., & Noonan, C. W. (2009). Randomized trial assessing the effectiveness of a pharmacist-delivered program for smoking cessation. *The Annals of Pharmacotherapy, 43*(2), 194–201. http://dx.doi.org/10.1345/aph.1L556.

Fritz, M., Wallner, R., Grohs, U., Kemmler, G., Saria, A., & Zernig, G. (2010). Comparable sensitivities of urine cotinine and breath carbon monoxide at follow-up time points of three months or more in a smoking cessation trial. *Pharmacology, 85*(4), 234–240. http://dx.doi.org/10.1159/000280435.

Gaalema, D. E., Higgins, S. T., Bradstreet, M. P., Heil, S. H., & Bernstein, I. M. (2011). Using NicAlert strips to verify smoking status among pregnant cigarette smokers. *Drug and Alcohol Dependence, 119*(1–2), 130–133. http://dx.doi.org/10.1016/j.drugalcdep.2011.05.014.

Gariti, P., Alterman, A. I., Ehrman, R., Mulvaney, F. D., & O'Brien, C. P. (2002). Detecting smoking following smoking cessation treatment. *Drug and Alcohol Dependence, 65*(2), 191–196.

Haj Mouhamed, D., Ezzaher, A., Mabrouk, H., Sâadaoui, M. H., Neffati, F., Douki, W., … Najjar, M. F. (2012). Interference of tobacco smoke with immunochromatography assay for urinary drug detection. *Journal of Forensic and Legal Medicine, 19*(7), 369–372. http://dx.doi.org/10.1016/j.jflm.2012.04.010.

Hawkins, S. S., Dacey, C., Gennaro, S., Keshinover, T., Gross, S., Gibeau, A., … Aldous, K. M. (2014). Secondhand smoke exposure among nonsmoking pregnant women in New York city. *Nicotine & Tobacco Research: Official Journal of the Society for Research on Nicotine and Tobacco, 16*. http://dx.doi.org/10.1093/ntr/ntu034.

Hecht, S. S., Carmella, S. G., Murphy, S. E., Riley, W. T., Le, C., Luo, X., … Hatsukami, D. K. (2007). Similar exposure to a tobacco-specific carcinogen in smokeless tobacco users and cigarette smokers. *Cancer Epidemiology, Biomarkers & Prevention: A Publication of the American Association for Cancer Research, Cosponsored by the American Society of Preventive Oncology, 16*(8), 1567–1572. http://dx.doi.org/10.1158/1055-9965.EPI-07-0227.

Hilberink, S. R., Jacobs, J. E., Bottema, B. J. A.M., de Vries, H., & Grol, R. P. T. M. (2005). Smoking cessation in patients with COPD in daily general practice (SMOCC): six months' results. *Preventive Medicine, 41*(5–6), 822–827. http://dx.doi.org/10.1016/j.ypmed.2005.08.003.

Holl, R. W., Grabert, M., Heinze, E., & Debatin, K. M. (1998). Objective assessment of smoking habits by urinary cotinine measurement in adolescents and young adults with type 1 diabetes. Reliability of reported cigarette consumption and relationship to urinary albumin excretion. *Diabetes Care, 21*(5), 787–791.

Hukkanen, J., Jacob, P., 3rd, & Benowitz, N. L. (2005). Metabolism and disposition kinetics of nicotine. *Pharmacological Reviews, 57*(1), 79–115. http://dx.doi.org/10.1124/pr.57.1.3.

Iwai, M., Ogawa, T., Hattori, H., Zaitsu, K., Ishii, A., Suzuki, O., & Seno, H. (2013). Simple and rapid assay method for simultaneous quantification of urinary nicotine and cotinine using micro-extraction by packed sorbent and gas chromatography-mass spectrometry. *Nagoya Journal of Medical Science, 75*(3–4), 255–261.

Jung, H.-S., Kim, Y., Son, J., Jeon, Y.-J., Seo, H.-G., Park, S.-H., & Huh, B. R. (2012). Can urinary cotinine predict nicotine dependence level in smokers? *Asian Pacific Journal of Cancer Prevention: APJCP, 13*(11), 5483–5488.

Kehl, D., Thyrian, J. R., Lüdemann, J., Nauck, M., & John, U. (2010). A descriptive analysis of relations between parents' self-reported smoking behavior and infants' daily exposure to environmental tobacco smoke. *BMC Public Health, 10*, 424. http://dx.doi.org/10.1186/1471-2458-10-424.

Kralikova, E., Kozak, J. T., Rasmussen, T., Gustavsson, G., & Le Houezec, J. (2009). Smoking cessation or reduction with nicotine replacement therapy: a placebo-controlled double blind trial with nicotine gum and inhaler. *BMC Public Health*, *9*, 433. http://dx.doi.org/10.1186/1471-2458-9-433.

Krishnan-Sarin, S., Cavallo, D. A., Cooney, J. L., Schepis, T. S., Kong, G., Liss, T. B., ... Carroll, K. M. (2013). An exploratory randomized controlled trial of a novel high-school-based smoking cessation intervention for adolescent smokers using abstinence-contingent incentives and cognitive behavioral therapy. *Drug and Alcohol Dependence*, *132*(1–2), 346–351. http://dx.doi.org/10.1016/j.drugalcdep.2013.03.002.

Lai, H.-K., Hedley, A. J., Repace, J., So, C., Lu, Q.-Y., McGhee, S. M., ... Wong, C.-M. (2011). Lung function and exposure to workplace second-hand smoke during exemptions from smoking ban legislation: an exposure-response relationship based on indoor PM2.5 and urinary cotinine levels. *Thorax*, *66*(7), 615–623. http://dx.doi.org/10.1136/thx.2011.160291.

Lazcano-Ponce, E., Benowitz, N., Sanchez-Zamorano, L. M., Barbosa-Sanchez, L., Valdes-Salgado, R., Jacob, P., 3rd, ... Hernandez-Avila, M. (2007). Secondhand smoke exposure in Mexican discotheques. *Nicotine & Tobacco Research: Official Journal of the Society for Research on Nicotine and Tobacco*, *9*(10), 1021–1026. http://dx.doi.org/10.1080/14622200701495967.

Lindgren, T., Willers, S., Skarping, G., & Norbäck, D. (1999). Urinary cotinine concentration in flight attendants, in relation to exposure to environmental tobacco smoke during intercontinental flights. *International Archives of Occupational and Environmental Health*, *72*(7), 475–479.

Malcon, M. C., Menezes, A. M. B., Assunção, M. C. F., Neutzling, M. B., & Hallal, P. C. (2008). Agreement between self-reported smoking and cotinine concentration in adolescents: a validation study in Brazil. *The Journal of Adolescent Health: Official Publication of the Society for Adolescent Medicine*, *43*(3), 226–230. http://dx.doi.org/10.1016/j.jadohealth.2008.02.002.

Massadeh, A. M., Gharaibeh, A. A., & Omari, K. W. (2009). A single-step extraction method for the determination of nicotine and cotinine in Jordanian smokers' blood and urine samples by RP-HPLC and GCMS. *Journal of Chromatographic Science*, *47*(2), 170–177.

McClure, E. A., Vandrey, R. G., Johnson, M. W., & Stitzer, M. L. (2013). Effects of varenicline on abstinence and smoking reward following a programmed lapse. *Nicotine & Tobacco Research*, *15*(1), 139–148. http://dx.doi.org/10.1093/ntr/nts101.

Montalto, N. J., & Wells, W. O. (2007). Validation of self-reported smoking status using saliva cotinine: a rapid semiquantitative dipstick method. *Cancer Epidemiology, Biomarkers & Prevention: A Publication of the American Association for Cancer Research, Cosponsored by the American Society of Preventive Oncology*, *16*(9), 1858–1862. http://dx.doi.org/10.1158/1055-9965.EPI-07-0189.

Pacheco, S. A., Aguiar, F., Ruivo, P., Proença, M. C., Sekera, M., Penque, D., & Simões, T. (2012). Occupational exposure to environmental tobacco smoke: a study in Lisbon restaurants. *Journal of Toxicology and Environmental Health: Part A*, *75*(13–15), 857–866. http://dx.doi.org/10.1080/15287394.2012.690690.

Paek, Y. J., Kang, J. B., Myung, S.-K., Lee, D.-H., Seong, M.-W., Seo, H. G., ... Ko, J. A. (2009). Self-reported exposure to secondhand smoke and positive urinary cotinine in pregnant nonsmokers. *Yonsei Medical Journal*, *50*(3), 345–351. http://dx.doi.org/10.3349/ymj.2009.50.3.345.

Panaretto, K. S., Mitchell, M. R., Anderson, L., Gilligan, C., Buettner, P., Larkins, S. L., & Eades, S. (2009). Tobacco use and measuring nicotine dependence among urban indigenous pregnant women. *The Medical Journal of Australia*, *191*(10), 554–557.

Park, S. W., & Kim, J. Y. (2009). Validity of self-reported smoking using urinary cotinine among vocational high school students. *Journal of Preventive Medicine and Public Health—Yebang Ŭihakhoe Chi*, *42*(4), 223–230. http://dx.doi.org/10.3961/jpmph.2009.42.4.223.

Piller, M., Gilch, G., Scherer, G., & Scherer, M. (2014). Simple, fast and sensitive LC-MS/MS analysis for the simultaneous quantification of nicotine and 10 of its major metabolites. *Journal of Chromatography B: Analytical Technologies in the Biomedical and Life Sciences*, *951–952*, 7–15. http://dx.doi.org/10.1016/j.jchromb.2014.01.025.

Prochazka, A. V., Kick, S., Steinbrunn, C., Miyoshi, T., & Fryer, G. E. (2004). A randomized trial of nortriptyline combined with transdermal nicotine for smoking cessation. *Archives of Internal Medicine*, *164*(20), 2229–2233. http://dx.doi.org/10.1001/archinte.164.20.2229.

Rangiah, K., Hwang, W.-T., Mesaros, C., Vachani, A., & Blair, I. A. (2011). Nicotine exposure and metabolizer phenotypes from analysis of urinary nicotine and its 15 metabolites by LCMS. *Bioanalysis*, *3*(7), 745–761. http://dx.doi.org/10.4155/bio.11.42.

Riboli, E., Haley, N. J., De Waard, F., & Saracci, R. (1995). Validity of urinary biomarkers of exposure to tobacco smoke following prolonged storage. *International Journal of Epidemiology*, *24*(2), 354–358.

Schepis, T. S., Duhig, A. M., Liss, T., McFetridge, A., Wu, R., Cavallo, D. A., ... Krishnan-Sarin, S. (2008). Contingency management for smoking cessation: enhancing feasibility through use of immunoassay test strips measuring cotinine. *Nicotine & Tobacco Research: Official Journal of the Society for Research on Nicotine and Tobacco*, *10*(9), 1495–1501. http://dx.doi.org/10.1080/14622200802323209.

Scherer, G., Krämer, U., Meger-Kossien, I., Riedel, K., Heller, W.-D., Link, E., ... Behrendt, H. (2004). Determinants of children's exposure to environmental tobacco smoke (ETS): a study in southern Germany. *Journal of Exposure Analysis and Environmental Epidemiology*, *14*(4), 284–292. http://dx.doi.org/10.1038/sj.jea.7500323.

Schvartsman, C., Farhat, S. C. L., Schvartsman, S., & Saldiva, P. H. N. (2013). Parental smoking patterns and their association with wheezing in children. *Clinics (São Paulo, Brazil)*, *68*(7), 934–939. http://dx.doi.org/10.6061/clinics/2013(07)08.

Shin, H.-S., Kim, J.-G., Shin, Y.-J., & Jee, S. H. (2002). Sensitive and simple method for the determination of nicotine and cotinine in human urine, plasma and saliva by gas chromatography-mass spectrometry. *Journal of Chromatography B: Analytical Technologies in the Biomedical and Life Sciences*, *769*(1), 177–183.

Toraño, J. S., & van Kan, H. J. M. (2003). Simultaneous determination of the tobacco smoke uptake parameters nicotine, cotinine and thiocyanate in urine, saliva and hair, using gas chromatography-mass spectrometry for characterisation of smoking status of recently exposed subjects. *The Analyst*, *128*(7), 838–843.

Tung, K.-Y., Wu, K.-Y., Tsai, C.-H., Su, M.-W., Chen, C.-H., Lin, M.-H., ... Lee, Y. L. (2013). Association of time-location patterns with urinary cotinine among asthmatic children under household environmental tobacco smoke exposure. *Environmental Research*, *124*, 7–12. http://dx.doi.org/10.1016/j.envres.2013.03.002.

Vardavas, C. I., Fthenou, E., Patelarou, E., Bagkeris, E., Murphy, S., Hecht, S. S., ... Kogevinas, M. (2013). Exposure to different sources of secondhand smoke during pregnancy and its effect on urinary cotinine and tobacco-specific nitrosamine (NNAL) concentrations. *Tobacco Control*, *22*(3), 194–200. http://dx.doi.org/10.1136/tobaccocontrol-2011-050144.

Wang, X., Tager, I. B., Van Vunakis, H., Speizer, F. E., & Hanrahan, J. P. (1997). Maternal smoking during pregnancy, urine cotinine concentrations, and birth outcomes. A prospective cohort study. *International Journal of Epidemiology*, *26*(5), 978–988.

Watson, I. D. (1977). Rapid analysis of nicotine and cotinine in the urine of smokers by isocratic high-performance liquid chromatography. *Journal of Chromatography*, *143*(2), 203–206.

Wilson, S. R., Farber, H. J., Knowles, S. B., & Lavori, P. W. (2011). A randomized trial of parental behavioral counseling and cotinine feedback for lowering environmental tobacco smoke exposure in children with asthma: results of the LET'S manage asthma trial. *Chest*, *139*(3), 581–590. http://dx.doi.org/10.1378/chest.10-0772.

Wong, S. L., Malaison, E., Hammond, D., & Leatherdale, S. T. (2013). Secondhand smoke exposure among Canadians: cotinine and self-report measures from the Canadian health measures survey 2007–2009. *Nicotine & Tobacco Research: Official Journal of the Society for Research on Nicotine and Tobacco*, *15*(3), 693–700. http://dx.doi.org/10.1093/ntr/nts195.

Wu, C.-F., Feng, N.-H., Chong, I.-W., Wu, K.-Y., Lee, C.-H., Hwang, J.-J., ... Wu, M.-T. (2010). Second-hand smoke and chronic bronchitis in Taiwanese women: a health-care based study. *BMC Public Health*, *10*, 44. http://dx.doi.org/10.1186/1471-2458-10-44.

Yilmaz, G., Caylan, N., & Karacan, C. D. (2013). Brief intervention to preteens and adolescents to create smoke-free homes and cotinine results: a randomized trial. *Journal of Tropical Pediatrics*, *59*(5), 365–371. http://dx.doi.org/10.1093/tropej/fmt034.

Part III

Alcohol

Section A

General Aspects

Chapter 35

Ethanol Metabolism and Implications for Disease

Roshanna Rajendram[1], Rajkumar Rajendram[2,3], Victor R. Preedy[3]

[1]Department of Emergency Medicine, The Royal Woolverhampton NHS Trust, New Cross Hospital, Wolverhampton, UK; [2]Department of Anaesthesia and Intensive Care, College of Medicine, King Khalid University Hospital, King Saud University Medical City, Riyadh, Saudi Arabia; [3]Diabetes and Nutritional Sciences Research Division, Faculty of Life Science and Medicine, King's College London, London, UK

Abbreviations

ABV Alcohol by volume
ABW Percentage content of alcohol by weight
ADH Alcohol dehydrogenase
ALDH Aldehyde dehydrogenase
AUC Area under the curve (i.e., blood ethanol concentration curve)
AUCip Area under the curve after intraperitoneal administration
AUCiv Area under the curve after intravenous administration
AUCoral Area under the curve after oral administration of ethanol
AUD Alcohol use disorders
BEC Blood ethanol concentration
C_2H_5 Ethyl group of ethanol
CYP2E1 Cytochrome P4502E1
DALY Disability-adjusted life year
FAEE Fatty acid ethyl esters
FPM First-pass metabolism
GABA Gamma aminobutyric acid
GBD Global burden of disease
GI Gastrointestinal
GSH Glutathione
H_2O_2 Hydrogen peroxide
iv Intravenous
Km Half-maximal activity
MEOS Microsomal ethanol oxidizing system
mRNA Messenger ribonucleic acid
NAD+ Nicotinamide adenine dinucleotide
NADH Reduced form of nicotinamide adenine dinucleotide
NMDA N-methyl-D-aspartate
OH Hydroxyl group
SD Standard deviation
SNP Single-nucleotide polymorphism
YLL Years of life lost
WHO World Health Organization

INTRODUCTION

Ethanol, ethyl alcohol, or "alcohol" is one of the most commonly used recreational drugs and "drinking" colloquially describes the consumption of beverages containing ethanol. Alcoholic beverages have been produced since time immemorial, but it is widely believed that the word *alcohol* is derived from the Arabic word *al-kuḥl* (*al-* means "the" and *kohl* is a black powder used as eyeliner). However, the modern Arabic name for ethanol is *al-ġawl*, which can be translated as "spirit" or "demon," reflecting the potential harm caused by alcoholic beverages.

To a certain extent, the basic physical properties, pathways, and metabolism of ethanol have been known for decades (Rajendram, Hunter, & Preedy, 2013). However, the following aspects are emerging areas of alcohol-related research, some of which are mentioned in this review: (1) the enormous extent of alcohol's effects on the body and the wide-ranging consequences of alcohol misuse; (2) the role of acetaldehyde in damaging cells and organs; (3) the effects of free radical species generated by the metabolism of ethanol or arising as a consequence of impaired nutrition; (4) the genetics of alcohol misuse; (5) the molecular subcellular effects of alcohol, from nucleic acids to membranes and organelles; and (6) the societal effects of alcohol misuse. All of the aforementioned are interrelated to some degree.

THE PHYSICAL PROPERTIES OF ETHANOL

Ethanol is a relatively uncharged molecule that has a polar hydroxyl (OH) group and a nonpolar (C_2H_5) group. As a result, ethanol is highly water soluble, crosses cell membranes easily by passive diffusion, and dissolves lipids that can disrupt biological membranes.

Ethanol is ubiquitously present in the natural environment. Even fruit and freshly baked bread contain traces of ethanol. Ethanol is also produced during the digestion of food. The gastrointestinal (GI) tract contains millions of micro-organisms, which include yeasts that can produce ethanol from sugars within the GI tract. If ethanol was not metabolized, it would accumulate and cause constant intoxication. Almost all animals have therefore developed pathways to metabolize ethanol. In fact, the sheer number of pathways and enzymes involved in the metabolism and detoxification of ethanol highlights its potential for toxicity. Other alcohols (e.g., methanol and isopropyl alcohol) are rarely present in the natural environment. However, as we do not have enzymes pathways to metabolize them, they are extremely toxic if ingested.

Most people enjoy alcohol without harming themselves or others, and some gain pleasure from its effects (Hutton, 2012; Lin, Amodeo, Arthurs, & Reilly, 2012). Moderate alcohol consumption may even reduce the risk of certain conditions, such as ischemic heart disease (Goncalves et al., 2015; Preedy & Watson, 2004; Rehm, Sempos, & Trevisan, 2003a; Ronksley, Brien, Turner, Mukamal, & Ghali, 2011). However, in excess, alcohol can induce any of at least 60–200 different alcohol-related pathologies (Preedy & Watson, 2004; WHO, 2014). These conditions are wide ranging, including nutrient deficiencies (micro- and macronutrients) and frank diseases affecting almost all organ systems. As two "rules of thumb," (1) alcohol or its metabolites have the potential to affect virtually every single organ and biochemical pathway and (2) 50% of chronic alcohol misusers will have one or more organ pathology.

INTERNATIONAL VARIATIONS IN ALCOHOL CONSUMPTION

There are marked international variations in the consumption of alcohol (Rajendram, Lewison, & Preedy, 2006; WHO, 2014). For example, in Afghanistan, the per capita consumption of pure alcohol by the adult population as a whole is 0–0.07 liter per annum. On the other hand, consumption rates in Japan, USA, UK, and Zimbabwe are 7.2, 9.2, 11.6, and 5.7 liter per annum, respectively (WHO, 2014). If one takes into account those who drink, the corresponding values are 11.9, 19.4, 13.3, 13.8, and 14.6 liter per year, respectively (WHO, 2014). In other words, data for per capita drinking may mask the extent of alcohol misuse unless consideration is made of the proportion of those who drink and those who do not.

ALCOHOL AND THE BURDEN OF DISEASE

Alcohol consumption and the associated burden of disease are rife throughout the modern world (Rajendram et al., 2006; WHO, 2014). The global burden of disease (GBD) due to alcohol is increasing. Currently, alcohol misuse accounts for 5.9% of all deaths globally (WHO, 2014), while in the GBD 1990 study, only 1.5% was attributed to alcohol (Murray & Lopez, 1997). This difference may reflect variations in reporting alcohol misuse and associated harm, but there is no denying the fact that alcohol misuse has an important adverse impact on the health of individuals and communities. The number of disability-adjusted life years (DALYs) attributable to alcohol has also increased. In 2000, alcohol accounted for 4.0% of the worldwide total number of DALYs (Rehm, Room, et al., 2003b; WHO, 2002). Currently, the figure is 5.1% (WHO, 2014). The burden of disease caused by alcohol varies between countries, reflecting the patterns of drinking and average volume of consumption per person (WHO, 2014). In the Americas (North and South combined), the percentage of DALYs due to alcohol use exceeds that due to tobacco/smoking or being overweight/obese (WHO, 2007).

In an analysis of the regional impact of alcohol misuse, the percentage of total DALYs due to alcohol was approximately 8% for alcohol use compared to 13% for tobacco in America A (very low child and very low adult mortality; covering Canada, Cuba, and the United States). In America B (low child and low adult mortality; covering Antigua and Barbuda, Argentina, Bahamas, Barbados, Belize, Brazil, Chile, Colombia, Costa Rica, Dominica, Dominican Republic, El Salvador, Grenada, Guyana, Honduras, Jamaica, Mexico, Panama, Paraguay, Saint Kitts and Nevis, Saint Lucia, Saint Vincent and the Grenadines, Suriname, Trinidad and Tobago, Uruguay, and Venezuela), corresponding figures were 11% and 3% for alcohol and tobacco, respectively. For America D (high childhood and adult mortality; covering Bolivia, Ecuador, Guatemala, Haiti, Nicaragua, and Peru), the figure for alcohol was 6%, with no data for tobacco (WHO, 2007). In the United States, more detailed analysis of DALYs has suggested that the impact of alcohol on the burden of disease was much greater than previously thought (Rehm et al., 2014). For example, alcohol misuse in the United States accounts for 65,000 deaths annually and over 1.1 million years of life lost due to premature mortality (YLL) (Rehm et al., 2014). In the United Kingdom, it is estimated that alcohol consumption accounts for 31,000 deaths annually and 10% of all DALYs (Balakrishnan, Allender, Scarborough, Webster, & Rayner, 2009).

To limit the harm caused by alcohol in the United Kingdom, the Department of Health (1987) and Royal Colleges of Physicians, Psychiatrists and General Practitioners (1995) have recommended sensible limits for ethanol intake. Other countries also have guidelines for sensible drinking or limits to drinking; this has been reviewed to cover 30 countries including the USA (Furtwaengler & De Visser, 2013). The aforementioned authors reported that in the USA, a standard drink represents 14 g ethanol, with guideline daily maximal intakes at 56 and 42 g per day for men and women, respectively (Furtwaengler & De Visser, 2013). In the UK, the corresponding guideline daily maximal intakes are 32 and 24 g per day for men and women, respectively (Furtwaengler & De Visser, 2013).

There are also weekly maxima for consumption of ethanol. In the USA, the weekly maximal is 196 and 98 g per week for men and women, respectively (Furtwaengler & De Visser, 2013). In the UK, these figures correspond to 168 and 98 g per week for men and women, respectively. However, the amount of ethanol in one unit varies around the world (Table 1), which has the potential to cause confusion and prevents international comparisons, especially when examining epidemiological data on alcohol-related harm. Unless otherwise specified, all references to units of alcohol in this chapter are based on the definitions used in the UK, where a standard alcoholic drink or unit of alcohol contains 8 g of ethanol.

However, ethanol intake still varies significantly between individuals, although arguably some drinkers may not be aware of these guidelines (De Visser, 2015), many simply choose to ignore them. In fact, even among the medical profession, the understanding of the unit system or the amount of alcohol in a drink is poor (Das et al., 2014). An analysis of 586 doctors in training in the UK showed that 80% claimed to be knowledgeable about the unit system, although only one in six knew the precise details. The authors remarked that the extent of this lack of knowledge was similar to that 7 years previously (Das et al., 2014).

Most people enjoy the pleasant psycho-pharmacological effects of ethanol. This can be addictive, and those who consume excessive amounts of ethanol are known as "alcoholics." There are other terms that are also used (e.g., alcohol misuser, harmful drinker, hazardous drinker, etc.), which makes comparative alcohol intake on the health of people with normal ethanol metabolism are considered by some as controversial, excessive consumption of ethanol undoubtedly affects the nervous system and, as mentioned above, almost every other organ in the body.

The ethanol content of an alcoholic beverage is usually expressed as a percentage of alcohol by volume (ABV). The strength of an alcoholic beverage typically ranges from 3% to 50%. However, different brands of the same class of alcoholic beverage have different concentrations of ethanol. Therefore, the amount of ethanol in each type of alcoholic beverage and therefore the dose of ethanol from a drink varies significantly.

Many of the acute effects of alcohol correlate with the peak concentration of ethanol in the blood while drinking. It is therefore important to understand the factors that influence the blood ethanol concentration (BEC) achieved from a dose of ethanol (i.e., drinking an alcoholic beverage) because most of the acute effects correlate with the BEC.

ABSORPTION AND DISTRIBUTION OF ETHANOL

The fundamental principles of absorption of ethanol via the GI tract and the subsequent distribution of ethanol are well understood. These principles have been reviewed by Rajendram et al. (2013). The ethanol in alcoholic beverages enters the body through the mouth before passing down the esophagus into the stomach. Ethanol then continues to travel down the GI tract until absorbed (Halsted, Robles, & Mezey, 1973; Rajendram et al., 2013). The luminal concentration of ethanol therefore decreases down the GI tract. There is also a concentration gradient of ethanol from the lumen to the blood. The BEC is much lower than that in the lumen of the upper small intestine (Table 2; Halsted et al., 1973; Rajendram et al., 2013). Ethanol diffuses passively across the cell membranes of the mucosal surface into the submucosal space and then the submucosal capillaries (Kalant, 2004; Rajendram et al., 2013).

Uptake of ethanol occurs across all of the GI mucosa. However, absorption is fastest in the duodenum and jejunum. The main factor that determines the rate of absorption of ethanol is the rate of gastric emptying, as most ethanol is absorbed after leaving the stomach through the pylorus (Kalant, 2004; Rajendram et al., 2013). Ethanol probably enters the ileum and colon from the blood because concentrations of ethanol within the ileal lumen are similar to BEC (Bode & Bode, 2003). Luminal consequences of ethanol ingestion occur mainly in the upper small intestine.

Plasma proteins do not bind ethanol, so ethanol readily diffuses from the bloodstream, through the walls of capillaries, and into tissues, including the brain (Szabo & Lippai, 2014). Equilibration of ethanol within a tissue depends on the tissue mass, water content, and the rate of blood flow. Ethanol concentration equilibrates between blood and the extracellular fluid within a single pass. Equilibration between blood water and total tissue water can occur within minutes but can take up to an hour or more, depending on the area of the capillary bed and blood flow (Kalant, 2004). Ethanol enters most tissues, but its solubility in bone and fat is negligible. Therefore, after absorption, the volume of distribution of ethanol reflects total body water. Total body water correlates with lean body mass. Therefore, for a given dose of ethanol, BEC will reflect lean body mass (Rajendram et al., 2013).

TABLE 1 Variations in the Amount of Ethanol in Standard Drinks or Units

Country	Unit (g of ethanol)	Daily Limits of Ethanol (g per day)	
		Male	Female
USA	14	56	42
Mexico	–	48	36
Ireland	10	40	30
Estonia	10	40	20
Poland	10	40	20
Switzerland	10	40	20
Italy	12	36	24
South Africa	12	36	24
UK	8	32	24
Brazil	10	30	20
Bulgaria	10	30	20
France	10	30	20
Netherlands	10	30	20
New Zealand	10	30	20
Singapore	10	30	20
Spain	10	30	20
Slovakia	14	28	14
Canada	13	40	27
Austria	–	24	16
Czech Republic	–	24	16
Germany	12	24	12
Iceland	12	24	12
Australia	10	20	20
Portugal	10	20	20
Finland	12	20	10
Hong Kong	10	20	10
Slovenia	10	20	10
Denmark	12	–	–
Lithuania	10	–	–
Sweden	12	–	–

The unit system does not permit international comparisons.
Data from Furtwaengler and De Visser (2013).

analysis difficult. Those with genetic variants of the enzymes that metabolize ethanol may have adverse reactions if they drink and others are teetotalers (i.e., do not drink alcohol at all), usually for religious or health reasons. While the possible benefits of moderate

METABOLISM OF ETHANOL

The average rate of oxidation of ethanol is approximately 15 mg/dl blood/hour (Fisher, Simpson, & Kapur, 1987). Many factors influence this rate, which varies significantly between individuals. Three major enzyme pathways oxidize ethanol to acetaldehyde (Figure 1; Rajendram et al., 2013): alcohol dehydrogenase (ADH) in the cytosol, the microsomal ethanol oxidizing system (MEOS) in the endoplasmic reticulum, and peroxisomal catalase. However, metabolism of ethanol by ADH inhibits catalase, so catalase is usually of little significance in ethanol metabolism (Lieber, 2005). In addition, there are nonoxidative pathways, such as via the formation of ethyl glucuronide and fatty acid ethyl esters (FAEE) (Cabarcos et al., 2014).

Alcohol Dehydrogenase

Alcohol dehydrogenase oxidizes ethanol and reduces nicotinamide adenine dinucleotide (NAD+) to NADH (Höög & Ostberg, 2011; Kalant, 2004). Alcohol dehydrogenase is a zinc metalloprotein with five classes of isoenzymes that arise from the association of eight different subunits into dimers (Table 3; Kwo & Crabb, 2002). A genetic model accounts for seven ADH encoding genes on chromosome 4q22-23 (Wang, Kapoor, & Goate, 2012).

The class 1 ADH isoforms are the most relevant for oxidation of ethanol. Class 1 isoenzymes have a low Km, while class 2 isoenzymes have a higher Km. Class 3 ADH has a low affinity for ethanol and does not oxidize ethanol in the liver (Lieber, 2005). Class 4 ADH is present in the stomach (Moreno & Pares, 1991; Stone, Thomasson, Bosron, & Li, 1993) and class 5 is in the liver and the stomach (Yasunami, Chen, & Yoshida, 1991). While most of the metabolism of ethanol is hepatic, gastric ADH also oxidizes ethanol.

Microsomal Ethanol Oxidizing System

The MEOS has a higher Km for ethanol (8–10 mmol/l) than ADH (0.2–2.0 mmol/l). The MEOS is less involved than ADH at low

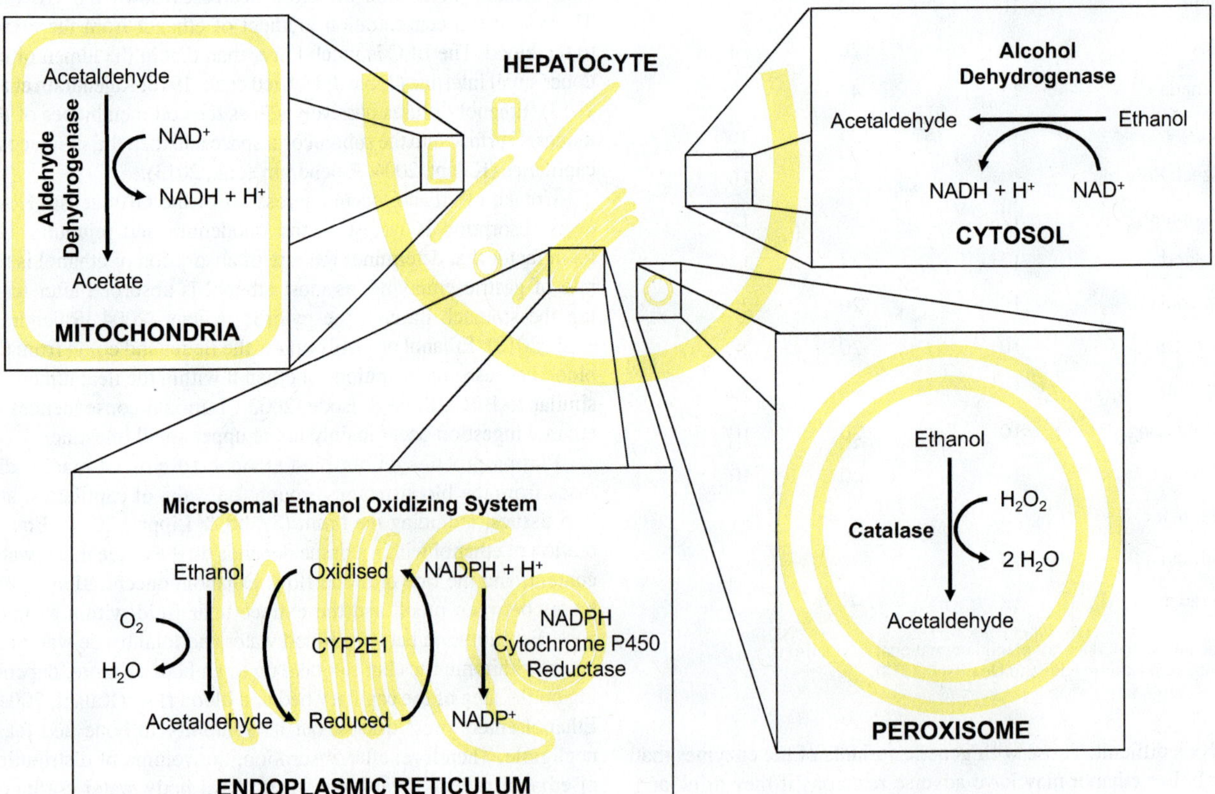

TABLE 2 Approximate Ethanol Concentrations in the GI Tract and in the Blood Following Oral Ethanol Administration (0.8 g/kg)

Site	Ethanol Concentration (g/dl)
Stomach	7–8
Proximal Jejunum	1–5
Ileum	0.1–0.2
Blood (15–120 min after dosage)	0.1–0.2

Ethanol appears in the blood as quickly as 5 min after ingestion and is rapidly distributed around the body. A dose of 0.8 g ethanol/kg body weight (56 g ethanol (7 Units) consumed by a 70 kg male) should result in a blood ethanol concentration of 100–200 mg/dl between 15 and 120 min after dosage. Highest concentrations occur after 30–90 min (Halsted et al., 1973; Rajendram et al., 2013).

FIGURE 1 **Major pathways of ethanol metabolism.** This figure illustrates the most important enzyme pathways that oxidize ethanol to acetaldehyde and acetate. *Adapted from Rajendram et al. (2013).*

BEC (Lieber, 2005). However, because the MEOS is induced by chronic ethanol exposure (Lieber, 2005), it is responsible for the increased clearance of ethanol from the blood in this setting. The most important enzyme of the MEOS is cytochrome P4502E1 (CYP2E1). Chronic ethanol use results in a 4- to 10-fold increase of CYP2E1 by increasing mRNA levels and the rate of translation of proteins (Lieber, 2005).

Metabolism of ethanol by CYP2E1 increases free radical and acetaldehyde production, which depletes intracellular defenses against oxidative stress (e.g., glutathione), further damaging hepatocytes (Konishi & Ishii, 2007; Lieber, 2005). Increased CYP2E1 activity accelerates lipid hydroperoxide production and contributes to the development of nonalcoholic fatty liver disease and nonalcoholic steatohepatitis. These diseases are commonly associated with obesity, type 2 diabetes, and hyperlipidemia.

Increased expression of CYP2E1 increases the metabolism of several medications (e.g., propranolol, warfarin, and diazepam), causing tolerance and reducing their effectiveness (Konishi & Ishii, 2007). Increased CYP2E1 also increases production of toxic metabolites, such as from paracetamol. This may exacerbate ethanol-induced disease.

As well as metabolizing ethanol and medications, CYP2E1 also has a role in normal physiology. CYP2E1 is involved in fatty acid oxidation and the diversion of ketones to gluconeogenesis (Konishi & Ishii, 2007; Lieber, 2005). CYP2E1 is involved in the conversion of acetone to acetol, which is then converted to methylglyoxal. Acetol and methylglyoxal both participate in gluconeogenesis.

The importance of understanding alcohol metabolism in relation to neuropathology pertains to the increased whole-body burden of free radicals, inflammatory mediators, and toxic by-products, which also affects the brain (Szabo & Lippai, 2014). Indeed, the concept of an alcohol-related liver–brain axis has some support (De La Monte, Derdak, & Wands, 2012; Mckillop, 2015; Murugan, Boyadjieva, & Sarkar, 2014); it is beyond the scope of this review but covered elsewhere (Preedy, 2016).

Acetaldehyde Metabolism

Acetaldehyde is a highly toxic product of ethanol metabolism. Acetaldehyde metabolism is discussed briefly here but is described in more detail in Rajendram et al. (2013).

Acetaldehyde is rapidly converted to acetate in a reaction that is catalyzed by aldehyde dehydrogenase (ALDH; Figure 1). There are several isoenzymes of ALDH (Table 4; Kwo & Crabb, 2002). The most important are ALDH1 (cytosolic) and ALDH2 (mitochondrial). The presence of ALDH in tissues that metabolize ethanol may reduce the toxic effects of acetaldehyde and thereby ethanol.

Several ALDH2 polymorphisms exist. For example, the ALDH2*2 allele results in a nearly inactive enzyme (Edenberg, 2007; Wang et al., 2012). Its expression is near dominant, so even heterozygotes have almost no detectable ALDH2 activity. The ALDH2*2 allele is common in East Asians but is essentially absent from Africans and Europeans (Edenberg, 2007; Wang et al., 2012).

TABLE 3 Alcohol Dehydrogenase (ADH) Isoenzymes

Class	Subunit	Location	Km (mmol/l)[a]	Vmax
Class 1				
ADH1	α	Liver	4	54
ADH2	β	Liver, lung	0.05–34	
ADH3	γ	Liver, stomach	0.6–1.0	
Class 2				
ADH4	π	Liver, cornea	34	40
Class 3				
ADH5	χ	Most tissues	1000	
Class 4				
ADH7	σ	Stomach, esophagus, other mucosae	40	1510
	μ		20	
Class 5				
ADH6	–	Liver, stomach	30	

This table describes the properties of the alcohol dehydrogenase (ADH) isoenzymes. ADH is a zinc metalloprotein with five classes of isoenzymes that arise from the association of eight different subunits into dimers. Km is the concentration of substrate that leads to half-maximal velocity. Class 1 isoenzymes generally require a low concentration of ethanol to achieve "half-maximal activity" (low Km), while class 2 isoenzymes have a relatively high Km. Class 3 ADH has a low affinity for ethanol and does not participate in the oxidation of ethanol in the liver (Lieber, 2005). Class 4 ADH is found in the human stomach (Moreno & Pares, 1991; Stone et al., 1993) and class 5 has been reported in liver and stomach (Yasunami et al., 1991).
[a]Km supplied is for ethanol; ADH also oxidizes other substrates.
Adapted from Kwo and Crabb (2002).

TABLE 4 Aldehyde Dehydrogenase (ALDH) Isoenzymes

Class	Structure	Location	Km (μmol/l)[a]
Class 1			
ALDH1	α4	**Cytosolic**	30
		Several tissues: Highest in liver	
Class 2			
ALDH2	α4	**Mitochondrial**	1
		Present in all tissues except red blood cells: Liver > kidney > muscle > heart	

This table describes the properties of the aldehyde dehydrogenase (ALDH) isoenzymes. Km is the concentration of substrate that leads to half-maximal velocity. Although there are several isoenzymes of ALDH the most important are ALDH1 (cytosolic) and ALDH2 (mitochondrial).
[a]Km supplied is for acetaldehyde; ALDH also oxidizes other substrates.
Adapted from Kwo and Crabb (2002).

ALDH2*2 reduces ethanol metabolism, so those with ALDH2*2 are easily intoxicated by ethanol (Peng et al., 2002). After exposure to alcohol, the blood acetaldehyde levels of these individuals rise rapidly and cause adverse reactions, including flushing, nausea, and tachycardia—effects that protect against alcohol dependence (Peng et al., 2002).

In chronic alcoholics, induction of MEOS increases oxidation of ethanol but reduces oxidation of acetaldehyde. In humans, hepatic acetaldehyde content increases with chronic ethanol consumption (Di Padova, Worner, Julkunen, & Lieber, 1987). A significant increase of acetaldehyde in hepatic venous blood reflects the high tissue level.

Metabolism of Acetate

Acetaldehyde is converted to acetate by ALDH. The serum acetate in human is usually under 0.2 mM. Serum acetate may be increased more than 20 times after ingestion of ethanol. Other common causes for increased serum acetate include prolonged starvation and lack of insulin (i.e., type 1 diabetes). Acetate metabolism in normal physiological conditions is not well described; the metabolic fate of acetate derived from alcohol is even less well understood. However, some important principles of acetate metabolism have been elucidated and reviewed by Cornier (2004) and Rajendram et al. (2013). Three important points are summarized here:

1. Most of the ethanol absorbed after oral intake is metabolized in the liver. This metabolism results in the production and release of acetate. Acetate readily crosses the blood–brain barrier and is metabolized in the brain. Cardiac and skeletal muscle also metabolize acetate.
2. The conversion of acetate, from any source, to acetyl-CoA is catalyzed by acetyl-CoA synthetase. This reaction requires adenosine triphosphate and results in the production of adenosine monophosphate. 5'nucleosidase catalyzes the production of adenosine from adenosine monophosphate.
3. The acetyl-CoA generated from acetate may be used to generate adenosine triphosphate via the Kreb's cycle. Acetyl-CoA may be converted to glycerol, glycogen, and lipid particularly, in the fed state. However, this only accounts for a small fraction of absorbed ethanol. The neurotransmitter acetylcholine is produced from acetyl-CoA in cholinergic neurons.

Fatty Acid Ethyl Esters

The location of fatty acid ethyl esters (FAEE) in neurological tissues has been reviewed previously (Zelner, Matlow, Natekar, & Koren, 2013). The authors concluded that they are found in human as well as rodent neurological tissues, including fetal brain after maternal ethanol exposure, gray and white matter, and various cell lines as a consequence of alcohol administration or consumption in vivo or in vitro (Zelner et al., 2013). The full importance and extent of these nonoxidative metabolites are not fully known, but there is evidence that they affect metabolic pathways in neurological tissues via a number of routes, including membrane disordering effects, mitochondrial dysfunction, and ion channels (Bora & Lange, 1993; Gubitosi-Klug & Gross, 1996; Laposata, Scherrer, Mazow, & Lange, 1987). Compared to our knowledge base of alcohol and acetaldehyde per se, there is a paucity of information on the neurological impact of nonoxidative metabolites.

Kinetics of Ethanol Elimination In Vivo

Alcohol elimination was originally thought to be a zero-order process—that is, a constant rate independent of the concentration of ethanol (Norberg, Jones, Hahn, & Gabrielsson, 2003). As the Km of most ADH isozymes for ethanol is low (about 1 mM), ADH is saturated even at low concentrations of ethanol; therefore, the rate of elimination is constant at maximal velocity and is independent of concentration (Norberg et al., 2003). However, elimination is not constant at very low concentrations when ADH is not saturated. Alcohol elimination follows Michaelis–Menten kinetics: the rate of elimination depends on the concentration of alcohol and the kinetic constants Km and Vmax (Matsumoto & Fukui, 2002).

In addition, because the metabolism by CYP2E1 and some ADH isozymes (e.g., ADH4) involves a high Km for alcohol, a concentration-dependent rate of ethanol elimination occurs, with higher rates of alcohol elimination at higher BEC. Because of concentration dependence, it is not possible to estimate a single rate of alcohol metabolism. Concentration-dependent metabolism of alcohol has been observed in some, but not all, studies on alcohol elimination (Ramchandani, Bostron, & Li, 2001).

Genetics of Ethanol Metabolism

Single nucleotide polymorphisms (SNPs) in ADH genes affect the characteristics of the translated proteins. This subject has been reviewed by Wang et al. (2012). The polymorphic forms of ADH (Class 1 ADH1B, ADH1C) vary between racial groups. There are three known polymorphisms in the ADH1B gene and two in the ADH1C gene. The reference ADH1 allele that encodes the β1 subunit is ADH1B*1 (Edenberg, 2007; Wang et al., 2012). This allele encodes arginine at positions 48 and 370. The ADH1B*2 allele encodes a subunit with histidine in position 48 (β2). The ADH1B*3 allele encodes a subunit with cysteine in position 370 (β3). These amino acid substitutions affect NAD+ binding, increasing turnover (Edenberg, 2007; Wang et al., 2012). The ADH1B*2 allele is common in Asians and the ADH1B*3 allele is common in Africans (Zintzaras, Stefanidis, Santos, & Vidal, 2006; Edenberg, 2007; Wang et al., 2012).

Studies of the association of alcoholism and alcohol-induced disease with the ADH2, ADH3, CYP2E1, and ALDH2 polymorphisms are inconclusive. A large meta-analysis found that the ADH2*1 and ADH3*2 alleles (less active ethanol metabolizing alcohol dehydrogenases) and the highly active ALDH2*1 allele increased the risk of alcoholism (Zintzaras et al., 2006). This may be because these polymorphisms reduce the accumulation of acetaldehyde. ALDH2*1 protects against liver disease by clearing acetaldehyde. Neither ADH2 nor the ADH3 polymorphisms are implicated in liver disease (Zintzaras et al., 2006). Variants of CYP2E1 do not affect the risk of alcoholism or alcoholic liver disease (Zintzaras et al., 2006).

This area of research is in its infancy. Much work still needs to be done to understand the relationships between these polymorphisms and alcohol-induced disease. However, the effects of these polymorphisms on alcohol metabolism and thereby blood ethanol concentration are highly relevant.

BLOOD ETHANOL CONCENTRATION

The lipid-to-water partition coefficient of ethanol is low. Therefore, because individuals have different body fat and water content, they may have very different BEC after the same dose of ethanol per kg body weight. However, if the same dose of ethanol is given, BEC is not significantly affected by the type of alcoholic beverage (although the rate of gastric emptying does).

Many effects correlate with the peak BEC or peak ethanol concentrations within organs during a drinking session. However, the relationship between the effects of ethanol and the BEC is extremely complex. It varies between individuals and also with drinking habits. Other effects of ethanol are due to the products of ethanol metabolism and the total dose of ethanol ingested over a period of time. These factors are linked as the ethanol concentration achieved while drinking determine which pathways of ethanol metabolism predominate.

The factors that increase the probability of higher maximum ethanol concentrations for any given level of consumption (e.g., drinking with an empty stomach) are therefore clinically relevant.

Effects of Food on Blood Ethanol Concentration

In a two-part crossover study, 10 healthy men were given ethanol to drink either after an overnight fast or immediately after standardized breakfast (Jones & Jönsson, 1994). The BEC was measured in samples taken at various times after drinking. The peak BEC was 67 mg/dl when a dose of 0.80 g/kg ethanol (standard deviation (SD) 9.5; 14.4 mmol/l, SD 2) was consumed after breakfast compared with 104 mg/dl (SD 14.5; 23.7 mmol/l, SD 3.1) if the ethanol was ingested after an overnight fast (p<0.001). The mean area under the BEC versus time curve (AUC; 0–6h) was 398 mg/dl per hour (SD 56) in fasting subjects compared with 241 mg/dl per hour (SD 34) in those who drank after breakfast (p<0.001). Metabolism of ethanol was approximately 2h faster if subjects drank before breakfast (Jones & Jönsson, 1994).

Therefore, the peak BEC is reduced if ethanol is consumed with or after food. Food delays gastric emptying into the duodenum and reduces the sharp early rise in BEC seen if ethanol is taken on an empty stomach. Food also increases elimination of ethanol from the blood. The area under the BEC versus time curve (AUC) is reduced. The mean clearance of blood ethanol is increased up to 50% after eating (Jones, 1993). However, food increases splanchnic blood flow, which maintains the ethanol diffusion gradient in the small intestine (Kalant, 2004). Food-induced impairment of gastric emptying may be partially offset by faster absorption of ethanol in the duodenum (Kalant, 2004).

In animal studies, ethanol is often administered with other nutrients in liquid diets. The AUC is less when ethanol is given in a liquid diet than with the same dose of ethanol in water (de Fiebre, de Fiebre, Booker, Nelson, & Collins, 1994). The different blood ethanol profile in these models could affect the expression of pathology in these models (Kalant, 2004).

First-Pass Metabolism of Ethanol

After oral administration of ethanol, the AUC (AUCoral) is significantly lower than after intravenous administration (AUCiv) or intraperitoneal administration (AUCip). The total iv ethanol dose is delivered to the systemic circulation. The difference between AUCoral and AUCiv is the amount of the oral dose that was either not absorbed or metabolized before entering the systemic circulation (i.e., first-pass metabolism; FPM). The ratio of AUCoral to AUCiv reflects the oral bioavailability of ethanol.

The study of ethanol metabolism has largely focused on the liver and its relationship to alcohol-induced liver disease. The role of the stomach in ethanol metabolism is controversial. However, several ADH isoenzymes are present in human gastric mucosa. This suggests that the stomach contributes to the FPM of ethanol (Lieber, 2005; Moreno & Pares, 1991; Yasunami et al., 1991).

Approximately 90% of all ethanol elimination occurs in the liver (Utne & Winkler, 1980). However, several factors including ethnicity (Dohmen et al., 1996), gender (Frezza et al., 1990), and alcoholism (Di Padova et al., 1987) alter overall FPM, despite having an opposite effect on hepatic metabolism of ethanol. The observations that their effects on FPM correlate with their effects on gastric alcohol dehydrogenase (ADH) activity (Di Padova et al., 1987; Dohmen et al., 1996; Frezza et al., 1990) is further evidence for ethanol metabolism in the stomach.

Gastric ADH activity is negligible compared to liver ADH activity when the BEC is below 100 mM. The gastric luminal cells are exposed to gastric juices, in which the ethanol concentration is much higher than BEC (Lieber, 2005). Increasing the ethanol concentration used to test for ADH activity increases the activity of gastric ADH (Lieber, 2005) but reduces that of hepatic ADH. This is the result of saturation and substrate inhibition of classes 1 and 2 ADH (Lieber, 2005).

Gastric ADH has a high activity and a Km around 40 mM (Kwo & Crabb, 2002). Thus, although hepatic oxidation of ethanol cannot increase once ADH is saturated, gastric ADH can metabolize ethanol significantly at the high concentrations in the stomach. Thus, gastric ADH can protect the body somewhat if excessive amounts of ethanol are consumed. If gastric emptying is delayed, prolonged contact with gastric ADH increases FPM of ethanol. Conversely, increasing the speed of gastric emptying reduces gastric FPM (Di Padova et al., 1987).

Contribution of First-Pass Metabolism to Overall Ethanol Metabolism

FPM increases with decreasing amounts of alcohol. For example, a dose of 0.15 g ethanol/kg body weight results in greater FPM than 0.3 or 0.8 g ethanol/kg body weight. When ethanol is given at a high dose, the presystemic contribution to its metabolism is relatively small, as hepatic metabolism dominates. In contrast, the significance of FPM to overall ethanol metabolism is relatively greater when small doses of alcohol are administered (Crabb, 1997; Gentry, Baraona, & Lieber, 1994).

Beverage Ethanol Content and Blood Ethanol Concentration

The ethanol concentration of the beverage consumed affects ethanol absorption and BEC. Absorption is fastest when the concentration is 10–30%. The concentration of ethanol in the GI tract also affects FPM. Gastric ADH requires a high ethanol concentration for optimal activity. The ethanol concentration in the beverage consumed affects oxidation of ethanol to acetaldehyde (Roine, Gentry, Lim, Baraona, & Lieber, 1991). After ingestion of equivalent amounts of ethanol, less FPM and higher blood levels occur after consumption of beer, which has a low ethanol concentration than whisky (Roine et al., 1993).

Quantification of Ethanol Absorption and First-Pass Metabolism

To define and quantify the absorption and FPM of ethanol, several groups have developed models of ethanol pharmacokinetics (for examples see Levitt, 2002). However, the accuracy of the models is poor, perhaps because there are several pathways of ethanol metabolism. Levitt (2002) analyzed human data with a computer program and suggested that the main factors influencing ethanol absorption and FPM are rate of gastric emptying and ethanol dose. When ethanol is ingested with food, absorption is slow and for a dose of 0.15 g/kg (12 g for an 80 kg man) the predicted fractional FPM was 36%, but only 7% for a dose of 0.3 g/kg (24 g for an 80 kg man; Levitt, 2002). For comparison, in fasting subjects, the absorption of oral ethanol is rapid and the predicted FPM was small.

Effects of Ethanol Metabolism on the Brain

Alcohol misuse causes structural and functional abnormalities of the nervous system and other organs. This has been identified clinically, with imaging and pathologically (Harper, 2009). For example, alcohol misusers lose neurons from several regions of the cerebral cortex, hypothalamus, and cerebellum.

Alcohol misusers can also develop liver disease and/or nutrient deficiency (e.g., thiamine deficiency) that also damage the nervous system. Although alcohol misusers with liver disease and/or nutritional deficiency are classified as "complicated" to differentiate them from the "uncomplicated misusers" without these complications, cognitively impaired uncomplicated alcohol misusers also have abnormalities (Harper, 2009).

The precise pathophysiological mechanisms of these effects of ethanol remain unclear. However, acetaldehyde has been implicated in the acute and chronic pharmacological, behavioral, and pathological effects of ethanol on the brain. Although blood levels of acetaldehyde are generally low and the blood–brain barrier prevents acetaldehyde entry, ethanol oxidation to acetaldehyde occurs within the brain itself (Hipólito, Sánchez, Polache, & Granero, 2007).

Ethanol metabolism generates reactive oxygen species (ROS) that cause lipid peroxidation. This is associated with the formation of malondialdehyde (MDA) and 4-hydroxy-2-nonenal (HNE), both of which form adducts with proteins. Acetaldehyde and MDA together can react with proteins to generate a stable MDA–acetaldehyde–protein adduct (MAA). Like acetaldehyde-amino acid adducts, peroxylipid-acetaldehyde adducts can induce immune responses and antibody formation. Importantly, MAA adducts can induce inflammatory processes in hepatic stellate and endothelial cells. Therefore, MDA and HNE production, formation of MAA adducts, and the development of liver disease are closely linked. Indeed, acetaldehyde adducts have been found in the brain of alcohol-dosed animals (Upadhya & Ravindranath, 2002) and alcoholics (Nakamura et al., 2003; Richards, Perry, Dodd, & Worrall, 2012).

Metabolism of ethanol by the MEOS also increases ROS production. This is associated with cancer, atherosclerosis, diabetes, inflammation, and aging. Cells regulate ROS with several defenses involving an array of antioxidants (e.g., glutathione, GSH). Intracellular ROS production and removal are normally balanced. This balance can be perturbed by ethanol.

Further research is required to more fully elucidate the pathophysiological role of ethanol metabolism in neurotoxicity. This is particularly important as in some cases ethanol-induced disease may be reversible.

The Contribution of Impaired Nutrition

Alcoholism can result in poor nutritional status as a consequence of one or more of the following:

- Poor or reduced intake: For example, this may be due to poor food selection or as a result of financial displacement (job losses or an overriding focus on buying alcoholic beverages).
- "Empty calories" of ethanol: Deficiencies will arise because alcoholic beverages contain very little micronutrients but are mainly comprised of metabolizable energy (ethanol provides 7.1 kcal/g compared to 9, 4, and 4 kcals/g for fats, proteins, and carbohydrates, respectively).
- Enhanced metabolic rate: There will be particularly increased energy requirements in alcohol-related diseases.
- Increased requirements: For example, there is an increased demand for antioxidants due to excessive free radical generation in the metabolism of alcohol and reduced tissue antioxidants (the reduced liver GSH is mentioned above).
- Maldigestion and malabsorption of micro- and macronutrients: This includes a cohort of contributing factors such as gastritis, altered gastric emptying, increased transit, microvillus injury, and pancreatic injury.
- Impaired metabolism of nutrients: For example, a functional liver is necessary to convert vitamin ergocalciferol (D2) and cholecalciferol (D3) to and 25-hydroxycholecaliferol and calcidiol respectively (Stokes, Volmer, Grunhage, & Lammert, 2013).
- Decreased hepatic storage: For example, vitamin A is stored in the liver and is reduced to maintain circulating levels.
- Increased urinary and fecal losses: This will include increased urinary phosphate, calcium and magnesium excretion, or increased excretion of fecal fats.

Nutritional aspects are important to consider for two reasons. The first relates to the fact that nutritional impairment will directly affect neurological tissue. These include Wernicke's encephalopathy and Korsakoff's psychosis induced by the deficiency of thiamine and pellagra due to niacin deficiency (Badawy, 2014). In pellagra, there is a wide range of neurological problems including stupor, delirium, catatonia, irritability, impaired concentration, anxiety, apathy, psychomotor defects, depression, axonal damage, chromatolysis in the pons, glial degeneration, and altered signal transmission in neurons (Lopez, Olivares, & Berrios, 2014). In addition to thiamine and niacin, there are other nutritional deficiencies occurring in alcoholics that may impact neurological tissues. These include deficiencies of vitamin A, other B vitamins, and vitamins C, D, and E. Mineral deficiencies in alcoholism (e.g., copper, selenium, zinc, iodine) also have the potential to impact on neurologic tissues. These effects of micronutrient deficiency on neurological tissues may be direct or indirect.

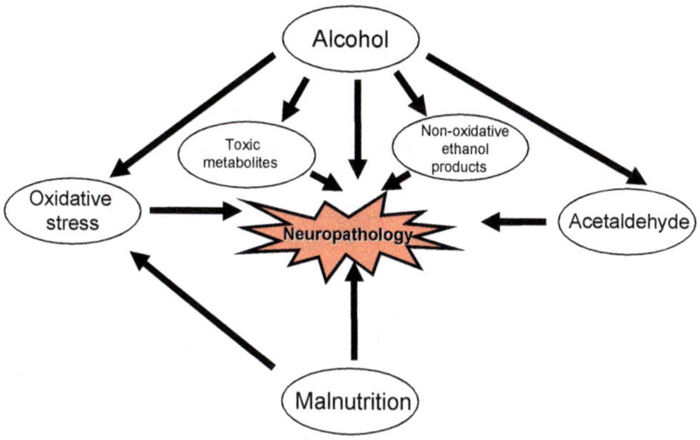

FIGURE 2 The interrelationship between alcohol, acetaldehyde, malnutrition, and oxidative stress on neuropathology. In this schematic diagram, the effects of alcohol, acetaldehyde, malnutrition, and oxidative stress are displayed. Nonoxidative ethanol products include fatty acid ethyl esters and ethyl glucuronide. Toxic metabolites include adducts of various kinds as well as inflammatory agents, etc. All of the aforementioned can target neurological tissues either at the molecular, cellular, structural, or functional levels. It is important to point out that oxidative stress will arise as a consequence of ethanol metabolism. For some micronutrients, such as selenium or vitamin E, deficiencies will increase oxidative stress. For example, selenium is an important component of the antioxidant enzyme glutathione peroxidase. There are other examples, such as the role of copper, zinc, and manganese in different forms of superoxide dismutase.

The second important point to consider is *metabolic superimposition*. This is when malnutrition occurs in the presence of ethanol- or acetaldehyde-induced neuronal toxicity. One must consider that in alcohol misuse there may be several pathways in which neurological tissues can be damaged, some of which are mentioned above (membrane disordering, biochemical effects, free radicals, inflammatory mediators, etc.). This is also illustrated in Figure 2.

APPLICATIONS TO OTHER ADDICTIONS AND SUBSTANCE MISUSE

Alcohol is unique because, in some individuals, it can cause addiction and/or organ damage if taken in excess amounts or chronically. The genetics of acetaldehyde formation and oxidation via ADHs, MEOS, and ALDHs means that some individuals are much more susceptible to disease, including neurological conditions. Of particular importance is the role of nutritional imbalance in alcohol misuse, which can play an important part in its overall impact on neurological conditions in some individuals. Up to 50% of chronic alcoholics may be malnourished with regards to one or more micro- and macronutrients (Gonzalez-Reimers et al., 2011; Knudsen et al., 2012; Manari, Preedy, & Peters, 2003; Ross, Wilson, Banks, Rezannah, & Daglish, 2012; Teixeira, Mota, & Fernandes, 2011; Wilkens et al., 2014).

In substance misuse other than alcohol, however, there are only a few studies in which malnutrition has been investigated as a focused area (e.g., Brock et al., 2010; Cemek et al., 2011; Kutan, Karatekin, & Nuhoglu, 2009; Saeland, Wandel, Bohmer, & Haugen, 2014). There are some other studies in which specific micronutrients have been examined for their impact on neurological tissue, such as vitamins C and E in relation to nicotine (Das, Gautam, Dey, Maiti, & Roy, 2009; Demiralay, Gursan, & Erdem, 2008; Naseer, Lee, & Kim, 2010). Examining the nutritional status of those exposed to the effects of addictions (either within fetal, adolescent, or adult life stages) will not directly prevent the addiction process. However, such studies may potentially unmask scenarios in which nutritional deficiencies contribute to the neuropathology in substance misuse.

SUMMARY

Ethanol is one of the most commonly used recreational drugs worldwide. Alcohol can damage almost every organ in the body and is responsible for at least 4% of the global burden of disease. When alcoholic beverages are consumed, ethanol is absorbed from the GI tract by diffusion and is rapidly distributed around the body in the blood before entering tissues by diffusion. Ethanol is metabolized to acetaldehyde mainly in the stomach and liver. Acetaldehyde is highly toxic and binds cellular constituents, generating harmful acetaldehyde adducts. Acetaldehyde is oxidized to acetate. Acetate metabolism and its role in the effects of ethanol are much less well understood. Ethanol and the products of its metabolism affect nearly every cellular structure or function and cause significant morbidity and mortality. The relationships between ethanol metabolism, blood ethanol concentration, and the harmful effects of ethanol are complex and vary between individuals and with drinking habits. Many acute effects correlate with the peak BEC or peak ethanol concentrations within organs during a drinking session. Other more chronic effects on the brain and other organs may be due to products of metabolism and the total dose of ethanol ingested over a period of time.

DEFINITION OF TERMS

Adducts A complex formed by a chemical binding to a biological molecule (e.g., protein or DNA).
Allele One of a number of variations of the same gene. Different alleles can result in the translation of slightly different proteins with different phenotypes. However, most alleles cause little or no observable differences.
Km The concentration of substrate that results in half-maximal velocity.
Oxidation Increase in oxidation state/loss of electrons by a molecule, atom, or ion.
Oxidative stress Imbalance between the production of reactive oxygen species and the ability to detoxify or repair the damage from these reactive molecules.
Reduction Decrease in oxidation state/gain of electrons by a molecule, atom, or ion.
Single-nucleotide polymorphism A common variation in a DNA sequence in which a single nucleotide in the genome (or other shared sequence) differs between members of a species. Therefore, if sequenced DNA fragments from two different individuals are ACTACT and ACTATT, which differ by only a single nucleotide, there are two alleles.

KEY FACTS OF ACETALDEHYDE

- Acetaldehyde is formed via the oxidation of ethanol and in turn acetaldehyde is converted to acetate.
- Acetaldehyde is an extremely reactive compound and a pure preparation of the chemical boils at room temperature.

- In the body, circulating levels of acetaldehyde are very low and bound to plasma proteins or red blood cells, thereby essentially reducing its toxicity.
- Acetaldehyde binds covalently with proteins, lipids, and nucleic acids.
- Acetaldehyde-protein adducts are toxic to cells.
- Acetaldehyde-protein adducts initiate the formation of neoantigens.

SUMMARY POINTS

- Alcohol can damage almost every organ in the body and in 2000 was responsible for at least 4% of the global burden of disease.
- The global burden of disease (GBD) due to alcohol is increasing.
- Absorbed ethanol is oxidized to acetaldehyde by three main pathways: alcohol dehydrogenase (ADH), microsomal ethanol oxidizing system (MEOS), or catalase.
- Acetaldehyde is highly toxic and forms harmful adducts by binding cellular constituents.
- Acetaldehyde is oxidized further to acetate.
- Acetate metabolism and its role in ethanol toxicity are not well understood.
- Approximately 90% of all ethanol elimination occurs in the liver.
- After ingestion of alcoholic beverages, ethanol travels along the GI tract until absorbed by diffusion.
- After consumption of alcoholic beverages, factors which affect the absorption, distribution, and first-pass metabolism of ethanol determine blood ethanol concentration.
- The factors that affect ethanol metabolism include gender, beverage ethanol content, time over which ethanol is consumed, and food.
- The relationships between ethanol metabolism, blood ethanol concentration, and the harmful effects of ethanol are complex and vary between individuals and with drinking habits.
- Many acute effects correlate with the peak BEC or peak ethanol concentrations within organs during a drinking session.
- More chronic effects of ethanol may be due to products of metabolism and the total dose of ethanol ingested over a period of time.

REFERENCES

Badawy, A. A. B. (2014). Pellagra and alcoholism: a biochemical perspective. *Alcohol and Alcoholism*, *49*, 238–250.

Balakrishnan, R., Allender, S., Scarborough, P., Webster, P., & Rayner, M. (2009). The burden of alcohol-related ill health in the United Kingdom. *Journal of Public Health*, *31*, 366–373.

Bode, C., & Bode, J. C. (2003). Effect of alcohol consumption on the gut. *Best Practice and Research Clinical Gastroenterology*, *17*, 575–592.

Bora, P. S., & Lange, L. G. (1993). Molecular mechanism of ethanol metabolism by human brain to fatty acid ethyl esters. *Alcoholism: Clinical and Experimental Research*, *17*, 28–30.

Brock, K. E., Graubard, B. I., Fraser, D. R., Weinstein, S. J., Stolzenberg-Solomon, R. Z., ... Albanes, D. (2010). Predictors of vitamin D biochemical status in a large sample of middle-aged male smokers in Finland. *European Journal of Clinical Nutrition*, *64*, 280–288.

Cabarcos, P., Tabernero, M. J., Otero, J. L., Miguez, M., Bermejo, A. M., Martello, S., ... Chiarotti, M. (2014). Quantification of fatty acid ethyl esters (FAEE) and ethyl glucuronide (EtG) in meconium for detection of alcohol abuse during pregnancy: Correlation study between both biomarkers. *Journal of Pharmaceutical and Biomedical Analysis*, *100*, 74–78.

Cemek, M., Buyukokuroglu, M. E., Hazman, O., Bulut, S., Konuk, M., & Birdane, Y. (2011). Antioxidant enzyme and element status in heroin addiction or heroin withdrawal in rats: effect of melatonin and vitamin E plus se. *Biological Trace Element Research*, *139*, 41–54.

Cornier, M.-A. (2004). Disposal of ethanol carbon atoms. In V. R. Preedy, & R. R. Watson (Eds.), *Comprehensive handbook of alcohol related pathology* (Vol. 1) (pp. 103–110). USA: Elsevier.

Crabb, D. W. (1997). First pass metabolism of ethanol: gastric or hepatic, mountain or molehill? *Hepatology*, *25*, 1292–1294.

Das, A. K., Corrado, O. J., Sawicka, Z., Haque, S., Anathhanam, S., Das, L., & West, R. (2014). Junior doctors' understanding of alcohol units remains poor. Clinical Medicine. *Journal of the Royal College of Physicians of London*, *14*, 141–144.

Das, S., Gautam, N., Dey, S. K., Maiti, T., & Roy, S. (2009). Oxidative stress in the brain of nicotine-induced toxicity: protective role of *Andrographis paniculata* Nees and vitamin E. *Applied Physiology, Nutrition, and Metabolism (Physiologie appliquee, nutrition et metabolisme)*, *34*, 124–135.

De La Monte, S., Derdak, Z., & Wands, J. R. (2012). Alcohol, insulin resistance and the liver-brain axis. *Journal of Gastroenterology and Hepatology (Australia)*, *27*(Suppl. 2), 33–41.

De Visser, R. O. (2015). Personalized feedback based on a drink-pouring exercise may improve knowledge of, and adherence to, government guidelines for alcohol consumption. *Alcoholism: Clinical and Experimental Research*, *39*, 317–323.

Demiralay, R., Gursan, N., & Erdem, H. (2008). The effects of erdosteine, N-acetylcysteine, and vitamin E on nicotine-induced apoptosis of hippocampal neural cells. *Journal of Cellular Biochemistry*, *104*, 1740–1746.

Department of Health. (1987). *Sensible drinking: The report of an Interdepartmental Working Group*. London: Department of Health.

Di Padova, C., Worner, T. M., Julkunen, R. J. K., & Lieber, C. S. (1987). Effects of fasting and chronic alcohol consumption on the first pass metabolism of ethanol. *Gastroenterology*, *92*, 1169–1173.

Dohmen, K., Baraona, E., Ishibashi, H., Pozzato, G., Moretti, M., Matsunaga, C., ... Lieber, C. S. (1996). Ethnic differences in gastric sigma-alcohol dehydrogenase activity and ethanol first-pass metabolism. *Alcohol Clinical and Experimental Research*, *20*, 1569–1576.

Edenberg, H. J. (2007). The genetics of alcohol metabolism: role of alcohol dehydrogenase and aldehyde dehydrogenase variants. *Alcohol Research and Health*, *30*, 5–13.

de Fiebre, N. C., de Fiebre, C. M., Booker, T. K., Nelson, S., & Collins, A. C. (1994). Bioavailability of ethanol is reduced in several commonly used liquid diets. *Alcohol*, *11*, 329–335.

Fisher, H. R., Simpson, R. I., & Kapur, B. M. (1987). Calculation of blood alcohol concentration (BAC) by sex, weight, number of drinks and time. *Canadian Journal of Public Health*, *78*, 300–304.

Frezza, M., Di Padova, C., Pozzato, G., Terpin, M., Baraona, E., & Lieber, C. S. (1990). High blood alcohol levels in women. The role of decreased gastric alcohol dehydrogenase activity and first-pass metabolism. *New England Journal of Medicine*, *322*, 95–99.

Furtwaengler, N. A. F. F., & De Visser, R. O. (2013). Lack of international consensus in low-risk drinking guidelines. *Drug and Alcohol Review*, *32*, 11–18.

Gentry, R. T., Baraona, E., & Lieber, C. S. (1994). Agonist: gastric first pass metabolism of alcohol. *Journal of Laboratory and Clinical Medicine, 123,* 32–33.

Goncalves, A., Claggett, B., Jhund, P. S., Rosamond, W., Deswal, A., Aguilar, D., ... Solomon, S. D. (2015). Alcohol consumption and risk of heart failure: the atherosclerosis risk in communities study. *European Heart Journal, 36,* 939–945.

Gonzalez-Reimers, E., Alvisa-Negrin, J., Santolaria-Fernandez, F., Martin-Gonzalez, M. C., Hernandez-Betancor, I., Fernandez-Rodriguez, C. M., ... Gonzalez-Diaz, A. (2011). Vitamin D and nutritional status are related to bone fractures in alcoholics. *Alcohol and Alcoholism, 46,* 148–155.

Gubitosi-Klug, R. A., & Gross, R. W. (1996). Fatty acid ethyl esters, nonoxidative metabolites of ethanol, accelerate the kinetics of activation of the human brain delayed rectifier K+ channel, Kv1.1. *Journal of Biological Chemistry, 271,* 32519–32522.

Halsted, C. H., Robles, E. A., & Mezey, E. (1973). Distribution of ethanol in the human gastrointestinal tract. *American Journal of Clinical Nutrition, 26,* 831–834.

Harper, C. (2009). The neuropathology of alcohol-related brain damage. *Alcohol and Alcoholism, 44,* 136–140.

Hipólito, L., Sánchez, M. J., Polache, A., & Granero, L. (2007). Brain metabolism of ethanol and alcoholism: an update. *Current Drug Metabolism, 8,* 716–727.

Höög, J. O., & Ostberg, L. J. (2011). Mammalian alcohol dehydrogenases-a comparative investigation at gene and protein levels. *Chemico-Biological Interactions, 191,* 2–7.

Hutton, F. (2012). Harm reduction, students and pleasure: an examination of student responses to a binge drinking campaign. *International Journal of Drug Policy, 23,* 229–235.

Jones, A. W. (1993). Disappearance rate of ethanol from the blood of human subjects: implications in forensic toxicology. *Journal of Forensic Science, 38,* 104–118.

Jones, A. W., & Jönsson, K.-Å. (1994). Food-induced lowering of blood-ethanol profiles and increased rate of elimination immediately after a meal. *Journal of Forensic Science, 39,* 1084–1093.

Kalant, H. (2004). Effects of food and body composition on blood alcohol levels. In V. R. Preedy, & R. R. Watson (Eds.), *Comprehensive handbook of alcohol related pathology* (Vol. 1) (pp. 87–102). London: Academic Press.

Knudsen, A. W., Jensen, J. E. B., Krag, A. A., Almdal, T. P., Nordgaard-Lassen, I., & Becker, U. (2012). Nutritional factors and bone metabolism in patients with alcohol dependence. *Clinical Nutrition Supplements, 7,* 168.

Konishi, M., & Ishii, H. (2007). Role of microsomal enzymes in development of alcoholic liver diseases. *Journal of Gastroenterology and Hepatology, 22*(Suppl. 1), S7–S10.

Kutan, F. A., Karatekin, G., & Nuhoglu, A. (2009). Fetal malnutrition in infants of smokers and passive smokers assessed by clinical assessment of nutritional status scoring. *Turkish Journal of Medical Sciences, 39,* 849–855.

Kwo, P. Y., & Crabb, D. W. (2002). Genetics of ethanol metabolism and liver disease. In D. I. N. Sherman, V. R. Preedy, & R. R. Watson (Eds.), *Ethanol and the liver. Mechanisms and Management* (pp. 95–129). London: Taylor and Francis.

Laposata, E. A., Scherrer, D. E., Mazow, C., & Lange, L. G. (1987). Metabolism of ethanol by human brain to fatty acid ethyl esters. *Journal of Biological Chemistry, 262,* 4653–4657.

Levitt, D. G. (2002). PKQuest: measurement of intestinal absorption and first pass metabolism - application to human ethanol pharmacokinetics. *BMC Clinical Pharmacology, 2,* 4.

Lieber, C. S. (2005). Alcohol metabolism: General aspects. In V. R. Preedy, & R. R. Watson (Eds.), *Comprehensive Handbook of alcohol related pathology* (Vol. 1) (pp. 15–26). Amsterdam: Elsevier.

Lin, J.-Y., Amodeo, L. R., Arthurs, J., & Reilly, S. (2012). Taste neophobia and palatability: the pleasure of drinking. *Physiology and Behavior, 106,* 515–519.

Lopez, M., Olivares, J. M., & Berrios, G. E. (2014). Pellagra encephalopathy in the context of alcoholism: review and case report. *Alcohol and Alcoholism, 49,* 38–41.

Manari, A. P., Preedy, V. R., & Peters, T. J. (2003). Nutritional intake of hazardous drinkers and dependent alcoholics in the UK. *Addiction Biology, 8,* 201–210.

Matsumoto, H., & Fukui, Y. (2002). Pharmacokinetics of ethanol: a review of the methodology. *Addiction Biology, 7,* 5–14.

Mckillop, I. H. (2015). Alcohol and the brain-liver axis: a further case of mind over matter? *Alcoholism: Clinical and Experimental Research, 39,* 405–407.

Moreno, A., & Pares, X. (1991). Purification and characterization of a new alcohol dehydrogenase from human stomach. *Journal of Biochemistry, 266,* 1128–1133.

Murray, C. J., & Lopez, A. D. (1997). Global mortality, disability, and the contribution of risk factors: Global Burden of Disease Study. *Lancet, 349*(9063), 1436–1442.

Murugan, S., Boyadjieva, N., & Sarkar, D. K. (2014). Protective effects of hypothalamic beta-endorphin neurons against alcohol-induced liver injuries and liver cancers in rat animal models. Alcoholism. *Clinical and Experimental Research, 38,* 2988–2997.

Nakamura, K., Iwahashi, K., Furukawa, A., Ameno, K., Kinoshita, H., Ijiri, I., ... Mori, N. (2003). Acetaldehyde adducts in the brain of alcoholics. *Archives of Toxicology, 77,* 591–593.

Naseer, M. I., Lee, H. Y., & Kim, M. O. (2010). Neuroprotective effect of vitamin C against the ethanol and nicotine modulation of GABAB receptor and PKA-alpha expression in prenatal rat brain. *Synapse, 64,* 467–477.

Norberg, A., Jones, W. A., Hahn, R. G., & Gabrielsson, J. L. (2003). Role of variability in explaining ethanol pharmacokinetics. *Clinical Pharmacokinetics, 42,* 1–31.

Peng, G. S., Yin, J. H., Wang, M. F., Lee, J. T., Hsu, Y. D., & Yin, S. J. (2002). Alcohol sensitivity in Taiwanese. *Association, 101,* 769–774.

Preedy, V. R. (Ed.). (2016). *The neuropathology of drug addictions and substance misuse* (Vol. 1) (1st ed.). USA: Elsevier.

Preedy, V. R., & Watson, R. R. (Eds.). (2004). *Handbook of alcohol-related pathology* (Vol. 1–3). London: Academic Press.

Rajendram, R., Hunter, R., & Preedy, V. R. (2013). Alcohol absorption, metabolism and physiological effects. In B. Caballero, L. Allen, & A. Prentice (Eds.), *Encyclopedia of human nutrition* (3rd ed.). UK: Elsevier.

Rajendram, R., Lewison, G., & Preedy, V. R. (2006). Worldwide alcohol-related research and the disease burden. *Alcohol and Alcoholism, 41,* 99–106.

Ramchandani, V. A., Bostron, W. F., & Li, T. K. (2001). Research advances in ethanol metabolism. *Pathologie Biologie, 49,* 676–682.

Rehm, J., Dawson, D., Frick, U., Gmel, G., Roerecke, M., Shield, K. D., & Grant, B. (2014). Burden of disease associated with alcohol use disorders in the United States. *Alcoholism, Clinical and Experimental Research, 38,* 1068–1077.

Rehm, J. T., Room, R., Monteiro, M., Gmel, G., Graham, K., Rehn, N., ... Jernigan, D. (2003). Alcohol as a risk factor for global burden of disease. *European Addiction Research, 9,* 157–164.

Rehm, J. T., Sempos, C. T., & Trevisan, M. (2003). Average volume of alcohol consumption, patterns of drinking and risk of coronary heart disease – a review. *Journal of Cardiovascular Risk, 10,* 15–20.

Richards, S., Perry, A., Dodd, P. R., & Worrall, S. (2012). Increased protein carbonyl content and elevated adduct formation in alcoholic cerebellar degeneration. *Alcoholism: Clinical and Experimental Research, 36*(Suppl. 2), 118A.

Roine, R. P., Gentry, R. T., Lim, R. T., Jr., Baraona, E., & Lieber, C. S. (1991). Effect of concentration of ingested ethanol on blood alcohol levels. *Alcoholism: Clinical and Experimental Research, 15,* 734–738.

Roine, R. P., Gentry, R. T., Lim, R. T., Jr., Heikkonen, E., Salaspuro, M., & Lieber, C. S. (1993). Comparison of blood alcohol concentrations after beer and whiskey. *Alcoholism: Clinical and Experimental Research, 17,* 709–711.

Ronksley, P. E., Brien, S. E., Turner, B. J., Mukamal, K. J., & Ghali, W. A. (2011). Association of alcohol consumption with selected cardiovascular disease outcomes: a systematic review and meta-analysis. *British Medical Journal, 342,* 479.

Ross, L. J., Wilson, M., Banks, M., Rezannah, F., & Daglish, M. (2012). Prevalence of malnutrition and nutritional risk factors in patients undergoing alcohol and drug treatment. *Nutrition, 28,* 738–743.

Royal College of Physicians, Royal College of Psychiatrists, Royal College of General Practitioners. (1995). *Alcohol and the heart in perspective, sensible limits reaffirmed.* Oxford: Oxprint.

Saeland, M., Wandel, M., Bohmer, T., & Haugen, M. (2014). Abscess infections and malnutrition–a cross-sectional study of polydrug addicts in Oslo, Norway. *Scandinavian Journal of Clinical and Laboratory Investigation, 74,* 322–328.

Stokes, C. S., Volmer, D. A., Grunhage, F., & Lammert, F. (2013). Vitamin D in chronic liver disease. *Liver International, 33,* 338–352.

Stone, C. L., Thomasson, H. R., Bosron, W. F., & Li, T. K. (1993). Purification and partial amino acid sequence of a high-activity human stomach alcohol dehydrogenase. *Alcoholism: Clinical and Experimental Research, 17,* 911–918.

Szabo, G., & Lippai, D. (2014). Converging actions of alcohol on liver and brain immune signaling. *International Review of Neurobiology, 118,* 359–380.

Teixeira, J., Mota, T., & Fernandes, J. C. (2011). Nutritional evaluation of alcoholic inpatients admitted for alcohol detoxification. *Alcohol and Alcoholism, 46,* 558–560.

Upadhya, S. C., & Ravindranath, V. (2002). Detection and localization of protein-acetaldehyde adducts in rat brain after chronic ethanol treatment. *Alcoholism: Clinical and Experimental Research, 26,* 856–863.

Utne, H. E., & Winkler, K. (1980). Hepatic and extrahepatic elimination of ethanol in cirrhosis. With estimates of intrahepatic shunts and Km for ethanol elimination. *Scandinavian Journal of Gastroenterology, 15,* 297–304.

Wang, J.-C., Kapoor, M., & Goate, A. M. (2012). The genetics of substance abuse. *Annual Review of Genomics and Human Genetics, 13,* 241–261.

Wilkens, K. A., Jensen, J.-E., Nordgaard-Lassen, I., Almdal, T., Kondrup, J., & Becker, U. (2014). Nutritional intake and status in persons with alcohol dependency: data from an outpatient treatment programme. *European Journal of Nutrition, 53,* 1483–1492.

World Health Organization. (2002). *World health report 2002. Reducing risks, promoting healthy life.* Geneva: WHO.

World Health Organisation. (2007). *Alcohol and public health in the Americas. A case for action.* Switzerland: World Health Organization.

World Health Organisation. (2014). *Global status report on alcohol and health 2014.* Switzerland: World Health Organization.

Yasunami, M., Chen, C. S., & Yoshida, A. (1991). A human alcohol dehydrogenase gene (ADH6) encoding an additional class of isozyme. *Proceedings of the National Academy of Science United States of America, 88,* 7610–7614.

Zelner, I., Matlow, J. N., Natekar, A., & Koren, G. (2013). Synthesis of fatty acid ethyl esters in mammalian tissues after ethanol exposure: a systematic review of the literature. *Drug Metabolism Reviews, 45,* 277–299.

Zintzaras, E., Stefanidis, I., Santos, M., & Vidal, F. (2006). Do alcohol-metabolizing enzyme gene polymorphisms increase the risk of alcoholism and alcoholic liver disease? *Hepatology, 43,* 352–361.

Chapter 36

Heavy Episodic Drinking or Binge Drinking: A Booming Consumption Pattern

Ana Adan[1,2], Irina Benaiges[1], Diego A. Forero[3]

[1]*Department of Psychiatry and Clinical Psychobiology, University of Barcelona, Barcelona, Spain;* [2]*Institute for Brain, Cognition and Behavior (IR3C), Barcelona, Spain;* [3]*Laboratory of NeuroPsychiatric Genetics, Biomedical Sciences Research Group, School of Medicine, Universidad Antonio Nariño, Bogotá, Colombia*

Abbreviations

AUD Alcohol use disorder
AUDIT Alcohol Use Disorders Identification Test
BAC Blood alcohol concentration
BD Binge drinking
BDNF Brain-derived neurotrophic factor
BMI Brief motivational intervention
CNS Central nervous system
ERP Event-related potentials
ESPAD European School Survey Project on Alcohol and Other Drugs
fMRI Functional magnetic resonance imaging
NIAAA National Institute on Alcohol Abuse and Alcoholism
PFC Prefrontal cortex
PM Prospective memory

INTRODUCTION

In recent decades, a new pattern of intermittent alcohol consumption has emerged involving binge intake, which happens in sessions lasting just a few hours, usually during weekend evenings and carried on by groups of peers. This pattern, known as heavy episodic drinking or binge drinking (BD), is found especially among adolescents and young adults in Western cultures.

BD involves the consumption of alcoholic beverages with the primary intention of reaching a marked intoxication, with low perception of risk and a tendency to equal intake in men and women (Parada et al., 2011a). Intoxication is characterized by a blood alcohol concentration (BAC) of at least 0.08 g/l. The practice of BD is increasing and expanding worldwide, and it has been recently labeled as a new culture of intoxication.

Epidemiological research has provided very variable data on the prevalence of BD according to the country and the sample studied. In the United States, where it has been most researched, BD is responsible for more than half of the deaths due to excessive alcohol intake, and it is the most prevalent alcohol use disorder (AUD). The data show that more than half of 12- to 17-year-olds consume alcohol with a BD pattern. Approximately 90% of the ethanol consumed under age 21 is as BD, and 70% of BD episodes involve adults aged 26 and older (Thiele, 2012).

A study conducted among 15- to 16-year-old adolescents in 35 European countries, the European School Survey Project on Alcohol and Other Drugs (ESPAD), showed that BD is found not only in Scandinavian countries, England, and Ireland—which are typically characterized as having BD problems (Pedersen & von Soest, 2013)—but in Mediterranean countries as well. The BD phenomenon has extended and appears to be on the rise in Mediterranean countries (Portugal, France, Italy, Spain), which had been traditionally characterized by a pattern of higher per capita consumption, most of which derived from daily consumption of wine during meals. On average, 43% of the ESPAD students reported heavy episodic drinking during the previous 30 days (Hibell et al., 2009).

In this chapter, we revise the concept of BD, which is still ambiguous nowadays, together with the risk and/or protection factors associated with the rise and maintenance of this pattern of alcohol intake. We highlight some consequences on health and its impact on neuropsychological functioning, in addition to possible early markers or endophenotypes for future use.

DEFINITION OF BINGE DRINKING

The Monitoring Future Study, carried on in the 1970s, proposed considering the pattern of BD as the consumption of five or more alcoholic drinks in one event, whether in men or women (Bachman, Johnston, & O'Malley, 1981). This has been and still is the criterion used in many studies on this topic.

The Harvard School of Public Health's College Alcohol Study, in the 1990s, proposed using different criteria according to sex. The concept of BD was established as the consumption of ≥5 alcoholic drinks for men and ≥4 drinks for women, in one single event and at least once in the previous 2 weeks (Wechsler, Davenport, Dowdall, Moeykens, & Castillo, 1994). This change of threshold in women was justified because women generally have a smaller stature and physiologic differences in the absorption and distribution of alcohol. In 2004, the National Institute on Alcohol Abuse and Alcoholism (NIAAA) Advisor Council endorsed the sex consideration in the definition of BD. The consumption of ≥5 drinks for men and ≥4 drinks for women in approximately 2h produces a BAC of 0.08 g/l or greater, taking into account that in the USA a standard drink has 14 g of alcohol. The new gender-specific definition increased the prevalence of women with BD pattern, due to the change in the criteria and not to actual changes in drinking behavior (Chavez, Nelson, Naimi, & Brewer, 2011).

There are significant differences in the amount of pure ethanol grams in a standard drink in different countries of the world, as published by the International Center for Alcohol Policies (2010). As Table 1 shows, the NIAAA criteria may be valid in Portugal, whereas in England the pattern of BD must be defined as the intake of ≥8 drinks for men and ≥6 drinks for women. In the countries where a standard drink contains 10 g of ethanol, the cutoff may be set to ≥6 drinks for men and ≥5 drinks for women (Parada et al., 2011a).

However, because teenagers may drink much larger amounts than those considered in the general definition, it has been proposed to consider several degrees of BD. Thus, the concept of extreme BD has emerged, with two subcategories: ≥10 drinks and ≥15 drinks (Patrick et al., 2013). In the USA, from the total percentage of high school seniors with BD, 10.5% were assigned to the first extreme category (≥10 drinks) and 5.6% to the second (≥15 drinks).

Regarding the timeframe to determine the presence of BD, there is also little agreement. Most of the studies set the frequency unit as 2 weeks, while others set it as 1 week and still others use time frames of 1–3 months or even 1 year. The timeframe of 1 or 2 weeks may be underestimating the prevalence of BD, while a long interval may overestimate it. The preferred and recommended criterion seems to be the occurrence of one episode of BD every 2–4 weeks.

The Alcohol Use Disorders Identification Test (AUDIT; Saunders, Aasland, Amundsen, & Grant, 1993) is the self-administered instrument most frequently used to detect a pattern of BD. Using its third question (How frequently do you drink six or more alcoholic beverages in one single event?), a cutoff line can be established different from that proposed by the NIAAA. However, no distinction between sexes is made, nor does it allow quantification of extreme intakes.

Another element to be taken into account is the BD trajectory in teenagers, from these possible four: accelerating (early onset and increased frequency), steep increase (delayed onset and rapid escalation), slow growth (delayed onset and gradual increase), and stable low (abstinence). High levels of alcohol intake and more physical health problems characterize the accelerating trajectory (Modecki, Barber, & Eccles, 2014).

CONSEQUENCES ON HEALTH AND POSSIBLE EARLY MARKERS

BD in the United States, Europe, and most developed countries is regarded as a major health and social concern that is associated with significant social and personal costs. BD elevates mortality risk (Plunk, Syed-Mohammed, Cavazos-Rehg, Bierut, & Grucza, 2014) and is related to physical injury, motor vehicle accidents, sexually transmitted diseases, unintended pregnancy, and medical complications, as pointed out by the NIAAA (2000). Table 2 summarizes the main effects on health that BD may cause.

The number of drinks and the BAC are good predictors of negative alcohol-related consequences (blackouts, physical fights, and getting physically sick) in first-year college students (Barnet et al., 2014). The BD adolescents show more risky sexual behavior (with the exception of non-condom use) and violent behavior (Stickley, Koyanagi, Koposov, Razvodovsky, & Ruchkin, 2013; Townshend, Kambouropoulos, Griffin, Hunt, & Milani, 2014; Xing, Ji, & Zhang, 2006), with there being a strong association with the number of BD days.

In healthy young binge drinkers, alterations have been found in macro- and microcirculation; these may represent early clinical manifestations of cardiovascular risk that should be considered in preventive actions (Goslawski et al., 2013). A study on rats shows that BD directly causes insulin resistance, which is a major risk factor to develop metabolic syndrome and type 2 diabetes (Lindtner et al., 2013). Alcohol appears to disrupt insulin-receptor signaling by causing inflammation in the hypothalamus, a nucleus of the central nervous system (CNS), which also participates in coronary artery disease and stroke.

Many studies have shown a link between psychological distress, such as anxiety and depression, and alcohol abuse (Balogun, Koyanagi, Stickley, Gilmour, & Shibuya, 2014; Cheng & Furnham, 2013; Wellman, Contreras, Dugas, O'Loughlin, & O'Loughlin, 2014). Alcohol BD is more severe among individuals with mental health problems and excessive alcohol consumption, which can in turn increase psychological distress. Suicide ideation and attempted suicide are also more prevalent in BD adolescents (Xing et al., 2006). Young BD adults had several sleep problems (trouble falling sleep, trouble staying asleep, and snoring/sleep apnea), independent of

TABLE 1 Grams of Pure Ethanol in a Standard Drink According to Country

Country	Ethanol (g)
England	8
Australia, Austria, France, Hungary, Ireland, Netherlands, New Zealand, Poland, Spain	10
Finland	11
Denmark, Italy, South Africa	12
Canada	13.6
Portugal (unofficial), United States	14
Japan	19.75

The definition of binge drinking based on standard drinks consumed is troublesome, because there is no international agreement on how many grams of pure ethanol there should be in a standard drink. Measures of standard units currently used in research can range from 8 g (England) to 19.75 g (Japan) of ethanol, making comparisons across studies difficult. Data were revised from the International Center for Alcohol Policies (2010), based on official government definitions.

psychiatric conditions, in a dose–response relationship (Popovici & French, 2013). While insomnia seems to be significantly associated in both sexes, the snoring/sleep apnea is related to BD only in males.

Ethanol also increases neuroinflammation in the CNS, and research has also observed persistent neuroimmune activation in BD, which could contribute to neurocognitive dysfunction in the prefrontal cortex (PFC) (Vetreno & Crews, 2012), as well as in myelination and white matter integrity (Jacobus, Squeglia, Bava, & Tapert, 2013). Brain-derived neurotrophic factor (BDNF) is a molecule identified as a major regulator of structural changes and inflammation processes in the brain related to alcohol use. The chronic BD pattern produces a decrease of BDNF, which correlates with lower survival and neuronal differentiation of cells in the hippocampus and the development of a depressive phenotype during the withdrawal period (Briones & Woods, 2013).

Moreover, a critical role of the corticotropin-releasing factor type 1 receptor in excessive ethanol intake has been observed related to stress response, in both models of BD and alcohol dependence (Kaur, Li, Stenzel-Poore, & Ryabynin, 2012). There is also evidence that BD upregulates the group 1 metabotropic glutamate receptor signaling throughout the amygdala (Cozzoli et al., 2014). Both mechanisms may be keys in the neurobiological mechanisms that underlie the transition from BD to ethanol dependence.

RISK AND PROTECTION FACTORS ASSOCIATED WITH BD

Drawing an explicative model of BD behavior is a highly complex task because there is a wide variety of interrelated factors in different domains, which in turn depend on the country, the sample, and the factors to be studied. Tables 3 and 4 show the main risk and protection factors associated with BD, respectively.

Most works have found a higher prevalence of BD and more drink intake in one single episode in males (Chavez et al., 2011; Hibell et al., 2009; Pedersen & von Soest, 2013). Moreover, there are differences between men and women in BD relating to their expectations towards drinking and to positive alcohol metacognitions. Women's expectations are related to improving their sociability and sexual ability, while men's are geared to reducing stress (Balodis, Potenza, & Olmstead, 2009). The positive alcohol metacognitions about cognitive self-regulation are a significant predictor of weekly levels of alcohol use in BD males (Clark et al., 2012). It is highly relevant to take into account the specific features associated with sex in the study of BD and the design of possible interventions.

Regarding age, the range and cutoff points used in different studies condition the data available. However alcohol intake

TABLE 2 Main Negative Consequences on Health and Neuropsychological Performance of Binge Drinking

Acute negative consequences	Death by ethylic intoxication Physical injury Motor vehicle accidents Physical fights Unplanned sex Unintended pregnancy Violent behavior
Cardiovascular and metabolic diseases	Coronary artery disease Stroke Insulin resistance Metabolic syndrome Type 2 diabetes
Psychological distress and mental disorders	Anxiety Depression Alcohol dependence Suicide (ideation and attempts) Insomnia (trouble falling sleep and staying asleep) Snoring/sleep apnea (only men)
Neurological diseases	Neuroinflammation Persistent neuroimmune activation Dysfunctions in myelination Alterations in the white matter integrity Maladaptive response to stress
Neuropsychological deficits	Sustained attention Memory (declarative, semantic, prospective and working) Associative learning Inhibitory control Decision making

The consequences are those evidenced by the research on adolescents, young adult, and adult participants with a consumption pattern of binge drinking.

TABLE 3 Risk Factors Associated to Binge Drinking (BD) Behavior in Adolescents and Young People

Individual characteristics	Sex (men) Adolescents and young Early pubertal timing Evening-type
Social and educational	Easy alcohol access Cheap alcohol Low socioeconomic status Adverse socioeconomic conditions
Family, parents, and peers	Parental alcohol consumption Parental BD Friends alcohol use
Other drug consumption	Smoking Marijuana use
Activities, beliefs, cognitions	Unaware of the risk of BD Unauthorized absence from the school (truancy) Sport involvement Use of products to enhance physique muscular
Personality traits	High impulsivity High novelty seeking High extraversion Low agreeableness Low conscientiousness

The factors are those obtained in the studies revised, related to the practice of binge drinking and/or which are predictors of sustained BD behavior.

TABLE 4 Protective Factors Associated to Binge Drinking (BD) Behavior in Adolescents and Young People

Individual characteristics	High intelligence morning-type
Social and educational	Educational achievement Higher school grades Occupational levels Stronger policy environment
Family, parents, and peers	Parental monitoring (control and supervision) Parental support Family oriented leisure Higher parental education (only for extreme BD)
Activities, beliefs, cognitions	School attachment Participation in school and non-school activities Religious involvement Belief that alcohol produces cognitive self-regulation (only men) Belief that alcohol enhances sociability and sexuality (only women)
Personality traits	Low impulsivity Low novelty seeking High conscientiousness

The factors are those obtained in the studies revised, related to the practice of non-binge drinking and/or not maintaining that behavior over time.

with BD behavior is a huge concern in all countries where there are statistics on the topic. In the United States and England, the prevalence among college students ranges from 43% to 58.5% in men and 32% to 54% in women (Howell et al., 2013). In Spain, a survey on this type of alcohol intake among teenagers and young adults (Spanish Observatory on Drugs, 2013) reported an increase of BD, which taking into account the previous 30 days was 35% in men and 20% in women aged 20–24.

The study of predictors of sustained BD in young adults shows similar determinants, being higher in males with no college/university studies (Pedersen & von Soest, 2013; Wellman et al., 2014). BD is related to parents and peers who also drink alcohol (Weitzman, Nelson, & Wechsler, 2003; Pedersen & von Soest, 2013; Strickly et al., 2013). Higher parental education is a protective factor for extreme BD, but it is a risk factor when considering BD for ≥5 drinks (Patrick et al., 2013).

College students more exposed to "wet" environments—defined as social, residential, and market surroundings where drinking is prevalent and alcohol is cheap and easily accessed—are more likely to engage in BD (Strickly et al., 2013; Weitzman et al., 2003). Moreover, girls and boys who practice BD are more prone to engage in all forms of substance use, with frequent polyconsumption. Tobacco and marijuana consumption are the most frequent (Stickley et al., 2013) and are predictors of the intensity of BD (Patrick et al., 2013). BD is a risk factor in the development of alcohol dependence (Balodis et al., 2009; White et al., 2011). Moreover, binge drinkers tend to be unaware of the risks of BD (Strickley et al., 2013).

It is very important to perform longitudinal studies to assess the correlates of adult BD. In this sense, we want to highlight a study by Cheng and Furnham (2013), who investigated the association of several variables as predictors of BD in 50-year-old adults from England. The results showed that individuals who come from lower parental social class with adverse socioeconomic conditions tend to develop more alcohol abuse behavior; in addition, childhood intelligence, educational achievement, and occupational levels were negatively associated with hazardous adult BD. Another variable related to BD in adolescents is school attachment, with truancy (unauthorized absence from the school) being a risk factor (Stickley et al., 2013). Frequent BD precedes long-term unemployment, but only in women; this result has little support for the social causation hypothesis because it appears after controlling for many potential confounder factors (Backhans, Lundin, & Hemmingsson, 2012). Social exclusion is also another consequence related to BD in the long term (Viner & Taylor, 2007).

A longitudinal study by Pedersen and von Soest (2013) in Norway found that the frequency of alcohol consumption and BD in parents predicted BD in their offspring at age 28, independently of sex. This study also observed that parental monitoring (control and supervision) and support with alcohol measures is of particular importance as a protective factor in adolescence.

In adolescents with accelerating BD trajectory, there is an early pubertal timing, low socioeconomic status, and more sport involvement compared to those classified as "stable low" (Modecki et al., 2014). However, participation in organized school or nonschool activities for early maturers protects against development of a "steep increase" trajectory. In addition, there is a higher probability for starting BD behavior in males who are highly concerned about muscularity and use supplements and other products to enhance their physique (Field et al., 2013).

Finally, it is important to highlight that a large majority of adolescents and young adults continue the BD behavior despite their knowledge of health risk (Zwaluw, Kleinjan, Lemmers, Spijkerman, & Engels, 2013). This may be explained by the theory of cognitive dissonance, which means that the drinkers' existing attitudes and cognitions will be modified to match the BD behavior. This suggests that, although knowledge and attitudes are treated in prevention, these may be inefficient on BD behavior.

PERSONALITY TRAITS, CIRCADIAN TYPOLOGY, AND BD

The study of personality traits associated with BD has confirmed consistently the presence of higher impulsivity (White et al., 2011; Townshend et al., 2014) and novelty seeking (Wellman et al., 2014) in youths, with both features being predictors of maintenance in the BD pattern. Both personality dimensions are also related to several nonadaptive behaviors and mental disorders, mainly substance use disorders (Marquez-Arrico & Adan, 2013).

Most personality questionnaires include an impulsivity dimension, although there are also specific instruments to measure it. The Dickman Impulsivity Inventory (DII, Dickman, 1990) may be highlighted because it proposes the existence of functional impulsivity as the tendency to make quick decisions when required by the situation to favor the person, in contrast to dysfunctional impulsivity, which is related to quick but irreflexive decisions that lead to negative consequences. Youths with BD, especially men,

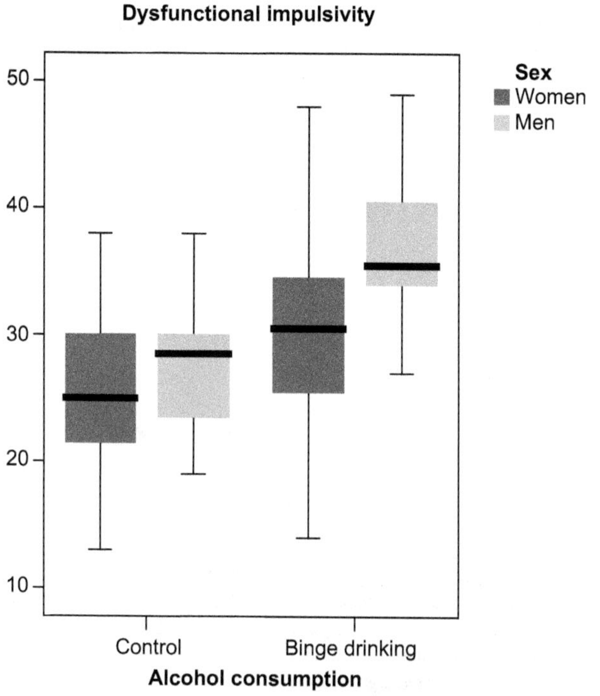

FIGURE 1 Dysfunctional impulsivity scores of binge drinking (BD) and control groups according to sex. The scores on dysfunctional impulsivity are higher in the BD group than in the control group. Men obtain higher scores in impulsivity, especially in the BD group. Dysfunctional impulsivity was measured using the Dickman Impulsivity Inventory (1990), with scores ranging from 12 to 60. *Data from Adan (2012), with permission from the publishers.*

obtain higher scores only in dysfunctional impulsivity compared to controls (Adan, 2012), as Figure 1 shows.

Impulsivity is the most consistent personality trait related to early use of drugs, repetition of intake, and progression to addiction, and it is a condition promoted with consumption that also favors relapse (Perry & Carroll, 2008). The data available on novelty seeking, defined as the need for adventure and excitement, preference for risk and strong emotions, boredom susceptibility, and disinhibition, point in the same direction (Castellanos-Ryan, Rubia, & Conrod, 2011). The disinhibition dimension is the best BAC predictor in both sexes (Legrand, Gomà-i-Freixanet, Kaltenbach, & Joli, 2007). Impulsivity and novelty seeking may explain the poor performance in executive tasks seen in teenagers (Castellanos-Ryan et al., 2011) and in addiction disorders (Dolan, Bechara, & Nathan, 2009).

Using the "Big Five" model of personality traits (extraversion, emotionality/neuroticism, conscientiousness, agreeableness, and intellect/openness) in a longitudinal birth cohort study, Cheng and Furnham (2013) observed that high extraversion and low agreeableness were significant predictors of BD. In other research, low conscientiousness also appears as a predictor of weekly levels of alcohol in BD university students (Clark et al., 2012). Perhaps not only one but the combination of several personality traits may configure in the future an endophenotype for the onset and maintenance of BD.

Circadian typology (morning-type, neither-type, and evening-type) is an individual difference that affects our biological and psychological functioning—not only in health, but also in disease (Adan et al., 2012 for a review). Morning-type subjects go to bed early and wake up early, and achieve their peak mental and physical performance in the early part of the day. In contrast, evening-type subjects go to bed and wake up late, and perform at their best toward the end of the day and evening hours. The evening-type, especially teenagers and young adults, show a personality pattern of higher impulsivity, novelty seeking, extraversion, and activity, and lower conscientiousness and harm avoidance. All the existing data also suggest that circadian typology is related to the consumption of all types of drugs and BD, with morning-type being a protective factor and evening-type a risk factor (Prat & Adan, 2011).

NEUROPSYCHOLOGICAL IMPACT OF BD

The study of the possible cognitive impact of BD is recent and the existing studies are difficult to compare due to methodological aspects that may influence the results (small samples, several definitions and duration of BD, use of different tasks, variable ages, etc.). The lack of control of the educational level, the consumption of other substances, the presence of psychopathological traits, and personality traits may greatly affect the results. The available data suggest that there are deficits in the cognitive performance in sustained attention tasks and memory and PFC activity tasks similar to those found in long-term alcohol abuse subjects, although to a lesser extent and with more mixed results.

In considering the neuropsychological impact of BD, we should distinguish between the effects of a single episode (next day) and those that may come from a sustained BD behavior over time (see Table 2). In both cases, the assessments are carried out with a zero BAC.

Effects of a Single Episode of BD

Neurocognitive impairment are found in subjective alertness, memory retrieval processes (delayed recall), and reaction time, while performance in attention tasks is not significantly impaired (McKinney, Coyle, & Verster, 2012; Verster, van Duin, Volkerts, Schreuder, & Verbaten, 2003). The pattern of impulsive acting, objectively assessed with inhibition tasks, also increases independently from individual differences in the impulsivity trait (McCarthy, Niculete, Treloar, Morris, & Bartholow, 2012).

Effects of BD on Learning and Memory Tasks

Studies have focused on retrospective memory in which learning, retention, and retrieval are evaluated. The existence of memory deficits under chronic alcohol consumption is well established, linked to alterations in the hippocampus and PFC. BD students perform worse in declarative memory tasks, both in immediate and delayed recall (Parada et al., 2011b). BD young adults present behavioral impairments in sustained attention, working memory, and associative learning (Parada et al., 2012; Scaife & Duka, 2009; Stephens and Duka, 2008).

For working memory, at an early age in young binge drinkers without AUD or mental comorbidity, the performance at a behavioral level is adequate. However, hypoactivation of the right anterior PFC was obtained using combined event-related potentials (ERP) and exact low-resolution brain electromagnetic tomography (Crego et al., 2010). Using functional magnetic resonance imaging (fMRI), higher bilateral activity in the supplementary premotor area was observed in a BD condition, as well as a positive correlation between the number of drinks consumed per event and the activity in the dorsomedial PFC and the cerebellar-thalamic-insular regions (Campanella et al., 2013).

The assessment of prospective memory (PM), conceptualized as the cognitive ability to remember to carry out some activity at some future point in time, is a crucial aspect of everyday cognitive function, and it is sensible to exhibit neuropsychological sequels associated with BD. Adolescents and young adults with BD showed impairments in a task of PM with fewer location–action combinations to recall (Heffernann, Clark, Bartholomew, Ling, & Stephens, 2010) and reduced function on time-based tasks (Heffernann & O'Neill, 2012) when compared with non-binge drinkers.

Effects of BD on Executive Functions

In BD university students, deficits in behavioral inhibition tasks have been observed, reflecting impulsivity and poorer decision making (Field, Schoenmakers, & Wiers, 2008; Scaife & Duka, 2009). This is more evident in high BD than in low BD subjects (Townshend et al., 2014), in men, and in those who engage in heavy BD at an early age (Goudriaan, Grekin, & Sher, 2011). The assessment of decision-making using the Iowa Gambling Task has found a pattern of "short-sightedness" for future consequences in BD, similar to that in addiction patients, and which implies impulsive behavior with lower inhibitory control (Johnson et al., 2008).

Heavy binge drinkers performed worse on a battery of tasks of executive functions compared with light social drinkers (Montgomery, Fisk, Murphy, Ryland, & Hilton, 2012). They generated significantly fewer words in a word fluency task (access to semantic memory), had worse random letter generation (inhibitory control), and had greater switch cost on a random letter task (switching), as shown in Figure 2.

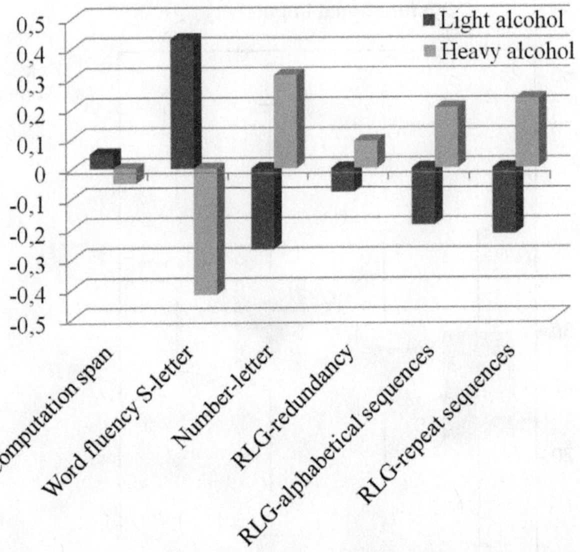

In the revision by Stephens & Duka's (2008), BD deficits in executive functions were found to underlie functional hyperactivity impairments in the PFC and the amygdala, which may cause neurotoxicity with time. Because the PFC and other brain areas involved are subject to development and maturation during adolescence and emerging adulthood, BD until the age of 24 poses a bigger harm to executive functioning than in later stages of life (Hermens et al., 2013).

The appearance of alterations in brain functions when effects in behavioral neurocognitive performance are not yet observable points at early cerebral disturbances, which should be studied in the future to check whether they persist when the BD habit ceases, as well as the possibility that they may be considered endophenotypes of the progression from BD to AUD. In this direction, the longitudinal study by Dager et al. (2013) using fMRI has identified that an amplified cue-elicited brain response was the most significant predictor of the emergence of BD and subsequent AUD. The overactivated regions were linked to habit formation, decision making, motivation, and attention.

FUTURE RESEARCH ON PREVENTION AND/OR TREATMENT ON BD

Countries affected by BD should set it as a priority to promote actions to reduce the number of persons practicing BD and the intensity of the associated intoxications. Some strategies that have

FIGURE 2 Executive functioning performance (standardized scores) in light and heavy alcohol social drinkers. Participants were 18–25 years old, had never used illicit substances, and had never been diagnosed with a substance use disorder or been advised to reduce their drinking. Heavy social drinkers generated significantly fewer words in a word fluency task (access to semantic memory), had worse random letter generation (RLG; inhibitory control), and greater switch cost on a random letter task (switching). *Data re-elaborated from Montgomery et al. (2012).*

proven effective in reducing alcohol consumption involve limiting the availability of alcohol, raising its price, promoting substance and alcohol-free environments (Naimi et al., 2014), and blocking legal access to alcohol to a later age. Although we might acknowledge that alcohol consumption may continue and that it is not realistic to advocate abstinence, promoting less harmful drinking patterns and their frequency is an advisable harm reduction strategy (Plunk et al., 2014).

Preventive and therapeutic approaches should take into account circadian rhythmicity. Involving chronobiology in prevention implies not only promoting healthy leisure activities or "what to do," but also placing such activities in appropriate daytime temporal moments or "when to do" them (Adan et al., 2012). Chronobiological therapeutic strategies such as to establish regular time patterns of sleep-wake, meals, and daily activity with a tendency towards a morning-type functioning, and exposition to light therapy and melatonin administration, may also be effective.

Few works have assessed the efficacy of interventions aimed at reducing dependence or intensity of BD in the population. The brief motivational intervention (BMI) is one of the few effective strategies for diminishing alcohol consumption. A single face-to-face session of BMI in a BD group was found to reduce alcohol use among subjects who experienced one or more alcohol-related adverse consequences (19% less drinking), whereas no preventive effect was obtained in control subjects (Daeppen et al., 2011). More research is needed in this area in the near future.

To conclude, research on pharmaceutical targets to avoid the health consequences of BD and the progression to AUD will hopefully offer knowledge to be applied in the clinic in the future, based on the promising findings from behavioral genetic approaches, such as the reduction of the BDNF or the upregulation of glutamate receptors in BD.

APPLICATIONS TO OTHER ADDICTIONS AND SUBSTANCE MISUSE

As we have seen in this chapter, the pattern of alcohol BD involves significant risks to both the immediate and short-term health outcomes of those who practice it. BD is more dangerous in the case of adolescents because the toxic amounts of alcohol in their bodies under maturation process aggravate the risks. The possible changes in gene expression and in a wide variety of neurotransmission systems should be noted, among which stands out the reinforcement dopaminergic pathway. These changes significantly increase biological vulnerability to progress towards the development of a substance use disorder, not just for alcohol or other substances with depressant pharmacological profile but in general. Similarly, these effects on the CNS can participate in the development of other mental disorders, such as depression and anxiety with and without comorbidity to the addictive disorder. The need to improve the knowledge about evidence-based treatments should be emphasized. However, efforts should prioritize developing preventive strategies starting in childhood in order to enhance the protective factors of BD practice.

DEFINITION OF TERMS

Attention Attention is a complex concept that represents a cohesive set of processes, which include sensory, motor, and cognitive processing.

Binge drinking Binge drinking is an intermittent pattern of heavy alcohol consumption of ≥5 drinks for men and ≥4 drinks for women, in one single event and at least once in the past two weeks.

Circadian typology This individual difference (morning-, neither- and evening-type) affects our biological and psychological functioning in health and in disease. Morning-types go to bed early and wake up early, and achieve their mental and physical peak performance in the early part of the day, in contrast to the evening-types.

Endophenotype This is an internal, inheritable, and measurable trait marker of vulnerability to develop a particular disease.

Executive functioning This set of cognitive skills is mainly linked to the functioning of the brain frontal lobe, which includes mental flexibility, planning and abstract reasoning abilities, self-regulation, and task monitoring, which determine goal-directed behavior.

Functional impulsivity Functional impulsivity is a tendency to make quick decisions to one's benefit when required by the situation.

Inhibition This mechanism prevents the entrance of nonpertinent information into working memory and suppresses information that has become irrelevant for a current task.

Impulsivity or dysfunctional impulsivity This personality trait is defined as the tendency to act with lack of prevision, making quick and irreflexive decisions that lead to negative consequences.

Novelty seeking This personality trait is characterized by a need for adventure and excitement, preference for risk, and strong emotions by the mere fact of living them, boredom susceptibility, and disinhibition.

Prospective memory This is the cognitive ability to remember to carry out some activity at some future point in time.

Working memory Working memory is the ability to hold and manipulate information in the mind over short periods of time—a mental workspace that is used to store important information in the course of our everyday lives.

KEY FACTS OF NEUROPSYCHOLOGY IN BINGE DRINKERS

- Assessing cognitive performance is more difficult than assessing biological aspects, because even in simple cognitive tasks there are several skills involved in their resolution (attention, motor control, etc.).
- BD behavior increases neuroinflammation in the CNS, which may contribute to dysfunction in PFC, myelination, and white matter integrity related to neuropsychological deficits.
- In healthy young binge drinkers, deficits have been found in sustained attention, learning, memory (declarative, semantic, prospective and working), and executive functions (inhibition control and decision making).
- Early cerebral disturbances measured by ERPs or fMRI in brain activity may be detected when effects on behavioral performance are not yet observable.
- Until the age of 24, BD behavior poses a larger harm than in later stages of life, because the brain areas implicated in the observed cognitive impairments (PFC, hippocampus and amygdala) are subject to development and maturation during adolescence and emerging adulthood.
- Future research should control factors that have a notorious influence on cognitive performance and that have been related to BD (level of education, consumption of other substances, personality traits, etc.) in order to reach more robust conclusions in this area.

SUMMARY POINTS

- This chapter focuses on the findings, mainly during the last decade, on the impact of BD.
- BD in the United States, Europe, and most developed countries is a major health and social concern, with significant social and personal costs.
- Mortality risk and comorbidity with physical (cardiovascular diseases, diabetes, metabolic syndrome, neurodegeneration) and mental (anxiety, depression, suicide, sleep disorders) pathologies are more prevalent in binge drinkers.
- Adolescents and young binge drinkers show neurocognitive impairments similar to those found in long-term alcohol abuse, although to a lesser extent and with more mixed results.
- The presence of high impulsivity and novelty seeking in young people are the best personality traits to predict the appearance and maintenance of the BD pattern.
- Circadian typology is a remarkable individual difference in BD and all addictive behaviors, with the morning-type being a protective factor and the evening-type being a risk factor.
- We need more systematic research on the effects of BD, preferably longitudinal studies incorporating biological and behavioral measures.
- Prevention and therapeutic interventions should give priority to BD, promoting protective factors and reducing risk factors known to affect its appearance and maintenance.

ACKNOWLEDGMENT

This work was supported by a grant from the Spanish Ministry of Economy and Competitiveness (PSI2012-32669).

REFERENCES

Adan, A. (2012). Functional and dysfunctional impulsivity in young binge drinkers. *Adicciones, 24*, 17–22.

Adan, A., Archer, S. N., Hidalgo, M. P., Di Milia, L., Natale, V., & Randler, C. (2012). Circadian typology: a comprehensive review. *Chronobiology International, 29*, 1153–1175.

Bachman, J. G., Johnston, L. D., & O'Malley, P. M. (1981). Smoking, drinking, and drug use among American high school students: correlates and trends, 1975–1979. *American Journal of Public Health, 71*, 59–69.

Backhans, M. C., Lundin, A., & Hemmingsson, T. (2012). Binge drinking – a predictor for or a consequence of unemployment? *Alcoholism: Clinical and Experimental Research, 36*, 1983–1990.

Balodis, I. M., Potenza, M. N., & Olmstead, M. C. (2009). Binge drinking in undergraduates: relationships with sex, drinking behaviours, impulsivity, and the perceived effects of alcohol. *Behavioral Pharmacology, 20*, 518–526.

Balogun, O., Koyanagi, A., Stickley, A., Gilmour, S., & Shibuya, K. (2014). Alcohol consumption and psychological distress in adolescents: a multi-country study. *Journal of Adolescent Health, 54*, 228–234.

Barnet, N. P., Clerkin, E. M., Wood, M., Monti, P. M., O'Leary Tewyan, T., Corriveau, D., ... Kahler, W. K. (2014). Description and predictors of positive and negative alcohol-related consequences in the first year of college. *Journal of Studies on Alcohol and Drugs, 75*, 103–114.

Briones, T. L., & Woods, J. (2013). Chronic binge-like alcohol consumption in adolescence causes depression-like symptoms possible mediated by the effects of BDNF on neurogenesis. *Neurocience, 254*, 324–334.

Campanella, S., Peigneux, P., Petit, G., Lallemand, F., Saeremans, M., Noël, X., ... Verbanck, P. (2013). Increased cortical activity in binge drinkers during working memory task: a preliminary assessment through a functional Magnetic Resonance Imaging. *PLoS One, 8*(4), e62260. http://dx.doi.org/10.1371/journal.pone.0062260.

Castellanos-Ryan, N., Rubia, K., & Conrod, P. J. (2011). Response inhibition and reward response bias mediate the predictive relationships between impulsivity and sensation seeking and common and unique variance in conduct disorder and substance misuse. *Alcoholism: Clinical and Experimental Research, 35*, 140–155.

Chavez, P. R., Nelson, D. E., Naimi, T. S., & Brewer, R. D. (2011). Impact of the new gender-specific definition for binge drinking on prevalence estimates for women. *American Journal of Preventive Medicine, 40*, 468–471.

Cheng, H., & Furnham, A. (2013). Correlates of adult binge drinking: evidence from a British cohort. *PLoS One, 8*(11), e78838. http://dx.doi.org/10.1371/journal.pone.007838.

Clark, A., Tran, C., Weiss, A., Caselli, G., Nikčević, A. V., & Spada, M. M. (2012). Personality and alcohol metacognitions as predictors of weekly levels of alcohol use in binge drinking university students. *Addictive Behaviors, 37*, 537–540.

Cozzoli, D. K., Courson, J., Wroten, M. G., Greentree, D. I., Lum, E. M., Campbell, R. N., ... Szumlinski, K. K. (2014). Binge alcohol drinking by mice requires intact group I metabotropic glutamate receptor signalling within the central nucleus of the amygdale. *Neuropsychopharmacology, 39*, 435–444.

Crego, A., Rodríguez-Holguín, S., Parada, M., Mota, N., Corral, M., & Cadaveira, F. (2010). Reduced anterior prefrontal cortex activation in young binge drinkers during a visual working memory task. *Drug and Alcohol Dependence, 109*, 45–56.

Daeppen, J.-B., Bertholet, N., Gaume, J., Fortini, C., Faouzi, M., & Gmel, G. (2011). Efficacy of brief motivational intervention in reducing binge drinking in young men: a randomized controlled trial. *Drug and Alcohol Dependence, 113*, 69–75.

Dager, A. D., Anderson, B. M., Rosen, R., Khadka, S., Sawyer, B., Jiantonio-Kelly, R. E., ... Pearlson, G. D. (2013). Functional magnetic resonance imaging (fMRI) response to alcohol pictures predicts subsequent transition to heavy drinking in college students. *Addiction, 109*, 585–595.

Dickman, S. J. (1990). Functional and dysfunctional impulsivity: personality and cognitive correlates. *Journal of Personality and Social Psychology, 58*, 95–102.

Dolan, S. L., Bechara, A., & Nathan, P. E. (2009). Executive dysfunction as a risk marker for substance abuse: the role of impulsive personality traits. *Behavioral Sciences and the Law, 26*, 799–822.

Field, A. E., Sonneville, K. R., Crosby, R. D., Swanson, S. A., Eddy, K. T., Camargo, C. A., ... Micali, N. (2013). Prospective associations of concerns about physique and the development of obesity, binge drinking, and drug use among adolescent boys and young adult men. *JAMA Pediatrics, 168*, 34–39.

Field, M., Schoenmakers, T., & Wiers, R. W. (2008). Cognitive processes in alcohol binges: a review and research agenda. *Current Drug Abuse Reviews, 1*, 263–279.

Goslawski, M., Piano, M. R., Bian, J.-T., Church, E. C., Szczurek, M., & Phillips, S. A. (2013). Binge drinking impairs vascular function in young adults. *Journal of the American College of Cardiology, 62*, 201–207.

Goudriaan, A. E., Grekin, E. R., & Sher, K. J. (2011). Decision making and response inhibition as predictor of heavy alcohol use: a prospective study. *Alcoholism: Clinical and Experimental Research, 35*, 1–8.

Heffernann, T., Clark, R., Bartholomew, J., Ling, J., & Stephens, S. (2010). Does binge drinking in teenagers affect their everyday prospective memory. *Drug and Alcohol Dependence, 109*, 73–78.

Heffernann, T., & O'Neill, T. (2012). Time based prospective memory deficits associated with binge drinking: evidence from the Cambridge Prospective Memory Test. *Drug and Alcohol Dependence, 123*, 207–212.

Hermens, D. F., Lagopoulos, J., Tobías-Webb, J., Tamara De Regt, T., Dore, G., Juckes, L., ... Hickie, I. B. (2013). Pathways to alcohol-induced brain impairment in young people: a review. *Cortex, 49*, 3–17.

Hibell, B., Guttormsson, U., Ahlström, S., Balakireva, O., Bjarnason, T., Kokkevi, A., & Kraus, L. (2009). *The 2007 ESPAD report. Substance use among students in 35 European countries*. Stockholm: The Swedish Council for Information on Alcohol and Other Drugs.

Howell, N. A., Worbe, Y., Lange, I., Tait, R., Irvine, M., Banca, P., ... Voon, V. (2013). Increased ventral striatal volume in college-aged binge drinkers. *PLoS One, 8*(9), e74164. http://dx.doi.org/10.1371/journal.pone.0074164.

International Center for Alcohol Policies (ICAP). (2010). *International drinking guidelines*. http://www.icap.org/Table/InternationalDrinkingGuidelines.

Jacobus, J., Squeglia, L. M., Bava, S., & Tapert, S. F. (2013). White matter characterization of adolescent binge drinking with and without co-occurring marijuana use: a 3-year investigation. *Psychiatry Research: Neuroimaging, 214*, 374–381.

Johnson, C. A., Xiao, L., Palmer, P., Sun, P., Wang, Q., Wei, Y., ... Bechara, A. (2008). Affective decision-making deficits, linked to dysfunctional ventromedial prefrontal cortex, revealed in 10th grade Chinese adolescent binge drinkers. *Neuropsychologia, 46*, 714–726.

Kaur, S., Li, J., Stenzel-Poore, M. P., & Ryabynin, A. E. (2012). Corticotropin-releasing factor acting on corticotropin-releasing factor receptor type I is critical for binger alcohol drinking in mice. *Alcoholism: Clinical and Experimental Research, 36*, 369–376.

Legrand, F. D., Gomà-i-Freixanet, M., Kaltenbach, M. L., & Joli, P. M. (2007). Association between sensation seeking and alcohol consumption in French college students: some ecological data collected in "open bar" parties. *Personality and Individual Differences, 43*, 1950–1959.

Lindtner, C., Scherer, T., Zielinski, E., Filatova, N., Fasshauer, M., Tonks, N. K., … Buettner, C. (2013). Binge drinking induces whole-body insulin resistance by impairing hypothalamic insulin action. *Science Translational Medicine, 5*, 170ra14 http://dx.doi.org/10.1126/scitranslmed.3005123.

Marquez-Arrico, J. E., & Adan, A. (2013). Dual diagnosis and personality traits: current situation and future research directions. *Adicciones, 25*(13), 195–202.

McCarthy, D. M., Niculete, M. E., Treloar, H. R., Morris, D. H., & Bartholow, B. D. (2012). Acute alcohol effects on impulsivity: associations with drinking and driving behavior. *Addiction, 107*, 2109–2114.

McKinney, A., Coyle, K., & Verster, J. (2012). Direct comparison of the cognitive effects of acute alcohol with the morning after a normal night's drinking. *Human Psychopharmacology: Clinical and Experimental Research, 27*, 295–304.

Modecki, K. L., Barber, B. L., & Eccles, J. S. (2014). Binge drinking trajectories across adolescents: for early maturing youth, extra-curricular activities are protective. *Journal of Adolescent Health, 54*, 61–66.

Montgomery, C., Fisk, J. E., Murphy, P. N., Ryland, I., & Hilton, J. (2012). The effects of heavy social drinking on executive function: a systematic review and meta-analytic study of existing literature and new empirical findings. *Human Psychopharmacology: Clinical and Experimental Research, 27*, 187–199.

Naimi, T. S., Blachette, J., Nelson, T. F., Nguyen, T., Oussayef, N., Heeren, T. C., … Xuan, Z. (2014). A new scale of the U.S. alcohol policy environment and its relationship to binge drinking. *American Journal of Preventive Medicine, 46*, 10–16.

National Institute on Alcohol Abuse and Alcoholism. (2000). *NIAAA: 10th special report to the US congress on alcohol and health*. Bethesda: National Institutes of Health.

National Institute on Alcohol Abuse and Alcoholism. (2004). *NIAAA council approves definition of binge drinking*. NIAAA Newsletter. No 3. Winter.

Observatorio Español sobre Drogas. Delegación del Gobierno para el Plan Nacional sobre Drogas. (2013). *Encuesta domiciliaria sobre alcohol y drogas de España (EDADES). 2011/2012*. Disponible en: http://www.pnsd.msc.es/Categoria2/observa/pdf/EDADES2011.pdf.

Parada, M., Corral, M., Caamaño-Isorna, F., Mota, N., Crego, A., Rodríguez-Olguín, S., & Cadaveira, F. (2011a). Definición del concepto de consumo intensivo de alcohol adolescente (binge drinking). *Adicciones, 23*, 53–63.

Parada, M., Corral, M., Caamaño-Isorna, F., Mota, N., Crego, A., Rodríguez-Olguín, S., & Cadaveira, F. (2011b). Binge drinking and declarative memory in university students. *Alcoholism, Clinical and Experimental Research, 35*, 1–10.

Parada, M., Corral, M., Mota, N., Crego, A., Rodríguez-Olguín, S., & Cadaveira, F. (2012). Executive functioning and alcohol binge drinking in university students. *Addictive Behaviors, 37*, 167–172.

Patrick, M. E., Schulenberg, J. E., Martz, M. E., Maggs, J. L., O'Malley, P. M., & Johnston, L. D. (2013). Extreme binge drinking among 12th-grade students in the United States. *JAMA Pediatrics, 167*, 1019–1025.

Pedersen, W., & von Soest, T. (2013). Socialization to binge drinking: a population-based, longitudinal study with emphasis on parental influences. *Drug and Alcohol Dependence, 133*, 587–592.

Perry, J. L., & Carroll, M. E. (2008). The role of impulsive behavior in drug abuse. *Psychopharmacology, 200*, 1–26.

Plunk, A. D., Syed-Mohammed, H., Cavazos-Rehg, P., Bierut, L. J., & Grucza, R. A. (2014). Alcohol consumption, heavy drinking, and mortality: rethinking the J-shaped curve. *Alcoholism: Clinical and Experimental Research, 38*, 471–478.

Popovici, I., & French, M. T. (2013). Binge drinking and sleep problems among young adults. *Drug and Alcohol Dependence, 132*, 207–215.

Prat, G., & Adan, A. (2011). Influence of circadian typology on drug consume, hazardous alcohol use and hangover symptoms. *Chronobiology International, 28*, 248–257.

Saunders, J. B., Aasland, G., Amundsen, A., & Grant, M. (1993). Alcohol consumption and related problems among primary health care patients: WHO collaborative project on early detection of person with harmful alcohol consumption I. *Addiction, 88*, 349–362.

Scaife, J. C., & Duka, T. (2009). Behavioural measures of frontal lobe function in a population of young social drinkers with binge drinking pattern. *Pharmacology, Biochemistry and Behavior, 93*, 354–362.

Stephens, D. N., & Duka, T. (2008). Cognitive and emotional consequences of binge drinking: role of amygdala and prefrontal cortex. *Philosophical Transactions of the Royal Society of Biological Sciences, 363*, 3169–3179.

Stickley, A., Koyanagi, A., Koposov, R., Razvodovsky, Y., & Ruchkin, V. (2013). Adolescent binge drinking and risky health behaviours: findings from northern Russia. *Drug and Alcohol Dependence, 133*, 838–844.

Thiele, T. E. (2012). Commentary: studies on binge-like ethanol drinking may help to identify the neurobiological mechanisms underlying the transition to dependence. *Alcohol Clinical and Experimental Research, 36*, 193–196.

Townshend, J. M., Kambouropoulos, N., Griffin, A., Hunt, F. J., & Milani, R. M. (2014). Binge drinking, reflection impulsivity, and unplanned sexual behavior: Impaired decision-making in young social drinkers. *Alcoholism: Clinical and Experimental Research, 30*, 1143–1150.

Verster, J. C., van Duin, D., Volkerts, E. R., Schreuder, A., & Verbaten, M. N. (2003). Alcohol hangover effects on memory functioning and vigilance performance after an evening of binge drinking. *Neuropsychopharmacology, 28*, 740–746.

Vetreno, R. P., & Crews, F. T. (2012). Adolescent binge drinking increases expression of the danger signal receptor agonist HMGB1 and Toll-like receptors in the adult prefrontal cortex. *Neuroscience, 226*, 475–488.

Viner, R. M., & Taylor, B. (2007). Adult outcomes of binge drinking in adolescence: findings from a UK national birth cohort. *Journal of Epidemiological Community Health, 61*, 902–907.

Wechsler, H., Davenport, A., Dowdall, G., Moeykens, B., & Castillo, S. (1994). Health and behavioral consequences of binge drinking in college. A national survey of students at 140 campuses. *Journal of the American Medical Association, 272*, 1672–1677.

Weitzman, E. R., Nelson, T. F., & Wechsler, H. (2003). Taking up binge drinking in college: the influences of person, social group, and environmental. *Journal of Adolescent Health, 32*, 26–35.

Wellman, R. J., Contreras, G. A., Dugas, E. N., O'Loughlin, E. K., & O'Loughlin, J. L. (2014). Determinants of sustained binge drinking in young adults. *Alcoholism: Clinical and Experimental Research, 38*, 1409–1415.

White, H. R., Marmorstein, N. R., Crews, F. T., Bates, M. E., Mun, E.-Y., & Loeber, R. (2011). Associations between heavy episodic drinking and changes in impulsive behavior among adolescent boys. *Alcoholism: Clinical and Experimental Research, 35*, 295–303.

Xing, Y., Ji, C., & Zhang, L. (2006). Relationship of binge drinking and other health-compromising behaviors among urban adolescents in China. *Journal of Adolescent Health, 39*, 495–500.

Zwaluw, C. S., Kleinjan, M., Lemmers, L., Spijkerman, R., & Engels, R. C. M. E. (2013). Longitudinal associations between attitudes towards binge drinking and alcohol free-drinks, and binge drinking behaviors in adolescents. *Addictive Behaviors, 38*, 2110–2114.

Section B

Molecular and Cellular Aspects

Chapter 37

Alcohol and Endogenous Opioids

Sara Palm, Ingrid Nylander
Neuropharmacology, Addiction and Behaviour, Department of Pharmaceutical Biosciences, Uppsala University, Uppsala, Sweden

Abbreviations

6-OHDA 6-Hydroxydopamine
ACTH Adrenocorticotropic hormone
AUD Alcohol use disorder
BEND Beta-endorphin
CNS Central nervous system
CLIP Corticotropin-like intermediate lobe peptide
CRF Corticotropin-releasing factor
DYN Dynorphin
ENK Enkephalin
GABA Gamma-aminobutyric acid
HPA Hypothalamic-pituitary-adrenal
MEAP Met-enkephalin-Arg^6-Phe^7
MSH Melanocyte-stimulating hormone
NMDA N-Methyl-D-aspartate
PDYN Prodynorphin
PENK Proenkephalin
POMC Proopiomelanocortin
SNP Single-nucleotide polymorphism
VTA Ventral tegmental area

Alcohol ingestion induces a number of effects, ranging from stimulant to sedative. Its reported subjective effects are also largely dependent on the individual. The multitude of actions induced by alcohol mirror the complex and not yet fully elucidated mechanism of action in the central nervous system (CNS). A wide range of targets has been proposed and alcohol has been shown to affect many of the neurotransmitter systems, such as dopamine, gamma-aminobutyric acid (GABA), glutamate, and acetylcholine (Söderpalm & Ericson, 2013). The neurobiological mechanisms underlying alcohol's addictive properties remain unclear, but long-term actions on synaptic transmission and plasticity in a number of neuronal networks are implicated (Vengeliene, Bilbao, Molander, & Spanagel, 2008). This review will focus on alcohol-induced effects on neuropeptides and the role of neuropeptides in alcohol use disorder (AUD).

NEUROPEPTIDES AS TARGET SYSTEM

There are a number of different proteins associated with peptide transmission, and they are all putative targets for alcohol. Neuropeptides are often co-localized with classical transmitters that are present in small vesicles; importantly, they differ from classical transmitters in how they are synthesized, stored, and released. Peptides are synthesized in the neuronal cell soma from precursor proteins and are then transported within the soma to dendrites or down the axon (Figure 1). The precursors are cleaved enzymatically into one or several bioactive peptides and this processing, as well as other modifications such as acetylation, occurs inside the large dense core vesicles. After release, the peptides act on specific receptor proteins that are localized postsynaptically on the presynaptic terminal or on other adjacent cells. Neuropeptides have also been shown to diffuse to reach cells more distant to the release site, which can explain the mismatch between expression of precursors and receptors. In addition, neuropeptides can be secreted directly into the blood circulation and act as hormones. A number of enzymes participate in peptide turnover; processing enzymes cleave the precursor protein into active peptides, converting enzymes participate in the conversion of one bioactive peptide into another peptide (that may have other effects through actions on another receptor), and finally, enzymes degrade peptides into inactive fragments. The processing of a particular precursor is tissue specific, and the peptide released from a neuron will be dependent on the available enzymes. As a consequence, neurons can change the content of peptides as an adaptation to new stimuli. These actions imply plastic systems that can change "on demand" and are well in line with the modulatory effects of neuropeptides in the CNS.

Alcohol may cause altered activity in the genes encoding the precursor proteins, the receptors, and/or the enzymes. Changes in any of these proteins may contribute to altered activity in neuronal circuits that utilize peptides as transmitters and in networks that are modulated by peptides. Studies of the involvement of neuropeptides in the development of AUD are therefore complex and encompass mRNA, peptide, enzyme, and receptor levels and functions as well as epigenetic mechanisms governing gene expression. Indeed, numerous studies have reported acute and/or chronic alcohol-induced effects on neuropeptide systems and peptide receptor agonists and antagonists have also been shown to modulate alcohol consumption in both humans and experimental animals. Furthermore, neuropeptides have been implicated in the development of AUD. A number of peptides are of interest, such as neuropeptide Y, corticotropin-releasing factor (CRF), substance P, ghrelin, and endogenous opioids (Söderpalm & Ericson, 2013; Vengeliene et al., 2008). In this chapter, the focus will be on endogenous opioid peptides.

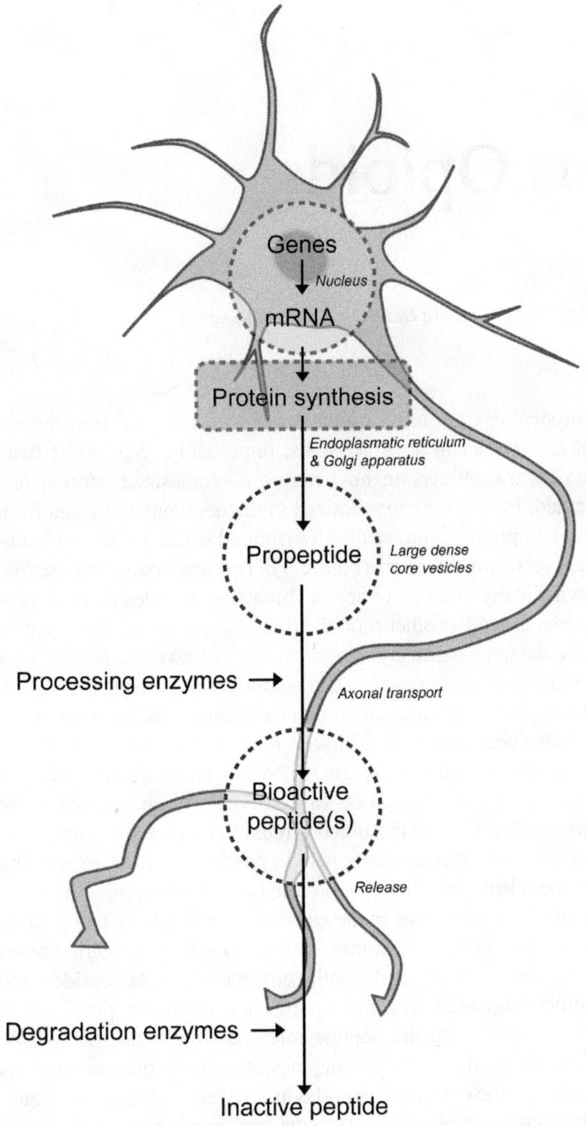

FIGURE 1 Neuropeptide synthesis, release, and degradation. Neuropeptides are synthesized in the neuronal cell soma from precursor proteins. The propeptides are processed enzymatically into one or several bioactive peptides inside large dense core vesicles. The large dense core vesicles are transported down the axon. Release, or secretion, can occur at the terminals, axons, dendrites, or cell soma. After release, the peptides are enzymatically degraded into inactive fragments.

THE ENDOGENOUS OPIOID SYSTEM

History

The pharmacological effects noted for morphine and similar substances started speculations about the presence of specific receptors for these substances. Evidence for an existing endogenous opioid system came in the late 1960s and early 1970s; in 1973, three independent research groups (led by Snyder, Simon, and Terenius, respectively) demonstrated the existence of stereospecific binding sites for opioids (Snyder & Pasternak, 2003). It was later found that there were different types of opioid receptors; they were termed mu-, delta-, and kappa-opioid receptors. Parallel studies also showed that opioid-like activity could be found in brain extracts, indicating that there were endogenous ligands for these receptors. The endogenous ligands were named endorphin (from "endogenous morphine"), enkephalin (from the Greek "in the head") and dynorphin (from the Greek word *dynamis*, meaning "power").

Later, the nociceptin/orphanin FQ receptor, previously referred to as the opioid receptor-like orphan receptor, was identified together with its ligand nociceptin/orphanin FQ. The nociceptin receptor was included among the opioid receptors due to the high sequence homology (Stevens, 2009). However, many opioid ligands lack affinity for the nociceptin receptor, and it is clear that the nociceptin/orphanin FQ receptor is distinct from mu-, delta- and kappa-receptors.

Furthermore, endomorphins have been identified as selective mu-receptor agonists; to date there are at least 10 endogenous opioids, from different families of compounds, that have been identified. The more recently discovered endogenous ligands differ from the three classical ligands in several ways; for instance, they do not have the amino acid sequence Tyr-Gly-Gly-Phe at the amino terminal (see below).

The endogenous opioid peptides and receptors are widely distributed in the brain and are involved in regulation of a number of physiological processes. The compiled data in the literature provide strong evidence for opioid involvement in natural reinforcement, in drug-induced reward, and also in adaptive processes induced by long-term drug intake (Le Merrer, Becker, Befort, & Kieffer, 2009).

Opioid Precursor Proteins

The endogenous opioid peptides are derived from the precursor proteins proopiomelanocortin, proenkephalin A, and prodynorphin (proenkephalin B), which are encoded by the genes proopiomelanocortin (*POMC*), proenkephalin (*PENK*), and prodynorphin (*PDYN*), respectively (Figure 2). Proopiomelanocortin, which gives rise to beta-endorphin and several nonopioid peptides (Figure 2(A)), is synthesized in the anterior lobe of the pituitary gland, arcuate nucleus of the hypothalamus, and nucleus tractus solitarius. Proenkephalin synthesis is widespread in the CNS and the protein gives rise to multiple enkephalins, including Leu-enkephalin, Met-enkephalin, Met-enkephalin-Arg^6-Phe^7, and Met-enkephalin-Arg^{6}Gly^{7}Leu^8 (Figure 2(B)). Prodynorphin synthesis is also widespread and generates dynorphin A, dynorphin B, neoendorphin, and Leu-enkephalin (Figure 2(C)).

The endogenous ligands have different amino acid sequences and different affinities for the receptors (Table 1), but the amino acid sequence Tyr-Gly-Gly-Phe at the amino terminal of the peptides is important for their binding to opioid receptors and is common to all three classes of endogenous opioid peptides (Le Merrer et al., 2009).

Opioid Receptors

Beta-endorphin has equal affinity for the mu- and delta-receptors, while the enkephalins are more selective for the delta-receptor and the dynorphins have higher affinity for the kappa-receptor (Yaksh & Wallace, 2011). Through autoradiography and in situ hybridization, receptor binding and expression have been mapped in different species and found to be widespread throughout the CNS. All three receptor types are highly abundant within the limbic regions and can, for

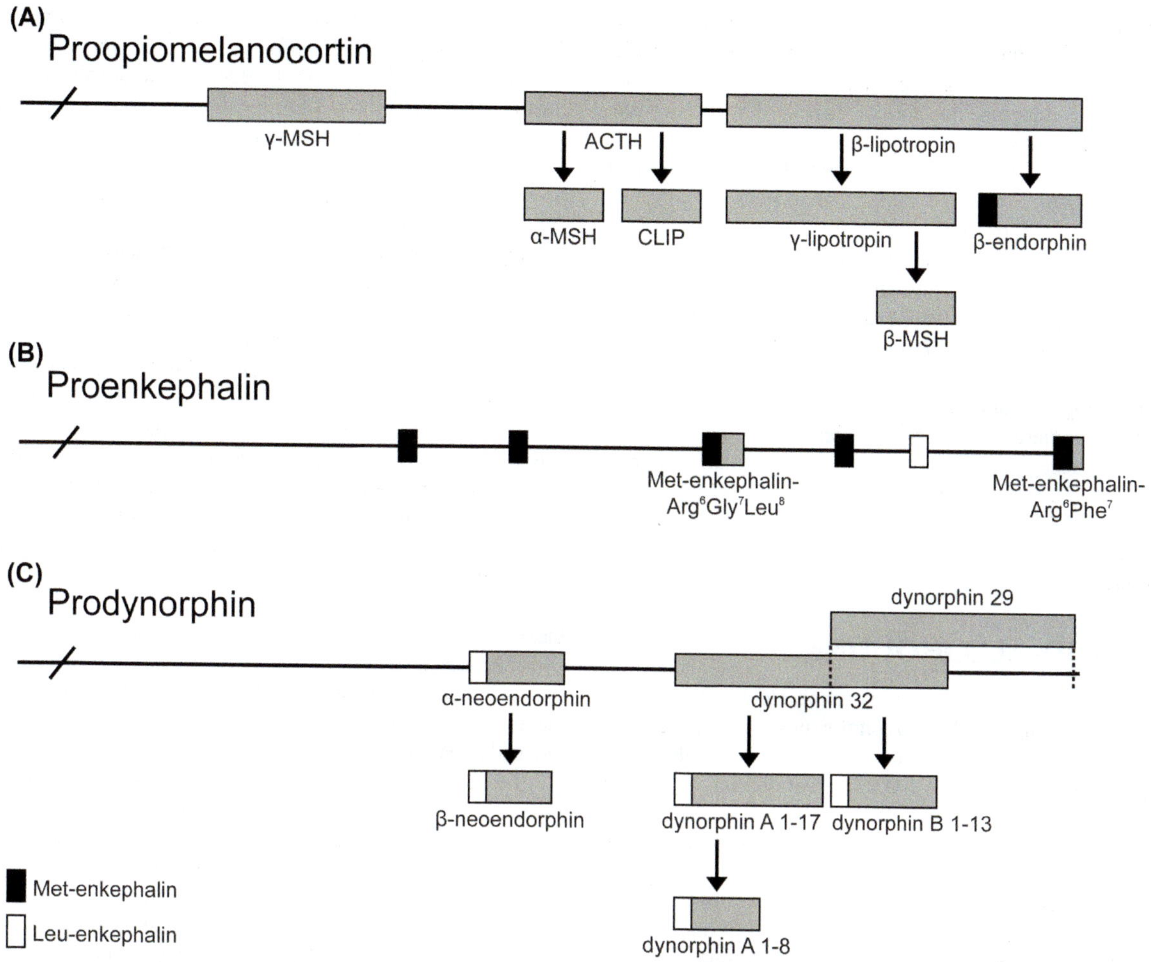

(A) Proopiomelanocortin

γ-MSH | ACTH | β-lipotropin
α-MSH | CLIP | γ-lipotropin | β-endorphin
β-MSH

(B) Proenkephalin

Met-enkephalin-Arg⁶Gly⁷Leu⁸ — Met-enkephalin-Arg⁶Phe⁷

(C) Prodynorphin

dynorphin 29
α-neoendorphin | dynorphin 32
β-neoendorphin | dynorphin A 1-17 | dynorphin B 1-13
dynorphin A 1-8

■ Met-enkephalin
□ Leu-enkephalin

FIGURE 2 **The endogenous opioid peptide families.** The precursors for the endogenous opioids and related products: (A) proopiomelanocortin, (B) proenkephalin, and (C) prodynorphin. Met-enkephalin, illustrated with black boxes, can be found within beta-endorphin and the octa and hepta Met-enkephalin peptides. Leu-enkephalin, illustrated with white boxes, can be found within many of the products of prodynorphin. ACTH, adrenocorticotropic hormone; CLIP, corticotropin-like intermediate lobe peptide; MSH, melanocyte-stimulating hormone.

TABLE 1 Amino Acid Sequences for Some of the Endogenous Opioid Peptides and Their Relative Receptor Affinities

Endogenous Peptide	Amino Acid Sequence	Receptor Affinity		
		Mu	Delta	Kappa
Beta-endorphin (rat)	Tyr-Gly-Gly-Phe-Met-Thr-Ser-Glu-Lys-Ser-Gln-Thr-Pro-Leu-Val-Thr-Leu-Phe-Lys-Asn-Ala-Ile-Ile-Lys-Asn-Ala-Tyr-Lys-Lys-Gly-Glu	+++	+++	
Met-enkephalin	Tyr-Gly-Gly-Phe-Met	++	+++	
Met-enkephalin-Arg⁶-Phe⁷	Tyr-Gly-Gly-Phe-Met-Arg-Phe	+++	+++	+++
Leu-enkephalin	Tyr-Gly-Gly-Phe-Leu	++	+++	
Dynorphin A	Tyr-Gly-Gly-Phe-Leu-Arg-Arg-Ile-Arg-Pro-Lys-Leu-Lys-Trp-Asp-Asn-Gln	++		+++
Dynorphin B	Tyr-Gly-Gly-Phe-Leu-Arg-Arg-Gln-Phe-Lys-Val-Val-Thr	+	+	+++

example, be found in high concentrations in the dorsal striatum, nucleus accumbens, amygdala, and hippocampus (Le Merrer et al., 2009). The various cell populations within these regions express different receptors and may thus be differentially regulated by the opioid peptides.

The opioid receptors are all G-protein coupled receptors that inhibit the production of cyclic AMP through adenylate cyclase and decrease neuronal excitability. Consequently, endogenous opioids commonly reduce transmitter release; however, through indirect, disinhibitory actions, the end result may be increased transmitter release. Mu- and delta-receptors generally mediate the positive, reinforcing effects in the limbic system, while kappa-receptors mediate negative, aversive effects. Mu-receptor agonists, such as morphine and heroin, are rewarding and can produce euphoric effects, while kappa-agonists produce dysphoria (Le Merrer et al., 2009). The involvement of the delta-receptor in the reward system is more complex, but agonists are generally associated with pleasurable effects.

Several subtypes of opioid receptors have been proposed based on pharmacological studies and this has led to some confusion over the terminology in the literature. Herein, the term "type" refers to the different receptor proteins arising from the different genes and "subtypes" refers to differences in binding based on pharmacological studies. This is in contrast to the terminology used for adrenergic receptors, for example, where subtypes of the adrenergic receptor, alpha 1 and alpha 2, are actually different gene products. Subtypes of opioid receptors may arise from alternative splicing of receptor mRNA (Stevens, 2009) and there are also reports of heterodimers of different types of opioid receptors (Stockton & Devi, 2012).

THE OPIOID–ALCOHOL LINK

Over the years, close interactions between alcohol and opioid networks have been described; alcohol exposure affects endogenous opioids and, vice versa, opioids modulate alcohol consumption (Oswald & Wand, 2004). However, the results are not congruent and the exact nature of these interactions remains to be elucidated. There are several reasons for the incoherent picture. For example, alcohol-induced effects may be different depending on the route of administration, timing of measurement, age, gender, and animal strain used. Furthermore, differences have been found even between animals from different suppliers (Palm, Roman, & Nylander, 2012). Generally, effects after voluntary alcohol consumption are not as well studied as after forced administration, but the effects may differ between individuals who are inclined to drink versus individuals who are not, such as between animals that are alcohol-preferring and nonpreferring. Individual differences in alcohol-induced effects are common and may involve different responses in opioid networks. For example, differences have been attributed to genotype and early life environment; see later sections on individual variation in vulnerability and response to treatment.

Acute Alcohol-Induced Effects on Opioids

Drugs of abuse commonly act on the reward system and increase extracellular levels of dopamine in the nucleus accumbens acutely after intake, and this is true for alcohol as well (Imperato & Di Chiara, 1986). The mechanisms behind the dopamine-releasing actions of alcohol are however not yet fully understood, but several systems are likely interacting. Opioids have been implicated in these mechanisms because opioid antagonists attenuate alcohol-induced dopamine release (Gonzales & Weiss, 1998), as described later.

Alcohol has been shown to increase the levels of beta-endorphin, enkephalin, and proenkephalin mRNA in the rat midbrain shortly after injection or gavage (Jarjour, Bai, & Gianoulakis, 2009; Mendez & Morales-Mulia, 2008). The peptides can bind to mu-opioid receptors on GABA interneurons in the ventral tegmental area (VTA) and inhibit the release of GABA upon the dopaminergic neurons (Trigo, Martin-Garcia, Berrendero, Robledo, & Maldonado, 2010). Because the dopaminergic neurons in the VTA are under tonic control by the GABA interneurons, this results in disinhibition of the dopaminergic neurons, which subsequently releases dopamine in the nucleus accumbens (Figure 3). It has also been shown that mu- and delta-opioid receptors facilitate dopamine release locally in the nucleus accumbens as well (Hirose et al., 2005), which could be another access point for alcohol–opioid interactions (Figure 3). However, the timings of peptide release are not entirely in line with timing of the dopamine increase, which in some studies have been shown to increase earlier than enkephalin and beta-endorphin. This suggests that the opioids may have a modulatory role in alcohol-induced dopamine release, but that the initial increase in dopamine may be due to actions on other targets. Studies provide evidence for $GABA_A$-, N-methyl-D-aspartate (NMDA)-, nicotine-, glycine-, and serotonin-receptors as primary targets for alcohol (Vengeliene et al., 2008). In support of this is a more detailed study of the timing of naltrexone's attenuating effects on dopamine release, which revealed that naltrexone (injected intravenously or locally into the VTA) did not prevent the initiation of release in the nucleus accumbens after alcohol but rather prevented the delayed alcohol-induced release (Valenta et al., 2013). This effect is different from the immediate attenuating effects of naltrexone on morphine-induced dopamine release.

However, there are also studies showing that activation of the mesolimbic dopamine pathway is not necessarily critical for the acute reinforcing effects of alcohol (Rassnick, Stinus, & Koob, 1993). Lesions of the dopaminergic pathway between the VTA and nucleus accumbens by 6-hydroxydopamine (6-OHDA) infusions in experimental animals led to depletion of dopamine in the nucleus accumbens. Despite the lack of dopamine, the animals continued to self-administer alcohol or heroin, which suggests that drugs that can activate opioid receptors may have other reinforcing mechanism beyond dopamine release in the nucleus accumbens.

Opioid levels in other brain regions are also acutely affected by alcohol. For example, proenkephalin mRNA and beta-endorphin levels increased in rat amygdala shortly after gavage administration of alcohol (Mendez & Morales-Mulia, 2008; Oliva, Ortiz, Pérez-Rial, & Manzanares, 2008).

Acute effects after voluntary alcohol drinking in animals are scarce, but there are studies showing increased levels of dynorphin in the amygdala and striatum, and decreased levels of enkephalin in the hypothalamus, medial prefrontal cortex, and hippocampus (Palm & Nylander, 2014). These effects are somewhat different from injection or gavage studies and may be explained by the timing of measurements. Microdialysis studies have revealed increases in enkephalin at 30–120 min (Marinelli, Lam, Bai, Quirion, & Gianoulakis, 2006; Mendez, Barbosa-Luna, Perez-Luna, Cupo, & Oikawa, 2010; Seizinger, Bovermann, Maysinger, Höllt, & Herz, 1983), beta-endorphin at 60–240 min, and dynorphin at 2–3 h (Jarjour et al., 2009; Lam, Marinelli, Bai, & Gianoulakis, 2008) after injection, with exact timing depending on the brain region. When peptide measurements are done with some delay after voluntary intake, degradation of released peptides can result in low levels of enkephalin after an initial release and compensatory release of dynorphin in response to increased dopamine can account for the increases in dynorphin.

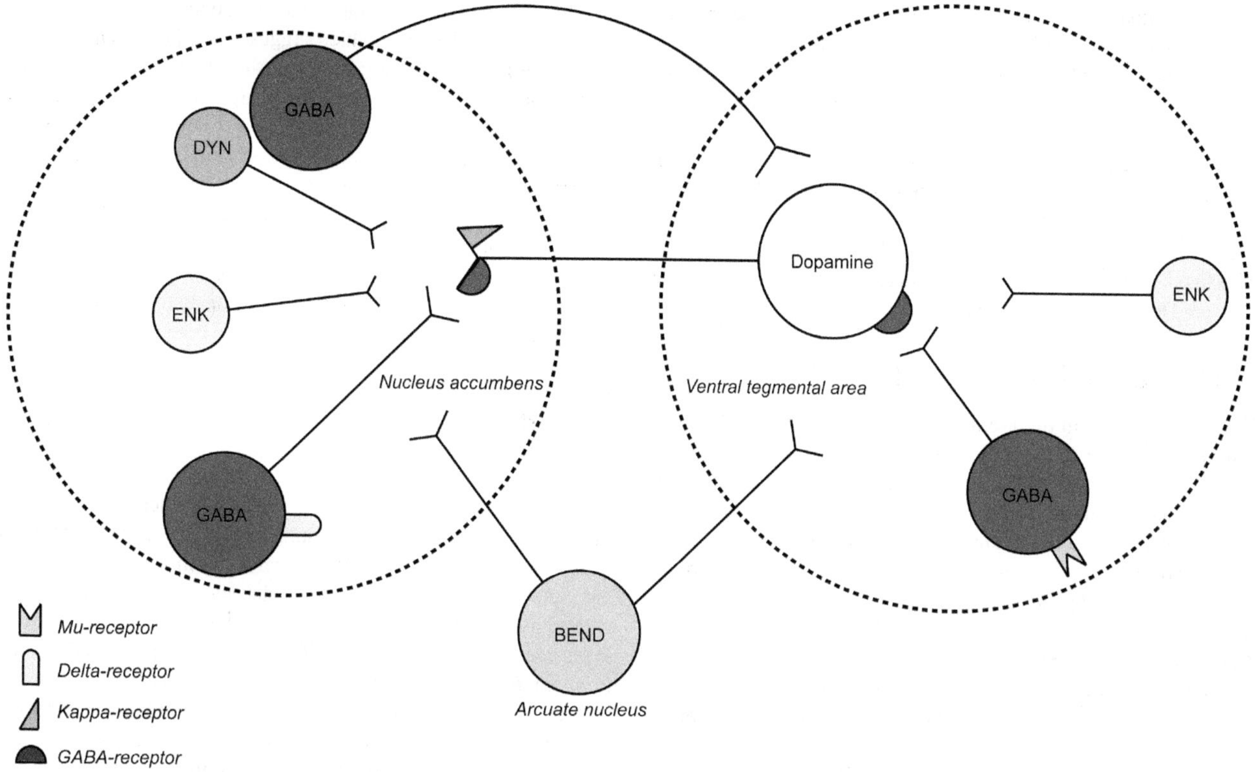

FIGURE 3 **Alcohol-induced dopamine release.** Schematic illustration of the proposed involvement of the endogenous opioids in alcohol-induced dopamine release in the nucleus accumbens. BEND, beta-endorphin; DYN, dynorphin; ENK, enkephalin; GABA, gamma-aminobutyric acid.

Chronic Effects

Chronic alcohol intake also modulates the opioid system, and these effects can be long lasting. Although many differences between studies exist and the results are far from conclusive, the general consensus is that prolonged alcohol exposure downregulates the positive aspects but upregulates the negative aspects of the endogenous opioid system. For example, prodynorphin mRNA and dynorphin B levels in the amygdala have been shown to increase after 3 or 4 weeks of voluntary alcohol intake in mice and rats (Chang et al., 2010; Ploj, Roman, Gustavsson, & Nylander, 2000). Chronic alcohol injections for 2 weeks also increase dynorphin B levels in the rat nucleus accumbens (Lindholm, Ploj, Franck, & Nylander, 2000). However, there are also studies that show increased proenkephalin mRNA, as well as enkephalin levels in the amygdala after 3–8 weeks of voluntary alcohol intake in rats (Chang et al., 2010; Cowen & Lawrence, 2001; Nylander & Roman, 2012). One study showed that delta-receptors in the amygdala become functional only after ethanol exposure, while in other areas, delta- and mu-receptors display decreased functional coupling after chronic alcohol (Trigo et al., 2010). Thus, the sensitivity to the effects of alcohol seems to be largely dependent on brain region.

Opioids Affect Alcohol Intake

Through studies of opioid receptor agonists and antagonists as well as genetic knockouts of the different receptors and precursor proteins, it has been shown that opioids can modulate alcohol intake. Most studies show that mu-opioid receptor antagonists decrease alcohol intake (Oswald & Wand, 2004) and mu-receptor knockouts drink less than their wild-type counterparts (Le Merrer et al., 2009). The effects of knocking out beta-endorphin are not as consistent as knocking out the receptor. There are studies on beta-endorphin knockouts that show increased (Grisel et al., 1999), decreased (Racz et al., 2008), or unchanged intake (Hayward, Hansen, Pintar, & Low, 2004).

Results on the involvement of the delta-receptor are also somewhat inconclusive. Delta-receptor antagonists administered systemically or locally into the VTA, dorsal striatum, or central nucleus of the amygdala decrease alcohol intake (Chu Sin Chung & Kieffer, 2013), but there are also studies showing no effect on alcohol intake (Honkanen et al., 1996). Delta-receptor agonists have been shown to stimulate alcohol self-administration, but paradoxically, delta-receptor knockouts have been shown to have an enhanced alcohol intake in comparison to wild-types. To explain these discrepancies, differences in response of different delta-receptor subtypes have been proposed, as well as the possibility of mu-/delta-receptor heterodimers. Delta-receptor knockouts also display more anxiety-like behaviors, and there is a possibility that this is what drives their increased intake. Knocking out enkephalin, on the other hand, seems to have no effect on alcohol intake (Chu Sin Chung & Kieffer, 2013).

Morphine locally injected into the nucleus accumbens or paraventricular nucleus (PVN) of the hypothalamus induced increased alcohol intake in animals. This effect has been attributed to the delta-receptors because a locally injected delta-agonist, but not a mu-agonist, increased alcohol intake (Barson et al., 2009, 2010).

Furthermore, kappa-opioid receptor agonists can reduce alcohol intake (Lindholm, Werme, Brene, & Franck, 2001). Further studies have shown that locally injected kappa-agonists reduced alcohol intake in the PVN but had no effect in the nucleus accumbens (Barson et al., 2009, 2010). Kappa-opioid receptor knockouts decrease their alcohol intake (Kovacs et al., 2005; van Rijn & Whistler, 2009), but dynorphin knockouts display increased (Femenia & Manzanares, 2012; Racz, Markert, Mauer, Stoffel-Wagner, & Zimmer, 2013), decreased (Blednov, Walker, Martinez, & Harris, 2006), or unchanged (Sperling, Gomes, Sypek, Carey, & McLaughlin, 2010) alcohol intake compared to their wild-type counterparts.

ENDOGENOUS OPIOIDS IN AUD

The compiled data in the literature clearly show that endogenous opioids are linked to the vulnerability to AUD, although many aspects of the opioid involvement in different phases of development of AUD, from the initiation of alcohol consumption to craving and relapse, are not yet understood. Early studies showed that alcohol-preferring rodents have innate basal differences in opioid networks (Jamensky & Gianoulakis, 1999; Nylander, Hyytia, Forsander, & Terenius, 1994; Ploj et al., 2000). Together with studies showing that endogenous opioids are critical for alcohol-induced rewards, a number of studies formed the basis for different hypotheses on the involvement of the endogenous opioid system in the vulnerability towards AUD (Oswald & Wand, 2004).

The opioid deficit theory hypothesizes that an inherent low basal opioid activity will reinforce alcohol consumption and lead to increased alcohol intake in order to compensate for the low activity. The opioid surfeit theory suggests that vulnerable individuals have inherent excess/surfeit opioid activity that stimulates alcohol intake, which in turn further increases opioid activity and maintains a high alcohol intake. A third hypothesis is that propensity for high alcohol intake is influenced by the individual sensitivity of the opioid system to alcohol-induced effects; for example, if an individual responds with a great activation of the opioid system upon alcohol intake, it will lead to more positive effects and subsequently to greater chances that the individual will want to experience those effects again. To support this are findings that individuals with risk for excessive alcohol consumption and AUD show greater alcohol-induced opioid activation than individuals with low risk (Gianoulakis, de Waele, & Thavundayil, 1996; Nylander & Roman, 2012).

Opioids are also implicated in adaptational processes occurring in the CNS as a response to long-term alcohol consumption. Alcohol-induced changes in neuronal networks can result in opioid dysfunction that, in turn, contributes to development of AUD. The neurobiological basis for the opponent process theory, which is used to describe adaptive processes in the brain underlying the transition into compulsive drug intake, has been suggested to involve opioids. It hypothesizes that a dysregulation of the endogenous opioid system forms during the addiction process, which leads to an upregulated dynorphin/kappa-receptor system, among other changes, and that the negative effect associated with this drives continued drinking despite little or no positive effects (Koob & Le Moal, 2008). Findings of an upregulated dynorphin/kappa-system in human alcoholics also give support for this theory (Sirohi, Bakalkin, & Walker, 2012). Furthermore, it has been shown that the kappa-receptor antagonist norbinaltorphimine selectively decreased alcohol intake in alcohol-dependent rats and not in nondependent rats, which suggests that there is an important difference in the dynorphin/kappa-system between addicted individuals and individuals with merely a high alcohol intake (Walker & Koob, 2008). An alternate explanation for these findings may be that the upregulated dynorphin/kappa-system is what underlies the vulnerability to AUD in these individuals. However, there are studies that suggest the opposite—that is, inherent high activity in the dynorphin/kappa-system may be protective against AUD (Gieryk, Ziolkowska, Solecki, Kubik, & Przewlocki, 2010). The protective effects may be caused by the ability of dynorphin peptides to reduce drug-induced reward and dopamine release.

Opioid Involvement in Individual Differences in Vulnerability to AUD

About half of the individual variations in vulnerability to AUD may be linked to genetics (Kimura & Higuchi, 2011). AUD has been associated with a number of gene variations, including opioid genes. For example, a single nucleotide polymorphism (SNP) at position 118 in the gene coding for the mu-receptor has been extensively studied (Ray et al., 2012). A substitution of adenosine to guanine (A118G) leads to an amino acid change (Asn40Asp) in the extracellular portion of the receptor. Individuals with the minor G allele have a greater hedonic response to alcohol, greater cue-elicited craving, and enhanced alcohol-induced dopamine response than individuals who are homozygous for the A allele. Whether this also means that G allele carriers have an increased risk for AUD remains controversial given the mixed evidence in studies of self-reported drinking and diagnostic status.

Environmental influence and its interaction with genetics have profound impact on vulnerability to AUD, and the early life environment plays a particularly important role. Early environmental factors may consist of psychological and physiological factors, as well as pharmacological and toxicological factors, during the pre- and postnatal period, childhood, and adolescence. The early environment can induce risk towards or protection against later adversities, such as psychiatric disorders, through interactions with genotype. In humans, early adversity is a robust predictor of a number of psychiatric disorders, including drug abuse, and there are a number of different types of early adversity, such as physical and emotional abuse, neglect, and parental drug use (Gershon, Sudheimer, Tirouvanziam, Williams, & O'Hara, 2013). Studies also show that early adversity shifts the onset of drug use to younger ages (Andersen & Teicher, 2009), which in itself is believed to be an important early environmental risk factor for later drug use and addiction (Grant et al., 2006). The underlying neurobiology behind the increased vulnerability towards AUD after early adversity and early alcohol use is as of yet poorly understood. However, several studies provide evidence for the involvement of epigenetic processes and, for example, increased DNA methylation has been reported in persons exposed to childhood trauma (Graff, Kim, Dobbin, & Tsai, 2011). In addition, emerging evidence shows that alcohol exposure can induce a number of effects on the epigenetic machinery (Shukla et al., 2008).

One hypothesis involves maladaptation of the stress response and derives from studies showing that early stressors have long-lasting effects on hypothalamic-pituitary-adrenal (HPA) axis activity (Gershon et al., 2013). For example, adults with a history of childhood abuse have flattened diurnal variability in cortisol and lower cortisol responses to stressors. However, there are also reports of increased reactivity of the HPA axis in response to stress in subjects with a history of early adversity. When functioning correctly, a stress response is activated by the release of CRF from the hypothalamus, which initiates the endocrine response to stress and subsequent cortisol release. Cortisol then inhibits further release through negative feedback mechanisms. In addition to their downstream peripheral effects, CRF and cortisol influence the limbic regions of the CNS, where they, together with central catecholamines, modulate cognitive and behavioral stress responses. Studies have shown that with increased levels of emotional or physiological stress, there is a decrease in behavioral control and an increase in impulsivity, possibly due to decreased prefrontal functioning and increased limbic-striatal responding (Sinha, 2008). Such changes are believed to promote addictive behaviors and could increase vulnerability to drug addiction and other psychiatric disorders. It has also been suggested that stress can sensitize the reward pathways to drugs of abuse.

In addition to effects on CRF, early life adversity has also been shown to affect the endogenous opioid system. The endogenous opioids are important for early social behavior and normal neuronal development, and it is not surprising that disturbances early in life can have long-term consequences. Animal studies of early environmental influence on opioid peptides have shown that particularly levels of Met-enkephalin-Arg^6-Phe^7 (MEAP) are sensitive to early life manipulations (Nylander & Roman, 2012). Animals with adverse early life experience have lower basal levels of MEAP, drink more alcohol, and show a more pronounced alcohol-induced response in MEAP levels compared to animals without adverse early life experience. In addition, studies have shown that alcohol-induced changes in beta-endorphin and dynorphin levels in the amygdala are dependent on the adolescent social environment (Figure 4) in animals given voluntary access to alcohol during adolescence (Palm & Nylander, 2014). These results show that the close connection between endogenous opioids and alcohol can be extended to a link between the early environment, alcohol, and opioids.

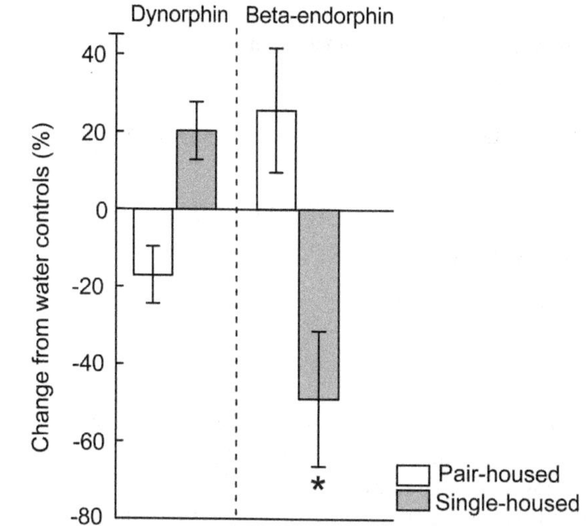

Opioids in Pharmacotherapy of AUD

The effects of opioid receptor antagonists, particularly mu-receptor antagonists, on alcohol intake have been utilized in the search for treatments of AUD (Ray et al., 2012). Two opioid receptor antagonists are currently used for treatment of AUD: naltrexone and nalmefene. Naltrexone is approved for therapeutic use in many countries worldwide. It has affinity for all three receptor types, but the effects have mainly been ascribed the mu-receptor antagonism. Nalmefene is approved for AUD treatment in Europe but not in the United States. It too has affinity for all the opioid receptors and favors the mu-receptor, but unlike naltrexone it has partial agonistic effects at the kappa-receptor.

Both drugs can reduce alcohol intake, and their effectiveness in doing so has been explained by their ability to attenuate alcohol reward through central actions on dopamine release (Gonzales & Weiss, 1998; O'Malley et al., 1992; Volpicelli, Alterman, Hayashida, & O'Brien, 1992). However, there are also studies indicating that the effectiveness of naltrexone in the treatment of AUD is due to its effects on the stress response, in addition to its

FIGURE 4 **Environmental impact on alcohol effects.** Alcohol-induced changes in dynorphin and beta-endorphin in the amygdala differ depending on social environment. The rats were pair- or single-housed during adolescence and given a choice to drink alcohol for 2 h per day, three times per week for six weeks. Levels are shown as change in percent (±SEM) from the water controls for the respective group (N = 10 per group). *$p < 0.05$ compared to water controls. *Adapted from Palm and Nylander (2014).*

attenuating effects on dopamine release. Individuals with AUD have an abnormal HPA axis, which may generate more cortisol during active drinking and withdrawal, but a blunted cortisol response following a period of abstinence (Lovallo, Dickensheets, Myers, Thomas, & Nixon, 2000). Stress and glucocorticoids are involved in reward through the mesolimbic dopamine system, where glucocorticoids can enhance dopamine release. After chronic drinking, it has been suggested that there is a development of tolerance in the HPA axis response and consequently a blunted cortisol response (Zhou & Kreek, 2014). The blunted cortisol response in abstinent patients with AUD may thus contribute to dysphoria and craving, and finally relapse during the early abstinence period. Naltrexone can disinhibit the effects of beta-endorphin on mu-receptors that tonically inhibit release of CRF. Naltrexone's ability to reduce craving may be due to the transient increase in ACTH and cortisol levels after administration, thus creating a normalized HPA axis response while at the same time blocking alcohol reward by blocking alcohol's activation of the HPA axis and central dopamine release.

Individual Differences in Response to Opioid Antagonists

Unfortunately, the effect sizes reported from clinical trials of both naltrexone and nalmefene are often small. The reasons for this may be attributed to the individual variations in neurobiology underlying AUD and involve genetic and/or environmental factors. Variations in the gene coding for the mu-receptor, in particular the A118G polymorphism described earlier in this chapter, have been related to variation in response to naltrexone (Kasai & Ikeda, 2011). Individuals with the minor G allele respond better to treatment with naltrexone. Furthermore, individuals with a family history of AUD responded with reduced total number of drinks after naltrexone as compared to controls (Krishnan-Sarin, Krystal, Shi, Pittman, & O'Malley, 2007).

In addition, environmental factors have been shown to affect naltrexone response. In an experimental study, it was shown that naltrexone treatment decreased alcohol intake only in animals with adverse early life experience; this may be explained by the more pronounced opioid activation in these animals (Daoura & Nylander, 2011).

APPLICATIONS TO OTHER ADDICTIONS AND SUBSTANCE MISUSE

The focus in this chapter is alcohol, but it is important to emphasize that endogenous opioids are not only implicated in drug-induced reward but also in natural reward (Le Merrer et al., 2009). In addition, endogenous opioids are involved in the mechanism of action of a number of drugs of abuse such as nicotine, psychostimulants, and cannabis (Trigo et al., 2010). Naturally, the endogenous opioid system is also involved in opioid addiction and the long-term changes seen in the balance between the euphoric and dysphoric effects of the endogenous opioids after alcohol can also be found after long-term use of opioids. Chronic treatment with nicotine, psychostimulants, and cannabis have been shown to have long-lasting effects on several aspects of the endogenous opioid system, such as increases in mRNA levels for the peptides and receptors. Furthermore, naltrexone blocks nicotine-, psychostimulant- and cannabis-induced dopamine overflow in the nucleus accumbens, most likely through interactions with the mu-opioid receptor. For example, nicotine's rewarding properties were blocked in knockouts lacking the mu-receptor or the proenkephalin gene. Interestingly, studies have also found that mu- and cannabinoid-1-receptors may form heterodimers that can control GABA and glutamate release in the nucleus accumbens. Furthermore, naltrexone has efficacy on smoking cessation and can attenuate self-administration and conditioned place preference to psychostimulants. Naltrexone also reduced craving and amphetamine use in a sample of amphetamine-dependent subjects (Jayaram-Lindstrom, Hammarberg, Beck, & Franck, 2008).

In conclusion, the endogenous opioid system is closely linked to natural reward as well as drug-induced reward. The modulatory role of the opioid receptors on dopamine release in the nucleus accumbens and dorsal striatum makes the endogenous opioid system important in drug addiction and addictive behavior. Furthermore, the long-term changes in the endogenous opioid system in response to many different classes of addictive drugs makes the system highly interesting, both as a neurobiological mediator of the behavioral changes that occur in the course of the addiction cycle and as a possible target for treatment.

DEFINITION OF TERMS

6-OHDA lesions 6-OHDA is a neurotoxic synthetic compound that is used to destroy dopaminergic and noradrenergic neurons. It is taken up through the reuptake transporters. In conjunction with selective noradrenaline reuptake inhibitors, it can be used to selectively destroy dopaminergic neurons.

A118G A SNP at position 118 in the gene coding for the mu-receptor. A substitution of adenosine to guanine (A118G) leads to an amino acid change (Asn40Asp) in the extracellular portion of the receptor.

Alcohol-preferring rodents These rats or mice have inherent high preference for alcohol. Many of the alcohol-preferring rat lines have been purposely bred for their high preference and often have a low-preferring counterpart.

Autoradiography The use of radiolabeled ligands to determine the tissue distribution of receptors. Ligands may be given in vivo or applied to sections in vitro.

Early life manipulations In this text, it refers to rodent experimental models of early life adversity, where pups are subjected to separation from the dam for shorter or longer periods of time during the first few weeks of life.

In situ hybridization Radiolabeled oligonucleotides or ribonucleic acids are used to determine tissue distribution of RNA transcripts.

Microdialysis In this sampling technique for continuous measurement of analyte concentrations in extracellular fluid, the microdialysis probe consists of a semipermeable membrane that surrounds inlet and outlet tubing. The probe is continuously perfused and analytes from the brain can diffuse over the semipermeable membrane and be transported to a sampler. The samples can then be analyzed further by, for example, high-performance liquid chromatography.

Voluntary alcohol intake Animals are given a free choice between alcohol and water. Alcohol percentage and time of access may be varied. Common procedures include a two-bottle free-choice paradigm with intermittent (e.g., every other day) access or access during the active (dark) period of the light/dark cycle.

KEY FACTS OF OPIOIDS

- The existence of binding sites for opioids was shown in 1973 by three independent research groups.
- There are three types of opioid receptors; mu-, delta-, and kappa-receptors.
- The classical endogenous opioids are endorphins, enkephalins, and dynorphins.
- To date, at least 10 other endogenous opioids have been identified, but they are different from the three classical endogenous opioids.
- The endogenous opioid system regulates a number of physiological processes, including reward mechanisms.
- The endogenous opioids are involved in the development of addiction.

SUMMARY POINTS

- The endogenous opioid peptides are involved in acute and chronic alcohol-induced effects and are critical for alcohol-induced rewards.
- Alcohol exposure induces a number of changes in endogenous opioid systems.
- Opioid agonists and antagonists modulate alcohol consumption.
- The mu-, delta-, and kappa-opioid receptors mediate different effects on reward and alcohol consumption.
- Opioid dysfunction may contribute to vulnerability to AUD.
- Opioid antagonists are widely used in pharmacotherapy of AUD.
- Genetic and environmental factors modify endogenous opioid function and affect vulnerability to AUD and the efficacy of treatment with opioid antagonists.
- Endogenous opioids are implicated in more general aspects of reward and addiction and are possible targets for treatment of addictive disorders.

REFERENCES

Andersen, S. L., & Teicher, M. H. (2009). Desperately driven and no brakes: developmental stress exposure and subsequent risk for substance abuse. *Neuroscience and Biobehavioral Reviews, 33*, 516–524.

Barson, J. R., Carr, A. J., Soun, J. E., Sobhani, N. C., Leibowitz, S. F., & Hoebel, B. G. (2009). Opioids in the nucleus accumbens stimulate ethanol intake. *Physiology and Behavior, 98*, 453–459.

Barson, J. R., Carr, A. J., Soun, J. E., Sobhani, N. C., Rada, P., Leibowitz, S. F., & Hoebel, B. G. (2010). Opioids in the hypothalamic paraventricular nucleus stimulate ethanol intake. *Alcoholism, Clinical and Experimental Research, 34*, 214–222.

Blednov, Y. A., Walker, D., Martinez, M., & Harris, R. A. (2006). Reduced alcohol consumption in mice lacking preprodynorphin. *Alcohol, 40*, 73–86.

Chang, G. Q., Barson, J. R., Karatayev, O., Chang, S. Y., Chen, Y. W., & Leibowitz, S. F. (2010). Effect of chronic ethanol on enkephalin in the hypothalamus and extra-hypothalamic areas. *Alcoholism, Clinical and Experimental Research, 34*, 761–770.

Chu Sin Chung, P., & Kieffer, B. L. (2013). Delta opioid receptors in brain function and diseases. *Pharmacology and Therapeutics, 140*, 112–120.

Cowen, M. S., & Lawrence, A. J. (2001). Alterations in central preproenkephalin mRNA expression after chronic free-choice ethanol consumption by fawn-hooded rats. *Alcoholism, Clinical and Experimental Research, 25*, 1126–1133.

Daoura, L., & Nylander, I. (2011). The response to naltrexone in ethanol-drinking rats depends on early environmental experiences. *Pharmacology, Biochemistry, and Behavior, 99*, 626–633.

Femenia, T., & Manzanares, J. (2012). Increased ethanol intake in prodynorphin knockout mice is associated to changes in opioid receptor function and dopamine transmission. *Addiction Biology, 17*, 322–337.

Gershon, A., Sudheimer, K., Tirouvanziam, R., Williams, L. M., & O'Hara, R. (2013). The long-term impact of early adversity on late-life psychiatric disorders. *Current Psychiatry Reports, 15*, 352.

Gianoulakis, C., de Waele, J. P., & Thavundayil, J. (1996). Implication of the endogenous opioid system in excessive ethanol consumption. *Alcohol, 13*, 19–23.

Gieryk, A., Ziolkowska, B., Solecki, W., Kubik, J., & Przewlocki, R. (2010). Forebrain PENK and PDYN gene expression levels in three inbred strains of mice and their relationship to genotype-dependent morphine reward sensitivity. *Psychopharmacology (Berlin), 208*, 291–300.

Gonzales, R. A., & Weiss, F. (1998). Suppression of ethanol-reinforced behavior by naltrexone is associated with attenuation of the ethanol-induced increase in dialysate dopamine levels in the nucleus accumbens. *Journal of Neuroscience, 18*, 10663–10671.

Graff, J., Kim, D., Dobbin, M. M., & Tsai, L. H. (2011). Epigenetic regulation of gene expression in physiological and pathological brain processes. *Physiological Reviews, 91*, 603–649.

Grant, J. D., Scherrer, J. F., Lynskey, M. T., Lyons, M. J., Eisen, S. A., Tsuang, M. T., ... Bucholz, K. K. (2006). Adolescent alcohol use is a risk factor for adult alcohol and drug dependence: evidence from a twin design. *Psychological Medicine, 36*, 109–118.

Grisel, J. E., Mogil, J. S., Grahame, N. J., Rubinstein, M., Belknap, J. K., Crabbe, J. C., & Low, M. J. (1999). Ethanol oral self-administration is increased in mutant mice with decreased beta-endorphin expression. *Brain Research, 835*, 62–67.

Hayward, M. D., Hansen, S. T., Pintar, J. E., & Low, M. J. (2004). Operant self-administration of ethanol in C57BL/6 mice lacking beta-endorphin and enkephalin. *Pharmacology, Biochemistry, and Behavior, 79*, 171–181.

Hirose, N., Murakawa, K., Takada, K., Oi, Y., Suzuki, T., Nagase, H., ... Koshikawa, N. (2005). Interactions among mu- and delta-opioid receptors, especially putative delta1- and delta2-opioid receptors, promote dopamine release in the nucleus accumbens. *Neuroscience, 135*, 213–225.

Honkanen, A., Vilamo, L., Wegelius, K., Sarviharju, M., Hyytia, P., & Korpi, E. R. (1996). Alcohol drinking is reduced by a mu 1- but not by a delta-opioid receptor antagonist in alcohol-preferring rats. *European Journal of Pharmacology, 304*, 7–13.

Imperato, A., & Di Chiara, G. (1986). Preferential stimulation of dopamine release in the nucleus accumbens of freely moving rats by ethanol. *Journal of Pharmacology and Experimental Therapeutics, 239*, 219–228.

Jamensky, N. T., & Gianoulakis, C. (1999). Comparison of the proopiomelanocortin and proenkephalin opioid peptide systems in brain regions of the alcohol-preferring C57BL/6 and alcohol-avoiding DBA/2 mice. *Alcohol, 18*, 177–187.

Jarjour, S., Bai, L., & Gianoulakis, C. (2009). Effect of acute ethanol administration on the release of opioid peptides from the midbrain including the ventral tegmental area. *Alcoholism, Clinical and Experimental Research, 33*, 1033–1043.

Jayaram-Lindstrom, N., Hammarberg, A., Beck, O., & Franck, J. (2008). Naltrexone for the treatment of amphetamine dependence: a randomized, placebo-controlled trial. *American Journal of Psychiatry, 165*, 1442–1448.

Kasai, S., & Ikeda, K. (2011). Pharmacogenomics of the human microopioid receptor. *Pharmacogenomics, 12*, 1305–1320.

Kimura, M., & Higuchi, S. (2011). Genetics of alcohol dependence. *Psychiatry and Clinical Neurosciences, 65*, 213–225.

Koob, G. F., & Le Moal, M. (2008). Review. Neurobiological mechanisms for opponent motivational processes in addiction. *Philosophical Transactions of the Royal Society of London. Series B, Biological Sciences, 363*, 3113–3123.

Kovacs, K. M., Szakall, I., O'Brien, D., Wang, R., Vinod, K. Y., Saito, M., ... Vadasz, C. (2005). Decreased oral self-administration of alcohol in kappa-opioid receptor knock-out mice. *Alcoholism, Clinical and Experimental research, 29*, 730–738.

Krishnan-Sarin, S., Krystal, J. H., Shi, J., Pittman, B., & O'Malley, S. S. (2007). Family history of alcoholism influences naltrexone-induced reduction in alcohol drinking. *Biological Psychiatry, 62*, 694–697.

Lam, M. P., Marinelli, P. W., Bai, L., & Gianoulakis, C. (2008). Effects of acute ethanol on opioid peptide release in the central amygdala: an in vivo microdialysis study. *Psychopharmacology (Berlin), 201*, 261–271.

Le Merrer, J., Becker, J. A., Befort, K., & Kieffer, B. L. (2009). Reward processing by the opioid system in the brain. *Physiological Reviews, 89*, 1379–1412.

Lindholm, S., Ploj, K., Franck, J., & Nylander, I. (2000). Repeated ethanol administration induces short- and long-term changes in enkephalin and dynorphin tissue concentrations in rat brain. *Alcohol, 22*, 165–171.

Lindholm, S., Werme, M., Brene, S., & Franck, J. (2001). The selective kappa-opioid receptor agonist U50,488H attenuates voluntary ethanol intake in the rat. *Behavioural Brain Research, 120*, 137–146.

Lovallo, W. R., Dickensheets, S. L., Myers, D. A., Thomas, T. L., & Nixon, S. J. (2000). Blunted stress cortisol response in abstinent alcoholic and polysubstance-abusing men. *Alcoholism, Clinical and Experimental Research, 24*, 651–658.

Marinelli, P. W., Lam, M. P., Bai, L., Quirion, R., & Gianoulakis, C. (2006). Microdialysis profile of dynorphin A1-8 release in the nucleus accumbens following alcohol administration. *Alcoholism, Clinical and Experimental Research, 30,* 982–990.

Mendez, M., Barbosa-Luna, I. G., Perez-Luna, J. M., Cupo, A., & Oikawa, J. (2010). Effects of acute ethanol administration on methionine-enkephalin expression and release in regions of the rat brain. *Neuropeptides, 44,* 413–420.

Mendez, M., & Morales-Mulia, M. (2008). Role of mu and delta opioid receptors in alcohol drinking behaviour. *Current Drug Abuse Reviews, 1,* 239–252.

Nylander, I., Hyytia, P., Forsander, O., & Terenius, L. (1994). Differences between alcohol-preferring (AA) and alcohol-avoiding (ANA) rats in the prodynorphin and proenkephalin systems. *Alcoholism, Clinical and Experimental Research, 18,* 1272–1279.

Nylander, I., & Roman, E. (2012). Neuropeptides as mediators of the early-life impact on the brain; implications for alcohol use disorders. *Frontiers in Molecular Neuroscience, 5,* 77.

Oliva, J. M., Ortiz, S., Pérez-Rial, S., & Manzanares, J. (2008). Time dependant alterations on tyrosine hydroxylase, opioid and cannabinoid CB1 receptor gene expressions after acute ethanol administration in the rat brain. *European neuropsychopharmacology: The Journal of the European College of Neuropsychopharmacology, 18,* 373–382.

O'Malley, S. S., Jaffe, A. J., Chang, G., Schottenfeld, R. S., Meyer, R. E., & Rounsaville, B. (1992). Naltrexone and coping skills therapy for alcohol dependence. A controlled study. *Archives of General Psychiatry, 49,* 881–887.

Oswald, L. M., & Wand, G. S. (2004). Opioids and alcoholism. *Physiology and Behavior, 81,* 339–358.

Palm, S., & Nylander, I. (2014). Alcohol-induced changes in opioid peptide levels in adolescent rats are dependent on housing conditions. *Alcoholism, Clinical and Experimental Research, 38,* 2978–2987.

Palm, S., Roman, E., & Nylander, I. (2012). Differences in basal and ethanol-induced levels of opioid peptides in Wistar rats from five different suppliers. *Peptides, 36,* 1–8.

Ploj, K., Roman, E., Gustavsson, L., & Nylander, I. (2000). Basal levels and alcohol-induced changes in nociceptin/orphanin FQ, dynorphin, and enkephalin levels in C57BL/6J mice. *Brain Research Bulletin, 53,* 219–226.

Racz, I., Markert, A., Mauer, D., Stoffel-Wagner, B., & Zimmer, A. (2013). Long-term ethanol effects on acute stress responses: modulation by dynorphin. *Addiction Biology, 18,* 678–688.

Racz, I., Schurmann, B., Karpushova, A., Reuter, M., Cichon, S., Montag, C., ... Zimmer, A. (2008). The opioid peptides enkephalin and beta-endorphin in alcohol dependence. *Biological Psychiatry, 64,* 989–997.

Rassnick, S., Stinus, L., & Koob, G. F. (1993). The effects of 6-hydroxydopamine lesions of the nucleus accumbens and the mesolimbic dopamine system on oral self-administration of ethanol in the rat. *Brain Research, 623,* 16–24.

Ray, L. A., Barr, C. S., Blendy, J. A., Oslin, D., Goldman, D., & Anton, R. F. (2012). The role of the Asn40Asp polymorphism of the mu opioid receptor gene (OPRM1) on alcoholism etiology and treatment: a critical review. *Alcoholism, Clinical and Experimental Research, 36,* 385–394.

van Rijn, R. M., & Whistler, J. L. (2009). The delta(1) opioid receptor is a heterodimer that opposes the actions of the delta(2) receptor on alcohol intake. *Biological Psychiatry, 66,* 777–784.

Seizinger, B. R., Bovermann, K., Maysinger, D., Höllt, V., & Herz, A. (1983). Differential effects of acute and chronic ethanol treatment on particular opioid peptide systems in discrete regions of rat brain and pituitary. *Pharmacology, Biochemistry, and Behavior, 18,* 361–369.

Shukla, S. D., Velazquez, J., French, S. W., Lu, S. C., Ticku, M. K., & Zakhari, S. (2008). Emerging role of epigenetics in the actions of alcohol. *Alcoholism, Clinical and Experimental Research, 32,* 1525–1534.

Sinha, R. (2008). Chronic stress, drug use, and vulnerability to addiction. *Annals of the New York Academy of Sciences, 1141,* 105–130.

Sirohi, S., Bakalkin, G., & Walker, B. M. (2012). Alcohol-induced plasticity in the dynorphin/kappa-opioid receptor system. *Frontiers in Molecular Neuroscience, 5,* 95.

Snyder, S. H., & Pasternak, G. W. (2003). Historical review: opioid receptors. *Trends in Pharmacological Sciences, 24,* 198–205.

Söderpalm, B., & Ericson, M. (2013). Neurocircuitry involved in the development of alcohol addiction: the dopamine system and its access points. *Current Topics in Behavioral Neurosciences, 13,* 127–161.

Sperling, R. E., Gomes, S. M., Sypek, E. I., Carey, A. N., & McLaughlin, J. P. (2010). Endogenous kappa-opioid mediation of stress-induced potentiation of ethanol-conditioned place preference and self-administration. *Psychopharmacology (Berlin), 210,* 199–209.

Stevens, C. W. (2009). The evolution of vertebrate opioid receptors. *Frontiers in Bioscience (Landmark Edition), 14,* 1247–1269.

Stockton, S. D., Jr., & Devi, L. A. (2012). Functional relevance of mu-delta opioid receptor heteromerization: a role in novel signaling and implications for the treatment of addiction disorders: from a symposium on new concepts in mu-opioid pharmacology. *Drug and Alcohol Dependence, 121,* 167–172.

Trigo, J. M., Martin-Garcia, E., Berrendero, F., Robledo, P., & Maldonado, R. (2010). The endogenous opioid system: a common substrate in drug addiction. *Drug and Alcohol Dependence, 108,* 183–194.

Valenta, J. P., Job, M. O., Mangieri, R. A., Schier, C. J., Howard, E. C., & Gonzales, R. A. (2013). mu-opioid receptors in the stimulation of mesolimbic dopamine activity by ethanol and morphine in Long-Evans rats: a delayed effect of ethanol. *Psychopharmacology (Berlin), 228,* 389–400.

Vengeliene, V., Bilbao, A., Molander, A., & Spanagel, R. (2008). Neuropharmacology of alcohol addiction. *British Journal of Pharmacology, 154,* 299–315.

Volpicelli, J. R., Alterman, A. I., Hayashida, M., & O'Brien, C. P. (1992). Naltrexone in the treatment of alcohol dependence. *Archives of General Psychiatry, 49,* 876–880.

Walker, B. M., & Koob, G. F. (2008). Pharmacological evidence for a motivational role of kappa-opioid systems in ethanol dependence. *Neuropsychopharmacology, 33,* 643–652.

Yaksh, T. L., & Wallace, M. S. (2011). Opioids, analgesia and pain management. In L. L. Brunton, B. A. Chabner, & B. C. Knollmann (Eds.), *Goodman & Gilman's the pharmacological basis of therapeutics* (12th ed.) (pp. 481–526). New York: McGraw-Hill.

Zhou, Y., & Kreek, M. J. (2014). Alcohol: a stimulant activating brain stress responsive systems with persistent neuroadaptation. *Neuropharmacology, 87,* 51–58.

Chapter 38

Effects of Alcohol on Nicotinic Acetylcholine Receptors and Impact on Addiction

Josephine Tarren, Masroor Shariff, Joan Holgate, Selena E. Bartlett

Addiction Neuroscience and Obesity Group, Institute of Health and Biomedical Innovation, School of Clinical Sciences, Faculty of Health, Queensland University of Technology, Translational Research Institute, Woolloongabba, QLD, Australia

Abbreviations

ACh Acetylcholine
AUD Alcohol use disorder
CNS Central nervous system
DA Dopamine
DHβE Dihydroxy beta-erythroidine
FDA Food and Drug Administration
GABA γ-Aminobutyric acid
MLA Methyllycaconitine
NAc Nucleus accumbens
nAChR Nicotinic acetylcholine receptor
PFC Prefrontal cortex
VTA Ventral tegmental area
-ergic Producing

INTRODUCTION

Alcohol and nicotine addictions commonly occur together, with 80–90% of alcoholics also recurrently using tobacco (Batel, Pessione, Maitre, & Rueff, 1995). Not only do smokers have a 10-fold risk for developing concurrent alcohol dependence (DiFranza & Guerrera, 1990), epidemiological studies indicate a higher risk of cancer and psychological illness with concomitant alcohol and nicotine abuse (Pelucchi, Gallus, Garavello, Bosetti, & La Vecchia, 2006). This striking observation correlates with neurobiological studies showing that nicotine exposure increases ethanol intake and induces reinstatement of ethanol seeking—an effect that is significantly reduced by neuronal nicotinic acetylcholine receptor (nAChR) antagonists (Clark, Lindgren, Brooks, Watson, & Little, 2001). Two of these antagonists, mecamylamine and varenicline, are currently marketed as smoking cessation aids. Studies by Steensland, Simms, Holgate, Richards, and Bartlett (2007) using rodent models have shown that varenicline significantly reduces ethanol consumption and seeking (Figure 1). This finding has been translated into humans with multisite clinical trials supporting the efficacy of varenicline in reducing alcohol consumption (Litten et al., 2013; Mitchell, Teague, Kayser, Bartlett, & Fields, 2012).

A whole host of disparate strategies are currently being employed to stem the many facets of alcohol and nicotine addiction. The discovery and unraveling of previously unknown or ill-understood neurobiological processes in the brain has contributed greatly towards developing novel pharmacotherapeutics, with the aim of improving patient outcomes in treatment of various mental illnesses, including addiction. The current and progressive research into the neuropathology of alcohol and the nAChRs is accelerating our understanding of addictions, to elucidate and enhance potential therapeutic targets.

Neuronal Nicotinic Acetylcholine Receptors

The neuronal nAChRs have long been regarded as a significant mediator of addiction, modulating glutamatergic, gamma-aminobutyric acid producing (GABAergic), and dopaminergic (DAergic) transmission within reward circuits in the brain. The nAChRs are pentameric ligand-gated ion channels made up of α and β subunits. There are a total of 11 genes that encode for nAChR α and β subunits in the human central nervous system (CNS), comprised of eight α subunits (α2-α7, α9, α10), and three β subunits (β2-β4; Sargent, 1993). The five subunits assemble together to form a channel–receptor complex with wide-ranging functional and pharmacological characteristics. The majority of the nAChRs assemble heteromerically, composed of both the α and β subunits usually in a 2:3 stoichiometric ratio (Cooper, Couturier, & Ballivet, 1991); research into native receptors by Gotti et al. (2007) identified the most common of these to be the α4β2 formation. The α7 subunit, although regularly assembling as a homopentamer, is increasingly found in combination with other α/β subunits (Khiroug et al., 2002). The α9 and α10 subunits assemble as α-only heteropentamers (Sgard et al., 2002), while the α5 subunit forms functional nAChRs only when co-assembled with another α subunit in an αxαyβ configuration (Ramirez-Latorre et al., 1996). Furthermore, the β3 forms functional heteropentameric nAChRs only in the presence of another β subunit in a αβxβy configuration (Boorman, Groot-Kormelink, & Sivilotti, 2000).

These nAChRs belong to a superfamily of ligand-gated ion channel receptors; when bound to an agonist, they allow movement

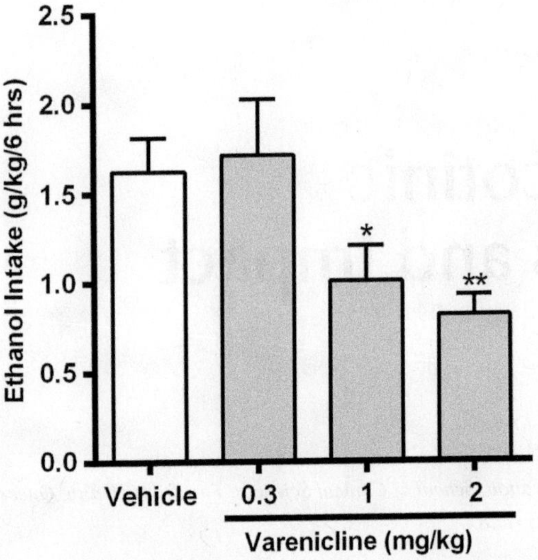

FIGURE 1 Varenicline significantly decreased ethanol consumption in rats that chronically consume low to moderate amounts of ethanol (continuous access to 10% ethanol). Varenicline (0.3–2 mg/kg s.c.) was administered 30 min before the start of the drinking session. Varenicline (1 and 2 mg/kg) significantly decreased ethanol consumption 6 h after the onset of drinking. The values are expressed as mean ethanol consumed (g/kg) ± SEM (repeated-measures analysis of variance followed by Newman–Keuls post-hoc test). *$p<0.05$; **$p<0.01$ compared with vehicle, n = 7. Adapted from Steensland et al. (2007). Copyright (2007) National Academy of Sciences, U.S.A.

of specific ions in and out of neurons. This transient change modulates transmission of nerve signals throughout the brain, via the release of neurotransmitters such as dopamine (Changeux, 2012). The relationship between alcohol and nAChRs has been unraveling since the late 1960s when Inoue and Frank (1967) highlighted ethanol's interactions with nAChRs in the peripheral nervous system. It was not until 1980, however, that the link between neuronal nicotinic receptors, addiction, and alcohol use disorders was made. Since then, a large body of work has been published to demonstrate ethanol's interactions with neuronal nAChRs, in vivo using animal models and in vitro via neuronal cell cultures.

Structural Diversity

Nicotinic receptors have important roles in the development of synaptic plasticity; they mediate activity-dependent mechanisms linked to learning, memory, and attention (Albuquerque et al., 1997). Electrophysiology indicates there are two acetylcholine (ACh) binding sites per receptor, with the agonist and protein interaction at the ligand-binding pocket limited by a defining amino acid of the subunit. These binding studies confirmed that both the α and β subunits are involved in ACh and nicotine binding. In the case of homomeric receptors, binding is determined by the adjacent positioning subunits (Galzi et al., 1990). The pharmacology of the binding properties of neuronal nAChR subunits is imperative to our understanding of alcohol's inflection on brain systems involved in addiction and to better tailor pharmacotherapeutics. Each subunit displays a distinct profile to agonists, antagonists, and modulators, ultimately determining the agonist sensitivity and calcium permeability of the receptor (Tapia, Kuryatov, & Lindstrom, 2007).

The subunit arrangement most widely accepted to be involved in alcohol addiction is the α4β2 configuration, with refined single channel measurements by Cooper et al. (1991) determining the ratio of subunit formation is $(\alpha 4)_2(\beta 2)_3$. Initially, work in *Xenopus* oocytes revealed that different expression ratios evoked altered agonist responses. For α4β2, a 1:1 ratio of subunits available for expression reduced the maximal current, while a 1:9 ratio increased ACh sensitivity it also reduced desensitization overall (Zwart & Vijverberg, 1998). Alcohol also alters nAChRs expression and desensitization. Previously, chronic drug use has been shown to downregulate and desensitize the receptors activated after extreme and excessive stimulation in an attempt to regulate the neural network and create homeostasis (Montastruc, Galitzky, Berlan, & Montastruc, 1993). However, nAChRs respond differently to long-term agonist exposure. Initially, there is a loss of receptor function, promoting an upregulation and increase in the ratio of high-affinity to low-affinity nAChRs. This compensates for the diminished signaling and reduced rewarding effects of the drug, and has been linked to the instigation of nicotine, and more recently alcohol addiction (Govind, Vezina, & Green, 2009). Ligand binding in M10 cells has shown that acute alcohol exposure blunts receptor signaling; this effect was reversed and expression enhanced with chronic exposure. This upregulation was proposed to be the result of a conformational change that decreases the degradation and removal of the receptor from the cell surface. The same study also indicates that chronic alcohol exposure even upregulates the actions of nicotine, an effect that may explain the co-abuse of these drugs (Dohrman & Reiter, 2003). Similarly, animal models have shown that long-term consumption of ethanol increases the expression of nAChRs in specific brain areas (e.g., hypothalamus and thalamus) but decreases receptor expression in others (hippocampus) (Peng, Gerzanich, Anand, Whiting, & Lindstrom, 1994).

Distribution

The extensive presynaptic, postsynaptic, and nonsynaptic locations of nAChRs underlie their modulatory roles throughout many areas of the brain linked to alcohol addiction. They can function through direct Ca^{2+} influx-mediated neurotransmitter release; Ca^{2+} induced Ca^{2+} release (CICR) from intracellular stores; and activation of presynaptic voltage-gated Ca^{2+} channels via neuronal depolarization (Sharma & Vijayaraghavan, 2003). Only recently has the availability of subunit-specific antibodies allowed the expression of nAChRs in the brain to be mapped (see Figure 2). The α4β2 nAChRs are expressed in numerous brain structures, importantly the cortico-mesolimbic pathway, while the α7 nAChRs are richly expressed in the hippocampus, cortex, and limbic regions (Dani & Bertrand, 2007).

The cortico-mesolimbic pathway is comprised of afferent and efferent neuronal projections of the ventral tegmental area (VTA), nucleus accumbens (NAc; part of the striatum), and the prefrontal cortex (PFC). The majority of the VTA (approximately 60%) is comprised of DAergic neurons that project mainly to the NAc, while gamma-aminobutyric acid producing (GABAergic) interneurons and projection neurons comprise the rest of the cell population in this area (Hendrickson, Guildford, & Tapper, 2013). There are several different nAChR subunits expressed in the VTA, some of which are the α3-α7, β2, and β3 subtypes. The α4 and β2 mRNAs are expressed in nearly all DAergic and GABAergic VTA neurons. VTA DAergic neurons contain several nAChR subtypes,

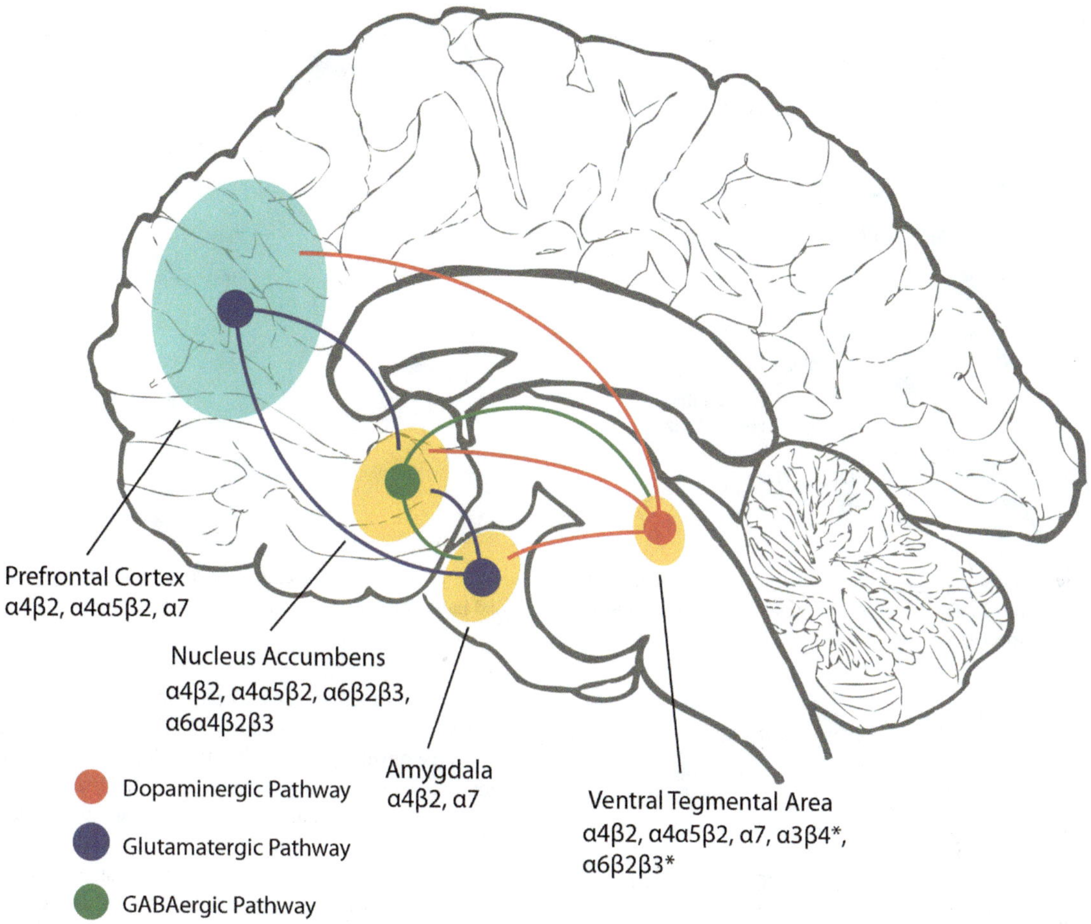

FIGURE 2 Distribution of neuronal nicotinic acetylcholine receptors (nAChRs) within the reward pathway. Sagittal view of the brain, highlighting brain regions pertaining to alcohol addiction and reward—the ventral tegmental area, amygdala, nucleus accumbens, and prefrontal cortex. Neural circuits are shown, highlighting vital DAergic, GABAergic, and glutamatergic innervation. Known neuronal nAChR subunit assembly and distribution is shown for each brain region (for review, see Gotti et al. (2007)).

including $\alpha 4\beta 2^*$, $\alpha 4\alpha 5\beta 2^*$, $\alpha 4\alpha 6\beta 2^*$, $\alpha 6\beta 2^*$, $\alpha 3\beta 2^*$, and $\alpha 7$ (Feduccia, Chatterjee, & Bartlett, 2012; Hendrickson et al., 2013). Insight into the role of nAChRs in modulating a neural response to ethanol has come from various perspectives such as pharmacology studies, using knockout models for various subunits of the nAChRs, gene expression studies, in vitro and in vivo analyses, as well as genome-wide association studies.

The Role of nAChRs in Alcohol Addiction

While the diagnostic criteria for alcohol addiction have been well characterized (see the *Diagnostic and Statistical Manual for Mental Disorders, 5th Edition*; APA, 2013), the molecular underpinnings for alcohol addiction, especially in relation to the role of neuronal nAChRs, are poorly understood. From the perspective of addiction, it is the role of the mesocorticolimbic dopamine system that has been implicated in reward and reinforcement to various drugs of abuse including alcohol (Koob, 1992). Alcohol consumption leads to an increase in ACh in the VTA. ACh then binds to nAChRs, facilitating the extracellular influx of dopamine into the NAc and producing the reinforcing effects of alcohol (Figure 3). While the VTA, NAc, and PFC play central roles in the positive reinforcing effects of alcohol, the hippocampus and amygdala are primarily involved in alcohol's negative effects, mediating memory and cue associations. These brain structures, in concert, provide the neural basis for encoding reward and reinforcement to natural rewards and when exposed to substances of abuse go awry, consequently leading to addiction (Table 1).

Gene Knockout Studies

Mouse gene knockout (KO) studies have offered some insight into the putative role of the various nAChR subunits. While α_4 KO studies have revealed the importance of α_4 in ethanol consumption and conditioned place preference (Bhutada et al., 2012), KO studies with other subunits are not as clear. For example, β_2 deletion abolishes α-conotoxin MII binding (Marubio et al., 2003) and β_2 antagonism modulates ethanol-related behaviors (Dawson, Miles, & Damaj, 2013), β_2 KO mice exhibit similar levels of ethanol consumption to their wild-type litter mates (Kamens, Andersen, & Picciotto, 2010). This also holds true for the α_6 and β_3 KOs (Kamens, Hoft, Cox, Miyamoto, & Ehringer, 2012). However, it is prudent to interpret these KO study results with caution because subunit compensation does occur in KO mouse models and other lines of corroborative

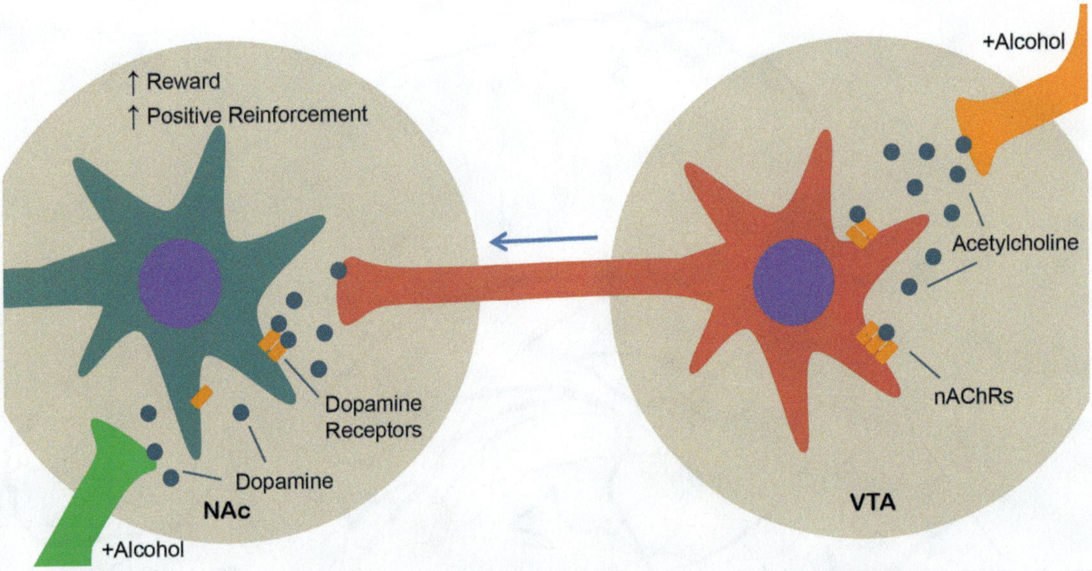

FIGURE 3 Schematic representation of the involvement of alcohol in the mesolimbic dopaminergic system. Self-administration of alcohol leads to an extracellular influx of dopamine into the nucleus accumbens (NAc) via direct and indirect mechanisms. This dopamine signal enforces environmental cues and reward expectations that form the basis of alcohol dependence.

TABLE 1 Key Facts of the Reward Pathway

- The reward pathway is a group of brain structures that regulate and govern behaviors by inducing pleasurable effects.
- It is comprised of the nucleus accumbens, ventral tegmental area, and prefrontal cortex.
- Self-administration of alcohol leads to an ACh increase in the VTA, facilitating the extracellular influx of dopamine into the NAc. This dopamine signal enforces environmental cues and reward expectations that form the basis of alcohol dependence.

This table lists the key facts of the reward pathway, including the function of the reward pathway, areas of the brain that comprise this network and the basic concept behind alcohol's modulation of this pathway.

evidence will be needed to substantiate these findings. Notably though, α_6 and α_5 KO mice exhibit increased ethanol-induced sedation (Kamens et al., 2012; Santos, Chatterjee, Henry, Holgate, & Bartlett, 2013). In the case of α_5, this observation is consistent with a study whose results suggest that α_5 does not bind agonists but rather occupies an auxiliary position in the pentameric nAChR (Marotta, Dilworth, Lester, & Dougherty, 2014). This is also supported by findings demonstrating that the α_5 subunit regulates the α_4 subunit in the VTA (Chatterjee et al., 2013). Furthermore, gene expression studies of the nAChR subunits have identified α_6 and β_3 mRNA present in the VTA (Larsson, Jerlhag, Svensson, Soderpalm, & Engel, 2004), a finding supported by work that shows the importance of the α_6 subunit in contributing to ethanol-mediated behaviors via the VTA (Powers, Broderick, Drenan, & Chester, 2013) as well as dopamine release in the NAc (Wang et al., 2014).

Stress, Alcohol Consumption, and Relapse

Stress has long been established to play a critical role in mediating irrational and compulsive ethanol intake and relapse in both humans and laboratory animals. Chronic ethanol intake leads to alteration in the allostasis and allostatic load of stress hormone responses, which lead to neuroadaptations that increase susceptibility to the development of chronic, relapsing AUDs (McEwen & Gianaros, 2011; Srinivasan, Shariff, & Bartlett, 2013). This profound stimulation of neurological stress systems interacts with, but is independent of, hormonal stress systems. These stress systems affect and produce the negative emotional state linked with drug seeking, and many posit its localization to be the circuitry of the amygdala (Koob, 2008). The amygdala is intrinsically linked with components of the mesolimbic DAergic pathway and its reaffirmation of alcohol dependence. It provides a connection between the mesolimbic pathway, the limbic system, and the hypothalamic-pituitary-adrenal axis, to modulate responses to environmental stimuli, sensory, and cognitive responses to addiction, as well as negative reinforcement and fear conditioning (Roberto, Gilpin, & Siggins, 2012). nAChR binding studies have shown a distinct population of these receptors located presynaptically in glutamatergic afferents, pyramidal neurons, and GABAergic interneurons (Hill, Zoli, Bourgeois, & Changeux, 1993; Perry et al., 2002). Refined work by Tang et al. (2011) supports the idea that the activation of alcohol-affected nAChRs in the amygdala during stress stimulates GABAergic and glutamatergic signaling to the NAc, reinforcing natural reward seeking. Dihydroxy beta-erythroidine (DHβE: an α4β2 antagonist), when microinfused into the basolateral nucleus of the amygdala in rats resulted in distinct memory deficits, indicating that β2 containing nAChRs are involved in this area (Addy, Nakijama, & Levin, 2003).

Pharmacotherapeutic Development Based on Nicotinic Receptors

Mecamylamine, a noncompetitive and nonspecific nAChR antagonist, was seminal in establishing the involvement of nAChRs in relation to effects of alcohol in the mesocorticolimbic pathway. Studies have demonstrated that systemic mecamylamine administration reduced ethanol consumption as well as ethanol-mediated dopamine release into the NAc (Blomqvist, Ericson, Engel, & Soderpalm, 1997; Blomqvist, Ericson, Johnson, Engel, & Soderpalm, 1996), a key signature for reward encoding (Schultz, 1998). Furthermore, when mecamylamine was infused directly into the VTA (but not the NAc), there was a concomitant decrease in ethanol-mediated dopamine release in the NAc (Blomqvist et al., 1997). Notably, reduced operant responding (lever pressing) for ethanol and its associated cues as well as a decrease in ethanol consumption during relapse was also observed (Kuzmin, Jerlhag, Liljequist, & Engel, 2009). The preceding cumulative evidence positioned the nAChR as potential therapeutic target for AUDs. Indeed, preclinical studies with mecamylamine using these models may hold predictive value because patients given mecamylamine report less pleasure from alcohol (Chi & de Wit, 2003). Mecamylamine, however, being a nonselective nAChR antagonist, does not aid in identifying the specific nAChR subtypes that are implicated in modulating a response to ethanol. Therefore, a suite of subunit-specific nAChR ligands have been used and findings are described below.

Dihydroxy beta-erythroidine (DHβE) is an antagonist at $\alpha_4\beta_2$ nAChRs. Also, methyllycaconitine (MLA) is an α_7 subunit-specific nAChR antagonist. It has been demonstrated that neither of these nAChR ligands are effective in reducing ethanol consumption nor the associated ethanol-mediated dopamine release in the NAc (Bito-Onon, Simms, Chatterjee, Holgate, & Bartlett, 2011). This suggests that neither the $\alpha_4\beta_2$ nor the α_7 nAChR play a role in response to an ethanol challenge. Interestingly, studies with α-conotoxin MII, a ligand specific for $\alpha_3\beta_2$*, β_3*, and α_6* (where * indicates the presence of other types of subunits), when infused into the VTA have shown reduced ethanol consumption and operant responding, as well as a decrease in dopamine release in the NAc (Kuzmin et al., 2009; Larsson et al., 2004). Surprisingly, a large percentage of the α-conotoxin MII-sensitive nAChRs also contains the α_4 subunit (Garcao, Oliveira, Cunha, & Agostinho, 2014). Furthermore, deletion of the β_2 subunit abolishes α-conotoxin MII binding in the VTA (Marubio et al., 2003), pointing to a relevant role for the α_4 and β_2 subunits. While seemingly conflicting with studies involving DHβE, there is evidence to suggest that the pentameric nAChR can be composed of more than just one α or β subunit type in a given nAChR. Furthermore, α-conotoxin PIA, which is specific for the α_6 subunit, does not cause a reduction in ethanol-modulated dopamine release into the NAc (Dowell et al., 2003), placing the α_3 subunit as an attractive candidate for ethanol-induced nAChR modulation.

Varenicline

One of the most promising advances in the treatment of AUDs comes from a partial agonist at $\alpha_4\beta_2$* nAChRs, varenicline. Based on the obvious co-addictive nature of nicotine and alcohol described above, this Food and Drug Administration (FDA)-approved smoking cessation aid (marketed as Champix) became the forefront of an extensive range of clinical and pharmaceutical research. Initial work by Ericson, Lof, Stomberg, and Soderpalm (2009) used in vivo microdialysis to study the effects of varenicline on extracellular dopamine in response ethanol in male Wistar rats. It was consequently shown that semichronic treatment with varenicline antagonized the stimulation and release of dopamine by systemic ethanol administration, as well as co-administration of combined nicotine and ethanol. The study also delineated the dose-dependent response of ethanol to varenicline, while effectively identifying nicotinic receptors, namely $\alpha_4\beta_2$*, in the instigation of AUDs.

In light of the pronounced preclinical evidence outlining the potential of this drug in alcohol use disorders (Ericson et al., 2009; Steensland et al., 2007), subsequent clinical studies were initiated. A double-blind placebo-controlled study was established to examine the effect of a clinically relevant dose of varenicline on alcohol self-administration. McKee et al. (2009) led a preliminary human study in 20 heavy-drinking smokers. Subjects underwent 7 days of pretreatment with 2 mg/day of varenicline, after which an alcohol priming dose (0.3 g/kg) was administered, and the physiologic activity of ethanol was assessed. Directly following was a 2 h self-administration period where the subjects could choose to consume up to eight additional drinks at 0.15 g/kg. Using this well-established self-administration paradigm, McKee and colleagues effectively illustrated that varenicline significantly reduced ethanol administration compared to the placebo, while also attenuating alcohol craving and positive reinforcement. Varenicline was also well tolerated within the group, with minimal side effects and no reactivity in response to the priming drink. This study not only justified varenicline as a treatment for alcohol and nicotine codependence, but as a primary treatment for AUDs.

Mitchell et al. (2012) completed a larger scale, double-blind, and randomized 16-week investigation, also in heavy drinking smokers. Outpatient subjects underwent 12 weeks of varenicline treatment at 2 mg/day or placebo, with a drug titration at onset and offset to mitigate side effects. During the course of treatment, subjects recorded the number of drinks and cigarettes they consumed during each 24-h period, alcohol craving, medication use, and any medication and nonmedication related illness. Varenicline was found to reduce cumulative consumption of alcohol and ongoing consumption as reported using ethyl glucuronide measurements. Although varenicline carries warnings for hostility, depression, and suicide-related events, subjects generally reported a low rate of side effects, the authors suggesting a connection between concurrent psychostimulant use and adverse side effects. Mitchell and colleagues were able to effectively attenuate alcohol consumption in heavy drinking smokers using varenicline.

In 2013, Litten and colleagues presented the results of a phase 2, randomized, double-blind placebo-controlled, parallel-group, multisite 13-week study. Patients were randomly assigned varenicline or a placebo at a titrated dose up to 1 mg per day (weeks 2–13). The primary efficacy endpoint was regarded as the percent of "heavy drinking days," defined as four or more drinks per day for females and five or more drinks per day for males. Secondary endpoints measured ranged from drinks per day and percent days abstinent to alcohol craving and cigarettes smoked per day. Litten et al. (2013) showed that varenicline significantly reduced all measures of alcohol use and craving. Unlike previous studies, the authors were also able to show that the effects of varenicline were independent of smoking status, suggesting that it may be a promising treatment for alcohol dependence with or without concurrent nicotine dependence. Vatsalya and colleagues have also shown that varenicline decreases fMRI blood oxygen level dependent (BOLD) activation in regions associated with motivation and incentive salience of alcohol reward (NAc, amygdala, and posterior insula) in heavy drinkers (unpublished data).

APPLICATIONS TO OTHER ADDICTIONS AND SUBSTANCE MISUSE

A picture is beginning to emerge that places various nAChRs as important modulators of DAergic reward and reinforcement pathway. The VTA plays a pivotal role in this process, but nAChRs in the NAc are also potent players for their role in responding to a drug challenge.

In the NAc specifically, a small population of cholinergic interneurons maintains and drive the cholinergic tone via broad arborization in this region. The presence of presynaptic nAChRs on the DAergic neurons projecting from the VTA places the NAc as a key modulator of the mesocorticolimbic dopamine pathway in response to ethanol. Indeed, it has been shown that the α_6 subunit plays an important role in the NAc (Wang et al., 2014). Furthermore, the small population of cholinergic interneurons in the NAc independently evoke spontaneous tonic firing that causes a continuous release of ACh (Bennett & Wilson, 1999). The released ACh, via presynaptic nAChR modulation on the DAergic afferents in the NAc, modulates further DA release (Zhou, Wilson, & Dani, 2002). Also, the inhibition of NAc cholinergic interneurons increases the firing rate in the majority of neighboring cells. This phenomenon produces an increase in conditioned place preference for cocaine (Witten et al., 2010). Interestingly, work has shown that varenicline, a partial agonist to nAChR, when injected into the NAc core and core–shell border, reduces ethanol consumption in rats, possibly via actions involving the modulatory effects of the cholinergic interneurons (Feduccia, Simms, Mill, Yi, & Bartlett, 2014). Taken together, this positions the cholinergic interneurons as pivotal modulators of reward-related behaviors in the NAc.

This places nAChRs with a prominent role in maintaining allostasis and allostatic load within the DAergic reward pathway and exposure to substances impacting on this pathway will also alter cholinergic function. Research into substance use disorders, outside of nicotine and alcohol addiction, is still in the early stages, but the evidence is growing supporting a role for nAChRs in addiction in general. To date, clinical studies have focused on cocaine and methamphetamine misuse with mixed results for and against the use of nAChR targeted pharmacotherapeutics. However, the findings of these studies consistently indicate that comorbid use of nicotine improves the outcomes suggesting the efficacy of the nAChR compound employed is dependent on the use of nicotine (for review, see Crunelle, Miller, Booij, & van den Brink, 2010). Early reports indicate varenicline may also prove useful for treating alcoholism and obesity (Cocores & Gold, 2008) and smoking and gambling (Austin, Duka, Rusted, & Jackson, 2014). Given the prevalence of alcohol and nicotine use in society, the high incidence of comorbid tobacco use within alcoholics and maturity of research in this area, we will focus on nicotine and alcohol codependence for the remainder of this chapter.

Nicotine and Alcohol Codependence

Alcohol is one of the most widely available drugs in today's society, and the co-occurrence of alcohol dependence with other drug addictions is not uncommon. While much research has elucidated nicotine's role in upregulating and reinstating ethanol consumption (Clark et al., 2001), ethanol's regulation of nicotine addiction is less understood. Through nAChRs, alcohol hijacks the natural reward circuitry and creates conditioned drug associations. This endows it with the power to sustain, prolong, and reinstate further drug-seeking behaviors—most significantly, nicotine seeking. Alcohol's modulation of nAChRs can impact nicotine addiction via several pathways, not limited to cross-tolerance and cross-sensitizations, upsetting conditioning mechanisms and by heightening psychosocial factors (Funk, Marinelli, & Le, 2006). The ability of alcohol to both escalate and inhibit many properties of nicotine may be important in mediating the strong relationship between these two drugs and the development of addiction. Using *Xenopus* oocytes, Cardoso et al. (1999) described that low doses of ethanol block nicotine-induced potentiation of $\alpha 7$ nAChR actions, which suggests that ethanol instigates rapid desensitization of these receptors. Ethanol is also known to elicit changes in receptor binding to nicotine (Booker & Collins, 1997). Dohrman and Reiter (2003) used M10 cells to illustrate ethanol's overall biphasic effect on nicotine regulation of nAChRs. Initially, ethanol served to blunt the upregulation of nAChRs originally seen by nicotine. However, by 96 h, cell receiving both alcohol and nicotine showed a significant upregulation of nAChR expression compared to either alcohol or nicotine alone. The same study also further delineated the stabilizing properties of these interactions. These cells exhibited a prolonged increase in expression, a key fact that may explain codependent individuals and increased rates of relapse (Hertling et al., 2005). It can be seen that alcohol acts directly and immediately when the drugs are taken in combination, while also chronically affecting neuronal plasticity over time with repeated insults of either one of both drugs. Finally, it should be noted that nAChR upregulation enhances neuronal excitability at glutamatergic and GABAergic synapses within many brain regions discussed above (Roberto et al., 2012; Tang et al., 2011), ultimately increasing currents directed throughout the mesocorticolimbic pathway, again insurmountably increasing activation of reward circuits (Table 2).

CONCLUSION

This chapter has summarized multiple different mechanisms that are ultimately delineating the role of neuronal nAChRs in alcohol use disorders. Clearly, the delineation of these cholinergic receptors is the key to advancing targeted therapeutics and reducing the comorbidity of nicotine addiction. This is further fueled by the compelling results seen in recent clinical trials of the partial nicotinic receptor agonist varenicline. More importantly, there is a growing need to address alterations in the allostatic functioning of stress hormone systems in the amygdala, leading to neuroadaptations that increase susceptibility to the development of chronic, relapsing addictions. This future area of research may elucidate novel therapeutics directed at drug-associated memories and ultimately aid in treating stress-related neuropsychiatric diseases, such as addiction.

TABLE 2 Key Facts of Alcohol and Tobacco Codependence

- Approximately 80–90% of individuals with an AUD are regular tobacco users.
- Both alcohol and tobacco exert their reinforcing actions via enhancing dopamine in the nucleus accumbens.
- Neuronal nicotinic acetylcholine receptors represent a common effector for the actions of nicotine and alcohol. Both drugs also modulate each other's actions via these receptors.

This table lists the key facts of alcohol and tobacco codependence, including statistics on the prevalence of co-abuse and the causal links between these dependencies.

DEFINITION OF TERMS

α-conotoxin MII This neurotoxic peptide from a group of conotoxins inhibits the function of ion channels. This particular conotoxin is specific for $\alpha_3\beta_2^*$, β_3^*, and α_6^* neuronal nicotinic acetylcholine receptors.

Agonist This drug binds to a receptor to induce a biological response.

Antagonist This ligand blocks or reduces agonist binding without eliciting a biological response.

Alcohol Dependence Alcohol dependence refers collectively to the behavioral, cognitive, and physiological disorders that develop after recurrent alcohol use.

Allostasis/Allostatic load This cascade of cause and effect relating to primary stress mediators leads to secondary and tertiary effects within stress systems and their ability to maintain homeostasis.

Cross-Sensitization In this phenomenon, the individual becomes sensitized to a drug different from the initial drug causal to the sensitization. These drugs usually have similar pharmacological or biological properties.

Dihydroxy beta-erythroidine (DHβE) This full antagonist at $\alpha_4\beta_2$ neuronal nicotinic receptors is used primarily to characterize subunit-specific activity related to addiction and fear conditioning.

Drug seeking Drug seeking is an uncontrollable desire and craving for the drug of interest, linked with extreme and conceited efforts to obtain the drug.

Mecamylamine First used as an antihypertensive drug, this noncompetitive and nonspecific neuronal nicotinic acetylcholine receptor antagonist is now known for its applications as a treatment for alcohol and nicotine addiction.

Pharmacotherapeutics Pharmacotherapeutics is the study of how drugs are employed to treat disease.

Synaptic Plasticity This is the ability of synapses to alter their activity in response to agonists or antagonists. Plastic changes can also result from an alteration in receptor number or expression around a synapse.

Tolerance Tolerance is a condition of adaptation to a pharmacologically active drug so that increasingly larger doses are required to produce the desired effect obtained earlier with smaller doses.

Varenicline This FDA-approved prescription medication is used primarily to treat nicotine addiction. It is a partial agonist at $\alpha_4\beta_2$ neuronal nicotinic receptors and effectively stimulates without causing a downstream release of dopamine.

SUMMARY POINTS

- Neuronal nAChRs are significant mediator of addiction, modulating glutamatergic, GABAergic, and DAergic transmission within reward circuits in the brain.
- Through neuronal nAChRs, alcohol hijacks the natural reward circuitry and creates conditioned drug associations with the power to sustain, prolong, and reinstate further alcohol and nicotine-seeking behaviors.
- Using pharmacological and gene knockout studies, nicotinic receptor subtypes involved in the reward pathway and alcohol dependence have been elucidated, leading to advances in targeted therapeutics.
- Additionally to the reward pathway, chronic ethanol intake has been linked to alterations in the homeostatic functioning of stress hormone systems in the amygdala, leading to neuroadaptations that increase susceptibility to the development of chronic, relapsing AUDs.
- While much research has elucidated nicotine's role in upregulating and reinstating ethanol consumption, ethanol's role in this relationship is less understood.
- The ability of alcohol to both escalate and inhibit many properties of nicotine may be important in mediating the strong relationship between these two drugs and the development of addiction.
- Clinical trials have been able to effectively attenuate alcohol consumption using the FDA-approved smoking cessation drug varenicline.
- The delineation of these cholinergic receptors may be the key to creating targeted therapeutics for alcohol use disorders and reducing the comorbidity of nicotine addiction.

ACKNOWLEDGMENT

This work was supported by funding from the National Health and Medical Research Council (1049427), the National Institutes of Health (R01AA017924-4), an Australian Research Council Future Fellowship (FT1110884), and by the Institute of Health and Biomedical Innovation—Queensland University of Technology.

REFERENCES

Addy, N. A., Nakijama, A., & Levin, E. D. (2003). Nicotinic mechanisms of memory: effects of acute local DHbetaE and MLA infusions in the basolateral amygdala. *Brain Research Cognitive Brain Research*, *16*(1), 51–57.

Albuquerque, E. X., Alkondon, M., Pereira, E. F., Castro, N. G., Schrattenholz, A., Barbosa, C. T., ... Maelicke, A. (1997). Properties of neuronal nicotinic acetylcholine receptors: pharmacological characterization and modulation of synaptic function. *Journal of Pharmacology and Experimental Therapeutics*, *280*(3), 1117–1136.

APA. (2013). *Diagnostic and statistical manual of mental disorders: Dsm-5*. Amer Psychiatric Pub Incorporated.

Austin, A. J., Duka, T., Rusted, J., & Jackson, A. (2014). Effect of varenicline on aspects of inhibitory control in smokers. *Psychopharmacology (Berlin)*, *231*(18), 3771–3785. http://dx.doi.org/10.1007/s00213-014-3512-7.

Batel, P., Pessione, F., Maitre, C., & Rueff, B. (1995). Relationship between alcohol and tobacco dependencies among alcoholics who smoke. *Addiction*, *90*(7), 977–980.

Bennett, B. D., & Wilson, C. J. (1999). Spontaneous activity of neostriatal cholinergic interneurons in vitro. *Journal of Neuroscience*, *19*(13), 5586–5596.

Bhutada, P., Mundhada, Y., Ghodki, Y., Dixit, P., Umathe, S., & Jain, K. (2012). Acquisition, expression, and reinstatement of ethanol-induced conditioned place preference in mice: effects of exposure to stress and modulation by mecamylamine. *Journal of Psychopharmacology*, *26*(2), 315–323. http://dx.doi.org/10.1177/0269881111431749.

Bito-Onon, J. J., Simms, J. A., Chatterjee, S., Holgate, J., & Bartlett, S. E. (2011). Varenicline, a partial agonist at neuronal nicotinic acetylcholine receptors, reduces nicotine-induced increases in 20% ethanol operant self-administration in Sprague-Dawley rats. *Addiction Biology*, *16*(3), 440–449. http://dx.doi.org/10.1111/j.1369-1600.2010.00309.x.

Blomqvist, O., Ericson, M., Engel, J. A., & Soderpalm, B. (1997). Accumbal dopamine overflow after ethanol: localization of the antagonizing effect of mecamylamine. *European Journal of Pharmacology*, *334*(2–3), 149–156.

Blomqvist, O., Ericson, M., Johnson, D. H., Engel, J. A., & Soderpalm, B. (1996). Voluntary ethanol intake in the rat: effects of nicotinic acetylcholine receptor blockade or subchronic nicotine treatment. *European Journal of Pharmacology, 314*(3), 257–267.

Booker, T. K., & Collins, A. C. (1997). Long-term ethanol treatment elicits changes in nicotinic receptor binding in only a few brain regions. *Alcohol, 14*(2), 131–140. http://dx.doi.org/10.1016/S0741-8329(96)00116-4.

Boorman, J. P., Groot-Kormelink, P. J., & Sivilotti, L. G. (2000). Stoichiometry of human recombinant neuronal nicotinic receptors containing the b3 subunit expressed in Xenopus oocytes. *Journal of Physiology, 529*(Pt 3), 565–577.

Cardoso, R. A., Brozowski, S. J., Chavez-Noriega, L. E., Harpold, M., Valenzuela, C. F., & Harris, R. A. (1999). Effects of ethanol on recombinant human neuronal nicotinic acetylcholine receptors expressed in Xenopus oocytes. *Journal of Pharmacology and Experimental Therapeutics, 289*(2), 774–780.

Changeux, J. P. (2012). The nicotinic acetylcholine receptor: the founding father of the pentameric ligand-gated ion channel superfamily. *Journal of Biological Chemistry, 287*(48), 40207–40215. http://dx.doi.org/10.1074/jbc.R112.407668.

Chatterjee, S., Santos, N., Holgate, J., Haass-Koffler, C. L., Hopf, F. W., Kharazia, V., ... Bartlett, S. E. (2013). The alpha5 subunit regulates the expression and function of alpha4*-containing neuronal nicotinic acetylcholine receptors in the ventral-tegmental area. *PLoS One, 8*(7), e68300. http://dx.doi.org/10.1371/journal.pone.0068300.

Chi, H., & de Wit, H. (2003). Mecamylamine attenuates the subjective stimulant-like effects of alcohol in social drinkers. *Alcoholism Clinical and Experimental Research, 27*(5), 780–786. http://dx.doi.org/10.1097/01.alc.0000065435.12068.24.

Clark, A., Lindgren, S., Brooks, S. P., Watson, W. P., & Little, H. J. (2001). Chronic infusion of nicotine can increase operant self-administration of alcohol. *Neuropharmacology, 41*(1), 108–117. http://dx.doi.org/10.1016/S0028-3908(01)00037-5.

Cocores, J. A., & Gold, M. S. (2008). Varenicline, appetite, and weight reduction. *Journal of Neuropsychiatry and Clinical Neuroscience, 20*(4), 497–498. http://dx.doi.org/10.1176/appi.neuropsych.20.4.497.

Cooper, E., Couturier, S., & Ballivet, M. (1991). Pentameric structure and subunit stoichiometry of a neuronal nicotinic acetylcholine receptor. *Nature, 350*(6315), 235–238. http://dx.doi.org/10.1038/350235a0.

Crunelle, C. L., Miller, M. L., Booij, J., & van den Brink, W. (2010). The nicotinic acetylcholine receptor partial agonist varenicline and the treatment of drug dependence: a review. *European Neuropsychopharmacology, 20*(2), 69–79. http://dx.doi.org/10.1016/j.euroneuro.2009.11.001.

Dani, J. A., & Bertrand, D. (2007). Nicotinic acetylcholine receptors and nicotinic cholinergic mechanisms of the central nervous system. *Annual Review of Pharmacology and Toxicology, 47*, 699–729. http://dx.doi.org/10.1146/annurev.pharmtox.47.120505.105214.

Dawson, A., Miles, M. F., & Damaj, M. I. (2013). The beta2 nicotinic acetylcholine receptor subunit differentially influences ethanol behavioral effects in the mouse. *Alcohol, 47*(2), 85–94. http://dx.doi.org/10.1016/j.alcohol.2012.12.004.

DiFranza, J. R., & Guerrera, M. P. (1990). Alcoholism and smoking. *Journal of Studies on Alcohol, 51*(2), 130–135.

Dohrman, D. P., & Reiter, C. K. (2003). Ethanol modulates nicotine-induced upregulation of nAChRs. *Brain Research, 975*(1–2), 90–98. http://dx.doi.org/10.1016/S0006-8993(03)02593-9.

Dowell, C., Olivera, B. M., Garrett, J. E., Staheli, S. T., Watkins, M., Kuryatov, A., ... McIntosh, J. M. (2003). Alpha-conotoxin PIA is selective for alpha6 subunit-containing nicotinic acetylcholine receptors. *Journal of Neuroscience, 23*(24), 8445–8452.

Ericson, M., Lof, E., Stomberg, R., & Soderpalm, B. (2009). The smoking cessation medication varenicline attenuates alcohol and nicotine interactions in the rat mesolimbic dopamine system. *Journal of Pharmacology and Experimental Therapeutics, 329*(1), 225–230. http://dx.doi.org/10.1124/jpet.108.147058.

Feduccia, A. A., Chatterjee, S., & Bartlett, S. E. (2012). Neuronal nicotinic acetylcholine receptors: neuroplastic changes underlying alcohol and nicotine addictions. *Frontiers in Molecular Neuroscience, 5*, 83. http://dx.doi.org/10.3389/fnmol.2012.00083.

Feduccia, A. A., Simms, J. A., Mill, D., Yi, H. Y., & Bartlett, S. E. (2014). Varenicline decreases ethanol intake and increases dopamine release via neuronal nicotinic acetylcholine receptors in the nucleus accumbens. *British Journal of Pharmacology, 171*(14), 3420–3431. http://dx.doi.org/10.1111/bph.12690.

Funk, D., Marinelli, P. W., & Le, A. D. (2006). Biological processes underlying co-use of alcohol and nicotine: neuronal mechanisms, cross-tolerance, and genetic factors. *Alcohol Research and Health, 29*(3), 186–192.

Galzi, J. L., Revah, F., Black, D., Goeldner, M., Hirth, C., & Changeux, J. P. (1990). Identification of a novel amino acid alpha-tyrosine 93 within the cholinergic ligands-binding sites of the acetylcholine receptor by photoaffinity labeling. Additional evidence for a three-loop model of the cholinergic ligands-binding sites. *Journal of Biological Chemistry, 265*(18), 10430–10437.

Garcao, P., Oliveira, C. R., Cunha, R. A., & Agostinho, P. (2014). Subsynaptic localization of nicotinic acetylcholine receptor subunits: a comparative study in the mouse and rat striatum. *Neuroscience Letters, 566*, 106–110. http://dx.doi.org/10.1016/j.neulet.2014.02.018.

Gotti, C., Moretti, M., Gaimarri, A., Zanardi, A., Clementi, F., & Zoli, M. (2007). Heterogeneity and complexity of native brain nicotinic receptors. *Biochemical Pharmacology, 74*(8), 1102–1111. http://dx.doi.org/10.1016/j.bcp.2007.05.023.

Govind, A. P., Vezina, P., & Green, W. N. (2009). Nicotine-induced upregulation of nicotinic receptors: underlying mechanisms and relevance to nicotine addiction. *Biochemical Pharmacology, 78*(7), 756–765. http://dx.doi.org/10.1016/j.bcp.2009.06.011.

Hendrickson, L. M., Guildford, M. J., & Tapper, A. R. (2013). Neuronal nicotinic acetylcholine receptors: common molecular substrates of nicotine and alcohol dependence. *Frontiers in Psychiatry, 4*, 29. http://dx.doi.org/10.3389/fpsyt.2013.00029.

Hertling, I., Ramskogler, K., Dvorak, A., Klingler, A., Saletu-Zyhlarz, G., Schoberberger, R., ... Lesch, O. M. (2005). Craving and other characteristics of the comorbidity of alcohol and nicotine dependence. *European Psychiatry, 20*(5–6), 442–450. http://dx.doi.org/10.1016/j.eurpsy.2005.06.003.

Hill, J. A., Jr., Zoli, M., Bourgeois, J. P., & Changeux, J. P. (1993). Immunocytochemical localization of a neuronal nicotinic receptor: the β2-subunit. *Journal of Neuroscience, 13*(4), 1551–1568.

Inoue, F., & Frank, G. B. (1967). Effects of ethyl alcohol on excitability and on neuromuscular transmission in frog skeletal muscle. *British Journal of Pharmacology and Chemotherapy, 30*(1), 186–193.

Kamens, H. M., Andersen, J., & Picciotto, M. R. (2010). Modulation of ethanol consumption by genetic and pharmacological manipulation of nicotinic acetylcholine receptors in mice. *Psychopharmacology (Berlin), 208*(4), 613–626. http://dx.doi.org/10.1007/s00213-009-1759-1.

Kamens, H. M., Hoft, N. R., Cox, R. J., Miyamoto, J. H., & Ehringer, M. A. (2012). The alpha6 nicotinic acetylcholine receptor subunit influences ethanol-induced sedation. *Alcohol, 46*(5), 463–471. http://dx.doi.org/10.1016/j.alcohol.2012.03.001.

Khiroug, S. S., Harkness, P. C., Lamb, P. W., Sudweeks, S. N., Khiroug, L., Millar, N. S., & Yakel, J. L. (2002). Rat nicotinic ACh receptor alpha7 and beta2 subunits co-assemble to form functional heteromeric nicotinic receptor channels. *Journal of Physiology*, *540*(Pt 2), 425–434.

Koob, G. F. (1992). Neural mechanisms of drug reinforcement. *Annals of the New York Academy of Sciences*, *654*, 171–191.

Koob, G. F. (2008). A role for brain stress systems in addiction. *Neuron*, *59*(1), 11–34. http://dx.doi.org/10.1016/j.neuron.2008.06.012.

Kuzmin, A., Jerlhag, E., Liljequist, S., & Engel, J. (2009). Effects of subunit selective nACh receptors on operant ethanol self-administration and relapse-like ethanol-drinking behavior. *Psychopharmacology (Berlin)*, *203*(1), 99–108. http://dx.doi.org/10.1007/s00213-008-1375-5.

Larsson, A., Jerlhag, E., Svensson, L., Soderpalm, B., & Engel, J. A. (2004). Is an alpha-conotoxin MII-sensitive mechanism involved in the neurochemical, stimulatory, and rewarding effects of ethanol? *Alcohol*, *34*(2–3), 239–250.

Litten, R. Z., Ryan, M. L., Fertig, J. B., Falk, D. E., Johnson, B., Dunn, K. E., ... Stout, R. (2013). A double-blind, placebo-controlled trial assessing the efficacy of varenicline tartrate for alcohol dependence. *Journal of Addiction Medicine*, *7*(4), 277–286. http://dx.doi.org/10.1097/ADM.0b013e31829623f4.

Marotta, C. B., Dilworth, C. N., Lester, H. A., & Dougherty, D. A. (2014). Probing the non-canonical interface for agonist interaction with an alpha5 containing nicotinic acetylcholine receptor. *Neuropharmacology*, *77*, 342–349. http://dx.doi.org/10.1016/j.neuropharm.2013.09.028.

Marubio, L. M., Gardier, A. M., Durier, S., David, D., Klink, R., Arroyo-Jimenez, M. M., ... Changeux, J. P. (2003). Effects of nicotine in the dopaminergic system of mice lacking the alpha4 subunit of neuronal nicotinic acetylcholine receptors. *European Journal of Neuroscience*, *17*(7), 1329–1337.

McEwen, B. S., & Gianaros, P. J. (2011). Stress- and allostasis-induced brain plasticity. *Annual Review of Medicine*, *62*, 431–445. http://dx.doi.org/10.1146/annurev-med-052209-100430.

McKee, S. A., Harrison, E. L., O'Malley, S. S., Krishnan-Sarin, S., Shi, J., Tetrault, J. M., ... Balchunas, E. (2009). Varenicline reduces alcohol self-administration in heavy-drinking smokers. *Biological Psychiatry*, *66*(2), 185–190. http://dx.doi.org/10.1016/j.biopsych.2009.01.029.

Mitchell, J. M., Teague, C. H., Kayser, A. S., Bartlett, S. E., & Fields, H. L. (2012). Varenicline decreases alcohol consumption in heavy-drinking smokers. *Psychopharmacology (Berlin)*, *223*(3), 299–306. http://dx.doi.org/10.1007/s00213-012-2717-x.

Montastruc, J. L., Galitzky, J., Berlan, M., & Montastruc, P. (1993). Mechanism of receptor regulation during repeated administration of drugs. *Therapie*, *48*(5), 421–426.

Pelucchi, C., Gallus, S., Garavello, W., Bosetti, C., & La Vecchia, C. (2006). Cancer risk associated with alcohol and tobacco use: focus on upper aero-digestive tract and liver. *Alcohol Research and Health*, *29*(3), 193–198.

Peng, X., Gerzanich, V., Anand, R., Whiting, P. J., & Lindstrom, J. (1994). Nicotine-induced increase in neuronal nicotinic receptors results from a decrease in the rate of receptor turnover. *Molecular Pharmacology*, *46*(3), 523–530.

Perry, D. C., Xiao, Y., Nguyen, H. N., Musachio, J. L., Davila-Garcia, M. I., & Kellar, K. J. (2002). Measuring nicotinic receptors with characteristics of alpha4beta2, alpha3beta2 and alpha3beta4 subtypes in rat tissues by autoradiography. *Journal of Neurochemistry*, *82*(3), 468–481.

Powers, M. S., Broderick, H. J., Drenan, R. M., & Chester, J. A. (2013). Nicotinic acetylcholine receptors containing alpha6 subunits contribute to alcohol reward-related behaviours. *Genes Brain and Behavior*, *12*(5), 543–553. http://dx.doi.org/10.1111/gbb.12042.

Ramirez-Latorre, J., Yu, C. R., Qu, X., Perin, F., Karlin, A., & Role, L. (1996). Functional contributions of alpha5 subunit to neuronal acetylcholine receptor channels. *Nature*, *380*(6572), 347–351. http://dx.doi.org/10.1038/380347a0.

Roberto, M., Gilpin, N. W., & Siggins, G. R. (2012). The central amygdala and alcohol: role of gamma-aminobutyric acid, glutamate, and neuropeptides. *Cold Spring Harbor Perspectives in Medicine*, *2*(12), a012195. http://dx.doi.org/10.1101/cshperspect.a012195.

Santos, N., Chatterjee, S., Henry, A., Holgate, J., & Bartlett, S. E. (2013). The alpha5 neuronal nicotinic acetylcholine receptor subunit plays an important role in the sedative effects of ethanol but does not modulate consumption in mice. *Alcoholism Clinical and Experimental Research*, *37*(4), 655–662. http://dx.doi.org/10.1111/acer.12009.

Sargent, P. B. (1993). The diversity of neuronal nicotinic acetylcholine receptors. *Annual Review of Neuroscience*, *16*, 403–443. http://dx.doi.org/10.1146/annurev.ne.16.030193.002155.

Schultz, W. (1998). Predictive reward signal of dopamine neurons. *Journal of Neurophysiology*, *80*(1), 1–27.

Sgard, F., Charpantier, E., Bertrand, S., Walker, N., Caput, D., Graham, D., ... Besnard, F. (2002). A novel human nicotinic receptor subunit, alpha10, that confers functionality to the alpha9-subunit. *Molecular Pharmacology*, *61*(1), 150–159.

Sharma, G., & Vijayaraghavan, S. (2003). Modulation of presynaptic store calcium induces release of glutamate and postsynaptic firing. *Neuron*, *38*(6), 929–939.

Srinivasan, S., Shariff, M., & Bartlett, S. E. (2013). The role of the glucocorticoids in developing resilience to stress and addiction. *Frontiers in Psychiatry*, *4*, 68. http://dx.doi.org/10.3389/fpsyt.2013.00068.

Steensland, P., Simms, J. A., Holgate, J., Richards, J. K., & Bartlett, S. E. (2007). Varenicline, an alpha4beta2 nicotinic acetylcholine receptor partial agonist, selectively decreases ethanol consumption and seeking. *Proceedings of the National Academy of Sciences of the United States of America*, *104*(30), 12518–12523. http://dx.doi.org/10.1073/pnas.0705368104.

Tang, A. H., Karson, M. A., Nagode, D. A., McIntosh, J. M., Uebele, V. N., Renger, J. J., ... Alger, B. E. (2011). Nerve terminal nicotinic acetylcholine receptors initiate quantal GABA release from perisomatic interneurons by activating axonal T-type (Cav3) Ca(2)(+) channels and Ca(2)(+) release from stores. *Journal of Neuroscience*, *31*(38), 13546–13561. http://dx.doi.org/10.1523/jneurosci.2781-11.2011.

Tapia, L., Kuryatov, A., & Lindstrom, J. (2007). Ca^{2+} permeability of the (alpha4)3(beta2)2 stoichiometry greatly exceeds that of (alpha4)2(beta2)3 human acetylcholine receptors. *Molecular Pharmacology*, *71*(3), 769–776. http://dx.doi.org/10.1124/mol.106.030445.

Wang, Y., Lee, J. W., Oh, G., Grady, S. R., McIntosh, J. M., Brunzell, D. H., ... Drenan, R. M. (2014). Enhanced synthesis and release of dopamine in transgenic mice with gain-of-function alpha6* nAChRs. *Journal of Neurochemistry*, *129*(2), 315–327. http://dx.doi.org/10.1111/jnc.12616.

Witten, I. B., Lin, S. C., Brodsky, M., Prakash, R., Diester, I., Anikeeva, P., ... Deisseroth, K. (2010). Cholinergic interneurons control local circuit activity and cocaine conditioning. *Science*, *330*(6011), 1677–1681. http://dx.doi.org/10.1126/science.1193771.

Zhou, F. M., Wilson, C. J., & Dani, J. A. (2002). Cholinergic interneuron characteristics and nicotinic properties in the striatum. *Journal of Neurobiology*, *53*(4), 590–605. http://dx.doi.org/10.1002/neu.10150.

Zwart, R., & Vijverberg, H. P. (1998). Four pharmacologically distinct subtypes of alpha4beta2 nicotinic acetylcholine receptor expressed in *Xenopus laevis* oocytes. *Molecular Pharmacology*, *54*(6), 1124–1131.

Chapter 39

Alcohol and Its Impact on Myelin

Consuelo Guerri, María Pascual
Department of Cellular Pathology, Príncipe Felipe Research Centre, Valencia, Spain

Abbreviations

CNP 2′:3′-Cyclic nucleotide 3′-phosphodiesterase
CNS Central nervous system
COX-2 Cyclooxygenase-2
IL Interleukin
iNOS Inducible nitric oxide synthase
KO mice Knockout mice
MAG Myelin-associated glycoprotein
MBP Myelin basic protein
MHC-II Major histocompatibility complex-II
MOG Myelin oligodendrocyte glycoprotein
NF-κB Nuclear factor-κB
NO-cGMP-PKG pathway Nitric oxide-cyclic GMP-protein kinase G pathway
PLP Proteolipid protein
TLR4 Toll-like receptor 4
TNF-α Tumor necrosis factor-α
WT mice Wild-type mice

INTRODUCTION

Myelin is an important component of brain white matter and is formed by specialized glial cells that ensheath axons, with a lipid-rich insulating membrane that allows the acceleration of the impulse conduction along nerve fibers. Continuous communication between neurons and glial cells is essential for myelin maintenance and axonal integrity (White & Kramer-Albers, 2014).

The myelination program is initiated in response to axon-glia recognition, mediated mostly by membrane-bound cell adhesion molecules to trigger the reorganization of the glial cytoskeleton and cell polarization (Simons & Trotter, 2007). In humans and rodents, myelination in the central nervous system (CNS) occurs during early postnatal development (Young et al., 2013). However, in certain brain areas, myelination takes place during the perinatal period, while in others (e.g., the prefrontal cortex), myelination is completed by the end of the second decade or even later (Toga, Thompson, & Sowell, 2006).

Within the CNS, white matter is found in many structures, but it concentrates particularly in areas where many signals must be sent over long distances. These areas include the thalamus and hypothalamus, which govern processes such as blood pressure and other essential life support functions that do not require conscious attention to be executed. Within the subcortex, myelinated axons pass signals between the two brain hemispheres (e.g., corpus callosum) and between different areas in the same hemisphere. These connections are necessary for proper and efficient communication pathways, which shape the integrated neural systems responsible for higher order functioning (Fornari, Knyazeva, Meuli, & Maeder, 2007).

Given myelin's critical role in brain communication, the processes that induce myelin dysfunction or disruption of myelin development may result in reduced brain connectivity and inefficient interneuronal communication. Consequently, white matter dysfunction and demyelination take place in many CNS disorders, and myelin is a hallmark of some neurodegenerative autoimmune diseases. The most common diseases in this group include multiple sclerosis, viral infection, and disseminated encephalomyelitis. These immune-mediated disorders are characterized by inflammation, demyelination, and axon degeneration in the CNS. Activation of macrophages and/or microglia with the release of oxygen radicals, cytokines (e.g., TNF-α), nitric oxide, and other toxic compounds that are harmful to oligodendrocytes (Merrill & Benveniste, 1996) participate in the demyelinating diseases associated with immune reactions (Martin & McFarland, 1995). Similarly, some chemical toxins (e.g., lead, cuprizone, lysolecithin, organotin) can also induce oligodendrocyte apoptosis and microglia activation, causing white matter dysfunction and cognitive disorders (Gemert & Killeen, 1998).

Alcohol is a neurotoxic compound whose abuse can cause structural and functional brain alterations (Harper & Matsumoto, 2005). Activation of the brain innate immune system and neuroinflammation has been suggested to contribute to the ethanol-induced brain damage and behavioral dysfunctions (Alfonso-Loeches, Pascual-Lucas, Blanco, Sanchez-Vera, & Guerri, 2010; Mayfield, Ferguson, & Harris, 2013; Pascual, Balino, Alfonso-Loeches, Aragon, & Guerri, 2011). Evidence from neuroimaging and postmortem studies have also revealed that alcohol abuse induces white matter loss (Wang et al., 2009), downregulates myelin-associated genes (Liu et al., 2006), and causes transcallosal white matter fiber degradation in human alcoholics (Pfefferbaum, Rosenbloom, Fama, Sassoon, & Sullivan, 2010; Schulte, Muller-Oehring, Rohlfing, Pfefferbaum, & Sullivan, 2010). Alcohol exposure during brain ontogeny also alters myelination processes (Creeley, Dikranian, Johnson, Farber, & Olney, 2013) and affects the structure of the white matter and fiber tract integrity in human adolescents who participate in binge drinking (Bava, Jacobus, Thayer, & Tapert, 2013). All of these findings indicate that myelin is a target of the actions of ethanol on adult and developing brains.

This chapter reviews the functional role of oligodendrocytes and myelin formation in developing and adult brains, as well as the functional cognitive consequences of myelin disruptions during brain ontogeny. It further discusses the potential cellular and molecular mechanisms of actions of ethanol on myelin dysfunctions observed in both adult alcoholics and children/adolescents exposed to alcohol. Finally, we comment on whether other drugs of abuse can affect myelin structure in correlation with behavioral dysfunction. Potential therapeutic approaches, which can reduce or even eliminate the deleterious effects of ethanol on myelin alterations, are proposed.

THE MYELINATION PROCESS AND BRAIN DEVELOPMENT

The myelin sheath of axons is a predominant white matter element composed of glial cells and capillaries and myelinated axons. Myelin accounts for the glistening white appearance, high lipid content, and relatively low water content of white matter. Gray matter, however, mainly contains nerve cell bodies with their extensive dendritic arborizations. The cells that supply the myelin to peripheral neurons are Schwann cells, whereas oligodendrocytes myelinate the axons of the CNS.

During the myelination process, oligodendrocytes synthesize large amounts of plasma membrane to form multiple myelin internodes that wrap around axons. Initially, the successive turns of the spiral of paired membrane sheets are separated by the cytoplasm. Eventually, however, the cytoplasm is extruded from between turns to result in a mature, compact myelin (Peters, 2009) (Figure 1(A)). An oligodendrocyte forms several internodal lengths of myelin, each on a different axon (Figure 1(B)). The turns of myelin terminating the sheath gradually become thinner and eventually end at the nodes of Ranvier, which separate the successive internodal lengths of myelin (Peters, 2009). Thus, oligodendrocytes form the internodal lengths of myelin in neuronal axons, which is very important for the nervous system function (Simons & Trotter, 2007). One main function of myelin is to insulate the axon and to cluster sodium channels into the nodes of Ranvier which, in turn, enables the action potential to jump from one node to another (Waxman, 2006). The development of myelin and saltatory nerve conduction provides the basis for fast information processing in a relatively small space and is fundamental for impulse conductions along nerve fibers in the CNS.

The significance of the integrity of the myelin structure and its maturation process for the connectivity between brain areas and for the development of cognitive, motor, and sensory functions has been shown in different studies and also in the developing brain (Gogtay et al., 2004; Paus et al., 1999). Age-dependent changes in white matter density, such as myelination or axon diameter (Paus et al., 1999), have been observed during childhood and adolescence and support motor and speech functions. Longitudinal magnetic resonance imaging studies performed in developing children and adolescent brains also confirmed that myelination is associated with maturation during human cortical development. The brain parts are associated with more basic functions, such as motor and sensory regions mature first, followed by areas involved in spatial orientation, speech, language development, and attention (upper and lower parietal lobes). Notably, the frontal cortex, involved in executive functions, is the last to mature and coincides with later myelination and dendrite pruning (Gogtay et al., 2004). White matter or myelin maturation not only changes during brain maturation but also occurs with sexual dimorphism. In a large longitudinal study on adolescent brain development, Giedd found differences between males and females in the rate at which white matter increases in various specific cortical brain regions (Giedd et al., 2010). Whereas males and females have been found to have similar frontal white matter volume during childhood, volume in males seems to increase more rapidly during adolescence than in females to result in greater frontal white matter volumes in males by young adulthood (Giedd et al., 2010). Hence, these changes are slightly associated with age in girls but more so in boys (Lenroot et al., 2007), and male adolescence is mediated by the effect of testosterone on myelination.

The importance of myelin structure during brain development has been demonstrated by neuroimaging studies, which have shown that an abnormal white matter microstructure in the posterior cerebral tracts strongly correlates with atypical unimodal and multisensory integration behavior in children with sensory processing (Owen et al., 2013). Similarly, an atypical white matter microstructure has also been observed in children with attention-deficit/hyperactivity disorder (Wang et al., 2012) characterized by the altered structural connectivity of the brain. These observations highlight the importance of correct myelin formation in the CNS and its role in cognitive and behavioral functions.

COMPOSITION OF MYELIN

The composition of the myelin sheath is very different from that of glial cell membranes and also differs from membranous outgrowths or extensions of the oligodendroglial cell body (Figure 1(B)), which provide a bridge of mixed composition. Myelin is characteristically richer in lipids than in proteins; thus, its dry mass is composed of 70–85% of lipids and 15–30% of proteins. The most typical lipid of myelin is cerebroside. Although its presence in myelin formation is not required, it plays key roles because of its insulating capacity and stability. In addition to cerebroside, the major lipids of myelin are cholesterol and ethanolamine-containing plasmalogens. Moreover, not only does the lipid composition of myelin show a high quality of this membrane, but the fatty acid composition of many individual lipids is also a typical characteristic of the myelin membrane.

Regarding the protein composition of myelin sheaths, the most abundant proteins of CNS myelin are myelin basic protein (MBP) and the proteolipid protein (PLP), whereas other proteins and glycoproteins are present but to a lesser extent. The MBP's function is to facilitate mature myelin sheath compaction in the CNS by maintaining its structural integrity, while the PLP is involved in the formation and maintenance of the multilamellar myelin structure. There are other higher-molecular-weight proteins present in myelin, such as 2′:3′-cyclic nucleotide 3′-phosphodiesterase (CNP), which is predominantly expressed in the myelin sheath. CNP contains phosphodiesterase activity and is essential for the normal interaction of oligodendrocytes with axons (Braun, Lee, & Gravel, 2004). Mice deficient in CNP suffer progressive axonal degeneration, which leads to premature death (Lappe-Siefke et al., 2003). Another important myelin protein is proteoglycan NG2. This protein is expressed in oligodendrocyte progenitor cells,

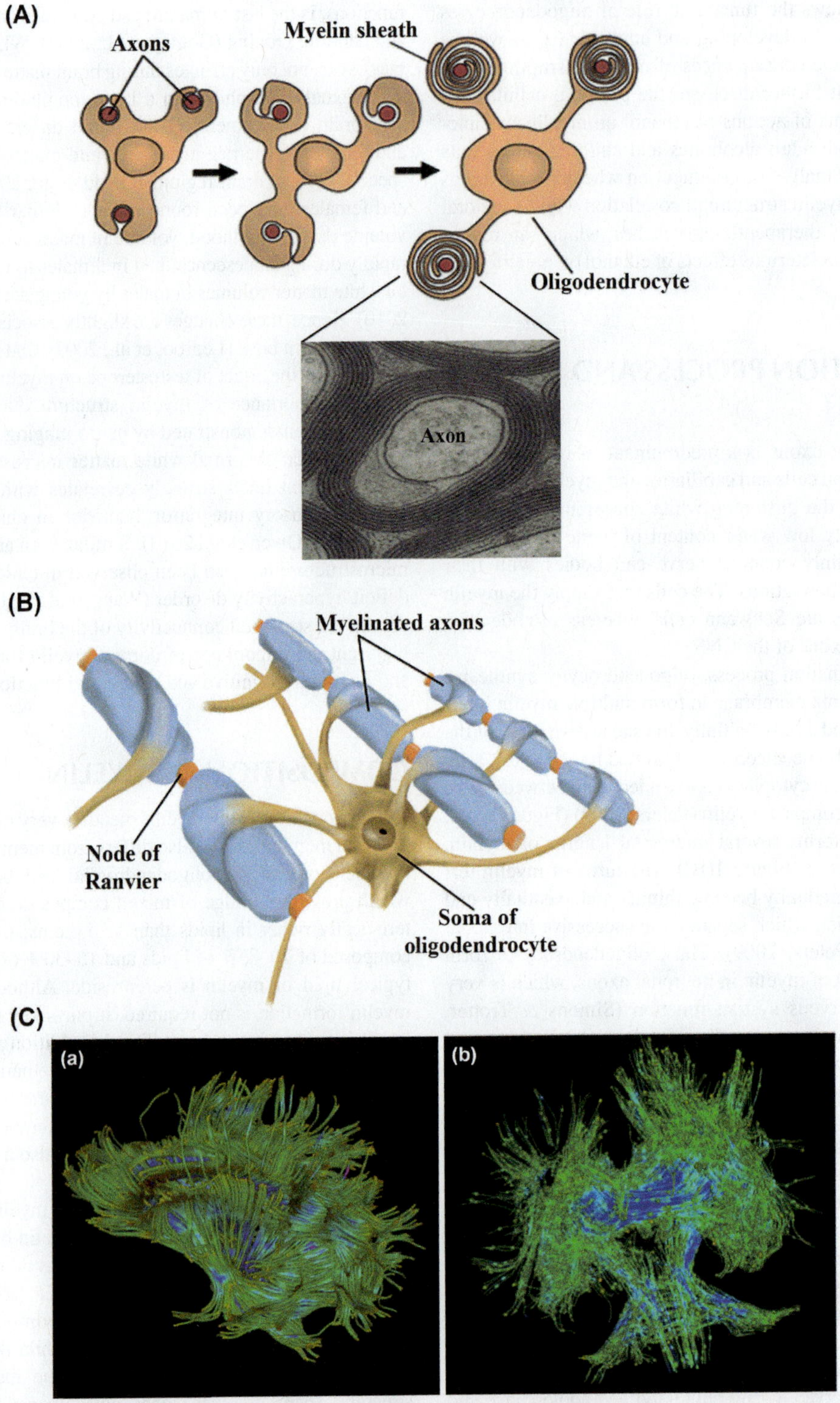

FIGURE 1 Myelin formation in the CNS. (A) An oligodendrocyte surrounds several axons. Although initially the external surfaces of the oligodendrocyte plasma membrane are not fused, the plasma membrane forms compacted spirals around the axon at the end of this process, leading to myelin sheaths. (B) The diagram shows how an oligodendrocyte wraps different axons to form multiple myelin internodes between the nodes of Ranvier. (C) A representative illustration of the white matter fiber tracts in a three-dimensional (3D) tract visualization of the human brain pathways using the diffusion tensor imaging technique and tractography. Diffusion tensor imaging is a noninvasive procedure to represent the organization of in vivo brain white matter pathways, while tractography is a 3D modeling technique to obtain virtual reconstruction of the trajectory of water molecules along white matter bundles, using data collected by diffusion tensor imaging. (a) Sagittal view with tube tracts display. (b) Coronal view with tensor glyphs (ellipsoids) and line tracts display, which depicted the corpus callosum white matter fiber tracts (blue). *These images have been provided by Dr. Maria de la Iglesia Vayá, with 3DSlider software, at the Systems Biology Department, Príncipe Felipe Research Centre, Valencia, Spain.*

promotes oligodendrocyte progenitor cell proliferation, and increases in response to injury and demyelination (Kang, Fukaya, Yang, Rothstein, & Bergles, 2010). Finally, the myelin-associated glycoprotein (MAG), an oligodendroglial membrane protein, plays a role in glia–axon interactions. However, this protein is not essential for myelin formation because MAG-null mice myelinate relatively normally. MAG also confers axonal protection from acute toxic insults, including inflammatory mediators (Nguyen et al., 2009).

PATHOPHYSIOLOGY OF THE DEMYELINATION AND REMYELINATION PROCESSES IN THE CNS

Myelination processes are relevant to the development of the CNS and the adult brain because they allow nerve impulses to propagate faster along myelinated fibers. Thus, both the demyelination and remyelination processes, as well as the neuroprotection of the myelin structures of nerve fibers, are the hallmark of diverse neurodegenerative or autoimmune diseases or brain damage induced by toxic conditions (Kadhim, De Prez, Gazagnes, & Sebire, 2003). One of the pathologic conditions associated with brain damage and demyelination is the activation of the brain innate immune system response, which can cause neuroinflammation.

Until quite recently, the CNS has been considered an immunologically privileged site. While the blood–brain barrier protects the CNS from peripheral immune and inflammatory activation, the CNS immune cells, such as microglia and astrocytes, are able to respond to infections or insults by releasing cytokines and toxic inflammatory compounds. However, although the acute inflammatory response is generally beneficial, sustained activation of glial cells and overproduction of inflammatory mediators, such as cytokines and chemokines (Donnelly & Popovich, 2008), can produce an inflammatory environment that enhances the recruitment and activation of immune cells, and amplifies the immune response (see Figure 2). Accumulation of inflammatory products can lead to myelin disruption and axonal damage. Indeed, the myelin debris generated after brain damage not only stimulates the inflammatory response but also inhibits axonal regeneration and brain remyelination (Kotter, Li, Zhao, & Franklin, 2006).

The regeneration of myelin sheaths, or remyelination, after CNS demyelination is important for restoring saltatory conduction and for preventing axonal loss. During remyelination, oligodendrocyte precursor cells, distributed throughout both the gray and white matter of the CNS, are activated in response to myelin injury and undergo proliferation, migration to the site of damage, and differentiation into mature myelinating oligodendrocytes to create new thinner myelin sheaths than the normal ones on demyelinated axons. This process helps to protect the axon from further damage and overall degeneration and, once again, increases conductance (Franklin & Ffrench-Constant, 2008). However, adult axons have a limited regrowth capacity after injury, mainly because of the presence of inhibitory molecules. For instance, myelin-associated proteins, such as Nogo, MAG, and OMgp, limit axonal outgrowth, while their blockage improves the regeneration of damaged fiber tracts, one of the main hindrances to neuronal regeneration (Geoffroy & Zheng, 2014). Yet under pathologic conditions, these myelin proteins (e.g., Nogo, MAG, or OMgp) display inhibitory effects for neuronal regeneration, and several findings have demonstrated that the genes associated with the immune response,

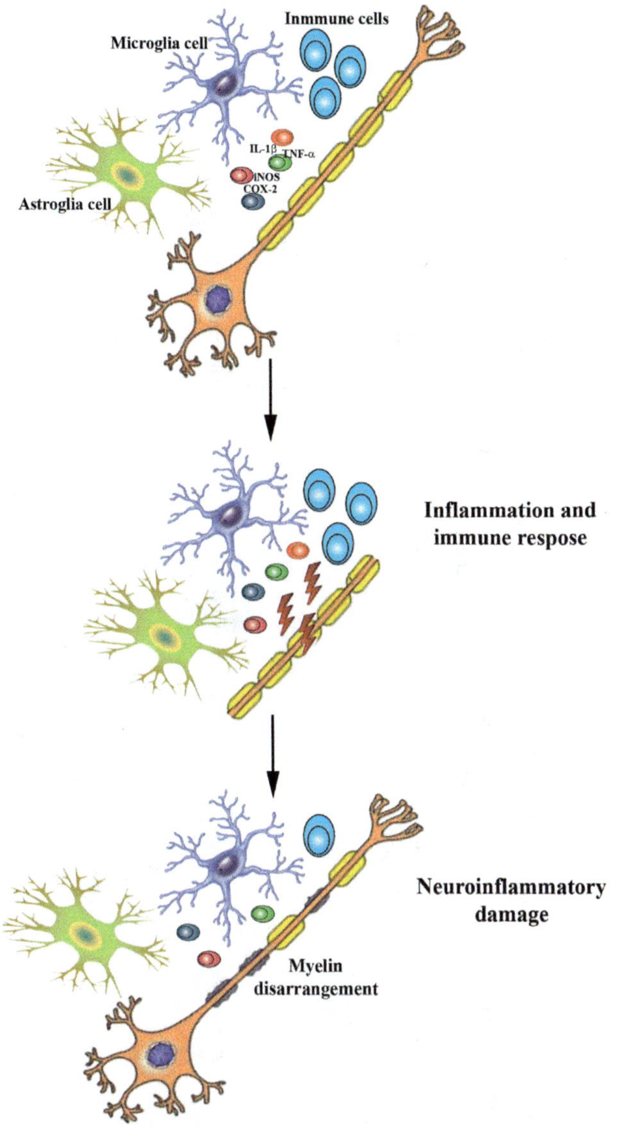

FIGURE 2 Immune response and myelin disruptions. It has been proposed that the activation of glial cells and the release of inflammatory mediators and cytokines, in addition to the direct attack by immune cells, can cause neuroinflammatory damage with myelin disarrangements.

such as TNF-α or MHC-II, can be involved in the regeneration of oligodendrocytes and remyelination (Arnett, Wang, Matsushima, Suzuki, & Ting, 2003). As the inflammatory response may also contribute to progressive axonal loss, it is necessary to understand CNS remyelination mechanisms in order to preserve axon integrity and to develop effective remyelination therapies.

ALCOHOL ABUSE AND WHITE MATTER DISTURBANCES

Myelin Dysfunction in Alcoholics

The first evidence that alcohol abuse can negatively affect white matter and disturb both myelin and axonal integrity in the alcoholic brain were mainly derived from both postmortem and neuroimaging studies (Table 1). Brain samples from postmortem

TABLE 1 White Matter and Myelin Alterations in Alcoholics

Material	Brain Regions	Effects	Study
Diffusion tensor imaging	Fronto-occipital fasciculus (FOF) fiber bundles	White matter (WM) integrity in FOF bundles might be related with to visual processing skills	Bagga et al. (2014)
Gene expression	Postmortem prefrontal and primary motor cortices	Alteration of WM in nonhepatic encephalopathy alcoholics	Sutherland, Sheedy, Sheahan, Kaplan, and Kril (2014)
Tract based spatial statistics and Iowa gambling task	Corpus callosum, and parietal, occipital and frontal regions	Decision-making related with abnormal WM integrity	Zorlu et al. (2013)
Gene expression	Postmortem hippocampal tissue	Myelination and neurogenesis and inflammation, hypoxia and stress	McClintick et al. (2013)
Diffusion tensor imaging	Inferior and superior longitudinal fasciculus in the left hemisphere	Exposure to alcohol and familial risk for alcohol abuse predicts WM	Hill, Terwilliger, and McDermott (2013)
Loss in connectivity	Brain reward system	WM alterations in alcohol dependence associated with brain reward system	Kuceyeski, Meyerhoff, Durazzo, and Raj (2013)
Immunohistochemistry	Postmortem cortical, striatal, and nigral regions	Superoxide dismutase-1 and metalloproteinase-9 associated with severe myelin damage	Skuja et al. (2013)
Diffusion tensor imaging	Brain reward system	Abnormalities in WM contribute to neurocognitive and executive functioning deficits	Durkee, Sarlls, Hommer, and Momenan (2013)
Magnetic resonance imaging	Whole brain	Additional liver disease and malnutrition induce osmotic demyelination	Malhotra and Ortega (2013)
Magnetic resonance imaging	Frontal, temporal, ventricular, and corpus callosum regions	Gender differences in WM volumes	Ruiz et al. (2012)
Magnetic resonance imaging	Corpus callosum region	WM abnormalities contribute to pathological gambling	Yip et al. (2013)
Diffusion tensor imaging	Frontal lobes	Abnormal frontal WM tract related with fractional anisotropy and radial diffusivity	Sorg et al. (2012)
Diffusion tensor imaging	Commissural region	WM associated with cognitive and motor impairment	Pfefferbaum et al. (2010)
Diffusion tensor imaging	Superior longitudinal fasciculus region	More severe WM disruption development in females than in males	Thatcher, Pajtek, Chung, Terwilliger, and Clark (2010)
Diffusion tensor imaging	Limbic fiber tracts	WM abnormalities associated with emotion and cognition areas	Schulte et al. (2010)
Western blot analysis	Postmortem cortical and subcortical regions	Myelin proteins in cirrhotic alcoholics, no changes in brain pathology or brain weight	Lewohl, Wixey, Harper, and Dodd (2005)

This table lists the various studies performed using different techniques, which show that alcohol abuse can cause myelin dysfunctions in several brain regions of alcoholics.

alcoholic patients have indicated that alcohol abuse induces white matter atrophy, reduces several myelin-related genes, and causes spreading demyelination (Lewohl et al., 2000; McClintick et al., 2013). More recently, the development of new neuroimaging techniques, such as magnetic resonance imaging and diffusion tensor imaging (e.g., Figure 1(C)) with fractional anisotropy, have allowed the analysis and quantification of white matter integrity in vivo.

Using these methodologies, different studies have confirmed that alcohol abuse alters white matter microstructure and myelin dysfunction in different brain regions, and this correlates with behavioral impairments (see Table 1). Thus, microstructural abnormalities in white matter have been found in several important fiber bundles pathways (e.g., corpus callosum, cortico-limbic frontal, superior fronto-occipital fasciculus), and these myelin alterations have been correlated with deficits in cognitive and executive functioning, information processing, and motor impairments in alcoholics (Bagga et al., 2014; Durkee et al., 2013; Pfefferbaum et al., 2010). Alterations in cortico-limbic myelin fibers have been proposed to contribute to emotion, cognition, and behavioral deficits (Schulte et al., 2010). Gender differences in regional white matter dysfunctions have also been observed, with the frontal and temporal white matter being more affected in alcoholic women, with corpus callosum fibers found among alcoholic men (Ruiz et al., 2012). Regarding confounding factors, such as hepatic complications with hepatic encephalopathy in alcoholics, some findings have indicated that although ethanol abuse can directly induce protein myelin changes in alcoholics without liver disease (Sutherland et al., 2014), alcohol abusers with additional liver disease and malnutrition can induce osmolar disturbances in the cerebral microenvironment, which lead to loss of the myelin sheath of neurons and demyelination (Malhotra & Ortega, 2013).

Myelin Dysfunctions in Adolescent Drinkers

Although alcoholism is more associated with adults, many alcoholics report having started alcohol drinking during adolescence. As mentioned in the introduction, the adolescence stage is a critical period of brain maturation, in which certain brain areas, such as the prefrontal cortex, myelination of axons, and pruning of synapses, play an important role for efficient communication between brain regions, higher order cognitive functioning, and complex behaviors. Studies conducted in human adolescents have provided evidence that binge alcohol drinking reduces prefrontal white matter (De Bellis et al., 2005; Medina et al., 2008), alters myelin integrity in the superior longitudinal fasciculus (Elofson, Gongvatana, & Carey, 2013), and reduces myelin fiber tracts with frontal connections (Bava et al., 2013). These events in myelin fiber tracts alterations, as observed with neuroimaging studies, have been correlated with long-term dysfunctions in the cognitive abilities of learning, memory, and executive functions reported in binge-drinking adolescents (Jacobus & Tapert, 2013). It is noteworthy that neuroimaging studies have also revealed that female adolescents who report binge drinking exhibit a more marked reduction in the prefrontal cortex with larger white fiber disruptions, such as corpus callosum, than male adolescents (De Bellis et al., 2005; Medina et al., 2008; Pfefferbaum et al., 2010).

Prenatal Alcohol Exposure and Alterations in White Matter

Alterations in white matter have also been observed in children exposed to alcohol prenatally. It is well-established that alcohol is a teratogen, and that in utero exposure can cause dysfunctions in cognitive and psychosocial functioning, as well as alterations in brain structure and function. These impairments are present in both children with fetal alcohol syndrome and in utero alcohol exposure, who do not present the facial dysmorphology called fetal alcohol spectrum disorders (Guerri, Bazinet, & Riley, 2009). Neuropsychological and brain-imaging studies have detected differences in brain structure relating to alcohol exposure in multiple brain systems and abnormalities in the white matter connecting different brain regions (see Nunez, Roussotte, & Sowell, 2011). Alcohol-induced apoptosis of oligodendrocytes in the fetal macaque brain has been proposed to be a mechanism of myelin disruption (Creeley et al., 2013). Notwithstanding, ethanol reduces growth factors and essential elements that maintain myelin structure during brain development, and this could also be involved.

MECHANISMS OF ALCOHOL-INDUCED MYELIN ALTERATIONS

Even though the neuropathological basis of ethanol-induced myelin disruption is presently uncertain, several mechanisms have been proposed. Some authors have suggested that ethanol, by inducing oxidative stress, can affect myelin structure because reduced superoxide dismutase-1 is accompanied by severe myelin damage in some brain regions (Skuja et al., 2013), indicating that oxidative stress is one of the widespread contributors to alcohol-induced brain damage. Other authors have proposed that by delaying c-Fos downregulation, ethanol can disrupt the normal timing for the expression of those genes involved in oligodendrocytes differentiation, and that MBP expression can alter the myelin sheath composition (Bichenkov & Ellingson, 2009).

It has been shown that by activating the innate immune system, and specifically toll-like receptor 4 (TLR4) in glial cells, ethanol can trigger the production of pro-inflammatory cytokines (IL-1β, TNF-α, IL-6) and inflammatory mediators (iNOS, COX-2) to induce neuroinflammation and brain damage (Alfonso-Loeches et al., 2010; Fernandez-Lizarbe, Pascual, & Guerri, 2009). Inhibition or elimination of the TLR4 signaling pathway, by using TLR4-KO (knockout) mice, abolishes the ethanol-induced activation of astroglia and microglial cells, the production of cytokines and inflammatory mediators, and neural death in the prefrontal cortex (Alfonso-Loeches et al., 2010; Fernandez-Lizarbe et al., 2009). Furthermore, ethanol-induced neuroinflammation and brain damage is accompanied by a downregulation in the protein and gene expression of several proteins associated with myelination (PLP, MBP, MOG, CNP, and MAG), while increasing NG2-proteoglycan in several brain regions of ethanol-treated wild-type (WT) mice. Chronic ethanol treatment also causes oligodendrocyte death and induces major ultrastructural myelin sheath disarrangements in the corpus callosum and cerebral cortex of chronic ethanol-treated WT mice.

Elimination of TLR4 (TLR4-KO) in mice abolishes ethanol-induced myelin disruptions, although small focal fiber disruptions have been noted in the same brain areas (Alfonso-Loeches et al., 2012) (see Figure 3). These results highlight the importance of TLR4 and the inflammatory environment on the myelin alterations induced by alcohol abuse. Based on these results, a possible mechanism is presented in Figure 4, in which ethanol, by activating TLR4 receptor signaling in astroglial and microglial cells, triggers the release of inflammatory mediators and cytokines (iNOS, COX-2, IL-1β, TNF-α), leading to myelin disruption, brain damage, and apoptotic neuronal death. This inflammatory response not only induces myelin disarrangements but can cause behavior and cognitive dysfunctions (Pascual et al., 2011), as well as alcohol consumption (Mayfield et al., 2013). These results suggest that the ethanol-induced activation of the

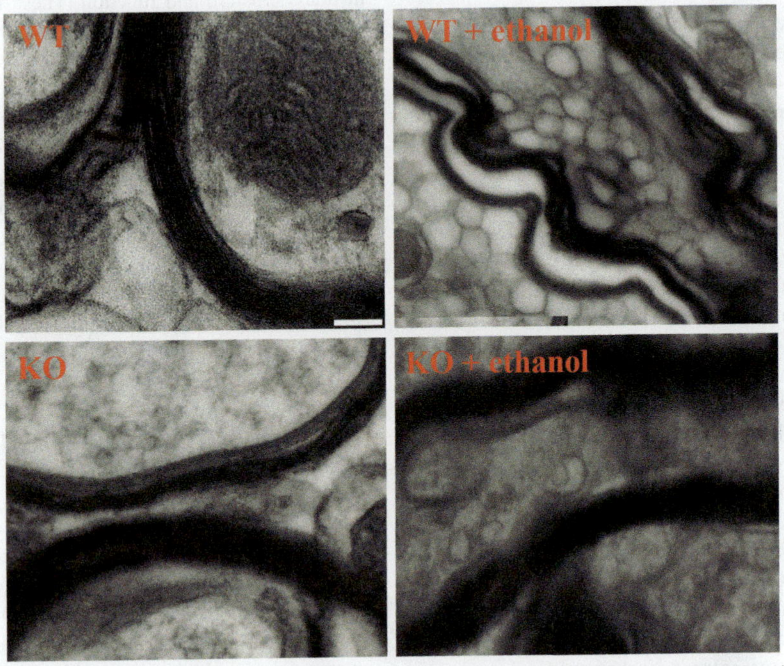

FIGURE 3 Electron microscopic examination of the corpus callosum myelinated axons of the wild-type (WT) and TLR4$^{-/-}$ knockout (KO) mice treated with or without ethanol for 5 months. The WT mice treated with chronic ethanol show myelin sheath disarrangements when compared with their control counterparts, while minimal changes in myelin disruptions are noted in the ethanol-treated TLR4$^{-/-}$ mice. The scale bar is 100 nm. *These images have been provided by Dr. Jaime Renau-Piqueras at the La Fe Research Hospital Centre, Valencia, Spain.*

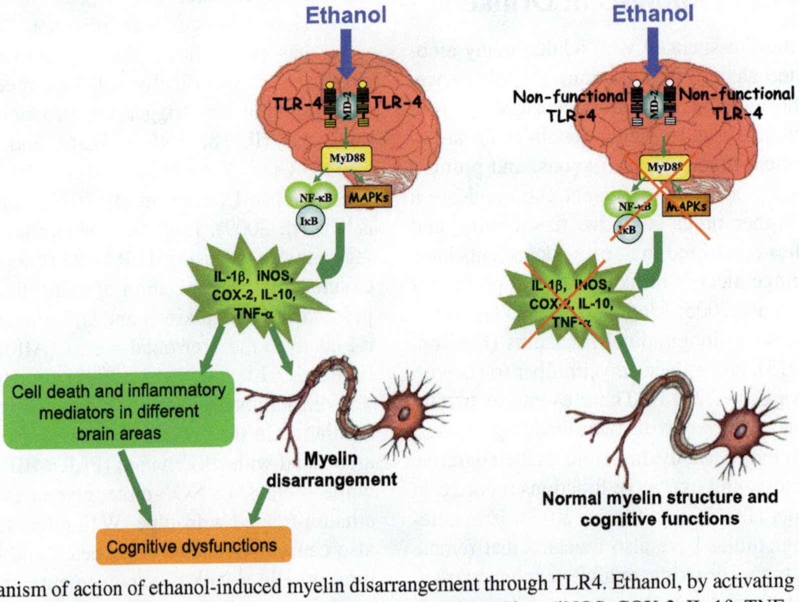

FIGURE 4 Possible mechanism of action of ethanol-induced myelin disarrangement through TLR4. Ethanol, by activating the TLR4 signaling pathway in astroglial and microglial cells, triggers the release of inflammatory mediators and cytokines (iNOS, COX-2, IL-1β, TNF-α), which leads to brain damage and apoptotic neuronal death, and also to myelin disarrangements and cognitive impairments. The TLR4 knockout is able to abolish ethanol-induced neuroinflammation and myelin disruptions.

neuroimmune system and TLR4 signaling contribute to white matter loss and behavioral dysfunctions associated with alcohol abuse.

Data have demonstrated that inflammatory damage can also contribute to the myelin alteration caused by binge drinking during adolescence. Studies done in experimental animals have demonstrated that binge-like ethanol treatment during adolescence upregulates the expression of TLR4 and TLR2, and the levels of inflammatory mediators, in the prefrontal cortex (Pascual, Pla, Minarro, & Guerri, 2014). These events have been associated with a sharp drop in several myelin proteins, such as MBP and myelin oligodendrocyte glycoprotein (MOG) (Pascual et al., 2014), suggesting that the activation of TLRs and inflammatory cytokines can induce myelin disruptions—effects that might participate in the long-term cognitive and behavioral impairments observed in rats exposed to alcohol during adolescence (Pascual, Blanco, Cauli, Minarro, & Guerri, 2007). It is noteworthy that despite low levels of cytokines having protective effects, overproduction of cytokines and inflammatory mediators by chronic ethanol abuse can cause neuroinflammation, demyelination, and brain damage. Some evidence suggests that in mice lacking TNF-α, remyelination decreases and neuronal damage increases following an insult, which indicates the neuroprotective role of TNF-α (Schmitz & Chew, 2008). Similarly, whereas high NO levels have toxic effects on differentiated oligodendroglia (Boullerne, Nedelkoska, & Benjamins, 2001), moderate iNOS levels in astroglia and microglia protect oligodendroglia against ethanol toxicity through the activation of the NO-cGMP-PKG pathway and NF-κB (Bonthius, Bonthius, Li, & Karacay, 2008).

In addition to the direct neurotoxic effects of ethanol within the brain, alcohol misuse may also establish a liver–brain axis of neurodegeneration mediated by toxic lipid trafficking across the blood–brain barrier. For example, when ceramides arrive at the CNS, they initiate a complex cascade of neurodegeneration mediated by insulin resistance, inflammation, and oxidative stress, leading to neuronal loss with progressive white matter degeneration and cognitive impairment. A study of human alcoholics has shown that white matter atrophy and degeneration with a reduced expression of myelin-associated genes, and increased levels of oxidative stress, are associated with an increased expression of pro-ceramide genes (de la Monte et al., 2008).

Finally, although different mechanisms have been proposed to explain the effects of ethanol on myelin disruption, data from experimental studies suggest that scavengers or anti-inflammatory compounds, which can prevent the production of oxygen radical species and cytokines and other inflammatory mediators, may prevent or ameliorate the effects of ethanol on myelin structure.

APPLICATIONS TO OTHER ADDICTIONS AND SUBSTANCE MISUSE

Evidence from clinical and experimental studies demonstrates that the use of different substances misuse can impair white matter in the CNS. We now go on to offer some examples.

Cocaine

Studies have shown that cocaine—one of the most frequently abused drugs in the world—induces dysfunction in white matter and in myelin structure. Postmortem studies of cocaine abusers have evidenced that the myelin PLP1 expression diminishes in the ventral and dorsal regions of the caudate, putamen, and internal capsule, a protein that is essential for maintaining stability of the myelin sheaths—effects that have been attributed to the adaptive changes that follow chronic cocaine abuse (Kristiansen, Bannon, & Meador-Woodruff, 2009). Microarray analyses have confirmed that cocaine abuse lowers the expression of various myelin-related genes (Albertson, Schmidt, Kapatos, & Bannon, 2006). According to these studies, neuroimaging studies also show an altered prefrontal white matter structure (Lim, Choi, Pomara, Wolkin, & Rotrosen, 2002) and abnormal white matter maturation in the frontal and temporal lobes (Bartzokis et al., 2002) in cocaine-dependent individuals. These findings indicate that cocaine causes a dysregulation of myelin in human cocaine abusers.

Amphetamine

Another illicit drug is amphetamine which, despite being an approved drug for the treatment of attention-deficit/hyperactivity disorder and narcolepsy, is also an addictive CNS stimulant and has been commonly used illegally by adolescents and young adults. Berman, O'Neill, Fears, Bartzokis, and London (2008) reviewed the association between amphetamine abuse structural changes in the brain and suggested that white matter abnormalities in amphetamine abusers occur more often than gray matter abnormalities. Indeed, frontal white matter integrity has been reported to be altered in methamphetamine abusers (Chung et al., 2007) and chronic methamphetamine abuse induces prominent occipital and temporal white matter hypertrophy, which may result from altered myelination and adaptive glial changes, including gliosis secondary to neuronal damage (Thompson et al., 2004). Similarly, adult methamphetamine abusers have presented microstructural abnormalities in the white matter fibers interconnecting prefrontal cortices, and also in the hippocampus in abstinent individuals, and these effects have been correlated to the psychiatric symptoms assessed during this period (Tobias et al., 2010).

Studies conducted with experimental animals have also shown that repeated administrations with a nonneurotoxic dose of amphetamine can cause microstructural changes in white matter in some brain regions—effects that have been associated with behavioral sensitization and impaired working memory. Notably, methamphetamine exposure in prenatal children leads to abnormalities in the white matter microstructures in the frontal lobe—effects that have been correlated with long-term executive dysfunction (Colby et al., 2012). These findings suggest that myelin structure is sensitive to amphetamine toxicity in the adult and developing brains.

Cannabis

Cannabis (marijuana) use typically begins in adolescence and early adulthood—a period when cannabinoid receptors are still abundant in white matter pathways across the brain. Findings have indicated that long-term cannabis abuse can be hazardous to the white matter in the developing brain. Delaying the age at which regular use begins may minimize the degree of severity of microstructural myelin impairments (Zalesky et al., 2012).

Using diffusion tensor imaging, differences in the mean diffusivity or fractional anisotropy in the corpus callosum and in frontal white matter fiber tracts have been observed in chronic cannabis subjects (Gruber, Silveri, Dahlgren, & Yurgelun-Todd, 2011). These findings suggest that chronic cannabis exposure can alter white matter structural integrity either by affecting demyelination or causing axonal damage, or even indirectly through delaying normal brain development. Notably, early marijuana use results in reduced fractional anisotropy in frontal white matter tracts—events that have been associated with increased impulsivity and that ultimately contribute to the initiation of marijuana use or the inability to discontinue use (Gruber et al., 2011).

The cerebellum is rich in cannabinoid receptors and is implicated in the neuropathology of schizophrenia. Long-term heavy cannabis use in healthy individuals has been associated with smaller cerebellar white matter volume, similar to that observed in schizophrenia. Reduced volumes have been reported to be even more pronounced in patients with schizophrenia who use cannabis, suggesting that cannabis use can alter the course of the brain maturational processes associated with schizophrenia (Solowij et al., 2011).

In summary, there is evidence to demonstrate that cannabis use in both early adolescence and adulthood is hazardous to white matter, leading to altered brain maturation processes and contributing to some behavioral dysfunctions associated with cannabis abuse and addiction.

Opioids

White matter damage and myelin pathology have been observed in opiate addiction, and these effects have been related to impaired neuronal connectivity and cognitive deficit in opiate addicts (Bora et al., 2012; Qiu et al., 2013). Alterations in the white matter of heroin-dependent subjects have been associated with the length of heroin dependence and correlated with decision-making impairments (Qiu et al., 2013). Similarly to alcohol, the disruption of normal white matter development from exposure to opiates is usually more severe in females than in males (De Bellis et al., 2005; Medina et al., 2008; Thatcher et al., 2010).

Nicotine

Human neuroimaging studies have shown an abnormal white matter structure among smokers (Cao et al., 2013; Lin, Wu, Zhu, & Lei, 2012). Myelination defects have also been observed in the brains of adolescent rats exposed to nicotine during the gestational period. Indeed, gestational nicotine exposure alters the expression of a number of genes involved in myelination processes in adolescent brains, with sex and brain region differences, indicating that gestational nicotine disturbs oligodendrocyte development and myelin formation. Alterations in the expression of several myelin genes in adolescent rats exposed to gestational nicotine have been related with changes in transcription and trophic factors, such as neuregulin receptors Erbb3 and aspartyl protease BACE1, which is involved in neuregulin proteolysis. These results suggest that abnormal brain myelination underlies several psychiatric disorders and drug abuse, including prenatal exposure to cigarette smoke (Cao et al., 2013).

Polydrug Use

Polydrug abuse has also been associated with brain matter alterations. Hence, some studies have analyzed the integrity of neuroanatomical pathways in adolescent marijuana users with concomitant alcohol use; they have demonstrated major alterations in the fronto-parietal circuitry, which leads to white matter disruptions and suggests that adolescent marijuana and alcohol users can present aberrant axonal and myelin maturation with compromised myelin fiber integrity (Bava et al., 2009).

A report evaluated the white matter integrity of pre- and post-marijuana and alcohol initiation in adolescence (Jacobus & Tapert, 2013). The findings demonstrate that, in most regions of the brain, teenagers who used both alcohol and marijuana show white matter integrity that is greater than or equal to those who initiated alcohol use only. The findings suggest poorer tissue integrity in association with the combined initiation of heavy alcohol drinking and marijuana use. Other studies have confirmed these findings, showing an abnormal white matter microstructure in adults who report substance use disorders (Baker, Yucel, Fornito, Allen, & Lubman, 2013).

However, the mechanisms participating in the white matter alterations induced by drugs of abuse are currently uncertain. Furthermore, magnetic resonance imaging in substance abusers (heroin, cocaine, and cannabis, but not alcohol) demonstrated reduced frontal lobe white matter volume compared with matched controls (Schlaepfer et al., 2006). This finding suggests that either polydrug abuse affects frontal lobe maturation leading to impaired judgment, or a preexisting condition leads individuals to substance misuse and consecutive abuse.

THERAPEUTIC APPROACHES

Therapies that can prevent or ameliorate the myelin dysfunction induced by ethanol abuse or other substance misuse are warranted. In this context, the administration of natural antioxidant and anti-inflammatory compounds can ameliorate the neuroinflammation (e.g., Pascual et al., 2007) and myelin dysfunctions induced by ethanol abuse. Transplantation of oligodendrocyte progenitor cells, derived from human-induced pluripotent stem cells, has been suggested to allow the treatment of demyelinated brain disorders (Wang et al., 2013).

DEFINITION OF TERMS

Nodes of Ranvier These are the gaps formed between the myelin sheath where the axons are left uncovered. Because the myelin sheath is largely composed of an insulating fatty substance, the nodes of Ranvier allow the generation of a fast electrical impulse along the axon. This rapid rate of conduction is called *saltatory conduction*.

Magnetic resonance imaging MRI is a type of scan that uses strong magnetic fields and radio waves to produce detailed images of the brain or other body organs.

Diffusion tensor imaging Also called diffusion magnetic resonance imaging, this is a method that provides a description of the diffusion of water through tissue, and can be used to highlight structural changes in tissue tracts. It is commonly used in brain research.

Fractional anisotropy This method is used to evaluate white matter fiber tracts.

Oligodendrocytes These are the myelinating cells of the CNS, which provide structural support to axons and produce the myelin sheath.

Neuroinflammation This is an early, nonspecific immune reaction to tissue damage or pathogen invasion in the CNS, characterized by increased glial activation, pro-inflammatory cytokine concentration, blood–brain-barrier permeability, and leukocyte invasion.

Toll-like receptors TLRs are receptors that play an essential role in the activation of innate immunity by recognizing specific patterns of microbial components. Their activation leads to an inflammatory response with the release of cytokines and inflammatory mediators.

Cytokines These small proteins include chemokines, interferons, interleukins, and tumor necrosis factor. They are produced by a wide range of cells, including immune cells such as macrophages, microglia, or astrocytes and act through receptors to modulate the immune response.

KEY FACTS

- Myelin is an important brain white matter component and is fundamental for impulse conduction along the nerve fibers in the CNS.
- Demyelination and remyelination processes are the hallmark of diverse neurodegenerative or autoimmune diseases or brain damage induced by toxic conditions.
- Activation of the brain's innate immune system response can cause neuroinflammation, a pathologic condition associated with demyelination.
- Alcohol consumption is the third largest risk factor for disease and disabilities worldwide, including those related to CNS dysfunctions (WHO, 2011, 2014).
- Alcohol abuse can induce white matter dysfunction in both the adult and developing brains, including adolescence. In all cases, alterations in white matter are associated with cognitive and behavioral dysfunction.
- One of the mechanisms involved in ethanol-induced brain damage and myelin dysfunction/disruptions is that through the activation of the innate immune response, which induces a brain inflammatory response, or neuroinflammation.
- The use and abuse of other drugs can also cause white matter dysfunction.

SUMMARY POINTS

- This chapter focuses on the effects of ethanol on the myelin structure observed in adult alcoholics and during brain development.
- New neuroimaging techniques show that alcohol abuse causes microstructural white matter abnormalities in several important fiber bundle pathways (e.g., corpus callosum, cortico-limbic frontal, superior fronto-occipital fasciculus).
- Myelin alterations in alcoholic brains have been correlated with deficits in cognitive and executive functioning, information processing, and motor impairments.
- Alcohol exposure during the prenatal period or during adolescence also induces white matter reduction and dysfunction—events associated with cognitive deficits.
- The mechanisms of ethanol-induced dysfunctions in CNS myelin are presently unknown.
- Recently, experimental evidence demonstrated that by activating the innate immune receptors and TLR4 signaling in glial cells, ethanol triggers inflammatory mediators and cytokines, which can damage myelin structure.
- Elimination of TLR4 abolishes most of the effects induced by ethanol.
- Clinical evidence has also shown impairments of the myelin and brain white matter structure in individuals who misuse different substances, such as cocaine, amphetamine, cannabis, opioids, or nicotine.
- Programs or therapies that can prevent or ameliorate myelin dysfunction induced by ethanol abuse or other substances misuse are warranted.

ACKNOWLEDGMENTS

This work has been supported by grants from the Spanish Ministry of Economics and Competitiveness (SAF2012-33747), the Spanish Ministry of Health: The Institute Carlos III and FEDER funds (RTA-Network RD12-0028-007), and PNSD (Ex. 20101037), GV-Consellería de Educación: PROMETEO ACOM 2014-062; ERAB(EA-13-08).

REFERENCES

Albertson, D. N., Schmidt, C. J., Kapatos, G., & Bannon, M. J. (2006). Distinctive profiles of gene expression in the human nucleus accumbens associated with cocaine and heroin abuse. *Neuropsychopharmacology, 31*(10), 2304–2312.

Alfonso-Loeches, S., Pascual, M., Gomez-Pinedo, U., Pascual-Lucas, M., Renau-Piqueras, J., & Guerri, C. (2012). Toll-like receptor 4 participates in the myelin disruptions associated with chronic alcohol abuse. *Glia, 60*(6), 948–964.

Alfonso-Loeches, S., Pascual-Lucas, M., Blanco, A. M., Sanchez-Vera, I., & Guerri, C. (2010). Pivotal role of TLR4 receptors in alcohol-induced neuroinflammation and brain damage. *Journal of Neuroscience, 30*(24), 8285–8295.

Arnett, H. A., Wang, Y., Matsushima, G. K., Suzuki, K., & Ting, J. P. (2003). Functional genomic analysis of remyelination reveals importance of inflammation in oligodendrocyte regeneration. *Journal of Neuroscience, 23*(30), 9824–9832.

Bagga, D., Sharma, A., Kumari, A., Kaur, P., Bhattacharya, D., Garg, M. L., ... Singh, N. (2014). Decreased white matter integrity in fronto-occipital fasciculus bundles: relation to visual information processing in alcohol-dependent subjects. *Alcohol, 48*(1), 43–53.

Baker, S. T., Yucel, M., Fornito, A., Allen, N. B., & Lubman, D. I. (2013). A systematic review of diffusion weighted MRI studies of white matter microstructure in adolescent substance users. *Neuroscience & Biobehavioral Reviews, 37*(8), 1713–1723.

Bartzokis, G., Beckson, M., Lu, P. H., Edwards, N., Bridge, P., & Mintz, J. (2002). Brain maturation may be arrested in chronic cocaine addicts. *Biological Psychiatry, 51*(8), 605–611.

Bava, S., Frank, L. R., McQueeny, T., Schweinsburg, B. C., Schweinsburg, A. D., & Tapert, S. F. (2009). Altered white matter microstructure in adolescent substance users. *Psychiatry Research, 173*(3), 228–237.

Bava, S., Jacobus, J., Thayer, R. E., & Tapert, S. F. (2013). Longitudinal changes in white matter integrity among adolescent substance users. *Alcoholism: Clinical & Experimental Research, 37*(Suppl. 1), E181–E189.

Berman, S., O'Neill, J., Fears, S., Bartzokis, G., & London, E. D. (2008). Abuse of amphetamines and structural abnormalities in the brain. *Annals of the New York Academy of Sciences, 1141,* 195–220.

Bichenkov, E., & Ellingson, J. S. (2009). Ethanol alters the expressions of c-Fos and myelin basic protein in differentiating oligodendrocytes. *Alcohol, 43*(8), 627–634.

Bonthius, D. J., Bonthius, N. E., Li, S., & Karacay, B. (2008). The protective effect of neuronal nitric oxide synthase (nNOS) against alcohol toxicity depends upon the NO-cGMP-PKG pathway and NF-kappaB. *Neurotoxicology, 29*(6), 1080–1091.

Bora, E., Yucel, M., Fornito, A., Pantelis, C., Harrison, B. J., Cocchi, L., ... Lubman, D. I. (2012). White matter microstructure in opiate addiction. *Addiction Biology, 17*(1), 141–148.

Boullerne, A. I., Nedelkoska, L., & Benjamins, J. A. (2001). Role of calcium in nitric oxide-induced cytotoxicity: EGTA protects mouse oligodendrocytes. *Journal of Neuroscience Research, 63*(2), 124–135.

Braun, P. E., Lee, J., & Gravel, M. (2004). 2′,3′-cyclic nucleotide 3′-phosphodiesterase: structure, biology and function. In R. A. Lazzarini (Ed.), *Myelin biology and disorders* (pp. 499–522). San Diego, CA: Elsevier Academic Press.

Cao, J., Wang, J., Dwyer, J. B., Gautier, N. M., Wang, S., Leslie, F. M., & Li, M. D. (2013). Gestational nicotine exposure modifies myelin gene expression in the brains of adolescent rats with sex differences. *Translational Psychiatry, 3*, e247.

Chung, A., Lyoo, I. K., Kim, S. J., Hwang, J., Bae, S. C., Sung, Y. H., ... Renshaw, P. F. (2007). Decreased frontal white-matter integrity in abstinent methamphetamine abusers. *International Journal of Neuropsychopharmacology, 10*(6), 765–775.

Colby, J. B., Smith, L., O'Connor, M. J., Bookheimer, S. Y., Van Horn, J. D., & Sowell, E. R. (2012). White matter microstructural alterations in children with prenatal methamphetamine/polydrug exposure. *Psychiatry Research, 204*(2–3), 140–148.

Creeley, C. E., Dikranian, K. T., Johnson, S. A., Farber, N. B., & Olney, J. W. (2013). Alcohol-induced apoptosis of oligodendrocytes in the fetal macaque brain. *Acta Neuropathologica Communications, 1*(1), 23.

De Bellis, M. D., Narasimhan, A., Thatcher, D. L., Keshavan, M. S., Soloff, P., & Clark, D. B. (2005). Prefrontal cortex, thalamus, and cerebellar volumes in adolescents and young adults with adolescent-onset alcohol use disorders and comorbid mental disorders. *Alcoholism: Clinical & Experimental Research, 29*(9), 1590–1600.

Donnelly, D. J., & Popovich, P. G. (2008). Inflammation and its role in neuroprotection, axonal regeneration and functional recovery after spinal cord injury. *Experimental Neurology, 209*(2), 378–388.

Durkee, C. A., Sarlls, J. E., Hommer, D. W., & Momenan, R. (2013). White matter microstructure alterations: a study of alcoholics with and without post-traumatic stress disorder. *PLoS One, 8*(11), e80952.

Elofson, J., Gongvatana, W., & Carey, K. B. (2013). Alcohol use and cerebral white matter compromise in adolescence. *Addictive Behaviors, 38*(7), 2295–2305.

Fernandez-Lizarbe, S., Pascual, M., & Guerri, C. (2009). Critical role of TLR4 response in the activation of microglia induced by ethanol. *Journal of Immunology, 183*(7), 4733–4744.

Fornari, E., Knyazeva, M. G., Meuli, R., & Maeder, P. (2007). Myelination shapes functional activity in the developing brain. *NeuroImage, 38*(3), 511–518.

Franklin, R. J., & Ffrench-Constant, C. (2008). Remyelination in the CNS: from biology to therapy. *Nature Reviews Neuroscience, 9*(11), 839–855.

Gemert, M., & Killeen, J. (1998). Chemically induced myelinopathies. *International Journal of Toxicology, 17*(3), 231–275.

Geoffroy, C. G., & Zheng, B. (2014). Myelin-associated inhibitors in axonal growth after CNS injury. *Current Opinion in Neurobiology, 27C*, 31–38.

Giedd, J. N., Stockman, M., Weddle, C., Liverpool, M., Alexander-Bloch, A., Wallace, G. L., ... Lenroot, R. K. (2010). Anatomic magnetic resonance imaging of the developing child and adolescent brain and effects of genetic variation. *Neuropsychology Review, 20*(4), 349–361.

Gogtay, N., Giedd, J. N., Lusk, L., Hayashi, K. M., Greenstein, D., Vaituzis, A. C., ... Thompson, P. M. (2004). Dynamic mapping of human cortical development during childhood through early adulthood. *Proceedings of the National Academy of Sciences of the United States of America, 101*(21), 8174–8179.

Gruber, S. A., Silveri, M. M., Dahlgren, M. K., & Yurgelun-Todd, D. (2011). Why so impulsive? White matter alterations are associated with impulsivity in chronic marijuana smokers. *Experimental and Clinical Psychopharmacology, 19*(3), 231–242.

Guerri, C., Bazinet, A., & Riley, E. P. (2009). Foetal alcohol spectrum disorders and alterations in brain and behaviour. *Alcohol and Alcoholism, 44*(2), 108–114.

Harper, C., & Matsumoto, I. (2005). Ethanol and brain damage. *Current Opinion in Pharmacology, 5*(1), 73–78.

Hill, S. Y., Terwilliger, R., & McDermott, M. (2013). White matter microstructure, alcohol exposure, and familial risk for alcohol dependence. *Psychiatry Research, 212*(1), 43–53.

Jacobus, J., & Tapert, S. F. (2013). Neurotoxic effects of alcohol in adolescence. *Annual Review of Clinical Psychology, 9*, 703–721.

Kadhim, H., De Prez, C., Gazagnes, M. D., & Sebire, G. (2003). In situ cytokine immune responses in acute disseminated encephalomyelitis: insights into pathophysiologic mechanisms. *Human Pathology, 34*(3), 293–297.

Kang, S. H., Fukaya, M., Yang, J. K., Rothstein, J. D., & Bergles, D. E. (2010). NG2+ CNS glial progenitors remain committed to the oligodendrocyte lineage in postnatal life and following neurodegeneration. *Neuron, 68*(4), 668–681.

Kotter, M. R., Li, W. W., Zhao, C., & Franklin, R. J. (2006). Myelin impairs CNS remyelination by inhibiting oligodendrocyte precursor cell differentiation. *Journal of Neuroscience, 26*(1), 328–332.

Kristiansen, L. V., Bannon, M. J., & Meador-Woodruff, J. H. (2009). Expression of transcripts for myelin related genes in postmortem brain from cocaine abusers. *Neurochemical Research, 34*(1), 46–54.

Kuceyeski, A., Meyerhoff, D. J., Durazzo, T. C., & Raj, A. (2013). Loss in connectivity among regions of the brain reward system in alcohol dependence. *Human Brain Mapping, 34*(12), 3129–3142.

Lappe-Siefke, C., Goebbels, S., Gravel, M., Nicksch, E., Lee, J., Braun, P. E., ... Nave, K. A. (2003). Disruption of Cnp1 uncouples oligodendroglial functions in axonal support and myelination. *Nature Genetics, 33*(3), 366–374.

Lenroot, R. K., Gogtay, N., Greenstein, D. K., Wells, E. M., Wallace, G. L., Clasen, L. S., ... Giedd, J. N. (2007). Sexual dimorphism of brain developmental trajectories during childhood and adolescence. *NeuroImage, 36*(4), 1065–1073.

Lewohl, J. M., Wang, L., Miles, M. F., Zhang, L., Dodd, P. R., & Harris, R. A. (2000). Gene expression in human alcoholism: microarray analysis of frontal cortex. *Alcoholism: Clinical & Experimental Research, 24*(12), 1873–1882.

Lewohl, J. M., Wixey, J., Harper, C. G., & Dodd, P. R. (2005). Expression of MBP, PLP, MAG, CNP, and GFAP in the Human Alcoholic Brain. *Alcoholism, Clinical and Experimental Research, 29*(9), 1698–1705.

Lim, K. O., Choi, S. J., Pomara, N., Wolkin, A., & Rotrosen, J. P. (2002). Reduced frontal white matter integrity in cocaine dependence: a controlled diffusion tensor imaging study. *Biological Psychiatry, 51*(11), 890–895.

Lin, F., Wu, G., Zhu, L., & Lei, H. (2012). Heavy smokers show abnormal microstructural integrity in the anterior corpus callosum: a diffusion tensor imaging study with tract-based spatial statistics. *Drug and Alcohol Dependence*, *129*(1–2), 82–87.

Liu, J., Lewohl, J. M., Harris, R. A., Iyer, V. R., Dodd, P. R., Randall, P. K., & Mayfield, R. D. (2006). Patterns of gene expression in the frontal cortex discriminate alcoholic from nonalcoholic individuals. *Neuropsychopharmacology*, *31*(7), 1574–1582.

Malhotra, K., & Ortega, L. (June 21, 2013). Central pontine myelinolysis with meticulous correction of hyponatraemia in chronic alcoholics. *BMJ Case Reports*, 1–4.

Martin, R., & McFarland, H. F. (1995). Immunological aspects of experimental allergic encephalomyelitis and multiple sclerosis. *Critical Reviews in Clinical Laboratory Sciences*, *32*(2), 121–182.

Mayfield, J., Ferguson, L., & Harris, R. A. (2013). Neuroimmune signaling: a key component of alcohol abuse. *Current Opinion in Neurobiology*, *23*(4), 513–520.

McClintick, J. N., Xuei, X., Tischfield, J. A., Goate, A., Foroud, T., Wetherill, L., ... Edenberg, H. J. (2013). Stress-response pathways are altered in the hippocampus of chronic alcoholics. *Alcohol*, *47*(7), 505–515.

Medina, K. L., McQueeny, T., Nagel, B. J., Hanson, K. L., Schweinsburg, A. D., & Tapert, S. F. (2008). Prefrontal cortex volumes in adolescents with alcohol use disorders: unique gender effects. *Alcoholism: Clinical & Experimental Research*, *32*(3), 386–394.

Merrill, J. E., & Benveniste, E. N. (1996). Cytokines in inflammatory brain lesions: helpful and harmful. *Trends in Neuroscience*, *19*(8), 331–338.

de la Monte, S. M., Tong, M., Cohen, A. C., Sheedy, D., Harper, C., & Wands, J. R. (2008). Insulin and insulin-like growth factor resistance in alcoholic neurodegeneration. *Alcoholism: Clinical & Experimental Research*, *32*(9), 1630–1644.

Nguyen, T., Mehta, N. R., Conant, K., Kim, K. J., Jones, M., Calabresi, P. A., ... Griffin, J. W. (2009). Axonal protective effects of the myelin-associated glycoprotein. *Journal of Neuroscience*, *29*(3), 630–637.

Nunez, C. C., Roussotte, F., & Sowell, E. R. (2011). Focus on: structural and functional brain abnormalities in fetal alcohol spectrum disorders. *Alcohol Research & Health*, *34*(1), 121–131.

Owen, J. P., Marco, E. J., Desai, S., Fourie, E., Harris, J., Hill, S. S., ... Mukherjee, P. (2013). Abnormal white matter microstructure in children with sensory processing disorders. *NeuroImage: Clinical*, *2*, 844–853.

Pascual, M., Balino, P., Alfonso-Loeches, S., Aragon, C. M., & Guerri, C. (2011). Impact of TLR4 on behavioral and cognitive dysfunctions associated with alcohol-induced neuroinflammatory damage. *Brain, Behavior, and Immunity*, *25*(Suppl. 1), S80–S91.

Pascual, M., Blanco, A. M., Cauli, O., Minarro, J., & Guerri, C. (2007). Intermittent ethanol exposure induces inflammatory brain damage and causes long-term behavioural alterations in adolescent rats. *European Journal of Neuroscience*, *25*(2), 541–550.

Pascual, M., Pla, A., Minarro, J., & Guerri, C. (2014). Neuroimmune activation and myelin changes in adolescent rats exposed to high-dose alcohol and associated cognitive dysfunction: a review with reference to human adolescent drinking. *Alcohol and Alcoholism*, *49*(2), 187–192.

Paus, T., Zijdenbos, A., Worsley, K., Collins, D. L., Blumenthal, J., Giedd, J. N., ... Evans, A. C. (1999). Structural maturation of neural pathways in children and adolescents: in vivo study. *Science*, *283*(5409), 1908–1911.

Peters, A. (2009). The effects of normal aging on myelinated nerve fibers in monkey central nervous system. *Frontiers in Neuroanatomy*, *3*, 11.

Pfefferbaum, A., Rosenbloom, M. J., Fama, R., Sassoon, S. A., & Sullivan, E. V. (2010). Transcallosal white matter degradation detected with quantitative fiber tracking in alcoholic men and women: selective relations to dissociable functions. *Alcoholism: Clinical & Experimental Research*, *34*(7), 1201–1211.

Qiu, Y., Jiang, G., Su, H., Lv, X., Zhang, X., Tian, J., & Zhuo, F. (2013). Progressive white matter microstructure damage in male chronic heroin dependent individuals: a DTI and TBSS study. *PLoS One*, *8*(5), e63212.

Ruiz, S. M., Oscar-Berman, M., Sawyer, K. S., Valmas, M. M., Urban, T., & Harris, G. J. (2012). Drinking history associations with regional white matter volumes in alcoholic men and women. *Alcoholism: Clinical & Experimental Research*, *37*(1), 110–122.

Schlaepfer, T. E., Lancaster, E., Heidbreder, R., Strain, E. C., Kosel, M., Fisch, H. U., & Pearlson, G. D. (2006). Decreased frontal white-matter volume in chronic substance abuse. *International Journal of Neuropsychopharmacology*, *9*(2), 147–153.

Schmitz, T., & Chew, L. J. (2008). Cytokines and myelination in the central nervous system. *Scientific World Journal*, *8*, 1119–1147.

Schulte, T., Muller-Oehring, E. M., Rohlfing, T., Pfefferbaum, A., & Sullivan, E. V. (2010). White matter fiber degradation attenuates hemispheric asymmetry when integrating visuomotor information. *Journal of Neuroscience*, *30*(36), 12168–12178.

Simons, M., & Trotter, J. (2007). Wrapping it up: the cell biology of myelination. *Current Opinion in Neurobiology*, *17*(5), 533–540.

Skuja, S., Groma, V., Ravina, K., Tarasovs, M., Cauce, V., & Teteris, O. (2013). Protective reactivity and alteration of the brain tissue in alcoholics evidenced by SOD1, MMP9 immunohistochemistry, and electron microscopy. *Ultrastructural Pathology*, *37*(5), 346–355.

Solowij, N., Yucel, M., Respondek, C., Whittle, S., Lindsay, E., Pantelis, C., & Lubman, D. I. (2011). Cerebellar white-matter changes in cannabis users with and without schizophrenia. *Psychological Medicine*, *41*(11), 2349–2359.

Sorg, S. F., Taylor, M. J., Alhassoon, O. M., Gongvatana, A., Theilmann, R. J., Frank, L. R., & Grant, I. (2012). Frontal white matter integrity predictors of adult alcohol treatment outcome. *Biological Psychiatry*, *71*(3), 262–268.

Sutherland, G. T., Sheedy, D., Sheahan, P. J., Kaplan, W., & Kril, J. J. (2014). Comorbidities, confounders, and the white matter transcriptome in chronic alcoholism. *Alcoholism: Clinical and Experimental Research*, *38*(4), 994–1001.

Thatcher, D. L., Pajtek, S., Chung, T., Terwilliger, R. A., & Clark, D. B. (2010). Gender differences in the relationship between white matter organization and adolescent substance use disorders. *Drug and Alcohol Dependence*, *110*(1–2), 55–61.

Thompson, P. M., Hayashi, K. M., Simon, S. L., Geaga, J. A., Hong, M. S., Sui, Y., ... London, E. D. (2004). Structural abnormalities in the brains of human subjects who use methamphetamine. *Journal of Neuroscience*, *24*(26), 6028–6036.

Tobias, M. C., O'Neill, J., Hudkins, M., Bartzokis, G., Dean, A. C., & London, E. D. (2010). White-matter abnormalities in brain during early abstinence from methamphetamine abuse. *Psychopharmacology (Berl)*, *209*(1), 13–24.

Toga, A. W., Thompson, P. M., & Sowell, E. R. (2006). Mapping brain maturation. *Trends in Neuroscience*, *29*(3), 148–159.

Wang, J. J., Durazzo, T. C., Gazdzinski, S., Yeh, P. H., Mon, A., & Meyerhoff, D. J. (2009). MRSI and DTI: a multimodal approach for improved detection of white matter abnormalities in alcohol and nicotine dependence. *NMR in Biomedicine*, *22*(5), 516–522.

Wang, S., Bates, J., Li, X., Schanz, S., Chandler-Militello, D., Levine, C., ... Goldman, S. A. (2013). Human iPSC-derived oligodendrocyte progenitor cells can myelinate and rescue a mouse model of congenital hypomyelination. *Cell Stem Cell, 12*(2), 252–264.

Wang, Y., Horst, K. K., Kronenberger, W. G., Hummer, T. A., Mosier, K. M., Kalnin, A. J., ... Mathews, V. P. (2012). White matter abnormalities associated with disruptive behavior disorder in adolescents with and without attention-deficit/hyperactivity disorder. *Psychiatry Research, 202*(3), 245–251.

Waxman, S. G. (2006). Axonal conduction and injury in multiple sclerosis: the role of sodium channels. *Nature Reviews Neuroscience, 7*(12), 932–941.

White, R., & Kramer-Albers, E. M. (2014). Axon-glia interaction and membrane traffic in myelin formation. *Frontiers in Cellular Neuroscience, 7*, 284.

WHO. (2011). *Global status report on alcohol and health.* Geneva: WHO Press.

WHO. (2014). *Global status report on alcohol and health.* Geneva: WHO Press.

Yip, S. W., Lacadie, C., Xu, J., Worhunsky, P. D., Fulbright, R. K., Constable, R. T., & Potenza, M. N. (2013). Reduced genual corpus callosal white matter integrity in pathological gambling and its relationship to alcohol abuse or dependence. *World Federation of the Societies of Biological Psychiatry, 14*(2), 129–138.

Young, K. M., Psachoulia, K., Tripathi, R. B., Dunn, S. J., Cossell, L., Attwell, D., ... Richardson, W. D. (2013). Oligodendrocyte dynamics in the healthy adult CNS: evidence for myelin remodeling. *Neuron, 77*(5), 873–885.

Zalesky, A., Solowij, N., Yucel, M., Lubman, D. I., Takagi, M., Harding, I. H., ... Seal, M. (2012). Effect of long-term cannabis use on axonal fibre connectivity. *Brain, 135*(Pt. 7), 2245–2255.

Zorlu, N., Gelal, F., Kuserli, A., Cenik, E., Durmaz, E., Saricicek, A., & Gulseren, S. (2013). Abnormal white matter integrity and decision-making deficits in alcohol dependence. *Psychiatry Research, 214*(3), 382–388.

Chapter 40

The Effects of Acute and Chronic Ethanol Exposure on GABAergic Neuroactive Steroid Immunohistochemistry: Relationship to Ethanol Drinking

Matthew C. Beattie, Antoniette Maldonado-Devincci, Jason B. Cook, A. Leslie Morrow
Department of Psychiatry & Pharmacology, Bowles Center for Alcohol Studies, University of North Carolina, Chapel Hill, NC, USA

Abbreviations

3α-HSD 3α-Hydroxysteroid dehydrogenase
3α,5α-THDOC 3α,21-Dihydroxy-5α-pregnan-20-one
3α,5α-THP 3α-Hydroxy-5α-pregnan-20-one
5α-R 5α-Reductase
BxD C57BL/6J × DBA/2J recombinant inbred mouse panel
CA Cornus ammonis
CeA Central nucleus of the amygdala
CIE Chronic intermittent ethanol
CNS Central nervous system
DOC Deoxycorticosterone
EtOH Ethanol
g/kg Grams per kilogram
GABA$_A$ γ-Aminobutyric acid type A
GABR$_A$2 γ-Aminobutyric acid type A receptor, α2
HPA Hypothalamic–pituitary–adrenal
IHC Immunohistochemistry
mRNA Messenger ribonucleic acid
NAc Nucleus accumbens
P450scc Cytochrome P450 side-chain cleavage
rAAV2 Recombinant adenoassociated virus vector
SNP Single-nucleotide polymorphism
TH Tyrosine hydroxylase
VTA Ventral tegmental area

INTRODUCTION

Ethanol consumption produces many well-known behavioral effects including anxiolysis, impaired motor coordination, impaired cognitive function, sedation, hypnosis, and anticonvulsant and proaggressive actions. While ethanol has many sites of action in the brain, the behavioral effects produced by ethanol consumption overlap with the effects of γ-aminobutyric acid type A (GABA$_A$) receptor agonists and can be altered by GABA$_A$ receptor modifiers. GABA$_A$ receptor agonists (benzodiazepines, muscimol) increase ethanol responses, whereas inverse agonists and antagonists (Ro15-4513, picrotoxin, bicuculline) decrease ethanol responses (see Grobin, Matthews, Devaud, & Morrow, 1998, for review). GABA$_A$ receptors are a family of chloride ion channels that serve as the primary inhibitory receptor family in the brain and mediate many of the behavioral effects of ethanol in the central nervous system. These mechanisms include direct and indirect effects on GABA$_A$ receptors, as well as effects on GABA release and the synthesis and availability of neuroactive steroids (see Kumar et al., 2009, for review).

Chronic ethanol exposure produces tolerance to ethanol and cross-tolerance to benzodiazepines as well as dependence upon ethanol in rodents and humans. Ethanol dependence is marked by increased anxiety and seizure susceptibility, impaired sleep and cognition, generalized dysphoria, and increased ethanol consumption. These changes are associated with alterations in GABA$_A$ receptor function and expression, as well as alterations in GABAergic neuroactive steroid levels in plasma and various brain regions (see Kumar et al., 2009, for review). This chapter focuses on GABAergic neuroactive steroid responses to acute and chronic ethanol exposure in monkeys, rats, and mice, with an emphasis on new information contributed by the application of immunohistochemical techniques.

Neuroactive steroids are endogenous steroids that rapidly alter neuronal excitability via membrane receptors. These steroids are derived from cholesterol and can be synthesized de novo in the brain and the adrenal glands. The biosynthetic pathway (Figure 1) for these compounds involves the conversion of cholesterol into pregnenolone, which is further metabolized into several steroid hormones that include progesterone, glucocorticoids, dehydroepiandrosterone (DHEA), estrogens, and their metabolites. GABAergic neuroactive steroids function as positive allosteric modulators of GABA$_A$ receptors. Among the most potent are derivatives of deoxycorticosterone (DOC) and progesterone, (3α,5α)-3,21-dihydroxypregnan-20-one (3α,5α-THDOC or allotetrahydrodeoxycorticosterone), and

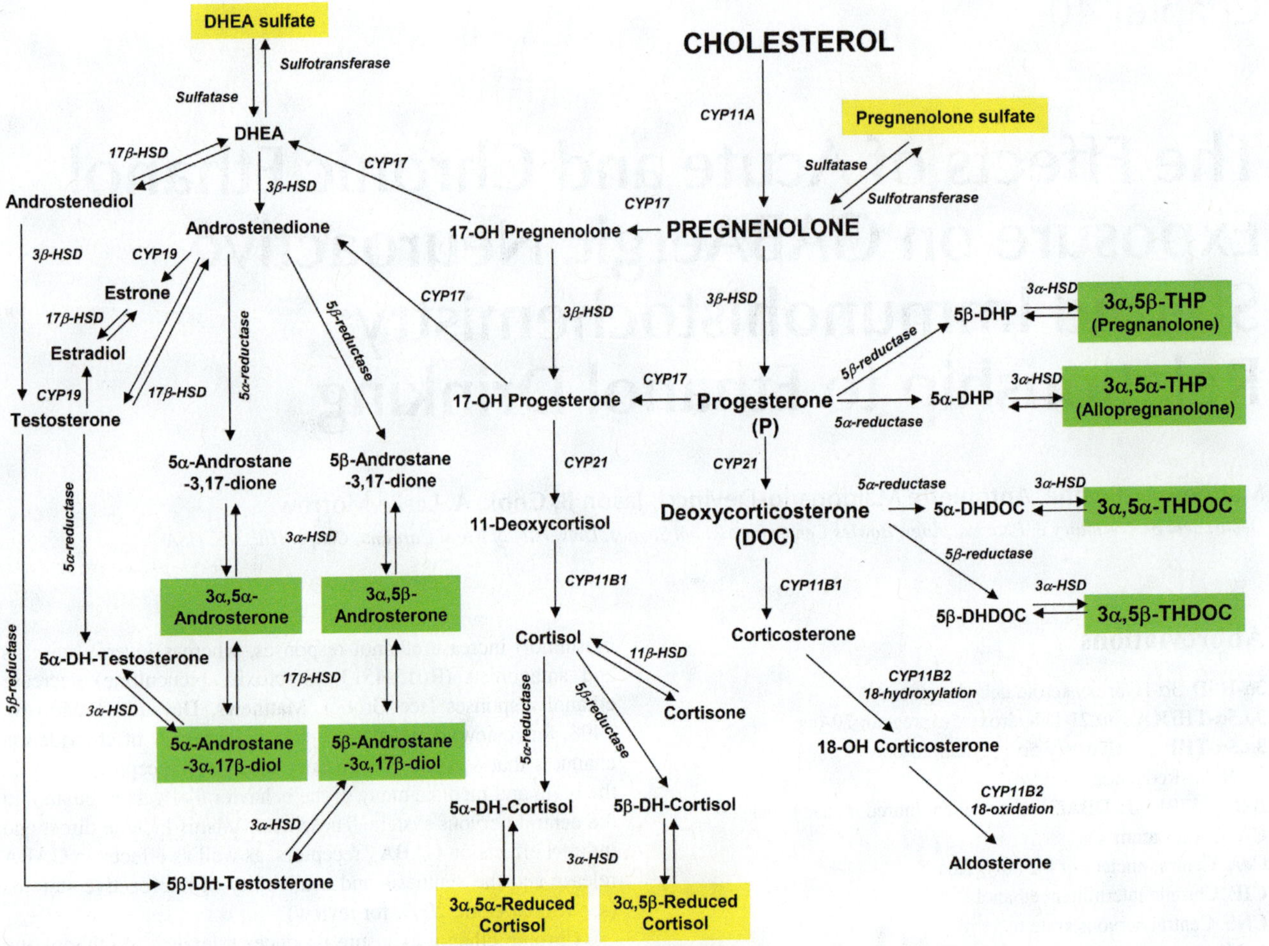

FIGURE 1 Neuroactive steroid biosynthesis pathway. The inhibitory neuroactive steroids with potent GABA$_A$ receptor-positive modulatory effects are highlighted in green, whereas the excitatory neuroactive steroids with weak GABA$_A$ receptor antagonist effects are highlighted in yellow.

($3\alpha,5\alpha$)-3-hydroxypregnan-20-one ($3\alpha,5\alpha$-THP or allopregnanolone). Systemic administration of GABAergic neuroactive steroids exerts a variety of pharmacological responses, including anxiolytic, antidepressant, anticonvulsant, sedative, anesthetic, and analgesic effects, in animal models and human studies (Belelli, Bolger, & Gee, 1989; Bitran, Hilvers, & Kellogg, 1991; Carl et al., 1990; Kavaliers, 1988; Khisti, Chopde, & Jain, 2000) that are consistent with their GABAergic actions. Furthermore, neuroactive steroids appear to influence ethanol drinking in various animal models of excessive drinking, although different effects are found in various mouse strains and rat models (Porcu & Morrow, 2014).

IMMUNOHISTOCHEMISTRY METHODOLOGY AND STRENGTHS IN NEUROACTIVE STEROID RESEARCH

Immunohistochemistry (IHC) has long been used to identify and quantify enzymes associated with neurosteroid biosynthesis and regulation; however, recent studies have utilized steroid antibodies (primarily $3\alpha,5\alpha$-THP) to visualize neuroactive steroid distribution and modulation by ethanol. Comparative studies between ethanol-exposed and control animals allow us to assess individual animal brain sections to quantify levels and distribution of neuroactive steroids. This is a very powerful approach, as it can allow for direct visualization of specific brain regions and cell types therein to provide an understanding of how ethanol affects regulation of neuroactive steroids in various paradigms.

GABAergic NEUROACTIVE STEROIDS AND ETHANOL INTERACTIONS IN RATS

Systemic administration of moderate ethanol doses (1–2.5 g/kg) increases plasma, cerebrocortical, and hippocampal levels of $3\alpha,5\alpha$-THP and $3\alpha,5\alpha$-THDOC in Sprague–Dawley rats and Sardinian alcohol-preferring rats (Barbaccia et al., 1999; VanDoren et al., 2000). In the Sprague–Dawley rat hippocampal slice, acute ethanol increased neurosteroidogenesis in CA1 pyramidal cells in the hippocampus, while functionally decreasing long-term potentiation (Sanna et al., 2004). However, chronic exposure to ethanol blunted the ethanol-induced increase in $3\alpha,5\alpha$-THP and other neuroactive steroids in ethanol-withdrawn rats (Khisti, Boyd, Kumar, & Morrow, 2005).

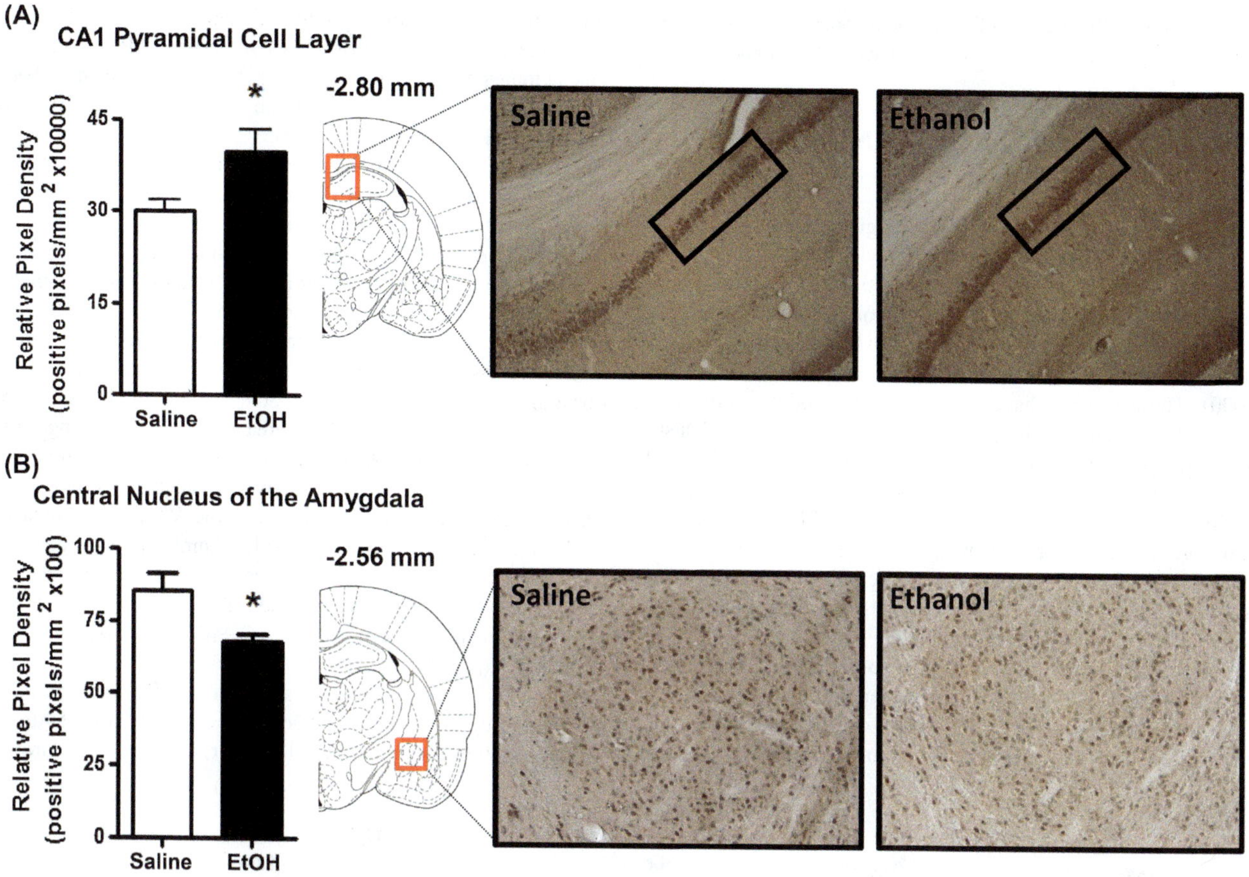

FIGURE 2 Effect of acute EtOH administration (2 g/kg, ip) on 3α,5α-THP immunoreactivity in the pyramidal cell layer of the CA1 hippocampus and CeA. (A) EtOH administration increased 3α,5α-THP immunoreactivity in the pyramidal cell layer of the CA1 hippocampus compared to saline controls. Representative photomicrographs (original magnification 10×) of 3α,5α-THP immunoreactivity in CA1 pyramidal cells (highlighted in rectangle, 2.80 mm relative to bregma) following saline (n=8) or EtOH (n=8) administration are shown. (B) EtOH administration reduced 3α,5α-THP immunoreactivity in the CeA compared to saline controls. Representative photomicrographs (original magnification 10×) of 3α,5α-THP immunoreactivity in the CeA (2.56 mm relative to bregma) following saline (n=8) or EtOH (n=8) administration are shown. EtOH (2 g/kg, ip) or saline was administered 60 min prior to tissue fixation and collection. Data are expressed as mean positive pixels/mm^2 ± SEM. *$p<0.01$ compared to saline administration. CeA, central nucleus of the amygdala; EtOH, ethanol; CA1, cornus ammonis area 1. *Adapted from Cook, Dumitru, et al. (2014).*

Immunohistochemical techniques have been used to determine ethanol-induced changes in cellular 3α,5α-THP across brain, showing both regional and cell-population specificity in this response in Wistar rats (Cook, Dumitru, O'Buckley, & Morrow, 2014). Ethanol increased 3α,5α-THP immunoreactivity in the medial prefrontal cortex, the hippocampal CA1 pyramidal cell layer, the polymorph cell layer of the dentate gyrus, the bed nucleus of the stria terminalis, and the paraventricular nucleus of the hypothalamus. In contrast, ethanol administration significantly reduced 3α,5α-THP immunoreactivity in the nucleus accumbens and the central nucleus of the amygdala (Figure 2). No changes were observed in the ventral tegmental area, dorsomedial striatum, granule cell layer of the dentate gyrus, or the lateral or basolateral amygdala (Cook, Dumitru, et al., 2014).

Ethanol, therefore, produces divergent brain-region- and cell-type-specific changes in 3α,5α-THP concentrations. The ethanol-induced increases in plasma neuroactive steroids were thought to be mediated by the hypothalamic–pituitary–adrenal (HPA) axis, since the effects were not observed in the cerebral cortex of hypophysectomized and adrenalectomized Sprague–Dawley or Wistar rats (O'Dell et al., 2004).

However, studies of ethanol's effects using IHC reveal that ethanol-induced elevations of 3α,5α-THP are independent of adrenal activation in the CA1 pyramidal cell layer, dentate gyrus polymorphic layer, bed nucleus of the stria terminalis, and paraventricular nucleus of the hypothalamus of Wistar rats. Furthermore, ethanol produced decreases in 3α,5α-THP labeling in the nucleus accumbens shore and central nucleus of the amygdala that also occurred independent of adrenal activation. However, in the medial prefrontal cortex, ethanol increased 3α,5α-THP immunoreactivity after sham surgery, but there was no change in 3α,5α-THP after adrenalectomy. These data indicate that ethanol dynamically regulates local 3α,5α-THP levels in several subcortical regions, but the adrenal glands contribute to 3α,5α-THP elevations in the medial prefrontal cortex (Cook, Nelli, et al., 2014).

Brain synthesis of 3α,5α-THP in response to ethanol seems to be isolated to specific cellular populations, as increases have been shown in the hippocampal CA1 pyramidal cell layer and the polymorph cell layer of the dentate gyrus, but no ethanol-induced changes in 3α,5α-THP were seen in the granule cell layer of the dentate gyrus. Therefore, the presence of this very specific effect of ethanol on cellular 3α,5α-THP in the hippocampus may underlie neuron-specific responses to ethanol across the brain.

NEUROACTIVE STEROIDS AND ETHANOL DRINKING IN RATS

Ethanol-induced elevations in GABAergic neuroactive steroids are thought to contribute to many behavioral effects of ethanol in rodents. Indeed, elevations in neuroactive steroids in response to ethanol are sufficient to reach physiologically relevant concentrations capable of enhancing GABAergic transmission (Barbaccia et al., 1999; Morrow et al., 1999; VanDoren et al., 2000). These steroids have been shown to modulate ethanol's anticonvulsant effects (VanDoren et al., 2000), sedation (Khisti, VanDoren, O'Buckley, & Morrow, 2003), impairment of spatial memory (Morrow, VanDoren, Fleming, & Penland, 2001), and anxiolytic-like (Hirani, Sharma, Jain, Ugale, & Chopde, 2005) and antidepressant-like (Hirani, Khisti, & Chopde, 2002) actions. This suggests that elevations in neuroactive steroids influence many of the GABAergic effects of ethanol in vivo and contribute to sensitivity to the behavioral effects of ethanol.

Neuroactive steroids alter both ethanol reinforcement and ethanol consumption in rodents. Both pregnenolone and the synthetic GABAergic neuroactive steroid (3,5)-20-oxo-pregnane-3-carboxylic acid (O'Dell et al., 2005) dose-dependently reduce ethanol self-administration without producing sedation (Besheer, Lindsay, O'Buckley, Hodge, & Morrow, 2010). 3α,5α-THP and ganaxolone, a longer acting synthetic analogue of 3α,5α-THP, have been shown to produce biphasic effects on ethanol self-administration (Ford, Nickel, Phillips, & Finn, 2005).

The synthesis of neuroactive steroids requires cholesterol conversion to pregnenolone by the mitochondrial cytochrome P450 side-chain cleavage (P450scc) enzyme, the rate-limiting enzymatic reaction in steroid synthesis (See Figure 1).

Studies in alcohol-preferring (P) rats using adenovirus-mediated delivery of the steroidogenic enzyme cytochrome P450scc utilized IHC techniques to determine local neuroactive steroid levels and distribution, as well as viral infection efficiencies. Local increases in P450scc led to increased 3α,5α-THP immunoreactivity in the ventral tegmental area (VTA), but no change was observed in the nucleus accumbens following infection (Figure 3). Overexpression of P450scc by recombinant adenoassociated virus vector (rAAV2)–P450scc infusion into the VTA was shown to reduce long-term operant ethanol self-administration (Figure 3). This reduction in ethanol reinforcement and consumption was associated with an increase in 3α,5α-THP-positive cells in the VTA of animals that received rAAV2–P450scc infusion (Cook, Werner, et al., 2014).

Immunohistochemical studies of neuroactive steroids have also uncovered colocalization of 3α,5α-THP with tyrosine hydroxylase (a putative dopaminergic cell marker) and NeuN (a neuronal

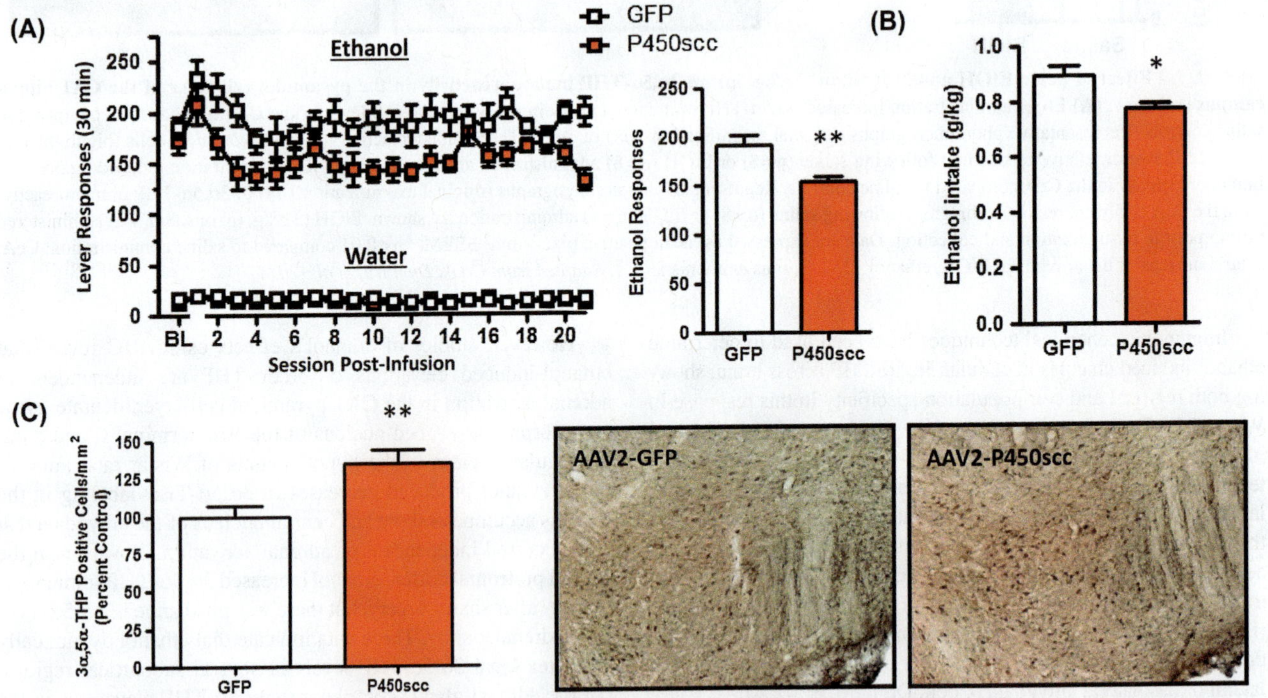

FIGURE 3 Recombinant adenoassociated virus vector (rAAV2)–P450scc transduction in the VTA produces long-term reductions in operant ethanol self-administration and increases 3α,5α-THP-positive cells. (A) rAAV2–P450scc transduction in the VTA reduced operant ethanol ($p < 0.005$) but not water responding over the 21 days of test sessions, compared with rAAV2–GFP (green fluorescent protein) controls. Mean ethanol responding over the 21 days of test sessions is collapsed in the bar graph. (B) rAAV2–P450scc transduction in the VTA reduced mean ethanol intake (g/kg; $p < 0.01$) over the 21 days of test sessions, compared with rAAV2–GFP controls. (C) Infusion of rAAV2–P450scc (2 μl) into the VTA increased 3α,5α-THP-positive cells in the VTA ($p < 0.005$) at 4 weeks postsurgery as shown by representative photomicrographs (original magnification 10×) of cellular 3α,5α-THP immunoreactivity in the VTA. Baseline (BL) represents 1 week average of ethanol responding during the week before surgery; *$p < 0.01$ and **$p < 0.005$ compared with control values. *Adapted from Cook, Werner, et al. (2014).*

marker) in the VTA of P rats. These data suggest that 3α,5α-THP is localized to neurons in the VTA.

The ability of rAAV2–P450scc transduction of VTA neurons to modulate ethanol self-administration is probably due to modulation of neural circuitry via $GABA_A$ receptor-mediated neuronal inhibition. The available data suggest that increasing 3α,5α-THP within a cell reduces the excitability of that particular cell (Tokuda, O'Dell, Izumi, & Zorumski, 2010). Therefore, rAAV2–P450scc transduction of a cell most likely produces a presynaptic inhibitory effect. It is not clear how 3α,5α-THP accesses the neuroactive steroid transmembrane binding sites on $GABA_A$ receptors, but it has been proposed to occur via intracellular (i.e., presynaptic) lateral diffusion through the cell membrane (Akk et al., 2007) or by a paracrine or autocrine mechanism (Herd, Belelli, & Lambert, 2007), as no active release mechanism has been identified.

It has been suggested that ethanol-induced elevations in GABAergic neuroactive steroids protect against the risk for ethanol dependence (Morrow, Porcu, Boyd, & Grant, 2006). Diminished elevations in GABAergic neuroactive steroids following ethanol exposure would result in reduced sensitivity to the anxiolytic, sedative, anticonvulsant, cognitive-impairing, and discriminative stimulus properties of ethanol. Reduced sensitivity to ethanol is associated with greater risk for the development of alcoholism in individuals with genetic vulnerability to alcoholism (Schuckit, 2009).

GABAergic NEUROACTIVE STEROIDS AND ETHANOL INTERACTIONS IN MICE

Acute ethanol administration (2 g/kg) decreased plasma 3α,5α-THP levels in C57BL/6J mice (Porcu et al., 2010). Furthermore, acute ethanol administration failed to alter cerebrocortical and hippocampal levels of 3α,5α-THP in C57BL/6J mice (Porcu et al., 2014). Similarly, acute injection of 2 g/kg ethanol did not change whole-brain 3α,5α-THP levels in C57BL/6J mice (Finn, Sinnott, et al., 2004). These data indicate that acute ethanol administration does not alter brain levels of 3α,5α-THP, but decreases circulating 3α,5α-THP levels.

In contrast, chronic ethanol consumption, using a limited-access paradigm (2 h/day), resulted in elevated cerebral 3α,5α-THP levels in C57BL/6J mice (Finn, Long, Tanchuck, & Crabbe, 2004). Alternatively, 3α,5α-THP dose-dependently alters ethanol responses in mice. Low-dose 3α,5α-THP administration increased ethanol intake, whereas high-dose 3α,5α-THP administration decreased ethanol intake in C57BL/6J mice (Ford, Nickel, Phillips, et al., 2005). This effect appears to be time dependent, where the increase in ethanol consumption was observed only during the first hour, and not the second hour, of access to ethanol in males after 17 days of ethanol drinking (Sinnott, Phillips, & Finn, 2002). Furthermore, intracerebroventricular administration of 3α,5α-THP increased ethanol drinking in C57BL/6J mice (Ford, Mark, Nickel, Phillips, & Finn, 2007). Finally, inhibition of 5α-reduced neurosteroid synthesis appears to reverse the elevation of ethanol consumption following 3α,5α-THP administration. Specifically, acute treatment with finasteride, a 5α-reductase inhibitor, decreased ethanol intake (Ford, Nickel, & Finn, 2005). However, when finasteride treatment was extended over 7 days, the decrease in ethanol intake was partially reversed in C57BL/6J mice (Ford, Nickel, & Finn, 2005). This could indicate that tolerance developed to repeated finasteride treatment or that changes in neurosteroid modulation of GABAergic inhibitory tone may have occurred.

Chronic intermittent ethanol (CIE) exposure is a model that produces ethanol dependence in C57BL/6J mice. In this model, mice are exposed to cycles of vaporized ethanol to induce ethanol dependence, which is subsequently followed by a short withdrawal period. Repeated CIE exposure and withdrawal increases subsequent voluntary ethanol consumption in C57BL/6J mice (Lopez, Griffin, Melendez, & Becker, 2012). Repeated ethanol vapor exposure decreases the aversive properties associated with ethanol (Lopez et al., 2012). These data indicate that repeated exposure to CIE results in elevated ethanol consumption, which could be mediated by decreased aversion associated with ethanol exposure.

Many brain regions are likely to contribute to elevated ethanol consumption following CIE exposure in ethanol-dependent mice. Changes in gene expression are observed in various brain regions following CIE exposure in the C57BL/6J mouse (Melendez, McGinty, Kalivas, & Becker, 2012). However, the precise mechanisms mediating the changes in behavior and gene expression are not well understood.

Previously, our laboratory found that cerebral cortical 3α,5α-THP levels were elevated 72 h following CIE exposure in ethanol-dependent mice using radioimmunoassay (Morrow and Porcu, 2009). We conducted a subregional analysis on changes in 3α,5α-THP immunoreactivity in CIE-exposed vs. air-exposed C57BL/6J mice using IHC (Maldonado-Devincci et al., 2014). CIE exposure produced decreases in 3α,5α-THP immunoreactivity in various cortical and limbic subregions (Figure 4). Specifically, decreases in 3α,5α-THP immunoreactivity were observed 72 h following withdrawal in the medial prefrontal cortex (−25.0±9.3%), nucleus accumbens core (−29.9±6.6%), dorsolateral striatum (−18.5±6.0%), lateral amygdala (−27.5±12.4%), and VTA (−31.6±12.4%). IHC targets intracellular changes in 3α,5α-THP; therefore these changes would be expected to blunt GABAergic tone by decreasing the amount of intracellular 3α,5α-THP present in these brain regions and produce a state of hyperexcitability.

Conversely, in the CA3 pyramidal cell layer of the hippocampus, we observed increased (+42.8±19.5%) 3α,5α-THP immunoreactivity in CIE-exposed mice (Maldonado-Devincci et al., 2014). At physiologically relevant concentrations, 3α,5α-THP increases presynaptic glutamate release in the CA3 pyramidal cell layer in the hippocampus (Park et al., 2011). Greater neuronal activation was observed in the CA3 pyramidal cell layer of the hippocampus in ethanol-withdrawn C57BL/6J mice, as indexed by c-Fos immunoreactivity (Chen, Reilly, Kozell, Hitzemann, & Buck, 2009). This increase in 3α,5α-THP immunoreactivity may be related to altered pyramidal neuronal activation.

In general, the work we have conducted investigating changes in 3α,5α-THP levels following CIE in the C57BL/6J mouse model using IHC demonstrates decreased 3α,5α-THP immunoreactivity in several cortical and limbic brain regions, except in the CA3 pyramidal cell layer of the hippocampus, where we observed an increase in 3α,5α-THP immunoreactivity. Chronic ethanol exposure and withdrawal induces a switch in excitatory and inhibitory

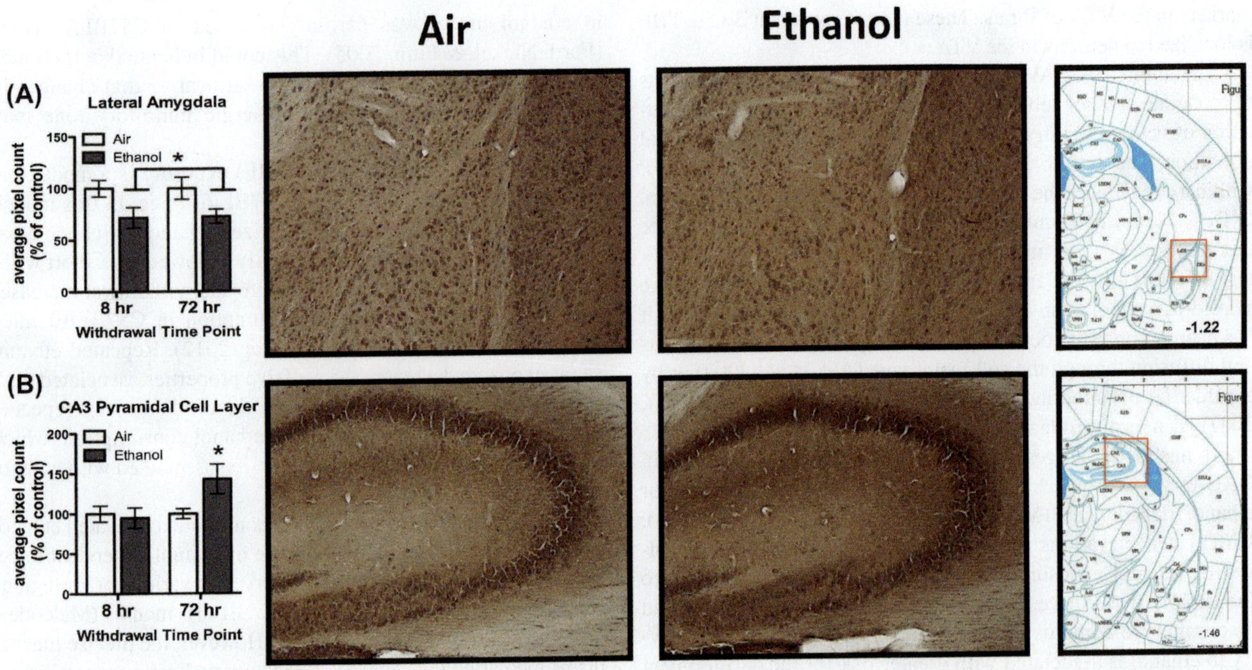

FIGURE 4 Effects of CIE exposure on 3α,5α-THP immunoreactivity in the amygdala and hippocampal pyramidal cell layer following withdrawal. The effects of CIE exposure on 3α,5α-THP immunoreactivity in the (A) lateral amygdala and (B) CA3 pyramidal cell layer 8 or 72 h after withdrawal in air-exposed (clear bars) or ethanol-exposed mice (gray bars). Data depicted are mean positive pixels/mm^2 ± SEM. Representative photomicrographs (original magnification 10×) of 3α,5α-THP immunoreactivity following 72 h withdrawal in air- (left; n = 10–12) or ethanol-exposed (right; n = 10–13) mice. Red box indicates coordinates relative to bregma depicted in photomicrographs. *p < 0.05 compared to respective air-exposed control. *Adapted from Maldonado-Devincci et al. (2014).*

tone, with increased excitation and decreased inhibition (Kumar et al., 2009; Olsen & Spigelman, 2012), manifested as a loss of GABAergic tone. Overall, the decrease in 3α,5α-THP immunoreactivity observed following 72-h withdrawal from CIE exposure may indicate reduced GABAergic neuronal inhibition, which would be expected to increase output firing of projection neurons. Maintaining a reduction in 3α,5α-THP levels may decrease GABAergic neurotransmission during ethanol withdrawal (Finn et al., 2006), which would probably produce a hyperexcitable state during ethanol withdrawal. The likely mechanism for decreased 3α,5α-THP levels within the brain may be a change in the biosynthetic pathway, with increased deoxycorticosterone and corticosterone levels produced in the C57BL/6J mouse (Porcu & Morrow, 2014).

Individual differences in vulnerability to alcoholism have a genetic component (Schuckit, 2009). Studies in rodents indicate a shared genetic sensitivity to ethanol, anxiety, and stress/HPA axis response (Crabbe, Phillips, Buck, Cunningham, & Belknap, 1999). Significant genetic variation in neuroactive steroid levels can be found across the C57BL/6J (B6) × DBA/2J (D2) (BxD) recombinant inbred mouse strains, a reference population to study networks of phenotypes and their modulation by gene variants (Williams, Gu, Qi, & Lu, 2001). Across the BxD mouse population, basal cerebral cortical 3α,5α-THP levels across strains ranged between 1.81 and 3.72 ng/g, equivalent to a 2.0-fold genetic variation and heritability of 0.40. The ethanol-induced changes in cerebral cortical 3α,5α-THP levels ranged between +4% and +63%. Both basal and ethanol-induced cerebral cortical 3α,5α-THP levels were correlated with some ethanol phenotypes previously determined in the BxD strains and available in GeneNetwork. Basal 3α,5α-THP levels were negatively correlated with consumption of 10% ethanol at 2 h (Spearman r = −0.82, p = 0.02, n = 7, GeneNetwork ID 12733), that is, those strains with low basal 3α,5α-THP levels consumed more alcohol. Interestingly, the ethanol-induced changes in 3α,5α-THP levels were negatively correlated with both 3% (Spearman r = −0.82, p = 0.02, n = 7, GeneNetwork ID 10474 (Phillips, Crabbe, Metten, & Belknap, 1994)) and 10% (Spearman r = −0.82, p = 0.02, n = 7, GeneNetwork ID 10582 (Rodriguez, Plomin, Blizard, Jones, & McClearn, 1994)) ethanol consumption. Those strains with increased 3α,5α-THP levels in response to acute ethanol consumed less alcohol. Basal and ethanol-induced 3α,5α-THP levels differ across BxD strains, indicating that genetic variation may contribute to the differences in the neuroactive steroid responses to stress or ethanol challenges. These studies have mainly been conducted in mouse strains, but it is likely that genetic variation could also explain the different neuroactive steroid sensitivity observed in rats and humans, compared to C57BL/6J mice.

NEUROACTIVE STEROIDS AND ETHANOL DRINKING IN NONHUMAN PRIMATES

Long-term ethanol exposure is difficult to study in rodent models because of their short life spans. It is also very difficult to train rodents to self-administer large doses of ethanol and establish daily drinking patterns that are similar to those achieved by human alcoholics. Nonhuman primates are important for the study of complex biomedical disease processes, including alcoholism.

Anatomical, physiological, genetic, and behavioral similarities to humans provide unique translational research opportunities. Cynomolgus macaques (*Macaca fascicularis*) will freely self-administer intoxicating levels of ethanol with drinking patterns similar to those seen in humans (Vivian et al., 2001), making them a good model to study the effects of chronic ethanol consumption. In this capacity, acute ethanol effects in nonhuman primates have not been extensively reported on. However, in contrast to what has been shown in the rodent, acute ethanol administration did not alter the GABAergic neuroactive steroids measured in monkeys (Porcu et al., 2010).

Chronic ethanol self-administration by the cynomolgus macaque significantly shifts GABA potency, but not efficacy, for basolateral amygdala $GABA_A$ receptors (Floyd et al., 2004). Long-term ethanol self-administration selectively reduces expression of α2, α3, and α1 mRNAs without substantially influencing α4 expression in the basolateral amygdala (Floyd et al., 2004). Similarly, prolonged intermittent ethanol drinking decreases GABAergic synaptic transmission while concomitantly increasing glutamatergic transmission in the putamen (Cuzon Carlson et al., 2011). These studies indicate that chronic ethanol leads to changes in GABA receptor structure and function, potentially altering neuroactive steroid modulation.

The GABAergic neuroactive steroids 3α,5α-THP, 3α,5β-THP, and 3α,5α-androsterone produce a discriminative stimulus effect that is similar to those of both relatively low (1.0 g/kg) and higher (2.0 g/kg) doses of ethanol. In addition, increased circulating progesterone as a result of menstrual phase produces increased sensitivity to ethanol, suggesting additive effects of circulating neuroactive steroids with administered ethanol. A negative correlation has been established between dexamethasone suppression of DOC responses and average daily ethanol consumption during a 12-month period (Porcu, Grant, Green, Rogers, & Morrow, 2006). Since the monkeys had no alcohol exposure prior to the HPA axis challenges, the correlation between subsequent alcohol drinking and suppression of the DOC response to dexamethasone (suppression of the HPA axis) may indicate a trait marker of propensity to consume ethanol. We have observed that changes in plasma DOC levels following dexamethasone administration to ethanol-naïve monkeys were correlated with 3α,5α-THP in the lateral amygdala and CA1 of the hippocampus following subsequent ethanol intake. Similarly, the lateral amygdala and hippocampus CA1 levels of 3α,5α-THP following ethanol intake correlate with pregnenolone responses to dexamethasone challenge conducted prior to ethanol exposure. These responses to dexamethasone challenge may represent a trait marker for alcohol drinking risk that is related to 3α,5α-THP levels in the amygdala and hippocampus. Further, ongoing work using IHC indicates that 3α,5α-THP levels in the amygdala are reduced by chronic ethanol administration (Beattie et al., 2015). Lateral and basolateral amygdala levels of 3α,5α-THP were inversely correlated with average daily drinking across a cohort of cynomolgus monkeys that had been induced to drink ethanol and were given free access for 12 months (Table 1). Conversely, 3α,5α-THP levels in the hippocampus CA1 region were positively correlated with average daily drinking (Table 1). These data indicate a reciprocal relationship between voluntary ethanol drinking and brain neuroactive steroid levels in limbic brain regions (Beattie et al., 2015).

TABLE 1 Correlations between 3α,5α-THP Immunoreactivity and Voluntary Ethanol Consumption in Cynomolgus Monkeys

Limbic Brain Subregion	Correlation between 3α,5α-THP and Voluntary Ethanol Consumption
Lateral amygdala	Negative
Basolateral amygdala	Negative
Basomedial amygdala	No correlation
Hippocampus CA1	Positive

Correlations between 3α,5α-THP immunoreactivity and voluntary ethanol consumption in limbic brain subregions of cynomolgus monkeys.

NEUROACTIVE STEROIDS AND ETHANOL DRINKING IN HUMANS

Low- and moderate-dose ethanol administration in humans decreases or does not change plasma 3α,5α-THP levels (Pierucci-Lagha et al., 2006; Porcu et al., 2010). However, following severe intoxication, circulating levels of 3α,5α-THP were elevated in both male and female human adolescents (Torres & Ortega, 2003, 2004). It is likely that the dose of alcohol administered to the human mediates the discrepancies in the literature. Higher doses may be required to observe systematic changes in 3α,5α-THP in the laboratory, such as those observed in the work by Torres and Ortega (2003, 2004).

Finasteride has also been shown to moderate drinking in heavy drinkers. Men who took low-dose finasteride for treatment of male pattern baldness reported a decrease in alcohol consumption, which was greatest for those individuals who consumed the highest level of alcohol (Irwig, 2013). Likewise, dutasteride pretreatment decreased heavy drinking days up to 2 weeks after the laboratory session, while no change in drinking was observed in the light drinkers (Covault et al., 2014). It is unclear if these effects involve neuroactive steroids, however, since 3α,5α-THP levels were not measured in either study and low-dose finasteride would not be expected to alter 3α,5α-THP levels, since it primarily affects type 2 5α-reductase activity. Nonetheless, finasteride has also been shown to reduce drinking in C57BL/6J mice (Ford et al., 2008). Further investigation of these phenomena is warranted.

GABRA2 GENETIC POLYMORPHISMS

Low level of responding to alcohol has been associated with higher alcohol consumption (Hinckers et al., 2006) and alcohol dependence (Schuckit, 1994). One of the major genes that have been examined in humans with its relevance to alcohol dependence and drinking is the *GABRA2* gene. In general, genetic variations in the *GABRA2* gene are associated with decreased subjective responses to alcohol and changes in alcohol drinking, which may influence the development of alcohol use disorders (Bauer et al., 2007). There are differences in individuals with different alleles that may promote drinking or contribute to differences in subjective responses to alcohol.

In general, individuals with the G allele appear to be protected from developing changes in drinking patterns (Milivojevic, Feinn, Kranzler, & Covault, 2014). Specifically, heavy and light drinkers who possess the protective G allele showed no differences in sedation in response to alcohol (Milivojevic et al., 2014). Among alcohol-dependent people, the G allele appears to be overrepresented (Covault, Gelernter, Hesselbrock, Nellissery, & Kranzler, 2004), which may indicate that those with this allele can continue to drink more alcohol because they experience fewer subjective effects associated with alcohol consumption.

Individuals homozygous for the A allele on the *GABRA2* gene reported greater subjective effects of alcohol compared to individuals with one or more G alleles (Pierucci-Lagha et al., 2005). Postmortem human prefrontal cortex brain samples that possessed the AA genotype of the *GABRA2* gene showed greater GABRA2 subunit mRNA levels compared to those with the heterozygous AG genotype (Haughey et al., 2008). However, there were no differences in α2 subunit mRNA levels between controls and alcohol-dependent individuals. Individuals with a homozygous genotype (GG or AA) may be at greater risk for experiencing greater rewarding effects of alcohol, which may promote further drinking (Haughey et al., 2008). These data suggest drinking level and genetic variation moderate excessive alcohol drinking, while those who are heavy drinkers and possess the risk-allele (C) will be more likely to drink even more.

Individuals who possess the C allele experience changes in their subjective response to alcohol, which could promote heavier drinking. Specifically, heavy and light drinkers homozygous for the C allele show different subjective responses to alcohol. Heavy drinkers with the C allele show low sedative effects after alcohol administration, while light drinkers with the C allele showed higher sedation in response to alcohol (Milivojevic et al., 2014).

GENETIC POLYMORPHISMS AND NEUROACTIVE STEROIDS

Polymorphic variations in the biosynthetic enzymes (5α-reductase and 3α-hydroxysteroid dehydrogenase (3α-HSD)) that produce 3α,5α-THP have been shown to be associated with risk of alcohol dependence. The minor C allele for SRD5A1 (the gene encoding 5α-reductase) exon 1 SNP rs248793 and the minor G allele for AKR1C3 (the gene encoding 3α-HSD) exon 1 SNP rs12529 were found to be more frequent in control subjects compared to alcohol-dependent subjects. These results suggest that the minor allele for 5α-reductase type 1 and 3α-HSD type 2 genes may decrease the risk of developing alcohol dependence (Milivojevic, Kranzler, Gelernter, Burian, & Covault, 2011). These data provide indirect evidence that neuroactive steroids may play a role in alcohol effects in humans.

Other work has been conducted to show the importance of GABAergic neuroactive steroids in ethanol's actions in humans. Finasteride, a 5α-reductase inhibitor, reduced the stimulatory and sedative subjective responses to alcohol in humans (Pierucci-Lagha et al., 2005). The ability of finasteride to reduce the subjective effects of alcohol was not observed in individuals carrying the GABA$_A$ α2-subunit polymorphism G allele associated with alcoholism, and individuals carrying this polymorphism had reduced sensitivity to alcohol as well as finasteride (Pierucci-Lagha et al., 2005). The type 1 5α-reductase inhibitor dutasteride also reduced the sedative and anesthetic effects of alcohol (Covault et al., 2014). These data are consistent with the idea that GABAergic neuroactive steroid responses contribute to ethanol sensitivity in humans and may be indicative of risk for alcoholism.

CONCLUSIONS

Different species show divergent effects of ethanol on GABAergic neuroactive steroids in both plasma and brain. The molecular mechanisms by which ethanol alters neuroactive steroid modulation are not well understood. Endocrine differences between rodents and primates, including the ovarian cycle or the predominant type of neuroactive steroid, must be taken into account. Moreover, the amount and the chronic nature of alcohol exposure that underlie addiction to alcohol also differ between rodents and primates. Immunohistochemical techniques have been used to understand the regulation of neuroactive steroids by ethanol in various species. Previously, very little was known about neuroactive steroid localization in the brain. Studies using IHC have been successful at positively labeling individual cells differentially across various brain regions to identify neuroactive steroid responses to ethanol. Dual-labeling immunofluorescence has allowed us to identify specific cell types in which neuroactive steroids are present or absent. Utilizing these techniques and others can improve our understanding of how genetic differences play a role in neuroactive steroids levels, as well as their effects on cell signaling, and could help us develop therapeutic interventions for alcohol use disorders (Figure 5).

APPLICATIONS TO OTHER ADDICTIONS AND SUBSTANCE MISUSE

Neuroactive steroids exert potent effects at GABA$_A$ receptors, which play a critical role in the brain's reward system. Chronic ethanol exposure leads to significant alterations in neuroactive steroid levels in various models, including human alcoholics. As mentioned in the previous section, local increases in neuroactive steroids in the rat VTA reduce ethanol self-administration. Similarly, in nonhuman primates high levels of drinking are associated with low limbic neuroactive steroid levels. Based on these findings neuroactive steroids may act as a therapeutic intervention for alcohol use disorders.

GABAergic transmission is the major inhibitory input in the brain and interactions between drugs of abuse and dopaminergic and GABAergic systems may underlie drug-induced alterations in neuronal activity and brain neurotransmitter levels. 3α,5α-THP administration has been shown to decrease both cue- and stress-induced reinstatement of extinguished cocaine-seeking behavior (Anker & Carroll, 2010; Schmoutz, Runyon, & Goeders, 2014). This suggests that neuroactive steroids play a key role in the neurobiology of conditioned cocaine reward. Similarly 3α,5α-THP administration blocks the escalation of cocaine self-administration (Anker, Zlebnik, & Carroll, 2010) and leads to reduced methamphetamine-primed reinstatement in rats (Holtz, Lozama, Prisinzano, & Carroll, 2012). Therapeutically, administration of GABAergic neuroactive steroids may be effective at decreasing

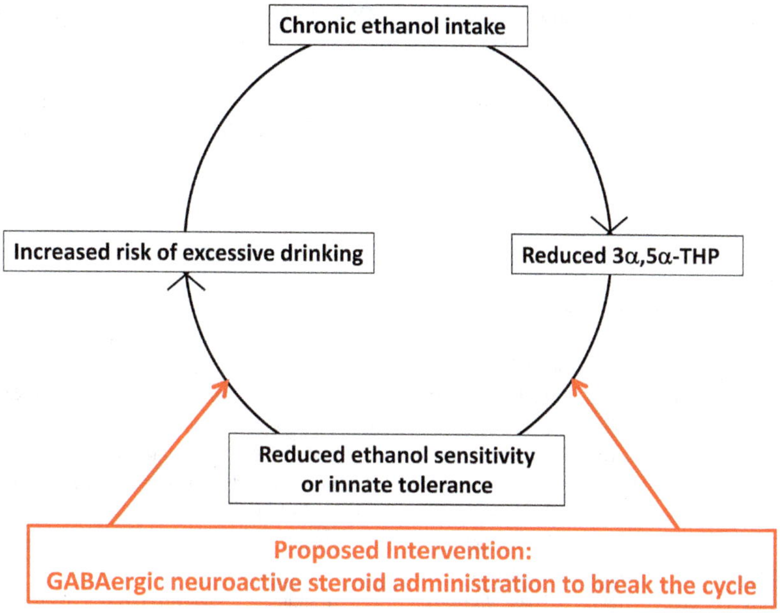

FIGURE 5 Model of chronic ethanol exposure cycle with proposed GABAergic neurosteroid therapeutic intervention to reduce excessive drinking.

relapse to some drugs of abuse in humans. Utilizing immunohistochemical techniques to further understand the relationship between neuroactive steroids and drugs of abuse could lead to potential therapeutic interventions for relapse prevention and drug addiction.

DEFINITION OF TERMS

Immunohistochemistry This is a technique by which antigens in cells of a tissue section are detected by the use of antibodies designed to bind specifically to the antigen.

BxD This is a panel of recombinant inbred mice derived from C57BL/6J and DBA/2J strains that serves as a genetic reference mouse model.

($3\alpha,5\alpha$)-3-Hydroxypregnan-20-one Commonly referred to as allopregnanolone, it is among the more potent GABAergic neuroactive steroids.

Recombinant adenoassociated virus vector 2 This is a method for delivering transgenes via viral infection.

Chronic intermittent ethanol This refers to consistent or repeated exposures to ethanol and withdrawal.

KEY FACTS

Key Facts of Neuroactive Steroids

- Neuroactive steroids, particularly $3\alpha,5\alpha$- and $3\alpha,5\beta$-reduced metabolites of progesterone, deoxycorticosterone, dihydroepiandrosterone, and testosterone, exert inhibitory actions that are mediated by synaptic and extrasynaptic $GABA_A$ receptors.
- GABAergic neuroactive steroids exert a variety of pharmacological responses including anxiolytic, antidepressant, anticonvulsant, sedative, anesthetic, and analgesic effects.
- Neuroactive steroid concentrations vary across physiological conditions mostly associated with reproduction and development such as puberty, ovarian cycling, pregnancy, or aging.
- Neuroactive steroids are thought to play a role in stress response, homeostasis, and allostasis through modulation of the HPA axis.

Key Facts of Ethanol

- Effects of ethanol involve various GABAergic mechanisms that contribute to many of its behavioral effects, including anxiolytic, anticonvulsant, sedative–hypnotic, cognitive-impairing, and motor-incoordinating actions.
- Ethanol has effects on GABA release and the synthesis and availability of endogenous neuroactive steroids.

SUMMARY POINTS

- This chapter focuses on neuroactive steroids, which rapidly alter neuronal excitability via membrane receptors.
- Neuroactive steroids are endogenously synthesized and primarily function at the $GABA_A$ receptors.
- In rats, acute ethanol administration leads to significant alterations in neuroactive steroids in various brain regions, including amygdalar reductions and hippocampal increases in $3\alpha,5\alpha$-THP levels.
- Neuroactive steroids appear to influence ethanol drinking in various animal models of excessive drinking, though different effects are found in various mouse strains and rat models.
- Local increases in $3\alpha,5\alpha$-THP in the VTA of rats reduces ethanol self-administration.
- CIE exposure produces decreases in $3\alpha,5\alpha$-THP immunoreactivity in various cortical and limbic subregions as well as increases in subregions of the hippocampus.
- Chronic ethanol administration in nonhuman primates reduces amygdalar $3\alpha,5\alpha$-THP levels that are negatively correlated with drinking behavior.

- Immunohistochemical techniques have proven to be a powerful approach to visualizing neuroactive steroid distribution and modulation by ethanol.
- Neuroactive steroids may be a target for therapeutic interventions for alcohol use disorders.

REFERENCES

Akk, G., Covey, D. F., Evers, A. S., Steinbach, J. H., Zorumski, C. F., & Mennerick, S. (2007). Mechanisms of neurosteroid interactions with GABA(A) receptors. *Pharmacology and Therapeutics, 116*, 35–57.

Anker, J. J., & Carroll, M. E. (2010). Reinstatement of cocaine seeking induced by drugs, cues, and stress in adolescent and adult rats. *Psychopharmacology (Berlin), 208*, 211–222.

Anker, J. J., Zlebnik, N. E., & Carroll, M. E. (2010). Differential effects of allopregnanolone on the escalation of cocaine self-administration and sucrose intake in female rats. *Psychopharmacology (Berlin), 212*, 419–429.

Barbaccia, M. L., Affricano, D., Trabucchi, M., Purdy, R. H., Colombo, G., Agabio, R., & Gessa, G. L. (1999). Ethanol markedly increases "GABAergic" neurosteroids in alcohol-preferring rats. *European Journal of Pharmacology, 384*, R1–R2.

Bauer, L. O., Covault, J., Harel, O., Das, S., Gelernter, J., Anton, R., & Kranzler, H. R. (2007). Variation in GABRA2 predicts drinking behavior in project MATCH subjects. *Alcoholism Clinical and Experimental Research, 31*, 1780–1787.

Beattie, M. C., Maldonado-Devincci, A. M., Porcu, P., O'Buckley, T. K., Daunais, J., Grant, K. A., & Morrow, A. L. (2015). Voluntary ethanol consumption reduces GABAergic neuroactive steroid (3α,5α) 3-hydroxypregnan-20-one (3α,5α-THP) in the amygdala of the cynomolgus monkey. *Addiction Biology*. http://dx.doi.org/10.1111/adb.12326. (Epub ahead of print).

Belelli, D., Bolger, M. B., & Gee, K. W. (1989). Anticonvulsant profile of the progesterone metabolite 5α-pregnan-3α-ol-20-one. *European Journal of Pharmacology, 166*, 325–329.

Besheer, J., Lindsay, T. G., O'Buckley, T. K., Hodge, C. W., & Morrow, A. L. (2010). Pregnenolone and ganaxolone reduce operant ethanol self-administration in alcohol-preferring P rats. *Alcoholism Clinical and Experimental Research, 34*, 2044–2052.

Bitran, D., Hilvers, R. J., & Kellogg, C. K. (1991). Anxiolytic effects of 3α-hydroxy-5α[β]-pregnan-20-one: endogenous metabolites of progesterone that are active at the $GABA_A$ receptor. *Brain Research, 561*, 157–161.

Carl, P., Hogskilde, S., Nielsen, J. W., Sorensen, M. B., Lindholm, M., Karlen, B., & Bäckstrøm, T. (1990). Pregnanolone emulsion. A preliminary pharmacokinetic and pharmacodynamic study of a new intravenous anaesthetic agent. *Anaesthesia, 45*, 189–197.

Chen, G., Reilly, M. T., Kozell, L. B., Hitzemann, R., & Buck, K. J. (2009). Differential activation of limbic circuitry associated with chronic ethanol withdrawal in DBA/2J and C57BL/6J mice. *Alcohol, 43*, 411–420.

Cook, J. B., Dumitru, A. M., O'Buckley, T. K., & Morrow, A. L. (2014). Ethanol administration produces divergent changes in GABAergic neuroactive steroid immunohistochemistry in the rat brain. *Alcoholism Clinical and Experimental Research, 38*, 90–99.

Cook, J. B., Nelli, S. M., Neighbors, M. R., Morrow, D. H., O'Buckley, T. K., Maldonado-Devincci, A. M., & Morrow, A. L. (2014). Ethanol alters local cellular levels of (3alpha,5alpha)-3-hydroxypregnan-20-one (3alpha,5alpha-THP) independent of the adrenals in subcortical brain regions. *Neuropsychopharmacology, 39*, 1978–1987.

Cook, J. B., Werner, D. F., Maldonado-Devincci, A. M., Leonard, M. N., Fisher, K. R., O'Buckley, T. K., ... Morrow, A. L. (2014). Overexpression of the steroidogenic enzyme cytochrome P450 side chain cleavage in the ventral tegmental area increases 3alpha,5alpha-THP and reduces long-term operant ethanol self-administration. *Journal of Neuroscience, 34*, 5824–5834.

Covault, J., Gelernter, J., Hesselbrock, V., Nellissery, M., & Kranzler, H. R. (2004). Allelic and haplotypic association of GABRA2 with alcohol dependence. *American Journal of Medical Genetics Part B: Neuropsychiatric Genetics, 129*, 104–109.

Covault, J., Pond, T., Feinn, R., Arias, A. J., Oncken, C., & Kranzler, H. R. (2014). Dutasteride reduces alcohol's sedative effects in men in a human laboratory setting and reduces drinking in the natural environment. *Psychopharmacology (Berlin), 231*, 3609–3618.

Crabbe, J. C., Phillips, T. J., Buck, K. J., Cunningham, C. L., & Belknap, J. K. (1999). Identifying genes for alcohol and drug sensitivity: recent progress and future directions. *Trends in Neurosciences, 22*, 173–179.

Cuzon Carlson, V. C., Seabold, G. K., Helms, C. M., Garg, N., Odagiri, M., Rau, A. R., ... Grant, K. A. (2011). Synaptic and morphological neuroadaptations in the putamen associated with long-term, relapsing alcohol drinking in primates. *Neuropsychopharmacology, 36*, 2513–2528.

Finn, D. A., Long, S. L., Tanchuck, M. A., & Crabbe, J. C. (2004). Interaction of chronic ethanol exposure and finasteride: sex and strain differences. *Pharmacology Biochemistry and Behavior, 78*, 435–443.

Finn, D. A., Sinnott, R. S., Ford, M. M., Long, S. L., Tanchuck, M. A., & Phillips, T. J. (2004). Sex differences in the effect of ethanol injection and consumption on brain allopregnanolone levels in C57BL/6 mice. *Neuroscience, 123*, 813–819.

Finn, D. A., Douglass, A. D., Beadles-Bohling, A. S., Tanchuck, M. A., Long, S. L., & Crabbe, J. C. (2006). Selected line difference in sensitivity to a GABAergic neurosteroid during ethanol withdrawal. *Genes, Brain and Behavior, 5*, 53–63.

Floyd, D. W., Friedman, D. P., Daunais, J. B., Pierre, P. J., Grant, K. A., & McCool, B. A. (2004). Long-term ethanol self-administration by cynomolgus macaques alters the pharmacology and expression of $GABA_A$ receptors in basolateral amygdala. *Journal of Pharmacology and Experimental Therapeutics, 311*, 1071–1079.

Ford, M. M., Mark, G. P., Nickel, J. D., Phillips, T. J., & Finn, D. A. (2007). Allopregnanolone influences the consummatory processes that govern ethanol drinking in C57BL/6J mice. *Behavioural Brain Research, 179*, 265–272.

Ford, M. M., Nickel, J. D., & Finn, D. A. (2005). Treatment with and withdrawal from finasteride alter ethanol intake patterns in male C57BL/6J mice: potential role of endogenous neurosteroids? *Alcohol, 37*, 23–33.

Ford, M. M., Nickel, J. D., Phillips, T. J., & Finn, D. A. (2005). Neurosteroid modulators of $GABA_A$ receptors differentially modulate ethanol intake patterns in male C57BL/6J mice. *Alcoholism Clinical and Experimental Research, 29*, 1630–1640.

Ford, M. M., Yoneyama, N., Strong, M. N., Fretwell, A., Tanchuck, M., & Finn, D. A. (2008). Inhibition of 5α-Reduced steroid biosynthesis impedes acquisition of ethanol drinking in male C57BL/6J mice. *Alcoholism Clinical and Experimental Research, 32*, 1408–1416.

Grobin, A. C., Matthews, D. B., Devaud, L. L., & Morrow, A. L. (1998). The role of $GABA_A$ receptors in the acute and chronic effects of ethanol. *Psychopharmacology, 139*, 2–19.

Haughey, H. M., Ray, L. A., Finan, P., Villanueva, R., Niculescu, M., & Hutchison, K. E. (2008). Human γ-aminobutyric acid A receptor α2 gene moderates the acute effects of alcohol and brain mRNA expression. *Genes, Brain and Behavior, 7*, 447–454.

Herd, M. B., Belelli, D., & Lambert, J. J. (2007). Neurosteroid modulation of synaptic and extrasynaptic GABA$_A$ receptors. *Pharmacology and Therapeutics, 116*, 20–34.

Hinckers, A. S., Laucht, M., Schmidt, M. H., Mann, K. F., Schumann, G., Schuckit, M. A., & Heinz, A. (2006). Low level of response to alcohol as associated with serotonin transporter genotype and high alcohol intake in adolescents. *Biological Psychiatry, 60*, 282–287.

Hirani, K., Khisti, R. T., & Chopde, C. T. (2002). Behavioral action of ethanol in Porsolt's forced swim test: modulation by 3α-hydroxy-5α-pregnan-20-one. *Neuropharmacology, 43*, 1339–1350.

Hirani, K., Sharma, A. N., Jain, N. S., Ugale, R. R., & Chopde, C. T. (2005). Evaluation of GABAergic neuroactive steroid 3α-hydroxy-5α-pregnane-20-one as a neurobiological substrate for the anti-anxiety effect of ethanol in rats. *Psychopharmacology, 180*, 267–278.

Holtz, N. A., Lozama, A., Prisinzano, T. E., & Carroll, M. E. (2012). Reinstatement of methamphetamine seeking in male and female rats treated with modafinil and allopregnanolone. *Drug and Alcohol Dependence, 120*, 233–237.

Irwig, M. S. (2013). Decreased alcohol consumption among former male users of finasteride with persistent sexual side effects: a preliminary report. *Alcoholism Clinical and Experimental Research, 37*, 1823–1826.

Kavaliers, M. (1988). Inhibitory influences of the adrenal steroid, 3α,5α-tetrahydroxycorticosterone on aggression and defeat-induced analgesia in mice. *Psychopharmacology, 95*, 488–492.

Khisti, R. T., Boyd, K. N., Kumar, S., & Morrow, A. L. (2005). *Differential effects of acute and chronic ethanol exposures on plasma and brain deoxycorticosterone levels* Vol. Program No. 798.7, p Online. 2005 Abstract Viewer/Itinerary: Society for Neuroscience.

Khisti, R. T., Chopde, C. T., & Jain, S. P. (2000). Antidepressant-like effect of the neurosteroid 3α-hydroxy-5α-pregnan-20-one in mice forced swim test. *Pharmacology Biochemistry and Behavior, 67*, 137–143.

Khisti, R. T., VanDoren, M. J., O'Buckley, T. K., & Morrow, A. L. (2003). Neuroactive steroid 3α-hydroxy-5α-pregnan-20-one modulates ethanol-induced loss of righting reflex in rats. *Brain Research, 980*, 255–265.

Kumar, S., Porcu, P., Werner, D. F., Matthews, D. B., Diaz-Granados, J. L., Helfand, R. S., & Morrow, A. L. (2009). The role of GABA$_A$ receptors in the acute and chronic effects of ethanol: a decade of progress. *Psychopharmacology (Berlin), 205*, 529–564.

Lopez, M. F., Griffin, W. C., 3rd, Melendez, R. I., & Becker, H. C. (2012). Repeated cycles of chronic intermittent ethanol exposure leads to the development of tolerance to aversive effects of ethanol in C57BL/6J mice. *Alcoholism Clinical and Experimental Research, 36*, 1180–1187.

Maldonado-Devincci, A. M., Beattie, M. C., Morrow, D. H., McKinley, R. E., Cook, J. B., O'Buckley, T. K., & Morrow, A. L. (2014). Reduction of circulating and selective limbic brain levels of (3alpha,5alpha)-3-hydroxy-pregnan-20-one (3alpha,5alpha-THP) following forced swim stress in C57BL/6J mice. *Psychopharmacology (Berlin), 231*, 3281–3292.

Melendez, R. I., McGinty, J. F., Kalivas, P. W., & Becker, H. C. (2012). Brain region-specific gene expression changes after chronic intermittent ethanol exposure and early withdrawal in C57BL/6J mice. *Addiction Biology, 17*, 351–364.

Milivojevic, V., Feinn, R., Kranzler, H. R., & Covault, J. (2014). Variation in AKR1C3, which encodes the neuroactive steroid synthetic enzyme 3alpha-HSD type 2 (17beta-HSD type 5), moderates the subjective effects of alcohol. *Psychopharmacology (Berlin), 231*, 3597–3608.

Milivojevic, V., Kranzler, H. R., Gelernter, J., Burian, L., & Covault, J. (2011). Variation in genes encoding the neuroactive steroid synthetic enzymes 5alpha-reductase type 1 and 3alpha-reductase type 2 is associated with alcohol dependence. *Alcoholism Clinical and Experimental Research, 35*, 946–952.

Morrow, A. L., Janis, G. C., VanDoren, M. J., Matthews, D. B., Samson, H. H., Janak, P. H., & Grant, K. A. (1999). Neurosteroids mediate pharmacological effects of ethanol: a new mechanism of ethanol action? *Alcoholism, Clinical and Experimental Research, 23*, 1933–1940.

Morrow, A. L., & Porcu, P. (2009). Neuroactive steroid biomarkers of alcohol sensitivity and alcoholism risk. edn. In M. Ritsner (Ed.), *Neuropsychiatric biomarkers, endophenotypes, and genes* (pp. 47–57). Dordrecht: Springer Science+Business Media B.V.

Morrow, A. L., Porcu, P., Boyd, K. N., & Grant, K. A. (2006). Hypothalamic-pituitary-adrenal axis modulation of GABAergic neuroactive steroids influences ethanol sensitivity and drinking behavior. *Dialogues in Clinical NeuroSciences, 8*, 463–477.

Morrow, A. L., VanDoren, M. J., Fleming, R., & Penland, S. (2001). Ethanol and neurosteroid interactions in the brain. edn. In G. Biggio, & R. H. Purdy (Eds.), *International Review of Neurobiology* (Vol. 46) (pp. 349–377). San Diego: Academic Press.

O'Dell, L. E., Alomary, A. A., Vallee, M., Koob, G. F., Fitzgerald, R. L., & Purdy, R. H. (2004). Ethanol-induced increases in neuroactive steroids in the rat brain and plasma are absent in adrenalectomized and gonadectomized rats. *European Journal of Pharmacology, 484*, 241–247.

O'Dell, L. E., Purdy, R. H., Covey, D. F., Richardson, H. N., Roberto, M., & Koob, G. F. (2005). Epipregnanolone and a novel synthetic neuroactive steroid reduce alcohol self-administration in rats. *Pharmacology Biochemistry and Behavior, 81*, 543–550.

Olsen, R. W., & Spigelman, I. (2012). GABAA receptor plasticity in alcohol withdrawal. In J. L. Noebels, M. Avoli, M. A. Rogawski, R. W. Olsen, & A. V. Delgado-Escueta (Eds.), *Jasper's basic mechanisms of the epilepsies* (4th ed.). Bethesda (MD).

Park, H. M., Choi, I. S., Nakamura, M., Cho, J. H., Lee, M. G., & Jang, I. S. (2011). Multiple effects of allopregnanolone on GABAergic responses in single hippocampal CA3 pyramidal neurons. *European Journal of Pharmacology, 652*, 46–54.

Phillips, T. J., Crabbe, J. C., Metten, P., & Belknap, J. K. (1994). Localization of genes affecting alcohol drinking in mice. *Alcoholism Clinical and Experimental Research, 18*, 931–941.

Pierucci-Lagha, A., Covault, J., Feinn, R., Khisti, R. T., Morrow, A. L., Marx, C. E., … Kranzler, H. R. (2006). Subjective effects and changes in steroid hormone concentrations in humans following acute consumption of alcohol. *Psychopharmacology, 186*, 451–461.

Pierucci-Lagha, A., Covault, J., Feinn, R., Nellissery, M., Hernandez-Avila, C., Oncken, C., … Kranzler, H. R. (2005). GABRA2 alleles moderate the subjective effects of alcohol, which are attenuated by finasteride. *Neuropsychopharmacology, 30*, 1193–1203.

Porcu, P., Grant, K. A., Green, H. L., Rogers, L. S., & Morrow, A. L. (2006). Hypothalamic-pituitary-adrenal axis and ethanol modulation of deoxycorticosterone levels in cynomolgus monkeys. *Psychopharmacology, 186*, 293–301.

Porcu, P., Locci, A., Santoru, F., Beretti, R., Morrow, A. L., & Concas, A. (2014). Failure of acute ethanol administration to alter cerebrocortical and hippocampal allopregnanolone levels in C57BL/6J and DBA/2J mice. *Alcoholism Clinical and Experimental Research, 38*, 948–958.

Porcu, P., & Morrow, A. L. (2014). Divergent neuroactive steroid responses to stress and ethanol in rat and mouse strains: relevance for human studies. *Psychopharmacology (Berlin), 17*, 3257–3272.

Porcu, P., O'Buckley, T. K., Alward, S. E., Song, S. C., Grant, K. A., de Wit, H., & Morrow, A. L. (2010). Differential effects of ethanol on serum GABAergic 3α,5α/3α,5β neuroactive steroids in mice, rats, cynomolgus monkeys and humans. *Alcoholism Clinical and Experimental Research, 34*, 432–442.

Rodriguez, L. A., Plomin, R., Blizard, D. A., Jones, B. C., & McClearn, G. E. (1994). Alcohol acceptance, preference, and sensitivity in mice. I. Quantitative genetic analysis using BXD recombinant inbred strains. *Alcoholism Clinical and Experimental Research, 18*, 1416–1422.

Sanna, E., Talani, G., Busonero, F., Pisu, M. G., Purdy, R. H., Serra, M., & Biggio, G. (2004). Brain steroidogenesis mediates ethanol modulation of $GABA_A$ receptor activity in rat hippocampus. *Journal of Neuroscience, 24*, 6521–6530.

Schmoutz, C. D., Runyon, S. P., & Goeders, N. E. (2014). Effects of inhibitory GABA-active neurosteroids on cocaine seeking and cocaine taking in rats. *Psychopharmacology (Berlin), 231*, 3391–3400.

Schuckit, M. A. (1994). Low level of response to alcohol as a predictor of future alcoholism. *American Journal of Psychiatry, 151*, 184–189.

Schuckit, M. A. (2009). An overview of genetic influences in alcoholism. *Journal of Substance Abuse Treatment, 36*, S5–S14.

Sinnott, R. S., Phillips, T. J., & Finn, D. A. (2002). Alteration of voluntary ethanol and saccharin consumption by the neurosteroid allopregnanolone in mice. *Psychopharmacology, 162*, 438–447.

Tokuda, K., O'Dell, K. A., Izumi, Y., & Zorumski, C. F. (2010). Midazolam inhibits hippocampal long-term potentiation and learning through dual central and peripheral benzodiazepine receptor activation and neurosteroidogenesis. *Journal of Neuroscience, 30*, 16788–16795.

Torres, J. M., & Ortega, E. (2003). Alcohol intoxication increases allopregnanolone levels in female adolescent humans. *Neuropsychopharmacology, 28*, 1207–1209.

Torres, J. M., & Ortega, E. (2004). Alcohol intoxication increases allopregnanolone levels in male adolescent humans. *Psychopharmacology, 172*, 352–355.

VanDoren, M. J., Matthews, D. B., Janis, G. C., Grobin, A. C., Devaud, L. L., & Morrow, A. L. (2000). Neuroactive steroid 3α-hydroxy-5α-pregnan-20-one modulates electrophysiological and behavioral actions of ethanol. *Journal of Neuroscience, 20*, 1982–1989.

Vivian, J. A., Green, H. L., Young, J. E., Majerksy, L. S., Thomas, B. W., Shively, C. A., ... Grant, K. A. (2001). Induction and maintenance of ethanol self-administration in cynomolgus monkeys (*Macaca fascicularis*): long-term characterization of sex and individual differences. *Alcoholism Clinical and Experimental Research, 25*, 1087–1097.

Williams, R. W., Gu, J., Qi, S., & Lu, L. (2001). The genetic structure of recombinant inbred mice: high-resolution consensus maps for complex trait analysis. *Genome Biology, 2*, 1–46.

Chapter 41

Ethanol and Its Impact on the Brain's Electrical Activity

Rubem Carlos Araújo Guedes[1], Ranilson de Souza Bezerra[2], Ricardo Abadie-Guedes[3]

[1]*Departamento de Nutrição, Universidade Federal de Pernambuco, Recife, Pernambuco, Brazil;* [2]*Departamento de Bioquímica, Universidade Federal de Pernambuco, Recife, Pernambuco, Brazil;* [3]*Departamento de Fisiologia e Farmacologia, Universidade Federal de Pernambuco, Recife, Pernambuco, Brazil*

Abbreviations

CNS Central nervous system
CSD Cortical spreading depression
DC Direct current
ECoG Electrocorticogram
EEG Electroencephalogram
MEG Magnetoencephalogram
ROS Reactive oxygen species
TMS Transcranial magnetic stimulation

ALCOHOL ABUSE AND ITS NEUROLOGICAL CONSEQUENCES

Alcoholic beverages constitute a class of substances with neural actions that are among the most consumed by diverse populations worldwide. In human history, the production and consumption of fermented beverages occurred as early as thousands of years BC, and the habit of drinking beer is likely to have preceded bread consumption (Mody, 2008). For thousands of years, fermented beverages such as beer and wine were used daily by western populations to mitigate thirst, because at that time the society considered water inappropriate for human consumption. This occurred in a time when water sterilizing processes were unknown (Vallee, 1994). At that period of the human history, pure water existed only at the mountain sources, and this was available only to a small number of people. With the development of techniques for water sterilizing and alcoholic beverage distillation, the pattern of alcohol intake changed from low alcohol level (fermented) beverages to high alcohol levels (distilled) drinks, which enabled the ingestion of much higher alcohol amounts in smaller volumes and demonstrated the negative effects of alcohol.

Regular, frequent, and gradually increasing consumption of alcohol is a condition defined as alcoholism. From a public health point of view, alcoholism is a matter of great concern. This is one of the main reasons that led to substantially increase the worries of the health authorities about the deleterious effects of alcohol consumption (Vallee, 1994). Nowadays, alcoholism is considered a serious social problem, which is associated to the incidence of several diseases, such as cancer (Boffetta, Hashibe, La Vecchia, Zatonski, & Rehm, 2006; Parkin, 2011), liver insufficiency (Lu, Wu, Wang, Ward, & Cederbaum, 2010), and stroke (Patra et al., 2010). Alcohol can permeate almost all brain tissue; the alcohol-induced disruption of the biochemical, morphological, and physiological organization of the brain depends directly or indirectly on the production of free radicals (reactive oxygen- or nitrogen species; see Abadie-Guedes, Guedes, & Bezerra, 2012; Abadie-Guedes, Santos, Cahú, Guedes, & Bezerra, 2008; Tapiero, Townsend, & Tew, 2004). In its initial stage, alcohol intake provokes sensations of pleasure and well-being. As time passes, the organism becomes dependent on alcohol and a regular pattern of alcohol ingestion is established; the amount of alcohol required to obtain pleasant sensations augments with time, increasing the collateral, deleterious effects on the nervous system. The neurological impact of such a condition has been the object of much investigation. The social impact and the economic costs represented by alcoholism are very substantial regarding the health assistance to the affected persons (Bouchery, Harwood, Sacks, Simon, & Brewer, 2011), and these factors justify the great importance of studying the brain's effects in alcoholism.

Basic brain processes responsible for the neural analysis of sensory information and perception, as well as the efficient production and control of motor activity, can be affected by excessive alcohol consumption, which can also alter more complex brain functions responsible for learning, memory, consciousness, cognition, and emotion. Such alcohol-dependent disorders sometimes can also permanently affect the nervous system, leading to the establishment of more or less disabling diseases (Evren et al., 2008). Experimental evidence indicates that alcohol intake can also interfere with brain development and function (Schindler, Tsutsui, & Clark, 2014). All of these facts have motivated a number of clinical and experimental researchers to study neural susceptibility to the lasting negative effects of alcohol consumption.

In this chapter, we review experimental results that demonstrate the effects of acute and chronic ethanol consumption on the brain's electrophysiological activity, using the phenomenon known as *cortical spreading depression* (CSD) as a well-established experimental model. We also comment on how CSD can be used in studies on animals under several other conditions in addition to alcoholism, which is of interest for human health (see Guedes, 2011 for an example).

INFLUENCE OF ALCOHOL ON THE ELECTRICAL ACTIVITY OF THE BRAIN

The production of electrical activity by the brain constitutes the basic mechanism underlying the normal execution of the numerous brain functions. Currently, different techniques enable the recording of the brain's electrical activity, providing relevant information on the brain's physiological functioning. The noninvasive recording of the human brain's electrical activity is demonstrated by electroencephalogram (EEG). In order to perform the EEG, several pairs of electrodes are placed at distinct regions of the scalp; each pair enables the recording of differences of electrical potential between the two recording points. EEG is very frequently employed in human patients to help in the diagnosis of certain neurological diseases, such as epilepsy. The use of EEG is extensive because the recording does not have great ethical concerns as it is a noninvasive technique.

The analysis of the EEG allows the identification of normal and pathological EEG patterns. Such analysis has evolved from visual inspection by trained professionals (in the 1940s) into an analysis based on computation techniques (in the 1970s). While visual inspection was good for analyzing the qualitative aspects of the EEG, a computation-based technique enables the employment of spectral analysis and fast Fourier transform algorithms to quantify and classify features such as amplitude and frequency patterns of the EEG waves (see Morgane et al., 1978). With this technique, several authors characterized the ontogeny of EEG patterns in children (Schulte & Bell, 1973) and animals (Gramsbergen, 1976). More recently, the combination of EEG and magnetoencephalography techniques proved to be useful in providing insight into the neural basis of changes in the oscillatory patterns of the brain's electrical activity associated with alcohol consumption (Rosen et al., 2014).

The hypothesis that alcohol-dependent humans can develop epilepsy is supported by some EEG findings (Sand et al., 2010), although this issue is still controversial. The confirmation of the causal relationship between human alcoholism and epileptogenesis still requires solid scientific investigation, including studies using electrophysiological techniques. In adult rats that were previously exposed to alcohol during brain development, Bonthius et al. (2001) found electrophysiological signs of hippocampal seizures, kindling, and spreading depression (see below), suggesting that exposure to alcohol during brain development can permanently alter the electrophysiological activity of the hippocampal formation. In this context, we have electrophysiologically investigated alcohol-induced changes in the rat neocortex using cortical spreading depression (CSD) as a model (which is presented in the next part of this chapter). Some key features on the electrophysiological effects of ethanol are in Table 1, which presents studies in humans and laboratory animals on neural electrophysiological processes that can be affected by alcohol consumption.

CSD AS A TOOL TO STUDY THE EFFECTS OF ALCOHOL ON THE BRAIN

CSD was first described in the rabbit cerebral cortex by the Brazilian neuroscientist Aristides Leão as a reduction (depression) of the spontaneous and evoked electrical activity of one previously stimulated point of the cerebral cortex (Leão, 1944), with a simultaneous direct current (DC) slow potential change in the tissue (Leão, 1947). The CSD slow potential change is of an "all-or-none" nature, and it is very convenient for calculation of the CSD velocity of propagation. Compared with the velocity with which neuronal action potentials propagate (in the order of meters per second), the velocity of propagation of CSD is significantly lower, ranging from 2 to 5 mm/min, as evaluated in all vertebrate species so far studied (Guedes, 2005). Once elicited at a certain stimulated cortical point, CSD concentrically propagated to remote cortical regions, while the initially depressed point started to recover, characterizing a slow propagating and fully reversible phenomenon. In contrast to the neuronal action potentials (whose propagation has an ion-based mechanism), CSD propagation is postulated to be based on a humoral process. According to the current postulations, the cells under CSD would release one or more chemical factors, which would "contaminate" the neighbor cells, depressing them; these newly depressed cells would also release the CSD chemical factors, which would recruit more cells into the CSD state, generating a feedback loop that would render the CSD propagation in an autoregenerative mode (Martins-Ferreira, Oliveira Castro, Stuchiner, & Rodrigues, 1974). CSD induction usually requires a perturbation of the brain homeostasis. The application of electrical, chemical, or mechanical stimuli can disrupt the brain homeostasis and therefore can experimentally elicit CSD when applied to one point of the cortical surface, as evidenced in several animal species (see Gorji, 2001; Guedes, 2011 for a review). Figure 1 provides an example of the electrophysiological recording of a CSD episode elicited by chemical stimulation (KCl) of one point of the rat cerebral cortex. In experiments on animals, based on CSD recordings on at least two cortical points, the calculation of the CSD velocity of propagation is very easy, based on the time spent by a CSD wave to pass the distance between the two recording points. Therefore, animal studies involving CSD recording and quantification of parameters such as propagation velocity, under several paradigms of alcohol consumption, may be helpful to understand the mechanisms of ethanol electrophysiological effects on the brain, as we intend to demonstrate in this chapter.

The CSD phenomenon has also been recorded in the human brain, both in vitro (Gorji & Speckmann, 2004) and in vivo (Fabricius et al., 2008). CSD causes a brain electrical silence, as a consequence of neuronal and glial depolarization. This electrical silence is completely reversed after a few minutes. Some authors use the term "spreading depolarization" to designate the causal event of spreading EEG depression (Dreier, 2011). Considering that CSD has already been demonstrated in the human brain, as stated above, it becomes important to mention that CSD is involved in the production of the sensorial (usually visual) aura that precedes the painful sensations in the migraine (Lauritzen et al., 2011). CSD is also causally related to other important neurological diseases, such as stroke and epilepsy (Dreier et al., 2012; Lauritzen et al., 2011; Leão, 1944). During CSD, Leão described the appearance of "epileptiform waves," similar to those found in the epileptic EEG (Leão, 1944). This abnormal EEG activity appeared while the spontaneous activity was depressed and led to the postulation that CSD- and excitability-related brain disorders would share at least some common mechanisms (Leão, 1944, 1972). Therefore, understanding CSD mechanisms could help in comprehending the mechanisms of excitability-dependent neurological diseases, and this raises the hope for future therapeutic strategies with better outcomes than the current ones.

TABLE 1 Examples of Key Features on the Brain Electrophysiological Effects of Ethanol

Animal	References	Condition	Effect
Humans	Kähkönen, Wilenius, Nikulin, Ollikainen, and Ilmoniemi (2003)	EEG and TMS in nine healthy subjects under alcohol.	Alcohol reduces the excitability in the prefrontal cortex.
Humans	Conte et al. (2008)	The motor evoked potential (MEP) amplitude and the cortical silent period (CSP) duration were measured during repetitive TMS before and after acute ethanol in 10 healthy subjects and in 13 patients with chronic ethanol.	The acute ethanol intake mainly acts on GABAergic neurotransmission, whereas chronic ethanol abuse alters glutamate-dependent mechanisms of short-term cortical plasticity.
Humans	Hertle et al. (2012)	Recording with surface electrode strip, during neurosurgery.	Ketamine reduced "spreading depolarizations." Midazolam anesthesia increased number of "spreading depolarization" clusters.
Humans	Dreier et al. (2012)	Brain electrophysiological recordings in vivo and in vitro.	GABA-mediated inhibition protected from "spreading convulsions." The authors also reported on "spreading convulsions" in monopolar subdural recordings.
Humans	Rosen et al. (2014)	EEG and MEG recordings in healthy social drinkers who participated in both alcohol and placebo conditions.	Changes in spontaneous brain oscillations regarding their spectral and spatial characteristics.
Rats	Marty and Spigelman (2012)	Whole-cell patch-clamp recordings in brain slices.	Chronic intermittent ethanol increased excitability parameters in vitro.
Rats	Huang, Yen, Tsai, Valenzuela, and Huang (2012)	Electrophysiological recordings from microwire arrays implanted in the anterior cerebellum of freely moving rats.	Increased firing and oscillations in cerebellar Golgi cells in vivo.
Rats	Granato, Palmer, De Giorgio, Tavian, and Larkum (2012)	In vitro electrophysiology on pyramidal neurons in neocortical slices.	Neurons in layer 5 from ethanol-treated animals displayed a lower number and a shorter duration of dendritic spikes.
Mice (GAD67-GFP Knock-in)	Wadleigh and Valenzuela (2012)	Whole-cell patch-clamp of cerebellar slices, to measure $GABA_A$ receptor-mediated spontaneous and miniature inhibitory postsynaptic currents.	Ethanol increased GABAergic transmission and excitability in cerebellar interneurons.

A considerable amount of information on CSD phenomenology has been accumulated during the 70 years that elapsed since the CSD initial description. However, the complete knowledge of the CSD mechanisms is still a task that requires further investigation. Most discussions about CSD mechanisms often highlight the possible involvement of either certain ions (Guedes & Do Carmo, 1980; Torrente et al., 2014), neurotransmitter activity (Gorelova, Koroleva, Amemori, Pavlík, & Bureš, 1987; Guedes, Amancio-dos-Santos, Manhães-de-Castro, & Costa-Cruz, 2002; Guedes, Rocha-de-Melo, Lima, Albuquerque, & Francisco, 2013), or free radicals produced in the nervous tissue (El-Bachá, Lima-Filho, & Guedes, 1998; Guedes, Amorim, & Teodósio, 1996). The increase in the generation of reactive oxygen species (ROS) in the brain deserves particular attention because this process is present in the organisms that consume alcohol chronically (Tapiero et al., 2004). The cortical tissue naturally offers a relative resistance to CSD propagation (Abadie-Guedes et al., 2008) and experimental treatments can decrease or increase this resistance, resulting in augmented or reduced CSD propagation velocities, respectively, compared to nontreated controls (Guedes, 2011). Therefore, experimental procedures that modify the brain's ability to propagate CSD may provide valuable clues to the understanding of the phenomenon and the diseases related to them.

ETHANOL INFLUENCES CSD PROPAGATION: THE ROLE OF ANTIOXIDANT MOLECULES

To investigate the influence of ethanol on CSD, we first compared the CSD propagation in the cortical surface of adult rats whose dams had been treated per gavage during the whole gestation or lactation period with either ethanol or distilled water (control group). The ethanol treatment during the lactation, but not during

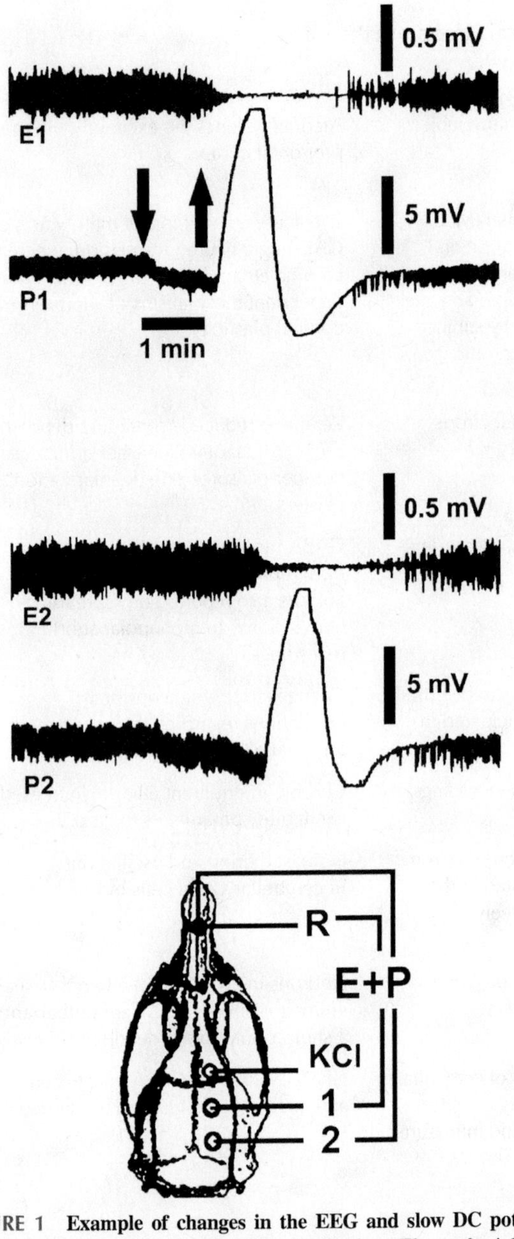

FIGURE 1 Example of changes in the EEG and slow DC potential recorded during cortical spreading depression. Electrophysiological recordings showing the reduction of the spontaneous cortical electrical activity (E) and the appearance of a negative slow potential change (P) during cortical spreading depression (CSD) in two cortical points (1 and 2) on the parietal cortical surface of an anesthetized rat. CSD was elicited by applying a cotton ball (1–2 mm diameter) soaked in 2% KCl solution (approximately 270 mM) for 1 min on the frontal cortex. Upward and downward arrows indicate respectively the beginning and the end of KCl stimulation. The skull diagram shows the recording positions 1 and 2 (on the parietal cortex) and the position of the reference electrode (R), on the nasal bones, as well as the place of KCl stimulus (on the frontal cortex). Vertical calibration bars (negativity upward) equal −0.5 mV for the ECoG- and −5 mV for the P-recordings. *Unpublished record from our laboratory.*

gestation, was associated with higher CSD propagation velocities compared with the water-treated controls (Figure 2(A)), suggesting a higher vulnerability of the lactation period compared with the gestation. Next, we demonstrated that ethanol treatment of adult rats for 7 days also facilitated CSD propagation compared with water-treated controls. Figure 2(B) summarizes these findings from our laboratory (Guedes & Frade, 1993). Evidence from others confirmed the facilitatory effect of ethanol treatment for 7 days on the CSD elicitation (Bonthius et al., 2001).

In contrast with the CSD facilitating effects of chronic ethanol, the acute intravenous infusion of ethanol in rats impairs CSD propagation (Sonn & Mayevsky, 2001), suggesting different mechanisms of action for acute and chronic effects of ethanol on the brain. This suggestion is also supported by evidence from motor-evoked potential measurements during repetitive transcranial magnetic stimulation (TMS) in humans (Conte et al., 2008). Alcohol consumption increases lipid peroxidation, and this enhances the formation of ROS in the brain. The intensification of ROS production represents a deleterious factor to the cerebral tissue. Under physiological conditions, the brain antioxidant substances neutralize the negative effects of ROS (Bondy, 1992). One important class of antioxidants is represented by the carotenoids, which are present in a number of vegetable- and animal-derived foods. Among the foods from animal origin, the crustaceans, such as shrimps, are important sources of carotenoids. Shrimps contain large amounts of carotenoids present as a carotenoprotein in the legs, blood, eyes, eggs, hepatopancreas, ovary, and the carapace (Kuo, Lee, Chichester, & Simpson, 1976). To test the hypothesis that shrimp carotenoids influence CSD propagation, we treated rat dams during pregnancy or lactation with a shrimp carotenoid ethanolic extract (30 μg/kg/d, per gavage) and examined CSD propagation in their progenies, comparing them with pups from mothers treated either with the vehicle (ethanol) or with distilled water. Compared with the distilled water group, the ethanol-treated rats displayed higher ($p<0.05$) CSD velocities, which is coherent with previous data from our laboratory (Guedes & Frade, 1993). In contrast, ethanol + carotenoid-treated rats displayed lower CSD velocities ($p<0.05$) compared with the ethanol-treated group. The CSD antagonizing effect associated with the carotenoids was more intense in the groups treated during lactation compared with the groups treated during the gestation period (Bezerra et al., 2005). The data, suggesting a protective action of shrimp carotenoids against the effects of chronic ethanol on CSD, are in Figure 3.

Among the shrimp carotenoids, astaxanthin is the most abundant. Because of that, we considered the possibility that the effects of the shrimp carotenoid extract on CSD were mainly caused by the large amount of astaxanthin present in the extract. We investigated this possibility by treating adult rats per gavage for 18 days with pure astaxanthin (from Sigma Co., St Louis, MO, USA) in doses of 2.5, 10, or 90 μg/kg/d. Figure 4 illustrates three CSD recordings representative from the chronic ethanol, control (water-treated), and ethanol + astaxanthin groups. Compared with the ethanol-treated animals, the water- and astaxanthin-treated groups presented with longer latencies for a CSD wave to cross the interelectrode distance, indicating lower CSD velocities. This astaxanthin action against the ethanol-induced CSD acceleration was dose dependent and followed an exponential decay model, which is represented by the equation $y = 2.9048 + 1.1988e^{(-x/9.0217)}$; ($r^2 = 0.9998$), in which the y-variable is the CSD velocity of propagation and the x-variable is astaxanthin dose in μg/kg/d (for details, see Abadie-Guedes et al., 2008). The results are compatible with the conclusion that the similar effects in the rats previously treated with the shrimp carotenoid extract are probably due to the high proportion of astaxanthin present in that extract.

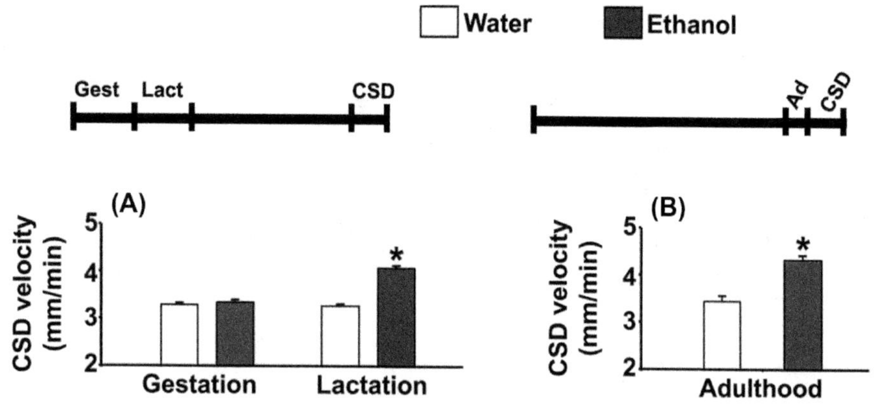

FIGURE 2 **CSD propagation in rats chronically treated with ethanol during gestation, lactation, or adulthood.** Mean ± SEM CSD velocities of propagation in adult rats after ethanol or water (control group) treatment. (A) Rats whose dams had been treated with ethanol during gestation (Gest) or lactation (Lact), compared with controls treated with distilled water during the same periods ($n=15$ for each group). (B) Rats that received ethanol or water for 7 days at adulthood (Ad) just before the CSD recording ($n=6$ for each group). The horizontal black lines indicate the time points of ethanol treatment in relation to the CSD recording period. *$p<0.05$ compared with the corresponding water-treated control group (ANOVA plus Holm–Sidak test). *Unpublished figure from our laboratory.*

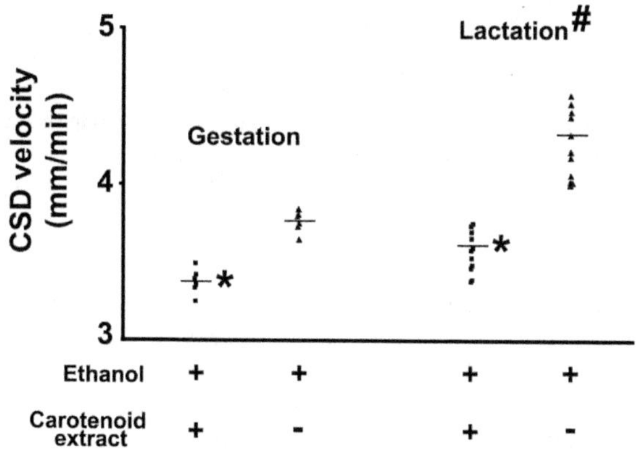

FIGURE 3 **Antagonizing action of an extract of shrimp carotenoids against the CSD facilitating effect of chronic ethanol in the rat brain.** CSD velocities of propagation in four groups of adult rats whose dams had been treated with ethanol, or ethanol plus a shrimp carotenoid extract during the whole gestation ($n=7$ for each treatment), or during the whole lactation period ($n=11$ for each treatment). Data are presented as individual values for each animal (points) and mean values for each group (horizontal bars). *$p<0.05$ compared with the corresponding ethanol-treated group. #$p<0.05$ compared with the respective group whose dams had been treated during gestation (ANOVA plus Holm–Sidak test). *Modified from a figure of our previous publication (Bezerra et al., 2005).*

It is interesting that acute exposure to ethanol reduces the level of glutamate and aspartate in the cerebral cortex (Tiwari, Veeraiah, Subramaniam, & Patel, 2014) and influences cortical excitability (Conte et al., 2008). Brain effects of acute ethanol also include impairment of CSD (Sonn & Mayevsky, 2001). Because of that, we decided to investigate whether acute astaxanthin could also antagonize the CSD effects of acute ethanol.

We analyzed CSD propagation in young and adult Wistar rats (respectively 60–80 days old and 150–180 days old), treated per gavage with a single dose of ethanol (3 g/kg), vehicle (water or olive oil), or ethanol plus 10 µg/kg astaxanthin (Abadie-Guedes et al., 2012). As can be observed in Figure 5, these experiments confirmed two types of effects previously suggested by our group and by others: (1) acute ethanol decelerates CSD (Sonn & Mayevsky, 2001) and (2) aging also decelerates CSD (Batista-de-Oliveira, Lopes, Mendes-da-Silva, & Guedes, 2012; Guedes et al., 1996). Furthermore, the acute treatment with ethanol plus astaxanthin resulted in CSD velocities comparable to the controls, indicating that astaxanthin antagonized the impairing action of acute ethanol on CSD (Abadie-Guedes et al., 2012).

APPLICATIONS TO OTHER ADDICTIONS AND SUBSTANCE MISUSE

The ability of the brain in exerting its normal function by means of generation and propagation of electrical activity constitutes its main physiological property. Therefore, the use of electrophysiologically based animal models, such as the CSD phenomenon, constitutes a very interesting strategy for the study of brain alterations under ethanol abuse, and it can be useful also for investigating brain effects of other drug addictions. The electrophysiological data from humans and laboratory animals presented in this chapter documents the relevance of the CSD in studying the nervous system, as previously stressed (Guedes, 2011). The importance of the lactation period as the developmental time point for the execution of the brain structural and functional maturation program is now well established in the rat. The brain alterations induced by ethanol (and probably other drugs; Gulley & Juraska, 2013; Rocha-de-Melo, Lima, Albuquerque, Oliveira, & Guedes, 2008) that occur during this period probably result from some degree of disruption of brain development programming. Processes such as dendrite development, synapse formation, cell migration, and myelination are certainly implicated in the neurophysiologic alterations found in the progeny of the rat dams that consumed ethanol early in life (Bezerra et al., 2005).

Adequate data extrapolation from the rat to the human species still depends upon further clinical and experimental investigation. This includes the growing utilization of animal models such as the CSD phenomenon, as presently described. In this context, CSD can be considered as a supporting tool to understand the reciprocal relationship between neural development and function on one hand and the abuse of drugs on the other.

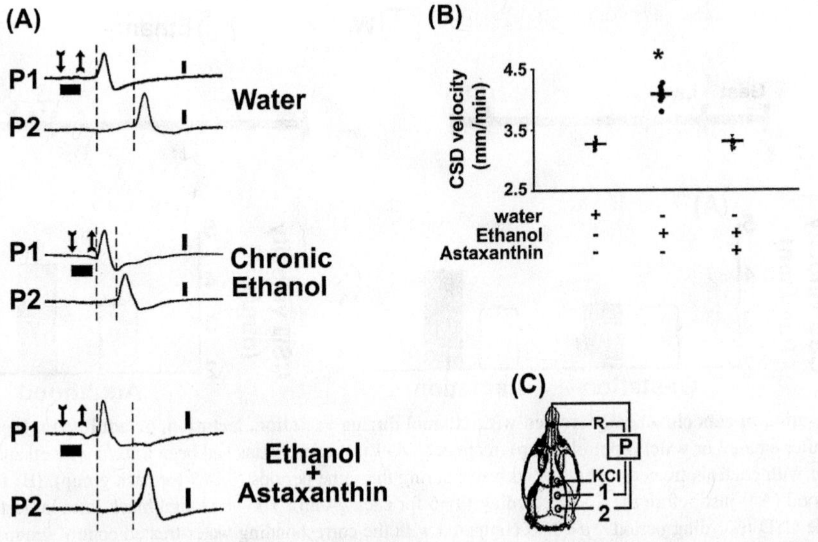

FIGURE 4 Antagonizing effect of astaxanthin against the CSD facilitating action of chronic ethanol in rats. (A) Electrophysiological recording (slow DC potential variation) showing CSD propagation in three adult rats (150–180 days old) treated during 18 days per gavage with water (control), or 3 g/kg/d ethanol, or 3 g/kg/d ethanol plus 10 μg/kg/d astaxanthin. The shorter CSD propagation latency (delimited by the dashed vertical lines) in the ethanol-treated rat indicates higher CSD velocity, compared with the control and ethanol + astaxanthin rats, as quantified in (B). The time (1 min) between down and up arrows corresponds to the application of a cotton ball (1–2 mm diameter) soaked with 2% KCl solution to one point on the frontal cortex to elicit CSD. The horizontal bars under the P1 traces indicate 1 min. The vertical solid bars equal 5 mV (negative upward). (B) CSD velocities of propagation in the three groups of adult rats exemplified in (A) ($n=7$ for each group). Data are presented as individual values for each animal (points) and mean values for each group (horizontal bars). $*p<0.05$ compared with the water and ethanol + astaxanthin groups (ANOVA plus Holm–Sidak test). (C) Skull diagram showing the recording positions 1 and 2 (on the right parietal cortex) and the position of the reference electrode (R), on the nasal bones, as well as the place of KCl stimulus (on the right frontal cortex). *Unpublished picture from the authors' laboratory, based on data of our previous publication (Abadie-Guedes et al., 2008).*

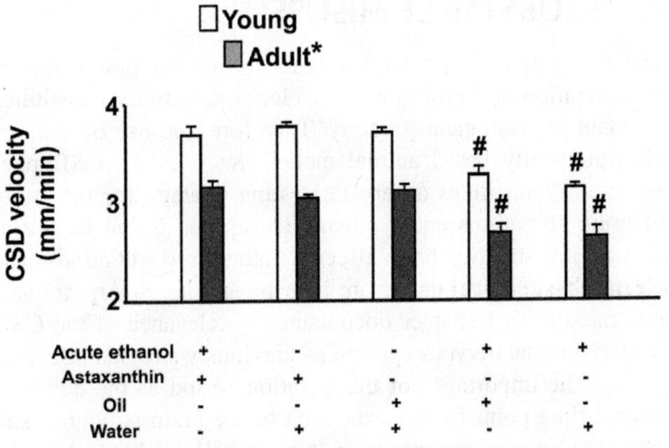

FIGURE 5 Antagonizing effect of astaxanthin against the CSD impairing action of acute ethanol in young and adult rats. CSD velocities (mean ± standard error) in young and adult Wistar rats (respectively 60–80 days old and 150–180 days old), treated per gavage with a single dose of ethanol (3 g/kg), vehicle (water or olive oil), or ethanol plus 10 μg/kg astaxanthin. The number of animals in each age group ranged from 9 to 10 in the young groups and from 8 to 11 in the adult condition. $*p<0.05$ compared with the corresponding young groups. $\#p<0.05$ compared with the water + oil, water + astaxanthin, and ethanol + astaxanthin groups (ANOVA plus Holm–Sidak test). *Modified from a figure of our previous publication (Abadie-Guedes et al., 2012).*

Considering that adverse conditions represented by the abuse of several drugs, including ethanol, currently affect an expressive part of the human population, it can be concluded that the investigation of electrophysiological features of phenomena such as CSD represents a valuable experimental approach to understand the underlying functional brain alterations. We believe that clarifying the mechanisms by which the abuse of several drugs affect CSD propagation may shed light on the role of this drug in modulating important human neurological diseases. Furthermore, the complete understanding of CSD generation and propagation mechanisms might be very helpful in developing better treatment of brain disorders such as stroke, epilepsy, and migraine (Dohmen et al., 2008; Fabricius et al., 2008; Margineanu & Klitgaard, 2009).

DEFINITIONS AND EXPLANATIONS OF KEY TERMS OR WORDS USED IN THE CHAPTER

Alcoholism Alcoholism is defined as frequent and intense alcohol consumption, which characterizes a status of drug dependency and affects increasing human contingents in many parts of the world. This condition affects seriously several organs, including the brain, generating physiological and psychological disturbances in the individual and also influencing social interaction.

Carotenoids These organic pigment molecules may have antioxidant action exerting scavenging effects in the brain, protecting it against reactive oxygen species produced in certain pathological states, such as alcoholism. Animals are not biochemically capable of biosynthesizing carotenoids, but they can accumulate them and/or convert their precursors, which they obtain from the diet.

Cortical spreading depression (CSD) CSD is a slow and reversible brain response that consists of the reduction (depression) of electrical activity, usually elicited by mechanical, electrical, or chemical stimulation of a point of the brain tissue. Current information on the

CSD points to the existence of causal association between this phenomenon and human neurological diseases such as brain ischemia, migraine, and epilepsy.

Electroencephalogram (EEG) EEG is a noninvasive technique largely used in the neurologic clinic for diagnosis of brain disorders, such as epilepsy. EEG consists in the recording of the spontaneous brain electrical activity. Formerly performed in chart-paper, ink-writer machines, currently EEG is mostly digitalized in computer-based devices.

Epilepsy This is a serious neurological disease caused by the abnormal functioning of a certain brain neuronal population, which starts to produce its electrical activity in an abnormally intense and uncontrolled manner. A generalized epileptic seizure usually presents with uncontrolled muscle contractions (convulsions) and loss of consciousness, as the main clinical signs.

Fast Fourier Transform This mathematical, algorithm-based technique is used to decompose the "spectrum of EEG-waves" into its distinct component frequencies, enabling the detection of disease-related alterations in one or more of such frequencies.

Free radicals This is a chemistry term that describes an atomic or molecular species with unpaired electrons on an otherwise open shell configuration. See also "reactive oxygen species" below.

Kindling Kindling is an experimental model of producing epileptic seizures in animals. It has been described by Goddard, McIntyre, and Leech (1969) and consists of the repeated excitatory stimulation of certain brain structures such as hippocampus, which initially results in partial seizures. With the repetition of the same stimulation paradigm, the seizures become permanently more severe and longer lasting. Kindling is considered a model of the human temporal lobe epilepsy.

Magnetoencephalography (MEG) MEG is a noninvasive, functional neuroimaging technique that, after detection and amplification, can record the field associated with the brain electrical activity. This technique is suitable for mapping brain activity using arrays of specific detectors known as "superconducting quantum interference devices," abbreviated as SQUIDs.

Reactive oxygen species (ROS) ROS are products derived from metabolic O_2 that are partially reduced and possess higher reactivity compared with molecular O_2 (e.g., the superoxide anion $[O_2 \cdot^-]$ and hydrogen peroxide $[H_2O_2]$, formed respectively by one- and two-electron reductions of O_2). The presence of ROS in excess in the brain has been associated with neurodegenerative and excitability-related diseases.

KEY FACTS OF ELECTROPHYSIOLOGY

- The primary function of the brain is to control, directly or indirectly, the function of all physiologic systems in the organism.
- The brain function is realized via the generation and propagation of electrical activity by the neurons.
- EEG is the noninvasive recording of the spontaneous brain electrical activity via electrodes placed on the scalp.
- EEG is the usual abbreviation for electroencephalogram.
- In human beings, alterations in the pattern of the EEG waves can help in diagnosing certain neurological diseases, such as epilepsy.
- In laboratory animals, the brain electrical activity can be recorded by placing the electrodes directly over the cortex surface, via skull openings.
- The recording of electrical activity directly from the cortical surface is denominated electrocorticogram (abbreviated as ECoG).
- In laboratory animals, the ECoG can help in understanding the mechanisms of certain disorders of the nervous system.
- EEG findings in humans and ECoG analyses in animals suggest that alcohol abuse can predispose to epilepsy.
- CSD is a brain response that can be evoked and propagated over the cortical tissue, and can be studied in animals via ECoG.
- Electrophysiological findings from the rat cortex on the effects of ethanol on CSD are reviewed in this chapter.

SUMMARY POINTS

- An organism under intense and repeated alcohol consumption can suffer developmental and physiological alterations.
- Depending on the intensity and duration of the alcohol consumption, neurological disturbances can appear, including changes in the brain electrical activity.
- Clinical data suggest that some alcohol-addicted humans have a higher propensity to develop epileptic manifestations compared with nonalcoholic controls. However, further studies are needed to definitely confirm this causal relationship.
- We reviewed the neurophysiological impact of alcohol consumption, analyzing in laboratory animals the alcohol-associated changes in the brain electrical activity.
- The use of the phenomenon of CSD in studies on brain ethanol effects was reviewed. Electrophysiological investigation using the CSD phenomenon is very helpful to get knowledge on the mechanisms of certain human neurological diseases including migraine, epilepsy, and stroke.
- We report on previously published experimental data on the CSD effects of acute and chronic alcohol consumption, and discuss possible underlying mechanisms for alcohol's action on brain electrophysiological activity.
- The propagation velocity of CSD constitutes a good indicator of electrophysiological alterations induced by alcohol, as indicated by a growing body of data from our laboratory and from others.

ACKNOWLEDGMENTS

The authors thank the Brazilian agencies CAPES, FINEP/IBN-Net (No. 01.06.0842-00), Instituto Nacional de Neurociência Translacional (INCT no. 573604/2008-8), FINEP/RECARCINA (1650/10), CNPq (558258/2009-3, 475787/2009-9), CAPES (Procad/2007 and Ciências doMar/2009), for the financial support. RCAG and RSB are Research Fellows from CNPq (No. 301190/2010-0 and 303570/2009-1, respectively).

REFERENCES

Abadie-Guedes, R., Guedes, R. C. A., & Bezerra, R. S. (2012). The impairing effect of acute ethanol on spreading depression is antagonized by astaxanthin in rats of 2 young-adult ages. *Alcoholism: Clinical and Experimental Research, 36*, 1563–1567.

Abadie-Guedes, R., Santos, S. D., Cahú, T. B., Guedes, R. C. A., & Bezerra, R. S. (2008). Dose-dependent effects of astaxanthin on cortical spreading depression in chronically ethanol-treated adult rats. *Alcoholism: Clinical and Experimental Research, 32*, 1417–1421.

Batista-de-Oliveira, M., Lopes, A. A. C., Mendes-da-Silva, R. F., & Guedes, R. C. A. (2012). Aging-dependent brain electrophysiological effects in rats after distinct lactation conditions, and treadmill exercise: a spreading depression analysis. *Experimental Gerontology, 47*, 452–457.

Bezerra, R. S., Abadie-Guedes, R., Melo, F. R. M., Paiva, A. M. A., Amâncio-dos-Santos, A., & Guedes, R. C. A. (2005). Shrimp carotenoids protect the developing rat cerebral cortex against the effects of ethanol on cortical spreading depression. *Neuroscience Letters, 391*, 51–55.

Boffetta, P., Hashibe, M., La Vecchia, C., Zatonski, W., & Rehm, J. (2006). The burden of cancer attributable to alcohol drinking. *International Journal of Cancer, 119*, 884–887.

Bondy, S. C. (1992). Reactive oxygen species: relation to aging and neurotoxic damage. *Neurotoxicology, 13*, 87–100.

Bonthius, D. J., Pantazis, N. J., Karacay, B., Bonthius, N. E., Taggard, D. A., & Lothman, E. W. (2001). Alcohol exposure during the brain growth spurt promotes hippocampal seizures, rapid kindling, and spreading depression. *Alcoholism: Clinical and Experimental Research, 25*, 734–745.

Bouchery, E. E., Harwood, H. J., Sacks, J. J., Simon, C. J., & Brewer, R. D. (2011). Economic costs of excessive alcohol consumption in the U.S., 2006. *American Journal of Preventive Medicine, 41*, 516–524.

Conte, A., Attilia, M. L., Gilio, F., Iacovelli, E., Frasca, V., Marini Bettolo, C., ... Inghilleri, M. (2008). Acute and chronic effects of ethanol on cortical excitability. *Clinical Neurophysiology, 119*, 667–674.

Dohmen, C., Sakowitz, O. W., Fabricius, M., Bosche, B., Reithmeier, T., Ernestus, R. I., ... Co-Operative Study of Brain Injury Depolarization (COSBID) (2008). Spreading depolarizations occur in human ischemic stroke with high incidence. *Annals of Neurology, 63*, 720–728.

Dreier, J. P. (2011). The role of spreading depression, spreading depolarization and spreading ischemia in neurologyical disease. *Nature Medicine, 17*, 439–447.

Dreier, J. P., Major, S., Pannek, H. W., Woitzik, J., Scheel, M., Wiesenthal, D., ... COSBID study group (2012). Spreading convulsions, spreading depolarization and epileptogenesis in human cerebral cortex. *Brain, 135*(Pt. 1), 259–275.

El-Bachá, R. S., Lima-Filho, J. L., & Guedes, R. C. A. (1998). Dietary antioxidant deficiency facilitates cortical spreading depression induced by photo-activated riboflavin. *Nutritional Neuroscience, 1*, 205–212.

Evren, C., Sar, V., Evren, B., Semiz, U., Dalbudak, E., & Cakmak, D. (2008). Dissociation and alexithymia among men with alcoholism. *Psychiatry and Clinical Neurosciences, 62*, 40–47.

Fabricius, M., Fuhr, S., Willumsen, L., Dreier, J. P., Bhatia, R., Boutelle, M. G., ... Lauritzen, M. (2008). Association of seizures with cortical spreading depression and peri-infarct depolarisations in the acutely injured human brain. *Clinical Neurophysiology, 119*, 1973–1984.

Goddard, G. V., McIntyre, D. C., & Leech, C. K. (1969). A permanent change in brain function resulting from daily electrical stimulation. *Experimental Neurology, 25*, 295–330.

Gorelova, N. A., Koroleva, V. I., Amemori, T., Pavlík, V., & Bureš, J. (1987). Ketamine blockade of cortical spreading depression in rats. *Electroencephalography and Clinical Neurophysiology, 66*, 440–447.

Gorji, A. (2001). Spreading depression: a review of the clinical relevance. *Brain Research Reviews, 38*, 33–60.

Gorji, A., & Speckmann, E. J. (2004). Spreading depression enhances the spontaneous epileptiform activity in human neocortical tissues. *European Journal of Neurosciences, 19*, 3371–3374.

Gramsbergen, A. (1976). The development of the EEG in the rat. *Developmental Psychobiology, 9*, 501–515.

Granato, A., Palmer, L. M., De Giorgio, A., Tavian, D., & Larkum, M. E. (2012). Early exposure to alcohol leads to permanent impairment of dendritic excitability in neocortical pyramidal neurons. *Journal of Neuroscience, 32*, 1377–1382.

Guedes, R. C. A. (2005). Electrophysiological methods: application in nutritional neuroscience. In H. Liebermann, R. Kanarek, & C. Prasad (Eds.), *Nutritional neurosciences: Overview of an emerging field* (pp. 39–54). New York: CRC Press.

Guedes, R. C. A. (2011). Cortical spreading depression: a model for studying brain consequences of malnutrition. In V. R. Preedy, R. R. Watson, & C. R. Martin (Eds.), *Handbook of behavior, food and nutrition* (pp. 2343–2355). Berlin: Springer.

Guedes, R. C. A., Amancio-dos-Santos, A., Manhães-de-Castro, R., & Costa-Cruz, R. R. G. (2002). Citalopram has an antagonistic action on cortical spreading depression in well-nourished and early-malnourished adult rats. *Nutritional Neuroscience, 5*, 115–123.

Guedes, R. C. A., Amorim, L. F., & Teodósio, N. R. (1996). Effect of aging on cortical spreading depression. *Brazilian Journal of Medical and Biological Research, 29*, 1407–1412.

Guedes, R. C. A., & Do Carmo, R. J. (1980). Influence of ionic alterations produced by gastric washing on cortical spreading depression. *Experimental Brain Research, 39*, 341–349.

Guedes, R. C. A., & Frade, S. F. (1993). Effect of ethanol on cortical spreading depression. *Brazilian Journal of Medical and Biological Research, 26*, 1241–1244.

Guedes, R. C. A., Rocha-de-Melo, A. P., Lima, K. R., Albuquerque, J. M. S., & Francisco, E. S. (2013). Early malnutrition attenuates the impairing action of naloxone on spreading depression in young rats. *Nutritional Neuroscience, 16*, 142–146.

Gulley, J. M., & Juraska, J. M. (2013). The effects of abused drugs on adolescent development of corticolimbic circuitry and behavior. *Neuroscience, 249*, 3–20.

Hertle, D. N., Dreier, J. P., Woitzik, J., Hartings, J. A., Bullock, R., Okonkwo, D. O., ... Cooperative Study of Brain Injury Depolarizations (COSBID) (2012). Effect of analgesics and sedatives on the occurrence of spreading depolarizations accompanying acute brain injury. *Brain, 135*, 2390–2398.

Huang, J. J., Yen, C. T., Tsai, M. L., Valenzuela, C. F., & Huang, C. (2012). Acute ethanol exposure increases firing and induces oscillations in cerebellar Golgi cells of freely moving rats. *Alcoholism: Clinical and Experimental Research, 36*, 2110–2116.

Kähkönen, S., Wilenius, J., Nikulin, V. V., Ollikainen, M., & Ilmoniemi, R. J. (2003). Alcohol reduces prefrontal cortical excitability in humans: a combined TMS and EEG study. *Neuropsychopharmacology, 28*, 747–754.

Kuo, H. C., Lee, T. C., Chichester, C. O., & Simpson, K. L. (1976). The carotenoids in the deep sea red crab, *Geryon quinquedens*. *Comparative Biochemistry and Physiology B, 54*, 387–390.

Lauritzen, M., Dreier, J. P., Fabricius, M., Hartings, J. A., Graf, R., & Strong, A. J. (2011). Clinical relevance of cortical spreading depression in neurological disorders: migraine, malignant stroke, subarachnoid and intracranial hemorrhage, and traumatic brain injury. *Journal of Cerebral Blood Flow and Metabolism, 31*, 17–35.

Leão, A. A. P. (1944). Spreading depression of activity in the cerebral cortex. *Journal of Neurophysiology, 7*, 359–390.

Leão, A. A. P. (1947). Further observations on the spreading depression of activity in the cerebral cortex. *Journal of Neurophysiology, 10*, 409–414.

Leão, A. A. P. (1972). Spreading depression. In D. P. Purpura, K. Penry, D. B. Tower, D. M. Woodbury, & R. D. Walter (Eds.), *Experimental models of epilepsy* (pp. 173–195). New York: Raven Press.

Lu, Y., Wu, D., Wang, X., Ward, S. C., & Cederbaum, A. I. (2010). Chronic alcohol-induced liver injury and oxidant stress are decreased in cytochrome P4502E1 knockout mice and restored in humanized cytochrome P4502E1 knock-in mice. *Free Radical Biology & Medicine, 49*, 1406–1416.

Margineanu, D. G., & Klitgaard, H. (2009). Brivaracetam inhibits spreading depression in rat neocortical slices in vitro. *Seizure, 18*, 453–456.

Martins-Ferreira, H., Oliveira Castro, G., Stuchiner, C. J., & Rodrigues, P. S. (1974). Liberation of chemical factors during spreading depression in isolated retina. *Journal of Neurophysiology, 37*, 785–791.

Marty, V. N., & Spigelman, I. (2012). Effects of alcohol on the membrane excitability and synaptic transmission of medium spiny neurons in the nucleus accumbens. *Alcohol, 46*, 317–327.

Mody, I. (2008). Extrasynaptic GABAA receptors in the crosshairs of hormones and ethanol. *Neurochemistry International, 52*, 60–64.

Morgane, P. J., Miller, M., Kemper, T., Stern, W., Forbes, W., Hall, R., ... Resnick, O. (1978). The effects of protein malnutrition on the developing nervous system in the rat. *Neuroscience and Biobehavioral Reviews, 2*, 137–230.

Parkin, D. M. (2011). Cancers attributable to consumption of alcohol in the UK in 2010. *British Journal of Cancer, 105*, S14–S18.

Patra, J., Taylor, B., Irving, H., Roerecke, M., Baliunas, D., Mohapatra, S., ... Rehm, J. (2010). Alcohol consumption and the risk of morbidity and mortality for different stroke types - a systematic review and meta-analysis. *BioMed Central Public Health, 10*(258), 1–12. http://dx.doi.org/10.1186/1471-2458-10-258.

Rocha-de-Melo, A. P., Lima, K. R., Albuquerque, J. M. S., Oliveira, A. K. P., & Guedes, R. C. A. (2008). Chronic neonatal exposure of rats to the opioid antagonist naloxone impairs propagation of cortical spreading depression in adulthood. *Neuroscience Letters, 441*, 315–318.

Rosen, B. Q., O'Hara, R., Kovacevic, S., Schulman, A., Padovan, N., & Marinkovic, K. (2014). Oscillatory spatial profile of alcohol's effects on the resting state anatomically-constrained MEG. *Alcohol, 48*, 89–97.

Sand, T., Bjørk, M., Bråthen, G., Michler, R. P., Brodtkorb, E., & Bovim, G. (2010). Quantitative EEG in patients with alcohol-related seizures. *Alcoholism: Clinical and Experimental Research, 34*, 1751–1758.

Schindler, A. G., Tsutsui, K. T., & Clark, J. J. (2014). Chronic alcohol intake during adolescence, but not adulthood, promotes persistent deficits in risk-based decision making. *Alcoholism: Clinical and Experimental Research, 38*(6), 1622–1629. http://dx.doi.org/10.1111/acer.12404.

Schulte, F. J., & Bell, E. F. (1973). Bioelectric brain development. An atlas of EEG power spectra in infants and young children. *Neuropädiatrie, 4*, 30–45.

Sonn, J., & Mayevsky, A. (2001). The effect of ethanol on metabolic, hemodynamic and electrical responses to cortical spreading depression. *Brain Research, 908*, 174–186.

Tapiero, H., Townsend, D. M., & Tew, K. D. (2004). The role of carotenoids in the prevention of human pathologies. *Biomedicine & Pharmacotherapy, 58*, 100–110.

Tiwari, V., Veeraiah, P., Subramaniam, V., & Patel, A. B. (2014). Differential effects of ethanol on regional glutamatergic and GABAergic neurotransmitter pathways in mouse brain. *Journal of Neurochemistry, 128*, 628–640.

Torrente, D., Mendes-da-Silva, R. F., Lopes, A. A. C., González, J., Barreto, G. E., & Guedes, R. C. A. (2014). Increased calcium influx triggers and accelerates cortical spreading depression in vivo in male adult rats. *Neuroscience Letters, 558*, 87–90.

Vallee, B. L. (1994). Alcohol in human history. In B. Jansson, H. Jörnvall, U. Rydberg, L. Terenius, & B. L. Vallee (Eds.), *Toward a molecular basis of alcohol use and abuse* (pp. 1–8). Basel: Birkhäuser.

Wadleigh, A., & Valenzuela, C. F. (2012). Ethanol increases GABAergic transmission and excitability in cerebellar molecular layer interneurons from GAD67-GFP knock-in mice. *Alcohol and Alcoholism, 47*, 1–8.

Chapter 42

Alcohol and Neuroimmune Interactions

Cynthia J.M. Kane[1], Susan E. Bergeson[2], Paul D. Drew[1]

[1]Department of Neurobiology and Developmental Sciences, University of Arkansas for Medical Sciences, Little Rock, AR, USA; [2]Department of Pharmacology and Neuroscience, Texas Tech University Health Sciences Center, Lubbock, TX, USA

Abbreviations

AUD Alcohol use disorders
CCL2 Chemokine (C—C motif) ligand 2; MCP-1
CNS Central nervous system
CX3CL1 Chemokine (C—X_3—C motif) ligand 1; fractalkine
CX3CR1 Chemokine (C—X_3—C motif) receptor 1
FASD Fetal alcohol spectrum disorders
FDA Food and Drug Administration
HMGB High-mobility group box
IL Interleukin
IFN Interferon
NF-κB Nuclear factor-κB
TGF Transforming growth factor
TLR Toll-like receptor
TNF Tumor necrosis factor

INTRODUCTION

The number of people who consume alcohol in the United States and throughout the world is high. In the United States, alcohol use disorders (AUD) occur in approximately 7% of adults and binge drinking occurs in 24.5% and 12.5% of adult men and women, respectively (Center for Disease Control and Prevention, 2013; Substance Abuse and Mental Health Services Administration, 2014). The cost of alcohol consumption among adults in the United States is estimated at $223.5 billion yearly. Adolescent and college-age alcohol abuse is a serious problem. Drinking among adolescents is prevalent and AUD occur in 3.4% of adolescents (Substance Abuse and Mental Health Services Administration, 2014). Binge drinking occurs in 40% of college students and 20% have AUD (Blanco, Okuda, et al., 2008). At least 12% of pregnant women drink alcohol (Floyd, Weber, Denny, & O'Connor, 2009). As a result, 2–5% of children in the United States are born with fetal alcohol spectrum disorders (FASD), which are the leading cause of mental retardation (May et al., 2009). Clearly, AUD and FASD exert a tremendous toll on individuals, families, and society.

Alcohol consumption across the lifespan, including fetal exposure during pregnancy, produces marked pathological changes in the central nervous system (CNS) that manifest as neurological, psychiatric, behavioral, cognitive, and motor defects. Chronic heavy drinking is associated with marked CNS pathology, including loss of brain volume and loss of neurons, resulting in neurological and cognitive dysfunction. A binge-drinking pattern in which high consumption is condensed into short periods of time appears to produce the most serious CNS pathology. Such a pattern of drinking is especially prevalent in adolescents and college-age individuals. Maternal consumption of alcohol during pregnancy is particularly damaging to the developing brain and long-term CNS function. The pathological effects of alcohol in the fetal brain produce a variety of lifelong disabilities, including increased risk of AUD later in life (Fryer, McGee, Matt, Riley, & Mattson, 2007; Streissguth et al., 2004).

AUD is the medical term currently used to describe the spectrum of clinically relevant, problematic alcohol consumption as defined by the *Diagnostic Statistical Manual of Mental Disorders, Fifth Edition*. The new diagnosis replaces the previously used bimodal terms of *alcohol abuse* and *alcoholism*, as a spectrum from mild to severe. As alcohol abuse and alcoholism have been generally accepted as complex traits (polygenic with multifaceted environmental influences), the new description as a spectrum fits. Evidence of alcohol effects on neuroimmune function—the focus of this review—also seems to fit a spectrum-associated hypothesis that lifespan timing (age of exposure), dose, chronicity, and drinking pattern affect outcomes and may explain some experimental discrepancies as biologically significant rather than technical differences. Data from several laboratories studying human and/or animal models clearly shows that alcohol consumption or exposure leads to neuroimmune changes across the entire lifetime, from in utero through adulthood. Emerging evidence shows that neuroinflammation contributes to behavioral, cognitive, psychiatric, and neurological problems associated with alcohol misuse. Fetal and adult brains exhibit neuroinflammatory events in response to alcohol exposure, including microglial activation and induction of pro-inflammatory cytokines and chemokines. The early adolescent brain may be differentially vulnerable; initial insult and lower doses do not produce a pro-inflammatory response, while high doses that are usually associated with neural damage and degeneration have been shown to do so. Particularly important, advances toward targeting alcohol-related neuroimmune changes show pharmacotherapeutic promise.

MICROGLIA

The CNS is composed principally of glia (microglia, astrocytes, and oligodendrocytes) and neurons. Microglia comprise

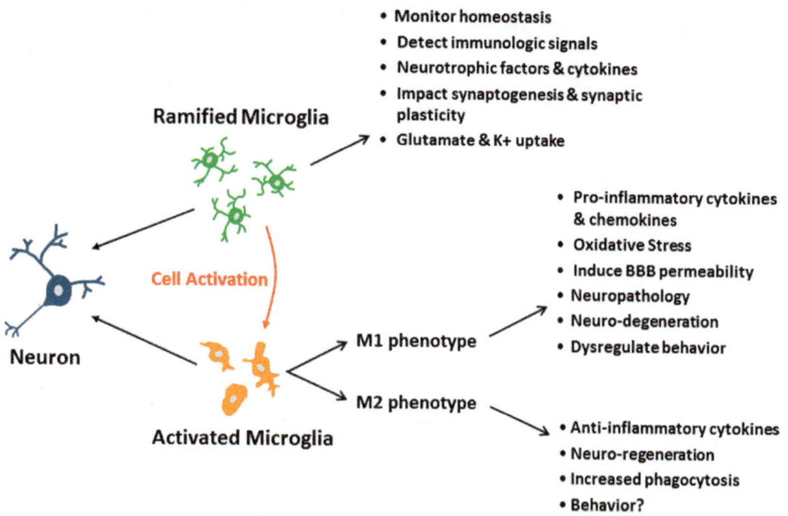

FIGURE 1 **Function and phenotypes of microglia.** Microglia normally have a ramified morphology and are very active maintaining CNS homeostasis and protecting neurons. Upon insult to the CNS, microglia undergo activation to either an M1 or M2 phenotype with distinct pro- or anti-inflammatory functions.

approximately 10% of the cells in the CNS. Normally, microglia play a critical role in maintaining CNS homeostasis (Figure 1). For example, microglia protect neurons through production of neurotrophic factors as well as removal of neurotoxic molecules. Distinct from other glia and neurons, which are derived from neuroectoderm, microglia are of hematopoietic origin (Saijo & Glass, 2011). Microglia are generated in the primitive yolk sac and move into the CNS prior to formation of the blood–brain barrier. Molecules including the growth factor colony stimulating factor 1 and the transcription factor PU.1 are required for differentiation of microglia and peripheral monocytes, supporting a common myeloid linage for these cells and microglia. Further support for a common linage of these cells comes from observations that microglia, peripheral monocytes, and peripheral macrophages express many of the same molecules and perform similar functions. Controversy has existed concerning whether monocytes are capable of replenishing microglia following formation of the blood–brain barrier. Early bone marrow chimera studies suggested that monocytes enter the CNS and form microglia. However, interpretation of these studies was complicated by irradiation used to produce the chimeras, which compromised the blood–brain barrier. More recent studies in which donor and recipient blood supplies were connected by parabiosis indicate that, in the absence of irradiation, few peripheral monocytes enter the CNS and these cells do not differentiate into microglia (Ajami, Bennett, Krieger, Tetzlaff, & Rossi, 2007). Studies also indicate that microglia are long-lived cells that can be generated under conditions of CNS pathology, likely from CNS microglial progenitors (Glass, Saijo, Winner, Marchetto, & Gage, 2010).

In the healthy CNS, microglia commonly exhibit a ramified appearance with long, branched processes and a relatively small cell body. The processes are believed to occupy defined territories, which do not overlap with adjacent microglia. The relative abundance of microglia varies regionally and is higher in gray than in white matter. Morphology can also change regionally. Collectively, this suggests disparate local functions for microglia. In the healthy CNS, microglia were classically defined as being in a "resting" state, which was distinguished from "activated" microglia present in the infected or inflamed CNS. However, microglia in the normal CNS are far from resting or inactive. Phagocytic microglia remove cellular debris, including cells that normally undergo apoptosis during development. Microglia in the healthy brain can also be highly motile. Two-photon microscopy studies demonstrated that microglia are highly dynamic in vivo, particularly at the synapse (Nimmerjahn, Kirchhoff, & Helmchen, 2005). There, microglia extend and retract processes rapidly and repeatedly, presumably to assess and modify the synapse. During development, microglia modulate the synapse through activity-dependent pruning as well as by modulating plasticity of developing and mature synapses (Boulanger, 2009; Tremblay, Lowery, & Majewska, 2010). Furthermore, microglial products play an important role in microglia-neuron communication. For example, chemokine (C—X$_3$—C motif) receptor 1 (CX3CR1) is believed to be expressed exclusively on microglia in the CNS and to interact with chemokine (C—X$_3$—C motif) ligand 1 (CX3CL1/fractalkine) expressed by neurons. Studies utilizing CX3CR1 knockout mice exhibited transient decreases in microglia early in development and altered synaptic plasticity in the hippocampus, supporting a role for microglial-neuron communication in these processes (Paolicelli et al., 2011).

In addition to functioning to maintain homeostasis in the healthy CNS, microglia function in innate immunity in the infected or inflamed CNS. Under these conditions, microglia change shape from a ramified morphology and small soma to an activated state, characterized by short, blunted processes and hypertrophy of the soma or an amoeboid morphology (Figure 2). Activated microglia produce a variety of cytokines and chemokines as well as cell surface molecules, including major histocompatibility (MHC) proteins critical to antigen presentation to T cells that control the adaptive immune response (Ransohoff & Perry, 2009; Saijo & Glass, 2011). Expression of pro-inflammatory molecules by microglia can play a protective role, such as in clearing pathogens from the infected CNS. However, these same pro-inflammatory

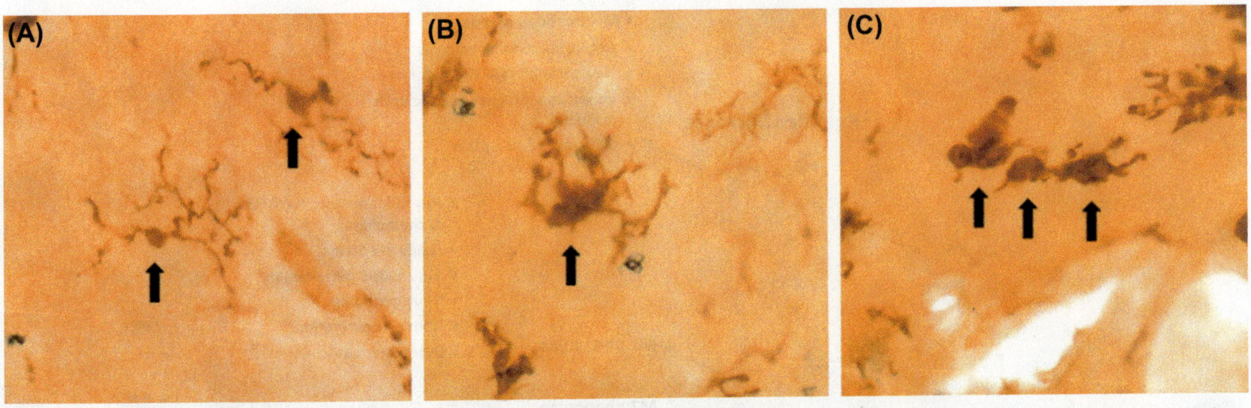

FIGURE 2 Microglial morphology. (A) Cells are normally ramified in the unperturbed CNS. (B) Cells hypertrophy and the processes become shorter and broader with partial activation. (C) Cells become amoeboid with further activation. *Reproduced from Drew and Kane (2014).*

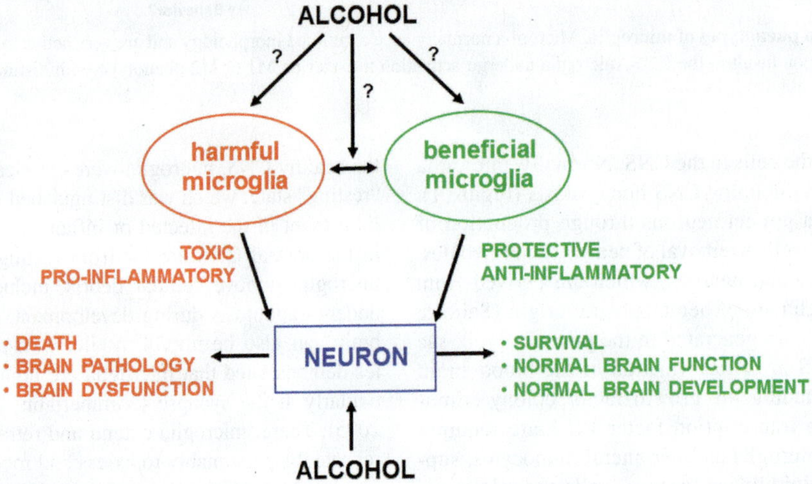

FIGURE 3 Model outlining the hypothesis that alcohol effects on microglia shift the balance between neuron survival with normal brain function and neuron death with neuropathology and brain dysfunction.

molecules can result in damage to CNS cells including neurons, particularly when microglia are chronically activated. Similar to peripheral macrophages, microglia can be activated by distinct signals to form classically activated M1 microglia that exhibit a pro-inflammatory phenotype or can also be alternatively activated to an anti-inflammatory and protective M2 phenotype (Ransohoff & Perry, 2009). Classical activation of microglia to the M1 state occurs in response to pro-inflammatory cytokines such as interferon (IFN)-γ or engagement of toll-like receptors (TLRs) on the surface of microglia. TLRs interact with exogenous ligands including conserved motifs present on the surface of pathogens and endogenous ligands such as high mobility group box (HMGB) 1 and heat shock protein 70. Anti-inflammatory cytokines, including interleukin (IL)-4 and transforming growth factor (TGF)-β are known to trigger the alternative activation of M2 microglia. The relative abundance of M1 and M2 microglia is suggested to mediate the balance between a destructive and a neuroprotective environment in the CNS (Hu et al., 2014). Microglia are capable of reverting back to a resting phenotype in the absence of activating stimuli. However, the reverted cells do not appear to be truly naïve, as they are believed to be primed for subsequent activation (Hanisch & Kettenmann, 2007).

FETAL ALCOHOL SPECTRUM DISORDERS

The effect of alcohol on microglia in the developing brain was essentially unknown until recently but was hypothesized to directly impact neuronal survival and function (Syapin, Hickey, & Kane, 2005) (Figure 3). The impact of alcohol on microglial activation in the fetal CNS has been investigated by us and others in rodent models of FASD. Our findings in a neonatal mouse model demonstrated that alcohol treatment reduces the number of microglia in the cerebellum (Kane et al., 2011); unknown is whether other brain regions show changes in microglial cell number. Microglial function in the developing brain is particularly important because they are neurotrophic, phagocytose debris from naturally occurring neuron death, mediate synaptic pruning, and affect synaptic plasticity as described previously. Importantly, our studies in the neonatal mouse also demonstrate that microglial activation is induced by alcohol exposure, including in the hippocampus, cerebellum, and cerebral cortex as measured by morphological change and increased intensity of Iba-1 immunostaining (Drew, Johnson, Douglas, Phelan, & Kane, 2015; Kane et al., 2011). Compared to microglia in control mice, cells in alcohol-treated mice exhibited shorter and broader processes and hypertrophy of the soma.

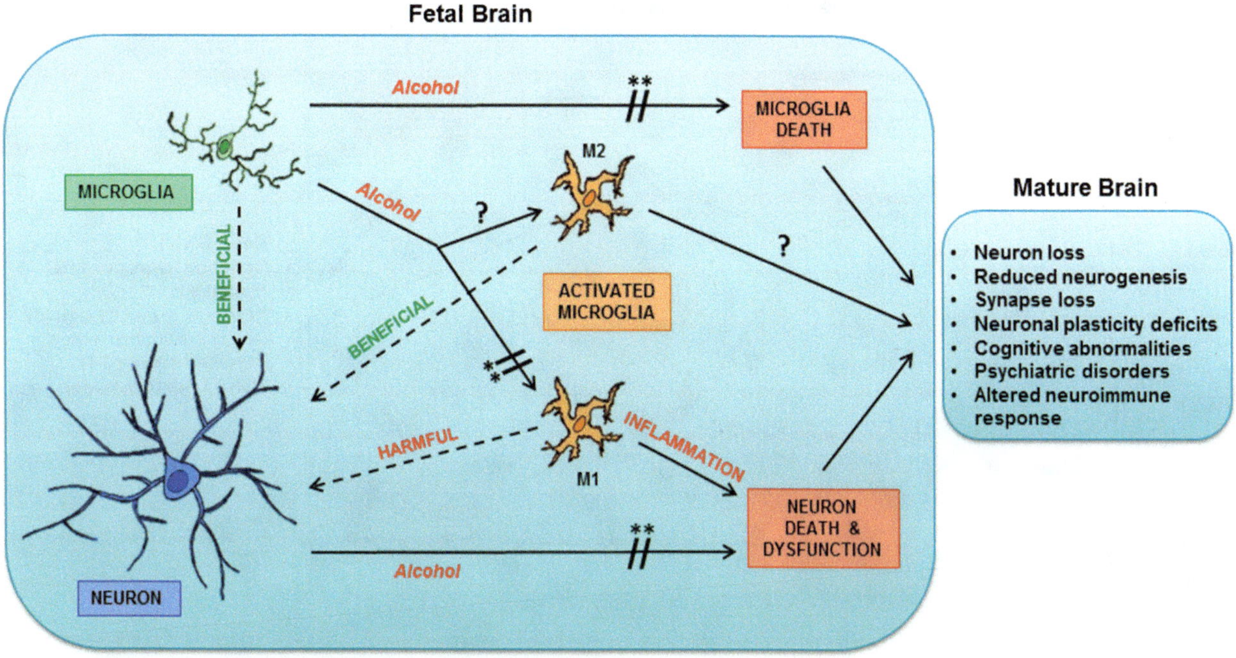

FIGURE 4 **Alcohol effects on microglia in the fetal brain.** Normally beneficial microglia die due to alcohol exposure. Surviving microglia can be activated to an M1 phenotype, leading to inflammation and neuron death, with consequential CNS dysfunction in the mature brain. Anti-inflammatory therapeutics may protect against alcohol pathogenesis by intervention in these processes. Alcohol can also activate microglia to an M2 neuroprotective phenotype. Asterisks (**) indicate sites of anti-inflammatory therapeutic intervention in the neuropathology and long-term consequences of AUD and FASD. *Modified from Drew and Kane (2014).*

The morphological phenotype of microglia in the alcohol-exposed mice suggests that alcohol produced partial microglial activation in these key areas of the developing brain that are linked to behavioral deficits in FASD.

Alteration of neuroimmune gene expression is also a hallmark of neuroinflammatory processes and is evident in rodent models of FASD. Studies by us and others demonstrate alcohol-induced increases in pro-inflammatory cytokines, chemokines, and transcription factors. Our studies in neonatal mice revealed that expression of IL-1β, tumor necrosis factor (TNF)-α, and chemokine (C—C motif) ligand 2 (CCL2) are induced in the hippocampus, cerebellum, and cerebral cortex following 6 days of alcohol treatment (Drew et al., 2015). Other studies in neonatal rats showed that alcohol treatment led to expression of IL-1β, TGF-β1, TNF-α, and nuclear factor-κB (NF-κB) in the hippocampus and cerebral cortex following 3 days of alcohol treatment (Tiwari & Chopra, 2011).

Collectively, these findings are significant for several reasons. Neuroinflammation with expression of these molecules is classically associated with neuropathology and neurodegeneration. Particularly relevant, fetal neuroinflammation has been shown to be associated with psychiatric and cognitive defects in adulthood (Green & Nolan, 2014), similar to behaviors in individuals with FASD (Fryer et al., 2007; Streissguth et al., 2004). Thus, it is hypothesized that alcohol-induced microglial loss and activation is pathogenic in the developing brain and that these changes produce persistent behavioral and cognitive deficits in children and adults (Drew et al., 2015; Drew & Kane, 2013, 2014; Kane, Phelan, & Drew, 2012; Kane et al., 2011) (Figure 4).

ADOLESCENT AND COLLEGE-AGE ALCOHOL CONSUMPTION

Adolescent and college-age drinking causes serious concern because of the vulnerability of the developing CNS at these ages to alcohol pathogenesis and associated long-term consequences. The pressing problem with alcohol consumption at these ages rests on the facts that drinking in both populations is high and occurs primarily in a binge pattern of consumption, compounded by high resistance to the sedative effects of alcohol; the CNS continues to develop critical cognitive functions during this period; a very high proportion of college-aged students drink; and adolescent drinking significantly increases the risk of adult AUD and psychiatric disorders. Continuing development is a very active process in the adolescent and young adult brain, with a high level of neurogenic activity and dynamic synaptic remodeling.

Rodent studies have begun to elucidate the mechanisms of alcohol-induced damage in the adolescent brain (Figure 5). Neurogenesis appears to be highly vulnerable to alcohol inhibition in the adolescent (Broadwater, Liu, Crews, & Spear, 2014; McClain, Hayes, Morris, & Nixon, 2011). It is intriguing, then, that neuroimmune activation and neuroinflammation have detrimental effects on neurogenesis and neuroprogenitor cell proliferation (Russo, Barlati, & Bosetti, 2011). For example, renewal of the stem cell population and proliferation of progenitors is inhibited by pro-inflammatory molecules including IFN-γ, IL-1β, IL-6, TNF-α, and reactive oxygen and nitrogen species (Sierra et al., 2014). Furthermore, neurogenesis is suppressed by microglial activation, although it should be noted that normal microglial function is

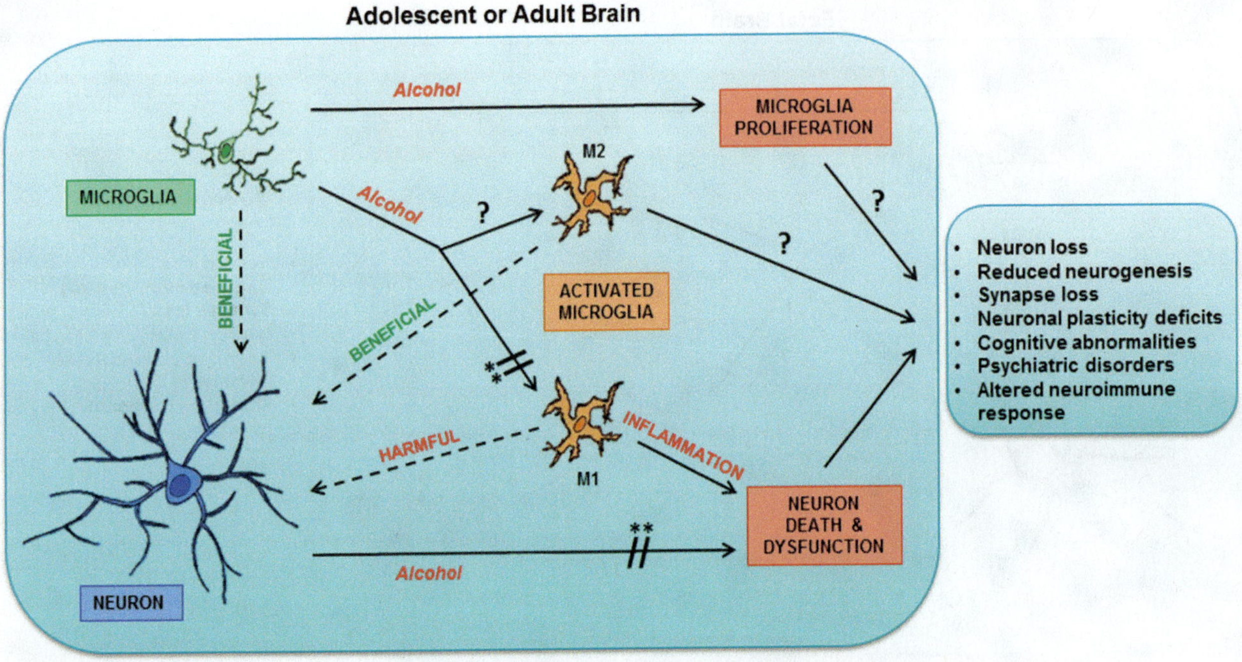

FIGURE 5 **Alcohol effects on microglia in adolescent and adult brain.** Alcohol can alter the proliferation and activation state of adolescent and adult microglia. Depending on the activation status of microglia, this can lead to neuron loss or protection. Anti-inflammatory therapeutics may protect against alcohol pathogenesis. Asterisks (**) indicate sites of anti-inflammatory therapeutic intervention in the neuropathology and long-term consequences of AUD and FASD.

important because it provides protection and a beneficial environment for neurogenesis (Hanisch & Kettenmann, 2007; Ziv et al., 2006). Studies in the hippocampus of adolescent rats treated with alcohol for 4 days demonstrated a change in microglial morphology suggestive of partial microglial activation (McClain, Morris, et al., 2011).

The finding that morphological change together with protracted microglial proliferation occurred without an increase in TNF-α protein or the activation markers ED1 or MHC-II further supported the conclusion that alcohol induced partial microglial activation in the hippocampus of the adolescent brain (McClain, Morris, et al., 2011). Binge alcohol exposure resulted in increased TLR2, TLR4, IL-1β, and TNF-α expression in the prefrontal cortex of adolescent mice (Pascual, Pla, Minarro, & Guerri, 2014). Our studies in the hippocampus, cerebellum, and cerebral cortex of adolescent mice did not identify a change in expression of CCL-2, IL-6, or TNF-α (Kane et al., 2014), which is consistent with limited microglial activation. In our global gene expression studies, genomic analysis of the pattern of alcohol-induced neuroimmune gene expression in adolescent rats suggested that neuroimmune pathways were not significantly activated by alcohol treatment (Agrawal et al., 2014). The neuroinflammatory relevance of this observation was supported by our demonstration that the anti-inflammatory pharmaceutical minocycline failed to reduce adolescent drinking behavior (Agrawal et al., 2014), although we have shown minocycline reduces drinking behavior in adults (Agrawal, Hewetson, George, Syapin, & Bergeson, 2011), as further discussed below. Due to the serious long-term consequences of adolescent drinking on adult behavior and risk of AUD, investigation of the role of neuroimmune activity in alcohol pathogenesis in the adolescent brain warrants additional study.

ADULT ALCOHOL CONSUMPTION

Microglial changes and expression of neuroimmune genes have been widely investigated in human alcoholics and adult rodent models of AUD and are hypothesized to contribute to the long-term consequences of neuropathology and cognitive and behavioral deficits (Figure 5). Investigation of human brain tissue from individuals with chronic alcohol use has revealed that expression of neuroinflammatory genes and proteins were increased in the frontal cortex (Liu et al., 2006). Other studies show that expression of CCL2 and NF-κB were elevated in human alcoholics (He & Crews, 2008; Yakovleva, Bazov, Watanabe, Hauser, & Bakalkin, 2011). CCL2 expression is interesting because CCL2-deficient mice have reduced drinking behavior and lower levels of neuroinflammation (Blednov et al., 2005). Overexpression of CCL2 in mice leads to altered hippocampal synaptic transmission (Nelson, Hao, Manos, Ransohoff, & Gruol, 2011), suggesting that increased CCL2 expression in alcoholics contributes to decreased synaptic plasticity and cognitive dysfunction.

Early reports indicated an increase in microglial cell number in the cerebellum of adult rats treated with alcohol for several months (Riikonen et al., 2002). Other studies also demonstrated increased microglial proliferation and an increased number of microglia in additional brain regions, specifically the hippocampus and motor and somatosensory cortex (Marshall et al., 2013; Nixon, Kim, Potts, He, & Crews, 2008). However, no change was seen in the number of microglia in studies of aged adult rats treated with alcohol (Dlugos & Pentney, 2001). Studies have analyzed the morphology and antigenic expression of microglia to assess alcohol-induced microglial activation in adult rats (Marshall et al., 2013). Expression of the activation marker OX42 was consistent with microglial activation, but the activation appeared to be only partial because the cells did not express elevated levels of OX6 and ED1 antigens, expression of IL-6 and TNF-α was not increased, and microglial morphology was suggestive of partial activation. However, in aged adult rats where microglia constitutively expressed the inflammatory marker OX42, expression of the OX42 antigen was not increased by alcohol and there was no suggestion of morphological change indicative of activation in the cells (Dlugos & Pentney, 2001). Microglial activation is a key component of neuroimmune processes in the CNS and further analyses of the microglial activation phenotype and neuroinflammatory processes in different brain regions associated with alcohol consumption in the adult are important (Figure 5).

Multiple studies in various adult rodent models of alcohol consumption have revealed increased expression of neuroinflammatory molecules in diverse brain regions. For example, studies by us and others in which adult mice were treated with alcohol for 10 days demonstrated elevation of IL-1β, TNF-α, and CCL2 (Kane et al., 2013, 2014; Qin et al., 2008). In two of these studies, expression of these cytokines and chemokines was induced in the hippocampus, cerebral cortex, and cerebellum, but not all cytokines and chemokines were induced in all brain regions (Kane et al., 2013, 2014). In the other study, not all of these molecules were induced, perhaps because expression in specific brain regions was diluted by analysis of the whole brain (Qin et al., 2008). In contrast, following 4-day alcohol exposure in rats, expression of the pro-inflammatory molecules TNF-α, IL-1β, IFN-γ, and CXCL1 were not altered in the hippocampus, frontal cortex, cingulate cortex, anterior cerebellar vermis, hypothalamus, or striatum (Zahr, Luong, Sullivan, & Pfefferbaum, 2010). The observation was supported by lack of pro-inflammatory TNF-α and IL-6 expression in the hippocampus and entorhinal cortex in the same model (Marshall et al., 2013). No expression of the anti-inflammatory cytokines IL-4, IL-5, and IL-13 was seen in the Zahr et al. (2010) study. Interestingly, there was a small but significant increase in anti-inflammatory IL-10 and TGF-β cytokines in the Marshall et al. (2013) study, which may represent partial or M2 activation of microglia as discussed previously.

Expression of various levels and markers of neuroimmune activation appears to be dependent on the brain region analyzed, the dose and pattern of alcohol administration, the age of the animal, and perhaps the species. Although neuroinflammatory molecules have potent effects on neuroimmune processes, neuropathology, neuron survival, and neuron function, the results of alcohol exposure to date are also consistent with the emerging understanding that neuroinflammatory molecules have additional roles in the CNS independent of their inflammatory function, which may be differentially altered by alcohol.

NEUROIMMUNE SIGNALING

Microglia and astrocytes are the principal glial cells mediating innate immune responses in the CNS. In addition to responding to pathogens, glia can respond to endogenous danger signals released following tissue injury. Studies suggest that alcohol elicits the release of endogenous danger signals including HMGB1, which triggers innate immune responses by glia through binding to TLR4 (Vetreno, Qin, & Crews, 2013) (Figure 6(A)). Alcohol is believed to activate transcription factors including NF-κB, AP-1, and CREB (Blanco, Valles, Pascual, & Guerri, 2005; Davis & Syapin, 2004; Zou & Crews, 2006), which increase the expression of pro-inflammatory molecules including cytokines, chemokines, nitric oxide, and cyclooxygenase-2 (Blanco et al., 2005), consistent with an M1 state of microglial activation.

Studies utilizing TLR4 neutralizing antibodies or TLR4 knockout mice demonstrated that TLR4 plays a critical role in alcohol-induced neuroinflammation. For example, alcohol-induced expression of pro-inflammatory molecules was blocked by TLR4 neutralizing antibodies (Blanco et al., 2005). Alcohol also induced expression of pro-inflammatory molecules in wild-type but not TLR4 knockout mice (Alfonso-Loeches, Pascual-Lucas, Blanco, Sanchez-Vera, & Guerri, 2010). The role of TLR4 on microglia in modulating neurodegeneration was supported by studies showing conditioned media from alcohol-treated wild-type, but not TLR4 knockout, microglia triggered apoptosis of cortical neurons. Alcohol induction of neuroinflammatory molecules was associated with TLR4-dependent changes in histone acetylation, suggesting that alcohol induced alterations in neuroinflammation by altering chromatin configuration through epigenetic changes (Pascual, Balino, Alfonso-Loeches, Aragon, & Guerri, 2011). TLR4 moved to lipid rafts in response to alcohol, suggesting that lipid rafts also may play a role in alcohol-induced neuroinflammation (Blanco, Perez-Arago, Fernandez-Lizarbe, & Guerri, 2008). TLR4 signaling can result in the activation of two distinct downstream signaling pathways, the MyD88-dependent and the MyD88-independent/TRIF-dependent pathways. Future studies are needed to determine if alcohol effects on neuroinflammation and neurodegeneration are mediated through MyD88-dependent and/or MyD88-independent pathways.

Inflammasomes are protein complexes that function as part of the innate immune response. A variety exists, but caspase-1 activating inflammasomes are some of the best characterized and are responsible for cleaving immature caspase-1 into a functionally active form. Caspase-1, in turn, is required to process a variety of molecules, most notably pro-IL-1β, to an active form. Studies demonstrated that IL-1β expression was increased in response to alcohol as discussed above and plays a significant role in alcohol-induced neuroinflammation (Alfonso-Loeches et al., 2010). More recent studies demonstrated that the caspase-1 activating NLRP3 inflammasome was critical to alcohol-induced neuroinflammation and alcohol increased the expression of neuroinflammatory molecules in wild-type but not NLRP3 knockout mice (Lippai, Bala, Petrasek, et al., 2013). Future studies are needed to determine if alcohol induction of neuroinflammation is mediated by other inflammasomes in addition to the NLRP3 inflammasome.

DNA methylation status and chromatin structure have been shown to be modified by alcohol consumption, both of which have direct effects on gene expression, including for noncoding genes. Although a direct role in neuroimmune function has not been established for epigenetic mechanisms, it is reasonable to hypothesize that they play an upstream role in modulation of alcohol-mediated regulation of neuroimmune function. Of the noncoding RNAs, several alcohol-responsive miRNAs have been identified for all life stages. However, the mechanistic specifics remain understudied, and most studies report ratio changes and presume a repression of gene expression via translational blockage.

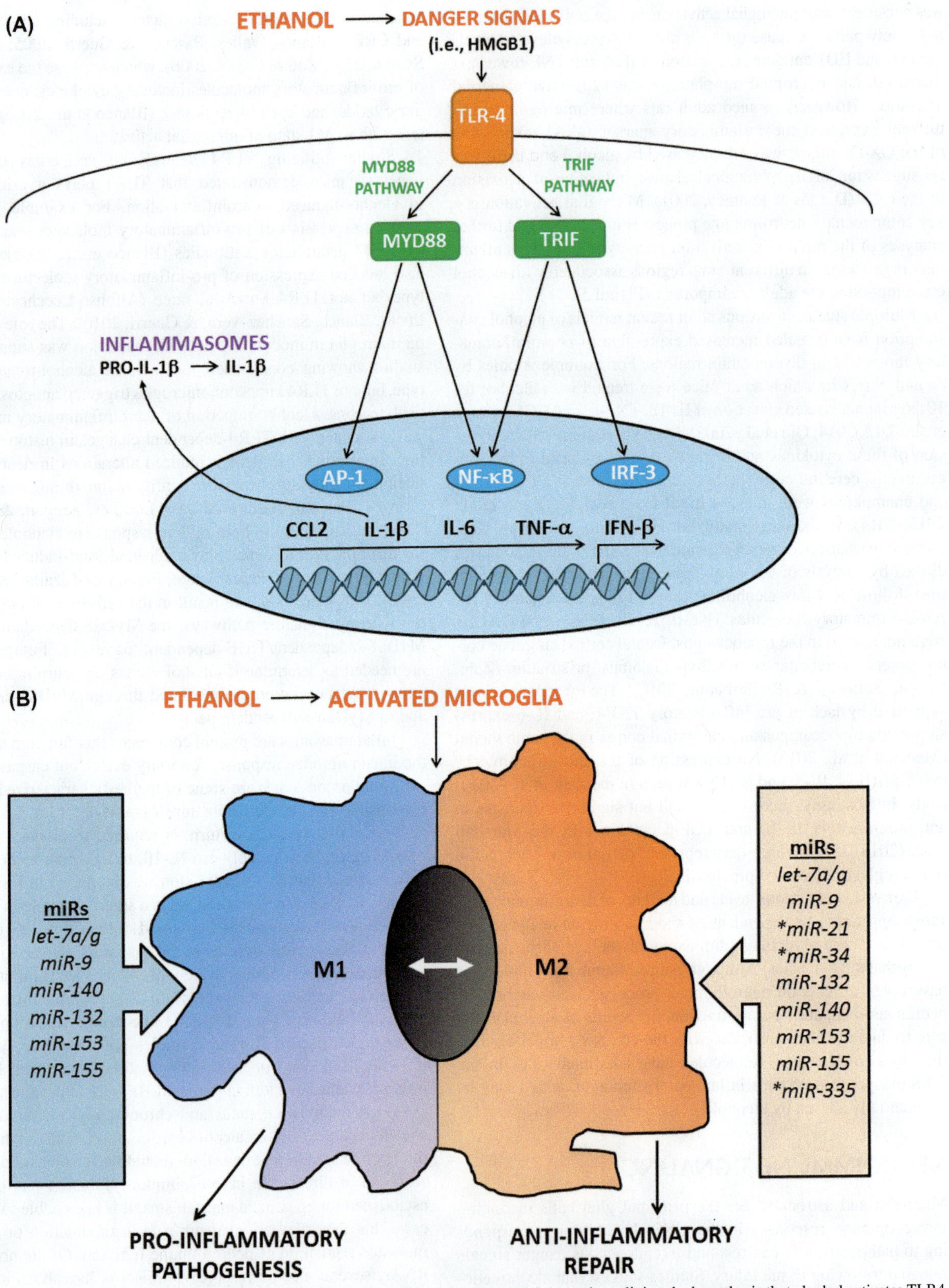

FIGURE 6 Alcohol effects on neuroimmune signaling and microglial activation. (A) Schematic outlining the hypothesis that alcohol activates TLR4 signaling, which can result in the activation of MyD88-dependent and/or TRIF-dependent signaling pathways. Inflammasomes also may be important in regulation of alcohol-induced neuroimmune signaling. (B) Schematic outlining the hypothesis that alcohol-induced miRNAs may alter the balance of M1 and M2 microglia. Asterisk (*) indicates miRNAs that are believed to act on M2 but not M1 microglia.

Although most miRNAs have been studied with respect to the ability to either degrade mRNA or block translation by binding the transcript and inhibiting processivity, general activation and repression of expression by a variety of mechanisms have now been shown. In fact, miRNAs have also been shown to direct DNA methylation and chromatin structural changes and most likely the story of miRNA mechanisms has not been completely written, especially with respect to alcohol and its effect on neuroinflammation. Currently, in mammals, miRNAs are known to be produced from miRNA genes, and also from intragenic and exonic regions. Effects are produced by forming hairpin structures, which are cleaved to double-stranded structures and further processed in the cytosol to single stranded ~22-mers. The mature miRNAs then most often bind to protein coding mRNAs in a mismatched fashion to block translation via at least four mechanisms including: a change in three-dimensional structure that inhibits 5′-cap and small ribosomal binding, inhibition of large ribosome binding, blockade of initiation of elongation, and growing peptide/ribosome premature termination due to miRNA/complex occlusion of processivity. In addition, nascent protein degradation and sequestration in P-bodies can occur. mRNA decay and cleavage can occur via miRNA-RISC associated exo- and endonuclease action.

In some cases, transcription can also be increased or decreased via miRNA-directed DNA methylation and chromatin reorganization or by binding to response elements. In most cases, miRNA action leads to a decrease in protein product, but it is important to note that is not always the case, and in response to alcohol, the direct action of miRNAs remains poorly understood. Across the lifespan, mammalian alcohol-responsive miRNAs with significant overrepresentation of neuroimmune targeting have been found, including in AUD postmortem brain. Figure 6(B) shows miRNAs modulated by alcohol in humans and animal models of AUD and FASD, which control target genes that are overrepresented in neuroimmune pathways. The miRNAs were identified from published literature (Balaraman, Tingling, Tsai, & Miranda, 2013; Lippai, Bala, Csak, Kurt-Jones, & Szabo, 2013; Robinson et al., 2014) and were analyzed using the Wikipathways function in WebGestalt by comparing the miRNA target gene lists from the TargetScan to the whole genome. The analyses also indicated that some of the miRNAs (miR-21, miR-34, and miR-335) may differentially regulate M2 but not M1 associated molecules, suggesting alcohol-induced miRNAs may shift the balance between M1 and M2 microglia. Collectively, the results suggest that miRNAs may modulate AUD and FASD. Although it is unknown whether the miRNA responses are pro-inflammatory or an attempt to recover, the results suggest that alcohol-related neuroimmune changes are impacted by miRNAs. Additional studies are needed to clarify alcohol-mediated epigenetic effects on neuroimmune functioning.

ANTI-INFLAMMATORY THERAPEUTICS FOR TREATMENT OF ALCOHOL USE DISORDERS

As presented, alcohol is known to induce neuroinflammation, which is believed to contribute to alcohol-induced neuropathology (Figures 4–6). Furthermore, neuroinflammation has been suggested to contribute to the development of AUD and FASD, suggesting that anti-inflammatory pharmaceuticals may be effective in the treatment of these disorders (Figures 4 and 5).

Pioglitazone is an agonist of peroxisome proliferator-activated receptor (PPAR)-γ, a member of the nuclear receptor family of proteins. Pioglitazone was approved by the US Food and Drug Administration (FDA) for the treatment of type II diabetes. We and others have shown that PPAR-γ agonists were potent suppressors of microglial activation in vitro and also suppressed the development of disease in a variety of animal models characterized by neuroinflammation and neurodegeneration (Mandrekar-Colucci, Sauerbeck, Popovich, & McTigue, 2013). Recently, we demonstrated that pioglitazone protected cerebellar neurons and microglia from the toxic effects of alcohol in an animal model of FASD (Kane et al., 2011). We also demonstrated that pioglitazone suppressed alcohol induction of pro-inflammatory cytokines and chemokines in the developing cerebellum, hippocampus, and cortex (Drew et al., 2015). Neuroinflammation has been suggested to contribute to increased alcohol consumption, with PPAR-α and PPAR-γ mutations having AUD trait associations in humans, and PPAR agonist treatment showing decreased alcohol consumption and relapse in alcohol-seeking behavior in rodents (Blednov et al., 2015; Stopponi et al., 2011). The studies support the idea that PPAR-α/γ agonists may be effective in the treatment of AUD and FASD.

Minocycline is an FDA-approved antibiotic and has anti-inflammatory properties. It suppressed microglial activation both in vitro and in animal models of disease associated with neuroinflammation and neurodegeneration (Garrido-Mesa, Zarzuelo, & Galvez, 2013). Studies by us and others demonstrated minocycline also altered alcohol consumption and behavioral responses to alcohol (Agrawal et al., 2011; Wu et al., 2011). Interestingly, minocycline suppressed short-term binge alcohol consumption in adult but not adolescent rodents (Agrawal et al., 2014). However, under high alcohol treatment, neuroinflammation has been seen during adolescence (Qin & Crews, 2012). Together with additional studies (Kane et al., 2014; McClain, Morris, et al., 2011), the data suggest that alcohol induces neuroinflammation, although more strongly in adult than adolescent mice, and support a link between neuroinflammation, alcohol dose, age, and the AUD spectrum.

Naltrexone was FDA-approved for the treatment of AUD and is believed to act by modulating opioid receptors. In addition to suppressing alcohol consumption, naltrexone suppressed alcohol-induced microglial activation in adult rodents (Qin & Crews, 2012). The anti-inflammatory properties of naltrexone were not surprising because naltrexone is a TLR4 antagonist and, as discussed above, many of the effects of alcohol on neuroinflammation are mediated through TLR4.

Collectively, the studies outlined above suggest anti-inflammatory pharmaceuticals, including PPAR agonists, minocycline, and naltrexone, may be effective in the treatment of AUD and FASD and may protect against the neuropathology associated with these disorders. The suppressive effects of naltrexone on alcohol consumption were demonstrated to be potentiated by pioglitazone, suggesting possible distinct mechanisms of action of these therapeutics (Stopponi et al., 2013). Due to the complex nature of the alcohol-mediated neuroimmune interactions, it is possible that a combination of anti-inflammatory agents acting at distinct sites may be most effective in blocking the neuroinflammation and neurodegeneration associated with alcohol use.

SUMMARY

Alcohol stimulates neuroimmune responses in humans and animal models. Alcohol effects on neuroimmune activity depend on a number of factors including age, dose of alcohol, and pattern of exposure. The mechanisms by which alcohol induces neuroinflammation are just beginning to be elucidated, and additional studies are required to better define these mechanisms. However, early studies suggest that anti-inflammatory therapeutics hold promise to block the long-term neuropathological consequences of alcohol.

APPLICATIONS TO OTHER ADDICTIONS AND SUBSTANCE MISUSE

- Other drugs of abuse, such as methamphetamine and cocaine, stimulate neuroinflammation in a manner similar to alcohol.
- Neuroinflammation may contribute to addiction to alcohol as well as other drugs of abuse.
- Anti-inflammatory pharmaceuticals may be effective in the treatment of addiction to alcohol and other drugs of abuse.

DEFINITION OF TERMS

Alcohol use disorder Alcohol use disorder is a clinically relevant disorder characterized by a spectrum of harmful consequences related to repeated harmful alcohol use. It is medially defined by the *Diagnostic Statistical Manual of Mental Disorders, Fifth Edition.*

Chemokine This protein affects the movement of responding cells, including immune cells.

Cytokine This protein alters the function of cells, including inflammatory cytokines, which alter the function of immune cells.

Fetal alcohol spectrum disorders (FASD) FASD occurs in some individuals whose mothers consumed alcohol during pregnancy. The defects associated with FASD can include facial dysmorphology, small size, brain damage, and neurological, psychiatric, behavioral, and cognitive problems.

Glia These are three types of cells in the brain that are not neurons, including microglia, astrocytes, and oligodendrocytes.

Microglial activation These changes in the morphology, antigen expression, and function of microglia occur in response to pathogens or noninfectious insults to the CNS.

miRNA A microRNA is a small, noncoding RNA molecule, which functions in RNA silencing or occasionally activation, and transcriptional regulation of gene expression or translational regulation of protein production.

Neuroimmune response These immune responses occur in the CNS, including innate immune responses by CNS glia.

Neuroinflammation This is an inflammatory response to infectious agents or noninfectious insults in the CNS. Alcohol is an example of a noninfectious insult that causes neuroinflammation.

Cell signaling pathways This system allows cells to respond to surrounding signals. The result is changes in the molecules produced by and the function of the responding cell.

Therapeutics Therapeutics are pharmaceutical medications that can prevent or treat disorders or diseases.

KEY FACTS

Alcohol Use Disorders

- The economic consequences of AUD in the United States is more than $200 billion yearly.
- Binge drinking causes the most severe damage to the brain.
- Among adults, approximately 25% of men and 13% of women binge-drink alcohol. Approximately 40% of college students binge-drink alcohol.
- At least 12% of pregnant women drink alcohol and 2–5% of the children in the United States are born with FASD.
- A significant level of brain pathology is present in individuals with AUD and in babies with FASD. The consequences of alcohol consumption include loss and dysfunction of neurons and inflammation in the brain.

Microglial Function

- The brain is composed of neurons and nonneuronal cells called glia. Glia include microglia, astrocytes, and oligodendrocytes.
- Microglia are responsible for maintaining homeostasis and protecting neurons.
- Microglia are the principal cells in the brain that produce and respond to immune molecules.
- Microglia can become "activated" by insult to the brain and become either pro-inflammatory (detrimental to neurons) or anti-inflammatory (protective of neurons).
- Microglia are activated by alcohol, which produces inflammation in the brain (neuroinflammation).

Neuroinflammation

- Neuroinflammatory events are primarily mediated by microglia.
- Alcohol consumption or exposure of the fetus to alcohol causes neuroinflammatory responses in the brain.
- Alcohol consumption or exposure of the fetus to alcohol stimulates microglial activation and production of pro-inflammatory cytokines and chemokines, thought to contribute to alcohol pathology in the brain.
- Neuroinflammation induced by alcohol is suggested to cause loss and dysfunction of neurons and contribute to the neurological, psychiatric, behavioral, and cognitive problems that are a consequence of alcohol consumption.

SUMMARY POINTS

- The fetal, adolescent, and adult brain exhibit neuroinflammatory events in response to alcohol exposure, including microglial activation and induction of pro-inflammatory cytokines and chemokines.
- Emerging evidence suggests that neuroinflammation contributes to the neurological, psychiatric, behavioral, and cognitive problems associated with alcohol consumption in the adult or fetal alcohol exposure.

- The signaling pathways controlling neuroimmune responses provide targets for development of therapeutics against alcohol pathogenesis.
- FDA-approved anti-inflammatory therapeutics hold promise for suppression of alcohol-induced neuroinflammation and the consequential neurological, psychiatric, behavioral, and cognitive problems, including AUD and FASD.

ACKNOWLEDGMENT

This work was supported by National Institutes of Health grants AA021142 (SEB), NIGMS IdeA Program award P30 GM110702 (CJMK), and AA18834 (CJMK).

REFERENCES

Agrawal, R. G., Hewetson, A., George, C. M., Syapin, P. J., & Bergeson, S. E. (2011). Minocycline reduces ethanol drinking. *Brain Behavior and Immunity*, 25(Suppl. 1), S165–S169. http://dx.doi.org/10.1016/j.bbi.2011.03.002.

Agrawal, R. G., Owen, J. A., Levin, P. S., Hewetson, A., Berman, A. E., Franklin, S. R., ... Bergeson, S. E. (2014). Bioinformatics analyses reveal age-specific neuroimmune modulation as a target for treatment of high ethanol drinking. *Alcoholism Clinical and Experimental Research*, 38(2), 428–437. http://dx.doi.org/10.1111/acer.12288.

Ajami, B., Bennett, J. L., Krieger, C., Tetzlaff, W., & Rossi, F. M. (2007). Local self-renewal can sustain CNS microglia maintenance and function throughout adult life. *Nature Neuroscience*, 10(12), 1538–1543. http://dx.doi.org/10.1038/nn2014 pii:nn2014.

Alfonso-Loeches, S., Pascual-Lucas, M., Blanco, A. M., Sanchez-Vera, I., & Guerri, C. (2010). Pivotal role of TLR4 receptors in alcohol-induced neuroinflammation and brain damage. *Journal of Neuroscience*, 30(24), 8285–8295. http://dx.doi.org/10.1523/JNEUROSCI.0976-10.2010.

Balaraman, S., Tingling, J. D., Tsai, P. C., & Miranda, R. C. (2013). Dysregulation of microRNA expression and function contributes to the etiology of fetal alcohol spectrum disorders. *Alcohol Research*, 35(1), 18–24.

Blanco, A. M., Perez-Arago, A., Fernandez-Lizarbe, S., & Guerri, C. (2008). Ethanol mimics ligand-mediated activation and endocytosis of IL-1RI/TLR4 receptors via lipid rafts caveolae in astroglial cells. *Journal of Neurochemistry*, 106(2), 625–639. http://dx.doi.org/10.1111/j.1471-4159.2008.05425.x.

Blanco, A. M., Valles, S. L., Pascual, M., & Guerri, C. (2005). Involvement of TLR4/type I IL-1 receptor signaling in the induction of inflammatory mediators and cell death induced by ethanol in cultured astrocytes. *Journal of Immunology*, 175(10), 6893–6899.

Blanco, C., Okuda, M., Wright, C., Hasin, D. S., Grant, B. F., Liu, S. M., & Olfson, M. (2008). Mental health of college students and their non-college-attending peers: results from the National Epidemiologic Study on Alcohol and Related Conditions. *Archives of General Psychiatry*, 65(12), 1429–1437. http://dx.doi.org/10.1001/archpsyc.65.12.1429.

Blednov, Y. A., Benavidez, J. M., Black, M., Ferguson, L. B., Schoenhard, G. L., Goate, A. M., ... Harris, R. A. (2015). Peroxisome proliferator-activated receptors alpha and gamma are linked with alcohol consumption in mice and withdrawal and dependence in humans. *Alcoholism Clinical and Experimental Research*, 39(1), 136–145.

Blednov, Y. A., Bergeson, S. E., Walker, D., Ferreira, V. M., Kuziel, W. A., & Harris, R. A. (2005). Perturbation of chemokine networks by gene deletion alters the reinforcing actions of ethanol. *Behavioural Brain Research*, 165(1), 110–125. http://dx.doi.org/10.1016/j.bbr.2005.06.026 pii:S0166-4328(05)00273-1.

Boulanger, L. M. (2009). Immune proteins in brain development and synaptic plasticity. *Neuron*, 64(1), 93–109. http://dx.doi.org/10.1016/j.neuron.2009.09.001 pii:S0896-6273(09)00678-3.

Broadwater, M. A., Liu, W., Crews, F. T., & Spear, L. P. (2014). Persistent loss of hippocampal neurogenesis and increased cell death following adolescent, but not adult, chronic ethanol exposure. *Developmental Neuroscience*, 36(3–4), 297–305. http://dx.doi.org/10.1159/000362874.

Center for Disease Control and Prevention. (2013). *Morbidity and mortality weekly report*, 62, 1–187.

Davis, R. L., & Syapin, P. J. (2004). Ethanol increases nuclear factor-kappa B activity in human astroglial cells. *Neuroscience Letters*, 371(2–3), 128–132. http://dx.doi.org/10.1016/j.neulet.2004.08.051 pii:S0304-3940(04)01062-6.

Dlugos, C. A., & Pentney, R. J. (2001). Quantitative immunocytochemistry of glia in the cerebellar cortex of old ethanol-fed rats. *Alcohol*, 23(2), 63–69.

Drew, P. D., Johnson, J. W., Douglas, J. C., Phelan, K. D., & Kane, C. J. M. (2015). Pioglitazone blocks ethanol induction of microglial activation and immune responses in the hippocampus, cerebellum, and cerebral cortex in a mouse model of fetal alcohol spectrum disorders. *Alcoholism Clinical and Experimental Research*, 39(3), 445–454.

Drew, P. D., & Kane, C. J. M. (2013). Neuroimmune mechanisms of glia and their interplay with alcohol exposure across the lifespan. In L. Grandison, C. Cui, & A. Noronha (Eds.), *Neural-immune interactions in brain function and alcohol related disorders* (pp. 359–386). New York, NY: Springer.

Drew, P. D., & Kane, C. J. M. (2014). Fetal alcohol spectrum disorders and neuroimmune changes. *International Review of Neurobiology*, 118, 41–80. http://dx.doi.org/10.1016/B978-0-12-801284-0.00003-8.

Floyd, R. L., Weber, M. K., Denny, C., & O'Connor, M. J. (2009). Prevention of fetal alcohol spectrum disorders. *Developmental Disabilities Research Reviews*, 15(3), 193–199. http://dx.doi.org/10.1002/ddrr.75.

Fryer, S. L., McGee, C. L., Matt, G. E., Riley, E. P., & Mattson, S. N. (2007). Evaluation of psychopathological conditions in children with heavy prenatal alcohol exposure. *Pediatrics*, 119(3), e733–741. http://dx.doi.org/10.1542/peds.2006-1606.

Garrido-Mesa, N., Zarzuelo, A., & Galvez, J. (2013). Minocycline: far beyond an antibiotic. *British Journal of Pharmacology*, 169(2), 337–352. http://dx.doi.org/10.1111/bph.12139.

Glass, C. K., Saijo, K., Winner, B., Marchetto, M. C., & Gage, F. H. (2010). Mechanisms underlying inflammation in neurodegeneration. *Cell*, 140(6), 918–934. http://dx.doi.org/10.1016/j.cell.2010.02.016 pii:S0092-8674(10)00168-6.

Green, H. F., & Nolan, Y. M. (2014). Inflammation and the developing brain: consequences for hippocampal neurogenesis and behavior. *Neuroscience and Biobehavioral Reviews*, 40, 20–34. http://dx.doi.org/10.1016/j.neubiorev.2014.01.004.

Hanisch, U. K., & Kettenmann, H. (2007). Microglia: active sensor and versatile effector cells in the normal and pathologic brain. *Nature Neuroscience*, 10(11), 1387–1394. http://dx.doi.org/10.1038/nn1997 pii:nn1997.

He, J., & Crews, F. T. (2008). Increased MCP-1 and microglia in various regions of the human alcoholic brain. *Experimental Neurology*, 210(2), 349–358. http://dx.doi.org/10.1016/j.expneurol.2007.11.017 pii:S0014-4886(07)00422-0.

Hu, X., Leak, R. K., Shi, Y., Suenaga, J., Gao, Y., Zheng, P., & Chen, J. (2014). Microglial and macrophage polarization-new prospects for brain repair. *Nature Reviews Neurology.* http://dx.doi.org/10.1038/nrneurol.2014.207.

Kane, C. J. M., Phelan, K. D., Douglas, J. C., Wagoner, G., Johnson, J. W., Xu, J., & Drew, P. D. (2013). Effects of ethanol on immune response in the brain: region-specific changes in aged mice. *Journal of Neuroinflammation, 10,* 66. http://dx.doi.org/10.1186/1742-2094-10-66.

Kane, C. J. M., Phelan, K. D., Douglas, J. C., Wagoner, G., Johnson, J. W., Xu, J., ... Drew, P. D. (2014). Effects of ethanol on immune response in the brain: region-specific changes in adolescent versus adult mice. *Alcoholism Clinical and Experimental Research, 38*(2), 384–391. http://dx.doi.org/10.1111/acer.12244.

Kane, C. J. M., Phelan, K. D., & Drew, P. D. (2012). Neuroimmune mechanisms in fetal alcohol spectrum disorder. *Developmental Neurobiology, 72*(10), 1302–1316. http://dx.doi.org/10.1002/dneu.22035.

Kane, C. J. M., Phelan, K. D., Han, L., Smith, R. R., Xie, J., Douglas, J. C., & Drew, P. D. (2011). Protection of neurons and microglia against ethanol in a mouse model of fetal alcohol spectrum disorders by peroxisome proliferator-activated receptor-gamma agonists. *Brain Behavior and Immunity, 25*(Suppl. 1), S137–S145. http://dx.doi.org/10.1016/j.bbi.2011.02.016.

Lippai, D., Bala, S., Csak, T., Kurt-Jones, E. A., & Szabo, G. (2013). Chronic alcohol-induced microRNA-155 contributes to neuroinflammation in a TLR4-dependent manner in mice. *PLoS One, 8*(8), e70945. http://dx.doi.org/10.1371/journal.pone.0070945.

Lippai, D., Bala, S., Petrasek, J., Csak, T., Levin, I., Kurt-Jones, E. A., & Szabo, G. (2013). Alcohol-induced IL-1beta in the brain is mediated by NLRP3/ASC inflammasome activation that amplifies neuroinflammation. *Journal of Leukocyte Biology, 94*(1), 171–182. http://dx.doi.org/10.1189/jlb.1212659.

Liu, J., Lewohl, J. M., Harris, R. A., Iyer, V. R., Dodd, P. R., Randall, P. K., & Mayfield, R. D. (2006). Patterns of gene expression in the frontal cortex discriminate alcoholic from nonalcoholic individuals. *Neuropsychopharmacology, 31*(7), 1574–1582. http://dx.doi.org/10.1038/sj.npp.1300947 pii:1300947.

Mandrekar-Colucci, S., Sauerbeck, A., Popovich, P. G., & McTigue, D. M. (2013). PPAR agonists as therapeutics for CNS trauma and neurological diseases. *ASN Neuro, 5*(5), e00129. http://dx.doi.org/10.1042/AN20130030.

Marshall, S. A., McClain, J. A., Kelso, M. L., Hopkins, D. M., Pauly, J. R., & Nixon, K. (2013). Microglial activation is not equivalent to neuroinflammation in alcohol-induced neurodegeneration: the importance of microglia phenotype. *Neurobiology of Disease, 54,* 239–251. http://dx.doi.org/10.1016/j.nbd.2012.12.016.

May, P. A., Gossage, J. P., Kalberg, W. O., Robinson, L. K., Buckley, D., Manning, M., & Hoyme, H. E. (2009). Prevalence and epidemiologic characteristics of FASD from various research methods with an emphasis on recent in-school studies. *Developmental Disabilities Research Reviews, 15*(3), 176–192. http://dx.doi.org/10.1002/ddrr.68.

McClain, J. A., Hayes, D. M., Morris, S. A., & Nixon, K. (2011). Adolescent binge alcohol exposure alters hippocampal progenitor cell proliferation in rats: effects on cell cycle kinetics. *Journal of Comparative Neurology, 519*(13), 2697–2710. http://dx.doi.org/10.1002/cne.22647.

McClain, J. A., Morris, S. A., Deeny, M. A., Marshall, S. A., Hayes, D. M., Kiser, Z. M., & Nixon, K. (2011). Adolescent binge alcohol exposure induces long-lasting partial activation of microglia. *Brain Behavior and Immunity, 25*(Suppl. 1), S120–S128. http://dx.doi.org/10.1016/j.bbi.2011.01.006.

Nelson, T. E., Hao, C., Manos, J., Ransohoff, R. M., & Gruol, D. L. (2011). Altered hippocampal synaptic transmission in transgenic mice with astrocyte-targeted enhanced CCL2 expression. *Brain Behavior and Immunity, 25*(Suppl. 1), S106–S119. http://dx.doi.org/10.1016/j.bbi.2011.02.013 pii:S0889-1591(11)00063-8.

Nimmerjahn, A., Kirchhoff, F., & Helmchen, F. (2005). Resting microglial cells are highly dynamic surveillants of brain parenchyma in vivo. *Science, 308*(5726), 1314–1318. http://dx.doi.org/10.1126/science.1110647 pii:1110647.

Nixon, K., Kim, D. H., Potts, E. N., He, J., & Crews, F. T. (2008). Distinct cell proliferation events during abstinence after alcohol dependence: microglia proliferation precedes neurogenesis. *Neurobiology of Disease, 31*(2), 218–229. http://dx.doi.org/10.1016/j.nbd.2008.04.009.

Paolicelli, R. C., Bolasco, G., Pagani, F., Maggi, L., Scianni, M., Panzanelli, P., ... Gross, C. T. (2011). Synaptic pruning by microglia is necessary for normal brain development. *Science, 333*(6048), 1456–1458. http://dx.doi.org/10.1126/science.1202529 pii:science.1202529.

Pascual, M., Balino, P., Alfonso-Loeches, S., Aragon, C. M., & Guerri, C. (2011). Impact of TLR4 on behavioral and cognitive dysfunctions associated with alcohol-induced neuroinflammatory damage. *Brain Behavior and Immunity, 25*(Suppl. 1), S80–S91. http://dx.doi.org/10.1016/j.bbi.2011.02.012.

Pascual, M., Pla, A., Minarro, J., & Guerri, C. (2014). Neuroimmune activation and myelin changes in adolescent rats exposed to high-dose alcohol and associated cognitive dysfunction: a review with reference to human adolescent drinking. *Alcohol and Alcoholism, 49*(2), 187–192. http://dx.doi.org/10.1093/alcalc/agt164.

Qin, L., & Crews, F. T. (2012). Chronic ethanol increases systemic TLR3 agonist-induced neuroinflammation and neurodegeneration. *Journal of Neuroinflammation, 9,* 130. http://dx.doi.org/10.1186/1742-2094-9-130.

Qin, L., He, J., Hanes, R. N., Pluzarev, O., Hong, J. S., & Crews, F. T. (2008). Increased systemic and brain cytokine production and neuroinflammation by endotoxin following ethanol treatment. *Journal of Neuroinflammation, 5,* 10. http://dx.doi.org/10.1186/1742-2094-5-10.

Ransohoff, R. M., & Perry, V. H. (2009). Microglial physiology: unique stimuli, specialized responses. *Annual Review of Immunology, 27,* 119–145. http://dx.doi.org/10.1146/annurev.immunol.021908.132528.

Riikonen, J., Jaatinen, P., Rintala, J., Porsti, I., Karjala, K., & Hervonen, A. (2002). Intermittent ethanol exposure increases the number of cerebellar microglia. *Alcohol and Alcoholism, 37*(5), 421–426.

Robinson, G., Most, D., Ferguson, L. B., Mayfield, J., Harris, R. A., & Blednov, Y. A. (2014). Neuroimmune pathways in alcohol consumption: evidence from behavioral and genetic studies in rodents and humans. *International Review of Neurobiology, 118,* 13–39. http://dx.doi.org/10.1016/B978-0-12-801284-0.00002-6.

Russo, I., Barlati, S., & Bosetti, F. (2011). Effects of neuroinflammation on the regenerative capacity of brain stem cells. *Journal of Neurochemistry, 116*(6), 947–956. http://dx.doi.org/10.1111/j.1471-4159.2010.07168.x.

Saijo, K., & Glass, C. K. (2011). Microglial cell origin and phenotypes in health and disease. *Nature Reviews Immunology, 11*(11), 775–787. http://dx.doi.org/10.1038/nri3086 pii:nri3086.

Sierra, A., Beccari, S., Diaz-Aparicio, I., Encinas, J. M., Comeau, S., & Tremblay, M. E. (2014). Surveillance, phagocytosis, and inflammation: how never-resting microglia influence adult hippocampal neurogenesis. *Neural Plasticity, 2014,* 610343. http://dx.doi.org/10.1155/2014/610343.

Stopponi, S., de Guglielmo, G., Somaini, L., Cippitelli, A., Cannella, N., Kallupi, M., ... Ciccocioppo, R. (2013). Activation of PPARgamma by pioglitazone potentiates the effects of naltrexone on alcohol drinking and relapse in msP rats. *Alcoholism Clinical and Experimental Research, 37*(8), 1351–1360. http://dx.doi.org/10.1111/acer.12091.

Stopponi, S., Somaini, L., Cippitelli, A., Cannella, N., Braconi, S., Kallupi, M., ... Ciccocioppo, R. (2011). Activation of nuclear PPARgamma receptors by the antidiabetic agent pioglitazone suppresses alcohol drinking and relapse to alcohol seeking. *Biological Psychiatry, 69*(7), 642–649. http://dx.doi.org/10.1016/j.biopsych.2010.12.010.

Streissguth, A. P., Bookstein, F. L., Barr, H. M., Sampson, P. D., O'Malley, K., & Young, J. K. (2004). Risk factors for adverse life outcomes in fetal alcohol syndrome and fetal alcohol effects. *Journal of Developmental and Behavioral Pediatrics, 25*(4), 228–238.

Substance Abuse and Mental Health Services Administration. (2014). *2012 National survey on drug use and health*.

Syapin, P. J., Hickey, W. F., & Kane, C. J. M. (2005). Alcohol brain damage and neuroinflammation: is there a connection? *Alcoholism Clinical and Experimental Research, 29*, 1080–1089.

Tiwari, V., & Chopra, K. (2011). Resveratrol prevents alcohol-induced cognitive deficits and brain damage by blocking inflammatory signaling and cell death cascade in neonatal rat brain. *Journal of Neurochemistry, 117*(4), 678–690. http://dx.doi.org/10.1111/j.1471-4159.2011.07236.x.

Tremblay, M. E., Lowery, R. L., & Majewska, A. K. (2010). Microglial interactions with synapses are modulated by visual experience. *PLoS Biology, 8*(11), e1000527. http://dx.doi.org/10.1371/journal.pbio.1000527.

Vetreno, R. P., Qin, L., & Crews, F. T. (2013). Increased receptor for advanced glycation end product expression in the human alcoholic prefrontal cortex is linked to adolescent drinking. *Neurobiology of Disease, 59*, 52–62. http://dx.doi.org/10.1016/j.nbd.2013.07.002.

Wu, Y., Lousberg, E. L., Moldenhauer, L. M., Hayball, J. D., Robertson, S. A., Coller, J. K., ... Hutchinson, M. R. (2011). Attenuation of microglial and IL-1 signaling protects mice from acute alcohol-induced sedation and/or motor impairment. *Brain Behavior and Immunity, 25* (Suppl. 1), S155–S164. http://dx.doi.org/10.1016/j.bbi.2011.01.012.

Yakovleva, T., Bazov, I., Watanabe, H., Hauser, K. F., & Bakalkin, G. (2011). Transcriptional control of maladaptive and protective responses in alcoholics: a role of the NF-kappaB system. *Brain Behavior and Immunity, 25*(Suppl. 1), S29–S38. http://dx.doi.org/10.1016/j.bbi.2010.12.019 pii:S0889-1591(10)00592-1.

Zahr, N. M., Luong, R., Sullivan, E. V., & Pfefferbaum, A. (2010). Measurement of serum, liver, and brain cytokine induction, thiamine levels, and hepatopathology in rats exposed to a 4-day alcohol binge protocol. *Alcoholism Clinical and Experimental Research, 34*(11), 1858–1870. http://dx.doi.org/10.1111/j.1530-0277.2010.01274.x.

Ziv, Y., Ron, N., Butovsky, O., Landa, G., Sudai, E., Greenberg, N., ... Schwartz, M. (2006). Immune cells contribute to the maintenance of neurogenesis and spatial learning abilities in adulthood. *Nature Neuroscience, 9*(2), 268–275. http://dx.doi.org/10.1038/nn1629.

Zou, J. Y., & Crews, F. T. (2006). CREB and NF-kappaB transcription factors regulate sensitivity to excitotoxic and oxidative stress induced neuronal cell death. *Cellular and Molecular Neurobiology, 26*(4–6), 385–405. http://dx.doi.org/10.1007/s10571-006-9045-9.

Chapter 43

Role of Glutamate Transport in Alcohol Withdrawal

Osama A. Abulseoud[1,2], Christina L. Ruby[3], Victor Karpyak[2]
[1]National Institute on Drug Abuse, National Institutes of Health, Baltimore, MD, USA; [2]Department of Psychiatry and Psychology, Mayo Clinic, Rochester, MN, USA; [3]Department of Biology, Indiana University of Pennsylvania, Indiana, PA, USA

Abbreviations

AMPA α-Amino-3-hydroxy-5-methylisoxazole-4-propionic acid
AWS Alcohol withdrawal syndrome
CaMKII Ca^{2+}-calmodulin dependent protein kinase II
CNS Central nervous system
CSF Cerebrospinal fluid
EAAT2 Excitatory amino acid transporter 2
ERK1/2 Extracellular signal regulated kinase
FDA Food and Drug Administration
GABA γ-Amino butyric acid
Glx Glutamate + glutamine
i.p. Intraperitoneal
KA Kainic acid or kainate
LTD Long-term depression
LTP Long-term potentiation
mGluR2/3 Metabotropic glutamate receptor 2/3
mGluR7 and mGluR8 Glutamate receptors 7 and 8
mPFC Medial prefrontal cortex
MRS Magnetic resonance spectroscopy
NAC Nucleus accumbens
NMDA N-methyl-D-aspartate
PKA Protein kinase A
PKC Protein kinase C
SLC1 Solute carrier family 1
TCA Tricarboxylic acid cycle
α-KG α-Ketoglutarate

INTRODUCTION

Alcohol withdrawal syndrome (AWS) is a serious medical condition characterized by anxiety, sleep disturbance, hyperexcitability, autonomic instability, and in severe cases delirium tremens, seizures, and death (Mayo-Smith et al., 2004; Schuckit, 2009, 2014). AWS is a significant public health concern. As much as one-fifth of the total national health care expenditure in the United States is spent on disorders related to excessive alcohol consumption (Bouchery, Harwood, Sacks, Simon, & Brewer, 2011), with a significant portion of that expenditure utilized annually to manage approximately 500,000 episodes of acute AWS (Kosten & O'Connor, 2003). Many of these episodes occur during postoperative recovery and specifically in trauma patients (16% and 31% incidence respectively; Spies & Rommelspacher, 1999).

Alcohol is a central nervous system (CNS) depressant that enhances γ-amino butyric acid (GABA) function and attenuates glutamate receptor activity (Erdozain & Callado, 2014). Prolonged heavy alcohol intake results in neuroadaptive changes to mitigate the continued presence of ethanol in the brain. These neuroadaptive changes attempt to regain the disrupted GABA/glutamate balance by enhancing glutamatergic and attenuating GABAergic neurotransmission (Becker & Mulholland, 2014). Enhanced glutamate signaling and activation of N-methyl-D-aspartate (NMDA) glutamate receptors result in extracellular calcium influx and mobilization of additional calcium from endoplasmic reticulum that can cause neuronal toxicity (Ye, Shi, & Yin, 2013). Abrupt cessation of alcohol intake eliminates alcohol from the brain, uncovering the disrupted GABA/glutamate balance that manifests as AWS. Despite intense research efforts, only a handful of glutamatergic genetic factors seem to play a role in the withdrawal process. Moreover, little is known about the effects of genetic variation on vulnerability or symptom severity during withdrawal.

Despite a study demonstrating the superior efficacy of three antiglutamatergic strategies in ameliorating withdrawal severity in human subjects (Krupitsky et al., 2007), benzodiazepines remain the criterion standard when treating AWS in clinical practice. Benzodiazepines are GABA receptor agonists that have proven effective in reducing withdrawal severity and mortality rates (Mayo-Smith, 1997). However, these drugs boost the reduced inhibitory GABAergic neurotransmission without affecting the elevation in glutamate levels. Leaving this hyperglutamatergic state uncorrected has been associated with more intense alcohol craving (Bauer et al., 2013) and higher chance of postwithdrawal relapse (Malcolm, 2003), and may have neurotoxic consequences. In this chapter, we review the glutamatergic mechanisms and associated genetic factors underlying withdrawal, with special emphasis on the newly discovered role of excitatory amino acid transporter type 2 (EAAT2) in the pathogenesis and potential treatment of AWS.

GLUTAMATE SIGNALING UNDER NORMAL PHYSIOLOGICAL CONDITIONS

Synthesis, Release, and Receptors

Glutamate is the major excitatory neurotransmitter in the brain. Glutamatergic neurons can synthesize glutamate from α-ketoglutarate, a product of the tricarboxylic acid (TCA) cycle, by the enzyme glutamate dehydrogenase, or from glutamine by the enzyme glutaminase. Synthesized glutamate is packaged into synaptic vesicles in the presynaptic terminal by vesicular glutamate transporters and released into synaptic space to bind to three different types of ionotropic glutamate receptors located on the head of the postsynaptic dendritic spine: the N-methyl-D-aspartate (NMDA) receptor, the α-amino-3-hydroxy-5-methylisoxazole-4-propionic acid (AMPA) receptor, and the kainic acid (kainate, KA) receptor (Figure 1) (Olive, 2009).

The NMDA receptor is a heterotetramer containing two NR1 subunits and two NR2 subunits (Cioffi, 2013). Each subunit is organized of four domains: (1) an extracellular amino-terminal domain, (2) an extracellular ligand-binding domain, (3) a transmembrane domain, and (4) an intracellular carboxy-terminal domain (Vyklicky et al., 2014). NMDA receptor activation requires two molecules of glutamate (agonist) and two molecules of glycine (coagonist) to bind, causing magnesium ions (Mg^{2+}) to exit the channel, which allows an influx of calcium (Ca^{2+}) and sodium (Na^+) followed by efflux of potassium (K^+), creating membrane depolarization (Cioffi, 2013). NMDA receptors interact with numerous different intracellular signaling pathways through several kinases, including protein kinase A (PKA), protein kinase C (PKC), extracellular signal regulated kinase 1/2 (ERK1/2), and Ca^{2+}-calmodulin dependent protein kinase II (CaMK II), which directly interact with C-terminal domain of the NR2B subunit (Contractor, Mulle, & Swanson, 2011; Szczurowska & Mares, 2015) to promote long-term potentiation (LTP) and long-term depression (LTD) and neuronal plasticity (Figure 2).

AMPA receptors mediate most of the fast excitatory synaptic transmission, while KA receptors are located presynaptically where they can directly and indirectly modulate GABAergic synaptic transmission and neuronal excitability (Gass & Olive, 2008). In addition to inotropic receptors, glutamate also binds to metabotropic glutamate receptors, which are located either in the perisynaptic annulus or on presynaptic terminals to mediate slower, modulatory glutamate transmission (Olive, 2009).

Maintaining Synaptic Glutamate Concentration at Low Physiological Levels

Once glutamate dissociates from its binding site at the glutamatergic receptor, it is cleared quickly from synaptic space by a family of glutamate transporters. These high affinity transporters, called excitatory amino acid transporters (EAATs), keep extracellular levels of glutamate tightly regulated (reviewed by Beart & O'Shea, 2007; Benarroch, 2010; Danbolt, 2001; Kanai & Hediger, 2003).

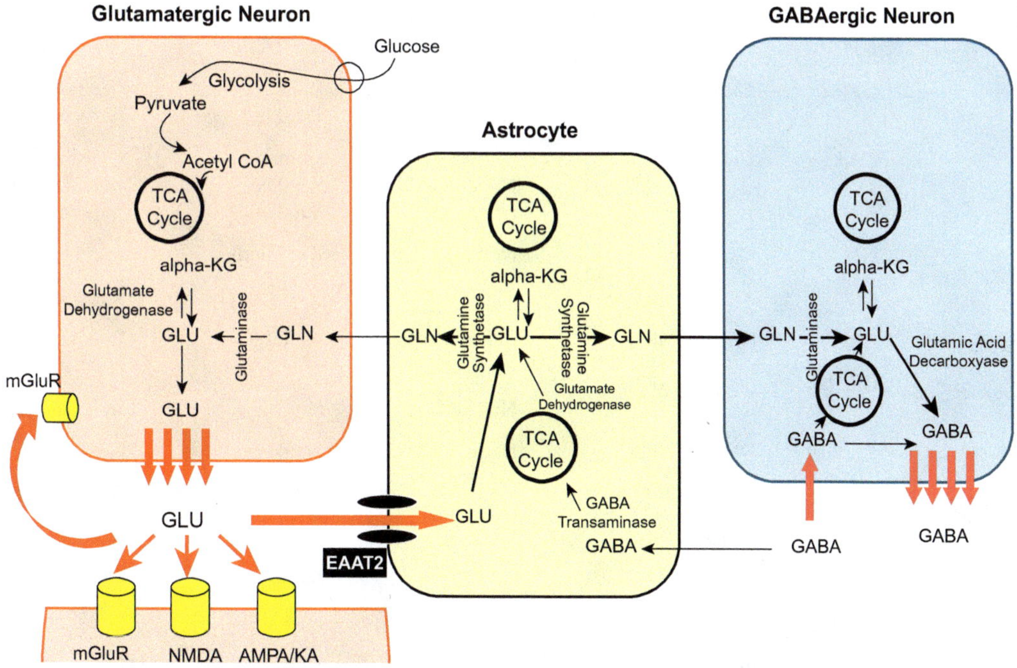

FIGURE 1 **Astrocytes maintain the balance between glutamatergic and GABAergic neurons.** Within glutamatergic neurons (left), glutamate is synthesized from alpha-ketoglutarate by the enzyme glutamate dehydrogenase or from glutamine by the enzyme glutaminase. Glutamate is released into synaptic space, where it binds to glutamate receptors. Following dissociation from receptor binding, glutamate is picked up by astrocytes (center) via EAAT2 and is converted to glutamine by the enzyme glutamine synthetase, then shuttled back to glutamatergic neurons or to GABAergic neurons (right) to produce GABA by the enzyme glutamic acid decarboxylase. (TCA, tricarboxylic acid; alpha-KG, alpha-ketoglutarate; EAAT2, excitatory amino acid transporter type 2; GLU, glutamate; GLN, glutamine; AMPA, α-amino-3-hydroxy-5-methyl-4-isoxazolepropionic acid; NMDA, N-methyl-D-aspartate; mGluR, metabotropic glutamate receptor.)

EAATs are members of the solute carrier family 1 (SLC1), which also includes two neutral amino acid transporters: alanine serine cysteine transporters 1 and 2 (ASC 1 and 2; reviewed by Grewer, Gameiro, & Rauen, 2014; Kanai & Hediger, 2003). Brain glutamate transporters are expressed in both glial cells and neurons (Danbolt, 2001; Schmitt, Asan, Puschel, Jons, & Kugler, 1996). Five subtypes of excitatory amino acid transporters have been identified (Table 1). Glial EAATs transport one glutamate molecule together with three Na^+ ions and one H^+ ion from the synaptic space into the astrocyte and countertransport one K^+ (Figure 3) (Zerangue & Kavanaugh, 1996). EAAT2 alone transports an estimated 90% of synaptic glutamate into astrocytes (Haugeto et al., 1996).

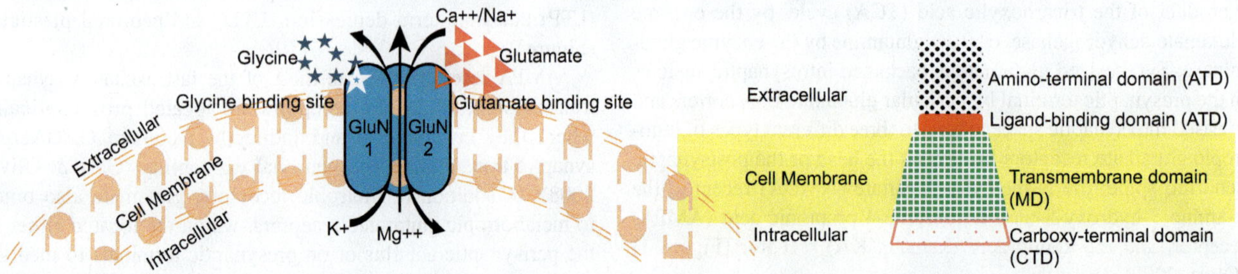

FIGURE 2 NMDA receptor structure. The heterotetrameric structure of the NMDA receptor (left) and the four domains in which each subunit is organized (right).

TABLE 1 Excitatory Amino Acid Transporter (EAAT) Subtypes in Humans with Their Corresponding Nomenclatures in the Solute Carrier Family 1, in Rodents, and Cellular Localization

Gene Set	Gene	Protein	Correlation	Species	References
Glutamate receptors	GRIA1	AMPAR GluR1	+	Rat (DMS)	Wang et al. (2012)
			None	Rat (Hipp)	Ferreira et al. (2001)
	GRIA2	AMPAR GluR2	+	Rat (DMS)	Wang et al. (2012)
			None	Rat (Hipp)	Ferreira et al. (2001)
	GRIA3	AMPAR GluR3	None	Rat (Hipp)	Ferreira et al. (2001)
	GRIK1	GluR5	None	Rat (Hipp)	Ferreira et al. (2001)
	GRIK2	GluR6	+	Rat (Hipp)	Carta et al. (2002)
			None	Rat (Hipp)	Ferreira et al. (2001)
	GRIK3	GluR7	+	Human	Preuss et al. (2006)
			None	Rat (DVC)	Freeman et al. (2012)
			None	Rat (Hipp)	Ferreira et al. (2001)
	GRIK5	KA2	None	Rat (Hipp)	Ferreira et al. (2001)
	GRIN1	NMDAR NR1	+	Human	Rujescu et al. (2005)
	GRIN2A	NMDAR NR2A	− (early), + (late)	Rat (DVC)	Freeman et al. (2012)
	GRIN2B	NMDAR NR2B	+	Human	Biermann et al. (2009)
			None	Human	Tadic et al. (2005)
	GRIN2C	NMDAR NR2C	None	Rat (Hipp)	Ferreira et al. (2001)
	GRIN2D	NMDAR NR2D	None	Rat (Hipp)	Ferreira et al. (2001)
	GRIN3A	NMDAR NR3A	− (early), + (late)	Rat (DVC)	Freeman et al. (2012)
	GRM5	mGluR5	+	Mouse	Blednov and Harris (2008)
	GRM7	mGluR7	None	Human	Preuss et al. (2002)
	GRM8	mGluR8	None	Human	Preuss et al. (2002)
Glutamate transporters	SLC1A2	EAAT2	−	Rat (NAC, mPFC)	Abulseoud et al. (2014)
			None	Human	Sander et al. (2000)
Glutamate metabolism	GLUL	Glutamine synthetase	−	Mouse	Letwin et al. (2006)

DMS, dorsomedial striatum; Hipp, hippocampus; DVC, dorsal vagal complex; NAC, nucleus accumbens; mPFC, medial prefrontal cortex.

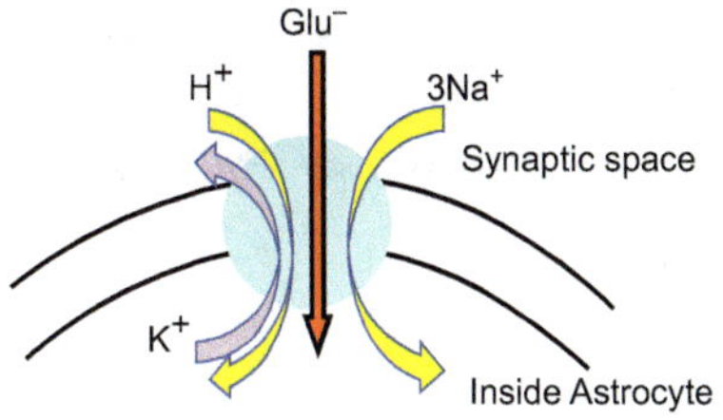

FIGURE 3 **Transportation of glutamate by EAAT2.** AAT2 transports one glutamate (Glu⁻) molecule from the synaptic space to the inside of astrocyte together with three sodium (Na$^+$) ions and one hydrogen (H$^+$) ion, and countertransports one potassium (K$^+$) ion from inside the astrocyte to the synaptic space.

Astrocytic Glutamate

Once inside the astrocyte, glutamate serves several important functions. First, it acts as the substrate for the production of glutamine by the enzyme glutamine synthetase, which rids the cell of toxic ammonia (Rose, Verkhratsky, & Parpura, 2013). Glutamine is then shuttled to glutamatergic neurons to be converted back to glutamate by glutaminase (Laake, Slyngstad, Haug, & Ottersen, 1995; Nicklas, Zeevalk, & Hyndman, 1987), termed the glutamate–glutamine cycle (McKenna, 2007). Glutamate also provides a substrate for GABA production in GABAergic neurons, catalyzed by the enzyme glutamic acid decarboxylase (Sze, 1979). Glutamate also functions in the glutamate/cystine antiporter system (Lewerenz, Maher, & Methner, 2012), which is critical for glutathione production. Finally, glutamate can enter the TCA cycle to maintain cellular energy production (Schousboe, Scafidi, Bak, Waagepetersen, & McKenna, 2014).

GLUTAMATE SIGNALING DURING ALCOHOL WITHDRAWAL

Alcohol is a CNS depressant that potentiates inhibitory GABAergic signaling and inhibits excitatory glutamatergic signaling. During heavy and prolonged alcohol consumption, the brain compensates by downregulating GABAergic neurotransmission and upregulating glutamatergic neurotransmission. Upon abrupt termination of alcohol drinking, these neuroadaptations manifest clinically as CNS hyperexcitability, which is the hallmark of AWS (Crews et al., 2005; Tsai et al., 1998). Several lines of evidence show that disruptions in glutamatergic signaling at every level underlie that pathophysiology of AWS.

Reduced Glutamate Synthesis

Under normal conditions, the brain derives its energy from glycolysis, which provides acetyl-CoA for the TCA cycle which, in turn, produces α-ketoglutarate (α-KG) and ATP. α-KG is then converted into glutamate by the enzyme glutamate dehydrogenase. Ethanol is metabolized into acetate, which provides a direct source of acetyl-CoA to the TCA cycle. During alcohol dependence, the TCA cycle becomes dependent on this source of acetyl-CoA, while during withdrawal, these substrates become relatively unavailable. The production of ATP through ethanol-derived acetate suppresses glycolysis enzymes and the conversion of pyruvate to acetyl-CoA. Abrupt cessation of ethanol intake halts the acetate-driven acetyl-CoA supply to the TCA cycle, jeopardizing α-KG production and, hence, glutamate synthesis. Likewise, ATP production is impaired (Derr, 1984; Schreiber, 1979).

Although both cell types are ultimately affected by the disruption in normal energy production in the brain, there is a differential effect on astrocytes versus neurons. While neurons can take up only glucose, astrocytes can take up both glucose and acetate (de Graaf, Mason, Patel, Behar, & Rothman, 2003). Hence, it is primarily astrocytes that are affected, while the glucose-driven TCA cycle and ATP production in neurons remains relatively stable during drinking and AWS. Because EAAT2-mediated uptake of glutamate is ATP-dependent, the impairment in astrocytic production of ATP during AWS reduces the ability of astrocytes to clear glutamate from the synapse (Derr, 1984; Schreiber, 1979). Decreased intracellular glutamate in astrocytes compromises the glutamate–glutamine cycle, ultimately reducing astrocytic replenishment of glutamine as a substrate for glutamate production in neurons (Figure 4; (Marx, Billups, & Billups, 2015)). As GABAergic neurons also require glutamine from astrocytes to synthesize GABA, reduced production of GABA may also result. Although the effect of alcohol withdrawal on ammonia levels has not yet been investigated, it is reasonable to speculate that the decrease in intracellular glutamate would disrupt the ability of astrocytes to detoxify ammonia, leading to further neurotoxicity.

Enhanced Glutamate Release

Despite the decrease in glutamate production in neurons, glutamate release is enhanced during alcohol withdrawal. Central to this imbalance are altered expression of group II metabotropic glutamate receptor 2/3 (mGluR2/3) by chronic exposure to alcohol (Meinhardt et al., 2013) and altered neuroglial interactions. mGluR2/3 receptors are located on presynaptic terminals of glutamatergic neurons, where their main function is to inhibit the release of glutamate (Ferraguti & Shigemoto, 2006; Holmes, Spanagel, & Krystal, 2013). These receptors are activated by nonvesicular, extrasynaptic glutamate that is released by astrocytes in exchange for cystine (Lutgen et al., 2013). Activated mGluR2/3 receptors then inhibit neuronal release of glutamate into the synapse (Moran, McFarland, Melendez, Kalivas, & Seamans, 2005; Tsai, Scott, Lewis, & Dodd, 2005). Alcohol withdrawal compromises the ability of astrocytes to take up synaptic glutamate due to TCA cycle dysfunction and poor ATP production (discussed previously), as well as downregulation of EAAT2 expression (discussed later). Combined, these changes lead to reduced astrocytic glutamate concentration and, hence, less glutamate released into extrasynaptic space by astrocytes. This attenuated glutamatergic tone at mGluR2/3 receptors during alcohol withdrawal results in excessive release of synaptic glutamate in response to neuronal activation (Figure 5). It could be argued that the elevated glutamate concentration in the synaptic space could "spill over" to the extrasynaptic space and activate the mGluR2/3 receptors. However, a study by Griffin, Haun, Hazelbaker, Ramachandra, and Becker (2014) showed that microinjection of the mGluR2/3 agonist LY379268 in ethanol-dependent mice was associated with significant reduction in synaptic glutamate concentration and ethanol drinking behavior.

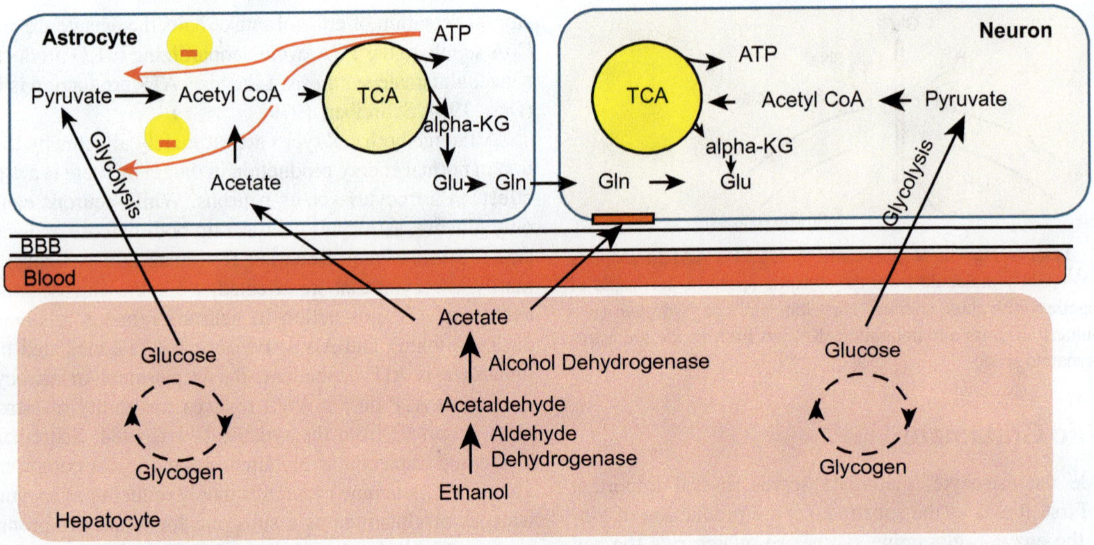

FIGURE 4 Ethanol withdrawal impacts astrocyte functionality. Ethanol is metabolized in the liver into acetate which acts as a substrate for the astrocytic, but not neuronal, TCA cycle and suppresses glycolysis. During acute withdrawal, the absence of ethanol and, hence, acetate deprives the astrocytic TCA cycle of its substrate, leading to astrocytic dysfunction. (TCA, tricarboxylic acid; alpha-KG, alpha-ketoglutarate; GLU, glutamate; GLN, glutamine; BBB, blood–brain barrier.)

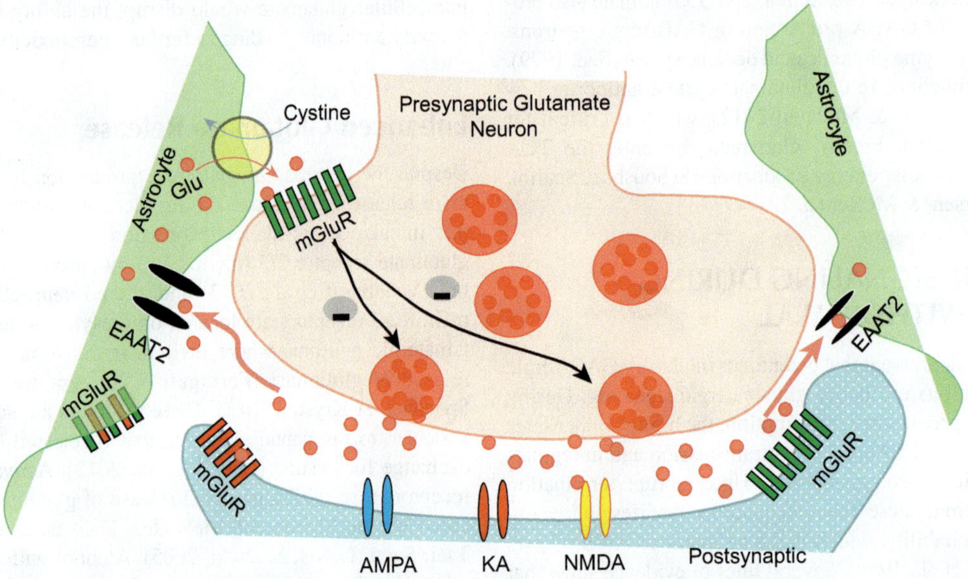

FIGURE 5 Alcohol has a significant impact on NMDA and mGluR2/3 receptors. Chronic alcohol consumption upregulates NMDA receptor expression and functionality, which manifest during alcohol withdrawal as hyperexcitability and seizure. (NMDA, N-methyl-D-aspartate receptor; AMPA, α-amino-3-hydroxy-5-methylisoxazole-4-propionic acid receptor; KA, kainic acid receptor; mGluR, metabotropic glutamate receptor; EAAT2, excitatory amino acid transporter type 2.)

Overactivated Glutamate Receptors

Another important change in glutamatergic signaling that is unmasked during alcohol withdrawal involves the upregulation of glutamate receptors. Acute exposure to alcohol inhibits NMDA receptors in several brain regions, such as the hippocampus and striatum (Gass & Olive, 2008; Holmes et al., 2013; Tsai & Coyle, 1998). Region-specific NMDA receptor inhibition impairs long-term potentiation and neuronal plasticity, and as a consequence, memory, cognition, and other functions that depend upon these basic processes (Tsai et al., 1998). Glutamatergic neurons respond to this alcohol-induced inhibition by increasing the functionality of existing NMDA receptors and upregulating their expression. During acute alcohol withdrawal, upregulated NMDA receptor expression and function lead to overstimulation of postsynaptic neurons, manifesting as CNS hyperexcitability and seizure susceptibility (Gass & Olive, 2008). Upregulation of NMDA receptors on noradrenergic locus coeruleus neurons is also thought to contribute to the anxiety and autonomic instability that characterizes acute alcohol withdrawal (Tsai et al., 1998).

Impaired Synaptic Glutamate Clearance

Several lines of evidence support the impairment of glutamate clearance as an important contributing factor to alcohol-induced disruption of glutamatergic neurotransmission. We have shown a significant reduction in EAAT2 protein expression in the striatum and medial prefrontal cortex (mPFC) in alcohol-preferring (P) and Wistar rats 7 days following severe alcohol withdrawal (Abulseoud et al., 2014). Likewise, Othman et al. (Othman, Sinclair, Haughey, Geiger, & Parkinson, 2002) found that 30-min exposure of cultured rat astrocytes to ethanol (50 mM) inhibited glutamate uptake. Melendez, Hicks, Cagle, and Kalivas (2005) found that the extracellular glutamate concentration was elevated and glutamate uptake was reduced at 24 h, but not at 14 days, after a 7-day, moderate-dose ethanol administration paradigm (1 g/kg/day, i.p.). Interestingly, however, total NAC EAAT2 content, as measured by Western blot, was not altered by ethanol exposure in this study.

Alele and Devaud (2005) also found no significant changes in glutamate transporter levels or function in the cortex of male rats after 3 days of ethanol withdrawal. Several potential explanations exist for the lack of congruency between some of the above studies, including differences in species, ethanol doses, routes of administration, and postwithdrawal time at which glutamate uptake measures were assessed. Alternatively, these inconsistencies could be understood in the light of reports that under certain pathological conditions, EAAT2 may remain functional, but the directionality of transporting glutamate is reversed; instead of clearing glutamate from synaptic space, EAAT2 "spits out" glutamate (reviewed by Billups et al., 1998). Although this pattern of reverse transport has not been reported specifically in AWS, it has been shown in several neurological conditions involving hyperglutamatergic signaling, such as severe ischemic strokes (Rossi, Oshima, & Attwell, 2000), amyotrophic lateral sclerosis (Foran & Trotti, 2009), epilepsy (Jabs, Seifert, & Steinhauser, 2008), Huntington's disease (Estrada-Sanchez & Rebec, 2012), Alzheimer disease (Lauderback et al., 1999), and Parkinson's disease (Assous et al., 2014).

Increased Extracellular Glutamate Concentration

Elevated extracellular glutamate levels during acute alcohol withdrawal have been documented in animal models as well as human subjects (Krystal, Petrakis, Mason, Trevisan, & D'Souza, 2003; Tsai & Coyle, 1998; Tsai, Gastfriend, & Coyle, 1995). Direct measurement of extracellular glutamate concentrations during acute alcohol withdrawal shows robust increases in the forebrain, hippocampus, and NAC in rodents (Buckman, Meshul, Finn, & Janowsky, 1999; Dahchour et al., 1998; Melendez et al., 2005; Rossetti & Carboni, 1995). In agreement with these studies, it has been reported increased glutamate + glutamine (Glx) levels in the NAC of mice during ethanol withdrawal, with the antiglutamatergic drug acamprosate normalizing Glx levels (Hinton et al., 2012). Another preclinical research group reported increased medial prefrontal glutamate and decreased glutamine as measured by 9.4T magnetic resonance spectroscopy (MRS) in acute withdrawal, both of which resolved within several weeks (Hermann et al., 2012). The high spectroscopic resolution in this study made it the first to provide support that the increase in the composite Glx signal during acute alcohol withdrawal was driven by glutamate.

Human studies also report increased glutamate levels during alcohol withdrawal. A 1994 study by Aliyev et al. assayed plasma levels of excitatory and inhibitory amino acid neurotransmitters in 20 male patients following alcohol withdrawal, finding that excitatory amino acids were higher in withdrawing patients than in normal controls, while inhibitory amino acids were lower (Aliyev, Aliyev, & Aliguliyev, 1994). In addition, this group found a positive correlation between the level of excitatory amino acids and the rating of subjective discomfort, and a negative correlation of this measure with inhibitory amino acids. Tsai and Coyle (Tsai & Coyle, 1998) found a significant increase in cerebrospinal fluid (CSF) glutamate concentration that correlated positively with the severity of alcohol dependence, as measured by the Alcohol Dependence Severity Scale (Gossop et al., 1995; Umhau et al., 2010). Using noninvasive proton MRS to quantify the glutamate concentration during withdrawal, Hermann et al. (Hermann et al., 2012) showed increased glutamate levels in the anterior cingulate of patients in withdrawal (versus controls) that normalize by day 14. Another group found an increased Glx-to-creatine ratio (Glx/Cr) in the anterior cingulate in patients at day 15 after the last drink compared to controls (Lee et al., 2007). Likewise, Bauer et al. (2013) quantified glutamate levels immediately after detoxification in alcohol-dependent patients versus healthy control subjects and found significantly elevated glutamate levels in the ventral striatum that correlated positively with craving. In addition, patients randomized to acamprosate at the initiation of abstinence showed a trend toward reduced glutamate from day 4 to day 25, while no change was observed in controls randomized to placebo (Umhau et al., 2010). One study (Thoma et al., 2011) reported *lower* glutamate in withdrawing patients, and *higher* glutamine in remitted patients versus controls in the medial cingulate. The basis for the disparity between this and the other studies is unknown.

CONTRIBUTION OF GENETIC VARIATION IN GLUTAMATE SIGNALING TO ALCOHOL WITHDRAWAL RISK AND SEVERITY

Despite the large body of evidence indicating the involvement of glutamatergic mechanisms in AWS, a review of the genetics of alcohol withdrawal included no evidence of the association between alcohol withdrawal and variation in genes governing glutamate neurotransmission (Schmidt & Sander, 2000). A handful of studies since 2000 have examined whether glutamatergic gene variants and/or expression patterns influence risk for and severity of AWS in humans and animal models. We summarize these findings in the following sections and in Table 2.

Human Studies

Human studies have revealed that genetic variation or expression patterns in three glutamatergic signaling genes—one encoding a subunit of the kainate receptor and the other two encoding subunits of the NMDA receptor—are associated with AWS severity. A functional Ser310Ala polymorphism of the GRIK3 gene, which encodes kainate receptor subunit GluR7, was associated with delirium tremens in alcoholics (Preuss et al., 2006). Rujescu et al. (2005) found that the 2108A allele and A-containing genotypes of the GRIN1 gene, which encodes the NR1 subunit of the NMDA receptor,

TABLE 2 Genes Encoding Proteins Involved in Glutamate Signaling, Transport, and Metabolism and Their Associations with AWS

EAAT Subtype in Human	Solute Carrier Family 1 Subtype	Homologue in Rodents	Cellular Localization	References
EAAT1	SLC1A3	L-glutamate L-aspartate transporter (GLAST)	Glial cells Bergman glial cells of the cerebellum	Berger, DeSilva, Chen, and Rosenberg, (2005)
EAAT2	SLC1A2	Glutamate transporter 1 (GLT1)	Almost exclusively glial Hippocampal neurons	Danbolt (2001) and Danbolt, Storm-Mathisen, and Kanner (1992) Chen et al. (2004)
EAAT3	SLC1A1	EAAC1	Neurons	Schmitt et al. (1996)
EAAT4	SLC1A6	EAAC4	Neurons Cerebellar Purkinje cells	Schmitt et al. (1996) Takayasu, Iino, Takatsuru, Tanaka, and Ozawa (2009)
EAAT5	SLC1A7		Cone photoreceptors and bipolar cells of the retina	Rauen and Kanner, (1994)

were associated with a history of withdrawal-induced seizures. Polymorphisms in the GRIN2B gene, encoding NR2B subunit of NMDA receptor were not associated with alcohol dependence, alcohol withdrawal-induced seizures, or delirium tremens (Tadic et al., 2005). However, GRIN2B gene expression was increased in blood samples from alcoholics during day 1 (P=0.001) and day 3 (P=0.029) of alcohol withdrawal (Biermann et al., 2009). The authors believe this finding may be explained by the negative correlation between alcohol drinking severity and the methylation of a cluster of five CPG-sites within the GRIN2B promoter.

Two other human studies reported that genes encoding metabotropic glutamate receptor subunits and glutamate transporter EAAT2 appear to be unrelated to AWS symptoms or severity. There was no association between polymorphisms in metabotropic glutamate receptors 7 and 8 (mGluR7 and mGluR8) and withdrawal seizures or delirium tremens (Preuss et al., 2002). Genotype frequencies of a silent G603A nucleotide exchange in exon 5 of the EAAT2 gene were assessed in 342 alcohol-dependent subjects, including 112 with a history of alcohol withdrawal seizures or delirium tremens and 54 with an antisocial personality disorder, compared with 223 control subjects (all of German descent; Sander et al., 2000). The frequency of the G603A polymorphism did not differ significantly between control subjects and either the entire sample of alcoholics or the alcoholics with severe physiological withdrawal symptoms. However, there is strong preclinical evidence that EAAT2 expression can be neuroprotective against AWS manifestations (discussed below).

Animal Studies

A handful of experimental studies have examined glutamatergic signaling genes and proteins using various animal models of alcohol withdrawal, with similarly mixed results as the human studies. As with the preclinical research summarized in the previous section, discrepancies between studies are likely reflective of differences in species/strain, alcohol administration/withdrawal paradigms, brain regions of interest, and time at which gene or protein expression was assessed. Ferreira et al. (2001) found that alcohol withdrawal induced no changes in the expression of several glutamate receptor subunits in the rat hippocampus, including AMPA receptor subunits GluR1, GluR2, GluR3, kainate receptor subunits GluR5, GluR6, GluR7, KA2, and NMDA receptor subunits NR2C and NR2D. However, Carta, Olivera, Dettmer, and Valenzuela (2002) reported an increase in kainate GluR6 expression in the rat hippocampus, and Haugbol, Ebert, and Ulrichsen (2005) reported upregulation of NMDA and AMPA glutamate receptor subtypes during alcohol withdrawal in rats.

Likewise, other studies by Blednov and Harris (2008), Freeman et al. (2012), Wang, Liu, Zhang, Wu, and Zeng (2012), and Abulseoud et al. (2014) have also revealed significant changes in glutamatergic gene and/or protein expression in animal models of AWS. NMDA receptor subunit NR2A and NR3A are downregulated during early withdrawal and upregulated during late withdrawal (Freeman et al., 2012). Blednov and Harris (2008) showed that mice lacking mGluR5 had reduced alcohol withdrawal. Similarly, decreased expression of genes coding for glutamine synthetase was linked to increased glutamate-mediated neurotransmission, increased ethanol-induced locomotion, and seizure susceptibility (Letwin et al., 2006). Wang et al. (2012) demonstrated that AMPA receptor subunits GluR1 and GluR2 were upregulated after alcohol withdrawal following intermittent alcohol exposure. Finally, in agreement with studies showing that mice lacking EAAT2 (GLT1) show lethal spontaneous seizures (Tanaka et al., 1997), our group demonstrated that upregulation of EAAT2 in the NAC and mPFC was associated with amelioration of AWS manifestations in rats (Abulseoud et al., 2014), a finding that may have significant clinical importance (discussed below).

PRECLINICAL EVIDENCE FOR THE USE OF CEFTRIAXONE TO TREAT ACUTE AWS

Ceftriaxone is an FDA-approved β-lactam antibiotic with good central nervous system penetration. The safety and side effect profile of ceftriaxone in medical settings has been well documented

(Bai, Bao, Cheng, Yang, & Li, 2012). Ceftriaxone has been shown to reduce synaptic glutamate levels in animal models of drug addiction (Abulseoud, Miller, Wu, Choi, & Holschneider, 2012; Knackstedt, Melendez, & Kalivas, 2010), an action presumably resulting from its ability to upregulate central EAAT2 expression (Rothstein et al., 2005).

We have demonstrated compelling evidence for the use of ceftriaxone as an effective method to treat acute AWS (Abulseoud et al., 2014). In this study, we developed a novel method for rating ethanol withdrawal manifestations based on quantifying individual behaviors over the complete duration of ethanol withdrawal using continuous video recording. An important advantage of the method we described is its superior temporal resolution for the appearance and cessation of individual withdrawal manifestations. The enhanced temporal resolution provided a more complete picture of withdrawal than previous studies, which relied upon short-term observations (15s) at various time points after cessation of alcohol administration (Lal et al., 1988; Waller, McBride, Lumeng, & Li, 1982). We also focused on objectively defined, clinically relevant measures of withdrawal rather than more subjective measures such as hyperactivity. Finally, we did not collapse all withdrawal signs into one score (Majchrowicz & Hunt, 1976). Instead, each withdrawal manifestation was scored individually, allowing for more robust clinical translatability.

We showed that ceftriaxone effectively reduced or abolished all manifestations of ethanol withdrawal in two different rat strains (alcohol-preferring P rats and Wistar rats) that are thought to resemble alcohol-dependent and social drinking animal models, respectively. We also found that the blockade of AWS manifestations was associated with increased EAAT2 protein expression in two key brain regions, the NAC and the mPFC. Furthermore, we observed that ceftriaxone blocked AWS-induced increases in alcohol consumption and preference. Because ceftriaxone appears to work by upregulating EAAT2, it is reasonable to speculate that not only does it reduce problematic extracellular glutamate levels, but it may also restore the deficits in astrocytic glutamate (discussed previously), effectively "resetting" glutamatergic signaling alterations induced by long-term, heavy alcohol consumption. This unique mechanism positions ceftriaxone as an ideal candidate for the treatment of AWS.

APPLICATIONS TO OTHER ADDICTIONS AND SUBSTANCE MISUSE

The role of glutamatergic neurotransmission in other addictive disorders is becoming more evident. Although there is much variability between different drugs of abuse in terms of the mechanisms by which glutamate is disrupted, the brain region(s) involved, and the overall effect on glutamate transmission (reviewed by Gass & Olive, 2008; Reissner & Kalivas, 2010), increasing EAAT2-mediated glutamate uptake appears to have potential therapeutic value for addiction to substances other than alcohol. In this regard, preclinical studies have demonstrated that ceftriaxone is effective in attenuating relapse-like behavior for cocaine (Knackstedt et al., 2010; Sari, Smith, Ali, & Rebec, 2009), reducing methamphetamine reinstatement (Abulseoud et al., 2012), decreasing development of morphine tolerance (Rawls, Baron, & Kim, 2010), and reducing nicotine withdrawal manifestations (Alajaji, Bowers, Knackstedt, & Damaj, 2013).

All of these studies also confirmed that ceftriaxone upregulated EAAT2 levels and/or function. Thus, ceftriaxone or other agents that increase glutamate uptake may be promising treatment options for substance use disorders in general.

SUMMARY AND CONCLUSIONS

A large body of evidence suggests that disruption in glutamatergic signaling at all levels contributes greatly to AWS and relapse to heavy drinking after abstinence. Although studies examining the contribution of individual molecular players involved in glutamatergic signaling are not entirely consistent, there is compelling evidence that astrocytic glutamate uptake transporter EAAT2 may represent an important, novel target for the development of therapeutics against the often life-threatening manifestations of alcohol withdrawal. In this regard, the FDA-approved antibiotic ceftriaxone may be well suited to treatment of acute AWS and prevention of relapse.

DEFINITION OF TERMS

Hyperglutamatergic state This is an elevation in brain glutamate concentration that has been reported in alcohol withdrawal.
Excitatory amino acid transporter This protein transports glutamate from synaptic space into astrocytes.
Upregulation This refers to an increase in the number or function of a cellular component, such as a receptor, transporter, or signaling molecule.

KEY FACTS ON THE TRICARBOXYLIC ACID CYCLE

- The TCA cycle is also known as the Krebs Cycle (named for its discoverer, Hans Adolf Krebs) and the citric acid cycle (named after the intermediate citric acid, or citrate).
- The TCA cycle metabolizes acetate derived from carbohydrates, proteins, and fats to form adenosine triphosphate (ATP), the body's energy currency.
- Oxidation of one glucose molecule through the combined action of glycolysis, the TCA cycle, and oxidative phosphorylation yields an estimated 30–38 ATP molecules.
- TCA cycle intermediates oxaloacetate and α-ketoglutarate give rise to the amino acids aspartate and glutamate, respectively—both of which act as excitatory neurotransmitters in the brain.
- There are eight enzymes in the TCA cycle that oxidize acetyl-coenzyme A (acetyl-CoA) into two molecules of carbon dioxide.
- Citrate produced in the TCA cycle can also be used in fatty acid synthesis.

SUMMARY POINTS

- AWS is a serious global health concern with deleterious effects on individuals and society.
- Current treatment strategies for AWS are lacking. GABAergic medications (benzodiazapines) are effective at reducing symptom severity, but they do not address the primary pathology of abnormally elevated glutamate.

- Astrocytes play several key roles in glutamatergic signaling, including glutamate production, metabolism, and synaptic clearing, ultimately maintaining extracellular glutamate concentrations within a narrow physiological range.
- Alcohol disrupts glutamatergic signaling at every level, leading to CNS hyperexcitability that underlies acute AWS, as well as postwithdrawal craving and relapse.
- Despite the glutamatergic basis of alcohol withdrawal manifestations, the contribution of genetic variability in glutamate signaling molecules to AWS is relatively understudied and poorly understood.
- We now have an animal model of severe AWS that meets face validity criteria. This model can help us explore other withdrawal-related phenomena, such as repeated withdrawal-induced kindling and even delirium tremens.
- Strong preclinical evidence exists for the use of the FDA approved β-lactam antibiotic, ceftriaxone, to treat acute AWS. The efficacy of ceftriaxone in reducing withdrawal manifestations in animal models of AWS appears to be based upon its ability to upregulate astrocytic glutamate transporter EAAT2 and, thus, clear excessive glutamate from the synapse.

REFERENCES

Abulseoud, O. A., Camsari, U. M., Ruby, C. L., Kasasbeh, A., Choi, S., & Choi, D. S. (2014). Attenuation of ethanol withdrawal by ceftriaxone-induced upregulation of glutamate transporter EAAT2. *Neuropsychopharmacology*, *39*(7), 1674–1684.

Abulseoud, O. A., Miller, J. D., Wu, J., Choi, D. S., & Holschneider, D. P. (2012). Ceftriaxone upregulates the glutamate transporter in medial prefrontal cortex and blocks reinstatement of methamphetamine seeking in a condition place preference paradigm. *Brain Research*, *1456*, 14–21.

Alajaji, M., Bowers, M. S., Knackstedt, L., & Damaj, M. I. (2013). Effects of the beta-lactam antibiotic ceftriaxone on nicotine withdrawal and nicotine-induced reinstatement of preference in mice. *Psychopharmacology (Berlin)*, *228*, 419–426.

Alele, P. E., & Devaud, L. L. (2005). Differential adaptations in GABAergic and glutamatergic systems during ethanol withdrawal in male and female rats. *Alcoholism Clinical and Experimental Research*, *29*, 1027–1034.

Aliyev, N. A., Aliyev, Z. N., & Aliguliyev, A. R. (1994). Amino acid neurotransmitters in alcohol withdrawal. *Alcohol and Alcoholism*, *29*, 643–647.

Assous, M., Had-Aissouni, L., Gubellini, P., Melon, C., Nafia, I., Salin, P., … Kachidian, P. (2014). Progressive Parkinsonism by acute dysfunction of excitatory amino acid transporters in the rat substantia nigra. *Neurobiology of Disease*, *65*, 69–81.

Bai, Z. G., Bao, X. J., Cheng, W. D., Yang, K. H., & Li, Y. P. (2012). Efficacy and safety of ceftriaxone for uncomplicated gonorrhoea: a meta-analysis of randomized controlled trials. *International Journal of STD and AIDS*, *23*, 126–132.

Bauer, J., Pedersen, A., Scherbaum, N., Bening, J., Patschke, J., Kugel, H., … Ohrmann, P. (2013). Craving in alcohol-dependent patients after detoxification is related to glutamatergic dysfunction in the nucleus accumbens and the anterior cingulate cortex. *Neuropsychopharmacology*, *38*, 1401–1408.

Beart, P. M., & O'Shea, R. D. (2007). Transporters for L-glutamate: an update on their molecular pharmacology and pathological involvement. *British Journal of Pharmacology*, *150*, 5–17.

Becker, H. C., & Mulholland, P. J. (2014). Neurochemical mechanisms of alcohol withdrawal. *Handbook of Clinical Neurology*, *125*, 133–156.

Benarroch, E. E. (2010). Glutamate transporters: diversity, function, and involvement in neurologic disease. *Neurology*, *74*, 259–264.

Berger, U. V., DeSilva, T. M., Chen, W., & Rosenberg, P. A. (2005). Cellular and subcellular mRNA localization of glutamate transporter isoforms GLT1a and GLT1b in rat brain by in situ hybridization. *The Journal of Comparative Neurology*, *492*, 78–89.

Biermann, T., Reulbach, U., Lenz, B., Frieling, H., Muschler, M., Hillemacher, T., … Bleich, S. (2009). N-methyl-D-aspartate 2b receptor subtype (NR2B) promoter methylation in patients during alcohol withdrawal. *Journal of Neural Transmission*, *116*, 615–622.

Billups, B., Rossi, D., Oshima, T., Warr, O., Takahashi, M., Sarantis, M., … Attwell, D. (1998). Physiological and pathological operation of glutamate transporters. *Progress in Brain Research*, *116*, 45–57.

Blednov, Y. A., & Harris, R. A. (2008). Metabotropic glutamate receptor 5 (mGluR5) regulation of ethanol sedation, dependence and consumption: relationship to acamprosate actions. *The International Journal of Neuropsychopharmacology/Official Scientific Journal of the Collegium Internationale Neuropsychopharmacologicum*, *11*, 775–793.

Bouchery, E. E., Harwood, H. J., Sacks, J. J., Simon, C. J., & Brewer, R. D. (2011). Economic costs of excessive alcohol consumption in the U.S., 2006. *American Journal of Preventive Medicine*, *41*, 516–524.

Buckman, J. F., Meshul, C. K., Finn, D. A., & Janowsky, A. (1999). Glutamate uptake in mice bred for ethanol withdrawal severity. *Psychopharmacology (Berlin)*, *143*, 174–182.

Carta, M., Olivera, D. S., Dettmer, T. S., & Valenzuela, C. F. (2002). Ethanol withdrawal upregulates kainate receptors in cultured rat hippocampal neurons. *Neuroscience Letters*, *327*, 128–132.

Chen, W., Mahadomrongkul, V., Berger, U. V., Bassan, M., DeSilva, T., Tanaka, K., … Rosenberg, P. A. (2004). The glutamate transporter GLT1a is expressed in excitatory axon terminals of mature hippocampal neurons. *Journal of Neuroscience*, *24*, 1136–1148.

Cioffi, C. L. (2013). Modulation of NMDA receptor function as a treatment for schizophrenia. *Bioorganic and Medicinal Chemistry Letters*, *23*, 5034–5044.

Contractor, A., Mulle, C., & Swanson, G. T. (2011). Kainate receptors coming of age: milestones of two decades of research. *Trends in Neurosciences*, *34*, 154–163.

Crews, F. T., Buckley, T., Dodd, P. R., Ende, G., Foley, N., Harper, C., … Sullivan, E. V. (2005). Alcoholic neurobiology: changes in dependence and recovery. *Alcoholism Clinical and Experimental Research*, *29*, 1504–1513.

Dahchour, A., De Witte, P., Bolo, N., Nedelec, J. F., Muzet, M., Durbin, P., & Macher, J. P. (1998). Central effects of acamprosate: part 1. Acamprosate blocks the glutamate increase in the nucleus accumbens microdialysate in ethanol withdrawn rats. *Psychiatry Research*, *82*, 107–114.

Danbolt, N. C. (2001). Glutamate uptake. *Progress in Neurobiology*, *65*, 1–105.

Danbolt, N. C., Storm-Mathisen, J., & Kanner, B. I. (1992). An [Na++K+] coupled L-glutamate transporter purified from rat brain is located in glial cell processes. *Neuroscience*, *51*, 295–310.

Derr, R. F. (1984). The ethanol withdrawal syndrome: a consequence of lack of substrate for a cerebral Krebs-cycle. *Journal of Theoretical Biology, 106*, 375–381.

Erdozain, A. M., & Callado, L. F. (2014). Neurobiological alterations in alcohol addiction: a review. *Adicciones, 26*, 360–370.

Estrada-Sanchez, A. M., & Rebec, G. V. (2012). Corticostriatal dysfunction and glutamate transporter 1 (GLT1) in Huntington's disease: interactions between neurons and astrocytes. *Basal Ganglia, 2*, 57–66.

Ferraguti, F., & Shigemoto, R. (2006). Metabotropic glutamate receptors. *Cell and Tissue Research, 326*, 483–504.

Ferreira, V. M., Frausto, S., Browning, M. D., Savage, D. D., Morato, G. S., & Valenzuela, C. F. (2001). Ionotropic glutamate receptor subunit expression in the rat hippocampus: lack of an effect of a long-term ethanol exposure paradigm. *Alcoholism Clinical and Experimental Research, 25*, 1536–1541.

Foran, E., & Trotti, D. (2009). Glutamate transporters and the excitotoxic path to motor neuron degeneration in amyotrophic lateral sclerosis. *Antioxidants and Redox Signaling, 11*, 1587–1602.

Freeman, K., Staehle, M. M., Gumus, Z. H., Vadigepalli, R., Gonye, G. E., Nichols, C. N., … Schwaber, J. S. (2012). Rapid temporal changes in the expression of a set of neuromodulatory genes during alcohol withdrawal in the dorsal vagal complex: molecular evidence of homeostatic disturbance. *Alcoholism Clinical and Experimental Research, 36*, 1688–1700.

Gass, J. T., & Olive, M. F. (2008). Glutamatergic substrates of drug addiction and alcoholism. *Biochemical Pharmacology, 75*, 218–265.

Gossop, M., Darke, S., Griffiths, P., Hando, J., Powis, B., Hall, W., & Strang, J. (1995). The Severity of Dependence Scale (SDS): psychometric properties of the SDS in English and Australian samples of heroin, cocaine and amphetamine users. *Addiction, 90*, 607–614.

de Graaf, R. A., Mason, G. F., Patel, A. B., Behar, K. L., & Rothman, D. L. (2003). In vivo 1H-[13C]-NMR spectroscopy of cerebral metabolism. *NMR in Biomedicine, 16*, 339–357.

Grewer, C., Gameiro, A., & Rauen, T. (2014). SLC1 glutamate transporters. *Pflugers Archiv: European Journal of Physiology, 466*, 3–24.

Griffin, W. C., 3rd, Haun, H. L., Hazelbaker, C. L., Ramachandra, V. S., & Becker, H. C. (2014). Increased extracellular glutamate in the nucleus accumbens promotes excessive ethanol drinking in ethanol dependent mice. *Neuropsychopharmacology, 39*, 707–717.

Haugbol, S. R., Ebert, B., & Ulrichsen, J. (2005). Upregulation of glutamate receptor subtypes during alcohol withdrawal in rats. *Alcohol and Alcoholism, 40*, 89–95.

Haugeto, O., Ullensvang, K., Levy, L. M., Chaudhry, F. A., Honore, T., Nielsen, M., … Danbolt, N. C. (1996). Brain glutamate transporter proteins form homomultimers. *Journal of Biological Chemistry, 271*, 27715–27722.

Hermann, D., Weber-Fahr, W., Sartorius, A., Hoerst, M., Frischknecht, U., Tunc-Skarka, N., … Sommer, W. H. (2012). Translational magnetic resonance spectroscopy reveals excessive central glutamate levels during alcohol withdrawal in humans and rats. *Biological Psychiatry, 71*, 1015–1021.

Hinton, D. J., Lee, M. R., Jacobson, T. L., Mishra, P. K., Frye, M. A., Mrazek, D. A., … Choi, D. S. (2012). Ethanol withdrawal-induced brain metabolites and the pharmacological effects of acamprosate in mice lacking ENT1. *Neuropharmacology, 62*, 2480–2488.

Holmes, A., Spanagel, R., & Krystal, J. H. (2013). Glutamatergic targets for new alcohol medications. *Psychopharmacology (Berlin), 229*, 539–554.

Jabs, R., Seifert, G., & Steinhauser, C. (2008). Astrocytic function and its alteration in the epileptic brain. *Epilepsia, 49*(Suppl. 2), 3–12.

Kanai, Y., & Hediger, M. A. (2003). The glutamate and neutral amino acid transporter family: physiological and pharmacological implications. *European Journal of Pharmacology, 479*, 237–247.

Knackstedt, L. A., Melendez, R. I., & Kalivas, P. W. (2010). Ceftriaxone restores glutamate homeostasis and prevents relapse to cocaine seeking. *Biological Psychiatry, 67*, 81–84.

Kosten, T. R., & O'Connor, P. G. (2003). Management of drug and alcohol withdrawal. *New England Journal of Medicine, 348*, 1786–1795.

Krupitsky, E. M., Rudenko, A. A., Burakov, A. M., Slavina, T. Y., Grinenko, A. A., Pittman, B., … Krystal, J. H. (2007). Antiglutamatergic strategies for ethanol detoxification: comparison with placebo and diazepam. *Alcoholism Clinical and Experimental Research, 31*, 604–611.

Krystal, J. H., Petrakis, I. L., Mason, G., Trevisan, L., & D'Souza, D. C. (2003). N-methyl-D-aspartate glutamate receptors and alcoholism: reward, dependence, treatment, and vulnerability. *Pharmacology and Therapeutics, 99*, 79–94.

Laake, J. H., Slyngstad, T. A., Haug, F. M., & Ottersen, O. P. (1995). Glutamine from glial cells is essential for the maintenance of the nerve terminal pool of glutamate: immunogold evidence from hippocampal slice cultures. *Journal of Neurochemsitry, 65*, 871–881.

Lal, H., Harris, C. M., Benjamin, D., Springfield, A. C., Bhadra, S., & Emmett-Oglesby, M. W. (1988). Characterization of a pentylenetetrazol-like interoceptive stimulus produced by ethanol withdrawal. *Journal of Pharmacology and Experimental Therapeutics, 247*, 508–518.

Lauderback, C. M., Harris-White, M. E., Wang, Y., Pedigo, N. W., Jr., Carney, J. M., & Butterfield, D. A. (1999). Amyloid beta-peptide inhibits Na+-dependent glutamate uptake. *Life Sciences, 65*, 1977–1981.

Lee, E., Jang, D. P., Kim, J. J., An, S. K., Park, S., Kim, I. Y., … Namkoong, K. (2007). Alteration of brain metabolites in young alcoholics without structural changes. *Neuroreport, 18*, 1511–1514.

Letwin, N. E., Kafkafi, N., Benjamini, Y., Mayo, C., Frank, B. C., Luu, T., … Elmer, G. I. (2006). Combined application of behavior genetics and microarray analysis to identify regional expression themes and gene-behavior associations. *Journal of Neuroscience, 26*, 5277–5287.

Lewerenz, J., Maher, P., & Methner, A. (2012). Regulation of xCT expression and system x (c) (-) function in neuronal cells. *Amino Acids, 42*, 171–179.

Lutgen, V., Qualmann, K., Resch, J., Kong, L., Choi, S., & Baker, D. A. (2013). Reduction in phencyclidine induced sensorimotor gating deficits in the rat following increased system xc(-) activity in the medial prefrontal cortex. *Psychopharmacology (Berlin), 226*, 531–540.

Majchrowicz, E., & Hunt, W. A. (1976). Temporal relationship of the induction of tolerance and physical dependence after continuous intoxication with maximum tolerable doses of ethanol in rats. *Psychopharmacology (Berlin), 50*, 107–112.

Malcolm, R. J. (2003). GABA systems, benzodiazepines, and substance dependence. *Journal of Clinical Psychiatry, 64*(Suppl. 3), 36–40.

Marx, M. C., Billups, D., & Billups, B. (2015). Maintaining the presynaptic glutamate supply for excitatory neurotransmission. *Journal of Neuroscience Research, 93*(7), 1031–1044.

Mayo-Smith, M. F. (1997). Pharmacological management of alcohol withdrawal. A meta-analysis and evidence-based practice guideline. American Society of Addiction Medicine Working Group on Pharmacological Management of Alcohol Withdrawal. *JAMA, 278*, 144–151.

Mayo-Smith, M. F., Beecher, L. H., Fischer, T. L., Gorelick, D. A., Guillaume, J. L., Hill, A., … Working Group on the Management of Alcohol Withdrawal Delirium, Practice Guidelines Committee, American Society of Addiction Medicine (2004). Management of alcohol withdrawal delirium. An evidence-based practice guideline. *Archives of Internal Medicine, 164*, 1405–1412.

McKenna, M. C. (2007). The glutamate-glutamine cycle is not stoichiometric: fates of glutamate in brain. *Journal of Neuroscience Research*, 85, 3347–3358.

Meinhardt, M. W., Hansson, A. C., Perreau-Lenz, S., Bauder-Wenz, C., Stahlin, O., Heilig, M., ... Sommer, W. H. (2013). Rescue of infralimbic mGluR2 deficit restores control over drug-seeking behavior in alcohol dependence. *Journal of Neuroscience*, 33, 2794–2806.

Melendez, R. I., Hicks, M. P., Cagle, S. S., & Kalivas, P. W. (2005). Ethanol exposure decreases glutamate uptake in the nucleus accumbens. *Alcoholism Clinical and Experimental Research*, 29, 326–333.

Moran, M. M., McFarland, K., Melendez, R. I., Kalivas, P. W., & Seamans, J. K. (2005). Cystine/glutamate exchange regulates metabotropic glutamate receptor presynaptic inhibition of excitatory transmission and vulnerability to cocaine seeking. *Journal of Neuroscience*, 25, 6389–6393.

Nicklas, W. J., Zeevalk, G., & Hyndman, A. (1987). Interactions between neurons and glia in glutamate/glutamine compartmentation. *Biochemical Society Transactions*, 15, 208–210.

Olive, M. F. (2009). Metabotropic glutamate receptor ligands as potential therapeutics for addiction. *Current Drug Abuse Reviews*, 2, 83–98.

Othman, T., Sinclair, C. J., Haughey, N., Geiger, J. D., & Parkinson, F. E. (2002). Ethanol alters glutamate but not adenosine uptake in rat astrocytes: evidence for protein kinase C involvement. *Neurochemical Research*, 27, 289–296.

Preuss, U. W., Koller, G., Bahlmann, M., Zill, P., Soyka, M., & Bondy, B. (2002). No association between metabotropic glutamate receptors 7 and 8 (mGlur7 and mGlur8) gene polymorphisms and withdrawal seizures and delirium tremens in alcohol-dependent individuals. *Alcohol and Alcoholism*, 37, 174–178.

Preuss, U. W., Zill, P., Koller, G., Bondy, B., Hesselbrock, V., & Soyka, M. (2006). Ionotropic glutamate receptor gene GRIK3 SER310ALA functional polymorphism is related to delirium tremens in alcoholics. *The Pharmacogenomics Journal*, 6, 34–41.

Rauen, T., & Kanner, B. I. (1994). Localization of the glutamate transporter GLT-1 in rat and macaque monkey retinae. *Neuroscience Letters*, 169, 137–140.

Rawls, S. M., Baron, D. A., & Kim, J. (2010). Beta-Lactam antibiotic inhibits development of morphine physical dependence in rats. *Behavioural Pharmacology*, 21, 161–164.

Reissner, K. J., & Kalivas, P. W. (2010). Using glutamate homeostasis as a target for treating addictive disorders. *Behavioural Pharmacology*, 21, 514–522.

Rose, C. F., Verkhratsky, A., & Parpura, V. (2013). Astrocyte glutamine synthetase: pivotal in health and disease. *Biochemical Society Transactions*, 41, 1518–1524.

Rossetti, Z. L., & Carboni, S. (1995). Ethanol withdrawal is associated with increased extracellular glutamate in the rat striatum. *European Journal of Pharmacology*, 283, 177–183.

Rossi, D. J., Oshima, T., & Attwell, D. (2000). Glutamate release in severe brain ischaemia is mainly by reversed uptake. *Nature*, 403, 316–321.

Rothstein, J. D., Patel, S., Regan, M. R., Haenggeli, C., Huang, Y. H., Bergles, D. E., ... Fisher, P. B. (2005). Beta-lactam antibiotics offer neuroprotection by increasing glutamate transporter expression. *Nature*, 433, 73–77.

Rujescu, D., Soyka, M., Dahmen, N., Preuss, U., Hartmann, A. M., Giegling, I., ... Szegedi, A. (2005). GRIN1 locus may modify the susceptibility to seizures during alcohol withdrawal. *American Journal of Medical Genetics. Part B, Neuropsychiatric Genetics: The Official Publication of the International Society of Psychiatric Genetics*, 133B, 85–87.

Sander, T., Ostapowicz, A., Samochowiec, J., Smolka, M., Winterer, G., & Schmidt, L. G. (2000). Genetic variation of the glutamate transporter EAAT2 gene and vulnerability to alcohol dependence. *Psychiatric Genetics*, 10, 103–107.

Sari, Y., Smith, K. D., Ali, P. K., & Rebec, G. V. (2009). Upregulation of GLT1 attenuates cue-induced reinstatement of cocaine-seeking behavior in rats. *Journal of Neuroscience*, 29, 9239–9243.

Schmidt, L. G., & Sander, T. (2000). Genetics of alcohol withdrawal. *European Psychiatry: The Journal of the Association of European Psychiatrists*, 15, 135–139.

Schmitt, A., Asan, E., Puschel, B., Jons, T., & Kugler, P. (1996). Expression of the glutamate transporter GLT1 in neural cells of the rat central nervous system: non-radioactive in situ hybridization and comparative immunocytochemistry. *Neuroscience*, 71, 989–1004.

Schousboe, A., Scafidi, S., Bak, L. K., Waagepetersen, H. S., & McKenna, M. C. (2014). Glutamate metabolism in the brain focusing on astrocytes. *Advances in Neurobiology*, 11, 13–30.

Schreiber, R. A. (1979). Sources of energy for the brain and physical dependence on ethanol. *Medical Hypotheses*, 5, 629–634.

Schuckit, M. A. (2009). Alcohol-use disorders. *Lancet*, 373, 492–501.

Schuckit, M. A. (2014). Recognition and management of withdrawal delirium (delirium tremens). *New England Journal of Medicine*, 371, 2109–2113.

Spies, C. D., & Rommelspacher, H. (1999). Alcohol withdrawal in the surgical patient: prevention and treatment. *Anesthesia and analgesia*, 88, 946–954.

Szczurowska, E., & Mares, P. (2015). Different action of a specific NR2B/NMDA antagonist Ro 25-6981 on cortical evoked potentials and epileptic afterdischarges in immature rats. *Brain Research Bulletin*, 111, 1–8.

Sze, P. Y. (1979). L-Glutamate decarboxylase. *Advances in Experimental Medicine and Biology*, 123, 59–78.

Tadic, A., Dahmen, N., Szegedi, A., Rujescu, D., Giegling, I., Koller, G., ... Soyka, M. (2005). Polymorphisms in the NMDA subunit 2B are not associated with alcohol dependence and alcohol withdrawal-induced seizures and delirium tremens. *European Archives of Psychiatry and Clinical Neuroscience*, 255, 129–135.

Takayasu, Y., Iino, M., Takatsuru, Y., Tanaka, K., & Ozawa, S. (2009). Functions of glutamate transporters in cerebellar Purkinje cell synapses. *Acta Physiologica*, 197, 1–12.

Tanaka, K., Watase, K., Manabe, T., Yamada, K., Watanabe, M., Takahashi, K., ... Wada, K. (1997). Epilepsy and exacerbation of brain injury in mice lacking the glutamate transporter GLT-1. *Science*, 276, 1699–1702.

Thoma, R., Mullins, P., Ruhl, D., Monnig, M., Yeo, R. A., Caprihan, A., ... Gasparovic, C. (2011). Perturbation of the glutamate-glutamine system in alcohol dependence and remission. *Neuropsychopharmacology*, 36, 1359–1365.

Tsai, G., & Coyle, J. T. (1998). The role of glutamatergic neurotransmission in the pathophysiology of alcoholism. *Annual Review of Medicine*, 49, 173–184.

Tsai, G., Gastfriend, D. R., & Coyle, J. T. (1995). The glutamatergic basis of human alcoholism. *American Journal of Psychiatry*, 152, 332–340.

Tsai, G. E., Ragan, P., Chang, R., Chen, S., Linnoila, V. M., & Coyle, J. T. (1998). Increased glutamatergic neurotransmission and oxidative stress after alcohol withdrawal. *American Journal of Psychiatry*, 155, 726–732.

Tsai, V. W., Scott, H. L., Lewis, R. J., & Dodd, P. R. (2005). The role of group I metabotropic glutamate receptors in neuronal excitotoxicity in Alzheimer's disease. *Neurotoxicity Research*, 7, 125–141.

Umhau, J. C., Momenan, R., Schwandt, M. L., Singley, E., Lifshitz, M., Doty, L., ... Heilig, M. (2010). Effect of acamprosate on magnetic resonance spectroscopy measures of central glutamate in detoxified alcohol-dependent individuals: a randomized controlled experimental medicine study. *Archives of General Psychiatry, 67*, 1069–1077.

Vyklicky, V., Korinek, M., Smejkalova, T., Balik, A., Krausova, B., Kaniakova, M., ... Vyklicky, L. (2014). Structure, function, and pharmacology of NMDA receptor channels. *Physiological Research/Academia Scientiarum Bohemoslovaca, 63*(Suppl. 1), S191–S203.

Waller, M. B., McBride, W. J., Lumeng, L., & Li, T. K. (1982). Induction of dependence on ethanol by free-choice drinking in alcohol-preferring rats. *Pharmacology Biochemistry and Behavior, 16*, 501–507.

Wang, K. S., Liu, X., Zhang, Q., Wu, L. Y., & Zeng, M. (2012). Genome-wide association study identifies 5q21 and 9p24.1 (KDM4C) loci associated with alcohol withdrawal symptoms. *Journal of Neural Transmission, 119*, 425–433.

Ye, H. B., Shi, H. B., & Yin, S. K. (2013). Mechanisms underlying taurine protection against glutamate-induced neurotoxicity. *The Canadian Journal of Neurological Sciences. Le journal canadien des sciences neurologiques, 40*, 628–634.

Zerangue, N., & Kavanaugh, M. P. (1996). Flux coupling in a neuronal glutamate transporter. *Nature, 383*, 634–637.

Chapter 44

Ethanol, Vitamins, and Brain Dysfunction

Emilio González-Reimers[1], Camino Fernández-Rodríguez[1], Emilio González-Arnay[2], Francisco Santolaria-Fernández[1]

[1]*Servicio de Medicina Interna, Hospital Universitario de Canarias, Universidad de La Laguna, La Laguna, Tenerife, Canary Islands, Spain;*
[2]*Departamento de Anatomía, Anatomía Patológica e Histología, Hospital Universitario de Canarias, Universidad de La Laguna, La Laguna, Tenerife, Canary Islands, Spain*

Abbreviations

cAMP Cyclic adenosine monophosphate
CD Cluster determinant
CREB cAMP responsive element binding protein
CRMP-2 Collapsin response mediator protein 2
DNA Deoxyribonucleic acid
EGF Epidermal growth factor
GABA Gamma-amino-butyric acid
HMBG-1 High mobility group box 1
IL Interleukin
MAP-LC3 Microtubule-associated protein-light chain 3
MCP Monocyte chemoattractant protein
MRI Magnetic resonance imaging
NFκB Nuclear factor kappa B
NADP Nicotinamide adenine dinucleotide phosphate
NADPH Reduced nicotinamide adenine dinucleotide phosphate
NGF Nerve growth factor
NMDA *N*-Methyl-D-aspartate
NOX Nicotinamide adenine dinucleotide phosphate oxidase
ROS Reactive oxygen species
TLR Toll-like receptor
TNF Tumor necrosis factor

Brain alterations are frequent among alcoholics and include several distinct clinicopathological pictures (Table 1). Some of them are uncommon, such as central pontine myelinolysis or Marchiafava-Bignami disease. Prevalence of Wernicke encephalopathy associated with thiamine deficiency shows important geographic variations (Harper, Fornes, Duyckaerts, Lecomte, & Hauw, 1995), but the most frequent central nervous system alterations are brain atrophy, accompanied by cognitive dysfunction and several neuropsychological alterations, and cerebellar atrophy, related to motor alterations and gait disturbance, and possibly also with cognitive impairment. The true prevalence of brain atrophy is not well known. In a study by Torvik and Torp (1986), cerebellar atrophy was found in 42% of alcoholics under 70 years—a prevalence even higher than that observed among nonalcoholic individuals over 70 years. This figure roughly parallels that of cognitive dysfunction, which affects 50–80% of alcoholics. A link exists between atrophy and altered cognition, although deranged blood flow, reversible with prolonged abstinence (Gansler et al., 2000), may contribute. In addition, several associated conditions, such as liver dysfunction, protein–calorie malnutrition, and deficiency in some antioxidant vitamins and trace elements, also damage the brain. This chapter reviews the mechanisms leading to brain atrophy in adolescent or adult alcoholics, and the potential role of vitamin alterations on these changes.

BRAIN ATROPHY AFFECTS BOTH WHITE MATTER AND GRAY MATTER

Gray Matter Atrophy

Gray matter atrophy can be viewed as a result of an imbalance between altered neurogenesis and increased neuron degeneration. Brain tissue, including cortical neurons, derives from the neuroepithelium of the neural tube. This neuroepithelium is formed by pluripotent stem cells that are able to differentiate both into neurons and glia; it is located in an area known as the ventricular zone. In this area, stem cells evolve to form immature neurons and the radial glia, which, in turn, generate cortical astrocytes but also preserve the ability to transform into new neurons. After the seventh week, both neurons and glia from the ventricular zone migrate to the subventricular zone—an area that progressively evolves into the main neurogenetic area (Meyer, Perez-Garcia, Abraham, & Caput, 2002). Cells of this area give raise to the six-layered neocortex of the mammalian brain through radial migration—a process disrupted by in utero ethanol exposure. Ethanol also affects dendritic arborization of developing hippocampal neurons (Kumada, Jiang, Cameron, & Komuro, 2007). The Cajal-Retzius neurons produce reelin, an essential matrix protein that organizes the normal structure of the cortex (Lambert de Rouvroit & Goffinet, 1998). This protein becomes altered by ethanol in vitro (Mooney, Siegenthaler, & Miller, 2004).

These changes, of undisputed importance in the development of the so-called fetal alcohol syndrome, may be also relevant in the

TABLE 1 Brain Alterations in Alcoholic Patients
Brain atrophy (frontal lobes, hippocampi)
Cerebellar atrophy
Wernicke–Korsakoff syndrome
Centropontine myelinolisis
Marchiafava-Bignami disease (necrosis of corpus callosum)
Ethanol intoxication
Witdrawal syndrome
Trauma (subdural + parenchymatous hematoma)
Stroke
Optic neuritis
Frontal atrophy and/or cerebellar atrophy are observed in 50–70% of heavy alcoholics.

adulthood. Indeed, hippocampus is an extraordinarily important neurogenetic area, with activity persisting until late adulthood. A normal hippocampus increases in size and myelination from childhood to adulthood (Suzuki et al., 2005). Studies of these authors performed on 23 younger adolescents, aged 13–14 years, compared with 30 older ones (19–21 years) revealed a marked volume increase among the latter, especially among males. Progressive myelination was also observed by Benes, Turtle, Khan, and Farol (1994) in 164 individuals; the myelination process was especially intense in the first two to three decades, especially among women, although it persisted during the lifetime. Indeed, it is currently accepted that, in humans, neuronogenesis continues in the subventricular zone of the wall of the lateral ventricles and in the hippocampal subgranular zone of the dentate gyrus throughout life (Knoth et al., 2010). These authors found doublecortin expression—a marker of neuronogenesis—in the granular zone of the dentate gyrus in brains of 54 individuals whose age at death ranged from birth to 100 years. Maximal expression was observed, however, during adolescence and young adulthood, until 30–40 years of age. In addition to these active neurogenetic areas, specialized forms of astrocytes from the subventricular zone migrate to the olfactory bulb, where they evolve to mature neurons (Lim & Alvarez-Buylla, 2014). Therefore, adolescence and young adulthood are important age periods for the full maturation and development of the hippocampus. However, in this life period, binge drinking is most common.

Ethanol heavily affects hippocampal neuronogenesis both during embryonic development (possibly underlying the cognitive deficits observed in the fetal alcohol syndrome, as commented) and in adulthood (as discussed below). Hippocampus affectation is especially important among adolescent binge drinkers, who frequently show a striking hippocampal volume shrinkage, especially in the right hippocampus. Agartz, Momenan, Rawlings, Kerich, and Hommer (1999) in a study of 52 excessive drinkers (age range 27–53 years) compared with 36 controls, found a decrease in both hippocampi volume among alcoholic women; however, no differences were observed in left hippocampi among men but were found in right ones. De Bellis et al. (2000) reported that both right and left hippocampi were atrophied in 12 alcoholics compared with 24 controls. Nagel, Schweinsburg, Phan, and Tapert (2005) found decreased left hippocampal volumes in 14 alcoholics aged 15–17 years compared to 17 controls, although hippocampal volume did not correlate with alcohol consumption. From these and other studies, it can be concluded that the adolescent hippocampus is especially sensitive to the deleterious effects of ethanol. These changes are in a certain way similar to those observed in degenerative situations. In patients with Alzheimer disease, a reduction in the left hippocampal volume was related to brain dysfunction.

Although it is clear that hippocampal atrophy is typical in heavy alcoholics, there is controversy about the relative importance of blunted neuronogenesis or increased neuron loss in its pathogenesis. Several authors, using animal models, have shown decreased neuronogenesis both in the hippocampus (Crews & Nixon, 2009; He, Nixon, Shetty, & Crews, 2005) and subventricular zone, but others have challenged this view. A study by Sutherland et al. (2013) of the brains of 15 chronic alcoholics and 16 age-matched controls failed to find differences in the number of proliferative cells in the olfactory bulb (however, mean age was around 55 years, far from the adolescent period). This result is challenging because, in line with the active migration of neuron precursor cells from the subventricular zone to the olfactory bulb, hyposmia has been reported as an early predictor of dementia in degenerative brain diseases and it has been proposed as a manifestation of brain atrophy in alcoholics (Rupp et al., 2003).

Ethanol also affects cerebellar development. Cerebellar neurons are especially sensitive to the noxious effects of ethanol, partly dependent on the altered signaling mechanisms mediated by the retinoic acid receptors and NMDA receptors and partly dependent on altered growth factors (Kumar, LaVoie, DiPette, & Singh, 2013).

White Matter Atrophy

White matter atrophy is also striking among alcoholics. The prefrontal white matter, corpus callosum, and cerebellum (Harper, 2009) are especially affected areas. However, some controversy exists: postmortem examination revealed that Purkinje cell volume was reduced among 10 alcoholics aged 45.5 years compared with 10 age-matched controls, but no changes were observed regarding white matter cerebellar atrophy (Andersen, 2004).

In part, white matter atrophy results from axonal injury secondary to neuronal death, but alterations in myelination have been also described (Harper, 2009). Repeated binge-drinking episodes lead to demyelination in the prefrontal cortex of adolescent rats but not in adult ones (Pascual, Pla, Miñarro, & Guerri, 2014), reinforcing the greater vulnerability of the brain during adolescence and early youth. Also, a single binge-drinking episode by a mother may lead to oligodendrocyte apoptosis in the fetus, and, consequently, decreased myelin production. This was shown in an experimental model in which macaque mothers were exposed to ethanol (Creeley, Dikranian, Johnson, Farber, & Olney, 2013) in an amount necessary to achieve a blood ethanol concentration in the range of 300–400 mg/dl (similar to those achieved during a binge-drinking episode) during 8 h. Also, in rat models, chronic ethanol exposure reduces the dendritic growth of newborn neurons (He et al., 2005).

The functional nature of white matter atrophy was shown by Bartsch et al. (2007), who used magnetic resonance imaging (MRI) to study 15 alcoholic patients who completed abstinence, and repeated the exploration 6–7 weeks after sobriety. This population was compared with 10 controls. Overall, brain mass showed a 2% increase after abstinence, especially in the superior vermis, perimesencephalic, supratentorial and infratentorial periventricular borders, and frontomesial and frontoorbital edges. These morphologic changes were accompanied by biochemical ones, suggesting a recovery from white matter lesions consistent with remyelination, and neuropsychological improvement (attention and concentration). Therefore, white matter damage seems to occur in relation to alcohol intake and is reversible after abstinence.

White matter lesions are also heavily dependent on vascular disease. Brain MRI studies allow the assessment of white matter hyperintensity, a parameter related to reduced blood flow. However, MRI studies using automated methods to assess white matter signal hyperintensity or diffusion tensor imaging (which allows examination of random mobility of tissue water, informing about the integrity of white matter fiber bundles) have yielded conflicting results in alcoholics (Cardenas et al., 2013).

FUNCTIONAL CONSEQUENCES

Brain morphologic alterations in alcoholics lead to functional changes. For instance, repeated withdrawal syndromes are associated with impaired cognitive function, but it seems that also acute ethanol ingestion deranges cognitive function (Duka, Townshend, Collier, & Stephens, 2003). Binge drinking impairs learning; interestingly, memory impairment is more intense when alcohol is consumed during adolescence than when it is consumed in adulthood, in parallel with anatomic and biochemical changes, which are also more intense when alcohol is consumed during adolescence. Moreover, one of the neuropsychological effects of ethanol-mediated organic brain damage may be the altered ability to abandon drinking habitus, which may be associated with a higher risk for resumption of ethanol drinking. In a study of 75 alcoholics treated against alcohol dependence, future relapsers showed smaller brain volumes in the mesocorticolimbic system than nonrelapsers (Cardenas et al., 2011). Cerebellar alteration may lead to ataxia and gait disturbance, although it has been also shown that it may also lead to cognitive dysfunction (Fitzpatrick, Jackson, & Crowe, 2008).

PATHOGENESIS OF NEUROTOXICITY IN ALCOHOLICS

Many data support the hypothesis that proinflammatory cytokines and oxidative stress contribute to brain damage. Ethanol promotes activation of the transcription factor nuclear factor kappa B (NFκB). Enhanced transcription of this factor is associated with increases in proinflammatory cytokines, especially tumor necrosis factor (TNF)-α and interleukin (IL)-1β and chimiokines, such as monocyte chemoattractant protein 1 (MCP-1). These cytokines are produced by activated glia, and numerous conditions, including ischemia, Alzheimer disease, brain injury, infection, or toxic exposure are potent inductors of microglia activation. In addition, proinflammatory cytokines produced elsewhere can reach the central nervous system, cross the blood–brain barrier, and secondarily activate microglia. Therefore, systemic inflammatory response due to ethanol consumption is involved in brain atrophy (Crews & Nixon, 2009).

Moreover, the duration of raised TNF values in brain is long lasting. A single intraperitoneal lipopolysaccharide injection provokes raised cytokine levels during 10 months, and activation of toll-like receptors (TLR) in microglia leads to a more prolonged secretion of cytokines in brain than in peripheral tissues. TLR-4 plays a pivotal role in this process. Alfonso-Loeches, Pascual-Lucas, Blanco, Sanchez-Vera, and Guerri (2010) have shown that neuroinflammation is blunted in mice lacking TLR-4 receptors. Ethanol also increases brain expression of TLR-3 and high mobility group box 1 (HMGB-1), a cytokine-like protein that functions as a TLR coagonist. Microglial activation, by multiple pathways, ultimately leads to microglia synthesis of proinflammatory cytokines, induction of NFκB, induction of NOX, and increased reactive oxygen species (ROS) production. ROS can activate neurons and cause neuronal induction of NOX leading to neuronal death (Qin & Crews, 2012). Therefore, microglial activation occurs by several different pathways, including ethanol itself, a local production of proinflammatory cytokines, and an induction of inflammation by cytokines produced in distant organs that reach microglia via a saturable blood to brain transport system.

Microglial activation of NFκB constitutes a key step in neuronal damage. It opposes to transcription of the cAMP responsive element binding protein (CREB). This transcription factor can be viewed as a neuron survival factor, because it protects neurons from apoptosis and excitotoxicity, and its expression keeps an inverse relation with the presence of ethanol. In fact, ethanol reduces levels of DNA binding to CREB in hippocampal entorhinal cortex slice cultures (Crews & Nixon, 2009), coincident with the enhancement in oxidative enzymatic pathways and NFκB transcription. Because NFκB induces NOX and enhances ROS production, ultimately, oxidative stress is an important mediator of the effects of ethanol on neurodegeneration. This fact raises questions about the role played by antioxidants in brain alterations. In other settings, Patel, Rogers, and Huang (2008) showed that flavonoids protect patients with Alzheimer disease, and Mehlig et al. (2008) showed that ingestion of wine in moderate amounts may protect against dementia. In addition, several studies suggest that antioxidants could protect the brain from binge ethanol-induced damage (Crews & Nixon, 2009). These findings support the importance of analyzing the role of antioxidant vitamins on brain alterations in the alcoholic patient.

LIPOSOLUBLE VITAMINS

As discussed below, the role of vitamin A or vitamin E deficiency on brain alterations and altered cognition has been well described for decades; more recently, the knowledge about vitamin D deficiency and brain alteration has suffered an explosive expansion. We comment on some relevant features regarding the effects of deficiency of these vitamins in alcoholics in the following sections. Recent research has pointed out that vitamin K deficiency seems to play a role in brain alterations, participating in sphingolipid metabolism and also as a brain antioxidant (Ferland, 2012).

Vitamin E Deficiency

Vitamin E (tocopherol) was discovered about 90 years ago, as a product contained in lettuce that prevented fetal loss in animals fed a rancid lard diet. Its function was previously thought to be restricted to the reproductive area. Later, it became evident that its principal effect was the protection of polyunsaturated fatty acids from oxidation. This effect was also evident in membrane phospholipids; therefore, there are theoretical basis that support a protective role of vitamin E on damaged myelin and neuronal cells. This has been experimentally demonstrated in murine models in which vitamin E deficiency was associated with cognitive dysfunction (Nishida et al., 2006). In vitamin E deficient animals, there is apoptosis and deposition of β-amyloid proteins, but before these histological changes occur, there is an alteration of collapsin response mediator protein 2 (CRMP-2), a cytoplasmatic protein involved in normal axonal function. These changes take place in the face of an increased expression of microtubule-associated protein-light chain 3 (MAP-LC3), an autophagy-related protein, possibly enhanced by increased oxidative damage (Fukui et al., 2012). Axonal dysfunction took place before any changes were detectable in the cell bodies of the hippocampal neurons. Vitamin E deficiency also causes ataxia, mainly due to its deleterious action on Purkinje cells.

Chronic ethanol feeding results in decreased serum vitamin E levels, partly due to an increased vitamin E demand by the liver due to increased transformation into α-tocopheryl quinone after scavenging of ROS (although in a murine model, ethanol-treated rats showed increased liver α-tocopherol concentration; Reilly, Patel, Peters, & Preedy, 2000). In alcoholics, poor nutrition and malabsorption probably contribute to decreased vitamin E levels. Therefore, less vitamin E remains available for performing antioxidant functions in tissues other than the liver, and possibly, vitamin E deficiency contributes to brain damage. Bondy, Guo, and Adams (1996) found that rats treated with α-tocopherol (200 mg/kg body weight by daily intraperitoneal injection for 15 days) showed raised levels of glutathione in both brain and liver. Simultaneous treatment with α-tocopherol prevented ethanol-induced decrease in liver and brain glutathione levels. Ethanol-treated rats also showed hyperhomocysteinemia and DNA damage, which were reverted by vitamin E supplementation (Shirpoor, Salami, Khadem-Ansari, Minassian, & Yegiazarian, 2009).

Vitamin E is a liposoluble vitamin whose absorption may be impaired in liver diseases. Therefore, theoretically, the development of liver cirrhosis should aggravate vitamin E deficiency in alcoholics—a result confirmed in a recent study (Figure 1). However, relationships among vitamin E levels and brain alterations are poor.

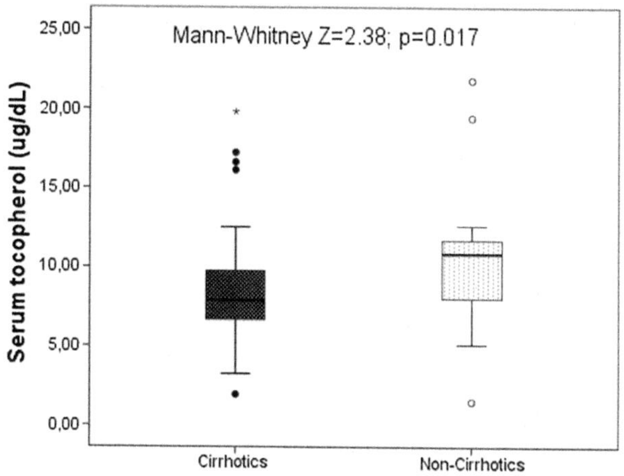

Vitamin A Deficiency

Vitamin A is usually ingested as beta carotene, which is transformed in the intestinal mucosa and (especially) in the liver into vitamin A. Carotenoids, in a similar way to α-tocopherol, protect unsaturated fatty acids from oxidative damage. Malnutrition and associated malabsorption contribute to vitamin A deficiency in alcoholics (Halsted, 2004). The situation becomes aggravated by ethanol-mediated microsomal cytochrome P-450 induction, which leads to vitamin A depletion. Indeed, liver vitamin A levels are reduced in alcoholics, and increased liver vitamin A breakdown has been reported in these patients (Clugston & Blaner, 2012).

In the brain, retinoic acid is involved in cognition and neuronal development. As with other vitamins, experimental data support a role for decreased vitamin A levels on brain affectation. In adult

FIGURE 1 Serum vitamin E levels in cirrhotics and noncirrhotics. Lower values are observed in cirrhotics. Circles represent outliers; asterisks represent extreme values.

rats, it has been clearly shown that memory and spatial learning become severely impaired with vitamin A deprivation. Vitamin A deficiency reduces hippocampal neuronogenesis and leads also to a marked reduction in retinoic acid receptors, leading to memory impairment (Etchamendy et al., 2003). The decrease in the number of hippocampal neurons and impaired brain function associated with vitamin A deprivation can be reversed by vitamin A supplementation (Bonnet et al., 2008). However, vitamin A supplementation may be also harmful, as it has been associated with increased oxidant capacity (Behr et al., 2012).

In a study performed in Kuala Lumpur on 333 individuals aged 60 or more years, the prevalence of mild cognitive impairment was 21.1%. Binary logistic regression indicated that the predictors of cognitive impairment were being married, being overweight or obese, and having vitamin A deficiency (Shahar et al., 2013). However, in a study on centenarians, the relationship between vitamin A levels (in brain or serum) and cognitive performance was weak (Johnson et al., 2013).

Several authors have reported reduced levels of vitamin A (or its metabolites) in serum and liver of alcoholics (Halsted, 2004). Being a liposoluble compound, the same reasons argued for a more intense depletion of vitamin E in cirrhotics are also valid for vitamin A. In accordance, we found that vitamin A levels were markedly decreased in cirrhotics (Figure 2), keeping a relationship with deranged liver function. Also, vitamin A levels kept an independent relationship with cerebellar atrophy.

Vitamin D Deficiency

It has been shown that systemic effects of vitamin D lay far beyond bone and calcium homeostasis. Vitamin D exerts positive effects on neuronal differentiation, migration, and proliferation, and it inhibits apoptosis. Individuals older than 60 with cognitive impairment show lower vitamin D levels than individuals without cognitive impairment (Chei et al., 2014). The results of the study by Chei et al. confirm those reported by many others, with few exceptions. Meta-analysis of five cross-sectional and two longitudinal studies including 7688 participants showed an increased risk of cognitive impairment in those with low vitamin D compared

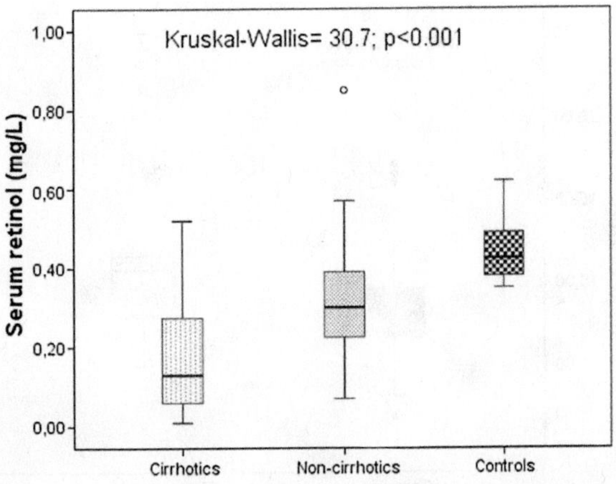

FIGURE 2 Serum vitamin A in cirrhotics, noncirrhotics, and controls. Lower values are observed in cirrhotics. Circles represent outliers.

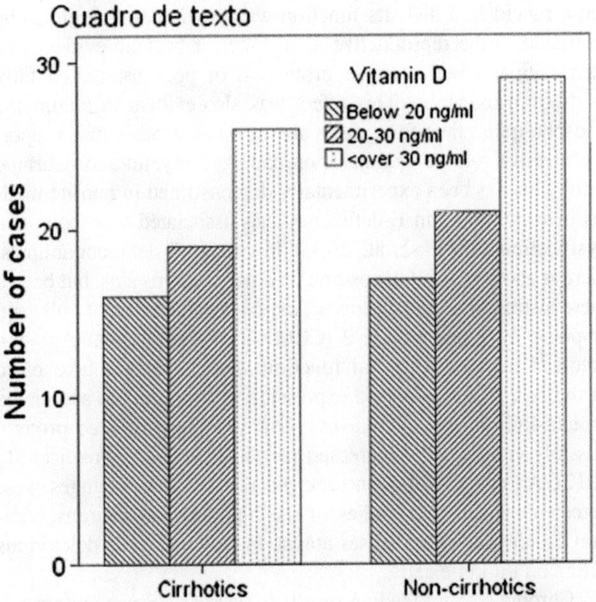

FIGURE 3 Prevalence of normal vitamin D, vitamin D insufficiency (below 30 ng/ml), and vitamin D deficiency (below 20 ng/ml) in cirrhotics and noncirrhotics. Vitamin D insufficiency was equally frequent among cirrhotics and noncirrhotics.

with normal vitamin D (OR 2.39, 95% CI 1.91–3.00; p<0.0001, Etgen, Sander, Bickel, Sander, & Förstl, 2012). Although heterogeneity of these population-based studies and the many confounding variables inherent to these studies oblige one to interpret these results with caution, several brain functions are clearly influenced by vitamin D. Vitamin D receptor is present in the brain, especially in the hippocampus. Vitamin D is involved in brain development and may be considered as a potent antioxidant in the adult brain (Briones & Darwish, 2014).

Vitamin D deficiency is common among alcoholics. It is usual to define vitamin D deficiency as serum vitamin D values below 20 ng/ml, and vitamin D insufficiency when vitamin D levels are below 30 ng/ml. Most studies in alcoholics show that the prevalence of vitamin D deficiency is about 40–60%, and that of vitamin D insufficiency reaches 60–90%, although normal values have been reported by other authors (e.g., Santori et al., 2008). We recently found vitamin D insufficiency in 73 out of 128 alcoholics, and vitamin D deficiency in 36 of them (Figure 3). Several factors may contribute to vitamin D alterations, including dietary habits, latitude, sun exposure, malabsorption due to pancreatic insufficiency, or portal hypertension. In addition, chronic ethanol consumption may alter renal metabolism of 25 OH D3, inducing the synthesis of the inactive metabolite 24–25 (OH)2 D3 (Shankar et al., 2008).

The deficient vitamin D levels observed in alcoholics may not only impair the antioxidant defense in brain, leading to increased neurodegeneration. Vitamin D deficiency is also related to increased blood pressure both by renin-dependent and renin-independent mechanisms, leading to the development of lacunar infarcts and vascular dementia. In addition, vitamin D deficiency is associated with insulin resistance.

HYDROSOLUBLE VITAMINS

Vitamin B12, Folic Acid, and Homocysteine

Cobalamin deficiency leads to several neurological syndromes, including myelopathy with ataxia, neuropsychiatric syndromes, neuropathy, optic neuritis, and cognitive deficits. Although in several studies vitamin B12 levels are related to dementia and judgment impairment, and the decrease of B12 levels over time is also related to incident dementia, it is also clear that cognitive changes already occur, even when B12 levels are in the lowest quartile of normality (Hin et al., 2006). Equally important, it has been also disentangled that serum B12 levels may be not a good marker to assess the status of the vitamin in tissues; plasma holotranscobalamin, methylmalonic acid, or homocysteine are better indicators, and variations of the serum levels of these parameters are associated with significant changes in cognitive performance. In parallel with the functional changes, a decrease in B12 is also related to the intensity of brain atrophy.

These alterations are due in part to the defective myelin synthesis associated with B12 deficiency. Indeed, it was shown that B12 in the form of adenosylcobalamine is an essential cofactor of myelin synthesis (Thakkar & Billa, 2014). However, demyelination associated with B12 deficiency is also heavily influenced by increased cytokine levels, such as nerve growth factor (NGF), TNF-α, and the soluble complex CD40:CD40 ligand, in the face of decreased neurotrophic molecules, such as epidermal growth factor (EGF) and IL-6 (Scalabrino, Veber, & Mutti, 2008). These findings establish a link between B12 deficiency and neuroinflammation.

S-adenosylmethionine is a key product in central nervous system metabolism because it acts as a methyl donor. Its levels are dramatically dependent both on B12 and folate, because these substances are required in the methylation of homocysteine to methionine, with the latter being the precursor of S-adenosylmethionine. Therefore, it is not surprising that both folate and vitamin B12 deficiency may cause similar brain alterations, including dementia, and a demyelinating myelopathy. Cobalamin and/or folate deficiency lead to increased homocysteine levels. Homocysteine can be degraded to cystathionine and cysteine by vitamin B6 (Figure 4).

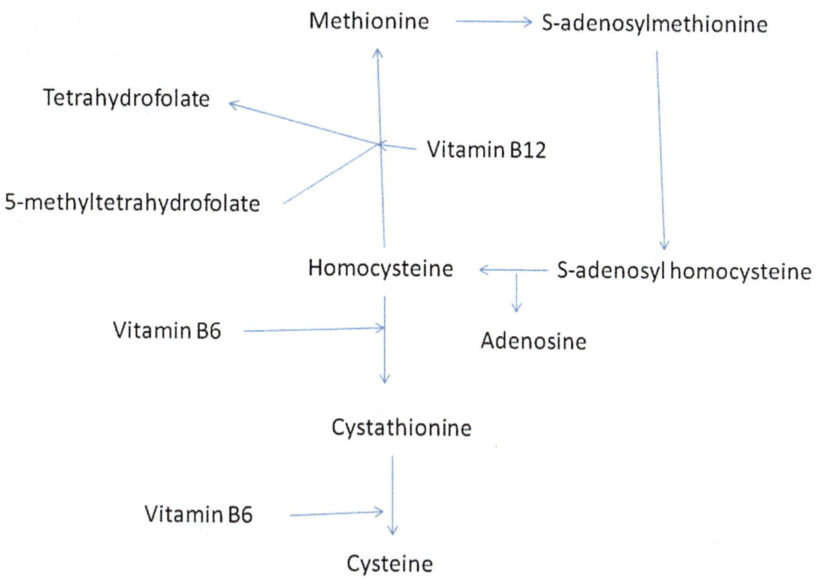

FIGURE 4 Role of B12, B6, and folate in homocysteine metabolism. B12 or folate deficiency blocks the transformation of homocysteine into methionine, whereas B6 deficiency blocks the catabolism of homocysteine into cystathionine.

Hyperhomocysteinemia is a classic factor associated with vascular risk and cognitive deficiency, with the hippocampus being especially sensitive to hyperhomocysteinemia. Possibly, oxidative damage is the underlying mechanism of the deleterious effect of hyperhomocysteinemia. Homocysteine downregulates glutathione peroxidase, therefore impairing the antioxidant machinery. There are several reports showing increased levels of homocysteine in alcoholics. In a study on 52 individuals who showed high levels of homocysteine, low levels of folate and B6 but normal B12 levels were observed. Hyperhomocysteinemia was related to hippocampal atrophy (Bleich et al., 2003). Heese et al. (2012), in another study on 168 alcoholics, found high homocysteine levels (median = 15.3 µmol/l), which decreased after abstinence (median 10.7 µmol/l at the 11th day), keeping a relation with riboflavin and folate, but not with cobalamin levels.

Results of these studies are in accordance with observations performed on otherwise healthy elderly subjects and in cases of dementia. In a study on 1092 subjects aged 76 years without dementia at inclusion, 111 developed dementia (83 of them due to Alzheimer disease) over an 8-year study period. The adjusted relative risk of dementia was 1.4 (95% confidence interval, 1.1–1.9) for each increase of one standard deviation of the log-transformed homocysteine levels. Plasma homocysteine levels higher than 14 µmol/l were associated with a double risk for developing Alzheimer disease (Seshadri et al., 2002). In another prospective study of 107 individuals aged 61–87 years without cognitive impairment at enrollment, vitamin B12 in the lowest tertile (<308 pmol/l) was associated with increased rate of brain volume loss along a 5-year study period (Vogiatzoglou et al., 2008) and more rapid cognitive decline. However, precise pathogenetic mechanisms, as well as the role of vitamin B12 in brain atrophy of alcoholics, are largely unknown. Moreover, although folate deficiency is a common finding in alcoholics, there is a trend to raised serum B12 levels, especially in cirrhotics. Preliminary data on 114 alcoholic patients showed that B12 levels were significantly higher among Child C patients compared with Child A and Child B ones (KW = 27.9; p < 0.001; Figure 5). However, neither folic acid, homocysteine, nor B12 were related with the intensity of frontal atrophy.

Vitamin C Deficiency

Vitamin C acts as a scavenger of ROS: ascorbic acid oxidizes to monodehydroascorbic acid, which is later deoxidized by the glutathione reductase activity, linking the beneficial effect of vitamin C to selenium stores. Several studies have reported decreased vitamin C in patients with dementia, supporting the importance of oxidative stress on brain damage. However, in the review performed by Crichton, Bryan, and Murphy (2013), there is no conclusive evidence that dietary antioxidants protect against the development of dementia. One year of treatment with vitamin C and E did not modify the course of Alzheimer disease, although it was accompanied by an antioxidant effect, as assessed by cerebrospinal fluid analysis (Aarlt et al., 2012). In contrast, several studies point out a beneficial effect for ethanol-induced hippocampal neurodegeneration, both in the developing rat fetus and in adult animals. Among alcoholics, vitamin C deprivation has been described, especially among smokers (Guequen et al., 2003).

Vitamin B3 Deficiency

Deficiency of some vitamins may provoke more or less acute clinical neurologic syndromes and response to specific therapy, but it complicates the clinical course of the alcoholics. Niacin deficiency is an example. Both reduced intake and impaired absorption may lead to vitamin B3 deficiency, which, in severe cases, may present with the full-blown picture of diarrhea, dermatitis, and dementia; the latter is more a delirium/confusion state than a true dementia. Given the lack of stores in the body, this clinical picture may ensue as soon as 60 days of dietary deprivation.

Niacin plays a role in the redox reaction involving reduced nicotinamide adenine dinucleotide and nicotinamide adenine dinucleotide phosphate (NADPH). The possibility exists that niacin is derived from tryptophan metabolism, but the conversion from tryptophan to niacin requires riboflavin, thiamine, and pyridoxine, probably equally deficient in alcoholics deprived

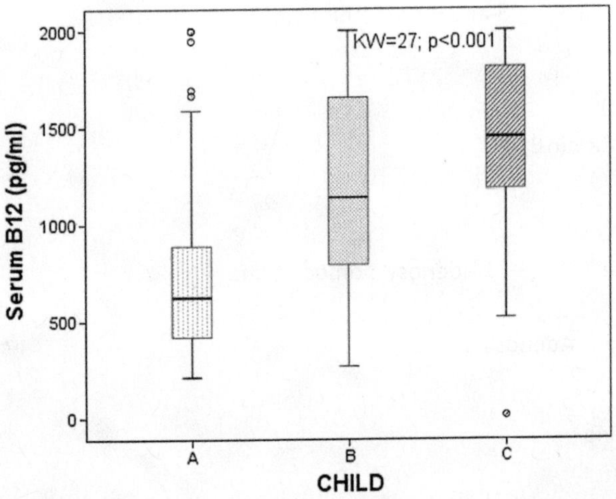

FIGURE 5 Serum vitamin B12 levels according to Child's classification. Vitamin B12 increases among Child C patients. Circles represent outliers.

from niacin. Therefore, malnourished alcoholics may develop the clinical-pathological features of niacin deficiency.

In a study of 20 necropsies, widespread central neuronal chromatolysis affecting cranial nerve and pontine nuclei and Betz neurons (Ishii & Nishihara, 1981) was observed in patients with niacin deficiency who had presented with confusion, hallucination, tremor, and extrapyramidal rigidity. Extrapyramidal signs, oppositional hypertonus, confusion, hallucinations, and insomnia are cardinal, although nonspecific, features of the so-called alcoholic pellagra encephalopathy. Niacin deficiency may provoke acute delirium in the alcoholic patient, mimicking withdrawal syndrome (Oldham & Ivkovic, 2012). Often, associated vitamin deficiencies usually complicate the clinical picture of pellagra (López, Olivares, & Berrios, 2014).

Thiamine Deficiency and Wernicke Encephalopathy

In many alcoholics, multiple vitamin deficiencies coexist, obscuring differential diagnosis of the neurologic disturbance that prompted admission to a hospital. Previously mentioned cases (López et al., 2014; Oldham & Ivkovic, 2012) and others (e.g., Ishii and Nishihara, 1981) constitute paradigmatic examples of a chronic alcoholic patient, like many attended in our unit, in whom severe impairment of protein–calorie nutritional status, which confers a poor prognosis, is associated with deficits of several micronutrients and vitamins. The Wernicke–Korsakoff syndrome is due to thiamine deficiency, which may occur in the alcoholic patient from one or more of the following mechanisms: inadequate intake, impaired absorption, a reduced liver storage, and decreased transformation of thiamine in its active form. It consists mainly of a constellation of focal neurological symptoms, including ophthalmoplegia, stupor or coma, and cerebellar dysfunction. These symptoms typically improve after appropriate treatment with thiamine; however, in many patients (nearly 80% in some series) the Korsakoff psychosis ensues, characterized by a more or less permanent cognitive impairment, confabulation, anterograde amnesia, visuospatial alteration, reduced affect, and impaired problem-solving capacity.

Lesions are more prominent in mammillary bodies and the dorsomedial and anterior nuclei of the thalamus. Pathogenesis includes impaired oxidative metabolism, alteration of mitochondrial function, and reduction of thiamine-dependent enzyme activity, such as transketolase, α-ketoglutarate dehydrogenase, and pyruvate dehydrogenase—all leading to neuronal death. Studies searching for a genetic predisposition for Wernicke–Korsakoff encephalopathy have yielded inconclusive results (Guerrini, Thomson, & Gurling, 2009). Likely, thiamine and ethanol exert synergistic effects, at least regarding white matter shrinkage: brain atrophy and atrophy of corpus callosum are observed in relation with Wernicke's encephalopathy in cases related to ethanol misuse. Cerebellar shrinkage is another feature observed both in thiamine deficiency conditions and ethanol.

CONCLUSIONS

In conclusion, ethanol exerts several effects on brain function; the most commonly observed are brain atrophy, cerebellar atrophy, ventricular enlargement, and white matter shrinkage. Neurogenesis is severely impaired, and neuronal death is increased. Underlying mechanisms include cytokine-derived inflammatory response. These cytokines are either locally produced or have a systemic origin, but in either case they generate inflammation accompanied by increased ROS production. The increased lipid peroxidation is aggravated by the fact that antioxidant defense is impaired, particularly those mechanisms that depend on some antioxidant vitamins, such as vitamin A, vitamin E, vitamin D, and vitamin C, among others, whose levels are usually depressed in alcoholics.

Despite this, and despite the abundance of experimental studies apparently confirming this hypothesis, results in human beings are more controversial. In general, trials adding vitamin supplements have yielded, in the best of cases, only modest results, with the exceptions of the well-known therapeutic effect of thiamine on Wernicke encephalopathy and niacin on pellagra-associated dementia. However, ethanol withdrawal leads to a marked improvement of brain performance and reversal of atrophy, thus remaining as the only effective therapy for this situation.

APPLICATIONS TO OTHER ADDICTIONS AND SUBSTANCE MISUSE

In addition to ethanol, other drugs damage the brain, although the underlying pathogenic mechanisms differ. Many studies support the role of ethanol in neuroinflammation, oxidative stress, and ultimately neuronal death. These features are also observed in polydrug abusers. Findings include neurodegenerative alterations, neuronal loss, axonal damage, and microglial activation. Oxidative stress may underlie the toxic effects of opiates, heroin, or morphine. Consumption of these drugs leads to depletion of antioxidant mechanisms, including cellular antioxidant systems and antioxidant vitamins; however, antioxidant therapy, although efficacious in some experimental models, fails to exert protection in humans. Cocaine, amphetamine, and synthetic derivatives severely impair brain blood perfusion. Stroke is a relatively common consequence. Opiates may also cause hypoxic brain damage due to respiratory depression. Studies have also shown depletion of the total antioxidant capacity among cocaine and/or methamphetamine consumers. Cocaine consumption is also associated with extensive white matter damage. Interestingly, chronic cocaine use alters the response to avoid punishment, enhances perseverative responding, and possibly, by increasing ROS production, contributes to the reinforcing effect of the drug; at the same time, oxidative stress impairs learning and memory. Consumption of marijuana is associated with smaller hippocampi and altered frontal white matter, and a greater impulsivity. Therefore, most drugs damage the brain by mechanisms related to neuroinflammation and excessive ROS production. Although there are theoretical bases for antioxidant vitamins supplementation and some experimental data support their efficacy, clinical studies have yielded poor results.

DEFINITION OF TERMS

Hippocampus This gyrated region of the temporal lobe is mainly composed of two interrelated regions of the archicortex called dentate gyrus and *Cornu Ammonis*. It is heavily involved in learning, behavior, and memory acquisition.

Archicortex This part of brain cortical tissue is organized in three layers (instead of six, as most of the remaining cortex). In the human brain, it includes the hippocampus and olfactory cortex.

Microglia The phagocytes of the central nervous system share many properties with other cells of the reticuloendothelial system, including cytokine production and oxygen free radical generation.

Oxidative stress This alteration of the structure of diverse molecules occurs by reaction with highly reactive oxygen species.

Antioxidants These enzymatic pathways or cofactors transform reactive oxidant species into less reactive ones.

Excitotoxicity The mechanism of neuronal cell death is derived from intense and prolonged activation of excitatory neurotransmitter receptors. This activation alters intracellular enzymatic pathways, ultimately leading to cell death. This mechanism is especially important during ethanol withdrawal.

Toll-like receptors These cell surface receptors are usually activated by-products derived from microorganisms, such as membrane structures from gram-negative bacteria. In the brain, toll-like receptors are present in microglia cells.

Apoptosis In this process, cells suffer a series of modifications affecting membranes, cytoplasma, and the nucleus, which ultimately lead to cell death, typically without accompanying inflammatory reaction.

Cytokines These small protein molecules are of importance in the immune response. They usually act on transcription factors that activate genes. Their functions are varied, mainly inducing the synthesis of inflammatory mediators and activating several kinds of cells, especially endothelial cells, immune cells, and phagocytes, among others.

Transcription factors These intracellular proteins bind to DNA sequences in the nucleus and modify the transcription of the DNA genome into messenger RNA, therefore orchestrating the synthesis of certain proteins.

KEY FACTS

Key Facts of Ethanol-Mediated Brain Damage among Adolescents

- Binge drinking is defined by the ingestion of at least five drinks (for men) or four (for women) during the same drinking episode.
- Binge drinking is more common during adolescence.
- This episodic pattern of heavy drinking in a short time is associated with episodic increases in glutamate release, which can cause excitotoxicity and neuronal death.
- During youth, the hippocampus is more sensitive to the noxious effects of alcohol-derived neuroinflammation and oxidative damage. This explains why this pattern of consumption is more damaging to the brain than regular excessive consumption.
- Possibly, hippocampal damage during binge drinking may predispose an individual to alcohol addiction.

Key Facts of Gray Matter

- Gray matter is composed of neuronal bodies.
- Significant neuronogenesis takes place during infancy, adolescence, and youth.
- Ethanol inhibits neuronogenesis.
- Ethanol increases neuronal death.
- The deficiency of some vitamins, such as thiamine, also leads to neuronal death and the effects are potentiated by ethanol.

SUMMARY POINTS

- Brain atrophy, especially affecting the frontal lobes and hippocampi, and cerebellar atrophy are major manifestations of heavy alcoholism.
- Both gray matter and white matter are affected.
- The secretion of proinflammatory cytokines and oxidative damage constitute the main pathogenic mechanisms involved. Therefore, there is a theoretical basis for the therapeutic use of antioxidants.
- Decreased levels of vitamin E, vitamin A, vitamin D, and vitamin C have been reported in patients or experimental models of ethanol-mediated brain atrophy.
- Experimental data are promising regarding the therapeutic effect of vitamins supplementation, but clinical trials show no definite benefits. The best therapy is alcohol abstention.
- Clinical syndromes due to specific vitamins deficiency may coexist in the same patient, and with a superimposed ethanol withdrawal syndrome, obscuring the diagnosis and making difficult the implementation of adequate therapy.

REFERENCES

Agartz, I., Momenan, R., Rawlings, R. R., Kerich, M. J., & Hommer, D. W. (1999). Hippocampal volume in patients with alcohol dependence. *Archives of General Psychiatry, 56*, 356–363.

Alfonso-Loeches, S., Pascual-Lucas, M., Blanco, A. M., Sanchez-Vera, I., & Guerri, C. (2010). Pivotal role of TLR4 receptors in alcohol-induced neuroinflammation and brain damage. *The Journal of Neuroscience, 30*, 8285–8295.

Andersen, B. B. (2004). Reduction of Purkinje cell volume in cerebellum of alcoholics. *Brain Research, 1007*, 10–18.

Arlt, S., Müller-Thomsen, T., Beisiegel, U., & Kontush, A. (2012). Effect of one-year vitamin C- and E-supplementation on cerebrospinal fluid oxidation parameters and clinical course in Alzheimer's disease. *Neurochemical Research, 37*, 2706–2714.

Bartsch, A. J., Homola, G., Biller, A., Smith, S. M., Weijers, H. G., Wiesbeck, G. A., ... Bendszus, M. (2007). Manifestations of early brain recovery associated with abstinence from alcoholism. *Brain, 130* (Pt 1), 36–47.

Behr, G. A., Schnorr, C. E., Simões-Pires, A., da Motta, L. L., Frey, B. N., & Moreira, J. C. (2012). Increased cerebral oxidative damage and decreased antioxidant defenses in ovariectomized and sham-operated rats supplemented with vitamin A. *Cell Biology and Toxicology, 28,* 317–330.

Benes, F. M., Turtle, M., Khan, Y., & Farol, P. (1994). Myelination of a key relay zone in the hippocampal formation occurs in the human brain during childhood, adolescence, and adulthood. *Archives of General Psychiatry, 51,* 477–484.

Bleich, S., Bandelow, B., Javaheripour, K., Müller, A., Degner, D., Wilhelm, J., ... Kornhuber, J. (2003). Hyperhomocysteinemia as a new risk factor for brain shrinkage in patients with alcoholism. *Neuroscience Letters, 335,* 179–182.

Bondy, S. C., Guo, S. X., & Adams, J. D. (1996). Prevention of ethanol-induced changes in reactive oxygen parameters by alpha-tocopherol. *Alcohol and Alcoholism, 31,* 403–410.

Bonnet, E., Touyarot, K., Alfos, S., Pallet, V., Higueret, P., & Abrous, D. N. (2008). Retinoic acid restores adult hippocampal neurogenesis and reverses spatial memory deficit in vitamin A deprived rats. *PLoS One, 3,* e3487.

Briones, T. L., & Darwish, H. (2014). Decrease in age-related tau hyperphosphorylation and cognitive improvement following vitamin D supplementation are associated with modulation of brain energy metabolism and redox state. *Neuroscience, 262,* 143–155.

Cardenas, V. A., Durazzo, T. C., Gazdzinski, S., Mon, A., Studholme, C., & Meyerhoff, D. J. (2011). Brain morphology at entry into treatment for alcohol dependence is related to relapse propensity. *Biological Psychiatry, 70,* 561–567.

Cardenas, V. A., Greenstein, D., Fouche, J. P., Ferrett, H., Cuzen, N., Stein, D. J., & Fein, G. (2013). Not lesser but greater fractional anisotropy in adolescents with alcohol use disorders. *NeuroImage. Clinical, 2,* 804–809.

Chei, C. L., Raman, P., Yin, Z. X., Shi, X. M., Zeng, Y., & Matchar, D. B. (October 3, 2014). Vitamin D levels and cognition in elderly adults in China. *Journal of the American Geriatrics Society, 62*(11), 2125–2129.

Clugston, R. D., & Blaner, W. S. (2012). The adverse effects of alcohol on vitamin A metabolism. *Nutrients, 4,* 356–371.

Creeley, C. E., Dikranian, K. T., Johnson, S. A., Farber, N. B., & Olney, J. W. (2013). Alcohol-induced apoptosis of oligodendrocytes in the fetal macaque brain. *Acta Neuropathologica Communications, 1,* 23.

Crews, F. T., & Nixon, K. (2009). Mechanisms of neurodegeneration and regeneration in Alcoholism. *Alcohol and Alcoholism, 44,* 115–127.

Crichton, G. E., Bryan, J., & Murphy, K. J. (2013). Dietary antioxidants, cognitive function and dementia–a systematic review. *Plant Foods for Human Nutrition, 68,* 279–292.

De Bellis, M. D., Clark, D. B., Beers, S. R., Soloff, P. H., Boring, A. M., Hall, J., ... Keshavan, M. S. (2000). Hippocampal volume in adolescent-onset alcohol use disorders. *American Journal of Psychiatry, 157,* 737–744.

Duka, T., Townshend, J. M., Collier, K., & Stephens, D. N. (2003). Impairment in cognitive functions after multiple detoxifications in alcoholic inpatients. *Alcoholism Clinical and Experimental Research, 27,* 1563–1572.

Etchamendy, N., Enderlin, V., Marighetto, A., Pallet, V., Higueret, P., & Jaffard, R. (2003). Vitamin A deficiency and relational memory deficit in adult mice: relationships with changes in brain retinoid signalling. *Behavioural Brain Research, 145,* 37–49.

Etgen, T., Sander, D., Bickel, H., Sander, K., & Förstl, H. (2012). Vitamin D deficiency, cognitive impairment and dementia: a systematic review and meta-analysis. *Dementia and Geriatric Cognitive Disorders, 33,* 297–305.

Ferland, G. (2012). Vitamin K, an emerging nutrient in brain function. *Biofactors, 38,* 151–157.

Fitzpatrick, L. E., Jackson, M., & Crowe, S. F. (2008). The relationship between alcoholic cerebellar degeneration and cognitive and emotional functioning. *Neuroscience and Biobehavioral Reviews, 32,* 466–485.

Fukui, K., Kawakami, H., Honjo, T., Ogasawara, R., Takatsu, H., Shinkai, T., ... Urano, S. (2012). Vitamin E deficiency induces axonal degeneration in mouse hippocampal neurons. *Journal of Nutritional Science and Vitaminology, 58,* 377–383.

Gansler, D. A., Harris, G. J., Oscar-Berman, M., Streeter, C., Lewis, R. F., Ahmed, I., & Achong, D. (2000). Hypoperfusion of inferior frontal brain regions in abstinent alcoholics: a pilot SPECT study. *Journal of Studies on Alcohol, 61,* 32–37.

Gueguen, S., Pirollet, P., Leroy, P., Guilland, J. C., Arnaud, J., Paille, F., ... Herbeth, B. (2003). Changes in serum retinol, alpha-tocopherol, vitamin C, carotenoids, zinc and selenium after micronutrient supplementation during alcohol rehabilitation. *Journal of the American College of Nutrition, 22,* 303–310.

Guerrini, I., Thomson, A. D., & Gurling, H. M. (2009). Molecular genetics of alcohol-related brain damage. *Alcohol and Alcoholism, 44,* 166–170.

Halsted, C. H. (2004). Nutrition and alcoholic liver disease. *Seminars in Liver Disease, 24,* 289–304.

Harper, C. (2009). The neuropathology of alcohol-related brain damage. *Alcohol and Alcoholism, 44,* 136–140.

Harper, C., Fornes, P., Duyckaerts, C., Lecomte, D., & Hauw, J. J. (1995). An international perspective on the prevalence of the Wernicke-Korsakoff syndrome. *Metabolic Brain Disease, 10,* 17–24.

Heese, P., Linnebank, M., Semmler, A., Muschler, M. A., Heberlein, A., Frieling, H., ... Hillemacher, T. (2012). Alterations of homocysteine serum levels during alcohol withdrawal are influenced by folate and riboflavin: results from the German Investigation on Neurobiology in Alcoholism (GINA). *Alcohol and Alcoholism, 47,* 497–500.

He, J., Nixon, K., Shetty, A. K., & Crews, F. T. (2005). Chronic alcohol exposure reduces hippocampal neurogenesis and dendritic growth of newborn neurons. *European Journal of Neuroscience, 21,* 2711–2720.

Hin, H., Clarke, R., Sherliker, P., Atoyebi, W., Emmens, K., Birks, J., ... Evans, J. G. (2006). Clinical relevance of low serum vitamin B12 concentrations in older people: the Banbury B12 study. *Age and Ageing, 35,* 416–422.

Ishii, N., & Nishihara, Y. (1981). Pellagra among chronic alcoholics: clinical and pathologic study of 20 necropsy cases. *Journal of Neurology, Neurosurgery, and Psychiatry, 44,* 209–215.

Johnson, E. J., Vishwanathan, R., Johnson, M. A., Hausman, D. B., Davey, A., Scott, T. M., ... Poon, L. W. (2013). Relationship between serum and brain carotenoids, α-tocopherol, and retinol concentrations and cognitive performance in the oldest old from the Georgia Centenarian Study. *Journal of Aging Research, 2013,* 951786.

Knoth, R., Singec, I., Ditter, M., Pantazis, G., Capetian, P., Meyer, R. P., ... Kempermann, G. (2010). Murine features of neurogenesis in the human hippocampus across the lifespan from 0 to 100 years. *PLoS One, 5*(1), e8809.

Kumada, T., Jiang, Y., Cameron, D. B., & Komuro, H. (2007). How does alcohol impair neuronal migration? *Journal of Neuroscience Research, 85,* 465–470.

Kumar, A., LaVoie, H. A., DiPette, D. J., & Singh, U. S. (2013). Ethanol neurotoxicity in the developing cerebellum: underlying mechanisms and implications. *Brain Sciences, 3*, 941–963.

Lambert de Rouvroit, C., & Goffinet, A. M. (1998). The reeler mouse as a model of brain development. *Advances in Anatomy Embryology and Cell Biology, 150*, 1–106.

Lim, D. A., & Alvarez-Buylla, A. (September 12, 2014). Adult neural stem cells stake their ground. *Trends in Neurosciences, 37*(10), 563–571 pii:S0166-2236(14) 00149–0.

López, M., Olivares, J. M., & Berrios, G. E. (2014). Pellagra encephalopathy in the context of alcoholism: review and case report. *Alcohol and Alcoholism, 49*, 38–41.

Mehlig, K., Skoog, I., Guo, X., Schütze, M., Gustafson, D., Waern, M., … Lissner, L. (2008). Alcoholic beverages and incidence of dementia: 34-year follow-up of the prospective population study of women in Goteborg. *American Journal of Epidemiology, 167*, 684–691.

Meyer, G., Perez-Garcia, C. G., Abraham, H., & Caput, D. (2002). Expression of p73 and reelin in the developing human cortex. *The Journal of Neuroscience, 22*, 4973–4986.

Mooney, S. M., Siegenthaler, J. A., & Miller, M. W. (2004). Ethanol induces heterotopias in organotypic cultures of rat cerebral cortex. *Cerebral Cortex, 14*, 1071–1080.

Nagel, B. J., Schweinsburg, A. D., Phan, V., & Tapert, S. F. (2005). Reduced hippocampal volume among adolescents with alcohol use disorders without psychiatric comorbidity. *Psychiatry Research, 139*, 181–190.

Nishida, Y., Yokota, T., Takahashi, T., Uchihara, T., Jishage, K., & Mizusawa, H. (2006). Deletion of vitamin E enhances phenotype of Alzheimer disease model mouse. *Biochemical and Biophysical Research Communications, 350*, 530–536.

Oldham, M. A., & Ivkovic, A. (2012). Pellagrous encephalopathy presenting as alcohol withdrawal delirium: a case series and literature review. *Addiction Science and Clinical Practice, 7*, 12.

Pascual, M., Pla, A., Miñarro, J., & Guerri, C. (2014). Neuroimmune activation and myelin changes in adolescent rats exposed to high-dose alcohol and associated cognitive dysfunction: a review with reference to human adolescent drinking. *Alcohol and Alcoholism, 49*, 187–192.

Patel, A. K., Rogers, J. T., & Huang, X. (2008). Flavanols, mild cognitive impairment, and Alzheimer's dementia. *International Journal of Clinical and Experimental Medicine, 1*, 181–191.

Qin, L., & Crews, F. T. (2012). Chronic ethanol increases systemic TLR3 agonist-induced neuroinflammation and neurodegeneration. *Journal of Neuroinflammation, 9*, 130.

Reilly, M. E., Patel, V. B., Peters, T. J., & Preedy, V. R. (2000). In vivo rates of skeletal muscle protein synthesis in rats are decreased by acute ethanol treatment but are not ameliorated by supplemental alpha-tocopherol. *Journal of Nutrition, 130*, 3045–3049.

Rupp, C. I., Kurz, M., Kemmler, G., Mair, D., Hausmann, A., Hinterhuber, H., & Fleischhacker, W. W. (2003). Reduced olfactory sensitivity, discrimination, and identification in patients with alcohol dependence. *Alcoholism Clinical and Experimental Research, 27*, 432–439.

Santori, C., Ceccanti, M., Diacinti, D., Attilia, M. L., Toppo, L., D'Erasmo, E., … Minisola, S. (2008). Skeletal turnover, bone mineral density, and fractures in male chronic abusers of alcohol. *Journal of Endocrinological Investigation, 31*, 321–326.

Scalabrino, G., Veber, D., & Mutti, E. (2008). Experimental and clinical evidence of the role of cytokines and growth factors in the pathogenesis of acquired cobalamin-deficient leukoneuropathy. *Brain Research Reviews, 59*, 42–54.

Seshadri, S., Beiser, A., Selhub, J., Jacques, P. F., Rosenberg, I. H., D'Agostino, R. B., … Wolf, P. A. (2002). Plasma homocysteine as a risk factor for dementia and Alzheimer's disease. *The New England Journal of Medicine, 346*, 476–483.

Shahar, S., Lee, L. K., Rajab, N., Lim, C. L., Harun, N. A., Noh, M. F., … Jamal, R. (2013). Association between vitamin A, vitamin E and apolipoprotein E status with mild cognitive impairment among elderly people in low-cost residential areas. *Nutritional Neuroscience, 16*, 6–12.

Shankar, K., Liu, X., Singhal, R., Chen, J. R., Nagarajan, S., Badger, T. M., & Ronis, M. J. (2008). Chronic ethanol consumption leads to disruption of vitamin D homeostasis associated with induction of renal 1,25 dihydroxyvitamin D3 24 hydroxylase (CYP24A1). *Endocrinology, 149*, 1748–1756.

Shirpoor, A., Salami, S., Khadem-Ansari, M. H., Minassian, S., & Yegiazarian, M. (2009). Protective effect of vitamin E against ethanol-induced hyperhomocysteinemia, DNA damage, and atrophy in the developing male rat brain. *Alcoholism Clinical and Experimental Research, 33*, 1181–1186.

Sutherland, G. T., Sheahan, P. J., Matthews, J., Dennis, C. V., Sheedy, D. S., McCrossin, T., … Kril, J. J. (2013). The effects of chronic alcoholism on cell proliferation in the human brain. *Experimental Neurology, 247*, 9–18.

Suzuki, M., Hagino, H., Nohara, S., Zhou, S. Y., Kawasaki, Y., Takahashi, T., … Kurachi, M. (2005). Male-specific volume expansion of the human hippocampus during adolescence. *Cerebral Cortex, 15*, 187–193.

Thakkar, K., & Billa, G. (August 13, 2014). Treatment of vitamin B12 deficiency-Methylcobalamine? Cyancobalamine? Hydroxocobalamin?-clearing the confusion. *European Journal of Clinical Nutrition, 69*(1), 1–2.

Torvik, A., & Torp, S. (1986). The prevalence of alcoholic cerebellar atrophy. A morphometric and histological study of an autopsy material. *Journal of Neurological Sciences, 75*, 43–51.

Vogiatzoglou, A., Refsum, H., Johnston, C., Smith, S. M., Bradley, K. M., de Jager, C., … Smith, A. D. (2008). Vitamin B12 status and rate of brain volume loss in community-dwelling elderly. *Neurology, 71*, 826–832.

Chapter 45

Phospholipase D: A Central Switch in the Development of Fetal Alcohol Spectrum Disorder

Ute Burkhardt, Jochen Klein
Department of Pharmacology, Goethe University Frankfurt, Frankfurt, Germany

Abbreviations

APP β-Amyloid precursor protein
ARF ADP-Ribosylation factor
DAG Diacylglycerol
EGFR Epidermal growth factor receptor
ER Endoplasmic reticulum
FAE Fetal alcohol embryopathy
FAS Fetal alcohol syndrome
FASD Fetal alcohol spectrum disorder
FCS Fetal calf serum
GABA γ-Aminobutyric acid
GnRH Gonadotropin-releasing hormone
HKD motif H, histidine; K, lysine; D, aspartic acid
IGF-1 Insulin-like growth factor one
JAK3 Janus kinase 3
LPA Lyso-phosphatidic acid
mTOR Mammalian target of rapamycin
NAA/Cho N-acetylaspartate/choline ratio
NAA/Cr N-acetylaspartate/creatine ratio
NMDA N-methyl-D-aspartate
NPC Neural progenitor cells
NSC Neuronal stem/precursor cell
PA Phosphatidic acid
PAE Prenatal alcohol exposure
PAT Passive avoidance test
PC Phosphatidylcholine
PDB Phorbol 12,13-dibutyrate
PDGF Platelet-derived growth factor
PE Phosphatidylethanolamine
PEth Phosphatidylethanol
PH Pleckstrin homology
PIP$_2$ Phosphatidylinositol-4,5-bis-phosphate
PKC Protein kinase C
PLD Phospholipase D
PMA Phorbol-12-myristate-13-acetate
PSer Phosphatidylserine
PX Phox homology
RTK Receptor tyrosine kinase
SGZ Subgranular zone
SIDS Sudden infant death syndrome
SLO Streptolysin O
SM Sphingomyelin
SMase Sphingomyelinase
SVZ Subventricular zone
TM Transmembrane domain

Phospholipases D (PLDs) are a group of enzymes that were identified for the first time in plants (Hanahan & Chaikoff, 1947). It took more than 20 years until Saito and Kanfer (1973) described PLD in mammalian tissue, specifically in the rat brain. Since then, multiple PLD isoforms were found and characterized. PLDs and their substrates have been identified as key players in signal transduction, endo- and exocytosis, vesicular trafficking and reorganization, development, and cell death. These PLD functions, together with the high activity in mammalian brain and the unique transphosphatidylation reaction with primary alcohols that will be described later, make PLDs an interesting object of study for the pathology of fetal alcohol spectrum disorder (FASD; Klein, 2005).

FETAL ALCOHOL SPECTRUM

The fetal alcohol spectrum (FAS) occurs when pregnant women drink alcohol in large quantities. In an international prospective study in the Western part of the world, Abel and Sokol (1991) reported that 1 out of 100 children showed some symptoms of FAS, and 1 out of 1000 showed signs of fetal alcohol embryopathy (FAE). The symptoms of FAE include delayed development and cranial deformations in the children of binge-drinking women. Fetal alcohol syndrome is diagnosed when both chronic abuse of alcohol by the mother is known and three symptoms occur in children, such as craniofacial abnormalities, prenatal and postnatal growth retardation, and cognitive or behavioral dysfunctions.

There is a range of symptoms of varying extent, because the severity of FAS depends on the amount of ethanol consumption

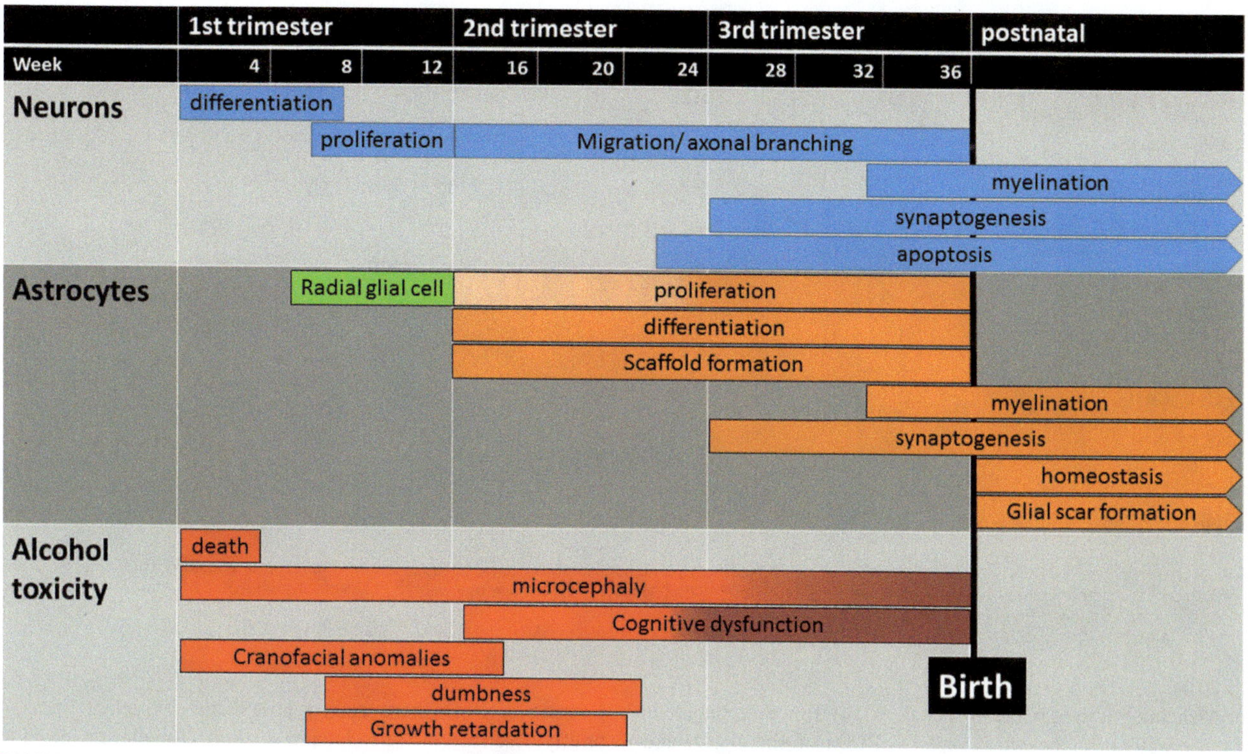

FIGURE 1 **Timeline of embryogenesis and ethanol toxicity.** Neurons differentiate and proliferate during the first trimester. Later on, they migrate to the final position in the brain and neurites grow out to form synapses. This process is accompanied by massive apoptosis of unused neurons and myelination of the axons by glial cells. Between weeks 9 and 12, the neuron-glia switch turns neuronal proliferation to glial proliferation from neuronal precursor cells (NPCs) and radial glial cells. The maximum of glial proliferation is at the end of the third trimester. Through the second and third trimester, glial cells differentiate and form the brain scaffold. Glial cells are involved in myelination, but they are also very important for the formation of synapses, regulation of the homeostasis, and repair processes throughout life. The effect of ethanol intoxication during embryogenesis is time dependent. In the early stages of pregnancy, ethanol toxicity leads to death of the embryo. The craniofacial characteristics only occur with alcohol toxicity during gastrulation (first trimester), while dumbness can occur because of toxicity during the development of the ears (week 8–20), in parallel to general growth retardation. The most vulnerable time is the brain growth spurt during the third trimester (i.e., during synaptogenesis and gliogenesis). Ethanol intoxication during this time leads to microcephaly and lifelong learning and memory deficits. *Modified from Andersen (2003).*

(Clarren, Bowden, & Astley, 1987) and the timing of ethanol consumption, which is a critical factor for the outcome of FAS as well. The craniofacial characteristics only occur when high peak alcohol concentrations are present during the embryonic stage of gastrulation—that is, in the early stage of pregnancy (Sulik, 2005). The second critical period is between the 7th and 20th weeks in humans (G12 to G20 in rats) during neuroepithelial cell proliferation and migration; that is the time of differentiation of almost all cells, and disturbances during this period lead to growth reduction (Suzuki, 2007). The last and most vulnerable time is the brain growth spurt, the time of glial development, myelination, and synaptogenesis. In humans, the brain growth spurt is during the third trimester (during fetogenesis), but in rats it occurs postnatally on days P1 to P10. Alcohol abuse in this period leads to microcephaly and neuronal loss, especially in the hippocampus (Guerri, 2002). This results in learning and memory deficits and long-term neurobehavioral dysfunctions (Figure 1).

PHOSPHOLIPASES D

In general, PLDs catalyze the head group exchange on phosphodiester bonds of glycerolipids (Jenkins & Frohman, 2005). They are membrane-bound proteins and belong to the group of HKD enzymes, which all share a conserved HKD motif (H, histidine; K, lysine; D, aspartic acid) in the active site. Most PLDs have two HKD motifs, which form the catalytic center (Selvy, Lavieri, Lindsley, & Brown, 2011). Unfortunately, no three-dimensional (3D) PLD structures are available for study due to difficulties in expression and purification of recombinant eukaryotic PLDs, but a 3D structural model has been postulated (Mahankali, Alter, & Gomez-Cambronero, 2014). Enzymes with PLD activity were identified in viruses, prokaryotic organisms, and eukaryotic organisms.

PLD Isoforms

In the 1990s, two mammalian phosphatidylcholine (PC)-specific PLD isoforms were cloned: PLD1 and PLD2. Both share 50% sequence homology and differ mostly in the length of the sequence between the HKD motifs. PLD1 has an extended loop and therefore is bigger than PLD2. PLD1 and PLD2 have N-terminal, highly conserved phox homology (PX) and pleckstrin homology (PH) domains, which are binding domains for their cofactor phosphatidylinositol-bis-4,5-phosphate (PIP_2). Another member of the PLD group was described recently. PLD3, a non-PC-specific PLD, was found to be involved in amyloid precursor protein (APP) processing, and mutations in PLD3 gene apparently increase the risk

TABLE 1 Overview of PLD Binding Partners and Their Functions

Binding Partner	PLD1	PLD2	Function
ARF1 and ARF6	+	(+)	Vesicular trafficking, mitogenic effects
RhoA	+	n.d.	Actin cytoskeleton organization and axonal signaling
Cdc42	+	n.d.	Differentiation
Rheb	+	n.d.	mTOR activation
mTOR/raptor		+	mTOR activation
PKC	+	+	GPCR signaling
EGFR	n.d.	+	Proliferation
JAK3	n.d.	+	Glial cell differentiation
Scr	n.d.	+	Proliferation
β-actin	−	−	Negative regulation of PLD

Binding partners of PLD1 and PLD2 are protein kinases (Cdc42, Rheb, PKCs, Scr), small GTPases (ARF1, 6, and RhoA), receptor tyrosine kinases (EGFR, JAK3), protein complexes (mTOR/raptor) or cytoskeletal proteins (β-actin). Most of them activate one or both PLD isoforms while β-actin is an inhibitory regulator. n.d., not determined (reviewed by Jang et al., 2012).

for Alzheimer disease (Cruchaga et al., 2014). However, the focus of this chapter is on the two PC-specific PLD isoforms (PLD1 and PLD2) that are critical for signal transduction during development.

PLD Location

PLD1 and PLD2 are ubiquitously expressed in all tissues, although the expression level differs. They are highly expressed in the heart, spleen, and brain. Both are especially enriched in the white matter of the brain. Although there are only slight changes in the expression level of PLD1 throughout development in rats (E15-P49), PLD2 is upregulated in gray matter during the early postnatal stage, in addition to an enhanced expression in white matter (Saito, Sakagami, & Kondo, 2000). This is an indication of the potential relevance of PLDs, and especially PLD2, for brain development.

Although both PLD isoforms share the same substrate and have similar tissue distribution, they differ in their subcellular location. PLD1 is mainly located in the perinuclear region, such as in Golgi apparatus and the endoplasmic reticulum, while PLD2 is located at the cellular membrane. However, both can translocate upon stimulation. PLD1 in particular translocates to late endosomes and the plasma membrane upon stimulation with fetal calf serum (FCS), and phorbol esters such as PMA stimulate translocation of PLD1 to the plasma membrane (Du, Huang, Liang, & Frohman, 2004). The translocation is used to recycle vesicles upon agonist stimulation and for desensitization of receptors. In some cases, both PLDs co-localize in the perinuclear region and plasma membrane and coordinately regulate cellular processes (reviewed by Jang, Lee, Hwang, & Ryu, 2012).

REGULATION OF PLDs

In addition to differences in the subcellular localization, PLD1 and PLD2 also differ in the way they are regulated. Many binding partners of PLDs have been reported and important ones are briefly described in the following sections (Table 1, reviewed by Jang et al., 2012).

Protein Kinases and Small GTPases

PLD2 has a high basal activity, and it is generally accepted that PLD2 is regulated by dissociation of inhibiting proteins. In addition, PLD2 can also be activated by receptor tyrosine kinases (RTK), which leads to proliferation and differentiation. PLD2 directly interacts with the intracellular part of the EGF receptor (EGFR), and EGFR stimulation leads to cell proliferation via PLD and protein kinase C (PKC) signaling (Sung et al., 2001). Other RTK-PLD2 signaling pathways involve JAK3 or Scr, which enhance cell proliferation.

In contrast, PLD1 activity is mainly stimulated by small GTPases, such as members of the ARF and Rho family. Activation of PLD1 by the small GTPases is associated with vesicle transport and neurite formation, but also with cell proliferation. ARF mainly stimulates PLD1 and, to a much lesser extent, PLD2. ARF1 interacts with PLD1 in the Golgi apparatus while ARF6 stimulates PLD1 located at the plasma membrane (Caumont, Galas, Vitale, Aunis, & Bader, 1998). While ARF1 is more important in proliferation processes, ARF6 regulates membrane trafficking via PLD. Members of the Rho family (RhoA, Rc1, and Cdc42) enhance PLD1 activity, which is important for mitogenic signals, phagosome formation, neurite outgrowth, and cytoskeletal reorganization. Direct protein–protein interactions are described for RhoA and PLD1. Furthermore, PLD1 was identified as the link between CdC42 and mTOR.

Protein Kinase C Isoforms

Protein kinase C isoforms are known to regulate both PLD isoforms upon GPCR activation. There is also a synergistic stimulation of PLD observed in combination with ARF6 or members of the Rho family. While conventional PKCs (such as PKCα) are mainly responsible for stimulation of PLD activity, such as by phorbol esters, atypical PKCs such as PKCζ are downstream targets of PLD2 (reviewed by Jang et al., 2012).

Cytoskeletal Proteins

Cytoskeletal proteins, such as actins and tubulins, generally inhibit PLD activity. The close interaction between PLD and cytoskeletal proteins is evidently important for cytoskeletal reorganization and vesicular transport.

ENZYMATIC REACTIONS OF PLDs

Hydrolysis

Under physiological conditions, PLDs use water as nucleophile to hydrolyze PC to phosphatidic acid (PA) and choline. The resulting PA accounts for only 1–2% of the total lipid amount of the cell. There are several pathways to generate PA: in particular, it is an important precursor for other phospholipids such as diacylglycerol (DAG) as well as for phosphatidylethanolamine (PE), phosphatidylserine (PSer), and triacylglycerols. Only a small part of the total PA pool (less than 10%) is used for signaling. Most of this pool of fast-processed PA is likely produced by PLDs.

Transphosphatidylation

Transphosphatidylation is a unique reaction of PLDs in the presence of primary alcohols. PLDs catalyze the head group exchange from PC to phosphatidylethanol (PEth) even at low concentrations of alcohols, because short-chain alcohols such as ethanol or butanol are much better nucleophiles than water (Figure 2). Choline is released and PEth is formed at the expense of PA. It is important to note that ethanol is not a "PLD inhibitor" as it is sometimes described in the literature, but it interrupts the pathways that PLDs are involved in by modifying the products of the catalytic reaction of PLDs, forming PEth instead of PA. This process has two consequences. First, PEth, a nonphysiological lipid, accumulates in the cell due to its slow breakdown. The half-life of PEth in the brain of ethanol-fed rats was estimated at approximately 8–10 h. The accumulation of PEth may lead to disturbance of membrane function; for example, a reduced binding of (^{3}H)-inositol 1,4,5-trisphosphate to rat cerebellar membranes was observed in the presence of PEth (Rodriguez, Lundqvist, Alling, & Gustavsson, 1996). However, there is no convincing evidence that PEth formation is involved in ethanol toxicity in vivo.

As a second and more relevant consequence, the reduced formation of PA in the presence of ethanol disrupts the PLD-PA signaling pathway. PA is a lipid second messenger that targets more than 50 proteins (Jang et al., 2012) involved in vesicular trafficking, cytoskeletal reorganization, proliferation, cell migration, apoptosis, and neuronal differentiation. PA directly enhances the activity of certain protein kinases, small GTPases, and others. In addition, PA is a precursor for other lipid second messengers, such as diacylglycerol (DAG) and lyso-phosphatidic acid (LPA). All three lipid second messengers regulate proliferation, cytoskeletal organization, and survival.

BRAIN CELLS AND PLD

Neurons

Approximately 50% of the cells (around 86 billion) in adult human brains are neurons. Proliferation and differentiation from neuronal precursor cells occurs almost entirely during early embryogenesis. During adulthood, only two regions of neurogenesis in the subventricular zone (SVZ) and subgranular zone (SGZ) remain. In the later stages of embryogenesis, immature neurons migrate to their final position and neurites grow out to form the neuronal network with multiple axons and synapses; one neuron in the adult brain may interact with 10,000 other neurons. The axons are myelinated by oligodendrocytes for faster signal transduction. Programmed cell death of neurons, but also of glial cells, occurs extensively during synaptogenesis to optimize the neuronal network (Kristiansen & Ham, 2014). An overview of the processes during neurogenesis and gliogenesis is given in Figure 1. PLDs are involved in neuronal differentiation and neurite outgrowth but also in signal transduction via vesicle trafficking, exocytosis, and signal termination via endocytosis and internalization of receptors.

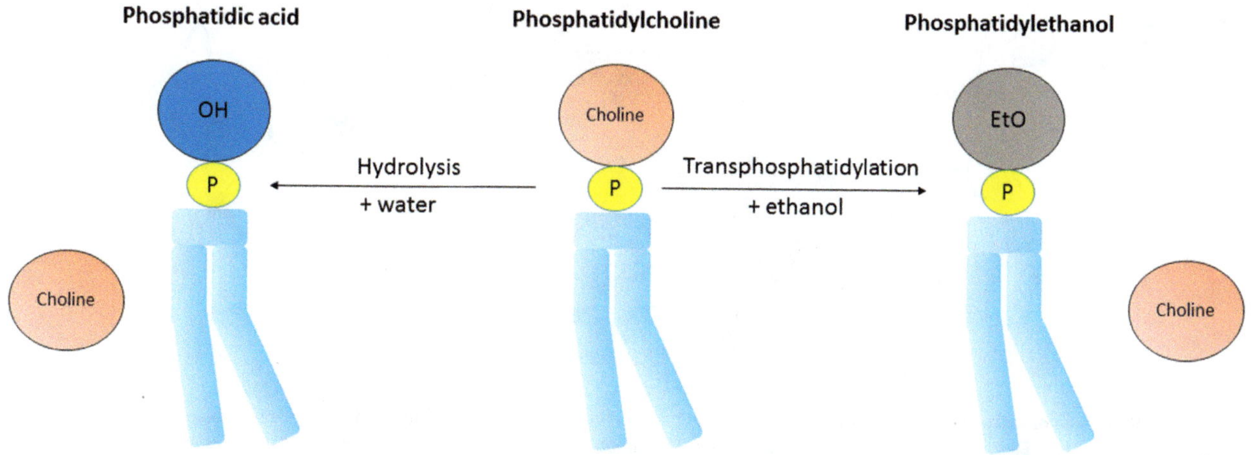

FIGURE 2 Reactions of phospholipases D (PLDs). Under physiological conditions, PLDs catalyze the hydrolysis of phosphatidylcholine (PC) to choline and phosphatidic acid (PA). In the presence of ethanol, phosphatidylethanol (PEth) is produced at the expense of PA. *Modified from Klein et al. (1995).*

Astrocytes

Beside neurons, glial cells are the second most important group of cells in the brain and are key players in brain development and maintenance of brain function. Glial cells are a heterogeneous group of cells consisting of astrocytes, oligodendrocytes, and microglia. They have different morphology and functions. All glial cells originate from neural progenitor cells (NPC), which generate radial glial cells and neuronal precursor cells. During early embryogenesis (between week 9 and 12), the production of neurons stops and glial cells start to proliferate; this is the so-called neuron-glia switch (Kessaris, Pringle, & Richardson, 2001). The highest peak of astrocyte proliferation, however, accompanies synaptogenesis. Astrocytes are further divided into fibrous astrocytes, which populate the white matter and form long fiber-like domains, and protoplasmic astrocytes, which populate the gray matter and have more irregular processes. The branching processes guide the migration of neurons and neurite outgrowth during mid and late embryogenesis. The branching processes of astrocytes that enclose synaptic clefts are crucial for synaptogenesis and, later, for termination of neuronal signaling via uptake of neurotransmitters such as glutamate or GABA. Astrocytes also form the blood–brain barrier by surrounding the blood vessels with end feet of their processes. Moreover, astrocytes are important for energy utilization and regulation of homeostasis in the adult brain. Upon injury of the brain, they also form the glial scar to protect other brain regions. The functions of astrocytes are summarized in Figure 3.

PLD FUNCTION

Proliferation

PLD activation is required for cell proliferation during brain development. This is supported by several findings. (1) In a neuronal stem/precursor cell (NSC), which can generate all cells of the brain, PLD activity is upregulated by EGF and FGF2 (Fujita et al., 2008). (2) Astroglial PLD was constantly found to be activated by mitogens such as FCS, PDGF, EGF, or IGF-1. This activation was accompanied by increased cell proliferation (Burkhardt, Stegner, et al., 2014; Kötter & Klein, 1999). (3) Several neurotransmitters increase PLD activity. For example, muscarinic agonists (Guizzetti, Costa, Peters, & Costa, 1996) stimulate PLD activity and, concomitantly, lead to astroglial proliferation. Interestingly, the activation of PLD by neurotransmitters depends on the timing during development. In the early postnatal stage (up to P14), glutamate leads to increased PLD activity, with the highest response in hippocampal slices taken from 8-day-old rats (Klein, Chalifa, Liscovitch, & Löffelholz, 1995). In adult rats, the basal PLD activity is high and glutamate induced delayed PLD activation (Klein et al., 1998). Muscarinic activation was only demonstrated in immature but not in adult tissue (reviewed Klein, 2005). (4) Addition of exogenous PLD and PA to streptolysin O (SLO)-permeabilized astrocytes led to astroglial cell growth (Schatter, Walev, & Klein, 2003). (5) Inhibition of PLD activity by downregulation of PLD isoforms using PLD-specific siRNA or PLD inhibitors reduced cell proliferation

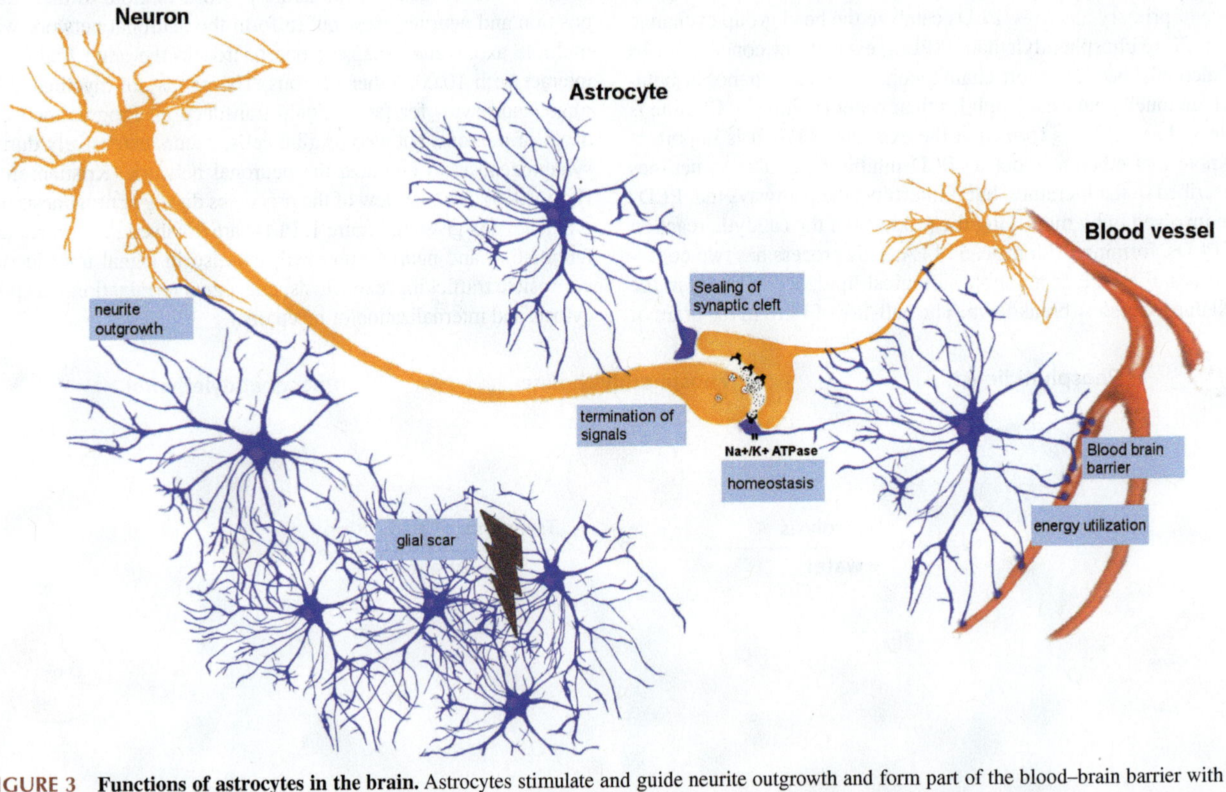

FIGURE 3 **Functions of astrocytes in the brain.** Astrocytes stimulate and guide neurite outgrowth and form part of the blood–brain barrier with the end feet of their branches. They are crucial for the energy utilization of neurons via the connection to blood vessels. Astrocytes are also strongly involved in synaptogenesis and, with their branches, close the synaptic cleft. In addition, they terminate transmission signals via uptake of neurotransmitters. The Na^+/K^+-ATPase activity of astrocytes is necessary for potassium homeostasis.

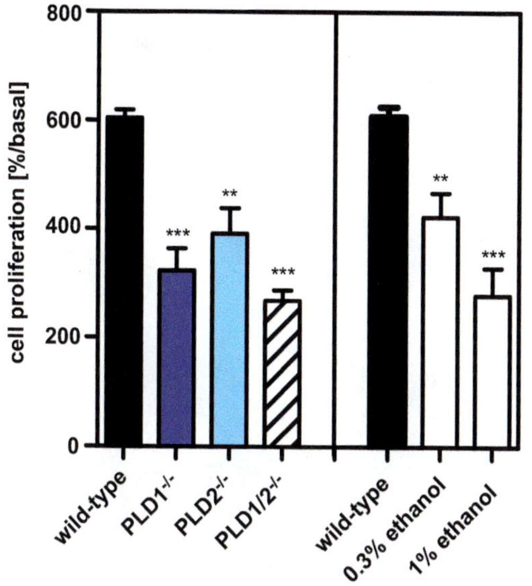

FIGURE 4 Effect of PLD ablation and ethanol on PDB-stimulated proliferation in astrocytes. Wild-type, PLD1$^{-/-}$, PLD2$^{-/-}$, and PLD1/2$^{-/-}$ astrocytes were stimulated with 1 µM PDB. Wild-type astrocytes were stimulated with 1 µM PDB and treated with 0.3% and 1% ethanol. The figure demonstrates that cell proliferation induced by the strongly mitogenic phorbol ester PDB is reduced in PLD-deficient cells. Similarly, ethanol causes a strong reduction of stimulated DNA synthesis. Data is shown as means ± SEM ($n=6$). Statistics: ***$p<0.001$ and **$p<0.01$ versus wild-type cells (one-way ANOVA + Dunnett's post-test).

in astrocytes (Burkhardt, Wojcik, Zimmermann, & Klein, 2014). (6) Recently, PLD-deficient mice strains were described (Elvers et al., 2010; Thielmann et al., 2012), and astrocyte cultures derived from PLD1$^{-/-}$, PLD2$^{-/-}$ or PLD1/2$^{-/-}$ mice proliferated less upon stimulation by growth factors (FCS, IGF-1, or phorbol ester PDB) than cells from wild-type mice (Figure 4; also see Burkhardt, Wojcik, et al., 2014).

Apoptosis

Besides pro-proliferative effects, PLD and PA are also involved in antiapoptotic processes. For example, Schatter, Jin, Löffelholz, and Klein (2005) reported that PA counteracts apoptosis. There are two lines of action to prevent apoptosis. On the one hand, PLD suppresses the pro-apoptotic sphingomyelinase (SMase) pathway. The product of SMase, ceramide, is a second messenger of pro-apoptotic pathways targeting kinases and phosphatases that are required for programmed cell death. Overexpression of PLD in PC12 cells showed reduced ceramide production (Kim, Lee, Kim, Park, & Han, 2003) and treatment of transiently SLO-permeabilized astrocytes with exogenous PA reduced ceramide levels and apoptosis (Schatter et al., 2005). Conversely, ceramide suppressed PLD activity in astrocytes, but the target of ceramide in the PLD signaling pathway is not known. These findings suggest that apoptosis and cell proliferation may be controlled by the balance of PLD and SMase activities (Klein, 2005). On the other hand, Kim et al. (2006) demonstrated that PLD terminates apoptosis by suppression of pro-apoptotic proteins and upregulation of antiapoptotic genes in fibroblasts.

Neurite Outgrowth

PLDs are known to play a key role in neurite outgrowth, especially in axonal growth (Kanaho, Funakoshi, & Hasegawa, 2009). Increased expression of PLD1 and PLD2 induced hippocampal mossy fiber sprouting and may modulate neuronal plasticity (Zhang et al., 2004), whereas inactive mutants of PLD inhibited axonal sprouting (Sung et al., 2001). The neuronal cell adhesion molecule L1 is important for cell migration, axonal outgrowth, and fasciculation of cerebral granule neurons (CGNs), and it was reported that PLD2 and the resulting PA are downstream targets of the L1-mediated MAP kinase pathway that induces neurite outgrowth (Watanabe et al., 2004). In addition, stimulation of astrocytes by carbachol, a muscarinic receptor agonist, led to neurite outgrowth of hippocampal neurons and elongation of axons. The neurite outgrowth is mediated by PLD-dependent synthesis and release of fibronectin, laminin, and plasminogen in astrocytes (Guizzetti, Moore, Giordano, & Costa, 2008).

Signaling

PLDs also play a role in neurotransmitter release. PLD1 is required for Ca^{2+}-dependent exocytosis in neuroendocrine cells by recruiting secretory granules to the exocytosis site (Humeau et al., 2001). In this study, PLD activity was increased by GnRH in immortalized gonadotropin-releasing hormone (GnRH) neurons, and vice versa, GnRH release was reduced when PLD signaling was interrupted. Therefore, PLD regulates intracellular signaling and neurosecretion in GT1 cells via a coupling to the GnRH receptor (Zheng, Krsmanovic, Vergara, Catt, & Stojilkovic, 1997). In cholinergic neurons, it was shown that PLD generates choline for acetylcholine synthesis and that this process is regulated by PKC-activating neurotransmitters (Zhao, Frohman, & Blusztajn, 2001). In addition to exocytosis, PLDs regulate signal termination via endocytosis of receptors, such as EGF, angiotensin II, or µ-opioid receptors (Du et al., 2004). The commonly postulated process is that PLD produces PA, a cone-shaped lipid that forces the membrane to bend inward and to form vesicles.

EFFECTS OF ETHANOL IN VITRO

Proliferation

The inhibitory effect of ethanol on astroglial cell growth has been repeatedly described in the literature. In these studies, glial cells were more vulnerable to ethanol than neurons (Guerri, 2002). Ethanol has been proposed to interact specifically with mitogenic signaling pathways (Costa, Vitalone, & Guizzetti, 2004; DeVito, Stone, & Mori, 1997; Guerri, 2002), including PLD signaling. As described above, the formation of PA by PLD is suppressed by primary alcohols, because phosphatidylalcohols are formed at the expense of PA (Klein et al., 1995). Isoforms of butanol (1-butanol and t-butanol) were often used to study the interruption of the PLD signaling pathway by alcohols. In contrast to 1-butanol, t-butanol is not a substrate in the transphosphatidylation reaction due to steric hindrance and can serve as inactive control. Using comparative studies of the butanol isoforms, it was shown that alcohols inhibit serum-, PDGF-, PDB-, and IGF-1-induced cell proliferation of rat cortical astrocytes, apparently by disrupting PLD signaling (Figure 4; Burkhardt,

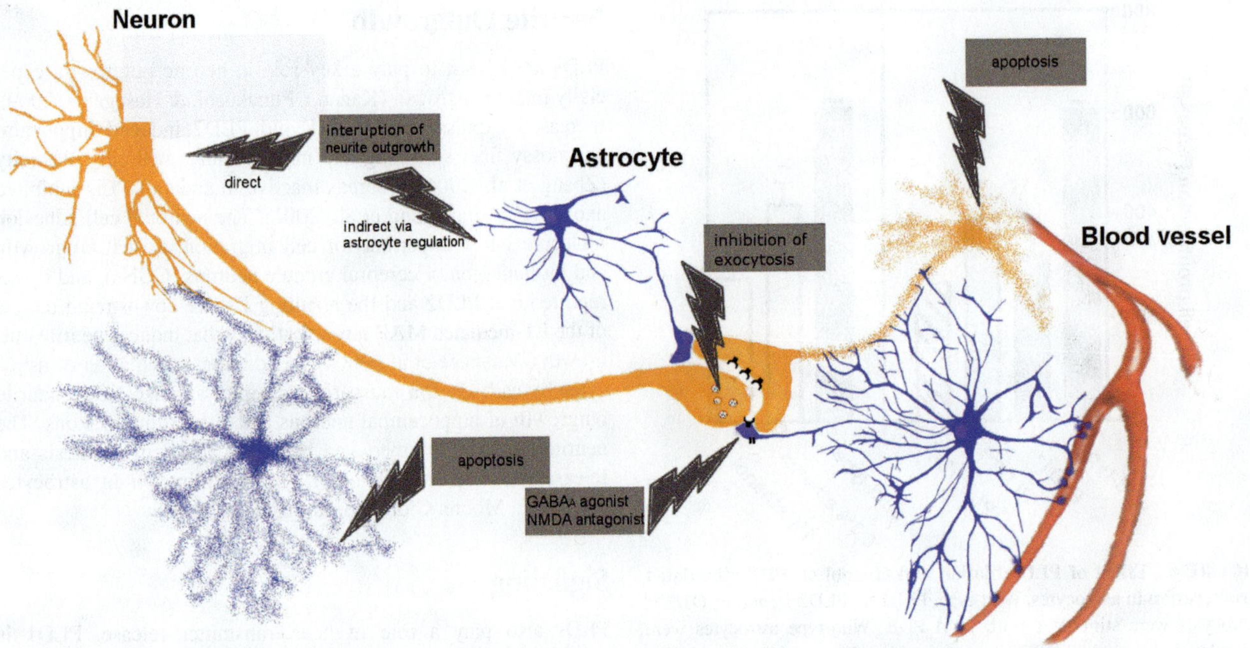

FIGURE 5 **Sites of ethanol toxicity in the brain.** Ethanol inhibits neurite formation directly in neurons but also inhibits regulation of neurite growth by astrocytes. Ethanol induces massive apoptosis of neurons and astrocytes, especially during brain development. Beside inhibition of exocytosis in synaptic clefts, ethanol acts as a $GABA_A$ agonist and NMDA antagonist. Except for these receptor interactions, the inhibition of PLD signaling pathway is involved in all of the ethanol-mediated processes illustrated above.

Wojcik, et al., 2014; Kötter & Klein, 1999). Exogenous addition of PLD and PA using transient permeabilization by SLO confirmed that ethanol inhibits PLD-stimulated astroglial proliferation but has no effect on PA-stimulated cell growth (Schatter et al., 2003). Fujita et al. (2008) also demonstrated that exogenous addition of PA to ethanol-inhibited neural stem cells (NSCs) rescued NSC proliferation. Moreover, PLD-deficient astrocytes are less vulnerable to ethanol than wild-type cells (Burkhardt, Stegner, et al., 2014).

Apoptosis

Alcohol exposure increases apoptosis during the development of the neocortex. Proteins regulating apoptosis (p53 and bcl-2) are elevated in ethanol-treated embryos (Lee et al., 2005), but ethanol-induced apoptosis is also triggered by the ceramide pathway in astrocytes (Pascual, Valles, Renau-Piqueras, & Guerri, 2003) and in neuronal crest cells (E10) (Wang & Bieberich, 2010). As mentioned before, there seems to be cross talk between PLD and ceramide. While ethanol reduced PA formation in astrocytes, ceramide production was increased. However, exogenous PA, added to transiently permeabilized astrocytes, suppressed ethanol-induced ceramide formation (Schatter et al., 2005).

Neurite Outgrowth

Ethanol exposure of neurons taken from 7-day-old rats inhibited neurite formation in CGNs; this depended on the activity of Rho GTPases (Joshi et al., 2006). Furthermore, neurite outgrowth depends on PLD and Rho activation by PDGF (Sung et al., 2001). The L1-MAPK-PLD neurite outgrowth cascade is inhibited by ethanol (Bearer, 2001; Watanabe et al., 2004). Finally, ethanol interfered with carbachol-induced axonal growth of prenatal rat hippocampal pyramidal neurons, while neuronal viability was not affected (VanDemark, Guizzetti, Giordano, & Costa, 2009).

Signaling

Ethanol alters synaptic function and signal transduction. Inhibition of the PLD pathway by ethanol caused a concomitant drop of GnRH release (Zheng et al., 1997). Ethanol also disturbed the endosomal recycling of dopamine transporters (Methner & Mayfield, 2010). In contrast to t-butanol, ethanol and 1-butanol inhibited exocytosis of catecholamines in chromaffin cells, and exocytosis was dependent on PLD and ARF6 (Caumont et al., 1998). In addition, ethanol suppresses neurotransmission by its $GABA_A$ agonistic and NDMA antagonistic properties (Ikonomidou et al., 2000). Olney (2004) described apoptosis in 7-day-old rats treated with $GABA_A$ agonists (benzodiazepines and barbiturates) and NDMA antagonists, which is caspase 3-dependent. Taken together, these findings explain ethanol toxicity combining features of massive neurodegeneration and apoptosis.

Sites of action of ethanol that correlate with PLD function are summarized in Figure 5.

EFFECTS OF ETHANOL AND PLD DELETION IN VIVO

Microcephaly

Brain imaging and neurobehavioral tests demonstrate the devastating consequences of prenatal alcohol exposure on the developing brain, with long-lasting effects on cognition and behavior (Guerri, Bazinet, & Riley, 2009). Neuroimaging techniques provide insight into structural and functional alterations of different brain regions caused by ethanol. Clarren and Smith (1978) reported microcephaly and anomalies of brain structures in offspring exposed to alcohol. In particular, white matter seems to be altered as the decrease of white matter is higher than the decrease of gray matter in humans (Archibald et al., 2001). In functional magnetic resonance imaging (MRI) studies, it was reported that the *N*-acetylaspartate/choline (NAA/Cho) ratio and the *N*-acetylaspartate/creatine (NAA/Cr) ratio were decreased, especially in the parietal and frontal cortices, which has been interpreted as an impairment of glial rather than neuronal cells (Fagerlund et al., 2006). In an in vivo MRI study with newborn mice, PLD1-deficient mice developed normally and similarly to wild-type mice while PLD2$^{-/-}$ mice and PLD1/2$^{-/-}$ mice had a delayed body and brain development (Burkhardt, Stegner, et al., 2014). Interestingly, brains of PLD2$^{-/-}$ mice and double knockout mice look less structured than brains of wild-type or PLD1$^{-/-}$ mice (Figure 6); the formation of the ventricle seems to be delayed and differentiation between cortices and striatum is difficult to see in prenatal stages. In PLD1$^{-/-}$ mice, the ventricles (especially the lateral ventricles) were strongly enlarged, while total brain volume was not changed compared to wild-type mice. Thus, the size of brain tissue may be smaller in PLD-deficient mice, and this may be due to disturbance of white matter development.

The above mentioned differences in brain development were observed in immature mouse brains. As adults, however, PLD-deficient mice had caught up with wild-type mice with respect to brain volume. Some MRI studies in rodents, which were exposed to ethanol as fetuses, were reported in the literature. So far, however, all of them were done postmortem. In rats exposed to high doses of ethanol, reduced body weight and brain volume with reduced cortical thickness were found (Coleman, Oguza, Lee, Styner, & Crews, 2012; Leigland, Ford, Lerch, & Kroenke, 2013). Others reported that prenatal alcohol exposure (PAE) in mice, after a dose regimen of 4.0 g ethanol/kg body weight throughout gestation, caused delayed body development, but mice caught up until adulthood; a higher dose of ethanol (6.0 g ethanol/kg) resulted in a lifelong smaller body size (Abel & Dintcheff, 1978).

Cognitive Dysfunction

Studies on the behavior in PLD-deficient mice using social recognition task (SRT), object recognition task (ORT), and passive avoidance test (PAT) showed that long-term and short-term memory as well as recognition of novelty and sociability were affected by PLD deletion (Burkhardt, Stegner, et al., 2014). The SRT depends on the short-term recognition of mates and the curiosity of mice. In contrast to wild-type mice, the short-term memory is disturbed in PLD1$^{-/-}$, PLD2$^{-/-}$, and PLD1/2$^{-/-}$-deficient mice. The recognition of novelty in the ORT requires more cognitive skills,

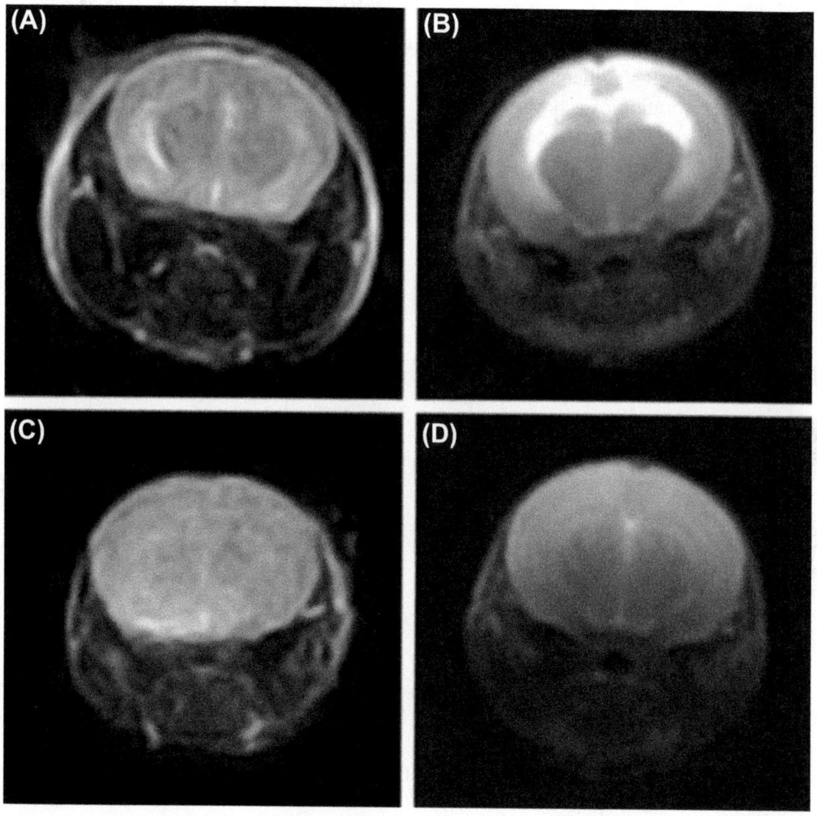

FIGURE 6 Microcephaly and structural disturbance of PLD-deficient mice. Exemplary MRI scan of (A) wild-type, (B) PLD1$^{-/-}$, (C) PLD2$^{-/-}$, and (D) PLD1/2$^{-/-}$ mouse embryos (embryonal age: P18–P22 days). The MRI scans were performed in a Siemens Magnetom® Trio with an eight-channel phase array coil usually used for wrist detection. The brains of the pups were measured in 19–20 transversal slices with a spatial resolution of 0.2×0.2×0.9 mm (cf. Burkhardt, Stegner et al., 2014).

relative to exploration of novel environments or a single novel object. While PLD2$^{-/-}$- and PLD1/2$^{-/-}$-deficient mice still failed to remember the objects and spent their time randomly with all objects, PLD1$^{-/-}$ mice learned better, but not as well as wild-type mice, in this test for long-term memory. In contrast to the recognition tests, the fear-driven PA test showed no difference between the mouse strains.

In PAE in rodents, the timing and concentration of ethanol treatment is critical for the change in behavior. In a mild dose regimen, juvenile mice had deficits in memory and learning, but they behaved normally as adults (Schneider, Moore, & Adkins, 2011). In particular, inhibitory control, including reversal learning and passive avoidance, were altered in juveniles (P18 to P60), but this effect did not persist into adulthood (Abel & Sokol, 1991); evidently, there are compensatory mechanisms in the adult animal.

Neurochemistry

PLDs catalyze the hydrolysis of PC to PA and choline. Part of the free choline in nerve tissue serves for the synthesis of ACh (Zhao et al., 2001), the neurotransmitter of cholinergic neurons. Central cholinergic neurons project axons from the forebrain to the hippocampus and cortical brain regions to regulate global functions such as attention, motivation, and memory (Wolff, 1996). As mentioned before, the hippocampus plays a key role in social and object recognition and in learning and memory in general. Neurochemical studies using microdialysis in the ventral hippocampus of PLD-deficient mice, therefore, were used to investigate cholinergic activity in PLD-deficient mice (Burkhardt, Stegner, et al., 2014). These studies showed that percentage ACh release was significantly decreased in PLD-deficient mice during the exploration of a novel environment, called the open field. Pharmacological stimulation of ACh release by scopolamine (M$_2$/M$_4$ inhibitor), in contrast, revealed no difference between wild-type and PLD-deficient mice (Figure 7).

Changes of neurotransmitter release were also suggested by previous studies. Humeau et al. (2001) already showed that PLD1, which is located and activated at the plasma membrane, is important for neurotransmitter release. They injected an inactive PLD1 mutant (K898R) into Aplysia neurons, thereby inhibiting ACh release. This inhibition was not caused by impaired vesicular trafficking or stimulus-secretion coupling, but by reducing the number of active presynaptic releasing sites. Zhao et al. (2001) prepared PLD2-overexpressing murine basal forebrain cholinergic SN56 cells and measured increased ACh levels. Overexpression of PLD1 had the same effect, but PLD1 mRNA was originally not found in these cells. Hence, they suggested PLD2 as the enzyme that generates choline for ACh synthesis in SN56 cells, but PLD1 may evidently also be used for this purpose. Therefore, PLD2 may be the main isoform to generate choline for ACh synthesis upon stimulation, and a lack of choline may lead to reduced ACh synthesis and release.

In our hands, however, ACh release was only reduced upon physiological stimulation, not during scopolamine infusion. This suggests that not just lack of choline, but differential activation of cholinergic neurons, underlies our findings (Burkhardt, Stegner, et al., 2014). PLD1 deletion may have had a smaller effect than PLD2-deletion because there is still enough ACh in the neuron, but the presynaptic releasing sites may be reduced. However, it is not clear how PLD, and especially PLD2, is involved in learning processes. PLD deletion in vivo elucidated the importance of PLD for the development of memory. Axonal guidance by astrocytes, neurotransmitter release, and/or LTP may be affected by PLD deletion.

APPLICATIONS TO OTHER ADDICTIONS AND SUBSTANCE MISUSE

Ethanol is the most fetotoxic substance among the commonly used addictive drugs, such as nicotine, heroin, cannabis, and benzodiazepines. In contrast to the severe malformations and cognitive dysfunctions caused by maternal ethanol consumption, other addictive drugs lead to less severe and usually transient symptoms. Smoking during pregnancy increases the risk of abortion and reduced birth weight; in addition, evidence is accumulating that nicotine increases the risk of sudden infant death syndrome (SIDS) postnatally. Cannabis has no obvious impact of the development of the

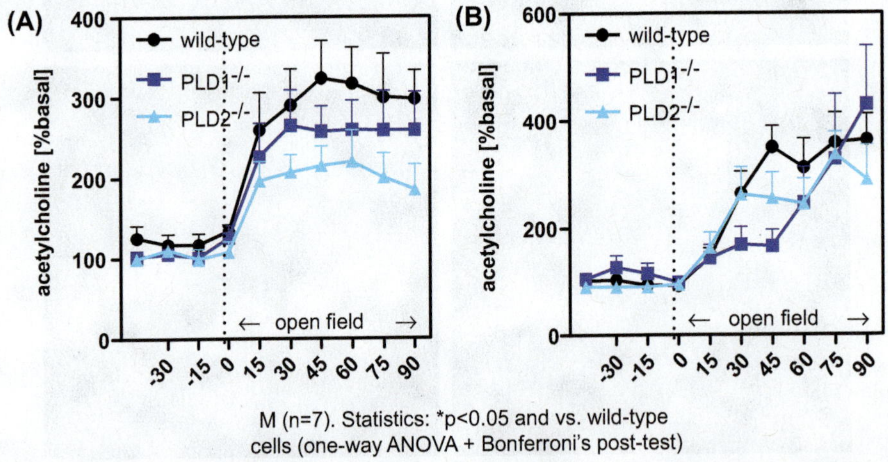

FIGURE 7 **Acetylcholine release in the hippocampus of PLD-deficient mice as measured by microdialysis.** Effect of PLD-gene deletion on acetylcholine (ACh) release after (A) exposure of the mouse in the open field and (B) local administration of scopolamine (M2/M4 inhibitor) in wild-type and PLD1−/−and PLD2−/− mice. Exposure to the open field or administration of scopolamine (0.1 μM) was initiated for 90min at time point zero. Data are shown as mean±SEM (n=7). M (n=7). Statistics: *p<0.05 and versus wild-type cells (one-way ANOVA+Bonferroni's post-test).

fetus, and few or no withdrawal symptoms were observed postpartum. In contrast, babies exposed to opiates (e.g., heroin or methadone) during fetal development showed symptoms of neonatal abstinence syndrome but achieved normal growth and development after 6 months. Benzodiazepine consumption during pregnancy has no impact on embryonic and postnatal development, but infants exposed to benzodiazepines during the third trimester exhibit the "floppy" infant syndrome. Symptoms of floppy babies are sedation, hypotonia, cyanosis, and reluctance to suck. However, benzodiazepines lack long-term effects on development, growth, and cognitive functions. Prenatal cocaine exposure also has remarkably little effect on infants but is associated with premature birth and subtly disturbed behavior, such as impaired cognitive and attention skills in later life. There is no evidence that PLD signaling is involved in any of these processes. Only one publication suggests that cocaine-impaired synaptic plasticity and behavior of rats depends on PLD-linked metabotropic glutamate receptors in the amygdala (Krishnan et al., 2011). However, this finding requires further study.

CONCLUSION

Considering the documented relevance of PLD in glial cell proliferation, neurite outgrowth, and neurotransmitter release, and the unique interference of ethanol with the PLD-catalyzed transphosphatidylation reaction, PLD is certainly one important target of ethanol pathology in the FAS besides suppressed neurotransmission by its $GABA_A$ agonistic and NDMA antagonistic properties (Ikonomidou et al., 2000). Nevertheless, there are still open questions regarding PLD function in the developing and aging brain. One further target of interest is certainly the potential role of PLD in neurodegenerative diseases, such as Alzheimer disease (Cruchaga et al., 2014; Oliviera et al., 2010). Whether ethanol interference with PLD function plays a role in aging also has to be investigated further. Clearly, the interference of ethanol with the PLD signaling pathway deserves study both in immature and mature (aging) brain.

DEFINITION OF TERMS

Astrocytes Astrocytes belong to the class of glial cells and are important for the formation of the brain scaffold during development. They guide neurite outgrowth and synaptogenesis. They also build up the blood–brain barrier and supply energy metabolites for neurons. In addition, they are important for brain homeostasis, signal termination, and nervous system repair.

Phospholipases D Phospholipases D belong to the large group of enzymes hydrolyzing phospholipids. The catalytic site consists of two HKD motifs. Mammalian PLD1 and PLD2 are phosphatidylcholine-specific. The PLD products are phosphatidic acid and choline.

Phosphatidic acid Phosphatidic acid (PA) is a lipid second messenger and important for cell growth and survival, as well as for endo- and exocytosis and cytoskeletal reorganization.

Brain growth spurt Brain growth spurt is the period of massive synaptogenesis, glial cell proliferation, and myelination in the third trimester of embryogenesis. In rats, the brain growth spurt is postnatally between day 1 and day 14.

Transphosphatidylation Transphosphatidylation is the unique reaction of PLDs. In the presence of primary alcohols such as ethanol, the head group (choline) is exchanged with alcohol. The resulting phosphatidylethanol (PEth) has a long half-life and no further function in the cell, but the PLD signaling pathway is interrupted because PA formation is inhibited.

Prenatal alcohol exposure (PAE) Mice are a common model to study the effects of fetal alcohol exposure via prenatal alcohol exposure.

Fetal alcohol syndrome Fetal alcohol syndrome was described for the first time in 1973 by Jones and Smith. Criteria for FAS are chronic alcohol abuse by the mother plus three symptoms such as craniofacial abnormalities, prenatal and postnatal growth retardation, and cognitive or behavioral dysfunction.

Fetal alcohol spectrum disorder For a diagnosis of fetal alcohol spectrum disorder (FASD), in contrast to FAS, the alcohol consumption of the mother must not be established. The term FASD was introduced by Riley and McGee (2005) and better describes the wide range of symptoms of prenatal alcohol toxicity in children because the severity depends on the amount and timing of alcohol consumption.

Neurite outgrowth Neurite outgrowth is the elongation and organization of neuronal branches. Neurite outgrowth is a central step on the way to synaptogenesis and network formation during brain development.

Microdialysis Microdialysis is a minimally invasive method to continuously collect samples of soluble molecules in the extracellular fluid of almost every tissue over a period of time.

Insulin-like growth factor 1 (IGF-1) Insulin-like growth factor 1 (IGF-1) shows characteristics of a peripheral peptide hormone that regulates energy supply and tissue remodeling, but also has characteristics of a growth factor with paracrine and autocrine function. It is very important for brain development.

Social recognition test A social recognition test depends on the sociability of species and the individual recognition of mates. Two mice not familiar with each other, once placed in a neutral area, will interact and make intensive contacts. Familiar mice will show less frequent contacts than unfamiliar mice.

Object recognition test An object recognition test depends on the curiosity of mice. Placed in a neutral area and confronted with a new and a familiar object, the mice will investigate the new object more intensively than the familiar object.

KEY FACTS OF PHOSPHOLIPASES D

- PLDs have ubiquitous expression in all species (bacteria, fungi, plants, and mammals) and all mammalian tissues.
- PLDs catalyze the head group exchange on phosphodiester bonds of glycerolipids.
- Transphosphatidylation is a unique reaction of PLDs in the presence of primary alcohols.
- Of the five isoforms in mammals, only PLD1 and PLD2 are PC-specific and are active in the transphosphatidylation reaction.
- Functions of PLDs include cell proliferation, migration, cytoskeletal reorganization, and endo- and endocytosis.
- PLDs are relevant in development, neurodegeneration, cancer, inflammation, and cardiovascular diseases.

SUMMARY POINTS

- This chapter discusses the inhibitory effect of ethanol on PLD signaling in astrocytes.
- PLDs are highly expressed during brain development.

- In the presence of ethanol, PLD catalyze the unique transphosphatidylation reaction. PEth is produced on the expense of PA.
- Interruption of PA signaling leads to inhibition of cell growth and apoptosis. In addition, vesicular transport and endo- and exocytosis is disturbed.
- Inhibition of PLD signaling in astrocytes by PLD isoform-specific inhibitors, siRNA against PLDs and PLD ablation leads to reduced cell proliferation in astrocytes.
- PLD-deficient mice had microcephaly during development and showed behavioral deficits that are comparable to PAE mice.

REFERENCES

Abel, E., & Dintcheff, B. (1978). Effects of prenatal alcohol exposure on growth and development in rats. *The Journal of Pharmacology and Experimental Therapeutics, 207*(3), 916–921.

Abel, E., & Sokol, R. (1991). A revised estimate of the economic impact of fetal alcohol syndrome. *Recent Developments in Alcoholism, 9*, 117–125.

Andersen, S. (2003). Trajectories of brain development: point of vulnerability or window of opportunity? *Neuroscience and Biobehavioral Reviews, 27*(1–2), 3–18.

Archibald, S., Fennema-Notestine, C., Gamst, A., Riley, E., Mattson, S., & Jernigan, T. (2001). Brain dysmorphology in individuals with severe prenatal alcohol exposure. *Developmental Medicine and Child Neurology, 43*(3), 148–154.

Bearer, C. (2001). L1 cell adhesion molecule signal cascades: targets for ethanol developmental neurotoxicity. *Neurotoxicology, 22*(5), 625–633.

Burkhardt, U., Stegner, D., Hattingen, E., Beyer, S., Nieswandt, B., & Klein, J. (2014). Impaired brain development and reduced cognitive function in phospholipase D-deficient mice. *Neuroscience Letters, 572*, 48–52.

Burkhardt, U., Wojcik, B., Zimmermann, M., & Klein, J. (2014). Phospholipase D is a target for inhibition of astroglial proliferation by ethanol. *Neuropharmacology, 79*, 1–9.

Caumont, A., Galas, M., Vitale, N., Aunis, D., & Bader, M. (1998). Regulated exocytosis in chromaffin cells. Translocation of ARF6 stimulates a plasma membrane-associated phospholipase D. *The Journal of Biological Chemistry, 273*(3), 1373–1379.

Clarren, S., Bowden, D., & Astley, S. (1987). Pregnancy outcomes after weekly oral administration of ethanol during gestation in the pig-tailed macaque (*Macaca nemestrina*). *Teratology, 35*(3), 345–354.

Clarren, S., & Smith, D. (1978). The fetal alcohol syndrome. *The New England Journal of Medicine, 298*(19), 1063–1067.

Coleman, L. J., Oguza, I., Lee, J., Styner, M., & Crews, F. (2012). Postnatal day 7 ethanol treatment causes persistent reductions in adult mouse brain volume and cortical neurons with sex specific effects on neurogenesis. *Alcohol, 46*(6), 603–612.

Costa, L., Vitalone, A., & Guizzetti, M. (2004). Signal transduction mechanisms involved in the antiproliferative effects of ethanol in glial cells. *Toxicology Letters, 149*, 67–73.

Cruchaga, C., Karch, C., Jin, S., Benitez, B., Cai, Y., Guerreiro, R., ... Goate, A. (2014). Rare coding variants in the phospholipase D3 gene confer risk for Alzheimer's disease. *Nature, 505*(7484), 550–554.

DeVito, W., Stone, S., & Mori, K. (1997). Low concentrations of ethanol inhibits prolactin-induced mitogenesis and cytokine expression in cultured astrocytes. *Endocrinology, 138*, 922–928.

Du, G., Huang, P., Liang, B., & Frohman, M. (2004). Phospholipase D2 localizes to the plasma membrane and regulates angiotensin II receptor endocytosis. *Molecular Biology of the Cell, 15*, 1024–1030.

Elvers, M., Stegner, D., Hagedorn, I., Kleinschnitz, C., Braun, A., Kuijpers, M., ... Nieswandt, B. (2010). Impaired αIIbβ3 integrin activation and shear-dependent thrombus formation in mice lacking phospholipase D1. *Science Signaling, 3*(103), 1–10.

Fagerlund, A., Heikkinen, S., Autti-Rämö, I., Korkman, M., Timonen, M., Kuusi, T., ... Lundbom, N. (2006). Brain metabolic alterations in adolescents and young adults with fetal alcohol spectrum disorders. *Alcoholism, Clinical and Experimental Research, 30*(12), 2097–2104.

Fujita, Y., Hiroyama, M., Sanbe, A., Yamauchi, J., Murase, S., & Tanoue, A. (2008). ETOH inhibits embryonic neural stem/precursor cell proliferation via PLD signaling. *Biochemical and Biophysical Research Communications, 370*, 169–173.

Guerri, C. (2002). Mechanisms involved in central nervous system dysfunctions induced by prenatal ethanol exposure. *Neurotoxicity Research, 4*(4), 327–335.

Guerri, C., Bazinet, A., & Riley, E. (2009). Foetal alcohol spectrum disorders and alterations in brain and behaviour. *Alcohol and Alcoholism, 44*, 108–114.

Guizzetti, M., Costa, P., Peters, J., & Costa, L. (1996). Acetylcholine as a mitogen: muscarinic receptor-mediated proliferation of rat astrocytes and human astrocytoma cells. *European Journal of Pharmacology, 297*, 265–273.

Guizzetti, M., Moore, N., Giordano, G., & Costa, L. (2008). Modulation of neuritogenesis by astrocyte muscarinic receptors. *The Journal of Biological Chemistry, 283*, 31884–31897.

Hanahan, D., & Chaikoff, I. (1947). A new phospholipid-splitting enzyme specific for the ester linkage between the nitrogenous base and the phosphoric acid group. *The Journal of Biological Chemistry, 169*(3), 699–705.

Humeau, Y., Vitale, N., Chasserot-Golaz, S., Dupont, J., Du, G., Frohman, M., ... Poulain, B. (2001). A role for phospholipase D1 in neurotransmitter release. *Proceedings of the National Academy of Sciences of the United States of America, 98*(26), 15300–15305.

Ikonomidou, C., Bittigau, P., Ishimaru, M., Wozniak, D., Koch, C., Genz, K., ... Olney, J. (2000). Ethanol-induced apoptotic neurodegeneration and fetal alcohol syndrome. *Science, 287*(5455), 1056–1060.

Jang, J.-H., Lee, C. S., Hwang, D., & Ryu, S. H. (2012). Understanding of the roles of phospholipase D and phosphatidic acid through their binding partners. *Cellular Signaling, 51*, 71–81.

Jenkins, G., & Frohman, M. (2005). Phospholipase D: a lipid centric review. *Cellular and Molecular Life Sciences, 62*(19–20), 2305–2316.

Joshi, S., Guleria, R., Pan, J., Bayless, K., Davis, G., Dipette, D., & Singh, U. (2006). Ethanol impairs Rho GTPase signaling and differentiation of cerebellar granule neurons in a rodent model of fetal alcohol syndrome. *Cellular and Molecular Life Sciences, 63*(23), 2859–2870.

Kanaho, Y., Funakoshi, Y., & Hasegawa, H. (2009). Phospholipase D signalling and its involvement in neurite outgrowth. *Biochimica et Biophysica Acta, 1791*(9), 898–904.

Kessaris, N., Pringle, N., & Richardson, W. (2001). Ventral neurogenesis and the neuron-glial switch. *Neuron, 31*(5), 677–680.

Kim, J., Lee, Y., Kwon, T., Chang, J., Chung, K., & Min, D. (2006). Phospholipase D prevents etoposide-induced apoptosis by inhibiting the expression of early growth response-1 and phosphatase and tensin homologue deleted on chromosome 10. *Cancer Research, 66*(2), 784–793.

Kim, K., Lee, K., Kim, Y., Park, S., & Han, J. (2003). Anti-apoptotic role of phospholipase D isozymes in the glutamate-induced cell death. *Experimental and Molecular Medicine, 28*, 38–45.

Klein, J. (2005). Functions and pathophysiological roles of phospholipase D in the brain. *Journal of Neurochemistry, 94*, 1473–1487.

Klein, J., Chalifa, V., Liscovitch, M., & Löffelholz, K. (1995). Role of phospholipase D activation in nervous system physiology and pathophysiology. *Journal of Neurochemistry, 65*, 1445–1455.

Klein, J., Vakil, M., Bergmann, F., Holler, T., Iovino, M., & Löffelholz, K. (1998). Glutamatergic activation of hippocampal phospholipase D: postnatal fading and receptor desensitization. *Journal of Neurochemistry, 70*(4), 1679–1685.

Kötter, K., & Klein, J. (1999). Ethanol inhibits astroglial cell proliferation by disruption of phospholipase D-mediated signaling. *Journal of Neurochemistry, 74*, 2517–2523.

Krishnan, B., Genzer, K., Pollandt, S., Lui, J., Gallagher, J., & Shinnick-Gallagher, P. (2011). Dopamine-induced plasticity, phospholipase D (PLD) activity and cocaine-cue behavior depend on PLD-linked metabotropic glutamate receptors in amygdala. *PLoS One, 6*(9), e25639.

Kristiansen, M., & Ham, J. (2014). Programmed cell death during neuronal development: the sympathetic neuron model. *Cell Death and Differentiation, 21*(7), 1025–1035.

Lee, R., An, S., Kim, S., Rhee, G., Kwack, S., Seok, J., … Park, K. (2005). Neurotoxic effects of alcohol and acetaldehyde during embryonic development. *Journal of Toxicology and Environmental Health, 68*, 23–24.

Leigland, L., Ford, M., Lerch, J., & Kroenke, C. (2013). The influence of fetal ethanol exposure on subsequent development of the cerebral cortex as revealed by magnetic resonance imaging. *Alcoholism, Clinical and Experimental Research, 37*(6), 924–932.

Mahankali, M., Alter, G., & Gomez-Cambronero, J. (2014). Mechanism of enzymatic reaction and protein–protein interactions of PLD from a 3D structural model. *Cellular Signalling, 27*. pii:S0898–6568(14)00312–X.

Methner, D., & Mayfield, R. (2010). Ethanol alters endosomal recycling of human dopamine transporters. *The Journal of Biological Chemistry, 285*(14), 10310–10317.

Oliveira, T., Chan, R., Huasong, T., Laredo, M., Shui, G., Staniszewski, A., … Paolo, G. D. (2010). Phospholipase D2 ablation ameliorates Alzheimer's disease-linked synaptic dysfunction and cognitive deficits. *The Journal of Neuroscience, 30*, 16419–16428.

Olney, J. (2004). Fetal alcohol syndrome at the cellular level. *Addiction Biology, 9*, 137–149.

Pascual, M., Valles, S., Renau-Piqueras, J., & Guerri, C. (2003). Ceramide pathways modulate ethanol-induced cell death in astrocytes. *Journal of Neurochemistry, 87*, 1535–1545.

Riley, E., & McGee, C. (2005). Fetal alcohol spectrum disorders: an overview with emphasis on changes in brain and behavior. *Experimental Biology and Medicine, 230*(6), 357–365.

Rodriguez, F., Lundqvist, C., Alling, C., & Gustavsson, L. (1996). Ethanol and phosphatidylethanol reduce the binding of (^{3}H)inositol 1,4,5-trisphosphate to rat cerebellar membranes. *Alcohol and Alcoholism, 31*(5), 453–461.

Saito, M., & Kanfer, J. (1973). Solubilization and properties of a membrane-bound enzyme from rat brain catalyzing a base-exchange reaction. *Biochemical and Biophysical Research Communications, 538*(2), 391–398.

Saito, S., Sakagami, H., & Kondo, H. (2000). Localization of mRNAs for phospholipase D type 1 and 2 in the brain of developing and mature rat. *Developmental Brain Research, 120*, 41–47.

Schatter, B., Jin, S., Löffelholz, K., & Klein, J. (2005). Cross-talk between phosphatidic acid and ceramide during ethanol-induced apoptosis in astrocytes. *BMC Pharmacology, 5*, 1–11.

Schatter, B., Walev, I., & Klein, J. (2003). Mitogenic effects of phospholipase D and phosphatidic acid in transiently permeabilized astrocytes: effects of ethanol. *Journal of Neurochemistry, 87*, 95–100.

Schneider, M., Moore, C., & Adkins, M. (2011). The effects of prenatal alcohol exposure on behavior: rodent and primate studies. *Neuropsychology Review, 21*.

Selvy, P., Lavieri, R., Lindsley, C., & Brown, H. (2011). Phospholipase D: Enzymology, functionality, and chemical modulation. *Chemical Reviews, 111*, 6064–6119.

Sulik, K. (2005). Genesis of alcohol-induced craniofacial dysmorphism. *Experimental Biology and Medicine, 230*(6), 366–375.

Sung, J., Lee, S., Min, D., Eom, T., Ahn, Y., Choi, M., … Chung, K. C. (2001). Differential activation of phospholipases by mitogenic EGF and neurogenic PDGF in immortalized hippocampal stem cell lines. *Journal of Neurochemistry, 78*, 1044–1053.

Suzuki, K. (2007). Neuropathology of developmental abnormalities. *Brain and Development, 29*(3), 129–141.

Thielmann, I., Stegner, D., Kraft, P., Hagedorn, I., Krohne, G., Kleinschnitz, C., … Nieswandt, B. (2012). Redundant functions of phospholipases D1 and D2 in platelet α-granule release. *Journal of Thrombosis and Haemostasis, 10*(11), 2361–2372.

VanDemark, K., Guizzetti, M., Giordano, G., & Costa, L. (2009). Ethanol inhibits muscarinic receptor-induced axonal growth in rat hippocampal neurons. *Alcoholism, Clinical and Experimental Research, 33*(11), 1945–1955.

Wang, G., & Bieberich, E. (2010). Prenatal alcohol exposure triggers ceramide-induced apoptosis in neural crest-derived tissues concurrent with defective cranial development. *Cell Death and Disease, 1*(5).

Watanabe, H., Yamazaki, M., Miyazaki, H., Arikawa, C., Itoh, K., Sasaki, T., … Kanaho, Y. (2004). Phospholipase D2 functions as a downstream signaling molecule of MAP kinase pathway in L1-stimulated neurite outgrowth of cerebellar granule neurons. *The Journal of Neuroscience, 89*(1), 142–151.

Wolff, N. (1996). Global and serial neurons form a hierarchically arranged interface proposed to underlie memory and cognition. *Neuroscience, 74*(3), 625–651.

Zhang, Y., Huang, P., Du, G., Kanaho, Y., Frohman, M., & Tsirka, S. (2004). Increased expression of two phospholipase D isoforms during experimentally induced hippocampal mossy fiber outgrowth. *Glia, 46*(1), 74–83.

Zhao, D., Frohman, M., & Blusztajn, J. (2001). Generation of choline for acetylcholine synthesis by phospholipase D isoforms. *BMC Neuroscience, 2*, 16.

Zheng, L., Krsmanovic, L., Vergara, L., Catt, K., & Stojilkovic, S. (1997). Dependence of intracellular signaling and neurosecretion on phospholipase D activation in immortalized gonadotropin-releasing hormone neurons. *Proceedings of the National Academy of Sciences of the United States of America, 94*(4), 1573–1578.

Chapter 46

N-Methyl-D-Aspartate Receptor Subunits and Alcohol

Christina N. Nona[1,2], José N. Nobrega[1,2,3,4]

[1]*Department of Pharmacology and Toxicology, University of Toronto, Toronto, ON, Canada;* [2]*Campbell Family Mental Health Research Institute, Behavioural Neurobiology Laboratory, Research Imaging Centre, Centre for Addiction and Mental Health, Toronto, ON, Canada;* [3]*Department of Psychiatry, University of Toronto, Toronto, ON, Canada;* [4]*Department of Psychology, University of Toronto, Toronto, ON, Canada*

Abbreviations

CTD Carboxy-terminal domain
GIIβ Glucosidase II beta subunit
NMDA N-Methyl-D-aspartate
NTD Amino-terminal domain
STEP Striatal enriched protein tyrosine phosphatase
STK Src tyrosine kinase
TM Transmembrane domain
tPA Tissue plasminogen activator

INTRODUCTION

Ethanol (EtOH) is the component in alcoholic beverages responsible for their psychoactive properties. EtOH is a very small molecule and is of low potency; large amounts, typically millimolar concentrations, are needed for it to produce its biological effects. For these reasons, EtOH is known to interact with a multitude of biological targets in the central nervous system (CNS), unlike many other addictive agents whose effects arise due to their actions primarily at a single site of action. Of all EtOH's possible targets, it is a widely held view that neurotransmitter-gated ion channel receptors are among the most important, with the N-methyl-D-aspartate receptor (NMDAR) being a key target (Moykkynen & Korpi, 2012).

The NMDAR is a ligand-gated, voltage-dependent ion channel that requires activation by both glutamate and glycine or D-serine (Figure 1). At resting membrane potentials, the receptor is inactive due to ion channel blockade by extracellular Mg^{2+}. Upon depolarization of the neuronal membrane, Mg^{2+} is expelled, and binding of glutamate and glycine triggers opening of the ion channel, allowing for the influx of Na^+ and Ca^{2+} and the expulsion of K^+. Due to their large Ca^{2+} permeability and the subsequent signaling mechanisms that become activated via Ca^{2+} entry, NMDARs are important in CNS excitability, cognition, motor coordination, and synaptic plasticity processes underlying learning and memory (Paoletti, Bellone, & Zhou, 2013). Given their role in synaptic plasticity, it is not surprising that NMDARs have been heavily implicated in addiction, which many investigators consider to be an aberrant form of learning and memory.

NMDARs are heteromeric protein complexes with three families of subunits that aggregate in different combinations to form a functional receptor: the obligatory NR1 subunit, needed to form functional channels, consists of eight splice variants; the NR2 subunits have four distinct isoforms (A–D); and the NR3 subunit exists in two isoforms (A and B). The structural organization of a single subunit is shown in Figure 2. The conventional NMDAR exists as a heteromer of two NR1 and two NR2 subunits, while complexes containing the NR3 subunit form either diheteromer (NR1/NR3) or triheteromer (NR1/NR2/NR3) receptors (Paoletti, 2011). The NR1 and NR3 subunits contain the glycine binding site, while the NR2 subunits contain the glutamate binding site. The subunit composition of NMDARs has been known to vary during development and across different brain areas, thus providing functional diversity to the receptor complex and explaining differential brain region sensitivities to EtOH.

EtOH dose-dependently and noncompetitively antagonizes NMDAR function (Figure 3). The onset of inhibition is rapid and includes reduction in the mean open time and open frequency of the channel but not in single channel conductance. This has led to the hypothesis that EtOH directly interacts with NMDAR subunits to regulate channel gating in mediating its antagonist effects, likely by interacting with domains regulating channel gating (Moykkynen & Korpi, 2012). Furthermore, external factors influence the receptor's sensitivity to EtOH inhibition, such as presence of cofactors, intracellular signaling and scaffolding proteins, and posttranslational receptor modifications (Ron, 2004). The precise mechanism by which ethanol inhibits channel activity as well as the subunits and specific amino acids involved are still in the early stages of discovery. However, as discussed in this chapter, exciting new data are beginning to unravel this mystery.

Acutely, EtOH inhibits NMDAR function, whereas chronic treatment results in an upregulation of NMDAR expression and activity as a result of repeated blockade (Krystal, Petrakis, Limoncelli, et al., 2003). It has been hypothesized that this enhanced receptor function may serve as a mechanism for EtOH

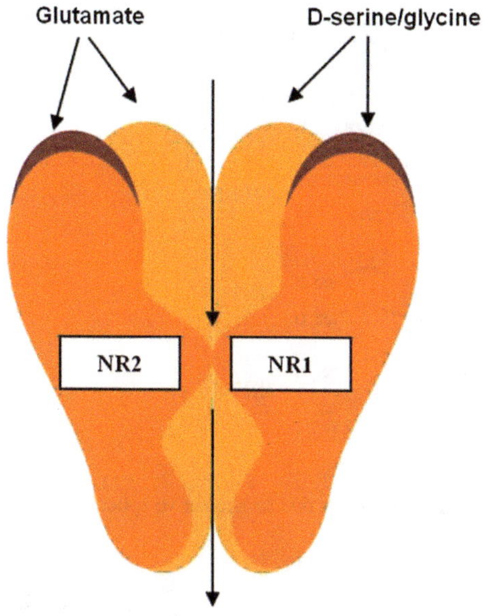

FIGURE 1 **The NMDA receptor.** The conventional N-methyl-D-aspartate receptor (NMDAR) is composed of two obligatory NR1 subunits that bind glycine or D-serine, and two NR2 subunits that bind glutamate. Membrane depolarization removes the magnesium ion in the channel (not shown) and binding of the co-agonists leads to receptor activation and the influx of calcium (Ca^{2+}) and sodium (Na^+) ions, as well as the expulsion of potassium ions.

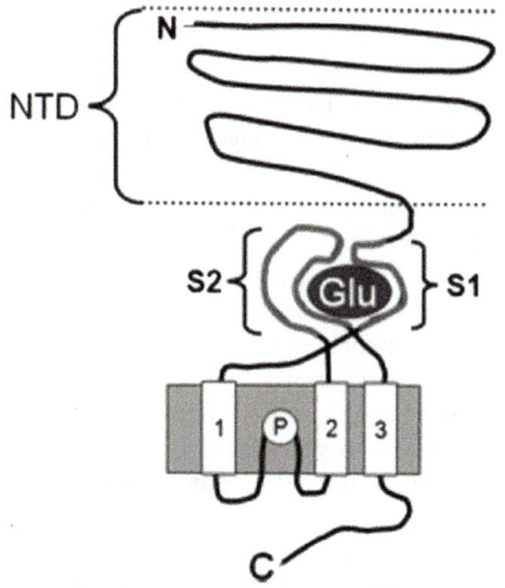

FIGURE 2 **The basic composition of an NMDA receptor subunit.** N-methyl-D-aspartate receptor (NMDAR) subunits consist of an amino-terminal domain (NTD), ligand binding domain (S1-S2), a transmembrane domain (TM 1–3), a membrane loop that forms part of the channel pore (P), and the carboxy-terminal domain (CTD). The NTD binds a variety of endogenous ligands such as zinc ions, phenylethanolamines, polyamines, and protons, while the CTD binds postsynaptic proteins, including scaffolding proteins, kinases, and phosphatases, as well as intracellular signaling molecules. *Reprinted with permission from Moykkynen and Korpi (2012).*

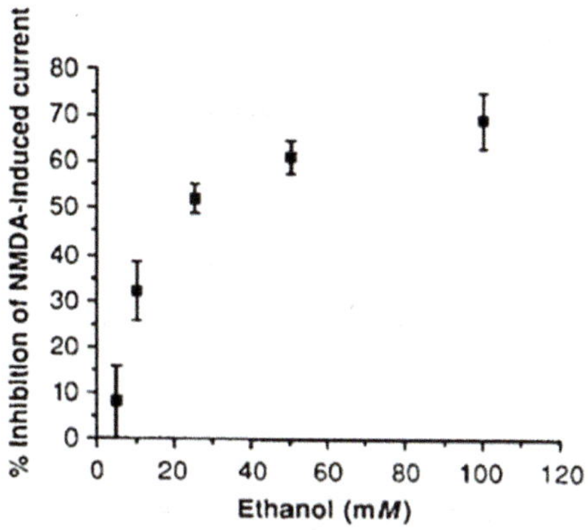

FIGURE 3 **Dose-dependent inhibition of NMDA receptors by ethanol.** Ethanol rapidly and dose-dependently inhibits NMDA-activated currents in hippocampal neurons. Values are mean ± SEM. *Reprinted with permission from Lovinger, White, and Weight (1989).*

tolerance by decreasing receptor sensitivity to EtOH's effects (Krystal, Petrakis, Limoncelli, et al., 2003). This would suggest that EtOH's actions at the NMDAR may contribute to the acquisition and maintenance of alcohol use disorders. Therefore, the focus of this chapter is on ethanol's actions at the NMDAR and its subunits, as well as the modulatory role of the subunits in mediating EtOH's actions.

ETHANOL'S ACTIONS AT THE NMDAR: ROLE IN THE SUBJECTIVE AND BEHAVIORAL EFFECTS OF ALCOHOL

A plethora of evidence indicates that EtOH's behavioral and subjective effects are due to its actions at the NMDAR. For example, NMDA antagonists mimic the behavioral and intoxicating effects of EtOH in multiple animal species (Grant & Colombo, 1993; Grant, Knisely, Tabakoff, Barrett, & Balster, 1991). That this is evident in various species solidifies the hypothesis that EtOH's behavioral effects are attributable at least in part to NMDAR antagonism. Furthermore, alcohol-dependent patients and individuals at risk for alcohol use disorders have altered NMDAR responses to antagonists of this receptor (Krystal, Petrakis, Krupitsky, et al., 2003; Petrakis et al., 2004; Phelps et al., 2009). These studies support an important role for NMDAR antagonism in EtOH's effects and suggest that altered NMDAR responses may arise prior to the development of alcoholism and/or be caused by long-term alcohol use.

NMDAR SUBUNITS AND ETHANOL

The NR1 Subunit

Effect of Splice Variants on EtOH's Ability to Inhibit NMDARs

The NR1 subunit is encoded for by a single gene that contains three areas of alternative splicing. One site is present in the N-terminus (the amino-terminal N1 cassette; exon 5) and two in the C-terminus (the carboxyl terminal C1 and C2 cassettes; exon

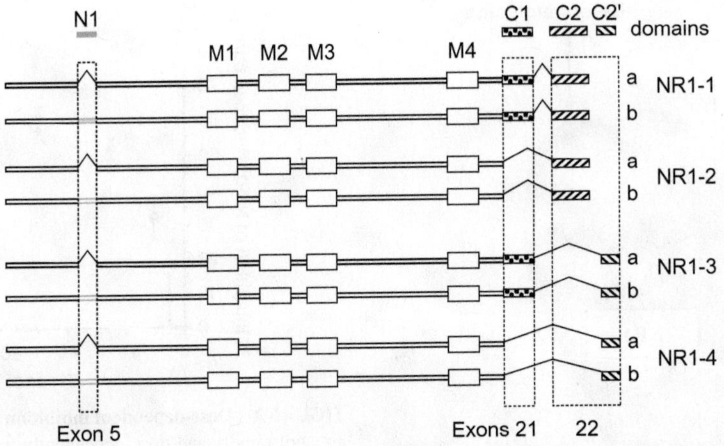

FIGURE 4 Splice variants of the NR1 transcript. The NR1 subunit can have eight possible isoforms due to alternative splicing of the NR1 gene. Reprinted with permission from Dingledine, Borges, Bowie, and Traynelis (1999).

21 and exon 22, respectively), resulting in a total of eight possible isoforms referred to as NR1-4a and NR1-4b (Figure 4; Paoletti, et al., 2013).

The NR1 splice variants confer differing degrees of channel sensitivity to EtOH inhibition. The NR2C and NR2D NMDARs that contain either an NR1-3b or NR1-4b splice variant show only a 15–18% inhibition by EtOH, whereas the level of inhibition is over 50% for the NR1-2B/NR2C combination (Jin & Woodward, 2006). Interestingly, no NR1 splice variant is consistent in regulating NMDAR inhibition by EtOH, emphasizing the fact that the effects of EtOH on an individual NMDAR are dependent upon the combination of NR1 and NR2 subunits. As will be discussed below, the NR1 splice variant also influences EtOH response in NR3-containing receptors (see NR3 section). Although not systemically studied, it is likely that different NR1 splice variants alter the 3D conformation of the NR1 protein, thereby modulating its interactions with the other subunits and with EtOH.

Does the NR1 Subunit Contain an EtOH Binding Site?

The N-terminal domain and C-terminal domain modulate the receptor's inhibition of EtOH, but they do not contain an EtOH binding site. In contrast, studies have strongly suggested that residues in TM3 and TM4 domains, which also appear to regulate channel gating, may contain an EtOH binding site (den Hartog et al., 2013; Ren, Zhao, Dwyer, & Peoples, 2012; Ronald, Mirshahi, & Woodward, 2001). These amino acids have been identified as glycine (Gly-638), phenylalanine (Phe-639), methionine (Met-818), and leucine (Leu-819) and interact with another specific set of residues in the NR2A subunit at the intersubunit interface to regulate EtOH binding (see NR2 section; Figure 5(A), Ren et al., 2012; Ronald et al., 2001).

Modulation of Subunit Expression by Chronic EtOH Exposure and Withdrawal

Repeated EtOH treatment increases NR1 protein levels, with most reports indicating no change in total mRNA levels but an upregulation of certain splice variants (Kumari & Anji, 2005; Nagy, 2008).

In vitro and in vivo, EtOH has been shown to increase the stability of the NR1 mRNA and this stability appears to be due to EtOH-induced interactions between the 3′UTR of the NR1 gene and the RNA binding proteins GIIβ, annexin A2, and a third unidentified protein, which have been suggested to stabilize the mRNA (Anji & Kumari, 2006, 2011).

Posttranslational Modifications of NR1 Subunit in EtOH's Actions

Carboxy-terminal domain (CTD) phosphorylation of NMDARs regulates surface expression, synaptic delivery, and NMDAR functioning, thus altering synaptic strength (Nagy, 2008; Wang et al., 2006). Because synaptic plasticity at glutamatergic synapses has been heavily implicated in the aberrant learning that has been postulated to occur during the acquisition of drug dependence, NMDAR modulation by EtOH may be one step in a cascade of events that leads to long-term changes in the brain and the addicted phenotype (Luscher, 2013).

Mutagenesis studies have shown that NR1 CTD residues regulate EtOH's sensitivity. For example, replacing serine 890 with either an alanine or aspartate residue, and in doing so preventing phosphorylation, increased EtOH's inhibitory effects on the NMDAR when combined with the NR2A, but not the NR2B, subunit (Xu, Smothers, & Woodward, 2011). When the mutation was modified to span six residues, because multiple sites are likely to be phosphorylated in vivo, again there was greater EtOH inhibition when the NR1 subunit was paired with NR2A, not NR2B. This highlights the importance of NMDAR subunit composition in EtOH's actions and indicates that phosphorylation would increase channel activity and reduce the inhibitory actions of EtOH.

Although these results highlight the modulatory influence of CTD phosphorylation in EtOH inhibition, it is unclear what kinases might be responsible for this. Surprisingly, modulating PKA activity does not appear to influence the inhibitory effects of EtOH in HEK cells, nor in cultured cortical or hippocampal neurons (Xu & Woodward, 2006). Additionally, removal of the binding site for CaMKII in the NR1 subunit CTD had no effect on EtOH inhibition of NMDARs; this is also seen in mutants that mimic or

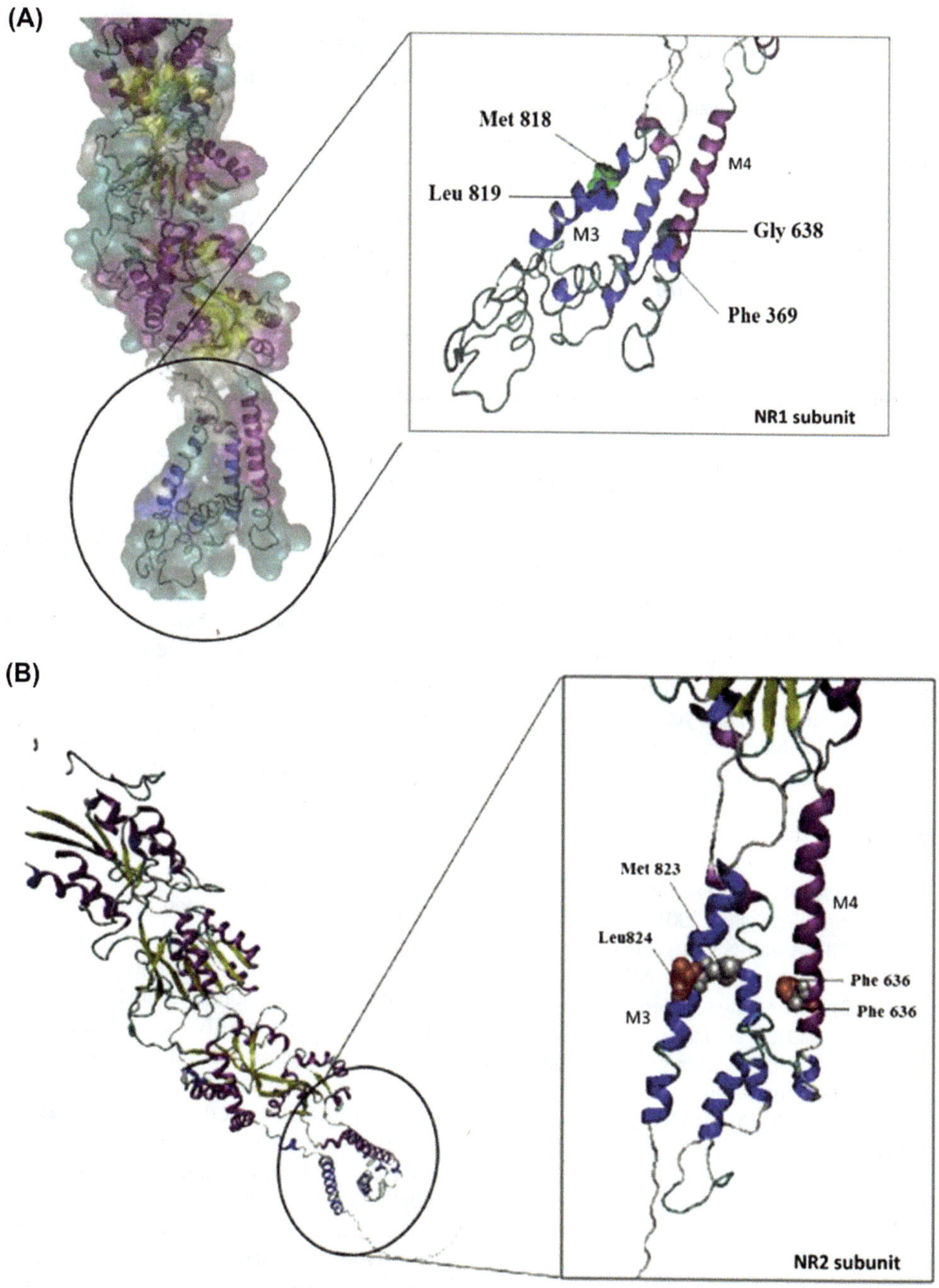

FIGURE 5 **Residues comprising the ethanol binding site.** Residues in the third and fourth transmembrane domain of the NR1 (A) and NR2A subunit (B) that interact with each other to make up the ethanol binding site. Abbreviations: Met, methionine; Leu, leucine; Gly, glycine; Phe, phenylalanine. *Reprinted with permission from Chandrasekar (2013).*

prevent CaMKII phosphorylation of the subunit (Xu, Chandler, & Woodward, 2008). Although these studies would seem to suggest that PKA and CaMKII do not regulate NMDAR responses to EtOH, studies using tissue slices have shown that EtOH treatment increases phosphorylation of the NR1 subunit via a mechanism involving PKA and PKC (Ferrani-Kile, Randall, & Leslie, 2003; Kumar, Lane, & Morrow, 2006). These discrepancies further highlight the fact that different results that can be obtained depending on methodology used and whether the NMDARs studied are native or recombinant.

Role of NR1 Subunit in EtOH's Behavioral Effects

The NR1 subunit has been implicated in a wide range of EtOH's behavioral effects. Reduced protein and gene expression levels have been shown to decrease EtOH consumption, reduce withdrawal severity, and are associated with sensitization to the locomotor stimulant effects of EtOH (Abrahao et al., 2013; Du, Elberger, Matthews, & Hamre, 2012; Nona, Li, & Nobrega, 2014). Transgenic mice with a mutation in this subunit that renders NMDARs insensitive to inhibition by EtOH are resistant to its acute stimulant, ataxic, and anxiolytic properties, suggesting that EtOH inhibition of NMDAR via NR1 is required for these properties of EtOH (den Hartog et al., 2013). However, it is important to be aware that transgenic mice may have compensatory developmental adaptations; thus, findings from such studies should be treated with caution.

The NR2 Subunits

Effect of Isoforms on EtOH's Ability to Inhibit NMDARs

Studies examining the role of the NR2 subunits in EtOH inhibition have shown that the overall sensitivity of a single NMDAR to EtOH is dependent upon the specific combinations of NR1 splice variants and NR2 subunits, as well as the expression system used (Jin & Woodward, 2006; Smothers, Clayton, Blevins, & Woodward, 2001). Studies using *Xenopus oocytes* to express recombinant NMDARs have consistently shown that NR1/2A and NR1/2B containing NMDARs are the most sensitive to being inhibited by EtOH (Chu, Anantharam, & Treistman, 1995; Mirshahi & Woodward, 1995). However, studies using HEK 293 cells have reported mixed findings. In some cases, NR1/2B containing NMDARs are the most sensitive to inhibition by EtOH; other studies suggest that it is the NR1/2B and NR1/2C receptors, and yet others indicate that EtOH exerts similar levels of inhibition among the NR1/NR2B and NR1/2A (Smothers & Woodward, 2003; Xu, Smothers, Trudell, & Woodward, 2012; Xu et al., 2011). Further evidence demonstrating an important role for the cell expression system used in NMDAR sensitivity to EtOH antagonism comes from a study showing that human NMDAR subunit sensitivity to EtOH differed across different cell lines (Smothers et al., 2001). Nonetheless, the current consensus is that NR2A and NR2B generally confer the greatest sensitivity to EtOH inhibition, while NR2C and NR2D do the least (Allgaier, 2002). Attempting to determine which combinations among the NR2A and 2B confer greater sensitivity to EtOH is likely to be elusive, as differences in cell systems and brain regions examined account for the discrepancies among studies that report that the 2A containing NMDARs are most sensitive to EtOH inhibition than 2B and vice versa.

Do the NR2 Subunits Contain an EtOH Binding Site?

As with the NR1 subunit, the CTD and amino-terminal domain (NTD) of NR2 subunits modulate EtOH responses without containing a binding site. Met 823, Leu 824, Phe 636, and Phe 637 in TM3 and TM4 of the NR2A subunit have been reported to contain an EtOH binding site (Ren et al., 2012; Ren, Zhao, Wu, & Peoples, 2013). Specifically, these four residues along with the four in the NR1 subunit are located in the intersubunit interface in TM3 and TM4, and they interact with each other to regulate EtOH sensitivity (Figure 5(B)).

Modulation of Subunit Expression by Chronic EtOH Exposure and Withdrawal

The majority of studies have shown that following withdrawal from chronic EtOH treatment, there is an upregulation of NR2A and NR2B mRNA and protein levels, with increases in the NR2B subunit and NR2B-containing NMDARs being the most consistently reported (Holmes, Spanagel, & Krystal, 2013; Nagy, 2008). Another factor contributing to the predominance of NR2B-containing NMDARs following EtOH treatment may be the internalization of the NR2A subunit via EtOH-mediated activation of H-Ras (Suvarna et al., 2005).

The increase in NR2B gene expression has been attributed to chronic EtOH-induced increases in demethylation of CpG islands in the NR2B gene (Biermann et al., 2009; Marutha Ravindran & Ticku, 2004, 2005; Qiang et al., 2010). This leads to increases in the binding of transcription factors to the gene promoter, increasing gene transcription, and elevating NR2B protein levels. Additionally, EtOH-induced increases in the protease, tPA, appear to be required for the observed increase in NR2B subunit expression following chronic EtOH treatment (Pawlak, Melchor, Matys, Skrzypiec, & Strickland, 2005).

Posttranslation Modifications in EtOH's Action

CTD phosphorylation of NR2 subunits enhances NMDAR activity and modifies response to bound ligands (Groveman et al., 2012; Ron, 2004). Interestingly, studies using recombinant NMDARs have generally shown that NR2 subunit phosphorylation does not influence EtOH inhibition of the channels. For example, modification of the Src, Fyn, and CaMKII phosphorylation sites in these studies did not change the degree of EtOH blockade of NMDAR function, except in one case where the largest EtOH concentration activated Fyn kinase and lead to phosphorylation of NR1/2A-, but not NR1/NR2B-, containing receptors, rendering them less sensitive to being inhibited by EtOH (Anders, Blevins, Sutton, Chandler, & Woodward, 1999; Anders, Blevins, Sutton, Swope, et al., 1999; Xu et al., 2008). Although such results would seem to suggest that NR2 subunit phosphorylation is not a major determinant of EtOH action at the NMDAR, this does not seem to be the case when examining native NMDARs from brain tissue.

In cortical neurons as well as slices of the ventral and dorsolateral striatum, EtOH did not alter the phosphorylation status of the NR2B subunit, but it did so in the hippocampus and dorsomedial striatum (Kalluri & Ticku, 1999; Miyakawa et al., 1997; Wang et al., 2007, 2010). Increases in NR2B phosphorylation reduced EtOH's antagonist actions at the receptor.

The increase in NR2B phosphorylation by EtOH is as a result of EtOH-induced increases in PKA activity, leading to the dissociation of the NR2B–binding scaffolding protein, RACK1, whose removal allows Fyn kinase to phosphorylate NR2B (Figure 6, Yaka, Phamluong, & Ron, 2003). In support of this model, STK inhibitors prevented EtOH-induced increases in NMDAR function during treatment and washout, suggesting that EtOH's potentiating effects on NMDAR activity are mediated by Fyn kinase activity (Wang et al., 2007; Yaka et al., 2003).

Not only does EtOH lead to NMDAR subunit phosphorylation, but, paradoxically, it has been shown to also decrease phosphorylation of both the NR2A and NR2B subunits in hippocampal and cortical slices (Ferrani-Kile et al., 2003; Hicklin et al., 2011). It has been suggested that EtOH may increase the activity of tyrosine phosphatases, as inhibiting their activity reduces the antagonistic effects of EtOH at the NMDAR (Ferrani-Kile et al., 2003). Indeed, it has been shown that the tyrosine phosphatase STEP is necessary for EtOH-induced

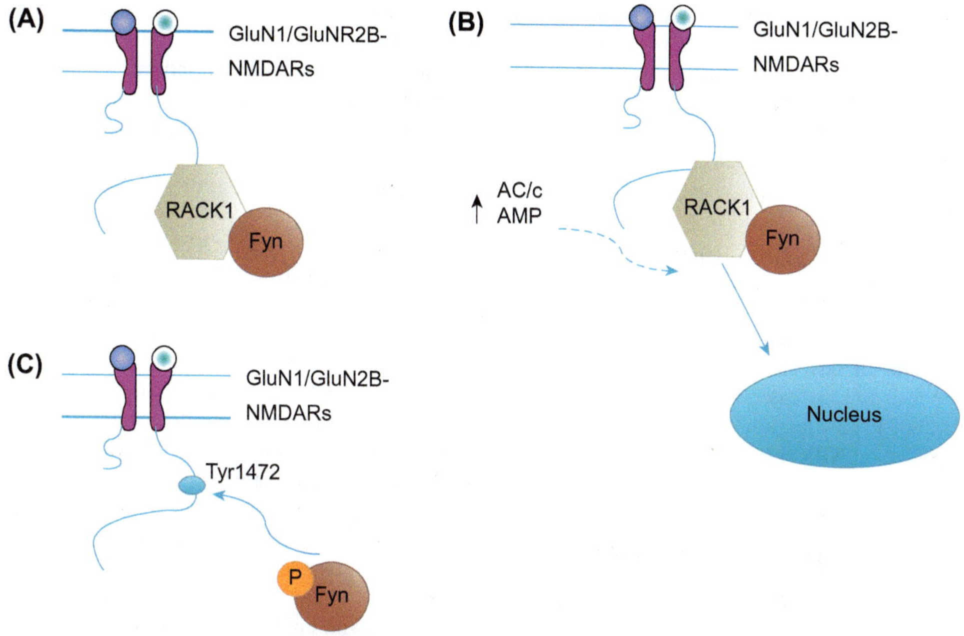

FIGURE 6 **Ethanol-induced phosphorylation of the NR2B subunit.** The scaffolding protein RACK1 binds to both the NR2B carboxy-terminal domain and Fyn kinase (A). EtOH-mediated increase in PKA activity leads to dissociation of the RACK1 from NR2B (B), permitting Fyn kinase to gain access to and phosphorylate the carboxy-terminal domain of NR2B at Tyr1472 (C), rendering the NMDAR less sensitive to EtOH inhibition. Abbreviations: NMDAR, N-methyl-D-aspartate receptor; RACK1, receptor for activated C-kinase; PKA, protein kinase A; pFyn, phosphorylated Fyn kinase; Tyr-1472, tyrosine 1472; cAMP, cyclic adenosine monophosphate; AC, adenylate cyclise. *Reprinted with permission from Trepanier, Jackson, and MacDonald (2012).*

dephosphorylation of NR2B and subsequent receptor inhibition (Hicklin et al., 2011). These findings highlight the dynamic regulation between kinase and phosphatase activity induced by EtOH exposure and demonstrate that the balance between the two will dictate phosphorylation status, and therefore sensitivity to EtOH inhibition.

Role of NR2 Subunits in EtOH's Behavioral Effects

The NR2A Subunit

The NR2A subunit appears to regulate many of EtOH's behavioral actions. Transgenic mice lacking this subunit show normal levels of EtOH consumption, but they have increased motor incoordination in response to EtOH, do not develop tolerance to its sedative/hypnotic effects, do not show withdrawal anxiety, and fail to demonstrate ethanol reward, as measured in a conditioned place preference (CPP) paradigm (Boyce-Rustay & Holmes, 2006; Daut et al., 2014). Furthermore, mice lacking the NR2A C terminus are more sensitive to EtOH's sedative/hypnotic effects (Gordey, Mekmanee, & Mody, 2001). Finally, sensitivity to EtOH's stimulant effects is associated with decreased NR2A protein expression, whereas mice resistant to sensitization have increased NR2A gene expression in the brain (Abrahao et al., 2013; Nona et al., 2014). Collectively, these findings indicate an important role for the NR2A subunit in EtOH reward, motor incoordination, sedation, and locomotor sensitization.

The NR2B Subunit

This subunit has been most implicated in excessive EtOH intake, relapse, tolerance to the sedative/hypnotic properties of EtOH, and aberrant plasticity, which may contribute to the mechanisms underlying drug intake. These effects appear to be regulated by Fyn kinase phosphorylation and subsequent activation of NR2B containing NMDARs. For example, pharmacological blockade of NR2B containing NMDARs potentiates the sedative/hypnotic effects of EtOH to levels seen in Fyn KO mice (Boyce-Rustay & Holmes, 2005; Miyakawa et al., 1997; Yaka et al., 2003). In the dorsomedial striatum, EtOH-induced Fyn activity and NR2B phosphorylation is necessary for EtOH self-administration and reinstatement of EtOH seeking (Wang et al., 2007, 2010). This subunit does not appear important in EtOH's stimulant actions, as the selective NR2B-containing NMDAR antagonist, ifenprodil, did not affect the expression of EtOH sensitization (Broadbent, Kampmueller, & Koonse, 2003).

The NR3 Subunits

Effect of Isoforms on EtOH's Ability to Inhibit NMDARs

Few studies have looked at the role of the NR3 subunits in EtOH's inhibitory response. NR3 subunits bind glycine and they can associate with NR1/NR2 receptors, where they decrease Ca^{2+} permeability. In addition, they can exist in an NR1/NR3 configuration where they form functional receptors that generate glycine-dependent currents (Paoletti, 2011; Smothers & Woodward, 2007). EtOH has been shown to inhibit glycine-activated currents in NR1/NR3A/NR3B receptors, with the NR1 splice variant influencing the degree of EtOH inhibition (Jin, Smothers, & Woodward, 2008; Smothers & Woodward, 2007).

Mg^{2+} increases EtOH-induced inhibition of NMDARs and co-expressing NR3A with NR1/NR2A, but not with NR1/NR2B receptors, has been shown to prevent this effect (Jin et al., 2008; Smothers & Woodward, 2003). Because the Mg^{2+} levels present in the media were similar to concentrations expressed under physiological conditions, these findings suggest that NR3 subunits might regulate EtOH sensitivity of NMDARs in vivo (Jin et al., 2008).

Modulation of Subunit Expression by Chronic EtOH Exposure and Withdrawal

To date, only one study has examined NR3 expression following chronic EtOH treatment. In rats chronically exposed to EtOH, NR3 subunit mRNA was not increased in the amygdala (Floyd, Jung, & McCool, 2003). Future studies should examine the effects of EtOH exposure on NR3 subunit expression in different brain areas.

APPLICATIONS TO OTHER ADDICTIONS AND SUBSTANCE MISUSE

An important role for NMDA receptors in other drugs of abuse has been demonstrated. For example, exposure to cocaine, amphetamine, and morphine has been shown to increase the phosphorylation levels of the NR1, NR2A, and NR2B subunits and modify subunit expression (Ma, Cepeda, & Cui, 2009). These biochemical changes appear to have important implications for behavior, as subunit knockdown and receptor antagonism block drug reward and sensitization to the locomotor stimulant effects of these drugs (Kao, Huang, & Tao, 2011; Ma et al., 2009; Tzschentke, 2007; Tzschentke & Schmidt, 2000). Clearly, EtOH's actions at NMDARs are not unique to this drug and instead appear to be a common action of other addictive agents. It is highly likely that modulation of NMDAR function and expression by drugs of abuse is a key step in the acquisition of drug dependence (Luscher, 2013).

CONCLUDING REMARKS

Research into understanding the mechanism of EtOH action within the nervous system and at the NMDAR complex has certainly come along way. In spite of the vast knowledge base acquired to date, there still remain limitations to what we currently know. To fully appreciate the effects of EtOH at this receptor site, such limitations must be kept in mind when interpreting data in the literature and when conducting future research in the hopes of gaining a greater insight into EtOH's mechanism of action.

An important fact to recognize is the effect of methodological aspects on the reported degree of receptor inhibition by EtOH in different studies. The degree of inhibition at a particular concentration can differ across studies largely due to different experimental protocols employed, including animal gender, the behavioral model used, developmental age, and the brain region examined. With use of in vitro assays comes the further complication that the type of cell system used can impact results and may not always reflect what occurs in vivo, since cofactors, scaffolding proteins, and culture milieu can be very different from the intact biological system. When it comes to human alcohol consumption, an added source of variability in EtOH's effects on NMDARs is the presence of congeners present in alcoholic beverages, which can alter EtOH's ability to inhibit NMDAR in a subunit-dependent manner. Indeed, such factors can explain the observation that EtOH concentrations, which significantly impair an individual's behavior (in the 4–68 mM range), have been shown in some studies to only modestly inhibit NMDAR function, questioning the extent of NMDAR involvement in EtOH's behavioral effects.

The identification of EtOH binding sites should also be treated with caution. Because such studies were carried out in vitro in cell expression systems, they are unable to assess the roles of external factors present in vivo in EtOH actions at the NMDAR, such as scaffolding proteins, intracellular signaling molecules, and NMDAR cofactors, which, although not yet a fully explored research area, are known to influence EtOH at this receptor. Additionally, such studies seem to suggest that EtOH's ability to modulate channel function arises from a direct action of EtOH on the protein. While this may be true, and indeed is the view supported by one theory of EtOH's action in the nervous system (the protein theory), it may not be entirely accurate. It has been argued in the past that one way EtOH alters protein function is by a direct interaction with lipid membranes, which would subsequently disrupt the lipid matrix surrounding the protein, thereby indirectly effecting protein function (referred to as the lipid theory of EtOH action). As residues within NMDAR subunits that regulate EtOH action are also known to orient and interact with the surrounding lipid membrane, it is possible that EtOH's action at the NMDAR might be regulated not only by a direct interaction with the protein but also by the surrounding lipid environment—a hypothesis that can be tested in intact tissue. Furthermore, the high concentrations of EtOH used in in vitro studies (millimolar range) may have profoundly different effects in vivo, not only altering NMDAR protein but potentially even the surrounding lipid environment. Indeed, with the use of cell expression systems, we only have a piece of the puzzle by which EtOH acts at the NMDAR, which is likely to be different from an intact biological system.

The recent discovery of functionally different NR3 subunits and the finding that they may play an important role in regulation of EtOH sensitivity of NMDARs in vivo adds impetus to study this subunit in EtOH's actions to a much greater extent than is currently the case. Indeed, only a paucity of data currently exists, examining the effects of EtOH exposure on this subunit as well as how NR3 may affect EtOH's actions at the NMDAR. Certainly, more focused studies on this receptor subunit will be a fruitful avenue to explore.

In spite of these limitations and the need for further research, it should be clear from the evidence reviewed in this chapter that EtOH actions at NMDARs play an important role in the behavioral effects of alcohol. Future work in this field holds promise in elucidating alcohol's actions in the CNS with the hopes of using such knowledge to develop more effective medication for the treatment of alcohol use disorders.

DEFINITION OF TERMS

Neuroplasticity These experience-induced changes in the brain can involve changes in synapses (i.e., synaptic plasticity) or changes in neural pathways, leading to larger scale structural brain changes at the cortical level.

Synaptic plasticity Synaptic plasticity is the strengthening or weakening of synapses in response to the degree of activity. Synaptic strengthening includes increased probably of neurotransmitter release, increase in the synaptic response of a postsynaptic cell, thereby enhancing the probably of generating an action potential. Synaptic weakening involves reduction in the neurotransmitter release probability and a decrease in the postsynaptic cell's response.

Single channel currents Single channel currents are the measurement of current flow through an ion channel over time. Channels fluctuate between open and closed states and stimuli can affect transition between the two states, such as change in membrane voltage or binding of ligand, while phosphorylation status of channel regulates channel activity.

Phosphorylation/Dephosphorylation These are the addition (phosphorylation) or removal (dephosphorylation) of a phosphate group, PO_4^{3-} from a biomolecule, thereby altering protein activity and function. Kinases add PO_4^{3-} group on a specific amino acid, thereby phosphorylating their target, whereas phosphatases remove the PO_4^{3-} group, dephosphorylating the target.

DNA methylation This is the addition of a methyl group ($-CH_4$) to a cytosine nucleotide, thereby suppressing gene expression.

CpG islands CpG islands are regions where a cytosine nucleotide is located beside a guanine nucleotide in a linear sequence along the DNA strand. CpG is shorthand for cytosine-phosphate-guanine. The cytosine nucleotide can be methylated.

Gene promoter This region of DNA, upstream the start of a gene, initiates transcription by the binding of transcription factors.

Transcription factor Transcription factor is a protein that binds to the gene promoter and regulates the degree of transcription from DNA to mRNA.

Site-directed mutagenesis In this technique, changes are made to the DNA sequence of a gene to investigate its function. In the studies described in this chapter, residues in the CTD of NMDAR subunits were mutated in such a way to prevent or mimic phosphorylation, in order to study of the effects of subunit phosphorylation on channel inhibition by EtOH.

H-Ras H-Ras is an enzyme that breaks down guanosine triphosphate, thereby stimulating the kinase Raf and leading to activation of the MAPK/ERK signal transduction pathway

KEY FACTS OF GLUTAMATE

- Glutamate is the most abundant excitatory neurotransmitter in the brain and plays a critical role in learning, memory, and neuroplasticity.
- Glutamate is produced in neurons, where it is converted from the amino acid glutamine by the enzyme glutaminase.
- The newly synthesized glutamate is then packaged in the neuronal terminal for subsequent release.
- Maintaining appropriate concentrations of glutamate in the synapse is critical, as too low or too high amounts can be harmful.
- What constitutes an "appropriate" concentration depends on many factors, such as the synaptic location being discussed and the level of activity at that particular region at a particular point in time.
- Too much glutamate can lead to seizures and neuronal death, while too little can cause psychosis, coma, and death.
- As a result, extracellular glutamate levels are tightly regulated by glutamate transporters, membrane proteins located on neurons, and glial cells which uptake glutamate,
- Glutamate binds to two classes of receptors: ionotropic and metabotropic.
- The ionotropic receptors are ligand-gated ion channels, allowing the passage of sodium, potassium, and in some cases small amounts of calcium ions. They mediate fast excitatory transmission and there are three such receptors.
- The metabotropic receptors are coupled to intracellular signaling mechanisms and mediate slow excitatory transmission. There are eight such receptors.

SUMMARY POINTS

- The NMDAR is a key target of EtOH.
- NMDARs are heteromeric protein complexes, with the conventional NMDAR consisting of two NR1 and two NR2 subunits.
- EtOH antagonizes NMDAR channel function in many brain regions, which may be important in EtOH's intoxicating and anesthetic properties.
- Chronic EtOH modifies NMDAR subunit phosphorylation status and expression, thereby regulating NMDAR sensitivity to EtOH inhibition.
- In general, subunit phosphorylation reduces NMDAR sensitivity to EtOH inhibition, whereas dephosphorylation enhances the inhibitory actions of EtOH.
- Studies examining EtOH inhibition of NMDARs are heavily influenced by methodological aspects.
- A putative EtOH binding site exists in the NR1/2A subunit interface in the third and fourth transmembrane domains.

REFERENCES

Abrahao, K. P., Ariwodola, O. J., Butler, T. R., Rau, A. R., Skelly, M. J., Carter, E., ... Weiner, J. L. (2013). Locomotor sensitization to ethanol impairs NMDA receptor-dependent synaptic plasticity in the nucleus accumbens and increases ethanol self-administration. *Journal of Neuroscience, 33*, 4834–4842.

Allgaier, C. (2002). Ethanol sensitivity of NMDA receptors. *Neurochemistry International, 41*, 377–382.

Anders, D. L., Blevins, T., Sutton, G., Chandler, L. J., & Woodward, J. J. (1999). Effects of c-Src tyrosine kinase on ethanol sensitivity of recombinant NMDA receptors expressed in HEK 293 cells. *Alcoholism: Clinical & Experimental Research, 23*, 357–362.

Anders, D. L., Blevins, T., Sutton, G., Swope, S., Chandler, L. J., & Woodward, J. J. (1999). Fyn tyrosine kinase reduces the ethanol inhibition of recombinant NR1/NR2A but not NR1/NR2B NMDA receptors expressed in HEK 293 cells. *Journal of Neurochemistry, 72*, 1389–1393.

Anji, A., & Kumari, M. (2006). A novel RNA binding protein that interacts with NMDA R1 mRNA: regulation by ethanol. *European Journal of Neuroscience, 23*, 2339–2350.

Anji, A., & Kumari, M. (2011). A cis-acting region in the N-methyl-d-aspartate R1 3′-untranslated region interacts with the novel RNA-binding proteins beta subunit of alpha glucosidase II and annexin A2–effect of chronic ethanol exposure in vivo. *European Journal of Neuroscience, 34*, 1200–1211.

Biermann, T., Reulbach, U., Lenz, B., Frieling, H., Muschler, M., Hillemacher, T., ... Bleich, S. (2009). N-methyl-D-aspartate 2b receptor subtype (NR2B) promoter methylation in patients during alcohol withdrawal. *Journal of Neural Transmission, 116*, 615–622.

Boyce-Rustay, J. M., & Holmes, A. (2005). Functional roles of NMDA receptor NR2A and NR2B subunits in the acute intoxicating effects of ethanol in mice. *Synapse, 56*, 222–225.

Boyce-Rustay, J. M., & Holmes, A. (2006). Ethanol-related behaviors in mice lacking the NMDA receptor NR2A subunit. *Psychopharmacology (Berlin), 187*, 455–466.

Broadbent, J., Kampmueller, K. M., & Koonse, S. A. (2003). Expression of behavioral sensitization to ethanol by DBA/2J mice: the role of NMDA and non-NMDA glutamate receptors. *Psychopharmacology (Berlin), 167*, 225–234.

Chandrasekar, R. (2013). Alcohol and NMDA receptor: current research and future direction. *Frontiers in Molecular Neuroscience, 6*, 14.

Chu, B., Anantharam, V., & Treistman, S. N. (1995). Ethanol inhibition of recombinant heteromeric NMDA channels in the presence and absence of modulators. *Journal of Neurochemistry, 65*, 140–148.

Daut, R. A., Busch, E. F., Ihne, J., Fisher, D., Mishina, M., Grant, S. G., ... Holmes, A. (2014). Tolerance to ethanol intoxication after chronic ethanol: role of GluN2A and PSD-95. *Addiction Biology, 20*(2), 259–262.

Dingledine, R., Borges, K., Bowie, D., & Traynelis, S. F. (1999). The glutamate receptor ion channels. *Pharmacological Reviews, 51*, 7–61.

Du, X., Elberger, A. J., Matthews, D. B., & Hamre, K. M. (2012). Heterozygous deletion of NR1 subunit of the NMDA receptor alters ethanol-related behaviors and regional expression of NR2 subunits in the brain. *Neurotoxicology and Teratology, 34*, 177–186.

Ferrani-Kile, K., Randall, P. K., & Leslie, S. W. (2003). Acute ethanol affects phosphorylation state of the NMDA receptor complex: implication of tyrosine phosphatases and protein kinase A. *Brain Research Molecular Brain Research, 115*, 78–86.

Floyd, D. W., Jung, K. Y., & McCool, B. A. (2003). Chronic ethanol ingestion facilitates N-methyl-D-aspartate receptor function and expression in rat lateral/basolateral amygdala neurons. *Journal of Pharmacology and Experimental Therapeutics, 307*, 1020–1029.

Gordey, M., Mekmanee, L., & Mody, I. (2001). Altered effects of ethanol in NR2A(DeltaC/DeltaC) mice expressing C-terminally truncated NR2A subunit of NMDA receptor. *Neuroscience, 105*, 987–997.

Grant, K. A., & Colombo, G. (1993). Discriminative stimulus effects of ethanol: effect of training dose on the substitution of N-methyl-D-aspartate antagonists. *Journal of Pharmacology and Experimental Therapeutics, 264*, 1241–1247.

Grant, K. A., Knisely, J. S., Tabakoff, B., Barrett, J. E., & Balster, R. L. (1991). Ethanol-like discriminative stimulus effects of non-competitive n-methyl-d-aspartate antagonists. *Behavioural Pharmacology, 2*, 87–95.

Groveman, B. R., Feng, S., Fang, X. Q., Pflueger, M., Lin, S. X., Bienkiewicz, E. A., & Yu, X. (2012). The regulation of N-methyl-D-aspartate receptors by Src kinase. *FEBS Journal, 279*, 20–28.

den Hartog, C. R., Beckley, J. T., Smothers, T. C., Lench, D. H., Holseberg, Z. L., Fedarovich, H., ... Woodward, J. J. (2013). Alterations in ethanol-induced behaviors and consumption in knock-in mice expressing ethanol-resistant NMDA receptors. *PLoS One, 8*, e80541.

Hicklin, T. R., Wu, P. H., Radcliffe, R. A., Freund, R. K., Goebel-Goody, S. M., Correa, P. R., ... Browning, M. D. (2011). Alcohol inhibition of the NMDA receptor function, long-term potentiation, and fear learning requires striatal-enriched protein tyrosine phosphatase. *Proceedings of the National Academy of Sciences of the United States of America, 108*, 6650–6655.

Holmes, A., Spanagel, R., & Krystal, J. H. (2013). Glutamatergic targets for new alcohol medications. *Psychopharmacology (Berlin), 229*, 539–554.

Jin, C., Smothers, C. T., & Woodward, J. J. (2008). Enhanced ethanol inhibition of recombinant N-methyl-D-aspartate receptors by magnesium: role of NR3A subunits. *Alcoholism: Clinical & Experimental Research, 32*, 1059–1066.

Jin, C., & Woodward, J. J. (2006). Effects of 8 different NR1 splice variants on the ethanol inhibition of recombinant NMDA receptors. *Alcoholism: Clinical & Experimental Research, 30*, 673–679.

Kalluri, H. S., & Ticku, M. K. (1999). Effect of ethanol on phosphorylation of the NMDAR2B subunit in mouse cortical neurons. *Brain Research Molecular Brain Research, 68*, 159–168.

Kao, J. H., Huang, E. Y., & Tao, P. L. (2011). NR2B subunit of NMDA receptor at nucleus accumbens is involved in morphine rewarding effect by siRNA study. *Drug and Alcohol Dependence, 118*, 366–374.

Krystal, J. H., Petrakis, I. L., Krupitsky, E., Schutz, C., Trevisan, L., & D'Souza, D. C. (2003). NMDA receptor antagonism and the ethanol intoxication signal: from alcoholism risk to pharmacotherapy. *Annals of the New York Academy of Sciences, 1003*, 176–184.

Krystal, J. H., Petrakis, I. L., Limoncelli, D., Webb, E., Gueorgueva, R., D'Souza, D. C., ... Charney, D. S. (2003). Altered NMDA glutamate receptor antagonist response in recovering ethanol-dependent patients. *Neuropsychopharmacology, 28*, 2020–2028.

Kumar, S., Lane, B. M., & Morrow, A. L. (2006). Differential effects of systemic ethanol administration on protein kinase cepsilon, gamma, and beta isoform expression, membrane translocation, and target phosphorylation: reversal by chronic ethanol exposure. *Journal of Pharmacology and Experimental Therapeutics, 319*, 1366–1375.

Kumari, M., & Anji, A. (2005). An old story with a new twist: do NMDAR1 mRNA binding proteins regulate expression of the NMDAR1 receptor in the presence of alcohol? *Annals of the New York Academy of Sciences, 1053*, 311–318.

Lovinger, D. M., White, G., & Weight, F. F. (1989). Ethanol inhibits NMDA-activated ion current in hippocampal neurons. *Science, 243*, 1721–1724.

Luscher, C. (2013). Drug-evoked synaptic plasticity causing addictive behavior. *Journal of Neuroscience, 33*, 17641–17646.

Ma, Y. Y., Cepeda, C., & Cui, C. L. (2009). The role of striatal NMDA receptors in drug addiction. *International Review of Neurobiology, 89*, 131–146.

Marutha Ravindran, C. R., & Ticku, M. K. (2004). Changes in methylation pattern of NMDA receptor NR2B gene in cortical neurons after chronic ethanol treatment in mice. *Brain Research Molecular Brain Research, 121*, 19–27.

Marutha Ravindran, C. R., & Ticku, M. K. (2005). Role of CpG islands in the up-regulation of NMDA receptor NR2B gene expression following chronic ethanol treatment of cultured cortical neurons of mice. *Neurochemistry International, 46*, 313–327.

Mirshahi, T., & Woodward, J. J. (1995). Ethanol sensitivity of heteromeric NMDA receptors: effects of subunit assembly, glycine and NMDAR1 Mg(2+)-insensitive mutants. *Neuropharmacology, 34*, 347–355.

Miyakawa, T., Yagi, T., Kitazawa, H., Yasuda, M., Kawai, N., Tsuboi, K., & Niki, H. (1997). Fyn-kinase as a determinant of ethanol sensitivity: relation to NMDA-receptor function. *Science, 278*, 698–701.

Moykkynen, T., & Korpi, E. R. (2012). Acute effects of ethanol on glutamate receptors. *Basic & Clinical Pharmacology & Toxicology, 111*, 4–13.

Nagy, J. (2008). Alcohol related changes in regulation of NMDA receptor functions. *Current Neuropharmacology, 6*, 39–54.

Nona, C. N., Li, R., & Nobrega, J. N. (2014). Altered NMDA receptor subunit gene expression in brains of mice showing high vs. low sensitization to ethanol. *Behavioural Brain Research, 260*, 58–66.

Paoletti, P. (2011). Molecular basis of NMDA receptor functional diversity. *European Journal of Neuroscience, 33*, 1351–1365.

Paoletti, P., Bellone, C., & Zhou, Q. (2013). NMDA receptor subunit diversity: impact on receptor properties, synaptic plasticity and disease. *Nature Reviews Neuroscience, 14*, 383–400.

Pawlak, R., Melchor, J. P., Matys, T., Skrzypiec, A. E., & Strickland, S. (2005). Ethanol-withdrawal seizures are controlled by tissue plasminogen activator via modulation of NR2B-containing NMDA receptors. *Proceedings of the National Academy of Sciences of the United States of America, 102*, 443–448.

Petrakis, I. L., Limoncelli, D., Gueorguieva, R., Jatlow, P., Boutros, N. N., Trevisan, L., ... Krystal, J. H. (2004). Altered NMDA glutamate receptor antagonist response in individuals with a family vulnerability to alcoholism. *American Journal of Psychiatry, 161*, 1776–1782.

Phelps, L. E., Brutsche, N., Moral, J. R., Luckenbaugh, D. A., Manji, H. K., & Zarate, C. A., Jr. (2009). Family history of alcohol dependence and initial antidepressant response to an N-methyl-D-aspartate antagonist. *Biological Psychiatry, 65*, 181–184.

Qiang, M., Denny, A., Chen, J., Ticku, M. K., Yan, B., & Henderson, G. (2010). The site specific demethylation in the 5′-regulatory area of NMDA receptor 2B subunit gene associated with CIE-induced upregulation of transcription. *PLoS One, 5*, e8798.

Ren, H., Zhao, Y., Dwyer, D. S., & Peoples, R. W. (2012). Interactions among positions in the third and fourth membrane-associated domains at the intersubunit interface of the N-methyl-D-aspartate receptor forming sites of alcohol action. *Journal of Biological Chemistry, 287*, 27302–27312.

Ren, H., Zhao, Y., Wu, M., & Peoples, R. W. (2013). A novel alcohol-sensitive position in the N-methyl-D-aspartate receptor GluN2A subunit M3 domain regulates agonist affinity and ion channel gating. *Molecular Pharmacology, 84*, 501–510.

Ron, D. (2004). Signaling cascades regulating NMDA receptor sensitivity to ethanol. *Neuroscientist, 10*, 325–336.

Ronald, K. M., Mirshahi, T., & Woodward, J. J. (2001). Ethanol inhibition of N-methyl-D-aspartate receptors is reduced by site-directed mutagenesis of a transmembrane domain phenylalanine residue. *Journal of Biological Chemistry, 276*, 44729–44735.

Smothers, C. T., Clayton, R., Blevins, T., & Woodward, J. J. (2001). Ethanol sensitivity of recombinant human N-methyl-D-aspartate receptors. *Neurochemistry International, 38*, 333–340.

Smothers, C. T., & Woodward, J. J. (2003). Effect of the NR3 subunit on ethanol inhibition of recombinant NMDA receptors. *Brain Research, 987*, 117–121.

Smothers, C. T., & Woodward, J. J. (2007). Pharmacological characterization of glycine-activated currents in HEK 293 cells expressing N-methyl-D-aspartate NR1 and NR3 subunits. *Journal of Pharmacology and Experimental Therapeutics, 322*, 739–748.

Suvarna, N., Borgland, S. L., Wang, J., Phamluong, K., Auberson, Y. P., Bonci, A., & Ron, D. (2005). Ethanol alters trafficking and functional N-methyl-D-aspartate receptor NR2 subunit ratio via H-Ras. *Journal of Biological Chemistry, 280*, 31450–31459.

Trepanier, C. H., Jackson, M. F., & MacDonald, J. F. (2012). Regulation of NMDA receptors by the tyrosine kinase Fyn. *FEBS Journal, 279*, 12–19.

Tzschentke, T. M. (2007). Measuring reward with the conditioned place preference (CPP) paradigm: update of the last decade. *Addiction Biology, 12*, 227–462.

Tzschentke, T. M., & Schmidt, W. J. (2000). Blockade of behavioral sensitization by MK-801: fact or artifact? A review of preclinical data. *Psychopharmacology (Berlin), 151*, 142–151.

Wang, J., Carnicella, S., Phamluong, K., Jeanblanc, J., Ronesi, J. A., Chaudhri, N., ... Ron, D. (2007). Ethanol induces long-term facilitation of NR2B-NMDA receptor activity in the dorsal striatum: implications for alcohol drinking behavior. *Journal of Neuroscience, 27*, 3593–3602.

Wang, J., Lanfranco, M. F., Gibb, S. L., Yowell, Q. V., Carnicella, S., & Ron, D. (2010). Long-lasting adaptations of the NR2B-containing NMDA receptors in the dorsomedial striatum play a crucial role in alcohol consumption and relapse. *Journal of Neuroscience, 30*, 10187–10198.

Wang, J. Q., Liu, X., Zhang, G., Parelkar, N. K., Arora, A., Haines, M., ... Mao, L. (2006). Phosphorylation of glutamate receptors: a potential mechanism for the regulation of receptor function and psychostimulant action. *Journal of Neuroscience Research, 84*, 1621–1629.

Xu, M., Chandler, L. J., & Woodward, J. J. (2008). Ethanol inhibition of recombinant NMDA receptors is not altered by coexpression of CaMKII-alpha or CaMKII-beta. *Alcohol, 42*, 425–432.

Xu, M., Smothers, C. T., Trudell, J., & Woodward, J. J. (2012). Ethanol inhibition of constitutively open N-methyl-D-aspartate receptors. *Journal of Pharmacology and Experimental Therapeutics, 340*, 218–226.

Xu, M., Smothers, C. T., & Woodward, J. J. (2011). Effects of ethanol on phosphorylation site mutants of recombinant N-methyl-D-aspartate receptors. *Alcohol, 45*, 373–380.

Xu, M., & Woodward, J. J. (2006). Ethanol inhibition of NMDA receptors under conditions of altered protein kinase A activity. *Journal of Neurochemistry, 96*, 1760–1767.

Yaka, R., Phamluong, K., & Ron, D. (2003). Scaffolding of Fyn kinase to the NMDA receptor determines brain region sensitivity to ethanol. *Journal of Neuroscience, 23*, 3623–3632.

Chapter 47

Alcohol Dehydrogenase Alleles and Impact on Neuropathology

Neil C. Dodge[1], Joseph L. Jacobson[1,2], Sandra W. Jacobson[1,2]
[1]Department of Psychiatry and Behavioral Neurosciences, Wayne State University School of Medicine, Detroit, MI, USA; [2]Departments of Human Biology and of Psychiatry and Mental Health, Faculty of Health Sciences, University of Cape Town, Detroit, MI, USA

Abbreviations

ADH Alcohol dehydrogenase
ADH1B β Subunit of the class I alcohol dehydrogenase
ALDH Aldehyde dehydrogenase
BAC Blood alcohol concentration
FAS Fetal alcohol syndrome
FASD Fetal alcohol spectrum disorders
MDI Bayley Mental Development Index

INTRODUCTION

Alcohol use disorders are exceptionally common, with an estimated lifetime prevalence of nearly 5% across the world's population, and in high-income countries treatment for alcohol-related illnesses accounts for an estimated 12.8% of health care costs (Rehm et al., 2009). According to the US Centers for Disease Control, excessive alcohol consumption, defined as binge drinking, heavy drinking, and any drinking by pregnant women or people younger than age 21, led to approximately 88,000 deaths and 2.5 million years of potential life lost per year in the United States from 2006 to 2010 (Stahre, Roeber, Kanny, Brewer, & Zhang, 2014). Excessive alcohol use has a number of associated short-term health risks, including motor vehicle accidents and falls, violence, and alcohol poisoning, as well as long-term health risks, including high blood pressure, heart disease, stroke, cancer, mental health problems, dementia, and substance dependence (National Institute of Alcohol Abuse and Alcoholism, 2000). In addition, alcohol adversely affects the offspring of pregnant heavy and moderate drinkers resulting in a range of fetal alcohol spectrum disorders (FASD). Despite more than 4 decades of research and published prevention guidelines, a high proportion of women continue to drink heavily during pregnancy, and 6.6% report binge-drinking episodes (Substance Abuse and Mental Health Services Administration, 2013). The prevalence of FASD has been estimated at 2–5% of the school-age population in the United States, Canada, and western Europe (Chudley et al., 2005; May et al., 2009) and 13.6–20.9% in the Cape Colored (mixed ancestry) community in South Africa (May et al., 2013). Fetal alcohol syndrome (FAS), the most severe of the FASD and most common preventable form of mental retardation (Abel & Sokol, 1991), occurs in the United States at 0.2–3.0 cases per 1000 live births—rates comparable to Down syndrome or spina bifida (Bertrand et al., 2004), 9–10 per 1000 in some Native American tribes (May, Hymbaugh, Aase, & Samet, 1983), and 5.9–9.1% in the Cape Colored South African community (May et al., 2013). FAS is a disorder characterized by a distinctive pattern of craniofacial dysmorphic features and growth retardation (Hoyme et al., 2005). Cognitive and behavioral impairment and growth deficits are also seen in nonsyndromal children lacking fetal alcohol-related dysmorphology whose mothers consume moderate-to-heavy levels of alcohol during pregnancy (Carter et al., 2012; Hoyme et al., 2005; Suttie et al., 2013) and in low-to-moderately exposed children (Day et al. 2002).

In the United States, only 4–10% of the children born to women who drink heavily during pregnancy develop full FAS (Abel, 1995). It has, therefore, long been of interest to find biomarkers that may help identify risk factors that place exposed children at the greatest likelihood of being affected by their exposure (J. Jacobson et al., 2011; S. Jacobson et al., 2011) and thus may lead to improved differential diagnosis and development of innovative treatments. Although maternal age has been identified as a moderator of fetal alcohol effects (Chiodo et al., 2010; Jacobson, Jacobson, Sokol, Chiodo, & Corobana, 2004; May, 1991), there has been relatively little systematic empirical investigation of other factors that may determine which children are affected.

The alcohol metabolism enzyme alcohol dehydrogenase (ADH) plays a prominent role in the elimination of alcohol from the body shown to have an impact on FASD vulnerability. Research on polymorphisms of this enzyme has added considerably to the understanding of genetic contributions to vulnerability both to alcohol use disorders and to fetal alcohol teratogenesis.

ALCOHOL METABOLISM

The major pathway of alcohol elimination from the body is its oxidation to acetaldehyde in the liver, primarily by the enzyme ADH (Figure 1; Bosron & Li, 1987). Acetaldehyde is then oxidized to acetate by the enzyme aldehyde dehydrogenase (ALDH). Because oxidation of alcohol by ADH is the slowest component of this metabolic pathway, it constitutes the rate-limiting step; the enzyme ADH, therefore, determines the rate at which alcohol is eliminated from the body.

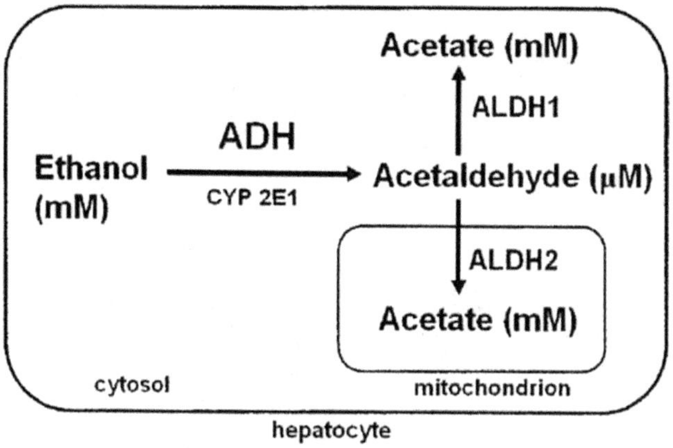

FIGURE 1 Ethanol metabolism. Ethanol is metabolized in liver hepatocytes by the enzyme alcohol dehydrogenase (ADH) to acetaldehyde, a molecule that is highly toxic. Acetaldehyde is then rapidly metabolized by the enzyme aldehyde dehydrogenase (ALDH) to acetate, which can then be excreted by the body. *From Warren and Li (2005). Reprinted with permission from John Wiley & Sons.*

Three functional single-nucleotide polymorphisms in the locus encoding the β subunit of the class I ADH (*ADH1B*) have been found to alter rates of alcohol metabolism, and each has distinct pharmacokinetic properties (Table 1). These polymorphisms differ in amino acid composition at two positions at the *ADH1B* locus. *ADH1B*1*, the most common allele in Caucasian and African populations, has the amino acid arginine at positions 47 and 369 (Brennan et al., 2004). The *ADH1B*2* allele substitutes histidine at position 47. *ADH1B*3* substitutes cysteine at position 369. Alcohol is cleared much more rapidly in individuals who are homozygous or heterozygous for either of these allele variants owing to larger maximal velocities (i.e., faster turnover rates), compared with individuals who are homozygous for *ADH1B*1* (Bosron & Li, 1987; Neumark et al., 2004; Thomasson, Beard, & Li, 1995). *ADH1B*3* also has a much larger K_m for ethanol and is, therefore, slower than the *ADH1B*1* allele at low ethanol concentrations, but at high concentrations it is more than 10-fold faster (Lee, Hoog, & Yin, 2004).

The *ADH1B*2* allele is most prevalent in Asian populations (Table 2) (e.g., Brennan et al., 2004). This polymorphism is also relatively prevalent in Jewish populations in the United States and Israel (e.g., Shea, Wall, Carr, & Li, 2001) and has also been found in the Cape Colored (mixed ancestry) population in the Western Cape Province of South Africa (Viljoen et al., 2001). The Cape Colored population, composed mainly of descendants of white European, Malaysian, and Khoi African ancestors, has historically comprised the large majority of workers in the wine-producing Western Cape. The *ADH1B*3* allele is found primarily in individuals of West African descent and occurs at an allele frequency of approximately 15–20% in African-American samples (Bosron & Li, 1987; Brennan et al., 2004; Jacobson et al., 2006).

ADH ALLELES AND ALCOHOL USE DISORDERS

Studies of Asian, European, and Jewish populations have shown that individuals with at least one *ADH1B*2* allele consume less alcohol on average and are less likely to develop alcohol use disorders than those who are homozygous for *ADH1B*1* (Beirut et al., 2012; Carr et al., 2002; Thomasson et al., 1991). Similarly, the *ADH1B*3* allele has been shown to have a protective effect against alcohol use disorders in African-Americans (Edenberg et al., 2006), Trinidadians of African descent (Ehlers et al., 2007), and Mission Indians in southern California (Wall, Carr, & Ehlers, 2003).

The proposed mechanism by which the *ADH1B*2* and *ADH1B*3* alleles protect against development of alcohol use disorders involves greater susceptibility by individuals with these polymorphisms to the aversive effects of a transient elevation in acetaldehyde, which is the initial by-product of alcohol metabolism in the liver. Acetaldehyde is toxic, and its accumulation is associated with nausea, increased heart rate, and facial redness, known as flushing. A study of Jewish college-aged men and women found that those men with at least one *ADH1B*2* allele reported more unpleasant reactions, including flushing, feelings of anxiety, itching, palpitation or fast heartbeat, headaches, and breathlessness, after the consumption of alcohol (Carr et al., 2002). Similarly, in a sample of Caucasian college students, those with an *ADH1B*2* allele reported a greater response to alcohol, were more likely to have experienced alcohol-induced headaches following one or two drinks, and reported more severe hangovers (Wall, Shea, Luczak, Cook, & Carr, 2005). In another study, participants with at least one *ADH1B*2* allele reported higher levels of subjective feelings of intoxication and increased body sway in response to an alcohol challenge (Duranceaux et al., 2006). Participants with the *ADH1B*3* allele had a higher pulse rate and stronger feelings of sedation than those who lacked the allele in response to a moderate alcohol challenge (McCarthy, Pedersen, Lobos, Todd, & Wall, 2010).

ADH ALLELES AND EFFECTS ON NEUROPATHOLOGY

The effects of the ADH alleles on neuropathology have been examined in adults with Wernicke–Korsakoff syndrome. This syndrome, in which Wernicke encephalopathy and Korsakoff syndrome resulting from vitamin B1 deficiency are both present, is often observed in cases of long-term severe alcoholism. Wernicke encephalopathy is characterized by dementia, ocular disturbances, and ataxia. Korsakoff syndrome is defined as the presence of anterograde amnesia, variable presentation of retrograde amnesia, and one of aphasia, apraxia, agnosia, or executive function deficit occurring when the individual is not in a state of intoxication, withdrawal, or delirium, as described in the *Diagnostic and Statistical Manual of Mental Disorders,* fourth edition. In a Japanese study comparing 47 male alcoholics with Wernicke–Korsakoff syndrome with 342 alcoholics without the syndrome and 175 nonalcoholics, Matsushita, Kato, Muramatsu, and Higuchi (2000) found a significantly higher allele frequency of *ADH1B*1* among those with Wernicke–Korsakoff syndrome compared to nonsyndromal alcoholic and control subjects. Thus, those with Wernicke–Korsakoff syndrome were less likely to have the rapid metabolizing *ADH1B*2* allele commonly found in Asian populations.

FETAL ALCOHOL SPECTRUM DISORDERS

The impact of ADH alleles on neuropathology has also been studied in the context of the teratogenicity of alcohol. As noted above, the FASD comprise a broad range of physical and neurobehavioral

TABLE 1 Characteristics of *ADH1B* Variants

Name	Amino Acid Differences	Enzyme Subunit	K_m for Ethanol (mM)	Turnover Rate (min^{-1})
*ADH1B*1*	Arg 47, Arg 369	β1	0.05	4
*ADH1B*2*	His 47, Arg 369	β2	0.9	350
*ADH1B*3*	Arg 47, Cys 369	β3	40.0	300

This table details the differences of the three variants of the β subunit of the class I alcohol dehydrogenase gene (*ADH1B*). The most common variant, *ADH1B*1*, has the amino acid arginine at positions 47 and 369 and has a slower turnover rate compared to the other less common variants. Note that the K_m for *ADH1B*3* is much higher than for the other variants. This indicates that much higher concentrations of ethanol are required to achieve maximal velocity. Adapted from Warren and Li (2005).

TABLE 2 *ADH1B* Allele Frequencies by Ethnicity (Neumark Friedlander, Thomasson, and Li, 1998; Thomasson et al., 1991, 1995)

	*ADH1B*1*	*ADH1B*2*	*ADH1B*3*
Caucasian (European and American)	>95%	<5%	<5%
Jewish (Israeli and American)	80%	20%	<5%
African American	85%	<5%	22%
Asian (Chinese, Japanese, and Korean)	35%	65%	<5%

This table indicates the frequency of the β subunit of the class I alcohol dehydrogenase gene (*ADH1B*) alleles in different ethnic populations. *ADH1B*1* is the most prevalent in all populations, with the exception of Asians, among whom the *ADH1B*2* allele is the most prevalent. The *ADH1B*3* allele is most common in those of African descent. Adapted from Warren and Li (2005).

TABLE 3 Key Facts Relating to Fetal Alcohol Spectrum Disorders

- Fetal alcohol spectrum disorder (FASD) is the clinical term used to describe the range of cognitive, behavioral, and growth outcomes associated with prenatal alcohol exposure (Hoyme et al., 2005).
- Fetal alcohol syndrome (FAS) is the most severe of the FASD and is characterized by a distinctive pattern of craniofacial dysmorphic features, small head circumference, and pre- and/or postnatal growth retardation.
- Partial FAS occurs when some but not all of the alcohol-related craniofacial anomalies are present.
- Alcohol-related neurodevelopmental disorder is used to characterize children who lack the distinctive alcohol-related craniofacial features but exhibit significant cognitive and behavioral impairment.
- The prevalence of FASD has been estimated at 2–5% of the school-age population in the United States, Canada, and Western Europe (May et al., 2009).

This table lists key information relating to FASD.

outcomes associated with prenatal alcohol exposure (Hoyme et al., 2005). FASD range from nonsyndromal individuals with known histories of prenatal exposure who exhibit subtle neurobehavioral impairment to the most severely affected individuals, who meet the criteria for a diagnosis of FAS (Table 3). Neurobehavioral outcomes associated with prenatal alcohol exposure include lower intelligence quotient (Jacobson et al., 2004; Mattson et al., 1997), slower information processing speed (Burden, Jacobson, & Jacobson, 2005; Coles, Platzman, Lynch, & Freides, 2002; Jacobson, Jacobson, Sokol, & Ager, 1993), and poor attention and executive function (Burden, Jacobson, Sokol, & Jacobson, 2005; Coles et al., 1997; Kodituwakku, Handmaker, Cutler, Weathersby, & Handmaker, 1995), verbal learning and memory (Lewis et al., 2015; Mattson, Riley, Delis, Stern, & Jones, 1996; Willford, Richardson, Leech, & Day, 2004), and arithmetic skills (Goldschmidt, Richardson, Stoffer, Geva, & Day, 1996; Howell, Lynch, Platzman, Smith, & Coles, 2006; Jacobson, Dodge, et al., 2011, Jacobson, Jacobson, et al., 2011; Streissguth et al., 1994).

Children and adolescents prenatally exposed to alcohol are also more likely to exhibit parent- and teacher-reported behavioral problems (Brown et al., 1991; Carmichael Olson et al., 1997; Jacobson et al., 2006; Larkby, Goldschmidt, Hanusa, & Day, 2011) and problems in social functioning (Lynch, Coles, Corley, & Falek, 2003). Using the Achenbach Child Behavior Checklist, Mattson and Riley (2000) found that children with a history of prenatal alcohol exposure exhibited more problems in internalizing and externalizing problems compared with nonexposed children. Other studies have found associations of prenatal alcohol exposure primarily with more externalizing behavior problems, which are seen in both childhood (e.g., Sood et al., 2001) and adolescence (Disney, Iacono, McGue, Tully, & Legrand, 2008). Larkby et al. (2011) found that an average of one or more drinks per day during the first trimester was associated with an increased rate of conduct disorder in adolescents.

Although FASD is associated with a broad range of adverse outcomes, not all children born to mothers who drink during pregnancy are affected (Abel, 1995). Experimental studies with laboratory animals have demonstrated that dose and timing of exposure are among the factors that determine vulnerability or severity of outcome associated with prenatal alcohol exposure (Bonthius & West, 1991; Goodlett, Horn, & Zhou, 2005). In addition, four maternal factors have been identified as potentially significant moderators of severity of outcome—older maternal age at delivery (Abel & Dintcheff, 1985; Chiodo et al., 2010; Jacobson et al., 2004; May, 1991), home environment, alcohol abuse history (Jacobson et al., 2004; Majewski, 1993), and maternal *ADH1B* status.

ADH ALLELES AND NEUROPATHOLOGY IN FASD

McCarver, Thomasson, Martier, Sokol, and Li (1997) were the first to study the effects of the *ADH1B*3* allele on outcomes in infants exposed to alcohol in utero. In an African-American sample, they found that fetal alcohol exposure was associated with reduced birth weight and lower scores on the Mental Development Index (MDI) from the Bayley Scales of Infant Development, the most extensively used standardized assessment of infant developmental status, in exposed infants whose mothers were homozygous for the *ADH1B*1* allele. These effects were not seen in alcohol-exposed infants of mothers with at least one *ADH1B*3*, whose birth weight and Bayley MDI performance were similar to those of the nonexposed infants. Similarly, Das, Cronk, Martier, Simpson, and McCarver (2004) found protective effects of the *ADH1B*3* allele on alcohol-related facial dysmorphology when both the mother and the child had at least one copy of the *ADH1B*3* allele. In a South African Cape Colored sample, Viljoen et al. (2001) found that children with FAS and their mothers were significantly less likely to carry an *ADH1B*2* allele compared to non-FAS children and their mothers, suggesting that the *ADH1B*2* allele may also be protective against prenatal alcohol exposure.

Stoler, Ryan, and Holmes (2002) was the only paper to report that affected infants, defined as infants with four of six FAS facial features and/or growth deficits greater than 2 standard deviations below the mean, were more likely to have an *ADH1B*3* allele than unaffected infants. Additionally, the presence of a maternal *ADH1B*3* allele was associated with a greater risk of having an affected infant after controlling for cigarette use and weight gain during pregnancy. In this sample, more women with the *ADH1B*3* allele reported heavy alcohol use (≥1 drink per day) during pregnancy than those without the *ADH1B*3* allele (70% vs 44%). However, as noted by Warren and Li (2005), very few of the mothers in the sample (10 of 108) reported heavy alcohol use (≥1 drink per day) during pregnancy, and 80 were abstainers, reducing the likelihood of detecting an effect of prenatal alcohol exposure. Thus, not surprisingly, alcohol consumption was not related to infant outcome after controlling for maternal genotype, cigarette use, and weight gain.

In the most extensive investigation of the moderating effects of the *ADH1B*3* allele as of this writing, Jacobson et al. (2006) presented findings from the Detroit Longitudinal Prenatal Alcohol Exposure Cohort collected during infancy and at 7.5 years. Confirming McCarver et al. (1997), the effects of prenatal alcohol on Bayley MDI scores were markedly less severe in children born to mothers with at least one copy of the *ADH1B*3* allele (Table 4). The presence of a maternal *ADH1B*3* allele was also protective against effects of alcohol exposure on infant head circumference, symbolic development assessed using the elicited play measure, and infant reaction time assessed in the visual expectancy paradigm. At 7.5 years, protective effects of the maternal *ADH1B*3* allele were seen on measures of attention, working memory, and executive function. In addition, prenatal alcohol exposure was associated with increased social problems, inattention, aggressive behaviors, and hyperactivity problems as reported on the Achenbach Teacher Report Form in children whose mothers were homozygous for the *ADH1B*1* allele but not in those born to mothers with at least one *ADH1B*3* allele.

More recently, Dodge, Jacobson, and Jacobson (2014) found that the presence of a maternal *ADH1B*3* allele continued to protect against the effects of prenatal alcohol exposure on behavior problems when the Detroit Longitudinal Prenatal Alcohol Exposure cohort was assessed during adolescence (Table 5). In the sample as a whole, and among adolescents born to mothers lacking the *ADH1B*3* allele, prenatal alcohol exposure was associated with increased teacher-reported attention problems, aggression, and externalizing problems, whereas prenatal alcohol was not related to any of these outcomes among those born to mothers with at least one *ADH1B*3* allele.

The mechanism by which the maternal *ADH1B*3* allele conveys this protective effect is not clearly understood. One possible explanation is that the presence of the *ADH1B*3* allele may cause mothers to drink less because of increased sensitivity to the transient increases in acetaldehyde associated with alcohol metabolism. However, no consistent differences in the amount or frequency of alcohol consumption across pregnancy were seen in mothers with and without the *ADH1B*3* allele in the Detroit Longitudinal Prenatal Alcohol Exposure cohort, suggesting that the protective effects of the *ADH1B*3* allele observed in that sample are probably not due to reductions in pregnancy drinking. Alternatively, although the amounts of alcohol consumed may be similar, peak blood alcohol concentrations (BACs) are likely to be lower in mothers with the *ADH1B*3* allele compared to those homozygous for the *ADH1B*1* allele owing to the more rapid metabolism of alcohol conveyed by the *ADH1B*3* allele. With lower peak BAC, lower concentrations of alcohol would be expected to reach the fetus in mothers with the *ADH1B*3* allele, thus reducing the impact of fetal exposure to alcohol. Experimental animal studies have demonstrated that peak BAC is an important determinant of extent of fetal alcohol damage. Ingestion of a given dose of alcohol over a short time period generates a higher peak BAC and greater neuronal (Bonthius & West, 1990) and behavioral impairment (Goodlett, Kelly, & West, 1987) than a larger dose ingested more gradually over several days.

ROLE OF THE ADH ALLELES IN THE FETUS

Given that the *ADH1B*3* allele is more likely to be present in the offspring of women with that allele, it can be difficult to discriminate the influences of the maternal and filial alleles. For example, in the Viljoen et al. (2001) study, there was complete overlap, in that all four of the FAS children with the *ADH1B*2* allele had mothers with the same allele, and in the Detroit cohort, 70% of the children with the *ADH1B*3* allele had mothers with that allele (Dodge et al., 2014). It is, therefore, not surprising that similar protective effects are often observed in alcohol-exposed offspring with the *ADH1B*3* allele. For example, McCarver et al. (1997) found that alcohol-exposed infants with the *ADH1B*3* allele had MDI scores similar to those of the nonexposed infants; only the exposed infants lacking the *ADH1B*3* allele had lower MDI scores than the other two groups. Dodge et al. (2014) found increased teacher-reported behavior problems in relation to prenatal alcohol exposure primarily in adolescents lacking the *ADH1B*3* allele, a pattern that was similar to that seen in relation to the maternal allele, although less pronounced. Moreover, the data relating to the filial allele are less consistent than those for the maternal allele. At 7.5 years, Jacobson et al. (2006) found increased teacher-reported behavior problems in relation to prenatal alcohol exposure in children with both genotypes on some end points (aggressive behavior and impulsivity), but the effects on others, particularly inattention

TABLE 4 Relation of Prenatal Alcohol Exposure to Infant and Child Outcomes in Children Born to Mothers with and without the *ADH1B*3* Allele

	Absent			Present		
	n	r	β	n	r	β
Infant Outcomes						
Birth size						
Weight	140	−0.22**	−0.11	77	−0.26*	−0.19
Head circumference	136	−0.31***	−0.23**	76	−0.05	0.04
Bayley Scales						
MDI	110	−0.26**	−0.24**	57	−0.12	−0.10
Elicited symbolic play	98	−0.26**	−0.18†	52	−0.06	−0.06
Processing Speed						
FTII	88	0.03	0.03	52	0.23†	0.31*
Cross-modal	106	0.34***	0.26**	59	−0.09	−0.11
Visual Expectancy Paradigm						
Reaction time	30	0.49**	0.46*	22	0.06	0.17
Proportion quick responses	30	−0.57***	−0.55**	22	−0.19	−0.26
7.5 Years						
Cognitive Outcomes						
Freedom from distractibility	118	−0.12	−0.16†	71	0.05	−0.10
Digit cancellation (interference)	119	0.28**	0.23*	71	0.02	−0.09
Category fluency	118	−0.19*	−0.23*	71	−0.07	−0.09
CPT RT (AX task)	108	0.15	0.18†	65	0.01	0.07
Magnitude Estimation (Slope)						
Number	106	−0.30**	−0.30**	65	−0.06	−0.06
Arrows	115	0.22*	0.22*	69	0.04	0.02
Achenbach Teacher Report Form						
Social problems	88	0.25*	0.24*	58	−0.04	−0.01
Attention problems	88	0.26*	0.21*	58	0.03	0.06
Aggressive behavior	88	0.25*	0.22*	58	−0.03	−0.06
Delinquent behavior	88	0.18†	0.14	58	−0.02	−0.04
Externalizing	88	0.25*	0.23*	58	−0.03	−0.02
Total problems	88	0.41***	0.38***	58	−0.00	0.03
ADHD Rating Scale						
Impulsivity	87	0.32**	0.29**	57	−0.03	−0.06
Inattention	87	0.28**	0.22*	57	−0.01	−0.04
ADHD classification	92	0.30**	0.30**	60	0.10	0.14

†$p<0.10$, *$p<0.05$, **$p<0.01$, ***$p<0.001$.
This table shows the zero-order correlation (r) and the standardized regression coefficient adjusted for potential confounders (β) for the relation between prenatal alcohol exposure (measured as average ounces of absolute alcohol/day) and developmental outcomes. Control variables related to each outcome at $p<0.10$ were included as potential confounders in the regression analyses for that outcome. Note that prenatal alcohol exposure is significantly associated with poorer outcomes in those born to mothers without the β subunit of the class I alcohol dehydrogenase gene *ADH1B*3* allele, while those born to mothers with the allele are protected. MDI, Bayley Mental Development Index; FTII, Fagan Test of Infant Intelligence; CPT RT, Continuous Performance Test Reaction Time; ADHD, attention deficit hyperactivity disorder.
Adapted from Jacobson et al. (2006).

TABLE 5 Relation of Prenatal Alcohol Exposure to Teacher-Reported Behavior Problems in Adolescents Born to Mothers with and without the ADH1B*3 Allele

	Total Sample		Maternal ADH1B*3						
			Absent				Present		
	r	β	n	r	β	n	r	β
Achenbach Teacher Report Form (TRF) (n=135)									
Attention problems	0.20*	0.13	65	0.33**	0.25*	47	0.04	−0.08
Inattention	0.17*	0.12	65	0.29*	0.23†	47	0.06	−0.04
Hyperactivity	0.22**	0.14	65	0.39**	0.30*	47	−0.01	−0.15
Social problems	0.17*	0.11	65	0.25*	0.15	47	0.12	−0.01
Thought problems	0.02	0.01	65	0.13	0.15	47	−0.04	−0.06
Anxious/depressed	0.06	0.03	65	0.19	0.16	47	−0.04	−0.15
Somatic complaints	0.11	0.03	65	0.09	−0.01	47	0.14	0.21
Withdrawn	0.07	0.07	65	−0.06	−0.06	47	0.07	0.07
Aggression	0.25**	0.20*	65	0.41***	0.39**	47	0.06	−0.02
Delinquency	0.23**	0.18*	65	0.22†	0.15	47	0.20	0.17
Internalizing problems	0.08	0.06	65	0.06	0.07	47	0.05	0.09
Externalizing problems	0.25**	0.23**	65	0.38**	0.35**	47	0.09	0.08
TRF total	0.22**	0.16†	65	0.23**	0.29*	47	0.07	−0.02
Disruptive Behavior Disorders Rating Scale (n=132)									
Inattention	0.21*	0.19*	66	0.33**	0.31**	45	0.10	0.09
Hyperactivity	0.17*	0.13	66	0.39**	0.33**	45	−0.04	0.08
Oppositional defiant disorder	0.06	0.06	66	0.12	0.12	45	0.05	0.05
Conduct disorder	0.06	0.05	66	0.20†	0.17	45	−0.08	−0.09

†p<0.10, * p<0.05, ** p<0.01, *** p<0.001.
This table shows the correlation (r) and the standardized regression coefficient adjusted for confounders (β) for the relation between prenatal alcohol exposure and teacher-reported behavior problems for adolescents born to mothers with (present) and without (absent) the ADH1B*3 allele. Control variables related to each outcome at p<0.10 were included as potential confounders in the regression analyses for that outcome. Note that prenatal alcohol exposure is significantly associated with poorer outcomes in those born to mothers without the ADH1B*3 allele, while those born to mothers with the allele are protected.
Adapted from Dodge et al. (2014).

and delinquency, were actually stronger in the children with at least one copy of the *ADH1B*3* allele.

Several mechanistic considerations suggest that the apparent protective effect of the filial *ADH1B* allele may be spurious, that is, attributable to the likelihood that mother and child will share the same allele variant. Unlike the maternal *ADH1B*3* allele, which would be expressed throughout the pregnancy, the fetal *ADH1B* genes are not expressed until the beginning of the second trimester (Smith, Hopkinson, & Harris, 1971). However, the heaviest drinking typically occurs during the first trimester (Cornelius et al., 1994), particularly during the period prior to pregnancy recognition (Streissguth et al., 1994). In the Detroit cohort, the mothers who reported drinking ≥1 oz absolute alcohol (AA) or the equivalent of two standard drinks per day around the time of conception reduced their drinking to an average of <0.3 AA/day across pregnancy (Jacobson, Chiodo, Sokol, & Jacobson, 2002). Moreover, many of the most severe effects of prenatal alcohol exposure—including craniofacial dysmorphology—have been linked to first-trimester exposure in experimental animal studies (Sulik, 2005). Thus, given that the fetal *ADH1B*3* allele is not expressed until later in the pregnancy after alcohol consumption has in most cases decreased, it would not be expected to provide an extensive protective effect on developmental outcome.

APPLICATIONS TO OTHER ADDICTIONS AND SUBSTANCE MISUSE

The application of our understanding of how ADH alleles influence neuropathology to other addictions is limited because, to our knowledge, the enzymes encoded by the ADH alleles lack the ability to act on other known drugs of abuse. However, given the evidence in this chapter indicating how enzymes involved in the metabolism of ethanol can have an impact on alcohol-related neuropathology, it seems possible that polymorphisms in key enzymes involved in the metabolism of other substances of abuse might also have an impact on substance abuse-related neuropathology. Studies have identified genetic modifiers of drug metabolism that affect the development of nicotine dependence (e.g., Pianezza, Sellers, & Tyndale, 1998) and opiate dependence (e.g., Tyndale, Droll, & Sellers, 1997). However, none to date have examined whether and how these genetic metabolism modifiers may influence neuropathology.

DEFINITION OF TERMS

ADH1B This is a gene that encodes the β subunit of ADH.
Alcohol dehydrogenase This is an enzyme that catalyzes the first step in the oxidative metabolism of ethanol, in which ethanol is oxidized to acetaldehyde.
Aldehyde dehydrogenase This is an enzyme that catalyzes the second step in the oxidative metabolism of ethanol, in which acetaldehyde is converted to acetic acid.
Acetaldehyde This is a molecule produced by ADH in the metabolism of ethanol.
Ethanol This is the psychoactive molecule in alcoholic beverages that provides the feeling of intoxication and the associated side effects.
Fetal alcohol spectrum disorders This is the term used to describe the range of outcomes resulting from prenatal alcohol exposure, including FAS, partial FAS, and alcohol-related neurodevelopmental disorder.
Functional polymorphisms These are changes in gene structure that are known to influence gene function.
K_m Known as the Michaelis constant, this is the substrate concentration at which the rate of reaction is half of V_{max}.
Rate-limiting step This is the step in a metabolic pathway that determines the overall rate of the reaction. Often, this is the slowest step of the pathway.
Teratogenicity This is the ability of a substance to adversely affect a fetus.
V_{max} In Michaelis–Menten kinetics, V_{max} represents the maximum reaction rate under saturating substrate conditions.
Wernicke–Korsakoff syndrome This is a disorder commonly seen in alcoholic patients related to thiamine deficiency. Its symptoms include dementia, ataxia, and amnesia.

SUMMARY POINTS

- ADH is a critical enzyme in the metabolism of alcohol.
- The expression of three alleles at the *ADH1B* locus results in enzymes that differ in turnover rate and affinity for alcohol.
- *ADH1B*2*, which is primarily found in Asian and Jewish populations, and *ADH1B*3*, which is primarily observed in those of African descent, are associated with faster alcohol metabolism, and individuals who carry at least one of these two alleles tend to drink less and have a reduced risk for developing alcohol use disorders.
- These ADH alleles also protect against alcohol-related neuropathology.
- Asian men who carry the *ADH1B*2* allele are less likely to develop Wernicke–Korsakoff syndrome.
- Children and adolescents prenatally exposed to alcohol whose mothers carry the *ADH1B*3* allele are less likely to exhibit the effects of prenatal alcohol exposure, including reduced growth, cognitive deficits, and behavioral abnormalities.
- The protective effect of the *ADH1B*3* allele on prenatal alcohol exposure may be due to the rapid metabolism of ethanol associated with this allele.

REFERENCES

Abel, E. L. (1995). An update on incidence of FAS—FAS is not an equal opportunity birth defect. *Neurotoxicology and Teratology, 17*, 437–443.

Abel, E. L., & Dintcheff, B. A. (1985). Factors affecting the outcome of maternal alcohol exposure: II. Maternal age. *Neurobehavioral Toxicology and Teratology, 7*, 263–266.

Abel, E. L., & Sokol, R. J. (1991). A revised conservative estimate of the incidence of FAS and its economic impact. *Alcoholism: Clinical and Experimental Research, 15*, 514–524.

Bertrand, J., Floyd, R. L., Weber, M. K., O'Connor, M., Riley, E. P., Johnson, K. A., ... National Task Force on FAS/FAE (2004). *Fetal alcohol syndrome: Guidelines for referral and diagnosis*. Atlanta, GA: Center for Disease Control and Prevention.

Bierut, L. J., Goate, A. M., Breslau, N., Johnson, E. O., Bertelsen, S., Fox, L., ... Edenberg, H. J. (2012). ADH1B is associated with alcohol dependence and alcohol consumption in populations of European and African ancestry. *Molecular Psychiatry, 17*, 445–450.

Bonthius, D. J., & West, J. R. (1990). Alcohol-induced neuronal loss in developing rats: increased brain damage with binge exposure. *Alcoholism: Clinical and Experimental Research, 14*, 107–118.

Bonthius, D. J., & West, J. R. (1991). Permanent neuronal deficits in rats exposed to alcohol during the brain growth spurt. *Teratology, 44*, 147–163.

Bosron, W. F., & Li, T.-K. (1987). Catalytic properties of human liver alcohol dehydrogenase isoenzymes. *Enzyme, 37*, 19–28.

Brennan, P., Lewis, S., Hashibe, M., Bell, D. A., Boffetta, P., Bouchardy, C., ... Benhamou, S. (2004). Pooled analysis of alcohol dehydrogenase genotypes and head and neck cancer: a HuGE review. *American Journal of Epidemiology, 159*, 1–16.

Brown, R. T., Coles, C. D., Smith, I. E., Platzman, K. A., Silverstein, J., Erickson, S., & Falek, A. (1991). Effects of prenatal alcohol exposure at school age. II. Attention and behavior. *Neurotoxicology and Teratology, 13*, 369–376.

Burden, M. J., Jacobson, S. W., & Jacobson, J. L. (2005). The relation of prenatal alcohol exposure to cognitive processing speed and efficiency in childhood. *Alcoholism: Clinical and Experimental Research, 29*, 1473–1483.

Burden, M. J., Jacobson, S. W., Sokol, R. J., & Jacobson, J. L. (2005). Effects of prenatal alcohol exposure on attention and working memory at 7.5 years of age. *Alcoholism: Clinical and Experimental Research, 29*, 443–452.

Carmichael Olson, H., Streissguth, A. P., Sampson, P. D., Barr, H. M., Bookstein, F. L., & Thiede, K. (1997). Association of prenatal alcohol exposure with behavioral and learning problems in early adolescence. *Journal of the American Academy of Child and Adolescent Psychiatry, 36*, 1187–1194.

Carr, L. G., Foroud, T., Stewart, T., Castelluccio, P., Edenberg, H. J., & Li, T. K. (2002). Influence of ADH1B polymorphism on alcohol use and its subjective effects in a Jewish population. *American Journal of Medical Genetics, 112*, 138–143.

Carter, R. C., Jacobson, J. L., Molteno, C. D., Jiang, H., Meintjes, E. M., Jacobson, S. W., & Duggan, C. (2012). Effects of heavy prenatal alcohol exposure and iron deficiency anemia on child growth and body composition through age 9 years. *Alcoholism: Clinical and Experimental Research, 36*, 1973–1982.

Chiodo, L. M., Da Costa, D. E., Hannigan, J. H., Covington, C. Y., Sokol, R. J., Janisse, J., ... Delaney-Black, V. (2010). The impact of maternal age on the effects of prenatal alcohol exposure on attention. *Alcoholism: Clinical and Experimental Research, 34*, 1813–1821.

Chudley, A. E., Conry, J., Cook, J. L., Loock, C., Rosales, T., & LeBlanc, N. (2005). Fetal alcohol spectrum disorder: Canadian guidelines for diagnosis. *Canadian Medical Association Journal, 172*, S1–S21.

Coles, C. D., Platzman, K. A., Lynch, M. E., & Freides, D. (2002). Auditory and visual sustained attention in adolescents prenatally exposed to alcohol. *Alcoholism: Clinical and Experimental Research, 26*, 263–271.

Coles, C. D., Platzman, K. A., Raskind-Hood, C. L., Brown, R. T., Falek, A., & Smith, I. E. (1997). A comparison of children affected by prenatal alcohol exposure and attention deficit, hyperactivity disorder. *Alcoholism: Clinical and Experimental Research, 21*, 150–161.

Cornelius, M. D., Richardson, G. A., Day, N. L., Cornelius, J. R., Geva, D., & Taylor, P. M. (1994). A comparison of prenatal drinking in two recent samples of adolescents and adults. *Journal of Studies on Alcohol and Drugs, 55*, 412.

Das, U. G., Cronk, C. E., Martier, S. S., Simpson, P. M., & McCarver, D. (2004). Alcohol dehydrogenase 2*3 affects alterations in offspring facial morphology associated with maternal ethanol intake in pregnancy. *Alcoholism: Clinical and Experimental Research, 28*, 1598–1606.

Day, N. L., Leech, S. L., Richardson, G. A., Cornelius, M. D., Robles, N., & Larkby, C. (2002). Prenatal alcohol exposure predicts continued deficits in offspring size at 14 years of age. *Alcoholism: Clinical and Experimental Research, 26*, 1584–1591.

Disney, E. R., Iacono, W., McGue, M., Tully, E., & Legrand, L. (2008). Strengthening the case: prenatal alcohol exposure is associated with increased risk for conduct disorder. *Pediatrics, 122*, e1225–e1230.

Dodge, N. C., Jacobson, J. L., & Jacobson, S. W. (2014). Protective effects of the alcohol dehydrogenase-ADH1B* 3 allele on attention and behavior problems in adolescents exposed to alcohol during pregnancy. *Neurotoxicology and Teratology, 41*, 43–50.

Duranceaux, N. C., Schuckit, M. A., Eng, M. Y., Robinson, S. K., Carr, L. G., & Wall, T. L. (2006). Associations of variations in alcohol dehydrogenase genes with the level of response to alcohol in non-Asians. *Alcoholism: Clinical and Experimental Research, 30*, 1470–1478.

Edenberg, H. J., Xuei, X., Chen, H. J., Tian, H., Wetherill, L. F., Dick, D. M., ... Foroud, T. (2006). Association of alcohol dehydrogenase genes with alcohol dependence: a comprehensive analysis. *Human Molecular Genetics, 15*, 1539–1549.

Ehlers, C. L., Montane-Jaimem, K., Moore, S., Shafe, S., Joseph, R., & Carr, L. G. (2007). Association of the ADH1B*3 allele with alcohol-related phenotypes in Trinidad. *Alcoholism: Clinical and Experimental Research, 31*, 216–220.

Goldschmidt, L., Richardson, G. A., Stoffer, D. S., Geva, D., & Day, N. L. (1996). Prenatal alcohol exposure and academic achievement at age six: a nonlinear fit. *Alcoholism: Clinical and Experimental Research, 20*, 763–770.

Goodlett, C. R., Horn, K. H., & Zhou, F. C. (2005). Alcohol teratogenesis: Mechanisms of damage and strategies for intervention. *Experimental Biology and Medicine, 230*, 394–406.

Goodlett, C. R., Kelly, S. J., & West, J. R. (1987). Early postnatal alcohol exposure that produces high blood alcohol levels impairs development of spatial navigation learning. *Psychobiology, 15*, 64–74.

Howell, K. K., Lynch, M. E., Platzman, K. A., Smith, G. H., & Coles, C. D. (2006). Prenatal alcohol exposure and ability, academic achievement, and school functioning in adolescence: a longitudinal follow-up. *Journal of Pediatric Psychology, 311*, 16–126.

Hoyme, H. E., May, P. A., Kalberg, W. O., Kodituwakku, P., Gossage, J. P., Trujillo, P. M., ... Robinson, L. K. (2005). A practical clinical approach to diagnosis of fetal alcohol spectrum disorders: clarification of the 1996 Institute of Medicine criteria. *Pediatrics, 115*, 39–47.

Jacobson, J. L., Dodge, N. C., Burden, M. J., Klorman, R., & Jacobson, S. W. (2011). Number processing in adolescents with prenatal alcohol exposure and ADHD: differences in the neurobehavioral phenotype. *Alcoholism: Clinical and Experimental Research, 35*, 431–442.

Jacobson, S. W., Carr, L. G., Croxford, J., Sokol, R. J., Li, T. K., & Jacobson, J. L. (2006). Protective effects of the alcohol dehydrogenase-ADH1B allele in African American children exposed to alcohol during pregnancy. *Journal of Pediatrics, 148*, 30–37.

Jacobson, S. W., Chiodo, L. M., Sokol, R. J., & Jacobson, J. L. (2002). Validity of maternal report of prenatal alcohol, cocaine, and smoking in relation to neurobehavioral outcome. *Pediatrics, 109*, 815–825.

Jacobson, S. W., Jacobson, J. L., Sokol, R. J., & Ager, J. W. (1993). Prenatal alcohol exposure and infant information processing ability. *Child Development, 64*, 1706–1721.

Jacobson, S. W., Jacobson, J. L., Sokol, R. J., Chiodo, L. M., & Corobana, R. (2004). Maternal age, alcohol abuse history, and quality of parenting as moderators of the effects of prenatal alcohol exposure on 7.5-year intellectual function. *Alcoholism: Clinical and Experimental Research, 28*, 1732–1745.

Jacobson, S. W., Jacobson, J. L., Stanton, M. E., Meintjes, E. M., & Molteno, C. D. (2011). Biobehavioral markers of adverse effect in fetal alcohol spectrum disorders. *Neuropsychology Review, 21*, 148–166.

Kodituwakku, P. W., Handmaker, N. S., Cutler, S. K., Weathersby, E. K., & Handmaker, S. D. (1995). Specific impairments in self-regulation in children exposed to alcohol prenatally. *Alcoholism: Clinical and Experimental Research, 19*, 1558–1564.

Larkby, C. A., Goldschmidt, L., Hanusa, B. H., & Day, N. L. (2011). Prenatal alcohol exposure is associated with conduct disorder in adolescence: findings from a birth cohort. *Journal of the American Academy of Child and Adolescent Psychiatry, 50*, 262–271.

Lee, S. L., Hoog, J. O., & Yin, S. J. (2004). Functionality of allelic variations in human alcohol dehydrogenase gene family: assessment of a functional window for protection against alcoholism. *Pharmacogenetics, 14*, 725–732.

Lewis, C. E., Thomas, K. G. F., Dodge, N. C., Molteno, C. D., Meintjes, E. M., Jacobson, J. L., & Jacobson, S. W. (2015). Verbal learning and memory impairment in children with fetal alcohol spectrum disorders. *Alcoholism: Clinical and Experimental Research, 39*(4), 724–732.

Lynch, M. E., Coles, C. D., Corley, T., & Falek, A. (2003). Examining delinquency in adolescents differentially prenatally exposed to alcohol: the role of proximal and distal risk factors. *Journal of Studies on Alcohol and Drugs, 64*, 678.

Majewski, F. (1993). Alcohol embryopathy: experience in 200 patients. *Developmental Brain Dysfunction, 6*, 248–265.

Matsushita, S., Kato, M., Muramatsu, T., & Higuchi, S. (2000). Alcohol and aldehyde dehydrogenase genotypes in Korsakoff syndrome. *Alcoholism: Clinical and Experimental Research, 24*, 337–340.

Mattson, S. N., & Riley, E. P. (2000). Parent ratings of behavior in children with heavy prenatal alcohol exposure and IQ-matched controls. *Alcoholism: Clinical and Experimental Research, 24*, 226–231.

Mattson, S. N., Riley, E. P., Delis, D. C., Stern, C., & Jones, K. L. (1996). Verbal learning and memory in children with fetal alcohol syndrome. *Alcoholism: Clinical and Experimental Research, 20*, 810–816.

Mattson, S. N., Riley, E. P., Gramling, L., Delis, D. C., Jones, K. L., & of Dysmorphology, T. D. (1997). Heavy prenatal alcohol exposure with or without physical features of fetal alcohol syndrome leads to IQ deficits. *The Journal of Pediatrics, 131*, 718–721.

May, P. A. (1991). Fetal alcohol effects among North American Indians: evidence and implications for society. *Alcohol Health and Research World, 15*, 239–247.

May, P. A., Blankenship, J., Marais, A. S., Gossage, J. P., Kalberg, W. O., Barnard, R., ... Seedat, S. (2013). Approaching the prevalence of the full spectrum of fetal alcohol spectrum disorders in a South African population-based study. *Alcoholism: Clinical and Experimental Research, 37*, 818–830.

May, P. A., Gossage, J. P., Kalberg, W. O., Robinson, L. K., Buckley, D., Manning, M., & Hoyme, H. E. (2009). Prevalence and epidemiologic characteristics of FASD from various research methods with an emphasis on recent in-school studies. *Developmental Disabilities Research Reviews, 15*, 176–192.

May, P. A., Hymbaugh, K. J., Aase, J. M., & Samet, J. M. (1983). Epidemiology of fetal alcohol syndrome among American Indians of the Southwest. *Biodemography and Social Biology, 30*, 374–387.

McCarthy, D. M., Pedersen, S. L., Lobos, E. A., Todd, R. D., & Wall, T. L. (2010). ADH1B*3 and response to alcohol in African-Americans. *Alcoholism: Clinical and Experimental Research, 34*, 1274–1281.

McCarver, D. G., Thomasson, H. R., Martier, S. S., Sokol, R. J., & Li, T. K. (1997). Alcohol dehydrogenase-2*3 allele protects against alcohol-related birth defects among African Americans. *Journal of Pharmacology and Experimental Therapeutics, 283*, 1095–1101.

National Institute of Alcohol Abuse and Alcoholism. (2000). *Tenth special report to the U.S. Congress on alcohol and health*. Bethesda, MD: National Institute of Health.

Neumark, Y. D., Friedlander, Y., Durst, R., Leitersdorf, E., Jaffe, D., Ramchandani, V. A., ... Li, T. K. (2004). Alcohol dehydrogenase polymorphisms influence alcohol–elimination rates in a male Jewish population. *Alcoholism: Clinical and Experimental Research, 28*, 10–14.

Neumark, Y. D., Friedlander, Y., Thomasson, H. R., & Li, T. K. (1998). Association of the ADH2*2 allele with reduced ethanol consumption in Jewish men in Israel: a pilot study. *Journal of studies on alcohol, 59*(2), 133–139.

Pianezza, M. L., Sellers, E. M., & Tyndale, R. F. (1998). Nicotine metabolism defect reduces smoking. *Nature, 393*, 750.

Rehm, J., Mathers, C., Popova, S., Thavoncharoensap, M., Teerawattananon, Y., & Patra, J. (2009). Global burden of disease and injury and economic cost attributable to alcohol use and alcohol-use disorders. *The Lancet, 373*, 2223–2233.

Shea, S. H., Wall, T. L., Carr, L. G., & Li, T. K. (2001). ADH2 and alcohol-related phenotypes in Ashkenazic Jewish American college students. *Behavior Genetics, 31*, 231–239.

Smith, M., Hopkinson, D. A., & Harris, H. (1971). Developmental changes and polymorphism in human alcohol dehydrogenase. *Annals of Human Genetics, 34*, 251–271.

Sood, B., Delaney-Black, V., Covington, C., Nordstrom-Klee, B., Ager, J., Templin, T., ... Sokol, R. J. (2001). Prenatal alcohol exposure and childhood behavior at age 6 to 7 years: I. Dose-response effect. *Pediatrics, 108*, e34.

Stahre, M., Roeber, J., Kanny, D., Brewer, R. D., & Zhang, X. (2014). Contribution of excessive alcohol consumption to deaths and years of potential life lost in the United States. *Preventing Chronic Disease, 11*, 130293. http://dx.doi.org/10.5888/pcd11.130293.

Stoler, J. M., Ryan, L. M., & Holmes, L. B. (2002). Alcohol dehydrogenase 2 genotypes, maternal alcohol use, and infant outcome. *Journal of Pediatrics, 141*, 780–785.

Streissguth, A. P., Barr, H. M., Carmichael-Olson, H., Sampson, P. D., Bookstein, F. L., & Burgess, D. M. (1994). Drinking during pregnancy decreases word attack and arithmetic scores on standardized tests: adolescent data from a population-based prospective study. *Alcoholism: Clinical and Experimental Research, 18*, 248–254.

Substance Abuse and Mental Health Services Administration. (2013). *The NSDUH report: 18 percent of pregnant women drink alcohol during early pregnancy*. NSDUH Report.

Sulik, K. K. (2005). Genesis of alcohol-induced craniofacial dysmorphism. *Experimental Biology and Medicine, 230*, 366–375.

Suttie, M., Foroud, T., Wetherill, L., Jacobson, J. L., Molteno, C. D., Meintjes, E. M., ... Hammond, P. (2013). Facial dysmorphism across the fetal alcohol spectrum. *Pediatrics, 131*, e779–e788.

Thomasson, H. R., Beard, J. D., & Li, T. K. (1995). ADH2 gene polymorphisms are determinants of alcohol pharmacokinetics. *Alcoholism: Clinical and Experimental Research, 19*, 1494–1499.

Thomasson, H. R., Edenberg, H. J., Crabb, D. W., Mai, X. L., Jerome, R. E., Li, T. K., ... Yin, S. J. (1991). Alcohol and aldehyde dehydrogenase genotypes and alcoholism in Chinese men. *American Journal of Human Genetics, 48*, 677.

Tyndale, R. F., Droll, K. P., & Sellers, E. M. (1997). Genetically deficient CYP2D6 metabolism provides protection against oral opiate dependence. *Pharmacogenetics and Genomics, 7*, 375–379.

Viljoen, D. L., Carr, L. G., Foroud, T. M., Brooke, L., Ramsey, M., & Li, T. K. (2001). Alcohol dehydrogenase-2*2 allele is associated with decreased prevalence of fetal alcohol syndrome in the mixed-ancestry population of the Western Cape Province, South Africa. *Alcoholism: Clinical and Experimental Research, 25*, 1719–1722.

Wall, T. L., Carr, L. G., & Ehlers, C. L. (2003). Protective association of genetic variation in alcohol dehydrogenase with alcohol dependence in Native American Mission Indians. *American Journal of Psychiatry, 160*, 41–46.

Wall, T. L., Shea, S. H., Luczak, S. E., Cook, T. A., & Carr, L. G. (2005). Genetic associations of alcohol dehydrogenase with alcohol use disorders and endophenotypes in white college students. *Journal of Abnormal Psychology, 114*, 456.

Warren, K. R., & Li, T. K. (2005). Genetic polymorphisms: impact on the risk of fetal alcohol spectrum disorder. *Birth Defects Research Part A: Clinical and Molecular Teratology, 73*, 195–203.

Willford, J. A., Richardson, G. A., Leech, S. L., & Day, N. L. (2004). Verbal and visuospatial learning and memory function in children with moderate prenatal alcohol exposure. *Alcoholism: Clinical and Experimental Research, 28*, 497–507.

Chapter 48

ADH Cluster Genes, Genome-Wide Association Studies, and Alcohol Dependence

Byung Lae Park[1], Hyoung Doo Shin[2]
[1]Department of Genetic Epidemiology, SNP Genetics, Inc., Seoul, Republic of Korea; [2]Department of Life Science, Sogang University, Seoul, Republic of Korea

Abbreviations

AA African Americans
AD Alcohol dependence
ADH Alcohol dehydrogenase
ALDH Aldehyde dehydrogenase
ASAM American Society of Addiction Medicine
EA European Americans
GWAS Genome-wide association study
NCADD National Council on Alcoholism and Drug Dependence

INTRODUCTION

Alcohol dependence (AD; MIM# 103780) is a common and multifactorial disorder characterized by (1) continuous or periodic impaired control over drinking; (2) preoccupation with alcohol; (3) use of alcohol despite adverse consequences; (4) distortion in thinking, according to the National Council on Alcoholism and Drug Dependence (NCADD) and the American Society of Addiction Medicine (ASAM); and (5) often being accompanied by chronic consumption of hazardous levels of ethanol. The lifetime prevalence of AD has been estimated to be about 12.5% (Grant et al., 2004; Hasin, Stinson, Ogburn, & Grant, 2007). In general, the morbidity rate for men with AD is higher than for women, because men tend to drink more heavily and more frequently, putting them at increased risk for disease and death (Lim et al., 2012; WHO, 2011). The World Health Organization (WHO) Global Status Report on Alcohol and Health and the Global Burden of Disease Study 2010 both list alcohol as the third leading risk factor for death and disability (Lim et al., 2012; WHO, 2011). WHO (2011) estimates that alcohol consumption causes about 2.5 million deaths per year—nearly 4% of the total deaths worldwide.

Twin and family studies have demonstrated that genetic predisposition is one of the most important factors for the risk of AD, estimated to be in the range of 40–60% (Dick & Bierut, 2006; Enoch & Goldman, 2002; Kendler, Heath, Neale, Kessler, & Eaves, 1992; Kendler, Neale, Heath, Kessler, & Eaves, 1994; Kessler et al., 1997; Prescott & Kendler, 1999). Monozygotic twins of alcoholics exhibit greater risk for alcoholism, whereas dizygotic twins of alcoholics are at approximately the same risk as full siblings. However, genetic studies have explained that only 2–3% of the genetic variances affect the diagnosis of AD, which may be due to the complexity and heterogeneity of the AD phenotype (Dick & Bierut, 2006).

Alcoholism is believed to be a multifactorial and polygenic disorder involving complex gene-to-gene and gene-to-environment interactions. For this reason, a large number of candidate gene studies on the risk loci of AD have looked at the various gene pathways, including the ethanol metabolic pathway. The alcohol dehydrogenase (ADH) gene family on 4q22-23 has been the most extensively studied due to its important function in the pathway of alcohol metabolism, and associations of the gene family's polymorphisms with the risk of AD has been thoroughly identified by multiple studies (Choi et al., 2005; Edenberg et al., 2006; Kim et al., 2008; Li, Yin, Crabb, O'Connor, & Ramchandani, 2001; Osier et al., 1999, 2002; Shen et al., 1997; Thomasson et al., 1991; Tolstrup, Nordestgaard, Rasmussen, Tybjaerg-Hansen, & Gronbaek, 2008). However, it is still unclear which specific genes or their variants contribute to the risk of AD and how they function. These research areas continue to be very active, in the broad effort to identify new genes and their variants.

ALCOHOL METABOLISM AND MAJOR GENETIC FACTORS

The conversion of alcohol to acetaldehyde, a reactive and toxic molecule, is catalyzed by ADH. This is the rate-limiting step in the elimination of ethanol in humans. Ethanol is oxidized to acetaldehyde in a reversible reaction primarily in the cytosol, and the acetaldehyde is further oxidized to acetate, principally by the mitochondrial aldehyde dehydrogenase ALDH2 with some contribution from cytosolic aldehyde dehydrogenases (Hurley & Edenberg, 2012). Many studies on the genetic component of ADHs and ALDH2 genes have demonstrated that ADH1B*47His (rs1229984; exon 3) and ALDH2*487Glu (rs671; exon 12) protect individuals from developing AD through either faster production or slower removal of acetaldehyde, a metabolite that triggers aversive reactions (Thomasson et al., 1991). In particular, ADH1B*47His encodes enzymes with V_{max} values, resulting in both higher pH optimum and lower turnover number of the atypical enzyme. This gene has consistently been found at significantly higher frequencies in individuals with AD than in controls in East Asian samples (Chen et al., 1999; Li et al., 2001; Osier et al., 1999, 2002; Shen et al., 1997; Thomasson et al., 1991). Individuals carrying the ALDH2 487*Lys gene exhibit a distinct reaction, such as prominent facial flushing, tachycardia, and nausea, even if they consume small amounts of alcohol, due to their poor conversion rate of toxic acetaldehyde to acetate. ALDH2 Glu487Lys is fairly common in East Asia, where up to 15–40% of nonalcoholics among Korean, Han Chinese, and Japanese populations carry at least one copy. However, the gene is extremely rare in those of non-Asian descent, with almost no individuals of European or African descent carrying this allele (Kim et al., 2008; Li, Zhao, & Gelernter, 2012; Luczak, Glatt, & Wall, 2006; Oota et al., 2004).

In the study by Kim et al. (2008), it was demonstrated that two major genes for alcohol metabolism, ADH1B and ALDH2, showed a dramatic combined effect on the risk of AD in the Korean population. The protective allele of ADH1B (ADH1B*47His) encodes for a rapid ethanol-metabolizing enzyme, and the susceptible allele of ALDH2 (ALDH2*487Lys) is strongly associated with a decreased rate of metabolizing acetaldehyde (Table 1). The combined analysis of two polymorphisms suggested that individuals bearing susceptible alleles at both loci have a 91 times greater risk for AD, and individuals bearing one susceptible and one protective allele at either loci have an 11 times greater risk compared with subjects who have both protective alleles (Table 2). About 5% of disease and population groups have susceptible/susceptible genotypes in Koreans, and up to 25% of East Asians also have protective genotypes at both loci, although considerable variation in allele frequency exists among East Asian populations. However, very low numbers among European and African populations have been reported to have these protective genotypes (Li et al., 2007; Oota et al., 2004). In other words, some East Asians, especially Koreans and Japanese, have high frequencies of the protective ADH1B and ALDH2 alleles (Matsuo et al., 2006), whereas most European and African populations have susceptible alleles. Although various social and cultural factors may be involved, these different genetic backgrounds regarding alcohol metabolism could very likely explain the lower incidence of alcohol abuse in some East Asian populations (Grant et al., 2004).

Several hypotheses for the natural selection of this allele in Korean, Japanese, and Han Chinese have been suggested to explain the high frequencies of ADH1B*47His and ALDH2*487Lys in East Asian populations (Han et al., 2007; Li et al., 2007). Among them, the hypothesis of selection pressure for higher ADH1B (mediated by ADH1B*47His) and lower ALDH2 (mediated by ALDH2*487Lys) against some endemic disease(s), including parasitism, in areas with high frequencies of those alleles might be the most plausible explanation, although the nature of the selective pressure and the time period during which it operates are still unknown. There is also convincing evidence that the frequency of ADH1B*47His increased independently in West and East Asia after humans migrated across Eurasia (Li et al., 2007). Similarly, natural selection of ALDH2*487Lys in East Asian populations has also been suggested (Oota et al., 2004). East Asians show similar genetic effects across their populations due to similar ethnic origin and regional characteristics, and there are only minute differences in their responses to alcohol.

GENOME-WIDE ASSOCIATION STUDY AND ADH GENE CLUSTER

The ADH gene cluster is known to be composed of seven ADH genes categorized into five classes, based on the structural similarity as well as the kinetic properties. The class I enzymes, such as ADH1A, ADH1B, and ADH1C, have high affinities for ethanol and contribute to over 70% of the total ethanol oxidizing ability. The class II ADH4 and class V ADH6 enzymes participate in the metabolism of a wide variety of substrates (Hurley, Edenberg, & Li, 2002; Lee, Hoog, & Yin, 2004). The class III ADH5 enzyme is a glutathione-dependent formaldehyde dehydrogenase and has only been detected in the brain (Kaiser, Holmquist, Vallee, & Jornvall, 1991). Unlike other classes of ADH enzymes, which are mainly expressed in the liver and account for about 80% of postabsorptive alcohol metabolism, the class IV ADH7 is mainly expressed in the upper digestive tract, where it oxidizes ethanol at high concentrations and acts early in the timeline of alcohol metabolism in the stomach mucosa (Edenberg, 2007; Farres et al., 1994). The class II and class V enzymes demonstrate greater efficiency in ethanol metabolism than that of class IV. Early studies on identification

TABLE 1 Combined Genetic Effect of ADH1B rs1229984 (A>G; His47Arg) and ALDH2 rs671 (G>A; Glu487Lys) on the Risk of Alcohol Dependence among Korean Male Subjects (n=1032)

Loci	Genotype	Amino Acid	AD, n (%)	NC, n (%)	Genetic Effect
ADH1B rs1229984 (A>G; His47Arg)	AA	His/His	217 (39.5)	298 (61.7)	Protective
	AG	His/Arg	145 (26.4)	155 (32.1)	Protective
	GG	Arg/Arg	187 (34.1)	30 (6.2)	Susceptible
ALDH2 rs671 (G>A; Glu487Lys)	GG	Glu/Glu	530 (96.5)	346 (73.6)	Susceptible
	AG	Glu/Lys	19 (3.5)	122 (25.3)	Protective
	AA	Lys/Lys	0 (0.0)	15 (3.1)	Protective

TABLE 2 Combined Association Analysis of ADH1B rs1229984 (A>G; His47Arg) and ALDH2 rs671 (G>A; Glu487Lys) with the Risk of Alcohol Dependence in Korean Male Subjects (n = 1416)

Genetic Effect		% of Subjects							
ADH1B rs1229984 (A>G; His47Arg)	ALDH2 rs671 (G>A; Glu487Lys)	AD, n=549 (%)	NC, n=483 (%)	PC[a], n=384 (%)	OR	p value	OR	p value	Global p value
Protective	Protective	2.2	26.5	24.8	1	–	1	–	2.2×10^{-31}
Protective	Susceptible	63.7	67.3	68.5	11.49	4.7×10^{-15}	11.40	3.5×10^{-15}	2.6×10^{-19}
Susceptible	Protective	1.3	1.9	2.3	8.30	0.0003			
Susceptible	Susceptible	33.8	4.3	4.4	91.43	1.4×10^{-32}	91.43	1.4×10^{-32}	16.14

[a]Population controls (n=384). Population controls were used for attributable fraction (AF, 87.7%) calculation (OR (10.53) and frequency (75.2% of PC) of subjects with one or two susceptible alleles for either loci).

of the genetic factors in the risk of AD were focused on the class I genes because they play major functions in alcohol metabolism. Studies have also shown that ADH4 and ADH7 gene variants are a plausible genetic component (Birley et al., 2008; Luo et al., 2005; Osier et al., 2002; Whitfield, 2002). These functional and genetic evidences aid to explain why numerous studies have focused on the ADH gene cluster to identify genetic components not only for the risk of AD, but also for other related diseases, such as upper aerodigestive tract (UAT) cancer and squamous cell carcinoma of the head and neck (SCCHN) (Ji et al., 2011; Oze et al., 2009).

Genome-wide association study (GWAS) has enabled the systematic identification of risk loci for many diseases, including AD. Large-scale GWASs offer considerable promise and have revolutionized the search for common genetic variants that influence individual risk for complex diseases. Subsequent replication studies of an independent population may then provide the substantiation of their genetic influences. A fine-mapping study as a follow-up is also a useful tool to identify a causative polymorphism in the region and to finely localize the signal. Several GWASs have explored candidate genes potentially related to AD (Bierut et al., 2010; Edenberg et al., 2010; Frank et al., 2011; Lind et al., 2010; Park et al., 2013; Treutlein et al., 2009; Wang et al., 2013; Zlojutro et al., 2011; Zuo et al., 2013, 2011). Highlights of these studies are summarized in Table 3.

GWASs and replication studies of African Americans (AA), European Americans (EA), Australians, and three HapMap subjects indicated that a 90-Mb region around the PHF3-PTP4A1 locus was associated with AD in AA (Zuo et al., 2011). The initial GWAS was conducted with 681 AA alcoholics and 508 AA controls. The top-ranked single-nucleotide polymorphisms (SNPs) from the GWAS were then retested in a primary replication sample of 1409 EA cases and 1518 EA controls. The replicable associations were subjected to secondary replication in a sample of 6438 Australian family subjects. A functional expression quantitative trait locus (eQTL) analysis of these replicable SNPs was subsequently done in order to explore their cis-acting regulatory effects on gene expression. Another GWAS of AAs and EAs, combining evidence from the case–control and follow-up study, suggested the association of a cluster of genes on chromosome 11 with AD (Edenberg et al., 2010). Edenberg et al. conducted a GWAS in 1192 AD cases and 692 controls of EA and AA ancestry from the COGA study. They followed up on the top-ranked SNPs in 262 pedigrees. Although no single SNP met genome-wide criteria for significance, there was strong support for the association of a cluster of genes, such as SLC22A18, PHLDA2, NAP1L4, SNORA54, CARS, and OSBPL5, with AD. Bierut et al. (2010) also explored genetic influences on AD in 1897 EA and AA subjects with AD and 1932 unrelated, alcohol-exposed, nondependent controls. Association analysis of GWAS data and two independent replication studies showed that no SNP passed a replication threshold.

In a GWAS of subjects of German descent, rs1789891, which is located between ADH1B and ADH1C and is tightly linked with the function of ADH1C R272Q, was observed to be significantly associated with the risk of AD (Frank et al., 2011). The German subjects comprised 1333 males with severe AD and 2168 controls. The rs1789891, which is located between the ADH1B and ADH1C genes, was detected to be the top-ranked SNP and achieved genome-wide significance. Other markers from this region were also associated with AD, and conditional analyses indicated that these made a partially independent contribution. The SNP rs1789891 is in complete linkage disequilibrium with the functional variant of the ADH1C gene, Arg272Gln, which has been reported to modify the rate of ethanol oxidation to acetaldehyde in vitro. This study may be the first GWAS of AD to provide genome-wide significant support for the role of the ADH gene cluster. A study by Treutlein et al. (2009) conducted an initial GWAS involving 476 patients and 1358 controls. The top SNPs were then followed up in a sample of 1024 patients and 996 controls. The GWAS produced 121 SNPs with nominal $p<10^{-4}$. These nominally associated SNPs from the GWAS and 19 additional SNPs from homologues of rat genes showing differential expression were analyzed in the follow-up study. Fifteen SNPs showed significant association with the same allele as in the GWAS, and two closely linked intergenic SNPs on 2q35 met the genome-wide significance criteria in the combined analysis (rs7590720, $p=9.72\times10^{-9}$; rs1344694, $p=1.69\times10^{-8}$).

In the GWAS in a Korean population, it was found that two major regions, 4q22-23 and 12q24, were significantly associated with the risk of AD, even after Bonferroni correction (Park et al., 2013). When the GWAS of Koreans was compared with the GWASs of other ethnic populations, several genetic loci in chromosomes 1, 2, and 4 showed significant associations with the risk of AD in all the GWASs, and most of the loci were on the ADH gene cluster, especially the ADH1B-ADH1C-ADH7 gene region. Although the strength of significance differed among GWASs, the ADH gene cluster might be an important region for AD in East Asians. In a subsequent fine-mapping study in a Korean population based on the GWA results, 72 successfully genotyped SNPs, including ADH1B rs1229984 (H47R; previously determined to be an important risk factor for AD in Koreans; Choi et al., 2005; Kim et al., 2008), were used for association analyses; ADH1B rs1229984 (H47R) showed the most significant association in the GWAS. The rest of the SNPs also showed a series of different association strengths and magnitudes of risk for AD across the ADH gene cluster region. It was previously demonstrated that the genetic effects of ADH1B and ADH1C regions on the risk of AD might come from the ADH1B*47Arg/*47Arg genotype, and that positive signals from other sites of ADH1B and ADH1C could be a result of tracking the genetic effect of ADH1B His47Arg (Choi et al., 2005). Investigating the effect of ADH1B rs1229984 (H47R) across the ADH gene cluster through conditional analysis by controlling for odds ratios derived from referent analysis of ADH1B rs1229984 (H47R) revealed that all positive signals on the adjacent ADH genes disappeared (Figure 1). These results indicate that ADH1B rs1229984 (H47R) may be the sole genetic marker tracking the genetic effects of the risk of AD in the ADH gene cluster region. However, the tracking effect of ADH1B rs1229984 (H47R) may be limited to several East Asian populations, such as Korean, Japanese, and Han Chinese, due to their geographic distribution and ethnic similarity (Li et al., 2007; Osier et al., 2002; Shen et al., 1997). There has been no evidence suggesting whether ADH1B rs1229984 (H47R) could act as the sole marker for AD in the ADH gene cluster in Japanese and Han Chinese populations. Nonetheless, the important role of ADH1B rs1229984 (H47R) in AD is supported by the cumulative results of Japanese and Han Chinese studies. ADH1B H47R in these two ethnic populations showed a similar effect on the pattern of genotype distribution, magnitude of risk, and strength of association to that of the Korean

TABLE 3 Summary of Published GWAS and Replication Study of AD

PMID	Author (Year)	Ethnic Group	GWAS Case	GWAS Control	Gene	SNP	AA Change	OR	p-value	Replication Study Case	Replication Study Control	OR	p-value
23455491	Zuo et al. (2013)	Caucasian	1409	1518	SERINC2	rs1039630		1.32	2.6×10^{-7}	1645	4793	1.25	0.049
					SERINC2	rs4478858		1.31	4.4×10^{-7}			1.26	0.021
					SERINC2	rs2275436		1.32	2.4×10^{-7}			1.34	0.043
19581569	Treutlein et al. (2009)	German	487	1358	Intergenic	rs1344694		0.7	5.57×10^{-6}	1024	996	0.82	3.36×10^{-3}
					PECR	rs7590720		0.69	5.7×10^{-6}			0.79	6.68×10^{-4}
					PECR	rs705648		1.41	2.68×10^{-5}			1.19	1.24×10^{-2}
					ADH1C	rs1614972		1.37	2.84×10^{-4}			1.16	3.62×10^{-2}
					CAST	rs13362120		0.71	5.72×10^{-5}			0.84	1.25×10^{-2}
					ERAP1	rs13160562		0.7	2.31×10^{-5}			0.85	2.47×10^{-2}
					PPP2R2B	rs1864982		1.49	9.71×10^{-5}			1.3	4.47×10^{-3}
					ESR1	rs6902771		1.35	9.31×10^{-5}			1.2	3.15×10^{-3}
					Intergenic	rs729302		0.73	6.19×10^{-5}			0.85	1.84×10^{-2}
					GATA4	rs13273672		1.27	2.18×10^{-3}			1.15	3.61×10^{-2}
					Intergenic	rs1487814		1.37	3.58×10^{-5}			1.15	2.48×10^{-2}
					CEP83	rs7138291		1.61	2.46×10^{-5}			1.23	4.1×10^{-2}
					Intergenic	rs36563		1.54	6.21×10^{-6}			1.19	3.83×10^{-2}
					CDH13	rs11640875		1.32	5.25×10^{-4}			1.19	9.85×10^{-3}
					Intergenic	rs12388359		0.47	1.21×10^{-5}			0.72	1.25×10^{-2}
23456092	Park et al. (2013)	Korean	117	279	ADH1B	rs1229984	His47Arg	–	–	504	471	2.35	2.63×10^{-21}
					ADH7	rs1442492		2.45	6.28×10^{-8}				
					ADH7	rs10516441		2.73	6.46×10^{-8}				
					ALDH2	rs671	Glu504Lys	0.22	8.42×10^{-8}				
					BRAP	rs3782886		0.31	4.65×10^{-6}				
					PRMT8	rs876594		1.96	1.77×10^{-5}				
22096494	Zuo et al. (2011)	African	681	508	Intergenic	rs9449291		1.54	3.8×10^{-5}	1409	1518	0.85	3.1×10^{-3}
					Intergenic	rs9449312		1.18	0.05			0.84	1.3×10^{-3}
					Intergenic	rs6942342		1.56	2.0×10^{-5}			0.85	1.8×10^{-3}
					Intergenic	rs9353016		1.54	4.1×10^{-5}			0.85	2.2×10^{-3}

				Intergenic	rs429811	1.54	3.0×10^{-5}	0.85	2.5×10^{-3}
				Intergenic	rs457499	1.52	4.5×10^{-5}	0.85	2.2×10^{-3}
				Intergenic	rs2758259	1.52	5.0×10^{-5}	0.85	1.8×10^{-3}
				Intergenic	rs1744134	0.82	0.032	1.14	0.027
				Intergenic	rs1744140	1.52	4.9×10^{-5}	0.85	3.0×10^{-3}
				Intergenic	rs2984458	1.52	4.4×10^{-5}	0.85	3.0×10^{-3}
				Intergenic	rs1681957	1.54	3.8×10^{-5}	0.85	3.3×10^{-3}
				Intergenic	rs1197905	1.54	3.8×10^{-5}	0.85	3.0×10^{-3}
				PTP4A1	rs2622274	1.27	5.6×10^{-3}	0.85	2.7×10^{-3}
				PTP4A1	rs1322416	1.54	3.0×10^{-5}	0.86	5.2×10^{-3}
				Intergenic	rs9294269	1.56	1.6×10^{-5}	1.17	2.4×10^{-3}
				PHF3	rs6932538	1.39	9.4×10^{-4}	0.88	0.016
				PHF3	rs10485358	1.45	3.2×10^{-4}	0.88	0.014
				PHF3	rs10755432	1.43	3.4×10^{-4}	0.88	0.022
				Intergenic	rs1057530	1.3	5.7×10^{-3}	0.88	0.022
				EYS	rs12205302	1.33	2.6×10^{-3}	0.88	0.021
				EYS	rs319924	1.41	6.4×10^{-4}	0.85	3.0×10^{-3}
				EYS	rs319920	1.43	2.9×10^{-4}	0.85	2.7×10^{-3}
				EYS	rs756274	1.34	5.0×10^{-4}	0.89	0.038
				EYS	rs6921058	1.37	8.0×10^{-4}	0.85	3.4×10^{-3}
				EYS	rs12205984	1.33	1.6×10^{-3}	0.85	2.6×10^{-3}
				EYS	rs321498	1.22	0.019	0.86	7.7×10^{-3}
				EYS	rs321494	1.39	1.1×10^{-3}	0.85	3.5×10^{-3}
				EYS	rs729291	1.3	5.2×10^{-3}	0.85	3.7×10^{-3}
				EYS	rs1482451	0.81	0.017	1.19	2.0×10^{-3}
				EYS	rs3003672	0.79	0.016	2.86	7.9×10^{-3}
20202923	Bierut et al. (2010)	Caucasian/African	1897	1932	PKNOX2	rs10893366	1.39	1.93×10^{-7}		
				CC2D2B	rs2039617	0.69	5.95×10^{-7}		
				Intergenic	rs9302534	0.78	2.73×10^{-6}		
				SH3BP5	rs1318937	1.35	3.54×10^{-6}		
				Intergenic	rs2700648	1.29	3.99×10^{-6}		
				Intergenic	rs10803574	1.28	4.41×10^{-6}		
				GRM5	rs6483362	1.42	4.53×10^{-6}		

Continued

TABLE 3 Summary of Published GWAS and Replication Study of AD—cont'd

PMID	Author (Year)	Ethnic Group	GWAS Case	GWAS Control	Gene	SNP	AA Change	OR	p-value	Replication Study Case	Replication Study Control	OR	p-value
					ZNF285A	rs2722650		0.76	7.14×10^{-6}				
					PKNOX2	rs10893365		1.31	7.2×10^{-6}				
					Intergenic	rs1386449		1.98	7.29×10^{-6}				
					PKNOX2	rs750338		1.28	7.61×10^{-6}				
					Intergenic	rs1505846		1.28	8.01×10^{-6}				
					Intergenic	rs9636231		1.3	9.13×10^{-6}				
					Intergenic	rs1363605		0.76	9.62×10^{-6}				
					TPK1	rs10224675		0.46	9.75×10^{-6}				
22004471	Frank et al. (2011)	German	1333	2168	Intergenic	rs1789891		1.46	1.27×10^{-8}				
23089632	Wang et al. (2013)	Caucasian	684	1638	ADH1C	rs1693482	Arg272Gln	1.31	1.24×10^{-7}				
21529783	Heath et al. (2011)	Caucasian	2062	6692	C15orf53	rs12903120			1.09×10^{-6}				
					TMEM108	rs10935045		1.18	1.7×10^{-6}				
					ANKS1A	rs1737727			5.5×10^{-5}				
					ANKS1A	rs2140418			4.4×10^{-5}				
20201924	Edenberg et al. (2010)	Caucasian	1192	692	SLC22A18/PHLDA2	rs2583442			1.2×10^{-4}				
					NAP1L4	rs4758621			3.7×10^{-4}				
					NAP1L4	rs12805661			1.5×10^{-4}				
					NAP1L4	rs729662			8.5×10^{-5}				
					SNORA54	rs3814964			7.4×10^{-5}				
					CARS	rs7481584			9.5×10^{-5}				
					CARS	rs377765			5.2×10^{-4}				
					CARS	rs369461			5.7×10^{-4}				
					OSBPL5	rs4758533			4.8×10^{-5}				
					OSBPL5	rs4468331			6.9×10^{-4}				
20158304	Lind et al. (2010)	Caucasian	1224	1162	MARK1	rs7530302			1.9×10^{-9}				
					DDX6	rs1784300			2.6×10^{-9}				
					KIAA1409	rs12882384			4.86×10^{-8}				

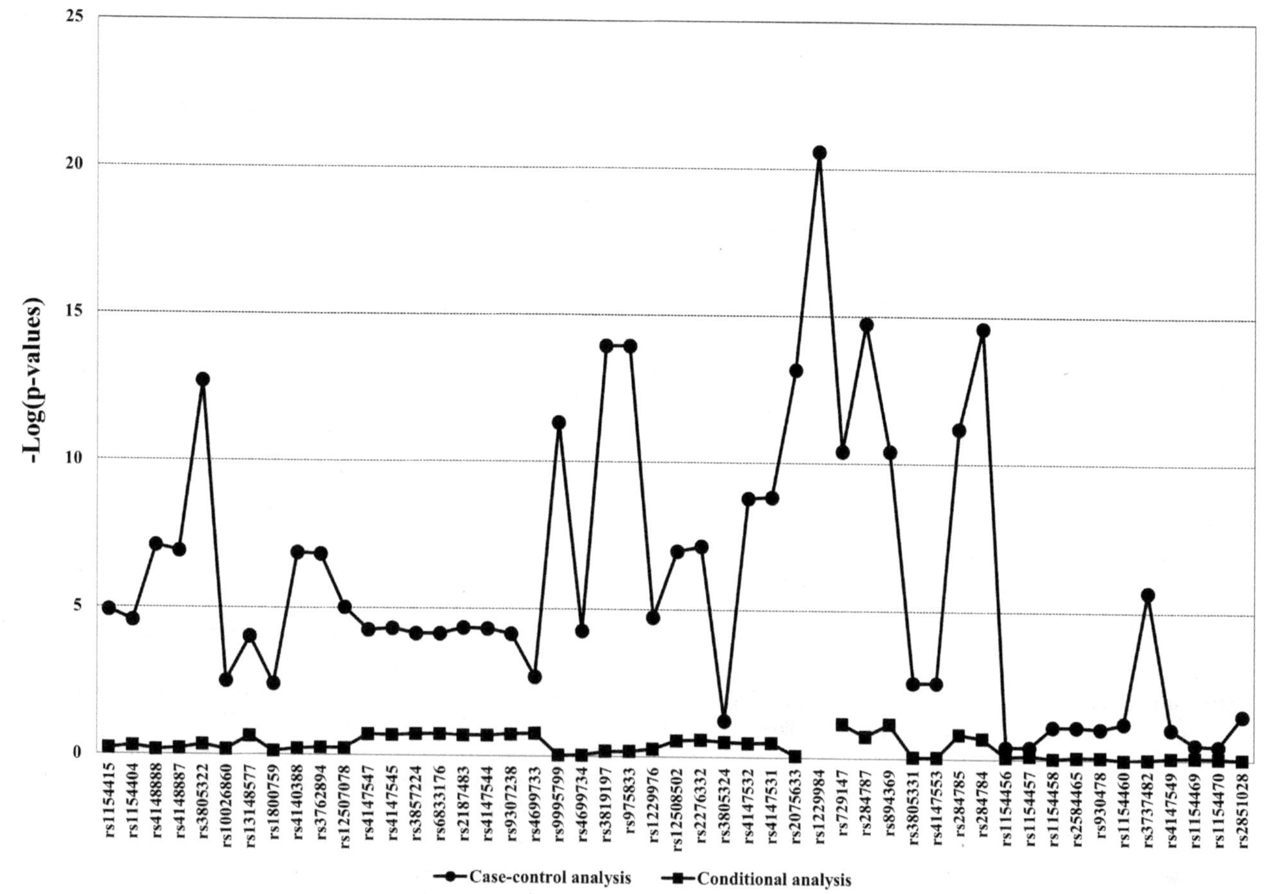

FIGURE 1 **Conditional analysis of genetic variants on *ADH* gene cluster with risk of AD.** Conditional analysis was corrected by *ADH1B H48R* referent OR value (HH, 1; HR, 1.35; RR, 8.205, respectively). Circular symbol means the –log (*p* value) of initial case–control analysis with the risk of AD, and square symbol means the –log (*p* value) after conditional analysis.

population (Higuchi, 1994; Nakamura et al., 1996; Whitfield, 2002). However, unlike in the three noted East Asian populations, rs1789891, located between ADH1B and ADH1C on the ADH gene cluster, showed the most powerful genetic effect on AD in the German male, and conditional analyses indicated that the variant made a partially independent contribution. These results suggest that the etiological characteristics of AD in the German male might be polygenic (Frank et al., 2011).

CONCLUSIONS

Family and twin studies have demonstrated that genetic factors play a substantial role in the risk of AD, and extensive study of candidate genes involved in alcohol metabolism and alcohol-related pathways has demonstrated their important function in disease risk. Additional genes also have been identified by GWAS and subsequent replication studies using high-density microarrays. However, the number of findings to date is still modest, and there have been a few conflicting results due to ethnic differences. A GWAS in a Korean alcoholic cohort showed that two major gene regions (the ADH gene cluster on 4q22-23 and ALDH2 on 12q24) were significantly associated with the risk of AD. The ADH gene cluster was focused on to find major genetic factors, and a fine-mapping study of the cluster revealed

that ADH1B rs1229984 (H47R) might be the sole genetic marker tracking the genetic effects of the risk of AD in the ADH gene cluster region in East Asians. However, multiple genetic markers of ADH genes for the risk of AD have been identified in various ethnic populations.

APPLICATIONS TO OTHER ADDICTIONS AND SUBSTANCE MISUSE

Drug dependence, including cocaine and opioid dependence, is one of the most common phenotypes comorbid with AD. Drug dependence has been shown to have a number of characteristics in common with AD, such as response to specific treatments. In fact, drug dependence may share susceptibility genes with AD. Several studies have demonstrated that polymorphisms of OPRM1 may moderate the risk of AD and/or drug dependence (Hoehe et al., 2000; Luo, Kranzler, Zhao, & Gelernter, 2003; Schinka et al., 2002). Luo et al. (2006, 2005) showed that ADH4 may play an important role in AD as well as drug dependence based on association studies of the relationship between the ADH gene cluster and drug dependence. Both the Hardy–Weinberg equilibrium and a case–control comparison revealed that the association of ADH4 gene with drug dependence reached levels of statistical significance that were at least as great as those for AD. Overall, the ADH gene cluster may predominantly contribute to the risk of AD, and ADH4 on the ADH gene cluster may similarly contribute to drug dependence, such as cocaine and opioid.

DEFINITION OF TERMS

ADH gene cluster The ADH gene cluster is known to be composed of seven ADH genes divided into five classes, based on structural similarity and kinetic properties.

Alcohol dehydrogenase (ADH) ADH refers to a family of enzymes that catalyze the reversible oxidation of primary or secondary alcohols to aldehydes or ketones, respectively. ADH has many roles in the body. One major function is to catalyze the oxidation of ethanol to acetaldehyde as the first step in ethanol metabolism by the liver.

Alcohol dependence (AD) AD is a substance-related disorder in which an individual is physically or psychologically dependent upon the consumption of alcohol.

Aldehyde dehydrogenase (ALDH) ALHDs are a group of enzymes that catalyze the oxidation (dehydrogenation) of aldehydes. Nineteen ALDH genes have been identified within the human genome, and they participate in a wide variety of biological processes, including the detoxification of exogenously and endogenously generated aldehydes.

Collaborative Studies on Genetics of Alcoholism (COGA) COGA has been funded by the NIAAA since 1989 to learn how genes affect vulnerability to alcoholism. The goal of the studies is to identify specific genes that impact the likelihood of developing alcoholism.

Fine mapping Fine mapping refers to the process of searching a gene region identified by GWAS for possible causal alleles.

Genome-wide association study (GWAS) GWAS is an examination of common genetic variants in different individuals to see whether any variant is associated with particular traits.

Twin study Twin studies are a key tool in behavioral genetics and in content fields from biology to psychology. Studies of twins show the absolute and relative importance of environmental and genetic influences on individuals.

KEY FACTS ABOUT ALCOHOL DEPENDENCE

- AD is characterized by uncontrolled and compulsive consumption of alcoholic beverages.
- The physical effects of AD may include cirrhosis of the liver, epilepsy, pancreatitis, peptic ulcers, sexual dysfunction, and nutritional deficiencies.
- AD can cause a wide range of mental health problems, such as brain damage, dementia, anxiety, panic disorder, depression, confusion, and psychosis.
- AD affects various parts of the body, including the gastrointestinal tract, respiratory system, cardiovascular system, and genitourinary system. It also accelerates the aging process.
- Twin and family studies have demonstrated that genetic predisposition is one of the most important factors for the risk of AD, estimated to be in the range of 40–60%.

SUMMARY

- AD is a multifactorial and polygenic disorder involving complex gene-to-gene and gene-to-environment interactions.

- ADH1B and ALDH2 gene variants play an important genetic role in the risk of AD.
- GWASs may help identify common genetic variants influencing individual risk for complex diseases, including AD.
- The ADH gene cluster may be a potential genetic region for the risk of AD in East Asian populations.
- ADH1B His47Arg may be the sole functional genetic marker across the ADH gene cluster in East Asian populations.

REFERENCES

Bierut, L. J., Agrawal, A., Bucholz, K. K., Doheny, K. F., Laurie, C., Pugh, E., … Gene, Environment Association Studies Consortium (2010). A genome-wide association study of alcohol dependence. *Proceedings of the National Academy of Sciences of the United States of America*, 107(11), 5082–5087.

Birley, A. J., James, M. R., Dickson, P. A., Montgomery, G. W., Heath, A. C., Whitfield, J. B., & Martin, N. J. (2008). Association of the gastric alcohol dehydrogenase gene ADH7 with variation in alcohol metabolism. *Human Molecular Genetics*, 17(2), 179–189.

Chen, C. C., Lu, R. B., Chen, Y. C., Wang, M. F., Chang, Y. C., Li, T. K., & Yin, S. J. (1999). Interaction between the functional polymorphisms of the alcohol-metabolism genes in protection against alcoholism. *The American Journal of Human Genetics*, 65(3), 795–807.

Choi, I. G., Son, H. G., Yang, B. H., Kim, S. H., Lee, J. S., Chai, Y. G., … Shin, H. D. (2005). Scanning of genetic effects of alcohol metabolism gene (ADH1B and ADH1C) polymorphisms on the risk of alcoholism. *Human Mutation*, 26(3), 224–234.

Dick, D. M., & Bierut, L. J. (2006). The genetics of alcohol dependence. *Current Psychiatry Reports*, 8(2), 151–157.

Edenberg, H. J. (2007). The genetics of alcohol metabolism: role of alcohol dehydrogenase and aldehyde dehydrogenase variants. *Alcohol Research and Health*, 30(1), 5–13.

Edenberg, H. J., Koller, D. L., Xuei, X., Wetherill, L., McClintick, J. N., Almasy, L., … Foroud, T. (2010). Genome-wide association study of alcohol dependence implicates a region on chromosome 11. *Alcoholism, Clinical and Experimental Research*, 34(5), 840–852.

Edenberg, H. J., Xuei, X., Chen, H. J., Tian, H., Wetherill, L. F., Dick, D. M., … Foroud, T. (2006). Association of alcohol dehydrogenase genes with alcohol dependence: a comprehensive analysis. *Human Molecular Genetics*, 15(9), 1539–1549.

Enoch, M. A., & Goldman, D. (2002). Problem drinking and alcoholism: diagnosis and treatment. *American Family Physician*, 65(3), 441–448.

Farres, J., Moreno, A., Crosas, B., Peralba, J. M., Allali-Hassani, A., Hjelmqvist, L., … Parés, X. (1994). Alcohol dehydrogenase of class IV (sigma sigma-ADH) from human stomach: cDNA sequence and structure/function relationships. *European Journal of Biochemistry*, 224(2), 549–557.

Frank, J., Cichon, S., Treutlein, J., Ridinger, M., Mattheisen, M., Hoffmann, P., … Rietschel, M. (2011). Genome-wide significant association between alcohol dependence and a variant in the ADH gene cluster. *Addiction Biology*, 17(1), 171–180.

Grant, B. F., Dawson, D. A., Stinson, F. S., Chou, S. P., Dufour, M. C., & Pickering, R. P. (2004). The 12-month prevalence and trends in DSM-IV alcohol abuse and dependence: United States, 1991–1992 and 2001–2002. *Drug and Alcohol Dependence*, 74(3), 223–234.

Han, Y., Gu, S., Oota, H., Osier, M. V., Pakstis, A. J., Speed, W. C., ... Kidd, K. K. (2007). Evidence of positive selection on a class I ADH locus. *The American Journal of Human Genetics, 80*(3), 441–456.

Hasin, D. S., Stinson, F. S., Ogburn, E., & Grant, B. F. (2007). Prevalence, correlates, disability, and comorbidity of DSM-IV alcohol abuse and dependence in the United States: results from the National Epidemiologic Survey on Alcohol and Related Conditions. *Archives of General Psychiatry, 64*(7), 830–842.

Heath, A. C., Whitfield, J. B., Martin, N. G., Pergadia, M. L., Goate, A. M., Lind, P. A., ... Montgomery, G. W. (2011). A quantitative-trait genome-wide association study of alcoholism risk in the community: findings and implications. *Biological Psychiatry, 70*(6), 513–518.

Higuchi, S. (1994). Polymorphisms of ethanol metabolizing enzyme genes and alcoholism. *Alcohol and Alcoholism Supplement, 2*, 29–34.

Hoehe, M. R., Kopke, K., Wendel, B., Rohde, K., Flachmeier, C., Kidd, K. K., ... Church, G. M. (2000). Sequence variability and candidate gene analysis in complex disease: association of mu opioid receptor gene variation with substance dependence. *Human Molecular Genetics, 9*(19), 2895–2908.

Hurley, T. D., & Edenberg, H. J. (2012). Genes encoding enzymes involved in ethanol metabolism. *Alcohol Research, 34*(3), 339–344.

Hurley, T. D., Edenberg, H. J., & Li, T. K. (2002). Pharmacogenomics of alcoholism. In J. Licinio, & M. L. Wong (Eds.), *Pharmacogenomics: The search for individualized therapies* (pp. 417–441). Weinheim, Germany: Wiley-VCH.

Ji, Y. B., Tae, K., Ahn, T. H., Lee, S. H., Kim, K. R., Park, C. W., ... Shin, H. D. (2011). ADH1B and ALDH2 polymorphisms and their associations with increased risk of squamous cell carcinoma of the head and neck in the Korean population. *Oral Oncology, 47*(7), 583–587.

Kaiser, R., Holmquist, B., Vallee, B. L., & Jornvall, H. (1991). Human class III alcohol dehydrogenase/glutathione-dependent formaldehyde dehydrogenase. *Journal of Protein Chemistry, 10*(1), 69–73.

Kendler, K. S., Heath, A. C., Neale, M. C., Kessler, R. C., & Eaves, L. J. (1992). A population-based twin study of alcoholism in women. *JAMA, 268*(14), 1877–1882.

Kendler, K. S., Neale, M. C., Heath, A. C., Kessler, R. C., & Eaves, L. J. (1994). A twin-family study of alcoholism in women. *The American Journal of Psychiatry, 151*(5), 707–715.

Kessler, R. C., Crum, R. M., Warner, L. A., Nelson, C. B., Schulenberg, J., & Anthony, J. C. (1997). Lifetime co-occurrence of DSM-III-R alcohol abuse and dependence with other psychiatric disorders in the National Comorbidity Survey. *Archives of General Psychiatry, 54*(4), 313–321.

Kim, D. J., Choi, I. G., Park, B. L., Lee, B. C., Ham, B. J., Yoon, S., ... Shin, H. D. (2008). Major genetic components underlying alcoholism in Korean population. *Human Molecular Genetics, 17*(6), 854–858.

Lee, S. L., Hoog, J. O., & Yin, S. J. (2004). Functionality of allelic variations in human alcohol dehydrogenase gene family: assessment of a functional window for protection against alcoholism. *Pharmacogenetics, 14*(11), 725–732.

Li, D., Zhao, H., & Gelernter, J. (2012). Strong protective effect of the aldehyde dehydrogenase gene (ALDH2) 504lys (*2) allele against alcoholism and alcohol-induced medical diseases in Asians. *Human Genetics, 131*(5), 725–737.

Li, H., Mukherjee, N., Soundararajan, U., Tarnok, Z., Barta, C., Khaliq, S., ... Kidd, K. K. (2007). Geographically separate increases in the frequency of the derived ADH1B*47His allele in eastern and western Asia. *The American Journal of Human Genetics, 81*(4), 842–846.

Li, T. K., Yin, S. J., Crabb, D. W., O'Connor, S., & Ramchandani, V. A. (2001). Genetic and environmental influences on alcohol metabolism in humans. *Alcoholism, Clinical and Experimental Research, 25*(1), 136–144.

Lim, S. S., Vos, T., Flaxman, A. D., Danaei, G., Shibuya, K., Adair-Rohani, H., ... Memish, Z. A. (2012). A comparative risk assessment of burden of disease and injury attributable to 67 risk factors and risk factor clusters in 21 regions, 1990–2010: a systematic analysis for the Global Burden of Disease Study 2010. *The Lancet, 380*(9859), 2224–2260.

Lind, P. A., Macgregor, S., Vink, J. M., Pergadia, M. L., Hansell, N. K., de Moor, M. H., ... Madden, P. A. (2010). A genomewide association study of nicotine and alcohol dependence in Australian and Dutch populations. *Twin Research and Human Genetics, 13*(1), 10–29.

Luczak, S. E., Glatt, S. J., & Wall, T. L. (2006). Meta-analyses of ALDH2 and ADH1B with alcohol dependence in Asians. *Psychological Bulletin, 132*(4), 607–621.

Luo, X., Kranzler, H. R., Zhao, H., & Gelernter, J. (2003). Haplotypes at the OPRM1 locus are associated with susceptibility to substance dependence in European-Americans. *American Journal of Medical Genetics Part B: Neuropsychiatric Genetics, 120B*(1), 97–108.

Luo, X., Kranzler, H. R., Zuo, L., Lappalainen, J., Yang, B. Z., & Gelernter, J. (2006). ADH4 gene variation is associated with alcohol dependence and drug dependence in European Americans: results from HWD tests and case-control association studies. *Neuropsychopharmacology, 31*(5), 1085–1095.

Luo, X., Kranzler, H. R., Zuo, L., Yang, B. Z., Lappalainen, J., & Gelernter, J. (2005). ADH4 gene variation is associated with alcohol and drug dependence: results from family controlled and population-structured association studies. *Pharmacogenetics and Genomics, 15*(11), 755–768.

Matsuo, K., Wakai, K., Hirose, K., Ito, H., Saito, T., & Tajima, K. (2006). Alcohol dehydrogenase 2 His47Arg polymorphism influences drinking habit independently of aldehyde dehydrogenase 2 Glu487Lys polymorphism: analysis of 2,299 Japanese subjects. *Cancer Epidemiology, Biomarkers and Prevention, 15*(5), 1009–1013.

Nakamura, K., Iwahashi, K., Matsuo, Y., Miyatake, R., Ichikawa, Y., & Suwaki, H. (1996). Characteristics of Japanese alcoholics with the atypical aldehyde dehydrogenase 2*2. I. A comparison of the genotypes of ALDH2, ADH2, ADH3, and cytochrome P-4502E1 between alcoholics and nonalcoholics. *Alcoholism, Clinical and Experimental Research, 20*(1), 52–55.

Oota, H., Pakstis, A. J., Bonne-Tamir, B., Goldman, D., Grigorenko, E., Kajuna, S. L., ... Kidd, K. K. (2004). The evolution and population genetics of the ALDH2 locus: random genetic drift, selection, and low levels of recombination. *Annals of Human Genetics, 68*(Pt 2), 93–109.

Osier, M., Pakstis, A. J., Kidd, J. R., Lee, J. F., Yin, S. J., Ko, H. C., ... Kidd, K. K. (1999). Linkage disequilibrium at the ADH2 and ADH3 loci and risk of alcoholism. *The American Journal of Human Genetics, 64*(4), 1147–1157.

Osier, M. V., Pakstis, A. J., Soodyall, H., Comas, D., Goldman, D., Odunsi, A., ... Kidd, K. K. (2002). A global perspective on genetic variation at the ADH genes reveals unusual patterns of linkage disequilibrium and diversity. *The American Journal of Human Genetics, 71*(1), 84–99.

Oze, I., Matsuo, K., Suzuki, T., Kawase, T., Watanabe, M., Hiraki, A., ... Tanaka, H. (2009). Impact of multiple alcohol dehydrogenase gene polymorphisms on risk of upper aerodigestive tract cancers in a Japanese population. *Cancer Epidemiology, Biomarkers and Prevention, 18*(11), 3097–3102.

Park, B. L., Kim, J. W., Cheong, H. S., Kim, L. H., Lee, B. C., Seo, C. H., ... Choi, I. G. (2013). Extended genetic effects of ADH cluster genes on the risk of alcohol dependence: from GWAS to replication. *Human Genetics*, *132*(6), 657–668.

Prescott, C. A., & Kendler, K. S. (1999). Genetic and environmental contributions to alcohol abuse and dependence in a population-based sample of male twins. *The American Journal of Psychiatry*, *156*(1), 34–40.

Schinka, J. A., Town, T., Abdullah, L., Crawford, F. C., Ordorica, P. I., Francis, E., ... Mullan, M. (2002). A functional polymorphism within the mu-opioid receptor gene and risk for abuse of alcohol and other substances. *Molecular Psychiatry*, *7*(2), 224–228.

Shen, Y. C., Fan, J. H., Edenberg, H. J., Li, T. K., Cui, Y. H., Wang, Y. F., ... Xia, G. Y. (1997). Polymorphism of ADH and ALDH genes among four ethnic groups in China and effects upon the risk for alcoholism. *Alcoholism, Clinical and Experimental Research*, *21*(7), 1272–1277.

Thomasson, H. R., Edenberg, H. J., Crabb, D. W., Mai, X. L., Jerome, R. E., Li, T. K., ... Yin, S. J. (1991). Alcohol and aldehyde dehydrogenase genotypes and alcoholism in Chinese men. *The American Journal of Human Genetics*, *48*(4), 677–681.

Tolstrup, J. S., Nordestgaard, B. G., Rasmussen, S., Tybjaerg-Hansen, A., & Gronbaek, M. (2008). Alcoholism and alcohol drinking habits predicted from alcohol dehydrogenase genes. *The Pharmacogenomics Journal*, *8*(3), 220–227.

Treutlein, J., Cichon, S., Ridinger, M., Wodarz, N., Soyka, M., Zill, P., ... Rietschel, M. (2009). Genome-wide association study of alcohol dependence. *Archives of General Psychiatry*, *66*(7), 773–784.

Wang, J. C., Foroud, T., Hinrichs, A. L., Le, N. X., Bertelsen, S., Budde, J. P., ... Goate, A. M. (2013). A genome-wide association study of alcohol-dependence symptom counts in extended pedigrees identifies C15orf53. *Molecular Psychiatry*, *18*(11), 1218–1224.

Whitfield, J. B. (2002). Alcohol dehydrogenase and alcohol dependence: variation in genotype-associated risk between populations. *The American Journal of Human Genetics*, *71*(5), 1247–1250.

WHO. (2011). *Global status report on alcohol and health*. WHO.

Zlojutro, M., Manz, N., Rangaswamy, M., Xuei, X., Flury-Wetherill, L., Koller, D., ... Almasy, L. (2011). Genome-wide association study of theta band event-related oscillations identifies serotonin receptor gene HTR7 influencing risk of alcohol dependence. *American Journal of Medical Genetics Part B: Neuropsychiatric Genetics*, *156B*(1), 44–58.

Zuo, L., Wang, K., Zhang, X. Y., Krystal, J. H., Li, C. S., Zhang, F., ... Luo, X. (2013). NKAIN1-SERINC2 is a functional, replicable and genome-wide significant risk gene region specific for alcohol dependence in subjects of European descent. *Drug and Alcohol Dependence*, *129*(3), 254–264.

Zuo, L., Zhang, C. K., Wang, F., Li, C. S., Zhao, H., Lu, L., ... Luo, X. (2011). A novel, functional and replicable risk gene region for alcohol dependence identified by genome-wide association study. *PLoS One*, *6*(11), e26726.

Chapter 49

Neuropathology of Gene Expression during Alcohol Withdrawal

Harinder Aujla
Department of Psychology, University of Winnipeg, Winnipeg, MB, Canada

Abbreviations

5-HT 5-Hydroxytryptamine (serotonin)
BDNF Brain-derived neurotrophic factor
BLA Basolateral amygdala
CeA Central nucleus of the amygdala
CRH Corticotropin-releasing hormone
DAT Dopamine transporter
DNA Deoxyribonucleic acid
DTs Delirium tremens
GABA γ-Aminobutyric acid
mRNA Messenger ribonucleic acid
NMDA *N*-methyl-D-aspartate
NOP Nociceptin receptor
NPS Neuropeptide S
PAG Periaqueductal gray
SNP Single-nucleotide polymorphism
TLR-4 Toll-like receptor 4
VTA Ventral tegmental area

INTRODUCTION

Cessation of chronic alcohol use is associated with persistent neurologic dysregulation that contributes to the expression of withdrawal and subsequent vulnerability to relapse (Becker, Lopez, & Doremus-Fitzwater, 2011). Withdrawal from alcohol is characterized by heightened stress responsivity, dysregulation of reward processing, and executive dysfunction—particularly inhibitory control (Koob et al., 2004). In severe cases, withdrawal from alcohol is also associated with the manifestation of seizures (Eyer et al., 2011). For a thorough review of the cognitive consequences of alcohol use, please refer to Chapter 57, "Neuropathobiology of Alcohol-Induced Cognitive Deficits," in the present text. The broad range of consequences resulting from alcohol use suggests alterations spanning multiple functional systems and possible damage to general processing capabilities within the nervous system.

Investigations including gene linkage and gene microarray studies in alcohol-dependent individuals (Chamorro et al., 2012; Dahlgren et al., 2011; Ducat et al., 2013; Edenberg et al., 2004; Grzywacz et al., 2012; He & Crews, 2008; Hendershot, Claus, & Ramchandani, 2014; Hillemacher et al., 2009) or animal models of alcoholism (Aujla et al., 2012; Covault, Gelernter, Hesselbrock, Nellissery, & Kranzler, 2004; D'Addario et al., 2013; Freeman, Staehle, Gümüş, et al., 2012; Freeman, Staehle, Vadigepalli, et al., 2013; Jee et al., 2013) have revealed important clues on polymorphisms that may be related to alcohol withdrawal. However, it is important to note that the identification of single-nucleotide polymorphisms (SNPs) from some approaches may reflect alterations that confer vulnerability to alcohol use rather than those that result from alcohol use or withdrawal. Thus, it is important to consider findings from investigations that have assessed changes in DNA expression utilizing groups randomly assigned to alcohol intake conditions to provide direct evidence for alcohol-induced alterations in gene expression during withdrawal.

The remainder of this chapter is organized around relating features of alcohol withdrawal to functional alterations in gene expression of selected neurotransmitter systems and neuroinflammation-related factors.

Glutamate

Alcohol reduces glutamatergic transmission by acting as a functional antagonist at ionotropic and metabotropic glutamate receptors and, with chronic administration, alters the expression of glutamate receptors during withdrawal in a compensatory fashion (Holmes, Spanagel, & Krystal, 2013). One consequence is increased excitatory *N*-methyl-D-aspartate (NMDA)-mediated transmission, which has been linked to withdrawal symptoms, including seizures (Eyer et al., 2011), that are reduced with the application of antiglutamatergic agents such as topiramate and acamprosate (Gass & Foster Olive, 2008)—the latter of which has also been shown to be effective in preventing relapse in abstinent users (Mann, Lehert, & Morgan, 2004).

Delirium tremens (DTs), a constellation of severe alcohol-withdrawal symptoms, reflect increased autonomic activity and psychotic features that have been linked to increased activity of ionotropic glutamate receptors (Gass & Foster Olive, 2008). Utilizing the Munich Gene Data Bank, Preuss et al. (2005) reported that the Ser 310 polymorphism of the *GRIK3* gene, which encodes the GluR7 subunit of the kainate receptor, was more frequently expressed in alcoholic patients with DTs versus those without DTs or nonalcoholic controls. Furthermore, Rujescu et al. (2005)

discovered alterations in the 2108A polymorphism of *GRIN1*, which encodes the NR1 subunit of the NMDA receptor, in alcoholic patients with seizures versus alcoholic patients without seizures or versus nondependent controls. However, no significant difference in polymorphism expression of *GRIN1* was observed between controls and alcoholic patients without seizures, including those with DTs, suggesting that the alteration at *GRIN1* was strongly associated specifically with seizure activity during alcohol withdrawal, rather than DTs per se (Rujescu et al., 2005).

In a 2012 report examining the consequences of withdrawal from chronic alcohol in rats on gene expression in the dorsal vagal complex, Freeman et al. (2012) reported a transient downregulation of *Grin2a* and *Grik3*, which code for the NMDAR2A and the GluR7 subunits of the kainate receptor, respectively, at 4h of withdrawal followed by an upregulation of the same genes beginning at 18h of withdrawal. While there is a relative paucity of data on whether alcohol withdrawal produces alterations in gene expression of metabotropic glutamate receptors, this family of receptors may play an important role in alcohol withdrawal. Preuss et al. (2005) did not observe any polymorphisms related to mGluR7 or mGluR8 between alcoholic patients with DTs and those without DTs or between alcoholic patients with withdrawal seizures and those without seizures. However, in a 2013 investigation, Kupila et al. (2013) examined mGluR1/5 binding in Cloninger type I alcoholic versus nonalcoholic participants and found increased receptor binding in the cornu ammonis (CA) 2 region of the hippocampus, a structure that is associated with seizures (Eyer et al., 2011), but not in CA1, CA3, the dentate gyrus, or three striatal structures—the caudate, putamen, and nucleus accumbens.

Alcohol-related alterations in gene expression have also been noted in systems related to motivation and stress. In a microarray analysis of gene expression in the nucleus accumbens and amygdala of alcohol-preferring rats following intake of alcohol or a control solution, Rodd et al. (2008) noted a significant decrease in accumbens expression of Homer1, a member of the Homer family of proteins, which serve to functionally regulate NMDA receptors, as well as amygdalar expression of *GRM1*, which codes for mGluR1, indicating a role for alcohol in modifying glutamate-related gene expression. Further evidence for alcohol-induced changes in amygdalar glutamatergic transmission comes from a 2013 report from Freeman et al. (2013) that noted a transient decrease in the expression of *GRIN2C*, which codes the NR2c subunit of NMDA, at 4h followed by a return to baseline by 18h of withdrawal from an alcohol liquid diet. These findings are consistent with previous findings that strongly implicate the amygdala in mediating the expression of heightened anxiety and stress reactivity associated with withdrawal from alcohol (Koob et al., 2004).

Decreased reward function, reflected, in part, by diminished activity in the ventral tegmental area (VTA) and nucleus accumbens, is associated with chronic alcohol administration (Fitzgerald, Liu, & Morzorati, 2012; Szumlinski, Ary, Lominac, Klugmann, & Kippin, 2007) and may contribute to relapse risk (Koob et al., 2004). Persistent modification of Homer2 in the nucleus accumbens following chronic alcohol administration has been reported by Szumlinski et al. (2007) up to 2 months following the cessation of alcohol. Interestingly, upregulation of NR2b and mGluR1, regulated by Homer2, was also observed during withdrawal, but returned to normal levels after 2 weeks (Szumlinski et al., 2007) (see Figure 1).

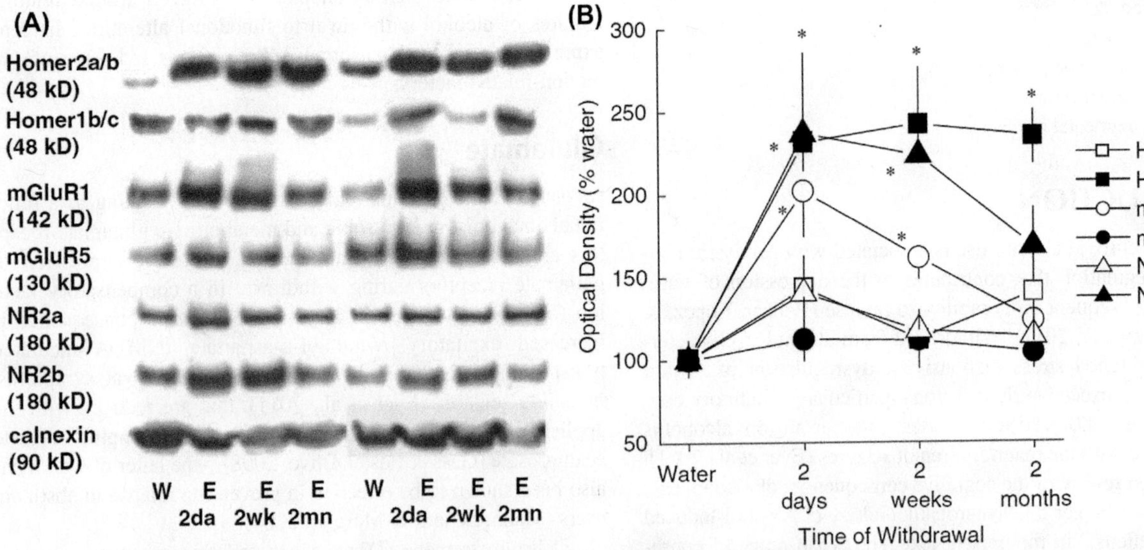

FIGURE 1 **Chronic voluntary alcohol intake produces enduring Homer2, but transient NR2b and mGluR1, elevations in the nucleus accumbens (NAC).** (A) Representative immunoblots for the total protein levels of Homer2a/b, Homer1b/c, NR2a, NR2b, mGluR1, mGluR5, and calnexin (loading control) in the NAC of groups of mice killed at 2 days (da), 2 weeks (wk), and 2 months (mn) withdrawal from 3 months of water (W) or alcohol consumption (E; mean daily intake 11.271.5 g/kg). (B) Summary of the changes in protein expression following withdrawal from 3 months of continuous alcohol consumption. Compared to water-drinking mice, chronic alcohol consumption elevated NAC Homer2a/b levels at all withdrawal time points ($F(3,33)=14.0$, $p<0.0001$), but did not affect significantly NAC Homer1b/c levels ($p=0.25$). Alcohol withdrawal did not affect NR2a levels ($p=0.24$) or the levels of mGluR5 ($p=0.35$), but elicited a rise in NR2b and mGluR1a that persisted for at least 2 weeks (for NR2b, $F(3,34)=2.7$, $p=0.06$; for mGluR1a, $F(3,34)=8.1$, $p<0.0001$). Data in (B) represent the mean±SEM of 7–9 animals per time point. *$p<0.05$ versus water control. *Reproduced from Szumlinski et al. (2007) with permission from the Nature Publishing Group.*

Thus, in addition to dysregulation within homeostatic and stress-related regions, glutamate-related alterations in gene expression in reward-related brain regions are produced following chronic alcohol use.

γ-Aminobutyric Acid

Alcohol acts as an agonist at γ-aminobutyric acid A (GABA$_A$) receptors and chronic alcohol use has been shown to produce tolerance to GABA-mediated receptor function (Ticku & Burch, 1980), an outcome associated with the manifestation of withdrawal symptoms, including heightened anxiety (Koob et al., 2004), seizure activity (Eyer et al., 2011) and, through disinhibition of dopaminergic neurons, DTs (Heinz et al., 1996).

In a 2014 report, examination of the GABA$_A$ receptor complex in the central nucleus of the amygdala (CeA) of human alcoholics revealed decreased gene expression of the α2 subunit—the most abundant subunit in the human CeA (Jin et al., 2014). Notably, the α2 subunit is particularly important in mediating anxiety-like behavior (Low, 2000) and reward-related actions of GABA agonists, such as benzodiazepines, in the nucleus accumbens (Engin et al., 2014). Further support for a role of the α2 subunit in alcoholism comes from the identification of SNPs in *GABRA2*, which codes for α2, in alcoholics (Covault et al., 2004; Edenberg et al., 2004; Lappalainen et al., 2005).

In the rat CeA, Freeman et al. (2013) reported decreased expression of *GABRD*, which codes for the δ receptor subunit of the GABA$_A$ receptor, at 4 and 32h, but not at 48h, of withdrawal from an alcohol liquid diet. Moreover, in a previous report, Freeman et al. (2012) noted altered expression of *GABRA1*, which codes for the α1 subunit of GABA$_A$, in the dorsal vagal complex—a source of homeostatic input into the CeA—at 4 and 8h of withdrawal. These findings are similar to those for genes encoding a variety of ionotropic glutamate receptor subunits—suggesting that certain GABA and glutamate receptor units are functionally linked, perhaps by the common transcription factor PAX-8 (Freeman et al., 2012).

In an in vitro examination of rat hippocampal neurons during alcohol exposure and 3h of withdrawal, Sanna et al. (2003) found dysregulation of several GABA$_A$ subunits. Alcohol exposure decreased mRNA expression of α1, γ2L, and γ2S and increased expression of α2, α3, and α4 (Sanna et al., 2003). At 3h of withdrawal, no increase in expression was observed for any GABA$_A$ subunit, while the decreased expression of α1, γ2L, and γ2S persisted, and decreased expression was also observed in α3—which represents a reversal from the increased expression seen during alcohol exposure (Sanna et al., 2003) (see Figure 2). These findings are consistent with previous observations of hippocampal hyperexcitability during alcohol withdrawal—a contributing factor for seizures (Eyer et al., 2011).

The periaqueductal gray (PAG) has been shown to play an important role in alcohol withdrawal-related audiogenic seizures, which are attenuated by intra-PAG administration of the NMDA receptor antagonist AP7 in alcohol-withdrawn rats (Yang, 2003). However, while intra-PAG administration of the GABA$_A$ agonist muscimol was able to reduce fear-potentiated startle in controls, it had no effect on reducing the enhanced fear-potentiated startle response in alcohol-withdrawn animals (Silva & Nobre, 2014). Moreover, increased excitability in the PAG has also been reported in mice following administration of chronic intermittent alcohol (Lowery-Gionta et al., 2014). Taken together, these findings suggest possible downregulation of PAG GABA$_A$ receptors during alcohol withdrawal. Thus, alcohol-induced alterations in GABA transmission contribute to a wide range of symptoms associated with alcohol withdrawal, including autonomic dysfunction as well emotional and reward-related processing dysregulation.

Opioids

Alcohol acts as an agonist at the μ opioid receptor to promote dopamine release from the VTA, and chronic use is associated with cross-tolerance to opiates (Zhou & Kreek, 2014). Withdrawal from alcohol has been shown to decrease the anxiolytic efficacy of dorsal PAG infusions of the μ opioid receptor agonist morphine in the fear-potentiated startle test (Silva & Nobre, 2014), pointing toward possible alteration in the gene that encodes the μ opioid receptor—*OPRM1*. However, while in humans the A118 polymorphism of *OPRM1* is related to a predisposition to initiate alcohol use, and also to levels of alcohol drinking (Hendershot et al., 2014) and cue-induced craving (Ray, 2011), there is little information on whether withdrawal from alcohol, or chronic alcohol administration per se, results in alterations in *OPRM1*, *OPRD1*, or *OPRK1*—the genes that encode the μ, δ, and κ opioid receptors, respectively.

It is notable that the effectiveness of naltrexone, a μ opioid receptor antagonist, in treating alcohol withdrawal is related to a polymorphism of *OPRM1* (Chamorro et al., 2012). In a meta-analysis, Chamorro et al. (2012) found lower rates of relapse in naltrexone-treated alcoholic patients possessing the G rather than the AA allele of the A118G SNP of *OPRM1*. They suggest that the effect of this polymorphism on relapse is related to the role of the A118 allele in mediating hypothalamic–pituitary–adrenal axis activity, which is related to stress-induced relapse (Ducat et al., 2013). Thus, as of this writing, findings support a role for altered opioid receptor function as a predisposing factor to alcohol use and relapse treatment, rather than as a consequence of alcohol use.

Dopamine

Dopamine release from the VTA projection to the nucleus accumbens is a common consequence of administration of drugs of abuse, and chronic activation of this pathway has been strongly implicated in withdrawal-related depression of reward function (Koob et al., 2004). Although alcohol does not function as a direct agonist at dopamine receptors, its actions at μ opioid receptors (Zhou & Kreek, 2014) and 5-hydroxytryptamine 3 (5-HT3) receptors (Engleman, Rodd, Bell, & Murphy, 2008) ultimately increase dopaminergic activity and, with chronic use, produce dopaminergic dysregulation during withdrawal (Koob et al., 2004).

Grzywacz et al. (2012) investigated polymorphisms of the *DRD2* gene, which codes for the dopamine D2 receptor, in alcohol-dependent individuals with or without seizures and/or DTs. They found that with respect to *DRD2* exon 8, individuals with the A/A+ genotype exhibited a significantly higher rate of seizures versus individuals with the A/A− genotype. Moreover, the A/G+ genotype conferred protection, as carriers exhibited a significantly lower rate of seizures versus those with the A/G− genotype. No significant differences were reported between the previous

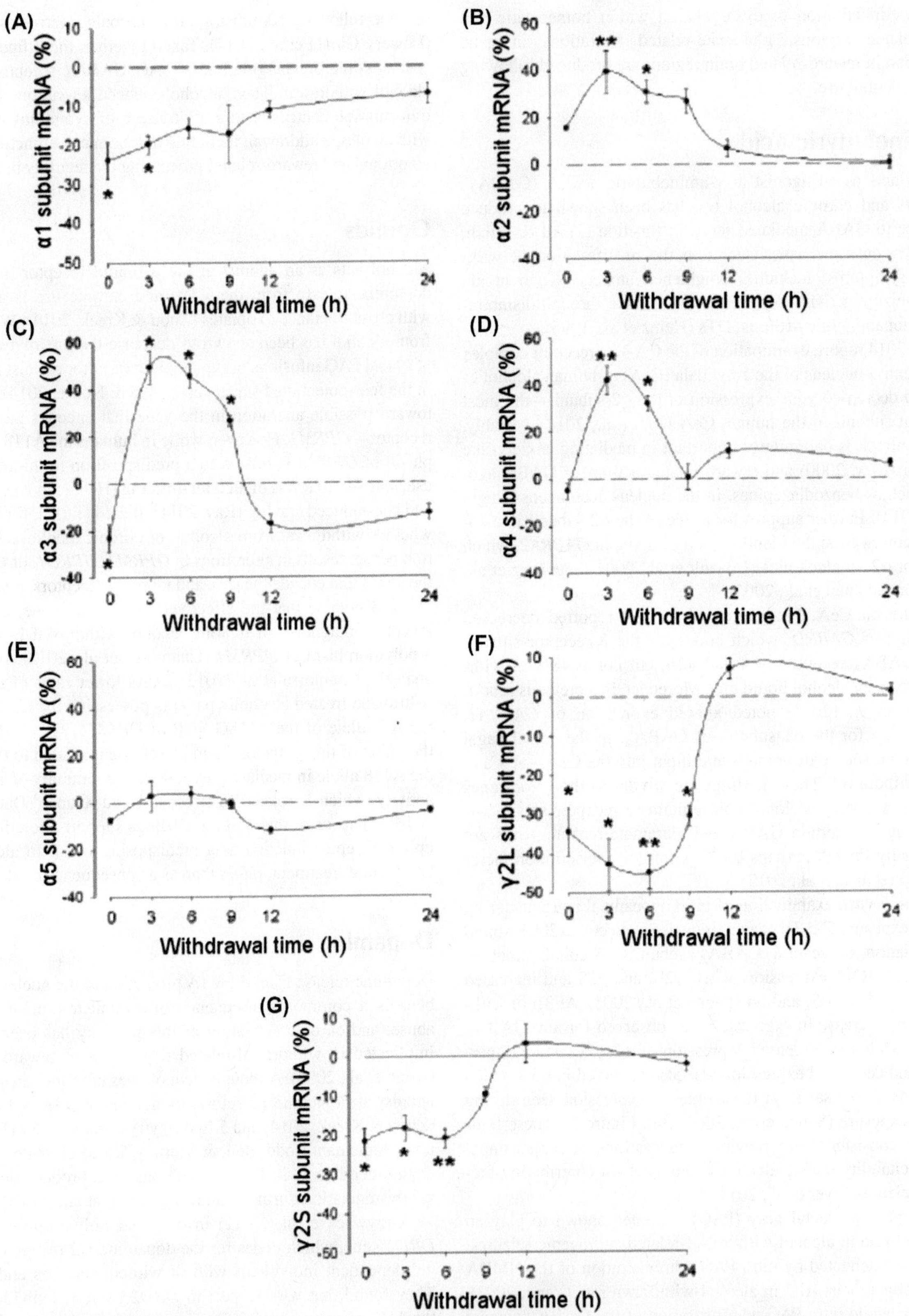

FIGURE 2 Time course of the effects of ethanol withdrawal on the abundance of GABA$_A$ receptor subunit mRNAs in hippocampal cells. Cells were incubated first for 5 days with 100 mM ethanol and then for the indicated times in ethanol-free medium. The amounts of GABA$_A$ receptor α1 (A), α2 (B), α3 (C), α4 (D), α5 (E), γ2L (F), and γ2S (G) subunit mRNAs were then determined by RNase protection assay. Data are means ± SE of 6–13 values from three independent experiments and are expressed as a percentage of the corresponding value for control cultures incubated in the absence of ethanol for 5 days. *$p<0.05$, **$p<0.001$ versus control. *Reproduced from Sanna et al. (2003) with permission from the Society for Neuroscience.*

genotypes on presentation of DTs during withdrawal (Grzywacz et al., 2012). Further examination of the *DRD2* gene at the A1 allele by Dahlgren et al. (2011) revealed that the A1+ (i.e., A1/A1 or A1/A2) polymorphism of *ANKK* Taq1A, which is functionally related to the *DRD2* gene (Noble, 2003) and may be included in the *DRD2* gene region, was significantly associated with increased relapse rates in alcohol-dependent individuals (Dahlgren et al., 2011). Interestingly, Grzywacz et al. (2012) did not observe any difference in presentation of withdrawal-related seizures or DTs as a function of A1+ polymorphisms. This indicates that relapse risk may stem from reward dysregulation in addition to alcohol seeking related to the alleviation of withdrawal.

The dopamine transporter (DAT) has been shown to be sensitive to alcohol exposure in rats (Szot, White, Veith, & Rasmussen, 1999) and human alcoholics (Tiihonen et al., 1995). In an investigation of alcohol-related changes in the epigenetic regulation of DAT, Hillemacher et al. (2009) reported increased methylation of the DAT promoter in alcohol-dependent individuals versus nondependent controls. Moreover, in alcohol-dependent individuals, DAT-promoter methylation levels were negatively correlated with craving scores on the Obsessive–Compulsive Drinking Scale—possibly owing to the protective effect of increased dopamine availability resulting from decreased DAT activity (Hillemacher et al., 2009). However, in a 2014 report, Nieratschker et al. (2014) did not observe any difference in DAT-promoter methylation as a function of alcohol history, nor did they observe a significant relationship between DAT-promoter methylation and craving. Thus, further investigation is warranted to resolve whether epigenetic modification of DAT results from alcohol use.

Dopamine transmission has also been implicated in regulating increased mRNA expression of ryanodine receptors during withdrawal from alcohol (Kurokawa, Mizuno, & Ohkuma, 2013). Kurokawa et al. (2013) reported elevated expression of ryanodine R1 and R2 receptors in the frontal cortex, limbic forebrain, and nucleus accumbens of mice following 9 days of alcohol vapor exposure—an effect that was reversed with the administration of the dopamine D1 receptor antagonist SCH23390 (see Figure 3), but not the dopamine D2 receptor antagonist sulpiride. However, both SCH23390 and sulpiride were similarly efficacious in reducing signs of physical withdrawal (Kurokawa et al., 2013), which suggests that transmission at D2 receptors may regulate withdrawal in a ryanodine-independent manner or within an alternate brain region.

Corticotropin-Releasing Hormone

Corticotropin-releasing hormone (CRH) has long been known to play a critical role in regulating the hypothalamic–pituitary–adrenal axis response to stress (Koob et al., 2004). More recent investigations have shed light on the role of extrahypothalamic CRH in stress and anxiety-like phenomena, including anxiogenic consequences associated with alcohol withdrawal (Menzaghi et al., 1994). Furthermore, Lê et al. (2000) have demonstrated that antagonism of CRH receptors attenuates foot-shock-induced reinstatement of responding to alcohol 5–8 days following the cessation of alcohol self-administration, which suggests an important contribution of CRH dysregulation to stress-related relapse during withdrawal.

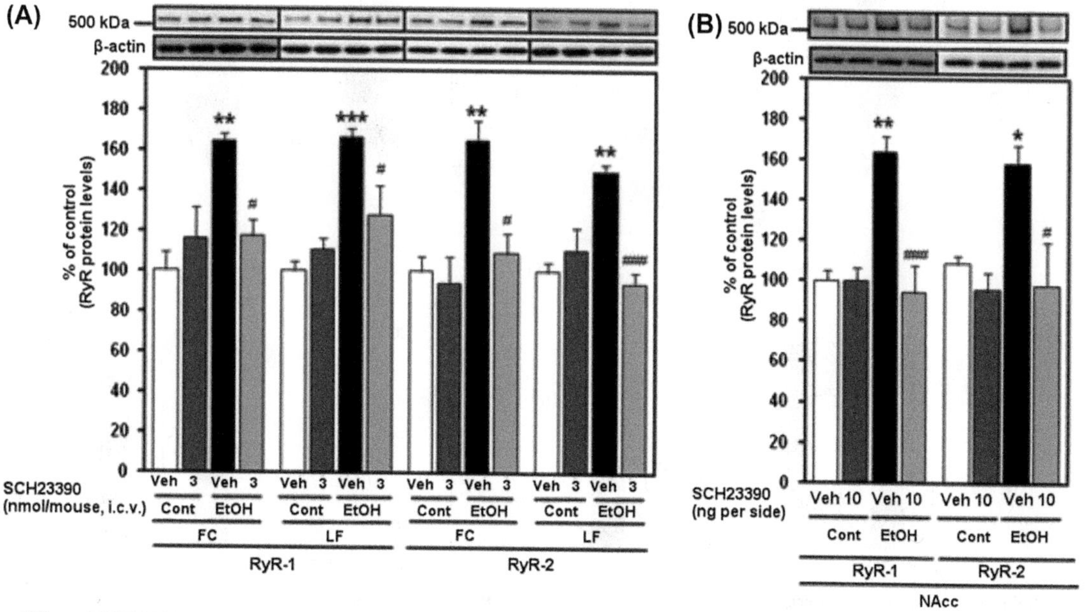

FIGURE 3 Effect of SCH23390 on increased expression of ryanodine receptor (RyR)-1 and RyR-2 protein in the frontal cortex (FC), limbic forebrain (LF), and nucleus accumbens (NAcc) obtained from mice with ethanol (EtOH) physical dependence. (A) SCH23390 (3 nmol/mouse) or vehicle (Veh) was intracerebroventricularly (i.c.v.) administered once a day during continuous EtOH vapor inhalation for 9 days. Samples used for measuring RyR protein were prepared after the continuous exposure to EtOH vapor for 9 days. **$p<0.01$, ***$p<0.001$ versus control (Cont); #$p<0.05$, ###$p<0.001$ versus Veh–EtOH group (Bonferroni multiple comparison test). (B) Effects of intra-nucleus accumbens (NAcc) administration of SCH23390 on increased expression of RyR-1 and RyR-2 in the NAcc obtained from mice with EtOH physical dependence. SCH23390 (10 ng per side) was intra-NAcc administered once a day during continuous EtOH vapor inhalation for 9 days. Each column represents the mean ± SEM of four mice. *$p<0.05$ and **$p<0.01$ versus vehicle control group; #$p<0.05$ and ###$p<0.001$ versus Veh–EtOH. *Reproduced from Kurokawa et al. (2013) with permission from John Wiley & Sons.*

Consistent with the preclinical literature on CRH–stress–alcohol interactions, in human alcoholics, Schmid et al. (2010) found that SNPs of the *CRHR1* gene, which codes for the CRH1 receptor, are associated with stress-induced onset of drinking and alcohol intake. Furthermore, Ray (2011) reported that an SNP of the CRH-binding protein gene is associated with alcohol-withdrawal-related phenomena including stress-induced craving, subjective tension, and negative mood. In a direct examination of withdrawal-related CRH dysregulation, Sommer et al. (2008) evaluated appetitive- and anxiety-related performance, as well as gene expression of *CRHR1* and *CRHR2*, which code for CRH receptors 1 and 2, respectively, during withdrawal from chronic alcohol. In comparison to nondependent controls, rats with 7 weeks of voluntary alcohol administration exhibited increased self-administration of alcohol, potentiated forced swim stress-induced alcohol intake, and heightened sensitivity to punishment-induced reduction of responding to alcohol in a conflict test—an effect that was completely blocked by administration of the CRHR1 antagonist MTIP (Sommer et al., 2008). Furthermore, elevated CRHR1 mRNA expression was observed at 3 months of withdrawal from alcohol in the basolateral and medial nuclei of the amygdala, while a reduction in expression of CRHR2 mRNA was observed in the basolateral amygdala (BLA) at the same time point (Sommer et al., 2008). Elevation of CRH mRNA expression was also observed in the central nucleus of the amygdala (Sommer et al., 2008) (see Figure 4). Thus, these findings demonstrate persistent dysregulation of CRH function that may play a critical role in long-term relapse risk in human alcoholics.

CRH-like regulation of stress is well conserved from an evolutionary perspective. Jee et al. (2013) demonstrated that loss-of-function and gain-of-function mutations of the gene *SEB-3*, which codes for a CRH-like receptor in *Caenorhabditis elegans*, mediated both heat-stress- and alcohol-withdrawal-induced locomotion, behavioral arousal, and tremor activity. Specifically, SEB-3 gain-of-function mutants exhibited increased sensitivity to the avoidance of, and decreased latency to respond to, 1-octanol. SEB-3 gain-of-function mutants also exhibited increased baseline tremor as well as increased heat- or alcohol-withdrawal-induced tremors versus wild type, whereas SEB-3 loss-of-function mutants exhibited performance changes in the opposite direction (Jee et al., 2013).

Nociceptin

Nociceptin is a neuropeptide that has received attention for its antianxiety actions in vivo (Aujla et al., 2012) as well as its efficacy in reducing alcohol administration and alcohol-cue-induced reinstatement (Ciccocioppo, Economidou, Fedeli, & Massi, 2003).

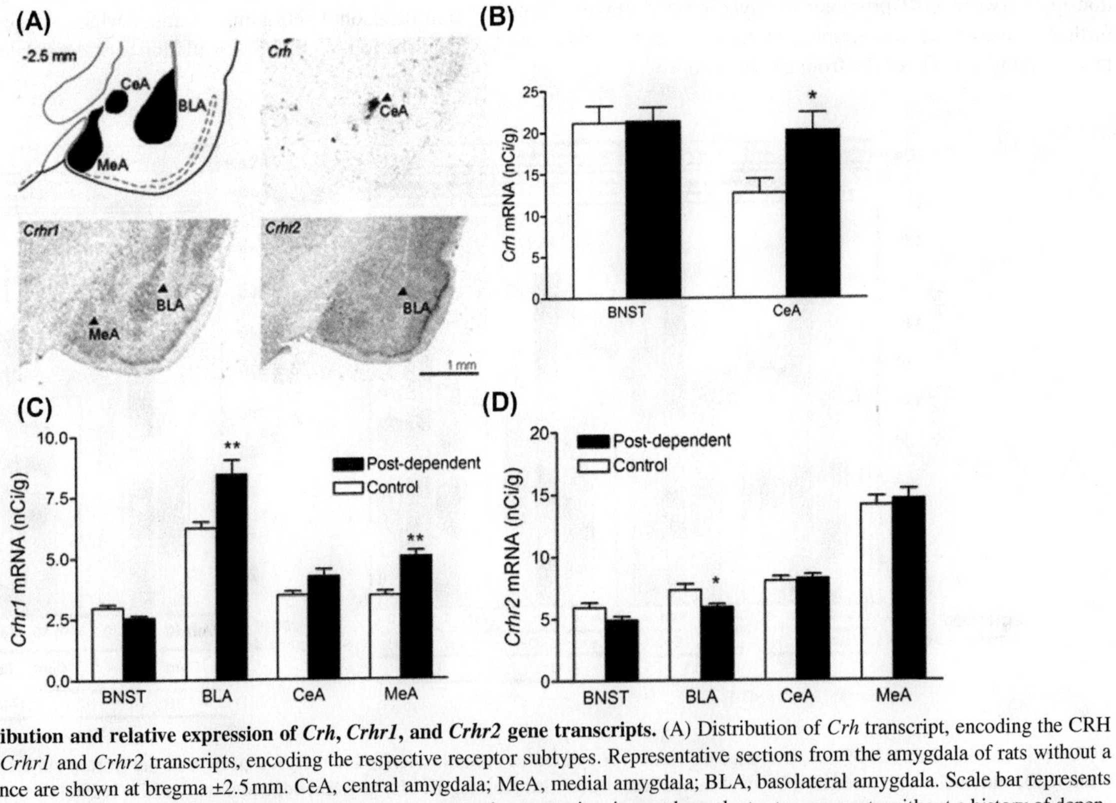

FIGURE 4 **Distribution and relative expression of *Crh*, *Crhr1*, and *Crhr2* gene transcripts.** (A) Distribution of *Crh* transcript, encoding the CRH precursor, and the *Crhr1* and *Crhr2* transcripts, encoding the respective receptor subtypes. Representative sections from the amygdala of rats without a history of dependence are shown at bregma ±2.5 mm. CeA, central amygdala; MeA, medial amygdala; BLA, basolateral amygdala. Scale bar represents 1 mm. Quantification of expression levels (nCi/g, mean ± SEM) for the respective transcripts in postdependent rats versus rats without a history of dependence is shown in (B) through (D). (B) *Crh* expression was upregulated within the CeA, which was the only amygdala region where measurable levels of this transcript were present. (C) *Crhr1* message was robustly upregulated within the BLA and MeA but not in the CeA or BNST. (D) Expression of the *Crhr2* transcript was unaffected within the extended amygdala, with the exception of the BLA, where a moderate decrease was seen. For all panels, $*p<0.05$, $**p<0.01$, corrected for multiple tests. BNST, bed nucleus of stria terminalis; CRH, corticotropin-releasing hormone. *Reproduced from Sommer et al. (2008) with permission from Elsevier.*

Nociceptin transmission is closely related to that of CRH, and nociceptin receptor (NOP) agonists reverse the anxiogenic actions of CRH (Ciccocioppo, Cippitelli, Economidou, Fedeli, & Massi, 2004). Examination of SNPs of *OPRLM1*, which codes for NOP, has revealed significant alterations in both type I and type II Scandinavian alcohol-dependent individuals versus nondependent controls at SNP rs6010718 (Huang, Young, Pletcher, Heilig, & Wahlestedt, 2008), suggesting a role for altered NOP function in alcohol dependence.

In a preclinical study, functional changes in NOP and prepronociceptin—the precursor for nociceptin—were investigated by Aujla et al. (2012) in rats with or without a history of chronic alcohol administered over 6 days via intragastric intubation. They found elevated gene expression of NOP in the bed nucleus of the stria terminalis in alcohol-dependent rats versus nondependent controls at both 1 and 3 weeks of withdrawal, while elevated NOP gene expression was observed in the lateral hypothalamus in alcohol-dependent versus nondependent controls only at 3 weeks of withdrawal. In addition, elevated gene expression of prepronociceptin in alcohol-dependent versus nondependent controls was observed in the CeA at 3 weeks but not 1 week of withdrawal (Aujla et al., 2012). D'Addario et al. (2013) also examined levels of gene expression for prepronociceptin as well as NOP in animals that had received alcohol via intragastric intubation for 1 or 5 days versus controls. They found elevated expression of prepronociceptin in the amygdala at 1 or 5 days of abstinence following 1 day of alcohol administration and at 1 day after 5 days of alcohol administration. However, no difference between alcohol and alcohol-naïve groups were observed with respect to gene expression of NOP. Following 13 days of chronic alcohol administration in rats, Lindholm, Ploj, Franck, and Nylander (2002) observed decreased expression of nociceptin 30 min, but not 5 or 21 days, following cessation of alcohol in the cingulate cortex, while decreased expression of nociceptin was observed in the hippocampus at 5 but not 21 days of abstinence.

The functional significance of nociceptin transmission in the CeA is highlighted by the findings of Cruz, Herman, Kallupi, and Roberto (2012), who observed alterations in the functional relationship between nociceptin and CRH resulting from chronic alcohol exposure. Specifically, they noted decreased basal GABA release and enhancement of nociceptin-mediated blockade of CRH-induced GABA release in alcohol-exposed versus alcohol-naïve animals—a mechanism that the authors suggest may account for nociceptin's anti-alcohol and anti-CRH actions in vivo (Cruz et al., 2012). Moreover, Economidou et al. (2011) established that previously discussed alterations in nociceptin/NOP transmission corresponded to alcohol-withdrawal-induced anxiety-like behavior. One week after 6 days of intragastric alcohol intubation, nociceptin was found to reverse alcohol-withdrawal-induced decreases in open-arm exploration in alcohol-dependent animals at doses that did not produce anxiolytic actions in nondependent controls (Economidou et al., 2011). This finding was later replicated by Aujla et al. (2012) and extended with the observation that, at 3 weeks, alcohol-dependent animals displayed similar levels of open-arm exploration versus nondependent controls when receiving vehicle, but exhibited anxiogenic-like performance when receiving nociceptin at doses that produced anxiolytic-like actions after 1 week of withdrawal in both groups—paralleling elevated gene expression of prepronociceptin in the CeA of alcohol-dependent animals at 1 versus 3 weeks of withdrawal (see Figure 5).

These alterations may reflect increased sensitivity to nociceptin's actions on anxiety, rather than a "switch" from anxiolytic to anxiogenic, as Aujla and Nedjadrasul (2015) demonstrated anxiolytic actions of low doses of nociceptin 3 weeks following cessation of alcohol administration in alcohol-dependent rats but not in alcohol-naïve controls.

Neuropeptide S

Neuropeptide S (NPS) has received considerable attention as a putative substrate for alcohol-related behaviors and, consequently, as a potential therapeutic target in the treatment of alcohol

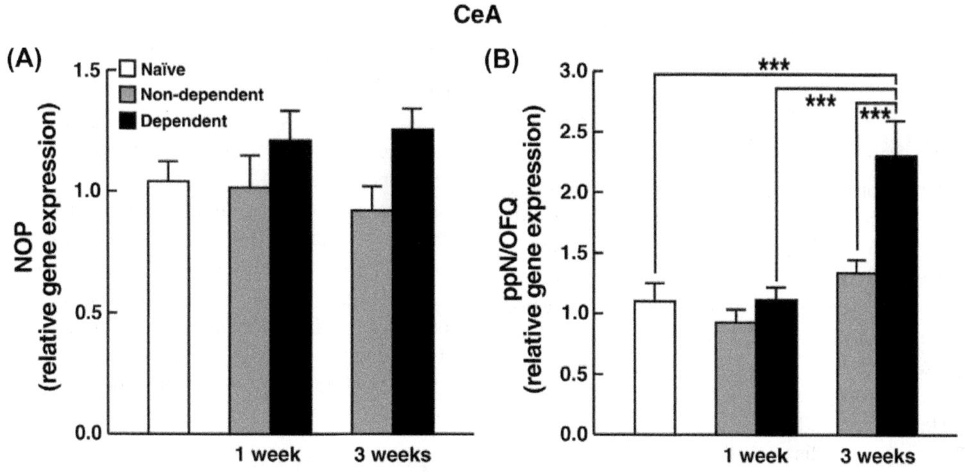

FIGURE 5 **Relative gene expression levels *NOP* and *ppN/OFQ* in the CeA of rats following termination of chronic ethanol or vehicle administration.** Relative gene expression levels of (A) *NOP* or (B) *ppN/OFQ* in the CeA of rats measured 1 or 3 weeks following termination of chronic ethanol (ethanol-dependent) or vehicle (nondependent) administration are shown. Data are expressed as mean (±SEM) values relative to mRNA levels in experimentally naïve age-matched controls. ***$p<0.001$. CeA, central nucleus of the amygdala; NOP, nociceptin/orphanin FQ-opioid peptide receptor; ppN/OFQ, prepronociceptin/orphanin FQ. *Reproduced from Aujla et al. (2012) with permission from Taylor & Francis Ltd.*

dependence (Reinsheid & Xu, 2005). Moreover, in humans, a polymorphism of the NPS receptor (NPS-R) has been shown to be associated with alcohol use disorders (Laas et al., 2014).

In rats, gene expression of the NPS-R has been found to be elevated during acute withdrawal and protracted abstinence in animals with a history of alcohol intake (Ruggeri et al., 2010). Ruggeri et al. (2010) reported that at 12h of abstinence, elevated expression of NPS-R was observed in alcohol-dependent versus nondependent controls in the primary somatosensory cortex, BLA, lateral hypothalamus, and paraventricular nucleus of the hypothalamus. However, at 7 days of abstinence, elevated gene expression in alcohol-dependent animals emerged in the endopiriform nucleus, motor cortex 2, and anterior cortical amygdaloid nucleus, while expression in the primary somatosensory cortex and endopiriform nucleus returned to levels similar to those observed in nondependent controls (Ruggeri et al., 2010). Furthermore, alcohol-dependent animals exhibited greater sensitivity to the anxiolytic actions of intracerebroventricular NPS in the shock-probe defensive burying test, suggesting that altered NPS-R gene expression produced functional differences in withdrawal-related anxiety (Ruggeri et al., 2010). Enquist, Ferwerda, Madhavan, Hok, and Whistler (2012) also demonstrated a similar differential sensitivity to the anxiolytic actions of NPS as a function of alcohol exposure following intra-BLA administration of NPS.

Myelin

Decreased brain white matter has been observed in postmortem tissue from alcohol-dependent individuals (refer to Chapter 39, "Alcohol and Its Impact on Myelin," for a review of these findings). Lewohl et al. (2000) utilized a microarray to examine gene expression changes in the frontal cortices of alcohol-dependent individuals and identified a subset of myelin-related genes that exhibited downregulated expression in alcohol-dependent individuals—myelin-associated glycoprotein, myelin and T cell differentiation protein, and apolipoprotein D. Furthermore, Mon et al. (2013) examined the promyelination neurotrophin brain-derived neurotrophic factor (BDNF) and identified alterations in the expression of the BDNF Val66Met (rs6265) polymorphism that predicted levels of gray and white matter recovery during abstinence from alcohol.

Immune/Proinflammatory Factors

Growing interest in the contribution of alcohol-stimulated immune response and consequent neuroinflammation promises to provide novel insights in the understanding of the neuropathology underlying features of alcohol withdrawal. Postmortem analysis of tissue from human alcoholic-dependent individuals has revealed increased expression of the proinflammatory cytokine monocyte chemoattractant protein 1 in the VTA, substantia nigra, hippocampus, and amygdala in alcohol-dependent individuals versus nondependent controls (He & Crews, 2008).

Alcohol is an activator of glial Toll-like receptor 4 (TLR-4), which produces downstream activation of immune factors that facilitate cell death (Crews, Zou, & Qin, 2011). Alfonso-Loeches, Pascual-Lucas, Blanco, Sanchez-Vera, and Guerri (2010) demonstrated that TLR-4-knockdown mice failed to exhibit chronic alcohol-induced increases in cortical levels of immune factors including CD11b (microglial marker), glial fibrillary acidic protein, and cytokines such as interleukin (IL)-1β, tumor necrosis factor-α, and IL-6, as well as the immune/inflammation-related enzymes cyclooxygenase 2 and inducible nitric oxide synthase (see Figure 6).

Crews et al. (2011) provide a thorough review of addiction from the perspective of dysregulated gene expression of immune factors.

CONCLUSION

Observations of altered gene expression during alcohol withdrawal provide important insights into the functioning of the "alcoholic brain" as well as potential leads on treatment targets. Symptoms that are associated with alcohol withdrawal, including hypofrontality-related cognitive deficits, presence of seizures, dysphoria, and heightened stress reactivity, generally correspond well to findings from the gene expression studies discussed in this chapter. Thus, treatment of symptoms associated with withdrawal from alcohol may be achieved through modification of gene expression and epigenetic factors.

APPLICATIONS TO OTHER ADDICTIONS AND SUBSTANCE MISUSE

Most drugs of abuse, despite different proximal sites of action, ultimately increase neurotransmission in the mesolimbic dopamine system (Koob et al., 2004). Consequently, alterations in gene expression, similar to those reviewed with respect to alcohol in this chapter, may occur following withdrawal from a variety of drugs of abuse. In a 2015 review, Bühler et al. (2015) describe alterations in gene expression in nicotine-, alcohol-, cannabis-, cocaine-, and heroin-dependent individuals. Common genes that appear to be similarly affected across drugs include those in the cholinergic cluster as well as *DRD2* and *ANKK1* (Bühler et al., 2015). Thus, overlap in altered gene expression taken together with the established efficacy of current treatments, such as naltrexone against both alcohol and heroin use, suggests that newly emerging gene therapies may also provide some efficacy across abused substances.

A common form of polydrug use—cocaine and alcohol—results in the production of cocaethylene, a potently reinforcing substance that may exacerbate many of the gene expression changes described in this chapter. The relative paucity of information on the genetics of cocaine dependence presents a significant challenge to better understanding of how polydrug use specifically alters gene expression.

Research is now beginning to draw parallels in gene expression changes that result from drug use and problem gambling. In a 2014 report, Lobo et al. (2014) identified changes in *DRD3* and *CAMK2D* that result from disordered gambling in a preclinical model of gambling—the rat gambling task, suggesting a commonality in disruption of gene expression in dopaminergic systems with drugs of abuse and behavioral addiction.

DEFINITION OF TERMS

Gene This is a functional section of a DNA molecule that codes for a protein product.

Neuroreceptor This is a protein embedded in the membrane of a cell (typically a neuron or glial cell) that is activated when bound by a neurotransmitter or drug.

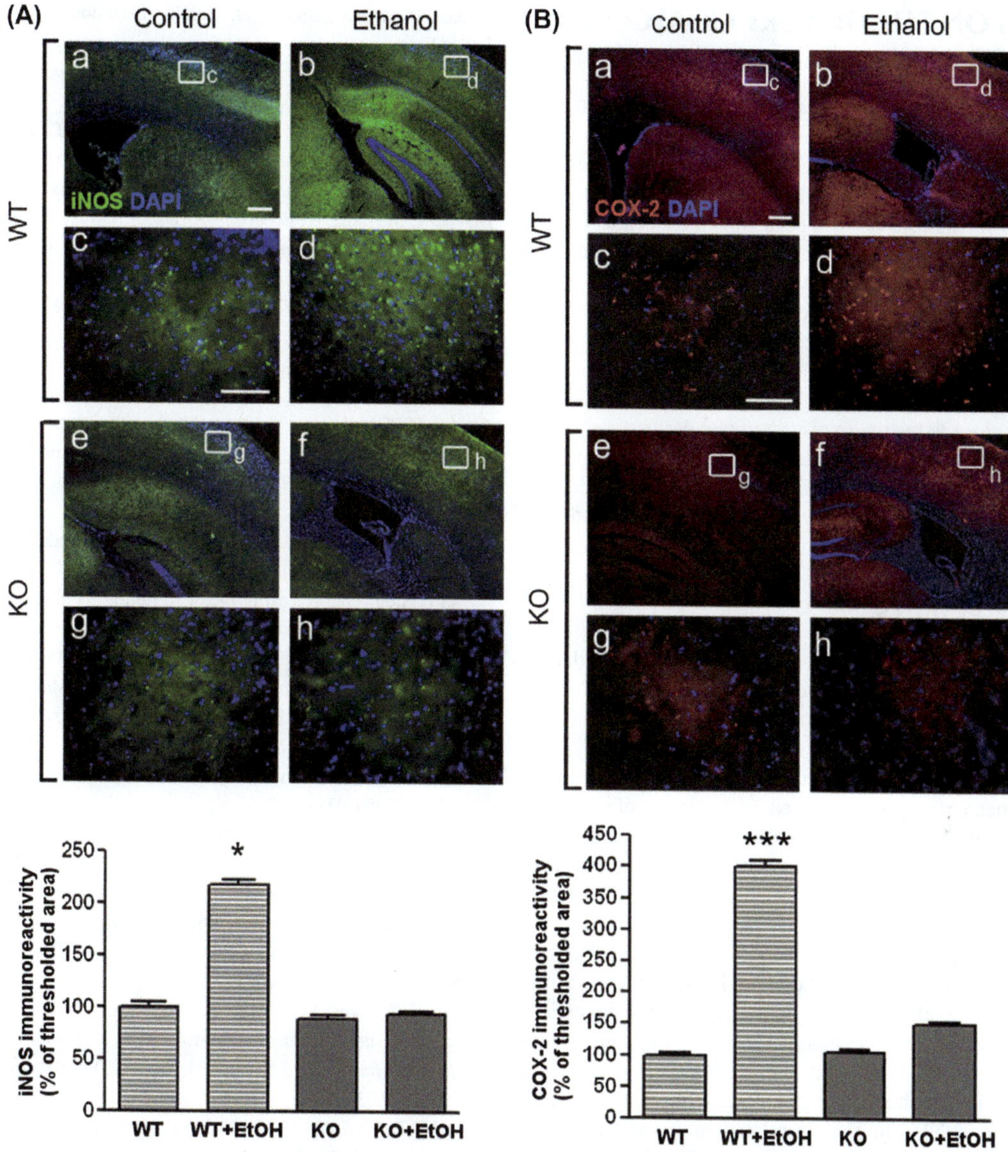

FIGURE 6 Role of the TLR-4 receptors in the ethanol-induced upregulation of the inducible nitric oxide synthase (iNOS) and cyclooxygenase 2 (COX-2) expression in the cerebral cortices. (A and B) iNOS and COX-2 immunostaining was performed in brain sections from the medial frontal cortex (scale bars: a, b, e, f, 200 μm; c, d, g, h, 50 μm). Bars represent the values of the quantification of iNOS and COX-2 immunoreactivity expressed as the percentage of the thresholded area occupied by the specific staining in relation to the whole area versus wild-type (WT) control. For quantification, five to eight high-power fields were analyzed per coverslip. Values represent the mean ± SD of at least three animals per group; $*p<0.05$ and $***p<0.005$ versus the WT control animals. KO, knockout; EtOH, ethanol. *Reproduced from Alfonso-Loeches et al. (2010) with permission from the Society for Neuroscience.*

Single-nucleotide polymorphism This is a difference of one nucleotide base pair in an otherwise shared DNA sequence between individuals of a species. Each variant is referred to as an allele.

Epigenetics These are alterations in the structure of DNA that do not alter the nucleotide sequence. Epigenetic modifications may result from experience and alter the expression of gene products.

Gene microarray This is a technology that permits simultaneous examination of the activity across thousands of genes. A modern microarray typically contains approximately 21,000 spots, each of which contains multiple copies of specific DNA strands, which are mapped out spatially in a computer database.

Delirium tremens This is a severe form of alcohol withdrawal that is associated with a number of psychological and overt physical symptoms including tremors, anxiety, agitation, confusion, and disorientation. In some cases, expression of DTs may be accompanied by seizures.

Messenger ribonucleic acid mRNA is single stranded and created from a DNA template. It carries genetic information from the nucleus of a cell to the ribosome where it undergoes translation into a protein.

Small interfering ribonucleic acid This is a double-stranded RNA that prevents the translation of genes that contain the complementary nucleotide sequence. siRNAs are commonly used to selectively examine how specific gene products contribute to a phenotype.

Anxiolysis This is a reduction in anxiety.

Elevated-plus maze This is a rodent model of anxiety that contains two enclosed arms and two arms without walls (i.e., open). An increased proportion of time spent exploring the open versus the enclosed arms is inferred to reflect a reduction in anxiety.

KEY FACTS ON THE HISTORY OF ALCOHOL CONSUMPTION

- Recent archaeological evidence suggests that the earliest consumption of alcohol may have occurred during the Neolithic period 9000 years ago in the form of fermented honey or berries.
- Although the ability to create beverages with high alcohol concentrations (i.e., spirits) had been known for centuries, the consumption of distilled spirits did not begin to increase in popularity in Britain until the seventeenth century.
- Preferences among alcoholic beverages vary between countries, with North Americans and the British generally preferring beer, while wine is preferred in Portugal, France, and Spain.
- Wine consumption has steadily increased in the United States and is now close to beer with respect to popularity.
- Men generally tend to consume larger amounts of alcohol than women and also exhibit higher rates of alcohol dependence. There is, however, significant variation in the magnitude of this gender difference across countries.

SUMMARY POINTS

- Clinical and preclinical studies on alcohol misuse have identified a number of genetic alterations that are related to alcohol consumption and that are functionally related to alcohol withdrawal symptoms.
- Alterations in genes that code for glutamate and GABA receptors have been strongly implicated in a variety of alcohol-withdrawal-related phenomena including DTs, anxiety, and homeostatic function, while glutamate receptors have also been implicated in dysregulated reward processing.
- Dopamine-related genetic changes resulting from alcohol withdrawal have been shown in reward-related brain regions, and gene expression of dopamine-related ryanodine receptors is altered in frontal cortex and limbic forebrain, suggesting that dopamine may be important in mediating cognitive and anxiety-related alcohol withdrawal symptoms.
- Consistent with the pharmacology literature implicating CRH, nociceptin, and NPS in anxiety-related processes, alcohol-withdrawal-induced alterations in genes coding for the receptors of these neuropeptides, particularly in the amygdala, are associated with anxiety-like behaviors.
- An emerging literature points toward alterations in gene expression of immune and proinflammation factors that are associated with alcohol withdrawal.
- Modification of gene products holds therapeutic promise in the treatment of alcohol misuse.

REFERENCES

Alfonso-Loeches, S., Pascual-Lucas, M., Blanco, A. M., Sanchez-Vera, I., & Guerri, C. (2010). Pivotal role of TLR4 receptors in alcohol-induced neuroinflammation and brain damage. *Journal of Neuroscience*, *30*(24), 8285–8295. http://dx.doi.org/10.1523/JNEUROSCI.0976-10.2010.

Aujla, H., Cannarsa, R., Romualdi, P., Ciccocioppo, R., Martin-Fardon, R., & Weiss, F. (2012). Modification of anxiety-like behaviors by nociceptin/orphanin FQ (N/OFQ) and time-dependent changes in N/OFQ-NOP gene expression following ethanol withdrawal. *Addiction Biology*, *18*(3), 467–479. http://dx.doi.org/10.1111/j.1369-1600.2012.00466.x.

Aujla, H., & Nedjadrasul, D. (2015). Low-dose Nociceptin/Orphanin FQ reduces anxiety-like performance in alcohol-withdrawn, but not alcohol-naïve, Male Wistar rats. *Neuropharmacology*. http://dx.doi.org/10.1016/j.neuropharm.2015.01.006.

Becker, H. C., Lopez, M. F., & Doremus-Fitzwater, T. L. (2011). Effects of stress on alcohol drinking: a review of animal studies. *Psychopharmacology*, *218*(1), 131–156. http://dx.doi.org/10.1007/s00213-011-2443-9.

Bühler, K.-M., Giné, E., Echeverry-Alzate, V., Calleja-Conde, J., de Fonseca, F. R., & López-Moreno, J. A. (2015). Common single nucleotide variants underlying drug addiction: more than a decade of research: SNPs and drug addiction. *Addiction Biology*, *20*(5), 845–871. http://dx.doi.org/10.1111/adb.12204.

Chamorro, A.-J., Marcos, M., Mirón-Canelo, J.-A., Pastor, I., González-Sarmiento, R., & Laso, F.-J. (2012). Association of µ-opioid receptor (OPRM1) gene polymorphism with response to naltrexone in alcohol dependence: a systematic review and meta-analysis. *Addiction Biology*, *17*(3), 505–512. http://dx.doi.org/10.1111/j.1369-1600.2012.00442.x.

Ciccocioppo, R., Cippitelli, A., Economidou, D., Fedeli, A., & Massi, M. (2004). Nociceptin/orphanin FQ acts as a functional antagonist of corticotropin-releasing factor to inhibit its anorectic effect. *Physiology & Behavior*, *82*(1), 63–68. http://dx.doi.org/10.1016/j.physbeh.2004.04.035.

Ciccocioppo, R., Economidou, D., Fedeli, A., & Massi, M. (2003). The nociceptin/orphanin FQ/NOP receptor system as a target for treatment of alcohol abuse: a review of recent work in alcohol-preferring rats. *Physiology & Behavior*, *79*(1), 121–128. http://dx.doi.org/10.1016/S0031-9384(03)00112-4.

Covault, J., Gelernter, J., Hesselbrock, V., Nellissery, M., & Kranzler, H. R. (2004). Allelic and haplotypic association of GABRA2 with alcohol dependence. *American Journal of Medical Genetics. Part B, Neuropsychiatric Genetics: The Official Publication of the International Society of Psychiatric Genetics*, *129B*(1), 104–109. http://dx.doi.org/10.1002/ajmg.b.30091.

Crews, F. T., Zou, J., & Qin, L. (2011). Induction of innate immune genes in brain create the neurobiology of addiction. *Brain, Behavior, and Immunity*, *25*(Suppl 1), S4–S12. http://dx.doi.org/10.1016/j.bbi.2011.03.003.

Cruz, M. T., Herman, M. A., Kallupi, M., & Roberto, M. (2012). Nociceptin/Orphanin FQ blockade of corticotropin-releasing factor-induced gamma-aminobutyric acid release in central amygdala is enhanced after chronic ethanol exposure. *Biological Psychiatry*, *71*(8), 666–676. http://dx.doi.org/10.1016/j.biopsych.2011.10.032.

D'Addario, C., Caputi, F. F., Rimondini, R., Gandolfi, O., Del Borrello, E., Candeletti, S., & Romualdi, P. (2013). Different alcohol exposures induce selective alterations on the expression of dynorphin and nociceptin systems related genes in rat brain. *Addiction Biology*, *18*(3), 425–433. http://dx.doi.org/10.1111/j.1369-1600.2011.00326.x.

Dahlgren, A., Wargelius, H.-L., Berglund, K. J., Fahlke, C., Blennow, K., Zetterberg, H., ... Balldin, J. (2011). Do alcohol-dependent individuals with DRD2 A1 allele have an increased risk of relapse? A pilot study. *Alcohol and Alcoholism (Oxford, Oxfordshire)*, *46*(5), 509–513. http://dx.doi.org/10.1093/alcalc/agr045.

Ducat, E., Ray, B., Bart, G., Umemura, Y., Varon, J., Ho, A., & Kreek, M. J. (2013). Mu-opioid receptor A118G polymorphism in healthy volunteers affects hypothalamic-pituitary-adrenal axis adrenocorticotropic hormone stress response to metyrapone. *Addiction Biology*, *18*(2), 325–331. http://dx.doi.org/10.1111/j.1369-1600.2011.00313.x.

Economidou, D., Cippitelli, A., Stopponi, S., Braconi, S., Clementi, S., Ubaldi, M., ... Ciccocioppo, R. (2011). Activation of brain NOP receptors attenuates acute and protracted alcohol withdrawal symptoms in the rat. *Alcoholism, Clinical and Experimental Research*, 35(4), 747–755. http://dx.doi.org/10.1111/j.1530-0277.2010.01392.x.

Edenberg, H. J., Dick, D. M., Xuei, X., Tian, H., Almasy, L., Bauer, L. O., ... Begleiter, H. (2004). Variations in GABRA2, encoding the α2 subunit of the GABAA receptor, are associated with alcohol dependence and with brain oscillations. *The American Journal of Human Genetics*, 74(4), 705–714. http://dx.doi.org/10.1086/383283.

Engin, E., Bakhurin, K. I., Smith, K. S., Hines, R. M., Reynolds, L. M., Tang, W., ... Rudolph, U. (2014). Neural basis of benzodiazepine reward: requirement for α2 containing GABAA receptors in the nucleus accumbens. *Neuropsychopharmacology*, 39(8), 1805–1815. http://dx.doi.org/10.1038/npp.2014.41.

Engleman, E. A., Rodd, Z. A., Bell, R. L., & Murphy, J. M. (2008). The role of 5-HT3 receptors in drug abuse and as a target for pharmacotherapy. *CNS & Neurological Disorders Drug Targets*, 7(5), 454–467.

Enquist, J., Ferwerda, M., Madhavan, A., Hok, D., & Whistler, J. L. (2012). Chronic ethanol potentiates the effect of neuropeptide S in the basolateral amygdala and shows increased anxiolytic and anti-depressive effects. *Neuropsychopharmacology: Official Publication of the American College of Neuropsychopharmacology*, 37(11), 2436–2445. http://dx.doi.org/10.1038/npp.2012.102.

Eyer, F., Schuster, T., Felgenhauer, N., Pfab, R., Strubel, T., Saugel, B., & Zilker, T. (2011). Risk assessment of moderate to severe alcohol withdrawal–predictors for seizures and delirium tremens in the course of withdrawal. *Alcohol and Alcoholism (Oxford, Oxfordshire)*, 46(4), 427–433. http://dx.doi.org/10.1093/alcalc/agr053.

Fitzgerald, G. J., Liu, H., & Morzorati, S. L. (2012). Decreased sensitivity of NMDA receptors on dopaminergic neurons from the posterior ventral tegmental area following chronic nondependent alcohol consumption. *Alcoholism, Clinical and Experimental Research*, 36(10), 1710–1719. http://dx.doi.org/10.1111/j.1530-0277.2012.01762.x.

Freeman, K., Staehle, M. M., Gümüş, Z. H., Vadigepalli, R., Gonye, G. E., Nichols, C. N., ... Schwaber, J. S. (2012). Rapid temporal changes in the expression of a set of neuromodulatory genes during alcohol withdrawal in the dorsal vagal complex: molecular evidence of homeostatic disturbance. *Alcoholism, Clinical and Experimental Research*, 36(10), 1688–1700. http://dx.doi.org/10.1111/j.1530-0277.2012.01791.x.

Freeman, K., Staehle, M. M., Vadigepalli, R., Gonye, G. E., Ogunnaike, B. A., Hoek, J. B., & Schwaber, J. S. (2013). Coordinated dynamic gene expression changes in the central nucleus of the amygdala during alcohol withdrawal. *Alcoholism, Clinical and Experimental Research*, 37(Suppl. 1), E88–E100. http://dx.doi.org/10.1111/j.1530-0277.2012.01910.x.

Gass, J. T., & Foster Olive, M. (2008). Glutamatergic substrates of drug addiction and alcoholism. *Biochemical Pharmacology*, 75(1), 218–265. http://dx.doi.org/10.1016/j.bcp.2007.06.039.

Grzywacz, A., Jasiewicz, A., Małecka, I., Suchanecka, A., Grochans, E., Karakiewicz, B., ... Samochowiec, J. (2012). Influence of DRD2 and ANKK1 polymorphisms on the manifestation of withdrawal syndrome symptoms in alcohol addiction. *Pharmacological Reports: PR*, 64(5), 1126–1134.

He, J., & Crews, F. T. (2008). Increased MCP-1 and microglia in various regions of the human alcoholic brain. *Experimental Neurology*, 210(2), 349–358. http://dx.doi.org/10.1016/j.expneurol.2007.11.017.

Heinz, A., Schmidt, K., Baum, S. S., Kuhn, S., Dufeu, P., Schmidt, L. G., & Rommelspacher, H. (1996). Influence of dopaminergic transmission on severity of withdrawal syndrome in alcoholism. *Journal of Studies on Alcohol*, 57(5), 471–474.

Hendershot, C. S., Claus, E. D., & Ramchandani, V. A. (2014). Associations of *OPRM1* A118G and alcohol sensitivity with intravenous alcohol self-administration in young adults: *OPRM1* and self-administration. *Addiction Biology*. http://dx.doi.org/10.1111/adb.12165.

Hillemacher, T., Frieling, H., Hartl, T., Wilhelm, J., Kornhuber, J., & Bleich, S. (2009). Promoter specific methylation of the dopamine transporter gene is altered in alcohol dependence and associated with craving. *Journal of Psychiatric Research*, 43(4), 388–392. http://dx.doi.org/10.1016/j.jpsychires.2008.04.006.

Holmes, A., Spanagel, R., & Krystal, J. H. (2013). Glutamatergic targets for new alcohol medications. *Psychopharmacology*, 229(3), 539–554. http://dx.doi.org/10.1007/s00213-013-3226-2.

Huang, J., Young, B., Pletcher, M. T., Heilig, M., & Wahlestedt, C. (2008). Association between the nociceptin receptor gene (OPRL1) single nucleotide polymorphisms and alcohol dependence. *Addiction Biology*, 13(1), 88–94. http://dx.doi.org/10.1111/j.1369-1600.2007.00089.x.

Jee, C., Lee, J., Lim, J. P., Parry, D., Messing, R. O., & McIntire, S. L. (2013). SEB-3, a CRF receptor-like GPCR, regulates locomotor activity states, stress responses and ethanol tolerance in *Caenorhabditis elegans*. *Genes, Brain, and Behavior*, 12(2), 250–262. http://dx.doi.org/10.1111/j.1601-183X.2012.00829.x.

Jin, Z., Bhandage, A. K., Bazov, I., Kononenko, O., Bakalkin, G., Korpi, E. R., & Birnir, B. (2014). Expression of specific ionotropic glutamate and GABA-A receptor subunits is decreased in central amygdala of alcoholics. *Frontiers in Cellular Neuroscience*, 8. http://dx.doi.org/10.3389/fncel.2014.00288.

Koob, G. F., Ahmed, S. H., Boutrel, B., Chen, S. A., Kenny, P. J., Markou, A., ... Sanna, P. P. (2004). Neurobiological mechanisms in the transition from drug use to drug dependence. *Neuroscience & Biobehavioral Reviews*, 27(8), 739–749. http://dx.doi.org/10.1016/j.neubiorev.2003.11.007.

Kupila, J., Kärkkäinen, O., Laukkanen, V., Tupala, E., Tiihonen, J., & Storvik, M. (2013). mGluR1/5 receptor densities in the brains of alcoholic subjects: a whole-hemisphere autoradiography study. *Psychiatry Research*, 212(3), 245–250. http://dx.doi.org/10.1016/j.pscychresns.2012.04.003.

Kurokawa, K., Mizuno, K., & Ohkuma, S. (2013). Dopamine D1 receptor signaling system regulates ryanodine receptor expression in ethanol physical dependence. *Alcoholism, Clinical and Experimental Research*, 37(5), 771–783. http://dx.doi.org/10.1111/acer.12036.

Laas, K., Reif, A., Akkermann, K., Kiive, E., Domschke, K., Lesch, K.-P., ... Harro, J. (2014). Neuropeptide S receptor gene variant and environment: contribution to alcohol use disorders and alcohol consumption. *Addiction Biology*. http://dx.doi.org/10.1111/adb.12149.

Lappalainen, J., Krupitsky, E., Remizov, M., Pchelina, S., Taraskina, A., Zvartau, E., ... Gelernter, J. (2005). Association between alcoholism and gamma-amino butyric acid alpha2 receptor subtype in a Russian population. *Alcoholism, Clinical and Experimental Research*, 29(4), 493–498.

Lê, A. D., Harding, S., Juzytsch, W., Watchus, J., Shalev, U., & Shaham, Y. (2000). The role of corticotrophin-releasing factor in stress-induced relapse to alcohol-seeking behavior in rats. *Psychopharmacology*, 150(3), 317–324.

Lewohl, J. M., Wang, L., Miles, M. F., Zhang, L., Dodd, P. R., & Harris, R. A. (2000). Gene expression in human alcoholism: microarray analysis of frontal cortex. *Alcoholism, Clinical and Experimental Research*, 24(12), 1873–1882.

Lindholm, S., Ploj, K., Franck, J., & Nylander, I. (2002). Nociceptin/orphanin FQ tissue concentration in the rat brain. Effects of repeated ethanol administration at various post-treatment intervals. *Progress in Neuro-Psychopharmacology and Biological Psychiatry*, 26(2), 303–306.

Lobo, D. S. S., Aleksandrova, L., Knight, J., Casey, D. M., el-Guebaly, N., Nobrega, J. N., & Kennedy, J. L. (2014). Addiction-related genes in gambling disorders: new insights from parallel human and pre-clinical models. *Molecular Psychiatry*. http://dx.doi.org/10.1038/mp.2014.113.

Low, K. (2000). Molecular and neuronal substrate for the selective attenuation of anxiety. *Science*, 290(5489), 131–134. http://dx.doi.org/10.1126/science.290.5489.131.

Lowery-Gionta, E. G., Marcinkiewcz, C. A., & Kash, T. L. (2014). Functional alterations in the dorsal raphe nucleus following acute and chronic ethanol exposure. *Neuropsychopharmacology*. http://dx.doi.org/10.1038/npp.2014.205.

Mann, K., Lehert, P., & Morgan, M. Y. (2004). The efficacy of acamprosate in the maintenance of abstinence in alcohol-dependent individuals: results of a meta-analysis. *Alcoholism: Clinical & Experimental Research*, 28(1), 51–63. http://dx.doi.org/10.1097/01.ALC.0000108656.81563.05.

Menzaghi, F., Rassnick, S., Heinrichs, S., Baldwin, H., Pich, E. M., Weiss, F., & Koob, G. F. (1994). The role of corticotropin-releasing factor in the anxiogenic effects of ethanol withdrawal. *Annals of the New York Academy of Sciences*, 739, 176–184.

Mon, A., Durazzo, T. C., Gazdzinski, S., Hutchison, K. E., Pennington, D., & Meyerhoff, D. J. (2013). Brain-derived neurotrophic factor genotype is associated with brain gray and white matter tissue volumes recovery in abstinent alcohol-dependent individuals. *Genes, Brain, and Behavior*, 12(1), 98–107. http://dx.doi.org/10.1111/j.1601-183X.2012.00854.x.

Nieratschker, V., Grosshans, M., Frank, J., Strohmaier, J., von der Goltz, C., El-Maarri, O., … Rietschel, M. (2014). Epigenetic alteration of the dopamine transporter gene in alcohol-dependent patients is associated with age. *Addiction Biology*, 19(2), 305–311. http://dx.doi.org/10.1111/j.1369-1600.2012.00459.x.

Noble, E. P. (2003). D2 dopamine receptor gene in psychiatric and neurologic disorders and its phenotypes. *American Journal of Medical Genetics Part B: Neuropsychiatric Genetics*, 116B(1), 103–125. http://dx.doi.org/10.1002/ajmg.b.10005.

Preuss, U. W., Zill, P., Koller, G., Bondy, B., Hesselbrock, V., & Soyka, M. (2005). Ionotropic glutamate receptor gene GRIK3 SER310ALA functional polymorphism is related to delirium tremens in alcoholics. *The Pharmacogenomics Journal*, 6(1), 34–41.

Ray, L. A. (2011). Stress-induced and cue-induced craving for alcohol in heavy drinkers: preliminary evidence of genetic moderation by the OPRM1 and CRH-BP Genes: stress and cue-induced alcohol craving. *Alcoholism, Clinical and Experimental Research*, 35(1), 166–174. http://dx.doi.org/10.1111/j.1530-0277.2010.01333.x.

Reinscheid, R. K., & Xu, Y.-L. (2005). Neuropeptide S as a novel arousal promoting peptide transmitter: NPS produces arousal and anxiolysis. *FEBS Journal*, 272(22), 5689–5693. http://dx.doi.org/10.1111/j.1742-4658.2005.04982.x.

Rodd, Z. A., Kimpel, M. W., Edenberg, H. J., Bell, R. L., Strother, W. N., McClintick, J. N., … McBride, W. J. (2008). Differential gene expression in the nucleus accumbens with ethanol self-administration in inbred alcohol-preferring rats. *Pharmacology Biochemistry and Behavior*, 89(4), 481–498. http://dx.doi.org/10.1016/j.pbb.2008.01.023.

Ruggeri, B., Braconi, S., Cannella, N., Kallupi, M., Soverchia, L., Ciccocioppo, R., & Ubaldi, M. (2010). Neuropeptide S receptor gene expression in alcohol withdrawal and protracted abstinence in postdependent rats. *Alcoholism, Clinical and Experimental Research*, 34(1), 90–97. http://dx.doi.org/10.1111/j.1530-0277.2009.01070.x.

Rujescu, D., Soyka, M., Dahmen, N., Preuss, U., Hartmann, A. M., Giegling, I., … Szegedi, A. (2005). GRIN1 locus may modify the susceptibility to seizures during alcohol withdrawal. *American Journal of Medical Genetics Part B: Neuropsychiatric Genetics*, 133B(1), 85–87. http://dx.doi.org/10.1002/ajmg.b.30112.

Sanna, E., Mostallino, M. C., Busonero, F., Talani, G., Tranquilli, S., Mameli, M., … Biggio, G. (2003). Changes in GABAA receptor gene expression associated with selective alterations in receptor function and pharmacology after ethanol withdrawal. *The Journal of Neuroscience*, 23(37), 11711–11724.

Schmid, B., Blomeyer, D., Treutlein, J., Zimmermann, U. S., Buchmann, A. F., Schmidt, M. H., … Laucht, M. (2010). Interacting effects of CRHR1 gene and stressful life events on drinking initiation and progression among 19-year-olds. *The International Journal of Neuropsychopharmacology*, 13(06), 703–714. http://dx.doi.org/10.1017/S1461145709990290.

Silva, L. B. C., & Nobre, M. J. (2014). Impaired fear inhibitory properties of GABAA and µ opioid receptors of the dorsal periaqueductal grey in alcohol-withdrawn rats. *Acta Neurobiologiae Experimentlis*, 74, 54–66.

Sommer, W. H., Rimondini, R., Hansson, A. C., Hipskind, P. A., Gehlert, D. R., Barr, C. S., & Heilig, M. A. (2008). Upregulation of voluntary alcohol intake, behavioral sensitivity to stress, and amygdala Crhr1 expression following a history of dependence. *Biological Psychiatry*, 63(2), 139–145. http://dx.doi.org/10.1016/j.biopsych.2007.01.010.

Szot, P., White, S. S., Veith, R. C., & Rasmussen, D. D. (1999). Reduced gene expression for dopamine biosynthesis and transport in midbrain neurons of adult male rats exposed prenatally to ethanol. *Alcoholism, Clinical and Experimental Research*, 23(10), 1643–1649.

Szumlinski, K. K., Ary, A. W., Lominac, K. D., Klugmann, M., & Kippin, T. E. (2007). Accumbens Homer2 overexpression facilitates alcohol-induced neuroplasticity in C57BL/6J mice. *Neuropsychopharmacology*, 33(6), 1365–1378.

Ticku, M. K., & Burch, T. (1980). Alterations in gamma-aminobutyric acid receptor sensitivity following acute and chronic ethanol treatments. *Journal of Neurochemistry*, 34(2), 417–423.

Tiihonen, J., Kuikka, J., Bergström, K., Hakola, P., Karhu, J., Ryynänen, O. P., & Föhr, J. (1995). Altered striatal dopamine re-uptake site densities in habitually violent and non-violent alcoholics. *Nature Medicine*, 1(7), 654–657.

Yang, L. (2003). Neurons in the periaqueductal gray are critically involved in the neuronal network for audiogenic seizures during ethanol withdrawal. *Neuropharmacology*, 44(2), 275–281. http://dx.doi.org/10.1016/S0028-3908(02)00367-2.

Zhou, Y., & Kreek, M. J. (2014). Alcohol: a stimulant activating brain stress responsive systems with persistent neuroadaptation. *Neuropharmacology*. http://dx.doi.org/10.1016/j.neuropharm.2014.05.044.

Chapter 50

The Quest of Candidate Gene in Alcohol Dependence: The Dopamine D2 Receptor (*DRD2*) Gene

Sheng-Yu Lee[1,3], Yun-Hsuan Chang[2,3], Ru-Band Lu[3,4,5,6]
[1]*Department of Psychiatry, Kaohsiung Veteran's Hospital, Kaohsiung, Taiwan;* [2]*Department of Psychology, College of Medical and Health Science, Asia University, Taichung, Taiwan;* [3]*Department of Psychiatry, College of Medicine, National Cheng Kung University;* [4]*Department of Psychiatry, National Cheng Kung University Hospital, Tainan, Taiwan;* [5]*Institute of Behavioral Medicine, College of Medicine, National Cheng Kung University, Tainan, Taiwan;* [6]*Addiction Research Center, National Cheng Kung University, Tainan, Taiwan*

Abbreviations

ALC Alcoholism, alcohol dependence
ALC + BP Alcoholism comorbid with bipolar disorder
ALDH2 Aldehyde dehydrogenase 2
Antisocial ALC Antisocial alcoholism
Antisocial non-ALC Antisocial nonalcoholism
ANX/DEP ALC Anxiety/depression alcoholism
ASPD Antisocial personality disorder
COMT Catechol-*o*-methyltransferase
DOPAC 3,4-Dihydroxyphenylacetic acid
DOPAL 3,4-Dihydroxyphenyl-acetaldehyde
DRD2 Dopamine D2 receptor
HA Harm avoidance
MAOA Monoamine oxidase A
Mixed ALC Mixed alcoholism
NS Novelty seeking
Pure ALC Pure alcoholism

INTRODUCTION

Alcohol abuse and dependence are multifactorial mental disorders that are confounded by heterogeneity and sociocultural factors (Merikangas, 1990). Family, twin, and adoption studies suggest a strong hereditary component in alcohol dependence (Cloninger, 1987). Some have suggested that the heritability of alcoholism is at least 50% (Reich, Hinrichs, Culverhouse, & Bierut, 1999). Alcoholism is three to five times more frequent in the parents, siblings, and children of alcoholics than in the general population. In twin studies, Pickens et al. (1991) reported a significant difference between monozygotic and dizygotic twin concordance for alcohol dependency in 886 male same-sex twin pairs. An investigation (Kendler, Heath, Neale, Kessler, & Eaves, 1992) of 1030 female–female twin pairs in Virginia found a significant genetic component for alcoholism. Moreover, a Danish study (Goodwin, Schulsinger, Hermansen, Guze, & Winokur, 1973) of 133 male adoptees separated from their parents by 6 weeks of age found that 18% of those with a biological father who was an alcoholic developed alcoholism compared with only 5% of the adoptees who did not have a biological father with a history of alcoholism. Taking all these findings together suggests a genetic component in alcoholism.

Many candidate genes that contribute to alcohol dependence susceptibility have been proposed, yet few of the published results leading to the proposal of these candidate genes can be replicated (Dick & Foroud, 2003). Being a multifactorial mental disorder, the typology of alcoholism to reduce heterogeneity is therefore important and critical in both areas of research and clinical practice. However, exploration of genetic underpinning after the subtyping alcoholism is largely overlooked.

MOLECULAR GENETICS OF ALCOHOLISM

The dopamine (DA) system was thought to be involved in pathogenesis of alcohol dependence because alcohol may stimulate the nucleus accumbens via the mesolimbic dopaminergic system to arouse self-rewarding and continuous drinking (Gessa, Muntoni, Collu, Vargiu, & Mereu, 1985). The dopamine D2 receptor (*DRD2*) gene has been considered a candidate gene for alcohol dependence (Blum et al., 1990). Animal studies have shown that ethanol stimulates the dopaminergic neurons of the ventral tegmental area (Brodie, 2002), and alcohol-preferring rats have lower densities of *DRD2* in the limbic system than do alcohol-nonpreferring rats (McBride et al., 1993). Alcohol-abusing and alcohol-dependent humans likewise have a reduced number of striatal *DRD2* than do controls (Hietala et al., 1994). Brain-imaging study shows that healthy controls with an *A1* allele of the *DRD2 TaqIA* gene have fewer *DRD2* receptors than do those without the *A1* allele (Jonsson et al., 1999). Individuals with at least one *A1* allele appear to have up to 40% fewer striatal *DRD2* receptors than do those carrying the *A2* allele (Ritchie & Noble, 2003). The *DRD2 TaqIA A1* allele was associated with fewer *DRD2* in the striatum and, hence, a lower dopaminergic function. In personality traits, the *DRD2 TaqIA A1* allele was associated with both novelty seeking and alcoholism (Noble, 1998).

Blum et al. (1990) firstly reported a positive association between alcoholism and the *Taq*IA polymorphism of the *DRD2* gene. However, the association remains inconsistent. One comprehensive literature review concluded that neither studies using the transmission disequilibrium test nor those using the affected family-based association test found any association between the *DRD2* gene and alcoholism (Edenberg et al., 1998). On the contrary, subsequent independent studies using the same Collaborative Studies on Genetics of Alcoholism data supported an association (Curtis, Shirk, & Fall, 1999). A large-sample study with 1217 Swedish participants (Berggren et al., 2006) and a meta-analysis study of case control (Munafo, Matheson, & Flint, 2007) also showed an association. One possible explanation for such a discrepancy may be due to different definitions of control groups, different racially or ethnically mixed study populations, and the phenotypic heterogeneity of alcoholism, in part, which confound the effect of the *DRD2 TaqIA* polymorphism on alcohol dependence. An important confounding factor in a genetic association study is that different populations may have different allele frequencies at many loci and different prevalence rates of disease (Bowcock et al., 1987). The prevalence rates for alcoholism in the two aboriginal groups in Taiwan were 20.7% in Atayal and 17.1% in Ami—considerably higher than the rate reported for the Han Chinese (1.5%) group (Chen, Huang, Osio, Fitzpatrick, & Cohen, 1993). However, our study shows no association between severe alcoholism and alleles at both TaqI A and B polymorphisms, either individually or with *DRD2* haplotypes in the samples of Chinese Han, Atayal, and Ami males (Lu et al., 1996). These results were inconsistent with the conclusions reached by Blum et al. (1990) and Blum, Roman, and Martin (1993), even after controlling for ethnic factors.

Dopamine is metabolized through oxidative deamination by monoamine oxidase A (MAOA), an isozyme of monoamine oxidase, or via O-methylation catalyzed by catechol-*O*-methyltransferase. In a rat model, 90% of the metabolism was via the deamination route in the corpus striatum to form 3,4-dihydroxyphenyl-acetaldehyde (DOPAL) (Westerink & de Vries, 1985). DOPAL is subsequently oxidized to 3,4-dihydroxyphenylacetic acid (DOPAC) by aldehyde dehydrogenase (ALDH) (Figure 1). It is possible that inhibiting the activity of the *ALDH2* enzyme suppresses alcohol intake by inhibiting DA metabolism (Keung & Vallee, 1998). The foregoing observations led us to hypothesize that the *MAOA* and *ALDH2* genes, along with the *DRD2* gene (all three are related to DA metabolism), and possibly their interaction, may influence drinking behavior. The association of the *MAOA*, *ALDH2*, and *DRD2* genes with alcohol dependence thus warrants further studies.

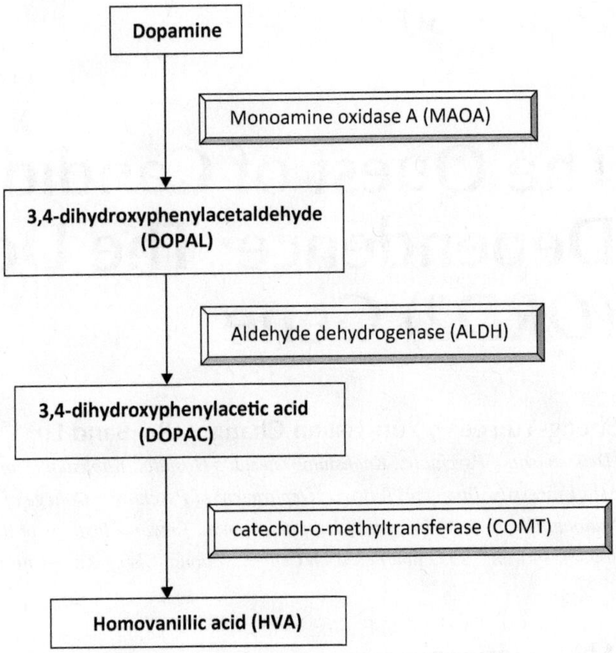

SUBTYPES OF ALCOHOLISM IDENTIFIED IN HAN CHINESE

The clinical heterogeneity of alcoholism may be related to genetic heterogeneity (Cloninger, 1987). The typology of alcoholism is, therefore, important for reducing inconsistencies both in research and in clinical practice. Individual subtypes of alcoholism may highlight genetic perturbations within particular domains of psychopathology that might not be revealed when studying alcoholism as a whole. Therefore, differentiating subtypes of alcoholism may reduce contradictory factors and uncover a more powerful association between genes and specific subtypes of alcoholism.

Alcohol dependence is a complex behavioral disorder that is often comorbid with other psychiatric disorders, such as anxiety, depression, bipolar, and antisocial personality disorder (Merikangas, Risch, & Weissman, 1994). The high comorbidity, possibly at the genetic level, makes differentiating the phenotype of alcoholism in association studies extremely important. Most patients diagnosed with alcohol dependence meet almost all of the DSM-IV-TR diagnostic criteria, suggesting that the disorder is homogeneous except with heterogenous comorbidity. The DSM-IV criteria have been considered more reliable for clinical use and have also been used in clinical research (Samochowiec et al., 1999), and most patients diagnosed with alcohol dependence meet most of the DSM-IV-TR diagnostic criteria, which suggests that the disorder is more homogeneous except for the comorbidity. Our research team previously categorized alcoholism (ALC) into four subtypes according to DSM-IV comorbidities (Huang et al., 2004): pure alcoholism (pure ALC), anxiety/depression alcoholism (ANX/DEP ALC), antisocial alcoholism (antisocial ALC), and mixed alcoholism (mixed ALC). We attempted to use dopaminergic genes and dopaminergic related genes to set up the genetic validation of subtypes of alcoholism (Huang et al., 2004; Lee, Chen, Chang, & Lu, 2014; Lee et al., 2010; Lu et al., 2012, 2005). The detailed descriptions of each type of alcoholism are listed below:

FIGURE 1 The metabolic pathway of dopamine.

1. *Pure ALC*: onset is usually in middle adulthood; associated with no comorbid mental illnesses, and associated with fewer alcohol-related problems. The patient might have alcohol withdrawal syndrome when stopping drinking.
2. *ANX/DEP ALC*: comorbid with anxiety or depression; onset is during late adolescence or early adulthood. The patient almost always has comorbid anxiety or depression before, during, or after drinking; this subtype is associated with more severe alcohol use.
3. *Antisocial ALC*: comorbid with antisocial personality disorder; onset is usually during early or middle adolescence; associated with more alcohol-related problems, social consequences, and physical consequences. Alcohol abuse might be a symptom of antisocial behavior.

4. *Mixed ALC*: comorbid with other major mental illnesses or multiple substance disorders. One common comorbid major mental illness is bipolar disorder (BD). The prevalence of alcoholism comorbid with the abuse of other substances, such as morphine, is relatively lower in Han Chinese than in Westerners. Established studies (Kidorf et al., 2004) have reported that about 70% of heroin-dependent patients had alcohol use disorder comorbidities. In the Han Chinese population, our previous studies found the comorbidity rate of alcohol use disorder was only 16.0% in patients with opioid dependence (Wang et al., 2013).

ASSOCIATION STUDIES OF THE *DRD2* GENE WITH ALCOHOLISM SUBTYPES

The *DRD2* Gene is Associated with Alcoholism with Conduct Disorder

In 1999, we first examined whether there is evidence for an association between alcoholism with conduct disorder and the *DRD2* *Taq*I A polymorphisms by categorizing participants into alcoholics with conduct disorder, alcoholics without conduct disorder, and nonalcoholics. Alcoholism with conduct disorder had shown a significantly higher frequency of the *DRD2* *Taq*I A polymorphisms compared to the controls. This is the first report on the positive association between subcategorized alcoholism (Lee et al., 1999).

A common comorbidity with conduct disorders is the presence of an anxiety disorder. Between 22% and 33% of children from the community and 60–75% in clinic-referred children with conduct disorder also have an anxiety disorder (Russo & Beidel, 1994). Moreover, evidence for an association between alcoholism and anxiety has emerged from clinical studies of patients with alcoholism (Weiss, 1985), and those of patients with anxiety disorders (Wesner, 1990). Family studies have confirmed that there is an increased risk of alcoholism among relatives of anxiety neurotics (Harris, Noyes, Crowe, & Chaudhry, 1983). Therefore, if the *DRD2* locus is linked to a predisposition to conduct disorder and alcoholism, its relation with anxiety disorder and alcoholism is worth further examination. Knutson et al. (1998) proposed a molecular explanation that a low serotonin turnover rate and aggressive behavior are mediated by negative emotions, such as insecurity and anxiety. Some studies also suggest that conduct disorder might lead to alcoholism because of the tendency for a person to be impulsive (Wiers, Sergeant, & Gunning, 1994) and to exhibit behavior disinhibition (McGue, 1997). Besides the association of *DRD2* gene with anxiety alcoholism, we also set forth to determine the possible temperament dimensional involvement in conduct disorder and alcoholism, and aggressiveness (possibly antisocial personality disorder) to examine any association between *DRD2* and these special dimensions.

The *DRD2* Gene Associated with Anxiety/Depression Alcoholism after Stratification of the *ALDH2* Gene

Several alcohol detoxification studies found that the *DRD2* gene may be associated with high scores of anxiety and depression in alcohol dependence. Alcohol dependence that was comorbid with anxiety and/or depressive disorders represented the greatest risk of relapse (Finckh et al., 1997; Lucht et al., 2001). Moreover, in a double-blind bromocriptine treatment study, individuals with alcohol dependence and the *DRD2 A1* allele showed the greatest improvement in craving and anxiety (Lawford et al., 1995). These observations are consistent with the notion that the *DRD2* gene is associated with anxiety-depressive alcohol dependence (or abuse). To elucidate the association between the *DRD2* gene and alcohol dependence in the Han Chinese population with attempts to overcome the possible confounding effects and to reduce false-positive or false-negative results, we designed a comparison between individuals with solely anxiety/depression (ANX/DEP), individuals with both alcohol dependence and anxiety/depression (ANX/DEP ALC), and individuals with pure alcohol dependence and normal controls (Huang et al., 2004).

A strong linkage disequilibrium between the *Taq*I A and B polymorphisms of the *DRD2* gene was evident. We found that the frequency of the *A1/B1* haplotype was significantly higher in the ANX/DEP ALC group than that of controls. There was no association between the *DRD2* haplotype and pure alcohol dependence or ANX/DEP when compared to controls. A comparison of ANX/DEP ALC with ANX/DEP also revealed that the frequency of the *DRD2 A1/B1* haplotype was significantly higher in ANX/DEP ALC. Initially, these results were consistent with the idea that the homozygous *Taq*IA *A1/A1* and *B1/B1* genotypes and the *A1/B1* haplotype are risk factors for alcoholism. This possibility was subsequently excluded by the lack of a statistically significant difference between pure ALC and controls in the frequency of the *DRD2* polymorphisms. Moreover, the frequencies of the polymorphisms were not significantly different for ANX/DEP versus controls or for ANX/DEP ALC versus pure ALC, indicating that the *DRD2* gene is not associated with ANX/DEP. Therefore, our result supports the notions that 1) the *DRD2* gene is indeed associated with alcohol dependence, but maybe not pure ALC; 2) the ANX/DEP ALC might be a specific subtype of alcohol dependence in the Han Chinese population; and 3) phenotype definition is an important issue in candidate gene association studies of complex disorders, such as alcohol dependence. If an important candidate gene may increase the propensity for mental illness has been studied and reported on in several hundred or thousands of scientific articles but still remain controversial, we suggest that this candidate gene is indeed involved in the pathogenesis of the mental illness. The most commonly seen reason for past inconsistencies in the association between the *DRD2* gene and alcoholism in genetic studies may be due to different compositions of the control groups used, different racially or ethnically mixed study populations, and the phenotypic heterogeneity of alcoholism (Amadeo et al., 1993; Lee & Humphreys, 2014; Noble, Blum, Ritchie, Montgomery, & Sheridan, 1991; Parsian et al., 1991). We have studied patients from same ethnic groups, classified the subtypes of alcoholism according to clear definitions (using comorbidity), and used a well-selected control group (super controls). We found that the *DRD2* gene is related only to ANX/DEP alcoholism. This finding may explain the history of conflicting findings about the association between the *DRD2* gene and the subtypes of alcoholism (Huang et al., 2004).

Because *ALDH2* is crucial enzyme for ethanol catabolism which might also play an important role in DA catabolism and risk for alcoholism, the involvement of the *ALDH2* gene with the association of the *DRD2* gene and ANX/DEP ALC was further investigated. It is shown that the *DRD2* gene is associated with ANX/DEP ALC only after controlling for the *ALDH2*1/*1* genotype, supporting the contention that the *DRD2* gene may interact with the *ALDH2* genes in ANX/DEP ALC.

The *DRD2* Gene Was Not Associated with Antisocial Alcoholism, But Was Associated with Antisocial Personality Disorder after Interaction with the *ALDH2* Gene

A strong hereditary component in both alcohol dependence and antisocial personality disorder (ASPD) has been revealed by family, twin, and adoption studies (Reich et al., 1999). ASPD comorbid with alcoholism (antisocial ALC) is characterized by the early-onset of alcohol-related problems and increases the severity of alcohol dependence (Cloninger, 1987). The prevalence rate of ASPD and other adult antisocial behaviors among individuals with alcohol use disorders ranges from 20% to 33% (Goldstein et al., 2007). The high comorbidity of ASPD and antisocial behaviors with alcoholism suggests a shared genetic influence that hinders differentiating their categorical diagnoses in association studies. Past inconsistent findings on an association between the *DRD2* gene and alcoholism comorbid or not comorbid with ASPD might be attributable to a failure to separate the effects of ASPD from those of alcoholism and to recruiting nonalcoholic ASPD subjects only (van den Bree, Svikis, & Pickens, 1998). In an attempt to elucidate whether past positive genetic associations of the *DRD2* gene on antisocial ALC were contributed by ASPD or whether alcoholism is merely a symptom of ASPD, our past study recruited a combined ASPD group—antisocial ALC and antisocial non-ALC participants—to clarify the nature of these disorders (Lu et al., 2012).

Enzymes that function in the metabolic breakdown of acetaldehyde are considered a major biological factor that influences both drinking behavior and the development of alcoholism. The *ALDH2* gene has a functional polymorphism in exon 12 and shows two variant alleles: *ALDH2*1* and *ALDH2*2*. The *ALDH2*1/*1*-encoded enzyme is active in the metabolism of acetaldehyde, whereas the enzymes encoded by the *ALDH2*1/*2* and *ALDH2*2/*2* are partially and totally inactive, respectively. It is believed that the *ALDH2*2* allele, with reduced enzyme activity, provides protection against the risk of developing alcoholism (Chen et al., 1999). Approximately half of the East Asian population has the *ALDH2*2* allele variant (Agarwal & Goedde, 1992), including Han Chinese in Taiwan (Chen et al., 1999; Thomasson et al., 1991), but this allele is rarely found in other ethnic populations. It is therefore comparatively easier to recruit individuals with antisocial non-ALC in the Han Chinese populations.

No association was found between the genotype for the *TaqIA* polymorphisms at the *DRD2* locus in the antisocial ALC, antisocial non-ALC, or healthy control groups. However, a significant interaction between the *DRD2 A1/A1* and *ALDH2*1/*1* genotypes was associated with the antisocial non-ALC groups using healthy controls as the reference group. Only a borderline significant gene-to-gene interaction effect was found in the total ASPD (antisocial ALC and antisocial non-ALC) groups. The study demonstrated that neither the *DRD2 TaqIA* polymorphism by itself nor its interaction with the *ALDH2* polymorphisms were associated with antisocial ALC (Lu et al., 2012).

As we stated previously, a positive association was found between ANX/DEP ALC and the interaction of the *DRD2 TaqIA A1/A1* and *ALDH2*1/*1* genotypes (Huang et al., 2004). Combining the finding with antisocial non-ALC, we hypothesize that a gene-to-gene interaction of the homozygous *A1* allele at the *DRD2* gene and **1* allele at the *ALDH2* gene coexists in the antisocial non-ALC and ANX/DEP ALC groups. The *ALDH2*1/*1* genotype is associated with *ALDH2* activity, which results in an active form during the fast catabolism of acetaldehyde (Chen et al., 1999). The *A1* allele of the *DRD2* gene has been associated with low *DRD2* density (Jonsson et al., 1999), which may also indicate a low DA function. Logically, the interaction between the *A1+* allele (*A1/A1* or *A1/A2*) of the *DRD2* gene and the *ALDH2*1/*1* polymorphism may also cause a lower DA function, which is known to be associated with alcoholism (Volkow, Fowler, Wang, & Swanson, 2004). Theoretically, the combination of both liabilities, alcoholism comorbid with ASPD (antisocial ALC), should lead to the lowest DA level compared with the antisocial non-ALC group. However, the current study result supports that the antisocial non-ALC group, instead of the antisocial ALC group, had the lowest DA level. One possible explanation is that the lower DA level may not be explained by the *ALDH2* and *DRD2* genes only because other enzymes—MAO and COMT (catechol-*o*-methyltransferase)—are also involved in the metabolism of DA. The interaction of these genes should be investigated in further studies to elucidate the influence of DA level in subtypes of alcoholism, such as ANX/DEP ALC as well as in ASPD.

The *DRD2* Gene Associated with Alcoholism Comorbid with Bipolar Disorder after Interaction with *MAOA* Gene

BP is a severe, chronic mood disorder characterized by episodes of mania and depression at rates that vary markedly between and within individuals. It is well-known that BP patients are heterogeneous with regard to underlying genetic risk factors. The lifetime prevalence rate of alcoholism is about 39–60.7% in BP patients (Kessler, Rubinow, Holmes, Abelson, & Zhao, 1997), and alcoholism has a negative effect on the prognosis of BP. We hypothesize that alcoholism comorbid with bipolar disorder (ALC+BP) is a genetically distinct subtype of alcoholism and investigated whether a common genetic vulnerability exists that contributes to risk of ALC+BP.

When looking for candidate genes for alcoholism, the *DRD2* and *MAOA* genes are logical choices due to their central roles in dopaminergic pathways. The *DRD2 TaqIA* polymorphism was never considered as a risk factor for alcoholism and BD. On the other hand, monoamine oxidase A (MAOA), a mitochondrial enzyme, is involved in the degradation of DA, serotonin, and norepinephrine. The *MAOA* gene is therefore a strong candidate locus for neuropsychiatric disorders, including alcoholism and BP. Because individual genes may have only a limited effect on the development of alcoholism comorbid with BP, a gene–gene interaction approach was done to detect weak gene contributions to this complex subtyped alcoholism. The association between the *MAOA-uVNTR* and *DRD2 TaqIA* genes and their interaction in ALC+BP was investigated (Hu et al., 2013).

Interaction between the *DRD2 TaqI A1/A2* genotype and the *MAOA-uVNTR* 3-repeat polymorphism in ALC+BP patients was found. The *DRD2* gene alone was not associated with ALC+BP. However, a significant association between *DRD2* genes and ALC+BP was found only after stratification for the *MAOA-uVNTR* 3-repeat polymorphism, but not after the stratification of *MAOA-uVNTR* 4-repeat polymorphism. We also found that the *MAOA-uVNTR* 3-repeat polymorphism by itself may protect BP patients against alcoholism; however, its interaction with the *DRD2 TaqI A1/A2* genotype nullified the protection. The function of DA in the interaction between the *DRD2 TaqI A1/A2* and the *MAOA-uVNTR* 3-repeat genotypes needs additional studies.

The genetic interaction in ALC+BP in this study is similar to that in Huang et al. (2007), who reported the association of the *DRD2* gene and ANX/DEP ALC persisted only after stratification of the *MAOA-uVNTR* 3-repeat polymorphism. Huang et al. (2007) concluded that the *MAOA* gene may modify the association between the *DRD2* gene and ANX/DEP ALC phenotype. Anxiety disorders are the most commonly found comorbid illnesses of all patients with BP. The lifetime prevalence rate of BP comorbid with any kind of anxiety disorder ranges from 51.2% to 92.9% (Bauer et al., 2005). This shared genetic underpinning may account for the high comorbidity between BP and anxiety disorder. Further investigation of other candidate genes for ANX/DEP ALC in ALC+BP may be warranted in order to pursue the similarities and differences between these two subtypes. After all, current findings of genetic interaction provide an empirical genetic characterization of ALC+BP. In addition, classification of ALC+BP as a special subtype using other dopaminergic genes of alcoholism should be considered.

THE *DRD2* GENE AND TEMPERAMENT INTERACTION IN SUBTYPED ALCOHOLISM

Cloninger (1987) proposed a neurobiological learning model of alcoholism that described two genetic subtypes: type I (milieu-limited) and type II (male-limited) alcoholism. Type I alcohol use disorders included late-onset drinking behavior, more psychological dependence, high harm avoidance, and low novelty seeking, whereas type II alcohol use disorders included early-onset drinking behavior, more behavioral disturbances, low harm avoidance, and high novelty seeking. Each of the personality dimensions was postulated to be associated with a particular neurotransmitter system. Specifically, novelty seeking (NS) was mediated by the dopaminergic system and harm avoidance (HA) by the serotoninergic system (Cloninger, 1987). Evidence supported that dopaminergic and serotonergic neurotransmission might interact at the molecular level (Kapur & Remington, 1996). A relationship between NS and serotonin (5-HT), as well as HA and DA had also been demonstrated (Kremer et al., 2005). A significant relationship was demonstrated between the presence of *DRD2 TaqI A1* allele and the higher NS scores (Noble, 1998).

Lesch et al. (1996) postulated an existing relationship between the serotonin transporter gene-linked polymorphic region (*5-HTTLPR*) and HA among healthy subjects. Individuals with the short form allele may have higher anxiety-related personality traits, such as HA (Samochowiec et al., 2001). On the other hand, the *5-HTTLPR* S/S was also found to be related with NS-related traits such as impulsiveness, antisocial behavior, and aggressive behavior (Gerra et al., 2005). One of the major possibilities for the inconsistent results in the above-mentioned gene–personality association studies might be due to the heterogeneity of alcoholism proposed by Cloninger (1987). Meanwhile, the use of Cloninger's Tridimensional Personality Questionnaire (TPQ), which lacks a cutoff point to distinguish various subtypes of alcoholism, might cause such discrepancies when subdividing alcoholism into different categories, and the obscure definition of high or low score personality traits may fluctuate according to social cultural differences. A well-defined phenotype or a refined quantitative phenotype can lead to an adequate statistical analysis.

Interaction of Personality Traits and Genes in Anxiety–Depressive Alcoholism

In Cloninger's hypothesis, type I alcoholism might have lower NS and higher HA scores when compared with healthy volunteers. Huang et al. (2004, 2007) successfully proved anxiety-depressive alcohol dependence (ANX/DEP ALC) to be a genetically well-defined subtype of alcoholism which was linked to *DRD2*, having similar clinical characteristics as type I alcoholism such as a late-onset and more anxious/depressed traits. Moreover, subjects of this subtype may suffer from anxiety or depression, before, during, or after heavy drinking. Because ANX/DEP ALC diagnosis was constructed by the DSM-IV diagnostic criteria, we might be able to make a more efficient clinical classification of alcoholism from it to avoid possible confounding variables when adopting Cloninger's model. We therefore examined the specific personality traits of ANX/DEP ALC by TPQ and whether or not the *DRD2* and *5-HTTLPR* genes contributed to NS and HA personality traits, respectively, in ANX/DEP ALC among Han Chinese in Taiwan.

We found that ANX/DEP ALC scored higher in both NS and HA than pure ALC. Thus, our results might be in support of previous findings, which suggested that ANX/DEP ALC might belong to a specific subtype of alcoholism in comparison with normal volunteers and pure ALC (Huang et al., 2004). Furthermore, we found that ANX/DEP ALC was associated with NS only in subjects with *DRD2 TaqIA A1+* allele (including *A1/A1* or *A1/A2* genotype). Moreover, the difference in NS between ANX/DEP ALC and Pure ALC existed in subjects with S/S genotype of *5-HTTLPR*. Thus, it can be postulated that the elevation of NS scores and the development of ANX/DEP ALC might relate to *DRD2 TaqIA A1+* allele and other potential genes or environmental factors, which might include various pathogenesis marks that were frequently affiliated with ANX/DEP ALC (Huang et al., 2004).

After stratification of the *DRD2 TaqIA A1/A1* or *A1/A2* genotype subjects, the difference in NS scores was only found in subjects with *5-HTTLPR* S/S genotype. We found that ANX/DEP ALC was associated with HA only in subjects with *5-HTTLPR* S/L and L/L genotypes. This study suggested that the personality traits of type I alcoholism in Cloninger's model might need modification. For instance, *5-HTTLPR* polymorphism involved in both NS and HA could make us infer that personality traits might be associated with multiple genes, or that a single gene might influence several personality traits, syndromes, or mental illnesses. Therefore, the consideration of multiple genes in the study design might provide more interpretations for the gene–personality association (Lin et al., 2007).

Interaction of Personality Traits and Genes in Antisocial Alcoholism

We recruited antisocial nonalcoholics as an appropriate control to examine (1) the specific personality traits in antisocial alcoholism, which were similar to Cloninger's type II alcoholism; (2) the frequency differences in candidate genes related to the dopaminergic and serotonergic functions between antisocial alcoholism and antisocial nonalcoholism; and (3) the association between gene and specific personality trait, as well as between gene-to-gene interaction and specific personality traits in antisocial alcoholism.

We found that antisocial alcoholics scored higher than antisocial nonalcoholics on NS, but no difference was observed on the HA dimension. We could not support Cloninger's hypothesis, which assumed that type II alcoholism had higher NS and lower HA personality traits than normal controls (mimicking antisocial alcoholism). The inconsistent findings with Cloninger's hypothesis might result from the use of different comparator groups. Previous studies indicated that individuals with antisocial personality disorder may have specific personality traits and gene expressions (Soyka et al., 2004) that are considerably different from normal subjects. If we compare antisocial alcoholics with normal subjects, several confounding factors and variations could occur (Lu et al., 2005). Thus, the use of antisocial nonalcoholics as a comparator group might minimize the effects of confounding variables caused by a biasing of TPQ and the discrepancies between normal controls and antisocial alcoholism.

We found that the difference in novelty-seeking scores between antisocial alcoholism and antisocial nonalcoholism could only be detected in subjects who carry *DRD2 TaqIA A1+* allele (including *DRD2 TaqIA A1/A1* and *A1/A2* genotypes) and *5-HTTLPR* S/S genotype. Our results were consistent with previous studies that *DRD2 TaqIA A1* allele was related to high novelty-seeking behavior traits (Noble, 1998) and the *5-HTTLPR* S-allele was highly correlated with novelty seeking, impulsivity, and antisocial behavior in type II alcoholism (Hallikainen et al., 1999). For as much as the interaction between DA and serotonin was documented (Lu, Ko, Lin, Lin, & Ho, 1989), the DA- and serotonin-related genes, such as the *TaqIA* at *DRD2* gene and *5-HTTLPR* polymorphism, might also be associated with mental illnesses as well as related hereditable personality traits. In this study, we did not find either gene to be related to harm avoidance personality traits. Although Cloninger (1987) suggested that 5-HT function might be related to the harm avoidance personality trait, our study findings could not support the association between *5-HTTLPR*, the serotonin-related gene, and harm avoidance personality traits.

In conclusion, we found that Han Chinese antisocial alcoholics have high novelty-seeking behavior traits, but no associations with harm avoidance behavior traits were present. Current results suggest that Cloninger's type II alcoholism hypotheses may need further modification. As to the relationship between specific personality traits and other genes or their interactions, the *DRD2* gene and the *5-HTTLPR* polymorphism might be unable to comprehensively explain the pathogenesis of specific alcoholism subtypes and related personality traits, only illustrating one of the important factors (Wu et al., 2008).

SUMMARY

The association of the *DRD2* gene to several subtypes of alcoholism in Han Chinese was reviewed. Different subtypes of alcoholism might have different pathogenic mechanisms, even in the *DRD2* gene, which was first known for its effect on susceptibility for unsubtyped alcoholism. This chapter elucidates possible reasons for past inconsistent findings between the *DRD2* gene and unsubtyped alcoholism and highlights that phenotype definition is an important issue in candidate gene association studies. Furthermore, some identified genetic interaction with *DRD2* gene is associated only with certain subtypes. Different subtypes manifest similar genetic factors, but the genetic modulation effects vary by subtype. Because alcoholism is a complex behavioral disorder, the typology of alcoholism to reduce heterogeneity is important for research and clinical practice. This review of the journey to pursue the association between the *DRD2* gene and subtyped alcoholism indicates that continued research is needed to dissect the genetic underpinnings of different subtypes of alcoholism.

DEFINITION OF TERMS

Acetaldehyde dehydrogenase 2 (*ALDH2*) gene The *ALDH2* gene is mapped to chromosome 12q24. This gene has a functional SNP in exon 12, resulting in a glutamic acid/lysine exchange at position 487 and shows two variant alleles: *ALDH2*1* and *ALDH2*2*. This polymorphism causes a reduction in the aldehyde dehydrogenase enzyme's activity. The *ALDH2*1/*1*-encoded enzyme is active in the metabolism of acetaldehyde, whereas the enzymes encoded by the *ALDH2*1/*2* and *ALDH2*2/*2* are partially and totally inactive, respectively.

Dopamine D2 receptor (*DRD2*) gene The *DRD2 TaqIA* (rs1800497) restriction fragment length polymorphism is linked to the density of *DRD2*. *DRD2* density is 30–40% lower in *A1* allele carriers than in homozygotic *A2* allele carriers.

Monoamine oxidase A (MAOA) This mitochondrial enzyme is involved in the degradation of dopamine, serotonin, and norepinephrine.

Monoamine oxidase A upstream variable number tandem repeat (*MAOA-uVNTR*) gene A functional 30-base pair (bp) upstream variable number tandem repeat in the promoter region of the *MAOA* gene may alter transcriptional efficiency. The polymorphism consists of 2, 3, 3.5, 4, and 5 repeats. The alleles with 3.5 and 4 repeats seem to be more active than those with 3, while the activities of the alleles with 2 and 5 repeats are still unclear.

Antisocial Alcoholism (Antisocial ALC) For antisocial personality disorder comorbid with alcohol dependence, the onset is usually during early or middle adolescence; it is associated with more alcohol-related problems, social consequences, and physical consequences.

Anxiety/depression comorbid with alcoholism (ANX/DEP ALC) For alcoholism comorbid with anxiety or depression, onset is during late adolescence or early adulthood. The patient almost always has comorbid anxiety or depression before, during, or after drinking.

Pure alcoholism (Pure ALC) Its onset is usually in middle adulthood; it is associated with no comorbid mental illnesses and fewer alcohol-related problems.

Bipolar disorder comorbid with alcoholism (ALC+BP) A common genetic vulnerability for the comorbidity has been proposed. According to clinical observations, whether BP patients are in the manic, hypomanic, or depressive state may all greatly aggravate the risk of binge drinking. Long-term alcohol drinking will also worsen the course of illness of BP.

Tridimensional Personality Questionnaire (TPQ) TPQ has been widely used in many studies of mental illness. In Cloninger's Tridimensional Theory of Personality, the unified neurobiological model of personality was divided into three independent heritable dimensions of temperament—novelty seeking, harm avoidance, and reward dependence—which were proposed in relation to dopamine, serotonin, and norepinephrine, respectively.

KEY FACTS OF SUBTYPED ALCOHOLISM

- The clinical heterogeneity of alcoholism may be related to genetic heterogeneity.
- The typology of alcoholism is important for reducing inconsistencies both in research and clinical practice.
- Individual subtypes of alcoholism may highlight genetic perturbations within particular domains of psychopathology, which might not be revealed when studying alcoholism as a whole.
- Differentiating subtypes of alcoholism may reduce contradictory factors and uncover a more powerful association between genes and specific subtypes of alcoholism.
- Researchers assume that the many differences between alcoholics are associated with biological, psychological, and sociocultural factors.
- Since 1960, researchers have used a biopsychosocial approach to subtype alcoholism.
- Studies that have tried to genetically validate different subtypes of alcoholism have yielded only limited findings.
- Most subtypes are difficult to use clinically because of cultural differences.
- Because alcoholic patients frequently have other comorbid mental illnesses, using the comorbidity of other mental illnesses to reclassify the subtypes of alcoholism has been investigated.

SUMMARY POINTS

- Phenotype definition is an important issue in candidate gene association studies.
- Past inconsistent findings between the *DRD2* gene and unsubtyped alcoholism may be attributed to the heterogeneity of alcoholism, a complex behavioral disorder.
- The association of the *DRD2* gene in several subtypes of alcoholism in Han Chinese was reviewed.
- The *DRD2* gene is associated with alcoholism with conduct disorder.
- In subtyped alcoholism, the *DRD2* gene alone is associated with anxiety/depression alcoholism, but not antisocial alcoholism or pure alcoholism.
- Furthermore, the *DRD2* gene may interact with the *ALDH2* genes in anxiety/depression alcoholism.
- A gene-to-gene interaction of the *DRD2* and the *ALDH2* gene exists in antisocial nonalcoholism but not antisocial alcoholism.
- The *DRD2* gene alone was not associated with ALC+BP. However, an interaction between the *DRD2* and the *MAOA-uVNTR* polymorphism in ALC+BP patients has been identified.

REFERENCES

Agarwal, D. P., & Goedde, H. W. (1992). Medicobiological and genetic studies on alcoholism. Role of metabolic variation and ethnicity on drinking habits, alcohol abuse and alcohol-related mortality. *Clinical Investigation, 70*, 465–477.

Amadeo, S., Abbar, M., Fourcade, M. L., Waksman, G., Leroux, M. G., Madec, A., ... Mallet, J. (1993). D2 dopamine receptor gene and alcoholism. *Journal of Psychiatric Research, 27*, 173–179.

Bauer, M. S., Altshuler, L., Evans, D. R., Beresford, T., Williford, W. O., Hauger, R., & Team, V. A. C.S. (2005). Prevalence and distinct correlates of anxiety, substance, and combined comorbidity in a multi-site public sector sample with bipolar disorder. *Journal of Affective Disorders, 85*, 301–315.

Berggren, U., Fahlke, C., Aronsson, E., Karanti, A., Eriksson, M., Blennow, K., ... Balldin, J. (2006). The taqI DRD2 A1 allele is associated with alcohol-dependence although its effect size is small. *Alcohol and Alcoholism, 41*, 479–485.

Blum, K., Noble, E. P., Sheridan, P. J., Montgomery, A., Ritchie, T., Jagadeeswaran, P., ... Cohn, J. B. (1990). Allelic association of human dopamine D2 receptor gene in alcoholism. *JAMA, 263*, 2055–2060.

Blum, T. C., Roman, P. M., & Martin, J. K. (1993). Alcohol consumption and work performance. *Journal of Studies on Alcohol, 54*, 61–70.

Bowcock, A. M., Hebert, J. M., Christiano, A. M., Wijsman, E., Cavalli-Sforza, L. L., & Boyd, C. D. (1987). The pro alpha 1 (IV) collagen gene is linked to the D13S3 locus at the distal end of human chromosome 13q. *Cytogenetics and Cell Genetics, 45*, 234–236.

van den Bree, M. B. M., Svikis, D. S., & Pickens, R. W. (1998). Genetic influences in antisocial personality and drug use disorders. *Drug and Alcohol Dependence, 49*, 177–187.

Brodie, M. S. (2002). Increased ethanol excitation of dopaminergic neurons of the ventral tegmental area after chronic ethanol treatment. *Alcoholism: Clinical & Experimental Research, 26*, 1024–1030.

Chen, C. C., Lu, R. B., Chen, Y. C., Wang, M. F., Chang, Y. C., Li, T. K., & Yin, S. J. (1999). Interaction between the functional polymorphisms of the alcohol-metabolism genes in protection against alcoholism. *American Journal of Human Genetics, 65*, 795–807.

Chen, L. H., Huang, C. Y., Osio, Y., Fitzpatrick, E. A., & Cohen, D. A. (1993). Effects of chronic alcohol feeding and murine AIDS virus infection on liver antioxidant defense systems in mice. *Alcoholism: Clinical & Experimental Research, 17*, 1022–1028.

Cloninger, C. R. (1987). Neurogenetic adaptive mechanisms in alcoholism. *Science, 236*, 410–416.

Curtis, A. J., Shirk, M. C., & Fall, R. (1999). Allylic or benzylic stabilization is essential for catalysis by bacterial benzyl alcohol dehydrogenases. *Biochemical and Biophysical Research, 259*, 220–223.

Dick, D. M., & Foroud, T. (2003). Candidate genes for alcohol dependence: a review of genetic evidence from human studies. *Alcoholism: Clinical & Experimental Research, 27*, 868–879.

Edenberg, H. J., Foroud, T., Koller, D. L., Goate, A., Rice, J., Van Eerdewegh, P., ... Begleiter, H. (1998). A family-based analysis of the association of the dopamine D2 receptor (DRD2) with alcoholism. *Alcoholism: Clinical & Experimental Research, 22*, 505–512.

Finckh, U., Rommelspacher, H., Kuhn, S., Dufeu, P., Otto, G., Heinz, A., ... Rolfs, A. (1997). Influence of the dopamine D2 receptor (DRD2) genotype on neuroadaptive effects of alcohol and the clinical outcome of alcoholism. *Pharmacogenetics, 7*, 271–281.

Gerra, G., Garofano, L., Castaldini, L., Rovetto, F., Zaimovic, A., Moi, G., ... Donnini, C. (2005). Serotonin transporter promoter polymorphism genotype is associated with temperament, personality traits and illegal drugs use among adolescents. *Journal of Neural Transmission, 112*, 1397–1410.

Gessa, G. L., Muntoni, F., Collu, M., Vargiu, L., & Mereu, G. (1985). Low doses of ethanol activate dopaminergic neurons in the ventral tegmental area. *Brain Research, 348*, 201–203.

Goldstein, R. B., Compton, W. M., Pulay, A. J., Ruan, W. J., Pickering, R. P., Stinson, F. S., & Grant, B. F. (2007). Antisocial behavioral syndromes and DSM-IV drug use disorders in the United States: results from the national epidemiologic survey on alcohol and related conditions. *Drug and Alcohol Dependence, 90*, 145–158.

Goodwin, D. W., Schulsinger, F., Hermansen, L., Guze, S. B., & Winokur, G. (1973). Alcohol problems in adoptees raised apart from alcoholic biological parents. *Archives of General Psychiatry, 28*, 238–243.

Hallikainen, T., Saito, T., Lachman, H. M., Volavka, J., Pohjalainen, T., Ryynanen, O. P., ... Tiihonen, J. (1999). Association between low activity serotonin transporter promoter genotype and early onset alcoholism with habitual impulsive violent behavior. *Molecular Psychiatry, 4*, 385–388.

Harris, E. L., Noyes, R., Jr., Crowe, R. R., & Chaudhry, D. R. (1983). Family study of agoraphobia. Report of a pilot study. *Archives of General Psychiatry, 40*, 1061–1064.

Hietala, J., West, C., Syvalahti, E., Nagren, K., Lehikoinen, P., Sonninen, P., & Ruotsalainen, U. (1994). Striatal D2 dopamine receptor binding characteristics in vivo in patients with alcohol dependence. *Psychopharmacology (Berlin), 116*, 285–290.

Hu, M. C., Lee, S. Y., Wang, T. Y., Chen, S. L., Chang, Y. H., Chen, S. H., ... Lu, R. B. (2013). Association study of DRD2 and MAOA genes with subtyped alcoholism comorbid with bipolar disorder in Han Chinese. *Progress in Neuro-Psychopharmacology & Biological Psychiatry, 40*, 144–148.

Huang, S. Y., Lin, W. W., Ko, H. C., Lee, J. F., Wang, T. J., Chou, Y. H., ... Lu, R. B. (2004). Possible interaction of alcohol dehydrogenase and aldehyde dehydrogenase genes with the dopamine D2 receptor gene in anxiety-depressive alcohol dependence. *Alcoholism: Clinical & Experimental Research, 28*, 374–384.

Huang, S. Y., Lin, W. W., Wan, F. J., Chang, A. J., Ko, H. C., Wang, T. J., ... Lu, R. B. (2007). Monoamine oxidase-A polymorphisms might modify the association between the dopamine D2 receptor gene and alcohol dependence. *Journal of Psychiatry Neuroscience, 32*(3), 185–192.

Jonsson, E. G., Nothen, M. M., Grunhage, F., Farde, L., Nakashima, Y., Propping, P., & Sedvall, G. C. (1999). Polymorphisms in the dopamine D2 receptor gene and their relationships to striatal dopamine receptor density of healthy volunteers. *Molecular Psychiatry, 4*, 290–296.

Kapur, S., & Remington, G. (1996). Serotonin-dopamine interaction and its relevance to schizophrenia. *American Journal of Psychiatry, 153*, 466–476.

Kendler, K. S., Heath, A. C., Neale, M. C., Kessler, R. C., & Eaves, L. J. (1992). A population-based twin study of alcoholism in women. *JAMA, 268*, 1877–1882.

Kessler, R. C., Rubinow, D. R., Holmes, C., Abelson, J. M., & Zhao, S. (1997). The epidemiology of DSM-III-R bipolar I disorder in a general population survey. *Psychological Medicine, 27*, 1079–1089.

Keung, W. M., & Vallee, B. L. (1998). Daidzin and its antidipsotropic analogs inhibit serotonin and dopamine metabolism in isolated mitochondria. *Proceedings of the National Academy of Sciences of the United States of America, 95*, 2198–2203.

Kidorf, M., Disney, E. R., King, V. L., Neufeld, K., Beilenson, P. L., & Brooner, R. K. (2004). Prevalence of psychiatric and substance use disorders in opioid abusers in a community syringe exchange program. *Drug and Alcohol Dependence, 74*, 115–122.

Knutson, B., Wolkowitz, O. M., Cole, S. W., Chan, T., Moore, E. A., Johnson, R. C., ... Reus, V. I. (1998). Selective alteration of personality and social behavior by serotonergic intervention. *American Journal of Psychiatry, 155*, 373–379.

Kremer, I., Bachner-Melman, R., Reshef, A., Broude, L., Nemanov, L., Gritsenko, I., ... Ebstein, R. P. (2005). Association of the serotonin transporter gene with smoking behavior. *American Journal of Psychiatry, 162*, 924–930.

Lawford, B. R., Young, R. M., Rowell, J. A., Qualichefski, J., Fletcher, B. H., Syndulko, K., ... Noble, E. P. (1995). Bromocriptine in the treatment of alcoholics with the D2 dopamine receptor A1 allele. *Nature Medicine, 1*, 337–341.

Lee, J. F., Lu, R. B., Ko, H. C., Chang, F. M., Yin, S. J., Pakstis, A. J., & Kidd, K. K. (1999). No association between DRD2 locus and alcoholism after controlling the ADH and ALDH genotypes in Chinese Han population. *Alcoholism: Clinical & Experimental Research, 23*, 592–599.

Lee, S. S., & Humphreys, K. L. (2014). Interactive association of dopamine receptor (DRD4) genotype and ADHD on alcohol expectancies in children. *Experimental and Clinical Psychopharmacology, 22*, 100–109.

Lee, S. Y., Chen, S. L., Chang, Y. H., & Lu, R. B. (2014). Variation of types of alcoholism: review and subtypes identified in Han Chinese. *Progress in Neuro-Psychopharmacology & Biological Psychiatry, 48*, 36–40.

Lee, S. Y., Hahn, C. Y., Lee, J. F., Huang, S. Y., Chen, S. L., Kuo, P. H., ... Lu, R. B. (2010). MAOA interacts with the ALDH2 gene in anxiety-depression alcohol dependence. *Alcoholism: Clinical & Experimental Research, 34*, 1212–1218.

Lesch, K. P., Bengel, D., Heils, A., Sabol, S. Z., Greenberg, B. D., Petri, S., ... Murphy, D. L. (1996). Association of anxiety-related traits with a polymorphism in the serotonin transporter gene regulatory region. *Science, 274*, 1527–1531.

Lin, S.-C., Wu, P.-L., Ko, H.-C., Wu, J. Y.-W., Huang, S.-Y., Lin, W.-W., & Lu, R.-B. (2007). Specific personality traits and dopamine, serotonin genes in anxiety-depressive alcoholism among Han Chinese in Taiwan. *Progress in Neuro-Psychopharmacology and Biological Psychiatry, 31*, 1526–1534.

Lu, R. B., Ko, H. C., Chang, F. M., Castiglione, C. M., Schoolfield, G., Pakstis, A. J., ... Kidd, K. K. (1996). No association between alcoholism and multiple polymorphisms at the dopamine D2 receptor gene (DRD2) in three distinct Taiwanese populations. *Biological Psychiatry, 39*(6), 419–429.

Lu, R. B., Ko, H. C., Lee, J. F., Lin, W. W., Huang, S. Y., Wang, T. J., ... Chou, Y. H. (2005). No alcoholism-protection effects of ADH1B*2 allele in antisocial alcoholics among Han Chinese in Taiwan. *Alcoholism: Clinical & Experimental Research, 29*, 2101–2107.

Lu, R. B., Ko, H. C., Lin, W. L., Lin, Y. T., & Ho, S. L. (1989). CSF neurochemical study of tardive dyskinesia. *Biological Psychiatry, 25*, 717–724.

Lu, R. B., Lee, J. F., Huang, S. Y., Lee, S. Y., Chang, Y. H., Kuo, P. H., ... Ko, H. C. (2012). Interaction between ALDH2*1*1 and DRD2/ANKK1 TaqI A1A1 genes may be associated with antisocial personality disorder not co-morbid with alcoholism. *Addiction Biology, 17*, 865–874.

Lucht, M. J., Kuehn, K. U., Schroeder, W., Armbruster, J., Abraham, G., Schattenberg, A., ... Freyberger, H. J. (2001). Influence of the dopamine D2 receptor (DRD2) exon 8 genotype on efficacy of tiapride and clinical outcome of alcohol withdrawal. *Pharmacogenetics, 11*, 647–653.

McBride, W. J., Murphy, J. M., Gatto, G. J., Levy, A. D., Yoshimoto, K., Lumeng, L., & Li, T. K. (1993). CNS mechanisms of alcohol self-administration. *Alcohol and Alcoholism Supplement, 2*, 463–467.

McGue, M. (1997). A behavioral-genetic perspective on children of alcoholics. *Alcohol Health & Research World, 21*, 210–217.

Merikangas, K. R. (1990). The genetic epidemiology of alcoholism. *Psychological Medicine, 20*, 11–22.

Merikangas, K. R., Risch, N. J., & Weissman, M. M. (1994). Comorbidity and co-transmission of alcoholism, anxiety and depression. *Psychological Medicine, 24*, 69–80.

Munafo, M. R., Matheson, I. J., & Flint, J. (2007). Association of the DRD2 gene Taq1A polymorphism and alcoholism: a meta-analysis of case-control studies and evidence of publication bias. *Molecular Psychiatry, 12*, 454–461.

Noble, E. P. (1998). DRD2 gene and alcoholism. *Science, 281*, 1287–1288.

Noble, E. P., Blum, K., Ritchie, T., Montgomery, A., & Sheridan, P. J. (1991). Allelic association of the D2 dopamine receptor gene with receptor-binding characteristics in alcoholism. *Archives of General Psychiatry, 48*, 648–654.

Parsian, A., Todd, R. D., Devor, E. J., O'Malley, K. L., Suarez, B. K., Reich, T., & Cloninger, C. R. (1991). Alcoholism and alleles of the human D2 dopamine receptor locus. Studies of association and linkage. *Archives of General Psychiatry, 48*, 655–663.

Pickens, R. W., Svikis, D. S., McGue, M., Lykken, D. T., Heston, L. L., & Clayton, P. J. (1991). Heterogeneity in the inheritance of alcoholism. A study of male and female twins. *Archives of General Psychiatry, 48*, 19–28.

Reich, T., Hinrichs, A., Culverhouse, R., & Bierut, L. (1999). Genetic studies of alcoholism and substance dependence. *American Journal of Human Genetics, 65*, 599–605.

Ritchie, T., & Noble, E. P. (2003). Association of seven polymorphisms of the D2 dopamine receptor gene with brain receptor-binding characteristics. *Neurochemical Research, 28*, 73–82.

Russo, M. F., & Beidel, D. C. (1994). Comorbidity of childhood anxiety and externalizing disorders: prevalence, associated characteristics, and validation issues. *Clinical Psychology Review, 14*, 199–221.

Samochowiec, J., Lesch, K. P., Rottmann, M., Smolka, M., Syagailo, Y. V., Okladnova, O., … Sander, T. (1999). Association of a regulatory polymorphism in the promoter region of the monoamine oxidase A gene with antisocial alcoholism. *Psychiatry Research, 86*, 67–72.

Samochowiec, J., Rybakowski, F., Czerski, P., Zakrzewska, M., Stepien, G., Pelka-Wysiecka, J., … Hauser, J. (2001). Polymorphisms in the dopamine, serotonin, and norepinephrine transporter genes and their relationship to temperamental dimensions measured by the temperament and character inventory in healthy volunteers. *Neuropsychobiology, 43*, 248–253.

Soyka, M., Preuss, U. W., Koller, G., Zill, P., Hesselbrock, V., & Bondy, B. (2004). No association of CRH1 receptor polymorphism haplotypes, harm avoidance and other personality dimensions in alcohol dependence: results from the Munich gene bank project for alcoholism. *Addiction Biol, 9*, 73–79.

Thomasson, H. R., Edenberg, H. J., Crabb, D. W., Mai, X. L., Jerome, R. E., Li, T. K., … Yin, S. J. (1991). Alcohol and aldehyde dehydrogenase genotypes and alcoholism in Chinese men. *American Journal of Human Genetics, 48*, 677–681.

Volkow, N. D., Fowler, J. S., Wang, G. J., & Swanson, J. M. (2004). Dopamine in drug abuse and addiction: results from imaging studies and treatment implications. *Molecular Psychiatry, 9*, 557–569.

Wang, T. Y., Lee, S. Y., Chen, S. L., Huang, S. Y., Chang, Y. H., Tzeng, N. S., … Lu, R. B. (2013). Association between DRD2, 5-HTTLPR, and ALDH2 genes and specific personality traits in alcohol- and opiate-dependent patients. *Behavioural Brain Research, 250*, 285–292.

Weiss, W. (1985). Social representations of alcohol in children and adolescents. *Soz Praventivmed, 30*, 329–332.

Wesner, R. B. (1990). Alcohol use and abuse secondary to anxiety. *Psychiatric Clinics of North America, 13*, 699–713.

Westerink, B. H., & de Vries, J. B. (1985). On the origin of dopamine and its metabolite in predominantly noradrenergic innervated brain areas. *Brain Research, 330*, 164–166.

Wiers, R. W., Sergeant, J. A., & Gunning, W. B. (1994). Psychological mechanisms of enhanced risk of addiction in children of alcoholics: a dual pathway? *Acta Paediatrica Supplement, 404*, 9–13.

Wu, C. Y., Wu, Y. S., Lee, J. F., Huang, S. Y., Yu, L., Ko, H. C., & Lu, R. B. (2008). The association between DRD2/ANKK1, 5-HTTLPR gene, and specific personality trait on antisocial alcoholism among Han Chinese in Taiwan. *American Journal of Medical Genetics, Part B: Neuropsychiatric Genetics, 147B*, 447–453.

Chapter 51

Acetaldehyde: A Reactive Metabolite

Roshanna Rajendram[1], Rajkumar Rajendram[2], Victor R. Preedy[2]
[1]Department of Emergency Medicine, The Royal Woolverhampton NHS Trust, New Cross Hospital, Wolverhampton, UK; [2]Diabetes and Nutritional Sciences Research Division, Faculty of Life Science and Medicine, King's College London, London, UK

INTRODUCTION

The interrelationship between acetaldehyde and neurotoxicity is still largely an unexplored area compared to other toxic agents. Much of the knowledge base pertaining to the cytotoxic effects of acetaldehyde generated from nonneurological tissues is pertinent to neuropathology, as some fundamental processes are the same. The main purpose here is not so much to describe the effects of acetaldehyde on the brain per se, but also to emphasize its reactivity.

Acetaldehyde is an aldehyde with the systematic name ethanal and the formula CH_3CHO (Figure 1). It is formed by the oxidation of ethanol via three pathways: alcohol dehydrogenases, catalase, or the microsomal ethanol-oxidizing system (MEOS) (Figure 2). The acetaldehyde is then converted into acetate via aldehyde dehydrogenases (ALDHs) (Figure 3).

Acetaldehyde is mutagenic and genotoxic (CEPA, 2000). It is classified by the International Agency for Research on Cancer and the World Health Organization as a Class 1 agent, possibly carcinogenic to humans.

At room temperature acetaldehyde is a colorless gas with a fruity odor that is flammable and unstable in air. It is highly reactive and toxic, causing damage at the cellular and genomic levels. The chemical properties of acetaldehyde are listed in Table 1.

Acetaldehyde occurs naturally in coffee, bread, and ripe fruit and is produced by plants and is manufactured on a large scale in industry. It is also produced by the partial oxidation of ethanol. Routes of exposure include air, water, land, or groundwater, as well as drinking and smoking. The sources of human exposure to acetaldehyde may be endogenous or exogenous.

ENDOGENOUS ALDEHYDES

Cytochrome P450-Catalyzed Metabolic Activation

Endogenous aldehydes with one less carbon atom than the initial alcohol substrate (e.g., acetaldehyde) are formed by cytochrome P450 (P450) 2E1-catalyzed oxidation of glycerol, ethylene glycol, 1,2-propane diols, or polyhydroxylated alcohols containing vicinal diols, probably mediated by a ferryl-type oxidant species involving a radical mechanism or the homolytic cleavage of a dioxetane intermediate (Clejan & Cederbaum, 1992; Kukielka & Cederbaum, 1995).

Myeloperoxidase-Catalyzed Metabolic Activation

Plasma contains myeloperoxidase and hydrogen peroxide (H_2O_2) (released from activated neutrophils at sites of inflammation), as well as chloride (0.1 M) and α-amino acids (4–5 mM). Neutrophils and chloride incubated with alanine form acetaldehyde (Hazen, Hsu, d'Avignon, & Heinecke, 1998).

Lipid Peroxidation

The background levels of exocyclic propano/etheno–DNA adducts in tissues from unexposed humans arise from the endogenous alkenals, hydroxyalkenals, dialdehydes, and alkanal products of lipid peroxidation (LPO) decomposition (Bartsch, 1999).

Endogenous carbonylated proteins are increased in the liver during LPO induced by ethanol. Using malondialdehyde antibodies, the major adduct was identified as a malondialdehyde adduct with cytochrome oxidase subunit IV (Chen, Petersen, Schenker, & Henderson, 2000). Cytochrome oxidase was also inactivated and could contribute to subsequent cytotoxicity.

Doxorubicin administered to rats increased aldehyde levels in the heart and plasma before cardiotoxicity developed. The increase in aldehyde levels and cardiotoxicity were both prevented by carnitine (Luo, Reichetzer, Trines, Benson, & Lehotay, 1999). Doxorubicin (10 mg/kg) increased the urinary

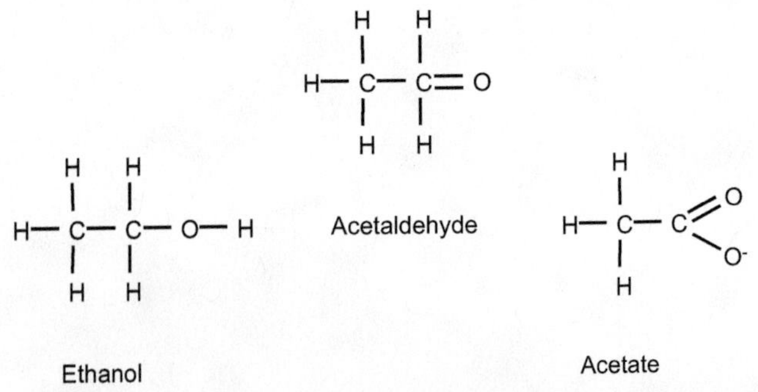

FIGURE 1 The chemical structures of acetaldehyde, ethanol, and acetate. *Figure adapted from Rajendram, Hunter, and Preedy (2005) with permission from Elsevier.*

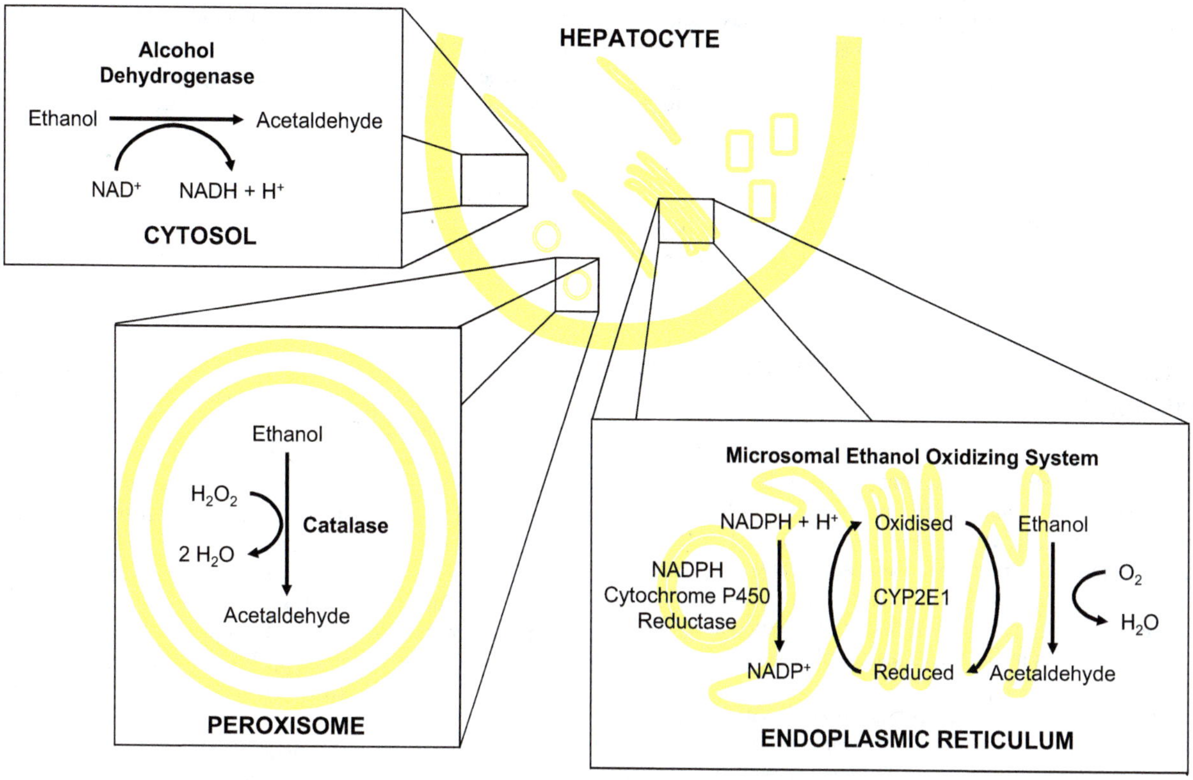

FIGURE 2 The pathways of acetaldehyde production from ethanol metabolism. *Figure adapted from Rajendram et al. (2005) with permission from Elsevier.*

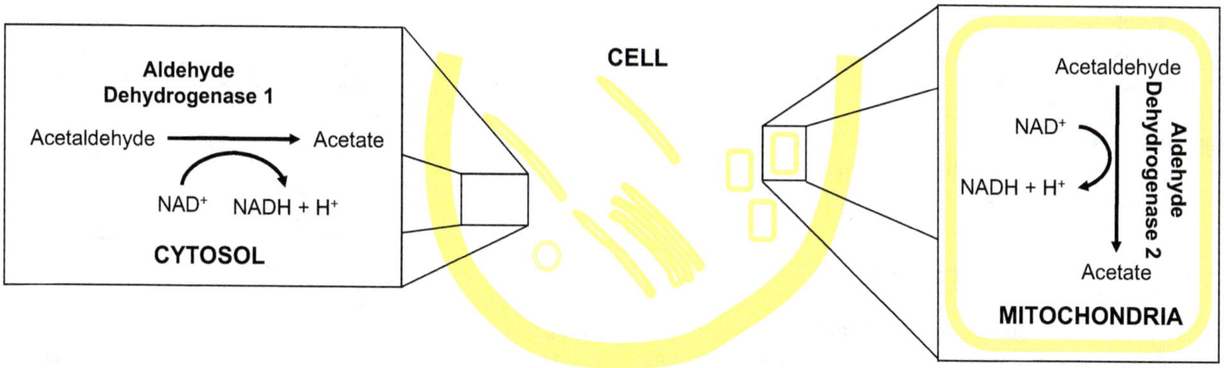

FIGURE 3 The pathways of acetate production from acetaldehyde metabolism. *Figure adapted from Rajendram et al. (2005) with permission from Elsevier.*

TABLE 1 The Chemical Properties of Acetaldehyde (Ethanal)

	Property
Chemical formula	CH_3CHO
Class	Aldehyde
Molecular weight	44.0
Melting point	−124 °C
Boiling point	20.2 °C
Description	Colorless liquid with a fruity aroma

excretion of acetaldehyde from 2 to 14 nM/kg 6 h later. Aldehyde levels were still high 72 h later (Bagchi, Bagchi, Hassoun, Kelly, & Stohs, 1995). Plasma cytotoxic alkenals (*trans*-2-heptenal, 4-hydroxynonenal (4-HNE), *trans*-2-nonenal) were increased up to 3.5 times (Luo et al., 1999).

EXOGENOUS ACETALDEHYDES

Dietary Acetaldehyde

The greatest source of human exposure to acetaldehyde is the diet. Alcoholic beverages contain traces of acetaldehyde and substantial amounts of ethanol (Feron et al., 1991). Metabolism of this ethanol generates acetaldehyde in vivo (Figure 2).

Acetaldehyde is naturally present in many foods including fruits and vegetables (e.g., peas; Feron et al., 1991). Acetaldehyde concentrations in vegetables have been reported to be 0.2–400 μg/g and in wine to be from 0.7 to 290 μg/g. Different alcoholic beverages have different amounts of acetaldehyde; one study showed the amount of acetaldehyde consumed by cider drinkers was greater than that by vodka drinkers (Lachenmeier, Gill, Chick, & Rehm, 2015). Acetaldehyde has been used to add flavor to food and beverages and is produced when fat-containing foods are cooked (Lane & Smathers, 1991).

Natural and Manufactured Aldehydes: Urban, Rural, and Indoor

The use of alternate automobile fuels is increasing aldehyde emissions. Adding ethanol to automobile fuels as an oxygenate increases the acetaldehyde emissions in exhaust fumes (Kirchstetter, Singer, & Harly, 1996). Grosjean, Grosjean, and Moreira (2002) attributed the increase in ambient acetaldehyde in Brazil to a change in reliance on ethanol as a vehicle fuel.

Acetaldehyde also has many industrial uses, such as in alkyd resin production. It is released as a reactor off-gas during the production of acetic acid.

Aldehydes are present in homes and workplaces and are detectable in confined spaces such as airplane cabins (James, 1997; NRC, 2002). Major indoor sources include furniture, carpets, fabrics, and paints (Kelley, Sadola, & Smith, 1966). Cooking fumes contain acetaldehyde, and it is also emitted from wood-burning stoves and fireplaces (Cao et al., 2007; Cerqueira, Gomes, Tarelho, & Pio, 2013; Lane & Smathers, 1991; Svensson et al., 1999). Acetaldehyde is generated indoors by the action of ozone on anthropogenic hydrocarbons (Weschler & Shields, 1997). Cigarette smoke is an important indoor source of acetaldehyde (Cao et al., 2007; Rickert, Robinson, & Young, 1980). Analysis of cigarette smoke for aldehydes showed that acetaldehyde was the major component at 709 μg/cigarette (Smith & Hansch, 2000), though values will vary between brands.

ACETALDEHYDE REACTIVITY

Acetaldehyde is implicated in the development of many diseases and particularly as a consequence of alcohol ingestion and misuse. However, the steps between alcohol exposure, acetaldehyde formation, and the development of biochemical and functional complications have not been fully elucidated. Acetaldehyde, a highly reactive metabolite, is potentially toxic via the formation of adducts with DNA and proteins (i.e., acetaldehyde–DNA or acetaldehyde–protein adducts) (Figure 4). This will have implications not only for liver disease (the liver is where most of the acetaldehyde is formed after alcohol consumption) but also for neurological tissues. Certainly alcohol-derived protein adducts have been found in neurological tissues (Nakamura et al., 2003, 2000; Niemela, 2007; Upadhya & Ravindranath, 2002) (Figure 5).

The mutagenic potency of aldehydes is listed in Table 2 (Marnett et al., 1985). Aldehydes are more effective at inhibiting cell proliferation than at causing cell death. The ability of aldehydes to inhibit the cell growth of mouse fibroblasts is listed in Table 3 (Wieslander et al., 1995).

There is increasing evidence that formation of protein adducts may be important in the development of alcohol- and acetaldehyde-induced disease. Protein adducts are posttranslationally modified proteins formed by covalent linkage of reactive nucleophiles (e.g., aldehydes) to parent proteins (Worrall & Thiele, 2001).

Protein adducts may inactivate proteins (Chen et al., 2000) and/or render them more susceptible to proteolysis (Nicholls, Fowles, Worrall, de Jersey, & Wilce, 1994). This may induce neoantigen formation and thereby antibody generation (Worrall & Thiele, 2001).

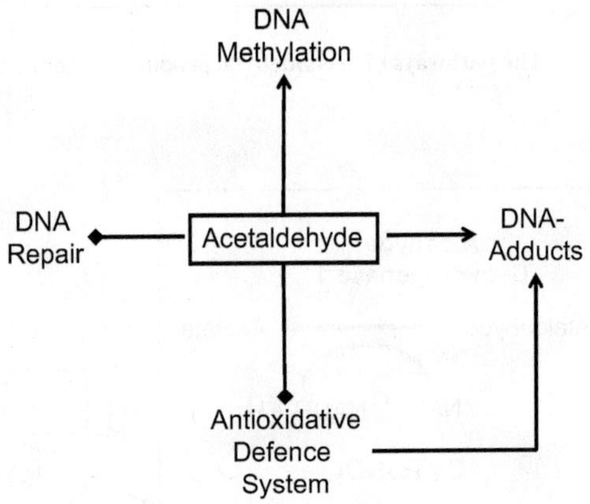

FIGURE 4 **The pathogenic effects of acetaldehyde.** Acetaldehyde causes DNA adducts, inhibits DNA repair and DNA methylation, and damages the antioxidative defense system. These effects are toxic, mutagenic, and carcinogenic.

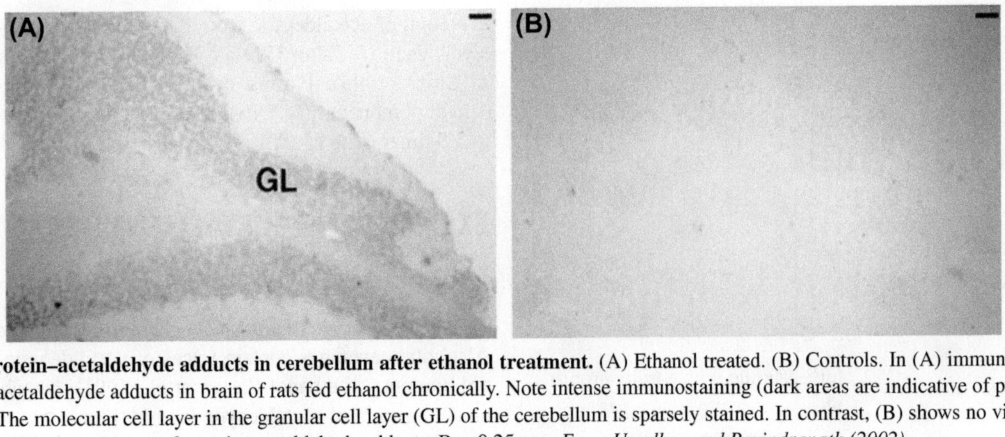

FIGURE 5 **Protein–acetaldehyde adducts in cerebellum after ethanol treatment.** (A) Ethanol treated. (B) Controls. In (A) immunohistochemistry shows protein–acetaldehyde adducts in brain of rats fed ethanol chronically. Note intense immunostaining (dark areas are indicative of protein–acetaldehyde adducts). The molecular cell layer in the granular cell layer (GL) of the cerebellum is sparsely stained. In contrast, (B) shows no visible staining at this level, indicative of an absence of protein–acetaldehyde adducts. Bar, 0.25 mm. *From Upadhya and Ravindranath (2002).*

TABLE 2 The Mutagenic Potency of Aldehydes
Methylglyoxal
Glutaraldehyde
Glyoxal
Formaldehyde
Acrolein
Crotonaldehyde
Acetaldehyde
Malondialdehyde

This table lists the mutagenic potency of aldehydes in descending order (Marnett et al., 1985).

TABLE 3 The Ability of Aldehydes to Inhibit Cell Growth
Aldehyde
Formaldehyde
Methylglyoxal
2-Furaldehyde
Glyoxal
Acetaldehyde

This table lists the ability of aldehydes to inhibit the cell growth of mouse fibroblasts in descending order (Wieslander et al., 1995).

While the mutagenic potency of acetaldehyde is relatively low, exposure is high in alcohol misuse and so the potential risk from acetaldehyde is greater than that from many other aldehydes. However, in the long term, there is very little evidence that malignancy of nervous tissues arises in alcohol misuse, though other cancers are elevated. In a large meta-analysis of 19 studies there was no overt risk of glioma with alcohol consumption; however, a reduced risk of glioma was seen in North American studies but not other geographical areas (Qi, Shao, Yang, Wang, & Hui, 2014). Another meta-analysis of other cancer types affecting the nervous system of adults was also inconclusive, though there were indications of increased risk with the consumption of higher volumes of alcohol and spirits (Galeone et al., 2013).

PROTEIN ADDUCT FORMATION

Acetaldehyde can form protein adducts in two ways (Freeman et al., 2005):

1. formation of protein adducts and
2. oxidative stress.

Acetaldehyde can initiate the formation of protein adducts (i.e., acetaldehyde–protein adducts) under both reducing and nonreducing conditions. That is, the acetaldehyde binds directly to proteins. There are also ethanol-derived adducts such as the hydroxyethyl radical–protein adducts.

Acetaldehyde can also act indirectly as oxidative stress occurs as a consequence of acetaldehyde metabolism. This results in the formation of volatile species such as those derived from LPO (i.e., malondialdehyde– or 4-HNE–protein adducts). When there is simultaneous acetaldehyde and malondialdehyde formation, a hybrid adduct is formed, namely the malondialdehyde–acetaldehyde–protein adduct; MAA. The major protein adduct found in the liver of alcoholic patients is a cyclic 2:1 MAA epitope condensation product. It is a 4-methyl-1,4-dihydropyridine-3,5-dicarbaldehyde derivative of protein amino groups.

Adduct formation may

1. impair the function of proteins,
2. form neoantigens that trigger autoimmune responses, and
3. be toxic to cells.

The importance of protein adduct formation in the etiology of alcoholic organ damage is well described by Freeman et al. (2005).

DAMAGE TO DNA

Damage to DNA and accumulation of DNA adducts are thought to play a causative role in several diseases, including cancer. Many aldehydes form adducts or modify DNA, resulting in mutagenicity (Figure 4). A variety of aldehyde–DNA lesions occur.

One DNA lesion caused by acetaldehyde is N^2-ethyl-2′-deoxyguanosine (N^2-ethyl-dG), which is also derived from N^2-ethylidene-2′-deoxyguanosine (N^2-ethylidene-dG). Essentially these are DNA adducts caused by the reaction of acetaldehyde with the 2′-deoxyguanosine in DNA (Garcia et al., 2011). The reaction of acetaldehyde with deoxyguanosine initially forms an unstable imine. A reduction step is required to form the stable N^2-ethyl-dG lesion (Vaca, Fang, & Schweda, 1995).

Fang and Vaca (1995) developed a highly sensitive postlabeling assay to detect N^2-ethyl-dG in cellular DNA. This detected low levels (one lesion per 10^8 nucleotides) in the liver DNA of rats given 10% ethanol in their drinking water. N^2-ethyl-dG was not present in the DNA of the control animals. The number of lesions in human alcohol abusers was an order of magnitude higher but basal levels of this lesion were seen in some nondrinking human controls. The white blood cell DNA of alcohol abusers contained two or three lesions per 10^7 nucleotides. This was significantly higher than the levels in nondrinking control subjects.

Using liquid chromatography–mass spectrometry, Matsuda et al. (1999) were able to detect N^2-ethyl-dG in urine samples from individuals tested after abstaining from alcohol for 1 week. This suggests that the lesion in urine results either from acetaldehyde formed endogenously or from an endogenous ethylating agent.

There is a role for ALDHs in preventing the formation of the DNA adducts N^2-ethyl-dG and N^2-ethylidene-dG. In esophageal tissue of ALDH2-knockout mice greater amounts of N^2-ethylidene-dG are found compared to tissue from those that express ALDH2 (Yukawa et al., 2014). These studies have not been extended to neurological tissues.

Acetaldehyde can cause many other DNA lesions (Garcia et al., 2011; Wang et al., 2000). Of these, two are of particular significance: DNA interstrand cross-links and 1,N^2-propano-2′-deoxyguanosine (1,N^2-PdG, or sometimes abbreviated to 1,N^2-propanodGuo).

1,N^2-PdG is an endogenous DNA adduct (Nath, Ocando, & Chung, 1996) It is responsible for the mutagenic, genotoxic, and carcinogenic properties of crotonaldehyde (Eder & Budiawan, 2001; Garcia et al., 2011).

Wang et al. (2000) found that formation of 1,N^2-PdG required 40 mM acetaldehyde. Such levels would not occur from ethanol metabolism in vivo. However, histones can facilitate the formation of 1,N^2-PdG from acetaldehyde and DNA (Sako, Inagaki, Esaka, & Deyashiki, 2003). Formation of 1,N^2-PdG from acetaldehyde and deoxyguanosine does not require reduction, unlike the formation of N^2-ethyl-dG.

PROTEIN CARBONYLATION/OXIDATION AND CYTOTOXICITY

It would be imprudent to consider that adduct formation arises as a direct consequence of just the binding of acetaldehyde per se to proteins or indeed other macromolecular compounds. The text below discusses other major forms of protein modifications that will arise as a consequence of oxidative stress and/or alcohol.

Endogenous carbonylated proteins (also termed protein carbonyls) increase with age and in various pathological states, including premature diseases, muscular dystrophy, rheumatoid arthritis, metal storage diseases, and atherosclerosis (Chevion, Berenshtein, & Stadtman, 2000), and neuropathologies (Rommer, Greilberger, Herwig, Auff, & Leutmezer, 2013), including those due to traumatic brain injury (Lazarus, Buonora, Jacobowitz, & Mueller, 2015) or malnutrition (Dkhar & Sharma, 2014).

Oxidative stress is an unavoidable consequence of using oxygen. Aging may be due to the accumulation of oxidatively damaged biomolecules. During aging the structure of the mitochondria changes and energy production declines. Lysosomes accumulate lipofuscin, particularly in postmitotic cells such as neurons and cardiac myocytes (Brunk & Terman, 2002). Carbonylation affects approximately 1 in 10 protein molecules and markedly increases during the last third of life, such that eventually 1 in 3 protein molecules are carbonylated and functionally impaired (Stadtman & Levine, 2000). Senescent cell cultures also show increased protein cross-linking and lysosomal lipofuscin formation, possibly resulting from lysosomal dysfunction (Sitte, Merker, Von Zglinicki, Davies, & Grune, 2000).

Carbonyls increase cellular oxidative stress by inactivating antioxidant enzymes (e.g., glutathione reductase, glutathione peroxidase, and glutathione transferases). Nicotinamide adenine dinucleotide phosphate (NADPH) partially protects glutathione reductase (Boggaram & Mannervik, 1982). This observation suggests that the target is the NADPH binding site.

Protein carbonylation effectively damages proteins and they form large, covalently cross-linked hydrophobic aggregates, which threaten cell viability and function. While carbonylated/oxidized nucleic acids are repaired by highly efficient excision/insertion mechanisms, the carbonylated/oxidatively damaged proteins are degraded by a highly regulated and complex process involving proteases. Replacements must then be synthesized and relocated (Shringarpure, Grune, & Davies, 2001).

Protein oxidative damage and protein covalent binding by the metabolites of xenobiotics are highly selective (Cohen et al., 1997). For example, mitochondrial adenine nucleotide translocase and aconitase are targets of oxidative damage in houseflies during aging or hyperoxia (Yan & Sohal, 1998). Proteins with a transition metal binding site may be targets for reactive oxygen species carbonylation.

Protein carbonyls have been reported to increase in brains of pregnant mice and their offspring as a consequence of ethanol exposure (Akhtar, Rouse, & Maffi, 2015). In a postmortem study, it was reported that the cerebella of alcoholics had higher levels of protein carbonyls (Richards, Perry, Dodd, & Worrall, 2012). A preliminary study has also suggested that protein carbonyls in blood are higher in alcoholics with psychiatric disorders and that there is a relationship with the severity of psychopathology (Verbenko, Malev, & Zakharova, 2012).

More modern techniques are able to identify specific proteins in neurological tissues that are carbonylated (Oikawa et al., 2014; Shen et al., 2015). However, as far as we know, these methods have not been applied to examine the specific proteins carbonylated as a consequence of alcohol toxicity. Potentially, proteomics can also be used to determine the specific proteins adapted with acetaldehyde per se. However, such studies have not yet been documented either.

MOLECULAR CYTOTOXIC MECHANISMS OF ALDEHYDES

The cytotoxicity of aldehydes is thought to involve protein carbonylation and protein glycation. The extent of protein carbonylation depends on the aldehyde concentration and the activity and intracellular location of the carbonyl-metabolizing enzymes (which may be inhibited by the aldehydes). The proteins targeted for glycation may differ from those targeted for carbonylation. Dicarbonyls react rapidly with arginine and lysine, whereas monocarbonyls react slowly with lysine. Xenobiotic-induced oxidative stress cytotoxicity can also result in protein carbonylation as a result of the intracellular formation of aldehydic decomposition products. In this case the oxidative stress and extent of protein carbonylation probably reflect the balance between pro- and antioxidant systems.

The susceptibility of specific cells to different aldehydes varies considerably. The activity and intracellular location of detoxifying enzymes within these cells are relevant. The antioxidant enzymes glutathione reductase and glutathione transferase and enzymes associated with energy production were also inactivated by aldehydes and are expected to determine the cells' susceptibility to aldehydes (Morgan, Dean, & Davies, 2002).

The cytotoxic effectiveness of aldehydes for inducing cell death as measured by accelerated cytotoxic mechanism screening is listed in Table 4 (Niknahad, Shuhendler, et al., 2003; Niknahad, Siraki, et al., 2003).

The cytotoxicity of aldehydes is related to their hydrophobicity and electrophilicity (Niknahad, Shuhendler, et al., 2003; Niknahad, Siraki, et al., 2003). Hydrophobic and electrophilic aldehydes bind most easily to macromolecules and inactivate the carbonyl-metabolizing enzymes that detoxify them. Decreasing protein carbonylation (e.g., with dithiothreitol) prevents cytotoxicity.

Acetaldehyde administered intragastrically to rats is oxidized by the liver, as methyl radicals can be detected by spin traps in bile (Nakao, Kadiiska, Mason, Grijalba, & Augusto, 2000). The methyl radicals are probably formed by decarbonylation of acetyl radicals generated when acetaldehyde is oxidized by hydroxyl radicals (Nakao et al., 2000).

TABLE 4 The Cytotoxic Effectiveness of Aldehydes

Aldehyde	LD_{50} (mM)
Acrolein	0.08
Crotonaldehyde	0.5
Salicylaldehyde	0.6
Formaldehyde	2.5
Glyoxal	5.0
Acetaldehyde	15

This table lists the cytotoxic effectiveness of aldehydes for inducing cell death as measured by accelerated cytotoxic mechanism screening in descending order (Niknahad, Shuhendler, et al., 2003; Niknahad, Siraki, et al., 2003). The individual dose required to kill 50% of a population is the LD_{50}.

A distinction must be made, however, between acetaldehydes per se and the aldehydes that may be generated endogenously by LPO.

Mitochondrial toxicity is also involved in the molecular cytotoxic mechanism with most aldehydes studied. Mitochondrial respiration is inhibited, membrane potential collapses, mitochondrial morphology changes, oxidative phosphorylation is inhibited, and cellular ATP is depleted before cytotoxicity occurs. Furthermore, cytotoxicity is prevented by glycolytic substrates (Sood & O'Brien, 1993).

ACETALDEHYDE OR ETHANOL?

The important question that is often raised in alcohol-related research is whether acetaldehyde mediates the effects of ethanol administration in the experimental setting (or indeed alcohol misuse in the clinical setting, especially in the scenario of addictions).

Stoichiometrically, one molecule of alcohol will generate one molecule of acetaldehyde. Put another way, 1 mol of ethanol (46.07 g) will be oxidized to produce 1 mol (44.05 g) of acetaldehyde if there were no losses via the breath, urine, or sweat or no oxidative metabolism via fatty acid ethyl esters. Thus, a standard UK unit (10.0 ml) will contain 7.89 g of ethanol, which will be converted to 7.54 g of acetaldehyde.

Acetaldehyde is extremely toxic in vivo, however, owing to its extreme volatility and chemical reactivity (though in some studies acetaldehyde has been added directly to in vitro systems). In a 2015 study on oligodendrocytes in 24-h exposure studies in vitro, ethanol at concentrations ≤120 mmol/l did not cause cell death, but at 240 mmol/l cell death was observed (Coutts & Harrison, 2015). Acetaldehyde on the other hand was extremely toxic at 0.5 mmol/l and above (Coutts & Harrison, 2015). The authors explain the results of these studies in terms of the white matter loss in the brain in alcohol toxicity.

To address the toxicity of acetaldehyde in vivo, metabolic tools have been used. Cyanamide inhibits acetaldehyde dehydrogenase and 4-methylpyrazole inhibits alcohol dehydrogenase. Essentially the use of cyanamide increases acetaldehyde. The use of 4-methylpyrazole reduces acetaldehyde. Acetaldehyde can also be sequestered with D-penicillamine, thereby reducing its concentration.

Catalase activity is an important pathway for the conversion of alcohol to acetaldehyde in the brain (Hamby-Mason, Chen, Schenker, Perez, & Henderson, 1997). The contribution of catalase is determined by the use of inhibitors such as 3-amino-1,2,4-triazole (also called aminotriazole) or azide. α-Lipoic acid will scavenge hydrogen peroxide, which is a component of the catalase pathway (Figure 2) (Melis, Carboni, Caboni, & Acquas, 2015).

Inhibition of MEOS can be carried out by use of 1-butanol or metyrapone. In some studies knock-in or knockout rodents have been used and in other cell studies, gene silencing has been employed.

For example, treatment of rats with ethanol or ethanol plus cyanamide (the ALDH inhibitor) increases the amount of acetaldehyde-derived protein adducts in skeletal muscle (Niemela, Parkkila, Koll, & Preedy, 2002). Another example of the use of inhibitors to demonstrate the role of acetaldehyde in alcohol-induced toxicity relates to the fetal alcohol syndrome, which may result from alcohol use during pregnancy. Hemoglobin–acetaldehyde adducts are increased in pregnant women and in those who deliver infants with fetal alcohol effects (Niemela, 2007). Rat studies showed that cyanamide (increasing acetaldehyde) markedly increased fetal alcohol effects in pregnant rats treated with ethanol from day 9 to 12 of gestation. This suggests that acetaldehyde causes the deleterious effects of ethanol (Ali & Persaud, 1988).

In one study, looking at the actions of acetaldehyde, the authors examined the role of acetaldehyde on brain cAMP-protein kinase A (PKA) signaling (Tarragon, Balino, & Aragon, 2014). They showed that ethanol markedly increased brain PKA activation in the brain of experimental animals but this was blocked with D-penicillamine pretreatment (Figure 6). This suggests that the acetaldehyde caused the cAMP-PKA activation.

In brain tissue, the use of cyanamide dosing showed that the increases in brain nitric oxide levels were dependent on both alcohol and acetaldehyde (Finnerty, O'Riordan, Klamer, Lowry, & Palsson, 2015). In another study ethanol-induced increases in motor activity were measured (Marti-Prats et al., 2013). Reducing acetaldehyde formation with D-penicillamine and azide reduced motor activity. In contrast, increasing acetaldehyde with cyanamide increased motor activity (Marti-Prats et al., 2013). Other studies using these inhibitors have included examination of cholinergic function (Jamal et al., 2007) and dopamine activity (Melis et al., 2015).

It is important to reemphasize the fact that adducts may be formed by LPO and by acetaldehyde from ethanol metabolism. Effectively this is a two-pronged attack on cells and is illustrated schematically in Figure 7. Malnutrition will also cause oxidative stress and increase formation of aldehydes such as malondialdehyde (Figure 7).

HUMAN HEALTH RISKS

Alzheimer Disease and Parkinson Disease

Acetaldehyde may be involved in the pathogenesis of Alzheimer disease (AD), as brain aldehyde-detoxifying enzymes are affected in AD patients (D'Souza, Elharram, Soon-Shiong, Andrew, & Bennett, 2015; Li et al., 2009).

Aldoketoreductase 7A2 (AKR7A2) is present in the glia and neurons in the substantia nigra and is elevated in senile plaques in the cortex and hippocampus by microglial activation and influx (Picklo, Olson, Hayes, Markesbery, & Montine, 2001). ALDH2

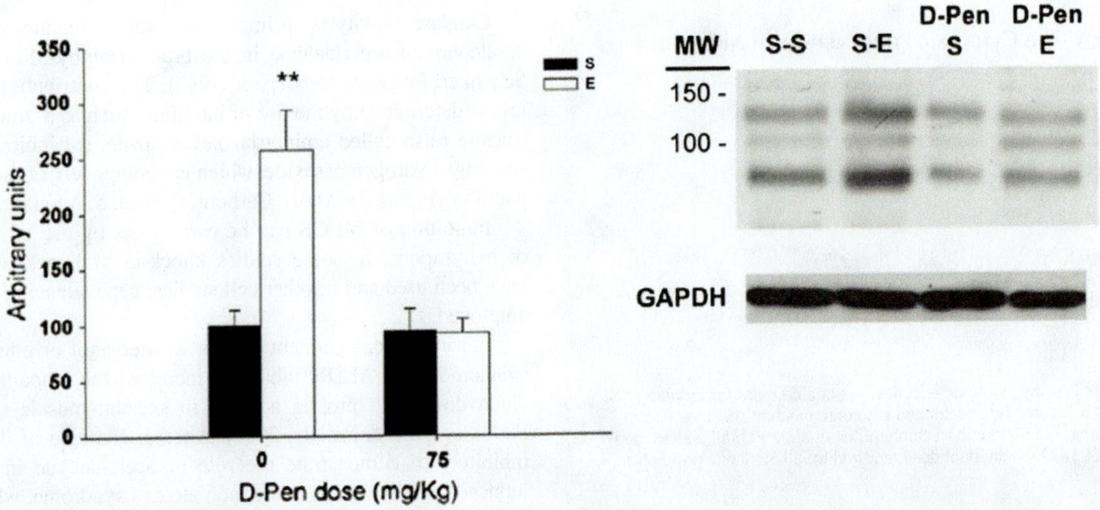

FIGURE 6 The use of D-penicillamine to study the role of acetaldehyde on brain cAMP-PKA signaling. In these studies mice were predosed with D-penicillamine (D-Pen; 75 mg/kg body weight) or vehicle (0 dose) and then treated 30 min afterward with either ethanol (E; clear bars; 2.5 g/kg body weight) or saline (filled bars; S). The histograms represent summary densitometry in arbitrary units after Western blotting of extracts from the striatum. Representative blots using glyceraldehyde-3-phosphate dehydrogenase (GAPDH) as the housekeeping protein are on the right-hand side. As no effect is seen in D-penicillamine plus ethanol-dosed mice we can assume that acetaldehyde is necessary for the activation of cAMP-PKA. *From Tarragon et al. (2014)*.

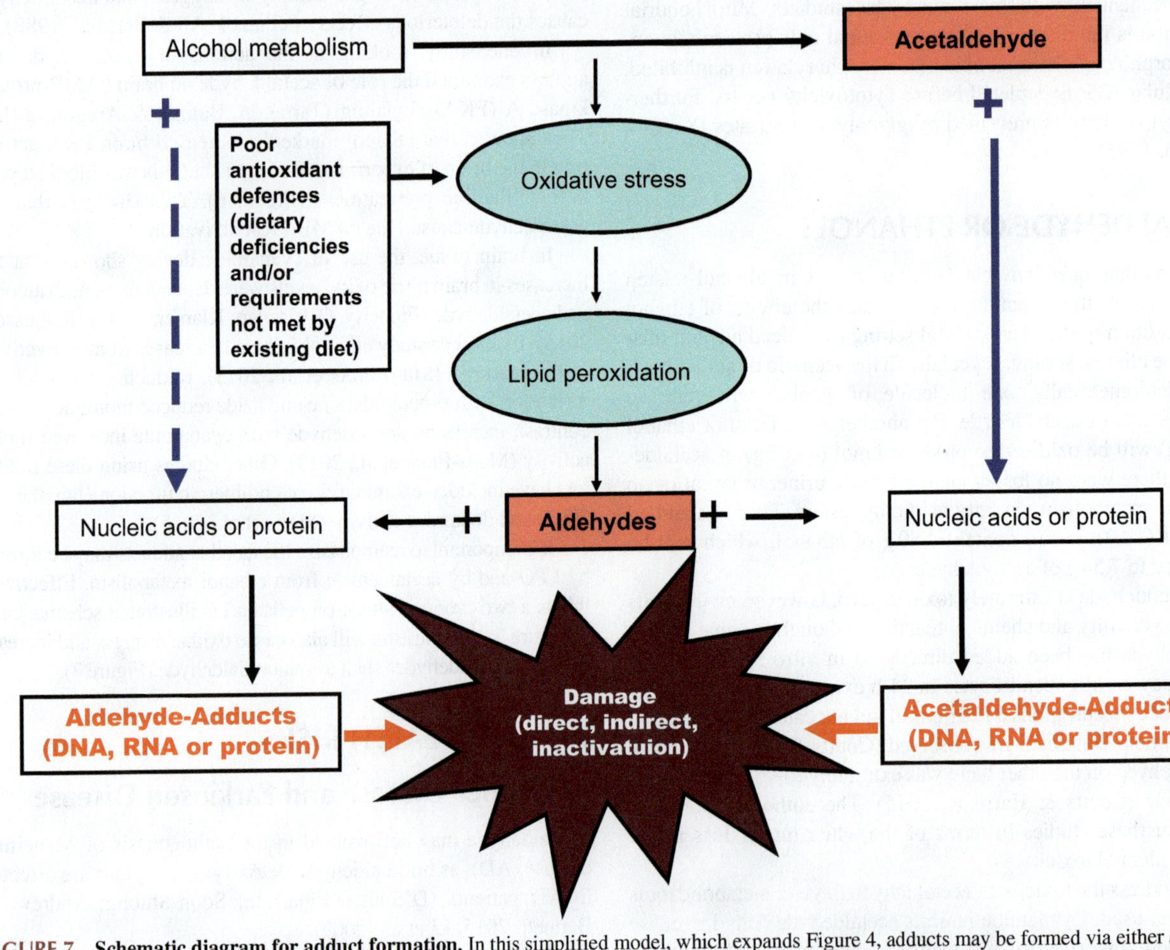

FIGURE 7 Schematic diagram for adduct formation. In this simplified model, which expands Figure 4, adducts may be formed via either aldehydes or the specific directive metabolite acetaldehyde. The acetaldehydes react with proteins, RNA, and DNA to form adducts, as well as with phospholipids. These adducts may be toxic or cause structural changes to cell membranes (direct), initiate antibody formation (indirect processes), or even inactivate proteins. Nutritional impairment in alcoholism will also exacerbate the formation of adducts. The dashed line represents the formation of ethanol-derived (i.e., hydroxyethyl radical) adducts.

but not ALDH1 is present in glia but not in neurons. This suggests that the neuronal mitochondria have no defense against reactive aldehydes. This could explain the vulnerability of the AD brain. ALDH2 was elevated in senile plaques from the influx of reactive astrocytes and microglia. Interestingly, the risk of AD is higher in Japanese people with the less active ALDH2*2 allele. Furthermore, overexpression of ALDH3, AKR1A1, and glutathione S-transferase Mu in PC12 cells (a cell line derived from rat pheochromocytoma), neuroblastoma cells, or neuronal cultures protected them from 4-HNE. 4-HNE cytotoxicity was also prevented by N-acetylcysteine or aminoguanidine (Picklo et al., 2001).

Oxidative damage to neuronal nucleic acids and mitochondrial dysfunction increases with age in the rat hippocampus. The hippocampus is important for memory. The age-related loss of spatial memory and mitochondrial dysfunction were partially reversed in rats fed acetylcarnitine and/or lipoic acid (Liu et al., 2002).

These observations suggest many potential therapeutic targets for the treatment and prevention of AD.

Wernicke Encephalopathy

This is an acquired neuropsychiatric disorder caused by thiamine deficiency. It develops in patients with impaired nutrition associated with gastrointestinal disease and chronic alcoholism. It is treated with thiamine. Thiamine deficiency inhibits the thiamine-dependent pyruvate dehydrogenase and α-ketoglutarate dehydrogenase activities. This slows down the citric acid cycle, resulting in mitochondrial uncoupling and ATP depletion. Furthermore, inhibition of cytosolic transketolase probably decreases NADPH, compromising the antioxidant defenses of the cell. This cytotoxicity was prevented by α-tocopherol or butylated hydroxyanisole. The cytotoxic mechanism may therefore involve LPO (Pannunzio, Hazell, Pannunzio, Rao, & Butterworth, 2000). Cerebral free radicals were also increased in thiamine-deficient rats (Langlais, Anderson, Guo, & Bondy, 1997). Parkhomenko et al. (2014) studied neuronal preparations in vitro and showed that acetaldehyde reduced the activity of both thiamine triphosphatase and thiamine pyrophosphate kinase.

CONCLUDING REMARKS

Various diseases, particularly those associated with aging, have been associated with the accumulation of acetaldehyde, including cancer, drug toxicity, and neurodegenerative diseases. Some of these effects have been associated with aldehyde-metabolizing enzyme polymorphisms, for example, mitochondrial ALDH2 and AKR.

Understanding the potential adverse effects and pathogenesis of acetaldehyde is a pressing need. The sheer number of enzymes involved in the metabolism and detoxication of acetaldehyde is a testament to the impact of its reactivity. Metabolic or autoxidation pathways lead to the formation of endogenous acetaldehyde, and a large number of genes function in the metabolic detoxication of acetaldehyde. Paradoxically, acetaldehyde may even be one of the body's defenses against bacteria, viruses, parasites, and tumor cells. Most ALDHs, reductases, and P450s detoxify aldehydes, except that P450 may oxidize aldehydes to toxic radicals. P450 often also generates formaldehyde while catalyzing the O-demethylation of xenobiotics, which could contribute to xenobiotic toxicity. Inhibition of ALDHs and reductases markedly increases aldehyde cytotoxicity, whereas P450 inhibitors, radical traps, and antioxidants can prevent cytotoxicity.

In environmental health, more toxicological studies and risk assessments of acetaldehyde are needed, as it is ubiquitously present and increased levels are associated with various diseases and pathological conditions.

APPLICATIONS TO OTHER ADDICTIONS AND SUBSTANCE MISUSE

There is clear evidence that acetaldehyde is an extremely toxic agent. While acetaldehyde is the immediate metabolite of ethanol metabolism, it does have its relevance in terms of tobacco. Tobacco leaf may also be smoked by itself or with cannabis. The acetaldehyde is formed via the heat-imposed decomposition of plant carbohydrates contained in the tobacco leaf (i.e., smoking; Lin, Ye, Deng, & Zhang, 2008).

Acetaldehyde may also contribute to the addictive process in tobacco usage, possibly by the formation of acetaldehyde-biogenic amine adducts (Talhout, Opperhuizen, & van Amsterdam, 2007) or another mechanism that may depend on the catalase-derived formation of acetaldehyde (Karahanian et al., 2011). Depending on the commercial brand, the amount of acetaldehyde can vary between 0.16 and 1.34 mg per cigarette (Lin et al., 2008).

It has been strongly argued that acetaldehyde in cigarette smoke is harmful to the brain, for example, affecting midbrain dopamine metabolism (Hoffman & Evans, 2013). Cannabis cigarette smoke and acetaldehyde per se can both interact with DNA to form N^2-ethylidene-dG adducts (Singh et al., 2009). Dissecting out the pathogenic mechanisms due to ethanol-derived acetaldehyde may help in a more detailed understanding of the harm due to tobacco-derived acetaldehyde.

DEFINITION OF TERMS

Acetaldehyde This is an aldehyde with the systematic name ethanal and the formula CH_3CHO.

Adducts These are complexes formed by a chemical binding to a biological molecule (e.g., protein or DNA).

Carbonylation This refers to oxidation of protein side chains or introduction of carbon monoxide into organic and inorganic substrates.

Glycation This is the covalent bonding of a sugar molecule with protein or lipid.

Oxidation This is an increase in oxidation state/loss of electrons by a molecule, atom, or ion.

Oxidative stress This describes an imbalance between the production of reactive oxygen species and the ability to detoxify or repair the damage from these reactive molecules.

Reduction This is a decrease in oxidation state/gain of electrons by a molecule, atom, or ion.

KEY FACTS OF BRAIN METABOLISM

- The adult brain utilizes 20% of the total oxygen consumed by the whole body.
- It oxidizes 120 g of glucose per day via glycolysis, the citric acid cycle, and oxidative phosphorylation as ATP-generating steps.

- Glucose is the main fuel, although in starvation it switches to ketone bodies, namely 3-hydroxybutyrate and acetoacetate. The citric acid cycle and oxidative phosphorylation are ATP-generating steps.
- Cell lines are a particularly useful tool in in vitro studies, which can be carried out using defined influencing factors and at the same time excluding others (cytokines, adducts, cellular mediators, and metabolites).
- More sophisticated tools have been used to understand brain metabolism in the clinical setting or in experimental animals in vivo.
- Analytical tools used in vivo include magnetic resonance spectroscopy (^{1}H or ^{13}C, for example) or positron emission tomography with fluorodeoxyglucose.
- Different brain regions and different cells have different rates of metabolism.
- Some researchers have used modeling tools to understand the metabolism of the brain (for example see Amaral, Alves, & Teixeira, 2014).

SUMMARY POINTS

- Acetaldehyde (ethanal) is highly reactive and toxic.
- Acetaldehyde causes damage at the cellular and genomic levels.
- Acetaldehyde is implicated in the development of many diseases, including those caused by alcohol, AD, and stroke.
- The greatest source of human exposure to acetaldehyde is the diet.
- Alcoholic beverages contain traces of acetaldehyde and substantial amounts of ethanol. Metabolism of this ethanol generates acetaldehyde in vivo.

REFERENCES

Akhtar, F. F., Rouse, C. A., & Maffi, S. K. (2015). Modulation of maternal oxidative stress causes inflammation and alters fetal neurodevelopment: alcoholism: clinical and experimental research. In *Conference: 38th Annual scientific meeting of the research society on alcoholism, San Antonio, TX, United States* (p. 230A).

Ali, F., & Persaud, T. V. (1988). Mechanisms of fetal alcohol effects: role of acetaldehyde. *Experimental Pathology*, 33, 17–21.

Amaral, A. I., Alves, M., & Teixeira, A. P. (2014). Metabolic flux analysis tools to investigate brain metabolism in vitro. *Neuromethods*, 90, 107–144.

Bagchi, D., Bagchi, M., Hassoun, E. A., Kelly, J., & Stohs, S. J. (1995). Adriamycin-induced hepatic and myocardial lipid peroxidation and DNA damage, and enhanced excretion of urinary lipid metabolites in rats. *Toxicology*, 95, 1–9.

Bartsch, H. (1999). Keynote address: exocyclic adducts as new risk markers for DNA damage in man. *IARC Scientific Publication*, 150, 1–16.

Boggaram, V., & Mannervik, B. (1982). Essential arginine residues in the pyridine nucleotide binding sites of glutathione reductase. *Biochimica et Biophysica Acta*, 701, 119–126.

Brunk, U. T., & Terman, A. (2002). Lipofuscin: mechanisms of age-related accumulation and influence on cell function. *Free Radical Biology & Medicine*, 33, 611–619.

Cao, J., Belluzzi, J. D., Loughlin, S. E., Keyler, D. E., Pentel, R., & Leslie, F. M. (2007). Acetaldehyde, a major constituent of tobacco smoke, enhances behavioral, endocrine, and neuronal responses to nicotine in adolescent and adult rats. *Neuropsychopharmacology*, 32(9), 2025–2035.

CEPA. (2000). *Acetaldehyde. Canadian Environmental Protection Act. Priority substances list assessment report*. Environ. Canada & Health Canada En40-215/50E.

Cerqueira, M., Gomes, L., Tarelho, L., & Pio, C. (2013). Formaldehyde and acetaldehyde emissions from residential wood combustion in Portugal. *Atmospheric Environment*, 72, 171–176.

Chen, J., Petersen, D. R., Schenker, S., & Henderson, G. I. (2000). Formation of malondialdehyde adducts in livers of rats exposed to ethanol: role in ethanol-mediated inhibition of cytochrome c oxidase. *Alcoholism: Clinical and Experimental Research*, 24, 544–552.

Chevion, M., Berenshtein, E., & Stadtman, E. R. (2000). Human studies related to protein oxidation: protein carbonyl content as a marker of damage. *Free Radical Research*, 33(Suppl.), S99–S108.

Clejan, L. A., & Cederbaum, A. I. (1992). Role of cytochrome P450 in the oxidation of glycerol by reconstituted systems and microsomes. *The FASEB Journal*, 6, 765–770.

Cohen, S. D., Pumford, N. R., Khairallah, E. A., Boekelheide, K., Pohl, L. R., Amouzadeh, H. R., & Hinson, J. A. (1997). Selective protein covalent binding and target organ toxicity. *Toxicology and Applied Pharmacology*, 143, 1–12.

Coutts, D. J. C., & Harrison, N. L. (2015). Acetaldehyde, not ethanol, impairs myelin formation and viability in primary mouse oligodendrocytes. *Alcoholism: Clinical and Experimental Research*, 39(3), 455–462.

Dkhar, P., & Sharma, R. (2014). Late-onset dietary restriction modulates protein carbonylation and catalase in cerebral hemispheres of aged mice. *Cellular and Molecular Neurobiology*, 34(2), 307–313.

D'Souza, Y., Elharram, A., Soon-Shiong, R., Andrew, R. D., & Bennett, B. M. (2015). Characterization of Aldh2$^{-/-}$ mice as an age-related model of cognitive impairment and Alzheimer's disease. *Molecular Brain*, 8, 1.

Eder, E., & Budiawan (2001). Cancer risk assessment for the environmental mutagen and carcinogen crotonaldehyde on the basis of TD50 and comparison with 1,N^2-propanodeoxyguanosine adduct levels. *Cancer Epidemiology Biomarkers and Prevention*, 10, 883–888.

Fang, J. L., & Vaca, C. E. (1995). Development of a 32P-postlabelling method for the analysis of adducts arising through the reaction of acetaldehyde with 2'-deoxyguanosine-3'-monophosphate and DNA. *Carcinogenesis*, 16, 2177–2185.

Feron, V. J., Til, H. P., de Vrijer, E., Woutersen, R. A., Cassee, F. R., & van Bladeren, P. J. (1991). Aldehydes: occurrence, carcinogenic potential, mechanism of action and risk assessment. *Mutation Research*, 259, 363–385.

Finnerty, N., O'Riordan, S. L., Klamer, D., Lowry, J., & Palsson, E. (2015). Increased brain nitric oxide levels following ethanol administration. *Nitric Oxide – Biology and Chemistry*, 47, 52–57.

Freeman, T. L., Tuma, D. J., Thiele, G. M., Klassen, L. W., Worrall, S., Niemela, O., ... Preedy, V. R. (2005). Recent advances in alcohol-induced adduct formation. *Alcoholism: Clinical and Experimental Research*, 29, 1310–1316.

Galeone, C., Malerba, S., Rota, M., Bagnardi, V., Negri, E., Scotti, L., ... Pelucchi, C. (2013). A meta-analysis of alcohol consumption and the risk of brain tumours. *Annals of Oncology*, 24(2), 514–523.

Garcia, C. C. M., Angeli, J. P. F., Freitas, F. P., Gomes, O. F., De Oliveira, T. F., Loureiro, A. P. M., ... Medeiros, M. H. (2011). [$^{13}C_2$]-acetaldehyde promotes unequivocal formation of 1,N^2-propano-2′-deoxyguanosine in human cells. *Journal of the American Chemical Society, 133*(24), 9140–9143.

Grosjean, D., Grosjean, E., & Moreira, L. F. (2002). Speciated ambient carbonyls in Rio de Janeiro, Brazil. *Environmental Science and Technology, 36*, 1389–1395.

Hamby-Mason, R., Chen, J. J., Schenker, S., Perez, A., & Henderson, G. I. (1997). Catalase mediates acetaldehyde formation from ethanol in fetal and neonatal rat brain. *Alcoholism: Clinical and Experimental Research, 21*(6), 1063–1072.

Hazen, S. L., Hsu, F. F., d'Avignon, A., & Heinecke, J. W. (1998). Human neutrophils employ myeloperoxidase to convert alpha-amino acids to a battery of reactive aldehydes: a pathway for aldehyde generation at sites of inflammation. *Biochemistry, 37*, 6864–6873.

Hoffman, A. C., & Evans, S. E. (2013). Abuse potential of non-nicotine tobacco smoke components: acetaldehyde, nornicotine, cotinine, and anabasine. *Nicotine and Tobacco Research, 15*(3), 622–632.

Jamal, M., Ameno, K., Ikuo, U., Kumihashi, M., Wang, W., & Ijiri, I. (2007). Ethanol and acetaldehyde: in vivo quantitation and effects on cholinergic function in rat brain. *Novartis Foundation Symposium, 285*, 137–141 discussion 141–144, 198–199.

James, J. T. (1997). Carcinogens in spacecraft air. *Radiation Research, 148*(Suppl. 5), S11–S16.

Karahanian, E., Quintanilla, M. E., Tampier, L., Rivera-Meza, M., Bustamante, D., Gonzalez-Lira, V., ... Israel, Y. (2011). Ethanol as a prodrug: brain metabolism of ethanol mediates its reinforcing effects. *Alcoholism: Clinical and Experimental Research, 35*(4), 606–612.

Kelley, T. J., Sadola, J. R., & Smith, D. L. (1966). Emission rates of formaldehyde and other carbonyls from consumer and industrial products found in California homes. In *Proc. Int. Spec. Conf. Air Waste Manage. Assoc.* (pp. 521–526).

Kirchstetter, T. W., Singer, B. C., & Harly, R. A. (1996). Impact of oxygenated gasoline use on California light-duty vehicle emissions. *Environmental Science and Technology, 30*, 661–670.

Kukielka, E., & Cederbaum, A. I. (1995). Increased oxidation of ethylene glycol to formaldehyde by microsomes after ethanol treatment: role of oxygen radicals and cytochrome P450. *Toxicology Letters, 78*, 9–15.

Lachenmeier, D. W., Gill, J. S., Chick, J., & Rehm, J. (2015). The total margin of exposure of ethanol and acetaldehyde for heavy drinkers consuming cider or vodka. *Food and Chemical Toxicology, 83*, 210–214.

Lane, R. H., & Smathers, J. L. (1991). Monitoring aldehyde production during frying by reversed-phase liquid chromatography. *Journal of Association of Official Analytical Chemists, 74*, 957–960.

Langlais, P. J., Anderson, G., Guo, S. X., & Bondy, S. C. (1997). Increased cerebral free radical production during thiamine deficiency. *Metabolic Brain Disease, 12*, 137–143.

Lazarus, R. C., Buonora, J. E., Jacobowitz, D. M., & Mueller, G. P. (2015). Protein carbonylation after traumatic brain injury: cell specificity, regional susceptibility, and gender differences. *Free Radical Biology and Medicine, 78*, 89–100.

Li, H., Borinskaya, S., Yoshimura, K., Kal'ina, N., Marusin, A., Stepanov, V. A., ... Kidd, K. K. (2009). Refined geographic distribution of the oriental ALDH2* 504Lys (nee 487Lys) variant. *Annals of Human Genetics, 73*(3), 335–345.

Lin, H., Ye, Q., Deng, C., & Zhang, X. (2008). Field analysis of acetaldehyde in mainstream tobacco smoke using solid-phase microextraction and a portable gas chromatograph. *Journal of Chromatography A, 1198–1199*(1–2), 34–37.

Liu, J., Head, E., Gharib, A. M., Yuan, W., Ingersoll, R. T., Hagen, T. M., ... Ames, B. N. (2002). Memory loss in old rats is associated with brain mitochondrial decay and RNA/DNA oxidation: partial reversal by feeding acetyl-L-carnitine and/or R-alpha-lipoic acid. *Proceedings of National Academy of Sciences of the United States of America, 99*, 2356–2361.

Luo, X., Reichetzer, B., Trines, J., Benson, L. N., & Lehotay, D. C. (1999). L-Carnitine attenuates doxorubicin-induced lipid peroxidation in rats. *Free Radical Biology and Medicine, 26*, 1158–1165.

Marnett, L. J., Hurd, H. K., Hollstein, M. C., Levin, D. E., Esterbauer, H., & Ames, B. N. (1985). Naturally occurring carbonyl compounds are mutagens in Salmonella tester strain TA104. *Mutation Research, 148*, 25–34.

Marti-Prats, L., Sanchez-Catalan, M. J., Orrico, A., Zornoza, T., Polache, A., & Granero, L. (2013). Opposite motor responses elicited by ethanol in the posterior VTA: the role of acetaldehyde and the non-metabolized fraction of ethanol. *Neuropharmacology, 72*, 204–214.

Matsuda, T., Terashima, I., Matsumoto, Y., Yabushita, H., Matsui, S., & Shibutani, S. (1999). Effective utilization of N^2-ethyl-2′-deoxyguanosine triphosphate during DNA synthesis catalyzed by mammalian replicative DNA polymerases. *Biochemistry, 38*, 929–935.

Melis, M., Carboni, E., Caboni, P., & Acquas, E. (2015). Key role of salsolinol in ethanol actions on dopamine neuronal activity of the posterior ventral tegmental area. *Addiction Biology, 20*(1), 182–193.

Morgan, P. E., Dean, R. T., & Davies, M. J. (2002). Inactivation of cellular enzymes by carbonyls and protein-bound glycation/glycoxidation products. *Archives of Biochemistry and Biophysics, 403*, 259–269.

Nakamura, K., Iwahashi, K., Furukawa, A., Ameno, K., Kinoshita, H., Ijiri, I., ... Mori, N. (2003). Acetaldehyde adducts in the brain of alcoholics. *Archives of Toxicology, 77*(10), 591–593.

Nakamura, K., Iwahashi, K., Itoh, M., Ameno, K., Ijiri, I., Takeuchi, Y., & Suwaki, H. (2000). Immunohistochemical study on acetaldehyde adducts in alcohol-fed mice. *Alcoholism: Clinical and Experimental Research, 24*(Suppl. 4), 93S–96S.

Nakao, L. S., Kadiiska, M. B., Mason, R. P., Grijalba, M. T., & Augusto, O. (2000). Metabolism of acetaldehyde to methyl and acetyl radicals: in vitro and in vivo electron paramagnetic resonance spin-trapping studies. *Free Radical Biology and Medicine, 29*, 721–729.

Nath, R. G., Ocando, J. E., & Chung, F.-L. (1996). Detection of 1,N^2-propanodeoxyguanosine adducts as potential endogenous DNA lesions in rodent and human tissues. *Cancer Research, 56*, 452–456.

National Research Council (NRC). (2002). *The airliner cabin environment and the health of passengers and crews*. Washington, DC: National Research Council, National Academic Press.

Niemela, O. (2007). Acetaldehyde adducts in circulation. *Novartis Foundation symposium, 285*, 183–192 discussion 193–197.

Niemela, O., Parkkila, S., Koll, M., & Preedy, V. R. (2002). Generation of protein-adducts with malondialdehyde and acetaldehyde in muscles with predominantly type I or type II fibers in rats exposed to ethanol and the acetaldehyde dehydrogenase inhibitor cyanamide. *American Journal of Clinical Nutrition, 76*(3), 668–674.

Niknahad, H., Shuhendler, A., Galati, G., Siraki, A. G., Easson, E., Poon, R., & O'Brien, P. J. (2003). Modulating carbonyl cytotoxicity in intact rat hepatocytes by inhibiting carbonyl metabolizing enzymes. II. Aromatic aldehydes. *Chemico Biological Interaction, 143–144*, 119–128.

Niknahad, H., Siraki, A. G., Shuhendler, A., Khan, S., Teng, S., Galati, G., ... O'Brien, P. J. (2003). Modulating carbonyl cytotoxicity in intact rat hepatocytes by inhibiting carbonylmetabolizing enzymes. I. Aliphatic alkenals. *Chemico Biological Interaction, 143–144*, 107–117.

Nicholls, R. M., Fowles, L. F., Worrall, S., de Jersey, J., & Wilce, P. A. (1994). Distribution and turnover of acetaldehyde-modified proteins in liver and blood of ethanol-fed rats. *Alcohol and Alcoholism, 29*, 149–157.

Oikawa, S., Kobayashi, H., Kitamura, Y., Zhu, H., Obata, K., Minabe, Y., … Yamashima, T. (2014). Proteomic analysis of carbonylated proteins in the monkey substantia nigra after ischemia-reperfusion. *Free Radical Research, 48*(6), 694–705.

Pannunzio, P., Hazell, A. S., Pannunzio, M., Rao, K. V., & Butterworth, R. F. (2000). Thiamine deficiency results in metabolic acidosis and energy failure in cerebellar granule cells: an in vitro model for the study of cell death mechanisms in Wernicke's encephalopathy. *Journal of Neuroscience Research, 62*, 286–292.

Parkhomenko, Y. M., Donchenko, G. V., Chornyi, S. A., Yanchiy, O. R., Strokina, A. O., Stepanenko, S. P., … Pogorelaya, N. K. (2014). Thiamine metabolism in neurons and their vital capacity upon the action of ethanol and acetaldehyde. *Neurophysiology, 46*(1), 1–9.

Picklo, M. J., Sr., Olson, S. J., Hayes, J. D., Markesbery, W. R., & Montine, T. J. (2001). Elevation of AKR7A2 (succinic semialdehyde reductase) in neurodegenerative disease. *Brain Research, 916*, 229–238.

Qi, Z.-Y., Shao, C., Yang, C., Wang, Z., & Hui, G.-Z. (2014). Alcohol consumption and risk of glioma: a meta-analysis of 19 observational studies. *Nutrients, 6*(2), 504–516.

Rajendram, R., Hunter, R., & Preedy, V. R. (2005). Alcohol absorption, metabolism and physiological effects. In B. Caballero, L. Allen, & A. Prentice (Eds.), *Encyclopedia of human nutrition* (2nd ed.). Oxford, UK: Elsevier.

Richards, S., Perry, A., Dodd, R., & Worrall, S. (2012). Increased protein carbonyl content and elevated adduct formation in alcoholic cerebellar degeneration: alcoholism: clinical and experimental research. In *Conference: 2012 International Society for Biomedical Research on Alcoholism world congress, (ISBRA 2012) Sapporo, Japan* (p. 118A).

Rickert, W. S., Robinson, J. C., & Young, J. C. (1980). Estimating the hazards of "less hazardous" cigarettes. I. Tar, nicotine, carbon monoxide, acrolein, hydrogen cyanide, and total aldehyde deliveries of Canadian cigarettes. *Journal of Toxicology and Environmental Health, 6*, 351–365.

Rommer, P., Greilberger, J., Herwig, R., Auff, E., & Leutmezer, F. (2013). Carbonyl proteins as marker of oxidative stress derived protein damage in neuroinflammatory and neurodegenerative diseases. *Journal of the Neurological Sciences*, e366 Conference: 21st World Congress of Neurology, Vienna, Austria.

Sako, M., Inagaki, S., Esaka, Y., & Deyashiki, Y. (2003). Histones accelerate the cyclic 1,N²-propanoguanine adduct-formation of DNA by the primary metabolite of alcohol and carcinogenic crotonaldehyde. *Bioorganic and Medical Chemistry Letters, 13*, 3497–3498.

Shen, L., Chen, C., Yang, A., Chen, Y., Liu, Q., & Ni, J. (2015). Redox proteomics identification of specifically carbonylated proteins in the hippocampi of triple transgenic Alzheimer's disease mice at its earliest pathological stage. *Journal of Proteomics, 123*, 101–113.

Shringarpure, R., Grune, T., & Davies, K. J. (2001). Protein oxidation and 20S proteasome-dependent proteolysis in mammalian cells. *Cellular and Molecular Life Sciences, 58*, 1442–1450.

Singh, R., Sandhu, J., Kaur, B., Juren, T., Steward, W. P., Segerback, D., & Farmer, P. B. (2009). Evaluation of the DNA damaging potential of cannabis cigarette smoke by the determination of acetaldehyde derived N²-ethyl-2′-deoxyguanosine adducts. *Chemical Research in Toxicology, 22*(6), 1181–1188.

Sitte, N., Merker, K., Von Zglinicki, T., Davies, K. J., & Grune, T. (2000). Protein oxidation and degradation during cellular senescence of human BJ fibroblasts: part II–aging of nondividing cells. *The FASEB Journal, 14*, 2503–2510.

Smith, C. J., & Hansch, C. (2000). The relative toxicity of compounds in mainstream cigarette smoke condensate. *Food and Chemical Toxicology, 38*, 637–646.

Sood, C., & O'Brien, P. J. (1993). Molecular mechanisms of chloroacetaldehyde-induced cytotoxicity in isolated rat hepatocytes. *Biochemical Pharmacology, 46*, 1621–1626.

Stadtman, E. R., & Levine, R. L. (2000). Protein oxidation. *Annals New York Academy of Sciences, 899*, 191–208.

Svensson, S., Some, M., Lundsjo, A., Helander, A., Cronholm, T., & Hoog, J. O. (1999). Activities of human alcohol dehydrogenases in the metabolic pathways of ethanol and serotonin. *European Journal of Biochemistry, 262*, 324–329.

Talhout, R., Opperhuizen, A., & van Amsterdam, J. G. C. (2007). Role of acetaldehyde in tobacco smoke addiction. *European Neuropsychopharmacology, 17*(10), 627–636.

Tarragon, E., Balino, P., & Aragon, C. M. G. (2014). Centrally formed acetaldehyde mediates ethanol-induced brain PKA activation. *Neuroscience Letters, 580*, 68–73.

Upadhya, S. C., & Ravindranath, V. (2002). Detection and localization of protein-acetaldehyde adducts in rat brain after chronic ethanol treatment. *Alcoholism: Clinical and Experimental Research, 26*(6), 856–863.

Vaca, C. E., Fang, J. L., & Schweda, E. K. (1995). Studies of the reaction of acetaldehyde with deoxynucleosides. *Chemico Biological Interaction, 98*, 51–67.

Verbenko, A., Malev, A. L., & Zakharova, A. N. (2012). The diagnostic role of carbonyl derivatives, as a marker of oxidative stress in schizophrenia and alcoholic psychiatric disorders. *European Psychiatry* Conference: 20th European Congress of Psychiatry, EPA 2012 Prague Czech Republic.

Wang, M., McIntee, E. J., Cheng, G., Shi, Y., Villalta, W., & Hecht, S. S. (2000). Identification of DNA adducts of acetaldehyde. *Chemical Research in Toxicology, 13*, 1149–1157.

Weschler, C. J., & Shields, H. C. (1997). Potential reactions among indoor pollutants. *Atmospheric Environment, 31*, 3487–3495.

Wieslander, A. P., André, A. H., Nilsson-Thorell, C., Muscalu, N., Kjellstrand, P. T., & Rippe, B. (1995). Are aldehydes in heat-sterilized peritoneal dialysis fluids toxic in vitro? *Peritoneal Dialysis International, 15*, 348–352.

Worrall, S., & Thiele, G. M. (2001). Protein modification in ethanol toxicity. *Adverse Drug Reactions and Toxicology Reviews, 20*, 133–159.

Yan, L. J., & Sohal, R. S. (1998). Mitochondrial adenine nucleotide translocase is modified oxidatively during aging. *Proceedings of the National Academy of Sciences of the United States of America, 95*, 12896–12901.

Yukawa, Y., Ohashi, S., Amanuma, Y., Nakai, Y., Tsurumaki, M., Kikuchi, O., Miyamoto, S., … Muto, M. (2014). Impairment of aldehyde dehydrogenase 2 increases accumulation of acetaldehyde-derived DNA damage in the esophagus after ethanol ingestion. *American Journal of Cancer Research, 4*(3), 279–284.

Section C

Structural and Functional Aspects

Chapter 52

Cortical Morphology in Fetal Alcohol Spectrum Disorders: Insights from Computational Neuroimaging

François De Guio[1], Ernesta Meintjes[2], Jean-François Mangin[3], David Germanaud[4,5,6]

[1]Université Paris Diderot, Sorbonne Paris Cité, UMR-S 1161 INSERM, Paris, France; [2]MRC/UCT Medical Imaging Research Unit, Faculty of Health Sciences, University of Cape Town, Cape Town, South Africa; [3]CEA, NeuroSpin Center, UNATI, Gif-sur-Yvette, France; [4]AP-HP, Robert-Debré Hospital, DHU PROTECT "Department of Child Neurology", Paris, France; [5]CEA, NeuroSpin Center, UNIACT, Gif-sur-Yvette, France; [6]INSERM, UMR1129, Paris, France

INTRODUCTION

Prenatal alcohol exposure is a leading cause of neurodevelopmental disorders. Both brain growth and structural organization can be strongly affected (Guerri, Bazinet, & Riley, 2009), resulting in fetal alcohol spectrum disorders (FASD). Numerous neuroimaging studies using in particular structural magnetic resonance imaging (MRI) have been undertaken to describe the morphological alterations in the exposed brain and its consequences on behavior or cognition (Moore, Migliorini, Infante, & Riley, 2014).

These studies notably raise two important questions. The first concerns the diagnosis of certain clinical forms of FASD. Indeed, in the most typical and severe form, which is the fetal alcohol syndrome (FAS), the clinical phenotype is enough to ascertain the etiological diagnosis (Astley, 2013). However, in other forms called alcohol-related neurodevelopmental disorders (ARND), the lack of a specific element gives a more probabilistic value to the link between alcohol exposure and disorder, to the point at which less explicit names have been proposed, such as static encephalopathy or neurobehavioral disorder in the context of alcohol exposure (Astley & Clarren, 2000). Conventional radiology has not been shown to be accurate enough to improve diagnosis specificity. By providing new specific phenotypic elements (or combination of elements) otherwise hidden to the physician, MRI-based computational morphology of the brain may turn out to be very valuable for FASD diagnosis. The other key question concerns the understanding of the underlying pathophysiology. The search for functional correlates of anatomical MRI markers, and the other way round, would certainly help in the understanding of FASD-related disabilities.

T1-weighted millimetric three-dimensional (3D) MRI provides good tissue contrast and high spatial resolution allowing for accurate rendering of the brain anatomy. Many structural parameters of the brain can be measured through more or less complex computational treatments of this type of neuroimaging data. Be they measured at the whole-brain level or at a more local one, those parameters can be roughly classified into two categories, size-related (quantity) and shape-related (geometry), without precluding an interaction between size and shape. Thus, whole-brain volume has been found redundantly reduced in children or adults with FASD (Lebel, Roussotte, & Sowell, 2011). Tissue-specific volumetry has also been used in various cohorts, for instance, to assess longitudinal cortical volume change in children and young adults with FASD (Lebel et al., 2012) or to follow the development of white matter volume in relation with cognitive functions (Gautam, Nunez, Narr, Kan, & Sowell, 2014). As well, region-specific volumetry of deep gray matter structures such as the hippocampus (Willoughby, Sheard, Nash, & Rovet, 2008) or the caudate nucleus (Fryer et al., 2012) has been conducted in relation to cognitive performance (Nardelli, Lebel, Rasmussen, Andrew, & Beaulieu, 2011; Roussotte et al., 2012). Otherwise, concurrent methods not assessing directly the volume of a given region but rather estimating a nonlinear deformation to match structures of individual subjects to a specific atlas (voxel-based morphometry, tensor-based morphometry) have been applied to look for local volumetric excesses and deficits (Meintjes et al., 2014; Sowell et al., 2010; Sowell, Thompson, et al., 2001). Considering now shape beyond volume variations, brain shape abnormalities and asymmetries have notably been examined through surface-based analysis in which surface anatomies across individuals were matched by sulcal landmarks (Sowell, Thompson, Mattson, et al., 2002; Sowell, Thompson, Peterson, et al., 2002). The shape of the corpus callosum (Bookstein, Streissguth, Sampson, Connor, & Barr, 2002; Sowell, Mattson, et al., 2001) or the hippocampal or the caudate nucleus has also been examined with surface deformation-based analysis (Joseph et al., 2014).

In addition to whole-brain and subcortical structure parameters, the cortex is by itself a good proxy for the developmental history of the brain and some functional characteristics (Dubois et al., 2008; Mangin, Jouvent, & Cachia, 2010). Indeed, the characterization of the amount (or dimensions) and geometry of the cortex in neurological disorders by means of computational neuroimaging has proved to be a valuable tool (Mangin et al., 2010). Therefore, the aims of this chapter are (1) to briefly explain the methods and concepts behind quantitative cortical morphology (morphometry), encompassing here cortical thickness and cortical shape analysis; (2) to gather and present the results in the context of FASD; and (3) to gauge the capacity of these neuroimaging tools to provide new biomarkers improving both the diagnosis of nonsyndromal forms and the understanding of the pathophysiology in FASD.

WHAT IS CORTICAL MORPHOLOGY?

Concepts

In this chapter, we propose to distinguish between two ways of quantitative investigations of cortical morphology from structural MRI: (1) *corticometry*, which is the measurement of the radiological amount of cortex through cortical thickness and cortical surface area, these parameters providing complementary information (Winkler et al., 2012); (2) *cortical morphometry*, which is the study of the cortical geometry and folding shape revealing folding intensity, folding pattern, fold morphotypes, and their variability. Both corticometry and cortical morphometry rely on the reconstruction of cortical surfaces from structural MR images, typically 3D T1-weighted images (Figure 1). Indeed, the starting point is to generate accurate models of both the gray/white interface and the pial (gray/cerebrospinal fluid (CSF)) surfaces from the volumetric segmentation of white matter and cortical gray matter. We will focus on methods and tools that have been used in the field of FASD.

Corticometry

Several computational tools that are broadly used and validated allow one to estimate the cortical thickness and the extension of the cortical layer, such as FreeSurfer (http://surfer.nmr.mgh.harvard.edu) (Fischl & Dale, 2000) or CIVET (http://mcin-cnim.ca/neuroimagingtechnologies/civet/) (Lerch & Evans, 2005). Basically, the FreeSurfer pipeline first delineates a unitary white matter volume, which is used to tessellate an initial gray–white surface. This surface is then optimized and further expanded to the gray–CSF interface to finally get a 3D mesh model of both the inner and the outer cortical surfaces, composed of about 150,000 paired points. The distance between these two surfaces gives the thickness of the cortical gray matter at any point (Figure 2). The measurement of the gray/white interface area or any derived surface gives the cortical surface extension, either globally or regionally, after atlas-based segmentation of the brain, for instance. Used as a "black box" tool, it takes a 3D T1-weighted image as an input and releases a large set of measurements (e.g., global and atlas-based volumes, surface area, and thickness). Other methods have been developed to measure the cortical thickness, such as

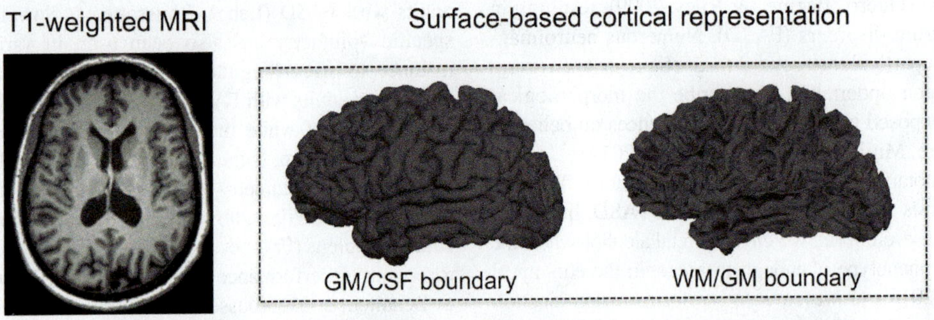

FIGURE 1 Surface-based cortical representation from high-resolution MRI. T1-weighted MRI is the typical anatomical input image for the various image-processing techniques developed for cortical morphology assessment. Two surfaces can be extracted to model the cortical surface: the interface between gray matter (GM) and cerebrospinal fluid (CSF) and the interface between white matter (WM) and GM.

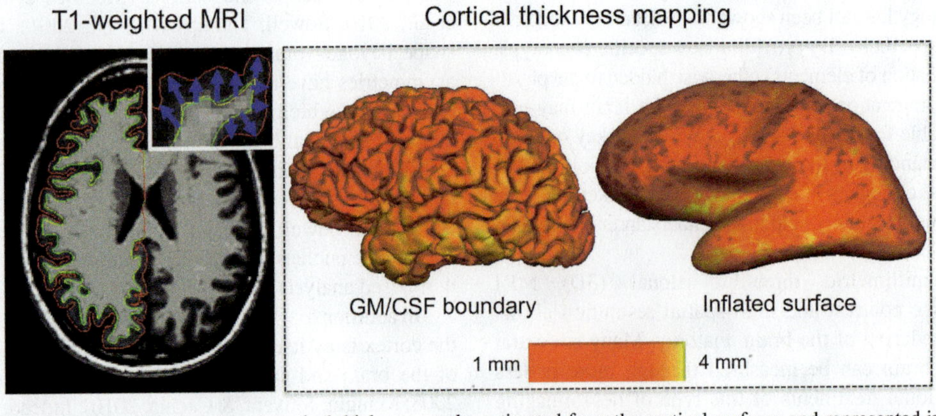

FIGURE 2 Cortical thickness estimation. Local cortical thickness can be estimated from the cortical surfaces and represented in the native space of the subject, or on an inflated surface as shown, or in a common template for group analyses.

voxel-based methods (Hutton, De Vita, Ashburner, Deichmann, & Turner, 2008) or methods dedicated to high-resolution images acquired with recent MR technology (Lusebrink, Wollrab, & Speck, 2013).

Cortical Morphometry

Surface-Based Morphometry

Cortical folding intensity reflects the relative amount of cortical surface buried in the folds and is measured through gyrification indices. Basically, it compares the observed cortical surface area to the one of a smoothed brain of the same bulk volume through a ratio that may be either locally or globally defined (Mangin et al., 2010). Historically, the gyrification index has been computed on 2D slices (Zilles, Armstrong, Schleicher, & Kretschmann, 1988). There are many further implementations of the concept of gyrification indices. Global and local gyrification indices have been transposed to three dimensions using surface-based models (Schaer et al., 2008) and consist, for instance, in measuring locally the extent of a cortical surface enclosed in a small sphere (Figure 3(C)) (Toro et al., 2008). A similar characterization of the surface may also be obtained through the local integration of the cortical curvature (Luders et al., 2006).

These measures of intensity are first estimates of the folding geometry but are rather blind to the way the cortical surface is buried, meaning the folding pattern and the shape of each fold. To better assess the geometry of the cortex, it is possible to map on the surface scalar functions that describe characteristics that are defined at each point and are good proxies of the geometric variations. A broadly used function is the cortical curvature (Figure 3(A)) (Luders et al., 2006). Hence, the folding geometry can be studied through the analysis of the variations of these functions. For all these local indices, comparison and group analyses are coordinate-based and performed after spatial normalization.

Apart from the direct analyses on the surface, various spectral approaches have been proposed to explore the spatial frequency compound of a folded surface (Chung, Hartley, Dalton, & Davidson, 2008; Seo & Chung, 2011). Recently, a spectral Fourier-like analysis of cortical curvature using the Laplace-Beltrami operator has been described (Figure 3(D)) (Germanaud et al., 2012). The computation is directly performed on native cortical meshes and achieves a quantitative analysis of the frequency compound of the cortical folding pattern, both in the frequency domain (band spectrum of the curvature) and in the image (spectral segmentation of the cortical folding pattern). By the same token, other new strategies to assess folding complexity have been developed, such as fractal analysis (Goni et al., 2013).

Object-Based Morphometry

Apart from these strictly surface-based analyses, object-based strategies have been developed that rely on computation from the cortical surface of models of the folds (Mangin et al., 2004a,b). Indeed, an alternative to coordinate-based analysis using spatial normalization is to apply a pattern recognition system to extract local shape-based features. Thus, mathematical models of sulci have been proposed as proxies of cortical folding (Im, Lee, Won Seo, et al., 2008; Mangin et al., 2004a,b) in a way that a 3D object that can be both identified and measured to get its geometric characteristics represents each individual sulcus. Basically, the Morphologist pipeline of the BrainVisa software (http://brainvisa.info) enables the automatic segmentation of the brain sulci by molding

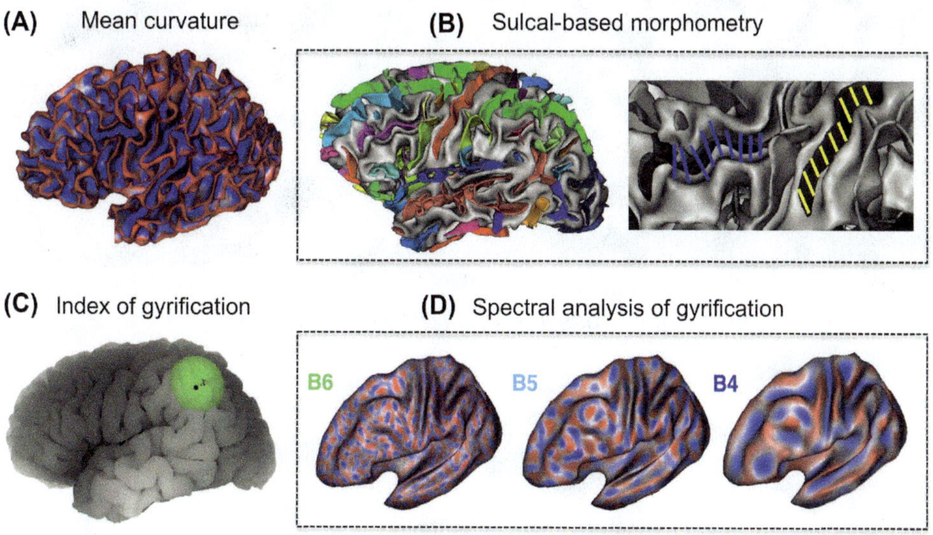

FIGURE 3 Computational tools for cortical morphometry. (A) Computation of the mean curvature mapped onto the WM/GM surface (gyri with positive curvatures (red) and sulci with negative curvatures (blue)). (B) In sulcal-based morphometry, sulci are automatically reconstructed and identified, enabling the measurement of associated features for each sulcus such as length, depth (represented in yellow on the right image), and opening (represented in blue). (C) Index of gyrification consists, for example, in measuring a surface ratio to quantify the local degree of folding from the surface-based representation. (D) Using a spectral-based analysis of gyrification, one can infer the contribution of each spectral band (e.g., B4–B6) on the local curvature (A) of the cortical surface.

of their "footprint" on the gray/white interface and their further recognition by a virtual anatomist algorithm. The basic features related to each individual sulcus, such as length, area, depth, and fold opening (the distance between the two gyral banks), are then estimated (Figure 3(B)) (Mangin et al., 2004a,b). More sophisticated features quantifying the complete spectrum of 3D shape variation have also been proposed (Sun et al., 2012).

CORTICAL MORPHOLOGY IN FETAL ALCOHOL SPECTRUM DISORDERS

Corticometry: Interesting Discrepant Results

In 1986, long before MRI studies in subjects with FASD, it was reported that ethanol-treated rats presented with a reduced cortical thickness by comparison to age-matched controls (Lopez-Tejero, Ferrer, M, & Herrera, 1986). Similarly, in 2013, Leigland and colleagues found a reduced cortical thickness and surface area using ex vivo MRI in a rat model of prenatal ethanol exposure (Leigland, Ford, Lerch, & Kroenke, 2013).

Sowell et al. (2008) carried on the first study of cortical thickness in human subjects with FASD in 2008 using an original method for cortical thickness assessment on T1-weighted MRI. Twenty-one subjects with heavy prenatal exposure (14 of which had the FAS characteristic facial appearance) were enrolled (mean age 12.6 years, range 8–22 years). In comparison to 21 control subjects, they found significant cortical excesses of up to 1.2 mm in temporal, parietal, and frontal regions. However, a couple of years later, Zhou et al. (2011) reported no cortical thickness increases in 33 participants (mean age 12.3 years, range 6–30) with FASD (only three with full FAS) versus 33 age- and sex-matched controls (Figure 4). In contrast, FASD presented with cortical thinning in several lobes. Meanwhile, another group assessed a greater cortical thickness in 20 individuals (mean age 10.9 years) with FAS and diagnosed with attention deficit hyperactivity disorder in comparison to individuals with attention deficit hyperactivity disorder but without FAS or in comparison to a control group (Fernandez-Jaen et al., 2011). Similarly, Yang et al. found a thicker cortex in 69 children and adolescents with FASD from three different sites and mixed diagnosis (mean age 13.4 years, range 8–16) versus 58 nonexposed controls (Figure 4) (Yang et al., 2012). In 2014, there was the first paper measuring cortical thickness longitudinally in FASD, but unfortunately in a small number of patients (Treit et al., 2014). Cortical thickness trajectories were compared between 11 children with FASD (mean age at first scan 8.9 years, range 6–12 years) and 21 controls undergoing two MRI scans 2–4 years apart. The FASD group had significantly lower mean cortical thickness at scan 1 but not at scan 2. Interestingly, FASD subjects showed less developmental thinning than controls in several brain regions. Finally, in a 2014 study, cortical volume, cortical thickness, and cortical surface area were computed in 36 participants diagnosed with ARND (mean age 11.4 years, range 8–15) and compared to 52 controls (Rajaprakash, Chakravarty, Lerch, & Rovet, 2014). They found no difference in cortical thickness but rather a decreased surface area, explaining the cortical volume deficit in alcohol-exposed subjects.

Overall, from the six transversal studies found in the literature on cortical thickness in FASD, three demonstrated greater cortical

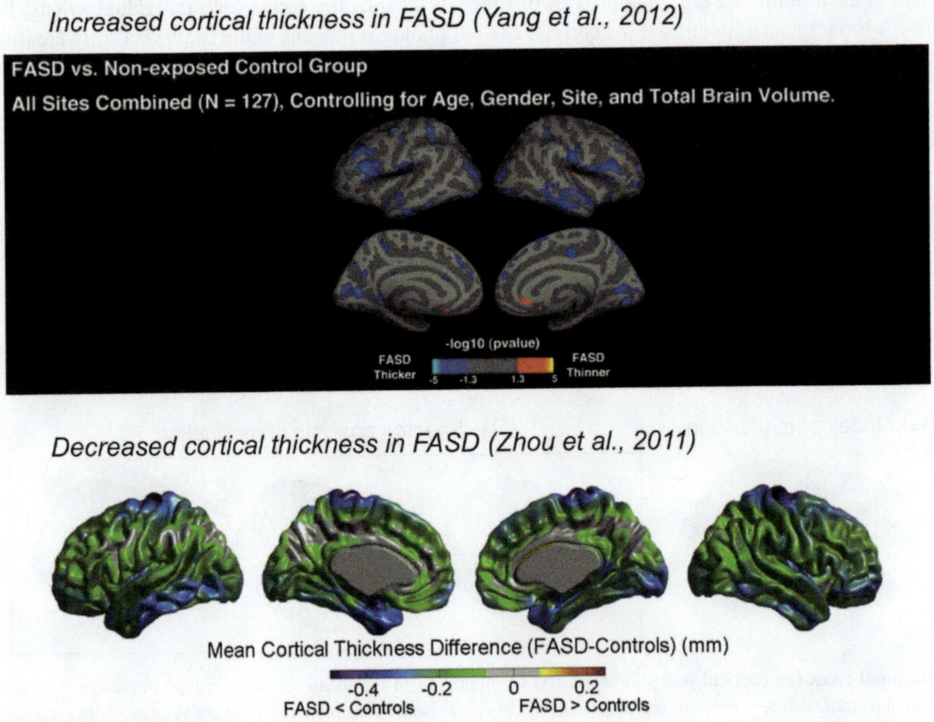

FIGURE 4 **Opposite results in cortical thickness analyses in FASD.** Top: Representation of a thicker cortical thickness in subjects with FASD ($n=69$, mean age ± SD 13.4 ± 1.9 years) compared to nonexposed subjects ($n=58$, mean age ± SD 13.0 ± 2.0 years) (Yang et al., 2012). Bottom: Illustration of the widespread cortical thinning observed in FASD children ($n=33$, mean age ± SD 12.3 ± 6.0) compared to controls ($n=33$, mean age ± SD 12.7 ± 6.0) (Zhou et al., 2011).

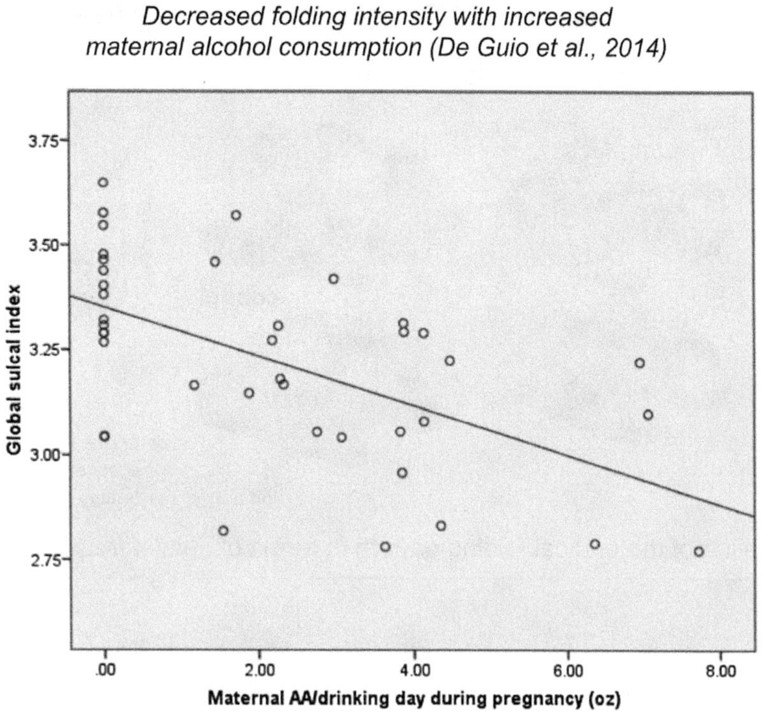

Decreased folding intensity with increased maternal alcohol consumption (De Guio et al., 2014)

FIGURE 5 Dose-dependent relationship between prenatal alcohol exposure and folding intensity. The g-SI as a function of maternal alcohol consumption expressed in absolute alcohol (AA) per drinking day during pregnancy (De Guio et al., 2014). The g-SI was defined as the percentage ratio between the total sulcal area and the outer cortex area and was negatively correlated with alcohol intake.

thickness, two suggested a decreased cortical thickness, and one established no difference. There may be some reasons for those opposite results, such as differences in the sample compositions (large proportion of FAS subjects for thickening results, mostly FASD or ARND subjects for conclusions of cortical thinning or absence of alterations) or various age ranges, while mean ages were rather similar between studies. Among other confounding factors may be the variation in global brain volume, since its consistent reduction in FASD may result in nonstrictly proportional variations in cortical thickness (Im, Lee, Lyttelton, et al., 2008). Ultimately, these small group studies remain prone to sampling bias. Nevertheless, the study of Treit et al. may bring an interesting explanation to those discrepant results, as it shows that the developmental trajectory itself is impaired (Treit et al., 2014). One can imagine that at an early time point in childhood cortical thickness is rather similar between groups and, later in adolescence, owing to an impaired cortical thinning, FASD subjects present with greater cortical thickness.

An Altered Cortical Morphometry?

Apart from corticometry, i.e., the measurement of cortical thickness or surface area, very few studies have assessed the cortical morphology using other methods. In 2014, sulcal-based morphometry was applied in 40 nine-year-old children (9 diagnosed with FAS, 15 heavily exposed (HE), and 16 controls) for whom maternal alcohol consumption was estimated (De Guio et al., 2014). At the whole-brain level, they reported a lower global sulcal index (g-SI; a gyrification index defined as the ratio between the total sulcal area and the outer cortex area) in the FAS group compared to the control group. This result is not surprising given that, even in the typically developing population, smaller brains are disproportionately less folded than larger ones because of the allometric scaling of cortical surface (Germanaud et al., 2012; Im, Lee, Lyttelton, et al., 2008; Toro et al., 2008). But interestingly, the HE group presented also with a significantly lower g-SI compared to controls, while both groups did not differ in brain or tissue volumes. In addition, there was a dose-dependent relationship between prenatal alcohol exposure and g-SI (Figure 5). Considering features extracted from sulcal-based morphometry, regional sulcal index and sulcal depth were not relevant, but fold opening (the mean distance between the two gyral banks of a sulcus) was found to be highly sensitive to prenatal alcohol exposure. There was an increased magnitude of fold opening with increased alcohol exposure that was significantly correlated in almost all regions of the brain. Results on cortical surface area (Rajaprakash et al., 2014) are in line with these observations, as a reduced cortical surface area is in agreement with a reduced g-SI. These results together describe the cortical folding pattern in alcohol-exposed children as less complex (wide folds, less total surface area, less sulcal surface area, or fewer folds).

As cortical morphometry is strongly dependent on brain volume (Germanaud et al., 2012) and given that brain volume reduction is a key feature of FASD (Lebel et al., 2011), the question of controlling for brain volume in such studies is challenging. A 2014 study combining spatial (conventional) and spectral analysis of gyrification with allometric modeling notably addresses this question by distinguishing even (nonspecific, related to brain size reduction) and uneven gyral simplifications (Germanaud et al., 2014). This distinction relied on the comparison of the

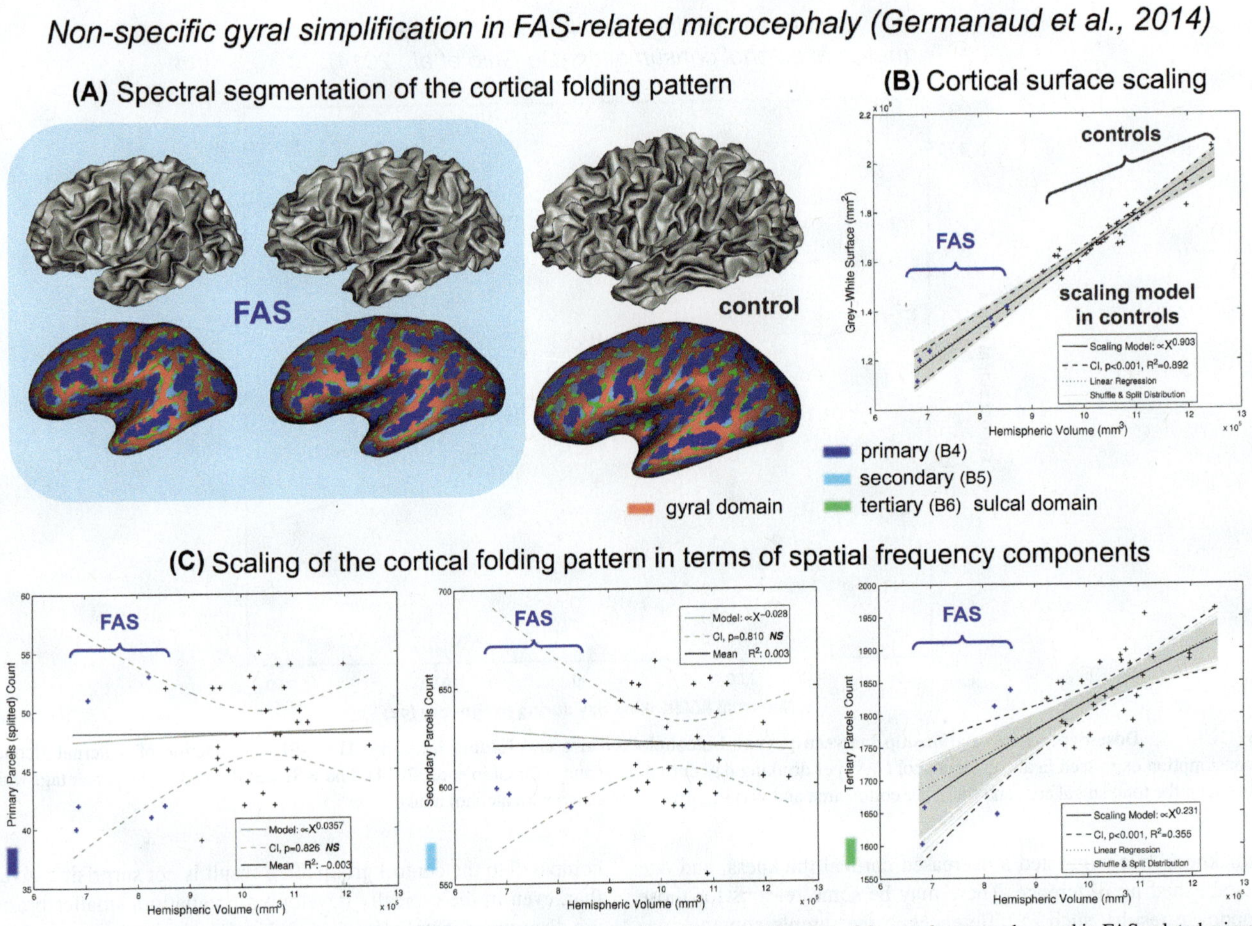

Non-specific gyral simplification in FAS-related microcephaly (Germanaud et al., 2014)

(A) Spectral segmentation of the cortical folding pattern

(B) Cortical surface scaling

■ primary (B4)
■ secondary (B5)
■ gyral domain   ■ tertiary (B6) sulcal domain

(C) Scaling of the cortical folding pattern in terms of spatial frequency components

FIGURE 6 **Nonspecific simplified gyral pattern in FAS-related microcephaly.** The simplification of the gyral pattern observed in FAS-related microcephaly seems nonspecific, i.e., totally explained by brain size reduction. (A) Evolution of cortical folding complexity along microcephalic patients and controls. Gray–white interface mesh (top) and spectral segmentation of folding pattern presented on totally smoothed mesh (bottom). Two FAS-related microcephalies (12 years, male, 681-ml brain; 15 years, male, 826-ml brain) and one control (13 years, male, 1070-ml brain). (B) Cortical surface area scaling in FAS-related microcephaly is as expected according to the scaling model fitted on controls. (C) Scaling of the cortical folding pattern in terms of spatial frequency components is as expected according to the scaling models fitted on controls. The spatial frequency components are the primary (low frequency), secondary (medium frequency), and tertiary (high frequency) folds given by the spectral segmentation of the cortical folding pattern. *Adapted from Germanaud et al. (2014).*

folding complexity in three different developmental diseases causing severe microcephaly, among which two were of genetic origin and the third was FAS (Germanaud et al., 2014). The FAS group consisted of six subjects with severe microcephaly (mean age 10.3 years, range 7–15) and was compared to a group of 30 controls with the same age range. While in the two genetic diseases a small part of the gyral simplification remained unexplained by the reduction of brain volume, FAS patient brains showed no significant deviation from the cortical surface area, gyrification index, and spectral characteristic of folding pattern expected for their brain volume (Figure 6). Although this result has to be taken cautiously because of a small sample size, it has to be kept in mind while looking for specific features in FASD cortical morphology without addressing the allometric scaling of the cortical layer.

Other studies are needed to examine the relevance of all these cortical markers in various groups of alcohol-exposed subjects, with different types of FASD and of different ages. Indeed, spanning the whole developmental range of age is important since there is a strong effect of age on myelin development (Dean et al., 2014) that may be influenced by a long-lasting effect of fetal alcohol exposure (Cao et al., 2014). Differences in myelination would not have to be mistaken for primary differences in cortical morphology as it relies on the segmentation of gray/white interface.

Relation to Diagnosis, Pathophysiology, and Perspectives

The total number of studies aimed at characterizing the cortical morphology in FAS and FASD is rather small. This may be due to a relative lack of interest in the scientific community to the matter of fetal alcohol exposure by comparison to other neurodevelopmental disorders (autistic spectrum disorders, attention deficit hyperactivity disorder, etc.), despite the heavy burden it represents in terms of cognitive and behavioral disability. Yet, recruiting large groups of FASD-affected participants for neuroimaging studies is indeed difficult, especially if one takes into account the peculiar context of fetal alcohol exposure. In addition, if corticometry is a standard in the field of neurodevelopmental disorders studies, cortical morphology is much less in use, even if cortical geometry can be a good record of the developmental history of the brain (Dubois et al., 2008; Mangin et al., 2010).

In the case of full FAS, clinical features have proved to be sufficient to ascertain the etiological diagnosis underlying the syndromic one, that is to say, the link between fetal alcohol exposure and symptoms (Astley, 2013). But in nonsyndromic FASD, often called ARND, the medical community lacks consistent phenotypic features that may help with the diagnosis and reinforce the probabilistic link between fetal alcohol exposure and symptoms. The search for a specific cognitive or behavioral phenotype has not been so far successful (Moore et al., 2014). As for a specific neuroradiologic phenotype, no attempt of diagnostic categorization or clustering has been published yet, neither with anatomical features nor with functional ones, while several reports exist in the field of autism spectrum disorders or schizophrenia, for instance (Ecker et al., 2010; Ingalhalikar et al., 2012). There are many difficulties in fulfilling this objective, such as the absence or imprecision of data reporting maternal alcohol consumption during pregnancy (quantities and periods of consumption during pregnancy) (Muggli, Cook, O'Leary, Forster, & Halliday, 2014); the large genetic variability of both the mother and the alcohol-exposed fetus, also explaining the variability of the teratogenic effects of alcohol (McCarthy & Eberhart, 2014); and the co-occurrence of other factors (other drugs or substance) possibly influencing brain morphology (see next section). Nevertheless, despite the limitations due to the small number of reported studies and their small sample size, it seems that cortical morphology could help in identifying abnormal neuroanatomical features in this population (De Guio et al., 2014; Rajaprakash et al., 2014; Zhou et al., 2011) to be tested as biomarkers in the search for such a compound neuroradiological-specific phenotype.

Apart from the diagnostic use, cortical biomarkers of FASD may be useful in the search for anatomical correlates of the pathophysiological process of fetal alcohol teratogenicity and the understanding of the observed cognitive dysfunctions. Recent results on cortical thickness and surface area or sulcal-based morphometry abnormalities, while having a low repeatability as of this writing, are encouraging and complementary. Indeed, measures of cortical thickness and cortical surface area are thought to reflect different processes occurring at different times in view of the radial-unit hypothesis of cortical development (Rakic, 1995). It has been shown in a large number of longitudinal MRI scans that cortical thickness and surface area provide independent information about the development of the cortex (Raznahan et al., 2011). A rough causal explanation of the impaired cortical morphology is the alcohol-responsible failure in the normal migration of radial cells. Minor changes in relative production of progenitors and neurons could produce dramatic changes in cortical surface area (Rakic, 2005) and could lead to a decreased number of radial columns and thus a reduced expansion of the cerebral cortex (Chenn & Walsh, 2002). On a different pathophysiological pathway, fetal alcohol exposure has been proven toxic not only for early proliferative and migratory processes but also for glial cell differentiation and maturation, also involved in cortical extension.

In conclusion, while numbers of fine computational tools exist to quantify in vivo the cortical morphology, few studies have searched for these promising structural markers in FASD. Future studies are needed to replicate or deepen the findings described in this chapter. In particular, studies with large samples, well-defined groups according to the various diagnoses under the FASD umbrella, and longitudinal design using MRI, from fetuses and neonates to adults, will help to better clarify the teratogenic effects of prenatal alcohol intake on brain cortical morphology.

APPLICATIONS TO OTHER ADDICTIONS AND SUBSTANCE MISUSE

There is no reason for which the described markers of cortical morphology would not be relevant for other addictions and substance misuse having proven neurodevelopmental toxicity or teratogenicity. For example, studies in humans have demonstrated an association between cortical thickness and the striatal dopamine response to drugs of abuse (Casey et al., 2013), heroin exposure (Li et al., 2014), or alcohol dependence (Momenan et al., 2012). Cortical thickness and gray matter volumetry methodologies have been applied to subjects prenatally exposed to tobacco (El Marroun et al., 2014), cocaine (Grewen et al., 2014), or opiates (Walhovd et al., 2007). It is worth noting that the main abnormalities in these three studies of prenatal substance exposure resemble those described for FASD: reduction of brain volume and thinner cortices. This highlights the importance of controlling for confounding substance intake when focusing on the effect of a single substance, properly modelling size effects in the analysis, and perhaps including other disorders impairing brain growth in the comparison groups.

DEFINITION OF TERMS

Cortical morphology This is the study of the structure and shape of the cortex, including the corticometry (measurement of the cortical thickness or surface area) and the cortical morphometry (study of the cortical geometry and folding shape, revealing folding intensity, folding pattern, fold morphotypes, and their variability).

Sulcal-based morphometry This is the extraction of a model of sulci from the cortical surface representation and the measurement of associated features (length, depth, opening, etc.).

Spectral analysis of gyrification This is the spectral decomposition of a proxy of cortical folding (e.g., curvature) of the cortical surface representation to study its spatial frequency compound.

KEY FACTS OF FETAL ALCOHOL SYNDROME DISORDERS

- FAS is the most common cause of preventable mental retardation.
- FAS is characterized by a distinctive pattern of craniofacial features, small head circumference (or other structural brain impairment), and growth retardation.
- The prevalence of FASD is estimated to be 2–5% of school-age children in the United States and Western Europe.
- There is no cure for FASD.

SUMMARY POINTS

- This chapter focuses on the study of cortical morphology in FASD.
- Various reliable markers of cortical morphology can be computed from high-resolution MRI and provide independent and complementary information.
- Overall, few studies have assessed the structural abnormalities in participants exposed prenatally to alcohol through the use of these markers.
- Cortical thickness alterations in FASD are challenging to explain given the discrepant results and its trajectories during development.
- Cortical morphometry reveals altered cortical folding in children with FASD without microcephaly with less cortical surface area, lower folding complexity, and wider folds.
- Future studies of cortical morphology are needed in large and well-defined diagnosed groups to repeat preliminary results and deepen the understanding of the effects of maternal alcohol consumption on the brain structure.

REFERENCES

Astley, S. J. (2013). Validation of the fetal alcohol spectrum disorder (FASD) 4-Digit Diagnostic Code. *Journal of Popular Therapeutics and Clinical Pharmacology*, 20(3), e416–467.

Astley, S. J., & Clarren, S. K. (2000). Diagnosing the full spectrum of fetal alcohol-exposed individuals: introducing the 4-digit diagnostic code. *Alcohol and Alcoholism*, 35(4), 400–410.

Bookstein, F. L., Streissguth, A. P., Sampson, P. D., Connor, P. D., & Barr, H. M. (2002). Corpus callosum shape and neuropsychological deficits in adult males with heavy fetal alcohol exposure. *NeuroImage*, 15(1), 233–251. http://dx.doi.org/10.1006/nimg.2001.0977.

Cao, W., Li, W., Han, H., O'Leary-Moore, S. K., Sulik, K. K., Allan Johnson, G., & Liu, C. (2014). Prenatal alcohol exposure reduces magnetic susceptibility contrast and anisotropy in the white matter of mouse brains. *NeuroImage*, 102(Pt 2), 748–755. http://dx.doi.org/10.1016/j.neuroimage.2014.08.035.

Casey, K. F., Cherkasova, M. V., Larcher, K., Evans, A. C., Baker, G. B., Dagher, A., ... Leyton, M. (2013). Individual differences in frontal cortical thickness correlate with the d-amphetamine-induced striatal dopamine response in humans. *Journal of Neuroscience*, 33(38), 15285–15294. http://dx.doi.org/10.1523/jneurosci.5029-12.2013.

Chenn, A., & Walsh, C. A. (2002). Regulation of cerebral cortical size by control of cell cycle exit in neural precursors. *Science*, 297(5580), 365–369. http://dx.doi.org/10.1126/science.1074192.

Chung, M. K., Hartley, R., Dalton, K. M., & Davidson, R. J. (2008). Encoding cortical surface by spherical harmonics. *Statistica Sinica*, 18(4), 1269–1291.

De Guio, F., Mangin, J. F., Riviere, D., Perrot, M., Molteno, C. D., Jacobson, S. W., ... Jacobson, J. L. (2014). A study of cortical morphology in children with fetal alcohol spectrum disorders. *Human Brain Mapping*, 35(5), 2285–2296. http://dx.doi.org/10.1002/hbm.22327.

Dean, D. C., 3rd, O'Muircheartaigh, J., Dirks, H., Waskiewicz, N., Lehman, K., Walker, L., ... Deoni, S. C. (2014). Modeling healthy male white matter and myelin development: 3 through 60 months of age. *NeuroImage*, 84, 742–752. http://dx.doi.org/10.1016/j.neuroimage.2013.09.058.

Dubois, J., Benders, M., Borradori-Tolsa, C., Cachia, A., Lazeyras, F., Ha-Vinh Leuchter, R., ... Hüppi, P. S. (2008). Primary cortical folding in the human newborn: an early marker of later functional development. *Brain*, 131(Pt 8), 2028–2041. http://dx.doi.org/10.1093/brain/awn137.

Ecker, C., Rocha-Rego, V., Johnston, P., Mourao-Miranda, J., Marquand, A., Daly, E. M., ... Murphy, D. G. (2010). Investigating the predictive value of whole-brain structural MR scans in autism: a pattern classification approach. *NeuroImage*, 49(1), 44–56. http://dx.doi.org/10.1016/j.neuroimage.2009.08.024.

El Marroun, H., Schmidt, M. N., Franken, I. H., Jaddoe, V. W., Hofman, A., van der Lugt, A., ... White, T. (2014). Prenatal tobacco exposure and brain morphology: a prospective study in young children. *Neuropsychopharmacology*, 39(4), 792–800. http://dx.doi.org/10.1038/npp.2013.273.

Fernandez-Jaen, A., Fernandez-Mayoralas, D. M., Quinones Tapia, D., Calleja-Perez, B., Garcia-Segura, J. M., Arribas, S. L., & Munoz Jareno, N. (2011). Cortical thickness in fetal alcohol syndrome and attention deficit disorder. *Pediatric Neurology*, 45(6), 387–391. http://dx.doi.org/10.1016/j.pediatrneurol.2011.09.004.

Fischl, B., & Dale, A. M. (2000). Measuring the thickness of the human cerebral cortex from magnetic resonance images. *Proceedings of the National Academy of Sciences of the United States of America*, 97(20), 11050–11055. http://dx.doi.org/10.1073/pnas.200033797.

Fryer, S. L., Mattson, S. N., Jernigan, T. L., Archibald, S. L., Jones, K. L., & Riley, E. P. (2012). Caudate volume predicts neurocognitive performance in youth with heavy prenatal alcohol exposure. *Alcoholism, Clinical and Experimental Research*, 36(11), 1932–1941. http://dx.doi.org/10.1111/j.1530-0277.2012.01811.x.

Gautam, P., Nunez, S. C., Narr, K. L., Kan, E. C., & Sowell, E. R. (2014). Effects of prenatal alcohol exposure on the development of white matter volume and change in executive function. *Neuroimage: Clinical*, 5, 19–27. http://dx.doi.org/10.1016/j.nicl.2014.05.010.

Germanaud, D., Lefevre, J., Fischer, C., Bintner, M., Curie, A., des Portes, V., ... Hertz-Pannier, L. (2014). Simplified gyral pattern in severe developmental microcephalies? New insights from allometric modeling for spatial and spectral analysis of gyrification. *NeuroImage*, 102(P2), 317–331. http://dx.doi.org/10.1016/j.neuroimage.2014.07.057.

Germanaud, D., Lefevre, J., Toro, R., Fischer, C., Dubois, J., Hertz-Pannier, L., & Mangin, J. (2012). Larger is twistier: spectral analysis of gyrification (SPANGY) applied to adult brain size polymorphism. *NeuroImage*, 63(3), 1257–1272. http://dx.doi.org/10.1016/j.neuroimage.2012.07.053.

Goni, J., Sporns, O., Cheng, H., Aznarez-Sanado, M., Wang, Y., Josa, S., ... Pastor, M. A. (2013). Robust estimation of fractal measures for characterizing the structural complexity of the human brain: optimization and reproducibility. *NeuroImage*, 83, 646–657. http://dx.doi.org/10.1016/j.neuroimage.2013.06.072.

Grewen, K., Burchinal, M., Vachet, C., Gouttard, S., Gilmore, J. H., Lin, W., ... Gerig, G. (2014). Prenatal cocaine effects on brain structure in early infancy. *NeuroImage*, 101, 114–123. http://dx.doi.org/10.1016/j.neuroimage.2014.06.070.

Guerri, C., Bazinet, A., & Riley, E. P. (2009). Foetal Alcohol Spectrum Disorders and alterations in brain and behaviour. *Alcohol and Alcoholism*, 44(2), 108–114. http://dx.doi.org/10.1093/alcalc/agn105.

Hutton, C., De Vita, E., Ashburner, J., Deichmann, R., & Turner, R. (2008). Voxel-based cortical thickness measurements in MRI. *NeuroImage*, 40(4), 1701–1710. http://dx.doi.org/10.1016/j.neuroimage.2008.01.027.

Im, K., Lee, J. M., Lyttelton, O., Kim, S. H., Evans, A. C., & Kim, S. I. (2008). Brain size and cortical structure in the adult human brain. *Cerebral Cortex*, *18*(9), 2181–2191. http://dx.doi.org/10.1093/cercor/bhm244.

Im, K., Lee, J., Won Seo, S., Hyung Kim, S., Kim, S., & Na, D. (2008). Sulcal morphology changes and their relationship with cortical thickness and gyral white matter volume in mild cognitive impairment and Alzheimer's disease. *NeuroImage*. http://dx.doi.org/10.1016/j.neuroimage.2008.07.016.

Ingalhalikar, M., Smith, A. R., Bloy, L., Gur, R., Roberts, T. P., & Verma, R. (2012). Identifying sub-populations via unsupervised cluster analysis on multi-edge similarity graphs. *Medical Image Computing and Computer Assisted Intervention*, *15*(Pt 2), 254–261.

Joseph, J., Warton, C., Jacobson, S. W., Jacobson, J. L., Molteno, C. D., Eicher, A., ... Meintjes, E. M. (2014). Three-dimensional surface deformation-based shape analysis of hippocampus and caudate nucleus in children with fetal alcohol spectrum disorders. *Human Brain Mapping*, *35*(2), 659–672. http://dx.doi.org/10.1002/hbm.22209.

Lebel, C., Mattson, S. N., Riley, E. P., Jones, K. L., Adnams, C. M., May, P. A., ... Sowell, E. R. (2012). A longitudinal study of the long-term consequences of drinking during pregnancy: heavy in utero alcohol exposure disrupts the normal processes of brain development. *Journal of Neuroscience*, *32*(44), 15243–15251. http://dx.doi.org/10.1523/jneurosci.1161-12.2012.

Lebel, C., Roussotte, F., & Sowell, E. R. (2011). Imaging the impact of prenatal alcohol exposure on the structure of the developing human brain. *Neuropsychology Review*, *21*(2), 102–118. http://dx.doi.org/10.1007/s11065-011-9163-0.

Leigland, L. A., Ford, M. M., Lerch, J. P., & Kroenke, C. D. (2013). The influence of fetal ethanol exposure on subsequent development of the cerebral cortex as revealed by magnetic resonance imaging. *Alcoholism, Clinical and Experimental Research*, *37*(6), 924–932. http://dx.doi.org/10.1111/acer.12051.

Lerch, J. P., & Evans, A. C. (2005). Cortical thickness analysis examined through power analysis and a population simulation. *NeuroImage*, *24*(1), 163–173. http://dx.doi.org/10.1016/j.neuroimage.2004.07.045.

Li, M., Tian, J., Zhang, R., Qiu, Y., Wen, X., Ma, X., ... Huang, R. (2014). Abnormal cortical thickness in heroin-dependent individuals. *NeuroImage*, *88*, 295–307. http://dx.doi.org/10.1016/j.neuroimage.2013.10.021.

Lopez-Tejero, D., Ferrer, I., Llobera, M., & Herrera, E. (1986). Effects of prenatal ethanol exposure on physical growth, sensory reflex maturation and brain development in the rat. *Neuropathology and Applied Neurobiology*, *12*(3), 251–260.

Luders, E., Thompson, P. M., Narr, K. L., Toga, A. W., Jancke, L., & Gaser, C. (2006). A curvature-based approach to estimate local gyrification on the cortical surface. *NeuroImage*, *29*(4), 1224–1230. http://dx.doi.org/10.1016/j.neuroimage.2005.08.049.

Lusebrink, F., Wollrab, A., & Speck, O. (2013). Cortical thickness determination of the human brain using high resolution 3T and 7T MRI data. *NeuroImage*, *70*, 122–131. http://dx.doi.org/10.1016/j.neuroimage.2012.12.016.

Mangin, J.-F., Jouvent, E., & Cachia, A. (2010). In-vivo measurement of cortical morphology: means and meanings. *Current Opinion in Neurology*, *23*(4), 359–367. http://dx.doi.org/10.1097/WCO.0b013e32833a0afc.

Mangin, J.-F., Rivière, D., Cachia, A., Duchesnay, E., Cointepas, Y., Papadopoulos-Orfanos, D., ... Régis, J. (2004a). A framework to study the cortical folding patterns. *NeuroImage*, *23*(Suppl. 1), S129–S138. http://dx.doi.org/10.1016/j.neuroimage.2004.07.019.

Mangin, J. F., Rivière, D., Cachia, A., Duchesnay, E., Cointepas, Y., Papadopoulos-Orfanos, D., ... Régis, J. (2004b). Object-based morphometry of the cerebral cortex. *IEEE Transactions on Medical Imaging*, *23*(8), 968–982. http://dx.doi.org/10.1109/tmi.2004.831204.

McCarthy, N., & Eberhart, J. K. (2014). Gene-ethanol interactions underlying fetal alcohol spectrum disorders. *Cellular and Molecular Life Sciences*, *71*(14), 2699–2706. http://dx.doi.org/10.1007/s00018-014-1578-3.

Meintjes, E. M., Narr, K. L., der Kouwe, A. J., Molteno, C. D., Pirnia, T., Gutman, B., ... Jacobson, S. W. (2014). A tensor-based morphometry analysis of regional differences in brain volume in relation to prenatal alcohol exposure. *Neuroimage: Clinical*, *5*, 152–160. http://dx.doi.org/10.1016/j.nicl.2014.04.001.

Momenan, R., Steckler, L. E., Saad, Z. S., van Rafelghem, S., Kerich, M. J., & Hommer, D. W. (2012). Effects of alcohol dependence on cortical thickness as determined by magnetic resonance imaging. *Psychiatry Research*, *204*(2-3), 101–111. http://dx.doi.org/10.1016/j.pscychresns.2012.05.003.

Moore, E. M., Migliorini, R., Infante, M. A., & Riley, E. P. (2014). Fetal alcohol spectrum disorders: recent neuroimaging findings. *Current Developmental Disorders Reports*, *1*(3), 161–172. http://dx.doi.org/10.1007/s40474-014-0020-8.

Muggli, E., Cook, B., O'Leary, C., Forster, D., & Halliday, J. (2014). Increasing accurate self-report in surveys of pregnancy alcohol use. *Midwifery*. http://dx.doi.org/10.1016/j.midw.2014.11.003.

Nardelli, A., Lebel, C., Rasmussen, C., Andrew, G., & Beaulieu, C. (2011). Extensive deep gray matter volume reductions in children and adolescents with fetal alcohol spectrum disorders. *Alcoholism, Clinical and Experimental Research*, *35*(8), 1404–1417. http://dx.doi.org/10.1111/j.1530-0277.2011.01476.x.

Rajaprakash, M., Chakravarty, M. M., Lerch, J. P., & Rovet, J. (2014). Cortical morphology in children with alcohol-related neurodevelopmental disorder. *Brain and Behavior*, *4*(1), 41–50. http://dx.doi.org/10.1002/brb3.191.

Rakic, P. (1995). A small step for the cell, a giant leap for mankind: a hypothesis of neocortical expansion during evolution. *Trends in Neuroscience*, *18*(9), 383–388.

Rakic, P. (2005). Less is more: progenitor death and cortical size. *Nature and Neuroscience*, *8*(8), 981–982. http://dx.doi.org/10.1038/nn0805-981.

Raznahan, A., Shaw, P., Lalonde, F., Stockman, M., Wallace, G. L., Greenstein, D., ... Giedd, J. N. (2011). How does your cortex grow? *Journal of Neuroscience*, *31*(19), 7174–7177. http://dx.doi.org/10.1523/jneurosci.0054-11.2011.

Roussotte, F. F., Sulik, K. K., Mattson, S. N., Riley, E. P., Jones, K. L., Adnams, C. M., ... Sowell, E. R. (2012). Regional brain volume reductions relate to facial dysmorphology and neurocognitive function in fetal alcohol spectrum disorders. *Human Brain Mapping*, *33*(4), 920–937. http://dx.doi.org/10.1002/hbm.21260.

Schaer, M., Cuadra, M. B., Tamarit, L., Lazeyras, F., Eliez, S., & Thiran, J. P. (2008). A surface-based approach to quantify local cortical gyrification. *IEEE Transactions on Medical Imaging*, *27*(2), 161–170. http://dx.doi.org/10.1109/tmi.2007.903576.

Seo, S., & Chung, M. K. (2011). Laplace-beltrami eigenfunction expansion of cortical manifolds. In *8th IEEE International Symposium on Biomedical Imaging: From Nano to Macro* (pp. 372–375). New York: IEEE.

Sowell, E. R., Leow, A. D., Bookheimer, S. Y., Smith, L. M., O'Connor, M. J., Kan, E., ... Thompson, P. M. (2010). Differentiating prenatal exposure to methamphetamine and alcohol versus alcohol and not methamphetamine using tensor-based brain morphometry and discriminant analysis. *Journal of Neuroscience*, *30*(11), 3876–3885. http://dx.doi.org/10.1523/jneurosci.4967-09.2010.

Sowell, E. R., Mattson, S. N., Kan, E., Thompson, P. M., Riley, E. P., & Toga, A. W. (2008). Abnormal cortical thickness and brain-behavior correlation patterns in individuals with heavy prenatal alcohol exposure. *Cerebral Cortex*, *18*(1), 136–144. http://dx.doi.org/10.1093/cercor/bhm039.

Sowell, E. R., Mattson, S. N., Thompson, P. M., Jernigan, T. L., Riley, E. P., & Toga, A. W. (2001). Mapping callosal morphology and cognitive correlates: effects of heavy prenatal alcohol exposure. *Neurology*, *57*(2), 235–244.

Sowell, E. R., Thompson, P. M., Mattson, S. N., Tessner, K. D., Jernigan, T. L., Riley, E. P., & Toga, A. W. (2001). Voxel-based morphometric analyses of the brain in children and adolescents prenatally exposed to alcohol. *Neuroreport*, *12*(3), 515–523.

Sowell, E. R., Thompson, P. M., Mattson, S. N., Tessner, K. D., Jernigan, T. L., Riley, E. P., & Toga, A. W. (2002). Regional brain shape abnormalities persist into adolescence after heavy prenatal alcohol exposure. *Cerebral Cortex*, *12*(8), 856–865.

Sowell, E. R., Thompson, P. M., Peterson, B. S., Mattson, S. N., Welcome, S. E., Henkenius, A. L., ... Toga, A. W. (2002). Mapping cortical gray matter asymmetry patterns in adolescents with heavy prenatal alcohol exposure. *NeuroImage*, *17*(4), 1807–1819.

Sun, Z. Y., Kloppel, S., Riviere, D., Perrot, M., Frackowiak, R., Siebner, H., & Mangin, J. F. (2012). The effect of handedness on the shape of the central sulcus. *NeuroImage*, *60*(1), 332–339. http://dx.doi.org/10.1016/j.neuroimage.2011.12.050.

Toro, R., Perron, M., Pike, B., Richer, L., Veillette, S., Pausova, Z., & Paus, T. (2008). Brain size and folding of the human cerebral cortex. *Cerebral Cortex (New York, NY: 1991)*, *18*(10), 2352–2357. http://dx.doi.org/10.1093/cercor/bhm261.

Treit, S., Zhou, D., Lebel, C., Rasmussen, C., Andrew, G., & Beaulieu, C. (2014). Longitudinal MRI reveals impaired cortical thinning in children and adolescents prenatally exposed to alcohol. *Human Brain Mapping*, *35*(9), 4892–4903. http://dx.doi.org/10.1002/hbm.22520.

Walhovd, K. B., Moe, V., Slinning, K., Due-Tonnessen, P., Bjornerud, A., Dale, A. M., ... Fischl, B. (2007). Volumetric cerebral characteristics of children exposed to opiates and other substances in utero. *NeuroImage*, *36*(4), 1331–1344. http://dx.doi.org/10.1016/j.neuroimage.2007.03.070.

Willoughby, K. A., Sheard, E. D., Nash, K., & Rovet, J. (2008). Effects of prenatal alcohol exposure on hippocampal volume, verbal learning, and verbal and spatial recall in late childhood. *Journal of the International Neuropsychological Society*, *14*(6), 1022–1033. http://dx.doi.org/10.1017/s1355617708081368.

Winkler, A. M., Sabuncu, M. R., Yeo, B. T., Fischl, B., Greve, D. N., Kochunov, P., ... Glahn, D. C. (2012). Measuring and comparing brain cortical surface area and other areal quantities. *NeuroImage*, *61*(4), 1428–1443. http://dx.doi.org/10.1016/j.neuroimage.2012.03.026.

Yang, Y., Roussotte, F., Kan, E., Sulik, K. K., Mattson, S. N., Riley, E. P., ... Sowell, E. R. (2012). Abnormal cortical thickness alterations in fetal alcohol spectrum disorders and their relationships with facial dysmorphology. *Cerebral Cortex*, *22*(5), 1170–1179. http://dx.doi.org/10.1093/cercor/bhr193.

Zhou, D., Lebel, C., Lepage, C., Rasmussen, C., Evans, A., Wyper, K., ... Beaulieu, C. (2011). Developmental cortical thinning in fetal alcohol spectrum disorders. *NeuroImage*, *58*(1), 16–25. http://dx.doi.org/10.1016/j.neuroimage.2011.06.026.

Zilles, K., Armstrong, E., Schleicher, A., & Kretschmann, H. J. (1988). The human pattern of gyrification in the cerebral cortex. *Anatomy and Embryology (Berl)*, *179*(2), 173–179.

Chapter 53

Abnormalities of Cerebellar Structure and Function in Alcoholism and Other Substance Use Disorders

Jessica W. O'Brien[1,2], Shirley Y. Hill[1,2,3]

[1]*Department of Psychiatry, University of Pittsburgh School of Medicine, Pittsburgh, PA, USA;* [2]*Department of Psychology, University of Pittsburgh, Pittsburgh, PA, USA;* [3]*Department of Human Genetics, Graduate School of Public Health, University of Pittsburgh, Pittsburgh, PA, USA*

Abbreviations

AUD Alcohol use disorder
BOLD Blood oxygen level-dependent
CB1 Cannabinoid receptor type 1
DSM *Diagnostic and Statistical Manual*
DTI Diffusion tensor imaging
FA Fractional anisotropy
fMRI Functional magnetic resonance imaging
GABA γ-Aminobutyric acid
MDMA 3,4-Methylenedioxymethamphetamine, "ecstasy"
MRI Magnetic resonance imaging
NAA *N*-Acetylaspartate
PET Positron emission tomography
rCBF Regional cerebral blood flow
SUD Substance use disorder

INTRODUCTION

Historically, the cerebellum has been viewed as a brain region principally involved in motor control and coordination. Studies going back to the nineteenth century have established that ablation of portions of the cerebellum of animals results in a wide range of motor impairments (Schmahmann, 1991). In addition, patients with cerebellar lesions have been observed to exhibit a wide range of motor disturbances including ataxia, dysarthria, dysmetria, and oculomotor dysfunction (Schmahmann, 1991). In the past two decades there has been increased attention paid to a broad range of neuropsychological sequelae associated with cerebellar dysfunction (O'Halloran, Kinsella, & Storey, 2012). Localized damage to the cerebellum has been associated with a specific constellation of cognitive symptoms, including deficits in attention, working memory, executive functioning, language functions, and visuospatial skills, as well as altered personality and affective behavior (Schmahmann, 1991). In addition, structural and functional abnormalities of the cerebellum have been implicated in a number of neurodevelopmental and neuropsychiatric conditions, including autism spectrum disorders, attention-deficit hyperactivity disorder, and schizophrenia (O'Halloran et al., 2012). Also, there is an emerging literature suggesting that use of a variety of commonly abused substances may be associated with altered cerebellar structure and function.

PREVALENCE OF SUBSTANCE USE DISORDERS

Substance use disorders (SUDs) are characterized by a cluster of cognitive, behavioral, and physiological symptoms indicating that an individual continues using alcohol, cannabis, hallucinogens, inhalants, opioids, sedatives/anxiolytics, stimulants, and/or tobacco despite significant substance-related problems (*Diagnostic and Statistical Manual 5* (DSM-5); American Psychiatric Association, 2013).

Alcohol Use Disorder

The DSM-5 diagnosis of alcohol use disorder (AUD) encompasses individuals with both alcohol abuse and alcohol dependence, classifications used in previous versions of the DSM. AUD is a highly prevalent form of SUD, affecting 8.5% of adults over the age of 18 and 4.6% of individuals between 12 and 17 years of age in the United States, with higher prevalence rates among men and individuals ages 18–24 (DSM-5; American Psychiatric Association, 2013; Hasin, Stinson, Ogburn, & Grant, 2007). Chronic use of alcohol in high doses may affect the gastrointestinal tract, cardiovascular system, and central and peripheral nervous systems and is associated with persistent cognitive deficits, including impaired judgment, blunted affect, poor insight, social withdrawal, reduced motivation, distractibility, cognitive rigidity, inattention, and perseveration (Oscar-Berman & Marinkovic, 2007).

Tobacco Use Disorder

Tobacco smoking is the most prevalent substance dependence worldwide and is associated with numerous public health problems (World Health Organization, 2008). In the United States, the 12-month prevalence of tobacco use disorder is 13% among individuals 18 and older (DSM-5; American Psychiatric Association, 2013). Tobacco use disorder is often observed in individuals with AUD and other SUDs (Durazzo, Gazdzinski, Banys, & Meyerhoff, 2004).

Cannabis Use Disorder

Cannabis is the most widely used illicit psychoactive substance in the United States, with cannabis use disorder affecting 3.4% of adolescents and 1.5% of adults (DSM-5; American Psychiatric Association, 2013; Compton, Thomas, Stinson, & Grant, 2007). Cannabis use has been associated with numerous acute and chronic mental health problems, including anxiety, depression, neuropsychological deficits, and increased risk of psychotic symptoms and disorders (Batalla et al., 2013).

Stimulant Use Disorder

Stimulant use disorders are characterized by a pattern of amphetamine-type substance (amphetamine, dextroamphetamine, and methamphetamine), cocaine, or other stimulant use leading to clinically significant impairment or distress. Amphetamine-type stimulant use disorder affects 0.2% of adolescents and 0.2% of adults, and cocaine-type stimulant use disorder affects 0.2% of adolescents and 0.3% of adults (DSM-5; American Psychiatric Association, 2013; Compton et al., 2007). Stimulant use disorder often co-occurs with other SUDs, especially those involving sedative properties such as alcohol or cannabis (DSM-5; American Psychiatric Association, 2013).

Opioid Use Disorder

Opioid use disorder affects approximately 0.4% of the population; heroin is the most commonly abused opioid, although this disorder also characterizes abuse of morphine, hydrocodones, oxycodone, codeine, and methadone (DSM-5; American Psychiatric Association, 2013; Compton et al., 2007). Acute opioid use affects μ, κ, and δ opioid receptors, and chronic opioid use is associated with changes in cortical blood flow, aberrant neural biochemistry, and cognitive deficits (Miquel, Toledo, Garcia, Coria-Avila, & Manzo, 2009).

THE CEREBELLUM AND ADDICTION CONNECTION

The central nervous system is one of the primary targets of drugs of abuse, and over the past two decades, converging evidence has implicated the cerebellum in addiction. Although classically considered to be a neural structure primarily responsible for motor functions, a growing body of research has implicated cerebellar involvement in a broad range of neuropsychological functions. Lesion and neuroimaging studies indicate that the cerebellum is involved in attention, working memory, learning, executive functioning, emotion, and affective states (O'Halloran et al., 2012). The cerebellum may also play a role in brain processes related to addiction, including sensitivity to reward, motivation, saliency, inhibitory control, and insight. Accordingly, structural and functional neuroimaging studies have demonstrated cerebellar abnormalities in individuals with SUDs. This chapter aims to review the existing literature on cerebellar dysfunction in SUDs. Additionally, we provide preliminary evidence that cerebellar dysfunction in addiction reflects a complex interaction between premorbid risk factors for SUDs and the consequences of the neurotoxic effects of addictive substances on the brain.

STRUCTURE AND FUNCTION OF THE CEREBELLUM

The cerebellum is located in the inferior posterior portion of the skull, superior to the brain stem (Figure 1). Owing to its heavily convoluted structure, it contains more than 50% of all neurons in the brain despite constituting only around 10% of the total brain volume (Sultan, Mock, & Thier, 2000). The anterior and posterolateral fissures divide the cerebellum into three anatomically distinct lobes: the anterior, posterior, and flocculonodular lobes (Figure 2). The cerebellum can also be divided into three sagittal zones: the vermis is located in the center of

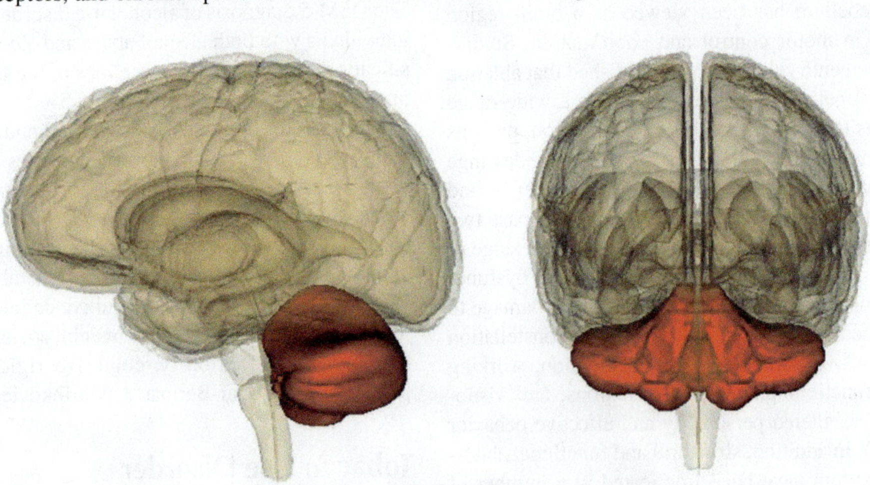

FIGURE 1 Localization of the cerebellum in the human brain. Sagittal (left) and coronal (right) views of the human brain illustrating the cerebellum (in red).

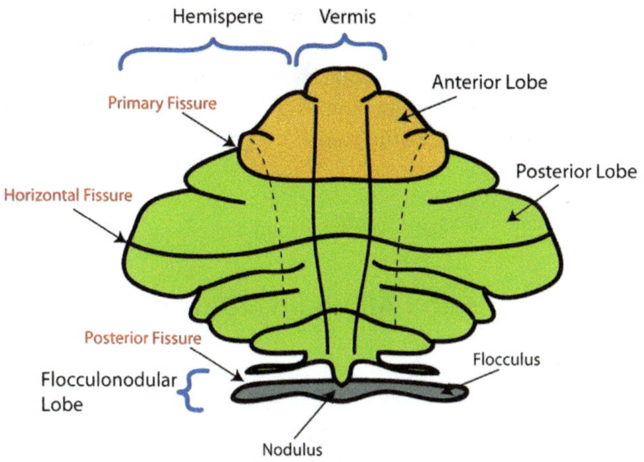

FIGURE 2 Anatomical divisions of the cerebellum. A diagram of the major cerebellum lobes is presented.

the cerebellum, spanning the anterior and posterior lobes. The intermediate zone is immediately lateral to the vermis, with the cerebellar hemispheres composing the most lateral aspects of the cerebellum.

Research with nonhuman primates indicates that the cerebellar cortex is not functionally homogeneous and that the cerebellum contains localized regions that are interconnected with specific motor or nonmotor areas of the cerebral cortex (Strick, Dum, & Fiez, 2009). In humans, lesion and neuroimaging studies indicate a general fractionation of the cerebellum such that anterior cerebellar regions appear to subserve motor and sensory functions, lateral posterior regions mediate cognitive abilities, and the medial cerebellar is probably important in emotional processing (Stoodley & Schmahmann, 2009). Furthermore, the right cerebellum appears to be more related to verbal memory and language, while the left cerebellum plays a role in visuospatial processing (O'Halloran et al., 2012). Although the distinct localization or circuitry of neuropsychological functions is not well understood at present, it is clear that the cerebellum is strongly interconnected to the cerebral cortex via afferent and efferent loops and plays an important role in cognition.

DEVELOPMENT OF THE CEREBELLUM

The use of addictive substances is exceedingly common during adolescence, a period of development during which the brain undergoes extensive morphometric and functional maturation (Gogtay et al., 2004; Johnston, O'Malley, Miech, Bachman, & Schulenberg, 2014). Cerebellar damage in childhood is associated with more conspicuous cognitive and affective changes than damage acquired in adulthood, indicating that disruptions in cerebellar activity during development may affect the organization and function of remote cerebral areas (Wang, Kloth, & Badura, 2014). Accordingly, to understand the possible effects of substance use on cerebellar morphology requires consideration of the normal developmental trajectory of the cerebellum.

Total cerebellar volume follows an inverted U-shaped trajectory between the ages of 5 and 24 years, and maximum cerebellar volume occurs later in development than does the point of maximal cerebral volume (ages 13.7 and 12.5, respectively), indicating a relatively prolonged developmental course (Tiemeier et al., 2010; Wierenga et al., 2014). Phylogenetically and ontogenetically diverse subregions of the cerebellum show differing developmental trajectories, such that areas of the cerebellum linked to later developing regions of the cerebrum also show prolonged maturation.

Developmental trajectories of the cerebellum are sexually dimorphic (Tiemeier et al., 2010; Wierenga et al., 2014). Maximum cerebellar volume occurs earlier in girls (age 11.8) than it does in boys (age 15.5), although total cerebellar volume is larger in males than in females after correcting for total cerebral volume. Volumes of the inferior posterior, anterior, and superior posterior lobes also peak earlier in girls (ages 11.1, 13.5, and 15.8, respectively) than in boys (ages 13.8, 15.7, and 18.2, respectively), although volumetric sex differences are least apparent in the anterior lobe (Tiemeier et al., 2010).

ALCOHOL USE DISORDERS AND THE CEREBELLUM

Classical postmortem neuropathology studies have demonstrated that chronic alcoholism is associated with notable shrinkage of the cerebellum (Harper, 1998; Kril & Harper, 1989). Atrophy of the cerebellar vermis is apparent, especially in individuals with exceptionally high levels of alcohol consumption or thiamine deficiency (Victor, Adams, & Mancall, 1959). On the cellular level, Purkinje cells and cells in the granular and molecular layers of the cerebellar cortex are particularly affected, especially in alcoholics with a history of thiamine deficiency (Baker, Harding, Halliday, Kril, & Harper, 1999; Phillips, Harper, & Kril, 1987). Neuropathological studies also indicate specific white matter reductions associated with chronic alcoholism, especially in the frontal cortex and cerebellum (Harper, 1998). In vivo imaging studies have also implicated damage to frontocerebellar circuitry as a consequence of alcoholism.

Magnetic Resonance Imaging Studies of the Cerebellum in Alcohol-Dependent Individuals

Magnetic resonance imaging (MRI) studies have revealed volumetric reductions in the cerebellum in alcoholics and these findings have not been limited to alcoholics with nutritional deficiencies (i.e., thiamine) or hepatic disorders (Oscar-Berman & Marinkovic, 2007). Importantly, reduced cerebellar volumes have been associated with both motor and cognitive deficits in alcoholics. Reduced volumes of the anterior superior cerebellar vermis and infratentorial tissue are associated with ataxia and greater postural sway (Sullivan, Deshmukh, Desmond, Lim, & Pfefferbaum, 2000; Sullivan, Rose, & Pfefferbaum, 2006) in alcohol-dependent individuals. Decreases in cerebellar gray matter volume are related to poorer performance on letter–number sequencing, a measure of working memory (Chanraud et al., 2007); higher rates of perseverative errors on the Wisconsin Card Sorting Task, a measure of executive functioning and cognitive flexibility (Chanraud et al., 2007); and deficits in the Frontal Assessment Battery, which assesses a number of executive functions (Nakamura-Palacios et al., 2014). Strikingly, cerebellar vermian volume is a stronger predictor of performance on the Wisconsin Card Sorting Task than volume of the prefrontal cortex (Sullivan, 2003), and cerebellar hemisphere white matter volume is a stronger predictor of visuospatial deficits in alcohol-dependent individuals than volume of the parietal cortex (Sullivan, 2003).

Cerebellar White Matter Structure in Alcoholism

Diffusion tensor imaging (DTI) studies also demonstrate white matter abnormalities of frontocerebellar circuits associated with alcoholism. The most commonly reported scalar measure derived from DTI is fractional anisotropy (FA), which is believed to be related to white matter integrity. Although the majority of DTI studies in individuals with AUD have not analyzed the cerebellum, one study has reported reduced fiber integrity (i.e., reduced FA) in the cerebellum (Yeh, Simpson, Durazzo, Gazdzinski, & Meyerhoff, 2009). Furthermore, corticocerebellar tracts connecting the midbrain and pons are characterized by fewer white matter fibers in alcoholics, and the number of fibers per voxel in this region is associated with performance on Part B of the Trail Making Test, which assesses visual search, working memory, and cognitive flexibility (Chanraud et al., 2009).

Cerebellar Neuronal Integrity in Alcoholism

Magnetic resonance spectroscopy is a noninvasive approach to the identification, visualization, and quantification of specific brain metabolites and neurotransmitters. In the cerebellum, recently detoxified alcoholics show abnormally low levels of *N*-acetylaspartate (NAA), a metabolite considered to be a marker of neuronal integrity (Bendszus et al., 2001; Parks et al., 2002), as well as lower levels of choline-containing compounds, which are associated with cell membrane synthesis, turnover, and metabolism (Bendszus et al., 2001). Furthermore, compared to individuals with AUD who remained abstinent, those who relapsed show lower baseline levels of cerebellar NAA, as well as total creatine, a metabolite influenced by the state of high-energy phosphate metabolism (Durazzo et al., 2004; Parks et al., 2002). Biochemical changes in the cerebellum in AUDs have been found to correlate with performance on tasks of visuospatial learning and memory (Durazzo et al., 2004) as well as auditory–verbal learning (Bendszus et al., 2001).

Functional Neuroimaging in Alcoholics

Further support for corticocerebellar dysfunction in alcoholics comes from functional MRI (fMRI) studies. fMRI utilizes the blood oxygen level-dependent (BOLD) signal to determine neural regions associated with task-based performance. Compared to unaffected individuals, alcoholic subjects have been shown to recruit additional neural regions, including the vermis of the cerebellum, when performing a self-paced finger-tapping task, potentially indicating compensatory alterations in frontocerebellar circuits to achieve adequate task performance (Parks et al., 2010). Similarly, alcohol-dependent subjects demonstrated unique neural activation in the left prefrontal cortex and right superior cerebellum during a verbal working-memory task (Desmond et al., 2003) and in the left cerebellum, thalamus, and pallidum during a visuospatial task (Schulte, Müller-Oehring, Rohlfing, Pfefferbaum, & Sullivan, 2010). While neuropathological and structural neuroimaging studies have interpreted structural abnormalities of the cerebellum as evidence of alcohol-induced damage, functional imaging studies indicate that the cerebellum may also play a compensatory role by improving performance on cognitive tasks typically dependent on intact frontal lobe functioning in substance-dependent individuals.

Alcohol Cues and Alcohol-Dependent Individuals

fMRI studies have also implicated activation of the cerebellum in response to alcohol-related cues. Alcohol-dependent individuals demonstrate heightened craving for alcohol when attending to alcohol-related environmental cues, and cue-induced craving is believed to play an important role in the development and persistence of addiction (Field & Cox, 2008). Schneider et al. (2001) assessed the neural response to ethanol odor in recently (less than 1 week) detoxified alcohol-dependent men and found that affected individuals showed unique BOLD activation in the cerebellum and amygdala in response to ethanol odor, compared to healthy controls. After a 3-week intervention involving cognitive behavioral group therapy and daily doxepin administration, stimulation with ethanol no longer elicited activation in these regions, implicating a role for the cerebellum in craving (Schneider et al., 2001). Similarly, across individuals in varying stages of alcohol dependence (current drinkers, recent abstainers undergoing treatment for alcohol dependence, and long-term abstainers with sustained remission from AUD), greater cerebellar activation to alcohol distractors during a visual oddball task (compared to nonalcohol distractors) was correlated with self-reported alcohol craving (Fryer et al., 2013), although group differences in activation were observed only in cortical regions. In a large sample of individuals with AUDs, an alcohol taste cue elicited greater bilateral cerebellar activation than did a nonalcoholic beverage, and there was a significant positive correlation between left cerebellar activation and self-reported level of alcohol dependence (Claus, Ewing, Filbey, Sabbineni, & Hutchison, 2011). In subsequent analyses of these data, Monnig and colleagues assessed correlations between BOLD response to alcohol taste cue and FA in 18 white matter tracts selected a priori (Monnig et al., 2014). Greater cerebellar BOLD activation to alcohol taste was associated with lower FA in the cingulate gyrus, external capsule, and fornix. Additionally, lower FA in the fornix and cingulate gyrus was associated with AUDs of greater severity. These converging data suggest that white matter connections among cortical and subcortical structures are involved in the development and maintenance of alcohol abuse and dependence.

Resting State Functional Connectivity in Alcoholics

Functional connectivity studies have also been used to investigate the neural consequences of alcoholism. Functional connectivity studies assess coherent signal fluctuations measured by fMRI that imply functional relationships between neural regions either at rest or during tasks completed in an MRI scanner

(Fox & Raichle, 2007). Various intrinsic connectivity networks have been identified in the resting brain, including the default mode network. The default mode network comprises brain regions (i.e., posterior cingulate, lateral temporoparietal areas, and medial and dorsolateral prefrontal cortices) that are more active during wakeful rest than during goal-directed activity. In addition, functional connectivity has been established in networks related to executive control, salience, attention, reward, emotion, somatosensation, hearing, and vision (Buckner, Andrews-Hanna, & Schacter, 2008). In healthy adults, the cerebellum demonstrates functional connectivity with neural regions implicated in executive control, attention, and somatosensation (Buckner, Krienen, Castellanos, Diaz, & Yeo, 2011). Functional connectivity studies with alcohol-dependent patients have documented network-specific patterns of functional connectivity differing in strength and spatial extent from controls in these systems.

Within the default-mode network, at rest, alcoholics show less connectivity between the left posterior cingulate cortex and the cerebellum than controls, and lesser local efficiency (Chanraud, Pitel, Pfefferbaum, & Sullivan, 2011). Lesser local efficiency in the cerebellum appears to be a marker of recency of AUD, with lesser local efficiency associated with a shorter duration of abstinence from alcohol. During wakeful rest, individuals with AUD also show spatially expanded and greater connectivity between the cerebellum and the postcentral gyrus, a connection believed to underlie somatosensation, as well as restricted connectivity between the superior parietal lobe and the cerebellum, a connection associated with attention (Müller-Oehring, Jung, Pfefferbaum, Sullivan, & Schulte, 2015). Restricted connectivity between the parietal lobe and the cerebellum appears to be associated with faster visuospatial coordination in alcoholics, indicating that tracts demonstrating aberrant patterns of functional connectivity may reflect the need for compensatory neural mechanisms owing to alcohol-related deficits in other regions (Müller-Oehring et al., 2015).

Task-Related Functional Connectivity in Alcoholism

Functional connectivity in alcohol-dependent individuals has been assessed during motor and cognitive tasks. Rogers and colleagues assessed functional connectivity during a finger-tapping task, a paradigm that has previously been reported to have a unique pattern of cerebellar activation in alcohol-dependent individuals compared to healthy controls (Parks et al., 2010), finding that frontocerebellar functional connectivity during task performance was lower in alcohol-dependent patients (Rogers, Parks, Nickel, Katwal, & Martin, 2012). Contrastingly, during a task of working memory typically associated with reduced connectivity between regions associated with the default mode network, alcoholics showed increased connectivity between the left posterior cingulate and the left anterior/medial cerebellar lobules IV/V (Chanraud et al., 2011). Although the cerebellum shows aberrant functional interactions with cortical regions during cognitive tasks, it remains unclear to what extent these connections underlie observed neuropsychological deficits associated with AUD or reflect compensatory neural mechanisms that may have developed as a result of alcohol-related damage to other neural regions and connections.

Conclusions—Structural and Functional Changes in Alcohol Dependence

Converging evidence from neuropathological, structural, and functional neuroimaging studies indicate numerous abnormalities of the cerebellum in AUDs. Importantly, structural and functional characteristics of the cerebellum are associated with measures of executive functioning, AUD severity, and treatment outcomes for affected individuals.

CEREBELLAR ABNORMALITIES IN SUD: CAUSE OR CONSEQUENCE?

The cerebellum demonstrates a protracted period of development, with marked changes occurring during adolescence for both males and females, particularly in subregions of the cerebellum associated with higher order cognition. Furthermore, although variations in total and regional brain volumes are genetically influenced, the cerebellum shows less heritability and more sensitivity to environmental effects than most other brain structures, especially in childhood and adolescence (Peper, Brouwer, Boomsma, Kahn, & Hulshoff Pol, 2007). As such, the cerebellum may be especially vulnerable to environmental effects in adolescence, including the neurotoxic effects of alcohol and other drugs. The earlier the onset of alcohol use during adolescence the greater is the likelihood that the individual will develop an SUD in adulthood (Grant & Dawson, 1997). It is reasonable to expect that the longer the individual is exposed to alcohol during adolescence the greater is his or her chance of experiencing cerebellar damage. It is possible that such damage may in turn contribute to the development of an SUD through the role that the cerebellum plays in executive functioning.

On a molecular and cellular level, drugs of abuse and dependence can induce long-term changes in cerebellar receptors, neurotransmitters, and intracellular signaling transduction pathways that may affect functional networks between the cerebellum and cortical areas (Miquel et al., 2009). Additionally, it has been shown that longer duration and greater severity of a variety of SUDs (alcohol, cannabis, cocaine, methamphetamine, heroin, tobacco) are associated with more pronounced structural and functional abnormalities of the cerebellum (Azizian, Monterosso, O'Neill, & London, 2009; Claus et al., 2011; Cousijn et al., 2012; Fryer et al., 2013; Monnig et al., 2014; Moreno-Lopez et al., 2015; Yalachkov, Kaiser, Gorres, Seehaus, & Naumer, 2013). Accordingly, the prevailing thought has been that cerebellar abnormalities observed in individuals with SUDs are a direct consequence of substance use itself. However, cerebellar abnormalities have also been observed in substance-naïve individuals with a family history of SUDs, indicating that cerebellar abnormalities may contribute to genetically mediated risk factors for SUD.

Premorbid Risk Factors in High-Risk Offspring

Twin, adoption, and family studies have provided strong evidence for a genetic mediation of alcohol dependence susceptibility within families (Kendler, Schmitt, Aggen, & Prescott, 2008). Individuals with a family history of alcoholism are at increased risk for alcohol and drug dependence in young adulthood, with those from families with multiple alcohol-dependent relatives at highest risk (Hill et al., 2008; Hill, Tessner, & McDermott, 2011). In addition to alcohol

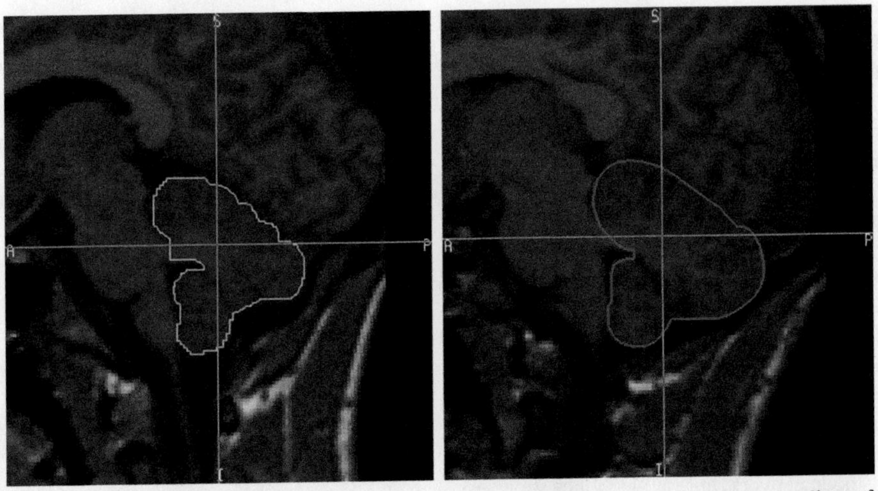

FIGURE 3 Region of interest tracings of the cerebellum in high- and low-risk individuals. Region of interest outlines of a typical low-risk male participant (blue outline) and of a typical high-risk male participant (red). Both male participants were 16 years of age at the time the scans were obtained. *From Hill, Wang, et al. (2011). The copyright permission is held by Psychiatry Research: Neuroimaging (Elsevier) and has not been requested for use in this publication because Academic Press is affiliated with Elsevier. Used with permission from Elsevier.*

dependence, genetic mediation of risk for developing nicotine and cannabis dependence has also been demonstrated (Kendler et al., 2008). Premorbid risk factors have been observed in children and adolescents with a family history of SUDs that are predictive of subsequent SUD, including neuropsychological deficits in response inhibition, emotion regulation, and cognitive control, comorbid psychiatric conditions, and abnormal neurobiological processes. Accordingly, studies examining substance-naïve youth with a family history of SUDs provide a powerful tool for determining potential premorbid risk factors for SUDs.

Cerebellum Structure and Familial Risk

Converging evidence from research examining high-risk offspring indicates that cerebellar abnormalities may be an important premorbid risk factor for SUDs. A number of studies that have examined cerebellar abnormalities and their relationship to familial risk for alcoholism have focused on structural morphology. Reduction in total cerebellar gray matter volume has been reported in a study of 20 high-risk male offspring with a family history of alcohol dependence who were contrasted with 21 controls (ages 8–24) using an automated imaging methodology. In addition to risk-group differences being seen, the authors noted that greater volumetric reductions were associated with higher levels of externalizing symptoms (Benegal, Antony, Venkatasubramanian, & Jayakumar, 2007). In contrast, two studies of adolescents and young adults from multiplex, alcohol-dependent families have shown greater total cerebellar volume and gray volume in comparison to controls (Hill et al., 2007; Hill, Wang, et al., 2011). In both of these studies manual tracing of the cerebellum was performed by trained individuals with high reliability (>90% agreement). Although manual tracing is quite labor intensive, it continues to be the gold standard against which automated algorithms are validated (Morey et al., 2009). Typical tracings from a high- and a low-risk subject may be seen in Figure 3. In one of these studies, 71 high-risk subjects were contrasted with 60 controls between the ages of 8 and 29 years with analyses focused on the entire sample and then on a subsample without an SUD (Hill, Wang, et al., 2011). The tendency for gray matter to be increased in high-risk individuals was accentuated when 24 of the 131 subjects with an SUD, a condition that may reduce volume, were removed before analysis (Figure 4).

The cerebellum has been shown to demonstrate an inverted U-shaped developmental trajectory between the ages of 7 and 24 in normal individuals, with males showing a delay in gray matter pruning (Wierenga et al., 2014)(see Figure 5). Because high-risk offspring show differing patterns of gray matter volume decline with age, it would appear that the normal pattern of gray matter pruning is disrupted. Studies of familial risk for other SUDs have also found a tendency for greater cerebellar gray matter volume among those with higher familial risk. Unaffected siblings of stimulant-dependent individuals show increased gray matter volume in the cerebellum compared to healthy controls with no family history (Ersche et al., 2013), and non-substance-abusing young adults with a family history of drug and alcohol abuse have a thicker lingula of the anterior cerebellar vermis compared to individuals with no family history (Anderson, Rabi, Lukas, & Teicher, 2010).

Cerebellar Lobes and Functional Differences

In addition to studies focusing on total cerebellar volume, there has been an increased emphasis on the study of functional differences of the individual cerebellar lobes, with the posterior lobe preferentially involved in cognition (Stoodley & Schmahmann, 2010). The distinct regions of the cerebellum can be identified by major fissures (Figure 2). Data (Hill et al., unpublished) indicate that high-risk offspring show especially pronounced volumetric differences in the corpus and inferior posterior lobe of the cerebellum, regions associated with higher-order cognitive and emotional processes (O'Halloran et al., 2012).

Cerebellar Functional Differences Associated with Family History Using Functional MRI

Individuals with a family history of alcoholism also show functional abnormalities of the cerebellum that converge with findings

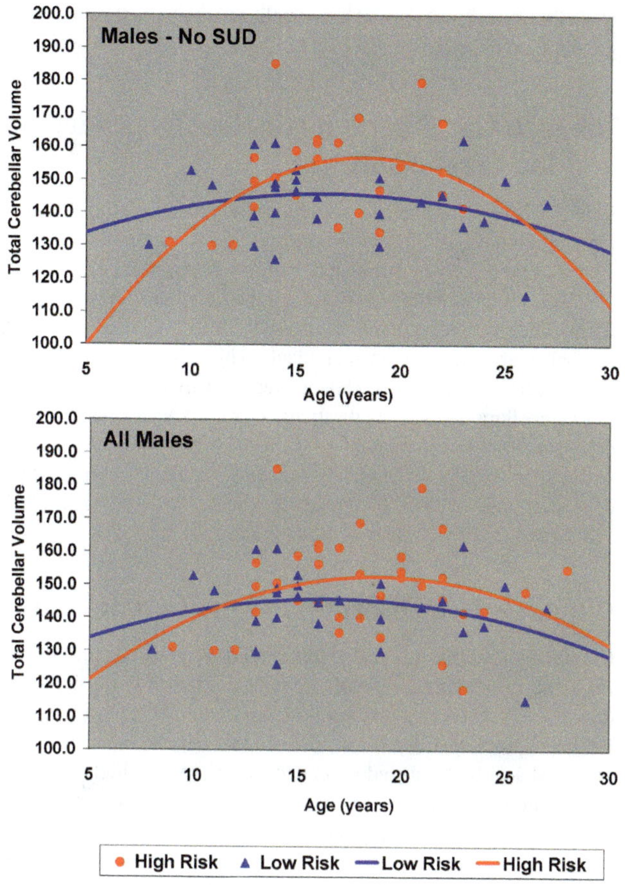

FIGURE 4 Developmental trajectories of cerebellar volume in high- and low-risk individuals with and without SUD. Total cerebellar volume by age is shown for high- and low-risk male subjects. The top shows the trajectories obtained for the 52 males without SUD prior to the MRI scan. The bottom shows all 65 male subjects and includes those with SUD ($n=13$). *From Hill, Wang, et al. (2011). The copyright permission is held by Psychiatry Research: Neuroimaging (Elsevier) and has not been requested for use in this publication because Academic Press is affiliated with Elsevier. Used with permission from Elsevier.*

in alcohol-dependent individuals. During a task of risky decision-making, individuals with a family history of AUD showed less BOLD response in the right dorsolateral prefrontal cortex and right cerebellum compared to individuals without a family history of AUD (Cservenka & Nagel, 2012). These findings suggest that atypical activity related in regions associated with executive functioning may contribute to risky choices regarding alcohol use, potentially explaining the higher rates of AUD observed in offspring with a family history of alcoholism. Similarly, during a task of spatial working memory, the density of a family history of alcoholism correlated with greater activation of the right superior parietal cortex but lesser activation of the right cerebellum (Mackiewicz Seghete, Cservenka, Herting, & Nagel, 2013), indicating that alterations in top-down cognitive control may be a general risk factor for individuals with a family history of AUD.

Functional Connectivity

Individuals with a family history of alcoholism show abnormal functional connectivity between the cerebellum and the cortical regions both at rest and during cognitive tasks. A reduction in resting state

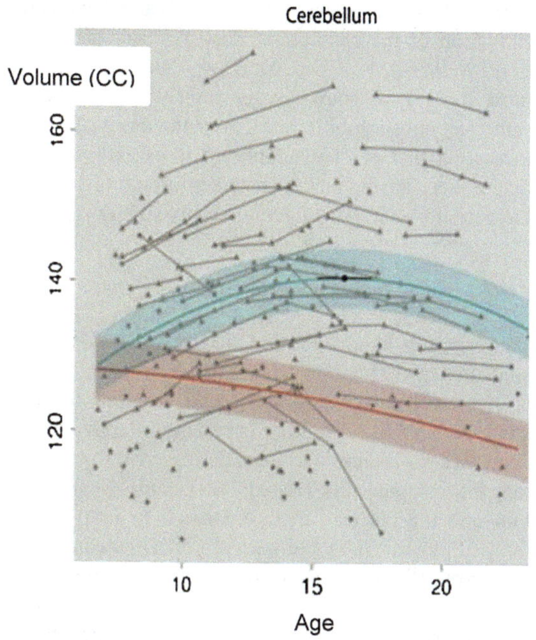

FIGURE 5 Trajectories of cerebellar volumes in healthy subjects. Trajectories of cerebellar volume in cubic centimeter (cc) in normally developing male and female subjects between the ages of 7 and 24. *From Wierenga et al. (2014). The copyright permission is held by NeuroImage (Elsevier) and has not been requested for use in this publication because Academic Press is affiliated with Elsevier. Used with permission from Elsevier.*

functional connectivity between the nucleus accumbens and the right cerebellum has been reported for individuals with a family history of AUD (Cservenka, Casimo, Fair, & Nagel, 2014), with greater density of family history associated with less connectivity between these regions. During tasks of cognitive interference, face perception, and reward-based decision-making, individuals with a family history show significantly reduced functional connectivity between bilateral anterior prefrontal cortices and contralateral cerebellar seed regions (Herting, Fair, & Nagel, 2011). Taken together, these findings indicate that the premorbid abnormalities in frontocerebellar circuitry may heighten the risk of developing an AUD through atypical cognitive processing during tasks requiring response inhibition, decision-making, and social cognition.

Cerebellar Volume and Substance Use Outcomes

Several studies have documented a relationship between cerebellar dysfunction and substance use outcomes. In a large prospective study, Manzardo and colleagues found that several measures of childhood motor development that rely on intact cerebellar functioning, including the age to first begin sitting and walking, predicted alcohol dependence at age 30 (Manzardo et al., 2005). Similarly, individuals with a family history of alcoholism demonstrate greater postural sway than their low-risk peers and smaller decreases in sway with age (Hill et al., 2000). Greater postural sway at age 15 is a particularly strong predictor of SUDs in young adulthood (Hill, Steinhauer, Locke-Wellman, & Ulrich, 2009). Adolescents who later transition to heavy drinking have less cerebellar white matter at baseline than adolescents who remain abstinent (Squeglia et al., 2014) and demonstrate reduced activation of the cerebellum during a task of response inhibition (Wetherill, Squeglia, Yang, & Tapert, 2013). Interestingly, these individuals show greater cerebellar activation during the same response-inhibition task after the onset of heavy drinking, indicating that cerebellar abnormalities associated with SUDs may reflect an interaction between premorbid risk factors and the neurotoxic effects of personal exposure to alcohol and drugs.

Conclusions—Familial Risk

Individuals at heightened genetic risk for SUDs show structural and functional abnormalities of the cerebellum before the initiation of alcohol and drug use. These high-risk offspring demonstrate abnormalities in gross cerebellar morphology and volume, decreased cerebellar activation during cognitive tasks, and aberrant intrinsic and task-related connectivity between the cerebellum and the cortical areas compared to individuals with low familial risk for SUD. Furthermore, SUD outcomes are related to premorbid behavioral measures of cerebellar function, cerebellar structure, and task-dependent cerebellar neural activation. These studies provide converging evidence that cerebellar abnormalities may be a premorbid risk factor involved in the etiology of SUDs. Thus, cerebellar abnormalities observed in individuals with SUDs may partially reflect preexisting risk factors for SUD, but may also be exacerbated by the neurotoxic effects of alcohol and drugs of abuse on the brain.

CEREBELLAR STRUCTURE AND FUNCTION IN OTHER SUBSTANCE USE DISORDERS

Structural and functional abnormalities of the cerebellum have also been documented in a number of other SUDs, including stimulant, cocaine, and opioid use disorders. Several studies have documented that cocaine use disorder is associated with reduced gray and white matter in the cerebellum, and greater severity and duration of cocaine use are associated with greater volumetric reductions (Moreno-Lopez et al., 2015; O'Neill, Cardenas, & Meyerhoff, 2001). Volumetric reductions in cerebellar gray matter also correlate with deficits in reversal learning in cocaine-dependent individuals (Moreno-Lopez et al., 2015), implicating relationships between cerebellar structure and cognition. Studies assessing cerebellar volume in methamphetamine-dependent adults have not documented differences compared to controls, although users of 3,4-methylenedioxymethamphetamine (MDMA; also known as "ecstasy") have reduced densities of cerebellar gray matter (Cowan et al., 2003). Furthermore, occasional users of cocaine and/or amphetamine-type stimulants (e.g., methamphetamine, methylphenidate, and MDMA) show significantly less gray matter in the right dorsolateral cerebellum, and greater lifetime consumption is associated with greater volumetric reductions in this area (Mackey, Stewart, Connolly, Tapert, & Paulus, 2014). Heroin-dependent individuals show reduced white matter integrity (i.e., lower FA) in the cerebellar peduncles and anterior vermis, as well as smaller cerebellar gray matter volumes (Bora et al., 2012; Lin, Chou, Chen, Huang, Chen, et al., 2012; Lin, Chou, Chen, Huang, Lu, et al., 2012). These structural abnormalities are associated with higher levels of depression, poorer memory, and impaired executive functioning (Lin, Chou, Chen, Huang, Chen, et al., 2012; Lin, Chou, Chen, Huang, Lu, et al., 2012), and more severe abnormalities are associated with a longer duration of opiate use (Bora et al., 2012).

Tobacco Use Disorder, Nicotine Dependence, and the Cerebellum

Individuals with tobacco use disorder show reduced gray matter density and volume of the cerebellum, and these reductions have been shown to be negatively correlated with the degree of nicotine dependence as well as lifetime exposure to tobacco smoke (Azizian et al., 2009). Both task-related and resting-state fMRI studies also indicate cerebellar abnormalities in individuals with tobacco use disorders. At rest, affected individuals show increased activity in the left posterior cerebellum but decreased activity in the left anterior cerebellum, compared to nonsmokers (Chu et al., 2014). Using task-related fMRI with either visually or haptically presented smoking cues, nicotine-dependent individuals show greater cerebellar activity than controls (Yalachkov, Kaiser, & Naumer, 2009; Yalachkov et al., 2013).

Cannabis Use Disorder and the Cerebellum

Cannabis acts on the endocannabinoid system, which modulates the neuronal activity of other neurotransmitters, including dopamine, through its actions on cannabinoid receptor 1 (CB1). CB1 receptors have a protracted period of development, reaching maximal levels during adolescence, the time at which cannabis use disorders are most prevalent. CB1 receptors are particularly concentrated in neural regions related to executive functioning, including the cerebellum. Accordingly, individuals with cannabis use disorder show deficits on psychomotor tasks dependent on corticocerebellar connections, including motor coordination, sensory integration, and eyeblink conditioning (Dervaux, Bourdel, Laqueille, & Krebs, 2013; Steinmetz et al., 2012). Because endogenous cannabinoids play an important role in the cellular aspects of cerebellar development, including synaptic pruning (Bossong & Niesink, 2010), exposure to exogenous cannabinoids from cannabis use during adolescence and young adulthood may disrupt normative cerebellar development.

Structural neuroimaging studies have shown that chronic cannabis use is associated with increased cerebellar gray matter volume in both adolescent and adult users (Batalla et al., 2013; Cousijn et al., 2012; Medina, Nagel, & Tapert, 2010). Increased volume of the cerebellar vermis is also associated with poorer executive functioning in cannabis users (Medina et al., 2010). Cerebellar white matter volume has been reported to be reduced in long-term, heavy cannabis users, with a longer history of cannabis use being associated with more severe reductions (Cousijn et al., 2012).

Functional abnormalities of the cerebellum have been observed in individuals affected by cannabis use disorder. Frequent cannabis users show lesser regional cerebral blood flow in the posterior cerebellum compared to other cortical areas and demonstrate aberrant connectivity between cerebellar hemispheres and between the cerebellum and other cortical regions (for a review, see Batalla et al., 2013). Furthermore, several fMRI and positron emission tomography studies have reported atypical cerebellar activation during the Iowa Gambling Task, a measure of probabilistic reasoning and decision-making under conditions of uncertain outcome (Batalla et al., 2013). Thus, abnormal intrinsic activity within the cerebellum and between the cerebellum and cortical structures may contribute to the cognitive and affective consequences of cannabis use.

CONCLUSIONS

Over the past two decades, it has become increasingly clear that the cerebellum plays an important role in both motor and cognitive functioning. Cerebellar activity has been implicated in cognitive processes known to be impaired in individuals with SUDs, including executive functioning, emotional processing, and affective states. Accordingly, individuals with SUD show both structural and functional abnormalities of the cerebellum. Although drugs of abuse demonstrate differing acute effects on the central nervous system, alcohol, cannabis, tobacco, opioids, cocaine, and other stimulants have all been shown to affect cerebellar structure and function. These cerebellar abnormalities have been found to correlate with cognitive abilities, duration and severity of SUD, and treatment outcomes, underlying the functional importance of intact cerebellar functioning. Interestingly, atypical cerebellar activation during cognitive tasks in individuals with SUD has been shown to correlate with better cognitive performance in some cases, suggesting that cerebellar abnormalities may also reflect neural reorganization that compensates for other affected regions and circuits. Cerebellar abnormalities have also been observed in substance-naïve individuals with a family history of SUDs, and measures of cerebellar functioning relate to future SUD outcomes. Thus, atypical structure and function of the cerebellum may be an important risk factor in both the etiology and the course of SUDs.

APPLICATIONS TO OTHER ADDICTIONS AND SUBSTANCE MISUSE

Abnormalities of cerebellar structure and function are associated with AUDs. Individuals with AUD show reduced cerebellar gray and white matter volumes, reduced white matter integrity in cerebellar tracts, atypical functional connectivity between the cerebellum and cortical structures, and differential activation of the cerebellum during cognitive tasks.

Cerebellar abnormalities have been noted in a number of other SUDs, including tobacco, cannabis, cocaine, stimulant, and opioid use disorders. Individuals affected by drug use disorders also show reductions in cerebellar volume, aberrant cerebellar functional connectivity, and atypical cerebellar activation during fMRI tasks.

Importantly, in both alcohol and drug use disorders, cerebellar dysfunction is correlated with measures of higher-order cognition as well as SUD severity and treatment outcomes. Although drugs of abuse demonstrate differing acute effects on the central nervous system, alcohol, cannabis, tobacco, opioids, cocaine, and other stimulants have all been shown to affect cerebellar structure and function.

DEFINITION OF TERMS

Ataxia This refers to atypical and uncoordinated movement, often considered a neurological sign of central nervous system dysfunction.
Cerebellar vermis This is the medial portion of the cerebellum connecting the right and left hemispheres.
Decision-making This is the cognitive process requiring the evaluation and selection of one option among several alternate possibilities; often assessed with the Iowa Gambling Task.
Default-mode network This is the neural network comprising brain regions (i.e., posterior cingulate, lateral temporoparietal areas, and medial and dorsolateral prefrontal cortices) that are more active during wakeful rest than during goal-directed activity.
Executive functioning This is an umbrella term referring to a number of higher-order cognitive processes related to goal-directed behavior, including response inhibition, working memory, and planning; it is historically considered to rely on intact functioning of the prefrontal cortex.
Frontocerebellar circuitry This is the functional neural network with nodes in the cerebellum and prefrontal cortex; it is implicated in executive functioning.
Functional connectivity This refers to the functional relationships between neural regions inferred from coherent signal fluctuations measured by fMRI, either at rest or during cognitive tasks.
Multiplex family study This is a research design utilizing individuals considered to be at ultrahigh risk for a disease of interest given the presence of multiple affected family members.
Postural sway This refers to the variation in the ability to control the body's position in space involving motor, sensory, and cognitive processes.
Purkinje cells This is a classification of γ-aminobutyric acid-ergic cells in the cerebellum defined by their large size and elaborate dendritic arbor; they appear to be especially sensitive to the neurotoxic effects of alcohol.

KEY FACTS OF THE CEREBELLUM

- The cerebellum is located in the inferior posterior portion of the skull, superior to the brainstem.
- Animal studies from the nineteenth century demonstrated that surgical removal of the cerebellum was associated with motor impairments.
- Clinical observations in humans indicated that cerebellar lesions were also associated with deficits in cognition and behavior.
- A growing body of research has implicated cerebellar involvement in a broad range of neuropsychological functions, including attention, working memory, learning, executive functioning, emotion, and affective states.
- The cerebellar cortex is not functionally homogeneous, and the cerebellum contains localized regions that are interconnected with specific motor or nonmotor areas of the cerebral cortex.
- The cerebellum may also play a role in brain processes related to addiction, including sensitivity to reward, motivation, saliency, inhibitory control, and insight.

SUMMARY POINTS

- This chapter focuses on cerebellar structure and function in AUD and other SUDs.
- The cerebellum is implicated in a broad range of neuropsychological functions, including attention, working memory, executive functioning, emotion, and affect.
- Alcohol and other drugs of abuse may damage the cerebellum at the molecular and cellular levels.

- Individuals with AUD show a reduced volume of the cerebellum, atypical neural activation in the cerebellum during cognitive tasks, and abnormal connectivity between the cerebellum and other brain structures.
- Individuals with other SUDs (cannabis, stimulants, opioids, and tobacco) also show structural and functional abnormalities of the cerebellum.
- Individuals with a family history of SUDs, who are at increased risk for drug and alcohol problems, also show structural and functional abnormalities of the cerebellum.
- Thus, atypical cerebellar structure and function may represent a premorbid risk factor for SUDs.

REFERENCES

American Psychiatric Association. (2013). *Diagnostic and statistical manual of mental disorders* (5th ed.). Washington, DC.

Anderson, C. M., Rabi, K., Lukas, S. E., & Teicher, M. H. (2010). Cerebellar lingula size and experiential risk factors associated with high levels of alcohol and drug use in young adults. *Cerebellum*, 9(2), 198–209. http://dx.doi.org/10.1007/s12311-009-0141-5.

Azizian, A., Monterosso, J., O'Neill, J., & London, E. D. (2009). Magnetic resonance imaging studies of cigarette smoking. In J. E. Henningfield, E. D. London, & S. Pogun (Eds.), *Nicotine psychopharmacology* (pp. 113–143). Berlin, Heidelberg: Springer.

Baker, K. G., Harding, A. J., Halliday, G. M., Kril, J. J., & Harper, C. G. (1999). Neuronal loss in functional zones of the cerebellum of chronic alcoholics with and without Wernicke's encephalopathy. *Neuroscience*, 91(2), 429–438.

Batalla, A., Bhattacharyya, S., Yucel, M., Fusar-Poli, P., Crippa, J. A., Nogue, S., ... Martin-Santos, R. (2013). Structural and functional imaging studies in chronic cannabis users: a systematic review of adolescent and adult findings. *PLoS One*, 8(2), e55821. http://dx.doi.org/10.1371/journal.pone.0055821.

Bendszus, M., Weijers, H. G., Wiesbeck, G., Warmuth-Metz, M., Bartsch, A. J., Engels, S., ... Solymosi, L. (2001). Sequential MR imaging and proton MR spectroscopy in patients who underwent recent detoxification for chronic alcoholism: correlation with clinical and neuropsychological data. *AJNR American Journal of Neuroradiology*, 22(10), 1926–1932.

Benegal, V., Antony, G., Venkatasubramanian, G., & Jayakumar, P. N. (2007). Gray matter volume abnormalities and externalizing symptoms in subjects at high risk for alcohol dependence. *Addiction Biology*, 12(1), 122–132. http://dx.doi.org/10.1111/j.1369-1600.2006.00043.x.

Bora, E., Yucel, M., Fornito, A., Pantelis, C., Harrison, B. J., Cocchi, L., ... Lubman, D. I. (2012). White matter microstructure in opiate addiction. *Addiction Biology*, 17(1), 141–148. http://dx.doi.org/10.1111/j.1369-1600.2010.00266.x.

Bossong, M. G., & Niesink, R. J. (2010). Adolescent brain maturation, the endogenous cannabinoid system and the neurobiology of cannabis-induced schizophrenia. *Progress in Neurobiology*, 92(3), 370–385.

Buckner, R. L., Andrews-Hanna, J. R., & Schacter, D. L. (2008). The brain's default network: anatomy, function, and relevance to disease. *Annals of the New York Academy of Sciences*, 1124, 1–38. http://dx.doi.org/10.1196/annals.1440.011.

Buckner, R. L., Krienen, F. M., Castellanos, A., Diaz, J. C., & Yeo, B. T. (2011). The organization of the human cerebellum estimated by intrinsic functional connectivity. *Journal of Neurophysiology*, 106(5), 2322–2345. http://dx.doi.org/10.1152/jn.00339.2011.

Chanraud, S., Martelli, C., Delain, F., Kostogianni, N., Douaud, G., Aubin, H. J., ... Martinot, J. L. (2007). Brain morphometry and cognitive performance in detoxified alcohol-dependents with preserved psychosocial functioning. *Neuropsychopharmacology*, 32(2), 429–438. http://dx.doi.org/10.1038/sj.npp.1301219.

Chanraud, S., Pitel, A. L., Pfefferbaum, A., & Sullivan, E. V. (2011). Disruption of functional connectivity of the default-mode network in alcoholism. *Cerebral Cortex*, 21(10), 2272–2281. http://dx.doi.org/10.1093/cercor/bhq297.

Chanraud, S., Reynaud, M., Wessa, M., Penttila, J., Kostogianni, N., Cachia, A., ... Martinot, J. L. (2009). Diffusion tensor tractography in mesencephalic bundles: relation to mental flexibility in detoxified alcohol-dependent subjects. *Neuropsychopharmacology*, 34(5), 1223–1232. http://dx.doi.org/10.1038/Npp.2008.101.

Chu, S., Xiao, D., Wang, S., Peng, P., Xie, T., He, Y., & Wang, C. (2014). Spontaneous brain activity in chronic smokers revealed by fractional amplitude of low frequency fluctuation analysis: a resting state functional magnetic resonance imaging study. *China Medical Journal*, 127(8), 1504–1509.

Claus, E. D., Ewing, S. W., Filbey, F. M., Sabbineni, A., & Hutchison, K. E. (2011). Identifying neurobiological phenotypes associated with alcohol use disorder severity. *Neuropsychopharmacology*, 36(10), 2086–2096. http://dx.doi.org/10.1038/npp.2011.99.

Compton, W. M., Thomas, Y. F., Stinson, F. S., & Grant, B. F. (2007). Prevalence, correlates, disability, and comorbidity of DSM-IV drug abuse and dependence in the United States: results from the National Epidemiologic Survey on Alcohol and Related Conditions. *Archives of General Psychiatry*, 64(5), 566–576.

Cousijn, J., Wiers, R. W., Ridderinkhof, K. R., van den Brink, W., Veltman, D. J., & Goudriaan, A. E. (2012). Grey matter alterations associated with cannabis use: results of a VBM study in heavy cannabis users and healthy controls. *NeuroImage*, 59(4), 3845–3851. http://dx.doi.org/10.1016/j.neuroimage.2011.09.046.

Cowan, R. L., Lyoo, I. K., Sung, S. M., Ahn, K. H., Kim, M. J., Hwang, J., ... Renshaw, P. F. (2003). Reduced cortical gray matter density in human MDMA (Ecstasy) users: a voxel-based morphometry study. *Drug and Alcohol Dependence*, 72(3), 225–235.

Cservenka, A., Casimo, K., Fair, D. A., & Nagel, B. J. (2014). Resting state functional connectivity of the nucleus accumbens in youth with a family history of alcoholism. *Psychiatry Research*, 221(3), 210–219. http://dx.doi.org/10.1016/j.pscychresns.2013.12.004.

Cservenka, A., & Nagel, B. J. (2012). Risky decision-making: an FMRI study of youth at high risk for alcoholism. *Alcoholism, Clinical and Experimental Research*, 36(4), 604–615. http://dx.doi.org/10.1111/j.1530-0277.2011.01650.x.

Dervaux, A., Bourdel, M. C., Laqueille, X., & Krebs, M. O. (2013). Neurological soft signs in non-psychotic patients with cannabis dependence. *Addiction Biology*, 18(2), 214–221. http://dx.doi.org/10.1111/j.1369-1600.2010.00261.x.

Desmond, J. E., Chen, S. H. A., DeRosa, E., Pryor, M. R., Pfefferbaum, A., & Sullivan, E. V. (2003). Increased frontocerebellar activation in alcoholics during verbal working memory: an fMRI study. *NeuroImage*, 19(4), 1510–1520. http://dx.doi.org/10.1016/S1053-8119(03)00102-2.

Durazzo, T. C., Gazdzinski, S., Banys, P., & Meyerhoff, D. J. (2004). Cigarette smoking exacerbates chronic alcohol-induced brain damage: a preliminary metabolite imaging study. *Alcoholism, Clinical and Experimental Research*, 28(12), 1849–1860.

Ersche, K. D., Jones, P. S., Williams, G. B., Smith, D. G., Bullmore, E. T., & Robbins, T. W. (2013). Distinctive personality traits and neural correlates associated with stimulant drug use versus familial risk of stimulant dependence. *Biological Psychiatry*, 74(2), 137–144. http://dx.doi.org/10.1016/j.biopsych.2012.11.016.

Field, M., & Cox, W. M. (2008). Attentional bias in addictive behaviors: a review of its development, causes, and consequences. *Drug and Alcohol Dependence*, 97(1–2), 1–20. http://dx.doi.org/10.1016/j.drugalcdep.2008.03.030.

Fox, M. D., & Raichle, M. E. (2007). Spontaneous fluctuations in brain activity observed with functional magnetic resonance imaging. *Nature Reviews Neuroscience*, 8(9), 700–711. http://dx.doi.org/10.1038/nrn2201.

Fryer, S. L., Jorgensen, K. W., Yetter, E. J., Daurignac, E. C., Watson, T. D., Shanbhag, H., ... Mathalon, D. H. (2013). Differential brain response to alcohol cue distractors across stages of alcohol dependence. *Biological Psychology*, 92(2), 282–291. http://dx.doi.org/10.1016/j.biopsycho.2012.10.004.

Gogtay, N., Giedd, J. N., Lusk, L., Hayashi, K. M., Greenstein, D., Vaituzis, A. C., ... Thompson, P. M. (2004). Dynamic mapping of human cortical development during childhood through early adulthood. *Proceedings of the National Academy of Sciences of the United States of America*, 101(21), 8174–8179. http://dx.doi.org/10.1073/pnas.0402680101.

Grant, B. F., & Dawson, D. A. (1997). Age at onset of alcohol use and its association with DSM-IV alcohol abuse and dependence: results from the National Longitudinal Alcohol Epidemiologic Survey. *Journal of Substance Abuse*, 9, 103–110.

Harper, C. (1998). The neuropathology of alcohol-specific brain damage, or does alcohol damage the brain? *Journal Neuropathology and Experimental Neurology*, 57(2), 101–110.

Hasin, D. S., Stinson, F. S., Ogburn, E., & Grant, B. F. (2007). Prevalence, correlates, disability, and comorbidity of DSM-IV alcohol abuse and dependence in the United States: results from the National Epidemiologic Survey on Alcohol and Related Conditions. *Archives of General Psychiatry*, 64(7), 830–842.

Herting, M. M., Fair, D., & Nagel, B. J. (2011). Altered fronto-cerebellar connectivity in alcohol-naive youth with a family history of alcoholism. *NeuroImage*, 54(4), 2582–2589. http://dx.doi.org/10.1016/j.neuroimage.2010.10.030.

Hill, S. Y., Muddasani, S., Prasad, K., Nutche, J., Steinhauer, S. R., Scanlon, J., ... Keshavan, M. (2007). Cerebellar volume in offspring from multiplex alcohol dependence families. *Biological Psychiatry*, 61(1), 41–47. http://dx.doi.org/10.1016/j.biopsych.2006.01.007.

Hill, S. Y., Shen, S., Locke, J., Lowers, L., Steinhauer, S., & Konicky, C. (2000). Developmental changes in postural sway in children at high and low risk for developing alcohol-related disorders. *Biological Psychiatry*, 47(6), 501–511.

Hill, S. Y., Shen, S., Lowers, L., Locke-Wellman, J., Matthews, A. G., & McDermott, M. (2008). Psychopathology in offspring from multiplex alcohol dependence families with and without parental alcohol dependence: a prospective study during childhood and adolescence. *Psychiatry Research*, 160(2), 155–166. http://dx.doi.org/10.1016/j.psychres.2008.04.017.

Hill, S. Y., Steinhauer, S. R., Locke-Wellman, J., & Ulrich, R. (2009). Childhood risk factors for young adult substance dependence outcome in offspring from multiplex alcohol dependence families: a prospective study. *Biological Psychiatry*, 66(8), 750–757. http://dx.doi.org/10.1016/j.biopsych.2009.05.030.

Hill, S. Y., Tessner, K. D., & McDermott, M. D. (2011). Psychopathology in offspring from families of alcohol dependent female probands: a prospective study. *Journal Psychiatric Research*, 45(3), 285–294. http://dx.doi.org/10.1016/j.jpsychires.2010.08.005.

Hill, S. Y., Wang, S., Carter, H., Tessner, K., Holmes, B., McDermott, M., ... Stiffler, S. (2011). Cerebellum volume in high-risk offspring from multiplex alcohol dependence families: association with allelic variation in GABRA2 and BDNF. *Psychiatry Research*, 194(3), 304–313. http://dx.doi.org/10.1016/j.pscychresns.2011.05.006.

Johnston, L. D., O'Malley, P. M., Miech, R. A., Bachman, J. G., & Schulenberg, J. E. (2014). *Monitoring the future national results on drug use: 1975-2013: Overview, key findings on adolescent drug use*. Ann Arbor, MI: Institute for Social Research, The University of Michigan.

Kendler, K. S., Schmitt, E., Aggen, S. H., & Prescott, C. A. (2008). Genetic and environmental influences on alcohol, caffeine, cannabis, and nicotine use from early adolescence to middle adulthood. *Archives of General Psychiatry*, 65(6), 674–682. http://dx.doi.org/10.1001/archpsyc.65.6.674.

Kril, J. J., & Harper, C. G. (1989). Neuronal counts from four cortical regions of alcoholic brains. *Acta Neuropathologica*, 79(2), 200–204.

Lin, W. C., Chou, K. H., Chen, C. C., Huang, C. C., Chen, H. L., Lu, C. H., ... Lin, C. P. (2012). White matter abnormalities correlating with memory and depression in heroin users under methadone maintenance treatment. *PLoS One*, 7(4), e33809. http://dx.doi.org/10.1371/journal.pone.0033809.

Lin, W. C., Chou, K. H., Chen, H. L., Huang, C. C., Lu, C. H., Li, S. H., ... Chen, C. C. (2012). Structural deficits in the emotion circuit and cerebellum are associated with depression, anxiety and cognitive dysfunction in methadone maintenance patients: a voxel-based morphometric study. *Psychiatry Research*, 201(2), 89–97. http://dx.doi.org/10.1016/j.pscychresns.2011.05.009.

Mackey, S., Stewart, J. L., Connolly, C. G., Tapert, S. F., & Paulus, M. P. (2014). A voxel-based morphometry study of young occasional users of amphetamine-type stimulants and cocaine. *Drug and Alcohol Dependence*, 135, 104–111. http://dx.doi.org/10.1016/j.drugalcdep.2013.11.018.

Mackiewicz Seghete, K. L., Cservenka, A., Herting, M. M., & Nagel, B. J. (2013). Atypical spatial working memory and task-general brain activity in adolescents with a family history of alcoholism. *Alcoholism, Clinical and Experimental Research*, 37(3), 390–398. http://dx.doi.org/10.1111/j.1530-0277.2012.01948.x.

Manzardo, A. M., Penick, E. C., Knop, J., Nickel, E. J., Hall, S., Jensen, P., & Gabrielli, W. F., Jr. (2005). Developmental differences in childhood motor coordination predict adult alcohol dependence: proposed role for the cerebellum in alcoholism. *Alcoholism, Clinical and Experimental Research*, 29(3), 353–357.

Medina, K. L., Nagel, B. J., & Tapert, S. F. (2010). Abnormal cerebellar morphometry in abstinent adolescent marijuana users. *Psychiatry Research*, 182(2), 152–159. http://dx.doi.org/10.1016/j.pscychresns.2009.12.004.

Miquel, M., Toledo, R., Garcia, L. I., Coria-Avila, G. A., & Manzo, J. (2009). Why should we keep the cerebellum in mind when thinking about addiction? *Current Drug Abuse Reviews*, 2(1), 26–40.

Monnig, M. A., Thayer, R. E., Caprihan, A., Claus, E. D., Yeo, R. A., Calhoun, V. D., & Hutchison, K. E. (2014). White matter integrity is associated with alcohol cue reactivity in heavy drinkers. *Brain and Behavior*, 4(2), 158–170. http://dx.doi.org/10.1002/brb3.204.

Moreno-Lopez, L., Perales, J. C., van Son, D., Albein-Urios, N., Soriano-Mas, C., Martinez-Gonzalez, J. M., ... Verdejo-Garcia, A. (2015). Cocaine use severity and cerebellar gray matter are associated with reversal learning deficits in cocaine-dependent individuals. *Addiction Biology*, 20(3), 546–556. http://dx.doi.org/10.1111/adb.12143.

Morey, R. A., Petty, C. M., Xu, Y., Hayes, J. P., Wagner, H. R., II, Lewis, D. V., ... McCarthy, G. (2009). A comparison of automated segmentation and manual tracing for quantifying hippocampal and amygdala volumes. *NeuroImage*, 45, 855–866.

Müller-Oehring, E. M., Jung, Y. C., Pfefferbaum, A., Sullivan, E. V., & Schulte, T. (2015). The resting brain of alcoholics. *Cerebral Cortex*, 25(11), 4155–4168. http://dx.doi.org/10.1093/cercor/bhu134.

Nakamura-Palacios, E. M., Souza, R. S., Zago-Gomes, M. P., de Melo, A. M., Braga, F. S., Kubo, T. T., & Gasparetto, E. L. (2014). Gray matter volume in left rostral middle frontal and left cerebellar cortices predicts frontal executive performance in alcoholic subjects. *Alcoholism, Clinical and Experimental Research, 38*(4), 1126–1133. http://dx.doi.org/10.1111/acer.12308.

O'Halloran, C. J., Kinsella, G. J., & Storey, E. (2012). The cerebellum and neuropsychological functioning: a critical review. *Journal of Clinical and Experimental Neuropsychology, 34*(1), 35–56. http://dx.doi.org/10.1080/13803395.2011.614599.

O'Neill, J., Cardenas, V. A., & Meyerhoff, D. J. (2001). Separate and interactive effects of cocaine and alcohol dependence on brain structures and metabolites: quantitative MRI and proton MR spectroscopic imaging. *Addiction Biology, 6*(4), 347–361. http://dx.doi.org/10.1080/13556210020077073.

Oscar-Berman, M., & Marinkovic, K. (2007). Alcohol: effects on neurobehavioral functions and the brain. *Neuropsychology Review, 17*(3), 239–257. http://dx.doi.org/10.1007/s11065-007-9038-6.

Parks, M. H., Dawant, B. M., Riddle, W. R., Hartmann, S. L., Dietrich, M. S., Nickel, M. K., … Martin, P. R. (2002). Longitudinal brain metabolic characterization of chronic alcoholics with proton magnetic resonance spectroscopy. *Alcoholism Clinical and Experimental Research, 26*(9), 1368–1380. http://dx.doi.org/10.1097/01.ALC.0000029598.07833.2D.

Parks, M. H., Greenberg, D. S., Nickel, M. K., Dietrich, M. S., Rogers, B. P., & Martin, P. R. (2010). Recruitment of additional brain regions to accomplish simple motor tasks in chronic alcohol-dependent patients. *Alcoholism-Clinical and Experimental Research, 34*(6), 1098–1109. http://dx.doi.org/10.1111/j.1530-0277.2010.01186.x.

Peper, J. S., Brouwer, R. M., Boomsma, D. I., Kahn, R. S., & Hulshoff Pol, H. E. (2007). Genetic influences on human brain structure: a review of brain imaging studies in twins. *Human Brain Mapping, 28*(6), 464–473. http://dx.doi.org/10.1002/hbm.20398.

Phillips, S. C., Harper, C. G., & Kril, J. (1987). A quantitative histological study of the cerebellar vermis in alcoholic patients. *Brain, 110*(Pt 2), 301–314.

Rogers, B. P., Parks, M. H., Nickel, M. K., Katwal, S. B., & Martin, P. R. (2012). Reduced fronto-cerebellar functional connectivity in chronic alcoholic patients. *Alcoholism, Clinical and Experimental Research, 36*(2), 294–301. http://dx.doi.org/10.1111/j.1530-0277.2011.01614.x.

Schmahmann, J. D. (1991). An emerging concept. The cerebellar contribution to higher function. *Archives of Neurology, 48*(11), 1178–1187.

Schneider, F., Habel, U., Wagner, M., Franke, P., Salloum, J. B., Shah, N. J., … Zilles, K. (2001). Subcortical correlates of craving in recently abstinent alcoholic patients. *The American Journal of Psychiatry, 158*(7), 1075–1083.

Schulte, T., Müller-Oehring, E. M., Rohlfing, T., Pfefferbaum, A., & Sullivan, E. V. (2010). White matter fiber degradation attenuates hemispheric asymmetry when integrating visuomotor information. *Journal of Neuroscience, 30*(36), 12168–12178. http://dx.doi.org/10.1523/Jneurosci.2160-10.2010.

Squeglia, L. M., Rinker, D. A., Bartsch, H., Castro, N., Chung, Y., Dale, A. M., … Tapert, S. F. (2014). Brain volume reductions in adolescent heavy drinkers. *Developmental Cognitive Neuroscience, 9*, 117–125. http://dx.doi.org/10.1016/j.dcn.2014.02.005.

Steinmetz, A. B., Edwards, C. R., Vollmer, J. M., Erickson, M. A., O'Donnell, B. F., Hetrick, W. P., & Skosnik, P. D. (2012). Examining the effects of former cannabis use on cerebellum-dependent eyeblink conditioning in humans. *Psychopharmacology (Berl), 221*(1), 133–141. http://dx.doi.org/10.1007/s00213-011-2556-1.

Stoodley, C. J., & Schmahmann, J. D. (2009). Functional topography in the human cerebellum: a meta-analysis of neuroimaging studies. *NeuroImage, 44*(2), 489–501. http://dx.doi.org/10.1016/j.neuroimage.2008.08.039.

Stoodley, C. J., & Schmahmann, J. D. (2010). Evidence for topographic organization in the cerebellum of motor control versus cognitive and affective processing. *Cortex, 46*(7), 831–844. http://dx.doi.org/10.1016/j.cortex.2009.11.008.

Strick, P. L., Dum, R. P., & Fiez, J. A. (2009). Cerebellum and non-motor function. *Annual Review of Neuroscience, 32*, 413–434. http://dx.doi.org/10.1146/annurev.neuro.31.060407.125606.

Sullivan, E. V. (2003). Compromised pontocerebellar and cerebellothalamocortical systems: speculations on their contributions to cognitive and motor impairment in nonamnesic alcoholism. *Alcoholism-Clinical and Experimental Research, 27*(9), 1409–1419. http://dx.doi.org/10.1097/01.Alc.0000085586.91726.46.

Sullivan, E. V., Deshmukh, A., Desmond, J. E., Lim, K. O., & Pfefferbaum, A. (2000). Cerebellar volume decline in normal aging, alcoholism, and Korsakoff's syndrome: relation to ataxia. *Neuropsychology, 14*(3), 341–352. http://dx.doi.org/10.1037/0894-4105.14.3.341.

Sullivan, E. V., Rose, J., & Pfefferbaum, A. (2006). Effect of vision, touch and stance on cerebellar vermian-related sway and tremor: a quantitative physiological and MRI study. *Cerebral Cortex, 16*(8), 1077–1086. http://dx.doi.org/10.1093/cercor/bhj048.

Sultan, F., Mock, M., & Thier, P. (2000). Functional architecture of the cerebellar system. In T. Klockgether (Ed.), *Handbook of ataxia disorders* (pp. 1–52). New York: Marcel Dekker.

Tiemeier, H., Lenroot, R. K., Greenstein, D. K., Tran, L., Pierson, R., & Giedd, J. N. (2010). Cerebellum development during childhood and adolescence: a longitudinal morphometric MRI study. *NeuroImage, 49*(1), 63–70. http://dx.doi.org/10.1016/j.neuroimage.2009.08.016.

Victor, M., Adams, R. D., & Mancall, E. L. (1959). A restricted form of cerebellar cortical degeneration occurring in alcoholic patients. *AMA Archives of Neurology, 1*(6), 579–688.

Wang, S. S., Kloth, A. D., & Badura, A. (2014). The cerebellum, sensitive periods, and autism. *Neuron, 83*(3), 518–532. http://dx.doi.org/10.1016/j.neuron.2014.07.016.

Wetherill, R. R., Squeglia, L. M., Yang, T. T., & Tapert, S. F. (2013). A longitudinal examination of adolescent response inhibition: neural differences before and after the initiation of heavy drinking. *Psychopharmacology (Berl), 230*(4), 663–671. http://dx.doi.org/10.1007/s00213-013-3198-2.

Wierenga, L., Langen, M., Ambrosino, S., van Dijk, S., Oranje, B., & Durston, S. (2014). Typical development of basal ganglia, hippocampus, amygdala and cerebellum from age 7 to 24. *NeuroImage, 96*, 67–72. http://dx.doi.org/10.1016/j.neuroimage.2014.03.072.

World Health Organization. (2008). *WHO Report on the Global Tobacco Epidemic, 2008: The MPOWER Package.* Geneva: World Health Organization.

Yalachkov, Y., Kaiser, J., Gorres, A., Seehaus, A., & Naumer, M. J. (2013). Sensory modality of smoking cues modulates neural cue reactivity. *Psychopharmacology (Berl), 225*(2), 461–471. http://dx.doi.org/10.1007/s00213-012-2830-x.

Yalachkov, Y., Kaiser, J., & Naumer, M. J. (2009). Brain regions related to tool use and action knowledge reflect nicotine dependence. *Journal of Neuroscience, 29*(15), 4922–4929. http://dx.doi.org/10.1523/JNEUROSCI.4891-08.2009.

Yeh, P. H., Simpson, K., Durazzo, T. C., Gazdzinski, S., & Meyerhoff, D. J. (2009). Tract-based spatial statistics (TBSS) of diffusion tensor imaging data in alcohol dependence: abnormalities of the motivational neurocircuitry. *Psychiatry Research-Neuroimaging, 173*(1), 22–30. http://dx.doi.org/10.1016/j.pscychresns.2008.07.012.

Chapter 54

Brain Volumes in Adolescents with Alcohol Use Disorder

Samanth J. Brooks[1], Dan J. Stein[2]

[1]Department of Psychiatry and Mental Health, Groote Schuur Hospital, University of Cape Town, Cape Town, South Africa; [2]Department of Psychiatry and Mental Health, Groote Schuur Hospital, MRC Unit on Anxiety and Stress Disorders, University of Cape Town, Cape Town, South Africa

INTRODUCTION

Alcohol use disorder (AUD), a psychiatric condition defined in the latest version of the American Psychiatric Association's *Diagnostic and Statistical Manual*, version 5 (DSM-5; American Psychiatric Association, 2013) falls under the rubric of substance abuse disorders. Previously, the DSM-IV (American Psychiatric Association, 1994) differentiated substance abuse and dependence, with abuse being conceptualized as a mild or early phase and dependence as the more severe manifestation. However, clinicians often reported that patients meeting the abuse criteria had severe illness, and the revised substance use disorder may better accommodate the symptoms presented and experienced by patients. Furthermore, clinicians reported that it was often confusing and difficult to define and diagnose dependence in their patients based on the DSM-IV criteria. Moreover, given that the general public often views dependence synonymously with "addiction," the term was deemed misleading, as dependence can also be a normal bodily response to a substance. According to the DSM-5 definition of substance abuse disorder, patients must meet at least 2 of the 11 criteria[1], which are clustered into four sections, (1) impaired control, (2) social impairment, (3) risky use, and (4) pharmacological dependence, in order to be diagnosed. Further, the number of criteria met gauges the severity of AUD: mild (2 or 3 criteria), moderate (4 or 5 criteria), and severe (6 or more criteria).

Adolescent AUD is a pertinent issue, given that consumption of alcohol disrupts normal brain development (Rourke, 1996), with cortical thickening and thinning from childhood during adolescence up to early adulthood (Sowell et al., 2004). Moreover, the prefrontal cortex and hippocampus are particularly susceptible to major neurodevelopmental changes during adolescence (De Bellis et al., 1999; Giedd et al., 1996; Jernigan et al., 2005; Suzuki et al., 2005). Protracted alcohol abuse during adolescence disrupts the normal rapid development of specific brain functions and can be observed using modern structural brain imaging. Gaining better insight into the links between adolescent AUD and brain volume may help pinpoint which brain regions, and therefore which brain functions, are associated with AUD. Such knowledge may ultimately enable clinicians to better treat adolescents who abuse alcohol and adults who began abusing alcohol during adolescence. Here we review the main differences in brain volume between adolescents with AUD and healthy controls and explore moderators or mediators (e.g., gender, psychiatric comorbidity, family history of alcohol abuse, and experience of childhood trauma) of the relationship between adolescent AUD and brain volume (Figure 1).

VOXEL-BASED MORPHOMETRY AND RELATED METHODS OF STRUCTURAL BRAIN IMAGING

Some of the studies included in this review (Table 1) have utilized an automated technique called *voxel-based morphometry* (VBM), a statistical parametric mapping (SPM), automated software approach developed by John Ashburner to estimate differential brain volume in subject groups, voxel by voxel (Ashburner & Friston, 2000). Of note, a voxel is a three-dimensional pixel contained within the acquired brain image and is usually an isotropic 3×3-mm cube. The VBM process begins by coregistering an individual's brain image to a template (usually the Montreal Neurological Institute template, and sometimes applying other registration techniques such as DARTEL; see Ashburner & Friston, 2000), which helps to account for individual differences in brain anatomy that are not related to the experimental condition being tested. The images are then segmented by the automated sequence into gray matter, white matter, and cerebrospinal fluid and finally smoothed using a Gaussian filter to account for outlying voxel intensities. Probability maps are then parametrically constructed

---

1. Impaired control: (1) taking more or for longer than intended; (2) unsuccessful efforts to stop or cut down use; (3) spending a great deal of time obtaining, using, or recovering from use; (4) craving for substance. Social impairment: (5) failure to fulfill major obligations because of use, (6) continued use despite problems caused or exacerbated by use, (7) important activities given up or reduced because of substance use. Risky use: (8) recurrent use in hazardous situations, (9) continued use despite physical or psychological problems that are caused or exacerbated by substance use. Pharmacologic dependence: (10) tolerance to effects of the substance, (11) withdrawal symptoms when not using or using less.

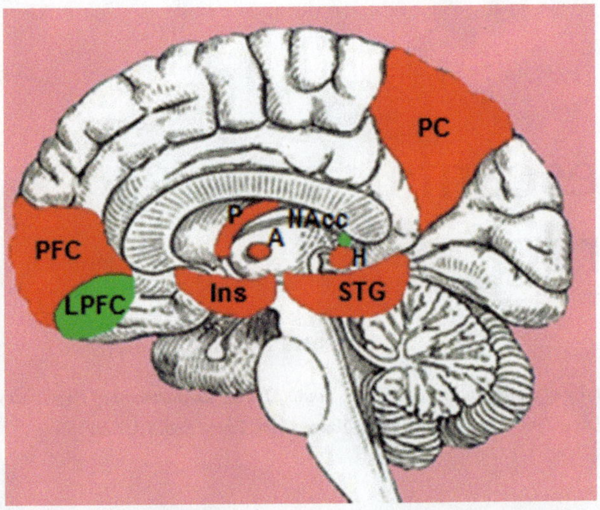

FIGURE 1 Schematic diagram of the brain regions involved in adolescent AUD. In red: regions that appear to be reduced; in green: regions that appear to be enlarged. PFC, prefrontal cortex; LPFC, lateral prefrontal cortex; Ins, insular cortex (although this region is on the lateral area of the cortex, beneath the temporal cortex); STG, superior temporal gyrus (although this region is on the lateral surface of the cortex); P, putamen (part of the dorsal striatum); A, amygdala; NAcc, nucleus accumbens; H, hippocampus; PC, parietal cortex.

using a general linear model approach to provide estimates of brain volume according to clusters of voxel intensity values, usually corrected using family-wise error (FWE) to account for multiple testing across thousands of voxels.

Other methods to measure brain structure of the human brain in vivo include semiautomated techniques and manual tracing. FreeSurfer, for example, measures both volume and cortical thickness, which can help to elucidate differences in the shape and structure of the cortex and its folding patterns, which may be altered in mental disorder. Manual tracing is another technique using a computer-interfaced tracing tool, often for subcortical regions such as the amygdala and substructures of the striatum that might be difficult for automated techniques to accurately measure, owing to their size and location.

HIPPOCAMPAL VOLUME IS REDUCED IN ADOLESCENTS WITH AUD

The hippocampus is associated with contextual learning and memory (see Table 2) (Eichenbaum, 1999), which are disrupted by protracted alcohol abuse during adolescence (Brown, Tapert, Granholm, & Delis, 2000). Basal ganglia, a group of limbic structures to which the hippocampi belong, as well as the prefrontal cortex, develop more rapidly during adolescence than childhood or adulthood. For example, the hippocampus increases in size, whereas the prefrontal cortex decreases and is perhaps a mark of synaptic pruning (De Bellis et al., 1999; Gied et al., 1996; Jernigan et al., 2005; Pfefferbaum et al., 1994). It is interesting to note that the first VBM study to be published demonstrated a significant increase in hippocampal volume associated with London taxi drivers completing the "Knowledge" test (a 5- to 7-year test of navigation around all the streets in London, required to gain a taxi license) (Maquire et al., 2000).

The first study to examine the effects of adolescent AUD on hippocampal and amygdala volumes was also conducted in 2000 (De Bellis et al., 2000) using semiautomated brain-tracing software. Previous data on the neurodevelopmental trajectories of the hippocampus prompted the authors to focus their inquiry on this region (as well as the amygdala), which forms part of the dopaminergic corticolimbic pathway associated with reward, motivation, and contextual memory. They specifically examined whether alcohol intake influences normal development in these regions in 12 male and female adolescents with AUD, comparing them to 24 healthy nondrinking adolescents. It was found that adolescents with AUD had significantly smaller bilateral hippocampal regions compared to controls, also after correction for gender and intracranial volume (Figure 2). Furthermore, hippocampal volume in the AUD adolescent group correlated positively with age of onset (e.g., the later onset corresponded to larger hippocampal volume) and negatively with duration of AUD (e.g., longer duration of alcohol abuse was associated with smaller hippocampal volumes). However, in the study by De Bellis et al., the amount of alcohol consumed did not correlate with hippocampal volume. The authors concluded that reduced hippocampal volume, which normally increases during adolescence, could be a symptom of alcohol-induced neurotoxicity and apoptosis.

Two other studies after the study by De Bellis et al. examined hippocampal volume differences between adolescents with AUD and a nondrinking group, using semiautomated software and manual tracing techniques (Medina, Schweinsburg, Cohen-Zion, Nagel, & Tapert, 2007; Nagel, Schweinsburg, Phan, & Tapert, 2005). Nagel and colleagues examined 14 AUD adolescents without psychiatric or other substance abuse comorbidities versus 17 healthy nondrinking adolescents. The authors found that AUD teens had significantly smaller left hippocampal volume than healthy controls, but again that levels of alcohol consumption did not correlate with hippocampal volume. Nagel and colleagues suggested therefore that smaller left hippocampal volume might be a premorbid and predisposing factor for AUD. Similarly, Medina and colleagues found that in 16 adolescents with AUD there was greater right to left hippocampal asymmetry (e.g., the left hippocampus was smaller) compared to 21 healthy controls and that the asymmetry was not influenced by gender. Medina and colleagues explored further whether severity of AUD, according to the DSM-IV classification of number of abuse/dependence criteria met, correlated with hippocampal asymmetry. They found that greater right than left hippocampal asymmetry did indeed correlate with number of criteria met (e.g., increased severity). Finally, Medina and colleagues discovered that hippocampal asymmetry in the healthy group predicted better verbal memory and total recall but that there was no such relationship between hippocampal asymmetry and neuropsychological measures in the adolescent AUD group.

A more recent VBM study by Brooks et al. (2014) examined the influence of early life adversity on brain volume in 58 AUD adolescents without psychiatric comorbidity versus 58 healthy controls and, in contrast to the previous studies, conducted a whole-brain, as opposed to a region-of-interest, analysis. A whole-brain analysis is potentially more robust given that region-of-interest analyses are biased by the accuracy of prior hypotheses. Brooks and colleagues found, in line with previous studies, that bilateral hippocampal volumes, but most significantly the left, were smaller in AUD

TABLE 1 Structural Brain Imaging Studies Examining the Effects of Alcohol on the Adolescent Brain, with a Brief Summary at the Bottom

Study (Year)	Title of Paper	Participant Characteristics	Comorbidities/Other Factors	Brain Imaging Tools	Summary of Findings (From Publication Abstracts)
Brooks et al. (2014)	Childhood adversity is linked to differential brain volumes in adolescents with alcohol use disorder: a voxel-based morphometry study.	Adolescent AUD (n=58; age 14.9±0.8) and with no other psychiatric comorbidities. Controls (n=58; age 14.7±0.8) age, gender, and protocol matched, light/nondrinking	Childhood trauma questionnaire (CTQ)	SPM/VBM	Reduced bilateral superior temporal gyrus. Negative correlation in the left hippocampus and right precentral gyrus with CTQ scores. Bilateral hippocampal volume negatively associated with CTQ scores.
Chung and Clark (2014)	Insula white matter volume linked to binge drinking frequency through enhancement motives in treated adolescents.	Adolescent AUD (age 16.6±1.1; n=30) baseline vs follow-up	Treated for substance abuse	FreeSurfer—cortical thickness	Enhancement motives associated with larger left insula white matter volume and frequency of binge drinking at baseline and 1-year follow-up. Right insula white matter volume was positively correlated with obsession/craving for alcohol.
Cservenka et al. (2015)	Family history density of alcoholism relates to left nucleus accumbens volume in adolescent girls.	Adolescents with high risk for AUD: 12–16 years (n=140). Mild family history of AUD (n=62, age 14.03±1.28). Positive family history of AUD (n=78, age 14.53±1.29)	No psychiatric or other comorbidities	FMRIB/FAST	Positive relationship between family history density (presence of AUD in first/second-degree relatives) and left NAcc volume/intra-cranial volume (ICV). Post hoc regressions indicated that this effect was significant only in adolescent females.
De Bellis et al. (2000)	Hippocampal volume in adolescent-onset alcohol use disorders.	Adolescent AUD (n=12, 17.2±2.2 years). Controls (n=24, 17.0±2.4 years) age, gender, height, weight, socioeconomic status, full-scale IQ, and handedness matched.	Cannabis use disorder, major depressive disorder, posttraumatic stress disorder, attention deficit hyperactivity disorder, conduct disorder, oppositional defiant disorder, generalized anxiety disorder.	IMAGE software, semiautomated segmentation approach	Bilateral hippocampal volumes were significantly smaller in subjects with AUD than in comparison subjects. Total hippocampal volume correlated positively with the age at onset and negatively with the duration of the AUD.
De Bellis et al. (2005)	Prefrontal cortex, thalamus, and cerebellar volumes in adolescents and young adults with adolescent onset alcohol use disorders and comorbid mental disorders.	Adolescent AUD (n=14, age 17.0±2.1). Controls (n=28, age 16.9±2.3), age, gender, height, weight, socioeconomic status, full-scale IQ, and handedness matched.	Cannabis use disorder, major depressive disorder, posttraumatic stress disorder, attention deficit hyperactivity disorder, conduct disorder, oppositional defiant disorder, generalized anxiety disorder.	IMAGE software, manual tracing	Adolescents with AUD had smaller PFC and PFC white matter volumes compared with control subjects. Right, left, and total thalamic; pons/brainstem; right and left cerebellar hemispheric; total cerebellar; and cerebellar vermis volumes did not differ between groups. There was a significant sex-by-group effect, indicating that males with an adolescent-onset AUD compared with control males had smaller cerebellar volumes, whereas the two female groups did not differ in cerebellar volumes. PFC volume variables significantly correlated with measures of alcohol consumption.

Continued

TABLE 1 Structural Brain Imaging Studies Examining the Effects of Alcohol on the Adolescent Brain, with a Brief Summary at the Bottom—cont'd

Study (Year)	Title of Paper	Participant Characteristics	Comorbidities/Other Factors	Brain Imaging Tools	Summary of Findings (From Publication Abstracts)
Doallo et al. (2014)	Larger mid-dorsolateral prefrontal gray matter volume in young binge drinkers revealed by voxel-based morphometry.	Adolescent alcohol use ($n=11$; age 22.43 ± 1.03), control subjects ($n=21$; age 22.18 ± 1.08) At least 3 years binge drinking in the alcohol group	No psychiatric or other comorbidities	SPM/VBM	Left middorsolateral PFC volume larger in alcohol use.
Fein et al. (2013)	Cortical and subcortical volumes in adolescents with alcohol dependence but without substance or psychiatric comorbidities.	Adolescent AUD ($n=64$, age 14.94 ± 0.79) Controls ($n=64$, age 14.71 ± 0.76), age and gender matched	No psychiatric or other comorbidities	FSL/ FSL imaging registration and segmentation tool (FIRST)/ VBM	Decreased gray matter density in all AUDs compared to controls located in the left lateral frontal, temporal, and parietal lobes, extending medially deep into the parietal lobe. Drinking boys had smaller thalamic and putamen volumes compared to nondrinking boys. AUD girls had larger thalamic and putamen volumes compared to nondrinking girls.
Medina et al. (2007)	Effects of alcohol and combined marijuana and alcohol use during adolescence on hippocampal volume and asymmetry	Adolescent AUD ($n=16$; age 16.9 ± 0.7) Control subjects ($n=21$; age 17.5 ± 1.1)	Marijuana use	AFNI, manual tracing of hippocampal ROIs	Adolescent alcohol users demonstrated a significantly different pattern of hippocampal asymmetry and reduced left hippocampal volume compared to nonusing controls. Increased alcohol abuse/dependence severity was associated with increased right over left asymmetry and smaller left hippocampal volumes. Greater right than left hippocampal volume was associated with superior verbal memory and total recall performance in controls.
Medina et al. (2008)	Prefrontal cortex volumes in adolescents with alcohol use disorders: Unique gender effects	Adolescent AUD ($n=14$; age 16.85 ± 0.65) Control subjects ($n=17$; age 16.55 ± 0.85)	Marijuana and other drug use	AFNI, manual tracing of PFC volumes	After controlling for conduct disorder, gender, and intracranial volume, AUD teens demonstrated marginally smaller anterior ventral PFC volumes than controls, and significant interactions between group and gender were observed. Compared with same-gender controls, females with AUD demonstrated smaller PFC volumes, while males with AUD had larger PFC volumes. The same pattern was observed for PFC white matter volumes.

Nagel et al. (2005)	Reduced hippocampal volume among adolescents with alcohol use disorders without psychiatric comorbidity.	Adolescent AUD ($n=14$; age 16.75 ± 0.68) Control subjects ($n=17$; age 16.46 ± 0.88), matched for gender, ethnicity, weight, socioeconomic status	No comorbidities	FMRIB/FAST for automated segmentation, AFNI for manual tracing	Adolescents with AUD had significantly smaller left hippocampal volumes than healthy teens, even after removal of teens with comorbid conduct disorder from the analyses. In contrast, the groups did not differ in right hippocampal, intracranial, or gray or white matter volumes or memory performance. Hippocampal volumes were not related to alcohol-consumption rates. These findings indicate that adolescents with AUD, but free from other psychiatric comorbidities, have reduced left hippocampal volume. Because hippocampal volume did not relate to alcohol use characteristics, it is possible that premorbid volumetric differences could account for some of the observed group differences in hippocampal volume.
Squeglia et al. (2012)	Binge drinking differentially affects adolescent male and female brain morphometry.	Adolescent binge drinkers ($n=29$, age 18.2 ± 0.79) Adolescent nondrinkers ($n=30$, age 18.0 ± 1.12) Matched for gender, age, pubertal development, and familial alcoholism	No comorbidities	FreeSurfer—cortical thickness	Binge × gender interactions were observed for cortical thickness in four left frontal regions: frontal pole, pars orbitalis, medial orbital frontal, and rostral anterior cingulate. For all interactions, female bingers had thicker cortices than female controls, while male bingers had thinner cortices than male controls. Thicker left frontal cortices corresponded to poorer visuospatial ability, inhibition, and attention performances for female bingers and worse attention for male bingers.

AUD, alcohol use disorder; HC, healthy control; CTQ, childhood trauma questionnaire; n, number of subjects; SPM, statistical parametric mapping; VBM, voxel-based morphometry; FMRI, functional magnetic resonance imaging of the brain; FAST, FMRIB Automated Segmentation Tool; FSL, FMRIB software library; AFNI, analysis of functional neuroimages; ROI, regions of interest.

TABLE 2 Summary of Structural Imaging Findings with Functional Significance in Adolescents with Alcohol Use Disorder

Brain Region	Structural Significance in Adolescents with AUD	Functional Significance of Brain Volume Differences
Bilateral hippocampal (but most significantly left)	Volumes are significantly smaller with some evidence that reduction is linked to AUD and duration of illness. Reduced volume in the bilateral hippocampus may be a premorbid effect predisposing adolescents to the risk of AUD, and may be associated with self-reported higher levels of childhood trauma. The effects might be more pronounced in the left hippocampus, and greater right than left hippocampal volumes (asymmetry) are associated with superior executive functioning.	Hippocampus function is associated with consolidation of contextual memory and fear learning. Reduced volume may be an indication of neurotoxic effects, for example, excessive exposure to alcohol and trauma during childhood. Reduced hippocampal volumes may adversely influence memory formation and goal-setting in adolescents with AUD. Reduction in the left hippocampus might be associated with specific deficits in memory, for example, immediate and delayed recall.
PFC	Volumes are reduced and are associated with amount of alcohol consumption, in particular the left lateral PFC, and smaller PFC white matter volume. Other studies show larger middorsolateral PFC volume in adolescents who abuse alcohol. There is some evidence for a gender interaction in the PFC, with females who abuse alcohol having smaller gray and white matter PFC volumes, and AUD males having larger PFC volumes. Females have thicker and males have thinner frontal cortices, which is associated with deficits in executive functioning.	Reduction in the left lateral PFC could be associated with deficits in cognitive control and other executive functions (e.g., planning, evaluation, self-referential thought) that hinder self-control over alcohol consumption in adolescents with AUD. Gender differences associated with volumes of the PFC might reflect differences in cognitive–affective responses to alcohol consumption.
Bilateral superior temporal gyrus	Reduced bilateral superior temporal gyrus in adolescents with AUD.	Superior temporal gyrus function is associated with comprehension and emotion processing, and reductions in this region may underlie deficits in affect regulation in adolescents with AUD.
Bilateral parietal cortex	Reduced in adolescents with AUD.	Parietal cortex is associated with sensory processing, especially involving embodiment, and reductions in this region may be related to difficulties in processing bodily responses associated with cognitive–affective processing.
Bilateral insular cortex	Reduced gray matter and white matter volume associated with frequency/amount of bingeing, as well as craving/obsession over alcohol consumption in adolescents with AUD.	Insular cortex function underlies interoceptive awareness, or the processing of bodily sensations, and has been linked to feelings of anxiety, hunger, and emotion. Reduced insula volume could be associated with deficits in affect regulation in those with AUD.
Basal ganglia	Left NAcc significantly larger in adolescents with a family history of alcohol abuse. Smaller thalamic and putamen volumes in adolescents with AUD, but more pronounced in males.	NAcc function is associated with the sensation of wanting and liking, and larger volume could be indicative of greater activation associated with craving for alcohol. The thalamus is associated with general arousal, and reduced volume in the thalamus could be associated with neurotoxic effects of hyperarousal.
Cerebellum	Smaller cerebellar volumes in males with AUD.	Cerebellum is associated with motor function, as well as appetitive processes. Reduced cerebellar volume in adolescents with AUD may be associated with deficits in these processes.

adolescents compared to controls. Furthermore, in the total group and not only the AUD group, levels of childhood trauma negatively correlated with bilateral hippocampal volume, suggesting premorbid effects and the influence of other factors such as genetic profile or socioeconomic status, rather than an interaction between early life adversity and AUD per se. In line with previous studies, Brooks and colleagues were also not able to find a correlation between hippocampal volume and amount of alcohol consumed.

In summary, four studies, as of this writing, have found hippocampal volume reduction in adolescents with AUD compared to healthy controls, most significantly on the left side. Age of onset and duration of AUD, but not amount of alcohol consumption, seem to influence the reduced hippocampal volume observed. There was also some evidence that reduced hippocampal volume indicates premorbid neuronal status and may be associated with the experience of early life adversity that may predispose an adolescent to develop a substance abuse disorder such as AUD. Additionally, gender effects and psychiatric comorbidity may also play roles in differential hippocampal volume in adolescents who abuse alcohol.

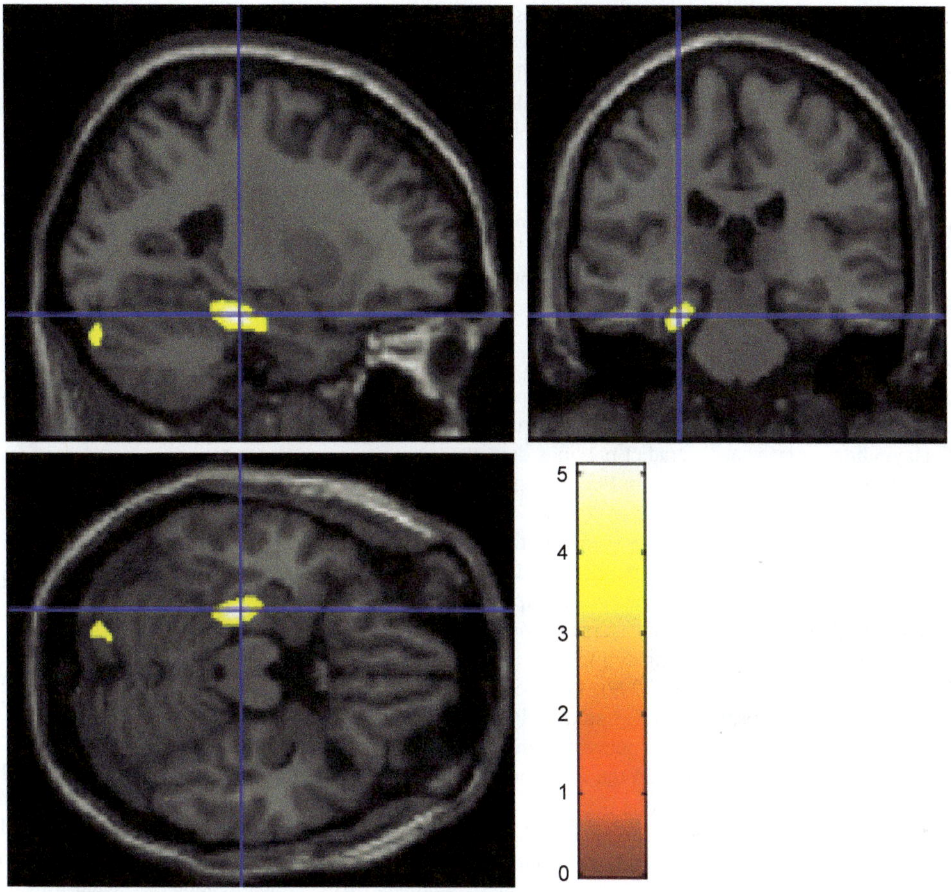

FIGURE 2 **Left hippocampal volume reduction in adolescents with AUD.** Brain map showing significantly reduced left hippocampal volume in adolescents with AUD compared to healthy age- and sex-matched controls *(Brooks et al., 2014)*. Color bar represents t-statistic ($p < 0.05$ FWE corrected).

PREFRONTAL CORTEX VOLUME IS DIFFERENTIAL IN ADOLESCENTS WITH ALCOHOL USE DISORDER

The prefrontal cortex (PFC) is associated with executive functioning, particularly those functions that pertain to self-regulation (Hall & Fong, 2015), such as goal-oriented cognitions, cognitive inhibition of behavior, and evaluation of salient stimuli, and as such aberrant PFC volume may underlie abnormalities in cognitive self-regulation in adolescents with AUD. Five structural imaging studies have reported reductions in PFC volume in adolescents with AUD (De Bellis et al., 2005; Doallo et al., 2014; Fein et al., 2013; Medina et al., 2008; Squeglia et al., 2012). The first study to show reduced PFC volume in adolescents with AUD was by De Bellis and colleagues, who examined 14 male and female AUD adolescents in comparison to 28 healthy adolescents with semiautomatic software and manual tracing techniques. While it must be noted that the AUD participants in this study also had various psychiatric comorbidities (e.g., cannabis use disorder, hallucinogen abuse, major depressive disorder, attention deficit disorder, posttraumatic stress disorder, conduct disorder, oppositional defiant disorder, generalized anxiety disorder, bipolar disorder), reduced gray and white matter volume in the PFC might be associated with adolescent AUD compared to healthy adolescents. Additionally, De Bellis and colleagues found that the average number of drinks per drinking episode, the number of drinks per maximum drinking episode, and the quantity and frequency of alcohol consumed all negatively correlated with the PFC gray matter volume.

Medina et al. (2008) were the next group to show PFC reductions in 14 adolescents with AUD compared to 17 healthy adolescents using semiautomated and manual tracing methods. Adolescents with AUD overall had marginally smaller anterior ventral PFC volumes. Additionally, Medina and colleagues found a gender interaction in that females had smaller, but males had larger (dorsolateral and orbitofrontal) PFC volumes. The authors suggest that this gender difference in PFC volume could be due to the observation that synaptic pruning of the PFC happens later in the adolescent development of boys (Lenroot & Giedd, 2006). Thus, female adolescents may be more susceptible to the neurotoxic effects of alcohol consumption on PFC development than males; conversely, alcohol consumption in males may interfere with healthy synaptic pruning during adolescence. Furthermore, the authors discuss evidence that rich concentrations of excitatory amino acid pathways in the PFC continue to develop during adolescence and so AUD during this life period may be associated with glutamate-mediated excitotoxicity, leading to cell shrinkage and axonal loss. Medina and colleagues go on to discuss the evidence that alcohol-related brain effects may also be due, in part, to an upregulation of inflammatory mediators (e.g., cyclo-oxygenase 2 and inducible nitric oxide synthase) resulting in cell death or damage as a result of protracted alcohol exposure during adolescence.

Squeglia et al. (2012) later reported significant differences in cortical thickness in the PFC of adolescents with AUD and were specifically interested in progressing research on gender differences. They measured 29 adolescent males and females with AUD and 29 age-, gender-, pubertal development-, and family alcohol history-matched controls using FreeSurfer automated software. While the authors note that their significant findings would not have survived strict Bonferroni correction, the effect sizes were within the medium range. Specifically, females with AUD had thicker cortices in the frontal pole, pars orbitalis, medial orbital frontal, and rostral anterior cingulate compared to female controls. Conversely, males with AUD had thinner cortices in these regions than male controls. Furthermore, thicker left frontal cortices corresponded to poorer visuospatial, inhibition, and attention performance for female binge drinkers and worse attention for male binge drinkers. Their findings somewhat contradict previous data that alcohol appears to be associated with smaller PFC volume; however, cortical thickness is not the same measure as cortical volume and might rather illustrate differences in cortical folding that underlie functional abnormalities, for example. As previous studies have also discussed, Squeglia and colleagues reiterate that synaptic refinement, which might be illustrated better by measures of cortical thickness, subserves efficient neural processing during early life stages (Giedd, 2004; Sowell et al., 2004; Spear, 2009). Therefore, cortical thickness in the female AUD sample in this study might indicate disruption to normal synaptic pruning and neural function. The gender differences in the effects of alcohol on the adolescent brain could also be due, in part, to pubertal hormone fluctuations.

Fein et al. (2013) more recently examined a unique cohort of AUD adolescent males and females without any psychiatric comorbidity, which can be a major confounding factor in structural brain imaging studies. Comparing 64 adolescents with AUD with 64 age- and gender-matched healthy adolescents, the authors found a large area of 12.5% decreased gray matter density in the left dorso- and ventrolateral PFC in the AUD group. Further, females had between 9% and 13% lower gray matter density compared to males in the left lateral and left anterior orbitofrontal. Additionally, higher gray matter volume in the left frontal cortex during adolescence was associated with worse psychomotor performance and more self-monitoring behaviors during neuropsychological testing. Also, lower gray matter density in the left lateral and left ventrolateral frontal regions was associated with a higher average number of monthly drinks, with a similar but less significant association on the right side. Thus, Fein and colleagues replicated previous findings but in a nonpsychiatric adolescent population, showing that alcohol consumption appears to be associated with reduced PFC volume and, perhaps, retardation of synaptic pruning in adolescent boys.

Finally, Doallo et al. (2014) were the latest to show differences in the PFC and were specifically interested in the relationship between brain volume in the dorsolateral PFC and executive function (e.g., working memory) in 11 adolescents with AUD compared to 21 healthy controls. Using VBM with region-of-interest (as opposed to whole-brain) analyses, the authors observed larger left middorsolateral PFC volume in the AUD group compared to the age- and gender-matched control group. Further, Doallo and colleagues reported that errors on a working memory task, as well as quantity and speed of alcohol intake, correlated positively with left middorsolateral PFC volume in the AUD group.

In summary, most studies report that PFC cortex gray and white matter volume is reduced in adolescents with AUD compared to healthy nondrinking adolescents and that reduced gray matter volume coincides with amount and frequency of alcohol consumption. However, caution must be applied to the interpretation of data from cohorts with other psychiatric comorbidities. Specifically, dorsolateral PFC and orbitofrontal cortices have been shown to be significantly smaller in adolescents with AUD and are linked to executive function deficits such as working memory, visuospatial ability, inhibition, and attention. Additionally, there was some evidence that boys may have larger PFC volumes, an indication of later synaptic pruning or upregulation of inflammatory mediators, but conversely thinner cortical thickness than girls, which may interact with large-scale hormonal changes and be associated with interference in cortical development.

OTHER BRAIN REGIONS SHOWING DIFFERENTIAL VOLUMES IN ADOLESCENT AUD COMPARED TO CONTROLS

Temporal Cortex

The superior temporal gyrus is involved in comprehension, language, and social cognition (Bigler et al., 2007), and structural alterations may disrupt these normal functions in adolescents with AUD. Two studies found significant reduction in bilateral superior temporal volumes of adolescents with AUD compared to nondrinking adolescents (Brooks et al., 2014; Fein et al., 2013). Brooks and colleagues suggested that reduced superior temporal gyri volume might be associated with Wernicke–Korsakoff syndrome (aphasia typified by memory and comprehension deficits) in adolescents who continue to abuse alcohol, which could be particularly detrimental for school-age individuals who are engaged in learning for formal examinations. Fein and colleagues observed a large area of decreased gray matter density in the left temporal cortex, extending to the frontal and parietal cortices, but these effects were not associated with any "drinking variable." Additionally, Fein and colleagues observed, in line with Squeglia et al. (2012), that larger volumes in the frontal/temporal/parietal region corresponded to worse neuropsychological performance (psychomotor and self-monitoring scores), which on first glance appears counterintuitive. However, as previously discussed, larger volumes may reflect a delay in synaptic pruning or upregulation of inflammatory mediators in this region.

Insular Cortex

The insular cortex, a region embedded in the temporal lobes, is involved in the processing of a variety of bodily sensations, including pleasant experiences during the consumption of substances such as alcohol, which are available to conscious perception (Craig, 2009; Duerden, Arsalidou, Lee, & Taylor, 2013). Pleasant bodily sensations can elicit motivated behaviors, such as craving for more alcohol, especially when self-regulation is impaired (Damasio, 1994; Verdejo-Garcia, Clark, & Dunn, 2012). Against this background, motivation for enhancement (e.g., "to get high") has been described as an incentive for substance abuse disorder (Naqvi & Bechara, 2010) and is supported by the hedonic bodily sensations associated with AUD, which might be represented in the insular cortex. Supported by the previous experimental data on the function of the insular cortex, Chung and colleagues conducted an examination of gray and white matter volumes in the insular cortex of 30 adolescents who were binge drinking, at baseline and 1 year later. Additionally, the authors divided the adolescent sample of binge drinkers into those with a DSM-IV AUD diagnosis ($n=14$) and those without ($n=16$) and found that those with a diagnosis had increased left white matter insula volume. Using a questionnaire to measure motivations for alcohol use, it was found that left insula white matter volume positively correlated with enhancement motivation in the adolescent AUD group. Furthermore, white matter volume in the right insula positively correlated with obsessions/craving for alcohol consumption. Gray matter insula effects were not observed.

Striatum

The striatum is an umbrella term for various substructures within the dopaminergic basal ganglia network, or the mesolimbic reward and motivation pathway of the midbrain. The internal capsule is the major white matter tract dividing the caudate nucleus and the putamen, which both together are often referred to as the dorsal striatum. Beneath these structures lies the nucleus accumbens (NAcc), often referred to as the ventral striatum. The dorsal striatum is generally involved in habitual, stimulus–response Pavlovian conditioning, whereas the ventral striatum is involved in motivated learning or operant conditioning (e.g., "wanting and liking," Berridge, 2007). As of this writing, two studies using automated segmentation techniques have reported alterations in striatal volume in adolescents with AUD (Cservenka et al., 2015; Fein et al., 2013). In the study by Fein et al., drinking boys had smaller putamen volumes than nondrinking boys, whereas drinking girls had larger putamen volumes than nondrinking girls. In the study by Cservenka et al., family history of alcohol use positively correlated with left NAcc volume in girls only. Thus, it appears that consumption of alcohol during adolescence, as well as a possible phenotypic risk, alters striatal volume differently in boys and girls. Larger volumes in the striatum might be indicative of neurotoxic effects of excessive dopamine release or a neuroinflammatory response to alcohol. Larger NAcc is also associated with greater adolescent reward seeking and risk taking and, within a family context of drinking, might exacerbate the onset of adolescent drinking behaviors.

MODERATING AND MEDIATING FACTORS—GENDER, PSYCHIATRIC COMORBIDITIES, FAMILY HISTORY OF ALCOHOL ABUSE, CHILDHOOD TRAUMA

Gender

Various differential relationships between AUD and brain volume were observed between boys and girls in five studies as of this writing (Cservenka et al., 2015; De Bellis et al., 2005; Fein et al., 2013; Medina et al., 2008; Squeglia et al., 2012). The first, by De Bellis and colleagues, observed that boys with AUD had smaller cerebellar volumes than nondrinking boys but that there was no difference in girls with AUD versus healthy girls. Medina and colleagues observed that girls with AUD compared to nondrinking girls had smaller PFC volumes, whereas AUD boys compared to nondrinking boys had larger PFC volumes. Squeglia et al. noted that four left frontal regions, namely the frontal pole, pars orbitalis, medial orbital frontal, and rostral anterior cingulate, were thicker in female bingers than in female controls. Conversely, male bingers had thinner cortices than male controls in these regions. Furthermore, thicker left frontal cortices corresponded with poorer visuospatial ability, inhibition, and attention performance for female bingers and worse attention for male bingers with thinner cortices. Fein and colleagues observed that drinking boys had smaller thalamic and putamen volumes compared to nondrinking boys. Conversely, AUD girls had larger thalamic and putamen volumes compared to nondrinking girls. Finally, Cservenka and colleagues reported a positive correlation between family history of alcohol abuse and volume of the left NAcc, but only in females.

To summarize, gender differences in the relationship between AUD status and brain volumes are intriguing. It could be that differences in the neurobiology between males and females may elucidate why the development and progression of AUD differs between the sexes (Keyes, Martins, Blanco, & Hasin, 2010; Schuckit, Daeppen, Tipp, Hesselbrock, & Bucholz, 1998) and could be due to hormonal differences in response to alcohol, variations in neurotransmitter systems between males and females, and neurotoxicity differences between genders (Ceylan-Isik, McBride, & Ren, 2010). Nevertheless, they provide some evidence that the functional effects of alcohol consumption on the brain may be different for boys and girls, suggesting that tailored neuropsychological intervention is necessary.

Psychiatric Comorbidities

Six of the 11 studies discussed in this chapter included adolescent samples with other comorbidities, including those with experience of childhood trauma (Brooks et al., 2014), those in treatment for other substance abuse (Chung & Clark, 2014), cannabis use disorder, major depressive disorder, posttraumatic stress disorder, attention deficit hyperactivity disorder, conduct disorder, oppositional defiant disorder, generalized anxiety disorder (De Bellis et al., 2000, 2005), and only marijuana use (Medina et al., 2007, 2008). Although it is difficult for researchers to recruit adolescent AUD samples without these comorbidities, it was argued by some authors, who were able to recruit nonpsychiatric samples for their studies, that volumetric brain changes could not be entirely linked to AUD status otherwise (Cservenka et al., 2015; Doallo et al., 2014; Fein et al., 2013; Nagel et al., 2005; Squeglia et al., 2012). Sampling nonpsychiatric cohorts revealed, in adolescents with AUD versus healthy nondrinking adolescents, reduced left hippocampal volume (Nagel et al., 2005); reduced left lateral frontal, temporal, and parietal lobes; smaller (AUD boys) and larger (AUD girls) thalamic and putamen volumes (Fein et al., 2013); larger left middorsolateral PFC volume (Doallo et al., 2014); a positive relationship between family history of AUD and left NAcc volume (Cservenka et al., 2015); and thicker frontal cortices (female bingers) and thinner frontal cortices (male bingers), which were linked to executive function deficits (Squeglia et al., 2012).

Summarizing the findings, a left-lateralized reduction in frontostriatal volume becomes more apparent in adolescents with AUD compared to nondrinking adolescents when other comorbidities are excluded. This is a pertinent finding, given that language function is predominantly supported by left-lateralized frontostriatal networks incorporating Wernicke's and Broca's areas (Tomasi & Volkow, 2012) and that Wernicke–Korsakoff syndrome (associated with deficits in encoding memory for new events, termed *anterograde amnesia*) is a classic symptom of protracted use of alcohol in adults (Brion et al., 2014). Thus, adolescents who continue to abuse alcohol may be at heightened risk of developing serious anterograde amnesia and other executive function deficits in adulthood, which would impinge on otherwise normal neural development. Indeed, some studies discussed here have shown that executive function deficits are already detectable in adolescents with AUD (Fein et al., 2013; Medina et al., 2007; Squeglia et al., 2012).

Childhood Trauma

One study noted that by excluding AUD adolescents with other comorbidities the prevailing finding of reduced hippocampal volume was no longer observed (Fein et al., 2013), which may suggest that there are other mediating factors in the relationship between adolescent AUD and hippocampal volume. Against this background, Brooks et al. (2014) examined how the experience of early life adversity interacts with brain volume in otherwise comorbidity-free adolescents with AUD. This line of inquiry appears to be relevant given that there is mounting evidence to link stress-by-gene interactions to hippocampal reduction (e.g., Frodl et al., 2014; Van Dam, Rando, Potenza, Tuit, & Sinha, 2014), which might in itself increase the risk for adolescent AUD. Brooks and colleagues found that childhood trauma scores negatively correlated with bilateral, but most significantly left, hippocampus, but were unable to demonstrate a link between childhood trauma, adolescent AUD status, and reduced hippocampal volume. Therefore, while it might be the case that early life stress has an impact on the development of the hippocampus, there are perhaps other factors (family history of AUD, genetic predisposition, socioeconomic status, gender) that may determine whether reduced hippocampal volume translates into adolescent AUD.

CONCLUSIONS

Eleven MRI studies as of this writing have compared brain volume and cortical thickness in adolescents who abuse alcohol versus nondrinking adolescents, and the most significant findings in the AUD group were that: (1) the bilateral hippocampus, but particularly the left, is reduced; (2) age of onset positively correlates, and duration of AUD negatively correlates, with hippocampal volume, but amount of alcohol consumed does not; (3) levels of childhood trauma positively correlate with hippocampal volume and may be a predisposing risk, along with other factors such as genetic predisposition, family history of drinking behavior, socioeconomic status, and psychiatric comorbidity; (4) PFC volume is reduced and negatively correlates with amount of alcohol consumed; (5) gender differences in PFC and striatum volumes are observed, in that AUD girls have smaller PFC and larger striatum, whereas boys have larger PFC and smaller striatum, which may be indicative of hormonal, neurotoxic, or synaptic pruning differences; (6) left temporal cortex is reduced in AUD and may be associated with Wernicke–Korsakoff aphasia, and in line with this, executive function deficits in AUD are observed in relation to temporal and prefrontal cortex volume differences; (7) increased white matter volume in the insular cortex may be indicative of increased craving for alcohol; (8) increased striatal volume may reflect neurotoxic responses to alcohol consumption and be related to family history of alcohol use; (9) moderating and mediating factors, such as gender, psychiatric comorbidities, family history of alcohol use, and experience of childhood trauma could be considered when examining the effects of AUD on the adolescent brain.

Together, these findings broadly implicate volumetric changes in the hippocampus and frontostriatal circuitry, which are associated with self-regulation, particularly with regard to emotion, reward, and motivation, as being most susceptible in adolescents who abuse alcohol. Additionally, mounting evidence suggests that the effects are most significantly lateralized to the left hemisphere, implicating learning, memory, and comprehension deficits (e.g., Wernike–Korsakoff aphasia). Moreover, executive function deficits, which are associated with degradation or retardation of the development of the PFC, are being detected by some studies of adolescent AUD. Since 2005, structural brain imaging studies on the neural effects of adolescent abuse of alcohol have truly honed in on the brain regions most associated with this life-damaging disorder. Combining structural brain imaging findings of future studies with neuropsychological measures and gene-by-environment analyses may further improve knowledge of the neurobiological impact of AUD.

APPLICATIONS TO OTHER ADDICTIONS AND SUBSTANCE MISUSE

Brain imaging of volumetric differences between adolescents with AUD and those who are healthy has implicated reduced bilateral hippocampal volume, particularly in the left hippocampus and the PFC, and increased volume in the basal ganglia. Volume reductions in the hippocampus and PFC may underlie problems with externalizing (e.g., extraversion and novelty seeking) in adolescents, which may be a phenotype that increases susceptibility for developing other substance misuse behaviors (Montigny et al., 2013). In line with this, other brain imaging studies have shown that duration of cocaine dependence, heavy cannabis use, and Internet addiction disorder are associated with reduced PFC and increased basal ganglia volumes (Cousijn et al., 2011; Ide et al., 2014; Yuan et al., 2011). Additionally, adolescents engaging in substance misuse show various executive function deficits that pertain to cognitive control, which could predispose to and exacerbate substance misuse (Wiers, Boelema, Nikolaou, & Gladwin, 2015). Finally, functional connectivity between the PFC and the basal ganglia (e.g., hippocampus, amygdala, striatum) develops significantly during adolescence, is involved in cognitive control of behavior and affect regulation, and may be damaged by significant trauma, such as early life adversity and/or substance misuse. Thus, it is becoming increasingly clear as more brain imaging studies are conducted that neural dysfunction between the PFC and the basal ganglia is associated with substance abuse disorder and is probably a biomarker for improvement following successful intervention.

DEFINITION OF TERMS

Adolescence This refers to the period following the onset of puberty, during which a young person develops from a child into an adult, approximately translating as the "teenage years" from 13 to 19. Striking brain changes occur during the adolescent period, which may contribute to a variety of mental health issues.

Alcohol use disorder This is a substance abuse disorder categorized by the DSM-5 and characterized by the harmful consequences of repeated alcohol use, a pattern of compulsive alcohol use, and (sometimes) physiological dependence on alcohol (i.e., tolerance and/or symptoms of withdrawal).

Amygdala This is a structure in the limbic system associated with emotional processing including fear responses and emotional memory, as well as the secretion of hormones, arousal, and the formation of emotional memories. The amygdala is a small, almond-shaped mass of nuclei located in the temporal lobes of the brain near the hippocampus. Researchers have also found that the amygdala plays an important role in emotional learning.

Hippocampus This is a small formation found symmetrically on both sides of the brain, in the shape of a seahorse, and also plays an important role in the limbic system. The hippocampus is involved in the formation of new memories and is also associated with contextual fear conditioning. When both sides of the hippocampus are damaged, the ability to create new memories can be altered, whereas memory is less impeded if only one side of the hippocampus is damaged. Age has an impact on the hippocampus, whereby during adolescence there is rapid growth of the hippocampus, and by the age of 80, up to 20% of the hippocampal volume may be lost.

Magnetic resonance imaging This is a method used to noninvasively image the living brain in vivo, by way of temporarily altering the resonance of protons in various brain tissues (gray matter, white matter, cerebrospinal fluid). By altering the resonance with a powerful magnet while the participant is in a supine position with his or her head inside the scanner, researchers can use the images derived to examine brain tissue alterations, usually across groups and/or in correlation with neuropsychological variables.

Neuroplasticity This is a generic term used to describe both synaptic plasticity and nonsynaptic plasticity—it refers to alterations in neural pathways and synapses due to changes in behavior, environment, neural processes, thinking, and emotions, as well as changes resulting from bodily injury.

Neuropsychology This is a branch of psychology that examines how cognition, emotion, learning, memory, movement, speech, vision, sensory processing, appetite, and other processes are associated with brain structure and function. Neuropsychological testing helps to establish whether brain injury has taken place and to chart improvements during recovery.

Parietal cortex This is the outer layer of one of the major lobes of the mammalian brain and is positioned between the frontal lobe and the occipital lobe and above the temporal cortex. Sensory information is processed by the parietal lobe and incorporates various modalities, including spatial sense and navigation (proprioception) and sense of touch (mechanoreception). The parietal cortex is also part of the dorsal visual stream, or the "where" pathway. Most sensory inputs from the skin (touch, temperature, and pain receptors) relay through the thalamus to the parietal lobe.

Prefrontal cortex This is the gray matter of the anterior part of the frontal lobe that is highly developed in humans and plays a role in the regulation of complex cognitive, emotional, and behavioral functioning. The PFC supports executive functioning in humans, such as cognitive inhibition, working memory, goal-oriented planning, stimulus evaluation, self-referential thought, conflict monitoring, and prediction error detection. Damage to the PFC can lead to alterations in personality while other vital functions remain intact.

Temporal cortex This is the outer layer of one of the major lobes of the mammalian brain and is positioned below the lateral fissure and the parietal lobe and between the occipital lobe and the frontal lobe. The temporal cortex is involved in visual memories, language comprehension, and emotion association and forms part of the ventral visual stream, or the "what" pathway.

Voxel-based morphometry This is a noninvasive, in vivo neuroimaging analysis technique utilizing MRI scans; it allows the measurement of focal differences in estimated brain volume, using the statistical approach of parametric mapping based on a general linear model.

KEY FACTS ABOUT BRAIN VOLUME DIFFERENCES IN ALCOHOL USE DISORDER

- **Bilateral hippocampal** volumes are significantly smaller in subjects with AUD, with some evidence that such reduction is linked to duration of illness.
- Reduced volume in the bilateral hippocampus may be a premorbid effect predisposing adolescents to the risk of AUD and may be associated with self-reported higher levels of childhood trauma.
- The effects might be more pronounced in the left hippocampus, and greater right than left hippocampal volume (asymmetry) has been associated with superior executive functioning.
- PFC volumes are significantly smaller in AUD and are associated with amount of alcohol consumption.
- There is some evidence for a gender interaction in the PFC, in that females who abuse alcohol may have smaller gray and white matter PFC volumes, and AUD males appear to have larger PFC volumes.
- Females may have thicker and males may have thinner frontal cortices, which has been linked to deficits in executive functioning.
- In other brain regions reduced bilateral superior temporal gyrus in AUD adolescents, and smaller parietal lobes, has been found.
- Reduced bilateral insula gray matter and white matter volume associated with frequency/amount of bingeing, as well as craving/obsession over alcohol consumption in adolescents with AUD, has been found.
- The left NAcc is significantly larger in adolescents with a family history of alcohol abuse.
- Smaller thalamic and putamen volumes have been found in adolescents with AUD, but are more pronounced in males.
- Smaller cerebellar volumes in males with AUD have been found.

SUMMARY POINTS

- This chapter focuses on a review of structural brain imaging studies that have examined differences in adolescents with AUD.
- Left hippocampal, prefrontal, temporal, and parietal cortex volumes are particularly susceptible to the effects of alcohol in adolescents.
- The neuropsychological link to brain volume differences in these regions is discussed.
- Other mediators/moderators of brain volume differences, such as the experience of early life adversity, comorbidity with psychiatric conditions, family history of alcohol abuse, and gender are discussed.
- Left-lateralized effects of AUD on brain volume may increase the risk for adolescents to develop Wernike–Korsakoff syndrome later in life.

REFERENCES

American Psychiatric Association. (1994). *Diagnostic and statistical manual of mental disorders: DSM-IV*. Washington, DC: American Psychiatric Association.

American Psychiatric Association. (2013). *Diagnostic and statistical manual of mental disorders: DSM-5*. Arlington, VA: American Psychiatric Association.

Ashburner, J., & Friston, K. J. (2000). Voxel-based morphometry—the methods. *NeuroImage, 11*(6), 805–821.

Berridge, K. C. (2007). The debate over dopamine's role in reward: the case for incentive salience. *Psychopharmacology, 191*(3), 391–431.

Bigler, E. D., Mortensen, S., Neeley, E. S., Ozonoff, S., Krasny, L., Johnson, M., ... Lainhart, J. E. (2007). Superior temporal gyrus, language function, and autism. *Developmental Neuropsychology, 31*(2), 217–238.

Brion, M., Pitel, A. L., Beaunieux, H., & Maurage, P. (2014). Revisiting the continuum hypothesis: toward an in-depth exploration of executive functions in korsakoff syndrome. *Frontiers in Human Neuroscience, 8*, 498.

Brooks, S. J., Dalvie, S., Cuzen, N. L., Cardenas, V., Fein, G., & Stein, D. J. (2014). Childhood adversity is linked to differential brain volumes in adolescents with alcohol use disorder: a voxel-based morphometry study. *Metabolic Brain Disease, 29*, 311–321.

Brown, S. A., Tapert, S. F., Granholm, E., & Delis, D. C. (2000). Neurocognitive functioning of adolescents: effects of protracted alcohol use. *Alcoholism: Clinical and Experimental Research, 24*, 164–171.

Ceylan-Isik, A. F., McBride, S. M., & Ren, J. (2010). Sex difference in alcoholism: who is at a greater risk for development of alcoholic complication? *Life Sciences, 87*, 133–138.

Chung, T., & Clark, D. B. (2014). Insula white matter volume linked to binge drinking frequency through enhancement motives in treated adolescents. *Alcoholism: Clinical and Experimental Research, 38*(7), 1932–1940.

Cousijn, J., Wiers, R. W., Ridderinkhof, K. R., van den Brink, W., Veltman, D. J., & Goudriaan, A. E. (February 15, 2012). Grey matter alterations associated with cannabis use: results of a VBM study in heavy cannabis users and healthy controls. *NeuroImage, 59*(4), 3845–3851. 2011.

Craig, A. D. (2009). How do you feel now? The anterior insula and human awareness. *Nature Reviews Neuroscience, 10*(1), 59–70.

Cservenka, A., Jones, S. A., & Nagel, B. J. (2015). Reduced cerebellar brain activity during reward processing in adolescent binge drinkers. *Developmental Cognitive Neuroscience*. pii: S1878-9293(15)00065-1.

Damasio, A. R. (1994). Descartes' error and the future of human life. *Scientific American, 271*, 144.

De Bellis, M. D., Clark, D. B., Beers, S. R., Soloff, P. H., Boring, A. M., Hall, J., ... Keshavan, M. S. (2000). Hippocampal volume in adolescent-onset alcohol use disorders. *American Journal of Psychiatry, 157*(5), 737–744.

De Bellis, M. D., Keshavan, M. S., Clark, D. B., Casey, B. J., Giedd, J., Boring, A. M., ... Ryan, N. D. (1999). AE Bennett research award: developmental traumatology, part II: brain development. *Biological Psychiatry, 45*, 1271–1284.

De Bellis, M. D., Narasimhan, A., Thatcher, D. L., Keshavan, M. S., Soloff, P., & Clark, D. B. (2005). Prefrontal cortex, thalamus, and cerebellar volumes in adolescents and young adults with adolescent-onset alcohol use disorders and comorbid mental disorders. *Alcoholism: Clinical and Experimental Research, 29*(9), 1590–1600.

Doallo, S., Cadaveira, F., Corral, M., Mota, N., López-Caneda, E., & Holguín, S. R. (May 2, 2014). Larger mid-dorsolateral prefrontal gray matter volume in young binge drinkers revealed by voxel-based morphometry. *PLoS One, 9*(5), e96380.

Duerden, E. G., Arsalidou, M., Lee, M., & Taylor, M. J. (2013). Lateralization of affective processing in the insula. *NeuroImage, 78*, 159–175.

Eichenbaum, H. (1999). The hippocampus and mechanisms of declarative memory. *Behavioural Brain Research, 103*(2), 123–133.

Fein, G., Greenstein, D., Cardenas, V. A., Cuzen, N. L., Fouche, J. P., Ferrett, H., ... Stein, D. J. (2013). Cortical and subcortical volumes in adolescents with alcohol dependence but without substance or psychiatric comorbidities. *Psychiatry Research: Neuroimaging, 214*, 1–8.

Frodl, T., Skokauskas, N., Frey, E. M., Morris, D., Gill, M., & Carballedo, A. (2014). BDNF Val66Met genotype interacts with childhood adversity and influences the formation of hippocampal subfields. *Human Brain Mapping, 35*(12), 5776–5783.

Giedd, J. N. (2004). Structural magnetic resonance imaging of the adolescent brain. *Annals of the New York Academy of Sciences, 1021*.

Giedd, J. N., Vaituzis, A. C., Hamburger, S. D., Lange, N., Rajapakse, J. C., Kaysen, D., ... Rapoport, J. L. (1996). Quantitative MRI of the temporal lobe, amygdala, and hippocampus in normal human development: ages 4–18. *Journal of Comparative Neurology, 366*, 223–230.

Hall, P. A., & Fong, G. T. (2015). Temporal self-regulation theory: a neurobiologically informed model for physical activity behavior. *Frontiers in Human Neuroscience, 25*(9), 117.

Ide, J. S., Zhang, S., Hu, S., Sinha, R., Mazure, C. M., & Li, C. S. (January 1, 2014). Cerebral gray matter volumes and low-frequency fluctuation of BOLD signals in cocaine dependence: duration of use and gender difference. *Drug and Alcohol Dependence, 134*, 51–62.

Jernigan, T. L., & Gamst, A. (2005). Changes in volume with age: consistency and interpretation of observed effects. *Neurobiology of Aging, 26*(9), 1271–1274.

Keyes, K. M., Martins, S. S., Blanco, C., & Hasin, D. S. (2010). Telescoping and gender differences in alcohol dependence: new evidence from two national surveys. *American Journal of Psychiatry, 167*, 969–976.

Lenroot, R. K., & Giedd, J. N. (2006). Brain development in children and adolescents: insights from anatomical magnetic resonance imaging. *Neuroscience & Biobehavioral Reviews, 30*, 718–729.

Maguire, E. A., Gadian, D. G., Johnsrude, I. S., Good, C. D., Ashburner, J., Frackowiak, R. S. J., & Frith, C. D. (2000). Navigation-related structural change in the hippocampi of taxi drivers. *Proceedings of the National Academy of Sciences, 97*(8), 4398–4403.

Medina, K. L., McQueeny, T., Nagel, B. J., Hanson, K. L., Schweinsburg, A. D., & Tapert, S. F. (2008). Prefrontal cortex volumes in adolescents with alcohol use disorders: unique gender effects. *Alcoholism: Clinical and Experimental Research, 32*(3), 386–394.

Medina, K. L., Schweinsburg, A. D., Cohen-Zion, M., Nagel, B. J., & Tapert, S. F. (2007). Effects of alcohol and combined marijuana and alcohol use during adolescence on hippocampal volume and asymmetry. *Neurotoxicology and Teratology, 29*(1), 141–152.

Montigny, C., Castellanos-Ryan, N., Whelan, R., Banaschewski, T., Barker, G. J., Büchel, C., ... IMAGEN Consortium. (November 14, 2013). A phenotypic structure and neural correlates of compulsive behaviors in adolescents. *PLoS One, 8*(11).

Nagel, B. J., Schweinsburg, A. D., Phan, V., & Tapert, S. F. (August 30, 2005). Reduced hippocampal volume among adolescents with alcohol use disorders without psychiatric comorbidity. *Psychiatry Research, 139*(3), 181–190.

Naqvi, N. H., & Bechara, A. (2010). The insula and drug addiction: an interoceptive view of pleasure, urges, and decision-making. *Brain Structure and Function, 214*, 435–450.

Pfefferbaum, A., Mathalon, D. H., Sullivan, E. V., Rawles, J. M., Zipursky, R. B., & Lim, K. O. (1994). A quantitative magnetic resonance imaging study of changes in brain morphology from infancy to late adulthood. *Archives of Neurology, 34*, 71–75.

Rourke, S. (1996). Neurobehavioral correlates of alcoholism. In *Neuropsychological assessment of neuropsychiatric disorders* (2nd ed.) (pp. 423–485).

Schuckit, M. A., Daeppen, J.-B., Tipp, J. E., Hesselbrock, M., & Bucholz, K. K. (1998). The clinical course of alcohol-related problems in alcohol dependent and nonalcohol dependent drinking women and men. *Journal of Studies on Alcohol and Drugs, 59*, 581–590.

Sowell, E. R., Thompson, P. M., Leonard, C. M., Welcome, S. E., Kan, E., & Toga, A. W. (2004). Longitudinal mapping of cortical thickness and brain growth in normal children. *Journal of Neuroscience, 24*, 8223–8231.

Spear, L. P. (2009). *The behavioral neuroscience of adolescence* (1st ed.). New York: W. W. Norton & Co Inc.

Squeglia, L. M., Sorg, S. F., Schweinsburg, A. D., Wetherill, R. R., Pulido, C., & Tapert, S. F. (2012). Binge drinking differentially affects adolescent male and female brain morphometry. *Psychopharmacology (Berlin), 220*(3), 529–539.

Suzuki, M., Hagino, H., Nohara, S., Zhou, S. Y., Kawasaki, Y., Takahashi, T., ... Kurachi, M. (2005). Male-specific volume expansion of the human hippocampus during adolescence. *Cerebral Cortex, 15*, 187–193.

Tomasi, D., & Volkow, N. D. (2012). Resting functional connectivity of language networks: characterization and reproducibility. *Molecular Psychiatry, 17*(8), 841–854.

Van Dam, N. T., Rando, K., Potenza, M. N., Tuit, K., & Sinha, R. (2014). Childhood maltreatment, altered limbic neurobiology, and substance use relapse severity via trauma-specific reductions in limbic gray matter volume. *JAMA Psychiatry, 71*(8), 917–925.

Verdejo-Garcia, A., Clark, L., & Dunn, B. D. (2012). The role of interoception in addiction: a critical review. *Neuroscience & Biobehavioral Reviews, 36*, 1857–1869.

Wiers, R. W., Boelema, S. R., Nikolaou, K., & Gladwin, T. E. (2015). On the development of implicit and control processes in relation to substance use in adolescence. *Current Addiction Reports, 2*(2), 141–155.

Yuan, K., Qin, W., Wang, G., Zeng, F., Zhao, L., Yang, X., ... Tian, J. (2011). Microstructure abnormalities in adolescents with internet addiction disorder. *PLoS One, 6*(6), e20708.

Chapter 55

Central Pontine Myelinolysis in Alcoholism: An Osmotic Demyelination Syndrome

Irena Dujmovic

Department for Immune-Mediated CNS Diseases, Clinic of Neurology, Clinical Centre of Serbia, Belgrade University School of Medicine, Belgrade, Serbia

Abbreviations

ADC Apparent diffusion coefficient
ADH Antidiuretic hormone
AQP1 Aquaporin-1
AQP4 Aquaporin-4
BBB Blood–brain barrier
CNS Central nervous system
CPM Central pontine myelinolysis
CT Computerized tomography
DWI Diffusion-weighted imaging
EPM Extrapontine myelinolysis
iNOS Inducible NO synthase
IVIG Intravenous immunoglobulins
LFB Luxol fast blue
MDMA 3,4-Methylenedioxymethamphetamine
MRI Magnetic resonance imaging
NO Nitric oxide
ODS Osmotic demyelination syndromes
TRH Thyrotropin-releasing hormone

INTRODUCTION

The first postmortem description of demyelinated lesions of the basis pontis dates back to 1932 and was reported by Lüthy in association with Wilson disease (Lüthy, 1932). Adams, Victor, and Mancall (1959) described a symmetrical butterfly-shaped patch of demyelination within the dorsal basis pontis in which an inflammatory component was absent. This feature prompted the authors to apply the term myelinolysis rather than demyelination and to introduce the concept of central pontine myelinolysis (CPM). In this original paper three of the four reported patients with CPM were alcoholics, whereas the fourth suffered only from severe malnutrition without alcoholism. As additional cases were identified, it became apparent that lesions were not confined to the pons, but could also occur at other sites of the central nervous system (CNS), which enabled the spectrum of CPM to be expanded to so-called extrapontine CPM or extrapontine myelinolysis (EPM) (Martin, 2004).

Is the contemporary concept of CPM far beyond its early descriptions? According to the current concept, CPM and EPM are defined as acquired, symmetrical, circumscribed, acute or subacute, noninflammatory CNS demyelinating lesions (Lupato et al., 2010). CPM and EPM belong to the spectrum of osmotic demyelination syndromes (ODS) (Lampl & Yazdi, 2002). CPM represents approximately one-half of all ODS cases, while in 20% of cases only EPM has been reported, and in the remaining 30% CPM was found to be associated with EPM (Martin, 2004).

CPM was reported to have a peak incidence between the ages of 30 and 50 years with a slight male preponderance (Lampl & Yazdi, 2002), but was also described in children (Norenberg, 2010). In a large number of cases and case series, CPM and EPM have been reported to be associated with various underlying diseases and conditions (Table 1). However, ever since the original descriptions of CPM (Adams et al., 1959) a striking feature has been the high association between ODS and chronic alcohol abuse (Lampl & Yazdi, 2002; Martin, 2004), observed in up to 40% of all ODS cases. For the cases associated with chronic alcoholism, many patients developed ODS during the terminal stage of binge drinking, although a few patients have been reported to develop ODS after alcohol withdrawal (An, Park, Han, & Song, 2010).

How could the chronic alcohol abuse contribute to the pathogenetic scenario of ODS?

PATHOPHYSIOLOGY

The pathophysiology of ODS is still poorly understood (Yoon, Shim, & Chung, 2008). In the original descriptions, Adams et al. (1959) postulated that the etiology must have been fundamentally biochemical since the observed pontine lesions were both symmetrical and constant in location. Tomlinson, Pierides, and Bradley (1976) reported two severely hyponatremic patients with a progressive neurological deterioration following the correction of hyponatremia, in whom the autopsy established the diagnosis of

TABLE 1 Disease States Associated with Osmotic Demyelination Syndromes (ODS)

Disease State/Condition

Chronic alcoholism, especially after alcoholic delirium

Electrolyte disturbances and disturbances of osmolality (hyponatremia, rapid correction of hyponatremia, hypernatremia, hypophosphatemia, hypokalemia, hypermagnesemia)

Malnutrition

Transplantation

Hepatic diseases of various origins

Hematologic disorders

Pulmonary infections

Diseases of the CNS (stoke, inflammation, tumors, craniocerebral trauma)

Malignant tumors (especially of the lung and the gastrointestinal tract) and chemotherapy

Renal failure, dialysis

Burns and complications

Sepsis

Diabetes mellitus, hyperglycemia, hypoglycemia

Acquired immune deficiency syndrome

Hyperemesis gravidarum, pregnancy without any special contributing factor

Acute porphyria

Wilson disease

Pancreatic failure

Immune-mediated diseases (celiac disease, systemic lupus erythematosus, Sjögren syndrome, Isaacs syndrome) and hyperinflammatory conditions (hemophagocytic syndrome)

Eating behavior disorder (anorexia nervosa, bulimia), persistent vomiting

Vitamin deficiency syndromes (Shoshin beriberi, B12 deficiency, folate depletion)

Intoxication (carbamate, lithium, ethylene glycol, toluene)

ODS may be associated with various underlying diseases and conditions. Often more than one association is present.

CPM. The precise sequence of events by which hyponatremia, or hypernatremia/hyperosmolarity following rapid correction of chronic hyponatremia, might lead to demyelination has still not been fully elucidated, but an intensive research over the past few decades has focused the pathogenetic role of severe osmotic stress in ODS development (Hurley, Filley, & Taber, 2011; Kumar, Fowler, Gonzalez-Toledo, & Jaffe, 2006; Martin, 2004; Norenberg, 2010). The prevailing hypothesis for the genesis of ODS implicates a reduced adaptive capacity of the neuroglia to large shifts in serum osmolarity (An et al., 2010; Hurley et al., 2011; Norenberg, 2010). The high concentration of ions is stressful to the cells as it interferes with the maintenance of proper protein structure and function, while organic osmolytes (i.e., myoinositol, taurine, glutamine, glutamate, creatine, phosphocreatine, glycerophosphorylcholine) tend to maintain normal protein structure and function (Hurley et al., 2011; Norenberg, 2010).

In alcoholics, alcohol itself could interfere with sodium/water regulation by interference with antidiuretic hormone (ADH) secretion, while an inadequate water intake, vomiting, and/or liver dysfunction might also contribute to electrolyte disturbances, osmotic stress, and the development of ODS (Martin, 2004; Ragland, 1990) (Figure 1). An acute ethyl alcohol ingestion suppresses endogenous ADH release and induces diuresis, while in the chronic state alcohol promotes retention of water and electrolytes due to increased ADH levels (Ragland, 1990). Additionally, organic osmolyte deficiency states in alcoholics due to liver disease (Bluml, Zuckerman, Tan, & Ross, 1998) or malnutrition (Zahr, Kaufman, & Harper, 2011) may disturb protective CNS mechanisms against osmotic stress (Norenberg, 1983, 2010; Yoon et al., 2008).

It has been suggested that osmotic stress can further contribute to the disruption of the blood–brain barrier (BBB) and the production of vasogenic edema (Norenberg, 1983, 2010). CNS endothelial tight junctions might be damaged mechanically by dehydration and shrinkage of endothelial cells, but also by agents released from damaged endothelial cells (Norenberg, 2010). Additionally, the BBB damage in ODS could also be mediated by nitric oxide (NO) (Tan, Harrington, Purcell, & Hurst, 2004) or cytokines such as tumor necrosis factor α (TNF-α) (Wiggins-Dohlvik et al., 2014), since an increased expression of inducible NO synthase (iNOS) and massive accumulations of microglia expressing TNF-α in brain demyelinating lesions were observed in the rat model of rapidly corrected severe hyponatremia (Takefuji et al., 2007). Popescu et al. (2013) suggested that aquaporin-1 (AQP1) and AQP4 CNS water channels might be involved in the pathogenesis of CPM.

We could assume that oxidative stress could damage the BBB in alcoholics (Figure 1) owing to the ability of ethanol to activate free radical-generating enzymes such as nicotinamide adenine dinucleotide phosphate oxidase and iNOS (Alikunju, Abdul Muneer, Zhang, Szlachetka, & Haorah, 2011). Alikunju et al. (2011) have shown evidence for the presence of alcohol-induced alterations of the tight-junction protein occludin in intact microvessels due to oxidative damage but also showed that alcohol-induced BBB damage enhanced immune cell adhesion at the BBB damage site (Figure 2). Additionally, acetaldehyde protein adduct formation in chronic alcoholics may induce a generalized immune response (Hoerner, Behrens, Worner, & Lieber, 1986) that could promote the BBB damage (Qin et al., 2008). He and Crews (2008) found positive signs of neuroinflammation in various brain regions in individuals with alcoholism and proposed an intriguing concept of alcohol-induced neuroinflammation. The mechanisms of alcohol-induced neuroinflammation (Table 2) are still only partially understood (Szabo & Lippai, 2014).

BBB damage could further contribute to demyelination in ODS. Some blood-derived factors such as complement could express oligodendrocyte toxicity (Wing, Zajicek, Seilly,

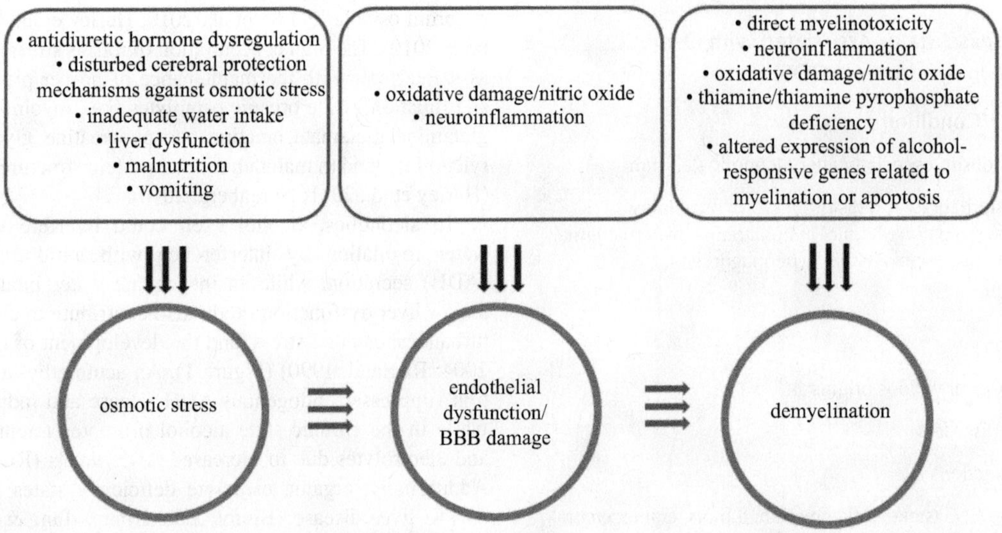

FIGURE 1 Alcohol-related pathogenetic mechanisms in ODS. The key pathogenetic steps in the development of demyelination in ODS (osmotic stress, endothelial dysfunction, and blood–brain barrier damage) may be influenced at multiple levels by alcohol-related pathogenetic mechanisms.

Compston, & Lachmann, 1992). In the rat model of rapid correction of chronic hyponatremia, Baker, Tian, Adler, and Verbalis (2000) showed marked increases in immunoglobulin G and C3d complement immunostaining in rat brains, further supporting the hypothesis that complement activation could be involved in the pathogenesis of demyelination in ODS. The concept of blood-derived myelinotoxic factors could also be supported by the typical location of ODS lesions in the admixture of gray and white matter, since blood-derived myelinotoxic factors derived from the richly vascular gray matter might be able to interact and damage the adjacent bundles of myelin-containing white matter (Norenberg, 1983). Additionally, the damage to endothelial cells could lead to the release of agents that could have toxic effects on oligodendrocytes and/or myelin sheets, such as proteases or cytokines (Norenberg, 2010), thus contributing to demyelination independent of the state of the BBB. Both osmotic stress and oligodendrocyte toxicity lead to the development of cytotoxic edema (An et al., 2010).

In alcoholics, a direct myelinotoxic effect of alcohol (Figure 1) might be supported by the results of Pons-Vázquez et al. (2011), who found that combined pre- and postnatal ethanol exposure disturbed myelination of optic axons in an experimental model of chronic ethanol exposure in rats. Malnourished alcoholics also have a high risk of thiamine deficiency (Zahr et al., 2011) and alcohol itself interferes with the conversion of thiamine to its metabolically active form thiamine pyrophosphate, which could negatively influence carbohydrate metabolism, lipid metabolism, and amino acid metabolism and could contribute not only to neuroaxonal damage but also to myelin damage, interrupting the production and maintenance of myelin (Zahr et al., 2011).

An additional mechanism of demyelination in ODS could be oligodendrocyte apoptosis (Ashrafian & Davey, 2001). The excess production of free radicals and the deranged NO metabolic effects in alcoholics may favor oligodendrocyte apoptosis (Ashrafian & Davey, 2001; Haider et al., 2011) (Figure 1). The myoinositol deficiency states in patients with liver disease might also favor apoptosis, since myoinositol also has antiapoptotic properties and could

modulate caspase-3 activity (Alfieri et al., 2002). DNA microarray investigations of human cortex from people with alcoholism have identified altered expression of several alcohol-responsive genes related to several metabolic functions, including myelination, apoptosis, and intracellular metabolism, thus potentially contributing to the myelin damage in alcoholics with ODS (Zahr et al., 2011).

PATHOLOGY

In gross specimens, ODS lesions present as areas of gray–tan softening well demarcated from the surrounding tissue (Jacob, Gupta, Nikolic, Gundogdu, & Ong, 2014). Luxol fast blue (LFB) combined with periodic acid–Schiff staining revealed pale, extensive, and sharply demarcated demyelinating lesions (Jacob et al., 2014; Kuhlmann, Lassmann, & Brück, 2008; Norenberg, 2010) (Figure 3).

The extent of axonal damage in ODS lesions is variable (Kuhlmann et al., 2008). Axon retraction bulbs and axonal fragmentation are recognized to occur in CPM although neurons are preserved (Ghosh, DeLuca, & Esiri, 2004). Within the lesions, numerous macrophages can be observed, while lymphocytic infiltrates are rare (Kuhlmann et al., 2008). Macrophages are widespread within the area of demyelination and show little or no increased abundance around veins and venules (Ghosh et al., 2004). It was suggested that the number of macrophages and the presence of LFB-positive myelin degradation products within the cytoplasm of macrophages depended on the age of the lesion (Kuhlmann et al., 2008). In 2013, astrocytic AQP1 and AQP4 loss, along with the presence of fewer and shorter astrocytic processes in lesional astrocytes, was reported in a subset of human CPM cases (Popescu et al., 2013). In chronic lesions, the proliferation of astrocytes in the areas of fibrillary gliosis has been observed (Thompson, Miller, Gledhill, & Rossor, 1989), which could lead to atrophy (Yuh, Simonson, D'Alessandro, Smith, & Hunsicker, 1995). Cystic (necrotic) cavitations may also occur in the center of older demyelinating lesions (Kuhlmann et al., 2008).

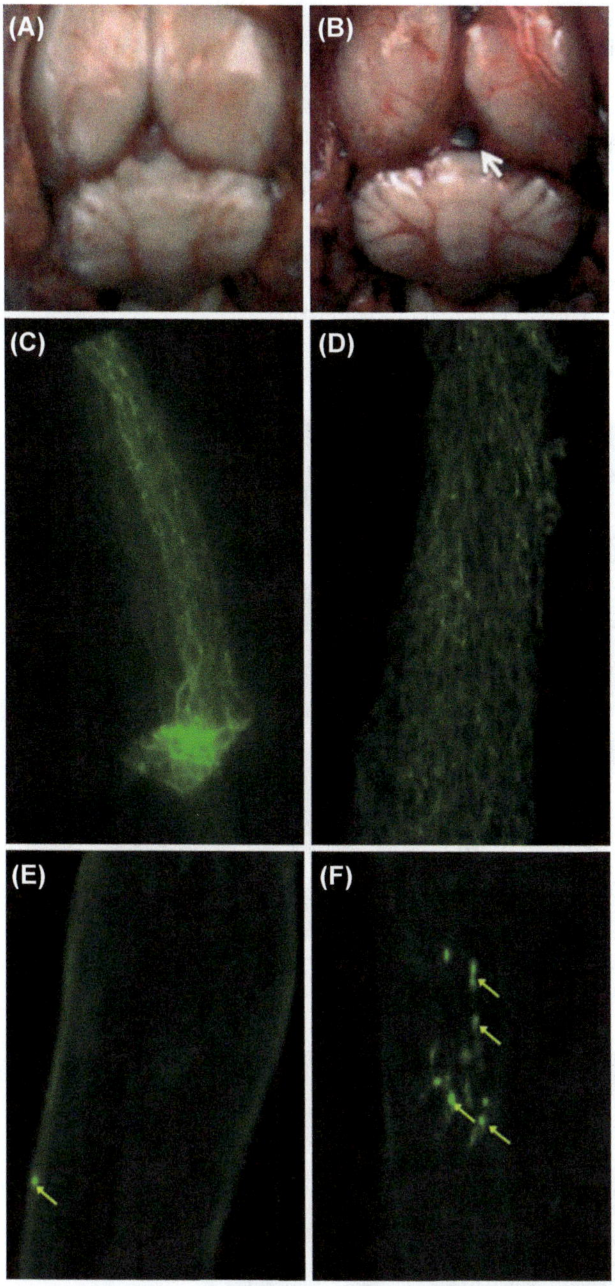

FIGURE 2 Alcohol-induced BBB damage enhances immune cell adhesion at the BBB damage site. Whole-brain structure of (A) control animal and (B) chronic alcohol ingestion animal indicating the breakage of brain microvessels and hemorrhage. The arrow shows the formation of a thrombus (B). (C and D) Immunohistochemical staining of the tight-junction protein occludin in intact brain microvessels of chronic alcohol ingestion compared with a pair-fed control animal. (E and F) The arrows indicate aggregations of macrophages at the BBB damage site. *Reprinted (minimally adapted) from Alikunju et al. (2011), Copyright—2011, with kind permission from Elsevier.*

NEUROIMAGING

Neuroimaging characteristics of ODS in alcoholics show no distinct characteristics compared to nonalcoholics. On brain computerized tomography (CT) scans ODS lesions typically present as

TABLE 2 Potential Mechanisms of Neuroinflammation in Chronic Alcoholics

- Adhesion of immune cells occurs at the site of alcohol-induced oxidative injury in the BBB.
- Alcohol activates microglia and astrocytes, cells that promote neuroinflammation.
- Alcohol-induced cell injury in the brain results in release of damage-associated molecular patterns, such as high-mobility group box 1, that trigger inflammatory changes through activation of PRRs.
- Alcohol consumption increases intestinal permeability and results in increased levels of pathogen-associated molecular patterns, such as endotoxin in the systemic circulation, that trigger PRRs and inflammation.
- The Toll-like receptor-4 pathway (which activates nuclear factor-κB and secretion of proinflammatory cytokines, TNF-α, IL-1β, and chemokines, including monocyte chemotactic protein-1) is activated.
- Alcohol-induced IL-1β secretion also requires Nod-like receptor-mediated inflammasome and caspase-1 activation.
- Delicate regulators of inflammatory gene expression are micro-RNAs (miRs) that have been identified in alcohol-related neuroinflammation.
- Alcohol induces miR-155, a regulator of inflammation in the brain, and deficiency of miR-155 in mice was protective against neuroinflammatory changes.

PRRs, pattern recognition receptors; IL-1β, interleukin-1β; RNA, ribonucleic acid.
According to Alikunju et al. (2011) and Szabo and Lippai (2014).

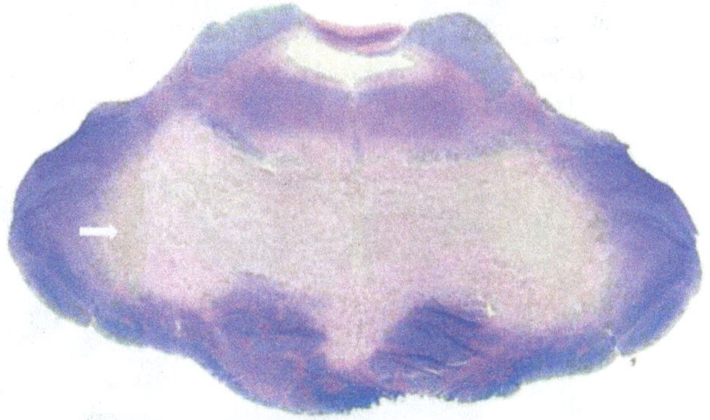

FIGURE 3 Classic histopathology of the pons in CPM. LFB/periodic acid–Schiff staining shows a symmetrical, central bat-wing area of demyelination affecting most of the pontine base. *Reprinted from Norenberg (2010), Figure 1, with kind permission from Springer Science and Business Media.*

low-density lesions in the pons or other affected regions that occasionally show enhancement (Zuccoli et al., 2010).

Nowadays the clinical suspicion of ODS is usually confirmed by magnetic resonance imaging (MRI), which is superior to CT, especially in the early phases of disease (Lampl & Yazdi, 2002). However, MRI performed in the early phase of the disease may also be unremarkable, and a repetition of neuroimaging after 10–14 days could help to confirm the clinical suspicion of ODS (Kumar, Mone, Gray, & Troost, 2000). On MRI, ODS typically present as nonenhancing ill-defined (Jacob et al., 2014) or well-defined (Dujmović, Vitas, Zlatarić, & Drulović, 2013) T1-weighted (T1W) hypointense and T2W and fluid-attenuated inversion recovery hyperintense lesions (Figures 4 and 5). In the early phases of ODS, contrast enhancement of T1W brain MRI lesions could be observed up to 4 weeks after the initial presentation, thus supporting the role of BBB damage in lesion development (Menger, Mackowski, Jörg, & Cramer, 1998). Pontine lesion in CPM frequently presents as a trident-shaped symmetrical central pontine abnormality (An et al., 2010; Dujmović et al., 2013) (Figure 4) that usually spares the tegmentum, ventrolateral pons, and corticospinal tracts (Thompson et al., 1989). However, in larger lesions, corticospinal, corticopontine, and pontocerebellar fibers may also be affected (Yuh et al., 1995; Liberatore et al., 2006). Extrapontine lesions (Figure 5) might involve various cerebral and/or cerebellar regions (in descending order of frequency: cerebellum, lateral geniculate body, external capsule, extreme capsule, hippocampus, putamen, cerebral cortex/subcortex, thalamus, caudate nucleus, claustrum, internal capsule, midbrain, internal medullary lamella, mammillary body, medulla oblongata) (Martin, 2004; Rego, Vieira, Correia, & Pereira, 2012), including the spinal cord, although spinal cord involvement in EPM is extremely rare. Jacob et al. (2014) have reported an autopsy CPM and EPM case with longitudinally extensive thoracic spinal cord lesion.

Diffusion-weighted imaging (DWI) can detect changes in water diffusion associated with cellular dysfunction (Chu, Kang,

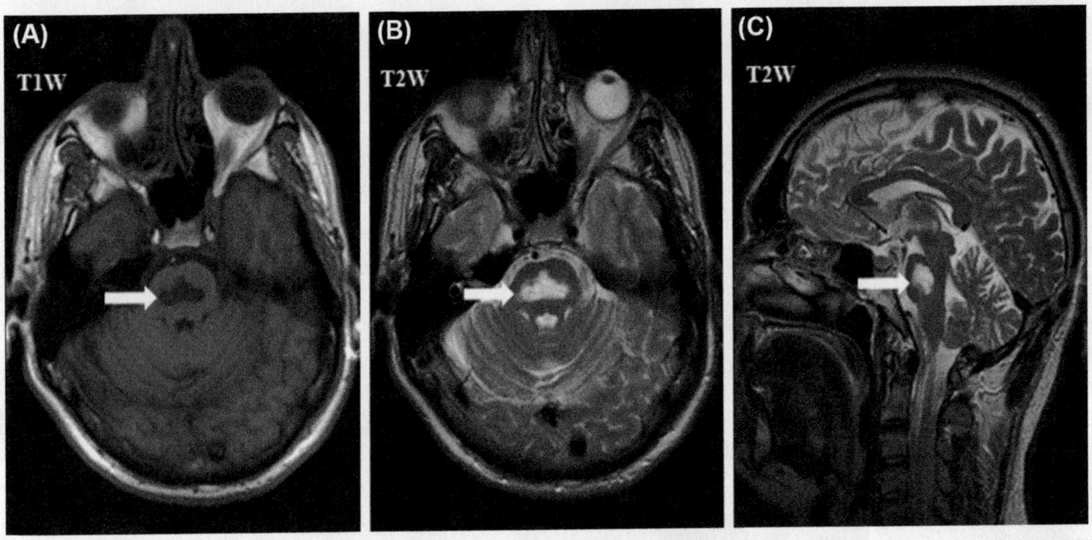

FIGURE 4 Brain MRI in a 30-year-old chronic alcoholic male with CPM. "Trident-shaped" pontine (A) T1W hypointensity and (B) T2W hyperintensity on axial brain MRI. (C) Well-defined T2W pontine hyperintensity on sagittal brain MRI.

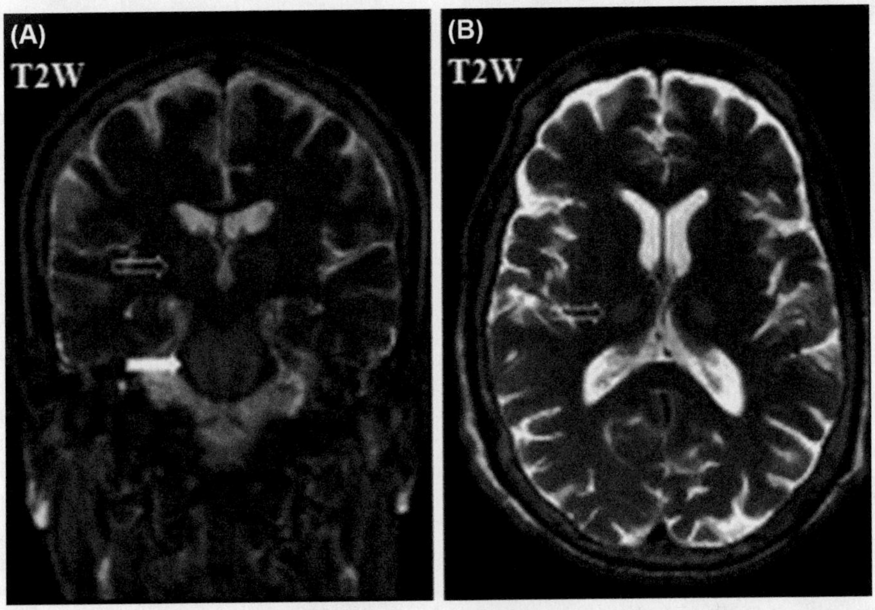

FIGURE 5 Brain MRI in a 47-year-old chronic alcoholic male with severe hypernatremia and multiple brain lesions suggestive of pontine and extrapontine myelinolysis. Hyperintense central pontine lesion (arrow) and symmetric bilateral extrapontine thalamic lesions (open arrows) on (A) coronal and (B) axial T2W brain image. *Reproduced from Rego et al. (2012), copyright notice 2012, with kind permission from BMJ Publishing Group Ltd.*

Ko, & Kim, 2001). Because the underlying process of ODS is the osmotic disturbance and water and electrolyte imbalance, DWI might be a better method for identifying early ODS lesions before the detection of any conventional brain MRI abnormality (Ruzek, Campeau, & Miller, 2004). DWI with an apparent diffusion coefficient (ADC) image is sensitive to cytotoxic and vasogenic edema (An et al., 2010). On DWI, vasogenic edema can be visualized as a high signal intensity and an increased ADC value (An et al., 2010; Rego et al., 2012), while cytotoxic edema is characterized by a high signal on DWI and a decrease in ADC value (An et al., 2010). In ODS cases with mixed vasogenic and cytotoxic edema, the ADC map shows an isointense signal (An et al., 2010).

Brain edema and demyelination are associated with a release of choline (Cho), a cell membrane and myelin marker, and therefore an increase in the Cho/creatine ratio could be found on brain magnetic resonance spectroscopy in patients with ODS (Nomoto, Arasaki, & Tamaki, 2004). Roh, Nam, and Lee (1998) reported CPM and EPM lesions on brain MRI after rapid correction of hyponatremia to show early hypermetabolism on [18F]-fluorodeoxyglucose positron emission tomography scan, potentially reflecting an increased glucose metabolism in activated microglia and astrocytes.

Brain MRI would be helpful to differentiate CPM/EPM from Marchiafava-Bignami disease (MBD), another demyelinating condition observed almost exclusively in alcoholics. CPM/EPM and MBD were shown to share a similar pathology, suggesting similar pathophysiology (Kuhlmann et al., 2008). However, extensive corpus callosum lesions are not typical for EPM and are more suggestive of MBD (Figure 6), although extracallosal and cortical lesions can be found in both MBD (Hillbom et al., 2014; Kuhlmann et al., 2008) and EPM (Martin, 2004).

Early CT/MRI abnormalities in patients with ODS are potentially reversible (Takei, Akahane, & Ikeda, 2003), most probably owing to resolution of acute edema and remyelination (Yuh et al., 1995). However, a time lag of several weeks has been observed between the disappearance of neurological manifestations and that of radiological abnormalities (Lampl & Yazdi, 2002). In the chronic phase, neuroimaging abnormalities are caused by chronic demyelinization and/or gliosis and are irreversible (Lampl & Yazdi, 2002). It has been shown that follow-up conventional MRI findings might not follow clinical recovery (Dujmović et al., 2013; Menger & Jörg, 1999).

CLINICAL PRESENTATION

The severity of clinical presentation in patients with ODS can be extremely variable, from asymptomatic cases in nonalcoholics (Lupato et al., 2010) or alcoholics (Nomoto et al., 2004) to severe neurological deterioration (Lampl & Yazdi, 2002).

CPM is usually characterized by spastic paraparesis and tetraparesis of varying degrees of severity; dysarthria; dysphagia; gaze and facial paralysis; bulbar and pseudobulbar paralysis; epileptic seizures; ataxia, in some cases with depressed or absent reflexes; lethargy; and arterial hypotension (Lampl & Yazdi, 2002; Lupato et al., 2010). Additionally, confusion (Lupato et al., 2010), somnolence (Dujmović et al., 2013), stupor (Wijdicks, Blue, Steers, & Wiesner, 1996), and coma (Spakowski, Nikolic, & Giambarba, 2009), as well as locked-in state (Sohn & Nam, 2014), have been reported in CPM cases. Consciousness disturbance in CPM could be misdiagnosed as alcohol withdrawal syndrome (Mochizuki et al., 2003). Psychiatric manifestations in CPM include agitated delirium, a pseudobulbar state with pathological laughing and crying, catatonia, and diverse neuropsychological deficits (Lampl & Yazdi, 2002).

Clinical presentation of EPM is highly variable (Kumar et al., 2006). Altered mental status, catatonia, emotional lability, akinetic mutism, gait disturbance, myoclonus (Kumar et al., 2006), parkinsonism or other extrapyramidal signs (Kumar et al., 2006), cognitive dysfunction (Yoon et al., 2008), and seizures (McNeill, Halpenny, Snow, Geoghegan, & Torreggiani, 2009), as well as other psychiatric, neuropsychological, and other CNS manifestations depending on the affected structures of the brain (Martin, 2004) or spinal cord (Jacob et al., 2014), have so far been observed as a clinical scenario in EPM.

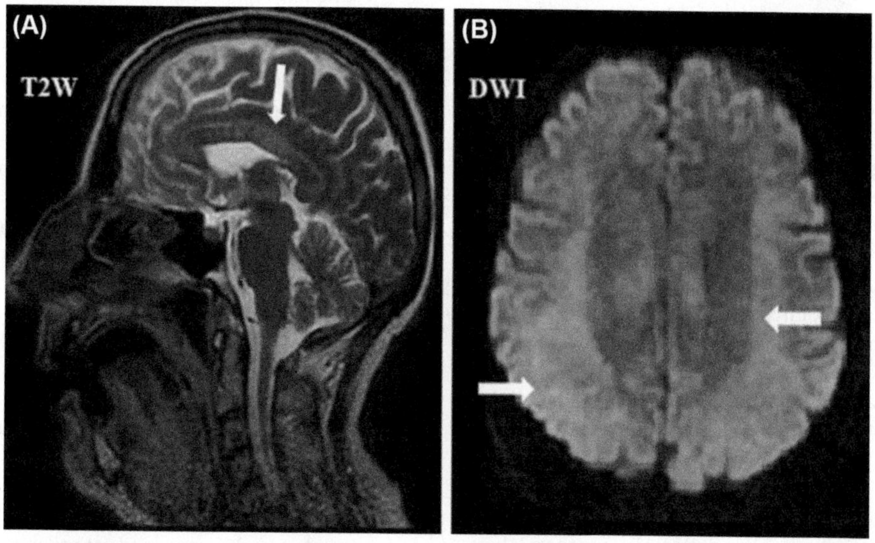

FIGURE 6 **Brain MRI in a 54-year-old chronic alcoholic female with MBD.** (A) T2W corpus callosum hyperintensity and diffuse swelling. (B) Bilateral cortical and white matter hyperintensity on DWI.

TREATMENT AND PROGNOSIS

The treatment of alcohol-related ODS is challenging, especially in severe cases. Prevention has been suggested to be the best treatment of ODS (Kumar et al., 2006). Alcoholics are at risk of hyponatremia (Ragland, 1990), which should be carefully corrected to minimize the risk of ODS development. Many authors agree that the correction of acute hyponatremia can be rapid (Martin, 2004). However, for chronic hyponatremia (duration of 1 week or longer) the blood sodium correction must be undertaken slowly to allow sufficient time for the brain cells to reacquire both electrolytes and organic osmolytes (Norenberg, 2010). Although the correction of not more than 10 mmol/l/day has been the "consensus" among neurologists, a 2004 recommendation suggests a correction not in excess of 8 mmol/l/day (Martin, 2004). Some authors also suggest stabilizing the patient in a mild hyponatremic state after the initial correction (Martin, 2004). In addition to the saline solutions and overall fluid restriction, intravenous and oral urea has been safely used to correct hyponatremia, since urea induces an osmotic diuresis and reduces renal sodium excretion by stimulating passive tubular sodium reabsorption (Kumar et al., 2006). More recently, the introduction of the vaptans, which are ADH receptor antagonists that induce an excretion of increased amounts of water without altered sodium or potassium excretion, could be of interest for ODS prevention in hyponatremic patients (Sterns, Silver, Kleinschmidt-DeMasters, & Rojiani, 2007).

There is no established effective treatment in alcoholics in whom ODS has already developed. A very careful intravenous water/electrolyte therapy should be undertaken with careful attention to the patient's nutritional status in addition to the treatment of other metabolic stressors (Kumar et al., 2006). Assuming the pathogenetic contribution of myelinotoxic factors, neuroinflammation, and BBB dysfunction, there have been several reports on the successful use of corticosteroids (Sakamoto et al., 2007), intravenous immunoglobulins (IVIG) (Saner et al., 2008), and plasmapheresis (Bibl et al., 1999; Grimaldi, Cavalleri, Vallone, Milanti, & Cortelli, 2005) in ODS. Corticosteroids may reduce brain edema, suppress demyelination, stabilize the BBB, and reduce neuroinflammation (Gold, Buttgereit, & Toyka, 2001) in ODS. A protective effect of corticosteroids on the development and size of lesions in animal models of osmotic-induced demyelination has been demonstrated (Rojiani, Prineas, & Cho, 1987). Sakamoto et al. (2007) reported a patient with a complete clinical recovery of CPM following high-dose pulse methylprednisolone treatment. Plasmapheresis was suggested to reduce high-molecular-weight myelinotoxic substances and could be considered as a safe and effective method to improve the clinical outcome of patients with severe clinical presentation of CPM (Bibl et al., 1999). Clinical improvement in patients with severe clinical presentation of CPM in the context of chronic alcohol abuse and rapid correction of hyponatremia was reported following therapeutic plasmapheresis by Grimaldi et al. (2005) (10 plasmapheresis sessions, two or three sessions per week over 1 month) and Bibl et al. (1999) (daily sessions over 4 days changing to twice a week over a further 3 weeks). However, Saner et al. (2008) could not observe a treatment effect of plasmapheresis in a patient with severe CPM that developed in association with liver transplantation owing to alcohol-induced cirrhosis, but reported an improvement following treatment with IVIG. Some other authors also supported the use of IVIG (0.4 g/kg body weight per day for 5 days) as a promising therapeutic option in ODS (Finsterer, Engelmayer, Trnka, & Stiskal, 2000), assuming that IVIG could reduce levels of myelinotoxic substances such as complement (Wing et al. 1992), modulate anti-myelin antibodies, and promote remyelination (Finsterer et al., 2000). An intriguing observation of the neurological improvement in patients with CPM following thyrotropin-releasing hormone (TRH) administration as reported by several authors deserves further study (Kumar et al., 2006). Although the exact mechanism of TRH action on the CNS has not been established, several interesting experiments supported a theory that TRH might act as a brain-stimulating substance (Wakui et al., 1991). The use of free-radical scavengers and antioxidant therapy was also proposed in ODS, but the effectiveness of these agents is still controversial (Kumar et al., 2006). Additionally, the symptomatic treatment can relieve symptoms until lesion normalization occurs (Kumar et al., 2006). However, owing to the complex alcohol-related pathogenetic mechanisms in the development of ODS, alcohol abstention is of great therapeutic importance (Dujmović et al., 2013).

Patient outcomes in ODS vary considerably and are difficult to predict. However, although there is usually a grave prognosis for the majority of ODS patients, a considerable number survive the acute phase and eventually show good recovery (Mochizuki et al., 2003). Recovery may be spontaneous, sometimes without any neurological sequelae (Brito et al., 2006). Mochizuki et al. (2003) suggested that CPM in alcoholics without iatrogenic correction of hyponatremia might be expected to follow a more benign clinical course. In a case series of 34 CPM patients (the vast majority were chronic alcoholics), Menger and Jörg (1999) reported that only 2 patients died, 11 completely recovered, 11 had some deficits but were independent, and 10 survived but were left dependent. In this study, the extent of the initial pontine lesion did not correlate with the clinical outcome. Similarly, in a series of 24 patients with CPM (18 were alcoholics), clinical outcome was not predicted by the T2W brain MRI lesion volume or the severity of hyponatremia (Graff-Radford, Fugate, Kaufmann, Mandrekar, & Rabinstein, 2011). However, Dervisoglu, Yegenaga, Anik, Sengul, and Turgut (2006) reported a rapid normalization of the brain ADC values during the first week and month associated with clinical improvement, suggesting that repeated DWI findings might be predictive in ODS.

Although mild or asymptomatic cases are possible, ODS in alcoholics might also be a devastating condition with considerable morbidity and/or mortality. A prompt recognition of ODS in alcoholics and initiation of intensive treatment and monitoring, in addition to an early multidisciplinary approach to care, may help modulate its progress and improve long-term outcome and prognosis. A better understanding of the pathophysiology of ODS should aid in the development of new causal therapeutic approaches and effective preventive strategies in chronic alcoholics with this condition.

APPLICATIONS TO OTHER ADDICTIONS AND SUBSTANCE MISUSE

1. Several drugs used to treat psychiatric symptoms or epilepsy in drug addicts may induce hyponatremia, such as antidepressants (tricyclic antidepressants, selective serotonin-reuptake inhibitors, monoamine oxidase inhibitors), antipsychotics (phenothiazines, butyrophenones), and antiepileptic drugs (carbamazepine, oxcarbazepine, sodium valproate, lamotrigine). The mechanisms

of drug-induced hyponatremia are mostly mediated through an increased hypothalamic production of ADH and/or potentiation of the ADH effect. Abuse of 3,4-methylenedioxymethamphetamine (MDMA, ecstasy) may induce hypotonic hyponatremia owing to ADH's secretagogue-like effects. Chronic hyponatremia might lead to ODS.

2. MDMA and other common and emerging drugs of abuse (anabolic androgenic steroids, cocaine and its levamisole-adulterated counterpart) may have nephrotoxic effects. Drug and substance abusers may suffer from liver abnormalities, respiratory compromise, hyperthermia with dehydration, various bacterial or fungal infections; are at risk of blood-borne viral infections (human immunodeficiency virus, hepatitis B and C, and others) in cases of injectable drug abuse; and often suffer from malnutrition, vitamin deficiency syndromes, nausea, and vomiting, while craniocerebral trauma in this population is also not rare. All these conditions might be associated with ODS.

3. Chronic glue sniffers are exposed to toluene, an aromatic hydrocarbon organic solvent that may cause acid–base and electrolyte disorders and CNS demyelinating lesions such as ODS.

4. In cases of cocaine-induced toxic encephalopathy, brain MRI could reveal T2W signal hyperintensity in the globus pallidi, splenium, and cerebral white matter that might resemble EPM.

DEFINITION OF TERMS

Antidiuretic hormone (ADH, vasopressin) This is produced in the hypothalamus and released by the posterior pituitary gland. ADH controls the permeability of the renal collecting ducts to water and decreases urine formation. Its secondary function is vasoconstriction.

Apoptosis This is a programmed cell death caused by complex mechanisms. An excessive apoptosis in the CNS leads to atrophy.

Cytotoxic edema This is an intracellular edema that results from uncompensated influx of cations, mainly sodium, and water into the cells. In the CNS it refers to the type of edema in which the BBB is intact.

Demyelination This is damage to the previously intact myelin sheath in the nervous system. It is distinguished from *dysmyelination*, which represents the defective primary structure and/or function of myelin sheaths.

Locked-in state This describes the neurological condition characterized by quadriplegia and anarthria with preserved consciousness.

Marchiafava-Bignami disease This is a rare condition mainly associated with alcoholism, characterized by extensive corpus callosum demyelination and necrosis in severe cases, although extracallosal and cortical lesions may also be present.

Osmotic stress This is the result of a sudden water and solute movement across a cell membrane due to rapid changes in osmolality.

Oxidative stress This is an imbalance of pro-oxidants and antioxidants that results in destructive free radical biochemical effects.

Plasmapheresis This is also known as therapeutic plasma exchange. It is a procedure that includes separating the blood, exchanging the plasma typically with donor plasma or albumin solution, and returning the blood cell components, primarily red blood cells, to the patient. It is an extracorporeal immunomodulation treatment strategy.

Vasogenic edema In the CNS this refers to the type of edema in which the BBB is damaged.

KEY FACTS

Key Facts of the Osmotic Demyelination Syndrome in Alcoholics

- The concept of CPM in alcoholics was introduced in 1959 but was expanded afterward to include extrapontine lesions, thus defining the spectrum of ODS.
- Osmotic stress, endothelial dysfunction, and BBB damage are the key pathogenetic steps in the development of demyelination in ODS.
- MRI changes may be delayed in ODS. Early MRI abnormalities are potentially reversible, but conventional MRI findings are not prognostic.
- Correction of chronic hyponatremia in alcoholics that is not in excess of 8 mmol/l/day minimizes the risk of ODS development.
- There is no established effective treatment for ODS, but corticosteroids, IVIG, or plasmapheresis may be considered as potential treatment options. Alcohol abstention is important in the treatment of ODS in alcoholics.
- ODS in alcoholics might be associated with a considerable morbidity and/or mortality, but some patients survive and eventually show good recovery, sometimes with no neurological sequelae.

Key Facts of the Blood–Brain Barrier

- The first direct evidence for the existence of the BBB was given at the dawn of the twentieth century by the Nobel Laureate Paul Ehrlich.
- The BBB has a surface area of nearly 20 m^2 and a total length of approximately 600 km.
- The BBB is a highly selective permeability barrier that consists of endothelial cells with tight junctions, a basement membrane, astrocytes with foot processes, and pericytes.
- The BBB separates circulating blood from the extracellular tissue in the CNS and protects the CNS from toxic and pathogenic agents in the blood.
- Alcohol passes the BBB.
- Disruption of the BBB is a critical event in the development and progression of several CNS disorders, being either a precipitating event or a consequence of the pathology.

SUMMARY POINTS

- This chapter aims to elucidate the association of alcoholism with ODS, the noninflammatory CNS demyelinating condition.
- Multiple alcohol-related pathogenetic mechanisms may contribute to the development of demyelination in ODS.
- Postmortem studies of CNS pathology in ODS have largely contributed to a better understanding of ODS pathogenesis.
- Neuroimaging characteristics of ODS in alcoholics show no distinct characteristics compared to nonalcoholics. The clinical suspicion of ODS is usually confirmed by MRI, which is superior to CT for the diagnosis of this condition.

- Clinical presentation in ODS can be extremely variable, varying from asymptomatic cases to severe neurological deterioration.
- Prevention of ODS alcoholics is of great importance. The treatment of alcohol-related ODS is challenging, especially in severe cases.
- Patient outcomes in ODS vary considerably and are difficult to predict.

REFERENCES

Adams, R. D., Victor, M., & Mancall, E. L. (1959). Central pontine myelinolysis: a hitherto undescribed disease occurring in alcoholic and malnourished patients. *A.M.A. Archives of Neurology and Psychiatry, 81*, 154–172.

Alfieri, R. R., Cavazzoni, A., Petronini, P. G., Bonelli, M. A., Caccamo, A. E., Borghetti, A. F., & Wheeler, K. P. (2002). Compatible osmolytes modulate the response of porcine endothelial cells to hypertonicity and protect them from apoptosis. *The Journal of Physiology, 540*, 499–508.

Alikunju, S., Abdul Muneer, P. M., Zhang, Y., Szlachetka, A. M., & Haorah, J. (2011). The inflammatory footprints of alcohol-induced oxidative damage in neurovascular components. *Brain, Behavior and Immunity, 25*(Suppl. 1), S129–S136.

An, J. Y., Park, S. K., Han, S. R., & Song, I. U. (2010). Central pontine and extrapontine myelinolysis that developed during alcohol withdrawal, without hyponatremia, in a chronic alcoholic. *Internal Medicine, 49*, 615–618.

Ashrafian, H., & Davey, P. (2001). A review of the causes of central pontine myelinosis: yet another apoptotic illness? *European Journal of Neurology, 8*, 103–109.

Baker, E. A., Tian, Y., Adler, S., & Verbalis, J. G. (2000). Blood–brain barrier disruption and complement activation in the brain following rapid correction of chronic hyponatremia. *Experimental Neurology, 165*, 221–230.

Bibl, D., Lampl, C., Gabriel, C., Jüngling, G., Brock, H., & Köstler, G. (1999). Treatment of central pontine myelinolysis with therapeutic plasmapheresis. *Lancet, 353*, 1155.

Bluml, S., Zuckerman, E., Tan, J., & Ross, B. D. (1998). Proton-decoupled 31P magnetic resonance spectroscopy reveals osmotic and metabolic disturbances in human hepatic encephalopathy. *Journal of Neurochemistry, 71*, 1564–1576.

Brito, A. R., Vasconcelos, M. M., Cruz Júnior, L. C., Oliveira, M. E., Azevedo, A. R., Rocha, L. G., & Mendonça, P. C. (2006). Central pontine and extrapontine myelinolysis: report of a case with a tragic outcome. *Jornal de Pediatria, 82*, 157–160.

Chu, K., Kang, D. W., Ko, S. B., & Kim, M. (2001). Diffusion-weighted MR findings of central pontine and extrapontine myelinolysis. *Acta Neurologica Scandinavica, 104*, 385–388.

Dervisoglu, E., Yegenaga, I., Anik, Y., Sengul, E., & Turgut, T. (2006). Diffusion magnetic resonance imaging may provide prognostic information in osmotic demyelination syndrome: report of a case. *Acta Radiologica, 47*, 208–212.

Dujmović, I., Vitas, J., Zlatarić, N., & Drulović, J. (2013). Central pontine myelinolysis in a chronic alcoholic: a clinical and brain magnetic resonance imaging follow-up. *Vojnosanitetski Pregled, 70*, 785–788.

Finsterer, J., Engelmayer, E., Trnka, E., & Stiskal, M. (2000). Immunoglobulins are effective in pontine myelinolysis. *Clinical Neuropharmacology, 23*, 110–113.

Ghosh, N., DeLuca, G. C., & Esiri, M. M. (2004). Evidence of axonal damage in human acute demyelinating diseases. *Journal of the Neurological Sciences, 222*, 29–34.

Gold, R., Buttgereit, F., & Toyka, K. V. (2001). Mechanism of action of glucocorticosteroid hormones: possible implications for therapy of neuroimmunological disorders. *Journal of Neuroimmunology, 117*, 1–8.

Graff-Radford, J., Fugate, J. E., Kaufmann, T. J., Mandrekar, J. N., & Rabinstein, A. A. (2011). Clinical and radiologic correlations of central pontine myelinolysis syndrome. *Mayo Clinic Proceedings, 86*, 1063–1067.

Grimaldi, D., Cavalleri, F., Vallone, S., Milanti, G., & Cortelli, P. (2005). Plasmapheresis improves the outcome of central pontine myelinolysis. *Journal of Neurology, 252*, 734–735.

Haider, L., Fischer, M. T., Frischer, J. M., Bauer, J., Höftberger, R., Botond, G., … Lassmann, H. (2011). Oxidative damage in multiple sclerosis lesions. *Brain, 134*, 1914–1924.

He, J., & Crews, F. T. (2008). Increased MCP-1 and microglia in various regions of the human alcoholic brain. *Experimental Neurology, 210*, 349–358.

Hillbom, M., Saloheimo, P., Fujioka, S., Wszolek, Z. K., Juvela, S., & Leone, M. A. (2014). Diagnosis and management of Marchiafava-Bignami disease: a review of CT/MRI confirmed cases. *Journal of Neurology, Neurosurgery and Psychiatry, 85*, 168–173.

Hoerner, M., Behrens, U. J., Worner, T., & Lieber, C. S. (1986). Humoral immune response to acetaldehyde adducts in alcoholic patients. *Research Communications in Chemical Pathology and Pharmacology, 54*, 3–12.

Hurley, R. A., Filley, C. M., & Taber, K. H. (2011). Central pontine myelinolysis: a metabolic disorder of myelin. *The Journal of Neuropsychiatry & Clinical Neurosciences, 23*, 369–374.

Jacob, S., Gupta, H., Nikolic, D., Gundogdu, B., & Ong, S. (2014). Central pontine and extrapontine myelinolysis: the great masquerader—an autopsy case report. *Case Reports in Neurological Medicine, 5*. Article ID 745347.

Kuhlmann, T., Lassmann, H., & Brück, W. (2008). Diagnosis of inflammatory demyelination in biopsy specimens: a practical approach. *Acta Neuropathologica, 115*, 275–287.

Kumar, S., Fowler, M., Gonzalez-Toledo, E., & Jaffe, S. L. (2006). Central pontine myelinolysis, an update. *Neurological Research, 28*, 360–366.

Kumar, S. R., Mone, A. P., Gray, L. C., & Troost, B. T. (2000). Central pontine myelinolysis: delayed changes on neuroimaging. *Journal of Neuroimaging, 10*, 169–172.

Lampl, C., & Yazdi, K. (2002). Central pontine myelinolysis. *European Neurology, 47*, 3–10.

Liberatore, M., Denier, C., Fillard, P., Petit-Lacour, M. C., Benoudiba, F., Lasjaunias, P., & Ducreux, D. (2006). Diffusion tensor imaging and tractography of central pontine myelinolysis. *Journal of Neuroradiology, 33*, 189–193.

Lupato, A., Fazio, P., Fainardi, E., Cesnik, E., Casetta, I., & Granieri, E. (2010). A case of asymptomatic pontine myelinolysis. *Neurological Sciences, 31*, 361–364.

Lüthy, F. (1932). Über die hepato-lentikuläre Degeneration. *Deutsche Zeitschrift für Nervenheilkunde, 123*, 102–181.

Martin, R. J. (2004). Central pontine and extrapontine myelinolysis: the osmotic demyelination syndromes. *Journal of Neurology, Neurosurgery and Psychiatry, 75*(Suppl. 3), 22–28.

McNeill, G., Halpenny, D., Snow, A., Geoghegan, A., & Torreggiani, W. C. (2009). Extrapontine myelinolysis after correction of hyponatremia presenting as generalised tonic seizures. *American Journal of Emergency Medicine, 27*, 243.

Menger, H., & Jörg, J. (1999). Outcome of central pontine and extrapontine myelinolysis (n=44). *Journal of Neurology, 246*, 700–705.

Menger, H., Mackowski, J., Jörg, J., & Cramer, B. M. (1998). Pontine and extra-pontine myelinolysis. Early diagnostic and prognostic value of cerebral CT and MRI. *Nervenarzt, 69*, 1083–1090.

Mochizuki, H., Masaki, T., Miyakawa, T., Nakane, J., Yokoyama, A., Nakamura, Y., ... Higuchi, S. (2003). Benign type of central pontine myelinolysis in alcoholism-clinical, neuroradiological and electrophysiological findings. *Journal of Neurology, 250*, 1077–1083.

Nomoto, N., Arasaki, K., & Tamaki, M. (2004). Central pontine myelinolysis in chronic alcoholism demonstrated by magnetic resonance imaging and spectroscopy. *European Neurology, 51*, 179–180.

Norenberg, M. D. (1983). A hypothesis of osmotic endothelial injury. A pathogenetic mechanism in central pontine myelinolysis. *Archives of Neurology, 40*, 66–69.

Norenberg, M. D. (2010). Central pontine myelinolysis: historical and mechanistic considerations. *Metabolic Brain Disease, 25*, 97–106.

Pons-Vázquez, S., Gallego-Pinazo, R., Galbis-Estrada, C., Zanon-Moreno, V., Garcia-Medina, J. J., Vila-Bou, V., ... Pinazo-Durán, M. D. (2011). Combined pre- and postnatal ethanol exposure in rats disturbs the myelination of optic axons. *Alcohol and Alcoholism, 46*, 514–522.

Popescu, B. F., Bunyan, R. F., Guo, Y., Parisi, J. E., Lennon, V. A., & Lucchinetti, C. F. (2013). Evidence of aquaporin involvement in human central pontine myelinolysis. *Acta Neuropathologica Communications, 1*, 40.

Qin, L., He, J., Hanes, R. N., Pluzarev, O., Hong, J. S., & Crews, F. T. (2008). Increased systemic and brain cytokine production and neuroinflammation by endotoxin following ethanol treatment. *Journal of Neuroinflammation, 5*, 10.

Ragland, G. (1990). Electrolyte abnormalities in the alcoholic patient. *Emergency Medicine Clinics of North America, 8*, 761–773.

Rego, I., Vieira, D., Correia, F., & Pereira, J. R. (March 9, 2012). Multiple brain lesions in a young man with hypernatraemia. *BMJ Case Reports.* http://dx.doi.org/10.1136/bcr.11.2011.5198.

Roh, J. K., Nam, H., & Lee, M. C. (1998). A case of central pontine and extrapontine myelinolysis with early hypermetabolism on 18FDG-PET scan. *Journal of Korean Medical Science, 13*, 99–102.

Rojiani, A. M., Prineas, J. W., & Cho, E. S. (1987). Protective effect of steroids in electrolyte-induced demyelination. *Journal of Neuropathology and Experimental Neurology, 46*, 495–504.

Ruzek, K. A., Campeau, N. G., & Miller, G. M. (2004). Early diagnosis of central pontine myelinolysis with diffusion-weighted imaging. *American Journal of Neuroradiology, 25*, 210–213.

Sakamoto, E., Hagiwara, D., Morishita, Y., Tsukiyama, K., Kondo, K., & Yamamoto, M. (2007). Complete recovery of central pontine myelinolysis by high dose pulse therapy with methylprednisolone. *Nihon Naika Gakkai Zasshi, 96*, 2291–2293.

Saner, F. H., Koeppen, S., Meyer, M., Kohnle, M., Herget-Rosenthal, S., Sotiropoulos, G. C., ... Broelsch, C. E. (2008). Treatment of central pontine myelinolysis with plasmapheresis and immunoglobulins in liver transplant patient. *Transplant International, 21*, 390–391.

Sohn, M. K., & Nam, J. H. (2014). Locked-in syndrome due to Central pontine myelinolysis: case report. *Annals of Rehabilitation Medicine, 38*, 702–706.

Spakowski, M., Nikolic, N., & Giambarba, C. (2009). Ethanol, water, salt and coma. *Praxis, 98*, 905–908.

Sterns, R. H., Silver, S., Kleinschmidt-DeMasters, B. K., & Rojiani, A. M. (2007). Current perspectives in the management of hyponatremia: prevention of CPM. *Expert Review of Neurotherapeutics, 7*, 1791–1797.

Szabo, G., & Lippai, D. (2014). Converging actions of alcohol on liver and brain immune signaling. *International Review of Neurobiology, 118*, 359–380.

Takefuji, S., Murase, T., Sugimura, Y., Takagishi, Y., Hayasaka, S., Oiso, Y., & Murata, Y. (2007). Role of microglia in the pathogenesis of osmotic-induced demyelination. *Experimental Neurology, 204*, 88–94.

Takei, Y., Akahane, C., & Ikeda, S. (2003). Osmotic demyelination syndrome: reversible MRI findings in bilateral cortical lesions. *Internal Medicine, 42*, 867–870.

Tan, K. H., Harrington, S., Purcell, W. M., & Hurst, R. D. (2004). Peroxynitrite mediates nitric oxide-induced blood–brain barrier damage. *Neurochemical Research, 29*, 579–587.

Thompson, P. D., Miller, D., Gledhill, R. F., & Rossor, M. N. (1989). Magnetic resonance imaging in central pontine myelinolysis. *Journal of Neurology, Neurosurgery and Psychiatry, 52*, 675–677.

Tomlinson, B. E., Pierides, A. M., & Bradley, W. G. (1976). Central pontine myelinolysis. Two cases with associated electrolyte disturbance. *The Quarterly Journal of Medicine, 45*, 373–386.

Wakui, H., Nishimura, S., Watahiki, Y., Endo, Y., Nakamoto, Y., & Miura, A. B. (1991). Dramatic recovery from neurological deficits in a patient with central pontine myelinolysis following severe hyponatremia. *Japanese Journal of Medicine, 30*, 281–284.

Wiggins-Dohlvik, K., Merriman, M., Shaji, C. A., Alluri, H., Grimsley, M., Davis, M. L., ... Tharakan, B. (2014). Tumor necrosis factor-α disruption of brain endothelial cell barrier is mediated through matrix metalloproteinase-9. *American Journal of Surgery, 208*(6), 954–960 pii:S0002-9610(14) 00450-4.

Wijdicks, E. F., Blue, P. R., Steers, J. L., & Wiesner, R. H. (1996). Central pontine myelinolysis with stupor alone after orthotopic liver transplantation. *Liver Transplantation and Surgery, 2*, 14–16.

Wing, M. G., Zajicek, J., Seilly, D. J., Compston, D. A., & Lachmann, P. J. (1992). Oligodendrocytes lack glycolipid anchored proteins which protect them against complement lysis. Restoration of resistance to lysis by incorporation of CD59. *Immunology, 76*, 140–145.

Yoon, B., Shim, Y. S., & Chung, S.-W. (2008). Central pontine and extrapontine myelinolysis after alcohol withdrawal. *Alcohol and Alcoholism, 43*, 647–649.

Yuh, W. T., Simonson, T. M., D'Alessandro, M. P., Smith, K. S., & Hunsicker, L. G. (1995). Temporal changes of MR findings in central pontine myelinolysis. *AJNR: American Journal of Neuroradiology, 16*(Suppl. 4), 975–977.

Zahr, N. M., Kaufman, K. L., & Harper, C. G. (2011). Clinical and pathological features of alcohol-related brain damage. *Nature Reviews: Neurology, 7*, 284–294.

Zuccoli, G., Siddiqui, N., Cravo, I., Bailey, A., Gallucci, M., & Harper, C. G. (2010). Neuroimaging findings in alcohol-related encephalopathies. *AJR: American Journal of Roentgenology, 195*, 1378–1384.

Chapter 56

The Role of Brain Glucocorticoid Systems in Alcohol Dependence

Annie M. Whitaker[1], Brittany M. Priddy[2], Scott Edwards[1], Leandro F. Vendruscolo[2]

[1]Department of Physiology, Alcohol and Drug Abuse Center of Excellence, LSU Health Sciences Center, New Orleans, LA, USA; [2]NIH/NIDA, IRP/INRB, Baltimore, MD, USA

Abbreviations

ACTH Adrenocorticotropic hormone
CeA Central amygdala
CORT Cortisol or corticosterone
CRF Corticotropin-releasing factor
DNA Deoxyribonucleic acid
FKBP FK506-binding protein
GR Glucocorticoid receptor
GRE Glucocorticoid response element
HPA Hypothalamic–pituitary–adrenal
HSP Heat shock protein
MAPK Mitogen-activated protein kinase
MW Molecular weight
NCoR1 Nuclear receptor corepressor 1
PVN Paraventricular nucleus
SRC1 Steroid receptor coactivator 1

INTRODUCTION

Alcoholism is characterized by compulsive drinking, an enduring propensity for relapse, and the emergence of negative emotional symptoms during abstinence (Koob et al., 2014). Currently, alcohol misuse is one of the leading causes of preventable deaths in the world and has an enormous cost to the individual, family, and society (World Health Organization, 2014). At the physiological level, alcohol dependence is associated with persistent changes in brain function that underlie excessive drinking and contribute to high rates of relapse after months or even years of alcohol abstinence. Further identification of causative brain changes linked to aberrant drinking behaviors will help with the discovery of new strategies to reduce and prevent alcohol-related morbidity and mortality.

Alcohol is initially consumed for its various gratifying effects (e.g., social facilitation and euphoria) via positive reinforcement processes. During the transition to dependence, neurobiological changes occur to adapt to repeated episodes of excessive alcohol intoxication and withdrawal, which lead to the emergence of negative emotional symptoms associated with withdrawal, such as anxiety, dysphoria, depression, irritability, and pain (Edwards & Koob, 2010). These symptoms create an internal state that drives compulsive alcohol drinking such that alcohol is ingested in increasing quantities in an attempt to cope with heightened negative emotions (i.e., negative reinforcement). This chapter covers preclinical models of alcohol dependence with a focus on the role of neuroendocrine/hypothalamic–pituitary–adrenal (HPA) and extrahypothalamic glucocorticoid systems in alcohol dependence, as these systems represent predominating stress-signaling mechanisms influenced by alcohol exposure.

ALCOHOL ACTIVATES THE HYPOTHALAMIC–PITUITARY–ADRENAL AXIS

The HPA axis is a major component of the neuroendocrine system that mediates bodily responses to environmental stressors (Myers, McKlveen, & Herman, 2014), facilitating a critical survival mechanism that prepares the body for fight-or-flight actions (McEwen, 2007). The HPA axis also regulates a variety of physiological processes such as reproduction, growth, metabolism, mood/emotions, and immune function (Molina, 2013). In response to environmental stressors, the paraventricular nucleus (PVN), a subregion of the hypothalamus, is responsible for the stimulation of the pituitary through the release of corticotropin-releasing factor (CRF). CRF binds to CRF1 receptors to signal the anterior part of the pituitary to release adrenocorticotropic hormone (ACTH). Through the general blood circulation, ACTH reaches the cortex of the adrenal glands, where binding of the ACTH receptor (also known as melanocortin receptor 2) by ACTH stimulates the rapid production of glucocorticoids (cortisol in primates or corticosterone in rodents; termed herein as CORT) and release into the bloodstream. Circulating CORT then acts on a variety of tissues to exert an array of vital physiological functions, often in a coordinated function to meet environmental demands and ultimately return body systems to homeostasis.

To exert its biological actions in the brain, CORT binds to two types of receptors: mineralocorticoid receptors (type I) and glucocorticoid receptors (GRs or type II). Mineralocorticoid receptors exhibit a high affinity for CORT and are almost fully occupied during resting conditions. GRs display a lower affinity for CORT and are activated predominantly at very high circulating CORT

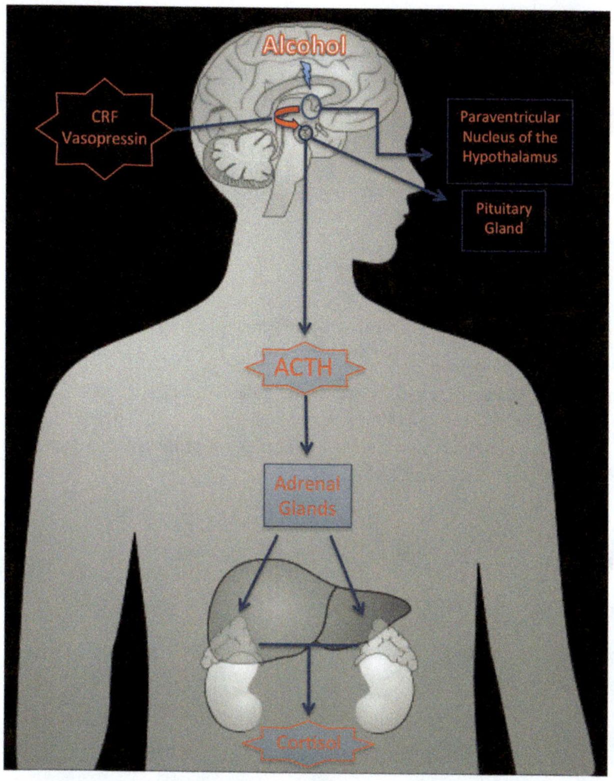

FIGURE 1 Alcohol-induced HPA axis activation. In response to alcohol intoxication, the paraventricular nucleus (PVN) of the hypothalamus releases corticotropin-releasing factor (CRF). CRF causes the release of adrenocorticotrophic hormone (ACTH) by the anterior pituitary. Following transit through the bloodstream, ACTH stimulates the release of glucocorticoids by the adrenal gland.

levels (McEwen, 2007), such as during particularly stressful situations. CORT negatively feeds back along different levels of the HPA axis to prevent further CORT release. Although moderate and sporadic HPA axis activation causes no enduring consequences to the body, intense and prolonged HPA axis activation may lead to persistent biological changes and psychiatric disease (Koob & Kreek, 2007; Sinha et al., 2011).

Similar to stress, acute alcohol exposure activates the HPA axis in humans (Ellis, 1966) and rodents (Richardson, Lee, O'Dell, Koob, & Rivier, 2008; for review, see Lu & Richardson, 2014) to significantly elevate CORT levels (Figure 1). In a nondependent state, CORT even facilitates alcohol's rewarding/reinforcing effects. Exogenous CORT has been shown to transiently increase alcohol drinking (Besheer, Fisher, Lindsay, & Cannady, 2013), whereas adrenalectomy reduces alcohol intake, an effect restored by CORT replacement (Fahlke, Hard, Eriksson, Engel, & Hansen, 1995; Fahlke, Hard, Henson, 1996).

CHRONIC ALCOHOL EXPOSURE DISRUPTS HPA AXIS FUNCTION

Excessive activation of the HPA axis following repeated or prolonged cycles of alcohol exposure and withdrawal can disrupt the balance of HPA axis activity (Adinoff et al., 1990; Adinoff, Ruether, Krebaum, Iranmanesh, & Williams, 2003; Lovallo, Dickensheets, Myers, Thomas, & Nixon, 2000; Rasmussen et al., 2000; Richardson et al., 2008; Sinha et al., 2011; Uhart & Wand, 2009; Zorrilla, Valdez, & Weiss, 2001). For example, Richardson et al. (2008) discovered that acute alcohol-induced activation of the HPA axis is decreased by chronic exposure to alcohol (i.e., neuroendocrine tolerance), an effect that may be partially due to downregulation of CRF in the PVN. In comparison, an upregulation of CRF occurs in extrahypothalamic brain regions such as the central nucleus of the amygdala. This apparent sensitization of brain stress systems represents a critical neuroadaptation that leads to compulsive drinking in alcohol dependence (for review, see Heilig & Koob, 2007). Moreover, previous exposure to chronic CORT has been shown to subsequently decrease alcohol-associated interoceptive effects (e.g., the capability of bodily information to reach conscious awareness; Besheer, Fisher, Grondin, Cannady, & Hodge, 2012; Paulus, Tapert, & Schulteis, 2009). Consequently, tolerance to alcohol's rewarding and other physiological effects might contribute to a perceived need to escalate alcohol drinking. Figure 2 illustrates the effect of alcohol on the HPA axis and extrahypothalamic stress systems in a nondependent state versus the transition to alcohol dependence.

It is important to note that the effects of alcohol on the neuroendocrine stress system are complex and modulated by several biological factors, including sex hormones and brain developmental phases (Allen, Lee, Koob, & Rivier, 2011; Przybycien-Szymanska, Gillespie, & Pak, 2012; Silva & Madeira, 2012). Environmental factors also regulate the effects of alcohol on neuroendocrine mechanisms. For instance, social interaction attenuates alcohol-induced changes in HPA axis function (Pang et al., 2013), indicating that these changes, although long-lasting, are not permanent but rather malleable.

ROLE OF BRAIN GLUCOCORTICOID RECEPTOR ACTIVITY IN THE TRANSITION TO ALCOHOL DEPENDENCE

Reports indicate that corticosteroid-dependent plasticity alters brain reward and stress systems. Expression levels of the GR are downregulated in several reward/stress-related brain regions (Vendruscolo et al., 2012). This finding can be interpreted as a compensatory mechanism for excessive circulating CORT caused by alcohol intoxication and withdrawal. Consistently, long-term alcohol exposure in mice produces a prolonged rise in glucocorticoid concentration in various brain regions and a downregulation of the GR in the prefrontal cortex (Little et al., 2008). These discoveries suggest that the GR system is significantly activated when alcohol intoxication and withdrawal occur repeatedly, and these changes might contribute to the development of alcoholism.

A functional role for the GR in compulsive alcohol drinking was examined through the use of the GR antagonist mifepristone (also known as RU-38486 or RU-486). In a preclinical model of alcoholism, rats that were exposed to repeated cycles of alcohol intoxication and withdrawal gradually displayed compulsive-like alcohol seeking. Importantly, the establishment of this condition was entirely blocked by ongoing mifepristone treatment (Vendruscolo et al., 2012). These findings indicate that GR activity is intimately associated with the development of compulsive alcohol drinking in

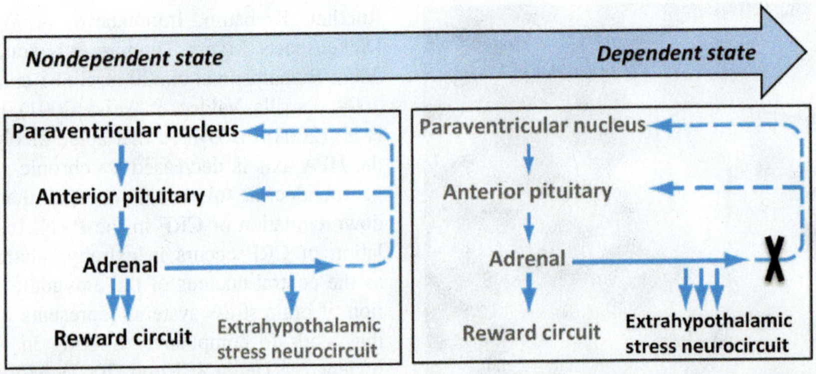

FIGURE 2 Alcohol-induced dysregulation of the HPA axis and extrahypothalamic stress systems. As indicated on the left, in a nondependent state, alcohol activates the HPA axis to release glucocorticoids, which facilitates the reinforcing effects of alcohol. As indicated on the right, repeated episodes of alcohol intoxication/withdrawal disrupt HPA axis function and "sensitize" extrahypothalamic stress systems. Sensitization of these stress systems is hypothesized to mediate compulsive alcohol drinking in alcohol dependence. *Modified from Vendruscolo et al. (2012).*

TABLE 1 Effects of Mifepristone on Alcohol-Related Behaviors

Author	Year	Species	Mifepristone Dose (route)	Outcome
Cippitelli et al.	2012	Rat	15 mg/kg (intraperitoneal)	Reduced neurotoxicity produced by binge-like exposure to alcohol.
Koenig et al.	2004	Rat	1, 5, and 20 mg/kg (intraperitoneal)	Decreased alcohol consumption under limited-access conditions.
O'Callaghan et al.	2005	Mice	100 mg/kg (intraperitoneal)	Prevented increases in alcohol consumption caused by moderate chronic stress.
Roberts et al.	1995	Mice	20 mg/kg (injection)	Attenuated stress and alcohol-induced behavioral sensitization.
Sharrett-Field et al.	2013	Rat	20 or 40 mg/kg (subcutaneous)	Significantly reduced the severity of alcohol withdrawal.
Simms et al.	2011	Rat	10 mg/side (intraperitoneal)	Reduced stress-induced reinstatement of alcohol seeking.
Vendruscolo et al.	2012	Rat	150 mg/21-day release (subcutaneous pellet)	Prevented the development of compulsive-like alcohol drinking.
Vendruscolo et al.	2015	Human	600 mg/day for 1 week	Decreased craving and drinking in alcoholics.

this particular model of alcohol dependence (for review, see Vendruscolo & Roberts, 2014). Additionally, mifepristone has been reported to decrease craving and drinking in alcoholics (Vendruscolo et al., 2015), prevent increases in alcohol consumption in mice (O'Callaghan, Craft, Jacquot, & Little, 2005), decrease alcohol drinking under limited-access conditions (Koenig & Olive, 2004), and reduce stress-induced reinstatement of alcohol seeking (Simms, Haass-Koffler, Bito-Onon, Li, & Bartlett, 2012), suggesting that mifepristone modulates alcohol-related behaviors under certain stressful conditions.

Mifepristone also attenuates other behavioral effects of alcohol that might contribute to compulsive alcohol drinking. For example, repeated restraint stress produces a sensitization of alcohol's activating (i.e., locomotor) effects in mice, and the administration of mifepristone attenuates this effect (Roberts, Lessov, & Phillips, 1995). Mifepristone also significantly reduced withdrawal severity (Sharrett-Field, Butler, Berry, Reynolds, & Prendergast, 2013) and neurotoxicity from alcohol in rats exposed to a binge-like pattern of alcohol intoxication (Cippitelli et al., 2014) or in mice after chronic alcohol intake (Jacquot et al., 2008). Table 1 summarizes several findings obtained with mifepristone treatment in animal models of alcohol dependence. A unique feature of mifepristone is its ability to differentially alter GR function at the molecular level based on its recruitment of coregulatory proteins in the nucleus (Schoch et al., 2010; Schulz et al., 2002). In the next section, we briefly summarize how nuclear trafficking and transcriptional activity of GR is regulated in target tissues.

GR STRUCTURE, CHAPERONE PROTEINS, AND NUCLEAR COFACTORS

At the molecular level, glucocorticoids are steroid hormones that mediate adaptive physiological processes by binding to GRs and subsequent glucocorticoid response elements (GREs) in the nucleus of target cells to regulate transcription of multiple genes (Kadmiel & Cidlowski, 2013). This system requires the coordination of several cofactors and cochaperones important for protein folding and nuclear trafficking (Figure 3). Because

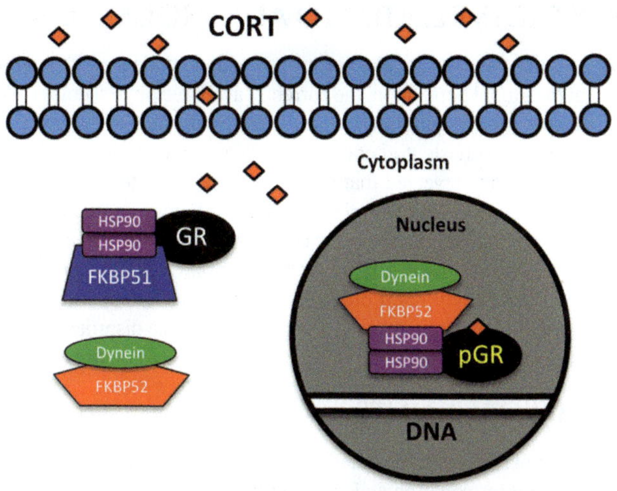

FIGURE 3 Gene transcriptional regulation by Glucocorticoid Receptor access and activity. Glucocorticoids such as cortisol and corticosterone (CORT) freely diffuse across lipid bilayer membranes to bind intracellular glucocorticoid receptors (GRs) that interact with a host of chaperone proteins including HSP90. Association with FKBP51 keeps GRs restricted to the cytoplasm. Upon ligand binding, FKBP52 facilitates GR translocation to the nucleus with assistance from the molecular motor protein dynein. Within the nucleus, GR phosphorylation (pGR) and association with additional coregulators (not shown) mediate the final balance of gene transcriptional activity.

glucocorticoids are lipophilic, they easily cross the plasma membrane and bind to intracellular GRs. Structurally, GRs consist of a DNA-binding domain, ligand-binding domain, and N-terminal transactivation domain and hinge region (Schaaf & Cidlowski, 2002). The DNA binding domain consists of two zinc fingers and an α-helix (Luisi et al., 1991). The N-terminal domain contains two transcription activation functions (AF1 and AF2) that are necessary for the recruitment of coregulators mediating transcription (Kumar & Thompson, 2003). These cofactors and coregulators are required for GR folding and gating of translocation to the nucleus upon ligand binding. For example, the inactive GR complex consists of heat shock protein (HSP) 90, HSP70, HSP56, HSP40, low-molecular-weight protein (p) 23 and p60, along with several immunophilins (Kumar & Thompson, 2003). HSP90 associates with the ligand-binding domain of the receptor and is critically important for GR folding.

In addition to protein folding and conformation, chaperone proteins are also heavily involved in GR trafficking into and out of the nucleus (Vandevyver, Dejager, & Libert, 2012). The immunophilin FK506-binding protein molecular weight (MW) 51 (FKBP51) is part of the GR complex that regulates GR trafficking, glucocorticoid sensitivity, and negative feedback (Wochnik et al., 2005). A role for FKBP51 dysfunction in the context of GR signal transduction has been implicated in disease states ranging from posttraumatic stress disorder to alcoholism (Binder, 2009; McClintick et al., 2013). Alterations in FKBP51 can lead to malfunction of GR trafficking and subsequent transcription of target genes, as binding of FKBP51 to HSP90 of the GR complex prevents nuclear translocation. In contrast, FK506-binding protein MW 52 (FKBP52) acts as a competitive inhibitor to FKBP51 (Wochnik et al., 2005). FKBP52 is bound to the molecular motor protein dynein. Binding of FKBP52 to the GR facilitates GR translocation to the nucleus,

transcription of target genes, and posttranslational GR phosphorylation events. In the absence of hormone, GR is phosphorylated primarily at serine 203 (Ser203); however, in the presence of glucocorticoids, GRs are phosphorylated at both Ser203 and Ser211 (Ser232 in rodents). Consequently, phosphorylation at Ser211/232 is closely associated with nuclear translocation and activation (Adzic et al., 2009).

INDUCTION OF TRANSCRIPTIONAL ACTIVATION BY GR

As described above, classical GR signal transduction involves glucocorticoids binding to GR to facilitate translocation to the nucleus. Here, the hormone/receptor complex associates with a GRE to regulate gene transcription. GREs are consensus DNA sequences located in the promoters of target genes. Binding of the GR hormone/receptor complex to GREs can prevent binding of other transcription factors, thus repressing gene activation. Alternatively, the complex can recruit key components of the transcriptional machinery to promote gene activation (Revollo & Cidlowski, 2009). Two crucial cofactors involved in this regulation are steroid receptor coactivator 1 and nuclear receptor corepressor 1 (NCoR1), which differentially associate with GR to respectively activate and inhibit transcription (Stevens et al., 2003). Importantly, the antagonistic properties of mifepristone are largely mediated by its recruitment of NCoR1 to the GR (Schulz et al., 2002).

Classic GR signaling typically involves genomic regulation. However, GRs can also produce rapid signaling events (i.e., within seconds to minutes) via nongenomic mechanisms (Kadmiel & Cidlowski, 2013). Nongenomic GR signaling mediates a multitude of physiological processes as well as the rapid glucocorticoid negative feedback at the level of the anterior pituitary (Tasker, Di, & Malcher-Lopes, 2006). Such rapid effects of glucocorticoid signaling are believed to result from interactions with plasma membrane-bound receptors (Dallman, 2005) and/or interaction with intracellular signal transduction processes such as the mitogen-activated protein kinase (MAPK) cascade (Haller, Mikics, & Makara, 2008). Indeed, along with GR activity, MAPK signaling in specific brain regions has been hypothesized to mediate the transition to alcohol dependence (Zamora-Martinez & Edwards, 2014).

APPLICATIONS TO OTHER ADDICTIONS AND SUBSTANCE MISUSE

Continued integration of behavioral and molecular experimentation to better understand GR activity in the context of excessive drug and alcohol exposure will no doubt shed light on the addiction process. A common feature of all substance use disorders is compulsivity, a loss of control over intake, and the appearance of negative emotional states when access to the substance is prevented. However, addiction to different substances also presents aspects that are specific for each agent (e.g., intensity of somatic withdrawal). The identification of commonalities and differences in addiction to different classes of drugs will help to develop more specific and individualized treatments for each disorder. For example, the role of the HPA axis in cocaine addiction and relapse has been repeatedly demonstrated (for review, see McReynolds, Pena, Blacktop, & Mantsch, 2014). Psychostimulants are strong activators of the HPA axis (Zuloaga, Johnson, Agam, & Raber, 2014) and GR function has been found to mediate several psychostimulant-induced behaviors. Inactivation of the GR in the whole brain or in dopaminoceptive cells decreases the rewarding/reinforcing properties and the locomotor-sensitizing effects of psychostimulants (Ambroggi et al., 2009; Barik et al., 2010; Deroche-Gamonet et al., 2003). Consistently, mifepristone injections reduce amphetamine or cocaine self-administration and sensitization (De Vries et al., 1996; Deroche-Gamonet et al., 2003; Fiancette, Balado, Piazza, & Deroche-Gamonet, 2010; Stairs, Prendergast, & Bardo, 2011). The effects of mifepristone on opioid abuse-related behaviors are less clear. Barik et al. (2010) have reported that inactivation of the GR in dopaminoceptive neurons has no effect on morphine reward or sensitization in mice, whereas mifepristone was shown to reduce morphine reward in rats (Dong et al., 2006). Furthermore, mifepristone attenuates somatic signs of opioid withdrawal (Navarro-Zaragoza, Hidalgo, Laorden, & Milanes, 2012) as well as opioid withdrawal-induced memory deficits (Mesripour, Hajhashemi, & Rabbani, 2008). Therefore, the HPA axis appears to be involved in the behavioral effects of multiple drugs of abuse, including alcohol, psychostimulants, and opioids. However, GRs appear to modulate different aspects of dependence-related behavior depending on the drug class. Further elucidation of the biological mechanisms by which GR regulates distinct components of drug addiction will improve the treatment of this devastating disorder.

DEFINITION OF TERMS

Adrenalectomy This is the surgical removal of the adrenal glands. The resultant loss of circulating glucocorticoids is associated with a decrease in alcohol drinking that is restored following CORT replacement.

Central amygdala This is a subcortical brain region that regulates innate and learned stress responses and contributes to the transition from alcohol use to dependence.

Ethanol Ethanol or ethyl alcohol is generally termed "alcohol." Ethanol is the primary type of alcohol in consumed alcoholic beverages. The chemical formula of ethanol is C_2H_5OH.

Homeostasis This is the tendency of an organism to maintain a constant internal state in the face of external challenges.

Negative reinforcement This is a process by which the removal of an aversive stimulus (e.g., negative emotional state of alcohol withdrawal) increases the probability of a response (drinking).

Positive reinforcement This is a process by which the presentation of a stimulus (alcohol) increases the probability of a response (drinking).

Receptor This is a large protein molecule (e.g., GR) that can be activated by a ligand (e.g., the glucocorticoid CORT) to produce a biological response.

Receptor agonist This refers to a substance/drug that binds to a receptor to produce a biological response.

Receptor antagonist This is a substance/drug (e.g., mifepristone) that blocks or reduces the effect of another substance/drug (e.g., the glucocorticoid CORT).

Stress From a biological perspective, it can be defined as the holistic physiological response of an organism to an environmental challenge.

KEY FACTS RELATED TO ALCOHOL MISUSE

- Despite the fact that alcohol misuse and dependence affect 6% of the global population, the currently available pharmacological treatments are quite ineffective. Alcohol dependence is associated with persistent alterations in brain stress systems. Animal models have revealed that GRs, which respond to the binding of glucocorticoids (the principal stress hormone in mammals), appear to be markedly activated during alcohol dependence. Chronic antagonism of GRs by mifepristone has been shown to block many symptoms of alcohol dependence and constitutes a potential novel pharmacotherapy for alcohol use disorders.

SUMMARY POINTS

- Alcohol abuse and dependence are major public health concerns.
- Alcohol intoxication activates the HPA axis, the primary neuroendocrine stress system, to release glucocorticoids.
- The effects of glucocorticoids are mediated through GR activation in close association with a host of coregulatory proteins.
- GRs are significantly activated during alcohol dependence.
- GR antagonism blocks compulsive alcohol drinking.
- The glucocorticoid system represents a potential target for the treatment of alcohol dependence.

ACKNOWLEDGMENT

Preparation of this review was supported by the Alcohol and Drug Abuse Center of Excellence at LSUHSC–New Orleans (A.M.W., S.E.) in addition to the National Institute on Drug Abuse, Intramural Research Program (B.M.P., L.F.V.) and the National Institute on Alcohol Abuse and Alcoholism (AA020839, S.E.).

REFERENCES

Adinoff, B., Martin, P. R., Bone, G. H., Eckardt, M. J., Roehrich, L., George, D. T., … Gold, P. W. (1990). Hypothalamic-pituitary-adrenal axis functioning and cerebrospinal fluid corticotropin releasing hormone and corticotropin levels in alcoholics after recent and long-term abstinence. *Archives of General Psychiatry*, 47(4), 325–330.

Adinoff, B., Ruether, K., Krebaum, S., Iranmanesh, A., & Williams, M. J. (2003). Increased salivary cortisol concentrations during chronic alcohol intoxication in a naturalistic clinical sample of men. *Alcoholism, Clinical and Experimental Research*, 27(9), 1420–1427. http://dx.doi.org/10.1097/01.ALC.0000087581.13912.64.

Adzic, M., Djordjevic, J., Djordjevic, A., Niciforovic, A., Demonacos, C., Radojcic, M., & Krstic-Demonacos, M. (2009). Acute or chronic stress induce cell compartment-specific phosphorylation of glucocorticoid receptor and alter its transcriptional activity in Wistar rat brain. *Journal of Endocrinology*, 202(1), 87–97. http://dx.doi.org/10.1677/JOE-08-0509.

Allen, C. D., Lee, S., Koob, G. F., & Rivier, C. (2011). Immediate and prolonged effects of alcohol exposure on the activity of the hypothalamic-pituitary-adrenal axis in adult and adolescent rats. *Brain, Behavior, and Immunity*, 25(Suppl. 1), S50–S60. http://dx.doi.org/10.1016/j.bbi.2011.01.016.

Ambroggi, F., Turiault, M., Milet, A., Deroche-Gamonet, V., Parnaudeau, S., Balado, E., ... Tronche, F. (2009). Stress and addiction: glucocorticoid receptor in dopaminoceptive neurons facilitates cocaine seeking. *Nature Neuroscience, 12*(3), 247–249. http://dx.doi.org/10.1038/nn.2282.

Barik, J., Parnaudeau, S., Saint Amaux, A. L., Guiard, B. P., Golib Dzib, J. F., Bocquet, O., ... Tronche, F. (2010). Glucocorticoid receptors in dopaminoceptive neurons, key for cocaine, are dispensable for molecular and behavioral morphine responses. *Biological Psychiatry, 68*(3), 231–239. http://dx.doi.org/10.1016/j.biopsych.2010.03.037.

Besheer, J., Fisher, K. R., Grondin, J. J., Cannady, R., & Hodge, C. W. (2012). The effects of repeated corticosterone exposure on the interoceptive effects of alcohol in rats. *Psychopharmacology, 220*(4), 809–822. http://dx.doi.org/10.1007/s00213-011-2533-8.

Besheer, J., Fisher, K. R., Lindsay, T. G., & Cannady, R. (2013). Transient increase in alcohol self-administration following a period of chronic exposure to corticosterone. *Neuropharmacology, 72*, 139–147. http://dx.doi.org/10.1016/j.neuropharm.2013.04.036.

Binder, E. B. (2009). The role of FKBP5, a co-chaperone of the glucocorticoid receptor in the pathogenesis and therapy of affective and anxiety disorders. *Psychoneuroendocrinology, 34*(Suppl. 1), S186–S195. http://dx.doi.org/10.1016/j.psyneuen.2009.05.021.

Cippitelli, A., Damadzic, R., Hamelink, C., Brunnquell, M., Thorsell, A., Heilig, M., & Eskay, R. L. (2014). Binge-like ethanol consumption increases corticosterone levels and neurodegneration whereas occupancy of type II glucocorticoid receptors with mifepristone is neuroprotective. *Addiction Biology, 19*(1), 27–36. http://dx.doi.org/10.1111/j.1369-1600.2012.00451.x.

Dallman, M. F. (2005). Fast glucocorticoid actions on brain: back to the future. *Frontiers in Neuroendocrinology, 26*(3–4), 103–108. http://dx.doi.org/10.1016/j.yfrne.2005.08.001.

De Vries, T. J., Schoffelmeer, A. N., Tjon, G. H., Nestby, P., Mulder, A. H., & Vanderschuren, L. J. (1996). Mifepristone prevents the expression of long-term behavioural sensitization to amphetamine. *European Journal of Pharmacology, 307*(2), R3–R4.

Deroche-Gamonet, V., Sillaber, I., Aouizerate, B., Izawa, R., Jaber, M., Ghozland, S., ... Piazza, P. V. (2003). The glucocorticoid receptor as a potential target to reduce cocaine abuse. *Journal of Neuroscience, 23*(11), 4785–4790.

Dong, Z., Han, H., Wang, M., Xu, L., Hao, W., & Cao, J. (2006). Morphine conditioned place preference depends on glucocorticoid receptors in both hippocampus and nucleus accumbens. *Hippocampus, 16*(10), 809–813. http://dx.doi.org/10.1002/hipo.20216.

Edwards, S., & Koob, G. F. (2010). Neurobiology of dysregulated motivational systems in drug addiction. *Future Neurology, 5*(3), 393–401. http://dx.doi.org/10.2217/fnl.10.14.

Ellis, F. W. (1966). Effect of ethanol on plasma corticosterone levels. *Journal of Pharmacology and Experimental Therapeutics, 153*(1), 121–127.

Fahlke, C., Hard, E., Eriksson, C. J., Engel, J. A., & Hansen, S. (1995). Consequence of long-term exposure to corticosterone or dexamethasone on ethanol consumption in the adrenalectomized rat, and the effect of type I and type II corticosteroid receptor antagonists. *Psychopharmacology, 117*(2), 216–224.

Fahlke, C., Hard, E., & Hansen, S. (1996). Facilitation of ethanol consumption by intracerebroventricular infusions of corticosterone. *Psychopharmacology, 127*(2), 133–139.

Fiancette, J. F., Balado, E., Piazza, P. V., & Deroche-Gamonet, V. (2010). Mifepristone and spironolactone differently alter cocaine intravenous self-administration and cocaine-induced locomotion in C57BL/6J mice. *Addiction Biology, 15*(1), 81–87. http://dx.doi.org/10.1111/j.1369-1600.2009.00178.x.

Haller, J., Mikics, E., & Makara, G. B. (2008). The effects of nongenomic glucocorticoid mechanisms on bodily functions and the central neural system. A critical evaluation of findings. *Frontiers in Neuroendocrinology, 29*(2), 273–291. http://dx.doi.org/10.1016/j.yfrne.2007.10.004.

Heilig, M., & Koob, G. F. (2007). A key role for corticotropin-releasing factor in alcohol dependence. *Trends in Neuroscience, 30*(8), 399–406. http://dx.doi.org/10.1016/j.tins.2007.06.006.

Jacquot, C., Croft, A. P., Prendergast, M. A., Mulholland, P., Shaw, S. G., & Little, H. J. (2008). Effects of the glucocorticoid antagonist, mifepristone, on the consequences of withdrawal from long term alcohol consumption. *Alcoholism, Clinical and Experimental Research, 32*(12), 2107–2116. http://dx.doi.org/10.1111/j.1530-0277.2008.00799.x.

Kadmiel, M., & Cidlowski, J. A. (2013). Glucocorticoid receptor signaling in health and disease. *Trends in Pharmacological Science, 34*(9), 518–530. http://dx.doi.org/10.1016/j.tips.2013.07.003.

Koenig, H. N., & Olive, M. F. (2004). The glucocorticoid receptor antagonist mifepristone reduces ethanol intake in rats under limited access conditions. *Psychoneuroendocrinology, 29*(8), 999–1003. http://dx.doi.org/10.1016/j.psyneuen.2003.09.004.

Koob, G., & Kreek, M. J. (2007). Stress, dysregulation of drug reward pathways, and the transition to drug dependence. *American Journal of Psychiatry, 164*(8), 1149–1159. http://dx.doi.org/10.1176/appi.ajp.2007.05030503.

Koob, G. F., Buck, C. L., Cohen, A., Edwards, S., Park, P. E., Schlosburg, J. E., ... George, O. (2014). Addiction as a stress surfeit disorder. *Neuropharmacology, 76*(Pt B), 370–382. http://dx.doi.org/10.1016/j.neuropharm.2013.05.024.

Kumar, R., & Thompson, E. B. (2003). Transactivation functions of the N-terminal domains of nuclear hormone receptors: protein folding and coactivator interactions. *Molecular Endocrinology, 17*(1), 1–10. http://dx.doi.org/10.1210/me.2002-0258.

Little, H. J., Croft, A. P., O'Callaghan, M. J., Brooks, S. P., Wang, G., & Shaw, S. G. (2008). Selective increases in regional brain glucocorticoid: a novel effect of chronic alcohol. *Neuroscience, 156*(4), 1017–1027. http://dx.doi.org/10.1016/j.neuroscience.2008.08.029.

Lovallo, W. R., Dickensheets, S. L., Myers, D. A., Thomas, T. L., & Nixon, S. J. (2000). Blunted stress cortisol response in abstinent alcoholic and polysubstance-abusing men. *Alcoholism, Clinical and Experimental Research, 24*(5), 651–658.

Lu, Y. L., & Richardson, H. N. (2014). Alcohol, stress hormones, and the prefrontal cortex: a proposed pathway to the dark side of addiction. *Neuroscience, 277*, 139–151. http://dx.doi.org/10.1016/j.neuroscience.2014.06.053.

Luisi, B. F., Xu, W. X., Otwinowski, Z., Freedman, L. P., Yamamoto, K. R., & Sigler, P. B. (1991). Crystallographic analysis of the interaction of the glucocorticoid receptor with DNA. *Nature, 352*(6335), 497–505. http://dx.doi.org/10.1038/352497a0.

McClintick, J. N., Xuei, X., Tischfield, J. A., Goate, A., Foroud, T., Wetherill, L., ... Edenberg, H. J. (2013). Stress-response pathways are altered in the hippocampus of chronic alcoholics. *Alcohol, 47*(7), 505–515. http://dx.doi.org/10.1016/j.alcohol.2013.07.002.

McEwen, B. S. (2007). Physiology and neurobiology of stress and adaptation: central role of the brain. *Physiological Reviews, 87*(3), 873–904. http://dx.doi.org/10.1152/physrev.00041.2006.

McReynolds, J. R., Pena, D. F., Blacktop, J. M., & Mantsch, J. R. (2014). Neurobiological mechanisms underlying relapse to cocaine use: contributions of CRF and noradrenergic systems and regulation by glucocorticoids. *Stress, 17*(1), 22–38. http://dx.doi.org/10.3109/10253890.2013.872617.

Mesripour, A., Hajhashemi, V., & Rabbani, M. (2008). Metyrapone and mifepristone reverse recognition memory loss induced by spontaneous morphine withdrawal in mice. *Basic and Clinical Pharmacology and Toxicology, 102*(4), 377–381. http://dx.doi.org/10.1111/j.1742-7843.2007.00183.x.

Molina, P. E. (2013). *Endocrine physiology* (4th ed.). McGraw-Hill.

Myers, B., McKlveen, J. M., & Herman, J. P. (2014). Glucocorticoid actions on synapses, circuits, and behavior: implications for the energetics of stress. *Frontiers in Neuroendocrinology, 35*(2), 180–196. http://dx.doi.org/10.1016/j.yfrne.2013.12.003.

Navarro-Zaragoza, J., Hidalgo, J. M., Laorden, M. L., & Milanes, M. V. (2012). Glucocorticoid receptors participate in the opiate withdrawal-induced stimulation of rats NTS noradrenergic activity and in the somatic signs of morphine withdrawal. *British Journal of Pharmacology, 166*(7), 2136–2147. http://dx.doi.org/10.1111/j.1476-5381.2012.01918.x.

O'Callaghan, M. J., Croft, A. P., Jacquot, C., & Little, H. J. (2005). The hypothalamopituitary-adrenal axis and alcohol preference. *Brain Research Bulletin, 68*(3), 171–178. http://dx.doi.org/10.1016/j.brainresbull.2005.08.006.

Pang, T. Y., Du, X., Catchlove, W. A., Renoir, T., Lawrence, A. J., & Hannan, A. J. (2013). Positive environmental modification of depressive phenotype and abnormal hypothalamic-pituitary-adrenal axis activity in female C57BL/6J mice during abstinence from chronic ethanol consumption. *Frontiers in Pharmacology, 4*, 93. http://dx.doi.org/10.3389/fphar.2013.00093.

Paulus, M. P., Tapert, S. F., & Schulteis, G. (2009). The role of interoception and alliesthesia in addiction. *Pharmacology, Biochemistry and Behavior, 94*(1), 1–7. http://dx.doi.org/10.1016/j.pbb.2009.08.005.

Przybycien-Szymanska, M. M., Gillespie, R. A., & Pak, T. R. (2012). 17beta-Estradiol is required for the sexually dimorphic effects of repeated binge-pattern alcohol exposure on the HPA axis during adolescence. *PLoS One, 7*(2), e32263. http://dx.doi.org/10.1371/journal.pone.0032263.

Rasmussen, D. D., Boldt, B. M., Bryant, C. A., Mitton, D. R., Larsen, S. A., & Wilkinson, C. W. (2000). Chronic daily ethanol and withdrawal: 1. Long-term changes in the hypothalamo-pituitary-adrenal axis. *Alcoholism, Clinical and Experimental Research, 24*(12), 1836–1849.

Revollo, J. R., & Cidlowski, J. A. (2009). Mechanisms generating diversity in glucocorticoid receptor signaling. *Annals of New York Academy of Sciences, 1179*, 167–178. http://dx.doi.org/10.1111/j.1749-6632.2009.04986.x.

Richardson, H. N., Lee, S. Y., O'Dell, L. E., Koob, G. F., & Rivier, C. L. (2008). Alcohol self-administration acutely stimulates the hypothalamic-pituitary-adrenal axis, but alcohol dependence leads to a dampened neuroendocrine state. *European Journal of Neuroscience, 28*(8), 1641–1653.

Roberts, A. J., Lessov, C. N., & Phillips, T. J. (1995). Critical role for glucocorticoid receptors in stress- and ethanol-induced locomotor sensitization. *Journal of Pharmacology and Experimental Therapeutics, 275*(2), 790–797.

Schaaf, M. J., & Cidlowski, J. A. (2002). Molecular mechanisms of glucocorticoid action and resistance. *Journal of Steroid Biochemistry and Molecular Biology, 83*(1–5), 37–48.

Schoch, G. A., D'Arcy, B., Stihle, M., Burger, D., Bar, D., Benz, J., ... Ruf, A. (2010). Molecular switch in the glucocorticoid receptor: active and passive antagonist conformations. *Journal of Molecular Biology, 395*(3), 568–577. http://dx.doi.org/10.1016/j.jmb.2009.11.011.

Schulz, M., Eggert, M., Baniahmad, A., Dostert, A., Heinzel, T., & Renkawitz, R. (2002). RU486-induced glucocorticoid receptor agonism is controlled by the receptor N terminus and by corepressor binding. *Journal of Biological Chemistry, 277*(29), 26238–26243. http://dx.doi.org/10.1074/jbc.M203268200.

Sharrett-Field, L., Butler, T. R., Berry, J. N., Reynolds, A. R., & Prendergast, M. A. (2013). Mifepristone pretreatment reduces ethanol withdrawal severity in vivo. *Alcoholism, Clinical and Experimental Research, 37*(8), 1417–1423. http://dx.doi.org/10.1111/acer.12093.

Silva, S. M., & Madeira, M. D. (2012). Effects of chronic alcohol consumption and withdrawal on the response of the male and female hypothalamic-pituitary-adrenal axis to acute immune stress. *Brain Research, 1444*, 27–37. http://dx.doi.org/10.1016/j.brainres.2012.01.013.

Simms, J. A., Haass-Koffler, C. L., Bito-Onon, J., Li, R., & Bartlett, S. E. (2012). Mifepristone in the central nucleus of the amygdala reduces yohimbine stress-induced reinstatement of ethanol-seeking. *Neuropsychopharmacology, 37*(4), 906–918. http://dx.doi.org/10.1038/npp.2011.268.

Sinha, R., Fox, H. C., Hong, K. I., Hansen, J., Tuit, K., & Kreek, M. J. (2011). Effects of adrenal sensitivity, stress- and cue-induced craving, and anxiety on subsequent alcohol relapse and treatment outcomes. *Archives of General Psychiatry, 68*(9), 942–952. http://dx.doi.org/10.1001/archgenpsychiatry.2011.49.

Stairs, D. J., Prendergast, M. A., & Bardo, M. T. (2011). Environmental-induced differences in corticosterone and glucocorticoid receptor blockade of amphetamine self-administration in rats. *Psychopharmacology, 218*(1), 293–301. http://dx.doi.org/10.1007/s00213-011-2448-4.

Stevens, A., Garside, H., Berry, A., Waters, C., White, A., & Ray, D. (2003). Dissociation of steroid receptor coactivator 1 and nuclear receptor corepressor recruitment to the human glucocorticoid receptor by modification of the ligand-receptor interface: the role of tyrosine 735. *Molecular Endocrinology, 17*(5), 845–859. http://dx.doi.org/10.1210/me.2002-0320.

Tasker, J. G., Di, S., & Malcher-Lopes, R. (2006). Minireview: rapid glucocorticoid signaling via membrane-associated receptors. *Endocrinology, 147*(12), 5549–5556. http://dx.doi.org/10.1210/en.2006-0981.

Uhart, M., & Wand, G. S. (2009). Stress, alcohol and drug interaction: an update of human research. *Addiction Biology, 14*(1), 43–64. http://dx.doi.org/10.1111/j.1369-1600.2008.00131.x.

Vandevyver, S., Dejager, L., & Libert, C. (2012). On the trail of the glucocorticoid receptor: into the nucleus and back. *Traffic, 13*(3), 364–374. http://dx.doi.org/10.1111/j.1600-0854.2011.01288.x.

Vendruscolo, L. F., & Roberts, A. J. (2014). Operant alcohol self-administration in dependent rats: focus on the vapor model. *Alcohol, 48*(3), 277–286. http://dx.doi.org/10.1016/j.alcohol.2013.08.006.

Vendruscolo, L. F., Barbier, E., Schlosburg, J. E., Misra, K. K., Whitfield, T. W., Jr., Logrip, M. L., ... Koob, G. F. (2012). Corticosteroid-dependent plasticity mediates compulsive alcohol drinking in rats. *Journal of Neuroscience, 32*(22), 7563–7571. http://dx.doi.org/10.1523/JNEUROSCI.0069-12.2012.

Vendruscolo, L. F., Estey, D., Goodell, V., Macshane, L. G., Logrip, . L., Schlosburg, J. E., ... Mason, B. J. (2015). Glucocorticoid receptor antagonism decreases alcohol seeking in alcohol-dependent individuals. *Journal of Clinical Investment, 125*(8), 3193–3197.

Wochnik, G. M., Ruegg, J., Abel, G. A., Schmidt, U., Holsboer, F., & Rein, T. (2005). FK506-binding proteins 51 and 52 differentially regulate dynein interaction and nuclear translocation of the glucocorticoid receptor in mammalian cells. *Journal of Biological Chemistry, 280*(6), 4609–4616. http://dx.doi.org/10.1074/jbc.M407498200.

World Health Organization. (2014). *Global status report on alcohol and health.*

Zamora-Martinez, E. R., & Edwards, S. (2014). Neuronal extracellular signal-regulated kinase (ERK) activity as marker and mediator of alcohol and opioid dependence. *Frontiers in Integrative Neuroscience, 8,* 24. http://dx.doi.org/10.3389/fnint.2014.00024.

Zorrilla, E. P., Valdez, G. R., & Weiss, F. (2001). Changes in levels of regional CRF-like-immunoreactivity and plasma corticosterone during protracted drug withdrawal in dependent rats. *Psychopharmacology, 158*(4), 374–381. http://dx.doi.org/10.1007/s002130100773.

Zuloaga, D. G., Johnson, L. A., Agam, M., & Raber, J. (2014). Sex differences in activation of the hypothalamic-pituitary-adrenal axis by methamphetamine. *Journal of Neurochemistry, 129*(3), 495–508. http://dx.doi.org/10.1111/jnc.12651.

Chapter 57

Neuropathobiology of Alcohol-Induced Cognitive Deficits

Vinod Tiwari[1], Kanwaljit Chopra[2]
[1]Department of Anesthesiology and Critical Care Medicine, Johns Hopkins University, School of Medicine, Baltimore, MD, USA; [2]Pharmacology Research Laboratory, University Institute of Pharmaceutical Sciences, UGC Centre of Advanced Study Panjab University, Chandigarh, India

Abbreviations

ACTH Adrenocorticotropic hormone
COX Cyclo-oxygenase
CREB cAMP response element-binding protein
DNA Deoxyribonucleic acid
FAS Fetal alcohol syndrome
GABA γ-Aminobutyric acid
GFAP Glial fibrillary acidic protein
GLUT Glucose transporter
IGF Insulin-like growth factor
IL-1β Interleukin-1β
iNOS Inducible-nitric oxide synthases
NADPH Nicotinamide adenine dinucleotide phosphate hydrogenase
NF-κB Nuclear factor-κB
NMDA N-Methyl-D-aspartate
ROS Reactive oxygen species
TLRs Toll-like receptors
TNF-α Tumor necrosis factor-α

INTRODUCTION

Cognition comprises a set of all the mental processes and abilities related to attention, judgment, evaluation, memory, problem solving, and decision making. Any barrier to all or any of these cognitive process leads to cognitive deficits. Chronic and excessive consumption of alcohol can lead to the development of a serious and potentially fatal brain damage, which may lead to the loss of specific brain functions including cognitive functioning (Weiss, Singewald, Ruepp, & Marksteiner, 2014). Approximately 50–75% of chronic alcoholics may show permanent cognitive impairment, making chronic alcoholism the second leading cause of dementia (Eckardt et al., 1998; Tiwari, Kuhad, & Chopra, 2009). Chronic alcoholics consistently show deficits in declarative and short-term memory along with frequent impairments in spatial learning and memory, which are indicative of hippocampal dysfunctioning (Parsons, 1998; Sullivan, Rosenbloom, & Pfefferbaum, 2000).

Atrophy of nerve cells and brain shrinkage in cortical and subcortical regions and hippocampus are also common features in alcoholics apart from behavioral and cognitive impairments, as indicated by imaging studies and postmortem analysis of brain structure (Harper, 1998; Sullivan & Pfefferbaum, 2005; Zahr, Kaufman, & Harper, 2011) The frontal lobes are the most affected region in the alcoholic brain, with significant neuronal loss (Kubota et al., 2001; Sullivan & Pfefferbaum, 2005). The frontal lobes are involved in the regulation of complex cognitive skills such as working memory, discrimination, and temporal ordering that underlie attention, risk taking, and judgment. Accordingly, chronic alcoholics demonstrate an inability to perform complex motor tasks, loss of planning and executive functions, impaired judgment, social withdrawal, reduced motivation, distractibility, and attention and impulse-control deficits (Parsons, 1998; Sullivan et al., 2000; Sullivan & Pfefferbaum, 2005). Both clinical (Sullivan et al. 2000) and preclinical studies suggest a direct relationship between chronic alcoholism and cognitive deficits (Wright & Taffe, 2014).

Alcohol consumption during pregnancy, on the other hand, is another significant public health problem, which is increasing at an alarming rate and results in adverse outcomes for the child. Heavy prenatal alcohol exposure results in severe neuropsychological deficits such as deficits in general intelligence, memory, language, attention, and learning and visuospatial abilities, along with impairment in motor skills and executive functioning (Mattson & Riley, 1998). Children with prenatal alcohol exposure have decreased academic scores and higher rates of learning disabilities than nonexposed children, which might be related to impairments in their verbal and nonverbal learning–memory skills (Roebuck-Spencer & Mattson, 2004). In the United States, England, and Canada, 20–32% of pregnant women drink and in some European countries the rate is even higher, exceeding 50%, making the situation more critical (May et al., 2005). Previous findings from our laboratory also suggested that chronic ethanol administration in rats leads to significant learning and memory impairments as evident from increased latencies in the Morris water maze (Figure 1) and elevated-plus maze (Figure 2) tasks. The clinical description of fetal alcohol syndrome (FAS) was first published in 1973 by Jones and Smith. Since then, FAS has been accepted as the leading identifiable cause of mental retardation and neurologic deficit in the Western world (Abel & Sokol, 1986). The economic cost of this disorder is huge and daunting, with direct costs in the United States estimated at $3.6–3.9 billion per year (Popova, Stade, Bekmuradov, Lange, & Rehm, 2011).

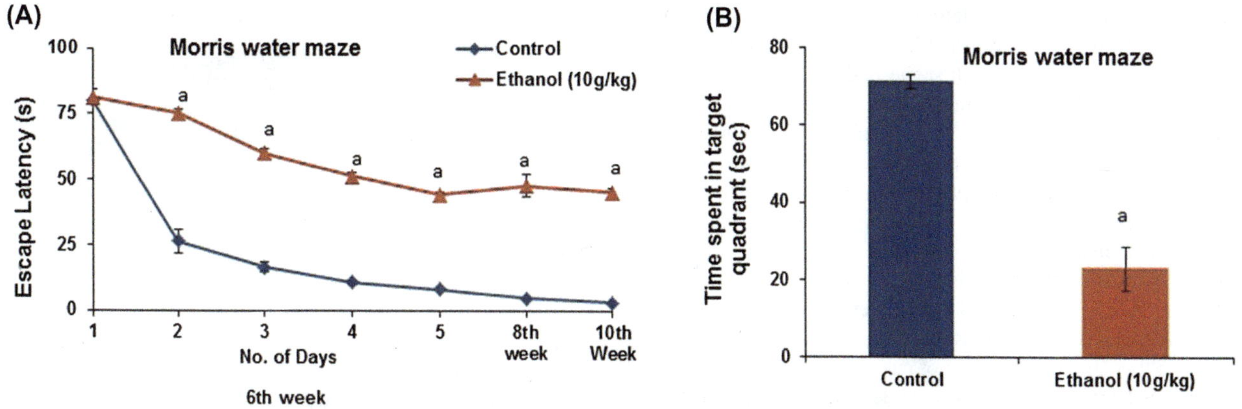

FIGURE 1 Effects of chronic administration of ethanol (10 g/kg; oral gavage) on the performance of rats in (A) the spatial memory acquisition phase (escape latency) and (B) time spent in the target quadrant in a probe trial of the Morris water maze task. Eight weeks (b.i.d.) of ethanol administration leads to a significant increase in escape latency and decrease in time spent in the target quadrant in the water maze task, suggesting impairment in learning and memory of ethanol-administered rats. Values are expressed as the mean ± SEM. $^a p<0.05$, different from control group.

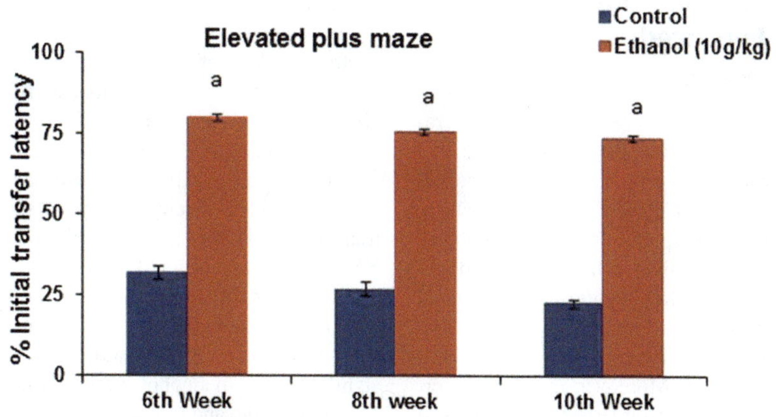

FIGURE 2 Effect of chronic administration of ethanol (10 g/kg; oral gavage) on percentage initial transfer latency of rats in the elevated-plus maze test. Percentage initial transfer latency was significantly increased after chronic ethanol administration, indicating memory deficits in rats. Values are expressed as the mean ± SEM. $^a p<0.05$, different from control group.

Therefore, understanding the neuropathological mechanisms involved in chronic alcohol-induced cognitive deficits is of great medical and economic importance so that new therapeutic arsenals can be developed against this devastating complication.

NEUROPATHOLOGICAL MECHANISMS ASSOCIATED WITH ALCOHOL-INDUCED COGNITIVE DEFICITS

In the past few decades, several preclinical and clinical reports have suggested diverse mechanisms involved in chronic alcohol-induced cognitive deficits, but the exact cellular, biochemical, and molecular mechanisms behind alcohol-induced neuronal damage and consequent cognitive deficit are still not fully understood. Alcohol enters the blood within 5 min after ingestion and its absorption peaks after 30–90 min. The key enzymes involved in the degradation of ethanol are alcohol dehydrogenase and acetaldehyde dehydrogenase, which converts ethanol into acetate followed by further metabolism. In alcoholics, there is a mutation in the acetaldehyde dehydrogenase enzyme, which leads to inefficient metabolism and accumulation of acetaldehyde levels up to 20 times higher than in individuals without the mutation (Kucera, Balaz, Varsik, & Kurca, 2002). A certain amount of acetaldehyde that is not metabolized through the usual pathways binds irreversibly to proteins, which results in the creation of cytotoxic proteins that adversely affect the function of nervous system cells (Achord, 1995). Some of the well-known mechanisms involved in alcohol-induced neuronal damage and cognitive deficits include free radical-induced damage (Crews et al., 2004; Tiwari, Kuhad, & Chopra, 2010), neuroinflammation (Alfonso-Loeches, Pascual-Lucas, Blanco, Sanchez-Vera, & Guerri, 2010), activation of the nuclear factor-κB (NF-κB) signaling pathway (Crews et al., 2006), N-methyl-D-aspartate (NMDA) receptor sensitization (Chandler, Sumners, & Crews, 1993), and inhibition of growth factors (Breese & Sonntag, 1995). Chronic alcoholism also causes new nerve cells to die, leading to inhibition of neurogenesis (Nixon & Crews, 2002). Some reports also suggest that alcohol may lead to interference in nerve transmission and thus the degradation of the brain's circuitry, which further leads to cognitive deficits (Sullivan & Pfefferbaum, 2005). Chronic alcoholism leads to a phenomenon known as "processing inefficiency." To compensate for alcohol-induced cognitive deficits, the brain reorganizes how it processes information and uses higher level functions for performing lower level tasks, which suggests that the brain needs to work harder to perform the same functions (Fama, Pfefferbaum, & Sullivan, 2004; Sullivan & Pfefferbaum, 2005). Women are especially more susceptible to alcohol's effects and experience the negative mental and physical consequences of alcoholism sooner than men, given the same number of years of drinking. For a long time this phenomenon, which is also known as "telescoping," has been known to account for relatively early onset of liver and heart damage among women alcoholics, but later findings suggest that the same is true with the brain as well (Mann et al., 2005).

Thus, although the prevalence of cognitive deficits and neurodegeneration is well supported by multiple studies, the neuropathological mechanisms involved are still poorly understood. Various neuropathological mechanisms that might play an important role in alcohol-induced cognitive deficits are discussed here.

Activation of Neuronal Free Radical Machinery Causing Enhanced Oxidative–Nitrosative Stress

Free radicals have been implicated in a variety of neurodegenerative disorders, including multiple sclerosis, Parkinson disease, and Alzheimer disease, and may also play a putative role in alcohol-induced cognitive deficits (Haorah et al., 2008; Jung, Gatch, & Simpkins, 2005). Oxidative stress results from an imbalance between the generation of free radicals and the body's endogenous antioxidant machinery. For maintaining normal biochemical processes, neurons are highly dependent on ATP generated from glucose metabolism, which leads to generation of free radicals as a by-product of glucose metabolism. Moreover, the central nervous system (CNS) is particularly vulnerable to reactive oxygen species (ROS)-induced damage because it is relatively deficient in oxidative defenses (poor catalase activity and moderate superoxide dismutase and glutathione peroxidase activities), requires a high amount of oxygen, and contains high levels of membrane polyunsaturated fatty acids, which are more susceptible to free radical attack (Halliwell, 1992). Enhanced oxidative stress attenuates the brain antioxidant defense system and activates secondary events leading to neuronal apoptosis by causing DNA damage and protein dysfunction. Ethanol enhances oxidative stress directly through generation of free radicals and causes lipid peroxidation along with suppression of endogenous antioxidant enzymes such as glutathione peroxidase/glutathione reductase (Siler-Marsiglio et al., 2004). Such compromised defense machinery may be adequate under normal circumstances, whereas during pro-oxidative conditions, such as chronic alcohol exposure, it can predispose the brain to oxidative damage. High and long-term exposure to alcohol may induce inducible nitric oxide synthase (iNOS) in the brain to further increase the release of nitric oxide (NO) and this excess amount of NO may suppress various physiological functions. This is further supported by findings suggesting increased iNOS induction in cerebellar cortical neurons of alcoholics (Konovko, Morozov, Kalinichenko, Dyuzen, & Motavkin, 2004). Peroxynitrite, a harmful oxidant formed by the reaction between superoxide and NO, interacts with lipids, DNA, and proteins via direct oxidative reactions or via indirect, radical-mediated mechanisms. These

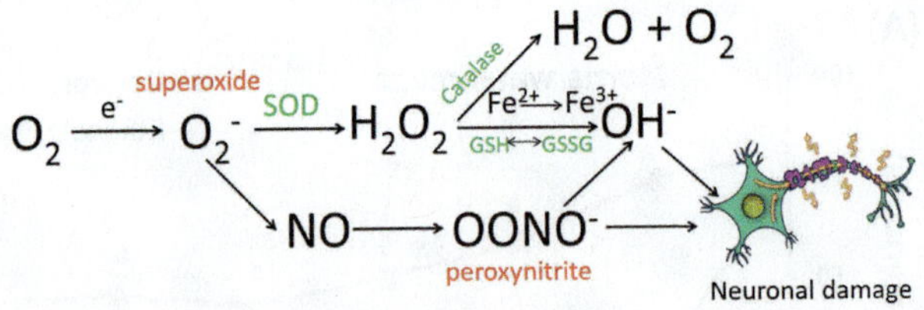

FIGURE 3 Generation and regulation of reactive oxygen and nitrogen species. SOD, superoxide dismutase; GSH, reduced glutathione; GSSG, oxidized glutathione.

reactions trigger cellular responses ranging from subtle modulations of cell signaling to overwhelming oxidative injury, committing cells to necrosis or apoptosis (Figure 3).

Aggravation of Neuroinflammatory Signaling and Cell Death Cascade

There are several clinical and preclinical findings that suggest the involvement of a neuroinflammatory signaling-mediated cell death cascade in ethanol-induced brain damage. Ethanol leads to changes in protein transcription, with increased DNA binding of NF-κB and reduced DNA binding of cAMP response element-binding protein (CREB). CREB is a family of transcription factors that, upon activation, promote neuronal survival and protect neurons from excitotoxicity and apoptosis by regulating prosurvival transcription factors (Lonze & Ginty, 2002). On the other hand, NF-κB is a transcriptional factor that is widely known for its ubiquitous roles in inflammation and the cell death cascade (Jung et al., 2005). Thus, the balance in expression and activation of these prosurvival versus proinflammatory transcription factors plays a very significant role in alcohol-induced brain damage (Figure 4).

Both in vivo and in vitro data suggest the involvement of a proinflammatory signaling-induced NF-κB induction as a key factor in alcohol-induced brain damage. Tumor necrosis factor-α is known to potentiate glutamate neurotoxicity by inhibiting glutamate uptake through NF-κB mechanisms (Zou & Crews, 2005). Ethanol leads to the induction of iNOS and enhances NF-κB–DNA binding in human astroglial cells, which normally regulate extracellular glutamate concentrations. Ethanol also induces several other enzymes of the free radical machinery, such as cyclo-oxygenase-2 (COX-2) and NADPH oxidase, that are downstream mediators of NF-κB (Knapp & Crews, 1999). ROS-producing enzymes, including NOS, COX-2, and NADPH oxidase, are all induced by NF-κB activation, suggesting that the ethanol-induced oxidoinflammatory cascade in the brain may be related to NF-κB activation (Chopra & Tiwari, 2013; Crews et al., 2006; Tiwari & Chopra, 2012). Alcohol-induced neurodegeneration as a result of enhanced neuroinflammatory response is further supported by findings from Crews et al. (2006), who reported that alcohol leads to NF-κB activation, microglial activation, and increased COX-2 immunoreactivity (Figure 4).

Findings from Jung et al. (2005) suggested a direct correlation between chronic ethanol consumption and neuronal apoptosis. Their report demonstrates that chronic ethanol leads to activation of protein kinase C, which subsequently phosphorylates IκB (the NF-κB inhibitor) present in NF-κB–IκB complexes. After phosphorylation, activated NF-κB, also known as a cell death signal, is released and translocates to the nucleus. Inside the nucleus NF-κB

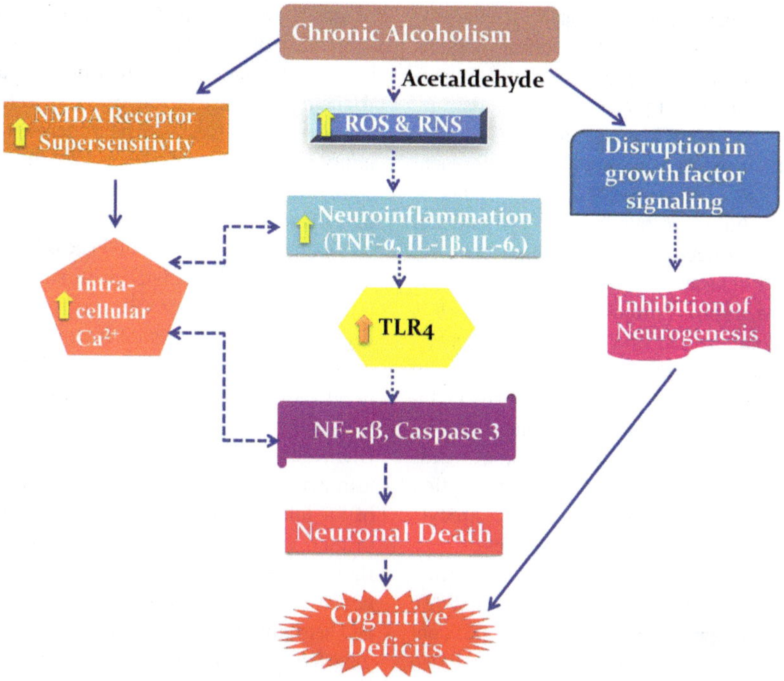

FIGURE 4 A schematic presentation of multiple pathways involved in alcohol-induced cognitive deficits.

binds to DNA and induces the expression of target genes, which results in DNA fragmentation and apoptosis through activation of caspases. Heavy alcohol exposure in rats during the period of brain development that is comparable to the human third trimester has been shown to produce the death of postmitotic neurons in the hypothalamus, cerebral cortex, cerebellum, and associated brainstem structures (Ikonomidou et al., 2000; Light, Belcher, & Pierce, 2002). Another finding in immature mice suggested that ethanol administration during the synaptogenesis period induces widespread apoptotic cell death in the developing brain (Ikonomidou et al., 2000), which is accompanied by caspase-3 activation and suggested to be responsible for cytological changes in the brain that characterize neuronal apoptosis. Thus, ethanol promotes a proinflammatory and antisurvival environment through the activation of proinflammatory transcription factors and the inhibition of prosurvival transcription factors, which finally leads to neuronal cell death (Figure 4).

Putative Role of N-Methyl-D-aspartate-Induced Excitotoxicity

Glutamate, the major excitatory neurotransmitter in brain, plays a putative role in a variety of neuropathological conditions including alcohol-induced cognitive deficits. Following neuronal damage there is increased glutamate release, which might lead to neuronal death via overstimulation of glutamate receptors and is referred to as excitotoxicity (Chandler et al., 1993). As of this writing four glutamate receptor subtypes have been identified, which include one metabotropic receptor and three ionotropic receptors (NMDA, kainate, and (S)-α-amino-4-bromo-3-hydroxy-5-isoxazolepropionic acid). Although activation of any of the glutamate receptors may contribute to glutamate neurotoxicity, studies have suggested NMDA receptors as being particularly important in excitotoxicity (Chandler et al., 1993). Ethanol has been known to have potent actions within the CNS for a long time, but the exact mechanisms by which it produces changes in behavioral and cognitive functions have not been fully elucidated. Ethanol-induced hyperexcitability is due to a combination of increased NMDA receptor activation, decreased γ-aminobutyric acid (GABA)$_A$ receptor activation, and increased function of voltage-activated calcium channels. Chronic and repetitive exposure to ethanol, particularly during withdrawal, is more dangerous to the CNS as it may destroy central neurons via excitotoxic mechanisms (Lovinger, 1993). Early in vitro studies suggested that acute ethanol exposure inhibited glutamatergic NMDA receptors, but long term exposure to ethanol resulted in NMDA receptor supersensitivity (Chandler et al., 1993). The interaction of glutamate with NMDA receptors causes calcium to flow into the signal-receiving neuron, and excessive activation of the NMDA glutamate receptor can lead to dangerously high calcium accumulation inside the neuron (Choi, 1995). This high intracellular calcium, if it persists for a prolonged duration, can lead to cell death by either apoptosis or necrosis (Choi, 1995).

Role of Astrocytes and Microglia in Alcohol-Induced Neurodegeneration

Glia are nonneuronal cells that form myelin, maintain homeostasis, and provide support and protection to neurons in both the peripheral nervous system and the CNS. Glial cells account for 70% of CNS cells and mainly comprise two types, the microglia and the macroglia (Watkins, Hutchinson, Milligan, & Maier, 2007). The macroglia are further divided into astrocytes, oligodendrocytes, and radial cells. Glial cells are also involved in the synthesis, release, and uptake of various neurotransmitters (Tiwari, Guan, & Raja, 2014;

Watkins et al., 2007). For over a century, it was believed that the glial cells do not have any role in neuronal growth and neurotransmission, but more recent studies have revealed that they indeed play important roles in various neurophysiological processes and in assisting neurons to form synaptic connections (Gourine et al., 2010).

Evidence of the effects of ethanol on astroglial development comes from both animal and clinical studies. Twenty days of alcohol intoxication (35% solution) in rats leads to a significant increase in the amount of the perineuronal and total glia along with a decrease in the number of neurons and the proliferation of the glia (Popova & Shchekalina, 1980). Chronic alcohol exposure is known to reduce the overall number of astrocytes in the cortex and interfere with the response to specific growth factors. Alcohol also causes astroglia to degenerate, leaving a void in trophic and metabolic support, which finally leads to degeneration of neurons. The loss of astroglia leads to a reduction in the ability to take up excess glutamate and eliminate free radicals along with dysregulation of ion homeostasis. These alcohol-induced changes in astrocyte development and function may have serious consequences on neuronal survival and migration and interfere with the correct formation of connections among neurons (Miguel-Hidalgo et al., 2002). Postmortem studies in human hippocampus suggest a significant loss of 37% of the glial cells in chronic alcoholics that included a reduction in astrocytes and oligodendrocytes (Korbo, 1999). Chronic alcohol exposure decreases glial fibrillary acidic protein (GFAP), which is an important neurofilament involved in many important CNS processes, including cell communication and the functioning of the blood–brain barrier, in the cerebellum of male and female rats (Rintala et al., 2001). The loss of GFAP expression is indicative of a loss of astrocyte function (Rintala et al., 2001) consistent with the finding that the number of astrocytes in the human hippocampus is reduced in alcoholics (Korbo, 1999). Prenatal exposure to alcohol may also have a profound impact on the development and proliferation of radial glial cells, which may result in the inhibition of neurogenesis and gliogenesis. A significant reduction in neuronal and glial production may result in drastically lower brain volumes (microcephaly) along with gross morphological and functional abnormalities including cognitive deficits within the developing CNS.

These clinical and experimental findings suggest that glial cells might be a primary target of ethanol-induced neurotoxicity. Alterations in neuron glial interactions can lead to several neuronal dysfunctions including cognitive deficits.

Activation of Toll-like Receptor-4 Induces Neuroinflammation and Brain Damage

Toll-like receptors (TLRs) are expressed by innate immune cells and play a significant role in their activation against foreign pathogens. Current knowledge suggests that at least 13 TLR genes and functional ligands exist in mammals. TLRs 1–9 are expressed in both humans and mice, whereas TLRs 10–13 are expressed only in mice (Trudler, Farfara, & Frenkel, 2010). TLRs play an important role in the cross talk between neurons and glial cells in the CNS. While TLR signaling was linked to neurogenesis, it was also found to be involved in the pathogenesis of CNS neurodegeneration and neural injury (Alfonso-Loeches et al., 2010). Activation of TLRs triggers the downstream stimulation of NF-κB and the induction of genes responsible for inflammation and neurodegeneration. A number of studies have demonstrated role of TLR-4 in brain injury. Elimination of TLR-4 protects against oxidative stress in Alzheimer disease, focal cerebral ischemia, human immunodeficiency virus-associated neurodegeneration. and ischemic brain injury (Alfonso-Loeches et al., 2010).

Chronic ethanol consumption is also associated with activation of TLR-4 signaling in astrocytes, microglia, and macrophages, indicating that activation of TLR-4 by ethanol could be an important mechanism associated with ethanol-induced neuroinflammation and brain damage (Blanco, Valles, Pascua, & Guerri, 2005; Figure 4). Chronic ethanol treatment increased the neuroinflammatory response, i.e., iNOS and COX-2 in the cerebral cortices of ethanol-treated wild-type mice, but the induction of these proteins did not take place in the cortices of TLR-4-knockout mice. Blanco et al. (2005) demonstrated that ethanol activates TLR-4 in astrocytes, triggers NF-κB activation, and leads to the induction of an inflammatory response, suggesting that TLR-4 activation in glial cells is a critical event during ethanol-induced neuroinflammation. In vivo studies with deficient TLR-4 function suggest a pivotal role for TLR-4 in the activation of both microglia and astroglia induced by ethanol, as it significantly reduced astroglial hypertrophy and completely abolished microglial activation (Alfonso-Loeches et al., 2010). Knocking down TLR-4 function significantly prevents ethanol-induced NF-κB activation and cytokine upregulation, suggesting the potential role of TLR-4/NF-κB in ethanol-induced neuroinflammation and cognitive decline.

Disruption in Growth-Factor Signaling

Growth factors are naturally occurring substances capable of stimulating cellular growth, differentiation, proliferation, and healing. Growth factors are important for regulating a variety of cellular processes important for physiological mechanisms. Growth factors typically act as signaling molecules that bind to specific receptors on the surface of their target cells to promote cell differentiation and maturation. Chronic alcohol exposure may interfere with the activity of growth factors that regulate neuronal cell proliferation and survival such as insulin-like growth factors (IGF) I and II. Alcohol exposure inhibits insulin and IGF signaling in both the liver and the brain by impairing the signaling cascade at multiple levels (de la Monte, Derdak, & Wands, 2012). This may further result in insulin resistance in the liver and CNS as the cells fail to adequately transmit information to the downstream signaling pathway comprising ERK/mitogen-activated protein kinase (needed for DNA synthesis) and phosphatidylinositol 3-kinase (promotes growth, survival, cell motility, glucose utilization, plasticity, and energy metabolism). Cytotoxic ceramides produced after liver damage may transfer from the liver to the blood and then to the brain, owing to their lipid-soluble nature, where they exert their neurodegenerative effects. Ceramides are capable of further activating proinflammatory cytokines and increasing lipid adducts and insulin resistance in the brain to impair motor and cognitive function (de la Monte et al., 2012). Alcohol may also induce cell death by inhibiting several other growth factors that support cells that have attained their final function and no longer divide (Zhang, Rubin, & Rooney, 1998). In another study, Cohen, Tong, Wands, and de la Monte (2007) also found that chronic ethanol exposure in adult rats may lead to insulin/IGF resistance with increased oxidative stress and impaired acetylcholine biosynthesis, which is responsible for ethanol-induced neurodegeneration and associated behavioral deficits.

Inhibition of Neurogenesis

Neurogenesis refers to the generation of new neurons and involves neuronal cell proliferation, differentiation, and migration. In most brain regions, neurogenesis declines as we approach young adulthood, except for two brain regions, i.e., the olfactory bulb and the hippocampus, which produce new neurons throughout the adult life (Gage, 2000). More than 30 years of research has shown that alcohol has determinant effects on neurogenesis in almost all regions of the developing brain (Campbell, Stipcevic, Flores, Perry, & Kippin, 2014). Binge alcohol exposure affects both cell proliferation and newborn cell survival and this is supported by evidence of cell death in the dorsal root ganglion following binge alcohol exposure and cell loss in the dentate gyrus following chronic alcohol exposure. Several mechanisms involved in alcohol-induced inhibition of neurogenesis include effects on NMDA receptors and a $GABA_A$ receptor (Ikonomidou et al., 2000). In cerebellar granule cell cultures, ethanol causes antagonism of NMDA receptors and leads to apoptotic cell death, which finally prevents the synthesis of brain-derived neurotrophic factor (Bhave, Ghoda, & Hoffman, 1999). There are other studies that suggest the involvement of corticotropin (ACTH)/corticosterone in alcohol-induced impairment of neurogenesis (McEwen, 1999). However, there is no evidence suggesting an increase in plasma ACTH/corticosterone levels of alcohol-fed animals, but some studies suggest blunted activity of the hypothalamic–pituitary–adrenal axis after alcohol intoxication (Ogilvie, Lee, Weiss, & Rivier, 1998).

Inhibiting neurogenesis has shown detrimental effects on hippocampus-based learning and memory. Several findings demonstrated that inhibition of neurogenesis has downstream effects on learning and memory. In a binge alcohol model of learning and memory, the performance of rats was impaired at 3 weeks following binge exposure at the same time point at which neurogenesis was found to be inhibited (Nixon & Crews, 2002). He, Nixon, Shetty, and Crews (2005) also demonstrated that ethanol treatment during adult neurogenesis blunted the growth of progenitors' dendritic arbor. Collectively these findings suggest that alcohol intoxication reduces neurogenesis, which is one of the most prominent causes of alcohol-induced cognitive deficits.

Alcohol-Induced Disruption in Neuronal Glucose Transport

Glucose is known to play a vital role in our body, including in the CNS. Apart from serving as the primary source of energy in all cells, it is also used in the production of various important molecules such as nucleic acids, lipids, hormones (steroids), and neurotransmitters. To cross the cell membrane and enter into cells glucose utilizes specific glucose transporter proteins designated GLUT-1 through GLUT-7. In the brain, the principal glucose transporter proteins include GLUT-1 and GLUT-3. Prolonged exposure to alcohol is known to be associated with reduced glucose uptake and GLUT-1 gene expression in rats (Abdul Muneer, Alikunju, Szlachetka, & Haorah, 2011; Singh, Pullen, Srivenugopal, Yuan, & Snyder, 1992). In cultured rat neurons, short-term alcohol exposure reduced cellular glucose uptake as well as the levels of glucose transporter proteins (Hu, Singh, & Snyder, 1995). Thus, alcohol-induced changes in glucose transport have broad implications and must be considered as an important potential contributor to CNS damage associated with alcohol exposure.

Possible Role of Alcohol-Induced Defects in Cell Adhesion

Cell-to-cell contact (cell adhesion) is very important for neurons to survive, migrate to their final destination, and develop appropriate connections with neighboring cells. Various cell adhesion molecules are known to play important roles in this process. Defects in one such cell adhesion molecule, known as L1, may lead to abnormal brain development in humans, characterized by mental retardation, absence of the corpus callosum, and abnormal development of the cerebellum (Fitzgerald, Charness, Leite-Morris, & Chen, 2011). Alcohol may interfere with normal brain development by reducing cell adhesion. Alcohol affects the L1 molecule in patients with FAS and thereby may contribute to several aspects of the FAS phenotype. This hypothesis is supported by findings that when cultured neurons are exposed to low levels of alcohol, the L1-mediated clumping of the cells is inhibited (Ramanathan, Wilkemeyer, Mittal, Perides, & Charness, 1996). These findings clearly suggest that the ethanol effect on cell adhesion is an important contributor to the harmful consequences of alcohol exposure.

FUTURE SCOPE AND CONCLUSIONS

Chronic alcohol consumption leads to severe physical, mental, and behavioral deficits, including cognitive dysfunction, and unfortunately no suitable therapeutic arsenal is available to prevent the sequela of neurodegeneration. Prolonged alcohol exposure in both adults and neonates leads to widespread neuropsychological deficits among several domains, including general intelligence, memory, language, attention, learning, executive functioning, motor skills, and social and adaptive functioning. Therefore, understanding how chronic alcohol exposure produces behavioral and neuropsychological deficits is of great medical importance.

Treatment of alcohol-induced cognitive deficits requires cessation of drinking, management of behavioral problems, and correction of nutritional deficiencies. Memantine is the only available drug known to produce a slight improvement in memory and cognitive functioning of alcoholics. Other psychiatric problems associated with chronic alcohol consumption are currently being treated with psychotropic medications such as antidepressants and/or antipsychotics. Impulsive or hostile behavior in alcoholics can be managed with anticonvulsants, antipsychotics. or lithium. Benzodiazepines may also be used sometimes to manage chronic alcohol-induced irritability or anxiety.

Thus, although considerable progress has been made in understanding the pathophysiology and pathways involved in alcohol-induced cognitive deficits in both adults and children exposed to alcohol, strategies regarding the therapeutic management of the disease need to be investigated so that novel therapeutic modalities targeted at disrupted molecular events can be developed for prevention as well as clinical management of alcohol-induced cognitive deficits.

DEFINITION OF TERMS

Apoptosis Apoptosis, also known as programmed cell death, is a series of biochemical events including cell blebbing, cell shrinkage, nuclear fragmentation, chromatin condensation, and chromosomal DNA fragmentation, which finally leads to cell death.

Astrocytes Astrocytes are a subtype of glial cell in the CNS and the most abundant cells in the human brain. They perform various vital functions such as biochemical support for endothelial cells, provision of nutrients to the nervous tissue, maintenance of extracellular ion balance, and help in the repair processes of the brain and spinal cord following injuries.

Dementia Dementia refers to defects in learning and memory, visuospatial ability, language, attention, and executive function (problem solving) of the individual. It may result from a wide variety of brain disorders that affect a person's ability to think, reason, and remember clearly.

Excitotoxicity It is the excessive stimulation of neurons by neurotransmitters such as glutamate that leads to neurodegeneration by allowing high levels of calcium ions (Ca^{2+}) to enter the cell.

Fetal alcohol syndrome FAS is permanent syndrome of birth defects comprising physical, behavioral, and cognitive defects that is caused by maternal consumption of alcohol during pregnancy.

Neurodegeneration It refers to progressive loss of structure or function of neurons, including neuronal cell death.

Neurogenesis Neurogenesis, also referred to as the birth of neurons, is the process by which neurons are generated from neural stem cells and progenitor cells.

Neuroinflammation This is the inflammation of neurons that is associated with various neurodegenerative disorders.

Oxidative stress It refers to an imbalance between the systemic manifestation of ROS and the body's antioxidant defense mechanisms.

Toll-like receptor-4 Toll-like receptor-4 is a protein that can detect lipopolysaccharide from gram-negative bacteria and thus is important in the activation of the innate immune system.

KEY FACTS ON ALCOHOL-INDUCED COGNITIVE DEFICIT AND ITS NEUROPATHOBIOLOGY

- Chronic alcoholism is the second leading cause of dementia after Alzheimer disease.
- Atrophy of nerve cells and brain shrinkage in the cortical and subcortical regions and the hippocampus are common features in alcoholics.
- Alcohol enhances oxidative stress directly through generation of free radicals along with suppression of endogenous antioxidant enzymes such as glutathione peroxidase/glutathione reductase.
- Alcohol leads to enhanced neuroinflammation along with NF-κB activation, microglial activation, and increased COX-2 immunoreactivity.
- Alcohol markedly increases neuronal hyperexcitability owing to a combination of increased NMDA receptor activation and decreased $GABA_A$ receptor activation.
- Alcohol may induce neuronal cell death by inhibiting several other growth factors that support cells that have attained their final function and no longer divide.

SUMMARY POINTS

- This chapter dwells on various pathophysiological cascades that are activated by chronic alcohol consumption and lead to long-term cognitive deficits and neurodegeneration in chronic alcoholics.
- Various biochemical, cellular, and molecular alterations associated with alcohol-induced cognitive deficits include free radical-induced damage, neuroinflammation, activation of the NF-κB signaling pathway, NMDA receptor sensitization, and inhibition of growth factors.
- Chronic alcoholism also causes new nerve cells to die, leading to inhibition of neurogenesis.
- Prolonged alcohol exposure in both adults and neonates leads to widespread neuropsychological deficits among several domains, including general intelligence, memory, language, attention, learning, executive functioning, motor skills, and social and adaptive functioning.
- Novel therapeutic modalities targeting disrupted molecular events need to be developed for prevention as well as clinical management of alcohol-induced cognitive deficits.

REFERENCES

Abel, E. L., & Sokol, R. J. (1986). Fetal alcohol syndrome is now leading cause of mental retardation. *Lancet, 2*, 1222.

Abdul Muneer, P. M., Alikunju, S., Szlachetka, A. M., & Haorah, J. (2011). Inhibitory effects of alcohol on glucose transport across the blood–brain barrier lead to neurodegeneration: preventive role of acetyl-L-carnitine. *Psychopharmacology, 214*, 707–718.

Achord, J. L. (1995). Alcohol and the liver. *Scientific American Science and Medicine, 2*, 16–25.

Alfonso-Loeches, S., Pascual-Lucas, M., Blanco, A. M., Sanchez-Vera, I., & Guerri, C. (2010). Pivotal role of TLR4 receptors in alcohol-induced neuroinflammation and brain damage. *Journal of Neuroscience, 30*, 8285–8295.

Bhave, S. V., Ghoda, L., & Hoffman, P. L. (1999). Brain-derived neurotrophic factor mediates the anti-apoptotic effect of NMDA in cerebellar granule neurons: signal transduction cascades and site of ethanol action. *Journal of Neuroscience, 19*, 3277–3286.

Blanco, A. M., Valles, S. L., Pascua, M., & Guerri, C. (2005). Involvement of TLR4/type I IL-1 receptor signaling in the induction of inflammatory mediators and cell death induced by ethanol in cultured astrocytes. *Journal of Immunology, 175*, 6893–6899.

Breese, C. R., & Sonntag, W. E. (1995). Effect of ethanol on plasma and hepatic insulin-like growth factor regulation in pregnant rats. *Alcoholism, Clinical and Experimental Research, 19*, 867–873.

Campbell, J. C., Stipcevic, T., Flores, R. E., Perry, C., & Kippin, T. E. (2014). Alcohol exposure inhibits adult neural stem cell proliferation. *Experimental Brain Research, 232*, 2775–2784.

Chandler, L. J., Sumners, C., & Crews, F. T. (1993). Ethanol inhibits NMDA receptor-mediated excitotoxicity in rat primary neuronal cultures. *Alcoholism, Clinical and Experimental Research, 17*, 54–60.

Choi, D. W. (1995). Calcium: still center-stage in hypoxic-ischemic neuronal death. *Trends in Neuroscience, 18*, 58–60.

Chopra, K., & Tiwari, V. (2013). Tocotrienol and cognitive dysfunction induced by alcohol. In R. R. Watson, V. R. Preedy, & S. Zibadi (Eds.), *Alcohol, nutrition, and health consequences* (pp. 181–202). New York: Humana Press, Springer.

Cohen, A. C., Tong, M., Wands, J. R., & de la Monte, S. M. (2007). Insulin and insulin-like growth factor resistance with neurodegeneration in an adult chronic ethanol exposure model. *Alcoholism, Clinical and Experimental Research, 31*, 1558–1573.

Crews, F., Nixon, K., Kim, D., Joseph, J., Shukitt-Hale, B., Qin, L., & Zou, J. (2006). BHT blocks NF-kappaB activation and ethanol-induced brain damage. *Alcoholism, Clinical and Experimental Research, 30*, 1938–1949.

Crews, F. T., Collins, M. A., Dlugos, C., Littleton, J., Wilkins, L., Neafsey, E. J., … Noronha, A. (2004). Alcohol-induced neurodegeneration: when, where and why? *Alcoholism, Clinical and Experimental Research, 28*, 350–364.

Eckardt, M. J., File, S. E., Gessa, G. L., Grant, K. A., Guerri, C., Hoffman, P. L., … Tabakoff, B. (1998). Effects of moderate alcohol consumption on the central nervous system. *Alcoholism, Clinical and Experimental Research, 22*, 998–1040.

Fama, R., Pfefferbaum, A., & Sullivan, E. V. (2004). Perceptual learning in detoxified alcoholic men: contributions from explicit memory, executive function, and age. *Alcoholism, Clinical and Experimental Research, 28*, 1657–1665.

Fitzgerald, D. M., Charness, M. E., Leite-Morris, K. A., & Chen, S. (2011). Effects of ethanol and NAP on cerebellar expression of the neural cell adhesion molecule L1. *PLoS One, 6*, e24364.

Gage, F. H. (2000). Mammalian neural stem cells. *Science, 287*, 1433–1438.

Gourine, A. V., Kasymov, V., Marina, N., Tang, F., Figueiredo, M. F., Lane, S., … Kasparov, S. (2010). Astrocytes control breathing through pH-dependent release of ATP. *Science, 329*(5991), 571–575.

Halliwell, B. (1992). Reactive oxygen species and the central nervous system. *Journal of Neurochemistry, 59*, 1609–1623.

Haorah, J., Ramirez, S. H., Floreani, N., Gorantla, S., Morsey, B., & Persidsky, Y. (2008). Mechanism of alcohol-induced oxidative stress and neuronal injury. *Free Radical Biology and Medicine, 45*, 1542–1550.

Harper, C. (1998). The neuropathology of alcohol-specific brain damage, or does alcohol damage the brain? *Journal of Neuropathology and Experimental Neurology, 57*(2), 101–110.

He, J., Nixon, K., Shetty, A. K., & Crews, F. T. (2005). Chronic alcohol exposure reduces hippocampal neurogenesis and dendritic growth of newborn neurons. *European Journal of Neuroscience, 21*, 2711–2720.

Hu, I. C., Singh, S. P., & Snyder, A. K. (1995). Effects of ethanol on glucose transporter expression in cultured hippocampal neurons. *Alcoholism, Clinical and Experimental Research, 19*, 1398–1402.

Ikonomidou, C., Bittigau, P., Ishimaru, M. J., Wozniak, D. F., Koch, C., Genz, K., … Olney, J. W. (2000). Ethanol-induced apoptotic neurodegeneration and fetal alcohol syndrome. *Science, 287*, 1056–1060.

Jones, K. L., & Smith, D. W. (1973). Recognition of the fetal alcohol syndrome in early infancy. *Lancet, 2*, 999–1001.

Jung, M. E., Gatch, M. B., & Simpkins, J. W. (2005). Estrogen neuroprotection against the neurotoxic effects of ethanol withdrawal: potential mechanisms. *Experimental Biology and Medicine, 230*, 8–22.

Knapp, D. J., & Crews, F. T., (1999). Induction of cyclooxygenase-2 in brain during acute and chronic ethanol treatment and ethanol withdrawal. *Alcoholism, Clinical and Experimental Research, 23*(4), 633–643.

Konovko, O. O., Morozov, Y. E., Kalinichenko, S. G., Dyuzen, I. V., & Motavkin, P. A. (2004). Induction of NO-synthase and acetaldehyde dehydrogenase in neurons of human cerebellar cortex during chronic alcohol intoxication. *Bulletin of Experimental Biology and Medicine, 137*, 211–214.

Korbo, L. (1999). Glial cell loss in the hippocampus of alcoholics. *Alcoholism, Clinical and Experimental Research, 23*, 164–168.

Kubota, M., Nakazaki, S., Hirai, S., Saeki, N., Yamaura, A., & Kusaka, T. (2001). Alcohol consumption and frontal lobe shrinkage: study of 1432 non-alcoholic subjects. *Journal of Neurology, Neurosurgery and Psychiatry, 71*, 104–106.

Kucera, P., Balaz, M., Varsik, P., & Kurca, E. (2002). Pathogenesis of alcoholic neuropathy. *Bratislavske Lekarske Listy, 103*, 26–29.

Light, K. E., Belcher, S. M., & Pierce, D. R. (2002). Time course and manner of Purkinje neuron death following a single ethanol exposure on postnatal day 4 in the developing rat. *Neuroscience, 114*, 327–337.

Lonze, B. E., & Ginty, D. D. (2002). Function and regulation of CREB family transcription factors in the nervous system. *Neuron, 35*, 605–623.

Lovinger, D. M. (1993). Excitotoxicity and alcohol-related brain damage. *Alcoholism, Clinical and Experimental Research, 17*, 19–27.

Mann, K., Ackermann, K., Croissant, B., Mundle, G., Nakovics, H., & Diehl, A. (2005). Neuroimaging of gender differences in alcohol dependence: are women more vulnerable? *Alcoholism, Clinical and Experimental Research, 29*, 896–901.

Mattson, S. N., & Riley, E. P. (1998). A review of the neurobehavioral deficits in children with fetal alcohol syndrome or prenatal exposure to alcohol. *Alcoholism, Clinical and Experimental Research, 22*, 279–294.

May, P. A., Gossage, J. P., Brooke, L. E., Snell, C. L., Marais, A. S., Hendricks, L. S., … Viljoen, D. L. (2005). Maternal risk factors for fetal alcohol syndrome in the Western Cape province of South Africa: a population-based study. *American Journal of Public Health, 95*, 1190–1199.

McEwen, B. S. (1999). Stress and hippocampal plasticity. *Annual Review of Neuroscience, 22*, 105–122.

Miguel-Hidalgo, J. J., Wei, J., Andrew, M., Overholser, J. C., Jurjus, G., Stockmeier, C. A., & Rajkowska, G. (2002). Glia pathology in the prefrontal cortex in alcohol dependence with and without depressive symptoms. *Biological Psychiatry, 52*(12), 1121–1133.

de la Monte, S., Derdak, Z., & Wands, J. R. (2012). Alcohol, insulin resistance and the liver-brain axis. *Journal of Gastroenterology and Hepatology, 27*(2), 33–41.

Nixon, K., & Crews, F. T. (2002). Binge ethanol exposure decreases neurogenesis in adult rat hippocampus. *Journal of Neurochemistry, 83*, 1087–1093.

Ogilvie, K., Lee, S., Weiss, B., & Rivier, C. (1998). Mechanisms mediating the influence of alcohol on the hypothalamic-pituitary-adrenal axis responses to immune and nonimmune signals. *Alcoholism, Clinical and Experimental Research, 22*, 243S–247S.

Parsons, O. A. (1998). Neurocognitive deficits in alcoholics and social drinkers: a continuum? *Alcoholism, Clinical and Experimental Research, 22*, 954–961.

Popova, E. N., & Shchekalina, G. A. (1980). Effect of alcohol on glial cells. *Zhurnal nevropatologii i psikhiatrii imeni S.S. Korsakova, 80*, 539–544.

Popova, S., Stade, B., Bekmuradov, D., Lange, S., & Rehm, J. (2011). What do we know about the economic impact of fetal alcohol spectrum disorder? A systematic literature review. *Alcohol and Alcoholism, 46*, 490–497.

Ramanathan, R., Wilkemeyer, M. F., Mittal, B., Perides, G., & Charness, M. E. (1996). Alcohol inhibits cell-cell adhesion mediated by human L1. *Journal of Cell Biology, 133*, 381–390.

Rintala, J., Jaatinen, P., Kiianmaa, K., Riikonen, J., Kemppainen, O., Sarviharju, M., & Hervonen, A. (2001). Dose-dependent decrease in glial fibrillary acidic protein immunoreactivity in rat cerebellum after lifelong ethanol consumption. *Alcohol, 23*, 1–8.

Roebuck-Spencer, T. M., & Mattson, S. N. (2004). Implicit strategy affects learning in children with heavy prenatal alcohol exposure. *Alcoholism, Clinical and Experimental Research, 28*, 1424–1431.

Siler-Marsiglio, K. I., Paiva, M., Madorsky, I., Serrano, Y., Neeley, A., & Heaton, M. B. (2004). Protective mechanisms of pycnogenol in ethanol-insulted cerebellar granule cells. *Journal of Neurobiology, 61*, 267–276.

Singh, S. P., Pullen, G. L., Srivenugopal, K. S., Yuan, X. H., & Snyder, A. K. (1992). Decreased glucose transporter 1 gene expression and glucose uptake in fetal brain exposed to ethanol. *Life Science, 51*, 527–536.

Sullivan, E. V., & Pfefferbaum, A. (2005). Neurocircuitry in alcoholism: a substrate of disruption and repair. *Psychopharmacology, 180*, 583–594.

Sullivan, E. V., Rosenbloom, M. J., & Pfefferbaum, A. (2000). Pattern of motor and cognitive deficits in detoxified alcoholic men. *Alcoholism, Clinical and Experimental Research, 24*, 611–621.

Tiwari, V., & Chopra, K. (2012). Attenuation of oxidative stress, neuroinflammation, and apoptosis by curcumin prevents cognitive deficits in rats postnatally exposed to ethanol. *Psychopharmacology, 224*(4), 519–535.

Tiwari, V., Guan, Y., & Raja, S. N. (2014). Modulating the delicate glial-neuronal interactions in neuropathic pain: promises and potential caveats. *Neuroscience and Biobehavioral Reviews, 45*, 19–27.

Tiwari, V., Kuhad, A., & Chopra, K. (2009). Suppression of neuro-inflammatory signaling cascade by tocotrienol can prevent chronic alcohol-induced cognitive dysfunction in rats. *Behavioural Brain Research, 203*, 296–303.

Tiwari, V., Kuhad, A., & Chopra, K. (2010). Epigallocatechin-3-gallate ameliorates alcohol-induced cognitive dysfunctions and apoptotic neurodegeneration in the developing rat brain. *International Journal of Neuropsychopharmacology, 13*, 1053–1066.

Trudler, D., Farfara, D., & Frenkel, D. (2010). Toll-like receptors expression and signaling in glia cells in neuro-amyloidogenic diseases: towards future therapeutic application. *Mediators of Inflammation, 2010*, 2010.

Watkins, L. R., Hutchinson, M. R., Milligan, E. D., & Maier, S. F. (2007). "Listening" and "talking" to neurons: implications of immune activation for pain control and increasing the efficacy of opioids. *Brain Research Reviews, 56*(1), 148–169.

Weiss, E., Singewald, E. M., Ruepp, B., & Marksteiner, J. (2014). Alcohol induced cognitive deficits. *Wiener Medizinische Wochenschrift, 164*, 9–14.

Wright, M. J., Jr., & Taffe, M. A. (2014). Chronic periadolescent alcohol consumption produces persistent cognitive deficits in rhesus macaques. *Neuropharmacology, 86*, 78–87.

Zahr, N. M., Kaufman, K. L., & Harper, C. G. (2011). Clinical and pathological features of alcohol-related brain damage. *Nature Reviews Neurology, 7*, 284–294.

Zhang, F. X., Rubin, R., & Rooney, T. A. (1998). Ethanol induces apoptosis in cerebellar granule neurons by inhibiting insulin-like growth factor 1 signaling. *Journal of Neurochemistry, 71*, 196–204.

Zou, J. Y., & Crews, F. T. (2005). TNF alpha potentiates glutamate neurotoxicity by inhibiting glutamate uptake in organotypic brain slice cultures: neuroprotection by NF kappa B inhibition. *Brain Research, 1034*, 11–24.

Chapter 58

Neuropathology in Neuropsychiatric Disorders Secondary to Alcohol Misuse

Jagdeo Prasad Rawat, Charles Pinto, Malay Dave, Kirti Yeshwant Tandel
Department of Psychiatry, Jagjivanram Western Railways Hospital, Mumbai, Maharashtra, India

Abbreviations

5-HT 5-hydroxytryptamine
ACD Alcoholic cerebellar degeneration
ARD Alcohol-related dementia
BA Brodmann's area
CA Cornu ammonis
CMRGlu Cerebral metabolic rate of glucose
CRF Corticotrophin-releasing factor
DSM Diagnostic and Statistical Manual of Mental Disorders
DT Delirium tremens
DTI Diffusion tensor imaging
FDG-PET Fluorodeoxyglucose positron emission tomography
FLAIR Fluid-attenuated inversion recovery
GABA γ-aminobutyric acid
ICD International Classification of Diseases
K_m Michaelis–Menten constant
KS Korsakoff syndrome
MBD Marchiafava–Bignami disease
NA Noradrenaline
NMDA N-methyl-D-aspartate
PDH Pyruvate dehydrogenase
PET Positron emission tomography
ROS Reactive oxygen species
SPECT Single-photon emission computed tomography
THDP Thiamine diphosphate
TK Transketolase
WE Wernicke encephalopathy
α-KGDH α-ketoglutarate dehydrogenase

kintu madyaṃ svabhāvena

yathaivānnaṃ tathā smṛitam ॥

ayuktiyuktaṃ rogāya,

yuktiyuktaṃ tathā'mṛitaṃ ॥

[But alcohol by nature is like food,

If taken improperly, it causes disease,

And if taken properly, it is like ambrosia]

The *shloka* above is quoted in Chapter VII of *Todaranda Ayurveda Saukhyam—Diagnosis and Treatment of Diseases in Ayurveda,* an ancient text on Ayurveda, a form of traditional medicine in India (Dash & Kashyap, 1984). It states that alcohol (*madyam*), when used in excess, can be dangerous (*rogāya*), but when used in moderate quantities can be as good as ambrosia (ā'*mṛitam*). It is believed that alcohol has been highly valued in various cultures throughout history. In ancient times, alcohol was seen as an important medicinal ingredient and an essential part of the diet. However, in the succeeding centuries, the irresistible desire of mankind to achieve an altered level of consciousness through the use of inebriants like alcohol has resulted in its misuse and subsequent health hazards. Alcohol use disorders are maladaptive patterns of alcohol consumption that include alcohol abuse and dependence (American Psychiatric Association, 2013). They are associated with economic cost, domestic violence, and loss of work productivity. The 2013 *Diagnostic and Statistical Manual of Mental Disorders* (DSM), fifth edition, integrates the two DSM-IV disorders, alcohol abuse and alcohol dependence, into a single disorder called alcohol use disorder, with mild, moderate, and severe subclassifications. The neuropathology behind various such neuropsychiatric disorders secondary to alcohol misuse will be discussed in this chapter.

WERNICKE ENCEPHALOPATHY

Wernicke encephalopathy (WE) is a neuropsychiatric disorder characterized by ataxia, ophthalmoplegia, and mental status disorders. It is frequently associated with alcohol misuse as a result of the impairment of thiamine-dependent enzymatic activity in susceptible brain cells. Thiamine diphosphate (THDP) works as a cofactor for several apoenzymes like transketolase (TK), pyruvate dehydrogenase (PDH), and α-ketoglutarate dehydrogenase (α-KGDH), which are essential in carbohydrate and amino acid metabolism. During the intracellular transfer of thiamine, THDP is formed as a result of thiamine pyrophosphokinase. During this transfer, magnesium acts as a vital cofactor for the enzyme. Magnesium and THDP are both required during the formation of TK, PDH, and α-KGDH for their proper function. Chronic alcoholics show deficiency in magnesium stores (Traviesa, 1974).

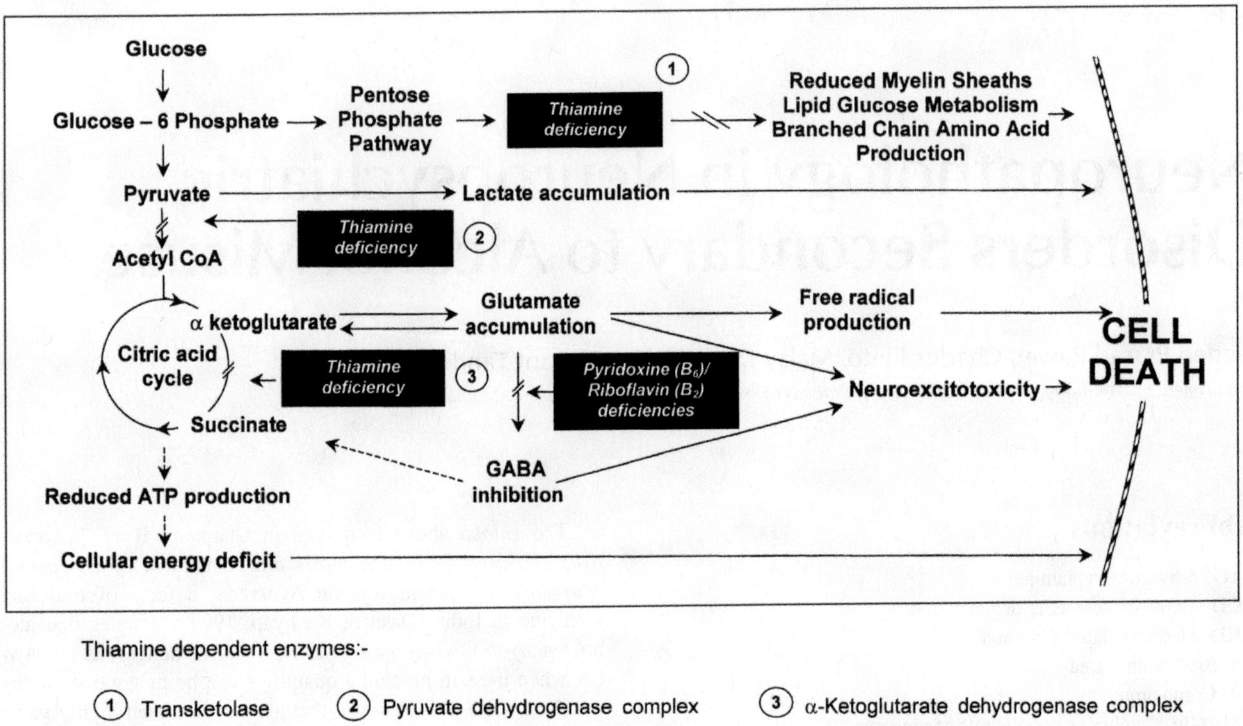

FIGURE 1 **Potential mechanisms leading to brain damage.** Thiamine-dependent enzymes: (1) TK; (2) PDH complex; (3) α-KGDH complex. *Thomson, A.D., Christopher Cook, C.H., Touquet, R., & Henry, J.A. (2002). The Royal College of Physicians report on alcohol: Guidelines for managing Wernicke's encephalopathy in the accident and emergency department. Alcohol and Alcoholism, 37(6), 513–521, by permission of Oxford University Press.*

Magnesium deficiency reduces the proper binding of THDP to the apoenzymes, causing reduced enzyme activity. In addition, a few animal studies have shown that increased concentration of magnesium reduces N-methyl-D-aspartate (NMDA) receptor-mediated excitotoxicity and severe deficiency increases it (Dodd, Beckmann, Davidson, & Wilce, 2000). Concurrent magnesium deficiency can inhibit the response to thiamine treatment. Therefore it may be important to correct magnesium deficiency before giving thiamine therapy in WE patients (Figure 1).

Nutritional deficiency along with vomiting and diarrhea and coexisting decrease in thiamine intake are commonly seen in chronic alcoholism. The gastrointestinal tract gradually loses its capacity to absorb a fixed dose of thiamine, especially in malnourished individuals. The liver, which stores a large part of the body's supplies of thiamine, also loses its capacity to store and handle thiamine. This reduced utilization of thiamine results in inhibition of the active thiamine-requiring enzymes. In the presence of thiamine deficiency, the NMDA receptors are upregulated in chronic alcoholics, causing the neuronal damage (Freund & Anderson, 1996). WE is a clinically underdiagnosed, acute or subacute illness that can cause permanent memory disturbance or death if proper treatment is not given in time. Thiamine deficiency can also cause cerebellar degeneration and neuropathy, due to either chronic alcoholism or nutritional deficiency. If the THDP-dependent enzymatic activity is not restored, irreversible structural damage ensues.

The histopathological changes in WE can be divided into acute and chronic changes. The acute changes consist of punctate hemorrhages around the third and fourth ventricles, the mammillary bodies, and the aqueduct. The presence of punctate hemorrhages in the mammillary bodies is specific for WE (Thorarinsson & Olafsson, 2011). The other regions involved are the hypothalamus, the dorsomedial thalamic nuclei, the oculomotor nuclei, the nuclei dorsalis of the vagus nerve, the abducent, and the vestibular nuclei. The thalamus may show a spongy or granular brown-grayish discoloration in >75% of the patients (Victor & Adams, 1971). Sometimes discoloration may be seen in the reticular formation. Microscopically, WE is characterized by symmetrical microhemorrhages and microglial proliferation in the regions involved. The endothelial hypertrophy and capillary proliferation render the capillaries prominent. The capillaries are surrounded by microhemorrhages and pericapillary edema with extravasations of red cells. The myelin destruction and axonal degeneration lead to astrocytic reaction and accumulation of macrophages with hemosiderin, especially in the brainstem and the thalamus.

The permanent or chronic changes in WE can be divided into macroscopic and microscopic changes. These findings are often seen mostly in the mammillary bodies and the dorsomedial nuclei of the thalami, unlike the acute changes, which are more extensive. Grossly the regions affected show gliotic changes with atrophy seen especially in the mammillary bodies and around the third and fourth ventricles. The lateral ventricles appear dilated in the late stages, with the corpus callosum showing atrophic changes. The histopathological changes in WE show a similar astrocytic reaction with gliosis and tissue destruction. However, there are no fresh pericapillary hemorrhages seen and the endothelium appears normal. In the cerebellum, WE is seen in the form of atrophy of the cerebellar vermis with thinning of the cerebellar cortex. The Purkinje cell layer is almost diminished, with a significant reduction in the number of granule cells and increase in the glial cells in the molecular layer.

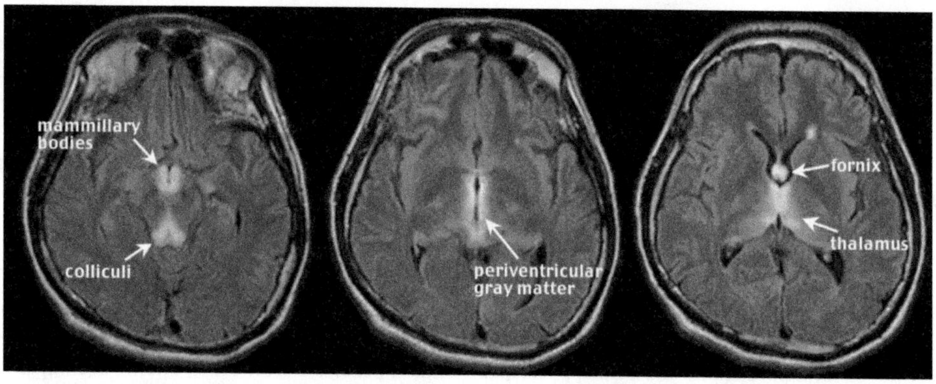

FIGURE 2 MRI brain findings in WE. Three contiguous FLAIR images (5 mm thick with a 2.5-mm skip) of an acute WE case. Note the hyperintense signal in the mammillary bodies and colliculi (left), periventricular gray matter (middle), and fornix and thalamus (right). *Sullivan, E.V., & Pfefferbaum A. (2009). Neuroimaging of the Wernicke–Korsakoff syndrome. Alcohol and Alcoholism, 44(2), 155–165, by permission of Oxford University Press.*

Studies (Sechi & Serra, 2007) have shown changes in the spinal cord in the form of a decrease in the anterior horn cells and involvement of the anterior and the posterior horn cells (Figure 2).

The selectivity of the topographical involvement in WE has been a matter of debate and research. The most affected brain regions are postulated to be those with higher metabolic demands and higher thiamine requirements. Nixon and Jordan (2008) reported that the periventricular areas are affected to a greater degree because of the consequences to the parenchyma of the high cerebrospinal fluid glutamate levels as a result of the incapacity of the choroid plexus to convert glutamate to other metabolites.

> **Case 1:**
>
> Mr A, a 48-year-old male, presented to a peripheral health center with acute confusion, disorientation, upward gaze palsy of the left eye, and ataxia for 5 days with a history of fall associated with injuries to the right lower limb. He had been a chronic alcoholic for 25 years, drinking 300–400 ml of alcohol daily. He was treated with intravenous thiamine 100 mg/day. The left upward gaze palsy and gait improved after 10 days of treatment. However, he remained agitated. After 3–4 weeks he started getting paranoid toward family members. He was referred to a tertiary psychiatry center. He showed marked confabulation during history with significant recent memory disturbances. His mini-mental state examination (MMSE) score was 20/30. Magnetic resonance imaging (MRI) of the brain showed multiple, hypointense lesions on T1-weighted imaging (T1WI) and hyperintensity on T2WI involving basal ganglias, thalami, midbrain, and periventricular regions with cerebral, cerebellar, and mammillary body atrophy. He was continued on multivitamins including thiamine. However, his memory deficit persisted.

KORSAKOFF SYNDROME

Russian neuropsychiatrist Sergei S. Korsakoff (1854–1900), during 1887–1889, described an illness similar to WE in a larger group of patients when he worked as the chief of the Moscow University Psychiatric Clinic. These patients presented with acute confusion, confabulations, pseudoreminiscences, and peripheral neuropathy with a variable degree of limb muscle atrophy. He described the memory loss as anterograde in nature. He concluded that agitation and prodromal confusion usually preceded the onset of memory disturbances. The patients' profound anterograde memory loss was characterized by an inability to learn new verbal and nonverbal information. Initially named as "cerebropathia psychica toxaemica" in 1889, it was renamed as "Korsakoff's disease" in 1897.

The locations of the lesions responsible for the memory impairment in Korsakoff syndrome (KS) remain controversial. Victor and Adams (1971) showed that the neuropathology in KS involves neuronal loss, microhemorrhages, and gliosis in the paraventricular and aqueductal gray matter. He concluded that damage to the medial dorsal nucleus was responsible for giving rise to the memory impairment. Another study contradicted the findings by reporting marked neuronal loss from the medial mammillary bodies and CA1 region of the hippocampus, along with gliosis in the medial thalamus adjacent to the wall of the third ventricle, a region known as the paratenial nucleus (Mayes, Meudell, Mann, & Pickering, 1988). However, the autopsy findings in the study did not reveal any lesions in the medial dorsal nuclei. Harding, Halliday, Caine, and Kril (2000) in their study compared the brains of five alcoholic patients with WE with absence of amnesia to eight other alcoholic patients with severe amnesia. They concluded that the neurodegeneration seen in the anterior principal thalamic nuclei was the only consistent lesion found in alcoholics with KS that differentiated them from those with WE.

Thiamine replacement shows remarkable improvement in confabulation and disorientation in the early stages of KS. The deficiency of thiamine along with the neurotoxic action of alcohol is understood to cause the persistent memory loss in KS. Nixon, Kaczmarek, Tate, and Kerr (1984) reported that a variant TK and thiamine deficiency together contribute to the pathogenesis of brain damage seen in KS. They found that the apparent Michaelis–Menten constant (K_m) values for the cofactor THDP were the same for patients and the control group. However, isoelectric focusing separated red cell TK into different isozymes characterized by p*I* values in the range 6.6–9.2. The isozyme pattern found in 39 of 42 patients with KS was present in only 8 of 36 controls. Coy et al. (1996) in their study showed the possibility that variation in the X-linked TK gene may account for a genetic predisposition to the development of Wernicke–Korsakoff syndrome.

Computed tomography (CT) and MRI demonstrate generalized cortical atrophy, with some studies (Shimamura, Jernigan, & Squire, 1988; Jacobson & Lishman, 1990) also indicating frontal lobe involvement. They suggested that KS results from two separate effects, the diencephalic damage, contributing to memory changes, and cortical atrophy. The former may result as a sequela to nutritional deficiency, whereas the direct toxic effects of alcohol result in cortical atrophy. Jacobson and Lishman (1987) showed the relation between the severity of memory impairment and third-ventricular enlargement. Colchester et al. (2001) showed significant atrophy in the thalamus, mammillary bodies, and frontal lobes in patients with KS. However, the study did not find significant differences in the anterolateral, medial temporal, hippocampal, and parahippocampal volumes. Relative hypermetabolism in white matter and hypometabolism in thalamic, orbitomedial, frontal, and retrosplenial regions by fluorodeoxyglucose positron emission tomography (FDG-PET) was shown by Reed et al. (2003) and interpreted as secondary metabolic effects within the diencephalic–limbic memory circuits.

The principal brain areas involved in episodic memory are the hippocampus, anterior thalamic nuclei, mammillary bodies, and cingulate cortex (Aggleton & Brown, 1999). The regions are connected by the fornix, mammillothalamic tract, and cingulum. The memory disturbances in KS primarily involve the semantic memory and the episodic memory. Semantic memory (also called generic memory) refers to the memory of meanings, understandings, and other concept-based knowledge and underlies the conscious recollection of factual information and general knowledge about the world. Episodic memory is a category of long-term memory that involves the recollection of specific events, situations, and experiences. It is the implicit memory that is relatively preserved in KS, the memory that uses past experiences to remember things without conscious awareness of them.

The basic deficit that is proposed in the memory disturbances is that the psychological processes involved in the encoding of information are impaired. The patients with KS are not able to process the meaningful aspects of the memory, i.e., the semantic memory, despite being able to encode the direct sensory impressions. There is also impairment in the process of "consolidation," or the storage of information in a permanent form. Some studies showed that Korsakoff patients display a disproportionate impairment in recalling the temporal aspects of learned information as a result of difficulty in associating two or more pieces of information. Thus there is a dissociation between the recall and the recognition memory. Kril and Harper (2012) suggested that damage to the anterior principal nucleus of the thalamus is responsible for the clinical manifestation of KS. However, it was not possible to determine whether it is the anterior nucleus alone or a combination of two or all three nuclei that is required for KS, as the mammillary bodies and the mediodorsal nucleus of the thalamus were also damaged in these patients (Table 1).

MARCHIAFAVA–BIGNAMI DISEASE

Marchiafava–Bignami disease (MBD) or syndrome is a rare progressive neurological disease characterized by selective demyelination

TABLE 1 Neuropathological Spectrum of Neuropsychiatric Disorders Secondary to Alcohol Misuse

Etiology	Brain Regions	Neuropathological Findings
Wernicke encephalopathy	Mammillary bodies, periaqueductal gray matter, periventricular regions	• MRI brain: hyperdensities in mammillary bodies and periaqueductal gray matter • abnormal enhancement on T1WI
Korsakoff syndrome	Mammillary bodies, hippocampus, thalamus, orbitofrontal cortices	• diencephalic damage • cortical atrophy • impairment in "consolidation" of memory
Marchiafava–Bignami disease	Corpus callosum	• hyperintensities on T2 and hypointensities on T1 in corpus callosum • necrosis and cyst formation • increased choline levels
Alcohol withdrawal syndrome	Nonspecific	• hyperexcitability of CNS • changes in the major neurotransmitters
Alcohol-related dementia	Frontal cortex	• glutamate excitotoxicity • disruption in neurogenesis • mitochondrial damage • oxidative stress
Alcoholic cerebellar degeneration	Cerebellum	• cerebellar atrophy especially superior cerebellar vermis
Alcoholic hallucinosis	Thalamus, dopamine pathways	• abnormal dopaminergic transmission • elevated levels of β-carbolines • thalamic dysfunction
Central pontine myelinolysis	Pons	• increased susceptibility to osmotic stress • MRI: hyperintensities on T2 in pontine region

Table 1 summarizes the clinical features, primary brain regions affected, and main neuropathological findings in the neuropsychiatric disorders described in this chapter.

of the corpus callosum and cortical laminar necrosis involving the frontal or temporal lobes and subsequently resulting in atrophy. It is usually seen in chronic alcohol consumption and rarely in nonalcoholic subjects. It was initially described by Carducci in 1898 and 5 years later by Italian pathologists Amico Bignami (1862–1919) and Ettore Marchiafava (1847–1935) in Italian red wine drinkers. Since then MBD has been demonstrated in alcoholics from all over the world, in subjects consuming alcoholic beverages varying from Chianti red wine from Italy to locally made illicit liquor (Mahuwa alcohol) in India (Rawat, Pinto, Kulkarni, Muthusamy, & Dave, 2014). Although the underlying etiology is still not understood, MBD is probably caused by the combined ill effects of chronic alcohol consumption with malnutrition (Figure 3).

Case 2:

Mr B, a 57-year-old male with a history of chronic alcohol abuse for 35 years, used to consume 400–500 ml of Mahuwa Daru (*Madhuca longifolia*) daily. Because of social pressure, the patient decided to stop drinking alcohol. He developed tremors, sleep disturbances, and disorientation along with fecal and urinary incontinence. His central nervous system (CNS) examination showed symmetrical pupillary. There were no lateralizing deficits and no signs of meningeal irritation. Liver and renal function tests were normal. He was started on injectable thiamine (100 mg) along with an antiwithdrawal line of management. He still had four episodes of seizures within 3 days of admission. However, his EEG was normal. Brain MRI showed hyperintense signal intensities involving the genu, body, and splenium of the corpus callosum on T2WI and fluid-attenuated inversion recovery (FLAIR). The patient gradually improved. On the 20th day he was symptom free and MSE was clear. His MMSE was 30/30 on discharge. After 1 month he could resume his work.

In the early years all case reports on MBD were essentially autopsy findings. However, owing to advances in neuroimaging in the past few decades, the diagnosis of MBD can be established antemortem by CT and MRI. MRI is currently the most sensitive diagnostic tool. MRI typically shows hyperintense lesions on T2-phase and FLAIR images, indicating edema and demyelination, with hypointense lesions on T1WI of the corpus callosum and sometimes even the genu and the splenium, signifying bleeding and hemosiderin deposits. In MBD, the lesions in the corpus callosum are symmetrical and the thin dorsal and ventral edges are spared ("sandwich sign"). This finding distinguishes it from other disorders with similar MRI findings, like ischemic stroke, multiple sclerosis, lymphoma, etc. Necrosis and cyst formation may also occur. Similar lesions may be found in the middle cerebellar peduncles and in the centrum semiovale. As lesions become chronic, cystic lesions are likely to develop. A few months later the signal intensity alterations become less evident. However, the residual atrophy of the involved structure usually remains. Heinrich, Runge, and Khaw (2004) described two subtypes: Type A is characterized by major impairment of consciousness, T2-hyperintense swelling of the entire corpus callosum on early MRI, and poor outcome. Type B is characterized by slight impairment of consciousness, partial callosal lesions on MRI, and a favorable outcome (Table 2).

A few PET studies (Pappata, Chabriat, Levasseur, Legault-Demare, & Baron, 1994) demonstrated markedly reduced cerebral metabolic rate of glucose in the frontal and temporoparieto-occipital association cortices. In their study of two patients, Nalini, Kovoor, Dawn, and Kallur (2009) showed similar findings by PET–CT. Glucose metabolism was reduced markedly, and MRI showed lesions essentially limited to the corpus callosum and white matter. These observations may suggest that in MBD, perfusion and metabolism defects may affect structures beyond the corpus callosum, like cerebral hemispheres, and the widespread cerebral functional involvement may account for the complex neurologic and cognitive deficits. The findings suggest that disconnection of not only the commissural fibers but also the extracallosal projection and association fibers

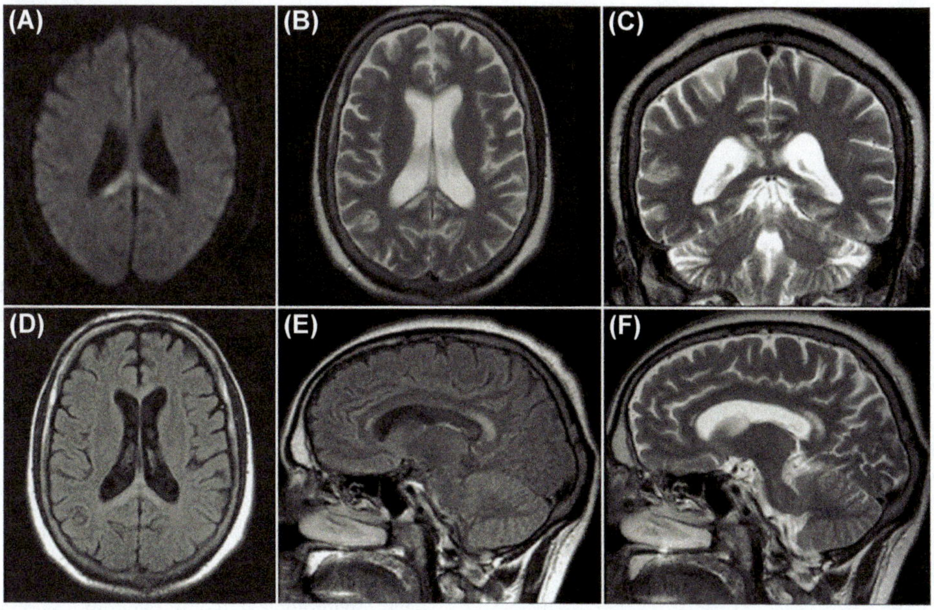

FIGURE 3 **MRI findings in MBD.** (A) Axial diffusion-weighted image shows symmetric hyperintensities in the splenium. T2-weighted (B) axial and (C) coronal and FLAIR (D) axial and (E and F) sagittal images show hyperintensities in the genu and splenium of callosum. *Data from Rawat et al. (2014), with permission from Medknow Publishers.*

TABLE 2 Subtypes of Marchiafava–Bignami Disease

	Subtypes of MBD	
	Type A	Type B
Clinical presentation	Coma, stupor, and pyramidal tract symptoms	Normal or mild impairment in mental status
Radiological signs	Entire corpus callosum is involved	Partial lesions are seen in corpus callosum
Mortality rate	21%	0%
Disability rate	86%	19%

The clinical subtypes of MBD are shown on the basis of both clinical and radiological signs.
Data published with permission of Medknow Publishers.

TABLE 3 Changes in the Activity of Various Neurotransmitters in Alcohol Withdrawal Syndrome

Increased Activity	Decreased Activity
Glutamate	GABA
Noradrenaline	Dopamine
Acetylcholine	Serotonin
CRF	Neuropeptide Y
Vasopressin	

Table 3 shows the activities of various neurotransmitters in the pathogenesis of alcohol withdrawal syndrome (GABA, γ-aminobutyric acid; CRF, corticotrophin-releasing factor).

involving the corticocortical and corticosubcortical connectivity can cause transneural depression of the cortical metabolism and function. MR spectroscopy has been used to detect patients with MBD. The initial findings show increased choline levels secondary to acute demyelination, low *N*-acetylaspartate secondary to neuronal damage, and the presence of lactate.

ALCOHOL WITHDRAWAL SYNDROME

Alcohol withdrawal syndrome is seen after cessation of a prolonged ingestion of alcohol. The symptoms usually appear within 6h to 2 days after the last consumption of alcohol. The alcohol withdrawal syndrome is characterized by tremors, nausea, vomiting, weakness, hyperreflexia, autonomic hyperactivity, diarrhea, tachycardia, sweating, hypertension, irritability, anxiety, and seizures. It may also be associated with transient auditory and visual hallucinations, which occur within the first 2 days of decreasing or discontinuing alcohol consumption. Delirium tremens presents within 2–4 days of the last drink. It is characterized by disorientation, persistent visual auditory and visual hallucinations, tremulousness, agitated behavior, and autonomic hyperactivity as a result of stress-related hormones (Table 3).

> **Case 3:**
>
> Mr C, a 46-year-old man, presented with altered mental status and agitation. He was repeatedly vomiting and passing highly colored urine. He had no significant past history of medical/surgical illness. He was a heavy alcoholic for the past 20 years, and 2 days back, he had stopped alcohol for religious reasons. He reported two instances of seizures in the past when he had abstained from alcohol. On examination, he had icterus with tremulousness. His blood pressure was 156/88 and pulse 106 beats/min. MSE revealed visual and auditory hallucinations. Abdominal examination was suggestive of hepatomegaly. Liver function tests were deranged. Thiamine was immediately given. His withdrawal symptoms were controlled by giving lorazepam 6 mg on day 1 and stepped up to 12 mg over next few days, after which his mental state improved and agitation decreased gradually. He was discharged after 2 weeks.

The pathophysiology behind this syndrome is explained by the inhibitory effects of γ-aminobutyric acid (GABA) on signal-receiving neurons, lowering the neuronal activity, which is enhanced by alcohol. More alcohol is required to produce the same inhibitory effect over time as tolerance develops, resulting in GABA receptors becoming less responsive to neurotransmitters. The sudden cessation of alcohol consumption from this adapted system, in the absence of GABA's inhibitory response, shifts the balance in favor of the excitatory neurotransmission mediated by glutamate (De Witte, 2004). Both the number and the function of NMDA receptors were reported to be increased during alcohol withdrawal (Haugbol, Ebert, & Ulrichsen, 2005). The overactivity of the noradrenergic nervous system and decreased serotonergic activity in several areas of the CNS also play major roles in alcohol withdrawal syndrome. Patkar et al. (2003) reported changes in the concentrations of both plasma noradrenaline (NA) and 5-hydroxytryptamine (5-HT) during alcohol withdrawal, of which NA activity was found to normalize by late withdrawal, whereas 5-HT activity seemed to be more persistent. A few studies (Shen, 2003) showed that there is a reduction in the activity of the mesolimbic dopaminergic system during alcohol withdrawal. The upregulation of L-type calcium channels during alcohol withdrawal (N'Gouemo & Morad, 2003) and differences in the cholinergic activity and postsynaptic sensitivity to cholinergic convulsants (Mark & Finn, 2002) were shown to be associated with seizure susceptibility. Alcohol withdrawal provokes the intense generation of reactive oxygen species and the activation of stress-responding protein kinases, along with suppression of mitochondrial enzymes, damaging the mitochondrial membrane potential and perturbing redox balance (Jung & Metzger, 2010). Decreased cellular expression of neuropeptide Y was reported in the central and medial nuclei of the amygdala as well as in the cortical and hypothalamic structures (Roy & Pandey, 2002). There is also increased activation of the hypothalamic–pituitary–adrenal axis after alcohol withdrawal, with increased plasma cortisol levels and increased cerebrospinal fluid corticotrophin-releasing factor levels (Zimmermann, Hundt, Spring, Grabner, & Holsboer, 2003). Vasopressin has been reported to increase during alcohol withdrawal (Linkola, Ylikahri, Fyhrquist, & Wallenius, 1978).

ALCOHOL-RELATED DEMENTIA

The neuropathological issues regarding alcohol-related dementia remain under debate. The cognitive decline seen as a result of chronic alcohol abuse and/or dependence can range from isolated amnesia or mild cognitive impairment to full-blown dementia. Glutamate excitotoxicity, disruption of neurogenesis, mitochondrial damage, and oxidative stress are a few ways through which the physiological processes are directly affected by chronic ethanol consumption. This forms the basis of the "neurotoxicity" hypothesis. The drinking patterns of repeated binge drinking and abstinence with withdrawal enhance neuronal injury through increased vulnerability of upregulated NMDA receptors to glutamate-induced excitotoxicity. Repeated exposures to toxic amounts of alcohol result in neural plasticity due to both reducing GABAergic inhibition and increasing glutamatergic excitation.

> **Case 4:**
>
> Mr D, a 40-year-old man with an 18-year history of alcoholism, was admitted for intoxication, forgetfulness, abnormal behavior, sleep disturbances, and inability to care for himself. He had no family history of Alzheimer disease and no history of head injury. His brain MRI showed diffuse cerebral atrophy with mild to moderate cerebellar atrophy. His initial MMSE on presentation was 19/30. He was given supportive treatment and neuroleptics to control his behavior. He was advised to strictly abstain from alcohol during this period. His MMSE started improving during the period of sobriety. After 3 months his MMSE was 28/30 with no behavioral abnormalities and significant improvement in memory disturbances.

Walker, Barnes, Zornetzer, Hunter, and Kubanis (1980), through a series of animal studies, showed that ethanol is neurotoxic as a result of various neuropathological changes leading to memory and learning impairment. The studies demonstrated loss of hippocampal CA1 and CA3 pyramidal neurons, mossy fiber–CA3 synapses, and dentate granules along with loss of cholinergic neurons in the basal forebrain. In human studies, the neuronal loss is also seen in the superior frontal association cortex, hypothalamus, basal ganglia, cerebellum, and reticular activating system. The prefrontal cortex is vulnerable in chronic alcoholics, leading to disinhibition, lack of insight, perseveration, and difficulties in planning, organization, abstract thinking, and problem solving. Bleich, Bandelow, and Javaheripour (2003) proposed a role for hyperhomocysteinemia in ethanol neurotoxicity. Their article reported that homocysteine is an excitatory amino acid and it markedly enhances the vulnerability of neuronal cells to excitotoxic and oxidative stress. Folate deficiency leads to hyperhomocysteinemia and since homocysteine acts as an agonist at the glutamate NMDA receptors, it increases NMDA receptor transmission and hence leads to glutamate excitotoxicity. The study also concluded that raised levels of homocysteine are associated with hippocampal brain atrophy in alcoholism. Ethanol also decreases the content of neurotrophin in the affected regions, which is important for neuronal survival, thus resulting in impaired intracellular signaling pathways. The damage to the DNA strands induced by ethanol might cause neuronal death. Animal studies showed that the ethanol metabolite acetaldehyde was found in frontal lobe cortex and white matter, causing protein adduct formation.

Alcohol in large quantities causes a decrease in brain size, widening of sulci, and enlargement of ventricles. Ethanol-related white matter loss is mostly due to demyelination and neuronal loss as a result of cellular degeneration. Rosenbloom, Sullivan, and Pfefferbaum (2003) stated that a newer MRI technique called diffusion tensor imaging (DTI) helps in detecting the degraded white matter, which is usually not seen in conventional MRI. Their studies using DTI showed that relative to age-matched control subjects, alcoholic men have lower regional fractional anisotropy in the anterior part of the corpus callosum and centrum semiovale. This macroscopic and microscopic brain damage is reflected in impaired performance on tests of cognition and motor skills.

ALCOHOLIC CEREBELLAR DEGENERATION

Alcoholic cerebellar degeneration (ACD) is caused by irreversible toxic degeneration of the Purkinje cells. It is thought to be caused by a combination of alcohol neurotoxicity and nutritional deficiency. Victor, Adams, and Mancall (1959) reported a clinical syndrome in 50 chronic alcoholics that manifested severe progressive truncal ataxia with sparing of upper extremities, without incoordination of the limbs, usually evolving over months to years; tremors; dysarthria; gaze-evoked nystagmus; and rarely titubation.

Cerebellar damage probably arises from the direct toxic effects of very high levels of alcohol rather than from a nutritional effect. There is atrophy in the superior cerebellar vermis, paramedian superior cerebellar hemispheres, and flocculi. Sullivan, Deshmukh, Desmond, Lim, and Pfefferbaum (2000) showed atrophy in the anterior part of the cerebellar vermis as well. In a study conducted in Japan (Yokota et al., 2006) on 1509 autopsies obtained from three series, the frequencies of memory impairment and ataxia in ACD cases were significantly higher than those in alcoholics without any alcohol-related pathologies. In this study, loss of Purkinje cells, narrowing of the width of the molecular layer, and tissue rarefaction in the granular layer were observed in the anterior and superior portions of the vermis. In adjacent regions, the Purkinje cell and molecular layers were more mildly affected. This study confirmed the frequency of asymptomatic cerebellar degeneration in alcoholics, suggesting that early intervention in alcoholism in the subclinical phase is important to prevent the development of cerebellar symptoms.

Baker, Harding, and Halliday (1999) reported the role of concomitant nutritional deficiency of thiamine. Changes in the levels of a number of thiamine-dependent enzymes were shown during protein expression profiles of cortical area BA9 gray and white matter and the cerebellar vermis (Alexander-Kaufman, Cordwell, Harper, & Matsumoto, 2007). More recent studies (Lisdahl, Thayer, Squeglia, McQueeny, & Tapert, 2013) reported smaller cerebellar volumes even in healthy teens after more intense binge drinking.

OTHER NEUROPSYCHIATRIC DISORDERS

Alcoholic Hallucinosis

The pathological basis of alcoholic hallucinosis closely resembles paranoid schizophrenia. (Taylor & Chandrasena, 2013). The possible mechanisms involved are changes in dopaminergic transmission systems and elevated levels of β-carbolines (Soyka, 1996). A 2000 FDG-PET study showed abnormally low thalamic functioning in alcoholics suffering from acute hallucinations (Soyka, Zetzsche, Dresel, & Tatsch, 2000). The thalamic dysfunction was postulated to be a functional rather than structural abnormality.

The regional blood flow in frontal lobes, left basal ganglia, and left thalamus was shown to decrease by using single-photon emission CT (Kitabayashi, Narumoto, Shibata, Ueda, & Fukui, 2007).

Central Pontine Myelinolysis

Central pontine myelinolysis, also called osmotic demyelination, is caused by nerve damage secondary to destruction of the myelin sheath in the pons. It is believed that the proapoptotic drive with inadequate energy provision seen in chronic alcoholism makes nerves susceptible to osmotic stress, decreasing their adaptive capacity in a hypo-osmotic hyponatremic environment (Ashrafian & Davey, 2001). The grid arrangement of the oligodendrocytes in the base of the pons limits their mechanical flexibility and their capacity to swell, leading to myelinolysis, making them prone to damage during sodium replacement. MRI findings on T1-weighted scans show hyperintensities in the pontine base. T2 images show round or triangular hyperintensities in the central pons.

APPLICATION TO OTHER ADDICTIONS AND SUBSTANCE MISUSE

Alcohol, nicotine, cannabis, and other psychostimulants, though having different chemical structures, seem to have similar neurochemical or neurobiological pathways through which they lead to addiction. Alcohol and tobacco use are highly correlated behaviors. Several common mechanisms may promote the combined use of alcohol and nicotine. There is a need for further research to find out the role of various neurobiological, genetic, learning, and conditioning processes and the underlying psychosocial factors contributing to the existence of this dual addiction. Also the increased prevalence of this concomitant behavior in mood disorders also leads to strategies to distinguish between independent mood episodes and substance-induced cognitive disturbances. Also gender differences should be studied in such cases. In the near future, individuals with substance-related problems will be characterized comprehensively by psychiatrists, psychotherapists, neurologists, and geneticists so as to plan the treatment process individually.

DEFINITION OF TERMS

Alcohol use disorder Alcohol use disorder is a condition characterized by the harmful consequences of repeated alcohol use, a pattern of compulsive alcohol use, and (sometimes) physiological dependence on alcohol (i.e., tolerance and/or symptoms of withdrawal).

Diffusion tensor imaging This is an advanced MRI technique that measures macroscopic axonal organization in the CNS.

Gliosis This is excessive proliferation of astroglial cells as a result of the healing process secondary to any injury in the CNS.

Glutamate excitotoxicity This refers to the death of central neurons after prolonged exposure to the toxic actions of glutamate or due to excessive influx of ions into the cell.

Neuropsychiatric disorder This is a psychiatric disorder attributable to an organic disease of the brain.

Neurotoxicity hypothesis This hypothesis states that chronic ethanol consumption results in various toxic effects by affecting various bodily physiological processes through glutamate excitotoxicity, disruption in neurogenesis, mitochondrial damage, and oxidative stress.

Neuropeptide Y This is a 36-amino-acid neuropeptide belonging to the pancreatic polypeptide family, with a potential role in the pathophysiology of mood and anxiety disorders, feeding behavior, alcohol use disorders, and sleep behavior.

Wallerian degeneration This is the process of antegrade degeneration of the axons and the myelin sheaths following proximal axonal or neuronal cell body lesions.

KEY FACTS

Key Facts about Wernicke Encephalopathy

- WE was first diagnosed in 1881 by German physician Carl Wernicke (1848–1905).
- Key features include ataxia, oculomotor abnormalities, and mental status disorder or confusion.
- Levels of THDP, the active form of intracellular thiamine are reduced. THDP is an essential cofactor of TK, PDH, and α-KGDH, which are key enzymes in glucose and amino acid metabolic pathways.
- Acute changes in the form of punctate hemorrhages are seen in mammillary bodies, third and fourth ventricles, and aqueduct of the midbrain.
- Chronic gliotic changes occur in the same areas with atrophy associated with widening of ventricles and the aqueduct.

Key Facts about Marchiafava–Bignami Disease

- MBD was first described by Carducci in 1898 and later by Italian pathologists Amico Bignami and Ettore Marchiafava in 1903.
- Symptoms include altered consciousness, emotional disturbances, depression, aggression, ataxia, apraxia, seizures, hemiparesis, and rarely death.
- There is selective demyelination of the corpus callosum and cortical laminar necrosis in the frontotemporal regions resulting in atrophy.
- MRI, the most sensitive diagnostic tool, shows hyperintense lesions on T2-phase and FLAIR images with hypointense lesions on T1WI of the corpus callosum and the genu and the splenium.
- Type A is characterized by altered consciousness, T2-hyperintense swelling of the entire corpus callosum on early MRI, and poor outcome. Type B is characterized by slight impairment of consciousness, partial callosal lesions on MRI, and a favorable outcome.

SUMMARY POINTS

- WE due to thiamine deficiency in chronic alcoholics and malnutrition is characterized by acute punctate hemorrhages seen in mammillary bodies, third and fourth ventricles, and aqueduct with capillary endothelial hypertrophy. Chronic changes include gliosis, atrophy, and ventricular dilatation.
- The main lesions seen in KS are in the periventricular and periaqueductal gray matter, the thalami, and the mammillary bodies. Neurodegeneration of the anterior principal nucleus of the thalamus differentiates Korsakoff patients from those with WE.
- Patients with MBD are characterized by selective demyelination of the corpus callosum and cortical laminar necrosis in frontotemporal regions.

- Alcohol facilitates GABA receptor function and blocks NMDA receptor activity. The adaptations within these two systems contribute to withdrawal-related symptoms, seizures, and neurotoxicity.
- In alcohol-related dementia, symptoms are seen as a result of the neuronal loss seen in the superior frontal association cortex, hypothalamus, basal ganglia, cerebellum, reticular activating system, and prefrontal cortex leading to memory disturbances and additional cognitive impairment.
- ACD is caused by irreversible toxic degeneration of the Purkinje cells in the anterior and superior portions of the vermis of the cerebellum.
- The possible mechanisms behind alcoholic hallucinosis are changes in dopaminergic transmission and elevated levels of β-carbolines.
- In central pontine myelinolysis, the proapoptotic drive with inadequate energy provisions seen in chronic alcoholism increases the susceptibility to osmotic stress.

REFERENCES

Aggleton, J. P., & Brown, M. W. (1999). Episodic memory, amnesia, and the hippocampal-anterior thalamic axis. *The Behavioral and Brain Sciences, 22*, 425–444.

Alexander-Kaufman, K., Cordwell, S., Harper, C., & Matsumoto, I. (2007). A proteome analysis of the dorsolateral prefrontal cortex in human alcoholic patients. *Proteomics. Clinical Applications, 1*, 62–72.

American Psychiatric Association (2013). Substance-related and addictive disorders. retrieved from http://www.dsm5.org/documents/substance%20use%20disorder%20fact%20sheet.pdf.

Ashrafian, H., & Davey, P. (2001). A review of the causes of central pontine myelinosis: yet another apoptic illness? *European Journal of Neurology, 8*, 103–109.

Baker, K., Harding, A., & Halliday, G. (1999). Neuronal loss in functional zones of the cerebellum of chronic alcoholics with and without Wernicke's encephalopathy. *Neuroscience, 91*, 429–438.

Bleich, S., Bandelow, B., & Javaheripour, K. (2003). Hyperhomocysteinemia as a new risk factor for brain shrinkage in patients with alcoholism. *Neuroscience Letters, 335*, 179–182.

Colchester, A., Kingsley, D., Lasserson, D., Kendall, B., Bello, F., Rush, C., ... Kopelman, M.D. (2001). Structural MRI volumetric analysis in patients with organic amnesia, 1: methods and findings, comparative findings across diagnostic groups. *Journal of Neurology, Neurosurgery and Psychiatry, 71*, 13–22.

Coy, J. F., Dübel, S., Kioschisa, P., Thomasb, K., Micklemb, G., Deliusa, H., & Poustkaa, A. (1996). Molecular cloning of tissue-specific transcripts of a transketolase-related gene: implications for the evolution of new vertebrate genes. *Genomics, 32*(3), 309–316.

Dash, V. B., & Kashyap, L. (1984). *Diagnosis and treatment of diseases in ayurveda*. New Delhi: Concept Publishing Company (Chapter VII, pp. 101).

De Witte, P. (2004). Imbalance between neuroexcitatory and neuroinhibitory amino acids causes craving for ethanol. *Addictive Behaviors, 29*, 1325–1339.

Dodd, P. R., Beckmann, A. M., Davidson, M. S., & Wilce, P. A. (2000). Glutamate-mediated transmission, alcohol and alcoholism. *Neurochemistry International, 37*, 509–533.

Freund, G., & Anderson, K. J. (1996). Glutamate receptors in the frontal cortex of alcoholics. *Alcoholism, Clinical and Experimental Research, 20*, 1165–1172.

N'Gouemo, P., & Morad, M. (2003). Ethanol withdrawal seizure susceptibility is associated with upregulation of L- and P-type Ca^{2+} channel currents in rat inferior colliculus neurons. *Neuropharmacology, 45*, 429–437.

Harding, A., Halliday, G., Caine, D., & Kril, J. J. (2000). Degeneration of anterior thalamic nuclei differentiates alcoholics with amnesia. *Brain, 123*(1), 141–154.

Haugbol, S. R., Ebert, B., & Ulrichsen, J. (2005). Upregulation of glutamate receptor subtypes during alcohol withdrawal in rats. *Alcohol and Alcoholism, 40*, 89–95.

Heinrich, A., Runge, U., & Khaw, A. V. (2004). Clinicoradiologic subtypes of Marchiafava-Bignami disease. *Journal of Neurology, 251*(9), 1050–1059.

Jacobson, R. R., & Lishman, W. (1987). Selective memory loss and global intellectual deficits in alcoholic Korsakoff's syndrome. *Psychological Medicine, 17*, 649–655.

Jacobson, R. R., & Lishman, W. A. (1990). Cortical and diencephalic lesions in Korsakoff's syndrome: a clinical and CT scan study. *Psychological Medicine, 20*, 63–75.

Jung, M. E., & Metzger, D. B. (2010). Alcohol withdrawal and brain injuries: beyond classical mechanisms. *Molecules, 15*(7), 4984–5011.

Kitabayashi, Y., Narumoto, J., Shibata, K., Ueda, H., & Fukui, K. (2007). Neuropsychiatric background of alcohol hallucinosis: a SPECT study. *Journal of Neuropsychiatry and Clinical Neuroscience, 19*, 85–89.

Kril, J. J., & Harper, C. G. (2012). Neuroanatomy and neuropathology associated with Korsakoff's syndrome. *Neuropsychology Review, 22*(2), 72–80.

Linkola, J., Ylikahri, R., Fyhrquist, F., & Wallenius, M. (1978). Plasma vasopressin in ethanol intoxication and hangover. *Acta Physiologica Scandinavica, 104*, 180–187.

Lisdahl, K. M., Thayer, R., Squeglia, L. M., McQueeny, T. M., & Tapert, S. F. (2013). Recent binge drinking predicts smaller cerebellar volumes in adolescents. *Psychiatry Research, 211*(1), 17–23.

Mark, G. P., & Finn, D. A. (2002). The relationship between hippocampal acetylcholine release and cholinergic convulsant sensitivity in withdrawal seizure-prone and withdrawal seizure-resistant selected mouse lines. *Alcoholism, Clinical and Experimental Research, 26*, 1141–1152.

Mayes, A. R., Meudell, P. R., Mann, D., & Pickering, A. (1988). Location of lesions in Korsakoff's syndrome: neuropsychological and neuropathological data on two patients. *Cortex; A Journal Devoted to the Study of the Nervous System and Behaviour, 24*, 367–388.

Nalini, A., Kovoor, J. M., Dawn, R., & Kallur, K. G. (2009). Marchiafava-Bignami disease: two cases with magnetic resonance imaging and positron emission tomography scan findings. *Neurology India, 57*, 644–648.

Nixon, P. F., & Jordan, L. (2008). Choroid plexus dysfunction: the initial event in the pathogenesis of Wernicke's encephalopathy and ethanol intoxication. *Alcoholism, Clinical and Experimental Research, 32*(8), 1513–1523.

Nixon, P. F., Kaczmarek, M. J., Tate, J., Kerr, R. A., & Price, J. (1984). An erythrocyte transketolase isoenzyme pattern associated with the Wernicke-Korsakoff syndrome. *European Journal of Clinical Investigation, 14*, 278–281.

Pappata, S., Chabriat, H., Levasseur, M., Legault-Demare, F., & Baron, J. C. (1994). Marchiafava-Bignami disease with dementia: severe cerebral metabolic depression revealed by PET. *Journal of Neural Transmission. Parkinson's Disease and Dementia Section, 8*, 131–137.

Patkar, A. A., Gopalakrishnan, R., Naik, P. C., Murray, H. W., Vergare, M. J., & Marsden, C. A. (2003). Changes in plasma noradrenaline and serotonin levels and craving during alcohol withdrawal. *Alcohol and Alcoholism, 38*, 224–231.

Rawat, J. P., Pinto, C., Kulkarni, K. S., Muthusamy, M. A. K., & Dave, M. D. (2014). Marchiafawa Bignami disease possibly related to consumption of a locally brewed alcoholic beverage: report of two cases. *Indian Journal of Psychiatry*, *56*(1), 76–78.

Reed, L. J., Lasserson, D., Marsden, P., Stanhope, N., Stevens, T., Bello, F., ... Kopelman, M. D. (2003). 18FDG-PET findings in the Wernicke-Korsakoff syndrome. *Cortex*, *39*, 1027–1045.

Rosenbloom, M., Sullivan, E. V., & Pfefferbaum, A. (2003). Using magnetic resonance imaging and diffusion tensor imaging to assess brain damage in alcoholics. *Alcohol Research and Health*, *27*(2), 146–152.

Roy, A., & Pandey, S. C. (2005). The decreased cellular expression of neuropeptide Y protein in rat brain structures during ethanol withdrawal after chronic ethanol exposure. *Alcoholism, Clinical and Experimental Research*, *26*, 796–803.

Sechi, G., & Serra, A. (2007). Wernicke's encephalopathy: new clinical settings and recent advances in diagnosis and management. *The Lancet Neurology*, *6*(5), 442–455.

Shen, R. Y. (2003). Ethanol withdrawal reduces the number of spontaneously active ventral tegmental area dopamine neurons in conscious animals. *The Journal of Pharmacology and Experimental Therapeutics*, *307*, 566–572.

Shimamura, A. P., Jernigan, T. L., & Squire, L. R. (1988). Korsakoff's syndrome: radiological (CT) findings and neuropsychological correlates. *Journal of Neuroscience*, *8*, 4400–4410.

Soyka, M. (November 1996). Alcohol induced hallucinosis. Clinical aspects, pathophysiology and therapy. *Nervenarzt*, *67*(11), 891–895.

Soyka, M., Zetzsche, T., Dresel, S., & Tatsch, K. (2000). FDG-PET and IBZM-SPECT suggest reduced thalamic activity but no dopaminergic dysfunction in chronic alcohol hallucinosis. *The Journal of Neuropsychiatry and Clinical Neurosciences*, *12*, 287–288.

Sullivan, E. V., Deshmukh, A., Desmond, J. E., Lim, K. O., & Pfefferbaum, A. (July 2000). Cerebellar volume decline in normal aging, alcoholism, and Korsakoff's syndrome: relation to ataxia. *Neuropsychology*, *14*(3), 341–352.

Taylor, J. E., & Chandrasena, R. D. (2013). Chronic alcoholic hallucinosis masquerading as schizophrenia: a case report. *The Primary Care Companion for CNS Disorders*, *15*(4), PCC.13l01523. http://doi.org/10.4088/PCC.13l01523.

Thorarinsson, B. L., & Olafsson, E. (2011). Wernicke's encephalopathy in chronic alcoholics. *Laeknabladid*, *97*(1), 21–29.

Traviesa, D. C. (1974). Magnesium deficiency: a possible cause of thiamine refractoriness in Wernicke-Korsakoff encephalopathy. *Journal of Neurology, Neurosurgery, and Psychiatry*, *37*(8), 959–962.

Victor, M., Adams, R. D., & Mancall, E. L. (1959). A restricted form of cerebellar cortical degeneration occurring in alcoholic patients. *AMA Archives of Neurology*, *1*(6), 579–688.

Victor, M., & Adams, R. D. (1971). The Wernicke-Korsakoff syndrome. A clinical and pathological study of 245 patients, 82 with post-mortem examinations. *Contemporary Neurology Series*, *7*, 1–206.

Walker, D. W., Barnes, D. E., Zornetzer, S. F., Hunter, B. E., & Kubanis, P. (1980). Neuronal loss in hippocampus induced by prolonged ethanol consumptions in rats. *Science*, *209*, 711–713.

Wernicke, C. (1881). Die akute haemorrhagische Polioencephalitissuperior. *Lehrbuch der Gehirnkrankheiten für Ärzte und Studierende* (Bd II. Kassel: Fischer Verlag, 229–242.

Yokota, O., Tsuchiya, K., Terada, S., Oshima, K., Ishizu, H., Matsushita, M., ... Akiyama, H. (2006). Frequency and clinicopathological characteristics of alcoholic cerebellar degeneration in Japan: a cross-sectional study of 1,509 postmortems. *Acta Neuropathologica*, *112*, 43–51.

Zimmermann, U., Hundt, W., Spring, K., Grabner, A., & Holsboer, F. (2003). Hypothalamic-pituitary-adrenal system adaptation to detoxification in alcohol-dependent patients is affected by family history of alcoholism. *Biological Psychiatry*, *53*, 75–84.

Chapter 59

Social Isolation and Ethanol Drinking: A Preclinical Model of Addiction Vulnerability in Males, but Not Females

Tracy R. Butler[1], Jeffrey L. Weiner[2]
[1]Department of Psychology, University of Dayton, Dayton, OH, USA; [2]Department of Physiology and Pharmacology, Wake Forest University School of Medicine, Winston–Salem, NC, USA

Abbreviations

ACTH Adrenocorticotropic hormone
CORT Corticosterone
DSM-IV *Diagnostic and Statistical Manual of Mental Disorders*, fourth edition
GABA γ-Aminobutyric acid
GH Group housed
HPA Hypothalamic–pituitary–adrenal
PND Postnatal day
SI Socially isolated

INTRODUCTION

Genetic and environmental factors influence the development of alcohol dependence, and in particular, early life stress in humans is an environmental condition that has been shown to affect gene expression and render individuals more vulnerable to developing addictive disorders (Enoch, 2011). Modeling this, adolescent social isolation in male rats recapitulates many important features of alcohol dependence and is emerging as an important model of addiction vulnerability in male rats. However, this model does not engender the same phenotype in female rats. Further examination of the clinical literature highlights myriad behavioral and neurobiological sex differences in models of alcohol dependence, early life stress, and hypothalamic–pituitary–adrenal (HPA) axis responsivity. Indeed, healthy men and women show differential responsivity to a pharmacological challenge (Lovallo, Farag, Sorocco, Cohoon, & Vincent, 2012), thus making it unsurprising that sex differences would also emerge in models of addiction vulnerability. For instance, adolescent social isolation in male rats results in enduring increases in a range of behaviors that have been associated with increased risk of alcoholism, such as anxiety-like behavior and voluntary ethanol self-administration, and is proving to be a useful model to study neurobiological correlates of addiction vulnerability in male subjects. Unfortunately, this model has shown equivocal results in female subjects and as of this writing, no studies have identified an effective model to study the impact of early life stress on addiction vulnerability in females. The focus of this chapter is to review sex differences in models of early life stress with the hope that providing a fuller understanding of these differences may provide a clearer path toward the development of an effective model in females.

SEX DIFFERENCES: FROM ACUTE RESPONSE TO ALCOHOL TO NEUROBIOLOGICAL CONSEQUENCES OF ALCOHOL DEPENDENCE

As of this writing, the majority of the preclinical research that has been devoted to studying the neural substrates of alcohol, affective disorders, and their comorbidity has utilized male subjects. Importantly, however, some human epidemiological studies report greater rates of comorbidity between alcohol use disorders and psychiatric disorders in females than in males (Merikangas et al., 1998). A number of sex/gender differences have also been identified in objective, subjective, and neurobiological consequences related to alcohol and alcohol use disorders (Nolen-Hoeksema, 2004). For instance, nondependent men and women show different levels of impairment under acute alcohol on objective laboratory measures (Weafer, Miller, & Fillmore, 2010), and heavily drinking women show greater alcohol craving and anxiety following exposure to stress cues versus neutral cues than do men (Hartwell & Ray, 2013). Women also show a shorter time period in the transition from alcohol abuse to dependence, a phenomenon termed "telescoping" (Piazza, Vrbka, & Yeager, 1989). Consequences of alcohol abuse may also be sex-dependent, as studies have demonstrated greater loss of brain matter and vulnerability to neurotoxicity in brains of female subjects (Butler, Smith, Self, Braden, & Prendergast, 2008; Butler, Smith, Berry, Sharrett-Field, & Prendergast, 2009; Hommer et al., 1996; Hommer, Momenan, Kaiser, & Rawlings, 2001). These converging lines of evidence suggest that factors that are related to vulnerability for alcoholism differ in males and females, which can precede greater neurobiological insult in women who develop alcoholism. Despite these differences and the significant number of women who meet criteria for lifetime prevalence of alcohol dependence

(Hasin & Grant, 2004), much less work exists using female rodents to study neurobiological correlates of behaviors related to vulnerability for alcohol-use disorders or persistent alcohol use.

EPIDEMIOLOGY BY SEX: ASSOCIATION WITH ALCOHOL USE DISORDERS

Approximately 12.5% of individuals sampled in the 2001–2002 US National Epidemiologic Survey on Alcohol and Related Conditions met the *Diagnostic and Statistical Manual of Mental Disorders*, fourth edition, criteria for lifetime alcohol dependence, with a current prevalence rate of 9%, which equals approximately 17.6 million Americans (Hasin & Grant, 2004). Rates of comorbidity between alcohol use disorders and mood or anxiety disorders are 20% and 18%, respectively, which is nearly twice that of individuals without alcohol use disorders (Grant et al., 2004). Even more striking, 78% of alcohol-dependent men and 86% of alcohol-dependent women have a comorbid psychiatric disorder (Castle, 2008; Kessler et al., 1997). Individuals with comorbid addiction and psychiatric diagnoses are less responsive to treatment and show poorer prognosis (Kranzler & Rosenthal, 2003). Diagnosis of a mood or anxiety disorder can contribute to risk, maintenance, and relapse in alcohol-dependent individuals. In particular, chronic early life stress has been shown to be a significant risk factor for alcohol dependence (Pilowsky, Keyes, & Hasin, 2009).

SEX DIFFERENCES: NEUROENDOCRINE RESPONSES IN VULNERABLE AND ADDICTED POPULATIONS

HPA axis dysfunction in alcohol-dependent men has been noted in several studies, hallmarked by hyporesponsiveness, even after prolonged abstinence (Adinoff, Iranmanesh, Veldhuis, & Fisher, 1998; Adinoff et al., 2005; Costa et al., 1996). Evidence suggests that HPA axis dysfunction predisposes and contributes to maintenance of alcohol use disorders (Eames et al., 2014). Indeed, cortisol levels have been shown to be elevated in still intoxicated subjects (during withdrawal) as well as in abstinent individuals. Using a similar paradigm in female alcohol-dependent individuals, however, shows no difference in HPA axis responsivity or basal cortisol/adrenocorticotropic hormone (ACTH) levels in alcohol-dependent women versus control subjects. Comparing three separate age groups and four levels of drinkers (none, light, heavy, and dependent) Gianoulakis, Dai, and Brown (2003) observed significant sex differences in HPA axis responses. Independent of age or drinking status, females had lower basal ACTH and β-endorphin than males, with no effect of phase of reproductive cycle on levels of ACTH, cortisol, or β-endorphin. Nondrinking and light-drinking females (ages 18–29) had higher cortisol levels than their male counterparts, but at ages 30–44, nondrinkers (independent of sex) had significantly lower cortisol than all drinking groups. These clinical data highlight significant effects of developmental stage, as well as sex and drinking status, on stress-related measures. Further, these sex-dependent differences in HPA axis functioning may have clinically significant implications, as alleviation of HPA axis dysfunction has been suggested as a treatment option for alcohol-dependent individuals (Prendergast & Mulholland, 2012). Most interestingly, in a large clinical study, Garbutt et al. (2005) showed that long-acting naltrexone, a US Food and Drug Administration-approved drug for alcohol dependence, was highly effective in reducing heavy drinking in alcohol-dependent men, but not women. This sex-dependent effect of naltrexone may be due, in part, to opioid modulation of HPA axis function. Indeed, Lovallo et al. (2012) showed that naltrexone elicited a significant cortisol response in women but not men. This suggests that women have greater opioid inhibition of basal HPA axis activity, and/or that the cortisol response elicited by naltrexone may mitigate potential therapeutic effects in women.

SOCIAL ISOLATION AS A MODEL OF ADDICTION VULNERABILITY IN MALE RODENTS

It has long been known that socially isolating rats during adolescence can result in profound and enduring behavioral and neurobiological alterations (for review see Hall, 1998; Robbins, Jones, & Wilkinson, 1996). Studies have begun to examine the utility of employing adolescent social isolation as a model of alcohol addiction vulnerability. For example, Whitaker, Degoulet, and Morikawa (2013) recently found that housing male Sprague–Dawley rats in single cages for a 3-week window in early adolescence significantly enhanced conditioned place preference for ethanol and amphetamine, relative to rats that were group housed during this critical developmental period. Social isolation was also associated with a significant enhancement of glutamatergic synaptic plasticity in the ventral tegmental area, a brain region known to play an integral role in the etiology of addictive disorders. Other studies have also shown increases in ethanol drinking and associated neurobiological changes in male mice in a model of adolescent social isolation (Lopez, Doremus-Fitzwater, & Becker, 2011), although female mice did not show the same effect of increased drinking in response to social isolation. Further, these data showed differential responses in male and female group-housed and socially isolated mice exposed to chronic variable stress, such that stressed male mice generally had greater ethanol intake versus control males, and stressed female mice generally had less ethanol intake versus control females.

Our model of social isolation uses male Long Evans rats procured from a supplier immediately postweaning (postnatal day (PND) 21), which are then housed in cohorts of four (group housed; GH) or housed individually (socially isolated; SI) for 6 weeks (PND 28–72) (Figure 1). This form of chronic adolescent stress results in numerous behavioral alterations associated with increased risk of alcoholism, including (1) hyperactivity in a novel open-field environment, (2) greater anxiety-like behavior on the elevated-plus maze, (3) impaired prepulse inhibition, (4) HPA axis dysfunction, (5) greater ethanol intake and preference in a home-cage, self-administration paradigm and an operant ethanol self-administration procedure, and (6) deficits in fear extinction (Figure 2) (Butler, Ariwodola, & Weiner, 2014; Chappell et al., 2013; McCool & Chappell, 2009; Skelly, Chappell, Carter, & Weiner, 2015). Notably, several of these alterations persist for months into adulthood, even when SI animals are compared with adolescent GH animals that were then isolated in adulthood (Chappell et al., 2013; Yorgason, Espana, Konstantopoulos, Weiner, & Jones, 2013). A recent study by Ravenelle, Neugebauer, Niedzielak, and Donaldson (2014) showed that the effect of SI or environmental enrichment (GH plus exposure to toys) on anxiety-like behavior in Long Evans rats differs in animals bred for a low- or high-anxiety phenotype;

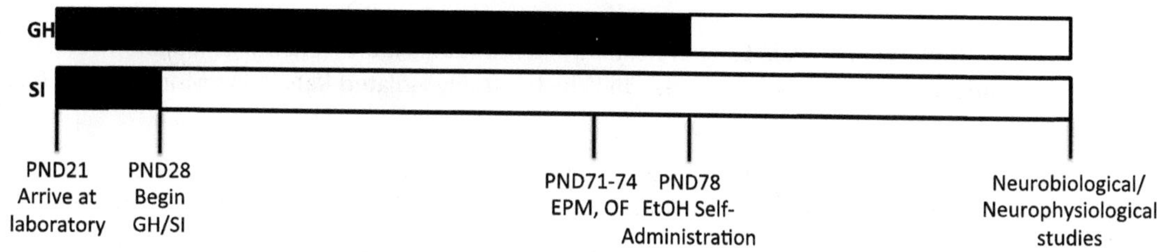

FIGURE 1 Typical experimental procedural timeline in social isolation model using male Long Evans rats (e.g., Butler, Ariwodola et al., 2014; Chappell, Carter, McCool, & Weiner, 2013). EPM, elevated-plus maze; OF, open field.

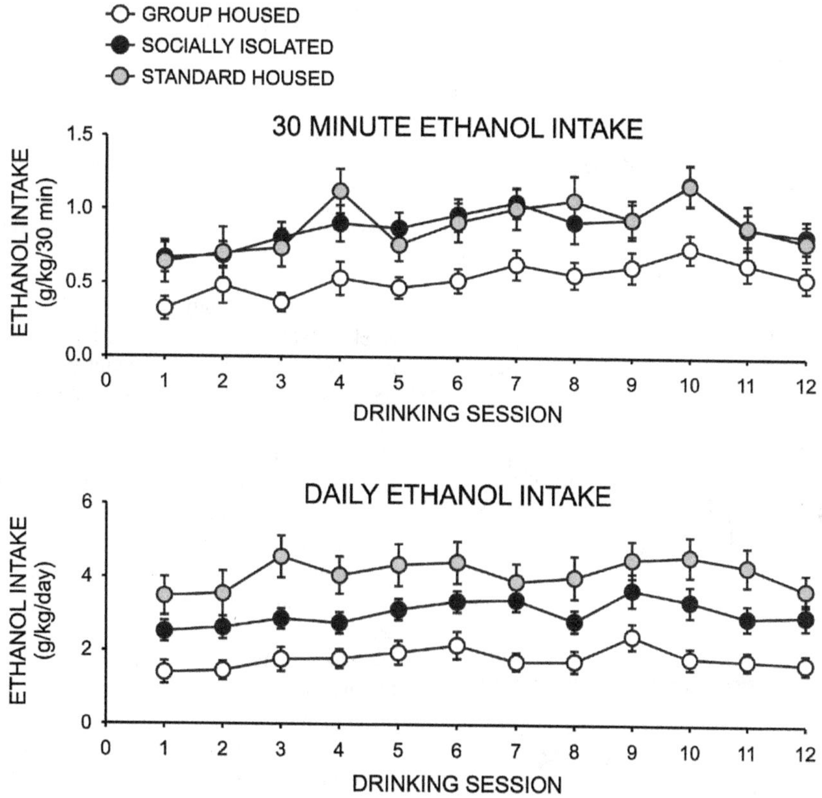

FIGURE 2 Twenty-four-hour ethanol intake in male Long Evans SI/GH rats over 4 weeks in a home-cage two-bottle choice intermittent-access ethanol-intake model. In this study, GH and SI rats were also compared to standard-housed rats that were procured from the supplier as adults before social isolation and ethanol intake. *Previously published in Chappell et al. (2013).*

that is, enrichment was successful in reducing anxiety-like behavior only in those rats that had high anxiety, with no further anxiolytic effect in low-anxiety rats. This behavioral phenotype of addiction vulnerability has proven to be robust and reproducible in male subjects, and it has provided a model in which to study neurobiological correlates of these behaviors (Table 1). To that end, SI rats have disrupted dopamine release kinetics in the nucleus accumbens (Yorgason et al., 2013), enhanced dopamine and norepinephrine release in response to ethanol administration in the nucleus accumbens (Karkhanis, Locke, McCool, Weiner, & Jones, 2014) and basolateral amygdala (Karkhanis, Alexander, McCool, Weiner, & Jones, 2015), and impaired plasticity and increased neuronal excitability in the basolateral amygdala (Rau et al., 2015; Yorgason et al., 2013). SI rats, but not GH rats, also show a correlation between ethanol intake and corticosterone level and between anxiety-like behavior and corticosterone level (Butler, Ariwodola, et al., 2014).

Surprisingly, using the identical experimental procedure, female Long Evans rats do not show the same behavioral phenotype (Figure 3) (Butler, Carter, & Weiner, 2014). That is, group differences are not present when GH/SI females are tested for locomotion in a novel open-field chamber or assessed for anxiety-like behavior in the elevated-plus maze. Although SI females exhibited transient increases in measures of ethanol intake and preference compared to GH rats early in the self-administration procedure, these differences do not persist. Moreover, whereas initial measures of anxiety-like behavior correlated with ethanol intake in male GH and SI rats, these measures were not correlated in females (Butler, Carter, et al., 2014; Chappell et al., 2013). Lukkes, Engelman, Zelin, Hale, and Lowry (2012) also failed to show hyperactivity in SI female Sprague–Dawley rats (21 days SI) or increased anxiety-like behavior in a social interaction test. Interestingly, Lopez et al. (2011) showed that SI male mice consistently drank more ethanol than their GH counterparts, whereas SI female mice did not. It should be noted, however, that another study did report greater anxiety-like behavior in female rats following adolescent social isolation (Pisu et al., 2013), although a different rat strain

TABLE 1 Representative Studies Using Male Long Evans Rats in a Model of Adolescent Social Isolation versus Group Housing as a Model of Chronic Early Life Stress, in Which Socially Isolated Rats Drink More Ethanol than Group-Housed Rats

Author (Year)	Rodent Strain	SI Period (PND)	Behavioral Test and Outcome for SI Rodents	Physiological Test and Outcome for SI Rodents
Butler, Arivodola et al. (2014)	Long Evans	6 weeks (28–72)	EPM: n.d. OF: Hyperactive EtOH SA: Greater intake and preference	Baseline CORT: n.d. DST: Failure to suppress CORT BLA fEPSP: n.d.
Chappell et al. (2013)	Long Evans	6 weeks (28–72)	EPM: Greater anxiety-like behavior OF: Hyperactive Response to novelty: n.d. EtOH: Greater intake	n.s.
Yorgason et al. (2013)	Long Evans	6–7 weeks (28–77)	EPM: Greater anxiety-like behavior	Voltammetry in NAc: increased DA release and uptake

BLA, basolateral amygdala; DA, dopamine; DST, dexamethasone suppression test; EPM, elevated-plus maze; EtOH, ethanol; OF, open field; fEPSP, field excitatory postsynaptic potential; NAc, nucleus accumbens; n.s., not studied; n.d., no difference SI versus GH; SA, self-administration.

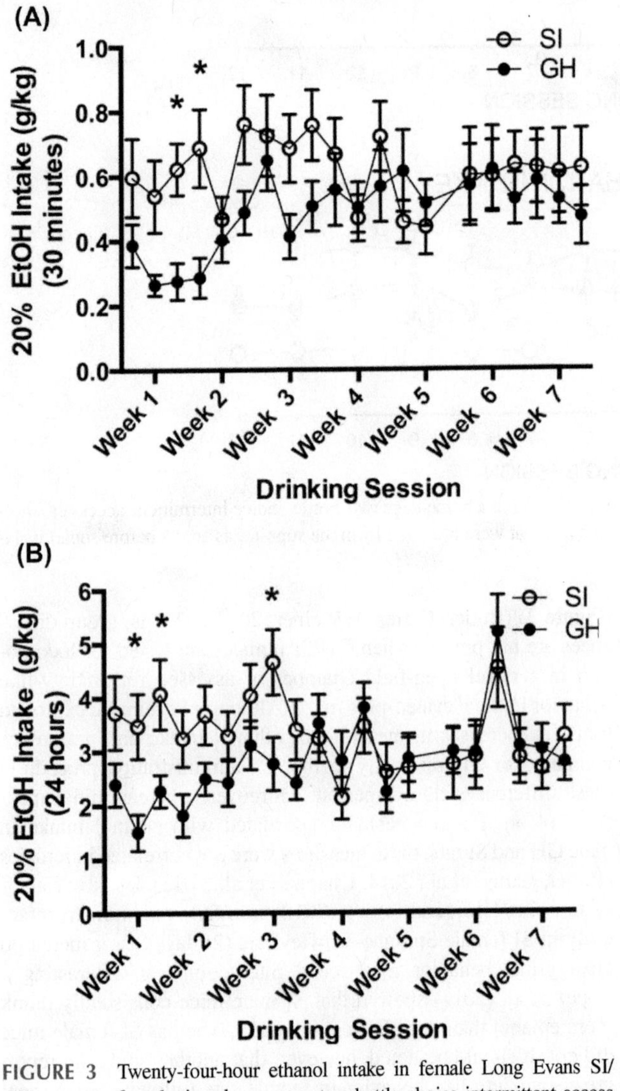

FIGURE 3 Twenty-four-hour ethanol intake in female Long Evans SI/GH rats over 6 weeks in a home-cage two-bottle choice intermittent-access ethanol-intake model. *Previously published in Butler, Ariwodola, et al. (2014).*

was used in that study (Sprague–Dawley). Advani, Hensler, and Koek (2007) showed increased ethanol preference in SI male mice, whereas increased ethanol preference in female mice was observed only in those females that were subjected to social isolation *and* maternal separation in early life. Female Wistar rats have also been shown to be resistant to the anxiogenic effects associated with chronic intermittent stress, whereas male Wistar rats do show greater anxiety-like behavior following the same protocol (Mitra, Vyas, Chatterjee, & Chattarji, 2005). Taken together, these data suggest that female subjects are markedly more resilient to a variety of chronic stressors that have a long-lasting impact on male rats. Importantly, the relationship between early life stress-induced increases in anxiety-like behavior and ethanol self-administration that has been consistently reported in male rats and mice is rarely observed in females.

One possible explanation for the observed sex differences in these models, however, is that they are not optimized for testing the behavioral parameter intended. For instance, the validity of the elevated-plus maze as a measure of anxiety-like behavior is based on many studies showing that anxiogenic drugs decrease open-arm time, and anxiolytic drugs (including the clinical gold standard benzodiazepine, diazepam) increase open-arm time in male rats (Pellow, Chopin, File, & Briley, 1985). However, when tested using a standard protocol that would be used in male rats, female Sprague–Dawley and Long Evans rats fail to show less anxiety-like behavior following diazepam administration (Ravenelle et al., 2014; Simpson, Ryan, Curley, Mulcaire, & Kelly, 2012). Nevertheless, the absence of robust and enduring increases in ethanol intake in female rats subjected to adolescent stress suggests that this model may have very limited utility to address the neurobiological substrates linking increases in anxiety measures and ethanol intake in females.

HORMONES IN MALE AND FEMALE RODENTS EXPOSED TO EARLY LIFE STRESS

Another factor that could contribute to sex differences is hormonal milieu, which not only differs between sexes, but also undergoes considerable changes during development. The stress hyporesponsive period occurs in rats immediately after birth, during which time HPA functioning has not fully matured to respond to stressors. It is also well accepted that adolescence is a period of profound changes in hormones, brain development, and behavior, in humans as well as in rodents. Particularly, negative feedback mechanisms that are important in physiological regulation of HPA axis function are not fully developed at the time of weaning (Vazquez, Morano, Lopez, Watson, & Akil, 1993), resulting in a different pattern of hormonal response following stress exposure compared to adults. A mature, or adult-like acute response to stress as measured by increases in ACTH and corticosterone (CORT) is present by the time of weaning (Goldman, Winget, Hollingshead, & Levine, 1973). However, development of the negative feedback system continues from weaning into adulthood (Goldman et al., 1973). Some data do suggest that ACTH and CORT levels poststress vary by stage of estrous cycle in adult female Long Evans rats (both higher during proestrus vs estrus and diestrus) (Viau & Meaney, 1991). Postpuberty, some behaviors may be influenced by sex hormones in females, although results are not consistent across laboratories. In adult rats, some studies have shown that female rats show less anxiety-like behavior during proestrus (Molina-Hernandez, Olivera-Lopez, Patricia Tellez-Alcantara, Perez-Garcia, & Teresa Jaramillo, 2006) and anxiolytic drugs are more effective during proestrus compared to metestrus/diestrus (Molina-Hernandez, Tellez-Alcantara, Olivera-Lopez, & Jaramillo, 2013), although several studies also report no differences in anxiety-like behavior across the stages of the estrous cycle (Bitran, Hilvers, & Kellogg, 1991; Nomikos & Spyraki, 1988) (Table 2).

Much evidence has shown that behaviors during adulthood can be influenced by stressful experiences in adolescence, which can be independent or dependent upon gonadal hormone modulation of HPA axis function. Wilkin, Waters, McCormick, and Menard (2012) showed that female rats exposed to intermittent stress during early adolescence exhibited greater anxiety-like behavior compared to female rats exposed to stress during mid-adolescence. Romeo, Lee, and McEwen (2004) showed that levels of stress hormones after restraint stress are independent of ovarian hormones in females (control vs ovariectomized rats), although CORT levels are higher poststress in adolescent than in adult females, which is probably explained by maturation of negative feedback mechanisms. Further, male prepubertal and adult rats have similar increases in levels of CORT and ACTH, but hormone levels take significantly longer to return to baseline in adolescent rats. It is suggested that this effect in males is attributable to increases in testosterone that are observed postpuberty and have been shown to exert inhibitory effects on HPA axis responses. Importantly, however, the adrenal glands are also a source of progesterone, as it has been shown that not only do progesterone levels increase in response to stress and drug challenge, but ovariectomized rats still have measureable levels of progesterone (Walker, Francis, Cabassa, & Kuhn, 2001).

SEX DIFFERENCES IN ALCOHOL MODELS: FEMALES DRINK MORE!

Across various strains of rats and mice, female subjects tend to consume more ethanol (as measured in grams per kilogram of body weight) and show greater ethanol preference (Lancaster, Brown, Coker, Elliott, & Wren, 1996). Females also tend to drink more postpuberty than as juveniles (Lancaster et al., 1996). The data are equivocal as to whether ethanol intake and/or preference varies by stage of estrous cycle. Female rats have been shown to be less sensitive to an acute, sedating dose of ethanol during diestrus and proestrus compared to males, although drinking levels were not less than those observed during other stages of the estrous cycle (Cha et al., 2006). Drinking pattern has been shown to vary across the estrous cycle without changing total ethanol (g/kg) intake (Ford et al., 2002b). For instance, intake in an operant self-administration paradigm has been shown to be similar in freely cycling female rats across the phases of the estrous cycle (Roberts et al., 1998). However, it cannot be completely ruled out that hormones can affect ethanol intake, as exogenous introduction of hormones in intact female rats has been shown to affect ethanol intake (Ford et al., 2002a).

MODELS OF AN ADDICTION-VULNERABLE PHENOTYPE IN FEMALE RATS

Although SI does not produce the same addiction-vulnerable phenotype in female rats as in male rats, there is some evidence that other types of stressors may engender such a phenotype in females. For instance, female rats appear more vulnerable to an addiction-like phenotype following chronic social instability stress during adolescence (Figure 4). That is, as adults, females that were exposed to social instability stress via multiple novel cage partner pairings during development showed greater anxiety-like behavior (McCormick, Smith, & Mathews, 2008). Importantly, this effect was not apparent immediately after the chronic social stress experience, but emerged later in adulthood. Social instability stress also produces locomotor sensitization to nicotine and amphetamines in female rats (Mathews, Mills, & McCormick, 2008; McCormick, Robarts, Gleason, & Kelsey, 2004). Social instability stress exposure during adolescence has also been shown to affect anxiety-like behavior in adulthood in an estrous-cycle stage-dependent manner. And interestingly, although female anxiety-like behavior was affected, CORT levels in females were not changed (McCormick et al., 2008). McCormick, Merrick, Secen, and Helmreich (2007) also showed that social instability stress does not alter CORT levels at PND 45 in females, although blunted CORT levels are observed in males. These data are consistent with the data from human alcoholics, in which males show blunted HPA axis responsivity but females do not. Thus, a model of chronic social instability in females holds promise as a potential model of behavioral correlates of addiction vulnerability, although much work must be done to evaluate important parameters such as the timeline of the critical period during development and when these behaviors emerge, effects of estrous cycle stage, and parameters of behavioral tests. It will also be critical to determine if these models engender long-lasting increases in measures of ethanol drinking.

SEX DIFFERENCES IN ADDICTION VULNERABILITY: FUTURE DIRECTIONS

In summary, a number of studies have demonstrated the utility of adolescent social isolation as a preclinical model of addiction vulnerability in male rats and have used this model to begin to identify key neurobiological substrates that may contribute to an increased risk of addiction. It is important to continue these lines of study; however, just as important is the need for developing equally effective preclinical models using female subjects. Clinical data clearly

TABLE 2 Effects of Stage of Estrous Cycle on Ethanol Self-Administration

Author (Year)	Rodent Strain (PND)	Stages Assessed	Comparison Group	Drug; Model (% EtOH)	Drug Outcome
Advani et al. (2007)	C57BL/6J mice (PND2–60)	None	Maternal separation/social isolation; males	At PND60; 2-bottle choice 4 days/week (5–20%)	EtOH preference in females increased only in rats exposed to both separation and isolation; EtOH intake was decreased by isolation in females
Bertholomey, Henderson, Badia-Elder, and Stewart (2011)	P rats (12–16 weeks)	None	1 h restraint stress during 3 weeks IA	Continuous free-choice access (15%) or IA (5 days on, 2 days off)	Stress did not alter EtOH intake or preference
Butler, Carter et al. (2014)	Long Evans (28–72)	None	Socially isolated versus group housed	Home-cage 2-bottle choice, intermittent access (20% EtOH)	GH>SI at early time points; later, n.d.
Cha, Li, Wilson, and Swartzwelder (2006)	Adult; Sprague–Dawley (60–75)	Swabs daily; proestrus, estrus, metestrus, diestrus	Males	Acute 5 g/kg	Sleep time: adult females less sleep time versus adult males; adult females less sleep time during proestrus and diestrus versus estrus and metestrus; fewer sIPSCs in females after EtOH versus males
Chester, Barrenha, Hughes, and Keuneke (2008)	HAP1 mice, adolescent and adult	None	Foot shock (30 min; 10 days)	2-bottle choice, continuous 30 days	Stress as adolescent led to greater intake and preference as adult versus control; stress as adult did not affect intake or preference
Ford, Eldridge, and Samson (2002b)	Long Evans (adult; 185–220 g)	Estrus, diestrus (early and late), proestrus	Cycle phase	Operant self-administration, voluntary access (10% EtOH; 10 weeks)	No effect of phase on intake
Ford, Eldridge, and Samson (2002a)	Long Evans (adult; 190–220 g)	–	Sham, OVX + E2, P	Operant self-administration (10%; 14 days)	Exogenous administration of estradiol increased ethanol intake in sham rats
Lancaster and Spiegel (1992)	Long Evans	None	Males	15 days (5–10%)	Females had greater g/kg intake and preference versus males
Lancaster et al. (1996)	Sprague–Dawley (30–60)	None	Males	Home-cage 2-bottle choice, free daily access	Females had greater intake and preference as adults than adolescents
Lopez et al. (2011)	C57BL/6 mice (adult)	None	Social isolation, chronic variable stress	Limited-access (2 h) 2-bottle choice	Isolated as adolescents consistently drink more than their GH counterparts; female mice isolated in adulthood drank significantly less than mice group housed in adulthood
Roberts, Smith, Weiss, Rivier, and Koob (1998)	Wistar (180–200 g)	Estrus, diestrus (early and late), proestrus	Cycle phase	Operant self-administration	Reduced EtOH intake during estrus and proestrus

E2, estradiol; HAP1, high-alcohol-preferring 1 line; IA, intermittent access; sIPSCs, spontaneous inhibitory postsynaptic currents; OVX, ovariectomy; P, alcohol-preferring.

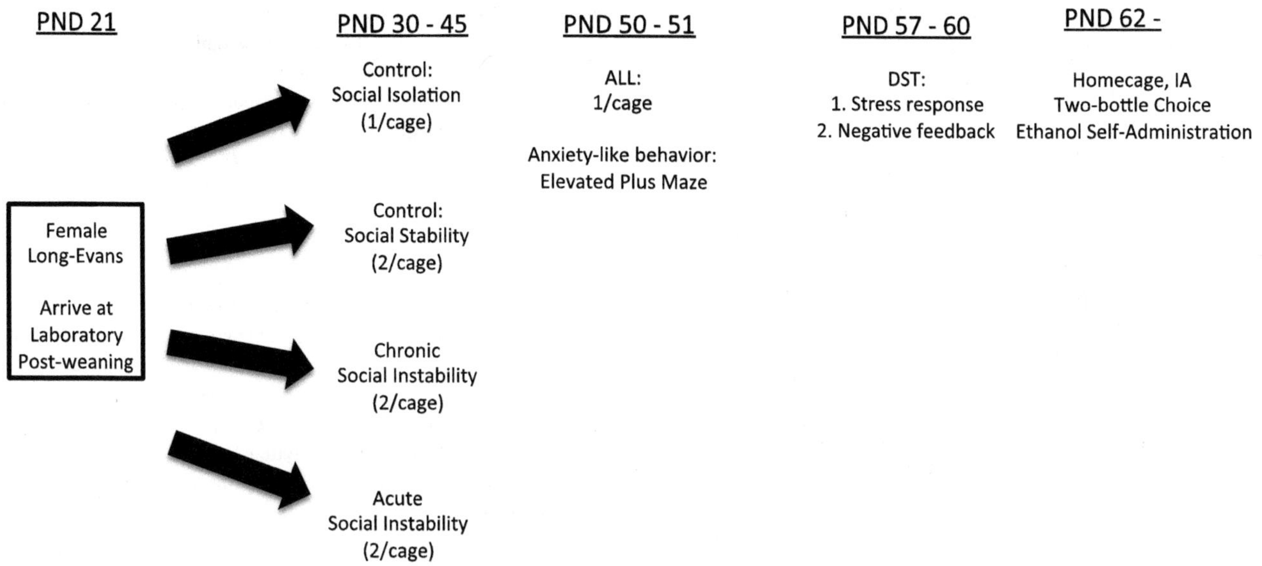

FIGURE 4 Proposed model of chronic social instability stress in female Long Evans rats. DST, dexamethasone suppression test. *Adapted with modifications from Chronic Social Instability Stress (after McCormick et al., 2008).*

support that high rates of comorbidity between addiction and affective disorders exist in females, demanding greater understanding of the neurobiological substrates responsible for this relationship. To that end, current studies are under way to optimize parameters of current models of anxiety-like behavior in female rats, as well as to explore models that engender greater anxiety-like behavior and stress responses in female rats to study their association with future alcohol intake, preference, and motivation for alcohol, as well as the neurobiological alterations that contribute to these behaviors. It will be important to determine what consistent relationships among anxiety-like behaviors, stress measures, and measures of ethanol drinking emerge using any of these models. Then, it will be important to determine if neurobiological correlates identified in male SI rats are also observed in females subjected to stressors that do result in increased anxiety and ethanol drinking. Such brain targets will include dopaminergic systems in the nucleus accumbens, glutamatergic systems in the ventral tegmental area, and glutamatergic and γ-aminobutyric acid-ergic systems in the amygdala.

APPLICATIONS TO OTHER ADDICTIONS AND SUBSTANCE MISUSE

Social isolation has also been shown to engender some behavioral attributes of an addiction-like phenotype to drugs of abuse other than alcohol (for review, see Robbins et al., 1996). For instance, psychostimulant-related behavior appears sensitive to rearing environment. SI rats have been shown to respond to and ingest more cocaine (Peitz et al., 2013) and show greater amphetamine-induced hyperactivity (Cain, Mersmann, Gill, & Pittenger, 2012); and a history of chronic social isolation and methylphenidate leads to greater alcohol intake and protracted increases in anxiety-like behavior (Gill, Chappell, Beveridge, Porrino, & Weiner, 2014).

DEFINITION OF TERMS

Addiction vulnerability This describes behavioral characteristics known to be correlated with greater risk for the development of addictive disorders in adulthood in the human population.
Adolescence Broadly defined in the rat, this critical period of development spans approximately PND 21–60.
Corticosterone This is the primary stress hormone in rodents; it is released following stress exposure in response to ACTH from the adrenal gland to affect peripheral and central processes.
Dexamethasone suppression test Dexamethasone is a synthetic glucocorticoid that upon administration engages negative feedback of the HPA axis. This results in reduced levels of circulating CORT in healthy subjects.
Estrous cycle This is the reproductive cycle in female rats, which lasts approximately 4–6 days and is divided into phases defined by hormone levels. In the laboratory, phases are most readily defined by the ratio of cell types observed from vaginal swabs.
Hypothalamic–pituitary–adrenal axis This is a system of central nuclei, peripheral organs, and hormones that constitutes the primary stress response in mammals.
HPA axis dysfunction The HPA axis responds to stress and is terminated via negative feedback. HPA axis dysfunction can refer to a blunted stress response to an acute stressor and inability to cease the stress response.
Group housing In relation to animal housing, this refers to living in a same-sex cohort in a much larger cage with two or three conspecifics.
Social instability In relation to animal housing, this refers to daily housing with a novel same-sex partner.
Social isolation In relation to animal housing, this refers to living alone in a standard-sized laboratory cage in the same colony room as other animals in the cohort.

KEY FACTS OF CONSEQUENCES FOLLOWING ADOLESCENT SOCIAL ISOLATION MODEL IN MALE VERSUS FEMALE RATS

- The elevated-plus maze is a well-validated model of anxiety-like behavior in male rats, with males showing greater time in the open arms of the maze after administration of benzodiazepines (e.g., diazepam). However, diazepam failed to show this same effect in female rats on the elevated-plus maze, suggesting that it may not be the most effective model for female anxiety-like behavior, or parameters on this model should be adjusted for use with females (Pellow et al., 1985).
- The most consistent finding across labs and various models of SI/GH is greater locomotion of SI rats in a novel environment relative to GH rats. However, female rats have not consistently shown this same response.
- Similar effects of SI have been noted in other rat strains as well as mice (males) (e.g., Gill et al., 2014).
- Greater alcohol intake in SI rats has been shown to result in pharmacologically relevant levels, which is not always observed in outbred male rats.
- Neurobiological consequences of SI include impaired plasticity and disrupted transporter kinetics, as well as slightly impaired negative feedback of the HPA axis.

SUMMARY POINTS

- Preclinical animal models are necessary to understand behavioral and neurobiological correlates of addiction vulnerability.
- Important sex differences related to alcohol exist in numerous laboratory, clinical, and neurobiological measures.
- Chronic stress is a common underlying factor in vulnerability, maintenance, and relapse of alcohol use disorders.
- Social isolation provides a model in male subjects that will continue to be useful for studying behavioral, physiological, and neurobiological correlates of addiction vulnerability, as well as consequences of long-term stress and alcohol exposure.
- Models using female subjects need to be explored for their utility in studying behavioral, physiological, and neurobiological correlates of addiction vulnerability for translation to the female human population.
- Understanding these relationships in male and female subjects will facilitate the advent of new pharmacotherapeutic options.

ACKNOWLEDGMENT

This work was supported by National Institutes of Health Grants AA17531, AA21099, AA10422, and F32 AA022270-01A1.

REFERENCES

Adinoff, B., Iranmanesh, A., Veldhuis, J., & Fisher, L. (1998). Disturbances of the stress response: the role of the HPA axis during alcohol withdrawal and abstinence. *Alcohol Health and Research World*, *22*(1), 67–72.

Adinoff, B., Krebaum, S. R., Chandler, P. A., Ye, W., Brown, M. B., & Williams, M. J. (2005). Dissection of hypothalamic-pituitary-adrenal axis pathology in 1-month-abstinent alcohol-dependent men, part 1: adrenocortical and pituitary glucocorticoid responsiveness. *Alcoholism, Clinical and Experimental Research*, *29*(4), 517–527.

Advani, T., Hensler, J. G., & Koek, W. (2007). Effect of early rearing conditions on alcohol drinking and 5-HT1A receptor function in C57BL/6J mice. *International Journal of Neuropsychopharmacology*, *10*(5), 595–607. http://dx.doi.org/10.1017/S1461145706007401.

Bertholomey, M. L., Henderson, A. N., Badia-Elder, N. E., & Stewart, R. B. (2011). Neuropeptide Y (NPY)-induced reductions in alcohol intake during continuous access and following alcohol deprivation are not altered by restraint stress in alcohol-preferring (P) rats. *Pharmacology, Biochemistry, and Behavior*, *97*(3), 453–461.

Bitran, D., Hilvers, R. J., & Kellogg, C. K. (1991). Ovarian endocrine status modulates the anxiolytic potency of diazepam and the efficacy of gamma-aminobutyric acid-benzodiazepine receptor-mediated chloride ion transport. *Behavioral Neuroscience*, *105*(5), 653–662.

Butler, T. R., Ariwodola, O. J., & Weiner, J. L. (2014). The impact of social isolation on HPA axis function, anxiety-like behaviors, and ethanol drinking. *Frontiers in Integrative Neuroscience*, *7*, 102. http://dx.doi.org/10.3389/fnint.2013.00102.

Butler, T. R., Carter, E., & Weiner, J. L. (2014). Adolescent social isolation does not lead to persistent increases in anxiety-like behavior or ethanol intake in female Long-Evans rats. *Alcoholism, Clinical and Experimental Research*, *38*(8), 2199–2207. http://dx.doi.org/10.1111/acer.12476.

Butler, T. R., Smith, K. J., Berry, J. N., Sharrett-Field, L. J., & Prendergast, M. A. (2009). Sex differences in caffeine neurotoxicity following chronic ethanol exposure and withdrawal. *Alcohol and Alcoholism*, *44*, 567–574.

Butler, T. R., Smith, K. J., Self, R. L., Braden, B. B., & Prendergast, M. A. (2008). Sex differences in the neurotoxic effects of adenosine A1 receptor antagonism during ethanol withdrawal: reversal with an A1 receptor agonist or an NMDA receptor antagonist. Alcoholism. *Clinical and Experimental Research*, *32*, 1260–1270.

Cain, M. E., Mersmann, M. G., Gill, M. J., & Pittenger, S. T. (2012). Dose-dependent effects of differential rearing on amphetamine-induced hyperactivity. *Behavioural Pharmacology*, *23*(8), 744–753. http://dx.doi.org/10.1097/FBP.0b013e32835a38ec.

Castle, D. J. (2008). Anxiety and substance use: layers of complexity. *Expert Review of Neurotherapeutics*, *8*(3), 493–501. http://dx.doi.org/10.1586/14737175.8.3.493.

Cha, Y. M., Li, Q., Wilson, W. A., & Swartzwelder, H. S. (2006). Sedative and GABAergic effects of ethanol on male and female rats. *Alcoholism, Clinical and Experimental Research*, *30*(1), 113–118. http://dx.doi.org/10.1111/j.1530-0277.2006.00005.x.

Chappell, A. M., Carter, E., McCool, B. A., & Weiner, J. L. (2013). Adolescent rearing conditions influence the relationship between initial anxiety-like behavior and ethanol drinking in male Long-Evans rats. *Alcoholism, Clinical and Experimental Research*, *37*(Suppl. 1), E394–E403. http://dx.doi.org/10.1111/j.1530-0277.2012.01926.x.

Chester, J. A., Barrenha, G. D., Hughes, M. L., & Keuneke, K. J. (2008). Age- and sexdependent effects of footshock stress on subsequent alcohol drinking and acoustic startle behavior in mice selectively bred for high-alcohol preference. *Alcoholism, Clinical and Experimental Research*, *32*, 1782–1794.

Costa, A., Bono, G., Martignoni, E., Merlo, P., Sances, G., & Nappi, G. (1996). An assessment of hypothalamo-pituitary-adrenal axis functioning in non-depressed, early abstinent alcoholics. *Psychoneuroendocrinology, 21*(3), 263–275.

Eames, S. F., Businelle, M. S., Suris, A., Walker, R., Rao, U., North, C. S., ... Adinoff, B. (2014). Stress moderates the effect of childhood trauma and adversity on recent drinking in treatment-seeking alcohol-dependent men. *Journal of Consulting and Clinical Psychology.* http://dx.doi.org/10.1037/a0036291.

Enoch, M. A. (2011). The role of early life stress as a predictor for alcohol and drug dependence. *Psychopharmacology (Berlin), 214*(1), 17–31. http://dx.doi.org/10.1007/s00213-010-1916-6.

Ford, M. M., Eldridge, J. C., & Samson, H. H. (2002a). Ethanol consumption in the female Long-Evans rat: a modulatory role of estradiol. *Alcohol, 26*(2), 103–113.

Ford, M. M., Eldridge, J. C., & Samson, H. H. (2002b). Microanalysis of ethanol self-administration: estrous cycle phase-related changes in consumption patterns. *Alcoholism, Clinical and Experimental Research, 26*(5), 635–643.

Garbutt, J. C., Kranzler, H. R., O'Malley, S. S., Gastfriend, D. R., Pettinati, H. M., Silverman, B. L., ... Vivitrex Study, G. (2005). Efficacy and tolerability of long-acting injectable naltrexone for alcohol dependence: a randomized controlled trial. *JAMA: The Journal of the American Medical Association, 293*, 1617–1625.

Gianoulakis, C., Dai, X., & Brown, T. (2003). Effect of chronic alcohol consumption on the activity of the hypothalamic-pituitary-adrenal axis and pituitary beta-endorphin as a function of alcohol intake, age, and gender. *Alcoholism, Clinical and Experimental Research, 27*(3), 410–423. http://dx.doi.org/10.1097/01.ALC.0000056614.96137.B8.

Gill, K. E., Chappell, A. M., Beveridge, T. J., Porrino, L. J., & Weiner, J. L. (2014). Chronic methylphenidate treatment during early life is associated with greater ethanol intake in socially isolated rats. *Alcoholism, Clinical and Experimental Research, 38*(8), 2260–2268. http://dx.doi.org/10.1111/acer.12489.

Goldman, L., Winget, C., Hollingshead, G. W., & Levine, S. (1973). Post-weaning development of negative feedback in the pituitary-adrenal system of the rat. *Neuroendocrinology, 12*(3), 199–211.

Grant, B. F., Stinson, F. S., Dawson, D. A., Chou, S. P., Dufour, M. C., Compton, W., ... Kaplan, K. (2004). Prevalence and co-occurrence of substance use disorders and independent mood and anxiety disorders: results from the national epidemiologic survey on alcohol and related conditions. *Archives of General Psychiatry, 61*(8), 807–816. http://dx.doi.org/10.1001/archpsyc.61.8.807.

Hall, F. S. (1998). Social deprivation of neonatal, adolescent, and adult rats has distinct neurochemical and behavioral consequences. *Critical Reviews in Neurobiology, 12*(1–2), 129–162.

Hartwell, E. E., & Ray, L. A. (2013). Sex moderates stress reactivity in heavy drinkers. *Addictive Behaviors, 38*, 2643–2646.

Hasin, D. S., & Grant, B. F. (2004). The co-occurrence of DSM-IV alcohol abuse in DSM-IV alcohol dependence: results of the national epidemiologic survey on alcohol and related conditions on heterogeneity that differ by population subgroup. *Archives of General Psychiatry, 61*(9), 891–896. http://dx.doi.org/10.1001/archpsyc.61.9.891.

Hommer, D., Momenan, R., Kaiser, E., & Rawlings, R. (2001). Evidence for a gender-related effect of alcoholism on brain volumes. *American Journal of Psychiatry, 158*(2), 198–204.

Hommer, D., Momenan, R., Rawlings, R., Ragan, P., Williams, W., Rio, D., & Eckardt, M. (1996). Decreased corpus callosum size among alcoholic women. *Archives of Neurology, 53*, 359–363.

Karkhanis, A., Alexander, N. J., McCool, B. A., Weiner, J. L., & Jones, S. R. (2015). Chronic social isolation during adolescence augments catecholamine response to acute ethanol in the basolateral amygdala. *Synapse, 69*(8), 385–395.

Karkhanis, A., Locke, J. L., McCool, B. A., Weiner, J. L., & Jones, S. R. (2014). Social isolation rearing increases nucleus accumbens dopamine and norepinephrine responses to acute ethanol in adulthood. *Alcoholism: Clinical and Experimental Research, 38*(11), 2270–2279.

Kessler, R. C., Crum, R. M., Warner, L. A., Nelson, C. B., Schulenberg, J., & Anthony, J. C. (1997). Lifetime co-occurrence of DSM-III-R alcohol abuse and dependence with other psychiatric disorders in the national comorbidity survey. *Archives of General Psychiatry, 54*(4), 313–321.

Kranzler, H. R., & Rosenthal, R. N. (2003). Dual diagnosis: alcoholism and co-morbid psychiatric disorders. *The American Journal on Addictions, 12*(Suppl. 1), S26–S40.

Lancaster, F. E., Brown, T. D., Coker, K. L., Elliott, J. A., & Wren, S. B. (1996). Sex differences in alcohol preference and drinking patterns emerge during the early postpubertal period. *Alcoholism, Clinical and Experimental Research, 20*(6), 1043–1049.

Lancaster, F. E., & Spiegel, K. S. (1992). Sex differences in pattern of drinking. *Alcohol, 9*(5), 415–420.

Lopez, M. F., Doremus-Fitzwater, T. L., & Becker, H. C. (2011). Chronic social isolation and chronic variable stress during early development induce later elevated ethanol intake in adult C57BL/6J mice. *Alcohol, 45*(4), 355–364. http://dx.doi.org/10.1016/j.alcohol.2010.08.017.

Lovallo, W. R., Farag, N. H., Sorocco, K. H., Cohoon, A. J., & Vincent, A. S. (2012). Naltrexone effects on cortisol secretion in women and men in relation to a family history of alcoholism: studies from the Oklahoma Family Health Patterns Project. *Psychoneuroendocrinology, 37*(12), 1922–1928.

Lukkes, J. L., Engelman, G. H., Zelin, N. S., Hale, M. W., & Lowry, C. A. (2012). Post-weaning social isolation of female rats, anxiety-related behavior, and serotonergic systems. *Brain Research, 1443*, 1–17. http://dx.doi.org/10.1016/j.brainres.2012.01.005.

Mathews, I. Z., Mills, R. G., & McCormick, C. M. (2008). Chronic social stress in adolescence influenced both amphetamine conditioned place preference and locomotor sensitization. *Developmental Psychobiology, 50*(5), 451–459. http://dx.doi.org/10.1002/dev.20299.

McCool, B. A., & Chappell, A. M. (2009). Early social isolation in male Long-Evans rats alters both appetitive and consummatory behaviors expressed during operant ethanol self-administration. *Alcoholism, Clinical and Experimental Research, 33*(2), 273–282. http://dx.doi.org/10.1111/j.1530-0277.2008.00830.x.

McCormick, C. M., Merrick, A., Secen, J., & Helmreich, D. L. (2007). Social instability in adolescence alters the central and peripheral hypothalamic-pituitary-adrenal response to a repeated homotypic stressor in male and female rats. *Journal of Neuroendocrinology, 19*(2), 116–126.

McCormick, C. M., Robarts, D., Gleason, E., & Kelsey, J. E. (2004). Stress during adolescence enhances locomotor sensitization to nicotine in adulthood in female, but not male, rats. *Hormones and Behavior, 46*(4), 458–466. http://dx.doi.org/10.1016/j.yhbeh.2004.05.004.

McCormick, C. M., Smith, C., & Mathews, I. Z. (2008). Effects of chronic social stress in adolescence on anxiety and neuroendocrine response to mild stress in male and female rats. *Behavioural Brain Research, 187*(2), 228–238. http://dx.doi.org/10.1016/j.bbr.2007.09.005.

Merikangas, K. R., Mehta, R. L., Molnar, B. E., Walters, E. E., Swendsen, J. D., Aguilar-Gaziola, S., ... Kessler, R. C. (1998). Comorbidity of substance use disorders with mood and anxiety disorders: results of the International Consortium in Psychiatric Epidemiology. *Addictive Behaviors*, *23*(6), 893–907.

Mitra, R., Vyas, A., Chatterjee, G., & Chattarji, S. (2005). Chronic-stress induced modulation of different states of anxiety-like behavior in female rats. *Neuroscience Letters*, *383*(3), 278–283. http://dx.doi.org/10.1016/j.neulet.2005.04.037.

Molina-Hernandez, M., Olivera-Lopez, J. I., Patricia Tellez-Alcantara, N., Perez-Garcia, J., & Teresa Jaramillo, M. (2006). Estrus variation in anxiolytic-like effects of intra-lateral septal infusions of the neuropeptide Y in Wistar rats in two animal models of anxiety-like behavior. *Peptides*, *27*(11), 2722–2730. http://dx.doi.org/10.1016/j.peptides.2006.05.017.

Molina-Hernandez, M., Tellez-Alcantara, N. P., Olivera-Lopez, J. I., & Jaramillo, M. T. (2013). Estrous cycle variation in anxiolytic-like effects of topiramate in Wistar rats in two animal models of anxiety-like behavior. *Pharmacology, Biochemistry and Behavior*, *103*(3), 631–636. http://dx.doi.org/10.1016/j.pbb.2012.11.002.

Nolen-Hoeksema, S. (2004). Gender differences in risk factors and consequences for alcohol use and problems. *Clinical psychology review*, *24*, 981–1010.

Nomikos, G. G., & Spyraki, C. (1988). Influence of oestrogen on spontaneous and diazepam-induced exploration of rats in an elevated plus maze. *Neuropharmacology*, *27*(7), 691–696.

Peitz, G. W., Strickland, J. C., Pitts, E. G., Foley, M., Tonidandel, S., & Smith, M. A. (2013). Peer influences on drug self-administration: an econometric analysis in socially housed rats. *Behavioural Pharmacology*, *24*(2), 114–123. http://dx.doi.org/10.1097/FBP.0b013e32835f1719.

Pellow, S., Chopin, P., File, S. E., & Briley, M. (1985). Validation of open:closed arm entries in an elevated plus-maze as a measure of anxiety in the rat. *Journal of Neuroscience Methods*, *14*(3), 149–167.

Piazza, N. J., Vrbka, J. L., & Yeager, R. D. (1989). Telescoping of alcoholism in women alcoholics. *The International Journal of the Addictions*, *24*, 19–28.

Pilowsky, D. J., Keyes, K. M., & Hasin, D. S. (2009). Adverse childhood events and lifetime alcohol dependence. *American Journal of Public Health*, *99*(2), 258–263. http://dx.doi.org/10.2105/AJPH.2008.139006.

Pisu, M. G., Garau, A., Olla, P., Biggio, F., Utzeri, C., Dore, R., & Serra, M. (2013). Altered stress responsiveness and hypothalamic-pituitary-adrenal axis function in male rat offspring of socially isolated parents. *Journal of Neurochemistry*, *126*(4), 493–502. http://dx.doi.org/10.1111/jnc.12273.

Prendergast, M. A., & Mulholland, P. J. (2012). Glucocorticoid and polyamine interactions in the plasticity of glutamatergic synapses that contribute to ethanol-associated dependence and neuronal injury. *Addiction Biology*, *17*(2), 209–223. http://dx.doi.org/10.1111/j.1369-1600.2011.00375.x.

Rau, A. R., Chappell, A. M., Butler, T. R., Ariwodola, O. J., & Weiner, J. L. (2015). Increased basolateral amygdala pyramidal cell excitability may contribute to the anxiogenic phenotype induced by chronic early life stress. *Journal of Neuroscience*, *35*(26), 9730–9740.

Ravenelle, R., Neugebauer, N. M., Niedzielak, T., & Donaldson, S. T. (2014). Sex differences in diazepam effects and parvalbumin-positive GABA neurons in trait anxiety Long-Evans rats. *Behavioural Brain Research*, *270*, 68–74. http://dx.doi.org/10.1016/j.bbr.2014.04.048.

Robbins, T. W., Jones, G. H., & Wilkinson, L. S. (1996). Behavioural and neurochemical effects of early social deprivation in the rat. *Journal of Psychopharmacology*, *10*(1), 39–47. http://dx.doi.org/10.1177/026988119601000107.

Roberts, A. J., Smith, A. D., Weiss, F., Rivier, C., & Koob, G. F. (1998). Estrous cycle effects on operant responding for ethanol in female rats. *Alcoholism, Clinical and Experimental Research*, *22*(7), 1564–1569.

Romeo, R. D., Lee, S. J., & McEwen, B. S. (2004). Differential stress reactivity in intact and ovariectomized prepubertal and adult female rats. *Neuroendocrinology*, *80*(6), 387–393. http://dx.doi.org/10.1159/000084203.

Simpson, J., Ryan, C., Curley, A., Mulcaire, J., & Kelly, J. P. (2012). Sex differences in baseline and drug-induced behavioural responses in classical behavioural tests. *Progress in Neuropsychopharmacology and Biological Psychiatry*, *37*(2), 227–236. http://dx.doi.org/10.1016/j.pnpbp.2012.02.004.

Skelly, M. J., Chappell, A. E., Carter, E., & Weiner, J. L. (2015). Adolescent social isolation increases anxiety-like behavior and ethanol intake and impairs fear extinction in adulthood: Possible role of noradrenergic signaling. *Neuropsychopharmacology*, *97*, 149–159.

Vazquez, D. M., Morano, M. I., Lopez, J. F., Watson, S. J., & Akil, H. (1993). Short-term adrenalectomy increases glucocorticoid and mineralocorticoid receptor mRNA in selective areas of the developing hippocampus. *Molecular and Cellular Neuroscience*, *4*(5), 455–471. http://dx.doi.org/10.1006/mcne.1993.1057.

Viau, V., & Meaney, M. J. (1991). Variations in the hypothalamic-pituitary-adrenal response to stress during the estrous cycle in the rat. *Endocrinology*, *129*(5), 2503–2511. http://dx.doi.org/10.1210/endo-129-5-2503.

Walker, Q. D., Francis, R., Cabassa, J., & Kuhn, C. M. (2001). Effect of ovarian hormones and estrous cycle on stimulation of the hypothalamo-pituitary-adrenal axis by cocaine. *Journal of Pharmacology and Experimental Therapeutics*, *297*(1), 291–298.

Weafer, J., Miller, M. A., & Fillmore, M. T. (2010). Response conflict as an environmental determinant of gender differences in sensitivity to alcohol impairment. *Current Drug Abuse Reviews*, *3*, 147–155.

Whitaker, L. R., Degoulet, M., & Morikawa, H. (2013). Social deprivation enhances VTA synaptic plasticity and drug-induced contextual learning. *Neuron*, *77*(2), 335–345. http://dx.doi.org/10.1016/j.neuron.2012.11.022.

Wilkin, M. M., Waters, P., McCormick, C. M., & Menard, J. L. (2012). Intermittent physical stress during early- and mid-adolescence differentially alters rats' anxiety- and depression-like behaviors in adulthood. *Behavioral Neuroscience*, *126*(2), 344–360. http://dx.doi.org/10.1037/a0027258.

Yorgason, J. T., Espana, R. A., Konstantopoulos, J. K., Weiner, J. L., & Jones, S. R. (2013). Enduring increases in anxiety-like behavior and rapid nucleus accumbens dopamine signaling in socially isolated rats. *The European Journal of Neuroscience*, *37*(6), 1022–1031. http://dx.doi.org/10.1111/ejn.12113.

Section D

Methods

Chapter 60

The Ultrarapid Alcohol, Smoking, and Substance Involvement Screening Test (ASSIST-Lite) and Implications for Neuropathology

Robert Ali[1], Linda Gowing[2], Jennifer Harland[1]

[1]*Discipline of Pharmacology, School of Medical Sciences, The University of Adelaide, Adelaide, SA, Australia;* [2]*Drug and Alcohol Services of South Australia, Discipline of Pharmacology, Medical School, The University of Adelaide, Adelaide, SA, Australia*

Abbreviations

ASSIST Alcohol, Smoking, and Substance Involvement Screening Test
BI Brief intervention
MI Motivational interviewing

INTRODUCTION

Globally, 2 billion people consume alcohol, 1.3 billion smoke tobacco, and up to 224 million people use cannabis (World Drug Report, 2012). Amphetamine-type stimulants and cocaine are used by 73 million people and 36 million people use illicit opioids (WHO, 2009). Nonmedical use of prescription stimulants, sedatives, and opioids is an established or emerging problem for many countries (World Drug Report, 2012).

Addictive behaviors, including addiction to licit and illicit drugs, are chronic relapsing brain disorders involving the neurological circuits that regulate reward, motivation, memory, and decision-making. Repeated exposure to addictive substances leads to egocentric behaviors with a focus on obtaining the drug by any means and on taking the drug under adverse psychosocial and medical conditions (Cadet, Bisagno, & Milroy, 2014).

Most people with clinically significant but low- to moderate-severity problems relating to their substance use will not seek help from specialist addiction treatment services (Humeniuk et al., 2012). They may seek assistance from a general practitioner, but it is more likely that low- to moderate-severity substance use will be identified during a presentation for some other reason. Hence frontline health care professionals can play a key role in the detection and prevention of hazardous alcohol consumption, smoking, and substance-related problems (Madras et al., 2009). Any health care visit presents an opportunity for screening and brief intervention, and routine practice should include asking about and documenting substance use.

The World Health Organization (WHO) Alcohol, Smoking, and Substance Involvement Screening Test (ASSIST) was developed in 1997 by the WHO and specialist addiction researchers in response to the overwhelming public health burden associated with psychoactive substance use worldwide. The ASSIST has undergone significant testing to ensure that it is feasible, reliable, valid, flexible, comprehensive, and cross-culturally relevant and able to be linked to brief interventions (Humeniuk, Henry-Edwards, Ali, Poznyak, & Monteiro, 2010). In 2012, the Ultrarapid Alcohol, Smoking, and Substance Involvement Screening Test (ASSIST-Lite) was developed as an ultrarapid screener; the version has been optimized for general medical settings (Ali, Meena, Eastwood, & Marsden, 2013), reflecting the context in which most screening for substance use is likely to occur.

There is strong evidence for the effectiveness of brief interventions for alcohol (Kaner et al., 2007) and tobacco (Stead et al., 2013) in primary care settings and growing evidence that brief interventions are effective for cannabis (Copeland & Swift, 2009; Copeland, Swift, Roffman, & Stephens, 2001), benzodiazepines (Bashir, King, & Ashworth, 1994; Heather et al., 2004), amphetamines (Baker, Boggs, & Lewin, 2001; Baker & Lee, 2003), opiates (Saunders, Wilkinson, & Phillips, 1995), and cocaine use (Stotts, Schmitz, Rhoades, & Grabowski, 2001). Research has shown that screening and brief intervention has led to a reduction in mortality (Cuijpers, Riper, & Lemmers, 2004), health care costs (Kaner et al., 2008), criminal justice involvement, and societal costs (Sullivan, Tetrault, Braithwaite, Turner, & Fiellin, 2011).

This chapter explores the relationship between use of the ASSIST-Lite and implications for neuropathology.

DEVELOPMENT OF THE ASSIST

In 1997 the WHO developed the ASSIST to:

- be faster to administer than existing diagnostic tests for substance use and substance use disorders;
- screen for all psychoactive substances, not just alcohol or tobacco;

- be able to be used in primary health care settings;
- have cross-cultural relevance; and
- be able to link easily into a brief intervention (Humeniuk et al., 2010).

The ASSIST has been through three main phases of testing to ensure that it is a reliable and valid instrument in international settings and able to be linked into a brief intervention. Phase I of the WHO ASSIST project was conducted in 1997 and 1998. It involved the development of the first version of the ASSIST (version 1.0). The draft questionnaire had 12 items (WHO ASSIST Working Group, 2002). The reliability and feasibility of the questionnaire items were assessed in a test–retest reliability study, which was carried out in Australia, Brazil, India, Ireland, Israel, the Palestinian Self-Rule Areas, Puerto Rico, the United Kingdom of Great Britain and Northern Ireland, and Zimbabwe. The sites were chosen to ensure that study participants would be culturally diverse and have different substance use patterns. The results showed that the ASSIST had good reliability and feasibility (Humeniuk et al., 2006). The ASSIST was revised to an eight-item questionnaire (version 2.0) on the basis of feedback from the study participants and to ensure that all items were easy to administer and understand (Humeniuk et al., 2010).

Phase II of the project was an international study to validate the ASSIST questionnaire in a variety of primary health care and drug treatment settings (Newcombe et al., 2005). The study took place during 2000 and 2002 and was carried out in Australia, Brazil, India, Thailand, the United Kingdom, the United States of America, and Zimbabwe. Participants were recruited from both primary care and alcohol and drug treatment services to ensure that individuals with different substance use patterns were adequately represented. The study demonstrated that the ASSIST had good concurrent construct, had predictive and discriminant validity, and included the development of cutoff scores for lower-, moderate-, and high-risk use for each type of substance. The resulting questionnaire, ASSIST version 3.0, was finally revised to the ASSIST version 3.1 for clinical use in health and welfare settings, and it is now advised that version 3.0 is used only for research purposes (Humeniuk et al., 2010).

A pilot study conducted at the same time as Phase II demonstrated that participants recruited from primary health care settings did reduce their substance use if given a brief intervention related to their ASSIST scores. This finding was investigated in more detail during Phase III of the ASSIST project, which consisted of a randomized controlled trial investigating the effectiveness of a brief intervention linked to ASSIST scores for moderate-risk cannabis, cocaine, amphetamine-type stimulant, or opioid use. Participants were recruited from primary health care settings and scored within the moderate-risk range for at least one of these substances (Humeniuk et al., 2010).

Phase III was conducted between 2003 and 2007 in Australia, Brazil, India, and the United States. The brief intervention lasted between 5 and 15 min and was based on the FRAMES model (Bien, Miller, & Tonigan, 1993) and incorporated motivational interviewing techniques (Miller & Rollnick, 2002). It focused on the delivery of personalized feedback regarding the participant's ASSIST scores and associated risk through the use of a purpose-designed ASSIST feedback report card. The brief intervention was bolstered with take-home self-help information (Humeniuk, Henry-Edwards, & Ali, 2003).

The results showed that participants receiving a brief intervention for illicit substances had significantly reduced ASSIST scores after 3 months compared with control participants who did not receive a brief intervention for their substance use. Moreover, over 80% of participants reported attempting to cut down on their substance use after receiving the brief intervention and also provided positive comments on the impact of the brief intervention (Humeniuk, Dennington, & Ali, 2008).

Despite the demonstrated effectiveness shown by Phase III, uptake of the ASSIST for routine use in primary health services has been lower than anticipated. Time constraints on staff appear to be the main barrier to adoption of systematic screening for substance use disorders in general medical settings (Smith, Schmidt, Davis, & Saitz, 2010). Perceptions that the time required to administer the ASSIST (approximately 15 min) was too long may have deterred its use (Mdege & Lang, 2011). This was the basis for development of the ASSIST-Lite in 2012 as an ultrarapid screener optimized for general medical settings (Ali et al., 2013).

ASSIST-LITE PROTOCOL

The ASSIST-Lite is an ultrarapid screener that has been optimized for general medical setting. The ASSIST-Lite asks about psychoactive substance use in the past 3 months, specifically, tobacco, alcohol, cannabis, amphetamine-type stimulants (or cocaine or a stimulant medication not as prescribed), sedatives (or sleeping medication not as prescribed), and street opioids (or an opioid containing medication not as prescribed). There is the opportunity to add in other substances, not specified, via the final question, "Did you use any other psychoactive altering substance?" A "yes" response then prompts: "What did you take?"

For each substance there are two (or three, for alcohol) questions to determine level of use, and the ASSIST-Lite scores them as follows:

- Not used in the past 3 months (0);
- Used, but no other questions positive (1);
- Used and either question positive (2);
- Used and both questions positive (3).

The cutoff score for a likely substance use disorder is 2. Any use of tobacco, cannabis, stimulants, sedatives, and opioids (including nonprescribed use of pharmaceutical products for the last three substance types) is a risk to health. However, alcohol is a logical exception. To capture hazardous drinking related to intoxication it is recommended that an additional criterion question be included. There is no international consensus on a level of alcohol consumption that is hazardous to health. As an example, the Australian national guideline (no more than four standard drinks (one standard drink=10 g alcohol)) on a single occasion for both men and women (NHMRC, 2009). Alternatively, the third question from the Alcohol Use Disorder Identification Test could be used, binary scored for ease of administration and scoring.

A respondent who has consumed alcohol in the past 3 months and screens negative on the alcohol ASSIST-Lite items, but has recently consumed more than four drinks on an occasion, could receive normative feedback and advice on health risks. With a hazardous consumption item included, the alcohol cutoff for the ASSIST-Lite increases to a score of 3. Optionally, including a final question to record another substance type, given local prevalence or emergent substance use patterns, can help to cue further evaluation in screening programs (Ali et al., 2013).

With an ultrabrief, multisubstance screener in hand (which it is estimated can be completed in less than 2 min), an electronic self-completion format using computer tablet and smartphone technologies is being developed. It is envisaged that this approach, which presents items based on previous responses, can underpin a screening that offers personalized feedback and personalized, brief psychological treatments.

Figure 1 shows the final simplified question structure and scoring for a single-page, interviewer-administered format for the ASSIST-Lite, using the Australian four standard drinks criterion as an example. To minimize false-positive reporting, the phrasing of the medications to refer to uses "not as prescribed" to avoid questioning patients who are taking a medication as instructed. This phrase captures use for the direct psychoactive effect and also whereby the patient self-medicates (i.e., takes more than prescribed or consumes via a different administration route).

NEUROPATHOLOGY AS A REASON FOR SCREENING AND INTERVENTION

Evidence from preclinical and clinical studies suggests that addiction represents sequential neuroadaptations. As a result, an initial impulsive action turns into a compulsive behavior and the addiction becomes chronic and relapsing. Imaging studies have provided evidence that this transition involves reprogramming of neuronal circuits that process:

1. Reward and motivation
2. Memory, conditioning, and habituation
3. Executive function and inhibitory control
4. Interoception and self-awareness
5. Stress reactivity (Koob & Volkow, 2010).

The transition from substance use to addiction is heavily influenced by genetic, developmental, and environmental factors and their dynamic interactions, which determine the course and severity of the addiction (see Figure 2). The abuse of substances is accompanied by moderate to severe neuropsychological impairments that appear to be secondary to functional and structural changes in various brain regions, including both cortical and subcortical regions of the human brain (Cadet et al., 2014).

Drug abuse initiates a cascade of interacting toxic, vascular, and hypoxic factors that finally result in widespread disturbances within the complex network of central nervous system cell–cell interactions (Buttner & Weis, 2006). As outlined by Koob and Volkow (2010), drug addiction has been conceptualized as a disorder that involves elements of both impulsivity and compulsivity that yield a composite addiction cycle composed of three stages:

1. binge/intoxication,
2. withdrawal/negative affect,
3. preoccupation/anticipation (craving).

Animal and human imaging studies have revealed discrete circuits that mediate the three stages of the addiction cycle. Elements of the ventral tegmental area and ventral striatum (the dopamine reward pathway) are a focal point for the binge/intoxication stage; hence binge/intoxication is driven by positive reinforcement. The extended amygdala plays a key role in the withdrawal/negative affect stage; negative reinforcement or substance use to alleviate negative affect is relevant at this stage.

The preoccupation/anticipation stage involves a widely distributed network including the orbitofrontal cortex–dorsal striatum, prefrontal cortex, basolateral amygdala, hippocampus, and insula; these areas are involved in craving and executive control. The transition to addiction involves neuroplasticity in all of these structures that may begin with changes in the mesolimbic dopamine system and a cascade of neuroadaptations from the ventral striatum to the dorsal striatum and orbitofrontal cortex and eventually dysregulation of the prefrontal cortex, cingulated gyrus, and extended amygdala (Koob & Volkow, 2010). It is the combination of positive and negative reinforcement and reduced executive function that is the basis of substance use becoming a compulsive behavior. In turn this is the basis of the loss of control over substance use that is a distinguishing feature of addiction.

The neuroadaptations associated with addiction are what makes substance use disorders, particularly dependence, difficult to treat. Preventing uptake of substance use is desirable but it is unrealistic to think that it is possible to prevent all use of substances for nonmedical purposes. The next preferred goal is to identify and intervene in substance use before neuroadaptation has occurred and before the behavior has become compulsive. This is the basis for promoting screening and brief intervention for substance use in a wide range of settings.

The neurobiological argument for screening and brief intervention is considered further in the following sections on each group of substances included in the ASSIST-Lite (tobacco, alcohol, cannabis, amphetamine-type stimulants, opioids, and sedatives).

Tobacco

The WHO (2011) reports that tobacco use is the second most significant risk factor for development of nontransmissible diseases. About 9% of global deaths are attributed to tobacco use, including direct consumption of tobacco and exposure to secondhand smoke (WHO, 2011).

The harmful effects of tobacco smoking are primarily due to tars, carcinogens, and carbon monoxide in smoke. The risks of ischemic heart disease; cerebrovascular disease; cancer of the trachea, bronchus, and lung; and chronic obstructive lung disease are all increased by tobacco smoking (WHO Global Report: Mortality Attributable to Tobacco, 2012).

The addictive component of tobacco is nicotine. Like most other drugs of dependence, nicotine increases extracellular levels of dopamine in the nucleus accumbens, but in the case of nicotine, the release of dopamine is not associated with euphoric effects (EMCDDA, 2009). Nicotine activates nicotinic acetylcholine receptors in the ventral tegmental area, nucleus accumbens, and amygdala, either directly or indirectly (Koob & Volkow, 2010).

Nicotine has been shown to reduce anxiety and relieve stress in smokers (although this property may be relief of

Alcohol, Smoking and Substance Involvement Test (ASSIST-Lite)

Instructions: These questions ask about psychoactive substances in the PAST 3 MONTHS ONLY

Question	Response	Notes
1 Did you smoke a cigarette containing tobacco?	Yes [1] No [0]	No: Skip to Q2
1a Did you usually smoke more than 10 cigarettes each day?	Yes [1] No [0]	
1b Did you usually smoke within 30 minutes after waking?	Yes [1] No [0]	Tobacco score: __ [0-3] Cut-off=2
2 Did you have a drink containing alcohol?	Yes [1] No [0]	No: Skip to Q3
2a On any one occasion, did you drink more than 4 standard drinks of alcohol?*	Yes [1] No [0]	
2b Have you ever tried and failed to control, cut down or stop drinking?	Yes [1] No [0]	
2c Has anyone expressed concern about your drinking?	Yes [1] No [0]	
*1 standard drink is about 1 small glass of wine, or one can of medium strength beer, or one single shot of spirits		Alcohol score:__ [0-4] Cut-off =3
3 Did you use cannabis?	Yes [1] No [0]	No: Skip to Q4
3a Have you had a strong desire or urge to use cannabis at least once a week or more often?	Yes [1] No [0]	
3b Has anyone expressed concern about your use of cannabis?	Yes [1] No [0]	Cannabis score: __ [0-3] Cut-off=2
4 Did you use an amphetamine-type stimulant, or cocaine, or a stimulant medication not as prescribed?	Yes [1] No [0]	No: Skip to Q5
4a Did you use a stimulant at least once a week or more often?	Yes [1] No [0]	
4b Has anyone expressed concern about your use of a stimulant?	Yes [1] No [0]	Stimulant score: __ [0-3] Cut-off=2
5 Did you use a sedative or sleeping medication not as prescribed?	Yes [1] No [0]	No: Skip to Q6
5a Have you had a strong desire or urge to use a sedative or sleeping medication at least once a week or more often?	Yes [1] No [0]	
5b Has anyone expressed concern about your use of a sedative or sleeping medication?	Yes [1] No [0]	Sedative score: __ [0-3] Cut-off=2
6 Did you use a street opioid (e.g. heroin), or an opiod-containing medication not as prescribed?	Yes [1] No [0]	No: Skip to Q7
6a Have you tried and failed to control, cut down or stop using an opiod?	Yes [1] No [0]	
6b Has anyone expressed concern about your use of an opiod?	Yes [1] No [0]	Opioid score: __ [0-3] Cut-off=2
7 Did you use any other psychoactive altering substance? What did you take _____		Not scored – but prompts further assessment

*Note that this is the Australian national guideline shown as an example.

FIGURE 1 Structure of the ASSIST-Lite, an ultrarapid screening tool based on the Alcohol, Smoking and Substance Involvement Test (ASSIST).

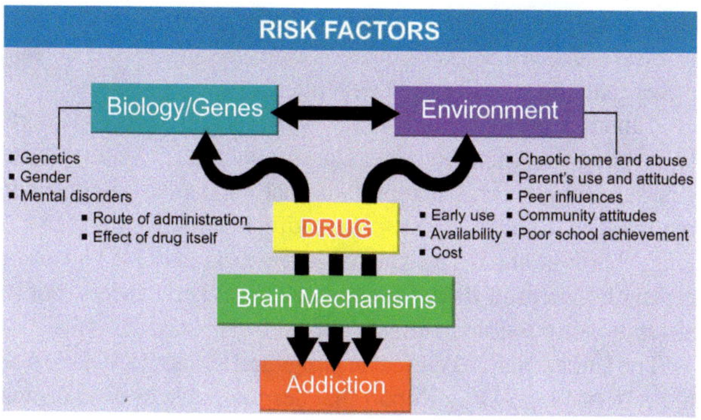

FIGURE 2 *Representation of risk factors for the development of addiction developed by the National Institute on Drug Abuse. http://www.drugabuse.gov/publications/drugs-brains-behavior-science-addiction/drug-abuse-addiction.*

anxiety caused by withdrawal and may not extend to nonsmokers). Nicotine also decreases food consumption and metabolism in humans and animals, making smoking a method of appetite control (Picciotto, 1998). The potential weight gain associated with smoking cessation is of importance to young people, particularly women, in an image-conscious society. Smoking also has an effect on vigilance and rapid information processing, which is beneficial for tasks requiring working memory. This is consistent with smokers reporting having a cigarette before doing a complex task that requires attention and arousal (Picciotto, 1998).

To identify the risk level of tobacco smoking, the ASSIST-Lite asks three questions related to tobacco use. If the answer to the first question, "Did you smoke a cigarette containing tobacco?" is "yes," two supplementary questions are asked. These are "Did you usually smoke more than 10 cigarettes per day" and "Did you usually smoke within 30 min after waking?" The supplementary questions identify likely nicotine dependence.

A score of 2 or above on tobacco smoking will prompt a brief intervention and/or referral to specialist intervention. This typically would include nine steps:

1. *Feedback* on the risks of smoking
2. *Advice* on how to reduce the risks
3. *Responsibility*, that is, the choice to quit is ultimately the client's
4. *Concern* about score
5. *Good things* about smoking
6. *Less good things* about smoking
7. *Summary* of the points raised
8. *Concern* about less good things
9. *Take-home information* and booklet

For a person identified as at risk from smoking, the brief intervention would include the known risks of smoking, advice on quitting (e.g., nicotine replacement therapy and supportive counseling), acknowledging that ultimately the choice to quit is his or hers, gaging the level of concern about the score and risks, exploration and discussion of the good and less good things of smoking, and take home advice. This approach has been shown to reduce tobacco smoking and some of the associated harms (Humeniuk et al., 2010).

Alcohol

Alcohol is consumed worldwide, but in some parts of the world there are high rates of abstinence from alcohol. The WHO estimated that in 2010, 61.7% of the global population ages 15 years and over had not drunk alcohol in the past 12 months. However, of those who drink alcohol, about 16% engage in heavy episodic drinking, and alcohol is nonetheless associated with significant health impacts. The WHO estimated that in 2012, about 3.3 million deaths, or 5.9% of all global deaths, were attributable to alcohol consumption (WHO, 2014).

The harmful effects of alcohol are related to the volume of alcohol consumed, the pattern of drinking, and the quality of alcohol consumed (WHO, 2014). With chronic, high-level alcohol consumption, the toxic effects of alcohol on organs and tissues cause increased risk of cardiovascular diseases (including stroke), cancers, and liver disease. With heavy episodic drinking (high-level consumption on a single occasion) the acute consequences include alcohol poisoning, injury, and violence. The quality of alcoholic beverages is a factor when homemade or illegally produced alcohol introduces the risk of contamination with methanol or other toxic substances (WHO, 2014).

The glutamate and γ-aminobutyric acid (GABA) pathways are the major control systems in the brain. GABA is inhibitory (GABA neurons make other neurons less likely to fire), while glutamate is excitatory (glutamate neurons make other neurons more likely to fire). Alcohol affects both the glutamate and the GABA pathways; it is the effect of alcohol on the control systems in the brain that helps to explain the broad effects of alcohol. As with other drugs of dependence, alcohol consumption triggers the release of dopamine in the nucleus accumbens, but this effect of alcohol is thought to be primarily through the indirect effects of alcohol on modulating neurotransmitters. Acute alcohol consumption suppresses the firing rate of ventral tegmental area GABA neurons, which leads to less suppression of dopamine neurons in the ventral tegmental area. This disinhibition leads to firing of the dopamine neurons and release of dopamine in the nucleus accumbens. Chronic alcohol sensitizes the system, leading to a relative dopamine deficiency (Petrakis, 2006).

Alcohol also has an effect on the glutamate system—alcohol is an N-methyl-D-aspartic acid glutamate receptor antagonist. Increased glutamatergic tone is associated with chronic alcohol consumption and is associated with the signs and symptoms of alcohol withdrawal (Petrakis, 2006).

One of the last areas of the brain to complete development is the prefrontal cortex (Gogtay et al., 2004), which is the area of the brain involved in decision-making and control of emotions and desires. This makes adolescents susceptible to the neurological effects of drugs of dependence but this susceptibility may be particularly marked with alcohol, because of its broad effects on many targets, including the glutamate and GABA systems, which have a critical role in the process of adult neurogenesis (Geil et al., 2014). The greater susceptibility of adolescents to the effects of alcohol is consistent with evidence that the commencement of alcohol drinking at an early age is associated with increased risk of problematic alcohol consumption in later life (Lee, Young, Kendler, & Prescott, 2012).

Alcohol appears to target some brain regions more than others, with a critical cluster of alcohol-induced impairments in behavior

FIGURE 3 Standard drinks chart (Australia). *Drug and Alcohol Services of Australia.*

control, learning, memory, mood, and decision-making attributed to the integrity of the hippocampus (Mechtcheriakov et al., 2007). Although there are many theories about the development of alcohol use disorders, all involve the fact that repeated bouts of excessive alcohol intake change the brain in a way that drives a loss of control over consumption (Koob & Le Moal, 1997).

The ASSIST-Lite can play a significant role in the identification of risky and hazardous alcohol use, with the aim of intervening before significant changes have occurred in the brain. It first establishes if the person has consumed alcohol by asking "Did you have a drink containing alcohol?" An affirmative response leads to three further questions to identify level of risk. These questions are:

"On any one occasion, did you drink more than four standard drinks?"
(This is based on the Australian national guidelines, in which the short-term risk associated with intoxication has been set at 40 g, see Figure 1, with a standard drink defined as containing 10 g of alcohol, see Figure 3).
"Have you tried and failed to control, cut down, or stop drinking?"
"Has anyone expressed concern about your drinking?"

A score of 3 or more suggests signs of alcohol use disorder and the person should receive a brief intervention (nine steps as outlined above) with a focus on a further assessment and/or referral to specialist services. A respondent who has consumed alcohol in the past 3 months and screens negative on the ASSIST-Lite items could receive normative feedback and advice on health risks (Ali et al., 2013).

Cannabis

Based on systematic reviews of epidemiological data, Degenhardt et al. (2013) estimated that in 2010 there were an estimated 13.1 million people globally who were cannabis dependent (around 0.2% of the adult population). Prevalence of cannabis dependence peaked in the 20- to 24-year age group, was higher in males than in females, and was higher in high income regions. It is estimated that approximately 10% of people who ever use cannabis become daily users (Anthony, Warner, & Kessler, 1994); hence it would be expected that the prevalence of cannabis use would be much higher than the prevalence of dependence.

Cannabis use is not associated with increased risk of mortality, but regular cannabis use is a risk factor for schizophrenia. It is as a trigger for schizophrenia that cannabis use contributes to global estimates of disability-adjusted life years (DALYs) (Degenhardt et al., 2013). Cannabis smoking may also be associated with increased risk of respiratory harm and cancers, but it is currently not possible to quantify this level of risk.

The United Nations Office on Drugs and Crime (UNODC) noted in the 2014 World Drug Report that despite stable or even decreasing prevalence of cannabis use, increasing numbers of people were seeking treatment for cannabis use each year. It would be expected that the group of cannabis users seeking treatment is likely to be those who are dependent and experiencing problems related to their cannabis use. A minority of cannabis users fit into this category; most cannabis users do not require structured treatment (Swift, Hall, & Teesson, 2001) but do constitute a group who might be expected to respond positively to screening and brief interventions.

The pharmacological effects of cannabis occur through stimulation of the cannabinoid receptors by Δ^9-tetrahydrocannabinol (Elsohly & Slade, 2005). These are the CB1 and CB2 receptors that are abundant in the brain's reward circuitries and implicated in addiction. CB1 receptors are also found in human peripheral tissue, but in much lower doses. CB2 is present in immune tissues and cells (Gailiègue et al., 1995).

The endocannabinoid system may have a role in various aspects of drug addiction including influencing the rewarding properties of drugs, drug-seeking behavior, and cravings and relapse (Maldonao et al., 2011). Cannabis users suffer from deficits in broad cognitive domains that include memory, attention, decision-making, and episodic memory, which may influence the degree to which cannabis users engage in risky behaviors with negative health consequences (Schuster, Crane, Mermelstein, & Gonzalez, 2012).

To identify current cannabis users the ASSIST-Lite asks "Did you use cannabis?" If the person responds "yes," they are prompted to answer two supplementary questions:

"Have you had a strong desire or urge to use cannabis at least once a week or more often?"
"Has anyone expressed concern about your use of cannabis?"

These two supplementary questions identify possible cannabis dependence and suggest the need for a brief intervention and possible referral for further assessment.

Brief interventions have shown promise among both adult and adolescent cannabis users, with research finding reductions in the quantity and frequency of use and the number of cannabis-associated problems at follow-up subsequent to brief interventions (Carroll et al., 2006; Copeland et al., 2001; Dennis et al., 2004; Kamon, Budney, & Stanger, 2005; Stephens, Roffman, & Curtin, 2000). These studies have typically compared a brief intervention (from one to six sessions) to a delayed-treatment control condition and are largely based upon cognitive behavioral therapy (CBT) and motivational enhancement therapy.

An Australian study by Martin and Copeland (2008) found that individuals who received a single assessment session combined with a motivational enhancement therapy-based feedback session had greater reductions in cannabis use and number of Diagnostic and Statistical Manual of Mental Disorders-IV criteria endorsed for cannabis dependence in comparison to those in a delayed-treatment control condition.

A meta-analysis of brief interventions for cannabis undertaken by the UK National Institute for Health and Clinical Excellence found a significant risk reduction (relative risk 3.33, 95% CI 1.99–5.56). Among emergency department samples of adolescents, a peer-delivered brief intervention with booster sessions was effective in preventing cannabis use and "days high" at 12 months (Bernstein et al., 2009). McCambridge, Strang, Platts, and Witton (2003) found that there was no difference in outcome between motivational interviewing and drug information and advice at either follow-up study interval. There was, however, substantial evidence of variability in outcomes by practitioner, regardless of the intervention being delivered, including in relation to the primary outcome measure of reduced frequency of cannabis use at the 6-month follow-up interval (McCambridge et al., 2003).

Earlier age of initiation of cannabis use increases the severity of psychosocial and substance use problems (Walton et al., 2013). Treatment engagement among young people has been suggested to be quite low, despite reported increases in treatment uptake among adults (Copeland, 2004). Time-limited, brief interventions may be efficacious in engaging young people resistant to long-term treatment (Martin & Copeland, 2008). Simply providing advice may be an effective brief intervention with young cannabis users (McCambridge et al., 2003).

Psychostimulants

Degenhardt, Baxter et al. (2014) estimated from a systematic review that in 2010, there were 24.1 million people dependent on psychostimulants, equating to around 0.1% of the global adult population dependent on cocaine and around 0.25% dependent on amphetamines. Cocaine dependence was most prevalent in the Americas, whereas amphetamine dependence is most prevalent in Asian regions (Degenhardt, Baxter et al., 2014).

Significant adverse health effects related to psychostimulant use, particularly dependent use, arise from injury or violence, psychosis, and other mental health complications (Degenhardt, Baxter et al., 2014). Injection use may also be associated with increased risk of blood-borne diseases such as human immunodeficiency virus (HIV) and hepatitis C.

Psychostimulants (cocaine, amphetamines, ecstasy) directly increase the amount of dopamine available for postsynaptic signaling in the nucleus accumbens (the dopamine reward pathway) either by increasing dopamine release or by reducing dopamine reuptake from the synapse (EMCDDA, 2009; Krasnova & Cadet, 2009). With chronic use, dopamine levels become depleted, with studies finding significantly lower dopamine in the brains of methamphetamine users (McCann, Szabo, Scheffel, Dannals, & Ricaurte, 1998; Wilson, Kalasinsky, & Levey, 1996).

Methamphetamine dependence is associated with complaints of cognitive dysfunction, memory problems, depressed mood, and self-reported deficits in everyday functioning (Dean, Groman, Morales, & London, 2013; Rendell, Mazur, & Henry, 2009). The ASSIST-Lite gages stimulant use by asking the question: "Did you use an amphetamine-type stimulant, or cocaine, or a stimulant medication not as prescribed?" The supplementary questions are:

"Did you use a stimulant at least once a week or more often?"
"Has anyone expressed concern about your use of a stimulant?"

The cutoff score for stimulant use disorder is 2. There is evidence supporting the effectiveness of psychological interventions in addressing methamphetamine use and dependence (Lee & Rawson, 2008). In a randomized controlled trial of regular amphetamine users, Baker et al. (2001) reported that brief interventions (consisting of motivational interviewing and CBT) were feasible and associated with better outcomes compared with a control group. The main finding was that more people in the intervention condition abstained from amphetamines at 6-month follow-up compared to the control group (Baker et al., 2001).

In a follow-up study, Baker et al. (2005) reported a significant increase in the likelihood of abstinence from amphetamines among those receiving two or more treatment sessions. Reduction in amphetamine use was accompanied by significant improvements in stage of change, benzodiazepine use, tobacco smoking, polydrug use, drug-injection risk behavior, criminal activity level, and psychiatric distress and depression level.

From their findings Baker et al. (2005) recommended a stepped-care approach. This would consist of a structured assessment of amphetamine use and related problems, self-help material, and regular monitoring of amphetamine use and related harms. Baker et al. (2005) suggested offering two sessions of CBT at the outset for regular amphetamine users, with further treatment offered depending on response.

Benefits have been shown after two sessions of motivational interviewing for cocaine users with low initial motivation to change (Rohsenow et al., 2004). Platt (1997), from a review of cocaine abuse, concluded that a nonconfrontational, empathic, and mutually respective therapeutic relationship is more likely to engage those more entrenched users who are unwilling to accept that they have a problem. The ASSIST-Lite clinical intervention is consistent with this approach.

Sedatives

Prescription drug abuse is increasing at alarming rates (UNODC, 2011). The number of prescriptions being written by medical professionals has increased dramatically since 2005. The nonmedical use or abuse of these prescription drugs is now seen as a serious and growing public health problem, with an estimated 52 million people having used prescription drugs for nonmedical reasons in their lifetime (NIDA, 2011).

Three classes of controlled prescription drugs are most commonly abused: opiates, such as oxycodone; sedative hypnotics, including benzodiazepines; and stimulants (UNODC, 2011). Responses to prescription drug abuse tend to focus on the prescribing practices of doctors, but, as has been noted by others (Brown, Swiggart, Dewey, & Ghulyan, 2012), the ability to identify and respond appropriately to substance abuse is an important component in changing prescribing practices.

The ASSIST-Lite helps identify this increasing problem by asking the question, "Did you use a sedative or sleeping medication not as prescribed?" A "yes" response prompts two supplementary questions:

"Have you had a strong desire or urge to use a sedative or sleeping medication at least once a week or more often?"
"Has anyone expressed concern about your use of a sedative or sleeping medication?"

Opioids

The United Nations Office on Drugs and Crime estimates that between 28.6 and 38 million people globally used opioids, including heroin and prescription painkillers, in one year. This equates to a global average prevalence of 0.7% of the population (World Drug Report, 2014).

Degenhardt, Charlson et al. (2014) estimated that there were 15.5 million opioid-dependent people globally in 2010, equating to 0.22% of the adult population. Prevalence was higher among males (0.3%) than among females (0.14%) and peaked at 25–29 years of age. Opioid dependence was estimated to account for 0.37% of global DALYs.

There are three types of opioid receptor: μ, δ, and κ, also referred to MOP, DOP, and KOP, respectively (Pathan & Williams, 2012). The MOP is responsible for most if not all the observed acute and chronic effects of opioid agonists such as morphine and heroin, including the addictive properties (Traynor, 2012). The effects of opioid drugs are determined by the distribution of opioid receptors in the central nervous system, gut, and other peripheral tissue (Pathan & Williams, 2012). The acute actions of μ-opioid agonists include analgesia, euphoria, constipation, and respiratory depression (Traynor, 2012).

The harmful effects of opioid drugs used for nonmedical purposes derive from injection as the primary route of administration and the effect of respiratory depression that is the basis of overdose. As opioid use occurs particularly in younger age groups, the years of life lost due to opioid overdose are substantial (Darke, Mills, Ross, & Teesson, 2011). Transmission of blood-borne diseases such as HIV/AIDS and hepatitis C is a major source of morbidity associated with use of opioid drugs (Degenhardt, Charlson et al., 2014).

The ASSIST-Lite identifies recent opioid use by asking: "Did you use a street opioid (e.g., heroin) or an opioid-containing medication not as prescribed?" If "yes," the respondent is prompted to answer two further questions:

"Have you failed to control, cut down, or stop using an opioid?"
"Has anyone expressed concern about your use of an opioid?"

A score of 2 or more suggests opioid dependence and the respondent should receive a further assessment with the view of referral to a specialist service.

Research into brief interventions with heroin users is limited. Saunders et al. (1995) concluded that brief motivational interventions were a useful adjunct to clients on methadone programs. They found that over a 6-month period clients receiving brief motivational interventions regarding their illicit opiate use demonstrated a greater and immediate commitment to abstinence. Compared to the control group, the intervention group also reported more positive expected outcomes for abstinence and fewer opiate-related problems, were initially more contemplative of change, complied with the methadone program longer, and relapsed less quickly (Saunders et al., 1995).

Other Drugs

It is important to recognize that illicit substances are not taken by sharply distinct population groups and two or more substances may be used concurrently or during the same use episode (Humeniuk et al., 2012). The ASSIST-Lite asks about other drug use via the questions, "Did you use any other psychoactive altering substance? What did you take?" Although these questions are not scored, they prompt further assessment.

CONDUCTING A BRIEF INTERVENTION WITH AN ILLICIT DRUG USER

To deliver an effective brief intervention, it is essential to understand the underlying principles of behavior change within the spirit of motivational interviewing. Motivational interviewing is a directive, client-centered style of interaction aimed at helping people to

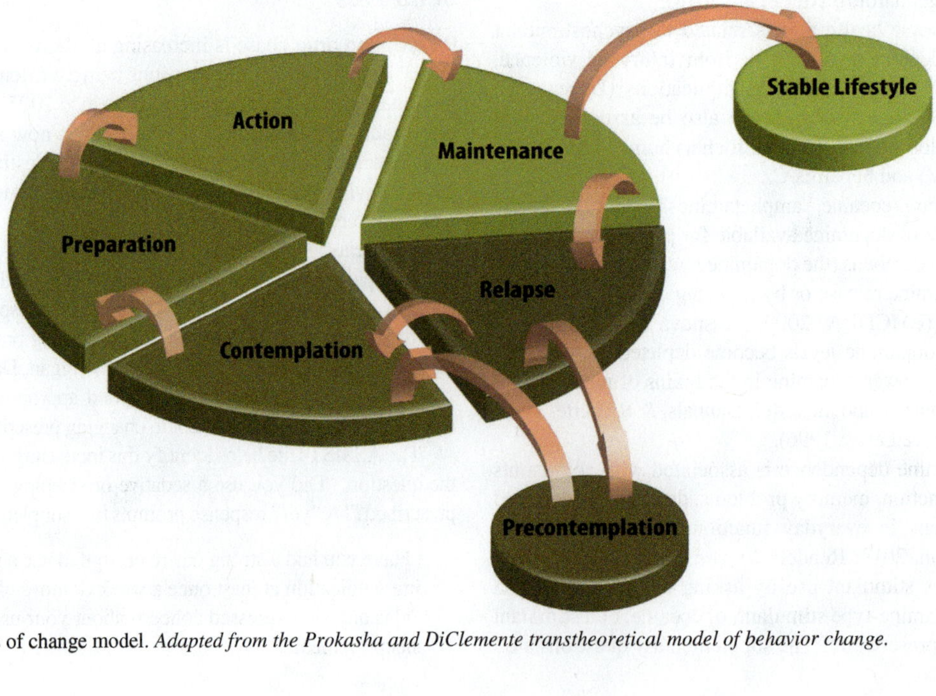

FIGURE 4 Stages of change model. *Adapted from the Prokasha and DiClemente transtheoretical model of behavior change.*

explore and resolve their ambivalence about their substance use and move through the stages of change (Miller & Rollnick, 2002). It is especially useful when working with patients in the precontemplation and contemplation stages but the principles and skills are important at all stages (Miller & Rollnick, 2013). The brief intervention approach suggested here is based on the motivational interviewing principles developed by Miller and Rollnick (2002, 2013).

Motivational interviewing makes use of five specific skills. These skills are used together to encourage patients to talk, to explore their ambivalence about their substance use, and to clarify their reasons for reducing or stopping their substance use. The first four skills are often known by the acronym OARS—open-ended questions, affirmation, reflective listening, and summarizing. The fifth skill is "eliciting change talk" and involves using the OARS to guide the patient to present the arguments for changing his or her substance use behavior (Miller & Rollnick, 2002, 2013).

The transtheoretical model offers an integrative framework for understanding the process of behavior change. The stages of change represent a key component of the transtheoretical model and describe a series of changes through which people pass as they change behavior (DiClemente & Velasquez, 2002). The model proposes that individuals in a change process tend to negotiate a cyclical pathway that traverses a number of recognizable change stages (DiClemente & Hughes, 1990). The stages of change model can be used to match interventions with a person's readiness to take in information and change their substance use (Figure 4).

CONCLUSION

Addictions are brain disorders that affect neural pathways that subsume reward, motivation, and memory. Only a small proportion of people who initiate substance use will go on to become addicted (dependent) and there is a continuum of use and associated risks between abstinence and dependence. Screening and brief intervention present the opportunity to identify and intervene in risky substance use. Early intervention, prior to the development of neurological changes, can prevent the development of dependence and encourage reductions in risk behaviors. People identified by screening as being at substantial risk of dependence can be encouraged to enter more structured treatment for their substance use. The ASSIST-Lite offers an approach to screening and brief intervention that is quick and suitable for use in generalist health settings, where it is likely that drug use might be detected prior to the development of dependence.

KEY FACTS

- Repeated exposure to addictive substances leads to egocentric behaviors with a focus on obtaining the drug by any means and on taking the drug under adverse psychosocial and medical conditions.
- Evidence from preclinical and clinical studies suggests that addiction represents sequential neuroadaptations.
- Most people with clinically significant but low- to moderate-severity problems relating to their substance use will not seek help from a specialist addiction treatment service.
- There is strong evidence for the effectiveness of brief interventions for alcohol and tobacco in primary care settings and growing evidence that brief interventions are effective for cannabis, benzodiazepines, amphetamines, opiates, and cocaine use.

SUMMARY POINTS

- Early intervention, prior to the development of neurological changes, can prevent the development of dependence and encourage reductions in risk behaviors.
- Addictive behaviors, including addiction to licit and illicit drugs, are chronic relapsing brain disorders involving the neurological circuits that regulate reward, motivation, memory, and decision-making.
- Frontline health care professionals can play a key role in the detection and prevention of hazardous alcohol consumption, smoking, and substance-related problems.
- The WHO ASSIST was developed in 1997 by the WHO and specialist addiction researchers in response to the overwhelming public health burden associated with psychoactive substance use worldwide.
- The ASSIST-Lite was developed as an ultrarapid screener; the version has been optimized for general medical settings reflecting the context in which most screening for substance use is likely to occur.
- Screening and brief intervention present the opportunity to identify and intervene in risky substance use.

REFERENCES

Ali, R., Meena, S., Eastwood, B., & Marsden, J. (2013). Ultra-rapid screening for substance-use disorders: the alcohol, smoking and substance involvement screening test (ASSIST-Lite). *Drug and Alcohol Dependence, 132,* 352–361.

Anthony, J. C., Warner, L. A., & Kessler, R. C. (1994). Comparative epidemiology of dependence on tobacco, alcohol, controlled substances, and inhalants: basic findings from the National Comorbidity Survey. *Experimental and Clinical Psychopharmacology, 2*(3), 244–268.

Baker, A., Boggs, T. G., & Lewin, T. J. (2001). Randomised controlled trial of brief cognitive-behavioural interventions among regular users of amphetamine. *Addiction, 96,* 1279–1287.

Baker, A., & Lee, N. (2003). A review of psychosocial interventions for amphetamine use. *Drug and Alcohol Review, 22,* 323–335.

Baker, A., Lee, N. K., Claire, M., Lewin, T. J., Grant, T., Pohlman, S., ... Carr, V. J. (2005). Brief cognitive behavioural interventions for regular amphetamine users: a step in the right direction. *Addiction, 100*(3), 367–378.

Bashir, K., King, M., & Ashworth, M. (1994). Controlled evaluation of brief intervention by general practitioners to reduce chronic use of benzodiazepines. *British Journal of General Practice, 44,* 408–412.

Bernstein, E., Edwards, E., Dorfman, D., Heeren, T., Bliss, C., & Bernstein, J. (2009). Screening and brief intervention to reduce marijuana use among youth and young. *Adults in a Pediatric Emergency Department, 16*(11), 1174–1185.

Bien, T. H., Miller, W. R., & Tonigan, S. (1993). Brief intervention for alcohol problems: a review. *Addiction, 88,* 315–336.

Brown, M. E., Swiggart, W. H., Dewey, C. M., & Ghulyan, M. V. (2012). Searching for answers: proper prescribing of controlled prescription drugs. *Journal of Psychoactive Drugs, 44*(1), 79–85.

Buttner, A., & Weis, S. (2006). Neuropathological alterations in drug abusers: the involvement of neurons, glial, and vascular systems. *Forensic Science, Medicine, and Pathology, 2*(2), 115–126.

Cadet, J. L., Bisagno, V., & Milroy, C. M. (2014). Neuropathology of substance use disorders. *Acta Neuropathology, 127*(1), 91–107.

Carroll, K. M., Easton, C. J., Nich, C., Hunkele, K. A., Neavins, T. M., Sinha, R., ... Rounsaville, B. J. (2006). The use of contingency management and motivational/skills-building therapy to treat young adults with marijuana dependence. *Journal of Consulting and Clinical Psychology, 74*(5), 955–966.

Copeland, J. (2004). Developments in the treatment of cannabis use disorder. *Current Opinion in Psychiatry, 17*(3), 161–167.

Copeland, J., & Swift, W. (2009). Cannabis use disorder: epidemiology and management. *International Review of Psychiatry, 21*, 96–103.

Copeland, J., Swift, W., Roffman, R., & Stephens, R. (2001). A randomised controlled trial of brief cognitive-behavioural interventions for cannabis use disorder. *Journal of Substance Abuse Treatment, 21*, 55–64.

Cuijpers, P., Riper, H., & Lemmers, L. (2004). The effects on mortality of brief interventions for problem drinking: a meta-analysis. *Addiction, 99*, 839–845.

Darke, S., Mills, K. L., Ross, J., & Teesson, M. (2011). Rates and correlates of mortality amongst heroin users: findings from the Australian Treatment Outcome Study (ATOS), 2001–2009. *Drug and Alcohol Dependence, 115*(3), 190–195.

Dean, A. C., Groman, S. M., Morales, A. M., & London, E. D. (2013). An evaluation of the evidence that methamphetamine abuse causes cognitive decline in humans. *Neuropsychopharmacology, 38*(2), 259–274.

Degenhardt, L., Baxter, A. J., Lee, Y. Y., Hall, W., Sara, G. E., Johns, N., ... Vos, T. (2014). The global epidemiology and burden of psychostimulant dependence: findings from the Global Burden of Disease Study 2010. *Drug and Alcohol Dependence, 137*, 36–47. http://dx.doi.org/10.1016/j.drugalcdep.2013.12.025.

Degenhardt, L., Charlson, F., Mathers, B., Hall, W. D., Flaxman, A. D., Johns, N., & Vos, T. (2014). The global epidemiology and burden of opioid dependence: results from the global burden of disease 2010 study. *Addiction, 109*, 1320–1333.

Degenhardt, L., Ferrari, A. J., Calabria, B., Hall, W. D., Norman, R. E., McGrath, J., ... Vos, T. (2013). The global epidemiology and contribution of cannabis use and dependence to the global burden of disease: results from the GBD 2010 study. *PloS One, 8*(10), e76635. http://dx.doi.org/10.1371/journal.pone.0076635.

Dennis, M., Godley, S., Diamond, G., Tims, F., Babor, T., & Donaldson, J. (2004). The Cannabis Youth Treatment Study: main findings from two randomized trials. *Journal of Substance Abuse Treatment, 27*, 197–213.

DiClemente, C. C., & Hughes, S. O. (1990). Stages of change profiles in outpatient alcoholism treatment. *Journal of Substance Abuse, 2*(2), 217–235.

DiClemente, C. C., & Velasquez, M. M. (2002). Motivational interviewing and the stages of change. In W. R. Miller, & S. Rollnick (Eds.), *Motivational interviewing: Preparing people for change* (2nd ed.). New York: Guilford.

Elsohly, M. A., & Slade, D. (2005). Chemical constituents of marijuana: the complex mixture of natural canninoids. *Life Science, 78*, 539–548.

European Monitoring Centre for Drugs and Drug Addiction (EMCDDA). (2009). Addiction neurobiology: ethical and social implications. In A. Carter, B. Capps, & W. Hall (Eds.), *EMCDDA monographs, no 9*. Luxembourg: Office for Official Publications of the European Communities.

Gailiegue, S., Mary, S., & Marchand, J. (1995). Expression of central and peripheral cannibinoid receptors in human immune tissue and luekocyte subpopulations. *Europe Journal of Biochemistry, 232*, 54–61.

Geil, C. R., Hayes, D. M., McClain, J. A., Liput, D. J., Marshall, A., Chen, K. Y., & Nixon, K. (2014). Alcohol and adult hippocampal neurogenisis: promiscuous drug, wanton effects. *Progress in Neuro-Pschopathology and Biological Psychiatry, 54*, 103–113.

Gogtay, N., Giedd, J. N., Lusk, L., Hayashi, K. M., Greenstein, D., Vaituzis, A. C., ... Thompson, P. M. (2004). Dynamic mapping of human cortical development during childhood through early adulthood. *PNAS, 101*(21), 8174–8179.

Heather, N., Bowie, A., Ashton, H., McAvoy, B., Spencer, I., Brodiw, J., & Giddings, D. (2004). Randomised controlled trial of two brief interventions against long-term benzodiazepine use: outcome of intervention. *Addict Res Theory, 12*, 141–154.

Humeniuk, R., Ali, R. L., & on behalf of the WHO ASSIST Phase II Study Group. (2006). *Validation of the alcohol, smoking and substance involvement screening test (ASSIST) and pilot brief intervention: A technical report of phase II findings of the WHO ASSIST project*. Geneva: World Health Organization.

Humeniuk, R., Ali, R., Babor, T., Souza-Formigoni, M. L. O., Boerngen de Lacerda, R., Ling, W., ... Vendetti, J. (2012). A randomized controlled trial of a brief intervention for illicit drugs linked to the Alcohol, Smoking and Substance Involvement Screening Test (ASSIST) in clients recruited from primary health-care settings in four countries. *Addiction, 107*, 957–966.

Humeniuk, R. E., Dennington, V., & Ali, R. L. (2008). *The effectiveness of a brief intervention for illicit drugs linked to the ASSIST screening test in primary health care settings: A technical report of phase III findings of the WHO ASSIST randomised controlled trial*. Geneva: World Health Organization.

Humeniuk, R. E., Henry-Edwards, S., & Ali, R. L. (2003). Self-help strategies for cutting down or stopping substance use: a guide. In *Draft version 1.1 for field testing*. Geneva: World Health Organization.

Humeniuk, R. E., Henry-Edwards, S., Ali, R. L., Poznyak, V., & Monteiro, M. (2010). *The alcohol, smoking and substance involvement screening test (ASSIST): Manual for use in primary care*. Geneva: World Health Organization.

Kamon, J., Budney, A., & Stanger, C. (2005). A contingency management intervention for adolescent marijuana abuse and conduct problems. *Journal of the American Academy of Child and Adolescent Psychiatry, 44*(6), 513–521.

Kaner, E. F., Beyer, F., Dickinson, H. O., Pienaar, E., Campbell, F., Schlesinger, C., ... Burnand, B. (2007). Effectiveness of brief alcohol interventions in primary care populations. *Cochrane Database of Systematic Reviews* (2), CD004148. http://dx.doi.org/10.1002/14651858.CD004148.pub3.

Kaner, E., Dickinson, H., Beyer, F., Pienaar, E., Schlesinger, C., Campbell, F., ... Heather, N. (2008). The effectiveness of brief alcohol intervention in primary care settings: a systematic review. *Drug and Alcohol Review, 28*, 301–323.

Koob, G. F., & Le Moal, M. (1997). Drug abuse: hedonic homeostatic dysregulation. *Science, 278*, 52–58.

Koob, G. F., & Volkow, N. D. (2010). Neurocircuitry of addiction. *Neuropsychopharmacology Reviews, 35*, 217–238.

Krasnova, I. N., & Cadet, J. L. (2009). Methamphetamine toxicity and messengers of death. *Brain Research Reviews, 60*, 379–407.

Lee, N. K., & Rawson, R. A. (2008). A systematic review of cognitive and behavioural therapies for methamphetamine dependence. *Drug and Alcohol Review, 27*(3), 309–317.

Lee, L. O., Young, K. C., Kendler, K. S., & Prescott, C. A. (2012). The effects of age at drinking onset and stressful life events on alcohol use in adulthood: a replication of the extension using a population-based twin study. *Alcoholism: Clinical and Experimental Research, 36*(4), 693–704.

Madras, B. K., Compton, W. M., Avula, D., Stegbauer, T., Stein, J. B., & Clark, H. W. (2009). Screening, brief interventions, referral to treatment (SBIRT) for illicit drug and alcohol use at multiple healthcare sites: comparison at intake and 6 months later. *Drug Alcohol Dependence, 99*, 280–295.

Maldonado, R., Berrendero, F., Ozaita, A., & Robledo, P. (2011). Neurochemical basis of cannabis addiction. *Neuroscience, 181*, 1–17.

Martin, G., & Copeland, J. (2008). The adolescent cannabis check-up: randomized trial of a brief intervention for young cannabis users. *Journal of Substance Abuse Treatment, 34*, 407–414.

McCambridge, J., Strang, J., Platts, S., & Witton, J. (2003). Cannabis use and the GP: brief motivational intervention increases clinical enquiry by GPs in a pilot study. *British Journal of General Practice, 53*, 637–639.

McCann, U. D., Szabo, Z., Scheffel, U., Dannals, R. F., & Ricaurte, G. A. (1998). Positron emission tomographic evidence of toxic effect of MDMA ("Ecstasy") on brain serotonin neurons in human beings. *Lancet, 352*(9138), 1433–1437.

Mechtcheriakov, S., Brenneis, C., Egger, K., Koppelstaetter, F., Schocke, M., & Marksteiner, J. (2007). A widespread distinct pattern of cerebral atrophy in patients with alcohol addiction revealed by voxel-based morphometry. *Journal of Neurology, Neurosurgery and Psychiatry, 78*, 610–614.

Mdege, N. D., & Lang, J. (2011). Screening instruments for detecting illicit drug use/abuse that could be useful in general hospital wards: a systematic review. *Addictive Behaviors, 36*, 1111–1119.

Miller, W., & Rollnick, S. (2002). *Motivational interviewing* (2nd ed.). New York, London: Guilford Press.

Miller, W. R., & Rollnick, S. (2013). *Motivational interviewing helping people change* (3rd ed.). Guilford Press. ISBN: 978-1-60918-227-4.

National Health and Medical Research Council (NHMRC). (2009). Australian guidelines to reduce health risks from drinking alcohol. Canberra, Commonwealth of Australia.

National Institute on Drug Abuse (NIDA). (2011). *Prescription drugs: Abuse and addiction*. Available at http://www.drugabuse.gov/publications/research-reports/prescription-drugs/director Accessed 20.01.16.

Newcombe, D., Humeniuk, R. E., & Ali, R. L. (2005). Validation of the World Health Organization alcohol smoking and substance involvement screening test (ASSIST): phase II study. Report from the Australian site. *Drug and Alcohol Review, 24*(3), 217–226.

Pathan, H., & Williams, J. (2012). Basic opioid pharmacology: an update. *British Journal of Pain, 6*(1), 11–16.

Petrakis, I. L. (2006). A rational approach to the pharmacotherapy of alcohol dependence. *Journal of Clinical Psychopharmacology, 26*(Suppl. 1), S3–S12.

Picciotto, M. R. (1998). Common aspects of the action of nicotine and other drugs of abuse. *Drug and Alcohol Dependence, 51*, 165–172.

Platt, J. J. (1997). *Cocaine Addiction: Theory, research and treatment*. Cambridge, MA: Harvard University Press.

Rendell, P. G., Mazur, M., & Henry, J. D. (2009). Prospective memory impairment in former users of methamphetamine. *Psychopharmacology, 203*, 609–616.

Rohsenow, D. J., Monti, P. M., Martin, R. A., Colby, S. M., Myers, M. G., Gulliver, S. B., ... Abrams, D. B. (2004). Motivational enhancement and coping skills training for cocaine abusers: effects on substance use outcomes. *Addiction, 99*(7), 862–874.

Saunders, B., Wilkinson, C., & Phillips, M. (1995). The impact of a brief motivational intervention with opiate users attending a methadone programme. *Addiction, 90*, 415–424.

Schuster, R. M., Crane, N. A., Mermelstein, R., & Gonzalez, R. (2012). The influence of inhibitory control and episodic memory on the risky sexual behavior of young adult cannabis users. *Journal of the International Neuropsychological Society, 18*(5), 827–833.

Smith, P. C., Schmidt, S. M., Davis, D. A., & Saitz, R. (2010). A single-question screeningtest for drug use in primary care. *Archives of Internal Medicine, 170*, 1155–1160.

Stead, L. F., Buitrago, D., Preciado, N., Sanchez, G., Hartmann-Boyce, J., & Lancaster, T. (2013). Physician advice for smoking cessation. *Cochrane Database of Systematic Reviews* (5), CD000165. http://dx.doi.org/10.1002/14651858.CD000165.pub4.

Stephens, R. S., Roffman, R. A., & Curtin, L. (2000). Comparison of extended versus brief treatments for marijuana use. *Journal of Consulting and Clinical Psychology, 68*, 898–908.

Stotts, A. L., Schmitz, J. M., Rhoades, H. M., & Grabowski, J. (2001). Motivational interviewing with cocaine-dependent patients: a pilot study. *Journal of Consulting and Clinical Psychology, 69*(5), 858–862.

Sullivan, L. E., Tetrault, J. M., Braithwaite, R. S., Turner, B. J., & Fiellin, D. A. (2011). A meta-analysis of the efficacy of nonphysician brief interventions for unhealthy alcohol use: implications for the patient-centered medical home. *American Journal on Addictions, 20*, 343–356.

Swift, W., Hall, W., & Teesson, M. (2001). Cannabis use and dependence among Australian adults: results from the National Survey of Mental Health and Wellbeing. *Addiction, 96*(5), 737–748.

Traynor, J. (2012). μ-Opioid receptors and regulators of G protein signaling (RGS) proteins: from a symposium on new concepts in mu-opioid pharmacology. *Drug and Alcohol Dependence, 121*(3), 173–180.

United Nations Office on Drugs and Crime (UNODC). (2011). *The Nonmedical use of prescription drugs: Policy direction issues*. New York: United Nations.

Walton, M. A., Bohnert, K., Resko, S., Barry, K., Chermack, S. T., Zucker, R. A., ... Blow, F. C. (2013). Computer and therapist based brief interventions among cannabis-using adolescents presenting to primary care: one year outcomes. *Drug and Alcohol Dependence, 132*(3), 646–653.

WHO ASSIST Working Group. (2002). The alcohol, smoking and substance involvement screening test (ASSIST): development, reliability and feasibility. *Addiction, 97*, 1183–1194.

Wilson, J. M., Kalasinsky, K. S., & Levey, A. I. (1996). Striatal dopamine nreve terminal makers in human, chronic methamphetamine users. *Nature Medicine, 2*, 699–703.

World drug report 2012. Vienna: United Nations Office on Drug and Crime.

World drug report 2014. United Nations Office on Drugs and Crime. Available from www.unodc.org.

World Health Organization. (2009). *Global health risks. Mortality and burden of diseases attributable to selected major risks*. Geneva: WHO Press, World Health Organization.

World Health Organization. (2011). *Global status report on non-communicable diseases 2010 – Description of the global burden of NCD's, their risk factors and determinants*. World Health Organization. 1-31. 4-20-2011.

World Health Organization. (2012). *WHO global report: Mortality attributable to tobacco*. Geneva: WHO Press, World Health Organization.

World Health Organization. (2014). *Global status report on alcohol and health 2014*. Geneva: WHO Press, World Health Organization.

Part IV

Cannabinoids

Section A

General Aspects

Chapter 61

Overview of Cannabis Use, Misuse, and Addiction

Cristina A.J. Stern, Leandro J. Bertoglio, Reinaldo N. Takahashi
Department of Pharmacology, Federal University of Santa Catarina, Florianópolis, Santa Catarina, Brazil

Abbreviations

CB1 receptor Cannabinoid type 1 receptor
CB2 receptor Cannabinoid type 2 receptor
CBD Cannabidiol
CPA Conditioned place aversion
CPP Conditioned place preference
CUD Cannabis use disorder
DSM-V *Diagnostic and Statistical Manual of Mental Disorders,* fifth edition
eCB Endocannabinoid
FAAH Fatty acid amide hydrolase
SA Self-administration
THC Δ^9-Tetrahydrocannabinol

INTRODUCTION

The earliest descriptions indicate that cannabis (marijuana) has been cultivated in China since 4000 BC. Its medicinal use was primarily documented in the Chinese pharmacopoeia around 2700 BC (Li, 1973). The Irish doctor O'Shaughnessy and the French psychiatrist Jacques-Joseph Moreau were responsible for spreading the medicinal properties of cannabis in the West during the nineteenth century (reviewed in Pamplona & Takahashi, 2012).

The scientific research focusing on cannabis and (endo)cannabinoids has grown since 1950 (Figure 1); for instance, some drugs, based on the acquired knowledge, have already been proven to be of value to patients with chronic pain and multiple sclerosis (Murray, Morrison, Henquet, & Di Forti, 2007). Meanwhile, a number of reports have demonstrated that long-term or heavy users have a risk of developing schizophrenia and cannabis use disorder (CUD; formerly known as cannabis abuse and dependence; Marshall, Gowing, Ali, & Le Foll, 2014; Volkow, Compton, & Weiss, 2014). This scenario has contributed to the dual view of cannabis use, as a potential medicine or a harmful drug. As a result, despite being illegal in some nations, there has been unprecedented interest in determining whether cannabis and its isolated compounds, particularly Δ^9-tetrahydrocannabinol (THC) and cannabidiol (CBD), are safe when used for medicinal and/or recreational purposes.

Cannabis has commonly been the most widely used illicit substance worldwide (UNODC, 2012). As shown in Table 1, the highest consumption is in Oceania, followed by North America, Europe, Africa, South America, and Asia. As shown in Figure 2, there seems to be a difference in cannabis consumption between men and women, at least in Europe and South America, for which data are available. Regardless of gender, however, its use decreases with aging, suggesting that most people may stop cannabis consumption (UNODC, 2012).

Cannabis leaves or flower tops can be smoked, eaten, or drunk as tea, and hashish, a product of the flowers' resin, is usually smoked. Depending on the route of administration, people will be exposed to different concentrations of THC, which causes the high from cannabis. Of note, the content of THC in cannabis has increased from nearly 3% in 1980 to 15% in 2012, and sometimes up to 25%, which is a phenomenon partially explained by the increased indoor cultivation of cannabis (van Amsterdam, Brunt, & van den Brink, 2015).

ACUTE AND LONG-TERM EFFECTS OF CANNABIS USE

Cannabis is used because of its pleasurable effects, which cause an experience of being relaxed, also known as being "high." However, depending on the amount consumed and/or the relative THC and CBD quantities, it can precipitate acute psychotic symptoms, auditory and visual hallucinations, and anxiety. The acute administration of cannabis also produces cognitive effects, impairing executive functions such as attention and working memory (D'Souza et al., 2004). However, how long these symptoms last is still a controversy, with studies reporting up to 4 weeks (Pope, Gruber, Hudson, Huestis, & Yurgelun-Todd, 2002). The long-term effects of cannabis consumption among heavy users are still a matter of debate. It has been shown that they present a decrease in cannabinoid type 1 (CB1) receptors in the cortical brain regions, which is correlated with the consumption period, although after 4 weeks of abstinence, the CB1 receptor density returned to a level similar to controls (Hirvonem et al., 2012). Moreover, no significant changes in the volume and/or morphology of the cortical and subcortical structures were found when samples of both adolescent and adult daily marijuana users versus nonusers were compared (Weiland et al., 2015).

The age of starting cannabis use seems to determine its long-term consequences. An onset before the age of 17 causes more

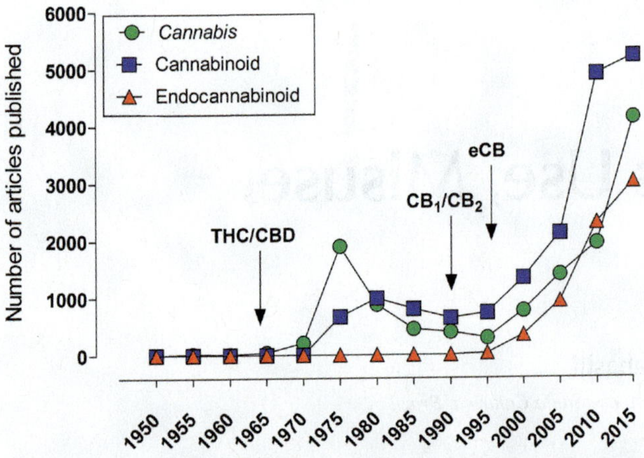

FIGURE 1 Number of published articles in which the terms cannabis, cannabinoid, and/or endocannabinoid are mentioned, according to a survey conducted in March 2015 through the PubMed Website (http://www.ncbi.nlm.nih.gov/pubmed). The period from 1950 to 2015 was binned by each 5-year period (the 2015 value was estimated based on its first trimester). The arrows indicate key discoveries in the history of cannabis research. For instance, by the 1960s, THC was discovered and CBD was isolated. They are considered the main psychotomimetic and nonpsychotomimetic compounds found in cannabis, respectively. The identification of cannabinoid type 1 and 2 (CB1 and CB2) receptors occurred in the 1990s. During the mid-1990s, the term endocannabinoid (eCB) system was suggested. *Adapted from Pamplona and Takahashi (2012).*

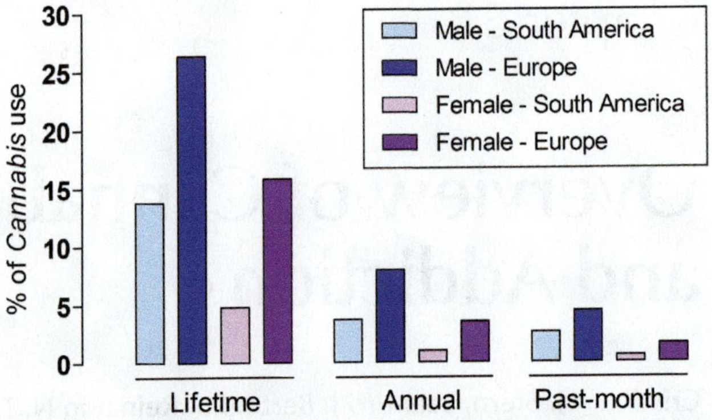

TABLE 1 Prevalence of Cannabis Use in 2010 or in the Latest Year (UNODC, 2012)

Region	Prevalence (%)
Oceania (Australia and New Zealand)	9.1–14.6
North America	10.8
Western and Central Europe	7.0
West and Central Africa	5.2–13.5
South America	2.9–3.0
Asia	1.0–3.4

FIGURE 2 Percentage of cannabis use in men and women from South America and Europe. Data were collected during 2005 and 2010 in 8 countries in South America and in 14 countries in Europe. The prevalence was evaluated for lifetime, annual, and past-month use of cannabis. *Adapted from UNODC (2012).*

CANNABIS MISUSE AND THE RISK OF ADDICTION

The risk of cannabis addiction depends on the onset age of its consumption and the frequency of use, with 25–50% of daily smokers becoming addicted (Volkow et al., 2014). It is also reported that cannabis use causes tolerance and a withdrawal syndrome. According to the *Diagnostic and Statistical Manual of Mental Disorders*, fifth edition, cannabis withdrawal is defined by the development of three or more of the following signs and symptoms within approximately 1 week of the cessation of heavy and prolonged use: (1) irritability, anger, or aggression; (2) nervousness or anxiety; (3) sleep difficulty; (4) decreased appetite or weight loss; (5) restlessness; (6) depressed mood; (7) at least one of the following physical symptoms causing significant discomfort: stomach pain, shakiness or tremors, sweating, fever, chills, or headache (American Psychiatric Association, 2013). Of note, most symptoms start about 24–48 h after the cessation of cannabis use, with a peak within 4–6 days, and the symptoms may persist for up to 1–4 weeks (Milin, Manion, Dare, & Walker, 2008). Estimates of the number of cannabis users experiencing withdrawal are variable, but the total number of cigarettes smoked predicts the intensity of withdrawal during abstinence from cannabis (McClure, Stitzer, & Vandrey, 2012).

In comparison with people who began cannabis use in adulthood, those who started during adolescence present two to four times more symptoms of dependence in the first 2 years subsequent to the first use (Volkow et al., 2014). An important topic of debate concerning cannabis addiction is the question of whether its use, mainly during adolescence, predicts an increased risk of the use of other illicit drugs, contributing to the view that cannabis could be a "gateway drug" (Kandel, 1975). Indeed, epidemiological surveys have shown that cannabis misuse predicts dependence on other illicit drugs (Fergusson, Boden, & Horwood, 2006). For instance, a positive correlation between cannabis use and the risk of abusing opiates and psychostimulants has been reported (Lynskey et al., 2003).

At the time of this writing, most attention has been focused on the effects of synthetic cannabinoids, also known as "spice." They were designed to mimic the effects of THC, but are up to 100 times more potent (Underwood, 2015). Spice products are a mixture of cannabis and synthetic CB1 receptor agonists, and thus, they can

cognitive and emotional deficits than later onset (Ehrenreich et al., 1999). These effects can be described as deficits in attention (Ehrenreich et al., 1999) and decision-making (Dougherty et al., 2013), poor intelligence quotient (Pope, Jacobs, Mialet, Yurgelun-Todd, & Gruber, 1997), deficits in executive functions (Fontes et al., 2011), and altered emotional states (Reilly, Didcott, Swift, & Hall, 1998). Moreover, there seems to be a correlation between the development of schizophrenia and cannabis use during adolescence. Indeed, some studies report that cannabis consumption during adolescence increases the risk of schizophrenia in adulthood four- to sixfold (Andreasson, Allebeck, Engstrom, & Rydberg, 1987; Arseneault et al., 2002), but others suggest that persons with the disposition to develop psychosis use more cannabis than healthy people, probably in an attempt to self-medicate and obtain relief from symptoms.

TABLE 2 Preclinical Studies Investigating the Rewarding/Reinforcing Effects of Δ^9-Tetrahydrocannabinol (THC)

Behavioral Model	Treatment	Species	Results	References
SA	THC 6.25 and 12.5 µg/kg iv SA	Food-restricted Wistar rats	↑SA	Takahashi and Singer (1979)
SA and CPP	Four days of THC 0.015–6 mg/kg. THC 0.01–1 µg/2 µl/infusion	Wistar rats	THC 0.075–0.75 mg/kg, ↑CPP; 1–3 mg/kg, ○; 6 mg/kg, ↑SA and CPA	Braida, Iosuè, Pegorini, and Sala (2004)
CPP	Five days of THC 1.0 and 10 mg/kg and 5 days of vehicle	ICR mice	THC 1.0 mg/kg, ↑CPP; THC 10 mg/kg, ↑CPA	Vann et al. (2008)
SA	THC 4 µg/kg in a fixed ratio 10 schedule	Squirrel monkeys	↑SA	Justinova et al. (2008)
SA	SA of THC (4 µg/kg/infusion)	Squirrel monkeys	↑SA	Justinova et al. (2013) and Justinová, Redhi, Goldberg, and Ferré (2014)

SA, self-administration; CPP, conditioned place preference; THC, Δ^9-tetrahydrocannabinol; ↑, increase; ↓, decrease; ○, no change; CPA, conditioned place aversion.

be smoked. They are sold with brand names like K2, Spice, Black Mamba, or Synthetic Marijuana, among others. The use of spice leads to more serious side effects than the use of cannabis. For instance, an increased risk of overdose, with heart attack, kidney failure, psychosis, and even death, has been reported in adolescents and adults (van Amsterdam et al., 2015). Contrary to cannabis, which contains CBD, which may provide protection against some THC-induced side effects, individuals using synthetic CB1 receptor agonists are probably more susceptible to developing psychosis.

TRANSLATIONAL RESEARCH ON CANNABIS USE IN ANIMAL MODELS

In rodents and primates, the reinforcing effects of THC can be measured directly with self-administration procedures and from the drug's ability to produce a conditioned preference for the environment paired with its administration. Although these animal models have considerable predictive validity for most psychotropic drugs, their results present some inconsistencies in the case of this cannabis constituent. As shown in Table 2, early preclinical studies of THC-induced reinforcement used high doses, unrelated to those that induce subjective effects in humans, with some studies suggesting an aversive profile for cannabis, which contrasts with the widespread notion that it has positive reinforcing effects in humans. Although THC may be the key dependence-inducing compound, animal models do not take into account certain differences, such as the oral and olfactory cues involved and the biological differences between animal models and human drug use.

The "gateway drug theory" has also been investigated in animal models, mainly using self-administration and locomotor sensitization procedures. For example, regarding the interaction of cannabis with psychostimulants, there are conflicting results, while for opioids, it appears that THC can prime the brain for enhanced response to opioids (Table 3). It is also worth mentioning that licit drugs, such as alcohol and nicotine, can be categorized as gateway drugs, as they also prime the brain for an enhanced response to other drugs. However, as pointed out in a review by Volkow et al. (2014), an alternative explanation is that people who are more susceptible to drug-taking are simply more likely to start with cannabis because of its accessibility, and their subsequent social interactions with peer drug users increase the probability of trying other drugs.

THERAPIES FOR CANNABIS USE AND WITHDRAWAL

Demand for treatment by patients with a long history of CUD has been increasing worldwide (Murray et al., 2007; Volkow et al., 2014). Currently, psychological approaches, such as cognitive–behavioral therapy and motivational enhancement therapy, are the mainstay (Danovitch & Gorelick, 2012). The available evidence supporting the effectiveness and safety of a pharmacological intervention to mitigate this psychiatric condition and/or the withdrawal symptoms of cannabis is still scarce. Indeed, it has been shown that fluoxetine, escitalopram, nefazodone, mirtazapine, venlafaxine, bupropion, buspirone, atomoxetine, lithium, divalproex sodium, olanzapine, risperidone, and quetiapine are probably of little value for the specific treatment of CUD (Balter, Cooper, & Haney, 2014; Marschal et al., 2014). However, it is worth remembering that some of the aforementioned drugs may be useful for the treatment of the anxiety and/or mood disorders commonly seen in cannabis users.

Preparations containing THC (or its synthetic form, dronabinol) alone and in conjunction with CBD are of potential value in reducing the withdrawal symptoms of cannabis (Allsop et al., 2014; Levin et al., 2011), although they are still considered experimental, because the appropriate dose ratio of these compounds and the duration of the treatment have not yet been determined. Of note, it is possible that an attenuating effect of these phytocannabinoids on cannabis use will be seen when they are combined with effective psychological therapies (Marschal et al., 2014). Moreover, preliminary positive evidence on gabapentin and N-acetylcysteine

TABLE 3 Effects of Previous Exposure to Δ⁹-Tetrahydrocannabinol (THC) on the Rewarding Effects of Psychostimulants and Opioids

Behavioral Model	Treatment	Species and Age of Onset of Treatment	Results	References
SA of amphetamine	Intercalated SA of THC and amphetamine	Adult Wistar rats	○ Amphetamine consumption	Takahashi and Singer (1981)
Amphetamine-induced stereotypy and locomotion	Chronic THC (14 days)	Adult Wistar rats	↑ Amphetamine-induced locomotion	Gorriti, Rodríguez de Fonseca, Navarro, and Palomo (1999)
Amphetamine-induced locomotion	Chronic THC (20 days)	Adult Wistar rats	↑ Amphetamine-induced locomotion	Lamarque, Taghzouti, and Simon (2001)
Amphetamine-induced stereotypy and locomotion	Pretreatment with THC at 28 and 32 postnatal days	Adolescent Sprague–Dawley rats	○ Locomotion and stereotypy	Ellgren, Hurd, and Franck (2004)
Amphetamine SA and psychostimulant-induced locomotion	Pretreatment with THC (15 days, 3 days between administrations)	Adult Sprague–Dawley rats	○ Consumption and locomotion	Cortright, Lorrain, Beeler, Tang, and Vezina (2011)
Operant SA of heroin	Pretreatment with THC at 28–49 postnatal days	Adult Long-Evans rats	↑ Heroin consumption; ↑ Expression of mRNA, density, and function of μ opioid receptors in limbic brain regions	Ellgren, Spano, and Hurd (2007)
Extinction and reinstatement of operant SA of heroin	Pretreatment with THC at 28–49 postnatal days	Adult Long-Evans rats	↑ Stress-induced heroin-seeking behavior	Spano, Ellgren, Wang, and Hurd (2007)

SA, self-administration; ↑, increase; ↓, decrease; ○, no change; THC, Δ⁹-tetrahydrocannabinol.
Adapted from De Carvalho and Takahashi (2014).

suggests they may also be promising for attenuating CUD and/or the withdrawal symptoms of cannabis (Balter et al., 2014), but further investigation is compulsory.

IS THERE A POTENTIAL THERAPEUTIC VALUE FOR MEDICAL CANNABIS?

The term medical cannabis may, in fact, encompass some of the different forms in which cannabinoids occur, namely: (1) standardized extracts of the *Cannabis* plant and (2) phytocannabinoids, such as THC and CBD, which were purified from the herbal extracts and are used in isolation or in conjunction (Pamplona & Takahashi, 2012; Pertwee, 2006).

The use of the standardized extracts of cannabis has been shown to relieve the symptoms of acquired immunodeficiency syndrome (AIDS), cancer chemotherapy, and multiple sclerosis (Ellis et al., 2009; Haney, Rabkin, Gunderson, & Foltin, 2005). There are also studies showing positive effects on chronic pain and glaucoma (e.g., Ware et al., 2010). The availability of various THC/CBD dose ratios is worth mentioning. For instance, Bedrocan® contains 22% THC and <1% CBD, Bediol® contains 6.5% THC and 8% CBD, while Bedrolite® contains <1% THC and 9% CBD (Fischedick, Van Der Kooy, & Verpoorte, 2010). However, which concentration of either of these phytocannabinoids is potentially the best one for each medical condition is still under investigation.

The CB1/CB2 receptor agonist THC (Marinol®) can also be prescribed for weight loss in AIDS. The synthetic form of THC is known as nabilone (Cesamet®) and is prescribed to suppress the vomiting and nausea induced by chemotherapy. Sativex® contains a similar dose ratio of THC and CBD, and it is used to treat spasticity in multiple sclerosis. Moreover, extracts of cannabis with high CBD concentrations have been used in the treatment of refractory epilepsy (Hughes, 2013).

Preclinical and preliminary clinical studies have suggested the use of cannabis-isolated compounds in the treatment of other medical conditions. For example, a study conducted in patients with posttraumatic stress disorder who received THC orally showed an improvement in the severity of some symptoms, such as sleep quality, and a reduction in nightmares (Roitman, Mechoulam, Cooper-Kazaz, & Shalev, 2014). In rodents, the efficacy of CBD, THC, and/or synthetic cannabinoids, which interfere with the endocannabinoid system in attenuating anxiety-like responses, morphine-related memory reconsolidation, and promoting neuroprotection, has been shown (Chiarlone et al., 2014; De Carvalho, Pamplona, Cruz, & Takahashi, 2014; Stern, Gazarini, Takahashi, Guimarães, & Bertoglio, 2012). The results also suggest that reducing the activity of fatty acid amide hydrolase, an enzyme that terminates anandamide signaling and may be inhibited by CBD, induces a gain in function in the regulation of fear and anxiety (Dincheva et al., 2015). Although effective when administered alone, preparations containing CBD are better tolerated than isolated THC

(Russo & Guy, 2006), probably because CBD counteracts some THC-induced side effects, such as psychotic and anxiety symptoms (Zuardi et al., 2012).

POLICY CHANGES IN CANNABIS REGULATION

As reviewed by Bostwick (2012), although recreational and "medical" users differ in how they consume cannabis, a significant overlap between them has been demonstrated. For example, in a Canadian study of 104 patients infected with human immunodeficiency virus, 43% reported cannabis use in the previous year. Although two-thirds had endorsed medical indications, ranging from anorexia to insomnia, emesis, and anxiety, nearly 80% of this group also used cannabis recreationally (Furler, Einarson, Millson, Walmsley, & Bendayan, 2004). Another study posited the notion that "typically medical cannabis use followed the recreational one and the majority of those interviewed were long-term and sometimes heavy recreational users" (Bostwick, 2012). As policy shifts toward the legalization of cannabis, it is reasonable to hypothesize that its use will increase and, by extension, the number of people with adverse health consequences will also increase.

In addition, there is considerable variation both in the prevalence of cannabis use and in the legal constraints on its use in various countries/states. For example, whereas the Netherlands has been closing coffee shops, Uruguay has legalized cannabis with the allegation that more harm comes from its criminalization than from its deleterious effects on mental health. Because laws of this kind have never before been implemented in a national or state jurisdiction, no previous case studies predictive of the changes to be expected are available. Thus, the impact of this legislation can be measured and evaluated only through reliable data and regular monitoring efforts.

CONCLUSIONS

In this overview, we briefly explored the controversies surrounding the use and misuse of cannabis, the most widely used illicit drug in the world for recreational purposes. In addition to its well-known acute effects on cognitive and motor functions, compelling evidence suggests that long-term cannabis users are at risk of developing psychosis and CUD. However, whether early cannabis use is related to a preexisting pathology, which is exacerbated by cannabis use, remains an open question. Moreover, we showed that studies using animal models are important for evaluating the reinforcing and gateway effects of cannabis. Additional preclinical research is necessary to investigate further the molecular and neurochemical mechanisms underlying the deleterious effects of cannabis. Not all users will need therapies to manage the withdrawal or support the cessation of cannabis use. However, it is important to establish effective pharmacotherapies for the treatment of cannabis withdrawal, especially in long-term or heavy cannabis users. The increasing knowledge regarding cannabinoids and the public approval of medicinal cannabis raises the possibility of many promising therapeutic applications. Finally, policy changes in cannabis regulation in the Americas have allowed the establishment of a licit supply chain, including licensing for production, personal cultivation, and retail commercialization of the market. Thus, it is reasonable to hypothesize that cannabis use will increase. Because the implementation of regulatory laws, for example, in Uruguay and in the states of Washington and Colorado (USA), may differ substantially, the legal status of recreational or medical cannabis will remain contradictory and complex.

APPLICATIONS TO OTHER ADDICTIONS AND SUBSTANCE MISUSE

Alcohol and nicotine have commonly been the two most widely used licit substances worldwide. Some authors agree that both may serve as a gateway to other drugs, such as cannabis, cocaine, and other illicit drugs. For example, among humans ages 18–34 years who used cocaine at least once in the United States, 90% used cigarettes before beginning cocaine use, and only 2.5% had never smoked (Levine et al., 2011). In rodents, nicotine increases the response to cocaine, promoting plastic changes in brain-rewarding areas, such as the striatum; however, reversing the drug order did not cause effects similar to those caused by nicotine, suggesting that, in this case, nicotine works as a gateway to cocaine misuse. If this condition is applied to humans, and considering the decrease in cigarette consumption around the world, one could expect a consequent reduction in cocaine addiction. However, the biological disposition to become addicted to any illicit drug should also be taken into account. Thus, more rigorous policies regarding nicotine use, for example, must be implemented, as well as educational programs on the risk of developing a drug addiction, even from using a licit drug.

DEFINITION OF TERMS

Cannabinoid This is a synthetic drug that usually potentiates the endocannabinoid system (e.g., by activating the cannabinoid receptors) and induces pharmacological effects similar to those of cannabis.
Endocannabinoids These are molecules that are endogenously synthesized, usually on demand, and that bind to (and activate the) cannabinoid receptors.
Conditioned place preference This is a form of Pavlovian conditioning used to assess the rewarding and motivational properties of drugs of abuse.
High This refers to the cannabis-induced rewarding effects.
Reward This is a stimulus that induces hedonic pleasure.
Relapse This is the resumption of drug misuse after a period of abstinence.
Tolerance This is the attenuation of the cannabis-induced effects associated with the repeated use of this substance.
Withdrawal syndrome This refers to a set of symptoms produced after a period in which the drug of addiction has not been used.

KEY FACTS OF THE CANNABINOID TETRAD TEST

- In the mid-1960s, at the same time that Raphael Mechoulam discovered THC, Billy Martin developed the cannabinoid tetrad test.
- The cannabinoid tetrad test consists of a series of behavioral tests in rodents aimed at the observation of four signs: hypolocomotion, hypothermia, analgesia, and catalepsy.
- The cannabinoid tetrad is used for screening drugs with cannabinoid-like activity.

- As expected, CB1 receptor agonists, such as THC and WIN 55,212-2, induce hypolocomotion, hypothermia, analgesia, and catalepsy.
- The tetrad effect of cannabinoids is due to the high expression of CB1 receptors in brain regions related to the tetrad signs.

SUMMARY POINTS

- THC, the main psychotomimetic compound of cannabis, binds to CB1 and CB2 receptors present in limbic and rewarding areas, promoting the effects of "high" and pleasure.
- Depending on the Cannabis strain and the amount smoked, it can precipitate acute psychotic symptoms, as well as visual and auditory hallucinations.
- The earlier the onset of cannabis use, the greater is the risk of becoming addicted to it.
- The risk of developing dependence on cannabis is smaller than the majority of other illicit drugs, but the number of users is significant.
- As with other drugs of abuse, the risk of developing an addiction is influenced by multiple factors, including environmental, genetic, and individual ones; the facility of obtaining the drug; impulsivity; and a risk-taking personality.
- The cannabis withdrawal syndrome is considered mild compared to those of heroin or ethanol, possibly because of the long half-lives of cannabinoids found in the herbal extract.

ACKNOWLEDGMENT

This work was supported by grants from the Conselho Nacional de Desenvolvimento Científico e Tecnológico (CNPq), Coordenação de Aperfeiçoamento de Pessoal de Nível Superior, Fundação de Amparo à Pesquisa e Inovação do Estado de Santa Catarina, and Programa de Apoio aos Núcleos de Excelência, all of Brazil. Dr C.A.J. Stern is supported by a scholarship from the CNPq and Professors R.N. Takahashi and L.J. Bertoglio are supported by research fellowships from the CNPq. We thank Dr Cristiane Ribeiro de Carvalho for comments on the manuscript.

REFERENCES

Allsop, D. J., Copeland, J., Lintzeris, N., Dunlop, A. J., Montebello, M., Sadler, C., ... McGregor, I. S. (2014). Nabiximols as an agonist replacement therapy during cannabis withdrawal: a randomized clinical trial. *JAMA Psychiatry, 71*, 281–291.

American Psychiatric Association. (2013). *Diagnostic and statistical manual of mental disorders: DSM-V*. Washington, DC.

van Amsterdam, J., Brunt, T., & van den Brink, W. (2015). The adverse health effects of synthetic cannabinoids with emphasis on psychosis-like effects. *Journal of Psychopharmacology, 29*, 254–263.

Andreasson, S., Allebeck, P., Engstrom, A., & Rydberg, U. (1987). Cannabis and schizophrenia. A longitudinal study of Swedish conscripts. *Lancet, 2*, 1483–1486.

Arseneault, L., Cannon, M., Poulton, R., Murray, R., Caspi, A., & Moffitt, T. E. (2002). Cannabis use in adolescence and risk for adult psychosis: longitudinal prospective study. *BMJ, 325*, 1212–1213.

Balter, R. E., Cooper, Z. D., & Haney, M. (2014). Novel pharmacologic approaches to treating cannabis use disorder. *Current Addiction Reports, 1*, 137–143.

Bostwick, J. M. (2012). Blurred boundaries: the therapeutics and politics of medical marijuana. *Mayo Clinic Proceedings, 87*, 172–186.

Braida, D., Iosuè, S., Pegorini, S., & Sala, M. (2004). Delta9-tetrahydrocannabinol-induced conditioned place preference and intracerebroventricular self-administration in rats. *European Journal of Pharmacology, 506*, 63–69.

Chiarlone, A., Bellocchio, L., Blázquez, C., Resel, E., Soria-Gómez, E., Cannich, A., ... Guzmán, M. (2014). A restricted population of CB1 cannabinoid receptors with neuroprotective activity. *Proceedings of the National Academy of Sciences of the United States of America, 111*, 8257–8262.

Cortright, J. J., Lorrain, D. S., Beeler, J. A., Tang, W. J., & Vezina, P. (2011). Previous exposure to $\Delta 9$-tetrahydrocannabinol enhances locomotor responding to but not self-administration of amphetamine. *The Journal of Pharmacology and Experimental Therapeutics, 337*, 724–733.

Danovitch, I., & Gorelick, D. A. (2012). State of the art treatments for cannabis dependence. *The Psychiatric Clinics of North America, 35*, 309–326.

De Carvalho, C. R., Pamplona, F. A., Cruz, J. S., & Takahashi, R. N. (2014). Endocannabinoids underlie reconsolidation of hedonic memories in Wistar rats. *Psychopharmacology, 231*, 1417–1425.

De Carvalho, C. R., & Takahashi, R. N. (2014). A maconha aumenta a vulnerabilidade a opioides em animais de laboratório. *Revista da Biologia USP, 13*, 24–27.

Dincheva, I., Drysdale, A. T., Hartley, C. A., Johnson, D. C., Jing, D., King, E. C., ... Lee, F. S. (2015). FAAH genetic variation enhances fronto-amygdala function in mouse and human. *Nature Communications, 6*, 6395.

Dougherty, D. M., Mathias, C. W., Dawes, M. A., Furr, R. M., Charles, N. E., Liguori, A., ... Acheson, A. (2013). Impulsivity, attention, memory, and decision-making among adolescent marijuana users. *Psychopharmacology, 226*, 307–319.

D'Souza, D. C., Perry, E., MacDougall, L., Ammerman, Y., Cooper, T., Wu, Y. T., ... Krystal, J. H. (2004). The psychotomimetic effects of intravenous delta-9-tetrahydrocannabinol in healthy individuals: implications for psychosis. *Neuropsychopharmacology, 29*, 1558–1572.

Ehrenreich, H., Rinn, T., Kunert, H. J., Moeller, M. R., Poser, W., Schilling, L., ... Hoehe, M. R. (1999). Specific attentional dysfunction in adults following early start of cannabis use. *Psychopharmacology, 142*, 295–301.

Ellgren, M., Hurd, Y. L., & Franck, J. (2004). Amphetamine effects on dopamine levels and behavior following cannabinoid exposure during adolescence. *European Journal of Pharmacology, 497*, 205–213.

Ellgren, M., Spano, S. M., & Hurd, Y. L. (2007). Adolescent cannabis exposure alters opiate intake and opioid limbic neuronal populations in adult rats. *Neuropsychopharmacology, 32*, 607–615.

Ellis, R. J., Toperoff, W., Vaida, F., van den Brande, G., Gonzales, J., Gouaux, B., ... Atkinson, J. H. (2009). Smoked medicinal cannabis for neuropathic pain in HIV: a randomized, crossover clinical trial. *Neuropsychopharmacology, 34*, 672–680.

Fergusson, D. M., Boden, J. M., & Horwood, L. J. (2006). Cannabis use and other illicit drug use: testing the cannabis gateway hypothesis. *Addiction, 101*, 556–569.

Fischedick, J., Van Der Kooy, F., & Verpoorte, R. (2010). Cannabinoid receptor 1 binding activity and quantitative analysis of *Cannabis sativa* L. smoke and vapor. *Chemical and Pharmaceutical Bulletin, 58*, 201–207.

Fontes, M. A., Bolla, K. I., Cunha, P. J., Almeida, P. P., Jungerman, F., Laranjeira, R. R., ... Lacerda, A. L. (2011). Cannabis use before age 15 and subsequent executive functioning. *British Journal of Psychiatry, 198*, 442–447.

Furler, M. D., Einarson, T. R., Millson, M., Walmsley, S., & Bendayan, R. (2004). Medicinal and recreational marijuana use by patients infected with HIV. *AIDS Patient Care STDS, 18,* 215–228.

Gorriti, M. A., Rodríguez de Fonseca, F., Navarro, M., & Palomo, T. (1999). Chronic (Δ)-9-tetrahydrocannabinol treatment induces sensitization to the psychomotor effects of amphetamine in rats. *European Journal of Pharmacology, 365,* 133–142.

Haney, M., Rabkin, J., Gunderson, E., & Foltin, R. W. (2005). Dronabinol and marijuana in HIV(+) marijuana smokers: acute effects on caloric intake and mood. *Psychopharmacology, 181,* 170–178.

Hirvonen, J., Goodwin, R. S., Li, C. T., Terry, G. E., Zoghbi, S. S., Morse, C., ... Innis, R. B. (2012). Reversible and regionally selective downregulation of brain cannabinoid CB(1) receptors in chronic daily cannabis smokers. *Molecular Psychiatry, 17,* 642–649.

Hughes, S. (2013). *FDA approves cannabis extract study in pediatric epilepsy* (Vol. 2014).

Justinova, Z., Mangieri, R. A., Bortolato, M., Chefer, S. I., Mukhin, A. G., Clapper, J. R., ... Goldberg, S. R. (2008). Fatty acid amide hydrolase inhibition heightens anandamide signaling without producing reinforcing effects in primates. *Biological Psychiatry, 64,* 930–937.

Justinova, Z., Mascia, P., Wu, H. Q., Secci, M. E., Redhi, G. H., Panlilio, L. V., ... Goldberg, S. R. (2013). Reducing cannabinoid abuse and preventing relapse by enhancing endogenous brain levels of kynurenic acid. *Nature Neuroscience, 16,* 1652–1661.

Justinová, Z., Redhi, G. H., Goldberg, S. R., & Ferré, S. (2014). Differential effects of presynaptic versus postsynaptic adenosine A_{2A} receptor blockade on Δ^9-tetrahydrocannabinol (THC) self-administration in squirrel monkeys. *The Journal of Neuroscience, 34,* 6480–6484.

Kandel, D. (1975). Stages in adolescent involvement in drug use. *Science, 190,* 912–914.

Lamarque, S., Taghzouti, K., & Simon, H. (2001). Chronic treatment with delta (9) tetrahydrocannabinol enhances the locomotor response to amphetamine and heroin. Implications for vulnerability to drug addiction. *Neuropharmacology, 41,* 118–129.

Levine, A., Huang, Y., Drisaldi, B., Griffin, E. A., Jr., Pollak, D. D., Xu, S., ... Kandel, E. R. (2011). Molecular mechanism for a gateway drug: epigenetic changes initiated by nicotine prime gene expression by cocaine. *Science Translational Medicine, 3,* 107ra109.

Levin, F. R., Mariani, J. J., Brooks, D. J., Pavlicova, M., Cheng, W., & Nunes, E. V. (2011). Dronabinol for the treatment of cannabis dependence: a randomized, double-blind, placebo-controlled trial. *Drug and Alcohol Dependence, 116,* 142–150.

Li, H. L. (1973). An archaeological and historical account of cannabis in China. *Economic Botany, 28,* 437–448.

Lynskey, M. T., Heath, A. C., Bucholz, K. K., Slutske, W. S., & Madden, P. A. (2003). Escalation of drug use in early-onset cannabis users vs co-twin controls. *JAMA, 289,* 427–433.

Marshall, K., Gowing, L., Ali, R., & Le Foll, B. (2014). Pharmacotherapies for cannabis dependence. *The Cochrane Database of Systematic Reviews, 12,* CD008940.

McClure, E. A., Stitzer, M. L., & Vandrey, R. (2012). Characterizing smoking topography of cannabis in heavy users. *Psychopharmacology, 220,* 309–318.

Milin, R., Manion, I., Dare, G., & Walker, S. (2008). Prospective assessment of cannabis withdrawal in adolescents with cannabis dependence: a pilot study. *Journal of the American Academy of Child and Adolescent Psychiatry, 47,* 174–178.

Murray, R. M., Morrison, P. D., Henquet, C., & Di Forti, M. (2007). Cannabis, the mind and society: the hash realities. *Nature Reviews Neuroscience, 8,* 885–895.

Pamplona, F. A., & Takahashi, R. N. (2012). Psychopharmacology of the endocannabinoids: far beyond anandamide. *Journal of Psychopharmacology, 26,* 7–22.

Pertwee, R. G. (2006). The pharmacology of cannabinoid receptors and their ligands: an overview. *International Journal of Obesity, 30*(Suppl. 1), S13–S18.

Pope, H. G., Jr., Gruber, A. J., Hudson, J. I., Huestis, M. A., & Yurgelun-Todd, D. (2002). Cognitive measures in long-term cannabis users. *Journal of Clinical Pharmacology, 42,* 41S–47S.

Pope, H. G., Jr., Jacobs, A., Mialet, J. P., Yurgelun-Todd, D., & Gruber, S. (1997). Evidence for a sex-specific residual effect of cannabis on visuospatial memory. *Psychotherapy and Psychosomatics, 66,* 179–184.

Reilly, D., Didcott, P., Swift, W., & Hall, W. (1998). Long-term cannabis use: characteristics of users in an Australian rural area. *Addiction, 93,* 837–846.

Roitman, P., Mechoulam, R., Cooper-Kazaz, R., & Shalev, A. (2014). Preliminary, open-label, pilot study of add-on oral Δ^9-tetrahydrocannabinol in chronic post-traumatic stress disorder. *Clinical Drug Investigation, 34,* 587–591.

Russo, E., & Guy, G. W. (2006). A tale of two cannabinoids: the therapeutic rationale for combining tetrahydrocannabinol and cannabidiol. *Medical Hypotheses, 66,* 234–246.

Spano, M. S., Ellgren, M., Wang, X., & Hurd, Y. L. (2007). Prenatal cannabis exposure increases heroin seeking with allostatic changes in limbic enkephalin systems in adulthood. *Biological Psychiatry, 61,* 554–563.

Stern, C. A., Gazarini, L., Takahashi, R. N., Guimarães, F. S., & Bertoglio, L. J. (2012). On disruption of fear memory by reconsolidation blockade: evidence from cannabidiol treatment. *Neuropsychopharmacology, 37,* 2132–2142.

Takahashi, R. N., & Singer, G. (1979). Self-administration of delta 9-tetrahydrocannabinol by rats. *Pharmacology Biochemistry and Behavior, 11,* 737–740.

Takahashi, R. N., & Singer, G. (1981). Cross self-administration of delta 9-tetrahydrocannabinol and D-amphetamine in rats. *Brazilian Journal of Medical and Biological Research, 14,* 395–400.

Underwood, E. (2015). A new drug war – as a growing wave of designer drugs hits the streets, researchers try to forecast which will prove to be most popular – and dangerous. *Science, 347,* 469–473.

UNODC. (2012). Available at http://www.unodc.org/.

Vann, R. E., Gamage, T. F., Warner, J. A., Marshall, E. M., Taylor, N. L., Martin, B. R., & Wiley, J. L. (2008). Divergent effects of cannabidiol on the discriminative stimulus and place conditioning effects of delta(9)-tetrahydrocannabinol. *Drug and Alcohol Dependence, 94,* 191–198.

Volkow, N. D., Compton, W. M., & Weiss, S. R. (2014). Adverse health effects of marijuana use. *The New England Journal of Medicine, 370,* 2219–2227.

Ware, M. A., Wang, T., Shapiro, S., Robinson, A., Ducruet, T., Huynh, T., ... Collet, J. P. (2010). Smoked cannabis for chronic neuropathic pain: a randomized controlled trial. *CMAJ, 182,* E694–E701.

Weiland, B. J., Thayer, R. E., Depue, B. E., Sabbineni, A., Bryan, A. D., & Hutchison, K. E. (2015). Daily marijuana use is not associated with brain morphometric measures in adolescents or adults. *The Journal of Neuroscience, 35,* 1505–1512.

Zuardi, A. W., Crippa, J. A., Hallak, J. E., Bhattacharyya, S., Atakan, Z., Martin-Santos, R., ... Guimaraes, F. S. (2012). A critical review of the antipsychotic effects of cannabidiol: 30 years of a translational investigation. *Current Pharmaceutical Design, 18,* 5131–5140.

Chapter 62

An Overview of Major and Minor Phytocannabinoids

Jahan P. Marcu
Americans for Safe Access, Washington, DC, USA; Green Standard Diagnostics, Inc., Las Vegas, NV, USA

Abbreviations

BCP β-Caryophyllene
CBC Cannabichromene
CBG Cannabigerol
ECS Endocannabinoid system
THC Δ^9-tetrahydrocannabinol
THCA Tetrahydrocannabinol acid
THCV Tetracannabivarin
TRP Transient receptor potential
TRPV Transient receptor potential vanilloid type

INTRODUCTION

For nearly five millennia, *Cannabis* has been documented as a medicine with unmatched applicability (Russo, 2011). The mechanism of action of *Cannabis* remained a mystery until fairly recently, with the discovery of phytocannabinoids, or plant cannabinoids, and the receptor system known as the endocannabinoid system (ECS). Phytocannabinoids are terpenophenolic compounds associated with the effects of the *Cannabis* plant and mimic the effects of endogenous cannabinoids. These phytocannabinoids are biosynthesized and secreted by glandular trichomes found on the flower tops of the *Cannabis* plant (Figure 1). In the 1960s a host of cannabinoids were discovered, including cannabigerol (CBG), tetracannabivarin (THCV), and cannabichromene (CBC) (Figure 2). More than 100 cannabinoids have been identified in *Cannabis* and reports are emerging of their occurrence in other plants (Appendino, Chianese, & Taglialatela-Scafati, 2011; Pertwee, 2014). The great bulk of clinical cannabinoid research focuses on the psychotropic Δ^9-tetrahydrocannabinol (THC), inadvertently marginalizing the roles of other cannabinoids in altering the pharmacological activities of *Cannabis* compounds. Many compounds in the plant can enhance or inhibit various aspects of THC pharmacology, for example, the inhibition of unwanted side effects such as with the coadministration of cannabidiol (CBD) (Russo & Guy, 2006). Major and minor phytocannabinoids can have remarkably positive effects in mammalian behavior related to anxiety and drug acquisition and may offer novel drug abuse treatment options.

The ratios of these major and minor compounds can vary greatly and some compounds are not often detected or tested for or reported. Furthermore, the *Cannabis* plant is not the only source of phytocannabinoids. Excellent in-depth reviews of the phytocannabinoids have been published (Pertwee, 2014; Pharmacopoeia, 2013). The following is an overview of the major and minor phytocannabinoids that can be found in *Cannabis*.

TETRAHYDROCANNABINOL ACID, TETRAHYDROCANNABIVARINIC ACID, AND CANNABIGEROL ACID

Cannabinoid acids are found as primary metabolites in *Cannabis* plants. For example, tetrahydrocannabinol acid (THCA) is synthesized in glandular trichomes of the *Cannabis* plant and forms THC after the parent compound is decarboxylated by UV exposure, prolonged storage, or heat (Figure 3). The cannabinoid acids do not produce any significant or documented psychotropic effects. THCA, tetrahydrocannabivarinic acid, and cannabigerol acid (CBGA) are the immediate natural precursors of THC, THCV, and CBG. THCA and CBGA are the primary phytocannabinoid metabolites and can cause apoptosis of insect cells (Figure 3) (Sirikantaramas, 2004)

Basic research has shown that these acidic compounds can efficiently activate TRPM8 channels and stimulate or desensitize a range of other transient receptor potential (TRP) cation channels. THCA and CBGA have been found to inhibit enzymes responsible for the breakdown of endocannabinoids, as well as cyclooxygenase-1 and -2, thus stimulating the ECS by increasing levels of endogenous cannabinoids.

CANNABIGEROL

This compound was the first cannabinoid purified from *Cannabis* (Gaoni & Mechoulam, 1964). CBG lacks the psychotropic effects of THC (Grunfeld & Edery, 1969a, 1969b, 1969c). This compound does not have a significant subjective psychotropic effect in humans but CBG may stimulate a range of receptors important for pain, inflammation, and heat sensitization. This compound can antagonize transient receptor potential vanilloid type (TRPV) 8 receptors and stimulates TRPV1, TRPV2, TRPA1, TRPV3, TRPV4, and α_2-adrenoceptor activity (Cascio, Gauson, Stevenson, Ross, & Pertwee, 2010; De Petrocellis & Di Marzo, 2009; De Petrocellis &

An Overview of Major and Minor Phytocannabinoids **Chapter | 62** 673

FIGURE 1 The *Cannabis* plant and its trichomes. Immature *Cannabis* plants are seen in (A), while in (B) there is an example of flowering plants. An example of a harvested and dried flower from a *Cannabis* plant is shown in (C), while (D and E) display close-ups of the trichomes, visible as pale stalks. A glandular cystolic trichome can be seen in (F); these are found in *Cannabis* flowers and biosynthesize cannabinoids. A cannabinoid-rich oil is secreted out of the top of trichomes in a waxy layer. *Photos A–C, courtesy of J.P.M., photos D–F, courtesy of E.R.*

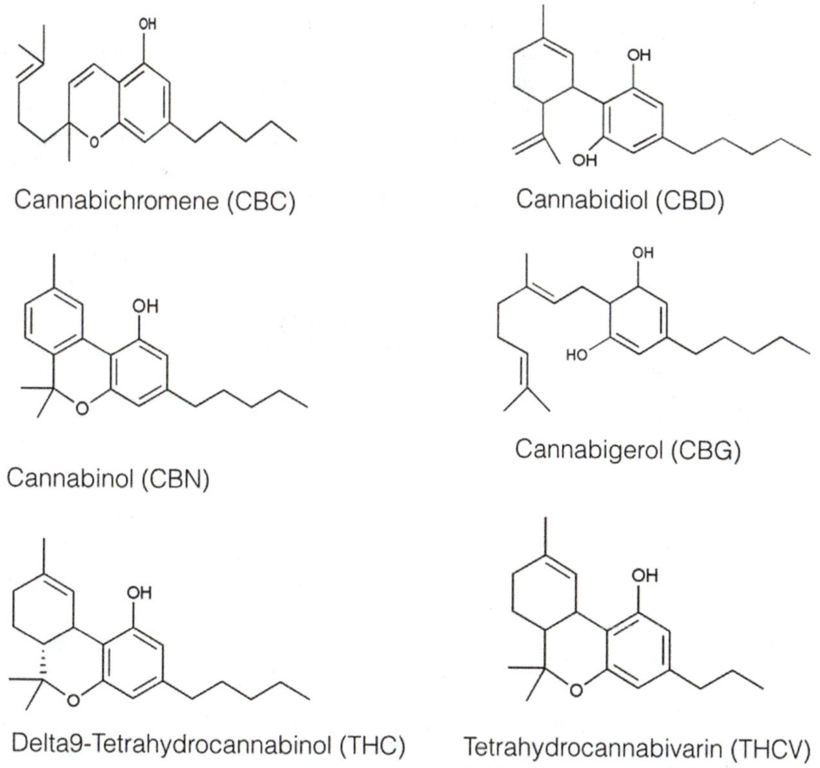

FIGURE 2 **Structures of cannabinoids found in the *Cannabis* plant.** These compounds are the secondary metabolites of *Cannabis*. These compounds are synthesized as acidic versions and decarboxylate over time or when heat is applied. *All images generated by J.P.M. using ChemDraw software.*

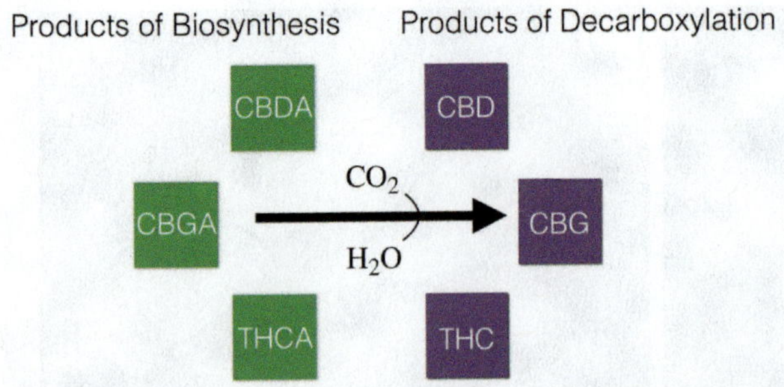

FIGURE 3 Products of biosynthesis and decarboxylation. Phytocannabinoids are biosynthesized as acidic precursors, such as THCA. Acidic phytocannabinoids are decarboxylated to form neutral cannabinoids. The conversion of acidic to neutral cannabinoid can occur from prolonged storage and when heat is applied, generating carbon dioxide and water. CBDA, cannabidiolic acid.

Di Marzo, 2010; De Petrocellis et al., 2011, 2012). CBG can also antagonize the stimulation of serotonin 5-HT1A and cannabinoid type 1 (CB1) receptors with significant efficiency.

Δ⁹-TETRAHYDROCANNABINOL

THC is the degradation product of nonpsychotropic THCA, which is synthesized from CBGA by THCA synthase. THC is a partial agonist at CB1 and CB2 receptors with a high affinity for both receptors. Stimulation of CB1 receptors by THC can lead to a tetrad of effects in assays with laboratory animals; these effects are documented as suppression of locomotor activity, hypothermia, catalepsy (ring test), and antinociceptive effects in tail flick test (Martin et al., 1991). THC can stimulate CB2 receptors, which may decrease the growth of some cancers and reduce arthritic pain and edema in models of arthritis. Perhaps most surprising is that direct stimulation of CB2 receptors can result in significantly reducing cocaine self-administration in animals (Gardner, 2013). Stimulation of CB2 receptors is not associated with the psychotropic effects of *Cannabis* use.

THC also has several non-CB receptor mechanisms that have been reported; these include inhibiting the 5-HT3A receptor, enhancing glycine receptor activation by allosteric modification, elevating calcium levels via TRPA1 or TRPV2, reducing elevated intracellular calcium levels from TRPM8 activity, stimulating nuclear receptors, and stimulating G Protein Receptor 18 (Barann et al., 2002; De Petrocellis et al., 2012, 2008; Hejazi et al., 2006; McHugh, Page, Dunn, & Bradshaw, 2012; O'Sullivan, Tarling, Bennett, Kendall, & Randall, 2005).

Oral THC administration can have significant effects on anxiety, depression, and mood. The effects of THC in humans can vary depending on the experience of the subject. The oral administration of pure THC to naïve subjects can induce anxiety but this is not reported with experienced users, and use of the drug is not significantly associated with developing anxiety and depressive disorders later in life (Bahi et al., 2014; Ballard, Bedi, & de Wit, 2012; Campos et al., 2013; Crippa et al., 2009). Oral administration of THC results in its metabolism to 11-hydroxy-THC, which possesses up to 10 times greater potency. This metabolism can explain some discrepancies between the observed effects in groups administered oral or inhaled forms of THC.

THC-predominant *Cannabis* is reported to be a commonly abused or misused substance. Street *Cannabis* can be contaminated or adulterated and this may underlie the negative health aspects of lifelong use. THC can cause temporary impairments to neuropsychomotor performance. All observable negative effects of THC administration on neurocognitive tasks disappear within 30 days regardless of the amount or length of use (Pope, Gruber, Hudson, Huestis, & Yurgelun-Todd, 2001). The proposed treatment for so-called *Cannabis* addiction or withdrawal is oral THC (Lichtman & Martin, 2005). CBD may also be considered an antidote for THC, as CBD and compounds in the plant tame or inhibit the psychotropic effects of THC (Russo, 2011). Excellent reviews are available covering numerous clinical trials with oral and inhaled THC for the treatment of over 10 pathologies (Ben Amar, 2006; Hazekamp & Grotenhermen, 2010; Pacher et al., 2006).

THC, the ECS, and the endorphin/opiate system can interact in remarkable ways (Table 1). Animal research has demonstrated a potential prophylactic effect on developing opiate dependence, as adolescent exposure to chronic THC blocks opiate dependence in maternally deprived rats (Morel, Giros, & Daugé, 2009). The ECS is proposed to interact with endorphins, through the release of opioid peptides from CB receptor activation and the synthesis of endocannabinoids induced by opiate receptor stimulation

TABLE 1 The ECS Is Proposed to Interact with Opioids through a Few Mechanisms

References	Finding
Morel et al. (2009)	Adolescent exposure to THC may impart resistance to opiate dependence in maternally deprived animals.
Abrams et al. (2011) and Russo et al. (2008)	CB receptor stimulation can result in endorphin release, and opioid receptor stimulation may increase synthesis of endogenous cannabinoids.
Abrams et al. (2011)	Clinical research demonstrates that THC can enhance the pain-relieving effects of suboptimal doses of opiates.
Bachhuber et al. (2014)	The state of Colorado has experienced a significant decrease in opiate-related deaths since the implementation of medical and commercial *Cannabis* laws.

The release of opioid peptides by CBs and the release of endocannabinoids by opioids may be one mechanism (Abrams et al., 2011; Russo, 2008). Clinically, THC may enhance the pain-relieving effects of opiates, lowering the amount of an opiate necessary for relief. Drug abuse studies demonstrate that adolescent exposure to chronic THC blocks opiate dependence in maternally deprived rats. There is evidence of the existence of a direct receptor–receptor interaction and cellular pathways, such as via allosteric modification of heterodimers.

(Abrams, Couey, Shade, Kelly, & Benowitz, 2011; Russo et al., 2008). Clinically, THC may enhance the pain-relieving effects of opiates, lowering the amount of an opiate necessary for relief. Surveys suggest *Cannabis* is used to decrease the use of other drugs (alcohol, nicotine, and opiates) (Reiman, 2009). In the United States, the state governments that have passed commercial *Cannabis*/marijuana laws report lower opiate overdose and related death statistics; these populations may reflect what has been observed in surveys and clinical studies of THC and opiates (Bachhuber, Saloner, Cunningham, & Barry, 2014).

TETRAHYDROCANNABIVARIN

THCV is a propyl analogue of THC and most often occurs as a small percentage of dried plant material, and THCV-rich plants that are 16% THCV by dry weight have been developed by a pharmaceutical company (Pharmacopoeia, 2013). Mechanistically speaking, THCV can behave as both an agonist and an antagonist at CB receptors depending on the concentration (Pertwee, 2008).

Antagonizing CB receptors can suppress appetite and the intoxicating effects of THC. However, caution must be emphasized when developing CB1 receptor antagonists. Clinical studies in human populations studying the antagonists of CB receptors with the drug rimonabant (SR141716A) led to depressive episodes and potentially worsened neurodegenerative disease outcomes, and ultimately this drug was withdrawn from the market (McLaughlin, 2012). Despite this setback, SR141716A remains a very important research tool for unlocking potential medical treatments targeting the CB receptors and deepening the understanding of the ECS.

CANNABIDIOL

The main nonpsychotropic phytocannabinoids are CBD and its acidic precursor cannabidiolic acid. These are the most abundant phytocannabinoids in European hemp (Pharmacopoeia, 2013). CBD has a very low affinity for CB receptors but may have significant CB1- and CB2-independent mechanisms of action. CBD is reported to be an agonist at TRPV1 and 5-HT1A receptors and to enhance adenosine receptor signaling (Russo, Burnett, Hall, & Parker, 2005). Exceptional tolerability of CBD in humans has been demonstrated (Mechoulam, Parker, & Gallily, 2002). CBD can produce a wide range of pharmacological activity including anticonvulsive, anti-inflammatory, antioxidant, and antipsychotic effects. These effects underlie the neuroprotective properties of this CB and support its role in the treatment of a number of neurological and neurodegenerative disorders, including epilepsy, seizures, Parkinson disease, amyotrophic lateral sclerosis, Huntington disease, Alzheimer disease, and multiple sclerosis (Hofmann & Frazier, 2013; de Lago & Fernández-Ruiz, 2007; Martin-Moreno et al., 2011; Scuderi et al., 2009).

CBD possesses the unique ability to counteract the intoxicating effects of *Cannabis* (Russo & Guy, 2006). The benefits of CBD include reducing the unwanted side effects of THC, a dynamic pharmacological effect that has been fairly well studied in clinical trials. CBD is included in a specific ratio of 1/1 in the medicinal *Cannabis* preparation and licensed pharmaceutical known as Sativex®, which has been studied in numerous properly controlled clinical trials representing some 30,000 patient years (Flachenecker, Henze, & Zettl, 2014; Rog, 2010; Sastre-Garriga, Vila, Clissold, & Montalban, 2014; Wade, Collin, Stott, & Duncombe, 2010).

CANNABICHROMENE

CBC can be one of the most abundant nonpsychotropic CBs found in strains or varieties of *Cannabis* (Brown & Harvey, 1990; Holley, Hadley, & Turner, 1975). CBC can cause strong anti-inflammatory effects in animal models of edema through non-CB receptor mechanisms (DeLong, Wolf, Poklis, & Lichtman, 2010). CBC has been shown to significantly interact with TRP cation channels, including TRPA1, TRPV1–4, and TRPV8 (Pertwee, 2014). CBC can also produce behavioral activity in the Billy Martin tetrad assay for the effects of CB administration. The effects of CBC can be augmented for additive results when THC is coadministered. CBC administration can induce nociception by itself and can potentiate the nociceptive effects of THC in animal models.

CANNABINOL

CBN is a degradation product of THC and its presence in the *Cannabis* plant may indicate the relative age of the harvested plant material or how well the material was stored. CBN binds less efficiently than THC to CB1 and CB2 receptors. CBN binds more tightly to CB2 receptors than to CB1 receptors. The metabolite of CBN is 11-hydroxy-CBN, which is reported to be more potent at CB1 receptors (Yamamoto et al., 2003).

A recent review of phytocannabinoids summarized the ability of CBN to inhibit the activity of a number of enzymes, including cyclooxygenase, lipoxygenase, and a host of cytochrome P450 (CYP) enzymes (e.g., CYP1A1, CYP1A2, CYP2B6, CYP2C9, CYP3A4, CYP3A5, CYP2A6, CYP2D6, CYP1B1, and CYP3A7) (Pertwee & Cascio, 2014). CBN may also stimulate the activity of phospholipases.

β-CARYOPHYLLENE

BCP is a volatile terpene with CB receptor activity, found ubiquitously throughout nature and in great abundance in *Cannabis*, cloves, and black pepper (Figure 4). This terpene is an efficient CB2 receptor agonist, is generally regarded as safe by the US Food and Drug Administration, and is available commercially (Gertsch et al., 2008). BCP binds and stimulates CB2 receptors, causing analgesic and anti-inflammatory activity without psychotropic effects. BCP has also been shown to reduce drug administration,

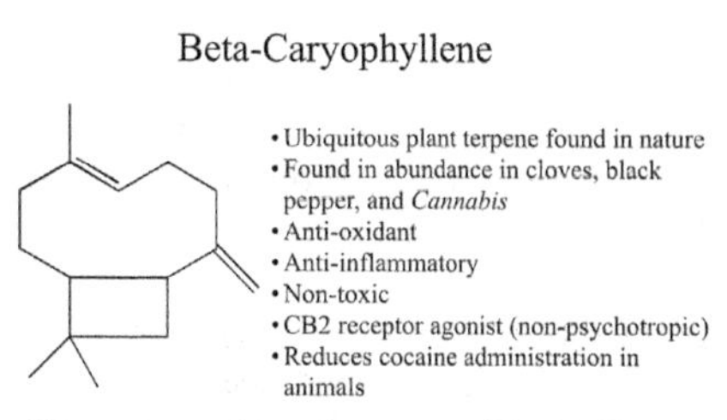

Beta-Caryophyllene

- Ubiquitous plant terpene found in nature
- Found in abundance in cloves, black pepper, and *Cannabis*
- Anti-oxidant
- Anti-inflammatory
- Non-toxic
- CB2 receptor agonist (non-psychotropic)
- Reduces cocaine administration in animals

FIGURE 4 Structure and key facts regarding BCP. BCP is a volatile terpene that activates CB2 receptors and may be a useful therapeutic compound. *Structure generated with ChemDraw by J.P.M.*

improving scores of depression and anxiety in mammals (Bahi et al., 2014; Onaivi et al., 2008; Xi et al., 2011).

APPLICATIONS TO OTHER ADDICTIONS AND SUBSTANCE MISUSE

Medical *Cannabis* and *Cannabis*-based medicines could potentially be developed as drug addiction disorder treatments or used as a substitute for alcohol and other drugs such as opiates and cocaine (Aggarwal, 2008; Otto, 2012; Reiman, 2009; Subbaraman, 2014). BCP and other CBs that activate the CB2 receptor may provide a safe treatment for drug addiction and withdrawal symptoms by providing anti-inflammatory effects and pain relief and improving mood, but without any intoxicating effects. CB1 receptor-based therapies may be appropriate for patients who have previous experience with *Cannabis*, as naïve patients have been shown to be less tolerant of the side effects of CB1 activation compared to experienced users in clinical settings.

DEFINITION OF TERMS

- Cannabinoids: This is a group of closely related compounds, similar to THC or other compounds found in plants.
- CB1 receptor: This is a G-protein-coupled receptor densely located in the brain and nervous tissue.
- CB2 receptor: This is a G-protein-coupled receptor densely located in immune tissue and also found in the brain.
- Endocannabinoid system: This is a mammalian biological system consisting of receptors (i.e., CB1, CB2), endogenously produced compounds (i.e., anandamide, 2-arachidonoylglycerol), and proteins responsible for the synthesis, breakdown, and transport of endogenous CBs.
- Phytocannabinoid: These are CB compounds that are found in plants.

SUMMARY

This chapter focuses on a group of terpenophenolic compounds found in the *Cannabis* plant (Table 2). *Cannabis* is a plant that has been documented as a medicine for millennia. Neurocognitive deficits related to *Cannabis* use are reversible regardless of the amount or duration of use over a lifetime. CBs such as BCP and CBD may offer novel therapeutic strategies to develop treatments for drug abuse-related disorders. BCP, CBD, and other phytocannabinoids are nonpsychotropic and do not cause intoxication. CBD is well tolerated in humans and can reduce anxiety. Furthermore, the administration of CB2 agonists reduces anxiety and depression in animal models, with supporting, but limited, evidence in humans.

THC and the ECS can interact with the opiate system. Clinically coadministration of THC with opiates allows the administration of significantly less opiate to reach the desired analgesic effects. Additionally, administration of CB2 agonists reduces drug-seeking behavior and signs of withdrawal in animal models. *Cannabis* and its pharmacological agents may have a potential role in drug abuse treatment programs; evidence exists from animal and human research suggesting clinical benefits related to cocaine and opiate pharmacodynamics.

TABLE 2 Chemical Types and Numbers of Cannabinoids

Chemical Class or Type	Number of Compounds
THC	18
CBG	17
Cannabidiol (CBD)	8
Other CBs	61
Non-CBs	>400

This table represents a simple breakdown of compounds found in *Cannabis*. There are several isomers of each class or type, as well as hundreds of non-CB components, including terpenes, flavonoids, vitamin K, fatty acids, nitrogenous compounds, etc.

REFERENCES

Abrams, D. I., Couey, P., Shade, S. B., Kelly, M. E., & Benowitz, N. L. (2011). Cannabinoid–opioid interaction in chronic pain. *Clinical Pharmacology and Therapeutics*, *90*(6), 844–851. http://dx.doi.org/10.1038/clpt.2011.188.

Aggarwal, S. K. (2008). *The medical geography of cannabinoid botanicals in Washington State: Access, delivery, and distress*. ProQuest.

Appendino, G., Chianese, G., & Taglialatela-Scafati, O. (2011). Cannabinoids: occurrence and medicinal chemistry. *Current Medicinal Chemistry*, *18*(7), 1085–1099.

Bachhuber, M. A., Saloner, B., Cunningham, C. O., & Barry, C. L. (2014). Medical cannabis laws and opioid analgesic overdose mortality in the United States, 1999–2010. *JAMA Internal Medicine*, *174*(10), 1668–1673. http://dx.doi.org/10.1001/jamainternmed.2014.4005.

Bahi, A., Mansouri, Al, S., Memari, Al, E., Ameri, Al, M., Nurulain, S. M., & Ojha, S. (2014). β-Caryophyllene, a CB2 receptor agonist produces multiple behavioral changes relevant to anxiety and depression in mice. *Physiology and Behavior*, *135C*, 119–124. http://dx.doi.org/10.1016/j.physbeh.2014.06.003.

Ballard, M. E., Bedi, G., & de Wit, H. (2012). Effects of delta-9-tetrahydrocannabinol on evaluation of emotional images. *Journal of Psychopharmacology*, *26*(10), 1289–1298. http://dx.doi.org/10.1177/0269881112446530.

Barann, M., Molderings, G., Brüss, M., Bönisch, H., Urban, B. W., & Göthert, M. (2002). Direct inhibition by cannabinoids of human 5-HT3A receptors: probable involvement of an allosteric modulatory site. *British Journal of Pharmacology*, *137*(5), 589–596. http://dx.doi.org/10.1038/sj.bjp.0704829.

Ben Amar, M. (2006). Cannabinoids in medicine: a review of their therapeutic potential. *Journal of Ethnopharmacology*, *105*(1–2), 1–25. http://dx.doi.org/10.1016/j.jep.2006.02.001.

Brown, N. K., & Harvey, D. J. (1990). In vitro metabolism of cannabichromene in seven common laboratory animals. *Drug Metabolism and Disposition*, *18*(6), 1065–1070.

Campos, A. C., Ortega, Z., Palazuelos, J., Fogaça, M. V., Aguiar, D. C., Díaz-Alonso, J., ... Galve-Roperh, I. (2013). The anxiolytic effect of cannabidiol on chronically stressed mice depends on hippocampal neurogenesis: involvement of the endocannabinoid system. *The International Journal of Neuropsychopharmacology*, *16*(06), 1407–1419. http://dx.doi.org/10.1017/S1461145712001502.

Cascio, M. G., Gauson, L. A., Stevenson, L. A., Ross, R. A., & Pertwee, R. G. (2010). Evidence that the plant cannabinoid cannabigerol is a highly potent α2–adrenoceptor agonist and moderately potent 5HT1A receptor antagonist. *British Journal of Pharmacology*, *159*(1), 129–141. http://dx.doi.org/10.1111/j.1476-5381.2009.00515.x.

Crippa, J. A., Zuardi, A. W., Martín-Santos, R., Bhattacharyya, S., Atakan, Z., McGuire, P., & Fusar-Poli, P. (2009). Cannabis and anxiety: a critical review of the evidence. *Human Psychopharmacology: Clinical and Experimental*, *24*(7), 515–523. http://dx.doi.org/10.1002/hup.1048.

De Petrocellis, L., & Di Marzo, V. (2009). An introduction to the endocannabinoid system: from the early to the latest concepts. *Best Practice and Research Clinical Endocrinology and Metabolism*, *23*(1), 1–15. http://dx.doi.org/10.1016/j.beem.2008.10.013.

De Petrocellis, L., & Di Marzo, V. (2010). Non-CB1, non-CB2 receptors for endocannabinoids, plant cannabinoids, and synthetic cannabimimetics: focus on G-protein-coupled receptors and transient receptor potential channels. *Journal of Neuroimmune Pharmacology*, *5*(1), 103–121. http://dx.doi.org/10.1007/s11481-009-9177-z.

De Petrocellis, L., Ligresti, A., Moriello, A. S., Allarà, M., Bisogno, T., Petrosino, S., ... Di Marzo, V. (2011). Effects of cannabinoids and cannabinoid-enriched Cannabis extracts on TRP channels and endocannabinoid metabolic enzymes. *British Journal of Pharmacology*, *163*(7), 1479–1494. http://dx.doi.org/10.1111/j.1476-5381.2010.01166.x.

De Petrocellis, L., Orlando, P., Moriello, A. S., Aviello, G., Stott, C., Izzo, A. A., & Di Marzo, V. (2012). Cannabinoid actions at TRPV channels: effects on TRPV3 and TRPV4 and their potential relevance to gastrointestinal inflammation. *Acta Physiologica*, *204*(2), 255–266. http://dx.doi.org/10.1111/j.1748-1716.2011.02338.x.

De Petrocellis, L., Vellani, V., Schiano-Moriello, A., Marini, P., Magherini, P. C., Orlando, P., & Di Marzo, V. (2008). Plant-derived cannabinoids modulate the activity of transient receptor potential channels of ankyrin type-1 and melastatin type-8. *The Journal of Pharmacology and Experimental Therapeutics*, *325*(3), 1007–1015. http://dx.doi.org/10.1124/jpet.107.134809.

DeLong, G. T., Wolf, C. E., Poklis, A., & Lichtman, A. H. (2010). Pharmacological evaluation of the natural constituent of *Cannabis sativa*, cannabichromene and its modulation by Δ^9-tetrahydrocannabinol. *Drug and Alcohol Dependence*, *112*(1–2), 126–133. http://dx.doi.org/10.1016/j.drugalcdep.2010.05.019.

Flachenecker, P., Henze, T., & Zettl, U. K. (2014). Nabiximols (THC/CBD oromucosal Spray, Sativex®) in clinical Practice - results of a Multicenter, non-Interventional Study (MOVE 2) in patients with multiple sclerosis spasticity. *European Neurology*, *71*(5–6), 271–279. http://dx.doi.org/10.1159/000357427.

Gaoni, Y., & Mechoulam, R. (1964). Structure+ synthesis of cannabigerol new hashish constituent. *Proceedings of the Chemical Society of London*, 82.

Gardner, E. L. (2013). CB2 agonist and antagonist effects on cocaine self-administration and other cocaine-induced actions. Substance Abuse.

Gertsch, J., Leonti, M., Raduner, S., Racz, I., Chen, J.-Z., Xie, X.-Q., ... Zimmer, A. (2008). Beta-caryophyllene is a dietary cannabinoid. *Proceedings of the National Academy of Sciences*, *105*(26), 9099–9104. http://dx.doi.org/10.1073/pnas.0803601105.

Grunfeld, Y., & Edery, H. (1969a). Psychopharmacological activity of some substances extracted from *Cannabis sativa* L. (hashish). *Electroencephalography and Clinical Neurophysiology*, *27*(2), 219–220.

Grunfeld, Y., & Edery, H. (1969b). Psychopharmacological activity of the active constituents of hashish and some related cannabinoids. *Psychopharmacologia*, *14*(3), 200–210. http://dx.doi.org/10.1007/BF00404218.

Grunfeld, Y., & Edery, H. (1969c). Psychopharmacological activity of the active constituents of hashish and some related cannabinoids. *Audio and Electroacoustics Newsletter, IEEE*, *14*(3), 200–210. http://dx.doi.org/10.1007/BF00404218.

Hazekamp, A., & Grotenhermen, F. (2010). Review on clinical studies with cannabis and cannabinoids 2005–2009. *Cannabinoids*, *5*(special), 1–21.

Hejazi, N., Zhou, C., Oz, M., Sun, H., Ye, J. H., & Zhang, L. (2006). Delta9-tetrahydrocannabinol and endogenous cannabinoid anandamide directly potentiate the function of glycine receptors. *Molecular Pharmacology*, *69*(3), 991–997. http://dx.doi.org/10.1124/mol.105.019174.

Hofmann, M. E., & Frazier, C. J. (2013). Marijuana, endocannabinoids, and epilepsy: potential and challenges for improved therapeutic intervention. *Experimental Neurology*, *244*, 43–50.

Holley, J. H., Hadley, K. W., & Turner, C. E. (1975). Constituents of *Cannabis sativa* L. XI: cannabidiol and cannabichromene in samples of known geographical origin. *Journal of Pharmaceutical Sciences*, *64*(5), 892–894.

de Lago, E., & Fernández-Ruiz, J. (2007). Cannabinoids and neuroprotection in motor-related disorders. *CNS and Neurological Disorders Drug Targets*, *6*(6), 377–387.

Lichtman, A. H., & Martin, B. R. (2005). Cannabinoid tolerance and dependence. *Handbook of Experimental Pharmacology*, *168*, 691–717.

Martin-Moreno, A. M., Reigada, D., Ramirez, B. G., Mechoulam, R., Innamorato, N., Cuadrado, A., & de Ceballos, M. L. (2011). Cannabidiol and other cannabinoids reduce microglial activation in vitro and in vivo: relevance to Alzheimer's disease. *Molecular Pharmacology*, *79*(6), 964–973. http://dx.doi.org/10.1124/mol.111.071290.

Martin, B. R., Compton, D. R., Thomas, B. F., Prescott, W. R., Little, P. J., Razdan, R. K., ... Ward, S. J. (1991). Behavioral, biochemical, and molecular modeling evaluations of cannabinoid analogs. *Pharmacology, Biochemistry, and Behavior*, *40*(3), 471–478. http://dx.doi.org/10.1016/0091-3057(91)90349-7.

McHugh, D., Page, J., Dunn, E., & Bradshaw, H. B. (2012). Δ(9)-Tetrahydrocannabinol and N-arachidonyl glycine are full agonists at GPR18 receptors and induce migration in human endometrial HEC-1B cells. *British Journal of Pharmacology*, *165*(8), 2414–2424. http://dx.doi.org/10.1111/j.1476-5381.2011.01497.x.

McLaughlin, P. J. (2012). Reports of the death of CB1 antagonists have been greatly exaggerated. *Behavioural Pharmacology*, *23*(5 and 6), 537–550. http://dx.doi.org/10.1097/FBP.0b013e3283566a8c.

Mechoulam, R., Parker, L. A., & Gallily, R. (2002). Cannabidiol: an overview of some pharmacological aspects. *Journal of Clinical Pharmacology*, *42*, 11S–19S.

Morel, L. J., Giros, B., & Daugé, V. (2009). Adolescent exposure to chronic delta-9-tetrahydrocannabinol blocks opiate dependence in maternally deprived rats. *Neuropsychopharmacology*, *34*, 2469–2476. http://dx.doi.org/10.1038/npp.2009.70.

O'Sullivan, S. E., Tarling, E. J., Bennett, A. J., Kendall, D. A., & Randall, M. D. (2005). Novel time-dependent vascular actions of Delta9-tetrahydrocannabinol mediated by peroxisome proliferator-activated receptor gamma. *Biochemical and Biophysical Research Communications*, *337*(3), 824–831. http://dx.doi.org/10.1016/j.bbrc.2005.09.121.

Onaivi, E. S., Ishiguro, H., Gong, J.-P., Patel, S., Meozzi, P. A., Myers, L., ... Uhl, G. R. (2008). Brain neuronal CB2 cannabinoid receptors in drug abuse and depression: from mice to human subjects. *PLoS One*, *3*(2), e1640. http://dx.doi.org/10.1371/journal.pone.0001640.

Otto, M. A. (2012). Medical marijuana often used as a prescription drug substitute. *Clinical Psychiatry News*, *40*(1), 33. http://dx.doi.org/10.1016/S0270-6644(12)70022-X.

Pacher, P., Bátkai, S., & Kunos, G. (2006). The endocannabinoid system as an emerging target of pharmacotherapy. *Pharmacological Reviews, 58*(3), 389–462. http://dx.doi.org/10.1124/pr.58.3.2.

Pertwee, R. G. (2008). The diverse CB 1 and CB 2 receptor pharmacology of three plant cannabinoids: Δ 9-tetrahydrocannabinol, cannabidiol and Δ 9-tetrahydrocannabivarin. *British Journal of Pharmacology, 153*(2), 199–215. http://dx.doi.org/10.1038/sj.bjp.0707442.

Pertwee, R. (2014). *Handbook of cannabis*. Oxford University Press.

Pertwee, R. G., & Cascio, M. G. (2014). Known pharmacological actions of delta-9-Tetrahydrocannabinol and of four other chemical constituents of cannabis that activate cannabinoid receptors. In *Handbook of cannabis* (pp. 115–136). http://dx.doi.org/10.1093/acprof:oso/9780199662685.003.0006.

Pharmacopoeia, A. H. (2013). *Cannabis inflorescence*.

Pope, H. G., Gruber, A. J., Hudson, J. I., Huestis, M. A., & Yurgelun-Todd, D. (2001). Neuropsychological performance in long-term cannabis users. *Archives of General Psychiatry, 58*(10), 909–915.

Reiman, A. (2009). Cannabis as a substitute for alcohol and other drugs. *Harm Reduction Journal, 6*, 35. http://dx.doi.org/10.1186/1477-7517-6-35.

Rog, D. J. (2010). Cannabis-based medicines in multiple sclerosis – a review of clinical studies. *Immunobiology, 215*(8), 658–672. http://dx.doi.org/10.1016/j.imbio.2010.03.009.

Russo, E. B. (2008). Cannabinoids in the management of difficult to treat pain. *Therapeutics and Clinical Risk Management, 4*(1), 245–259.

Russo, E. B. (2011). Taming THC: potential cannabis synergy and phytocannabinoid-terpenoid entourage effects. *British Journal of Pharmacology, 163*(7), 1344–1364. http://dx.doi.org/10.1111/j.1476-5381.2011.01238.x.

Russo, E. B., Burnett, A., Hall, B., & Parker, K. K. (2005). Agonistic properties of cannabidiol at 5-HT1a receptors. *Neurochemical Research, 30*(8), 1037–1043. http://dx.doi.org/10.1007/s11064-005-6978-1.

Russo, E., & Guy, G. W. (2006). A tale of two cannabinoids: the therapeutic rationale for combining tetrahydrocannabinol and cannabidiol. *Medical Hypotheses, 66*(2), 234–246. http://dx.doi.org/10.1016/j.mehy.2005.08.026.

Russo, E. B., Jiang, H. E., Li, X., Sutton, A., Carboni, A., del Bianco, F., ... Li, C. S. (2008). Phytochemical and genetic analyses of ancient cannabis from Central Asia. *Journal of Experimental Botany, 59*(15), 4171–4182. http://dx.doi.org/10.1093/jxb/ern260.

Sastre-Garriga, J., Vila, C., Clissold, S., & Montalban, X. (2014). THC and CBD oromucosal spray (Sativex®) in the management of spasticity associated with multiple sclerosis. *Expert Review of Neurotherapeutics, 11*(5), 627–637. http://dx.doi.org/10.1586/ern.11.47.

Scuderi, C., De Filippis, D., Iuvone, T., Blasio, A., Steardo, A., & Esposito, G. (2009). Cannabidiol in medicine: a review of its therapeutic potential in CNS disorders. *Phytotherapy Research, 23*(5), 597–602. http://dx.doi.org/10.1002/ptr.2625.

Sirikantaramas, S. (2004). The gene controlling marijuana psychoactivity: molecular cloning and heterologous expression of delta-1-tetrahydrocannabinolic acid synthase from *Cannabis sativa* L. *Journal of Biological Chemistry, 279*(38), 39767–39774. http://dx.doi.org/10.1074/jbc.M403693200.

Subbaraman, M. S. (2014). Can cannabis be considered a substitute medication for alcohol? *Alcohol and Alcoholism, 49*(3), 292–298. http://dx.doi.org/10.1093/alcalc/agt182.

Wade, D. T., Collin, C., Stott, C., & Duncombe, P. (2010). Meta-analysis of the efficacy and safety of Sativex (nabiximols), on spasticity in people with multiple sclerosis. *Multiple Sclerosis, 16*(6), 707–714.

Xi, Z.-X., Peng, X.-Q., Li, X., Song, R., Zhang, H.-Y., Liu, Q.-R., ... Gardner, E. L. (2011). Brain cannabinoid CB2 receptors modulate cocaine's actions in mice. *Nature Neuroscience, 14*(9), 1160–1166. http://dx.doi.org/10.1038/nn.2874.

Yamamoto, I., Watanabe, K., Matsunaga, T., Kimura, T., Funahashi, T., & Yoshimura, H. (2003). Pharmacology and toxicology of major constituents of Marijuana—On the metabolic activation of cannabinoids and its mechanism. *Journal of Toxicology: Toxin Reviews, 22*(4), 577–589. http://dx.doi.org/10.1081/TXR-120026915.

Chapter 63

Treating the Phenomenon of New Psychoactive Substances: Synthetic Cannabinoids and Synthetic Cathinones

Maurizio Coppola[1], Raffaella Mondola[2], Francesco Oliva[3], Rocco Luigi Picci[3], Daniele Ascheri[3], Federica Trivelli[3]

[1]*Department of Addiction, ASL CN2, Alba, Italy;* [2]*Department of Mental Health, ASL CN1, Saluzzo, Italy;* [3]*Department of Clinical and Biological Sciences, San Luigi Gonzaga Medical School, University of Turin, Torino, Italy*

INTRODUCTION

Despite some exceptions, information reported from international agencies confirms that global drug consumption has been substantially stable since the late 1990s. However, there has been a growing alert related to three phenomena: the consumption of prescription medicines, polydrug use, and the spread of new psychoactive substances. In particular, the rapid and large increase in the spread of new psychoactive substances among drug users observed since 2000 has determined a public health concern worldwide (United Nations Office on Drugs and Crime, 2014). These compounds, known by various names, including "smart drugs," "legal highs," and "research chemicals," are synthetic compounds and plant derivatives marketed as legal alternatives to traditional and illicit drugs of abuse (European Monitoring Centre for Drugs and Drug Addiction, 2014; United Nations Office on Drugs and Crime, 2013a, 2013b, 2014).

Although the commercialization of new drugs of abuse is not a novel phenomenon, the recent spread of new molecules within the recreational drug market has no precedent in terms of number and typology of the substances, with an increasing trend from year to year (United Nations Office on Drugs and Crime, 2013a, 2013b, 2014). In 2013, 81 new substances have been officially reported to the European Union via the Early Warning System (European Monitoring Centre for Drugs and Drug Addiction, 2014). The numbers of substances reported in 2009, 2010, 2011, and 2012 were 24, 41, 49, and 73, respectively. Overall, in the European Union more than 350 new substances are being monitored in terms of health risk and spread (European Monitoring Centre for Drugs and Drug Addiction, 2014). United Nations Office on Drug and Crime (2014) reports that about 90% of the member states have signaled a health risk related to the consumption of new psychoactive substances. Overall, 348 new substances have been identified on the recreational drug market by the United Nations Office on Drug and Crime member states, a number higher than that of the internationally controlled psychoactive substances, which corresponds to 234 (United Nations Office on Drug and Crime, 2014).

New psychoactive substances have been defined by the European Union as follows: "new narcotics or psychotropic drugs, in pure form or in a preparation, that are not scheduled under the Single Convention on Narcotic Drugs of 1961 or the Convention on Psychotropic Substances of 1971, but which may pose a public health threat comparable to that posed by substances listed in those conventions" (Council of the European Union Decision 2005/387/JHA) (European Monitoring Centre for Drugs and Drug Addiction, 2009, 2011a). Thus, the adjective "new" does not necessarily refer to drugs of new synthesis, but includes any substance not scheduled that has recently appeared on the recreational drug market. The popularity of these drugs appears to be related to three principal factors. First, they fall outside the international drug laws, reducing the risk of legal problems for sellers and buyers. Second, they are principally marketed via Web stores that are able to change the availability of products quickly in relation to changes in drug laws. Third, considering that these drugs are frequently labeled as plant derivatives, herbal blends, or bath salts, they can induce a false sense of safety in drug users (Coppola & Mondola, 2012a).

The epidemiology of this phenomenon is little known, but some information is available. In 2011, a flash Eurobarometer survey performed on more than 12,000 people ages 15–24 estimated that about 5% of young Europeans had used new psychoactive substances at least once. In the United Kingdom, Latvia, and Poland the percentage was higher (10%), while Ireland was the country in which the new psychoactive substances were most used (16%). Conversely, Finland, Greece, and Italy were the countries with the smallest percentage of use (3%). Overall, 2.9 million Europeans ages 15–64 have used new psychoactive substances at least once (Eurobarometer, 2011). In the United States, the number of calls received by poison control centers shows that, after cannabis, the new psychoactive substances are the most used recreational compounds among students (United Nations Office on Drug and Crime, 2013a, 2013b). A study performed on 654 Australian ecstasy users showed that 44% of them had used new psychoactive drugs in the past 6 months (in particular 2,5-dimethoxy-4-iodophenethylamine and 4-bromo-2,5-dimethoxyphenethylamine, which are best known as 2C-I and 2C-B, respectively) (Burns et al., 2014).

A survey conducted by the European Monitoring Centre for Drugs and Drug Addiction (2012) has shown that Web stores offering for sale at least one psychoactive drug quadrupled from January 2010 to January 2012, growing from 170 to 693. This study confirms that the Internet is the principal route of spread of these products. Kratom, *Salvia divinorum*, and hallucinogenic mushrooms are the plant derivatives most frequently offered online, while methoxetamine, 5,6-methylenedioxy-2-aminoindane, 6-(2-aminopropyl)benzofuran, 3,4-methylenedioxypyrovalerone (MDPV), 4-methylethcathinone, methiopropamine, and 5-iodo-2-aminoindane are the synthetic compounds most frequently offered on Web stores.

Poison control centers and emergency units have signaled cases of severe intoxication and deaths related to the recreational consumption of new psychoactive substances worldwide (Wood, Hill, Thomas, & Dargan, 2014). However, this information could be underestimated, since the new drugs are not identified by immunological screening tests (Favretto, Pascali, & Tagliaro, 2013). Acute intoxications are mainly characterized by cardiovascular, neurological, and psychopathological symptoms that require medical assistance (Papaseit, Farré, Schifano, & Torrens, 2014). In contrast, there is little information about the chronic use of these compounds. Tolerance, dependence, withdrawal symptoms, protracted psychosis, and increased risk of chronic psychosis have been reported, however (Zamengo, Frison, Bettin, & Sciarrone, 2014). On the whole, the phenomenon of new psychoactive substances represents both a public health concern across the world and a challenge for physicians. Generally, the pharmacological and toxicological characteristics of these compounds are unknown. There is very little information about doses, drug interactions, and mechanisms of action. In many cases, the new psychoactive substances are drugs synthesized for preclinical research or medicines abandoned because of their severe side effects. Furthermore, products sold online frequently contain a combination of new substances that can be different from those reported on the package (Maxwell, 2014). In addition, drug users can consume new drugs unknowingly, because these molecules may be added to the traditional substances of abuse (Giné, Espinosa, & Vilamala, 2014).

Numerous molecules have been analytically detected in biological samples of people intoxicated by products sold as legal alternatives to traditional drugs of abuse. However, the most popular and widespread new psychoactive substances belong to two groups of drugs: synthetic cannabinoids and synthetic cathinones. In this chapter we examine clinical, pharmacological, and toxicological information about these new substances of abuse appearing on the recreational drug market.

SYNTHETIC CANNABINOIDS

Synthetic cannabinoids, more correctly named synthetic cannabinoid receptor agonists, are a large group of synthetic molecules functionally similar to Δ^9-tetrahydrocannabinol. These substances were synthesized more than 40 years ago as potential medicines for the treatment of various forms of chronic pain. Since 2004, information derived from Internet forums has evidenced the use of these compounds as legal alternatives to natural cannabis. In 2008, the Austrian National Focal Point reported to the European Monitoring Centre for Drugs and Drug Addiction the presence of the synthetic cannabinoid JWH-018 in recreational products sold with the brand name "Spice." Subsequently, many seizures of synthetic cannabinoids have been performed by the police worldwide. Furthermore, numerous analytically confirmed cases of severe intoxication related to the consumption of these substances have been reported in many countries. On the whole, these compounds can be considered the most popular new psychoactive substances (European Monitoring Centre for Drugs and Drug Addiction, 2009, 2011b, 2012).

Chemistry

Synthetic cannabinoids are synthetic molecules acting as agonists of cannabinoid receptors CB1 and/or CB2. They mimic the action of both endocannabinoids (e.g., anandamide and 2-arachidonoylglycerol) and vegetable cannabinoids. Synthetic cannabinoids are a heterogeneous group of chemicals formed by molecules with various chemical structures (Figure 1). Generally, they

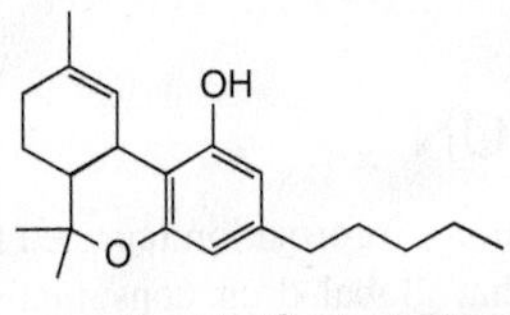

Classical cannabinoid: Δ^9-tetrahydrocannabinol

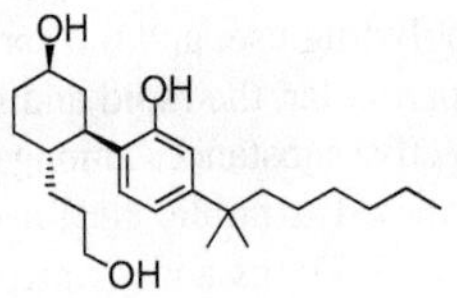

Nonclassical cannabinoid: CP-55,940

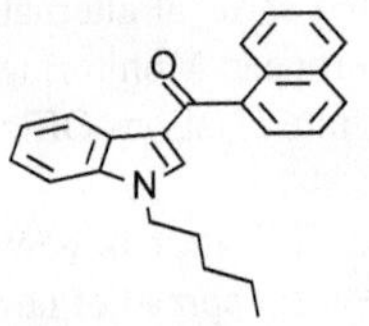

Aminoalkylindole: JWH-018

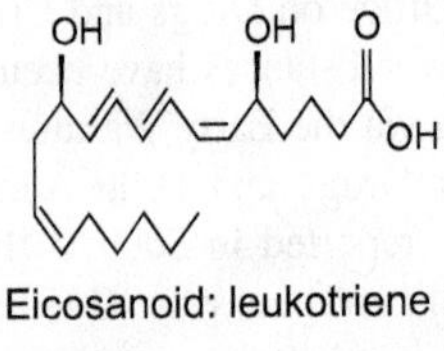

Eicosanoid: leukotriene B4

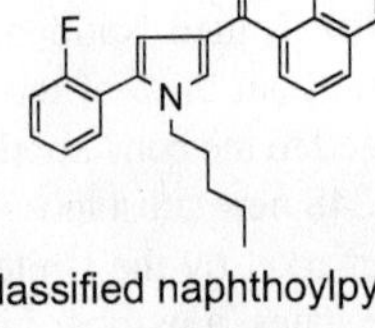

Unclassified naphthoylpyrrole: JWH-307

FIGURE 1 Chemical structures of cannabinoids. Cannabinoids show various chemical structures. Some examples are shown.

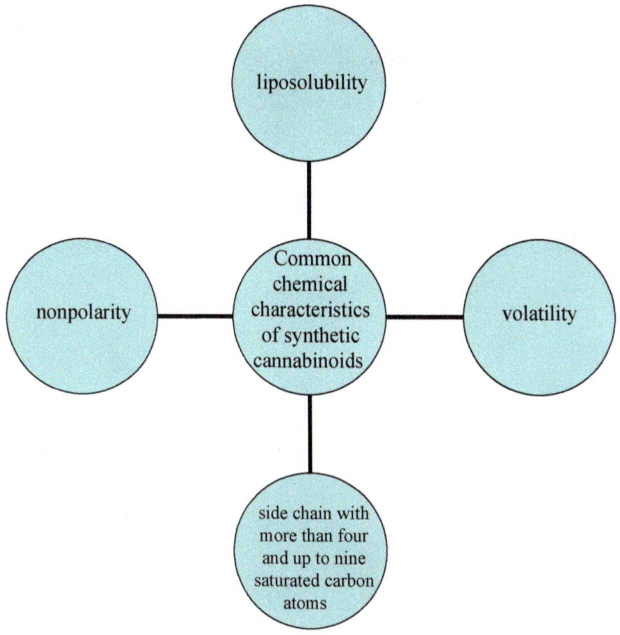

FIGURE 2 Common chemical characteristics of synthetic cannabinoids. Generally, synthetic cannabinoids share the chemical characteristics shown (European Monitoring Centre for Drugs and Drug Addiction, 2009).

have in common some characteristics, including liposolubility, nonpolarity, volatility, and a side chain with more than four and up to nine saturated carbon atoms (Figure 2) (European Monitoring Centre for Drugs and Drug Addiction, 2009). These substances have been classified in the following categories: classical cannabinoids, nonclassical cannabinoids, hybrid cannabinoids, aminoalkylindoles, eicosanoids, and others (Howlett et al., 2002). Classical cannabinoids are a subgroup of synthetic cannabinoids including Δ^9-tetrahydrocannabinol and other chemical constituents of the cannabis plant as well as their structurally related synthetic analogues. In contrast, the other subgroups are formed by molecules structurally different from the natural cannabinoids (Table 1) (European Monitoring Centre for Drugs and Drug Addiction, 2009; Howlett et al., 2002).

Pharmacology

To date, there is very little information about the pharmacological properties of these substances. They are agonists of CB1 and/or CB2 receptors with a potency that differs from molecule to molecule. In vitro studies have demonstrated that some compounds, such as HU-210, show high affinity for the CB1 receptor, while other molecules, such as HU-308 and JWH-015, display higher affinity for the CB2 receptor (Hanus et al., 1999). Unlike cannabis, synthetic cannabinoids are stably present in their active form, which is rapidly absorbed via inhalation. Oral administration is also reported; however, their low solubility in water can determine a delayed absorption. Little is known about their distribution and bioavailability; however, the lipophilic nature of these compounds is predictive of both a high volume of distribution and a high bioavailability after inhalation. Consumption via inhalation permits one to avoid the hepatic first pass and could promote the accumulation of these molecules in the fat tissues of the body. Monohydroxylated compounds and their glucuronide derivatives appear to be the principal metabolites of aminoalkylindoles; however, multiple hydroxylations, N-dealkylation, a combination of hydroxylation and dealkylation, and oxidation of the terminally hydroxylated N–alkyl moiety to the corresponding carboxylic acid have also been observed. Information on the metabolism of synthetic cannabinoids is available for only some molecules that have been studied in human liver microsomes or in urine samples extracted from the drug users. Finally, regarding the elimination of these cannabinoids and their metabolites, some evidence suggests that they can be excreted via the urine and in minimal part via the feces (European Monitoring Centre for Drugs and Drug Addiction, 2009; United Nations Office on Drugs and Crime, 2012).

Toxicology

In 2009, a self-experiment by Auwärter et al. (2009) showed that 0.3 g of an herbal blend marketed under the brand name "Spice Diamond" produced a cannabis-like effect for some hours. This work was the first controlled study to demonstrate the psychotropic effects of synthetic cannabinoids sold on the recreational drug market. Drug users who take synthetic cannabinoids desire the following effects: increased socialization skills, mild euphoria, relaxation, mild dissociation, and well-being. On the other hand, numerous cases of severe intoxication and death related to the consumption of these compounds have been reported across the world. Adverse effects reported include confusion, red or bloodshot eyes, dilated pupils, disorganized speech, nausea, sustained muscle contraction, vomiting, diarrhea, restlessness, fever, sweating, dryness of the mouth, tremor, insomnia, headache, seizures, tachycardia, tachypnea, hypertension, chest pain, weakness, anxiety, agitation, panic attack, depression, suicidality, dissociation, hallucinations, and delirium. ST elevation; high levels of troponin, creatinine kinase, and potassium; respiratory depression; bradycardia; and kidney damage have also been reported. Tolerance, dependence, and withdrawal symptoms have been observed after repeated use (European Monitoring Centre for Drugs and Drug Addiction, 2009; United Nations Office on Drugs and Crime, 2012). Finally, there have been described some analytically confirmed cases of people dying from myocardial infarction that developed after synthetic cannabinoid consumption (Behonick et al., 2014; Ibrahim, Al-Saffar, & Wannenburg, 2014). Another fatality was related to the consumption of the synthetic cannabinoid MAM-2201; however, no symptomatological details were available (Saito et al., 2013).

Synthetic Cannabinoids and Psychosis

The association between cannabis consumption and psychosis, in particular schizophrenia, is controversial and debated. Chronic consumption of cannabis can produce cognitive impairment, psychotic symptoms, and behavioral alterations similar to those present in schizophrenic patients. Risk of psychosis appears to be directly related to the potency of cannabis used, frequency of use, age of onset, and genetic vulnerability. Some genetic polymorphisms such as those involving both catechol-O-methyltransferase and the AKT1 gene may increase the risk of developing cannabis-related psychosis. On the other hand, the presence of cannabidiol

TABLE 1 Cannabinoid Classification

Natural Cannabinoids and Structurally Related Synthetic Analogues or Classical Cannabinoids	Cannabinoids Structurally Unrelated to Natural Cannabinoids	Molecules of Uncertain Classification
Δ^9-Tetrahydrocannabinol Δ^8-Tetrahydrocannabinol AM-411 AM-905 AM-906 AM-919 HU-210 O-823 O-1184	Nonclassical cannabinoids CP-47,497–C6 CP-47,497–C8 CP-47,497–C9 CP-55,940, CP-55,244 HU-308 HU-320	Oleamide Erucamide
	Hybrid cannabinoids AM-4030 AM-938	
	Aminoalkylindoles Divided into Naphthoylindoles: JWH-015 JWH-018 JWH-073 JWH-081 JWH-098 JWH-122 JWH-184 JWH-185 JWH-200 JWH-210 JWH-398 WIN-55,212 Naphthylmethylindoles JWH-175 Phenylacetylindoles RCS-8 JWH-167 JWH-203 JWH-250 JWH-251 Benzoylindoles AM-694 AM-1241 RCS-4 WIN 48,098	
	Eicosanoids Leukotrienes Prostaglandins Prostacyclins Thromboxanes O-1812	
	Cannabinoids unclassified elsewhere Divided into Diarylpyrazoles SR141716 Naphthoylpyrroles JWH-030 JWH-147 JWH-307 Naphthylmethylindenes JWH-176 1-Naphthalenyl[4-(pentylox)-1-naphthalenyl]-methanone	

Cannabinoids have been classified into various groups, including classical cannabinoids, nonclassical cannabinoids, hybrid cannabinoids, aminoalkylindoles, eicosanoids, and others (Howlett et al., 2002). The potential subdivision of example cannabinoids into subgroups is shown.

within the cannabis plant represents a protective factor because this molecule exerts antipsychotic and neuroprotective activity (Radhakrishnan, Wilkinson, & D'Souza, 2014). Generally, synthetic cannabinoids are potent CB1 receptor agonists and since they are synthetic compounds, they do not contain cannabidiol. Thus, theoretically, these substances could represent a significant risk factor for the development of psychotic symptoms. Literature data report some analytically confirmed cases of psychosis associated with the consumption of synthetic cannabinoids. Mainly, they have involved vulnerable patients in which the use of synthetic cannabinoids has caused a relapse or an exacerbation of a preexisting psychotic disorder. Clinical features reported include delusion, visual and auditory hallucinations, psychomotor agitation, dissociation, aggression, behavioral alterations, alterations in language and thought, and suicidal ideation (Every-Palmer, 2011, 2010; Hurst, Loeffler, & McLay, 2011; Tung, Chiang, & Lam, 2012).

SYNTHETIC CATHINONES

Synthetic cathinones are a large group of stimulants chemically related to cathinone, the most potent amphetamine-like compound naturally present in the khat plant (*Catha edulis*) (Figure 3) (Hassan, Gunaid, & Murray-Lyon, 2007). Synthetic cathinones, synthesized since 1928 (methcathinone), were tested as antidepressants in the former Soviet Union in the 1930s and 1940s and as stimulants for the treatment of chronic fatigue (pyrovalerone) in France and the United States in the 1970s (Goldberg, Gardos, & Cole, 1973). The recreational use of these stimulants was reported in the 1970s and 1980s in the former Soviet Union and in the 1990s in the United States. Moreover, between 1997 and 2004 synthetic cathinones abuse was reported in Germany (Coppola & Mondola, 2012b). Since 2005, numerous synthetic cathinones have been found in recreational products marketed as legal highs across the world. These compounds, frequently labeled as "plant food" or "bath salts," are popular as substitutes for traditional stimulants such as ecstasy and cocaine. Since their appearance on the recreational drug market, many cases of severe intoxication and death have been reported worldwide (Coppola & Mondola, 2012b; Paillet-Loilier, Cesbron, Le Boisselier, Bourgine, & Debruyne, 2014).

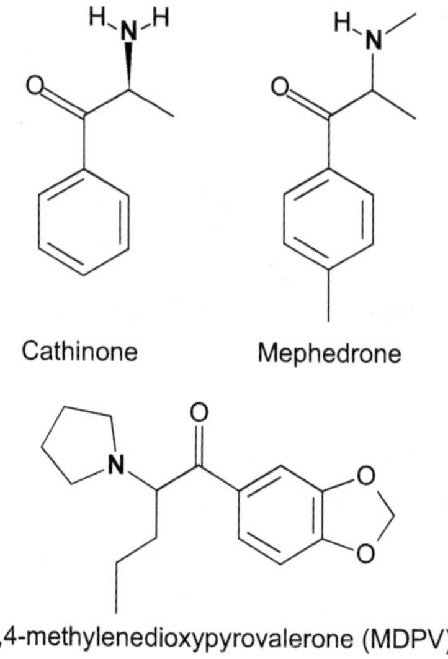

Chemistry

Synthetic cathinones are β-keto-phenethylamine derivatives structurally related to the natural cathinone, a monoamine alkaloid present in the khat plant (Valente, Guedes de Pinho, de Lourdes Bastos, Carvalho, & Carvalho, 2014). Cathinone is chemically similar to amphetamine, from which it differs by the presence of a ketone oxygen atom at the β position of the side chain (Figure 4). The molecular structure of cathinone can be modified in a number of foreseeable manners to produce a series of compounds chemically related to it. Many synthetic cathinones differ from related amphetamines only in the functional group attached at the β carbon on the aminoalkyl chain linked to the phenyl ring. This technique has been used to circumvent international drug laws (Coppola & Mondola, 2012b; Valente, et al., 2014). Cathinone derivatives show substitution patterns at four positions of the general cathinone skeleton (Figure 5). In particular, the chemical structure can be modified as follows: variation on the α-carbon substituent (frequently with an alkyl group) on the carbon atom linked to the carbon in the α position (R1), N-alkylation or addition of ring substituents (R2, R3), or inclusion of additional functional groups to the aromatic ring (R4). Groups used as substituents in R1, R2, R3, and R4 include methyl, methylenedioxy, alkyl, alkoxy, alkylenedioxy, haloalkyl, halide, and phenyl groups. In addition, a synthetic cathinone named 4-methyl-α-pyrrolidino-α-methylpropiophenone,

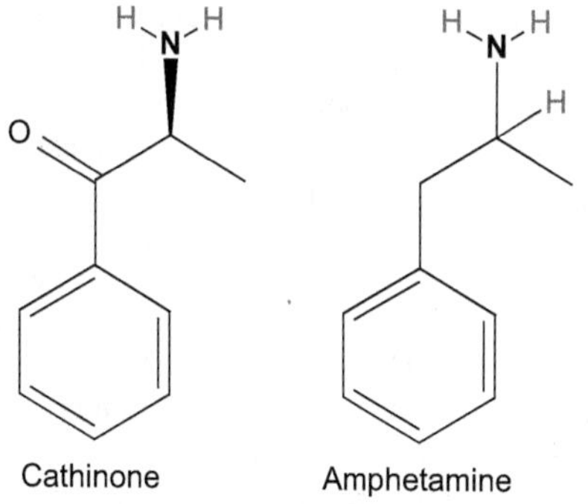

FIGURE 3 **Chemical structures of natural cathinone and some synthetic analogues.** Synthetic cathinones are synthetic derivatives of natural cathinone. The chemical structures of cathinone and two of the most popular synthetic cathinones, mephedrone and MDPV, are shown.

FIGURE 4 **Chemical structures of cathinone and amphetamine.** Cathinone is chemically similar to amphetamine, from which it differs by the presence of a ketone oxygen atom on the β position of the side chain (Coppola & Mondola, 2012a, 2012b; Valente, et al., 2014).

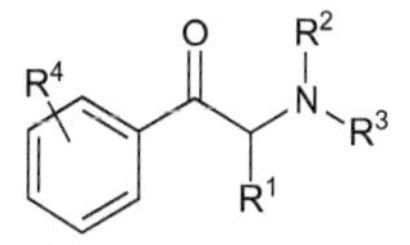

FIGURE 5 **General skeleton of synthetic cathinones.** Cathinone derivatives show substitution patterns at four positions of the general cathinone skeleton (Coppola & Mondola, 2012b; Paillet-Loilier et al., 2014; Valente et al., 2014).

derived from the α-pyrrolidinopropiophenone via a methylation in the *para* position on the phenyl ring (R5), has been mentioned. A single substituted cathinone can present multiple modifications. It is important to remember that cathinones presenting a unique structure have also been described. Particularly, the molecule named naphyrone shows a naphthyl ring that is not present in other cathinone derivatives. Furthermore, like other phenethylamines, substituted cathinones are chiral molecules and can exist in two stereoisomeric forms that may differ in their potency and receptor affinity (Coppola & Mondola, 2012b; Paillet-Loilier et al., 2014; Valente et al., 2014).

Pharmacology

Pharmacological properties of synthetic cathinones differ from molecule to molecule. Generally, the β-keto group increases the polarity of these substances if compared to related amphetamines. The increased polarity reduces the ability to cross the blood–brain barrier and, consequently, the potency of these compounds with respect to amphetamines. However, the presence of a pyrrolidine ring in some molecules reduces the polarity and increases the blood–brain barrier permeability (Coppola & Mondola, 2012a, 2012b; Paillet-Loilier et al., 2014; Valente et al., 2014). This subgroup of synthetic cathinones, including MDPV, shows high potency and neurotoxicity (Coppola & Mondola, 2012a, 2012b; Paillet-Loilier et al., 2014; Valente et al., 2014). Furthermore, considering the high blood–brain barrier permeability of some of these cathinones, the presence of a specific transporter is plausible (Simmler et al., 2013). Synthetic cathinones exert their effects by increasing the concentration of catecholamines such as dopamine, serotonin, and norepinephrine in the synaptic cleft. An increase in monoamine concentration is determined through dopamine, serotonin, and norepinephrine reuptake inhibition as well as via the release of monoamines from intracellular stores. Studies performed in rat synaptosomes and in HEK 293 cells that stably express human dopamine (DAT), serotonin (SERT), and norepinephrine (NET) transporters have shown that substituted cathinones can inhibit the monoamine transporters with different potencies and selectivities. Like amphetamines, they are also able to stimulate the release of monoamines from intracellular stores (Coppola & Mondola, 2012b; Paillet-Loilier et al., 2014; Valente et al., 2014). MDPV, pyrovalerone, flephedrone, and ephedrone show very high affinity for both DAT and NET and poor affinity for SERT. They are the most potent DAT inhibitors among cathinones and they are more potent than cocaine and amphetamines. Moreover, MDPV and pyrovalerone act predominantly as dopamine-reuptake inhibitors rather than dopamine releasers. 4-Fluoromethcathinone, methylone, ethylone, butylone, naphyrone, and mephedrone are nonselective monoamine-reuptake inhibitors (Eshleman et al., 2013; Simmler et al., 2013). Furthermore, except for naphyrone, they are also monoamine releasers (Eshleman et al., 2013).

Toxicology

Studies performed in rats and mice have shown that synthetic cathinones can induce a dose-dependent hyperlocomotion (Fantegrossi, Gannon, Zimmerman, & Rice, 2013; López-Arnau et al., 2013; López-Arnau, Martínez-Clemente, Pubill, Escubedo, & Camarasa, 2012). Studies performed in animal models have highlighted that these substances can reduce working memory performance (Wright, Vandewater, Angrish, Dickerson, & Taffe, 2012) as well as producing long-term memory impairment (den Hollander et al., 2013; Motbey et al., 2012). Mephedrone-induced neurotoxicity has been tested in mouse neuronal cultures. Additionally, this substance can produce a significant loss in dopamine-reuptake sites in mouse striatum and frontal cortex cells (Martinez-Clemente et al., 2014). Finally, animal studies have shown that synthetic cathinones can produce reinforcing properties and abuse liability (Watterson et al. 2012, 2014). A study performed in male Wistar rats trained to self-administer MDPV and D-methamphetamine showed that both compounds induced self-administration and that MDPV was more potent and effective than D-methamphetamine (Aarde, Huang, Creehan, Dickerson, & Taffe, 2013). Synthetic cathinones are used for their stimulant and entactogenic effects. Desired effects include mild euphoria, intensification of sensory perceptions, increase in sociability, increase in energy and capacity to work, and increase in sexual performance. However, the international literature reports severe adverse effects related to the consumption of these substances. Acute intoxication is characterized by numerous undesired effects, including psychomotor agitation, headache, nausea, vomiting, vertigo, dizziness, altered vision, hyperthermia, sweating, bruxism, mydriasis, abdominal pain, tachycardia, palpitation, arrhythmia, blood hypertension, chest pain, ST segment changes, vasoconstriction, seizures, tremors, extrapyramidal symptoms, insomnia, confusion, delirium, panic attack, aggression, suicidal ideation, delusion, and hallucinations. Cerebral edema, stroke, rhabdomyolysis, hyponatremia, acidosis, cardiorespiratory collapse, myocardial infarction, myocarditis, renal failure, and multiple organ failure have also been reported in cases of overdose (Coppola & Mondola, 2012b; Paillet-Loilier et al., 2014; Valente et al., 2014). Fatalities have been associated with the consumption of various cathinones such as mephedrone, MDPV, methedrone, butylone, and methcathinone (Coppola & Mondola, 2012b; Paillet-Loilier et al., 2014; Valente et al., 2014). Additionally, abuse, tolerance, dependence, and withdrawal symptoms have been described in chronic users (Coppola & Mondola, 2012b; Paillet-Loilier et al., 2014; Valente et al., 2014). Synthetic cathinones are generally nasally insufflated or orally ingested; however, intravenous or intramuscular injection, rectal administration, and inhalation have also been reported. These stimulants are not detected by routine immunoassays, but require the use of gas chromatography–mass spectrometry or liquid chromatography–tandem mass spectrometry (Coppola & Mondola, 2012b; Paillet-Loilier et al., 2014).

APPLICATIONS TO OTHER ADDICTIONS AND SUBSTANCE MISUSE

1. MDPV acts as a dopamine- and norepinephrine-reuptake inhibitor, increasing the monoamine concentration in the synaptic cleft.
2. This mechanism of action is similar to that of cocaine.
3. In animal models, MDPV is a complete substitute for the discriminative stimulus effects of cocaine.

4. Studies in animal models have shown that MDPV has the same abuse liability of cocaine.
5. MDPV could become a cocaine substitute among cocaine-dependent patients.

DEFINITION OF TERMS

Δ^9-Tetrahydrocannabinol This is the main active principle of the cannabis plant acting as a CB1 and CB2 receptor agonist.

Cannabidiol This is a nonpsychotropic phytochemical present in the cannabis plant. It exerts neuroprotective and antipsychotic effects.

CB1 receptor This is a G-protein-coupled receptor present in the central and peripheral nervous systems.

CB2 receptor This is a G-protein-coupled receptor present in the central and peripheral nervous systems, but principally in the immune and gastrointestinal systems.

Cannabis This is an annual, dioecious, flowering herb belonging to the Cannabaceae family. It contains 66 psychotropic principles and more than 400 substances.

Catha edulis This is a flowering plant belonging to the Celastraceae family. It contains some amphetamine-like compounds, in particular cathine and cathinone.

Cathine This is an amphetamine-like alkaloid present in the khat plant (*C. edulis*).

Cathinone This is an amphetamine-like alkaloid present in the khat plant (*C. edulis*).

Methcathinone This was the first synthetic cathinone synthesized, in 1928.

JWH-018 This was the first synthetic cannabinoid found in recreational products marketed as legal cannabis.

KEY FACTS OF SYNTHETIC CANNABINOIDS AND PSYCHOSIS

- Synthetic cannabinoids are potent CB1 receptor agonists.
- Because they are synthetic molecules, they do not contain cannabidiol.
- A potent CB1 receptor stimulation is related to the development of psychotic symptoms.
- Cannabidiol can produce antipsychotic and neuroprotective effects.
- The presence of potent CB1 stimulation and absence of cannabidiol can increase the risk of acute and chronic psychosis.
- Clinical evidence confirms the risk of psychosis associated with the consumption of products containing synthetic cannabinoids.

SUMMARY POINTS

- New psychoactive substances represent a public health concern worldwide.
- The spread of these products increases from year to year.
- Among these compounds, synthetic cathinones and synthetic cannabinoids represent the most popular substances.
- They have been associated with numerous cases of severe intoxication and death across the world.
- Chronic consumption of these new drugs could be associated with the development of chronic brain diseases, including psychotic disorders.

REFERENCES

Aarde, S. M., Huang, P. K., Creehan, K. M., Dickerson, T. J., & Taffe, M. A. (2013). The novel recreational drug 3,4-methylenedioxypyrovalerone (MDPV) is a potent psychomotor stimulant: self-administration and locomotor activity in rats. *Neuropharmacology, 71*, 130–140.

Auwärter, V., Dresen, S., Weinmann, W., Müller, M., Pütz, M., & Ferreirós, N. (2009). Spice and other herbal blends: harmless incense or cannabinoid designer drugs? *Journal of Mass Spectrometry, 44*, 832–837.

Behonick, G., Shanks, K. G., Firchau, D. J., Mathur, G., Lynch, C. F., Nashelsky, M., ... Meroueh, C. (2014). Four postmortem case reports with quantitative detection of the synthetic cannabinoid, 5F-PB-22. *Journal of Analytical Toxicology, 38*(8), 559–562 pii: bku048.

Burns, L., Roxburgh, A., Matthews, A., Bruno, R., Lenton, S., & Van Buskirk, J. (2014). The rise of new psychoactive substance use in Australia. *Drug Testing and Analysis, 6*(8), 846–849. http://dx.doi.org/10.1002/dta.1626.

Coppola, M., & Mondola, R. (2012a). 3,4-Methylenedioxypyrovalerone (MDPV): chemistry, pharmacology and toxicology of a new designer drug of abuse marketed online. *Toxicology Letters, 208*, 12–15.

Coppola, M., & Mondola, R. (2012b). Synthetic cathinones: chemistry, pharmacology and toxicology of a new class of designer drugs of abuse marketed as "bath salts" or "plant food". *Toxicology Letters, 211*, 144–149.

Eshleman, A. J., Wolfrum, K. M., Hatfield, M. G., Johnson, R. A., Murphy, K. V., & Janowsky, A. (2013). Substituted methcathinones differ in transporter and receptor interactions. *Biochemical Pharmacology, 85*, 1803–1815.

Eurobarometer. (2011). *Youth attitudes on drugs.* http://ec.europa.eu/public_opinion/flash/fl_330_en.pdf. Accessed 17.05.14.

European Monitoring Centre for Drugs and Drug Addiction (EMCDDA). (2009). *Understanding the 'Spice' phenomenon.* http://www.emcdda.europa.eu/attachements.cfm/att_80086_EN_EMCDDA_Understanding%20the%20%E2%80%98Spice%E2%80%99%20phenomenon_4Update%2020090813.pdf. Accessed 17.05.14.

European Monitoring Centre for Drugs and Drug Addiction (EMCDDA). (2011a). *Drugs in focus.* http://www.emcdda.europa.eu/attachements.cfm/att_145850_EN_EMCDDA_DiF22_EN.pdf. Accessed 09.07.09.

European Monitoring Centre for Drugs and Drug Addiction (EMCDDA). (2011b). *Synthetic cannabinoids and 'Spice'.* http://www.emcdda.europa.eu/publications/drug-profiles/synthetic-cannabinoids. Accessed 16.07.09.

European Monitoring Centre for Drugs and Drug Addiction (EMCDDA). (2012). *Annual report 2012: The state of the drugs problem in Europe.* http://www.emcdda.europa.eu/attachements.cfm/att_190854_EN_TDAC12001ENC_.pdf. Accessed 09.07.14.

European Monitoring Centre for Drugs and Drug Addiction (EMCDDA). (2014). *European drug report 2014: Trends and developments.* Available at http://www.emcdda.europa.eu/attachements.cfm/att_228272_EN_TDAT14001ENN.pdf. Accessed 09.07.14.

Every-Palmer, S. (2010). Warning: legal synthetic cannabinoid-receptor agonists such as JWH-018 may precipitate psychosis in vulnerable individuals. *Addiction, 105*, 1859–1860.

Every-Palmer, S. (2011). Synthetic cannabinoid JWH-018 and psychosis: an explorative study. *Drug and Alcohol Dependence, 117*, 152–157.

Fantegrossi, W. E., Gannon, B. M., Zimmerman, S. M., & Rice, K. C. (2013). In vivo effects of abused 'bath salt' constituent 3,4-methylenedioxypyrovalerone (MDPV) in mice: drug discrimination, thermoregulation, and locomotor activity. *Neuropsychopharmacology, 38*, 563–573.

Favretto, D., Pascali, J. P., & Tagliaro, F. (2013). New challenges and innovation in forensic toxicology: focus on the "New Psychoactive Substances". *Journal of Chromatography A., 1287*, 84–95.

Giné, C. V., Espinosa, I. F., & Vilamala, M. V. (2014). New psychoactive substances as adulterants of controlled drugs. A worrying phenomenon? *Drug Testing and Analysis*. http://dx.doi.org/10.1002/dta.1610.

Goldberg, J., Gardos, G., & Cole, J. O. (1973). A controlled evaluation of pyrovalerone in chronically fatigued volunteers. *International Pharmacopsychiatry, 8*, 60–69.

Hanus, L., Breuer, A., Tchilibon, S., Shiloah, S., Goldenberg, D., Horowitz, M., ... Fride, E. (1999). HU-308: a specific agonist for CB2, a peripheral cannabinoid receptor. *Proceedings of the National Academy of Sciences of the United States of America, 25*, 14228.

Hassan, N. A., Gunaid, A. A., & Murray-Lyon, I. M. (2007). Khat (*Catha edulis*): health aspects of khat chewing. *Eastern Mediterranean Health Journal, 13*, 706–718.

den Hollander, B., Rozov, S., Linden, A. M., Uusi-Oukari, M., Ojanperä, I., & Korpi, E. R. (2013). Long-term cognitive and neurochemical effects of "bath salt" designer drugs methylone and mephedrone. *Pharmacology, Biochemistry, and Behavior, 103*, 501–509.

Howlett, A. C., Barth, F., Bonner, T. I., Cabral, G., Casellas, P., Devane, W. A., ... Pertwee, R. G. (2002). International Union of Pharmacology. XXVII. Classification of cannabinoid receptors. *Pharmacological Reviews, 54*, 161–202.

Hurst, D., Loeffler, G., & McLay, R. (2011). Psychosis associated with synthetic cannabinoid agonists: a case series. *American Journal of Psychiatry, 118*, 1119.

Ibrahim, S., Al-Saffar, F., & Wannenburg, T. (2014). A unique case of cardiac arrest following K2 abuse. *Case Reports in Cardiology, 2014*, 3. http://dx.doi.org/10.1155/2014/120607.

López-Arnau, R., Martínez-Clemente, J., Carbò, M. I., Pubill, D., Escubedo, E., & Camarasa, J. (2013). An integrated pharmacokinetic and pharmacodynamic study of a new drug of abuse, methylone, a synthetic cathinone sold as "bath salts". *Progress in Neuropsychopharmacology and Biological Psychiatry, 45*, 64–72.

López-Arnau, R., Martínez-Clemente, J., Pubill, D., Escubedo, E., & Camarasa, J. (2012). Comparative neuropharmacology of three psychostimulant cathinone derivatives: butylone, mephedrone and methylone. *British Journal of Pharmacology, 167*, 407–420.

Martínez-Clemente, J., López-Arnau, R., Abad, S., Pubill, D., Escubedo, E., & Camarasa, J. (2014). Dose and time-dependent selective neurotoxicity induced by mephedrone in mice. *PLoS One, 9*, e99002.

Maxwell, J. C. (2014). Psychoactive substances-Some new, some old: a scan of the situation in the U.S. *Drug and Alcohol Dependence, 134*, 71–77.

Motbey, C. P., Karanges, E., Li, K. M., Wilkinson, S., Winstock, A. R., Ramsay, J., ... McGregor, I. S. (2012). Mephedrone in adolescent rats: residual memory impairment and acute but not lasting 5-HT depletion. *PLoS One, 7*, e45473.

Paillet-Loilier, M., Cesbron, A., Le Boisselier, R., Bourgine, J., & Debruyne, D. (2014). Emerging drugs of abuse: current perspectives on substituted cathinones. *Substance Abuse and Rehabilitation, 5*, 37–52.

Papaseit, E., Farré, M., Schifano, F., & Torrens, M. (2014). Emerging drugs in Europe. *Current Opinion in Psychiatry, 27*, 243–250.

Radhakrishnan, R., Wilkinson, S. T., & D'Souza, D. C. (2014). Gone to pot – a review of the association between cannabis and psychosis. *Frontiers in Psychiatry, 5*, 54.

Saito, T., Namera, A., Miura, N., Ohta, S., Miyazaki, S., Osawa, M., & Inokuchi, S. (2013). A fatal case of MAM-2201 poisoning. *Forensic Toxicology, 31*, 333–337.

Simmler, L. D., Buser, T. A., Donzelli, M., Schramm, Y., Dieu, L. H., Huwyler, J., ... Liechti, M. E. (2013). Pharmacological characterization of designer cathinones in vitro. *British Journal of Pharmacology, 168*, 458–470.

Tung, C. K., Chiang, T. P., & Lam, M. (2012). Acute mental disturbance caused by synthetic cannabinoid: a potential emerging substance of abuse in Hong Kong. *East Asian Archives of Psychiatry, 22*, 31–33.

United Nations Office on Drugs and Crime (UNODC). (2012). *Synthetic cannabinoids in herbal products*. https://www.unodc.org/documents/scientific/Synthetic_Cannabinoids.pdf. Accessed 16.07.14.

United Nations Office on Drugs and Crime (UNODC). (2013a). *The challenge of new psychoactive substances*. http://www.unodc.org/documents/scientific/NPS_Report.pdf. Accessed 09.07.14.

United Nations Office on Drugs and Crime (UNODC). (2013b). *World drugs report*. http://www.unodc.org/unodc/secured/wdr/wdr2013/World_Drug_Report_2013.pdf. Accessed 09.07.14.

United Nations Office on Drugs and Crime (UNODC). (2014). *World drugs report*. https://www.unodc.org/documents/data-and-analysis/WDR2014/World_Drug_Report_2014_web.pdf. Accessed 09.07.14.

Valente, M. J., Guedes de Pinho, P., de Lourdes Bastos, M., Carvalho, F., & Carvalho, M. (2014). Khat and synthetic cathinones: a review. *Archives of Toxicology, 88*, 15–45.

Watterson, L. R., Hood, L., Sewalia, K., Tomek, S. E., Yahn, S., Johnson, C. T., ... Olive, M. F. (2012). The reinforcing and rewarding effects of methylone, a synthetic cathinone commonly found in "bath salts". *Journal of Addiction Research and Therapy, 9*, 002.

Watterson, L. R., Kufahl, P. R., Nemirovsky, N. E., Sewalia, K., Grabenauer, M., Thomas, B. F., ... Olive, M. F. (2014). Potent rewarding and reinforcing effects of the synthetic cathinone 3,4-methylenedioxypyrovalerone (MDPV). *Addiction Biology, 19*, 165–174.

Wood, D. M., Hill, S. L., Thomas, S. H., & Dargan, P. I. (2014). Using poisons information service data to assess the acute harms associated with novel psychoactive substances. *Drug Testing and Analysis, 6*(7–8), 850–860. http://dx.doi.org/10.1002/dta.1671.

Wright, M. J., Jr., Vandewater, S. A., Angrish, D., Dickerson, T. J., & Taffe, M. A. (2012). *British Journal of Pharmacology, 167*, 1342–1352.

Zamengo, L., Frison, G., Bettin, C., & Sciarrone, R. (2014). Understanding the risks associated with the use of new psychoactive substances (NPS): high variability of active ingredients concentration, mislabelled preparations, multiple psychoactive substances in single products. *Toxicology Letters, 229*, 220–228.

Section B

Molecular and Cellular Aspects

Chapter 64

Role of Lipid Rafts and the Underlying Filamentous-Actin Cytoskeleton in Cannabinoid Receptor 1 Signaling

Dimitra Mangoura[1], Olga Asimaki[1], Emmanouella Tsirimonaki[1], Nikos Sakellaridis[2]
[1]Basic Research Center, Biomedical Research Foundation of the Academy of Athens, Athens, Greece; [2]Pharmacology Department, Medical School, University of Thessaly, Larissa, Greece

Abbreviations

CB1 Cannabinoid 1 receptor
DAG Diacyl Glycerol
EGFR Epidermal growth factor receptor
ERK1/2 Extracellular signal-regulated kinases 1/2
FGFR Fibroblast growth factor receptor
GPCR G-Protein-coupled receptor
HPTLC High-performance thin-layer chromatography
MCD Methyl-β-cyclodextrin
PKCε Protein kinase Cε
PLC Pospholipase C
R(+)-MA Methanandamide
RTK Receptor tyrosine kinase

CANNABINOID RECEPTOR 1 SIGNALING: A FORCEFUL REGULATOR OF NEURONAL DEVELOPMENT AND FUNCTION

The endocannabinoid system is a complex signal transduction system, comprising the G-protein-coupled cannabinoid receptors 1 and 2 (CB1 and CB2), the best characterized endogenous ligands 2-arachidonoylglycerol (2-AG) and N-arachidonoylethanolamine (anandamide, AEA), and the enzymes necessary for the biosynthesis and degradation of these ligands. Uniquely among neurotransmitters, endocannabinoids are produced from precursor phospholipid molecules in the plasma membrane (PM) upon demand, for example, elevation of intracellular calcium (Cadas, Gaillet, Beltramo, Venance, & Piomelli, 1996). AEA binds to CB1 as a partial agonist, while 2-AG is a full agonist of both CB1 and CB2. The second major difference between the two receptors is that CB1 is prominently found in the central nervous system (CNS), where it mediates the central pharmacological effects of Δ⁹-tetrahydrocannabinol, the main psychoactive ingredient of marijuana (Howlett et al., 2002), whereas CB2 is predominantly expressed in peripheral tissues. The abundance of CB1, particularly in the cerebral cortex, striatum, cerebellum, hippocampus, and hypothalamus, and the discovery of several endogenous ligands have further established the CB1 system as an extensive base for the regulation of CNS physiology and function. CB1 is expressed as early as the trophoblast (Sun & Dey, 2008), becomes abundant in the developing brain, and plays important roles in the self-renewal proliferation capacity of neural cells and then in neuronal differentiation, controlling important aspects such as dendritic branching, and synaptic density and transmission (Asimaki & Mangoura, 2011; Diaz-Alonso et al., 2012; Tagliaferro et al., 2006; Takahashi & Castillo, 2006; Xapelli et al., 2013). These neuromodulatory actions of CB1 highlight the effects of cannabinoids in a wide range of systemic responses such as motor activity, memory, fear extinction, analgesia, and euphoria. CB1 is also important for appetite and metabolism regulation. The gut–brain axis that acts to maintain energy homeostasis is controlled by CB1 expression in the hypothalamus, while persistence of higher body weight even after resection of the stomach curvature correlates well with persistently high CB1 expression (Fedonidis et al., 2014) and the fact that certain CB1 genotypes associate with human obesity (Russo et al., 2007). A third difference between CB1 and CB2 has been documented at the cellular and molecular level, which is their association with lipid rafts, that is, the frequent detection of CB1, but not CB2, in these specialized membrane subdomains. This chapter focuses on the importance of lipid rafts as a platform of CB1 signaling.

CB1 SIGNALING TO THE MAJOR EFFECTOR EXTRACELLULAR SIGNAL-REGULATED KINASE

The signaling pathways that elicit the diverse cellular and organismal effects of CB1 have been the focus of extensive research efforts. As with all G-protein-coupled receptors (GPCRs), initiation of signaling cascades upon agonist–receptor binding occurs at the plasma membrane (cell surface). The specific topology of a given GPCR in discrete plasma membrane domains allows access of a liganded receptor to signaling lipids and to specific effectors available at this topology and therefore instant activation and/or interactions of effector proteins that form signaling complexes. These topologies regulate how long these formed signaling complexes will be maintained and whether they would traverse the endocytic pathway, which additionally contributes to the biological action of a specific agonist (Moore, Milano, & Benovic, 2007). The whole process, as it possesses cell type- and cell state-specific effectors, allows for differences in the magnitude and duration of the specific GPCR signal and therefore of its downstream biological output. More specifically, the signal (1) may modify preexisting proteins and alter the physiological processes in which they are involved, e.g., phosphorylation of cytoskeleton proteins that control cellular process formation; (2) may be a transducer into the nucleus and alter transcription and therefore expression of proteins; and (3) in neurons, may do both at the periphery, where translation of proteins may occur from locally stored mRNAs. A typical paradigm for the importance of the magnitude and duration in the biological outcome of the signaling is the activation of the signaling kinase extracellular signal-regulated kinase (ERK). ERK is the last effector in the three-kinase module consisting of Raf (mitogen-activated protein kinase (MAPK) kinase kinase), MAPK kinase, and ERK (or MAPK). GPCRs activate the ERK cascade utilizing a number of effectors and protein–protein signaling interactions, including protein kinase C (PKC), protein kinase A, Src-family kinases, and the transactivation of membrane receptors with tyrosine kinase activity (RTKs), such as epidermal growth factor receptor (EGFR), to converge onto Ras and Raf activation (Daub, Weiss, Wallasch, & Ullrich, 1996; Jiang, Dong, Trujillo, Miller, & Eberhardt, 2001; Luttrell et al., 1996; Mangoura & Dawson, 1993). The ERK signal is passed on to modifications in the activity of transcription factors, as well as to modifications of peripheral proteins responsible for the acute responses of a cell. Indeed, the magnitude and duration of ERK activation have been causally linked to specific cellular responses in neurons and neural cells, as transient and strong activations (<10 min) associate with cell proliferation, while sustained and milder activations induce differentiation (e.g., Leondaritis, Petrikkos, & Mangoura, 2009) and correlate well with enhancement of memory (e.g., Kelleher, Govindarajan, Jung, Kang, & Tonegawa, 2004; Zisopoulou et al., 2013). ERK activation by CB1 has been demonstrated in a variety of cells in culture (Asimaki & Mangoura, 2011; Bouaboula et al., 1995; Galve-Roperh, Rueda, Gomez del Pulgar, Velasco, & Guzman, 2002; Rubovitch, Gafni, & Sarne, 2004) and in brain slices (Lin, Mao, Su, & Gean, 2009), and the signaling mechanism of the GPCR CB1 to ERK has become a specific research focus.

CB1 SIGNALING TO ERK EMANATES FROM LIPID RAFTS

For several GPCRs, and for CB1 in particular, signaling to ERK depends on lipid rafts, that is, lipid-driven, relatively ordered, fluid domains in plasma membranes (Figure 1). These fluctuating nanoscale membrane assemblies, enriched in lipids like cholesterol,

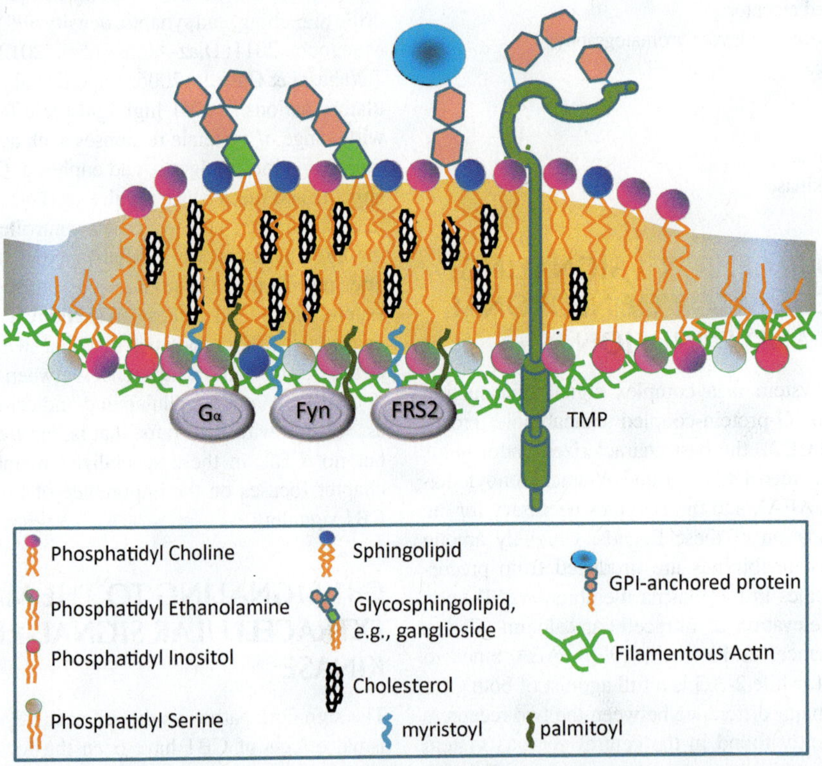

FIGURE 1 Schematic representation of the outer and inner leaflets of a lipid raft. Lipid rafts are nanoscale assemblies of sphingolipids, such as sphingomyelin, and glycosphingolipids, such as gangliosides, cholesterol, glycosylphosphatidylinositol (GPI)-anchored proteins, glycosylated transmembrane proteins, and lipid-modified, membrane-tethered proteins that contribute to membrane heterogeneity and constitute a base for cell type-specific signaling. The underlying cortical filamentous actin cytoskeleton and its associated proteins impose corral-like constraints that control the lateral diffusion of membrane proteins and lipids.

TABLE 1 Detection of CB1 in Lipid Rafts with Various Methodologies and the Effects of Agonism and/or Methyl-β-Cyclodextrin

Cell Type	Method of Lipid Raft Isolation	CB1 Detection in Rafts	Effect of Treatment	Effect of MCD	References
BV2 murine microglial cell line, caveolin 1-null	No detergent (OptiPrep fractionation or carbonate extraction)	Yes	Cannabidiol, no	ND	Rimmerman et al. (2011)
C6-2B rat glioma cell line	Carbonate extraction	Yes (caveolin rafts)	ND	ND	Bari et al. (2008)
Primary cortical neurons	No detergent OptiPrep fractionation	Yes	R(+)-MA, transiently into rafts	Blocks ERK activation	Asimaki et al. (2011)
MDA-MB-231, human breast cancer cell line	Detergent (Triton X-100-resistant membranes)	Yes	Long-term (24 h AEA), yes, away from rafts	CB1 moves away from rafts	Sarnataro et al. (2005)

glycolipids, or sphingolipids, and in glycosylphosphatidylinositol-anchored proteins and transmembrane receptors (Harder & Simons, 1999; Karnovsky, Kleinfeld, Hoover, & Klausner, 1982; del Pozo et al., 2005), are regulatory for both intracellular signaling and membrane trafficking. As raft lipids are also enriched in intracellular sorting vesicles intended for the PM (Orci et al., 1981), lipid-driven rafts may serve as a mechanism for sorting proteins with transmembrane domains along the secretory pathway from the trans-Golgi network to the PM and for endosomal sorting/recycling (Brown & Rose, 1992; Klemm et al., 2009; Lusa et al., 2001). What regulates the affinity of transmembrane proteins, as receptors are, to rafts is still only partially understood; their propensity to interact with raft lipids or raft-embedded proteins and their lipid modifications are certainly important. Among the lipid modifications, palmitoylation, the only reversible lipid modification and one posted for CB1 (Oddi et al., 2011), does regulate "raftophilicity" (Resh, 2006 and references therein). Studies have shown that, in addition to palmitoylation, transmembrane domain length and transmembrane sequence are also critical determinants of membrane raft association (Diaz-Rohrer, Levental, Simons, & Levental, 2014).

The cytoplasmic face of lipid rafts is enriched in molecules for intracellular signaling, such as Src-family kinases, G proteins, small GTPases, and adapter molecules, tethered there by their post-translational lipid modifications. Again, palmitoylation predicts, as for FRS2 (Kouhara et al., 1997), and double modification by both palmitoylation and myristoylation ascertains, as for G_α species (reviewed in Chen & Manning, 2001), Fyn (Alland, Peseckis, Atherton, Berthiaume, & Resh, 1994), or flotillin 1 (Neumann-Giesen et al., 2004), a raft residence (Figure 1). Moreover, several other signaling proteins or transmembrane receptors may conditionally flow into lipid rafts; for example, signaling phosphorylations and resulting protein–protein interactions recruit the adapter protein GRB2 (Ridyard & Robbins, 2003), the tyrosine kinases Src and fibroblast growth factor receptor (FGFR) (Asimaki, Leondaritis, Lois, Sakellaridis, & Mangoura, 2011), or the guanine nucleotide exchange factor SOS (Aronheim et al., 1994). As a consequence lipid rafts facilitate the formation of signaling complexes upon stimulation of the membrane receptors with ligands. In addition to the generation of relatively linear signaling cascades, the formed signaling complexes serve to generate positive or negative signaling feedback loops that may increase the efficiency of the signal by providing both amplification mechanisms and the necessary turn-off switches. These switches, through regulation of the magnitude and duration of the signaling cascade, contribute significantly to the specificity of the signal.

LIPID RAFTS, FILAMENTOUS-ACTIN NETWORKS, AND MICROTUBULES REGULATE CB1 SIGNALING AND LOCALIZATION

Effects of Lipid Raft Cholesterol

As with other cyclodextrins, the water-soluble, cyclic glucose heptamer methyl-β-cyclodextrin (MCD) sequesters cholesterol in its hydrophobic core into soluble inclusion complexes and blocks clathrin- or lipid raft (with or without caveolin)-dependent vesicle trafficking from (or to) the membrane (Puri et al., 2001), all disrupting GPCR signaling. A few studies, using Western blotting analysis of membrane or cell fractions, have postulated the localization of CB1 to rafts (Table 1). Despite the different methodologies used, that is, cell lysis with detergent (preparation of Triton X-100-resistant membranes), with detergent-free solutions, or by carbonate extraction and then fractionation of the lysates in OptiPrep™ or sucrose gradients, CB1 is detected in the rafts. At least for the detergent-free method of Macdonald and Pike, protein raft markers like flotillin 1 show the expected distribution. In addition, we validated the preservation of the native composition of lipid rafts with high-performance thin-layer chromatography of phospholipids to verify enrichment of sphingomyelin (Figure 2). The methodology may account in part for the differences in detection of CB1 movement; however, differences in the time of exposure to CB1 agonist, as well as the different cellular backgrounds, play important roles too. When

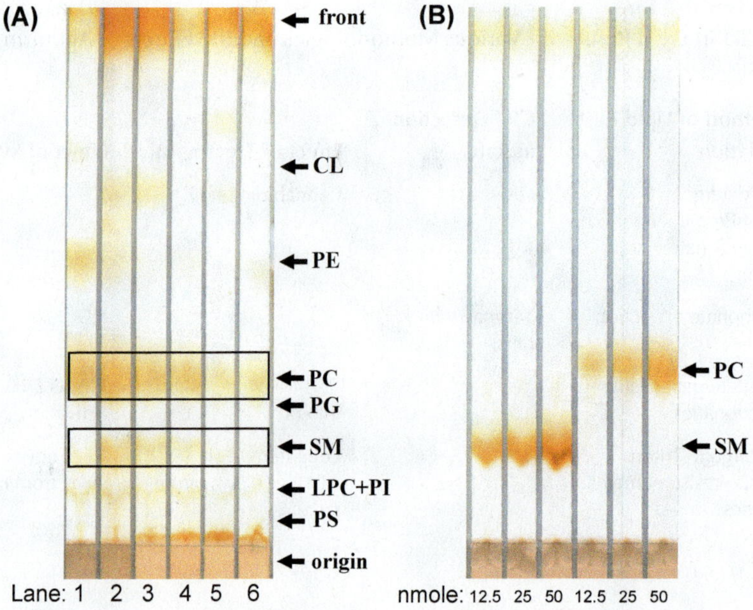

FIGURE 2 Detergent-free lipid-raft preparations preserve native lipid composition and highlight the enrichment in sphingomyelin. (A) Total lipids were extracted using a modified Bligh-Dyer method from the embryonic stem cell line E14T (lane 1) and OptiPrep fractions were combined as follows: lipid rafts, fractions 1–5 (lane 2); nonraft membranes, fractions 6–8 (lane 3); and endoplasmic reticulum–Golgi membranes, fractions 9 and 10 (lane 4), 11 (lane 5), and 12 (lane 6). Equal amounts of each extract were chromatographed on borate-impregnated thin-layer chromatography (TLC) plates with a solvent system consisting of $CHCl_3:CH_3OH:H_2O:NH_4OH$ (120:75:6:2) and lipids were visualized by iodine staining. Arrows denote the positions of authentic phospholipid standards chromatographed in separate lanes on the same TLC plate. PS, phosphatidylserine; LPC, lysophosphatidylcholine; PI, phosphatidylinositol; SM, sphingomyelin; PG, phosphatidylglycerol; PC, phosphatidylcholine; PE, phosphatidylethanolamine; CL, cardiolipin. (B) PC and SM standards (12.5–50 nM) were chromatographed as in A.

applied, MCD leads to the disappearance of CB1 from the raft fractions (Asimaki et al., 2011) or the Triton X-100-resistant membranes (Sarnataro et al., 2005).

The effects of MCD on the distribution of CB1 are readily seen with confocal microscopy. The upper row in Figure 3 presents the typical fluorescence confocal microscopic images of the CB1–GFP (green fluorescent protein) subcellular localization in COS-7 cells, which have an ample cytoplasm and a well-developed cell cortex. Seventy-two hours posttransfection and under resting conditions (control), fluorescence contributed by the tagged receptor at the level of the maximum nuclear diameter (equatorial section, column A) is seen at the PMs of lamellipodia (arrows) and filopodia (arrowheads); such distribution becomes evident closer to the adhering surface of the cell (column B). PM expression reflects proper receptor folding and stability, as well as intracellular movement and processing through the endoplasmic reticulum (ER) and Golgi organelles. Indeed the receptor is also seen in a juxtanuclear compartment and paranuclear compartments around the nucleus, consistent with an ER (asterisk) and Golgi cisternae and associated vesicular profile localization (asterisks), as expected for an overexpressed receptor that is constantly synthesized. Some endosomal origin of this intensity should be considered, attributable to cycling of the receptor between PM and endosomes after autocrine or paracrine stimulation of the CB1 itself or through its transactivation by other membrane receptors, again due to paracrine or autocrine stimulation, often characterized as constitutive endocytosis. When cells were incubated with the specific CB1 agonist methanandamide ($R(+)$-MA; 9 nM × 15 min; Figure 3, second row), PM localization becomes minimal, while the intensity of the intracellular juxta- and perinuclear pools of CB1 receptors significantly increases by threefold (inset in first versus second row). This enrichment may reflect the presence of the receptor on recycling endosomes or even a proteasomal localization, as proteasomes are denser in the vicinity of the microtubule organizing center (MTOC) by the nucleus and on the cytosolic side of the ER. Notably, a perinuclear, ring-like enrichment of CB1 appears, consistent with an additional distribution to the ER (inset second row). While this may further argue for a presence in recycling endosomes, it should be noted that endosomal EGFR interacts with the ER phosphatase PTP1B, in an apparently global mechanism of direct interactions between PM receptors and ER molecules (Eden, White, Tsapara, & Futter, 2010). With MCD treatment (10 mM × 15 min, at 37 °C), the receptor appears minimal at the PM and concentrated in some intracellular pools, including some irregular large vesicular structures (arrows), not seen in controls. When MCD incubation is followed by $R(+)$-MA, we do not observe any measurable difference in intracellular densities compared to MCD alone, nor does the overall pattern present with similarities to that with $R(+)$-MA alone, all indicating disruption of the $R(+)$-MA-induced intracellular flow of the receptor. Furthermore, CB1–GFP fluorescence did not show any significant colocalization with that of Lyso-Tracker Red DND-99, used as a marker of lysosomes (Figure 4). After treatment with $R(+)$-MA, and despite the increased intracellular signal, only a small pool of the receptor is seen to colocalize with lysosomes (Figure 4, arrows).

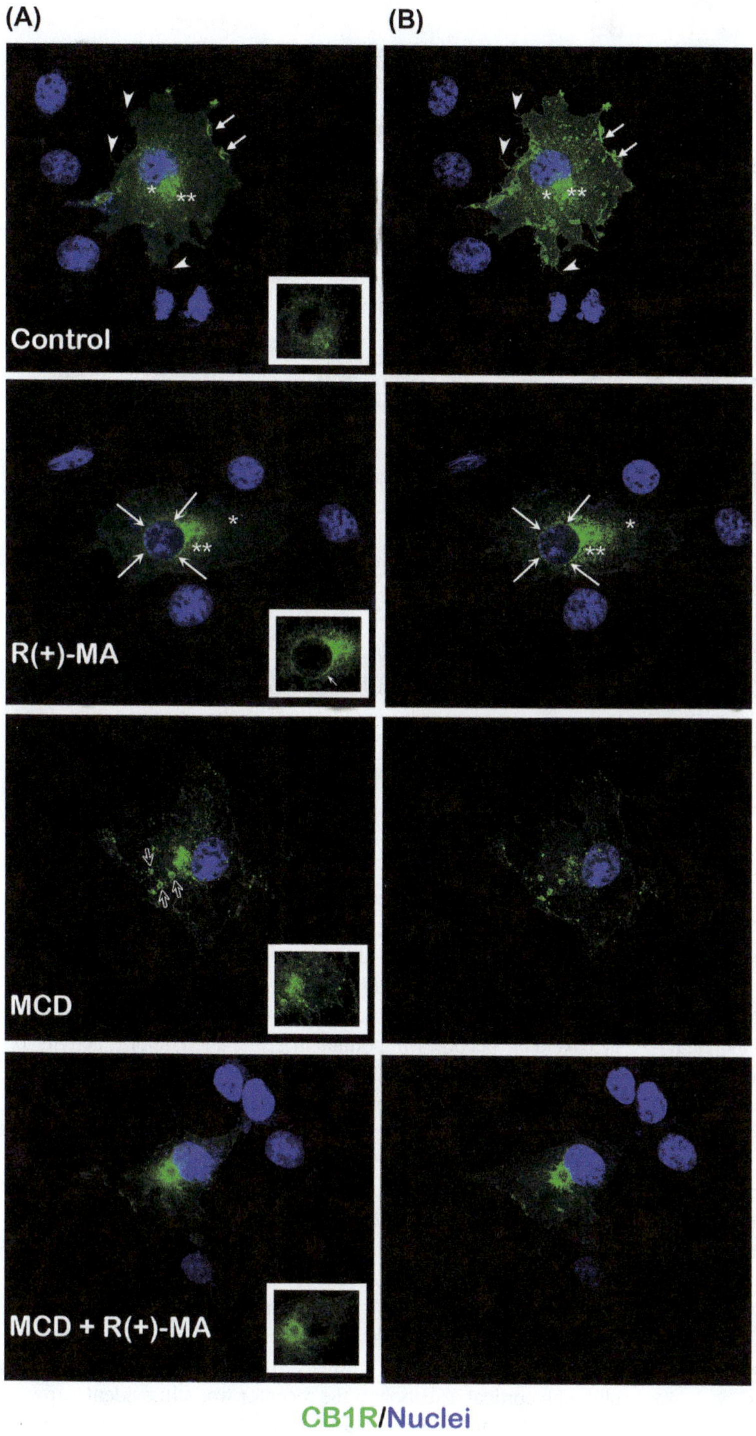

FIGURE 3 Subcellular distribution of CB1: effects of agonism and/or MCD. COS-7 cells were transfected with a CB1–GFP construct and were either left untreated (control) or treated with R(+)-MA (9 nM) for 15 min, with or without cholesterol extraction with 10 mM MCD, fixed with 4% paraformaldehyde, and stained with Hoechst 33258 to visualize chromatin (nuclei). Fluorescence was imaged using a Leica TCS SP5 inverted confocal microscope with a motorized stage, a 63× HC PL APO CS oil lens, and a tandem scanner (Biological Imaging Unit, Biomedical Research Foundation, Academy of Athens). Images are projections of four serial 0.25-μm-thick Z-stacks (~2 μm each panel) at the equatorial plane (left column) or at the adhering surface (right column). Detailed description is given in the text.

Effects of Microtubule Integrity

An integral functional partner of lipid rafts is the tubulins, resident proteins in raft microdomains (Palestini et al., 2000). Both α- and β-tubulin have been shown to physically interact with the lipid-raft markers flotillin 1, caveolin, and G-proteins (Dremina, Sharov, & Schoneich, 2005). Microtubules, the polymerized dimers of α- and β-tubulin, in the cell periphery are very dynamic and the rapid interchange between stable elongation and catastrophe is modulated by a number of molecules in or in close association with lipid rafts. Moreover, both internalizing endosomes and vesicles transported from the ER to the Golgi may utilize microtubules to translocate. Although the fixing conditions for visualizing receptors, in this case CB1–GFP, are not optimal for precisely preserving the architecture of microtubules and in particular peripheral tubulin dimers, a number of CB1 vesicles may be seen along the immunofluorescence of β-tubulin (Figure 5(A), arrows in inset), especially along thicker microtubule bundles

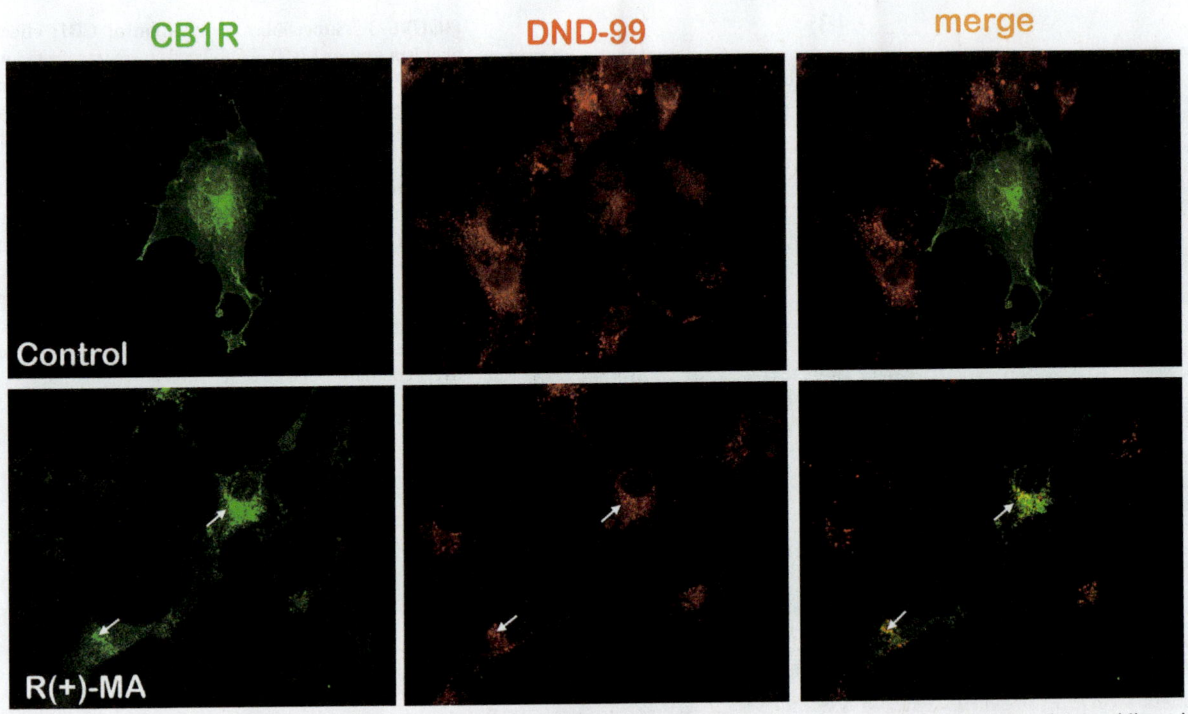

FIGURE 4 Limited localization of CB1 in lysosomes. CB1–GFP COS-7 cells were preincubated with LysoTracker DND-99 to detect acidic vesicular organelles (lysosomes) and then imaged under resting conditions (control) or after treatment with $R(+)$-MA. Intracellular patterns of CB1–GFP are quite distinct from those of DND-99 under either condition; arrows point to occasional colocalization upon receptor internalization. Images are projections of four serial 0.25-μm-thick Z-stacks (~2 μm each panel) at the equatorial plane.

(double arrows). This close association may also be seen in mitotic cells at telophase, in which internalization of the receptor after treatment with $R(+)$-MA may be readily seen (Figure 5(B)). The flow of CB1–GFP fluorescence appears to be in close association with microtubules emanating from the MTOC (yellow arrow), while there is an unexpected enrichment of small CB1 vesicles on the central spindle microtubules (double arrows) that promote ingression of the cleavage furrow and cytokinesis. When cells were mildly treated with nocodazole (660 nM × 1 h), a drug that binds to tubulins with a two-affinity Michaelis–Menten kinetics pattern and inhibits microtubule assembly, the cytoplasm collapses into few thick processes around the soma and the GFP-tagged receptors appear with a diffuse distribution in it, with no particular enrichments or PM presentation (Figure 6). Thus microtubules are essential for proper CB1 surface presentation and intracellular trafficking.

Effects of Filamentous-Actin Membrane Cytoskeleton Integrity

It is now adequately documented that lipid rafts coalesce within constraints applied by the underlying cortical filamentous-actin (F-actin) cytoskeleton and its associated proteins to form signaling platforms from where signaling cascades emanate. More specifically, constraints are imposed by the F-actin and cross-linking proteins that act as fences and by interacting proteins that act as pickets, namely by the cytosolic domains of transmembrane proteins or cytosolic proteins, tethered into the membrane through lipid posttranslational modifications (Kusumi & Suzuki, 2005; Morone et al., 2006). This extensive network of "fences" and "pickets" forms restraining "corrals" for the movement of lipid rafts through the entirety of the PM. F-actin-binding proteins are instantly modified by receptor-mediated signaling kinase activation and transiently assume new affinities and diffusion rates (Asimaki & Mangoura, 2011; Harder & Simons, 1999; Kusumi & Suzuki, 2005; Mangoura, 1997). Thus, a signal-dependent reorganization of the cortical actin cytoskeleton becomes important in the spatiotemporal activity of signaling pathways, because signaling complexes often internalize on pinched-off vesicles along cytoskeleton tracks, further propagating the signal intracellularly (del Pozo et al., 2005).

The importance of an intact F-actin cytoskeleton becomes apparent when imaging the subcellular localization of CB1 after the collapse of the F-actin cytoskeleton by cytochalasin D (10 μM × 1 h, at 37 °C) (Figure 7). More specifically, cytochalasin caused CB1–GFP fluorescence to appear as irregular, scattered foci, mostly throughout the cytoplasm (arrows), while some remained in the fine processes that are left as the cytoplasm retracts (arrowheads). Treatment with $R(+)$-MA resulted in further retraction of the fine processes, although cortical detection of the receptor was still evident (arrows, lower row), while some increase in the intracellular juxta- and perinuclear pools of CB1 was detectable, yet lower than with $R(+)$-MA alone (Figure 3, second row, vs Figure 7, lower row). It is thus possible that CB1 signaling and endocytosis may still take place when F-actin is disrupted, yet, the signal is modified as to (1) result in less intracellular trafficking of the receptor and (2) be "consumed" in the cell periphery, where more intense phosphorylation of F-actin-binding proteins, like MARCKS or p120 (Asimaki & Mangoura, 2011), leads to intense disengagement of the F-actin cytoskeleton from the PM and hence further collapsing of filopodia, as seen with CB1 agonism in hippocampal neurons (Roland et al., 2014).

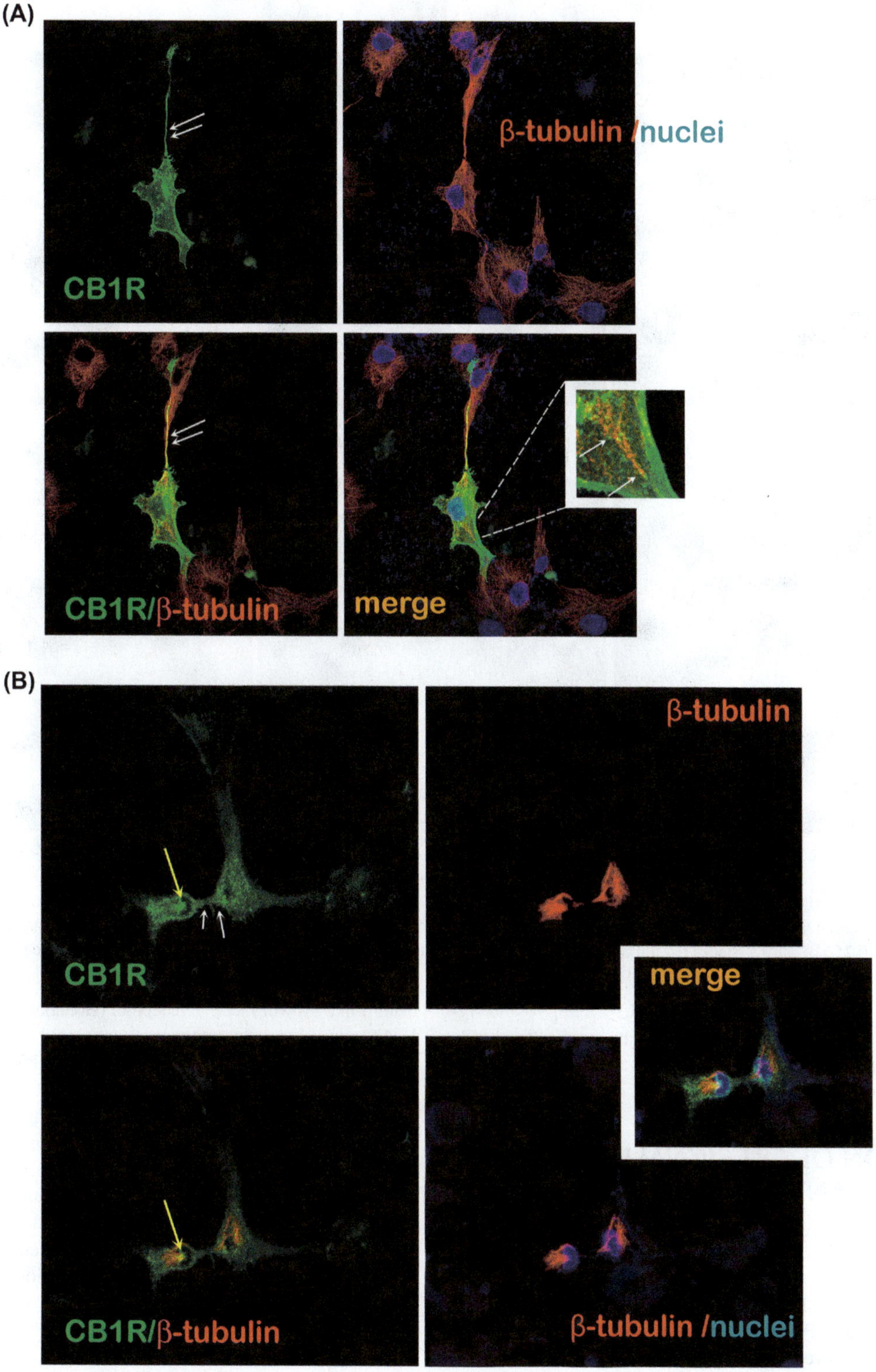

FIGURE 5 **CB1–GFP fluorescence partially localizes along microtubules.** CB1–GFP-expressing COS-7 cells were immunostained with β-tubulin, while nuclei are visualized with Hoechst 33258. (A) In control cells a number of CB1 vesicles may be seen along the microtubule tracks (arrows) detected with β-tubulin. (B) CB1–GFP internalization after treatment with R(+)-MA may be seen in cells at the last stage of mitosis, as deduced from chromatin (Hoechst 33258, blue) and β-tubulin patterns of staining. Yellow arrows point to a CB1 enrichment in the vicinity of the MTOC and double arrow to the central spindle microtubules. All images are projections of four serial 0.25-μm-thick Z-stacks (~2 μm each panel) at the equatorial plane.

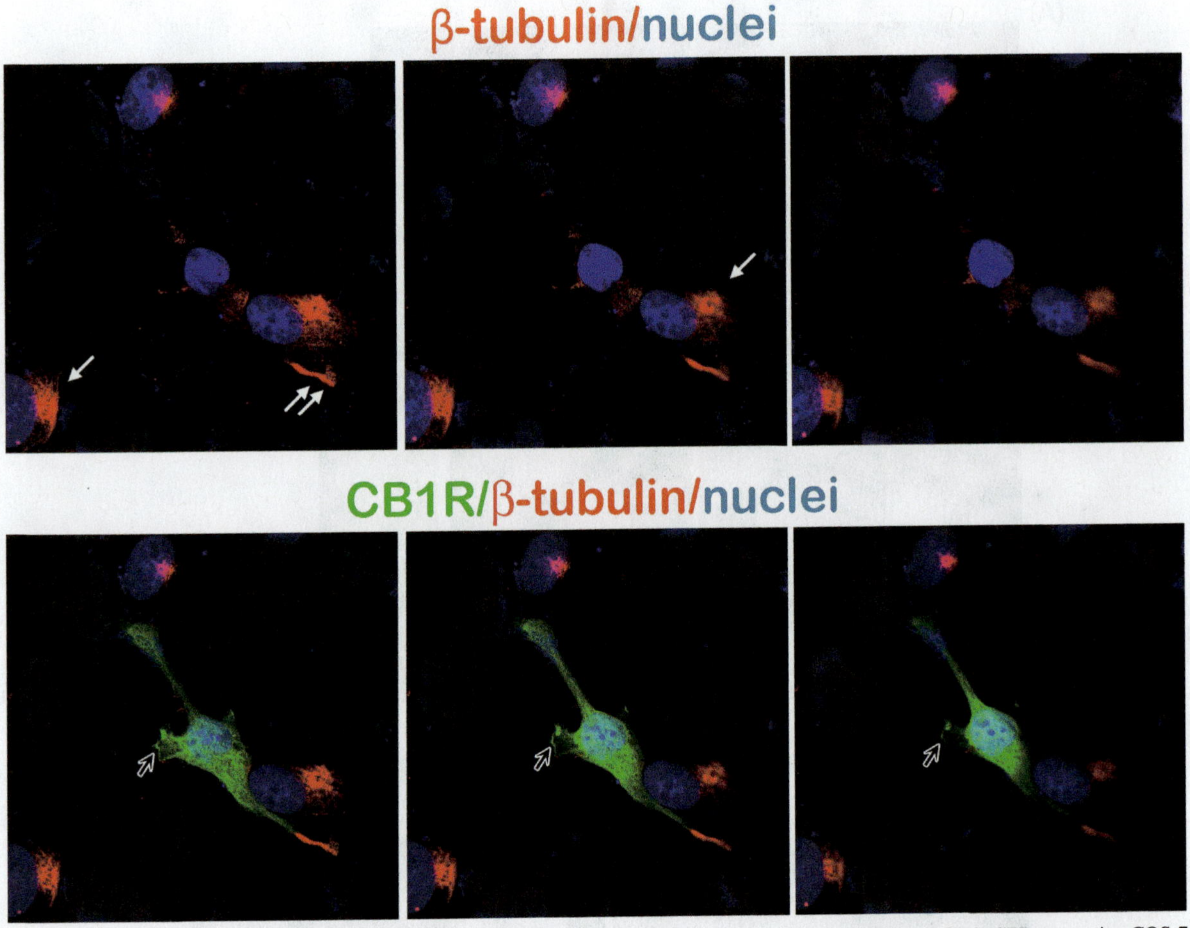

β-tubulin/nuclei

CB1R/β-tubulin/nuclei

FIGURE 6 Microtubule integrity is essential for proper CB1 surface presentation and intracellular trafficking. CB1–GFP-expressing COS-7 cells were treated with the microtubule disrupter nocodazole (660 nM, 1 h, 37 °C) and immunostained with β-tubulin (red), while nuclei were visualized with Hoechst 33258. Nocodazole causes microtubule collapse toward the MTOC (single arrows) and in the leading process of a migrating cell (double arrow), and CB1–GFP fluorescence appears scattered throughout the cytoplasm with only limited membrane presentation (empty arrows). All images are projections of four serial 0.25-μm-thick Z-stacks (~2 μm each panel), starting at the equatorial plane (left column) and moving toward the free surface of the cell.

These data collectively indicate that CB1 targeting to and signaling/trafficking from the PM is regulated by lipid rafts and its functional partners, the F-actin membrane and the intracellular microtubule cytoskeleton.

FORMATION OF CB1 SIGNALING COMPLEXES AND TRANSACTIVATION OF TYROSINE KINASE RECEPTORS IN LIPID RAFTS

Interreceptor cross talk has long been recognized as an important mechanism that links GPCRs to ERK1/2 activation. CB1s may form homotypic dimers or even heterotypic dimers with other GPCRs for both signal amplification and receptor internalization (e.g., Ellis, Pediani, Canals, Milasta, & Milligan, 2006). The distinctly different heterotypic complexes of GPCRs and RTKs are, however, less understood and characterized, except in that they are known to require scaffolding proteins (Maudsley et al., 2000; Pyne, Waters, Moughal, Sambi, & Pyne, 2003). Cross talk between CB1 and RTKs was first described in CHO cells cotransfected with CB1 and insulin or insulin-like growth factor 1 receptors (Bouaboula et al., 1997), then EGFR was shown to integrate cannabinoid signaling to activate ERK1/2 in tumor cell lines (Hart, Fischer, & Ullrich, 2004; Rubovitch et al., 2004), while a significant level of cross talk between CB1 and FGFRs, also prominent receptors in promoting neuritic outgrowth, was shown to regulate the axonal growth of rat cerebellar granule neurons (Williams, Walsh, & Doherty, 2003) or neuritic outgrowth in chick embryo cortical neurons (Asimaki & Mangoura, 2011).

In canonical pathways, GPCR-dependent transactivation of RTKs requires tyrosine kinases of the Src family, which activate RTKs through intermolecular phosphorylation (e.g., Sandilands et al., 2007), while dependency on Ca^{2+}, G proteins, or PKC activity appears to be receptor and cell type specific (Ma, Huang, Ali, Lowry, & Huang, 2000; Mangoura, 1997; Mangoura & Dawson, 1993). Ligand-induced organization of GPCR–RTK multiprotein signaling complex provides amplification of the signal through the engagement of additional downstream effectors and may also contribute to the endocytosis and intracellular trafficking of the receptor. This highly complex regulation of membrane receptors, serving neurotransmission, growth, or polarity, is considered pivotal for CNS development and synapse function.

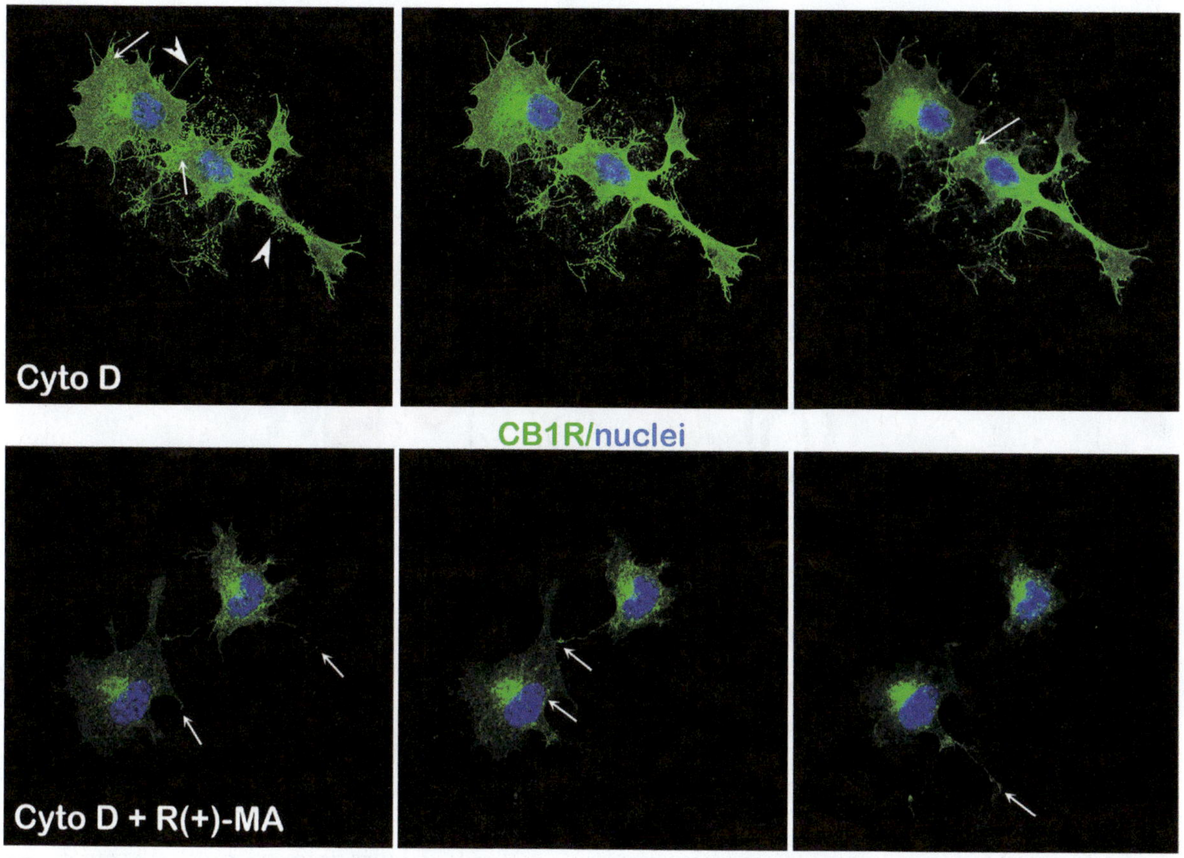

FIGURE 7 Integrity of the F-actin cytoskeleton is essential for proper CB1 surface presentation, intracellular trafficking, and endocytosis. CB1–GFP-expressing COS-7 cells were treated with 10 μM cytochalasin D (1 h at 37 °C) prior to addition of vehicle (upper row) or $R(+)$-MA for 15 min (lower row), fixed, stained with Hoechst 33258, and imaged. Detailed description is given in the text. All images are projections of four serial 0.25-μm-thick Z-stacks (~2 μm each panel), starting at the equatorial plane (left column) and moving toward the free surface of the cell.

Our investigations on CB1 proximal signaling in early cortical neurons in primary cultures have revealed that CB1 agonism with $R(+)$-MA induces a biphasic activation of ERK1/2 in a Src- and Fyn-dependent manner (Asimaki & Mangoura, 2011). Most interestingly this 90-min-long time course of ERK activation clearly exhibits two sequential increases: the first peak starts within seconds and reaches a plateau by 5 min, and the second, additional amplification at 15 min lasts for up to 30 min, decreasing slowly thereafter. This CB1-induced, sustained duration of ERK activation translates to induction of neuritic outgrowth in chick cortical neurons in the long term.

Among the earliest signaling events in this model of a CB1–ERK cascade is a $G_{q/11}$-mediated phospholipase C and PKC activation. With proper pharmacological analysis, we postulate that PKCε is the PKC isoform immediately activated. This isoform uniquely bears an actin-binding site, which allows it, upon activation, to bind to F-actin and modify F-actin-binding proteins, all leading to regulation of neurite outgrowth (Asimaki & Mangoura, 2011; Mangoura, 1997; Zeidman, Troller, Raghunath, Pahlman, & Larsson, 2002). Moreover, endogenous CB1 and PKCε physically associate and reciprocally coimmunoprecipitate, while stoichiometric cotransfections with appropriate constructs show that CB1 interacts directly with the regulatory domain of PKCε (Asimaki & Mangoura, 2011). Dissociation of the two molecules occurs probably owing to conformational changes in CB1 with ligand binding (Howlett et al., 2002) and to the phospholipase C (PLC)/diacyl glycerol-induced "opening" (activation) of PKCε; an open-conformation PKCε preferentially associates with Src-family kinases (Song et al., 2002). While activation of both Fyn and Src occurs upon $R(+)$-MA agonism in primary cortical neurons, their specific inhibition abolishes both peaks, suggesting that this CB1/G_q/PLC/PKCε/Src–Fyn/Ras/Raf/ERK pathway (Figure 8) induces a second, proximal signaling event that leads to intense amplification, possibly recruitment of an RTK.

Cross talk of CB1 and FGFR has been postulated in the axonal growth of rat cerebellar granule neurons (Williams et al., 2003) and, indeed, PD173074, an FGFR-specific inhibitor known to abolish neuritogenesis, significantly reduces the second phase of $R(+)$-MA-induced ERK1/2 activation, but not the first (Asimaki et al., 2011). As probed by immunoprecipitation, $R(+)$-MA-induced activation of CB1 indeed elicits tyrosine phosphorylation of FGFR (and therefore activation; Furdui, Lew, Schlessinger, & Anderson, 2006) in a time frame that correlates well with the second wave of ERK1/2 activation and constitutes an amplification of the first wave (Figure 8). Moreover, direct activation of PKCε (with the PKCε-specific activator peptide ψεRACK) increases tyrosine phosphorylation of FGFR (Asimaki et al., 2011). Further probing this transactivation, with analysis of CB1 immunoprecipitates during the second phase of ERK1/2 activation, shows that, upon CB1 agonism and $G_{i/o}$ stimulation, a complex containing CB1 and activated Src, Fyn, and FGFR is formed.

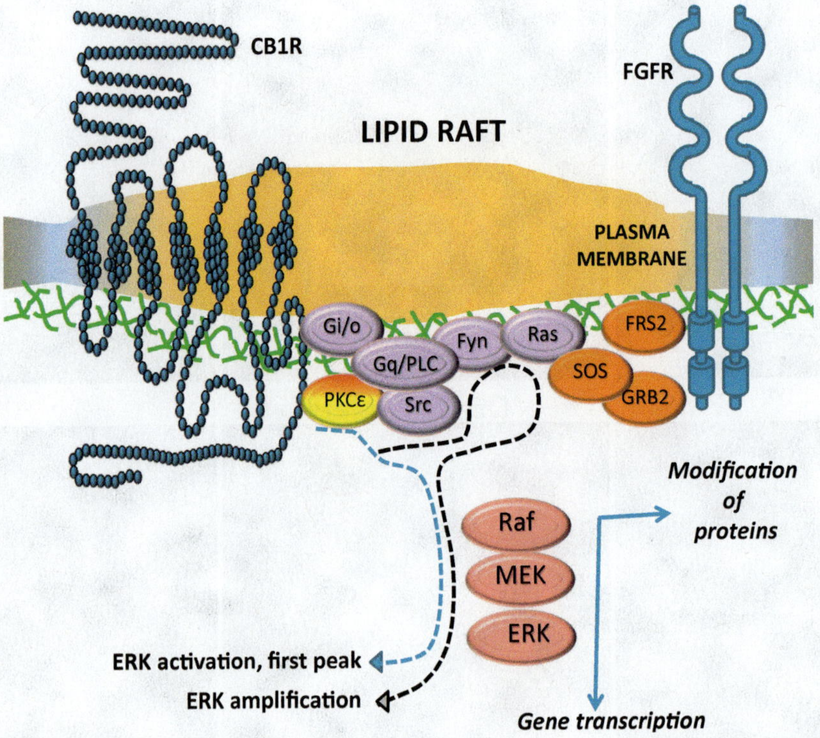

FIGURE 8 Schematic representation of CB1 signaling out of a lipid raft. CB1 causes a sustained and biphasic ERK activation in cortical neurons. A significant pool of CB1s preassociates with regulatory domains of PKCε, which upon ligand binding and sequential acute activation of $G_{q/11}$/PLC becomes activated and dissociates from CB1. The liganded receptor flows into lipid rafts, while the activated PKCε acutely forms transient signaling complexes with activated Src and Fyn kinases to elicit the first wave of ERK1/2 activation that lasts for 5 min (CB1/G_q/PLC/PKCε/Src–Fyn/Ras/Raf/ERK). Concurrently, several F-actin cytoskeleton proteins are modified by phosphorylation. A second pool of CB1s couples to activation of $G_{i/o}$ and, utilizing as effectors additional Src and Fyn molecules, transactivates FGFR to drive a second amplifying wave of ERK1/2 activation (CB1/$G_{i/o}$/Src–Fyn/FGFR/Ras/Raf/ERK).

Demonstrating specificity, specific CB1 antagonists, or potent inhibitors of PKCε and of Src or Fyn, abolish the cannabinoid-induced transactivation of FGFRs. This amplification of the magnitude and duration of the CB1-dependent activation of ERK through FGFR activation is of high functional significance, because inhibition of FGFR activation with PD173074 abolishes the CB1-dependent increase in the length of the major dendrites in primary cortical neurons.

More importantly, lipid-raft integrity is required for this CB1/G_i/PKCε/Src–Fyn/Ras/Raf/ERK pathway, because MCD diminishes the association of CB1 with Src, Fyn, and FGFR. Moreover, recruitment and activation of CB1 in lipid rafts constitutes an early event that precedes and is required for these functional CB1 multiprotein signaling complexes. As mentioned above, CB1 is widely distributed in raft and nonraft membranes of primary cortical neurons under basal conditions, when analyzed by a detergent-free lipid-raft preparation that provides sufficient resolution of genuine lipid rafts from nonraft PMs and intracellular ER and Golgi membranes (Figure 2). After 2 min of $R(+)$-MA, CB1 rapidly incorporates into the lipid-raft fractions and by 7 min it progressively traffics toward the nonraft membranes as well as the intracellular ER and Golgi membranes. After 15 min, the original pool of CB1 molecules in the lipid rafts has been restored. CB1 has evolved to bind lipid neurotransmitters, and at least for the prototypical agonist AEA, cholesterol is pivotal for insertion into the lipid bilayer (Di Pasquale, Chahinian, Sanchez, & Fantini, 2009). It is therefore possible that ligand binding induces conformational changes in CB1 that allow its flow in the lipid rafts, where it associates with other lipid-raft-prone signaling proteins. Indeed, Fyn kinase, a raft resident through its double lipid modification, is detected as activated in the lipid rafts within 2 min of treatment with $R(+)$-MA (Asimaki et al., 2011). Concurrently, that is by 2 min, $R(+)$-MA induces incorporation of an FGFR pool in the lipid rafts. FGFRs lack a GRB2 binding site and depend on recruiting the intermediate docking myristoylated phosphoprotein scaffolds FRS2 and FRS3, a function unique to vertebrates and available only to a limited repertoire of RTKs (Gotoh, 2008), in order to activate SOS and thus Ras, Raf, and ERK. Moreover, as ERK1/2 phosphorylates FRS2 to negatively affect FGF-induced tyrosine phosphorylation of FRS2 (Gotoh, 2008), the activation of FGFR provides another off switch in this signaling cascade. This is an additional mechanism to control the CB1-dependent ERK1/2 activation, the first being the PKC-dependent recruitment of off switches for the ERK pathway, such as neurofibromin (Mangoura et al., 2006). Taken together, FGFR1, known to signal mitogenic ERK1/2 signals from non-lipid-raft membranes, may also signal from lipid rafts in the case in which it is recruited as a tyrosine kinase that amplifies the CB1 signal in neurons.

CONCLUDING REMARKS

Collective knowledge has documented an important role for CB1 in a diverse array of physiological and pathophysiological processes ranging from normal development to control of appetite to pathogenesis of cancer and the development of addiction. Therefore understanding the mechanistic details of CB1 signaling is of critical importance, as it holds great therapeutic promise. In this chapter we have reviewed experimental evidence that shows that lipid rafts constitute the PM platform on which both early CB1 signaling events and subsequent ordered complex formation with receptor tyrosine kinases, in particular with FGFR, are organized and primed toward the regulated propagation of ERK1/2 activation. These observations may provide new insights into the pharmacological effects and cellular functions modulated by the CB1-induced signaling events, which additionally underlie several of the less understood acute effects of cannabinoid drugs.

APPLICATIONS TO OTHER ADDICTIONS AND SUBSTANCE MISUSE

CB1 has evolved as a major regulatory molecule for almost all known aspects of the development and function of the CNS, while it is capable of establishing addiction. It is therefore very important to understand the precise mechanisms and intracellular signal transduction pathways that CB1 utilizes for its effects. As elaborated in this chapter, the highly complex intracellular signaling of CB1 is initiated and organized at specialized membrane microdomains, the lipid rafts.

Lipid rafts are also regulatory to a number of other receptors that CB1 interacts with and which may also induce addiction, such as the opioid receptors. Given that the composition of lipid rafts may be altered with various approaches, pharmacological or nutritional, understanding the role of lipid rafts in the signaling of membrane receptors opens new ways to target addiction.

Moreover, several of the intracellular signaling molecules that are utilized by both opioid receptors and CB1 are mediators of the effects of alcohol, such as PKCε, and understanding their function may lead to unraveling the molecular basis of addiction and eventually lead to its treatment.

DEFINITION OF TERMS

Plasma membrane This is a lipid bilayer (two leaflets), rich in proteins, that encloses each cell and separates it from its environment, thus generating the extracellular and the intracellular space; the latter includes all the cellular organelles and the nucleus with the genetic information.

The cannabinoid 1 receptor This is a protein interlaced in the PM in a specific manner, namely, its initial domain along with three other more centrally exposed to the extracellular space, while three other domains plus the terminal domain are exposed to the intracellular space. Exogenous (synthetic, plant-derived) or endogenous (produced by cells in the organism) cannabinoid substances (CB1 agonists) bind specifically to CB1; CB1 then changes its three-dimensional form and starts an intracellular signaling cascade.

Intracellular signaling cascades or signal transduction pathways or simply signaling These are a series of interactions among proteins that transmit the event or message that CB1 agonists have been bound to the CB1. Intracellular signaling cascades transmit this message (1) to proteins in the intracellular/cytoplasmic space and change transiently their function in physiological processes, e.g., shape of the cell, and (2) to the nucleus to modify gene expression.

Phosphorylation This is the addition of a phosphate group to certain amino acids, the building blocks of proteins. It is catalyzed by enzymes called *kinases*, and most kinases become activated during signaling. Phosphorylation is a reversible modification of proteins.

Lipid rafts These are specially organized microdomains with a diameter of 10–200 nm and slightly thicker than the rest of the PM; hence they appear as rafts in water. Lipid rafts have a particular composition of lipids and proteins (including receptors). This particular composition along with their ability to coalesce allows them to function as a regulatory platform for receptor signaling.

Lipid raft asymmetry This refers to the fact that the outer lipid layer or outer leaflet (which faces the extracellular space) and the inner leaflet (which faces the intracellular space) differ in protein and lipid composition, further regulating receptor signaling.

F-actin cytoskeleton or membrane skeleton or cortical cytoskeleton This is a special cytoskeleton structure composed of F-actin and proteins that bind to it. This fishnet-like structure underlies the PM and creates corrals that restrain the movement of lipid rafts. Several of its proteins interact with signaling cascades (act as scaffolds that facilitate propagation of the signal).

KEY FACTS OF CB1

- CB1 is expressed very early in embryo development and is involved in the mediation of important aspects of cell and tissue growth.
- CB1 is a prominent receptor in the CNS.
- CB1 is particularly expressed in the cerebral cortex, striatum, cerebellum, hippocampus, and hypothalamus, CNS areas that control cognition, memory, emotion, and movement, as well as appetite and metabolism.
- CB1 modulatory actions on cells are mediated by its intracellular signaling, which may transiently modify both preexisting proteins and protein dosage in general, via signaling to the nucleus for changes in gene expression (changes in gene transcription are translated to changes in protein content: the effects may be up- or downregulation of a given protein).
- Aberrant CB1 signaling and aberrations in protein dosage of proteins in CNS cells are the molecular basis of addiction.

SUMMARY POINTS

- Lipid rafts are important for CB1 plasma presentation, intracellular trafficking, and signaling.
- Integrity of F-actin and microtubule cytoskeletons is required for the same processes.
- CB1 flows to lipid rafts upon ligand binding.
- The CB1 signaling to ERK is served by two additive pathways:
  - CB1/G_q/PLC/PKCε/Src–Fyn/Ras/Raf
  - CB1/$G_{i/o}$/Src–Fyn/FGFR/Ras/Raf

REFERENCES

Alland, L., Peseckis, S. M., Atherton, R. E., Berthiaume, L., & Resh, M. D. (1994). Dual myristylation and palmitylation of Src family member p59fyn affects subcellular localization. *Journal of Biological Chemistry*, 269(24), 16701–16705.

Aronheim, A., Engelberg, D., Li, N., al-Alawi, N., Schlessinger, J., & Karin, M. (1994). Membrane targeting of the nucleotide exchange factor Sos is sufficient for activating the Ras signaling pathway. *Cell*, 78(6), 949–961.

Asimaki, O., Leondaritis, G., Lois, G., Sakellaridis, N., & Mangoura, D. (2011). Cannabinoid 1 receptor-dependent transactivation of fibroblast growth factor receptor 1 emanates from lipid rafts and amplifies extracellular signal-regulated kinase 1/2 activation in embryonic cortical neurons. *Journal of Neurochemistry, 116*(5), 866–873.

Asimaki, O., & Mangoura, D. (2011). Cannabinoid receptor 1 induces a biphasic ERK activation via multiprotein signaling complex formation of proximal kinases PKCepsilon, Src, and Fyn in primary neurons. *Neurochemistry International, 58*(2), 135–144.

Bari, M., Oddi, S., De Simone, C., Spagnolo, P., Gasperi, V., Battista, N., ... Maccarrone, M. (2008). Type-1 cannabinoid receptors colocalize with caveolin-1 in neuronal cells. *Neuropharmacology, 54*(1), 45–50.

Bouaboula, M., Perrachon, S., Milligan, L., Canat, X., Rinaldi-Carmona, M., Portier, M., ... Casellas, P. (1997). A selective inverse agonist for central cannabinoid receptor inhibits mitogen-activated protein kinase activation stimulated by insulin or insulin-like growth factor 1. Evidence for a new model of receptor/ligand interactions. *Journal of Biological Chemistry, 272*(35), 22330–22339.

Bouaboula, M., Poinot-Chazel, C., Bourrie, B., Canat, X., Calandra, B., Rinaldi-Carmona, M., ... Casellas, P. (1995). Activation of mitogen-activated protein kinases by stimulation of the central cannabinoid receptor CB1. *Biochemical Journal, 312*(Pt 2), 637–641.

Brown, D. A., & Rose, J. K. (1992). Sorting of GPI-anchored proteins to glycolipid-enriched membrane subdomains during transport to the apical cell surface. *Cell, 68*(3), 533–544.

Cadas, H., Gaillet, S., Beltramo, M., Venance, L., & Piomelli, D. (1996). Biosynthesis of an endogenous cannabinoid precursor in neurons and its control by calcium and cAMP. *The Journal of Neuroscience, 16*(12), 3934–3942.

Chen, C. A., & Manning, D. R. (2001). Regulation of G proteins by covalent modification. *Oncogene, 20*(13), 1643–1652.

Daub, H., Weiss, F. U., Wallasch, C., & Ullrich, A. (1996). Role of transactivation of the EGF receptor in signalling by G-protein-coupled receptors. *Nature, 379*(6565), 557–560.

Diaz-Alonso, J., Aguado, T., Wu, C. S., Palazuelos, J., Hofmann, C., Garcez, P., ... Galve-Roperh, I. (2012). The CB(1) cannabinoid receptor drives corticospinal motor neuron differentiation through the Ctip2/Satb2 transcriptional regulation axis. *The Journal of Neuroscience, 32*(47), 16651–16665.

Diaz-Rohrer, B. B., Levental, K. R., Simons, K., & Levental, I. (2014). Membrane raft association is a determinant of plasma membrane localization. *Proceedings of the National Academy of Sciences of the United States of America, 111*(23), 8500–8505.

Di Pasquale, E., Chahinian, H., Sanchez, P., & Fantini, J. (2009). The insertion and transport of anandamide in synthetic lipid membranes are both cholesterol-dependent. *PLoS One, 4*(3), e4989.

Dremina, E. S., Sharov, V. S., & Schoneich, C. (2005). Protein tyrosine nitration in rat brain is associated with raft proteins, flotillin-1 and alpha-tubulin: effect of biological aging. *Journal of Neurochemistry, 93*(5), 1262–1271.

Eden, E. R., White, I. J., Tsapara, A., & Futter, C. E. (2010). Membrane contacts between endosomes and ER provide sites for PTP1B-epidermal growth factor receptor interaction. *Nature Cell Biology, 12*(3), 267–272.

Ellis, J., Pediani, J. D., Canals, M., Milasta, S., & Milligan, G. (2006). Orexin-1 receptor-cannabinoid CB1 receptor heterodimerization results in both ligand-dependent and -independent coordinated alterations of receptor localization and function. *Journal of Biological Chemistry, 281*(50), 38812–38824.

Fedonidis, C., Alexakis, N., Koliou, X., Asimaki, O., Tsirimonaki, E., & Mangoura, D. (2014). Long-term changes in the ghrelin-CB1 axis associated with the maintenance of lower body weight after sleeve gastrectomy. *Nutrition and Diabetes, 4*, e127.

Furdui, C. M., Lew, E. D., Schlessinger, J., & Anderson, K. S. (2006). Autophosphorylation of FGFR1 kinase is mediated by a sequential and precisely ordered reaction. *Molecular Cell, 21*(5), 711–717.

Galve-Roperh, I., Rueda, D., Gomez del Pulgar, T., Velasco, G., & Guzman, M. (2002). Mechanism of extracellular signal-regulated kinase activation by the CB(1) cannabinoid receptor. *Molecular Pharmacology, 62*(6), 1385–1392.

Gotoh, N. (2008). Regulation of growth factor signaling by FRS2 family docking/scaffold adaptor proteins. *Cancer Science, 99*(7), 1319–1325.

Harder, T., & Simons, K. (1999). Clusters of glycolipid and glycosylphosphatidylinositol-anchored proteins in lymphoid cells: accumulation of actin regulated by local tyrosine phosphorylation. *European Journal of Immunology, 29*(2), 556–562.

Hart, S., Fischer, O. M., & Ullrich, A. (2004). Cannabinoids induce cancer cell proliferation via tumor necrosis factor alpha-converting enzyme (TACE/ADAM17)-mediated transactivation of the epidermal growth factor receptor. *Cancer Research, 64*(6), 1943–1950.

Howlett, A. C., Barth, F., Bonner, T. I., Cabral, G., Casellas, P., Devane, W. A., ... Pertwee, R. G. (2002). International Union of Pharmacology. XXVII. Classification of cannabinoid receptors. *Pharmacological Reviews, 54*(2), 161–202.

Jiang, S. W., Dong, M., Trujillo, M. A., Miller, L. J., & Eberhardt, N. L. (2001). DNA binding of TEA/ATTS domain factors is regulated by protein kinase C phosphorylation in human choriocarcinoma cells. *Journal of Biological Chemistry, 276*(26), 23464–23470.

Karnovsky, M. J., Kleinfeld, A. M., Hoover, R. L., & Klausner, R. D. (1982). The concept of lipid domains in membranes. *Journal of Cell Biology, 94*(1), 1–6.

Kelleher, R. J. III., Govindarajan, A., Jung, H. Y., Kang, H., & Tonegawa, S. (2004). Translational control by MAPK signaling in long-term synaptic plasticity and memory. *Cell, 116*(3), 467–479.

Klemm, R. W., Ejsing, C. S., Surma, M. A., Kaiser, H. J., Gerl, M. J., Sampaio, J. L., ... Simons, K. (2009). Segregation of sphingolipids and sterols during formation of secretory vesicles at the trans-Golgi network. *Journal of Cell Biology, 185*(4), 601–612.

Kouhara, H., Hadari, Y. R., Spivak-Kroizman, T., Schilling, J., Bar-Sagi, D., Lax, I., & Schlessinger, J. (1997). A lipid-anchored Grb2-binding protein that links FGF-receptor activation to the Ras/MAPK signaling pathway. *Cell, 89*(5), 693–702.

Kusumi, A., & Suzuki, K. (2005). Toward understanding the dynamics of membrane-raft-based molecular interactions. *Biochimica et Biophysica Acta, 1746*(3), 234–251.

Leondaritis, G., Petrikkos, L., & Mangoura, D. (2009). Regulation of the Ras-GTPase activating protein neurofibromin by C-tail phosphorylation: implications for protein kinase C/Ras/extracellular signal-regulated kinase 1/2 pathway signaling and neuronal differentiation. *Journal of Neurochemistry, 109*(2), 573–583.

Lin, H. C., Mao, S. C., Su, C. L., & Gean, P. W. (2009). The role of prefrontal cortex CB1 receptors in the modulation of fear memory. *Cerebral Cortex, 19*(1), 165–175.

Lusa, S., Blom, T. S., Eskelinen, E. L., Kuismanen, E., Mansson, J. E., Simons, K., & Ikonen, E. (2001). Depletion of rafts in late endocytic membranes is controlled by NPC1-dependent recycling of cholesterol to the plasma membrane. *Journal of Cell Science, 114*(Pt 10), 1893–1900.

Luttrell, L. M., Hawes, B. E., van Biesen, T., Luttrell, D. K., Lansing, T. J., & Lefkowitz, R. J. (1996). Role of c-Src tyrosine kinase in G protein-coupled receptor- and Gbetagamma subunit-mediated activation of mitogen-activated protein kinases. *Journal of Biological Chemistry*, *271*(32), 19443–19450.

Ma, Y. C., Huang, J., Ali, S., Lowry, W., & Huang, X. Y. (2000). Src tyrosine kinase is a novel direct effector of G proteins. *Cell*, *102*(5), 635–646.

Mangoura, D. (1997). μ-Opioids activate tyrosine kinase focal adhesion kinase and regulate cortical cytoskeleton proteins cortactin and vinculin in chick embryonic neurons. *Journal of Neuroscience Research*, *50*(3), 391–401.

Mangoura, D., & Dawson, G. (1993). Opioid peptides activate phospholipase D and protein kinase C-epsilon in chicken embryo neuron cultures. *Proceedings of the National Academy of Sciences of the United States of America*, *90*(7), 2915–2919.

Mangoura, D., Sun, Y., Li, C., Singh, D., Gutmann, D. H., Flores, A., … Vallianatos, G. (2006). Phosphorylation of neurofibromin by PKC is a possible molecular switch in EGF receptor signaling in neural cells. *Oncogene*, *25*(5), 735–745.

Maudsley, S., Pierce, K. L., Zamah, A. M., Miller, W. E., Ahn, S., Daaka, Y., … Luttrell, L. M. (2000). The beta(2)-adrenergic receptor mediates extracellular signal-regulated kinase activation via assembly of a multi-receptor complex with the epidermal growth factor receptor. *Journal of Biological Chemistry*, *275*(13), 9572–9580.

Moore, C. A., Milano, S. K., & Benovic, J. L. (2007). Regulation of receptor trafficking by GRKs and arrestins. *Annual Review of Physiology*, *69*, 451–482.

Morone, N., Fujiwara, T., Murase, K., Kasai, R. S., Ike, H., Yuasa, S., … Kusumi, A. (2006). Three-dimensional reconstruction of the membrane skeleton at the plasma membrane interface by electron tomography. *Journal of Cell Biology*, *174*(6), 851–862.

Neumann-Giesen, C., Falkenbach, B., Beicht, P., Claasen, S., Luers, G., Stuermer, C. A., … Tikkanen, R. (2004). Membrane and raft association of reggie-1/flotillin-2: role of myristoylation, palmitoylation and oligomerization and induction of filopodia by overexpression. *Biochemical Journal*, *378*(Pt 2), 509–518.

Oddi, S., Dainese, E., Fezza, F., Lanuti, M., Barcaroli, D., De Laurenzi, V., … Maccarrone, M. (2011). Functional characterization of putative cholesterol binding sequence (CRAC) in human type-1 cannabinoid receptor. *Journal of Neurochemistry*, *116*(5), 858–865.

Orci, L., Montesano, R., Meda, P., Malaisse-Lagae, F., Brown, D., Perrelet, A., & Vassalli, P. (1981). Heterogeneous distribution of filipin–cholesterol complexes across the cisternae of the Golgi apparatus. *Proceedings of the National Academy of Sciences of the United States of America*, *78*(1), 293–297.

Palestini, P., Pitto, M., Tedeschi, G., Ferraretto, A., Parenti, M., Brunner, J., & Masserini, M. (2000). Tubulin anchoring to glycolipid-enriched, detergent-resistant domains of the neuronal plasma membrane. *Journal of Biological Chemistry*, *275*(14), 9978–9985.

del Pozo, M. A., Balasubramanian, N., Alderson, N. B., Kiosses, W. B., Grande-Garcia, A., Anderson, R. G., & Schwartz, M. A. (2005). Phospho-caveolin-1 mediates integrin-regulated membrane domain internalization. *Nature Cell Biology*, *7*(9), 901–908.

Puri, V., Watanabe, R., Singh, R. D., Dominguez, M., Brown, J. C., Wheatley, C. L., … Pagano, R. E. (2001). Clathrin-dependent and -independent internalization of plasma membrane sphingolipids initiates two Golgi targeting pathways. *Journal of Cell Biology*, *154*(3), 535–547.

Pyne, N. J., Waters, C., Moughal, N. A., Sambi, B. S., & Pyne, S. (2003). Receptor tyrosine kinase-GPCR signal complexes. *Biochemical Society Transactions*, *31*(Pt 6), 1220–1225.

Resh, M. D. (2006). Trafficking and signaling by fatty-acylated and prenylated proteins. *Nature Chemical Biology*, *2*(11), 584–590.

Ridyard, M. S., & Robbins, S. M. (2003). Fibroblast growth factor-2-induced signaling through lipid raft-associated fibroblast growth factor receptor substrate 2 (FRS2). *Journal of Biological Chemistry*, *278*(16), 13803–13809.

Rimmerman, N., Juknat, A., Kozela, E., Levy, R., Bradshaw, H. B., & Vogel, Z. (2011). The non-psychoactive plant cannabinoid, cannabidiol affects cholesterol metabolism-related genes in microglial cells. *Cellular and Molecular Neurobiology*, *31*(6), 921–930.

Roland, A. B., Ricobaraza, A., Carrel, D., Jordan, B. M., Rico, F., Simon, A., … Lenkei, Z. (2014). Cannabinoid-induced actomyosin contractility shapes neuronal morphology and growth. *eLife*, *3*, e03159.

Rubovitch, V., Gafni, M., & Sarne, Y. (2004). The involvement of VEGF receptors and MAPK in the cannabinoid potentiation of Ca^{2+} flux into N18TG2 neuroblastoma cells. *Brain Research. Molecular Brain Research*, *120*(2), 138–144.

Russo, P., Strazzullo, P., Cappuccio, F. P., Tregouet, D. A., Lauria, F., Loguercio, M., … Siani, A. (2007). Genetic variations at the endocannabinoid type 1 receptor gene (CNR1) are associated with obesity phenotypes in men. *Journal of Clinical Endocrinology & Metabolism*, *92*(6), 2382–2386.

Sandilands, E., Akbarzadeh, S., Vecchione, A., McEwan, D. G., Frame, M. C., & Heath, J. K. (2007). Src kinase modulates the activation, transport and signalling dynamics of fibroblast growth factor receptors. *EMBO Reports*, *8*(12), 1162–1169.

Sarnataro, D., Grimaldi, C., Pisanti, S., Gazzerro, P., Laezza, C., Zurzolo, C., & Bifulco, M. (2005). Plasma membrane and lysosomal localization of CB1 cannabinoid receptor are dependent on lipid rafts and regulated by anandamide in human breast cancer cells. *FEBS Letters*, *579*(28), 6343–6349.

Song, C., Vondriska, T. M., Wang, G. W., Klein, J. B., Cao, X., Zhang, J., … Ping, P. (2002). Molecular conformation dictates signaling module formation: example of PKCepsilon and Src tyrosine kinase. *American Journal of Physiology–Heart and Circulatory Physiology*, *282*(3), H1166–H1171.

Sun, X., & Dey, S. K. (2008). Aspects of endocannabinoid signaling in periimplantation biology. *Molecular and Cellular Endocrinology*, *286*(1–2 Suppl. 1), S3–S11.

Tagliaferro, P., Ramos, A. J., Onaivi, E. S., Evrard, S. G., Vega, M. D., & Brusco, A. (2006). Morphometric study on cytoskeletal components of neuronal and astroglial cells after chronic CB1 agonist treatment. *Methods in Molecular Medicine*, *123*, 91–104.

Takahashi, K. A., & Castillo, P. E. (2006). The CB1 cannabinoid receptor mediates glutamatergic synaptic suppression in the hippocampus. *Neuroscience*, *139*(3), 795–802.

Williams, E. J., Walsh, F. S., & Doherty, P. (2003). The FGF receptor uses the endocannabinoid signaling system to couple to an axonal growth response. *Journal of Cell Biology*, *160*(4), 481–486.

Xapelli, S., Agasse, F., Sarda-Arroyo, L., Bernardino, L., Santos, T., Ribeiro, F. F., … Malva, J. O. (2013). Activation of type 1 cannabinoid receptor (CB1) promotes neurogenesis in murine subventricular zone cell cultures. *PLoS One*, *8*(5), e63529.

Zeidman, R., Troller, U., Raghunath, A., Pahlman, S., & Larsson, C. (2002). Protein kinase Cepsilon actin-binding site is important for neurite outgrowth during neuronal differentiation. *Molecular Biology of the Cell*, *13*(1), 12–24.

Zisopoulou, S., Asimaki, O., Leondaritis, G., Vasilaki, A., Sakellaridis, N., Pitsikas, N., & Mangoura, D. (2013). PKC-epsilon activation is required for recognition memory in the rat. *Behavioural Brain Research*, *253*, 280–289.

Chapter 65

Cannabinoid Agonists

Ana Bagues, Carlos Goicoechea

A. Farmacología y Nutrición, Departamento de Ciencias Básicas de la Salud. Unidad Asociada de I+D+i al CSIC, Rey Juan Carlos University, Alcorcón, Madrid, Spain

Abbreviations

2-AG 2-Arachidonoylglycerol
CNS Central nervous system
COX Cyclooxygenase
DAGL Diacylglycerol lipase
DSE Depolarization-induced suppression of excitation
DSI Depolarization-induced suppression of inhibition
FAAH Fatty acid amide hydrolase
HPA Hypothalamic–pituitary–adrenocortical
LOX Lipoxygenase
LTD Long-term depression
MAG Monoacylglycerol
MAGL Monoacylglycerol lipase
NADA *N*-Arachidonoyldopamine
NAE *N*-Acylethanolamine
NAPE *N*-Acylphosphatidylethanolamine
NAPE-PLD *N*-Acylphosphatidylethanolamine-hydrolyzing phospholipase D
NArPE *N*-Arachidonoylphosphatidylethanolamine
NAT *N*-Acyltransferase
PLC Phospholipase C
PNS Peripheral nervous system
TRPV1 Transient receptor potential cation channel subfamily V member 1
Δ^9-THC Δ^9-Tetrahydrocannabinol

INTRODUCTION

Cannabis has been cultivated by humans for the longest time. It is not known when or where the first plant arose, although it is thought to have happened in Asia. All throughout history it has been used by humankind for several purposes: in industry, using its fiber and obtaining oil from its seeds, but also in medical and religious contexts.

Cannabinol was the first component of cannabis to be isolated at the end of the nineteenth century (Wood, Spivey, & Easterfield, 1899), but its full chemical structure was not clarified until the mid-twentieth century (Adams, Baker, et al., 1940). That same year (1940), cannabidiol was also purified (Adams, Hunt, et al., 1940).

In the 1940s and 1950s the first cannabinoid agonists were synthesized, but research on these molecules was ceased because of the contradicting results obtained from various studies and the psychoactive effects they induced.

The real revolution in cannabis research came in the 1960s with the characterization of Δ^9-tetrahydrocannabinol (Δ^9-THC), the cannabinoid agonist responsible for the psychoactive effects of cannabis (Gaoni & Mechoulam, 1964). This finding led to further research with the objectives of understanding the medical properties of cannabis, synthesis of new cannabinoid agonists, and discovery of the endocannabinoid system.

Depending on their origin, cannabinoid agonists can be classified as:

- Endogenous substances found in the animal organism that are part of the endocannabinoid system (endocannabinoids),
- Molecules that can be found in the plant (phytocannabinoids),
- Molecules that are synthesized in the laboratory (synthetic cannabinoids).

ENDOCANNABINOIDS

The endocannabinoid system is formed by several elements: at least two specific receptors named cannabinoid receptors; the endogenous ligands of these receptors, also called endocannabinoids; and several enzymes and proteins that regulate ligand concentration.

An Introduction to the Endocannabinoid System

The first component of the endocannabinoid system to be discovered was the CB1 receptor. This receptor was first cloned from the brain of the rat (Matsuda, Lolait, Brownstein, Young, & Bonner, 1990), sometime after from humans (Gerard, Mollereau, Vassart, & Parmentier, 1991), and 5 years later from the mouse (Chakrabarti, Onaivi, & Chadhuri, 1995). Overall they share 97–99% sequence identity (Howlett et al., 2002) and similar distributions throughout the species (Herkenham et al., 1990).

Three years after the CB1 receptor was cloned, a second cannabinoid receptor was identified from human promyelocytic cells and was named CB2 (Munro, Thomas, & Abushaar, 1993). It presents a slightly lower homology in the sequence of amino acids between humans, rats, and mice: 81% and 82%, respectively (Griffin, Tao, & Abood, 2000; Shire et al., 1996).

Both receptor types are members of the seven-transmembrane-domain G-protein-coupled family. They share 44% homology in their amino acid sequences and 68% when only the transmembrane region is considered (Munro et al., 1993). The main differences between these receptors are the way the signals are transduced to the interior of the cell and their distribution in the various tissues (Howlett et al., 2002).

When cannabinoid receptors were discovered it was thought that the CB1 receptor was exclusively located within the central nervous system (CNS) and CB2 in the periphery. Nowadays it is known that these differences in location are not so strict. CB1 is mainly expressed within the CNS; this receptor is the most abundant G-protein-coupled receptor in the brain, with especially high levels in the striatum, cerebellum, basal ganglia, cerebral cortex, and hippocampus (Herkenham et al., 1991). In the peripheral nervous system it is also found, but to a lesser extent. In peripheral nonneural tissues it has been described to be in adrenal gland, adipose tissue, heart, liver, lung, prostate, uterus, ovary, testis, bone marrow, thymus, and tonsils (Galiegue et al., 1995). The widespread distribution of the CB1 receptor is consistent with the multiplicity of effects of cannabinoid agonists, including hypomotility, increased food intake, disruption of short-term memory consolidation, antinociception, deficits of executive function, anxiety/anxiolysis, and psychotropic effects.

On the other hand, the CB2 receptor is mostly, though not exclusively, expressed in peripheral tissues, especially in immune tissues. This receptor is mainly expressed in spleen, tonsils, and thymus, tissues responsible for immune cell production and regulation. These immune cells include mast cells, B cells, T4 and T8 cells, microglial cells, macrophages, natural killer cells, and, to a lesser extent, monocytes and polymorphonuclear neutrophils (Howlett et al., 2004).

As stated before, the first studies indicated that CB2 receptors were not present in neurons of the CNS, although in past years they have been localized in several structures of the CNS: mRNA has been found in cortex, hippocampus, cerebellum, and brainstem (Van Sickle et al., 2005). A more detailed review of the distribution of CB1 and CB2 receptors in the CNS has been performed by Svizenska and coworkers (Svizenska, Dubovy, & Sulcova, 2008).

After the discovery of the CB1 receptor, Mechoulam and coworkers followed the reasoning that had led to the discovery of endogenous opioids: evolution could not have maintained in the organism a receptor for only a plant to stimulate; this led to the identification of the first endocannabinoid, N-arachidonoylethanolamine, which was named anandamide after the Sanskrit word *ananda*, which means bliss, joy, delight (Devane et al., 1992).

A few years later a second endocannabinoid was identified, 2-arachidonoylglycerol (2-AG), first in the intestine of the dog (Mechoulam et al., 1995) and then in the brain (Sugiura et al., 1995).

Since then other compounds have been isolated, such as 2-arachidonylglyceryl ether (noladin ether), O-arachidonoylethanolamine (virodhamine), and N-arachidonoyldopamine (NADA). However, the endogenous functions in physiological processes for all these compounds have not yet been established in detail and as of this writing, anandamide and 2-AG are the most studied endocannabinoids of the two major classes of endocannabinoids, N-acylethanolamines (NAEs) and monoacylglycerols, respectively. These compounds are frequently referred to as the "major" endocannabinoids (De Petrocellis & Di Marzo, 2009).

Endocannabinoid Agonists in the CNS

The levels of endocannabinoids vary across the structures of the CNS, 2-AG being the most abundant endocannabinoid, in both naïve rats and postmortem human brain (Buczynski & Parsons, 2010; Palkovits et al., 2008).

Anandamide is a partial agonist for CB1 and CB2 receptors, though with slightly higher affinity for CB1. In contrast, 2-AG behaves as a full agonist for CB1 and CB2 receptors, with higher receptor potency at CB1 and CB2 than anandamide. Both show less affinity for CB2 than CB1 receptor (Pertwee et al., 2010).

The available literature gives CB1 a predominant role as the major target for endogenous and exogenous cannabinoids at axon terminals within the CNS, although emerging data are demonstrating a possible role of CB2 receptors in synaptic transmission (den Boon et al., 2012). Future experiments are needed to determine the participation of CB2 in neuronal activity.

One of the peculiar characteristics of endocannabinoids is that, when modulating synaptic transmission, they act in a retrograde way (although not exclusively). So, endocannabinoids are synthesized from the postsynaptic neuron and bind to the presynaptic cannabinoid receptor. The result of this binding is a presynaptic inhibition of excitatory or inhibitory neurotransmitter release in the same or a neighboring synapse, which can be short or long term.

Another peculiar characteristic is that endocannabinoids cannot be stored in vesicles as other neurotransmitters. Because of their lipophilic structure, they are synthesized on demand once the postsynaptic neuron is stimulated. Several different stimuli, such as an increase in intracellular calcium, activation of group I metabotropic glutamate receptors, and/or $G_{q/11}$-protein-coupled receptor stimulation, lead to the cleavage of membrane phospholipids and synthesis of endocannabinoids (Pagotto, Marsicano, Cota, Lutz, & Pasquali, 2006).

Endocannabinoid Metabolism

Although both endocannabinoids are synthesized using phospholipids as precursors the synthesis is achieved through distinct mechanisms.

Membrane glycerophospholipids are the precursor molecules of anandamide and other NAEs. Their biosynthesis can be subdivided into two different phases: the first resulting in the respective N-acylphosphatidylethanolamine (NAPE) and the second giving the final NAE.

In the case of anandamide the first reaction consists of the synthesis of N-arachidonoylphosphatidylethanolamine (NArPE). This is obtained by two Ca^{2+}-dependent reactions, a first one in which diarachidonoylphosphatidylcholine is formed and a second to form NArPE. The second phase in the synthesis of anandamide consists of the hydrolysis of NArPE, resulting in anandamide and phosphatidic acid. The enzyme responsible for the first reaction is an N-acyltransferase (NAT) and the second reaction is catalyzed by the N-acylphosphatidylethanolamine-hydrolyzing phospholipase D (NAPE-PLD). While the pathways of NArPE synthesis seem to be selective for anandamide, although the molecular characterization of the NAT responsible for these reactions has yet not been achieved, NAPE-PLD does not distinguish between NAPEs.

Two other pathways for anandamide synthesis have been proposed; both start off with NArPE. In one of these pathways an

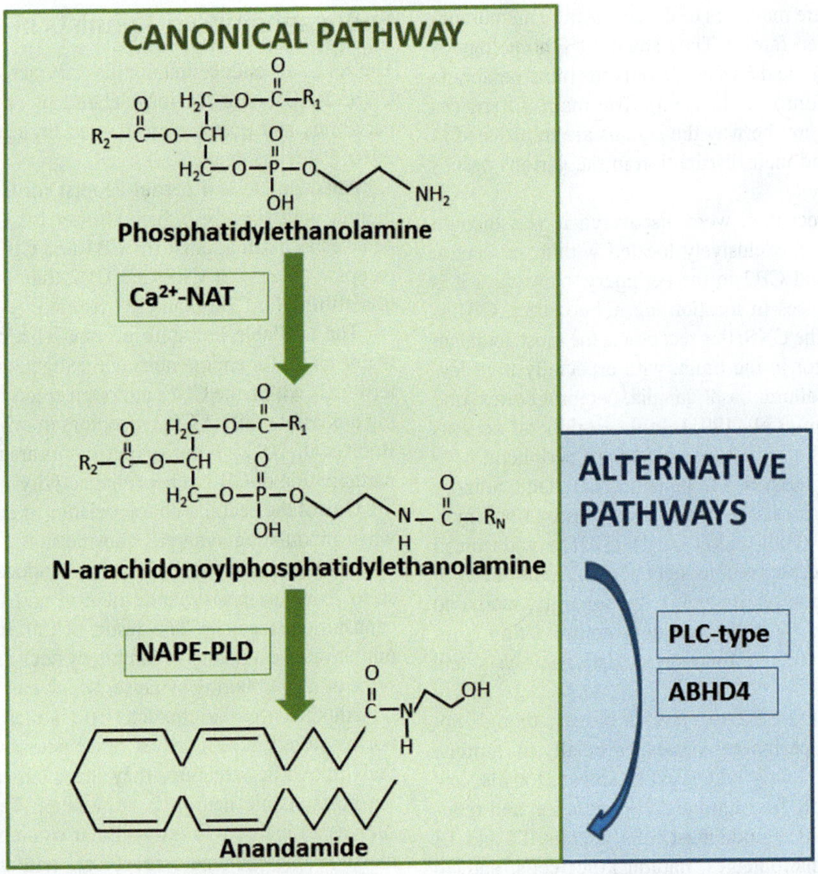

FIGURE 1 Canonical and alternative pathways for anandamide synthesis. The first reaction consists of the synthesis of NArPE by the enzyme NAT. Then NARPE is hydrolyzed by NAPE-PLD, releasing anandamide and phosphatidic acid. Two other alternative pathways have been proposed in which an enzyme with PLC activity and the enzyme ABHD4 are implicated.

enzyme with phospholipase C (PLC) activity converts NArPE into phosphoanandamide, and a phosphatase activity cleaves phosphoanandamide into anandamide and free phosphate. In the second pathway proposed, first glycerophosphoanandamide is catalyzed by the enzyme α/β hydrolase domain-containing protein 4 (ABHD4), which may be transformed into anandamide by cleavage of its phosphodiester bond, catalyzed by the phosphodiesterase GDE1, with release of glycerol phosphate (Piomelli, 2014; Rahman, Tsuboi, Uyama, & Ueda, 2014; Reisenberg, Singh, Williams, & Doherty, 2012) (Figure 1).

In contrast to the synthesis of anandamide, the routes by which 2-AG are formed seem to be better understood. The most important route by which 2-AG is synthesized is through the hydrolysis of 2-arachidonoylphosphatidylinositol by the PLC enzyme to form 2-arachidonoyldiacylglycerol, which is then hydrolyzed by diacylglycerol lipase (DAGL) to 2-AG (Ueda, Tsuboi, & Uyama, 2013).

Among PLC enzymes, PLC-β can be activated by several $G_{q/11}$-protein-coupled receptors. This suggests that activation of these receptors by neurotransmitters in postsynaptic neurons activates PLC-β, which will consequently synthesize 2-AG. DAGL is a membrane-associated enzyme, which can be stimulated by Ca^{2+}; two subtypes have been implicated in 2-AG formation: DAGLα and DAGLβ. Their contribution to 2-AG is tissue dependent; DAGLα appears to be more important in brain tissues and spinal cord (Piomelli, 2014; Rahman et al., 2014; Reisenberg et al., 2012; Ueda et al., 2013).

As with anandamide, an alternative pathway for 2-AG synthesis has been proposed: a first step mediated by the phospholipase A1, which gives 2-arachidonoyllysophosphatidylinositol, and a final reaction in which the enzyme lyso-PLC hydrolyzes 2-arachidonoyllysophosphatidylinositol into 2-AG (Piomelli, 2014; Rahman et al., 2014; Reisenberg et al., 2012; Ueda et al., 2013) (Figure 2).

Cannabinoid degradation occurs within the cell, so once endocannabinoids are shed to the synaptic cleft they must be transported to the intracellular space. Three different mechanisms have been proposed for the transportation of anandamide through the membrane: passive diffusion, facilitated transport, and internalization through caveolae. Once anandamide is internalized, evidence demonstrates the existence of intracellular binding proteins, which might allow its transport through the cytosol (Maccarrone, Dainese, & Oddi, 2010).

Anandamide is metabolized by the fatty acid amide hydrolase (FAAH) into arachidonic acid and ethanolamine (Cravatt et al., 1996). This enzyme is a membrane-bound serine hydrolase, although unlike other serine hydrolases it possesses an atypical catalytic triad consisting in Ser–Ser–Lys instead of the Ser–His–Asp.

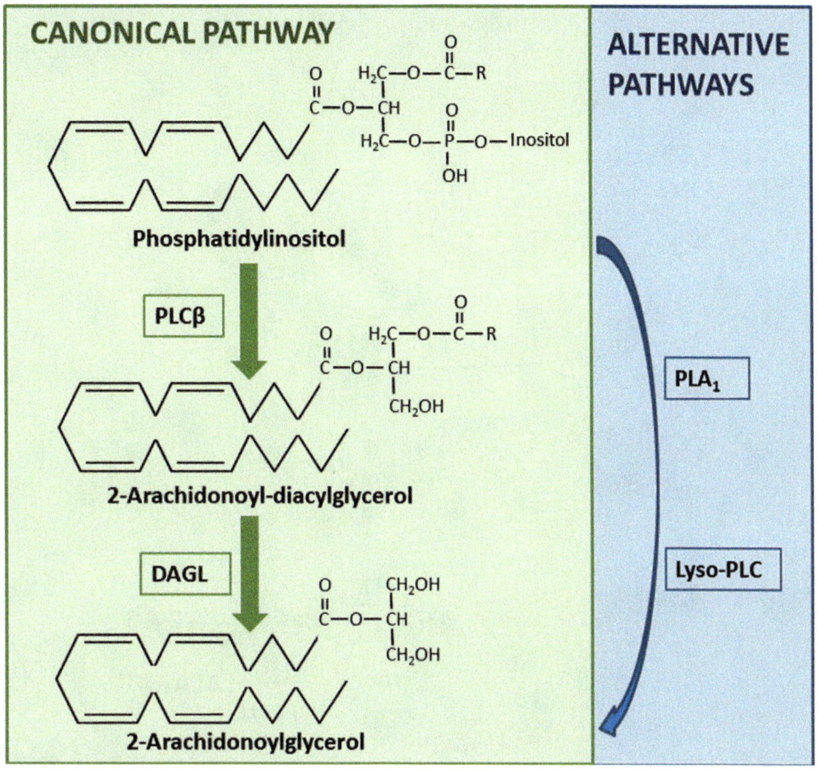

FIGURE 2 Canonical and alternative pathways for 2-AG synthesis. The most important route by which 2-AG is synthesized is through the hydrolysis of 2-arachidonoylphosphatidylinositol by the PLC enzyme to form 2-arachidonoyldiacylglycerol, which is then hydrolyzed by DAGL to 2-AG. An alternative pathway for 2-AG synthesis has been proposed: a first step mediated by the phospholipase A1 (PLA_1), which gives 2-arachidonoyllysophosphatidylinositol, and a final reaction in which the enzyme lyso-PLC hydrolyzes 2-arachidonoyllysophosphatidylinositol into 2-AG.

FAAH is formed by a large domain that anchors the enzyme to the membrane; a channel, which permits the entry of the substrate; a hydrophobic cavity, which interacts with the side chain of the substrate; and a cytosolic port, which interacts with the polar head of the substrate and is connected with the cytosol (Feledziak, Lambert, Marchand-Brynaert, & Muccioli, 2012).

A second isoform (FAAH-2) has been identified in mammals, but not in rodents. Both enzymes share 20% amino acid sequence identity, though their orientation in the membrane is different. Also their distribution patterns vary: FAAH-1 is preferentially expressed in brain, testis, and small intestine, while FAAH-2 is predominant in cardiac tissue (Yates & Barker, 2009).

The main enzyme responsible for the degradation of 2-AG is monoacylglycerol lipase (MAGL). MAGL breaks down 2-AG into glycerol and arachidonic acid, which is then incorporated into the membrane or transformed into the eicosanoid family of lipid mediators. It belongs to the serine hydrolase superfamily and is localized in the cytosol. 2-AG can also be broken down by other enzymes such as cyclooxygenase-2, lipoxygenase, or FAAH. It has a heterogeneous distribution in the rat brain with the highest expression in the regions where the CB1 receptor is most abundant, such as the hippocampus, cortex, and cerebellum (Ueda et al., 2013).

Endocannabinoids and Synaptic Modulation

Once endocannabinoids are synthesized, 2-AG and anandamide are shed to the synaptic cleft by a mechanism that is, as of this writing, unknown, and bind to the presynaptic CB1 receptor (Figure 3). This binding results in a short-term inhibition of γ-aminobutyric acid (GABA) or glutamate release, which persists for tens of seconds. This phenomenon was termed, depending on the neurotransmitter inhibited, depolarization-induced suppression of inhibition (DSI), when GABA release is inhibited, or depolarization-induced suppression of excitation (DSE), when glutamate release is inhibited (Castillo, Younts, Chavez, & Hashimotodani, 2012; Melis, Greco, & Tonini, 2014).

In addition to the short-term inhibition of neurotransmitter release, it has also been proven that endocannabinoids can mediate presynaptic forms of long-term depression (LTD) at both excitatory and inhibitory synapses. This can be induced by both anandamide and 2-AG. It is thought that the postsynaptic mechanisms leading to LTD are similar to DSE/DSI, although presynaptic proteins such as Rab3B/R1M1α or a reduction in P/Q-type voltage-gated channels may be involved. Also, in DSI/DSE there is an inhibition of presynaptic calcium influx through Ca^{2+} voltage-gated channels, probably via the βγ subunits, while in LTD, inhibition of adenylyl cyclase and downregulation of the cAMP/protein kinase A pathway via the $α_{i/o}$ limb are required (Castillo et al., 2012; Melis et al., 2014).

With the introduction of different pharmacological tools (such as FAAH or MAGL inhibitors) and genetically altered mice, the specific roles of these two major endocannabinoids in the CNS are being studied. Both endocannabinoids have been shown to be involved in neuroprotection, learning and memory, nociception,

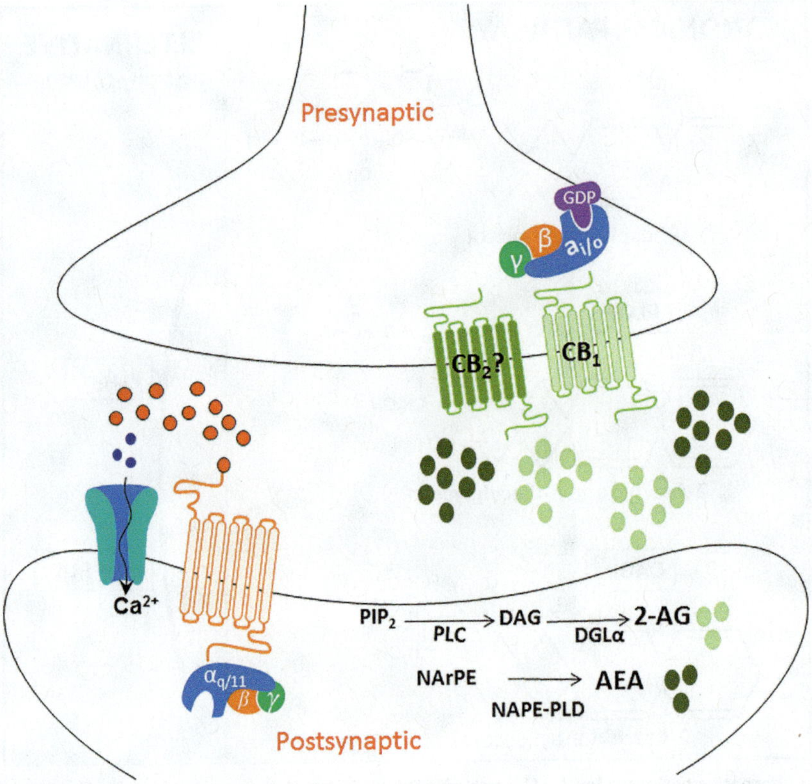

FIGURE 3 Schematic of endocannabinoid synthesis and retrograde signaling. Postsynaptic increase in Ca^{2+} influx or activation of $G_{q/11}$ induces endocannabinoid synthesis. Endocannabinoids are then shed to the synaptic cleft and bind to the CB1 receptor (the role of the CB2 receptor is yet to be determined), which consequently inhibit GABA and glutamate release.

anxiety, reward, and drug addiction, although the participation of either endocannabinoid in each function is different. 2-AG is the main endocannabinoid responsible for most forms of DSI/DSE and some forms of LTD, while anandamide mediates certain forms of synaptic homeostasis and plasticity. Anandamide and 2-AG take part in highly specialized compartments of physiological and pathophysiological functioning; even when they cooperate anandamide and 2-AG take part in different steps of the same function, although how these two endocannabinoids interact is yet fairly unknown. For a review, see Luchicchi and Pistis (2012).

An example of how anandamide and 2-AG interact is observed in the modulation of acute stress. Studies have demonstrated that there is a tonic endocannabinoid tone in the prefrontal cortex and amygdala, induced mainly by the binding of anandamide to CB1, inhibiting the hypothalamic–pituitary–adrenocortical (HPA) system. Under stimuli of acute stress there is a rapid reduction in anandamide levels due to an increased activity of FAAH. When the stimulus is ceased there is a rapid increase in 2-AG, which has been proposed to have an inhibitory effect over the HPA system and in consequence the recovery from the stressful stimuli (Gunduz-Cinar, Hill, McEwen, & Holmes, 2013).

Under conditions of chronic stress anandamide is persistently reduced. It is not clear if this reduction is due to an increase in FAAH activity or disruption in the synthesis of anandamide. This reduction in anandamide induces an impaired cannabinoid signaling hyperactivation of the HPA axis and consequently an increased vulnerability to stress-related illnesses (Gunduz-Cinar et al., 2013).

All the physiological effects of endocannabinoids cannot be fully understood through their binding to the CB1 and CB2 receptors, however. Anandamide can also activate different receptors, such as the peroxisome proliferator-activated receptors or the transient receptor potential cation channel subfamily V member 1 (TRPV1) receptor.

The TRPV1 receptor belongs to the transient receptor potential family, which is composed of six subfamilies. Receptors belonging to three of these subfamilies have been suggested to interact with cannabinoid agonists: TRPV1, TRPV2, and TRPV4 (belonging to the vanilloid subfamily), TRPM8 (belongs to the melastatin subfamily), and TRPA1 (belongs to the ankyrin subfamily).

TRPV1 is a nonselective cation channel that is activated by noxious heat (>43 °C), low pH (<6.9), and capsaicin (pungent agent of chili peppers) and consequently induces nociception, although it is also expressed in the brain, where it has been shown to participate in antinociception (Kim et al., 2012), locomotor control (Lee, Di Marzo, & Brotchie, 2006), and regulation of affective behaviors (Moreira, Aguiar, Terzian, Guimaraes, & Wotjak, 2012).

Anandamide, but not 2-AG, acts as a full agonist; its affinity is similar or slightly lower than that of capsaicin, and it seems to interact with TRPV1 at the same intracellular binding site as capsaicin (Pertwee et al., 2010).

TRPV1 colocalizes with CB1 and CB2 in sensory neurons and with CB1 in several brain areas and CB2 in osteoclasts. Anandamide can bind to both presynaptic and postsynaptic TRPV1 receptors. When binding to the postsynaptic TRPV1 receptor it will induce hyperpolarization of neurons, whereas when binding to the presynaptic TRPV1 it will facilitate glutamatergic signaling. But how these two receptors interplay to modulate synaptic transmission is, as of this writing, not completely clarified (Di Marzo & De Petrocellis, 2012).

Other endogenous cannabinoids that have been shown to bind to the TRPV1 receptor are NADA with a higher potency than anandamide or *N*-oleoyldopamine (Pertwee et al., 2010).

Less is known about the action that endocannabinoids exert over other TRP channels. Along this line anandamide and NADA have been shown to antagonize the effects of the TRPM8 agonists menthol and icilin and weakly activate the TRPA1 channel (De Petrocellis & Di Marzo, 2009; De Petrocellis et al., 2012), although future experiments are needed to determine the physiologic functions they exert.

The orphan G-protein-coupled receptor GPR55 has also been proposed to mediate some of the effects of endocannabinoids and Δ^9-THC. This protein is highly expressed in the brain and peripheral tissues, although it presents little homology to either cannabinoid receptor already cloned. The results are contradicting and future experiments are needed (De Petrocellis & Di Marzo, 2009).

PHYTOCANNABINOIDS: Δ^9-TETRAHYDROCANNABINOL

Nowadays it is known that the *Cannabis sativa* plant has around 400 compounds, and more than 60 different cannabinoids have been identified. The main cannabinoids present in the plant are Δ^9-THC, Δ^8-THC, cannabidiol, and cannabinol. Although all of them, except for cannabidiol, have psychoactive effects, Δ^9-THC is the molecule responsible for most of the psychoactive effects of cannabis. All are extremely hydrophobic molecules.

The pharmacokinetics of Δ^9-THC depends on the method of its administration. When smoked, Δ^9-THC is rapidly absorbed into the bloodstream, and maximal concentration is reached within a few minutes. In contrast, when orally administered, Δ^9-THC has a lower bioavailability, as it is sensitive to gastric acid and hepatic metabolism, reaching maximum plasma concentrations within 1–2 h. In plasma 60% of the drug is associated with the lipoprotein; the remainder of the drug appears to be bound by albumin and blood cells. Δ^9-THC is initially distributed to highly perfused organs such as the heart, kidneys, lung, placenta, or liver and then into adipose tissue, where it can remain and take several weeks to be totally eliminated, though its retention in this tissue reduces its penetration into the CNS (its presence in the CNS is usually around 1% of the maximum plasma concentration). Most metabolites are excreted through feces (68%) and urine (12%), although also through hair, saliva, and sweat.

Most Δ^9-THC metabolism occurs in the liver, where it is hydroxylated to 11-hydroxy-Δ^9-THC, although it can also be metabolized in the lung or intestine. Levels of this metabolite can be detected minutes after smoking Δ^9-THC and reach maximum plasma concentrations 30–60 min after smoking. As it is also a psychoactive metabolite, it could potentiate the effects of Δ^9-THC in the CNS (Gonzalez, Sagreso, et al., 2002; Huestis, Mazzoni, & Rabin, 2011).

Like anandamide, Δ^9-THC is a partial agonist for CB1 and CB2 receptors; it resembles anandamide in its CB1 affinity, although with less efficacy than anandamide and with less relative intrinsic activity at CB2 than at CB1 receptors. Cannabidiol displays a very low affinity for CB1 and CB2 receptors; despite this low affinity cannabidiol can interact with these receptors at low concentrations and block several of the in vivo effects of Δ^9-THC (Pertwee, 2008; Pertwee et al., 2010). This has important clinical implications, for example, the neurocognitive and psychological effects after sporadic or long-term use of cannabis are likely to vary depending on the proportion of Δ^9-THC and cannabidiol present in the drug. Currently the strains that are high in Δ^9-THC and low in cannabidiol dominate the market (Schoeler & Bhattacharyya, 2013).

When used as a drug of abuse, Δ^9-THC can alter short-term memory, sense of time, attention, problem solving, verbal fluency, reaction time, and psychomotor control and induce positive feelings as mild euphoria and relaxation or, on the contrary, anxiety, paranoia, and panic reaction. These opposing effects are dose and context dependent: in general, low doses of Δ^9-THC exert anxiolytic and mood-enhancing effects, whereas high doses are anxiogenic and dysphoric. However, these dose-dependent effects also vary with other factors, including environment and previous experience with the drug (Clapper, Mangieri, & Piomelli, 2009). But the molecular mechanisms by which Δ^9-THC induces these antagonistic effects are yet not completely known. Like endogenously released cannabinoid agonists, Δ^9-THC can act through neuronal presynaptic CB1 receptors to inhibit ongoing neurotransmitter release onto other neurotransmitter-releasing neurons, consequently having a mixed stimulatory–inhibitory effect on neurotransmitter release. On the other hand, in some assays, it has been found that Δ^9-THC can also behave as a CB1 and CB2 receptor antagonist of exogenous and endogenous cannabinoids at these receptors. The way Δ^9-THC can act either as an agonist or as an antagonist is not fully clear, although the hypothesis exists that because Δ^9-THC has a relatively low cannabinoid receptor efficacy, its capacity to activate these receptors will be influenced by the density and coupling efficiencies of these receptors (Pertwee, 2008).

The effects of Δ^9-THC vary depending on the individual's history of Δ^9-THC abuse, that is, chronic use of Δ^9-THC has shown to induce tolerance (in which either the potency or the efficacy of a drug changes such that the physiological and/or psychological consequences of the same drug dose are significantly diminished with repeated use), both in humans and in laboratory animals. Although tolerance to the various effects of cannabinoids does not occur at the same time, in animals it has been shown that tolerance to motor impairment or hypothermia appears within the first 3–7 days of treatment, while neuroendocrine or memory impairment needs a much longer time of treatment (Gonzalez, Cebeira, & Fernandez-Ruiz, 2005). As in animals, tolerance to the various effects of cannabinoids also appears in humans at different time points (D'Souza et al., 2008).

The mechanism by which tolerance occurs has not been completely elucidated. It has been shown that downregulation (loss of receptors) and desensitization (attenuated receptor-mediated G-protein and effector activity) of the CB1 receptor occur and contribute to tolerance after chronic Δ^9-THC administration in animals (Lazenka, Selley, & Sim-Selley, 2013). Findings in humans and in animals have shown that downregulation after chronic exposure to Δ^9-THC does not occur in equal ways along the various structures of the CNS; CB1 receptors in the hippocampus suffer adaptations faster, while changes in striatum are slower. This could explain, partly, why tolerance to the different effects induced by cannabis consumption appears at different time points (Hirvonen et al., 2012).

Previously it was thought that cannabis did not induce dependence, but it has been shown that approximately 10% of first-time users and 50% of cannabis users develop dependence. But the molecular basis responsible for the interindividual differences in becoming cannabis dependent is not known. Several factors, such as genetic variability, have been proposed. Interestingly, different genetic variants of the CB1 receptor and FAAH genes have been associated with an increase in susceptibility to drug addiction (Clapper et al., 2009; Oliere, Joliette-Riopel, Potvin, & Jutras-Aswad, 2013).

Ninety percent of cannabis dependents that go under treatment have a difficult time dealing with abstinence as they experience a withdrawal syndrome that is characterized by craving, irritability, anxiety, depressed mood, decreased appetite, and sleep difficulties and display similar scope and severity to the withdrawal associated with tobacco use. Interestingly, the affective and cognitive effects seem to have greater responsibility in cannabis relapse than craving. As of this writing, no specific therapeutic approaches have been developed for cannabis dependence and most interventions are based on those used for alcohol dependence. Studies in animals are investigating the use of anandamide transport and FAAH inhibitors for the treatment of cannabis dependence; the majority of evidence obtained so far suggests that indirect activation of CB1 receptors by increasing levels of synaptically available anandamide does not mimic the reinforcing effects typical of direct-acting cannabinoid agonists (Clapper et al., 2009; Maldonado, Berrendero, Ozaita, & Robledo, 2011).

SYNTHETIC CANNABINOIDS

Synthetic cannabinoids or cannabimimetic compounds were initially synthesized with the objective to target selectively CB1 or CB2 receptors for research purposes, although in the past few years there has been an increase in the intake of these drugs with an illicit purpose. Synthetic cannabinoids can be chemically classified into naphthoylindoles, benzoylindoles, phenylacetylindoles, adamantylindoles, cyclophenols, and a miscellaneous group (ElSohly, Gul, Wanas, & Radwan, 2014).

In the United States, synthetic cannabinoids first emerged on the drug market in 2008 and were marketed as "legal alternatives to marijuana," since the use of synthetic cannabinoids produces effects similar to those of cannabis. Since then 74 synthetic cannabinoids have been reported (United Nations Office on Drugs and Crime). This seems to be due to the manufacturer's intention to be one step ahead of legislation; as various cannabinoids are banned they are replaced by other compounds similar in structure and pharmacologically active. Unlike Δ^9-THC, the synthetic cannabinoids used are high-potency full agonists at the brain CB1 receptor (Spaderna, Addy, & D'Souza, 2013).

Hundreds of synthetic cannabinoids are categorized into the following structural groups: adamantoylindoles (e.g., AB-001), aminoalkylindoles (e.g., WIN 55,212-2), benzoylindoles (e.g., AM694), cyclohexylphenols (e.g., CP 55,940), dibenzopyrans (e.g., HU-210), naphthoylindoles (e.g., JWH-018), naphthylmethylindoles (e.g., JWH-175), naphthylmethylindenes (e.g., JWH-176), naphthoylpyrroles (e.g., JWH-307), phenylacetylindoles (e.g., JWH-250), tetramethylcyclopropyl ketone indoles (e.g., UR-144), quinolinyl ester indoles (e.g., PB-22), and indazole carboxamide compounds (Figure 1; Castaneto et al., 2014).

These synthetic cannabinoids have been found to be mixed with FAAH and MAGL enzyme inhibitors, cathinones, or tryptamines (Nakajima et al., 2013; Uchiyama, Kawamura, Kikura-Hanajiri, & Goda, 2013). For a review of emerging compounds used as drugs of abuse, see Lewin, Seltzman, Carroll, Mascarella, and Reddy (2014).

CANNABINOID AGONISTS IN THE CLINIC

As of this writing, three different cannabinoid drugs have been approved in different countries for the treatment of certain types of pain, anorexia, and nausea and vomiting.

The first was nabilone, a CB1/CB2 receptor agonist, which is a synthetic analogue of Δ^9-THC, indicated for the relief of nausea and vomiting produced by chemotherapy. For this purpose dronabinol was also approved some years later, although because 5-HT3 antagonists have higher effectiveness and better tolerability they are generally used instead. Dronabinol was introduced for the treatment of AIDS-related anorexia (Howard, Twycross, Shuster, Mihalyo, & Wilcock, 2013). Sativex® was approved in multiple countries for the treatment of neuropathic pain in patients with multiple sclerosis and patients with advanced cancer. Sativex® is composed of approximately equal amounts of dronabinol and cannabidiol.

A CB1 receptor antagonist, rimonabant, was introduced in the clinic for the treatment of obesity, although anxiety and depression symptoms were observed in a significant proportion of individuals, and a higher suicide risk, and it was removed from the market (Christensen, Kristensen, Bartels, Bliddal, & Astrup, 2007).

At present various strategies are being used to exploit the possible therapeutic effects of cannabinoid agonists, reducing the side effects mainly due to the activation of CNS CB1 receptors. These can be summarized as follows (Pertwee, 2009):

- Targeting cannabinoid receptors located outside the blood–brain barrier;
- Targeting cannabinoid receptors expressed by a particular tissue, which can be obtained by the local and topical administration of a cannabinoid agonist without reaching the CNS;
- Targeting upregulated cannabinoid receptors—some disorders trigger a "protective" upregulation of a subpopulation of CB1 and CB2 receptors that can mediate symptom relief or oppose disease progression;
- Multitargeting—this strategy relies on the capacity of cannabinoid agonists to act additively or synergistically with other systems (i.e., cannabinoid agonists are being studied in combined administration with opioids for the treatment of pain).

APPLICATION TO OTHER ADDICTIONS AND SUBSTANCE MISUSE

The role of cannabinoid receptors in cocaine addiction has been studied, but few studies have aimed to investigate the implication of cannabinoid agonists in cocaine addiction.

Endocannabinoids play a central role in several cognitive and physiological processes associated with addiction, such as reward, stress responsiveness, and drug-related synaptic plasticity.

Acute and chronic administration of drugs has been shown to alter the endocannabinoid levels, i.e., acute administration of cocaine increases levels of 2-AG but not anandamide in the limbic forebrain (Patel, Rademacher, & Hillard, 2003). On the other hand, chronic cocaine exposure decreases 2-AG content of the limbic forebrain but not in other areas such as the cerebral cortex striatum or midbrain nor does it alter anandamide levels.

Endocannabinoid levels are not modified depending only on the time of drug abuse but also depending on the drug being abused. Chronic alcohol exposure has been shown to increase or decrease, in midbrain and limbic forebrain, the contents of anandamide and 2-AG, and nicotine induces changes in endocannabinoid contents in other structures such as the limbic forebrain, brainstem, hippocampus, and cerebral cortex (Gonzalez, Cascio, et al., 2002).

As of this writing, various preclinical studies aim to investigate cannabinoid receptor antagonists and FAAH inhibitors for the treatment of different addictions. The results obtained from these studies are controversial. In genetically modified mice without FAAH enzyme (FAAH knockout), or upon administration in mice of a FAAH inhibitor, nicotine withdrawal is increased. Interestingly, different results have been observed in rats in which inhibitors of FAAH inhibited nicotine reward, decreased reinstatement of nicotine-seeking behavior, and reversed nicotine withdrawal anxiety (Maldonado, Robledo, & Berrendero, 2013; Muldoon, Lichtman, Parsons, & Damaj, 2013).

Similar to what happens with nicotine, the results obtained in studies using cocaine and opioids vary among species. A pharmacological increase in anandamide levels reduced morphine withdrawal and reinstatement of cocaine-seeking behavior in rats and mice and did not reinstate extinguished drug-seeking behavior in monkeys that had previously self-administered cocaine (Justinova et al., 2008; Maldonado et al., 2013).

It seems without doubt that endocannabinoids are implicated in some aspects of drug addiction, such as in reward. Modulating endocannabinoid agonist levels could be a useful pharmacological option in the treatment of various addictions although future experiments are necessary.

DEFINITION OF TERMS

Cannabis *Cannabis* is an annual, dioecious, flowering herb that includes three different species, *Cannabis sativa*, *Cannabis indica*, and *Cannabis ruderalis*.

Agonist This is a chemical that binds to a receptor and activates the receptor to produce a biological response.

Endocannabinoids These are endogenous substances found in animal organisms that are part of the endocannabinoid system and bind to cannabinoid receptors.

Phytocannabinoids These are molecules that can be found in the cannabis plant or other plants and bind to cannabinoid receptors.

Synthetic cannabinoids These are molecules that are synthesized in the laboratory to bind selectively or not to one or more cannabinoid receptors.

Δ^9-Tetrahydrocannabinol This is the phytocannabinoid responsible for the psychoactive effects of cannabis.

Anandamide (*N*-arachidonoylethanolamine) This was the first endocannabinoid to be identified. It is a partial agonist for CB1 and CB2 receptors, with a slightly higher affinity for CB1.

2-Arachidonoylglycerol This was the second endocannabinoid to be identified. It is a full agonist for CB1 and CB2 receptors with a slightly higher affinity for CB1.

Fatty acid amide hydrolase This is an enzyme belonging to the serine hydrolase family and is responsible for metabolizing anandamide.

Tolerance This refers to when the potency or efficacy of a drug changes such that the physiological and/or psychological consequences of the same drug dose are significantly diminished with repeated use.

KEY FACTS

Key Facts of Cannabis

The first written references to the use of marijuana for medical purposes are found in the Chinese pharmacopoeia of Shen Nung (2600 BC) and Shen Nung (2700 BC).

The first scientific studies on the therapeutic effect of cannabis were published by Sir William B. O'Shaughnessy in 1839. The article described the analgesic, antispasmodic, and muscle-relaxing properties of cannabis. O'Shaughnessy's article roused interest in cannabis research, which led on to active cannabis research by other researchers.

Key Facts of Agonists

Agonists can behave as full agonists or partial agonists. When an agonist binds to the receptor and elicits a maximal response it is a full agonist, while partial agonists are not able to induce a maximal response.

Key Facts of Receptors

Receptors are protein molecules localized in the cell that are directly and specifically in charge of the chemical inter- and intracellular signaling.

There are various types of receptors; two of these types are ligand-gated ion channels and G-protein-coupled receptors. Ligand-gated ion channels are those in which the interaction of the agonist with the receptor causes the opening and closing of an ion channel in another part of the same molecule. G-protein receptors have seven transmembrane domains. These receptors are connected to the effector systems through GTP-binding proteins.

Key Facts of Endocannabinoid System

The endocannabinoid system is formed by at least two specific receptors named cannabinoid receptors; the endogenous ligands of these receptors, also called endocannabinoids, and several enzymes and proteins that regulate ligand concentration.

The endocannabinoid system modulates neurotransmission; endocannabinoids act primarily (though not exclusively) upon presynaptic receptors, which in turn inhibit the release of GABA or glutamate.

SUMMARY POINTS

- Cannabinoid agonists can be classified depending on their origin as endocannabinoids, phytocannabinoids, and synthetic cannabinoids.
- The most studied endocannabinoids are anandamide and 2-AG.
- Anandamide and 2-AG bind to presynaptic cannabinoid receptors to regulate synaptic transmission by inhibiting neurotransmitter release.
- Anandamide and 2-AG are synthesized from membrane phospholipids on demand, through different mechanisms that are not well established as yet. FAAH is the enzyme responsible for the metabolism of anandamide, and MAGL is the main enzyme responsible for 2-AG degradation.
- Anandamide and 2-AG can bind to other noncannabinoid receptors.
- Δ^9-THC and cannabidiol are phytocannabinoids. Δ^9-THC is the main molecule responsible for the psychoactive effects of cannabis, while cannabidiol does not induce psychoactive effects and can have a protective effect.
- The effects induced after cannabis abuse will depend on the concentrations of Δ^9-THC and cannabidiol.
- As of this writing, three different cannabinoid drugs have been approved with different indications.
- Research on the development of various cannabinoid drugs is very active, following different strategies: targeting peripheral cannabinoid receptors, receptors expressed in a certain tissue, or cannabinoids combined with other drugs.

REFERENCES

Adams, R., Baker, B., & Wearn, R. (1940). Structure of cannabinol. III. Synthesis of cannabinol, 1-hydroxy-3-n-amyl-6,6,9-trimethyl-6-dibenzopyran. *Journal of the American Chemical Society*, 62, 2204–2207.

Adams, R., Hunt, M., & Clark, J. (1940). Structure of cannabidiol, a product isolated from the marihuana extract of Minnesota wild hemp I. *Journal of the American Chemical Society*, 62, 196–200.

den Boon, F. S., Chameau, P., Schaafsma-Zhao, Q., van Aken, W., Bari, M., Oddi, S., ... Werkman, T. R. (2012). Excitability of prefrontal cortical pyramidal neurons is modulated by activation of intracellular type-2 cannabinoid receptors. *Proceedings of the National Academy of Sciences of the United States of America*, 109(9), 3534–3539.

Buczynski, M. W., & Parsons, L. H. (2010). Quantification of brain endocannabinoid levels: methods, interpretations and pitfalls. *British Journal of Pharmacology*, 160(3), 423–442.

Castaneto, M. S., Gorelick, D. A., Desrosiers, N. A., Hartman, R. L., Pirard, S., & Huestis, M. A. (2014). Synthetic cannabinoids: epidemiology, pharmacodynamics, and clinical implications. *Drug and Alcohol Dependence*, 144C, 12–41.

Castillo, P. E., Younts, T. J., Chavez, A. E., & Hashimotodani, Y. (2012). Endocannabinoid signaling and synaptic function. *Neuron*, 76(1), 70–81.

Chakrabarti, A., Onaivi, E., & Chadhuri, G. (1995). Cloning, sequencing and characterization of mouse brain-type cannabinoid receptor gene. *Faseb Journal*, 9(3), A404.

Christensen, R., Kristensen, P. K., Bartels, E. M., Bliddal, H., & Astrup, A. (2007). Efficacy and safety of the weight-loss drug rimonabant: a meta-analysis of randomised trials. *Lancet*, 370(9600), 1706–1713.

Clapper, J. R., Mangieri, R. A., & Piomelli, D. (2009). The endocannabinoid system as a target for the treatment of cannabis dependence. *Neuropharmacology*, 56(Suppl. 1), 235–243.

Cravatt, B. F., Giang, D. K., Mayfield, S. P., Boger, D. L., Lerner, R. A., & Gilula, N. B. (1996). Molecular characterization of an enzyme that degrades neuromodulatory fatty-acid amides. *Nature*, 384(6604), 83–87.

De Petrocellis, L., & Di Marzo, V. (2009). An introduction to the endocannabinoid system: from the early to the latest concepts. *Best Practice & Research Clinical Endocrinology & Metabolism*, 23(1), 1–15.

De Petrocellis, L., Moriello, A. S., Imperatore, R., Cristino, L., Starowicz, K., & Di Marzo, V. (2012). A re-evaluation of 9-HODE activity at TRPV1 channels in comparison with anandamide: enantioselectivity and effects at other TRP channels and in sensory neurons. *British Journal of Pharmacology*, 167(8), 1643–1651.

Devane, W., Hanus, L., Breuer, A., Pertwee, R., Stevenson, L., Griffin, G., ... Mechoulam, R. (1992). Isolation and structure of a brain constituent that binds to the cannabinoid receptor. RID E-1312-2011 *Science*, 258(5090), 1946–1949.

Di Marzo, V., & De Petrocellis, L. (2012). Why do cannabinoid receptors have more than one endogenous ligand? *Philosophical Transactions of the Royal Society of London. Series B, Biological sciences*, 367(1607), 3216–3228.

D'Souza, D. C., Ranganathan, M., Braley, G., Gueorguieva, R., Zimolo, Z., Cooper, T., ... Krystal, J. (2008). Blunted psychotomimetic and amnestic effects of Delta-9-tetrahydrocannabinol in frequent users of cannabis. *Neuropsychopharmacology*, 33(10), 2505–2516.

ElSohly, M. A., Gul, W., Wanas, A. S., & Radwan, M. M. (2014). Synthetic cannabinoids: analysis and metabolites. *Life Sciences*, 97(1), 78–90.

Feledziak, M., Lambert, D. M., Marchand-Brynaert, J., & Muccioli, G. G. (2012). Inhibitors of the endocannabinoid-degrading enzymes, or how to increase endocannabinoid's activity by preventing their hydrolysis. *Recent Patents on CNS Drug Discovery*, 7(1), 49–70.

Galiegue, S., Mary, S., Marchand, J., Dussossoy, D., Carriere, D., Carayon, P., ... Casellas, P. (1995). Expression of central and peripheral cannabinoid receptors in human immune tissues and leukocyte subpopulations. *European Journal of Biochemistry*, 232(1), 54–61.

Gaoni, Y., & Mechoulam, R. (1964). Isolation structure + partial synthesis of active constituent of hashish. *Journal of the American Chemical Society*, 86(8), 1646–1647.

Gerard, C., Mollereau, C., Vassart, G., & Parmentier, M. (1991). Molecular-Cloning of a human cannabinoid receptor which is also expressed in testis. *Biochemical Journal*, 279, 129–134.

Gonzalez, S., Cascio, M. G., Fernandez-Ruiz, J., Fezza, F., Di Marzo, V., & Ramos, J. A. (2002). Changes in endocannabinoid contents in the brain of rats chronically exposed to nicotine, ethanol or cocaine. *Brain Research*, 954(1), 73–81.

Gonzalez, S., Cebeira, M., & Fernandez-Ruiz, J. (2005). Cannabinoid tolerance and dependence: a review of studies in laboratory animals. *Pharmacology Biochemistry and Behaviour*, 81(2), 300–318.

Gonzalez, S., Sagreso, O., Gomez, M., & Ramos, J. A. (2002). *Química y metabolismo de los cannabinoides. Guia Basica sobre los Cannabinoides* (pp. 13–21). Madrid: Sociedad Española de Investigación sobre Cannabinoides.

Griffin, G., Tao, Q., & Abood, M. (2000). Cloning and pharmacological characterization of the rat CB2 cannabinoid receptor. *Journal of Pharmacology and Experimental Therapeutics*, *292*(3), 886–894.

Gunduz-Cinar, O., Hill, M. N., McEwen, B. S., & Holmes, A. (2013). Amygdala FAAH and anandamide: mediating protection and recovery from stress. *Trends in Pharmacological Sciences*, *34*(11), 637–644.

Herkenham, M., Lynn, A., Johnson, M., Melvin, L., Decosta, B., & Rice, K. (1991). Characterization and localization of cannabinoid receptors in rat-brain – a quantitative invitro autoradiographic study. *Journal of Neuroscience*, *11*(2), 563–583.

Herkenham, M., Lynn, A., Little, M., Johnson, M., Melvin, L., Decosta, B., & Rice, K. (1990). Cannabinoid receptor localization in brain. *Proceedings of the National Academy of Sciences of the United States of America*, *87*(5), 1932–1936.

Hirvonen, J., Goodwin, R. S., Li, C., Terry, G. E., Zoghbi, S. S., Morse, C., ... Innis, R. B. (2012). Reversible and regionally selective downregulation of brain cannabinoid CB1 receptors in chronic daily cannabis smokers. *Molecular Psychiatry*, *17*(6), 642–649.

Howard, P., Twycross, R., Shuster, J., Mihalyo, M., & Wilcock, A. (2013). Cannabinoids. *Journal of Pain and Symptom Management*, *46*(1), 142–149.

Howlett, A., Barth, F., Bonner, T., Cabral, G., Casellas, P., Devane, W., ... Pertwee, R. (2002). International Union of Pharmacology. XXVII. Classification of cannabinoid receptors. RID B-7358-2011 RID E-1312-2011 *Pharmacological Reviews*, *54*(2), 161–202.

Howlett, A., Breivogel, C., Childers, S., Deadwyler, S., Hampson, R., & Porrino, L. (2004). Cannabinoid physiology and pharmacology: 30 years of progress. *Neuropharmacology*, *47*, 345–358.

Huestis, M. A., Mazzoni, I., & Rabin, O. (2011). Cannabis in sport anti-doping perspective. *Sports Medicine*, *41*(11), 949–966.

Justinova, Z., Mangieri, R. A., Bortolato, M., Chefer, S. I., Mukhin, A. G., Clapper, J. R., ... Goldberg, S. R. (2008). Fatty acid amide hydrolase inhibition heightens anandamide signaling without producing reinforcing effects in primates. *Biological Psychiatry*, *64*(11), 930–937.

Kim, Y. H., Back, S. K., Davies, A. J., Jeong, H., Jo, H. J., Chung, G., ... Oh, S. B. (2012). TRPV1 in GABAergic interneurons mediates neuropathic mechanical allodynia and disinhibition of the nociceptive circuitry in the spinal cord. *Neuron*, *74*(4), 640–647.

Lazenka, M. F., Selley, D. E., & Sim-Selley, L. J. (2013). Brain regional differences in CB1 receptor adaptation and regulation of transcription. *Life Sciences*, *92*(8–9), 446–452.

Lee, J., Di Marzo, V., & Brotchie, J. M. (2006). A role for vanilloid receptor 1 (TRPV1) and endocannabinoid signalling in the regulation of spontaneous and L-DOPA induced locomotion in normal and reserpine-treated rats. *Neuropharmacology*, *51*(3), 557–565.

Lewin, A. H., Seltzman, H. H., Carroll, F. I., Mascarella, S. W., & Reddy, P. A. (2014). Emergence and properties of spice and bath salts: a medicinal chemistry perspective. *Life Sciences*, *97*(1), 9–19.

Luchicchi, A., & Pistis, M. (2012). Anandamide and 2-arachidonoylglycerol: pharmacological properties, functional features, and emerging specificities of the two major endocannabinoids. *Molecular Neurobiology*, *46*(2), 374–392.

Maccarrone, M., Dainese, E., & Oddi, S. (2010). Intracellular trafficking of anandamide: new concepts for signaling. *Trends in Biochemical Sciences*, *35*(11), 601–608.

Maldonado, R., Berrendero, F., Ozaita, A., & Robledo, P. (2011). Neurochemical basis of cannabis addiction. *Neuroscience*, *181*, 1–17.

Maldonado, R., Robledo, P., & Berrendero, F. (2013). Endocannabinoid system and drug addiction: new insights from mutant mice approaches. *Current Opinion in Neurobiology*, *23*(4), 480–486.

Matsuda, L., Lolait, S., Brownstein, M., Young, A., & Bonner, T. (1990). Structure of a cannabinoid receptor and functional expression of the cloned cDNA. RID B-8609-2009 *Nature*, *346*(6284), 561–564.

Mechoulam, R., Benshabat, S., Hanus, L., Ligumsky, M., Kaminski, N., Schatz, A., ... Vogel, Z. (1995). Identification of an endogenous 2-monoglyceride, present in canine gut, that binds to cannabinoid receptors. *Biochemical Pharmacology*, *50*(1), 83–90.

Melis, M., Greco, B., & Tonini, R. (2014). Interplay between synaptic endocannabinoid signaling and metaplasticity in neuronal circuit function and dysfunction. *European Journal of Neuroscience*, *39*(7), 1189–1201.

Moreira, F. A., Aguiar, D. C., Terzian, A. L., Guimaraes, F. S., & Wotjak, C. T. (2012). Cannabinoid type 1 receptors and transient receptor potential vanilloid type 1 channels in fear and anxiety-two sides of one coin? *Neuroscience*, *204*, 186–192.

Muldoon, P. P., Lichtman, A. H., Parsons, L. H., & Damaj, M. I. (2013). The role of fatty acid amide hydrolase inhibition in nicotine reward and dependence. *Life Sciences*, *92*(8–9), 458–462.

Munro, S., Thomas, K. L., & Abushaar, M. (1993). Molecular characterization of a peripheral receptor for cannabinoids. *Nature*, *365*(6441), 61–65.

Nakajima, J., Takahashi, M., Seto, T., Kanai, C., Suzuki, J., Yoshida, M., ... Hamano, T. (2013). Analysis of azepane isomers of AM-2233 and AM-1220, and detection of an inhibitor of fatty acid amide hydrolase [3'-(aminocarbonyl)(1,1'-biphenyl)-3-yl]-cyclohexylcarbamate (URB597) obtained as designer drugs in the Tokyo area. *Forensic Toxicology*, *31*(1), 76–85.

Oliere, S., Joliette-Riopel, A., Potvin, S., & Jutras-Aswad, D. (2013). Modulation of the endocannabinoid system: vulnerability factor and new treatment target for stimulant addiction. *Frontiers in Psychiatry*, *4*, 109.

Pagotto, U., Marsicano, G., Cota, D., Lutz, B., & Pasquali, R. (2006). The emerging role of the endocannabinoid system in endocrine regulation and energy balance. *Endocrine Reviews*, *27*(1), 73–100.

Palkovits, M., Harvey-White, J., Liu, J., Kovacs, Z. S., Bobest, M., Lovas, G., ... Kunos, G. (2008). Regional distribution and effects of postmortal delay on endocannabinoid content of the human brain. *Neuroscience*, *152*(4), 1032–1039.

Patel, S., Rademacher, D. J., & Hillard, C. J. (2003). Differential regulation of the endocannabinoids anandamide and 2-arachidonylglycerol within the limbic forebrain by dopamine receptor activity. *The Journal of Pharmacology and Experimental Therapeutics*, *306*(3), 880–888.

Pertwee, R. G. (2008). The diverse CB1 and CB2 receptor pharmacology of three plant cannabinoids: Delta(9)-tetrahydrocannabinol, cannabidiol and Delta(9)-tetrahydrocannabivarin. *British Journal of Pharmacology*, *153*(2), 199–215.

Pertwee, R. G. (2009). Emerging strategies for exploiting cannabinoid receptor agonists as medicines. *British Journal of Pharmacology*, *156*(3), 397–411.

Pertwee, R. G., Howlett, A. C., Abood, M. E., Alexander, S. P. H., Di Marzo, V., Elphick, M. R., ... Ross, R. A. (2010). International Union of Basic and Clinical Pharmacology. LXXIX. Cannabinoid receptors and their ligands: beyond CB_1 and CB_2. *Pharmacological Reviews*, *62*(4), 588–631.

Piomelli, D. (2014). More surprises lying ahead. The endocannabinoids keep us guessing. *Neuropharmacology*, *76*, 228–234.

Rahman, I. A., Tsuboi, K., Uyama, T., & Ueda, N. (2014). New players in the fatty acyl ethanolamide metabolism. *Pharmacological Research: The Official Journal of the Italian Pharmacological Society, 86,* 1–10.

Reisenberg, M., Singh, P. K., Williams, G., & Doherty, P. (2012). The diacylglycerol lipases: structure, regulation and roles in and beyond endocannabinoid signalling. *Philosophical Transactions of the Royal Society of London. Series B, Biological Sciences, 367*(1607), 3264–3275.

Schoeler, T., & Bhattacharyya, S. (2013). The effect of cannabis use on memory function: an update. *Substance Abuse and Rehabilitation, 4,* 11–27.

Shire, D., Calandra, B., RinaldiCarmona, M., Oustric, D., Pessegue, B., BonninCabanne, O., … Ferrara, P. (1996). Molecular cloning, expression and function of the murine CB2 peripheral cannabinoid receptor. *Biochimica Et Biophysica Acta-Gene Structure and Expression, 1307*(2), 132–136.

Spaderna, M., Addy, P. H., & D'Souza, D. C. (2013). Spicing things up: synthetic cannabinoids. *Psychopharmacology, 228*(4), 525–540.

Sugiura, T., Kondo, S., Sukagawa, A., Nakane, S., Shinoda, A., Itoh, K., … Waku, K. (1995). 2-Arachidonoylgylcerol – a possible endogenous cannabinoid receptor-ligand in brain. *Biochemical and Biophysical Research Communications, 215*(1), 89–97.

Svizenska, I., Dubovy, P., & Sulcova, A. (2008). Cannabinoid receptors 1 and 2 (CB1 and CB2), their distribution, ligands and functional involvement in nervous system structures - a short review. *Pharmacology Biochemistry and Behaviour, 90*(4), 501–511.

Uchiyama, N., Kawamura, M., Kikura-Hanajiri, R., & Goda, Y. (2013). URB-754: a new class of designer drug and 12 synthetic cannabinoids detected in illegal products. *Forensic Science International, 227*(1–3), 21–32.

Ueda, N., Tsuboi, K., & Uyama, T. (2013). Metabolism of endocannabinoids and related N-acylethanolamines: canonical and alternative pathways. *The FEBS Journal, 280*(9), 1874–1894.

Van Sickle, M., Duncan, M., Kingsley, P., Mouihate, A., Urbani, P., Mackie, K., … Sharkey, K. (2005). Identification and functional characterization of brainstem cannabinoid CB2 receptors. RID B-7358-2011 *Science, 310*(5746), 329–332.

Wood, T. B., Spivey, W. T., & Easterfield, T. H. (1899). Cannabinol, Part 1. *Journal Chemistry Society, 75,* 20–36.

Yates, M. L., & Barker, E. L. (2009). Inactivation and biotransformation of the endogenous cannabinoids anandamide and 2-arachidonoylglycerol. *Molecular Pharmacology, 76*(1), 11–17.

Chapter 66

Molecular Pharmacology of CB1 and CB2 Cannabinoid Receptors

Jahan P. Marcu[1,2], Jason B. Schechter[3]
[1]Americans for Safe Access, Washington, DC, USA; [2]Green Standard Diagnostics, Inc., Las Vegas, NV, USA; [3]Cortical Systematics LLC, Tucson, AZ, USA

Abbreviations

CB1 Cannabinoid type 1 receptor
CB2 Cannabinoid type 2 receptor
EC Extracellular
ECS Endocannabinoid system
GPCR G-protein-coupled receptor
^{3}H Tritium
MAP Mitogen-activated protein
PLC Phospholipase C
TMH Transmembrane helix
WT Wild type

INTRODUCTION

In the 1990s, a number of surprising discoveries laid a strong, new foundation for both *Cannabis* and cannabinoid research. The identification and early characterization of the cannabinoid receptors (Console-Bram, Marcu, & Abood, 2012; Matsuda, Lolait, Brownstein, Young, & Bonner, 1990; Pacher, 2006) literally provided substrate for the future of cannabinoid science. The CB1 receptor is now understood to be one of the most abundant proteins in the mammalian brain. The CB2 receptor—sharing significant genetic sequence (homology) with the CB1 receptor—has been found to be expressed predominately in immune tissue and cells of immune origin. Because of this characteristic distribution, cannabinoid receptors provide an attractive therapeutic target. The discovery and subsequent availability of various synthetic cannabinoid analogues have allowed researchers to begin characterizing the various physiological roles of these abundant G-protein-coupled receptors (GPCRs). The sum of all knowledge of how these cannabinoid receptors and their endogenous ligands work together to modulate mammalian biology and behavior is referred to as the endocannabinoid system (ECS).

INTRODUCTION TO CANNABINOID RECEPTOR RESEARCH

The investigation of novel compounds acting through CB1 and CB2 receptors has revealed at least five structurally distinct classes of cannabinoid ligands, each possessing their own unique binding properties (Hurst et al., 2002; Song & Bonner, 1996). Known cannabinoid classes include tricyclic cannabinoids (i.e., Δ^9-tetrahydrocannabinol (Δ^9-THC)), bicyclic cannabinoids (CP 55,940), aminoalkylindole or indole derivatives (WIN 55,212), fatty acid derivatives or eicosanoids (anandamide), and antagonist/inverse agonists (SR141716A).

Structure–activity relationship studies of ligands such as SR141716A have aided in the design of more high-affinity and selective ligands (Hanuš, 2009). Many of these cannabinoid compounds have demonstrated selective binding properties (see Figure 1 for cannabinoid ligand descriptions and a review of K_i values). For instance, CP 55,940, Δ^9-THC, and anandamide activate both CB1 and CB2 receptors with similar efficacies, whereas compounds like WIN 55,212 possess a significantly higher affinity for CB2 over CB1. The synthesis and availability of novel and potent cannabinoid ligands has driven significant research gains in recent years.

Recovering useful data on the transmembrane helical clusters that surround and comprise the ligand-binding pocket of a membrane-bound receptor is crucial to understanding how that protein carries out its functions. Investigation represents significant technical challenges. Crystallization techniques, often useful in revealing important structural information of proteins, failed to successfully crystallize the cannabinoid receptors. Whereas mutational analysis of transfected cell lines—cells forced to express the CB1 receptor—serve as a reliable tool for studying both the structure of the CB1 receptor and the pharmacology of cannabinergic ligands (Abood, 2009; Matsuda et al., 1990), the molecular structure of known cannabinoid receptors was largely determined through experiments combining site-directed mutagenesis with molecular modeling techniques. Cannabinoid receptors collected and processed from cell cultures have been used to study GPCR activity through various methods, including, but not limited to, GTPγS assays to determine receptor activity and tritiated (using [^{3}H]) cannabinoid assays to determine binding affinities and distribution (Figure 2). The techniques developed for delineation of the ECS have been among the most innovative in all of scientific enquiry.

Structure	Description	Reported Binding Activity: K_i (nM)
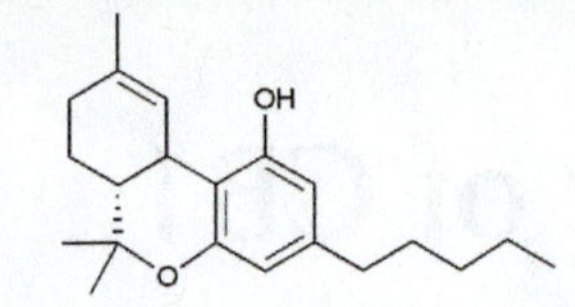	**THC** Botanical, Agonist	CB1: 5.05–80.3 CB2: 3.13–75.3
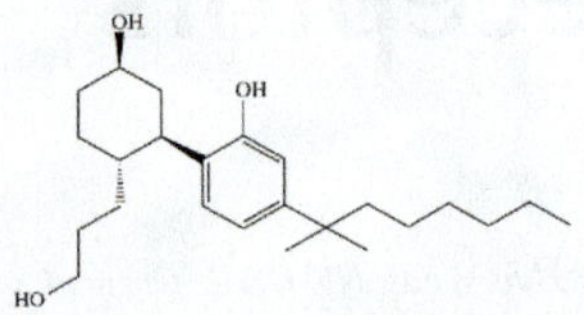	**CP55,940** Synthetic, Agonist	CB1: 0.06–0.073 CB2: 0.17–0.52
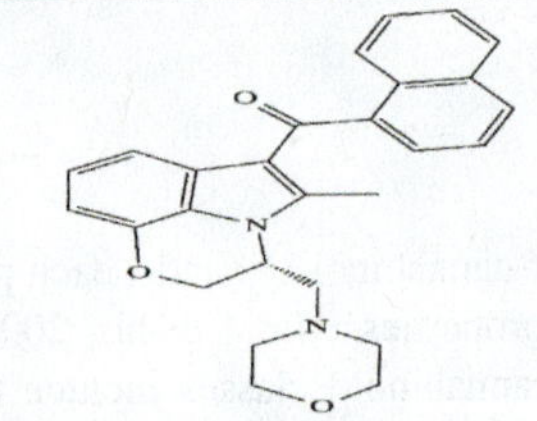	**WIN55,212** Synthetic, Indole or aminoalkylindole, Agonist	CB1: 1.89–124 CB2: 0.28–16.2
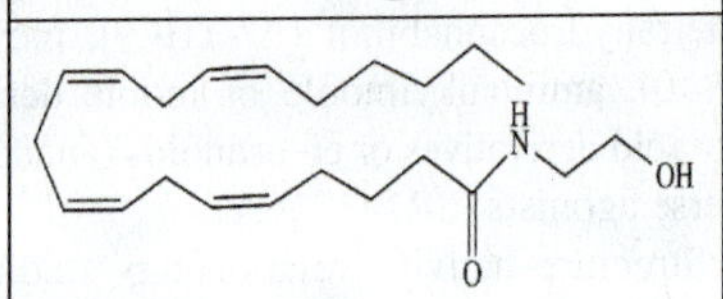	**Anandamide** Endocannabinoid, Fatty acid derivative, Agonist	CB:1 61–543 CB2: 279–1940
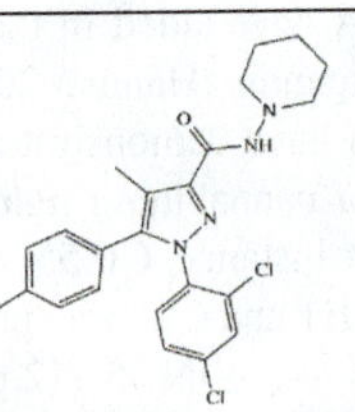	**SR141716A** Synthetic, Antagonist / Inverse agonist	CB1: 1.8–12.3 CB2: 5.14–13,2000

FIGURE 1 Structures and definitions of cannabinoid ligands. The structures of prototypical cannabinoid compounds from various structural classes, with binding activity values reported as a range of average K_i values for the displacement of a tritiated compound from CB1 and CB2 receptors (rat, human, or mouse) by a nonradioactive compound. In most cases, the tritiated compound used was [^{3}H]CP 55,940, but K_i values may also be derived from displacement of [^{3}H]SR141716A, R-(+)-[^{3}H]WIN55,212, or [^{3}H]HU-210. (For full references of reported values see Pertwee, 2014. All images were generated using ChemDraw by the author.)

OVERVIEW OF ENDOCANNABINOID SYSTEM MOLECULAR BIOLOGY

The CB1 receptor has a fundamental, perhaps unsung, role in the mammalian central nervous system (CNS). Dubbed "cannabinoid type 1," the CB1 is a member of the class A rhodopsin-like family of GPCRs found primarily in the CNS, where they are key to the regulation of neuronal activity. There is some evidence that the CB1 receptor is also expressed in peripheral tissues, albeit to a lesser extent, including the adrenal gland, bone marrow, heart, lung, and prostate (Howlett et al., 2002). The CB1 is a $G_{i/o}$-coupled GPCR that binds five structurally diverse classes of ligands; these include the previously mentioned endocannabinoids (typified by anandamide and 2-arachidonoylglycerol (2-AG)), the classical and nonclassical cannabinoids (typified by Δ^9-THC and CP 55,940, respectively), the indoles (typified by WIN 55,212-2), and the diarylpyrazole antagonists/inverse agonists (typified by SR141716A) (Picone, Fournier, & Makriyannis, 2002).

GPCRs are a superfamily of cell surface receptors with the principal role of transmitting information concerning the environment exterior to the cell to the inside of the cell. This transfer of information across the cellular membrane is relayed via heterotrimeric guanine nucleotide-binding regulatory proteins (i.e., G proteins) to other proteins (i.e., kinases). As cannabinoid receptors belong the "class A" family of GPCRs, they share a common membrane topology (Kristiansen, 2004). Class A receptors are generally characterized as having three distinct features: an extracellular N-terminus, an intracellular C-terminus, and seven transmembrane helices (TMHs), connected by loops (Ballesteros & Weinstein, 1995). Those regions crucial for ligand binding, receptor activation, and signal transduction lay within the bundles and stretches of amino acids throughout the seven TMHs (Abood, 2009).

Functional residues of the CB1 receptor can be divided into at least four groups by function; there are residues that affect ligand binding, receptor conformation, receptor activation, and G-protein

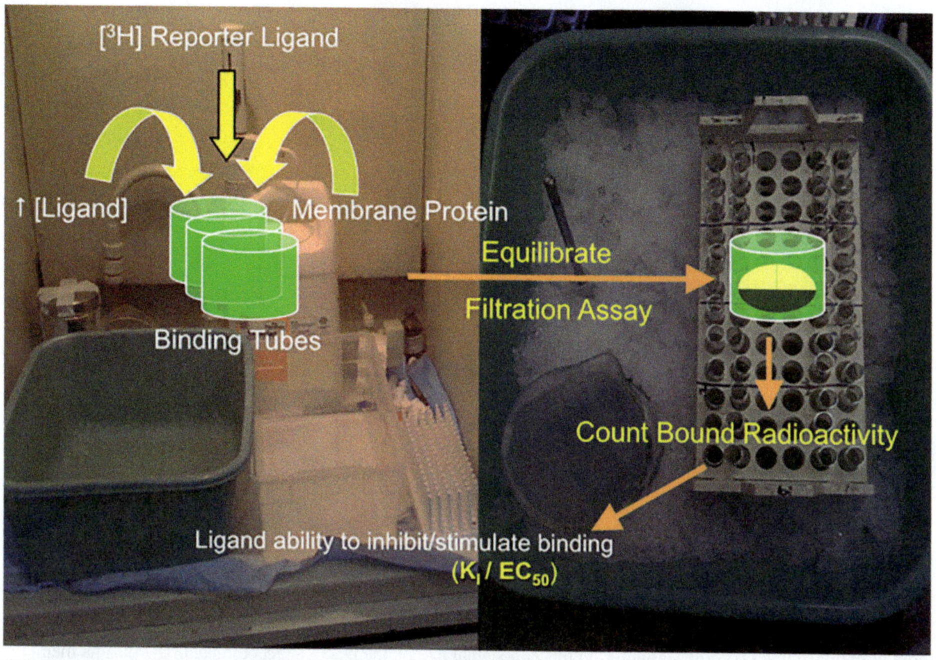

FIGURE 2 **Ligand binding experimental setup and work flow.** A reporter ligand, membrane protein, and salt solution are placed in siliconized tubes with increasing amounts of the reporter ligand. The assay is equilibrated and filtered. The radioactivity is collected on filter paper and counted via a scintillation counter. The count of bound radioactivity is used to generate binding and stimulation values. The background image on the left shows the setup for siliconizing glass tubes, the background image on the right shows the experimental setup prior to equilibration.

association. Alterations in binding affinities and conformations are usually associated with changes in basal and maximal stimulation. Mutations of residues that affect receptor confirmation may cause a failure of expression due to protein misfolding and/or a suppression of binding affinities and receptor activation. For instance, on the extracellular portion of TMHs 3–4–5–6, there is an aromatic cluster consisting of tryptophan, tyrosine, and phenylalanine residues. The residues F3.25, F3.36, W4.64, Y5.39, W5.43, and W6.48 constitute a "microdomain" of aromatic clusters that face the binding pocket (numbering is based on location and conservation of the residue across species Ballesteros & Weinstein, 1995). Study of this microdomain revealed that a tyrosine, Y5.39, is conserved between both CB1 and CB2 receptors. A mutation of this Y5.39 to another aromatic residue, phenylalanine (Y5.39F), resulted in only minor differences in affinity and transduction compared to the wild-type (WT) CB1 receptor. However, an isoleucine (Y5.39I) mutation results in a loss of ligand binding along with pronounced topological changes in TMHs 3–4–5. Such systematic analysis of aromatic microdomain demonstrates that the cannabinoids SR141716A and WIN 55,212 require aromatic stacking interactions of TMHs 3–4–5–6 to activate the CB1 receptor (McAllister et al., 2003). This study also revealed a selective binding region for aminoalkylindoles and diaryl pyrazole cannabinoids, but not for endogenous (anandamide) and bicyclic (CP 55,940) cannabinoids.

Aromatic residues F3.36 and W6.48 form a "toggle switch," and mutational analysis revealed that these residues are important for both binding and activation (McAllister et al., 2004). A mutation of F3.36 to alanine resulted in an increase in constitutive activity, confirmed by using the inverse agonist SR141716A to reduce basal levels of GTPγS binding. Mutation of W6.48 did not cause a rise in basal levels but, rather, an increase in ligand affinities.

Together, the F3.36 and W6.48 residues form an interaction that exists in the inactive state of the CB1 receptor. As a ligand is recognized by the receptor, these "bridges" are broken and the protein conformation changes—operating as a toggle switch mechanism.

Other residues may have more dramatic effects on protein folding or conformation. For example, tryptophan 4.64 (W4.64) is conserved between CB1 and CB2 receptors and mutational analysis suggests this aromatic residue is important for ligand binding and signaling (McAllister et al., 2003; Rhee, 2002). Substituting W4.64 with the nonaromatic residues leucine (W4.64L) and alanine (W4.64A) resulted in drastic reductions—W4.64L and W4.64A mutants displayed a significant loss of ligand binding and signal transduction. The W4.64A mutant did not localize to the cell surface and this may be the result of protein misfolding. (In many GPCRs, W4.64 is known for its role in proper protein folding. See Wess, Nanavati, Vogel, & Maggio, 1993). As another example, mutations in the N-terminus of the CB1 second extracellular loop (EC-2) resulted in retention of the receptor in the endoplasmic reticulum (Ahn, Bertalovitz, Mierke, & Kendall, 2009).

In contrast, isoleucine 2.62 (I2.62) and aspartate 2.63 (D2.63) are located in a nonbinding region of the cannabinoid receptor, but may be an important structural requirement (Kapur, Samaniego, Thakur, Makriyannis, & Abood, 2008). A mutation of I2.62 to threonine (I2.62T)—and substituting D2.63 for the (likewise similarly charged) glutamate residue D2.63E—did not significantly alter either binding or activation compared to WT. However, substituting the neutrally charged asparagine (D2.63N) reduced the potency of CP 55,940, WIN 55,212, HU-210, and AM4056. The double mutation (of I2.62T–D2.63N) resulted in a synergistic increase in the EC_{50} required for activation. Accordingly, D2.63A is hypothesized to form a "salt bridge" with a lysine (K) residue in

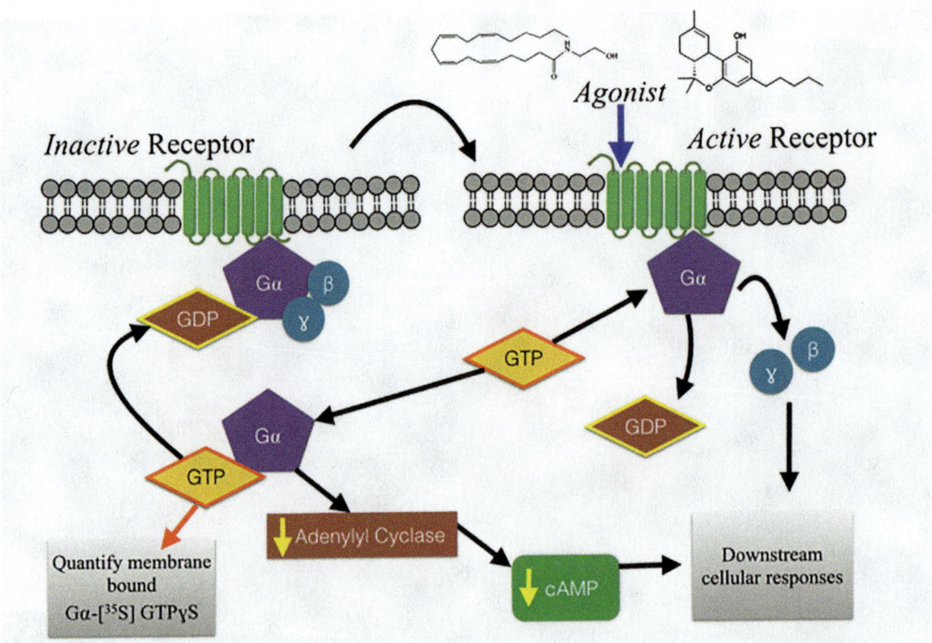

FIGURE 3 Basic model of cannabinoid receptor signaling. In this conceptual model, the inactive receptor sits in the plasma membrane with the G-protein complex. Upon the ligand binding, the active receptor undergoes a conformational change, which releases GDP for GTP, and the β and γ subunits separate to affect downstream cellular responses, such as increasing MAP kinase activity. The Gα–GTP complex can be used to measure receptor stimulation in bioassays using radiometric techniques, in this case adding Gα-[^{35}S]GTPγS, which irreversibly binds to Gα and can then be collected and measured.

the third EC loop (K373). Salt bridges between charged residues are the result of proximity and favorable electrostatic attractions and contribute to protein structure and the specificity of interaction of proteins with various biomolecules (Bosshard, Marti, & Jelesarov, 2004). Studies on GPCRs suggest these interactions are between charged residues from spatially distinct domains, i.e., a salt bridge between TMH2 and TMH7 modulates receptor function. Salt bridges are generally composed of negative charges from Asp, Glu, Tyr, Cys, and the C-terminal carboxylate group and of positive charges from His, Lys, Arg, and the N-terminal amino group (Kumar & Nussinov, 2002). In consideration of facts such as D2.63A–K373A producing dramatic reductions in receptor activation, salt-bridge interactions may be of particular importance for maintaining the structure of the CB1 receptor (see Figure 4 for schematic representation of important amino acids).

The CB1 and CB2 receptors share a significant degree of sequence homology, despite being predominately located in the CNS and immune system, respectively. The high degree of homology between the amino acid residue sequences from TMHs of various GPCRs has led to the identification of conserved residues shown to be crucial for receptor function (Kapur et al., 2008; Roche, Hoare, & Parker, 1992; Tao et al., 1999). For example, in the third TMH of both CB1 and CB2 receptors, there is a conserved lysine residue, Lys 192 (K3.28), which plays an important role in selective ligand binding.

In addition, charged interactions between amino acid residues from different TMH domains have been shown to be essential for either ligand binding or receptor function (Marcu et al., 2013; Sealfon et al., 1995; Xu et al., 1999; Zhou et al., 1994). Residues from the extracellular loops demonstrate a comparatively low sequence homology and were initially thought to connect the TMH domains rather than having a direct role in receptor functioning. However, more recent studies have demonstrated the critical role of the EC loops in ligand binding and receptor signaling (Marcu et al., 2013; Shore et al., 2014). Mutation studies have demonstrated that the first EC loop (EC-1) is important to the activation of the adenosine A_{2B} receptor (Peeters et al., 2011). The EC-2 loop has been shown to be important in ligand binding and activation at the V_{1a} vasopressin receptor (Conner et al., 2007), helix movement in rhodopsin (Ahuja et al., 2009), and the binding of allosteric modulators at the M_2 acetylcholine receptor (Avlani et al., 2007). Considerably less is known about the EC-3 loop. However, a key salt bridge between the EC-3 and the EC-2 loops has been observed to influence ligand binding and receptor activation at the $β_2$-adrenergic receptor (Bokoch et al., 2010).

While both the EC-1 and the EC-2 loops of the CB1 receptor have, thus far, been better characterized than its EC-3 loop (Ahn et al., 2009; Ahn, Mahmoud, & Kendall, 2012; Bertalovitz, Ahn, & Kendall, 2010; Murphy & Kendall, 2003), modeling studies suggest that EC-3 residue K373 may form a functionally important ionic interaction with a transmembrane residue, D2.63[176]. Previous D2.63[176] mutational studies demonstrated that the negative charge of D2.63[176] is critical for agonist efficacy, but not for ligand binding at the CB1 receptor (Kapur et al., 2008). This functional requirement (of a negatively charged residue at 2.63[176]) might be due to the residue's participation in an ionic interaction with K373—a relationship that is necessary for signal transduction. The efficacy of CP 55,940 and WIN 55,212-2 was markedly reduced by alanine-substitution mutations at D2.63 and K373, while the charge reversal mutation led to a partial rescue of WT levels of efficacy. Computational results indicate that the D2.63[176]–K373 ionic interaction strongly influences the conformation of the EC-3 loop, providing a structure-based rationale for the importance of the EC-3 loop to signal transduction in CB1. This putative ionic

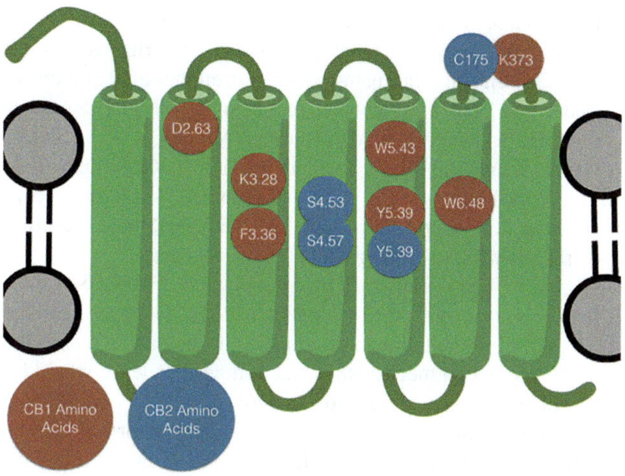

FIGURE 4 Important amino acids for CB1/CB2 receptors. The amino acid residues important for proper functioning of the CB1 and CB2 receptor are noted in red and blue spheres, respectively. The placement of the circles is a conceptual approximate location of the amino acids on this generic diagram of a seven-TMH GPCR.

interaction results in the EC-3 loop pulling over the top (EC side) of the receptor, resulting in an EC-3 loop conformation that may serve both protective and mechanistic roles (Figure 4).

Finally, EC loops may be important for allosteric interactions with cannabinoid receptors (Ahn et al., 2012). Stimulating allosteric sites on cannabinoid receptors can alter binding affinity and efficacy of orthosteric ligands. This offers a unique approach to cannabinoid receptor-based strategies, promising a more precise control of the effects of cannabinoid-based treatments by controlling the conformation of the receptor and the selectivity of the ligand it can interact with (Price, 2005; Shore et al., 2014).

OVERVIEW OF CANNABINOID SIGNALING IN CELLS

Cannabinoid receptors are members of the rhodopsin subfamily of GPCRs, meaning their signal transduction mechanisms are affected by G-protein coupling (Figure 3). As of this writing, cannabinoid receptor activation is known to involve regulation of adenylyl cyclase, alterations in mitogen-activated protein (MAP) kinases, modulation of ion channels, and modulation of intracellular Ca^{2+}. Cannabinoid signaling cascades are linked to processes of cell proliferation, cell differentiation, cell movement, and cell death. Understanding the molecular pathways involved in, and coupled to, cannabinoid receptor signaling is key to elucidating the mechanisms and control of mammalian physiology.

CB1 and CB2 receptors are coupled to $G_{i/o}$ proteins. Cannabinoid receptor signaling events are usually pertussis toxin sensitive, implicating the involvement of $G_{i/o}$-coupled proteins (Howlett, Qualy, & Khachatrian, 1986). Additional evidence from experiments that sequester $G_{i/o}$ proteins from being used by cannabinoid receptors has shown that CB1 can also associate with G_s. CB2 receptors are not known to associate with G_s (Calandra et al., 1999; Glass & Felder, 1997). In contrast, experiments have revealed CB1 receptor coupling with G_q proteins in HEK cells (Lauckner, Hille, & Mackie, 2005). Both CB1 and CB2 receptor stimulation can induce G_q-mediated Ca^{2+} release in insulinoma cells (De Petrocellis et al., 2007). The $G_{i/o}$ and G_s proteins can modulate cAMP and Ca^{2+} currents, while G_q is mainly involved with intracellular Ca^{2+} release. CB1 and CB2 receptors are known to associate with $G_{i/o}$ and G_q, but only the CB1 receptor is known to associate with G_s proteins.

Upon receptor stimulation—by synthetic, phytological, or endocannabinoid means—there is generally an inhibition of adenylyl cyclase and activation of the MAP kinase signaling pathway (Howlett, 1984; Howlett et al., 2002; Pertwee, 2005). The CB1 receptor when coupled to $G_{i/o}$ can also modulate A-type and inwardly rectifying potassium currents and cause an inhibition of N- and P/Q-type calcium currents.

The CB1 receptor-mediated inhibition of adenylyl cyclase was the first cannabinoid receptor mechanism to be characterized (Howlett, 1984). More than a decade later, it was discovered that CB2 receptors can also inhibit adenylyl cyclase activity (Bayewitch et al., 1995). The ability of cannabinoid receptors to interact with $G_{i/o}$ is exploited in the adenylyl cyclase assay (and the [^{35}S]GTPγS receptor stimulation assay). Functional inhibition of adenylyl cyclase, and thus cAMP has been demonstrated in many other experiments (Demuth & Molleman, 2006). This cannabinoid-mediated effect is sensitive to both pertussis toxin and CB1 antagonists such as SR141716A and leads to an attenuation of cAMP accumulation. Such alterations in cAMP concentrations can regulate protein kinase A (PKA) phosphorylation, resulting in significant changes in cellular activity (Childers & Deadwyler, 1996; Mu, Zhuang, Hampson, & Deadwyler, 2000). As one example, PKA activity can affect gene expression and regulate potassium channel activity (Hampson et al., 1995).

MAP kinase pathways are often activated by GPCRs. MAP kinase cascades can lead to the activation of extracellular signal-regulated kinase 1/2 (ERK1/2), c-Jun N-terminal kinase (JNK), p38 MAP kinase, and/or ERK5 proteins. Stimulation of CB1 and CB2 receptors, both in vivo and in vitro, has been shown to activate ERK1/2, phosphoinositide 3-kinase, and PKC through mechanisms that involve the activation of $G_{i/o}$ proteins and inhibition of adenylyl cyclase and alterations in the activity of protein kinases (Davis, Ronesi, & Lovinger, 2003). Via these mechanisms, cannabinoid receptor signaling can lead to alterations in the levels of p38 MAP kinase and JNK. The stimulation of various signaling molecules can also be dependent on the cell type or tissue.

Ion channels can also be modulated by CB1 and CB2 receptor stimulation (Atwood, Wager-Miller, Haskins, Straiker, & Mackie, 2012; McAllister, Griffin, Satin, & Abood, 1999). CB1 receptors can stimulate G-protein-coupled inwardly rectifying potassium channels, which are activated through $G_{i/o}$ proteins, and this stimulation may be induced by a variety of CB1 agonists (Turu & Hunyady, 2010). Experiments using Xenopus oocytes transfected with CB2 and G-protein-coupled inwardly rectifying potassium channels did not demonstrate CB2 modulation of ion channel function. However, CB2 cannabinoid receptors inhibit synaptic transmission when expressed in cultured autaptic neurons. CB1 receptors can also affect the function of L-type, N-type, and P/Q-type calcium channels (Mackie, Devane, & Hille, 1993; Mackie & Hille, 1992; Mackie, Lai, Westenbroek, & Mitchell, 1995).

Cannabinoids can modulate levels of intracellular calcium. Both 2-AG and WIN 55212-2 were able to increase intracellular Ca^{2+} in NG108-15 cells, a neuroblastoma–glioma hybrid cell line (Sugiura et al., 1996). These effects were inhibited by pertussis toxin and SR141716A, again suggesting the effect is mediated by $G_{i/o}$ proteins via the CB1 receptor. A phospholipase C (PLC) inhibitor was also able to block the response, implicating the involvement of IP_3 in the release of Ca^{2+} (Sugiura et al., 1997). Additional evidence supports cannabinoid modulation of Ca^{2+} in endothelial, cerebellar granule, canine kidney, and smooth muscle cells (Demuth & Molleman, 2006). Anandamide has also been shown to modulate intracellular Ca^{2+} in endothelial cells by activation of PLC, the effect of which was blocked by the CB2 antagonist SR144528 but not the CB1 antagonist SR141716A (Zoratti, Kipmen-Korgun, Osibow, Malli, & Graier, 2003). Collectively, these data support a CB1- and CB2-receptor-mediated release of intracellular Ca^{2+} through $G_{i/o}$ proteins and increases in PLC activity.

The CB1 receptor-coupled G proteins are also associated with other GPCRs, and these G proteins have been demonstrated to affect important signaling molecules in osteoblasts. $G_{i/o}$ proteins may regulate osteoblast proliferation by interacting with the MAP kinase cascade and by affecting downstream ERK activation. Antiapoptotic signals can result from AKT activation, which can induce a caspase cascade under the regulation of G_i proteins. The effect on such downstream mediators has been explored in cells, with promising results in osteoclasts (Ross, 2005).

G_q may regulate both osteoblast proliferation and differentiation. In vitro assays have revealed that G_q signaling promotes osteoblast proliferation (Wu, Deng, Zhu, & Li, 2010). In an apparent contradiction, transgenic mice with a constitutively activated G_q in osteoblasts appear to have reduced numbers of osteoblasts and lower bone marrow density (Ogata, Kawaguchi, Chung, Roth, & Segre, 2007). Cultures from these G_q-transgenic mice show impaired osteoblast differentiation confirmed by a decrease in activity of the differentiation marker alkaline phosphatase. This study was conducted with FVB/N mice and MC3T3 cells—it will be interesting to see if future studies find a consistent G_q-associated inhibition of osteoblastic differentiation across different mouse backgrounds such as C57/BL6 or CD1 mouse strains.

G_s proteins have a demonstrated role in bone remodeling as well. Transgenic animals with an induced G_s deficiency exhibit significant skeletal malformations, which surely supports the role played by G_s proteins in endochondral bone development (Castrop et al., 2007). A G_s deficiency in chondrocytes is lethal in mice, as the animals displayed severe growth plate defects, while osteoblast-specific G_s deficiency in mice reduced bone turnover, thickened cortical bone, and produced a narrow bone marrow cavity (Sakamoto, Chen, Kobayashi, Kronenberg, & Weinstein, 2005; Sakamoto, Chen, Nakamura, et al., 2005). G_s promotes bone formation in vivo, and osteoblast differentiation and proliferation are promoted in vitro.

SUMMARY AND APPLICATIONS TO OTHER ADDICTIONS AND SUBSTANCE MISUSE

While this biochemical saga continues to be elucidated, at the same time thousands of research articles have reported that cannabinoid receptor stimulation clearly mitigates numerous pathologies in humans. Positive outcomes have been evidenced in pathologies including, but clearly not limited to, Alzheimer disease, cancer, obesity, and pain (Pacher, 2006; Pertwee, 2014; Pertwee, 2009; Russo, 2008a; Russo, 2011). Additionally, cannabinoid receptor activity may decrease the untoward effects of drug withdrawal in animals, enhance the pain-attenuating effects of morphine in humans, and have a prophylactic effect on developing opiate or other drug addiction-related disorders (Abrams, Couey, Shade, Kelly, & Benowitz, 2011; Bachhuber, Saloner, Cunningham, & Barry, 2014; Ledent et al., 1999). Unfortunately, many attempts at harnessing the therapeutic potential of the cannabinoid receptors have failed owing to overregulation and "unacceptable" subjective CNS-related side effects, such as euphoria and, with far less incidence, depression and suicidal fixation (Christopoulou & Kiortsis, 2011). Clearly, a better understanding of the cannabinoid receptor signal transduction mechanism(s), at a molecular level, would be useful in realizing the therapeutic potential of modulating the ECS.

Mutational analysis of the cannabinoid receptor will surely contribute to our base of knowledge, building an increasingly accurate molecular model of the CB1 receptor. Tailoring future cannabinoid ligands to the receptor will herald the rise of potentially useful synthetic/modified cannabinergic therapeutic drugs and will very probably serve as future treatment modalities for symptoms of drug withdrawal and addiction. One tantalizing possibility, on the forefront of pharmacogenomics research, is to create compounds that may activate cannabinoid receptors when there is a loss of endogenous ligand activity or ligand recognition due to a mutation of those genes encoding the cannabinoid receptors.

Many studies have found associations with ECS polymorphisms in human diseases such as depression, drug addiction, and abnormal mental health (Hillard, Weinlander, & Stuhr, 2012). Hypothetically, a patient might possess a polymorphism (or acquire a mutation) that hinders efficient interactions between cannabinoid receptors and endocannabinoids. In this scenario, a drug or treatment might be chosen based on the patient's genotype. Creation of a different class of bespoke, neo-cannabinoids might be able to replace lost endogenous compounds or perhaps increase the efficiency of existing endogenous receptor–ligand interactions. Such a personalized approach would require either selecting a compound that has structurally distinct binding requirements from extant endogenous compounds (i.e., an indole such as WIN 55,212) or a compound that would enhance baseline endocannabinoid activity at the target. A "clinical endocannabinoid deficiency syndrome"—resulting from defects in the ECS—has already been proposed to underlie certain pathologies, including otherwise treatment-resistant conditions, and cannabinoid receptor polymorphisms are already associated with several diseases (Russo, 2008b). To date, a mutation has yet to be identified in the human cannabinoid receptor that results in conclusive alteration of ligand–receptor interactions. The future of *Cannabis*- and cannabinoid-based clinical trials could be enhanced by developing methods for screening patient polymorphisms or mutations in those genes associated with the ECS.

DEFINITION OF TERMS

Cannabinoids These are a group of closely related compounds, similar to Δ^9-THC or other compounds found in plants.
CB1 receptor This is a GPCR densely located in the brain and nervous tissue.
CB2 receptor This is a GPCR densely located in immune tissue and also found in the brain.

Endocannabinoid system This is a mammalian biological system consisting of receptors (i.e., CB1, CB2), endogenously produced compounds (i.e., anandamide, 2-AG), and proteins responsible for the synthesis, breakdown, and transport of endogenous cannabinoids.

Mutational analysis This is an experimental process of mutating amino acids of a protein and investigating the mutant protein's function.

SUMMARY POINTS

- The cannabinoid receptors are highly expressed in nearly all mammalian tissues.
- The various classes of cannabinoid receptors share a significant amount of sequence homology.
- The CB1 and CB2 receptors have distinct expression profiles; CB1 is localized in neuronal tissue, while CB2 is found on cells of immune origin.
- There exist at least five distinct classes of cannabinoid ligands.
- Genomics studies have revealed strong associations between substance abuse and cannabinoid receptor polymorphisms.
- Developing modern therapeutic cannabinergic agents will depend on how well the ECS is understood.
- Cannabinoid receptor mutational analysis has provided insight into how these receptors can influence drug addiction-related disorders and disease progression.

REFERENCES

Abood, M. E. (2009). Molecular biology of cannabinoid receptors: mutational analyses of the CB receptors. *The Cannabinoid Receptors*. http://dx.doi.org/10.1007/978-1-59745-503-9-8.

Abrams, D. I., Couey, P., Shade, S. B., Kelly, M. E., & Benowitz, N. L. (2011). Cannabinoid–opioid interaction in chronic pain. *Clinical Pharmacology & Therapeutics*, *90*(6), 844–851. http://dx.doi.org/10.1038/clpt.2011.188.

Ahn, K., Johnson, D. S., Mileni, M., Beidler, D., Long, J. Z., McKinney, M. K., ... Cravatt, B. F. (2009). Discovery and characterization of a highly selective FAAH inhibitor that reduces inflammatory pain. *Chemistry & Biology*, *16*(4), 411–420. http://dx.doi.org/10.1016/j.chembiol.2009.02.013.

Ahn, K. H., Bertalovitz, A. C., Mierke, D. F., & Kendall, D. A. (2009). Dual role of the second extracellular loop of the cannabinoid receptor 1: ligand binding and receptor localization. *Molecular Pharmacology*, *76*(4), 833–842. http://dx.doi.org/10.1124/mol.109.057356.

Ahn, K. H., Mahmoud, M. M., & Kendall, D. A. (2012). Allosteric modulator ORG27569 induces CB1 cannabinoid receptor high affinity agonist binding state, receptor internalization, and G_i protein-independent ERK1/2 kinase activation. *The Journal of Biological Chemistry*, *287*(15), 12070–12082. http://dx.doi.org/10.1074/jbc.M111.316463.

Ahuja, S., Hornak, V., Yan, E. C. Y., Syrett, N., Goncalves, J. A., Hirshfeld, A., ... Eilers, M. (2009). Helix movement is coupled to displacement of the second extracellular loop in rhodopsin activation. *Nature Structural & Molecular Biology*, *16*(2), 168–175. http://dx.doi.org/10.1038/nsmb.1549.

Atwood, B. K., Wager-Miller, J., Haskins, C., Straiker, A., & Mackie, K. (2012). Functional selectivity in CB(2) cannabinoid receptor signaling and regulation: implications for the therapeutic potential of CB(2) ligands. *Molecular Pharmacology*, *81*(2), 250–263. http://dx.doi.org/10.1124/mol.111.074013.

Avlani, V. A., Gregory, K. J., Morton, C. J., Parker, M. W., Sexton, P. M., & Christopoulos, A. (2007). Critical role for the second extracellular loop in the binding of both orthosteric and allosteric G protein-coupled receptor ligands. *The Journal of Biological Chemistry*, *282*(35), 25677–25686. http://dx.doi.org/10.1074/jbc.M702311200.

Bachhuber, M. A., Saloner, B., Cunningham, C. O., & Barry, C. L. (2014). Medical cannabis laws and opioid analgesic overdose mortality in the United States, 1999–2010. *JAMA Internal Medicine*, *174*(10), 1668–1673. http://dx.doi.org/10.1001/jamainternmed.2014.4005.

Ballesteros, J. A., & Weinstein, H. (1995). [19] Integrated methods for the construction of three-dimensional models and computational probing of structure-function relations in G protein-coupled receptors. In *Receptor molecular biology* (Vol. 25) (pp. 366–428). Elsevier. http://dx.doi.org/10.1016/S1043-9471(05)80049-7.

Bayewitch, M., Avidor-Reiss, T., Levy, R., Barg, J., Mechoulam, R., & Vogel, Z. (1995). The peripheral cannabinoid receptor: adenylate cyclase inhibition and G protein coupling. *FEBS Letters*, *375*(1–2), 143–147.

Bertalovitz, A. C., Ahn, K. H., & Kendall, D. A. (2010). Ligand binding sensitivity of the extracellular loop two of the cannabinoid receptor 1. *Drug Development Research*, *71*(7), 404–411. http://dx.doi.org/10.1002/ddr.20388.

Bokoch, M. P., Zou, Y., Rasmussen, S. G. F., Liu, C. W., Nygaard, R., Rosenbaum, D. M., ... Kobilka, B. K. (2010). Ligand-specific regulation of the extracellular surface of a G-protein-coupled receptor. *Nature*, *463*(7277), 108–112. http://dx.doi.org/10.1038/nature08650.

Bosshard, H. R., Marti, D. N., & Jelesarov, I. (2004). Protein stabilization by salt bridges: concepts, experimental approaches and clarification of some misunderstandings. *Journal of Molecular Recognition*, *17*(1), 1–16. http://dx.doi.org/10.1002/jmr.657.

Calandra, B., Portier, M., Kernéis, A., Delpech, M., Carillon, C., Le Fur, G., ... Shire, D. (1999). Dual intracellular signaling pathways mediated by the human cannabinoid CB1 receptor. *European Journal of Pharmacology*, *374*(3), 445–455. http://dx.doi.org/10.1016/S0014-2999(99)00349-0.

Castrop, H., Oppermann, M., Mizel, D., Huang, Y., Faulhaber-Walter, R., Weiss, Y., ... Schnermann, J. (2007). Skeletal abnormalities and extraskeletal ossification in mice with restricted Gsalpha deletion caused by a renin promoter-Cre transgene. *Cell and Tissue Research*, *330*(3), 487–501. http://dx.doi.org/10.1007/s00441-007-0491-6.

Childers, S. R., & Deadwyler, S. A. (1996). Role of cyclic AMP in the actions of cannabinoid receptors. *Biochemical Pharmacology*, *52*(6), 819–827.

Christopoulou, F. D., & Kiortsis, D. N. (2011). An overview of the metabolic effects of rimonabant in randomized controlled trials: potential for other cannabinoid 1 receptor blockers in obesity. *Journal of Clinical Pharmacy and Therapeutics*, *36*(1), 10–18. http://dx.doi.org/10.1111/j.1365-2710.2010.01164.x.

Conner, M., Hawtin, S. R., Simms, J., Wootten, D., Lawson, Z., Conner, A. C., ... Wheatley, M. (2007). Systematic analysis of the entire second extracellular loop of the V(1a) vasopressin receptor: key residues, conserved throughout a G-protein-coupled receptor family, identified. *The Journal of Biological Chemistry*, *282*(24), 17405–17412. http://dx.doi.org/10.1074/jbc.M702151200.

Console-Bram, L., Marcu, J., & Abood, M. E. (2012). Cannabinoid receptors: nomenclature and pharmacological principles. *Progress in Neuropsychopharmacology & Biological Psychiatry*, *38*(1), 4–15. http://dx.doi.org/10.1016/j.pnpbp.2012.02.009.

Davis, M. I., Ronesi, J., & Lovinger, D. M. (2003). A predominant role for inhibition of the adenylate cyclase/protein kinase A pathway in ERK activation by cannabinoid receptor 1 in N1E-115 neuroblastoma cells. *The Journal of Biological Chemistry, 278*(49), 48973–48980. http://dx.doi.org/10.1074/jbc.M305697200.

De Petrocellis, L., Marini, P., Matias, I., Moriello, A. S., Starowicz, K., Cristino, L., … Di Marzo, V. (2007). Mechanisms for the coupling of cannabinoid receptors to intracellular calcium mobilization in rat insulinoma beta-cells. *Experimental Cell Research, 313*(14), 2993–3004. http://dx.doi.org/10.1016/j.yexcr.2007.05.012.

Demuth, D. G., & Molleman, A. (2006). Cannabinoid signalling. *Life Sciences, 78*(6), 549–563. http://dx.doi.org/10.1016/j.lfs.2005.05.055.

Glass, M., & Felder, C. C. (1997). Concurrent stimulation of cannabinoid CB1 and dopamine D2 receptors augments cAMP accumulation in striatal neurons: evidence for a Gs linkage to the CB1 receptor. *The Journal of Neuroscience: the Official Journal of the Society for Neuroscience, 17*(14), 5327–5333.

Hampson, R. E., Evans, G. J., Mu, J., Zhuang, S. Y., King, V. C., Childers, S. R., & Deadwyler, S. A. (1995). Role of cyclic AMP dependent protein kinase in cannabinoid receptor modulation of potassium "A-current" in cultured rat hippocampal neurons. *Life Sciences, 56*(23–24), 2081–2088.

Hanuš, L. O. (2009). Pharmacological and therapeutic secrets of plant and brain (endo)cannabinoids. *Medicinal Research Reviews, 29*(2), 213–271. http://dx.doi.org/10.1002/med.20135.

Hillard, C. J., Weinlander, K. M., & Stuhr, K. L. (2012). Contributions of endocannabinoid signaling to psychiatric disorders in humans: genetic and biochemical evidence. *Neuroscience, 204*(C), 207–229. http://dx.doi.org/10.1016/j.neuroscience.2011.11.020.

Howlett, A. C. (1984). Inhibition of neuroblastoma adenylate cyclase by cannabinoid and nantradol compounds. *Life Sciences, 35*(17), 1803–1810. http://dx.doi.org/10.1016/0024-3205(84)90278-9.

Howlett, A. C., Barth, F., Bonner, T. I., Cabral, G., Casellas, P., Devane, W. A., … Pertwee, R. G. (2002). International Union of Pharmacology. XXVII. Classification of cannabinoid receptors. *Pharmacological Reviews, 54*(2), 161–202. http://dx.doi.org/10.1021/jm00038a020.

Howlett, A. C., Qualy, J. M., & Khachatrian, L. L. (1986). Involvement of Gi in the inhibition of adenylate cyclase by cannabimimetic drugs. *Molecular Pharmacology, 29*(3), 307–313.

Hurst, D. P., Lynch, D. L., Barnett-Norris, J., Hyatt, S. M., Seltzman, H. H., Zhong, M., … Reggio, P. H. (2002). N-(piperidin-1-yl)-5-(4-chlorophenyl)-1-(2,4-dichlorophenyl)-4-methyl-1H-pyrazole-3-carboxamide (SR141716A) interaction with LYS 3.28(192) is crucial for its inverse agonism at the cannabinoid CB1 receptor. *Molecular Pharmacology, 62*(6), 1274–1287.

Kapur, A., Samaniego, P., Thakur, G. A., Makriyannis, A., & Abood, M. E. (2008). Mapping the structural requirements in the CB1 cannabinoid receptor transmembrane Helix II for signal transduction. *The Journal of Pharmacology and Experimental Therapeutics, 325*(1), 341–348. http://dx.doi.org/10.1124/jpet.107.133256.

Kristiansen, K. (2004). Molecular mechanisms of ligand binding, signaling, and regulation within the superfamily of G-protein-coupled receptors: molecular modeling and mutagenesis approaches to receptor structure and function. *Pharmacology and Therapeutics, 103*(1), 21–80. http://dx.doi.org/10.1016/j.pharmthera.2004.05.002.

Kumar, S., & Nussinov, R. (2002). Close-range electrostatic interactions in proteins. *Chembiochem: A European Journal of Chemical Biology, 3*(7), 604–617. http://dx.doi.org/10.1002/1439-7633(20020703)3:7<604::CBI C604>3.0.CO;2-X.

Lauckner, J. E., Hille, B., & Mackie, K. (2005). The cannabinoid agonist WIN55,212-2 increases intracellular calcium via CB1 receptor coupling to Gq/11 G proteins. *Proceedings of the National Academy of Sciences of the United States of America, 102*(52), 19144–19149. http://dx.doi.org/10.1073/pnas.0509588102.

Ledent, C., Valverde, O., Cossu, G., Petitet, F., Aubert, J. F., Beslot, F., … Parmentier, M. (1999). Unresponsiveness to cannabinoids and reduced addictive effects of opiates in CB1 receptor knockout mice. *Science (New York, NY), 283*(5400), 401–404. http://dx.doi.org/10.1126/science.283.5400.401.

Mackie, K., Devane, W. A., & Hille, B. (1993). Anandamide, an endogenous cannabinoid, inhibits calcium currents as a partial agonist in N18 neuroblastoma cells. *Molecular Pharmacology, 44*(3), 498–503.

Mackie, K., & Hille, B. (1992). Cannabinoids inhibit N-type calcium channels in neuroblastoma-glioma cells. *Proceedings of the National Academy of Sciences of the United States of America, 89*(9), 3825–3829. http://dx.doi.org/10.1073/pnas.89.9.3825.

Mackie, K., Lai, Y., Westenbroek, R., & Mitchell, R. (1995). Cannabinoids activate an inwardly rectifying potassium conductance and inhibit Q-type calcium currents in AtT20 cells transfected with rat brain cannabinoid receptor. *The Journal of Neuroscience: The Official Journal of the Society for Neuroscience, 15*(10), 6552–6561.

Marcu, J., Shore, D. M., Kapur, A., Trznadel, M., Makriyannis, A., Reggio, P. H., & Abood, M. E. (2013). Novel insights into CB1 cannabinoid receptor signaling: a key interaction identified between the extracellular-3 loop and transmembrane helix 2. *The Journal of Pharmacology and Experimental Therapeutics, 345*(2), 189–197. http://dx.doi.org/10.1124/jpet.112.201046.

Matsuda, L. A., Lolait, S. J., Brownstein, M. J., Young, A. C., & Bonner, T. I. (1990). Structure of a cannabinoid receptor and functional expression of the cloned cDNA. *Nature, 346*(6284), 561–564. http://dx.doi.org/10.1038/346561a0.

McAllister, S. D., Griffin, G., Satin, L. S., & Abood, M. E. (1999). Cannabinoid receptors can activate and inhibit G protein-coupled inwardly rectifying potassium channels in a xenopus oocyte expression system. *The Journal of Pharmacology and Experimental Therapeutics, 291*(2), 618–626.

McAllister, S. D., Hurst, D. P., Barnett-Norris, J., Lynch, D., Reggio, P. H., & Abood, M. E. (2004). Structural mimicry in class A G protein-coupled receptor rotamer toggle switches: the importance of the F3.36(201)/W6.48(357) interaction in cannabinoid CB1 receptor activation. *The Journal of Biological Chemistry, 279*(46), 48024–48037. http://dx.doi.org/10.1074/jbc.M406648200.

McAllister, S. D., Rizvi, G., Anavi-Goffer, S., Hurst, D. P., Barnett-Norris, J., Lynch, D. L., … Abood, M. E. (2003). An aromatic microdomain at the cannabinoid CB(1) receptor constitutes an agonist/inverse agonist binding region. *Journal of Medicinal Chemistry, 46*(24), 5139–5152. http://dx.doi.org/10.1021/jm0302647.

Murphy, J. W., & Kendall, D. A. (2003). Integrity of extracellular loop 1 of the human cannabinoid receptor 1 is critical for high-affinity binding of the ligand CP 55,940 but not SR 141716A. *Biochemical Pharmacology, 65*(10), 1623–1631. http://dx.doi.org/10.1016/S0006-2952(03)00155-2.

Mu, J., Zhuang, S. Y., Hampson, R. E., & Deadwyler, S. A. (2000). Protein kinase-dependent phosphorylation and cannabinoid receptor modulation of potassium A current (IA) in cultured rat hippocampal neurons. *Pflügers Archiv: European Journal of Physiology, 439*(5), 541–546.

Ogata, N., Kawaguchi, H., Chung, U.-I., Roth, S. I., & Segre, G. V. (2007). Continuous activation of G alpha q in osteoblasts results in osteopenia through impaired osteoblast differentiation. *The Journal of Biological Chemistry, 282*(49), 35757–35764. http://dx.doi.org/10.1074/jbc.M611902200.

Pacher, P. (2006). The endocannabinoid system as an emerging target of pharmacotherapy. *Pharmacological Reviews, 58*(3), 389–462. http://dx.doi.org/10.1124/pr.58.3.2.

Peeters, M. C., van Westen, G. J. P., Guo, D., Wisse, L. E., Müller, C. E., Beukers, M. W., & Ijzerman, A. P. (2011). GPCR structure and activation: an essential role for the first extracellular loop in activating the adenosine A2B receptor. *FASEB Journal, 25*(2), 632–643. http://dx.doi.org/10.1096/fj.10-164319.

Pertwee, R. (2014). *Handbook of cannabis*. Oxford University Press.

Pertwee, R. G. (2005). *Cannabinoids. Handbook of experimental pharmacology*.

Pertwee, R. G. (2009). Emerging strategies for exploiting cannabinoid receptor agonists as medicines. *British Journal of Pharmacology, 156*(3), 397–411. http://dx.doi.org/10.1111/j.1476-5381.2008.00048.x.

Picone, R. P., Fournier, D. J., & Makriyannis, A. (2002). Ligand based structural studies of the CB1 cannabinoid receptor. *The Journal of Peptide Research: Official Journal of the American Peptide Society, 60*(6), 348–356.

Price, M. R. (2005). Allosteric modulation of the cannabinoid CB1 receptor. *Molecular Pharmacology, 68*(5), 1484–1495. http://dx.doi.org/10.1124/mol.105.016162.

Rhee, M.-H. (2002). Functional role of serine residues of transmembrane dopamin VII in signal transduction of CB2 cannabinoid receptor. *Journal of Veterinary Science, 3*(3), 185–191.

Roche, P. J., Hoare, S. A., & Parker, M. G. (1992). A consensus DNA-binding site for the androgen receptor. *Molecular Endocrinology (Baltimore, MD), 6*(12), 2229–2235. http://dx.doi.org/10.1210/mend.6.12.1491700.

Ross, F. P. (2005). Lipid links to better bone: a hypothesis. *Cell Metabolism, 2*(1), 2–4. http://dx.doi.org/10.1016/j.cmet.2005.06.003.

Russo, E. B. (2008a). Cannabinoids in the management of difficult to treat pain. *Therapeutics and Clinical Risk Management, 4*(1), 245–259.

Russo, E. B. (2008b). Clinical endocannabinoid deficiency (CECD): can this concept explain therapeutic benefits of cannabis in migraine, fibromyalgia, irritable bowel syndrome and other treatment-resistant conditions? *Neuro Endocrinology Letters, 29*(2), 192–200.

Russo, E. B. (2011). Taming THC: potential cannabis synergy and phytocannabinoid-terpenoid entourage effects. *British Journal of Pharmacology, 163*(7), 1344–1364. http://dx.doi.org/10.1111/j.1476-5381.2011.01238.x.

Sakamoto, A., Chen, M., Kobayashi, T., Kronenberg, H. M., & Weinstein, L. S. (2005). Chondrocyte-specific knockout of the G protein G(s) alpha leads to epiphyseal and growth plate abnormalities and ectopic chondrocyte formation. *Journal of Bone and Mineral Research, 20*(4), 663–671. http://dx.doi.org/10.1359/JBMR.041210.

Sakamoto, A., Chen, M., Nakamura, T., Xie, T., Karsenty, G., & Weinstein, L. S. (2005). Deficiency of the G-protein alpha-subunit G(s)alpha in osteoblasts leads to differential effects on trabecular and cortical bone. *The Journal of Biological Chemistry, 280*(22), 21369–21375. http://dx.doi.org/10.1074/jbc.M500346200.

Sealfon, S. C., Chi, L., Ebersole, B. J., Rodic, V., Zhang, D., Ballesteros, J. A., & Weinstein, H. (1995). Related contribution of specific helix 2 and 7 residues to conformational activation of the serotonin 5-HT2A receptor. *The Journal of Biological Chemistry, 270*(28), 16683–16688.

Shore, D. M., Baillie, G. L., Hurst, D. H., Navas, F., Seltzman, H. H., Marcu, J. P., ... Reggio, P. H. (2014). Allosteric modulation of a cannabinoid G protein-coupled receptor: binding site elucidation and relationship to G protein signaling. *The Journal of Biological Chemistry, 289*(9), 5828–5845. http://dx.doi.org/10.1074/jbc.M113.478495.

Song, Z. H., & Bonner, T. I. (1996). A lysine residue of the cannabinoid receptor is critical for receptor recognition by several agonists but not WIN55212-2. *Molecular Pharmacology, 49*(5), 891–896.

Sugiura, T., Kodaka, T., Kondo, S., Tonegawa, T., Nakane, S., Kishimoto, S., ... Waku, K. (1996). 2-Arachidonoylglycerol, a putative endogenous cannabinoid receptor ligand, induces rapid, transient elevation of intracellular free Ca2+ in neuroblastoma×glioma hybrid NG108-15 cells. *Biochemical and Biophysical Research Communications, 229*(1), 58–64. http://dx.doi.org/10.1006/bbrc.1996.1757.

Sugiura, T., Kodaka, T., Kondo, S., Tonegawa, T., Nakane, S., Kishimoto, S., ... Waku, K. (1997). Inhibition by 2-arachidonoylglycerol, a novel type of possible neuromodulator, of the depolarization-induced increase in intracellular free calcium in neuroblastoma x glioma hybrid NG108-15 cells. *Biochemical and Biophysical Research Communications, 233*(1), 207–210. http://dx.doi.org/10.1006/bbrc.1997.6425.

Tao, Q., McAllister, S. D., Andreassi, J., Nowell, K. W., Cabral, G. A., Hurst, D. P., ... Abood, M. E. (1999). Role of a conserved lysine residue in the peripheral cannabinoid receptor (CB2): evidence for subtype specificity. *Molecular Pharmacology, 55*(3), 605–613.

Turu, G., & Hunyady, L. (2010). Signal transduction of the CB1 cannabinoid receptor. *Journal of Molecular Endocrinology, 44*(2), 75–85. http://dx.doi.org/10.1677/JME-08-0190.

Wess, J., Nanavati, S., Vogel, Z., & Maggio, R. (1993). Functional role of proline and tryptophan residues highly conserved among G protein-coupled receptors studied by mutational analysis of the m3 muscarinic receptor. *The EMBO Journal, 12*(1), 331–338.

Wu, M., Deng, L., Zhu, G., & Li, Y.-P. (2010). G Protein and its signaling pathway in bone development and disease. *Frontiers in Bioscience (Landmark Edition), 15*, 957–985.

Xu, H., Lu, Y. F., Partilla, J. S., Zheng, Q. X., Wang, J. B., Brine, G. A., ... Rothman, R. B. (1999). Opioid peptide receptor studies, 11: involvement of Tyr148, Trp318 and His319 of the rat mu-opioid receptor in binding of mu-selective ligands. *Synapse (New York, NY), 32*(1), 23–28. http://dx.doi.org/10.1002/(SICI)1098-2396(199904)32:1<23::SYN3>3.0.CO;2-N.

Zhou, W., Flanagan, C., Ballesteros, J. A., Konvicka, K., Davidson, J. S., Weinstein, H., ... Sealfon, S. C. (1994). A reciprocal mutation supports helix 2 and helix 7 proximity in the gonadotropin-releasing hormone receptor. *Molecular Pharmacology, 45*(2), 165–170.

Zoratti, C., Kipmen-Korgun, D., Osibow, K., Malli, R., & Graier, W. F. (2003). Anandamide initiates Ca(2+) signaling via CB2 receptor linked to phospholipase C in calf pulmonary endothelial cells. *British Journal of Pharmacology, 140*(8), 1351–1362. http://dx.doi.org/10.1038/sj.bjp.0705529.

Chapter 67

Does Agonist Efficacy Alter the In Vivo Effects of Cannabinoids: Girl, Could We Get Much "Higher"?

Roger S. Gifford[1,2], Jimit G. Raghav[1,2], Torbjörn U.C. Järbe[1,2]

[1]Center for Drug Discovery (CDD), Northeastern University, Boston, MA, USA; [2]Department of Pharmaceutical Sciences, Northeastern University, Boston, MA, USA

Abbreviations

2-AG 2-Arachidonoylglycerol
11-OH 11-Hydroxy
AEA Arachidonoylethanolamine/anandamide
CB1R Cannabinoid 1 receptor
CB2R Cannabinoid 2 receptor
DEA Drug Enforcement Agency
ED$_{50}$ Median effective dose
G$_i$ G-protein that inhibits cAMP
G$_s$ G-protein that stimulates cAMP
GABA γ-Aminobutyric acid
GPCR G-protein-coupled receptor
GTPγS Guanosine 5'-O-[γ-thio]triphosphate
ICSS Intracranial self-stimulation
NIDA National Institute of Drug Abuse
SAMHSA Substance Abuse and Mental Health Services Administration
SAR Structure–activity relationship
Δ^8-THC (−)-*trans*-Δ^8-Tetrahydrocannabinol
Δ^9-THC (−)-*trans*-Δ^9-Tetrahydrocannabinol

INTRODUCTION

(−)-*trans*-Δ^9-Tetrahydrocannabinol (Δ^9-THC), the main psychoactive ingredient in marijuana, has potent psychoactive effects on users, with sedation and altered cognition produced by high doses. Marijuana has been used for millennia, and Substance Abuse and Mental Health Services Administration (SAMHSA) survey data stated that marijuana is the most widely used illicit drug in the United States, with 7.5% of the population over 12 reporting use in the past month (SAMHSA, 2014). But new synthetic recreational drugs that target the cannabinoid 1 receptor (CB1R) have sprung up on the market, and their effects are still not well understood. Such recreational compounds will collectively be referred to as "Spice" in this chapter, irrespective of their chemical class or pharmacological profile once the compounds were detected and made illegal by law enforcement measures. Most of marijuana's effects are mediated through the CB1R, a G-protein-coupled receptor found throughout much of the brain. Most of these Spice compounds act as full agonists at CB1R, and Δ^9-THC is a partial agonist at CB1R, producing roughly 40% of the in vitro maximal effect of a full agonist such as CP-55,940 (Brents et al., 2012). This has raised concern that the Spice compounds may produce effects much stronger than those of Δ^9-THC, and case reports have been made of kidney failure, seizures, and death, symptoms rarely if ever seen with marijuana or Δ^9-THC (Behonick et al., 2014; Bhanushali, Jain, Fatima, Leisch, & Thornley-Brown, 2013; Harris & Brown, 2013). In the laboratory, though, most in vivo experiments have not found clear differences regarding partial versus full CB1R agonists beyond potency, i.e., the effects of full agonists could be replicated by a partial agonist at high enough doses (Fan, Compton, Ward, Melvin, & Martin, 1994). Before the rise of Spice abuse around 2005, very little was known about the recreational use of CB1R full agonists. But since then, Spice use has quickly grown into one of the most commonly abused drugs worldwide, with 8% of American high school seniors reporting using it in 2012 (Johnston, O'Malley, Bachman, & Schulenberg, 2013). To evaluate whether Spice's higher efficacy is responsible for its potential toxicity we need to review the differences in vivo between full and partial CB1R agonists.

The endocannabinoid system (ECS) is composed of two receptors, CB1R and CB2R. Of these, CB1R is responsible for the majority of the psychotropic effects of cannabinoid drugs. CB2R is mainly involved with immune function and will not be covered in this chapter. CB1R predominantly works via the G$_i$ pathway, leading to inhibition of cAMP. It is at this level that we define partial and full agonist, based upon their maximal effect upon activation of CB1R. Partial agonists are purportedly limited in their maximal effect compared to full agonists (Childers, 2006). A hypothetical example is shown in Figure 1.

CANNABINOID AGONIST CLASSIFICATION

According to the International Union of Basic and Clinical Pharmacology (Pertwee et al., 2010), cannabinoid agonists currently

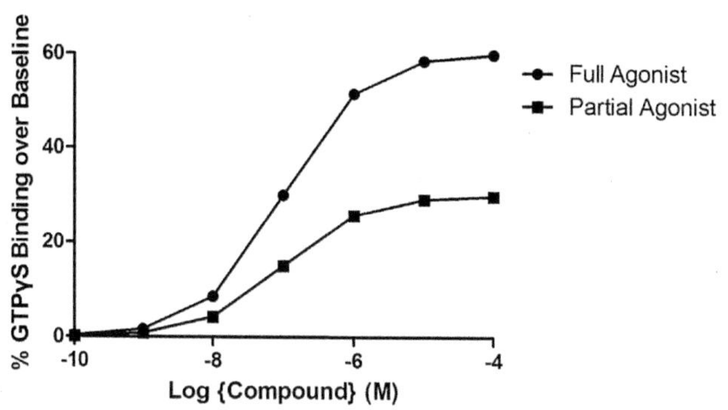

FIGURE 1 Example of efficacy differences at CB1R. Hypothetical representation of outcomes of full and partial agonists using an in vitro guanosine 5'-O-[γ-thio]triphosphate (GTPγS) receptor binding assay. Represented compounds have the same potency but different efficacies.

used for medicinal research and recreational use can be classified into four main groups as follows:

1. Classical cannabinoids: The compounds falling into this category have a dibenzopyran ring system. Δ^9-THC, the major psychoactive constituent of the marijuana plant, is the standard example. Other examples include Δ^8-THC, a minor isomer of marijuana that produces effects similar to but less potent than those of Δ^9-THC. The Δ^8-THC isomer is more chemically stable than the Δ^9-THC isomer and hence often used as the template for creating new ligands affecting the ECS (Jarbe & Gifford, 2014). HU-210 is a synthetic Δ^8-THC-derived cannabinoid that also falls in this category. HU-210 was the most potent cannabinoid compound synthesized at its creation, being 100–800 times more potent than Δ^9-THC, with a slow onset of effect but a long duration of action (Jarbe, Hiltunen, & Mechoulam, 1989; Mechoulam et al., 1988). HU-210 was classified as a Schedule I compound in 2008 after the US Drug Enforcement Agency found HU-210 in confiscated batches of Spice across the United States.

2. Nonclassical cannabinoids: Compounds in this class do not carry pyran rings and most of them are bicyclic ring compounds. CP-55,940 was the first compound synthesized in this class by the drug company Pfizer in 1974 (Melvin & Johnson, 1987), and it is still one of the most widely used compounds in the field of cannabinoid research. CP-47,497, a compound structurally similar to CP-55,940, was also synthesized by Pfizer (Weissman, Milne, & Melvin, 1982). This compound is more readily synthesized and later found its way into clandestine laboratories used for Spice production. CP-47,497 was placed on Schedule I in 2008.

3. Cannabinergic indoles: The compounds in this class have chemical structures different from the above cannabinoid agonists. WIN 55,212-2 was one of the first CB1R agonists to be synthesized in this class (Ward et al., 1990). Following this discovery, the National Institute of Drug Abuse sponsored further structure–activity relationship exploration of this class of compounds in collaboration with various investigators, resulting in a plethora of potent aminoalkylindoles and related indoles. JWH-018 (also known as AM678) was also an early compound in the aminoalkylindole series, and because of its potency at CB1R and the ease of synthesis, the compound made its way onto the streets as one of the earliest discovered Spice compounds (Atwood, Huffman, Straiker, & Mackie, 2010; Wiley et al., 1998). This eventually led to its ban in most countries of the world. Other potent CB1R agonists in this class that are now classified as Schedule I include JWH-073, AM2201, XLR-11, and UR-144.

4. Eicosanoids and derivatives: The two best known compounds in this class are arachidonoylethanolamine/anandamide (AEA) and 2-arachidonoylglycerol (2-AG), representing some of the body's endogenous cannabinoids. These compounds will not be discussed much in this chapter as, while they do activate CB1R (with AEA acting as a partial agonist and 2-AG as a full agonist), their pharmacological profiles do not completely match that of other cannabinoid drugs, and this extends to synthetic AEA analogues such as AM356 and AM1346 (Jarbe, Li, Liu, & Makriyannis, 2009; McMahon, Ginsburg, & Lamb, 2008).

The chemical structures of representative cannabinoid ligands from the above four classes are illustrated in Figure 2.

HERBAL INCENSE: NOT YOUR PARENTS' MARIJUANA?

While there has been extensive work, both clinical and observational, on the effects of marijuana and Δ^9-THC, very little was known about the compounds found in Spice until after reports of emergency room visits started surfacing. As such compounds had not originally been intended for use in clinical trials, no scientific data on the effects in humans were known and initially only a few in vivo animal studies had been performed. This lack of laboratory data precluded easy comparisons with marijuana and forced the study of Spice in humans to heavily rely on anecdotal reports of individuals who have used Spice, usually after their admission to a rehabilitation facility or a hospital.

Spice products are marketed in head shops all over the world, labeled as "not for human consumption." Various case studies reveal that the psychotropic effects of Spice can involve cases in which people experienced extreme paranoia along with panic attacks, while other anecdotal studies reported that the "high" obtained from Spice products is fairly similar to that obtained from smoking marijuana (Fattore & Fratta, 2011). Many of the users of Spice report that the onset of action of most Spice concoctions is faster than that of marijuana and the duration of the high is typically shorter compared to the marijuana-induced high. This may have some bearing on the sequence of tolerance development and dependence.

One 2013 survey concluded that there is a difference in the intensity of the highs caused by marijuana and Spice. In this study ($n=14,966$), 17% of the sample reported using Spice, and 99% of all the subjects using Spice in the study reported being regular users of marijuana. The study concluded that all the surveyed marijuana users prefer marijuana over Spice (Winstock & Barratt, 2013). The main reason for their use of Spice was that it may go unnoticed in drug tests, whereas marijuana/Δ^9-THC does not. Finally, rarely do the users or the treatment providers know what dose, let alone what drug, is actually being taken, as labeling is scant to nonexistent. Overall, the anecdotal reports highlight side effects rarely seen with marijuana. Since Spice compounds are

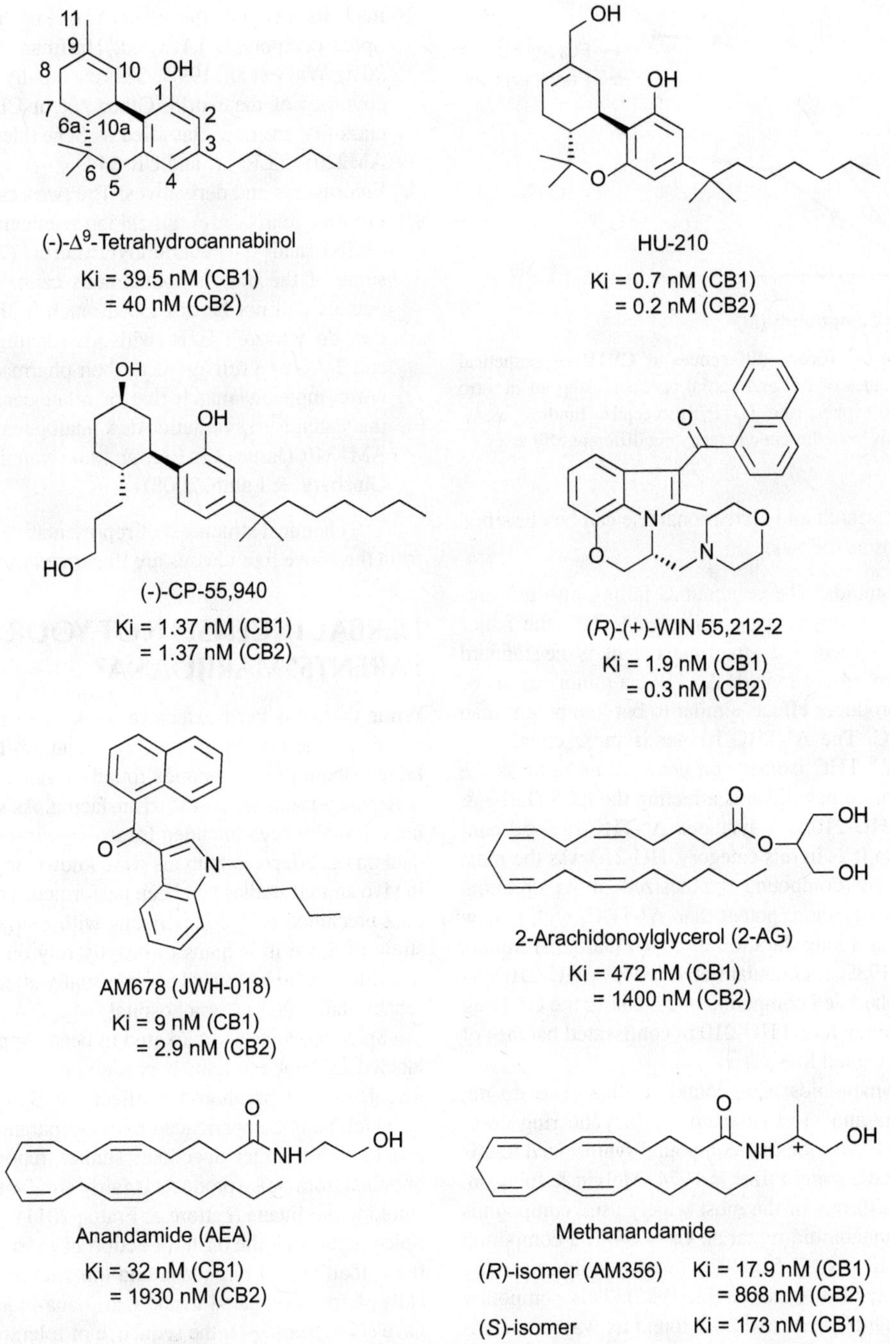

FIGURE 2 Structures of various cannabinoids from different classes. Chemical structures of Δ^9-THC and other ligands with affinity (Ki) for CB1R. AEA and 2-AG represent endogenous cannabinoids and AM356/methanandamide is a more stable analog of AEA.

almost entirely full CB1R agonists, it has been speculated that this may be the reason for the discrepancy between Spice products and marijuana/Δ^9-THC.

PRECLINICAL STUDIES

Almost all information about Spice use in humans is gained from recreational users' personal experiences, so objective data on Spice have to rely upon animal models. Very few studies have used Spice compounds though, as alternative synthetic full agonists were already in use in laboratories. Thus, most research on cannabinoid full agonists has used compounds that were not necessarily ever found in Spice. A common model used to test cannabinoid drugs is the "tetrad." By testing hypothermia, locomotor effects, analgesia, and catalepsy, a general picture of a cannabinoid's effects can be gained. Many early studies with full agonists, though, failed to find much difference between full agonists and Δ^9-THC (Fan et al., 1994). Paronis, Nikas, Shukla, and

Makriyannis (2012), however, reported that Δ^9-THC's maximal effect in a hypothermia assay was less than that of the full CB1R agonist AM2389. More importantly, the researchers also found that Δ^9-THC attenuated the hypothermia produced by AM2389 when the animals were pretreated with Δ^9-THC (Paronis et al., 2012). This antagonism could be overcome by increasing the AM2389 dose, thus displaying a theoretical hallmark effect of partial agonists combined with full agonists. One has to be careful, though, as other experiments failed to find this difference, and our own laboratory (unpublished) has found that the maximal effect of Δ^9-THC in hypothermia was on par with those of two full agonists (CP-55,940 and AM5983) and that problems of solubility limit the effectiveness of Δ^9-THC at the highest doses. In other tests of the tetrad, efficacy does not produce reliable differences of effect (Fan et al., 1994; Marshell et al., 2014). That Δ^9-THC will act inconsistently as a partial agonist in vivo may be due to uncontrolled variables, but what these conditions are is not known. Differences in species, handling, and previous drug history can all affect the results, and so what differences exist have not been ascertained. Since most of the experiments were carried out with varying methods, direct comparisons are difficult but suggestive that efficacy may be important under certain (unknown) conditions.

MARIJUANA DEPENDENCE STUDIES IN HUMANS

Marijuana, unlike other drugs of abuse such as morphine, barbiturates, and alcohol, is considered lacking a strong physical dependence component. For much of the early twentieth century it was widely believed that marijuana ingestion did not result in physical dependence. However, studies starting in the 1970s reported that people who consume marijuana chronically did indeed seek treatment for dependence (Haney, Ward, Comer, Foltin, & Fischman, 1999a). A few characteristics of marijuana dependence shared with other drugs of abuse is compulsive seeking behavior, limited self-control, and relapsing back to marijuana consumption despite the knowledge of the potential harm done by such dependence behavior (Maldonado, Berrendero, Ozaita, & Robledo, 2011). The dependence produced by marijuana is considered relatively mild compared to other drugs of abuse. Anxiety, insomnia, anorexia, and increased irritability are some of the symptoms associated with marijuana dependence (Gonzalez, Cebeira, & Fernandez-Ruiz, 2005). Some reports based on Δ^9-THC studies in humans strongly support the existence of such symptoms. A study with marijuana smokers was performed to elucidate withdrawal symptoms. In this study, subjects showed symptoms of withdrawal, which included increased anxiety, increased aggression, and reduction of appetite along with stomach pains (Haney, Ward, Comer, Foltin, & Fischman, 1999b). The general consensus about the time course of withdrawal seems to be that the withdrawal symptoms appear 24h post-marijuana cessation, and the peak of these withdrawal symptoms lasts 1–4 days. The symptoms seem to dissipate within 1–2 weeks of abstinence (Cooper & Haney, 2008). There have also been case reports of people experiencing withdrawal from Spice, with symptoms appearing similar to those of marijuana withdrawal (Zimmermann et al., 2009).

DEPENDENCE AND WITHDRAWAL STUDIES IN ANIMALS

Δ^9-THC and most other cannabinoids to date are fat soluble, which is responsible for a "depot effect," wherein the drug is stored in fat tissues of the body and slowly released over time, providing a sustained release that makes withdrawal symptoms difficult to observe (Tanda & Goldberg, 2003). A widely used technique to elicit withdrawal and examine dependence in animals is by administering a CB1R antagonist (e.g., rimonabant) to animals chronically exposed to a cannabinoid agonist and thereby producing a state of withdrawal (Rodriguez de Fonseca, Carrera, Navarro, Koob, & Weiss, 1997). As of this writing there is no study available for cannabinoid dependence and withdrawal using Spice compounds such as JWH-018 or AM2201, although they have been studied with the aminoalkylindole WIN 55,212-2, wherein it produced effects similar to those of Δ^9-THC withdrawal (Aceto, Scates, & Martin, 2001).

TOLERANCE

Tolerance to cannabinoid agonists is well documented and can be formed in less than a week and may produce tolerance necessitating 50× more drug to achieve the same effect (Desai et al., 2013). While all reported cannabinoid agonists produce tolerance, one reason partial vs full CB1R agonists may differ in their effects lies in whether they produce the same amount of tolerance. One theory for why full agonists do not differ from partial agonists is that the receptor pool for CB1R is so large that maximal effect is achieved before saturation of all receptors (Gifford et al., 1999). As one cause of tolerance is downregulation of available receptors, the pool of receptors may decrease to the point at which high and low efficacy agonists may exert different maximal effects. Figure 3 shows an example of what this situation may look like. Data from Hruba, Ginsburg, and McMahon (2012) showed that chronic Δ^9-THC for 14 days significantly shifted the ED_{50} of Δ^9-THC drug discrimination for Δ^9-THC, CP-55,940, JWH-018, and JWH-073 by a factor of 9.2, 3.6, 4.3, and 5.6, respectively. Of these, both CP-55,940 and JWH-018's potency ratios were significantly different from that of Δ^9-THC. While the JWH-073 potency ratio difference did not reach significance compared to Δ^9-THC, the data are still supportive that tolerance for Δ^9-THC increased more than for the full agonist. This is not definitive, though, as the stronger tolerance to Δ^9-THC may just be due to Δ^9-THC being used to induce the tolerance, and other partial CB1R agonists were not examined. In support of this suggestion, other researchers (Desai et al., 2013) found that chronic administration of AM411, a partial CB1R agonist, did not produce differences in tolerance for Δ^9-THC, AM4054, and WIN 55,212, the last two of which are full CB1R agonists. These discrepancies highlight that induction of tolerance is highly dependent on both the protocol and the compound used, as tolerance is really a combination of various factors, from receptor downregulation to psychological mechanisms.

DRUG DISCRIMINATION

One way of measuring the subjective effects of drugs is through drug discrimination. Drug discrimination is a procedure by which animals are trained to perform a certain response to receive

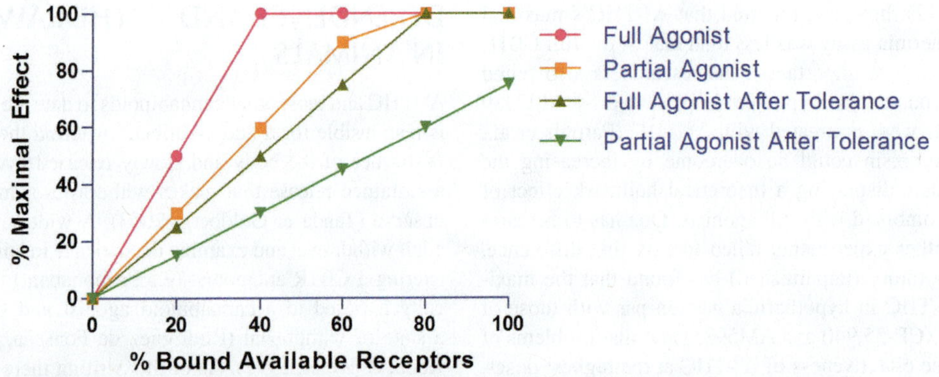

FIGURE 3 Receptor pool interactions with full and partial agonists. Example of how a large pool of receptors may cause a full agonist and a partial agonist to have the same maximal effect. The figure also includes an example of "tolerance" in which half the receptors have been lost, demonstrating how the partial agonist may differ from the full agonist under such circumstances.

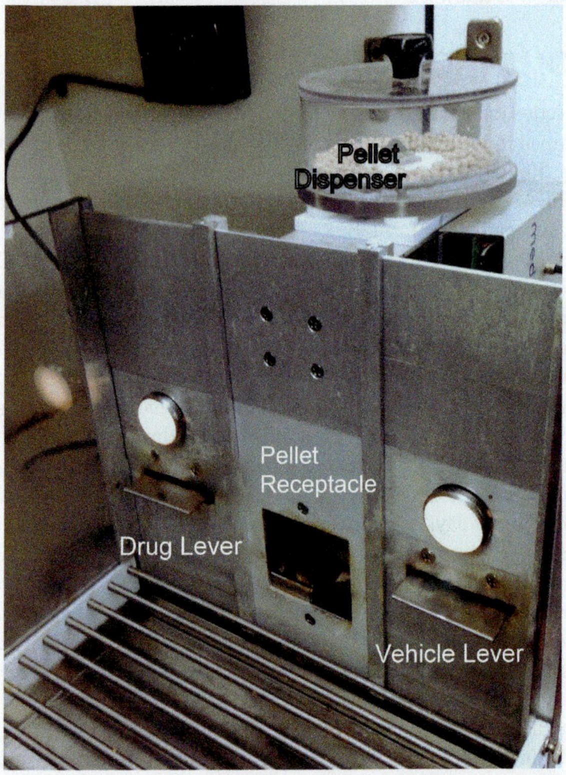

FIGURE 4 Example drug discrimination chamber. Pictured is an operant chamber used for drug discrimination. By pressing one of the two levers during training sessions and being rewarded with food pellets, animals learn to associate the respective lever with a certain condition (in this case left is the training drug and right is the placebo).

reinforcement after presentation of a specific stimulus, in this case either the presence or the lack of a certain drug. By training animals to recognize the effects of a specific drug, their response to another one can be measured to determine if it differs. Figure 4 shows an operant chamber used in such drug discrimination tests. Tests using both partial and full agonists as the training drug have demonstrated that animals generalize their response to the training drug to include other cannabinoid agonists, indicative that the subjective effects of CB1R agonists are all relatively similar (Jarbe, Deng, Vadivel, & Makriyannis, 2011; Jarbe, Li, Vadivel, & Makriyannis, 2010; Rodriguez & McMahon, 2014). To examine efficacy, Jarbe et al. (2012) took this a step further by using animals trained to discriminate between doses of AM5983, a full CB1R agonist. Theorizing that different doses of a full agonist may produce differing levels of tolerance, they found that while the EC_{50} for all compounds increased proportionally with training dose, the EC_{50} of the partial agonists Δ^9-THC and AM356 increased more rapidly with a higher training dose in comparison to the full agonist aminoalkylindole. This was attributed to the increased tolerance generated by regular exposure to a full agonist, which may help to whittle down the pool of CB1R available, requiring that partial agonists be given in proportionally greater doses to achieve similar levels of effect. Other researchers reported a similar outcome comparing the ED_{50} dose generalization curves for Δ^9-THC and CP-55,940 at two training doses of the full agonist CP-55,940 (De Vry & Jentzsch, 2003).

REINFORCEMENT

Research into the reinforcing value of cannabinoid drugs has been somewhat ambiguous, as many studies have reported inconsistent findings, some reporting no effect or negative reinforcement, while others found them to provide reinforcement. While the exact effect seems to be species/strain, dose, and prior history specific, there is strong evidence that cannabinoids play a role in reward processing and that they can be reinforcing under certain conditions. By reviewing the literature on reward and cannabinoids, Panagis, Mackey, and Vlachou (2014) found that, while there is discrepancy in the literature, the overall trends seem to indicate some differences between Δ^9-THC and synthetic full CB1R agonist compounds. For one, Δ^9-THC self-administration in naïve animals was accomplished only in 2013, as reviewed elsewhere (Justinova et al., 2013). By using low-dose intravenous perfusions into squirrel monkeys, reliably increased rates of Δ^9-THC self-administration were achieved compared to placebo sessions. So far, though, this feat has been accomplished only using squirrel monkeys and also in only one laboratory thus far. All other studies that reported Δ^9-THC self-administration included histories of either other drug exposure or severe food restriction. By using a synthetic full agonist like WIN 55,212-2, other laboratories have managed to produce synthetic cannabinoid self-administration in both rats and mice, although it still remains difficult. Interestingly, rats self-administering WIN 55,212-2 did not sustain drug self-administration when THC was substituted for WIN 55,212-2 (Lefever, Marusich, Antonazzo, & Wiley, 2014). The failure to produce Δ^9-THC self-administration in rats, and the apparent difficulty involved in getting Δ^9-THC self-administration in any nonhuman model, is suggestive that full agonist synthetic cannabinoids may possess more reinforcing qualities than partial agonists, although discrepancies with other reinforcement models suggest something else may be occurring.

Conditioned place preference/avoidance assays are purported to reflect the inherent reinforcing qualities of a drug, such that rewarding drugs produce conditioned place preference and unpleasant ones produce aversion. Studies of cannabinoids find both outcomes, with both Δ^9-THC and full agonists producing place preference and aversion, depending on the exact protocol used (Panagis et al., 2014). For example, one research group (Hyatt & Fantegrossi, 2014) was able to produce conditioned place preference with Δ^9-THC at low doses and aversion at higher doses, but JWH-018 produced only conditioned place aversion at all doses tested. This could be overcome, though, by having the animals receive repeated exposure to Δ^9-THC. This caused low doses of JWH-018 to instead produce conditioned place preference. This is important, as it indicates that cannabinoid-naïve individuals may find Spice overtly strong and unpleasant and that only those with history of marijuana use may find it enjoyable.

One other approach to studying reward is by intracranial self-stimulation (ICSS), wherein an electrode is placed into the reward circuits of the brain and animals learn to perform a certain action to receive an electrical burst from the electrode. Typically, administration of rewarding drugs acts to lower the threshold of electrical stimulation necessary to act as reinforcement, while unpleasant/aversive drugs raise the threshold. Studies using Δ^9-THC show variable action, with it often acting to lower the threshold at low doses, but will instead increase the threshold at higher doses. Studies using full CB1R agonist compounds, though, were able only to raise the threshold for ICSS, with no documented case of actually lowering the threshold (Panagis et al., 2014). This appears counterintuitive to the fact that full agonists more readily act as reinforcers in self-administration studies.

IF NOT EFFICACY, WHAT, THEN?

While we have focused on efficacy as the underlying reason Spice may be more dangerous, there are numerous differences between Δ^9-THC/marijuana and Spice products that may be responsible for the discrepancies between the reported effects. One possibility is off-target effects, including those of possible metabolites, which have only lately begun to be investigated. For example, research (Brents et al., 2012) has shown that JWH-073 has multiple active metabolites that act as CB1R partial agonists, as well as one that acts as a CB1R antagonist. These partial agonists and the antagonist would muddy the effects in vivo, making comparisons difficult. This is true even of Δ^9-THC, whose main metabolite, 11-OH-Δ^9-THC, has been shown to be a full agonist (Matsuda, Lolait, Brownstein, Young, & Bonner, 1990). Whether this is the reason for problems separating Δ^9-THC's effects from those of full agonists is not known. Possibly the best way to study this would be to use controlled deactivation cannabinoids, where CB1R agonists are specifically designed to have inactive metabolites, allowing for a better separation of the effects of full and partial agonists (Nikas et al., 2015; Sharma et al., 2013).

In addition to metabolites, functional selectivity of compounds may play a role. For example, Laprairie and colleagues tested six distinct cannabinoids from various chemical classes and found differing patterns of second-messenger recruitment for each, with CP-55,940 even producing some G_s activity, which usually produces a reaction opposite the standard G_i activity. In addition to second messengers, each ligand was also found to promote various amounts of arrestin to associate with CB1R, which is important in tolerance development and may explain some differences in tolerance production (Laprairie, Bagher, Kelly, Dupre, & Denovan-Wright, 2014).

In addition to differences in possible pharmacological targets/effects, the pharmacokinetics of the drug, including its method of administration, is very important. Nabilone is one of the two approved cannabinoid drugs in the United States, and since its approval, nabilone has been shown to act as a full agonist at CB1R. Its side effect profile typically mirrors that of dronabinol (i.e., Δ^9-THC), and there does not appear to be much difference in their safety profile. A major difference between nabilone and recreational synthetic cannabinoids, though, is its mode of administration. Interestingly, a review of potential abuse of nabilone found recreational use of nabilone to be rare (Ware & St Arnaud-Trempe, 2010). Users found nabilone to have more side effects and longer onset of effect than smoked marijuana. Possibly the side effects are due to the higher efficacy, but the lack of reported abuse is also likely to be related to its mode of administration. For further demonstration that mode of administration is important, see Marshell et al. (2014), who found that inhaled versus intraperitoneal (IP) injections of doses of JWH-018, JWH-073, and Δ^9-THC displayed different effects. Ignoring potency differences, one sees that inhalation produces none of the catalepsy observed after IP administration. Inhalation also reduced the maximal amount of analgesia produced, regardless of the agonist examined.

One other reason it may be so hard to detect partial CB1R agonists is that it may depend on the exact system used. For example Laaris, Good, & Lupica (2010), using γ-aminobutyric acid (GABA)/CB1R-expressing hippocampal axon terminals, found that both Δ^9-THC and WIN 55,212-2 had the same maximal effect on inhibitory postsynaptic currents. This is in contrast to glutamate/CB1R-expressing neurons from the same area, which showed Δ^9-THC to act as a partial agonist. The authors suggested that this is because the GABA neurons have higher levels of CB1R. This highlights the importance of the specific assay in defining full and partial agonists.

The main reason comparing marijuana to Spice is difficult is that marijuana is not composed solely of Δ^9-THC, but contains a myriad of other related compounds, which may have other pharmacological effects that may modulate the effects of marijuana (see Mechoulam, Fride, & Di Marzo, 1998; Russo, 2011). Furthermore, 11-OH-Δ^9-THC, a main metabolite of Δ^9-THC, has been studied for its cannabimimetic activity. Drug discrimination studies done with 11-OH-Δ^9-THC showed that it is generally more potent than Δ^9-THC in generalization studies done with Δ^9-THC as the training drug. 11-OH-Δ^9-THC, as well as 11-OH-Δ^8-THC, produces effects similar to those of Δ^9-THC also in humans, as summarized elsewhere (Balster & Prescott, 1992).

Other than Δ^9-THC, the only partial agonists studied to some extent in the literature are AEA, AM356, AM411, some other phytocannabinoids, and the compound BAYER 59-3074. While AEA and its analogue AM356 have shown some differences from other cannabinoid agonists, such as not being fully antagonized by rimonabant and failing to affect animals at 100 mg/kg after chronic AM411 (Desai et al., 2013), these differences are not shared with Δ^9-THC. As mentioned previously, endocannabinoids and their analogues display some differences compared to other classes, regardless of efficacy (Jarbe, Lamb, Lin, & Makriyannis, 2001; Jarbe, Lamb, Liu, & Makriyannis, 2003). This lack of partial agonists tested means most assays compare only one partial agonist with a full agonist and thus cannot reliably indicate that efficacy is the reason for any difference in effect.

CONCLUSION

Overall, it seems that high/low efficacy may not be the most effective way to differentiate the important effects of various cannabinoids. While there definitely appears to be some discrepancies between Δ^9-THC and high-efficacy Spice compounds, the sparse research on partial agonists other than Δ^9-THC and AEA does not allow for a strong argument that CB1R efficacy is vital to understanding the effects of cannabinoid agonists. Instead, it is suggested that other differences between cannabinoids may be responsible for the differing reports of marijuana- and Spice-induced effects. Off-target effects, pharmacokinetics, metabolism, and ligands with biased agonism may all contribute to the effects of specific Spice compounds, and their inherent higher efficacy may be important only under specific circumstances, and whether the recreational use of Spice may produce these is still being determined.

APPLICATIONS TO OTHER ADDICTIONS AND SUBSTANCE MISUSE

While this research focuses on the differences among cannabinoid drugs, aspects of it may be helpful in guiding future research with other classes of drugs' partial and full agonists. It is important to note how efficacy can be altered depending on the assay/effect studied. These may also be relevant to research looking into the differences between inverse agonists and neutral antagonists, which also do not consistently produce different effects.

DEFINITION OF TERMS

Spice "Spice" was an early brand of herbal incense found to contain synthetic cannabinoids. The term has been generalized to refer to any herbal incense blend and/or the synthetic cannabinoid compounds found within it sold recreationally.
Drug discrimination This a behavioral technique for assessing "subjective" effects of drugs in animals or humans.
Cannabinoid This is a ligand that binds to and affects CB1R and/or CB2R. It may also refer to bioactive compounds of cannabis that do not necessarily bind to CB1R/CB2R.
Cannabinoid 1 receptor This was the first cannabinoid receptor discovered, responsible for the psychoactive effects of CB1R agonists and most observed effects of Δ^9-THC/marijuana.
Efficacy This refers to the maximal effect of a drug on a specific test.
Potency This refers to the dose of drug needed to achieve a certain level of effect.
Phytocannabinoid This is a cannabinoid drug derived from plants, especially marijuana.
Schedule I This is the top tier of compounds in the Controlled Substances Act. Placement in Schedule I means that there is a high potential of abuse and no established medical use.
Structure–activity relationship This refers to the relation between a molecule's chemical structure and its biological activity.
Conditioned place preference/avoidance This is a behavioral assay that utilizes two distinct environments wherein an animal learns to associate one environment with a specific stimulus, often a drug. The animal will then either prefer or avoid the associated environment based on the specific stimulus's reinforcing properties.
Intracranial self-stimulation This is a behavioral experiment wherein an electrode is inserted into certain brain areas where electrical stimulation acts as reinforcement.

KEY FACTS

Key Facts of Spice/Synthetic Marijuana

- Spice is a product typically sold as "herbal incense" or as a legal alternative to marijuana that was found to contain synthetic cannabinoid drugs.
- Spice was originally a specific brand of herbal incense and now is a generic term for any herbal incense.
- The exact contents of most Spice blends are not typically known and the drugs/dosage used are variable.
- Spice has been linked to various health problems not seen before with marijuana.
- As governments ban the use of certain compounds found in Spice, new synthetic cannabinoids have taken their place.

Key Facts of Cannabinoid Agonists

- Cannabinoid agonists of the CB1R produce similar effects in a dose-dependent manner, such as mood changes, altered memory, sedation, hypothermia, and analgesia.
- Spice compounds detected so far have all been full CB1R agonists; Δ^9-THC is a partial agonist.
- Most in vivo assays do not find a difference between partial and full agonists.
- Δ^9-THC is marketed medically as dronabinol. Nabilone is another approved cannabinoid medication and is a full agonist for CB1R.

SUMMARY POINTS

- Cannabinoid agonists can be differentiated between those with high and those with low efficacy.
- Spice is typically composed of synthetic full agonists for CB1R in comparison to marijuana/Δ^9-THC, a natural partial agonist for CB1.
- Case reports have found profound side effects from Spice never seen before from marijuana/Δ^9-THC.
- Most lab in vivo assays, though, do not find much difference between full and partial CB1R agonists.
- Some assays find differences only after long-term use of the drugs.
- Synthetic CB1R full agonists have many other differences with Δ^9-THC in addition to efficacy.

ACKNOWLEDGMENT

Preparation of this chapter was defrayed in part by National Institute on Drug Abuse Grant 5RO1DA 009064-19. We thank Dr K. Vemuri for providing the chemical structures displayed in Figure 2. The authors declare that the study sponsor did not have any role in study design; in the collection, analysis, and interpretation of data; in the writing of the report; or in the decision to submit the chapter for publication.

REFERENCES

Aceto, M. D., Scates, S. M., & Martin, B. B. (2001). Spontaneous and precipitated withdrawal with a synthetic cannabinoid, WIN 55212-2. *European Journal of Pharmacology, 416*(1–2), 75–81.

Atwood, B. K., Huffman, J., Straiker, A., & Mackie, K. (2010). JWH018, a common constituent of 'Spice' herbal blends, is a potent and efficacious cannabinoid CB receptor agonist. *British Journal of Pharmacology, 160*(3), 585–593. http://dx.doi.org/10.1111/j.1476-5381.2009.00582.x.

Balster, R. L., & Prescott, W. R. (1992). Delta 9-tetrahydrocannabinol discrimination in rats as a model for cannabis intoxication. *Neuroscience and Biobehavioral Reviews, 16*(1), 55–62.

Behonick, G., Shanks, K. G., Firchau, D. J., Mathur, G., Lynch, C. F., Nashelsky, M., ... Meroueh, C. (2014). Four postmortem case reports with quantitative detection of the synthetic cannabinoid, 5F-PB-22. *Journal of Analytical Toxicology, 38*(8), 559–562. http://dx.doi.org/10.1093/jat/bku048.

Bhanushali, G. K., Jain, G., Fatima, H., Leisch, L. J., & Thornley-Brown, D. (2013). AKI associated with synthetic cannabinoids: a case series. *Clinical Journal of the American Society of Nephrology, 8*(4), 523–526. http://dx.doi.org/10.2215/cjn.05690612.

Brents, L. K., Gallus-Zawada, A., Radominska-Pandya, A., Vasiljevik, T., Prisinzano, T. E., Fantegrossi, W. E., ... Prather, P. L. (2012). Monohydroxylated metabolites of the K2 synthetic cannabinoid JWH-073 retain intermediate to high cannabinoid 1 receptor (CB1R) affinity and exhibit neutral antagonist to partial agonist activity. *Biochemical Pharmacology, 83*(7), 952–961. http://dx.doi.org/10.1016/j.bcp.2012.01.004.

Childers, S. R. (2006). Activation of G-proteins in brain by endogenous and exogenous cannabinoids. *AAPS Journal, 8*(1), E112–E117. http://dx.doi.org/10.1208/aapsj080113.

Cooper, Z. D., & Haney, M. (2008). Cannabis reinforcement and dependence: role of the cannabinoid CB1 receptor. *Addiction Biology, 13*(2), 188–195. http://dx.doi.org/10.1111/j.1369-1600.2007.00095.x.

De Vry, J., & Jentzsch, K. R. (2003). Intrinsic activity estimation of cannabinoid CB_1 receptor ligands in a drug discrimination paradigm. *Behavioural Pharmacology, 14*(5–6), 471–476.

Desai, R. I., Thakur, G. A., Vemuri, V. K., Bajaj, S., Makriyannis, A., & Bergman, J. (2013). Analysis of tolerance and behavioral/physical dependence during chronic CB1 agonist treatment: effects of CB1 agonists, antagonists, and noncannabinoid drugs. *Journal of Pharmacology and Experimental Therapeutics, 344*(2), 319–328. http://dx.doi.org/10.1124/jpet.112.198374.

Fan, F., Compton, D. R., Ward, S., Melvin, L., & Martin, B. R. (1994). Development of cross-tolerance between delta 9-tetrahydrocannabinol, CP 55,940 and WIN 55,212. *Journal of Pharmacology and Experimental Therapeutics, 271*(3), 1383–1390.

Fattore, L., & Fratta, W. (2011). Beyond THC: the new generation of cannabinoid designer drugs. *Frontiers in Behavioral Neuroscience, 5*, 60. http://dx.doi.org/10.3389/fnbeh.2011.00060.

Gifford, A. N., Bruneus, M., Gatley, S. J., Lan, R., Makriyannis, A., & Volkow, N. D. (1999). Large receptor reserve for cannabinoid actions in the central nervous system. *Journal of Pharmacology and Experimental Therapeutics, 288*(2), 478–483.

Gonzalez, S., Cebeira, M., & Fernandez-Ruiz, J. (2005). Cannabinoid tolerance and dependence: a review of studies in laboratory animals. *Pharmacology, Biochemistry, and Behavior, 81*(2), 300–318. http://dx.doi.org/10.1016/j.pbb.2005.01.028.

Haney, M., Ward, A. S., Comer, S. D., Foltin, R. W., & Fischman, M. W. (1999a). Abstinence symptoms following oral THC administration to humans. *Psychopharmacology (Berlin), 141*(4), 385–394.

Haney, M., Ward, A. S., Comer, S. D., Foltin, R. W., & Fischman, M. W. (1999b). Abstinence symptoms following smoked marijuana in humans. *Psychopharmacology (Berlin), 141*(4), 395–404.

Harris, C. R., & Brown, A. (2013). Synthetic cannabinoid intoxication: a case series and review. *Journal of Emergency Medicine, 44*(2), 360–366. http://dx.doi.org/10.1016/j.jemermed.2012.07.061.

Hruba, L., Ginsburg, B. C., & McMahon, L. R. (2012). Apparent inverse relationship between cannabinoid agonist efficacy and tolerance/cross-tolerance produced by Delta(9)-tetrahydrocannabinol treatment in rhesus monkeys. *Journal of Pharmacology and Experimental Therapeutics, 342*(3), 843–849. http://dx.doi.org/10.1124/jpet.112.196444.

Hyatt, W. S., & Fantegrossi, W. E. (2014). Delta9-THC exposure attenuates aversive effects and reveals appetitive effects of K2/'Spice' constituent JWH-018 in mice. *Behavioural Pharmacology, 25*(3), 253–257. http://dx.doi.org/10.1097/fbp.0000000000000034.

Jarbe, T. U., Deng, H., Vadivel, S. K., & Makriyannis, A. (2011). Cannabinergic aminoalkylindoles, including AM678=JWH018 found in 'Spice', examined using drug (Delta(9)-tetrahydrocannabinol) discrimination for rats. *Behavioural Pharmacology, 22*(5–6), 498–507. http://dx.doi.org/10.1097/FBP.0b013e328349fbd5.

Jarbe, T. U., & Gifford, R. S. (2014). "Herbal incense": designer drug blends as cannabimimetics and their assessment by drug discrimination and other in vivo bioassays. *Life Sciences, 97*(1), 64–71. http://dx.doi.org/10.1016/j.lfs.2013.07.011.

Jarbe, T. U., Hiltunen, A. J., & Mechoulam, R. (1989). Stereospecificity of the discriminative stimulus functions of the dimethylheptyl homologs of 11-hydroxy-delta 8-tetrahydrocannabinol in rats and pigeons. *Journal of Pharmacology and Experimental Therapeutics, 250*(3), 1000–1005.

Jarbe, T. U., Lamb, R. J., Lin, S., & Makriyannis, A. (2001). (R)-methanandamide and Delta 9-THC as discriminative stimuli in rats: tests with the cannabinoid antagonist SR-141716 and the endogenous ligand anandamide. *Psychopharmacology (Berlin), 156*(4), 369–380.

Jarbe, T. U., Lamb, R. J., Liu, Q., & Makriyannis, A. (2003). (R)-Methanandamide and delta9-tetrahydrocannabinol-induced operant rate decreases in rats are not readily antagonized by SR-141716A. *European Journal of Pharmacology, 466*(1–2), 121–127.

Jarbe, T. U., Li, C., Liu, Q., & Makriyannis, A. (2009). Discriminative stimulus functions in rats of AM1346, a high-affinity CB1R selective anandamide analog. *Psychopharmacology (Berlin), 203*(2), 229–239. http://dx.doi.org/10.1007/s00213-008-1199-3.

Jarbe, T. U., Li, C., Vadivel, S. K., & Makriyannis, A. (2010). Discriminative stimulus functions of methanandamide and delta(9)-THC in rats: tests with aminoalkylindoles (WIN55,212-2 and AM678) and ethanol. *Psychopharmacology (Berlin), 208*(1), 87–98. http://dx.doi.org/10.1007/s00213-009-1708-z.

Jarbe, T. U., Tai, S., LeMay, B. J., Nikas, S. P., Shukla, V. G., Zvonok, A., & Makriyannis, A. (2012). AM2389, a high-affinity, in vivo potent CB1-receptor-selective cannabinergic ligand as evidenced by drug discrimination in rats and hypothermia testing in mice. *Psychopharmacology (Berlin)*, *220*(2), 417–426. http://dx.doi.org/10.1007/s00213-011-2491-1.

Johnston, L. D., O'Malley, P. M., Bachman, J. G., & Schulenberg, J. E. (2013). *Monitoring the Future National Results on Drug Use: 2012 Overview, Key Findings on Adolescent Drug Use*. Ann Arbor: Institute for Social Research, The University of Michigan.

Justinova, Z., Mascia, P., Wu, H. Q., Secci, M. E., Redhi, G. H., Panlilio, L. V., … Goldberg, S. R. (2013). Reducing cannabinoid abuse and preventing relapse by enhancing endogenous brain levels of kynurenic acid. *Nature Neuroscience*, *16*(11), 1652–1661. http://dx.doi.org/10.1038/nn.3540.

Laaris, N., Good, C. H., & Lupica, C. R. (2010). Delta9-tetrahydrocannabinol is a full agonist at CB1 receptors on GABA neuron axon terminals in the hippocampus. *Neuropharmacology*, *59*(1–2), 121–127. http://dx.doi.org/10.1016/j.neuropharm.2010.04.013.

Laprairie, R. B., Bagher, A. M., Kelly, M. E., Dupre, D. J., & Denovan-Wright, E. M. (2014). Type 1 cannabinoid receptor ligands display functional selectivity in a cell culture model of striatal medium spiny projection neurons. *Journal of Biological Chemistry*, *289*(36), 24845–24862. http://dx.doi.org/10.1074/jbc.M114.557025.

Lefever, T. W., Marusich, J. A., Antonazzo, K. R., & Wiley, J. L. (2014). Evaluation of WIN 55,212-2 self-administration in rats as a potential cannabinoid abuse liability model. *Pharmacology, Biochemistry, and Behavior*, *118*, 30–35. http://dx.doi.org/10.1016/j.pbb.2014.01.002.

Maldonado, R., Berrendero, F., Ozaita, A., & Robledo, P. (2011). Neurochemical basis of cannabis addiction. *Neuroscience*, *181*, 1–17. http://dx.doi.org/10.1016/j.neuroscience.2011.02.035.

Marshell, R., Kearney-Ramos, T., Brents, L. K., Hyatt, W. S., Tai, S., Prather, P. L., & Fantegrossi, W. E. (2014). In vivo effects of synthetic cannabinoids JWH-018 and JWH-073 and phytocannabinoid Delta9-THC in mice: inhalation versus intraperitoneal injection. *Pharmacology, Biochemistry, and Behavior*, *124*, 40–47. http://dx.doi.org/10.1016/j.pbb.2014.05.010.

Matsuda, L. A., Lolait, S. J., Brownstein, M. J., Young, A. C., & Bonner, T. I. (1990). Structure of a cannabinoid receptor and functional expression of the cloned cDNA. *Nature*, *346*(6284), 561–564. http://dx.doi.org/10.1038/346561a0.

McMahon, L. R., Ginsburg, B. C., & Lamb, R. J. (2008). Cannabinoid agonists differentially substitute for the discriminative stimulus effects of Delta(9)-tetrahydrocannabinol in C57BL/6J mice. *Psychopharmacology (Berlin)*, *198*(4), 487–495. http://dx.doi.org/10.1007/s00213-007-0900-2.

Mechoulam, R., Feigenbaum, J. J., Lander, N., Segal, M., Jarbe, T. U., Hiltunen, A. J., & Consroe, P. (1988). Enantiomeric cannabinoids: stereospecificity of psychotropic activity. *Experientia*, *44*(9), 762–764.

Mechoulam, R., Fride, E., & Di Marzo, V. (1998). Endocannabinoids. *European Journal of Pharmacology*, *359*(1), 1–18.

Melvin, L. S., & Johnson, M. R. (1987). Structure-activity relationships of tricyclic and nonclassical bicyclic cannabinoids. *NIDA Research Monograph*, *79*, 31–47.

Nikas, S. P., Sharma, R., Paronis, C. A., Kulkarni, S., Thakur, G. A., Hurst, D., … Makriyannis, A. (2015). Probing the carboxyester side chain in controlled deactivation (−)-delta(8)-tetrahydrocannabinols. *Journal of Medicinal Chemistry*, *58*(2), 665–681. http://dx.doi.org/10.1021/jm501165d.

Panagis, G., Mackey, B., & Vlachou, S. (2014). Cannabinoid regulation of brain reward processing with an emphasis on the role of CB1 receptors: a step back into the future. *Frontiers in Psychiatry*, *5*, 92. http://dx.doi.org/10.3389/fpsyt.2014.00092.

Paronis, C. A., Nikas, S. P., Shukla, V. G., & Makriyannis, A. (2012). Delta(9)-Tetrahydrocannabinol acts as a partial agonist/antagonist in mice. *Behavioural Pharmacology*, *23*(8), 802–805. http://dx.doi.org/10.1097/FBP.0b013e32835a7c4d.

Pertwee, R. G., Howlett, A. C., Abood, M. E., Alexander, S. P., Di Marzo, V., Elphick, M. R., … Ross, R. A. (2010). International Union of Basic and Clinical Pharmacology. LXXIX. Cannabinoid receptors and their ligands: beyond CB(1) and CB(2). *Pharmacological Reviews*, *62*(4), 588–631. http://dx.doi.org/10.1124/pr.110.003004.

Rodriguez de Fonseca, F., Carrera, M. R., Navarro, M., Koob, G. F., & Weiss, F. (1997). Activation of corticotropin-releasing factor in the limbic system during cannabinoid withdrawal. *Science*, *276*(5321), 2050–2054.

Rodriguez, J. S., & McMahon, L. R. (2014). JWH-018 in rhesus monkeys: differential antagonism of discriminative stimulus, rate-decreasing, and hypothermic effects. *European Journal of Pharmacology*, *740*, 151–159. http://dx.doi.org/10.1016/j.ejphar.2014.06.023.

Russo, E. B. (2011). Taming THC: potential cannabis synergy and phytocannabinoid-terpenoid entourage effects. *British Journal of Pharmacology*, *163*(7), 1344–1364. http://dx.doi.org/10.1111/j.1476-5381.2011.01238.x.

SAMHSA, (2014). *Results from the 2013 National Survey on Drug Use and Health: Summary of National Findings, NSDUH Series H-48, HHS Publication No. (SMA) 14-4863*. Rockville, MD: Substance Abuse and Mental Health Services Administration.

Sharma, R., Nikas, S. P., Paronis, C. A., Wood, J. T., Halikhedkar, A., Guo, J. J., … Makriyannis, A. (2013). Controlled-deactivation cannabinergic ligands. *Journal of Medicinal Chemistry*, *56*(24), 10142–10157. http://dx.doi.org/10.1021/jm4016075.

Tanda, G., & Goldberg, S. R. (2003). Cannabinoids: reward, dependence, and underlying neurochemical mechanisms–a review of recent preclinical data. *Psychopharmacology (Berlin)*, *169*(2), 115–134. http://dx.doi.org/10.1007/s00213-003-1485-z.

Ward, S. J., Baizman, E., Bell, M., Childers, S., D'Ambra, T., Eissenstat, M., … Pacheco, M. (1990). Aminoalkylindoles (AAIs): a new route to the cannabinoid receptor? *NIDA Research Monograph*, *105*, 425–426.

Ware, M. A., & St Arnaud-Trempe, E. (2010). The abuse potential of the synthetic cannabinoid nabilone. *Addiction*, *105*(3), 494–503. http://dx.doi.org/10.1111/j.1360-0443.2009.02776.x.

Weissman, A., Milne, G. M., & Melvin, L. S., Jr. (1982). Cannabimimetic activity from CP-47,497, a derivative of 3-phenylcyclohexanol. *Journal of Pharmacology and Experimental Therapeutics*, *223*(2), 516–523.

Wiley, J. L., Compton, D. R., Dai, D., Lainton, J. A., Phillips, M., Huffman, J. W., & Martin, B. R. (1998). Structure-activity relationships of indole- and pyrrole-derived cannabinoids. *Journal of Pharmacology and Experimental Therapeutics*, *285*(3), 995–1004.

Winstock, A. R., & Barratt, M. J. (2013). Synthetic cannabis: a comparison of patterns of use and effect profile with natural cannabis in a large global sample. *Drug and Alcohol Dependence*, *131*(1–2), 106–111. http://dx.doi.org/10.1016/j.drugalcdep.2012.12.011.

Zimmermann, U. S., Winkelmann, P. R., Pilhatsch, M., Nees, J. A., Spanagel, R., & Schulz, K. (2009). Withdrawal phenomena and dependence syndrome after the consumption of "spice gold". *Deutsches Arzteblatt International*, *106*(27), 464–467. http://dx.doi.org/10.3238/arztebl.2009.0464.

Chapter 68

Cannabis, Cannabinoid Receptors, and Stress-Induced Excitotoxicity

Borja García-Bueno, Javier R. Caso

Department of Pharmacology, Faculty of Medicine, University Complutense, Madrid, Spain

Abbreviations

2-AG 2-Arachidonoylglycerol
7TM Seven transmembrane domains
Δ^9-THC Δ^9-Tetrahydrocannabinol
AEA Anandamide
AP-1 Activator protein 1
BDNF Brain-derived neurotrophin receptor
CNS Central nervous system
COX-2 Cyclooxygenase 2
DAG-L Diacylglycerol lipase
EAAT-2 Excitatory amino acid transporter 2
ECS Endocannabinoid system
FAAH Fatty acid amide hydrolase
GPR55 G-protein-coupled receptor 55
GR Glucocorticoid receptor
HPA axis Hypothalamus–pituitary–adrenal axis
IL-1β Interleukin 1β
iNOS Inducible NO synthase
MAG-L Monoacylglycerol lipase
MAPK Mitogen-activated protein kinases
MCP-1 Monocyte chemotactic protein 1
NAPE-PLD *N*-acylphosphatidylethanolamine-hydrolyzing phospholipase D
NF-κB Nuclear factor κB
NMDA receptor *N*-methyl-D-aspartic acid receptor
nNOS Neuronal NO synthase
NO· Nitric oxide
OEA *N*-oleoylethanolamine
ONOO⁻ Peroxynitrite anion
PEA *N*-palmitoylethanolamine
PGE$_2$ Prostaglandin E$_2$
PPARα Peroxisome proliferator-activated receptor α
PPARγ Peroxisome proliferator-activated receptor γ
PPI Prepulse inhibition test
PVN Paraventricular nucleus of hypothalamus
ROS Reactive oxygen species
TNF-α Tumor necrosis factor α
TPVR1 Transient receptor potential vanilloid 1
VCAM-1 Vascular cell adhesion protein 1

INTRODUCTION

Cannabis

"Cannabinoids" are a family of psychoactive compounds present in the plant *Cannabis sativa*, both in the leaf/flowered sprout and in the resin. Preparations of this plant have been much used throughout history, for therapeutic reasons or as part of social/religious rituals in various civilizations. However, the use of cannabis in Western societies began at the start of the nineteenth century.

C. sativa contains more that 460 different chemical compounds, and more than 70 of them are grouped under the term cannabinoids (with a common carbocyclic structure of 21 atoms). In 1964, Δ^9-tetrahydrocannabinol, the main psychoactive component of the plant, was isolated and determined.

The Endocannabinoid System

The endocannabinoid system (ECS) consists of the cannabinoid receptors, their endogenous ligands (known as endocannabinoids; eCB), the enzymes involved in the synthesis and degradation of these eCBs, and their uptake/reuptake mechanisms. The two main eCB $G_{i/o}$-protein-coupled receptors CB1R and CB2R were characterized in the early 1990s. CB1R is predominantly located in the central nervous system (CNS) but is also expressed in several peripheral tissues. CB2R is primarily expressed in peripheral immune cells, but also in some brain areas, predominantly in microglia and neural stem cells. Other receptors related to the ECS are the nuclear receptor peroxisome proliferator-activated receptor α (PPARα), which interacts with the eCB-like molecules *N*-palmitoylethanolamine (PEA) and *N*-oleoylethanolamine (OEA), regulating ingestion, pain, and inflammation; the transient receptor potential vanilloid 1 (TPVR1); and the GPR55 orphan receptor. eCBs belong to a family of polyunsaturated fatty acid derivatives that function as lipid signaling molecules released "on demand." The main eCBs are *N*-arachidonoylethanolamine (anandamide, AEA) and 2-arachidonoylglycerol (2-AG). AEA is produced by immune cells and neurons and it is more selective for CB1R than CB2R, whereas 2-AG acts as a full agonist for both receptors. AEA has several possible routes of synthesis, the most important being constituted by a two-step enzyme reaction catalyzed by N-acylphosphatidylethanolamine-hydrolyzing phospholipase D, and the major degradation enzyme for AEA is fatty acid amide hydrolase. On the other hand, the synthesis of 2-AG is catalyzed by diacylglycerol lipase. Several enzymes can hydrolyze 2-AG, but this reaction appears to be principally catalyzed by monoacylglycerol lipase.

Cannabinoid Receptors

CB1R is predominantly located in the CNS, but it is also expressed in several peripheral tissues (e.g., adrenal glands, heart, liver, spleen, gut, and testis). The main function of CB1R is the regulation of neurotransmission, and its stimulation is responsible of the psychoactive effects derived from *C. sativa* consumed in its various preparations. At the CNS level, CB1R is mostly expressed in the basal ganglia, cerebral cortex, cerebellum, and hippocampus, but also in the brainstem, thalamic nuclei, amygdalar complex, and dorsal horn of the medulla. This distribution suggests a fundamental function for CB1R in the regulation of movement, memory, and pain. CB1R is mainly expressed in neurons and also in astrocytes, microglia, and oligodendroglia, although to a lesser extent. Neuronal CB1R expression is mainly presynaptic, in axons and nerve endings, where it regulates the release of neurotransmitters.

CB2R is primarily expressed in peripheral immune cells and organs, but also in other organic systems, such as muscle, liver, gut, testis, and adipose tissue. CB2R is also present in some types of cells of the brain, predominantly in microglia and neural stem cells. In general the expression of CB2R in the CNS is low but it is increased in glial cells in response to various pathological stimuli and conditions (e.g., dopaminergic neurotoxins, stroke, β-amyloid deposition, glioma, etc.).

The main intracellular pathways activated by CB1/CB2R lead to the inhibition of the enzyme adenylate cyclase, the control of different ionic channels (K^+ and Ca^{2+} dependent), the activation of the mitogen-activated protein kinases, and the production of nitric oxide (NO·; mainly through the regulation of the enzyme neuronal NO· synthase) and arachidonic acid (by means of the stimulation of the enzymes phospholipase D and A_2).

Other receptors related to the ECS are the nuclear receptor PPARα, which interacts with the eCB-like molecules PEA and OEA, regulating ingestion, pain, and inflammation; the TPVR1; and the GPR55 orphan receptor.

Functions of Cannabinoids

A brief description of the more relevant functions of cannabinoids, in terms of neuroprotection, follows:

Antioxidant

Mainly because of their chemical structure, some cannabinoids afford neuroprotection through the regulation of the oxidative status of a cell in potential neuropathological situations of oxidative stress (i.e., neurodegenerative pathologies). Oxidative stress could be produced by a marked increase in reactive oxygen species and/or a deficiency in the antioxidant endogenous systems, and some cannabinoids act at both levels. The cannabinoids with a higher antioxidant potential are the phytocannabinoids cannabidiol and cannabinol, or their analogues nabilone, levonantradol, and dexanabinol.

Anti-inflammatory Profile

Some cannabinoid agonists present an anti-inflammatory profile in diverse neurologic/neurodegenerative pathologies such as stroke, Parkinson, Alzheimer, and Huntington disease, mainly by the regulation of activated microglia, astroglia, and macrophages. Specifically, some CB1R and CB2R agonists such as arachidonyl-2'-chloroethylamide (ACEA), JWH-133, or WIN 55,212-2, among others, are capable of downregulating proinflammatory molecules such as cytokines (interleukin 1β (IL-1β) or tumor necrosis factor-α (TNF-α)), chemokines (monocyte chemotactic protein 1), the adhesion molecule vascular cell adhesion protein 1, nuclear factors (nuclear factor κB (NF-κB), activator protein 1), enzymes (inducible NO synthase (iNOS), cyclooxygenase-2 (COX-2)), etc.

Regulation of the Intracellular Concentration of Ca^{2+}

The rapid and massive increase in intracellular Ca^{2+} levels is a key event in the neuronal damage observed in some chronic neurodegenerative pathologies. This increase produces the depolarization of the cellular and mitochondrial membranes, precipitating their failure and potentiating the oxidative stress. These events trigger programmed neuronal death, or apoptosis. The cannabinoid agonists, acting on postsynaptic CB1R, are able to block voltage-dependent Ca^{2+} channels, inhibiting the activation of multiple intracellular cascades correlated with cellular damage/death.

Regulation of Brain Vasculature

Some CB1R agonists improve neuronal degeneration by increasing the blood supply in an episode of cerebral stroke. This vasodilatation is mediated by the inhibition of some endothelin-derived molecules, such as endothelin-1 or NO. This effect is mainly mediated by CB1R, but other authors speculate about the existence of a specific "endothelial cannabinoid receptor."

Finally, two of the most relevant functions of cannabinoids are the regulation of the activation of the hypothalamus–pituitary–adrenal (HPA) axis after stress exposure and the control of neurotransmitter metabolism (specifically glutamate). These functions will be discussed in detail in the following subsections.

ENDOCANNABINOID SYSTEM AND STRESS: A BIDIRECTIONAL RELATIONSHIP

Since 2005, increasing evidence has supported the existence of a decisive role for the ECS as a regulator of the stress response. The ECS controls the release of stress-related neurotransmitters (catecholamines, dopamine, glutamate, and γ-aminobutyric acid (GABA)) in various neuronal populations associated with stress and emotional responses. Several alterations in the elements of the ECS have been demonstrated after stress and, in contrast, there are also changes in the stress response following modulation of eCB signaling. As an illustrative example, it has been shown that acute stress produces transient changes in the content of eCBs in the amygdala and the prefrontal cortex and, on the other hand, the activation of cannabinoid receptors prevents the activation of the HPA stress axis, as occurs in the case of the administration of the dual CB1/CB2R agonist WIN 55,212-2 in the basolateral amygdala.

CB1R and Stress

The relationship between cannabis consumption, the ECS, and the mechanisms governing the stress response are complex.

Cannabis abuse is classically described as an acute stressor, able to activate the HPA stress axis. In fact, several epidemiological studies have identified that cannabinoids affect anxiety and stress responsivity.

On the other hand, there is growing evidence showing that the activation of the HPA axis by exposure to physical, psychological, or mixed stress produces the immediate release of eCBs in a mechanism dependent on the activation of the glucocorticoid receptors. This release of eCBs activates the CB1R expressed in glutamatergic and GABAergic neurons in specific brain areas such as the prefrontal cortex, amygdala, striatum, and hypothalamus, in an attempt to restore the homeostasis of the HPA axis after stress.

Alterations in HPA axis activity under basal conditions and after stress exposure in CB1R-knockout (KO) mice have been also shown. CB1R-KO mice present increased levels of corticotrophin-releasing factor mRNA in the hypothalamic paraventricular nucleus (PVN) and an overactivity of the HPA axis negative feedback mechanism. In addition to HPA axis dysregulation, chronic excitotoxicity, and neuroinflammation, CB1R-KO mice present defective adult neurogenesis, increased anxiety, and deficiency in neuronal plasticity from decreasing brain-derived neurotrophin receptor (BDNF) levels in the hippocampus, diminished locomotor activity, reduced expression of the serotonin receptor 5-HT2C, and reactivity of raphe nuclei to stress, suggesting a possible role for CB1R in the pathophysiology of stress-related depression and its pharmacological manipulation as a potential therapeutic strategy. Based on all these alterations, CB1R-KO mice have been proposed as a suitable animal model to study depression. From a pharmacological point of view, CB1R seems to be implicated in the regulation of GABAergic neurotransmission and in the anxiolytic profile of benzodiazepines. CB1R-KO mice are also less responsive to treatment with antidepressant drugs.

CB2R and Stress

The role of CB2R in the regulation of the stress response is more controversial. Using a protocol of immobilization and acoustic stress, the increase in plasma corticosterone induced by stress is not modified by genetic (transgenic CB2xP mice overexpressing CB2R in neurons and CB2R-KO mice) or pharmacological manipulations (CB2R agonist JWH-133) of CB2R. In agreement with this, increased plasma corticosterone content elicited by systemic endotoxin administration did not change after the pharmacological modulation of CB2R.

The expression of these receptors in stress-responsive neural circuits, however, such as the hippocampus, amygdala, and hypothalamus, indirectly suggests that CB2R activation could regulate the neuroendocrine response. In fact, CB2xP mice submitted to 30 min of restraint stress presented lower levels of pro-opiomelanocortin mRNA in the arcuate nuclei than their wild-type (WT) counterparts, as well as a complete block of the stress-induced increase in the mRNA for corticotropin-releasing factor in the PVN of the hypothalamus.

Thus, more detailed neuroendocrine studies regarding the time course of synthesis and release of corticosterone and other stress hormones in stress models of different nature and duration are needed.

CB2R-KO mice presented an enhanced neuroinflammatory response in the frontal cortex after stress exposure, and deletion of the CB2R also induced schizophrenia (deficits in the prepulse inhibition test) and depression-like behaviors in mice. In addition, transgenic mice overexpressing CB2R (CB2xP) have increased levels of the neurotrophin BDNF in the hippocampus and a better response to depression behavioral tests.

STRESS AND CELLULAR DAMAGE/DEATH BY EXCITOTOXICITY

One of the initial processes that take place in the stress response is the release of excitatory amino acids (glutamate and aspartate) in some brain areas. As early as 20 min after the onset of immobilization (used as an experimental model of psychological stress in rodents), there is an immediate and sustained release of glutamate and aspartate into the synaptic cleft that reaches excitotoxic levels, indicating a role for excitatory amino acids in excitotoxic necrotic brain injury. Interestingly, in contrast to apoptotic cell death, necrotic cells swell and burst, releasing proinflammatory mediators. The increased extracellular glutamate binds to its ionotropic N-methyl-D-aspartic acid (NMDA) receptor, whose overactivation causes a continuous excitation of neurons. This overactivation induces further glutamate release, ATP depletion, and a dramatic increase in intracellular Ca^{2+} levels, which then activates Ca^{2+}-dependent enzymes (e.g., proteases, lipases, peroxidases) and eventually leads to neuronal death. In the particular case of restraint/immobilization stress exposure, this massive release of glutamate induces the release of proinflammatory cytokines such as TNF-α or IL-1β. Stress also activates the NF-κB pathway in a TNF-α-dependent mechanism. NF-κB activation elicits the expression and activity of proinflammatory enzymatic sources, such as iNOS and COX-2, among others. The result of this sequence of events is the accumulation of oxidative and nitrosative mediators, which can attack membrane phospholipids and cause cell damage in a process known as lipid peroxidation. Despite stress-induced production and accumulation of potentially cytotoxic and/or proinflammatory mediators like glutamate, NO$^{\bullet}$, peroxynitrite anion (ONOO$^-$), or prostaglandin E$_2$, various authors have discussed the possibility that some of the many changes caused by stress response effectors are not damaging to the neurons, but in fact are predominantly beneficial to their structure and function.

This process is especially relevant, considering that uncontrolled excitotoxicity and neuroinflammation contribute to cell death and damage in neurological and neuropsychiatric diseases, including some that are related to stress exposure (neurodegenerative diseases, depression, posttraumatic stress disorder, and schizophrenia).

An important issue to consider when excitotoxicity is analyzed is the contribution of uptake mechanisms. The excitatory amino acid transporters (EAATs) are responsible for the uptake of glutamate from the synaptic cleft and, specifically, astrocyte glutamate uptake is a high-affinity process regulated by EAAT-2. Immobilization stress in rats induces a decrease in the expression of EAAT-2 in the cortex and it is plausible that stress-induced regulation of the expression of EAAT-2 may be involved in the excitotoxic effects of stress in the brain. Interestingly, this change seems to be dependent on the duration of the stress exposure and the brain area studied. Using a repeated-stress (21–40 days) model, an increase in EAAT-2 expression has been found in the hippocampus, which could be considered a counterregulatory mechanism.

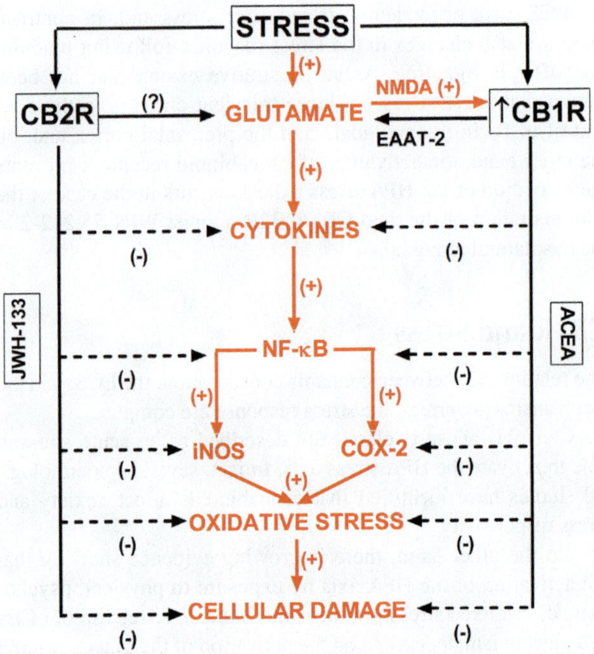

CANNABINOID RECEPTOR ACTIVATION AND STRESS-INDUCED EXCITOTOXICITY

CB1R and Excitotoxicity

A CB1R upregulation in the mouse prefrontal cortex exposed to subchronic restraint/acoustic stress in a mechanism dependent on NMDA glutamate receptors has been demonstrated. In addition, the daily administration of the potent and highly selective CB1R agonist ACEA prevented stress-induced upregulation of CB1R mRNA and protein and presented an antiexcitotoxic profile, regulating glutamate transport and uptake after stress exposure, at least in part by restoring protein expression of EAAT-2. This effect could explain, to some extent, the anti-inflammatory profile of ACEA against stress-induced NF-κB-dependent neuroinflammation (see Figure 1).

Stress-induced CB1R upregulation could represent a compensatory response aimed at controlling the massive release of glutamate into the synaptic cleft that might produce cellular damage and even death by excitotoxicity after stress exposure. CB1R activation with ACEA enhances the main glutamate uptake mechanism, by restoring protein expression of the glutamate transporter EAAT-2. Astrocyte uptake of glutamate is a high-affinity process regulated by EAAT-2; the fact that stress decreases EAAT-2 expression in the cerebral cortex indicates a specific damaging effect at this level and makes the effect of ACEA more interesting. CB1R effects on EAAT-2 have also been demonstrated in other experimental settings, such as an in vitro model of multiple sclerosis or after prenatal exposure to the dual CB1R/CB2R agonist WIN 55,212-2. It has been widely demonstrated that presynaptic CB1R inhibits glutamate release, but glutamatergic neurotransmission is also

FIGURE 1 **Schematic view of the regulatory role of CB1R and CB2R on stress-induced excitotoxicity, neuroinflammation, and oxidative stress damage.** Stress exposure induces excitotoxicity through the activation of proinflammatory cytokines, nuclear factors, proinflammatory enzymes, and oxidative/nitrosative stress mediators in the rat prefrontal cortex. This excitotoxic/proinflammatory/oxidant pathway could be regulated by the endogenous/exogenous activation of CB1R and CB2R. NMDA, N-methyl-D-aspartic acid ionotropic receptor; EAAT-2, excitatory amino acid transporter 2; NF-κB, nuclear factor κB; iNOS, inducible nitric oxide synthase; COX-2, cyclooxygenase-2.

regulated by CB1R at the level of postsynaptic reuptake clearance mechanisms.

How neuronal CB1R activation enhances glutamate glial uptake may be explained by neuron–astrocyte communication that regulates the PPARγ activity in astrocytes. Notably, the EAAT-2 glutamate transporter is a target gene of PPARγ in excitotoxic-related models, such as stress exposure or stroke. As well, a study identified eCBs as PPARγ agonists affording neuroprotection targeting EAAT-2 in an in vitro model of multiple sclerosis.

In addition, several in vivo and in vitro studies have demonstrated that some cannabinoids can block neuronal death by excitotoxicity through a decrease in the release of glutamate in the presynaptic neuron. This mechanism is CB1R-dependent and affects several processes, such as motor activity, memory, and the stress response.

CB2R and Excitotoxicity

The other cannabinoid receptor type, CB2R, has been recognized as a major regulator of the immune system in the periphery, as it is highly expressed in a wide range of immune cells. However, it is now accepted that CB2R is also expressed in the CNS, by microglia, astrocytes, and subpopulations of neurons present in brain areas related to the HPA axis, although the extent of CB2R expression in neurons remains controversial. Furthermore, CB2R is inducible in microglia under neuroinflammatory conditions, suggesting that such upregulation could be a common pattern of response against various types of chronic human neurodegenerative and neurological pathologies. Although there is still debate regarding the role of CB2R in the brain, specific functions for CB2R in neuropsychiatric conditions are emerging.

In synaptosomes, samples from the prefrontal cortex of mice exposed to subchronic restraint/acoustic stress, the administration of the selective CB2R agonist JWH-133, or the overexpression of CB2R in the transgenic mice CB2xP showed increased control levels of glutamate uptake, which were reduced by stress back to control levels. The absence of CB2R (CB2R-KO mice) did not modify the glutamate uptake by synaptosomes, compared with WT mice, under either control or stress conditions (see Figure 1). These effects were not due to changes in the protein expression of EAAT-2. However, other authors have demonstrated a role for CB2R in the regulation of α-amino-3-hydroxy-5-methyl-4-isoxazolepropionic acid (AMPA) excitotoxicity in in vivo and in vitro models of multiple sclerosis, through the upregulation of EAAT-2. Under stress conditions, other approaches, such as the determination of glutamate levels in the tissue, will help to show whether the effects of CB2R manipulation on glutamate uptake are due to inflammation-related actions on EAAT-2 activity or to a direct effect of the CB2R in neurons.

THERAPEUTIC POTENTIAL IN STRESS-RELATED NEUROPSYCHOPATHOLOGIES

There is increasing evidence supporting the idea that cellular damage/death produced by excitotoxicity/neuroinflammation is a main factor in the pathophysiology of neurologic/neurodegenerative diseases such as stroke, Alzheimer, Parkinson's, and Huntington's diseases. Thus, the use of cannabinoid-based compounds (i.e., Sativex©) is now in the clinical area in diverse ongoing clinical trials.

On the other hand, it is worth mentioning that the use of stress-based animal models is relevant to study possible mechanisms involved in the pathophysiology of schizophrenia and depression, bearing in mind that stress is a major contributor to the etiology and progression of psychiatric illness in its multiple clinical and subclinical manifestations. Thus, the most used animal models for depression are based on exposure to chronic stress. It has been suggested that excitotoxicity/neuroinflammation could be a cause of the pathophysiological alterations observed in particular brain structures of patients diagnosed with psychotic disease, bipolar disorder, and schizophrenia. If this is finally the case, the therapeutic use of cannabinoid-based compounds in the treatment of psychiatric diseases is promising and deserves further investigation.

Finally, "cannabinoid" treatment affords neuroprotection in various stress- and non-stress-related pathological experimental settings (see a detailed summary in Table 1).

DEFINITION OF TERMS

Excitotoxicity This is the process, by means of glutamate or related excitatory amino acids, which mediates the death of central neurons under specific conditions (e.g., stress exposure).

Neuroinflammation This is the specific immune response of the CNS against endogenous/exogenous threatening stimuli.

Oxidative stress This is a deregulation of the balance between the production of reactive oxygen species (free radicals) and the endogenous cellular antioxidant defenses

Hypothalamic–pituitary–adrenal axis This is a multifaceted set of homeostatic interactions between the hypothalamus, the pituitary gland, and the suprarenal glands aimed at regulating the response of an organism against stress exposure.

N-Methyl-D-aspartic acid ionotropic receptor This is a glutamate-gated ion channel widely expressed in the CNS.

Excitatory amino acid transporter 2 This is a membrane-bound protein that is mainly localized in astroglial cells placed in presynaptic glutamatergic nerve endings, whose function is the termination of the glutamatergic neurotransmission.

Synaptosome This is the isolated synaptic terminal from a neuron that allows ex vivo studies of neurotransmitter transport.

KEY FACTS OF EXCITOTOXICITY

- Excitotoxicity is a key event in the pathophysiology of neurologic/neurodegenerative diseases such as amyotrophic lateral sclerosis, Huntington disease, or stroke.
- The "excitotoxic hypothesis" of schizophrenia has received an increasing interest in past years.
- The use of drugs capable of regulating glutamate transport and metabolism is emerging as a putative therapeutic strategy for the treatment of depression and other stress-related psychopathologies.
- In January 2015, the term "excitotoxicity" retrieved 6196 entries in PubMed and "neuroinflammation," a hot topic in biomedical sciences, a very similar number, 6770.
- In the US National Institutes of Health Web page, www.ClinicalTrials.gov, the term "glutamate" retrieved 306 clinical trials.

SUMMARY POINTS

- Excitotoxicity and neuroinflammation contribute to the pathophysiology of both neurological/neurodegenerative and neuropsychiatric diseases.
- Stress exposure elicits excitotoxicity and neuroinflammation in limbic brain areas such as the prefrontal cortex.
- The ECS is a homeostatic system that regulates the organic stress response.
- CB1R and CB2R agonists present a multifaceted neuroprotective profile regulating stress-induced excitotoxicity, neuroinflammation, and oxidative stress.
- The pharmacological manipulation of the cannabinoid receptors is a promising therapeutic strategy for the treatment of CNS pathologies that deserves further elucidation.

736 PART | IV Cannabinoids

TABLE 1 Effects of "Cannabinoid" Treatment on Excitotoxicity

Main Effect	"Cannabinoid" Treatment	Model	References
Prevention of mitochondrial dysfunction and oxidative stress	Dual CB1R/CB2R agonist WIN 55,212-2	Quinolinic acid-induced excitotoxicity in rat striatal cultured cells and rat brain synaptosomes	Rangel-López et al. (2015)
More susceptibility to cell lost due to excitotoxic damage	—	Mutant mice lacking CB1R in glutamatergic neurons injected in the striatum with quinolinic acid	Chiarlone et al. (2014)
Increased control levels of glutamate uptake	CB2R agonist JWH-133	Prefrontal cortex synaptosomes of rats exposed to immobilization and acoustic stress	Zoppi et al. (2014)
Protected dentate gyrus granule cells and reduced the number of activated microglia	G-protein-coupled receptor 55 ligand L-α-lysophosphatidylinositol	Rat organotypic hippocampal slice cultures exposed to NMDA	Kallendrusch et al. (2013)
Decreased excitotoxicity (glutamate/N-acetylaspartate ratio)	CB2R activation with cannabidiol	Hypoxic-ischemic newborn pigs	Pazos et al. (2013)
Protected neuronal cells from excitotoxicity	Cannabigerol quinone (VCE-003)	Theiler murine encephalomyelitis virus (TMEV) model of multiple sclerosis (MS)	Granja et al. (2012)
Protected neuronal cells from NMDA excitotoxicity	Cannabinoid agonist HU-210	Ethanol withdrawal increased NMDA-evoked neuronal death by excitotoxicity	Rubio et al. (2011)
Prevented stress-induced decrease in glutamate uptake and EAAT-2 protein levels	CB1R agonist ACEA	Prefrontal cortex synaptosomes of rats exposed to immobilization and acoustic stress	Zoppi et al. (2011)
Reduced AMPA-induced excitotoxicity both in vivo and in vitro	Cannabinoid agonist HU-210 through CB1R and CB2R	TMEV model of MS	Docagne et al. (2007)
Increased neuronal viability	Δ⁹-tetrahydrocannabinol	Hippocampal neurons in culture exposed to reduced extracellular concentrations of Mg^{2+}	Gilbert, Kim, Waataja, and Thayer (2007)
Reduced lactate dehydrogenase efflux and increased glutamate extracellular levels	ACEA, JWH-133, WIN 55,212-2	Brain slices from 7-day-old Wistar rats were exposed to oxygen–glucose deprivation for 30 min	Fernández-Lopez et al. (2006)
Increased neuronal viability	WIN 55,212-2	Murine cerebrocortical neurons in vitro and mouse cerebral cortex in vivo exposed to NMDA toxicity	Kim, Won, Mao, Jin, and Greenberg (2006)
More susceptibility to cell lost due to excitotoxicity in hippocampal pyramidal neurons	—	CB1R-KO mice exposed to kainic acid	Marsicano et al. (2003)
Attenuated cytotoxic edema	Anandamide, arvanil acting at both CB1R and VR1 receptor	Ouabain-induced in vivo excitotoxicity in rats	Veldhuis et al. (2003)
Reducing the infarct area and the number of cortical degenerating neurons	WIN 55,212-2	Unilateral intrastriatal microinjection of NMDA	Hansen et al. (2002)
Increased cellular viability	Palmitoylethanolamide	Excessive NMDA receptor stimulation of cultured mouse cerebellar granule cells	Skaper et al. (1996)

The pharmacological modulation of CB1R/CB2R reduces excitotoxic cellular damage/death in various in vivo and in vitro models of neuropathologies.

REFERENCES

Chiarlone, A., Bellocchio, L., Blázquez, C., Resel, E., Soria-Gómez, E., Cannich, A., ... Guzmán, M. (2014). A restricted population of CB1 cannabinoid receptors with neuroprotective activity. *Proceedings of the National Academy of Sciences of the United States of America*, *111*(22), 8257–8262.

Docagne, F., Muñetón, V., Clemente, D., Ali, C., Loría, F., Correa, F., ... Guaza, C. (2007). Excitotoxicity in a chronic model of multiple sclerosis: neuroprotective effects of cannabinoidsthrough CB1 and CB2 receptor activation. *Molecular and Cellular Neuroscience*, *34*(4), 551–561.

Fernández-López, D., Martínez-Orgado, J., Nuñez, E., Romero, J., Lorenzo, P., Moro, M. A., & Lizasoain, I. (2006). Characterization of the neuroprotective effect of the cannabinoid agonist WIN-55212 in an in vitro model of hypoxic-ischemic brain damage in newborn rats. *Pediatric Research*, *60*(2), 169–173.

Gilbert, G. L., Kim, H. J., Waataja, J. J., & Thayer, S. A. (2007). Delta9-tetrahydrocannabinol protects hippocampal neurons from excitotoxicity. *Brain Research*, *1128*(1), 61–69.

Granja, A. G., Carrillo-Salinas, F., Pagani, A., Gómez-Cañas, M., Negri, R., Navarrete, C., ... Muñoz, E. (2012). A cannabigerol quinone alleviates neuroinflammation in a chronic model of multiple sclerosis. *Journal of Neuroimmune Pharmacology*, *7*(4), 1002–1016.

Hansen, H. H., Azcoitia, I., Pons, S., Romero, J., García-Segura, L. M., Ramos, J. A., ... Fernández-Ruiz, J. (2002). Blockade of cannabinoid CB(1) receptor function protects against in vivo disseminating brain damage following NMDA-induced excitotoxicity. *Journal of Neurochemistry*, *82*(1), 154–158.

Kallendrusch, S., Kremzow, S., Nowicki, M., Grabiec, U., Winkelmann, R., Benz, A., ... Koch, M. (2013). The G protein-coupled receptor 55 ligand l-α-lysophosphatidylinositol exerts microglia-dependent neuroprotection after excitotoxic lesion. *Glia*, *61*(11), 1822–1831.

Kim, S. H., Won, S. J., Mao, X. O., Jin, K., & Greenberg, D. A. (2006). Molecular mechanisms of cannabinoid protection from neuronal excitotoxicity. *Molecular Pharmacology*, *69*(3), 691–696.

Marsicano, G., Goodenough, S., Monory, K., Hermann, H., Eder, M., Cannich, A., & Lutz, B. (2003). CB1 cannabinoid receptors and on-demand defense against excitotoxicity. *Science*, *302*(5642), 84–88.

Pazos, M. R., Mohammed, N., Lafuente, H., Santos, M., Martínez-Pinilla, E., Moreno, E., ... Martínez-Orgado, J. (2013). Mechanisms of cannabidiol neuroprotection in hypoxic-ischemic newborn pigs: role of 5HT(1A) and CB2 receptors. *Neuropharmacology*, *71*, 282–291.

Rangel-López, E., Colín-González, A. L., Paz-Loyola, A. L., Pinzón, E., Torres, I., Serratos, I. N., ... Santamaría, A. (2015). Cannabinoid receptor agonists reduce the short-term mitochondrial dysfunction and oxidative stress linked to excitotoxicity in the rat brain. *Neuroscience*, *285*, 97–106.

Rubio, M., Villain, H., Docagne, F., Roussel, B. D., Ramos, J. A., Vivien, D., ... Ali, C. (2011). Pharmacological activation/inhibition of the cannabinoid system affects alcohol withdrawal-induced neuronal hypersensitivity to excitotoxic insults. *PLoS One*, *6*(8), e23690.

Skaper, S. D., Buriani, A., Dal Toso, R., Petrelli, L., Romanello, S., Facci, L., & Leon, A. (1996). The ALIAmide palmitoylethanolamide and cannabinoids, but not anandamide, are protective in a delayed postglutamate paradigm of excitotoxic death in cerebellar granule neurons. *Proceedings of the National Academy of Sciences of the United States*, *93*(9), 3984–3989.

Veldhuis, W. B., van der Stelt, M., Wadman, M. W., van Zadelhoff, G., Maccarrone, M., Fezza, F., ... Di Marzo, V. (2003). Neuroprotection by the endogenous cannabinoid anandamide and arvanil against in vivo excitotoxicity in the rat: role of vanilloid receptors and lipoxygenases. *Journal of Neuroscience*, *23*(10), 4127–4133.

Zoppi, S., Pérez Nievas, B. G., Madrigal, J. L., Manzanares, J., Leza, J. C., & García-Bueno, B. (2011). Regulatory role of cannabinoid receptor 1 in stress-induced excitotoxicity and neuroinflammation. *Neuropsychopharmacology*, *36*(4), 805–818.

Zoppi, S., Madrigal, J. L., Caso, J. R., García-Gutiérrez, M. S., Manzanares, J., Leza, J. C., & García-Bueno, B. (2014). Regulatory role of the cannabinoid CB2 receptor in stress-induced neuroinflammation in mice. *British Journal of Pharmacology*, *171*(11), 2814–2826.

Chapter 69

Cannabis, Endocannabinoid CB1 Receptors, and the Neuropathology of Vision

Miguel Dasilva[1], Kenneth L. Grieve[1], Casto Rivadulla[2]

[1]Faculty of Life Sciences, University of Manchester, Manchester, UK; [2]NEUROcom, Depto de Medicina, Univ da Coruña, Campus de Oza, A Coruña, Spain; The Institute for Biomedical Research of Coruña (INIBIC), A Coruña, Spain

Abbreviations

2-AG 2-Arachidonoylglycerol
5-HTTP L-5-Hydroxytryptophan
CB Cannabinoid
CB1 Cannabinoid receptor type 1
EEG Electroencephalogram
LFP Local field potential
mRNA Messenger ribonucleic acid
SR141716A N-(piperidin-1-yl)-5-(4-chlorophenyl)-1-(2,4-dichlorophenyl)-4-methyl-1H-pyrazole-3-carboxamide hydrochloride
SSVEP Steady-state visual evoked potential
THC Δ^9-Tetrahydrocannabinol
WIN 55,212-2 (R)-(+)-[2,3-dihydro-5-methyl-3-(4-morpholinylmethyl) pyrrolo[1,2,3-de]-1,4-benzoxazin-6-yl]-1-naphthalenylmethanone

INTRODUCTION

Pathology of vision is known to occur as the result of drug action and side effects in a number of well-known cases (Li, Tripathi, & Tripathi, 2008). While anecdotal and apocryphal descriptions of changes in visual experiences as a result of cannabis ingestion abound, hard evidence for pathologies resulting from, for example, cellular firing changes in the early visual system, which might underpin these reported effects, is much rarer. Indeed, many descriptions of cannabis-induced visual changes use emotive phrases involving "perception," "interpretation," and "appreciation" rather than scientifically measurable disturbances of physiology. The discovery of the endocannabinoid system came about from the characterization of the chemical structure of Δ^9-tetrahydrocannabinol (THC), the main active component of the Cannabis sativa plant (Mechoulam & Shvo, 1963). This allowed the identification and cloning of the two main membrane receptors with which THC interacts: cannabinoid receptor type 1 (CB1) and cannabinoid receptor type 2 (CB2) (Devane, Dysarz, Johnson, Melvin, & Howlett, 1988; Matsuda, Lolait, Brownstein, Young, & Bonner, 1990). Later the two main endogenous ligands for these receptors were discovered: anandamide (AEA) and 2-arachidonoylglycerol (2-AG) (Devane et al., 1992; Sugiura et al., 1995). Since then, our knowledge of the endocannabinoid system has grown quickly. Today, it is widely accepted that endocannabinoid system components—both ligands and receptors—are broadly distributed throughout the central nervous system (Matias, Bisogno, & Di Marzo, 2006; Tsou, Brown, Sañudo-Peña, Mackie, & Walker, 1998), where endocannabinoids play important roles in various aspects of its physiology, development, and organization. However, even though CBs have been shown to mediate elements of cognitive processes such as attention and memory (Kanayama, Rogowska, Pope, Gruber, & Yurgelun-Todd, 2004), few studies have explored the effects of the endocannabinoid system on sensory processing, especially vision.

Several factors have contributed to this lack of research. First, cannabis is simply the name for a genus of plants (Cannabis spp.) from which an active substance or substances (cannabis, marijuana, hashish) have been extracted—little information was available about these substances and the mechanisms by which they affected behavior and cognition before the endocannabinoid system was discovered and characterized. The effects of marijuana on general state and attention and the difficulty of establishing appropriate controls had also made it difficult to obtain and interpret psychophysical data related to vision (Hollister, 1971; Weil, Zinberg, & Nelsen, 1968). The reputation of marijuana as a drug of abuse and its legal status had also for a long time hampered a systematic and concise study of the psychophysical and physiological mechanisms involved in its effects on brain activity. Finally, histological studies aimed at detecting the expression of CB receptors in the brain had classically shown weak or no expression in primary sensory pathways (Herkenham et al., 1991; Moldrich & Wenger, 2000).

More recently, however, studies have found contrasting results in key components of sensory systems with regard to the distribution of the components of the endocannabinoid system, suggesting a higher incidence of CB ligands and receptors (Sun, Wu, Lu, & Beierlein, 2011), along with the machinery involved in their synthesis and degradation (Felder et al., 1996). Furthermore, studies specifically aimed at studying the role of endocannabinoids in various parts of the visual system have greatly supported this conclusion (Dasilva, Grieve, Cudeiro, & Rivadulla, 2012, 2014; Ohiorhenuan et al., 2014; Sales-Carbonell et al., 2013; Sun et al.,

2011). Indeed, a careful scrutiny of the previous literature offers numerous direct and indirect insights on the close relationship between CBs and vision. Thus, the evidence now supports the existence of a significant relationship between the components of the endocannabinoid system and visual function, a relationship that can be observed from a psychophysical and electrophysiological point of view.

PSYCHOPHYSICAL EFFECTS OF CANNABINOIDS ON VISION

Early psychophysical studies on the effect of marijuana on visual perception in human subjects found no significant results when orally administering cannabis and assessing behavior (Clark & Nakashima, 1968) or during visual brightness discrimination tasks after smoking marijuana (Caldwell, Myers, Domino, & Merriam, 1969). However, these conclusions were challenged by a contemporary study in animals, which found cannabis to affect color but not form perception when administered intraperitoneally to pigeons trained in a visual discrimination task (Siegel, 1969). These results were corroborated soon after in humans by studies demonstrating an acute effect of smoked marijuana on color discrimination (Domino, Rennick, & Pearl, 1976; Kiplinger, Manno, Rodda, & Forney, 1971) and were extended further showing a decrement in the ability to detect geometric shapes in human subjects intoxicated with marijuana (Carlin, Bakker, Halpern, & Post, 1972).

An effect of marijuana on peripheral vision was also demonstrated (Moskowitz, Sharma, & McGlothlin, 1972); in this study participants were tested under control conditions and after the administration of marijuana during a peripheral light detection task. Interestingly, peripheral vision decreased up to 50% after the consumption of marijuana. The authors correctly demonstrated that these effects were not the simple consequence of the induction of a general drowsiness, as reaction times were not significantly affected.

Another effect of marijuana on vision was revealed by studying the phenomenon of apparent visual motion (Sharma & Moskowitz, 1972). In this study human subjects were instructed to report whether they observed any movement of a centrally presented flashing stationary light. After smoking marijuana subjects reported a higher index of apparent visual motion during the presentation of the stimulus and reported this to happen with complex patterns like zigzag, elliptical, or circular movements and changes in apparent intensity (closer/farther movements). However, this early study did not measure eye movements.

Around the same time, a set of studies focused on the general effects of marijuana in humans reported a decrease in Vernier visual acuity (Agurell, 1976) and eye accommodation (Thomas & Chester, 1973) after smoking marijuana, results that were replicated by another study on the ocular manifestations of CBs, which observed a decrease in Snellen visual acuity (Shapiro, 1974).

However, even though those preliminary studies had suggested an effect of marijuana on various aspects of vision, it was not until the study by Dawson, Jiménez-Antillon, Perez, and Zeskind (1977), that a systematic approach to the effects of marijuana on vision was accomplished. Here the authors studied the effects of chronic marijuana use on various eye functions in human subjects. They found that marijuana users presented consistently smaller pupil diameters compared to nonusers. They also reported, in agreement with previous studies (Agurell, 1976; Shapiro, 1974), a decrease in visual acuity during monocular vision, measured by the ability to read Snellen letters. However, they did not find significant differences in this parameter between luminance levels when comparing users with nonusers. The authors also analyzed color and brightness matching. For this purpose subjects were presented with stimuli of different tones and intensities of red and green in each visual hemifield and were asked to change either parameter until both stimuli appeared equal. Marijuana users presented values similar to those of nonusers in terms of midpoint color match and brightness, but users showed a significant decrease in the range of color and brightness at which they could successfully perform the task, confirming the previous findings in pigeons (Siegel, 1969). In addition, Dawson and colleagues also reported less complete dark adaptation and increased photosensitivity in marijuana users compared to nonusers (Dawson et al., 1977), results that were subsequently partially corroborated (Adams, Brown, Haegerstrom-Portnoy, Flom, & Jones, 1978), showing that marijuana had a significant acute effect, decreasing the time delay to recover from an intense light exposure (measured as the ability to detect low-contrast stimuli).

In line with these previous studies, more recent research has reported the presence of visual disturbances in human subjects both acutely after smoking marijuana and as flashbacks after ceasing cannabis use (Lerner, Goodman, Rudinski, & Bleich, 2011). In this study subjects claimed the presence of seven types of visual disturbances including blurred perception, distorted perception of distance, movement illusions, color intensification, color dimness, distortions in size perception, and blending of patterns. These disturbances were perceived up to 12 weeks after the cessation of cannabis use, becoming less intense and frequent over this period until completely disappearing. Thus, although not tested, the authors argued the possibility of a downregulation in serotonergic neurotransmission by marijuana, as cannabis has been previously seen to affect serotonin neurotransmission, and various medicines acting on the mechanism of action of this neurotransmitter have been shown to induce visual disturbances (Lauterbach, Abdelhamid, & Annandale, 2000).

Marijuana has a potential facilitatory action on night vision. The phenomenon was first reported by West in 1991 in a short communication (West, 1991). In this, the author reflected on his observations about the astonishing improvement in night vision of Jamaican fishermen who could navigate with their boats through difficult coral reefs during the night after smoking or ingesting marihuana. A similar study some years later by Merzouki and Molero Messa (1999) reported the same effects observed by West, but in Moroccan fishermen. These fishermen attributed the ability of seeing during the night to the consumption of "kif," a mixture of tobacco and marijuana. However, neither of these authors attempted a systematic measurement of their observations. It was not until 2004 that Russo and colleagues objectively tested this phenomenon (Russo, Merzouki, Mesa, Frey, & Bach, 2004). They studied a group of fishermen of northern Morocco and reported a significant decrease in scotopic sensitivity and dark adaptation after smoking kif (Figure 1(A)). However, and even though these effects seemed to be dose dependent, they did not find any alteration in Snellen chart visual acuity as previously reported (Dawson et al., 1977). When trying to explain their results the authors suggested an action of marijuana on the retina as the likeliest mechanism, basing this assertion on the fact that the retina expresses

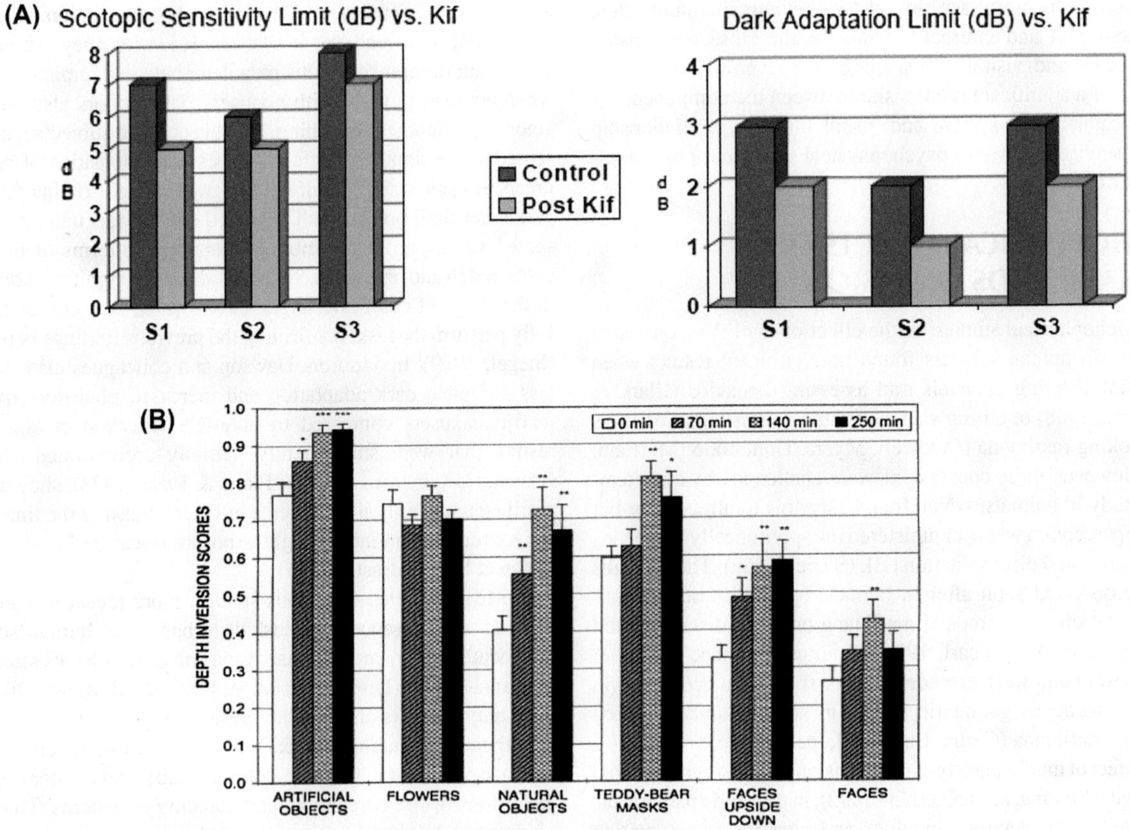

FIGURE 1 Effects of cannabis on visual function. (A) Scotopic sensitivity limit (left) and dark adaptation limit (right) before and after smoking cannabis in three subjects. A decrease of at least 1 dB occurred in all subjects for both conditions. *(Adapted from Russo et al. (2004).)* (B) Depth inversion scores during the presentation of various classes of objects to human subjects before and after the administration of THC. Scores were decreased for all objects but flowers. Asterisks indicate the error probability revealed by respective Wilcoxon tests (*$p \leq 0.05$; **$p \leq 0.01$; ***$p \leq 0.001$). *Adapted from Leweke et al. (1999).*

CB receptors in higher concentrations than other parts of the central nervous system. They also suggested a stimulatory effect of marijuana on rods to be responsible for the improvement in night vision, but they did not go further into the issue.

Another phenomenon that has thrown some light on our current knowledge of the effects of CBs on vision is binocular depth inversion (an illusion created by the brain to make sense of visual irrational experiences generated when the visual information that should normally reach one eye is sent to the other, e.g., an "inside-out face"). In this situation, the information provided by bottom-up mechanisms must be modified or disregarded by top-down influences, as otherwise the system would not be able to properly represent the erroneous information provided by bottom-up pathways during the illusion. Using this tool, Emrich et al. (1991), found a strong impairment of depth inversion in human subjects after smoking marijuana. They attributed these results to an action of marijuana as a modulator in the interactions between visual sensation and perceptual awareness.

Leweke, Schneider, Thies, Münte, and Emrich (1999), went a step further in the study of this relationship by investigating the effect of oral administration of the synthetic THC dronabinol on the binocular depth inversion of various classes of natural and artificial objects. In this study dronabinol was chosen to provide a less contaminated source of THC than in previous studies (Figure 1(B)).

In agreement with Emrich et al. (1991), the authors found that dronabinol induced a strong impairment of binocular depth inversion for both types of stimuli—natural and artificial—with the interesting exception of flowers. Accordingly, and considering that binocular depth inversion results from a domination of top-down over bottom-up signals, the authors assumed that the fact that binocular depth inversion is reduced by THC indicates an effect of this compound on the top-down processing of visual information. They support this assumption by the observation that binocular depth inversion for objects with high levels of familiarity showed less reduction than artificial ones, which is in agreement with the observation that the influence of top-down processing on the depth inversion of objects with high familiarity is higher than for objects with lower familiarity. So, if THC is affecting top-down processing, the stronger this path, the less affected by the drug it should be. The hypothesis was further supported by the observation of a time decline on the effects of THC, which suggested a dose-dependent effect.

Another study led by Leweke, Schneider, Radwan, Schmidt, and Emrich (2000) aimed to test the theory that the disruption of binocular depth inversion induced by cannabis is due to an impairment of the top-down mechanisms involved in interpreting and contextualizing visual information. To this end the authors tested the effect of psychotropic (nabilone) and nonpsychotropic (cannabidiol) CBs on binocular depth inversion. As they predicted, they found that binocular depth inversion was significantly affected only during the administration of the psychotropic CB nabilone and concluded that psychotropic CBs weakened the correction effect of top-down pathways during the presentation of erroneous visual information (binocular depth inversion). These results were further confirmed by two other works (Koethe et al., 2006; Semple, Ramsden, & McIntosh, 2003) in healthy individuals and schizophrenic patients, respectively.

Also interesting are two case studies reporting an improvement in nystagmus in an individual with a congenital manifestation of this condition (Pradeep, Thomas, Roberts, Proudlock, & Gottlob, 2008) and in a patient suffering from multiple sclerosis (Schon et al., 1999). In both cases smoking marijuana decreased amplitude, frequency, and intensity of nystagmus, which was reflected in an improvement in visual acuity. Interestingly, a previous study by Brown, Adams, Haegerstrom-Portnoy, and Jones (1975), reported a decrease in visual acuity induced by marijuana only when targets were in motion and required coordinated eye movements, and Adams, Brown, Flom, Jones, and Jampolsky (1975), found no alterations in static visual acuity after consumption of marijuana.

Several studies on different pathological conditions have also offered support for the role of CBs on vision. In some cancer patients the analgesic effect induced by the administration of THC is often accompanied by alterations in vision (Noyes, Brunk, Baram, & Canter, 1975). It has also been reported while assessing the effects of smoked cannabis on multiple sclerosis that patients not only had improved motor symptoms or reduced pain, but also experienced a decrease in the incidence of double and blurred vision associated with the illness (Consroe, Musty, Rein, Tillery, & Pertwee, 1997). Patients have also reported previously irrelevant visual patterns to become potent distracters after the intravenous injection of THC in a study relating the CB system with psychosis (D'Souza et al., 2004). And very recently Lax, Esquiva, Altavilla, and Cuenca (2014) demonstrated a protective effect of the synthetic CB HU210 on retinal structure and function against retinitis pigmentosa.

Although less often, the effect of marijuana on vision has also been studied in animals from a psychophysical point of view. Here results are even more scarce and contradictory and would need further investigation. Apart from the aforementioned study by Siegel on pigeons (Siegel, 1969), Hockman, Perrin, and Kalant (1971), observed strange behaviors in cats after intraperitoneal injections of THC, reporting that these animals usually stared into space and followed nonexistent stimuli with their eyes. In contrast, a pair of studies in behaving monkeys (Schulze et al., 1988) found no significant effects on a task involving color discrimination or remembrance–identification of geometric shapes before and after the intravenous administration of THC or the breathing of marijuana smoke. However, Aigner found a significant drop in performance of as much as 20% during a visual recognition task in monkeys after either acute or chronic oral administration of THC (Aigner, 1988). More recently, a study in behaving rats performing a visual discrimination task found an impairment in brightness detection after the administration of the synthetic CB1 agonist WIN 55,212-2 (Arguello & Jentsch, 2004). Interestingly, these effects were abolished by the coadministration of SR141716A, a specific CB1 antagonist, which had no action when administered alone.

ELECTROPHYSIOLOGICAL EFFECTS OF CANNABINOIDS ON VISION

Electroencephalography (EEG) has been one of the main electrophysiological techniques used to assess the relationship between CBs and visual perception. Early studies found no significant changes in EEG of human subjects after smoking marijuana. However, Tinklenberg, Kopell, Melges, and Hollister (1972), reported an increase in the amplitude and latency of components within visually evoked responses. These results were confirmed later (Lewis, Dustman, Peters, Straight, & Beck, 1973) in a study in which visual evoked responses to flashing stimuli were recorded on occipital, central, and frontal scalp areas in healthy subjects, before and after the oral administration of THC. Here, a significant increase in the latency of primary and secondary components of visual evoked responses was observed at all recorded regions, although the effects were more pronounced on the occipital scalp. In addition, effects on the magnitude of visual evoked responses were much less evident, showing a general decremental trend, which achieved significance only at the occipital area. The authors suggested that this result was unusual, as most drug-induced alterations in EEG evoked responses, like those induced by barbiturates, affected only the amplitude and not the latency of the responses. They proposed that this lack of action of marijuana on the amplitude of visual evoked responses along with the consistently increased latencies reflected a lack of drug action on brainstem centers, since brainstem centers were known to be greatly affected by drugs modulating the amplitude of late components. Finally they also suggested a general increase in the firing threshold of all neuronal elements involved in the circuit underpinning visual evoked responses as a possible mechanism to explain their results.

More recently Patrick et al. (1995), found no significant differences in the magnitude of P300 in chronic marijuana smokers. However, P300 is an EEG waveform that appears symmetrically around the central–parietal cortex and reflects complex information processing and attentional vigilance rather than a purely sensory signal. Thus, it is likely that the authors did not see any effect because of the nature of the signal itself, which seems unlikely to be highly modulated by visual processing or any other modulation of primary sensory perception.

Another important EEG study investigating the effects of marijuana on vision was accomplished by Skosnik, Krishnan, Vohs, and O'Donnell (2006). In this study the authors measured occipital steady-state visual evoked potentials (SSVEPs) during passive photic stimulation with flickering lights on habitual cannabis users versus normal human controls (Figure 2(A)). SSVEPs are an EEG response to visual periodic stimulation generated by the tuning of the primary visual cortex to both the frequency and the phase of the stimulus, so it is an indicator of the functional state of the neural circuits in the visual cortex. The authors observed a decrease in the magnitude of SSVEPs in the cannabis users and this observation was positively correlated with the age of first cannabis use. Thus, these results clearly point to an alteration in sensory function in V1 of human subjects, probably due to an alteration in the oscillatory activity of the neurons in this cortical area.

Böcker et al. (2010), studied the effect of smoked marijuana with high levels of THC on visual evoked response potentials during a visual selective attention task in which subjects had to

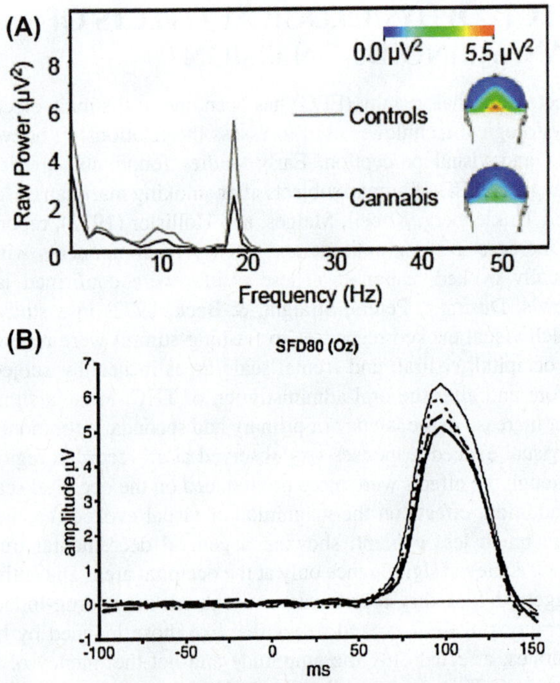

FIGURE 2 Cannabis exposure alters EEG. (A) Averaged power spectra recorded on the occipital area of the skull in human cannabis users and controls. Topographic maps from EEGs recorded on the same subjects in response to periodic photic stimuli at 18 Hz are also shown. *(Adapted from Sosnik et al. (2006).)* (B) Averaged evoked response potentials to gratings recorded on the occipital area of the skull in groups of human subjects exposed to various concentrations of CBs. Placebo (continuous thin line), low CB concentration (thin dashed line), medium CB concentration (thick dashed line), high CB concentration (continuous thick line). The amplitude of the response decreased with increasing exposure. *Adapted from Böcker et al. (2010).*

discriminate between spatial frequency and orientation of square-wave gratings (Figure 2(B)). They found that marijuana induced a dose-dependent decrease in the amplitude of occipital SDF80, an evoked response potential that represents sensory processing that is not affected by attention. As a consequence, this study corroborated earlier findings (Sknosnik et al., 2006) and reflected a dose-dependent effect of THC on brain correlates of sensory processing affecting bottom-up pathways involved in spatial frequency discrimination. Similar decrements in evoked response potentials related to visual sensory processing have also been reported for both acute (Ilan, Gevins, Coleman, ElSohly, & de Wit, 2005) and chronic cannabis use (Patrick, Straumanis, Struve, Fitzgerald, & Manno, 1997).

There is less available published work on the effects of cannabis at a cellular, neurophysiological level within the sensory systems. The presence of CB receptors at the level of the retina was first demonstrated by detecting the presence of mRNAs for these receptors in the rat retina (Buckley, Hansson, Harta, & Mezey, 1998; Porcella, Casellas, Gessa, & Pani, 1998). A detailed study of the cellular distribution of CB1 receptors came soon after in two contemporaneous studies (Straiker, Maguire, Mackie, & Lindsey, 1999; Yazulla, Studholme, McIntosh, & Deutsch, 1999). These studies found CB1 receptors in retinal inner and outer plexiform layers of rhesus monkeys, mice, rats,

chicken, goldfish, and tiger salamander, especially in the synaptic terminals of photoreceptors, although amacrine, horizontal, and ganglion cells were also positive in the majority of these species.

The physiological role of CBs in the retina has been a subject of extensive research. It has been demonstrated that the activation of CB1 receptors inhibits the release of noradrenaline, dopamine, and glutamate (Opere et al., 2006; Schlicker, Timm, & Göthert, 1996) and disrupts γ-aminobutyric acid (GABA)-ergic inhibitory neurotransmission (Warrier & Wilson, 2007). In addition, CB agonists have also been seen to inhibit the presence of high-voltage calcium currents in retinal ganglion cells, effectively modulating retinal output (Lalonde, Jollimore, Stevens, Barnes, & Kelly, 2006). Suppression of calcium and potassium currents in bipolar cells has also been reported, suggesting a retrograde CB-mediated inhibition by retinal ganglion cells (Straiker et al., 1999). Interestingly, all *on* bipolar cells are affected by CBs, while only one-third of *off* cells are modulated. It is under debate, thus, whether endocannabinoids experience a bias toward the *on* pathway in the retina, which would imply ganglion cell CB-mediated retrograde inhibition of bipolar cells as present mainly during increments rather than decrements of light intensity.

With regard to the effect of CBs on photoreceptors, it has been reported that bipolar cells can exert a retrograde inhibitory action on cone excitability by synthesizing and releasing CBs in a calcium-dependent manner (Fan & Yazulla, 2007). This inhibition may follow both a voltage-dependent and a voltage-independent mechanism. Thus the former would allow bipolar cells to act as a positive feedback, amplifying the reduction of glutamate release during increments in light intensity, while the latter would act as a negative feedback enhancing the effect of light increments (Fan & Yazulla, 2007; Klooster, Studholme, & Yazulla, 2001).

The presence of CB receptors in the elements of the visual pathway beyond the retina has been seldom studied. General immunohistochemical localization of CB1 receptors in the rat and mouse brain originally reported only sparse staining in the dorsal thalamus, including the dorsal lateral geniculate nucleus and visual cortex (Herkenham et al., 1991; Moldrich & Wenger, 2000). Nonetheless, high levels of AEA have been found in the thalamus (Felder et al., 1996) along with high levels of fatty acid amide hydrolase, its degradative enzyme (Thomas & Chester, 1973). These results have been supported by studies specifically targeting the localization of CB1 receptors in the lateral geniculate nucleus and visual cortex, which have found high levels of these receptors in both areas of the mouse (Sun et al., 2011; Yoneda et al., 2013).

Perhaps the fact that CB1 receptors were traditionally considered to be absent from central brain structures involved in visual processing hampered a systematic study on the effects of CBs in these areas from a cellular/system neurophysiological point of view. There are, however, a handful of studies that have thrown some light on the issue. The first report found in the literature relating the visual effects of CBs from a cellular perspective was the study by Bieger and Hockman (1973). In this work the authors studied the response of lateral geniculate nucleus neurons in the hooded rat during the intravenous administration of THC. They found a consistent THC-induced inhibition in the firing rate of cells activated by light, while cells inhibited by light either remained unchanged or experienced an

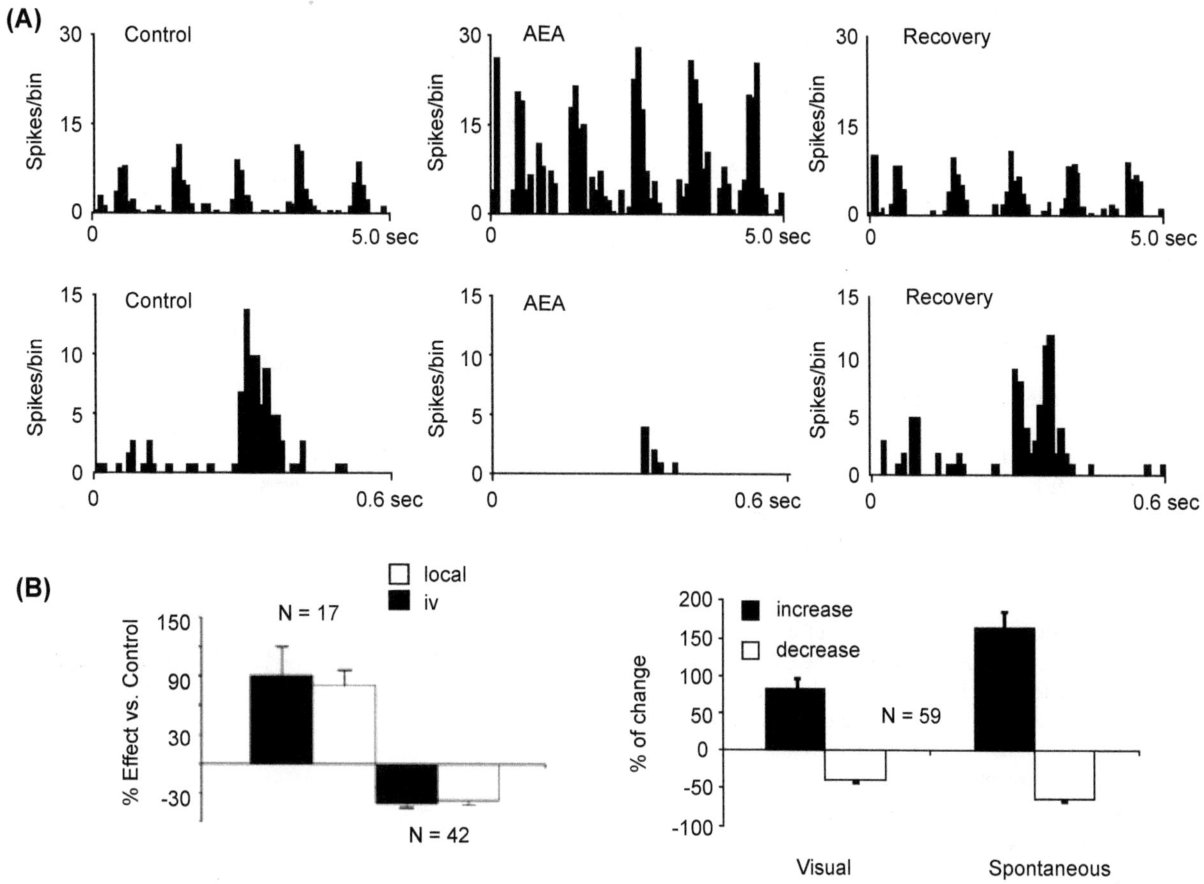

FIGURE 3 **Endocannabinoids modulate thalamic neuronal responses.** (A) Responses of two well-isolated neurons in the dorsolateral geniculate nucleus (dLGN) of the rat to a grating (upper traces) and spot (lower traces) of optimal characteristics before and after local administration of AEA by pressure ejection. The response of the first neuron to the visual stimulus increases during the administration of AEA, reverting to control conditions during recovery. The contrary happens to the other neuron. (B) Bar graphs showing averaged population responses of single cells in the dLGN of the rat during local and systemic administration of endocannabinoid agonists. Both routes of administration induced similar results in sign and magnitude (left), although effects were more prominent during spontaneous than during visual responses (right), effectively increasing signal-to-noise ratio. *Adapted from Dasilva et al. (2012).*

increase in firing. When stimulating the optic nerve and recording evoked field potentials in the lateral geniculate nucleus, the authors found that THC affected only field potential components due to intrinsic geniculate activity and not to the activation by the optic nerve. Further, when the biogenic amine precursor L-5-hydroxytryptophan was administered after THC, the original effect of cannabis on light-sensitive neurons was reversed. Accordingly, the authors proposed THC to have a direct action on lateral geniculate nucleus neurons, probably by interfering with serotonergic midbrain projections, which could result in blurred vision and distortions of perception.

While Bieger and Hockman's (1973) study was the first attempt to relate visual function and neuronal activity in terms of CB modulation, this was very preliminary work. Only multiunit activity and local field potentials were studied, without a detailed characterization of neuronal responses. Consequently neither receptive fields nor response properties were fully characterized. Visual stimulation consisted of a broad illumination of the entire eye, which raises the question of whether the recorded responses were the result of groups of rather than single neurons. Further, owing to the intravenous administration of THC, the authors expressed doubt about whether the observed effects were the result of a direct action on lateral geniculate nucleus neurons or whether they were the consequence of a more general alteration in brain function at higher levels.

A new study came about more recently attempting a systematic characterization of the effect of CBs in the visual thalamus (Dasilva et al., 2012). Here the extracellular activity of well-isolated single neurons was recorded in the lateral geniculate nucleus of the hooded rat while manipulating the activation of CB1 receptors (Figure 3(A) and (B)). In agreement with the earlier study of Bieger and Hockman (1973), the intravenous administration of various CB agonists induced an inhibition of visual responses in the majority of the cells tested, while a minority of the cells expressed an excitatory effect. However, in contrast to that previous study, Dasilva and coworkers did not find any relationship between the observed effects and the physiological characterization of cells (on vs off, sustained vs transient). Similar effects were observed when the CB agonists were administered locally into the lateral geniculate nucleus by iontophoresis, indicating a local effect of CBs in the thalamus (Figure 3(A) and (B)), thereby further validating the observations of Bieger and Hockman. In addition, the observed effects were demonstrated to be specifically mediated by CB1 receptors and to affect also spontaneous activity. Spontaneous activity, like the visual responses, increased or decreased in magnitude during the administration of CB1 agonists, although the magnitude of the effect was higher than during visual responses. This work confirmed that CBs can act locally in the visual thalamus and that the endocannabinoid system modulates the processing of information by this nucleus, altering how this information is transmitted to the cortex. In this regard, the authors described how CBs effectively increased the signal-to-noise ratio and decreased the variability of neuronal responses in their population of cells and suggested these results to support a putative role of CBs in sending clearer and less variable information to the visual cortex. The results corroborate previous proposals on the "filtering" of information by CBs in the visual system (D'Souza et al., 2004; Solowij, Michie, & Fox, 1991).

As of this writing only one study has been published testing the effect of CBs in the visual cortex at the cellular level (Ohiorhenuan et al., 2014). In this study (Figure 4(A)) the intravenous administration of a synthetic CB receptor agonist (CP 55,940) induced an alteration in network dynamics consisting of a decrease in EEG and local field potential (LFP) power at the low gamma range (25–50 Hz), with a decrease in LFP–LFP coherence. Interestingly, the authors were unable to correlate these obvious changes in the firing pattern of individual neurons with alterations at the population level, the only exception being a decrease in firing rate in neurons that experienced an increase in latency and duration of visual responses. This CB-induced decrease in firing rate is in agreement with Dasilva et al. (2012), who, as mentioned above, observed a general decrease in the firing rate of thalamic neurons. Taken together, these results raise the intriguing possibility of a synergistic effect of CBs at different levels of the thalamocortical circuit and beyond.

In support of this hypothesis, a 2012 study (Figure 4(B)) in the somatosensory system (Sales-Carbonell et al., 2013) has found that the activation of CB1 receptors present at glutamatergic cortical neurons decreased cortical synchrony at relatively high frequencies (>12 Hz), while, in the same study, increased synchrony of thalamocortical oscillations in the spindle-like range (4–8 Hz) was also reported as a consequence of the activation of CB1 receptors present at thalamic striatonigral synapses. Similar results have been found in the visual system at the level of the thalamus (Dasilva et al., 2014), where the activation of CB1 receptors specifically induced the appearance of delta-like oscillations (1–4 Hz) as a consequence of a general hyperpolarizing effect of CBs in this nucleus. This hyperpolarizing effect of endocannabinoids in the visual thalamus could thus result from the effect of these neuromodulators in the thalamic reticular nucleus, where 2-AG inhibits intrinsic inhibitory synapses (Sun et al., 2011). These synapses limit oscillatory activity in thalamic circuits (Figure 4(C)) and as a consequence, its inhibition by endocannabinoids would enhance intrinsic thalamic synchronicity in the spindle and delta-like ranges as demonstrated by Sales-Carbonell et al., 2013, and Dasilva et al. (2014). Thus, these results, along with the findings of Ohiorhenuan et al. (2014), in the cortex, seem to confirm a dynamic role of endocannabinoid signaling in regulating synchronous activity at various levels of the thalamocortical circuit with obvious implications for visual function. While these

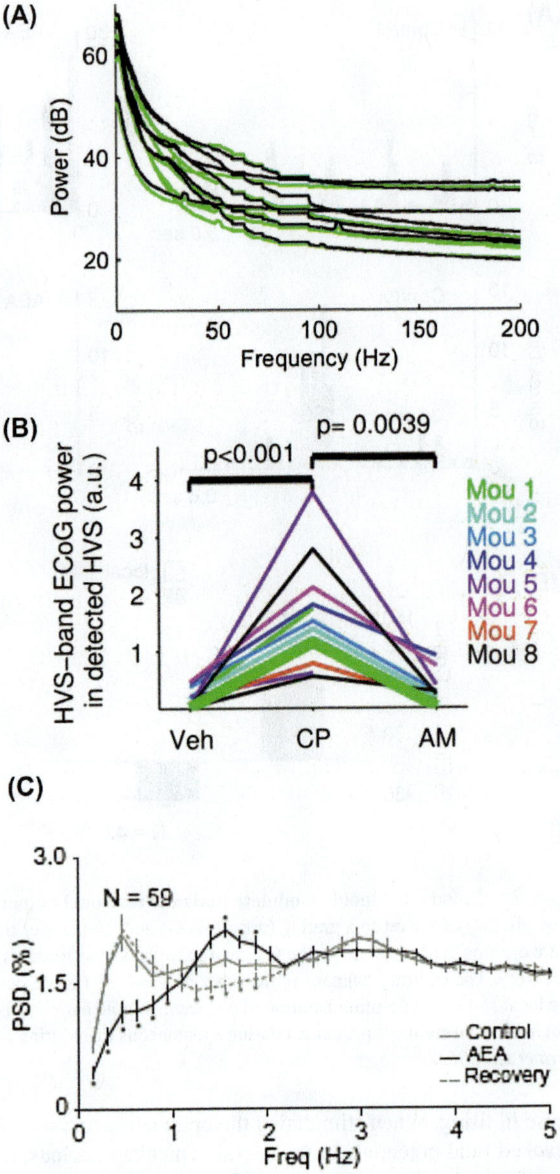

FIGURE 4 **Endocannabinoids play a role in neuronal rhythmicity and synchrony.** (A) LFP power of seven recording protocols in the visual cortex of two anesthetized monkeys before (black) and after (green) the intravenous administration of the CB agonist CP 55,940. The agonist decreased the power of low-gamma frequencies in both animals. (B) Systemic administration of CP 55,940 induces high-voltage spindles (HVS) in the cortex of mice. This effect is abolished by the administration of the CB antagonist AM251. Each line represents an experiment and each color an animal. *(Adapted from Sales-Carbonell et al. (2013).)* (C) Power spectral density (PSD) graph showing the effect induced by the intravenous administration of AEA on the oscillatory activity of simultaneously recorded groups of rat dLGN neurons. Note how cells start to oscillate at delta-like frequencies (~1.5 Hz) in the presence of AEA. *$p \leq 0.05$ vs control. *Adapted from Dasilva et al. (2014).*

basic cellular mechanisms may underpin some of the visual shifts seen at the psychophysical level, they will require further work at higher cognitive levels to fully understand the pathology associated with cannabis misuse.

CONCLUSIONS AND FUTURE DIRECTIONS

The role of endogenous and exogenous CBs in the visual system has been little studied. However, evidence from the literature suggests that the endocannabinoid system is an important component of visual processing. It is important to understand the basic physiology to fully understand the potentially pathological effects of exogenous substances acting on this system, which may have complex effects at various levels in visual processing.

Psychophysical studies both in humans and in animals have demonstrated several effects of CBs, administered both as pure compounds and as the drug marijuana, on color, form, and brightness perception; peripheral vision and visual acuity; night vision; perception of motion; and reconstruction of illusory and erroneous visual information. In addition, direct and indirect evidence has been brought about from EEG studies observing different alterations in visual evoked responses after acute or chronic consumption of various CB agonists during the performance of a wide range of behavioral tasks. Finally, although less intensively studied, firm evidence on the role of CBs in visual processing has been shown at various hierarchical levels from a neurophysiological perspective/cellular level. Not only the internal processing of visual information by the retina, but also further processing and computation of visual signals by the thalamus and cortex have been demonstrated to be modulated by CBs at both single-cell and population levels. However, multiple questions remain unanswered with regard to the exact mechanisms by which endocannabinoids modulate the processing of visual information in the retina and different levels of the thalamocortical circuit. New in vitro studies in the retina and beyond will be needed to clarify the exact synaptic mechanisms and neurotransmitters involved in the modulation of visual information processing by endocannabinoids and new and more detailed in vivo studies will also be necessary to integrate the mechanism of action of endocannabinoids into the physiology of the visual system. Only once these basic issues are resolved will we be fully able to explain the effects of cannabis on vision, as mediated via the endocannabinoid CB1 receptor.

APPLICATIONS TO OTHER ADDICTIONS AND SUBSTANCE MISUSE

The endocannabinoid system utilizes its own receptors through which it regulates the activity of the main neurotransmitter systems of the brain. Hence it has the capability of influencing the action of many other substances. For instance, CBs modulate the release of GABA in many brain areas; these receptors are the target receptors for ethanol or barbiturates.

Furthermore, the CBs can alter the activity of the dopamine-mediated process of reward, with a prominent role in the neurobiology of addiction.

The CB system has been implicated in the pharmacological and behavioral effects of ethanol, including tolerance development. Blocking the CB1 receptor has been shown to reduce voluntary ethanol intake in rats.

Similarly, a relationship between the endocannabinoid system and cocaine, morphine, and amphetamine abuse has been shown.

On the other hand, CBs demonstrate neuroprotective properties against methamphetamine-induced damage and interact with opioid agonists for inducing an enhanced antinociceptive response.

In the visual system, the interaction between CBs and other substances has not been studied in detail; however, it is clear that the effect of simultaneous consumption of several substances could produce interactions between them, since they act in similar processes, such as visual memory, attention, and ocular movements.

DEFINITION OF TERMS

Apparent motion This is a phenomenon that occurs when visual stimuli that are successively presented at different locations are perceived as a continuous motion.

Peripheral vision This is the ability to see objects and movement outside the fixation point, in contrast to central vision, which refers to vision at the fixation spot.

Vernier visual acuity This is a visual acuity test that measures the ability to detect a misalignment between two line segments or gratings.

Bottom-up and top-down processing This refers to different mechanisms of processing sensory inputs. In the bottom-up system each station of the process analyzes the input and sends the result to the next level of processing. In the top-down mechanism, the information from higher levels is sent to the previous ones, influencing, even controlling, the information processing.

Multiple sclerosis This is a syndrome in which the insulating myelin layer of nerve cells in the brain and spinal cord is damaged. It can cause visual deficits such as optic neuritis, diplopia, or nystagmus.

Neural oscillatory activity This is a pattern of rhythmic or repetitive activity of neurons in the central nervous system. This activity can be recorded at the cellular level or the system level. Different frequencies have been associated with different levels of consciousness.

Retinitis pigmentosa This is an inherited, degenerative eye disease that causes vision impairment and often progresses to blindness.

KEY FACTS OF THE CANNABINOID SYSTEM

- The endocannabinoid receptors were discovered at the beginning of the 1990s.
- They contain two main receptors broadly distributed through the central nervous system.
- The CB system has been implicated in processes such as regulation of appetite, memory, analgesia, or sleep.
- Research interest on the system has grown with the development of molecules targeting the CB receptors.
- It has been described in many animals.
- The endocannabinoid system is an important component of visual processing.
- The CB system shows promising therapeutic possibilities for the treatment of addiction to many substances other than cannabis.
- Medical disorders such as multiple sclerosis or anorexia are also potential targets for CB-mediated therapies.

SUMMARY POINTS

- Endocannabinoid system components—both ligands and receptors—are broadly distributed throughout the central nervous system.
- Studies reflect a critical role for CBs in visual perception from the retina to higher cortical centers.

- CBs affect color, form, and brightness perception; peripheral vision and visual acuity; night vision; perception of motion; and reconstruction of illusory and erroneous visual information.
- Marijuana facilitates night vision. Jamaican fishermen navigate with their boats through difficult coral reefs during the night after smoking or ingesting marijuana.
- Knowing the synaptic mechanisms and neurotransmitters involved in the modulation of visual information processing by endocannabinoids is an open question to be answered in the future.
- Not only vision, but also other sensory systems are influenced by CB-mediated activity.
- CBs mediate the reward system of the brain, hence they are implicated in the abuse of many other substances.

REFERENCES

Adams, A. J., Brown, B., Flom, M. C., Jones, R. T., & Jampolsky, A. (1975). Alcohol and marijuana effects on static visual acuity. *American Journal of Optometry and Physiological Optics*, 52(11), 729–735.

Adams, A. J., Brown, B., Haegerstrom-Portnoy, G., Flom, M. C., & Jones, R. T. (1978). Marijuana, alcohol, and combined drug effects on the time course of glare recovery. *Psychopharmacology (Berlin)*, 56(1), 81–86.

Agurell, S. (1976). Pharmacokinetics of delta-9-THC in man after smoking: relations to physiological and psychological effects. In M. C. Braude, & S. Szara (Eds.), *The pharmacology of marihuana* (Vol. I) (pp. 49–61).

Aigner, T. G. (1988). Delta-9-tetrahydrocannabinol impairs visual recognition memory but not discrimination learning in rhesus monkeys. *Psychopharmacology (Berlin)*, 95(4), 507–511.

Arguello, P. A., & Jentsch, J. D. (2004). Cannabinoid CB1 receptor-mediated impairment of visuospatial attention in the rat. *Psychopharmacology (Berlin)*, 177(1–2), 141–150.

Bieger, D., & Hockman, C. H. (1973). Differential effects produced by delta 1-tetrahydrocannabinol on lateral geniculate neurones. *Neuropharmacology*, 12(3), 269–273.

Böcker, K. B., Gerritsen, J., Hunault, C. C., Kruidenier, M., Mensinga, T. T., & Kenemans, J. L. (2010). Cannabis with high δ9-THC contents affects perception and visual selective attention acutely: an event-related potential study. *Pharmacology, Biochemistry, and Behavior*, 96(1), 67–74.

Brown, B., Adams, A. J., Haegerstrom-Portnoy, G., & Jones, R. T. (1975). Effects of alcohol and marijuana on dynamic visual acuity: I. Threshold measurements. *Perception and Psychophysics*, 18(6), 441–446.

Buckley, N. E., Hansson, S., Harta, G., & Mezey, E. (1998). Expression of the CB1 and CB2 receptor messenger RNAs during embryonic development in the rat. *Neuroscience*, 82(4), 1131–1149.

Caldwell, D. F., Myers, S. A., Domino, E. F., & Merriam, P. E. (1969). Auditory and visual threshold effects of marihuana in man. *Perceptual and Motor Skills*, 29(3), 755–759.

Carlin, A. S., Bakker, C. B., Halpern, L., & Post, R. D. (1972). Social facilitation of marijuana intoxication: impact of social set and pharmacological activity. *Journal of Abnormal Psychology*, 80(2), 132–140.

Clark, N. D., & Nakashima, E. N. (1968). Experimental studies on marihuana. *The American Journal of Psychiatry*, 125(3), 135–140.

Consroe, P., Musty, R., Rein, J., Tillery, W., & Pertwee, R. (1997). The perceived effects of smoked cannabis on patients with multiple sclerosis. *European Neurology*, 38(1), 44–48.

Dasilva, M. A., Grieve, K. L., Cudeiro, J., & Rivadulla, C. (2012). Endocannabinoid CB1 receptors modulate visual output from the thalamus. *Psychopharmacology (Berlin)*, 219(3), 835–845.

Dasilva, M., Grieve, K. L., Cudeiro, J., & Rivadulla, C. (2014). Anandamide activation of CB1 receptors increases spontaneous bursting and oscillatory activity in the thalamus. *Neuroscience*, 265, 72–82.

Dawson, W. W., Jiménez-Antillon, C. F., Perez, J. M., & Zeskind, J. A. (1977). Marijuana and vision–after ten years' use in Costa Rica. *Investigative Ophthalmology and Visual Science*, 16(8), 689–699.

Devane, W. A., Dysarz, F. A., 3rd, Johnson, M. R., Melvin, L. S., & Howlett, A. C. (1988). Determination and characterization of a cannabinoid receptor in rat brain. *Molecular Pharmacology*, 34(5), 605–613.

Devane, W. A., Hanus, L., Breuer, A., Pertwee, R. G., Stevenson, L. A., Griffin, G., ... Mechoulam, R. (1992). Isolation and structure of a brain constituent that binds to the cannabinoid receptor. *Science*, 258(5090), 1946–1949.

Domino, E., Rennick, P., & Pearl, J. H. (1976). Short-term neuropsychopharmacological effects of marijuana smoking in experienced male users. In M. C. Braude, & S. Szara (Eds.), *The pharmacology of marijuana* (Vol. I) (pp. 393–412).

D'Souza, D. C., Perry, E., MacDougall, L., Ammerman, Y., Cooper, T., Wu, Y. T., ... Krystal, J. H. (2004). The psychotomimetic effects of intravenous delta-9-tetrahydrocannabinol in healthy individuals: implications for psychosis. *Neuropsychopharmacology*, 29(8), 1558–1572.

Emrich, H. M., Weber, M. M., Wendl, A., Zihl, J., von Meyer, L., & Hanisch, W. (1991). Reduced binocular depth inversion as an indicator of cannabis-induced censorship impairment. *Pharmacology, Biochemistry, and Behavior*, 40(3), 689–690.

Fan, S. F., & Yazulla, S. (2007). Retrograde endocannabinoid inhibition of goldfish retinal cones is mediated by 2-arachidonoyl glycerol. *Visual Neuroscience*, 24(3), 257–267.

Felder, C. C., Nielsen, A., Briley, E. M., Palkovits, M., Priller, J., Axelrod, J., ... Becker, G. W. (1996). Isolation and measurement of the endogenous cannabinoid receptor agonist, anandamide, in brain and peripheral tissues of human and rat. *FEBS Letters*, 393, 231–235.

Herkenham, M., Lynn, A. B., Johnson, M. R., Melvin, L. S., de Costa, B. R., & Rice, K. C. (1991). Characterization and localization of cannabinoid receptors in rat brain: a quantitative in vitro autoradiographic study. *Journal of Neuroscience*, 11(2), 563–583.

Hockman, C. H., Perrin, R. G., & Kalant, H. (1971). Electroencephalographic and behavioral alterations produced by delta-1-tetrahydrocannabinol. *Science*, 172(3986), 968–970.

Hollister, L. E. (1971). Marihuana in man: three years later. *Science*, 172, 21–29.

Ilan, A. B., Gevins, A., Coleman, M., ElSohly, M. A., & de Wit, H. (2005). Neurophysiological and subjective profile of marijuana with varying concentrations of cannabinoids. *Behavioural Pharmacology*, 2005(16), 487–497.

Kanayama, G., Rogowska, J., Pope, H. G., Gruber, S. A., & Yurgelun-Todd, D. A. (2004). Spatial working memory in heavy cannabis users: a functional magnetic resonance imaging study. *Psychopharmacology (Berlin)*, 176(3–4), 239–247.

Kiplinger, G. F., Manno, J. E., Rodda, B. E., & Forney, R. B. (1971). Dose-response analysis of the effects of tetrahydrocannabinol in man. *Clinical Pharmacology and Therapeutics*, 12(4), 650–657.

Klooster, J., Studholme, K. M., & Yazulla, S. (2001). Localization of the AMPA subunit GluR2 in the outer plexiform layer of goldfish retina. *The Journal of Comparative Neurology*, 441(2), 155–167.

Koethe, D., Gerth, C. W., Neatby, M. A., Haensel, A., Thies, M., Schneider, U., ... Leweke, F. M. (2006). Disturbances of visual information processing in early states of psychosis and experimental delta-9-tetrahydrocannabinol altered states of consciousness. *Schizophrenia Research, 88*(1–3), 142–150.

Lalonde, M. R., Jollimore, C. A., Stevens, K., Barnes, S., & Kelly, M. E. (2006). Cannabinoid receptor-mediated inhibition of calcium signaling in rat retinal ganglion cells. *Molecular Vision, 12*, 1160–1166.

Lauterbach, E. C., Abdelhamid, A., & Annandale, J. B. (2000). Posthallucinogen-like visual illusions (palinopsia) with risperidone in a patient without previous hallucinogen exposure: possible relation to serotonin 5HT2a receptor blockade. *Pharmacopsychiatry, 33*(1), 38–41.

Lax, P., Esquiva, G., Altavilla, C., & Cuenca, N. (2014). Neuroprotective effects of the cannabinoid agonist HU210 on retinal degeneration. *Experimental Eye Research, 120*, 175–185.

Lerner, A. G., Goodman, C., Rudinski, D., & Bleich, A. (2011). Benign and time-limited visual disturbances (flashbacks) in recent abstinent high-potency heavy cannabis smokers: a case series study. *Israel Journal of Psychiatry and Related Sciences, 48*(1), 25–29.

Leweke, F. M., Schneider, U., Radwan, M., Schmidt, E., & Emrich, H. M. (2000). Different effects of nabilone and cannabidiol on binocular depth inversion in man. *Pharmacology, Biochemistry, and Behavior, 66*(1), 175–181.

Leweke, F. M., Schneider, U., Thies, M., Münte, T. F., & Emrich, H. M. (1999). Effects of synthetic delta9-tetrahydrocannabinol on binocular depth inversion of natural and artificial objects in man. *Psychopharmacology (Berlin), 142*(3), 230–235.

Lewis, E. G., Dustman, R. E., Peters, B., Straight, R. C., & Beck, E. C. (1973). The effects of varying doses of delta 9-tetrahydrocannabinol on the human visual and somatosensory evoked response. *Electroencephalography and Clinical Neurophysiology, 35*(4), 347–354.

Li, J., Tripathi, R. C., & Tripathi, B. J. (2008). Drug-induced ocular disorders. *Drug Safety, 31*(2), 127–141.

Matias, I., Bisogno, T., & Di Marzo, V. (2006). Endogenous cannabinoids in the brain and peripheral tissues: regulation of their levels and control of food intake. *International Journal of Obesity (London), 30*(Suppl. 1), S7–S12.

Matsuda, L. A., Lolait, S. J., Brownstein, M. J., Young, A. C., & Bonner, T. I. (1990). Structure of a cannabinoid receptor and functional expression of the cloned cDNA. *Nature, 346*(6284), 561–564.

Mechoulam, R., & Shvo, Y. (1963). The structure of cannabidiol. *Tetrahedron, 19*, 2073–2078.

Merzouki, A., & Molero Mesa, J. (1999). Le chanvre (*Cannabis sativa* L.) dans la pharmacopee traditionnelle du Rif (Nord du Maroc). *ARS Pharmaceutica, 4*, 233–240.

Moldrich, G., & Wenger, T. (2000). Localization of the CB1 cannabinoid receptor in the rat brain. An immunohistochemical study. *Peptides, 21*(11), 1735–1742.

Moskowitz, H., Sharma, S., & McGlothlin, W. (1972). Effect of marihuana upon peripheral vision as a function of the information processing demands in central vision. *Perceptual and Motor Skills, 35*(3), 875–882.

Noyes, R., Jr., Brunk, S. F., Baram, D. A., & Canter, A. (1975). Analgesic effect of delta-9-tetrahydrocannabinol. *Journal of Clinical Pharmacology, 15*(2–3), 139–143.

Ohiorhenuan, I. E., Mechler, F., Purpura, K. P., Schmid, A. M., Hu, Q., & Victor, J. D. (2014). Cannabinoid neuromodulation in the adult early visual cortex. *PLoS One, 9*(2), e87362.

Opere, C. A., Zheng, W. D., Zhao, M., Lee, J. S., Kulkarni, K. H., & Ohia, S. E. (2006). Inhibition of potassium- and ischemia-evoked [3H] D-aspartate release from isolated bovine retina by cannabinoids. *Current Eye Research, 31*(7–8), 645–653.

Patrick, G., Straumanis, J. J., Struve, F. A., Nixon, F., Fitz-Gerald, M. J., & Soucair, M. (1995). Auditory and visual P300 event related potentials are not altered in medically and psychiatrically normal chronic marihuana users. *Life Sciences, 56*(23–24), 2135–2140.

Patrick, G., Straumanis, J. J., Struve, F. A., Fitzgerald, M. J., & Manno, J. E. (1997). Early and middle latency evoked potentials in medically and psychiatrically normal daily marihuana users: a paucity of significant findings. *Clinical Electroencephalography, 28*, 26–31.

Porcella, A., Casellas, P., Gessa, G. L., & Pani, L. (1998). Cannabinoid receptor CB1 mRNA is highly expressed in the rat ciliary body: implications for the antiglaucoma properties of marihuana. *Brain Research. Molecular Brain Research, 58*(1–2), 240–245.

Pradeep, A., Thomas, S., Roberts, E. O., Proudlock, F. A., & Gottlob, I. (2008). Reduction of congenital nystagmus in a patient after smoking cannabis. *Strabismus, 16*(1), 29–32.

Russo, E. B., Merzouki, A., Mesa, J. M., Frey, K. A., & Bach, P. J. (2004). Cannabis improves night vision: a case study of dark adaptometry and scotopic sensitivity in kif smokers of the Rif mountains of northern Morocco. *Journal of Ethnopharmacology, 93*(1), 99–104.

Sales-Carbonell, C., Rueda-Orozco, P. E., Soria-Gómez, E., Buzsáki, G., Marsicano, G., & Robbe, D. (2013). Striatal GABAergic and cortical glutamatergic neurons mediate contrasting effects of cannabinoids on cortical network synchrony. *Proceedings of the National Academy of Sciences of the United States of America, 110*(2), 719–724.

Schlicker, E., Timm, J., & Göthert, M. (1996). Cannabinoid receptor-mediated inhibition of dopamine release in the retina. *Naunyn Schmiedebergs Archives of Pharmacology, 354*(6), 791–795.

Schon, F., Hart, P. E., Hodgson, T. L., Pambakian, A. L., Ruprah, M., Williamson, E. M., & Kennard, C. (1999). Suppression of pendular nystagmus by smoking cannabis in a patient with multiple sclerosis. *Neurology, 53*(9), 2209–2210.

Schulze, G. E., McMillan, D. E., Bailey, J. R., Scallet, A., Ali, S. F., Slikker, W., Jr., & Paule, M. G. (1988). Acute effects of delta-9-tetrahydrocannabinol in rhesus monkeys as measured by performance in a battery of complex operant tests. *The Journal of Pharmacology and Experimental Therapeutics, 245*(1), 178–186.

Semple, D. M., Ramsden, F., & McIntosh, A. M. (2003). Reduced binocular depth inversion in regular cannabis users. *Pharmacology, Biochemistry, and Behavior, 75*(4), 789–793.

Shapiro, D. (1974). The ocular manifestations of cannabinols. *Ophthalmologica, 168*, 366.

Sharma, S., & Moskowitz, H. (1972). Effect of marihuana on the visual autokinetic phenomenon. *Perceptual and Motor Skills, 35*(3), 891–894.

Siegel, R. K. (1969). Effects of *Cannabis sativa* and lysergic acid diethylamide on a visual discrimination task in pigeons. *Psychopharmacologia, 15*(1), 1–8.

Skosnik, P. D., Krishnan, G. P., Vohs, J. L., & O'Donnell, B. F. (2006). The effect of cannabis use and gender on the visual steady state evoked potential. *Clinical Neurophysiology, 117*(1), 144–156.

Solowij, N., Michie, P. T., & Fox, A. M. (1991). Effects of long-term cannabis use on selective attention: an event-related potential study. *Pharmacology, Biochemistry, and Behavior, 40*(3), 683–688.

Straiker, A. J., Maguire, G., Mackie, K., & Lindsey, J. (1999). Localization of cannabinoid CB1 receptors in the human anterior eye and retina. *Investigative Ophthalmology and Visual Science, 40*(10), 2442–2448.

Sugiura, T., Kondo, S., Sukagawa, A., Nakane, S., Shinoda, A., Itoh, K., ... Waku, K. (1995). 2-Arachidonoylglycerol: a possible endogenous cannabinoid receptor ligand in brain. *Biochemical and Biophysical Research*, 215(1), 89–97.

Sun, Y. G., Wu, C. S., Lu, H. C., & Beierlein, M. (2011). Target-dependent control of synaptic inhibition by endocannabinoids in the thalamus. *The Journal of Neuroscience*, 31(25), 9222–9230.

Thomas, R., & Chester, G. (1973). The pharmacology of marihuana. *Medical Journal of Australia*, 2, 229.

Tinklenberg, J. R., Kopell, B. S., Melges, F. T., & Hollister, L. E. (1972). Marihuana and alcohol, time production and memory functions. *Archives of General Psychiatry*, 27(6), 812–815.

Tsou, K., Brown, S., Sañudo-Peña, M. C., Mackie, K., & Walker, J. M. (1998). Immunohistochemical distribution of cannabinoid CB1 receptors in the rat central nervous system. *Neuroscience*, 83(2), 393–411.

Warrier, A., & Wilson, M. (2007). Endocannabinoid signaling regulates spontaneous transmitter release from embryonic retinal amacrine cells. *Visual Neuroscience*, 24(1), 25–35.

Weil, A. T., Zinberg, N. E., & Nelsen, J. M. (1968). Clinical and psychological effects of marihuana in man. *Science*, 162, 1234–1242.

West, M. E. (1991). Cannabis and night vision. *Nature*, 351(6329), 703–704.

Yazulla, S., Studholme, K. M., McIntosh, H. H., & Deutsch, D. G. (1999). Immunocytochemical localization of cannabinoid CB1 receptor and fatty acid amide hydrolase in rat retina. *The Journal of Comparative Neurology*, 415(1), 80–90.

Yoneda, T., Kameyama, K., Esumi, K., Daimyo, Y., Watanabe, M., & Hata, Y. (2013). Developmental and visual input-dependent regulation of the CB1 cannabinoid receptor in the mouse visual cortex. *PLoS One*, 8(1), e53082.

Chapter 70

Cannabidiol and 5-HT$_{1A}$ Receptors

Manoela V. Fogaça, Alline C. Campos, Francisco S. Guimarães
Department of Pharmacology, Medical School of Ribeirão Preto, University of São Paulo (FMRP-USP), São Paulo, Brazil

Abbreviations

5-HT Serotonin
5-HT$_{1A}$ 5-Hydroxytryptamine-1A receptor
5-HT$_{2A}$ 5-Hydroxytryptamine-2A receptor
8-OH-DPAT 8-Hydroxy-2-(di-*n*-propylamino)tetralin hydrobromide
BNST Bed nucleus of stria terminalis
cAMP Cyclic adenosine monophosphate
CB$_1$ Cannabinoid receptor type 1
CB$_2$ Cannabinoid receptor type 2
CBD Cannabidiol
CFC Contextual fear conditioning
CHO Chinese hamster ovary cells
CNS Central nervous system
dlPAG Dorsolateral periaqueductal gray matter
DRN Dorsal raphe nucleus
EPM Elevated-plus maze
ETM Elevated-T maze
FAAH Fatty acid amide hydrolase
GABA γ-Aminobutyric acid
GPR55 G-protein-coupled receptor 55
ip Intraperitoneal
PL Prelimbic medial prefrontal cortex
PPARγ Peroxisome proliferator-activated receptor γ
PTSD Posttraumatic stress disorder
THC Δ^9-Tetrahydrocannabinol
TRPA$_1$ Transient receptor potential ankyrin type 1
TRPM$_8$ Transient receptor potential melastin type 8
TRPV$_1$ Transient receptor potential vanilloid type 1
TRPV$_2$ Transient receptor potential vanilloid type 2

INTRODUCTION

CANNABIDIOL: A BRIEF HISTORY

Cannabidiol (CBD) is the major nonpsychotomimetic compound present in the *Cannabis sativa* plant, which has more than 70 substances called phytocannabinoids. It may constitute around 40% of the plant's extract (Campos, Moreira, Gomes, Del Bel, & Guimarães, 2012). Even though CBD was first isolated in 1940 by Adams et al. and Todd et al. (for a detailed revision see Adams, 1941; Todd, 1946) its chemical structure and stereochemistry were characterized only in the 1960s by Professor Mechoulam's group (Mechoulam & Shvo, 1963), who also isolated other *Cannabis* constituents such as cannabinol, cannabigerol, and Δ^9-tetrahydrocannabinol (THC). Although the pharmacology of cannabinoid compounds has been investigated since the 1960s and 1970s, the interest in this field greatly increased in the 1990s after the discovery of the endocannabinoid system. Nowadays, CBD is commercially available as an oromucosal spray solution combined with THC (in a 1:1 proportion) for the management of spasticity in patients with multiple sclerosis. In addition, several studies have shown promising CBD results for the treatment of various pathologies, as will be discussed in this chapter.

MULTIPLE EFFECTS OF CANNABIDIOL AND ITS THERAPEUTIC POTENTIAL

CBD was, from the beginning, a good candidate for therapeutic use, since it presents a remarkable lack of psychoactive effects and is well tolerated in humans. The first studies of CBD's pharmacological activity were conducted in the early 1970s, showing that it exerts anticonvulsive effects in rodents. In addition to epilepsy, there is a huge literature reporting CBD potential therapeutic effects on several conditions such as stroke, inflammatory diseases, pain, cancer, nausea, neurodegeneration, and psychiatric disorders. Although multiple mechanisms of action seem to be responsible for this wide range of therapeutic effects, several pieces of evidence suggest that some of them are mediated by serotonin 5HT$_{1A}$ receptors.

CANNABIDIOL AND 5HT$_{1A}$ RECEPTORS

In 1957, almost 10 years after the discovery of serotonin, a monoamine derived from the amino acid tryptophan, serotonin (5-HT) receptors were identified in the periphery. Almost 20 years later, 5-HT receptor heterogeneity in the central nervous system was better elucidated by binding studies that indicate the existence of two 5-HT receptors in the rat brain, named 5-HT$_1$ and 5-HT$_2$ (Peroutka & Snyder, 1979). From that date the number of 5-HT receptors has greatly expanded. As of this writing, seven 5-HT receptor classes have been identified, one ionotropic and six G-protein coupled. Specifically, the 5-HT$_1$ class, which is coupled to a G$_{i/o}$ protein that reduces the formation of cyclic adenosine monophosphate (cAMP), includes five subtypes: 5HT$_{1A}$, 5-HT$_{1B}$, 5-HT$_{1D}$, 5-HT$_{1E}$, and 5-HT$_{1F}$. 5HT$_{1A}$ receptors are present in presynaptic membranes as autosomic receptors in the raphe nuclei but are also found postsynaptically. Once these receptors inhibit adenylate cyclase, the activation of autosomic $5HT_{1A}$ receptors decreases the release of 5-HT from neuron terminals in the projection areas (Bevilaqua et al., 1997).

The first study showing that CBD could exert part of its effects through $5HT_{1A}$ receptor activation was conducted by Russo, Burnett, Hall, and Parker (2005). In this study CBD displaced [^{3}H]8-hydroxy-DPAT hydrobromide ([^{3}H]8-OH-DPAT), a known $5HT_{1A}$ receptor agonist, from Chinese hamster ovary (CHO) cell membranes transfected with human and rat $5HT_{1A}$ receptors. Also, since the $5HT_{1A}$ G-protein-coupled receptor system is negatively coupled to cAMP formation, CBD increased [^{35}S]GTPγS incorporation into the cell membranes and decreased cAMP concentration. These effects were similar to those induced by 5-HT (Russo et al., 2005). Following this study, the role of $5HT_{1A}$ receptors in the in vivo effects of CBD started to be investigated. As will be discussed below, there is now evidence that these receptors are involved in the anxiolytic- and antidepressive-like, neuroprotective, antinausea, and motor effects of CBD (Table 1).

Generalized Anxiety and Depression

One of the first works suggesting that CBD produces anxiolytic effects was conducted in normal volunteers by Zuardi, Shirakawa, Finkelfarb, and Karniol (1982). The authors showed that CBD

TABLE 1 CBD Effects Mediated by $5HT_{1A}$ Receptors Evaluated in Distinct Animal Models

Predictive Model	Animal Model	Schedule of CBD Administration	Effective CBD Dose	Reference
Anxiety/panic	EPM	Intra-dlPAG	30 nmol	Campos and Guimarães (2008)
	EPM/VCT	Intra-BNST	30–60 nmol	Gomes et al. (2011)
	CFC	Intra-BNST	30–60 nmol	Gomes et al. (2012)
	EPM/CFC	Intra-PL	30 nmol	Fogaça et al. (2014)
	ETM/electrical stimulation	Intra-dPAG	30–60 nmol	Soares et al. (2010)
	Restraint stress and EPM	ip	10 mg/kg	Resstel et al. (2009)
	ETM	ip/21 days	5 mg/kg	Campos et al. (2013)
	Wild snake exposure	ip	3 mg/kg	Twardowschy et al. (2013)
	Cat exposure	ip/7 days	5 mg/kg	Campos et al. (2012)
Depression	FST	ip	30 mg/kg	Zanelati et al. (2010)
Neuroprotection	Middle cerebral artery occlusion	ip	3 mg/kg	Mishima et al. (2005) and Hayakawa et al. (2007)
	Hypoxic–ischemic brain injury	ip	1 mg/kg	Pazos et al. (2013)
Nausea/vomiting	Lithium-induced conditioned gaping	ip/intra-DRN	5 mg/kg–10 µg	Rock et al. (2012)
Motor control	Rearing behavior	ip	20 mg/kg	Espejo-Porras et al. (2013)
	Hepatic encephalopathy	ip/4 weeks	5 mg/kg	Magen et al. (2010)
	Catalepsy	ip	30 mg/kg	Gomes et al. (2013)
Cardiovascular responses	Baroreflex measurement	Intra-BNST	60 nmol	Alves et al. (2010)
	Blood pressure/heart rate measurements	ip	10 mg/kg	Resstel et al. (2009)
	Blood pressure/heart rate measurements	Intra-BNST/restraint stress	30 nmol	Gomes et al. (2012)
Addiction	Intracranial self-stimulation	ip	0.5 mg/kg	Katsidoni et al. (2013)

EPM, elevated-plus maze; *VCT*, Vogel conflict test; *CFC*, contextual fear conditioning; *ETM*, elevated-T maze; *FST*, forced swimming test; *dlPAG*, dorsolateral periaqueductal gray matter; *BNST*, bed nucleus of the stria terminalis; *PL*, prelimbic medial prefrontal cortex; *dPAG*, dorsal periaqueductal gray matter; *DRN*, dorsal raphe nucleus; *ip*, intraperitoneal.

blocked the increase in anxiety induced by THC, the major psychotomimetic compound of marijuana. Soon after, a preclinical study reported that acute intraperitoneal (ip) injection of CBD induces anxiolytic-like effects in rats tested in the elevated-plus maze test (EPM) (Guimarães, Chiaretti, Graeff, & Zuardi, 1990). This study also showed that CBD produces an inverted dose–response U-shaped curve, being anxiolytic at low, but ineffective at higher doses. Following these initial results, several other studies reported anxiolytic-like effects of CBD in various animal models. For example, systemic administration of CBD reduced the increase in blood pressure and heart rate and prevented the delayed anxiogenic-like effects induced by acute restraint stress (Resstel et al., 2009). Similar anxiolytic effects were detected when the drug was microinjected into brain structures related to defensive responses, such as the dorsolateral periaqueductal gray matter (dlPAG) (Campos & Guimarães, 2008), the bed nucleus of the stria terminalis (BNST) (Gomes, Resstel, & Guimarães, 2011; Gomes et al., 2012), and the prelimbic medial prefrontal cortex (PL) (Fogaça, Reis, Campos, & Guimarães, 2014). In the last structure CBD induced either anxiolytic- or anxiogenic-like effects, depending on the animal model employed, i.e., the EPM or contextual fear conditioning (CFC), respectively (Fogaça et al., 2014). All these effects seem to depend on $5HT_{1A}$ receptor activation, since they were abolished by pretreatment with the specific receptor antagonist WAY 100635 (Campos & Guimarães, 2008; Fogaça et al., 2014; Gomes et al., 2011) (Figures 1 and 2). Also, the antidepressive-like effect of CBD in mice submitted to the forced swimming model was blocked by this antagonist (Zanelati, Biojone, Moreira, Guimarães, & Joca, 2010).

In humans, although no study so far has investigated the involvement of $5HT_{1A}$ receptors on CBD effects, the anxiolytic effect of this drug in volunteers submitted to a simulated public speaking model was very similar to the effect in those treated with ipsapirone, a $5HT_{1A}$ partial agonist (Zuardi, Cosme, Graeff, & Guimarães, 1993).

Panic and Posttraumatic Stress Disorder

CBD injected directly into the dlPAG induced antiaversive effects in the elevated-T maze (ETM) and attenuated flight reactions induced by local electrical stimulation, animal models used to investigate panic-like responses (Soares et al., 2010). Also, repeated but not acute ip injections of CBD for 21 days decreased escape responses in the ETM, without altering $5HT_{1A}$ and $5-HT_{1C}$ receptor mRNA expression and 5-HT extracellular concentrations (Campos, de Paula Soares, et al., 2013). The panicolytic-like effect of CBD was also observed in an animal model of innate fear-induced responses evoked by the presence of a wild snake (Twardowschy et al., 2013). The intra-dlPAG and ip effects of CBD were abolished by WAY 100635 (Campos, de Paula Soares, et al., 2013; Soares et al., 2010; Twardowschy et al., 2013). In addition to panic disorder, CBD also interfered with models associated with posttraumatic stress disorder (PTSD). Repeated daily injections of CBD in rats that started 1 h after a single exposure to a live cat blocked the long-lasting anxiogenic consequences observed in these animals (Campos, Ferreira, & Guimarães, 2012). Pretreatment with WAY 100635 attenuated the CBD anti-predator-induced anxiety effect, indicating that this response also depends on facilitation of $5HT_{1A}$ neurotransmission (Campos, Ferreira, et al., 2012).

Neuroprotection

CBD has potent neuroprotective properties, preventing neurodegeneration induced by glutamate β-amyloid 6-hydroxydopamine or cerebral ischemia (Campos, Moreira et al., 2012). CBD reduced the infarct volume induced by middle cerebral artery occlusion

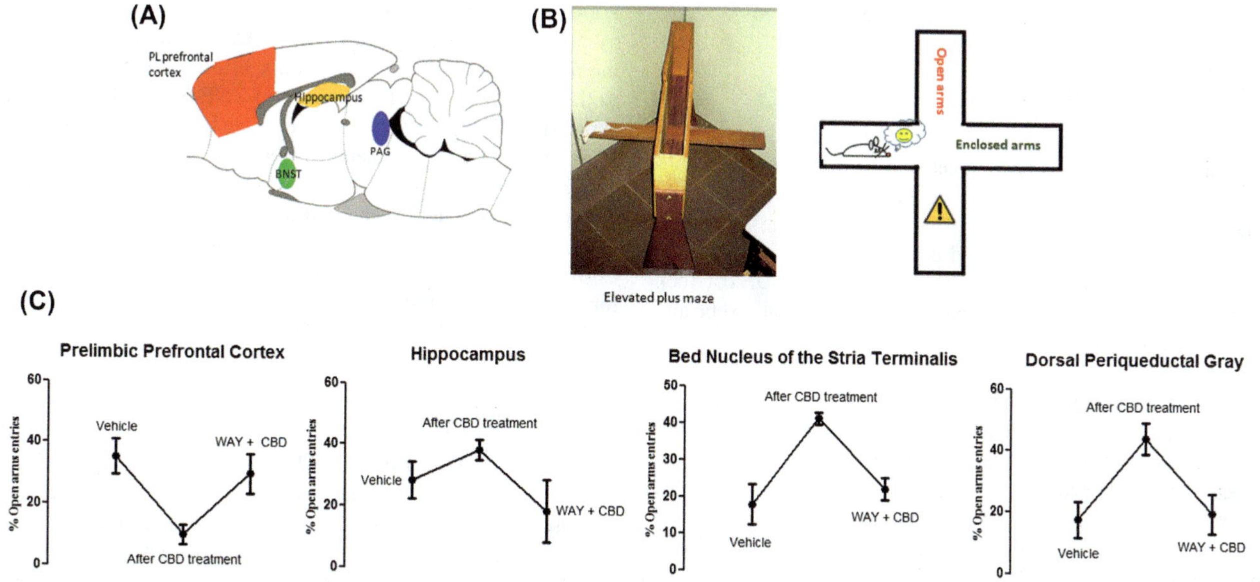

FIGURE 1 CBD controls anxiety-like behaviors recruiting $5HT_{1A}$ receptors in various brain structures. (A) Sagittal representation of the CBD site of injections in the rat brain. (B) The EPM, an animal model used to test the effects of CBD on anxiety-like behaviors (left, representative image of the EPM; right, when in the enclosed arms, animals feel safer than when in the open arms, therefore increased levels of open-arms exploration indicate less anxiety). (C) Representative graphics of the results obtained from the EPM test. Results are shown as the mean ± SEM. *Used with permission from Fogaça et al. (2014), Campos and Guimarães (unpublished observation), Gomes et al. (2011), and Campos and Guimarães (2008) respectively.*

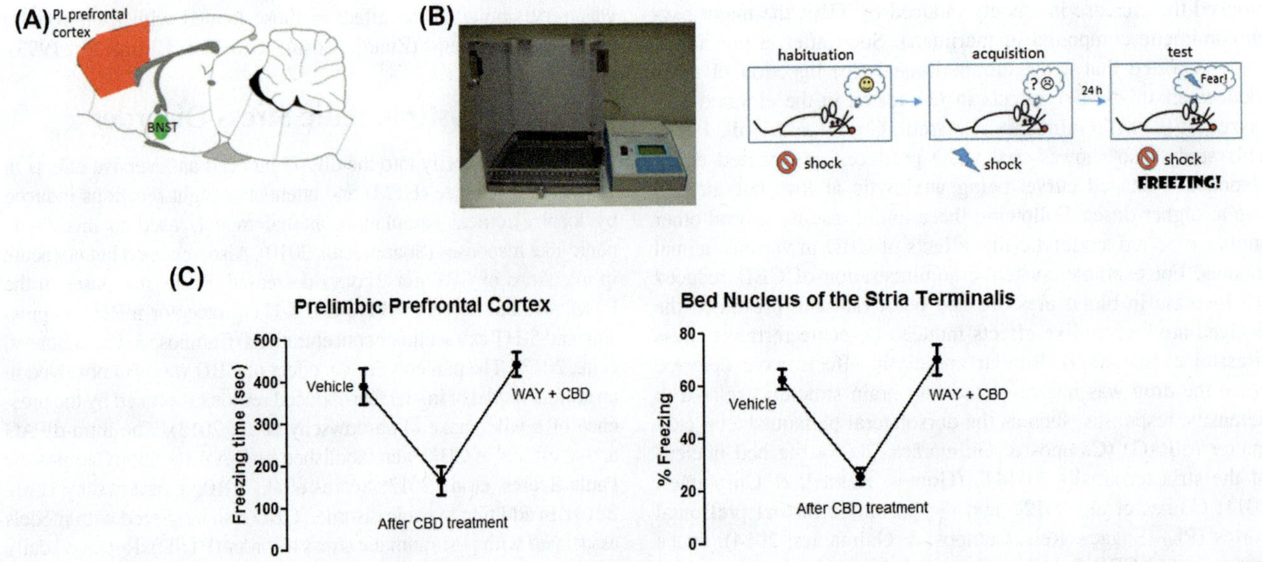

FIGURE 2 CBD decreases fear conditioning behavior by activating 5HT$_{1A}$ receptors in various brain structures. (A) Sagittal representation of the site of injections of CBD in the rat brain. (B) CFC paradigm was used to test the effects of CBD on behavioral responses in rats (left, representative photograph of the apparatus; right, schematic representation of the CFC protocol; after receiving a mild foot shock in the acquisition session and being reexposed to the same environment, animals will express freezing behaviors, which indicate increased anxiety). (C) Representative graphics of the results obtained in the CFC. Results are shown as the mean ± SEM. *Used with permission from Fogaça et al. (2014) and Gomes et al. (2012).*

and increased cerebral blood flow without inducing tolerance after repetitive treatment (Hayakawa et al., 2007; Mishima et al., 2005). These effects were attenuated by pretreatment with WAY 100635, but not by cannabinoid type 1 (CB$_1$) and transient receptor potential vanilloid type 1 (TRPV$_1$) receptor antagonists. Similarly, acute injection of CBD prevented the alterations induced by hypoxic–ischemic brain injury in newborn pigs. These changes were also abolished by previous injection of WAY 100635 or a cannabinoid type 2 (CB$_2$) receptor antagonist (Pazos et al., 2013).

Nausea and Vomiting

CBD presented antinausea effects in an animal model based on lithium chloride-induced conditioned gaping to a flavor or a context (Parker, Kwiatkowska, & Mechoulam, 2006). A pioneer study relating the antinausea effects of CBD to the serotonergic system indicated that the drug suppressed conditioned gaping in rats and vomiting in shrews similar to the 5HT$_{1A}$ agonist 8-OH-DPAT (Rock et al., 2011). Soon after, it was demonstrated that the antiemetic and antinausea effect induced by systemic administration of CBD was abolished by pretreatment with WAY 100135 (a 5-HT$_{1A}$ receptor antagonist that also has affinity for 5-HT$_{1B, 1C, 2}$, α1, α2, and D2 receptors) and/or WAY 100635 (Rock et al., 2012). Also, the antinausea effects of CBD (ip) were abolished by WAY 100635 injected directly into the dorsal raphe nucleus (DRN), indicating that these effects may be mediated by 5HT$_{1A}$ autoreceptors. The same occurred when an inverse schedule of administration was used (i.e., CBD into the DRN and WAY 100635 systemically (Rock et al., 2012)).

Motor Control

CBD can influence motor behavior via the serotonergic system producing, however, a U-shaped dose–response curve (Espejo-Porras, Fernández-Ruiz, Pertwee, Mechoulam, & García, 2013). Higher doses of this compound (>10 mg/kg) decreased rearing (vertical activity) behavior. The 5HT$_{1A}$ agonist 8-OH-DPAT induced a similar response but with a higher potency. Both drug effects were abolished by pretreatment with WAY 100635 but not by a CB$_1$ receptor antagonist. However, 8-OH-DPAT- but not CBD-induced motor activity was associated with increased 5-HT levels in the basal ganglia, a brain region involved in motor control (Espejo-Porras et al., 2013). In a model of hepatic encephalopathy induced by bile-duct ligation, repeated injections of CBD (over 4 weeks) improved cognition and locomotion, effects that were accompanied by a reversion in hippocampal TNF-α1 and BDNF gene expression. Except for the last one, these effects were abolished by WAY 100635 pretreatment (Magen et al., 2010). Finally, CBD attenuated catalepsy induced by the D2 receptor antagonist haloperidol, a classic antipsychotic drug used to treat schizophrenia, without inducing catalepsy by itself. This effect was prevented by WAY 100635. This result suggests that CBD could be useful for the treatment of striatal disorders such as Parkinson disease (Gomes, Del Bel, & Guimarães, 2013).

Other 5HT$_{1A}$-Related Effects

Injection of CBD or 8-OH-DPAT into the BNST facilitated the baroreflex response, inducing bradycardia after arterial pressure increase. Pretreatment with WAY 100635 abolished these responses (Alves et al., 2010). In an animal model of hyperphagia, CBD prevented the increase in food intake induced by 8-OH-DPAT (Scopinho, Guimarães, Corrêa, & Resstel, 2011). Finally, CBD inhibited the reward-facilitating effects of morphine but not cocaine in the intracranial self-stimulation paradigm through activation of 5HT$_{1A}$ receptors located in the DRN (Katsidoni, Anagnostou, & Panagis, 2013).

HOW DOES CANNABIDIOL INTERACT WITH 5HT$_{1A}$ RECEPTORS?

As reported above, Russo et al. (2005) observed that CBD, in the micromolar range, is able to displace the 5HT$_{1A}$ receptor agonist 8-OH-DPAT from cloned human 5-HT1A receptors expressed in CHO cells, acting as an agonist at these receptors. However, regardless of the innumerous papers reviewed above that have shown successful blockage of CBD effects by 5HT$_{1A}$ receptor antagonists, findings using brainstem preparations suggested that CBD does not act at these receptors as a full agonist (Rock et al., 2011, 2012). It remains unclear, therefore, how this phytocannabinoid facilitates 5HT$_{1A}$ receptor function. The 5HT$_{1A}$ receptor is a G$_i$-coupled receptor that, when activated, enhances K$^+$ currents and inhibits adenylyl cyclase activity, resulting in hyperpolarization, inhibition of neuronal firing, and changes in the levels of cAMP and protein kinase A (PKA) (Bevilaqua et al., 1997). In addition, stimulation of the mitogen-activated protein kinase (MAPK) and protein kinase B/Akt pathways has also been described after the activation of 5HT$_{1A}$ receptors (Cloëz-Tayarani, Kayyali, Fanburg, & Cavaillon, 2004). Thus, another possible mechanism of CBD facilitation of 5HT$_{1A}$-mediated transmission is the interaction with intracellular signaling transduction pathways (Figure 3). Unpublished observations from our group suggest that CBD may increase neural precursor cell proliferation in vitro by activating extracellular signal-regulated kinase (ERK) 1/2 signaling. This intracellular pathway is recruited after cannabinoid or serotonergic receptor activation and has been associated with progenitor/stem cell proliferation (Fogaça, Galve-Roperh, Guimarães, & Campos, 2013). Additional studies, however, are needed to verify the involvement of this last mechanism on CBD 5HT$_{1A}$-mediated effects.

Since CBD does not seem to be a 5HT$_{1A}$ full agonist, it could facilitate 5HT$_{1A}$-mediated signaling by targeting allosteric sites in the receptor and enhancing the ability of 5HT$_{1A}$ agonists to stimulate [^{35}S]GTPγS binding (Rock et al., 2011). Also, interactions between CB$_1$ and 5HT$_{1A}$ receptors have been observed in the prefrontal cortex of rats chronically treated with fluoxetine (Marco et al., 2011) and cannabinoid receptor agonists are proposed to enhance 5-HT receptor activity via MAPK–ERK1/2 signaling (Franklin & Carrasco, 2013). Cannabinoid receptors can make heterodimers with several other receptors such as those for dopamine, orexin, and opioids (Hojo et al., 2008; Przybyla & Watts, 2010). CBD shows low affinity for CB$_1$/CB$_2$ receptors but can facilitate endocannabinoid neurotransmission by inhibiting anandamide metabolism/uptake (Bisogno et al., 2001; Campos, Ortega, et al., 2013). Endocannabinoids can modulate not only serotonergic neurotransmission but also the brain expression of 5-HT subtypes 1A and 2A/2C receptors (Cassano et al., 2011). Genetic deletion of fatty acid amide hydrolase (FAAH), which is responsible for endocannabinoid/endovanilloid degradation, increases the firing of serotonergic neurons located in the DRN. Consequently, 5-HT release is increased in limbic areas such as prefrontal cortex (Bambico et al., 2010).

In the 1970s and 1980s several papers came out suggesting that phytocannabinoids could facilitate monoaminergic neurotransmission in synaptosome preparations. The capacity of CBD to indirectly activate 5HT$_{1A}$ by facilitating serotonergic neurotransmission was tested by Campos, de Paula Soares, et al. (2013). However, at least in the dlPAG, acute or chronic treatment with CBD failed to increase extracellular 5-HT measured by microdialysis (Campos, de Paula Soares, et al., 2013). The capacity of CBD to enhance 5-HT levels in other brain areas and biological systems is still open for investigation.

Changes in 5HT$_{1A}$ receptor expression are believed to play an important role in the action of antidepressant drugs. Repeated administration of 5HT$_{1A}$ agonists or antidepressants induces internalization of 5HT$_{1A}$ autoreceptors in the raphe neurons but not of postsynaptic

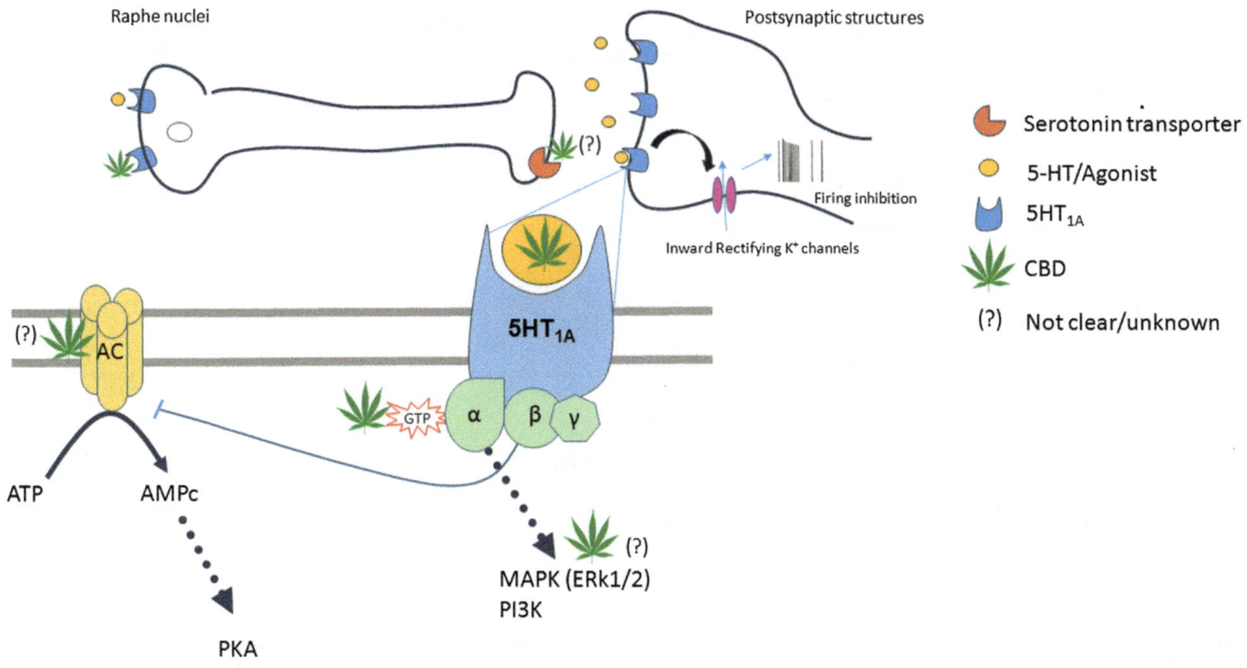

FIGURE 3 Possible mechanisms involved in the facilitatory effects of CBD on 5-HT$_{1A}$-mediated neurotransmission. CBD modulates 5HT$_{1A}$ functions in the brain. Although CBD was originally proposed to act as a 5-HT$_{1A}$ receptor agonist, more recent evidence favors mechanisms that involve allosteric modulation or changes in intracellular pathways. AC, Adenylyl Ciclase; PI3K, Phosphoinositide 3-kinase.

5HT$_{1A}$ receptors in the hippocampus or prefrontal cortex (Blier, Pineyro, el Mansari, Bergeron, & de Montigny, 1998). However, we failed to find any change in 5HT$_{1A}$ receptor mRNA expression after 21 days of treatment with CBD in the dorsal portions of the PAG of rats (Campos, de Paula Soares, et al., 2013). Despite this negative finding, new studies aiming at investigating CBD effects on 5HT$_{1A}$ receptor expression in other brain areas are clearly needed.

ARE ALL CANNABIDIOL EFFECTS MEDIATED BY 5-HT$_{1A}$ RECEPTORS?

In addition to the complex interaction of CBD with 5HT$_{1A}$ receptors, this phytocannabinoid produces several other important effects over a wide range of doses and concentrations in a 5HT$_{1A}$ receptor-independent manner (Table 2) (Campos, Moreira, et al., 2012).

TABLE 2 Multiple CBD Mechanisms of Action Assessed in In Vitro Studies

Biological System	Mechanism	Methodology	Concentration Range	Reference
Endocannabinoid- and endovanilloid-related mechanisms	CB$_1$ receptor antagonist	Whole mouse brain membranes	1 μM	Thomas et al. (2007)
	CB$_2$ receptor inverse agonist	CHO transfected cells	1 μM	Thomas et al. (2007)
	TRPV$_1$ receptor agonist	HEK-293 transfected cells	10 μM 1 nM–25 μM	Bisogno et al. (2001) De Petrocellis et al. (2011)
	TRPA$_1$ receptor agonist	HEK-293 transfected cells	1 nM–25 μM	De Petrocellis et al. (2011)
	TRPM$_8$ receptor antagonist	HEK-293 transfected cells	1 nM–25 μM	De Petrocellis et al. (2011)
	TRPV$_2$ receptor agonist	HEK-293 transfected cells	1 nM–25 μM 0.01–100 μM	De Petrocellis et al. (2011) Qin et al. (2008)
	FAAH enzyme inhibitor	HEK-293 transfected cells Membranes from rat brain	30 μM 1–100 μM	Bisogno et al. (2001) De Petrocellis et al. (2011)
	Anandamide transporter inhibitor	HEK-293 transfected cells	30 μM	Bisogno et al. (2001)
	MAGL/DAGLα/NAAA enzyme inhibitor	COS-7 or HEK-293 transfected cells	1–100 μM	De Petrocellis et al. (2011)
Serotonergic-related mechanisms	5-HT$_{1A}$ receptor agonist	CHO transfected cells	16 μM	Russo et al. (2005)
	5-HT$_{2A}$ receptor agonist	CHO transfected cells	32 μM	Russo et al. (2005)
	5-HT$_3$ receptor antagonist	*Xenopus laevis* oocytes	10–30 μM	Xiong et al. (2011)
	Tryptophan degradation inhibitor	Human blood cells	3 μg/ml	Jenny et al. (2009)
Others	Intracellular [Ca^{2+}] modulator	Hippocampal cell cultures/hippocampal preparations	~1 μM	Ryan et al. (2006)
	Allosteric modulator of μ and δ opioid receptors	Cerebral cortex preparations	100 μM	Kathmann et al. (2006)
	PPARγ receptor agonist	Aorta preparations	5 μM	O'Sullivan et al. (2009)
	Adenosine uptake/A2 receptors	Migroglia and macrophage cell cultures	120 nM	Liou et al. (2008)
	α7-nACh receptor	*Xenopus* oocytes Hippocampal slices	100 μM 10 μM	Mahgoub et al. (2013)

HEK-293, human embryonic kidney 293 cells; *CHO*, Chinese hamster ovary cells; *FAAH*, fatty acid amide hydrolase; *MAGL*, monoacylglycerol lipase; *DAGL*, diacylglycerol lipase; *NAA*, N-arachidonoylphosphatidylethanolamine; *TRPV*, transient receptor potential vanilloid; *TRPA*, transient receptor potential ankyrin; *TRPM*, transient receptor potential melastin; *PPAR*, peroxisome proliferators-activated receptor; *A2 receptors*, adenosine type 2 receptors; *α7-nACh receptor*, α7-nicotinic acetylcholine receptors.

CBD might act on or modify the function of several receptors, including CB_1, CB_2, G-protein-coupled receptor 55, $TRPV_1$, and $TRPV_2$, transient receptor potential melastin type 8 ($TRPM_8$), and ankyrin type 1 ($TRPA_1$) (Bisogno et al., 2001; De Petrocellis et al., 2011; Russo et al., 2005; Thomas et al., 2007). It could also inhibit the FAAH enzyme and the adenosine transporter (Bisogno et al. 2001; Pandolfo et al., 2011), indirectly increasing the levels of endocannabinoids or adenosine. Additionally, some CBD effects involve intracellular pathways or receptors, such as peroxisome proliferator-activated receptor γ (PPARγ), that play fundamental roles in cell survival (Esposito et al., 2011).

Endocannabinoid System and Cannabidiol Interactions

Several in vitro studies suggest that CBD shows low affinity for CB_1/CB_2 receptors (Campos, Moreira, et al., 2012). Despite this finding, some important in vitro and in vivo effects of CBD are mediated by CB_1 or CB_2 receptors. For instance, CBD interferes with aversive memories by facilitating endocannabinoid-mediated neurotransmission. CBD effects on fear memory extinction and reconsolidation, important features related to PTSD, were prevented by previous treatment with CB_1 but not $5HT_{1A}$ receptor antagonists (Stern, Gazarini, Takahashi, Guimarães, & Bertoglio, 2012). Similar CB_1-dependent effects were observed in the marble-burying model, an animal model proposed to evaluate obsessive–compulsive behaviors (Casarotto, Gomes, Ressel, & Guimarães, 2010). Because of the low affinity of CBD for CB_1 receptors, the actions of this phytocannabinoid on cannabinoid receptors are most probably mediated by inhibition of endocannabinoid degradation/uptake.

Endocannabinoids can reduce the release of several neurotransmitters, including glutamate, the major excitatory neurotransmitter (Takahashi & Castillo, 2006). An indirect antiglutamatergic action via increased endocannabinoid neurotransmission may be involved in some central effects of CBD such as anticonvulsant effects. Epileptic patients present a significant reduction in CB_1 receptors located on glutamatergic terminals leading to increased neural excitability and consequently seizures (Ludanyi et al., 2008).

Cannabinoid signaling also contributes to CBD effects on neurogenesis. Wolf et al. (2010) showed that the proneurogenic effect of CBD was absent in CB1-knockout mice. This result suggested that CBD's effect was mediated by an indirect activation of these receptors, possibly by inhibition of anandamide metabolism/uptake (Bisogno et al., 2001). Our group demonstrated that CBD increases proliferation of hippocampal progenitor cells in vivo and in vitro by increasing the production of the endocannabinoid anandamide, an effect mimicked by CB_1 or CB_2 receptor agonists and prevented by antagonists of these receptors (Campos, Ortega, et al., 2013; Fogaça et al., unpublished data).

$TRPV_1$ Channels and Cannabidiol Actions

The $TRPV_1$ receptor was one of the first identified members of this family, being a nonselective cation channel with a preference for calcium. It is activated by noxious stimuli, heat, protons (pH < 5.9), and various, mostly noxious, natural products such as capsaicin (Tominaga et al., 1998). CBD can also act as an agonist at these receptors (Bisogno et al., 2001). $TRPV_1$ receptors are present in the brain and can be activated by the endocannabinoid/endovanilloid anandamide (Zygmunt et al., 1999). Also, different from CB_1 receptor activation, agonist activity at these receptors increases glutamate release (Xing & Li, 2007). $TRPV_1$ activation seems to be related to the inverted U-shaped dose–response curves commonly observed after cannabinoid injection into various brain structures (Campos & Guimarães, 2009; Fogaça, Aguiar, Moreira, & Guimarães, 2012; Fogaça, Gomes, Moreira, Guimarães, & Aguiar, 2013). Accordingly, using intradorsal PAG injections, our group demonstrated that local pretreatment with an ineffective dose of the $TRPV_1$ antagonist capsazepine is able to turn a higher, ineffective dose (60 nmol) of CBD into an anxiolytic one (Campos & Guimarães, 2009). In addition to $TRPV_1$, CBD could also interfere with other members of the TRP family, activating $TRPV_2$ and $TRPA_1$ channels and antagonizing $TRPM_8$ channels (De Petrocellis et al., 2008, 2011; Qin et al., 2008). The importance of these mechanisms for CBD's effects remains to be further investigated.

Other Relevant Mechanisms Related to Cannabidiol Effects

CBD increases intracellular calcium concentrations via mitochondrial uptake and release, associated with activation of type L voltage-gated calcium channels. CBD also inhibits oxidative and nitrosative stress, a mechanism that has been related to its action on PPARγ receptors, with promising implications for the treatment of Alzheimer, Huntington, and Parkinson diseases. CBD decreases the neuronal damage promoted by β-amyloid protein deposits and attenuates the depletion of tyrosine hydroxylase, dopamine, and γ-aminobutyric acid levels induced by neurotoxic stimuli, modulating the expression of the inducible isoform of nitric oxide synthase and reducing the production of reactive oxygen species-generating NADPH oxidases. Moreover, CBD treatment attenuated high-glucose-induced mitochondrial superoxide generation and nuclear factor-κB activation, along with expression of the adhesion molecules ICAM-1 and VCAM-1. Although the exact mechanisms involved in these CBD effects remain unclear, they might be due to CBD agonist activity at PPAR receptors. CBD could also act at the level of the mitochondria or inside the nucleus to oppose oxidative/nitrosative stress (Fernandez-Ruiz et al., 2013).

CONCLUSION

CBD is a mysterious and fascinating drug. Although it is the phytocannabinoid with the largest range of potential therapeutic use, its mechanisms are still largely unknown. Evidence reviewed here indicates that part of CBD's effects depend on facilitation of $5HT_{1A}$-receptor mediated neurotransmission. How this is accomplished, however, is unclear (see Figure 3). Despite the initial claim that it acts as a receptor agonist, results indicate that this does not seem to be the case. Allosteric mechanisms and/or changes in intracellular signaling transduction pathways are good candidates to explain this effect, but further studies are needed to clarify this problem.

Moreover, it is becoming increasingly clear that the unique CBD drug profile depends on multiple mechanisms of action. This notion challenges the common belief that drugs aimed at single pharmacological targets are usually better therapeutic candidates. Actually, CBD adds to other examples, such as clozapine, suggesting that the so-called "dirty" drugs might be better therapeutic options, particularly for complex disorders.

APPLICATION TO OTHER ADDICTIONS AND SUBSTANCE MISUSE

In this chapter, we reported that CBD exerts several therapeutic activities through $5HT_{1A}$ receptor activation. Specifically in drug addiction, various studies have reported altered function of the serotonergic system in individuals with alcohol dependence. For example, $5HT_{1A}$ receptor expression was elevated in the prefrontal cortex tissues from alcohol-dependent subjects, probably as a compensatory response to the hyposerotonergic function induced by alcohol abuse, which was correlated with the occurrence of depressive symptoms (Thompson, Cruz, Olukotun, & Delgado, 2012). In rodents, drugs acting both as agonists and as antagonists of $5HT_{1A}$ receptors can decrease ethanol intake. In a study, the antagonism of $5HT_{1A}$ receptors enhanced ethanol-induced conditioned place preference, suggesting that the drug increases sensitivity to ethanol reward and then decreases ethanol intake (Risinger & Boyce, 2002). As CBD is suggested to act as an allosteric modulator of $5HT_{1A}$ receptors, it is expected that the compound could be effective to treat drug addiction. In fact, clinical studies showed that CBD was effective for the treatment of *Cannabis* withdrawal syndrome (Crippa et al., 2013) and reduced cigarette consumption in tobacco smokers (Morgan, Das, Joye, Curran, & Kamboj, 2013). Also, preclinical studies showed that CBD attenuated cue-induced heroin-seeking behavior (Ren, Whittard, Higuera-Matas, Morris, & Hurd, 2009). Moreover, low doses of THC and CBD potentiated the extinction of cocaine- and amphetamine-induced conditioned place preference learning, an effect that was not dependent on CB_1 receptor activation (Parker, Burton, Sorge, Yakiwchuk, & Mechoulam, 2004). CBD inhibited the reward-facilitating effect of morphine by $5HT_{1A}$ receptor activation located in the dorsal raphe (Katsidoni et al., 2013). Although only a few studies relating CBD to $5HT_{1A}$ receptors in drug addiction are available in the literature, the evidence described above suggests that the action of CBD on $5HT_{1A}$ receptors could be a therapeutic target to treat addiction.

DEFINITION OF TERMS

Allosteric modulation This refers to indirect modulation of the receptor response by the binding of a drug to a site distinct from that of the agonist.
Bed nucleus of the stria terminalis This is a limbic structure closely associated with the amygdaloid complex and involved in the regulation of autonomic, neuroendocrine, and behavioral responses to stressful and/or threatening stimuli.
Dorsal raphe nucleus This is a midbrain structure from which, together with the median raphe nucleus, originate most of the 5-HT prosencephalic projections.
Dorsolateral periaqueductal gray matter This is the dorsolateral column of the periaqueductal gray matter, a structure that surrounds the Sylvius aqueduct that connects the third and fourth ventriculi. Electrical or chemical stimulation of this region in humans and laboratory animals induces strong defensive responses.
Endocannabinoid system This is a neurotransmitter system discovered in the 1990s that comprises the endogenous agonists anandamide and 2-arachidonoylglycerol, the enzymes responsible for their synthesis and metabolism, and the cannabinoid receptors CB1 and CB2.
G_i-coupled receptor This is a G-protein-coupled receptor that, when activated, inhibits the formation of the second-messenger cyclic AMP.
Prelimbic medial prefrontal cortex This is part of the ventral medial prefrontal cortex, the anterior pole of the mammalian brain that receives projections from the mediodorsal thalamic nucleus. Together with the adjacent infralimbic cortex, this region plays an important role in the modulation of defensive responses.
Receptor agonist This refers to a chemical compound that can bind with higher affinity to a cellular specialized structure called a receptor and, once bound, is able to activate this receptor and induce a physiological and/or pharmacological effect.
Receptor antagonist This is a chemical compound that can bind with higher affinity to a cellular specialized structure called a receptor but, once bound, is unable to activate this receptor, preventing or decreasing the physiological and/or pharmacological effects induced by a receptor agonist.
Serotonin This is a monoamine neurotransmitter first identified in 1945 as a vasoconstrictor substance. Later studies confirmed its important role as a neurotransmitter in the central nervous system, being chemically characterized as 5-hydroxytryptamine (5-HT).

KEY FACTS

Key Facts of Cannabidiol

- CBD is the major nonpsychotomimetic compound present in the C. sativa plant, constituting around 40% of the plant's extract.
- CBD was first isolated in 1940 but its chemical structure and stereochemistry were characterized only in the 1960s.
- Nowadays, CBD is commercially available as an oromucosal spray solution, combined with THC (in a 1:1 proportion), for the management of spasticity in patients with multiple sclerosis.

Key Facts of 5-HT Receptors

- In 1957, almost 10 years after the discovery of 5-HT, 5-HT receptors were identified in the periphery.
- The $5HT_{1A}$ is a G_i-coupled receptor that, when activated, enhances K^+ currents and inhibits adenylyl cyclase activity, resulting in hyperpolarization.
- Seven 5-HT receptor classes have been identified so far, one ionotropic and six G-protein coupled.

SUMMARY POINTS

- CBD is a promising drug for treating several conditions such as epilepsy, stroke, inflammatory diseases, pain, cancer, nausea, neurodegeneration, and psychiatric disorders.
- CBD's effects seem to involve various neurotransmitter and transduction signaling systems.
- One of the main CBD target is the serotonergic system, particularly $5HT_{1A}$ receptors.

- 5HT$_{1A}$ receptors are involved in the anxiolytic- and antidepressive-like, neuroprotective, antinausea, and motor effects of CBD.
- Despite the initial claim that it acts as a receptor agonist, more recent results indicate that this does not seem to be the case.
- Allosteric mechanisms and/or changes in intracellular signaling transduction pathways could explain CBD effects, but further studies are needed to clarify this problem.

REFERENCES

Adams, R. (1941). Marihuana. *Harvey Lectures, 37*, 168–197.

Alves, F. H., Crestani, C. C., Gomes, F. V., Guimarães, F. S., Correa, F. M., & Resstel, L. B. (2010). Cannabidiol injected into the bed nucleus of the stria terminalis modulates baroreflex activity through 5-HT$_{1A}$ receptors. *Pharmacological Research, 62*(3), 228–236.

Bambico, F. R., Cassano, T., Dominguez-Lopez, S., Katz, N., Walker, C. D., Piomelli, D., & Gobbi, G. (2010). Genetic deletion of fatty acid amide hydrolase alters emotional behavior and serotonergic transmission in the dorsal raphe, prefrontal cortex, and hippocampus. *Neuropsychopharmacology, 35*(10), 2083–2100.

Bevilaqua, L., Ardenghi, P., Schroder, N., Bromberg, E., Schmitz, P. K., Schaeffer, E., ... Izquierdo, I. (1997). Drugs acting upon the cyclic adenosine monophosphate/protein kinase A signalling pathway modulate memory consolidation when given late after training into rat hippocampus but not amygdala. *Behavioural Pharmacology, 8*(4), 331–338.

Bisogno, T., Hanus, L., De Petrocellis, L., Tchilibon, S., Ponde, D. E., Brandi, I., ... Di Marzo, V. (2001). Molecular targets for cannabidiol and its synthetic analogues: effect on vanilloid VR1 receptors and on the cellular uptake and enzymatic hydrolysis of anandamide. *British Journal of Pharmacology, 134*(4), 845–852.

Blier, P., Pineyro, G., el Mansari, M., Bergeron, R., & de Montigny, C. (1998). Role of somatodendritic 5-HT autoreceptors in modulating 5-HT neurotransmission. *Annals of the New York Academy of Sciences, 861*, 204–216.

Campos, A. C., & Guimarães, F. S. (2008). Involvement of 5HT$_{1A}$ receptors in the anxiolytic-like effects of cannabidiol injected into the dorsolateral periaqueductal gray of rats. *Psychopharmacology (Berlin), 199*(2), 223–230.

Campos, A. C., & Guimarães, F. S. (2009). Evidence for a potential role for TRPV1 receptors in the dorsolateral periaqueductal gray in the attenuation of the anxiolytic effects of cannabinoids. *Progress in Neuro-Psychopharmacology & Biological Psychiatry, 33*(8), 1517–1521.

Campos, A. C., Ferreira, F. R., & Guimarães, F. S. (2012). Cannabidiol blocks long-lasting behavioral consequences of predator threat stress: possible involvement of 5HT$_{1A}$ receptors. *Journal of Psychiatric Research, 46*(11), 1501–1510.

Campos, A. C., Moreira, F. A., Gomes, F. V., Del Bel, E. A., & Guimarães, F. S. (2012). Multiple mechanisms involved in the large-spectrum therapeutic potential of cannabidiol in psychiatric disorders. *Philosophical Transactions of the Royal Society, Series B, Biological Sciences, 367*(1607), 3364–3378.

Campos, A. C., Ortega, Z., Palazuelos, J., Fogaça, M. V., Aguiar, D. C., Diaz-Alonso, J., ... Guimarães, F. S. (2013). The anxiolytic effect of cannabidiol on chronically stressed mice depends on hippocampal neurogenesis: involvement of the endocannabinoid system. *International Journal of Neuropsychopharmacology, 16*(6), 1407–1419.

Campos, A. C., de Paula Soares, V., Carvalho, M. C., Ferreira, F. R., Vicente, M. A., Brandão, M. L., ... Guimarães, F. S. (2013). Involvement of serotonin-mediated neurotransmission in the dorsal periaqueductal gray matter on cannabidiol chronic effects in panic-like responses in rats. *Psychopharmacology (Berlin), 226*(1), 13–24.

Casarotto, P. C., Gomes, F. V., Resstel, L. B., & Guimarães, F. S. (2010). Cannabidiol inhibitory effect on marble-burying behaviour: involvement of CB$_1$ receptors. *Behavioural Pharmacology, 21*(4), 353–358.

Cassano, T., Gaetani, S., Macheda, T., Laconca, L., Romano, A., Morgese, M. G., ... Piomelli, D. (2011). Evaluation of the emotional phenotype and serotonergic neurotransmission of fatty acid amide hydrolase-deficient mice. *Psychopharmacology (Berlin), 214*(2), 465–476.

Cloëz-Tayarani, I., Kayyali, U. S., Fanburg, B. L., & Cavaillon, J. M. (2004). 5-HT activates ERK MAP kinase in cultured-human peripheral blood mononuclear cells via 5-HT$_{1A}$ receptors. *Life Science, 76*(4), 429–443.

Crippa, J. A., Hallak, J. E., Machado-de-Sousa, J. P., Queiroz, R. H., Bergamaschi, M., Chagas, M. H., & Zuardi, A. W. (2013). Cannabidiol for the treatment of cannabis withdrawal syndrome: a case report. *Journal of Clinical Pharmacy and Therapeutics, 38*(2), 162–164.

De Petrocellis, L., Ligresti, A., Moriello, A. S., Allarà, M., Bisogno, T., Petrosino, S., ... Di Marzo, V. (2011). Effects of cannabinoids and cannabinoid-enriched Cannabis extracts on TRP channels and endocannabinoid metabolic enzymes. *British Journal of Pharmacology, 163*(7), 1479–1494.

De Petrocellis, L., Vellani, V., Schiano-Moriello, A., Marini, P., Magherini, P. C., Orlando, P., & Di Marzo, V. (2008). Plant-derived cannabinoids modulate the activity of transient receptor potential channels of ankyrin type-1 and melastatin type-8. *Journal of Pharmacology and Experimental Therapeutics, 325*(3), 1007–1015.

Espejo-Porras, F., Fernández-Ruiz, J., Pertwee, R. G., Mechoulam, R., & García, C. (2013). Motor effects of the non-psychotropic phytocannabinoid cannabidiol that are mediated by 5-HT$_{1A}$ receptors. *Neuropharmacology, 75*, 155–163.

Esposito, G., Scuderi, C., Valenza, M., Togna, G. I., Latina, V., De Filippis, D., ... Steardo, L. (2011). Cannabidiol reduces Aβ-induced neuroinflammation and promotes hippocampal neurogenesis through PPAR-gamma involvement. *PLoS One, 6*(12), e28668.

Fernandez-Ruiz, J., Sagredo, O., Pazos, M. R., Garcia, C., Pertwee, R., Mechoulam, R., & Martinez-Orgado, J. (2013). Cannabidiol for neurodegenerative disorders: important new clinical applications for this phytocannabinoid? *British Journal of Clinical Pharmacology, 75*(2), 323–333.

Fogaça, M. V., Aguiar, D. C., Moreira, F. A., & Guimarães, F. S. (2012). The endocannabinoid and endovanilloid systems interact in the rat prelimbic medial prefrontal cortex to control anxiety-like behavior. *Neuropharmacology, 63*(2), 202–210.

Fogaça, M. V., Galve-Roperh, I., Guimarães, F. S., & Campos, A. C. (2013). Cannabinoids, neurogenesis and antidepressant drugs: is there a link? *Current Neuropharmacology, 11*(3), 263–275.

Fogaça, M. V., Gomes, F. V., Moreira, F. A., Guimarães, F. S., & Aguiar, D. C. (2013). Effects of glutamate NMDA and TRPV1 receptor antagonists on the biphasic responses to anandamide injected into the dorsolateral periaqueductal grey of Wistar rats. *Psychopharmacology (Berlin), 226*(3), 579–587.

Fogaça, M. V., Reis, F. M., Campos, A. C., & Guimarães, F. S. (2014). Effects of intra-prelimbic prefrontal cortex injection of cannabidiol on anxiety-like behavior: involvement of 5HT$_{1A}$ receptors and previous stressful experience. *European Neuropsychopharmacology, 24*(3), 410–419.

Franklin, J. M., & Carrasco, G. A. (2013). Cannabinoid receptor agonists upregulate and enhance serotonin 2A (5-HT$_{2A}$) receptor activity via ERK1/2 signaling. *Synapse, 67*(3), 145–159.

Gomes, F. V., Del Bel, E. A., & Guimarães, F. S. (2013). Cannabidiol attenuates catalepsy induced by distinct pharmacological mechanisms via 5-HT$_{1A}$ receptor activation in mice. *Progress in Neuro-Psychopharmacology & Biological Psychiatry, 46*, 43–47.

Gomes, F. V., Reis, D. G., Alves, F. H., Corrêa, F. M., Guimarães, F. S., & Resstel, L. B. (2012). Cannabidiol injected into the bed nucleus of the stria terminalis reduces the expression of contextual fear conditioning via 5-HT$_{1A}$ receptors. *Journal of Psychopharmacology, 26*(1), 104–113.

Gomes, F. V., Resstel, L. B., & Guimarães, F. S. (2011). The anxiolytic-like effects of cannabidiol injected into the bed nucleus of the stria terminalis are mediated by 5-HT$_{1A}$ receptors. *Psychopharmacology (Berlin), 213*(2–3), 465–473.

Guimarães, F. S., Chiaretti, T. M., Graeff, F. G., & Zuardi, A. W. (1990). Antianxiety effect of cannabidiol in the elevated plus-maze. *Psychopharmacology (Berlin), 100*(4), 558–559.

Hayakawa, K., Mishima, K., Nozako, M., Ogata, A., Hazekawa, M., Liu, A. X., ... Fujiwara, M. (2007). Repeated treatment with cannabidiol but not delta9-tetrahydrocannabinol has a neuroprotective effect without the development of tolerance. *Neuropharmacology, 52*(4), 1079–1087.

Hojo, M., Sudo, Y., Ando, Y., Minami, K., Takada, M., Matsubara, T., & Uezono, Y. (2008). mu-Opioid receptor forms a functional heterodimer with cannabinoid CB$_1$ receptor: electrophysiological and FRET assay analysis. *Journal of Pharmaceutical Sciences, 108*(3), 308–319.

Jenny, M., Santer, E., Pirich, E., Schennach, H., & Fuchs, D. (2009). Delta9-tetrahydrocannabinol and cannabidiol modulate mitogen-induced tryptophan degradation and neopterin formation in peripheral blood mononuclear cells in vitro. *Journal of Neuroimmunology, 207*(1–2), 75–82.

Kathmann, M., Flau, K., Redmer, A., Tränkle, C., & Schlicker, E. (2006). Cannabidiol is an allosteric modulator at mu- and delta-opioid receptors. *Naunyn-Schmiedeberg's Archives of Pharmacology, 372*(5), 354–361.

Katsidoni, V., Anagnostou, I., & Panagis, G. (2013). Cannabidiol inhibits the reward-facilitating effect of morphine: involvement of 5-HT$_{1A}$ receptors in the dorsal raphe nucleus. *Addiction Biology, 18*(2), 286–296.

Liou, G. I., Auchampach, J. A., Hillard, C. J., Zhu, G., Yousufzai, B., Mian, S., ... Khalifa, Y. (2008). Mediation of cannabidiol anti-inflammation in the retina by equilibrative nucleoside transporter and A2A adenosine receptor. *Investigative Ophthalmology & Visual Science, 49*(12), 5526–5531.

Ludanyi, A., Eross, L., Czirjak, S., Vajda, J., Halasz, P., Watanabe, M., ... Katona, I. (2008). Downregulation of the CB$_1$ cannabinoid receptor and related molecular elements of the endocannabinoid system in epileptic human hippocampus. *Journal of Neuroscience, 28*(12), 2976–2990.

Magen, I., Avraham, Y., Ackerman, Z., Vorobiev, L., Mechoulam, R., & Berry, E. M. (2010). Cannabidiol ameliorates cognitive and motor impairments in bile-duct ligated mice via 5-HT$_{1A}$ receptor activation. *British Journal of Pharmacology, 159*(4), 950–957.

Mahgoub, M., Keun-Hang, S. Y., Sydorenko, V., Ashoor, A., Kabbani, N., Al Kury, L., ... Oz, M. (2013). Effects of cannabidiol on the function of α7-nicotinic acetylcholine receptors. *European Journal of Pharmacology, 720*(1–3), 310–319.

Marco, E. M., Garcia-Gutierrez, M. S., Bermudez-Silva, F. J., Moreira, F. A., Guimarães, F., Manzanares, J., & Viveros, M. P. (2011). Endocannabinoid system and psychiatry: in search of a neurobiological basis for detrimental and potential therapeutic effects. *Frontiers in Behavioral Neuroscience, 5*, 63.

Mechoulam, R., & Shvo, Y. (1963). The structure of cannabidiol. *Tetrahedron, 19*, 2073–2078.

Mishima, K., Hayakawa, K., Abe, K., Ikeda, T., Egashira, N., Iwasaki, K., & Fujiwara, M. (2005). Cannabidiol prevents cerebral infarction via a serotonergic 5-hydroxytryptamine1A receptor-dependent mechanism. *Stroke, 36*(5), 1077–1082.

Morgan, C. J., Das, R. K., Joye, A., Curran, H. V., & Kamboj, S. K. (2013). Cannabidiol reduces cigarette consumption in tobacco smokers: preliminary findings. *Addictive Behaviors, 38*(9), 2433–2436.

O'Sullivan, S. E., Sun, Y., Bennett, A. J., Randall, M. D., & Kendall, D. A. (2009). Time-dependent vascular actions of cannabidiol in the rat aorta. *European Journal of Pharmacology, 612*(1–3), 61–68.

Pandolfo, P., Silveirinha, V., dos Santos-Rodrigues, A., Venance, L., Ledent, C., Takahashi, R. N., ... Kofalvi, A. (2011). Cannabinoids inhibit the synaptic uptake of adenosine and dopamine in the rat and mouse striatum. *European Journal of Pharmacology, 655*(1–3), 38–45.

Parker, L. A., Burton, P., Sorge, R. E., Yakiwchuk, C., & Mechoulam, R. (2004). Effect of low doses of delta9-tetrahydrocannabinol and cannabidiol on the extinction of cocaine-induced and amphetamine-induced conditioned place preference learning in rats. *Psychopharmacology (Berlin), 175*(3), 360–366.

Parker, L. A., Kwiatkowska, M., & Mechoulam, R. (2006). Delta-9-tetrahydrocannabinol and cannabidiol, but not ondansetron, interfere with conditioned retching reactions elicited by a lithium-paired context in Suncus murinus: an animal model of anticipatory nausea and vomiting. *Physiology & Behavior, 87*(1), 66–71.

Pazos, M. R., Mohammed, N., Lafuente, H., Santos, M., Martínez-Pinilla, E., Moreno, E., ... Martínez-Orgado, J. (2013). Mechanisms of cannabidiol neuroprotection in hypoxic-ischemic newborn pigs: role of 5HT$_{1A}$ and CB$_2$ receptors. *Neuropharmacology, 71*, 282–291.

Peroutka, S. J., & Snyder, S. H. (1979). Multiple serotonin receptors: differential binding of [3H]5-hydroxytryptamine, [3H]lysergic acid diethylamide and [3H]spiroperidol. *Molecular Pharmacology, 16*(3), 687–699.

Przybyla, J. A., & Watts, V. J. (2010). Ligand-induced regulation and localization of cannabinoid CB$_1$ and dopamine D2L receptor heterodimers. *Journal of Pharmacology and Experimental Therapeutics, 332*(3), 710–719.

Qin, N., Neeper, M. P., Liu, Y., Hutchinson, T. L., Lubin, M. L., & Flores, C. M. (2008). TRPV2 is activated by cannabidiol and mediates CGRP release in cultured rat dorsal root ganglion neurons. *Journal of Neuroscience, 28*(24), 6231–6238.

Ren, Y., Whittard, J., Higuera-Matas, A., Morris, C. V., & Hurd, Y. L. (2009). Cannabidiol, a nonpsychotropic component of cannabis, inhibits cue-induced heroin seeking and normalizes discrete mesolimbic neuronal disturbances. *Journal of Neuroscience, 29*(47), 14764–14769.

Resstel, L. B., Tavares, R. F., Lisboa, S. F., Joça, S. R., Corrêa, F. M., & Guimarães, F. S. (2009). 5-HT$_{1A}$ receptors are involved in the cannabidiol-induced attenuation of behavioural and cardiovascular responses to acute restraint stress in rats. *British Journal of Pharmacology, 156*(1), 181–188.

Risinger, F. O., & Boyce, J. M. (2002). 5-HT$_{1A}$ receptor blockade and the motivational profile of ethanol. *Life Science, 71*(6), 707–715.

Rock, E. M., Bolognini, D., Limebeer, C. L., Cascio, M. G., Anavi-Goffer, S., Fletcher, P. J., ... Parker, L. A. (2012). Cannabidiol, a non-psychotropic component of cannabis, attenuates vomiting and nausea-like behaviour via indirect agonism of 5-HT$_{1A}$ somatodendritic autoreceptors in the dorsal raphe nucleus. *British Journal of Pharmacology*, *165*(8), 2620–2634.

Rock, E. M., Goodwin, J. M., Limebeer, C. L., Breuer, A., Pertwee, R. G., Mechoulam, R., & Parker, L. A. (2011). Interaction between non-psychotropic cannabinoids in marihuana: effect of cannabigerol (CBG) on the anti-nausea or anti-emetic effects of cannabidiol (CBD) in rats and shrews. *Psychopharmacology (Berlin)*, *215*(3), 505–512.

Russo, E. B., Burnett, A., Hall, B., & Parker, K. K. (2005). Agonistic properties of cannabidiol at 5-HT$_{1A}$ receptors. *Neurochemical Research*, *30*(8), 1037–1043.

Ryan, D., Drysdale, A. J., Pertwee, R. G., & Platt, B. (2006). Differential effects of cannabis extracts and pure plant cannabinoids on hippocampal neurones and glia. *Neuroscience Letters*, *408*(3), 236–241.

Scopinho, A. A., Guimarães, F. S., Corrêa, F. M., & Resstel, L. B. (2011). Cannabidiol inhibits the hyperphagia induced by cannabinoid-1 or serotonin-1A receptor agonists. *Pharmacology Biochemistry and Behavior*, *98*(2), 268–272.

Soares Vde, P., Campos, A. C., Bortoli, V. C., Zangrossi, H., Jr., Guimarães, F. S., & Zuardi, A. W. (2010). Intra-dorsal periaqueductal gray administration of cannabidiol blocks panic-like response by activating 5-HT$_{1A}$ receptors. *Behavioural Brain Research*, *213*(2), 225–229.

Stern, C. A., Gazarini, L., Takahashi, R. N., Guimarães, F. S., & Bertoglio, L. J. (2012). On disruption of fear memory by reconsolidation blockade: evidence from cannabidiol treatment. *Neuropsychopharmacology*, *37*(9), 2132–2142.

Takahashi, K. A., & Castillo, P. E. (2006). The CB$_1$ cannabinoid receptor mediates glutamatergic synaptic suppression in the hippocampus. *Neuroscience*, *139*(3), 795–802.

Thomas, A., Baillie, G. L., Phillips, A. M., Razdan, R. K., Ross, R. A., & Pertwee, R. G. (2007). Cannabidiol displays unexpectedly high potency as an antagonist of CB$_1$ and CB$_2$ receptor agonists in vitro. *British Journal of Pharmacology*, *150*(5), 613–623.

Thompson, P. M., Cruz, D. A., Olukotun, D. Y., & Delgado, P. L. (2012). Serotonin receptor, SERT mRNA and correlations with symptoms in males with alcohol dependence and suicide. *Acta Psychiatrica Scandinavica*, *126*(3), 165–174.

Todd, A. R. (1946). Hashish. *Experientia*, *2*, 55–60.

Tominaga, M., Caterina, M. J., Malmberg, A. B., Rosen, T. A., Gilbert, H., Skinner, K., ... Julius, D. (1998). The cloned capsaicin receptor integrates multiple pain-producing stimuli. *Neuron*, *21*(3), 531–543.

Twardowschy, A., Castiblanco-Urbina, M. A., Uribe-Mariño, A., Biagioni, A. F., Salgado-Rohner, C. J., Crippa, J. A., & Coimbra, N. C. (2013). The role of 5-HT$_{1A}$ receptors in the anti-aversive effects of cannabidiol on panic attack-like behaviors evoked in the presence of the wild snake *Epicrates cenchria crassus* (Reptilia, Boidae). *Journal of Psychopharmacology*, *27*(12), 1149–1159.

Wolf, S. A., Bick-Sander, A., Fabel, K., Leal-Galicia, P., Tauber, S., Ramirez-Rodriguez, G., ... Kempermann, G. (2010). Cannabinoid receptor CB$_1$ mediates baseline and activity-induced survival of new neurons in adult hippocampal neurogenesis. *Cell Communication and Signaling*, *8*, 12.

Xing, J., & Li, J. (2007). TRPV1 receptor mediates glutamatergic synaptic input to dorsolateral periaqueductal gray (dl-PAG) neurons. *Journal of Neurophysiology*, *97*(1), 503–511.

Xiong, W., Koo, B. N., Morton, R., & Zhang, L. (2011). Psychotropic and nonpsychotropic cannabis derivatives inhibit human 5-HT(3A) receptors through a receptor desensitization-dependent mechanism. *Neuroscience*, *16*(184), 28–37.

Zanelati, T. V., Biojone, C., Moreira, F. A., Guimarães, F. S., & Joca, S. R. (2010). Antidepressant-like effects of cannabidiol in mice: possible involvement of 5-HT$_{1A}$ receptors. *British Journal of Pharmacology*, *159*(1), 122–128.

Zuardi, A. W., Cosme, R. A., Graeff, F. G., & Guimarães, F. S. (1993). Effects of ipsapirone and cannabidiol on human experimental anxiety. *Journal of Psychopharmacology*, *7*(Suppl. 1), 82–88.

Zuardi, A. W., Shirakawa, I., Finkelfarb, E., & Karniol, I. G. (1982). Action of cannabidiol on the anxiety and other effects produced by delta 9-THC in normal subjects. *Psychopharmacology (Berlin)*, *76*(3), 245–250.

Zygmunt, P. M., Petersson, J., Andersson, D. A., Chuang, H., Sorgard, M., Di Marzo, V., ... Hogestatt, E. D. (1999). Vanilloid receptors on sensory nerves mediate the vasodilator action of anandamide. *Nature*, *400*(6743), 452–457.

Chapter 71

CB1 Receptor-Mediated Signaling Mechanisms in the Deleterious Effects of Spice Abuse

Balapal S. Basavarajappa

Division of Analytical Psychopharmacology, Nathan Kline Institute for Psychiatric Research, Orangeburg, NY, USA; Department of Psychiatry, College of Physicians and Surgeons, New York State Psychiatric Institute, Columbia University, New York, NY, USA

INTRODUCTION

Marijuana is the most commonly abused illegal drug (Figure 1). The rate of abuse increased from 14.5 million in 2007 to 18.1 million in 2011, with an estimated ~5 million adult daily cannabis users. Regular marijuana abuse appears to start in the eighth grade for 1.3% of children and between 12 and 17 years of age for 7.9% of children (Johnston, O'Malley, Bachman, & Schulenberg, 2012). Moreover, the average content of Δ^9-tetrahydrocannabinol (Δ^9-THC), a major psychoactive molecule of many bioactive phytocannabinoids found in the *Cannabis sativa* plant (Gaoni & Mechoulam, 1964), in marijuana increased from 3.4% in 1993 to 8.8% in 2008 (Mehmedic et al., 2010). Furthermore, an increase in the concentration of Δ^9-THC (13.8%) in highly potent plant varieties (sinsemilla, "skunk") (Mehmedic et al., 2010) was also observed in this period. The potent psychological changes observed after Δ^9-THC consumption are similar to those following recreationally consumed cannabis (Pertwee, 1988). Most effects of Δ^9-THC are mediated through the endocannabinoid system. The endocannabinoid system includes receptors for Δ^9-THC known as cannabinoid receptors type 1 and 2 (CB1R and CB2R), endogenous receptor ligands ("endocannabinoids," ECs), and EC-synthesizing and -degrading enzymes (Basavarajappa, 2014). CB1R is predominately expressed in the brain, particularly in areas such as the hippocampus, basal ganglia, cortex, amygdala, and cerebellum, all of which are areas associated with the behavioral effects of Δ^9-THC (Herkenham, Lynn, de Costa, & Richfield, 1991). The EC system has a homeostatic function, but its dysfunction can contribute to pathological conditions (Basavarajappa, 2007a, 2007b; Battistella et al., 2014).

Synthetic cannabinoids were developed as research tools to explore the function of the EC system and as potential therapeutic agents (Huffman, Dong, Martin, & Compton, 1994). However, since 2005, synthetic cannabinoids have been sprayed onto plant material and subsequently packaged and sold under brand names such as "Spice" or "K2" to mimic the effects of marijuana (Vardakou, Pistos, & Spiliopoulou, 2010). Although these packages are labeled "not for human consumption," these products are often smoked, resulting in a marijuana-like high, as well as other physiological effects. Many synthetic cannabinoids are now classified as Schedule I drugs under the US Controlled Substance Act (US Drug Enforcement Administration, 2014). Because synthetic cannabinoids are now illegal and are classified in the most dangerous category of scheduled drugs, more structurally diverse cannabimimetic compounds are emerging, which might not be listed under scheduled drug regulations. In addition, "medical" marijuana is being legalized in many states across the United States. Therefore, individuals who previously would not have risked the consequences of procuring an illegal drug might now consider exposing themselves to marijuana and synthetic cannabinoids. Synthetic cannabinoids are classified (Table 1) based on the chemical structure of the molecule (Howlett et al., 2002). The pharmacological effects of many synthetic cannabinoids vary widely and have been reviewed (Castaneto et al., 2014). The most frequently studied synthetic cannabinoids are the aminoalkylindole WIN 55,212-2 (134 studies), cyclohexylphenol CP 55,940 (54), and HU-210 (39). Only a few studies have included the naphthoylindoles (JWH-018 and analogs) or the newer synthetic structural families of cannabinoids that are now dominating the market (Castaneto et al., 2014). Synthetic cannabinoids and their metabolites have been found to possess a higher binding affinity for cannabinoid receptors than marijuana, which implies greater potency, greater adverse effects, and perhaps a longer duration of action. Although most Spice drugs are widely known to be potent CB1R agonists, knowledge of the exact signaling mechanisms that are essential for their deleterious effects is largely limited. In this chapter, the author presents an integrative overview of the current research on the signaling mechanisms involved in the harmful effects of Spice compounds.

CANNABINOID RECEPTORS

Evidence of the existence of the marijuana receptor has been available since the 1980s (Devane, Dysarz, Johnson, Melvin, & Howlett, 1988). It has now been shown that cannabinoids have two specific receptor subtypes, CB1 and CB2, which are expressed in

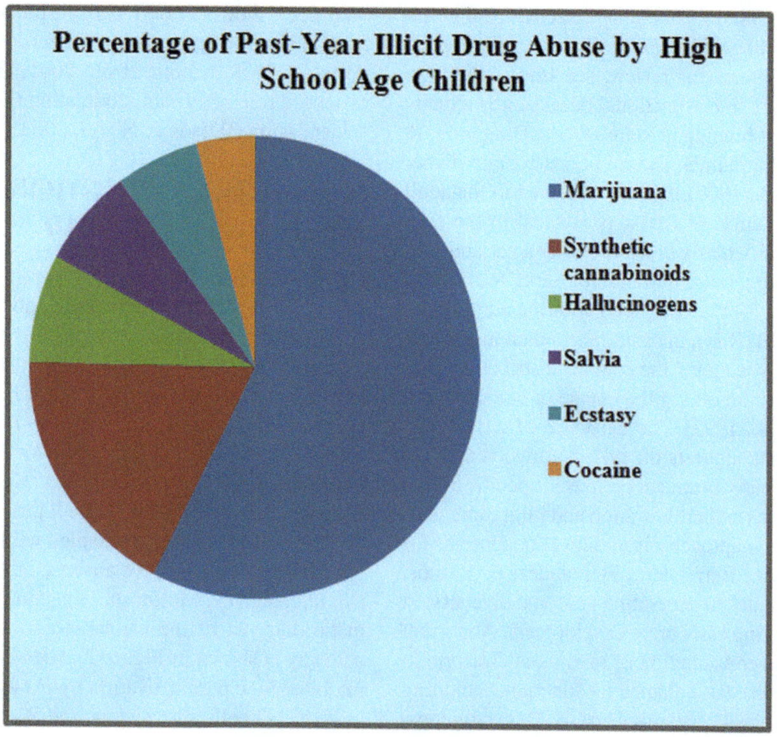

FIGURE 1 Pie diagram of illicit drug abuse rates among high-school-age children. Synthetic cannabinoid abuse is most popular among young people; of the illicit drugs most used by high-school seniors, they are second only to marijuana (http://www.drugabuse.gov/publications/drugfacts/spice-synthetic-marijuana). Easy availability and the misperception that Spice products are safe to get high and therefore harmless have probably contributed to their popularity. Another important point is that the synthetic cannabinoids sprayed in Spice products are not easily detected in standard drug tests.

TABLE 1 Classification of Synthetic Cannabinoids (Howlett et al., 2002)

Class	Examples
Classical cannabinoids	Δ^9-THC, HU-210, AM906, AM411, O-1184
Nonclassical cannabinoids	CP 47,497-C8, CP 55,940, CP 55,244
Hybrid cannabinoids	AM4030
Aminoalkylindoles	JWH-018, JWH-073, JWH-398, JWH-015, JWH-122, JWH-210, JWH-081, JWH-200, WIN 55,212, JWH-250, JWH-251, pravadoline, AM694, RSC-4
Eicosanoids	AEA and methanandamide
Others	Diarylpyrazoles (SR141716A), naphthoylpyrroles (JWH-307), naphthylmethylindenes, or derivatives of naphthalene-1-yl-(4-pentyloxynaphthalen-1-yl)methanone (CRA-13)

all vertebrates and have been cloned. CB1R predominantly functions in the nervous system but is expressed in many cells throughout the body. CB2R is primarily associated with cells with immune function such as splenocytes, macrophages, monocytes, microglia, and B and T cells. Evidence has suggested that CB2R is present in other cells and is often upregulated under pathological conditions (reviewed in Pertwee et al., 2010). The CB1 and CB2 receptors belong to the large superfamily of heptahelical G-protein-coupled receptors (GPCRs) and couple to $G_{i/o}$ proteins (for more details, see Basavarajappa, 2007a, 2007b; Basavarajappa, 2014). CB1R is among the most abundant GPCRs in the brain, with densities similar to the levels of γ-aminobutyric acid (GABA)- and glutamate-gated ion channels (Herkenham, Lynn, Johnson, et al., 1991). Functional CB2R is also present in limited amounts and in distinct locations in the brains of several animal species, including humans (Onaivi et al., 2008). However, the functional significance of this receptor in the brain is slowly emerging (Onaivi et al., 2008), and its involvement in many pathological conditions has also been reported.

CB1R has a wide expression pattern in the developing nervous system, and its expression follows neuronal differentiation in the embryo from the earliest stages. Several studies have described the expression pattern of CB1R mRNA and the distribution of CB1Rs in the fetal and neonatal rodent brain (Cristino & Di Marzo, 2014).

CB1R mRNA and receptor binding could be detected from gestational day 14 in rats, coinciding with the time of phenotypic expression of most neurotransmitters (for review, see Insel, 1995). At this fetal age, CB1R appears to be functional because it is already found to be coupled to GTP-binding proteins (Glass, Dragunow, & Faull, 1997). The developing human and rat brains contain higher levels of CB1R (Glass et al., 1997) than are observed in the adult brain. However, the distribution of CB1R is atypical in the fetal and early neonatal brain, particularly in white matter areas and subventricular zones of the forebrain (Berrendero, Sepe, Ramos, Di Marzo, & Fernandez-Ruiz, 1999), compared with the adult brain. This atypical location of CB1R is a transient phenomenon because the receptors progressively acquire the classic distribution pattern observed in the adult brain during the course of late postnatal development (Romero et al., 1997). The existence of CB1R during early brain development suggests the possible involvement of CB1R during the fetal and early postnatal periods in specific events of brain development such as cell proliferation and migration, axonal elongation, and, later, synaptogenesis and myelinogenesis (for review, see Fernandez-Ruiz, Berrendero, Hernandez, & Ramos, 2000). Thus, CB1R contributes to generating neuronal diversity, in particular, brain regions during early brain development. Abnormal CB1R activation due to the consumption of marijuana (Tortoriello et al., 2014) or other agents that enhance its function, including ethanol (Subbanna, Nagaraja, Umapathy, Pace, & Basavarajappa, 2014; Subbanna, Shivakumar, Psychoyos, Xie, & Basavarajappa, 2013) and synthetic cannabinoids (Tree, Scotto di Perretolo, Peyronnet, & Cayetanot, 2014), has long-lasting deleterious consequences on neurobehavioral outcomes in adulthood.

Consistent with their role during early brain development, there is evidence that perinatal exposure to marijuana or synthetic cannabinoids modifies the maturation of neurotransmitter systems, as well as their related behaviors (Fernandez-Ruiz et al., 2000). These effects are caused through CB1R activation. The administration of synthetic cannabinoids in doses similar to those attained by marijuana consumption was shown to alter the development of the neurotransmitter system, resulting in neurobehavioral disturbances. Thus, adult animals that were perinatally exposed to synthetic cannabinoids exhibited neurobehavioral abnormalities (Fernandez-Ruiz et al., 2000). Apparently, most of these neurobehavioral abnormalities originate through the developmental alteration of several neurotransmitter systems caused by exposure to marijuana or synthetic cannabinoids and probably through CB1R activation during critical prenatal and early postnatal periods of brain development (Goldschmidt, Richardson, Willford, Severtson, & Day, 2012; Shabani, Mahnam, Sheibani, & Janahmadi, 2014; Silva, Zhao, Popp, & Dow-Edwards, 2012).

The distribution of CB1R in the adult brain is highly heterogeneous, with the highest densities of the receptor present in the basal ganglia, substantia nigra, pars reticulata, and globus pallidus. In addition, very high levels of binding are found in the hippocampus, particularly within the dentate gyrus, as well as in the molecular layer of the cerebellum. In contrast, little CB1R is present in the brainstem (Howlett, 2002). A similar CB1R distribution is found in humans (Biegon & Kerman, 2001; Glass et al., 1997). The highest densities are found in association with the limbic cortices, with much lower levels within the primary sensory and motor regions, suggesting an important role for CB1R in motivational (limbic) and cognitive (association) information processing. CB1R has also been shown to be localized presynaptically on GABAergic interneurons and glutamatergic neurons (Basavarajappa, Ninan, & Arancio, 2008; Katona et al., 2001). This is consistent with the proposed role of ECs in modulating GABA and glutamate neurotransmission (Wilson & Nicoll, 2002).

SIGNAL TRANSDUCTION MECHANISM OF CB1R

CB1R activation promotes its interaction with G proteins, resulting in guanosine diphosphate/guanosine triphosphate exchange and the subsequent dissociation of the α and $\beta\gamma$ subunits. These subunits regulate the activity of multiple downstream effector proteins to elicit biological functions. CB1Rs are coupled with G_i or G_o proteins. However, their affinity for G_i or G_o proteins might vary, as revealed by several receptor ligand and receptor ligand-stimulated GTPγS-binding studies (Kearn, Greenberg, DiCamelli, Kurzawa, & Hillard, 1999). CB1R differs from many other GPCR–G protein pairs, as it is precoupled with G proteins and is therefore constitutively active in the absence of exogenously added agonists (Mukhopadhyay, McIntosh, Houston, & Howlett, 2000). A schematic diagram of the cannabinoid-mediated signal transduction pathway is shown in Figure 2. Adenylate cyclase (AC) activity is inhibited by CB1R activation by R-(+)-methanandamide and ECs in N18TG2 cells (for review, see Basavarajappa, 2007a, 2007b). In some cases, the upregulation of AC activity was reported without $G_{i/o}$ coupling (pertussis toxin-sensitive), probably through the activation of G_s proteins (Glass & Felder, 1997). The in vitro expression of specific isoforms of AC (I, III, V, VI, or VIII) with the coexpression of CB1R was associated with the inhibition of cyclic AMP (cAMP) accumulation. However, the expression of other AC isoforms (II, IV, or VII) with the coexpression of CB1R was associated with the stimulation of cAMP accumulation (Rhee, Bayewitch, Avidor-Reiss, Levy, & Vogel, 1998). The questions of whether CB1R coupling to G_s proteins has physiological importance and whether this coupling increases after G_i or G_o protein sequestration by colocalized noncannabinoid $G_{i/o}$-protein-coupled receptors have yet to be answered. Further characterization of the mechanism by which CB1R activation leads to the accumulation of GαGTP$\beta\gamma$ heterotrimers would enhance our understanding of the mechanism of CB1R signal transduction. A mechanism has been proposed for other GPCRs (Lambert, 2008). It is also important to characterize whether these heterotrimers (Gα, G$\beta\gamma$, and GαGTP$\beta\gamma$) can interact with specific downstream effector molecules to provide specificity to the signaling events.

The inhibition of N-type voltage-gated channels by some synthetic cannabinoids has been demonstrated in several neuronal cells (for review, see Basavarajappa, 2007a, 2007b). L-type Ca^{2+} channels are inhibited by synthetic cannabinoids in cat brain arterial smooth muscle cells that express CB1R (Gebremedhin, Lange, Campbell, Hillard, & Harder, 1999). This effect is blocked by pertussis toxin and SR141716A, suggesting a CB1R/$G_{i/o}$-mediated signaling mechanism (Mechoulam & Parker, 2013). CB1Rs are also coupled to ion channels through $G_{i/o}$ proteins and activate A-type and inwardly rectifying potassium channels. The coupling to A-type and D-type potassium channels is thought to occur through AC. These effects are also stimulated by the inhibition of AC by cannabinoids. Because of the decrease in cAMP accumulation, cAMP-dependent protein kinase (PKA) is inhibited by CB1R

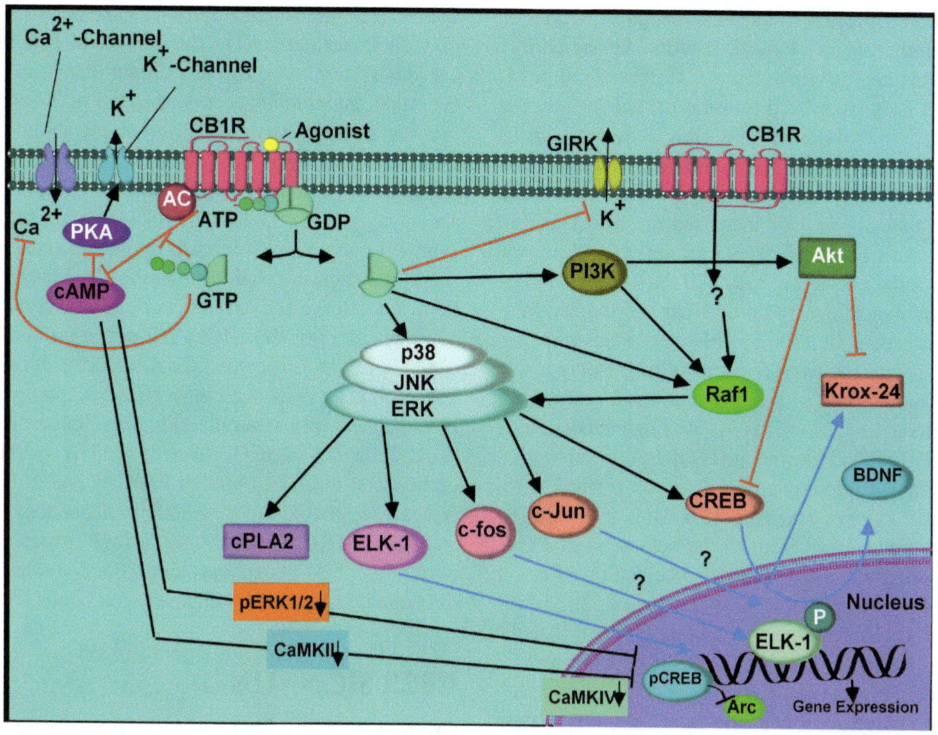

FIGURE 2 Schematic outline of the CB1R signaling pathway. Δ^9-THC and other synthetic cannabinoids elicit their effects by binding to CB1R. CB1R is a seven-transmembrane-domain GPCR located in the cell membrane. The Ca^{2+} channels inhibited by CB1R activation include N-, P/Q-, and L-type channels. The actions on Ca^{2+} channels and AC are thought to be mediated by the α subunits of the G protein, and the actions on G-protein-gated inwardly rectifying potassium (GIRK) channels and inositol 1,4,5-triphosphate kinase (PI3K) are thought to be mediated by the βγ subunits. AC inhibition and the subsequent decrease in cAMP lead to decreased activation of cAMP-dependent protein kinase (PKA). This leads to decreased K^+ channels and Ca^{2+}/calmodulin-dependent protein kinase IV (CaMKIV) and cAMP-response element-binding protein (CREB) phosphorylation, which might lead to gene expression inhibition. Stimulatory effects are shown by (→) symbols and inhibitory effects by (⊥) symbols. BDNF, brain-derived neurotrophic factor; cPLA2, cytoplasmic phospholipase A2.

activation (for review, see Basavarajappa, 2007a, 2007b). Without synthetic cannabinoids, PKA phosphorylates the potassium channel protein by exerting decreased outward potassium current. In the presence of cannabinoids, however, phosphorylation of the channel by PKA is reduced, which leads to an increased outward potassium current (Childers & Deadwyler, 1996). Cannabinoids inhibit N-type and P/Q-type calcium channels and D-type potassium channels (Howlett et al., 2002). In addition, synthetic cannabinoids can close sodium channels, but whether this effect is receptor mediated has yet to be proven. There is also evidence from experiments with rat hippocampal CA1 pyramidal neurons that CB1R is negatively coupled to M-type potassium channels (Schweitzer, 2000). CB1R might also mobilize arachidonic acid (AA), close 5-HT3 receptor ion channels, and under certain conditions, couple with G_s proteins to activate AC (Calandra et al., 1999) to reduce the outward potassium K current, possibly through AA-mediated stimulation of protein kinase C (Hampson, Mu, & Deadwyler, 2000). CB1R has also been reported to activate phospholipase C (PLC) through G proteins in COS-7 cells cotransfected with CB1R and G_α subunits (Ho, Uezono, Takada, Takase, & Izumi, 1999). In cultured cerebellar granule neurons, CB1R works through a PLC-sensitive mechanism to increase N-methyl-D-aspartate (NMDA)-elicited calcium release from inositol 1,4,5-triphosphate-gated intracellular stores (Netzeband, Conroy, Parsons, & Gruol, 1999). CB1R activation by synthetic cannabinoid agonists evokes a rapid, transient increase in intracellular free Ca^{2+} in N18TG2 and NG108-15 cells (Sugiura, Kishimoto, Oka, & Gokoh, 2006). These targets might have significant roles in the regulation of neurotransmitter release, but this requires further investigation.

The regulation of cellular growth has usually been associated with tyrosine kinase receptors. However, studies suggest that GPCRs can stimulate the mitogen-activated protein kinase (MAPK) pathway and thereby induce cellular growth. After the first observation of MAPK cascade activation by ECs and anandamide (AEA) (Wartmann, Campbell, Subramanian, Burstein, & Davis, 1995), several studies using both ECs and synthetic cannabinoids have implicated the involvement of this pathway in both in vivo and in vitro models. The activation of two isoforms (p42/p44) of MAPK mediated by CB1R/$G_{i/o}$ was observed in nonneuronal U373MG astrocytoma cells and in host cells expressing recombinant CB1R (Bouaboula et al., 1995). Similarly, the activation of CB1R/$G_{i/o}$ by Δ^9-THC and synthetic cannabinoid (HU-210) led to p42/p44 MAPK activation in C6 glioma cell and primary astrocyte cultures (Guzman, Sanchez, & Galve-Roperh, 2002). In WI-38 fibroblasts, AEA, via CB1R/$G_{i/o}$, promoted tyrosine phosphorylation of extracellular signal-regulated kinase 2 (ERK2 or p44) and increased MAPK activity (Wartmann et al., 1995). In some cells, CB1R-mediated MAPK activation was mediated through the inositol 1,4,5-triphosphate kinase (PI3K) pathway (Wartmann et al., 1995). AEA and synthetic cannabinoids

(CP 55,940 and WIN 55,212-2) increased the phosphorylation of FAK+6,7, a neural isoform of focal adhesion kinase (FAK), in hippocampal slices and cultured neurons (Derkinderen et al., 2001). Δ^9-THC and ECs stimulated tyrosine phosphorylation of the Tyr 397 residue, which is crucial for FAK activation, in the hippocampus (Derkinderen et al., 2001). Synthetic cannabinoids increased the phosphorylation of p130-Cas, a protein associated with FAK in the hippocampus. ECs increased the association of Fyn but not of Src with FAK+6,7. These effects were mediated through inhibition of the cAMP pathway. CB1R-stimulated FAK autophosphorylation was shown to be upstream of the Src-family kinases (Derkinderen et al., 2001). This new mechanism for cannabinoid regulation of the MAPK pathway might play a role in the EC-induced modulation of synaptic plasticity, cell migration, and neurite remodeling, as well as in pathological conditions (Subbanna et al., 2013). Δ^9-THC promoted Raf-1 phosphorylation and its subsequent translocation to the membrane in cortical astrocytes (Sanchez, Galve-Roperh, Rueda, & Guzman, 1998). CB1R-mediated release of βγ subunits leads to the activation of PI3K, resulting in the tyrosine phosphorylation and activation of Raf-1 and finally MAPK phosphorylation. The activation of p38 MAPK was observed in CHO cells expressing recombinant CB1R (Rueda, Galve-Roperh, Haro, & Guzman, 2000) and in human vascular endothelial cells with endogenous CB1R (Liu et al., 2000). Δ^9-THC was shown to induce activation of c-Jun N-terminal kinase (JNK1 and JNK2) in CHO cells expressing recombinant CB1R (Rueda et al., 2000). The pathway for JNK activation involves the $G_{i/o}$ protein, PI3K, and Ras (Rueda et al., 2000).

Activation of the Na^+/H^+ exchanger in CHO cells stably expressing CB1R was shown to be mediated through MAPK and CB1R. EC-stimulated activation of MAPK activity was shown to lead to the phosphorylation of cytoplasmic phospholipase A2 and the release of AA, resulting in the synthesis of prostaglandin E2, in WI-38 cells. MAPK activation by synthetic cannabinoids was shown to induce immediate early gene expression (*krox*-24) in U373MG human astrocytoma cells. Δ^9-THC induced the expression of *krox*-24, brain-derived neurotrophic factor, and c-Fos in the mouse hippocampus. The CB1R- and MAPK/ERK-mediated activation of *krox*-24 is negatively regulated through PI3K/protein kinase B, also known as Akt, in neuro2a cells. The suppression of prolactin receptor and tyrosine kinase nerve growth factor receptor synthesis by EC was shown to be associated with a CB1R-mediated decrease in PKA and an increase in MAPK activities. Synthetic cannabinoid exposure induces the expression of c-Fos and c-Jun in the brain (for review, see Basavarajappa, 2007a, 2007b); whether this is mediated by CB1R-activated MAPK is not known. The Δ^9-THC-induced phosphorylation of the transcription factor ELK-1 is mediated by MAPK/ERK (Valjent, Caboche, & Vanhoutte, 2001). Intracerebroventricular injection of EC evoked an increase in c-Fos protein in the rat brain with a distribution generally similar to that of CB1R (Patel, Moldow, Patel, Wu, & Chang, 1998). The activation of protein kinase B/Akt (isoform IB) by synthetic cannabinoids is mediated by $G_{i/o}$ and PI3K in U373MG astrocytoma and CHO cells expressing recombinant CB1R (Gomez del Pulgar, Velasco, & Guzman, 2000). CB1R-mediated gene regulation through MAPK activation is an important physiological mechanism by which cannabinoids and ECs regulate synaptic plasticity.

In addition, CB1R activation by synthetic cannabinoids has been found to lead to the generation of ceramide (Guzman et al., 2002). This widespread lipid secondary messenger is known to play an important role in the control of cell fate in the central nervous system. Studies have shown that cannabinoid-dependent ceramide generation occurs through a G-protein-independent process and involves two different metabolic pathways: sphingomyelin hydrolysis and de novo ceramide synthesis. In turn, ceramide mediates synthetic cannabinoid-induced apoptosis, as shown by in vitro and in vivo studies. Activation of CB1Rs by synthetic cannabinoids induces apoptosis via ceramide accumulation, p38 phosphorylation, mitochondrial membrane depolarization, and caspase activation in both Mantle cell lymphoma (MCL) cell lines and primary MCLs but not in normal B cells (Guzman et al., 2002). In astrocytes, CB1R has been shown to be coupled to sphingomyelin hydrolysis through the adapter protein FAN, which is associated with neutral sphingomyelinase activation (Guzman et al., 2002). In 2010, CB1R activation by synthetic cannabinoids or ECs was shown to reduce ERK (Atwood, Huffman, Straiker, & Mackie, 2010), Ca^{2+}/calmodulin-dependent protein kinases II and IV (CaMKII, CaMKIV) and cAMP-response element-binding protein (CREB) phosphorylation (Basavarajappa, Nagre, Xie, & Subbanna, 2014; Basavarajappa & Subbanna, 2014).

CB2R SIGNALING

CB2R belongs to the seven-transmembrane-domain class of GPCRs. CB2R is coupled to $G_{i/o}$ proteins; thus, its activation is associated with the inhibition of AC and the cAMP/PKA-dependent pathway, as has been observed for CB1R. CB2R stimulation activates MAPK cascades, specifically the ERK and the p38 MAPK cascades (Herrera et al., 2006). In addition, the activation of CB2R has also been linked to the stimulation of additional intracellular pathways, including the PI3K/Akt pathway (Herrera et al., 2006; Figure 3). These pathways have been associated with prosurvival effects, as well as with the de novo synthesis of the sphingolipid messenger ceramide (Herrera et al., 2006), which has been linked to the proapoptotic effects of synthetic cannabinoids.

SPICE SIGNALING

Spice products contain several synthetic cannabinoids, and their potency differs at CB1R and CB2R. Therefore, their signaling mechanisms may also differ. In 2014, a list of synthetic cannabinoids found in Spice products, as well as the deleterious biological effects of these cannabinoids, was reviewed (Castaneto et al., 2014). Some known physiological effects of synthetic cannabinoids are listed in Table 2. Synthetic cannabinoids with higher affinity for either CB1R or CB2R were shown to elicit deleterious neurobehavioral effects. The majority of synthetic cannabinoids found in Spice products such as AM5983, AM678, AM2233, AM2389, SDB-001, AM4054, UR-144, XLR-11, JWH-081, and JWH-073 were found to have higher affinity for CB1R than CB2R, are assumed to be similar to Δ^9-THC in their mechanism of action, and appear to be more potent than Δ^9-THC, with some of them exhibiting additional symptoms (Hermanns-Clausen, Kneisel, Szabo, & Auwarter, 2013). Few synthetic cannabinoids detected in Spice products such as JWH-018, AM1710, and JWH-133 appear to have higher affinity for CB2R than for CB1R (for references, see https://www.drugs-forum.com/forum/showthread.php?t=117873).

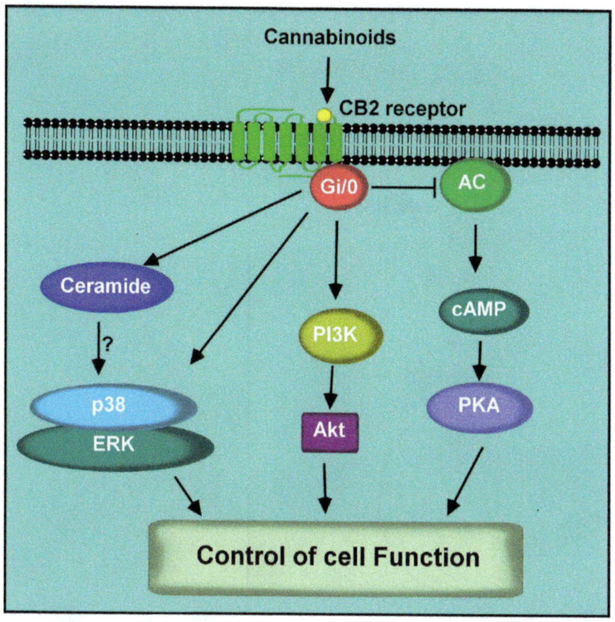

FIGURE 3 Summary of CB2R signaling. Δ^9-THC and synthetic cannabinoids also bind to CB2R. CB2Rs are also seven-transmembrane-domain GPCRs located in the cell membrane. The activation of CB2R is coupled to several different cellular pathways such as AC, cAMP, PKA, ERK, p38 MAPK, and Akt pathways, as well as the pathway for the de novo synthesis of ceramide. Stimulatory effects are shown by ($\rightarrow$) symbols and inhibitory effects by ($\perp$) symbols.

TABLE 2 Some Known Physiological Effects of Synthetic Cannabinoids (Castaneto et al., 2014)

Increase heart rate and blood pressure
Altered state of consciousness
Mild euphoria and relaxation
Perceptual alterations (time distortion)
Intensification of sensory experiences
Pronounced cognitive effects
Impaired short-term memory
Increase in reaction times

The signaling mechanisms used by these synthetic cannabinoids are not well characterized.

JWH-081 causes acute toxicity, as experienced by emergency patients, possibly through strong CB1R stimulation (Hermanns-Clausen et al., 2013). JWH-081 binds to CB1R with high affinity (1.2 nM) (Aung et al., 2000; Huffman et al., 2005). Acute bath application of JWH-081 to hippocampal slices inhibited long-term potentiation (LTP). This effect was absent in slices derived from CB1R-knockout (KO) mice. Thus, the deleterious effects of JWH-081 on LTP are mediated by CB1R. JWH-081 treatment impairs spontaneous alternation in the Y maze, as well as spatial and object recognition memory, in adult mice. The involvement of pCaMKIV, pCREB, and pERK1/2, which are key regulators of synaptic plasticity, learning, and memory (Martin, Grimwood, & Morris, 2000), in the acute effects of JWH-081 have also been studied. JWH-081 reduced CaMKIV and CREB phosphorylation in the hippocampus of CB1R-wild-type (WT) but not KO mice. CaMKIV phosphorylation was reduced by JWH-081 in a dose-dependent manner, whereas the inhibition of CREB phosphorylation was observed only at a higher dose of JWH-081 (1.25 mg/kg, 30 min). JWH-081 failed to alter ERK phosphorylation and total ERK protein in the hippocampus of CB1R-WT or -KO mice. In these studies, preadministration of the CB1R antagonist SR141716A 30 min before JWH-081 treatment rescued both CaMKIV and CREB phosphorylation. JWH-081 apparently elicited greater in vitro and in vivo responses relative to Δ^9-THC, the well-known, classical cannabinoid that is present in marijuana (Uchiyama, Kikura-Hanajiri, & Goda, 2011). In general, the CaMKIV-mediated phosphorylation of CREB at Ser 133 is essential for the transcriptional activation of the CREB/CRE-mediated signaling pathway (Bito, Deisseroth, & Tsien, 1996) and has been thought to play a central role in memory consolidation and LTP (Martin et al., 2000). Δ^9-THC significantly reduced the phosphorylation of CREB (Rubino & Parolaro, 2008) and another calmodulin kinase-related molecule such as CaMKII in a CB1R-dependent manner (Rubino et al., 2007). All Δ^9-THC metabolites except one are inactivated by oxidative metabolism (Maurer, Sauer, & Theobald, 2006), which prevents further CB1R activation. The higher affinity, potency, and efficacy of JWH-081 (Razdan, Vemuri, Makriyannis, & Huffman, 2009), coupled with its potential metabolism into other metabolites (Wohlfarth, Scheidweiler, Chen, Liu, & Huestis, 2013), suggests that both the acute and the chronic effects of JWH-081 might be greater than those of a similar dose of Δ^9-THC. Therefore, JWH-081 impairs CaMKIV and CREB phosphorylation via a signaling mechanism downstream of CB1R to induce its potent deleterious neurobehavioral effects in mice (Figure 4).

Another derivative of JWH, JWH-018, exhibits agonistic activity at CB1R (9 nM) and CB2R (2.94 nM) (Seely et al., 2012) and also produces the tetrad of behaviors classically associated with cannabinoids in rodent models (analgesia, catalepsy, hypomotility, and hypothermia) (Brents et al., 2011; Wiebelhaus et al., 2012), but is less potent compared with JWH-081. Chronic JWH-018 treatment also induces deficits in spatial memory in adolescent mice (Compton et al., 2012). JWH-018 inhibits forskolin-stimulated cAMP production (Chin, Murphy, Huffman, & Kendall, 1999) and has been shown to reduce ERK phosphorylation in cultured hippocampal neurons (Atwood et al., 2010). Although the effects of JWH-081 and JWH-018 are clearly due to CB1R activation, the potential role of CB2R in the effects of compounds present in Spice and K2 requires further study.

In summary, research on the psychoactive components of marijuana, as well as the role of the EC system in humans and its relationship to various brain disorders, has received much attention since the identification of cannabinoid receptors and their endogenous ligands. Moreover, in addition to the well-known symptoms of euphoria and elation, synthetic cannabinoid binding to CB1R can also cause anxiety, short-term memory loss, and attention deficits and can have many other cognitive, affective, and psychomotor effects. The extent to which brain development and functions are disturbed remains unknown in young synthetic cannabinoid users, who have been shown to develop an increased risk of

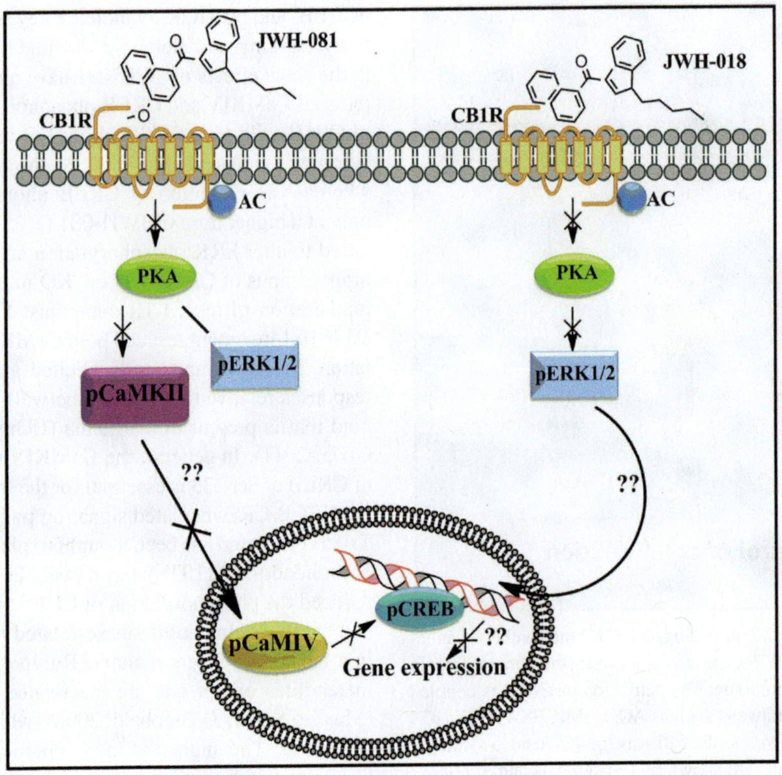

FIGURE 4 Schematic diagram depicting the CB1R signaling of the synthetic cannabinoids found in Spice products. JWH-081 and JWH-018 both bind to CB1R but elicit different CB1R signaling pathways. JWH-018, which has a stronger affinity for CB2R than for CB1R, decreases ERK1/2 phosphorylation. JWH-081, which has stronger affinity for CB1R than for CB2R, does not affect ERK1/2 phosphorylation but does impair the activation of CaMKIV and CREB, which are associated with Arc expression.

cannabinoid dependence. Although studies on the acute effects of synthetic cannabinoids are under way, studies on the chronic abuse of synthetic cannabinoid drugs are still insufficient; thus, this area requires further research. The immediate effects of acute synthetic cannabinoids are probably related to the activation of presynaptic CB1R-mediated signaling cascades and inhibition of the release of a number of neurotransmitters in the brain. In chronic synthetic cannabinoid users, it can be assumed that adaptive changes in the CB1R-mediated signaling cascades and related neurotransmitter system result in severe adverse effects. It is likely that the additional compounds identified in Spice/K2 preparations might also contribute, through CB1R or CB2R signaling, to the behavioral effects produced by smoking Spice/K2. Furthermore, their different pharmacologies might cause the different preparations of Spice/K2 to vary in their effects. Further investigation into these additional synthetic cannabinoids is required. Spice/K2 is marketed as a "natural" herbal blend but contains at least one very potent synthetic cannabinoid that acts through CB1R signaling, which probably accounts for the deleterious effects produced upon smoking Spice/K2.

APPLICATIONS TO OTHER ADDICTIONS AND SUBSTANCE MISUSE

1. Cannabinoid receptor signaling is also involved in other drugs of abuse such as ethanol, nicotine, and cocaine.
2. The EC system plays a major role in modulating motivation and reward processes.
3. CB1R antagonists are commonly applied in the treatment for alcohol and nicotine motivational processes.
4. The cannabinoid receptors modulate NMDA and GABA functions. Acute or chronic use of cannabinoids leads to cognitive impairments.
5. Chronic exposure to cannabinoids, as well as to alcohol, alters GABA and NMDA receptor functioning.
6. Cannabis or marijuana abuse leads to other drug use, which in turn leads to increased adult attention deficit/hyperactivity disorder symptoms.
7. Synthetic cannabinoids or marijuana may exacerbate existing substance use disorders and therefore should be avoided in cases of such diagnoses.

DEFINITION OF TERMS

Spice This product is a foil pack with herbal products sprayed with synthetic cannabinoids.
Marijuana This is the most commonly used illegal drug, and it contains the major psychoactive compound Δ^9-THC.
Endocannabinoid system The EC system is the cannabinoid system found in all vertebrate animals. The system comprises two receptors for Δ^9-THC, known as CB1R and CB2R, endogenous receptor ligands (ECs), and EC-metabolizing enzymes.
Schedule I US Controlled Substance Act This is a law that was passed by the US government to control the manufacturing, distribution, and use of certain psychotropic substances (e.g.,

synthetic cannabinoids) for nonscientific and nonmedical purposes.

Cannabinoid receptors These are the seven-transmembrane-domain proteins that are present on cell surfaces in all vertebrate animals and humans. Many compounds present in the *C. sativa* plant, such as Δ^9-THC, and the majority of synthetic cannabinoids bind to this protein to elicit their biological functions.

Neurotransmitter system Neurotransmitters are chemicals produced by neurons that bind to specialized receptors on target neurons to communicate among a large number of neurons. This system is necessary for the normal functioning of the nervous system.

Signal transduction Signal transduction occurs when an extracellular signaling molecule binds to a specific receptor located on the cell surface or inside of the cell. In turn, this receptor elicits a series of biochemical events inside of the cell, creating an amplified response. Depending on the cell, the signaling molecule, and the receptor, the response alters the function of the cell.

KEY FACTS OF SPICE PRODUCTS

- Numerous herbal products termed "Spice" and "K2" have been made and are readily sold on the Internet, in head shops, and in convenience stores. Although these products are labeled "not for human consumption," they are, nevertheless, typically synthetic cannabinoids sprayed onto plant material to mimic the effects of marijuana.
- The majority of Spice product users are males from 13 to 59 years of age, many of whom have a history of polydrug use, including cannabis, alcohol, and nicotine. These users perceive synthetic cannabinoids as a safer alternative to noncannabinoid illegal drugs and a favorable alternative to marijuana to elicit a marijuana-like "high" while avoiding detection by standard drug screening methods.
- Most of the synthetic cannabinoids that have been identified from Spice product samples are structurally distinct from the Δ^9-THC contained in marijuana.
- Many synthetic cannabinoids are Schedule I drugs under the US Controlled Substance Act. However, as new synthetic cannabinoids are scheduled, more structurally diverse cannabimimetic compounds are appearing, which may not be covered under the current regulations.
- High doses or chronic exposure to Spice products can lead to dangerous medical consequences, including psychosis, violent behaviors, tachycardia, hyperthermia, and death. These potent psychoactive effects are attributed to the synthetic cannabinoids. The lack of sensitive methods to detect these compounds in routine urine drug tests, the documented serious adverse effects, the limited animal studies and human pharmacology data, and, until recently, the legal status of these compounds in most jurisdictions all make synthetic cannabinoid consumption a significant public health problem and a serious safety concern.
- The majority of synthetic cannabinoids found in Spice products appear to have higher affinity and lower K_i values than Δ^9-THC for CB1R. In addition, some synthetic cannabinoid metabolites may be active at CB1R and prolong the psychoactive and physiological effects of the parent compound and contribute to the severity of intoxication.

SUMMARY POINTS

- This chapter focuses on the signaling mechanisms involved in the deleterious effects of synthetic cannabinoids found in Spice products.
- Spice/K2 products are simply plant materials sprayed with synthetic cannabinoids to mimic the effects of marijuana.
- Synthetic cannabinoids are potent CB1R or CB2R agonists. Thus, exposure to these compounds has the potential to cause neurobehavioral abnormalities.
- Because cannabinoids are coupled to G proteins, they activate a wide variety of downstream signaling events to regulate homeostatic functions; however, their dysfunction can contribute to pathological conditions.
- Data suggest that synthetic cannabinoids found in Spice products such as JWH-081 and JWH-018 possess different affinities for CB1R and CB2R and exhibit different CB1R signaling events in neurons.
- JWH-018 reduces ERK1/2 phosphorylation, whereas JWH-081 does not affect ERK1/2 phosphorylation but does impair CaMKIV and CREB phosphorylation. Both compounds inhibit synaptic plasticity and spatial learning and memory.
- Understanding the specific cannabinoid signaling events mediated by the various synthetic cannabinoids present in Spice products will be helpful to develop a therapeutic agent to prevent the deleterious effects of Spice abuse.

ACKNOWLEDGMENT

This work was supported in part by NIH/NIAAA Grant AA019443 (B.S.B.).

REFERENCES

Atwood, B. K., Huffman, J., Straiker, A., & Mackie, K. (2010). JWH018, a common constituent of "Spice" herbal blends, is a potent and efficacious cannabinoid CB receptor agonist. *British Journal of Pharmacology, 160*(3), 585–593.

Aung, M. M., Griffin, G., Huffman, J. W., Wu, M., Keel, C., Yang, B., ... Martin, B. R. (2000). Influence of the N-1 alkyl chain length of cannabimimetic indoles upon CB(1) and CB(2) receptor binding. *Drug and Alcohol Dependence, 60*(2), 133–140.

Basavarajappa, B. S. (2007a). The endocannabinoid signaling system: a potential target for next-generation therapeutics for alcoholism. *Mini-Reviews in Medicinal Chemistry, 7*, 769–779.

Basavarajappa, B. S. (2007b). Neuropharmacology of the endocannabinoid signaling system-Molecular mechanisms, biological actions and synaptic plasticity. *Current Neuropharmacology, 5*, 81–97.

Basavarajappa, B. S. (2014). Major enzymes of endocannabinoid metabolism. In A.-ur Rahman (Ed.), *Frontiers in protein and peptide sciences* (Vol. 1). Oak Park, IL, USA: Bentham Science Publishers (in press).

Basavarajappa, B. S., Nagre, N. N., Xie, S., & Subbanna, S. (2014). Elevation of endogenous anandamide impairs LTP, learning, and memory through CB1 receptor signaling in mice. *Hippocampus, 24*(7), 808–818.

Basavarajappa, B. S., Ninan, I., & Arancio, O. (2008). Acute ethanol suppresses glutamatergic neurotransmission through endocannabinoids in hippocampal neurons. *Journal of Neurochemistry, 107*(4), 1001–1013.

Basavarajappa, B. S., & Subbanna, S. (2014). CB1 receptor-mediated signaling underlies the hippocampal synaptic, learning and memory deficits following treatment with JWH-081, a new component of Spice/K2 preparations. *Hippocampus, 24*(2), 178–188.

Battistella, G., Fornari, E., Annoni, J. M., Chtioui, H., Dao, K., Fabritius, M., ... Giroud, C. (2014). Long-term effects of cannabis on brain structure. *Neuropsychopharmacology, 39*.

Berrendero, F., Sepe, N., Ramos, J. A., Di Marzo, V., & Fernandez-Ruiz, J. J. (1999). Analysis of cannabinoid receptor binding and mRNA expression and endogenous cannabinoid contents in the developing rat brain during late gestation and early postnatal period. *Synapse, 33*(3), 181–191.

Biegon, A., & Kerman, I. A. (2001). Autoradiographic study of pre- and postnatal distribution of cannabinoid receptors in human brain. *NeuroImage, 14*(6), 1463–1468.

Bito, H., Deisseroth, K., & Tsien, R. W. (1996). CREB phosphorylation and dephosphorylation: a Ca(2+)- and stimulus duration-dependent switch for hippocampal gene expression. *Cell, 87*(7), 1203–1214.

Bouaboula, M., Poinot-Chazel, C., Bourrie, B., Canat, X., Calandra, B., Rinaldi-Carmona, M., ... Casellas, P. (1995). Activation of mitogen-activated protein kinases by stimulation of the central cannabinoid receptor CB1. *The Biochemical Journal, 312*(Pt 2), 637–641.

Brents, L. K., Reichard, E. E., Zimmerman, S. M., Moran, J. H., Fantegrossi, W. E., & Prather, P. L. (2011). Phase I hydroxylated metabolites of the K2 synthetic cannabinoid JWH-018 retain in vitro and in vivo cannabinoid 1 receptor affinity and activity. *PLoS One, 6*(7), e21917.

Calandra, B., Portier, M., Kerneis, A., Delpech, M., Carillon, C., Le Fur, G., ... Shire, D. (1999). Dual intracellular signaling pathways mediated by the human cannabinoid CB1 receptor. *European Journal of Pharmacology, 374*, 445–455.

Castaneto, M. S., Gorelick, D. A., Desrosiers, N. A., Hartman, R. L., Pirard, S., & Huestis, M. A. (2014). Synthetic cannabinoids: epidemiology, pharmacodynamics, and clinical implications. *Drug and Alcohol Dependence, 144*.

Childers, S. R., & Deadwyler, S. A. (1996). Role of cyclic AMP in the actions of cannabinoid receptors. *Biochemical Pharmacology, 52*, 819–827.

Chin, C. N., Murphy, J. W., Huffman, J. W., & Kendall, D. A. (1999). The third transmembrane helix of the cannabinoid receptor plays a role in the selectivity of aminoalkylindoles for CB2, peripheral cannabinoid receptor. *The Journal of Pharmacology and Experimental Therapeutics, 291*(2), 837–844.

Compton, D. M., Seeds, M., Pottash, G., Gradwohl, B., Welton, C., & Davids, R. (2012). Adolescent exposure of JWH-018 "Spice" produces subtle effects on learning and memory performance in adulthood. *Journal of Behavioral and Brain Science, 2*, 146–155.

Cristino, L., & Di Marzo, V. (2014). Fetal cannabinoid receptors and the "dis-joint-ed" brain. *The EMBO Journal, 33*(7), 665–667.

Derkinderen, P., Toutant, M., Kadare, G., Ledent, C., Parmentier, M., & Girault, J. A. (2001). Dual role of Fyn in the regulation of FAK+6,7 by cannabinoids in hippocampus. *The Journal of Biological Chemistry, 276*, 38289–38296.

Devane, W. A., Dysarz, F. A. I., Johnson, M. R., Melvin, L. S., & Howlett, A. C. (1988). Determination and characterization of a cannabinoid receptor in rat brain. *Molecular Pharmacology, 34*, 605–613.

Fernandez-Ruiz, J., Berrendero, F., Hernandez, M. L., & Ramos, J. A. (2000). The endogenous cannabinoid system and brain development. *Trends in Neuroscience, 23*(1), 14–20.

Gaoni, Y., & Mechoulam, R. (1964). Isolation, structure and partial synthesis of an active constituent of hashish. *Journal of the American Chemical Society, 86*, 1646–1647.

Gebremedhin, D., Lange, A. R., Campbell, W. B., Hillard, C. J., & Harder, D. R. (1999). Cannabinoid CB1 receptor of cat cerebral arterial muscle functions to inhibit L-type Ca^{2+} channel current. *The American Journal of Physiology, 276*, H2085–H2093.

Glass, M., Dragunow, M., & Faull, R. L. (1997). Cannabinoid receptors in the human brain: a detailed anatomical and quantitative autoradiographic study in the fetal, neonatal and adult human brain. *Neuroscience, 77*, 299–318.

Glass, M., & Felder, C. C. (1997). Concurrent stimulation of cannabinoid CB1 and dopamine D2 receptors augments cAMP accumulation in striatal neurons: evidence for a Gs linkage to the CB1 receptor. *The Journal of Neuroscience, 17*(14), 5327–5333.

Goldschmidt, L., Richardson, G. A., Willford, J. A., Severtson, S. G., & Day, N. L. (2012). School achievement in 14-year-old youths prenatally exposed to marijuana. *Neurotoxicology and Teratology, 34*(1), 161–167.

Gomez del Pulgar, T., Velasco, G., & Guzman, M. (2000). The CB1 cannabinoid receptor is coupled to the activation of protein kinase B/Akt. *The Biochemical Journal, 347*, 369–373.

Guzman, M., Sanchez, C., & Galve-Roperh, I. (2002). Cannabinoids and cell fate. *Pharmacology & Therapeutics, 95*(2), 175–184.

Hampson, R. E., Mu, J., & Deadwyler, S. A. (2000). Cannabinoid and kappa opioid receptors reduce potassium K current via activation of G(s) proteins in cultured hippocampal neurons. *Journal of Neurophysiology, 84*, 2356–2364.

Herkenham, M., Lynn, A. B., de Costa, B. R., & Richfield, E. K. (1991). Neuronal localization of cannabinoid receptors in the basal ganglia of the rat. *Brain Research, 547*, 267–274.

Herkenham, M., Lynn, A. B., Johnson, M. R., Melvin, L. S., de Cost, B. R., & Rice, K. C. (1991). Characterization and localization of cannabinoid receptors in rat brain: a quantitative in vitro autoradiographic study. *The Journal of Neuroscience, 16*, 8057–8066.

Hermanns-Clausen, M., Kneisel, S., Szabo, B., & Auwarter, V. (2013). Acute toxicity due to the confirmed consumption of synthetic cannabinoids: clinical and laboratory findings. *Addiction, 108*(3), 534–544.

Herrera, B., Carracedo, A., Diez-Zaera, M., Gomez del Pulgar, T., Guzman, M., & Velasco, G. (2006). The CB2 cannabinoid receptor signals apoptosis via ceramide-dependent activation of the mitochondrial intrinsic pathway. *Experimental Cell Research, 312*(11), 2121–2131.

Ho, B. Y., Uezono, Y., Takada, S., Takase, I., & Izumi, F. (1999). Coupling of the expressed cannabinoid CB1 and CB2 receptors to phospholipase C and G protein-coupled inwardly rectifying K^+ channels. *Receptors & Channels, 6*, 363–374.

Howlett, A. C. (2002). The cannabinoid receptors. *Prostaglandins & Other Lipid Mediators, 68–69*, 619–631.

Howlett, A. C., Barth, F., Bonner, T. I., Cabral, G., Casellas, P., Devane, W. A., ... Pertwee, R. G. (2002). International Union of Pharmacology. XXVII. Classification of cannabinoid receptors. *Pharmacological Reviews, 54*, 161–202.

Huffman, J. W., Dong, D., Martin, B. R., & Compton, D. R. (1994). Design, synthesis and pharmacology of cannabimimetic indoles. *Bioorganic & Medicinal Chemistry Letters, 4*, 563–566.

Huffman, J. W., Zengin, G., Wu, M. J., Lu, J., Hynd, G., Bushell, K., ... Martin, B. R. (2005). Structure-activity relationships for 1-alkyl-3-(1-naphthoyl) indoles at the cannabinoid CB(1) and CB(2) receptors: steric and electronic effects of naphthoyl substituents. New highly selective CB(2) receptor agonists. *Bioorganic & Medicinal Chemistry, 13*(1), 89–112.

Insel, T. R. (1995). The development of brain and behavior. In F. E. Bloom, & D. J. Kupfer (Eds.), *Psychopharmacology: The four generation of progress* (pp. 683–694). New York: Raven Press.

Johnston, L. D., O'Malley, P. M., Bachman, J. G., & Schulenberg, J. E. (2012). *Marijuana use continues to rise among U.S. teens, while alcohol use hits historic lows.* University of Michigan News Service.

Katona, I., Rancz, E. A., Acsady, L., Ledent, C., Mackie, K., Hajos, N., ... Freund, T. F. (2001). Distribution of CB1 cannabinoid receptors in the amygdala and their role in the control of GABAergic transmission. *The Journal of Neuroscience, 21*(23), 9506–9518.

Kearn, C. S., Greenberg, M. J., DiCamelli, R., Kurzawa, K., & Hillard, C. J. (1999). Relationships between ligand affinities for the cerebellar cannabinoid receptor CB1 and the induction of GDP/GTP exchange. *Journal of Neurochemistry, 72*, 2379–2387.

Lambert, N. A. (2008). Dissociation of heterotrimeric G proteins in cells. *Science Signaling, 1*(25), re5.

Liu, J., Gao, B., Mirshahi, F., Sanyal, A. J., Khanolkar, A. D., Makriyannis, A., ... Kunos, G. (March 15, 2000). Functional CB1 cannabinoid receptors in human vascular endothelial cells. *The Biochemical Journal, 346*(Pt 3), 835–840.

Martin, S. J., Grimwood, P. D., & Morris, R. G. (2000). Synaptic plasticity and memory: an evaluation of the hypothesis. *Annual Review of Neuroscience, 23*, 649–711.

Maurer, H. H., Sauer, C., & Theobald, D. S. (2006). Toxicokinetics of drugs of abuse: current knowledge of the isoenzymes involved in the human metabolism of tetrahydrocannabinol, cocaine, heroin, morphine, and codeine. *Therapeutic Drug Monitoring, 28*(3), 447–453.

Mechoulam, R., & Parker, L. A. (2013). The endocannabinoid system and the brain. *Annual Review of Psychology, 64*, 21–47.

Mehmedic, Z., Chandra, S., Slade, D., Denham, H., Foster, S., Patel, A. S., ... ElSohly, M. A. (2010). Potency trends of Δ^9-THC and other cannabinoids in confiscated cannabis preparations from 1993 to 2008. *Journal of Forensic Sciences, 55*(5), 1209–1217.

Mukhopadhyay, S., McIntosh, H. H., Houston, D. B., & Howlett, A. C. (2000). The CB(1) cannabinoid receptor juxtamembrane C-terminal peptide confers activation to specific G proteins in brain. *Molecular Pharmacology, 57*(1), 162–170.

Netzeband, J. G., Conroy, S. M., Parsons, K. L., & Gruol, D. L. (1999). Cannabinoids enhance NMDA-elicited Ca^{2+} signals in cerebellar granule neurons in culture. *The Journal of Neuroscience, 19*, 8765–8777.

Onaivi, E. S., Ishiguro, H., Gong, J. P., Patel, S., Meozzi, P. A., Myers, L., ... Uhl, G. R. (2008). Brain neuronal CB2 cannabinoid receptors in drug abuse and depression: from mice to human subjects. *PLoS One, 3*(2), e1640.

Patel, N. A., Moldow, R. L., Patel, J. A., Wu, G., & Chang, S. L. (1998). Arachidonylethanolamide (AEA) activation of FOS proto-oncogene protein immunoreactivity in the rat brain. *Brain Research, 797*(2), 225–233.

Pertwee, R. G. (1988). The central neuropharmacology of psychotropic cannabinoids. *Pharmacology & Therapeutics, 36*(2–3), 189–261.

Pertwee, R. G., Howlett, A. C., Abood, M. E., Alexander, S. P., Di Marzo, V., Elphick, M. R., ... Ross, R. A. (2010). International Union of Basic and Clinical Pharmacology. LXXIX. Cannabinoid receptors and their ligands: beyond CB(1) and CB(2). *Pharmacological Reviews, 62*(4), 588–631.

Razdan, R. K., Vemuri, V. K., Makriyannis, A., & Huffman, J. W. (2009). *Cannabinoid receptor ligands and structure–activity relationships.* Totowa: Humana Press.

Rhee, M. H., Bayewitch, M., Avidor-Reiss, T., Levy, R., & Vogel, Z. (1998). Cannabinoid receptor activation differentially regulates the various adenylyl cyclase isozymes. *Journal of Neurochemistry, 71*, 1525–1534.

Romero, J., Garcia-Palomero, E., Berrendero, F., Garcia-Gil, L., Hernandez, M. L., Ramos, J. A., & Fernández-Ruiz, J. J. (1997). Atypical location of cannabinoid receptors in white matter areas during rat brain development. *Synapse, 26*(3), 317–323.

Rubino, T., & Parolaro, D. (2008). Long lasting consequences of cannabis exposure in adolescence. *Molecular and Cellular Endocrinology, 286*(1–2 Suppl. 1), S108–S113.

Rubino, T., Sala, M., Vigano, D., Braida, D., Castiglioni, C., Limonta, V., ... Parolaro, D. (2007). Cellular mechanisms underlying the anxiolytic effect of low doses of peripheral Δ^9-tetrahydrocannabinol in rats. *Neuropsychopharmacology, 32*(9), 2036–2045.

Rueda, D., Galve-Roperh, I., Haro, A., & Guzman, M. (2000). The CB(1) cannabinoid receptor is coupled to the activation of c-Jun N-terminal kinase. *Molecular Pharmacology, 58*(4), 814–820.

Sanchez, C., Galve-Roperh, I., Rueda, D., & Guzman, M. (1998). Involvement of sphingomyelin hydrolysis and the mitogen-activated protein kinase cascade in the Δ^9-tetrahydrocannabinol-induced stimulation of glucose metabolism in primary astrocytes. *Molecular Pharmacology, 54*, 834–843.

Schweitzer, P. (2000). Cannabinoids decrease the K(+) M-current in hippocampal CA1 neurons. *The Journal of Neuroscience, 20*, 51–58.

Seely, K. A., Brents, L. K., Radominska-Pandya, A., Endres, G. W., Keyes, G. S., Moran, J. H., & Prather, P. L. (2012). A major glucuronidated metabolite of JWH-018 is a neutral antagonist at CB1 receptors. *Chemical Research in Toxicology, 25*(4), 825–827.

Shabani, M., Mahnam, A., Sheibani, V., & Janahmadi, M. (2014). Alterations in the intrinsic burst activity of Purkinje neurons in offspring maternally exposed to the CB1 cannabinoid agonist WIN 55212-2. *The Journal of Membrane Biology, 247*(1), 63–72.

Silva, L., Zhao, N., Popp, S., & Dow-Edwards, D. (2012). Prenatal tetrahydrocannabinol (THC) alters cognitive function and amphetamine response from weaning to adulthood in the rat. *Neurotoxicology and Teratology, 34*(1), 63–71.

Subbanna, S., Nagaraja, N. N., Umapathy, N. S., Pace, B. S., & Basavarajappa, B. S. (2014). Ethanol exposure induces neonatal neurodegeneration by enhancing CB1R Exon1 histone H4K8 acetylation and up-regulating CB1R function causing neurobehavioral abnormalities in adult mice. *International Journal of Neuropsychopharmacology* (in press).

Subbanna, S., Shivakumar, M., Psychoyos, D., Xie, S., & Basavarajappa, B. S. (2013). Anandamide-CB1 receptor signaling contributes to postnatal ethanol-induced neonatal neurodegeneration, adult synaptic and memory deficits. *Journal of Neuroscience, 33*(15), 6350–6366.

Sugiura, T., Kishimoto, S., Oka, S., & Gokoh, M. (2006). Biochemistry, pharmacology and physiology of 2-arachidonoylglycerol, an endogenous cannabinoid receptor ligand. *Progress in Lipid Research, 45*(5), 405–446.

Tortoriello, G., Morris, C. V., Alpar, A., Fuzik, J., Shirran, S. L., Calvigioni, D., ... Harkany, T. (2014). Miswiring the brain: Δ^9-tetrahydrocannabinol disrupts cortical development by inducing an SCG10/stathmin-2 degradation pathway. *The EMBO Journal, 33*(7), 668–685.

Tree, K. C., Scotto di Perretolo, M., Peyronnet, J., & Cayetanot, F. (2014). In utero cannabinoid exposure alters breathing and the response to hypoxia in newborn mice. *European Journal of Neuroscience, 40*(1), 2196–2204.

Uchiyama, N., Kikura-Hanajiri, R., & Goda, Y. (2011). Identification of a novel cannabimimetic phenylacetylindole, cannabipiperidiethanone, as a designer drug in a herbal product and its affinity for cannabinoid CB(1) and CB(2) receptors. *Chemical & Pharmaceutical Bulletin (Tokyo), 59*(9), 1203–1205.

US Drug Enforcement Administration. (2014). Schedules of controlled substances: temporary placement of four synthetic cannabinoids into schedule I. *Notice of Intent to Federal Register, 79*, 7577–7582.

Valjent, E., Caboche, J., & Vanhoutte, P. (2001). Mitogen-activated protein kinase/extracellular signal-regulated kinase induced gene regulation in brain: a molecular substrate for learning and memory? *Molecular Neurobiology, 23*, 83–99.

Vardakou, I., Pistos, C., & Spiliopoulou, C. (2010). Spice drugs as a new trend: mode of action, identification and legislation. *Toxicology Letters, 197*(3), 157–162.

Wartmann, M., Campbell, D., Subramanian, A., Burstein, S. H., & Davis, R. J. (1995). The MAP kinase signal transduction pathway is activated by the endogenous cannabinoid anandamide. *FEBS Letters, 2–3*, 133–136.

Wiebelhaus, J. M., Poklis, J. L., Poklis, A., Vann, R. E., Lichtman, A. H., & Wise, L. E. (2012). Inhalation exposure to smoke from synthetic "marijuana" produces potent cannabimimetic effects in mice. *Drug and Alcohol Dependence, 126*(3), 316–323.

Wilson, R. I., & Nicoll, R. A. (2002). Endocannabinoid signaling in the brain. *Science, 296*, 678–682.

Wohlfarth, A., Scheidweiler, K. B., Chen, X., Liu, H. F., & Huestis, M. A. (2013). Qualitative confirmation of 9 synthetic cannabinoids and 20 metabolites in human urine using LC-MS/MS and library search. *Analytical Chemistry, 85*(7), 3730–3738.

Section C

Structural and Functional Aspects

Chapter 72

Cannabis Use Disorders and Brain Morphology

Valentina Lorenzetti[1,2], Janna Cousijn[3]

[1]Monash Clinical & Imaging Neuroscience, Monash University, Melbourne, VIC, Australia; [2]Melbourne Neuropsychiatry Centre, The University of Melbourne and Melbourne Health, Melbourne, VIC, Australia; [3]Department of Developmental Psychology, University of Amsterdam, The Netherlands

Abbreviations

ACC Anterior cingulate cortex
CUD Cannabis use disorder
DLPFC Dorsolateral prefrontal cortex
DSM *Diagnostic and Statistical Manual of Mental Disorders*
NAcc Nucleus accumbens
OFC Orbitofrontal cortex
PFC Prefrontal cortex
SUD Substance use disorder
THC Δ^9-Tetrahydrocannabinol
VTA Ventral tegmental area

INTRODUCTION

The past decades have been marked by an increase in the awareness of the addictive properties of cannabis, which parallels the rise in prevalence rates of cannabis use disorders (CUDs) (Degenhardt et al., 2013). CUDs currently have the highest burden of disease in drug treatment services in Oceania and Africa and the second highest in Europe and South America (UNODC, 2010). Approximately 30% of all cannabis users develop a CUD (Swift, Hall, & Teesson, 2001), but less than one-third of individuals with a CUD seek help (Stinson, Ruan, Pickering, & Grant, 2006). Moreover, relapse rates of CUDs range between 52% and 70% and are comparable to those of other substance use disorders (SUDs) (Budney, Vandrey, Hughes, Thostenson, & Bursac, 2008; Chauchard, Septfons, & Chabrol, 2013). The significant societal and personal harms associated with CUDs are alarming and warrant the development of new treatment and intervention strategies. These harms may be ascribed to the detrimental impact of addiction-related processes and of long-term cannabinoid exposure on the brain. Elucidating the brain correlates of CUDs is an important step that may help identify new treatment targets. However, in contrast to other SUDs like alcohol and cocaine, relatively little is known about the neurobiological mechanisms underlying CUDs. Most neuroimaging studies that investigate the effects of cannabis on the brain examine groups of heavy cannabis users rather than groups with a diagnosed CUD specifically. This chapter presents a narrative review of structural neuroimaging findings on gray matter abnormalities associated with CUDs. We summarize the existing studies on gray matter morphology in CUDs in an attempt to dissociate the potential adverse effects of cannabis use versus CUDs on gray matter morphology. Moreover, we discuss caveats of existing studies and offer suggestions for future work in this area. To provide a theoretical background for the discussion of the structural brain correlates of CUDs we start this chapter with a description of the nosology of CUDs and the neurocognitive processes associated with CUDs, including the related brain systems.

NOSOLOGY OF CANNABIS USE DISORDER

CUDs are similar to other SUDs in that they are characterized by compulsive substance use despite awareness of its harmful consequences (Leshner, 1997). In the *Diagnostic and Statistical Manual of Mental Disorders*, fifth edition (DSM-V), a CUD is diagnosed when the individual meets at least 2 of 11 criteria within a period of 12 months (APA, 1996). Diagnostic criteria of CUDs are similar to those of other SUDs and are clustered alongside four dimensions: loss of control, social problems, pharmacological consequences of cannabis use, and high-risk use (see Table 1).

Given the introduction of the DSM-V in May 2013, the structural neuroimaging studies discussed in this review investigated cannabis abuse and dependence as defined in the DSM-IV. In contrast to the DSM-IV, the DSM-V does not distinguish between cannabis abuse and dependence. Moreover, the DSM-V includes craving and cannabis withdrawal as diagnostic criteria and three stages of severity based on the number of diagnostic criteria that are met (mild, two or three criteria; moderate, four or five criteria; severe, six or more). These changes in diagnostic criteria raise questions on how the reviewed neurobiological findings on CUD as defined by the DSM-IV can be related to CUD as defined by the DSM-V. In this regard, Bailey, DuPont, and Teitelbaum (2014) noted that a moderate or severe CUD DSM-V diagnosis approximates cannabis dependence in the DSM-IV, while a mild CUD DSM-V diagnosis approximates cannabis abuse in the DSM-IV. Additional features of CUD include its gender-dependent clinical profile, psychiatric comorbidity, and low psychosocial outcomes. Males have a higher

TABLE 1 DSM-V Criteria for Cannabis Use Disorders

At least two of the following symptoms

Loss of control
- Cannabis is often taken in large amounts or over a longer period than was intended;
- Persistent desire to stop or cut down cannabis use or unsuccessful efforts in doing so;
- Craving or the strong urge or desire to use cannabis;
- Spending a great deal of time obtaining, using, or recovering from the use of cannabis.

Social problems
- Recurrent cannabis use resulting in a failure to fulfill obligations at home, school, or work;
- Recurrent cannabis use despite having persistent or recurrent social or interpersonal problems due to cannabis use;
- Important social, work, or recreational activities are given up or reduced owing to cannabis use.

Pharmacological consequences of cannabis use
- Withdrawal from the effects of cannabis, which is taken to relieve or avoid withdrawal symptoms;
- Tolerance to the effects of cannabis: more cannabis is needed to achieve the desired effect.

High-risk use
- Recurrent cannabis use in physically hazardous situations;
- Cannabis use is continued despite knowledge of physical and psychological problems.

chance of developing a CUD (UNODC, 2010), but craving appears to be more severe in females (King et al., 2011). Comorbid substance use and psychiatric disorders including depression, anxiety, and psychosis are common among individuals with a CUD (van der Pol et al., 2013). Finally, an earlier onset of cannabis use is associated with more severe outcomes with regard to psychiatric health, neurocognitive functioning, socioeconomic status, and academic achievements (Stinson et al., 2006; Swift et al., 2001).

NEUROCOGNITIVE ASPECTS OF CANNABIS USE DISORDER

Contemporary addiction models highlight the role of an imbalance between strong automatically triggered motivations to use and compromised cognitive control (disinhibition) in the transition from impulsive and recreational substance use toward compulsive substance use (e.g., Everitt & Robbins, 2005; Koob & Volkow, 2010; Robinson & Berridge, 2003; Wiers et al., 2007). Automatically triggered motivations—including craving, attention, and approach action tendencies—are thought to reflect sensitized and conditioned responses toward substance-related stimuli that develop during the development of SUDs. Exposure to cannabis-related versus neutral cues can induce craving (Gray, LaRowe, Watson, & Carpenter, 2011; Lundahl & Johanson, 2011), automatically capture attention (Asmaro, Carolan, & Liotti, 2014; Cousijn, Watson, et al., 2013), and activate approach tendencies (Cousijn, Goudriaan, & Wiers, 2011; Field, Eastwood, Bradley, & Mogg, 2006) in individuals with a CUD compared to non-cannabis-using controls (henceforth referred to as controls). Moreover, individuals with a CUD show impairments in executive functions like planning, organizing, problem solving, decision-making, memory, and emotional control (Solowij & Battisti, 2008). Cognitive impairments may already (mildly) emerge in recreational cannabis users and exacerbate as the CUD progresses (Martin-Santos et al., 2010). Some cognitive impairment may predate the onset of cannabis use and/or CUD, constituting a risk factor for CUD (Cousijn, Wiers, et al., 2013).

Cognitive deficits in CUD overlap with those observed in other SUDs (Fernández-Serrano, Pérez-García, & Verdejo-García, 2011). However, compared to other SUDs, neurocognitive and neurobiological research in CUD is still in its infancy. Yet, preliminary findings support similar neurobiological mechanisms underlying CUD and other SUDs. Brain areas that play a prominent role in the development of SUDs include subcortical regions such as the ventral tegmental area (VTA), nucleus accumbens (NAcc), amygdala, hippocampus, dorsal striatum, and regions of the prefrontal cortex (PFC) (see Figure 1; Everitt & Robbins, 2005; Koob & Volkow, 2010; Wilson, Sayette, & Fiez, 2004). SUD-associated brain areas (striatum, amygdala, hippocampus, and PFC (Burns et al., 2007)) are rich in endogenous cannabinoid receptor 1 (CB1), which mediates the psychoactive effects of cannabinoids including Δ^9-tetrahydrocannabinol (THC).

The VTA modulates the hedonic response to drug cues through the firing threshold of its dopamine neurons. The hedonic response to drug cues is thought to change as the addiction progresses, resulting in increased firing in response to substance-related cues (Koob & Volkow, 2010). VTA dopamine neurons project to the ventral striatum (e.g., NAcc), which mediates reward seeking by connecting motivational aspects of salient stimuli to motor actions (Everitt & Robbins, 2005). Neurobiological changes in the VTA and ventral striatum occur during recreational substance use, before the development of SUDs (Koob & Volkow, 2010). The dorsal striatum mediates the formation of habitual and compulsive substance use (Belin & Everitt, 2008; Everitt & Robbins, 2005). Notably, there is a shift from ventral to dorsal striatum involvement during the transition from controlled to compulsive substance use, but this evidence remains to be replicated in CUD (Everitt & Robbins, 2013).

The hippocampus and amygdala are also important in the development of cue-induced conditioned responses, such as craving and attention. The amygdala is involved in attributing emotional salience to cues and mediating approach and avoidance behavior (Schneider et al., 2001). The PFC is one of the main substrates involved in cognitive control and is therefore crucially involved in SUDs. Key components of the PFC that have been linked to cognitive control deficits in SUDs include the dorsolateral PFC (DLPFC), anterior cingulate cortex (ACC), and orbitofrontal cortex (OFC) (Koob & Volkow, 2010; Mansouri, Tanaka, & Buckley, 2009). More specifically, the ACC is involved in attention, conflict monitoring, and assessing salience of motivational information (Ridderinkhof, Ullsperger, Crone, & Nieuwenhuis, 2004). The OFC is key for reward evaluation and, together with the ACC, in the integration of motivational information into cognitive processes (Koob & Volkow, 2010; Mansouri et al., 2009).

STRUCTURAL NEUROIMAGING STUDIES INVESTIGATING CANNABIS USE DISORDER

Morphological alterations within the aforementioned brain regions have been observed in various SUDs including alcohol, opiate, and

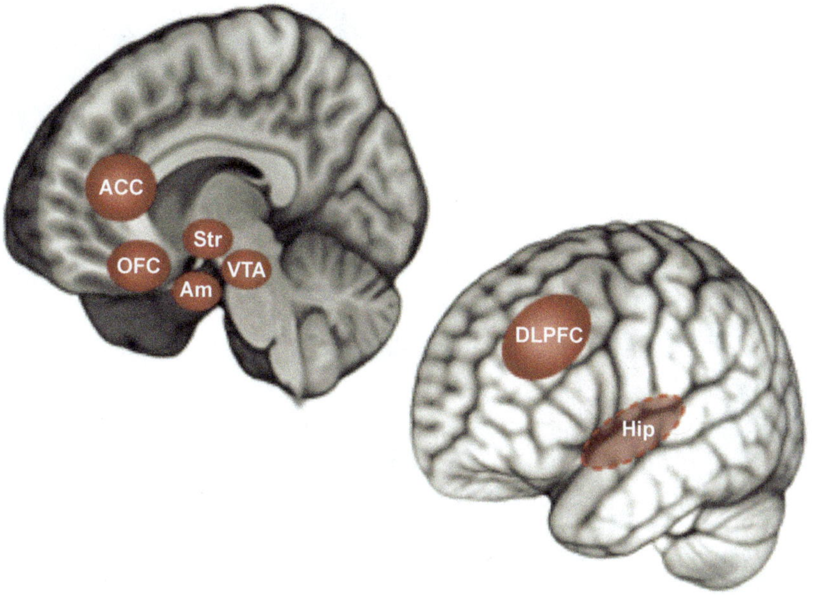

FIGURE 1 Brain regions implicated in addiction. ACC, anterior cingulate cortex; OFC, orbitofrontal cortex; DLPFC, dorsolateral prefrontal cortex; Str, striatum; Am, amygdala; VTA, ventral tegmental area; Hip, hippocampus.

cocaine use disorders (Everitt & Robbins, 2005; Koob & Volkow, 2010; Wilson et al., 2004). Given that CUD and other SUDs share neurocognitive deficits, the morphology of SUD-associated regions such as the VTA, striatum, hippocampus, amygdala, ACC, DLPFC, and OFC may also be affected in CUD. Morphological abnormalities of the ventral striatum (implicated in early stages of addiction) may already be evident in regular cannabis users before the onset of CUD, whereas morphological abnormalities of the dorsal striatum (implicated in compulsive, chronic substance use) may be evident only in individuals with a CUD. Similarly, given the role of disinhibition in SUDs, morphological abnormalities of the PFC may be prominent in individuals with a CUD (Cousijn et al., 2012).

Over the past decade, a growing number of neuroimaging studies examined gray matter morphology in regular cannabis users, but not specifically in individuals with a CUD (for reviews see Batalla et al., 2013; Lorenzetti, Solowij, Fornito, Lubman, & Yucel, 2014; Rocchetti et al., 2013). Notably, a history of regular cannabis use does not necessarily differentiate between cannabis users with and without a CUD (van der Pol et al., 2013). While the structural neuroimaging evidence in regular cannabis users is critical to characterize brain alterations associated with chronic cannabinoid exposure, neuroanatomical alterations specific to CUD remain largely unknown. This section summarizes structural neuroimaging findings on gray matter morphology in CUD specifically. First, we describe sociodemographic characteristics of the reviewed samples and the methods that were employed to measure CUDs. We then review the existing findings on the association between gray matter morphology and CUD and discuss (1) group differences between individuals with a CUD diagnosis and controls, (2) associations with CUD problem severity, (3) the role of abstinence and treatment, and (4) the role of gender. We identified a total of 16 studies in regular cannabis users that reported using instruments to assess diagnosis and/or severity of CUD (see Table 2). These studies were selected from those identified in a systematic literature review on brain morphology in regular cannabis users (Lorenzetti et al., 2014) and include additional studies that were published since January 2013.

Sample Characteristics and Diagnostic Instruments

Sample Size, Age, and Gender Distributions

Table 2 provides an overview of the characteristics of the reviewed studies. Most investigations used relatively small samples of regular cannabis users: only three studies included >30 participants (Cousijn et al., 2012; Mata et al., 2010; McQueeny et al., 2011). Moreover, most studies investigated young adults or adolescents. Only two investigations examined adults between 30 and 40 years of age (Tzilos et al., 2005; Yücel et al., 2008). CUDs are more prevalent in males (UNODC, 2010), which is reflected in the characteristics of the reviewed samples: most studies investigated samples that were composed of mostly males or exclusively of males.

Recruitment Sites of Regular Cannabis Users

The study of non-treatment-seeking regular cannabis users dominates the literature to date. Most investigations ($n=11$, see Table 2) recruited regular cannabis users from the general community, educational institutions, or Dutch coffee shops (outlets where legal cannabis can be bought). Only three studies recruited treatment-seeking, regular cannabis users from treatment services (Ashtari et al., 2011; Kumra et al., 2012; Yip et al., 2014).

Assessment of CUD Diagnosis and Symptoms

Assessment of CUD according to the DSM-IV (APA, 1996) was reported in all reviewed investigations. Specifically, the studies reported screening for any current or past psychiatric disorder and not excluding participants who endorsed cannabis abuse or

TABLE 2 Structural Neuroimaging Studies of Gray Matter Morphology in Regular Cannabis Users Assessed for Cannabis Use Disorder

Author (Year)	Sample Total n (Males)	Age, years (SD) CB	Age, years (SD) HC	Recruitment Site	In Treatment	Abuse Assessed	Abuse Reported	Dependence Assessed	Dependence Reported	Results Group Comparison	Association with Abuse/Dependence Symptoms
Battistella et al. (2014)	25 regular vs 22 occasional users. All males.	25 (2)	23 (2)	General community and university	No	Diagnosis, symptoms	No	No	N/A	Regular < occasional users for gray matter volume in temporal cortex/pole, parahippocampus, insula, OFC. Regular > occasional for cerebellar gray matter volume.	None performed (with abuse)
Gilman et al. (2014)	20 (9) CB vs 20 (9) HC	21 (2)	21 (2)	Not specified	No	Diagnosis	No	Diagnosis	None met criteria	CB > HC for left NAcc volume (trend) and density. Shape difference in left NAcc and right amygdala.	N/A
Yip et al. (2014)	20 CUD, of which 13 abstained for 21 days, vs 20 HC. All males.	27 (2)	29 (2)	Treatment service (outpatients)	Yes	No	N/A	Diagnosis, symptoms	All dependent	CUD > HC and abstinent > nonabstinent CUD for caudate volume. CUD = HC for putamen volume.	None performed (with abstinence days and addiction severity)
McQueeny et al. (2011)	35 (27) CB after 28 days abstinence vs 47 (36) HC	18 (1)	18 (1)	Local schools	No	Diagnosis, symptoms	Diagnosis endorsed in 65% of sample. Symptoms not reported.	Diagnosis	Endorsed in 65% of sample.	CB = HC and CB female > HC female, CB male > HC male for amygdala volume.	None performed (with withdrawal symptoms, substance-related life problems).
Kumra et al. (2012)	16 (8) CUD vs 51 (26)	17 (2)	16 (2)	Treatment service	Yes	Diagnosis	All CUDs	Diagnosis	All CUDs	CUD < HC for superior parietal cortex GM volume.	N/A
Cousijn et al. (2011)	33 (12) CUD vs 42 (16) HC	21 (1)	22 (2)	Coffee shops	No	Diagnosis, symptoms	Symptom severity exceeds abuse cutoff	Diagnosis, symptoms	Symptom severity suggests dependence	CB > HC for anterior cerebellum. CB = HC for volume of hippocampus, amygdala, OFC, ACC, striatum.	Neg. corr. between problem use and amygdala volume. No corr. between problem use and hippocampus, OFC, ACC, cerebellum.
Ashtari et al. (2011)	14 CUD after 7 months abstinence vs 14 HC. All males.	19 (1)	19 (1)	Therapeutic community (inpatients)	Yes	No	N/A	Diagnosis	Dependence early full remission	CUD < HC for hippocampus volume, CUD = HC for amygdala volume.	None performed (with abstinence duration).

Study	Sample	Age CB	Age HC	Recruitment	Col6	Col7	Col8	Col9	Findings	Correlations
Churchwell, Lopez-Larson, and Yurgelun-Todd (2010)	18 (16) CUD vs 18 (12) HC	17 (1)	17 (1)	General community	No	Diagnosis	All endorsed	Diagnosis	No	CUD<HC for right OFC. CUD=HC for left OFC.	N/A
Mata et al. (2010)	30 (23) CB vs 44 (25) HC	26 (5)	26 (6)	General community	No	Diagnosis	No	No	N/A	CB<HC for frontal sulcal concavity. CB>HC for frontal thickness.	N/A
Medina, Nagel, and Tapert (2010)	16 (12) CB 28 day abstinence vs 16 (12) HC	18 (1)	18 (1)	High school, colleges	No	Diagnosis, symptoms	No	Diagnosis, symptoms	No	CB>HC for cerebellar vermis and inferior posterior lobules XIII–X.	N/A
Medina et al. (2009)					No		$n = 2.2$ (3.1) DSM symptoms		$n = 2.2$ (3.1) DSM symptoms	CB = HC for PFC volume. CB female>HC female for PFC volume. CB male<HC controls for PFC volume.	No corr. between abuse/dependence symptoms and PFC (anterior, posterior, total gray and white matter, total ICV)
Yücel et al. (2008)	15 CUD vs 16 HC. All males	40 (9)	36 (10)	General community	No	Diagnosis	All endorsed	Diagnosis	N/A	CB<HC for hippocampus and amygdala volumes.	N/A
Medina, Schweinsburg, et al. (2007)	26 (19) CB vs 21 (14) HC	18 (1)	18 (1)	General community, high schools, university	No	Diagnosis, symptoms	$n = 2.2$ (3.1) DSM symptoms	Diagnosis, symptoms	$n = 2.2$ (3.1) DSM symptoms	CB = HC for hippocampal volume.	Pos. corr. between abuse/dependence symptoms and hippocampal volume and left>right asymmetry.
Medina, Nagel, et al. (2007)	16 (12) CB following 28-day abstinence vs 16 (11) HC	18 (1)	18 (1)		No		No		No	CB = HC for hippocampal volume.	N/A
Jager et al. (2007)	20 (13) CB abstinent for 7 days vs 20 (16) HC	25 (5)	24 (4)	General community, coffee shops, university	No	Diagnosis	No	No	N/A	CB = HC for parahippocampus gray and white matter density.	N/A
Tzilos et al. (2005)	22 (16) CUD vs 26 (19) HC	38 (6)	30 (9)	Not specified	No	No	N/A	Diagnosis	All endorsed	CUD=HC for hippocampus volume.	N/A

CB, regular cannabis users; CUD, participants diagnosed with a CUD (i.e., abuse or dependence); HC, non-cannabis-using controls; Neg, negative; corr., correlation; Pos., positive; ICV, intracranial volume. N/A, not applicable; OFC, orbitofrontal cortex; NAcc, nucleus accumbens.

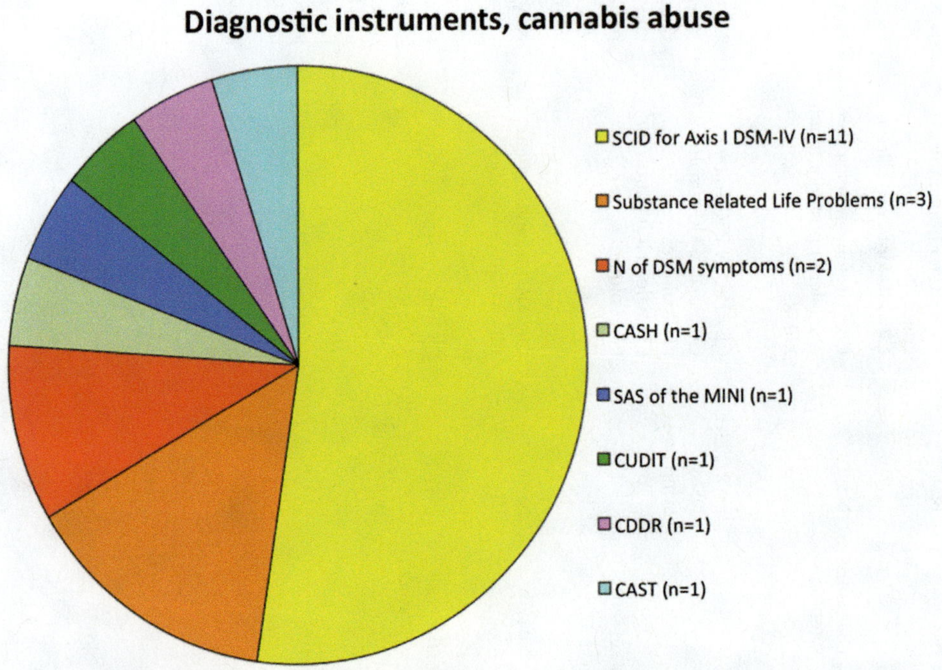

Diagnostic instruments, cannabis abuse

- SCID for Axis I DSM-IV (n=11)
- Substance Related Life Problems (n=3)
- N of DSM symptoms (n=2)
- CASH (n=1)
- SAS of the MINI (n=1)
- CUDIT (n=1)
- CDDR (n=1)
- CAST (n=1)

FIGURE 2 **Summary of instruments employed to diagnose cannabis abuse.** SCID, Structured Clinical Interview for Axis I DSM-IV Disorders (Ashtari et al., 2011; Churchwell et al., 2010; Gilman et al., 2014; Kumra et al., 2012; McQueeny et al., 2011; Medina et al., 2009, 2010; Medina, Nagel, et al., 2007; Medina, Schweinsburg, et al., 2007; Yip et al., 2014; Yücel et al., 2008); Substance-related life problems (McQueeny et al., 2011; Medina et al., 2009; Medina, Nagel, et al., 2007); number of DSM-endorsed symptoms (Medina et al., 2009; Medina, Schweinsburg, et al., 2007); CASH, Comprehensive Assessment of Symptoms History screening (Jager et al., 2007); SAS of the MINI, Substance Abuse Scales of the Mini International Neuropsychiatric Interview of the DSM (Jager et al., 2007); CUDIT, diagnostic threshold of the Cannabis Use Disorder Identification Test (Cousijn et al., 2012); CDDR, Customary Drinking and Drug Use Record (McQueeny et al., 2011); CAST, Cannabis Abuse Screening Test (both diagnosis and symptoms; Battistella et al., 2014).

dependence (see Table 2). The three investigations that examined treatment-seeking cannabis users, however, included mostly participants with comorbid psychopathologies and medication (Ashtari et al., 2011; Kumra et al., 2012; Yip et al., 2014). Only five studies were composed exclusively of individuals with a CUD diagnosis (Ashtari et al., 2011; Churchwell et al., 2010; Kumra et al., 2012; Tzilos et al., 2005; Yip et al., 2014), and four samples were composed of cannabis users with and without CUDs (Cousijn et al., 2012; McQueeny et al., 2011; Medina et al., 2009; Medina, Schweinsburg, Cohen-Zion, Nagel, & Tapert, 2007).

Thirteen investigations assessed cannabis abuse with a variety of instruments (see Table 2 and Figure 2), but only four reported the outcome of the diagnostic assessment. Seven studies assessed severity of cannabis abuse (Battistella et al., 2014; Cousijn et al., 2012; McQueeny et al., 2011; Medina et al., 2009; Medina, Nagel, Park, McQueeny, & Tapert, 2007; Medina et al., 2010; Medina, Schweinsburg, et al., 2007). Eleven studies examined cannabis dependence (see Figure 3) and seven studies the severity of dependence (see Figure 4; Cousijn et al., 2012; McQueeny et al., 2011; Medina et al., 2009, 2010; Medina, Nagel, et al., 2007; Medina, Schweinsburg, et al., 2007; Yip et al., 2014).

Cannabis Abstinence and Withdrawal

A total of six CUD samples were studied following prolonged abstinence from cannabinoids (>21 days, see Table 2). Most of the other samples abstained from cannabis for approximately 12 to 3 h. Several studies measured cannabis withdrawal as a criterion of CUD (n=5; McQueeny et al., 2011; Medina et al., 2009, 2010; Medina, Nagel, et al., 2007; Medina, Schweinsburg, et al., 2007), but the severity of withdrawal was not reported.

Cannabis Use Disorders Compared to Controls

This section summarizes the findings of studies that compared gray matter morphology between CUD and control participants. We reviewed findings separately for the striatum, amygdala, hippocampus, PFC, and cerebellum, all regions in which an association with CUD was reported.

Striatum

Striatal gray matter morphology was examined in three studies that showed mixed findings. Yip et al. (2014) found a smaller volume of some (i.e., caudate), but not all, parts of the striatum in cannabis-dependent participants compared to controls. In contrast, Cousijn et al. (2012) found no differences in striatal volume in regular cannabis users (of which most endorsed a CUD diagnosis) compared to controls. Interestingly, Gilman et al. (2014) found larger gray matter density in the NAcc (but no significant volume differences) in nondependent regular cannabis users. It is

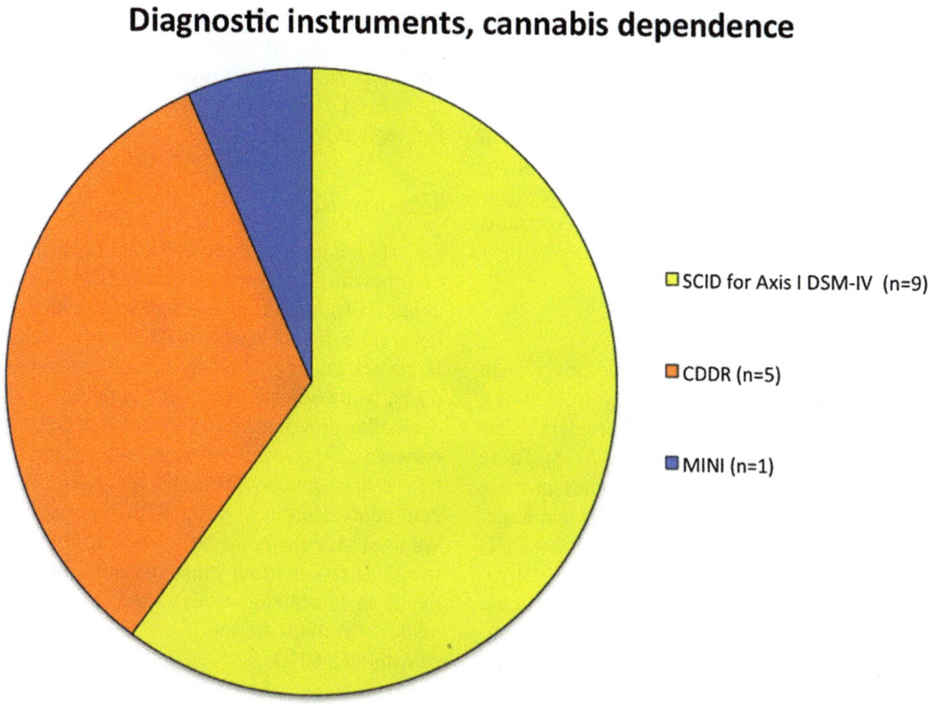

Diagnostic instruments, cannabis dependence

- SCID for Axis I DSM-IV (n=9)
- CDDR (n=5)
- MINI (n=1)

FIGURE 3 Summary of instruments employed to diagnose cannabis dependence. SCID, Structured Clinical Interview for Axis I DSM-IV Disorders (Ashtari et al., 2011; Churchwell et al., 2010; Gilman et al., 2014; McQueeny et al., 2011; Medina et al., 2009, 2010; Medina, Nagel, et al., 2007; Medina, Schweinsburg, et al., 2007; Yip et al., 2014); CDDR, Customary Drinking and Drug Use Record (McQueeny et al., 2011; Medina et al., 2009, 2010; Medina, Nagel, et al., 2007; Medina, Schweinsburg, et al., 2007); MINI, Mini International Neuropsychiatric Interview of the DSM (Cousijn et al., 2012).

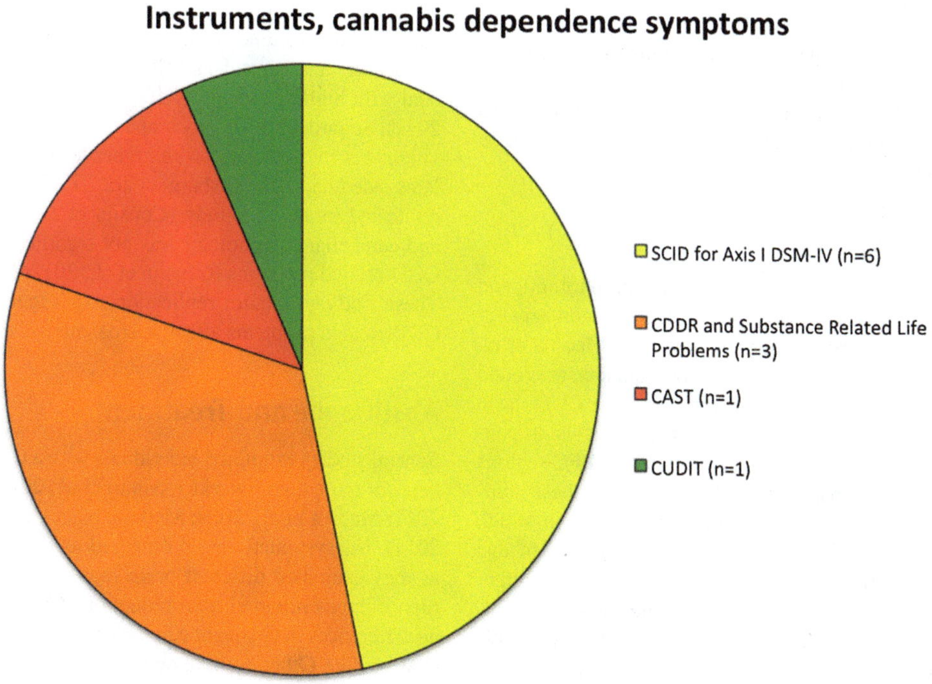

Instruments, cannabis dependence symptoms

- SCID for Axis I DSM-IV (n=6)
- CDDR and Substance Related Life Problems (n=3)
- CAST (n=1)
- CUDIT (n=1)

FIGURE 4 Summary of instruments employed to measure symptoms of cannabis dependence. SCID, Structured Clinical Interview for Axis I DSM IV Disorders (Churchwell et al., 2010; Gilman et al., 2014; McQueeny et al., 2011; Medina et al., 2009, 2010; Medina, Nagel, et al., 2007); Substance-related life problems and CDDR, Customary Drinking and Drug Use Record (McQueeny et al., 2011; Medina et al., 2009, 2010; Medina, Nagel, et al., 2007); CAST, Comprehensive Assessment of Symptoms History screening (Yip et al., 2014); CUDIT, Cannabis Use Disorder Identification Test (Cousijn et al., 2012).

notable that the NAcc was affected in nondependent cannabis users (Gilman et al., 2014), whereas the dorsal striatum (caudate) was affected in dependent users (Yip et al., 2014). These findings are in line with the model of Koob and Volkow (2010), which postulates a transition from ventral striatum (e.g., NAcc) to dorsal striatum (e.g., caudate) as drug use becomes compulsive (Koob & Volkow, 2010). However, the paucity of neuroimaging studies investigating striatal morphology in dependent and nondependent cannabis users prevents drawing conclusions about this issue.

Amygdala

Four studies examined gray matter morphology of the amygdala in samples endorsing cannabis dependence (McQueeny et al., 2011), cannabis abuse (Yücel et al., 2008), mixed levels of problem severity (Cousijn et al., 2012), and no CUD (Gilman et al., 2014). These studies showed mixed findings, with smaller volumes in male cannabis abusers vs male controls (Yücel et al., 2008) and larger volumes in CUD females compared to control females and CUD males (McQueeny et al., 2011). Gilman et al.'s (2014) investigation of nondependent regular cannabis users found no volumetric reduction, but a significant shape alteration of the amygdala, suggesting subtle detrimental effects of regular cannabis use predating CUD onset.

Hippocampus

Five studies examined hippocampal gray matter morphology (Ashtari et al., 2011; Cousijn et al., 2012; Medina, Schweinsburg, et al., 2007; Tzilos et al., 2005; Yücel et al., 2008). While hippocampal volume reduction is the most consistently reported finding in regular cannabis users (Rocchetti et al., 2013), only two studies found reduced volume in chronic cannabis abusers and in dependent individuals in early remission (Ashtari et al., 2011; Yücel et al., 2008). Hippocampal alterations are probably related to neurotoxic effects of chronic cannabinoids exposure rather than specific to CUD (Ashtari et al., 2011; Cousijn et al., 2012).

Prefrontal Cortex

Three studies investigated PFC gray matter morphology and showed mixed results. Churchwell et al. (2010) found smaller PFC volumes in CUD versus control participants. Medina et al. (2009) replicated this finding in male CUD participants, but found an effect in the opposite direction in females with CUD, who showed larger PFC volumes compared to female controls. In contrast, Cousijn et al. (2012) observed no alteration in PFC regions (ACC and OFC) in participants with CUD. While all these studies investigated samples of young adults, the samples in which group effects were found were between 16 and 18 years of age (Churchwell et al., 2010; Medina et al., 2009), slightly younger than the 21- to 22-year-old CUD participants who showed no PFC alterations (Cousijn et al., 2012). Other investigations of cannabis users between 23 and 25 years of age (Battistella et al., 2014) and around 26 years (Mata et al., 2010) also found PFC alterations, but failed to report whether cannabis users endorsed a CUD diagnosis. While it is unclear if PFC alterations relate to either chronic cannabis exposure or CUD, these findings preliminarily suggest that adolescent- versus adult-onset CUD may more strongly affect PFC morphology. The PFC plays a prominent role in cognitive control together with superior parietal brain regions. Interestingly, Kumra et al. (2012) showed smaller superior parietal cortex volumes in CUD, suggesting that impairments occur in cognitive control regions.

Cerebellum

Two studies investigated cerebellar gray matter and observed larger volumes in CUD versus controls (Cousijn et al., 2012; Medina et al., 2010), which is consistent with evidence in regular cannabis users (that did not report on CUD). Battistella et al. (2014) found larger cerebellar gray matter in regular versus occasional cannabis users, while Solowij et al., (2011) failed to observe differences in cerebellar gray matter (but found reduced white matter volume) between chronic cannabis users and controls. The cerebellum contains a high concentration of CB1 (Burns et al., 2007) and is not commonly associated with SUDs. Larger cerebellar gray matter volumes may therefore be specific to users of cannabis vs other substances. Reporting on diagnostic outcomes of CUD assessments in future studies will determine whether cerebellar gray matter alterations are specific to regular cannabis users either with or without a CUD.

Associations with Cannabis Use Disorder Severity

Three studies explored the linear association between symptoms of cannabis abuse/dependence and brain morphology (Cousijn et al., 2012; Medina, Nagel, et al., 2007; Medina, Schweinsburg, et al., 2007). Only one study investigated this in the amygdala, and although volume of this region did not differ between cannabis users and controls, smaller amygdala volumes were associated with higher levels of CUD-related problems (Cousijn et al., 2012). Regarding the hippocampus, one study observed that larger volumes were associated with higher levels of CUD-related problems (Medina, Schweinsburg, et al., 2007), but this effect was not replicated in another study (Cousijn et al., 2012). Striatal, PFC, and cerebellar morphology was not significantly associated with CUD-related problems (Cousijn et al., 2012; Medina et al., 2009). These findings require replication, but suggest that increases in CUD-related problems affect neuroanatomy.

Abstinence and Treatment

Several studies examined individuals with a CUD who were abstinent for more than 3 weeks (Ashtari et al., 2011; McQueeny et al., 2011; Medina et al., 2009; Medina, Nagel, et al., 2007; Yip et al., 2014). No systematic trend emerged across these investigations, as they examined different brain regions. Studies examining the hippocampus reported both reductions (Ashtari et al., 2011) and no differences in abstinent cannabis users versus controls (Medina et al., 2009). Other brain regions were investigated in single studies only, which demonstrated alterations in the caudate (Yip et al., 2014) and cerebellum (Medina et al., 2010), but not in the putamen (Yip et al., 2014), amygdala (McQueeny et al., 2011), and PFC (Medina et al., 2009). Notably, magnetic resonance imaging assessments were conducted prior to abstinence in one study

(Yip et al., 2014) but following prolonged abstinence in others (Ashtari et al., 2011; McQueeny et al., 2011; Medina et al., 2009). Thus, group differences may constitute preexisting vulnerabilities (Yip et al., 2014), subacute effects of cannabis use (Cousijn et al., 2012), or recovery after abstinence (Ashtari et al., 2011). For example, Ashtari et al. (2011) found reduced hippocampal but not amygdala volumes in 7-months-abstinent cannabis users. Thus, amygdala volume reductions, which were reported in current users (Yücel et al., 2008) and correlated with cannabis use-related problems (Cousijn et al., 2012), may recover following abstinence. In contrast, hippocampal volume reductions, the most consistently reported finding in regular cannabis users (Rocchetti et al., 2013), may persist after abstinence. Longitudinal studies are required to elucidate the neurobiological trajectories of change (and potentially recovery) after prolonged abstinence.

Patterns of neuroanatomical alterations associated with seeking treatment could not be noted, as different brain regions were examined by studies investigating treatment-seeking (Ashtari et al., 2011; Kumra et al., 2012; Yip et al., 2014) versus non-treatment-seeking cannabis users.

Gender Effects

A few studies examined gender effects and reported mixed findings. This is to no surprise given the predominance of small sample sizes and male dominance in the reviewed studies. However, two investigations reported group × gender effects, with larger brain volumes in females compared to males with a CUD (McQueeny et al., 2011; Medina et al., 2009). Medina et al. (2009) found larger PFC volumes in female cannabis users compared to female controls, while male cannabis users showed smaller volumes than male controls. Similarly, McQueeny et al. (2011) found larger amygdala volumes in female cannabis users compared to male cannabis users and female controls. Yet, two other studies did not replicate these effects (Cousijn et al., 2012; Gilman et al., 2014). No gender effect was reported on striatal (Cousijn et al., 2012; Gilman et al., 2014), PFC (Cousijn et al., 2012; Mata et al., 2010), medial temporal (Cousijn et al., 2012; Gilman et al., 2014), and superior parietal regions (Kumra et al., 2012). The scarcity of studies investigating gender effects contrast with strong preclinical evidence on the role of sex hormones in problematic cannabinoid consumption (Winsauer et al., 2011), warranting future investigations in larger CUD samples with a balanced male to female ratio.

DISCUSSION

This narrative review shows a lack of research and understanding of the neurobiological substrates underlying CUDs. The conducted studies often investigated relatively small samples of regular cannabis users with varying levels of CUD-related problems. Perhaps not surprisingly, these studies generally showed contradictory results that lack replication, hindering a proper and reliable integration of the findings to date. Yet, the reported significant associations between gray matter morphology and CUD are in line with contemporary addiction models and indicate structural alterations in different brain regions at different stages of CUD. Moreover, age of onset, gender, cumulative cannabis consumption, abstinence duration, CUD-related problem severity, and gender may be important moderators of the relation between CUD and gray matter morphology. In this section, we summarize the associations between gray matter morphology and CUDs and then discuss the implications of these finding for furthering the current understanding of the neurobiology of CUDs. Moreover, we will discuss important issues and limitations that became evident from the reviewed literature and suggest how these could be addressed by future studies.

Summary of Regional Alterations Associated with Cannabis Use Disorder

There are emerging trends for gray matter abnormalities in CUD samples within the striatum, amygdala, hippocampus, PFC, and cerebellum. Consistent with theoretical models of a ventral-to-dorsal striatal shift as the SUD progresses (Koob & Volkow, 2010), cannabis users without a CUD showed an altered ventral striatum (Gilman et al., 2014), while cannabis users with a CUD exhibited alterations within the dorsal striatum (Yip et al., 2014). Gender may moderate the effect of CUD on the brain, as the amygdala and OFC were differentially affected in male and female individuals with a CUD (McQueeny et al., 2011; Medina et al., 2009; Yücel et al., 2008). Hippocampal alterations in CUD participants are less consistently reported than in regular cannabis users (Rocchetti et al., 2013), suggesting that alterations within this region are related to neurotoxic effects of chronic cannabinoid exposure and may not exacerbate with increasing CUD-related problems. Similarly, larger cerebellar volume may be shared across regular cannabis users with and without a CUD (Cousijn et al., 2012; Medina et al., 2010). In contrast, more severe CUD symptoms may aggravate structural alterations in the amygdala (Cousijn et al., 2012). Within the PFC, alterations were apparent in late adolescents but not in young adults, suggesting that CUD detrimentally affects PFC morphology in earlier stages of adolescent neurodevelopment when significant PFC remodeling occurs (Gogtay et al., 2004). These region-dependent trends suggest that different brain regions are relevant for different aspects and stages of CUD. This notion remains speculative given the paucity of studies. Notably, the lack of neuroimaging studies in CUD is in sharp contrast with the growing body of studies in regular cannabis users (Lorenzetti et al., 2014). Individuals with similar levels of cannabis exposure (i.e., dosage, age of onset, frequency, and duration) can vary greatly in the use-related problems they experience (van der Pol et al., 2013). Studies in regular cannabis users lack CUD assessments and thereby cannot identify potential CUD-specific effects, emphasizing the lack of knowledge about the neurobiological mechanisms underlying CUDs.

Cannabis Use Disorder and Comorbid Psychopathologies

To study the neurobiology underlying CUD, one should preferably compare regular cannabis users with a CUD in treatment with regular cannabis users with nonsignificant psychological problems. These samples may even be matched on the level of cannabis use (van der Pol et al., 2013). Unfortunately, only three studies recruited participants from treatment services. While most investigations examined daily or almost daily cannabis users without comorbid diagnosable or subthreshold psychiatric problems, the three studies of treatment-seeking cannabis users examined samples that were composed mostly of participants with comorbid psychopathologies and medication (Ashtari et al., 2011; Kumra et al., 2012; Yip et al., 2014). This is consistent with epidemiological evidence of high comorbidity between CUD and other psychiatric disorders, including anxiety, depression, attention-deficit hyperactivity disorder, and psychotic disorders (Dorard, Berthoz, Phan, Corcos, & Bungener, 2008; Moore et al., 2007; Skinner, Conlon, Gibbons, & McDonald, 2011; van der Pol et al., 2013). To control for comorbid psychiatric symptoms, exclusion of participants with a diagnosed disorder other than CUD is common in the reviewed studies that investigated nonclinical samples. Findings in CUD without comorbid psychopathologies may not extend to both treatment- and non-treatment-seeking cannabis users, as a substantial part of them suffer from comorbid psychopathologies. Even though psychiatric comorbidity complicates the study of CUD, more investigations of ecologically valid clinical and subclinical groups are needed. To differentiate between abnormalities specific to CUD or general to multiple psychopathologies, studies could include clinical control groups matched on psychopathological symptoms other than CUD.

Limitations of Existing Studies and Recommendations for Future Studies

This review of structural neuroimaging studies highlights a number of limitations, including the lack of information about whether CUD was endorsed in the examined samples, the heterogeneous instruments used to assess CUD, the lack of studies in treatment-seeking cannabis users, the lack of longitudinal studies, and the small sample sizes with limited age ranges (mainly young adults and adolescents). First, most studies that screened for CUD in their samples did not report whether cannabis users endorsed a CUD diagnosis and did not examine the symptom severity. Thus, the literature on the neurobiology of regular cannabis users may be partly entrenched with CUD-related processes that do not occur in regular users who do not experience problems with their use. Moreover, the lack of studies in treatment-seeking cannabis users, and the concurrent abundance of studies in regular cannabis users with varying levels of CUD-related problems, may lead to an underestimation of the effects of CUD on brain morphology. As highlighted in the previous sections, the progress of addiction determines profound changes in brain functioning, independent of drug-specific neurotoxic effects, particularly within the striatum and PFC, with alterations shifting from the ventral to the dorsal portion of the striatum and from the medial to the lateral portion of the PFC (Koob & Volkow, 2010). Studying the commonalities and differences between regular cannabis users with and without a CUD, and between regular cannabis users and individuals with a similar psychopathological profile, will help to understand the neurobiological mechanisms underlying CUDs. This may especially prove fruitful in the development of new intervention and treatment strategies that target the functioning of specific brain systems.

A variety of instruments were employed to assess diagnosis and severity of CUD, which limits the direct comparability of findings across studies. Standardized and validated measures of the severity of cannabis dependence are currently lacking, and this issue highlights the need to develop objective measures of CUD (van der Pol et al., 2013). Moreover, differences between the new DSM-V criteria for CUD and older versions of the DSM should be noted, as the former no longer distinguishes between cannabis abuse and dependence, introduces craving and withdrawal as diagnostic criteria, and describes three stages of CUD severity. In light of these issues, future studies could benefit from investigating the association between brain morphology and CUD symptom severity as reported in the DSM-V (American Psychiatric Association, 2013). The DSM-V criteria of cannabis dependence can thereby provide a more standardized instrument to measure severity of dependence.

Finally, an important step in understanding the neurobiology of CUD is to dissociate causal and consequential effects and to determine potential neurobiological trajectories of recovery. Only a few investigations examined CUD participants after prolonged abstinence and no longitudinal studies are currently missing. The reviewed preliminary evidence suggests that the amygdala may recover to normal levels, while the hippocampus shows persistent reductions despite cannabis use cessation. Future studies may elucidate the neurobiological trajectories of change (and potentially recovery) after prolonged cannabinoid abstinence by performing magnetic resonance examinations of cannabis users prior to and following abstinence treatment.

APPLICATIONS TO OTHER ADDICTIONS AND SUBSTANCE MISUSE

The neurobiology of CUD may overlap with that of other SUDs. Yet neuroimaging and neurocognitive behavioral studies in long-time regular cannabis users show relatively mild to even absent neurocognitive deficits. While it is tempting to conclude that neurobiological and neurocognitive effects of CUD are less severe than those of other SUDs, such conclusion cannot be drawn (yet) given the paucity of neuroimaging studies in cannabis users with CUD. Moreover, it is essential to note that the potential mild neurocognitive effects and low addictive potential of cannabis compared to other substances like cocaine and heroin do not imply that the problems an individual can experience from a CUD are less severe than those experienced from another SUD.

CONCLUDING KEY FACTS ON THE ASSOCIATION BETWEEN GRAY MATTER AND CANNABIS USE DISORDER

- Our review highlights an urgent need for a better understanding of the neurobiological correlates of CUD.
- CUD may exacerbate adverse neurobiological outcomes of regular cannabis use, affecting additional brain regions.
- The few significant associations between gray matter morphology and CUD are consistent with contemporary addiction models.
- Age of onset, gender, cumulative cannabis consumption, abstinence, and CUD-associated problems may be important moderators in the association between CUD and brain morphology.

- Studying the commonalities and differences between ecologically valid samples of regular cannabis users with and without a CUD will be an important next step to understanding the neurobiological mechanisms underlying CUD and developing new treatment strategies.

DEFINITION OF TERMS

Cannabis use disorder This refers to the problematic use of cannabis and includes harms such as loss of control over use, social problems in relation to use, pharmacological consequences (tolerance and withdrawal), and high-risk use.

Substance use disorder This refers to problematic use of substances, including cannabis, that have progressive detrimental effects that pervade all aspects of the individual's personal, social, and work life.

Cannabinoids These are the chemical compounds of cannabis, including over 480 natural compounds, which have a variety of properties. Only a minority of cannabinoids are psychoactive.

Δ^9-Tetrahydrocannabinol This is the main psychoactive compound of cannabis and has been linked to the neurotoxic effects of cannabis on the central nervous system.

CB1 This refers to the cannabinoid type 1 receptor to which THC binds. CB1 receptors are located in the central nervous system, in presynaptic terminals. They are part of the endogenous cannabinoid system and have been described to mediate the effects of cannabis on brain and behavior.

Striatum This is an area of the brain implicated in processing reward. It is subdivided into a dorsal and a ventral portion, which mediate impulsive and compulsive behavior, respectively.

Prefrontal cortex This is a brain area involved in mediating cognitive control functions, including planning, decision-making, and conflict monitoring.

Hippocampus This is a region of the brain that mediates learning and memory.

Amygdala This is a brain area that mediates the regulation of emotion.

Diagnostic and Statistical Manual of Mental Disorders This manual is the standard used to classify psychopathologies by mental health professionals. It outlines diagnostic classifications, diagnostic criteria to endorse disorders, and their descriptions.

KEY FACTS OF THE STRIATUM

- The striatum is a deep-brain nucleus that links motivation to motor movements involved in the execution of simple motor tasks as well as more complex cognitive tasks, such as reward processing, decision-making, and social interactions.
- In medicine, the term "striatum" was initially used to refer to a variety of regions of the brain. The currently accepted definition of the term striatum has been used since 1941.
- The term "corpus striatum" originates from Latin and it means "striped mass" of gray and white matter.
- Gray matter regions of the striatum include the caudate nucleus and the putamen, which are separated by the white matter internal capsule and the NAcc.
- The dorsal striatum comprises the caudate and putamen, while the ventral striatum includes the NAcc and the ventromedial portions of the caudate and putamen.
- The main input regions of the striatum are the somatosensory and motor cortices, which project to the putamen; the PFC, which innervates the caudate; and the VTA, which projects to the NAcc.
- Drug and behavioral addictions affect the plasticity of the striatum. In recreational drug use and at the initial stages of addiction, impulsive reward seeking and the experience of pleasure are mediated by the ventral striatum. As addiction develops, habitual and compulsive use is mediated by the dorsal striatum.
- The striatum has also been implicated in mediating low levels of motivation in psychopathologies such as schizophrenia and depression.
- Alterations in the striatum are observed in neurodegenerative disorders such as Parkinson disease, characterized by degeneration of striatal dopaminergic innervations, and Huntington disease, which is accompanied by reduced striatal gray matter.

SUMMARY POINTS

- CUDs affect 13.1 million individuals worldwide and represent the most vulnerable portion of cannabis users.
- Neuroanatomical alterations may mediate the adverse outcome of CUD.
- This review summarizes findings from 16 neuroimaging studies of gray matter morphology in CUD.
- CUD-specific alterations emerged within the striatum, medial temporal lobe, PFC, and cerebellum.
- Age of onset, gender, cumulative cannabis consumption, abstinence, and CUD-associated problems may moderate the association between CUD and brain morphology.
- The paucity of conducted studies prevents drawing conclusions about CUD-specific alterations.
- Studying the commonalities and differences between cannabis users with and without a CUD is an important next step to understanding the neurobiological mechanisms underlying CUDs.

REFERENCES

APA. (1996). *DSM-IV diagnostic and statistical manual of mental disorders* (4th ed.).

APA, American Psychiatric Association D. S. M. Task Force. (2013). *Diagnostic and statistical manual of mental disorders: DSM-5*. From http://dsm.psychiatryonline.org/book.aspx?bookid=556.

Ashtari, M., Avants, B., Cyckowski, L., Cervellione, K. L., Roofeh, D., Cook, P., ... Kumra, S. (2011). Medial temporal structures and memory functions in adolescents with heavy cannabis use. *Journal of Psychiatry Research*, *45*(8), 1055–1066.

Asmaro, D., Carolan, P. L., & Liotti, M. (2014). Electrophysiological evidence of early attentional bias to drug-related pictures in chronic cannabis users. *Addictive Behaviors*, *39*(1), 114–121.

Bailey, J. A., DuPont, R. L., & Teitelbaum, S. A. (2014). Cannabis use disorder: Epidemiology, comorbidity, and pathogenesis. *UpToDate*. see http://www.uptodate.com/contents/cannabis-use-disorder-epidemiology-comorbidity-and-pathogenesis.

Batalla, A., Soriano-Mas, C., Lopez-Sola, M., Torrens, M., Crippa, J. A., Bhattacharyya, S., ... Martin-Santos, R. (2013). Modulation of brain structure by catechol-O-methyltransferase Val(158) Met polymorphism in chronic cannabis users. *Addiction Biology, 19*(4), 722–732.

Battistella, G., Fornari, E., Annoni, J. M., Chtioui, H., Dao, K., Fabritius, M., ... Giroud, C. (2014). Long-term effects of cannabis on brain structure. *Neuropsychopharmacology, 39*, 2041–2048.

Belin, D., & Everitt, B. J. (2008). Cocaine seeking habits depend upon dopamine-dependent serial connectivity linking the ventral with the dorsal striatum. *Neuron, 57*(3), 432–441.

Budney, A. J., Vandrey, R. G., Hughes, J. R., Thostenson, J. D., & Bursac, Z. (2008). Comparison of cannabis and tobacco withdrawal: severity and contribution to relapse. *Journal of Substance Abuse Treatment, 35*(4), 362–368.

Burns, H. D., Van Laere, K., Sanabria-Bohorquez, S., Hamill, T. G., Bormans, G., Eng, W. S., ... Hargreaves, R. J. (2007). [18F]MK-9470, a positron emission tomography (PET) tracer for in vivo human PET brain imaging of the cannabinoid-1 receptor. *Proceedings of the National Academy of Sciences of the United States of America, 104*(23), 9800–9805.

Chauchard, E., Septfons, A., & Chabrol, H. (2013). Motivations for cannabis cessation, coping and adaptation strategies, and perceived benefits: impact on cannabis use relapse and abstinence. *Encephale, 39*(6), 385–392.

Churchwell, J. C., Lopez-Larson, M., & Yurgelun-Todd, D. A. (2010). Altered frontal cortical volume and decision making in adolescent cannabis users. *Frontiers in Psychology, 1*, 225.

Cousijn, J., Goudriaan, A. E., & Wiers, R. W. (2011). Reaching out towards cannabis: approach-bias in heavy cannabis users predicts changes in cannabis use. *Addiction, 106*(9), 1667–1674.

Cousijn, J., Watson, P., Koenders, L., Vingerhoets, W. A., Goudriaan, A. E., & Wiers, R. W. (2013). Cannabis dependence, cognitive control and attentional bias for cannabis words. *Addictive Behavior, 38*(12), 2825–2832.

Cousijn, J., Wiers, R. W., Ridderinkhof, K. R., van den Brink, W., Veltman, D. J., & Goudriaan, A. E. (2012). Grey matter alterations associated with cannabis use: results of a VBM study in heavy cannabis users and healthy controls. *NeuroImage, 59*(4), 3845–3851.

Cousijn, J., Wiers, R. W., Ridderinkhof, K. R., van den Brink, W., Veltman, D. J., & Goudriaan, A. E. (2013). Effect of baseline cannabis use and working-memory network function on changes in cannabis use in heavy cannabis users: a prospective fMRI study. *Human Brain Mapping, 35*(5), 2470–2482.

Degenhardt, L., Ferrari, A. J., Calabria, B., Hall, W. D., Norman, R. E., McGrath, J., ... Vos, T. (2013). The global epidemiology and contribution of cannabis use and dependence to the global burden of disease: results from the GBD 2010 study. *PLoS One, 8*(10), e76635.

Dorard, G., Berthoz, S., Phan, O., Corcos, M., & Bungener, C. (2008). Affect dysregulation in cannabis abusers: a study in adolescents and young adults. *European Child and Adolescent Psychiatry, 17*(5), 274–282.

Everitt, B. J., & Robbins, T. W. (2005). Neural systems of reinforcement for drug addiction: from actions to habits to compulsion. *Nature Neuroscience, 8*(11), 1481–1489.

Everitt, B. J., & Robbins, T. W. (2013). From the ventral to the dorsal striatum: devolving views of their roles in drug addiction. *Neuroscience and Biobehavioral Reviews, 37*(9), 1946–1954.

Fernández-Serrano, M. J., Pérez-García, M., & Verdejo-García, A. (2011). What are the specific vs. generalized effects of drugs of abuse on neuropsychological performance? *Neuroscience & Biobehavioral Reviews, 35*(3), 377–406.

Field, M., Eastwood, B., Bradley, B. P., & Mogg, K. (2006). Selective processing of cannabis cues in regular cannabis users. *Drug and Alcohol Dependence, 85*(1), 75–82.

Gilman, J. M., Kuster, J. K., Lee, S., Lee, M. J., Kim, B. W., Makris, N., ... Breiter, H. C. (2014). Cannabis use is quantitatively associated with nucleus accumbens and amygdala abnormalities in young adult recreational users. *The Journal of Neuroscience, 34*(16), 5529–5538.

Gogtay, N., Giedd, J. N., Lusk, L., Hayashi, K. M., Greenstein, D., Vaituzis, A. C., ... Thompson, P. M. (2004). Dynamic mapping of human cortical development during childhood through early adulthood. *Proceedings of the National Academy of Sciences of the United States of America, 101*(21), 8174–8179.

Gray, K. M., LaRowe, S. D., Watson, N. L., & Carpenter, M. J. (2011). Reactivity to in vivo marijuana cues among cannabis-dependent adolescents. *Addictive Behaviors, 36*(1–2), 140–143.

Jager, G., Van Hell, H. H., De Win, M. M. L., Kahn, R. S., Van Den Brink, W., Van Ree, J. M., & Ramsey, N. F. (2007). Effects of frequent cannabis use on hippocampal activity during an associative learning memory task. *European Psychopharmacology, 17*, 289–297.

King, G. R., Ernst, T., Deng, W., Stenger, A., Gonzales, R. M. K., Nakama, H., & Chang, L. (2011). Altered brain activation during visuomotor integration in chronic active cannabis users: relationship to cortisol levels. *The Journal of Neuroscience, 31*(49), 17923–17931.

Koob, G. F., & Volkow, N. D. (2010). Neurocircuitry of addiction. *Neuropsychopharmacology, 35*(1), 217–238.

Kumra, S., Robinson, P., Tambyraja, R., Jensen, D., Schimunek, C., Houri, A., ... Lim, K. (2012). Parietal lobe volume deficits in adolescents with schizophrenia and adolescents with cannabis use disorders. *Journal of the American Academy of Child and Adolescent Psychiatry, 51*(2), 171–180.

Leshner, A. I. (1997). Addiction is a brain disease, and it matters. *Science, 278*(5335), 45–47.

Lorenzetti, V., Solowij, N., Fornito, A., Lubman, D. I., & Yucel, M. (2014). The association between regular cannabis exposure and alterations of human brain morphology: an updated review of the literature. *Current Pharmaceutical Design, 20*(13), 2138–2167.

Lundahl, L. H., & Johanson, C. E. (2011). Cue-induced craving for marijuana in cannabis-dependent adults. *Experimental and Clinical Psychopharmacology, 19*(3), 224–230.

Mansouri, F. A., Tanaka, K., & Buckley, M. J. (2009). Conflict-induced behavioural adjustment: a clue to the executive functions of the prefrontal cortex. *Nature Reviews Neuroscience, 10*(2), 141–152.

Martin-Santos, R., Fagundo, A. B., Crippa, J. A., Atakan, Z., Bhattacharyya, S., Allen, P., ... McGuire, P. (2010). Neuroimaging in cannabis use: a systematic review of the literature. *Psychological Medicine, 40*(3), 383–398.

Mata, I., Perez-Iglesias, R., Roiz-Santianez, R., Tordesillas-Gutierrez, D., Pazos, A., Gutierrez, A., ... Crespo-Facorro, B. (2010). Gyrification brain abnormalities associated with adolescence and early-adulthood cannabis use. *Brain Research, 1317*, 297–304.

McQueeny, T., Padula, C. B., Price, J., Medina, K. L., Logan, P., & Tapert, S. F. (2011). Gender effects on amygdala morphometry in adolescent marijuana users. *Behavioural Brain Research, 224*(1), 128–134.

Medina, K. L., McQueeny, T., Nagel, B. J., Hanson, K. L., Yang, T. T., & Tapert, S. F. (2009). Prefrontal cortex morphometry in abstinent adolescent marijuana users: subtle gender effects. *Addiction Biology, 14*(4), 457–468.

Medina, K. L., Nagel, B. J., Park, A., McQueeny, T., & Tapert, S. F. (2007). Depressive symptoms in adolescents: associations with white matter volume and marijuana use. *The Journal of Child Psychology and Psychiatry, 48*(6), 592–600.

Medina, K. L., Nagel, B. J., & Tapert, S. F. (2010). Abnormal cerebellar morphometry in abstinent adolescent marijuana users. *Psychiatry Research*, *182*(2), 152–159.

Medina, K. L., Schweinsburg, A. D., Cohen-Zion, M., Nagel, B. J., & Tapert, S. F. (2007). Effects of alcohol and combined marijuana and alcohol use hippocampal volume and asymmetry. *Neurotoxicology and Teratology*, *29*, 141–152.

Moore, T. H., Zammit, S., Lingford-Huges, A., Barnes, T. R., Jones, P. B., Burke, M., & Lewis, G. (2007). Cannabis use and risk of psychotic or affective mental health outcomes: a systematic review. *Lancet*, *370*, 319–328.

van der Pol, P., Liebregts, N., de Graaf, R., Ten Have, M., Korf, D. J., van den Brink, W., & van Laar, M. (2013). Mental health differences between frequent cannabis users with and without dependence and the general population. *Addiction*, *108*(8), 1459–1469.

Ridderinkhof, K. R., Ullsperger, M., Crone, E. A., & Nieuwenhuis, S. (2004). The role of the medial frontal cortex in cognitive control. *Science*, *306*(5695), 443–447.

Robinson, T. E., & Berridge, K. C. (2003). Addiction. *Annual Review of Psychology*, *54*, 25–53.

Rocchetti, M., Crescini, A., Borgwardt, S., Caverzasi, E., Politi, P., Atakan, Z., & Fusar-Poli, P. (2013). Is cannabis neurotoxic for the healthy brain? A meta-analytical review of structural brain alterations in non-psychotic users. *Psychiatry and Clinical Neurosciences*, *67*(7), 483–492.

Schneider, F., Habel, U., Wagner, M., Franke, P., Salloum, J. B., Shah, N. J., ... Zilles, K. (2001). Subcortical correlates of craving in recently abstinent alcoholic patients. *The American Journal of Psychiatry*, *158*(7), 1075–1083.

Skinner, R., Conlon, L., Gibbons, D., & McDonald, C. (2011). Cannabis use and non-clinical dimensions of psychosis in university students presenting to primary care. *Acta Psychiatrica Scandinavica*, *123*(1), 21–27.

Solowij, N., & Battisti, R. (2008). The chronic effects of cannabis on memory in humans: a review. *Current Drug Abusers Reviews*, *1*, 81–98.

Solowij, N., Yücel, M., Respondek, C., Whittle, S., Lindsay, E., Pantelis, C., & Lubman, D. I. (2011). Cerebellar white-matter changes in cannabis users with and without schizophrenia. *Psychological Medicine*, *41*(11), 2349–2359.

Stinson, F. S., Ruan, W. J., Pickering, R., & Grant, B. F. (2006). Cannabis use disorders in the USA: prevalence, correlates and co-morbidity. *Psychological Medicine*, *36*(10), 1447–1460.

Swift, W., Hall, W., & Teesson, M. (2001). Cannabis use and dependence among Australian adults: results from the National Survey of Mental Health and Wellbeing. *Addiction*, *96*(5), 737–748.

Tzilos, G. K., Cintron, C. B., Wood, J. B. R., DSimpson, N. S., Young, A. D., Pope, H. G. J., & Yurgelun-Todd, D. A. (2005). Lack of hippocampal volume change in long-term heavy cannabis users. *The American Journal on Addiction*, *14*, 64–72.

UNODC. (2010). World drug report 2010. In UNDOC (Ed.), *World drug report series 2010: United Nations Office on drugs and crime*.

Wiers, R. W., Bartholow, B. D., van den Wildenberg, E., Thush, C., Engels, R. C., Sher, K. J., ... Stacy, A. W. (2007). Automatic and controlled processes and the development of addictive behaviors in adolescents: a review and a model. *Pharmacology, Biochemistry, and Behavior*, *86*(2), 263–283.

Wilson, S. J., Sayette, M. A., & Fiez, J. A. (2004). Prefrontal responses to drug cues: a neurocognitive analysis. *Nature Neuroscience*, *7*(3), 211–214.

Winsauer, P. J., Daniel, J. M., Filipeanu, C. M., Leonard, S. T., Hulst, J. L., Rodgers, S. P., ... Sutton, J. L. (2011). Long-term behavioral and pharmacodynamic effects of delta-9-tetrahydrocannabinol in female rats depend on ovarian hormone status. *Addiction Biology*, *16*(1), 64–81.

Yip, S. W., DeVito, E. E., Kober, H., Worhunsky, P. D., Carroll, K. M., & Potenza, M. N. (2014). Pretreatment measures of brain structure and reward-processing brain function in cannabis dependence: an exploratory study of relationships with abstinence during behavioral treatment. *Drug and Alcohol Dependence*, *140*, 33–41.

Yücel, M., Solowij, N., Respondek, C., Whittle, S., Fornito, A., Pantelis, C., & Lubman, D. (2008). Regional brain abnormalities associated with heavy long-term cannabis use. *Archives of General Psychiatry*, *65*(6), 1–8.

Chapter 73

Imaging Dopamine Alterations in Cannabis Dependence

Jodi J. Weinstein[1], Anissa Abi-Dargham[2]

[1]Department of Psychiatry, New York State Psychiatric Institute and Columbia University Medical Center, New York, NY, USA; [2]Departments of Psychiatry and Radiology, Division of Translational Imaging, New York State Psychiatric Institute and Columbia University Medical Center, New York, NY, USA

Abbreviations

2TCM Two-tissue compartment model
AADC Amino acid decarboxylase
AST Associative striatum
BP$_{ND}$ Binding potential relative to nondisplaceable compartment
CB1 Cannabinoid-1 receptor
CU Cannabis user
D1 Dopamine-1 receptor
D2/3 Dopamine-2/3 receptor
DAT Dopamine transporter
ES Effect size
F-DOPA [^{18}F]Fluorodopa
HC Healthy control
IBZM [^{123}I]Iodobenzamide
K_i Influx rate constant
$K_{i(cer)}$ Influx rate constant relative to the cerebellum
LGA Logan graphical reference tissue model
MRTM Multilinear regression tissue model
PET Positron emission tomography
PHNO [^{11}C]PHNO
RAC [^{11}C]Raclopride
SMST Sensorimotor striatum
SPECT Single-photon emission computed tomography
SRTM Simplified reference tissue model
THC Δ^9-Tetrahydrocannabinol
UHR Individuals at ultrahigh risk for schizophrenia
VST Ventral striatum
ΔBP$_{ND}$ Percentage change in BP$_{ND}$ following a challenge

INTRODUCTION

Cannabis is the most commonly used drug in North America (SAMHSA, 2014), starting in adolescence and frequently leading to dependence (Budney, Moore, Vandrey, & Hughes, 2003), anxiety, and depression (Degenhardt, Hall, & Lynskey, 2001; Troisi, Pasini, Saracco, & Spalletta, 1998, but also see Moore et al., 2007). Associations with the onset of psychosis and schizophrenia have also been described (Henquet, Di Forti, Morrison, Kuepper, & Murray, 2008). It is suggested that the primary psychoactive component of cannabis may have long-lasting effects on the brain by interacting with normal maturational processes (Pistis et al., 2004). Abnormal brain structure, with smaller hippocampal and amygdala volumes (Yucel et al., 2008), and cognitive disturbances including deficits in verbal learning and working memory, particularly in adolescents, have been reported (Schweinsburg, Brown, & Tapert, 2008). Altogether, these reports suggest changes in brain function especially with adolescent use, which may lead to long-lasting effects on brain and behavior (Ghazzaoui & Abi-Dargham, 2014). Since deficits in striatal dopamine observed with other drugs of abuse predict poor outcome and higher rates of relapse (Martinez et al., 2011) and cannabis use is starting to receive approvals for medical and recreational use in certain countries, questions are being raised regarding its similarity to other drugs of abuse.

Dopamine, a neurotransmitter involved in reward prediction and salience attribution, plays a central role in mediating addictive behavior (Volkow, Fowler, Wang, & Swanson, 2004). For multiple substances of abuse, including cocaine, methamphetamine, heroin, alcohol, and nicotine, it has been consistently shown that substance-dependent subjects display low availability of dopamine-2/dopamine-3 receptors (D2/3) and blunted dopamine release in the striatum compared to demographically matched controls, when imaged in vivo using positron emission tomography (PET) and D2/3 radiotracer scans before and after a challenge with amphetamine or methylphenidate (Martinez et al., 2005, 2007; Volkow et al., 1997, see also review in Martinez & Narendran, 2010). Until recently, fewer imaging studies have examined these dopaminergic indices in cannabis abuse and dependence despite the well-known multiple interactions between the cannabinoid and the dopaminergic systems. While the direct psychopharmacological effects of the primary psychoactive component of cannabis, Δ^9-tetrahydrocannabinol (THC), are mediated by endocannabinoid (CB1) receptors, which are predominantly localized in the basal ganglia (Freund, Katona, & Piomelli, 2003), the CB1 receptor itself interacts with pre- and postsynaptic dopaminergic sites. Furthermore, cannabis does indirectly activate the same mesolimbic dopaminergic pathways as other drugs of abuse, producing similar actions in behavioral models of addiction. These observations support the potential of cannabis to induce dependence in humans (Gardner & Vorel, 1998; Maldonado, 2002; Maldonado & Rodriguez de Fonseca, 2002) and

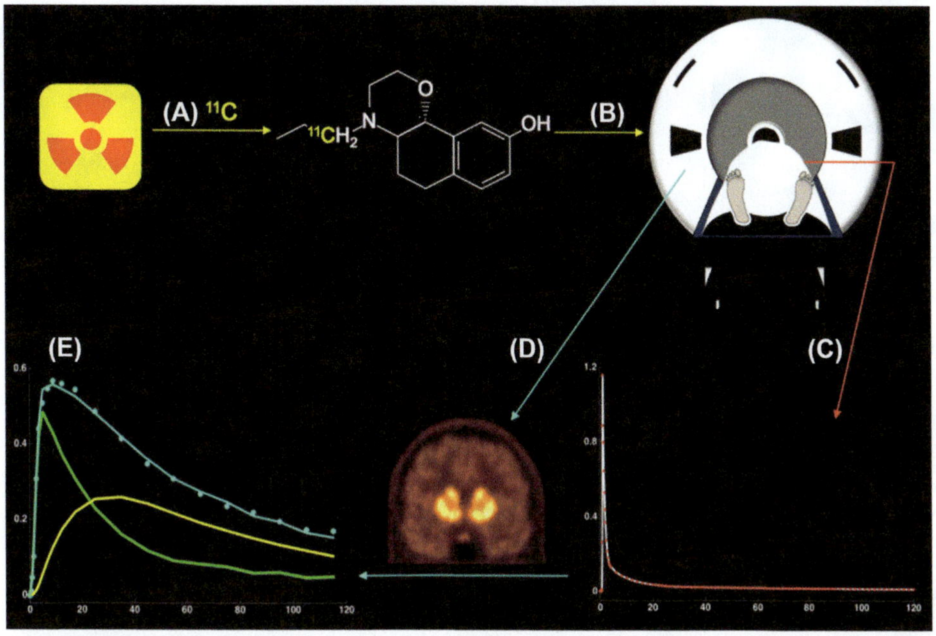

FIGURE 1 Flow diagram for typical PET neuroimaging study in cannabis dependence. Radioactive tracer is synthesized (A) and injected intravenously (B). The scan starts at the time of tracer injection. Tracer levels in the plasma are measured at multiple time points during the scan to generate an input function (C) for the pharmacokinetic model. The scanner output is a three-dimensional image of radioactive emission data (coronal slice shown in D). By fitting these data to a pharmacokinetic model (E), outcome measures such as BP_{ND} can be derived. This in vivo neuroimaging paradigm has been used in studies examining the dopamine system in individuals with cannabis dependence.

for this dependence to be mediated by dopaminergic mechanisms (for a more extensive review, see the chapter "Cannabis and the Mesolimbic System"). This chapter reviews the imaging methods used to study the dopaminergic system in humans and then summarizes the findings regarding acute and long-term effects of cannabis use on the dopaminergic system in humans.

IMAGING METHODS USED TO STUDY DOPAMINE IN HUMANS

PET and single-photon emission computed tomography (SPECT) are noninvasive imaging techniques that can be applied to measure brain neurochemistry in vivo using a radiolabeled ligand (radiotracer). Outcome measures such as a radiotracer's binding potential to target receptors relative to its nondisplaceable uptake (BP_{ND}) can be calculated by fitting radiotracer emission data to a pharmacokinetic model (Figure 1). Various data-fitting methods are in use for deriving BP_{ND}, including the two-tissue compartment model (2TCM; which requires measurement of the radiotracer concentration in arterial blood), simplified reference tissue model (SRTM), multilinear regression tissue model (MRTM), and Logan graphical reference tissue model (LGA).

Human imaging studies using these techniques have examined various aspects of the dopaminergic system, such as transporters, receptors, and the neurotransmitter itself. In particular, reversibly binding D2/3 radiotracers such as [^{11}C]raclopride and [^{11}C]PHNO ([^{11}C](+)-4-propyl-3,4,4a,5,6,10b-hexahydro-2H-naphtho[1,2-b][1,4]oxazin-9-ol) have been used to measure BP_{ND}, a measure of D2/3 availability, as well as the percentage change in BP_{ND} between scans as a result of a challenge (ΔBP_{ND}) to derive a measure of dopamine release in response to the challenge (Figure 2). Various challenges have been used: the drug itself (cannabis), a stress paradigm, and a psychostimulant challenge. The response to a psychostimulant challenge is considered a probe of dopamine storage and release capacity of the presynaptic dopaminergic neuron (Laruelle, 2000). Another tracer particularly useful for PET imaging of the dopamine system is [^{18}F]fluorodopa ([^{18}F]F-DOPA), a substrate for amino acid decarboxylase that can be measured as it partially follows the metabolic pathway of endogenous dopamine precursors and accumulates in presynaptic terminals. The quantification of its uptake rate (K_i) is interpreted as a combined indicator of presynaptic dopamine synthesis and storage capacity. Additional radiotracers have also been used to study the dopamine transporter (DAT).

EFFECT OF ACUTE CANNABINOID ADMINISTRATION ON THE DOPAMINE SYSTEM

Several imaging studies have examined the effect of acute THC administration on dopamine release in vivo and overall have shown a small but detectable dopamine release measured indirectly via displacement of D2/3 radiotracers (Table 1). These studies started with an initial anecdotal case report by Voruganti, Slomka, Zabel, Mattar, and Awad (2001), followed by a report, on nine healthy participants, of THC inhalation reducing the binding of the D2/3 radiotracer [^{11}C]raclopride in the ventral striatum (VST) and the precommissural dorsal putamen, compared to placebo, consistent with an increase in dopamine levels in these regions (Bossong et al., 2009). The synthetic cannabinoid dronabinol, compared to placebo (Stokes, Mehta, Curran, Breen, & Grasby, 2009), confirmed these findings and showed displacement of 0.6–3% in some striatal subregions, not reaching significance, but consistent in magnitude with the other reports. Administration of intravenous THC, in doses

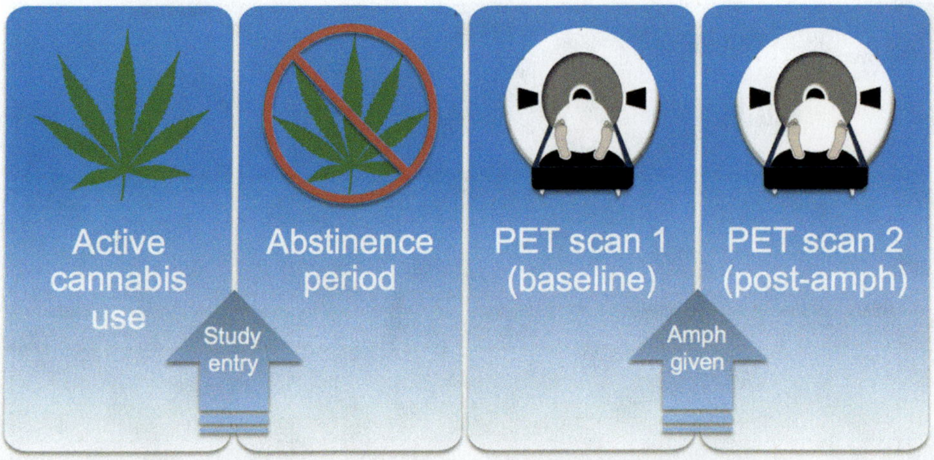

FIGURE 2 Typical study design for PET neuroimaging of cannabis dependence. Human PET neuroimaging studies of the dopamine system in cannabis dependence have been designed comparable to other studies of addiction. The typical study design consists of enrollment of individuals with cannabis dependence who agree to an abstinence period (duration and location are set by each study), followed by PET scans before (baseline) and after (e.g., post-amph) a challenge such as amphetamine (amph). The change in dopamine measures induced by this challenge (e.g., ΔBP_{ND}) is compared between individuals with cannabis dependence and healthy controls.

TABLE 1 Studies of Dopamine Release after Acute THC Administration

	Radiotracer	THC Administration	Dopamine Release Capacity
Voruganti et al. (2001)	[^{123}I]IBZM, $n=1$	Inhalation, dose unknown	20% displacement in striatum
Bossong et al. (2009)	[^{11}C]Raclopride, $n=9$	8 mg inhalation	Significant displacement in ventral striatum and precommissural putamen
Stokes et al. (2009)	[^{11}C]Raclopride, $n=13$	10 mg oral dronabinol	0.6–3% displacement in some striatal subregions not reaching significance, but magnitude consistent with other reports; correlation with plasma THC in associative striatum
Barkus et al. (2011)	[^{123}I]IBZM, $n=9$	2.5 mg intravenous	No difference from placebo for caudate or putamen

Several imaging studies have examined the effect of acute THC administration on dopamine release in vivo and, overall, have shown a small but detectable dopamine release measured indirectly via displacement (ΔBP_{ND}) of D2/3 radiotracers. IBZM, iodobenzamide.

sufficient to provoke psychotic symptoms, revealed no significant difference from placebo in stimulating dopamine release in the striatum (Barkus et al., 2011). There was also no apparent association between positive psychotic symptoms and dopamine release in this study. However, it is possible that this discrepant result arose from the pseudo-equilibrium analysis method used to derive the binding potential in this study. Pseudo-equilibrium designs for PET analyses are not considered fully quantitative methods, as they may lead to outcome measures that are affected by peripheral clearance and metabolism, or blood flow differences, rather than central receptor-related parameters (Ghazzaoui & Abi-Dargham, 2014).

EFFECT OF CHRONIC CANNABIS USE ON THE DOPAMINE SYSTEM

The effects of chronic cannabis use on the dopaminergic system are less clear-cut. Stemming from the common finding of blunted striatal dopamine transmission associated with dependence on other dopamine-enhancing substances of abuse (Martinez et al., 2005, 2007; Volkow et al., 1997), few studies have examined the effects of chronic cannabis use on dopaminergic indices using PET (Table 2). Early studies suggested no differences between cannabis user groups and healthy controls, although many of these studies were limited by sampling light cannabis users (i.e., those who could tolerate abstinence during the study period).

Imaging Studies of Stress Response

The response of the dopaminergic system in cannabis users to a stress paradigm in 13 cannabis users and 12 matched controls was investigated using the D3-preferring mixed D2/3 agonist radiotracer [^{11}C]PHNO and shown to be normal (Mizrahi et al., 2013).

Imaging Studies of the Dopamine Transporter

Striatal and extrastriatal DAT availabilities were measured using the PET tracer [^{11}C]PE2I (Leroy et al., 2012) in 11 healthy nonsmoker

TABLE 2 PET Studies of Dopaminergic Indices in Chronic Cannabis Users

Study	Radiotracer	Cannabis User (CU) and Healthy Control (HC) Groups	Comorbidity in CU and HC Groups	Prescan Abstinence	D2/3 Availability in CU, Compared to HC	Dopamine Release to Challenge in CU, Compared to HC
Sevy et al. (2008)	[11C]Raclopride	6 CU (chronic use, >3 joints, or 1.5g of cannabis/week)+6 HC	Mixed nicotine smokers in both groups (83% CU+17% HC)	15 weeks	= in striatum	n/a
Stokes et al. (2012)	[11C]Raclopride	10 CU (chronic use, 52 joints/lifetime year)+10 HC	Mixed current nicotine smokers in both groups (30% CU+20% HC), some with prior use of other drugs	18 months	= in striatum, controlling for age and nicotine-smoking status	n/a
Urban et al. (2012)	[11C]Raclopride	15 CU (current chronic use, 517±465 puffs[a] in past month; mild-moderate dependence)+16 HC	Mixed current nicotine smokers in both groups (13% CU+6% HC)	4 weeks	= in striatum	(A, intravenous) = in striatum
Albrecht et al. (2013)	[11C]Raclopride	10 CU (chronic use, ≥1 joint/week in past month, 13±12/week, 47±42/month)+8 HC, all males	Mixed nicotine smokers (50%), some with other drug use	20h	= in striatum Associated (−) with cannabis intake/day	n/a
Mizrahi et al. (2013)	[11C]PHNO	13 CU (chronic use, ≥3 times/week, 16 joints/week)+12 HC	Mixed nicotine smokers (23%) in CU, and past use of other drugs in 46% but no dependence	10h (range: 1–22h)	↑ in striatum (and in all subregions[b]) during sensorimotor control task = in pallidum = in midbrain	(S) = in striatum (but increased in VST when controlled for baseline BP$_{ND}$) ↓ in pallidum
Bloomfield et al. (2014)	[18F]F-DOPA[c]	19 CU (current, chronic use, ≥1 time/week, median: 26.3 g/month; 5 CU met criteria for dependence, 5 for abuse)+19 HC	Psychotic symptoms with cannabis use in CU. Both groups included nicotine smokers (79% CU+42% HC), and mixed use of other substances	14 h (median)	n/a	↓[c] in striatum, (and in AST and VST subregions) and associated with greater use (−) and age of onset of use (+)
Volkow et al. (2014)	[11C]Raclopride	24 CU (chronic cannabis abuse, ~5 days/week, ~5 joints/day)+24 HC	Mixed nicotine users (42% current in CU, 13% in HC), no other substance use, and no psychiatric comorbidities	Not reported	= in striatum	(M, intravenous) ↓ in striatum by measure of distribution volume, but not by ΔBP$_{ND}$
Weinstein et al. (2015)	[11C]PHNO	11 CU (severe cannabis dependence, chronic, daily use, 1503±947 puffs[a] in past month)+12 HC	No other substance use, including nicotine, and no psychiatric comorbidities	5–7 days	= in striatum, pallidum, midbrain, or thalamus	(A, oral) ↓ in striatum (and in AST and SMST) = in pallidum = in midbrain = in thalamus

With chronic daily cannabis use, in the absence of nicotine use or other psychiatric comorbidities, molecular imaging studies show a reduction in striatal and pallidal dopamine release capacity (ΔBP$_{ND}$), and, in cannabis users who experience psychotic symptoms, an overall decrease in dopamine synthesis rate. Earlier PET studies showing no group differences in dopaminergic indices sampled lighter cannabis users or included nicotine smokers in both patients and controls, thus minimizing group differences. These studies may also have lacked power to detect differences because of longer abstinence in users and imaging limitations related to use of D2/3 antagonist tracers such as [11C]raclopride with smaller signals. Values given for group characteristics, comorbidities, and duration of prescan abstinence reflect reported group means (or median, when noted). Arrows indicate direction of abnormality in the CU group for a given dopaminergic index, compared to the HC group, with (↓) indicating significantly reduced D2/3 availability (BP$_{ND}$) or, in the case of a challenge paradigm in which dopamine release is induced by amphetamine (A), methylphenidate (M), or stress (S), significantly decreased dopamine release (i.e., less displacement) in CU group compared to ΔBP$_{ND}$ in HC group. (=) indicates measure in CU group did not significantly differ from the HC group, (↑) indicates reported increase, and "n/a" denotes measure was not examined in the study. Directions of associations with clinical factors are indicated with (+) and (−).

[a]Subregions of striatum include associative (AST), sensorimotor (SMST), and ventral (VST).
[b]The number of "puffs" was estimated as follows: 1 pipe (5 puffs), "joint" (10 puffs), "bong" or "bowl" (12 puffs), and "blunt" (20 puffs).
[c]An indicator of presynaptic dopamine, often interpreted as a representation of the rate of dopamine synthesis; calculated using the rate constant of [18F]F-DOPA influx to a region of interest relative to the cerebellum reference tissue (K$_i^{cer}$).

subjects, 14 tobacco-dependent smokers, and 13 cannabis and tobacco smokers. DAT availability was significantly reduced in both of the substance-user groups, compared to controls. However, since both substance-user groups included nicotine, which has also been associated with lower DAT availability, it is not possible to know if cannabis on its own would have been associated with changes in DAT.

Imaging Studies of Capacity for Dopamine Release

A first report in six patients and six controls using [^{11}C]raclopride in chronic cannabis users showed no group differences in D2/3 receptor availability and no correlations between striatal D2/3 receptor availability and normalized glucose metabolism in any region of the frontal cortex or striatum (Sevy et al., 2008). This was followed by other negative reports using [^{11}C]raclopride, including one in a larger cohort (Stokes et al., 2012) and our own study examining the long-term effects of chronic use of cannabis on both D2/3 receptor availability and striatal dopamine release capacity in chronic cannabis-dependent, non-nicotine-smoking, participants ($n=15$) compared to matched controls ($n=16$) that found no differences between groups. We used the amphetamine challenge to induce dopamine release (Figure 2). Participants were imaged within similar time frames of abstinence, within a few weeks of last use, as confirmed by urine toxicology (Urban et al., 2012). A subsequent PET study focused on the acute withdrawal period also did not show between-group differences when comparing 10 individuals with chronic cannabis use (average last use was 20h prior to scan) to eight demographically matched controls (Albrecht et al., 2013; Ghazzaoui & Abi-Dargham, 2014).

Several considerations may explain these negative results. Severe cannabis dependence has a similarly high relapse rate and significant withdrawal symptoms comparable to other drugs of abuse (Budney et al., 2003; Kadden, Litt, Kabela-Cormier, & Petry, 2007). It is possible that alterations in the dopamine system are restricted to the more severe dependent subjects and may not be observed in lighter users. Stimulant-dependent participants with more blunted dopamine transmission are more likely to relapse into drug use than those with more preserved dopamine activity (Martinez et al., 2011; Wang et al., 2011), and studies of cannabis dependence that have required a few weeks of abstinence may have prevented heavier users from participating, possibly accounting for the negative results. Furthermore, normalization of the CB1 receptor in cannabis users within a few weeks of abstinence was reported using the CB1 receptor-selective radioligand [^{18}F]FMPEP-d$_2$. Chronic heavy users showed reduction of CB1 receptor binding in most brain regions that began to reverse after 4 weeks of abstinence (Hirvonen et al., 2012). Also cognitive impairment found in users immediately after cannabis cessation resolved by 28 days of monitored abstinence (Pope, Gruber, Hudson, Huestis, & Yurgelun-Todd, 2002), suggesting that the impact of THC on neurochemical networks subserving cognition, possibly including dopamine, may reverse during prolonged abstinence (Ghazzaoui & Abi-Dargham, 2014).

In contrast to these earlier negative studies, more recent data have provided more definitive evidence of a compromised integrity of the dopaminergic system in more severe cannabis dependence. In 2014, Volkow et al., using a methylphenidate challenge and [^{11}C]raclopride, reported group differences in dopamine release when comparing pre- to postdistribution volumes in the striatum.

We completed a PET study of 11 cannabis users and 12 matched controls using [^{11}C]PHNO that provided definitive evidence for a reduction in dopamine release capacity in the striatum as a whole ($p=0.002$, effect size (ES)$=1.48$), specifically in the associative striatum (AST; $p=0.003$, ES$=1.39$) and the sensorimotor striatum (SMST; $p=0.003$, ES$=1.41$) subregions, as well as in the pallidum ($p=0.012$, ES$=1.16$) in heavy cannabis users (Figure 3). There were no significant group differences in BP$_{ND}$ or ΔBP$_{ND}$ in midbrain or thalamus. The cannabis users in this study were carefully selected to represent the clinical population of severely cannabis-dependent users, excluding other comorbidities. In contrast to most previous studies, we included only cannabis-dependent participants, and their abstinence was facilitated with a 1-week inpatient stay. As a consequence, this cohort has more severely dependent subjects compared to some of the previous cohorts, which may account for the group differences in dopamine release.

The discrepancies between these later studies and the earlier ones reporting no group differences between cannabis users and controls could also be explained by a difference in duration of abstinence prior to scanning (Figure 2). Four weeks of abstinence, as in our first cohort (Urban et al., 2012), could result in a normalization of any dopamine deficit that may have been present earlier during the abstinence period. Reduced CB1 receptor availability in chronic heavy users begins to reverse after 4 weeks of abstinence (Hirvonen et al., 2012), and cerebral blood volume partially normalizes after 4 weeks of abstinence (Sneider et al., 2008). Finally, another difference from our previous study (Urban et al., 2012) is that we used [^{11}C]PHNO, a D3-preferring agonist tracer that is more sensitive to the dopamine-releasing effects of amphetamine than the D2/3 antagonist tracer [^{11}C]raclopride, which may have allowed more power to detect differences between groups that could have been missed with [^{11}C]raclopride (Figure 4).

The subregions of the striatum most affected were the AST and SMST, in contrast to other drug addictions, in which deficits in VST dopamine release are greatest and predict craving (Martinez et al., 2005, 2007; Volkow et al., 1997). A possible explanation could be the differential anatomical distribution of the CB1 receptor, which is less abundant in the ventral than in the dorsal striatum (Herkenham et al., 1990; Martin et al., 2008; Van Waes, Beverley, Siman, Tseng, & Steiner, 2012) and higher in the globus pallidum than in the thalamus (Herkenham et al., 1990). This would suggest a direct pharmacological effect of chronic, heavy cannabis use on dopamine release. Nevertheless, the possibility that striatal dopamine release is also reduced in the VST cannot be excluded.

Studies in Comorbid Cannabis Use and Psychosis

The effect of comorbid psychosis and cannabis use has also been investigated using PET imaging, in light of the observations of an association between these conditions (Henquet et al., 2008). A 2014 [^{11}C]PHNO PET study of participants with high risk for schizophrenia (UHR) showed reduced dopamine release on a stress task in UHR with heavy cannabis use compared to UHR without cannabis use (Mizrahi et al., 2014). Bloomfield et al. (2014) measured dopamine synthesis capacity in 19 cannabis users with

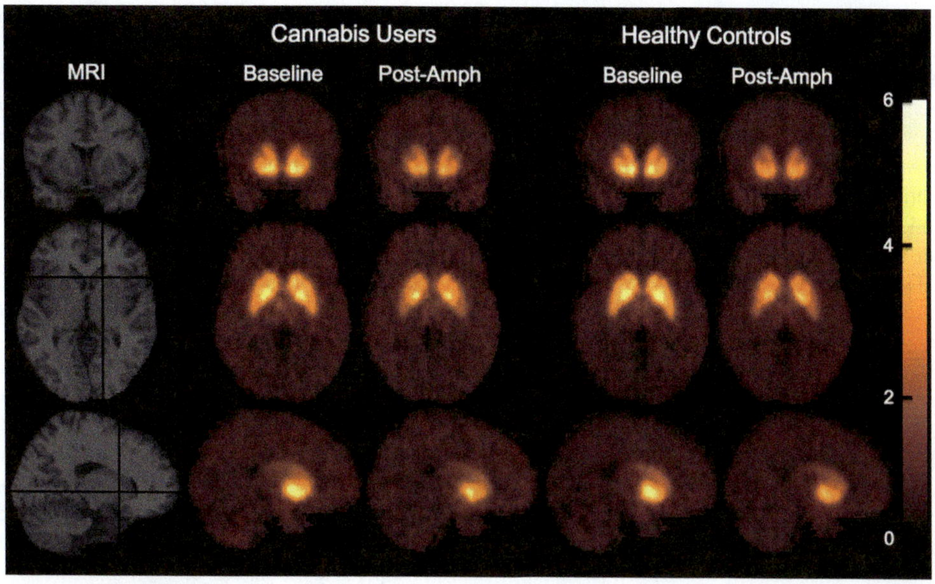

FIGURE 3 **Average [^{11}C]PHNO BP$_{ND}$ images before and after amphetamine administration in individuals with cannabis dependence and healthy control volunteers.** Once the tracer has been delivered to the brain tissue, it can bind reversibly to receptors. BP$_{ND}$ is proportional to the equilibrium ratio of specifically bound and free radiotracer and can be derived for each voxel in an image. These original images (unpublished) show [^{11}C]PHNO BP$_{ND}$ at baseline and after amphetamine challenge (post-amph) in individuals with cannabis dependence and healthy controls. The brightest areas (highest BP$_{ND}$) illustrate where binding to target receptors (D2/3) is greatest. Following amphetamine-induced dopamine release, BP$_{ND}$ decreases to an extent that reflects the capacity for dopamine release. Coronal, transverse, and sagittal slices are shown for coordinates (39, 54, 36) in normalized Montreal Neurologic Institute (MNI) template space.

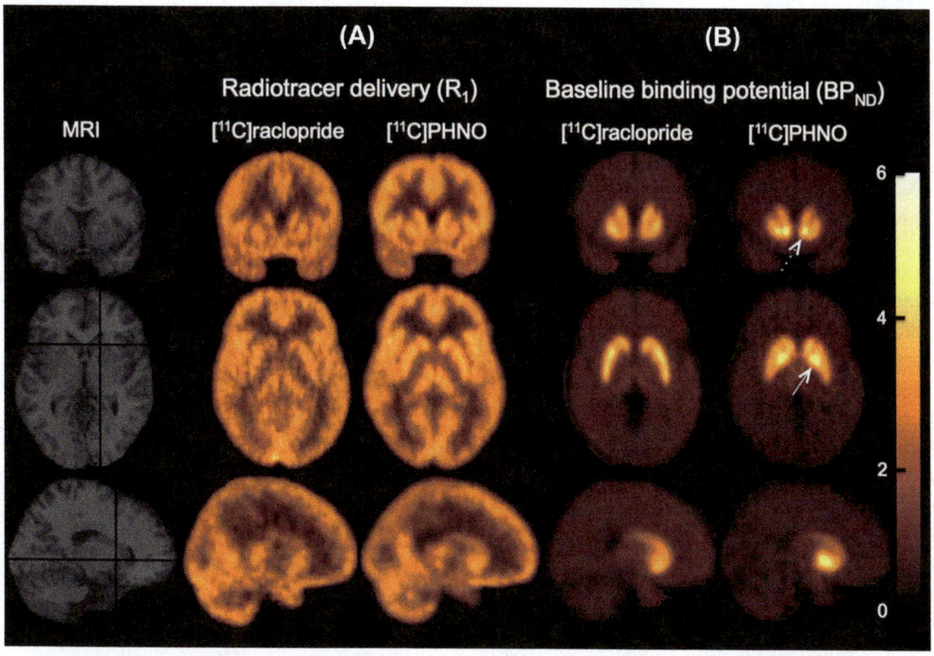

FIGURE 4 **Comparison of radiotracers used for imaging the dopamine system.** This figure illustrates the similar delivery but improved sensitivity of [^{11}C]PHNO over [^{11}C]raclopride for imaging the dopamine system. (A) Images of R$_1$, a parameter that is proportional to the product of blood flow and the extraction fraction of the radiotracer from capillary to brain tissue. (B) Images of BP$_{ND}$, an indicator of dopamine-D2/3 receptor availability. Note that R$_1$ is similar for both tracers, but greater uptake of [^{11}C]PHNO in dopamine-D3 receptor-rich regions enhances BP$_{ND}$ compared to that of [^{11}C]raclopride in regions such as the VST (dotted arrow) and globus pallidum (solid arrow). The original images (unpublished) are voxel-wise R$_1$ and BP$_{ND}$ values averaged across healthy controls at baseline after normalization into a common MNI template space. Coronal, transverse, and sagittal slices are shown for coordinates (39, 54, 36).

psychotic-like symptoms, compared with 19 nonusers, using PET and [^{18}F]F-DOPA. They found reduced striatal dopamine synthesis capacity in the cannabis users; however, a significant subset also used other substances of abuse. The results indicated that long-term cannabis use was correlated with a dose-dependent decrease in dopamine synthesis capacity in the striatum, especially in subjects meeting criteria for cannabis abuse or dependence (effect size 0.85; t36=2.54, $p=0.016$). Furthermore, dopamine synthesis capacity was negatively associated with magnitude of cannabis use ($r=-0.77$, $p<0.001$) and positively associated with age of onset of cannabis use ($r=0.51$, $p<0.027$), but not with cannabis-induced psychotic-like symptoms ($r=0.32$, $p=0.19$). The investigators concluded that the psychosis induced by cannabis use may not be mediated by dopamine transmission (Bloomfield et al., 2014). Alternatively, interactions between CB1 and D2 receptors as they colocalize on the D2-bearing medium spiny neurons in the striatum may result in a state of altered D2 signaling regardless of the magnitude of dopamine release and may contribute to the emergence of psychotic symptoms (Ghazzaoui & Abi-Dargham, 2014).

CONCLUSION

In summary, studies show a reduction in striatal and pallidal dopamine release capacity in chronic, daily cannabis users and an overall decrease in dopamine synthesis rate in cannabis users who experience psychotic symptoms. Earlier literature showing no group differences in dopaminergic indices sampled lighter cannabis users or included nicotine smokers in both patients and controls, thus minimizing group differences. These studies may also have lacked power to detect differences because of longer abstinence in users and imaging limitations related to the use of PET D2/3 antagonist tracers with smaller signals.

APPLICATIONS TO OTHER ADDICTIONS AND SUBSTANCE MISUSE

As in other addictions, such as alcohol, nicotine, heroin, cocaine, and methamphetamine (Martinez & Narendran, 2010; Volkow & Li, 2004), chronic daily cannabis use is associated with alterations in dopaminergic parameters. This suggests that striatal dopamine transmission, although with a slightly different topographic distribution than the striatal subregions typically thought to be involved in chronic drug addiction, does indeed play a major role in the neurochemical adaptations to chronic cannabis use.

The more recent data we presented here suggest that cannabis, when used heavily, has effects similar to those of other drugs of abuse on the dopamine system. Such dopaminergic changes are reported to be associated with poor functional outcomes (Martinez et al., 2011). These data have important public health implications, especially in light of the wider availability of cannabis, specifically to adolescents, who are more susceptible to experiencing negative behavioral and biological consequences.

DEFINITION OF TERMS

Positron emission tomography and single-photon emission computed tomography PET and SPECT are noninvasive imaging techniques that can be applied to measure brain neurochemistry in vivo using a radiolabeled ligand (radiotracer). PET and SPECT techniques include tracing metabolic pathways with tracers such as [^{18}F]F-DOPA (dopamine synthesis capacity) and [^{18}F]FDG (glucose metabolism), measuring target protein availability (e.g., receptors) with tracers that specifically bind to targets of interest, and indirectly measuring fluctuations in endogenous chemicals (e.g., dopamine) by measuring changes in (dopamine) receptor binding of competing, exogenous radioligands such as [^{11}C]raclopride or [^{11}C]PHNO.

BP_{ND} This is the binding potential of radiotracer relative to the nondisplaceable compartment (typically the reference tissue, such as the cerebellum), interpreted as a measure of receptor availability. Various data fitting methods are in use for deriving BP_{ND}, including 2TCM, which requires measurement of radiotracer concentration in arterial blood; SRTM; MRTM; and LGA.

ΔBP_{ND} This is the percentage change in BP_{ND} between two PET scans. In the context of PET imaging of the dopaminergic system, this is typically calculated using BP_{ND} values derived from scans before and after a challenge to derive a measure of dopamine release in response to the challenge.

Note about challenge paradigms For studies of dopamine transmission in substance dependence, various challenge paradigms have been used: the drug itself (e.g., cannabis for studies in cannabis dependence), a stress paradigm, and a psychostimulant challenge (e.g., amphetamine or methylphenidate administration).

Dopamine release capacity The response to a psychostimulant challenge, measured as ΔBP_{ND}, is considered a probe of dopamine storage and release capacity of the presynaptic dopaminergic neuron. When presynaptic stores of dopamine are released in response to a psychostimulant, synaptic concentrations of dopamine increase, correspondingly decreasing the availability of dopamine receptors for binding with the radiotracer ligand. Thus, a larger (i.e., more negative) ΔBP_{ND}, denoting more radiotracer displacement, implies a greater dopamine release capacity.

Striatum A part of the basal ganglia, this group of subcortical nuclei includes the caudate, putamen, and nucleus accumbens (also known as VST). Functional subdivisions of the striatum include the AST (comprising the caudate and the anterior putamen), the SMST (posterior putamen), and the limbic striatum (VST).

KEY FACTS ABOUT IMAGING THE DOPAMINE SYSTEM IN HUMANS USING PET

- PET imaging of the dopamine system has been particularly informative for addiction research since, for many substances of abuse, dependence has been associated with dopamine abnormalities in the striatum.
- Research PET scans typically last 1–2 h, starting when a radiotracer is injected intravenously.
- The PET tracer is a radioactively labeled molecule that either is a ligand for a receptor or transporter or partially traces the metabolic pathway of an endogenous compound such as dopamine.
- The tracer mass dose is kept very low and participants experience minimal, if any, effects.
- The scanner detects the amount of radioactive emission and generates a sequence of three-dimensional images.

- By fitting the emission data to a pharmacokinetic model, quantitative outcome measures can be derived.
- In the case of dopamine imaging, these outcome measures include:
  - BP_{ND} and V_T, which both reflect availability of dopamine receptors or transporters;
  - $K_{i(cer)}$, which is reported in [^{18}F]F-DOPA PET to reflect presynaptic dopamine synthesis and storage capacity.
- ΔBP_{ND} is an additional measure that reflects dopamine release capacity. Participants undergo PET scans before and after an intervention that transiently stimulates dopamine release, usually with a one-time dose of a psychostimulant. The percentage change in the BP_{ND} with intervention is reported as ΔBP_{ND} and is a quantitative measure of how much dopamine was released.

SUMMARY POINTS

Dopamine Alterations in Cannabis Dependence

- Molecular imaging studies show a reduction in striatal and pallidal dopamine release capacity in chronic, daily cannabis users and an overall decrease in dopamine synthesis rate in cannabis users who experience psychotic symptoms.
- Earlier literature showing no group differences in dopaminergic indices sampled lighter cannabis users or included nicotine smokers in both patients and controls, thus minimizing group differences. These studies may also have lacked power to detect differences because of longer abstinence in users and imaging limitations related to use of PET dopamine-D2/3 receptor antagonist tracers with smaller signals.
- As in other addictions, such as alcohol, nicotine, heroin, cocaine, and methamphetamine, chronic cannabis abuse or dependence is associated with alterations in dopaminergic parameters.
- This suggests that striatal dopamine transmission, although with a slightly different topographic distribution than the striatal subregions typically thought to be involved in chronic drug addiction, plays a major role in the neurochemical adaptations to chronic cannabis use.

ACKNOWLEDGMENT

This work was funded by the National Institute on Drug Abuse (R01: DA022455) and the National Institute of Mental Health (T32 MH018870). The authors thank Xiaoyan Xu and Mark Slifstein for assistance with image generation and content review.

REFERENCES

Albrecht, D. S., Skosnik, P. D., Vollmer, J. M., Brumbaugh, M. S., Perry, K. M., Mock, B. H., ... Yoder, K. K. (2013). Striatal D(2)/D(3) receptor availability is inversely correlated with cannabis consumption in chronic marijuana users. *Drug and Alcohol Dependence, 128*(1–2), 52–57.

Barkus, E., Morrison, P. D., Vuletic, D., Dickson, J. C., Ell, P. J., Pilowsky, L. S., ... Murray, R. M. (2011). Does intravenous Delta 9-tetrahydrocannabinol increase dopamine release? A SPET study. *Journal of Psychopharmacology, 25*(11), 1462–1468.

Bloomfield, M. A., Morgan, C. J., Egerton, A., Kapur, S., Curran, H. V., & Howes, O. D. (2014). Dopaminergic function in cannabis users and its relationship to cannabis-induced psychotic symptoms. *Biological Psychiatry, 75*(6), 470–478.

Bossong, M. G., van Berckel, B. N., Boellaard, R., Zuurman, L., Schuit, R. C., Windhorst, A. D., ... Kahn, R. S. (2009). Delta 9-tetrahydrocannabinol induces dopamine release in the human striatum. *Neuropsychopharmacology, 34*(3), 759–766.

Budney, A. J., Moore, B. A., Vandrey, R. G., & Hughes, J. R. (2003). The time course and significance of cannabis withdrawal. *Journal of Abnormal Psychology, 112*(3), 393–402.

Degenhardt, L., Hall, W., & Lynskey, M. (2001). Alcohol, cannabis and tobacco use among Australians: a comparison of their associations with other drug use and use disorders, affective and anxiety disorders, and psychosis. *Addiction, 96*(11), 1603–1614.

Freund, T. F., Katona, I., & Piomelli, D. (2003). Role of endogenous cannabinoids in synaptic signaling. *Physiological Reviews, 83*(3), 1017–1066.

Gardner, E. L., & Vorel, S. R. (1998). Cannabinoid transmission and reward-related events. *Neurobiology of Disease, 5*(6 Pt B), 502–533.

Ghazzaoui, R., & Abi-Dargham, A. (2014). Imaging dopamine transmission parameters in cannabis dependence. *Progress in Neuro-Psychopharmacology & Biological Psychiatry, 52*, 28–32.

Henquet, C., Di Forti, M., Morrison, P., Kuepper, R., & Murray, R. M. (2008). Gene-environment interplay between cannabis and psychosis. *Schizophrenia Bulletin, 34*(6), 1111–1121.

Herkenham, M., Lynn, A. B., Little, M. D., Johnson, M. R., Melvin, L. S., de Costa, B. R., & Rice, K. C. (1990). Cannabinoid receptor localization in brain. *Proceedings of the National Academy of Sciences of the United States of America, 87*(5), 1932–1936.

Hirvonen, J., Goodwin, R. S., Li, C. T., Terry, G. E., Zoghbi, S. S., Morse, C., ... Innis, R. B. (2012). Reversible and regionally selective downregulation of brain cannabinoid CB1 receptors in chronic daily cannabis smokers. *Molecular Psychiatry, 17*(6), 642–649.

Kadden, R. M., Litt, M. D., Kabela-Cormier, E., & Petry, N. M. (2007). Abstinence rates following behavioral treatments for marijuana dependence. *Addictive Behaviors, 32*(6), 1220–1236.

Laruelle, M. (2000). Imaging synaptic neurotransmission with in vivo binding competition techniques: a critical review. *Journal of Cerebral Blood Flow and Metabolism, 20*(3), 423–451.

Leroy, C., Karila, L., Martinot, J. L., Lukasiewicz, M., Duchesnay, E., Comtat, C., ... Trichard, C. (2012). Striatal and extrastriatal dopamine transporter in cannabis and tobacco addiction: a high-resolution PET study. *Addiction Biology, 17*(6), 981–990.

Maldonado, R. (2002). Study of cannabinoid dependence in animals. *Pharmacology & Therapeutics, 95*(2), 153–164.

Maldonado, R., & Rodriguez de Fonseca, F. (2002). Cannabinoid addiction: behavioral models and neural correlates. *Journal of Neuroscience, 22*(9), 3326–3331.

Martin, A. B., Fernandez-Espejo, E., Ferrer, B., Gorriti, M. A., Bilbao, A., Navarro, M., ... Moratalla, R. (2008). Expression and function of CB1 receptor in the rat striatum: localization and effects on D1 and D2 dopamine receptor-mediated motor behaviors. *Neuropsychopharmacology, 33*(7), 1667–1679.

Martinez, D., Carpenter, K. M., Liu, F., Slifstein, M., Broft, A., Friedman, A. C., ... Nunes, E. (2011). Imaging dopamine transmission in cocaine dependence: link between neurochemistry and response to treatment. *The American Journal of Psychiatry, 168*(6), 634–641.

Martinez, D., Gil, R., Slifstein, M., Hwang, D. R., Huang, Y., Perez, A., ... Abi-Dargham, A. (2005). Alcohol dependence is associated with blunted dopamine transmission in the ventral striatum. *Biological Psychiatry, 58*(10), 779–786.

Martinez, D., & Narendran, R. (2010). Imaging neurotransmitter release by drugs of abuse. *Current Topics in Behavioral Neurosciences, 3*, 219–245.

Martinez, D., Narendran, R., Foltin, R. W., Slifstein, M., Hwang, D. R., Broft, A., ... Laruelle, M. (2007). Amphetamine-induced dopamine release: markedly blunted in cocaine dependence and predictive of the choice to self-administer cocaine. *The American Journal of Psychiatry, 164*(4), 622–629.

Mizrahi, R., Kenk, M., Suridjan, I., Boileau, I., George, T. P., McKenzie, K., ... Rusjan, P. (2014). Stress-induced dopamine response in subjects at clinical high risk for schizophrenia with and without concurrent cannabis use. *Neuropsychopharmacology, 39*(6), 1479–1489.

Mizrahi, R., Suridjan, I., Kenk, M., George, T. P., Wilson, A., Houle, S., & Rusjan, P. (2013). Dopamine response to psychosocial stress in chronic cannabis users: a PET study with [(11)C]-(+)-PHNO. *Neuropsychopharmacology, 38*(4), 673–682.

Moore, T. H., Zammit, S., Lingford-Hughes, A., Barnes, T. R., Jones, P. B., Burke, M., & Leis, G. (2007). Cannabis use and risk of psychotic or affective mental health outcomes: a systematic review. *Lancet, 370*(9584), 319–328.

Pistis, M., Perra, S., Pillolla, G., Melis, M., Muntoni, A. L., & Gessa, G. L. (2004). Adolescent exposure to cannabinoids induces long-lasting changes in the response to drugs of abuse of rat midbrain dopamine neurons. *Biological Psychiatry, 56*(2), 86–94.

Pope, H. G., Jr., Gruber, A. J., Hudson, J. I., Huestis, M. A., & Yurgelun-Todd, D. (2002). Cognitive measures in long-term cannabis users. *Journal of Clinical Pharmacology, 42*(Suppl. 11), 41S–47S.

SAMHSA. (2014). *Results from the 2013 National Survey on Drug Use and Health: Summary of national findings. (HHS publication no. (SMA) 14–4863). Center for behavioral health statistics and quality.* Rockville, MD: Substance Abuse and Mental Health Services Administration.

Schweinsburg, A. D., Brown, S. A., & Tapert, S. F. (2008). The influence of marijuana use on neurocognitive functioning in adolescents. *Current Drug Abuse Reviews, 1*(1), 99–111.

Sevy, S., Smith, G. S., Ma, Y., Dhawan, V., Chaly, T., Kingsley, P. B., ... Eidelberg, D. (2008). Cerebral glucose metabolism and D2/D3 receptor availability in young adults with cannabis dependence measured with positron emission tomography. *Psychopharmacology (Berl), 197*(4), 549–556.

Sneider, J. T., Pope, H. G., Jr., Silveri, M. M., Simpson, N. S., Gruber, S. A., & Yurgelun-Todd, D. A. (2008). Differences in regional blood volume during a 28-day period of abstinence in chronic cannabis smokers. *European Neuropsychopharmacology: The Official Journal of the European College of Neuropsychopharmacology, 18*(8), 612–619.

Stokes, P. R., Egerton, A., Watson, B., Reid, A., Lappin, J., Howes, O. D., ... Lingford-Hughes, A. R. (2012). History of cannabis use is not associated with alterations in striatal dopamine D2/D3 receptor availability. *Journal of Psychopharmacology, 26*(1), 144–149 http://dx.doi.org/10.1177/0269881111414090 PMID: 21890594. (Epub 2011 Sep 2).

Stokes, P. R., Mehta, M. A., Curran, H. V., Breen, G., & Grasby, P. M. (2009). Can recreational doses of THC produce significant dopamine release in the human striatum? *NeuroImage, 48*(1), 186–190.

Troisi, A., Pasini, A., Saracco, M., & Spalletta, G. (1998). Psychiatric symptoms in male cannabis users not using other illicit drugs. *Addiction, 93*(4), 487–492.

Urban, N. B., Slifstein, M., Thompson, J. L., Xu, X., Girgis, R. R., Raheja, S., ... Abi-Dargham, A. (2012). Dopamine release in chronic cannabis users: a [11c]raclopride positron emission tomography study. *Biological Psychiatry, 71*(8), 677–683.

Van Waes, V., Beverley, J. A., Siman, H., Tseng, K. Y., & Steiner, H. (2012). CB1 cannabinoid receptor expression in the striatum: association with corticostriatal circuits and developmental regulation. *Frontiers in Pharmacology, 3*, 21.

Volkow, N. D., Fowler, J. S., Wang, G. J., & Swanson, J. M. (2004). Dopamine in drug abuse and addiction: results from imaging studies and treatment implications. *Molecular Psychiatry, 9*(6), 557–569.

Volkow, N. D., & Li, T. K. (2004). Drug addiction: the neurobiology of behaviour gone awry. *Nature Reviews. Neuroscience, 5*(12), 963–970.

Volkow, N. D., Wang, G. J., Fowler, J. S., Logan, J., Gatley, S. J., Hitzemann, R., ... Pappas, N. (1997). Decreased striatal dopaminergic responsiveness in detoxified cocaine-dependent subjects. *Nature, 386*(6627), 830–833.

Volkow, N. D., Wang, G. J., Telang, F., Fowler, J. S., Alexoff, D., Logan, J., ... Tomasi, D. (2014). Decreased dopamine brain reactivity in marijuana abusers is associated with negative emotionality and addiction severity. *Proceedings of the National Academy of Sciences of the United States of America, 111*(30), E3149–E3156.

Voruganti, L. N., Slomka, P., Zabel, P., Mattar, A., & Awad, A. G. (2001). Cannabis induced dopamine release: an in-vivo SPECT study. *Psychiatry Research, 107*(3), 173–177.

Wang, G. J., Smith, L., Volkow, N. D., Telang, F., Logan, J., Tomasi, D., ... Fowler, J. S. (2011). Decreased dopamine activity predicts relapse in methamphetamine abusers. *Molecular Psychiatry, 17*(9), 918–925.

Weinstein, J., van de Giessen, E., Cassidy, C., Haney, M., Slifstein, M., & Abi-Dargham, A. (2015). Blunted striatal dopamine release in cannabis dependence. *Journal of Nuclear Medicine, 56*, 32.

Yucel, M., Solowij, N., Respondek, C., Whittle, S., Fornito, A., Pantelis, C., & Lubman, D. I. (2008). Regional brain abnormalities associated with long-term heavy cannabis use. *Archives of General Psychiatry, 65*(6), 694–701.

Chapter 74

Cannabis and the Mesolimbic System

Carla Cannizzaro[1], Marco Diana[2]

[1]Department of Sciences for Health Promotion and Maternal Care, School of Medicine, University of Palermo, Palermo, Italy; [2]Department of Drug Sciences, School of Pharmacy, University of Sassari, Sassari, Italy

Abbreviations

CB1 Cannabinoid 1 subtype
CBD Cannabidiol
CPP Conditioned place preference
DA Dopamine
DSE Depolarization-induced suppression of excitation
DSI Depolarization-induced suppression of inhibition
ICSS Intracranial self-stimulation
LTD Long-term depression
LTP Long-term potentiation
mPFC Medial prefrontal cortex
MSN Medium spiny neurons
Nacc Nucleus accumbens
RMTg Rostromedial tegmental nucleus
VS Ventral striatum
VTA Ventral tegmental area
Δ^9-**THC** Δ^9-Tetrahydrocannabinol

THE DOPAMINERGIC MESOLIMBIC SYSTEM

Various classes of chemicals, including nitrogenous compounds, amino acids, hydrocarbons, sugar, terpenes, and simple fatty acids, together contribute to the unique pharmacological and toxicological properties of cannabis (Svizenska, Dubovy, & Sulcova, 2008). Among them Δ^9-tetrahydrocannabinol (Δ^9-THC) and cannabidiol, the two main ingredients of the *Cannabis sativa* plant, have distinct symptomatic and behavioral effects (Iversen, 2003). Δ^9-THC is the primary psychoactive constituent and is believed to be primarily responsible for the resulting cognitive effects, psychotic symptoms, and anxiety, as well as the addictive potential, of smoked cannabis (Leweke & Koethe, 2008). Although many factors contribute to addiction, cannabis and other drugs of abuse act upon the mesolimbic reward circuitry in the mammalian brain, thus promoting repetition of drug taking, a necessary precondition for the neuronal adaptations that underlie addiction. Indeed dopamine (DA) in the mesocorticolimbic system has long been considered essential for translating motivations into goal-directed actions. The reinforcing effects of major classes of abused drugs are produced through actions in the mesolimbic system, which originates from A10 DAergic neurons in the ventral tegmental area (VTA) and innervates the limbic component of the basal ganglia, to influence motivational, emotional, contextual, and affective components of behavior (Ikemoto, Qin, & Liu, 2005). Drug self-administration and electrical self-stimulation studies have shown that DAergic projections from the VTA are crucial in reward but DA is not a sole mediator. Major glutamatergic projections are provided by the amygdala, hippocampus, and medial prefrontal cortex (mPFC) to the nucleus accumbens (Nacc), which consists of anatomically heterogeneous elements classically represented by the core and shell. The Nacc has two main outputs, which are γ-aminobutyric acid (GABA)-ergic projections to the ventral pallidum (so-called indirect pathway) and the VTA/substantia nigra pars reticulata (direct pathway). Both the ventral pallidum and the VTA send GABAergic efferents to the medial dorsal thalamus. Glutamatergic projections from the medial dorsal thalamus to the mPFC close this circuitry (Figure 1). Studies suggested that within the Nacc shell, the medial portion is more responsive to psychomotor stimulants than its lateral counterpart; moreover, the medial shell, but not the core, supports self-administration of Δ^9-THC. Anatomically correlated, the medial olfactory tubercle and medial Nacc shell form the medial portion of the ventral striatum (VS), which appears to be more responsive to the rewarding effects of DAergic drugs than its lateral counterparts. Indeed DA neurons projecting to the medial VS are localized posteromedially in the VTA in relation to those projecting to the lateral VS, indicating that the medial part of the VTA–VS DA system is particularly important for reward and arousal and that the lateral portion is more closely involved in specific conditioned responses (Ikemoto et al., 2005). In 2009, a new nucleus was identified and named the rostromedial tegmental nucleus (RMTg) (Jhou, Fields, Baxter, Saper, & Holland, 2009), located close to the VTA so that the posterior end of the VTA overlaps with the anterior part of the RMTg. It contains predominantly GABAergic neurons, whose major projections target DAergic neurons in the VTA and substantia nigra. Together with the lateral habenula, the RMTg may represent a site of tonic inhibition over VTA DAergic neurons and may be a key site through which deprivation-type signals are conveyed over access-type messages in the mesolimbic system. One of the hallmarks of abused drugs is the ability to increase DA signaling from VTA neurons (Di Chiara & Imperato, 1988). These well-characterized cells fire in two distinct modes: low-frequency (1–5 Hz) tonic activity, which produces a steady DAergic "tone" on high-affinity inhibitory D2-like DA receptors of the mesolimbic system, and high-frequency (>20 Hz within bursts) phasic activity, which results in increased synaptic DA content

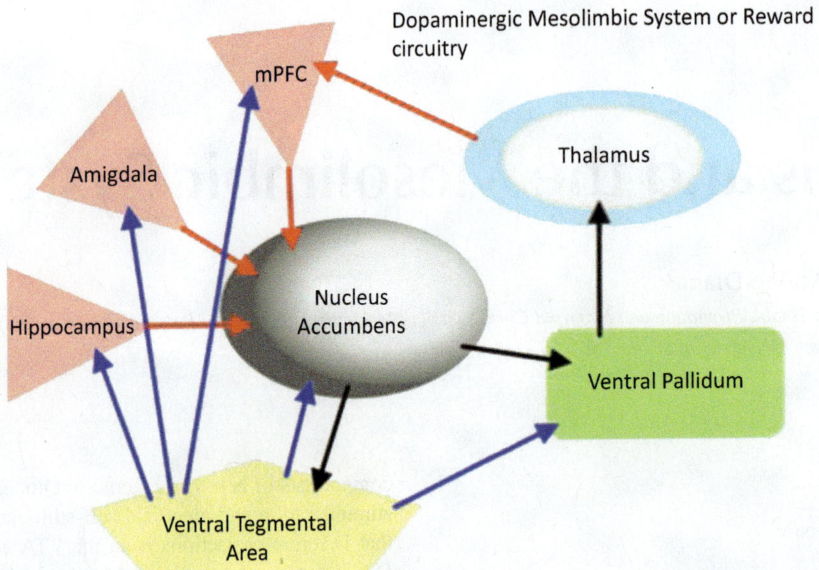

FIGURE 1 Schematic illustration of the mesolimbic system. Red arrows indicate glutamatergic pathways; black arrows indicate GABAergic pathways; blue arrows indicate DAergic pathways. See text for additional details. *Modified from Piercea and Kumaresana (2006).*

sufficient to occupy low-affinity D1-like receptors. The baseline DAergic tone is believed to facilitate long-term depression (LTD) at corticostriatal synapses and suppress activity of the indirect pathway of the basal ganglia (Shen, Flajolet, Greengard, & Surmeier, 2008); on the other hand D1 receptor activation following reward-related stimuli is coupled to an enhancement in long-term potentiation (LTP) of the excitatory transmission and to the activation of the striatal direct pathway, thereby motivating behavior (Grace, Floresco, Goto, & Lodge, 2007). Stimuli promoting burst activity of DAergic neurons in the VTA also produce transient increases in extracellular DA concentration at the terminal fields of the mesolimbic system, such as the Nacc (Diana, 1998). A shift in midbrain DAergic neuron activation from tonic low-frequency to phasic high-frequency firing results from changes in synaptic input from glutamate and GABA afferents to VTA DAergic cells. Indeed excitatory glutamatergic afferents project from sensory and cognitive regions, including the prefrontal cortex and the extended amygdala, whereas GABAergic input comes from the basal ganglia and the RMTg. Thus, to activate the "conditional" burst firing of DAergic neurons, glutamatergic input must activate *N*-methyl-D-aspartate (NMDA) glutamate receptors localized at synapses impinging on midbrain DA neurons (Overton & Clark, 1997). Conversely GABAergic input to the same cells dampens burst firing and returns the neurons to baseline pacemaker-like activity. Notably, the maintenance of midbrain DAergic firing patterns requires a balance between excitatory and inhibitory VTA afferent signals, and a key role in the "governance" of these modulatory networks is strategically played by the endocannabinoid system.

CANNABINOIDS IN THE REWARD CIRCUITRY

Although the euphorigenic properties of cannabis preparations have been appreciated by humans for centuries, only lately have we acquired the experimental tools to evaluate cannabinoid reward and abuse liability in experimental animals. Historically, intracranial self-stimulation (ICSS) has been utilized in rodents to study how pharmacological or molecular manipulations affect brain reward function: ICSS is an operant behavioral paradigm in which animals would work to obtain intracranial stimulation through electrodes implanted into discrete brain areas (such as the mesolimbic brain reward areas). The conditioned place preference (CPP) paradigm is based on the assumption that animals learn to approach stimuli paired with rewards and to avoid stimuli paired with aversive agents. In this procedure, the animal develops an association between the subjective state produced by the drug (e.g., a heightened feeling of euphoria comparable to pleasure in humans) and the environmental cues present during the drug state. Intravenous drug self-administration has been one of the most direct approaches to studying the rewarding properties of drugs of abuse in experimental animals, such as rodents or primates. In this behavioral paradigm, based on operant conditioning, animals learn to make an operant response, such as pressing a lever in an operant chamber or inserting their nose into a hole, to self-administer a reinforcer (e.g., a drug of abuse) after the completion of the reinforcement schedule requirement. The rewarding properties of Δ^9-THC are clearly shown by a decrease in brain-stimulation reward thresholds and an increase in self-administration behavior and CPP in a dose-dependent manner. It is now clear that cannabinoids exert emotional and motivational effects and produce drug reinforcement/drug-seeking behavior. To characterize the actions of cannabinoids on reward circuitry, the ability of Δ^9-THC to modulate DA transmission in the mesolimbic system has been largely investigated. Cannabinoid systemic administration enhances extracellular DA concentration in the Nacc shell, by increasing both the baseline firing rates and the bursting frequency of midbrain DAergic neurons. Smoked marijuana produces subjective feelings of well-being and euphoria in humans that are inhibited by the concomitant administration of rimonabant, a cannabinoid 1 subtype (CB1) receptor antagonist/inverse agonist; similarly DA enhancement in the Nacc, together with the related phenotypes, is

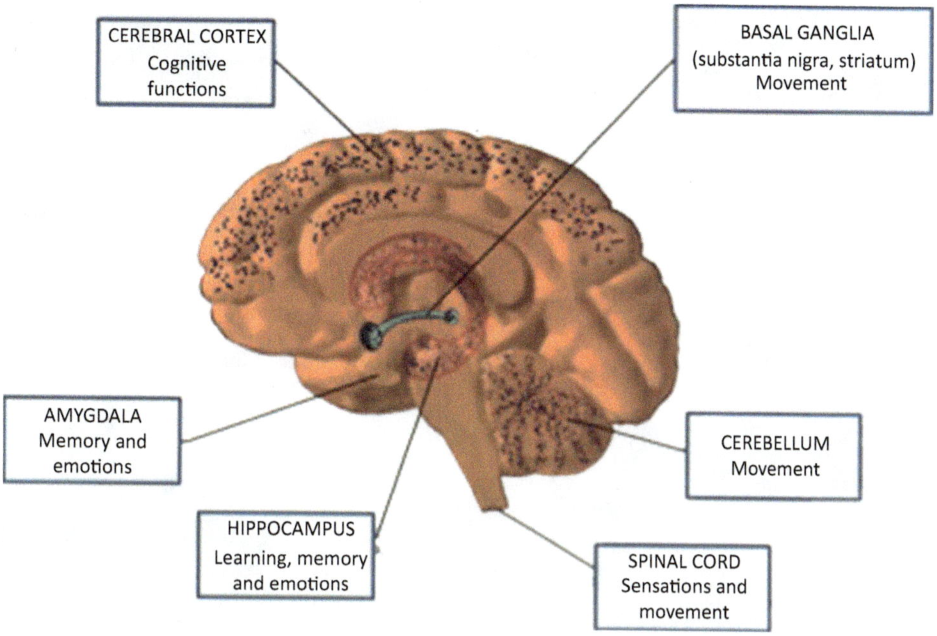

FIGURE 2 CB1 receptor distribution in the brain: correlation with functions that are affected by cannabinoid activity. *From Serpelloni, Diana, Gomma, and Rimondo (2011).*

abolished by blocking the same receptors, suggesting that many, if not all, psychological and neurochemical effects of cannabis in the mesolimbic system depend on the activation of CB1 receptors (Figure 2). Interestingly, while VTA DAergic cells release endocannabinoids during phasic activation, they do not express CB1 receptors. This suggests that cannabinoids can produce an enhancement in DA release in the mesolimbic system from VTA neurons via an indirect mechanism: indeed the midbrain contains GABAergic neurons that modulate DAergic cell activity, through $GABA_A$ receptors on DA somata (Riegel & Lupica, 2004). The prior application of the $GABA_A$ receptor antagonist bicuculline to VTA-containing slices blocked the excitatory effect exerted by cannabinoids on DA release. In other words, cannabinoids can enhance DA transmission through the inhibition of the inhibitory GABAergic tone, thus disinhibiting the VTA DAergic neurons to discharge. In line with this evidence Lupica and Riegel (2005) proposed a model of reciprocal interaction between DA and endocannabinoids in the mesolimbic system. Indeed, endocannabinoids released as a consequence of enhanced activation of VTA neurons upon depolarization act as "retrograde messengers," activate presynaptic CB1 receptors, and inhibit GABA release from the terminals (Figure 3). The majority of the evidence suggests that 2-arachidonoylglycerol mediates most forms of CB1-mediated retrograde regulation of synaptic transmission, i.e., depolarization-induced suppression of inhibition or excitation. Much more interactions occur, however, between DA and cannabinoids in the mesolimbic system, and it is noteworthy that the direct infusion of Δ^9-THC into the VTA does not increase DA accumulation in the Nacc. CB1 receptors are also located on GABAergic terminals originating from Nacc medium spiny output neurons that target $GABA_B$ receptors on DA neurons in the VTA, suggesting a second possible mechanism of disinhibition (Heimer, Zahm, Churchill, Kalivas, & Wohltmann, 1991). Thus, cannabinoids acting at CB1 receptors can control the release of GABA in the VTA that is derived from both intrinsic and extrinsic sources, indicating that the inputs from the Nacc to the VTA may represent a critical pathway for the expression of cannabinoid-induced reward in the mesolimbic system. Synthetic cannabinoid agonists and endocannabinoids, acting as retrograde messengers, can also inhibit glutamate release onto neurons in the mesolimbic circuitry and in the VTA in particular (Melis et al., 2004). This puzzle is made more complicated by the finding that glutamate release onto Nacc medium spiny neurons (MSNs) is also inhibited by cannabinoid agonists, apparently as a result of the activation of CB1 receptors coupled to voltage-dependent Kþ channels in the glutamatergic nerve terminals. The reduction of glutamatergic inputs that arise from neurons in the prefrontal cortex to the Nacc by CB1 receptor activation would reduce the excitation of GABAergic Nacc MSNs projecting to DA neurons in the VTA, and thereby would decrease the inhibition of DA neurons. Again, the relative contribution of the CB1 receptor modulation of GABAergic and glutamatergic synaptic transmission to the output of the Nacc might rely upon the functional setting under which each system transmits. The activation of CB1 receptors by endocannabinoids can modify the strength of LTP and can also initiate long-term forms of synaptic plasticity, including LTD (Kortleven, Fasano, Thibault, Lacaille, & Trudeau, 2011), in the mesolimbic system. Indeed, CB1-mediated long-term reduction of transmitter release at the same synapse defines endocannabinoid LTD, which emerged for the first time in 2002 at excitatory synapses in dorsal striatum. Since then, endocannabinoid LTD has been reported in several other brain structures such as the Nacc, amygdala and hippocampus, somatosensory cortex, prefrontal cortex, and VTA. Endocannabinoids and functional CB1 receptors are required to observe the LTD of glutamatergic cortical synaptic inputs to striatal MSNs; contextually, presynaptic activity is required in terms of presynaptic NMDA receptor recruitment. Endocannabinoid LTD is a widely expressed phenomenon in the brain that can be observed also

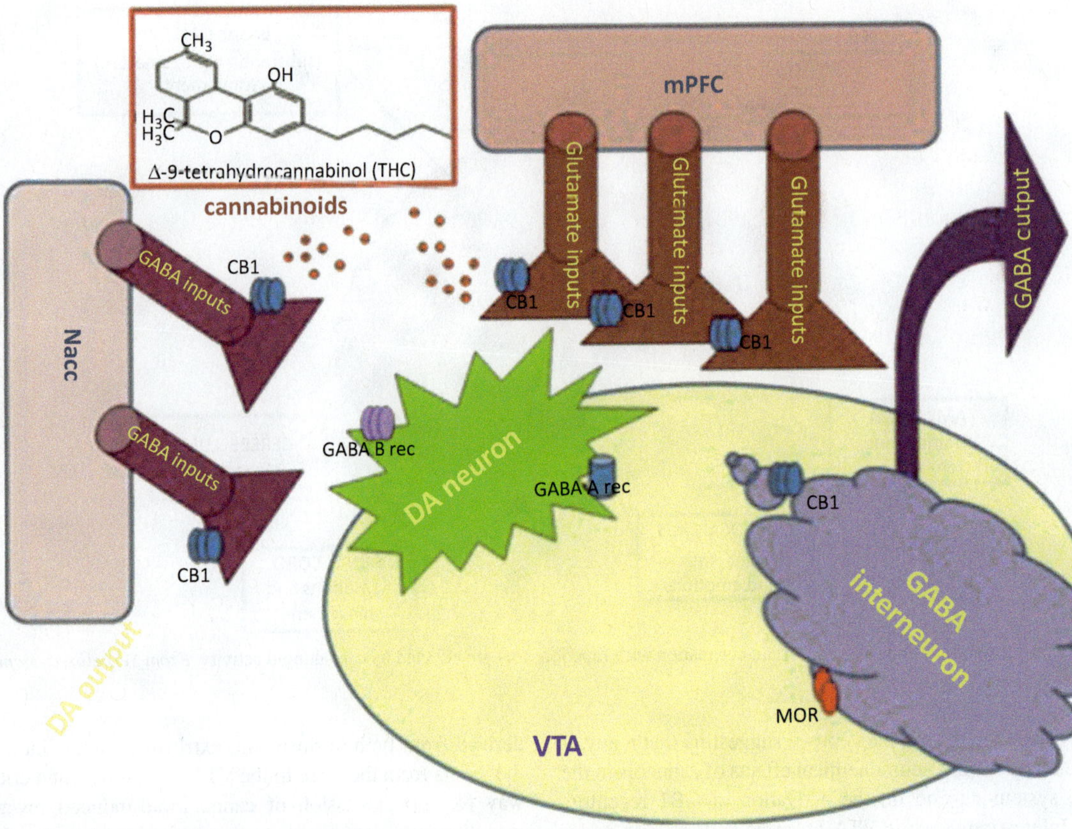

FIGURE 3 Cellular elements mediating cannabinoid actions in the mesolimbic system, through activation of presynaptic CB1 receptors. Note the extrinsic and intrinsic GABAergic inputs to the VTA. MOR, μ opioid receptor; DA, dopamine; Nacc, nucleus accumbens. *Modified from Lupica, Riegel, and Hoffman (2004).*

in inhibitory synapses. Although the precise role that LTD plays in regulating the function of the dorsal and ventral striatum is not known as of this writing, it has been speculated that the long-term changes in synaptic efficacy mediated by this system may be involved in communicating enduring information as to reward salience and in the establishment of motor habits associated with compulsive drug use, too (Gerdeman & Lovinger, 2003). CB1 receptors are located on both GABAergic and glutamatergic axon terminals in the Nacc, where they inhibit the release of each of these neurotransmitters. However, following in vivo treatment with Δ^9-THC for 1 week, the inhibition of glutamatergic and GABAergic synaptic transmission by WIN 55,212-2, a potent cannabinoid agonist, is greatly diminished, confirming CB1 receptor tolerance (Hoffman, Riegel, & Lupica, 2003). Synaptic remodeling of mesolimbic circuitries by long-term plasticity may represent a mechanism through which drug use may progress from casual to impulsive. Notably, short- or long-term exposure to Δ^9-THC can limit the degree to which Nacc glutamate synapses undergo LTD, and this appears to result from a downregulation of CB1 receptor function and the ability of cannabinoids to initiate this form of synaptic plasticity (Hoffman et al., 2003). Such a phenomenon amounts to metaplasticity, which refers to enduring changes in the ability of neurons to generate synaptic plasticity (Abraham, 2008). Considerable evidence using behavioral genetic approaches has demonstrated several important functional interrelationships between cannabinoids and opiate-related reward processing, both in terms of opiate-related reward behaviors and in terms of CB1 receptor-mediated modulation of neural regions critical for opiate reward processing, such as the mesocorticolimbic system. The mammalian VTA is a critical neural substrate for the processing of the primary rewarding properties of opiates, through DA-dependent and non-DA neuronal substrates (Laviolette, Gallegos, Henriksen, & van der Kooy, 2004; Nader & Van der Kooy, 1997), and a large body of evidence has demonstrated that cannabinoid compounds can acutely activate both the opiate and the DAergic signaling systems. For example, the acute administration of Δ^9-THC has been reported to increase levels of β-endorphins and enkephalinergic peptides directly in the mesolimbic DA pathway, including the VTA and Nacc. As a consequence, cannabinoids would alter motivational processes mediated by the Nacc and also play a role in modulating the reinforcing properties of other abused drugs, ultimately making long-term changes to neural circuitries and behavior.

CANNABIS DEPENDENCE

The effects sought by cannabis users seem to be produced primarily by Δ^9-THC in a dose-related manner so that a higher THC content leads to an increase in the subjective effects of this drug, probably enhancing the risk of dependence and psychotic symptoms of cannabis users (Hall & Degenhardt, 2009). Indeed compelling evidence suggests that exposure to marijuana during critical, adolescent windows of neurodevelopment is positively correlated with an increased propensity to develop schizophrenia-related psychoses in early adulthood. Functional interactions between cannabinoid signaling and mesolimbic DAergic transmission stand as a nexus point to understanding how dysregulation of the endocannabinoid system may relate to the well-established aberrations in DAergic transmission, as a core neuropathological feature of both addiction and schizophrenia (Tan, Ahmad, Loureiro, Zunder, & Laviolette, 2014). In epidemiological studies about 9% of those who ever use this drug become daily users, rising to 16% when the consumption is initiated during adolescence; heavy or "regular" cannabis use is usually defined as daily or near-daily use of up to three to five joints of potent cannabis a day. This pattern of consumption, when continued over years and decades, predicts increased risk of many of the adverse health effects attributed to cannabis. Both tolerance and dependence can develop with chronic use, as well as withdrawal signs and symptoms, as reported below (Diana, Melis, Muntoni, & Gessa, 1998; Fattore, Fadda, Spano, Pistis, & Fratta, 2008). In 1993 the first definition of cannabis dependence in the *Diagnostic and Statistical Manual of Mental Disorders* (DSM), third edition, included impaired control over cannabis use and difficulty ceasing use despite harms caused by it. In many countries such as Australia, Canada, and the United States, cannabis dependence is the most commonly treated type of drug dependence after alcohol and tobacco. This is not to be attributed to an increase in the number of people smoking marijuana, but to the greater potency of the marijuana that individuals are exposed to. Indeed, the average potency of Δ^9-THC in seized marijuana has increased from 3% in 1992 to 13.2% in 2012, and some samples of marijuana analyzed displayed a Δ^9-THC content of more than 30% (UNODC, 2014). Despite the common belief that cannabis preparations do not produce withdrawal, clinical reports on chronic consumers of even low daily doses of cannabis derivatives described the onset of physical dependence, defined by a withdrawal response that occurs upon cessation of drug administration. Abstinence from marijuana smoke or oral Δ^9-THC produces indeed symptoms such as irritability, anger, anxiety, decreased appetite, weight loss, restlessness, craving, and disrupted sleep. These symptoms usually occur 24h after last use, peak in 2–3 days, and last about 2–3 weeks (Budney & Hughes, 2006). Approximately 60–90% of individuals experiencing withdrawal during abstinence use marijuana to alleviate the symptoms, indicating that Δ^9-THC plays an essential role in the development of dependence and in the expression of withdrawal. Dependence and withdrawal can be induced and precipitated, respectively, in the animal model (Diana et al., 1998). Behavioral signs observed during precipitated withdrawal include whirling, wet-dog shakes, sniffing, front-paw tremor, genital licking, erection, ataxia, diarrhea, mastication, and piloerection (Diana et al., 1998; Wilson, Varvel, Harloe, Martin, & Lichtman, 2006). Because withdrawal is precipitated by a CB1 antagonist and alleviated by Δ^9-THC, it is clear that cannabinoid dependence is largely mediated by CB1 receptors in rodents. Disturbances in cannabinoid transmission within the mesocorticolimbic pathway may underlie DAergic dysregulation linked to a variety of neuropsychiatric disorders such as addiction and schizophrenia. Within the prefrontal cortex, CB1 receptor transmission appears to control emotional salience, in the context of both aversive, negative events and rewarding, appetitive, motivational stimuli. The underlying mechanisms are still to be elucidated. Moreover, cannabis use is associated with impairments in cognitive functions, including learning and memory, attention, and decision-making. Clinical findings have demonstrated significantly diminished DA synthesis capacity in regular marijuana users compared with nonusers and evidence of attenuated DA receptor expression levels in current or recently abstinent marijuana users (Bloomfield et al., 2014; Urban et al., 2012). In humans some studies have shown volume reductions in the hippocampus, amygdala, and cerebellum; greater gray matter density in marijuana users than in control participants in the left Nacc extending to the subcallosal cortex, hypothalamus, sublenticular extended amygdala, and left amygdala; volume increase in the left Nacc; and significant shape differences in the left Nacc and right amygdala. These observations suggest that marijuana exposure, even in young recreational users, is associated with exposure-dependent alterations of the neural matrix of core reward structures that are consistent with animal evidence of changes in dendritic arborization in the mesolimbic system (Gilman et al., 2014). The appearance of overt abstinence signs upon cessation of drug administration is paralleled by structural and functional neuroimaging studies on cannabis use (Budney & Hughes, 2006; Iversen, 2003). Moreover, work has shown that withdrawal from a regimen of chronic cannabinoid administration profoundly affects the mesolimbic system, causing a shrinkage of DA neurons in the VTA and a reduction in the spine density of dendrites of MSNs of the Nacc shell (Spiga, Lintas, Migliore, & Diana, 2010; Spiga et al., 2014; Figure 4); this represents a morphological correlate of the functional deficits detected by electrophysiological and neurochemical means that ultimately may contribute to the negative emotional state that characterizes withdrawal (Koob & Le Moal, 2001). In this regard, the spines' loss, in addition to contributing to the reduction in the already abated DA transmission, could explain the downregulation of CB1 receptors after Δ^9-THC withdrawal as well as the CB1-mediated inhibition of excitatory synaptic transmission between the prefrontal cortex and the Nacc. These findings suggest that cannabis affecting dendritic spines provides another example of drug-induced aberrant neural plasticity with marked reflections on the physiology of synapses, system structural organization, and neuronal circuitry remodeling that promote the transition from chronic drug intake to "addiction."

APPLICATIONS TO OTHER ADDICTIONS AND SUBSTANCE MISUSE

In this chapter we have approached the topic describing cannabis activity in the mesolimbic system. Cannabis is able, through its principal active ingredient Δ^9-THC, to increase DA release in the mesolimbic system, paralleling the activity of most abused drugs. Nevertheless, beyond this fact, exogenous and endogenous cannabinoids can alter synaptic process in both the Nacc and the VTA, suggesting that the regulation of inner and outer neural circuits to the mesolimbic system is relevant if not essential for the expression of value-attribution processing and in the modulation of reward-seeking behavior for different drugs of abuse. Chronic use of cannabis can therefore exert a continued regulation of these processes, thus contributing to addiction to several classes of drugs, such as alcohol. In particular, CB1 receptor manipulation is reported to affect

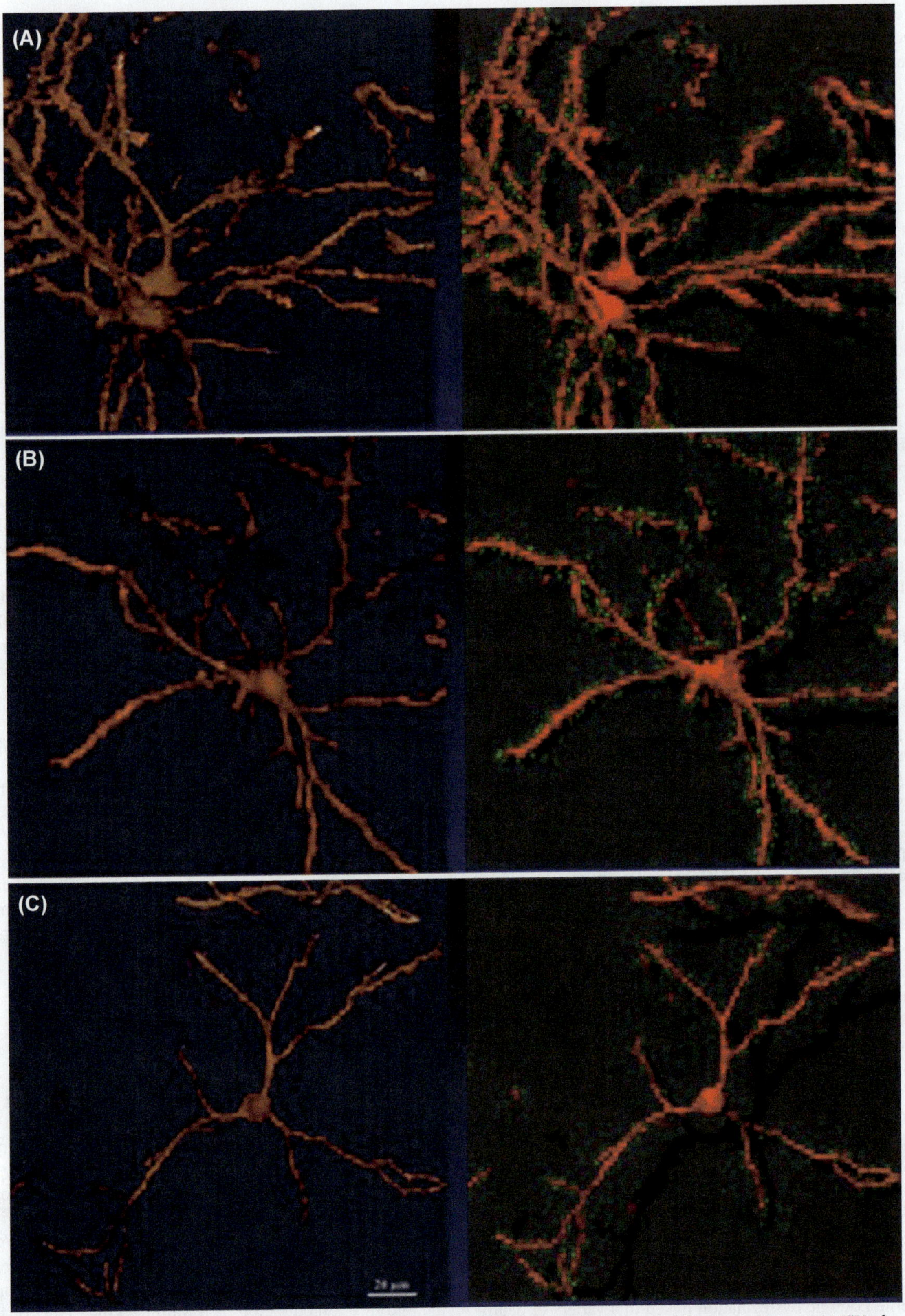

FIGURE 4 Representative confocal reconstructions synthesizing major findings on Golgi–Cox-stained MSNs in the Nacc. (A) MSNs from the core control group, (B) MSN from the shell control group, and (C) MSN from the shell in the Δ^9-THC-treated group. *From Spiga et al. (2010).*

ethanol-related behavior, and in fact CB1 antagonism decreases both voluntary ethanol intake and relapse to ethanol in several experimental models. Consistently, publications have shown that administration of a pure antagonist at the CB1 receptor, such as AM281, is able to reduce the motivational properties of the neuroactive metabolite of alcohol, acetaldehyde, in an operant-conflict self-administration paradigm. In particular, the administration of AM281 decreases drug seeking when the drug is not available, drug taking after a period of forced abstinence, and drug-related compulsive behavior, in an experimental model of drug use despite negative consequences (Plescia et al., 2014). The manipulation of the endocannabinoid system might represent a useful therapeutic strategy to treat addiction-related behaviors.

DEFINITION OF TERMS

Cannabis *C. sativa* L. (hemp) is an annual flowering plant that gives origin to marijuana, from dried leaves and female flowering tops, and hashish, from the resin, which originates on these female flowering tops.

Δ^9-Tetrahydrocannabinol Δ^9-THC is the main psychoactive component of *Cannabis* that was first isolated in 1964. It binds to G-protein-coupled cannabinoid receptors in the nervous system and in the periphery.

CB1 receptors These receptors are the most abundant G-protein-coupled receptors in the brain: they are highly expressed in the basal ganglia nuclei in the hippocampus, cerebellum, and neocortex. They modulate neurotransmitter release.

Retrograde messenger It is a signaling molecule that is produced by the postsynaptic terminal, travels "backward" across the synaptic cleft, and binds to the presynaptic membrane where it exerts its main physiological function.

Depolarization-induced suppression of inhibition or excitation These are short-term forms of synaptic plasticity induced by CB1 activation, which result in prevention of transmitter release through the inhibition of voltage-gated Ca^{2+} channels and/or increase in K^+ conductance.

Reward system The reward circuitry of the brain consists of subcortical nuclei that are synaptically interconnected. It classically consists of the VTA and the Nacc, which play key roles in the processing of rewarding environmental stimuli and in the setup of addictive behaviors.

Reinforcers Reinforcers refers to those stimuli—substances or acts—that are able to activate the DAergic mesolimbic system and produce a sense of euphoria and well-being, thus increasing the probability of repeating the behaviors paired with them.

Dependence This is a form of receptor and postreceptor plasticity, which establishes as a physiological response to repeated doses of many medications, including opioids, antidepressants, and β-blockers, and which is followed by a withdrawal syndrome upon abrupt cessation of administration.

Cannabis addiction It occurs with heavy chronic use in individuals who report difficulties in controlling their use and who continue cannabis consumption despite the personal experience of adverse consequences.

Cannabis use disorder This is the term that the DSM-5 work group has recommended to subsume cannabis abuse/dependence, indicating the main clinical criteria for the diagnosis.

KEY FACTS OF THE "ADDICTIVE" SPINES IN THE NUCLEUS ACCUMBENS

- Dendritic spines are the main postsynaptic compartments of excitatory synapses in the brain with peculiar and distinctive morphological features consisting of a bulbous head and a thinner neck that connects the spine to the dendritic shaft.
- At the ultrastructural level, the spine head is characterized by an electron-dense matrix of receptors and supporting proteins collectively known as the postsynaptic density, which dynamically changes its structure and composition during development and in response to synaptic activity.
- A significant subpopulation of spines on the distal dendrites of the MSNs in the Nacc possesses a particular synaptic architecture, called "striatal microcircuit" or "synaptic triad," that involves both DAergic and glutamatergic axons.
- All addictive drugs commonly abused by humans evoke variations on DA concentrations within the Nacc and this may have a role in spine density, morphology, and synaptic strength in the entire neuronal mesolimbic pathway.
- The withdrawal syndrome after chronic drug administration seems to be a crucial point of the addictive process that is manifested by the induction of rapid changes in dendritic spine density and morphology.
- Confocal analysis of Golgi–Cox-stained MSNs of the Nacc revealed a decrease in spine density in the shell during morphine, alcohol, and cannabis withdrawal, while no changes in the number of spines were observed during chronic treatments.
- As long as the drug is "on-board" it supports spine persistence and function, whereas abrupt withdrawal discloses spine pruning and synaptic dysfunction.
- Withdrawal after chronic drug intake induces major changes at the neural level, which, in turn, will elicit behavioral changes, thus representing the "driving force" in the transition from chronic drug intake to addiction.

SUMMARY POINTS

- This chapter focuses on cannabis, the most widely produced and most frequently used illegal plant-based drug in Europe, whose main active ingredient, Δ^9-THC, is able to heavily affect the mesolimbic reward circuitry.
- The regional distribution of CB1 receptors corresponds to the behavioral effects of cannabinoids, including those on mood, motor coordination, autonomic function, memory, sensation, and cognition.
- Cannabis as much as other drugs of abuse functions as a reinforcer in the brain through its actions on the mesolimbic system, where it produces an increase in the release of DA.
- In the mesolimbic system, the medial portions of the VS and the VTA appear to be more responsive to the rewarding effects of DAergic drugs than their lateral counterparts.
- Actions of cannabis derivatives on DA transmission are mediated by the modulation of GABA and glutamate release.
- Cannabis-induced subjective feelings of well-being and euphoria in humans are inhibited by the concomitant administration of CB1 receptor antagonists, suggesting that many, if not all, psychological and neurochemical effects of cannabis in the mesolimbic system depend on the activation of CB1 receptors.

- Chronic consumers of even low daily doses of cannabis derivatives display, upon cessation of drug administration, the appearance of overt abstinence signs paralleled by structural anomalies related to cannabis use.
- Withdrawal from a regimen of chronic cannabinoid administration profoundly affects the mesolimbic system, causing a shrinkage of DA neurons in the VTA, a reduction in spine density of the Nacc shell, and overall reduction in DA transmission that ultimately may contribute to the negative emotional state that characterizes withdrawal.

ACKNOWLEDGMENT

A special acknowledgment goes to Miss Carlotta Vita for her much appreciated graphic support.

REFERENCES

Abraham, W. C. (2008). Metaplasticity: tuning synapses and networks for plasticity. *Nature Reviews Neuroscience, 9*(5), 387.

Bloomfield, M. A., Morgan, C. J., Egerton, A., Kapur, S., Curran, H. V., & Howes, O. D. (2014). Dopaminergic function in cannabis users and its relationship to cannabis- induced psychotic symptoms. *Biological Psychiatry, 75*, 470–478.

Budney, A. J., & Hughes, J. R. (2006). The cannabis withdrawal syndrome. *Current Opinion in Psychiatry, 19*, 233–238.

Di Chiara, G., & Imperato, A. (1988). Drugs abused by humans preferentially increase synaptic dopamine concentrations in the mesolimbic system of freely moving rats. *Proceedings of the National Academy of Sciences of the United States of America, 85*, 5274–5278.

Diana, M. (1998). Drugs of abuse and dopamine cell activity. *Advances in Pharmacology, 42*, 998–1001.

Diana, M., Melis, M., Muntoni, A. L., & Gessa, G. L. (August 18, 1998). Mesolimbic dopaminergic decline after cannabinoid withdrawal. *Proceedings of the National Academy of Sciences of the United States of America, 95*(17), 10269–10273.

Fattore, L., Fadda, P., Spano, M. S., Pistis, M., & Fratta, W. (2008). Neurobiological mechanisms of cannabinoid addiction. *Molecular and Cellular Endocrinology, 286S*, S97–S107.

Gerdeman, G. L., & Lovinger, D. M. (2003). Emerging roles for endocannabinoids in long-term synaptic plasticity. *British Journal of Pharmacology, 140*(5), 781–789.

Gilman, J. M., Kuster, J. K., Lee, S., Lee, M. J., Kim, B. W., Makris, N., … Breiter, H. C. (August 16, 2014). Cannabis use is quantitatively associated with nucleus accumbens and amygdala abnormalities in young adult recreational users. *The Journal of Neuroscience, 34*(16), 5529–5538.

Grace, A., Floresco, S. B., Goto, Y., & Lodge, D. J. (2007). Regulation offering of dopaminergic neurones and control of goal-directed behaviors. *Trends in Neurosciences, 30*(5), 220–227.

Hall, W., & Degenhardt, L. (2009). Adverse health effects of non-medical cannabis use. *Lancet, 374*, 1383–1391.

Heimer, L., Zahm, D. S., Churchill, L., Kalivas, P. W., & Wohltmann, C. (1991). Specificity in the projection patterns of accumbal core and shell in the rat. *Neuroscience, 41*, 89–125.

Hoffman, A. F., Riegel, A. C., & Lupica, C. R. (2003). Functional localization of cannabinoid receptors and endogenous cannabinoid production in distinct neuron populations of the hippocampus. *European Journal of Neuroscience, 18*(3), 524–534.

Ikemoto, S., Qin, M., & Liu, Z. H. (2005). The functional divide for primary reinforcement of D-amphetamine lies between the medial and lateral ventral striatum: is the division of the accumbens core, shell and olfactory tubercle valid. *The Journal of Neuroscience, 25*, 5061–5065.

Iversen, L. (2003). Cannabis and the brain. *Brain, 126*, 1252–1270.

Jhou, T. C., Fields, H. L., Baxter, M. G., Saper, C. B., & Holland, P. C. (2009). The rostromedial tegmental nucleus (RMTg), a major GABAergic afferent to midbrain dopamine neurones, selectively encodes aversive stimuli and promotes behavioral inhibition. *Neuron, 61*(5), 786–800.

Koob, G. F., & Le Moal, M. (2001). Drug addiction, dyes regulation of reward, and allostasis. *Neuropsychopharmacology, 24*, 97129.

Kortleven, C., Fasano, C., Thibault, D., Lacaille, J. C., & Trudeau, L. E. (May 2011). The endocannabinoid 2 arachidonoylglycerol inhibits long-term potentiation of glutamatergic synapses onto ventral tegmental area dopamine neurones in mice. *European Journal of Neuroscience, 33*(10), 1751–1760.

Laviolette, S. R., Gallegos, R. A., Henriksen, S. J., & van der Kooy, D. (2004). Opiate state controls bi-directional reward signaling via GABAA receptors in the ventral tegmental area. *Nature Neuroscience, 7*, 160–169.

Leweke, F., & Koethe, D. (2008). Cannabis and psychiatric disorders: it is not only addiction. *Addiction Biology, 13*, 264–275.

Lupica, C. R., Riegel, A. C., & Hoffman, A. F. (2004). Marijuana and cannabinoid regulation of brain reward circuits. *British Journal of Pharmacology, 143*, 227–234.

Lupica, C. R., Riegel, A. C. (June 2005). Endocannabinoid release from midbrain dopamine neurons: a potential substrate or cannabinoid receptor antagonist treatment of addiction. *Neuropharmacology, 48*(8), 1105–1116.

Melis, M., Pistis, M., Perra, S., Muntoni, A. L., Pillola, G., & Gessa, G. L. (January 7, 2004). Endocannabinoids mediate presynaptic inhibition of glutamatergic transmission in rat ventral tegmental area dopamine neurones through activation of CB1 receptors. *The Journal of Neuroscience, 24*(1), 53–62.

Nader, K., & Van der Kooy, D. (1997). Deprivation state switches the neurobiological sub- strates mediating opiate reward in the ventral tegmental area. *The Journal of Neuroscience, 17*, 383–390.

Overton, P. G., & Clark, D. (1997). Burst firing in midbrain dopaminergic neurones. *Brain Research Reviews, 25*(3), 312–334.

Piercea, R. C., & Kumaresana, V. (2006). The mesolimbic dopamine system: the final common pathway for the reinforcing effect of drugs of abuse? *Neuroscience and Biobehavioral Reviews, 30*, 215–223.

Plescia, F., Brancato, A., Marino, R. A., Vita, C., Navarra, M., & Cannizzaro, C. (2014). Effect of acetaldehyde intoxication and withdrawal on NPY expression: focus on endocannabinoidergic system involvement. *Frontiers in Neuroscience, 7*(64), 1–9.

Riegel, A. C., & Lupica, C. R. (December 8, 2004). Independent presynaptic and postsynaptic mechanisms regulate endocannabinoid signaling at multiple synapses in the ventral tegmental area. *The Journal of Neuroscience, 24*(49), 11070–11078.

Serpelloni, G., Diana, M., Gomma, M., & Rimondo, C. (2011). *Cannabis and harm to health*. Italy: Department of Anti-Drug Policy, Presidenza del Consiglio dei Ministri.

Shen, W., Flajolet, M., Greengard, P., & Surmeier, D. J. (August 8, 2008). Dichotomous dopaminergic control of striatal synaptic plasticity. *Science, 321*(5890), 848–851.

Spiga, S., Lintas, A., Migliore, M., & Diana, M. (2010). Altered architecture and functional consequences of the mesolimbic dopamine system in cannabis dependence. *Addiction Biology, 15*(3), 266–276.

Spiga, S., Talani, G., Mulas, G., Licheri, V., Fois, G. R., Muggironi, G., ... Diana, M. (September 2, 2014). Hampered long-term depression and thin spine loss in the nucleus accumbens of ethanol-dependent rats. *Proceedings of the National Academy of Sciences of the United States of America, 111*(35), E3745–E3754.

Svizenska, I., Dubovy, P., & Sulcova, A. (2008). Cannabinoid receptors 1 and 2 (CB1 and CB2), their distribution, ligands and functional involvement in nervous system structures: a short review. *Pharmacology, Biochemistry, and Behavior, 90*, 501–511.

Tan, H., Ahmad, T., Loureiro, M., Zunder, J., & Laviolette, S. R. (2014). The role of cannabinoid transmission in emotional memory formation: implications for addiction and schizophrenia. *Frontiers in Psychiatry, 5*, 73. http://dx.doi.org/10.3389/fpsyt.2014.00073.

UNODC. (2014). *World drug report 2014*. Vienna: United Nations Office on Drug and Crime.

Urban, N. B. L., Slifstein, M., Thompson, J. L., Xu, X., Girgis, R. R., Raheja, S., ... Abi-Dargham, A. (2012). Dopamine release in chronic cannabis users: a [11C]raclopride positron emission tomography study. *Biological Psychiatry, 71*, 677–683.

Wilson, D. M., Varvel, S. A., Harloe, J. P., Martin, B. R., & Lichtman, A. H. (2006). SR 141716 (Rimonabant) precipitates withdrawal in marijuana-dependent mice. *Pharmacology, Biochemistry, and Behavior, 85*(1), 105–113.

Chapter 75

Role of the Endocannabinoid System and Major *Cannabis* Constituents in the Reconsolidation and Extinction of Rewarding Drug-Associated Memories

Cristiane R. de Carvalho, Cristina A.J. Stern, Leandro J. Bertoglio, Reinaldo N. Takahashi
Department of Pharmacology, Federal University of Santa Catarina, Florianópolis, Santa Catarina, Brazil

Abbreviations

2-AG 2-Arachidonoylglycerol
5-HT1A receptor Serotonin type-1A receptor
AEA Anandamide
AM251 N-(Piperidin-1-yl)-5-(4-iodophenyl)-1-(2,4-dichlorophenyl)-4-methyl-1H-pyrazole-3-carboxamide
AM630 6-Iodo-2-methyl-1-[2-(4-morpholinyl)ethyl]-1H-indol-3-yl-(4 methoxyphenyl) methanone
cAMP Cyclic adenosine monophosphate
CB1 receptor Cannabinoid type-1 receptor
CB2 receptor Cannabinoid type-2 receptor
CBD Cannabidiol
CNS Central nervous system
CPP Conditioned place preference
CS Conditioned stimulus
eCB Endocannabinoid
FAAH Fatty acid amide hydrolase
GPCR G-protein-coupled receptor
GPR55 Orphan G-protein-coupled receptor
LTD Long-term depression
LTP Long-term potentiation
MAGL Monoacylglycerol lipase
NAc Nucleus accumbens
PFC Prefrontal cortex
PPAR Peroxisome proliferator-activated receptor
PTSD Posttraumatic stress disorder
SR141716A Rimonabant hydrochloride
THC Δ^9-Tetrahydrocannabinol
TRPV1 Transient receptor potential vanilloid type-1
URB597 Cyclohexylcarbamic acid 3'-(aminocarbonyl)-[1,1'-biphenyl]-3-yl ester
US Unconditioned stimulus
VTA Tegmental ventral area
WAY 100635 N-[2-[4-(2-methoxyphenyl)-1-piperazinyl]ethyl]-N-2-pyridinylcyclohexanecarboxamide maleate salt
WIN 55,212 (R)-(+)-WIN 55,212

INTRODUCTION

Addiction has been defined as repetitive and compulsive drug use, despite its severe and negative consequences. Although methadone has been used medically to treat heroin (opiate) addiction and therapeutic interventions for nicotine withdrawal (smoking cessation) have been developed, very few medications are available and effective in attenuating permanently addiction to a given drug. Moreover, a high rate of drug relapse is a pervasive problem because, even after years of abstinence, it can occur when an addict encounters cues, including people, objects, or places, associated with prior drug experience. The formation and maintenance of strong associative memories that develop among drug-paired contextual cues and rewarding stimuli or withdrawal-associated aversive feelings have been suggested to contribute to this and other addiction-related outcomes (Foltin & Haney, 2000).

Accumulating evidence now indicates that drugs of abuse are able to induce a dysfunctional synaptic plasticity in brain regions responsible for processing reinforcement and reward aspects. For this reason, addiction may represent inappropriate and maladaptive learning and memory processing (Torregrossa, Corlett, & Taylor, 2011), similar to that underlying other psychiatric conditions such as posttraumatic stress disorder (PTSD) (Parsons & Ressler, 2013). Based on this fact, it would be interesting to establish the effectiveness of therapeutic interventions targeting the maintenance (reconsolidation) and/or suppression (extinction) of aberrant and enduring emotional memories.

The aim of this chapter is to provide an updated overview of the role of the endocannabinoid (eCB) system in learning and memory aspects relevant to drug addiction and traumatic events, and compare potentially promising findings from preclinical pharmacological studies targeting reconsolidation with those that have shown interference with the extinction of emotional memories.

ESSENTIAL ASPECTS ABOUT LEARNING AND MEMORY PROCESSING

Acquisition and Consolidation

More than a century has passed since the memory consolidation theory was proposed by Müller and Pilzecker (1900). On such occasion, it was found, in humans, that recently acquired information could be disrupted by learning new information immediately after the previous acquisition. Based on that evidence, the authors suggested that at the beginning memories are fragile, but become stable over time. They referred to this period of stabilization as consolidation (McGaugh, 2000). Therefore, whereas acquisition refers to the first stage of learning, consolidation refers to the progressive postacquisition stabilization of the memory trace (Roedirger, Dudai, & Fitzpatrick, 2007). Memory consolidation depends on the activity of the dorsal hippocampus, the amygdala nuclei, and the neocortex (Einarsson & Nader, 2012; McGaugh, 2000), where protein synthesis is required to induce the synaptic plasticity necessary to reinforce the neuronal circuit used for learning (McGaugh, 2000). Of note, activation of neural components of the mesocorticolimbic dopaminergic pathway by drugs of abuse enhances memory consolidation (Blaiss & Janak, 2006). Moreover, one of the most accepted mechanisms proposed to support memory consolidation is long-term potentiation (LTP), which increases synaptic efficacy after the delivery of a relevant stimulus (Bliss & Collingridge, 1993). This premise is supported by results showing that interventions that affect LTP also disrupt memory consolidation (Martin & Shapiro, 2000).

Retrieval

Accessing information stored as memory is a two-step process. The first is reactivation, which represents the passage from a stable to a labile form of the memory trace (Lewis, 1979). Memory reactivation can be succeeded by the expression of a behavior (and/or by verbalization in humans). Retrieval of a previously consolidated memory in the absence of reinforcement can lead the memory to undergo two distinct and independent processes, namely reconsolidation and extinction. The duration of the reexposure to the conditioned stimulus is one of the factors governing the occurrence of memory reconsolidation or extinction (Torregrossa & Taylor, 2013) (Figure 1).

Reconsolidation

As mentioned earlier, memory reactivation renders the memory trace labile, opening an opportunity to modify the original content of the memory. This opportunity is time-limited and is followed by a new restabilization phase known as reconsolidation (Alberini & LeDoux, 2013). Importantly, although consolidation and reconsolidation share some molecular mechanisms, they are considered different memory phases. It has been proposed that the reconsolidation process maintains a memory over time and/or allows its updating (Tronson & Taylor, 2007). The latter aspect is of particular interest owing to the therapeutic potential of disrupting the reconsolidation of cue-drug memories as an anti-relapse treatment for drug addiction (Milton & Everitt, 2012; Tronson & Taylor, 2007).

Extinction

Extinction is a form of inhibitory learning that suppresses the expression of the original memory. The extinction process does not generally modify the original conditioned stimulus–unconditioned stimulus (CS–US) association. It is used by therapists exposing patients to drug or threat cues, promoting the formation of a new and neutral associated memory (Conklin & Tiffany, 2002). In laboratory animals, memory extinction is induced by long retrieval sessions without the presentation of the CS. Despite extinction training, spontaneous recovery of the original memory, as well as its reinstatement and/or renewal, can still occur (Conklin & Tiffany, 2002).

OVERVIEW OF THE ENDOCANNABINOID SYSTEM

As illustrated in Figure 2, the eCB system comprises two different cannabinoid receptors, their endogenous ligands, and the enzymes involved in the synthesis and degradation of these eCBs (Di Marzo, 2009). Cannabinoid type-1 (CB1) and type-2 (CB2)

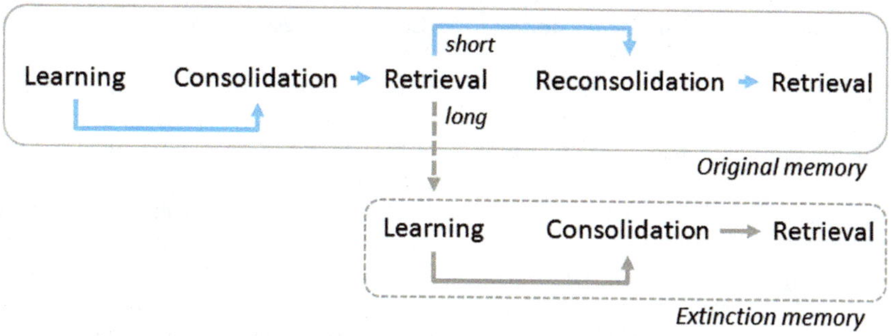

FIGURE 1 Phases of emotional (rewarding or aversive) memory processing and its extinction.

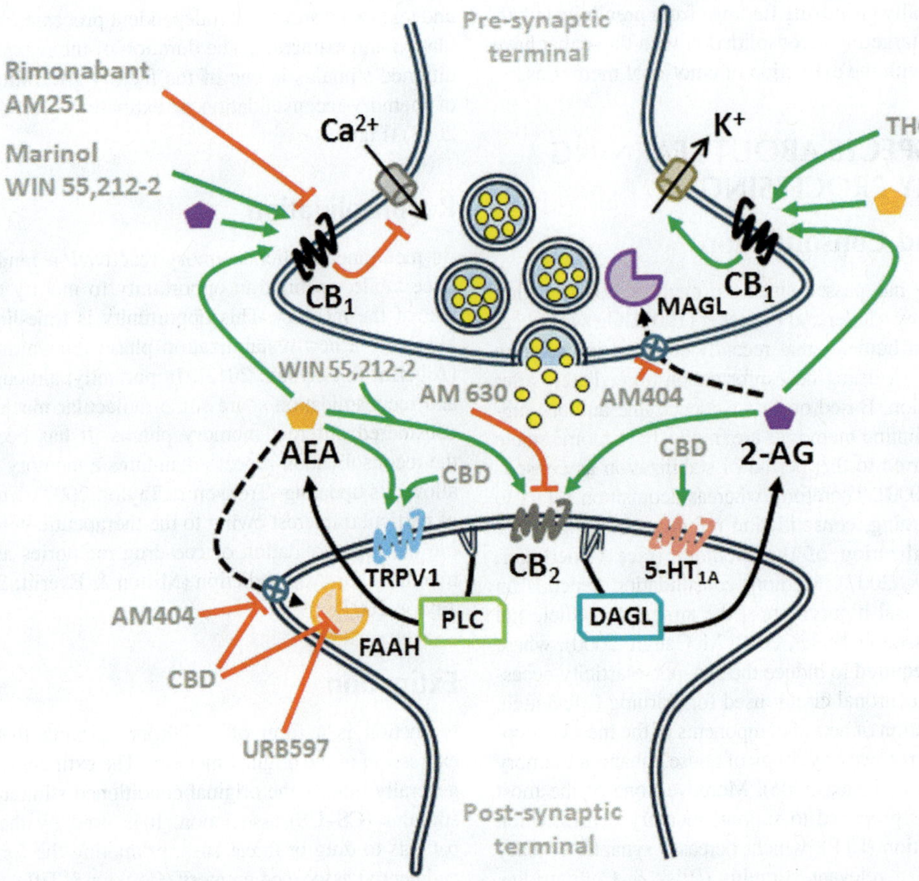

FIGURE 2 The brain eCB system and its main pharmacological targets. The eCBs are produced on demand in postsynaptic terminals: anandamide (AEA) is generated by phospholipase C (PLC), and 2-arachidonoylglycerol (2-AG) is generated by diacylglycerol lipase (DAGL). AEA and 2-AG diffuse retrogradely toward the synaptic cleft where, similar to certain phytocannabinoids (e.g., Δ^9-tetrahydrocannabinol, THC) and synthetic cannabinoids (e.g., WIN 55,212-2 and marinol), they activate presynaptic metabotropic CB1 receptors and/or postsynaptic metabotropic CB2 receptors, which are significantly less abundant than CB1 receptors in the brain. Stimulation of presynaptic CB1 receptors reduces the release of stored neurotransmitter by opening inwardly rectifying K^+ channels and closing Ca^{2+} channels. AEA and 2-AG may reenter post- or presynaptic nerve terminals, possibly through a specialized transporter that can be blocked by AM404, after which they are respectively catabolized by fatty acid amide hydrolase (FAAH) and monoacylglycerol lipase (MAGL). AM251 and rimonabant (SR141716) are CB1 receptor antagonists and AM630 is a CB2 receptor antagonist. Cannabidiol (CBD) is a phytocannabinoid that acts through multiple mechanisms of action, including inhibition of FAAH activity and AEA reuptake and activation of serotonin type-1A (5-HT1A) receptors and transient receptor potential vanilloid 1 (TRPV1) channels. See the text for additional details.

receptors have different densities and distributions in the central nervous system (CNS) and peripheral tissues. CB1 receptors are expressed by central and peripheral neurons, while CB2 receptors are expressed mostly by immune cells, although they are also expressed in the brain (Van Sickle et al., 2005). Initially, CB2 receptors were found in the brain only under pathological conditions (Ibrahim et al., 2003). However, further investigation has revealed expression of CB2 receptors in several brain regions under physiological conditions (Van Sickle et al., 2005), including those related to reward processing, such as the nucleus accumbens (NAc), which may suggest they contribute to drug addiction (Aracil-Fernández et al., 2012; Xi et al., 2011).

The two major endogenous ligands for CB1 and CB2 receptors are anandamide (AEA) and 2-arachidonylglycerol (2-AG). These eCBs are synthesized on demand, mainly postsynaptically, and act as retrograde messengers regulating the presynaptic release of various neurotransmitters (Castillo, Younts, Chávez, & Hashimotodani, 2012). AEA acts as a partial agonist at both CB1 and CB2 receptors (selectivity: CB1 >> CB2) and activates transient receptor potential vanilloid type 1 (TRPV1) channels (Starowicz, Nigam, & Di Marzo, 2007). 2-AG is the most abundant eCB in the CNS and nonselectively activates CB1 and CB2 receptors (Castilo et al., 2012; Di Marzo, 2009). It has been shown that AEA, 2-AG, and Δ^9-tetrahydrocannabinol (THC), the main psychotomimetic constituent of *Cannabis*, exert their effects mainly through activation of CB1 receptors (Castilo et al., 2012). In the case of eCBs, the effects are rapidly terminated through carrier-mediated uptake followed by intracellular enzymatic degradation. 2-AG and AEA are metabolized by monoacylglycerol lipase and fatty acid amide hydrolase (FAAH), respectively (Wilson & Nicoll, 2002).

The eCBs function as retrograde signaling messengers, regulating neuronal activity and plasticity by depolarization-induced suppression of inhibition or excitation (Wilson & Nicoll, 2002). Both phenomena are forms of short-term synaptic plasticity that contribute to the regulation of a number of physiological functions, including motivation, memory, and emotions (Di Marzo,

2009; Maldonado, Valverde, & Berrendero, 2006). Additionally, eCBs appear to modulate the memory process by changing synaptic plasticity and mediating more persistent forms of synaptic plasticity (LTP and long-term depression (LTD)) in several addiction- and memory-related brain areas (Maldonado et al., 2006; Sidhpura & Parsons, 2011).

USING THE CONDITIONED PLACE PREFERENCE PARADIGM TO INVESTIGATE THE ROLE OF THE ENDOCANNABINOID SYSTEM IN MEMORY RECONSOLIDATION AND EXTINCTION

Conditioned place preference (CPP), the most widely used paradigm to investigate hedonic memory reconsolidation and extinction, is a form of Pavlovian conditioning used to measure the reward and motivational aspects of drug abuse. The CPP apparatus has two distinct compartments or chambers. When a given chamber of the CPP apparatus is paired with a rewarding drug (e.g., amphetamine, cocaine, or morphine), the animal will show a preference for the associated compartment (Tzschentke, 2007). It is still not clear which psychological processes underlie place preference. Furthermore, the CPP task does not exactly reflect the extent of Pavlovian conditioning that occurs between environmental CS and the effects of a drug in human addicts who have an extensive history of drug use and therefore many pairings of CS and US. Despite these translational constraints, preclinical studies focusing on reconsolidation of rewarding drug memory have used the CPP procedure. Importantly, studies of memory require a change in animal behavior. Thus, the formation and storage of associative memories can be attributed to the expression of CPP (Tzschentke, 2007).

As illustrated in Figure 3, a typical protocol for studying memory reconsolidation or extinction using CPP involves three phases: training (or conditioning), retrieval, and testing, which is usually conducted within days or weeks postretrieval. In the first phase, the animals are trained to acquire drug-associative memories and then, following either extinction or forced abstinence, they are subjected to a retrieval session. Memory retrieval is triggered by reexposure to a CS, commonly without a US presentation. Thus, memories that were already consolidated can be retrieved and reactivated, becoming susceptible to disruption, if a reconsolidation blocker is present at the time of reactivation, or to facilitation, if an extinction facilitator is present (Nader, Schafe, & LeDoux, 2000).

ROLE OF THE ENDOCANNABINOID SYSTEM IN LEARNING AND MEMORY PROCESSING

CB1 receptors are highly abundant in brain areas, such as the NAc, ventral tegmental area (VTA), prefrontal cortex (PFC), amygdala, hippocampus, and dorsal striatum (Maldonado et al., 2006), responsible for learning and memory and drug-related behaviors (Robbins, Ersche, & Everitt, 2008). Importantly, brain-imaging studies in humans have revealed a similar neuroanatomical involvement in cue-elicited craving in drug addicts (Volkow, Fowler, & Wang, 2004).

The eCB system is involved in the primary rewarding effects of various drugs of abuse (alcohol, cannabinoids, nicotine, and opioids), possibly by increasing the firing rate of dopaminergic neurons of the mesolimbic pathway. For instance, the activation of CB1 receptors located on axon terminals of γ-aminobutyric acid (GABA)-ergic neurons in the VTA inhibits GABA transmission, removing the inhibitory input on dopaminergic neurons and leading to increased dopamine activity (Maldonado et al., 2006). Similarly, glutamatergic neurotransmission from neurons of the NAc is also modulated by the activation of CB1 receptors (Maldonado et al., 2006; Sidhpura & Parsons, 2011). The inhibition of glutamate release attenuates the excitation of GABAergic neurons from the NAc that project to the VTA, thus indirectly activating VTA dopaminergic neurons.

The impact of CB2 receptors on the reinforcing effects of drugs has also been described (Aracil-Fernández et al., 2012; Xi et al., 2011). Overall, these preclinical studies report that brain CB2 receptors interfere with drug-rewarding effects. Their activation within dopamine terminals could potentially inhibit dopamine release and, thus, reduce the reinforcing properties of cocaine (Aracil-Fernández et al., 2012; Xi et al., 2011). If so, these findings raise several new hypotheses to be tested subsequently, including whether this regulation of dopamine release by CB2 receptors could modulate the intake of other drugs of abuse and whether this action is evident in other animal models.

The eCB system also modulates the motivation to drug-seeking behavior, which is predictive of drug relapse, by a mechanism unrelated to the release of dopamine in the NAc (Maldonado et al., 2006). The PFC is a brain area that integrates sensory information, emotional processing, and hedonic

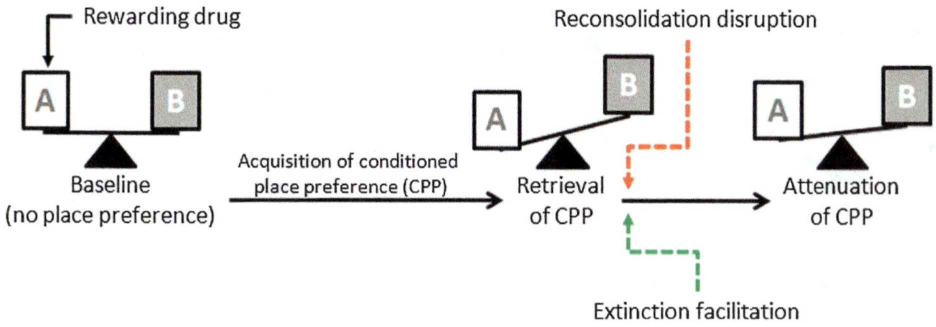

FIGURE 3 Postretrieval (pharmacological and/or behavioral) strategies to attenuate drug rewarding-related memories using the CPP paradigm in laboratory animals.

experience. It seems that activation of CB1 receptors in the PFC could explain the eCB system's involvement in the motivation to seek a drug (Kringelbach, 2005).

The eCB system has been implicated in relapse and drug-seeking behavior, probably by influencing the synaptic plasticity underlying drug-related memories. Convergent evidence has shown that the eCB-mediated LTD in the NAc, amygdala, hippocampus, PFC, and VTA is critical in preparing excitatory synapses for subsequent induction of LTP, which in turn contributes to consolidation of the reward-driven behavior required to establish the addictive process (Maldonado et al., 2006; Sidhpura & Parsons, 2011).

Although the involvement of the eCB system in emotional learning and memory is complex, there is agreement in the literature about the eCB system's role as a promoter of memory extinction, at least for aversive memories (Pamplona, Bitencourt, & Takahashi, 2008; Suzuki et al., 2004). Accordingly, either genetic or pharmacologic blockade of CB1 receptors impairs extinction of classical fear conditioning (Pamplona et al., 2008; Suzuki et al., 2004). On the other hand, these experimental interventions did not affect those memories produced by appetitive or drug-related stimuli (Harloe, Thorpe, & Lichtman, 2008; Manwell et al., 2009). Scarce evidence is currently available about the effects of cannabinoids on the reconsolidation of hedonic and aversive memories. Since impairing the reconsolidation produces longer lasting effects than facilitating the extinction process, interest in the mechanisms of reconsolidation has increased since 2005.

THE ENDOCANNABINOID SYSTEM AND RECONSOLIDATION OF REWARDING DRUG-RELATED MEMORIES

So far, studies have shown that systemic administration of the CB1 receptor antagonist rimonabant disrupts drug reward-associated memory. Indeed, it was demonstrated in rats that antagonism of CB1 receptors persistently impaired the reconsolidation of methamphetamine- (Yu et al., 2009), nicotine- (Fang et al., 2011), and morphine-induced CPP (De Carvalho, Pamplona, Cruz, & Takahashi, 2014) (Table 1). Activation of CB1 and CB2 receptors by WIN 55,212-2 did not alter morphine-induced CPP, but potentiating eCB signaling by inhibiting AEA metabolism promoted a transient and CB1 receptor-dependent increase in morphine-induced CPP (De Carvalho et al., 2014). It was also shown that systemic injection of AM630, a selective CB2 receptor antagonist, had no effect on the reconsolidation of morphine-induced CPP (De Carvalho et al., 2014). Of note, it is possible that the lack of effect of AM630 may be attributed to its relatively poor pharmacokinetic properties and/or blood–brain barrier passage (Xi et al., 2011). Some studies have shown that CB2 receptor activation reduces drug-related behavior in rodents (Aracil-Fernández et al., 2012; Xi et al., 2011). Altogether, these findings indicate that pharmacological manipulation of CB2 receptors could represent a potential target for drug addiction-related memory mitigation, even though further investigation is needed to corroborate this premise.

TABLE 1 Cannabinoid Effects on Rewarding Drug and Appetitive Memory Extinction and Reconsolidation

Drug/Route	Reconsolidation Effect	Extinction Effect	Species	Learning Task	References
Rimonabant (CB1 antagonist, systemic)	↓	O	Rat	Amphetamine; morphine; nicotine CPP	Manwell et al. (2009), Yu et al. (2009), Fang et al. (2011), and De Carvalho et al. (2014)
Rimonabant (CB1 antagonist, systemic)	—	O	Mice	Appetitively motivated operant conditioning task	Niyuhire et al. (2007)
AM630 (CB2 antagonist, systemic)	O	—	Rat	Morphine CPP	De Carvalho et al. (2014)
WIN 55,212-2 (CB1/CB2 agonist, systemic)	O	—	Rat	Morphine CPP	De Carvalho et al. (2014)
URB597 (FAAH inhibitor, systemic)	↑	↑	Rat	Naloxone-precipitated morphine withdrawal; morphine CPP	Manwell et al. (2009) and De Carvalho et al. (2014)
CBD (not CB1/CB2, systemic)	—	↑	Rat	Cocaine or amphetamine CPP; Intravenous heroin self-administration	Parker, Burton, Sorge, Yakiwchuk, and Mechoulam (2004) and Ren, Whittard, Higuera-Matas, Morris, and Hurd (2009)
THC (not CB1/CB2, systemic)	—	↑	Rat	Cocaine or amphetamine CPP	Parker et al. (2004)

CPP, conditioned place preference; ↑, facilitating; ↓, impairing; O, no effect; —, not evaluated.

THE ENDOCANNABINOID SYSTEM AND RECONSOLIDATION OF AVERSIVE MEMORIES

PTSD is frequently induced by exposure of individuals to an intense traumatic situation. Patients suffering from this psychiatric condition frequently and spontaneously retrieve the traumatic memory in an intrusive manner, which causes reexperience of the trauma and the expression of hyperarousal, avoidance, increased stress responses, and generalization. This has a deep impact on the individual's life and, to relieve these symptoms, PTSD patients tend to increase their use of addictive drugs (Müller et al., 2014). In this regard, it becomes clear that the reconsolidation of trauma-related memories could be a target for pharmacological interventions able to disrupt this memory restabilization phase and, consequently, ameliorate their corresponding symptoms (Alberini & LeDoux, 2013).

It has been shown in mice previously conditioned to contextual cues that activation of CB1 receptors is necessary to induce memory labilization (Suzuki, Mukawa, Tsukagoshi, Frankland, & Kida, 2008). However, the role of AEA and 2-AG in memory labilization is still unknown. Of note, PTSD patients exhibit low levels of AEA and present an upregulation of CB1 receptors (Neumeister, 2013).

Activation of CB1 receptors by AEA in the dorsal hippocampus of rats disrupts the reconsolidation of contextual fear memory (de Oliveira Alvares, Pasqualini Genro, Diehl, Molina, & Quillfeldt, 2008). Also in rats, infusing the agonist WIN 55,212-2 into the amygdala blocks the reconsolidation of fear-potentiated startle (Lin, Mao, & Gean, 2006). The drug also blocked the reconsolidation of conditioned taste aversion when infused into the intrainsular cortex (Kobilo, Hazvi, & Dudai, 2007). WIN 55,212-2 is a potent CB1/CB2 agonist and, although CB2 receptors are less expressed than CB1 receptors in the brain, some participation of CB2 receptors cannot be ruled out. There have been only a few other studies investigating the role of the eCB system in fear memory reconsolidation (Table 2). Overall, their results are consistent, suggesting that activation of this modulatory system could control the intensity of fear memories by blocking the reconsolidation step each time they are retrieved.

EFFECTS OF Δ⁹-TETRAHYDROCANNABINOL AND CANNABIDIOL ON RECONSOLIDATION OF DRUG, APPETITIVE, AND AVERSIVE MEMORIES

The two main phytocannabinoids of *Cannabis* are the psychotomimetic THC and the nonpsychotomimetic CBD. Since their discovery, the pharmacological effects of THC and CBD have been extensively characterized in laboratory animals and humans.

Preclinical studies investigating the effects of these two phytocannabinoids on memory extinction and reconsolidation are still scarce. One study (Parker, Burton, Sorge, Yakiwchuk, & Mechoulam, 2004) investigated the effects of a low dose of THC (0.5 mg/kg) on extinction of cocaine- and amphetamine-induced CPP in rats. The authors reported a facilitated extinction that was not blocked by antagonism of CB1 receptors with rimonabant, suggesting a mechanism not dependent on the eCB system. Regarding THC's effects on aversive memories, a study with rats that received a dose of 10 mg/kg THC once per day for 6 days reported impaired extinction of an auditory fear memory (Ashton, Smith, & Darlington, 2008). In contrast, a study with healthy humans subjected to classical fear conditioning reported that 7.5 mg/kg THC induced a better recall of extinction than controls. This effect was accompanied by increased activity in brain regions that process fear extinction, suggesting that the eCB system, or at least CB1 receptors, may be a potential target for pharmacological interventions to attenuate PTSD and other memory-related disorders (Rabinak et al., 2013). Altogether, the data about THC and extinction point to a possible use of low doses of THC for rewarding drug- or aversive-associated memories. A recent preclinical study has reported an impairing effect of THC (0.3-10 mg/kg i.p.) on fear memory reconsolidation in rats. This effect was prevented by pharmacological antagonism of CB1 receptors located in prelimbic subregion of the medial prefrontal cortex (Stern et al., 2015). There has still been no study aimed at investigating the effects of THC on the reconsolidation of drug rewarding-related memories.

In rodents, a low dose of CBD (5.0 mg/kg) has been shown to facilitate the extinction of both cocaine- and amphetamine-induced CPP (Parker et al., 2004). Like THC, this CBD effect was unrelated to CB1 receptor activation. A lack of hedonic effects of CBD was also shown, demonstrating that it does not induce addiction (Parker et al., 2004). This result is reinforced by the study of Katsidoni, Anagnostou, and Panagis (2013) showing that CBD did not affect the reinforcing effect of brain stimulation in rats. Rather, it prevented the rewarding effects of morphine, but not cocaine, an effect mediated through serotonin type-1A (5-HT1A) receptors (Katsidoni et al., 2013). The same dose of CBD also reduced heroin-seeking behavior reinstated in animals exposed to a conditioned cue, but did not change their self-intake of heroin, the extinction behavior, or the drug seeking induced by a prime heroin injection, suggesting CBD as an antirelapse drug (Ren et al., 2009).

A randomized double-blind placebo-controlled study demonstrated that the inhalation of CBD during 1 week reduced cigarette consumption in a long-lasting manner (Morgan, Das, Joye, Curran, & Kamboj, 2013). Although the authors did not investigate the mechanisms underlying this effect, they suggested a possible impairing effect of CBD on the reconsolidation process. Reinforcing the potential therapeutic effects of CBD, a study demonstrated that humans exposed to a transdermal gel containing 5% of CBD presented nearly 50% less neurodegeneration induced by alcohol binge consumption than controls (Liput, Hammell, Stinchcomb, & Nixon, 2013).

Infusing CBD into the brain ventricles (2.0 μg/μl) also facilitates contextual fear-conditioning extinction in rats, an effect blocked by the CB1 receptor antagonist rimonabant (Bitencourt et al., 2008). Another preclinical study reported an impairing effect of CBD (10 mg/kg) on fear memory reconsolidation in rats. This effect was blocked by the CB1 receptor antagonist AM251, but not by the 5-HT1A antagonist WAY 100635 (Stern et al., 2012). Although the effect of CBD in both extinction and reconsolidation could be prevented by CB1 receptor blockade, it is known that CBD is not a direct CB1 receptor agonist. Thus, it is thought to inhibit the activity of FAAH, with a consequent increase in the level of AEA (Bisogno et al., 2001). As AEA activates CB1 receptors and TRPV1 channels (Starowickz et al., 2007), a possible

TABLE 2 Cannabinoid Effects on Aversive Memory Extinction and Reconsolidation

Drug/Route	Reconsolidation Effect	Extinction Effect	Species	Learning Task	References
CB1 deletion	—	↓	Mice	Auditory fear conditioning	Marsicano et al. (2002), Kamprath et al. (2006), and Plendl and Wotjak (2010)
Rimonabant (CB1 antagonist, systemic)	↑	↓	Mice	Auditory fear conditioning; contextual fear conditioning; fear-potentiated startle; inhibitory avoidance	Marsicano et al. (2002), Suzuki et al. (2004), Chhatwal, Myers, Ressler, and Davis (2005), Niyuhire et al. (2007), Pamplona, Prediger, Pandolfo, and Takahashi (2006), and Pamplona et al. (2008)
Rimonabant (CB1 antagonist, insular cortex)	O	↓	Rat	Conditioned-taste aversion	Kobilo et al. (2007)
AM251 (CB1 antagonist, infralimbic cortex, amygdala, dorsal hippocampus)	↓↑	↓	Rat	Fear-potentiated startle; contextual fear conditioning	Bucherelli, Baldi, Mariottini, Passani, and Blandina (2006), de Oliveira Alvares et al. (2008), and Lin et al. (2006)
WIN 55,212-2 (CB1/CB2 agonist, systemic, dorsal hippocampus, amygdala, insular cortex)	↓	↑	Rat	Contextual fear conditioning; inhibitory avoidance; fear potentiated Startle; conditioned-taste aversion	Pamplona et al. (2006, 2008), Lin et al. (2006), Kobilo et al. (2007), and Abush and Akirav (2010)
Anandamide (endogenous CB1 agonist, dorsal hippocampus)	↓	↑	Rat	Contextual fear conditioning	de Oliveira Alvares et al. (2008)
AM404 (Anandamide uptake inhibitor, systemic, icv, infralimbic cortex, intra CA1)	—	↑	Rat	Contextual fear conditioning; inhibitory avoidance, fear-potentiated startle	Bitencourt, Pamplona, and Takahashi (2008), Pamplona et al. (2008), Abush and Akirav (2010), Chhatwal et al. (2005), and Lin et al. (2006)
URB597 (FAAH inhibitor, infralimbic cortex)	—	↑	Rat	Fear-potentiated startle	Lin et al. (2006)
CBD (CB1 indirect agonist, icv, systemic)	↓	↑	Rat	Contextual fear conditioning	Bitencourt et al. (2008) and Stern, Gazarini, Takahashi, Guimarães, and Bertoglio (2012)
CBD (mechanism not evaluated, oral)	—	↑	Human	Fear conditioning	Das et al. (2013)
THC (mechanism not evaluated, systemic for six days)	—	↓	Rat	Auditory fear conditioning	Ashton, Smith, and Darlington (2008)
Marinol (Synthetic THC, mechanism not evaluated, oral)	—	↑	Human	Fear conditioning	Rabinak et al. (2013)
THC (mechanism not evaluated, oral)	—	O	Human	Fear conditioning	Klumpers et al. (2012)

↑, facilitating, ↓, impairing, O, no effect, —, not evaluated, icv, intracerebroventricularly.

contribution to the latter cannot be excluded in advance. Despite the convergent evidence mentioned above, as yet no clinical study has investigated whether CBD can impair the reconsolidation (or facilitate the extinction) of drug- and trauma-related memories.

The studies reviewed herein indicate that more than one mechanism of action may be recruited by THC and/or CBD to induce their cognitive effects, which depend at least in part on the nature (rewarding vs aversive) of the memory evaluated. A mechanism relying on indirect CB1/CB2 activation could explain CBD's effects on extinction and reconsolidation of aversive memories, probably by an increase in AEA induced by FAAH inhibition. In support of this premise, increasing the endogenous levels of AEA in the brain facilitates the extinction and impairs the reconsolidation of fear memories (Bitencourt et al., 2008; Lin et al., 2006; de Oliveira Alvares et al., 2008; Pamplona et al., 2008). As mentioned earlier, however, the effects of THC and CBD on the extinction of drug-related memories appear to be unrelated to CB1 receptor-mediated signaling mechanisms. Possible candidates to be investigated are the 5-HT1A receptor and TRPV1 channels. Regarding THC, there are few and contradictory results and, like CBD, there appear to be multiple mechanisms of action mediating its effects on memory extinction and reconsolidation.

To summarize, whereas the available preclinical data provide evidence that CBD, which does not induce addiction by itself and can counteract the negative effects of THC (Niesink & van Laar, 2013), is able to attenuate rewarding drug- and fear-related memories through extinction facilitation and/or reconsolidation disruption, the effects of THC on these matters require further investigation.

CONCLUDING REMARKS

At present, numerous studies indicate that the reward caused by consuming drugs can pathologically usurp neural mechanisms of learning and memory. The CPP paradigm has been extensively used in drug addiction research to investigate reward-related learning and memory processes. Preclinical research focusing on the reconsolidation of drug rewarding-associated memories, particularly through manipulation of the eCB system, is still scarce, but evidence provided so far supports their relationship. The data reviewed herein also indicate that disruption of reconsolidation can be achieved by direct or indirect activation of brain CB1 receptors. Altogether, these findings support the notion that the eCB system mediates relapse episodes following cessation of drug exposure, highlighting the view that reconsolidation blockade, rather than extinction facilitation, may indicate a new direction for the treatment for drug- and context-induced relapse to drug seeking. However, further studies are necessary to elucidate the exact mechanisms by which the eCB signaling system modulates these processes.

APPLICATIONS TO OTHER ADDICTIONS AND SUBSTANCE MISUSE

Accumulating evidence has also pointed to a role for the eCB system in other types of addiction, such as those associated with food intake and gambling. Indeed, a dysfunctional eCB signaling seems to contribute not only to food addiction, by controlling appetite and food preference, but also to gambling, by controlling risk behaviors. Overall, it is thought these nondrug addictions also usurp the brain natural reward system and involve aberrant learning and memory processing. For that reason, the potential relevance of targeting the eCB system at specific time points, such as reconsolidation, to their treatment should be further investigated.

Current drug addiction treatments effectively relieve craving in the clinical context, but not when addicts return to their usual environment, as exposure to stimuli associated with the effects of the drug trigger the addict's habitual response of using the drug once again. Thus, disrupting reconsolidation of drug cues may boost conventional cue-exposure therapy's effectiveness, holding promise for a more effective treatment to addiction. In this context, whereas studies addressing the misuse of CNS stimulants are more numerous, those focusing on CNS depressants with potential of abuse and addiction are scarcer. In either case, however, modulating the activity of the eCB system may have a potential therapeutic value to the treatment of drug addiction, presenting promising applicability to the clinical setting once it theoretically targets the core of the problem, the formation and maintenance of an aberrant and enduring rewarding memory.

DEFINITION OF TERMS

Aversive memory This refers to a memory from a stressful life experience.
Conditioned stimulus This is an early neutral stimulus that is associated with a US or a reinforcer.
Conditioned place preference This is a form of Pavlovian conditioning used for assessing rewarding and motivational properties of drug abuse.
Extinction This refers to a new and inhibitory memory that suppresses the original one.
Memory This is the retention of internal representations generated by one or several experiences that can subsequently drive behavior.
Memory consolidation This refers to the progressive postacquisition stabilization phase of a memory trace.
Pavlovian conditioning This is a type of associative learning in which a previously motivationally neutral (conditioned) stimulus is paired in space and time with a motivationally relevant US (aversive or hedonic). The behavior of the individual does not affect the contingency between the presentations of both stimuli.
Reconsolidation This represents the restabilization phase of a memory that may occur after its retrieval and reactivation.
Retrieval This refers to the time point at which an established memory is accessed.
Retrograde messenger This is a neurotransmitter released by the postsynapse that travels backward and activates presynaptic receptors.
Reward This is a stimulus that induces hedonic pleasure.
Unconditioned stimulus This is a stimulus that is motivationally relevant to the individual (e.g., food, sex).

KEY FACTS OF MEMORY RECONSOLIDATION

- Around 1965 it was demonstrated that previously consolidated memories can be impaired by amnesic interventions when they are in a labile state again, as is the case during retrieval and

reactivation. These reactivated memories may undergo a process of restabilization, known as reconsolidation, to be maintained. For historical reasons, however, research on memory reconsolidation remained scarce for several decades.
- Since 2000, this process has been demonstrated for various types of memory, principally those with negative or positive emotional valence. Moreover, convergent evidence obtained from studies using laboratory animals and humans has implicated several brain regions (e.g., amygdala, hippocampus, and PFC) and neurotransmitters/neuromodulators (e.g., glutamate, GABA, noradrenaline, glucocorticoids, and eCBs) in memory reconsolidation.
- Accumulating evidence supports the theory that reconsolidation may offer an opportunity to update aberrant memories, such as those underlying addiction and PTSD. The capacity for plastic changes in memory strength or content following memory retrieval/reactivation is currently being investigated as a new target to improve the therapeutic interventions into these neuropsychiatric disorders.
- At least at the preclinical level, several boundary conditions may constrain memory reconsolidation. Overall, this process takes place when there is a prediction error (a discrepancy between actual and expected events) during its retrieval/reactivation. Moreover, it is favored when retrieval of the memory trace is short and when it is not so remote and/or strong.
- These and related issues have started to be addressed in human studies, thus increasing our understanding of the reconsolidation phenomenon and, in particular, its potential value in the permanent treatment of dysfunctional memory processes in mental disorders.

SUMMARY POINTS

- Drug addiction is a chronic, relapsing disorder, at least in part owing to the formation and maintenance of a strong memory that associates drugs and environmental cues.
- Interfering with the reconsolidation process has been proposed as a potential therapeutic strategy to attenuate and perhaps even erase aberrant memories underlying drug addiction and PTSD.
- eCBs are retrograde messengers that mediate the reinforcing effects of the drug of abuse and enhance synaptic plasticity in brain regions involved in the etiology of drug addiction.
- The eCB system participates in the common neurobiological mechanisms that underlie learning, memory, and drug-related behaviors.
- Pharmacologically manipulating the eCB system may be a promising therapeutic approach to disrupting the reconsolidation of hedonic or aversive memories, such as those experienced by people suffering from drug addiction or PTSD, respectively.
- CBD is a nonpsychotomimetic phytocannabinoid of potential relevance in attenuating cue-induced drug-seeking behavior and fear-related memories through extinction facilitation or reconsolidation disruption.

ACKNOWLEDGMENTS

This work was supported by grants from the Conselho Nacional de Desenvolvimento Científico e Tecnológico (CNPq), Coordenação de Aperfeiçoamento de Pessoal de Nível Superior, Fundação de Amparo à Pesquisa e Inovação do Estado de Santa Catarina, and Programa de Apoio aos Núcleos de Excelência, all of Brazil. Dr C.R. De Carvalho and Dr C.A.J. Stern are supported by scholarships from CNPq and Professor R.N. Takahashi and Professor L.J. Bertoglio are supported by research fellowships from CNPq.

REFERENCES

Abush, H., & Akirav, I. (2010). Cannabinoids modulate hippocampal memory and plasticity. *Hippocampus, 20*, 1126–1138.

Alberini, C. M., & LeDoux, J. E. (2013). Memory reconsolidation. *Current Biology, 23*, 746–750.

Aracil-Fernández, A., Trigo, J. A., García-Gutiérre, M. S., Ortega-Álvaro, A., Ternianov, A., Navarro, D., ... Manzanares, J. (2012). Decreased cocaine motor sensitization and self-administration in mice overexpressing cannabinoid CB_2 receptors. *Neuropsychopharmacology, 37*, 1749–1763.

Ashton, J. C., Smith, P. F., & Darlington, C. L. (2008). The effect of delta 9-tetrahydrocannabinol on the extinction of an adverse associative memory. *Pharmacology, 81*, 18–20.

Bisogno, T., Hanus, L., De Petrocellis, L., Tchilibon, S., Ponde, D. E., Brandi, I., ... Di Marzo, V. (2001). Molecular targets for cannabidiol and its synthetic analogues: effect on vanilloid VR1 receptors and on the cellular uptake and enzymatic hydrolysis of anandamide. *British Journal of Pharmacology, 134*, 845–852.

Bitencourt, R. M., Pamplona, F. A., & Takahashi, R. N. (2008). Facilitation of contextual fear memory extinction and anti-anxiogenic effects of AM404 and cannabidiol in conditioned rats. *European Neuropsychopharmacology, 18*, 849–859.

Blaiss, C. A., & Janak, P. H. (2006). Post-training and post-reactivation administration of amphetamine enhances morphine conditioned place preference. *Behavioural Brain Research, 171*, 329–337.

Bliss, T. V., & Collingridge, G. L. (1993). A synaptic model of memory: long-term potentiation in the hippocampus. *Nature, 361*, 31–39.

Bucherelli, C., Baldi, E., Mariottini, C., Passani, M. B., & Blandina, P. (2006). Aversive memory reactivation engages in the amygdala only some neurotransmitters involved in consolidation. *Learning & Memory, 13*, 426–430.

Castillo, P. E., Younts, T. J., Chávez, A. E., & Hashimotodani, Y. (2012). Endocannabinoid signaling and synaptic function. *Neuron, 76*, 70–81.

Chhatwal, J. P., Myers, K. M., Ressler, K. J., & Davis, M. (2005). Regulation of gephyrin and GABAA receptor binding within the amygdala after fear acquisition and extinction. *Journal of Neuroscience, 25*, 502–506.

Conklin, C. A., & Tiffany, S. T. (2002). Applying extinction research and theory to cue-exposure addiction treatments. *Addiction, 97*, 155–167.

Das, R. K., Kamboj, S. K., Ramadas, M., Yogan, K., Gupta, V., Redman, E., ... Morgan, C. J. (2013). Cannabidiol enhances consolidation of explicit fear extinction in humans. *Psychopharmacology (Berlin), 226*, 781–792.

De Carvalho, C. R., Pamplona, F. A., Cruz, J. S., & Takahashi, R. N. (2014). Endocannabinoids underlie reconsolidation of hedonic memories in Wistar rats. *Psychopharmacology (Berlin), 231*, 1417–1425.

Di Marzo, V. (2009). The endocannabinoid system: its general strategy of action, tools for its pharmacological manipulation and potential therapeutic exploitation. *Pharmacological Research: The Official Journal of the Italian Pharmacological Society, 60*, 77–84.

Einarsson, E. O., & Nader, K. (2012). Involvement of the anterior cingulate cortex in formation, consolidation, and reconsolidation of recent and remote contextual fear memory. *Learning & Memory, 19*, 449–452.

Fang, Q., Li, F. Q., Li, Y. Q., Xue, Y. X., He, Y. Y., Liu, J. F., ... Wang, J. S. (2011). Cannabinoid CB_1 receptor antagonist rimonabant disrupts nicotine reward-associated memory in rats. *Pharmacology, Biochemistry & Behavior, 99*, 738–742.

Foltin, R. W., & Haney, M. (2000). Conditioned effects of environmental stimuli paired with smoked cocaine in humans. *Psychopharmacology (Berlin), 149*, 24–33.

Harloe, J. P., Thorpe, A. J., & Lichtman, A. H. (2008). Differential endocannabinoid regulation of extinction in appetitive and aversive Barnes maze tasks. *Learning & Memory, 15*, 806–809.

Ibrahim, M. M., Deng, H., Zvonok, A., Cockayne, D. A., Kwan, J., Mata, H. P., ... Malan, T. P., Jr. (2003). Activation of CB_2 cannabinoid receptors by AM1241 inhibits experimental neuropathic pain: pain inhibition by receptors not present in the CNS. *Proceedings of the National Academy of Sciences of the United States of America, 100*, 10529–10533.

Kamprath, K., Marsicano, G., Tang, J., Monory, K., Bisogno, T., Di Marzo, V., ... Wotjak, C. T. (2006). Cannabinoid CB_1 receptor mediates fear extinction via habituation-like processes. *Journal of Neuroscience, 26*, 6677–6686.

Katsidoni, V., Anagnostou, I., & Panagis, G. (2013). Cannabidiol inhibits the reward-facilitating effect of morphine: involvement of 5-HT1A receptors in the dorsal raphe nucleus. *Addiction Biology, 18*, 286–296.

Klumpers, F., Denys, D., Kenemans, J. L., Grillon, C., van der Aart, J., & Baas, J. M. (2012). Testing the effects of Δ9-THC and D-cycloserine on extinction of conditioned fear in humans. *Journal of Psychopharmacology, 26*, 471–478.

Kobilo, T., Hazvi, S., & Dudai, Y. (2007). Role of cortical cannabinoid CB_1 receptor in conditioned taste aversion memory. *The European Journal of Neuroscience, 25*, 3417–3422.

Kringelbach, M. L. (2005). The human orbitofrontal cortex: linking reward to hedonic experience. *Nature Reviews. Neuroscience, 6*, 691–702.

Lewis, D. J. (1979). Psychobiology of active and inactive memory. *Psychology Bulletin, 86*, 1054–1083.

Lin, H. C., Mao, S. C., & Gean, P. W. (2006). Effects of intra-amygdala infusion of CB_1 receptor agonists on the reconsolidation of fear-potentiated startle. *Learning & Memory, 13*, 316–321.

Liput, D. J., Hammell, D. C., Stinchcomb, A. L., & Nixon, K. (2013). Transdermal delivery of cannabidiol attenuates binge alcohol-induced neurodegeneration in a rodent model of an alcohol use disorder. *Pharmacology, Biochemistry & Behavior, 111*, 120–127.

Maldonado, R., Valverde, O., & Berrendero, F. (2006). Involvement of the endocannabinoid system in drug addiction. *Trends in Neurosciences, 29*, 225–232.

Manwel, L., Satvat, E., Lang, S. T., Allen, C. P., Leri, F., & Parker, L. A. (2009). FAAH inhibitor, URB-597, promotes extinction and CB_1 antagonist, SR141716, inhibits extinction of conditioned aversion produced by naloxone-precipitated morphine withdrawal, but not extinction of conditioned preference produced by morphine in rats. *Pharmacology, Biochemistry & Behavior, 94*, 154–162.

Marsicano, G., Wotjak, C. T., Azad, S. C., Bisogno, T., Rammes, G., Cascio, M. G., ... Lutz, B. (2002). The endogenous cannabinoid system controls extinction of aversive memories. *Nature, 418*, 530–534.

Martin, P. D., & Shapiro, M. L. (2000). Disparate effects of long-term potentiation on evoked potentials and single CA1 neurons in the hippocampus of anesthetized rats. *Hippocampus, 10*, 207–212.

McGaugh, J. L. (2000). Memory a century of consolidation. *Science, 287*, 248–251.

Milton, A. L., & Everitt, B. J. (2012). The persistence of maladaptive memory: addiction, drug memories and anti-relapse treatments. *Neuroscience and Biobehavioral Reviews, 36*, 1119–1139.

Morgan, C. J., Das, R. K., Joye, A., Curran, H. V., & Kamboj, S. K. (2013). Cannabidiol reduces cigarette consumption in tobacco smokers: preliminary findings. *Addictive Behaviors, 38*, 2433–2436.

Müller, G. E., & Pilzecker, A. (1900). Experimentelle Beiträge zur Lehre vom Gedächtnis. *Zeitschrift für Psychologie. Ergänzungsband, 1*, 1–300.

Müller, M., Vandeleur, C., Rodgers, S., Rössler, W., Castelao, E., Preisig, M., & Ajdacic-Gross, V. (2014). Factors associated with comorbidity patterns in full and partial PTSD: findings from the PsyCoLaus study. *Comprehensive Psychiatry, 55*, 837–848.

Nader, K., Schafe, G. E., & LeDoux, J. E. (2000). The labile nature of consolidation theory. *Nature Reviews Neuroscience, 1*, 216–219.

Neumeister, A. (2013). The endocannabinoid system provides an avenue for evidence-based treatment development for PTSD. *Depression & Anxiety, 30*, 93–96.

Niesink, R. J., & van Laar, M. W. (2013). Does cannabidiol protect against psychological effects of THC? *Frontiers in Psychiatry, 4*, 130.

Niyuhire, F., Varvel, S. A., Thorpe, A. J., Stokes, R. J., Wiley, J. L., & Lichtman, A. H. (2007). The disruptive effects of the CB_1 receptor antagonist rimonabant on extinction learning in mice are task-specific. *Psychopharmacology (Berlin), 191*, 223–231.

de Oliveira Alvares, L., Pasqualini Genro, B., Diehl, F., Molina, V. A., & Quillfeldt, J. A. (2008). Opposite action of hippocampal CB_1 receptors in memory reconsolidation and extinction. *Neurosciece, 154*, 1648–1655.

Pamplona, F. A., Bitencourt, R. M., & Takahashi, R. N. (2008). Short- and long-term effects of cannabinoids on the extinction of contextual fear memory in rats. *Neurobiology of Learning & Memory, 90*, 290–293.

Pamplona, F. A., Prediger, R. D., Pandolfo, P., & Takahashi, R. N. (2006). The cannabinoid receptor agonist WIN 55,212-2 facilitates the extinction of contextual fear memory and spatial memory in rats. *Psychopharmacology (Berlin), 188*, 641–649.

Parker, L. A., Burton, P., Sorge, R. E., Yakiwchuk, C., & Mechoulam, R. (2004). Effect of low doses of delta9-tetrahydrocannabinol and cannabidiol on the extinction of cocaine-induced and amphetamine-induced conditioned place preference learning in rats. *Psychopharmacology (Berlin), 175*, 360–366.

Parsons, R. G., & Ressler, K. J. (2013). Implications of memory modulation for post-traumatic stress and fear disorders. *Nature Neuroscience, 16*, 146–153.

Plendl, W., & Wotjak, C. T. (2010). Dissociation of within- and between-session extinction of conditioned fear. *Journal of Neuroscience, 30*, 4990–4998.

Rabinak, C. A., Angstadt, M., Sripada, C. S., Abelson, J. L., Liberzon, I., Milad, M. R., & Phan, K. L. (2013). Cannabinoid facilitation of fear extinction memory recall in humans. *Neuropharmacology, 64*, 396–402.

Ren, Y., Whittard, J., Higuera-Matas, A., Morris, C. V., & Hurd, Y. L. (2009). Cannabidiol, a nonpsychotropic component of cannabis, inhibits cue-induced heroin-seeking and normalizes discrete mesolimbic neuronal disturbances. *The Journal of Neuroscience, 29*, 14764–14769.

Robbins, T. W., Ersche, K. D., & Everitt, B. J. (2008). Drug addiction and the memory systems of the brain. *Annals of the New York Academy of Sciences, 1141*, 1–21.

Roedirger, H. L., Dudai, Y., & Fitzpatrick, S. M. (2007). *Science of memory concepts*. New York: Oxford University Press.

Sidhpura, N., & Parsons, L. H. (2011). Endocannabinoid-mediated synaptic plasticity and addiction-related behavior. *Neuropharmacology, 61*, 1070–1087.

Starowicz, K., Nigam, S., & Di Marzo, V. (2007). Biochemistry and pharmacology of endovanilloids. *Pharmacology & Therapeutics, 114*, 13–33.

Stern, C. A., Gazarini, L., Takahashi, R. N., Guimarães, F. S., & Bertoglio, L. J. (2012). On disruption of fear memory by reconsolidation blockade: evidence from cannabidiol treatment. *Neuropsychopharmacology, 37*, 2132–2142.

Stern, C. A., Gazarini, L., Vanvossen, A. C., Zuardi, A. W., Galve-Roperh, I., Guimaraes, F. S., … Bertoglio, L. J. (2015). Δ^9-Tetrahydrocannabinol alone and combined with cannabidiol mitigate fear memory through reconsolidation disruption. *European Neuropsychopharmacology, 25*, 958–965.

Suzuki, A., Josselyn, S. A., Frankland, P. W., Masushige, S., Silva, A. J., & Kida, S. (2004). Memory reconsolidation and extinction have distinct temporal and biochemical signatures. *The Journal of Neuroscience, 24*, 4787–4795.

Suzuki, A., Mukawa, T., Tsukagoshi, A., Frankland, P. W., & Kida, S. (2008). Activation of LVGCCs and CB_1 receptors required for destabilization of reactivated contextual fear memories. *Learning & Memory, 15*, 426–433.

Torregrossa, M. M., Corlett, P. R., & Taylor, J. R. (2011). Aberrant learning and memory in addiction. *Neurobiology of Learning & Memory, 96*, 609–623.

Torregrossa, M. M., & Taylor, J. R. (2013). Learning to forget: manipulating extinction and reconsolidation processes to treat addiction. *Psychopharmacology (Berlin), 226*, 659–672.

Tronson, N. C., & Taylor, J. R. (2007). Molecular mechanisms of memory reconsolidation. *Nature Reviews. Neuroscience, 8*, 262–275.

Tzschentke, T. M. (2007). Measuring reward with the conditioned place preference (CPP) paradigm: update of the last decade. *Addiction Biology, 12*, 227–462.

Van Sickle, M. D., Duncan, M., Kingsley, P. J., Mouihate, A., Urbani, P., Mackie, K., … Sharkey, K. A. (2005). Identification and functional characterization of brainstem cannabinoid CB_2 receptors. *Science, 310*, 329–332.

Volkow, N. D., Fowler, J. S., & Wang, G. J. (2004). The addicted human brain viewed in the light of imaging studies: brain circuits and treatment strategies. *Neuropharmacology, 47*, 3–13.

Wilson, R. I., & Nicoll, R. A. (2002). Endocannabinoid signaling in the brain. *Science, 296*, 678–682.

Xi, Z. X., Peng, X. Q., Li, X., Song, R., Zhang, H. Y., Liu, Q. R., … Gardner, E. L. (2011). Brain cannabinoid CB_2 receptors modulate cocaine's actions in mice. *Nature Neuroscience, 14*, 1160–1166.

Yu, L. L., Wang, X. Y., Zhao, M., Liu, Y., Li, Y. Q., Li, F. Q., … Lu, L. (2009). Effects of cannabinoid CB_1 receptor antagonist rimonabant in consolidation and reconsolidation of methamphetamine reward memory in mice. *Psychopharmacol (Berlin), 204*, 203–211.

Chapter 76

Effects of Δ⁹-Tetrahydrocannabinol, Synthetic Cannabinoids, and Fatty Acid Amide Hydrolase Inhibitors on Mood and Serotonin Neurotransmission

Gabriella Gobbi[1], Nicolas Nuñez[1,2], Ryan McLaughlin[1,3], Francis Bambico[1,4]

[1]*Neurobiological Psychiatry Unit, McGill University, Montreal, QC, Canada;* [2]*Hospital Neurospsiquiatrico de Agudos y Cronicos Dr. Alejandro Korn/ Universidad, Nacional de La Plata, Buenos Aires, Argentina;* [3]*Department of Integrative Physiology and Neuroscience, College of Veterinary Medicine, Washington State University, Pullman, WA, USA;* [4]*Centre for Addiction and Mental Health, University Toronto, Toronto, ON, USA*

Abbreviations

5-HT 5-Hydroxytryptamine, serotonin
CB1R Cannabinoid type 1 receptor
DR Dorsal raphe
FAAH Fatty acid amide hydrolase
FST Forced swim test
GABA γ-Aminobutyric acid
ip Intraperitoneal
iv Intravenous
mPFCv Ventromedial prefrontal cortex
PFC Prefrontal cortex
SSRI Selective serotonin reuptake inhibitor
Δ⁹-THC (−)*trans*-Δ⁹-tetrahydrocannabinol

INTRODUCTION

For many centuries, the *Cannabis sativa* plant has been known to possess psychotropic effects linked to the regulation of mood. It was early in the seventeenth century when Robert Burton considered cannabis as a possible treatment for depression in his book: *The Anatomy of Melancholy*. The idea that the mood-elevating properties of cannabis could treat depression was soon set aside because of multiple adverse side effects and inconsistent efficacy, leaving a controversy over its mechanisms of action.

Certainly, one of the most salient effects of cannabinoids is related to elevation of mood: subjective effects include a sensation of "high," anxiety relief, and the feeling of being "relaxed" (Earleywine, 2005; Iversen, 2003). This subjective experience is highly variable depending on the dose of drug, the environment, and the experience and expectations of the drug user. The high produced by cannabis is a complex experience, characterized by a quickening of mental associations and a sharpened sense of humor, sometimes described as a state of "fatuous euphoria."

A breakthrough discovery in the twentieth century was the acknowledgment of a multifaceted endogenous cannabinoid system (i.e., endocannabinoid system) that plays a vital role in the etiology and pathophysiology of brain dysfunctions. Despite this knowledge, very little is known about the mechanism(s) that underlies the mood-elevating effects of cannabis, but it is very likely mediated by the central cannabinoid receptor CB1 (CB1R) since the CB1R antagonist rimonabant blocks this effect (Huestis et al., 2001).

When our laboratory started this line of research in 2002, only a few articles were published on the link between cannabis and serotonin (5-HT) function. At that time, it was known that CB1R was present at the level of the nucleus of the dorsal raphe (DR) (Matsuda, Bonner, & Lolait, 1993; Moldrich & Wenger, 2000; Tsou et al., 1998), the major source of 5-HT neurons in the brain, and that fatty acid amide hydrolase (FAAH), the enzyme responsible for the degradation of the endocannabinoid anandamide, was present in its oligodendrocytes (Egertova, Cravatt, & Elphick, 2003). These anatomical data were consistent with an electrophysiological study reporting that postsynaptic orexin receptors can modulate glutamatergic synaptic transmission to DR 5-HT neurons through retrograde endocannabinoid signaling (Haj-Dahmane & Shen, 2005). It was known that 5-HT1B, 5-HT2A (Devlin & Christopoulos, 2002), and 5-HT3 receptor subtypes were coexpressed with CB1R (Hermann, Marsicano, & Lutz, 2002) in γ-aminobutyric acid (GABA)-ergic interneurons (Morales, Wang, Diaz-Ruiz, & Jho, 2004), as was the inhibitory functional interaction between CB1R and 5-HT2A receptors (Darmani, 2001; Kimura, Ohta, Watanabe, Yoshimura, & Yamamoto, 1998) and between CB1R and 5-HT3 receptors (Barann et al., 2002; Fan, 1995). Moreover, it was reported that somatodendritic 5-HT1A receptors are involved in the hypothermic effects of Δ⁹-tetrahydrocannabinol (Δ⁹-THC) administration (Malone & Taylor, 2001), suggesting an interaction between these two systems at the hypothalamic level. At this time, the Gorzalka lab at the University of British Columbia showed that chronic treatment with the CB1R agonist HU-210 alters pharmacological responses to 5-HT2A and 5-HT1A receptor agonists (Hill, Sun, Tse, & Gorzalka, 2006).

However, the extent to which activation of the endocannabinoid system could directly modulate 5-HT firing properties was unknown. For these reasons, our laboratory started to better dissect the possible involvement of these compounds in modulating 5-HT firing activity. The activity of a putative antidepressant drug on 5-HT firing is essential for predicting its potential therapeutic effect on mood regulation.

HOW TO ASSESS IF A PUTATIVE DRUG MODULATES MOOD USING ANIMAL MODELS

Major depression is a heterogeneous disorder, but despite its complexity, several animal models able to mimic symptoms of human depression have been developed. Importantly, these tests, such as the forced swim test (FST), tail suspension test (TST), learned helplessness model, and chronic unpredictable stress paradigm, have been repeatedly validated and are currently the most popular models for detecting antidepressant-like activity because of their ease of use, reliability, and high predictive validity (see our review, Bambico & Gobbi, 2008).

It is well known that people suffering from mood disorders show impaired function of the 5-HT and/or norepinephrine system and that different classes of currently available antidepressants mostly act by modulating, directly or indirectly, these two neurotransmitter systems. However, compelling evidence indicates that another monoamine, dopamine, plays a role in the pathophysiology and treatment of depression. Using in vivo electrophysiology, it is possible to examine the modulation of these monoamines by clinically prescribed antidepressants. All conventional antidepressants increase/decrease monoaminergic electrical activity as final pathways, albeit through different mechanisms (for more details, see Gobbi & Blier, 2005; Bambico & Gobbi, 2008). Consequently, testing whether a novel compound modulates monoaminergic neurotransmission using in vivo electrophysiology represents a valid framework to assess its potential for antidepressant action.

Δ^9-TETRAHYDROCANNABINOL AND SEROTONIN FIRING ACTIVITY

The primary pharmacologically active component of cannabis, Δ^9-THC, probably mediates most of its psychoactive and mood-related effects. Although heavy or high-dose cannabis use has been associated with an elevated risk for developing mood disorders, anxiety, psychosis, and cognitive impairment, especially among teenagers, cannabis has been routinely used for self-medicating depressive symptoms, suggesting that it could have therapeutic benefits in primary and secondary depression (for review see Bambico & Gobbi, 2008). Nabilone (Cesamet), a synthetic Δ^9-THC derivative commercialized in Canada, has been reported to increase mood in 38% of people and to induce euphoria in another 14% (from *Compendium of Pharmaceuticals and Specialties*, Canada, 2012).

Similarly in rodents, antidepressant-like activity following Δ^9-THC administration has been reported in preclinical animal models such as the FST (Bambico, Hattan, Garant, & Gobbi, 2012; El-Alfy et al., 2010; Moreira, Grieb, & Lutz, 2004), the olfactory bulbectomy model (Elbatsh, Spicer, Marsden, Fone, & Kendall, 2009; Rodriguez-Gaztelumendi, Rojo, Pazos, & Diaz, 2009), and the TST (El-Alfy et al., 2010). On the other hand, Egashira et al. (2008) reported that Δ^9-THC administration increased, rather than decreased, immobility in the FST, which is indicative of behavioral despair. These conflicting results suggest that Δ^9-THC may also exert depressogenic effects under certain conditions, similar to cannabis use in humans.

Given the dual effects of cannabis and cannabis constituents on mood and antidepressant-like behavior, respectively, we examined whether Δ^9-THC modulates single-unit 5-HT firing in vivo. The intravenous administration of various doses of Δ^9-THC (0–1.6mg/kg) yielded a complex response profile. In fact, we identified three different groups of 5-HT neurons on the basis of their response to Δ^9-THC within this dose range: 25.58% ($n=22$) were excited ($\geq 10\%$ of baseline), 32.56% ($n=28$) were inhibited (<10% of baseline), and 41.86% ($n=36$) were nonresponding ($\chi^2=3.5$, $p=0.17$). These neurons, regardless of their response, exhibited identical electrophysiological characteristics and were all recorded from the rostrocaudal midline extent of the DR nucleus. The excitatory responses were mainly produced by doses >0.45mg/kg and were maximal at 1.0mg/kg. The inhibitory responses were mainly produced by doses <0.45mg/kg and were maximal at 0.4mg/kg. Inert responses were equally distributed between low and high dose ranges. In the brain, Δ^9-THC binds mostly to CB1R, but at higher doses may also bind to the transient receptor potential vanilloid type 1 (TRPV1) channel, also known as the capsaicin receptor or vanilloid receptor 1. In an attempt to determine which receptor subsystems were responsible for the contrasting excitatory and inhibitory responses, the CB1R antagonist rimonabant (1.0mg/kg) or the TRPV1 channel antagonist capsazepine (0.01 mg/kg) was intravenously administered following a Δ^9-THC-induced increase or decrease in DR 5-HT firing. We found that the excitatory response was attenuated by rimonabant, but not by capsazepine, suggesting that Δ^9-THC-induced increases in 5-HT firing are mediated by CB1R activation. On the other hand, the inhibitory response was only partially reversed by rimonabant in one of three neurons and was not at all sensitive to capsazepine, indicating a non-CB1R- and a non-TRPV1-mediated mechanism (Bambico et al., 2012).

Acute intraperitoneal (ip) administration of the dose of Δ^9-THC shown earlier to evoke the maximal excitatory response from 5-HT neurons (1.0mg/kg) produced a modest but nonsignificant enhancement of spontaneous 5-HT firing. Increasing the dose to 2 and 4mg/kg similarly yielded nonsignificant elevations in the mean 5-HT firing rate. However, repeated administration (once daily for 5days) of Δ^9-THC (1.0mg/kg, ip) produced a significant increase in the mean spontaneous DR 5-HT firing rate ($p<0.05$), which was blocked by the coapplication of rimonabant (1.0mg/kg, ip), which is suggestive of a CB1R-mediated effect (Figure 1). The reason for this discrepancy between acute vs chronic Δ^9-THC treatment is unknown, although it may be due to the pharmacodynamic/pharmacokinetic properties of Δ^9-THC or possibly a

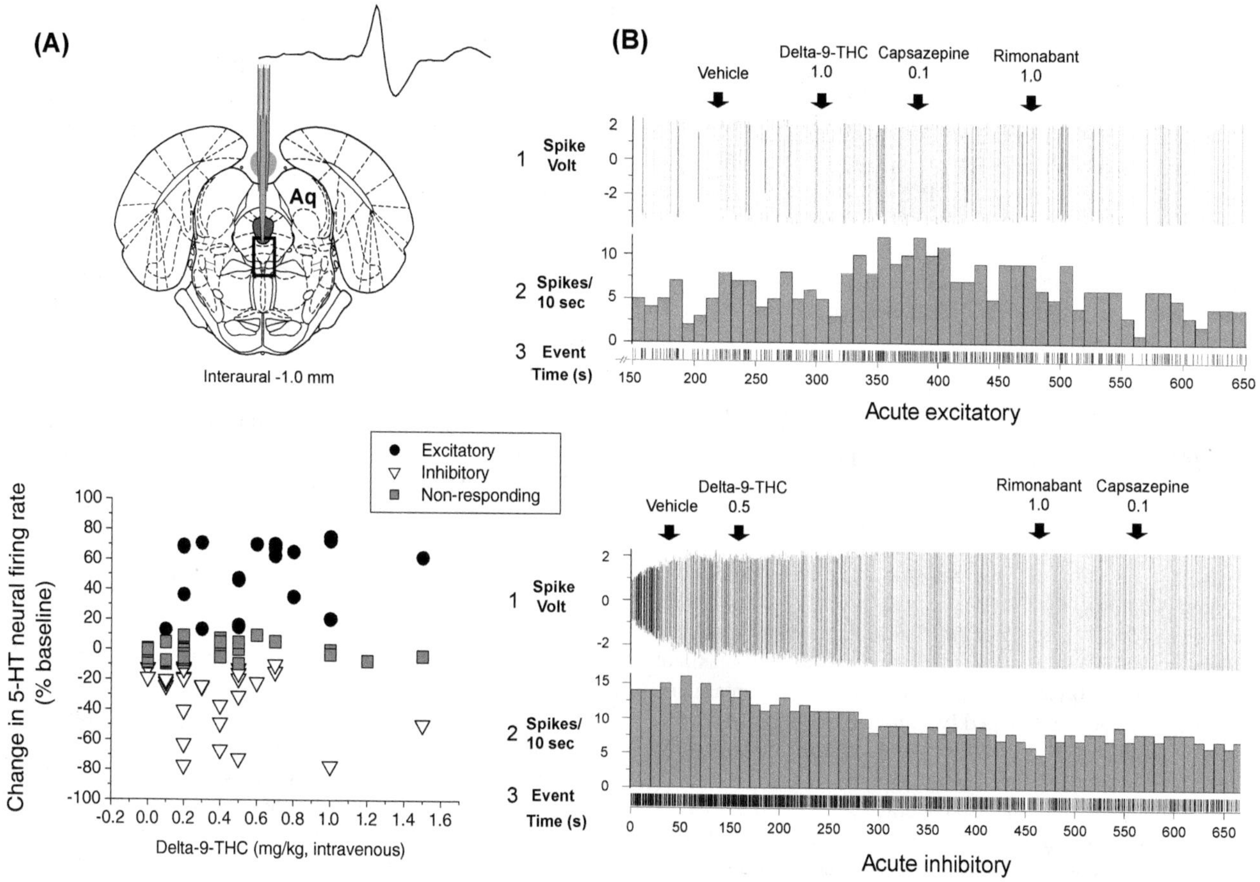

FIGURE 1 (A) Acute intravenous (iv) administrations of variable doses of Δ^9-THC elicited a complex response profile from DR 5-HT neurons, with neurons exhibiting either excitatory or inhibitory responses, as well as inert ones. A plate from Paxinos and Watson (2007) illustrating a brain coronal section through the DR nucleus is shown. The boxed area shows where putative 5-HT neurons were recorded. An exemplary spike waveform of a putative 5-HT neuron is shown above. (B, top panel) A representative 5-HT neural firing rate histogram showing that a Δ^9-THC-evoked excitatory response was reversed by the CB1R antagonist rimonabant but not by the vanilloid (TRPV1) antagonist capsazepine. (B, bottom panel) A representative 5-HT neural firing rate histogram showing that a Δ^9-THC-evoked inhibitory response was reversed neither by rimonabant nor capsazepine. Numbers above arrows indicate the iv dose administered. *From Bambico et al. (2012), Progress in Neuropsychopharmacology and Biological Psychiatry, with permission.*

long-term neuroplastic or synaptic modification produced by prolonged CB1R activation (Bambico et al., 2012).

Similar data were found in the FST, a popular preclinical paradigm used to assess the putative antidepressant-like properties of a drug (Castagne, Moser, Roux, & Porsolt, 2011). Following FST exposure, we found that acute Δ^9-THC administration (1.0 mg/kg, ip) was not sufficient to elicit differences in either the total duration or the frequency of active vs passive coping behaviors compared to control-treated rats. Conversely, a repeated administration schedule (5 days) elicited an antidepressant-like response, characterized by an increase in swimming duration and a decrease in immobility, similar to the conventional selective 5-HT reuptake inhibitor (SSRI) citalopram (Figure 1, Bambico et al., 2012). Moreover, chronic Δ^9-THC, but not citalopram, significantly increased climbing duration ($p=0.007$), suggesting that the antidepressant-like effects of Δ^9-THC administration are mediated by the noradrenergic system.

Last, we showed that repeated (5 days), but not acute, administration of Δ^9-THC (1.0 mg/kg, ip) increased the excitatory response of hippocampal CA3 pyramidal neurons to the 5-HT1A receptor antagonist WAY 100635 in a manner similar to 5-day treatment with citalopram (Bambico et al., 2012). These findings suggest that Δ^9-THC, similar to other classes of antidepressants, increases tonic 5-HT1A postsynaptic receptor activity, but only following a long-term treatment regimen (Besson, Haddjeri, Blier, & de Montigny, 2000; Haddjeri, Blier, & de Montigny, 1998). In summary, these electrophysiological experiments revealed that administration of various doses of Δ^9-THC produces a complex multimodal response, with excitation, inhibition, and nonresponse among different populations of 5-HT neurons. Additional research examining the pharmacological effects of Δ^9-THC on monoaminergic transmission and emotional processes is warranted. Evidence suggests that Δ^9-THC could act both as a partial and as a full CB1R agonist, depending on whether the receptor is localized to GABAergic or glutamatergic synapses (Laaris, Good, & Lupica, 2010).

Repeated Δ^9-THC treatment (1.0 mg/kg) yielded a significant elevation in the mean discharge rate of spontaneously active DR 5-HT neurons, an effect that was reversed by coadministration with the CB1R antagonist rimonabant. Similar results have also been obtained with the synthetic CB1R agonist WIN 55,212-2 ex vivo (Mendiguren & Pineda, 2009). Moreover, these researchers showed that both rimonabant and the CB1R inverse agonist AM251 decreased DR 5-HT neural activity in approximately 50% of neurons recorded (Mendiguren & Pineda, 2009).

Thus, there appears to be a general consensus that the endocannabinoid system modulates 5-HT neurotransmission, with CB1R activation resulting in enhanced DR 5-HT firing and increased postsynaptic 5-HT1A receptor activation. However, the response profile of CB1R agonists and antagonists is far from uniform, which could be due to the ubiquitous expression profile of CB1Rs and their ability to modulate both excitatory and inhibitory transmission in an on-demand fashion.

THE EFFECT OF WIN 55,212-2 ON 5-HT FIRING AND ANTIDEPRESSANT-LIKE ACTIVITY

In our laboratory, we have assessed the spontaneous single-unit firing activity of DR 5-HT neurons after cumulative intravenous (iv) administration of the CB1R agonist WIN 55,212-2. Increasing doses of WIN 55,212-2 (0.05–0.2 mg/kg) evoked a dose-dependent increase in 5-HT unit firing activity, which was half-maximal (ED_{50}) at a dose of 0.1 mg/kg and was not blocked by capsazepine (20 µg/kg, iv), but was blocked by rimonabant (1 mg/kg, iv) in 100% of neurons tested (Figure 2), (Bambico, Katz, Debonnel, & Gobbi, 2007).

WIN 55,212-2 treatment also increased 5-HT burst activity, a pattern that is associated with antidepressant-like activity (Gobbi et al., 2005). The maximal increase in burst frequency and in the mean number of spikes in a burst from baseline was recorded following the 0.2 mg/kg dose (Figure 2(E); Bambico et al., 2007). Among all neurons recorded, 66.67% of 5-HT neurons responded to increasing dose injections of WIN 55,212-2, while 33.33% of neurons were nonresponding. All responding and nonresponding neurons showed the same electrophysiological characteristics, were inhibited by the $GABA_B$ agonist baclofen, and were localized to the DR, thus indicating that not all DR neurons are activated by CB1R stimulation.

Importantly, cumulative doses higher than 0.2 mg/kg of WIN 55,212-2 injected iv produced a general decline in neuronal excitation that was significant at both 0.3 and 0.4 mg/kg and achieved a maximal level 45% below baseline (i.e., vehicle) following 0.4 mg/kg WIN 55,212-2. A waning of stimulatory effects was also observed with different parameters of burst activity: burst frequency, mean number of spikes in a burst, and mean burst length. In two of three neurons tested, capsazepine reversed the decrease induced by high doses of WIN 55,212-2, suggesting that TRPV1 receptors, but not CB1Rs, are involved in the inhibitory effects of high doses of WIN 55,212-2 on DR 5-HT firing. Interestingly, neither rimonabant (1 mg/kg, iv) nor capsazepine (20 µg/kg, iv) alone had a significant effect on 5-HT single-unit firing activity, meaning that at low doses these two drugs may act as inert antagonists or that such doses are unable to unmask a putative tonic cannabinoid- or vanilloid-mediated regulation of 5-HT firing.

We found that in the FST, WIN 55,212-2 at the dose of 0.1–0.2 mg/kg administered 23, 5, and 0.75 h before the test produced antidepressant-like effects characterized by decreased immobility and increased swimming activity (Bambico et al., 2007). However, the antidepressant-like activity disappeared when higher doses were injected (Bambico et al., 2007).

Incremental doses of WIN 55,212-2 and the mean spontaneous firing rate of 5-HT neurons showed a biphasic relationship. One-way analysis of variance revealed a dose-dependent increase with lower doses of WIN 55,212-2, while coadministration of rimonabant prevented this increase. A dose of 0.2 mg/kg WIN 55,212-2 yielded a maximal 126.32% increase in neuronal activity. On the other hand, a much higher dose of WIN 55,212-2 (2.0 mg/kg) yielded a significant 64% decrease in 5-HT firing compared to vehicle. We also calculated the mean number of neurons per electrode descent, which serves as an indirect measure of spontaneously active neurons (Gobbi et al., 2007). In comparison with vehicle injections, there were 28% more spontaneously active 5-HT neurons encountered after treatment with 0.1 mg/kg WIN 55,212-2 and 33.33% more active neurons with 0.2 mg/kg WIN 55,212-2 ($p<0.01$), while a high dose of WIN 55,212-2 (2.0 mg/kg) had 48.8% fewer active neurons than the control ($p<0.01$). The number of spontaneously active neurons in rats treated with a low dose of WIN 55,212-2 (0.2 mg/kg) coapplied with rimonabant (1.0 mg/kg) did not significantly differ from that of those treated with vehicle (Bambico et al., 2007).

We also asked if the WIN 55,212-2-induced stimulation of DR 5-HT neurons is mediated by CB1Rs located in neurons of the prefrontal cortex (PFC) that project to the DR. Indeed, DR 5-HT neurons receive important excitatory inputs from pyramidal (glutamatergic) cells of the PFC (Jankowski & Sesack, 2004). We performed systematic transections of the PFC–DR pathway prior to conducting electrophysiological recordings. Three types of transections were performed: a total bilateral PFC transection (tPFC), a selective transection of PFC focusing on the medial PFC (mPFC; areas transected included the dorsal peduncular, infralimbic, prelimbic Cg3, and cingulate Cg1 cortices), and a transection of the lateral aspect of the PFC (latPFC): mainly the lateral prefrontal/agranular insular, but also the frontal Fr2, Fr1, and Fr3; the ventrolateral and lateral orbital cortices; and some parts of parietal area 1 (modified after Hajós, Hajós-Korcsok, & Sharp, 1999). 5-HT single-unit recordings were conducted 1.5–2 h after each transection.

Following tPFC transection, WIN 55,212-2 failed to increase 5-HT single-unit firing activity at otherwise stimulatory doses in intact brains. To pinpoint the specific subregion of the PFC that is critical in mediating the modulation of 5-HT single-unit activity, we compared transection of the mPFC with that of the latPFC. The response of 5-HT single units to the latPFC transection did not significantly differ from the control, but on the other hand, mPFC transection produced an effect similar to tPFC transection and was significantly different from the control, thus indicating that the medial, but not lateral, subregions of the PFC contribute to the enhanced 5-HT firing activity following CB1R activation (Bambico et al., 2007). To further corroborate this hypothesis we also injected WIN 55,212-2 directly into the mPFC and then performed the FST; and we observed that the injection of the CB1 agonist into the mPFC was sufficient to induce an antidepressant-like effect and increase 5-HT firing activity of DR neurons (Bambico et al., 2007).

Altogether these results indicate that the ventromedial PFC (mPFCv) plays an instrumental role in mediating DR 5-HT firing and that local CB1R activation within this region exerts antidepressant-like effects via enhanced mPFCv-driven stimulation of DR 5-HT neuronal firing.

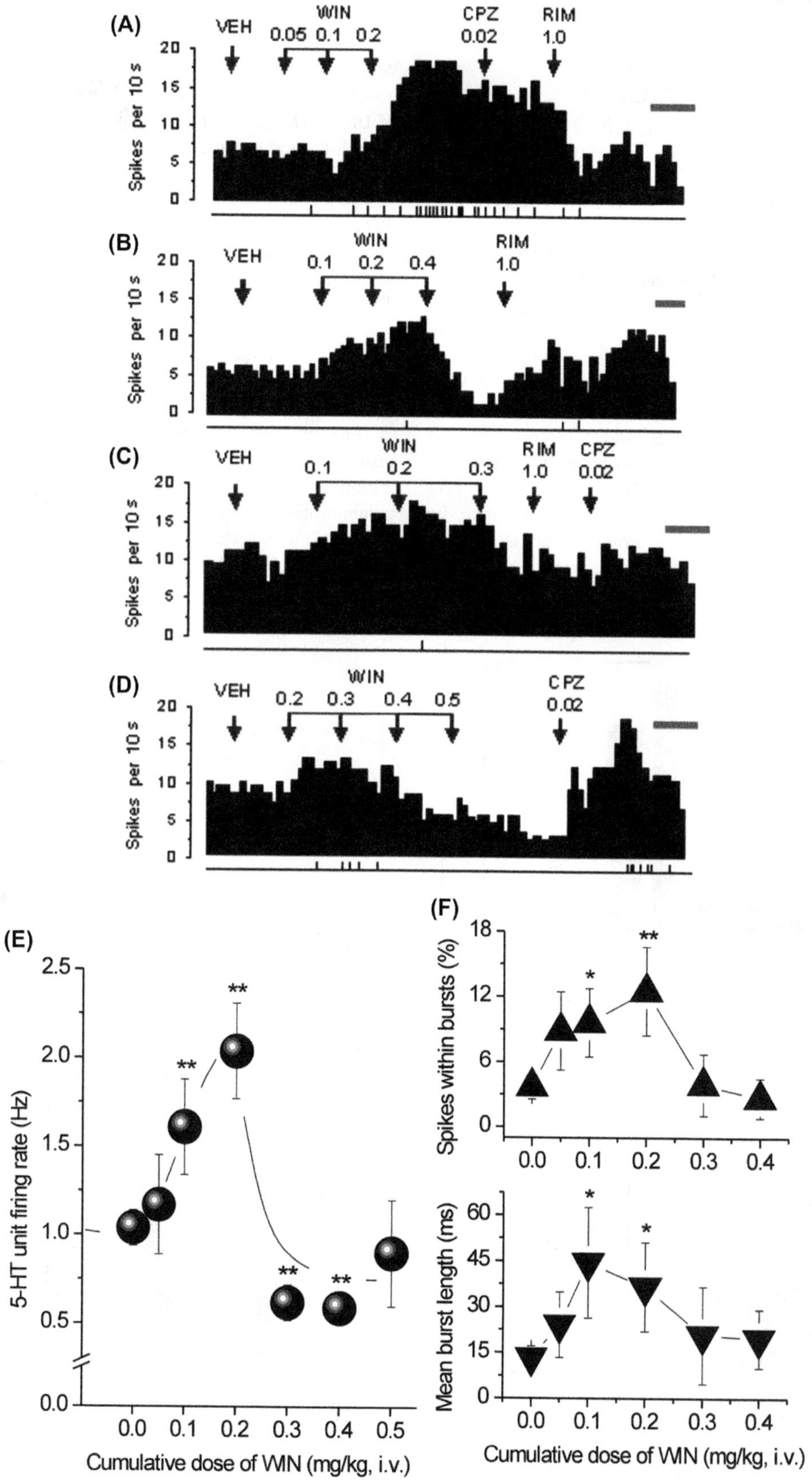

FIGURE 2 Effect of iv administration of cumulative doses of WIN 55,212-2 on DR 5-HT neurons. (A–D) Integrated firing rate histograms of 5-HT neurons illustrating that low doses of WIN (0.1–0.2 mg/kg, iv) rapidly increased single-unit firing activity. (A) This effect was reversed by rimonabant (RIM) (1.0 mg/kg, iv; $n=4$) but not by capsazepine (CPZ) (0.02 mg/kg, iv; $n=4$). (B–D) High dose of WIN (0.30–0.50 mg/kg, iv) rapidly decreased 5-HT single-unit firing activity. This effect was reversed by CPZ (0.02 mg/kg, iv) in two of three neurons (D) and partially reversed (B) or not reversed (C) by RIM (1 mg/kg, iv) in one and three neurons, respectively. 5-HT neuronal firing rate in each histogram is plotted as spikes per 10 s. Calibration bar on the right side of each histogram, 1 min. The vertical lines depicted below each histogram represent the frequency of neuronal burst activity such that each tick corresponds to a burst discharge event. (E) WIN (0.05–0.5 mg/kg, iv) produced a biphasic response profile in 5-HT single-unit activity. (F) Line graphs showing that cumulative doses of WIN modulated 5-HT neuronal burst activity measured as percentage of spikes within bursts and mean burst length (bottom). $*p<0.05$ or $**p<0.01$ vs baseline (vehicle). *From Bambico et al. (2007), Journal of Neuroscience, with permission.*

FATTY ACID AMIDE HYDROLASE INHIBITION, 5-HT FIRING, AND ANTIDEPRESSANT-LIKE ACTIVITY

The selective inhibitor of the enzyme FAAH, which catalyzes the intracellular hydrolysis of the endocannabinoid anandamide, has been proposed to be a useful alternative to direct CB1R agonists because of their capacity to increase endogenous CB1R signaling without inducing the typical cannabis-related side effects such as dependence and sedation (see Bambico & Gobbi, 2008 for review).

One of the first experiments carried out in our laboratory was to test whether the FAAH inhibitor URB 597 influences DR 5-HT transmission. We first measured spontaneous activity of 5-HT neurons in the DR of anesthetized rats. Single injections of URB 597 (0.03–0.3 mg/kg, iv) evoked a slow increase in 5-HT neuron firing activity, which was half-maximal at a dose of ≈0.06 mg/kg and was blocked by pretreatment with rimonabant (1 mg/kg, iv; Figure 3). Interestingly, the increase in firing was not as immediate as observed with the CB1R direct agonist, occurring only after 15–20 min; this delay is compatible with the pharmacodynamics of URB 597, which after passing the blood–brain barrier, inhibits FAAH, leading to a gradual accumulation in anandamide content, which subsequently activates CB1Rs (Gobbi et al., 2005).

This response was recapitulated following subchronic URB 597 treatment. Indeed, 4-day treatment with URB 597 (0.1 mg/kg, ip, once daily for 4 days) evoked an even stronger response, which was also reversed by rimonabant (1 mg/kg, ip) (Gobbi et al., 2005). This sustained 5-HT increase after a subchronic treatment regimen was also associated with an increase in 5-HT bursting activity and sustained 5-HT release in the hippocampus, but not in the PFC, as assessed by in vivo microdialysis in awake rats (Gobbi et al., 2005). A single injection of URB 597 had no such effect.

Finally, 4-day treatment with URB 597 did not affect the responsiveness of 5-HT neurons to local iontophoretic administration of the 5-HT1A receptor agonist 8-hydroxy-2-(di-n-propylamino)tetralin hydrobromide (8-OH-DPAT), suggesting that URB 597, unlike classical antidepressants (Artigas, Romero, de Montigny, & Blier, 1996; Gobbi & Blier, 2005), did not produce

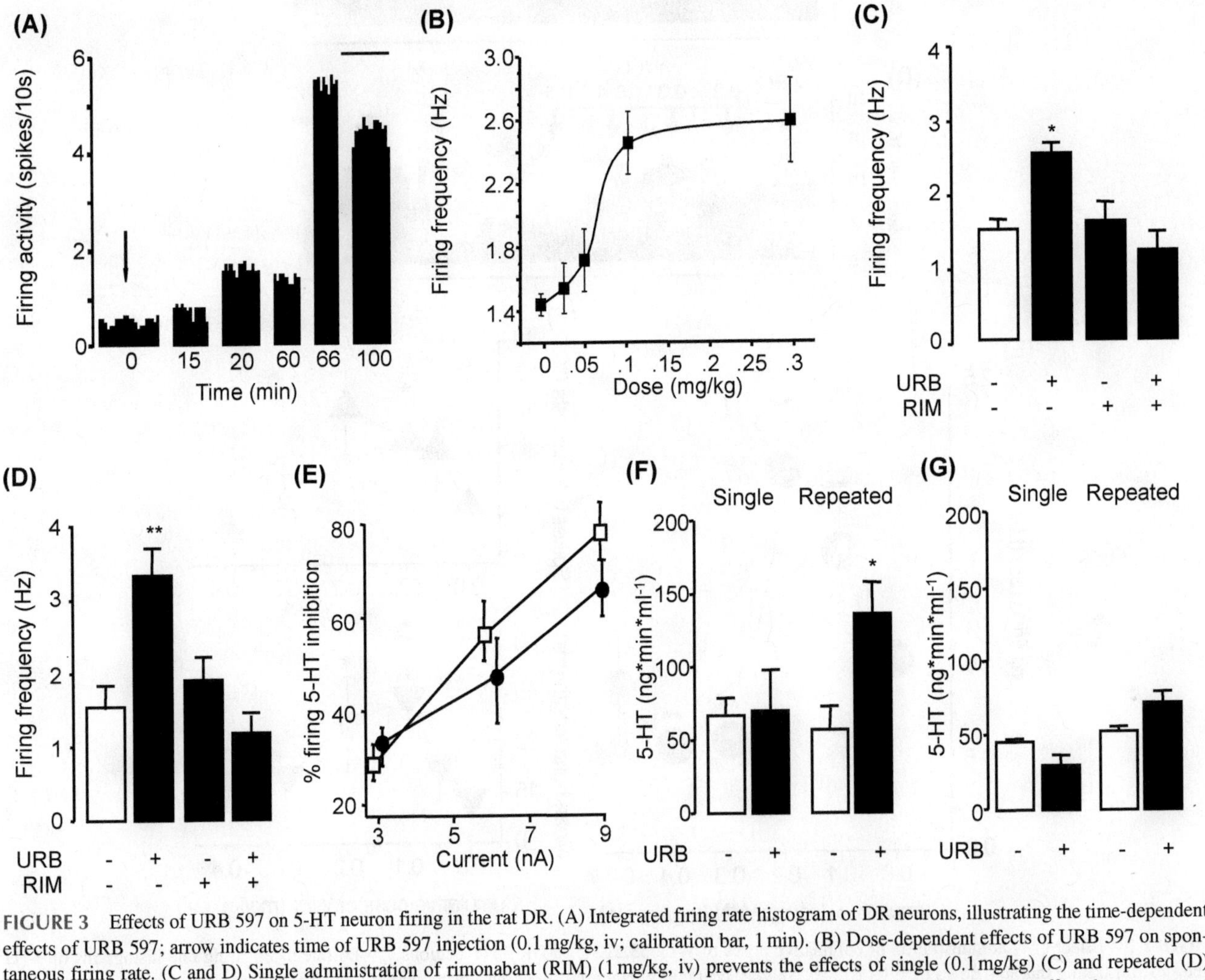

FIGURE 3 Effects of URB 597 on 5-HT neuron firing in the rat DR. (A) Integrated firing rate histogram of DR neurons, illustrating the time-dependent effects of URB 597; arrow indicates time of URB 597 injection (0.1 mg/kg, iv; calibration bar, 1 min). (B) Dose-dependent effects of URB 597 on spontaneous firing rate. (C and D) Single administration of rimonabant (RIM) (1 mg/kg, iv) prevents the effects of single (0.1 mg/kg) (C) and repeated (D) URB 597 injections (0.1 mg/kg, ip, once daily for 4 days) on 5-HT neuron firing. (E) Repeated URB 597 administration does not affect the response of 5-HT neurons to 8-hydroxy-2-(di-n-propylamino)tetralin, expressed as percentage inhibition of 5-HT-neuron firing rate. Open symbols represent vehicle. (F and G) Effects of single or repeated URB 597 injections on 5-HT outflow over 3 h in hippocampus (F) and PFC (G) of awake rats. $*p < 0.05$ vs vehicle; $**p < 0.01$ vs vehicle. *From Gobbi et al. (2005), Proceedings of the National Academy of Sciences, with permission.*

desensitization of 5-HT1A autoreceptors. Importantly, at the same doses (0.1 and 0.3 mg/kg) and duration (once a day for 4 days), URB 597 induced antidepressant-like effects in the mouse TST and the rat FST. This effect was more robust after repeated injections (4 days) and was reversed by the preadministration of the CB1R antagonist rimonabant (Figure 3; Gobbi et al., 2005).

Since we have previously shown that intra-mPFCv administration of WIN 55,212-2 elicits antidepressant-like responses as well as increases in DR 5-HT firing, we next examined whether *endogenous* medial prefrontocortical cannabinoid signaling modulates coping behaviors in the FST using a combination of behavioral, pharmacological, biochemical, and electrophysiological approaches. We first examined how FST exposure affects endocannabinoid ligand content in the mPFC using mass spectrometry and revealed that anandamide content experiences a rapid and robust decline immediately following the first FST exposure session. Anandamide content was partially (but not fully) restored when examined 24 h later, but was subject to an even greater decline following a second FST exposure session, and this was accompanied by behavioral despair (i.e., increased immobility) (McLaughlin et al., 2012). Thus, fluctuations in anandamide signaling in the mPFC were suspected to mediate the transition between active and passive coping strategies. To support this claim, we next demonstrated that local inhibition of FAAH in the mPFCv with URB 597 reduced the expression of passive, despair-like coping responses in the FST and consequently augmented the expression of a subset of active coping responses (i.e., swimming) that are known to be 5-HT mediated (McLaughlin et al., 2012). The enhancement of swimming cannot be attributed to a general increase in locomotion, since previous reports have demonstrated that URB 597 does not significantly affect basal locomotor activity (Adamczyk, Mccreary, & Filip, 2008). This effect in the FST was blocked by coadministration of the CB1R inverse agonist AM251, as well as by global pharmacological depletion of 5-HT precursors, suggesting that the ability of FAAH inhibition within the mPFCv to promote active coping strategies in the FST is both CB1R dependent and 5-HT mediated. Finally, using in vivo single-unit extracellular recordings, we demonstrated that local inhibition of FAAH within the mPFCv enhanced the firing rate of DR 5-HT neurons on a time course that mirrors the behavioral effects in the FST (McLaughlin et al., 2012). Together, these studies argue that anandamide/CB1R activity in the mPFCv mediates the expression of active coping responses in the FST via an enhancement of DR 5-HT neuronal firing, which is in line with previous findings from our group following intra-mPFCv administration of exogenous CB1R agonists (Bambico et al., 2007).

We were also able to recapitulate these electrophysiological and behavioral findings in FAAH-knockout mice (FAAH$^{-/-}$), in which we observed a marked increase (+34.68%) in DR 5-HT neural firing compared to their littermates, which was reversed by rimonabant (Bambico et al., 2010). This effect was particularly significant in a subset of neurons exhibiting high firing rates (33.15% mean decrease). FAAH$^{-/-}$ mice also showed a resilience profile characterized by reduced immobility in the FST and TST, predictive of antidepressant activity, which again was attenuated by rimonabant administration.

The delay in therapeutic onset of antidepressants has been attributed to gradual neuroplastic adaptations at the presynaptic and postsynaptic levels that result from the progressive augmentation of 5-HT activity. These modifications include desensitization of autoinhibitory 5-HT1A receptors and sensitization or increased tonic activation of postsynaptic 5-HT1A receptors (Besson et al., 2000; Haddjeri et al., 1998; Szabo & Blier, 2001). The hippocampal pyramidal response to the 5-HT1A receptor antagonist WAY 100635 indicates enhanced tonus on the hippocampal 5-HT1A heteroreceptors, a hallmark of antidepressant-like action. FAAH$^{-/-}$ mice, compared to their wild-type littermates, showed increased tonic activity of 5-HT1A receptors, as tested with ip administration of WAY 100635 (0.5 mg/kg), which potently disinhibited hippocampal pyramidal neural activity (Bambico et al., 2010). Together, these results suggest that genetic deletion or pharmacological inhibition of FAAH (systemically or intra-mPFCv) elicits antidepressant-like effects, paralleled by increased 5-HT transmission and augmented postsynaptic 5-HT1A and 5-HT2A/2C receptor function.

CHRONIC ADMINISTRATION OF CB1R AGONISTS IN ADOLESCENCE AND ADULTHOOD: IMPLICATIONS FOR 5-HT FIRING ACTIVITY

Cannabis remains the most abused illicit substance worldwide. During the period of 2010–2012, 11.69% of Americans age 12 and older had abused marijuana at least once in the year prior to being surveyed (Substance Abuse and Mental Health Services Administration, 2014). It is important to mention that there have been studies that suggest that heavy and/or prolonged use of cannabis early in life increases the risk for neuropsychiatric disorders such as anxiety, depression, and schizophrenia (Bambico, Duranti, Tontini, Tarzia, & Gobbi, 2009; Bambico & Gobbi, 2008; Howlett et al., 2004) independent of whether the individual uses other illicit drugs (Hayatbakhsh et al., 2007). The limited neurobiological data scrutinizing this association are somewhat inconsistent, with reports of both increased (O'Shea, McGregor, & Mallet, 2006; O'Shea, Singh, McGregor, & Mallet, 2004) and decreased (Biscaia et al., 2003; Rubino et al., 2008) emotional reactivity following adolescent cannabis use. The impact of adolescent vs adult cannabis exposure, as well as the neural mechanisms underlying the increased predisposition for developing mood disorders, deserves more attention, especially given the recent changes in laws governing recreational cannabis use in some US states.

The neurobiological impact of drug use is especially compounded during the critical adolescence period when brain development is punctuated by constant neuroplastic shaping, synaptic reorganization, and extensive neurochemical changes, along with reckless or impulsive behaviors, novelty and sensation seeking, risk taking, and partial anhedonia (Spear, 2000). During this stage, corticolimbic CB1R density is at its peak, undergoing gradual pruning thereafter (Belue, Howlett, Westlake, & Hutchings, 1995), which may well relate to the crucial role of endocannabinoids in brain developmental processes including neurogenic control, neural progenitor proliferation, lineage segregation, and the migration and phenotypic specification of immature neurons (Harkany, Keimpema, Barabas, & Mulder, 2008). Several lines of evidence support the notion that prolonged aberrations in CB1R signaling may dramatically alter CB1R density in ways that potentially disrupt the development of the monoaminergic systems, thereby influencing mood and anxiety control (Ellgren et al., 2008).

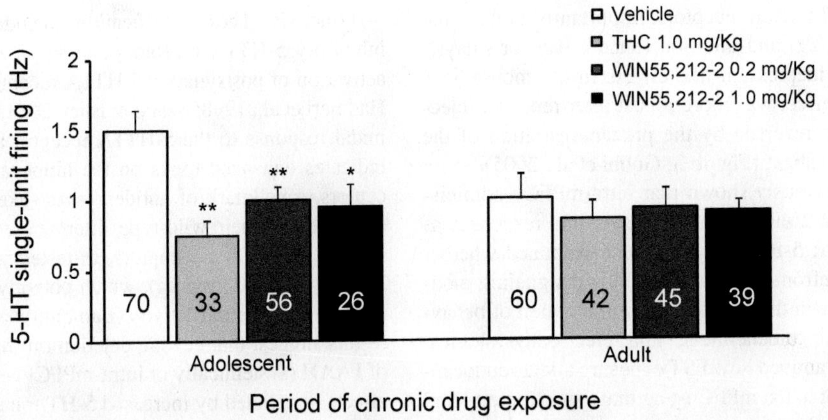

FIGURE 4 Alterations in 5-HT neurotransmission following chronic adolescent cannabinoid exposure. Chronic daily treatment with WIN 55,212-2 (0.2 or 1.0 mg/kg, ip) or THC (1.0 mg/kg, ip) when administered during adolescence but not when administered during adulthood resulted in a significant decrease in DR 5-HT spontaneous single-unit firing. Values within each bar denote the number of neurons recorded. $*p<0.05$, $**p<0.01$. Modified after Bambico et al. (2010), with permission.

In light of this evidence, we studied the impact of exogenous CB1R agonists on 5-HT firing activity after long-term exposure during adolescence or adulthood, hypothesizing that the adolescent brain would be subject to more detrimental, insidious effects of chronic CB1R agonism compared to that of mature subjects. Adolescent rats were treated with WIN 55,212-2 from postnatal day (PND) 30 to 50 (0.1 or 1 mg/kg, once a day) or with Δ^9-THC (1 mg/kg, once a day) or vehicle. At PND 70, they were tested using behavioral assays for emotional behavior or in vivo electrophysiology for assessing changes in 5-HT neurotransmission. Four distinct groups of adult rats were similarly treated for 20 days (from PND 70 to 90) and tested 20 days later (PND 110).

In the adolescent-exposed groups, all drug treatments significantly attenuated spontaneous 5-HT single-spike activity in adulthood (Bambico et al., 2010, Figure 4). Further analyses of neural activity revealed a modest nonsignificant decrease in burst firing activity (decreased number of spikes per burst and burst length, increased burst interspike interval, and decreased ratio (%) of spikes within bursts to the total number of spikes) following exposure to the low dose of WIN 55,212-2 (Bambico et al., 2010). Opposite effects (nonsignificant *increase* in burst activity) were observed after exposure to the high dose of WIN 55,212-2 as well as Δ^9-THC, which corresponded to positively skewed interspike interval distributions, further indicating irregular and burst-like neural firing (Bambico et al., 2010). Drug exposure during adulthood did not yield significant changes in 5-HT single-spike and burst firing activity.

From a behavioral point of view, chronic adolescent exposure, but not adult exposure, to WIN 55,212-2 (low 0.2 mg/kg and high 1.0 mg/kg) and Δ^9-THC (1 mg/kg) led to depressive-like behavior in the FST and sucrose preference test, while the high dose also induced anxiety-like responses in the novelty-suppressed feeding test.

Together these data suggest that the 5-HT system in adolescents is particularly sensitive to chronic consumption of CB1R agonists leading to a net decrease in 5-HT electrical activity and behavioral consequences such as vulnerability to depression and anxiety. Adult chronic cannabinoid exposure failed to elicit profound or persistent effects on emotional behavior and 5-HT neurotransmission. These findings are particularly relevant from a translational perspective in human populations, although more studies are needed to understand the reasons for these plastic changes induced by exogenous CB1R agonists. Moreover, greater efforts are needed in the prevention and treatment of mental health consequences induced by heavy and/or prolonged cannabis use during adolescence.

CONCLUSION

Major depression is a complex multifaceted disease whose etiology and pathophysiology is poorly understood. It is estimated to afflict 10–15% of the general population, resulting in profound alterations in emotional, cognitive, and neuroendocrine functioning (Lépine & Briley, 2011). The World Heath Organization predicts that major depression will become the most prevalent cause of illness-induced disability by the year 2030, leading to premature fatalities and disabilities. It is one of the most prevalent psychiatric disorders. The STAR-D (largest prospective randomized antidepressant trial) showed that 16% of study dropouts were due to intolerance of medication. Also 38.6% developed moderate/severe impairments due to an adverse event. It was also noted that the remission rate of first-line treatment with the SSRI Citalopram over a 12-week period was relatively modest (36.8%) (Rush et al. 2006). Therefore, the discovery and validation of new targets for the treatment of depression are consequently of extreme importance.

From our studies, we have demonstrated that direct activation of CB1R via Δ^9-THC or synthetic cannabinoid agonists and indirect activation of CB1R via inhibition of FAAH both exert putative antidepressant-like effects at low doses by modulating 5-HT neurotransmission, thereby producing effects on mood and behavior via a mechanism similar to that of conventional antidepressants. However, with respect to Δ^9-THC and other synthetic cannabinoid agonists, the opposite effect is observed at higher doses and the side effects induced by these cannabinoids prevent them from being considered as viable antidepressant compounds.

Another major concern of cannabinoid use for therapeutic purposes is the potentially harmful effects on brain plasticity and development following adolescent consumption. We revealed that 5-HT neurotransmission during adolescence is particularly sensitive to chronic consumption of CB1R agonists, leading to a net decrease in 5-HT activity, with adverse behavioral consequences such as increased vulnerability to developing depression and anxiety. Even if data from our laboratory and others suggest that the association between cannabis and depression during adulthood is restricted to early onset (i.e., adolescent) of cannabis use, it cannot be ruled out that vulnerable adult populations may also be sensitive to the adverse consequences of heavy and/or prolonged cannabis consumption.

The development of FAAH inhibitors might instead represent a valid alternative to cannabis and cannabinoid-based compounds, since they increase endogenous cannabinoids in a natural fashion, thereby avoiding many of the side effects of exogenous cannabinoids (Gobbi et al., 2005). However, more studies are clearly needed to better understand the neurobiological underpinnings of cannabis use and depression and to test novel FAAH inhibitors or other cannabis-derived medicines for the treatment of mood disorders in clinical settings.

APPLICATIONS TO OTHER ADDICTIONS AND SUBSTANCE MISUSE

1. The cannabinoid receptors can be stimulated by either endogenous or exogenous cannabinoids orchestrating a well-defined signaling response with multiple points of interaction with the mesocorticolimbic reward/stress circuit.
2. The mesocorticolimbic reward/stress circuit is also implicated in cocaine dependence, amphetamine dependence, opioid dependence, and alcohol dependence.
3. Cannabis abuse is associated with the abuse of other substances, such as nicotine and alcohol, and possibly harder drugs such as heroin, cocaine, and ecstasy.
4. Activation of CB1R affects mood regulation through indiscriminate modulation of brain structures that are crucial for stress responsivity and reward seeking (e.g., DR, hippocampus, amygdala, hypothalamus, ventral striatum, mPFC, ventral tegmental area). As such, dysfunctional CB1R activity may stimulate drug seeking and enhance vulnerability for developing affective disorders.
5. Our studies showed that chronic use of cannabis during adolescence, similar to chronic use of amphetamine or nandrolone, can have serious and irreversible effects on 5-HT and norepinephrine neurotransmission.

DEFINITION OF TERMS

Depression This is a complex heterogeneous disease clinically characterized by a constellation of symptoms impairing an individual's ability to function and cope with daily life activities. Its main alterations are in the cognitive, affective, somatic, and vegetative domains. The symptoms must be present mainly for 2 weeks, representing a change in previous functioning. Core symptoms are anhedonia and depressed mood.

Serotonin (5-hydroxytryptamine) This is the principal monoamine neurotransmitter that it is implicated in mood, sleep, and appetite. It has been also implicated in learning processes as well as in memory functions. Most conventional antidepressants act by modulating the amount of this neurotransmitter.

Δ⁹-Tetrahydrocannabinol This is the main psychoactive constituent of cannabis, with multiple functions and effects including changes in emotionality, behavior, and cognitive functions. Moreover, its analgesic, antinausea, hypnotic, and hyperphagic properties have been exploited for medicinal purposes for centuries. However, its potential for abuse and dependence and its association with many neuropsychiatric disorders make it necessary to better understand its complex neurobiological mechanism of action.

Forced swim test This is a behavioral test to examine antidepressant-like activity of putative antidepressant drugs.

Dorsal raphe This is the main source of 5-HT neurons in the brain. It has two subdivisions, the rostral and caudal regions, with many projections, mainly to the amygdala. The rostral subdivision has been associated with depression, as individuals suffering from depression display morphological abnormalities (i.e., smaller size) within this brain area.

Agonist This refers to a molecule that activates a receptor to produce a maximal biological response.

Antagonist This refers to a molecule that blocks a biological response by binding to a receptor.

Partial agonist This is a molecule that binds to a receptor, producing a biological response that is less than that of a full agonist. In the presence of a high concentration of endogenous agonists, the partial agonist acts like an antagonist.

KEY FACTS

Key Facts about Cannabis

- Cannabis remains one of the most abused illicit drugs for recreational purposes.
- Its main psychoactive component, Δ⁹-THC, elicits a complex subjective experience that depends on the potency of Δ⁹-THC, the user, and the environment in which the drug is consumed.
- There has been a steady rise in the use of cannabis in recent years. We are still trying to unravel its complex physiological and behavioral effects.
- Cannabis use has been shown to produce alterations in emotional and cognitive processes and has been linked to the onset of many neuropsychiatric disorders, most notably schizophrenia, in individuals with a genetic predisposition, as well as depression, possibly in nonpredisposed individuals.
- There is an ongoing debate regarding the therapeutic effects and detrimental consequences of cannabis consumption; accordingly, more studies are warranted.

Key Facts about Depression

- Depression is a relatively common debilitating mental disease, affecting almost 350 million people according to the World Health Organization.
- Epidemiological studies have reported that up to 9.5% of adults in the United States suffer from varying degrees of depression, while in Canada the prevalence of depression is estimated to be between 3.2% and 6%.

- Depression is characterized by persistent symptoms of depressed mood and anhedonia, feelings of worthlessness, alterations in somatic and vegetative functions, cognitive deficits, sleep disturbances, change in appetite, fatigue, anxiety, and psychomotor agitation.
- Numerous studies have revealed a greater likelihood for substance abusers to develop mood disorders, especially depression.
- Only one-third of people with depression respond to antidepressants, one-third have a partial response, and one-third have no response at all. Mental health researchers are still searching for novel treatments for drug-resistant depression and working toward a better comprehension of its pathophysiological underpinnings.

SUMMARY POINTS

- Psychoactive constituents of cannabis, synthetic cannabinoids, and endocannabinoid enhancers (FAAH inhibitors) regulate mood and affective states by modulating 5-HT neurons located in the DR nuclei.
- Δ^9-THC modulates 5-HT firing activity, as repeated administration of low doses (1 mg/kg) of Δ^9-THC elicits an increase in DR 5-HT single-unit activity and enhances tonic activity of 5-HT1A receptors in the dorsal hippocampus, resulting in an antidepressant-like effect.
- Δ^9-THC produces a mixed response on 5-HT firing activity, with 26% of neurons showing an increase, 33% showing a decrease, and 42% showing no response. However, after 4 days of ip injections, Δ^9-THC (1 mg/kg) produces a significant elevation of 5-HT firing.
- The increase in firing following WIN 55,212 and Δ^9-THC administration is prevented by the CB1R antagonist rimonabant, providing empirical support for a CB1R-mediated mechanism of action.
- The mPFCv plays an instrumental role in mediating the increase in 5-HT firing and the antidepressant-like effects of CB1R agonists.
- Chronic administration of CB1R agonists during adolescence leads to a net decrease in 5-HT electrical activity and impairments in emotional behavior during adulthood.
- Genetic and pharmacological inhibition of FAAH produces anxiolytic and antidepressant-like effects.
- CB1R activation strongly interacts with the 5-HT system to regulate mood; however, the use of such compounds during adolescence could lead to later impairments in 5-HT neurotransmission and emotional behavior.
- More studies are warranted to understand and untangle the neuroplastic modifications provoked by cannabis consumption, and future research is needed to design treatments targeting the endocannabinoid system that may prevent the development of major depression in vulnerable populations.

REFERENCES

Adamczyk, P., Mccreary, A. C., & Filip, M. (2008). Activation of endocannabinoid transmission induces antidepressant-like effects in rats. *Journal of Physiology and PharmacologyE, 59*(2), 217–228.

Artigas, F., Romero, L., de Montigny, C., & Blier, P. (1996). Acceleration of the effect of selected antidepressant drugs in major depression by 5-HT1A antagonists. *Trends in Neurosciences, 19*, 378–383.

Bambico, F. R., Cassano, T., Dominguez-Lopez, S., Katz, N., Walker, C. D., Piomelli, D., & Gobbi, G. (2010). Genetic deletion of fatty acid amide hydrolase alters emotional behavior and serotonergic transmission in the dorsal raphe, prefrontal cortex, and hippocampus. *Neuropsychopharmacology, 35*, 2083–2100.

Bambico, F. R., Duranti, A., Tontini, A., Tarzia, G., & Gobbi, G. (2009). Endocannabinoids in the treatment of mood disorders: evidence from animal models. *Current Pharmaceutical Design, 15*, 1623–1646.

Bambico, F. R., & Gobbi, G. (2008). The cannabinoid CB1 receptor and the endocannabinoid anandamide: possible antidepressant targets. *Expert Opinions on Therapeutic Targets, 12*, 1347–1366.

Bambico, F. R., Hattan, P. R., Garant, J. P., & Gobbi, G. (2012). Effect of delta-9-tetrahydrocannabinol on behavioral despair and on pre- and postsynaptic serotonergic transmission. *Progress in Neuro-psychopharmacology and Biological Psychiatry, 38*, 88–96.

Bambico, F. R., Katz, N., Debonnel, G., & Gobbi, G. (2007). Cannabinoids elicit antidepressant-like behavior and activate serotonergic neurons through the medial prefrontal cortex. *Journal of Neuroscience, 27*, 11700–11711.

Barann, M., Molderings, G., Bruss, M., Bonisch, H., Urban, B. W., & Gothert, M. (2002). Direct inhibition by cannabinoids of human 5-HT3A receptors: probable involvement of an allosteric modulatory site. *British Journal of Pharmacology, 137*, 589–596.

Belue, R. C., Howlett, A. C., Westlake, T. M., & Hutchings, D. E. (1995). The ontogeny of cannabinoid receptors in the brain of postnatal and aging rats. *Neurotoxicology and Teratology, 17*, 25–30.

Besson, A., Haddjeri, N., Blier, P., & de Montigny, C. (2000). Effects of the co-administration of mirtazapine and paroxetine on serotonergic neurotransmission in the rat brain. *European Neuropsychopharmacology: The Journal of the European College of Neuropsychopharmacology, 10*, 177–188.

Biscaia, M., Marin, S., Fernandez, B., Marco, E. M., Rubio, M., Guaza, C., … Viveros, M. P. (2003). Chronic treatment with CP 55,940 during the peri-adolescent period differentially affects the behavioural responses of male and female rats in adulthood. *Psychopharmacology, 170*, 301–308.

Castagne, V., Moser, P., Roux, S., & Porsolt, R. D. (2011). Rodent models of depression: forced swim and tail suspension behavioral despair tests in rats and mice. *Current Protocols in Neuroscience/Editorial Board* Jacqueline N Crawley et al., Chapter 8:Unit 8 10A.

CPHA. (2012). *CPS: Compendium of pharmaceuticals and specialties*. Canada: Canadian Pharmacists Assoc.

Darmani, N. A. (2001). Cannabinoids of diverse structure inhibit two DOI-induced 5-HT(2A) receptor-mediated behaviors in mice. *Pharmacology, Biochemistry, and Behavior, 68*, 311–317.

Devlin, M. G., & Christopoulos, A. (2002). Modulation of cannabinoid agonist binding by 5-HT in the rat cerebellum. *Journal of Neurochemistry, 80*, 1095–1102.

Earleywine, M. (2005). *Understanding marijuana: a new look at the scientific evidence*. USA: Oxford Univ. Press.

Egashira, N., Matsuda, T., Koushi, E., Higashihara, F., Mishima, K., Chidori, S., … Fujiwara, M. (2008). Delta(9)-tetrahydrocannabinol prolongs the immobility time in the mouse forced swim test: involvement of cannabinoid CB(1) receptor and serotonergic system. *European Journal of Pharmacology, 589*, 117–121.

Egertova, M., Cravatt, B. F., & Elphick, M. R. (2003). Comparative analysis of fatty acid amide hydrolase and cb(1) cannabinoid receptor expression in the mouse brain: evidence of a widespread role for fatty acid amide hydrolase in regulation of endocannabinoid signaling. *Neuroscience, 119*, 481–496.

El-Alfy, A. T., Ivey, K., Robinson, K., Ahmed, S., Radwan, M., Slade, D., ... Ross, S. (2010). Antidepressant-like effect of delta9-tetrahydrocannabinol and other cannabinoids isolated from Cannabis sativa L.. Pharmacology, Biochemistry, and Behavior, 95, 434–442.

Elbatsh, M., Spicer, C., Marsden, C., Fone, K., & Kendall, D. (2009). Antidepressant-like effects of chronic cannabinoid agonist and antagonist treatments in the olfactory bulbectomy model of depression in rats. British Pharmacological Society.

Ellgren, M., Artmann, A., Tkalych, O., Gupta, A., Hansen, H. S., Hansen, S. H., ... Hurd, Y. L. (2008). Dynamic changes of the endogenous cannabinoid and opioid mesocorticolimbic systems during adolescence: THC effects. European Neuropsychopharmacology : The Journal of the European College of Neuropsychopharmacology, 18, 826–834.

Fan, P. (1995). Cannabinoid agonists inhibit the activation of 5-HT3 receptors in rat nodose ganglion neurons. Journal of Neurophysiology, 73, 907–910.

Gobbi, G., Bambico, F. R., Mangieri, R., Bortolato, M., Campolongo, P., Solinas, M., ... Piomelli, D. (2005). Antidepressant-like activity and modulation of brain monoaminergic transmission by blockade of anandamide hydrolysis. PNAS, 102, 18620–18625.

Gobbi, G., & Blier, P. (2005). Effect of neurokinin-1 receptor antagonists on serotoninergic, noradrenergic and hippocampal neurons: comparison with antidepressant drugs. Peptides, 26, 1383–1393.

Gobbi, G., Cassano, T., Radja, F., Morgese, M. G., Cuomo, V., Santarelli, L., ... Blier, P. (2007). Neurokinin 1 receptor antagonism requires norepinephrine to increase serotonin function. European Neuropsychopharmacology : The Journal of the European College of Neuropsychopharmacology, 17, 328–338.

Haddjeri, N., Blier, P., & de Montigny, C. (1998). Long-term antidepressant treatments result in a tonic activation of forebrain 5-HT1A receptors. Journal of Neuroscience, 18, 10150–10156.

Haj-Dahmane, S., & Shen, R. Y. (2005). The wake-promoting peptide orexin-B inhibits glutamatergic transmission to dorsal raphe nucleus serotonin neurons through retrograde endocannabinoid signaling. Journal of Neuroscience, 25, 896–905.

Hajós, M., Hajós-Korcsok, É., & Sharp, T. (1999). Role of the medial prefrontal cortex in 5–HT1A receptor-induced inhibition of 5–HT neuronal activity in the rat. British journal of pharmacology, 126(8), 1741–1750.

Harkany, T., Keimpema, E., Barabas, K., & Mulder, J. (2008). Endocannabinoid functions controlling neuronal specification during brain development. Molecular and Cellular Endocrinology, 286, S84–S90.

Hayatbakhsh, M. R., Najman, J. M., Jamrozik, K., Mamun, A. A., Alati, R., & Bor, W. (2007). Cannabis and anxiety and depression in young adults: a large prospective study. Journal of the American Academy of Child and Adolescent Psychiatry, 46, 408–417.

Hermann, H., Marsicano, G., & Lutz, B. (2002). Coexpression of the cannabinoid receptor type 1 with dopamine and serotonin receptors in distinct neuronal subpopulations of the adult mouse forebrain. Neuroscience, 109, 451–460.

Hill, M. N., Sun, J. C., Tse, M. T., & Gorzalka, B. B. (2006). Altered responsiveness of serotonin receptor subtypes following long-term cannabinoid treatment. International Journal of Neuropsychopharmacology, 9, 277–286.

Howlett, A. C., Breivogel, C. S., Childers, S. R., Deadwyler, S. A., Hampson, R. E., & Porrino, L. J. (2004). Cannabinoid physiology and pharmacology: 30 years of progress. Neuropharmacology, 47(Suppl. 1), 345–358.

Huestis, M. A., Gorelick, D. A., Heishman, S. J., Preston, K. L., Nelson, R. A., Moolchan, E. T., & Frank, R. A. (2001). Blockade of effects of smoked marijuana by the CB1-selective cannabinoid receptor antagonist SR141716. Archives of General Psychiatry, 58, 322–328.

Iversen, L. (2003). Cannabis and the brain. Brain : A Journal of Neurology, 126, 1252–1270.

Jankowski, M. P., & Sesack, S. R. (2004). Prefrontal cortical projections to the rat dorsal raphe nucleus: ultrastructural features and associations with serotonin and gamma-aminobutyric acid neurons. Journal of Comparative Neurology, 468, 518–529.

Kimura, T., Ohta, T., Watanabe, K., Yoshimura, H., & Yamamoto, I. (1998). Anandamide, an endogenous cannabinoid receptor ligand, also interacts with 5-hydroxytryptamine (5-HT) receptor. Biological and Pharmaceutical Bulletin, 21, 224–226.

Laaris, N., Good, C. H., & Lupica, C. R. (2010). Delta9-tetrahydrocannabinol is a full agonist at CB1 receptors on GABA neuron axon terminals in the hippocampus. Neuropharmacology, 59, 121–127.

Lépine, J., & Briley, M. (2011). The increasing burden of depression. Neuropsychiatric Disease Treatment, 7, 3–7.

Malone, D. T., & Taylor, D. A. (2001). Involvement of somatodendritic 5-HT(1A) receptors in Delta(9)-tetrahydrocannabinol-induced hypothermia in the rat. Pharmacology, Biochemistry, and Behavior, 69, 595–601.

Matsuda, L. A., Bonner, T. I., & Lolait, S. J. (1993). Localization of cannabinoid receptor mRNA in rat brain. Journal of Comparative Neurology, 327, 535–550.

McLaughlin, R. J., Hill, M. N., Bambico, F. R., Stuhr, K. L., Gobbi, G., Hillard, C. J., & Gorzalka, B. B. (2012). Prefrontal cortical anandamide signaling coordinates coping responses to stress through a serotonergic pathway. European Neuropsychopharmacology, 22(9), 664–671.

Mendiguren, A., & Pineda, J. (2009). Effect of the CB(1) receptor antagonists rimonabant and AM251 on the firing rate of dorsal raphe nucleus neurons in rat brain slices. British Journal of Pharmacology, 158, 1579–1587.

Moldrich, G., & Wenger, T. (2000). Localization of the CB1 cannabinoid receptor in the rat brain. An immunohistochemical study. Peptides, 21, 1735–1742.

Morales, M., Wang, S. D., Diaz-Ruiz, O., & Jho, D. H. (2004). Cannabinoid CB1 receptor and serotonin 3 receptor subunit A (5-HT3A) are co-expressed in GABA neurons in the rat telencephalon. Journal of Comparative Neurology, 468, 205–216.

Moreira, F. A., Grieb, M., & Lutz, B. (2004). Role of serotonin on the antidepressant-like effect induced by delta-9-tetrahydrocannabinol, but not by rimonabant. In 18th Symposium of the International Cannabinoid Research Society Avemore, Scotland.

O'Shea, M., McGregor, I. S., & Mallet, P. E. (2006). Repeated cannabinoid exposure during perinatal, adolescent or early adult ages produces similar longlasting deficits in object recognition and reduced social interaction in rats. Journal of Psychopharmacology, 20, 611–621.

O'Shea, M., Singh, M. E., McGregor, I. S., & Mallet, P. E. (2004). Chronic cannabinoid exposure produces lasting memory impairment and increased anxiety in adolescent but not adult rats. Journal of Psychopharmacology, 18, 502–508.

Paxinos, G. & Watson, C. (2007). The rat brain in stereotaxis coordinates. New York, NY: Academic Press.

Rodriguez-Gaztelumendi, A., Rojo, M. L., Pazos, A., & Diaz, A. (2009). Altered CB receptor-signaling in prefrontal cortex from an animal model of depression is reversed by chronic fluoxetine. Journal of Neurochemistry, 108, 1423–1433.

Rubino, T., Vigano, D., Realini, N., Guidali, C., Braida, D., Capurro, V., … Parolaro, D. (2008). Chronic delta 9-tetrahydrocannabinol during adolescence provokes sex-dependent changes in the emotional profile in adult rats: behavioral and biochemical correlates. *Neuropsychopharmacology, 33*, 2760–2771.

Rush, A., Trivedi, M., Wisniewski, S., Nierenberg, A., Stewart, J., Warden, D., … Fava, M. (2006). Acute and longer-term outcomes in depressed outpatients requiring one or several treatment steps: a STAR* D report. *American Journal of Psychiatry, 163*(11), 1905–1917.

Spear, L. P. (2000). Neurobehavioral changes in adolescence. *Current Directions in Psychology Science, 9*, 111–114.

Substance Abuse and Mental Health Services Administration. (2014). *Substate estimates of substance use and mental disorders from the 2010–2012 national surveys on drug use and health: results and detailed tables*. Rockville, MD http://www.samhsa.gov/data/NSDUH/substate2k12/toc.aspx.

Szabo, S. T., & Blier, P. (2001). Effects of the selective norepinephrine reuptake inhibitor reboxetine on norepinephrine and serotonin transmission in the rat hippocampus. *Neuropsychopharmacology, 25*, 845–857.

Tsou, K., Nogueron, M. I., Muthian, S., Sañudo-Peña, M. C., Hillard, C. J., Deutsch, D. G., & Walker, J. M. (1998). Fatty acid amide hydrolase is located preferentially in large neurons in the rat central nervous system as revealed by immunohistochemistry. *Neuroscience letters, 254*(3), 137–140.

Chapter 77

CB1 Cannabinoid Receptors and Aggression: Relationship to Cannabis Use

Marta Rodríguez-Arias, José Miñarro, M. Carmen Arenas, María A. Aguilar
Unidad de Investigación Psicobiología de las Drogodependencias, Departamento de Psicobiología, Facultad de Psicología, Universitat de València, Valencia, Spain

Abbreviations

2-AG 2-Arachidonoylglycerol
5-HT Serotonin
AA Arachidonic acid
AEA *N*-Arachidonoylethanolamine or anandamide
AMT/AT Anandamide and 2-arachidonoylglycerol membrane transporter
ASPD Antisocial personality disorder
CB1-KO Mice lacking the CB1 receptor
DAG Diacylglycerol
DAGL Diacylglycerol lipase
ECS Endogenous cannabinoid system
FAAH Fatty acid amide hydrolase
GABA γ-Aminobutyric acid
GPR G-Protein-coupled receptor
MA Monoamine
MAGL Monoacylglycerol lipase
MAPK Mitogen-activated protein kinase
MAR Monoamine receptor
NADA *N*-Arachidonoyldopamine
NAPE *N*-Arachidonylphosphatidylethanolamine
NAT *N*-Acyltransferase
OA Oleic acid
OEA Oleoylethanolamide
PA Palmitic acid
PEA Palmithylethanolamide
Ph Chol Phosphatidylcholine
Ph Eth Phosphatidylethanolamine
PKA Protein kinase A
PK$^{-/-}$/TauVLW mice Parkin-null, human tau-overexpressing PK$^{-/-}$/TauVLW
PLC Phospholipase C
PLD Phospholipase D
PPAR Peroxisome proliferator-activated receptor
PPARα Peroxisome proliferator-activated receptor α
SUD Substance use disorder
TAG Triacylglycerol
THC Δ^9-Tetrahydrocannabinol
TRPA1 Transient receptor potential ankyrin 1
TRPV1 Transient receptor potential channel type V1

INTRODUCTION

The relationship between drug use and violent behavior has been firmly established through a great number of studies (see reviews Boles & Miotto, 2003; Chermack et al., 2010; Friedman, 1998). Substance use disorder is considered a common risk factor for perpetration of violence (Barrett, Teesson, & Mills, 2014), since it contributes more to violent behavior than any other mental health disorder (Pulay et al., 2008). Violent behavior can be necessary to gain access to drugs, to acquire the resources to purchase a drug, or to resolve disputes in the illegal and unregulated drug market. Violence and drug use can be results of the same factor, including personality traits such as high novelty seeking. Finally, drugs can increase the likelihood of aggression by exerting a direct effect on the subject (Hoake & Steward, 2003). In this line, drugs of abuse induce a state of intoxication (pharmacological effect), can provoke brain damage after prolonged use (neurotoxic effects), and induce extreme discomfort when abruptly discontinued after chronic use (withdrawal effects).

Aggression can be generally defined as behavior that inflicts harm and injury or threatens to do so (Berkowitz, 1993). According to Baron's definition (Baron & Richardson, 1994) aggression is any form of behavior whose goal is the harming or injuring of another living being who is motivated to avoid such treatment. In the literature, animal studies involving drug administration and laboratory measures tend to use the term aggression. On the other hand, research concerning individuals who have come into contact with law enforcement after drug consumption tend to use the term "violence" to denote aggression between two or more human beings involving physical harm or injury. Human aggression can be classified as defensive, premeditated, or impulsive–hostile (Vitiello & Stoff, 1997). This last type is linked to biological and environmental causes and is thus the most easily induced by consumption of drugs of abuse (Coccaro, Lee, & Kavoussi, 2010).

Human and nonhuman aggressive behaviors have some common features, but animal aggression is considered less complex. Nevertheless, animal aggression is a broad category of behavior rather than a single behavioral pattern and encompasses various origins, motivations, expressions, and functions (Miczek, Faccidomo, Fish, & DeBold, 2007). In the behavioral repertoire of a given species, various acts can be classified as aggressive and have varying ethological significance depending on their functionality in the survival and reproduction of the species. Blanchard and Brain classified aggressive behavior in animals as offensive or defensive depending on the distal and proximal conditions that produce it, the topography of the behavior, and its consequences (Blanchard & Blanchard, 1977; Brain & Benton, 1979).

Despite the great deal of controversy concerning the relationship between marijuana use and aggressive behavior in humans (Harris et al., 2010), relatively few studies have specifically examined the association between cannabis use and violence in humans (e.g., Dembo et al., 1987; Moore & Stuart, 2005; Myerscough & Taylor, 1985; Smith, Homish, Leonard, & Collins, 2013; Taylor et al., 1976). The acute effects of Δ^9-tetrahydrocannabinol (THC) (the primary psychoactive component of cannabis) on aggressive behavior have been the most studied aspect with respect to this area, but the information obtained is controversial. Results show a biphasic effect of THC; while low doses of THC can slightly increase aggression, moderate and high doses can suppress or even eliminate this type of behavior (Myerscough & Taylor, 1985; Taylor et al., 1976). However, in all of these studies, variables that could have influenced the data obtained should be taken into consideration, such as rapid eye movement sleep deprivation, social seclusion, or pretreatment with another drug.

The animal literature also largely fails to support the cannabis–violence relationship, as cannabis administration has been shown to enhance submissive behaviors and suppress attack behaviors. However, some animal studies, specifically those using Wistar rats, have associated cannabis administration with increased aggression.

This chapter aims to summarize the most important results concerning the relationship between the CB1 cannabinoid type 1 receptor and aggression. First, we offer a brief review of the history and current epidemiology of cannabis consumption and an explanation of the main components and functions of the endocannabinoid system. We then review studies of the relationship between cannabis and aggression in humans and discuss the main findings regarding the effects of acute or chronic cannabis consumption or withdrawal on aggressive behavior. Finally, we focus on the effects of cannabinoids in animal models of aggression.

THE ENDOCANNABINOID SYSTEM

Although cannabis has been used for thousands of years, its neurobiological mechanism of action was not discovered until 1964, when the chemical structure of its main psychoactive constituent, THC, was identified (Gaoni & Mechoulam, 1964). About 20 years later, the first cannabinoid receptor (CB1) was identified (Devane, Dysarz, Johnson, Melvin, & Howlett, 1988). In 1992, the first endogenous ligand on CB receptors—arachidonoylethanolamine (AEA; anandamide), its name originating from "ananda," meaning "the bliss" in Sanskrit—was identified (Devane et al., 1992). These findings opened the way for research that led to the discovery of the endogenous cannabinoid system (ECS), a widely distributed modulatory system involved in multiple physiological processes (pain, growth and development, immune function), behaviors (motor control, reproduction, sleep, eating, emotional homeostasis), learning and memory, and mental disorders (Mechoulam & Parker, 2013; Rodriguez de Fonseca, 2008; Vinod & Hungund, 2006).

Endocannabinoids, together with their receptors and enzymes involved in synthesis and metabolism, constitute the ECS (see Figure 1). Endocannabinoids are signaling molecules (amides, esters, and ethers of long-chain polyunsaturated fatty acids) found in abundance in the cerebral cortex, basal ganglia, and limbic structures (Matias, Bisogno, & Di Marzo, 2006). The best known are anandamide and 2-arachidonoylglycerol (2-AG) (Mechoulam et al., 1995), but several others have been identified: N-arachidonylglycine, N-arachidonoyldopamine, 2-arachidonoylglyceryl ether (noladin ether), O-arachidonoylethanolamine (virodhamine), and 9-octadecenoamide (oleamide). The ECS centers on G-protein-coupled receptors (GPR) (Pertwee et al., 2010; Rodriguez de Fonseca, 2008). Two cannabinoid receptor subtypes have been cloned: CB1 (Matsuda, Lolait, Brownstein, Young, & Bonner, 1990), which is expressed in the brain and many peripheral tissues, and CB2 (Munro, Thomas, & Abu-Shaar, 1993), which is expressed predominantly on immune cells and damaged tissues and at low levels in the brain (Mechoulam & Parker, 2013). A yet-to-be-cloned CB3 receptor has been identified in the hippocampus. CB1 is probably the most abundant GPR to exist in the brain (Pertwee, 2010). It is presynaptically located at nerve terminals where endocannabinoids act as modulators of synaptic transmission (Howlett, 2002). These receptors are coupled negatively to adenylyl cyclase and N- and P/Q-type Ca^{2+} channels, and positively to A-type and inwardly rectifying K^+ channels and mitogen-activated protein kinases by $G_{i/o}$ proteins (Howlett, 2002). CB1 receptors are located in the cerebral cortex, basal ganglia, hippocampus, anterior cingulate cortex, and cerebellum (Herkenham, Lynn, Little, Johnson, & Melvin, 1990), where they inhibit the release of several neurotransmitters (e.g., glutamate and γ-aminobutyric acid) and stimulate dopamine release in the nucleus accumbens (Mechoulam & Parker, 2013). Synthetic CB1 receptor ligands have been developed as research tools to explore the functions of the ECS system. The hundreds of CB1 receptor agonists include WIN 55,212, HU-210, and CP 55,940. The main CB1 receptor antagonists are rimonabant (SR141716A), AM251, and AM281. Genetic knockout mice that lack CB1 receptors (or other proteins related to the ECS system, e.g., CB2, fatty acid amide hydrolase (FAAH), monoacylglycerol lipase, etc.) have also been developed with this objective in mind.

CANNABIS AND AGGRESSION IN HUMANS

Cannabis was historically suspected of instigating a wide variety of aggressive behaviors, which led to it being considered a social "menace" and to its prohibition (Hoaken & Stewart, 2003). In 1993, after reviewing the scientific literature, the US National Research Council recognized that short-term use of cannabis inhibited aggressive behavior in humans, while long-term use could promote violence (Friedman, 1998). Today, the findings of observational research in humans are largely mixed, with studies showing positive (Arseneault, Moffitt, Caspi, Taylor, & Silva, 2000; Barrett et al., 2014; Harris et al., 2010; Moore et al., 2008; Morris, TenEyck, Barnes, & Kovandzic, 2014), negative (Arendt et al., 2007), or nonexistent (Lejoyeux et al., 2013; Macdonald, Erickson, Wells, Hathaway, & Pakula, 2008) associations between marijuana and violence.

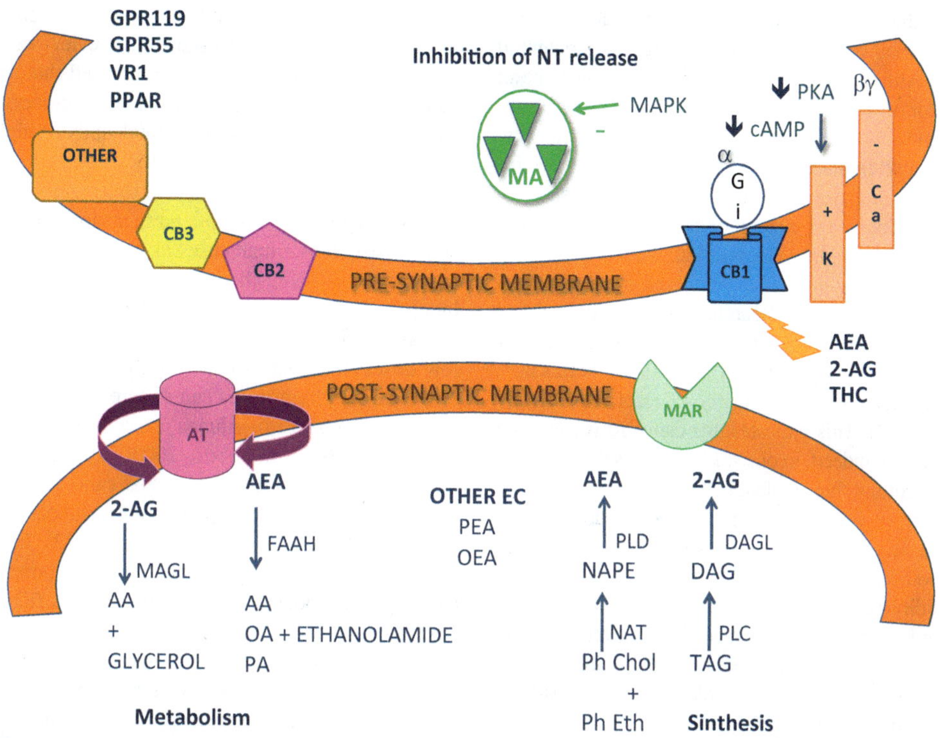

FIGURE 1 **The endocannabinoid system.** The two main endocannabinoids, anandamide and 2-AG, are synthesized in the postsynaptic nerve terminal. Anandamide is produced by hydrolysis of the membrane phospholipid *N*-arachidonylphosphatidylethanolamine, which is catalyzed by the enzyme phospholipase D. The synthesis of *N*-arachidonylphosphatidylethanolamine is mediated by an uncloned enzyme—*N*-acyltransferase—which detaches an arachidonate moiety from phospholipids (e.g., phosphatidylcholine) and transfers it to the primary amino group of phosphatidylethanolamine. 2-AG is produced by the metabolism of diacylglycerol by specific diacylglycerol lipases. The second messenger diacylglycerol is produced by hydrolysis of phospholipids or triglycerides such as triacylglycerol by phospholipase C. Anandamide and 2-AG, as well as other endocannabinoids such as palmithylethanolamide (PEA) and oleoylethanolamide (OEA), act mainly through the CB1 and CB2 receptors, although other receptors have been identified as targets of endocannabinoids (a new uncloned CB3 receptor, the GPR119 and GPR55 orphan receptors, the vanilloid VR1 receptor, and the peroxisome proliferator-activated receptor α). Endocannabinoids function as retrograde signaling molecules that inhibit the release of classical anterograde neurotransmitters by presynaptic terminals (such as monoamines) and bind with their receptor (monoamine receptor). After the activation of presynaptic CB1 receptors (by anandamide, 2-AG, or THC), different signal transduction mechanisms are stimulated. Via G inhibitory proteins, endocannabinoids inhibit adenylyl cyclase activity and subsequently reduce cAMP, leading to reduced activity of protein kinases such as protein kinase A and mitogen-activated protein kinase (MAPK) and to the modulation of ion channels (stimulation of potassium and inhibition of calcium channels) and inhibition of neurotransmitter release. The activity of endocannabinoids is limited by a transporter that takes up AEA and 2-AG into the postsynaptic cell. AEA, PEA, and OEA are degraded by the enzyme fatty acid amidohydrolase to ethanolamide plus arachidonic, oleic, or palmitic acids. 2-AG is degraded by the enzyme monoacylglycerol lipase to arachidonic acid plus glycerol.

The main reason for these discrepant results is the heterogeneity of the studies performed. First, as previously stated, human aggression is a multidimensional construct that hinders a simple definition. We have found research demonstrating an association between marijuana and delinquent behavior based on reports of marijuana consumers committing violent crimes (Arseneault et al., 2000) and exhibiting a significantly higher level of delinquent behaviors (Dembo et al., 1987) than subjects testing negative for marijuana. Conversely, White, Loeber, Stouthamer-Loeber, and Farrington (1999) found that the association between marijuana use and aggression was nonsignificant after controlling for history of aggression and alcohol use in a longitudinal study of substance use and aggression. Other studies evaluating the relation between marijuana use and violence have distinguished between nonintimate interpersonal and intimate partner violence (see review by Moore & Stuart, 2005). While mixed results were documented when nonintimate interpersonal violence was evaluated (Moore & Stuart, 2005), studies examining the relationship between drug abuse and aggression between intimate partners revealed a significant association with marijuana, considering it an important factor in the incidence of partner aggression (Moore et al., 2008).

Another reason for these relatively inconsistent findings could be the heterogeneity of the subjects assessed, which have included college students (Taylor et al., 1976), delinquents in a detention center (Dembo et al., 1987), and psychiatric patients (Arseneault et al., 2000; Barrett et al., 2014; Becker et al., 2012; Lejoyeux et al., 2013). Serious aggression in first-episode psychosis was associated with regular cannabis use (Harris et al., 2010), and posttraumatic stress disorder violence and trait aggression were associated with higher levels of cannabis use (Barrett et al., 2014). However, a lack of association between cannabis use and aggressive behavior was reported in a population of schizophrenics (Lejoyeux et al., 2013).

Correlational studies support a link between cannabis use and violent behavior, but their results do not offer insight into potential causal relationships. Although cannabis use has been shown to be an independent predictor of violence at a community level (Arseneault et al., 2000), it is unclear whether it is a direct cause of violence; whether people who use cannabis are violent as a result of a third factor, such as a personality type; or whether cannabis dependence is an indirect cause of violence as a result of criminal association or crime to obtain money for the drug (Harris et al., 2010).

This review principally examines studies based on psychopharmacological explanations. Multiple researchers have considered that cannabis in moderate-to-high doses reduces aggressive behavior (Hoaken & Stewart, 2003; Myerscough & Taylor, 1985). In this way, cannabis-intoxicated subjects are less likely to react aggressively (Hoaken & Stewart, 2003). This indicates that cannabis is used as a means of self-medication; indeed, in one study subjects that reported having problems controlling their violent behavior were much more likely to use cannabis to decrease aggression, as the drug helped them to relax and to decrease suspiciousness (Arendt et al., 2007). In favor of this hypothesis, a longitudinal study to assess the long-term predictive validity of antisocial personality disorder (ASPD) on criminal behavior in samples of substance abusers over a 30-year period reported a lower propensity for crime in cannabis users than in users of other stimulant drugs. Moreover, violence in ASPD patients was predicted by the absence of cannabis addiction (Fridell, Hesse, Jæger, & Kühlhorn, 2008). Marijuana use has also been related to decreased likelihood of violence with injury in interpersonal conflict incidents among men and women in substance use disorder treatment (Chermack et al., 2010). Moreover, Becker et al. (2012) observed that delinquent behavior predicted subsequent marijuana use, whereas marijuana use did not predict subsequent delinquent behavior.

Other studies reported no correlation with violent crime when marijuana was the only drug consumed (for review see Friedman, 1998).

However, other studies have reported that marijuana use predicts later delinquent behavior, while delinquency did not predict later marijuana use in a longitudinal study of high-school students (Becker et al., 2012). Additionally, consistent marijuana use during adolescence has been related to an increase in the likelihood of being involved in intimate partner violence in young adulthood and consistent marijuana use has been associated with an increase in the odds of being the perpetrator of intimate partner violence, independent of whether there is alcohol use (Morris et al., 2014).

As previously highlighted, researchers in general agree that low doses of cannabis can slightly increase aggression, but that moderate and high doses can suppress or even eliminate aggressive behavior, since marijuana use has been found to depress mental activity (Boles & Miotto, 2003; Myerscough & Taylor, 1985; Taylor et al., 1976). It is likely that cannabis temporarily inhibits aggression in the general population, whereas it increases aggression in some individuals (Smith et al., 2013). In some cases, when consumed in high doses or in an extremely potent form, marijuana can have psychoactive effects that are difficult to differentiate from those of hallucinogens such as LSD (Boles & Miotto, 2003). Additionally, other factors are likely to account for associations between marijuana use and violence, such as personality, expectancies about drug effects, and/or social contextual factors associated with general marijuana use (Chermack et al., 2010).

While cannabis intoxication seems to reduce the likelihood of violence, mounting evidence associates withdrawal with an increase of aggression (Budney & Hughes, 2006; Hoaken & Stewart, 2003; Kouri, Pope, & Lukas, 1999; Moore & Stuart, 2005; Smith et al., 2013) (see Figure 2). Anger, aggression, and

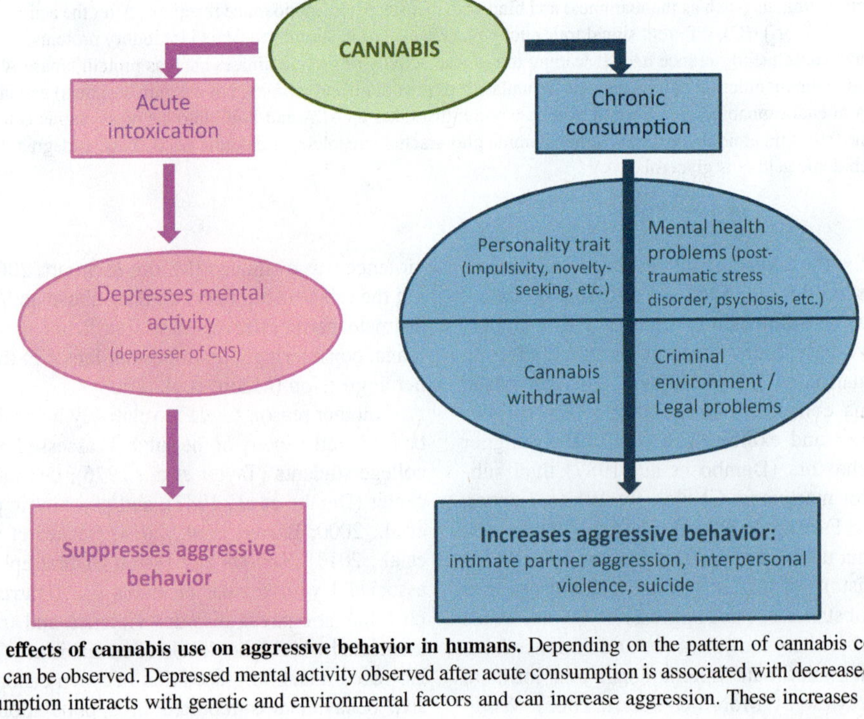

FIGURE 2 **Different effects of cannabis use on aggressive behavior in humans.** Depending on the pattern of cannabis consumption, increases or decreases in aggression can be observed. Depressed mental activity observed after acute consumption is associated with decreased aggression. Conversely, chronic cannabis consumption interacts with genetic and environmental factors and can increase aggression. These increases are observed particularly during withdrawal syndrome.

irritability are common symptoms observed during cannabis withdrawal (Ramesh, Schlosburg, Wiebelhaus, & Lichtman, 2011; Vandrey, Budney, Hughes, & Liguori, 2008), and can persist for weeks (Budney, Moore, Vandrey, & Hughes, 2003; Kouri et al., 1999). The greatest risk of violence seems to be within the first week of abstinence and is associated with different variables. In highly controlled experimental settings, it has been observed that heavy chronic THC users are significantly more aggressive than controls and with respect to their preabstinence aggression levels at 3 and 7, but not at 28 days, of abstinence (Kouri et al., 1999). In a laboratory experiment with dependent cannabis users, withdrawal-related angry outbursts were associated with high levels of distress (Allsop, Norberg, Copeland, Fu, & Budney, 2011). Finally, in one observational study, marijuana withdrawal symptoms were linked with current relationship aggression (but not with general aggression) among users with a history of aggression, but not among those without such a history (Smith et al., 2013). According to the authors, one explanation for the differential findings between relationship and general aggression stems from the transitive nature of marijuana withdrawal. Withdrawal symptoms peak after 2–6 days of abstinence and may last for up to 2 weeks, during which time users tend to interact more frequently with their partner than others. Moreover, as the authors of the study in question stated, these results do not exclude marijuana withdrawal as a contributing factor to general aggression in particular individuals. Chronic marijuana users may run an increased risk of aggression during periods of abstinence. Moreover, individuals who have a tendency to be aggressive are more likely to act aggressively during periods of marijuana withdrawal than other periods. This needs to be confirmed by research employing alternative approaches, such as longitudinal studies.

Finally, it has been hypothesized that sensitization of CB1-receptor-mediated G-protein signaling in the prefrontal cortex is one of the etiological or neuroadaptive factors in the pathophysiology of suicide, a less common form of aggression that is self-directed (Vinod & Hungund, 2006).

RELATION BETWEEN AGGRESSION AND CB1 RECEPTOR IN ANIMAL STUDIES

Numerous experimental studies have highlighted the involvement of the endocannabinoid system in the control of emotional behavior (for review see Valverde & Torrens, 2012). However, few studies have used animal models to specifically address the relationship between CB1 receptors and aggression. Most of those that have done so have shown that stimulation of this receptor decreases aggression, although blockade or absence of the CB1 receptor can also decrease aggression in several circumstances.

In a series of studies performed in the 1970s, Miczek and coworkers studied the effects of acute or chronic THC administration in mice and rats. Acute THC administration decreased offensive aggression in dominant animals, but also altered the submissive reaction of subordinate subjects (Miczek, 1978; Miczek & Barry, 1977). However, after chronic administration of THC for 5–8 weeks, 25–70% of previously "nonkiller" rats presented mouse-killing (Miczek, 1976).

Sativex®, a mixture of THC (a stimulant of both CB1 and CB2 receptors) and cannabidiol, has been approved for the treatment of spasticity in multiple sclerosis. In comparison with vehicle-treated PK$^{-/-}$/TauVLW mice (parkin-null, human tau-overexpressing PK$^{-/-}$/TauVLW, a model of complex frontotemporal dementia, parkinsonism, and lower motor neuron disease), Sativex®-treated animals show a significant improvement in abnormal behaviors related to stress, such as auto- and heteroaggressive behavior and stereotypes (Casarejos et al., 2013). In the study in question, there was a significant reduction in self-injury facial masks, which was very severe in vehicle-treated PK$^{-/-}$/TauVLW mice and very mild in Sativex®-treated littermates.

Repeated administration of psychostimulants and cannabinoids can elicit so-called behavioral sensitization, a gradually increased behavioral response to a drug. Landa, Slais, and Sulcova (2006) demonstrated that the CB1 receptor was involved in the development of sensitization to the antiaggressive effects of methamphetamine. Pretreatment with the CB1 agonist methanandamide resulted in cross-sensitization to this methamphetamine antiaggressive effect, whereas pretreatment with a CB2 agonist, JWH-015, did not. Combined pretreatment with methamphetamine plus a CB1 antagonist (AM251) suppressed this sensitization.

Among the countless consequences of stress in animals and in humans, evidence suggests that endocannabinoid transmission in the brain is altered. Indeed, stress has been shown to alter endocannabinoid content in several brain areas, including the amygdala, striatum, and prefrontal cortex (Rademacher et al., 2008). Activation of the endocannabinoid system during stress modulates complex responses such as stress-induced analgesia, escaping behavior, suppression of reproductive behavior, and sensitivity to natural reward. A social defeat stress paradigm was shown to cause a dramatic rearrangement of the CB1 receptor in the striatum (Rossi et al., 2008). However, Moise Eisenstein, Astarita, Piomelli, and Hohmann (2008) observed that neither unconditioned nor conditioned social defeat in the Syrian hamster depended on CB1 receptor activation, blockade of CB1, or inhibition of FAAH-altered conditioned defeat behavior. In line with this, Griebel, Stemmelin, and Scatton (2005) observed that the CB1 antagonist rimonabant did not affect flight or risk assessment when mice were able to escape from the oncoming rat. In contrast, when escape was not possible, the CB1 antagonist decreased defensive threat and attack reactions. The forced-contact test is thought to be particularly stressful for animals because of the impossibility of escape and unavoidable confrontation with the threat stimulus. Similar to the acute blockade of CB1 receptors by rimonabant, permanent deletion of the CB1 receptor gene (CB1$^{-/-}$) in mice led to a profile of reduced defensiveness, suggesting that said receptor plays an important role in the expression of this particular set of behaviors.

Genetically modified mice lacking the CB1 receptor behave normally under basal conditions, but can display altered behavior under adverse environmental conditions (for review see Valverde & Torrens, 2012). Experiments with CB1-knockout (CB1-KO) mice have revealed anxiogenic- and depressive-like phenotypes. To date, only two studies have evaluated the social and aggressive profile of CB1-KO mice. Martin, Ledent, Parmentier, Maldonado, and Valverde (2002) used the resident–intruder paradigm, in which aggressive behavior by a resident toward an intruder represents the species-typical repertoire of offensive aggressive acts and postures. In this paradigm, males living in a pair with a female or isolated for several weeks attack an intruder/opponent that is placed in their home cage (Rodriguez-Arias, Aguilar, & Simon, 2005). In Martin's study, resident CB1-KO mice were significantly more

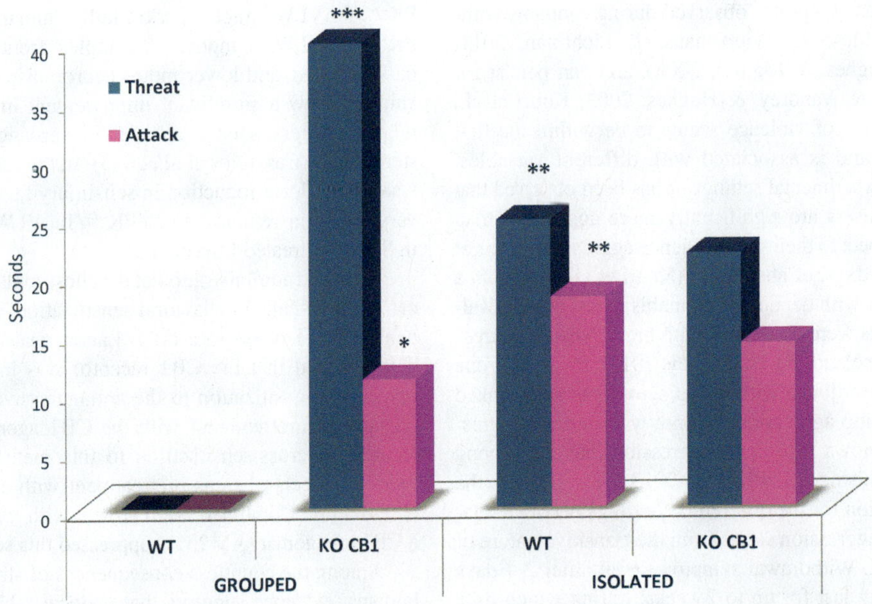

FIGURE 3 Aggressive behavior displayed by mice lacking the CB1 receptor in a social interaction test. Means ± SEM of accumulated times (in seconds) for threat (blue) and attack (pink) behaviors of group- or singly housed CB1-KO and WT adult mice during the social interaction test ($n = 10$–15 mice per group). Differences with respect to group-housed WT mice, $*p < 0.05$, $**p < 0.01$, $***p < 0.001$. *Modified with permission from Rodriguez-Arias et al. (2013).*

aggressive toward intruders than wild-type animals, although this difference disappeared after successive encounters.

In a more recent study performed in our laboratory we studied the aggressive behavior of CB1-KO mice by means of another paradigm, namely, intermale aggression (Rodriguez-Arias et al., 2013). We employed an ethological model of intermale aggression in which an experimental adult mouse is group housed or isolated for 28 days and is then confronted with a conspecific that has been housed in groups and rendered anosmic with zinc sulfate 1 day before the test. This opponent mouse induces an attack reaction in its opponent but does not outwardly provoke or defend itself, since it cannot perceive a pheromone that is present in the urine of the experimental animals. However, the opponent mouse elicits aggressive behavior in mice with a normal sense of smell. The agonistic encounter takes place in a neutral area, and both offensive and defensive behaviors are evaluated. Our results suggested that the CB1 receptor played a relevant role in the regulation of aggressive behavior. First, we observed that grouped CB1-KO mice showed more aggression than their wild-type (WT) counterparts, spending more time in threat and attack, needing less time to perform the first aggressive behavior, and showing longer aggressive interactions during the social interaction test (see Figure 3). However, when WT or CB1-KO mice were housed in isolation, a procedure that has been shown to facilitate the display of aggression in laboratory mice (Rodriguez-Arias et al., 2005), no differences were observed. These KO mice presented differences in serotonin (5-HT), a key neurotransmitter system involved in aggression control. 5-HT in the mammalian central nervous system is derived mainly from dorsal and medial raphe. Inhibition of the metabolism of monoamines renders 5-HT and other monoamines more available in the brain. However, CB1-KO mice seem to better metabolize 5-HT, as they show higher levels of catechol-O-methyltransferase (COMT) in the raphe nucleus and amygdala. Gene expression of monoamine oxidase A was also increased in the amygdala. This may have reduced 5-HT levels, which could have been related with the elevated concentration of 5-HT1Br observed in these mice. In the same study, we tested the antiaggressive effect of the CB1 agonist Arachidonyl-2'-chloroethylamide (ACEA), and observed how it significantly decreased aggressive behaviors in isolated, highly aggressive mice (see Figure 4).

CONCLUSIONS

The relationship between cannabis use and aggressive behavior, although widely studied in humans and in animal models, is a subject that requires clarification, as there are many confounding variables that need to be taken into consideration. As a general conclusion, we can affirm that acute cannabis use does not lead to increases in aggressive behavior, but rather has the opposite effect. However, under different circumstances, such as high stress, contrasting effects can be seen; for example, withdrawal syndrome after discontinuation of chronic cannabis use is one of the most strongly associated with heightened aggression. Pharmacological studies in animal models do not generally support the hypothesis that cannabis induces aggression. Studies performed in mice lacking the CB1 receptor confirm this lack of a relation.

APPLICATIONS TO OTHER ADDICTIONS AND SUBSTANCE MISUSE

- Alcohol is a drug that has been associated with heightened aggression and is often consumed with cannabis among adolescents and young people.
- The increase in aggression observed after cannabis withdrawal must be considered when seeking to abstain from multiple drugs simultaneously (e.g., tobacco or opioids).

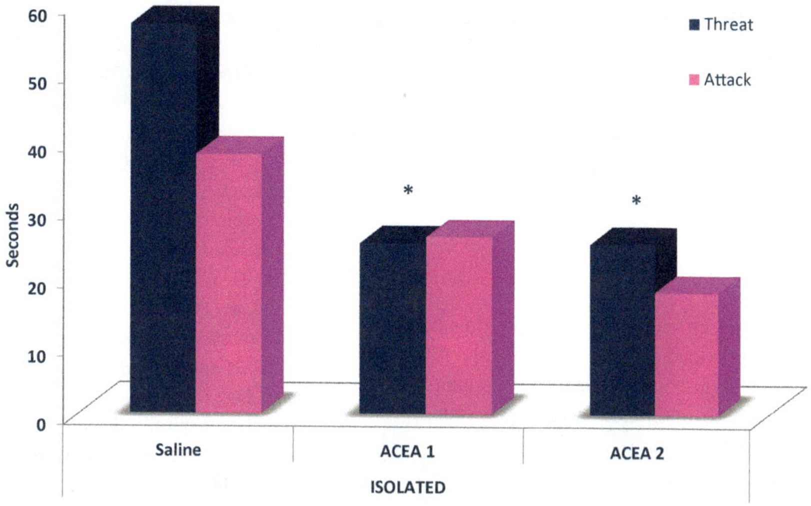

FIGURE 4 Effect of a single dose of the CB1 agonist ACEA. Means ± SEM of accumulated times (in seconds) for threat (blue) and attack (pink) behaviors during the social interaction test in highly aggressive isolated adult OF1 mice treated with saline or the CB1 agonist ACEA (1 or 2 mg/kg) ($n=9–12$ mice per group). Differences with respect to saline-treated group-housed mice, $*p<0.05$. *Modified with permission from Rodriguez-Arias et al. (2013).*

- The CB1 receptors are involved in feeding behavior and mood change.
- Among adolescents, cannabis use may be a means of forming relationships.
- Cannabis consumption is closely associated with psychiatric illness (dual pathology). Knowledge of the relation between cannabis and aggression could be of relevance for treatment of this comorbidity.

DEFINITION OF TERMS

Aggression This refers to any form of behavior directed toward the goal of harming or injuring another living being that is motivated to avoid such treatment.

Animal models of aggression An animal model allows for the study of one or several aspects of a human condition. Its purpose is the study of a given phenomenon found in humans. Animal models are essential because they have a heuristic, hypothesis-generating function and provide important parallels to human aggression.

CB1 receptor This is the most abundant G-protein-coupled receptor presynaptically located at nerve terminals, where endocannabinoids act as modulators of synaptic transmission.

CB1 receptor agonists These are compounds that induce a change in the configuration of the CB1 receptor, producing a biological response. As example is WIN 55,212.

CB1 receptor antagonists These are compounds that induce a change in the configuration of the CB1 receptor, without inducing any biological response. Examples are rimonabant and SR141716A.

Correlational study This is a scientific study in which researchers investigate associations between variables. It allows us to determine which variables are related. However, the fact that two variables are related or correlated does not mean there is a causal relationship.

Endocannabinoid system This is the retrograde neurotransmitter system formed by the cannabinoid receptors, endogenously produced compounds displaying significant affinity for these receptors (endocannabinoids), and enzymes involved in the synthesis and metabolism of endocannabinoids.

Longitudinal study This is a study in which researchers conduct several observations of the same subjects over a period of time, sometimes lasting many years. It is an observational study, as researchers do not interfere with their subjects.

Offensive and defensive aggression The distinction between offensive and defensive aggression is based upon the distal and proximal antecedent conditions that precipitate aggression, the topography of the behavior, and its consequences. Offensive aggressive behavior between conspecifics is ritually organized, and the attack is usually directed toward less vulnerable body areas such as the back and flanks of the opponent. Defensive aggression involves attack in defense of the self in response to threatening or fear-inducing stimuli and is often accompanied by escape.

Sensitization In the context of the use of drugs, sensitization refers to the increased effectiveness of a given drug with repeated administration.

Substance use disorder The *Diagnostic and Statistical Manual of Mental Disorders,* fifth edition (DSM-5), defines a substance use disorder as a cluster of cognitive, behavioral, and physiological symptoms indicating that the individual continues using the substance despite significant substance-related problems. An important characteristic of substance use disorders is an underlying change in brain circuits that may persist beyond detoxification, particularly in individuals with severe disorders

Withdrawal syndrome The DSM-5 defines this as a syndrome that occurs when blood or tissue concentrations of a substance decline in an individual who has maintained prolonged heavy use of the substance. After developing withdrawal symptoms, the individual is likely to consume the substance to relieve the symptoms.

KEY FACTS OF AGGRESSION

- Aggressive behavior is influenced by genetic and environmental factors.
- The amygdala, anterior cingulate cortex, and regions of the prefrontal cortex are associated with aggression control.
- Aggression is elicited by excessive reactivity in the amygdala plus inadequate prefrontal regulation.

- Insufficient serotonergic activity can enhance aggression.
- Functional polymorphisms in the monoamine oxidase A and 5-HT transporter may be of particular importance.
- Rodent models of aggression are not comparable to aggression in humans.
- Similar molecular mechanisms are found in aggression in mice and humans.

SUMMARY POINTS

- Studies of the relation between cannabis consumption and abnormal aggression have provided contrasting results.
- Cannabinoids act on a G-protein-coupled receptor presynaptically located at nerve terminals that modulate synaptic transmission, called the CB1 receptor.
- There is general agreement that low doses of cannabis can slightly increase aggression and that moderate and high doses can suppress or even eliminate aggressive behavior in human beings.
- Increasing evidence associates withdrawal after chronic cannabis use with an increase in aggression.
- Animal studies confirm a role for the CB1 receptor in controlling aggressive behavior.
- Several variables should be taken into consideration when studying cannabis and aggression, such as previous aggressive profile or stressful circumstances.

REFERENCES

Allsop, D. J., Norberg, M. M., Copeland, J., Fu, S., & Budney, A. J. (2011). The Cannabis Withdrawal Scale development: patterns and predictors of cannabis withdrawal and distress. *Drug and Alcohol Dependence*, *119*, 123–129.

Arendt, M., Rosenberg, R., Fjordback, L., Brandholdt, J., Foldager, L., Sher, L., & Munk-jørgensen, P. (2007). Testing the self-medication hypothesis of depression and aggression in cannabis-dependent subjects. *Psychological Medicine*, *37*, 935–945.

Arseneault, L., Moffitt, T. E., Caspi, A., Taylor, P. J., & Silva, P. A. (2000). Mental disorders and violence in a total birth cohort: results from the Dunedin Study. *Archives of General Psychiatry*, *57*, 979.

Baron, R. A., & Richardson, D. R. (1994). *Human aggression* (2nd ed.). New York: Plenum Press.

Barrett, E. L., Teesson, M., & Mills, K. L. (2014). Associations between substance use, post-traumatic stress disorder and the perpetration of violence: a longitudinal investigation. *Addictive Behaviors*, *39*, 1075–1080.

Becker, S. J., Nargiso, J. E., Wolff, J. C., Uhl, K. M., Simon, V. A., Spirito, A., & Prinstein, M. J. (2012). Temporal relationship between substance use and delinquent behavior among young psychiatrically hospitalized adolescents. *Journal of Substance Abuse Treatment*, *43*, 251–259.

Berkowitz, L. (1993). *Aggression: its causes, consequences and control*. Philadelphia: Temple University Press.

Blanchard, R. J., & Blanchard, D. C. (1977). Aggressive behavior in the rat. *Behavioral Biology*, *21*, 197–224.

Boles, S. M., & Miotto, K. (2003). Substance abuse and violence: a review of the literature. *Aggression and Violent Behavior*, *8*, 155–174.

Brain, P., & Benton, D. (1979). The interpretation of physiological correlates of differential housing in laboratory rats. *Life Sciences*, *24*, 99–115.

Budney, A. J., & Hughes, J. R. (2006). The cannabis withdrawal syndrome. *Current Opinion on Psychiatry*, *19*, 233–238.

Budney, A. J., Moore, B. A., Vandrey, R. G., & Hughes, J. R. (2003). The time course and significance of cannabis withdrawal. *Journal of Abnormal Psychology*, *112*, 393–402.

Casarejos, M. J., Perucho, J., Gomez, A., Muñoz, M. P., Fernandez-Estevez, M., Sagredo, O., ... Mena, M. A. (2013). Natural cannabinoids improve dopamine neurotransmission and tau and amyloid pathology in a mouse model of tauopathy. *Journal of Alzheimer Disease*, *35*, 525–539.

Chermack, S. T., Grogan-Taylor, A., Perron, B. E., Murray, R. L., De Chavez, P., & Walton, M. A. (2010). Violence among men and women in substance use disorder treatment: a multi-level event-based analysis. *Drug and Alcohol Dependence*, *112*, 194–200.

Coccaro, E. F., Lee, R., & Kavoussi, R. J. (2010). Aggression, suicidality, and intermittent explosive disorder: serotonergic correlates in personality disorder and healthy control subjects. *Neuropsychopharmacology*, *35*, 435–444.

Dembo, R., Walshburn, M., Wish, E., Yeung, H., Getreu, A., Berry, E., & Blount, W. R. (1987). Heavy marijuana use and crime among youths entering a juvenile detention center. *Journal of Psychoactive Drugs*, *19*, 47–56.

Devane, W. A., Dysarz, F. A., Johnson, M. R., Melvin, L. S., & Howlett, A. C. (1988). Determination and characterization of a cannabinoid receptor in rat brain. *Molecular Pharmacology*, *34*, 605–613.

Devane, W. A., Hanus, L., Breuer, A., Pertwee, R. G., Stevenson, L. A., Griffin, G., ... Mechoulam, R. (1992). Isolation and structure of a brain constituent that binds to the cannabinoid receptor. *Science*, *258*, 1946–1949.

Fridell, M., Hesse, M., Jæger, M. M., & Kühlhorn, E. (2008). Antisocial personality disorder as a predictor of criminal behavior in a longitudinal study of a cohort of abusers of several classes of drugs: relation to type of substance and type of crime. *Addictive Behaviors*, *33*, 799–811.

Friedman, A. S. (1998). Substance use/abuse as a predictor to illegal and violent behavior: a review of the relevant literature. *Aggression and Violent Behavior*, *3*, 339–355.

Gaoni, Y., & Mechoulam, R. (1964). Isolation, structure and partial synthesis of an active constituent of hashish. *Journal of American Chemical Society*, *86*, 1646–1647.

Griebel, G., Stemmelin, J., & Scatton, B. (2005). Effects of the cannabinoid CB1 receptor antagonist rimonabant in models of emotional reactivity in rodents. *Biological Psychiatry*, *57*, 261–267.

Harris, A. W. F., Large, M. M., Redoblado-Hodge, A., Nielssen, O., Anderson, J., & Brennan, J. (2010). Clinical and cognitive associations with aggression in the first episode of psychosis. *Australian and New Zealand Journal of Psychiatry*, *44*, 85–93.

Herkenham, M., Lynn, A. B., Little, M. D., Johnson, M. R., & Melvin, L. S. (1990). Cannabinoid receptor localization in brain. *Proceedings of the National Academy of Sciences of the United States of America*, *87*, 1932–1936.

Hoaken, P. N., & Stewart, S. H. (2003). Drugs of abuse and the elicitation of human aggressive behavior. *Addiction Behavior*, *28*, 1533–1554.

Howlett, A. C. (2002). The cannabinoid receptors. *Prostaglandins and Other Lipids Mediators*, *68–69*, 619–631.

Kouri, E. M., Pope, H. G., & Lukas, S. E. (1999). Changes in aggressive behavior during withdrawal from long-term marijuana use. *Psychopharmacology*, *143*, 302–308.

Landa, L., Slais, K., & Sulcova, A. (2006). Impact of cannabinoid receptor ligands on behavioural sensitization to antiaggressive methamphetamine effects in the model of mouse agonistic behaviour. *Neuroendocrinology Letters*, *27*, 703–710.

Lejoyeux, M., Nivoli, F., Basquin, A., Petit, A., Chalvin, F., & Embouazza, H. (2013). An investigation of factors increasing the risk of aggressive behavior among schizophrenic inpatients. *Frontiers in Psychiatry, 4*, 97.

Macdonald, S., Erickson, P., Wells, S., Hathaway, A., & Pakula, B. (2008). Predicting violence among cocaine, cannabis, and alcohol treatment clients. *Addictive Behaviors, 33*, 201–205.

Martin, M., Ledent, C., Parmentier, M., Maldonado, R., & Valverde, O. (2002). Involvement of CB1 cannabinoid receptors in emotional behaviour. *Psychopharmacology, 159*, 379–387.

Matias, I., Bisogno, T., & Di Marzo, V. (2006). Endogenous cannabinoids in the brain and peripheral tissues: regulation of their levels and control of food intake. *International Journal of Obesity, 30*, 7–12.

Matsuda, L. A., Lolait, S. J., Brownstein, M. J., Young, A. C., & Bonner, T. I. (1990). Structure of a cannabinoid receptor and functional expression of the cloned cDNA. *Nature, 346*, 561–564.

Mechoulam, R., Ben-Shabat, S., Hanus, L., Ligumsky, M., Kaminiski, N. E., Schatz, A. R., ... Vogel, Z. (1995). Identification of an endogenous 2-monoglyceride, present in canine gut, that binds to cannabinoid receptors. *Biochemical Pharmacology, 50*, 83–90.

Mechoulam, R., & Parker, L. A. (2013). The endocannabinoid system and the brain. *Annual Review of Psychology, 64*, 21–47.

Miczek, K. A. (1976). Mouse-killing and motor activity: effects of chronic delta9-tetrahydrocannabinol and pilocarpine. *Psychopharmacology, 47*, 59–64.

Miczek, K. A. (1978). Delta9-tetrahydrocannabinol: antiaggressive effects in mice, rats, and squirrel monkeys. *Science, 199*(4336), 1459–1461.

Miczek, K. A., & Barry, H., 3rd (1977). Comparison of the effects of alcohol, chlordiazepoxide, and delta9-tetrahydrocannabinol on intraspecies aggression in rats. *Advances in Experimental and Medical Biology, 85B*, 251–264.

Miczek, K. A., Faccidomo, S., Fish, E. W., & DeBold, J. F. (2007). Neurochemistry and molecular neurobiology of aggressive behavior. In J. Blaustein (Ed.), *Behavioral neurochemistry, neuroendocrinology and molecular neurobiology* (3rd ed.) (pp. 285–336). New York: Springer.

Moise, A. M., Eisenstein, S. A., Astarita, G., Piomelli, D., & Hohmann, A. G. (2008). An endocannabinoid signaling system modulates anxiety-like behavior in male syrian hamsters. *Psychopharmacology, 200*, 333–346.

Moore, T. M., & Stuart, G. L. (2005). A review of the literature on marijuana and interpersonal violence. *Aggressive and Violent Behavior, 10*, 171–192.

Moore, T. M., Stuart, G. L., Meehan, J. C., Rhatigan, D. L., Hellmuth, J. C., & Keen, S. M. (2008). Drug abuse and aggression between intimate partners: a meta-analytic review. *Clinical Psychology Review, 28*, 247–274.

Morris, R. G., TenEyck, M., Barnes, J. C., & Kovandzic, T. V. (2014). The effect of medical marijuana laws on crime: evidence from state panel data, 1990–2006. *PLoS One, 9*(3), e92816.

Munro, S., Thomas, K. L., & Abu-Shaar, M. (1993). Molecular characterization of a peripheral receptor for cannabinoids. *Nature, 365*, 61–65.

Myerscough, R., & Taylor, S. (1985). The effects of marijuana on human physical aggression. *Journal of Personality and Social Psychology, 49*, 1541–1546.

Pertwee, R. G. (2010). Receptors and channels targeted by synthetic cannabinoid receptor agonists and antagonists. *Current Medical Chemistry, 17*, 1360–1381.

Pertwee, R. G., Howlett, A. C., Abood, M. E., Alexander, S. P., Di Marzo, V., Elphick, M. R., ... Ross, R. A. (2010). International Union of Basic and Clinical Pharmacology. LXXIX. Cannabinoid receptors and their ligands: beyond CB_1 and CB_2. *Pharmacological Reviews, 62*, 588–631.

Pulay, A. J., Dawson, D. A., Hasin, D. S., Goldstein, R. B., Ruan, W. J., Pickering, R. P., & Grant, B. F. (2008). Violent behavior and DSM-IV psychiatric disorders: results from the national epidemiologic survey on alcohol and related conditions. *Journal of Clinical Psychiatry, 69*, 12–22.

Rademacher, D. J., Meier, S. E., Shi, L., Ho, W. S., Jarrahian, A., & Hillard, C. J. (2008). Effects of acute and repeated restraint stress on endocannabinoid content in the amygdala, ventral striatum, and medial prefrontal cortex in mice. *Neuropharmacology, 54*, 108–116.

Ramesh, D., Schlosburg, J. E., Wiebelhaus, J. M., & Lichtman, A. H. (2011). Marijuana dependence: not just smoke and mirrors. *Journal of the Institute for Laboratory Animal Research, 52*, 295–308.

Rodriguez-Arias, M., Aguilar, M. A., & Simon, V. M. (2005). Drugs of abuse and aggression: a review in animal models. *Current Topics in Pharmacology, 9*, 1–27.

Rodriguez-Arias, M., Navarrete, F., Daza-Losada, M., Navarro, D., Aguilar, M. A., Berbel, P., ... Manzanares, J. (2013). CB1 cannabinoid receptor-mediated aggressive behavior. *Neuropharmacology, 75*, 172–180.

Rodriguez de Fonseca, F. (2008). The endogenous cannabinoid system and drug addiction: 20 years after the discovery of the CB1 receptor. *Addiction Biology, 13*, 143–146.

Rossi, S., De Chiara, V., Musella, A., Kusayanagi, H., Mataluni, G., Bernardi, G., ... Centonze, D. (2008). Chronic psychoemotional stress impairs cannabinoid- receptor-mediated control of GABA transmission in the striatum. *The Journal of Neuroscience, 28*, 7284–7292.

Smith, P. H., Homish, G. G., Leonard, K. E., & Collins, R. L. (2013). Marijuana withdrawal and aggression among a representative sample of U.S. marijuana users. *Drug and Alcohol Dependence, 132*, 63–68.

Taylor, S. P., Vardaris, R. M., Rawtich, A. B., Gammon, C. B., Cranston, J. W., & Lubetkin, A. I. (1976). The effects of alcohol and delta-9-tetrahydrocannabinol on human physical aggression. *Aggressive Behavior, 2*, 153–161.

Valverde, O., & Torrens, M. (2012). CB1 receptor-deficient mice as a model for depression. *Neuroscience, 204*, 193–206.

Vandrey, R. G., Budney, A. J., Hughes, J. R., & Liguori, A. (2008). A within-subject comparison of withdrawal symptoms during abstinence from cannabis, tobacco, and both substances. *Drug and Alcohol Dependence, 92*, 48–54.

Vinod, K. Y., & Hungund, B. L. (2006). Role of the endocannabinoid system in depression and suicide. *Trends in Pharmacological Sciences, 27*, 539–545.

Vitiello, B., & Stoff, D. M. (1997). Subtypes of aggression and their relevance to child psychiatry. *Journal of American Academy of Child and Adolescent Psychiatry, 36*, 307–315.

White, H. R., Loeber, R., Stouthamer-Loeber, M., & Farrington, D. P. (1999). Developmental associations between substance use and violence. *Developmental Psychopathology, 11*, 785–803.

Chapter 78

Cannabis and Bipolar Manic Episodes

Jean-Michel Aubry
Mood Disorders Unit, Division of Psychiatric Specialties, Department of Mental Health and Psychiatry, Geneva University Hospitals, Geneva, Switzerland

Abbreviations

AUD Alcohol use disorder
BD Bipolar disorder
fMRI Functional magnetic resonance imaging
MDD Major depressive disorder
SUD Substance use disorder

SUBSTANCE USE AND BIPOLAR DISORDER

Among mood spectrum disorders, bipolar disorder bears the highest rate of substance abuse, with about 50% of subjects having a lifetime comorbidity of an alcohol use disorder and 41% having any other substance use disorder (Regier et al., 1990; Strakowski et al., 2007). The association between cannabis use and mood disorders is well recognized and cannabis is the most commonly used illegal drug among patients with bipolar disorders. Around one-third of bipolar patients have a diagnosis of cannabis abuse or dependence and lifetime prevalence rates from 26% to 46% have been reported (Agrawal, Nurnberger, & Lynskey, 2011; Cerullo & Strakowski, 2007; Leweke & Koethe, 2008).

Substance abuse and increased symptom severity have also been shown (Lagerberg et al., 2011).

CHILDHOOD TRAUMA, CANNABIS, AND BIPOLAR DISORDERS

A link has been reported between childhood trauma and increased risk of substance abuse in adulthood (Fergusson, Boden, & Horwood, 2008; Shin, Edwards, Heeren, & Amodeo, 2009). In a 2014 study exploring the influence of childhood trauma and substance abuse on the clinical picture of 587 patients with bipolar disorder (BD), Aas et al. (2014) showed significant additive effects between childhood abuse and cannabis abuse on increased clinical expression of BD.

IMPACT OF CANNABIS USE ON MANIA PHENOMENOLOGY

It has been proposed that manic episodes can be classified as predominantly euphoric, which is usually called "classic" or "pure" mania, or dysphoric mania, characterized by inner tension, irritability, and anger with verbal and/or physical aggressiveness (Dilsaver, Chen, Shoaib, & Swann, 1999; Swann et al., 2013).

Higher rates of substance use disorders (SUDs) have been reported in patients with dysphoric mania compared with euphoric mania (Feinman & Dunner, 1996). In a 2015 study with bipolar hospitalized manic patients, Güclü, Senormanci, Aydin, Erkiran, and Kokturk (2015) evaluated distinct clusters of subtypes and the relationship between the clinical features such as SUDs and the various clusters. They found that 39% of the patients in the psychomotor elevation cluster (with features such as increased motor activity, enhanced rate of speech, disruptive behavior, elevated mood, increased sexual interest, and diminished sleep) had an alcohol use disorder. In the dysphoric–psychotic cluster, 31.6% presented an alcohol and a cannabis use comorbidity. The authors raise the hypothesis that patients with classical or pure mania may use the sedative effect of alcohol as a self-medication. They also point to a causal relationship between cannabis and psychosis (Güclü et al., 2015).

Moreover, in the dysphoric–psychotic cluster, 47% of patients had made one suicide attempt during their lifetime. This result is in agreement with previous studies showing that dysphoria is associated with suicide attempts (Dalton, Cate-Carter, Mundo, Parikh, & Kennedy, 2003).

In one study, the duration of cannabis abuse was closely related with the duration of manic episodes (Cerullo & Strakowski, 2007). The impact of cannabis on mania phenomenology is summarized in Table 1.

IMPACT OF CANNABIS USE ON BIPOLAR DISORDER OUTCOME

In the field of mood disorders, the association of cannabis use and BDs has been less investigated than the association with unipolar depression (Feingold, Weiser, Rehm, & Lev-Ran, 2015). Nevertheless, several studies have reported an association between cannabis use and BD (Cerullo & Strakowski, 2007; Etain et al., 2013). As pointed out by Feingold et al. (2015), longitudinal studies have suggested a relationship between baseline cannabis use and the diagnosis of BD at follow-up (Baethge et al., 2008; Tijssen et al., 2010). However, methodological limitations imply that these findings must be taken with caution.

TABLE 1 Impact of Cannabis on Mania Phenomenology

- Psychotic symptoms
- Dysphoric mania
- More lifetime suicide attempts
- Longer episodes

The characteristics listed above have been linked to cannabis use.

In a longitudinal investigaion drawn from the National Epidemiology Survey on Alcohol and Related Conditions, Feingold et al. (2015) explored the association between cannabis use and major depressive disorder (MDD) and BD over a 3-year prospective study. For unipolar disorder, they found that there was no difference in the incidence of MDD between cannabis users compared to nonusers, which is in agreement with several previous studies. For BD, their results suggest that the association between cannabis use and increased incidence of BD was not confirmed after adjusting for very frequent (daily or almost daily) users as well as for subjects using cannabis on a less-than-weekly basis. The authors hypothesize that additional factors may mediate the association between cannabis use and BD (Feingold et al., 2015).

Some studies have shown a poorer medication adherence and a more severe course of illness for BD individuals with cannabis use (Agrawal et al., 2011). It has also been reported that bipolar patients with cannabis use spend more time in affective episodes (Strakowski et al., 2007) and that the number of manic or depressive episodes is higher (De Hert et al., 2011; Lev-Ran, Le Foll, McKenzie, George, & Rehm, 2013a).

A study that did not differentiate between cannabis use alone or cannabis plus other substances reported a higher rate of substance abuse for BD type I than BD type II (Regier et al., 1990). However, when types I and II BD were compared for the rate of cannabis abuse comorbidity and the course of illness, no significant difference was found (Bega, Schaffer, Goldstein, & Levitt, 2012). Stone et al. (2014) also showed that cannabis users who diminished or stopped the use of cannabis after a first hospitalization had a greater improvement in symptomatology at 1-year follow-up compared to subjects who continued to use cannabis at the same frequency. Based on data from a 2-year prospective observational study in individuals with a manic or mixed episode (EMBLEM), Zorrilla et al. (2015) reported that patients who stopped using cannabis during a manic/mixed episode had similar functional outcome compared to patients who never used cannabis. In contrast, continuous use was associated with poorer functioning and a higher risk of recurrence (Zorrilla et al., 2015).

Agrawal et al. (2011) have reported more frequent mixed episodes in patients with cannabis use disorders. Lev-Ran, Le Foll, McKenzie, George, and Rehm (2013b) showed a more severe course of illness and poorer treatment outcomes in BD patients with cannabis abuse.

In 2015, Kvitland et al. (2015) examined the association between recent onset BD type I and course of illness with cannabis use. They reported that continuous cannabis use was significantly associated with elevated mood and diminished global functioning at follow-up. Moreover, inferior global functioning was linked to elevated mood.

CANNABIS USE AND ONSET MANIC EPISODE

There are sufficient data showing that the risk of psychotic disorders increases with the frequency and intensity of cannabis use (Bally, Zullino, & Aubry, 2014). The question of whether cannabis also increases the risk of developing BD has been less investigated.

Using a retrospective design, Agrawal et al. (2011) reported that around half of bipolar patients identified their mood disorders prior to cannabis use. Some studies have also suggested that people with BD may use cannabis as self-medication for their (hypo)manic and or depressive mood episodes (Bizzarri et al., 2009, 2007) and that BD increases the risk of developing any future substance abuse (Merikangas et al., 2008).

However, there is now accumulating evidence suggesting that the age of onset of first manic or depressive episode is younger when there is co-occurring cannabis abuse (Duffy et al., 2012; Lagerberg et al., 2011). In a study with 101 subjects with first-treatment BD type I, Kvitland et al. (2014) reported that recent cannabis use was significantly associated with lower age at onset of first manic and psychotic episode.

In contrast, there was no association with first depressive episode.

In an 8-year prospective study with 705 young subjects ages 14–24 years, Tijssen et al. (2010) found that cannabis use was significantly associated with onset of (hypo)manic symptoms. In this study, subjects who reported past cannabis use were clearly more likely to develop manic symptomatology.

These data are in agreement with De Hert et al. (2011) and Strakowski et al. (1998, 2007), suggesting that substance abuse including cannabis can trigger the onset of BD and the occurrence of new manic episodes (Strakowski, DelBello, Fleck, & Arndt, 2000). They also support the hypothesis that cannabis may increase the risk of manic and psychotic episodes through an interaction with genetic vulnerability and other environmental stressors (Kvitland et al., 2014; Lev-Ran et al., 2013a). Major factors increasing the risk of bipolar onset episodes or psychosis in subjects using cannabis are shown (Figure 1).

In view of the published evidence and as stated in a 2014 review on cannabis and first manic episode (Bally et al., 2014), a causal relationship between cannabis use and onset (hypo)mania is building up. Nevertheless, further investigations are still needed to confirm the association between cannabis use and onset of BD.

CANNABIS USE AND PSYCHOSIS

A large enough body of data has now accumulated to point to a causal relationship between cannabis use and onset of psychosis (Burns, 2013). For example, the age at onset of positive symptoms in schizophrenia was younger for subjects using cannabis (Donoghue et al., 2014). Koskinen, Lohonen, Koponen, Isohanni, and Miettunen (2010) reported in a meta-analysis that young male subjects with first psychotic episode used cannabis more often. Moreover, in bipolar patients, cannabis use was shown to be associated with a higher risk of psychotic symptoms (van Rossum, Boomsma, Tenback, Reed, & van Os, 2009).

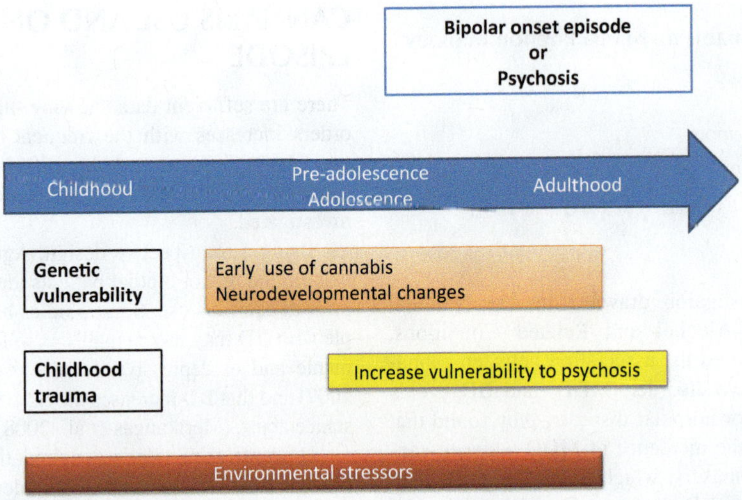

FIGURE 1 Longitudinal overview of cannabis use and bipolar onset episode or psychosis. Major factors increasing the risk of bipolar onset episodes or psychosis in subjects using cannabis are shown.

MECHANISMS UNDERLYING CANNABIS USE AND MANIC SYMPTOMS

Burns (2013) proposed two major pathways from cannabis use to psychosis. One pathway refers to the hypothesis that early use of cannabis during adolescence as well as lifetime use in people genetically vulnerable leads to neurodevelopmental changes. This in turn would end up in disruptions of normal endocannabinoid, γ-aminobutyric acid-ergic, and dopaminergic systems. Individuals would thus become more vulnerable to psychosis.

The second pathway applies to subjects without a lifetime history of cannabis use but who start cannabis use a short time before the onset of psychosis. Burns proposes that although cannabis use is starting only shortly before the episode, these subjects are already genetically and developmentally vulnerable to psychosis.

The neurobiological mechanisms underlying the relationship between cannabis use and mania are still unknown but some common factors with the above hypothesis could partially explain the pathway from cannabis to mania.

Interestingly, in a functional magnetic resonance imaging study comparing adolescents with BD, adolescents with cannabis use, adolescents with co-occurring cannabis use and BD, and healthy adolescents, Bitter et al. (2014) found that adolescents with comorbid BD and cannabis use do not show the same overactivation of the regions involved in emotional processing as seen in adolescents with BD alone. The authors propose that these results suggest that subjects with comorbid BD and cannabis use may have a unique endophenotype of BD. Alternatively, brain activation may be altered specifically in BD subjects who use cannabis (Bitter et al., 2014).

APPLICATIONS TO OTHER ADDICTIONS AND SUBSTANCES MISUSE

In healthy subjects, cannabis can induce a variety of experiences. They are usually temporary and include euphoric and mild psychotic experiences (D'Souza et al., 2004). However, there is now sufficient evidence that heavy cannabis use can participate in the development of a psychotic disorder (see review in Gibbs et al., 2015). Longitudinal studies suggest a causal link between cannabis use and onset of psychosis (Burns, 2013). For example, the age at onset of positive symptoms in schizophrenia was younger for subjects using cannabis (Donoghue et al., 2014).

Regarding other substances misuse, cocaine and amphetamines are well-known inducers of manic episodes (Table 2).

DEFINITION OF TERMS

Manic episode The main feature of a manic episode is an elevated mood associated with increased energy and physical as well as mental activity. Other features are also part of the clinical picture such as diminished need for sleep, increased libido, and increased risk taking. Delusions and hallucinations can be present in the severe forms of mania.
It must last at least several days to satisfy diagnosis criteria but can last for weeks or even for months before mood and behavior go back to normality. Manic episodes are usually followed by a depressive episode.

Onset mania About half of BDs start with a manic episode. The peak of onset episode is between 15 and 25 years of age.

Euphoric and dysphoric mania Euphoric, or classical, mania refers to a manic episode with elated mood and exaggerated optimism. This is in contrast to dysphoric mania, which manifests itself with inner tension, anxiety, and aggressive verbal and physical behaviors.

KEY FACTS OF MANIA AND BIPOLAR DISORDERS

- BD has a prevalence of around 2–3% worldwide.
- BD type I is characterized by manic and depressive episodes.
- Manic episodes can be triggered mainly by acute or chronic stress, substance abuse, antidepressants, and various somatic treatments.
- A manic episode can last from several days to several months and is usually followed by a depressive episode.
- The number of manic episodes is highly variable from one subject suffering from BD to another.

TABLE 2 Medications and Substances That May Induce Mania

Medication Category
Cardiology
Angiotensin converting enzyme inhibitors
Psychotropic medications
All antidepressants
Antidepressant discontinuation
Antiepileptic drugs
Anti-infectives
Various antibiotics
Metabolism
Anabolic steroids
Corticosteroids
Thyroid hormones
Varia
Antihistamines
Caffeine
Cyclosporin
Immunomodulators
Sympathomimetics (phenylpropanolamine)
Substance of Abuse
Amphetamine
Alcohol
Cocaine
Heroin
Morphine
Cannabis

Medications and substances of abuse that have been reported to induce manic symptomatology. For a complete description, see Aubry, Ferrero, and Schaad (2013).

SUMMARY POINTS

- Among mood spectrum disorders, BD bears the highest rate of substance abuse.
- Cannabis is the most commonly used illegal drug among patients with BD.
- The age of onset of first manic or depressive episode is younger when there is co-occurring cannabis abuse. There is a poorer medication adherence, a more severe course of illness, and a higher number of manic or depressive episodes with cannabis use.
- Mania phenomenology seems to be influenced by cannabis, found more frequently in dysphoric than in euphoric manic patients.
- Data published as of this writing suggest that there may be a causal relationship of cannabis use to onset mania and also to the occurrence of new manic episodes. However, further investigations are still necessary to confirm this causal link.

REFERENCES

Aas, M., Etain, B., Bellivier, F., Henry, C., Lagerberg, T., Ringen, A., ... Melle, I. (2014). Additive effects of childhood abuse and cannabis abuse on clinical expressions of bipolar disorders. *Psychological Medicine*, 44(8), 1653–1662. http://dx.doi.org/10.1017/S0033291713002316.

Agrawal, A., Nurnberger, J. I., Jr., & Lynskey, M. T. (2011). Cannabis involvement in individuals with bipolar disorder. *Psychiatry Research*, 185(3), 459–461. http://dx.doi.org/10.1016/j.psychres.2010.07.007.

Aubry, J. M., Ferrero, F., & Schaad, N. (2013). *Psychopharmacologie des troubles bipolaires*. Chêne-Bourg (GE): Médecine et Hygiène.

Baethge, C., Hennen, J., Khalsa, H. M., Salvatore, P., Tohen, M., & Baldessarini, R. J. (2008). Sequencing of substance use and affective morbidity in 166 first-episode bipolar I disorder patients. *Bipolar Disorders*, 10(6), 738–741. http://dx.doi.org/10.1111/j.1399-5618.2007.00575.x.

Bally, N., Zullino, D., & Aubry, J. M. (2014). Cannabis use and first manic episode. *Journal of Affective Disorders*, 165, 103–108. http://dx.doi.org/10.1016/j.jad.2014.04.038.

Bega, S., Schaffer, A., Goldstein, B., & Levitt, A. (2012). Differentiating between Bipolar Disorder Types I and II: results from the National Epidemiologic Survey on Alcohol and Related Conditions (NESARC). *Journal of Affective Disorders*, 138(1–2), 46–53. http://dx.doi.org/10.1016/j.jad.2011.12.032.

Bitter, S. M., Adler, C. M., Eliassen, J. C., Weber, W. A., Welge, J. A., Burciaga, J., ... DelBello, M. P. (2014). Neurofunctional changes in adolescent cannabis users with and without bipolar disorder. *Addiction*, 109(11), 1901–1909. http://dx.doi.org/10.1111/add.12668.

Bizzarri, J. V., Rucci, P., Sbrana, A., Miniati, M., Raimondi, F., Ravani, L., ... Cassano, G. B. (2009). Substance use in severe mental illness: self-medication and vulnerability factors. *Psychiatry Research*, 165(1–2), 88–95. http://dx.doi.org/10.1016/j.psychres.2007.10.009.

Bizzarri, J. V., Sbrana, A., Rucci, P., Ravani, L., Massei, G. J., Gonnelli, C., ... Cassano, G. B. (2007). The spectrum of substance abuse in bipolar disorder: reasons for use, sensation seeking and substance sensitivity. *Bipolar Disorders*, 9(3), 213–220. http://dx.doi.org/10.1111/j.1399-5618.2007.00383.x.

Burns, J. K. (2013). Pathways from cannabis to psychosis: a review of the evidence. *Frontiers in Psychiatry*, 4, 128. http://dx.doi.org/10.3389/fpsyt.2013.00128.

Cerullo, M. A., & Strakowski, S. M. (2007). The prevalence and significance of substance use disorders in bipolar type I and II disorder. *Substance Abuse Treatment, Prevention, and Policy*, 2, 29. http://dx.doi.org/10.1186/1747-597X-2-29.

Dalton, E. J., Cate-Carter, T. D., Mundo, E., Parikh, S. V., & Kennedy, J. L. (2003). Suicide risk in bipolar patients: the role of co-morbid substance use disorders. *Bipolar Disorders*, 5(1), 58–61. http://dx.doi.org/10.1034/j.1399-5618.2003.00017.x.

De Hert, M., Wampers, M., Jendricko, T., Franic, T., Vidovic, D., De Vriendt, N., ... van Winkel, R. (2011). Effects of cannabis use on age at onset in schizophrenia and bipolar disorder. *Schizophrenia Research*, 126(1–3), 270–276. http://dx.doi.org/10.1016/j.schres.2010.07.003.

Dilsaver, S. C., Chen, Y. R., Shoaib, A. M., & Swann, A. C. (1999). Phenomenology of mania: evidence for distinct depressed, dysphoric, and euphoric presentations. *The American Journal of Psychiatry*, 156(3), 426–430.

Donoghue, K., Doody, G. A., Murray, R. M., Jones, P. B., Morgan, C., Dazzan, P., ... Maccabe, J. H. (2014). Cannabis use, gender and age of onset of schizophrenia: data from the AESOP study. *Psychiatry Research*, *215*(3), 528–532. http://dx.doi.org/10.1016/j.psychres.2013.12.038.

D'Souza, D. C., Perry, E., MacDougall, L., Ammerman, Y., Cooper, T., Wu, Y. T., ... Krystal, J. H. (2004). The psychotomimetic effects of intravenous delta-9-tetrahydrocannabinol in healthy individuals: implications for psychosis. *Neuropsychopharmacology*, *29*(8), 1558–1572. http://dx.doi.org/10.1038/sj.npp.1300496.

Duffy, A., Horrocks, J., Milin, R., Doucette, S., Persson, G., & Grof, P. (2012). Adolescent substance use disorder during the early stages of bipolar disorder: a prospective high-risk study. *Journal of Affective Disorders*, *142*(1–3), 57–64. http://dx.doi.org/10.1016/j.jad.2012.04.010.

Etain, B., Mathieu, F., Liquet, S., Raust, A., Cochet, B., Richard, J. R., ... Bellivier, F. (2013). Clinical features associated with trait-impulsiveness in euthymic bipolar disorder patients. *Journal of Affective Disorders*, *144*(3), 240–247. http://dx.doi.org/10.1016/j.jad.2012.07.005.

Feingold, D., Weiser, M., Rehm, J., & Lev-Ran, S. (2015). The association between cannabis use and mood disorders: a longitudinal study. *Journal of Affective Disorders*, *172*, 211–218. http://dx.doi.org/10.1016/j.jad.2014.10.006.

Feinman, J. A., & Dunner, D. L. (1996). The effect of alcohol and substance abuse on the course of bipolar affective disorder. *Journal of Affective Disorders*, *37*(1), 43–49. http://dx.doi.org/10.1016/0165-0327(95)00080-1.

Fergusson, D. M., Boden, J. M., & Horwood, L. J. (2008). Exposure to childhood sexual and physical abuse and adjustment in early adulthood. *Child Abuse and Neglect*, *32*(6), 607–619. http://dx.doi.org/10.1016/j.chiabu.2006.12.018.

Gibbs, M., Winsper, C., Marwaha, S., Gilbert, E., Broome, M., & Singh, S. P. (2015). Cannabis use and mania symptoms: a systematic review and meta-analysis. *Journal of Affective Disorders*, *171*, 39–47. http://dx.doi.org/10.1016/j.jad.2014.09.016.

Güçlü, O., Senormanci, O., Aydin, E., Erkiran, M., & Kokturk, F. (2015). Phenomenological subtypes of mania and their relationships with substance use disorders. *Journal of Affective Disorders*, *174*, 569–573. http://dx.doi.org/10.1016/j.jad.2014.11.016.

Koskinen, J., Lohonen, J., Koponen, H., Isohanni, M., & Miettunen, J. (2010). Rate of cannabis use disorders in clinical samples of patients with schizophrenia: a meta-analysis. *Schizophrenia Bulletin*, *36*(6), 1115–1130. http://dx.doi.org/10.1093/schbul/sbp031.

Kvitland, L. R., Melle, I., Aminoff, S. R., Demmo, C., Lagerberg, T. V., Andreassen, O. A., & Ringen, P. A. (2015). Continued cannabis use at one year follow up is associated with elevated mood and lower global functioning in bipolar I disorder. *BMC Psychiatry*, *15*, 11. http://dx.doi.org/10.1186/s12888-015-0389-x.

Kvitland, L. R., Melle, I., Aminoff, S. R., Lagerberg, T. V., Andreassen, O. A., & Ringen, P. A. (2014). Cannabis use in first-treatment bipolar I disorder: relations to clinical characteristics. *Early Intervention in Psychiatry*. http://dx.doi.org/10.1111/eip.12138.

Lagerberg, T. V., Sundet, K., Aminoff, S. R., Berg, A. O., Ringen, P. A., Andreassen, O. A., & Melle, I. (2011). Excessive cannabis use is associated with earlier age at onset in bipolar disorder. *European Archives of Psychiatry and Clinical Neuroscience*, *261*(6), 397–405. http://dx.doi.org/10.1007/s00406-011-0188-4.

Lev-Ran, S., Le Foll, B., McKenzie, K., George, T. P., & Rehm, J. (2013a). Bipolar disorder and co-occurring cannabis use disorders: characteristics, co-morbidities and clinical correlates. *Psychiatry Research*, *209*(3), 459–465. http://dx.doi.org/10.1016/j.psychres.2012.12.014.

Lev-Ran, S., Le Foll, B., McKenzie, K., George, T. P., & Rehm, J. (2013b). Cannabis use and cannabis use disorders among individuals with mental illness. *Comprehensive Psychiatry*, *54*(6), 589–598. http://dx.doi.org/10.1016/j.comppsych.2012.12.021.

Leweke, F. M., & Koethe, D. (2008). Cannabis and psychiatric disorders: it is not only addiction. *Addiction Biology*, *13*(2), 264–275. http://dx.doi.org/10.1111/j.1369-1600.2008.00106.x.

Merikangas, K. R., Herrell, R., Swendsen, J., Rossler, W., Ajdacic-Gross, V., & Angst, J. (2008). Specificity of bipolar spectrum conditions in the comorbidity of mood and substance use disorders: results from the Zurich cohort study. *Archives of General Psychiatry*, *65*(1), 47–52. http://dx.doi.org/10.1001/archgenpsychiatry.2007.18.

Regier, D. A., Farmer, M. E., Rae, D. S., Locke, B. Z., Keith, S. J., Judd, L. L., & Goodwin, F. K. (1990). Comorbidity of mental disorders with alcohol and other drug abuse. Results from the Epidemiologic Catchment Area (ECA) Study. *JAMA*, *264*(19), 2511–2518.

Shin, S. H., Edwards, E., Heeren, T., & Amodeo, M. (2009). Relationship between multiple forms of maltreatment by a parent or guardian and adolescent alcohol use. *The American Journal on Addictions*, *18*(3), 226–234. http://dx.doi.org/10.1080/10550490902786959.

Stone, J. M., Fisher, H. L., Major, B., Chisholm, B., Woolley, J., Lawrence, J., ... MiData Consortium (2014). Cannabis use and first-episode psychosis: relationship with manic and psychotic symptoms, and with age at presentation. *Psychological Medicine*, *44*(3), 499–506. http://dx.doi.org/10.1017/S0033291713000883.

Strakowski, S. M., DelBello, M. P., Fleck, D. E., Adler, C. M., Anthenelli, R. M., Keck, P. E., Jr., ... Amicone, J. (2007). Effects of co-occurring cannabis use disorders on the course of bipolar disorder after a first hospitalization for mania. *Archives of General Psychiatry*, *64*(1), 57–64. http://dx.doi.org/10.1001/archpsyc.64.1.57.

Strakowski, S. M., DelBello, M. P., Fleck, D. E., & Arndt, S. (2000). The impact of substance abuse on the course of bipolar disorder. *Biological Psychiatry*, *48*(6), 477–485. http://dx.doi.org/10.1016/S0006-3223(00)00900-8.

Strakowski, S. M., Sax, K. W., McElroy, S. L., Keck, P. E., Jr., Hawkins, J. M., & West, S. A. (1998). Course of psychiatric and substance abuse syndromes co-occurring with bipolar disorder after a first psychiatric hospitalization. *The Journal of Clinical Psychiatry*, *59*(9), 465–471.

Swann, A. C., Suppes, T., Ostacher, M. J., Eudicone, J. M., McQuade, R., Forbes, A., & Carlson, B. X. (2013). Multivariate analysis of bipolar mania: retrospectively assessed structure of bipolar I manic and mixed episodes in randomized clinical trial participants. *Journal of Affective Disorders*, *144*(1–2), 59–64. http://dx.doi.org/10.1016/j.jad.2012.05.061.

Tijssen, M. J., Van Os, J., Wittchen, H. U., Lieb, R., Beesdo, K., & Wichers, M. (2010). Risk factors predicting onset and persistence of subthreshold expression of bipolar psychopathology among youth from the community. *Acta Psychiatrica Scandinavica*, *122*(3), 255–266. http://dx.doi.org/10.1111/j.1600-0447.2010.01539.x.

van Rossum, I., Boomsma, M., Tenback, D., Reed, C., & van Os, J. (2009). Does cannabis use affect treatment outcome in bipolar disorder? A longitudinal analysis. *The Journal of Nervous and Mental Disease*, *197*(1), 35–40. http://dx.doi.org/10.1097/NMD.0b013e31819292a6.

Zorrilla, I., Aguado, J., Haro, J. M., Barbeito, S., Lopez Zurbano, S., Ortiz, A., ... Gonzalez-Pinto, A. (2015). Cannabis and bipolar disorder: does quitting cannabis use during manic/mixed episode improve clinical/functional outcomes? *Acta Psychiatrica Scandinavica*, *131*(2), 100–110. http://dx.doi.org/10.1111/acps.12366.

Chapter 79

White Matter, Schizophrenia, and Cannabis

Candice E. Crocker[1,2], Philip G. Tibbo[1]

[1]*Department of Psychiatry, Dalhousie University, Halifax, NS, Canada;* [2]*Division of Neurology, Department of Medicine, Dalhousie University, Halifax, NS, Canada*

Abbreviations

2-AG 2-Arachidonoylglycerol
AK Axial kurtosis
CB1R Cannabinoid receptor-1
CB2R Cannabinoid receptor-2
CNPase 2′,3′-Cyclic nucleotide 3′-phosphodiesterase
DTI Diffusion tensor imaging
FA Fractional anisotropy
MAG Myelin-associated glycoprotein
MBP Myelin basic protein
MD Mean diffusivity
RD Radial diffusivity
SLF Superior longitudinal fasciculus
TBSS Tract-based spatial statistics
THC Tetrahydrocannabinol
WM White matter

INTRODUCTION

Schizophrenia can be a chronic severe disorder that has an estimated prevalence worldwide of 0.5–1.0% and an average age of onset in late adolescence and young adulthood. Much investigation has been done to try to carefully identify pathological changes that suggest possible biological mechanisms underlying the disease; however, the definitive causes of schizophrenia are currently not fully known. One of the most reproducible neurobiological abnormalities reported in schizophrenia is structural dysconnectivity, seen functionally as cognitive deficits and structurally as abnormalities in brain white matter (WM). Multiple lines of evidence have indicated that cannabis use, in particular during the important phase of adolescent brain development, may be a risk factor for the development of schizophrenia in those individuals sensitive to a gene × environment interaction. The focus of this chapter is on reviewing the evidence for WM changes in schizophrenia in relation to cannabis use.

The human brain is composed of two main neuronal tissue types—WM and gray matter. WM, so-called for its pale appearance in fixed postmortem tissue, comprises the main fiber tracts within the brain. The white color derives from the lipid content of the myelin wrapping of the fiber bundles, which provides insulation similar to the plastic coating on copper wiring and speeds conductance of nerve impulses along the fiber tracts. Myelin is produced by oligodendrocytes, a type of neuroglial cell that extends many processes, which repeatedly wrap an axon, forming this multilayer membrane of myelin (Baumann & Pham-Dinh, 2001). WM is predominantly a tissue involved in efficiently conducting nerve impulses throughout the brain.

WM can be grouped into three main classes of tracts. The first are projection fibers. These tracts are generally arranged vertically, allowing signals to be passed from lower brain regions (and the spinal cord) to higher ones and vice versa. The corticobulbar tract joins projections from many areas of the cortex to the brainstem. The corticospinal tract projects from the pre- and postcentral gyri and travels to the pyramidal tract at the pons. All thalamic radiations converge into the internal capsule, located between the putamen and the thalamus–caudate nucleus regions. The corticobulbar and corticospinal tracts and the thalamic fibers all penetrate the internal capsule, where the cortex–brainstem connection occupies the more lateral regions. The second group of fibers are the association fibers. These are intrahemispheric fiber pathways connecting either different lobes of that hemisphere (long association fibers) or signals from one gyrus to the next (short association fibers). The four best characterized association tracts are the superior longitudinal fasciculus, inferior longitudinal fasciculus, inferior fronto-occipital fasciculus, and uncinate fasciculus. The third and final group of fibers are the commissural or callosal fibers. These fibers connect the two hemispheres of the brain from the various gyri and nuclei within a single cerebral hemisphere to corresponding locations in the opposite hemisphere. The best known is the massive array of projections across the corpus callosum that forms the so-called callosal radiation. The projections from the genu of the corpus callosum form the forceps minor; those from the splenium form the forceps major. There are also strong projections from the splenium that sweep inferiorly along the lateral margin of the posterior horn of the lateral ventricle and project into the temporal lobes; these projections are known collectively as the tapetum. See Figure 1 for an illustration.

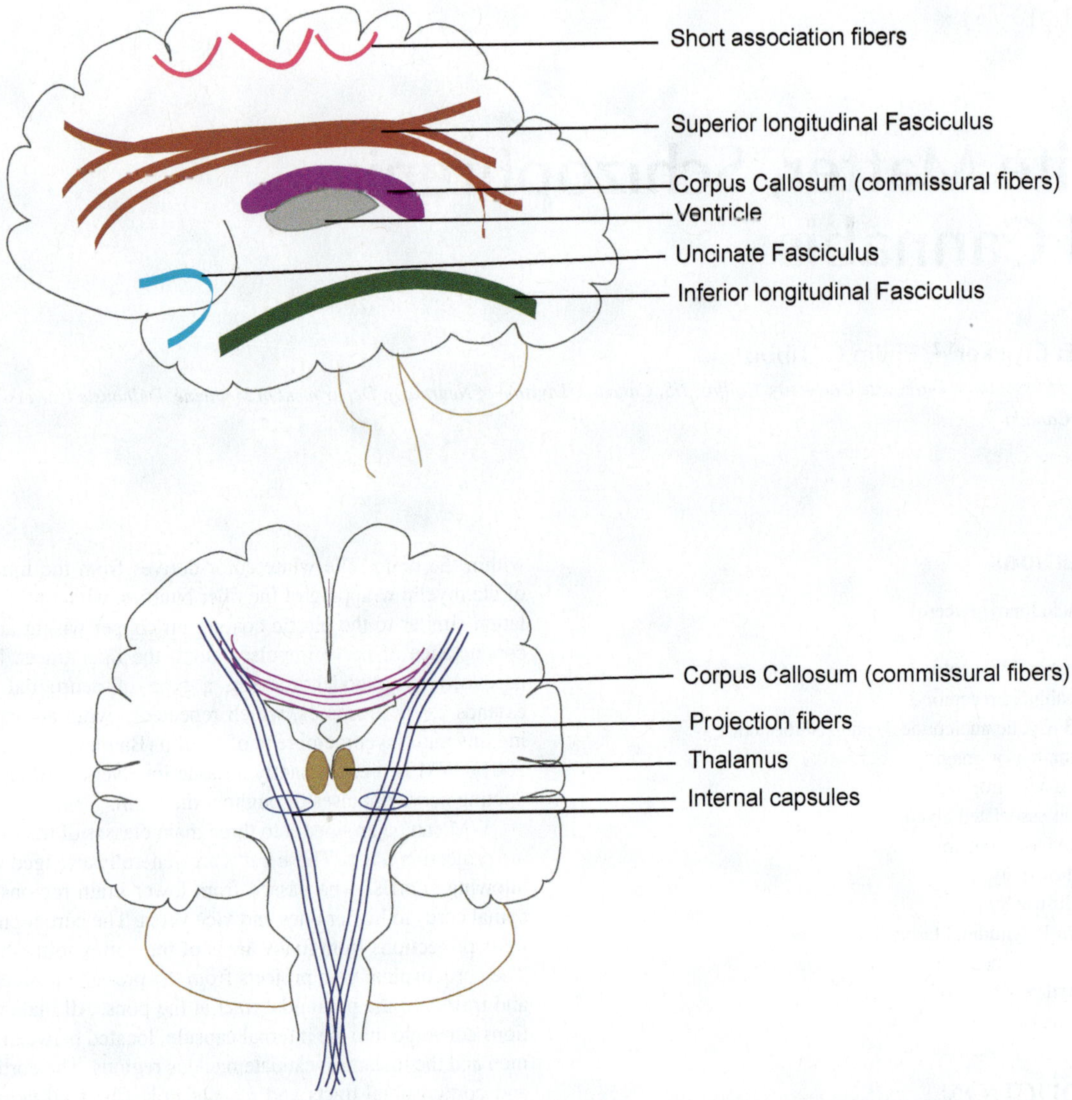

FIGURE 1 Diagram of the major WM tracts in the human brain. Cartoon of the location of various association, projection, and commissural WM tracts in the human brain seen in both a sagittal (top) and a coronal (bottom) view.

White Matter and the Endocannabinoid System

An interesting and important point in this discussion is that WM contains receptors for cannabis. *Cannabis sativa* (cannabis) contains greater than 70 different cannabinoid compounds, with Δ^9-tetrahydrocannabinol (THC) being the best characterized. The psychoactive properties of THC are the result of an interaction of THC with the body's endogenous cannabinoid system. The endogenous endocannabinoid ligands, 2-arachidonoylglycerol and anandamide being the two best characterized, activate receptors that are involved in intercellular communication and intracellular metabolism. This system is used by the human body for a variety of functions, including cognitive, emotional, and pain processing. Additionally, the endocannabinoid system is involved in brain development and maturation, in particular axon guidance and growth, and neuroplasticity. We currently know of two G-protein-coupled receptors, cannabinoid receptor-1 (CB1 receptor) and cannabinoid receptor-2 (CB2 receptor), that interact with these endogenous ligands. CB2 receptors are predominantly expressed on cells of the immune system. The CB1 receptor is 68% homologous to the CB2 receptor, but in contrast is found predominantly on cells of the peripheral and central nervous system. CB1 receptors are found in a particularly high concentration in dopamine receptor-rich areas of the brain, such as the hippocampus and basal ganglia, and expressed on nervous system tissues from the early embryonic period onward. Interestingly, there are temporal and regional variations of CB1 receptor distribution from the fetal period to adulthood, with greater WM expression during the fetal period that slowly changes to a greater receptor density in gray matter in adulthood (Chadwick, Miller, & Hurd, 2013).

The endocannabinoid system is active during adolescence, so it is important to appreciate the potential effects of cannabis use on the adolescent developing brain. In normal brain development and maturation, after a burst of synapse building in the preadolescent period, synaptic density decreases by 40% in adolescents (between the ages of 7 and 15), demonstrating that active synaptic remodeling is taking place during this phase of development.

Brain regions such as the hippocampus and frontal cortex are continuing to develop and mature during this period and levels of endocannabinoids are increased in these brain regions during adolescence compared to adulthood. Significant levels of endocannabinoid receptors have also been found in WM tracts of the adolescent brain, for example, on the glial cells responsible for the production and maintenance of WM (e.g., astrocytes and oligodendrocytes). As with the endocannabinoids, the distribution of CB1 receptors is not in its adult configuration until the end of adolescence, a significant difference between adult and adolescent brain (Chadwick et al., 2013). Thus adolescence represents a period of heightened CB1 receptor density and possibly functionality, meaning that the effects of cannabis during this time could be fundamentally different from the effects on a mature brain. It has therefore been postulated that early adolescent cannabis exposure interacts with the endocannabinoid processes, adversely affecting the trajectory of WM development, ultimately triggering psychosis in individuals who are vulnerable to the illness (Bava et al., 2009).

Measurement of White Matter Structural Connectivity

Structural connectivity refers to neuroanatomy that can be seen or measured at a macroscopic level by magnetic resonance imaging and microscopically by staining of postmortem tissue. In vivo structural connectivity of WM can be measured by diffusion tensor imaging (DTI). This magnetic resonance imaging technique uses a series of gradient pulses that are measured in multiple directions at once, typically 25 to 55. This allows the detection of diffusion of water molecules. Diffusion that is restricted to a particular direction is referred to as being anisotropic (such as is seen in axons) and the measure of freely moving water is referred to as isotropic (as is seen in ventricles). The amount of anisotropy in a measured region is quantified as the fractional anisotropy, or FA, and is a value between 0 (isotropic) and 1 (highly restricted). Another common measure is mean diffusivity, or MD, and it reflects the magnitude of the water movement but without directionality. Tractography can also be performed on the DT images, which can illustrate and examine the connections between regions of interest. The comparison of tracts between individuals is commonly done using tract-based spatial statistics (TBSS). WM can also be quantified by measuring the volume of WM tissue using voxel-based morphometry. Overall, DTI allows for characterization of WM in living individuals (see Figure 2 for illustration).

WM can also be studied in postmortem tissue. Postmortem investigations can be completed with either light or electron microscopy to look for abnormalities of the myelin sheath surrounding axons and of the oligodendrocytes that produce the myelin sheath. The study of WM in postmortem tissue by either method involves staining of the tissue for WM- or oligodendrocyte-associated proteins such as myelin basic protein (MBP), myelin-associated glycoprotein (MAG), and 2′,3′-cyclic nucleotide 3′-phosphodiesterase (CNPase). Each of these proteins has a particular profile in the healthy brain. MBP is found only in mature oligodendrocytes and is involved in the process of compacting myelin (see Figure 3 for an example of MBP staining) (Baumann & Pham-Dinh, 2001). MAG is found in the periaxonal membrane of the mature oligodendrocyte.

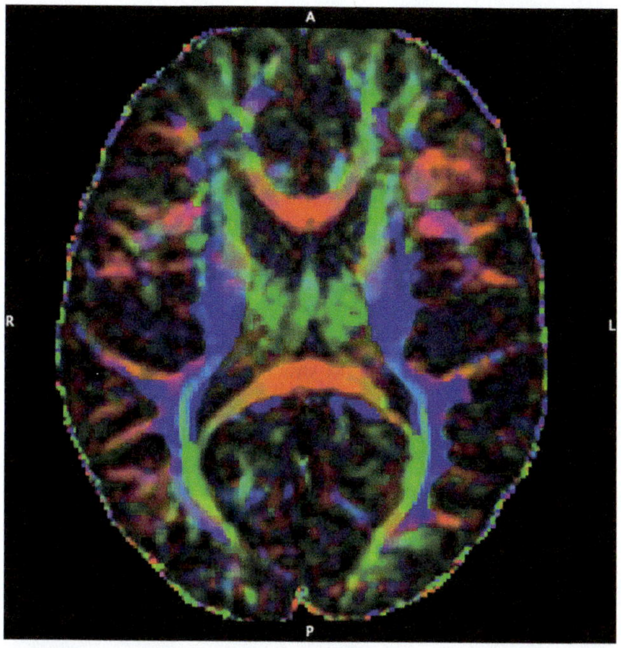

FIGURE 2 **Diffusion tensor FA map.** Human DT image showing prominent diffusion tensor direction by FA mapping. DTI was performed on a healthy subject at a field strength of 1.5T with 3-mm-thick slices (General Electric Healthcare, Fairfield, CT, USA). FA mapping was done using FMRIB's Diffusion Toolbox (Functional MRI of the Brain, FMRIB Analysis Group, Oxford, UK). The color convention is as follows: red for fibers crossing from left to right, green for fibers traversing in the anteroposterior direction, and blue for fibers going from superior to inferior portions of the brain.

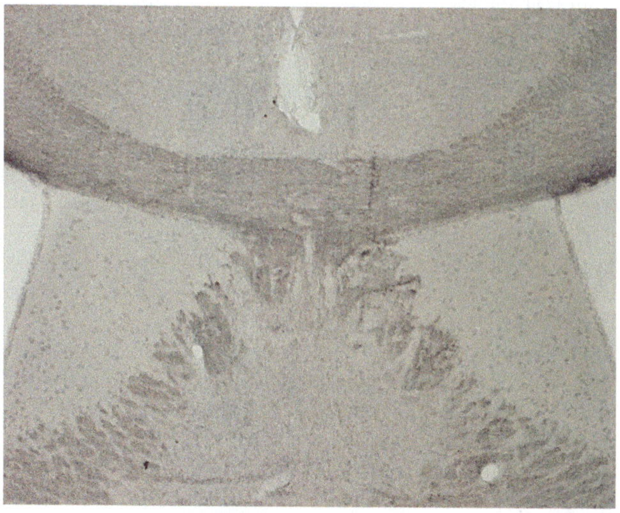

FIGURE 3 **Myelin staining.** Murine cerebral cortex tissue stained MBP by immunohistochemistry (formalin/paraformaldehyde-fixed section). Samples were blocked with 2.5% serum for 1 h followed by incubation with the primary antibody at a 1/200 dilution for 18 h. A biotin-conjugated rabbit anti-rat polyclonal antibody was used as the secondary antibody at a 1/500 dilution.

CNPase has a different distribution; it is absent from compacted myelin and instead is found in oligodendrocyte cytoplasm. CNPase can be detected early in development and at several stages of cell maturation in the precursor cells to oligodendrocytes (Baumann & Pham-Dinh, 2001).

WHITE MATTER CHANGES IN SCHIZOPHRENIA

Dysconnectivity is the most common abnormality reported in schizophrenia. While there is debate about whether this observation is a primary factor in causing schizophrenia or the result of other cortical dysfunction is not clear at this time. However, the end result is failed or at least inadequate information transfer between neurons and brain regions in affected individuals. There is a large body of literature examining WM in chronic schizophrenia patients, which is complicated by the "chicken or egg"-type discussion, that is, do WM abnormalities lead to schizophrenia or does schizophrenia affect WM structure itself? Most studies use a voxel-based morphometry or TBSS approach; however, there is variability among the studies, including replication difficulties, accounted for in part by lack of systematic control for specific variables such as correction for age and medication (Fitzsimmons, Kubicki, & Shenton, 2013). Age is an important consideration as WM integrity and oligodendrocyte apoptosis vary with age, especially during the adolescent years (Engelter, Provenzale, Petrella, DeLong, & MacFall, 2000; Lebel, Caverhill-Godkewitsch, & Beaulieu, 2010).

Findings of reduced FA in schizophrenia have been reported for the internal capsule, the anterior thalamic radiation, the left inferior longitudinal fasciculus, the left inferior frontal-occipital fasciculus, and portions of the uncinate fasciculus (Kubicki & Shenton, 2014). There are a number of DTI studies that extend the finding of decreased FA and correlate decreases in WM with negative symptom severity (Bracht et al., 2014). An argument for damage to WM resulting in the development of schizophrenia arises from DTI studies that include individuals with first-episode psychosis. These firmly establish that WM abnormalities exist before medication treatment begins and at the very early stages of illness (Andreasen et al., 2011; Kubicki & Shenton, 2014; Zhang et al., 2014). Even more evidence for causation versus epiphenomenon comes from work examining individuals who are currently healthy but who have childhood and adolescent risk factors for psychosis. These individuals show decreased FA values in the superior longitudinal fasciculus, which suggests further weight to the "two-hit" hypothesis of schizophrenia development (DeRosse et al., 2014). This hypothesis suggests that it takes both a genetic predisposition (the so-called first hit) and a second environmental or sociological insult (the second hit), such as cannabis use, to enable the development of schizophrenia (Figure 4) (Bayer, Falkai, & Maier, 1999).

Another method of more closely examining the cell packing of WM is diffusion kurtosis imaging (DKI). DKI has exhibited improved sensitivity and specificity in detecting developmental and pathological changes in neural tissues compared to conventional DTI. Zhu et al. (2015) examined by DKI the corpus callosum and internal capsule of chronic schizophrenia patients. They found significant abnormalities in two measures, axial kurtosis (AK) and radial diffusivity (RD). In schizophrenia patients, a measurement of decreased AK suggests axonal damage; however, the increased RD indicates myelin impairment. These findings suggest that diffusion and kurtosis parameters could provide complementary information and they should be jointly used to further elucidate pathological changes in schizophrenia (Zhu et al., 2015).

Postmortem studies also illustrate WM abnormalities in schizophrenia. Again, this work has not always been replicated. A possible reason for this inconsistency in both the neuroimaging and the postmortem work is the possible heterogeneity of the schizophrenic patient population used, and there is some evidence of antipsychotic drugs affecting WM (Ozcelik-Eroglu et al., 2014). With this caution in mind, there is postmortem evidence of increased cell density in deep WM along with maldistribution of neurons in deeper WM prefrontal cortex (Akbarian et al., 1996). There is also evidence of problems with myelin sheath formation, increased numbers of abnormal fibers, and oligodendroglial cell ultrastructure (Uranova, Vikhreva, Rachmanova, & Orlovskaya, 2011). Reductions in deep-WM, but not gyral, myelin staining have been seen with light microscopy in postmortem tissue from patients with schizophrenia compared to healthy controls, but the results were not statistically significant (Regenold et al., 2007). Another avenue of postmortem exploration relevant to our discussion here is a study that examined the cytoarchitecture of the substantia nigra by examining dopaminergic neurons, cells that are affected directly by THC. In this 2014 study, astrocyte density was decreased, but not oligodendrocyte density, in this region in schizophrenic patients compared to healthy controls and another patient group affected by depression (Williams et al., 2014). It has also been reported that cross-sectional WM decreases were more significant in the left superior gyrus of older (65–84 years) patients compared to younger (30–54 years) patients and age-matched healthy controls (Torii et al., 2012). This is in agreement with a longitudinal imaging study showing progressive losses in frontal lobe WM volume seen over a 3-year period, and these losses correlated with negative symptoms (Ho et al., 2003). However, what is not known is if significant WM changes early in the illness result in even more negative changes over the life span.

WHITE MATTER CHANGES WITH CANNABIS USE

Cannabis is generally considered not to be neurotoxic, so one might expect not to find differences in WM between users and nonusers. However, this assumption of lack of neurotoxicity is based primarily on animal studies of cannabis use and there are several reasons this may not be valid in humans. First, there is significant molecular divergence between the sequences for the CB1 receptor in rodents and humans (McPartland, Matias, Di Marzo, & Glass, 2006). Second, the receptor affinity for THC is significantly different between the species, and third, the distribution of CB1 receptors is different, with denser CB1 receptor expression in cognitive regions such as cerebral cortex in humans compared with rats (McPartland, Glass, & Pertwee, 2007). The assumption of cannabis being harmless based on animal research is therefore suspect; the animal work cannot be translated directly to humans. Hence, it is not surprising that all papers in this small body of literature show damage to WM in cannabis users. We considered all papers with greater than 10 subjects to allow for a sufficient breadth of literature to discuss here (see Table 1).

DTI studies show deficits in WM structural integrity with cannabis use in comparison to nonusing control subjects. Most of the studies examining DTI in cannabis users had a subject population that was either adolescent or young adult. Heavy cannabis use was associated with damage to the corpus callosum that is reflected in an increased MD value (Arnone et al., 2008). Damage with heavy

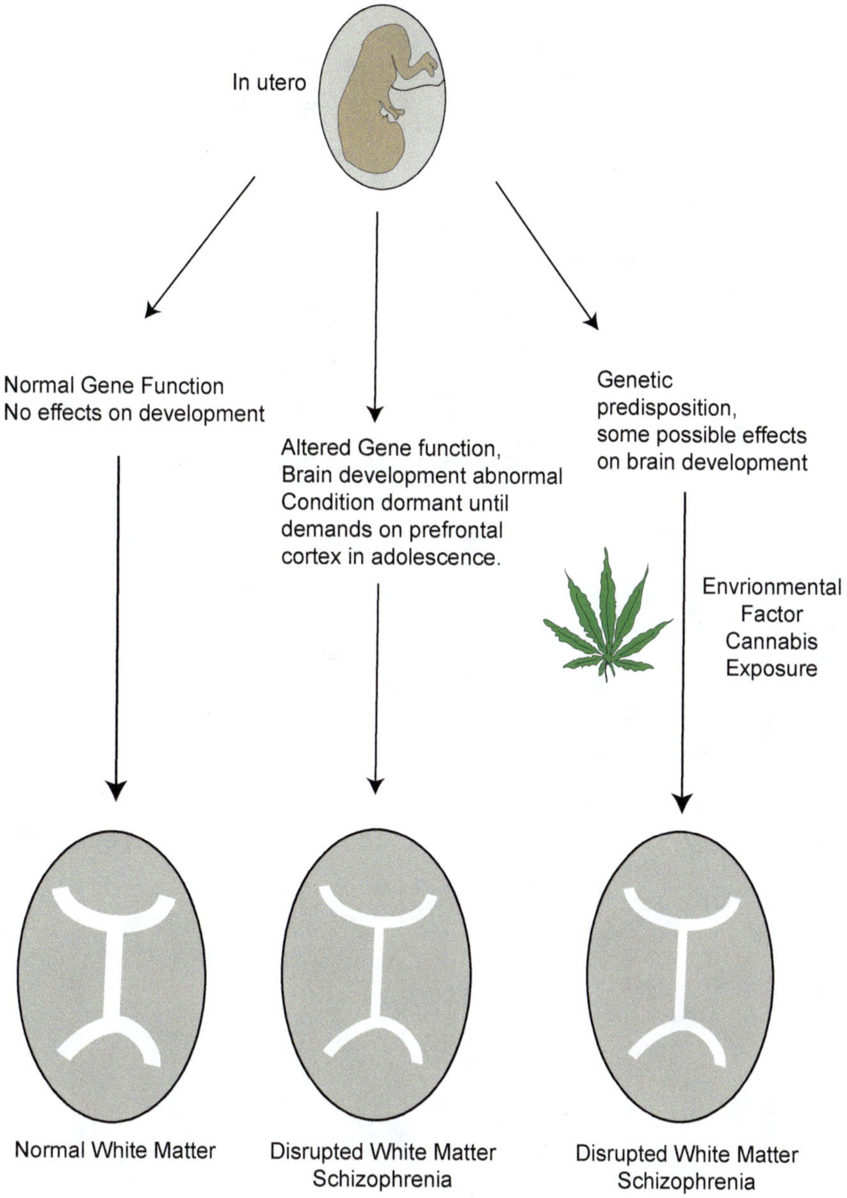

FIGURE 4 Illustration of two possible models of WM disruption leading to the development of schizophrenia. The two most common models of schizophrenia development are the genetic and the two-hit hypotheses. Here we show a modeling of the genetic and the two-hit hypotheses leading to disruptions in WM development, disorganization, and reduced myelination in a timeline fashion from conception to disease onset.

use was also noted to the frontotemporal region and the superior longitudinal fasciculus (SLF) as measured by tractography and with decreased FA values (Ashtari et al., 2009). Another group confirmed these findings in the SLF and found decreased FA in the hippocampus as well (Yucel et al., 2010). Decreased FA in the SLF was a definitive finding in two other studies that examined a variety of substance-use-dependent subjects with a cannabis subanalysis (Bava et al., 2009; Clark, Chung, Thatcher, Pajtek, & Long, 2012). Decreased FA was also measured in regions of the corpus callosum and the internal capsule in users in another study of cannabis-using adults (Gruber, Dahlgren, Sagar, Gonenc, & Lukas, 2014). What is interesting about this study is a subanalysis that showed that individuals who began cannabis smoking prior to age 16 had greater decrements in FA than those who began after age 16 (Gruber et al., 2014). This again is supportive of age of cannabis use onset effects on WM development, which parallels the epidemiological studies reporting age of regular use effects on risk of psychosis development.

Measurements of FA and MD both suggest changes in axonal connections and integrity of fiber tracts with cannabis use in an otherwise healthy population. Tractography and mapping of nerve cell connections are other methods to investigate WM changes. Synaptic remodeling, as measured by axonal connectivity analyses (by diffusion-weighted magnetic resonance imaging and mapping), was impaired in a measurable way in cannabis users compared to nonusers, in the right fimbria, the commissural fibers, and the corpus callosum, structures that contain abundant levels of cannabinoid receptors in the developing brain (Zalesky et al.,

TABLE 1 Diffusion Tensor Imaging Studies of White Matter with Cannabis Use

First Author (year)	Group (n)	Age, Years, Mean (SD)	Controlled for Comorbid Alcohol Use	Controlled for Other Illicit Substances	Results
Clark et al. (2012)	CAN+ (31) HC (20)	17 (1) 16 (1)	No	Group of 31 a subset of other substance use disorders	CAN+<HC in FA in SLF. Significant gender effect.
Kim et al. (2011)	CAN+ (12) HC (13)	19 (1) 22 (4)	Excluded more than 21 drinks per week in the past month or more than 5 drinks per occasion	No other illicit drugs previous 3 months	Decreased regional connectivity in CAN+. Tractography showed global network decreases in efficiency and altered regional connectivity.
Gruber et al. (2014)	CAN+ (25) HC (18)	23 (6) 23 (4)	Groups matched for alcohol consumption	Exclusion of other illicit drugs	CAN+<HC in FA. Subanalysis with earlier start (14 vs 17 years) showed greater FA decrease.
Zalesky et al. (2012)	CAN+ (59) HC (33)	33 (11) 32 (12)	Groups matched for alcohol consumption	Exclusion of other illicit drugs	CAN+<HC in FA, axonal connectivity reduced in callosal and limbic fiber tract families.
Ashtari et al. (2009)	CAN+ (14) HC (14)	19 (1) 19 (1)	5 CAN+ had concurrent alcohol abuse	Exclusion of other illicit drugs	CAN+<HC in FA, axonal connections increased in frontotemporal area.
Arnone et al. (2008)	CAN+ (11) HC (11)	25 (3) 23 (3)	Excluded more than 21 drinks per week in the past month	No other illicit drugs previous 6 months	No differences in FA in the corpus callosum. MD significantly increased in corpus callosum between prefrontal lobes.
Yucel et al. (2010)	CAN+ (11) HC (8)	19 (2) 20 (3)	No	Excluded if inhalants were used for 6 months or more.	CAN+<HC in FA in both hippocampi and right SLF.
Bava et al. (2009)	CAN+ (36) HC (36)	18 (1) 18 (1)	Subjects used both alcohol and cannabis	Excluded use of psychoactive drugs	CAN+<HC in FA in left SLF, left postcentral gyrus, and inferior frontotemporal tracts.
Clark et al. (2012)	CAN+ (31) HC (20)	17 (1) 16 (1)	Alcohol use significantly different between control and substance use group	None	CAN+<HC in FA in prefrontal and parietal volumes.

The minimal threshold for significance is $p<0.05$ (uncorrected for comparisons); HC, healthy control; CAN+, cannabis using; FA, fractional anisotropy; MD, mean diffusivity; SLF, superior longitudinal fasciculus.

2012). Another method of examining axonal connectivity is to use a graph theoretical model in conjunction with DTI. This method showed that brain networks in cannabis-using subjects were significantly less integrated and have altered regional connectivity, for example, in the cingulate region, compared to healthy nonusing controls (Kim et al., 2011). There is one other 2014 study to mention, even though it does not use DTI, that suggests even irregular cannabis use may be detrimental to the human brain. This study used magnetic resonance imaging to examine the size, shape, and density of the nucleus accumbens and the amygdala in 20 cannabis-using adolescents and 20 nonusers. Larger abnormalities in both brain structures were associated with greater cannabis intake (Gilman et al., 2014). All of these studies show that WM connectivity and integrity are reduced with cannabis use in an otherwise healthy population.

WHITE MATTER CHANGES WITH CANNABIS USE IN PEOPLE WITH SCHIZOPHRENIA

There is very little literature on the combined influences of cannabis consumption and schizophrenia on WM structure (see Table 2). This is a field of research that would benefit from longitudinal studies on this population. A discussion of cannabis use in schizophrenia is important; as previously mentioned, there is strong epidemiological evidence of a link between adolescent cannabis use and the development of schizophrenia (Andreasson, Allebeck, Engstrom, & Rydberg, 1987; Burns, 2013). The effect of cannabis on WM is one possible mechanism by which cannabis may promote the development of psychosis. The few DTI studies done as of this writing have limited conclusions owing to the inclusion of comorbid drug and alcohol use, and there are conflicting results from these studies. An intriguing finding from one of these studies suggests that cannabis use prior to age 17 creates a subgroup of patients with hyperconnectivity. This is counter to the majority of DTI research into schizophrenia, which shows decreased FA and hypoconnectivity. This study showed increased FA in the left posterior corpus callosum and right occipital and left temporal lobes in cannabis-using patients compared to healthy controls but no significant difference from cannabis-naïve patients (Peters et al., 2009). However, these same patient populations had a significant increase in FA values due to hard drug use as well, a significant confound to any conclusion about what the source of increased connectivity might be (Peters et al., 2009). James et al. (2011), showed decreased FA in cannabis-using schizophrenia patients in brainstem, internal capsule, corona radiata, and superior and inferior longitudinal fasciculus. However, another study using TBSS performed on using and nonusing first-episode psychosis patients showed no difference in WM parameters (Haller et al., 2013). This study is notable for its prospective design; however, no healthy control users or nonusers were included, making the

TABLE 2 Diffusion Tensor Imaging Studies of White Matter with Cannabis and Schizophrenia

First Author, Year of Publication	Group (n)	Age, Years, Mean (SD)	Controlled for Comorbid Alcohol Abuse	Controlled for Comorbid Stimulant Use	Results
Peters et al. (2009)	FEP CAN+ (35) HC (21)	22 (4) 23 (3)	Excluded alcohol abuse or dependence in HC	Exclusion of other illicit drugs in HC	FEP with cannabis use prior to age 17 had increased FA in bilateral uncinate fasciculus, anterior internal capsule, and frontal WM. FA highest in CAN+ FEP+ subgroup with history of other illicit drug use.
Epstein et al. (2014)	FEP (53) UHR (19) CAN+ (28) HC (53)	17 (2) 16 (3) 18 (2) 17 (3)	No	Exclusion of other illicit drugs	Corticospinal and left ILF and left inferior fronto-occipital FAs were all significantly different between FEP and controls. Left ILF in CAN+ had significantly decreased FA relative to HC. Corticospinal significantly different between CAN+ and FEP. 21 of the 55 FEP had cannabis use disorder diagnosis
Haller et al. (2013)	FEP CAN+ (33) FEP CAN− (17)	23 (4) 24 (4)	No	Exclusion of other illicit drugs	No difference between CAN− and CAN+ SCZ.
James et al. (2011)	FEP (32) CAN+ SCZ (16) CAN− SCZ (16) HC (28)	16 (1) 16 (1)	No heavy alcohol use reported by subjects	Exclusion of other illicit drugs	FEP<controls in FA, in association, brainstem, callosal and projection fibers. CAN+ FEP<CAN− FEP in FA, in brainstem, projection, and association fiber tracts

The minimal threshold for significance is p<0.05 (uncorrected for comparisons). FEP, first-episode psychosis; HC, healthy control; UHR, high risk for psychosis; SCZ, schizophrenia; CAN+, cannabis using; CAN−, not cannabis using; FA, fractional anisotropy; ILF, inferior longitudinal fasciculus.

impact of disease impossible to gauge (Haller et al., 2013). There is one other paper that examined schizophrenia and cannabis use in the same study but not in a separate group of cannabis-using schizophrenia subjects. Epstein et al. found decreased FA in cannabis users relative to healthy controls in the left inferior longitudinal fasciculus and the corticospinal tracts. First-episode psychosis patients studied in the same manner showed lower FA values in the corticospinal tracts than both the cannabis users and the healthy controls (Epstein et al., 2014). Twenty-one of the fifty-five psychosis subjects had cannabis use histories, but only eight had a positive urinalysis during the study (Epstein et al., 2014).

No postmortem papers on the state of WM in cannabis-using schizophrenia patients were found. However, some postmortem work has been done on cannabinoid receptor expression. CB1 receptors were reported to be downregulated in human brain with chronic use (Villares, 2007). Binding experiments in cannabis-using schizophrenia patients showed increased binding to CB1 receptors in the prefrontal cortex that was independent of recent cannabis consumption compared to cannabis-using controls (Dean, Sundram, Bradbury, Scarr, & Copolov, 2001). This absence of downregulation in schizophrenia could be another piece of the cannabis and schizophrenia puzzle.

APPLICATIONS TO OTHER ADDICTIONS AND SUBSTANCE MISUSE

Cannabis is not the only recreational substance to affect WM integrity. This is one of the major confounders to cannabis research, as there are very few users that are purely cannabis using. In other words, cannabis use is frequently comorbid with other drug use (Tsuang et al., 1998). Published studies presented here struggled particularly with exclusion of alcohol use. There are some studies that examine alcohol and cannabis use together. One study examined eight different WM regions in binge-drinking adolescents with or without cannabis use and healthy controls. This study showed significant differences in FA in all eight WM regions between the binge-drink-only group and controls, including the inferior longitudinal fasciculus, inferior fronto-occipital fasciculus, and SLF (Jacobus et al., 2009). Similar to one of the schizophrenia and cannabis papers, in four of these same regions, binge drinkers who were also heavy cannabis users had higher FA than nonusing binge drinkers (Jacobus et al., 2009; Peters et al., 2009). A prospective follow-up DTI study done by the same group on 16 adolescents (ages 16–18 years) with minimal alcohol and marijuana use who were scanned again 3 years later showed FA decreased in those individuals who initiated heavy alcohol and cannabis use but not for alcohol use alone (Jacobus, Squeglia, Infante, Bava, & Tapert, 2013). Another paper cited in the cannabis section examined cannabis users, inhalant users, and healthy controls and found lower FA values for both user groups but greater decreases in the inhalant group (Yucel et al., 2010). Another study showed WM disorganization of the SLF in all substance-dependent individuals, but the examination of specific substance-related variables to changes in the prefrontal cortex and parietal lobe FA showed a strong association with cannabis-related symptoms but not alcohol-related symptoms (Clark et al., 2012). The SLF, in addition to the inferior frontal and temporal lobe WM tracts, was also noted to have decreased FA in alcohol and marijuana users in another study comparing 36 users to 36 demographically similar healthy control subjects (Bava et al., 2009). Reduced FA has also been reported in the frontal lobe with cocaine use (Lim, Choi, Pomara, Wolkin, & Rotrosen, 2002). To further understand the effects of cannabis on WM structure and function, future studies thus have to take into consideration, and control for, the comorbidity of other substances of abuse.

DEFINITION OF TERMS

Schizophrenia This is a chronic long-term mental disorder that causes people to interpret reality abnormally. Schizophrenia is characterized as having core symptom clusters of positive, such as hallucinations, and negative symptoms, such as loss of personality, and cognitive deficits such as working memory abnormalities.

Early phase psychosis The first time someone experiences psychotic symptoms or a psychotic episode and the subsequent critical 2–5 years of illness are known as the early phase.

White matter This is a component of the central nervous system that is mostly composed of glial cells and myelinated axons. This tissue is pale in color and when formalin-fixed appears white.

White matter tracts These are bundles of myelinated nerve axons. The tracts form groups of known WM fibers called projection, association, and commissural tracts in the cerebral hemispheres.

Gray matter This is the other major component of the central nervous system, primarily composed of nerve cell bodies, unmyelinated axons, astroglia, and branching dendrites. This tissue appears darker in color than WM.

Diffusion tensor imaging This is the magnetic resonance spectroscopy method for characterizing neuropathology by measuring the movement of water within the brain.

Fractional anisotropy This is the amount of anisotropy measured in a region by DTI and quantified as a directional value between 0 (isotropic) and 1 (highly restricted).

Mean diffusivity This is the magnitude of the water movement measured by DTI but without directionality.

Myelin This is a mixture of proteins and phospholipids formed by oligodendrocytes that create an insulating sheath around many nerve fibers.

Oligodendrocyte This is a type of neuroglia that wraps axons and produces myelin.

KEY FACTS

Key Facts about White Matter in Schizophrenia

- WM abnormalities exist at disease onset.
- WM abnormalities may be the result of life experiences, such as cannabis use, that are risk factors for developing psychosis.
- There are genes affecting WM development that are associated with schizophrenia.
- WM loss may be accelerated with aging in people with schizophrenia.
- Greater WM deficits are associated with worse negative symptoms and cognitive problems.

Key Facts about the Association between White Matter, Adolescent Cannabis Use, and Schizophrenia

- The endocannabinoid system is important for brain development and maturation, particularly during adolescence.
- There is a temporal pattern associated with CB1 receptor and endocannabinoid ligands—higher expression during adolescence than adulthood.
- Adolescent use of cannabis is associated with higher risk of schizophrenia development as seen in large epidemiologic studies.
- Cannabis can affect the trajectory of WM development in an otherwise healthy population.
- Cannabis plus genetic liability can alter the trajectory of WM development toward the development of psychosis.

SUMMARY POINTS

- WM is a type of brain tissue that speeds conductance of electrical signals through the body.
- Changes in WM can be measured in vivo by DTI and in vitro by postmortem staining.
- Studies of WM brain tissue in patients with schizophrenia show agreement when the studies are corrected for age and other comorbidities.
- Studies of WM in cannabis users are in good agreement that WM integrity is damaged with chronic and heavy use and possibly also with recreational use.
- Studies of WM in cannabis-using people with schizophrenia are lacking as the studies that have been done are all complicated by lack of healthy control subjects or lack of controls for other substance use.

REFERENCES

Akbarian, S., Kim, J. J., Potkin, S. G., Hetrick, W. P., Bunney, W. E., Jr., & Jones, E. G. (1996). Maldistribution of interstitial neurons in prefrontal white matter of the brains of schizophrenic patients. *Archives of General Psychiatry, 53*(5), 425–436.

Andreasen, N. C., Nopoulos, P., Magnotta, V., Pierson, R., Ziebell, S., & Ho, B. C. (2011). Progressive brain change in schizophrenia: a prospective longitudinal study of first-episode schizophrenia. *Biological Psychiatry, 70*(7), 672–679. http://dx.doi.org/10.1016/j.biopsych.2011.05.017. pii:S0006-3223(11)00546-4.

Andreasson, S., Allebeck, P., Engstrom, A., & Rydberg, U. (1987). Cannabis and schizophrenia. A longitudinal study of Swedish conscripts. *Lancet, 2*(8574), 1483–1486.

Arnone, D., Barrick, T. R., Chengappa, S., Mackay, C. E., Clark, C. A., & Abou-Saleh, M. T. (2008). Corpus callosum damage in heavy marijuana use: preliminary evidence from diffusion tensor tractography and tract-based spatial statistics. *NeuroImage, 41*(3), 1067–1074. http://dx.doi.org/10.1016/j.neuroimage.2008.02.064.

Ashtari, M., Cervellione, K., Cottone, J., Ardekani, B. A., Sevy, S., & Kumra, S. (2009). Diffusion abnormalities in adolescents and young adults with a history of heavy cannabis use. *Journal of Psychiatric Research, 43*(3), 189–204. http://dx.doi.org/10.1016/j.jpsychires.2008.12.002.

Baumann, N., & Pham-Dinh, D. (2001). Biology of oligodendrocyte and myelin in the mammalian central nervous system. *Physiological Reviews, 81*(2), 871–927.

Bava, S., Frank, L. R., McQueeny, T., Schweinsburg, B. C., Schweinsburg, A. D., & Tapert, S. F. (2009). Altered white matter microstructure in adolescent substance users. *Psychiatry Research, 173*(3), 228–237. http://dx.doi.org/10.1016/j.pscychresns.2009.04.005.

Bayer, T. A., Falkai, P., & Maier, W. (1999). Genetic and non-genetic vulnerability factors in schizophrenia: the basis of the "two hit hypothesis". *Journal of Psychiatric Research, 33*(6), 543–548.

Bracht, T., Horn, H., Strik, W., Federspiel, A., Razavi, N., Stegmayer, K., … Walther, S. (2014). White matter pathway organization of the reward system is related to positive and negative symptoms in schizophrenia. *Schizophrenia Research, 153*(1–3), 136–142. http://dx.doi.org/10.1016/j.schres.2014.01.015.

Burns, J. K. (2013). Pathways from cannabis to psychosis: a review of the evidence. *Front Psychiatry, 4*, 128. http://dx.doi.org/10.3389/fpsyt.2013.00128.

Chadwick, B., Miller, M. L., & Hurd, Y. L. (2013). Cannabis use during adolescent development: susceptibility to psychiatric illness. *Front Psychiatry, 4*, 129. http://dx.doi.org/10.3389/fpsyt.2013.00129.

Clark, D. B., Chung, T., Thatcher, D. L., Pajtek, S., & Long, E. C. (2012). Psychological dysregulation, white matter disorganization and substance use disorders in adolescence. *Addiction, 107*(1), 206–214. http://dx.doi.org/10.1111/j.1360-0443.2011.03566.x.

Dean, B., Sundram, S., Bradbury, R., Scarr, E., & Copolov, D. (2001). Studies on [3H]CP-55940 binding in the human central nervous system: regional specific changes in density of cannabinoid-1 receptors associated with schizophrenia and cannabis use. *Neuroscience, 103*(1), 9–15.

DeRosse, P., Ikuta, T., Peters, B. D., Karlsgodt, K. H., Szeszko, P. R., & Malhotra, A. K. (2014). Adding insult to injury: childhood and adolescent risk factors for psychosis predict lower fractional anisotropy in the superior longitudinal fasciculus in healthy adults. *Psychiatry Research, 224*(3), 296–302. http://dx.doi.org/10.1016/j.pscychresns.2014.09.001.

Engelter, S. T., Provenzale, J. M., Petrella, J. R., DeLong, D. M., & MacFall, J. R. (2000). The effect of aging on the apparent diffusion coefficient of normal-appearing white matter. *AJR American Journal of Roentgenology, 175*(2), 425–430. http://dx.doi.org/10.2214/ajr.175.2.1750425.

Epstein, K. A., Cullen, K. R., Mueller, B. A., Robinson, P., Lee, S., & Kumra, S. (2014). White matter abnormalities and cognitive impairment in early-onset schizophrenia-spectrum disorders. *Journal of the American Academy of Child Adolescent Psychiatry, 53*(3), 362–372. e361–e362. http://dx.doi.org/10.1016/j.jaac.2013.12.007.

Fitzsimmons, J., Kubicki, M., & Shenton, M. E. (2013). Review of functional and anatomical brain connectivity findings in schizophrenia. *Current Opinion in Psychiatry, 26*(2), 172–187. http://dx.doi.org/10.1097/YCO.0b013e32835d9e6a.

Gilman, J. M., Kuster, J. K., Lee, S., Lee, M. J., Kim, B. W., Makris, N., … Breiter, H. C. (2014). Cannabis use is quantitatively associated with nucleus accumbens and amygdala abnormalities in young adult recreational users. *The Journal of Neuroscience, 34*(16), 5529–5538. http://dx.doi.org/10.1523/jneurosci.4745-13.2014.

Gruber, S. A., Dahlgren, M. K., Sagar, K. A., Gonenc, A., & Lukas, S. E. (2014). Worth the wait: effects of age of onset of marijuana use on white matter and impulsivity. *Psychopharmacology (Berlin), 231*(8), 1455–1465. http://dx.doi.org/10.1007/s00213-013-3326-z.

Haller, S., Curtis, L., Badan, M., Bessero, S., Albom, M., Chantraine, F., ... Merlo, M. (2013). Combined grey matter VBM and white matter TBSS analysis in young first episode psychosis patients with and without cannabis consumption. *Brain Topography*, 26(4), 641–647. http://dx.doi.org/10.1007/s10548-013-0288-8.

Ho, B. C., Andreasen, N. C., Nopoulos, P., Arndt, S., Magnotta, V., & Flaum, M. (2003). Progressive structural brain abnormalities and their relationship to clinical outcome: a longitudinal magnetic resonance imaging study early in schizophrenia. *Archives of General Psychiatry*, 60(6), 585–594. http://dx.doi.org/10.1001/archpsyc.60.6.58560/6/585.

Jacobus, J., McQueeny, T., Bava, S., Schweinsburg, B. C., Frank, L. R., Yang, T. T., & Tapert, S. F. (2009). White matter integrity in adolescents with histories of marijuana use and binge drinking. *Neurotoxicology and Teratology*, 31(6), 349–355. http://dx.doi.org/10.1016/j.ntt.2009.07.006.

Jacobus, J., Squeglia, L. M., Infante, M. A., Bava, S., & Tapert, S. F. (2013). White matter integrity pre- and post marijuana and alcohol initiation in adolescence. *Brain Sciences*, 3(1), 396–414. http://dx.doi.org/10.3390/brainsci3010396.

James, A., Hough, M., James, S., Winmill, L., Burge, L., Nijhawan, S., ... Zarei, M. (2011). Greater white and grey matter changes associated with early cannabis use in adolescent-onset schizophrenia (AOS). *Schizophrenia Research*, 128(1–3), 91–97. http://dx.doi.org/10.1016/j.schres.2011.02.014.

Kim, D. J., Skosnik, P. D., Cheng, H., Pruce, B. J., Brumbaugh, M. S., Vollmer, J. M., ... Newman, S. D. (2011). Structural network topology revealed by white matter tractography in cannabis users: a graph theoretical analysis. *Brain Connection*, 1(6), 473–483. http://dx.doi.org/10.1089/brain.2011.0053.

Kubicki, M., & Shenton, M. E. (2014). Diffusion tensor imaging findings and their implications in schizophrenia. *Current Opinion in Psychiatry*, 27(3), 179–184. http://dx.doi.org/10.1097/yco.0000000000000053.

Lebel, C., Caverhill-Godkewitsch, S., & Beaulieu, C. (2010). Age-related regional variations of the corpus callosum identified by diffusion tensor tractography. *NeuroImage*, 52(1), 20–31. http://dx.doi.org/10.1016/j.neuroimage.2010.03.072.

Lim, K. O., Choi, S. J., Pomara, N., Wolkin, A., & Rotrosen, J. P. (2002). Reduced frontal white matter integrity in cocaine dependence: a controlled diffusion tensor imaging study. *Biological Psychiatry*, 51(11), 890–895.

McPartland, J. M., Glass, M., & Pertwee, R. G. (2007). Meta-analysis of cannabinoid ligand binding affinity and receptor distribution: interspecies differences. *British Journal of Pharmacology*, 152(5), 583–593. http://dx.doi.org/10.1038/sj.bjp.0707399.

McPartland, J. M., Matias, I., Di Marzo, V., & Glass, M. (2006). Evolutionary origins of the endocannabinoid system. *Gene*, 370, 64–74. http://dx.doi.org/10.1016/j.gene.2005.11.004.

Ozcelik-Eroglu, E., Ertugrul, A., Oguz, K. K., Has, A. C., Karahan, S., & Yazici, M. K. (2014). Effect of clozapine on white matter integrity in patients with schizophrenia: a diffusion tensor imaging study. *Psychiatry Research*, 223(3), 226–235. http://dx.doi.org/10.1016/j.pscychresns.2014.06.001.

Peters, B. D., de Haan, L., Vlieger, E. J., Majoie, C. B., den Heeten, G. J., & Linszen, D. H. (2009). Recent-onset schizophrenia and adolescent cannabis use: MRI evidence for structural hyperconnectivity? *Psychopharmacology Bulletin*, 42(2), 75–88.

Regenold, W. T., Phatak, P., Marano, C. M., Gearhart, L., Viens, C. H., & Hisley, K. C. (2007). Myelin staining of deep white matter in the dorsolateral prefrontal cortex in schizophrenia, bipolar disorder, and unipolar major depression. *Psychiatry Research*, 151(3), 179–188. http://dx.doi.org/10.1016/j.psychres.2006.12.019.

Torii, Y., Iritani, S., Sekiguchi, H., Habuchi, C., Hagikura, M., Arai, T., ... Ozaki, N. (2012). Effects of aging on the morphologies of Heschl's gyrus and the superior temporal gyrus in schizophrenia: a postmortem study. *Schizophrenia Research*, 134(2–3), 137–142. http://dx.doi.org/10.1016/j.schres.2011.10.024.

Tsuang, M. T., Lyons, M. J., Meyer, J. M., Doyle, T., Eisen, S. A., Goldberg, J., ... Eaves, L. (1998). Co-occurrence of abuse of different drugs in men: the role of drug-specific and shared vulnerabilities. *Archives of General Psychiatry*, 55(11), 967–972.

Uranova, N. A., Vikhreva, O. V., Rachmanova, V. I., & Orlovskaya, D. D. (2011). Ultrastructural alterations of myelinated fibers and oligodendrocytes in the prefrontal cortex in schizophrenia: a postmortem morphometric study. *Schizophrenia Research Treatment*, 2011, 325789. http://dx.doi.org/10.1155/2011/325789.

Villares, J. (2007). Chronic use of marijuana decreases cannabinoid receptor binding and mRNA expression in the human brain. *Neuroscience*, 145(1), 323–334. http://dx.doi.org/10.1016/j.neuroscience.2006.11.012.

Williams, M. R., Galvin, K., O'Sullivan, B., MacDonald, C. D., Ching, E. W., Turkheimer, F., ... Maier, M. (2014). Neuropathological changes in the substantia nigra in schizophrenia but not depression. *European Archives of Psychiatry and Clinical Neuroscience*, 264(4), 285–296. http://dx.doi.org/10.1007/s00406-013-0479-z.

Yucel, M., Zalesky, A., Takagi, M. J., Bora, E., Fornito, A., Ditchfield, M., ... Lubman, D. I. (2010). White-matter abnormalities in adolescents with long-term inhalant and cannabis use: a diffusion magnetic resonance imaging study. *Journal of Psychiatry and Neuroscience*, 35(6), 409–412. http://dx.doi.org/10.1503/jpn.090177.

Żalesky, A., Solowij, N., Yucel, M., Lubman, D. I., Takagi, M., Harding, I. H., ... Seal, M. (2012). Effect of long-term cannabis use on axonal fibre connectivity. *Brain*, 135(Pt 7), 2245–2255. http://dx.doi.org/10.1093/brain/aws136.

Zhang, R., Wei, Q., Kang, Z., Zalesky, A., Li, M., Xu, Y., ... Huang, R. (2014). Disrupted brain anatomical connectivity in medication-naive patients with first-episode schizophrenia. *Brain Structure and Function*, 220(2), 1145–1159. http://dx.doi.org/10.1007/s00429-014-0706-z.

Zhu, J., Zhuo, C., Qin, W., Wang, D., Ma, X., Zhou, Y., & Yu, C. (2015). Performances of diffusion kurtosis imaging and diffusion tensor imaging in detecting white matter abnormality in schizophrenia. *NeuroImage: Clinical*, 7, 170–176. http://dx.doi.org/10.1016/j.nicl.2014.12.008.

Chapter 80

Electroencephalography and Cannabis: From Event-Related Potentials to Oscillations

Patrick D. Skosnik, Jose A. Cortes-Briones

Psychiatry Service, VA Connecticut Healthcare System, West Haven, CT, USA; Abraham Ribicoff Research Facilities, Connecticut Mental Health Center, New Haven, CT, USA; Department of Psychiatry, Yale University School of Medicine, New Haven, CT, USA

Abbreviations

ASSR Auditory steady-state response
CB1R Central cannabinoid receptor
CCK Cholecystokinin
EEG Electroencephalography
ERP Event-related potential
GABA γ-Aminobutyric acid
IRF Impulse response function
ITC Intertrial coherence
iv Intravenous
LZC Lempel–Ziv complexity
MMN Mismatch negativity
THC Δ^9-Tetrahydrocannabinol

INTRODUCTION

Cannabis continues to be the most commonly used illicit drug by adults (Johnston, O'Malley, Bachman, & Schulenberg, 2012). However, both in the United States and in other parts of the world, there is a growing trend toward decriminalization of the recreational use of cannabis (Bly, 2012). This trend continues despite the fact that there is now compelling evidence for the existence of a cannabis dependence syndrome (Clapper, Mangieri, & Piomelli, 2009). Furthermore, the concept of "medical" cannabis is being more readily accepted (Kleber & DuPont, 2012). Whether in the context of recreational or medicinal use, chronic and acute cannabis exposure alters several areas of neural function including sensory, perceptual, motor, and higher cognitive processes (Mechoulam & Parker, 2012). This highlights the importance of fully understanding the neural effects of both chronic and acute exposure to cannabis.

THE BRAIN CANNABINOID SYSTEM

Plant-derived cannabinoids (phytocannabinoids), most notably Δ^9-tetrahydrocannabinol (THC) (Mechoulam & Parker, 2012), exert their synaptic, psychotropic, and electrophysiological effects through the activation of central cannabinoid receptors (CB1Rs) (Pertwee et al., 2010). Interestingly, the CB1R is the most abundant G-protein-coupled receptor in the central nervous system (Eggan & Lewis, 2007). The CB1R is most abundant in areas such as the cerebral cortex, hippocampus, basal ganglia, and cerebellum (Eggan & Lewis, 2007). CB1Rs are mostly located presynaptically, and their activation (by either endogenous or exogenous cannabinoids) inhibits the release of other neurotransmitters such as γ-aminobutyric acid (GABA) and glutamate by decreasing Ca^{2+} influx via the inhibition of adenylate cyclase and N-type Ca^{2+} channels (Freund, Katona, & Piomelli, 2003; Twitchell, Brown, & Mackie, 1997). In the cerebral cortex and hippocampus, this neuromodulation principally occurs in networks of cholecystokinin-containing GABAergic interneurons (Eggan & Lewis, 2007). CB1Rs therefore function as molecular "brakes," primarily regulating the timing and release of GABA (Farkas et al., 2010). Perturbation of normal CB1R function via chronic or acute cannabis exposure would therefore be expected to alter the excitatory/inhibitory balance of neural networks, which would affect sensory, perceptual, and cognitive processes. One particularly powerful and noninvasive tool with which to study the neural effects of exogenous cannabinoids is electroencephalography (EEG).

ELECTROENCEPHALOGRAPHY

Neural activity as captured on the scalp by EEG results from the summed extracellular electric potential fields generated by hundreds of millions of pyramidal cells (Niedermeyer & da Silva, 2005). This makes EEG one of the only noninvasive techniques capable of directly capturing the electrical activity of neurons (Luck et al., 2011).

To be detectable by EEG, neural activity needs to be strong enough to cross the tissue and fluid (e.g., skull, scalp, cerebrospinal fluid) separating it from the surface electrodes. This relies on the ability of the brain to synchronize, either spontaneously or in response to stimulation, the activity of large populations of pyramidal cells (unsynchronized neurons would cancel out their individual contributions to the electric potential fields captured by the EEG) (Lachaux, Axmacher, Mormann, Halgren, & Crone, 2012).

The architecture of the connectivity between pyramidal cells and GABAergic interneurons, along with the specific characteristics of these cells (e.g., excitability, firing patterns, morphology, receptors, neurotransmitter dynamics, etc.), leads pyramidal cells to engage in transient states of synchronized activation (Lachaux et al., 2012). Thus, substances (e.g., cannabinoids) capable of interfering with these variables will tend to disrupt the electric potential fields resulting from the synchronization of pyramidal cells, which will appear on the EEG as abnormalities in event-related potentials (ERPs) or neural oscillations.

Numerous studies have examined the effects of chronic and acute cannabinoid exposure utilizing a number of EEG measures. While early studies primarily focused on resting/baseline EEG (Struve & Straumanis, 1990), more recent research has focused on transient ERPs such as the P50, N100, and P300 and mismatch negativity (MMN). Further, contemporary work has also begun to focus on the effects of cannabinoids on neural oscillatory activity using such paradigms as the auditory steady-state response (ASSR) and others. Each of these types of ERP components and neural oscillation paradigms will be described below, along with a review of the literature summarizing how both chronic and acute cannabinoids affect each of these measures. For a review of EEG studies on cannabinoids prior to 1990, please see Struve and Straumanis (1990).

CANNABINOIDS AND ERPs

ERPs are time-locked, electrical potentials detected at the scalp in response to specific stimuli (hence the term event-related, which serves to distinguish ERPs from resting or spontaneous EEG). According to Luck and Kappaenman (2012), an ERP waveform can be broadly defined as: "a depiction of the changes in scalp-recorded voltage over time that reflect the sensory, cognitive, affective, and motor processes elicited by a stimulus." Further, they define an ERP peak as "a reliable local positive or negative maximum in the observed ERP waveform" (Luck & Kappenman, 2012). These transient evoked potentials can be either positive (denoted with a "P") or negative (denoted "N") in polarity and can occur at different latencies poststimulus (the numerical value after the P or N signifies peak latency in milliseconds). Hence, the well-studied P300 is a positive voltage ERP typically observed around 300 ms poststimulus. A schematic of prototypical ERPs can be seen in Figure 1 (Picton, Hillyard, Krausz, and Galambos, 1974). In general, ERPs that occur prior to 100 ms are considered automatic/preattentional and are probably related to basic sensory processing. By contrast, those that occur at 100 ms and beyond are thought to be more involved in higher perceptual, affective, and cognitive operations. ERPs are ideally suited to the study of the effects of drugs on the brain, as the high temporal resolution of EEG (tens to hundreds of milliseconds) corresponds to the temporal scale in which human sensation, perception, motor behavior, affect, and cognition are thought to occur. Numerous ERPs have been studied, with each thought to correspond to varying aspects of neural processing (for a thorough, contemporary review of the various types of ERPs and their roles in normal and abnormal brain function, please see Luck & Kappenman, 2012). A brief description of several of the more well-studied evoked potentials follows, along with an examination of how both chronic and acute cannabis/cannabinoids affect these ERPs. Of note, while a treatment of

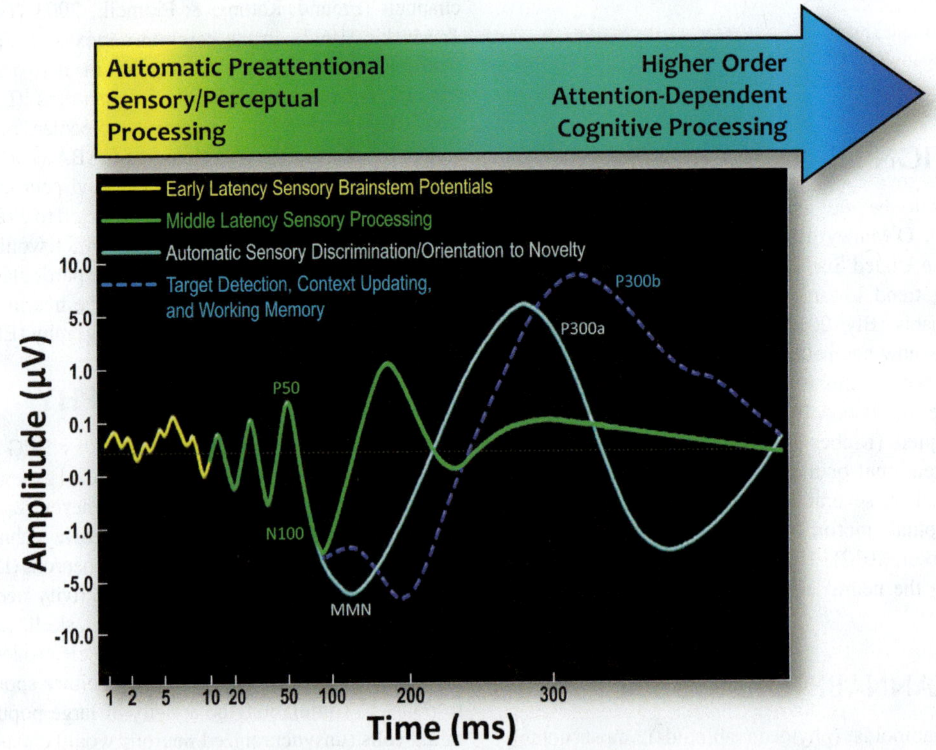

FIGURE 1 Illustration demonstrating prototypical ERPs. Time is shown using a logarithmic scale. Stages of information processing from automatic preattentional to higher order attentional processes are shown from left to right, respectively. *Adapted from Picton et al. (1974), generated de novo by the authors.*

EEG and ERP recording and analysis methods is beyond the scope of this chapter, the interested reader in encouraged to refer to Luck (2005).

The *P50* is a positive voltage, early to middle latency (which peaks around 50 ms), preattentive ERP component elicited by discrete auditory stimuli (e.g., brief white noise clicks). In the prototypical P50 sensory gating or "dual-click" procedure, the amplitude of the P50 to a second paired click (S2) is attenuated relative to the P50 amplitude to the first click (S1). In the optimum P50 sensory gating paradigm, a 500-ms interstimulus interval is utilized. While the P50 to S1 is thought to be related to the capacity of the brain to register salient stimuli, the reduction of the P50 amplitude to S2 (compared to S1) reflects the automatic suppression or inhibitory gating of redundant and irrelevant stimuli.

P50 suppression deficits are commonly observed in psychotic disorders such as schizophrenia. They are thought to be mediated by the hippocampus, temporoparietal region, auditory cortex, and prefrontal cortex (Grunwald et al., 2003; Mayer et al., 2009), all areas dense in CB1Rs (Eggan & Lewis, 2007). Chronic cannabis exposure has been associated with disruptions in P50 suppression (Patrick et al., 1999; Patrick & Struve, 2000) and this effect correlates with the magnitude of cannabis exposure (Edwards, Skosnik, Steinmetz, O'Donnell, & Hetrick, 2009). Another study performed after 28 days of abstinence demonstrated P50 gating deficits that correlated with the number of years of cannabis consumption (Rentzsch et al., 2007). More recently, Broyd et al. (2013) found that long-term cannabis users had significantly greater disruption in both P50 grating ratios and difference scores compared to short-term users and controls. While P50 gating deficits could predate the onset of cannabis use, Broyd et al. (2013) further demonstrated that the longer an individual used cannabis, the worse the gating deficit, partially ruling out this possibility. While no studies have measured the effect of acute cannabis or cannabinoid administration on sensory gating in humans, preclinical studies suggest that cannabinoid agonists disrupt sensory gating in animal analogues of the P50 (Hajos, Hoffmann, & Kocsis, 2008). Taken together, these findings suggest that cannabinoids may interfere with the inhibitory networks involved in the "gating out" of irrelevant or redundant sensory information.

The *P300* is a late positive, postattentional ERP component thought to be related to directed attention, contextual updating of working memory, and the attribution of salience to deviant or novel stimuli (Polich & Criado, 2006). This response is typically elicited utilizing standard "oddball" paradigms in which low-probability target tones (~10-20%) are embedded within a repeating sequence of high-probability standard tones (~80-90%) differing in some physical dimension (e.g., frequency, duration, amplitude, etc.). Rather than being generated by a single neural source, it this thought that the P300, particularly the P300b component, reflects activity from a distributed neural network including such areas as the thalamus, hippocampus, inferior parietal lobe, superior temporal gyrus, and frontal cortex (Kiehl, Laurens, Duty, Forster, & Liddle, 2001).

Studies assessing the effect of chronic cannabis use on the P300 have produced mixed results. Solowij, Michie, and Fox (1991) reported decreased P300 amplitudes in a small sample of recently abstinent cannabis users (Solowij et al., 1991). However, in a subsequent larger study, they failed to replicate the P300 amplitude deficits but observed slower P300 latencies. Furthermore, the latency deficits correlated with frequency of cannabis use (Solowij, Michie, & Fox, 1995). While Kempel, Lampe, Parnefjord, Hennig, and Kunert (2003) reported reduced P300 amplitudes (Kempel et al., 2003), Skosnik, Park, Dobbs, and Gardner (2008) reported increased P300 amplitudes, and de Sola et al. (2008), Patrick et al. (1995) were unable to detect P300 amplitude differences in cannabis users. While the reasons for these discrepant results are unclear, they may be related to differences in samples and the cognitive load of the task such that P300 is impaired in studies using cognitively challenging tasks (Kempel et al., 2003; Solowij et al., 1991, Solowij et al., 1995), but unimpaired in simple tasks (Skosnik, Park, et al., 2008; de Sola et al., 2008). Interestingly, a polymorphism of the CB1R gene has been associated with decreased P300 amplitude (Johnson et al., 1997; Stadelmann et al., 2011), suggesting that CB1R function may play a role in the regulation of P300 amplitude.

Regarding the effect of acute cannabinoids on the P300, both oral and smoked THC has been reported to reduce P300 amplitude (Ilan, Gevins, Coleman, ElSohly, & de Wit, 2005; Roser et al., 2008). Interestingly, P300 amplitude reductions have even been reported in heavy cannabis users after smoking THC, indicating that these THC-induced deficits may be resistant to the effects of tolerance (Theunissen et al., 2012). The above-described studies of P300 deficits after acute cannabinoid administration have been further substantiated by D'Souza et al. (2012) in the context of the intravenous (iv) administration of THC. Intravenous THC administration is ideal, as it reduces the intra- and intersubject variability of plasma THC typical of oral and smoked routes of administration. Further, iv THC eliminates the confound of potential "entourage" effects from other chemicals released during the pyrolyzation of cannabis (D'Souza et al., 2004).

In a double-blind, placebo-controlled, counterbalanced, and crossover study of the dose-related effects of THC on the P300 (utilizing a classic three-stimulus oddball task), D'Souza et al. (2012) demonstrated that both P300a and P300b amplitudes were disrupted in a dose-dependent manner (Figure 2) (D'Souza et al., 2012). Further, these deficits were not observed during early sensory-related ERPs, suggesting that THC specifically affects neural ensembles involved in the processing of novelty and context updating of working memory. Taken together, the P300 appears to be reliably disrupted by acute cannabinoids, which is consistent with the known attention- and memory-altering effects of cannabis intoxication.

MMN is an automatic, preattentive, negative voltage ERP component that is generated primarily in the superior temporal and prefrontal cortex (Naatanen & Alho, 1995; Rinne, Alho, Ilmoniemi, Virtanen, & Naatanen, 2000). It occurs approximately 100-200 ms after an auditory stimulus that deviates in frequency or duration from a sequence of standard auditory stimuli. It is thought to index basic auditory processing and sensory memory and is relatively independent of attention.

To date, two studies have examined auditory MMN in relation to chronic cannabinoid exposure. Roser et al. (2010) studied the effect of chronic cannabis exposure on MMN utilizing frequency and duration deviance. Initial analysis showed that cannabis users exhibited decreased MMN amplitudes at electrode Cz in the frequency deviance condition. More striking was the fact that both long-term and heavier users of cannabis had significantly lower MMN amplitudes compared to short-term or light users. Moreover,

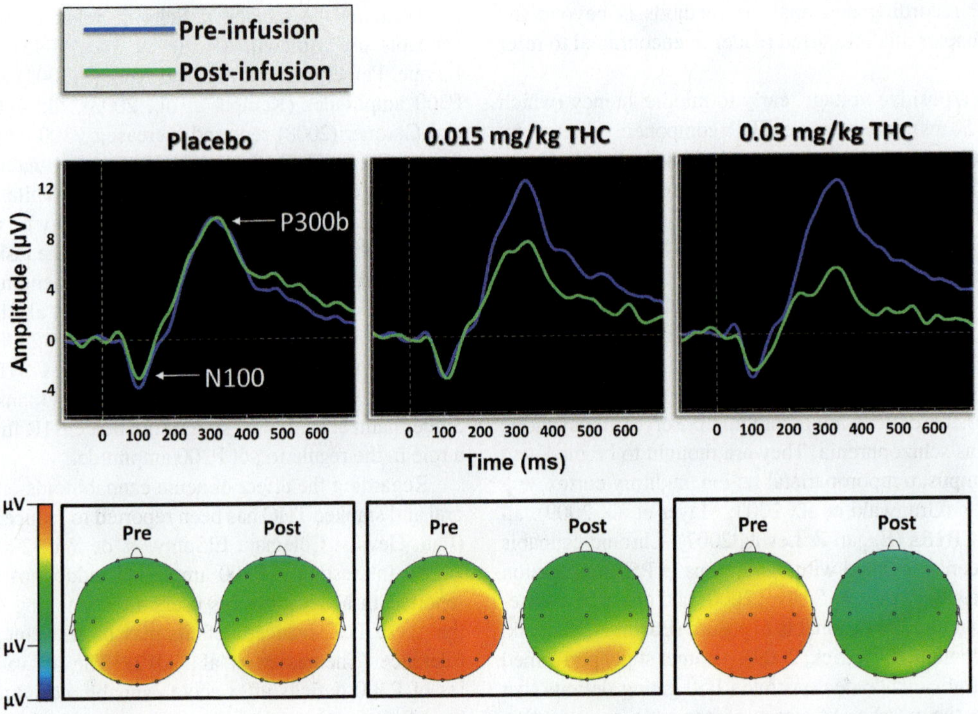

FIGURE 2 **The dose-related effects of THC on the auditory P300b.** The grand-averaged target P300b waveforms at electrode Pz across THC dose conditions are shown by topographic voltage maps from the peak grand-averaged P300b. *Reproduced with permission from D'Souza et al. (2012). Reproduced with permission from previous articles written by the authors in the journal Neuropsychopharmacology (please see: http://www.nature.com/reprints/permission-requests.html).*

duration of cannabis exposure was negatively correlated with MMN amplitudes in the frequency deviance condition at frontal electrode sites (Fz and F4) (Roser et al., 2010). In other words, those showing smaller amplitudes had greater overall chronicity of cannabis use. More recently, Greenwood et al. (2014) demonstrated similar deficits in MMN amplitudes to frequency deviants in a large sample of both short- and long-term cannabis users, which they suggest may be related to interactions between the cannabinoid and the glutamatergic transmitter systems. These converging data therefore suggest that chronic, heavy use of cannabis is associated with deficits in sensory memory as assessed via the MMN paradigm.

In terms of the acute effects of cannabinoids on MMN, Juckel, Roser, Nadulski, Stadelmann, and Gallinat (2007) examined the effects of oral THC or cannabis extract containing both THC and the nonpsychoactive constituent cannabidiol (CBD) on MMN using frequency deviance. Surprisingly, oral THC did not alter MMN amplitude compared to placebo (Juckel et al., 2007). Further, THC plus CBD was actually shown to increase the amplitude of the MMN ERP. The authors postulated that cannabis extract, with the addition of CBD, enhanced MMN by virtue of CBD's reputed antipsychotic effects. The lack of an MMN effect with pure THC could be related to the dose of THC chosen (10 mg orally) or due to the inter- and intraindividual variability inherent in oral routes of administration.

While less well studied than the EEG biomarkers discussed above, an additional psychophysiological index that may prove useful in the study of cannabinoids is the *N100* ERP. The N100 component is a large exogenous ERP that exists regardless of task demands (although it can be modulated by attention). It is thought to be related to basic sensory processing and, in the auditory domain, is probably generated by auditory and frontal cortices (Naatanen & Picton, 1987).

As of this writing, only a handful of studies have explicitly assessed the N100 in relation to heavy cannabis use. In a study using repetitive photic stimuli, Skosnik, Krishnan, Vohs, and O'Donnell (2006) demonstrated robust differences in the visual N160 response in chronic cannabis users tested after 24 h of abstinence. No differences in N160 latencies were shown. This effect was further demonstrated in the auditory modality for discrete 1000-Hz tones during an associative learning task (N100) (Skosnik, Edwards, et al., 2008). However, a subsequent study utilizing the same auditory stimuli with a new sample of cannabis users failed to replicate this finding (Edwards et al., 2008). Further, in a study examining the dose-related effects of exogenous cannabinoids on ERPs using an auditory oddball task (see discussion of D'Souza et al. (2012) above), THC failed to produce any deficits in the N100 response (Figure 2). Hence, the effects of chronic and acute cannabinoids on the N100 sensory ERP remain equivocal.

CANNABINOIDS AND NEURAL OSCILLATIONS

Complex mental processes such as perception, language, affect, and memory rely on the integrity of long-range functional networks consisting of transiently formed ensembles of brain areas (nodes), which process information in a coordinated manner (Varela, Lachaux, Rodriguez, & Martinerie, 2001). It has been proposed that the way in which the brain coordinates and "binds" the activity of the areas involved in a network is through synchronized neural oscillations.

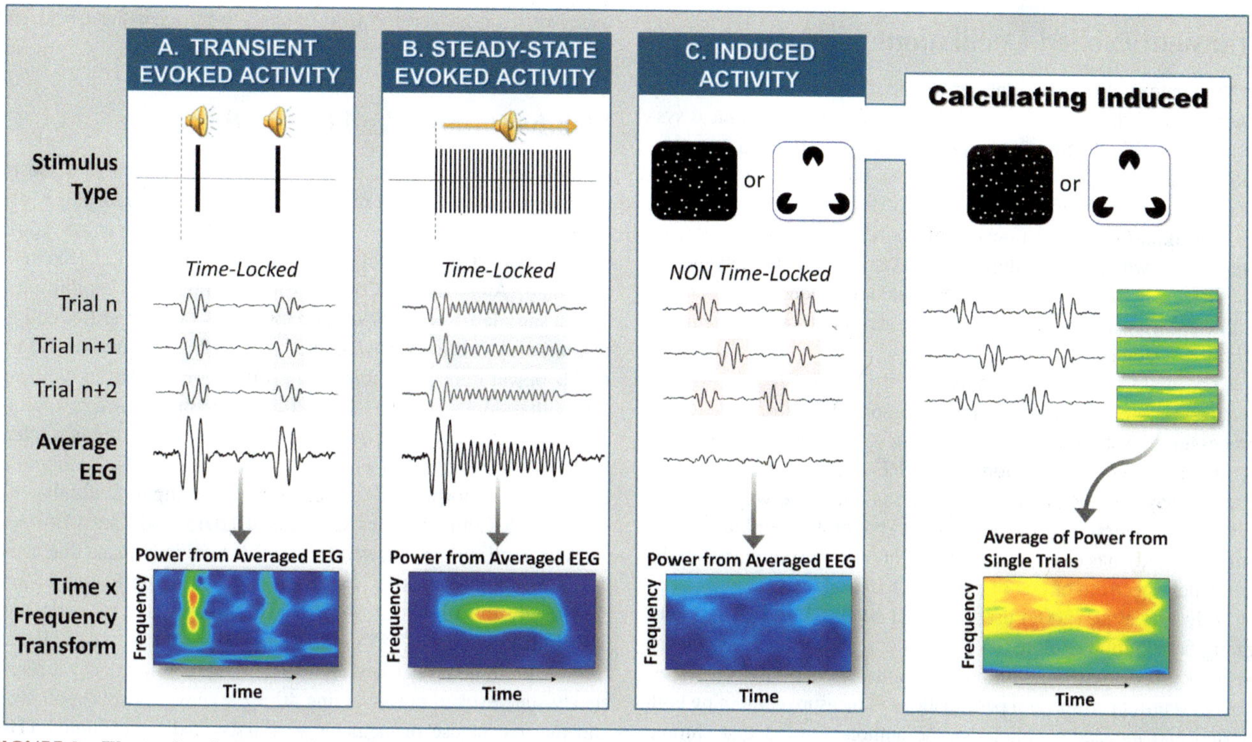

FIGURE 3 Illustration demonstrating various types of quantified EEG neural oscillations. Transient evoked and evoked steady-state activities are phase- and time-locked oscillations (A and B) elicited by discrete or steady-state sensory stimuli. For both of these types of activity, averaging across trials has the effect of canceling out non-phase-locked oscillations (owing to destructive interference), leaving the evoked response intact and magnified. Hence, time × frequency analysis can be performed on the averaged EEG (i.e., after averaging all the individual trials). By contrast, induced activity is typically elicited by higher perceptual stimuli during coherent motion or the perception of illusory shapes (see (C)). These responses are often non-time-locked (jittered across time from trial to trial). Hence, averaging the trials cancels out the response of interest due to destructive interference and is not visible in the time × frequency transform (bottom of (C), left). However, these responses can be detected if time × frequency analysis is performed on the single trials, and the time × frequency transforms of individual trials are then averaged ((C), right). *Generated de novo by the authors.*

In other words, the activity of disparate brain areas engaged in a specific functional network tends to oscillate in a synchronized way at particular frequencies (Singer, 1999). Different frequencies have typically been associated with different types of function. For example, gamma oscillations (30–80 Hz) have been implicated in higher perception and conscious awareness, whereas theta oscillations (3–7 Hz) are involved in various memory processes. If such a mechanism of neural coordination and synchrony were to go awry (e.g., as with chronic or acute cannabinoid exposure), one would expect disruptions in attention and working memory, the loosening of associations, dysregulated sensory–motor integration, and many other behavioral effects observed in relation to cannabinoids (Skosnik, Krishnan, Aydt, Kuhlenshmidt, & O'Donnell, 2006).

Types of Quantified Neural Oscillations

Neural oscillations in the EEG are quantified via time × frequency analysis utilizing tools such as the Fourier, wavelet, or Hilbert transform (Roach & Mathalon, 2008). The types of quantified neural oscillations are often differentiated into four categories: transient evoked, steady-state evoked, induced, and resting state. *Transient evoked,* or phase-locked, oscillations are a type of oscillatory response to discrete stimuli (e.g., tones or clicks) that are time-locked to stimulus onset (Figure 3(A)). These are not unlike *impulse response functions* as utilized in digital signal processing (see next section). Averaging across trials has the effect of canceling out non-phase-locked oscillations (due to destructive interference), leaving the evoked response intact. In fact, these transient evoked oscillations are likely the temporal–spectral equivalents of the traditional ERPs described above. *Steady-state evoked* oscillations result from periodic rhythmic stimuli presented at specific frequencies, which can be used to entrain the EEG to a particular phase and frequency, thus allowing the testing of oscillatory responses at functionally relevant brain rhythms (e.g., gamma (30–80 Hz)). These are commonly referred to as ASSRs and visual steady-state responses. Like transiently evoked oscillations, steady-state evoked responses are also phase- or time-locked with respect to stimulus onset (Figure 3(B)). In contrast to these two types of evoked oscillations, *induced* or non-phase-locked oscillations occur at variable phases or latencies with respect to stimulus onset, particularly during perceptual and cognitive tasks (Figure 3(C), left). These non-time-locked responses greatly diminish or disappear altogether during standard EEG trial averaging owing to destructive inference. Hence, analysis of induced oscillations require time × frequency analysis on single trials (Figure 3(C), right). Finally, *resting-state* oscillations refer to the spontaneous oscillatory activity of the brain occurring in the absence of stimulation (not shown). Similar to induced oscillations, resting-state oscillations are non-phase-locked, and hence spectral analysis of resting/spontaneous EEG requires analysis of single trials or epochs. Each of these types of oscillation in relation to both chronic and acute cannabinoids is discussed below.

Transient Evoked Oscillations

In the context of digital signal processing, one of the most widely used ways of characterizing the behavior of certain dynamical systems is to obtain the impulse response function (IRF) of the system, that is, the unique and stable way in which the system responds to a brief pulse of stimulation. The IRF provides information about the specific input/output relationship of a system, which results from the way in which the system processes stimuli. Thus, changes in the IRF (e.g., peak amplitude, peak latency, frequency spectrum, etc.) can be used to detect and identify aberrant processes occurring within a system.

The brain's response to pulses of stimulation (e.g., P50) has been considered by some authors as a sort of IRF of the circuits processing the stimuli (David, Harrison, & Friston, 2005). In view of this, a full characterization of the IRF-like responses of neural circuits may provide valuable information about the way in which drugs affect neural function. However, to the best of our knowledge, only one study has explored the effects of (chronic) cannabis on the frequency spectrum of one of the brain's IRF-like responses.

Edwards et al. (2009) used a paired-click (S1 and S2) sensory gating paradigm to explore the effect of chronic cannabis use on the frequency spectrum of the brain's response to the auditory clicks. Using event-related spectral perturbations (a measure combining both evoked and induced oscillations), the authors showed that chronic cannabis users had an attenuated response to the first click (S1) in the high-beta band (21–29 Hz) and to the second click (S2) in the gamma band (30–50 Hz). Correlation analyses revealed an inverse relationship between cannabis use and the gamma response after the second click. Furthermore, it was observed that theta-band (4–7 Hz) intertrial coherence (ITC) was also reduced in the response to the second click in chronic cannabis users. These findings suggest that heavy cannabis use disrupts the capacity of neural circuits to engage in theta, beta, and gamma oscillations in response to stimulation.

Steady-State Evoked Oscillations

In the first-ever study of the effects of chronic cannabis use on steady-state neural oscillations using the ASSR paradigm, it was demonstrated that 20 and 40-Hz harmonic EEG spectral power was decreased during beta-band auditory stimulation (Skosnik, Krishnan, Aydt, et al., 2006). A follow-up study examining a larger sample of confirmed cannabis users (via urine toxicology testing) and additional frequencies of stimulation found decreased 40-Hz power in the cannabis group (Figure 4). Interestingly, it was found that an earlier age of onset of cannabis use was related to lower 40-Hz steady-state responses. However, it remains unclear whether these changes were due to residual THC, withdrawal effects, or premorbid differences in drug-seeking individuals.

In 2015, Cortes-Briones, Cahill, et al. (2015) and Cortes-Briones, Skosnik, et al. (2015) showed for the first time in humans that acute THC disrupts steady-state evoked gamma-band oscillations. More specifically, this study demonstrated a dose-dependent (placebo, low, and high) reduction in ITC for 40 Hz stimulation during an ASSR task (Figure 5). In addition, they reported that THC (high dose) reduced steady-state evoked power during 40-Hz stimulation at a trend level and that recent use of cannabis was associated with blunted THC effects on ITC and evoked power. No significant effects were reported for the 20- or 30-Hz ASSRs, suggesting that cannabinoids selectively affected time-locked gamma-band activity. Interestingly, inverse relationships between ITC and the psychosis-relevant effects of THC were observed, indicating that THC's effect on gamma may underlie some of the psychotomimetic effects of cannabinoids. Considering that ITC is a measure of the consistency (nonrandomness) of the

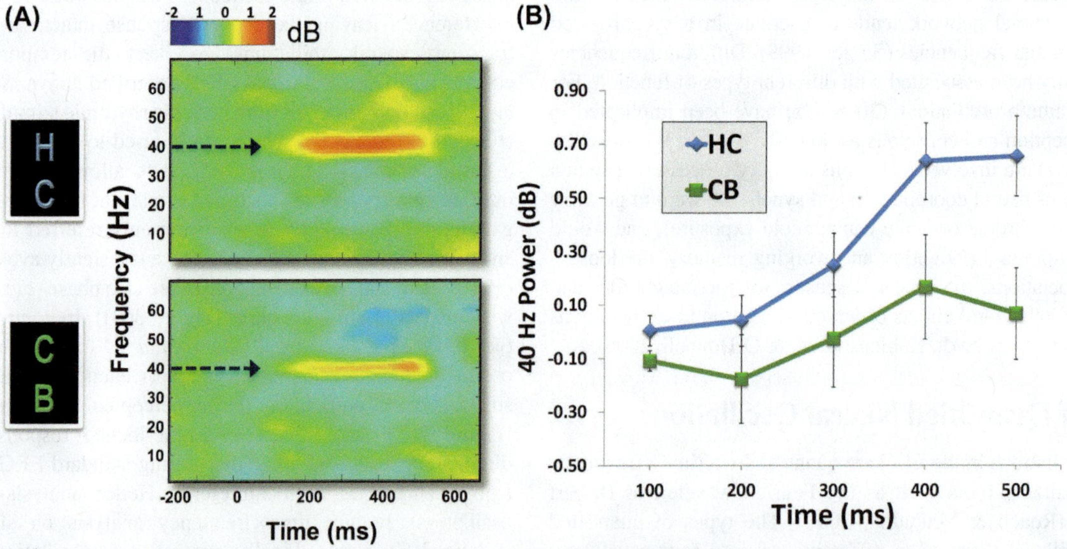

FIGURE 4 The effect of cannabis use on the gamma-band (40 Hz) auditory steady-state response. (A) Grand-averaged time × frequency plots of power during gamma-band (40 Hz) auditory stimulation at electrode FCz for healthy controls (HC; top; $n=24$) and cannabis users (CB; bottom; $n=22$). Greater 40-Hz power was seen in control subjects compared to cannabis users. (B) Average power values in 100-ms intervals during the 500-ms window after onset of the 40-Hz click trains for control subjects (blue line) and cannabis users (green line) at FCz (error bars indicate ± SE). *Reproduced with permission from Skosnik et al. (2012). Reproduced with permission from previous articles written by the authors in the journal Neuropsychopharmacology (please see: http://www.nature.com/reprints/permission-requests.html).*

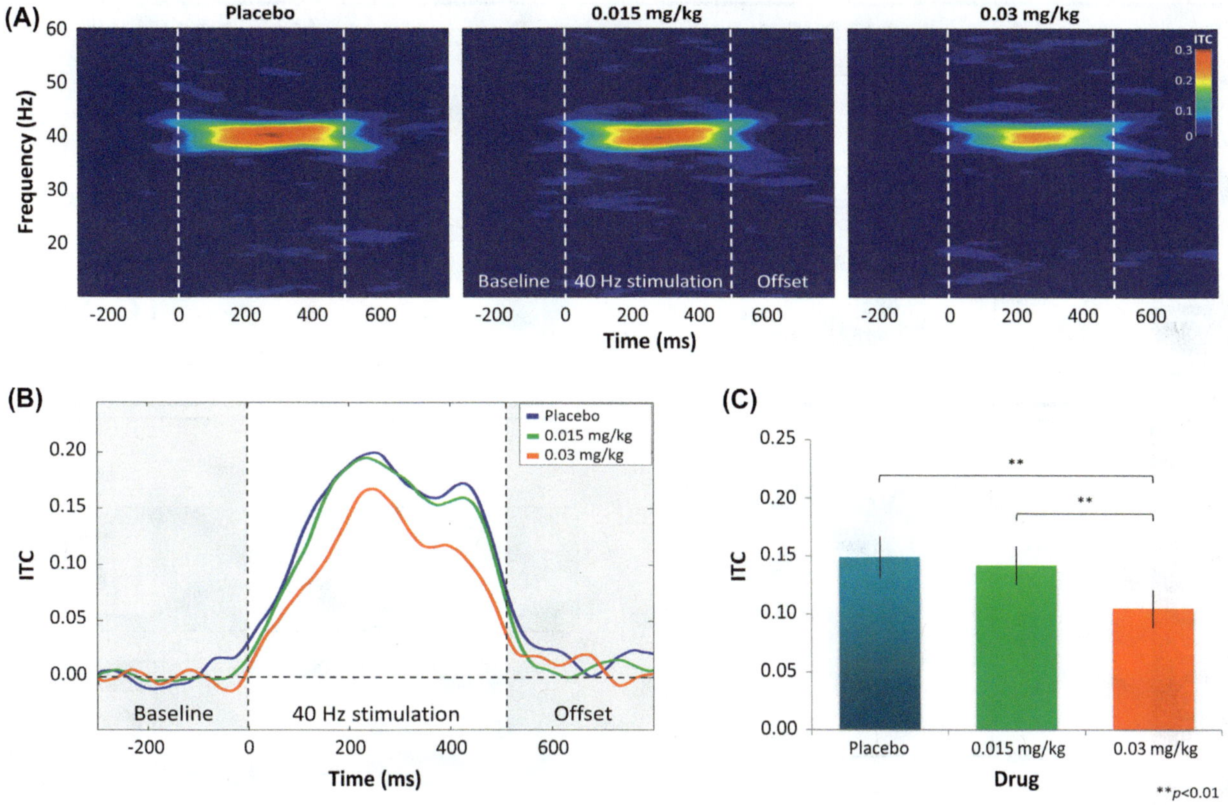

FIGURE 5 **40-Hz ITC per drug condition at electrode Cz.** (A) Hilbert transform-based time–frequency plots for ITC. From left to right, the plots show the group-averaged time–frequency plots of the placebo, 0.015 mg/kg, and 0.03 mg/kg conditions during 40-Hz stimulation at electrode Cz. (B) Time course of the group-averaged ITC per drug condition at electrode Cz. The blue line represents the ITC of the placebo condition, while the green and red lines represent the ITC of the 0.015 and 0.03 mg/kg doses, respectively. (C) ITC (mean ± SEM) by drug condition at electrode Cz during the time interval 50–550 ms after stimulation onset. Significant differences (**adjusted $p < 0.01$) between drug conditions are marked. *Reproduced with permission from previous articles written by the authors in the journal Neuropsychopharmacology (please see: http://www.nature.com/reprints/permission-requests.html).*

brain's response to a stimulus/event across trials, the authors hypothesized that these findings hint at a possible relationship between the psychosis-relevant effects of THC and the randomness (intraindividual variability) of the brain's response to gamma-range stimulation.

Induced Oscillations

Induced oscillations are commonly observed during higher perceptual processes such as coherent motion perception or the perception of illusory shapes (Figure 3(C)). Skosnik, Krishnan, D'Souza, Hetrick, and O'Donnell (2014) examined induced gamma oscillations during coherent motion perception in heavy cannabis users. The results showed that cannabis use was associated with decreased induced power in the gamma range compared to cannabis-naïve controls (Figure 6). No difference in induced power was observed in the incoherent motion or static conditions. These data suggest that cannabis use may interfere with the generation of induced gamma-band neural oscillations, which could in part mediate the perception-altering effects of exogenous cannabinoids.

Regarding acute studies, in a placebo-controlled study in which participants received iv THC, significant reductions in induced theta-band (3.5–7 Hz) power and bifrontal interelectrode coherence were observed during an *n*-back working-memory task. Furthermore, coherence deficits were positively correlated with psychotic-like symptoms (Morrison et al., 2011). The authors suggested that the psychotomimetic effects of THC could be related to impaired network dynamics as shown by a disruption in the communication between the right and the left frontal lobes. Associating the psychotomimetic effects of THC with a disruption in neural communication is a thought-provoking idea that is consistent with findings using measures of neural noise (see below).

Resting State

Few contemporary studies have examined the effects of chronic or acute cannabinoid exposure on resting-state oscillations in humans. In general terms, studies in chronic cannabis users have reported reductions in resting-state power. In a study that recorded resting-state EEG with the eyes closed in abstinent chronic cannabis users, Herning, Better, Tate, and Cadet (2003) reported a reduction in theta (4–7 Hz) and alpha-1 (8–10 Hz) power in users compared to healthy controls after both 72 h and 28 days of controlled abstinence. In a similar study from the same group (Herning, Better, & Cadet, 2008), the authors reported that after 72 h of controlled abstinence, long-term users (>8 years of use) had less alpha-2 (10–13.9 Hz) and beta-2 (25–40 Hz) power at posterior sites compared to short-term users

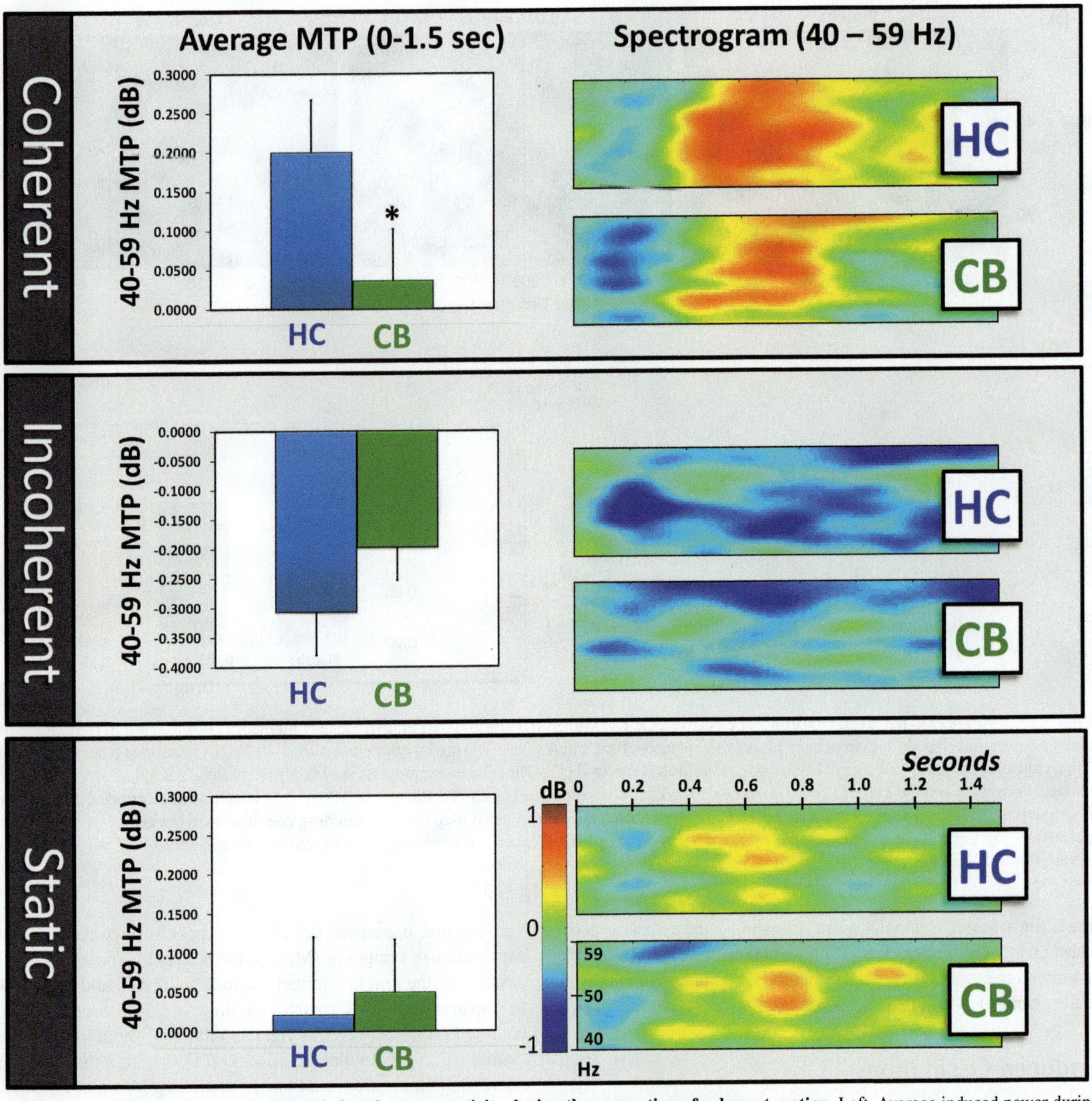

FIGURE 6 The effect of cannabis use on induced gamma activity during the perception of coherent motion. Left, Average induced power during the period of stimulation (0–1.5 s) for each of the three experimental conditions for the healthy control (HC) and cannabis (CB) groups (data from electrode Cz). Cannabis users demonstrated decreased gamma power (40–59 Hz) in the coherent motion condition compared to healthy controls (top). Right, Time × frequency spectrograms illustrating induced power from 40 to 59 Hz in each experimental condition for the control and cannabis groups (electrode Cz). Reproduced with permission from Skosnik et al. (2014). Reproduced with permission from previous articles written by the authors in the journal Neuropsychopharmacology (please see: http://www.nature.com/reprints/permission-requests.html).

(<8 years of use) and healthy controls. An issue with these studies is that they did not report baseline measures of resting-state power, that is, obtained before abstinence. The deficits observed after 72 h of abstinence could be partly explained by the acute state of withdrawal and not by a long-term abnormality associated with cannabis use. Instead, the deficits observed after 28 days of abstinence are more likely to reflect a long-term deficit associated with cannabis use, especially in view of their relationship with years of use.

Acute studies in animals have shown that CB1R agonists modulate resting-state oscillations in the cerebral cortex and hippocampus (Hajos et al., 2008). For example, systemic administration of the CB1R agonist CP 55,940 has been shown to decrease the amplitude of gamma-band oscillations in mice as measured via neocortical electrocorticograms (Sales-Carbonell et al., 2013). Similar findings have been observed via local field potential (LFP) recordings in rats (Hajos et al., 2008), in which CP 55,940 induced theta- and gamma-band deficits in the hippocampus and gamma deficits in the entorhinal cortex.

Only two studies in humans, as of this writing, have explored the acute effects of cannabinoids on resting-state EEG. In a placebo-controlled study in which participants smoked cigarettes containing different doses of THC, it was observed that THC reduced theta (4–8 Hz) and beta (12–30 Hz) closed-eye resting-state power compared to placebo (Böcker et al., 2009). Furthermore, the authors reported an inverse correlation between resting-state theta power and performance on a working-memory task. A number of limitations in the statistical methods (e.g., lack of correction for multiple comparisons) of this article, and the fact that the working-memory test was administered more than an hour after the EEG recordings, allow some caution regarding the interpretation and significance of the results.

In another placebo-controlled study in which participants received iv THC, a significant condition (placebo, THC) per frequency interval (21–27, 27–35, 35–45 Hz) interaction was reported for resting-state EEG recorded with the eyes closed (Nottage et al., 2015). Based on the visual inspection of the data, the authors concluded that THC induced a shift in the resting-state power spectra toward an increase in the higher frequency component (high-beta to low-gamma band).

CONCLUSION AND FUTURE DIRECTIONS

In this chapter, a review of the literature on the chronic and acute effects of cannabinoids on neural activity as measured by EEG was undertaken. A brief review of the cannabinoid system and the neural mechanisms of EEG was also carried out to provide the reader with some basic concepts useful later in the chapter. In general terms, the review of the ERP (P50, P300, MMN, and N100) literature suggests that both chronic and acute exposure to cannabinoids is associated with deficits in peak amplitude, latency, or both. However, a number of studies reported either no effects or inconsistent results (e.g., P300 in chronic users, MMN in acute exposure). In contrast, the literature on the effects of cannabinoids on neural oscillations (transient evoked, steady-state evoked, induced, and resting state) is more consistent in showing deficits or disruptions. This makes some theoretical sense, as neural oscillations are thought to be generated by GABAergic interneurons, which are the primary cell type containing CB1Rs. However, this consistency should be taken with caution considering the small number of studies addressing the effects of cannabinoids on oscillations for certain paradigms (e.g., transient evoked oscillations).

ERPs and spectral analyses of neural oscillations are among the oldest and more reliable measures of brain activity obtained from EEG signals. During the past decades, these measures, in combination with a relatively small number of experimental paradigms (e.g., P300, ASSR, etc.), have provided valuable insights into the chronic and acute effects of cannabinoids on the electrical activity of the brain. Furthermore, deficits in ERPs and spectral analyses have been shown to correlate with functional abnormalities induced by cannabinoids such as psychotic symptoms and working-memory deficits, suggesting that these EEG measures index core processes of the brain. However, paraphrasing Werner Heisenberg, it is important to remember that what we observe is not brain activity in itself, but brain activity exposed to our method of questioning. In other words, the analysis techniques we use to interrogate EEG signals (e.g., averaging across trials, Fourier and wavelet transforms, etc.) determine the kind of phenomena (e.g., time-locked changes in voltage, oscillations, frequency bands, etc.) that we find in our scientific explorations and, therefore, the way in which we understand neural processes.

Let us illustrate this point with an example: consider we conduct a series of analyses on a multichannel audio recording of a cocktail party, in which each channel has the signal registered by one of several microphones (sensors) located throughout the party room. A time × frequency analysis might reveal several periods of intersensor synchronization in a number of frequency bands, suggesting that widespread episodes of interaction between the signals' sources (people talking) may be an important characteristic of this cocktail party. By contrast, a semiotic analysis (i.e., an analysis of the meaning of words and sentences) conducted on the same data set might reveal that the periods of synchronization corresponded to toasts ("Cheers!") to celebrate the birthday of the person hosting the party. Both analyses captured a "real" part of what was going on at the party; however, the understanding and the hypotheses resulting from each analysis will probably lead us to very different conclusions.

There are an enormous variety of analyses and methods originated in the fields of information theory, telecommunications, and chaos theory, among others, available for studying EEG data. An increasing number of researchers have used these approaches for purposes such characterizing symptomatic exacerbation in schizophrenia, predicting seizures in epilepsy, and determining the level of consciousness in brain-injured patients, with very promising results (Casali et al., 2013). However, the implementation of these techniques to study the effects on EEG activity of psychotropic drugs in general, and THC in particular, has been slow.

As of this writing, only one study has used a nontraditional approach to characterize the effects of THC on EEG and the relationship between these effects and behavior. Cortes-Briones et al. (2015) used Lempel–Ziv complexity (LZC) to test the hypothesis that neural noise, that is, the randomness of neural activity, is involved in the psychotomimetic effects of THC. LZC is a nonlinear measure from information theory first developed to characterize the level of randomness of signals. The authors reported for the first time that THC increases neural noise in a dose-dependent manner (placebo, low, high) and that there is a strong positive relationship between neural noise and the psychosis-like positive and disorganization symptoms induced by THC (Figure 7). Interestingly, there was no relationship between neural noise and negative-like symptoms. Furthermore, these findings were independent of the changes in signal power induced by THC. Considering that random noise can interfere with and distort the information circulating within a network, the authors hypothesized that by increasing neural noise, THC may be inducing a *disconnection* (aberrant connection) (Stephan, Friston, & Frith, 2009) between brain areas and that this may be one of the mechanisms underlying the psychotomimetic effects of THC.

These results argue in favor of exploring new ways of analyzing EEG data to boost our understanding of the effects of cannabinoids and other psychotropic drugs on the brain. However, despite their advantages, some of these techniques could be misleading if not used correctly. Artifacts derived from muscle activity and other sources can easily lead to spurious results if not taken care of properly. In view of this, we propose that to improve our understanding of the effects of cannabinoids on the brain, it is a necessity to catch up with other areas of EEG research and incorporate alternative approaches, making sure that we do so in a careful and well-planned manner.

APPLICATIONS TO OTHER ADDICTIONS AND SUBSTANCE MISUSE

EEG has yielded important insights into the chronic and acute effects of cannabis on human brain function. Thus, this technique

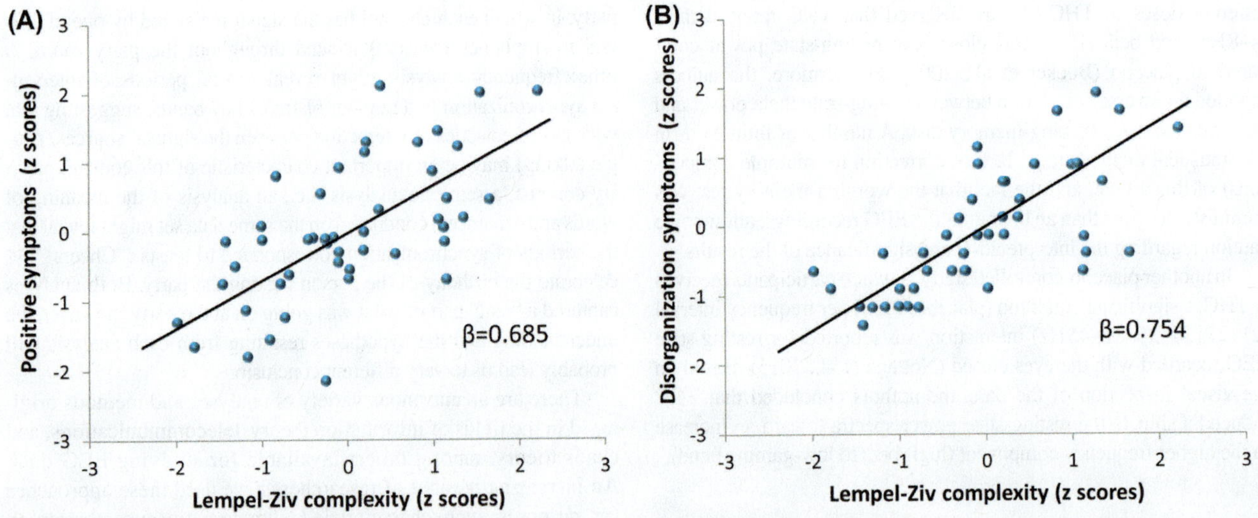

FIGURE 7 **THC-associated positive and disorganization symptoms versus LZC corrected for signal power.** The figure shows the regression lines and standardized coefficients of the regressions of positive (A) and disorganization (B) symptom factors of the Positive and Negative Syndrome Scale on LZC corrected for signal power. Positive and Negative Syndrome Scale scores and LZC values are presented in Z scores. *Reproduced with permission from a previous article written by the authors in the journal Biological Psychiatry (please see: http://www.biologicalpsychiatryjournal.com/content/bps-authorinfo#cop).*

and the advances in novel signal processing described above suggest that it may be useful in the study of other drugs of abuse. Further, since cannabis is commonly used concurrent with other substances of abuse, researchers studying noncannabinoid drug classes should consider the way in which cannabis interacts with other substances.

DEFINTION OF TERMS

Δ^9-Tetrahydrocannabinol This is the primary psychoactive constituent found in the drug cannabis.

Central cannabinoid receptor 1 This is a metabotropic, G-protein-coupled receptor found in the brain that mediates the effects of the drug cannabis (and cannabis's primary psychoactive constituent, THC). It is the most abundant metabotropic receptor found in the brain.

Electroencephalography Literally, "to write the electrical activity of the brain," this is a technique used in behavioral neuroscience to noninvasively record and study the electrical activity of the brain using electrodes placed on the scalp.

Event-related potential This is a time-locked, electrical potential detected at the scalp in response to a specific stimulus (hence, the term event-related, which serves to distinguish an ERP from resting or spontaneous EEG). ERPs are observed in the ongoing EEG as a positive or negative deflection in voltage.

Time×frequency transform This is an analysis technique used to extract the instantaneous spectral decomposition of EEG signals. Instantaneous decompositions are usually displayed in time×frequency plots.

Lempel–Ziv complexity This is a nonlinear measure of the level of randomness (noise) of signals. It reflects the minimum number of distinct patterns of activity that are necessary to reconstruct the signal. The higher the randomness of a signal is, the larger is the minimum number of patterns necessary to reconstruct it and, therefore, the higher is the LZC.

Neural oscillations As registered by EEG and LFPs, neural oscillations reflect periodic (recurrent) states of synchronized activation of large populations of pyramidal cells. It has been proposed that neural oscillations are involved in "binding" the activity of neurons to form functional networks.

KEY FACTS OF CANNABIS AND ELECTROENCEPHALOGRAPHY

- Chronic cannabis use is associated with disruptions in the P50 ERP during dual-click tasks, indicating that chronic use disrupts "sensory gating."
- Studies examining the P300 ERP, which is an index of attention, context updating, and working memory, have consistently shown disruptions in P300 amplitude after acute cannabinoid administration. Experiments examining the chronic effects of cannabis on the P300 have been mixed.
- Chronic cannabis use disrupts MMN, an index of sensory memory.
- Both chronic and acute cannabinoids disrupt various types of neural oscillations, suggesting that an intact endocannabinoid system is necessary for the generation and maintenance of synchronized neural activity, particularly in the gamma frequency range (30–80 Hz).
- Acute THC increases neural noise, suggesting that THC may be inducing a disconnection (aberrant connection) between brain areas, which could contribute to THC's psychotomimetic effects.

SUMMARY POINTS

- Cannabis continues to be one of the most commonly used drugs worldwide.
- The primary psychoactive constituent in cannabis is THC.
- The effect of THC is mediated by central cannabinoid receptors in the brain.
- In parallel with advances in our understanding of the endocannabinoid system, numerous studies have begun to examine the effects of exogenous cannabinoids on human brain function.

- EEG is one of the few noninvasive techniques that can directly record neural activity in humans.
- This chapter reviews studies that have utilized EEG in an attempt to examine the chronic and acute effects of cannabinoids on various aspects of neural function.
- A brief discussion of the nature of EEG is included, followed by sections describing the effects of cannabinoids on ERPs and neural oscillations.
- Last, a brief discussion of potential future directions for studies on the electrophysiological correlates of cannabinoids is included.

REFERENCES

Bly, L. S. (2012). Colorado, Washington OK recreational marijuana use. *USA Today*. Retrieved from http://www.usatoday.com/story/dispatches/2012/11/07/colorado-washington-legalize-recreational-marijuana-tourism/1689269/.

Böcker, K. B. E., Hunault, C. C., Gerritsen, J., Kruidenier, M., Mensinga, T. T., & Kenemans, J. L. (2009). Cannabinoid modulations of resting state EEG theta power and working memory are correlated in humans. *Journal of Cognitive Neuroscience*, 22(9), 1906–1916.

Broyd, S. J., Greenwood, L. M., Croft, R. J., Dalecki, A., Todd, J., Michie, P. T., … Solowij, N. (2013). Chronic effects of cannabis on sensory gating. *International Journal of Psychophysiology*, 89(3), 381–389.

Casali, A. G., Gosseries, O., Rosanova, M., Boly, M., Sarasso, S., Casali, K. R., … Massimini, M. (2013). A theoretically based index of consciousness independent of sensory processing and behavior. *Science Translational Medicine*, 5(198), 198ra105.

Clapper, J. R., Mangieri, R. A., & Piomelli, D. (2009). The endocannabinoid system as a target for the treatment of cannabis dependence. *Neuropharmacology*, 56(Suppl. 1), 235–243.

Cortes-Briones, J. A., Cahill, J. D., Skosnik, P. D., Mathalon, D. H., Williams, A., Sewell, R. A., … D'Souza, D. C. (2015). The psychosis-like effects of Δ^9-tetrahydrocannabinol are associated with increased cortical noise in healthy humans. *Biological Psychiatry*, 78(11), 805–813.

Cortes-Briones, J. A., Skosnik, P. D., Mathalon, D., Cahill, J., Pittman, B., Williams, A., … D'Souza, D. C. (2015). Δ9-THC disrupts gamma (γ)-Band neural oscillations in humans. *Neuropsychopharmacology*, 40(9), 2124–2134.

D'Souza, D. C., Fridberg, D. J., Skosnik, P. D., Williams, A., Roach, B., Singh, N., … Mathalon, D. (2012). Dose-related modulation of event-related potentials to novel and target stimuli by intravenous Delta(9)-THC in humans. *Neuropsychopharmacology*, 37, 1632–1646.

D'Souza, D. C., Perry, E., MacDougall, L., Ammerman, Y., Cooper, T., Wu, Y. T., … Krystal, J. H. (2004). The psychotomimetic effects of intravenous delta-9-tetrahydrocannabinol in healthy individuals: implications for psychosis. *Neuropsychopharmacology*, 29(8), 1558–1572.

David, O., Harrison, L., & Friston, K. J. (2005). Modelling event-related responses in the brain. *NeuroImage*, 25(3), 756–770.

Edwards, C. R., Skosnik, P. D., Steinmetz, A. B., O'Donnell, B. F., & Hetrick, W. P. (2009). Sensory gating impairments in heavy cannabis users are associated with altered neural oscillations. *Behavioral Neuroscience*, 123(4), 894–904.

Edwards, C. R., Skosnik, P. D., Steinmetz, A. B., Vollmer, J. M., O'Donnell, B. F., & Hetrick, W. P. (2008). Assessment of forebrain-dependent trace eyeblink conditioning in chronic cannabis users. *Neuroscience Letters*, 439(3), 264–268.

Eggan, S. M., & Lewis, D. A. (2007). Immunocytochemical distribution of the cannabinoid CB1 receptor in the primate neocortex: a regional and laminar analysis. *Cerebral Cortex*, 17(1), 175–191.

Farkas, I., Kallo, I., Deli, L., Vida, B., Hrabovszky, E., Fekete, C., … Liposits, Z. (2010). Retrograde endocannabinoid signaling reduces GABAergic synaptic transmission to gonadotropin-releasing hormone neurons. *Endocrinology*, 151(12), 5818–5829.

Freund, T. F., Katona, I., & Piomelli, D. (2003). Role of endogenous cannabinoids in synaptic signaling. *Physiological Reviews*, 83(3), 1017–1066.

Greenwood, L. M., Broyd, S. J., Croft, R., Todd, J., Michie, P. T., Johnstone, S., … Solowij, N. (2014). Chronic effects of cannabis use on the auditory mismatch negativity. *Biological Psychiatry*, 75(6), 449–458.

Grunwald, T., Boutros, N. N., Pezer, N., von Oertzen, J., Fernandez, G., Schaller, C., & Elger, C. E. (2003). Neuronal substrates of sensory gating within the human brain. *Biological Psychiatry*, 53(6), 511–519.

Hajos, M., Hoffmann, W. E., & Kocsis, B. (2008). Activation of cannabinoid-1 receptors disrupts sensory gating and neuronal oscillation: relevance to schizophrenia. *Biological Psychiatry*, 63(11), 1075–1083.

Herning, R. I., Better, W., & Cadet, J. L. (2008). EEG of chronic marijuana users during abstinence: relationship to years of marijuana use, cerebral blood flow and thyroid function. *Clinical Neurophysiology*, 119(2), 321–331.

Herning, R. I., Better, W., Tate, K., & Cadet, J. L. (2003). EEG deficits in chronic marijuana abusers during monitored abstinence. *Annals of the New York Academy of Sciences*, 993(1), 75–78.

Ilan, A. B., Gevins, A., Coleman, M., ElSohly, M. A., & de Wit, H. (2005). Neurophysiological and subjective profile of marijuana with varying concentrations of cannabinoids. *Behavioral Pharmacology*, 16(5–6), 487–496.

Johnson, J. P., Muhleman, D., MacMurray, J., Gade, R., Verde, R., Ask, M., … Comings, D. E. (1997). Association between the cannabinoid receptor gene (CNR1) and the P300 event-related potential. *Molecular Psychiatry*, 2(2), 169–171.

Johnston, L. D., O'Malley, P. M., Bachman, H. G., & Schulenberg, J. E. (2012). *Monitoring the future national results on adolescent drug use: Overview of key findings, 2011*. Ann Arbor: Institute for Social Research, The University of Michigan.

Juckel, G., Roser, P., Nadulski, T., Stadelmann, A. M., & Gallinat, J. (2007). Acute effects of Delta9-tetrahydrocannabinol and standardized cannabis extract on the auditory evoked mismatch negativity. *Schizophrenia Research*, 97(1–3), 109–117.

Kempel, P., Lampe, K., Parnefjord, R., Hennig, J., & Kunert, H. J. (2003). Auditory-evoked potentials and selective attention: different ways of information processing in cannabis users and controls. *Neuropsychobiology*, 48(2), 95–101.

Kiehl, K. A., Laurens, K. R., Duty, T. L., Forster, B. B., & Liddle, P. F. (2001). Neural sources involved in auditory target detection and novelty processing: an event-related fMRI study. *Psychophysiology*, 38(1), 133–142.

Kleber, H. D., & DuPont, R. L. (2012). Physicians and medical marijuana. *The American Journal of Psychiatry*, 169(6), 564–568.

Lachaux, J.-P., Axmacher, N., Mormann, F., Halgren, E., & Crone, N. E. (2012). High-frequency neural activity and human cognition: past, present and possible future of intracranial EEG research. *Progress in Neurobiology*, 98(3), 279–301.

Luck, S. J. (2005). *An introduction to the event-related potential technique*. Cambridge, Mass: MIT Press.

Luck, S. J., & Kappenman, E. S. (2012). *Oxford handbook of event-related potential components*. Oxford: Oxford University Press.

Luck, S. J., Mathalon, D. H., O'Donnell, B. F., Hamalainen, M. S., Spencer, K. M., Javitt, D. C., & Uhlhaas, P. J. (2011). A roadmap for the development and validation of event-related potential biomarkers in schizophrenia research. *Biological Psychiatry*, 70(1), 28–34.

Mayer, A. R., Hanlon, F. M., Franco, A. R., Teshiba, T. M., Thoma, R. J., Clark, V. P., & Canive, J. M. (2009). The neural networks underlying auditory sensory gating. *Neuroimage, 44*(1), 182–189.

Mechoulam, R., & Parker, L. A. (2012). The endocannabinoid system and the brain. *Annual Review of Psychology, 64*, 21–47.

Morrison, P. D., Nottage, J., Stone, J. M., Bhattacharyya, S., Tunstall, N., Brenneisen, R., ... Ffytche, D. H. (2011). Disruption of frontal theta coherence by [Delta]9-Tetrahydrocannabinol is associated with positive psychotic symptoms. *Neuropsychopharmacology, 36*(4), 827–836.

Naatanen, R., & Alho, K. (1995). Generators of electrical and magnetic mismatch responses in humans. *Brain Topography, 7*(4), 315–320.

Naatanen, R., & Picton, T. (1987). The N1 wave of the human electric and magnetic response to sound: a review and an analysis of the component structure. *Psychophysiology, 24*(4), 375–425.

Niedermeyer, E., & da Silva, F. L. (2005). *Electroencephalography: Basic principles, clinical applications, and related fields*. Lippincott Williams & Wilkins.

Nottage, J., Stone, J., Murray, R., Sumich, A., Bramon-Bosch, E., Ffytche, D., & Morrison, P. D. (2015). Delta-9-tetrahydrocannabinol, neural oscillations above 20 Hz and induced acute psychosis. *Psychopharmacology, 232*(3), 519–528.

Patrick, G., Straumanis, J. J., Struve, F. A., Fitz-Gerald, M. J., Leavitt, J., & Manno, J. E. (1999). Reduced P50 auditory gating response in psychiatrically normal chronic marihuana users: a pilot study. *Biological Psychiatry, 45*(10), 1307–1312.

Patrick, G., Straumanis, J. J., Struve, F. A., Nixon, F., Fitz-Gerald, M. J., Manno, J. E., & Soucair, M. (1995). Auditory and visual P300 event related potentials are not altered in medically and psychiatrically normal chronic marihuana users. *Life Sciences, 56*(23–24), 2135–2140.

Patrick, G., & Struve, F. A. (2000). Reduction of auditory P50 gating response in marihuana users: further supporting data. *Clinical EEG (Electroencephalography), 31*(2), 88–93.

Pertwee, R. G., Howlett, A. C., Abood, M. E., Alexander, S. P., Di Marzo, V., Elphick, M. R., ... Ross, R. A. (2010). International union of basic and clinical pharmacology. LXXIX. Cannabinoid receptors and their ligands: beyond CB and CB. *Pharmacological Reviews, 62*(4), 588–631.

Picton, T. W., Hillyard, S. A., Krausz, H. I., & Galambos, R. (1974). Human auditory evoked potentials. I. Evaluation of components. *Electroencephalography and Clinical Neurophysiology, 36*(2), 179–190.

Polich, J., & Criado, J. R. (2006). Neuropsychology and neuropharmacology of P3a and P3b. *Internation Journal of Psychophysiology, 60*(2), 172–185.

Rentzsch, J., Penzhorn, A., Kernbichler, K., Plockl, D., Gomez-Carrillo de Castro, A., Gallinat, J., & Jockers-Scherübl, M. C. (2007). Differential impact of heavy cannabis use on sensory gating in schizophrenic patients and otherwise healthy controls. *Experimental Neurology, 205*(1), 241–249.

Rinne, T., Alho, K., Ilmoniemi, R. J., Virtanen, J., & Naatanen, R. (2000). Separate time behaviors of the temporal and frontal mismatch negativity sources. *Neuroimage, 12*(1), 14–19.

Roach, B. J., & Mathalon, D. H. (2008). Event-related EEG time-frequency analysis: an overview of measures and an analysis of early gamma band phase locking in schizophrenia. *Schizophrenia Bulletin, 34*(5), 907–926.

Roser, P., Della, B., Norra, C., Uhl, I., Brune, M., & Juckel, G. (2010). Auditory mismatch negativity deficits in long-term heavy cannabis users. *European Archives of Psychiatry and Clinical Neuroscience, 260*(6), 491–498.

Roser, P., Juckel, G., Rentzsch, J., Nadulski, T., Gallinat, J., & Stadelmann, A. M. (2008). Effects of acute oral Delta9-tetrahydrocannabinol and standardized cannabis extract on the auditory P300 event-related potential in healthy volunteers. *European Neuropsychopharmacol, 18*(8), 569–577.

Sales-Carbonell, C., Rueda-Orozco, P. E., Soria-Gomez, E., Buzsaki, G., Marsicano, G., & Robbe, D. (2013). Striatal GABAergic and cortical glutamatergic neurons mediate contrasting effects of cannabinoids on cortical network synchrony. *Proceedings of the National Academy of Sciences of the United States of America, 110*(2), 719–724.

Singer, W. (1999). Neuronal synchrony: a versatile code for the definition of relations? *Neuron, 24*(1), 49–65, 111–125.

Skosnik, P. D., D'Souza, D. C., Steinmetz, A. B., Edwards, C. R., Vollmer, J. M., Hetrick, W. P., & O'Donnel, B. F. (2012). The effect of chronic cannabinoids on broadband EEG neural oscillations in humans. *Neuropsychopharmacology, 37*(10), 2184–2193.

Skosnik, P. D., Edwards, C. R., O'Donnell, B. F., Steffen, A., Steinmetz, J. E., & Hetrick, W. P. (2008). Cannabis use disrupts eyeblink conditioning: evidence for cannabinoid modulation of cerebellar-dependent learning. *Neuropsychopharmacology, 33*(6), 1432–1440.

Skosnik, P. D., Krishnan, G. P., Aydt, E. E., Kuhlenshmidt, H. A., & O'Donnell, B. F. (2006). Psychophysiological evidence of altered neural synchronization in cannabis use: relationship to schizotypy. *American Journal of Psychiatry, 163*(10), 1798–1805.

Skosnik, P. D., Krishnan, G. P., D'Souza, D. C., Hetrick, W. P., & O'Donnell, B. F. (2014). Disrupted gamma-band neural oscillations during coherent motion perception in heavy cannabis users (Original article). *Neuropsychopharmacology, 39*(13), 3087–3099.

Skosnik, P. D., Krishnan, G. P., Vohs, J. L., & O'Donnell, B. F. (2006). The effect of cannabis use and gender on the visual steady state evoked potential. *Clinical Neurophysiology, 117*(1), 144–156.

Skosnik, P. D., Park, S., Dobbs, L., & Gardner, W. L. (2008). Affect processing and positive syndrome schizotypy in cannabis users. *Psychiatry Research, 157*(1–3), 279–282.

de Sola, S., Tarancon, T., Pena-Casanova, J., Espadaler, J. M., Langohr, K., Poudevida, S., ... de la Torre, R. (2008). Auditory event-related potentials (P3) and cognitive performance in recreational ecstasy polydrug users: evidence from a 12-month longitudinal study. *Psychopharmacology (Berlin), 200*(3), 425–437.

Solowij, N., Michie, P. T., & Fox, A. M. (1991). Effects of long-term cannabis use on selective attention: an event-related potential study. *Pharmacology Biochemistry and Behavior, 40*(3), 683–688.

Solowij, N., Michie, P. T., & Fox, A. M. (1995). Differential impairments of selective attention due to frequency and duration of cannabis use. *Biological Psychiatry, 37*(10), 731–739.

Stadelmann, A. M., Juckel, G., Arning, L., Gallinat, J., Epplen, J. T., & Roser, P. (2011). Association between a cannabinoid receptor gene (CNR1) polymorphism and cannabinoid-induced alterations of the auditory event-related P300 potential. *Neuroscience Letters, 496*(1), 60–64.

Stephan, K. E., Friston, K. J., & Frith, C. D. (2009). Dysconnection in schizophrenia: from abnormal synaptic plasticity to failures of self-monitoring. *Schizophrenia Bulletin*. http://dx.doi.org/10.1093/schbul/sbn176.

Struve, F. A., & Straumanis, J. J. (1990). Electroencephalographic and evoked-potential methods in human marijuana research—historical review and future-trends. *Drug Development Research, 20*(3), 369–388.

Theunissen, E. L., Kauert, G. F., Toennes, S. W., Moeller, M. R., Sambeth, A., Blanchard, M. M., & Ramaekers, J. G. (2012). Neurophysiological functioning of occasional and heavy cannabis users during THC intoxication. *Psychopharmacology (Berlin), 220*(2), 341–350.

Twitchell, W., Brown, S., & Mackie, K. (1997). Cannabinoids inhibit N- and P/Q-type calcium channels in cultured rat hippocampal neurons. *Journal of Neurophysiology, 78*(1), 43–50.

Varela, F., Lachaux, J. P., Rodriguez, E., & Martinerie, J. (2001). The brainweb: phase synchronization and large-scale integration. *Nature Reviews Neuroscience, 2*(4), 229–239.

Part V

Opioids

Section A

General Aspects

Chapter 81

Street Level Heroin, an Overview on Its Components and Adulterants

Maryam Akhgari[1], Afshar Etemadi-Aleagha[2], Farzaneh Jokar[1]

[1]Department of Forensic Toxicology, Legal Medicine Research Center, Legal Medicine Organization, Tehran, Iran; [2]Department of Anesthesiology, Tehran University of Medical Sciences (TUMS), Amir Alam Hospital, Tehran, Iran

INTRODUCTION

There are many definitions for drug abuse in different societies. Drug or substance use for any reason other than therapeutic purposes to change physical or mental functions (euphoria, sedation, etc.) is termed drug or substance abuse. Substance abuse or illicit drug abuse has negative consequences, which may produce problems for abusers (Shannon et al., 2008). Abused substances are classified into various groups. Opium and its natural, semisynthetic, and synthetic derivatives constitute a big category of abused substances. Heroin is a semisynthetic derivative of morphine originated from opium. Street heroin contains many adulterants in addition to its main proposed active ingredient (heroin, diacetylmorphine, also known as diamorphine). Some of these adulterants are inert and others have pharmacological action. Adulterants can vary street heroin purity. The occurrence of fatalities with this highly addictive substance can be associated with the effects of adulterants, microbial contamination, and variation in heroin purity (McLauchlin et al., 2002).

This chapter provides an overview of the constituents in street heroin samples. It starts with the categorization of opioids followed by an explanation of the history of heroin production and marketing and finally its pulling out of the drug market. The production process for illicit heroin is explained in Sections A Focus on Heroin Production Process, Definition of Controlled Substances, and Controlled Substances Act Schedules. Then we cover the Controlled Substances Act and Controlled Substances Act Schedules. Opioid pharmacology and the effects of opioids on receptors are reviewed, followed by a discussion of the physical appearance of heroin, routes of heroin administration, and body packers of heroin. Street names or slang terms used for heroin, its combination with other drugs, and the presence of adulterants in heroin samples are discussed in Sections Street Names and Slang Terms for Heroin and Component Analysis of Heroin. The last, important, part of the chapter explains the medical consequences of street heroin abuse.

With the reviews contained in this chapter the reader will have a general knowledge about heroin pharmacology, its manufacturing processes, the adulterants added to it, and finally the health consequences of its use.

AN OVERVIEW OF OPIATES

Opium is a highly addictive natural and nonsynthetic opioid, which is obtained from the poppy plant, *Papaver somniferum*. Opium is the base substance for the synthesis of various medications.

Classification of Opioids According to Source

Opioid-derived drugs are categorized into three subgroups:

1. Natural opiates: morphine, codeine, and opium are natural opiates. These are alkaloids found naturally in poppy seeds.
2. Semisynthetic opiates: heroin, oxycodone, oxymorphone, and buprenorphine are half-natural substances. Opium is used as a base for the synthesis of these drugs.
3. Synthetic opiates: fentanyl, pethidine, and dextropropoxyphene are not found naturally and are manufactured in laboratories.

Classification of Opioids According to Chemical Structure

There are numerous chemical structures for opioids:

1. Phenanthrenes: morphine, codeine, heroin, hydromorphone, and oxycodone.
2. Benzomorphans: pentazocine and phenazocine
3. Diphenylpropylamines: propoxyphene, methadone, levo-α-acetylmethadol, loperamide
4. Phenylpiperidines: meperidine, also known as pethidine
5. Anilidopiperidines: fentanyl, alfentanil, and sufentanil
6. Oripavine derivatives: etorphine, dihydroetorphine, and buprenorphine
7. Morphinan derivatives: levorphanol and butorphanol
8. Other types of opioids: some types of synthetic opioids do not belong to the mentioned categories. One of these drugs is tramadol, a synthetic analogue of codeine (Stolberg, 2011).

Classification of Opioids According to Their Effects on Opioid Receptors

The opioid receptors are named μ, κ, δ, σ, and ϵ. Opioids can act as agonists, antagonists, agonists/antagonists, or partial agonists (Contet, Kieffer, & Befort, 2004).

HISTORICAL REVIEW OF HEROIN

The word heroin is derived from the German word "heroisch" (heroic) owing to its producing a heroic feeling in users. Heroin was first synthesized by chemist Charles Romley Alder Wright in 1874 in England to find a nonaddictive alternative for morphine (Wright). Twenty-three years later heroin was first marketed by Bayer Pharmaceutical Company in 1897. Heroin was sold as a cough medicine in a variety of dosage forms, tablets, mixed into cough syrups, mixed with glycerin to make an elixir, and as heroin salts mixed with water, and was exported to 23 countries as a cough medicine. As a result of physicians' reports of their patients' addiction to this medicine within months of its widespread prescription and use, Bayer pulled its heroin cough medicine from the market in 1913. Heroin was regulated by international authorities to restrict its production, distribution, and use. The ban on heroin production converted it to an illegal drug. Illicit manufacturing and trafficking of this highly addictive substance were started in many countries. The quantity of heroin seizures underwent a tenfold increase from 1970 to 2014 (Heroin Wikipedia, 2014).

A FOCUS ON THE HEROIN PRODUCTION PROCESS

One of the Asian businesses is illegal cultivation of *P. somniferum*. Western countries such as Mexico and Colombia produce opium poppy. Afghanistan, Pakistan, Turkey, Iran, and India are other countries that produce opium. According to a United Nations Office on Drugs and Crime report Afghanistan is the world's largest supplier of opium poppy. *P. somniferum* is a flowering plant that is the main source for heroin manufacturing. Flowers of the plant convert to round grayish-green fruits, which develop into capsules also called the seedpod or poppy head. After the pods are fully mature they are ready to be incised by iron or glass blades. Incisions cause a white latex to drip onto the surface of the pods. This white latex oxidizes, darkens, and thickens overnight to produce opium. Opium contains over 40 different alkaloids. Most of these alkaloids are salts of meconic acid. The most important alkaloid of opium is morphine. The *P. somniferum* morphine content is about 9–14%. The next prominent alkaloid is codeine. Other alkaloids of opium include noscapine, papaverine, thebaine, narceine, protopine, laudanine, codamine, and cryptopine, among others. Morphine can be extracted and purified from opium to be a source of illicit heroin production. However, clandestine laboratories are not able to extract and purify morphine ideally and some of the opium alkaloids remain as impurities of origin in illegally synthesized heroin.

Acetylation of morphine converts it to heroin. The most important chemical used in the acetylation process of heroin is acetic anhydride. Morphine is mixed with acetic anhydride and heated. The product is heroin (3,6-diacetylmorphine) with other impurities (Cannabis, Coca & Poppy; Nature's Addictive Plants, Production & Distribution, 2014; Zerell, Ahrens, & Gerz, 2005).

DEFINITION OF CONTROLLED SUBSTANCES

The abuse potential of a drug or substance can be estimated by some items; they include:

1. There is evidence that a drug or substance is used in sufficient amounts to become a health hazard for the abuser or other people in the community.
2. The drug was diverted significantly from a legal drug channel or its original purpose to an illegal use.
3. The drug is used by individuals without any medical advice from a practitioner.
4. The drug is new and its action is similar to that of other drugs already listed as having a potential for abuse.

CONTROLLED SUBSTANCES ACT SCHEDULES

Manufacturing, importation, possession, use, and distribution of certain drugs such as narcotics, stimulants, etc., and other chemicals are regulated by US federal drug policy. All substances that are regulated under federal law have been placed by the Controlled Substances Act into one of five schedules.

Schedule I

1. Has high abuse potential;
2. Has no accepted medical use in treatment in the United States or its use lacks accepted safety in treatment under medical supervision:

    Heroin, 3,4-methylenedioxymethamphetamine (ecstasy), methaqualone, lysergic acid diethylamide (LSD), marijuana, mescaline, psilocybin, tetrahydrocannabinol

Schedule II

1. The substance has high abuse potential;
2. The substance has currently accepted medical use in treatment in the United States or severe restrictions needed to be used in treatment;
3. Abuse of the substance induces severe psychological or physical dependence:

    Cocaine/crack, raw opium, codeine, hydromorphone, morphine, oxycodone (oxycontin), oxymorphone, methadone, pethidine, amphetamine, methamphetamine, phenmetrazine, phencyclidine

Schedule III

1. The potential for substance abuse is less than for the substances listed in Schedules I and II;
2. The substance has currently accepted medical use in treatment in the United States;
3. Abuse of the substance may induce moderate or low physical dependence or high psychological dependence:

    Anabolic steroid, ketamine

Schedule IV

1. The substance has a low potential for abuse in comparison with substances in Schedule III;
2. The substance has currently accepted medical use in treatment in the United States;
3. Abuse of the substance may induce limited physical dependence or psychological dependence in comparison to the substances in Schedule III:

   Dextropropoxyphene (Darvocet), alprazolam (Xanax), clonazepam (Klonopin), diazepam (Valium), lorazepam (Ativan), pentazocine, phentermine, zolpidem (Ambien)

Schedule V

1. The substance has low potential for abuse in comparison to the substances in Schedule IV;
2. The substance has currently accepted medical use in treatment in the United States;
3. The substance has limited physical dependence or psychological dependence problems in comparison to the controlled substances in Schedule IV:

   Cough medicines with codeine

As is seen in this section, heroin is categorized as a Schedule I substance with high potential for abuse (Controlled Substance Schedules, 2012).

HEROIN PHARMACOLOGY

After intravenous (iv) injection, smoking, or snorting, heroin enters the bloodstream and crosses the blood–brain barrier (BBB) rapidly, but in oral administration heroin undergoes extensive first-pass metabolism. Plasma butyrylcholinesterase is the primary enzyme responsible for the activation of heroin (Qiao, Han, & Zhan, 2013). Hydrolysis converts heroin to its active metabolite, 6-monoacetylmorphine (6-MAM), and its inactive one, 3-MAM, followed by transformation to morphine. Heroin is more lipid soluble, more potent, and 100 times faster acting than morphine. The difference between heroin and morphine lies in the two acetyl groups in heroin's structure, which increase the solubility of heroin in lipid tissues and make for an easy passage through the BBB (Zovko & Criscuolo, 2009). The affinity of heroin for μ opioid receptors is very low (Selley et al., 2001). Heroin exerts its analgesic effect by two active metabolites, 6-MAM and morphine. These two metabolites have an agonistic effect on μ opioid receptors in the brain and spinal cord of the central nervous system (CNS). Also, morphine is a weak agonist at the δ and κ opioid receptor subtypes. Morphine binding to δ and κ receptors reduces the inhibitory effect of γ-aminobutyric acid (GABA) on dopaminergic neurons by inhibiting the release of GABA from the nerve terminals (Katzung, 2001).

Morphine (heroin metabolite) is metabolized by conjugation with glucuronic acid and by other minor routes of metabolism, including deamination, to produce normorphine (Trescot, Datta, Lee, & Hansen, 2008). Conjugation yields two metabolites, morphine-3-glucuronide (M3G) and M6G. M6G has an analgesic effect and respiratory depressant activity (Peat, Hanna, Woodham, Knibb, & Ponte, 1991). The μ opioid receptors mediate some pharmacological actions of heroin, such as respiratory depression, euphoria, and physical dependence (Hosztafi, 2003). Street heroin has variable composition and purity and can cause death in heroin abusers through respiratory arrest.

WHAT DOES HEROIN LOOK LIKE?

Heroin is a fine white powder with a bitter taste in its purest form (Figure 1) (What is heroin cut with?, 2014). It contains up to 85–95% diacetylmorphine. Nonpure heroin varies in chemical and physical appearance. It contains additives and adulterants, which influence its appearance. Additives make heroin appear as gray, pink/beige, brown, or black granular lumps or black sticky material (black tar) (Figures 2 and 3). Asian heroin is a brown coarse powder with poor water solubility (Ciccarone, 2009).

FIGURE 1 Heroin is a fine white powder in its purest form.

FIGURE 2 Additives can make heroin appear as gray, pink/beige, brown, or black granular lumps.

FIGURE 3 Additives can also make heroin appear as a black sticky material (black tar).

Black tar is a product of postprocessing hydrolysis of heroin. Hydrolysis can be mediated by water or excess acid in heroin samples. 6-MAM content above 10% can be an indicator of postprocessing hydrolysis of heroin (Drugs of abuse, Heroin, 2014).

ROUTES OF ADMINISTRATION FOR HEROIN

The route of administration is an important factor influencing the onset of heroin effects. Heroin can be nasally insufflated, or "snorted," smoked, or injected into veins, into muscles, or under the skin. Oral or rectal routes are other ways of using heroin.

Heroin blood concentration rises quickly after iv injection. Using heroin by other routes of administration such as smoking, insertion into the vagina or anus, snorting, and oral routes can produce slower effects in comparison to the iv method (Klous, Van den Brink, Van Ree, & Beijnen, 2005). For reaching a quick and potent "high," heroin is most often used by iv injection to easily accessible arm veins or the femoral vein in the groin (Senbanjo, Tipping, Hunt, & Strang, 2012). As iv injection has the risk of sharing needles and spreading infectious diseases such as human immunodeficiency virus, hepatitis B and C, and other blood-borne diseases, some heroin abusers choose other routes of administrations such as insufflation or snorting (Clatts, Giang, Goldsamt, & Yi, 2007). In this method heroin is crushed into fine particles. This fine powder can be inhaled through a rolled-up paper into the nose. Liquefied heroin may be sniffed by nasal spray bottles, a practice known as "shabanging" (A Dictionary of Slang Drug Terms, Trade Names, and Pharmacological Effects and Uses, 1997; Heroin CESAR, 2014). Abusers who have the experience of iv routes do not use other methods of drug administration, because the "rush" encourages them to use the iv route rather than smoking, snorting, or eating (Hosztafi, 2011). Abusers of heroin experience little to no "rush" after an oral dose, also the first-pass metabolism decreases heroin bioavailability after oral administration (Klous, Van den Brink, Van Ree, & Beijnen, 2005).

Heroin may be used as suppository (anal insertion) or pessary (vaginal insertion) by pushing dissolved heroin in a water base into the anus or vagina (Heroin Abuse and Suppositories, 2014).

BODY PACKING OF HEROIN

Internal concealment of wrapped packets of illicit drugs (heroin, amphetamines, marijuana, and other substances) in the body to transport them is called body packing. These packets can either be swallowed or inserted into the rectum or vagina. Accidental deaths due to the rupture of drug packs have been reported in some studies (Njau, Raikos, Spagou, Tzikas, & Tsoukali, 2010). Figure 4 shows street heroin packets extracted from the stomach of a body packer. As is shown, one packet was punctured.

STREET NAMES AND SLANG TERMS FOR HEROIN

Definition of Street Drug Names

Street drugs or illegal substances often have slang names. These names are chosen by abusers to describe the drug or its action on the body or to confuse police agents who are looking for heroin suppliers. Many geographic areas have their specific slang names

FIGURE 4 Street heroin packets extracted from the stomach of a body packer. As is shown, one packet was punctured.

for illicit drugs; also, in police stations and courtrooms they are called by their own selected names. Another reason for using slang names is to disguise the activity of abusers.

Common Street Names for Heroin

Heroin has many common street names, for example, brown sugar, boy, black, black tar, black pearl, black stuff, black eagle, brown, brown crystal, brown tape, brown Rhine, big bag, blue bag, blue star, brick gum, China white, chiba or chiva, dope, dragon, H, Iranian crack, junk, Mexican brown, Mexican mud, Mexican horse, mud, Number 3, Number 4, Number 8, skag, smack, snow, snowball, scat, sack, skunk, tar, white, white nurse, white lady, white horse, white girl, white boy, white stuff.

Street Names for "Mixed" Drugs

Sometimes heroin is mixed with other drugs to obtain a certain pharmacologic effect. Slang terms for these mixtures are shown in Table 1 (Akhgari, Jokar, Bahmanabadi, & Etemadi-Aleagha, 2012; Casa Palmera Staff, 2010; Kazemifar, Solhi, & Badakhshan, 2011).

COMPONENT ANALYSIS OF HEROIN

Why is it important to analyze the components of heroin?

Analysis of seized street heroin is important for legal purposes. Analysis of heroin shows that two or more samples are linked to one another (i.e., they came from an identical source, from one batch or from the same clandestine laboratory, characterized by inert and life-threatening adulterants (Lurie, Driscoll, Cathapermal, & Panicker, 2013)).

Adulterants and Diluents in Street Heroin

The terms adulterant and diluent are used for substances added to illicitly distributed controlled drugs in addition to the active ingredients. Adulterants may have mild to serious health consequences and sometimes be fatal in abusers. While "adulterants" and "diluents" have differences, they share a common characteristic in that they refer to all additional substances added to illicit drugs

TABLE 1 Street Names for Drug Combinations with Street Heroin

Active Ingredients	Street Name
Heroin and marijuana	Atom bomb, canade, woola, woolie, woo-woo
Heroin and cold medicine	Cheese
Heroin and ecstasy	Chocolate chip cookies, H bomb
Heroin and alprazolam	Bars
Heroin and Ritalin	Pineapple
Heroin and cocaine	Belushi, boy-girl, he-she, dynamite, goofball, H&C, primo, snowball, Murder 1
Heroin and LSD	Beast, LBJ, neon nod
Heroin and crack cocaine	Eightball
Heroin and crack	Dragon rock, moon rock
Heroin, cocaine, marijuana, and PCP	El diablito
Heroin and cocaine	Goofball
Heroin and cocaine	H & C
Ecstasy and heroin	H-bomb
Heroin, phenobarbital, and methaqualone	Karachi
Heroin plus LSD and PCP	LBJ
Methamphetamine and heroin mixed in one syringe	Methball
Crack and heroin	Moonrock
Heroin and cocaine	Murder one
LSD and heroin	Neon Nod
Ecstasy particles added to a bag of heroin	On the ball
Heroin and dimenhydrinate	Polo
Heroin plus PCP	Poro
Heroin, sleeping pills, strychnine, and caffeine	Red rock opium/redrum
Two layers of cocaine with a layer of heroin in the middle	Sandwich
Low-purity heroin plus crack cocaine	Scramble
Heroin and methamphetamine	Screwball

This table shows the slang names for drugs mixed with heroin. These mixtures are produced to obtain a certain "high."

intentionally or synthesized during manufacturing, distribution, or storage steps.

Some of the reasons for adding adulterants to street drugs are:

1. Dilution or bulking of the substance of abuse (sugar);
2. Enhancing or mimicking the pharmacological effects of the active ingredient (fentanyl in heroin);
3. Facilitating the administration of the substance of abuse (caffeine in heroin).

There are many studies from various countries indicating the presence of adulterants in multiple samples of illicit drugs (Behrman, 2008; Chaudron-Thozet, Girard, & David, 1992; Cole et al., 2010; Janhunen & Cole, 1999; O'Neal, Poklis, & Lichtman, 2001; Wong, Curtis, & Wingert, 2008). A summary of the published studies on drug adulterants found in illicit heroin, the potential reason for their presence, and their neurological effects and neuropathy is provided in Table 2. As is shown in Table 2, acetylcodeine is one of the most notable impurities of origin in most heroin products as a result of the failure to remove codeine during the purification process of morphine. One of the key markers in signature analysis of street heroin is acetylcodeine because the ratio of heroin/acetylcodeine varies between sources. Acetylcodeine is a toxic by-product and potentiates the convulsant effect of diacetylmorphine (Johnston & King, 1998). Unreacted morphine and codeine are other components of street heroin and account for some adverse reactions in iv heroin abusers. Noscapine and papaverine are found in smaller amounts but have pharmacologic effects in street heroin. MAM is found in illicitly produced heroin as an impurity; this derivative could be a result of heroin hydrolysis and spontaneous deacetylation under humid conditions (Atasoy, Biçer, Açikkol, & Bilgiç, 1988; Gomez & Rodriguez, 1989).

Microbial Contamination of Heroin

Street heroin can cause bacterial infections. Bacterial infections are common in iv heroin abusers. As a result of poor or unsterile manufacturing techniques, nonstandard packaging, and distribution and storage conditions, bacteria, fungi, and viruses can contaminate street heroin. There are many reports demonstrating microbial contamination of street heroin. McLauchlin et al. (2002) studied 58 heroin samples and identified 17 species of bacteria in these samples. Brett, Hood, Brazier, Duerden, and Hahné (2005) showed that there was a dramatic increase in soft-tissue infections with spore-forming bacteria (*Clostridium* and *Bacillus* spp.) in injecting drug users. Dancer, McNair, Finn, and Kolsto (2002) studied a case of cellulitis. Aspirate samples from tissue and the patient's own heroin were positive for *Bacillus cereus*. Kalka-Moll, Aurbach, Schaumann, Schwarz, and Seifert (2007) recognized 12 clinical cases of iv drug abusers with abscess in various parts of the body in the metropolitan area of Cologne, Germany. Abscess samples of a number of cases showed positive results for botulinum toxin.

Adulteration of Heroin with Heavy Metals

One of the heavy metals detected in street heroin is lead. Parras, Patier, and Ezpeleta (1987) reported one case of lead poisoning from using lead-adulterated heroin. This heavy metal can be added to street drugs intentionally to increase weight. Another source of lead in heroin samples is lead pots used in the illicit manufacturing of street drugs (Zerell et al., 2005).

TABLE 2 Typical Adulterants Used to Cut Street Heroin

Adulterant	Licit Pharmacological Effect	Potential Reason for the Presence of Adulterant	Neurological Effects and Neuropathology	References
Procaine	Local anesthetic	Facilitate smoking Anesthetic property relieves pain of injection	Ischemic nerve injury	Atasoy, Biçer, Açikkol, and Bilgiç (1988), Cole et al. (2011), and Kalichman and Lalonde (1991)
Fentanyl	Potent synthetic opioid analgesic	Produces more powerful opiate effect	Bradykinesia	Wong et al. (2008) and Zesiewicz et al. (2009)
Clenbuterol	β2-Adrenergic agonist	Unknown	Tremor, agitation	Behman et al. (2008) and Brett, Dawson, and Brown (2014)
Diphenhydramine	Antihistamine	Relieves symptoms of allergy	Hallucinatory psychosis, encephalitis	Behman et al. (2008) and Jones, Dougherty, and Cannon (1986)
Acetaminophen	Over-the-counter pain reliever	Disguises poor-quality heroin with its bitter taste	Oxidative stress, neurotoxicity	Ghanizadeh et al. (2012) and Chaudron-Thozet et al. (1992)
Acetylcodeine	Synthetic by-product	Impurity of origin	Convulsion	Soltaninejad, Faryadi, Akhgari, and Bahmanabadi (2007)
6-Monoacetylmorphine	Degradation or synthetic by-product	Impurity of origin	CNS depression	Johnston and King (1998) and Hosztafi (2003)
Caffeine		Causes heroin to vaporize at a lower temperature	Neurodegenerative disease	Chaudron-Thozet et al. (1992) and Luong and Nguyen (2015)
Papaverine	Opium alkaloid	Impurity of origin	CNS depression	Johnston and King (1998)
Noscapine	Opium alkaloid	Impurity of origin	Hallucination	Johnston and King (1998)
Dextromethorphan	Antitussive	Induces "high" effect	CNS depression	Barbera, Busardò, Indorato, and Romano (2013)
Codeine	Opium alkaloid	Impurity of origin	CNS depression	Akhgari, Jokar, Bahmanabadi, and Etemadi-Aleagha (2012)
Morphine	Opium alkaloid	Impurity of origin	Neurodegeneration	Cunha-Oliveira, et al. (2007)
Phenobarbital	Anticonvulsant	Has sedative and hypnotic effects and facilitates heroin smoking	Impaired cognition	Akhgari et al. (2012) and Hong (2011)
Acetylthebaol	Synthetic by-product	Impurity of origin		Akhgari et al. (2012)
Diazepam	Antianxiety	Probably used for its antianxiety and sedative effects	CNS depression	Akhgari et al. (2012)
Chloroquine	Antimalarial and amebicidal drug	With no effect but widespread availability, low price, and color and crystalline structure that can be used to bulk heroin	Coma, convulsion	Messant, Jérémie, Lenfant, and Freysz (2004)
Tramadol	Synthetic opioid drug	Shares many effects with heroin	Seizure, serotonin syndrome	Randall et al. (2014)
Sugars (sucrose, lactose)	Sugars	Used to bulk and dilute heroin		Chaudron-Thozet et al. (1992)
Quinine	Antimalarial drug	Disguises poor-quality heroin with its bitter taste, provides a "rush feeling"	Brown-Séquard syndrome	Cole et al. (2010) and Krause (1983)
Lead	Metal	Increases weight	Peripheral neuropathy	Parras et al. (1987) and Krigman (1978)

Adulterants found in street heroin samples, their accepted medical use, the reason for their use as cutting agent, and their neurological effects are shown in Table 2.

Diluents or Bulking Agents Added to Heroin

Sugars (sucrose, fructose, glucose) are common cutting agents in heroin samples. Other diluents for cutting heroin samples are chalk, brick dust, powdered milk, and starch, which do not have any pharmacological effects and are benign substances but can cause health consequences (Coomber, 1997a, 1997b).

MEDICAL CONSEQUENCES OF STREET HEROIN ABUSE

Chronic consumption of street drugs can lead to altered mental status, damage to various organs in the body, and finally multiple organ failure. Adulterated street heroin contains other drugs in addition to heroin that cannot be detected by laboratory screening tests, but they can cause health problems (Radovanović, Milovanović, Ignjatović-Ristić, & Radovanović, 2012).

Neurological Effects and Neuropathology of Adulterants and Diluents in Street Heroin

A broad spectrum of morphofunctional and neuropathologic changes has been reported in the brain of heroin abusers (Neri et al., 2013). These lesions are caused by the active ingredient, heroin, and other substances added as adulterants (Büttner, Mall, Penning, & Weis, 2000). Postmortem investigations on the brains of drug-addicted individuals show biochemical and ultrastructural abnormalities. Neuropathologic changes in the brain can be precipitated by direct effects of heroin such as respiratory depression or by other reasons (infections and adulterants) (Büttner et al., 2000).

Cerebral edema with increased brain weight, decreased neuronal densities in the globus pallidus, bilateral and symmetric ischemic lesions, and hypodensities in the basal ganglia are caused by hypoxia related to heroin abuse (Andersen & Skullerud, 1999). As indicated by Niehaus and Meyer (1998) heroin abuse has caused focal neurological deficits and stroke. Flaccid paraparesis and paraplegia along with sensory loss in the legs are clinical presentations of heroin addiction-induced myelopathy. Neurotoxic effects of heroin and its adulterants, allergic reaction to "cutting agents," and adulterants, and embolism are some causes of myelopathy in heroin abusers (Büttner et al., 2000). Inhalation of preheated heroin can cause spongiform leukoencephalopathy due to a lipophilic toxin-induced process by contaminants or cerebral hypoxia (Büttner et al., 2000).

Bacterial infections are a common problem among injecting drug users (Cole et al., 2010). Unsterile preparation and distribution processes in combination with poor health conditions and use of contaminated injection equipment contribute to inducing bacterial, fungal, and viral infections (McLauchlin et al., 2002). Septic foci in the brain can be produced as a result of bacterial or fungal endocarditis. Other studies have described intracranial mycotic aneurysms and development of subarachnoid hemorrhage in drug abusers with endocarditis (Gilroy, Andaya, & Thomas, 1973).

Direct toxic effects of heroin together with the action of adulterants have been hypothesized in the production of neurologic complications in drug abusers. Embolism from heroin adulterants has been proposed in some studies (Büttner et al., 2000).

There is a case report concerning Brown–Séquard syndrome, characterized by right-sided hemiparalysis with a contralateral sensory loss of touch and pain and vasculitis in the cervical region following quinine-adulterated heroin ingestion. Toxic effects of heroin and quinine caused vasculitis, cellulitis, and arachnoiditis in this heroin abuser (Krause, 1983).

Some lesions in the peripheral nervous system have been attributed to heavy metal adulterants such as lead in heroin. These include polyradiculopathy, brachial and lumbosacral plexitis, Guillain–Barré syndrome, and mononeuropathy (Antonini, Palmieri, Spagnoli, & Millefiorini, 1989).

Side Effects on Kidney

A broad spectrum of pathologic changes in kidney was reported in previous studies (Buettner et al., 2014; Milroy & Parai, 2011). Atraumatic rhabdomyolysis, acute renal failure, and electrolytic disorders are reported in iv heroin abusers. The reason for this problem could be the drugs and toxins in injected street drugs, immune system disorders, muscle ischemia, and other causes. Signs of glomerular ischemia, interstitial inflammation, and arteriosclerosis were seen in postmortem studies on the kidneys of chronic drug abusers (Buettner et al., 2014; Radovanovis et al., 2012).

Side Effects on Liver

Liver diseases are common in iv drug abusers but usually unrecognized. Hepatitis C virus infection can be acquired by iv drug abusers and subsequently cause liver fibrosis. Liver autopsy samples of iv heroin addicts have shown fat changes, chronic hepatitis, cirrhosis, and Kupffer cell hypertrophy (Ilic, Karadzic, Kostic-Banovic, Stojanovic, & Antovic, 2010).

Side Effects on Lungs

Adulterants in street heroin enter the bloodstream as small particles. These particles can clog blood vessels in the lungs, liver, kidneys, and brain (Wilson et al., 2006). Lung diseases such as abscesses, pneumonia, and tuberculosis are usual among heroin abusers (Hind, 1990). Inhalation of low concentrations of talc dust for a long period of time or acute exposure to high concentrations of talc powder can produce talcosis or talc pneumoconiosis (Davis, 1983).

Arterial obstruction of the lung was reported to be due to the injection of drugs containing talc (Arnett, Battle, Russo, & Roberts, 1976). Talc is a general filler for producing tablets in the pharmaceutical industry. Talc serves as a foreign body and can produce arterial obstruction after injection of water-dissolved tablets intended for oral use. Ischemic necrosis in some parts of the distal extremities is produced by injection of drugs containing talc (Arnett et al., 1976).

Side Effects on Heart

Bacteria and other microorganisms are usual components of street heroin. These microorganisms have various sources; they can come from unsterile needles, be incorporated into heroin samples for dividing or distribution processes, or appear after filtering heroin for injection using unsterile cotton. However, from any source, these microorganisms attack many organs in the body. Endocarditis is one of the complications in iv heroin abusers (Panduranga, Al-Abri, & Al-Lawati, 2013). Particulate matters in street drugs such as talc can produce tricuspid valve damage (Frontera & Gradon, 2000). Infective endocarditis may be the result of contaminated drugs with large bacterial load. Bacteria can be introduced into the body of iv drug abusers through injection of drugs with unsterile or shared syringes (Frontera & Gradon, 2000).

CONCLUDING REMARKS

Heroin is one of the illicit drugs that are banned by international drug control treaties. Illicit use of street heroin is a significant cause of social problems, economic costs, and premature mortality from drug overdose, violence, and infectious diseases, especially in young age groups. There are some interventions for preventing the use of street heroin. Controlling the source countries of street heroin supply; legal prohibition by the criminal justice system and police force of manufacturing, possession, and use of opioids; and finally educational programs are methods for controlling street drug abuse.

APPLICATION TO OTHER ADDICTIONS AND SUBSTANCE MISUSE

Street heroin abuse neuropathology extends to addiction to other drugs and substances as well. There are reports that heroin elevates dopamine (DA) level. Oxidative metabolism of DA leads to the formation of reactive oxygen species (ROS). Carbohydrate, amino acid, phospholipid, and nucleic acid damage cause neuronal cell injury and neurodegeneration. The brain is the organ most sensitive to the oxidative stress induced by heroin (Xu et al., 2006).

The effect of heroin on ROS-induced neurotoxicity is the same as for amphetamine-type stimulants and ethanol.

Amphetamines are members of the phenylethylamine chemical structure family with psychoactive properties. It is known that amphetamines cause structural and functional alterations in the brain. They can easily diffuse through cellular membranes, especially in the brain, and interact with monoamine transporter sites. In animal models amphetamines show major neurotoxic action such as long-term deficits in dopaminergic and serotonergic systems, depletion of monoamine brain level, degeneration of neuronal fibers, and neuronal death. Some of the mechanisms for methamphetamine-induced neurotoxicity are DA release and subsequent autoxidation and enzymatic oxidation of DA. Also aberrant release of DA can induce oxidative stress (Davidson, Gow, Lee, & Ellinwood, 2001).

It is supposed that the first oxidative metabolite of ethanol, acetaldehyde, induces its neurotoxic effects via the formation of adducts with brain macromolecules such as proteins. Also ethanol can be metabolized by cytochrome P450 2E1 in the brain and produce ROS, thus contributing to neurodegeneration (Brocardo, Gil-Mohapel, & Christie, 2011).

The effects of drug abuse on the brain, however, are crucial for CNS toxicity. Several studies support a role for ROS in the neurotoxicity and neurodegeneration induced by many drugs of abuse in various organs, especially brain.

KEY FACTS

Key Facts of Street Heroin

- Heroin was first synthesized as a nonaddictive alternative for morphine.
- Owing to heroin's high addictive property, it is classified as a Schedule I drug under the Controlled Substances Act of 1970.
- Manufacturing, possession, distribution, and selling of heroin are illegal in many countries.
- Heroin can be diluted at each stage of the chain of distribution with pharmaceutical and nonpharmaceutical additives.
- Street heroin is used as a recreational drug on the black market.
- Street heroin is not a pure substance at all and it may contain many adulterants, which can have health consequences to abusers, their families, and the community.
- Heroin and its adulterants exert various neuropathologic changes and neurotoxic effects on the brain.
- Some neurotoxicology and neuropathology features associated with street heroin abuse are gray matter loss, neuronal apoptosis, oxidative cell damage, neurodegeneration, myelopathy, spongiform leukoencephalopathy, cerebral edema, ischemia, and stroke.

Key Points of the Personality Profile of Persons Vulnerable to Developing Addiction

- Clients with *paranoid personality disorder* can be attracted to the dominance drugs (alcohol, cocaine, and amphetamines), because they enhance the need for control that is central to the disorder (Benjamin, 1993). These drugs allow individuals with this personality to feel more powerful in a world that seems dangerous and hostile.
- Clients with *schizoid personality disorder* may be attracted to psychedelic drugs and become addicted to the state of arousal and satisfaction involved in facilitated fantasy (Milkman & Sunderwirth, 1987).
- Drugs such as marijuana and LSD may replicate the digressive, tangential quality of thought patterns already present in individuals with *schizotypal personality disorder*, and mere drug use can be enough to precipitate a psychiatric crisis. Psychoeducation is vital with these clients (Ekleberry, 1996).
- Clients with *antisocial personality disorders,* perhaps due to low neurological arousal, often seek thrills and are likely to be most attracted to stimulants. Their use of alcohol and drugs bothers them only in terms of the pressure they receive from employers, family, or the criminal justice system (Ekleberry, 1996).
- Clients with *borderline personality disorder* are the best candidates of all those with personality disorders for developing addictive disorders; they will use almost any drug of choice to worst advantage.
- Clients with *histrionic personality disorder* may value drugs and alcohol or compulsive behaviors for social enhancement. Antianxiety drugs are often sought; but stimulants provide them with dramatic mood boosts.

- Clients with *avoidant personality disorder* (AvPD) are vulnerable to substance use that can reduce interpersonal vulnerability or ease social paralysis. Drugs that will make a difference include sedative–hypnotics that calm anxiety and stimulants that provide a sense of strength or reduced vulnerability. Mild hallucinogens facilitate escape into fantasy and distract the AvPD client from the pain of his or her own self-absorption (Stone, 1993).
- Clients with *dependent personality disorder* may use alcohol and other substances as a passive way to escape from problems (Beck, 1993).

REFERENCES

A dictionary of slang drug terms, trade names, and pharmacological effects and uses. (1997). Retrieved from the Texas Commission on Alcohol and Drug Abuse website http://www.tcada.state.tx.us/research/slang/terms.pdf.

Akhgari, M., Jokar, F., Bahmanabadi, L., & Etemadi-Aleagha, A. (2012). Street-level heroin seizures in Iran: a survey of components. *Journal of Substance Use, 17*(4), 348–355.

Andersen, S. N., & Skullerud, K. (1999). Hypoxic/ischaemic brain damage, especially pallidal lesions, in heroin addicts. *Forensic Science International, 102*(1), 51–59.

Antonini, G., Palmieri, G., Spagnoli, L. G., & Millefiorini, M. (1989). Lead brachial neuropathy in heroin addiction. A case report. *Clinical Neurology and Neurosurgery, 91*(2), 167–170.

Arnett, E. N., Battle, W. E., Russo, J. V., & Roberts, W. C. (1976). Intravenous injection of talc-containing drugs intended for oral use. A cause of pulmonary granulomatosis and pulmonary hypertension. *American Journal of Medicine, 60*, 711–718.

Atasoy, S., Biçer, F., Açikkol, M., & Bilgiç, Z. (1988). Illicit drug abuse in the Marmara region of Turkey. *Forensic Science International, 38*(1–2), 75–81.

Barbera, N., Busardò, F. P., Indorato, F., & Romano, G. (2013). The pathogenetic role of adulterants in 5 cases of drug addicts with a fatal outcome. *Forensic Science International, 227*(1–3), 74–76.

Beck, A. T. (1993). Cognitive therapy: past, present, and future. *Journal of Consulting and Clinical Psychology, 61*, 194–198.

Behrman, A. D. (2008). Luck of the draw: common adulterants found in illicit drugs. *Journal of Emergency Nursing, 34*(1), 80–82.

Benjamin, L. S. (1993). *Interpersonal diagnosis and treatment of personality disorder* (1st ed.). NY: Guilford Press.

Brett, J., Dawson, A. H., & Brown, J. A. (2014). Clenbuterol toxicity: a NSW poisons information centre experience. *The Medical Journal of Australia, 200*(4), 219–221.

Brett, M. M., Hood, J., Brazier, J. S., Duerden, B. I., & Hahné, S. J. (2005). Soft tissue infections caused by spore-forming bacteria in injecting drug users in the United Kingdom. *Epidemiology & Infection, 133*(4), 575–582. Review.

Brocardo, P. S., Gil-Mohapel, J., & Christie, B. R. (2011). The role of oxidative stress in fetal alcohol spectrum disorders. *Brain Research Reviews, 67*(1–2), 209–225.

Buettner, M., Toennes, S. W., Buettner, S., Bickel, M., Allwinn, R., Geiger, H., … Jung, O. (2014). Nephropathy in illicit drug abusers: a postmortem analysis. *American Journal of Kidney Diseases, 63*(6), 945–953.

Büttner, A., Mall, G., Penning, R., & Weis, S. (2000). The neuropathology of heroin abuse. *Forensic Science International, 113*(1–3), 435–442. Review.

Cannabis, coca & poppy; nature's addictive plants, production & distribution. (2014). Retrieved from the Drug Enforcement Administration Museum & Visitor's Center website http://www.deamuseum.org/ccp/opium/production-distribution.html.

Casa Palmera Staff. (2010). *Nicknames, street names and slang for heroin.* Retrieved from the Casa Palmera, Del Mar California website http://casapalmera.com/nicknames-street-names-and-slang-for-heroin.

Chaudron-Thozet, H., Girard, J., & David, J. J. (1992). Analysis of heroin seized in France. *Bulletin on Narcotics, 44*(1), 29–33.

Ciccarone, D. (2009). Heroin in brown, black and white: structural factors and medical consequences in the US heroin market. *International Journal of Drug Policy, 20*, 277–282.

Clatts, M. C., Giang, L. M., Goldsamt, L. A., & Yi, H. (2007). Novel heroin injection practices: implications for transmission of HIV and other blood borne pathogens. *American Journal of Preventive Medicine, 32*(6 Suppl.), S226–S233.

Cole, C., Jones, L., McVeigh, J., Kicman, A., Syed, Q., & Bellis, M. A. (2010). *A guide to adulterants, bulking agents and other contaminants found in illicit drugs (Review).* Retrieved from the Liverpool: Centre for public health, Faculty of Health and Applied Social Sciences, Liverpool John Moores University website www.cph.org.uk/showPublication.aspx?pubid=632.

Contet, C., Kieffer, B. L., & Befort, K. (2004). Mu opioid receptor: a gateway to drug addiction. *Current Opinion in Neurobiology, 14*(3), 370–378. Review.

Controlled Substance Schedules. (2012). *Drug enforcement administration (DEA) diversion control.* Retrieved from the U. S. Department of justice website http://www.aids2012.org/WebContent/File/Controlled_Substance_Schedules_DEA.PDF.

Coomber, R. (1997a). Dangerous drug adulteration – an international survey of drug dealers using the Internet and the World Wide Web (WWW). *International Journal of Drug Policy, 8*(2), 71–78.

Coomber, R. (1997b). How often does the adulteration/dilution of heroin actually occur: an analysis of 228 "Street" samples across the UK (1995–1996) and discussion of monitoring policy. *International Journal of Drug Policy, 8*(4), 178–186.

Cunha-Oliveira, T., Rego, A. C., Garrido, J., Borges, F., Macedo, T., & Oliveira, C. R. (2007). Street heroin induces mitochondrial dysfunction and apoptosis in rat cortical neurons. *Journal of Neurochemistry, 101*(2), 543–554.

Dancer, S. J., McNair, D., Finn, P., & Kolsto, A. B. (2002). Bacillus cereus cellulitis from contaminated heroin. *Journal of Medical Microbiology, 51*(3), 278–281.

Davidson, C., Gow, A. J., Lee, T. H., & Ellinwood, E. H. (2001). Methamphetamine neurotoxicity: necrotic and apoptotic mechanisms and relevance to human abuse and treatment. *Brain Research Reviews, 36*(1), 1–22.

Davis, L. L. (1983). Pulmonary 'mainline' granulomatosis: talcosis secondary to intravenous heroin abuse with characteristic x-ray findings of asbestosis. *Journal of the National Medical Association, 75*(12), 1225–1228.

Drugs of abuse, Heroin. (2014). Retrieved from the National Institute on Drug Abuse (NIH) website http://www.drugabuse.gov/drugs-abuse/heroin.

Ekleberry, S. C. (1996). Dual diagnosis: addiction and axis II personality disorders. *The Counselor*, March/April, 7–13.

Frontera, J. A., & Gradon, J. D. (2000). Right-side endocarditis in injection drug users: review of proposed mechanisms of pathogenesis. *Clinical Infectious Diseases, 30*(2), 374–379.

Ghanizadeh, A. (2012). Acetaminophen may mediate oxidative stress and neurotoxicity in autism. *Medical Hypotheses*, 78(2), 351.

Gilroy, J., Andaya, L., & Thomas, V. J. (1973). Intracranial mycotic aneurysms and subacute bacterial endocarditis in heroin addiction. *Neurology* (11), 1193–1198.

Gomez, J., & Rodriguez, A. (1989). An evaluation of the results of a drug sample analysis. *Bulletin on Narcotics*, 41(1 2), 121–126.

Heroin abuse and suppositories. (2014). Retrieved from the Addiction Search, The latest addiction information website http://www.addictionsearch.com/treatment_articles/article/heroin-abuse-and-suppositories_173.html.

Heroin. (2014a). Retrieved from the Center for Substance Abuse Research (CESAR) website http://www.cesar.umd.edu/cesar/drugs/heroin.asp.

Heroin. (2014b). Retrieved from Wikipedia, the free encyclopedia website http://en.wikipedia.org/wiki/Heroin.

Hind, C. R. K. (1990). Pulmonary complications of intravenous drug misuse. 1. Epidemiology and non-infective complications. *Thorax*, 45, 891–898.

Hong, Z. (2011). The impact on cognition by phenobarbital in epilepsy treatment. *Neurology Asia*, 16(Suppl. 1), 65–66.

Hosztafi, S. (2003). Heroin, part III: the pharmacology of heroin. *Acta Pharmaceutica Hungarica*, 73(3), 197–205.

Hosztafi, S. (2011). Heroin addiction. *Acta Pharmaceutica Hungarica*, 81(4), 173–183.

Ilic, G., Karadzic, R., Kostic-Banovic, L., Stojanovic, J., & Antovic, A. (2010). Ultrastructural changes in the liver of intravenous heroin addicts. *Bosnian Journal of Basic Medical Sciences*, 10(1), 38–43.

Janhunen, K., & Cole, M. D. (1999). Development of a predictive model for batch membership of street samples of heroin. *Forensic Science International*, 102(1), 1–11.

Johnston, A., & King, L. A. (1998). Heroin profiling: Predicting the country of origin of seized heroin. *Forensic Science International*, 95(1), 47–55.

Jones, J., Dougherty, J., & Cannon, L. (1986). Diphenhydramine-induced toxic psychosis. *American Journal of Emergency Medicine*, 4(4), 369–371.

Kalichman, M. W., & Lalonde, A. W. (1991). Experimental nerve ischemia and injury produced by cocaine and procaine. *Brain Research*, 565(1), 34–41.

Kalka-Moll, W. M., Aurbach, U., Schaumann, R., Schwarz, R., & Seifert, H. (2007). Wound botulism in injection drug users. *Emerging Infectious Diseases*, 13(6), 942–943.

Katzung, B. G. (2001). Opioid analgesics and antagonists. In *Basic and clinical pharmacology* (8th ed.) (pp. 512–531). USA: The McGraw Hill Companies Inc.

Kazemifar, A. M., Solhi, H., & Badakhshan, D. (2011). Crack in Iran: Is it really cocaine? *Journal of Addiction Research & Therapy*, 2, 107.

Klous, M. G., Van den Brink, W., Van Ree, J. M., & Beijnen, J. H. (2005). Development of pharmaceutical heroin preparations for medical co-prescription to opioid dependent patients. *Drug and Alcohol Dependence*, 80(3), 283–295. Review.

Krause, G. S. (1983). Brown-Sequard syndrome following heroin injection. *Annals of Emergency Medicine* (9), 581–583.

Krigman, M. R. (1978). Neuropathology of heavy metal intoxication. *Environmental Health Perspectives*, 26, 117–120. Review.

Luong, K. V., & Nguyen, L. T. (2015). The role of caffeine in neurodegenerative diseases: possible genetic and cellular signaling mechanisms. In R. R. Watson, & V. R. Preedy (Eds.), *Bioactive nutraceuticals and dietary supplements in neurological and brain disease* (pp. 261–279). Academic Press.

Lurie, I. S., Driscoll, S. E., Cathapermal, S. S., & Panicker, S. (2013). Determination of heroin and basic impurities for drug profiling by ultra-high-pressure liquid chromatography. *Forensic Science International*, 231(1–3), 300–305.

McLauchlin, J., Mithani, V., Bolton, F. J., Nichols, G. L., Bellis, M. A., Syed, Q., ... Ashton, J. R. (2002). An investigation into the microflora of heroin. *Journal of Medical Microbiology*, 51, 1001–1008.

Messant, I., Jérémie, N., Lenfant, F., & Freysz, M. (2004). Massive chloroquine intoxication: importance of early treatment and pre-hospital treatment. *Resuscitation*, 60(3), 343–346.

Milkman, H. B., & Sunderwirth, S. G. (1987). *Craving for ecstasy: the consciousness and chemistry of escape*. Lexington, Mass: Lexington Books.

Milroy, C. M., & Parai, J. L. (2011). The histopathology of drugs of abuse. *Histopathology*, 59, 579–593.

Neri, M., Panata, L., Bacci, M., Fiore, C., Riezzo, I., Turillazzi, E., & Fineschi, V. (2013). Cytokines, chaperones and neuroinflammatory responses in heroin-related death: what can we learn from different patterns of cellular expression? *International Journal of Molecular Sciences*, 14(10), 19831–19845.

Niehaus, L., & Meyer, B. U. (1998). Bilateral borderzone brain infarctions in association with heroin abuse. *Journal of the Neurological sciences*, 160(2), 180–182.

Njau, S. N., Raikos, N., Spagou, K., Tzikas, A., & Tsoukali, H. (2010). Heroin body Packer's death in Greece. *The Open Forensic Science Journal*, 3, 53–56.

O'Neal, C. L., Poklis, A., & Lichtman, A. H. (2001). Acetylcodeine, an impurity of illicitly manufactured heroin, elicits convulsions, antinociception, and locomotor stimulation in mice. *Drug and Alcohol Dependence*, 65, 37–43.

Panduranga, P., Al-Abri, S., & Al-Lawati, J. (2013). Intravenous drug abuse and tricuspid valve endocarditis: growing trends in the Middle East Gulf region. *World Journal of Cardiology*, 5(11), 397–403.

Parras, F., Patier, J. L., & Ezpeleta, C. (1987). Lead-contaminated heroin as a source of inorganic-lead intoxication. *The New England Journal of Medicine*, 316(12), 755.

Peat, S. J., Hanna, M. H., Woodham, M., Knibb, A. A., & Ponte, J. (1991). Morphine-6-glucuronide: effects on ventilation in normal volunteers. *Pain*, 45(1), 101–104.

Qiao, Y., Han, K., & Zhan, C. G. (2013). Fundamental reaction pathway and free energy profile for butyrylcholinesterase-catalyzed hydrolysis of heroin. *Biochemistry*, 52(37), 6467–6479.

Radovanović, M. R., Milovanović, D. R., Ignjatović-Ristić, D., & Radovanović, M. S. (2012). Heroin addict with gangrene of the extremities, rhabdomyolysis and severe hyperkalemia. *Vojnosanitetski Pregled*, 69(10), 908–912.

Randall, C., & Crane, J. (2014). Tramadol deaths in Northern Ireland: a review of cases from 1996 to 2012. Review Article. *Journal of Forensic and Legal Medicine*, 23, 32–36.

Selley, D. E., Cao, C. C., Sexton, T., Schwegel, J. A., Martin, T. J., & Childers, S. R. (2001). Mu Opioid receptor-mediated G-protein activation by heroin metabolites: evidence for greater efficacy of 6-monoacetylmorphine compared with morphine. *Biochemical Pharmacology*, 62(4), 447–455.

Senbanjo, R., Tipping, T., Hunt, N., & Strang, J. (2012). Injecting drug use via femoral vein puncture: preliminary findings of a point-of-care ultrasound service for opioid-dependent groin injectors in treatment. *Harm Reduction Journal*, 20(9), 6.

Shannon, K., Rusch, M., Morgan, R., Oleson, M., Kerr, T., & Tyndall, M. W. (2008). HIV and HCV prevalence and gender-specific risk profiles of crack cocaine smokers and dual users of injection drugs. *Substance Use & Misuse, 43*(3–4), 521–534.

Soltaninejad, K., Faryadi, M., Akhgari, M., & Bahmanabadi, L. (2007). Chemical profile of counterfeited buprenorphine vials seized in Tehran, Iran. *Forensic Science International, 172*(2–3), e4–e5.

Stolberg, V. (2011). Synthetic narcotics. In M. Kleiman, & J. Hawdon (Eds.), *Encyclopedia of drug policy* (pp. 756–761). Thousand Oaks, CA: SAGE Publications, Inc. http://dx.doi.org/10.4135/9781412976961.n333.

Stone, M. H. (1993). Long-term outcome in personality disorders. *British Journal of Psychiatry, 162*, 299–313.

Trescot, A. M., Datta, S., Lee, M., & Hansen, H. (2008). Opioid pharmacology. *Pain Physician: Opioid Special Issue, 11*, S133–S153. ISSN:1533–3159.

What is heroin cut with? (2014). Retrieved from the Rehabs.com, the nation's best rehabs website http://luxury.rehabs.com/heroin-addiction/what-is-it-cut-with.

Wilson, L. E., Torbenson, M., Astemborski, J., Faruki, H., Spoler, C., Rai, R., ... Thomas, D. L. (2006). Progression of liver fibrosis among injection drug users with chronic hepatitis C. *Hepatology, 43*(4), 788–795.

Wong, S. C., Curtis, J. A., & Wingert, W. E. (2008). Concurrent detection of heroin, fentanyl, and xylazine in seven drug related deaths reported from the Philadelphia medical examiner's office. *Journal of Forensic Sciences, 53*(2), 495–498.

Wright, C. R. A. Retrieved from Wikipedia, the free encyclopedia website http://en.wikipedia.org/wiki/Charles_Romley_Alder_Wright.

Xu, B., Wang, Z., Li, G., Li, B., Lin, H., Zheng, R., & Zheng, Q. (2006). Heroin-administered mice involved in oxidative stress and exogenous antioxidant-alleviated withdrawal syndrome. *Basic & Clinical Pharmacology & Toxicology, 99*(2), 153–161.

Zerell, U., Ahrens, B., & Gerz, P. (2005). Documentation of a heroin manufacturing process in Afghanistan. *Bulletin on Narcotics, LVII*(1–2).

Zesiewicz, T. A., Hauser, R. A., Freeman, A., Sullivan, K. L., Miller, A. M., & Halim, T. (2009). Fentanyl-induced bradykinesia and rigidity after deep brain stimulation in a patient with Parkinson disease. *Clinical Neuropharmacology, 32*(1), 48–50.

Zovko, A., & Criscuolo, C. L. (2009). *The pharmacological effects of diacetylmorphine (heroin) after diffusion through the blood–brain barrier.* Retrieved from the website: http://www.ebookbrowse.com/the-pharmacological-effects-of-diacetylmorphine.

Chapter 82

Opioid Prescription Drug Abuse and Its Relation to Heroin Trends

Dessa Bergen-Cico[1], Susan Scholl[1], Nato Ivanashvili[2], Rachael Cico[3]
[1]Department of Public Health, Syracuse University, Syracuse, NY, USA; [2]Department of Public Health, Syracuse University, Tbilisi, Republic of Georgia; [3]Columbia University, New York, NY, USA

INTRODUCTION

This chapter focuses on the role that prescription opioids may play as a predictor of the development of opioid and opiate addiction in some countries. In the United States prescription opioid abuse has escalated in recent years, and the abuse of these drugs serves as a gateway to heroin use and addiction. In the United States, Estonia, and Finland synthetic and prescription opioids are increasingly being replaced by heroin use. This phenomenon is driven in part by increased availability, and a drop in price, of heroin. For people who are dependent on synthetic prescription opioids, it is often much less expensive to maintain their addiction with heroin rather than synthetic opioids (UNODC, 2014).

Opiates are powerful drugs that are naturally derived from the dried sap of the opium poppy, whereas synthetic and semisynthetic opioids are manufactured in chemical laboratories that yield compounds with similar chemical structures. Semisynthetic opiates include morphine, heroin, and codeine; examples of synthetically manufactured opioids include fentanyl, Vicodin, Subutex, Oxycontin, methadone, pethidine, desomorphine, and many others. Both synthetic opioids and plant opiates operate similarly in the body and attach to the same specific receptor sites in the brain, spinal cord, and gastrointestinal tract. Opiates and opioids block the transmission of pain messages and operate as central nervous system (CNS) depressants, thereby decreasing breathing/respiration and heart rate. At high levels, opiates and opioids reduce consciousness and may result in overdose and/or death due to the slowing or cessation of breathing and heart rate. Drugs in this category can also induce a sense of euphoria as well as sensations of warmth, drowsiness, and contentment. These secondary effects relieve stress and physical discomfort by creating a relaxed detachment from pain and desires. The use of these drugs has increased for three primary reasons: (1) availability, (2) legality, and (3) perceived safety. In certain countries, they are currently widely available. Until recently, possession of these medications, unlike possession of heroin, was safe from legal repercussions. Moreover these prescription opioids are erroneously perceived as being safe because of their status as pharmaceutical medications.

TRENDS

Based on worldwide data collected and reported by the United Nations Office on Drugs and Crime (UNODC), there were between 12 and 29 million opiate users worldwide in 2009. Since 2009, use has steadily increased when prescription opioid abuse is included alongside heroin and opium in the tracking of these trends (UNODC, 2012, 2014). Global analysis of past-year use of opioids, including heroin and prescription painkillers, is estimated at between 28 and 38 million people (UNODC, 2014), whereas heroin and opium use, excluding prescription opiates, is estimated at 16.5 million people, 0.4% of the world's population ages 15–64 (UNODC, 2014). This global analysis does not present as complete a picture as when data are analyzed regionally, with higher prevalence rates found in regions of Asia (southwest and central Asia), Australia, eastern Europe, Russia, and North America (UNODC, 2012). In Europe, heroin is the main opiate of abuse, with limited nonmedical prescription opioid abuse, whereas most recreational prescription opioid users are in North America and Australia (UNODC, 2014).

HEROIN AVAILABILITY ON THE WORLD MARKET

The exponential growth in opium cultivation in Afghanistan following the start of Operation Enduring Freedom has resulted in devastating rates of addiction in Afghanistan and neighboring Iran and Pakistan, as well as in its target markets in Russia and Europe. The rampant addiction within Afghanistan and Russia began in the 1980s during the Afghan Soviet War. Today Afghanistan produces 90% of the world's illicit heroin, which has flooded Afghanistan, Iran, Pakistan, Russia, and Europe, driving down the price of heroin on the world market while simultaneously increasing rates of addiction. There have also been major heroin seizures in Russia and central Asia, which has impeded heroin availability and is believed to have stimulated demand for opioids such as desomorphine and Subutex as replacements for heroin.

GLOBAL TRENDS IN OPIOID AND OPIATE ABUSE

In the late 1990s Purdue Pharma began marketing Oxycontin (time-released oral oxycodone) as a nonaddictive opioid for pain management. Since then, the marketing and prescribing of opioid-based medications has skyrocketed in North America. Concomitant with the proliferation of opioids in the medical market, opioid-related deaths have grown more than fivefold in North America (Dhalla et al., 2009).

The maps presented here illustrate the global prevalence by country for: (1) prescription opioid abuse (Figure 1), (2) all opioid abuse (all semisynthetic and synthetic opiates including prescription drugs) (Figure 2), and (3) opiate abuse (Figure 3). The size of the circle in each country represents the percentage of the population using the corresponding category of drugs. In Figure 1 the sizes of the circles range from a low of 0.01% to 5.7% for prescription opioids. In Figure 2 they range from a low of 0.2% to 6.1% of the population for all opioid abuse. Figure 3 illustrates opiate abuse; the smallest size of the circles represents the low end of the range at 0.01% with increasing prevalence by size of circle up to the higher rate of 2% of the population.

The source of the data in Table 1 is the UNODC 2014 World Drug Report using the UNODC Tableau database interface and the table represent all countries for which data are available. The countries included in Table 1 encompass only those in which 1% or greater of their population abuses prescription opioids. The only exception to this 1% threshold is the inclusion of Northern Ireland, England, and Wales, where the prevalence rates are lower. These countries are included in the table because they are part of the United Kingdom, whereas Scotland, also part of the United Kingdom, has an opioid abuse rate of 1.7% of the population. Unfortunately prevalence rates of heroin and opiate use for the United Kingdom cannot be reported because the data are unavailable. The far right column of Table 1 presents information on the percentage of population in these countries using heroin and/or opium. In each of these countries where the prevalence of prescription opioid abuse is greater than 1% of the population, the rate of opioid abuse is equal to or exceeds the rate of opium and heroin use, with the exception of Afghanistan. The difference between the percentage of the Afghan population abusing opioids (2.9%) and those using opiates (3%) is negligible, which is remarkable considering the fact that Afghanistan produces 90% of the world's illicit heroin (UNODC, 2014). In the narrative following Tables 1 and 2 we highlight three countries that have particularly high rates of opioid prescribing (Australia, Canada, United States) and the Republic of Georgia, which is recovering from an opioid epidemic stemming from abuse and diversion of Subutex.

OPIATE AND OPIOID OVERDOSE DEATHS

In 2011 there were 8457 overdose deaths in Europe and opioids were involved in the majority of these deaths (EMCDDA, 2013). In 2012 there were 13,504 opiate and opioid overdose deaths in the United States; of these the vast majority (73%, $n=9869$) were due

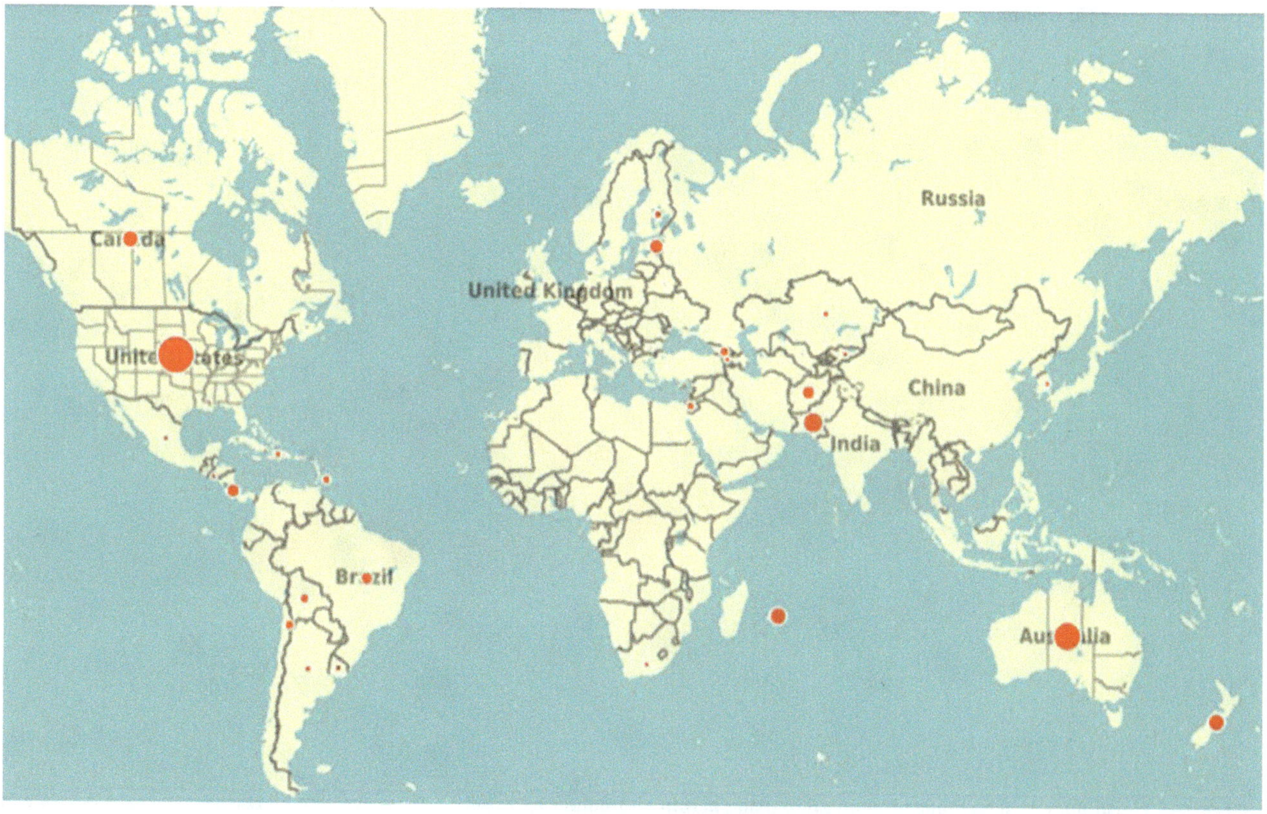

FIGURE 1 **World map of percentage of population abusing prescription opioids by country.** The global prevalence of prescription opioid drug abuse is illustrated by country. The size of the circle in each country represents the percentage of the population abusing prescription opioids. These range from the smallest circle and lowest rate of 0.01% to the largest circle representing 5.7% of the population. *Data Source: UNODC 2014 World Drug Report via Tableau.*

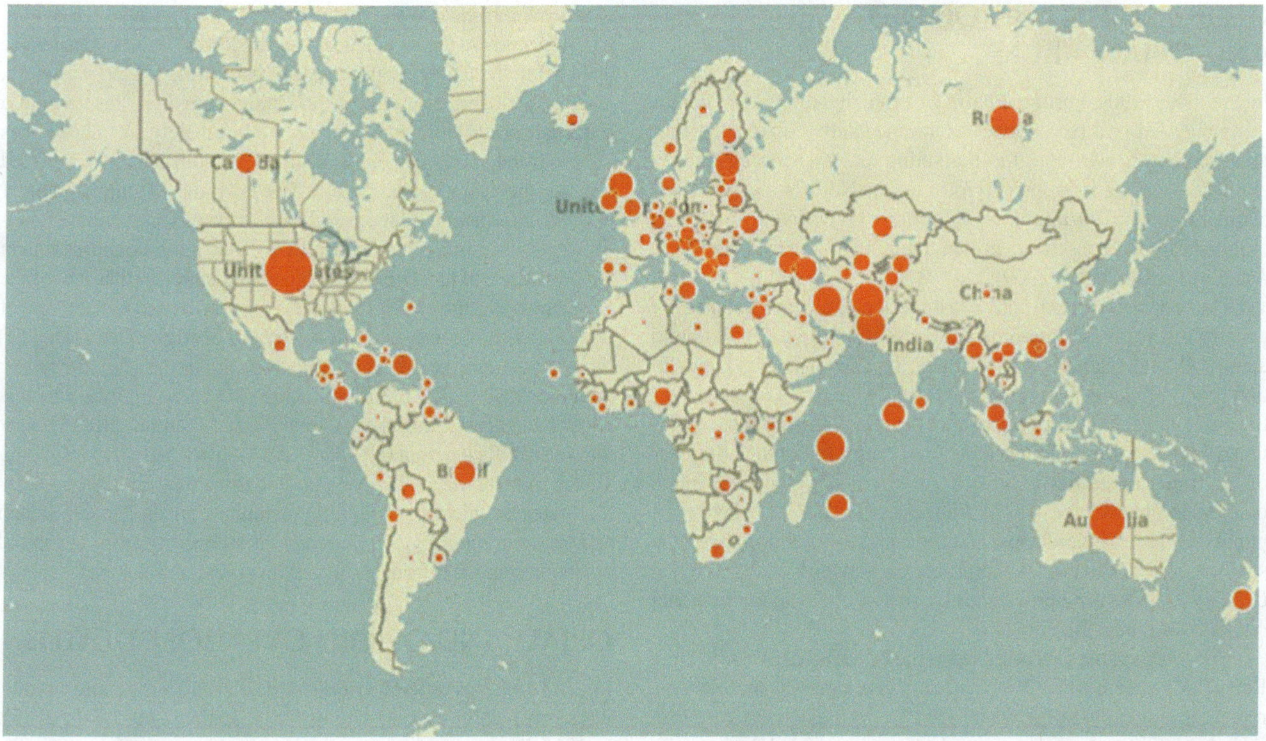

FIGURE 2 **World map of percentage of population abusing opioids by country.** The global prevalence of opioids (synthetic opiates and prescription drugs) is illustrated by country. The size of the circle in each country represents the percentage of the population using opioids, ranging from the smallest circle and lowest rate of 0.2% of the population to the largest circle representing 6.1% of the population. *Data Source: UNODC 2014 World Drug Report via Tableau.*

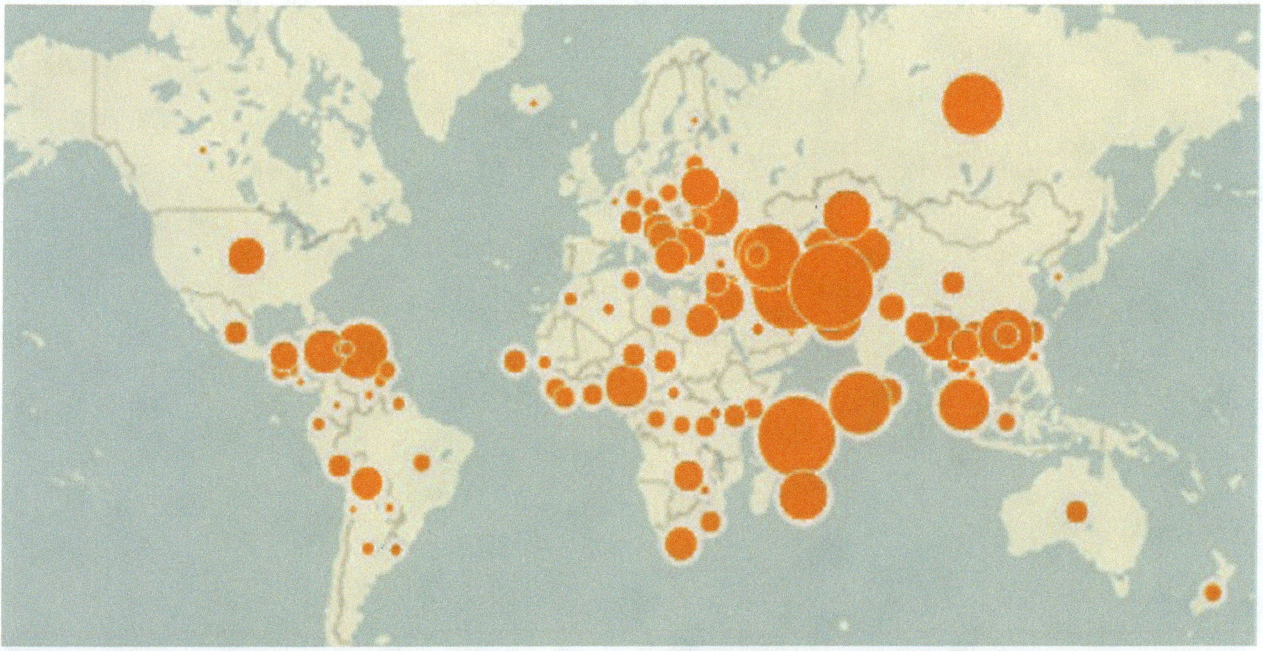

FIGURE 3 **World map of percentage of population using heroin and opium by country.** The prevalence by country of opiate (heroin and opium) use is illustrated. The size of the circle in each country represents the percentage of the population using opiates, ranging from the smallest circle and lowest rate of 0.01% of the population to the largest circle representing 2% of the population. *Data Source: UNODC 2014 World Drug Report via Tableau.*

TABLE 1 Percentage of Population Abusing Opioids and Heroin by Country

Country	Percentage of Population Abusing Opioids	Percentage of Population Using Opium and Heroin
Afghanistan	2.9	**3.0**
Australia	2.4	0.2
Azerbaijan	1.5	1.5
Brazil	1.45	0.1
Canada	1.0	N.A.
Georgia	1.36	N.A.
Islamic Republic of Iran	2.3	2.3
Jamaica	1	0.7
Kazakhstan	1	0.9
Macao	1.2	1.1
Maldives	1.46	1.3
New Zealand	1.1	0.1
Pakistan	2.4	1
Russian Federation	2.3	1.4
United Kingdom (England and Wales)	0.76	N.A.
United Kingdom (Northern Ireland)	0.13	N.A.
United Kingdom (Scotland)	1.7	N.A.
United States	**6.1**	0.6

Countries with the highest rate in each column are indicated by bold text.
Data Source: UNODC World Drug Report 2014.

TABLE 2 Number of Opiate and Opioid Overdose Deaths by Country

Country	Number Deaths
Austria	192
Belgium	118
Croatia	54
Cyprus	6
Czech Republic	6
Estonia	118
Finland	172
France	338
Hungary	10
Ireland	158
Italy	282
Latvia	9
Lithuania	42
Luxembourg	5
Malta	3
Norway	217
Portugal	17
Romania	11
Slovakia	12
Slovenia	10
Sweden	216
Turkey	86
United Kingdom	2250
United States[a]	13,504

[a]SAMHSA (2013).
Data Source: EMCDDA unless noted.

to prescription opioid overdose. Table 2 shows the numbers of opiate and opioid deaths by country for the most recent year for which data are available for each country. The total number of deaths in Europe was 4333, whereas the total number of deaths in the United States was more than three times higher, at 13,504.

UNITED STATES

There are an estimated 5.3 million opiate abusers in the United States, the majority of whom abuse prescription opioids, notably codeine cough syrup, prescription pain relievers (e.g., fentanyl, hydrocodone (Vicodin), oxycodone (Oxycontin, Percocet), and replacement therapy opiates (e.g., methadone, Subutex, buprenorphine hydrochloride). In 2012, these semisynthetic and synthetic prescription opiates comprised the second largest category of recreational drug initiates among Americans age 12 and older (estimated 1.8 million), second only to the 2.4 million cannabis initiates that year. In contrast, there were 156,000 new heroin users the same year (SAMHSA, 2013).

Opioid overdose is a leading cause of injury-related mortality in the United States. In just 2 years, from 2010 to 2012, the death rate from heroin overdose doubled in the United States, increasing from 1.0 per 100,000 ($n=1779$) in 2010 to 2.1 per 100,000 ($n=3635$) deaths in 2012. Heroin now accounts for approximately 16% of all overdose deaths in the United States (Yokell et al., 2014). Concomitantly, the number of prescription opiate overdose deaths declined 6.6% from 2010 (10,427 deaths) to 2012 (9869 deaths), from a rate of 6.0 to 5.6 per 100,000, respectively (Rudd et al., 2014). Although there has been an increase in heroin overdose deaths and a decline in prescription opioid overdose deaths, prescription opioid overdose represents the majority of opiate/opioid overdose deaths in the United States (Rudd et al., 2014). Figure 4 illustrates the trends in heroin and opioid overdose deaths

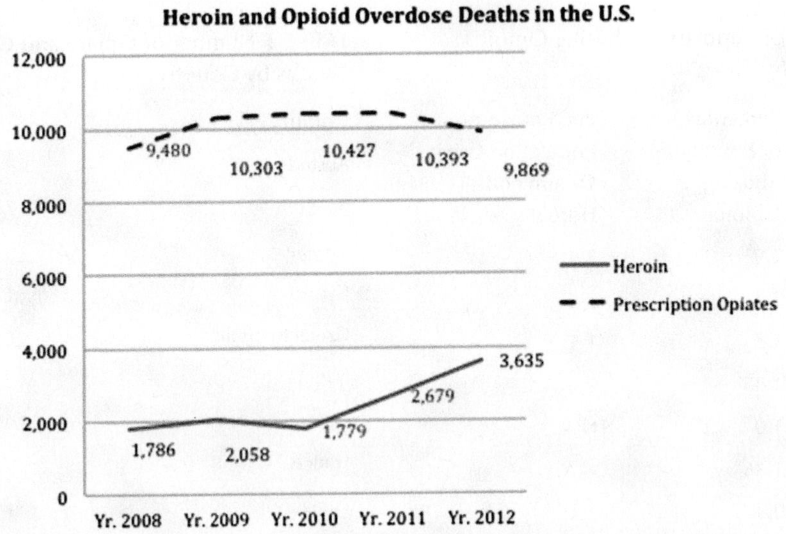

FIGURE 4 Number of overdose deaths by year for heroin and prescription opioids. *Data Source: US Substance Abuse and Mental Health Data Archive.*

in the United States from 2008 to 2012. More than two-thirds of emergency department visits for opioid overdoses involve prescription drugs; 84% of prescription opioid overdoses occur in urban areas, with 40% of all overdoses occurring in the southern part of the country. It is also notable that women account for the majority (53%) of prescription opioid overdoses (Yokell et al., 2014).

TRENDS IN TREATMENT FOR OPIATES AND OPIOIDS

The numbers of Americans age 12 and older entering treatment for opiate and opioid addiction increased nearly 275%, from 360,000 in 2002 to 973,000 in 2012 (SAMHSA, 2013). See Figure 5 for an illustration of the trend of heroin and opioid addiction admissions for treatment in the United States from 2002 to 2012.

Treatment for Prescription Opioid Abuse among Military and Veterans

As a subset of the American population, the US Army has noted a troubling trend in the illicit use of prescribed drugs stemming from the proliferation of prescription pain-management drugs, namely synthetic opiates. As such, the US Army has become increasingly vigilant about the use of opioids, particularly in light of the increased prescribing of synthetic opiates like oxycodone (Oxycontin) (Bergen-Cico, 2012). To assess the extent of possible abuse and addiction, the Army measures the level of authorized (legal as prescribed) and unauthorized (illegal abuse) use of prescription opioids. Alongside heroin, prescription opioids are now included in the Army's routine drug screening for active service members. Heroin use among veterans seeking treatment had been on the rise from 2000 to 2006, but then began to decline, whereas the numbers of veterans entering treatment for prescription opioid abuse have steadily increased. Following alcohol, the most commonly used substances of abuse in the military for recreational purposes are prescription drugs (U.S. Army, 2010). Figure 6 demonstrates the precipitous increase in military veterans being admitted to treatment for synthetic opiates (pharmaceuticals such as Vicodin, Oxycontin, Dilaudid) and the declining trend in veterans seeking treatment for heroin, morphine, and opium.

Republic of Georgia: Subutex and Heroin Drug Use Trends in Georgia from 2000 to 2014

Opiate and opioid use often increases in response to war and sociopolitical upheaval (Bergen-Cico, 2012). Following the dissolution of the Soviet Union in 1990, there was a breakdown in the antidrug and prohibition efforts of the totalitarian state, coupled with opening of the borders and increased drug trafficking in the Republic of Georgia. Social values were in flux and there were concomitant sociopolitical and economic crises, which contributed to rapid increases in drug use (Todadze & Lezhava, 2008). According to the International Narcotics Control Board (INCB) there was an 80% increase in the number of drug abusers in the Republic of Georgia between 2003 and 2005. The INCB attributed much of this increase to the import and illegal sale of Subutex (Parfitt, 2006). Subutex (buprenorphine) is a prescription medication approved for the treatment of opiate dependence. It contains the active ingredient buprenorphine hydrochloride, which binds to opiate receptor sites in the brain and body thereby reducing symptoms of opiate withdrawal and dependence. However, Subutex can also be used recreationally and abused.

The abuse of Subutex has been most notable in major urban areas like Tbilisi, (Kirtadze, Otiashvili, O'Grady, & Jones, 2012; Otiashvili et al., 2014). Journalist Gaeme Wood described Tbilisi in 2006 this way "…it had all the wrecked majesty of an ex-beauty queen with 6 years of track-marks down her arms. It was a great European capital in decay: crumbling bridges, refugees from war, and—most of all—cast-off syringes everywhere" (Wood, 2013). Georgia is in relatively close proximity to Afghanistan and it has been flooded with heroin in recent decades, as it has served as a transit point for drugs between Asia and Europe. The increased availability and use of heroin in Georgia, and the region, resulted in escalated rates of addiction; and people in Georgia began using Subutex for self-treatment of heroin addiction. However, Subutex has not alleviated the heroin problem, instead it became a new

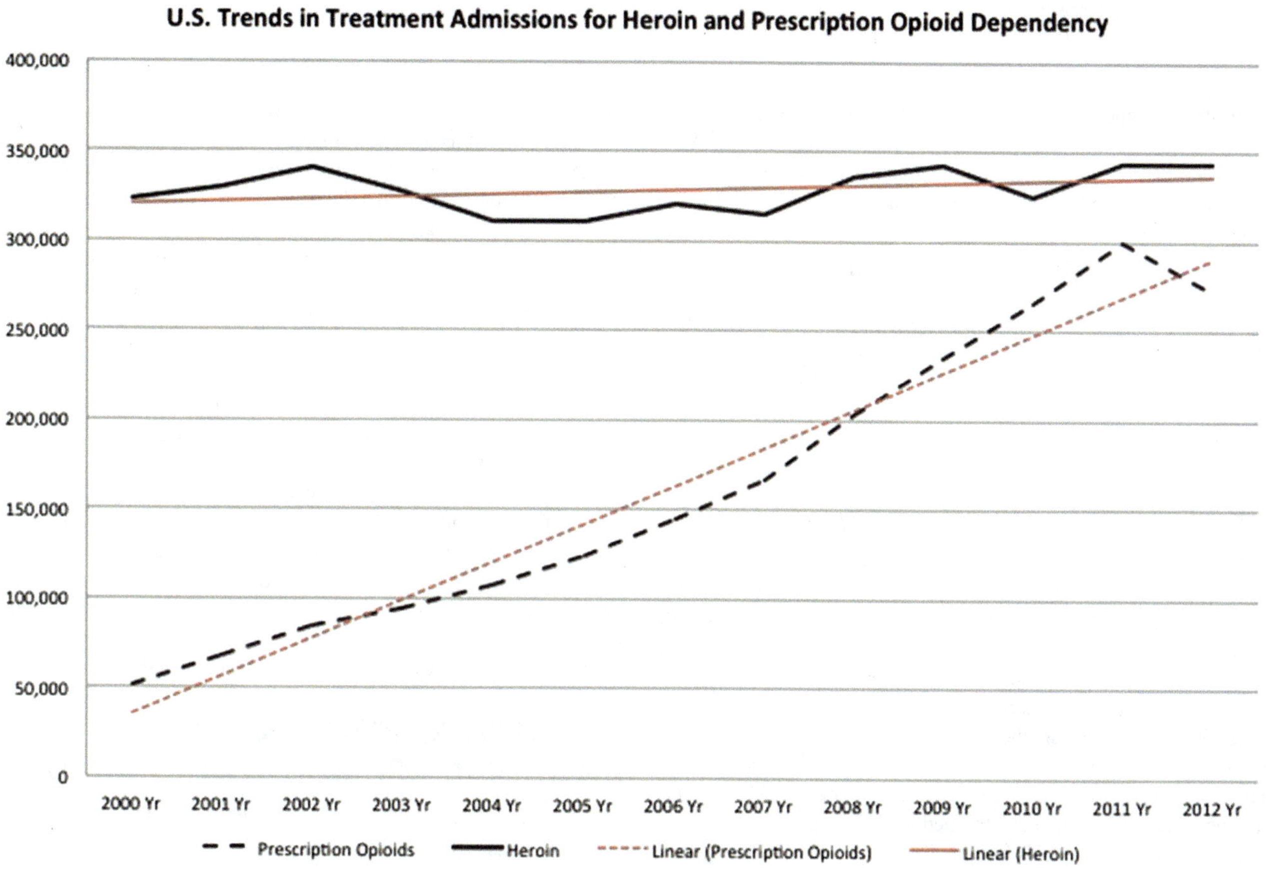

FIGURE 5 Trends in admissions for heroin and opioid treatment by year in the United States. *Data Source: US Substance Abuse and Mental Health Data Archive.*

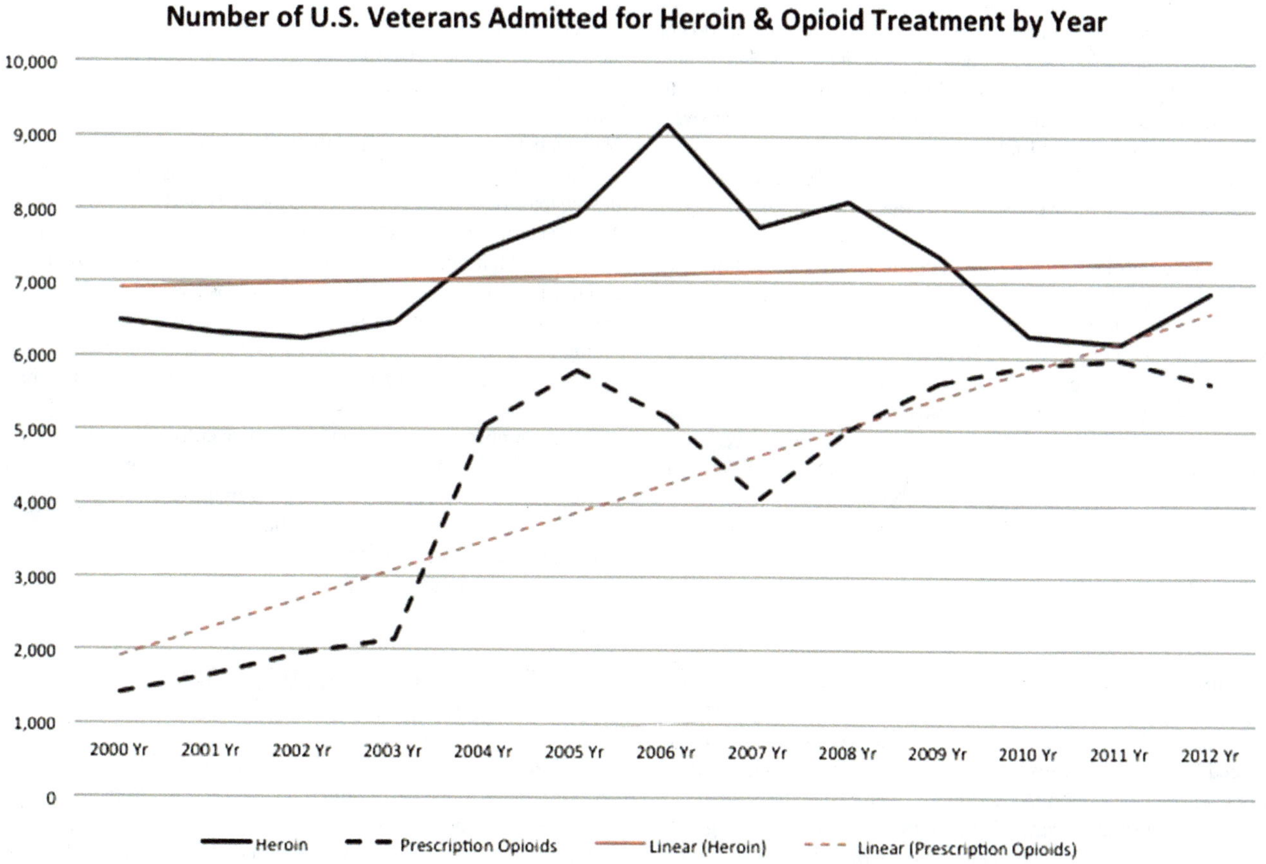

FIGURE 6 Trends in treatment among the US military veterans for heroin and opioid abuse by year. *Data Source: US Substance Abuse and Mental Health Data Archive.*

drug of abuse and it was reported that Georgians had been using heroin to get off Subutex (Rimple, 2006; Vaisman, 2006). As a result self-medication and self-management of addiction emerged out of necessity in Georgia owing to insufficient treatment for the estimated 275,000 opiate addicts in a country of just 4.6 million (Vaisman, 2006). Around 2010, crocodile, or krokodil, surfaced in Georgia and eastern Europe; krokodil is a desomorphine-based homemade injectable opioid (Sikharulidze, Kapanadze, Otiashvili, Poole, & Woody, 2014). The rates of morbidity and mortality for krokodil are high and its name is due to the discolored, ulcerated, and scale-like skin that users develop (Grund, Latypov, & Harris, 2013).

Canada

North America has the highest rate of opioid use in the world; this encompasses the United States and Canada (population approximately 35 million). From 2006 to 2011, the prescribing of opioids in Canada rose 23% (Gomes, Mamdani, Paterson, Dhalla, & Juurlink, 2014), with 17% of Canadians using prescription opioids and an estimated 1% rate of abuse among the total population (Canada Health, 2012, 2013). Of interest, the per capita rate of high-dose (doses exceeding 200 mg oral morphine equivalent—MEQ) opioid dispensing increased steadily in Canada between 2006 and 2008 before stabilizing around 2010 after the introduction of Canadian and American guidelines regarding MEQ prescription recommendations (Chou et al., 2009). Issues having an impact on the prescribing and use of these powerful drugs in Canada include aging, chronic pain, marketing, and media coverage. The oxycodone prescribing volume varies greatly from one province to another. Fentanyl dispensing appears as a lower volume choice. In certain provinces where restrictions do not exist on the prescribing of high-dose opioids, there are higher rates of these drugs being dispensed (Gomes et al., 2014). This seems to suggest that policies limiting high-dose prescribing may reduce overprescribing of opioids and related abuse. Interestingly, however, in Saskatchewan, high-dose opioid provision is not restricted for purposes of prescription coverage, and yet this province has the lowest rate of high-dose oxycodone dispensing. Great diversity exists in Canada in terms of urban versus rural opiate habitation. Even in provinces with larger cities, such as Toronto, Montreal, Calgary, Ottawa, Edmonton, Winnipeg, and Vancouver, where one may assume greater potential dispensing and use of high-dose opioids, significant variation can be seen in terms of the volume dispensed. While Canada may not be a rabid consumer of opioids like its American neighbor, trends in use should be tracked, as this country experiences very high rates of abuse in some regions. For example, among the First Nations people of the Sioux Lookout Zone it has been estimated that 80% of the adult population have used prescription drugs illicitly (Timpson & O'Gorman, 2010, p. iv).

Australia

As a nation, Australia is also part of the dialogue regarding escalating opioid use. In a 15-year period beginning in 1997, the supply/demand of oxycodone and fentanyl increased dramatically, to exceed 22% and 46%, respectively (Dobbins, 2014). The demand arena warrants attention. As is true in developed countries with an increased life span, Australia's demographic is aging. Along with aging come many issues, chief among them physical illness, injury, and chronic disease. Chronic pain, which can be experienced at any age, is cited as a reason for the prescribing of opioids as per typical medical protocol. Hence, a greater potential demand for prescription opioids from medical practitioners exists. Australia has seen increases among all opioid drugs, notably oxycodone, tramadol, and morphine (Leong, Murnion, & Haber, 2009).

Not surprisingly, since oxycodone had become the most widely prescribed opioid, it is this drug that was often reported as the most abused, diverted to the streets and sold in dispensed tablet form, then typically crushed and snorted or injected by users. The number of users injecting heroin had decreased compared to those injecting oxycodone (Kirwan, Dietze, & Lloyd, 2012). The introduction in early 2014 of Oxycontin in Australia, the new formulation of which prevents snorting or injecting, may require new monitoring and education efforts on the part of physicians, treatment agencies, and public health officials. Despite strict prescribing guidelines by Australia's Pharmaceutical Benefits Scheme, use of these drugs has increased. Oxycodone is the seventh most prescribed drug in Australia (Dobbins, 2014). Increased access to and use of these drugs both legally and illegally will require ongoing monitoring and research.

HARM REDUCTION RESPONSE TO INCREASING OPIOID AND HEROIN ABUSE

Naloxone (Narcan) is increasingly available in the United States to first responders (firefighters, paramedics, police) and the family members and peers of known opiate-dependent users. Naloxone is a safe response to opiate and opioid overdose, almost immediately reversing the overdose and the opiate/opioid's life-threatening effects on the CNS (e.g., decreased respiration and heart rate). The effects of naloxone last a relatively short time; however, this time is usually sufficient to get the person the medical intervention needed to save his or her life. In some parts of the United States, naloxone can be purchased without prescription. In Canada paramedics carry Narcan, but not police and firefighters. Naloxone is tightly controlled in Canada and citizens do not have access to it, as of the writing of this chapter (Canadian Harm Reduction Network). In Australia and North America there are efforts to make naloxone available over the counter through pharmacies, take-home naloxone (THN) programs, and needle and syringe exchange programs, as it is in many European countries. The THN programs combine overdose prevention education and first aid training for drug users, their family members, and their peers, with the distribution of the antagonist naloxone. Figure 7 is a photograph of a naloxone nasal spray kit.

The numbers of opiate and opioid overdose deaths may be greatly reduced through broader availability of naloxone and appropriate training in its use. Any meaningful discussion about prevention of opioid overdose and abuse must also include examination of the relationship between high rates of opioid prescriptions and rising rates of opioid abuse in countries where opioids like Oxycontin and Subutex proliferate.

APPLICATIONS TO OTHER ADDICTIONS AND SUBSTANCE MISUSE

There is evidence of a relationship between nonmedical use of prescription opioids and heroin in the United States and the Republic of Georgia, coupled with increasing prevalence of opioid abuse and overdose across the globe. Beyond the bidirectional abuse

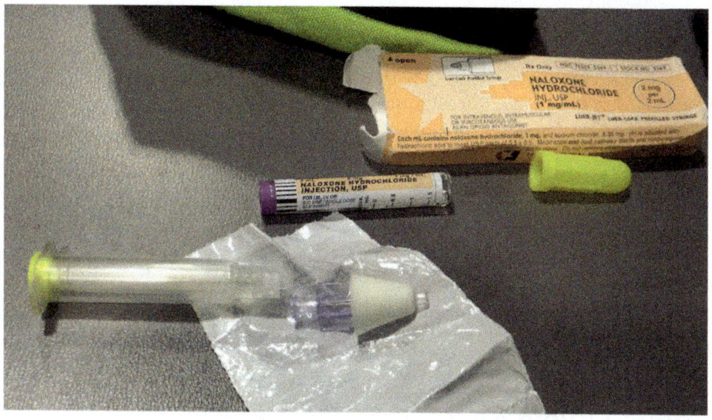

FIGURE 7 Naloxone nasal administration unit. *Photo by Brian J. Kievit.*

between heroin and prescription opioids, there are concerns about how the proliferation of prescription opioids may also negatively influence the effectiveness and access to these powerful medications for pain relief among those who are suffering. The prolonged use and abuse of opioids and opiates increase tolerance to these drugs and other depressant drugs, while simultaneously reducing the user's pain threshold. Combined, these factors increase opioid and opiate users' risk for cross-tolerance to other drugs categorized as CNS depressants, including alcohol.

CONCLUSION

The proliferation of prescription opioids presents public health challenges to ensuring prescription opioids are available for medical necessity, while also focusing on the misuse and diversion of these drugs (UNODC, 2014). Drug manufacturers must work with medical and public health personnel to provide the safest formulations possible to meet medical demand. Physicians and prescribing agents require current, research-driven information to work with patients to treat conditions for which these drugs are most utilized and to avoid addiction and diversion activity. Aggressive marketing by pharmaceutical companies has increased the social acceptability of the use of these medications, which has led to increases in dispensing of these prescriptions. These factors coupled with the flood of raw opium production in Afghanistan have created the perfect storm of environmental availability of opioids and opiates.

DEFINITION OF TERMS

Methadone Methadone is a prescription medication intended for opiate dependence treatment. Methadone can be used recreationally and abused, therefore its access is generally limited to being dispensed in clinics that specialize in addiction treatment.

Opiates Opiates are drugs derived from the opium poppy (opium) and semisynthetic drugs derived from an opium base (heroin, morphine, codeine). Semisynthetic and plant opiates operate similarly in the body and attach to the same specific receptor sites in the brain, spinal cord, and gastrointestinal tract. Opiates are CNS depressants and block the transmission of pain messages. When used improperly they can cause serious harm, including overdose and death.

Opioids Opioids are chemicals that are synthetically manufactured drugs similar in structure to opium. Synthetic opioids operate similarly in the body compared to plant-based opiates and attach to the same specific receptor sites in the brain, spinal cord, and gastrointestinal tract. Opioids are CNS depressants and they block the transmission of pain messages. When used improperly, or for recreational purposes, they can lead to addiction and cause serious harm, including overdose and death.

Subutex Subutex is a prescription medication approved for the treatment of opioid and opiate dependence. It contains the active ingredient buprenorphine hydrochloride, which reduces symptoms of opiate dependence. Subutex can be used recreationally and abused.

Suboxone Suboxone is a prescription medication approved for the treatment of opiate and opioid dependence. Suboxone contains the active ingredient buprenorphine hydrochloride and an additional ingredient called naloxone, the latter of which guards against misuse of this drug. Because Suboxone is formulated to minimize abuse potential individuals may be given take-home doses and prescriptions for management of opiate addiction symptoms.

Morphine equivalent This is a milligram measure used for standardized comparison of opioid strength across medications.

KEY FACTS

Key Facts about Opioids

Opioids are synthetic forms of opiates (opium, heroin, morphine). Opioids are powerful drugs that are synthetically manufactured (e.g., Oxycontin, Percocet, Subutex, Vicodin) to mimic opiates (morphine, heroin, codeine, opium). Synthetic opioids and plant-based opiates operate similarly in the body and attach to the same specific receptor sites in the brain, spinal cord, and gastrointestinal tract. Opioids are CNS depressants and they block the transmission of pain messages. However, when used improperly or for recreational purposes, they can cause serious harm, including overdose and death. Opioids slow down the CNS, which in turn decreases breathing/respiration and heart rate. At high levels opioids reduce consciousness and may result in overdose and death due to the slowing or cessation of breathing and heart rate. Opioids can also induce a sense of euphoria and sensations of warmth, drowsiness, and contentment. Opioids relieve stress and physical discomfort by creating a relaxed detachment from pain and desires. Natural opiates are derived from the dried sap of the opium poppy, whereas synthetic opiates are manufactured in chemical laboratories with a similar chemical structure.

Key Facts about Overdose

An overdose occurs when the body has more drugs in its system than it can process and handle, which can cause life-threatening dysfunction. People can overdose on many different substances, including alcohol and other drugs, including prescription drugs. In the case of an opioid overdose, the level of opioids or combination of opioids and other drugs, including alcohol, in the body decreases CNS functioning such that breathing and heart rate may stop. High doses of opioids and the combination of opioids with other CNS depressants such as alcohol and/or benzodiazepines greatly increase the risk of overdose.

Key Facts about Prescription Opioid Abuse

Prescription opioid drug abuse refers to the use of prescription medications that are: (1) not prescribed for the individual taking the prescription drug, (2) taken in a way other than the prescribed medication was intended to be taken as directed by one's physician, and/or (3) intentionally used as a recreational drug. Abuse of prescription opioids presents a high risk for addiction potential and abuse, including overdose similar to that of heroin.

Key Facts about the Prevalence of Prescription Opioids

Opioids such as Oxycontin, hydrocodone, and Vicodin are increasingly prescribed for populations with severe and chronic pain. For such populations these opioids are effective in managing debilitating pain and appropriate with proper oversight and management by qualified health care providers. However, the increased circulation of these narcotics presents increased risk of misuse and abuse by people seeking to use the drugs recreationally.

SUMMARY POINTS

- Opioids are powerful synthetically manufactured opiates. Synthetic opioids and plant-based opiates (opium, heroin, morphine) operate similarly in the body and attach to the same specific receptor sites in the brain, spinal cord, and gastrointestinal tract.
- Opioids and opiates are CNS depressants and they block the transmission of pain messages.
- Abuse of prescription opioids, including synthetic opioids such as fentanyl, Vicodin, Subutex, Oxycontin, methadone, pethidine, and desomorphine, has been increasing in certain areas of the world.
- The United States has the highest population percentage (6%) abusing pharmaceutical opioids in the world.
- Demand for opioids must be examined within the context of aging populations, who may experience chronic pain, injury, and illness disproportionate to younger peers.
- Prescription opioid drug abuse refers to the use of prescription medications that are: (1) not prescribed for the individual taking the prescription drug, (2) taken in a way other than the prescribed medication was intended to be taken as directed by one's physician, and/or (3) intentionally used as a recreational drug.
- Opioids such as Oxycontin, hydrocodone, and Vicodin are increasingly prescribed for populations with severe and chronic pain; however, dissimilarity in prescribing of opioids among physicians merits discussion.
- Between 2003 and 2005 there was an 80% increase in the number of drug abusers in the Republic of Georgia. This increase was largely attributed to increased importation and street diversion of the prescription opioid Subutex.
- The increased prescribing of opioids in recent decades has increased opioid abuse and dependence in some countries.
- Among countries in which the prevalence of prescription opioid abuse is greater than or equal to 1% of the population, opioid abuse is equal to or exceeds heroin and opium abuse.

REFERENCES

Bergen-Cico, D. (2012). *War and drugs: The role of military conflict in the development of substance abuse.* Boulder, CO: Paradigm Publishers.

Chou, R., Fanciullo, G. J., Fine, P. G., Adler, J. A., Ballantyne, J. C., Davies, P., ... American Pain Society-American Academy of Pain Medicine Opioids Guidelines Panel. (2009). Clinical guidelines for the use of chronic opioid therapy in chronic noncancer pain. *Journal of Pain, 10*(2), 113–130.

Dhalla, I. A., Mamdani, M. M., Sivilotti, M. L., Kopp, A., Qureshi, O., & Juurlink, D. N. (2009). Prescribing of opioid analgesics and related mortality before and after the introduction of long-acting oxycodone. *Canadian Medical Association Journal, 181*(12), 891–896.

Dobbins, M. (2014). Pharmaceutical drug misuse in Australia. *Australian Prescriber, 37*, 79–81.

European Monitoring Centre for Drugs and Drug Addiction (EMCDDA). (2013). *Data statistical bulletin 2013. Table DRD-1. Summary of characteristics of the deceased in drug-induced deaths according to national definitions.* Reitox National Reports 2012.

Gomes, T., Mamdani, M. M., Paterson, J. M., Dhalla, I. A., & Juurlink, D. N. (2014). Trends in high-dose opioid prescribing in Canada. *Canadian Family Physician, 60*(9), 826–832.

Grund, J. P. C., Latypov, A., & Harris, M. (2013). Breaking worse: the emergence of krokodil and excessive injuries among people who inject drugs in Eurasia. *International Journal on Drug Policy, 24*(4), 265–274.

Health Canada. (2012). *2011 Canadian alcohol and drug monitoring (CADUM) survey.* Accessed November 30, 2014 at www.hc-sc.gc.ca.

Health Canada. (2013). *2012 Canadian alcohol and drug use monitoring survey.* Ottawa, CA. Accessed November 30, 2014 at www.hc-sc.gc.ca.

Kirtadze, I., Otiashvili, D., O'Grady, K. E., & Jones, H. E. (2012). Behavioral treatment naltrexone reduces drug use and legal problems in the Republic of Georgia. *The American Journal of Drug and Alcohol Abuse, 38*(2), 171–175.

Kirwan, A., Dietze, P., & Lloyd, B. (2012). *Findings from the Illicit Drug Reporting System (IDRS).* Sydney: National Drug and Alcohol Research Centre, University of New South Wales.

Leong, M., Murnion, B., & Haber, P. S. (2009). Examination of opioid prescribing in Australia from 1992 to 2007. *Internal Medicine Journal, 39*, 676–681.

Otiashvili, D., Piralishvili, G., Sikharulidze, Z., Kamkamidze, G., Poole, S., & Woody, G. E. (2014). Methadone and Suboxone® for Subutex® injectors: primary outcomes of pilot RCT. *Drug and Alcohol Dependence, 140*, e166.

Parfitt, T. (2006). Designer drug Subutex takes its toll in Tbilisi. *The Lancet, 368*(9532), 273–274.

Rimple, P. (April 11, 2006). *Georgia fighting losing battle against designer drug*. Eurasia Insight: Latest News.

Rudd, R. A., Paulozzi, L. J., Bauer, M. J., Burleson, R. W., Carlson, R. E., Dao, D., & Zehner, A. M. (2014). Increases in heroin overdose deaths—28 states, 2010 to 2012. *MMWR: Morbidity and Mortality Weekly Report, 63*(39), 849–854.

Substance Abuse and Mental Health Services Administration (SAMHSA). (2013). *Results from the 2012 National Survey on Drug Use and Health: Summary of National Findings*. Rockville, MD: Substance Abuse and Mental Health Services Administration.

Sikharulidze, Z., Kapanadze, N., Otiashvili, D., Poole, S., & Woody, G. E. (2014). Desomorphine (crocodile) injection among in-treatment drug users in Tbilisi, Georgia. *Drug and Alcohol Dependence, 140*, e208.

Timpson, J., & O'Gorman, K. E. (2010). *Community-based options for addressing prescription drug abuse in remote northwestern Ontario first nations*. Thunder Bay, ON: Nishnawbe Aski Nation.

Todadze, K., & Lezhava, G. (2008). Implementation of drug substitution therapy in Georgia. *Central European Journal of Public Health, 16*(3), 121–123.

United Nations Office on Drugs and Crime (UNODC). (2012). *World drug report 2012*.

United Nations Office on Drugs and Crime (UNODC). (2014). *World drug report 2014*.

United States, Department of Health and Human Services, Substance Abuse and Mental Health Services Administration. (2012). *Monitoring the future survey*.

United States Army. (2010). *Army health promotion risk reduction suicide prevention report*.

Vaisman, D. (June 12, 2006). *Detox drug is Georgia's new habit*. Europe: The New York Times.

Yokell, M. A., Delgado, M., Zaller, N. D., Wang, N., McGowan, S. K., & Green, T. (2014). Presentation of prescription and nonprescription opioid overdoses to US emergency departments. *Journal of American Medical Association Internal Medicine, 174*(12), 2034–2037. http://dx.doi.org/10.1001/jamainternmed.2014.5413.

Section B

Molecular and Cellular Aspects

Chapter 83

Corticotropin-Releasing Factor Signaling at the Intersection of Pain and Opioid Addiction

M. Adrienne McGinn, Scott Edwards
Department of Physiology, Alcohol and Drug Abuse Center of Excellence, LSU Health Sciences Center, New Orleans, LA, USA

Abbreviations

ACC Anterior cingulate cortex
ACTH Adrenocorticotropic hormone
BLA Basolateral amygdala
CeA Central amygdala
CREB cAMP response element-binding protein
CRF Corticotropin-releasing factor
CRFR Corticotropin-releasing factor receptor
ERK Extracellular signal-regulated kinase
HPA Hypothalamic–pituitary–adrenal
ICSS Intracranial self-stimulation
MAPK Mitogen-activated protein kinase
NMDA *N*-Methyl-D-aspartate
PB Parabrachial area
PFC Prefrontal cortex
PKA Protein kinase A
PVN Paraventricular nucleus

INTRODUCTION

The opioid family of drugs consists of opiates, which are alkaloid substances derived from naturally occurring substances in the opium poppy (including morphine, named after the god of dreams, Morpheus; Huxtable & Schwarz, 2001) and the synthetic, psychoactive chemicals that resemble opiates (termed opioids, including heroin and most prescription opioid medications). Both groups represent exogenous ligands of the endogenous opioid receptors of the central and peripheral nervous system (Lutz & Kieffer, 2013). Opioids are among the world's oldest utilized drugs, with the opium poppy being cultivated as early as 3400 BC in Mesopotamia (Scott, 1969). At that time it was popular for both its analgesic effects and its ability to produce a strong sense of euphoria. Opium was recreationally used for centuries with few references to its addictive properties until the Chinese popularized the practice of smoking opium, thereby increasing its potency and availability to the brain (Scott, 1969). Following the development of the hypodermic syringe in 1856, dependence on highly refined opiate formulations quickly became a major public health concern (Levinthal, 2005).

Heroin was developed as a synthetic opioid in the late nineteenth century by the Bayer chemical company in an attempt to create a more potent analgesic (Brownstein, 1993). From 1895 to 1910, Bayer even marketed heroin as a nonaddictive morphine substitute until it was ultimately discovered that heroin is metabolized to morphine once reaching the brain.

THE ADDICTED PHENOTYPE

Opioid addiction, also known as opioid use disorder (DSM-5, 2013), is a chronic relapsing disease characterized by a compulsion to seek and take opioids, an escalation of opioid intake, and the manifestation of a negative emotional state typically observed during abstinence (Edwards & Koob, 2010). The emergence of such a dysphoric state with the transition to opioid addiction facilitates negative reinforcement processes and may underlie the final definitive stage of the addiction process. A greater understanding of the physiological basis of negative reinforcement may provide insight into the transition to opioid addiction, as well as revealing new treatment opportunities. However, there persists a gap in our knowledge regarding the relationship between the negative emotional conditions associated with dependence and the exacerbation of addiction, and the question remains whether treatment of such states themselves will be effective in reducing relapse or intake escalation.

As of 2012, there were 467,000 people in the United States who met the *Diagnostic and Statistical Manual of Mental Disorders* (DSM), fourth edition, criteria for heroin abuse or dependence, double the amount reported in 2002. The resurgence of heroin use is thought to be fueled by the current prescription-opioid abuse epidemic, with heroin being cheaper and sometimes more readily available to individuals dependent on prescription painkillers. Around 3.8 million patients legally receive a prescription opioid annually (Governale, 2010), and it is estimated that only 5% of patients who take their prescription as directed will develop an addiction-related disorder. However, approximately 2.4 million Americans use prescription drugs in a nonmedical fashion, and about 1.9 million people in the United States meet abuse or dependence criteria for prescription opioids

(Substance Abuse and Mental Health Services Administration, 2011). Thus, from ancient times to today, there appears to be an inevitable intersection between beneficial opioid use for pain management and potential risk for addiction. A more precise understanding of the biobehavioral mechanisms that delimit these categorizations is warranted to benefit those suffering from chronic pain conditions and addiction.

ANIMAL MODELS OF OPIOID INTAKE ESCALATION AND ADDICTION

A crucial advancement in our ability to test negative reinforcement theories of opioid use and addiction has been the development and refinement of reliable animal models of excessive opioid self-administration (Edwards & Koob, 2012). For example, prolonged access to heroin self-administration (6–23 h per day) leads to a significant escalation of heroin intake and manifestation of dependence-related symptoms in rats, while animals given restricted access (1 h per day) fail to escalate intake or exhibit withdrawal symptoms (Edwards, Graham, Whisler, & Self, 2009; Vendruscolo et al., 2011). Animals with a history of intake escalation go on to display a prolonged latency to extinguish operant self-administration of heroin (a measure of compulsive drug seeking) as well as an enhanced sensitivity to stress- and heroin-induced reinstatement, models of relapse (Ahmed, Walker, & Koob, 2000; Lenoir & Ahmed, 2007). Opioid withdrawal has also been demonstrated to heighten brain reward thresholds using the intracranial self-stimulation procedure (Schulteis, Markou, Gold, Stinus, & Koob, 1994). Such data provide the best evidence of anhedonia and affective system dysregulation following excessive opioid exposure, and this phenomenon is hypothesized to result from recruitment of a compensatory, antireward system opposing the drug-induced activation of the natural reward circuitry (Koob & Le Moal, 2008).

CHRONIC OPIOID EXPOSURE AND HYPERALGESIA

Somewhat paradoxically, chronic opioid exposure directed at alleviating pain often leaves individuals more sensitive to nociception, a condition known as opioid-induced hyperalgesia (Mehendale, Goldman, & Mehendale, 2013). Hyperalgesia is also prevalent in long-term opioid abusers, suggesting that this condition emerges with extended opioid use. In line with this hypothesis, opioid exposure also leads to a hyperalgesia in rodents that is exacerbated with chronic or excessive use (Simonnet & Rivat, 2003). Provided that chronic pain is known to generate both emotional distress and a negative emotional state (King et al., 2009), opioid-induced hyperalgesia would seem to fit diagnostic criteria as an endophenotype closely associated with addiction via facilitation of negative reinforcement processes. Shurman, Koob, and Gutstein (2010) have specifically hypothesized that targeted pain management with controlled doses of opioids that are strictly titrated toward effective analgesia represents the best treatment strategy from a clinical perspective, while overexposure to opioids in excessive amounts (i.e., above medical requirement, as seen in addicts) would be expected to promote an engagement of opponent motivational processes in terms of both pain (paradoxical hyperalgesia) and negative emotional states. Such opponent neuroadaptations manifest at even the cellular and molecular levels following opioid exposure, but may also progress toward the recruitment of multiple brain reinforcement circuits during the addiction process (Koob & Volkow, 2010).

OPIOID ADDICTION IN RELATION TO PAIN MANAGEMENT

Chronic pain affects over 100 million Americans (Institute of Medicine, 2011) and disrupts the lives of approximately 20% of the global population (Goldberg & McGee, 2011), a number that is likely to increase over the next several decades given an aging population in the United States and many Western countries. From a clinical view, controlled and clinically titrated exposure to an opioid with abuse potential is a condition entirely distinguishable from opioid diversion for nonmedical use and uncontrolled/excessive drug consumption (Volkow & McLellan, 2011). Unfortunately, a long-standing concern for medical practitioners is how to administer chronic opioid analgesics to pain patients without generating an addicted state (Fields, 2011). Importantly, opioid therapy rarely leads to abuse following drug discontinuation (Fishbain, Cole, Lewis, Rosomoff, & Rosomoff, 2008); however, the propensity for addiction is enhanced in patients with a history of illicit opioid use or abuse. Moreover, chronic pain patients tend to exhibit a significant craving for prescription opioids even in the absence of misuse risk (Wasan et al., 2012). Ren, Shi, Epstein, Wang, and Lu (2009) discovered that an enhanced nociceptive state is manifest for up to 5 months in abstinent opioid abusers, while those individuals with more pain sensitivity also displayed greater cue-induced opioid craving at this time point. These data suggest that pain sensitivity may promote or exacerbate addicted states (drug dependence and compulsive drug-seeking behaviors) in this more compromised population.

SHARED NEURAL MECHANISMS AND NEUROCIRCUITRY UNDERLYING PAIN AND OPIOID ADDICTION

Pain represents a uniquely subjective experience that may exert a powerful influence on reinforcement processes, possibly facilitating the transition to addiction. An association between chronic pain and a possibly enhanced motivation to obtain opioids has been demonstrated in rats (Martin et al., 2011; Martin, Kim, Buechler, Porreca, & Eisenach, 2007). In animals suffering from mechanical hyperalgesia of the hind paw after a spinal nerve ligation injury, only heroin doses that effectively alleviated hyperalgesia were able to sustain heroin self-administration, while subeffective doses were capable of maintaining heroin self-administration in control (uninjured) rats, suggesting that one driving motivation to self-administer ever larger doses of opioids in individuals with a sensitized nociceptive system may be to seek pain relief. Indeed, nerve-injured rats appear to require much higher doses of heroin to register its rewarding properties, and injured rats are also more sensitive to μ opioid receptor antagonism (Martin et al., 2011). Although these studies have proven valuable in understanding opioid efficacy in the presence of pain, a critical limitation of these studies was the use of animals with a limited history of opioid exposure. A more extensive investigation of the effects of chronic pain in the context of long-term opioid self-administration (and accompanying opioid-seeking behavior) is warranted.

Revelation of opioid-induced molecular neuroadaptations within brain reinforcement systems may provide valuable insights into potential mechanisms underlying the transition to drug addiction in vulnerable populations. Following excessive opioid exposure, reward neurotransmitter systems (including some endogenous opioid peptide systems) are compromised, while brain stress systems such as norepinephrine and corticotropin-releasing factor (CRF) signaling are recruited (Koob & Le Moal, 2008). Strong evidence suggests that the neural substrates associated with addiction may also overlap with substrates of emotional aspects of nociceptive processing in areas such as the prefrontal cortex (PFC) and amygdala (Egli, Koob, & Edwards, 2012), where ascending pain circuits terminate for processing the affective components of pain (Neugebauer, Li, Bird, & Han, 2004). Particularly interesting is the proposed role of the anterior cingulate cortex (ACC; part of the PFC) in organizing the execution of behavioral strategies (such as avoiding negative emotional conditions) following nociceptive input (Zhuo, 2014). Evidence for this function of the ACC has been demonstrated in rodent models, including a central role for mitogen-activated protein kinase (MAPK) signaling in the induction and expression of affective pain (Cao et al., 2009; Johansen, Fields, & Manning, 2001). Thus, it is conceivable that elevated MAPK activity within the ACC may link persistent pain-driven motivated behaviors with the urge to compulsively seek and take opioids.

Along with the ACC, pain-responsive neurons are also abundant in the laterocapsular division of the central amygdala (termed the nociceptive amygdala; Bernard, Huang, & Besson, 1990). The central amygdala (CeA) is critically important for opioid reward (Cai et al., 2013) and represents another possible neuroanatomical intersection of central pain modulation and drug reinforcement. Thus, opioids probably act directly within this area of ascending nociceptive circuitry to regulate neuronal plasticity related to the interaction of nociception and negative affect. Persistent pain produces multiple neurophysiological adaptations in the CeA, which receives functionally distinct projections from the pontine parabrachial area (nociceptive coding) and basolateral amygdala (sensory-affective coding) that are sensitized in the context of persistent pain (Ikeda et al., 2007; Neugebauer et al., 2004). Consequently, Neugebauer (2007) speculated that the CeA facilitates nociceptive signaling in chronic pain states. In relation to opioid use disorders, chronic pain-induced activation of the CeA is also accompanied by dysregulation of amygdala-driven PFC function and a resultant production of cognitive deficits (Ji et al., 2010). Importantly, executive system dysfunction is thought to play a central role in the heavily compromised decision-making that accompanies the transition from drug use to addiction (George & Koob, 2010), and via this mechanism individuals suffering from chronic pain may be more susceptible to opioid misuse and/or ineffective pain management.

ROLE OF CORTICOTROPIN-RELEASING FACTOR IN OPIOID-INDUCED HYPERALGESIA AND THE TRANSITION TO OPIOID DEPENDENCE

As a principal mediator of the sensitized stress response that manifests during opioid withdrawal, the functional potentiation of CRF signaling represents the recruitment of a central pharmacological system contributing to the establishment and maintenance of drug addiction (Zorrilla, Logrip, & Koob, 2014). CRF is a neuropeptide synthesized by the paraventricular nucleus of the hypothalamus, more famously known as the initial component of the hypothalamic–pituitary–adrenal axis response to stress. CRF stimulates synthesis of adrenocorticotropic hormone in the anterior pituitary, which then facilitates the secretion of glucocorticoids and mineralocorticoids from the adrenal gland to restore homeostasis. High concentrations of CRF are also present in the basal forebrain and brainstem (Swanson, Sawchenko, Rivier, & Vale, 1983), where the neuropeptide regulates autonomic and behavioral responses to chronic stress. Extrahypothalamic CRF1 receptors (CRF1Rs) are widely expressed throughout the brain, and activation of these receptors underlies a sensitization of stress responsiveness, while activation of complementary CRF2 receptors mediates a contrasting (but not necessarily opposing) effect on stress and anxiety-like behavior (Zhao et al., 2007).

Chronic opioid exposure leads to withdrawal-induced increases in brain CRF levels (Weiss et al., 2001). In turn, prophylactic administration of CRF1R antagonists prevents heroin intake escalation (Park et al., 2013). Decreases in excessive heroin intake via CRF1R blockade are believed to be due to an attenuation of negative reinforcement mechanisms that accompany heroin intake escalation (Kenny, Chen, Kitamura, Markou, & Koob, 2006). Importantly, administration of a CRF receptor antagonist into the CeA reduces the reward associated with negative reinforcement during morphine withdrawal (Heinrichs, Menzaghi, Schulteis, Koob, & Stinus, 1995), suggesting a possible link between negative affect and potentiated CRF signaling following chronic opioid exposure. McNally and Akil (2002) revealed a role for amygdala CRF receptor activity in regulating thermal hyperalgesia following precipitated morphine withdrawal in dependent animals. In this study, rats were made dependent via subcutaneous insertion of morphine pellets and withdrawal was induced by the opioid receptor antagonist naloxone. Under these conditions, dependent rats displayed a significantly reduced tail-flick latency compared to placebo-pelleted, naloxone-treated animals, and this index of thermal hyperalgesia was reduced via microinjection of a nonselective CRF receptor antagonist into the CeA.

Similar to morphine, heroin administration leads to a rapid reduction in nociceptive thresholds (Laulin, Larcher, Celerier, Le Moal, & Simonnet, 1998) that is worsened following chronic treatment, suggesting a recruitment or potentiation of pronociceptive physiological processing (Celerier, Laulin, Corcuff, Le Moal, & Simonnet, 2001; Shurman et al., 2010). In rodents, mechanical hyperalgesia development occurs in direct relation to amount of self-administered heroin (Edwards et al., 2012), while CRF1R blockade alleviates this hypersensitivity. In a similar set of studies, Park et al. (2013) measured the effects of CRF1R antagonism on hyperalgesia-like behavior during heroin withdrawal utilizing models of both acute and chronic heroin dependence. Consistent with previous studies, passive or self-administered heroin generated a robust reduction in mechanical nociceptive thresholds. In the context of acute opioid dependence induced via subcutaneous heroin injections, CRF1R antagonism alleviated withdrawal-induced mechanical hypersensitivity. In contrast, several functional adrenergic receptor antagonists that typically exhibit antihyperalgesic effects (including clonidine) failed to alter mechanical hypersensitivity under this condition. To extend these findings to a more chronic dependence model relevant to addiction, the authors next determined the efficacy of chronic CRF1R antagonism or clonidine treatment on heroin self-administration and revealed that CRF1R blockade (but not clonidine administration) reduced heroin intake escalation, while both treatments prevented chronic dependence-associated mechanical hypersensitivity. These results provide evidence that a rapid sensitization of CRF function occurs upon opioid exposure that may begin to promote both opioid-induced hyperalgesia and escalation of intake.

ROLE OF CORTICOTROPIN-RELEASING FACTOR SIGNALING IN CHRONIC PAIN STATES

Chronic pain produces a number of physiological changes throughout the body, including increases in CRF levels in the CeA (Rouwette et al., 2012). Furthermore, in addition to their role in alleviating opioid-induced pain, the antihyperalgesic effects of CRF1R antagonists have been demonstrated across multiple pain models. CRF would appear to play a part in the signaling of various pain modalities (e.g., mechanical, thermal, visceral), suggesting that its role is not modality specific but may generalize to all classes of pain. Importantly, CRF1R antagonists do not alter baseline measures of various nociception-related indices (e.g., mechanical or thermal paw withdrawal thresholds, audible or ultrasonic vocalizations) in uninjured animals, indicating the absence of a direct analgesic effect (e.g., Cohen et al., 2013; Fu & Neugebauer, 2008). CRF directly alters the electrophysiological properties of CeA neurons to drive pain sensitization. For example, rodents in a state of arthritic inflammatory pain exhibit enhanced CeA excitability that is attenuated by CRF1R antagonism (Ji & Neugebauer, 2007). The authors also found that the normally inhibitory role of CRF2R signaling in this region was lost in the arthritis condition, further demonstrating the complementary roles of CRF1 versus CRF2 receptors in neuronal physiology, including pain signaling (Ji & Neugebauer, 2008). An elevated anxiety-like behavior following arthritis induction is also alleviated following either systemic or intra-CeA CRF1R antagonism (Ji, Fu, Ruppert, & Neugebauer, 2007), suggesting that CRF signaling drives the intersection of pain and negative affect in the CeA (Figure 1; Egli et al., 2012; Neugebauer et al., 2004). Stress factors other than pain also produce an enhancement of CRF activity in the CeA, and it was demonstrated in 2013 that non-pain-related activation of CRF1R signaling in the CeA could augment nociceptive responsiveness (Ji, Fu, Adwanikar, & Neugebauer, 2013). Further investigation by Neugebauer and colleagues revealed that the interaction of CRF1Rs with N-methyl-D-aspartate receptor channels and/or downstream activation of protein kinase A (PKA) and extracellular signal-regulated kinase (ERK) pathways mediates nociceptive sensitization in the CeA (Fu et al., 2008; Ji & Neugebauer, 2009). Potentiation of CeA PKA and ERK signaling is associated with various indices of opioid addiction (Edwards et al., 2009; Zamora-Martinez & Edwards, 2014) and these pathways may enhance the levels of CeA CRF via activation of the transcription factor cAMP response element-binding protein, which is increased in the CRF-containing CeA neurons of morphine-dependent animals (Shaw-Lutchman et al., 2002).

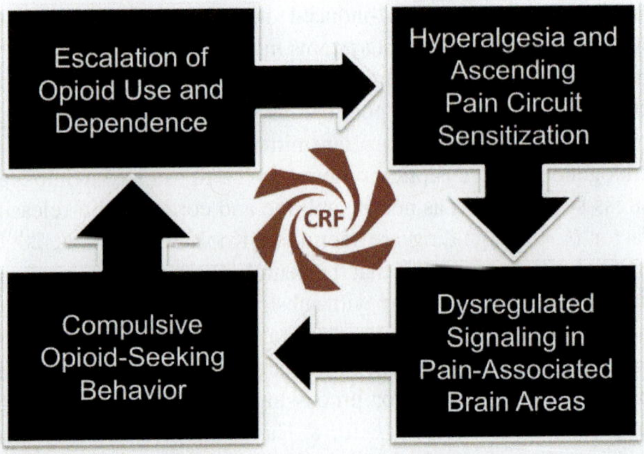

FIGURE 1 Interactive nature of CRF, pain, and opioid addiction. Studies have described an intersection of pain and drug-addiction-related behaviors. Excessive opioid-induced hyperalgesia in dependent animals may drive compulsive drug seeking and escalated intake via altered neuronal signaling as a negative reinforcement mechanism to alleviate pain. Blockade of this circuit via antagonism of CRF receptors in pain-associated brain areas (e.g., the CeA) is hypothesized to alleviate addiction-related behaviors, including hyperalgesia.

APPLICATIONS TO OTHER ADDICTIONS AND SUBSTANCE MISUSE

Koob and Le Moal (2008) have suggested that drug addiction can be modeled as a chronic emotional pain syndrome. The proposed sensitization of pain-driven negative affect (Ji et al., 2007) is thought to be mediated via brain reinforcement circuitry and to closely interact with neuronal mechanisms related to psychiatric disorder and potential addiction to multiple substances (Egli et al., 2012). As described above, the brain is hypothesized to respond to acute opioid exposure by activating neuronal and hormonal elements to promote homeostasis (Shurman et al., 2010). According to this conceptualization, opioids initially trigger a dual euphoric and analgesic response (reward and pain relief) followed by an opposing dysphoric and hyperalgesic condition during abstinence (Laulin et al., 1998; Park et al., 2013) that further sensitizes over time (Celerier et al., 2001; Edwards et al., 2012; Laulin et al., 1998). A common neurocircuitry probably promotes cross talk among a variety of psychiatric disorders, and this accounts for the commonly described comorbidities between affective disorders and both opioid abuse (Fischer, Lusted, Roerecke, Taylor, & Rehm, 2012) and chronic pain (Gerrits et al., 2012). In a similar fashion, drugs such as CRF1R antagonists that act on these shared systems would probably be effective for treating a range of affective pain-related conditions, including opioid use disorder. Moreover, dependence on other commonly abused drugs (e.g., alcohol and nicotine) is associated with a CRF1R-dependent hyperalgesia (Figure 2; Cohen et al., 2013; Edwards et al., 2012). As a result, patients with a history of alcohol or other substance use disorders may be susceptible to the activation of a common set of nociceptive factors that could drive relapse to seeking and further abuse of any or all of these substances. The CeA appears to play a central role in mediating the intersection of pain and affective-related behaviors expected to contribute to substance use disorders. In addition, pain may also drive cognitive

Animal Model or Condition	Hyperalgesia?	Responsive to Systemic CRF1R Antagonism?	Responsive to Intra-CeA CRF1R Antagonism?
Arthritis	Yes	Yes	Yes
Heroin Dependence	Yes	Yes	
Morphine Dependence	Yes		Yes
Alcohol Dependence	Yes	Yes	
Nicotine Dependence	Yes	Yes	Yes
Cocaine Dependence	No		

FIGURE 2 **Role of CRF signaling in chronic pain and drug dependence.** Animal models of chronic pain have revealed a key role for CeA CRF1R signaling in hyperalgesia. Similarly, dependence on a variety of abused substances also produces hyperalgesia that is alleviated via systemic or intra-CeA administration of CRF1R antagonists.

disruption via a CeA–PFC circuit (Ji et al., 2010), and such cognitive deficits may further exacerbate addiction-related behaviors including relapse propensity (Bossert et al., 2011). Importantly, the planned functional integration of addiction-related research objectives across the National Institutes of Health is expected to promote further collaboration among addiction and pain neuroscientists, leading to better therapeutic strategies for these devastating diseases.

DEFINITION OF TERMS

Anhedonia This is a state of being incapable of taking pleasure in activities normally considered rewarding.

Anterior cingulate cortex This is a subregion of the PFC that mediates the affective dimension of pain and also contributes to relapse vulnerability in the context of drug addiction.

Central amygdala This is a subcortical brain region that regulates anxiety, nociceptive sensitivity, and negative reinforcement processes that promote the transition to drug addiction.

Dysphoria This is a state of feeling intense anxiety, depression, and/or indifference.

Homeostasis This refers to the tendency for a physiological system to maintain a constant internal state, often achieved via the recruitment of opponent or negative feedback processes.

Hyperalgesia This refers to excessive pain sensitivity often measured as a reduction in nociceptive thresholds. Excessive opioid exposure leads to a paradoxical hyperalgesic state.

Negative reinforcement This is the process by which removal of an aversive stimulus (such as pain) increases the occurrence of an action (such as opioid use).

Opiate This refers to any of the naturally occurring alkaloid compounds derived from the opium poppy.

Opioid This refers to any substance that resembles opiates and binds to opioid receptors, including natural opiates, endogenous opioids (such as endorphins), and various synthetic chemicals that compose the prescription analgesic drug class.

Opioid use disorder In its most severe form, this is similar to opioid addiction or opioid dependence. The severity of opioid use disorder is defined according to the cumulative presence of several diagnostic criteria from the fifth edition of the DSM (DSM-5, 2013), including (but not limited to) taking more opioids than intended, craving opioids, and continuing opioid use despite the presence of significant social or psychological problems associated with their use.

KEY FACTS RELATED TO OPIOID-INDUCED HYPERALGESIA AND ADDICTION

Although opioids are highly effective analgesics, chronic or excessive opioid exposure leads to a paradoxical hyperalgesia. This heightened pain sensitivity may represent a negative emotional state that promotes opioid intake escalation in vulnerable individuals. Importantly, pain sensitivity continues to drive craving in former opioid addicts well into the abstinence stage. At the preclinical level, animal models of opioid addiction and pain hypersensitivity have revealed a causative role for CRF signaling in the CeA in both of these processes. Antagonism of CRF1Rs may thus prevent both opioid-induced hyperalgesia and the desire to escalate opioid intake.

SUMMARY POINTS

- Opioids are frequently used for the alleviation of pain, although the risk of dependence presents a challenge for the clinician.
- Neuroscientists have begun to conceptualize potential biobehavioral mechanisms that define the interactions between pain and addiction.
- Hyperalgesia resulting from chronic or excessive opioid exposure may facilitate the transition to dependence in vulnerable populations.
- The CeA is one critical neuroanatomical substrate that mediates both pain sensitivity and the transition to addiction.
- CRF signaling within the CeA is potentiated by both chronic pain and excessive opioid use, and blockade of CRF1Rs may represent a viable therapeutic strategy for hyperalgesia and prevention of opioid use disorder.

ACKNOWLEDGMENT

Preparation of this review was partially supported by funds from the National Institute on Alcohol Abuse and Alcoholism (AA020839, SE).

REFERENCES

Ahmed, S. H., Walker, J. R., & Koob, G. F. (2000). Persistent increase in the motivation to take heroin in rats with a history of drug escalation. *Neuropsychopharmacology*, *22*(4), 413–421. http://dx.doi.org/10.1016/S0893-133X(99)00133-5.

Bernard, J. F., Huang, G. F., & Besson, J. M. (1990). Effect of noxious somesthetic stimulation on the activity of neurons of the nucleus centralis of the amygdala. *Brain Research*, *523*(2), 347–350.

Bossert, J. M., Stern, A. L., Theberge, F. R., Cifani, C., Koya, E., Hope, B. T., & Shaham, Y. (2011). Ventral medial prefrontal cortex neuronal ensembles mediate context-induced relapse to heroin. *Nature Neuroscience*, *14*(4), 420–422. http://dx.doi.org/10.1038/nn.2758.

Brownstein, M. J. (1993). A brief history of opiates, opioid peptides, and opioid receptors. *Proceedings of the National Academy of Sciences of the United States of America*, *90*(12), 5391–5393.

Cai, Y. Q., Wang, W., Hou, Y. Y., Zhang, Z., Xie, J., & Pan, Z. Z. (2013). Central amygdala GluA1 facilitates associative learning of opioid reward. *The Journal of Neuroscience*, *33*(4), 1577–1588. http://dx.doi.org/10.1523/JNEUROSCI.1749-12.2013.

Cao, H., Gao, Y. J., Ren, W. H., Li, T. T., Duan, K. Z., Cui, Y. H., ... Zhang, Y. Q. (2009). Activation of extracellular signal-regulated kinase in the anterior cingulate cortex contributes to the induction and expression of affective pain. *The Journal of Neuroscience*, *29*(10), 3307–3321. http://dx.doi.org/10.1523/JNEUROSCI.4300-08.2009.

Celerier, E., Laulin, J. P., Corcuff, J. B., Le Moal, M., & Simonnet, G. (2001). Progressive enhancement of delayed hyperalgesia induced by repeated heroin administration: a sensitization process. *The Journal of Neuroscience*, *21*(11), 4074–4080.

Cohen, A., Treweek, J., Edwards, S., Leao, R. M., Schulteis, G., Koob, G. F., & George, O. (2013). Extended access to nicotine leads to a CRF receptor dependent increase in anxiety-like behavior and hyperalgesia in rats. *Addiction Biology*. http://dx.doi.org/10.1111/adb.12077.

DSM-5. (2013). *American Psychiatric Association: Diagnostic and statistical manual of mental disorders* (5th ed.). Arlington, VA.

Edwards, S., Graham, D. L., Whisler, K. N., & Self, D. W. (2009). Phosphorylation of GluR1, ERK, and CREB during spontaneous withdrawal from chronic heroin self-administration. *Synapse*, *63*(3), 224–235. http://dx.doi.org/10.1002/syn.20601.

Edwards, S., & Koob, G. F. (2010). Neurobiology of dysregulated motivational systems in drug addiction. *Future Neurology*, *5*(3), 393–401. http://dx.doi.org/10.2217/fnl.10.14.

Edwards, S., & Koob, G. F. (2012). Experimental psychiatric illness and drug abuse models: from human to animal, an overview. *Methods in Molecular Biology*, *829*, 31–48. http://dx.doi.org/10.1007/978-1-61779-458-2_2.

Edwards, S., Vendruscolo, L. F., Schlosburg, J. E., Misra, K. K., Wee, S., Park, P. E., ... Koob, G. F. (2012). Development of mechanical hypersensitivity in rats during heroin and ethanol dependence: alleviation by CRF(1) receptor antagonism. *Neuropharmacology*, *62*(2), 1142–1151. http://dx.doi.org/10.1016/j.neuropharm.2011.11.006.

Egli, M., Koob, G. F., & Edwards, S. (2012). Alcohol dependence as a chronic pain disorder. *Neuroscience and Biobehavioral Reviews*, *36*(10), 2179–2192. http://dx.doi.org/10.1016/j.neubiorev.2012.07.010.

Fields, H. L. (2011). The doctor's dilemma: opiate analgesics and chronic pain. *Neuron*, *69*(4), 591–594. http://dx.doi.org/10.1016/j.neuron.2011.02.001.

Fischer, B., Lusted, A., Roerecke, M., Taylor, B., & Rehm, J. (2012). The prevalence of mental health and pain symptoms in general population samples reporting nonmedical use of prescription opioids: a systematic review and meta-analysis. *The Journal of Pain*, *13*(11), 1029–1044. http://dx.doi.org/10.1016/j.jpain.2012.07.013.

Fishbain, D. A., Cole, B., Lewis, J., Rosomoff, H. L., & Rosomoff, R. S. (2008). What percentage of chronic nonmalignant pain patients exposed to chronic opioid analgesic therapy develop abuse/addiction and/or aberrant drug-related behaviors? A structured evidence-based review. *Pain Medicine*, *9*(4), 444–459. http://dx.doi.org/10.1111/j.1526-4637.2007.00370.x.

Fu, Y., Han, J., Ishola, T., Scerbo, M., Adwanikar, H., Ramsey, C., & Neugebauer, V. (2008). PKA and ERK, but not PKC, in the amygdala contribute to pain-related synaptic plasticity and behavior. *Molecular Pain*, *4*, 26. http://dx.doi.org/10.1186/1744-8069-4-26.

Fu, Y., & Neugebauer, V. (2008). Differential mechanisms of CRF1 and CRF2 receptor functions in the amygdala in pain-related synaptic facilitation and behavior. *The Journal of Neuroscience*, *28*(15), 3861–3876. http://dx.doi.org/10.1523/JNEUROSCI.0227-08.2008.

George, O., & Koob, G. F. (2010). Individual differences in prefrontal cortex function and the transition from drug use to drug dependence. *Neuroscience and Biobehavioral Reviews*, *35*(2), 232–247. http://dx.doi.org/10.1016/j.neubiorev.2010.05.002.

Gerrits, M. M., Vogelzangs, N., van Oppen, P., van Marwijk, H. W., van der Horst, H., & Penninx, B. W. (2012). Impact of pain on the course of depressive and anxiety disorders. *Pain*, *153*(2), 429–436. http://dx.doi.org/10.1016/j.pain.2011.11.001.

Goldberg, D. S., & McGee, S. J. (2011). Pain as a global public health priority. *BMC Public Health*, *11*, 770. http://dx.doi.org/10.1186/1471-2458-11-770.

Governale, L. (July 22, 2010). *Outpatient prescription opioid utilization in the U.S., years 2000–2009*. Available at: http://www.fda.gov. Accessed April 2014.

Heinrichs, S. C., Menzaghi, F., Schulteis, G., Koob, G. F., & Stinus, L. (1995). Suppression of corticotropin-releasing factor in the amygdala attenuates aversive consequences of morphine withdrawal. *Behavioral Pharmacology*, *6*(1), 74–80.

Huxtable, R. J., & Schwarz, S. K. (2001). The isolation of morphine—first principles in science and ethics. *Molecular Interventions*, *1*(4), 189–191.

Ikeda, R., Takahashi, Y., Inoue, K., & Kato, F. (2007). NMDA receptor-independent synaptic plasticity in the central amygdala in the rat model of neuropathic pain. *Pain*, *127*(1-2), 161–172. http://dx.doi.org/10.1016/j.pain.2006.09.003.

Institute of Medicine of the National Academies Report. (2011). *Relieving pain in America: A blueprint for transforming prevention, care, education, and research*. Washington, DC: The National Academies Press.

Ji, G., Fu, Y., Adwanikar, H., & Neugebauer, V. (2013). Non-pain-related CRF1 activation in the amygdala facilitates synaptic transmission and pain responses. *Molecular Pain*, *9*, 2. http://dx.doi.org/10.1186/1744-8069-9-2.

Ji, G., Fu, Y., Ruppert, K. A., & Neugebauer, V. (2007). Pain-related anxiety-like behavior requires CRF1 receptors in the amygdala. *Molecular Pain*, *3*, 13. http://dx.doi.org/10.1186/1744-8069-3-13.

Ji, G., & Neugebauer, V. (2007). Differential effects of CRF1 and CRF2 receptor antagonists on pain-related sensitization of neurons in the central nucleus of the amygdala. *Journal of Neurophysiology*, *97*(6), 3893–3904. http://dx.doi.org/10.1152/jn.00135.2007.

Ji, G., & Neugebauer, V. (2008). Pro- and anti-nociceptive effects of corticotropin-releasing factor (CRF) in central amygdala neurons are mediated through different receptors. *Journal of Neurophysiology*, *99*(3), 1201–1212. http://dx.doi.org/10.1152/jn.01148.2007.

Ji, G., & Neugebauer, V. (2009). Hemispheric lateralization of pain processing by amygdala neurons. *J Neurophysiol*, *102*(4), 2253–2264. http://dx.doi.org/10.1152/jn.00166.2009.

Ji, G., Sun, H., Fu, Y., Li, Z., Pais-Vieira, M., Galhardo, V., & Neugebauer, V. (2010). Cognitive impairment in pain through amygdala-driven prefrontal cortical deactivation. *The Journal of Neuroscience*, *30*(15), 5451–5464. http://dx.doi.org/10.1523/JNEUROSCI.0225-10.2010.

Johansen, J. P., Fields, H. L., & Manning, B. H. (2001). The affective component of pain in rodents: direct evidence for a contribution of the anterior cingulate cortex. *Proceedings of the National Academy of Sciences of the United States of America*, *98*(14), 8077–8082. http://dx.doi.org/10.1073/pnas.141218998.

Kenny, P. J., Chen, S. A., Kitamura, O., Markou, A., & Koob, G. F. (2006). Conditioned withdrawal drives heroin consumption and decreases reward sensitivity. *The Journal of Neuroscience*, *26*(22), 5894–5900. http://dx.doi.org/10.1523/JNEUROSCI.0740-06.2006.

King, T., Vera-Portocarrero, L., Gutierrez, T., Vanderah, T. W., Dussor, G., Lai, J., … Porreca, F. (2009). Unmasking the tonic-aversive state in neuropathic pain. *Nature Neuroscience*, *12*(11), 1364–1366. http://dx.doi.org/10.1038/nn.2407.

Koob, G. F., & Le Moal, M. (2008). Addiction and the brain antireward system. *Annual Review of Psychology*, *59*, 29–53. http://dx.doi.org/10.1146/annurev.psych.59.103006.093548.

Koob, G. F., & Volkow, N. D. (2010). Neurocircuitry of addiction. *Neuropsychopharmacology*, *35*(1), 217–238. http://dx.doi.org/10.1038/npp.2009.110.

Laulin, J. P., Larcher, A., Celerier, E., Le Moal, M., & Simonnet, G. (1998). Long-lasting increased pain sensitivity in rat following exposure to heroin for the first time. *European Journal of Neuroscience*, *10*(2), 782–785.

Lenoir, M., & Ahmed, S. H. (2007). Heroin-induced reinstatement is specific to compulsive heroin use and dissociable from heroin reward and sensitization. *Neuropsychopharmacology*, *32*(3), 616–624. http://dx.doi.org/10.1038/sj.npp.1301083.

Levinthal, C. F. (2005). *Drugs, behavior, and modern society* (4th ed.). Boston, MA: Pearson.

Lutz, P. E., & Kieffer, B. L. (2013). The multiple facets of opioid receptor function: implications for addiction. *Current Opinion in Neurobiology*, *23*(4), 473–479. http://dx.doi.org/10.1016/j.conb.2013.02.005.

Martin, T. J., Buechler, N. L., Kim, S. A., Ewan, E. E., Xiao, R., & Childers, S. R. (2011). Involvement of the lateral amygdala in the antiallodynic and reinforcing effects of heroin in rats after peripheral nerve injury. *Anesthesiology*, *114*(3), 633–642. http://dx.doi.org/10.1097/ALN.0b013e318209aba7.

Martin, T. J., Kim, S. A., Buechler, N. L., Porreca, F., & Eisenach, J. C. (2007). Opioid self-administration in the nerve-injured rat: relevance of antiallodynic effects to drug consumption and effects of intrathecal analgesics. *Anesthesiology*, *106*(2), 312–322.

McNally, G. P., & Akil, H. (2002). Role of corticotropin-releasing hormone in the amygdala and bed nucleus of the stria terminalis in the behavioral, pain modulatory, and endocrine consequences of opiate withdrawal. *Neuroscience*, *112*(3), 605–617.

Mehendale, A. W., Goldman, M. P., & Mehendale, R. P. (2013). Opioid overuse pain syndrome (OOPS): the story of opioids, prometheus unbound. *Journal of Opioid Management*, *9*(6), 421–438. http://dx.doi.org/10.5055/jom.2013.0185.

Neugebauer, V. (2007). The amygdala: different pains, different mechanisms. *Pain*, *127*(1–2), 1–2. http://dx.doi.org/10.1016/j.pain.2006.10.004.

Neugebauer, V., Li, W., Bird, G. C., & Han, J. S. (2004). The amygdala and persistent pain. *Neuroscientist*, *10*(3), 221–234. http://dx.doi.org/10.1177/1073858403261077.

Park, P. E., Schlosburg, J. E., Vendruscolo, L. F., Schulteis, G., Edwards, S., & Koob, G. F. (2013). Chronic CRF receptor blockade reduces heroin intake escalation and dependence-induced hyperalgesia. *Addiction Biology*. http://dx.doi.org/10.1111/adb.12120.

Ren, Z. Y., Shi, J., Epstein, D. H., Wang, J., & Lu, L. (2009). Abnormal pain response in pain-sensitive opiate addicts after prolonged abstinence predicts increased drug craving. *Psychopharmacology (Berlin)*, *204*(3), 423–429. http://dx.doi.org/10.1007/s00213-009-1472-0.

Rouwette, T., Vanelderen, P., de Reus, M., Loohuis, N. O., Giele, J., van Egmond, J., … Kozicz, T. (2012). Experimental neuropathy increases limbic forebrain CRF. *European Journal of Pain*, *16*(1), 61–71. http://dx.doi.org/10.1016/j.ejpain.2011.05.016.

Schulteis, G., Markou, A., Gold, L. H., Stinus, L., & Koob, G. F. (1994). Relative sensitivity to naloxone of multiple indices of opiate withdrawal: a quantitative dose-response analysis. *Journal of Pharmacology and Experimental Therapeutics*, *271*(3), 1391–1398.

Scott, J. M. (1969). *The white poppy: A history of opium*. London: Heinemann.

Shaw-Lutchman, T. Z., Barrot, M., Wallace, T., Gilden, L., Zachariou, V., Impey, S., … Nestler, E. J. (2002). Regional and cellular mapping of cAMP response element-mediated transcription during naltrexone-precipitated morphine withdrawal. *The Journal of Neuroscience*, *22*(9), 3663–3672.

Shurman, J., Koob, G. F., & Gutstein, H. B. (2010). Opioids, pain, the brain, and hyperkatifeia: a framework for the rational use of opioids for pain. *Pain Medicine*, *11*(7), 1092–1098. http://dx.doi.org/10.1111/j.1526-4637.2010.00881.x.

Simonnet, G., & Rivat, C. (2003). Opioid-induced hyperalgesia: abnormal or normal pain? *Neuroreport*, *14*(1), 1–7. http://dx.doi.org/10.1097/01.wnr.0000051540.96524.e7.

Substance Abuse and Mental Health Services Administration. (2011). *Results from the 2011 National Survey on Drug Use and Health: Summary of national findings*. Rockville, MD: SAMHSA.

Swanson, L. W., Sawchenko, P. E., Rivier, J., & Vale, W. W. (1983). Organization of ovine corticotropin-releasing factor immunoreactive cells and fibers in the rat brain: an immunohistochemical study. *Neuroendocrinology*, *36*(3), 165–186.

Vendruscolo, L. F., Schlosburg, J. E., Misra, K. K., Chen, S. A., Greenwell, T. N., & Koob, G. F. (2011). Escalation patterns of varying periods of heroin access. *Pharmacology Biochemistry and Behavior*, *98*(4), 570–574. http://dx.doi.org/10.1016/j.pbb.2011.03.004.

Volkow, N. D., & McLellan, T. A. (2011). Curtailing diversion and abuse of opioid analgesics without jeopardizing pain treatment. *JAMA*, *305*(13), 1346–1347. http://dx.doi.org/10.1001/jama.2011.369.

Wasan, A. D., Ross, E. L., Michna, E., Chibnik, L., Greenfield, S. F., Weiss, R. D., & Jamison, R. N. (2012). Craving of prescription opioids in patients with chronic pain: a longitudinal outcomes trial. *The Journal of Pain*, *13*(2), 146–154. http://dx.doi.org/10.1016/j.jpain.2011.10.010.

Weiss, F., Ciccocioppo, R., Parsons, L. H., Katner, S., Liu, X., Zorrilla, E. P., ... Richter, R. R. (2001). Compulsive drug-seeking behavior and relapse. Neuroadaptation, stress, and conditioning factors. *Annals of the New York Academy of Sciences, 937*, 1–26.

Zamora-Martinez, E. R., & Edwards, S. (2014). Neuronal extracellular signal-regulated kinase (ERK) activity as marker and mediator of alcohol and opioid dependence. *Frontiers in Integrative Neuroscience, 8*, 24. http://dx.doi.org/10.3389/fnint.2014.00024.

Zhao, Y., Valdez, G. R., Fekete, E. M., Rivier, J. E., Vale, W. W., Rice, K. C., ... Zorrilla, E. P. (2007). Subtype-selective corticotropin-releasing factor receptor agonists exert contrasting, but not opposite, effects on anxiety-related behavior in rats. *Journal of Pharmacology and Experimental Therapeutics, 323*(3), 846–854. http://dx.doi.org/10.1124/jpet.107.123208.

Zhuo, M. (2014). Long-term potentiation in the anterior cingulate cortex and chronic pain. *Philosophical Transactions of the Royal Society of London. Series B, Biological Sciences, 369*(1633), 20130146. http://dx.doi.org/10.1098/rstb.2013.0146.

Zorrilla, E. P., Logrip, M. L., & Koob, G. F. (2014). Corticotropin releasing factor: a key role in the neurobiology of addiction. *Frontiers in Neuroendocrinology, 35*(2), 234–244. http://dx.doi.org/10.1016/j.yfrne.2014.01.001.

Chapter 84

The Role of the δ Opioid Receptor Gene, *OPRD1*, in Addiction

Richard C. Crist, Wade H. Berrettini

Center for Neurobiology and Behavior, Department of Psychiatry, University of Pennsylvania School of Medicine, Philadelphia, PA, USA

Abbreviations

CPP Conditioned place preference
DOR δ Opioid receptor
GWAS Genome-wide association study
KOR κ Opioid receptor
MOR μ Opioid receptor
SNP Single-nucleotide polymorphism

INTRODUCTION

Drug abuse and dependence are complex disorders that are regulated by a vast interacting network of genes and pathways. The pathways involved control a variety of phenotypes, ranging from the direct reinforcing effects of the drug to learning and memory. Dissecting the roles of specific genes in addiction is a difficult task that requires parsing the contributions of individual genes and more complex gene–gene interactions. However, the potential benefits of understanding the mechanisms of addiction are significant: identification of at-risk individuals in the population, increased numbers of new therapeutic targets, and personalized selection of appropriate treatments. Genes of interest to the addiction field thus far have primarily been driven by clear hypotheses, rather than unbiased screens. The result is that addiction research frequently focuses on genes with clear links to drug dependence, including receptors directly activated by drugs of abuse. A prominent example of these receptors is the opioid receptor family.

A number of opioid receptors have been identified, but the primary members of the receptor family are the μ opioid receptor (MOR), the δ opioid receptor (DOR), and the κ opioid receptor (KOR). These three proteins are encoded by the *OPRM1*, *OPRD1*, and *OPRK1* genes, respectively. All three proteins are transmembrane G-protein-coupled receptors and are able to activate downstream effects after binding by endogenous ligands. β-Endorphin is the endogenous peptide for MOR, but the receptor is also the primary target of morphine, oxycodone, fentanyl, and a host of other opioid drugs. MOR regulates much of the euphoria and analgesia traditionally associated with opioid use. DOR is also responsible for mediating the euphoric and analgesic effects of both endogenous enkephalins and exogenous ligands. Since these rewarding effects underlie the addictive qualities of opioids, MOR and DOR are intimately involved in the mechanisms of opioid addiction. KOR serves as a counterpoint to the other opioid receptors, binding dynorphin and regulating the dysphoric and aversive effects of opioid use.

As the primary target of many illicit and prescription opioids, MOR has been extensively studied in the addiction field at both the genetic and the molecular levels. While DOR has been studied to a lesser degree, a wide range of reports since 1995 have heavily implicated DOR in addiction. This chapter summarizes the current understanding of the roles of *OPRD1* and DOR in addiction susceptibility and treatment.

OPRD1 AND HUMAN ADDICTION

Both candidate gene studies and genome-wide association studies (GWAS) have been used to identify genetic variation associated with dependence and substance abuse. Unfortunately addiction GWAS, including studies on alcoholism, smoking, heroin dependence, and cocaine dependence, have not identified any significant hits in *OPRD1*. It should be noted, however, that only a single GWAS has been performed for both heroin and cocaine dependence and the numbers of subjects were relatively low (Gelernter, Kranzler, et al., 2014; Gelernter, Sherva, et al., 2014). Increases in both the sample size and the number of studies may yield additional targets in the future. These GWAS also focus entirely on common genetic variants and do not necessarily preclude effects of rare variants or other polymorphisms in *OPRD1*.

Candidate genes studies focusing on *OPRD1* itself have had significantly more success in identifying relevant polymorphisms (Figure 1). *OPRD1* has two coding single-nucleotide polymorphisms (SNPs) with minor allele frequencies large enough to be relevant on a population level: rs2234918, a synonymous variant in exon 3, and rs1042114, a nonsynonymous variant in exon 1. The minor alleles of both polymorphisms have been associated with opioid dependence in individuals of European origin (Mayer et al., 1997; Zhang, Kranzler, Yang, Luo, & Gelernter, 2008). These SNPs are also part of a six-variant haplotype that was associated with drug dependence (Zhang et al., 2008). In addition to the coding variants, a larger number of intronic SNPs have also been linked with addiction. Cocaine dependence in African-Americans is associated with the rs678849 genotype, while rs2236861 is associated with opioid dependence in a population from western Europe (Beer et al., 2013; Crist, Ambrose-Lanci, et al., 2013). An additional study of Australians of European origin identified a third variant (rs2236857) that was associated with opioid dependence (Nelson et al., 2012). Dependence was also associated with a haplotype block of rs2236857 and rs581111 in this population (Nelson et al., 2012). SNPs in *OPRD1* may also affect alcohol and drug dependence in combination with genetic variants in other opioid receptor genes (Li & Zhang, 2013). Not all of the association studies have had positive findings though. No link between *OPRD1* genotype and methamphetamine dependence was seen in a Japanese population, for example (Kobayashi et al., 2006). The majority of the significant hits in *OPRD1* fall in intron 1, which is unsurprising considering the ~46-kb intron accounts for almost 90% of the gene. In total there is no apparent connection between the various significant SNPs that would implicate a specific region of the *OPRD1* gene in all of these phenotypes.

Some work has also been done on the pharmacogenetics of addiction treatment, although it has lagged far behind case–control studies (Table 1). This is probably due to both the relative youth of the pharmacogenetics field and the small number of effective addiction treatments to actually study. Opioid dependence has been effectively treated with both methadone, a MOR agonist, and buprenorphine, a MOR partial agonist and KOR antagonist. Our laboratory has described an association between the *OPRD1* variant rs678849 and both methadone and buprenorphine efficacy in African-Americans (Crist, Clarke, et al., 2013). We have also observed an association between two additional *OPRD1* intronic polymorphisms and buprenorphine treatment outcome in women of European descent (Clarke et al., 2014). An additional study of methadone by Crettol et al. (2008) focused on the synonymous SNP rs2234918, but did not find an association with outcome. Beyond opioid dependence the only other addiction touched on by *OPRD1* pharmacogenetics studies is alcoholism, which has several treatments of varying effectiveness. A genetic variant in the 3′ UTR (rs4654327) was associated with the ability of naltrexone, a MOR antagonist, to reduce alcohol craving in European-Americans (Ashenhurst, Bujarski, & Ray, 2012). However, a number of other polymorphisms in the gene showed no correlation between genotype and alcohol use in patients treated with naltrexone (Gelernter et al., 2007). An additional therapeutic option for alcoholism is nalmefene, which is an antagonist at MOR and DOR but a partial agonist at KOR. Neither of the two *OPRD1* SNPs analyzed in connection with nalmefene treatment were associated with therapeutic efficacy (Arias et al., 2008).

It is important to understand the limitations presented by genetic association studies and pharmacogenetic analyses. Most of the negative studies have examined only a small number of polymorphisms and do not suggest that *OPRD1* as a whole is unrelated to the phenotypes being studied. Several of the significant findings have also failed to replicate in subsequent studies, leaving questions about the validity of the associations (Franke et al., 1999; Loh el, Fann, Chang, Chang, & Cheng, 2004; Xuei et al., 2007). These discrepancies may indicate false positives caused by random chance in the analyzed

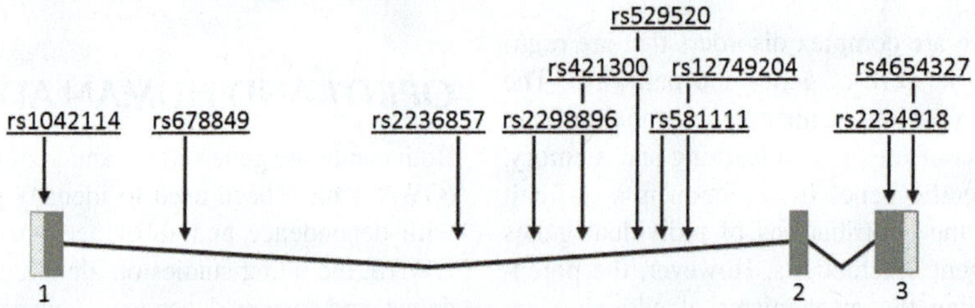

FIGURE 1 The locations of variants within the *OPRD1* gene that have been associated with addiction. The illustration shows the *OPRD1* gene at location 29,138,654–29,190,208 base pairs on chromosome 1. Gray boxes indicate the exons of the gene. The exon numbers are presented below the gene. Dotted boxes indicate the untranslated regions at the 5′ and 3′ ends. The gray lines represent intronic regions. The identification numbers and approximate locations of genetic variants associated with either addiction susceptibility or pharmacogenetics are shown.

TABLE 1 Associations between *OPRD1* Variants and Addiction Treatment Efficacy

Study	Dependence	Medication	SNP	Genotype	Effect
Crist, Clarke, et al. (2013)	Opioid	Buprenorphine	rs678849	C/C	Improved outcome in African-Americans
	Opioid	Methadone	rs678849	C/T and T/T	Improved outcome in African-Americans
Clarke et al. (2014)	Opioid	Buprenorphine	rs581111	G/G	Improved outcome in European-American women
	Opioid	Buprenorphine	rs529520	C/C	Improved outcome in European-American women
Ashenhurst et al. (2012)	Alcohol	Naltrexone	rs465327	A/G and A/A	Decreased alcohol stimulation and craving

The table shows the polymorphisms in the *OPRD1* gene that have statistically significant pharmacogenetic effects in addiction treatment. The addictive substance and the specific medication are included for each finding, as well as the polymorphism identification number and the relevant genotype(s). The genotype(s) for each finding is associated with the indicated effect.

cohort and demonstrate that even published associations should be considered suspect until confirmed in an independent population. Failures in replication studies may also be caused by ethnic differences in study samples and other confounding phenotypes such as comorbid addictions (Xu, Liu, Nagarajan, Gu, & Goldman, 2002). These potential problems highlight the need for detailed phenotyping and attention to confounds when designing association studies. Additionally, certain sets of genetic variants are inherited together in units known as a haplotype blocks. Polymorphisms in these blocks "tag" one another with varying degrees of correlation; for highly correlated SNPs, knowing the genotype at one locus in the block may permit imputation of the genotype at another locus. In this situation, it is difficult to determine whether the association with the phenotype is being driven by the variant that was genotyped or by a highly correlated allele in the same haplotype block. This can make it more difficult to design future experiments since tagged SNPs may be in different parts of the gene of interest or in other genes entirely. Despite these caveats, the growing volume of genetic associations suggests that there is likely to be some underlying connection between the *OPRD1* gene and human addictions.

RODENT MODELS AND ADDICTION

The associations between *OPRD1* polymorphisms and addiction phenotypes are potentially interesting but do not speak directly to a functional role for the gene. Virtually all drugs of abuse target receptors other than DOR and the few common opioids that do interact with the receptor are not thought to activate downstream signaling. However, the relevance of DOR to addiction is supported by substantial evidence in rodent model systems, which provide more direct methods of studying individual genes through the use of genetic manipulation.

A number of mouse lines have been generated with the mouse DOR gene, *Oprd1*, disabled, and these mice develop a range of drug-related phenotypes (Table 2). They do not develop tolerance to morphine, for example, although the analgesic effects of the drug are still present (Chefer & Shippenberg, 2009; Zhu et al., 1999). Further analyzing these knockout mice using traditional behavioral paradigms such as conditioned place preference (CPP), self-administration, and withdrawal has also revealed a number of phenotypic effects. Mice lacking *Oprd1* will self-administer morphine but do not develop place preference for the drug in CPP studies (Chefer & Shippenberg, 2009; Le Merrer et al., 2011). The presence of self-administration suggests that the rewarding effects of the drug are still present and any CPP differences are caused by other factors. More elaborate studies designed to dissect this discrepancy have provided evidence that the inability to develop place preference for morphine is caused by learning and memory deficits in the knockout mice (Le Merrer, Faget, Matifas, & Kieffer, 2012).

Rodent *Oprd1*-knockout models have also been used to study the effects on a variety of nonopioid drugs. *Oprd1*-null mutant mice show increased amounts of ethanol self-administration compared to control animals (Roberts et al., 2001). Loss of the *Oprd1* gene also abolishes nicotine CPP and decreases self-administration of a nicotine solution (Berrendero et al., 2012). *Oprd1*-knockout mice have also been studied in relation to Δ^9-tetrahydrocannabinol CPP and withdrawal; however, the loss of the gene had no significant effect (Ghozland et al., 2002).

Knocking out *Oprd1* in model organisms is not the only method for studying addiction in the absence of a functioning DOR. Treating animals with an antagonist of the receptor can accomplish the same effect, with the additional benefit of temporal control to avoid unforeseen issues caused by chronic loss of receptor function (Table 3). Disruption of DOR function by the antagonist naltrindole is able to inhibit morphine dependence as defined by naloxone-precipitated withdrawal, replicating the effects observed in *Oprd1*-knockout mice (Miyamoto, Portoghese, & Takemori, 1993a, 1993b). Antisense oligomers can be used to reduce DOR function as well, by preventing translation of *Oprd1* mRNA. Injection of antisense *Oprd1* RNA into the brains of mice

TABLE 2 Addiction Studies in *Oprd1*-Knockout Mice

Study	Drug	Paradigm	Effect of Knockout
Chefer and Shippenberg (2009)	Morphine	CPP	Decrease
	Morphine	Tolerance	Decrease
Zhu et al. (1999)	Morphine	Tolerance	Decrease
Le Merrer et al. (2011)	Morphine	CPP	Decrease
	Morphine	Self-administration	N/A
Le Merrer et al. (2012)	Morphine	Cue-associated CPP	N/A
Roberts et al. (2001)	Ethanol	Self-administration	Increase
Berrendero et al. (2012)	Nicotine	CPP	Decrease
	Nicotine	Self-administration	Decrease
Ghozland et al. (2002)	THC	CPP	N/A

The table lists studies of addiction-related phenotypes using *Oprd1*-knockout mouse lines. For each study, the specific drug and experimental paradigm are indicated. The effect observed in knockout mice compared to controls is also presented. N/A means that there was no observed effect of the knockout. CPP, conditioned place preference; Cue-associated CPP, conditioned place preference in which the drug was administered alongside audio and/or visual cues; THC, Δ^9-tetrahydrocannabinol.

TABLE 3 Nontransgenic Disruption of DOR in Rodent Models of Addiction

Study	Drug	Species	Method of Disruption	Paradigm	Effect
Miyamoto et al. (1993a)	Morphine	Mouse	Naltrindole	Precipitated withdrawal	Decrease
Miyamoto et al. (1993b)	Morphine	Mouse	Naltrindole	Precipitated withdrawal	Decrease
Kest et al. (1996)	Morphine	Mouse	Antisense mRNA	Tolerance	Decrease
Suzuki et al. (1997)	Morphine	Mouse	Antisense mRNA	CPP	Decrease
	Morphine	Mouse	Antisense mRNA	Precipitated withdrawal	Decrease
Billa et al. (2010)	Morphine	Rat	Naltriben	CPP	Decrease
Berrendero et al. (2012)	Nicotine	Mouse	Naltrindole	Self-administration	Decrease
	Nicotine	Mouse	Naltrindole	CPP	Decrease
Ismayilova and Shoaib (2010)	Nicotine	Rat	Naltrindole	Self-administration	N/A
Liu and Jernigan (2011)	Nicotine	Rat	Naltrindole	Self-administration	N/A
Margolis et al. (2008)	Alcohol	Rat	TIPP-Ψ	Self-administration	Decrease
Nielsen et al. (2012)	Alcohol	Rat	Naltrindole	Self-administration	Decrease
van Rijn and Whistler (2009)	Alcohol	Mouse	Naltriben	Self-administration	Decrease
Ward and Roberts (2007)	Cocaine	Rat	Naltrindole	Self-administration (VTA)	Increase
	Cocaine	Rat	Naltrindole	Self-administration (NA)	Decrease

The table shows studies of addiction-related phenotypes in rodents using disruption of *Oprd1* by means other than knockout models. For each study, the drug being studied, the model organism used, the method by which *Oprd1* or DOR is disrupted, the experimental paradigm used, and the observed effect compared to controls are indicated. N/A means there was no effect in treated rodents compared to controls. Where relevant, the specific location of self-administration is indicated. VTA, ventral tegmental area; NA, nucleus accumbens; CPP, conditioned place preference.

administered morphine resulted in a failure to develop tolerance or other symptoms of dependence (Kest, Lee, McLemore, & Inturrisi, 1996; Suzuki et al., 1997). Rodents injected with DOR antagonists also have decreased morphine CPP (Billa, Xia, & Moron, 2010). Both of these phenotypes were previously noted in *Oprd1*-knockout mice treated with morphine.

DOR antagonists also alter the effects of drugs outside the opioid family. Administering naltrindole to mice mirrors the effects of *Oprd1* knockout in regards to nicotine: the treated mice show decreases in self-administration and CPP (Berrendero et al., 2012). Similar treatments in rats do not result in the decreased self-administration seen in mice, although it is unclear if these differences are species-specific or simply strain-specific (Ismayilova & Shoaib, 2010; Liu & Jernigan, 2011). Naltrindole and other DOR antagonists also reduce alcohol self-administration in both rats and mice (Margolis, Fields, Hjelmstad, & Mitchell, 2008; Nielsen et al., 2012; van Rijn & Whistler, 2009). These findings are the opposite of the effects seen in *Oprd1*-knockout mice, suggesting that chemical antagonism of the receptor is not entirely functionally equivalent to complete loss of the receptor via knockout, perhaps due to developmental factors. Cocaine-seeking behavior is altered as well, with rats treated with antagonist performing significantly more lever presses for cocaine injected into the ventral tegmental area and significantly fewer level presses when the cocaine is instead injected into the nucleus accumbens (Ward & Roberts, 2007). These findings highlight the benefits of the spatial and temporal control provided by antagonist treatment versus constitutive knockout models.

A smaller number of studies have further analyzed DOR by administering DOR-specific agonists alongside illicit drugs (Table 4). Morphine CPP is increased by agonist treatment in mice (Suzuki et al., 1996). Activation of DOR also boosts the effects of two classes of stimulants: cocaine and amphetamines. Treatment of mice with cocaine under normal conditions results in increased locomotor activity, and repeated administration over time causes a sensitization effect in which each dose results in higher rates of locomotion than the previous doses (Kotlinska, Gibula-Bruzda, Witkowska, & Izdebski, 2013). The addition of deltorphins, endogenous agonists of DOR, enhances both the initial increase in movement caused by cocaine and the subsequent sensitization (Kotlinska et al., 2013). Amphetamine use causes dopamine efflux in both humans and rodents, resulting in the euphoria. Rats treated with the DOR agonist SNC80 show increased dopamine release following amphetamine injection (Bosse et al., 2014). Finally, several studies have examined the relationship between DOR agonists and ethanol dependence. Both SNC80 and DALA increased self-administration, but DPDPE treatment resulted in a decrease in alcohol consumption (Barson et al., 2010; Margolis et al., 2008; van Rijn, Brissett, & Whistler, 2010). The difference in results may reflect the location of agonist injection; DPDPE was injected into the ventral tegmental area while the other agonists were administered

TABLE 4 The Effects of DOR Agonists on Rodent Models of Addiction

Study	Drug	Species	Agonist	Paradigm	Effect
Suzuki et al. (1996)	Morphine	Mouse	TAN-67	CPP	Increase
Kotlinska et al. (2013)	Cocaine	Mouse	Deltorphin	Locomotor activity	Increase
	Cocaine	Mouse	Deltorphin	Cocaine sensitization	Increase
Bosse et al. (2014)	Amphetamine	Rat	SNC80	Dopamine efflux	Increase
Barson et al. (2010)	Alcohol	Rat	DALA	Self-administration (PVN)	Increase
Margolis et al. (2008)	Alcohol	Rat	DPDPE	Self-administration (VTA)	Decrease
van Rijn et al. (2010)	Alcohol	Rat	SNC80	Self-administration (IP)	Increase

The table lists the studies of addiction-related phenotypes in rodents using DOR agonists. The specific drug, species, chemical agonist, experimental paradigm, and observed effect are provided for each study. Where relevant, the specific location of self-administration is indicated. VTA, ventral tegmental area; PVN, paraventricular nucleus. IP, intraperitoneal; CPP, conditioned place preference.

intraperitoneally or to other regions of the brain (Barson et al., 2010; Margolis et al., 2008; van Rijn et al., 2010). Taken in aggregate, there is substantial evidence from rodent models to support a role for *Oprd1* and DOR in the drug dependence.

DRUGS AND δ OPIOID RECEPTOR FUNCTION

Human genetic association studies and animal models both suggest that DOR is involved in regulating addiction to a number of substances. Neither of these methods, however, indicates any specific mechanisms by which the receptor affects these phenotypes. Since no prominent drugs of abuse have DOR as their primary target, the potential relevance of the receptor is not readily apparent. Nonetheless, a number of groups have identified intriguing connections between opioids and DOR function, expression, and trafficking.

The addition of morphine to oral epithelial cell cultures results in cell migration and this migration is mediated by DOR (Charbaji, Schafer-Korting, & Kuchler, 2012). Morphine treatment in primary cultures of neurons also reduces internalization of DOR, resulting in increased levels of the receptor at the cell surface (Ong, Xue, Olmstead, & Cahill, 2014). The change in DOR trafficking is abolished by treatment with a DOR antagonist, which again supports some role for morphine in activating the receptor (Ong et al., 2014). An additional study, however, found that chronic administration of morphine in rats resulted in downregulation of DOR in the nucleus accumbens (Turchan et al., 1999). These varying results suggest that there may be differences in the physiological effects caused by chronic and acute morphine treatment. The data also suggest that morphine is directly or indirectly causing signaling through DOR despite being classified solely as a MOR agonist.

Methadone and buprenorphine may also have the ability to alter DOR trafficking and function. Levels of the DOR protein are significantly decreased in blood drawn from methadone maintenance patients compared with healthy controls (Toskulkao, Pornchai, Akkarapatumwong, Vatanatunyakum, & Govitrapong, 2010). This downregulation is mirrored in cell lines treated directly with the medication (Toskulkao et al., 2010). Cell lines exposed to methadone also have been shown to develop desensitization of DOR, which would have functional consequences similar to those of decreases in receptor expression (Liu, Liao, Gong, & Qin, 1999). Desensitization does not occur when morphine is used in lieu of methadone, suggesting that the drugs, while similar, have different downstream effects on DOR (Liu et al., 1999). As with morphine, methadone is thought to have limited affinity for the receptor (Kristensen, Christensen, & Christrup, 1995). The observed desensitization of DOR following methadone administration, therefore, suggests either that methadone indirectly affects DOR or that the medication has more of a direct effect on the receptor than previously appreciated. In support of the latter hypothesis, methadone has been shown to activate intracellular signaling through DOR, and not MOR, in mice treated with the drug for 3 days (Rady, Portoghese, & Fujimo, 2002).

Unlike methadone, buprenorphine has a relatively high affinity for DOR; however, conflicting publications have indicated a role for the medication as either an agonist or an antagonist of the receptor (Huang, Kehner, Cowan, & Liu-Chen, 2001; Kajiwara et al., 1986; Negus et al., 2002; Wood, Charleson, Lane, & Hudgin, 1981). Regardless of the compound's exact function, treatment of rats with buprenorphine has been shown to increase DOR expression in the brain (Belcheva et al., 1993, 1996). These data again suggest that some opioids may have downstream functions that affect DOR.

Changes in DOR function and expression have also been found in rodent models of addiction to substances other than opioids. Cocaine administration in rats causes downregulation of DOR in the nucleus accumbens, similar to that caused by morphine; however, this decrease in expression affects only a subtype of δ receptors known as δ1 and does not affect δ2 receptors (Turchan et al., 1999). Treatment of rats with clorazepate, a benzodiazepine, resulted in upregulation of DOR in the brain, and mice chronically treated with ethanol expressed increased levels of DOR in the spinal cord compared to controls (Quentin et al., 2005; van Rijn, Brissett, & Whistler, 2012). Levels of receptor in the ventral tegmental area were also associated with decreased self-administration of alcohol (Margolis et al., 2008; Mitchell, Margolis, Coker, & Fields, 2012). During withdrawal, nicotine-dependent mice displayed a decrease in functional DOR in the nucleus accumbens, again indicating that substances of abuse can affect DOR expression despite not directly targeting the receptor itself (McCarthy, Zhang, Neff, & Hadjiconstantinou, 2011).

The functional associations between DOR and addictive substances are still preliminary and significant work is required to expand on these early findings. However, the current data demonstrate two important points: (1) drugs from multiple classes can alter DOR expression or localization and (2) there are at least some cases in which drugs can activate downstream signaling via DOR. Both scenarios highlight the potential relevance of DOR in addiction onset and treatment.

HETERODIMERIZATION OF μ AND δ OPIOID RECEPTORS

The studies discussed in this chapter so far have focused on DOR or *OPRD1* alone. Often the experiments involved have treated each of the opioid receptor family members as independent entities, knocking out individual genes or using receptor-specific ligands to study phenotypic effects. There is substantial evidence, however, that indicates opioid receptors can form functional heterodimers in addition to monomers and homodimers (reviewed in Fujita, Gomes, & Devi, 2015). While DOR–KOR and MOR–KOR heterodimers have been observed, major efforts are currently focused on dissecting the relevance of MOR–DOR heterodimers. The heterodimers are known to have different signaling properties compared to the individual receptor types alone. For example, methadone, morphine, and other MOR agonists have been shown to have significantly higher efficacy at MOR–DOR heterodimers than at MOR alone (Yekkirala et al., 2012). MOR–DOR heterodimers also have different downstream signaling mechanisms compared to either of the receptors individually. The heterodimers bind different classes of G proteins compared to the individual receptors, constitutively recruit β-arrestin, and cause altered phosphorylation of extracellular signal-regulated kinase proteins compared to MOR alone (Rozenfeld & Devi, 2007).

In addition to the growing body of evidence for MOR–DOR heterodimers altering the effects of opioid ligands, there are data suggesting that opioids can alter the abundance of heterodimers. Chronic treatment of both mice and primary cultured neurons with morphine results in increased MOR–DOR heterodimers at the cell surface (Gupta et al., 2010; Ong et al., 2014). This increase appears to be the result of changes in trafficking of the heterodimer complex compared to the component receptors alone (Ong et al., 2014). In vitro experiments with methadone have also found differences in heterodimer trafficking. Internalization of MOR following methadone treatment primarily results in the receptor being trafficked back to the cell surface (Milan-Lobo & Whistler, 2011). Heterodimers, however, are degraded in response to methadone-induced internalization (Milan-Lobo & Whistler, 2011).

The discovery of MOR–DOR heterodimers is relatively recent but the field has already offered potential explanations for the role of DOR in addiction. The changes in receptor activation and downstream signaling caused by heterodimerization allow DOR to affect signaling through MOR by acting as a sink for the MOR protein. Upregulation of DOR, which can be caused by various factors, including administration of certain drugs, probably increases the chances of MOR–DOR heterodimerization. This scenario forces a shift in downstream activation from MOR-specific signaling to heterodimer-specific signaling, despite DOR alone never directly interacting with traditional MOR ligands (Figure 2). The observed effects of opioids on heterodimer prevalence at the cell surface and the higher efficacy of many of the compounds at the heterodimer complex may also create a type of feedback loop. This feedback could quickly shift the prevailing downstream activation toward heterodimer-dominant signaling rather than the MOR-dominant signaling traditionally believed to be associated with most opioid drugs.

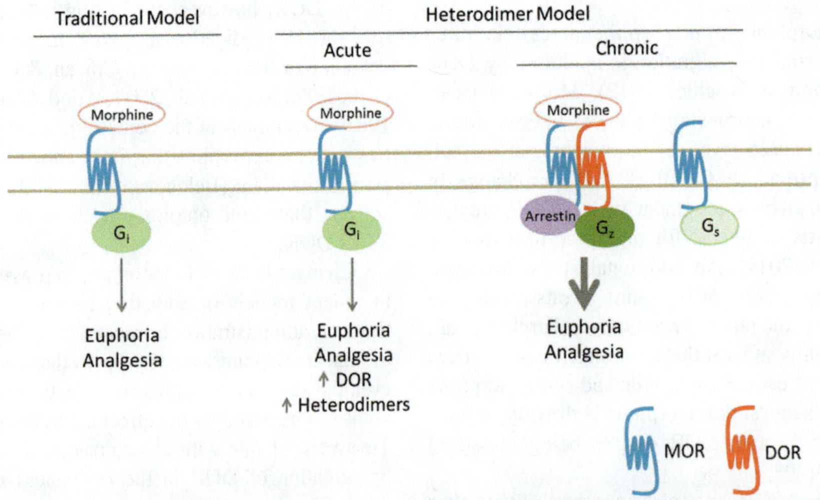

FIGURE 2 The role of MOR–DOR heterodimers in morphine use. The illustration compares the traditional model of morphine use with a newer model incorporating more recent data on MOR–DOR heterodimers. In the traditional model, morphine activates downstream signaling through MOR and G_i proteins. Newer data have demonstrated that morphine use causes upregulation of DOR and MOR–DOR heterodimers, as well as potentially switching MOR signaling to G_s proteins (Wang, Friedman, Olmstead, & Burns, 2005). Morphine has higher efficacy at the heterodimers, which differ from the individual receptors by constitutively interacting with β-arrestin and signaling through G_z proteins. Therefore, chronic morphine use may cause a shift from MOR-dominant signaling to heterodimer-dominant signaling.

CONCLUSION

The opioid receptor field is still fleshing out the role of DOR in addiction onset and treatment. Numerous sources of evidence, including human case–control analyses, rodent models, and in vitro studies, now indicate that *OPRD1* and DOR are intimately involved in dependence on a number of drug classes. However, several questions still remain to be answered. The most notable one is how DOR mediates these phenotypes. The relatively recent discovery of MOR–DOR heterodimers may be a key piece of information in answering these questions. Activation of MOR–DOR heterodimers has different downstream effects compared to activation of MOR or DOR alone, providing a potential mechanism for DOR to alter the effects of drugs that do not target DOR monomers directly. The connection between DOR and learning and memory may also link the receptor to addiction, which is a learned habit caused by dysfunctional reinforcement mechanisms. Mutations, polymorphisms, and environmental factors will almost certainly create a gradient of DOR expression levels in the population, and expression of the receptor may correlate with the ability to learn addictive behavior.

Beyond the underlying biomedical research connecting DOR and addiction is the question of how best to apply these findings to improve patient treatment. Confirmation of *OPRD1* polymorphisms associated with addiction may allow the identification of individuals with a higher susceptibility to dependence. Pharmacogenetics findings have even more benefit, potentially allowing clinicians to prescribe treatments that are more likely to have successful outcomes. Increased appreciation of the involvement of DOR also labels the receptor as a potential therapeutic target, either through heterodimer-specific ligands or DOR-specific agonists and antagonists. It is hoped that ongoing work on *OPRD1* and DOR will fill in the current knowledge gaps and put new and future findings into clinical practice.

APPLICATIONS TO OTHER ADDICTIONS AND SUBSTANCE MISUSE

The vast majority of human research on DOR and *OPRD1* has focused on opioid dependence, which is a logical extension of the known function of DOR and the receptor's similarity to MOR. However, this primary focus on opioid dependence may have resulted in important connections between the receptor and other addictions being overlooked. Extensive work in rodent models has demonstrated that disruption of *Oprd1* by transgenic manipulation or DOR by chemical antagonism alters response to several different drugs of abuse. For example, these models have shown *Oprd1*-knockout mice to self-administer more alcohol and less nicotine than controls. The mice also have changes in place preference and tolerance to morphine. Treatment of rodents with DOR agonists and antagonists further alters response to morphine, cocaine, amphetamines, nicotine, and alcohol. These data suggest that DOR is relevant to a wide variety of addictions and not simply opioid addiction. In support of this hypothesis, research has found associations between *OPRD1* variants and both cocaine dependence and the efficacy of naltrexone in reducing cravings for alcohol. Despite these results, the study of DOR in human research is still dominated by research on opioid dependence. The human and rodent findings highlight the need for the field to expand and better incorporate other addictions into DOR research. The receptor is likely to be relevant to a wide range of addictions, but that relevancy will be revealed only when researchers shift their focus to better include substances like cocaine, nicotine, and alcohol.

DEFINITION OF TERMS

Antisense oligomers These are single-stranded DNA molecules designed to complement a specific mRNA and form a DNA–RNA duplex.

Comorbid This refers to phenotypes that occur at the same time in a single patient.

Conditioned place preference This is a rodent model of addiction using a box with two distinct sides. Rodents are repeatedly administered saline on one side and the study drug on the other. If test animals subsequently prefer the drug side of the box, it indicates the drug has addictive qualities.

Desensitization This describes internalization of a cell surface receptor following activation of the receptor. This is used as a negative feedback mechanism.

Efflux This is the active transport of compounds through the cell membrane and into the extracellular space.

Endogenous This describes a substance produced internally by the organism.

Exogenous This describes a substance produced externally and then introduced into the organism.

Haplotype This is a group of genetic variants that are inherited together more often than predicted by random chance.

Knockout This is a transgenic model organism in which a specific gene has been made entirely nonfunctional by disruption at the DNA level.

Minor allele frequency This is the rate at which the less common of two alleles at a particular genetic variant is present in a population.

Tagged SNP This is a genetic variant whose genotype can be predicted if the genotype of another variant is already known, because the two variants are inherited together.

KEY FACTS OF PHARMACOGENETICS OF ADDICTION

- Even the best current addiction treatments have poor efficacy or negative side effects in many patients.
- These phenotypes are the result of a combination of environmental and genetic factors.
- Pharmacogenetics is the study of how genetic variation in patients affects these phenotypes.
- Identification of relevant polymorphisms could allow prospective genotyping and prescription of medications with the best chance of successful outcomes.
- Pharmacogenetics of addiction has been hampered by relatively small sample sizes, which reduce the statistical power to identify relevant variants.
- A lack of quality treatments for many addictions (e.g., cocaine dependence) has also limited the growth of the field thus far.

SUMMARY POINTS

- This chapter focuses on the DOR, which is a G-protein-coupled receptor activated by endogenous and exogenous opioids.
- The DOR is encoded by the *OPRD1* gene and is part of a family of receptors including the MOR and the KOR.

- Activation of DOR results in euphoria and analgesia and also has effects on learning and memory.
- Genetic association studies and pharmacogenetic studies have linked numerous *OPRD1* variants to human addiction.
- Disruption of the orthologous genes in mice and rats has also been shown to alter a wide variety of addiction phenotypes, including CPP, self-administration, and withdrawal.
- Administration of many drugs of abuse can alter the function and cellular trafficking of DOR, indicating a connection despite the receptor not being the primary target of the drugs.
- DOR can form heterodimers with MOR and these heterodimers have altered signaling properties compared to the individual receptors.
- Research indicates that DOR is intimately involved in the molecular processes of addiction but further research is required to elucidate the specific role of the receptor.

REFERENCES

Arias, A. J., Armeli, S., Gelernter, J., Covault, J., Kallio, A., Karhuvaara, S., ... Kranzler, H. R. (2008). Effects of opioid receptor gene variation on targeted nalmefene treatment in heavy drinkers. *Alcoholism, Clinical and Experimental Research*, 32(7), 1159–1166.

Ashenhurst, J. R., Bujarski, S., & Ray, L. A. (2012). Delta and kappa opioid receptor polymorphisms influence the effects of naltrexone on subjective responses to alcohol. *Pharmacology Biochemistry and Behavior*, 103(2), 253–259.

Barson, J. R., Carr, A. J., Soun, J. E., Sobhani, N. C., Rada, P., Leibowitz, S. F., & Hoebel, B. G. (2010). Opioids in the hypothalamic paraventricular nucleus stimulate ethanol intake. *Alcoholism, Clinical and Experimental Research*, 34(2), 214–222.

Beer, B., Erb, R., Pavlic, M., Ulmer, H., Giacomuzzi, S., Riemer, Y., & Oberacher, H. (2013). Association of polymorphisms in pharmacogenetic candidate genes (OPRD1, GAL, ABCB1, OPRM1) with opioid dependence in European population: a case-control study. *PLoS One*, 8(9), e75359.

Belcheva, M. M., Barg, J., McHale, R. J., Dawn, S., Ho, M. T., Ignatova, E., & Coscia, C. J. (1993). Differential down- and up-regulation of rat brain opioid receptor types and subtypes by buprenorphine. *Molecular Pharmacology*, 44(1), 173–179.

Belcheva, M. M., Ho, M. T., Ignatova, E. G., Jefcoat, L. B., Barg, J., Vogel, Z., ... Coscia, C. J. (1996). Buprenorphine differentially alters opioid receptor adaptation in rat brain regions. *Journal of Pharmacology and Experimental Therapeutics*, 277(3), 1322–1327.

Berrendero, F., Plaza-Zabala, A., Galeote, L., Flores, A., Bura, S. A., Kieffer, B. L., & Maldonado, R. (2012). Influence of delta-opioid receptors in the behavioral effects of nicotine. *Neuropsychopharmacology*, 37(10), 2332–2344.

Billa, S. K., Xia, Y., & Moron, J. A. (2010). Disruption of morphine-conditioned place preference by a delta2-opioid receptor antagonist: study of mu-opioid and delta-opioid receptor expression at the synapse. *European Journal of Neuroscience*, 32(4), 625–631.

Bosse, K. E., Jutkiewicz, E. M., Schultz-Kuszak, K. N., Mabrouk, O. S., Kennedy, R. T., Gnegy, M. E., & Traynor, J. R. (2014). Synergistic activity between the delta-opioid agonist SNC80 and amphetamine occurs via a glutamatergic NMDA-receptor dependent mechanism. *Neuropharmacology*, 77, 19–27.

Charbaji, N., Schafer-Korting, M., & Kuchler, S. (2012). Morphine stimulates cell migration of oral epithelial cells by delta-opioid receptor activation. *PLoS One*, 7(8), e42616.

Chefer, V. I., & Shippenberg, T. S. (2009). Augmentation of morphine-induced sensitization but reduction in morphine tolerance and reward in delta-opioid receptor knockout mice. *Neuropsychopharmacology*, 34(4), 887–898.

Clarke, T. K., Crist, R. C., Ang, A., Ambrose-Lanci, L. M., Lohoff, F. W., Saxon, A. J., ... Berrettini, W. H. (2014). Genetic variation in OPRD1 and the response to treatment for opioid dependence with buprenorphine in European-American females. *The Pharmacogenomics Journal*, 14(3), 303–308.

Crettol, S., Besson, J., Croquette-Krokar, M., Hammig, R., Gothuey, I., Monnat, M., ... Eap, C. B. (2008). Association of dopamine and opioid receptor genetic polymorphisms with response to methadone maintenance treatment. *Progress in Neuro-Psychopharmacology and Biological Psychiatry*, 32(7), 1722–1727.

Crist, R. C., Ambrose-Lanci, L. M., Vaswani, M., Clarke, T. K., Zeng, A., Yuan, C., ... Berrettini, W. H. (2013). Case-control association analysis of polymorphisms in the delta-opioid receptor, OPRD1, with cocaine and opioid addicted populations. *Drug and Alcohol Dependence*, 127(1–3), 122–128.

Crist, R. C., Clarke, T. K., Ang, A., Ambrose-Lanci, L. M., Lohoff, F. W., Saxon, A. J., ... Berrettini, W. H. (2013). An intronic variant in OPRD1 predicts treatment outcome for opioid dependence in African-Americans. *Neuropsychopharmacology*, 38(10), 2003–2010.

Franke, P., Nothen, M. M., Wang, T., Neidt, H., Knapp, M., Lichtermann, D., ... Maier, W. (1999). Human delta-opioid receptor gene and susceptibility to heroin and alcohol dependence. *American Journal of Medical Genetics*, 88(5), 462–464.

Fujita, W., Gomes, I., & Devi, L. A. (2015). Mu-Delta opioid receptor heteromers: new pharmacology and novel therapeutic possibilities. *British Journal of Pharmacology*, 172(2), 375–387.

Gelernter, J., Gueorguieva, R., Kranzler, H. R., Zhang, H., Cramer, J., Rosenheck, R., ... VA Cooperative Study #425 Study Group (2007). Opioid receptor gene (OPRM1, OPRK1, and OPRD1) variants and response to naltrexone treatment for alcohol dependence: results from the VA Cooperative Study. *Alcoholism, Clinical and Experimental Research*, 31(4), 555–563.

Gelernter, J., Kranzler, H. R., Sherva, R., Koesterer, R., Almasy, L., Zhao, H., & Farrer, L. A. (2014). Genome-wide association study of opioid dependence: multiple associations mapped to calcium and potassium pathways. *Biological Psychiatry*, 76(1), 66–74.

Gelernter, J., Sherva, R., Koesterer, R., Almasy, L., Zhao, H., Kranzler, H. R., & Farrer, L. A. (2014). Genome-wide association study of cocaine dependence and related traits: FAM53B identified as a risk gene. *Molecular Psychiatry*, 19(6), 717–723.

Ghozland, S., Matthes, H. W., Simonin, F., Filliol, D., Kieffer, B. L., & Maldonado, R. (2002). Motivational effects of cannabinoids are mediated by mu-opioid and kappa-opioid receptors. *The Journal of Neuroscience*, 22(3), 1146–1154.

Gupta, A., Mulder, J., Gomes, I., Rozenfeld, R., Bushlin, I., Ong, E., ... Devi, L. A. (2010). Increased abundance of opioid receptor heteromers after chronic morphine administration. *Science Signaling*, 3(131), ra54.

Huang, P., Kehner, G. B., Cowan, A., & Liu-Chen, L. Y. (2001). Comparison of pharmacological activities of buprenorphine and norbuprenorphine: norbuprenorphine is a potent opioid agonist. *Journal of Pharmacology and Experimental Therapeutics*, 297(2), 688–695.

Ismayilova, N., & Shoaib, M. (2010). Alteration of intravenous nicotine self-administration by opioid receptor agonist and antagonists in rats. *Psychopharmacology (Berlin)*, 210(2), 211–220.

Kajiwara, M., Aoki, K., Ishii, K., Numata, H., Matsumiya, T., & Oka, T. (1986). Agonist and antagonist actions of buprenorphine on three types of opioid receptor in isolated preparations. *The Japanese Journal of Pharmacology, 40*(1), 95–101.

Kest, B., Lee, C. E., McLemore, G. L., & Inturrisi, C. E. (1996). An antisense oligodeoxynucleotide to the delta opioid receptor (DOR-1) inhibits morphine tolerance and acute dependence in mice. *Brain Research Bulletin, 39*(3), 185–188.

Kobayashi, H., Hata, H., Ujike, H., Harano, M., Inada, T., Komiyama, T., ... Sora, I. (2006). Association analysis of delta-opioid receptor gene polymorphisms in methamphetamine dependence/psychosis. *American Journal of Medical Genetics Part B Neuropsychiatric Genetics, 141B*(5), 482–486.

Kotlinska, J. H., Gibula-Bruzda, E., Witkowska, E., & Izdebski, J. (2013). Involvement of delta and mu opioid receptors in the acute and sensitized locomotor action of cocaine in mice. *Peptides, 48*, 89–95.

Kristensen, K., Christensen, C. B., & Christrup, L. L. (1995). The mu1, mu2, delta, kappa opioid receptor binding profiles of methadone stereoisomers and morphine. *Life Sciences, 56*(2), PL45–PL50.

Le Merrer, J., Faget, L., Matifas, A., & Kieffer, B. L. (2012). Cues predicting drug or food reward restore morphine-induced place conditioning in mice lacking delta opioid receptors. *Psychopharmacology (Berlin), 223*(1), 99–106.

Le Merrer, J., Plaza-Zabala, A., Del Boca, C., Matifas, A., Maldonado, R., & Kieffer, B. L. (2011). Deletion of the delta opioid receptor gene impairs place conditioning but preserves morphine reinforcement. *Biological Psychiatry, 69*(7), 700–703.

Li, Z., & Zhang, H. (2013). Analyzing interaction of mu-, delta- and kappa-opioid receptor gene variants on alcohol or drug dependence using a pattern discovery-based method. *Journal of Addiction Research and Therapy* (Suppl. 7), 007.

Liu, J. G., Liao, X. P., Gong, Z. H., & Qin, B. Y. (1999). Methadone-induced desensitization of the delta-opioid receptor is mediated by uncoupling of receptor from G protein. *European Journal of Pharmacology, 374*(2), 301–308.

Liu, X., & Jernigan, C. (2011). Activation of the opioid mu1, but not delta or kappa, receptors is required for nicotine reinforcement in a rat model of drug self-administration. *Progress in Neuro-Psychopharmacology and Biological Psychiatry, 35*(1), 146–153.

Loh el, W., Fann, C. S., Chang, Y. T., Chang, C. J., & Cheng, A. T. (2004). Endogenous opioid receptor genes and alcohol dependence among Taiwanese Han. *Alcoholism, Clinical and Experimental Research, 28*(1), 15–19.

Margolis, E. B., Fields, H. L., Hjelmstad, G. O., & Mitchell, J. M. (2008). Delta-opioid receptor expression in the ventral tegmental area protects against elevated alcohol consumption. *The Journal of Neuroscience, 28*(48), 12672–12681.

Mayer, P., Rochlitz, H., Rauch, E., Rommelspacher, H., Hasse, H. E., Schmidt, S., & Höllt, V. (1997). Association between a delta opioid receptor gene polymorphism and heroin dependence in man. *Neuroreport, 8*(11), 2547–2550.

McCarthy, M. J., Zhang, H., Neff, N. H., & Hadjiconstantinou, M. (2011). Desensitization of delta-opioid receptors in nucleus accumbens during nicotine withdrawal. *Psychopharmacology (Berlin), 213*(4), 735–744.

Milan-Lobo, L., & Whistler, J. L. (2011). Heteromerization of the mu- and delta-opioid receptors produces ligand-biased antagonism and alters mu-receptor trafficking. *Journal of Pharmacology and Experimental Therapeutics, 337*(3), 868–875.

Mitchell, J. M., Margolis, E. B., Coker, A. R., & Fields, H. L. (2012). Alcohol self-administration, anxiety, and cortisol levels predict changes in delta opioid receptor function in the ventral tegmental area. *Behavioral Neuroscience, 126*(4), 515–522.

Miyamoto, Y., Portoghese, P. S., & Takemori, A. E. (1993a). Involvement of delta 2 opioid receptors in acute dependence on morphine in mice. *Journal of Pharmacology and Experimental Therapeutics, 265*(3), 1325–1327.

Miyamoto, Y., Portoghese, P. S., & Takemori, A. E. (1993b). Involvement of delta 2 opioid receptors in the development of morphine dependence in mice. *Journal of Pharmacology and Experimental Therapeutics, 264*(3), 1141–1145.

Negus, S. S., Bidlack, J. M., Mello, N. K., Furness, M. S., Rice, K. C., & Brandt, M. R. (2002). Delta opioid antagonist effects of buprenorphine in rhesus monkeys. *Behavioral Pharmacology, 13*(7), 557–570.

Nelson, E. C., Lynskey, M. T., Heath, A. C., Wray, N., Agrawal, A., Shand, F. L., ... Montgomery, G. W. (2012). Association of OPRD1 polymorphisms with heroin dependence in a large case-control series. *Addiction Biology, 19*(1), 111–121.

Nielsen, C. K., Simms, J. A., Li, R., Mill, D., Yi, H., Feduccia, A. A., ... Bartlett, S. E. (2012). Delta-opioid receptor function in the dorsal striatum plays a role in high levels of ethanol consumption in rats. *The Journal of Neuroscience, 32*(13), 4540–4552.

Ong, E. W., Xue, L., Olmstead, M. C., & Cahill, C. M. (2014). Prolonged morphine treatment alters delta opioid receptor post-internalisation trafficking. *British Journal of Pharmacology, 172*(2), 615–629.

Quentin, T., Debruyne, D., Lelong-Boulouard, V., Poisnel, G., Barre, L., & Coquerel, A. (2005). Clorazepate affects cell surface regulation of delta and kappa opioid receptors, thereby altering buprenorphine-induced adaptation in the rat brain. *Brain Research, 1063*(1), 84–95.

Rady, J. J., Portoghese, P. S., & Fujimo, J. M. (2002). Methadone and heroin antinociception: predominant delta-opioid-receptor responses in methadone-tolerant mice. *The Japanese Journal of Pharmacology, 88*(3), 319–331.

van Rijn, R. M., Brissett, D. I., & Whistler, J. L. (2010). Dual efficacy of delta opioid receptor-selective ligands for ethanol drinking and anxiety. *Journal of Pharmacology and Experimental Therapeutics, 335*(1), 133–139.

van Rijn, R. M., Brissett, D. I., & Whistler, J. L. (2012). Emergence of functional spinal delta opioid receptors after chronic ethanol exposure. *Biological Psychiatry, 71*(3), 232–238.

van Rijn, R. M., & Whistler, J. L. (2009). The delta(1) opioid receptor is a heterodimer that opposes the actions of the delta(2) receptor on alcohol intake. *Biological Psychiatry, 66*(8), 777–784.

Roberts, A. J., Gold, L. H., Polis, I., McDonald, J. S., Filliol, D., Kieffer, B. L., & Koob, G. F. (2001). Increased ethanol self-administration in delta-opioid receptor knockout mice. *Alcoholism, Clinical and Experimental Research, 25*(9), 1249–1256.

Rozenfeld, R., & Devi, L. A. (2007). Receptor heterodimerization leads to a switch in signaling: beta-arrestin2-mediated ERK activation by mu-delta opioid receptor heterodimers. *The FASEB Journal, 21*(10), 2455–2465.

Suzuki, T., Ikeda, H., Tsuji, M., Misawa, M., Narita, M., & Tseng, L. F. (1997). Antisense oligodeoxynucleotide to delta opioid receptors attenuates morphine dependence in mice. *Life Sciences, 61*(11), PL165–PL170.

Suzuki, T., Tsuji, M., Mori, T., Misawa, M., Endoh, T., & Nagase, H. (1996). Effect of the highly selective and nonpeptide delta opioid receptor agonist TAN-67 on the morphine-induced place preference in mice. *Journal of Pharmacology and Experimental Therapeutics, 279*(1), 177–185.

Toskulkao, T., Pornchai, R., Akkarapatumwong, V., Vatanatunyakum, S., & Govitrapong, P. (2010). Alteration of lymphocyte opioid receptors in methadone maintenance subjects. *Neurochemistry International*, *56*(2), 285–290.

Turchan, J., Przewlocka, B., Toth, G., Lason, W., Borsodi, A., & Przewlocki, R. (1999). The effect of repeated administration of morphine, cocaine and ethanol on mu and delta opioid receptor density in the nucleus accumbens and striatum of the rat. *Neuroscience*, *91*(3), 971–977.

Wang, H. Y., Friedman, E., Olmstead, M. C., & Burns, L. H. (2005). Ultra-low-dose naloxone suppresses opioid tolerance, dependence and associated changes in mu opioid receptor-G protein coupling and Gbetagamma signaling. *Neuroscience*, *135*(1), 247–261.

Ward, S. J., & Roberts, D. C. (2007). Microinjection of the delta-opioid receptor selective antagonist naltrindole 5'-isothiocyanate site specifically affects cocaine self-administration in rats responding under a progressive ratio schedule of reinforcement. *Behavioral Brain Research*, *182*(1), 140–144.

Wood, P. L., Charleson, S. E., Lane, D., & Hudgin, R. L. (1981). Multiple opiate receptors: differential binding of mu, kappa and delta agonists. *Neuropharmacology*, *20*(12A), 1215–1220.

Xu, K., Liu, X. H., Nagarajan, S., Gu, X. Y., & Goldman, D. (2002). Relationship of the delta-opioid receptor gene to heroin abuse in a large Chinese case/control sample. *American Journal of Medical Genetics*, *110*(1), 45–50.

Xuei, X., Flury-Wetherill, L., Bierut, L., Dick, D., Nurnberger, J., Jr., Foroud, T., & Edenberg, H. J. (2007). The opioid system in alcohol and drug dependence: family-based association study. *American Journal of Medical Genetics Part B Neuropsychiatric Genetics*, *144B*(7), 877–884.

Yekkirala, A. S., Banks, M. L., Lunzer, M. M., Negus, S. S., Rice, K. C., & Portoghese, P. S. (2012). Clinically employed opioid analgesics produce antinociception via μ-δ opioid receptor heteromers in Rhesus monkeys. *ACS Chemical Neuroscience*, *3*(9), 720–727.

Zhang, H., Kranzler, H. R., Yang, B. Z., Luo, X., & Gelernter, J. (2008). The OPRD1 and OPRK1 loci in alcohol or drug dependence: OPRD1 variation modulates substance dependence risk. *Molecular Psychiatry*, *13*(5), 531–543.

Zhu, Y., King, M. A., Schuller, A. G., Nitsche, J. F., Reidl, M., Elde, R. P., ... Pintar, J. E. (1999). Retention of supraspinal delta-like analgesia and loss of morphine tolerance in delta opioid receptor knockout mice. *Neuron*, *24*(1), 243–252.

Chapter 85

Genome-Wide Association Studies and Human Opioid Sensitivity

Daisuke Nishizawa, Kazutaka Ikeda
Addictive Substance Project, Tokyo Metropolitan Institute of Medical Science, Tokyo, Japan

Abbreviations

ABC Adenosine triphosphate-binding cassette
cAMP Cyclic adenosine 3′,5′-monophosphate
EA European American
ED$_{50}$ Effective dose 50
EM Extensive metabolizer
EPOS European Pharmacogenetic Opioid Study
GWAS Genome-wide association study
HWE Hardy–Weinberg equilibrium
IM Intermediate metabolizer
LD Linkage disequilibrium
LOD Logarithm of odds
M6G Morphine 6-glucuronide
MEAC Minimum effective analgesic blood concentration
NPL Nonparametric linkage or nonparametric LOD
OR Odds ratio
PM Poor metabolizer
RD Reward dependence
SNP Single-nucleotide polymorphism
STR Short tandem repeat
TCI Temperament and Character Inventory
UM Ultrarapid metabolizer
VNTR Variable number of tandem repeat

INTRODUCTION

Opioids, such as morphine, heroin, and fentanyl, are widely used as effective analgesics for the treatment of acute and chronic pain because of their antinociceptive effects. The opioid system is also involved in the mechanism of the rewarding effects of morphine, ethanol (Sora et al., 2001), cocaine (Hall, Goeb, Li, Sora, & Uhl, 2004), and various other drugs (Berrendero, Kieffer, & Maldonado, 2002; Contarino et al., 2002; Shen et al., 2010) and plays an important role in the euphoric effects of these substances and other addictive behaviors.

Considerable differences in the sensitivity to opioids are widely known, which can differentiate the analgesic effects that are required for adequate pain relief and the rewarding effects associated with addictive drugs, including opioids. For example, the minimum effective analgesic blood concentration of pethidine was estimated to be 0.10–0.82 mg/l in 15 of 16 patients with intractable pain, with a median of 0.25 mg/l (Mather & Glynn, 1982), indicating that approximately an eight-times higher amount of opioid analgesics is required in patients with possibly less opioid sensitivity than patients with higher opioid sensitivity. Interindividual differences in the plasma concentrations of opioids are attributable to various factors, including absorption, distribution, metabolism, and elimination (McQuay et al., 1990). The required amount of clinically prescribed opioid analgesics is considered to reflect opioid responsiveness or sensitivity, depending on individual pharmacodynamic/phamacokinetic factors. It may also vary among patients with pain that is caused by malignant disease or surgery, depending on age, sex, weight, basal pain sensitivity (Pleym, Spigset, Kharasch, & Dale, 2003; Wilder-Smith, 2005), the type of surgery (Aoki et al., 2014), perceived pain during the perioperative period (Aubrun, Langeron, Quesnel, Coriat, & Riou, 2003), and genetic factors.

To separate genetic and environmental factors, the twin-study paradigm estimates heritability as the proportion of overall phenotypic variance that is accounted for by genetic factors by calculating variances in monozygotic and dizygotic twin pairs. Although twin studies of the effects of opioids are still in their nascent stage, Angst et al. (2012) reported the outcomes of a twin study that provided a global estimation of the genetic and environmental contributions to interindividual differences in pain sensitivity and the analgesic effects of opioid. Two years earlier, they had published a paper that reported the preliminary data (Angst et al., 2010). In their study, 81 monozygotic and 31 dizygotic twin pairs successfully received computer-controlled infusions of the μ opioid receptor agonist alfentanil on a single occasion in a randomized, double-blind, placebo-controlled design. Pain sensitivity and analgesic effects were assessed using experimental heat and cold pressor pain models, together with numerous covariates, including demographic factors, depression, anxiety, and sleep quality. The detected significant heritability was for cold pressor pain tolerance and opioid-mediated elevations in heat and cold pressor pain thresholds. Genetic effects were estimated to account for 12%, 60%, and 30% of the observed response variance after analgesic administration for heat pain threshold, cold-pressor pain threshold, and cold-pressor pain tolerance, respectively. The results suggested that alfentanil-induced analgesia is more highly heritable for cold pain than for heat pain. As of this writing, however, the heritability of the effects of opioids has not been precisely estimated for other pain conditions aside from experimentally induced pain. These findings from twin-study paradigms also do not convey information about the potential genes involved.

Two major approaches have been adopted to explore specific human genetic variations that are involved in the susceptibility to various common or rare diseases and other complex traits for which the involvement of genetic factors can be implicated. One is the candidate gene approach, which is known as a "hypothesis-driven" method that targets several specific genes that are related to the phenotypes of interest. The other is the genome-wide approach, which is known as a "hypothesis-free" method that targets all genes in the human genome, regardless of preexisting assumptions related to the phenotypes of interest. In most cases, whole-genome genotyping arrays are used in this approach to genotype hundreds of thousands or even millions of genetic variations simultaneously. In terms of statistical analyses, family-based and population-based designs are both commonly used in genetic association studies, although family-based designs (e.g., linkage analyses, which require samples of subjects with specific traits of interest) appear to have lost predominance over the past few years (Ott, Kamatani, & Lathrop, 2011). Population-based case–control or quantitative trait studies are relatively more prevalent. In this chapter, the findings of several studies that reported candidate genes are briefly reviewed, followed by genome-wide association studies (GWASs) that evaluated opioid sensitivity and a possibly related phenotype, opioid dependence. Although many more studies are required to discover the precise genetic factors that are involved in human opioid sensitivity, several genetic variations may aid future personalized treatment by utilizing genetic information regarding pain and substance use disorders.

ASSOCIATION STUDIES ON SPECIFIC CANDIDATE GENES FOR HUMAN OPIOID SENSITIVITY

Many candidate gene association studies have been conducted to identify relationships between genetic variations, mostly single-nucleotide polymorphisms (SNPs), and human opioid sensitivity or related phenotypes. They typically focus on genes that are involved in pharmacokinetic or pharmacodynamic pathways of opioids or pain-related genes of various modalities. Described below are examples of these studies, including the μ opioid receptor (*OPRM1*); cytochrome P450, family 2, subfamily D, polypeptide 6 (*CYP2D6*); and other genes. More details are shown in Table 1 and elsewhere (Branford, Droney, & Ross, 2012; Droney, Riley, & Ross, 2012; Vuilleumier, Stamer, & Landau, 2012).

For the *OPRM1* gene, which encodes the major μ opioid receptor, most studies have focused on the functional A118G SNP, in which the 118G variant receptor reportedly binds β-endorphin, an endogenous opioid that activates the μ opioid receptor approximately three times more tightly than the 118A common form of the receptor (Bond et al., 1998). In other studies, other polymorphisms were not analyzed, or significant associations were only partially identified for those polymorphisms. Most studies revealed that the analgesic effects of opioids were higher in A/G or G/G genotype carriers than in A/A genotype carriers (Table 1). A few reports demonstrated the elevated effects of the A/A genotype compared with other genotypes (Janicki et al., 2006; Landau, Kern, Columb, Smiley, & Blouin, 2008). Another study that used human brain tissues and Chinese hamster ovary cells found that mRNA and protein expression was decreased in the 118G allele compared with the 118A allele (Zhang, Wang, Johnson, Papp, & Sadee, 2005), suggesting that the allelic expression imbalance observed in this study may reflect the in vivo analgesic effects of opioids.

CYP2D6 is a member of the CYP superfamily of enzymes, which are known as monooxygenases that are responsible for reactions of various compounds. With regard to opioid metabolism, CYP2D6 catalyzes codeine and oxycodone to generate morphine and oxymorphine, respectively. Many genetic polymorphisms have been identified for the *CYP2D6* gene, and enzymatic activity is classified according to haplotype or diplotype, with the involvement of a number of SNPs. They are generally classified based on enzymatic activity: extensive metabolizer (EM), intermediate metabolizer (IM), and poor metabolizer (PM). Enzymes with especially higher activity are designated as ultrarapid metabolizers (UMs). A carrier of this UM enzyme reportedly experienced severe pain in the epigastrium after codeine intake (Dalen, Frengell, Dahl, & Sjoqvist, 1997). In EM carriers, a significant positive correlation was reported between the increase in pain threshold and the plasma concentration of codeine (Sindrup et al., 1990). Stamer et al. (2003) reported that the percentage of nonresponders was significantly higher in the PM group (46.7%) than in the EM group (21.6%).

ABCB1, whose official name is adenosine triphosphate-binding cassette (ABC), subfamily B (MDR/TAP), member 1, is a member of the superfamily of ABC transporters, which transport various molecules across extra- and intracellular membranes. Although many studies have focused on the genetic polymorphisms 3435C>>T and 2677G>>T/A, other studies have investigated the rs1128503 C/T polymorphism (Table 1). Coulbault et al. (2006) found that the *ABCB1* homozygous GG–CC diplotype, consisting of the 3435C>>T and 2677G>>T/A SNPs, was significantly associated with fewer morphine side effects, such as nausea and vomiting. Park et al. (2007) reported that patients who carried the linked 3435T and 2677T alleles exhibited a significant difference in the level of respiratory suppression. Patients with genotypes that are susceptible to fentanyl (1236TT, 2677TT, and 3435TT) showed early (2–3 min) and profound respiratory suppression (65–73% of initial respiration rate) compared with other resistant genotypes (83–85% of initial respiration rate in 1236CC, 2677GG, and 3435CC). Ross et al. (2008) showed that patients who carried the common G allele at position 2677 in exon 26 were less likely to experience drowsiness and confusion or hallucinations than patients who carried the variant T or A allele, which encode alternate amino acid substitutions.

Catechol-*O*-methyltransferase (COMT) is an enzyme that is involved in metabolizing catecholamines, such as dopamine, epinephrine, and norepinephrine, and is a key modulator of these neurotransmitters. A functional polymorphism of the *COMT* gene encodes the substitution of valine (Val) by methionine (Met) at codon 158 (Val-158Met), which is associated with a difference in thermostability that leads to a three- to fourfold reduction in the activity of the COMT enzyme (Lotta et al., 1995). Rakvag et al. (2005) and Reyes-Gibby et al. (2007) studied cancer patients and found that carriers of the Val/Val genotype required more morphine than patients with the Met/Met genotype. In further analyses of the joint effects of the *COMT* Val/Met and *OPRM1* A118G SNPs (Reyes-Gibby et al., 2007), they found that carriers of the *OPRM1* AA and *COMT* Met/Met genotype required the lowest morphine dose to achieve pain relief (87 mg/24 h). Patients with neither the Met/Met nor the AA genotype required the highest morphine dose (147 mg/24 h).

TABLE 1 Genetic Polymorphisms Identified as Possibly Associated with Human Opioid Sensitivity in Candidate Gene Association Studies

Gene	Reported Polymorphism	SNP Reference ID	Drug	Reported Effects Related to Opioid Analgesic Effects	References
OPRM1	118A>G	rs1799971 (rs17181017)	M6G	Low M6G potency in A/G or G/G genotypes	Lötsch et al. (2002)
			Morphine, M6G	Smaller potency of pupil-constricting or fewer side effects of M6G and morphine in A/G or G/G genotypes	Skarke, Darimont, Schmidt, Geisslinger, and Lotsch (2003)
			Morphine	More morphine requirement in G/G genotype	Klepstad et al. (2004)
			M6G	Smaller analgesic responses to M6G in A/G than A/A genotype	Romberg et al. (2005)
			Levomethadone	Lower miotic potency in A/G or G/G genotypes than A/A	Lötsch et al. (2006)
			Morphine	More morphine requirement in G/G genotype	Chou et al. (2006a)
			Morphine	More morphine requirement in G/G genotype than A/A genotype	Chou et al., (2006b)
			Morphine sulfate equivalents	Higher required opioid dose in high-quartile chronic pain patients with A/A genotype than other genotypes	Janicki et al. (2006)
			Morphine	More morphine requirement in G/G genotype than A/A genotype	Reyes-Gibby et al. (2007)
			Fentanyl	Higher A/A intrathecal fentanyl ED_{50} than other genotypes	Landau et al. (2008)
			Alfentanil	More alfentanil requirement in G/G genotype than A/A genotype	Oertel, Schmidt, Schneider, Geisslinger, and Lotsch (2006)
			Morphine	Less morphine requirement and pain and more nausea in A/A genotype	Sia et al. (2008)
			Several opioids	More analgesic requirement in G/G genotype than other genotypes	Hayashida et al. (2008)
			Fentanyl	Higher consumption of postoperative fentanyl in G/G genotype than A/A or A/G genotypes	Zhang et al. (2010a)
COMT	472G>A	rs4680	Morphine	More morphine requirement in Val/Val genotype	Rakvag et al. (2005)
			Morphine	More morphine requirement in Val/Val and Val/Met genotypes than Met/Met genotype	Reyes-Gibby et al. (2007)
MC1R	451C>T, 478C>T, 880G>C	rs1805007, rs1805008, rs1805009	M6G	Less pain sensitivity and larger M6G effect in nonfunctional MC1R carriers	Mogil et al. (2005)
			Pentazocine (only women)	Larger pentazocine effect in female nonfunctional MC1R carriers	Mogil et al. (2003)
CYP2D6	2549A>del		Codeine	Codeine increases the pricking pain thresholds in extensive metabolizers	Sindrup et al. (1990)
		–	Codeine	Greater respiratory, psychomotor, and pupillary effects of codeine in extensive metabolizers	Caraco, Sheller, and Wood (1996)

continued

TABLE 1 Genetic Polymorphisms Identified as Possibly Associated with Human Opioid Sensitivity in Candidate Gene Association Studies–cont'd

Gene	Reported Polymorphism	SNP Reference ID	Drug	Reported Effects Related to Opioid Analgesic Effects	References
	1846G>A	rs28371711, rs3892097	Tramadol	More tramadol nonresponders in poor metabolizers	Stamer et al. (2003)
	Gene duplication/ amplification	–	Codeine	Experience of severe pain in the epigastrium after codeine intake in an ultrarapid metabolizer	Dalen et al. (1997)
			Codeine	Life-threatening codeine intoxication in an ultrarapid metabolizer	Gasche et al. (2004)
CYP3A4	CYP3A4*1G	rs2242480	Fentanyl	Decreased postoperative dose requirements in CYP3A4*1G	Zhang et al. (2010b)
	CYP3A5*3 and CYP3A4*1G	rs776746 and rs2242480	Fentanyl	Decreased postoperative dose requirements in CYP3A5*3 and CYP3A4*1G	Zhang et al. (2011)
ABCB1	2677G>T/A	rs2032582	Morphine	Fewer side effects (drowsiness, confusion, and hallucinations) in G allele carriers	Ross et al. (2008)
	3435T-2677T (haplotype)		Fentanyl	Higher level of respiratory suppression in 3435T-2677T haplotype carriers	Park et al. (2007)
	3435CC-2677GG (diplotype)		Morphine	Fewer morphine side effects in 3435CC-2677GG diplotype carriers	Coulbault et al. (2006)
	1236C>T	rs1128503	Several opioids	Less morphine requirement in C/C genotype than other genotypes	Kobayashi et al. (2009)
	3435C>T	rs1045642	Several opioids	Decreased dose in a gene dose-dependent manner with the 3435C>T	Lötsch et al. (2009)
IL6	−174G>C	rs1800795	Several opioids	Increased dose requirements in C/C genotype	Reyes-Gibby et al. (2008)
IL1RN	86-bp VNTR	rs2234663	Morphine	Different genotype distribution in medium morphine consumers	Bessler et al. (2006)
TNF	−308G>A	rs1800629	Several opioids	Decreased pain relief in GA+AA genotype	Reyes-Gibby et al. (2008)

Summary of knowledge of associations between candidate gene polymorphisms and human opioid sensitivity. M6G, morphine 6-glucuronide; ED_{50}, effective dose 50; VNTR, variable number of tandem repeat.

Increasing evidence indicates that the uncontrolled activation of microglial cells under neuropathic pain conditions induces the release of proinflammatory cytokines (e.g., interleukin-1β (IL-1β), IL-6, and tumor necrosis factor-α (TNF-α)). The results of many studies support the hypothesis that modulating glial and neuroimmune activation may be a potential therapeutic approach for enhancing morphine analgesia (Mika, 2008), suggesting that genetic polymorphisms in the genes that are related to cytokines may affect morphine sensitivity. Bessler, Shavit, Mayburd, Smirnov, and Beilin (2006) reported that in the medium postoperative morphine consumer group, among 76 women who underwent transabdominal hysterectomy, the distribution of the variable number of tandem repeat polymorphism in intron 2 of the IL-1 receptor antagonist gene (*IL1RN*) was significantly different compared with the other two groups.

Although many genetic variations have been reported in studies that targeted specific genes of interest, most of the results have not been robustly replicated in subsequent similar studies or suggest small effect sizes of the outcomes caused by their identified variations that possibly contribute to opioid sensitivity/responsiveness or related phenotypes, such as undesirable side effects. Many more studies need to be conducted to resolve the "missing heritability" of opioid sensitivity (Manolio et al., 2009).

GENOME-WIDE ASSOCIATION STUDIES OF HUMAN OPIOID SENSITIVITY

Although GWASs have successfully discovered genetic variations related to many diseases or other human traits, they have rarely been applied to phenotypic traits, such as human opioid sensitivity or responsiveness. Genetic factors related to individual differences in the potency of opioids, which is proportional to both the affinity and the efficacy of the drug, still remain to be comprehensively explored in the entire human genome. Only a few studies have conducted such

TABLE 2 Genome-Wide Studies of Human Opioid Sensitivity and Opioid Dependence

Reported Study	Phenotype	Drug	Number of Markers	Suggested Locus or SNP	Related Gene	References
Linkage study	Opioid dependence	Opioids	409	Chromosomes 17 and 2		Gelernter et al. (2006)
Linkage study	Opioid dependence	Opioid (heroin)	404	17q11.2 and 4q31.21		Glatt et al. (2006)
Linkage study	Opioid dependence	Opioids	10,204	14q and 10q	NRXN3	Lachman et al. (2007)
GWAS	Opioid dependence	Opioids	5633	rs770124	NAV3	Yu et al. (2008)
Linkage study	Opioid dependence	Opioid (heroin)	~386	4q31.21		Glatt et al. (2008)
GWAS	Opioid dependence	Opioid (heroin)	~10,000	rs965972 (1q31.2), rs1986513 (4q28.1), rs1714984 (17p12), rs1867898 (2q21.2)	MYOCD	Nielsen et al. (2008)
GWAS	Opioid dependence	Opioid (heroin)	113,135 or 113,174	rs10494334 (1q23.3), rs950302 (1q24.1)	DUSP27, RIMS2, CMYA3	Nielsen et al. (2010)
GWAS	Opioid sensitivity	Opioids	1 million	rs12948783	RHBDF2	Galvan et al. (2011)
GWAS	Opioid dependence	Opioids	889,659	rs62103177, rs60349741, rs114070671, rs115368721, rs73411566	KCNG2, KCNC1, APBB2, PARVA	Gelernter et al. (2014)
GWAS	Opioid sensitivity	Opioid (fentanyl)	295,036	rs2952768 (2q33.3–q34)	METTL21A, CREB1	Nishizawa et al. (2014)

Summary of knowledge from GWASs (genome-wide association studies) and linkage studies of human opioid sensitivity. SNP, single-nucleotide polymorphism.

investigations, two of which are described below. The key results of the GWASs of opioid sensitivity and opioid dependence are summarized in Table 2.

One report was part of the European Pharmacogenetic Opioid Study (EPOS), which was a prospective cross-sectional multinational multicenter study of 2294 patients with cancer from 11 European countries that was conducted from July 2005 to April 2008 (Kurita et al., 2011). The study included cancer patients who were treated with opioids for moderate or severe pain and recruited from 17 centers in 11 European countries (Galvan et al., 2012, 2011). Opioids that were administered to more than 5% of the patients included morphine, oxycodone, and fentanyl. These opioid doses were converted to equivalent total daily oral morphine doses for the analysis. Pain relief was measured using an 11-point numerical rating scale, from 0% ("no pain relief") to 100% ("complete pain relief") in 10% increments. The cancer patients in the first series were defined as "good" or "poor" responders to opioid therapy, based on pain relief phenotype scores ≥90% or ≤40%, respectively. The second series of 570 patients was selected based on pain relief phenotype scores ≥90% or ≤50%, respectively, to extend the results of the GWAS. DNA pools of "good" ($n=293$; mean pain relief=94.1%) and "poor" ($n=145$; mean pain relief=27.0%) responders were created using equal amounts of fluorimetrically measured DNA from each sample. Association analyses between approximately 1 million SNPs and the pain relief phenotype (categorical variable, "poor" or "good" responders, or quantitative variable) were performed, which included analyses of Hardy–Weinberg equilibrium, linkage disequilibrium (LD) between SNPs, and population-based associations between phenotypes and genotype/allelotype. Analyses of chromosome counts that were obtained from allelic frequencies in the DNA pools of "good" and "poor" responders identified 486 SNPs at the statistical threshold of $p<1.0\times10^{-3}$. Among the candidate SNPs, 72 were associated with the pain relief phenotype (at $p<1.0\times10^{-5}$ threshold value) and selected for genotyping in the second series of 570 EPOS patients; 57 of the SNPs were successfully genotyped. Of these, the rs12948783 SNP, which maps to chromosome 17 upstream of the RHBDF2 gene, showed the strongest association ($p=1.1\times10^{-7}$). Analyses of the total sample size of 1008 cancer patients for the quantitative pain relief phenotype pointed to eight SNPs that were associated at the statistical threshold of $p<1.0\times10^{-3}$, with the strongest allelic association for the rs12948783 SNP (RHBDF2; $p=1.1\times10^{-8}$). Genotype–phenotype analyses of this SNP in 936 patients (72 genotypes were missing) showed that patients who were homozygous for

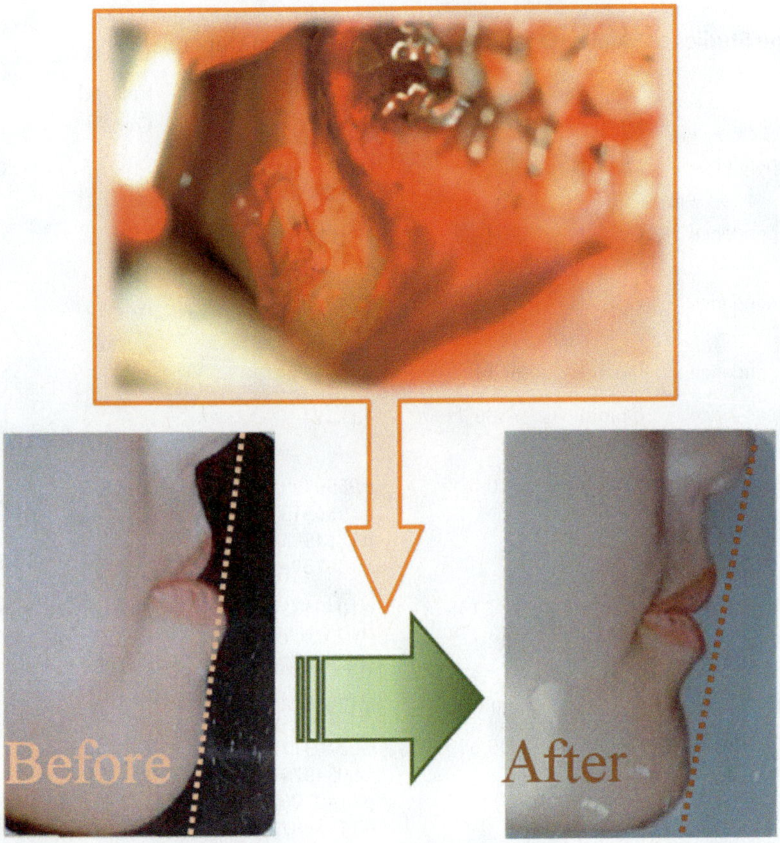

FIGURE 1 Schematic representation of cosmetic orthognathic surgery. Dotted line indicates Ricketts' esthetic line (E-line).

the rare allele (i.e., AA genotype) experienced a low normalized pain relief value (80 ± 9%; $n=23$) that did not statistically differ from patients with the GA genotype (88.5 ± 3%; $n=243$). Comparisons between subjects who carried the rare allele (i.e., AA or GA genotype, $n=266$) and patients who carried the common GG genotype (104 ± 1%; $n=670$) showed a highly statistically significant association ($p=8.1 \times 10^{-9}$) with patients who carried the GG genotype who had almost complete pain relief. These findings could represent a starting point for estimating the probability of being a poor responder to opioid treatment, thus leading to personalized cancer pain therapy in which, for example, individuals who are at genetic risk for a poor response would receive higher starting doses of opioids. Moreover, genes that are tagged by these polymorphisms may provide targets for new therapeutic strategies to control cancer pain perception.

Another study was published later that recruited 355 healthy subjects who were scheduled to undergo cosmetic orthognathic surgery for mandibular prognathism (Figure 1), which involves substantial pain uniformly in almost all subjects who had not incurred any invasion or pain before the surgery and thus could be considered ideal subjects for unveiling genetic contributions to individual differences in sensitivity to opioid analgesics (Fukuda et al., 2009). A multistage GWAS of a total of 353 subjects was conducted to investigate the association between 295,036 SNPs and requirements for an opioid analgesic, fentanyl, during the 24-h postoperative period, which was an end point for human opioid sensitivity (Nishizawa et al., 2014). As a result, 9, 12, and 10 SNPs were selected as the top candidates for additive, dominant, and recessive models for each minor allele, respectively, after the final stage (Figure 2). Among these, several SNPs that mapped to 2q33.3–q34 showed significant associations with 24-h postoperative fentanyl requirements after the final stage in the additive and recessive models (Figure 3). The genes that are located in this region include *METTL21A* (*FAM119A*) and *CREB1*, which encode methyltransferase-like 21A and cyclic adenosine 3′,5′-monophosphate response element binding protein 1, respectively. In the subsequent study, 112 patients were recruited who underwent major open abdominal surgery under combined general and epidural anesthesia (Hayashida et al., 2008), which involves modes of invasion different from that of the orthognathic surgery and might cause different pain modalities. Consequently, a significant difference in postoperative analgesic requirements was found between the subjects with the combined T/T and T/C genotype and subjects with the C/C genotype in the best candidate SNP in the 2q33.3–q34 region, the rs2952768 SNP. The subjects with the C/C genotype required significantly more analgesics than the subjects with the combined T/T and T/C genotype in the rs2952768 SNP, a pattern that was interestingly similar to the one observed in the subjects who underwent cosmetic orthognathic surgery (Figure 4). For this SNP, a significant difference was also found in the genotypic distribution between the absent and the present subgroups of polydrug use among the patients with methamphetamine dependence. Fewer polydrug abusers carried the C/C genotype compared with monodrug users. A similar result was found in patients with eating disorders, and a lower proportion of patients with drug dependence carried the C allele compared with patients without drug dependence. These results showed that carriers of the C allele among the patients with psychiatric disorders, especially the C/C genotype, tended not to abuse polydrugs and not to have comorbid alcohol or drug dependence, suggesting that carriers of the C allele in this SNP are less liable to the expression of symptoms of serious dependence. The association between this SNP and data from the Temperament and Character Inventory (TCI; a personality profiling questionnaire) in healthy volunteers was also investigated. Intriguingly, among the seven dimensions of the TCI, a significant association was found only for reward dependence (RD). The RD value decreased as the copy number of the C allele increased among the subjects. The mRNA expression levels of the *METTL21A* (*FAM119A*) and *CREB1* genes, which were included in the same LD block as the rs2952768 SNP, were additionally examined with RNA samples that were extracted from postmortem subject specimens. A significant association was found in the relative mRNA expression level of the *CREB1* gene between the combined T/T and T/C genotype subgroup and the C/C subgroup, and the C/C genotype of this SNP was significantly associated with elevated *CREB1* mRNA expression. Altogether, these results demonstrate that SNPs in this locus are the most potent genetic factors associated with human opioid sensitivity known to date, possibly affecting both the efficacy of opioid analgesics and the susceptibility to severe substance dependence. The results obtained in this study will provide valuable information for the personalized treatment of pain and substance dependence.

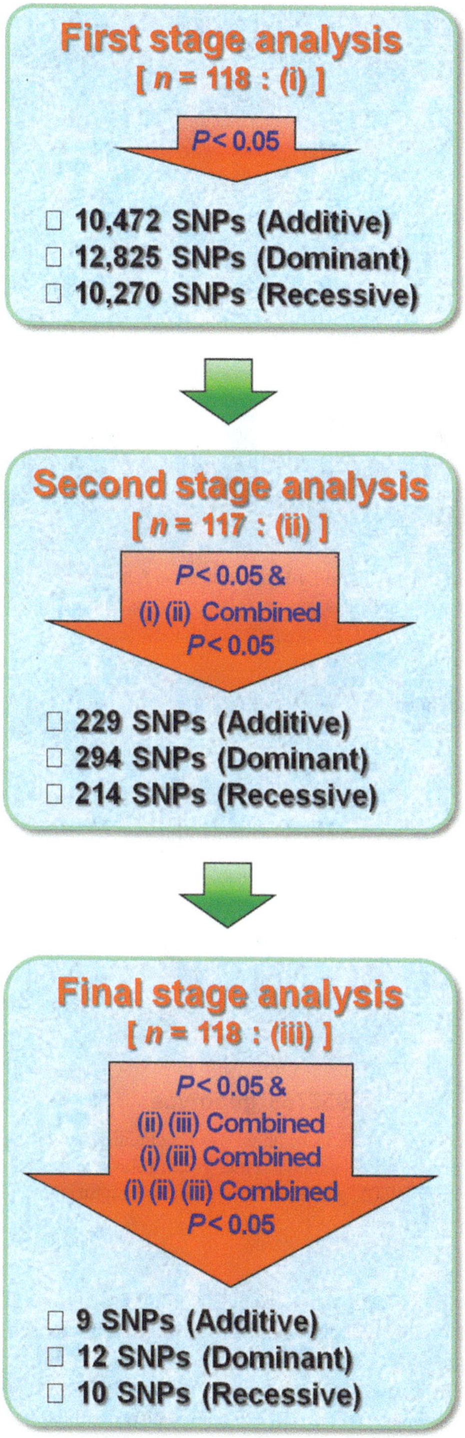

GENOME-WIDE ASSOCIATION OR LINKAGE STUDIES OF HUMAN OPIOID DEPENDENCE

To our knowledge, no studies of opioid dependence with a genome-wide scope were published before 2006. Initial studies that targeted the entire genomic region appeared as linkage studies that recruited a sample of hundreds of families with hundreds of short tandem repeat (STR) or microsatellite markers that spanned the genome.

Gelernter et al. (2006) used cluster analytic methods to identify opioid dependence-related symptom clusters, which were shown to be heritable, and then completed a genome-wide linkage scan (with 409 markers) for the opioid-dependence diagnosis and two cluster-defined phenotypes that were represented by 1250 families: the heavy-opioid-use cluster and non-opioid-use cluster. The statistically strongest results were seen for the cluster-defined traits. For the heavy-opioid-use cluster, the results showed a logarithm of odds (LOD) score of 3.06 on chromosome 17 ($p=0.0002$) for European-American (EA) and African-American subjects combined. For the non-opioid-use cluster, the results showed a LOD score of 3.46 elsewhere on chromosome 17 ($p=0.00002$, uncorrected for multiple traits studied) for EA subjects only. They also identified a possible linkage (LOD score of 2.43) between opioid dependence and chromosome 2 markers for African-American subjects.

Glatt et al. (2006) used samples from 194 fully independent affected sibling pairs from 192 Han Chinese families from Yunnan Province, China (near Asia's "Golden Triangle") for genotyping

FIGURE 2 Schematic illustration of the multistage GWAS. Potent candidate polymorphisms associated with human opioid sensitivity were selected in a three-stage GWAS.

with alcohol dependence, and a significantly lower proportion of drug abusers carried the C allele or C/C genotype compared with alcoholics without drug abuse. Furthermore, a significant difference in allelic distribution was found between the absent and the present subgroups of comorbid dependence among the patients

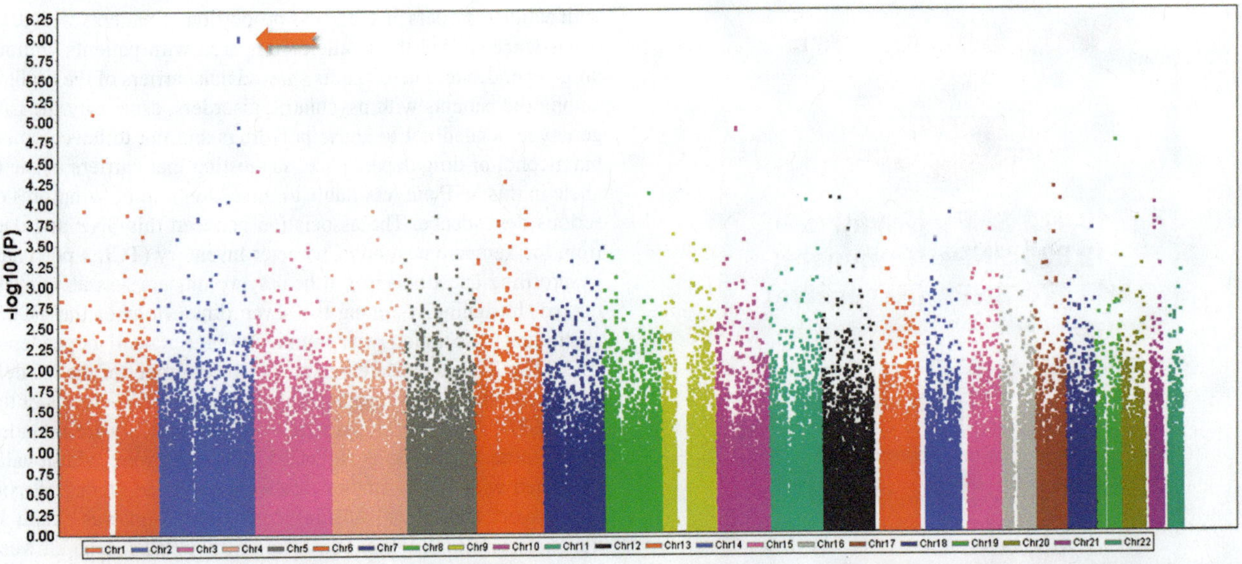

FIGURE 3 Genome-wide association for combined samples between polymorphism markers and 24-h postoperative fentanyl requirements in recessive model. The data are plotted for each chromosome, 1–22 (from left to right). Red arrow indicates genetic polymorphisms mapped to 2q33.3–q34.

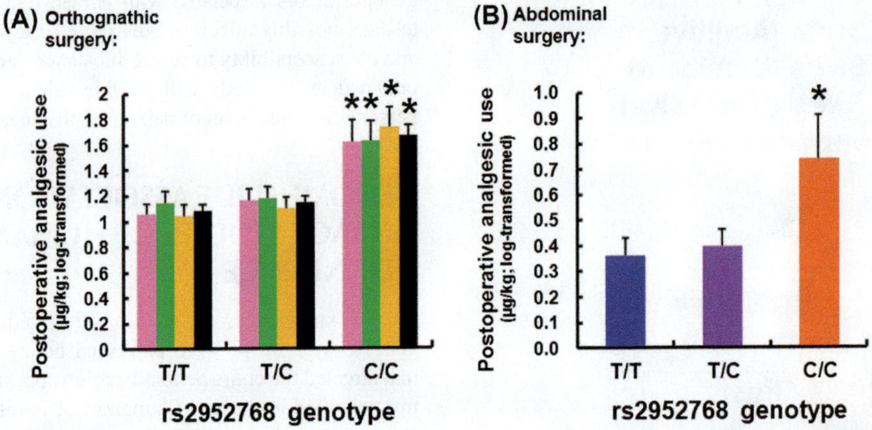

FIGURE 4 Association analysis between opioid analgesic requirements and the rs2952768 SNP. Dose of analgesics administered during the 24-h postoperative period after (A) cosmetic orthognathic surgery and (B) major open abdominal surgery is shown. Pink, green, yellow, and black bars represent doses of samples in first, second, final, and combined stage analyses, respectively. *$p<0.05$, greater dose of analgesic administered in the C/C genotype compared with the T/C and T/T genotypes. The data are expressed as the mean ± SEM.

404 STR markers spaced at an average intermarker distance of 9 cM. Although none of the findings achieved genome-wide significance, they found two regions with nonparametric linkage (NPL) Z scores > 2.0. An NPL Z score of 2.19 ($p=0.014$) was observed at D4S1644, located at 143.3 cM on chromosomal region 4q31.21. The highest NPL Z score of 2.36 ($p=0.009$) was observed at 53.4 cM on chromosomal region 17q11.2 at marker D17S1880. Two years later, these authors supplemented their samples with additional individuals and families, bringing the total number of genotyped individuals to 1513 and the number of independent sibling pairs to 397 (Glatt et al., 2008). Upon repeating similar analyses with this larger sample size, they found that the evidence for linkage at their most strongly implicated locus from the previous study (marker D17S1880, 53.4 cM on 17q11.2; NPL $Z = 2.36$; $p = 0.009$) was completely abolished ($Z = -1.13$; $p = 0.900$). In contrast, the evidence of linkage at the second-most strongly implicated locus from the previous study (D4S1644, 143.3 cM on 4q31.21; NPL $Z=2.19$; $p=0.014$) increased in magnitude and significance ($Z=2.64$; $p=0.004$), becoming the most strongly implicated locus overall in their full sample. Other loci on chromosomes 1, 2, 4, 12, 16, and X also displayed nominally significant evidence of linkage ($p \leq 0.05$), although these loci appeared to be entirely distinct from opioid-linked loci reported by other groups.

A high-density genome-wide linkage study of opioid dependence compared with other studies was reported by Lachman et al. (2007). They evaluated 305 opioid-dependent affected sibling pairs from an ethnically mixed population of methadone-maintained subjects and genotyped their DNA. The analysis identified a region on chromosome 14q with a nonparametric LOD (NPL) of 3.30. Their secondary analyses indicated that this locus was relatively specific to the self-identified Puerto Rican subset, in which the NPL increased from 3.30 to 5.00 ($NPL_{Caucasian} = 0.05$ and $NPL_{African-American} = 0.15$). The 14q peak encompasses the *NRXN3* gene (neurexin 3), which was previously identified as a potential candidate gene for addiction. Secondary analyses also identified several regions with gender-specific NPL scores >2.00. The most significant was a peak on 10q that increased from 0.90 to 3.22 when only males were considered ($NPL_{female} = 0.05$). Although these linkage studies identified several specific chromosomal loci that are possibly associated with opioid dependence, the GWAS appeared to have taken the place of linkage studies afterward, before extensive replication studies were conducted for the results obtained in those previous linkage studies.

An initial GWAS for opioid dependence was seemingly reported as late as 2008. Yu et al. (2008) genotyped 5633 tag SNP markers in 1699 subjects from 339 African-American families and 334 EA families ascertained through a sibling pair for either cocaine or opioid dependence. Associations between genetic markers and five substance dependence traits (cocaine dependence, opioid dependence, cocaine-induced paranoia, alcohol dependence, and nicotine dependence) were assessed. Eight SNPs were significant in both African-American and EA groups for opioid dependence. Three of these SNPs on chromosome 2 and one SNP on chromosome 20 were far from any known gene sequences. The most significant result among the gene-based SNPs was rs770124 in *NAV3* on chromosome 12 ($p = 0.0003$).

Nielsen et al. (2008) performed a GWAS to identify genes that may be associated with the vulnerability to develop heroin addiction using DNA from 104 individual former severe heroin addicts who met federal criteria for methadone maintenance and 101 individual control subjects, all Caucasian. Using separate analyses of autosomal and X-chromosomal variants, they found that the strongest associations between allele frequency and heroin addiction were with the autosomal variants rs965972 (located in the Unigene cluster Hs.147755; $Q = 0.053$) and rs1986513 ($Q = 0.187$). The three variants that exhibited the strongest association with heroin addiction by genotype frequency were rs1714984 (located in an intron of the gene for the transcription factor myocardin; $p = 0.000022$), rs965972 ($p = 0.000080$), and rs1867898 ($p = 0.000284$). One genotype pattern (AG–TT–GG) was significantly associated with developing heroin addiction (odds ratio (OR) = 6.25) and explained 27% of the population-attributable risk for heroin addiction in this cohort. Another genotype pattern (GG–CT–GG) of these variants was significantly associated with protection against developing heroin addiction (OR = 0.13). The lack of this genotype pattern explained 83% of the population-attributable risk for developing heroin addiction. Evidence of the involvement of five genes in heroin addiction was found: the genes that encode the μ opioid receptor, the metabotropic receptors mGluR6 and mGluR8, nuclear receptor NR4A2, and cryptochrome 1 (photolyase-like). This approach identified several new genes that are potentially associated with heroin addiction and confirmed the role of *OPRM1* in this disease. Nielsen et al. (2010) again conducted a GWAS using DNA from 325 methadone-stabilized, former severe heroin addicts and 250 control subjects. In this second study, DNA was pooled by ethnicity (Caucasian and African-American) and analyzed for 100,000 SNPs. The strongest association with the vulnerability to develop heroin addiction with experiment–wise significance ($p = 0.035$) was found in Caucasians with the variant rs10494334 (a variant in an unannotated region of the genome (1q23.3)). In African-American subjects, the variant that was most significantly associated with heroin addiction vulnerability was rs950302, which is found in the cytosolic dual-specificity phosphatase 27 gene ($p = 0.0079$). Furthermore, analyses of the top 500 variants with the most significant associations ($p \leq 0.0036$) in Caucasians showed that three of these variants were clustered in the regulating synaptic membrane exocytosis protein 2 gene *RIMS2*. Of the top 500 variants in African-Americans ($p \leq 0.0238$), three are in the cardiomyopathy-associated 3 gene *CMYA3*.

As of this writing, the most recent GWAS of opioid dependence was reported by Gelernter et al. (2014), for which two populations, African-American and EA, with opioid dependence in three sets of subjects were recruited. The design employed three phases. Phase 1 included their discovery GWAS data set, consisting of 5697 subjects (58% African-American) who were diagnosed with opioid and/or other substance dependence and control subjects. The subjects were genotyped, yielding 890,000 SNPs that were suitable for analysis. Additional genotypes were imputed with the 1000 Genomes reference panel. Top-ranked findings were further evaluated in phase 2 by incorporating information from the publicly available Study of Addiction: Genetics and Environment data set with GWAS data from 4063 subjects (32% African-American). In phase 3, the most significant SNPs from phase 2 were genotyped in 2549 independent subjects (32% African-American). Analyses were performed with case–control and ordinal trait designs. Consequently, the most significant results emerged from the African-American subgroup. Genome-wide significant associations ($p < 5.0 \times 10^{-8}$) were observed with SNPs from multiple loci. *KCNG2**rs62103177 was most significant after combining the results from data sets in every phase of the study. The most compelling results were obtained with genes involved in potassium signaling pathways (e.g., *KCNC1* and *KCNG2*). Pathway analysis also implicated genes that are involved in calcium signaling and long-term potentiation. This would be the first study that identified risk pathways for opioid dependence. Although the results that were obtained in these GWASs should be replicated, some of the outcomes may provide insights into novel therapeutic and prevention strategies for opioid dependence that target molecules related to the genes that GWASs identified as possibly involved in the etiology of the disease.

APPLICATIONS TO OTHER ADDICTIONS AND SUBSTANCE MISUSE

1. The opioid system is not only involved in the mechanism of the rewarding effects of its ligands, such as heroin and morphine, but also involved in the rewarding effects of other addictive substances, such as ethanol, cocaine, methamphetamine, and nicotine.
2. There is a common neurobiological mechanism that underlies addiction to various substances, and the opioid system is involved in addiction to various substances.
3. Genetic factors that affect human sensitivity to opioids may be at least partially shared with those that affect human sensitivity to other addictive substances.

4. Genetic variants that have been identified in candidate gene studies and GWASs associated with human opioid sensitivity can also influence interindividual differences in the vulnerability to addiction and misuse of other substances.
5. Personalized medicine, in which the treatment of opioid addiction or misuse is tailored to individual patients, depending on genetic information related to human opioid sensitivity, is also applicable to the personalized treatment of addiction or misuse of other substances.

DEFINITION OF TERMS

Opioids This is the collective designation for any narcotic drugs, which typically show high affinity for opioid receptors, such as μ opioid receptors.

Opioid sensitivity This is the degree of in vivo main or side effects of opioids, in which higher sensitivity represents higher effects of opioids.

Genome This is the complete set of genes to construct an organism and sustain its life.

Association study This is a statistical analysis to investigate associations between some specific traits, such as susceptibility to diseases or other qualitative or quantitative traits, and some genetic variations.

Genome-wide association study This describes an association study that targets genetic variations in all of the genes or chromosomal regions in the human genome.

Genetic polymorphisms These are genetic variations in which variant-type minor alleles exist at relatively high rates in the population, normally 1%.

Single-nucleotide polymorphism This is a genetic polymorphism of a single nucleotide, which usually denotes a polymorphism of one base difference at a specific site in the human genome.

Genome-wide linkage study This is a study, also called genome-wide linkage analysis, that is a family-based statistical analysis to identify chromosomal loci, including specific genes or markers within the whole human genome, that are more likely to be inherited by the subjects from their parents than other loci.

Pharmacogenomics This is the study of differences in the pharmacological actions of drugs caused by genetic variations throughout the whole genome.

Personalized medicine This refers to individualization of medical treatments, in which interindividual differences in constitutional predisposition caused by various factors, including genetic factors, are properly considered.

KEY FACTS

Key Facts of GWAS

- GWAS is the usual abbreviation for genome-wide association study. They can also be designated as GWA studies or WGA (whole-genome association) studies.
- Association studies are generally statistically conducted to investigate associations between some specific traits, such as susceptibility to disease or other qualitative or quantitative traits, and genetic variations that target several specific genes of interest.
- GWASs target genetic variations, mostly SNPs, in all of the genes or chromosomal regions in the human genome, regardless of preexisting assumptions related to the phenotypes of interest.
- Genotyping data for hundreds of thousands or millions of genetic variations are required to conduct GWASs. In most cases, whole-genome genotyping arrays are used to obtain such data simultaneously.
- GWASs often require special software to conduct statistical analyses for millions of SNPs at the same time, instead of normally used statistical software.
- In GWASs, corrections for multiple testing are often indispensable to determine significance of the whole tests because many genetic variations are analyzed simultaneously.

Key Facts of Opioid Sensitivity

- As with many other drugs, there is a wide range of interindividual differences in the main or side effects of opioids, in which some individuals are more sensitive and some individuals are less sensitive.
- The mechanism of opioid sensitivity has been more elaborately investigated in mice, including genetically modified mice and various wild mice, than in humans.
- Some mouse strains are known to be more sensitive to opioids, but other strains are less sensitive, such as CXBK mice, which harbor a longer untranslated region of the *Oprm1* mRNA than other mouse strains, leading to reduced levels of *Oprm1* mRNA expression and less sensitivity to opioids.
- A growing number of candidate gene studies and GWASs of human opioid sensitivity have been reported, but they are still at an early stage. Many more studies will be required to reveal unidentified heritability and the mechanisms that underlie interindividual differences in human opioid sensitivity.
- Given that the opioid system is involved in antinociception and the rewarding effects of various abused substances and possibly the neurobiological mechanisms that underlie the etiology of substance dependence, revealing the genetic factors of human opioid sensitivity by genetic analyses will provide valuable information for the appropriate individualization of opioid doses that are required for adequate pain control and the treatment of substance dependence.

SUMMARY POINTS

- This chapter focuses on GWASs applied to human opioid sensitivity.
- Opioids are commonly used as analgesics for the treatment of acute and chronic, moderate to severe pain, owing to their antinociceptive effects.
- The opioid system is also involved in the rewarding effects of various abused substances and plays a role in the neurobiological mechanisms of substance dependence.
- There is wide interindividual variability in the sensitivity to opioids, which has often been a critical problem in pain treatment that seeks to provide satisfactory pain relief for all patients. Interindividual variability in the sensitivity to opioids could also be associated with individual differences in the vulnerability to substance dependence.
- As of this writing, only a limited number of studies have addressed the relationship between human genetic variations and the sensitivity to opioids with a genome-wide scope.

- Comprehensively revealing such relationships by targeting variations in the entire human genome will provide valuable information for the appropriate individualization of opioid doses that are required for adequate pain control and the treatment of substance dependence.
- A growing number of genetic variations in candidate genes have been identified as possibly associated with human opioid sensitivity.
- Many more studies will be required to discover additional genetic factors that are involved in human opioid sensitivity to provide refined personalized treatment for pain and dependence.

REFERENCES

Angst, M. S., Phillips, N. G., Drover, D. R., Tingle, M., Galinkin, J. L., Christians, U., … Clark, J. D. (2010). Opioid pharmacogenomics using a twin study paradigm: methods and procedures for determining familial aggregation and heritability. *Twin Research and Human Genetics, 13*, 412–425.

Angst, M. S., Phillips, N. G., Drover, D. R., Tingle, M., Ray, A., Swan, G. E., … Clark, J. D. (2012). Pain sensitivity and opioid analgesia: a pharmacogenomic twin study. *Pain, 153*, 1397–1409.

Aoki, Y., Yoshida, K., Nishizawa, D., Kasai, S., Ichinohe, T., Ikeda, K., & Fukuda, K. (2014). Factors that affect intravenous patient-controlled analgesia for postoperative pain following orthognathic surgery for mandibular prognathism. *PLoS One, 9*, e98548.

Aubrun, F., Langeron, O., Quesnel, C., Coriat, P., & Riou, B. (2003). Relationships between measurement of pain using visual analog score and morphine requirements during postoperative intravenous morphine titration. *Anesthesiology, 98*, 1415–1421.

Berrendero, F., Kieffer, B. L., & Maldonado, R. (2002). Attenuation of nicotine-induced antinociception, rewarding effects, and dependence in μ-opioid receptor knock-out mice. *Journal of Neuroscience, 22*, 10935–10940.

Bessler, H., Shavit, Y., Mayburd, E., Smirnov, G., & Beilin, B. (2006). Postoperative pain, morphine consumption, and genetic polymorphism of IL-1β and IL-1 receptor antagonist. *Neuroscience Letters, 404*, 154–158.

Bond, C., LaForge, K. S., Tian, M., Melia, D., Zhang, S., Borg, L., … Yu, L. (1998). Single-nucleotide polymorphism in the human mu opioid receptor gene alters β-endorphin binding and activity: possible implications for opiate addiction. *Proceedings of the National Academy of Sciences of the United States of America, 95*, 9608–9613.

Branford, R., Droney, J., & Ross, J. R. (2012). Opioid genetics: the key to personalized pain control? *Clinical Genetics, 82*, 301–310.

Caraco, Y., Sheller, J., & Wood, A. J. (1996). Pharmacogenetic determination of the effects of codeine and prediction of drug interactions. *The Journal of Pharmacology and Experimental Therapeutics, 278*(3), 1165–1174.

Chou, W. Y., Yang, L. C., Lu, H. F., Ko, J. Y., Wang, C. H., Lin, S. H., … Hsu, C. J. (2006a). Association of μ-opioid receptor gene polymorphism (A118G) with variations in morphine consumption for analgesia after total knee arthroplasty. *Acta Anaesthesiologica Scandinavica, 50*(7), 787–792.

Chou, W. Y., Wang, C. H., Liu, P. H., Liu, C. C., Tseng, C. C., & Jawan, B. (2006b). Human opioid receptor A118G polymorphism affects intravenous patient-controlled analgesia morphine consumption after total abdominal hysterectomy. *Anesthesiology, 105*(2), 334–337.

Contarino, A., Picetti, R., Matthes, H. W., Koob, G. F., Kieffer, B. L., & Gold, L. H. (2002). Lack of reward and locomotor stimulation induced by heroin in μ-opioid receptor-deficient mice. *European Journal of Pharmacology, 446*, 103–109.

Coulbault, L., Beaussier, M., Verstuyft, C., Weickmans, H., Dubert, L., Tregouet, D., … Becquemont, L. (2006). Environmental and genetic factors associated with morphine response in the postoperative period. *Clinical Pharmacology and Therapeutics, 79*, 316–324.

Dalen, P., Frengell, C., Dahl, M. L., & Sjoqvist, F. (1997). Quick onset of severe abdominal pain after codeine in an ultrarapid metabolizer of debrisoquine. *Therapeutic Drug Monitoring, 19*, 543–544.

Droney, J., Riley, J., & Ross, J. (2012). Opioid genetics in the context of opioid switching. *Current Opinion in Supportive and Palliative Care, 6*, 10–16.

Fukuda, K., Hayashida, M., Ide, S., Saita, N., Kokita, Y., Kasai, S., … Ikeda, K. (2009). Association between *OPRM1* gene polymorphisms and fentanyl sensitivity in patients undergoing painful cosmetic surgery. *Pain, 147*, 194–201.

Galvan, A., Fladvad, T., Skorpen, F., Gao, X., Klepstad, P., Kaasa, S., & Dragani, T. A. (2012). Genetic clustering of European cancer patients indicates that opioid-mediated pain relief is independent of ancestry. *Pharmacogenomics Journal, 12*, 412–416.

Galvan, A., Skorpen, F., Klepstad, P., Knudsen, A. K., Fladvad, T., Falvella, F. S., … Dragani, T. A. (2011). Multiple loci modulate opioid therapy response for cancer pain. *Clinical Cancer Research, 17*, 4581–4587.

Gasche, Y., Daali, Y., Fathi, M., Chiappe, A., Cottini, S., Dayer, P., & Desmeules, J. (2004). Codeine intoxication associated with ultrarapid CYP2D6 metabolism. *The New England Journal of Medicine, 351*(27), 2827–2831.

Gelernter, J., Kranzler, H. R., Sherva, R., Koesterer, R., Almasy, L., Zhao, H., & Farrer, L. A. (2014). Genome-wide association study of opioid dependence: multiple associations mapped to calcium and potassium pathways. *Biological Psychiatry, 76*, 66–74.

Gelernter, J., Panhuysen, C., Wilcox, M., Hesselbrock, V., Rounsaville, B., Poling, J., … Kranzler, H. R. (2006). Genomewide linkage scan for opioid dependence and related traits. *American Journal of Human Genetics, 78*, 759–769.

Glatt, S. J., Lasky-Su, J. A., Zhu, S. C., Zhang, R., Zhang, B., Li, J., … Tsuang, M. T. (2008). Genome-wide linkage analysis of heroin dependence in Han Chinese: results from Wave Two of a multi-stage study. *Drug and Alcohol Dependence, 98*, 30–34.

Glatt, S. J., Su, J. A., Zhu, S. C., Zhang, R., Zhang, B., Li, J., … Tsuang, M. T. (2006). Genome-wide linkage analysis of heroin dependence in Han Chinese: results from Wave One of a multi-stage study. *American Journal of Medical Genetics B: Neuropsychiatric Genetics, 141B*, 648–652.

Hall, F. S., Goeb, M., Li, X. F., Sora, I., & Uhl, G. R. (2004). μ-Opioid receptor knockout mice display reduced cocaine conditioned place preference but enhanced sensitization of cocaine-induced locomotion. *Molecular Brain Research, 121*, 123–130.

Hayashida, M., Nagashima, M., Satoh, Y., Katoh, R., Tagami, M., Ide, S., … Ikeda, K. (2008). Analgesic requirements after major abdominal surgery are associated with *OPRM1* gene polymorphism genotype and haplotype. *Pharmacogenomics, 9*, 1605–1616.

Janicki, P. K., Schuler, G., Francis, D., Bohr, A., Gordin, V., Jarzembowski, T., … Mets, B. (2006). A genetic association study of the functional A118G polymorphism of the human μ-opioid receptor gene in patients with acute and chronic pain. *Anesthesia and Analgesia, 103*, 1011–1017.

Klepstad, P., Rakvag, T. T., Kaasa, S., Holthe, M., Dale, O., Borchgrevink, P. C., ... Skorpen, F. (2004). The 118 A > G polymorphism in the human μ-opioid receptor gene may increase morphine requirements in patients with pain caused by malignant disease. *Acta Anaesthesiologica Scandinavica, 48*(10), 1232–1239.

Kobayashi, D., Nishizawa, D., Kasai, S., Hasegawa, J., Nagashima, M., Katoh, R., ... Ikeda, K. (2009). Association between analgesic requirements after major abdominal surgery and polymorphisms of the opioid metabolism-related gene ABCB1. In F. Columbus (Ed.), *Acute pain* (pp. 101–110). New York: Nova Science Publishers.

Kurita, G. P., Sjogren, P., Ekholm, O., Kaasa, S., Loge, J. H., Poviloniene, I., & Klepstad, P. (2011). Prevalence and predictors of cognitive dysfunction in opioid-treated patients with cancer: a multinational study. *Journal of Clinical Oncology, 29*, 1297–1303.

Lachman, H. M., Fann, C. S., Bartzis, M., Evgrafov, O. V., Rosenthal, R. N., Nunes, E. V., ... Knowles, J. A. (2007). Genomewide suggestive linkage of opioid dependence to chromosome 14q. *Human Molecular Genetics, 16*, 1327–1334.

Landau, R., Kern, C., Columb, M. O., Smiley, R. M., & Blouin, J. L. (2008). Genetic variability of the μ-opioid receptor influences intrathecal fentanyl analgesia requirements in laboring women. *Pain, 139*, 5–14.

Lötsch, J., Skarke, C., Grosch, S., Darimont, J., Schmidt, H., & Geisslinger, G. (2002). The polymorphism A118G of the human μ-opioid receptor gene decreases the pupil constrictory effect of morphine-6-glucuronide but not that of morphine. *Pharmacogenetics, 12*(1), 3–9.

Lötsch, J., Skarke, C., Wieting, J., Oertel, B. G., Schmidt, H., Brockmoller, J., & Geisslinger, G. (2006). Modulation of the central nervous effects of levomethadone by genetic polymorphisms potentially affecting its metabolism, distribution, and drug action. *Clinical Pharmacology and Therapeutics, 79*(1), 72–89.

Lötsch, J., von Hentig, N., Freynhagen, R., Griessinger, N., Zimmermann, M., Doehring, A., ... Geisslinger, G. (2009). Cross-sectional analysis of the influence of currently known pharmacogenetic modulators on opioid therapy in outpatient pain centers. *Journal of Pharmacogenetics and Genomics, 19*(6), 429–436.

Lotta, T., Vidgren, J., Tilgmann, C., Ulmanen, I., Melen, K., Julkunen, I., & Taskinen, J. (1995). Kinetics of human soluble and membrane-bound catechol O-methyltransferase: a revised mechanism and description of the thermolabile variant of the enzyme. *Biochemistry, 34*, 4202–4210.

Manolio, T. A., Collins, F. S., Cox, N. J., Goldstein, D. B., Hindorff, L. A., Hunter, D. J., ... Visscher, P. M. (2009). Finding the missing heritability of complex diseases. *Nature, 461*, 747–753.

Mather, L. E., & Glynn, C. J. (1982). The minimum effective analgetic blood concentration of pethidine in patients with intractable pain. *British Journal of Clinical Pharmacology, 14*, 385–390.

McQuay, H. J., Carroll, D., Faura, C. C., Gavaghan, D. J., Hand, C. W., & Moore, R. A. (1990). Oral morphine in cancer pain: influences on morphine and metabolite concentration. *Clinical Pharmacology and Therapeutics, 48*, 236–244.

Mika, J. (2008). Modulation of microglia can attenuate neuropathic pain symptoms and enhance morphine effectiveness. *Pharmacological Reports, 60*, 297–307.

Mogil, J. S., Wilson, S. G., Chesler, E. J., Rankin, A. L., Nemmani, K. V., Lariviere, W. R., ... Fillingim, R. B. (2003). The melanocortin-1 receptor gene mediates female-specific mechanisms of analgesia in mice and humans. *Proceedings of the National Academy of Sciences of the United States of America, 100*(8), 4867–4872.

Mogil, J. S., Ritchie, J., Smith, S. B., Strasburg, K., Kaplan, L., Wallace, M. R., ... Dahan, A. S. (2005). Melanocortin-1 receptor gene variants affect pain and μ-opioid analgesia in mice and humans. *Journal of Medical Genetics, 42*(7), 583–587.

Nielsen, D. A., Ji, F., Yuferov, V., Ho, A., Chen, A., Levran, O., ... Kreek, M. J. (2008). Genotype patterns that contribute to increased risk for or protection from developing heroin addiction. *Molecular Psychiatry, 13*, 417–428.

Nielsen, D. A., Ji, F., Yuferov, V., Ho, A., He, C., Ott, J., & Kreek, M. J. (2010). Genome-wide association study identifies genes that may contribute to risk for developing heroin addiction. *Psychiatric Genetics, 20*, 207–214.

Nishizawa, D., Fukuda, K., Kasai, S., Hasegawa, J., Aoki, Y., Nishi, A., ... Ikeda, K. (2014). Genome-wide association study identifies a potent locus associated with human opioid sensitivity. *Molecular Psychiatry, 19*, 55–62.

Oertel, B. G., Schmidt, R., Schneider, A., Geisslinger, G., & Lotsch, J. (2006). The μ-opioid receptor gene polymorphism 118A>G depletes alfentanil-induced analgesia and protects against respiratory depression in homozygous carriers. *Pharmacogenetics and Genomics, 16*(9), 625–636.

Ott, J., Kamatani, Y., & Lathrop, M. (2011). Family-based designs for genome-wide association studies. *Nature Reviews Genetics, 12*, 465–474.

Park, H. J., Shinn, H. K., Ryu, S. H., Lee, H. S., Park, C. S., & Kang, J. H. (2007). Genetic polymorphisms in the *ABCB1* gene and the effects of fentanyl in Koreans. *Clinical Pharmacology and Therapeutics, 81*, 539–546.

Pleym, H., Spigset, O., Kharasch, E. D., & Dale, O. (2003). Gender differences in drug effects: implications for anesthesiologists. *Acta Anaesthesiologica Scandinavica, 47*, 241–259.

Rakvag, T. T., Klepstad, P., Baar, C., Kvam, T. M., Dale, O., Kaasa, S., ... Skorpen, F. (2005). The Val158Met polymorphism of the human catechol-O-methyltransferase (COMT) gene may influence morphine requirements in cancer pain patients. *Pain, 116*, 73–78.

Reyes-Gibby, C. C., Shete, S., Rakvag, T., Bhat, S. V., Skorpen, F., Bruera, E., ... Klepstad, P. (2007). Exploring joint effects of genes and the clinical efficacy of morphine for cancer pain: OPRM1 and COMT gene. *Pain, 130*, 25–30.

Reyes-Gibby, C. C., El Osta, B., Spitz, M. R., Parsons, H., Kurzrock, R., Wu, X., ... Bruera, E. (2008). The influence of tumor necrosis factor-alpha -308 G/A and IL-6 -174 G/C on pain and analgesia response in lung cancer patients receiving supportive care. *Cancer Epidemiology, Biomarkers & Prevention, 17*(11), 3262–3267.

Romberg, R. R., Olofsen, E., Bijl, H., Taschner, P. E., Teppema, L. J., Sarton, E. Y., ... Dahan, A. (2005). Polymorphism of μ-opioid receptor gene (OPRM1:c.118A>G) does not protect against opioid-induced respiratory depression despite reduced analgesic response. *Anesthesiology, 102*(3), 522–530.

Ross, J. R., Riley, J., Taegetmeyer, A. B., Sato, H., Gretton, S., du Bois, R. M., & Welsh, K. I. (2008). Genetic variation and response to morphine in cancer patients: catechol-O-methyltransferase and multidrug resistance-1 gene polymorphisms are associated with central side effects. *Cancer, 112*, 1390–1403.

Shen, X., Purser, C., Tien, L. T., Chiu, C. T., Paul, I. A., Baker, R., ... Ma, T. (2010). μ-Opioid receptor knockout mice are insensitive to methamphetamine-induced behavioral sensitization. *Journal of Neuroscience Research, 88*, 2294–2302.

Sia, A. T., Lim, Y., Lim, E. C., Goh, R. W., Law, H. Y., Landau, R., … Tan, E. C. (2008). A118G single nucleotide polymorphism of human μ-opioid receptor gene influences pain perception and patient-controlled intravenous morphine consumption after intrathecal morphine for post-cesarean analgesia. *Anesthesiology, 109*(3), 520–526.

Sindrup, S. H., Brosen, K., Bjerring, P., Arendt-Nielsen, L., Larsen, U., Angelo, H. R., & Gram, L. F. (1990). Codeine increases pain thresholds to copper vapor laser stimuli in extensive but not poor metabolizers of sparteine. *Clinical Pharmacology and Therapeutics, 48*, 686–693.

Skarke, C., Darimont, J., Schmidt, H., Geisslinger, G., & Lotsch, J. (2003). Analgesic effects of morphine and morphine-6-glucuronide in a transcutaneous electrical pain model in healthy volunteers. *Clinical Pharmacology and Therapeutics, 73*(1), 107–121.

Sora, I., Elmer, G., Funada, M., Pieper, J., Li, X. F., Hall, F. S., & Uhl, G. R. (2001). μ Opiate receptor gene dose effects on different morphine actions: evidence for differential in vivo μ receptor reserve. *Neuropsychopharmacology, 25*, 41–54.

Stamer, U. M., Lehnen, K., Hothker, F., Bayerer, B., Wolf, S., Hoeft, A., & Stuber, F. (2003). Impact of CYP2D6 genotype on postoperative tramadol analgesia. *Pain, 105*, 231–238.

Vuilleumier, P. H., Stamer, U. M., & Landau, R. (2012). Pharmacogenomic considerations in opioid analgesia. *Pharmgenomics and Personalized Medicine, 5*, 73–87.

Wilder-Smith, O. H. (2005). Opioid use in the elderly. *European Journal of Pain, 9*, 137–140.

Yu, Y., Kranzler, H. R., Panhuysen, C., Weiss, R. D., Poling, J., Farrer, L. A., & Gelernter, J. (2008). Substance dependence low-density whole genome association study in two distinct American populations. *Human Genetics, 123*, 495–506.

Zhang, Y., Wang, D., Johnson, A. D., Papp, A. C., & Sadee, W. (2005). Allelic expression imbalance of human mu opioid receptor (OPRM1) caused by variant A118G. *Journal of Biological Chemistry, 280*, 32618–32624.

Zhang, W., Chang, Y. Z., Kan, Q. C., Zhang, L. R., Lu, H., Chu, Q. J., … Zhang, J. (2010a). Association of human μ-opioid receptor gene polymorphism A118G with fentanyl analgesia consumption in Chinese gynaecological patients. *Anaesthesia, 65*(2), 130–135.

Zhang, W., Chang, Y. Z., Kan, Q. C., Zhang, L. R., Li, Z. S., Lu, H., … Zhang, J. (2010b). CYP3A4*1G genetic polymorphism influences CYP3A activity and response to fentanyl in Chinese gynecologic patients. *European Journal of Clinical Pharmacology, 66*(1), 61–66.

Zhang, W., Yuan, J. J., Kan, Q. C., Zhang, L. R., Chang, Y. Z., Wang, Z. Y., & Li, Z. S. (2011). Influence of CYP3A5*3 polymorphism and interaction between CYP3A5*3 and CYP3A4*1G polymorphisms on post-operative fentanyl analgesia in Chinese patients undergoing gynaecological surgery. *European Journal of Anaesthesiology, 28*(4), 245–250.

Chapter 86

Opiate Receptors and Gender and Relevance to Heroin Addiction

Maja Djurendic-Brenesel, Vladimir Pilija
Institute of Forensic Medicine, Clinical Center Vojvodina, Novi Sad, Serbia

Abbreviations

cAMP Cyclic adenosine monophosphate
DA Dopamine
DOP δ-Opiate receptors
GDP Guanosine diphosphate
GPCRs G-Protein-coupled receptors
GTP Guanosine triphosphate
KOP κ-Opiate receptors
MOP μ-Opiate receptors
NAc Nucleus accumbens
NOP Nociceptin orphanin FQ peptide receptor

INTRODUCTION

Opiates have been used since prehistoric times and opium was probably the first drug that was discovered. Excavations of the remains of Neolithic settlements in Switzerland (the Cortaillod culture, 3900–3500 BC) have shown that *Papaver* was already being cultivated. The writings of Theophrastus (third century BC) are the first known written source mentioning opium. The word opium derives from the Greek word "opos" meaning vegetable juice; after all, opium is prepared from the juice of the opium poppy, *Papaver somniferum* (Booth, 1996, chap. 2).

Opium contains a considerable number of various substances, and in the nineteenth century the most significant were isolated. In 1805, the German pharmacist Sertürner isolated and described the principal alkaloid and powerful active ingredient in opium. He named it morphine after Morpheus, the Greek god of dreams. This event was soon followed by the discovery of other opium alkaloids: noscapine (narcotine, 1817) and codeine (1832) by Robiquet, papaverine (1848) by Merck, and thebaine (1932) by Small and Lutz. By the 1850s these pure alkaloids, rather than the earlier crude opium preparations, were being commonly prescribed for the relief of pain, cough, and diarrhea. In this period, the invention of the hypodermic syringe (1853) led to widespread use of morphine intravenously as a painkiller.

The alkaloid content of opium is approximately 10–20%, with more than 40 individual alkaloids having been isolated, but the most significant are morphine, codeine, thebaine, papaverine, and noscapine. All but thebaine are used clinically as analgesics to reduce pain without a loss of consciousness. Thebaine possesses a mild ability to relieve pain, has stimulatory rather than depressant effects, is poisonous, and is primarily used to synthesize other drugs. Hundreds of semisynthetic derivatives have been synthesized from thebaine, such as hydrocodone (Vicodin), oxycodone (Percodan), hydromorphone (Dilaudid), oxymorphone (Opana), nalbuphine (Nubain), etorphine, naloxone (Narcan), naltrexone (Vivitrol), buprenorphine (Buprenex), and many of these have narcotic effects.

Morphine and morphine-like narcotic agonists act on both the central and the peripheral nervous systems and have the potential to produce euphoria, analgesia, nausea, vomiting, miosis, drowsiness, a sense of detachment, depressed cough reflex, hypothermia, and respiratory depression. Morphine is used for medical purposes in the relief of moderate to severe acute and chronic pain, as well as both preoperatively and intraoperatively in various anesthesia protocols. The main disadvantage of using morphine therapeutically is that the repeated administration leads to tolerance, psychological and physical dependence, and addiction as a consequence of compulsive drug use (Karch & Drummer, 2013, chap. 5; Rogers, Spector, & Trounce, 1981; Schiff, 2002, chap. 10).

Opium and morphine addiction problems prompted a scientific search for potent but nonaddictive painkillers for medical use. First synthesized from morphine in 1874 by Wright, the Bayer Company of Germany introduced diacetylmorphine (diamorphine), which was named heroin because of its "heroic" qualities as analgesic. However, soon after its introduction, heroin was recognized as having narcotic, toxic, and addictive properties far exceeding those of morphine. It is assumed that heroin is the most addictive drug, and its use has been forbidden since 1924. Nowadays, heroin abuse has risen to alarming levels and is the cause of a large number of deaths due to acute intoxication with a large variety of side effects (Schiff, 2002; Swerdlow, 2000, chap. 4).

Major advances have been made in understanding the mechanism of action of opiates. The rigid structural and stereochemical requirements essential for the analgesic actions of morphine and related opiate agonists led to the theory that they produce pharmacological effects directly by stimulating specific receptors. A most significant advance in opiate molecular biology was the cloning and characterization of these receptors. Also, increased knowledge of the cellular action of opiates and identification of their sites of action in the brain have made possible the characterization of the brain regions important for addiction and other effects, such as physical dependence (Nestler, 2004).

Investigations on both laboratory animals and humans have indicated differences between males and females in sensitivity to the effects following opiate administration and have deservedly received notable attention. Essential gender differences have been found in opiate-induced antinociception; hypothermia; respiratory, reinforcing, and discriminative stimulus effects; analgesia; tolerance; and dependence (Cicero, Ennis, Ogden, & Meyer, 2000; Cicero, Nock, & Meyer, 2000; Craft, Heideman, & Bartok, 1999; Kest, Adler, & Hopkins, 2000; Negus, Zuzga, & Mello, 2002; Peckham, Barkley, Divin, Cicero, & Traynor, 2005; Sarton et al., 2000). It is assumed that these gender-related differences are a consequence of the different organizational structures of the brain in males and females, associated with sex hormones (Cicero, Nock, O'Connor, & Meyer, 2002), and more precise studies may provide more tailored approaches to pain relief and treatment of drug abuse.

OPIATE RECEPTORS

Most recent advances stem from groundbreaking research first published in 1973, when opiate receptors were discovered in the brain, by applying radioligand binding assays. Opiate receptors are proteins located on the surfaces of nerve cells—neurons, the basic working units of the brain. Neurons communicate through an electrochemical process by releasing neurotransmitters, which bond with receptors on the target cell, setting off further biochemical processes within it.

The localization of opiate receptors in specific brain areas led to the discovery of endogenous opiate-like substances produced in the body, which appear to be responsible for the regulation of pain, immune responses, and other body functions. More recently, these endogenous opiate peptides have been hypothesized to be important in the regulation of normal mood states and the mediation of the action of abused substances other than opiates, such as alcohol and cocaine (Froehlich, 1997; Unterwald, Horne-King, & Kreek, 1992).

Opiate peptides that are produced in the body include endorphins, enkephalins, dynorphins, and endomorphins. Each family derives from a distinct precursor protein and has a characteristic anatomical distribution. β-Endorphins are the most powerful endogenous opiate peptide neurotransmitters and are found in the neurons of both the central and the peripheral nervous system. They are involved in many functions including pain modulation, nausea, vomiting, respiration, and hormonal regulation. Two pentapeptides possessing direct opiate activity are methionine-enkephalin (met-enkephalin) and leucine-enkephalin (leu-enkephalin). Dynorphins (dynorphin A and dynorphin B (rimorphin)) are produced in many different parts of the brain and the spinal cord and have many different physiological actions, but primarily act as modulators of pain response. The smallest endogenous opiate peptides possessing the highest known affinity and specificity for opiate receptors are two tetrapeptides, endomorphin-1 and endomorphin-2 (Koneru, Satyanarayana, & Rizwan, 2009; Sprenger, Berthele, Platzer, Boecker, & Tolle, 2005).

Exogenous opiate ligands, such as morphine and heroin, exert their analgesic effects by binding to and activating receptors that comprise part of an endogenous opiate system, mimicking and amplifying the actions of the above-mentioned endogenous neurotransmitters.

CLASSIFICATION AND LOCALIZATION OF OPIATE RECEPTORS

Classical pharmacological studies indicated the presence of multiple classes of opiate receptors long before their identification biochemically. The discovery of opiate receptor multiplicity was established through receptor binding studies and cloning experiments and implied the existence of multiple signaling pathways mediating a multiplicity of functions. Three major classes of opiate receptors are μ, δ, and κ, named by using the first letter of the first ligand that was found to bind to them. Morphine was the first ligand shown to bind to μ receptors, the drug ketocyclazocine was shown to attach to κ receptors, and δ receptors were named after the mouse vas deferens tissue in which receptors were first characterized. A fourth opiate receptor was identified in 1995 and has been included in the opiate receptor family, termed the nociceptin orphanin FQ peptide receptor (Dhawan, Cesselin, & Raghubir, 1996).

The International Union of Pharmacology has recommended appropriate terminology for opiate receptors: MOP (μ), KOP (κ), DOP (δ), and NOP for the nociceptin orphanin FQ peptide receptor (McDonald & Lambert, 2005).

Other receptor subtypes have been suggested, e.g., σ receptors (σ_1, σ_2), owing to the antitussive actions of opiates being mediated via these receptors. However, it was confirmed that σ receptors are nonopiate, because they are quite different in both function and gene sequence and bind diverse classes of psychotropic drugs. It was found that σ receptors have affinity for the several benzomorphans (pentazocine, cyclazocine), various structurally and pharmacologically distinct psychoactive chemicals such as haloperidol and cocaine, and neuroactive steroids like progesterone (Su & Hayashi, 2003).

In the scientific literature the existence of further opiate receptors, ε, ζ, ι, and λ, has been suggested based on pharmacological evidence of actions produced by endogenous opiate peptides, but these do not seem to be mediated through any of the known opiate receptors.

Pharmacological, physiological, and opiate receptor binding studies indicate the presence of an ε opiate receptor, stimulated by the endogenous opiate peptide β-endorphin, which induces the release of met-enkephalin, which, in turn, stimulates DOP for the production of antinociception. The ε opiate receptors in the brain are stimulated by the μ agonist etorphine and κ agonist bremazocine, as well as β-endorphin, while buprenorphine has been shown to act as an ε antagonist. It is noteworthy that the ε opiate receptor-mediated pain control system is different from the MOP, DOP, or KOP system, and there appears to be no ε-selective ligand currently available (Tseng, 2002).

The ζ opiate receptor has been cloned and classified as an opiate growth factor receptor. This receptor does not have any sequence homology with MOP, DOP, or KOP and has a quite different function. The ζ opiate receptor mediates the activity of the opiate growth factor met-enkephalin, a peptide that regulates developmental events in a variety of normal and tumorigenic tissues and cells, including the nervous system. Binding of met-enkephalin to the cloned ζ receptor was inhibited by a high concentration of naltrexone, but no additional binding affinity measures are available (Zagon, Verderame, & McLaughlin, 2002).

Other receptor subtypes described are ι, which preferentially binds enkephalins, being present in the ileum of rabbits, and λ, with affinity to epoxymorphine, being found in fresh preparations of cell membranes of rats. However, these receptors are poorly characterized, and wider acceptance of their existence awaits further experimental evidence relating to the identification of their respective genes. The unusual functional properties of these receptors might be associated with proteins encoded by nonhomologous genes or posttranslational modification of the protein products derived from the MOP, DOP, KOP, or NOP receptor genes (Martins, de Almeida, do Rego Monteiro, Kowacs, & Ramina, 2012).

Opiates express their most important pharmacological effects through their complex interactions with the three major classes of opiate receptors, μ, δ, and κ. Pharmacological evidence based on studies of activity with radioligands have suggested subtypes of these receptors: μ_1, μ_2, μ_3, δ_1, δ_2, κ_1, κ_2, and κ_3.

Opiate receptors are found in nerve cells located in various regions of the brain and the spinal cord, as well as in intramural nerve plexuses that are involved in the regulation of gastrointestinal and urogenital motility. A characteristic of the peripheral opiate system, which is very similar to that in the brain, is interaction with immune functions (Koneru et al., 2009; Martins et al., 2012; Rang, Dale, Ritter, & Moore, 2003, chap. 4). Immunohistochemical examinations of human brain tissue in cases of morphine and heroin overdose have shown that morphine localizes in the neuronal cytoplasm of the cortex, basal ganglia, brainstem, cerebellum,

TABLE 1 Location and Physiological Action of Opiate Receptors

Receptor	Subtypes	Location	Physiological Action		

MOP	μ_1, μ_2, μ_3	• Brain
		  • Cortex
		  • Thalamus
		  • Cerebellum
		  • Brainstem
		  • Limbic system
		  • Basal ganglia
		• Spinal cord
		  • Substantia gelatinosa
		• Peripheral
		• Sensory neurons
		• Intestinal tract
			μ_1:
			• Supraspinal and Spinal Analgesia
			• Hypothermia
			• Miosis
			• Catalepsy
			• Nociception
			• Prolactin release
			• Testosterone inhibition
			• Physical dependence
				μ_2:
				• Euphoria
				• Respiratory depression
				• Reduced gastrointestinal motility
				• Constipation
				• Bradycardia
				• Physical dependence
					μ_3:
					• Vasodilation

DOP	δ_1, δ_2	• Brain
		  • Cortex
		  • Basal ganglia
		  • Olfactory bulb
		  • Amygdala
		• Peripheral
		• Sensory neurons
			• Spinal analgesia
			• Euphoria
			• Hypotension
			• Hyperthermia
			• Convulsant effects
			• Brain reward
			• Antidepressant effects
			• Modulating cognitive functions
			• Physical dependence		

KOP	κ_1, κ_2, κ_3	• Brain
		  • Hypothalamus
		  • Claustrum
		  • Brainstem
		  • Midbrain Periaqueductal gray matter
		• Spinal cord
		  • Substantia gelatinosa
		• Peripheral sensory neurons
			κ_1, κ_3:
			• Peripheral and spinal analgesia
			• Stimulation
			• Nociception
			• Dissociative/hallucinogenic effects
			• Thermoregulation
			• Locomotor activity
			• Anticonvulsant effects
			• Miosis
			• Respiratory depression
			• Inhibition of antidiuretic hormone release
			• Neuroendocrine secretion function
				κ_2:
				• Ataxia
				• Sedation
				• Dysphoria	

NOP	NOP	• Brain
		  • Cortex
		  • Limbic system
		  • Basal ganglia
		• Spinal cord
			• Depression
			• Stress and anxiety
			• Learning and memory
			• Feeding
			• Tolerance to μ agonists
			• Reward/addiction
			• Urogenital activity		

This table lists the locations of opiate receptors in the central and peripheral nervous systems and physiological actions of the major classes of opiate receptors. Data from Vaughn et al. (1990), Calo', Guerrini, Rizzi, Salvadori, and Regoli (2000), and Rang et al. (2003, chap. 4).

and part of the limbic system—hippocampus and amygdala (Karch & Drummer, 2013, chap. 5). The location and physiological action of the major classes of opiate receptors are shown in Table 1.

Pharmacological studies have shown that certain endogenous and exogenous opiates act more on some receptor subtypes than others, which may lead to the development of new pain-relieving drugs that do not cause depressed breathing rate, addiction, or other side effects. Thus, endogenous opiate peptides, β-endorphins and endomorphins, interact preferentially with μ receptors, enkephalins with δ receptors, and dynorphins with κ receptors. Exogenous opiates, morphine and several other opiates, have high affinity for MOP, whereas other opiates have varying affinities for DOP and KOP. The opiate antagonist naloxone binds with high affinity to all opiate receptors, but its affinity for μ receptors is generally 10-fold greater. MOP mediates most of the pharmacological effects of opiates, and its regulation is of great importance to unraveling the molecular mechanisms underlying the physical responses to opiate treatment, such as tolerance and dependence (Koneru et al., 2009; Martins et al., 2012; Rang et al., 2003, chap. 4). Endogenous and exogenous opiate ligand specificity for the opiate receptors is shown in Table 2.

STRUCTURE OF OPIATE RECEPTORS AND SIGNAL TRANSDUCTION

Opiate receptor binding studies and cloning experiments confirmed that MOP, DOP, KOP, and NOP are lipoproteins from membranes of nerve cells and belong to the large family of integral membrane protein receptors, related to guanine nucleotide-binding proteins known as G proteins (GPCRs—G-protein-coupled receptors). GPCRs consist of seven transmembrane α-helices, an extracellular amino-terminal segment, an intracellular carboxy-terminal tail, and three extra- and intracellular loops. It has been found that cloned NOP is approximately 50% identical with MOP, DOP, and KOP, which possess the same general structure. GPCRs bind a wide range of extracellular ligands and transmit signals to intracellular G proteins, the most important components of the cell signaling cascade, i.e., they activate inside signal transduction pathways and ultimately cellular responses. G-protein-mediated signaling of GPCRs is intrinsically kinetic, including two key processes, G-protein activation and deactivation.

TABLE 2 Opiate Ligand Specificity for Opiate Receptors

Receptor	Agonists		Antagonists
	Endogenous Opiates	Exogenous Opiates	
MOP	β-Endorphins endomorphins	Morphine Codeine Oxymorphone Dextropropoxyphene Methadone Meperidine Etorphine Fentanyl Sufentanil Levorphanol Buprenorphine Norbuprenorphine	Naloxone Naltrexone Nalorphine Cyprodime
DOP	Enkephalins	Etorphine Levorphanol Norbuprenorphine	Naloxone Naltrexone Buprenorphine Naltrindole
KOP	Dynorphins	Etorphine Nalbuphine Butorphanol Levorphanol Bremazocine Ketazocine Pentazocine Nalorphine	Naloxone Naltrexone Buprenorphine Norbinaltorphimine
NOP	Nociceptin orphanin FQ peptide	Etorphine Buprenorphine Norbuprenorphine	

This table lists the specificity of endogenous and exogenous opiate agonists and antagonists for major classes of opiate receptors.
Data from Roth et al. (2002), Rang et al. (2003, chap. 4), Koneru et al. (2009), Djurendic-Brenesel and Pilija (2012), and Martins et al. (2012).

Heterotrimeric G proteins consist of distinct Gα, Gβ, and Gγ subunits. When the receptor is occupied, the Gα subunit is uncoupled and forms a complex, which interacts with cellular systems to produce an effect. In the inactive state the Gα subunit is bound to the nucleotide guanosine diphosphate (GDP) and is in complex with the Gβ and Gγ subunits. Agonist binding to an opiate receptor leads to a conformational change in the receptor and G-protein activation, causing dissociation of GDP from the Gα subunit, which is replaced by the nucleotide guanosine triphosphate (GTP), and then separation of the Gα–GTP from the two other G-protein subunits. In the active state Gα–GTP and Gβγ subunit complex interact with intracellular signaling proteins, such as potassium and voltage-sensitive calcium channels. At the postsynaptic membrane, opiate receptors mediate hyperpolarization by opening potassium channels and stimulate K^+ efflux, thereby preventing neuronal excitation. Activation of voltage-sensitive calcium channels suppresses Ca^{2+} influx, and thereby attenuates the excitability of neurons and/or reduces release of neurotransmitters such as substance P and calcitonin gene-related peptide (pronociceptive and proinflammatory neuropeptide). On the inner face of the plasma membrane Gα–GTP inhibit an enzyme, adenylate cyclase, which catalyzes the conversion of adenosine triphosphate into the cyclic adenosine monophosphate (cAMP), an intracellular signaling molecule, which may result in inhibition of neurotransmitter release (Figure 1). cAMP functions as a "second messenger" to relay extracellular signals to intracellular effectors, particularly protein kinase A, which controls numerous events from phosphorylation to regulation of glycogen, sugar, and lipid metabolism to gene transcription.

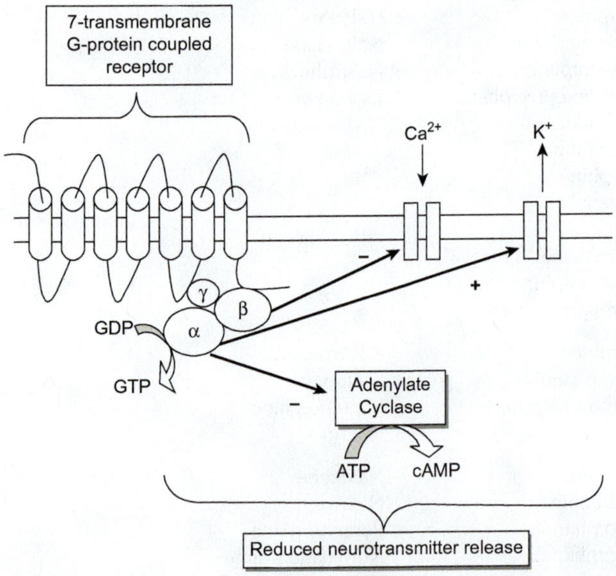

The intracellular signal is terminated by G proteins' intrinsic GTPase activity, which hydrolyzes the Gα-bound GTP to GDP and thus returns the G protein back to an inactive state, ready for a subsequent round of receptor activation (Bjarnadottir et al., 2006; McDonald & Lambert, 2005; Standifer & Pasternak, 1997; Traynor, 2012; Tuteja, 2009).

TOLERANCE, DEPENDENCE, AND ADDICTION

Tolerance, dependence, and addiction are manifestations of brain changes, induced by chronic exposure to morphine, heroin, and other opiates. The natural reward system, a collection of brain structures such as the ventral tegmental area (VTA), nucleus accumbens (NAc), and prefrontal cortex, is stimulated by the pleasure derived from opiates. By binding to opiate receptors, exogenous opiates generate signals in the VTA to produce the neurotransmitter dopamine (DA) and release it into the NAc, which causes feelings of pleasure. The information then travels to the prefrontal cortex, whose role is to evaluate reward impulses and determine whether they are safe to pursue. Opiate receptor activation by opiates results in feelings of reward and activates the pleasure circuit by causing greater amounts of DA to be released within the NAc, and that is a primary reason for repeated drug use. However, repeated exposure leads to tolerance and dependence, and prolonged use produces more long-lasting changes in the brain, which leads to compulsive drug-seeking behavior and addiction (Kosten & George, 2002).

Tolerance to the analgesic effects of opiates is a physiological state characterized by fading effects of opiates with chronic administration, which lead to progressive increases in the dose required to produce the desired effect. Tolerance usually develops after 1 to 3 weeks of opiate exposure. Repeated administration of an opiate may result in cross-tolerance, i.e., extension of the tolerance to other opiates, even to those that have not been previously administered. Although dependence usually accompanies tolerance, they are distinct phenomena.

Physical dependence is the expected response to continuous opiate administration and refers to the physiological adaptation of the body to the presence of an opiate. It occurs much more rapidly than tolerance, just after a few days of opiate administration, and is manifested by specific withdrawal symptoms or abstinence syndrome that can be produced by abrupt cessation of opiate administration, rapid dose reduction, and/or administration of an antagonist (e.g., naloxone). Withdrawal symptoms are physical and psychological (sweating, chills, abdominal cramping, nausea, vomiting, diarrhea), usually begin 6 to 8 h after the last opiate administration and peak after 2 days, and typically last 7 to 10 days. Some withdrawal symptoms such as insomnia, anxiety, and lack of interest can last 6 months or even longer. The withdrawal response is very complex and involves many brain regions.

Psychological dependence, also called emotional or psychic dependence, is characterized by emotional and mental preoccupation with the enjoyable effects of the opiate drug and by a persistent craving to take it again. The symptoms displayed are

FIGURE 1 Structure and activation of GPCRs and signal transduction. The structure of a GPCR, its activation, and the subsequent signal transduction are shown. Seven-transmembrane GPCR activation by opiate ligands leads to activation of the heterotrimeric G protein (αβγ), causing the exchange of GDP for GTP on the α subunit and its separation from βγ subunits, which initiate the intracellular transduction pathways that include stimulation of K^+ efflux, suppression of Ca^{2+} influx, and inhibition of adenylate cyclase, i.e., production of cAMP from ATP. Abbreviations: α, β, γ, G-protein subunits; GDP, guanosine diphosphate; GTP, guanosine triphosphate; ATP, adenosine triphosphate; cAMP, cyclic adenosine monophosphate. *From McDonald and Lambert (2005); with permission from Oxford University Press.*

not physical, as the drug is used to obtain relief from tension or emotional discomfort. Drug deprivation causes uneasiness, anxiety, and sometimes depression. Craving seems to be the most common withdrawal symptom. Psychological dependence is usually manifested by compulsive opiate use, but the pattern of use and frequency can differ considerably among individuals.

Addiction is often a consequence of tolerance, physical dependence, and withdrawal syndrome, but biochemical, social, and psychological factors are more important in its development. Addiction is a behavioral disorder characterized by impaired control over repeated drug use, compulsive drug seeking, drug craving, and continued use despite adverse social, psychological, and/or physical consequences. These disorder behaviors are extremely difficult to control, much more difficult than physical dependence, and they are responsible for the massive health and social problems caused by drug addiction (Adriaensen, Vissers, Noorduin, & Meer, 2003; Koob & Nestler, 1997; Ueda, Inoue, & Mizuno, 2003).

GENDER DIFFERENCES IN THE PHARMACOLOGICAL EFFECTS OF OPIATES

It has long been suspected that men and women have different sensitivity and responses to drugs. Investigations on laboratory animals (rats, mice, monkeys) and humans have consistently proven the existence of gender differences in a number of the aspects of the pharmacology of opiates. The most important gender differences have been found in opiate-induced antinociception; hypothermia; respiratory, reinforcing, and discriminative stimulus effects; analgesia; tolerance; and dependence.

Experimental results from studies with laboratory rodents and monkeys suggest the existence of gender differences in morphine-induced antinociception, with males consistently more sensitive than females. In assays of thermal nociception it has been observed that opiate μ and κ agonists have a higher potency or produce greater effects in males than in females. It was hypothesized that opiate ligands have a higher affinity for opiate receptors in males; however, studies with competitive opiate antagonists suggest that gender differences exist in the affinity of opiate ligands only for κ receptors and may contribute to gender differences in the effects of some opiates (Negus et al., 2002; Peckham et al., 2005).

Opiate agonists affect mammalian temperature regulation and can exhibit the hypothermic and/or hyperthermic effects on body temperature. These opiate effects are considered to involve both peripheral and central thermoregulatory alterations. Experiments with rodents show that females display greater sensitivity to the hypothermic effects of morphine relative to males. Kest et al. (2000) observed that tolerance to the thermoregulatory effects of morphine developed in both genders, but hypothermia was greater in females relative to males, which may contribute in part to gender differences in morphine analgesia.

A number of studies with rodents indicate that μ opiate agonists are more potent and in certain cases more efficacious in producing analgesia and sedation in males than in females (Sarton et al., 2000). On the other hand, females are more sensitive than males to reinforcing, discriminative, and locomotor stimulant effects of opiates (Cicero, Ennis, et al., 2000; Cicero, Nock, et al., 2000; Craft et al., 1999). It has been found that equivalent degrees of physical dependence were generated in males and females during chronic morphine administration, but expression of physical dependence was more severe in males than in females during spontaneous withdrawal, although it has been found that females self-administer much larger amounts of μ agonists than males (Cicero, Aylward, & Meyer, 2003; Cicero, Ennis, et al., 2000; Cicero, Nock, et al., 2000). These results suggest that studies of gender differences in drug-seeking behavior in humans should be of high priority.

Considering the complex pattern of gender differences in the pharmacology of opiates, investigations were carried out to examine gender differences in the pharmacokinetics of opiates, i.e., differences in their concentration in blood and in brain regions of male and female rats, at different times after treatment with seized heroin. As can be seen from Figures 2 and 3, opiate levels in blood samples attained the highest values 15 min after treatment in animals of both genders. In brain regions the highest opiate content

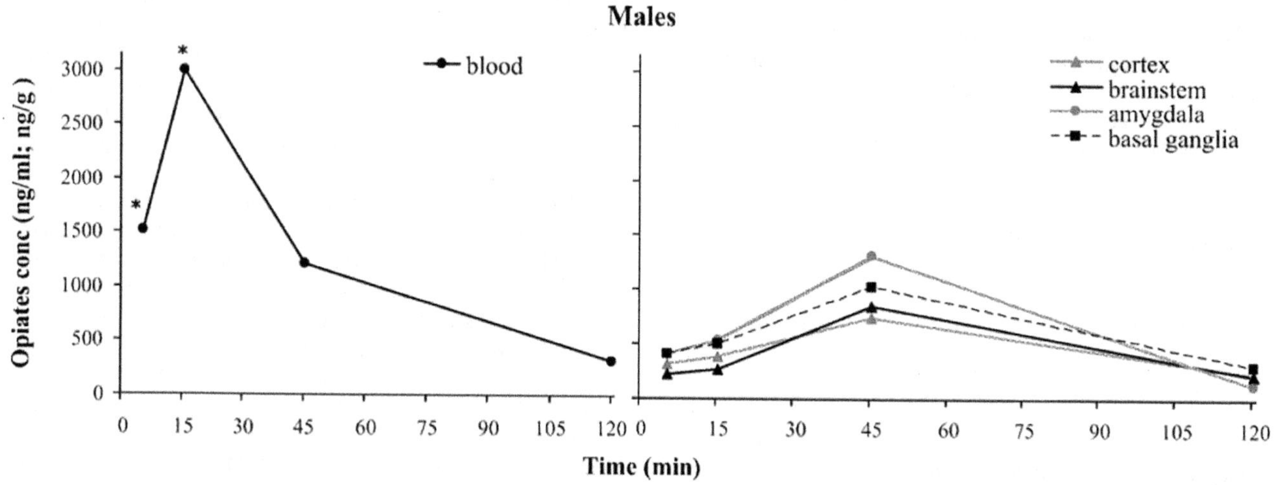

FIGURE 2 **Opiate content in blood and brain regions of male rats versus time.** Opiate content in blood and brain regions of male rats at various times postadministration is shown. Rats were treated intraperitoneally with seized heroin and sacrificed by decapitation after 5, 15, 45, and 120 min. Opiate content in blood and brain regions attained highest values after 15 and 45 min, respectively. Opiate content in blood after 5 and 15 min was significantly higher (*$p < 0.05$) compared to opiate content in brain regions. Each point is the mean of five rats. *From Djurendic-Brenesel et al. (2010); with permission from Elsevier.*

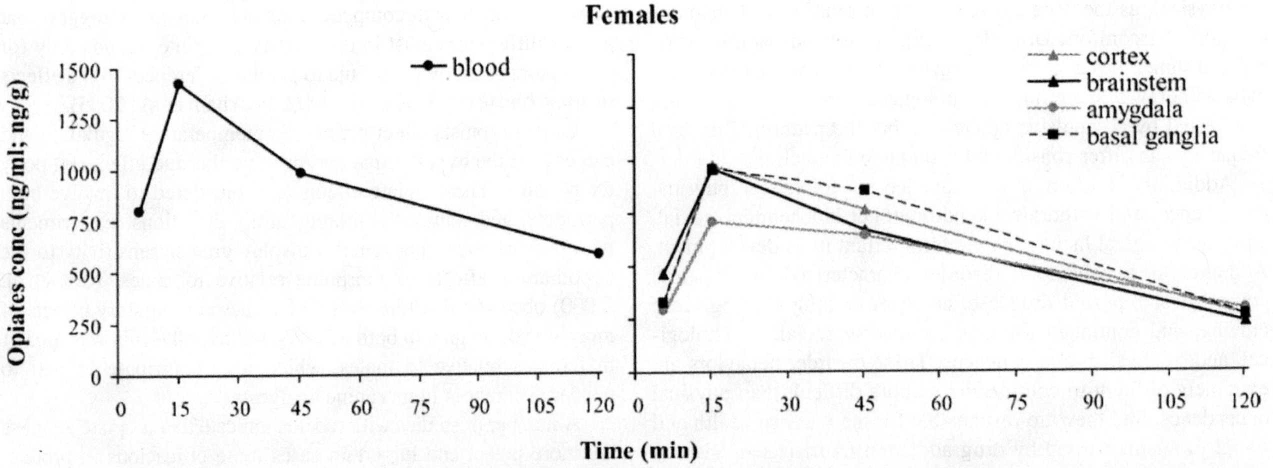

FIGURE 3 **Opiate content in blood and brain regions of female rats versus time.** Opiate content in blood and brain regions of female rats at various times postadministration is shown. Rats were treated intraperitoneally with seized heroin and sacrificed by decapitation after 5, 15, 45, and 120 min. Opiate content in blood and brain regions attained highest values after 15 min. Each point is the mean of five rats. *From Djurendic-Brenesel et al. (2010); with permission from Elsevier.*

was determined after 15 min in females and after 45 min in males. However, 15 min after treatment the values of opiate concentration in blood were significantly higher for males, whereas in brain regions they were significantly lower for males than for females (Table 3). These findings suggest a faster passage of opiates from blood to the brain regions in females than in males and could contribute to the large gender differences that have been observed in the acute pharmacological profiles of opiates (Djurendic-Brenesel, Mimica-Dukic, Pilija, & Tasic, 2010; Djurendic-Brenesel & Pilija, 2012; Djurendic-Brenesel, Pilija, Mimica-Dukic, Budakov, & Cvjeticanin, 2012; Đurendić-Brenesel, PIlija, Cvjetićanin, Ivetić, & Mimica-Dukić, 2012).

It is assumed that the observed gender differences in response to opiates are a consequence of different organizational structures in female and male brains, mediated by sex hormones such as estrogen and testosterone, which have a significant influence on the communication of brain neurons and are closely associated with the neurons' sensitivity to opiates (Cicero et al., 2002; Ji, Murphy, & Traub, 2006; Wiesenfeld-Hallin, 2005). On the other hand, investigations have provided evidence that different sensitivities to opiates in males and females are entirely genotype dependent, and genetic factors play an important role in the risk of the opiate addiction (Sternberg & Mogil, 2001).

GENDER DIFFERENCES IN HEROIN ADDICTION

Nowadays, heroin addiction is a major public health problem whose incidence rate has dramatically increased, especially among adolescents. A number of epidemiologic survey studies have indicated that the prevalence rates of heroin abuse are higher among men than among women. Some statistical data indicate that about 74% of heroin addicts are males, but this is in contrast to the animal and clinical studies indicating that females are more vulnerable to heroin abuse than males (Lee & Ho, 2013).

However, it is important to note that major discrepancies exist between human and animal data, because human females seem to have a greater response to μ agonists than human males, with differences in the magnitude and direction of response of the μ opiate system in several cortical and subcortical brain regions (Wiesenfeld-Hallin, 2005; Zubieta, Dannals, & Frost, 1999).

Gender differences have been observed in studies of clinical profiles of opiate-dependent individuals. Thus, it has been found that heroin (i.e., morphine)-induced depression was more severe in women than in men, women have a more rapid progression to opiate involvement and dependence than men, and craving for opiates was significantly higher among women (Back et al., 2011). Also, in one study among adolescent heroin users, females were about four times more likely to inject heroin than males (Wu & Howard, 2007).

As investigations have given apparent evidence that the KOP and MOP systems regulate the activity of DA neurons within the brain, studies have been carried out with κ agonists, such as nalbuphine, butorphanol, and pentazocine. It was found that they produce greater analgesia in women than in men (Gear et al., 1996), which indicates that robust analgesic gender differences in humans are not restricted to only μ opiate agonists. Furthermore, studies with experimental animals suggest that a decrease in the activity of the KOP system may be effective in the treatment of depression and drug addiction (Negus et al., 2002; Peckham et al., 2005).

It has been proposed that opiate abuse among women may be influenced by psychosocial and hormonal factors and estrogen-regulated neuroendocrine functions. Estrogen exerts various actions on the neurotransmitter systems (acetylcholine, serotonin, noradrenaline, DA); influences pain sensation, mood, and seizures; and decreases the secretion of β-endorphin, but direct evidence of interactions among estrogen and opiate receptors is lacking (McEwen, 2001).

Epidemiological studies indicate that the risk of opiate addiction is 50% genetic dependent, but the specific genes have not been identified as of this writing (Sternberg & Mogil, 2001). These facts indicate that it is necessary to investigate both functional interactions between sex hormones and opiates and genetic factors, to determine an individual's risk for addictive disorder and provide insight into gender differences in opiate addiction.

TABLE 3 Opiates' Contents in Blood and Brain Regions of Male and Female Rats in the Times When the Highest Contents Were Measured

Time (min)	Sample	Males		Females			
		Sum of Opiates (ng/ml; ng/g)	p1	Sum of Opiates (ng/ml; ng/g)	p2	p3	p4
15	Blood	**2994.5 ± 505.5**		**1423.1 ± 139.2**			
	Cortex	390.5 ± 18.1	0.043*	972.8 ± 147.1	0.133	0.030*	0.044*
	Brainstem	272.2 ± 25.6	0.044*	954.8 ± 169.5	0.129		0.034*
	Amygdala	532.7 ± 51.5	0.049*	706.8 ± 55.7	0.077		0.048*
	Basal ganglia	497.0 ± 37.2	0.048*	**956.1 ± 87.1**	0.100		0.039*
45	Blood	1209.7 ± 131.7		983.3 ± 113.5			
	Cortex	741.9 ± 53.1	0.066	759.7 ± 134.5	0.104	0.102	0.145
	Brainstem	848.5 ± 96.3	0.092	662.8 ± 68.5	0.204		0.092
	Amygdala	**138.7 ± 121.5**	0.128	641.1 ± 41.3	0.178		0.055
	Basal ganglia	**1025.3 ± 80.7**	0.152	850.8 ± 46.9	0.226		0.161

In this table are shown the highest measured opiate concentrations in blood and brain regions of male and female rats, 15 and 45 min after treatment. Rats were treated intraperitoneally with seized heroin and sacrificed by decapitation after 5, 15, 45, and 120 min. Opiate contents in blood were significantly higher after 15 min for males, whereas in brain regions were significantly lower in males compared to females. Values are means ± SD of five rats. The highest values are shown in bold. Significant difference: p1, males' blood and brain regions; p2, females' blood and brain regions; p3, males' and females' blood; p4, males' and females' brain regions; *p < 0.05.
From Djurendic-Brenesel et al. (2010); with permission from Elsevier.

A variety of treatments are available for heroin addiction, including both behavioral and pharmacological. Medications such as methadone, levo-α-acetylmethadol, buprenorphine, and naltrexone act on the same brain structures as addictive opiates, but with a relatively low success rate in addiction treatment (Kosten & George, 2002; Mijatović, Samojlik, Ajduković, Đurendić-Brenesel, & Petković, 2014).

Bearing in mind that treatment of heroin addiction involves a difficult struggle to overcome the effects of brain abnormalities, i.e., complex psychological, social, and biological disease aspects, pharmacological and appropriate psychosocial treatments must be applied together. Also, it is necessary to consider that heroin abuse by women is growing significantly, and therefore it is very important to understand how female biology influences the effects of opiates.

APPLICATIONS TO OTHER ADDICTIONS AND SUBSTANCE MISUSE

Until now there has been a deficiency of preclinical or clinical systematic data related to potential gender differences in opiate abuse, compared to the considerably earlier reports indicating the existence of gender differences in cocaine abuse, and the necessity for different prevention strategies and treatments for opposite genders (Griffin, Weiss, Mirin, & Lange, 1989). As well as opiates, cocaine stimulates the reward system in the brain, i.e., cells in the NAc, by increasing DA signaling and thereby producing feelings of pleasure and satisfaction. Cocaine is an extremely psychologically addictive drug, and therefore the complex treatment of cocaine addiction includes a variety of emotional, social, and behavioral problems (Nestler, 2005). Significant gender differences have been found in the initiation of cocaine use, the development of dependence, and the treatment-seeking behaviors in men and women. Thus, Wagner and Anthony (2006) indicated that females from the first use of cocaine progress more rapidly to dependence than males. Clinical studies indicate that addicted male individuals demonstrate greater neurobiological disruption relative to females, but expression of cocaine dependence is more severe in females (Adinoff et al., 2006). Addicted women express more craving and have greater depressive symptomatology and severity of physical, psychiatric, and social consequences compared to men. Despite these facts, women have more positive attitudes toward treatment of cocaine addiction (Najavits & Lester, 2008). Many pharmacological (medications, i.e., disulfiram, gabapentin, baclofen) and behavioral treatments such as cognitive behavioral therapy, motivational therapy, and 12-step facilitation therapy are available for treatment of cocaine addiction, but have a fairly low success rate (Palinkas, 1999). Owing to these facts, and the widespread abuse of cocaine in addition to heroin, extensive efforts are being invested in the development of the appropriate treatment programs considering social, neurobiological, and medical aspects of men's and women's drug abuse (Najavits & Lester, 2008).

DEFINITION OF TERMS

Opiate receptors These are lipoproteins located on the surface of nerve cells within the central and peripheral nervous systems that bind opiates. Major subtypes are μ, δ, and κ opiate receptors.

G proteins These are guanine nucleotide-binding cell membrane proteins, coupled to cell surface receptors, and binding of an extracellular ligand to the receptor affects biochemical actions within cells.

Endogenous opiates These are naturally occurring opiate-like substances within the body that bind to opiate receptors. These peptides are called endorphins, enkephalins, dynorphins, and endomorphins.

Exogenous opiates These are natural or synthetic opiate alkaloids, such as morphine and heroin, that bind to opiate receptors. They mimic and amplify the actions of endogenous opiates.

Reward system This is a collection of brain structures, such as the VTA, NAc, and prefrontal cortex, that are involved in feeling pleasure.

Dopamine This is a neurotransmitter in the brain that regulates emotional responses, motivation, movement, arousal, cognition, and reward, i.e., the feeling of pleasure. It is the primary neurotransmitter in the reward system.

Tolerance This refers to a physiological condition characterized by reduced reaction to a drug's effects owing to chronic administration.

Physical dependence This is a physiological neuroadaptation to the presence of a drug required for normal physiological functions, resulting in development of a withdrawal syndrome when drug use stops.

Psychological dependence This is the emotional need and craving for a drug's pleasurable effects helpful in alleviating emotional discomfort.

Addiction This is a complex chronic disease characterized by dysfunctional behavior, compulsive need, and craving for a drug despite adverse consequences.

KEY FACTS OF OPIATE ABUSE

- The production of opium in 2013 is estimated at about 7000 tons, of which 80% is produced in Afghanistan.
- According to the World Drug Report in 2013, about 16.5 million people are opiate abusers, of whom about 11 million abuse heroin.
- The number of opiate addicts is continuously growing, but the alarming fact is that very young addicts (ages 12–17 years) comprise about 26% of the total opiate abusers population.
- High mortality rates are associated with opiate use by injection, with males having higher mortality rates from drug overdose than females who inject opiates.
- The highest prevalence of injection opiate use and highest mortality rates among users are in high-income countries, particularly in western Europe.
- Methadone and buprenorphine are still used for substitution therapy, but in many countries this treatment is unavailable or illegal; e.g., in Russia, with the highest rates of opiate use in the world, this treatment is forbidden by law.

SUMMARY POINTS

- This chapter focuses on the pharmacological effects of opiates and their influence on males and females.
- Opiates produce pharmacological effects directly by stimulating opiate receptors, located on the surfaces of neurons in the various regions of brain.
- Three major classes of opiate receptors are μ, δ, and κ, of which μ receptors mediate most of the effects of opiates.
- Gender differences in the effects of opiates have been found in both laboratory animals and humans.

- Prevalence rates of heroin abuse are higher among men, but women are more vulnerable to heroin abuse and progress more rapidly to heroin dependence than men.
- Investigations indicate the necessity of developing appropriate addiction treatments considering social, neurobiological, and medical aspects of drug abuse by men and women.

REFERENCES

Adinoff, B., Williams, M. J., Best, S. E., Harris, T. S., Chandler, P., & Devous, M. D. (2006). Sex differences in medial and lateral orbitofrontal cortex hypoperfusion in cocaine-dependent men and women. *Gender Medicine, 3*(3), 206–222.

Adriaensen, H., Vissers, K., Noorduin, H., & Meer, T. (2003). Opioid tolerance and dependence: an inevitable consequence of chronic treatment? *Acta Anaesthesiologica Belgica, 54*(1), 37–47.

Back, S. E., Payne, R. L., Wahlquist, A. H., Carter, R. E., Stroud, Z., Haynes, L., … Ling, W. (2011). Comparative profiles of men and women with opioid dependence: results from a national multisite effectiveness trial. *American Journal of Drug and Alcohol Abuse, 37*, 313–323.

Bjarnadottir, T. K., Gloriam, D. E., Hellstrand, S. H., Kristiansson, H., Fredriksson, R., & Schioth, H. B. (2006). Comprehensive repertoire and phylogenetic analysis of the G protein-coupled receptors in human and mouse. *Genomics, 88*(3), 263–273.

Booth, M. (1996). *Opium: A history*. New York: St. Martin's Press.

Calo', G., Guerrini, R., Rizzi, A., Salvadori, S., & Regoli, D. (2000). Pharmacology of nociceptin and its receptor: a novel therapeutic target. *British Journal of Pharmacology, 129*(7), 1261–1283.

Cicero, T. J., Aylward, S. C., & Meyer, E. R. (2003). Gender differences in the intravenous self-administration of mu opiate agonists. *Pharmacology Biochemistry and Behavior, 74*(3), 541–549.

Cicero, T. J., Ennis, T., Ogden, J., & Meyer, E. R. (2000). Gender differences in the reinforcing properties of morphine. *Pharmacology Biochemistry and Behavior, 65*, 91–96.

Cicero, T. J., Nock, B., & Meyer, E. R. (2000). Gender-linked differences in the expression of physical dependence in the rat. *Pharmacology, Biochemistry and Behavior, 72*, 691–697.

Cicero, T. J., Nock, B., O'Connor, L., & Meyer, E. R. (2002). Role of steroids in sex differences in morphine-induced analgesia: activational and organizational effects. *Journal of Pharmacology and Experimental Therapeutics, 300*, 695–701.

Craft, R. M., Heideman, L. M., & Bartok, R. E. (1999). Effect of gonadectomy on discriminative stimulus effects of morphine in female versus male rats. *Drug and Alcohol Dependence, 53*, 95–109.

Dhawan, B. N., Cesselin, F., & Raghubir, R. (1996). Classification of opioid receptors. *Pharmacological Reviews, 48*, 567–592.

Djurendic-Brenesel, M., Mimica-Dukic, N., Pilija, V., & Tasic, M. (2010). Gender-related differences in the pharmacokinetics of opiates. *Forensic Science International, 194*(1–3), 28–33.

Djurendic-Brenesel, M., & Pilija, V. (2012). Opiate receptors and gender. In M. B. Guthrie, & B. M. Wooten (Eds.), *Heroin: Pharmacology, effects and abuse prevention* (pp. 1–27). New York: Nova Science Publishers Inc.

Djurendic-Brenesel, M., Pilija, V., Mimica-Dukic, N., Budakov, B., & Cvjeticanin, S. (2012). Distribution of opiate alkaloids in brain tissue of experimental animals. *Interdisciplinary Toxicology, 5*(4), 173–178.

Đurendić-Brenesel, M., Pilija, V., Cvjećianin, S., Ivetić, V., & Mimica-Dukić, N. (2012). Regional distribution of opiate alkaloids in experimental animals' brain tissue and blood. *Acta Veterinaria (Beograd), 62*(2–3), 137–149.

Froehlich, J. C. (1997). Opioid peptides. *Alcohol Health and Research World, 21*(2), 132–136.

Gear, R. W., Miaskowski, C., Gordon, N. C., Paul, S. M., Heller, P. H., & Levine, J. D. (1996). Kappa-opioids produce significantly greater analgesia in women than in men. *Nature Medicine, 2*, 1248–1250.

Griffin, M. L., Weiss, R. D., Mirin, S. M., & Lange, U. (1989). A comparison of male and female cocaine abusers. *Archives of General Psychiatry, 46*, 122–128.

Ji, Y., Murphy, A. Z., & Traub, R. J. (2006). Sex differences in morphine-induced analgesia of visceral pain are supraspinally and peripherally mediated. *American Journal of Physiology–Regulatory, Integrative and Comparative Physiology, 291*, 307–314.

Karch, S. B., & Drummer, O. (2013). *Karch's pathology of drug abuse* (5th ed.). New York: CRC Press.

Kest, B., Adler, M., & Hopkins, E. (2000). Sex differences in thermoregulation after acute and chronic morphine administration in mice. *Neuroscience Letters, 291*, 126–128.

Koneru, A., Satyanarayana, S., & Rizwan, S. (2009). Endogenous opioids: their physiological role and receptors. *Global Journal of Pharmacology, 3*(3), 149–153.

Koob, G. F., & Nestler, E. J. (1997). The neurobiology of drug addiction. *Journal of Neuropsychiatry and Clinical Neurosciences, 9*, 482–497.

Kosten, T. R., & George, T. P. (2002). The neurobiology of opioid dependence: implications for treatment. *Science and Practice Perspectives, 1*(1), 13–20.

Lee, C. W., & Ho, I. (2013). Sex differences in opioid analgesia and addiction: interactions among opioid receptors and estrogen receptors. *Molecular Pain, 9*, 45.

Martins, R. T., de Almeida, D. B., do Rego Monteiro, F. M., Kowacs, P. A., & Ramina, R. (2012). Opioid receptors to date. *Revista Dor Pesquisa Clínica e Terapêutica, 13*(1), 75–79.

McDonald, J., & Lambert, D. G. (2005). Opioid receptors. *Continuing Education in Anaesthesia, Critical Care & Pain, 5*(1), 22–25.

McEwen, B. S. (2001). Estrogens effects on the brain: multiple sites and molecular mechanisms. *Journal of Applied Physiology, 91*, 2785–2801.

Mijatović, V., Samojlik, I., Ajduković, N., Đurendić-Brenesel, M., & Petković, S. (2014). Methadone-related deaths–epidemiological, pathohistological, and toxicological traits in 10-year retrospective study in Vojvodina, Serbia. *Journal of Forensic Sciences, 59*(5), 1280–1285.

Najavits, L. M., & Lester, K. M. (2008). Gender differences in cocaine dependence. *Drug and Alcohol Dependence, 97*, 190–194.

Negus, S. S., Zuzga, D. S., & Mello, N. K. (2002). Sex differences in opioid antinociception in rhesus monkeys: antagonism of fentanyl and U50,488 by quadazocine. *The Journal of Pain, 3*(3), 218–226.

Nestler, E. J. (2004). Historical review: molecular and cellular mechanisms of opiate and cocaine addiction. *Trends in Pharmacological Sciences, 25*(4), 210–218.

Nestler, E. J. (2005). The neurobiology of cocaine addiction. *Science and Practice Perspectives, 5*, 4–10.

Palinkas, L. A. (1999). Cognitive behavioural therapy reduced cocaine abuse compared with 12 step facilitation. *Evidence Based Mental Health, 2*(2), 51.

Peckham, E. M., Barkley, L. M., Divin, M. F., Cicero, T. J., & Traynor, J. R. (2005). Comparison of the antinociceptive effect of acute morphine in female and male Sprague–Dawley rats using the long-lasting mu-antagonist methocinnamox. *Brain Research, 1058*, 137–147.

Rang, H. P., Dale, M. M., Ritter, J. M., & Moore, P. K. (2003). *Pharmacology* (5th ed.). Edinburgh: Churchill Livingstone.

Rogers, H., Spector, R. G., & Trounce, J. R. (1981). *A textbook of clinical pharmacology*. London: Hooder and Stoughton.

Roth, B. L., Baner, K., Westkaemper, R., Siebert, D., Rice, K. C., Steinberg, S., ... Salvinorin, A. (2002). A potent naturally occurring nonnitrogenous κ opioid selective agonist. *Proceedings of the National Academy of Sciences of the United States of America, 99*(18), 11934–11939.

Sarton, E., Olofsen, E., Romberg, R., den Hartigh, J., Kest, B., Nieuwenhuijs, D., ... Dahan, A. (2000). Sex differences in morphine analgesia: an experimental study in health volunteers. *Anesthesiology, 93*, 1245–1254.

Schiff, P. L., Jr. (2002). Opium and its alkaloids. *American Journal of Pharmaceutical Education, 66*(2), 186–194.

Sprenger, T., Berthele, A., Platzer, S., Boecker, H., & Tolle, T. R. (2005). What to learn from in vivo opioidergic brain imaging? *European Journal of Pain, 9*, 117–121.

Standifer, K. M., & Pasternak, G. W. (1997). G proteins and opioid receptor-mediated signalling. *Cellular Signalling, 9*(3/4), 237–248.

Sternberg, W. F., & Mogil, J. F. (2001). Genetic and hormonal basis of pain states. *Best Practice and Research Clinical Anaesthesiology, 15*, 229–245.

Su, T. P., & Hayashi, T. (2003). Understanding the molecular mechanism of sigma-1 receptors: towards a hypothesis that sigma-1 receptors are intracellular amplifiers for signal transduction. *Current Medicinal Chemistry, 10*(20), 2073–2080.

Swerdlow, J. L. (2000). *Nature's medicine.* Washington, DC: National Geographic.

Traynor, J. (2012). μ-Opioid receptors and regulators of G protein signaling (RGS) proteins: from a symposium on new concepts in mu-opioid pharmacology. *Drug and Alcohol Dependence, 121*, 173–180.

Tseng, L. F. (2002). Evidence for epsilon-opioid receptor-mediated beta-endorphin-induced analgesia. *Trends in Pharmacological Sciences, 22*(12), 623–630.

Tuteja, N. (2009). Signaling through G protein coupled receptors. *Plant Signaling & Behavior, 4*(10), 942–947.

Ueda, H., Inoue, M., & Mizuno, K. (2003). New approaches to study the development of morphine tolerance and dependence. *Life Sciences, 74*, 313–320.

Unterwald, E. M., Horne-King, J., & Kreek, M. J. (1992). Chronic cocaine alters brain mu opioid receptors. *Brain Research, 584*, 314–318.

Vaughn, L. K., Wire, W. S., Davis, P., Shimohigashi, Y., Toth, G., Knapp, R. J., ... Yamamura, H. I. (1990). Differentiation between rat brain and mouse vas deferens delta opioid receptors. *European Journal of Pharmacology, 177*(1–2), 99–101.

Wagner, F. A., & Anthony, J. C. (2006). Male-female differences in the risk of progression from first use to dependence upon cannabis, cocaine and alcohol. *Drug and Alcohol Dependence, 86*, 191–198.

Wiesenfeld-Hallin, Z. (2005). Sex differences in pain perception. *Gender Medicine, 2*, 137–145.

Wu, L. T., & Howard, M. O. (2007). Is inhalant use a risk factor for heroin and injection drug use among adolescents in the United States? *Addictive Behaviors, 32*, 265–281.

Zagon, I. S., Verderame, M. F., & McLaughlin, P. J. (2002). The biology of the opioid growth factor receptor (OGFr). *Brain Research Reviews, 38*(3), 351–376.

Zubieta, J. K., Dannals, R. F., & Frost, J. J. (1999). Gender and age influences on human brain mu-opioid receptor binding measured by PET. *American Journal of Psychiatry, 156*, 842–848.

Chapter 87

Transcriptional Effects of Heroin and Methamphetamine in the Striatum

Ryszard Przewlocki[1,2], Michal Korostynski[1], Marcin Piechota[1]

[1]*Department of Molecular Neuropharmacology, Institute of Pharmacology, Polish Academy of Sciences, Krakow, Poland;* [2]*Department of Neurobiology and Neuropsychology, Institute of Applied Psychology, Jagiellonian University, Krakow, Poland*

Abbreviations

GR Glucocorticoid receptor
HPA Hypothalamopituitary–adrenal
IEG Immediate early gene
VTA Ventral tegmental area

INTRODUCTION

Drug addiction is compulsive drug use despite adverse consequences and loss of control over drug use. Exposure to the drug of abuse initiates molecular and cellular alterations (Nugent, Penick, & Kauer, 2007; Ungless, Whistler, Malenka, & Bonci, 2001), while prolonged drug intake induces long-lasting changes in the brain reward system. The system consists of dopaminergic neuronal cells localized in the midbrain ventral tegmental area projecting to the mesocorticolimbic circuit, the dorsal and ventral (nucleus accumbens) striata, the prefrontal cortex, and the amygdala. The major neural target sites of the addictive drugs are the ventral and dorsal striata that control reward sensitivity, motor function, and habit learning (Gerdeman, Partridge, Lupica, & Lovinger, 2003). The dorsal striatum (caudate-putamen) is thought to underlie stimulus response and spatial learning and may be involved in longer term behavioral and molecular changes occurring in drug addictive states. The ventral striatum (nucleus accumbens) is involved in the initial rewarding effects of drugs of abuse, appetitive behavior, and reinforcement.

The mesocorticolimbic system is the neuronal substrate of rewarding effects of psychostimulants and opioids. Release of dopamine either directly, in the case of psychostimulants, or indirectly, for opioids, appears to be a common consequence of exposure to the drugs of abuse (Di Chiara & Imperato, 1988). Drug impact on the brain network induces neuroadaptations that lead to behavioral alterations and increased overall vulnerability to addiction with subsequent drug exposures (Hyman, Malenka, & Nestler, 2006). There are commonalities and differences in the ways opioids and psychostimulants affect behavior, neurobiology, and neurochemistry of the brain in animals and humans.

Common reinforcing and incentive effects are mediated, at least in part, by increasing dopamine release in the mesocorticolimbic system (Koob, 1992b). Self-administration of both classes of drugs has been demonstrated to enhance the release of dopamine in the nucleus accumbens (Di Chiara & Imperato, 1988). Furthermore, locomotor sensitization and conditioned place preference in rodents occur in response to both types of drugs. Both opioids and psychostimulants are self-administered in rodents and monkeys. Opioids, including heroin, act primarily as opioid receptor agonists, while psychostimulants, including methamphetamine, act by inhibition of dopamine reuptake. Opioids enhance dopamine release indirectly via the inhibition of γ-aminobutyric acidergic interneurons connected to dopaminergic cells in the ventral tegmental area (VTA), while psychostimulants enhance dopamine release by direct presynaptic inhibitory effects on the dopamine transporter. Long-term drug abuse leads to mood disorders and anhedonia in both humans and animals (Koob, 1992a; Newton, Kalechstein, Duran, Vansluis, & Ling, 2004).

Badiani et al. have argued that opioid and psychostimulant addictions are behaviorally and neurobiologically distinct (Badiani, Belin, Epstein, Calu, & Shaham, 2011). Dopamine receptor blockade decreases amphetamine but not heroin reward. Furthermore, repeated noncontingent injections of amphetamine increase while morphine self-administration decreases dendrite branching and spine density in the nucleus accumbens. The administration of opioids in humans produces a feeling of pleasure, satisfaction, and relaxation, while the administration of psychostimulants induces euphoria, increased attention, and a feeling of power. The effects of withdrawal from these substances are radically different in humans and mice. The molecular bases of these common and differential behavioral and neurobiological effects of psychostimulants and opioids are still unknown. However, the transcriptional effects within the reward system may contribute to the development of long-term drug-related alterations (Hyman et al., 2006; McClung & Nestler, 2008; Przewlocki, 2004).

The current evidence indicates that there are transient changes in gene transcription within the reward system after acute challenge as well as long-term changes after chronic treatment with drugs of abuse (Korostynski, Piechota, Kaminska, Solecki, & Przewlocki, 2007; Piechota et al., 2010, 2012). The drug-induced alterations result in the synthesis of new proteins, changing the morphology of certain neuronal cell types; plastic changes within neuronal networks; and finally changes in brain function. Therefore, one of the important goals in addiction research is to identify the drug-induced gene expression changes within specific brain structures that are related to the addictive properties of various drugs. It is important to estimate the extent of the molecular effects of opioids and psychostimulants and whether they are common or different. This knowledge will enhance our understanding of the cellular and molecular neurobiology of addiction, aid the development of novel effective therapies, and uncover the neuropathology of psychostimulant and opioid co-use. In fact, the neurobiological mechanisms of psychostimulant effects in opioid-dependent addicts or in patients on opioid maintenance therapy are poorly understood.

It is believed that drug-induced gene expression changes in the brain are related to the addictive properties of opioids and psychostimulants. The genes whose expression changes after the administration of opioids and/or psychostimulants could be associated with their common effects, such as the release of dopamine, behavioral sensitization, place conditioning, or anhedonia during withdrawal. Genes whose expression differs after administration of these substances could be related to or responsible for drug-specific effects, such as differences in behavioral response, independent of drug class neuroadaptations (Nestler, 2001), drug-induced synaptic remodeling (long-term potentiation or long-term depression) (Nugent et al., 2007; Ungless et al., 2001), or structural brain plasticity (Russo et al., 2010).

This issue generates several questions. Are there any differences in the transcriptional response to the drugs? Is the expression of those genes altered by opioids and/or psychostimulants? Which genes are activated and which are inhibited? What specific regulatory mechanisms may be involved? The next question is whether these changes are common or different for various drugs of abuse. Does chronic exposure to a drug of abuse cause long-lasting transcriptional changes in the brain that underlie the behavioral abnormalities of addiction? Exploring these dynamics in the gene expression profile of both classes of addictive drugs following acute administration, during chronic treatment, and during withdrawal could provide an answer for some of these questions. This review provides preclinical data and presents changes in gene transcription in the striatum of mice that occur following exposure to opioids or psychostimulants and discusses how they differ following repeated treatment and during periods of withdrawal.

PATTERNS OF GENE EXPRESSION REGULATED BY THE ACUTE ADMINISTRATION OF PSYCHOSTIMULANTS OR OPIOIDS

Multiple studies have used microarray analysis to study the molecular effects of chronic treatment with morphine (Korostynski et al., 2007; McClung, Nestler, & Zachariou, 2005), heroin (Kuntz-Melcavage, Brucklacher, Grigson, Freeman, & Vrana, 2009), cocaine (Newton et al., 2004), or methamphetamine (Yang et al., 2008) in the brain. These studies revealed many genes from various functional classes with expression altered by drugs of abuse (Kerns et al., 2005; Lemberger, Parkitna, Chai, Schutz, & Engblom, 2008; Treadwell & Singh, 2004). The majority of the results obtained were based on arbitrary significance or fold-change thresholds or focused only on individual substances or single time points. Therefore, determining when the transcriptional changes occur and whether the changes are transient or persistent is difficult. Exploring the dynamics in the gene expression profile following acute treatment, chronic treatment, and withdrawal could provide a better understanding of the molecular mechanisms underlying the drug action. The key question is whether gene expression changes are common for various drugs of abuse. While acute treatment with psychostimulants (cocaine vs methamphetamine) induces highly similar changes in the striatal transcriptome, opioids, a distinct pharmacological class, may have relatively distinct transcriptional profiles (methamphetamine vs heroin).

Our previous research focused on comparing the acute effects of various drugs of abuse. In this study, we aimed to define the sequence of changes in gene transcription in the mouse striatum in response to addictive drugs with various mechanisms of action (Piechota et al., 2010). Almost all of the altered genes were regulated in the form of two drug-responsive transcriptional modules, TM1 and TM2 (Figure 1). No other distinct transcriptional modules were identified, even at lower significance thresholds. This study depicted all the main patterns of drug-induced gene expression in the mouse striatum after administration of rewarding doses of opioids and psychostimulants.

DRUG-INDUCED REGULATION OF ACTIVITY-DEPENDENT GENES

The first group of genes (TM1) activated by both opioids and psychostimulants consisted of immediate early genes (IEGs), which are well-described markers of neuronal activation (Guzowski et al., 2005; Morgan, Cohen, Hempstead, & Curran, 1987). The IEGs are a group of genes that undergo a rapid (novel protein synthesis-independent) and transient induction at the transcriptional level as a result of synaptic activation of a neuron. They are therefore considered to be markers of neuronal activation known as activity-dependent genes (Sagar, Sharp, & Curran, 1988). Particular genes from this group that have been identified in the response to drugs of abuse include *Fos*, *Fosb*, and *Egr1*. Some IEGs (*Npas4*, *Homer1a*, and *Arc*) may be involved in protecting against neuronal overexcitability (Figure 2) (Carlezon, Duman, & Nestler, 2005). *Npas4* regulates inhibitory synapse development in an activity-dependent manner and diminishes the excitatory synaptic input neurons receive (Lin et al., 2008). *Homer1a* appears to participate in the attenuation of the gradual inhibition of glutamate receptor-dependent calcium mobilization, as well as in mitogen-activated protein kinase activation. Interestingly, *Homer1* is induced by psychostimulants at early time points (1–2 h) but is quite late (4–6 h) following opioid treatments (Piechota et al., 2010; Szumlinski, Ary, & Lominac, 2008). Finally, increased expression of *Arc* may play a role in reducing AMPA receptor-mediated synaptic transmission (Rial Verde, Lee-Osbourne, Worley, Malinow, & Cline, 2006).

Psychostimulants induce IEGs at early time points. The peak of TM1 transcriptional activation for cocaine was at 1 h, and methamphetamine peaked at 2 h after drug injection. In contrast, at early time points after opioid treatment (1–2 h), some genes from the TM1 group were silent. Furthermore, in situ hybridization and microarray studies demonstrated elevated levels of several IEG mRNAs (including *Fos*) in the mouse striatum at much later time points after the administration of morphine or heroin (Piechota et al., 2010; Ziolkowska,

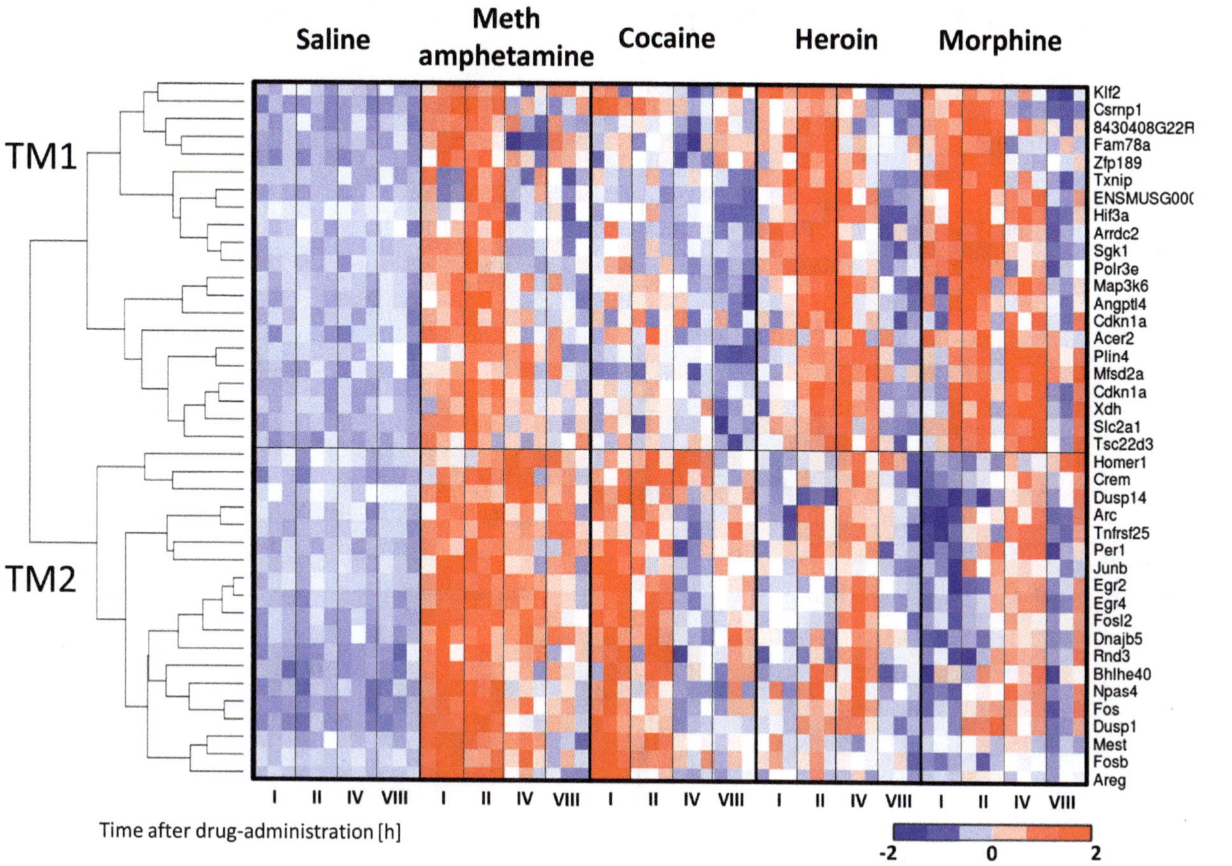

FIGURE 1 Hierarchical clustering of opioid and psychostimulant acute effects on the transcriptional profile in the mouse striatum. Microarray results are shown as a heat map and include genes with significant expression alterations. Colored rectangles represent transcript abundance 1, 2, 4, and 8 h after a single injection of the drug indicated at the top. The intensity of the color is proportional to the standardized values (between −2 and 2) from each microarray, as indicated on the bar below the heat map image. Clustering was performed using Euclidean distance according to the scale on the left. Major drug-responsive gene transcription patterns are arbitrarily described as TM1 and TM2.

Urbanski, Wawrzczak-Bargiela, Bilecki, & Przewlocki, 2005). A more detailed study demonstrated that morphine produces two episodes of IEG induction, which are separated in time and have different neuroanatomical distributions. At 30 min, IEGs were induced in a very limited area of the dorsal striatum and nucleus accumbens shell, cingulate cortex, and lateral septum. The late (4–6 h) IEG induction (of *Arc* and *Egr1*) was more widespread and involved most of the dorsal striatum and cortex. This delayed episode of opioid-produced IEG activation may have important functional consequences, mediating cellular neuroadaptations to opioid abuse.

ACTIVATION OF STRESS-RELATED GENES BY DRUGS OF ABUSE

The second group (TM2) of genes was identified as drug regulated (Kerns et al., 2005; Korostynski et al., 2007, 2013; Piechota et al., 2010; Treadwell & Singh, 2004). This set of genes consists of three subsequent clusters expressed at different time points. These genes are induced by opioids and, to a lesser extent, by methamphetamine, and only marginally by cocaine. The first cluster was induced at a relatively early time point (1–2 h). The expression of these genes appeared to depend on several regulatory proteins, for example, transcription factors of the FOX family. The next pattern was induced at later time points (4–8 h) and appeared to be regulated by steroid hormones that respond to morphine, heroin, and methamphetamine but not cocaine. This gene expression may be related to activation of the hypothalamopituitary–adrenal (HPA) axis after the administration of opioids (Ellis, 1966).

These genes modulate various aspects of cell functioning: hexose transport (*Slc2a1*) (Figure 3), lipid metabolism (*Angptl4*, *Pparg*), and the regulation of sodium channels, the actin cytoskeleton (*Sgk1*), and the cell cycle (*Cdkn1a*). Expression of these genes was also affected relatively weakly by methamphetamine compared to the much stronger effects of opioids. This group of genes also includes *Fkbp5*, a chaperone that inhibits glucocorticoid receptor (GR) translocation into the nucleus (Tatro, Everall, Kaul, & Achim, 2009). Expression of these genes depends on corticotropin-releasing factor and GR signaling as a key element of the neuroadaptive changes that are induced by opioids and, to some extent, by other drugs of abuse (de Kloet, Joels, & Holsboer, 2005; Marinelli, Aouizerate, Barrot, Le Moal, & Piazza, 1998). HPA activation and glucocorticoid release are two of the most recognized attributes of stress and common effects of various drugs of abuse (Ellis, 1966).

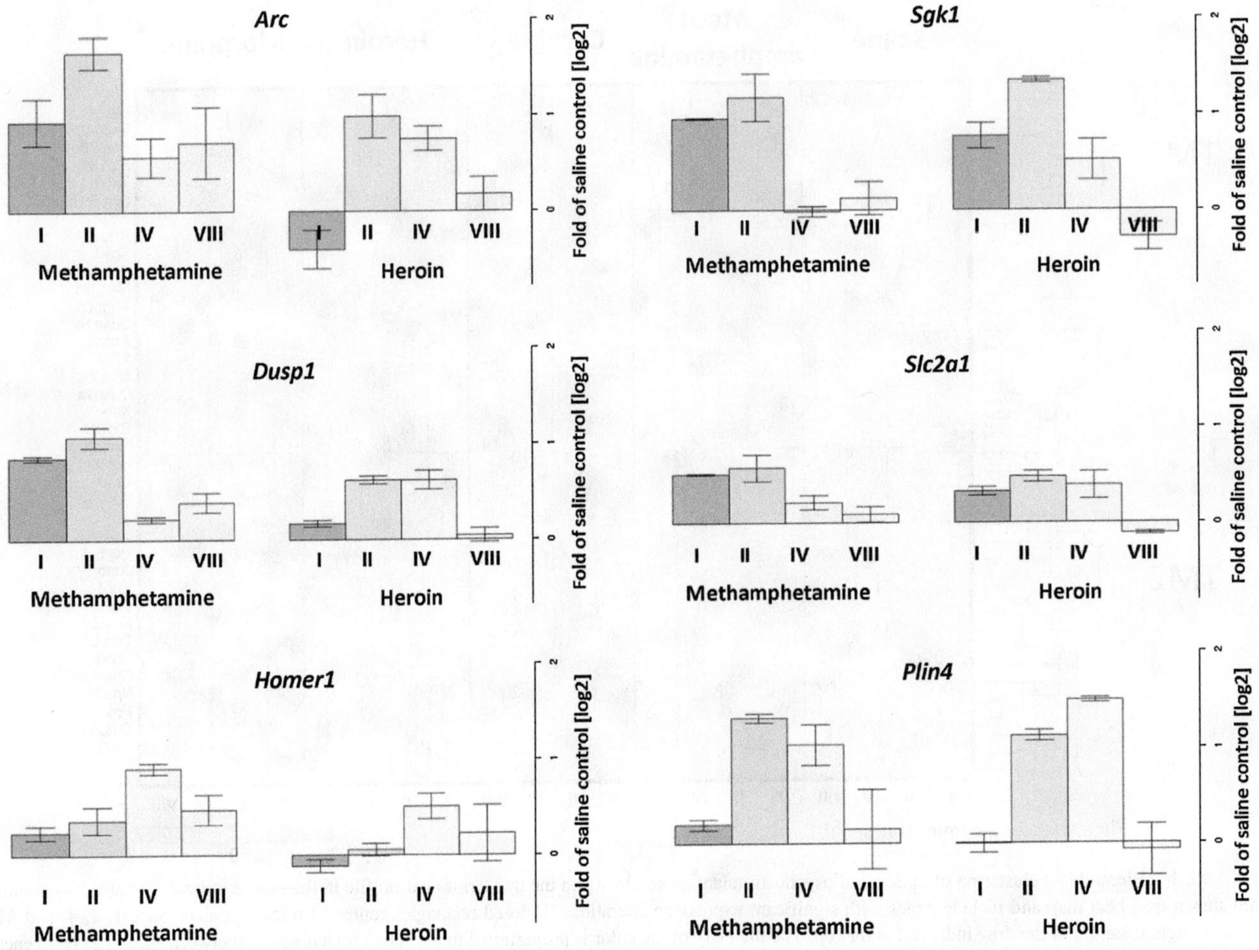

FIGURE 2 Differences and similarities in gene expression alterations in the striatum induced by methamphetamine and heroin. Bar graphs summarizing the microarray-based measurement of changes in selected gene expression after the indicated drug injection, presented as fold change over the saline control group with standard error. Transcript abundance levels were measured 1, 2, 4, and 8 h after a single drug injection.

Stress enhances drug self-administration in animal models (Le Moal, 2009), and chronic exposure to stress is associated with increased vulnerability to addiction (Piazza & Le Moal, 1998). On the other hand, stress, as well as glucocorticoids, sensitizes the rewarding properties of morphine, while adrenalectomy blocks the stress-induced potentiation of drug reward. The identification of the novel group of opioid-responsive genes may help to uncover a molecular mechanism linking stress and addiction (Goeders, 2003; Yachi, Inoue, Tanaka, Yoshikawa, & Tohyama, 2007). Moreover, Mantsch et al. demonstrated that corticosterone itself produces almost the same effects on drug use as stress. Therefore, corticosterone released after the administration of drugs of abuse enhances the rewarding properties of opioids in a manner similar to that of stress. The mechanism of corticosterone's contribution to addiction vulnerability is not well understood. Steroid-mediated enhancement of mesocorticolimbic dopamine neuron activity has been suggested to play a major role (Le Moal, 2009).

A correlation analysis between the induction of gene clusters and behavioral effects suggests that gene patterns may be associated with the rewarding effects of drugs (Piechota et al., 2010). The results suggested that some genes, such as *Sgk1* and *Tsc22d3*, might be associated with neuronal functions (Korostynski et al., 2007; Luca et al., 2009; Yachi et al., 2007) and neuroplastic changes after administration of drugs of abuse (Luca et al., 2009).

GLUCOCORTICOID RECEPTOR-DEPENDENT GENES INDUCED BY DRUGS OF ABUSE ARE EXPRESSED IN ASTROCYTES

Our studies identified opioid-induced transcriptional effects in the striatum expressed in glial cells rather than in neurons (Korostynski et al., 2007; Slezak et al., 2013). Moreover, several lines of evidence suggest that exposure to drugs of abuse leads to the activation of both astrocytes and microglia. Noncontingent administration of opioids and methamphetamine increases glial activation in multiple brain regions (Reichel, Ramsey, Schwendt, McGinty, & See, 2012). Interestingly, the vast majority of genes commonly activated by various drugs of abuse in vivo have also been activated in astrocytes in vitro by selective stimulation of the GR (Carter, Hamilton, & Thompson, 2013; Slezak et al., 2013).

Thus, a fraction of opioid-activated gene transcription occurs in glial cells and is mediated by glucocorticoids and regulated by the GR. This evidence further suggests that administration of

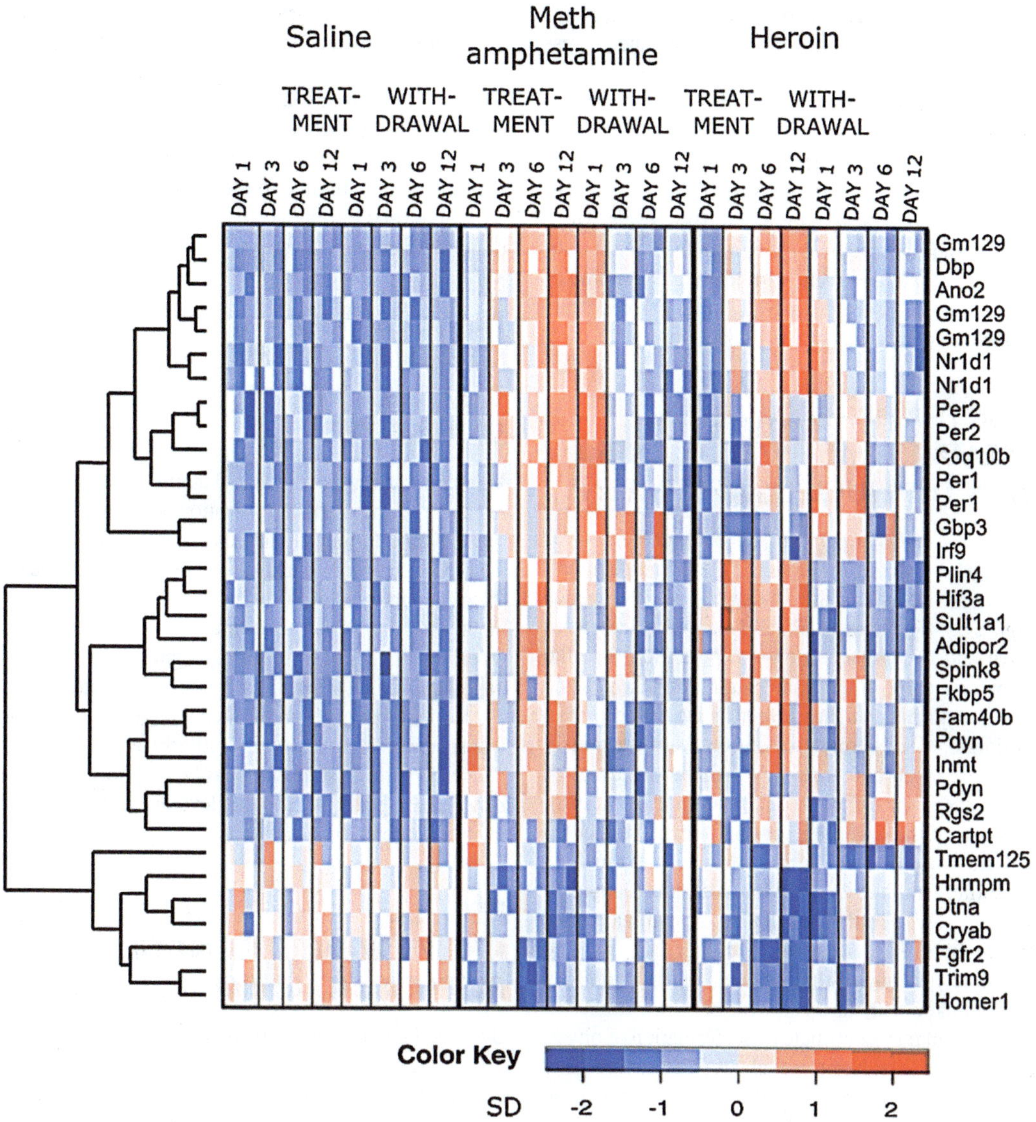

FIGURE 3 Gene expression changes during chronic heroin or methamphetamine treatment and withdrawal. Microarray results are shown as a heat map and include genes with genome-wide statistical significance. Colored rectangles represent transcript abundance of the gene as labeled on the right after 1, 3, 6, and 12 days of chronic treatment or withdrawal from the drug indicated at the top. The intensity of the color is proportional to the standardized values (between −2.5 and 2.5) from each microarray, as indicated on the bar below the heat map image.

various drugs of abuse indirectly activates glucocorticoids and may affect astrocytes in the brain reward circuit. The novel potential astrocyte contribution to the central effects of drugs of abuse opens a new direction for research focused on the role of glucocorticoids in astrocytes and a putative involvement of the GR in mediating the effects of drugs of abuse.

GENE EXPRESSION REGULATED BY CHRONIC ADMINISTRATION OF PSYCHOSTIMULANTS OR OPIOIDS

Drug addiction is a long-lasting state. Therefore, several studies suggest that genes with long-lasting changes in their expression may be the best candidates for explaining the persistence of addiction. Permanent changes in the gene expression profile have been suggested to be important factors in long-term alterations in behavior following treatment with drugs of abuse, such as psychomotor sensitization, dependence, and craving (Pierce, Bell, Duffy, & Kalivas, 1996). Multiple studies have used microarray analysis to study the molecular effects of chronic treatment with opioids (Korostynski et al., 2007; McClung et al., 2005) and psychostimulants (Newton et al., 2004; Yang et al., 2008). The majority of those studies focused only on the effects of drugs of abuse at single time points. Therefore, determining when the transcriptional changes occur and whether the changes are transient or persistent is difficult. Therefore, our study aimed to reveal permanent transcriptome changes in the striatum after chronic treatment of mice with methamphetamine and heroin with an increasing dose paradigm to avoid the effects of tolerance (Piechota et al., 2012).

We utilized whole-genome gene expression profiling to evaluate the common effects and reveal the time course of gene expression alterations in the striatum (including nucleus accumbens) following chronic treatment with methamphetamine or heroin (Figure 3). Changes in the transcriptome developed during treatment and returned to basal levels after a few days of withdrawal. In fact, during nearly 2 weeks of a protracted abstinence period, all altered levels of mRNA abundance returned to basal levels. Thus, our study failed to observe permanent changes in the striatal transcriptome following chronic treatment with methamphetamine or heroin. Therefore, changes in gene expression after chronic treatment with both methamphetamine and heroin are transient.

CHRONIC METHAMPHETAMINE AND HEROIN TREATMENT ACTIVATED GENES ENRICHED IN THE NUCLEUS ACCUMBENS

Small groups of genes (*Pdyn*, *Cartpt*, *Rgs2*, *Inmt*, and *Fam40b*) are induced by the repeated administration of both methamphetamine and heroin, as well as during early stages of spontaneous withdrawal (Figure 2) (Piechota et al., 2012). Two genes from this cluster (*Pdyn* and *Cartpt*) are known to be enriched in the striatum. More precisely, those genes are highly enriched in the nucleus accumbens and have higher levels in the nucleus than in the dorsal striatum. Genes from this cluster were previously shown to be regulated by chronic treatment with various drugs of abuse (Kuntz-Melcavage et al., 2009; Ziolkowska et al., 2006). Enhanced expression of these genes might be directly connected to the effects of chronic treatment with drugs of abuse, such as withdrawal, irritability, and anhedonia. Moreover, the changes in expression may be involved in the remodeling of neuronal structures and connections in the ventral striatum and, thus, may be indirectly responsible for long-lasting changes in behavior. The *Pdyn* gene has been linked with the emergence of sensitization and physical dependence and may contribute to the formation of physical dependence (Koob, 1992a; Newton et al., 2004). Another gene from this group, *Cartpt*, was previously linked to the regulation of food intake (Kristensen et al., 1998). However, in the nucleus accumbens, *Cartpt* was also linked to attenuated excessive excitation (Kim, Jung, Na, Hong, & Yoon, 2003).

Chronic methamphetamine and heroin probably activate the transcriptional program aimed at inhibiting dopamine excitatory input to the striatum. Therefore, a homeostatic response to repeated dopaminergic stimulation probably exists. A putative homeostatic response is also supported by evidence from the *Rgs2* gene, which appears to be involved in the action of chronically administered drugs of abuse (Burchett, Bannon, & Granneman, 1999; Piechota et al., 2012). Chronic exposure to psychostimulants increases the levels of *Rgs2* transcripts (Lomazzi, Slesinger, & Luscher, 2008). Rgs2 belongs to the family of G-protein-signaling regulators and has the potential to regulate dopamine D1 receptor signaling via direct inhibition of adenylyl cyclase or through the regulation of the coupling efficiency between G-protein-coupled receptors and G-protein-activated inwardly rectifying K^+ channels (Lomazzi et al., 2008).

The product of another gene from this group, indolethylamine N-methyltransferase (*Inmt*), catalyzes the N-methylation of tryptamine and structurally related compounds. In particular, *Inmt* can catalyze the transmethylation of tryptamine into dimethyltryptamine, which binds nonselectively to serotonin receptors. Thus, this induction may be responsible for some of the psychotic effects of the withdrawal from drugs of abuse (Zorick, Sugar, Hellemann, Shoptaw, & London, 2011). The function of another gene from this group, *Fam40b*, is still unknown. The distribution of the *Fam40b* gene is limited to the striatum, the CA2 field of the hippocampus, and Purkinje cells, which may imply that this gene has important neuronal functions.

TRANSCRIPTIONAL EFFECTS OF CHRONIC METHAMPHETAMINE AND HEROIN TREATMENT IN THE STRIATUM DISPLAY SIMILARITY

This observation is of particular interest because transcriptional responses to acute administration of these drugs overlap only partially and differ in their time dynamics (Piechota et al., 2010). Whole-genome microarray profiling revealed some common alterations and their time course in the striatum following chronic treatment with methamphetamine and heroin (Piechota et al., 2012). Overlap between lists of heroin- and methamphetamine-induced genes was highly significant and was related to drug-induced alterations in expression of circadian genes (Figure 4). The genes whose expression changed similarly after heroin and methamphetamine administration could be responsible for their common behavioral effects, such as the development of psychomotor sensitization or conditioned place preference.

REGULATION OF GLUCOCORTICOID RECEPTOR-DEPENDENT GENES IN RESPONSE TO CHRONIC TREATMENT

Both methamphetamine and heroin cause a dose-dependent induction of "stress" genes (e.g., *Fkbp5* and *Plin4*), which, as we have shown previously, are regulated by acute administration of drugs of abuse and glucocorticoids (Carter et al., 2013; Piechota et al., 2010; Slezak et al., 2013). The acute administration of heroin and methamphetamine activates the expression of glucocorticoid-dependent genes, which disappears within 8 h. After chronic treatment, even longer activation of the genes has been measured. However, the doses used in chronic treatment are much higher than the starting dose, and most likely, the levels of the glucocorticoid-regulated genes remain elevated for a longer period of time. In addition, few of those genes have a long mRNA half-life (*Fkbp5*, 11 h; *Sult1a1*, 18 h; and *Hif3a*, 21 h).

Therefore, chronic drug intake may lead to adaptive expression changes that alter the responsivity to stress. Drug withdrawal has a stimulatory effect on corticotropin-releasing factor secretion and HPA functioning similar to that of behavioral stress. Craving in humans (Sinha & Li, 2007) and reinstatement to self-administration of drugs of abuse in animals are a consequence of stress exposure (Piazza & Le Moal, 1998). Stress alters the proneness to drugs of abuse and, therefore, stress is considered an important factor contributing to drug-seeking behavior, relapse, and chronic states of addiction (Kreek & Koob, 1998). Interestingly, in human addicts, the HPA axis is dysregulated and even hyporesponsive to stress

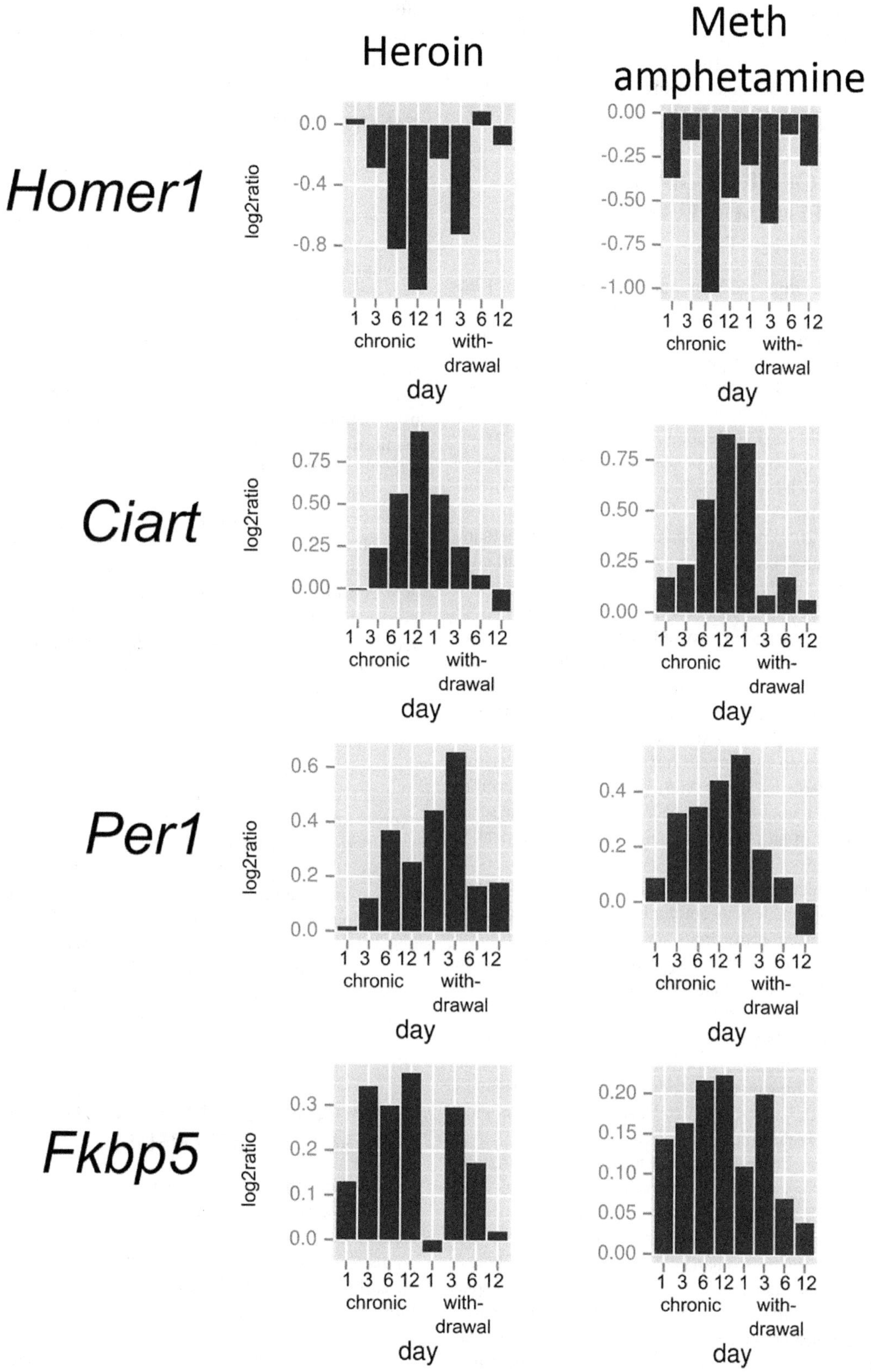

FIGURE 4 Validation of time-dependent drug-induced alterations in the gene expression profile. The bar graphs summarize the quantitative PCR-based changes in gene expression following the indicated treatment. Data are presented as the fold change over the saline control group.

(Lovallo, 2006). This altered response to stress is considered to be an important risk factor for relapse. Stress-induced HPA activity predicted relapse to drug use and amounts of subsequent use, indicating that stress not only elicits craving but also independently predicts relapse (Koob & Kreek, 2007).

ALTERATION OF CIRCADIAN GENES BY CHRONIC METHAMPHETAMINE AND HEROIN TREATMENT

Interaction between drug addiction and biological rhythms has evolved from both clinical studies and preclinical research. Chronic methamphetamine and heroin administration leads to changes in specific groups of genes (Perreau-Lenz & Spanagel, 2008; Piechota et al., 2012), which are widely known as circadian rhythm genes (Albrecht, Sun, Eichele, & Lee, 1997; Shearman, Zylka, Weaver, Kolakowski, & Reppert, 1997). Changes in expression levels of these genes return to baseline after a few days of drug withdrawal, which is most likely associated with the normalization of the circadian rhythm.

Approximately 20 genes are regulated by chronic methamphetamine and heroin. These genes are circadian genes, and eight of these genes are robustly regulated by circadian rhythms (*Per1, Per2, Nr1d1, Gm129, Dbp, Ano2, Cryab,* and *Coq10b*). *Gm129* has been renamed *Ciart* and is involved in the regulation of core clock genes and the response of the circadian clock to stress. Interestingly, Ciart may also form a complex with the GR and mediates the glucocorticoid response (Goriki et al., 2014). Circadian rhythm genes *Per1, Per2,* and *Nr1d1* (Albrecht et al., 1997; Shearman et al., 1997) are altered by chronic treatment with various substances of abuse (Perreau-Lenz & Spanagel, 2008). Furthermore, some of those genes may be involved in psychostimulant sensitization and reward in an opposite manner (Abarca, Albrecht, & Spanagel, 2002).

Treatment of mice with both methamphetamine and heroin during the light phase induces locomotor activation in mice, which are nocturnal. Therefore, if the drugs are administered during the diurnal phase, mice are forced to stay awake during their natural sleeping time. Such behavioral changes lead to a deregulation of the diurnal cycle following chronic drug administration causing alterations in the expression of circadian genes. Furthermore, a behavioral study demonstrated that if mice were treated during the nocturnal phase, immediately before the peaks of their natural locomotor activity, the alteration in the levels of circadian genes was not observed (Piechota et al., 2012).

Some researchers, however, consider the direct rather than the indirect effect of drugs of abuse on circadian rhythm (Falcon & McClung, 2009). The interaction is quite complex because not only chronic drug treatment but also acute treatment causes alterations in the levels of some circadian genes (i.e., *Per1*) (McClung & Nestler, 2003; Piechota et al., 2010). *Per1*-knockout mice display a lack of cocaine reward, whereas *Per2* mutants display a strong cocaine-induced place preference (Abarca et al., 2002). These two circadian genes, which are both induced by acute treatment with drugs of abuse, appear to have opposite functions in mediating the rewarding effects of cocaine, suggesting again that interplay between the circadian rhythm and drugs of abuse is complex. Interestingly, studies have shown that *Per1*-knockout mice did not differ from wild-type controls in self-administration and reinstatement of cocaine-seeking behavior (Halbout, Perreau-Lenz, Dixon, Stephens, & Spanagel, 2011).

CONCLUSIONS

Here we present evidence that the acute administration of psychostimulants and opioids induces different gene expression profiles in the mouse striatum. In general, acute administration of both opioids and psychostimulants activate two main groups of activity-dependent genes and glucocorticoid-regulated genes, but the effects differ in time course and magnitude. Psychostimulants induce activity-dependent genes early (1–2 h). Opioids appear to inhibit these genes at early time points and evoke activation peaks approximately 4 h after administration. Opioids profoundly stimulate the expression of glucocorticoid-dependent genes, and psychostimulants are much weaker inductors. Acute cocaine administration only slightly, if at all, influences the expression of a fraction of these genes. It is likely that the transcriptome alterations may be related to behaviorally and neurobiologically different opioid and psychostimulant features of addictive behavior.

The current theories suggest that opiate and psychostimulant addiction are similar or identical phenomena. There is growing evidence that the chronic intake of drugs of abuse induces prolonged alterations in brain plasticity, which have long-term consequences on behavior. Interestingly, chronic methamphetamine and heroin induce molecular alterations in similar sets of transcripts, which may be involved in the development of addiction symptoms, such as physical dependence and anhedonia during withdrawal. Moreover, the striatal transcriptome alterations are transient, persist only a few days after withdrawal, and do not correlate directly with behavioral sensitization. Therefore, gene expression alterations during chronic drug treatment and in the early period of withdrawal are involved in the establishment of persistent neuroplastic alterations at the cellular and molecular levels. Finally, a previous study emphasizes that there is a direct interaction between the circadian system and chronic drug use, which suggests that the alteration of circadian gene expression leads to molecular and behavioral changes that are associated with drug addiction. However, one study also emphasizes that there is a more complex interaction between the circadian system and drug response.

APPLICATIONS TO OTHER ADDICTIONS AND SUBSTANCE MISUSE

Gene expression changes in response to heroin and methamphetamine show some similarities, but also clear differences. Acute administration of both drugs activates IEGs but at different time points. Methamphetamine (and cocaine) induces early genes in the early period (1–2 h), heroin (and morphine) much later (4–6 h). Promoters of these genes are coregulated by two main activity-responsive transcription factors, cAMP-response element-binding protein and serum response factor (SRF). The activity-dependent gene expression is critical for certain forms of neuronal plasticity in the rodent nervous system. Both drugs also activate a group of genes regulated by steroids and the GR. This regulation appears to be common for ethanol, nicotine, opioids, and methamphetamine.

Therefore, the GR-dependent signaling system is emerging as a key element of the neuroadaptive changes that are induced by drugs of abuse. Interestingly, while expression of activity-dependent genes takes place in neurons, glucocorticoid-dependent genes appear to be activated in glial cells. The importance of glial activation and gliotransmitters acting on neuronal receptors and neuroplastic changes during drug taking, addiction, and withdrawal responses has been reported, but a role of glucocorticoid-dependent gene activation remains to be established. Chronic heroin and methamphetamine administration induces transient expression of circadian genes and genes coding for neuropeptides such as dynorphin. The expression persists for only a few days following prolonged chronic treatment and protractive withdrawal. Studies suggest that such changes may be common for opioids and psychostimulants.

DEFINITION OF TERMS

Activity-dependent transcription This refers to the expression of genes (including *Fos, Arc,* and *Npas4*) in neuronal cells induced in response to synaptic activity, which was found to be critical for certain forms of neuronal plasticity and survival in the mammalian nervous system.

Behavioral sensitization This is a nonassociative learning process in which repeated administrations of a stimulus (e.g., drug administration) results in the progressive amplification of a behavioral response (e.g., increase in locomotor response of mice to repeated treatment with psychostimulants).

Brain plasticity This is the ability of the brain and nervous system to change structurally and functionally in response to stimuli. Functional alterations related to neuroplasticity constitute the biological basis for phenomena such as memory, addiction, and recovery of brain function.

Circadian genes These are a group of genes that display transcriptional regulation in mammals by an endogenous, entrainable oscillation of ~24 h. Example genes under circadian regulation are *Per1, Per2,* and *Ciart*.

Drug-inducible gene expression This is the expression of genes regulated in the brain in response to drug treatment. These genes are induced in coexpressed patterns containing dozens of transcripts.

Glucocorticoid receptor-dependent regulation This refers to the control of gene expression level by GRs. Corticoids bind to the GR in the cytoplasm and the hormone–receptor complex is then translocated into the nucleus, where the GR binds to glucocorticoid response element and activates/inhibits transcription of target genes.

Polydrug abuse This is the use of multiple psychoactive drugs (e.g., psychostimulants with opioids) in combination to achieve a synergistic or supplementary effect.

Transcriptional regulator This is a factor that influences the level of transcript expression by promoting or blocking the recruitment of RNA polymerase to a specific gene. Several classes of transcription regulators are defined, including core promoter factors, enhancers, silencers, insulators, and various types of regulatory noncoding RNAs.

Reward circuit This consists of brain structures involved in mediating effects of reinforcement. The major neurochemical pathway of the reward system involves the mesolimbic pathway that goes from the VTA via the medial forebrain bundle to the nucleus accumbens.

KEY FACTS OF POLYDRUG ABUSE

- The first extensive use of methamphetamine occurred during World War II, by German soldiers because of its stimulant effects.
- Dextroamphetamine and morphine were found to be an excellent combination for pain relief during World War II.
- Speedball (intravenous use of cocaine and heroin mixed together in a single shot) produces the best rush effect described by addicts.
- Methamphetamine affects the brain by elevating levels of monoamine neurotransmitters such as dopamine, norepinephrine, and serotonin.
- Activation of opioid receptors in the VTA results in enhancing dopamine release in the nucleus accumbens.
- Elevated levels of monoamine produce a euphoric effect.
- The key brain structure for both methamphetamine and heroin actions is the striatum, in which dopamine is released.
- Activation of neurons by elevated levels of monoamines produces gene expression changes in specific brain areas.

SUMMARY POINTS

- This chapter focuses on gene expression changes in response to methamphetamine and heroin.
- Both classes of drug induce changes in gene expression in the brain reward center (e.g., striatum and nucleus accumbens), which presumably underlie their addictive properties.
- Acute administration of the drugs activates sets of activity-dependent genes in neurons.
- Interestingly, both methamphetamine and heroin activate glucocorticoid-dependent genes in nonneuronal cells such as glia and microglia.
- Chronic drug administration alters the expression of circadian genes and genes coding for neuropeptides such as dynorphin.
- However, the gene expression alterations are transient after acute administration and persist only for a few days after drug withdrawal following chronic treatment.
- The drug-induced gene expression changes are involved in drug-induced neuronal plasticity at the cellular and molecular levels and participate in establishing an addictive phenotype.

ACKNOWLEDGMENTS

This work was supported by Grant NCN 2013/08/A/NZ3/00848 Maestro and statutory activity of the IP PAS. The authors appreciate the assistance of Mrs. Lidia Radwan in the preparation of the manuscript.

REFERENCES

Abarca, C., Albrecht, U., & Spanagel, R. (2002). Cocaine sensitization and reward are under the influence of circadian genes and rhythm. *Proceedings of the National Academy of Sciences of the United States of America, 99,* 9026–9030.

Albrecht, U., Sun, Z. S., Eichele, G., & Lee, C. C. (1997). A differential response of two putative mammalian circadian regulators, mper1 and mper2, to light. *Cell, 91,* 1055–1064.

Badiani, A., Belin, D., Epstein, D., Calu, D., & Shaham, Y. (2011). Opiate versus psychostimulant addiction: the differences do matter. *Nature Reviews Neuroscience, 12*, 685–700.

Burchett, S. A., Bannon, M. J., & Granneman, J. G. (1999). RGS mRNA expression in rat striatum: modulation by dopamine receptors and effects of repeated amphetamine administration. *Journal of Neurochemistry, 72*, 1529–1533.

Carlezon, W. A., Jr., Duman, R. S., & Nestler, E. J. (2005). The many faces of CREB. *Trends in Neuroscience, 28*, 436–445.

Carter, B. S., Hamilton, D. E., & Thompson, R. C. (2013). Acute and chronic glucocorticoid treatments regulate astrocyte-enriched mRNAs in multiple brain regions in vivo. *Frontiers in Neuroscience, 7*, 139.

Di Chiara, G., & Imperato, A. (1988). Drugs abused by humans preferentially increase synaptic dopamine concentrations in the mesolimbic system of freely moving rats. *Proceedings of the National Academy of Sciences of the United States of America, 85*, 5274–5278.

Ellis, F. W. (1966). Effect of ethanol on plasma corticosterone levels. *Journal of Pharmacology and Experimental Therapeutics, 153*, 121–127.

Falcon, E., & McClung, C. A. (2009). A role for the circadian genes in drug addiction. *Neuropharmacology, 56*(Suppl. 1), 91–96.

Gerdeman, G. L., Partridge, J. G., Lupica, C. R., & Lovinger, D. M. (2003). It could be habit forming: drugs of abuse and striatal synaptic plasticity. *Trends in Neuroscience, 26*, 184–192.

Goeders, N. E. (2003). The impact of stress on addiction. *European Neuropsychopharmacology, 13*, 435–441.

Goriki, A., Hatanaka, F., Myung, J., Kim, J. K., Yoritaka, T., Tanoue, S., … Takumi, T. (2014). A novel protein, CHRONO, functions as a core component of the mammalian circadian clock. *PLoS Biology, 12*, e1001839.

Guzowski, J. F., Timlin, J. A., Roysam, B., McNaughton, B. L., Worley, P. F., & Barnes, C. A. (2005). Mapping behaviorally relevant neural circuits with immediate-early gene expression. *Current Opinion in Neurobiology, 15*, 599–606.

Halbout, B., Perreau-Lenz, S., Dixon, C. I., Stephens, D. N., & Spanagel, R. (2011). Per1(Brdm1) mice self-administer cocaine and reinstate cocaine-seeking behaviour following extinction. *Behavioural Pharmacology, 22*, 76–80.

Hyman, S. E., Malenka, R. C., & Nestler, E. J. (2006). Neural mechanisms of addiction: the role of reward-related learning and memory. *Annual Review of Neuroscience, 29*, 565–598.

Kerns, R. T., Ravindranathan, A., Hassan, S., Cage, M. P., York, T., Sikela, J. M., … Miles, M. F. (2005). Ethanol-responsive brain region expression networks: implications for behavioral responses to acute ethanol in DBA/2J versus C57BL/6J mice. *Journal of Neuroscience, 25*, 2255–2266.

Kim, J., Jung, J. I., Na, H. S., Hong, S. K., & Yoon, Y. W. (2003). Effects of morphine on mechanical allodynia in a rat model of central neuropathic pain. *Neuroreport, 14*, 1017–1020.

de Kloet, E. R., Joels, M., & Holsboer, F. (2005). Stress and the brain: from adaptation to disease. *Nature Reviews Neuroscience, 6*, 463–475.

Koob, G., & Kreek, M. J. (2007). Stress, dysregulation of drug reward pathways, and the transition to drug dependence. *American Journal of Psychiatry, 164*, 1149–1159.

Koob, G. F. (1992a). Drugs of abuse: anatomy, pharmacology and function of reward pathways. *Trends in Pharmacological Sciences, 13*, 177–184.

Koob, G. F. (1992b). Neural mechanisms of drug reinforcement. *Annals of New York Academy of Sciences, 654*, 171–191.

Korostynski, M., Piechota, M., Dzbek, J., Mlynarski, W., Szklarczyk, K., Ziolkowska, B., & Przewlocki, R. (2013). Novel drug-regulated transcriptional networks in brain reveal pharmacological properties of psychotropic drugs. *BMC Genomics, 14*, 606.

Korostynski, M., Piechota, M., Kaminska, D., Solecki, W., & Przewlocki, R. (2007). Morphine effects on striatal transcriptome in mice. *Genome Biology, 8*, R128.

Kreek, M. J., & Koob, G. F. (1998). Drug dependence: stress and dysregulation of brain reward pathways. *Drug and Alcohol Dependence, 51*, 23–47.

Kristensen, P., Judge, M. E., Thim, L., Ribel, U., Christjansen, K. N., Wulff, B. S., … Hastrup, S. (1998). Hypothalamic CART is a new anorectic peptide regulated by leptin. *Nature, 393*, 72–76.

Kuntz-Melcavage, K. L., Brucklacher, R. M., Grigson, P. S., Freeman, W. M., & Vrana, K. E. (2009). Gene expression changes following extinction testing in a heroin behavioral incubation model. *BMC Neuroscience, 10*, 95.

Le Moal, M. (2009). Drug abuse: vulnerability and transition to addiction. *Pharmacopsychiatry, 42*(Suppl. 1), S42–S55.

Lemberger, T., Parkitna, J. R., Chai, M., Schutz, G., & Engblom, D. (2008). CREB has a context-dependent role in activity-regulated transcription and maintains neuronal cholesterol homeostasis. *FASEB Journal, 22*, 2872–2879.

Lin, Y., Bloodgood, B. L., Hauser, J. L., Lapan, A. D., Koon, A. C., Kim, T. K., … Greenberg, M. E. (2008). Activity-dependent regulation of inhibitory synapse development by Npas4. *Nature, 455*, 1198–1204.

Lomazzi, M., Slesinger, P. A., & Luscher, C. (2008). Addictive drugs modulate GIRK-channel signaling by regulating RGS proteins. *Trends in Pharmacological Sciences, 29*, 544–549.

Lovallo, W. R. (2006). Cortisol secretion patterns in addiction and addiction risk. *International Journal of Psychophysiology, 59*, 195–202.

Luca, F., Kashyap, S., Southard, C., Zou, M., Witonsky, D., Di Rienzo, A., & Conzen, S. D. (2009). Adaptive variation regulates the expression of the human SGK1 gene in response to stress. *PLoS Genetics, 5*, e1000489.

Marinelli, M., Aouizerate, B., Barrot, M., Le Moal, M., & Piazza, P. V. (1998). Dopamine-dependent responses to morphine depend on glucocorticoid receptors. *Proceedings of the National Academy of Sciences of the United States of America, 95*, 7742–7747.

McClung, C. A., & Nestler, E. J. (2003). Regulation of gene expression and cocaine reward by CREB and DeltaFosB. *Nature Neuroscience, 6*, 1208–1215.

McClung, C. A., & Nestler, E. J. (2008). Neuroplasticity mediated by altered gene expression. *Neuropsychopharmacology, 33*, 3–17.

McClung, C. A., Nestler, E. J., & Zachariou, V. (2005). Regulation of gene expression by chronic morphine and morphine withdrawal in the locus ceruleus and ventral tegmental area. *Journal of Neuroscience, 25*, 6005–6015.

Morgan, J. I., Cohen, D. R., Hempstead, J. L., & Curran, T. (1987). Mapping patterns of c-fos expression in the central nervous system after seizure. *Science, 237*, 192–197.

Nestler, E. J. (2001). Molecular basis of long-term plasticity underlying addiction. *Nature Reviews Neuroscience, 2*, 119–128.

Newton, T. F., Kalechstein, A. D., Duran, S., Vansluis, N., & Ling, W. (2004). Methamphetamine abstinence syndrome: preliminary findings. *American Journal on Addiction, 13*, 248–255.

Nugent, F. S., Penick, E. C., & Kauer, J. A. (2007). Opioids block long-term potentiation of inhibitory synapses. *Nature, 446*, 1086–1090.

Perreau-Lenz, S., & Spanagel, R. (2008). The effects of drugs of abuse on clock genes. *Drug News and Perspectives, 21*, 211–217.

Piazza, P. V., & Le Moal, M. (1998). The role of stress in drug self-administration. *Trends in Pharmacological Sciences, 19*, 67–74.

Piechota, M., Korostynski, M., Sikora, M., Golda, S., Dzbek, J., & Przewlocki, R. (2012). Common transcriptional effects in the mouse striatum following chronic treatment with heroin and methamphetamine. *Genes, Brain, and Behavior, 11*, 404–414.

Piechota, M., Korostynski, M., Solecki, W., Gieryk, A., Slezak, M., Bilecki, W., ... Przewlocki, R. (2010). The dissection of transcriptional modules regulated by various drugs of abuse in the mouse striatum. *Genome Biology, 11*, R48.

Pierce, R. C., Bell, K., Duffy, P., & Kalivas, P. W. (1996). Repeated cocaine augments excitatory amino acid transmission in the nucleus accumbens only in rats having developed behavioral sensitization. *Journal of Neuroscience, 16*, 1550–1560.

Przewlocki, R. (2004). Opioid abuse and brain gene expression. *European Journal of Pharmacology, 500*, 331–349.

Reichel, C. M., Ramsey, L. A., Schwendt, M., McGinty, J. F., & See, R. E. (2012). Methamphetamine-induced changes in the object recognition memory circuit. *Neuropharmacology, 62*, 1119–1126.

Rial Verde, E. M., Lee-Osbourne, J., Worley, P. F., Malinow, R., & Cline, H. T. (2006). Increased expression of the immediate-early gene arc/arg3.1 reduces AMPA receptor-mediated synaptic transmission. *Neuron, 52*, 461–474.

Russo, S. J., Dietz, D. M., Dumitriu, D., Morrison, J. H., Malenka, R. C., & Nestler, E. J. (2010). The addicted synapse: mechanisms of synaptic and structural plasticity in nucleus accumbens. *Trends in Neurosciences, 33*, 267–276.

Sagar, S. M., Sharp, F. R., & Curran, T. (1988). Expression of c-fos protein in brain: metabolic mapping at the cellular level. *Science, 240*, 1328–1331.

Shearman, L. P., Zylka, M. J., Weaver, D. R., Kolakowski, L. F., Jr., & Reppert, S. M. (1997). Two period homologs: circadian expression and photic regulation in the suprachiasmatic nuclei. *Neuron, 19*, 1261–1269.

Sinha, R., & Li, C. S. (2007). Imaging stress- and cue-induced drug and alcohol craving: association with relapse and clinical implications. *Drug and Alcohol Review, 26*, 25–31.

Slezak, M., Korostynski, M., Gieryk, A., Golda, S., Dzbek, J., Piechota, M., ... Przewlocki, R. (2013). Astrocytes are a neural target of morphine action via glucocorticoid receptor-dependent signaling. *Glia, 61*, 623–635.

Szumlinski, K. K., Ary, A. W., & Lominac, K. D. (2008). Homers regulate drug-induced neuroplasticity: implications for addiction. *Biochemical Pharmacology, 75*, 112–133.

Tatro, E. T., Everall, I. P., Kaul, M., & Achim, C. L. (2009). Modulation of glucocorticoid receptor nuclear translocation in neurons by immunophilins FKBP51 and FKBP52: implications for major depressive disorder. *Brain Research, 1286*, 1–12.

Treadwell, J. A., & Singh, S. M. (2004). Microarray analysis of mouse brain gene expression following acute ethanol treatment. *Neurochemical Research, 29*, 357–369.

Ungless, M. A., Whistler, J. L., Malenka, R. C., & Bonci, A. (2001). Single cocaine exposure in vivo induces long-term potentiation in dopamine neurons. *Nature, 411*, 583–587.

Yachi, K., Inoue, K., Tanaka, H., Yoshikawa, H., & Tohyama, M. (2007). Localization of glucocorticoid-induced leucine zipper (GILZ) expressing neurons in the central nervous system and its relationship to the stress response. *Brain Research, 1159*, 141–147.

Yang, M. H., Jung, M. S., Lee, M. J., Yoo, K. H., Yook, Y. J., Park, E. Y., ... Park, J. H. (2008). Gene expression profiling of the rewarding effect caused by methamphetamine in the mesolimbic dopamine system. *Molecules and Cells, 26*, 121–130.

Ziolkowska, B., Stefanski, R., Mierzejewski, P., Zapart, G., Kostowski, W., & Przewlocki, R. (2006). Contingency does not contribute to the effects of cocaine self-administration on prodynorphin and proenkephalin gene expression in the rat forebrain. *Brain Research, 1069*, 1–9.

Ziolkowska, B., Urbanski, M. J., Wawrzczak-Bargiela, A., Bilecki, W., & Przewlocki, R. (2005). Morphine activates Arc expression in the mouse striatum and in mouse neuroblastoma Neuro2A MOR1A cells expressing mu-opioid receptors. *Journal of Neuroscience Research, 82*, 563–570.

Zorick, T., Sugar, C. A., Hellemann, G., Shoptaw, S., & London, E. D. (2011). Poor response to sertraline in methamphetamine dependence is associated with sustained craving for methamphetamine. *Drug and Alcohol Dependence, 118*, 500–503.

Section C

Structural and Functional Aspects

Chapter 88

Neurological Abnormalities in Opiate Addicts

Josef Finsterer[1], Claudia Stöllberger[2]

[1]Krankenanstalt Rudolfstiftung, Vienna, Austria; [2]Medical Department, Krankenanstalt Rudolfstiftung, Vienna, Austria

Abbreviations

AIDP Acute inflammatory demyelinating polyneuropathy
CCT Cerebral computed tomography scan
CNS Central nervous system
CSF Cerebrospinal fluid
HAND HIV-associated neurocognitive disorders
HIV Human immunodeficiency virus
MRI Magnetic resonance imaging
PNS Peripheral nervous system

INTRODUCTION

Opiate addiction is a central nervous system (CNS) disease caused by the use of opiate-based compounds (opioids—natural or synthetic substances with morphine-like effects, which act via opioid receptors), such as oxycontin (oxycodone), morphine, heroin, buprenorphine, opium, etc. (Campbell, 2003). Opiate addiction is characterized by psychological or physical dependence on opioids, which thus need to be regularly administered to avoid hardly bearable withdrawal symptoms. Opiate addicts carry an increased risk of developing secondary disease due to the toxicity of the drug, toxic ingredients in drug mixtures, the side effects of the drugs, or the lifestyle following the social decline associated with the progression of opiate addiction. Secondary disease in opiate addicts generally includes effects on the immune system leading to infections (Lucas, 2005), lung disease (Büttner, Mall, Penning, & Weis, 2000; Simonovska et al., 2012), renal disease (Simonovska et al., 2012), cardiac disease (Adle-Biassette, Marc, Benhaiem-Sigaux, Durigon, & Gray, 1996; Lucas, 2005), vascular disease (Büttner et al., 2000; Decet & Bianchin, 2005; Lucas, 2005; Simonovska et al., 2012), trauma (Lucas, 2005), orthopedic disease (Simonovska et al., 2012), or neurological disease (Lucas, 2005; Sadeghian et al., 2009; Table 1). Neurological disease is one of the most prevalent of the secondary diseases in opiate addicts (Lukacher, Vrublevskiĭ, & Laskova, 1987). Neurological abnormalities develop within the first months of drug abuse in almost all patients (Lukacher et al., 1987). This chapter deals with secondary neurological abnormalities caused by opiate addiction but also highlights preexisting neurological disease, which may promote opiate addiction.

CLASSIFICATION

Neurological disease in opiate addicts may be classified as primary (opiate addiction), secondary (direct neurological side effects of opiate addiction), or tertiary (neurological complications from nonneurological side effects, e.g., arrhythmias, endocarditis, atherosclerosis). Neurological disease in opiate addicts may be further classified according to various other aspects. It can be classified according to the duration of addiction into early or late neurological complications, according to the dosage into low-dosage or high-dosage neurological abnormalities, and according to the state of addiction into neurological disease under drug influence, during withdrawal, or under drug substitution.

CAUSES OF NEUROLOGICAL DISEASE IN OPIATE ADDICTS

Neurological disease in opiate addicts may result from the toxic effect of the drug, the combined use of opiates with other neurotoxic substances, infections, bacteremia from intravenous drug use, atherosclerosis from smoking, hyperlipidemia, arterial hypertension, trauma, reduced primitive reflexes during intoxication, compression of nerves, compression of arteries, rhabdomyolysis from muscle compression during prolonged sleep or coma, or rhythm abnormalities and heart failure (Paur, Wallner, Hermann, Stöllberger, & Finsterer, 2012).

NEUROLOGICAL ABNORMALITIES IN OPIATE ADDICTS

Neurological abnormalities may concern the CNS (Table 2), the peripheral nervous system (PNS) (Table 3), or both. CNS disease reported in opiate addiction includes ischemic stroke, cerebral bleeding, epilepsy, cerebellar dysfunction, extrapyramidal syndromes, migraine, CNS infections, or other rare CNS disorders (Table 2). PNS disease reported in opiate addicts includes muscle disease or effects on the peripheral nerves (Table 3).

TABLE 1 Secondary Disease in Opiate Addicts

Disease	References
Effects on the Immune System	
Infections of soft-tissue, solid-organ abscesses	Lucas (2005)
Lung Disease	
Opiate-induced respiratory depression	Simonovska et al. (2012) and Büttner et al. (2000)
Renal Disease	
Proteinuria	Simonovska et al. (2012)
Hematuria	Simonovska et al. (2012)
Cardiac Disease	
Endocarditis	Lucas (2005)
Arrhythmias	Lucas (2005)
Myocardial infarction	Adle-Biassette et al. (1996)
Vascular Disease	
Vasoconstriction	Lucas (2005)
Raynaud phenomenon	Simonovska et al. (2012)
Aneurysms	DiLuna et al. (2007)
Thrombosis	DiLuna et al. (2007)
Vasculitis	Simonovska et al. (2012) and Büttner et al. (2000)
Spontaneous carotid artery dissection	Decet and Bianchin (2005) and DiLuna et al. (2007)
Positional vascular compression	Lucas (2005), Simonovska et al. (2012), Büttner et al. (2000), and Decet and Bianchin (2005)
Trauma	Lucas (2005)
Orthopedic disease (Arthralgia)	Simonovska et al. (2012)
Neurological disease	Simonovska et al. (2012) and Sadeghian et al. (2009)

TABLE 2 CNS Disorders Induced by Opiate Addiction

Disorder	References
Ischemic stroke	Lucas (2005), Borne et al. (2005), Brust and Richter (1976), Pascual Calvet et al. (1989), and Sahni et al. (2008)
Subarachnoid bleeding	Quattrini et al. (1982)
Epilepsy	Quattrini et al. (1982), Pascual Calvet et al. (1989), and Sahni et al. (2008)
CNS infections	Borne et al. (2005)
Extrapyramidal disease	Heales et al. (2004)
Spongiform encephalopathy	Borne et al. (2005), Pascual Calvet et al. (1989), and Sahni et al. (2008)
Cerebellar dysfunction	Celius (1997)
Cerebral atrophy	Borne et al. (2005), Kivisaari et al. (2004), and Danos et al. (1998)
Cerebral edema	Jamshidi et al. (2013)
Transverse myelopathy (spinal stroke)	Büttner et al. (2000) and Pascual Calvet et al. (1989)
Respiratory depression	Adrish, Duncalf, Diaz-Fuentes, & Venkatram (2014)

ISCHEMIC STROKE IN OPIATE ADDICTS

Ischemic stroke is a frequent finding in heroin addicts and usually occurs within minutes or an hour after administration of the drug, but delayed-onset stroke also has been observed (Borne, Riascos, Cuellar, Vargas, Rojas, 2005; Neiman, Haapaniemi, & Hillbom, 2000). It may result from vasoconstriction (artery spasms), atherosclerosis (macroangiopathy, microangiopathy), thromboembolism (cardiac, aneurysmatic, septic), positional vascular compression, vasculitis, arrhythmias, arterial hypertension, arterial hypotension, dilated cardiomyopathy, foreign material injected with the diluents and contaminants, or heart failure (Decet & Bianchin, 2005). Focal ischemia during hypoventilation and shock after an overdose may be another cause (Brust and Richter, 1976). A more rare cause may be toxicity to the vessel wall resulting in arteritis (Brust and Richter, 1976). Rarely, bilateral border-zone infarctions result from vasculitis of the large cerebral arteries (Büttner et al., 2000). Occasionally, border-zone infarction results from vasospasm of the basal cerebral arteries (Niehaus, Röricht, Meyer, & Sander, 1997). A rare cause of ischemic stroke in opiate addicts is bilateral spontaneous carotid artery dissection during a withdrawal syndrome (Decet & Bianchin, 2005). A major cause of ischemic stroke in opiate addicts is fungal or bacterial endocarditis (Brust & Richter, 1976). Stroke may be found not only in intravenous heroin addicts but also in heroin sniffers (Adle-Biassette et al., 1996, Bartolomei, Nicoli, Swiader, & Gastaut, 1992). In these patients, stroke may occur a few hours after sniffing heroin (Bartolomei et al., 1992). Stroke in opiate addicts most commonly occurs in the territory of the medial cerebral artery but any other territory may be also affected (Bartolomei et al., 1992). In some cases, opiate addiction may simultaneously cause spinal border-zone infarction and cerebral border-zone infarction within the globus pallidus (Niehaus et al., 1997). Occasionally, ischemic stroke in opiate addicts may be associated with other neurological disorders, such as rhabdomyolysis (Hsu, Chiu, & Liao, 2009; Somala, 2009) or seizures (Celius, 1997). Rarely, opiate addicts may experience spinal stroke resulting in myelopathy (Büttner et al., 2000). Whether substitution of opiates by buprenorphine decreases the risk of opiate addicts to develop stroke, as has been proposed (Yulug & Ozan, 2009), needs further investigation.

TABLE 3 PNS Disorders Induced by Opiate Addiction	
Disease	References
Peripheral Nerve	
Plexopathy	Pascual Calvet et al. (1989) and Sahni et al. (2008)
Acute inflammatory demyelinating polyneuropathy	Pascual Calvet et al. (1989) and Sahni et al. (2008)
Demyelinating polyradiculoneuropathy	Sahni et al. (2008)
Mononeuropathy	Pascual Calvet et al. (1989) and Sahni et al. (2008)
Compression neuropathy	Paur et al. (2012)
Polyneuropathy	Paur et al. (2012) and Quattrini et al. (1982)
Skeletal Muscle	
Rhabdomyolysis	Somala (2009), Hsu et al. (2009), Pascual Calvet et al. (1989), and Sahni et al. (2008)
Compartment syndrome	Sahni et al. (2008)
Gluteal compartment syndrome	Adrish et al. (2014)
Acute bacterial myopathy	Pascual Calvet et al. (1989) and Sahni et al. (2008)
Fibrosing myopathy	Pascual Calvet et al. (1989) and Sahni et al. (2008)
Muscle ischemia	Lucas (2005)
Musculoskeletal syndrome	Pascual Calvet et al. (1989)

CEREBRAL BLEEDING IN OPIATE ADDICTS

Rarely, opiate addicts may experience intracerebral hemorrhage due to arterial hypertension, preexisting cavernoma, arteriovenous malformation, or small-vessel disease (Kivisaari et al., 2004, Knoblauch, Buchholz, Koller, & Kistler, 1983; Neiman et al., 2000). Cerebral bleeding (contusional bleeding) may also result from head trauma due to accidents, seizures, involvement in acts of violence, or falls of nonepileptic cause. In a single heroin addict, subarachnoid bleeding has been described, but conventional angiography did not reveal a causative aneurysm or other source of the bleeding (Quattrini et al., 1982). In a 54-year-old female with a history of 35 years of intravenous narcotic abuse and untreated arterial hypertension, subarachnoid bleeding occurred after vertebral puncture injury resulting from recurrent cervical intra-arterial injections of heroin. In this case, vertebral puncture injury was assumed to cause internal carotid artery dissection and fusiform aneurysm formation (Figure 1; DiLuna, Bydon, Gunel, & Johnson, 2007). Mural injury was attributed to either abscess formation or endarteritis resulting in local thrombosis or vasospasm (DiLuna et al., 2007). Bacterial endocarditis may cause intracranial mycotic aneurysms, which are at risk of rupture (Gilroy, Andaya, & Thomas, 1973).

EPILEPSY IN OPIATE ADDICTS

There are several reasons opiate addicts develop epilepsy. First, epilepsy may be a premorbid condition either due to a genetic defect or due to preaddiction cerebral lesions from hypoxia, infection, or trauma. Second, epilepsy may result from cerebral lesions acquired during drug addiction, such as hypoxia, infection, or trauma. Additionally, seizures occur frequently during withdrawal from heroin or poppy husk (Mattoo et al., 2009; Yousuf, Adjei, & Kinder, 2009). The frequency of epilepsy among opiate addicts is highly variable and reported as 0.4–15.1% (Basu, Banerjee, Harish, & Mattoo, 2009; Mattoo et al., 2009; Quattrini et al., 1982). Seizures seem to be more common in older patients with longer duration of substance abuse compared to young addicts (Mattoo et al., 2009). In a study of 33 heroin addicts, five patients developed seizures 1 to 3 years after the onset of addiction (Quattrini et al., 1982). Four of these five patients had developed focal seizures and one patient developed generalized seizures (Quattrini et al., 1982). Opiate addicts developing CNS infection are more likely to develop seizures compared to those without CNS infection (Débat Zoguéreh, Badiaga, & Girard, 1997).

The frequency of epilepsy is significantly increased in patients misusing dextropropoxyphene, a weak opioid often used as a psychoactive substance (Basu et al., 2009; Ng & Alvear, 1993). Among these patients, the frequency of epilepsy was reported as 20% (Basu et al., 2009). In a study of 73 dextropropoxyphene users, 53% developed epilepsy after starting with the drug, but only 17% of these cases were pure dextropropoxyphene abusers (Ng & Alvear, 1993). In the majority of cases (87%), generalized tonic–clonic seizures occurred (Basu et al., 2009). Seizures developed on the average 2 h after a higher than usual dosage of dextropropoxyphene. Occurrence of seizures was related to the duration of dextropropoxyphene abuse and the medical comorbidity (Basu et al., 2009).

TRAUMA

Head trauma or traumatic brain injury may result from falls due to inattention, seizures, or syncopes, or traffic accidents during compromised alertness or inattention, or involvement in acts of violence (Lucas, 2005). Although occasionally reported (Adrish et al., 2014), a systematic investigation of the risk in opiate addicts of experiencing traumatic brain injury, head trauma, or traumatic PNS lesions has not been carried out as of this writing. Compression neuropathy, rhabdomyolysis, or compartment syndrome usually do not result from direct trauma but prolonged abidance in a compromising limb position due to overdose can be regarded as an indirect trauma as well (Figure 2).

INFECTIONS

Various infections involving the CNS or the PNS have been described in opiate addicts (Borne et al., 2005). The most important infection with regard to outcome in opiate addicts is human immunodeficiency virus (HIV). HIV infection results from using

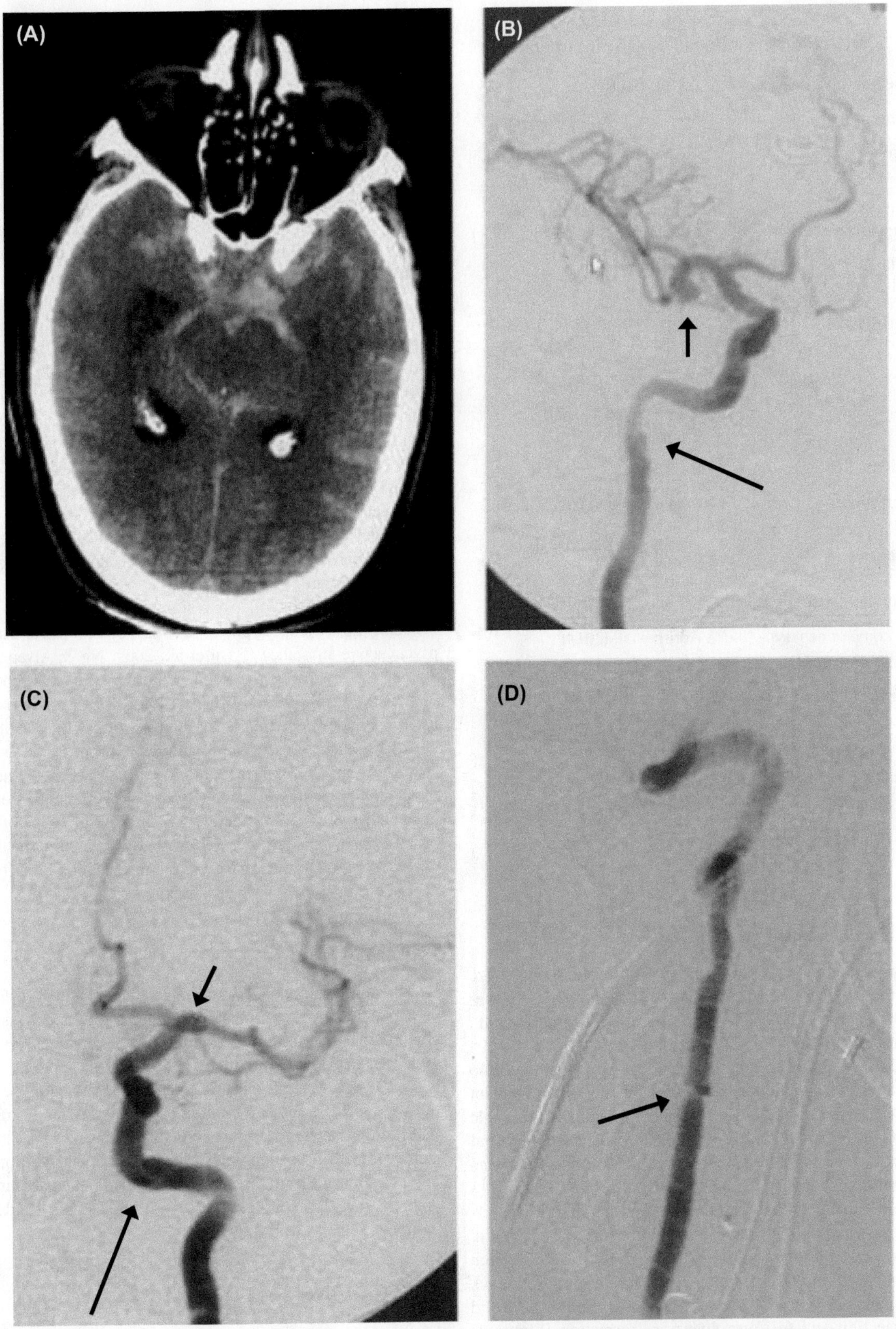

FIGURE 1 Complications from cervical intra-arterial heroin injection. (A) Axial noncontrast computed tomography through the basal cisterns reveals diffuse subarachnoid hemorrhage. (B) Anteroposterior (AP) view of the right internal carotid artery demonstrates a spiral dissection involving the cervical, petrous, and cavernous segments. Note the lobulated right middle cerebral artery aneurysm and absence of local vasospasm (*short arrow*). (C) AP view of the left internal carotid artery demonstrates a spiral dissection involving the distal cervical, petrous, and cavernous internal carotid segments, without narrowing or extraluminal contrast (*long arrow*). There is fusiform dilatation of the proximal (M1) segment of the middle cerebral artery on the left beginning just distal to the carotid summit (*short arrow*). There is mild fusiform dilatation of the proximal (A2) segment of the anterior cerebral artery. (D) Lateral view of the cervical left vertebral arteriogram revealed a focal dissection of the cervical vertebral artery with small opposing pseudoaneurysms, consistent with a puncture injury (*arrow*). With permission from DiLuna et al. (2007).

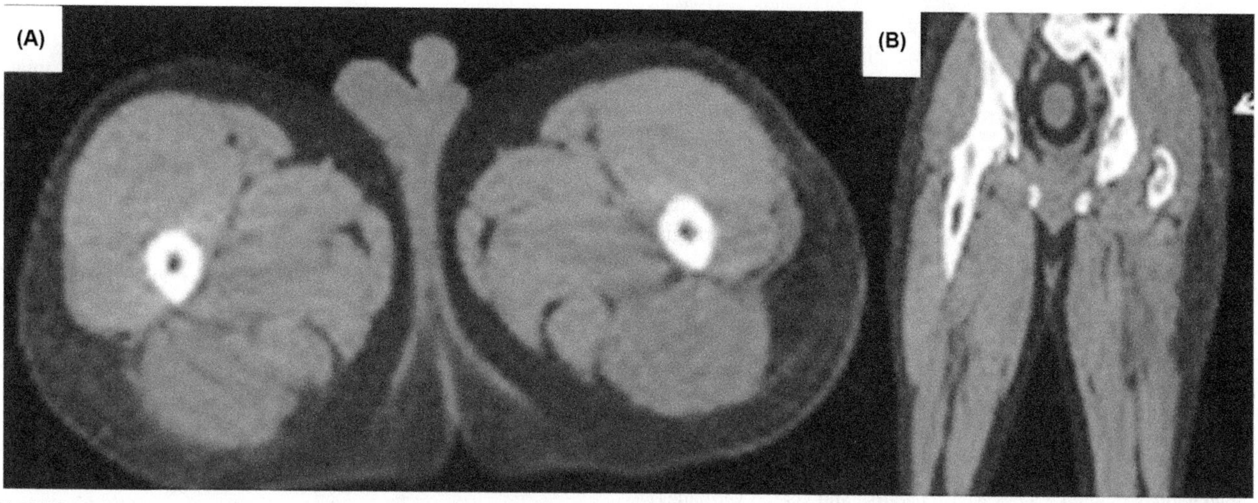

FIGURE 2 Computed tomography of lower extremities without contrast medium showing flank soft-tissue edema with prominence of the gluteal muscles on the left. There is apparent fascia edema of the lateral aspect of the vastus lateralis muscle along with edema on the lateral aspect of semitendinosus muscle bilaterally. *With permission from Adrish et al. (2014).*

contaminated needles or syringes or from the sexual behavior of these patients (sex selling, frequently changing partners). This is why opiate addicts represent one of the risk groups for HIV. HIV infection manifests in the CNS as HIV-associated neurocognitive disorders; HIV-associated opportunistic infections with viruses, bacteria, protozoa, or fungi; or HIV-associated malignancies, such as lymphoma or Kaposi sarcoma. In an 18-year-old heroin-addicted soldier, HIV infection manifested with specific subacute encephalopathy (Martini, Vion, Le Gangneux, Grandpierre, & Becquet, 1991). The combination of transverse myelitis, polyradiculoneuropathy, and leukoencephalopathy in an opiate addict is an indication of HIV infection (Pascual Calvet, Pou, Pedro-Botet, & Gutiérrez Cebollada, 1989). The frequent association between opiate addiction and HIV seems to result from the close relation between these two agents. According to previous investigations, opiates enhance the pathogenesis of HIV by selective activation of μ opioid receptors with downstream effects of disrupting glial homeostasis, increasing inflammation, and decreasing the threshold for proapoptotic events in neurons (Hauser, Fitting, Dever, Podhaizer, & Knapp, 2012). There are indications that opiates generally exhibit an immunosuppressive role in the CNS, since they decrease HIV-associated activation of markers that reflect neurotoxic pathways (Byrd, Murray, Safdieh, & Morgello, 2012). Patients who are hepatitis B, hepatitis C, or HIV positive are prone to acquire CNS infections more frequently than opiate addicts without a history of these infections. More rare CNS infections associated with opiate addiction include *Candida* meningitis manifesting with multiple cranial nerve lesions, weight loss (Carra Dalliere, Thouvenot, Baptista, Le Moing, & Charif, 2010) or with meningoencephalomyeloradiculitis (Pihet, Poulain, De Sèze, Camus, & Sendid, 2005), *Aspergillosis* manifesting as ventriculitis (Morrow, Wong, Finkelstein, Sternberg, & Armstrong, 1983), or cryptococcal meningitis manifesting with syringomyelia and hydrocephalus (McLone & Siqueira, 1976). Not only in heroin addicts but also in patients under substitution with buprenorphine, Candida meningitis or Candida retinitis may occur (Cazorla et al., 2005).

COGNITIVE IMPAIRMENT

Cognitive impairment is a well-recognized effect of opiate addiction (Anthony et al., 2010). The temporal and frontal cortices are most frequently implicated (Anthony et al., 2010). The cause of this effect is largely unknown but studies have shown that opiate users show early Alzheimer disease-related brain pathology and neuroinflammation (Anthony et al., 2010). Tau pathology could be the basis for cognitive impairment in opiate addicts (Anthony et al., 2010). Additionally, methadone maintenance patients exhibit not only an increased rate of psychiatric morbidity but also an increased rate of cognitive deficits (Darke, Sims, McDonald, & Wickes, 2000). Methadone maintenance patients perform more poorly on various neuropsychological domains, such as information processing, attention, short-term visual memory, delayed visual memory, short-term verbal memory, long-term verbal memory, or problem-solving, compared to controls (Darke et al., 2000). Generally, the number of nonfatal heroin overdoses is an independent predictor of poor cognitive performance (Darke et al., 2000).

OTHER CNS ABNORMALITIES IN OPIATE ADDICTS

Spongiform leukoencephalopathy is a rare CNS manifestation, particularly in abusers who inhale heroin vapors (Borne et al., 2005; Büttner et al., 2000). It is considered to result from a lipophilic toxin-induced process inducing or enhancing cerebral hypoxia, but a definite culprit has not yet been identified (Büttner et al., 2000). Rarely, transverse myelitis may occur (Sahni, Garg, Garg, Agarwal, & Singh, 2008). Single cases have been reported, which developed acute disabling cerebellar ataxia after intra-arterial injection of heroin (Celius, 1997). In patients who inhale heroin pyrolysate, temporary parkinsonism has been described (Heales, Crawley, & Rudge, 2004). In these patients there is evidence that tetrahydrobiopterin becomes deficient, causing altered dopamine metabolism (Heales et al., 2004). Acute parkinsonism with typical lesions in the basal ganglia may also occur in intravenous heroin addicts (Mätzler, Nägele, Gasser, & Krüger, 2007).

PNS DISEASE

PNS disease in opiate addicts may either affect the skeletal muscle or the peripheral nerves. Muscle disease induced by opiate addiction includes coma-induced rhabdomyolysis with quadriplegia, compartment syndrome, acute bacterial myopathy, muscle ischemia (Lucas, 2005), or fibrosing myopathy (Sahni et al., 2008). Disease of the peripheral nerves due to opiate addiction includes plexopathy, acute inflammatory demyelinating polyneuropathy, mononeuropathy, or compression neuropathy (Table 3; Hsu et al., 2009; Pascual Calvet et al., 1989; Paur et al., 2012; Sahni et al., 2008; Somala, 2009). Rarely, polyneuropathy has been described in opiate addicts (Paur et al., 2012; Quattrini et al., 1982). Compartment syndrome in opiate addicts may take a devastating course and lead to permanent disability if fasciotomy is delayed (Adrish et al., 2014). This is the case if the clinical manifestations of a compartment syndrome are falsely attributed to a preexisting neuropathy and why early nerve conduction studies and needle electromyography should be carried out in case muscle weakness remains unclear (Adrish et al., 2014). Compartment syndrome may manifest not only as mononeuropathy but also as plexopathy (Jost, Mayer, & Rossmanith, 2002). Rhabdomyolysis is presumably the most common of the neurological complications in opiate addicts. It may also occur in patients taking methadone (Valga-Amado et al., 2012). It occurs not only in patients with intravenous drug abuse but also in heroin sniffers. More frequently, however, rhabdomyolysis occurs in heroin users owing to the increased myotoxic effect of heroin (Kosmadakis, Michail, Filiopoulos, Papadopoulou, & Michail, 2011). In some cases rhabdomyolysis may lead to reflex sympathetic dystrophy (Lee, Chiu, Chen, Liu, & Yu, 2001). In other cases rhabdomyolysis may even be fatal, particularly if it goes undetected because of asymptomatic presentation on admission (Chan, Wong, & Chow, 1990).

NEUROLOGICAL WITHDRAWAL SYMPTOMS IN OPIATE ADDICTS AND IN NEONATES

Withdrawal from opiate addiction is known to go along with a number of neurological abnormalities (Gekht, Polunina, Briun, & Gusev, 2003). In a study of 79 patients with heroin addiction during 17 months and an average dosage of 0.5 g/day, hypomimia, hypokinesia, muscle hypotonia, reduced tendon reflexes, and facilitated nociceptive reflexes were found in more than 70% of the patients after the first week of abstinence (Gekht et al., 2003). Nonspecific signs, such as nystagmus, limited convergence, tremor, or dynamic ataxia were found in 35% of the patients (Gekht et al., 2003). Most of these microneurological signs disappeared within 10 days of abstinence (Gekht et al., 2003). These signs were most prevalent among heroin addicts who were addicted for more than 6 months (Gekht et al., 2003). Neurological withdrawal may also occur in neonates from pregnant opiate addicts. In a study of 20 pregnant heroin addicts the prenatal condition was stabilized by treating 16 of them with methadone (Vering, Seeger, Becker, Halberstadt, & Bender, 1992). Despite this measure and prophylactic treatment with barbiturates, neonates experienced severe withdrawal symptoms including seizures (Vering et al., 1992).

CNS IMAGING IN OPIATE ADDICTS

Imaging studies of the CNS may be carried out in opiate addicts with or without clinical neurologic manifestations. In opiate addicts without clinical manifestations of a neurological disease, cerebral magnetic resonance imaging (MRI) may show wider sylvian fissures and larger ventricles compared to controls (Kivisaari et al., 2004). Generally, the cerebrospinal fluid spaces seem to be enlarged in opiate addicts (Danos et al., 1998). Rarely, subcortical posttraumatic lesions may be found (Kivisaari et al., 2004). In a retrospective cerebral computed tomography (CCT) study of 71 opiate overdose patients, an abnormal CCT was found in 20% of the cases (Jamshidi et al., 2013). Eight patients had generalized cerebral edema and six patients had ischemic stroke (Jamshidi et al., 2013). The outcome of patients with CCT abnormalities was worse compared to patients with normal imaging (Jamshidi et al., 2013). Vascular lesions due to microvascular ischemia may be another finding on cerebral MRI in heroin addicts (Borne et al., 2005). In a small study of four patients, cerebral MRI showed areas of demyelination in the white matter (Volkow, Valentine, & Kulkarni, 1988). The cerebral blood flow was impaired on positron emission tomography (Volkow et al., 1988).

NEUROPATHOLOGY IN OPIATE ADDICTS

The major neuropathological findings in brains from opiate addicts are due to infections, such as spread from bacterial endocarditis, mycosis, or HIV infection (Büttner et al., 2000). Other neuropathological findings include ischemic or hypoxic lesions with cerebral edema or ischemic neuronal damage due to prolonged respiratory depression or stroke (Büttner et al., 2000). Other studies have shown cystic, cortical infarcts involving both hemispheres with an intralaminar pattern and a watershed distribution (Adle-Biassette et al., 1996). In heroin addicts who developed musculoskeletal syndrome, muscle biopsy revealed morphologic abnormalities similar to those found in alcohol myopathy (Pascual Calvet et al., 1989). In an autopsy study of 11 opiate addicts one presented with cerebral phycomycetosis (case 5) and three died from endocarditis resulting in mycotic aneurysm, septic embolism, and focal leptomeningitis (case 1); in disseminated cerebral microabscesses (case 2); or in leptomeningeal abscesses (case 3) (Adelman & Aronson, 1969). Another patient had hypoxic encephalomalacia (case 8), one had Wernicke encephalopathy (case 9), and one had tabes dorsalis (case 11) (Adelman & Aronson, 1969).

PREMORBID NEUROLOGICAL VULNERABILITIES FOSTERING THE DEVELOPMENT OF OPIATE ADDICTION

It is conceivable but not confirmed that premorbid neurological abnormalities may foster the development of opiate addiction. Among these conditions, epilepsy, traumatic brain injury, and cognitive impairment are the ones that have been confirmed to play a pathogenetic role in the development of opiate addiction. Traumatic brain injury can cause an organic personality disorder, which may cause decision-making deficits that increase the risk of drug abuse (Bjork & Grant, 2009). Although there is not much literature available (Genaïlo, 1990), it is conceivable that mental retardation, premorbid epilepsy, premorbid head trauma, or premorbid vascular or infectious CNS lesions play at least a contributing role in the development of opiate addiction.

DEPENDENCY OF NEUROLOGICAL ABNORMALITIES ON THE MODE OF OPIATE ADMINISTRATION, THE DURATION OF ADDICTION, AND THE OPIATE DOSAGE

Little is known about the dependency of neurological abnormalities on the type of opiate administration. It can be speculated, however, that intravenous administration of the drug bears a higher risk of developing neurological disease than oral or inhalational intake. With increasing duration of opiate addiction, affected patients sooner or later develop medical abnormalities in addition to their addiction or premorbid condition (Chen, Scheier, & Kandel, 1996; Lucas, 2005). The risk of developing neurological disease from opiate addiction increases with the duration of drug abuse and with the degree of social decline (Lucas, 2005). In heroin addicts it has been shown that an overdose may cause acute respiratory failure, spongiform leukoencephalopathy, seizures, stroke, toxic amblyopia, transverse myelopathy, mononeuropathy, plexopathy, acute inflammatory demyelinating polyradiculoneuropathy, rhabdomyolysis, compartment syndrome, fibrosing myopathy, or acute bacterial myopathy (Sahni et al., 2008). In a retrospective study of 452 heroin overdose patients, neurological complications were related to the total morphine levels but not those of conjugated morphine (Pascual Calvet et al., 1989). Overdosing was also related to high benzodiazepine levels (Pascual Calvet et al., 1989).

NEUROLOGICAL COMPLICATIONS DURING GENERAL ANESTHESIA

Noncardiac in-hospital complications after coronary artery bypass grafting do not seem to be increased among opiate addicts (Sadeghian et al., 2009).

SECONDARY DISEASE THAT CAUSES TERTIARY NEUROLOGICAL ABNORMALITIES

There are a number of nonneurological disorders in opiate addicts that could cause secondary neurological disease. Among these are endocarditis, atherosclerosis, arterial hypertension, heart failure, or general infection. Additionally opiate addiction may be associated with various autoantibodies and immunological abnormalities (Simonovska et al., 2012). In a study of 161 heroin addicts using heroin intravenously, serum IgA levels were significantly reduced (Simonovska et al., 2012). Compared to heroin addicts who inhaled heroin, those who used heroin intravenously had elevated serum IgG and serum IgM levels, elevated anti-β2GP1 cryoglobulins, elevated rheuma factors, and elevated anti-β2GP1 IgA and IgM levels (Simonovska et al., 2012). Cryoglobulin-positive heroin addicts presented more frequently with arthralgia, vasculitis, hematuria, respiratory problems, Raynaud phenomenon, proteinuria, renal insufficiency, or neurological disease compared to cryoglobulin-negative heroin addicts (Simonovska et al., 2012).

CONCLUSIONS

Although some of the frequent secondary or tertiary neurological side effects of opiate addiction have been addressed in single systemic studies, there is still a lack of knowledge about the risk of and presentation, treatment, and outcome of neurological abnormalities directly or indirectly resulting from opiate addiction. A main disadvantage of most of the available studies of neurological disease in opiate addicts is their poor design, which is frequently retrospective in nature or based on case studies or case reports. A further disadvantage is the frequent mixture of pure opiate addicts with those who are on a substitution therapy and those who take opiates in addition to cocaine, benzodiazepines, alcohol, or designer drugs. Even among the pure opiate abusers it is important to differentiate between the various types of opiates since they appear to carry an variable risk for neurological complications (e.g., buprenorphine, dextropropoxyphene). A further disadvantage of most of the studies is that patients were investigated clinically, with imaging techniques, or at autopsy. If neurological disease is investigated systematically in opiate addicts, a comprehensive investigation battery needs to be prospectively applied and group sizes need to be large enough to be statistically representative. Generally, opiate addicts are at risk to experience neurological disease more frequently than the general population. This risk increases with the duration of addiction, premorbid conditions, and number of comorbidities. Outcome of neurological disease in opiate addicts is poor not only because of lacking systematic studies but also because these patients are rarely routinely investigated by a neurologist and in case they are investigated, the compliance to neurological treatment is often deficient because of the inherent cognitive, social, or economic impairment.

APPLICATIONS TO OTHER ADDICTIONS AND SUBSTANCE MISUSE

Neurological side effects are not restricted to opiate addicts but may also occur in any other type of addiction or substance misuse. The spectrum and prevalence of neurological abnormalities, however, may vary considerably between the different addictions.

DEFINITION OF TERMS

Opiate-based compounds These are opioids (natural or synthetic substances) with morphine-like effects that act via opioid receptors.
Border-zone infarction This is a subtype of ischemic stroke located at the edge between two vascular territories, usually due to hypoperfusion.
Focal seizures This refers to synchronized activity of neuronal cells in a localized cortical area without spreading to other cortical regions.
Arteriovenous malformation This is an arteriovenous shunt in the absence of an in-between capillary bed.
Compartment syndrome This is a life-threatening condition after limb injury with consecutive increase in pressure within a muscular compartment and impaired blood supply to muscle and nerves.
Raynaud phenomenon This refers to excessively reduced blood flow due to vasospasms with decoloration of fingers or toes in response to cold or emotional stress.

Artery dissection This is a separation of muscular artery layers with consecutive narrowing of the artery diameter and reduction or discontinuation of the blood flow and ischemia in the appropriate vascular territory.

Nystagmus This is a condition of uncontrolled, rapid, jerky eye movements, most frequently from side to side but sometimes also up and down or circular, resulting in poor vision.

Tremor This is an involuntary rhythmic muscle contraction and relaxation involving to and fro movements and affecting hands, arms, eyes, face, head, trunk, vocal cords, or legs.

Embolism This is a blood clot, bacterial hump, fat globule, food particle, or gas bubble transported by the bloodstream and resulting in blockage of a blood vessel distal to the origin of embolus formation.

KEY FACTS

- A number of neurological disorders, such as stroke, bleeding, seizures, meningitis, compartment syndrome, or respiratory depression, in opiate addicts require emergency management so as not to miss the point at which neurological abnormalities become irreversible.
- Opiate addicts are at risk of experiencing neurological disease more frequently than the general population, a risk that increases with the duration of addiction, premorbid conditions, and number of comorbidities.
- Outcome of neurological disease in opiate addicts is poor because they are rarely routinely investigated by a neurologist and when they are investigated, the compliance with neurological treatment is often deficient owing to inherent cognitive, social, or economic impairment.
- The prevalence of neurological disease in opiate addicts increases if opiates are combined with other drugs or alcohol.

SUMMARY POINTS

- Neurological complications are frequent in opiate addicts.
- Neurological complications often remain unrecognized and untreated because opiate addicts do not visit a neurologist.
- Some neurological complications require emergency management and, if it is not provided, may result in severe long-term deficits or even death.
- The most frequent CNS disorders in opiate addicts include stroke, epilepsy, infections, or extrapyramidal disease.
- The most frequent PNS disorders in opiate addicts include compression neuropathy, polyradiculoneuropathy, plexopathy, rhabdomyolysis, or compartment syndrome.

REFERENCES

Adelman, L. S., & Aronson, S. M. (1969). The neuropathologic complications of narcotics addiction. *Bulletin of the New York Academy of Medicine, 45,* 225–234.

Adle-Biassette, H., Marc, B., Benhaiem-Sigaux, N., Durigon, M., & Gray, F. (1996). Cerebral infarctions in a drug addict inhaling heroin. *Archives Anatomie et de Cytologie Pathologiques, 44,* 12–17.

Adrish, M., Duncalf, R., Diaz-Fuentes, G., & Venkatram, S. (2014). Opioid overdose with gluteal compartment syndrome and acute peripheral neuropathy. *American Journal of Case Reports, 15,* 22–26.

Anthony, I. C., Norrby, K. E., Dingwall, T., Carnie, F. W., Millar, T., Arango, J. C., ... Bell, J. E. (2010). Predisposition to accelerated Alzheimer-related changes in the brains of human immunodeficiency virus negative opiate abusers. *Brain, 133,* 3685–3698.

Bartolomei, F., Nicoli, F., Swiader, L., & Gastaut, J. L. (1992). Ischemic cerebral vascular stroke after heroin sniffing. A new case. *La Presse Medicale, 21,* 983–986.

Basu, D., Banerjee, A., Harish, T., & Mattoo, S. K. (2009). Disproportionately high rate of epileptic seizure in patients abusing dextropropoxyphene. *American Journal on Addictions, 18,* 417–421.

Bjork, J. M., & Grant, S. J. (2009). Does traumatic brain injury increase risk for substance abuse? *Journal of Neurotrauma, 26,* 1077–1082.

Borne, J., Riascos, R., Cuellar, H., Vargas, D., & Rojas, R. (2005). Neuroimaging in drug and substance abuse part II: opioids and solvents. *Topics in Magnetic Resonance Imaging, 16,* 239–245.

Brust, J. C., & Richter, R. W. (1976). Stroke associated with addiction to heroin. *Journal of Neurology, Neurosurgery and Psychiatry, 39,* 194–199.

Büttner, A., Mall, G., Penning, R., & Weis, S. (2000). The neuropathology of heroin abuse. *Forensic Science International, 113,* 435–442.

Byrd, D., Murray, J., Safdieh, G., & Morgello, S. (2012). Impact of opiate addiction on neuroinflammation in HIV. *Journal of Neurovirology, 18,* 364–373.

Campbell, N. (Spring 2003). Dark Paradise: a history of opiate addiction in America (review). *Bulletin of the History of Medicine, 77*(1), 218–219.

Carra Dalliere, C., Thouvenot, E., Baptista, G., Le Moing, V., & Charif, M. (2010). Meningomyeloradiculitis in an immunocompetent patient. *Revue Neurologique (Paris), 166,* 741–744.

Cazorla, C., Grenier de Cardenal, D., Schuhmacher, H., Thomas, L., Wack, A., May, T., & Rabaud, C. (2005). Infectious complications and misuse of high-dose buprenorphine. *La Presse Medicale, 34,* 719–724.

Celius, E. G. (1997). Neurologic complications in heroin abuse. Illustrated by two unusual cases. *Tidsskr Nor Laegeforen, 117,* 356–357.

Chan, Y. F., Wong, P. K., & Chow, T. C. (1990). Acute myoglobinuria as a fatal complication of heroin addiction. *American Journal of Forensic Medicine and Pathology, 11,* 160–164.

Chen, K., Scheier, L. M., & Kandel, D. B. (1996). Effects of chronic cocaine use on physical health: a prospective study in a general population sample. *Drug and Alcohol Dependence, 43,* 23–37.

Danos, P., Van Roos, D., Kasper, S., Brömel, T., Broich, K., Krappel, C., ... Möller, H. J. (1998). Enlarged cerebrospinal fluid spaces in opiate-dependent male patients: a stereological CT study. *Neuropsychobiology, 38,* 80–83.

Darke, S., Sims, J., McDonald, S., & Wickes, W. (2000). Cognitive impairment among methadone maintenance patients. *Addiction, 95,* 687–695.

Débat Zoguéreh, D., Badiaga, S., & Girard, N. (1997). Periarteritis nodosa disclosed by epilepsy in a drug addict with hepatitis B and C virus carrier state. *La Revue de Médecine Interne, 18,* 311–315.

Decet, G., & Bianchin, A. (2005). Withdrawal syndrome in a drug addict caused diagnostic delay of spontaneous bilateral carotid artery dissection. *Journal of Neurosurgery and Anesthesiology, 17,* 59–60.

DiLuna, M. L., Bydon, M., Gunel, M., & Johnson, M. H. (2007). Neurological picture. Complications from cervical intra-arterial heroin injection. *Journal of Neurology, Neurosurgery and Psychiatry, 78,* 1198.

Gekht, A. B., Polunina, A. G., Briun, E. A., & Gusev, E. I. (2003). Neurological disturbances in heroin addicts in acute withdrawal and early post-abstinence periods. *Zhurnal Nevrologii Psikhiatrii Imeni S S Korsakova, 103,* 9–15.

Genaĭlo, S. P. (1990). Characteristics of premorbid conditions in drug addicts. *Zhurnal Nevropatologii Psikhiatrii Imeni S S Korsakova, 90*, 42–47.

Gilroy, J., Andaya, L., & Thomas, V. J. (1973). Intracranial mycotic aneurysms and subacute bacterial endocarditis in heroin addiction. *Neurology, 23*, 1193–1198.

Hauser, K. F., Fitting, S., Dever, S. M., Podhaizer, E. M., & Knapp, P. E. (2012). Opiate drug use and the pathophysiology of neuroAIDS. *Current HIV Research, 10*, 435–452.

Heales, S., Crawley, F., & Rudge, P. (2004). Reversible parkinsonism following heroin pyrolysate inhalation is associated with tetrahydrobiopterin deficiency. *Movement Disorders, 19*, 1248–1251.

Hsu, W. Y., Chiu, N. Y., & Liao, Y. C. (2009). Rhabdomyolysis and brain ischemic stroke in a heroin-dependent male under methadone maintenance therapy. *Acta Psychiatrica Scandinavica, 120*, 76–79.

Jamshidi, F., Sadighi, B., Aghakhani, K., Sanaei-Zadeh, H., Emamhadi, M., & Zamani, N. (2013). Brain computed tomographic scan findings in acute opium overdose patients. *American Journal of Emergency Medicine, 31*, 50–53.

Jost, U., Mayer, G., & Rossmanith, T. (2002). Complete brachial plexus paralysis caused by compartment syndrome in heroin intoxication. *Der Unfallchirurg, 105*, 392–394.

Kivisaari, R., Kähkönen, S., Puuskari, V., Jokela, O., Rapeli, P., & Autti, T. (2004). Magnetic resonance imaging of severe, long-term, opiate-abuse patients without neurologic symptoms may show enlarged cerebrospinal spaces but no signs of brain pathology of vascular origin. *Archives of Medical Research, 35*, 395–400.

Knoblauch, A. L., Buchholz, M., Koller, M. G., & Kistler, H. (1983). Hemiplegia following injection of heroin. *Schweizerische Medizinische Wochenschrift, 113*, 402–406.

Kosmadakis, G., Michail, O., Filiopoulos, V., Papadopoulou, P., & Michail, S. (2011). Acute kidney injury due to rhabdomyolysis in narcotic drug users. *International Journal of Artificial Organs, 34*, 584–588.

Lee, B. F., Chiu, N. T., Chen, W. H., Liu, G. C., & Yu, H. S. (2001). Heroin-induced rhabdomyolysis as a cause of reflex sympathetic dystrophy. *Clinical Nuclear Medicine, 26*, 289–292.

Lucas, C. E. (2005). The impact of street drugs on trauma care. *Journal of Trauma, 59*(Suppl.), S57–S60 discussion S67–75.

Lukacher, G. I., Vrublevskiĭ, A. G., & Laskova, N. B. (1987). Neurological aspects of opium addiction. *Zhurnal Nevropatologii Psikhiatrii Imeni S S Korsakova, 87*, 1653–1657.

Martini, L., Vion, P., Le Gangneux, E., Grandpierre, G., & Becquet, D. (1991). AIDS and myasthenia: an uncommon association. *Revue Neurologique (Paris), 147*, 395–397.

Mattoo, S. K., Singh, S. M., Bhardwaj, R., Kumar, S., Basu, D., & Kulhara, P. (2009). Prevalence and correlates of epileptic seizure in substance-abusing subjects. *Psychiatry and Clinical Neurosciences, 63*, 580–582.

Mätzler, W., Nägele, T., Gasser, T., & Krüger, R. (2007). Acute parkinsonism with corresponding lesions in the basal ganglia after heroin abuse. *Neurology, 68*, 414.

McLone, D. G., & Siqueira, E. B. (1976). Post-meningitic hydrocephalus and syringomyelia treated with a ventriculoperitoneal shunt. *Surgical Neurology, 6*, 323–325.

Morrow, R., Wong, B., Finkelstein, W. E., Sternberg, S. S., & Armstrong, D. (1983). Aspergillosis of the cerebral ventricles in a heroin abuser. Case report and review of the literature. *Archives in Internal Medicine, 143*, 161–164.

Neiman, J., Haapaniemi, H. M., & Hillbom, M. (2000). Neurological complications of drug abuse: pathophysiological mechanisms. *European Journal of Neurology, 7*, 595–606.

Ng, B., & Alvear, M. (1993). Dextropropoxyphene addiction–a drug of primary abuse. *American Journal of Drug and Alcohol Abuse, 19*, 153–158.

Niehaus, L., Röricht, S., Meyer, B. U., & Sander, B. (1997). Nuclear magnetic resonance tomography detection of heroin-associated CNS lesions. *Aktuelle Radiologie, 7*, 309–311.

Pascual Calvet, J., Pou, A., Pedro-Botet, J., & Gutiérrez Cebollada, J. (1989). Non-infective neurologic complications associated to heroin use. *Archivos De Neurobiologia (Madras), 52*(Suppl. 1), 155–161.

Paur, R., Wallner, C., Hermann, P., Stöllberger, C., & Finsterer, J. (2012). Neurological abnormalities in opiate addicts with and without substitution therapy. *American Journal of Drug and Alcohol Abuse, 38*, 239–245.

Pihet, M., Poulain, D., De Sèze, J., Camus, D., & Sendid, B. (2005). Candida albicans meningo-encephalo-myelo-radiculitis at an addict. *Annales de Biologie Clinique (Paris), 63*, 547–552.

Quattrini, A., Paggi, A., Ortenzi, A., Silvestri, R., Cianci, F., Ardito, S., ... Mancini, S. (1982). Neurological complications in drug dependence with special reference to the development of epileptic syndromes. *Rivista di Patologia Nervosa e Mentale, 103*, 262–270.

Sadeghian, S., Karimi, A., Dowlatshahi, S., Ahmadi, S. H., Davoodi, S., Marzban, M., ... Fathollahi, M. S. (2009). The association of opium dependence and postoperative complications following coronary artery bypass graft surgery: a propensity-matched study. *Journal of Opioid Management, 5*, 365–372.

Sahni, V., Garg, D., Garg, S., Agarwal, S. K., & Singh, N. P. (2008). Unusual complications of heroin abuse: transverse myelitis, rhabdomyolysis, compartment syndrome, and ARF. *Clinical Toxicology (Philadelphia), 46*, 153–155.

Simonovska, N., Bozinovska, C., Chibishev, A., Grchevska, L., Dimitrovski, K., & Neceva, V. (2012). Changes of some humoral immunologic indicators and clinical manifestations of cryoglubulinemia in heroin addicts. *Georgian Medical News, 212*, 45–53.

Somala, R. K. (2009). Rhabdomyolysis and brain ischemic stroke in a heroin-dependent male. Invited comment. *Acta Psychiatrica Scandinavica, 120*, 80–81.

Valga-Amado, F., Monzón-Vázquez, T. R., Hadad, F., Torrente-Sierra, J., Pérez-Flores, I., & Barrientos-Guzmán, A. (2012). Rhabdomyolysis with acute renal failure secondary to taking methadone. *Nefrologia, 32*, 262–263.

Vering, A., Seeger, J., Becker, S., Halberstadt, E., & Bender, H. G. (1992). Heroin abuse and methadone substitution in pregnancy. *Geburtshilfe und Frauenheilkunde, 52*, 144–147.

Volkow, N. D., Valentine, A., & Kulkarni, M. (1988). Radiological and neurological changes in the drug abuse patient: a study with MRI. *Journal of Neuroradiology, 15*, 288–293.

Yousuf, M. A., Adjei, S., & Kinder, B. A. (2009). 58-year-old woman with ST-segment elevation, seizures, and altered mental status in the setting of opiate withdrawal. *Chest, 135*, 1098–1101.

Yulug, B., & Ozan, E. (2009). Buprenorphine: a safe agent for opioid dependent patients who are under the increased risk of stroke? *Journal of Opioid Management, 5*, 134.

Chapter 89

Clinical and Preclinical Molecular Imaging in Chronic Pain—Implications for Analgesic Use and Misuse

Deepak Behera[1], Nida Ashraf[2]

[1]Molecular Imaging Program at Stanford, Department of Radiology, Stanford University School of Medicine, Stanford, CA, USA;
[2]Aga Khan University, Karachi, Pakistan

Abbreviations

BDNF Brain-derived neurotrophic factor
BOLD Blood oxygen level dependent
CGRP Calcitonin gene-related peptide
CT Computed tomography
DRG Dorsal root ganglion
DTI Diffusion tensor imaging
FDG Fluorodeoxyglucose
fMRI Functional magnetic resonance imaging
GABA γ-Aminobutyric acid
GTX Gonyautoxin
HVA High-voltage-activated calcium channels
IL-6 Interleukin 6
IL-1β Interleukin 1β
LVA Low-voltage-activated calcium channels
MAPK-1 Mitogen-activated protein kinase 1
MAPK-8 Mitogen-activated protein kinase 8
MRI Magnetic resonance imaging
MRS Magnetic resonance spectroscopy
NaV Voltage-gated sodium ion channel
NGF Nerve growth factor
NIRS Near-infrared fluorescence spectroscopy
NK-1 Neurokinin 1
NMDA *N*-methyl D-aspartate
NO Nitric oxide
PET Positron emission tomography
PGE$_2$ Prostaglandin E$_2$
rCBF Regional cerebral blood flow
RSN Resting state network
SNS Sensory neuron specific
STX Saxitoxin
S1R σ1 receptor
TTX-R Tetrodotoxin resistant
TTX-S Tetrodotoxin sensitive
VBM Voxel-based morphometry

INTRODUCTION

Pain is debilitating, not only for those suffering from it, but also for the economy as a whole: a 2011 report estimated the annual economic cost in the United States to be greater than $600 billion dollars, with analgesics being the most commonly prescribed medications (Committee on Advancing Pain Research Care and Education, 2011). A large proportion of those who experience pain suffer from chronic noncancer pain, including inflammatory joint diseases, neuropathic pain, and chronic pain syndromes. Nearly a third of the US population suffers from some kind of a chronic pain. According to some estimates, on average about half of the patients felt that their pain was not optimally controlled, whereas a third remained untreated (Bekkering et al., 2011; Harker et al., 2012).

Opioids are an effective therapy in the management of pain. Since 1995 there has been a liberalization of regulatory laws worldwide for the prescription of opioids, and physicians are now confronted with the challenge of trying to treat patients optimally without overprescribing these powerful medications (Atluri, Sudarshan, & Manchikanti, 2014; Hallinan, Osborn, Cohen, Dobbin, & Wodak, 2011; Ling, Mooney, & Hillhouse, 2011). The enormity of chronic pain burden in society together with a rise in prescription of opioids has resulted in an epidemic of opioid misuse and related economic impact (Ruetsch, 2010).

Recognizing that opioid prescription misuse is a growing health concern, various risk factors for addictive behavior have been hypothesized, both biological and environmental, in an attempt to identify potential corrective measures (Garland, Froeliger, Zeidan, Partin, & Howard, 2013). For instance, associations have been discovered between mental illnesses, including depression and anxiety, and increased risk of opioid misuse (Howe & Sullivan, 2014), while persistent pain has been implicated as a mediator of opioid reward by decreasing γ-aminobutyric acid (GABA) synaptic activity leading to neuronal disinhibition and, hence, misuse (Zhang et al., 2014). Similarly, the effects of variables such as age, gender, socioeconomic status, etc., on opioid misuse have also been investigated (Manchikanti, Fellows, Ailinani, & Pampati, 2010).

In addition, the inadequate identification of the cause or source of chronic intractable pain may contribute to the global potential for analgesic misuse. An increasing number of organizations are beginning to consider that chronic pain, which persists long after the cessation of the initial insult that caused it, evolves from being a mere symptom to a self-sustaining disease in its own right (Committee on Advancing Pain Research Care and Education, 2011). Not being able to recognize that chronic pain might be an independent condition often leads to a futile search for a separate cause. Current clinical and imaging practices, including computed tomography and magnetic resonance imaging (MRI), fall short of definitive localization of the site(s) of pain-generating pathology, primarily because they rely on subjective reports and/or anatomic changes. Clinical imaging identifies as many significant anatomic abnormalities in asymptomatic individuals as in chronic pain patients, casting a doubt on the specificity of such techniques (Jensen et al., 1994; Sher, Uribe, Posada, Murphy, & Zlatkin, 1995). Such abnormalities might be erroneously identified as the pain-causing culprit in the patient with chronic pain, sometimes leading to unsuccessful interventions and adding to the economic burden (Flynn, Smith, & Chou, 2011). Chronic pain management thus heavily relies on treatment of the "pain symptom" (hence the huge demand for analgesics), instead of treatment of the underlying "pain disease."

On the other hand, anatomic defects, as seen in current imaging methods, rarely correlate with the biological processes in chronic pain, which are undetectable via gross anatomic imaging. Molecular imaging techniques such as positron emission tomography (PET) and functional MRI (fMRI) draw upon the notion that prolonged nociceptive signals induce functional changes in neural structures, which in turn lead to chronic pain becoming a self-sustained condition even after the initial insult has healed. Several studies and reviews in the literature have documented molecular and morphological changes in the brain related to chronic pain (Apkarian, Hashmi, & Baliki, 2011; Tracey & Bushnell, 2009). Brain molecular imaging studies of chronic pain have opened up a Pandora's box—while they have improved our understanding of pain processing in the brain, there are still mysteries within the brain that leave much room for improvement in the specificity of such techniques (May, 2007).

Locating the source of pain can potentially allow local therapeutic interventions instead of systemic therapies that act on several physiological systems and lead to untoward effects. Molecular imaging has the potential for identifying and locating such sources of pain generation (in the nerves and spinal cord) and areas of pain perception and processing (in the brain) to guide targeted therapy. Also, identifying the pathological processes giving rise to self-sustaining pain can potentially open up an entire new field of drug development targeted toward interrupting such processes locally. As an example, there is a growing body of evidence showing involvement of microglial cells in the chronic neuropathic pain state, as well as potential for pain relief by manipulating these cells (Gosselin, Suter, Ji, & Decosterd, 2010). The promise of discovery of novel therapeutic approaches is exciting and may significantly alter current pain management therapy in the future. Targeted to deranged physiological processes and delivered at sites of derangement, such therapeutics may reduce the need for prolonged opioid use and potential for misuse.

PATHOPHYSIOLOGY RELEVANT TO PAIN IMAGING

Chronic pain can have an identifiable cause, as in the case of long-term inflammation, e.g., arthritis, or disease or injury of the nervous system, e.g., neuropathic pain, or the underlying cause may be unknown. It arises as a result of altered neuronal sensitivity, in which activation of low-threshold mechanoreceptors (normally generating innocuous sensations) gives rise to pain (Zimmermann, 2001). Various processes occurring in the spinal cord are accountable for this abnormal activity, including enhanced excitability, attenuated inhibition, and structural changes—these are collectively termed as central neuronal sensitization. Central sensitization manifests itself as tactile allodynia, temporal summation of pain, and secondary hyperalgesia (Woolf, 2011). It is to be distinguished from peripheral sensitization, which is primarily responsible for pain hypersensitivity in that it decreases the nociceptor's threshold to prolonged noxious stimuli and occurs only at the site where its terminals were exposed to inflammatory mediators (Chaban, 2010). Studies have identified the roles of inflammation, ion channels, astrocytes, microglia, gap junctions, membrane excitability, and gene transcription in the generation or perpetuation of pain (Scholz & Woolf, 2007).

Information is transmitted from the periphery to the spinal cord and brain by a variety of axons with myelin sheaths of varying degrees of thickness. The more myelinated the axons are, the more sensitive to changes in myelination they will be, as in the case of disease processes or nerve injury. When myelination is compromised there is a dysfunction of the mechanisms that transmit action potentials along axons owing to disruption of the protective and nourishing effect of the myelin sheath, and as a result, unnecessary or ectopic action potentials arise and augment nociceptive signals.

The vastly complex genetic infrastructure and metabolic machinery of nerve cells resides in the cell bodies located in the dorsal root ganglion (DRG). If a normally non-pain-mediating neuron begins to synthesize substance P, activity in this type of neuron may lead to the perception of pain. Inflammation of peripheral structures induces the production of chemical mediators like substance P and brain-derived neurotrophic factor in the DRG by exposing their afferent terminals to nerve growth factor and inflammatory mediators. Substance P acts via the neurokinin receptors on the dorsal horn neuron to augment the postsynaptic response and increase the activity of N-methyl D-aspartate (NMDA) receptors. Protein kinase C phosphorylates the NMDA receptor, altering its responsiveness. Under normal conditions, magnesium binding blocks the NMDA receptor but alteration in Mg^{2+} binding kinetics allows release of Mg^{2+} from the receptor and glutamate-induced activation and subsequent depolarization of the cell membrane (Kidd & Urban, 2001).

Another way in which gene activity might affect pain is in the expression or distribution of sodium channels, which are responsible for the generation of action potentials. Thus, a change in the way a cell produces or controls sodium channels may lead to a perception of persistent pain, despite the absence of a noxious stimulus. Two types of sodium channels exist—one is sensitive to puffer fish tetrodotoxin and the other is resistant (TTX-R). One of the subtypes of the TTX-R channels, namely, the sensory neuron-specific channel, is found within the DRG and its expression is increased in chronic inflammatory states (Rogers, Tang, Madge, & Stevens, 2006).

Calcium also plays a vital role in short- and long-term events triggered by synaptic transmission. Activation of these voltage-dependent calcium channels results in an increase in intracellular calcium, which in turn triggers the release of neurotransmitters, thereby setting up a cascade reaction including increased protein transcription (Bourinet et al., 2014). Calcium channels can be broadly classified into two types: low-voltage activated and high-voltage activated (HVA). The HVA can be further divided into L, N, P, Q, and R types. N, P, and Q types of calcium channels are primarily located in the presynaptic neurons and are involved in the release of neurotransmitters, including glutamate, calcitonin gene-related peptide, and substance P. Inhibiting these channels will result in decreased neurotransmission and, hence, analgesia. L-type channels are found in the muscles and heart as well as on the cell bodies of neurons, where they mediate activation of Ca-dependent enzymes and gene expression (Bourinet et al., 2014).

Direct neuronal signaling to other neurons in the pain pathway is not the only mechanism of persistent pain. Nonneuronal, immune cells, including microglia and astrocytes, are stimulated in the presence of inflammatory mediators, including protons, prostaglandins, substance P, and histamine, within the spinal cord and brain, or by injury to peripheral nerves, and contribute to disinhibitory mechanisms. They undergo structural and proliferative changes and begin the transcription of various genes, including the expression of purinergic cell surface receptors, ionotropic non-NMDA and NMDA receptors, as well as metabotropic glutamate receptors. Upon activation of astrocytes, the mitogen-activated protein kinase (MAPK)-1 and MAPK-8 signaling pathways come into play, and these increase the synthesis of interleukin 1β (IL-1β), IL-6, tumor necrosis factor-α, prostaglandin E_2, and nitric oxide, which consequently augment glutamate transporters (thereby increasing excitatory synaptic transmission) and further mediate astrocyte–astrocyte activation (Scholz & Woolf, 2007).

Another emerging biomarker associated with nerve injury and neuroinflammation is the σ1 receptor (S1R). Initially believed to be a subtype of opioid receptor, S1R modulates several ion channels. S1R ligands can potentiate the opioid analgesic effect. Studies are now examining its role in the mediation of chronic pain. It has been hypothesized that peripheral nerve injury induces several changes, one of which is the depletion of calcium, which subsequently activates S1R expression, leading to an increase in neuropathic pain. S1R is highly expressed in the central and peripheral nervous system and is now being targeted for the treatment of chronic pain (Zamanillo, Romero, Merlos, & Vela, 2013).

In addition to changes in concentration or expression of specific molecular targets, chronic pain produces nonspecific physiological changes, which are also exploited by molecular imaging methods. For instance, increased metabolic demand due to increased neuronal activity in pain centers of the brain may result in differentially increased blood flow and oxygen utilization, which can be mapped using certain MRI and PET techniques. Similarly, increased glucose utilization in neuronal tissue can be discerned using radiolabeled glucose.

Further exploitation of the processes involved in the generation of pathological pain through molecular imaging studies can identify targets for chronic pain management and can significantly alter its management in the years to come.

FUNCTIONAL AND MOLECULAR IMAGING OF THE BRAIN

Pain is a subjective experience and brain activity can be detected via imaging techniques in response to stimulation of cutaneous nerves. Studies have revealed activation of certain brain areas in response to thermal, mechanical, and chemical stimuli and they have been collectively termed as the "pain matrix." A 2000 review identified the roles of the anterior cingulate cortex, the primary and secondary somatosensory cortices, the insular area, the prefrontal cortex, and the caudate nucleus in pain processing (Peyron, Laurent, & García-Larrea, 2000) (Figure 1).

Functional and molecular neuroimaging are essential tools in revealing brain structures that are activated in pain. Probes that are able to detect the biological activity of enzymes, transporters, receptors, and other relevant target proteins are increasingly shaping the future of pain imaging (May, 2007). Imaging has fast become an essential tool in pain research (Apkarian et al., 2011). Various imaging modalities have been employed, which detect functional, chemical, and anatomical changes in the human brain in response to pain.

Early imaging attempts at studying the effects of pain in the human brain involved investigating hemodynamic effects using PET (Hsieh, Belfrage, Stone-Elander, Hansson, & Ingvar, 1995; Iadarola et al., 1995). PET requires the incorporation of radioactive atoms in relevant molecules and subsequently observing the emitted radioactivity using a network of spatially arranged detectors around the subject. These early studies primarily looked at

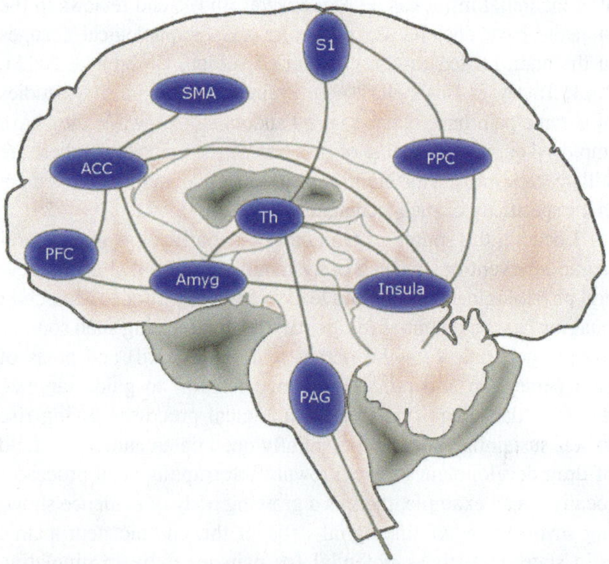

FIGURE 1 The "pain matrix." Imaging studies have identified a group of interconnected cortical and subcortical centers as consistently involved in chronic pain. These are the thalamus (Th), the amygdala (Amyg), the insular cortex (Insula), the supplementary motor area (SMA), the posterior parietal cortex (PPC), the prefrontal cortex (PFC), the cingulate cortex (ACC), the periaqueductal gray (PAG), the basal ganglia and cerebellar cortex (not shown), and the primary (S1) and secondary (not shown) sensory cortex (May, 2007).

regional cerebral blood flow changes in relation to pain. More recent studies using PET have begun to investigate the effects of treatment on radiotracer uptake and distribution in the brain (Brefel-Courbon et al., 2005).

However, PET techniques suffered from the limitations of low-resolution images and some amount of radiation dose, however small, to the patient. Also, blood flow measurements using PET required short-lived radioactive tracers that could be produced only in a cyclotron. Therefore the utility of such a technique is limited by the availability of such radiotracers. The possibility of using a variety of compounds to investigate specific pain-related biomarkers using PET still remains attractive. In a particularly interesting study PET was used to reveal opioid abuse potential (Love, Stohler, & Zubieta, 2009). In addition, the availability of PET tracers for dopamine, serotonin, cholinergic, GABA, adenosine, and opioid systems holds the potential for mapping these receptors under different pain conditions in the near future (Heiss & Herholz, 2006).

MRI-based techniques, owing to their better resolution and nonradioactive nature, have since taken the center stage of investigating pain processing in the brain (Figure 2).

fMRI is the most commonly used imaging technique, and it assesses human brain cortical function by discerning changes in blood flow. Referred to as blood oxygen level-dependent (BOLD) imaging, this technique uses magnetic signal changes from the paramagnetic deoxyhemoglobin in the blood. Regional variations between local concentrations of deoxyhemoglobin are dependent on blood flow and oxygen usage and reflect neural activity in the brain (Ogawa, Lee, Kay, & Tank, 1990). During inactivity, certain brain areas show higher BOLD signal than the rest of the brain. This consistently identifiable pattern of BOLD mapping of the brain is also known as the resting-state network (RSN) (Raichle & Snyder, 2007). The RSN is disrupted in several conditions, including chronic pain (Baliki, Geha, Apkarian, & Chialvo, 2008). fMRI is increasingly being used to understand brain processing of chronic pain and effects of analgesia on brain function (Borsook & Becerra, 2006).

Brain cortical thickness can be measured using an MRI technique known as voxel-based morphometry (Ashburner & Friston, 2000). Chronic pain affects neuronal density and connections within the brain cortex (Metz, Yau, Centeno, Apkarian, & Martina, 2009). This effect is seen on MRI scans as changes in cortical thickness in specific brain regions. For instance, patients suffering from chronic back pain show reduced cortical thickness in the bilateral dorsolateral prefrontal cortex and right anterior thalamus (Apkarian et al., 2004), while those with fibromyalgia show a reduction in the cingulate cortex, medial prefrontal cortex, parahippocampal gyrus, and insula (Robinson, Craggs, Price, Perlstein, & Staud, 2011).

Diffusion tensor imaging (DTI) is yet another MRI technique that measures the integrity of white matter tracts based on microstructural changes in water diffusion along these tracts (Alexander, Lee, Lazar, & Field, 2007). Measured as functional anisotropy, DTI can infer damage to normal neuronal fiber pathways. The use of DTI has been explored in an attempt to predict efficacy and guide surgical planning for deep brain stimulation for treating chronic pain (Owen et al., 2008). DTI studies in patients with migraine have suggested involvement of the brainstem in the initiation of an attack of migraine (Kara et al., 2013). In the healthy person, DTI has confirmed the functional neuronal tract connection between cortical and subcortical structures involved in nociception (Hadjipavlou, Dunckley, Behrens, & Tracey, 2006).

Magnetic resonance spectroscopy (MRS) is a noninvasive imaging technique that can study alterations in the chemical milieu within the brain, including neurotransmitter concentration and metabolites, thus revealing potential chemical processes involved in a pathological condition (Bertholdo, Watcharakorn, & Castillo, 2013). *N*-Acetylaspartate levels measured with MRS were found to correlate with clinical scores in patients with familial migraine (Dichgans, Herzog, Freilinger, Wilke, & Auer, 2005). A preliminary study reports that biochemical alterations in the prefrontal cortex, anterior cingulate cortex, and thalamus, identified with

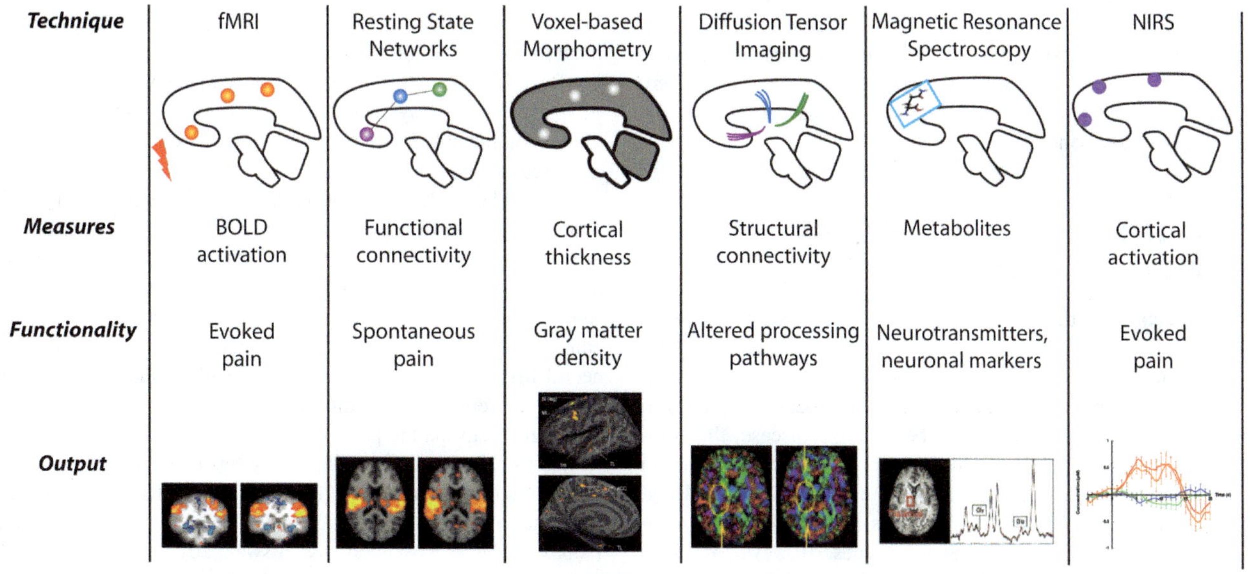

Technique	fMRI	Resting State Networks	Voxel-based Morphometry	Diffusion Tensor Imaging	Magnetic Resonance Spectroscopy	NIRS
Measures	BOLD activation	Functional connectivity	Cortical thickness	Structural connectivity	Metabolites	Cortical activation
Functionality	Evoked pain	Spontaneous pain	Gray matter density	Altered processing pathways	Neurotransmitters, neuronal markers	Evoked pain
Output						

FIGURE 2 **Brain imaging methods in pain research.** A list of techniques used for brain imaging in pain research, including the functional parameter that is measured, the quantitative information that is derived, and an illustrative example of the output image, is presented (Sava et al., 2009).

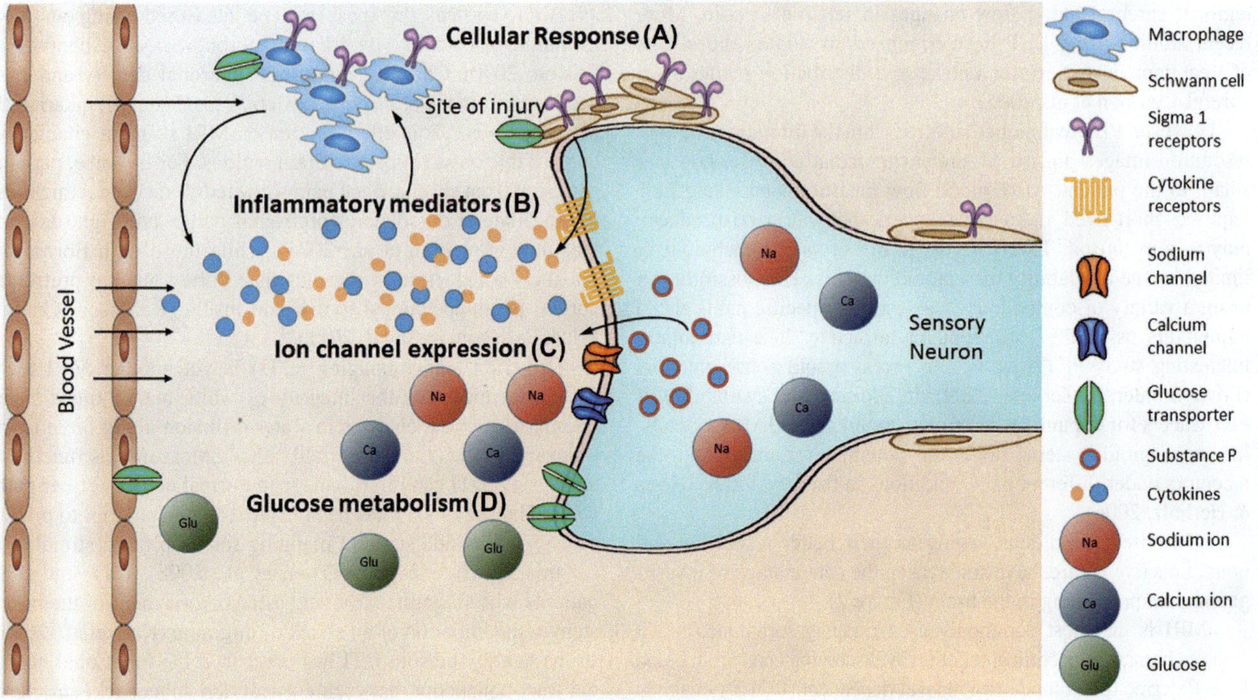

FIGURE 3 Molecular and cellular mechanisms of chronic pain visualized in imaging of the peripheral nervous system. Main mechanisms of peripheral neuropathic pain imaging. (A) Cellular response to neuropathy—Schwann cell and macrophage imaging. (B) Inflammatory mediator and receptor imaging. (C) Ion channel expression—sodium channel and calcium channel imaging. (D) Metabolic response—glucose utilization imaging (Tung et al., 2015).

MRS, could discriminate between patients with low back pain and healthy individuals (Siddall et al., 2006).

Cerebral near-infrared fluorescence spectroscopy (NIRS), similar to fMRI, detects hemodynamic and oxygenated hemoglobin alterations in relation to cortical processing of pain (Murkin & Arango, 2009; Owen-Reece, Smith, Elwell, & Goldstone, 1999). The basis of such imaging rests on the fact that stimulated and active areas of the brain receive increased blood flow and consume oxygen at an increased rate compared to the surrounding cortices. Few studies have used NIRS specifically for chronic pain. Painful stimulus evokes a distinct pattern of cortical activation as opposed to nonnoxious tactile stimulation (Becerra et al., 2008). A similar result was seen even in newborns (Bartocci, Bergqvist, Lagercrantz, & Anand, 2006).

FUNCTIONAL AND MOLECULAR IMAGING OF PERIPHERAL STRUCTURES

As stated earlier, part of the problem in finding a consistent therapeutic response for chronic pain is the inability of current technologies to identify the source of pain generation. While injury, surgery, or tissue damage might initiate the process, as pain progresses from being a symptom to becoming a disease, the chronicity of pain shifts to molecular and physiological changes in neuronal and associated cellular structures themselves (Tung, Behera, & Biswal, 2015). Such changes are impossible to identify via gross anatomic imaging techniques. However, imaging modalities targeted at identifying these deranged molecular and functional changes in neural structures may be able to identify the source of pain generation (Figure 3).

As of this writing, molecular imaging studies on peripheral structures in relation to chronic pain are rare in humans. One case report demonstrates increased radiolabeled glucose uptake on PET scan in the spinal cord and sciatic nerves ascribed to neuropathy (Cheng, Huang, & Zhuang, 2009). Another study reports that a radiolabeled fluoride PET scan is able to identify potential sources of pain in the bones of the foot, where conventional imaging failed. Further, the results of these PET scans affected clinical management in almost half of the patients studied (Fischer et al., 2010).

Despite the dearth of clinical studies, molecular imaging of the peripheral nervous system shows promising new targets for pain imaging in the preclinical imaging field. Some of these techniques are discussed here.

Fluorodeoxyglucose (FDG) is a metabolic marker used in PET imaging. Popular in oncology imaging, FDG follows glucose as it diffuses into the cell and becomes metabolically trapped within the glycolytic cycle. It is, thus, an excellent marker to identify tissues with increased metabolic demands. In patients with chronic pain, reduced nociceptive threshold causes frequent or continuous neural firing, which in turn is metabolically demanding. In addition, neurogenic inflammation is often an accompanying feature of chronic neuropathic pain, further increasing metabolic activity related to pain generation. Exploiting this biochemical process in a rat model of neuropathic pain, FDG PET/MRI hybrid imaging was able to identify the metabolically hyperactive nerve (Behera, Jacobs, Behera, Rosenberg, & Biswal, 2011). The same technique also demonstrated that long-term opioid administration in

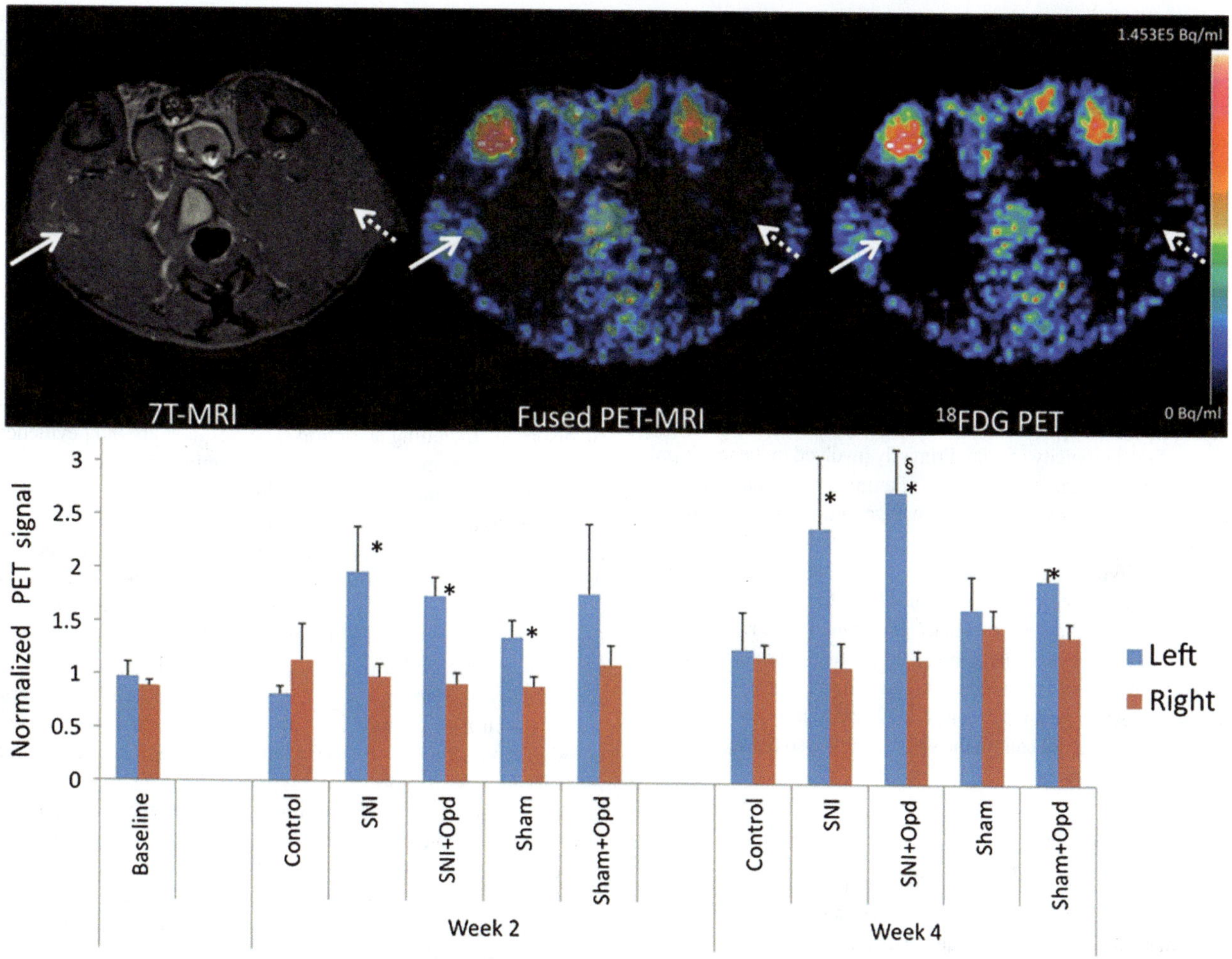

FIGURE 4 **PET/MRI of opioid-induced hyperalgesia.** Top: MRI, PET MRI, and PET images of a representative opioid-treated SNI rat, showing increased ^{18}FDG uptake in the injured nerve (solid arrows) but no uptake in the uninjured nerves (broken arrows) at 4 weeks. Bottom: Normalized ^{18}FDG PET signal in left and right sciatic nerves at baseline (presurgery) and 2 and 4 weeks after surgery. Progressively increasing PET signal is seen in the opioid-treated SNI group. (*$p < 0.05$ for PET signal in left nerve vs right nerve; §$p < 0.03$ for PET signal in left nerve at 4 weeks vs 2 weeks.) (Behera, Behera et al., 2011).

rats caused opioid-induced hyperalgesia, which was reflected in increasing signal from PET scans over time (Behera, Behera, & Biswal, 2011) (Figure 4).

Peripheral neuropathic pain is often the result of nerve damage. Identifying such sites of neural injury and repair is another strategy to locate pain generators. The role of Schwann cells in the healing response to nerve injury is widely acknowledged. Also, S1Rs, once believed to be a type of opioid receptor, are abundantly expressed in Schwann cells. These receptors are also implicated in chronic pain development in the central nervous system (Diaz et al., 2009). Using S1Rs as a target for molecular imaging probes can help identify areas of Schwann cell proliferation, which in turn can indicate sites of neural damage and accompanying repair process and inflammation. ^{18}F-FTC-146 is an S1R PET imaging agent, which has shown promise in identifying neural damage in rodents (Behera et al., 2012). Given the fact that S1R ligands are actively being investigated as potential analgesics, such a tool for noninvasive imaging of S1R density in the affected nerves of a patient with chronic pain could potentially be used to select patients who may benefit from such interventions.

Ion channels are also potential target mechanisms to use in molecular imaging of neural function, as ionic movement across neural membranes is responsible for action potential generation, maintenance of resting membrane potential, and signal transmission and propagation along the nerve pathway.

Studies of inflammatory and neuropathic pain pathology have noted increased levels of voltage-gated sodium ion channels (NaVs) in sensory and DRG neurons. In addition, activated macrophages and epidermal keratinocytes, which are implicated in the augmentation and perpetuation of chronic nociceptive activity, as well as intact afferents that neighbor the injured nerve also display amplified levels of NaV expression. Thus, the upregulation of NaVs is believed to play a major role in chronic and neuropathic pain symptomology (Black, Nikolajsen, Kroner, Jensen, & Waxman, 2008; Pertin et al., 2005). Guanidinium toxins, namely, saxitoxin (STX) and gonyautoxin III, bind to NaVs with high affinity and specificity. A radiolabeled STX derivative was used to image NaV expression in murine sciatic nerves, related to increased neural excitability and chronic pain (Hoehne et al., 2013). This mechanism is of particular interest as NaV blockade is one of the key mechanisms of action of some commonly used anesthetics, such as lidocaine. Similar to S1R ligands, blockade of pain-specific subtypes of NaV may herald a new generation of analgesics.

Among ion channels, calcium channels are also known to be upregulated in chronic pain. Manganese is a surrogate calcium flux marker, as it uses these channels to enter cells. Owing to its paramagnetic properties, manganese can be used as an MRI contrast agent. Hyperactive neurons need high calcium exchange across their membranes to generate action potentials. Therefore, they also accumulate manganese, when administered systemically, and can be highlighted. In a sciatic injury model, manganese-enhanced MRI highlighted the lumbar plexus and sciatic nerves. The MRI signals were correlated with pain behavior and neural manganese content measure in inductively coupled plasma spectrometry (Jacobs et al., 2012).

Macrophages and microglia are intimately associated with development of the chronicity of pain. Primarily involved in the repair process of nerve damage, they release inflammatory mediators, which add to the pathogenesis of chronic pain. Pinpointing areas of increased density of these cells in the vicinity of neural structures can be another way of identifying sources of chronic pain. These phagocytic cells engulf iron oxide nanoparticles introduced into the bloodstream. Iron, being a superparamagnetic substance, can thus be used to track the migration of phagocytic cells. Using MRI, a study was able to image accumulation of these cells around the site of nerve injury. More interestingly, minocycline, a macrophage inhibitor, significantly reduced the recruitment of these cells to the site of injury, also correlating with pain relief (Ghanouni et al., 2012).

CONCLUSION

Molecular imaging techniques indicate that chronic pain should be dealt with as a disease entity, rather than a symptom. Treatment approaches may have to be modified to include a holistic approach—instead of relying solely on symptomatic or sensory therapies, emotional and neurodegenerative aspects may have to be considered. Neuroprotective agents, for instance, could prevent or revert cerebral neuronal loss associated with chronic pain. In addition, high-risk populations, such as diabetics and postoperative patients, may benefit from prophylactic treatment for pain (Borsook, Moulton, Schmidt, & Becerra, 2007; Pergolizzi et al., 2013).

As these techniques evolve, they give us various new tools for drug development. They provide objective measures of pain unaffected by internal or external influences, of which currently there are none. Such measures can help in evaluating new drug development, as well as monitoring response to treatments.

Postmarketing monitoring may be performed for incremental long-term effects in brain function in the pain matrix, improving results of drug trials. Given the variability in the pain phenotype, such tools can be used for proper patient selection for a particular type of therapy. Indeed, some of these imaging agents are already similar to therapeutic agents, both current and new. Ion channel blockers, S1R antagonists, and macrophage inhibitors are all being explored as new therapeutic agents for chronic pain, potentially devoid of addiction potential.

Being able to identify pain generation at the source can potentially allow us to develop improved ways of drug delivery to these local sources, reducing the systemic side effects and potential for addiction and abuse.

To conclude, molecular and functional imaging techniques redefine our understanding of chronic pain as well as providing us with tools to better manage this expensive disease. Improving comprehension, new methods of targeted treatment, and novel holistic treatment strategies should reduce our dependence on opioids and other habit-forming drugs as the mainstay of pain therapy.

APPLICATIONS TO OTHER ADDICTIONS AND SUBSTANCE MISUSE

As discussed in this chapter, molecular imaging helps in identifying altered molecular and functional changes in neural structures. These imaging modalities are being increasingly used to study the neurobiological changes in the living neuronal tissue in psychiatric disorders, including addiction. There is now growing evidence that suggests the involvement of the dopaminergic system in drug addiction. Dopamine receptors and dopaminergic release is greatly attenuated in areas of the striatum in cocaine, methamphetamine, heroin, and alcohol dependence. Research pertaining to cocaine dependency is also in progress and, while still in its rudimentary stages, offers novel approaches to its management. Further improvements in molecular imaging techniques may also furnish insight into which substances may be abused and which may not. Additionally, this emerging technology can ascertain biomarkers/alterations in brain function, which may be associated with progression from drug use to abuse, and also objectively assess the recovery process.

Molecular and functional imaging techniques introduce a novel tool into the drug development process. As research into the area expands, molecular imaging biomarkers for various substance abuse drugs will hold an immense value to the industry as well as public health. These technologies sound promising and will contribute significantly toward clinical management of substance abuse disorders in the near future.

DEFINITION OF TERMS

Molecular imaging This is the visualization of molecular or functional derangements in pathological tissues using in vivo imaging of tagged molecules or cells.

Positron emission tomography This is a molecular imaging modality in which radioisotope-tagged molecules are injected into the body and their distribution is visualized. The radioisotopes decay by emitting positrons.

Pain matrix This refers to a group of interconnected regions in the brain that have been consistently shown by molecular imaging of the brain to be involved in pain processing.

Functional MRI This is a type of MRI that observes functional blood flow changes associated with brain activity in discrete regions.

Blood oxygen level-dependent MRI This is an fMRI technique that exploits the difference in magnetic properties of oxygenated and deoxygenated hemoglobin to measure relative oxygen utilization in various parts of the brain, thus indicating neural activity level.

Voxel-based morphometry This is an MRI-based technique that measures volume differences in parts of the brain by automated comparison of three-dimensional image units called voxels of the test brain with a standard template.

Diffusion tensor imaging This is the mapping of the diffusion process of water molecules using MRI. Linear diffusion is seen along nerve axons and nerve fibers. Detection of random diffusion (or anisotropy) is associated with microstructural axonal damage.

Magnetic resonance spectroscopy This studies alterations in the chemical profile of the tissue being imaged, including metabolite concentration, thus revealing abnormal chemical processes signifying a pathological condition.

Near-infrared fluorescence spectroscopy This is a high-resolution molecular imaging technique that uses fluorescent molecular probes that, when excited, emit light of near-infrared wavelength.

KEY FACTS OF MOLECULAR IMAGING

- Molecular imaging is a type of medical imaging that emanates from radiopharmacology.
- It combines molecular biology with in vivo imaging and allows for a noninvasive, spatial visualization of abnormalities in cellular function and/or the molecular processes in a pathological state.
- Molecular imaging utilizes biomarkers that help to observe functional changes in particular targets or pathways, as opposed to anatomical changes observed in traditional clinical imaging.
- Biomarkers are probes that interact with their surroundings within cells and tissues in vivo, thereby altering the image according to molecular/functional changes occurring within the area of interest.
- Molecular imaging complements traditional medical imaging by helping in earlier detection of disease, diagnosis of diseases that do not manifest with structural changes, stratification of patients or lesions for more accurate personalized therapy, and follow-up after treatment and as a preclinical in vivo visualization tool for drug discovery and development.

SUMMARY POINTS

- This chapter introduces the concept of molecular imaging in chronic pain and discusses its potential role in reducing analgesic misuse.
- Analgesic misuse is a major and expensive health problem.
- Inability to identify chronic pain generators may contribute to analgesic misuse.
- Maladaptive cellular and molecular changes in the nervous system, which may not be associated with any structural or anatomic defect, give rise to chronicity of pain.
- Current clinical imaging modalities lack specificity and sensitivity to identify causes of chronic pain because they rely on discovering pain-causing anatomic abnormalities.
- Molecular imaging uses molecular and functional changes in the nervous system to detect and characterize pain.
- The ability to "see" the pain process opens up avenues for new targeted treatment methods and allows development of novel drugs with low addiction potential.

REFERENCES

Alexander, A. L., Lee, J. E., Lazar, M., & Field, A. S. (2007). Diffusion tensor imaging of the brain. *Neurotherapeutics, 4*(3), 316–329.

Apkarian, A. V., Hashmi, J. A., & Baliki, M. N. (2011). Pain and the brain: specificity and plasticity of the brain in clinical chronic pain. *Pain, 152*(3), S49–S64.

Apkarian, A. V., Sosa, Y., Sonty, S., Levy, R. M., Harden, R. N., Parrish, T. B., & Gitelman, D. R. (2004). Chronic back pain is associated with decreased prefrontal and thalamic gray matter density. *The Journal of Neuroscience, 24*(46), 10410–10415.

Ashburner, J., & Friston, K. J. (2000). Voxel-based morphometry—the methods. *NeuroImage, 11*(6 Pt 1), 805–821.

Atluri, S., Sudarshan, G., & Manchikanti, L. (2014). Assessment of the trends in medical use and misuse of opioid analgesics from 2004 to 2011. *Pain Physician, 17*(2), E119–E128. Retrieved from http://www.ncbi.nlm.nih.gov/pubmed/24658483.

Baliki, M. N., Geha, P. Y., Apkarian, A. V., & Chialvo, D. R. (2008). Beyond feeling: chronic pain hurts the brain, disrupting the default-mode network dynamics. *The Journal of Neuroscience, 28*(6), 1398–1403.

Bartocci, M., Bergqvist, L. L., Lagercrantz, H., & Anand, K. J. S. (2006). Pain activates cortical areas in the preterm newborn brain. *Pain, 122*(1–2), 109–117.

Becerra, L., Harris, W., Joseph, D., Huppert, T., Boas, D. A., & Borsook, D. (2008). Diffuse optical tomography of pain and tactile stimulation: activation in cortical sensory and emotional systems. *NeuroImage, 41*(2), 252–259.

Behera, D., Behera, S., & Biswal, S. (2011). Metabolic peripheral nerve-related changes observed in opioid-induced hyperalgesia (OIH) can be detected with 18F-FDG PET-MRI. In *World molecular imaging congress* (p. P291). San Diego: World Molecular Imaging Society.

Behera, D., Jacobs, K. E., Behera, S., Rosenberg, J., & Biswal, S. (2011). (18)F-FDG PET/MRI can be used to identify injured peripheral nerves in a model of neuropathic pain. *Journal of Nuclear Medicine, 52*(8), 1308–1312.

Behera, D., Shen, B., James, M. L., Borgohain, P., Patankar, M., Andrews, L., ... Chin, F. T. (2012). Radiolabeled sigma-1 receptor ligand detects peripheral neuroinflammation in a neuropathic pain model using PET-MRI. In *World molecular imaging congress* (p. SS123). Dublin, Ireland: World Molecular Imaging Society.

Bekkering, G. E., Bala, M. M., Reid, K., Kellen, E., Harker, J., Riemsma, R., ... Kleijnen, J. (2011). Epidemiology of chronic pain and its treatment in The Netherlands. *The Netherlands Journal of Medicine, 69*(3), 141–153. Retrieved from http://www.ncbi.nlm.nih.gov/pubmed/21444943.

Bertholdo, D., Watcharakorn, A., & Castillo, M. (2013). Brain proton magnetic resonance spectroscopy. *Neuroimaging Clinics, 23*(3), 359–380.

Black, J. A., Nikolajsen, L., Kroner, K., Jensen, T. S., & Waxman, S. G. (2008). Multiple sodium channel isoforms and mitogen-activated protein kinases are present in painful human neuromas. *Annals of Neurology, 64*(6), 644–653.

Borsook, D., & Becerra, L. R. (2006). Breaking down the barriers: fMRI applications in pain, analgesia and analgesics. *Molecular Pain, 2*, 30.

Borsook, D., Moulton, E. A., Schmidt, K. F., & Becerra, L. R. (2007). Neuroimaging revolutionizes therapeutic approaches to chronic pain. *Molecular Pain, 3*, 25.

Bourinet, E., Altier, C., Hildebrand, M. E., Trang, T., Salter, M. W., & Zamponi, G. W. (2014). Calcium-permeable ion channels in pain signaling. *Physiological Reviews, 94*(1), 81–140.

Brefel-Courbon, C., Payoux, P., Thalamas, C., Ory, F., Quelven, I., Chollet, F., ... Rascol, O. (2005). Effect of levodopa on pain threshold in Parkinson's disease: a clinical and positron emission tomography study. *Movement Disorders, 20*(12), 1557–1563.

Chaban, V. V. (2010). Peripheral sensitization of sensory neurons. *Ethnicity and disease, 20*(1), S1–S3.

Cheng, G., Huang, S., & Zhuang, H. (2009). Elevated FDG activity in the spinal cord and the sciatic nerves due to neuropathy. *Clinical Nuclear Medicine, 34*(12), 950–951.

Committee on Advancing Pain Research Care and Education. (2011). *Relieving pain in America: A blueprint for transforming prevention, care, education, and research.* Washington, DC: Institute of Medicine.

Diaz, J. L., Zamanillo, D., Corbera, J., Baeyens, J. M., Maldonado, R., Pericas, M. A., ... Torrens, A. (2009). Selective sigma-1 (sigma1) receptor antagonists: emerging target for the treatment of neuropathic pain. *Central Nervous System Agents In Medicinal Chemistry, 9*(3), 172–183.

Dichgans, M., Herzog, J., Freilinger, T., Wilke, M., & Auer, D. P. (2005). 1H-MRS alterations in the cerebellum of patients with familial hemiplegic migraine type 1. *Neurology, 64*(4), 608–613.

Fischer, D. R., Maquieira, G. J., Espinosa, N., Zanetti, M., Hesselmann, R., Johayem, A., ... Strobel, K. (2010). Therapeutic impact of [18F]fluoride positron-emission tomography/computed tomography on patients with unclear foot pain. *Skeletal Radiology, 39*(10), 987–997.

Flynn, T. W., Smith, B., & Chou, R. (2011). Appropriate use of diagnostic imaging in low back pain: a reminder that unnecessary imaging may do as much harm as good. *The Journal of Orthopaedic and Sports Physical Therapy, 41*(11), 838–846.

Garland, E. L., Froeliger, B., Zeidan, F., Partin, K., & Howard, M. O. (2013). The downward spiral of chronic pain, prescription opioid misuse, and addiction: cognitive, affective, and neuropsychopharmacologic pathways. *Neuroscience and Biobehavioral Reviews, 37*(10 Pt 2), 2597–2607. http://dx.doi.org/10.1016/j.neubiorev.2013.08.006.

Ghanouni, P., Behera, D., Xie, J., Chen, X., Moseley, M., & Biswal, S. (2012). In vivo USPIO magnetic resonance imaging shows that minocycline mitigates macrophage recruitment to a peripheral nerve injury. *Molecular Pain, 8*(1), 49.

Gosselin, R.-D., Suter, M. R., Ji, R.-R., & Decosterd, I. (2010). Glial cells and chronic pain. *The Neuroscientist, 16*(5), 519–531.

Hadjipavlou, G., Dunckley, P., Behrens, T. E., & Tracey, I. (2006). Determining anatomical connectivities between cortical and brainstem pain processing regions in humans: a diffusion tensor imaging study in healthy controls. *Pain, 123*(1–2), 169–178.

Hallinan, R., Osborn, M., Cohen, M., Dobbin, M., & Wodak, A. (2011). Increasing the benefits and reducing the harms of prescription opioid analgesics. *Drug and Alcohol Review, 30*(3), 315–323. http://dx.doi.org/10.1111/j.1465-3362.2011.00294.x.

Harker, J., Reid, K. J., Bekkering, G. E., Kellen, E., Bala, M. M., Riemsma, R., ... Kleijnen, J. (2012). Epidemiology of chronic pain in Denmark and Sweden. *Pain Research and Treatment.*

Heiss, W.-D., & Herholz, K. (2006). Brain receptor imaging. *Journal of Nuclear Medicine, 47*(2), 302–312.

Hoehne, A., Behera, D., Parsons, W. H., James, M. L., Shen, B., Borgohain, P., ... Du Bois, J. (2013). A (18)F-Labeled saxitoxin derivative for in vivo PET-MR imaging of voltage-gated sodium channel expression following nerve injury. *Journal of the American Chemical Society, 135*(48), 18012–18015.

Howe, C. Q., & Sullivan, M. D. (2014). The missing "P" in pain management: how the current opioid epidemic highlights the need for psychiatric services in chronic pain care. *General Hospital Psychiatry, 36*(1), 99–104. http://dx.doi.org/10.1016/j.genhosppsych.2013.10.003.

Hsieh, J. C., Belfrage, M., Stone-Elander, S., Hansson, P., & Ingvar, M. (1995). Central representation of chronic ongoing neuropathic pain studied by positron emission tomography. *Pain, 63*(2), 225–236.

Iadarola, M. J., Max, M. B., Berman, K. F., Byas-Smith, M. G., Coghill, R. C., Gracely, R. H., & Bennett, G. J. (1995). Unilateral decrease in thalamic activity observed with positron emission tomography in patients with chronic neuropathic pain. *Pain, 63*(1), 55–64.

Jacobs, K. E., Behera, D., Rosenberg, J., Gold, G., Moseley, M., Yeomans, D., & Biswal, S. (2012). Oral manganese as an MRI contrast agent for the detection of nociceptive activity. *NMR in Biomedicine, 25*(4), 563–569.

Jensen, M. C., Brant-Zawadzki, M. N., Obuchowski, N., Modic, M. T., Malkasian, D., & Ross, J. S. (1994). Magnetic resonance imaging of the lumbar spine in people without back pain. *The New England Journal of Medicine, 331*(2), 69–73.

Kara, B., Kiyat Atamer, A., Onat, L., Ulusoy, L., Mutlu, A., & Sirvanci, M. (2013). DTI findings during spontaneous migraine attacks. *Clinical Neuroradiology, 23*(1), 31–36.

Kidd, B. L., & Urban, L. A. (2001). Mechanisms of inflammatory pain. *British Journal of Anaesthesia, 87*(1), 3–11.

Ling, W., Mooney, L., & Hillhouse, M. (2011). Prescription opioid abuse, pain and addiction: clinical issues and implications. *Drug and Alcohol Review, 30*(3), 300–305. http://dx.doi.org/10.1111/j.1465-3362.2010.00271.x.

Love, T. M., Stohler, C. S., & Zubieta, J.-K. (2009). Positron emission tomography measures of endogenous opioid neurotransmission and impulsiveness traits in humans. *Archives of General Psychiatry, 66*(10), 1124–1134.

Manchikanti, L., Fellows, B., Ailinani, H., & Pampati, V. (2010). Therapeutic use, abuse, and nonmedical use of opioids: a ten-year perspective. *Pain Physician, 13*(5), 401–435. Retrieved from http://www.ncbi.nlm.nih.gov/pubmed/20859312.

May, A. (2007). Neuroimaging: visualising the brain in pain. *Neurological Sciences, 28*(Suppl. 2), S101–S107.

Metz, A. E., Yau, H.-J., Centeno, M. V., Apkarian, A. V., & Martina, M. (2009). Morphological and functional reorganization of rat medial prefrontal cortex in neuropathic pain. *Proceedings of the National Academy of Sciences of the United States of America, 106*(7), 2423–2428.

Murkin, J. M., & Arango, M. (2009). Near-infrared spectroscopy as an index of brain and tissue oxygenation. *British Journal of Anaesthesia, 103*(Suppl. 1), i3–i13.

Ogawa, S., Lee, T., Kay, A., & Tank, D. (1990). Brain magnetic resonance imaging with contrast dependent on blood oxygenation. *Proceedings of the National Academy of Sciences, 87*(24), 9868–9872.

Owen, S. L., Heath, J., Kringelbach, M., Green, A. L., Pereira, E. A., Jenkinson, N., ... Aziz, T. Z. (2008). Pre-operative DTI and probabilisitic tractography in four patients with deep brain stimulation for chronic pain. *Journal of Clinical Neuroscience, 15*(7), 801–805.

Owen-Reece, H., Smith, M., Elwell, C. E., & Goldstone, J. C. (1999). Near infrared spectroscopy. *British Journal of Anaesthesia, 82*(3), 418–426.

Pergolizzi, J., Ahlbeck, K., Aldington, D., Alon, E., Coluzzi, F., Dahan, A., ... Varrassi, G. (2013). The development of chronic pain: physiological CHANGE necessitates a multidisciplinary approach to treatment. *Current Medical Research and Opinion, 29*(9), 1127–1135.

Pertin, M., Ji, R.-R., Berta, T., Powell, A. J., Karchewski, L., Tate, S. N., ... Decosterd, I. (2005). Upregulation of the voltage-gated sodium channel beta2 subunit in neuropathic pain models: characterization of expression in injured and non-injured primary sensory neurons. *The Journal of Neuroscience, 25*(47), 10970–10980.

Peyron, R., Laurent, B., & García-Larrea, L. (2000). Functional imaging of brain responses to pain. A review and meta-analysis. *Neurophysiologie Clinique (Clinical Neurophysiology), 30*(5), 263–288.

Raichle, M. E., & Snyder, A. Z. (2007). A default mode of brain function: a brief history of an evolving idea. *NeuroImage, 37*(4), 1083–1090.

Robinson, M. E., Craggs, J. G., Price, D. D., Perlstein, W. M., & Staud, R. (2011). Gray matter volumes of pain-related brain areas are decreased in fibromyalgia syndrome. *The Journal of Pain, 12*(4), 436–443.

Rogers, M., Tang, L., Madge, D. J., & Stevens, E. B. (2006). The role of sodium channels in neuropathic pain. *Seminars in Cell Developmental Biology, 17*(5), 571–581.

Ruetsch, C. (2010). Empirical view of opioid dependence. *Journal of Managed Care Pharmacy: JMCP, 16*(1 Suppl. B), S9–S13. Retrieved from http://www.ncbi.nlm.nih.gov/pubmed/20146549.

Sava, S., Lebel, A. A., Leslie, D. S., Drosos, A., Berde, C., Becerra, L., & Borsook, D. (2009). Challenges of functional imaging research of pain in children. *Molecular Pain, 5*, 30.

Scholz, J., & Woolf, C. J. (2007). The neuropathic pain triad: neurons, immune cells and glia. *Nature Neuroscience, 10*(11), 1361–1368.

Sher, J. S., Uribe, J. W., Posada, A., Murphy, B. J., & Zlatkin, M. B. (1995). Abnormal findings on magnetic resonance images of asymptomatic shoulders. *The Journal of Bone and Joint Surgery, 77*(1), 10–15.

Siddall, P. J., Stanwell, P., Woodhouse, A., Somorjai, R. L., Dolenko, B., Nikulin, A., ... Mountford, C. E. (2006). Magnetic resonance spectroscopy detects biochemical changes in the brain associated with chronic low back pain: a preliminary report. *Anesthesia and Analgesia, 102*(4), 1164–1168.

Tracey, I., & Bushnell, M. C. (2009). How neuroimaging studies have challenged us to rethink: is chronic pain a disease? *The Journal of Pain, 10*(11), 1113–1120.

Tung, K.-W., Behera, D., & Biswal, S. (2015). Neuropathic pain: mechanisms and treatments. *Seminars in Musculoskeletal Radiology, 19*, 103–111.

Woolf, C. J. (2011). Central sensitization: Implications for the diagnosis and treatment of pain. *Pain, 152*(Suppl. 3), S2–S15.

Zamanillo, D., Romero, L., Merlos, M., & Vela, J. M. (2013). Sigma 1 receptor: a new therapeutic target for pain. *European Journal of Pharmacology, 716*(1–3), 78–93.

Zhang, Z., Tao, W., Hou, Y.-Y., Wang, W., Lu, Y.-G., & Pan, Z. Z. (2014). Persistent pain facilitates response to morphine reward by downregulation of central amygdala GABAergic function. *Neuropsychopharmacology: Official Publication of the American College of Neuropsychopharmacology, 39*(9), 2263–2271. http://dx.doi.org/10.1038/npp.2014.77.

Zimmermann, M. (2001). Pathobiology of neuropathic pain. *European Journal of Pharmacology, 429*(1–3), 23–37.

Chapter 90

Single-Photon-Emission Computed Tomography Studies with Dopamine and Serotonin Transporters in Opioid Users

Shih-Hsien Lin, Kao Chin Chen, Yen Kuang Yang
Department of Psychiatry, National Cheng Kung University Hospital, College of Medicine, National Cheng Kung University, Tainan, Taiwan

Abbreviations

DAT Dopamine transporter
DRD2 Dopamine receptor D2
DRD4 Dopamine receptor D4
HIV Human immunodeficiency virus
PET Positron emission tomography
SERT Serotonin transporter
SPECT Single-photon-emission computed tomography
SSRI Selective serotonin reuptake inhibitor

INTRODUCTION

Addiction to opioids, such as morphine and heroin, is a global problem. Harm associated with illegal opioid use includes contraction of the human immunodeficiency virus (HIV), a poor quality of life, and a huge economic cost (Lin, Chen, Lee, Hsiao, et al., 2013; Mark, Woody, Juday, & Kleber, 2001). Several interacting biological mechanisms may be involved in opioid dependence. Monoamine neurotransmitters could play pivotal roles in opioid dependence, as well as other addiction behaviors (Howell & Kimmel, 2008; Kish et al., 2001). There are at least three important monoamine neurotransmitters: dopamine, serotonin, and noradrenaline.

A very large body of research has revealed the mechanism of the dopaminergic system in human behavior (Previc, 1999, 2009). Altered dopaminergic function was found to be associated with schizophrenia (Carlsson, 1988), as well as other psychiatric disorders, such as attention deficit hyperactivity disorder (Volkow, Wang, et al., 2009; Volkow et al., 2007). At least four major dopamine pathways exist in the central nervous system of humans: (1) the mesolimbic dopamine pathway, (2) the mesocortical dopamine pathway, (3) the nigrostriatal dopamine pathway, and (4) the tuberoinfundibular pathway (Stahl, 2000). The mesolimbic dopamine pathway, which is associated with the reward process, was found to be the common pathway for the reinforcing effect of drugs of abuse (Pierce & Kumaresan, 2006). In the past few decades, the role of dopamine in addiction has been well established (Goldstein & Volkow, 2002; Melis, Spiga, & Diana, 2005; Spanagel & Weiss, 1999; Volkow, Fowler, Wang, Baler, & Telang, 2009; Wise, 1996). Preclinical and clinical evidence also implies that opioid dependence, as an addiction behavior in nature, is also associated with the dopaminergic system (Jia, Wang, Liu, & Wu, 2005; Shi et al., 2008; Tjon et al., 1994; Xiao et al., 2006).

Meanwhile, it is also worthy of note that other monoamine neurotransmitters, such as serotonin, may also play a role (Gerra et al., 2005; Kish et al., 2001). Evidence from neuroimaging studies has indicated that the serotonergic system could be associated with alcoholism (Heinz et al., 1998), smoking (Staley et al., 2001), and cocaine dependence (Jacobsen et al., 2000). Although the association between serotonin and addiction remains to be elucidated, gaining an understanding of this mechanism may lead to valuable clinical implications, as the serotonergic system could be one of the most important sites for the treatment of mental disorders.

To understand the activities of neurotransmitters, positron emission tomography (PET) and single-photon-emission computed tomography (SPECT) can be used. Although the advantages of PET are well known, SPECT is considered more economical and more widely available than PET (Sharma & Ebadi, 2008). Several radioligands have been widely used for imaging of the receptors and transporters of dopamine and serotonin (Howes et al., 2007; Paterson, Kornum, Nutt, Pike, & Knudsen, 2013). In this chapter, we introduce the findings of associations between opioid dependence and dopamine and serotonin from SPECT studies.

DOPAMINE

Use of opioids may increase the dopaminergic activity, especially in the mesolimbic system, and when the activity of the dopaminergic system decreases, the addict will experience withdrawal, followed by craving (Volkow, Fowler, et al., 2009; Volkow, Fowler, Wang, & Swanson, 2004). This craving may induce bingeing. It is worth noting that this process, which is a dysregulation of reward, may result in allostasis of the hypodopamine activity (Koob & Le Moal, 2001). In other words, the dopamine function could be downwardly regulated.

Although the dopaminergic theory of opioid use is sound, evidence from neuroimaging studies is still being accumulated.

It is worth noting that the dopamine release induced by other drugs has been confirmed by SPECT studies (Voruganti, Slomka, Zabel, Mattar, & Awad, 2001). However, evidence supporting the induction of dopamine release by opioids remains to be found. In a SPECT study ($n=12$ abstinent opioid-dependent) with [^{123}I] IBZM, Zijlstra, Booij, van den Brink, and Franken (2008) found that exposure to opioid-related stimuli may induce dopamine release (with a lower dopamine receptor D2 (DRD2) availability in the right putamen) among those with opioid dependence.

Group Differences between Opioid-Dependent Subjects and Controls

The study of Zijistra et al. (2008) also indicated that those with opioid dependence have a lower DRD2 availability than controls ($n=18$) in the left caudate nucleus only (0.65 vs 0.79). The differences in DRD2 availability between groups in the right caudate nucleus and the left and right putamen were not significant. In another study using SPECT with [^{123}I]β-CIT, Cosgrove et al. (2010) reported that the difference (10%) in striatal dopamine transporter (DAT) availability between eight heroin users and eight healthy controls was not significant. However, in a SPECT study with [^{99m}Tc]TRODAT-1, Yeh et al. (2012) found that the difference in DAT availability between those with opioid dependence (DAT availability = 0.85 ± 0.28; $n=32$) and controls (DAT availability = 1.16 ± 0.26; $n=32$) was significant. The magnitude of the difference was larger (27%) than that reported in the study of Cosgrove et al. (2010). In the study of Yeh et al. (2012), 32 heroin users either were treated with a very low dose of methadone ($n=20$) or were undergoing a methadone-free abstinence period ($n=12$). The difference between the low-dose methadone users (0.78 ± 0.27) and the controls was greater than 30%. This finding may imply that chronic exposure to opioids may result in a lower availability of DAT, which could be a marker of hypodopaminergic activity. A similar result was also found in another independent study by Liu et al. (2013). This study also used [^{99m}Tc]TRODAT-1 and found a significantly lower DAT availability (28–34%) in the bilateral caudate and bilateral putamen in those with opioid dependence ($n=64$) compared with controls ($n=15$).

It is interesting that the studies reporting the two positive findings (Liu et al., 2013; Yeh et al., 2012) used the same radioligand and reported a similar magnitude of decrease in DAT availability, while in the study using [^{123}I]β-CIT (Cosgrove et al., 2010), the magnitude was found to be lower. The explanation for this phenomenon is unclear. However, we speculate that the statistical power of the study by Cosgrove et al. (2010) might not have been large enough to detect a 10% difference. We propose that the magnitude of difference (10% for [^{123}I]β-CIT and 27% for [^{99m}Tc] TRODAT-1) could be considered useful for planning the sample sizes of future studies.

The Effects of Treatment

It is controversial whether the altered dopaminergic system can be normalized by treatment. The findings of Yeh et al. (2012) indicated no differences between those receiving a very low dose of methadone ($n=20$) and those undergoing a period of methadone-free abstinence ($n=12$). However, it should be noted that the DAT availability before treatment (baseline) was not measured in the study of Yeh et al. (2012). Whether methadone treatment can increase the dopaminergic activity from baseline is unclear.

On the other hand, the findings of Liu et al. (2013) indicated that DAT availability was partially, but significantly, recovered (14–17%) after 6 months of treatment with Jitai tablets, a traditional Chinese medicine ($n=25$). This finding not only supported the effects of Jitai, but also implied that recovery of DAT availability might be an important surrogate marker for other treatment programs in the future. It is also worthy of note that the magnitude of recovery in the study by Yeh et al. (2012) was similar to that reported by Liu et al. (2013), as shown in Figure 1, but the statistical power might be too small to confirm the difference.

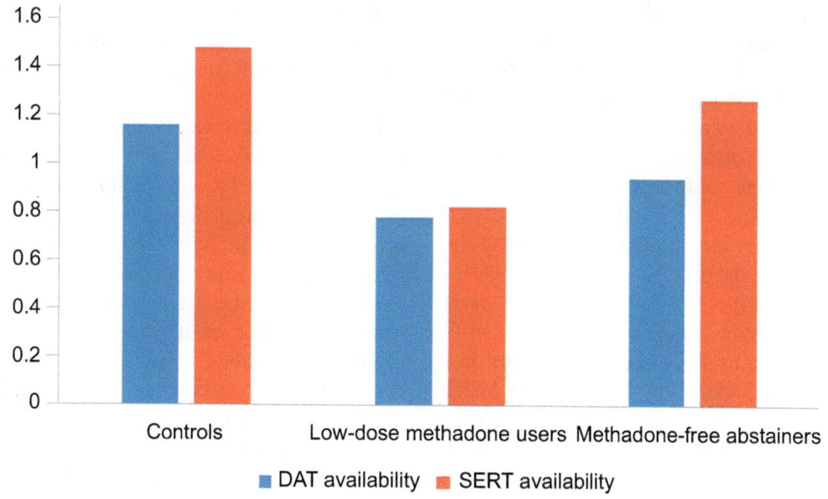

FIGURE 1 The mean differences in DAT and SERT availability between opioid users during low-dose methadone treatment, methadone-free subjects (after treatment), and controls. DAT, dopamine transporter; SERT, serotonin transporter. *Data have been published previously (Yeh et al., 2012).*

The Associations between Dopamine and Other Markers

Identification of the variables associated with dopaminergic activity could be important. In the study of Cosgrove et al. (2010), DAT availability was not associated with heroin use characteristics (total number of years of heroin use and quantity used per day). In the study of Liu et al. (2013), DAT availability among heroin users was associated not only with years of heroin use and heroin daily dosage, but also with depression tendency, as measured by the Hamilton depression rating scale (Hamilton, 1960). It is also worth noting that in the pilot study of Lin et al. (2015), a lower DAT availability was associated with greater heroin expenditure in 1 year.

The implications of these studies remain to be elucidated. However, the findings of Liu et al. (2013) and Lin et al. (2015) may confirm a downward regulation effect of chronic opioid use. These findings may also imply that low dopamine could cause heroin binge. The two explanations cannot be ruled out, as these findings were based on cross-sectional analysis.

SEROTONIN

The role of serotonin in opioid use has not yet been fully acknowledged, although studies have implied that the serotonergic system may play a role in the development of opioid dependence (Kish et al., 2001; Proudnikov et al., 2006; Xu et al., 2004). It should be noted that serotonin has been found to be associated with a wide range of affective disorders (Malison et al., 1998; Nikolaus, Hautzel, Heinzel, & Muller, 2012), which may be comorbid with opioid dependency. It would be very difficult to disentangle the complicated relationships between damage to the serotonergic system, opioid dependency, and the tendency toward affective disorders such as major depressive disorder and bipolar disorder.

Group Differences between Those with Opioid Dependence and Controls

Cosgrove et al. (2010) conducted a pilot study to examine the serotonergic activity in those with opioid dependence. Although the differences were not statistically significant, the binding potentials of [^{123}I]β-CIT in the diencephalon and brainstem among heroin users were lower than those in the controls (diencephalon, 21%; brainstem, 20%). In the dual isotopes SPECT study of Yeh et al. (2012), the group differences were confirmed by the highly selective radiotracer [^{123}I]ADAM. The availability of the midbrain serotonin transporter (SERT) was also significantly lower in the opioid users (0.99±0.42) than in the controls (1.48±0.55). Meanwhile, the difference between the low-dose methadone users ($n=20$; 0.82±0.24) and the methadone-free abstainers ($n=12$; 1.27±0.52) was also significant, as shown in Figure 1. Post hoc analysis confirmed that the methadone users had a lower SERT availability than both the healthy controls and the methadone-free users. No difference was found between the controls and the methadone-free abstainers. The findings of Yeh et al. (2012) could be striking, owing to: (1) the very large effect of the group difference between the patients and controls and (2) analysis implying that SERT availability is almost recovered in methadone-free abstainers. Nevertheless, replication is needed. The binding in the two-isotope study is illustrated in Figure 2.

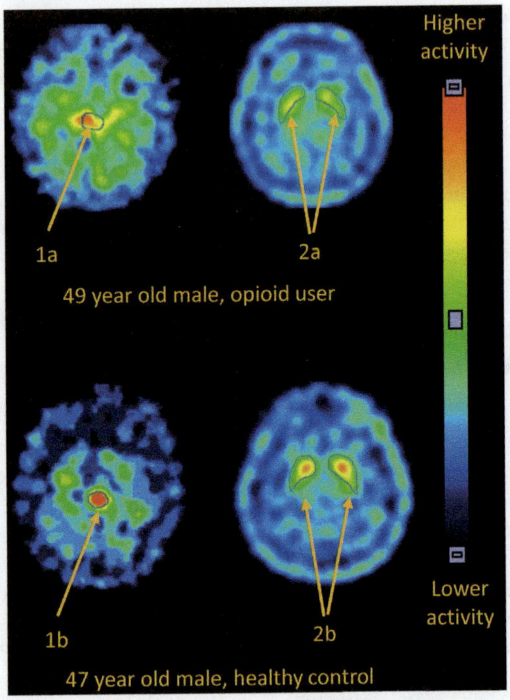

FIGURE 2 Examples of controls and opioid-dependent subjects to illustrate the binding areas of [^{99m}Tc]TRODAT-1 and [^{123}I]ADAM. The midbrain of the patient (1a) had a lower binding ratio (midbrain/cerebellum) of SERT than that of the healthy control (1b). The striatum of the patient (2a) had a lower binding ratio (striatum/occipital lobe) of DAT than that of the healthy control (2b). Levels of SPECT activity are color-coded from low (blue) to high (red). *Data have been published previously (Yeh et al., 2012).*

The Associations between Serotonin and Other Variables

Little is known about the associations between serotonergic activity and other biological or psychological variables among opioid users. It is worth noting that a significant association between SERT availability and depression tendency among those with opioid dependence was found in the study of Yeh et al. (2012). However, this association was suppressed after controlling for the group difference between methadone users and abstainers. This phenomenon may indicate that methadone use may also play a pivotal role in the causal chain between depression and altered serotonergic function among opioid users. Several possible underlying mechanisms could be proposed from this phenomenon. For example, whether the subject was using methadone, was experiencing a very stressful event (Yeh et al., 2009), or was already abstinent could be a common source that determines the association between SERT and depression. On the other hand, it is also plausible that a depressive tendency may influence opioid abstinence, leading to normalization of SERT functioning.

However, Lin, Chen, et al. (2012) found, controversially, that a higher availability of SERT after methadone treatment is associated with a quicker relapse. Although this study was a small-sample one ($n=9$), this finding may indicate that further investigation is needed to clarify the role of serotonin in opioid dependence.

DISCUSSION

Theoretical Implications of the Findings from SPECT Studies

In summary, evidence indicates that altered dopamine function can be found in those with opioid dependence. Meanwhile, the role of the serotonergic system is beginning to be revealed, but the current findings are controversial.

The current evidence is in agreement with the dopamine theory of addiction. The findings of Zijlstra et al. (2008) indicated that exposure to opioid-related stimuli may induce dopamine release (with a lower DRD2 availability in the right putamen) among those with opioid dependence. We speculate that the memory or conditioning process might play a role in this phenomenon. Although the patients did not receive opioids, the cue may activate the biological process that had been induced by the opioid. The opioid could induce pathological dopamine release that affects the mesolimbic pathway compared with controls (as shown in Figure 3). Prolonged and repeated opioid use may therefore result in a downward regulation of dopamine activity, resulting in a lower DAT availability (Liu et al., 2013; Yeh et al., 2012). The hypodopaminergic activity among opioid users may be a source of craving, and a dose-dependent response between the degree of hypodopaminergic activity and the money spent on heroin has also been noted, which might be a pathological economic behavior similar to bingeing (Lin et al., 2015). It is also worth noting that the study of Liu et al. (2013) confirmed that the altered dopaminergic function might be able to be recovered after long-term treatment. Although we have focused on the findings from SPECT studies in this chapter, it is worth noting that several PET studies have also reported comparable findings (Shi et al., 2008; Wang et al., 1997).

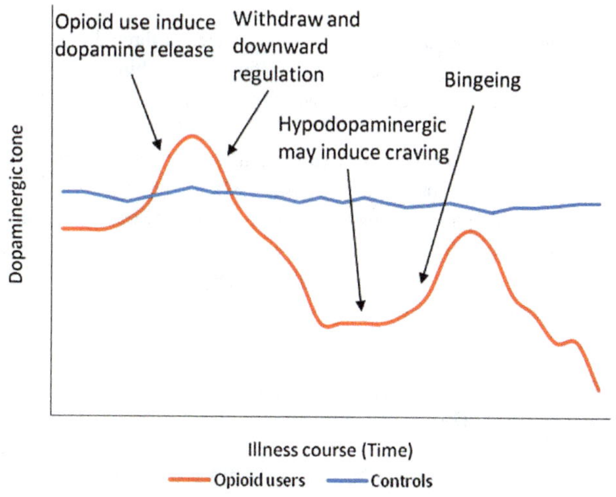

On the other hand, the role of serotonin remains to be identified. Although the finding was not significant statistically, the study of Cosgrove et al. (2010) shed light on the role of serotonergic activity in opioid dependence. A lower level of SERT among opioid users compared with controls was confirmed by Yeh et al. (2012). This study also indicated that depression tendency, opioid use, and serotonergic function could be intertwined. Meanwhile, the findings from Lin, Chen, et al. (2012) indicated that whether low serotonergic function is associated with a poor treatment outcome remains to be elucidated. In summary, the role of the serotonergic system in opioid dependence is unclear and might be controversial.

We speculate that the role of serotonin in addiction behavior is complex, and many other factors might also be involved in this relationship. It is well known that serotonergic activity could be associated with affective disorders (Malison et al., 1998; Nikolaus et al., 2012). Our study also found that the availability of SERT in vivo is associated with the stress hormone (Tsai et al., 2012), which also is predictive of the treatment outcome of methadone therapy (Lin, Chen, Chen, et al., 2013). On the other hand, assuming that individuals with the short allele of 5-HTTLPR have a lower in vivo SERT availability (Reimold et al., 2007), genetic study may help to explain the controversial findings regarding SERT. For example, the long allele was found to be associated with hazardous drinking behavior under psychosocial adversity (Laucht et al., 2009), unaffected by supportive social relationships (Way & Taylor, 2010) and risk preference (Homberg & Lesch, 2011). These findings from other areas might provide an explanation of the controversial findings of SPECT studies.

It should be noted that dopamine and serotonin are not independent systems. For instance, the serotonin neurons in the raphe nuclei project to dopamine neurons in the ventral tegmental area, which is an important site of the reward pathway (Di Giovanni, Di Matteo, & Esposito, 2008). Meanwhile, genetic study has also indicated that a statistical interaction exists between dopamine receptor D4 and SERT gene polymorphism in those with heroin dependence (Szilagyi et al., 2005). It is difficult to investigate the dopamine–serotonin interaction by SPECT study alone. Nevertheless, the interaction between dopamine and serotonin should not be neglected.

Clinical Implications

The findings from SPECT studies may confirm that the altered dopaminergic system could be the most important characteristic of opioid dependence. Some current treatments, such as methadone or buprenorphine, are also involved in the opioid system and may interact with the dopamine system, enhancing dopamine neurotransmission (Volkow, 2010). Good nutrition and a healthy lifestyle, which regulate the normal function of the dopaminergic system, might have some benefit for patients. Therefore, medicine-allied professions are needed to enhance the treatment effect. Some findings may also imply that in vivo DAT could be a surrogate marker of treatment (Liu et al., 2013; Yeh et al., 2012). Considering the high rate of relapse and the huge economic burden of opioid dependence (Lin, Chen, Lee, Hsiao, et al., 2013; Lin, Chen, et al., 2012), using SPECT as a clinical tool for decision-making and outcome evaluation might have future potential.

Although the role of serotonin is unclear from SPECT studies, it should be noted that a selective serotonin reuptake inhibitor (SSRI)

FIGURE 3 **Dopaminergic activity and addiction behavior.** The dopamine activity of healthy individuals may decrease slightly with age. Opioid use may increase the dopaminergic activity extensively, and then dopamine is downward regulated to a much lower level that leads to craving.

was found to be an effective add-on (Landabaso et al., 1998; Mannelli, Peindl, & Wu, 2011). However, its effect was also found to be diminished over time (Farren & O'Malley, 2002). Investigation of the role of serotonin in the treatment of opioid dependence is still needed.

APPLICATIONS TO OTHER ADDICTIONS AND SUBSTANCE MISUSE

The findings regarding dopamine and opioid use could be applied to other addiction behaviors and other types of substance abuse, as the mesolimbic system is the common reward pathway. Therefore, the findings reviewed in the section "Dopamine" could be considered general principles for addiction behaviors. However, the findings regarding serotonin are unclear as yet. We are therefore of a conservative view regarding whether the findings can be applied to other areas.

Other Potential Biological Mechanisms

In this chapter, we have introduced findings related to dopamine and serotonin from SPECT studies. It should be noted that other biological mechanisms might be associated with these monoamine neurotransmitters, which may have important theoretical and clinical implications. For example, oxytocin was found to be associated with the dopaminergic system (Baskerville & Douglas, 2010; Chang et al., 2014). Meanwhile, plasma cholesterol was also found to be a marker of the serotonin system (Hibbeln et al., 2000). Our findings indicated that the plasma levels of oxytocin and cholesterol are associated with craving for heroin (Lin, Chen, Lee, Lee, et al., 2013; Lin, Yang, et al., 2012). Considering the cost of SPECT, understanding the potential compact surrogate markers that may be associated with dopaminergic and serotonergic activity might be useful for clinical practice.

CONCLUSION

This chapter introduced the associations between opioid dependence and the dopamine and serotonin systems. The studies discussed are in agreement with the dopamine hypothesis of addiction. The findings imply that a low activity of dopamine neurons in the striatum might be a characteristic of the damage caused by opioid dependence. It is worth noting that the ratio of reduction in DAT availability is not small (10–34%). This neuropathology could induce heroin binge. A certain level of recovery after treatment has also been found. We speculate that the dopaminergic system could be a treatment target, and SPECT might be able to be used as a tool for diagnosis and prognosis.

SERT availability was also found to be reduced among heroin users. However, evidence is scarce. The effect (an approximate 20–30% reduction) could be used to plan future studies. The factors associated with serotonin are unclear and controversial. Considering the importance of the serotonergic system in treatment for psychiatric disorders, investigation of this issue is urgent.

In summary, we have introduced the findings from SPECT studies regarding the associations between opioid dependence and dopamine and serotonin in this chapter. Evidence is scarce, but nevertheless sheds light on the roles of dopamine and serotonin in opioid dependence. The findings may also provide hints for investigators and clinicians focused on other related biological mechanisms.

DEFINITION OF TERMS

Dopamine This is a hormone of the catecholamine family. Dopamine is one of the most important neurotransmitters in the human brain. Dopamine is associated with schizophrenia, attention deficit hyperactivity disorder, and Parkinson disease.

Heroin This is a synthesized opioid, often used as an illicit drug. One of the most harmful effects is the spreading of HIV by sharing needles for intravenous drug use.

Mesolimbic pathway This is one of the dopamine pathways. This pathway begins in the ventral tegmental area and connects to the limbic system via the nucleus accumbens (ventral striatum) and other areas. This pathway plays a pivotal role in the reward response, which is associated with addiction behavior.

Methadone maintenance treatment This is a treatment program applying long-term prescription of methadone as an alternative to illicit opioids. Methadone maintenance treatment could reduce the harm caused by opioid dependence.

Opioid This refers to any psychoactive chemical that resembles morphine or other opiates in terms of its pharmacological effects. Opioids are important and effective analgesics for acute pain. However, illicit opioid abuse could be one of the most harmful addiction behaviors.

Receptor (of a neurotransmitter) This is a membrane protein that is activated by a neurotransmitter. Receptors receive neurotransmitters released from presynaptic cells and then trigger an electrical signal.

Serotonin This is a monoamine neurotransmitter found throughout the human body. Low serotonergic activity is associated with depression and other affective disorders. The SERT is also one of the most important sites for the pharmacological treatment of psychiatric disorders.

Selective serotonin reuptake inhibitors This is a class of drugs that block the SERT and increase the extracellular level of serotonin. SSRIs are widely used as antidepressants.

Single-photon-emission computed tomography This is a nuclear medicine tomographic imaging technique. SPECT imaging is performed using a gamma camera and can yield a three-dimensional data set using a computer. Using radiotracers that bind to receptors or transporters of neurons, the activity of the neuron system can be assessed.

Transporter (of plasma membrane) This is a membrane-spanning protein that pumps the neurotransmitter dopamine out of the synapse. A complex homeostatic balance exists between the amount of neurotransmitter synthesized and the amount of reuptake via a transporter. Low (high) availability of transporters could be considered as a result of prolonged low (high) neuron activity.

KEY FACTS RELATED TO OPIOID DEPENDENCY

- Injection drug use (IDU) is strongly associated with HIV, accounting for 80% of cases in some countries in eastern Europe and central Asia. IDU has also been found to be the main route of hepatitis C infection, accounting for an estimated 90% of new hepatitis C infections.

- Heroin dependence could result in a poor quality of life. It was found that 72% of the income of heroin users is allocated to heroin purchase, while only 7% is spent on food and 5% on shelter and utilities.
- A meta-analysis reported that the crude mortality rate of opioid dependence was 2.09 per 100 person-years. The crude death rate in 2009 was 0.83 per 100 people in the general population worldwide.

SUMMARY POINTS

- This chapter focused on the dopaminergic and serotonergic functions among those with opioid dependence from studies using SPECT.
- The findings indicated that a low dopaminergic activity could be a characteristic of prolonged opioid use. This effect could be similar to the effects of other types of substance abuse.
- Limited evidence implies that low serotonergic activity might also be associated with opioid dependence. However, the role of serotonin remains to be investigated.
- The results of two independent studies may imply recovery of dopaminergic and serotonin activity after treatment.
- In vivo DAT availability might be a potential surrogate marker for the treatment outcome.
- SPECT may be able to be used as a tool to assist diagnosis and prognosis in the future.

REFERENCES

Baskerville, T. A., & Douglas, A. J. (2010). Dopamine and oxytocin interactions underlying behaviors: potential contributions to behavioral disorders. *CNS Neuroscience and Therapeutics, 16*, e92–123.

Carlsson, A. (1988). The current status of the dopamine hypothesis of schizophrenia. *Neuropsychopharmacology, 1*, 179–186.

Chang, W. H., Lee, I. H., Chen, K. C., Chi, M. H., Chiu, N. T., Yao, W. J., ... Chen, P. S. (2014). Oxytocin receptor gene rs53576 polymorphism modulates oxytocin-dopamine interaction and neuroticism traits-A SPECT study. *Psychoneuroendocrinology, 47*, 212–220.

Cosgrove, K. P., Tellez-Jacques, K., Pittman, B., Petrakis, I., Baldwin, R. M., Tamagnan, G., ... Staley, J. K. (2010). Dopamine and serotonin transporter availability in chronic heroin users: a [(123)I]beta-CIT SPECT imaging study. *Psychiatry Research, 184*, 192–195.

Di Giovanni, G., Di Matteo, V., & Esposito, E. (2008). *Serotonin-dopamine interaction: Experimental evidence and therapeutic relevance* (1st ed.). Amsterdam, Boston: Elsevier.

Farren, C. K., & O'Malley, S. (2002). A pilot double blind placebo controlled trial of sertraline with naltrexone in the treatment of opiate dependence. *American Journal on Addictions, 11*, 228–234.

Gerra, G., Garofano, L., Castaldini, L., Rovetto, F., Zaimovic, A., Moi, G., ... Donnini, C. (2005). Serotonin transporter promoter polymorphism genotype is associated with temperament, personality traits and illegal drugs use among adolescents. *Journal of Neural Transmission, 112*, 1397–1410.

Goldstein, R. Z., & Volkow, N. D. (2002). Drug addiction and its underlying neurobiological basis: neuroimaging evidence for the involvement of the frontal cortex. *American Journal of Psychiatry, 159*, 1642–1652.

Hamilton, M. (1960). A rating scale for depression. *Journal of Neurology, Neurosurgery and Psychiatry, 23*, 56–62.

Heinz, A., Ragan, P., Jones, D. W., Hommer, D., Williams, W., Knable, M. B., ... Linnoila, M. (1998). Reduced central serotonin transporters in alcoholism. *American Journal of Psychiatry, 155*, 1544–1549.

Hibbeln, J. R., Umhau, J. C., George, D. T., Shoaf, S. E., Linnoila, M., & Salem, N., Jr. (2000). Plasma total cholesterol concentrations do not predict cerebrospinal fluid neurotransmitter metabolites: implications for the biophysical role of highly unsaturated fatty acids. *American Journal of Clinical Nutrition, 71*, 331S–338S.

Homberg, J. R., & Lesch, K. P. (2011). Looking on the bright side of serotonin transporter gene variation. *Biological Psychiatry, 69*, 513–519.

Howell, L. L., & Kimmel, H. L. (2008). Monoamine transporters and psychostimulant addiction. *Biochemical Pharmacology, 75*, 196–217.

Howes, O. D., Montgomery, A. J., Asselin, M. C., Murray, R. M., Grasby, P. M., & McGuire, P. K. (2007). Molecular imaging studies of the striatal dopaminergic system in psychosis and predictions for the prodromal phase of psychosis. *British Journal of Psychiatry Supplement, 51*, s13–18.

Jacobsen, L. K., Staley, J. K., Malison, R. T., Zoghbi, S. S., Seibyl, J. P., Kosten, T. R., & Innis, R. B. (2000). Elevated central serotonin transporter binding availability in acutely abstinent cocaine-dependent patients. *American Journal of Psychiatry, 157*, 1134–1140.

Jia, S. W., Wang, W., Liu, Y., & Wu, Z. M. (2005). Neuroimaging studies of brain corpus striatum changes among heroin-dependent patients treated with herbal medicine, U'finer capsule. *Addiction Biology, 10*, 293–297.

Kish, S. J., Kalasinsky, K. S., Derkach, P., Schmunk, G. A., Guttman, M., Ang, L., ... Haycock, J. W. (2001). Striatal dopaminergic and serotonergic markers in human heroin users. *Neuropsychopharmacology, 24*, 561–567.

Koob, G. F., & Le Moal, M. (2001). Drug addiction, dysregulation of reward, and allostasis. *Neuropsychopharmacology, 24*, 97–129.

Landabaso, M. A., Iraurgi, I., Jimenez-Lerma, J. M., Sanz, J., Fernadez de Corres, B., Araluce, K., ... Gutierrez-Fraile, M. (1998). A randomized trial of adding fluoxetine to a naltrexone treatment programme for heroin addicts. *Addiction, 93*, 739–744.

Laucht, M., Treutlein, J., Schmid, B., Blomeyer, D., Becker, K., Buchmann, A. F., ... Banaschewski, T. (2009). Impact of psychosocial adversity on alcohol intake in young adults: moderation by the LL genotype of the serotonin transporter polymorphism. *Biological Psychiatry, 66*, 102–109.

Lin, S. H., Chen, K. C., Lee, S. Y., Hsiao, C. Y., Lee, I. H., Yeh, T. L., ... Yang, Y. K. (2013). The economic cost of heroin dependency and quality of life among heroin users in Taiwan. *Psychiatry Research, 209*, 512–517.

Lin, S. H., Chen, K. C., Lee, S. Y., Yao, W. J., Chiu, N. T., Lee, I. H., ... Yang, Y. K. (2012). The association between availability of serotonin transporters and time to relapse in heroin users: a two-isotope SPECT small sample pilot study. *European Neuropsychopharmacology, 22*, 647–650.

Lin, S. H., Chen, K. C., Lee, S. Y., Chiu, N. T., Lee, I. H., Chen, P. S., ... Yang, Y. K. (2015). The association between heroin expenditure and dopamine transporter availability—a single-photon emission computed tomography study. *Psychiatry Research, 231*, 292–297.

Lin, S.-H., Chen, P. S., Lee, L.-T., Lee, S.-Y., Tsai, C. H., Chen, K. C., ... Yang, Y. K. (2013). The association between level of plasma oxytocin and craving among heroin users. In *Paper presented at the 2014 NIDA International Forum, San Juan, Puerto Rico*.

Lin, S. H., Chen, W. T., Chen, K. C., Lee, S. Y., Lee, I. H., Chen, P. S., ... Yang, Y. K. (2013). Associations among hypothalamus-pituitary-adrenal axis function, novelty seeking, and retention in methadone maintenance therapy for heroin dependency. *Journal of Addiction Medicine, 7*, 335–341.

Lin, S. H., Yang, Y. K., Lee, S. Y., Hsieh, P. C., Chen, P. S., Lu, R. B., & Chen, K. C. (2012). Association between cholesterol plasma levels and craving among heroin users. *Journal of Addiction Medicine, 6*, 287–291.

Liu, Y., Han, M., Liu, X., Deng, Y., Li, Y., Yuan, J., ... Gao, J. (2013). Dopamine transporter availability in heroin-dependent subjects and controls: longitudinal changes during abstinence and the effects of Jitai tablets treatment. *Psychopharmacology, 230*, 235–244.

Malison, R. T., Price, L. H., Berman, R., van Dyck, C. H., Pelton, G. H., Carpenter, L., ... Charney, D. S. (1998). Reduced brain serotonin transporter availability in major depression as measured by [123I]-2 beta-carbomethoxy-3 beta-(4-iodophenyl)tropane and single photon emission computed tomography. *Biological Psychiatry, 44*, 1090–1098.

Mannelli, P., Peindl, K. S., & Wu, L. T. (2011). Pharmacological enhancement of naltrexone treatment for opioid dependence: a review. *Substance Abuse and Rehabilitation, 2011*, 113–123.

Mark, T. L., Woody, G. E., Juday, T., & Kleber, H. D. (2001). The economic costs of heroin addiction in the United States. *Drug and Alcohol Dependence, 61*, 195–206.

Melis, M., Spiga, S., & Diana, M. (2005). The dopamine hypothesis of drug addiction: hypodopaminergic state. *International Review of Neurobiology, 63*, 101–154.

Nikolaus, S., Hautzel, H., Heinzel, A., & Muller, H. W. (2012). Key players in major and bipolar depression–a retrospective analysis of in vivo imaging studies. *Behavioural Brain Research, 232*, 358–390.

Paterson, L. M., Kornum, B. R., Nutt, D. J., Pike, V. W., & Knudsen, G. M. (2013). 5-HT radioligands for human brain imaging with PET and SPECT. *Medicinal Research Reviews, 33*, 54–111.

Pierce, R. C., & Kumaresan, V. (2006). The mesolimbic dopamine system: the final common pathway for the reinforcing effect of drugs of abuse? *Neuroscience and Biobehavioral Reviews, 30*, 215–238.

Previc, F. H. (1999). Dopamine and the origins of human intelligence. *Brain and Cognition, 41*, 299–350.

Previc, F. H. (2009). *The dopaminergic mind in human evolution and history*. New York: Cambridge University Press.

Proudnikov, D., LaForge, K. S., Hofflich, H., Levenstien, M., Gordon, D., Barral, S., ... Kreek, M. J. (2006). Association analysis of polymorphisms in serotonin 1B receptor (HTR1B) gene with heroin addiction: a comparison of molecular and statistically estimated haplotypes. *Pharmacogenetics and Genomics, 16*, 25–36.

Reimold, M., Smolka, M. N., Schumann, G., Zimmer, A., Wrase, J., Mann, K., ... Heinz, A. (2007). Midbrain serotonin transporter binding potential measured with [11C]DASB is affected by serotonin transporter genotype. *Journal of Neural Transmission, 114*, 635–639.

Sharma, S., & Ebadi, M. (2008). SPECT neuroimaging in translational research of CNS disorders. *Neurochemistry International, 52*, 352–362.

Shi, J., Zhao, L. Y., Copersino, M. L., Fang, Y. X., Chen, Y., Tian, J., ... Lu, L. (2008). PET imaging of dopamine transporter and drug craving during methadone maintenance treatment and after prolonged abstinence in heroin users. *European Journal of Pharmacology, 579*, 160–166.

Spanagel, R., & Weiss, F. (1999). The dopamine hypothesis of reward: past and current status. *Trends in Neurosciences, 22*, 521–527.

Stahl, S. M. (2000). *Essential psychopharmacology: Neuroscientific basis and practical application* (2nd ed.). New York: Cambridge University Press.

Staley, J. K., Krishnan-Sarin, S., Zoghbi, S., Tamagnan, G., Fujita, M., Seibyl, J. P., ... Innis, R. B. (2001). Sex differences in [123I]beta-CIT SPECT measures of dopamine and serotonin transporter availability in healthy smokers and nonsmokers. *Synapse, 41*, 275–284.

Szilagyi, A., Boor, K., Szekely, A., Gaszner, P., Kalasz, H., Sasvari-Szekely, M., & Barta, C. (2005). Combined effect of promoter polymorphisms in the dopamine D4 receptor and the serotonin transporter genes in heroin dependence. *Neuropsychopharmacologia Hungarica, 7*, 28–33.

Tjon, G. H., De Vries, T. J., Ronken, E., Hogenboom, F., Wardeh, G., Mulder, A. H., & Schoffelmeer, A. N. (1994). Repeated and chronic morphine administration causes differential long-lasting changes in dopaminergic neurotransmission in rat striatum without changing its delta- and kappa-opioid receptor regulation. *European Journal of Pharmacology, 252*, 205–212.

Tsai, H. Y., Lee, I. H., Yeh, T. L., Yao, W. J., Chen, K. C., Chen, P. S., ... Yang, Y. K. (2012). Association between the dexamethasone suppression test and serotonin transporter availability in healthy volunteer: a SPECT with [123I] ADAM study. *European Neuropsychopharmacology, 22*, 641–646.

Volkow, N. D. (2010). Opioid-dopamine interactions: implications for substance use disorders and their treatment. *Biological Psychiatry, 68*, 685–686.

Volkow, N. D., Fowler, J. S., Wang, G. J., Baler, R., & Telang, F. (2009). Imaging dopamine's role in drug abuse and addiction. *Neuropharmacology, 56*(Suppl. 1), 3–8.

Volkow, N. D., Fowler, J. S., Wang, G. J., & Swanson, J. M. (2004). Dopamine in drug abuse and addiction: results from imaging studies and treatment implications. *Molecular Psychiatry, 9*, 557–569.

Volkow, N. D., Wang, G. J., Kollins, S. H., Wigal, T. L., Newcorn, J. H., Telang, F., ... Swanson, J. M. (2009). Evaluating dopamine reward pathway in ADHD: clinical implications. *JAMA, 302*, 1084–1091.

Volkow, N. D., Wang, G. J., Newcorn, J., Fowler, J. S., Telang, F., Solanto, M. V., ... Pradhan, K. (2007). Brain dopamine transporter levels in treatment and drug naive adults with ADHD. *NeuroImage, 34*, 1182–1190.

Voruganti, L. N., Slomka, P., Zabel, P., Mattar, A., & Awad, A. G. (2001). Cannabis induced dopamine release: an in-vivo SPECT study. *Psychiatry Research, 107*, 173–177.

Wang, G. J., Volkow, N. D., Fowler, J. S., Logan, J., Abumrad, N. N., Hitzemann, R. J., ... Pascani, K. (1997). Dopamine D2 receptor availability in opiate-dependent subjects before and after naloxone-precipitated withdrawal. *Neuropsychopharmacology, 16*, 174–182.

Way, B. M., & Taylor, S. E. (2010). Social influences on health: is serotonin a critical mediator? *Psychosomatic Medicine, 72*, 107–112.

Wise, R. A. (1996). Addictive drugs and brain stimulation reward. *Annual Review of Neuroscience, 19*, 319–340.

Xiao, Z. W., Cao, C. Y., Wang, Z. X., Li, J. X., Liao, H. Y., & Zhang, X. X. (2006). Changes of dopamine transporter function in striatum during acute morphine addiction and its abstinence in rhesus monkey. *Chinese Medical Journal, 119*, 1802–1807.

Xu, K., Lichtermann, D., Lipsky, R. H., Franke, P., Liu, X., Hu, Y., ... Goldman, D. (2004). Association of specific haplotypes of D2 dopamine receptor gene with vulnerability to heroin dependence in 2 distinct populations. *Archives of General Psychiatry, 61*, 597–606.

Yeh, T. L., Chen, K. C., Lin, S. H., Lee, I. H., Chen, P. S., Yao, W. J., ... Chiu, N. T. (2012). Availability of dopamine and serotonin transporters in opioid-dependent users-a two-isotope SPECT study. *Psychopharmacology, 220*, 55–64.

Yeh, T. L., Lee, I. H., Chen, K. C., Chen, P. S., Yao, W. J., Yang, Y. K., ... Lu, R. B. (2009). The relationships between daily life events and the availabilities of serotonin transporters and dopamine transporters in healthy volunteers–a dual-isotope SPECT study. *NeuroImage, 45*, 275–279.

Zijlstra, F., Booij, J., van den Brink, W., & Franken, I. H. (2008). Striatal dopamine D2 receptor binding and dopamine release during cue-elicited craving in recently abstinent opiate-dependent males. *European Neuropsychopharmacology, 18*, 262–270.

Chapter 91

Personality Dimensions, Impulsivity, and Heroin

Cuneyt Evren, Muge Bozkurt

Research, Treatment and Training Center for Alcohol and Substance Dependence (AMATEM), Bakirkoy Training and Research Hospital for Psychiatry, Neurology and Neurosurgery, Istanbul, Turkey

Abbreviations

AI Attentional impulsiveness
ASPD Antisocial personality disorder
BIS-11 Barrat Impulsiveness Scale-11
BPD Borderline personality disorder
C Cooperativeness
HA Harm avoidance
MG Maintenance group
MI Motor impulsiveness
NPI Nonplanning impulsiveness
NS Novelty seeking
P Persistence
RD Reward dependence
RG Relapse group
SD Self-directedness
ST Self-transcendence
SUD Substance use disorder
TCI Temperament and Character Inventory

INTRODUCTION

Since 2005, there has been an increasing interest in the role of impulsivity in substance use disorders (SUDs) (Bozkurt, Evren, Yilmaz, Can, & Cetingok, 2013). Impulsivity is a multifaceted construct that can be defined as a predisposition toward rapid, unplanned reactions to internal or external stimuli without regard to the negative consequences of these reactions to the impulsive individual or to others (Moeller, Barrat, Dougherty, Schmitz, & Swann, 2001). A growing number of studies have confirmed a strong association between impulsivity and SUDs; i.e., the level of impulsivity has been reported to be higher in SUDs, including alcohol (Mitchell, Fields, D'Esposito, & Boettiger, 2005) and heroin (Kirby, Petry, & Bickel, 1999) use disorders, than in non-substance-using populations.

Impulsivity has been related to risk taking, lack of planning, and quick decision-making (Eysenck & Eysenck, 1977). Definitions of impulsivity suggest that such behaviors tend to be committed without forethought or conscious judgment and are characterized by acting on the spur of the moment, the inability to focus on a specific task, and a lack of adequate planning (Moeller et al., 2001; Patton, Stanford, & Barratt, 1995). There is accumulating evidence from preclinical laboratory animal and clinical studies indicating that impulsive behavior might be causally linked to several distinct processes in drug addiction including onset, maintenance, related problems, and relapse (Pattij & De Vries, 2013).

High relative comorbidity is observed between SUDs and Axis I and Axis II psychiatric disorders from the impulse control spectrum, e.g., antisocial personality disorder (Verheul & van den Brink, 2000). Moeller et al. (2002) have found that the most marked difference in characteristics between substance users and control subjects is a high level of impulsivity in substance users. It has also been suggested that impulsivity is a temperamental risk factor for substance use and might be a fundamental mechanism in both the onset of excessive substance use and relapse (Lane, Cherek, Rhoades, Pietras, & Tcheremissine, 2003). Impulsivity has also been found to be a high-risk factor for early substance use and related to the severity of drug abuse and treatment retention (Tarter et al., 2003). Impulsivity may also serve as a moderator of the relationship between substance use behavior and substance use outcomes, such as substance use-related problems (Simons, 2003). Impulsivity might contribute significantly to the risk of suicide attempts in substance-dependent patients, may interrupt their outpatient or inpatient treatment, and may mediate the effects of substance use on aggression (Bozkurt et al., 2013).

Patients with an SUD are frequently associated with impulsivity that may underlie elevated levels of life-threatening types of behavior, including aggression. Impulsive aggression is characterized by an inability to regulate affect as well as aggressive impulses and is highly comorbid with other mental disorders, including substance abuse (Seo, Patrick, & Kennealy, 2008). Consistent with this, prisoners with substance abuse had both high impulsivity and aggressiveness compared to prisoners without substance abuse (Cuomo, Sarchiapone, Giannantonio, Mancini, & Roy, 2008). Studies also suggest that impulsive aggression may be associated with a certain subtype of substance abuse (i.e., early onset) (Seo et al., 2008). Some authors point out that impulsivity and aggression are expected to appear together on the phenotypic level, justifying the view that impulsive aggression is a single trait-like dimension (Seroczynski, Bergeman, & Coccaro, 1999). Others consider that impulsivity and aggression are two related but separate constructs and impulsivity should be accepted as a high-order phenotype preceding aggressive acts (García-Forero, Gallardo-Pujol, Maydeu-Olivares, & Andrés-Pueyo, 2009). In conclusion, there is a general consensus that impulsivity and aggression are closely related constructs but the nature of their relationship remains unclear (Bozkurt et al., 2013).

Individuals who have high levels of impulsivity are at increased risk for substance experimentation, problematic substance use, and inability to abstain from substance use. Additionally, impulsive individuals tend to begin using alcohol and other substances at earlier ages and illicit drug use is more common among them (Kollins, 2002). It has also been suggested that substance use might facilitate impulsivity by interfering with normal inhibitory controls (Verdejo-García, Lawrence, & Clark, 2008). Nevertheless, impulsive individuals are more likely to engage in behaviors that can be dangerous to themselves or others, including driving recklessly, starting fights, shoplifting, perpetrating domestic violence, and trying to hurt or kill themselves, and also they are exposed to a higher risk of lifetime trauma and to substantial physical and psychosocial impairment (Chamorro et al., 2012). Impulsivity has also been found to be related to craving (Evren, Durkaya, Evren, Dalbudak, & Cetin, 2012), suggested as a risk factor for relapse during an abstinence period (Evren, Durkaya, et al., 2012; De Wit, 2009; Mitchell et al., 2005) and which may have a negative effect on the treatment outcome of patients with SUD (Evren, Durkaya, et al., 2012; Moeller et al., 2001). Finally, studies have suggested that impulsivity is associated with suicide attempts and self-injurious behaviors (Evren, Cinar, Evren, & Celik, 2012).

Some studies have been conducted to evaluate the relationship of impulsivity with the preferred drug of abuse. Previous study using discount rates showed that heroin and cocaine abusers were more impulsive than alcoholics and non-substance-using controls (Kirby & Petry, 2004). Studies have also pointed out that the structure of impulsivity differed on the dimensional level depending on the type of substance dependence (Clark, Robbins, Ersche, & Sahakian, 2006), which may indicate the different psychopharmacological properties and behavioral functions of these substances. Although impulsivity showed no difference between heroin dependents with comorbid cocaine dependency and without (Nielsen et al., 2012), the presence of multiple-drug abuse may also affect the results.

In our previous study we evaluated impulsivity in two different groups of men with alcohol ($n=94$) or heroin ($n=78$) dependency and healthy controls ($n=63$) (Bozkurt et al., 2013). Impulsivity scores were higher among both the alcohol- and the heroin-dependent groups than in the healthy controls (Figure 1). Severity of impulsivity discriminated both alcohol dependents and heroin dependents from healthy controls. These findings were similar to previous studies showing higher impulsivity scores in alcohol dependents (Mitchell et al., 2005) and heroin dependents (Nielsen et al., 2012) compared to the general population. When subscales of impulsivity were taken as independent variables, together with current age and hostility, motor impulsiveness (MI) discriminated alcohol dependents, whereas, together with physical aggression, nonplanning impulsiveness (NPI) discriminated heroin dependents from healthy controls. Thus, although aggression and impulsivity did not discriminate alcohol and heroin dependents from each

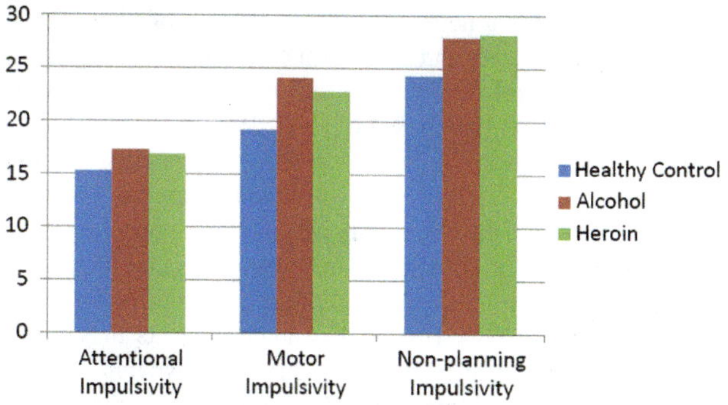

FIGURE 1 Comparison of impulsivity scores between groups. Alcohol and heroin groups > control group.

other, and impulsivity discriminated these groups from healthy controls, different dimensions of impulsivity discriminated these groups from healthy controls. The results suggest that both aggression and impulsivity are important constructs on which to focus in the treatment of substance dependents, but different dimensions might be the center of attention for patients with different substances of dependence. Consistent with the previous study (Clark et al., 2006), the NPI dimension of impulsivity, which can be defined as a tendency to choose a small, more immediate reward over a larger, more delayed reward and centered on the "present orientation" with a "lack of planning for the future and foresight," discriminated heroin dependents from controls in this study. These findings may suggest that NPI is characteristic of drug dependents rather than alcohol dependents. Finally, in a 2010 study, impulsivity was higher both in drug users and in their siblings than in controls, whereas, although drug users differed from controls on all three subscales of the Barrat Impulsiveness Scale-11 (BIS-11), siblings differed from controls only with respect to NPI (Ersche, Turton, Pradhan, Bullmore, & Robbins, 2010). This may indicate that NPI exists before drug exposure, while the MI and attentional impulsiveness (AI) components are exacerbated by drug exposure (Hogarth, 2011).

IMPULSIVITY AND HEROIN DEPENDENCE

Previous studies have shown that current heroin addicts have higher levels of impulsivity than nonaddicts do on measures of impulsivity (Bornovalova, Daughters, Hernandez, Richards, & Lejuez, 2005; Clark et al., 2006; Lejuez, Bornovalova, Daughters, & Curtin, 2005; Verdejo-García, Bechara, Recknor, & Pérez-García, 2007), while impulsivity measured with the BIS-11 distinguishes heroin users from nonusers (Dissabandara et al., 2014). Even the methadone-maintained patients scored higher on the BIS-11 subscales than did normal volunteers with no history of drug abuse or dependence (David et al., 2012). Previous studies did, however, show that methadone treatment reduces criminal and disruptive behaviors and that buprenorphine may produce even better results than methadone in patients with prominently violent suicidal and impulsive behaviors (Maremmani et al., 2011). The initiation of illicit drug use in some cases may be due to impaired impulse control, which may lead to drug addiction, with repeated episodes of drug use (Perry & Carroll, 2008). Drug addiction may facilitate impulsive acts by interfering with normal inhibitory controls (Verdejo-García et al., 2008). From a neuropsychological point of view, heroin use has implicit short- and long-term consequences. In particular, impulse control dysfunction and negative affective states have been reported (Koob & Le Moal, 1997). The continuous intake of this substance increases levels of impulsivity that return to baseline (preheroin) levels throughout abstinence; in heroin-dependent subjects, impulsivity therefore becomes more intense as a result of chronic heroin exposure, rather than being a vulnerability trait (Schippers, Binnekade, Schoffelmeer, Pattij, & De Vries, 2012). Heroin-dependent subjects may prefer this substance because of the self-medication dynamics, which have positive effects in managing the preexisting aggressiveness (Aharonovich, Nguyen, & Nunes, 2001) that is usually supported by impulsiveness (Seroczynski et al., 1999). It may therefore be true that individuals who become heroin users are more likely to show impulsiveness, not because of the drug itself, but because of a preexisting premorbid impulsive disposition that leads them to form ties selectively with heroin (Hoaken & Stewart, 2003).

The BIS-11 measures three subtypes of impulsiveness: cognitive (attentional) impulsiveness (inattention and cognitive instability), MI (motor disinhibition), and NPI (lack of self-control and intolerance of cognitive complexity) (Patton et al., 1995). MI seems to be an important dimension that may be linked with relapse at follow-up. In a 2012 study, impulsive personality was found to play a role in the onset of heroin use and, of the three factors, the motor factor appeared to have the greatest effect on the age of onset of drug use (Li et al., 2012). The motor factor was defined by Patton et al. (1995) as "acting on the spur of the moment." Substance-dependent individuals turn out to be impaired in carrying out tasks that measure motor impulsivity (Verdejo-Garcia & Perez-Garcia, 2007), as is manifested by their poor inhibitory control over overwhelming responses (Dougherty et al., 2003). Impulsivity has also been discussed as a potential mediator of the relationship of craving to relapse. In response to stress or environmental cues, an individual with substance abuse could use the substance to become capable of taking action rapidly, regardless of the consequences. Once the substance has been used, craving and withdrawal may lead to continued use or dependence (Jentsch & Taylor, 1999).

In our other study we evaluated the changes in impulsivity and aggression scores among male heroin-dependent patients using buprenorphine/naloxone as a maintenance treatment and those who relapsed within 12 months of their discharge from the hospital (Evren, Yilmaz, et al., 2014). Among 78 consecutively admitted male heroin dependents, 52 were available and examined by face-to-face interview 12 months after discharge from the hospital. Patients were investigated by using the BIS-11 both at baseline and at the end of 12 months. Among 52 heroin-dependent patients, 44.2% (n=23) were considered to have relapsed during the 12-month follow-up. Sociodemographic variables did not differ between the groups. The mean scores of impulsivity did not differ between the relapse group (RG) and the maintenance group (MG) at baseline, whereas impulsivity scores were higher in the RG than in the MG at the end (Figure 2). In the MG, impulsivity (motor and nonplanning) was lower at the end of 12 months, whereas impulsivity (attentional and nonplanning) was higher in the RG (Figure 3). In logistic regression models, MI evaluated both at baseline and after 12 months predicted relapse. MI seems

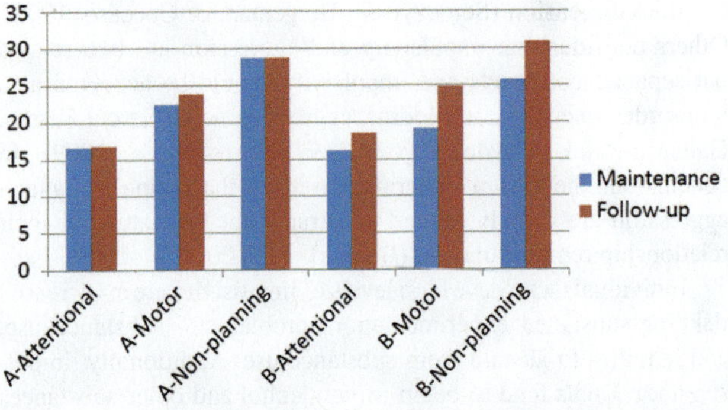

FIGURE 2 A comparison of impulsivity scores of patients in remission and patients that relapsed at baseline and at the end of the 12 month follow-up. A: Baseline evaluation. B: Follow-up evaluation.

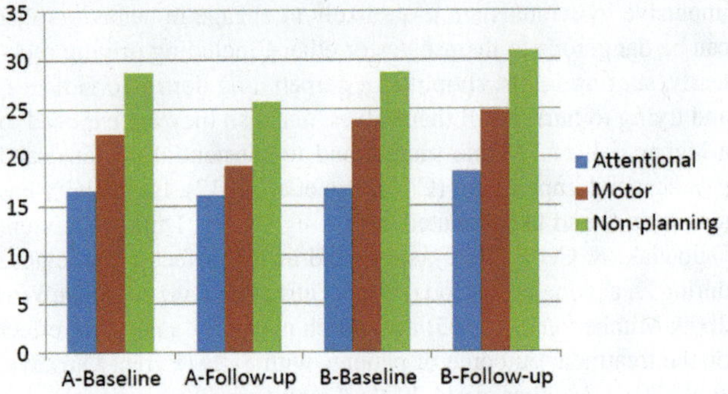

FIGURE 3 A comparison of impulsivity scores between baseline and the end of the 12-month follow-up in patients in remission and patients that relapsed. A: MG (n=29). B: RG (n=23).

to be an important dimension that may be related to relapse. While impulsivity scores increased in the RG during the 12-month follow-up, they decreased in the MG.

TEMPERAMENT AND CHARACTER DIMENSIONS OF PERSONALITY

Cloninger developed a dimensional, psychobiological model of personality that accounts for both normal and abnormal variations in two major components of personality: temperament and character. The Temperament and Character Inventory (TCI) is a self-administered dimensional questionnaire constructed to assess the seven basic dimensions of personality (Kose, 2003). The TCI was originally designed to explore genetic and environmental factors underlying normal and abnormal personality dimensions. The four temperament dimensions are assumed to be highly heritable and underlain by specific neurotransmission systems (Gourion, Pelissolo, & Lepine, 2003). Three character dimensions may be determined by genetic and biologic factors but are more influenced by environmental factors than temperament and are therefore less stable over time (Basiaux et al., 2001). The psychobiological model assumes interactions between temperament and character scales. These interactions elicit secondary emotions and are important in the development of personality (De la Rie, Duijsens, & Cloninger, 1998). According to Cloninger and Svrakic (1997), the presence and severity of personality disorders can be reliably assessed using interview or questionnaire versions of the TCI.

Individual differences in personality structure and development have a strong influence on the risk of all forms of psychopathology, including substance abuse (Cloninger & Svrakic, 1997). Substance abuse was considered an important clinical disorder in the evaluation of the TCI (Ball, Tennen, Poling, Kranzler, & Rounsaville, 1997), because Cloninger (1987) also subtyped alcoholism on the basis of some of these TCI personality dimensions. Evaluation of personality in substance-dependent patients may also be important in prediction, prevention, and treatment (Hosak, Preiss, Halir, Cermakova, & Csemy, 2004).

Ball et al. (1997) found that several TCI scales were associated with different personality disorders among substance-dependent patients, although not as strongly as the NEO Personality Inventory dimensions. In this study, the results did not support most of the predictions made for the TCI in earlier studies on substance abusers (Cloninger, Svrakic, & Przybeck, 1993). In contrast to this finding, Gourion et al. (2003) suggested that the results of their study highlighted the overall stability of the TCI in patients with opiate dependence and provided evidence for the usefulness of this questionnaire. Similarly, Basiaux et al. (2001) reported that the TCI data add to evidence concerning a higher probability of personality disorder in alcohol-dependent patients.

There are also studies that evaluated dimensions that would predict a different substance of choice. In one study, maturity and the drug of choice among female addicts were related to different TCI scales (Gerdner, Nordlander, & Pedersen, 2002). Indeed, an association between personality and substance choice has been reported by several studies (McCormick, Dowd, Quirk, & Zegarra, 1998). Le Bon et al. (2004) found that personality profiles were related to the drug of choice and suggested that personality screening might be of great value in preventive strategies.

The main finding of a 2007 study was that, among temperament dimensions, the novelty seeking (NS) score was higher and the reward dependence (RD) score lower in the drug-dependent group, whereas among character dimensions, self-directedness (SD) and cooperativeness (C) scores were lower in the drug-dependent group (Evren, Evren, Yancar, & Erkiran, 2007). Consistent with this study, the study of Le Bon et al. (2004) showed higher scores for NS in heroin-dependent patients than in alcohol-dependent patients, but in contrast, higher scores for SD in heroin-dependent patients. A possible explanation for this may be the heterogeneity of the drug-dependent group in the first study, in which only one-third consisted of opiate dependents and 77.7% of polysubstance abusers, compared to that of Le Bon et al. (2004). This difference might also be partly explained by the fact that the patients were from a developing country with a different culture and religion in the first study. Individuals with high NS scores tend to be quick-tempered, excitable, exploratory, curious, enthusiastic, ardent, easily bored, impulsive, and disorderly, whereas individuals with low RD scores are often described as practical, tough-minded, cold, and socially insensitive (Kose, 2003). Ball, Kranzler, Tennen, Poling, and Rounsaville (1998) found that severe substance abusers had high scores on the NS and RD dimensions. Although severity of substance abuse was not evaluated in our previous study (Evren et al., 2007), as most of the drug-dependent patients were polysubstance users they might be considered as severe abusers. In the literature, among substance abusers, alcoholics are characterized as less impulsive than cocaine or heroin abusers, and polysubstance abusers are known to have particularly high levels of impulsivity and sensation seeking (McCormick et al., 1998). The results of our study were similar (Evren et al., 2007). Research suggests that NS represents a vulnerability factor for substance abuse (Hosak et al., 2004). Gabel, Stallings, Schmitz, Young, and Fulker (1999) reported that NS made a significant contribution to the prediction of substance misuse. Our results showed that high NS was a predictor of drug dependency, as was young age. All categories of personality disorder are distinguished by low SD, regardless of the cluster or category of personality disorder (Cloninger et al., 1993). Low SD is defined by poor impulse control or weak ego strength and is described as being irresponsible, purposeless, immature, weak, fragile, blaming, destructive, ineffective, unreliable, helpless, poorly integrated, and low in self-acceptance. Basiaux et al. (2001) reported that alcohol dependents were characterized by higher NS and lower SD than nonpsychiatric control subjects. Gutierrez, Sangorrin, Martin-Santos, Torres, and Torrens (2002) reported that the character dimension SD was strongly associated with the presence and severity of all personality disorders, irrespective of subtype, in correctly classifying 77% of subjects among substance abusers. Patients with a low SD score are frequently described by clinicians as immature or having a personality disorder (Cloninger et al., 1993; De la Rie et al., 1998). Likewise, most individuals with personality disorders are low in C, which is characterized by poor interpersonal functioning and described as being intolerant, narcissistic, hostile or disagreeable, critical, unhelpful, revengeful, and opportunistic (Cloninger, 2000; Kose, 2003).

In our previous study we evaluated the differences in dimensions of temperament and character in Turkish alcohol- ($n=111$) and drug-dependent ($n=93$) inpatients and examined which dimensions would predict drug dependency (Evren et al., 2007). Among the temperament dimensions, NS score was higher and RD score was lower in drug-dependent patients than in alcohol-dependent patients. Among the character dimensions, SD and C scores were lower in drug-dependent patients (Figure 4). Low age and NS predicted drug dependency in a forward logistic regression model. Thus, as in previous studies, which indicate an association between personality and substance choice, in this study, TCI was shown to be an efficient tool in discriminating alcohol and drug dependents.

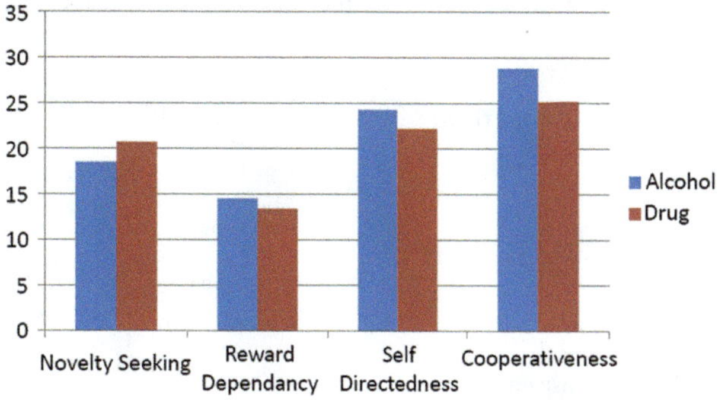

FIGURE 4 Comparing alcohol- and drug-dependent patients in terms of the TCI main dimensions. Only the main dimensions that showed differences between groups are shown.

RELATIONSHIPS OF PERSONALITY DIMENSIONS WITH IMPULSIVITY IN HEROIN DEPENDENTS

Trait impulsivity is an important determinant of substance use during development, and in adults momentary "state" increases in impulsive behavior may increase the likelihood of substance use, especially in individuals attempting to abstain (De Wit, 2009). Conversely, acute and chronic effects of substance use may increase impulsive behaviors, which may in turn facilitate further substance use (De Wit, 2009). Finally the association between impulsivity and substance abuse may be mediated through a common third factor (De Wit, 2009; Perry & Carroll, 2008), such as personality.

The TCI measures four basic temperament and three basic character dimensions based on Cloninger's model of personality development (Kose, 2003). Temperament consists of four traits, called harm avoidance (HA), NS, RD, and persistence (P). Character consists of three dimensions: SD, C, and self-transcendence (ST). The psychobiological model assumes interactions between temperament and character scales, eliciting secondary emotions, which are important in the development of personality (De la Rie et al., 1998). Individual differences in personality structure and development have a strong influence on the risk of all forms of psychopathology, including substance abuse (Cloninger & Svrakic, 1997). Each dimension is influenced by complex interactions among genetic and environmental variables because individual personalities develop as complex adaptive systems (Cloninger, 2000). Cloninger (1996) defines impulsive behavior as the coexistence of four heritable temperamental traits: high NS, low HA, low P, and, rarely, high RD. Among these traits, high NS was consistently found to be related with impulsivity in various populations such as patients with generalized anxiety disorder (GAD) (Pierò, 2010). Finally, it was suggested that the subjects most vulnerable to substance dependency may be those with high impulsivity and/or high NS (García-Forero et al., 2009). Thus, it is important to evaluate the relationship of personality dimensions with impulsivity in substance-dependent populations. Also, different from other populations, impulsivity is closely linked to substance use and abuse, both as a contributor to use and as a consequence of use (De Wit, 2009). Identification of personality predictors of impulsivity dimensions might allow us to differentiate among subgroups of patients with heroin dependency who, despite sharing the same diagnosis, may need different and specific approaches to the treatment of such dimension. Thus the personality dimensions that predict dimensions of impulsivity in heroin dependents may differ from those of other populations and from the model defined by Cloninger (1996).

The aim of our previous study was to evaluate the relationship of personality dimensions with impulsivity among men with heroin dependence (Evren, Bozkurt, et al., 2014). Also we wanted to control for the effects of depression and anxiety symptoms on this relationship. Participants were consecutively admitted male heroin-dependent ($n=78$) inpatients and healthy controls ($n=63$). Patients were investigated with the BIS-11, the TCI, and the Symptom Checklist–Revised. The severity of impulsivity and the dimensions of impulsivity were higher in heroin-dependent inpatients than in healthy controls (Table 1). Low SD, low P, and high NS scores predicted impulsivity in heroin-dependent male inpatients. Although depression predicted both attentional impulsiveness and MI, the personality dimensions that predict impulsivity dimensions differed (Table 2). This may suggest that when impulsivity is the problem, types of impulsivity and personality dimensions must be evaluated and the treatment should be shaped accordingly for heroin dependents.

Consistent with our previous study (Evren, Bozkurt, et al., 2014), in different populations, such as patients with GAD (Pierò, 2010) and a sample of Italian subjects (Martinotti et al., 2008), high NS and low SD (Martinotti et al., 2008; Pierò, 2010) were consistently related with impulsivity. It was suggested that a high risk-seeking predisposition (NS) in subjects with a low level of temperamental inhibition (low HA) may lead to an impulsive temperament (Cloninger, 1996). Nevertheless, consistent with the previous studies conducted in different populations (Martinotti et al., 2008; Pierò, 2010), low HA was not related with impulsivity in our previous study (Evren et al., 2014). This may suggest that combinations of temperament dimensions that define impulsivity may differ according to the population that is studied.

Whenever subjects present a personality weakness (low SD), the likelihood of impulsive personality disorders (cluster B of the *Diagnostic and Statistical Manual of Mental Disorders*, fourth edition) increases (Cloninger, 2000). Cluster B personality disorders (i.e., antisocial personality disorder—ASPD, borderline personality disorder—BPD) typically show a high degree of comorbidity with SUDs (Casillas & Clark, 2002; Verheul & van den Brink, 2000). A pathologic increase in impulsivity is at the core of impulsive personality disorders (BPD and ASPD). Among these personality disorders, ASPD has especially been reported to reflect high NS, low HA, and low RD traits, whereas "impulsive" BPD reflects high NS, high HA, and low RD traits (Kose, 2003). One study interestingly

TABLE 1 Comparison of BIS-11 and Subscales of BIS-11 between Healthy Control and Heroin-Dependent Groups

	Healthy Controls		Heroin Dependents		
	Mean	S.D.	Mean	S.D.	p
BIS-11	58.81	9.20	67.82	10.03	<0.001
Attentional impulsiveness	15.32	3.26	16.87	3.71	0.01
Motor impulsiveness	19.19	4.31	22.80	4.44	<0.001
Nonplanning impulsiveness	24.30	4.75	28.15	4.35	<0.001

TABLE 2 Predictors of Severity of Impulsivity, Attentional, Motor, and Nonplanning Impulsiveness in Heroin Dependents

BIS-11	Attentional	Motor	Nonplanning
Novelty seeking	Cooperativeness	Novelty seeking	Self-directedness
Self-directedness	Depression	Depression	Persistence
Persistence			

reported high HA in ASPD subjects with alcohol abuse (Tikkanen, Holi, Lindberg, & Virkkunen, 2007). These authors found that offenders with low HA committed less impulsive violence, whereas HA was high in subjects with either ASPD or BPD. A 2011 study also found that the NS and HA scores were higher, whereas RD and P scores were lower, in ASPD (Basoglu et al., 2011). Results indicate that impulsivity plays a significant role in cluster B personality disorders and SUDs, as well as in their association with each other (Casillas & Clark, 2002). Thus, consistent with the previous studies (Pierò, 2010), the findings of our previous study (Evren, Bozkurt, et al., 2014) may suggest that BPD and ASPD may be highly represented in our samples of substance dependents.

In response to stress or environmental cues, an individual with a substance abuse disorder could use the substance in a rapid unplanned action without regard to the consequences (Mitchell et al., 2005). Substance use may further diminish control over impulsivity regulation. Results suggest that although temperament dimensions of high NS and low P and a character dimension of low SD predict impulsivity, personality dimensions that predict impulsivity dimensions seem to differ. Measures of impulsivity dimensions seem therefore related to different measures of temperament; a combined approach to this issue could help to create more homogeneous diagnostic categories (or subtypes) and aid in the identification of at-risk individuals for specific symptoms, behaviors, or treatment responses. An adequate, tailored pharmacologic and psychologic treatment of impulsivity dimensions according to related temperament dimensions in subjects with heroin dependency could produce a better response to therapeutic efforts, as well as reducing the risk of dropout. This may suggest that when impulsivity is the problem, types of impulsivity and personality dimensions must be evaluated and the treatment should be shaped accordingly among heroin dependents. In particular, subjects with low SD may benefit from a specific psychotherapeutic and psychopharmacologic approach aimed at improving self-control and self-efficacy (Cloninger, 2000), which may reduce impulsivity.

APPLICATIONS TO OTHER ADDICTIONS AND SUBSTANCE MISUSE

It was mentioned in this chapter that high levels of impulsivity are associated with heroin dependency. Studies also confirmed a relationship between impulsivity and other SUDs. Impulsive behavior might be causally linked to several distinct processes in drug addiction, including onset, maintenance, related problems, and relapse (Pattij & De Vries, 2013). It has also been suggested that substance use might facilitate impulsivity by interfering with normal inhibitory controls (Verdejo-García et al., 2008). Impulsivity has also been found to be a high-risk factor for early substance use and related to the severity of drug abuse and treatment retention (Tarter et al., 2003) and it might contribute significantly to the risk of suicide attempts in substance-dependent patients, may interrupt their outpatient or inpatient treatment, and may mediate the effects of substance use on aggression (Bozkurt et al., 2013).

Studies have also pointed out that the structure of impulsivity differed on the dimensional level depending on the type of substance dependence (Clark et al., 2006), which may indicate the different psychopharmacological properties and behavioral functions of these substances.

Studies suggested that impulsivity can be considered as an endophenotype of addictive behaviors (Verdejo-Garcia et al., 2008). Consistent with this, studies also show associations between impulsivity and behavioral addictions, such as Internet addiction (Dalbudak, Evren, Aldemir, et al., 2013). Also, among university students, high NS scores, along with low character scores (SD and C), were related with Internet addiction (Dalbudak, Evren, Topcu, et al., 2013).

Impulsive behavior is defined as the coexistence of four heritable temperamental traits: high NS, low HA, low P, and, rarely, high RD (Cloninger, 1996). It was suggested that the subjects most vulnerable to substance dependency may be those with high impulsivity and/or high NS (García-Forero et al., 2009). Thus, it is important to evaluate the relationship of personality dimensions with impulsivity in substance-dependent populations.

Measures of impulsivity dimensions seem therefore related to various measures of temperament; a combined approach to this issue could help to create more homogeneous diagnostic categories (or subtypes) and aid in the identification of at-risk individuals for specific symptoms, behaviors, or treatment responses. An adequate, tailored pharmacologic and psychologic treatment of impulsivity dimensions according to related temperament dimensions in subjects with SUD could produce a better response to therapeutic efforts, as well as reducing the risk of dropout.

DEFINITION OF TERMS

Attentional impulsiveness It is a dimension of impulsivity that refers to inattention and cognitive instability.

Barrat Impulsiveness Scale-11 It is a scale that was designed to assess the personality trait of impulsiveness.

Character Character is self-concepts and individual differences in goals and values, which influence voluntary choices and the meaning and significance of what is experienced.

Cooperativeness C is a character dimension that quantifies the extent to which individuals conceive themselves as integral parts of human society.

Craving Craving is a strong desire to consume a particular substance and is a major factor in relapse and/or continued use after withdrawal.

Harm avoidance HA is a temperamental dimension, which refers to the inhibition of behavior in response to signals of punishment and frustrative nonreward.

Impulsivity Impulsivity is a predisposition toward rapid, unplanned reactions to internal or external stimuli without regard for the negative consequences of these reactions to the impulsive individual or to others.

Motor impulsiveness It is a dimension of impulsivity that refers to acting on the spur of the moment.

Nonplanning impulsiveness It is a tendency to choose a small, more immediate reward over a larger, more delayed reward and centered on the "present orientation" with a "lack of planning for the future and foresight."

Novelty seeking NS is a temperamental dimension that refers to the initiation of appetitive approach in response to novelty, approach to signals of reward, active avoidance of conditioned signals of punishment, and skilled escape from unconditioned punishment.

Persistence P is a temperamental dimension that refers to the maintenance of behavior despite frustration, fatigue, and intermittent reinforcement and is related to industriousness, determination, and perfectionism.

Reward dependence RD is a temperamental trait that refers to the maintenance of behavior in response to cues of social reward and is related to sentimentality, social sensitivity, attachment, and dependence on approval by others.

Self-directedness SD is a character dimension that quantifies the extent to which an individual is responsible, reliable, resourceful, goal-oriented, and self-confident.

Self-transcendence ST is a character dimension that quantifies the extent to which individuals conceive themselves as integral parts of the universe as a whole.

Temperament Temperament is heritable automatic emotional responses to experience and is thought to be stable throughout life.

Temperament and Character Inventory It is a self-administered dimensional questionnaire constructed to assess the seven basic dimensions of personality.

KEY FACTS OF IMPULSIVITY, TEMPERAMENT, AND CHARACTER

- There is accumulating evidence indicating that impulsive behavior might be causally linked to several distinct processes in drug addiction, including onset, maintenance, related problems, and relapse.
- Impulsivity is an important construct on which to focus in the treatment of substance dependents, but different dimensions might be the center of attention for patients with different substances of dependence.
- Individual differences in personality structure and development have a strong influence on the risk of substance abuse.
- Temperament consists of four traits: HA, NS, RD, and P. Character consists of three dimensions: SD, C, and ST.
- Impulsive behavior is defined as the coexistence of four heritable temperamental traits: high NS, low HA, low P, and, rarely, high RD. Among these traits, high NS was consistently found to be related with impulsivity in various populations.
- It was suggested that the subjects most vulnerable to substance dependency may be those with high impulsivity and/or high NS.

SUMMARY POINTS

- This chapter gives an overview of studies that evaluated the association of impulsivity with temperament and character dimension in SUDs and heroin dependents.
- Impulsivity is a temperamental risk factor for substance use and might be a fundamental mechanism in both the onset of excessive substance use and relapse.
- Previous studies have shown that current heroin addicts have higher levels of impulsivity than nonaddicts do on measures of impulsivity, while impulsivity measured by the BIS-11 distinguishes heroin users from nonusers.
- Impulsivity has also been found to be a high-risk factor for illicit and early substance use and related to the severity of drug abuse, craving, and treatment retention.
- Substance use might facilitate impulsivity by interfering with normal inhibitory controls.
- NS, which is a temperament dimension linked to impulsivity, was found to be higher in heroin dependents.
- Some studies indicate that the NPI dimension of impulsivity exists before drug exposure, while the MI and AI components are exacerbated by drug exposure.
- The continuous intake of heroin increases levels of impulsivity that return to baseline (preheroin) levels throughout abstinence; in heroin-dependent subjects, impulsivity therefore becomes more intense as a result of chronic heroin exposure.
- Heroin-dependent subjects may prefer this substance because of the self-medication dynamics, which have positive effects in managing the preexisting aggressiveness that is usually supported by impulsiveness.
- MI seems to play a major role in the age of onset of drug use and may be related to relapse in heroin dependents.
- Measures of impulsivity dimensions seem therefore related to various measures of temperament; a combined approach to this issue could help to create more homogeneous diagnostic categories and aid in the identification of at-risk individuals for specific symptoms, behaviors, or treatment responses.
- An adequate, tailored pharmacologic and psychologic treatment of impulsivity dimensions according to related temperament dimensions in subjects with heroin dependency could produce a better response to therapeutic efforts, as well as reducing the risk of dropout.

REFERENCES

Aharonovich, E., Nguyen, H. T., & Nunes, E. V. (2001). Anger and depressive states among treatment-seeking drug abusers: testing the psychopharmacological specificity hypothesis. *The American Journal on Addictions/American Academy of Psychiatrists in Alcoholism and Addictions, 10*, 327–334.

Ball, S. A., Kranzler, H. R., Tennen, H., Poling, J. C., & Rounsaville, B. J. (1998). Personality disorder and dimension differences between type A and type B substance abusers. *Journal of Personality Disorders, 12*, 1–12.

Ball, S. A., Tennen, H., Poling, J. C., Kranzler, H. R., & Rounsaville, B. J. (1997). Personality, temperament, and character dimensions and the DSMIV personality disorders in substance abusers. *Journal of Abnormal Psychology, 106*, 545–553.

Basiaux, P., le Bon, O., Dramaix, M., Massat, I., Souery, D., Mendlewicz, J., ... Verbanck, P. (2001). Temperament and Character Inventory (TCI) personality profile and sub-typing in alcoholic patients: a controlled study. *Alcohol and Alcoholism, 36*, 584–587.

Basoglu, C., Oner, O., Ates, A., Algul, A., Bez, Y., Ebrinc, S., & Cetin, M. (2011). Temperament traits and psychopathy in a group of patients with antisocial personality disorder. *Comprehensive Psychiatry, 52*, 607–612.

Bornovalova, M. A., Daughters, S. B., Hernandez, G. D., Richards, J. B., & Lejuez, C. W. (2005). Differences in impulsivity and risk-taking propensity between primary users of crack cocaine and primary users of heroin in a residential substance-use program. *Experimental and Clinical Psychopharmacology, 13*, 311–318.

Bozkurt, M., Evren, C., Yilmaz, A., Can, Y., & Cetingok, S. (2013). Aggression and impulsivity in different groups of alcohol and heroin dependent inpatient men. *Bulletin of Clinical Psychopharmacology, 23*, 335–344.

Casillas, A., & Clark, L. A. (2002). Dependency, impulsivity, and self-harm: traits hypothesized to underlie the association between cluster B personality and substance use disorders. *Journal of Personality Disorders, 16*, 424–436.

Chamorro, J., Bernardi, S., Potenza, M. N., Grant, J. E., Marsh, R., Wang, S., & Blanco, C. (2012). Impulsivity in the general population: a national study. *Journal of Psychiatric Research, 46*, 994–1001.

Clark, L., Robbins, T. W., Ersche, K. D., & Sahakian, B. J. (2006). Reflection impulsivity in current and former substance users. *Biological Psychiatry, 60*, 515–522.

Cloninger, C. R. (1987). Neurogenetic adaptive mechanisms in alcoholism. *Science, 24*, 410–416.

Cloninger, C. R. (1996). Assessment of the impulsive-compulsive spectrum of behavior by the seven-factor model of temperament and character. In J. M. Oldham, E. Hollander, & A. E. Skodol (Eds.), *Impulsivity and Compulsivity* (pp. 59–95). Washington: American Psychiatric Press.

Cloninger, C. R. (2000). A practical way to diagnosis personality disorder: a proposal. *Journal of Personality Disorders, 14*, 99–108.

Cloninger, C. R., & Svrakic, D. M. (1997). Integrative psychobiological approach to psychiatric assessment and treatment. *Psychiatry, 60*, 120–141.

Cloninger, C. R., Svrakic, D. M., & Przybeck, T. R. (1993). A psychobiological model of temperament and character. *Archives of General Psychiatry, 50*, 975–990.

Cuomo, C., Sarchiapone, M., Giannantonio, M. D., Mancini, M., & Roy, A. (2008). Aggression, impulsivity, personality traits, and childhood trauma of prisoners with substance abuse and addiction. *The American Journal of Drug and Alcohol Abuse, 34*, 339–345.

Dalbudak, E., Evren, C., Aldemir, S., Coskun, K. S., Ugurlu, H., & Yildirim, F. G. (2013). Relationship of internet addiction severity with depression, anxiety, and alexithymia, temperament and character in university students. *Cyberpsychology Behavior and Social Networking, 16*, 272–278.

Dalbudak, E., Evren, C., Topcu, M., Aldemir, S., Coskun, K. S., Bozkurt, M., ... Canbal, M. (2013). Relationship of Internet addiction with impulsivity and severity of psychopathology among Turkish university students. *Psychiatry Research, 210*, 1086–1091.

David, A., Nielsena, D. A., Ho, A., Bahl, A., Varma, P., Kellogg, S., ... Kreek, M. J. (2012). Former heroin addicts with or without a history of cocaine dependence are more impulsive than controls. *Drug and Alcohol Dependence, 124*, 113–120.

De la Rie, S. M., Duijsens, I. J., & Cloninger, C. R. (1998). Temperament, character, and personality disorders. *Journal of Personality Disorders, 12*, 362–372.

De Wit, H. (2009). Impulsivity as a determinant and consequence of drug use: a review of underlying processes. *Addiction Biology, 14*, 22–31.

Dissabandara, L. O., Loxton, N. J., Dias, S. R., Dodd, P. R., Daglish, M., & Stadlin, A. (2014). Dependent heroin use and associated risky behaviour: the role of rash impulsiveness and reward sensitivity. *Addictive Behaviors, 39*, 71–76.

Dougherty, D. M., Bjork, J. M., Harper, R. A., Marsh, D. M., Moeller, F. G., Mathias, C. W., & Swann, A. C. (2003). Behavioral impulsivity paradigms: a comparison in hospitalized adolescents with disruptive behavior disorders. *Journal of Child Psychology and Psychiatry, and Allied Disciplines, 44*, 1145–1157.

Ersche, K. D., Turton, A. J., Pradhan, S., Bullmore, E. T., & Robbins, T. W. (2010). Drug addiction endophenotypes: impulsive versus sensation-seeking personality traits. *Biological Psychiatry, 68*, 770–773.

Evren, C., Bozkurt, M., Evren, B., Can, Y., Yigiter, S., & Yılmaz, A. (2014). Relationships of personality dimensions with impulsivity in heroin dependent inpatient men. *Anadolu Psikiyatri Dergisi, 15*, 9–14.

Evren, C., Cinar, O., Evren, B., & Celik, S. (2012). Relationship of self-mutilative behaviours with severity of borderline personality, childhood trauma and impulsivity in male substance dependent inpatients. *Psychiatry Research, 200*, 20–25.

Evren, C., Durkaya, M., Evren, B., Dalbudak, E., & Cetin, R. (2012). Relationship of relapse with impulsivity, novelty seeking and craving in male alcohol-dependent inpatients. *Drug and Alcohol Review, 31*, 81–90.

Evren, E., Evren, B., Yancar, C., & Erkiran, M. (2007). Temperament and character model of personality profile of alcohol- and drug-dependent inpatients. *Comprehensive Psychiatry, 48*, 283–288.

Evren, C., Yilmaz, A., Can, Y., Bozkurt, M., Evren, B., & Umut, G. (2014). Severity of impulsivity and aggression in 12 month follow-up among male heroin dependent patients. *Bulletin of Clinical Psychopharmacology, 24*, 158–167.

Eysenck, S. B., & Eysenck, H. J. (1977). The place of impulsiveness in a dimensional system of personality description. *The British Journal of Social and Clinical Psychology, 16*, 57–68.

Gabel, S., Stallings, M. C., Schmitz, S., Young, S. E., & Fulker, D. W. (1999). Personality dimensions and substance misuse: relationships in adolescents, mothers and fathers. *The American Journal on Addictions/American Academy of Psychiatrists in Alcoholism and Addictions, 8*, 101–113.

García-Forero, C., Gallardo-Pujol, D., Maydeu-Olivares, A., & Andrés-Pueyo, A. (2009). Disentangling impulsiveness, aggressiveness and impulsive aggression: an empirical approach using self-report measures. *Psychiatry Research, 168*, 40–49.

Gerdner, A., Nordlander, T., & Pedersen, T. (2002). Personality factors and drug of choice in female addicts with psychiatric comorbidity. *Substance Use and Misuse, 37*, 1–18.

Gourion, D., Pelissolo, A., & Lepine, J. P. (2003). Test-retest reliability of the temperament and character inventory in patients with opiate dependence. *Psychiatry Research, 118*, 81–88.

Gutierrez, F., Sangorrin, J., Martin-Santos, R., Torres, X., & Torrens, M. (2002). Measuring the core features of personality disorders in substance abusers using the Temperament and Character Inventory (TCI). *Journal of Personality Disorders, 16*, 344–359.

Hoaken, P. N., & Stewart, S. H. (2003). Drugs of abuse and the elicitation of human aggressive behavior. *Addictive Behaviors, 28*, 1533–1554.

Hogarth, L. (2011). The role of impulsivity in the etiology of drug dependence: reward sensitivity versus automaticity. *Psychopharmacology (Berlin), 215*, 567–580.

Hosak, L., Preiss, M., Halir, M., Cermakova, E., & Csemy, L. (2004). Temperament and character inventory (TCI) personality profile in methamphetamine abusers: a controlled study. *European Psychiatry: The Journal of the Association of European Psychiatrists, 19*, 193–195.

Jentsch, J. D., & Taylor, J. R. (1999). Impulsivity resulting from frontostriatal dysfunction in drug abuse: implications for the control of behavior by reward-related stimuli. *Psychopharmacology (Berlin)*, *146*, 373–390.

Kirby, K. N., & Petry, N. M. (2004). Heroin and cocaine abusers have higher discount rates for delayed rewards than alcoholics or non-drug using controls. *Addiction*, *99*, 461–471.

Kirby, K. N., Petry, N. M., & Bickel, W. K. (1999). Heroin addicts have higher discount rates for delayed rewards than non-drug-using controls. *Journal of Experimental Psychology. General*, *128*, 78–87.

Kollins, S. H. (2002). Delay discounting is associated with substance use in college students. *Addictive Behaviors*, *28*, 1167–1173.

Koob, G. F., & Le Moal, M. (1997). Drug abuse: hedonic homeostatic dysregulation. *Science*, *278*, 52–58.

Kose, S. (2003). A psychobiological model of temperament and character: TCI. *New Symposium*, *41*, 86–97.

Lane, S. D., Cherek, D. R., Rhoades, H. M., Pietras, C. J., & Tcheremissine, O. V. (2003). Relationships among laboratory and psychometric measures of impulsivity: implications in substance abuse and dependence. *Addictive Disorders and Their Treatment*, *2*, 33–40.

Le Bon, O., Basiaux, P., Streel, E., Tecco, J., Hanak, C., Hansenne, M., ... Dupont, S. (2004). Personality profile and drug of choice; a multivariate analysis using Cloninger's TCI on heroin addicts, alcoholics, and a random population group. *Drug and Alcohol Dependence*, *73*, 175–182.

Lejuez, C. W., Bornovalova, M. A., Daughters, S. B., & Curtin, J. J. (2005). Differences in impulsivity and sexual risk behavior among inner-city crack/cocaine users and heroin users. *Drug and Alcohol Dependence*, *77*, 169–175.

Li, T., Du, J., Yu, S., Jiang, H., Fu, Y., Wang, D., ... Zhao, M. (2012). Pathways to age of onset of heroin use: a structural model approach exploring the relationship of the COMT gene, impulsivity and childhood trauma. *PLoS One*, *7*, e48735.

Maremmani, A. G., Rovai, L., Pani, P. P., Pacini, M., Lamanna, F., Rugani, F., ... Maremmani, I. (2011). Do methadone and buprenorphine have the same impact on psychopathological symptoms of heroin addicts? *Annals of General Psychiatry*, *10*, 17.

Martinotti, G., Mandelli, L., Di Nicola, M., Serretti, A., Fossati, A., Borroni, S., ... Janiri, L. (2008). Psychometric characteristic of the Italian version of the Temperament and Character Inventory–revised, personality, psychopathology, and attachment styles. *Comprehensive Psychiatry*, *49*, 514–522.

McCormick, R. A., Dowd, E. T., Quirk, S., & Zegarra, J. H. (1998). The relationship of NEO-PI performance to coping styles, patterns of use, and triggers for use among substance abusers. *Addictive Behaviors*, *23*, 497–507.

Mitchell, J. M., Fields, H. L., D'Esposito, M., & Boettiger, C. A. (2005). Impulsive responding in alcoholics. *Alcoholism, Clinical and Experimental Research*, *29*, 2158–2169.

Moeller, F. G., Barrat, E. S., Dougherty, D. M., Schmitz, J. M., & Swann, A. C. (2001). Psychiatric aspects of impulsivity. *American Journal of Psychiatry*, *158*, 1783–1793.

Moeller, F. G., Dougherty, D. M., Barratt, E. S., Oderine, V., Mathias, C. W., Harper, R. A., & Swann, A. C. (2002). Increased impulsivity in cocaine dependent subjects independent of antisocial personality disorder and aggression. *Drug and Alcohol Dependence*, *68*, 105–111.

Nielsen, D. A., Ho, A., Bahl, A., Varma, P., Kellogg, S., Borg, L., & Kreek, M. J. (2012). Former heroin addicts with or without a history of cocaine dependence are more impulsive than controls. *Drug and Alcohol Dependence*, *124*, 113–120.

Pattij, T., & De Vries, T. J. (2013). The role of impulsivity in relapse vulnerability. *Current Opinion in Neurobiology*, *23*, 700–705.

Patton, J. H., Stanford, M. S., & Barratt, E. S. (1995). Factor structure of the Barratt impulsiveness scale. *Journal of Clinical Psychology*, *51*, 768–774.

Perry, J. L., & Carroll, M. E. (2008). The role of impulsive behavior in drug abuse. *Psychopharmacology (Berlin)*, *200*, 1–26.

Pierò, A. (2010). Personality correlates of impulsivity in subjects with generalized anxiety disorders. *Comprehensive Psychiatry*, *51*, 538–545.

Schippers, M. C., Binnekade, R., Schoffelmeer, A. N., Pattij, T., & De Vries, T. J. (2012). Unidirectional relationship between heroin self-administration and impulsive decision-making in rats. *Psychopharmacology (Berlin)*, *219*, 443–452.

Seo, D., Patrick, C. J., & Kennealy, P. J. (2008). Role of serotonin and dopamine system interactions in the neurobiology of impulsive aggression and its comorbidity with other clinical disorders. *Aggression and Violent Behavior*, *13*, 383–395.

Seroczynski, A. D., Bergeman, C. S., & Coccaro, E. F. (1999). Etiology of the impulsivity/aggression relationship: genes or environment? *Psychiatry Research*, *86*, 41–57.

Simons, J. S. (2003). Differential prediction of alcohol use and problems: the role of biopsychological and social-environmental variables. *The American Journal of Drug and Alcohol Abuse*, *29*, 861–879.

Tarter, R. E., Kirisci, L., Mezzich, A., Cornelius, J. K., Pajer, K., Vanyukov, M., ... Clark, D. (2003). Neurobehavioral disinhibition in childhood predicts early age at onset of substance use disorder. *The American Journal of Psychiatry*, *160*, 1078–1085.

Tikkanen, R., Holi, M., Lindberg, N., & Virkkunen, M. (2007). Tridimensional Personality Questionnaire data on alcoholic violent offenders: specific connections to severe impulsive cluster B personality disorders and violent criminality. *BMC Psychiatry*, *7*, 36.

Verdejo-García, A., Bechara, A., Recknor, E. C., & Pérez-García, M. (2007). Negative emotion-driven impulsivity predicts substance dependence problems. *Drug and Alcohol Dependence*, *91*, 213–219.

Verdejo-García, A., Lawrence, A. J., & Clark, L. (2008). Impulsivity as a vulnerability marker for substance-use disorders: review of findings from high-risk research, problem gamblers and genetic association studies. *Neuroscience and Biobehavioral Reviews*, *32*, 777–810.

Verdejo-Garcia, A., & Perez-Garcia, M. (2007). Profile of executive deficits in cocaine and heroin polysubstance users: common and differential effects on separate executive components. *Psychopharmacology*, *190*, 517–530.

Verheul, R., & van den Brink, W. (2000). The role of personality pathology in the aetiology and treatment of substance use disorders. *Current Opinion in Psychiatry*, *13*, 163–169.

Chapter 92

fMRI of Heroin and Ketamine Addiction

Y.L. Jiang[1], Tony C.H. Chow[2], Sharon L. Wu[2], H.C. Tang[2], Maria S.M. Wai[2], D.T. Yew[2]
[1]*Department of Radiology, Kunming Medical University, Yunnan, China;* [2]*Schools of Biomedical Sciences and Chinese Medicine, The Chinese University of Hong Kong, Hong Kong, China*

Abbreviations

BOLD image Blood oxygen level-dependent image
fMRI Functional magnetic resonance imaging

INTRODUCTION

In functional magnetic resonance imaging (fMRI), for general studies, a stimulus or a movement is usually initiated and the normal BOLD (blood oxygen level-dependent) images are obtained. Subsequently images are compared between normal control individuals and addicts. If the investigator chooses to initiate a movement or muscular activity, the episode has to be simple, to avoid complications in interpretation. In our laboratory, we frequently ask the tested individuals to engage in movements like chewing and opposition of thumb and fingers.

The measurement of BOLD images is detailed as the search for an increase or decrease in blood oxygen level, which denotes the active sites of the functional activities. The idea is simple and direct. However, functional activities during MRI need to be continuous so that the neurons will require glucose and induce the vessels to continuously dilate. Termination of the continuous activity will arrest the functional response. In addition, there is usually a lag between the neuronal activity and the vascular response. Further, any ionic changes that can interrupt or enhance vascular activities will affect the accuracy of the functional recording. Often, the noise of the recording and the mental state of the individual can influence the recording. It is important to repeat the action several times and to enlist more specimens to obtain reliable results. fMRI can be performed on the developing brain (Fang et al., 2005).

APPLICATION TO TYPES OF ADDICTIONS

Our laboratory had performed fMRI on motor activities and upon cued movement of heroin addicts versus normal individuals (Jiang, Tian, Lu, & Rudd, 2014). The following is an example of an fMRI study on heroin addicts and normal individuals in the process of chewing.

When the individual is at rest and there is no motion, the superior parietal areas have very subtle activity, and BOLD images indicate a mixture of downregulated and moderated levels of oxygen usage. Blue colors on the figures indicate downregulated areas of oxygen. Moderate areas of oxygen are indicated by yellow, while high-oxygen-demand areas are in red in routine BOLD images (Figure 1). Upon movement, the superior parietal cortex displays a few more high-oxygen-demand sites in the BOLD image of the addict (Figure 2), while a lot of high-oxygen-demand areas are seen in the superior parietal areas of the normal individual doing the same chore or movement (Figure 3). Volumetrically, these high-oxygen-demand areas in the superior parietal cortex in the normal control are about three times the corresponding high-oxygen-demand areas in the addicted. Deeper slices of the brain of a normal individual at rest without motion again display many downregulated areas of oxygen demand in the BOLD image (Figure 4), while movement in these normal individuals elicits significant high-oxygen areas of demand in the BOLD image, in the premotor, motor, sensory, and association areas, as well as along the supplementary cortices (Figure 5). Although similar areas light up in the deeper brain of the addicts upon movement, the activation sites are much smaller in volume (Figure 6) than those of the normal controls. In the addicts who have been on heroin for less than 5 years (Figure 6), the activation is generally at least two times more in those areas than in those who have been on the drug for 10 years or more (Figure 7). In the brain slices through the diencephalon areas of the normal control (Figure 8), one can spot many activations around the thalamus and the internal capsule. The visual cortices light up as long as the tested individual is obviously still looking afar. Another two interesting areas that light up are the posterior part of the parahippocampal gyrus and the cingulate cortex in the normal controls. In the addicts, however, the activation of the parahippocampal gyrus is gone and the activation in the thalamus and the internal capsule is minimal (Figure 9). However, the activation of the cingulate cortex (posterior cingulated) persists in the addict as in the normal control (Figure 9). It is likely that this posterior cingulated gyrus, as an important emotional area, is not eliminated in the addict. The activation of the parahippocampal gyrus is likely to be related to memory and the limbic system in the normal control, while that in the heroin addict is downregulated. The cerebellum and the surrounding inferior temporal area are heavily labeled with high oxygen demand in the normal control (Figure 10), while some activities are still apparent here in the heroin addict, but such activities are greatly diminished (Figure 11). Although this general comparison was between movement activations in the brains of

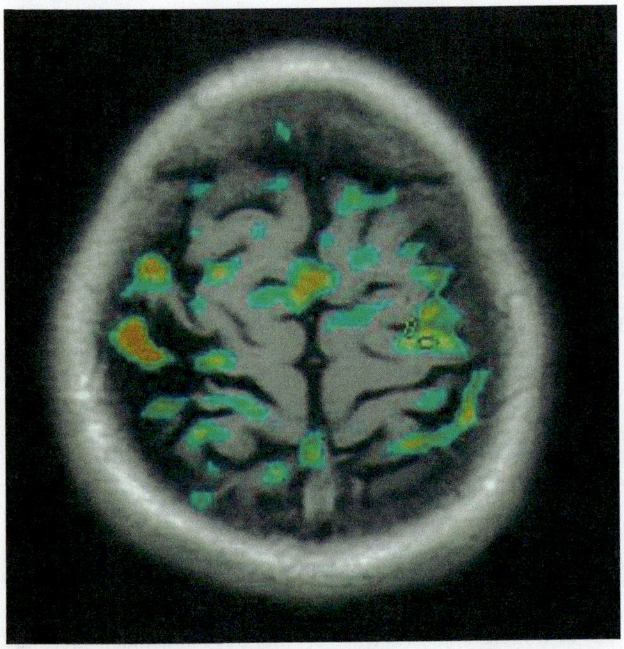

FIGURE 1 fMRI of a slice through the superior parietal cortex of a normal individual at rest, showing very few high-oxygen-demand active sites. Most sites are blue, meaning they are downregulated in oxygen demand.

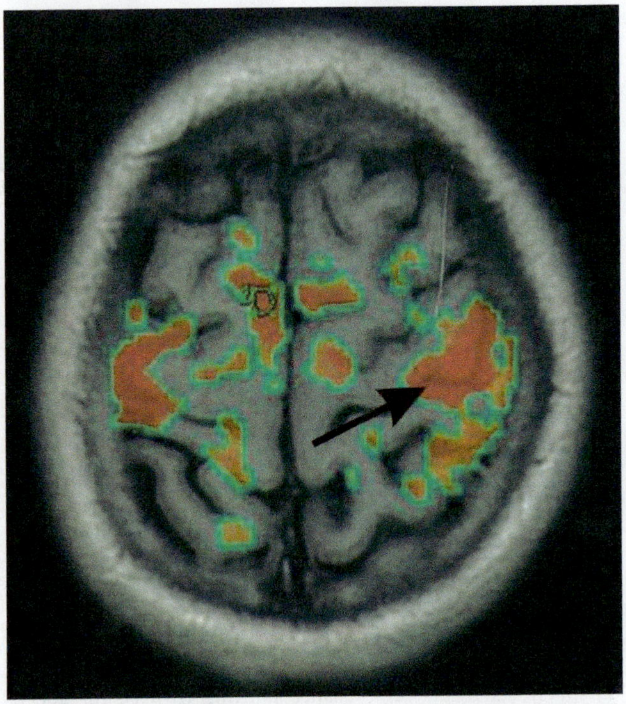

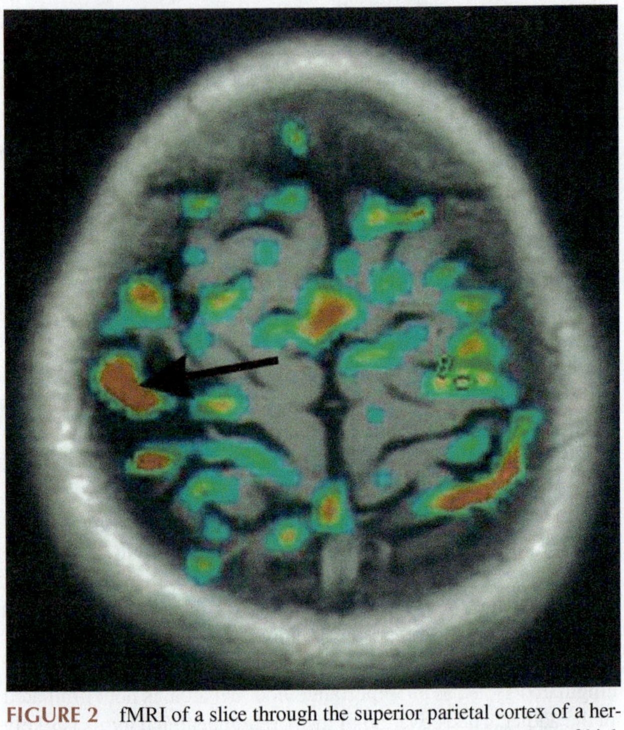

FIGURE 2 fMRI of a slice through the superior parietal cortex of a heroin addict in motion, showing an increase of a few activated sites of high oxygen demand (red color, arrow).

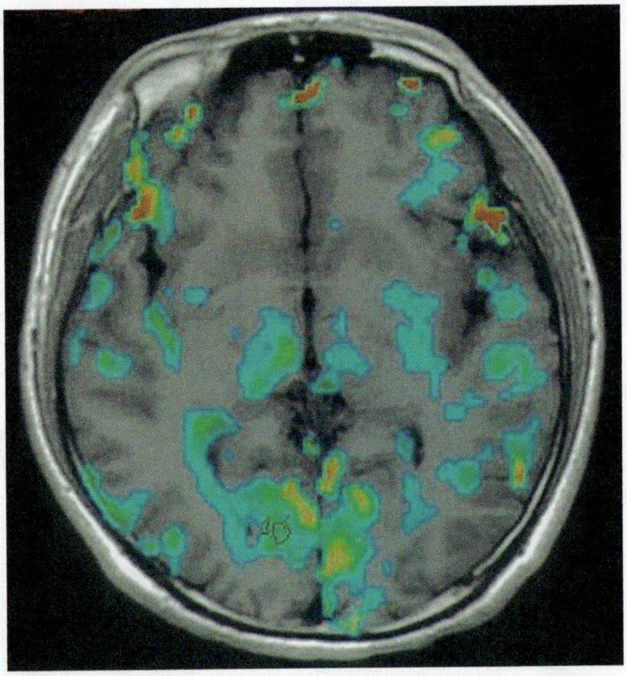

FIGURE 3 fMRI of a slice through the superior parietal cortex of a normal individual in motion, showing more active sites of high oxygen demand (arrow).

FIGURE 4 fMRI of a slice above the corpus callosum in a normal individual, showing downregulation at rest.

normal controls and of subjects who were addicted to opioids, our experience of humans and primates and their fMRI during movement under other abusive drugs is strikingly similar (e.g., ketamine). A general pattern of decreased activities in BOLD images, low activities of the diencephalon and memory areas, and underactivity of the cerebellum, along with decreases in muscular strength, is a feature of addicts on abusive drugs. However, noting the high activity in the cingulate and the prefrontal cortex of heroin addicts when shown a drug movie, one has to come to the conclusion that the many areas of the emotional and behavior systems of pathways are

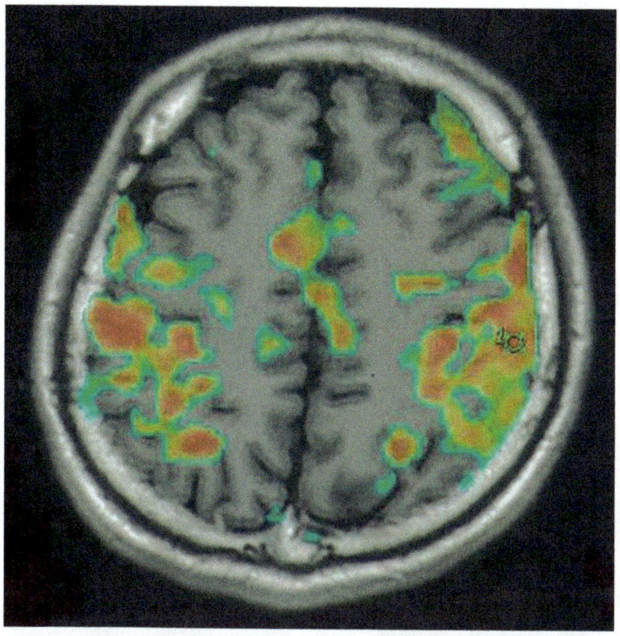

FIGURE 5 fMRI of a slice above the corpus callosum in a normal individual in motion, showing lots of active sites, including supplementary motor cortices.

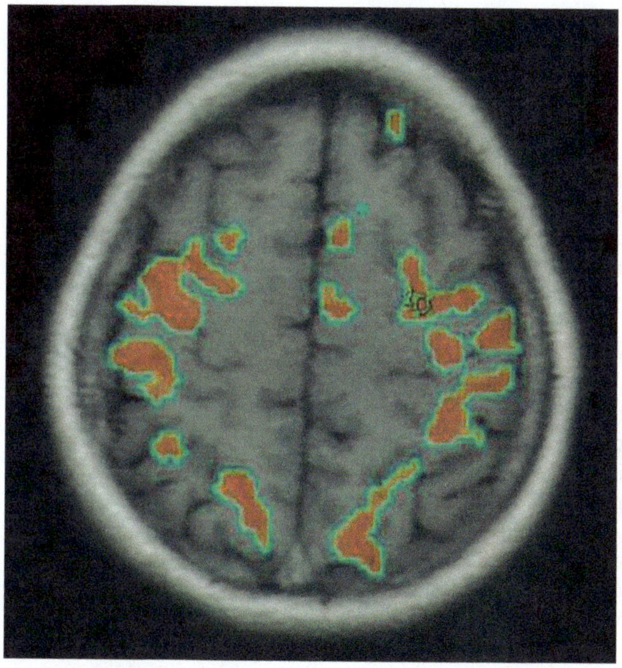

FIGURE 6 fMRI of a slice above the corpus callosum in a heroin addict of 5 years, in motion, showing fewer active sites than the normal control, shown in Figure 4.

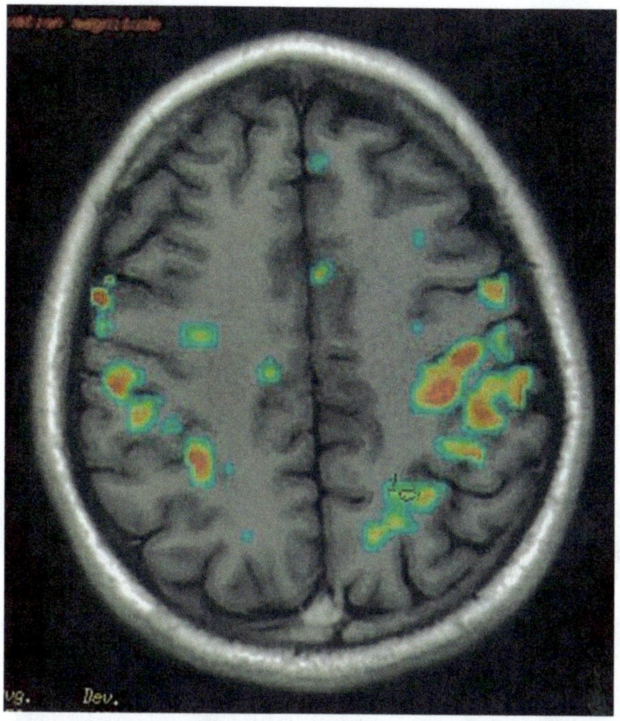

FIGURE 7 fMRI of a slice above the corpus callosum in a heroin addict of over 10 years, in motion, showing even fewer active sites than in the addict of 5 years.

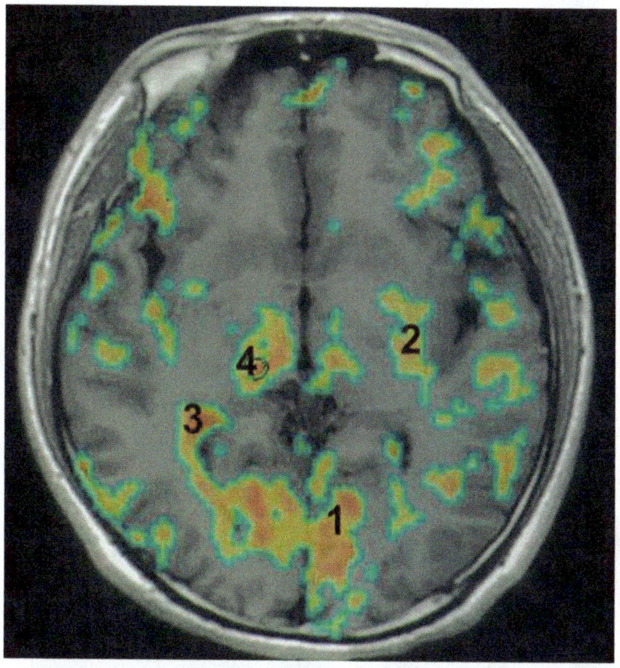

FIGURE 8 fMRI of a slice at the level of the diencephalon of a normal individual in motion. Note many active sites, including (1) the coagulate cortex, (2) the internal capsule, (3) the posterior parahippocampal gyrus, and (4) the thalamus.

still functioning in the addicts. The activated centers generated in this example of chewing in heroin addicts versus normal controls, on the whole, illustrate activated centers that are similar to those reported in opposition of thumb and fingers (Jiang et al., 2014).

In another experiments, we compared the prefrontal responses in heroin addicts when they were shown a drug puffing movie and a pornographic movie. As predicted, the prefrontal areas did not respond much to the pornographic movie as shown by the fMRI (at a much lower level than in a normal adult), but the response to the drug movie in the prefrontal cortex was several times higher than in controls. When we compared the fMRI of addicts of 5 years

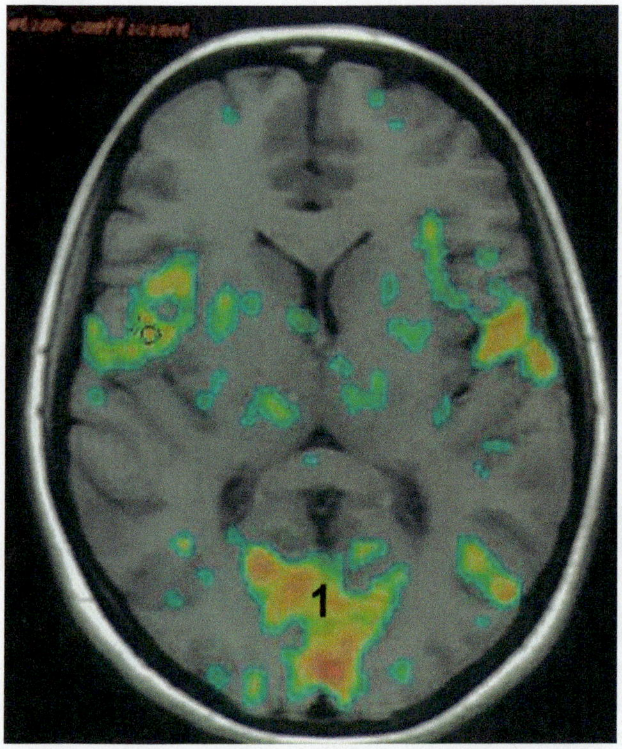

FIGURE 9 fMRI of a slice at the level of the diencephalon of a heroin addict in motion. Note fewer active sites than in the normal control, but the cingulate cortex (1) is still activated.

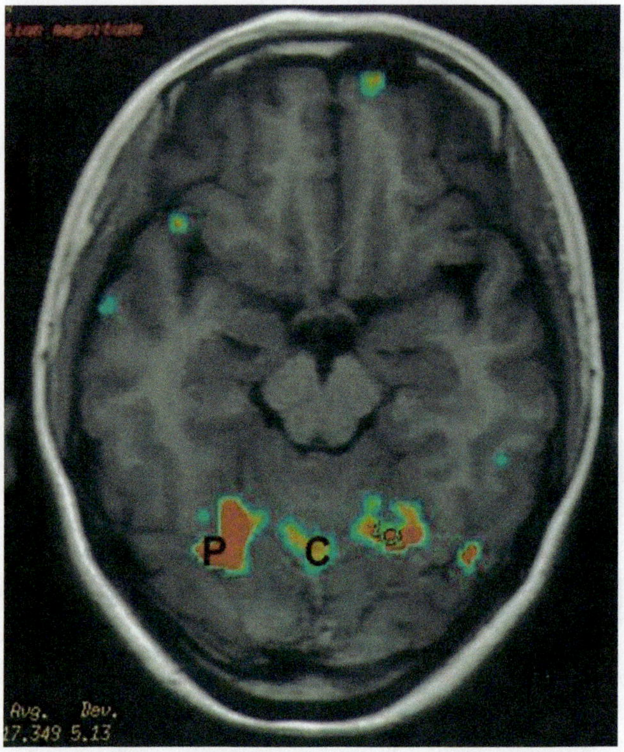

FIGURE 11 fMRI of a slice at the level of the cerebellum of a heroin addict in motion. Note active sites in the cerebellum (C) and the parahippocampal gyrus (P), which are less in volume than in the normal control.

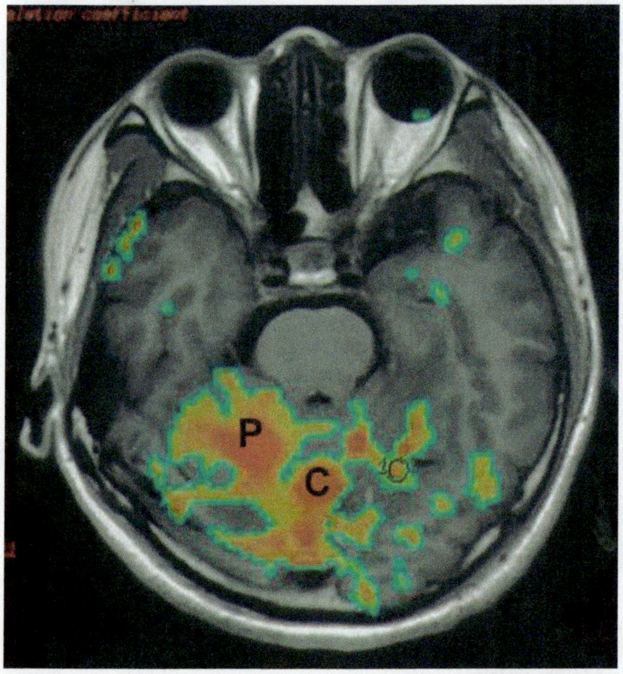

FIGURE 10 fMRI of a slice at the level of the cerebellum of a normal individual in motion. Note active sites in cerebellum (C) and the surrounding parahippocampal gyrus (P).

of addiction with those of 10 years of addiction, showing them the drug puffing movie, the activation volumes (in voxels) were not statistically different. This may indicate that the addiction potential was already firmly established after a few years of addiction. It must also be remembered that addiction to different abusive agents would take different durations. As an example, in our laboratory, we performed the conditioned place preference assay on our mice addicted to methamphetamine. We had mice that expressed behavioral and chemical addictive responses after only 2 weeks of treatment (preliminary studies). Of further interest, we have tried comparing addicts (on heroin or ketamine) watching a drug puffing movie and a movie of just cigarette smoking. The fMRI of the prefrontal cortices, to our surprise, showed equivalent volumes of activation sites. It may be that the puffing scene itself reminded the abusers of their addiction. There are as yet few studies using fMRI or MRI of humans addicted to different drugs of abuse. In Table 1, we present a comparison of brains between heroin addiction and ketamine addiction.

Brain damage in drug addiction, usually with abusive drugs, has been studied for years. The drugs frequently found and used by abusers in Asia include heroin, methamphetamine, lysergic acid diethylamide (LSD), cocaine, ecstasy, and ketamine. All of these agents in many ways affect the central nervous system. The drug ketamine, which was popular among abusers in Asia in the late 1990s and which is now gaining popularity in Europe and America, has been tested in various primate and mouse models. The brains of these animals are very susceptible to the effects of

TABLE 1 fMRI Changes after Motion

	Heroin	Ketamine
Motor cortex	↑	↓
Prefrontal cortex	↑	
Cerebellum	↑	
Brainstem		↓ (upregulated in mice)
Cingulate		↓
Parahippocampal	↓	↑
Striatum	↓	↑
Thalamus		↓

Yu et al. (2012), Sun et al. (2011, 2012), and Jiang et al. (2014).

ketamine and its metabolites. One of the first studies on the damage by ketamine was on cultured cells.

Neuroblastoma cells were used in those experiments. These cells can differentiate into neurons via the employment of retinoic acid. After differentiation, these neuron-like cells and the full assortment of axons and dendrites were infused with ketamine in the culture medium. Neuronal cells began to retract their processes and died (Mak, Lam, Lu, Wong, & Yew, 2010). Subsequently, the ketamine effects were tested on animals and were concluded to work on many organs in vivo, which included the heart, liver, and urogenital systems. Kidney malfunctioning initially resulted from tubular degeneration followed by glomerular damage and was accompanied by epithelial pain and fibrosis of the bladder, which might be related to an immune episode. In the brain of animals, focal lesions of cell loss, usually via apoptosis, were observed in both mice and primates (Yew, 2015). These lesions were not only present in the cortical areas, but were also observed subcortically, for example, in the cerebellum and the brainstem. Likewise, neuronal cells were not the only affected elements. The fibers from these cells started to degenerate early in addiction, with vesiculation of the myelin sheath and degeneration of the axon (Yew, 2015). In humans, MRI confirmed early fibrous track lesions as early as 1 year after addiction to ketamine, and by 6 to 7 years of addiction, the fiber track degenerations were so eminent that even computed tomographical scan could reveal degenerative fibers (Wang et al., 2013). The lesions in the brains of the human had impacts on the cell bodies of the cortex, again usually related to apoptosis, and severe atrophy began to appear after several years (usually 4–5 years) (Wang et al., 2013). The damage spread to the cerebellum and brainstem, and by 10 years of addiction, nearly all of the areas of the brain demonstrated lesions (Wang et al., 2013). In the experimental animals, mice and primates, brain damage went along with loss of muscle strength. In both animal models, immunohistochemistry tagged the presence of mutated tau protein in the neurons, suggesting a possibility of early neuronal cell damage akin to an episode of Alzheimer disease (Yeung et al., 2010).

Toxicity of other abusive drugs on the brain has been reported in the literature. Among those cited in this chapter, several important ones led to brain lesions. For example, LSD, which was tied to hallucinations, had been reported to induce recurrent cortical blindness in patients with migraine a few days after intake (Berhard & Ulrick, 2009). Methamphetamine has been known to bind to dopamine D3 receptors, particularly in the midbrain, and cause toxicity to the dopamine projections into the cortex, resulting in a decrease in gray matter volume (Heidbreder et al., 2005; Morales et al., 2015). The latter authors (Morales et al., 2015) reported that the mesolimbic system was especially targeted. In our studies with ketamine, we actually observed an increase in tyrosine hydroxylase neurons in the midbrain of animals treated long term with ketamine. It appears that dopaminergic enhanced firing would predispose target neurons to degenerate. Since, as of this writing, there is no proof that dopaminergic misfiring would lead to vascular changes, the degeneration of neurons via necrosis seems unlikely. It appears that cell death in the nervous system as a result of abusive drugs is related to apoptosis, as reported for ketamine.

Our fMRI studies discussed in this chapter and those MRI and fMRI studies in previous works pointed out that fiber tracts were affected. This was in agreement with studies on connectivity of the brain after abusive drug insult in the cocaine model. Ray, Gohel, and Biswal (2015) and Salzwedel et al. (2015) both reported decrease or increase in functional connectivity in the brain after cocaine intake. These results align with our findings on fiber degeneration as well as low activity in BOLD images of addicts. It seems that many of the abusive drugs, through different mechanisms, lead to common patterns of degeneration at the end.

SUMMARY POINTS

- Abusive drugs, of all drugs studied, cause cell death and nerve fiber changes in the central nervous system.
- Downregulation and upregulation of MRI activities were seen in various parts of the brain in heroin and ketamine addicts.
- The prefrontal area was important in craving.
- Motor and sensory cortices were downregulated in heroin and ketamine addicts.
- The cerebellum was also downregulated in heroin and ketamine addicts.

REFERENCES

Berhard, M. K., & Ulrick, K. (2009). Recurrent blindness after LSD intake. *Fortschritte der Neurologie Psychiatrie, 77*, 102–104.

Fang, M., Lorke, D. E., Li, J., Gong, X., Yew, J. C., & Yew, D. T. (2005). Postnatal changes in functional activities of the pig's brain: a combined functional magnetic resonance imaging and immunohistochemical study. *Neurosignals, 14*, 222–233.

Heidbreder, C. A., Gardner, E. L., Xi, Z. X., Thanos, P., Mugnaini, M., Hagen, J. J., & Ashby, C. R. (2005). The role of central D3 dopamine receptors in drug addiction: a review of pharmacological evidence. *Brain Research Reviews, 49*, 77–105.

Jiang, Y. L., Tian, W., Lu, G., & Rudd, J. A. (2014). Patterns of cortical activation following motor tasks and psychological induce movie cues in heroin uses: an fMRI study. *Internal Journal of Psychiatry in Medicine, 47*, 25–40.

Mak, Y. T., Lam, W. P., Lu, L., Wong, Y. W., & Yew, D. T. (2010). The toxic effect of ketamine on SH-SY5Y neuroblastoma cell line and human neuron. *Microscopy Research and Technique, 73*, 195–201.

Morales, A., Kohn, M., Robertson, C. L., Dean, A., Mandelkern, M. A., & London, E. (2015). Grey matter volume, midbrain dopamine D2/D3 receptors and drug craving in methamphetamine users. *Molecular Psychiatry, 20*, 764–771.

Ray, S., Gohel, S. R., & Biswal, B. B. (2015). Altered functional connectivity strength in abstinent cocaine smokers compared to healthy controls. *Brain Connectivity, 5*(8), 476–486. Epub 2015.

Salzwedel, A. P., Grewen, K. M., Vachat, C., Gerig, G., Lin, W., & Gao, W. (2015). Prenatal drug exposure affects neonatal functional connectivity. *The Journal of Neuroscience, 35*, 5860–5869.

Sun, L., Lam, W. P., Wong, Y. W., Lam, L. H., Tang, H. C., Wai, M. S., ... Yew, D. T. (2011). Permanent deficits in brain functions caused by long-term ketamine treatment in mice. *Human and Experimental Toxicology, 30*, 1287–1296.

Sun, L., Li, Q., Zhang, Y., Liu, D., Jiang, H., Pan, F., ... Yew, D. T. (2012). Chronic ketamine exposure induces permanent impairment of brain functions in adolescent cynomolgus monkeys. *Addiction Biology, 19*, 185–194.

Wang, C., Zheng, D., Xu, J., Lam, W., & Yew, D. T. (2013). Brain damages in ketamine addicts as revealed by magnetic resonance imaging. *Frontiers in Neuroanatomy, 7*, 23. http://dx.doi.org/10.33889/fnana.00023. Epub 2013.

Yeung, L. Y., Wai, M. S., Fan, M., Mak, Y. T., Lam, W. P., Li, Z., ... Yew, D. T. (2010). Hyperphosphorylated tau in brains of mice and monkeys with long-term administration of ketamine. *Toxicology Letters, 193*, 189–193.

Yew, D. T. (2015). Ketamine use and misuse: an impact on the nervous system. In D. T. Yew (Ed.), *Ketamine use and misuse* (pp. 1–12). Boca Raton, USA: CRC Press.

Yu, H., Li, Q., Wang, D., Shi, L., Lu, G., Sun, L., ... Yew, D. T. (2012). Mapping the central effects of chronic ketamine administration in an adolescent primate mode by functional magnetic imaging (fMRI). *Neurotoxicology, 33*, 70–77.

Chapter 93

Heroin-Induced Cerebrovascular Ischemia and Leukoencephalopathy: Role of Mitochondria

Yashar Yousefzadeh Fard[1], Seyyed A. Hashemi[2], Megan L. Fitzgerald[1]
[1]*Department of Psychiatry, New York State Psychiatry Institute/Columbia University, New York, NY, USA;* [2]*Immunogenetic Research Center, Traditional and Complementary Medicine Research Center, Medical School Mazandaran University of Medical Sciences, Sari, Iran*

Abbreviations

ACA/MCA/PCA Anterior/middle/posterior cerebral artery
CBF Cerebral blood flow
CSF Cerebrospinal fluid
CT Computed tomography
MRI Magnetic resonance imaging
MRS Magnetic resonance spectroscopy
NAA *N*-Acetylaspartate

There are few cases in the literature about cerebrovascular ischemic lesions and leukoencephalopathy following heroin addiction. Here, we present available reports of cases published in English in PubMed about lesions following heroin use, which have included magnetic resonance imaging (MRI) and/or computed tomography (CT) information, were published through December 27, 2014, and discuss possible etiologies of heroin-induced neurovascular accident and leukoencephalopathy. We used key words "heroin & stroke" and "heroin & leukoencephalopathy" separately for our searches. Because of the etiologic correlation between cerebrovascular accidents with both cocaine and amphetamine abuse, titles and abstracts were used to exclude simultaneous abuse of cocaine and amphetamine. Remaining case reports were screened for inclusion of MRI and/or CT scans. We have provided a chronologic review of the existing cases in the literature to make it easy for readers to find patterns of outbreaks and to get a sense of possible links between the epidemiology of heroin abuse and the disorders. A discussion of epidemiologic factors is important but beyond the scope of this chapter.

STROKE AND HEROIN: A CHRONOLOGICAL REVIEW OF CASES (TABLE 1)

The association of ischemic stroke and heroin is infrequent and there are few case reports about ischemic stroke following heroin use.

Two heroin addicts, ages 30 and 35 years, experienced severe ischemic stroke following intravenous injection of heroin. Extensive nonocclusive infarctions in the carotid territory were shown on arteriograms, CT scans, and technetium-99 scintigrams. Cerebral blood flow (CBF) investigation in one of these individuals showed severe hyperemia of the entire left carotid artery territory, including regions where there was no infarction. The etiology of stroke was concluded to be a combination of hypoxia with decreased perfusion pressure in the carotid territory on one side due to external compression of the carotid artery during the intoxicated comatose state (Jensen et al., 1990).

There is a report of two cases of vascular ballismus following heroin injection.

Two intravenous heroin abuser males, 34 and 19 years of age, were reported having ballistic movements following a period of impaired consciousness after injection of regular doses of heroin. They had been abusing heroin for 10 years and 1 year, respectively. Their MRI findings showed bilateral globus pallidus hyperintensities and ischemia in the right lenticulostriate and inferior branch of the right middle cerebral artery (MCA) vascular territory (Vila & Chamorro, 1997).

MRI findings of a heroin addict who presented with generalized epileptic seizure showed bilateral borderzone infarctions between territories of the anterior cerebral artery MCA, and posterior cerebral artery. Anatomical regions in the parieto-occiptal and frontal lobes were also involved. Cerebral angiogram showed basal arteritis. The lesions were related to heroin-associated vasculitis (Niehaus & Meyer, 1998).

A case of acute bilateral corona radiata and bilateral frontotemporoparietal white matter ischemia with chronic bilateral globus pallidus and cerebral peduncle lesions in a 33-year-old man was also reported. He had used heroin for 13 years and was on methadone maintenance therapy at the same time. He was diagnosed with heroin intoxication following injection (Hsu et al., 2009). It has been suggested that heroin and methadone had a synergistic impact and caused the neurological complication in this case (Somala, 2009).

A less common MRI finding in heroin-induced cerebrovascular ischemia-related literature was reported for the first time

TABLE 1 Case Reports of Stroke Following Heroin Exposure

First Author/Article	Imaging	Microscopic Tissue Study	Objective Findings in Brain Imaging and/or Biopsy/Autopsy	Suggested Reason	Country of the Report/Age (Range)/Gender/Route of Heroin Administration
Jensen, Olsen, and Winther (1990)/Severe nonocclusive ischemic stroke in young heroin addicts			Nonocclusive extensive infarction in carotid territory (CT, angiography)	Compression of carotid artery	Denmark/30 and 35/man and woman/intravenous (iv) injection
Vila and Chamorro (1997) Ballistic movements due to ischemic infarcts after intravenous heroin overdose: Report of two cases	+	−	Bilateral globus pallidus hyperintensities (MRI); basal ganglia hypodensities (CT)	Hypoxic–ischemic process and hypoperfusion following heroin intoxication	Spain/34/man/iv heroin user
			Caudate, putamen, and globus pallidus hyperintensities (MRI)	Embolic event or cardiac arrhythmia from myocardial hypoxia	Spain/19/man/iv heroin user
Niehaus and Meyer (1998) Bilateral borderzone brain infarctions in association with heroin abuse	+	−	Parieto-occipital and frontal hypodensities on CT, borderzone hyperintensities on MRI	Heroin-associated vasculitis	Germany/25/man/unknown route of heroin abuse
Hsu, Chiu, and Liao (2009) Rhabdomyolysis and brain ischemic stroke in a heroin-dependent male under methadone maintenance therapy	+	−	Acute bilateral corona radiata and frontoparietotemporal ischemia, old bilateral globus pallidus ischemia	Synergistic intoxication with heroin and methadone resulting in cerebral vasospasm, hypersensitivity reaction, and hypoventilation	Taiwan/33/man/chronic iv heroin and 6 months methadone maintenance
Benoilid, Collongues, de Seze, and Blanc (2013) Heroin inhalation-induced unilateral complete hippocampal stroke	+	−	Left hippocampus cortical laminar necrosis without vascular abnormality	Cerebral ischemic hypoxia or seizure	France/33/man/first-time heroin inhalation

The table shows identifying characteristics of each published report in PubMed (first author and title), imaging and/or tissue studies and their findings, suggested etiology by authors, and some epidemiologic characteristics of the studied population.
Cerebrovascular accident following heroin abuse and suggested pathologic reason.

by Benoilid et al. (2013). They found left hippocampal cortical laminar necrosis (CLN) 48 h after the initial heroin inhalation of a 33-year-old man. Presentation was with seizure, tongue bite, and amnesia. The authors could not definitively relate CLN with either a probable seizure disorder or heroin inhalation. No vascular abnormality was found, which was compatible with the MRI findings. A cardiac source of embolization was ruled out by echocardiography (Benoilid et al., 2013).

HEROIN-INDUCED LEUKOENCEPHALOPATHY: A CHRONOLOGICAL REVIEW OF CASES (TABLE 2)

Heroin-induced leukoencephalopathy presents with cerebellar and pyramidal symptoms including ataxia, gait disorders, spastic and flaccid paralysis, and, finally, death. Wolters et al. (1982) first described clinical features of heroin-induced leukoencephalopathy in 47 patients. They observed three main clinical stages in patients that ultimately present with spastic stretching in the terminal stage. All patients (11) with terminal symptoms finally died. All patients were heroin addicts who inhaled the white smoke of heated heroin (pyrolysate). The abused substance itself as well as patient serum and cerebrospinal fluid (CSF) findings were analyzed to find the responsible component. Nine patients' brain tissues were analyzed histochemically. CT scanning of all encephalopathy patients' brains, but not healthy heroin abusers, showed cerebellar and cerebral hypodensities. Postmortem histopathology revealed severe vacuolization in 100% of the cerebellar white matter examined and mild to severe spongiform degenerations in cerebral white matter. Epidemiologic investigations were conducted to find the source of heroin supply. All were directly or indirectly supplied from the same region in the Netherlands. The authors postulated that the toxic effect of the unknown substance in the abused heroin can result in leukoencephalopathy. However, chemical analysis of

TABLE 2 Case Reports of Leucoencephalopathy after Heroin Exposure

First Author/Article	Types of Studies Performed	Findings	Proposed Etiology	Country of Heroin Acquisition/Age/Gender/Route of Heroin Administration
Wolters et al. (1982)/Leucoencephalopathy after inhaling "heroin" pyrolysate	Extensive toxicology studies of heroin sample, serum, urine, CSF, and brain for known neurotoxins (using TLC, GC, MS) with similar leukoencephalopathogenetic character, post-heated heroin exposure animal behavior screening	—	Unknown toxic substance (ruling out effect of triethyltin, hexachlorophene, isonicotinic acid, and cycloleucine)	The Netherlands (Zeedijk, Amsterdam)/18–34/35 men, 12 women/inhaled pyrolysate
	Brain CT scanning	Symmetric cerebellar hypodensities of all patients, simultaneous cerebral hypodensities in 43 patients		
	Brainstem auditory evoked responses	Abnormal delay		
	Brain autopsy, light and electron microscopy/histochemical analysis	Spongiform degeneration of CNS white matter and reduced number of oligodendroglia, axoplasm and oligodendroglia vacuolization, and swollen mitochondria, thickened walls of capillaries, absent products of myelin breakdown		
Sempere, Posada, Ramo, and Cabello (1991)/Spongiform leucoencephalopathy after inhaling heroin	Toxicologic	—	Toxic effect of unknown adulterants (ruled out effects of caffeine, phenobarbital, methaqualone, procaine, piracetam, and lignocaine)	Spain (Madrid)/43/man/inhaled heroin smoke
	CT scanning	Symmetric cerebellar and cerebral hypodensities		
	Brain autopsy and microscopy	Cerebral and cerebellar spongiosis and vacuolization		
Celius and Andersson (1996)/Leucoencephalopathy after inhalation of heroin: A case report	Brain CT and MRI	CT hypodensities in cerebellum bilaterally, hyperintensities in T2 MRI	1) Toxic effect of a new compound from heating irregularly added substance(s) to heroin. 2) Inhalation delivers much higher doses of toxin to brain. 3) Individual predisposition	Norway/36/man/always inhaled the same batch of heroin
	Somatosensory and brainstem evoked potential	Normal		
Roulet Perez, Maeder, Rivier, and Deonna (1992)/Toxic leucoencephalopathy after heroin ingestion in a 2½-year-old child	Brain CT and MRI	Symmetrical hypodensities in cerebellar hemispheres and MRI T2 image hyperintensities	Neurotoxic adulterant widespread at the time (1982–1992) and place (Europe)	Switzerland/2.5/boy/unknown method of abuse (suggested ingestion) of heroin

Continued

TABLE 2 Case Reports of Leukoencephalopathy after Heroin Exposure—cont'd

First Author/Article	Types of Studies Performed	Findings	Proposed Etiology	Country of Heroin Acquisition/Age/Gender/Route of Heroin Administration
Weber, Henkes, Moller, Bade, and Kuhne (1998)/Toxic spongiform leucoencephalopathy after inhaling heroin vapor	Brain CT and MRI	Bilateral cerebellar and periventricular hypodensities and T2 image hyperintensities	Unknown neurotoxin inhaled with heroin	The Netherlands/35/man/inhalation
Tan, Algra, Valk, and Wolters (1994)/Toxic leukoencephalopathy after inhalation of poisoned heroin: MR findings	Brain CT and MRI	Bilateral cerebellar and periventricular cerebral white matter hypodensities (CT) and hyperintensities (T2 MRI), corticospinal tract and posterior limbs of internal capsule lesions	Unknown neurotoxin, similar myelin disorganization with TET administration in rats (proposed "poisoned heroin with triethyltin")	The Netherlands/33, 26, 34, 27 men/inhaled
Rizzuto et al.(1997)/Delayed spongiform leukoencephalopathy after heroin abuse	MRI	Diffuse hyperintensities in cerebellum, centrum ovale, and pallidum in T2 MRI	Hypoxia induced by hypotension following heroin overdose	Italy/30/man/intravenous
	Light microscopy	Edematous myelin spongiosis		
	Immunohistochemistry	Normal pattern of respiratory chain subunits of astrocytes and oligodendroglia		
Kriegstein et al. (1999)/Leukoencephalopathy and raised brain lactate from heroin vapor inhalation ("chasing the dragon")	Brain CT and MRI	Symmetric cerebellar, posterior cerebral, corticospinal tracts, hypodensities (CT) and white matter hyperintensities (T2 MRI)	Hypoxic ischemic injury induced by toxic effect of pyrolysate	USA (NY)/21/woman/40, 28/men/inhaled with different frequencies
	^{1}H MRSI	Elevated cerebellar and cortical lactate and decreased cerebellar NAA spectrum correlating with clinical presentation severity and substance use frequency		
	Biopsy and electron microscopy	Spongiform degeneration and vacuolization of myelin sparing axon		
Chen et al. (2000)/Heroin-induced spongiform leukoencephalopathy: Value of diffusion MR imaging	DW-MRI	Decreased white matter apparent diffusion coefficient	White matter pathology induced by neurotoxin	Taiwan/46/man/intravenous
Barnett et al. (2001)/Reversible delayed leukoencephalopathy following intravenous heroin overdose	Brain CT and MRI	Normal CT scanning, diffuse cerebral hyperintensity (FLAIR and T2)	Hypoxia induced by heroin overdose	Australia/45/man/intravenous

Reference/Title	Investigation	Findings	Proposed etiology	Epidemiology
Gacouin et al. (2003)/Reversible spongiform leucoencephalopathy after inhalation of heated heroin	Brain CT and MRI	Diffuse cerebral hyperdensities (CT) and hyperintensities (T2 MRI)	Inhalation of large quantity of toxic substance	France/21/man/inhalation
Vella et al. (2003)/Acute leukoencephalopathy after inhalation of a single dose of heroin	MRI	White matter hyperintensities (T2) and water diffusion disorder (DWI), sparing cerebellum	Mitochondrial injury as well as hypoxia	Switzerland/16/man/inhalation
	MRS	Decreased NAA, Chl, Glu, increased lactate		
Halloran, Ifthikharuddin, and Samkoff (2005)/Leukoencephalopathy after from "chasing the dragon"	MRI	Subcortical white matter hyperintensities (T2)	—	USA (NY)/49/man/inhalation
Chang, Lo, Kao, and Chen, 2006/MRI features of spongiform leukoencephalopathy following heroin inhalation	MRI	Symmetric hyperintensities in cerebellum, pons, and posterior limb of internal capsule (T2)	Mitochondrial dysfunction and neurotoxicity	Taiwan/26/man/inhalation
	¹H MRS	Decreased NAA/Cr ratio, doublet of lactate peak		
Gupta, Krishnan, and Sudhakar (2009)/Hippocampal involvement due to heroin inhalation—"chasing the dragon"	Brain MRI and CT	Bilateral cerebellar and hippocampal hyperdensities (CT) and hyperintensities (T2)	Nonspecific toxic demyelination	Bahrain/42/man/inhalation
Jee et al. (2009)/Heroin vapor inhalation-induced spongiform leukoencephalopathy	Brain MRI and CT	Bilateral hyperintensities in cerebellum, posterior limbs of internal capsule, pons, and splenium	Profound cellular damage due to unclear reason	Taiwan/26/man/inhalation
Villella, Iorio, Conte, Batocchi, and Bria (2010)/Toxic leukoencephalopathy after intravenous heroin injection: A case with clinical and radiological reversibility	MRI	Symmetric diffuse hyperintensities in periventricular white matter and semiovale centers (T2) and restricted diffusion (DWI)	Hypoxia hypoperfusion	Italy/32/man/intravenous
Kass-Hout et al. (2011)/"Chasing the dragon"—heroin-associated spongiform leukoencephalopathy: Case report	MRI	Diffuse cerebellar hyperintensities (T2)	Unknown toxin	USA/21/man/inhalation
Verma, Sharma, and Vidhate (2011)/A rare case of acute fatal leucoencephalopathy due to heroin exposure	MRI	Symmetric cerebral hyperintensities (T2)	—	India/25/man/inhalation
Cordova et al. (2014)/Chasing the dragon: New knowledge for an old practice	MRI	Increased intensity in both cerebral hemispheres	—	USA/46/woman/inhalation

The table shows identifying characteristics of each published report in PubMed (first author and title), types of investigations done in the article and their findings, proposed etiology, and some epidemiologic characteristics of the population.
TLC, thin-layer chromatography; GC, gas chromatography; MS, mass spectrometry; FLAIR, fluid-attenuated inversion recovery; Chl, choline; Glu, glutamate; DWI, diffusion weighted image; TET, triethyltin.

similar abused heroin to find the possible responsible neurotoxin and animal pyrolysate exposure studies were futile (Wolters et al., 1982).

The proposed toxin most likely appears during the last stages before heroin is inhaled (Buxton et al., 2011), because further toxicological investigations did not show any substance in the heroin batches likely to be responsible for the lesions (Sempere et al., 1991; Wolters et al., 1982). In addition, experimental exposure to heated heroin smoke did not reproduce similar presentations in animals (Wolters et al., 1982). This suggests the toxin is added or produced in the last step after heating and before inhalation by the abuser.

Autopsy and microscopic evaluation of brain tissue revealed extensive vacuolization, white matter spongiosis, and loss of oligodendroglia with almost normal axons and gray matter in the cerebrum and cerebellum of a 43-year-old heroin smoker. He presented with cerebellar and pyramidal symptoms. Cranial CT scan showed hypodensities of cerebral and cerebellar white matter (Sempere et al., 1991).

Neuroimaging findings of bilateral cerebellar and periventricular cerebral white matter lesions in four heroin addicts with cerebellar damage and encephalopathy presentations resulted in the suggestion by Tan et al. that the patients had inhaled heroin poisoned with triethyltin. The patients were all young men from the Netherlands who had inhaled pyrolysate for years and were diagnosed with toxic leukoencephalopathy (Tan et al., 1994).

There is a report of a 36-year-old man with similar symptoms who abused inhaled heated heroin. Brain CT scan revealed hypodense areas symmetrically in both cerebellar hemispheres. An MRI indicated corresponding lesions with decreased signal intensity on T1-weighted images and increased signal intensity similar to CSF on T2-weighted images. The patient inhaled heroin with his partner, but his partner never developed encephalopathy. It was suggested that, in addition to higher toxin delivery with the inhalation method, and possible toxic effects of new compounds produced by heating heroin, there may be individual predisposing factors (Celius & Andersson, 1996).

There is a case report of delayed leukoencephalopathy occurring 23 days after intravenous heroin injection. MRI findings showed bilateral cerebellar hyperintensities indicating leukoencephalopathy. The authors suggested hypoxia following hypotension after heroin overdose as a possible reason for spongiform leukoencephalopathy in the patient. Although immunohistochemical studies could not find any defects in proteins expressed by mitochondrial genes, they could not rule out defects in other levels of mitochondrial phosphorylation (Rizzuto et al., 1997).

Another case of toxic spongiform leukoencephalopathy after inhaling heroin vapor was reported in 1998. A 35-year-old male Caucasian presented with cerebellar symptoms. CT scan and MRI of brain showed symmetric hypodensity and hyperintensity in cerebellar hemispheres and both posterior limbs of the internal capsule. It was suggested to be due to the neurotoxic effect of an unknown substance inhaled with the heroin (Weber et al., 1998).

The first cases of heroin-induced leukoencephalopathy in the United States were reported in 1999 from New York. A 21-year-old woman and a 40-year-old man developed ataxia, dysdiadochokinesia, and abulia. They had inhaled pyrolysate for at least 6 months together but with different frequencies. They recovered partially after antioxidant administration. The third case was a 28-year-old male who had inhaled heroin with them for a few sessions. He was not symptomatic but showed milder symptoms in further neurological exam (dysdiadochokinesia on rapid alternating hand motion on left side only). The authors performed MRI and magnetic resonance spectroscopy (MRS), electrophysiologic studies, and electron microscopic pathologic investigation for the first patient. Although MRI findings and pathology studies were similar to those of previously reported outbreaks in Europe, the MRS showed higher lactate levels in the cerebellum of patients 1 and 2. In patient 3, who had minimal symptoms on examination, there was a slight increase in lactate level in one cerebellar voxel only. Lactate level is an indicator of anaerobic metabolism and its levels highlight hypoxic–ischemic injury in cerebellum. More importantly, lactate level correlated with severity of clinical symptoms of heroin-induced encephalopathy. The patients' electrophysiological studies did not show any white matter lesions. This encephalopathy had clear spongiform features in microscopic studies but MRS studies showed low levels of choline (a biomarker increased following white matter lesions) and high levels of N-acetylaspartate (NAA; increased following axonal injury and mitochondrial intoxication) in cerebellum. According to these findings, the authors believed the demyelination theory proposed by previous studies of similar outbreaks in Europe could not be true in these cases. They postulated a new theory based on a dose–response relation of inhaled heroin and recovery after antioxidant therapy. This theory proposes that it is hypoxic–ischemic injury that leads to reversible mitochondrial dysfunction (Kriegstein et al., 1999).

DISCUSSION OF POSSIBLE PATHOPHYSIOLOGIES

Cerebrovascular Accidents Following Heroin

Diacetylmorphine (traditionally called heroin) is the most commonly used street opioid. Although there are no confirmed statistics about the prevalence of opioid abuse and dependence worldwide (Degenhardt, Mathers, & Hall, 2014), evidence shows the prevalence of first-time heroin use and dependence is growing in the United States (Martino et al., 2010).

Cerebral perfusion is generally reduced after heroin consumption (Figure 1). This can result in cerebral hypoperfusion either locally or globally. Hypotension following histamine degranulation, parasympathetic stimulation, and adulterant embolization after heroin abuse can result in decreased cerebral perfusion (Ghuran & Nolan, 2000). A general decrease in CBF has been shown by single-photon-emission computed tomography in chronic heroin and nicotine addicts. The observed decreases were prominent in regions of the prefrontal cortex (Botelho et al., 2006). How this observed decrease in blood flow correlates with impaired judgment of heroin abusers is an interesting topic that is beyond the scope of this chapter.

Although the central nervous system (CNS) hypoperfusion–ischemic effect does not depend upon route of heroin administration, ischemic strokes are more common with intravenous heroin abuse (Benassi et al., 1996; Brust & Richter, 1976; Jensen et al., 1990; Rumbaugh, Bergeron, Fang, & McCormick, 1971).

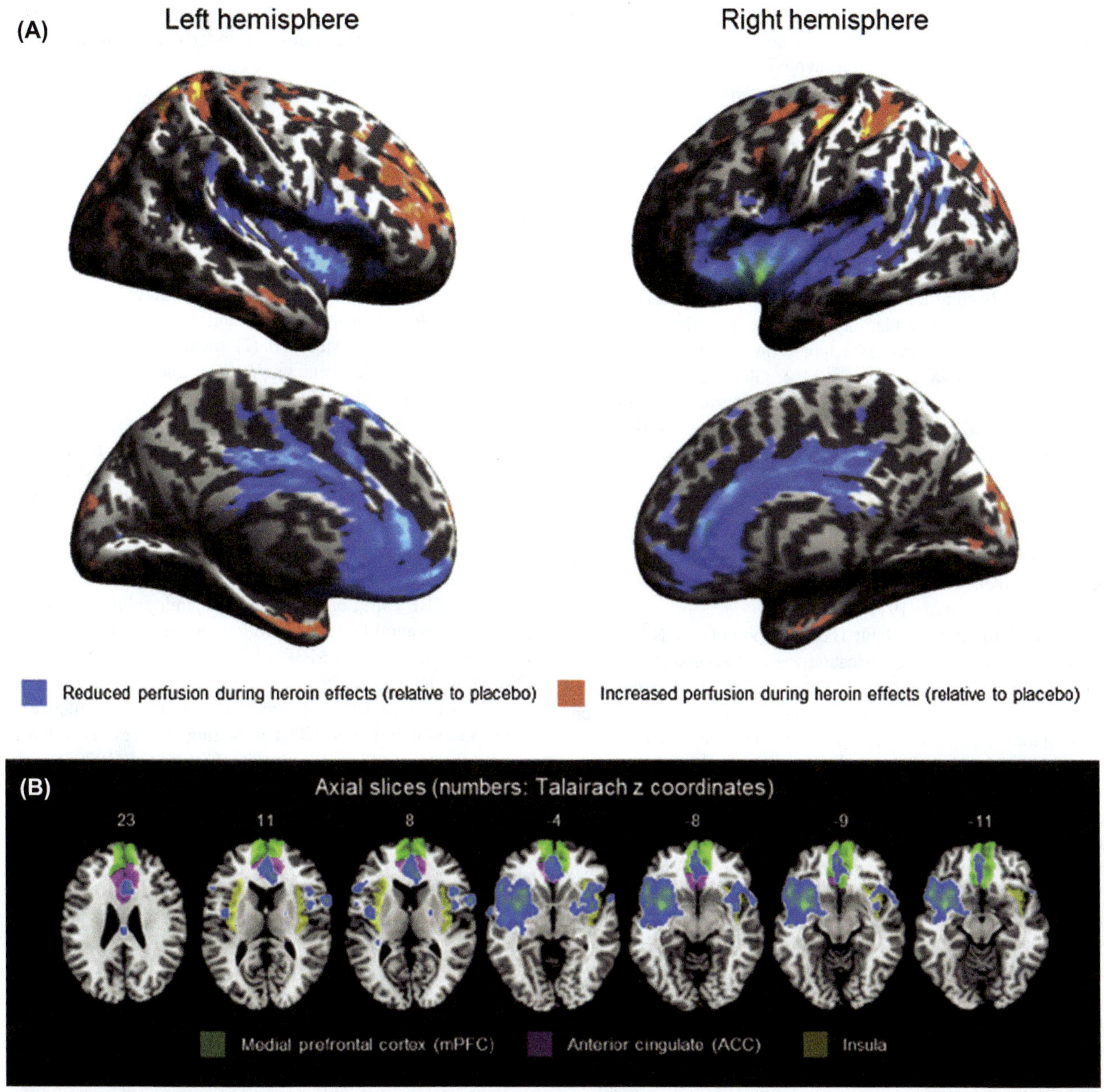

FIGURE 1 Cortical perfusion and relative thickness alterations in heroin users. (A) Altered perfusion during heroin effect. Relative hypoperfusion (blue) and hyperperfusion (red). (B) Correlation between regional cortical volume and perfusion (axial slices). *Permission granted by Denier et al. (2013).*

However, there are also reports of stroke due to intranasal heroin abuse (Adle-Biassette, Marc, Benhaiem-Sigaux, Durigon, & Gray, 1996; Zuckerman, Ruiz, Keller, & Brooks, 1996). The following paragraphs describe the pathogeneses by which heroin induces ischemic stroke. These are all more common with intravenous heroin abuse.

Embolization: Intravenous drug abusers are at high risk of infectious endocarditis, and subsequent septic emboli results in ischemic lesions in brain (Hagan & Burney, 2007). Another source of embolization is foreign bodies injected with impure heroin. There are many substances added to heroin to dilute it, such as talc, quinine, strychnine, lidocaine, and sugar (Lucas, 2005).

Following intravenous injection, they first embolize to the pulmonary artery and induce a granulomatous reaction in the lung parenchyma (Roberts, 2002). Ultimately, this results in pulmonary artery hypertension. The consequent right-to-left shunt will take the drug of abuse and impurities into systemic circulation including the brain (Bitar & Gomez, 1993; Esse, Fossati-Bellani, Traylor, & Martin-Schild, 2011).

Systemic blood pressure decrease: Overdose is more commonly seen among heroin injectors than inhalers (Swift, Maher, & Sunjic, 1999). Abrupt hypotension can result from a heroin overdose. A variety of pathologic mechanisms have been described for this, including arrhythmia, cardiac and pulmonary artery enlargement, low cardiac output, and eventually myocardial failure with no gross or microscopic change (Paranthaman & Khan, 1976). Hypotension can result in hypoxic–ischemic infarct in the globus pallidus and ischemia in the watershed areas of the MCA, with anterior and posterior cerebral artery supplementation (Adle-Biassette et al., 1996; Andersen & Skullerud, 1999).

Platelet count and hemostasis: One suggested mechanism for ischemia is platelet aggregation and increased clot formation in the entire vasculature, but particularly in cerebral arteries (Silvestrelli, Corea, Micheli, & Lanari, 2010). Studies have shown increased platelet count in heroin dependents, although this was not statistically significant compared to control subjects (Haghpanah, Afarinesh, & Divsalar, 2010). Also, significant thrombocytopenia or thrombocytosis happens in chronic heroin users (Zvetkova, Antonova, Ivanov, Savov, & Gluhcheva, 2010). More importantly, structural and morphologic destruction of platelets results in altered hemostatic status of heroin addicts (Zvetkova et al., 2010). All of these factors result in higher viscosity of blood in heroin addicts (Savov et al., 2006).

Hypersensitivity reaction: This mechanism depends upon patients having reexposure to the drug after a period of abstinence. The hypersensitivity results in histamine degranulation and subsequently hypotension and reduced cerebral perfusion (Citron et al., 1970; Rumbaugh et al., 1971).

Damage to mitochondria: The first report of a CLN lesion following heroin inhalation intoxication was by Beilinod et al. (2013). The lesion is a selective pan-necrosis of cortex, and it is described with no cerebral hypoxia or known mitochondrial disorders. Considering the unique finding in this case, which attributes no regional circulation abnormality, we suggest as a possibility mitochondrial impairment following heated heroin inhalation intoxication. The authors did not mention this possibility and instead suggested seizure-attributed CLN. A different case report described a link between status epilepticus with left hippocampal CLN (Heinrich, Runge, Kirsch, & Khaw, 2007). Hypoxia following multiple status epilepticus attacks can result in ischemia in specific brain regions, but there is no evidence in the case report by Beilinod et al. to show a history of several epileptic attacks. There is insufficient evidence for this case report to prove any causative link between the possible seizure and the lesion on MRI. In addition, there are reported cases of left hippocampal CLN concomitant with hereditary mitochondrial dysfunction in infantile onset spinocerebellar ataxia (IOSCA), suggesting the association of mitochondrial dysfunction with CLN (Lonnqvist, Paetau, Valanne, & Pihko, 2009).

It is therefore possible that mitochondrial damage following heated heroin inhalation is the pathology underlying the MRI finding in the case of Beilinod et al.

Heroin-Induced Spongiform Leukoencephalopathy

Heroin abuse in the long term results in decreased CBF (Botelho et al., 2006). This reduction in CBF results in cerebral myelin sheath damage. White matter vacuolization is a pathological finding in heroin-induced spongiform encephalopathy (Figure 2). Two proposed mechanisms for this white matter change are: (1) ischemic–hypoxic injury to oligodendrocytes and (2) myelin basic protein (MBP) hydrolysis in watershed regions (Yin, Lu, Chen, Fan, & Lu, 2013).

Postmortem studies have shown endothelial protrusions into vessel lumen in chronic ischemic stroke patients. It is therefore suggested that this microvascular damage may be the reason for white matter vacuolization in heroin abusers in whom there is chronic decreased CBF (Dziewulska, Rafalowska, Podlecka, & Szumanska, 2004; Yin et al., 2013). Ischemic–hypoxic microvascular damage results in calcium influx in myelinated axons. This can activate calcium-dependent proteolytic enzyme, which in turn degrades MBP (Miki et al., 2009). Ischemic injury to watershed regions in cerebral circulation following reduced CBF can result in myelin damage by MBP hydrolysis (Pernet, Joly, Christ, Dimou, & Schwab, 2008).

Toxin-related cerebral vasculitis: Heroin can induce systemic vascular diseases with necrotizing features. This can also involve the brain and result in stroke (Citron et al., 1970; Woods & Strewler, 1972). Jensen et al. (1990) showed normal angiograms and ruled out occlusive vascular disease as the cause of infarcts. However, their findings could not rule out ischemia since there was extensive hyperemia in the territory of carotid artery, which could indicate ischemia (Jensen et al., 1990).

Niehaus and Meyer (1998) presented Doppler sonography indicating pathologically increased mean blood flow velocities with disturbed flow pattern (high intensity of low-frequency components) in the supraclinoidal segments of both internal carotid arteries. These findings provide evidence that there was segmental vasospasm of basal cerebral arteries and exclude a hyperdynamic state of cerebral perfusion. However, high-dose corticosteroid prescription did work in this case and intensified the possibility of heroin-associated vasculitis (Niehaus & Meyer, 1998).

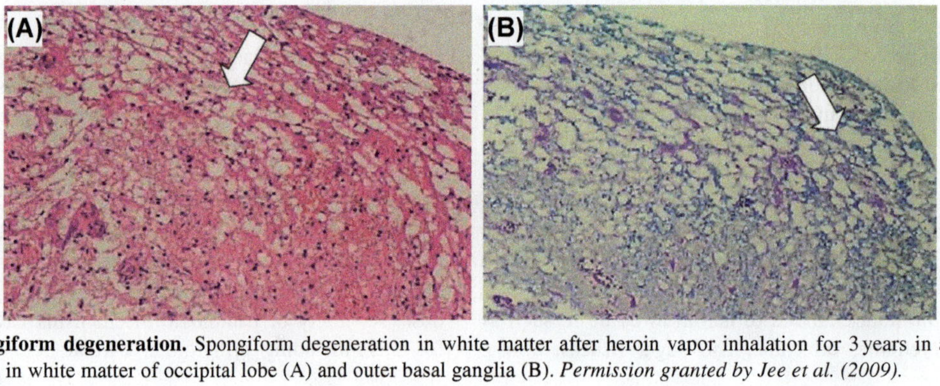

FIGURE 2 Spongiform degeneration. Spongiform degeneration in white matter after heroin vapor inhalation for 3 years in a 26-year old man. Vacuolar formation in white matter of occipital lobe (A) and outer basal ganglia (B). *Permission granted by Jee et al. (2009).*

Hypoventilation and hyperemia: Hyperemia develops after revascularization following ischemic lesions like spontaneous recanalization of occluded arteries in stroke patients. Intravenous heroin can result in intoxication, subsequent hypoventilation, and cerebral hypoxia. After normalization of ventilation, a hyperemic state develops in the entire brain. Jensen et al. (1990) showed that in one of their cases hyperemia was seen in the entire carotid territory and also outside the infarct area. However, the presence of a localized infarct in one hemisphere excludes hypoxia and global ischemia as the only cause of infarction. They suggested that a focal decrease in the perfusion pressure in the already hypoxic brain might be the cause of the infarction. Inappropriate placing of the neck during the comatose state may cause external compression or kinking of the carotid artery and may cause a critical lowering of perfusion pressure in the carotid territory. In a well-oxygenated state this is a very unlikely cause of decreased perfusion pressure in the carotid territory. However, in hypoxic states it might very well be the case.

Mitochondrial injury: As previously mentioned, heroin inhalation can result in mitochondrial injury, and there is evidence showing heroin inhalation can result in pathological findings of mitochondrial damage such as CLN (Benoilid et al., 2013). Mitochondrial damage in IOSCA can also result in clinical presentations of ataxia similar to what is seen in leukoencephalopathy and very similar findings on MRI (CLN) (Lonnqvist et al., 2009).

It was originally suggested by Celius and Andersson (1996) that there may be individual predisposing factors for heroin-induced leukoencephalopathy based on their dichotomous observation of symptoms in one heroin inhalation partner but not the other, who had abused in the same way and from the same batch of heroin. Although there are outbreak reports of heroin-induced encephalopathy in groups of people who have used heroin together (Kriegstein et al., 1999), it is still necessary to consider the possibility of individual predisposition. The observed difference in individual vulnerability on inhaling the same substance (Celius & Andersson, 1996) can be related to differences in mitochondrial susceptibility among subjects.

Some studies have shown the role of genes upstream of mitochondrial cytochrome c in heroin-induced neuronal apoptosis in cerebellar and cerebral cortex (Cunha-Oliveira et al., 2007; Lai, Pu, Cao, Jing, & Liu, 2011). Considering similar pathological features in mitochondrial DNA in inherited enzyme diseases like IOSCA (Lonnqvist et al., 2009), one hypothesis could be that an individual predisposition to heroin inhalation-induced leukoencephalopathy exists based on mitochondrial enzyme defects. There is evidence in case reports for pathological changes in oligodendroglial and mitochondrial defects (swelling) following heated heroin inhalation (Wolters et al., 1982). It has been suggested that inhaled heroin contains an unknown toxin or toxins that appear irregularly in the black market and result in outbreaks. We suggest that this toxin induces an opioid receptor overstimulation effect in spongiform leukoencephalopathy via mitochondrial toxicity. There might also be variation in mitochondrial susceptibility to hypoxic damage.

Kriegstein et al. (1999) showed for first time that, although there were spongiform degenerations visible through electron microscopy, no myelin breakdown products (choline) were detected by MRS imaging (MRSI). However, they found markers of hypoxic damage (lactate) in regions with leukoencephalopathy lesions (Figure 3). They hypothesized hypoxic ischemic injury as the underlying pathophysiology for heroin-induced leukoencephalopathy. This injury among their reported patients had a clear correlation with the severity of clinical findings. It is possible that pyrolysate results in mitochondrial toxicity and hypoxic injury in oligodendrocytes. There are reports of similar clinical pictures and neuroimaging findings in mitochondrial enzyme deficiency syndromes like IOSCA that further emphasize the possibility of an underlying mitochondrial susceptibility to pyrolysate-induced toxicity as the pathology of heroin-induced leukoencephalopathy (Lonnqvist et al., 2009).

Research has focused on metabolite accumulation assays in regions of heroin-induced leukoencephalopathy. They have shown an enormous increase in lactate, a product of hypoxic metabolism, and mitochondrial injury following heroin inhalation has been postulated. On the other hand, MRS has failed to show a rise in choline titers (Figure 4). Hypoxic injuries to oligodendroglia and astrocytes can result in mitochondrial damage, which in turn will further promote hypoxic metabolism in oligodendroglia and vacuole production through myelin sheaths consequently. These vacuoles can impair water diffusion through white matter (diffusion water image) (Vella, Kreis, Lovblad, & Steinlin, 2003).

HYPOXIA IS A COMMON PATHWAY TO INDUCE LEUKOENCEPHALOPATHY

There are some case reports of delayed-onset leukoencephalopathy following intravenous heroin overdose (Barnett, Miller, Reddel, & Davies, 2001; Chen et al., 2000; Rizzuto et al., 1997). Considering the fact that all these cases were comatose after intoxication and before developing leukoencephalopathy, cerebral hypoxia could be the common pathway between intravenous heroin-induced and pyrolysate inhalation-induced leukoencephalopathy. These cases were typically nonlethal and reversible, except one (Rizzuto et al., 1997). Injected adulterants can result in vasodilation and hypotension. Heroin can result in bradycardia that, in turn, adds to development of hypotension in a short time. The hypotension subsequently results in hypoxia–ischemia in CNS. The difference between heroin inhalation-induced and intravenous injection-induced leukoencephalopathy is the pathophysiology of the hypoxia. As mentioned before, the mitochondriotoxic effect of unknown adulterants or their compounds after heating is responsible for hypoxic injury in oligodendroglia and astrocytes. However, intravenous injection of heroin adds to the vasodilator effect of contaminating adulterants and results in hypotension–hypoventilation that, in turn, yields hypoxia. Depending on the severity and rate of this hypotension-induced hypoxia it could be accompanied by shock presentation or result in coma.

Further Clues to Mitochondrial Injury, and Cases in Minors

A dose–response relationship between symptom severity and inhaled heroin has been reported (Buxton et al., 2011; Kriegstein et al., 1999). Anatomical locations of the lesion on MRI findings vary from cerebellar prominent to diffuse cerebral. This can be related to a different suggested pathophysiology of the lesion, from hypotension-induced global cerebral hypoperfusion to heroin-induced oligodendroglial toxicity. Intravenous overdose potentially

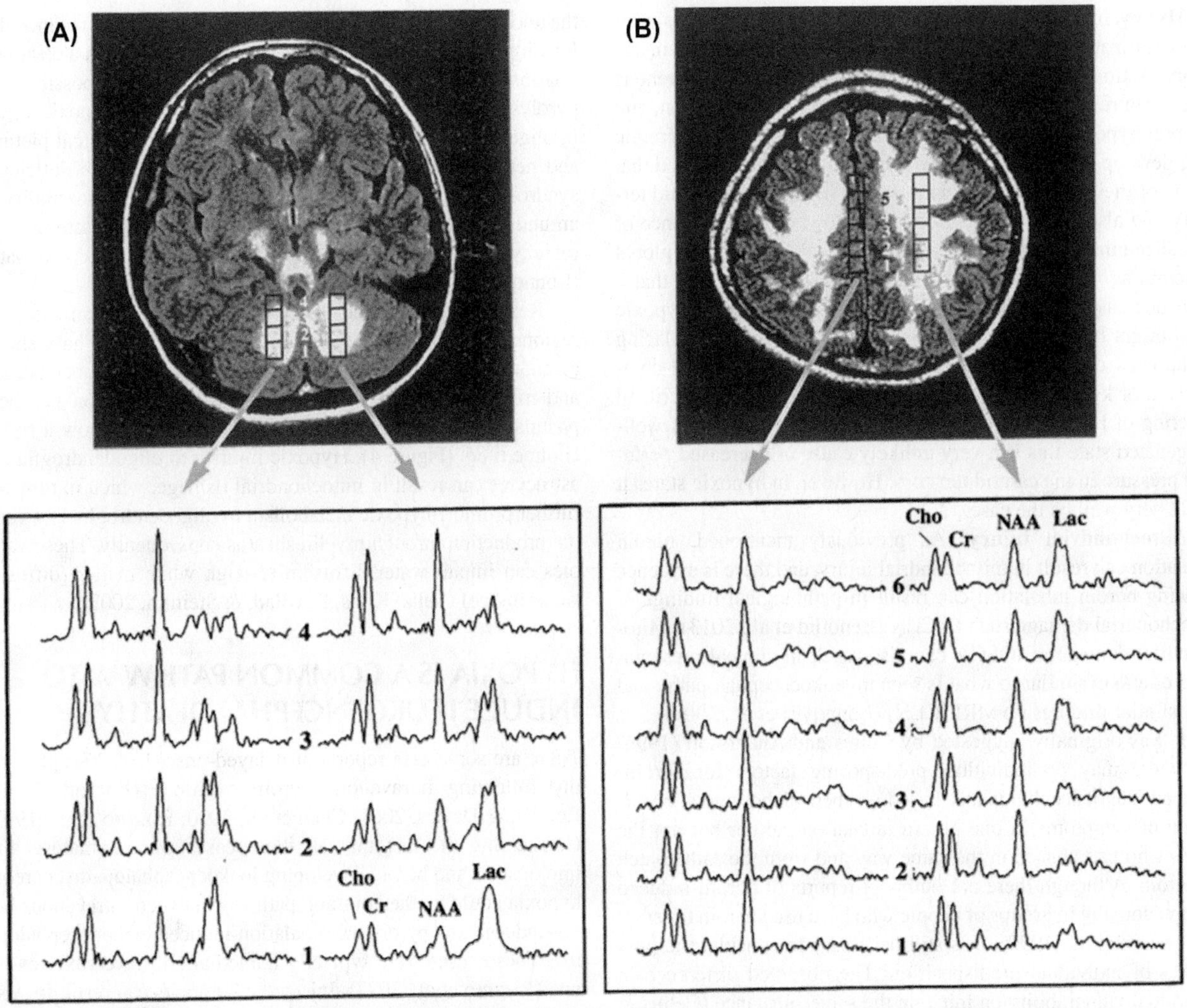

998 PART | V Opioids

FIGURE 3 **MRSI findings in heroin-induced leukoencephalopathy.** Fluid-attenuated inversion recovery images and corresponding ^{1}H MRSI spectra within several adjacent voxels in gray matter and white matter showing abnormal presence of choline (Cho), creatine (Cr), N-acetylaspartate (NAA) and lactate (Lac) in cerebellar (A) and cerebral white matter (B). *Permission granted by Kriegstein et al. (1999).*

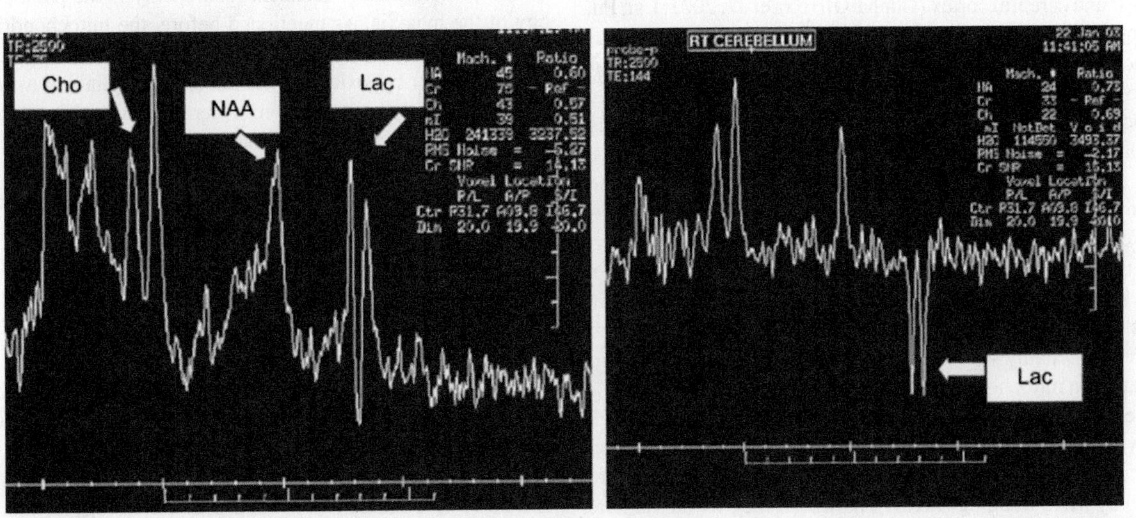

FIGURE 4 **MRS of a single voxel in cerebellum** indicating reduced choline (Cho) and N-acetylaspartate (NAA) (A) and increased lactate (Lac) in a different frequency of magnetic field (B). *Permission granted by Offiah and Hall (2008).*

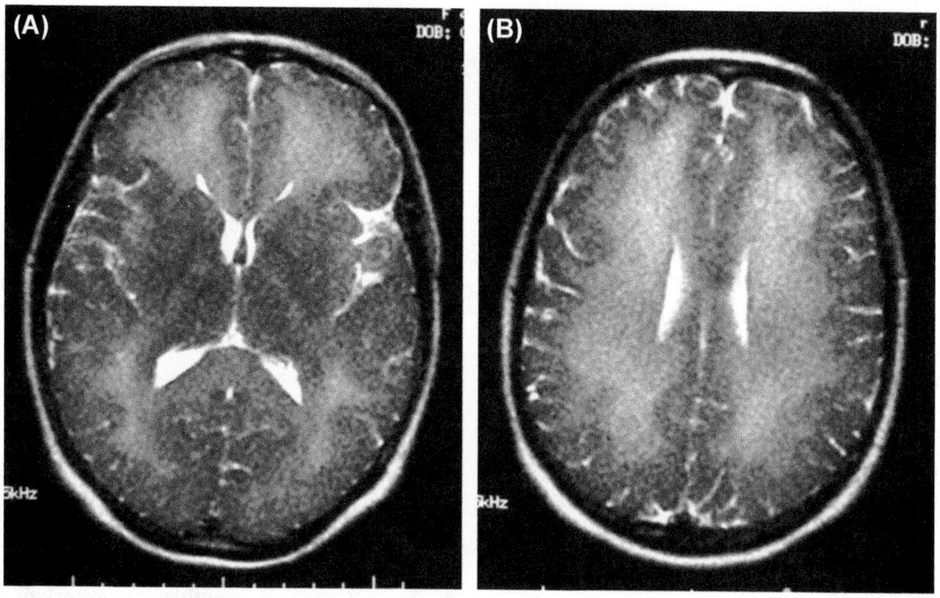

FIGURE 5 **MRI finding in hypoxic–ischemic leukoencephalopathy.** T2 weighted MRI in 22 days post encephalitis attack following morphine overdose shows diffuse subcortical (A) and supraventricular (B) white matter changes sparing basal ganglia. *Permission granted by Molley et al. (2006).*

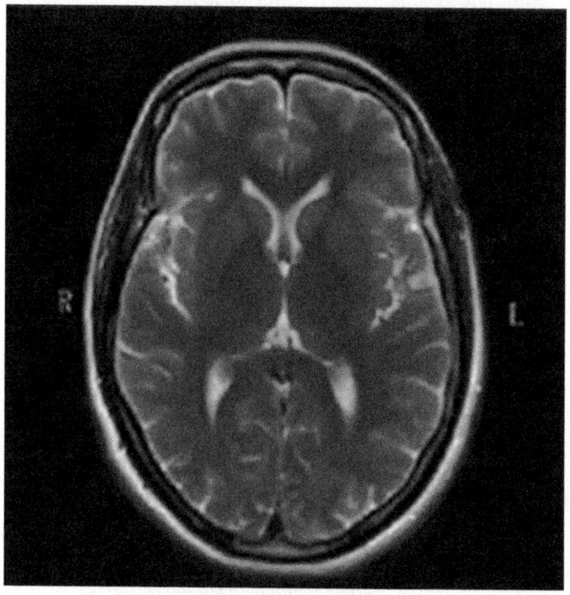

FIGURE 6 **MRI finding in heroin-induced leukoencephalopathy.** T2 MRI shows spongiform leukoencephalopathy in a patient addicted to heroin for 12 months. *With permission from Li, Deng, and Ye (2012).*

gives hypotension and global brain hypoperfusion. But other routes of administration, like inhalation, can result in a more local dissolution of heroin or other lipophilic toxins in white matter with local distribution of subsequent lesions.

MRI findings of hypoxic–ischemic leukoencephalopathy have been reported to have a pattern very similar to that of MRI findings in heroin-induced leukoencephalopathy (Molloy, Soh, & Williams, 2006; Figures 5 and 6). However, neuroimaging findings of heroin vapor-induced leukoencephalopathy have revealed cerebellar sparing and more cerebral lesions (Ryan, Molloy, Farrell, & Hutchinson, 2005; Vella et al., 2003). Pathologic studies with electron microscopy have also revealed extensive axonal injuries (Ryan et al., 2005). It has been postulated that pathologic differences in cases of heroin-induced leukoencephalopathy are probably due to different adulterants and/or the presence of cocaine in heroin mixtures (Blasel et al., 2010; Ryan et al., 2005).

There are a few case reports about similar leukoencephalopathies following ingestion of heroin and methadone in minors. These reports suggest that inhalation route-induced lesion is not a possible etiology—at least, not in those cases. Ingestion of opiates has resulted in similar clinical and imaging findings in these case reports (Anselmo et al., 2006; Roulet Perez et al., 1992). There are some physiological differences in minors that make them different from adult cases. Intestinal surface-to-body weight ratio in young children makes them more vulnerable to toxic effects of possible ingested neurotoxins in heroin compared to adult parents who have used the same amount of heroin. Portal circulation attenuates symptoms of possible toxins, however, relative to inhalation. This can explain the less severe symptoms in the child in the Anselmo et al. report after ingestion compared to leukoencephalopathy following heroin inhalation. However, because of the similarity in methadone-induced leukoencephalopathy in this case to heroin-induced lesions, it was suggested that the pathogenesis of leukoencephalopathy is opioid receptor overstimulation (Anselmo et al., 2006). Mixed adulterants may accentuate this effect. However, because the patient's ataxia and spasticity did not improve, another possibility that the authors did not address is that of a mitochondrial defect-induced encephalopathy presentation. Considering the fact that cerebellar symptoms did not resolve completely following treatment, it is possible that the MRI findings are actually signs of mitochondrial encephalopathy. Unfortunately, there is no information about metabolic measures that would help to clarify a possible mitochondrial etiology of the neuroimaging and clinical findings in these cases.

CONCLUSION

The likely mitochondriotoxic effect of heroin may be enhanced to a clinically symptomatic level by unknown and irregularly appearing adulterants. This effect is independent of the method of abuse, although the adulterants most probably are chemical compounds produced after heating in the last stage before administration by abusers. Leukoencephalopathy after heroin abuse has been successfully treated by antioxidants in some case reports (Bach et al., 2012; Gacouin et al., 2003; Kriegstein et al., 1999; Sempere et al., 1991; Wolters et al., 1982), which may alleviate mitochondrial injury.

APPLICATIONS TO OTHER ADDICTIONS AND SUBSTANCE MISUSE

Abused drugs can induce CNS white matter damage (leukoencephalopathy), including stimulants (methamphetamine, MDMA, and amphetamine), cocaine, narcotics, and heroin, with unknown toxin etiologies. There are mitochondriotoxic effects described for cocaine, stimulants, and heroin. Therefore, mitochondrial toxicity could be a common pathway of leukoencephalopathy caused by substance abuse for several abused substances.

DEFINITION OF TERMS

Pyrolysate This is the product of heating heroin, usually over a foil. The white smoke of the heroin is inhaled through a pipe.

Chasing the dragon This refers to the act of inhaling white smoke from heated heroin (pyrolysate) through a pipe.

Spongiform leukoencephalopathy This is a lesion in white matter characterized by vacuole formation in the myelin of neurons, which gives a spongiform appearance to the white matter under a microscope.

Magnetic resonance spectroscopy imaging This is used to measure the spectrum of MR signals of metabolites in the body after excitation. Each MR signal refers to a specific molecule.

Abulia This is a neurologic symptom, appearing as lack of will or initiative.

KEY FACTS OF SPONGIFORM LEUKOENCEPHALOPATHY

- It is also known as toxic spongiform leukoencephalopathy. The term refers to infrequent brain white matter damage induced by drugs of abuse, environmental toxins, or chemotherapeutic drugs.
- MRI is a common way to study and diagnose the lesion.
- The mechanism and pathophysiology underlying toxic leukoencephalopathy are unknown and vary between sources of toxicity.
- Clinical severity depends on the duration of exposure, concentration of the toxin, and its purity.
- Although toxic leukoencephalopathy is known to be reversible after removing the toxic agent, heroin-induced spongiform leukoencephalopathy is usually an irreversible and fatal lesion.
- Heroin inhalation is well known to induce leukoencephalopathy but there are few reports of leukoencephalopathy following injection or ingestion of heroin and many more sporadic reports of other recreational substances exposure.
- The first heroin-induced leukoencephalopathy was reported in 1982 from the Netherlands.
- The first report of heroin-induced leukoencephalopathy in United States was reported in 1999.
- Heroin inhalation (chasing the dragon) is a common route of heroin administration in Asia but far fewer reports of heroin-induced leukoencephalopathy exist in Asia than in Europe.
- There is no specific treatment for heroin-induced leukoencephalopathy. However, there are some reports of alleviated symptoms after antioxidant administration.

SUMMARY POINTS

- This chapter focuses on two CNS lesions induced by heroin exposure, namely stroke and leukoencephalopathy.
- Stroke and leukoencephalopathy following stroke could be caused by various pathologic mechanisms.
- Embolization of impurities, hypotension, external pressure, increased platelet count and hemostasis, and mitochondrial damage can result in stroke following heroin exposure.
- Pathologic mechanisms leading to leukoencephalopathy following heroin exposure are microvasculitis due to toxin, hypoventilation, and mitochondrial injury.
- All described pathologic mechanisms in both lesions are common in inducing hypoxia.
- Heroin has a mitochondriotoxic effect. However, this effect is enhanced by the unknown compound(s) of heated adulterants when "chasing the dragon."

ACKNOWLEDGMENT

Special thanks go to Dr. Salehi Sadaghiani and Dr. Gharedaghi for their help with this review. Both provided valuable suggestions.

REFERENCES

Adle-Biassette, H., Marc, B., Benhaiem-Sigaux, N., Durigon, M., & Gray, F. (1996). Infarctus cerebraux chez un toxicomane inhalant l'heroine [Cerebral infarctions in a drug addict inhaling heroin]. *Archives D'anatomie et de Cytologie Pathologiques*, *44*(1), 12–17.

Andersen, S. N., & Skullerud, K. (1999). Hypoxic/ischaemic brain damage, especially pallidal lesions, in heroin addicts. *Forensic Science International*, *102*(1), 51–59.

Anselmo, M., Campos Rainho, A., do Carmo Vale, M., Estrada, J., Valente, R., Correia, M., ... Barata, D. (2006). Methadone intoxication in a child: toxic encephalopathy? *Journal of Child Neurology*, *21*(7), 618–620.

Bach, A. G., Jordan, B., Wegener, N. A., Rusner, C., Kornhuber, M., Abbas, J., & Surov, A. (2012). Heroin spongiform leukoencephalopathy (HSLE). *Clinical Neuroradiology*, *22*(4), 345–349. http://dx.doi.org/10.1007/s00062-012-0173-y.

Barnett, M. H., Miller, L. A., Reddel, S. W., & Davies, L. (2001). Reversible delayed leukoencephalopathy following intravenous heroin overdose. *Journal of Clinical Neuroscience: Official Journal of the Neurosurgical Society of Australasia*, *8*(2), 165–167.

Benassi, G., Rinaldi, R., Azzimondi, G., Stracciari, A., DAlessandro, R., & Pazzaglia, P. (1996). Acute generalized dystonia due to a bilateral lesion of basal ganglia mainly affecting the nuclei pallidi. *Italian Journal of Neurological Sciences, 17*(1), 71–73. http://dx.doi.org/10.1007/Bf01995712.

Benoilid, A., Collongues, N., de Seze, J., & Blanc, F. (2013). Heroin inhalation-induced unilateral complete hippocampal stroke. *Neurocase, 19*(4), 313–315. http://dx.doi.org/10.1080/13554794.2012.667125.

Bitar, S., & Gomez, C. R. (1993). Stroke following injection of a melted suppository. *Stroke; A Journal of Cerebral Circulation, 24*(5), 741–743.

Blasel, S., Hattingen, E., Adelmann, M., Nichtweiss, M., Zanella, F., & Weidauer, S. (2010). Toxic leukoencephalopathy after heroin abuse without heroin vapor inhalation MR imaging and clinical features in three patients. *Clinical Neuroradiology, 20*(1), 48–53. http://dx.doi.org/10.1007/S00062-010-0022-9.

Botelho, M. F., Relvas, J. S., Abrantes, M., Cunha, M. J., Marques, T. R., Rovira, E., ... Macedo, T. (2006). Brain blood flow SPET imaging in heroin abusers. *Annals of the New York Academy of Sciences, 1074*, 466–477.

Brust, J. C., & Richter, R. W. (1976). Stroke associated with addiction to heroin. *Journal of Neurology, Neurosurgery and Psychiatry, 39*(2), 194–199.

Buxton, J. A., Sebastian, R., Clearsky, L., Angus, N., Shah, L., Lem, M., & Spacey, S. D. (2011). Chasing the dragon – characterizing cases of leukoencephalopathy associated with heroin inhalation in British Columbia. *Harm Reduction Journal, 8*. http://dx.doi.org/10.1186/1477-7517-8-3 Article No: 3.

Celius, E. G., & Andersson, S. (1996). Leucoencephalopathy after inhalation of heroin: a case report. *Journal of Neurology, Neurosurgery and Psychiatry, 60*(6), 694–695.

Chang, W.-C., Lo, C.-P., Kao, H.-W., & Chen, C.-Y. (2006). MRI features of spongiform leukoencephalopathy following heroin inhalation. *Neurology, 67*(3), 504.

Chen, C. Y., Lee, K. W., Lee, C. C., Chin, S. C., Chung, H. W., & Zimmerman, R. A. (2000). Heroin-induced spongiform leukoencephalopathy: value of diffusion MR imaging. *Journal of Computer Assisted Tomography, 24*(5), 735–737.

Citron, B. P., Halpern, M., McCarron, M., Lundberg, G. D., McCormick, R., Pincus, I. J., ... Haverback, B. J. (1970). Necrotizing angiitis associated with drug abuse. *New England Journal of Medicine, 283*(19), 1003–1011. http://dx.doi.org/10.1056/NEJM197011052831901.

Cordova, J. P., Balan, S., Romero, J., Korniyenko, A., Alviar, C. L., Paniz-Mondolfi, A., & Jean, R. (2014). "Chasing the dragon": New knowledge for an old practice. *American Journal of Therapeutics, 21*(1), 52–55. http://dx.doi.org/10.1097/MJT.0b013e31820b8856.

Cunha-Oliveira, T., Rego, A. C., Garrido, J., Borges, F., Macedo, T., & Oliveira, C. R. (2007). Street heroin induces mitochondrial dysfunction and apoptosis in rat cortical neurons. *Journal of Neurochemistry, 101*(2), 543–554. http://dx.doi.org/10.1111/j.1471-4159.2006.04406.x.

Degenhardt, L., Mathers, B., & Hall, W. D. (2014). Response to Hser et al. (2014): the necessity for more and better data on the global epidemiology of opioid dependence. *Addiction, 109*(8), 1335–1337. http://dx.doi.org/10.1111/add.12612.

Denier, N., Schmidt, A., Gerber, H., Schmid, O., Riecher-Rössler, A., Wiesbeck, G. A., ... Borgwardt, S. (2013). Association of frontal gray matter volume and cerebral perfusion in Heroin addiction: A multi modal neuroimaging study. *Frontiers in Psychiatry, 4*, 135.

Dziewulska, D., Rafalowska, J., Podlecka, A., & Szumanska, G. (2004). Remote morphological changes in the white matter after ischaemic stroke. *Folia neuropathologica/Association of Polish Neuropathologists and Medical Research Centre, Polish Academy of Sciences, 42*(2), 75–80.

Esse, K., Fossati-Bellani, M., Traylor, A., & Martin-Schild, S. (2011). Epidemic of illicit drug use, mechanisms of action/addiction and stroke as a health hazard. *Brain and Behavior, 1*(1), 44–54. http://dx.doi.org/10.1002/brb3.7.

Gacouin, A., Lavoue, S., Signouret, T., Person, A., Dinard, M. D., Shpak, N., & Thomas, R. (2003). Reversible spongiform leucoencephalopathy after inhalation of heated heroin. *Intensive Care Medicine, 29*(6), 1012–1015.

Ghuran, A., & Nolan, J. (2000). The cardiac complications of recreational drug use. *The Western Journal of Medicine, 173*(6), 412–415.

Gupta, P. K., Krishnan, P. R., & Sudhakar, P. J. (2009). Hippocampal involvement due to heroin inhalation-"Chasing the Dragon". *Clinical Neurology and Neurosurgery, 111*(3), 278–281. http://dx.doi.org/10.1016/J.Clineuro.2008.09.004.

Hagan, I. G., & Burney, K. (2007). Radiology of recreational drug abuse. *Radiographics, 27*(4), 919–940. http://dx.doi.org/10.1148/rg.274065103.

Haghpanah, T., Afarinesh, M., & Divsalar, K. (2010). A review on hematological factors in opioid-dependent people (opium and heroin) after the withdrawal period. *Addiction and Health, 2*(1–2), 9–16.

Halloran, O., Ifthikharuddin, S., & Samkoff, L. (2005). Leukoencephalopathy from "chasing the dragon". *Neurology, 64*(10), 1755.

Heinrich, A., Runge, U., Kirsch, M., & Khaw, A. V. (2007). A case of hippocampal laminar necrosis following complex partial status epilepticus. *Acta neurologica Scandinavica, 115*(6), 425–428.

Hsu, W. Y., Chiu, N. Y., & Liao, Y. C. (2009). Rhabdomyolysis and brain ischemic stroke in a heroin-dependent male under methadone maintenance therapy. *Acta Psychiatrica Scandinavica, 120*(1), 76–79. http://dx.doi.org/10.1111/j.1600-0447.2009.01378.x.

Jee, R. C., Tsao, W. L., Shyu, W. C., Yen, P. S., Hsu, Y. H., & Liu, S. H. (2009). Heroin vapor inhalation-induced spongiform leukoencephalopathy. *Journal of the Formosan Medical Association, 108*(6), 518–522.

Jensen, R., Olsen, T. S., & Winther, B. B. (1990). Severe non-occlusive ischemic stroke in young heroin addicts. *Acta Neurologica Scandinavica, 81*(4), 354–357.

Kass-Hout, O., Kass-Hout, T., Darkhabani, M. Z., Mehta, B., Mokin, M., & Radovic, V. (2011). "Chasing the dragon" – heroin-associated spongiform leukoencephalopathy: case report. *Neurology, 76*(9), A595.

Kriegstein, A. R., Shungu, D. C., Millar, W. S., Armitage, B. A., Brust, J. C., Chillrud, S., ... Lynch, T. (1999). Leukoencephalopathy and raised brain lactate from heroin vapor inhalation ("chasing the dragon"). *Neurology, 53*(8), 1765–1773.

Lai, B., Pu, H., Cao, Q., Jing, H., & Liu, X. (2011). Activation of caspase-3 and c-Jun NH2-terminal kinase signaling pathways involving heroin-induced neuronal apoptosis. *Neuroscience Letters, 502*(3), 209–213.

Li, X., Deng, L., & Ye, B. (2012). Home-based rehabilitation for heroin-induced spongiform leukoencephalopathy. *Neural Regeneration Research, 7*(7), 534–538.

Lonnqvist, T., Paetau, A., Valanne, L., & Pihko, H. (2009). Recessive twinkle mutations cause severe epileptic encephalopathy. *Brain, 132*(Pt 6), 1553–1562. http://dx.doi.org/10.1093/brain/awp045.

Lucas, C. E. (2005). The impact of street drugs on trauma care. *Journal of Trauma, 59*(Suppl. 3), S57–S60; discussion S67–S75.

Martino, S., Brigham, G. S., Higgins, C., Gallon, S., Freese, T. E., Albright, L. M., ... Condon, T. P. (2010). Partnerships and pathways of dissemination: the National Institute on Drug Abuse-Substance Abuse and Mental Health Services Administration Blending Initiative in the Clinical Trials Network. *Journal of Substance Abuse Treatment*, *38*(Suppl. 1), S31–S43. http://dx.doi.org/10.1016/j.jsat.2009.12.013.

Miki, K., Ishibashi, S., Sun, L., Xu, H., Ohashi, W., Kuroiwa, T., & Mizusawa, H. (2009). Intensity of chronic cerebral hypoperfusion determines white/gray matter injury and cognitive/motor dysfunction in mice. *Journal of Neuroscience Research*, *87*(5), 1270–1281.

Molloy, S., Soh, C., & Williams, T. L. (2006). Reversible delayed posthypoxic leukoencephalopathy. *AJNR American Journal of Neuroradiology*, *27*(8), 1763–1765.

Niehaus, L., & Meyer, B. U. (1998). Bilateral borderzone brain infarctions in association with heroin abuse. *Journal of the Neurological Science*, *160*(2), 180–182.

Offiah, C., & Hall, E. (2008). Heroin-induced leukoencephalopathy: characterization using MRI, diffusion-weighted imaging and MR-spectroscopy. *Clinical Radiology*, *63*(2), 146–152.

Paranthaman, S. K., & Khan, F. (1976). Acute cardiomyopathy with recurrent pulmonary edema and hypotension following heroin overdosage. *Chest*, *69*(1), 117–119.

Pernet, V., Joly, S., Christ, F., Dimou, L., & Schwab, M. E. (2008). Nogo-A and myelin-associated glycoprotein differently regulate oligodendrocyte maturation and myelin formation. *The Journal of Neuroscience: The Official Journal of the Society for Neuroscience*, *28*(29), 7435–7444.

Rizzuto, N., Morbin, M., Ferrari, S., Cavallaro, T., Sparaco, M., Boso, G., & Gaetti, L. (1997). Delayed spongiform leukoencephalopathy after heroin abuse. *Acta Neuropathologica*, *94*(1), 87–90.

Roberts, W. C. (2002). Pulmonary talc granulomas, pulmonary fibrosis, and pulmonary hypertension resulting from intravenous injection of talc-containing drugs intended for oral use. *Proceedings (Baylor University Medical Center)*, *15*(3), 260–261.

Roulet Perez, E., Maeder, P., Rivier, L., & Deonna, T. (1992). Toxic leucoencephalopathy after heroin ingestion in a 2 1/2-year-old child. *Lancet*, *340*(8821), 729.

Rumbaugh, C. L., Bergeron, R. T., Fang, H. C., & McCormick, R. (1971). Cerebral angiographic changes in the drug abuse patient. *Radiology*, *101*(2), 335–344. http://dx.doi.org/10.1148/101.2.335.

Ryan, A., Molloy, F. M., Farrell, M. A., & Hutchinson, M. (2005). Fatal toxic leukoencephalopathy: clinical, radiological, and necropsy findings in two patients. *Journal of Neurology, Neurosurgery and Psychiatry*, *76*(7), 1014–1016.

Savov, Y., Antonova, N., Zvetkova, E., Gluhcheva, Y., Ivanov, I., & Sainova, I. (2006). Whole blood viscosity and erythrocyte hematometric indices in chronic heroin addicts. *Clinical Hemorheology and Microcirculation*, *35*(1–2), 129–133.

Sempere, A. P., Posada, I., Ramo, C., & Cabello, A. (1991). Spongiform leucoencephalopathy after inhaling heroin. *Lancet*, *338*(8762), 320.

Silvestrelli, G., Corea, F., Micheli, S., & Lanari, A. (2010). Clinical pharmacology and vascular risk. *Open Neurology Journal*, *4*, 64–72. http://dx.doi.org/10.2174/1874205X01004020064.

Somala, R. K. (2009). Rhabdomyolysis and brain ischemic stroke in a heroin-dependent male. Invited comment. *Acta Psychiatrica Scandinavica*, *120*(1), 80–81. http://dx.doi.org/10.1111/j.1600-0447.2009.01384.x.

Swift, W., Maher, L., & Sunjic, S. (1999). Transitions between routes of heroin administration: a study of Caucasian and Indochinese heroin users in south-western Sydney, Australia. *Addiction (Abingdon, England)*, *94*(1), 71–82.

Tan, T. P., Algra, P. R., Valk, J., & Wolters, E. C. (1994). Toxic leukoencephalopathy after inhalation of poisoned heroin: MR findings. *AJNR American Journal of Neuroradiology*, *15*(1), 175–178.

Vella, S., Kreis, R., Lovblad, K. O., & Steinlin, M. (2003). Acute leukoencephalopathy after inhalation of a single dose of heroin. *Neuropediatrics*, *34*(2), 100–104. http://dx.doi.org/10.1055/s-2003-39604.

Verma, R., Sharma, P., & Vidhate, M. R. (2011). A rare case of acute fatal leucoencephalopathy due to heroin exposure. *Neurology India*, *59*(1), 127–128. http://dx.doi.org/10.4103/0028-3886.76863.

Vila, N., & Chamorro, A. (1997). Ballistic movements due to ischemic infarcts after intravenous heroin overdose: report of two cases. *Clinical Neurology and Neurosurgery*, *99*(4), 259–262.

Villella, C., Iorio, R., Conte, G., Batocchi, A. P., & Bria, P. (2010). Toxic leukoencephalopathy after intravenous heroin injection: a case with clinical and radiological reversibility. *Journal of Neurology*, *257*(11), 1924–1926. http://dx.doi.org/10.1007/S00415-010-5620-6.

Weber, W., Henkes, H., Moller, P., Bade, K., & Kuhne, D. (1998). Toxic spongiform leucoencephalopathy after inhaling heroin vapour. *European Radiology*, *8*(5), 749–755. http://dx.doi.org/10.1007/s003300050467.

Wolters, E. C., van Wijngaarden, G. K., Stam, F. C., Rengelink, H., Lousberg, R. J., Schipper, M. E., & Verbeeten, B. (1982). Leucoencephalopathy after inhaling "heroin" pyrolysate. *Lancet*, *2*(8310), 1233–1237.

Woods, B. T., & Strewler, G. J. (1972). Hemiparesis occurring six hours after intravenous heroin injection. *Neurology*, *22*(8), 863–866.

Yin, R., Lu, C., Chen, Q., Fan, J., & Lu, J. (2013). Microvascular damage is involved in the pathogenesis of heroin induced spongiform leukoencephalopathy. *International Journal of Medical Sciences*, *10*(3), 299–306.

Zuckerman, G. B., Ruiz, D. C., Keller, I. A., & Brooks, J. (1996). Neurologic complications following intranasal administration of heroin in an adolescent. *Annals of Pharmacotherapy*, *30*(7–8), 778–781.

Zvetkova, E., Antonova, N., Ivanov, I., Savov, Y., & Gluhcheva, Y. (2010). Platelet morphological, functional and rheological properties attributable to addictions. *Clinical Hemorheology and Microcirculation*, *45*(2–4), 245–251. http://dx.doi.org/10.3233/CH-2010-1305.

Chapter 94

The Hypothalamic–Pituitary–Adrenal Axis and Related Brain Stress-Response Systems and Heroin

Yan Zhou, Hilary Briggs, Mary Jeanne Kreek
The Laboratory of the Biology of Addictive Diseases, The Rockefeller University, New York, NY, USA

Abbreviations

AVP Arginine vasopressin
CRF Corticotropin-releasing factor
CRF-R1 CRF type I receptor
CPP Conditioned place preference
eGFP Enhanced green fluorescent protein
HPA Hypothalamic–pituitary–adrenal
KOP-r κ Opioid receptor
MC2R Adrenocorticotropic hormone receptor
MOP-r μ Opioid receptor
NAc Nucleus accumbens
OPRM1 μ Opioid receptor gene
POMC Proopiomelanocortin
PVN Paraventricular nucleus
pPVN Parvocellular division of PVN
V1b AVP type 1b receptor
VTA Ventral tegmental area.

INTRODUCTION

Opiate addiction is a major global public health problem and there remains a further need for the better utilization of effective medications (e.g., methadone and buprenorphine plus naltrexone maintenance treatment) and the development of novel medications for the treatment of excessive opiate use and opiate addiction. In humans, heroin addiction can be characterized by a drug user's initial use resulting in increased habituation that causes severe withdrawal symptoms during periods of involuntary abstinence. The reward pathway and withdrawal symptoms, and the desire to avoid them, often result in relapse and reescalation of drug use. The length of time a person spends in each of the developing stages of addiction varies by individual. There are three main factors that contribute to the development of heroin addiction: the environment (e.g., stress), the drug's reinforcing effects, and genetics. Our laboratory, as well as others, has implicated the dysregulation of the stress-responsive hypothalamic–pituitary–adrenal (HPA) axis and brain stress-response systems, at least in part, during the acquisition of preaddictive behavior and progression toward opiate addiction.

This review will discuss four important systems in current addiction research: (1) the HPA axis in human heroin addiction and rodent models, (2) the role of the arginine vasopressin (AVP) and AVP type 1b receptor (V1b) systems in rodent stress response and drug addiction, (3) the role of the endogenous μ and κ opioid systems in rodent stress response and drug addiction, and (4) orexin and dynorphin in the lateral hypothalamus and their roles in rodent stress response and drug addiction.

First, with regard to the HPA axis, there is an increasing body of literature demonstrating that opiates alter the HPA axis (Figure 1), and in turn the abnormal HPA activity contributes to opiate consumption, development of opiate addiction, and relapse to drug use. In our discussion of the HPA axis, we will provide an overview of research on opiate addiction, with specific emphasis on animal models for laboratory-based research on the HPA axis and basic clinical research to elucidate the biology of heroin addiction. We will also include a discussion of the studies on the genetic correlates of opiate addictive diseases.

Second is the AVP system (Figure 2(A)), which has been studied in neuroendocrinology and drug addiction. The HPA axis in rodents and humans is directly influenced by the AVP system. This neuropeptide system is profoundly altered by opiates in rodent models, which is discussed in detail in this chapter. Of interest is the impact of drugs of abuse, such as heroin and cocaine, on the AVP and the V1b receptor system in rodent addiction models. Different and interesting brain areas such as the medial amygdala, paraventricular nucleus (PVN) of the hypothalamus, and anterior pituitary localizations will be discussed as well as the functions of AVP and its V1b receptor, which have been further identified and elucidated (Zhou, Proudnikov, Yuferov, & Kreek, 2010).

The third relevant topic is the endogenous opioid system and its critical role in the control of the HPA axis. Animal and human studies have demonstrated that β-endorphin and dynorphin exert tonic inhibition and stimulation of HPA activity by acting on the μ opioid receptor (MOP-r) (Farren et al., 1999; Kreek, 1973a, 1973b; Kreek et al., 1983; Rosen, Kosten, & Kreek, 1999; Schluger et al., 1998; Volavka et al., 1990) and the κ opioid receptor (KOP-r) (Zhou, Leri, Grella, Aldrich, & Kreek, 2013), respectively.

In human genetic studies, polymorphisms in genes encoding the MOP-r have been found to be associated with heroin addiction.

Finally, the fourth topic discussed herein is the stress-responsive orexin (or hypocretin) system (Figure 2(B)). Most of the lateral hypothalamic orexin neurons coexpress dynorphin (Figure 2(C)). Orexin A and B are two orexin-derived peptides acting on the orexin 1 and 2 receptors, respectively. Orexin has actions in the reward areas of the brain, such as the nucleus accumbens (NAc) and ventral tegmental area (VTA) with implications in addiction and addiction-like behaviors (Harris, Wimmer, & Aston-Jones, 2005).

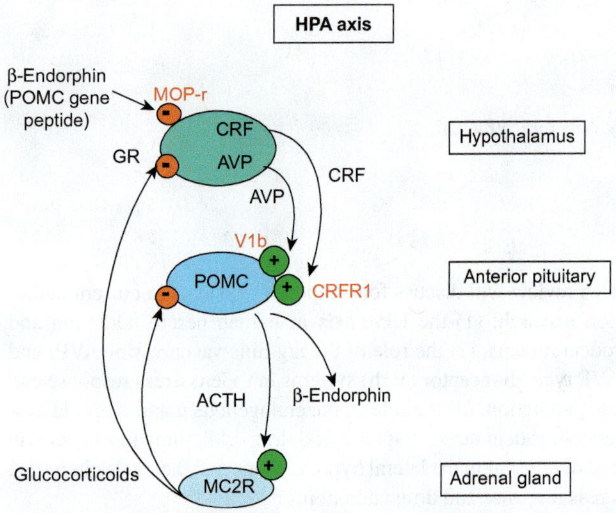

HPA AXIS IN HUMAN HEROIN ADDICTION AND RODENT MODELS

In humans, stress plays a major role in drug addiction and elevates drug craving (Kreek & Koob, 1998). Of importance, the stress-induced elevation of HPA activity predicts relapse to drug use and amounts of subsequent use, clearly indicating that stress not only elicits craving, but independently predicts drug relapse (e.g., Sinha, Garcia, Paliwal, Kreek, & Rounsaville, 2006). Vulnerability to drug abuse is enhanced by stress, and the HPA response to stress is one of the critical factors influencing individual vulnerability to drug abuse. Environmental stress modulates the effects of drugs of abuse on the acquisition of drug self-administration behavior, locomotor activity, and reinstatement of self-administration after extinction (Zhou et al., 2010).

The HPA axis is a well-studied stress-response system in animal models and also in humans (see Figure 1). Stress, triggered by either internal or external stimuli, increases corticotropin-releasing factor (CRF) and AVP (Figure 2(A)), causing their release into the pituitary portal circulation from terminals of the hypothalamic PVN. When CRF type 1 receptor (CRF-R1) and V1b receptor are activated, this drives the release of the proopiomelanocortin (POMC) peptides (Vale, Spiess, Rivier, & Rivier, 1981). Two major POMC gene products are adrenocorticotropic hormone (ACTH) and β-endorphin (39 and 31 amino acids, respectively).

Acting primarily on the adrenal cortex, ACTH releases corticosterone in rodents and cortisol in humans. Glucocorticoids, which are essential for life in mammals, act in a negative feedback mood by decreasing biosynthesis, release, and function of CRF and AVP in the hypothalamus and of CRF-R1 and V1b in the anterior pituitary. Glucocorticoids also act to reduce the processing and release of POMC peptides in the anterior pituitary. CRF, CRF-R1, CRF-R2, AVP, and V1b are all located in extrahypothalamic brain regions. Interestingly, POMC has also been identified in extrahypothalamic brain regions like the NAc, amygdala, and mesolimbic area (Zhou, Colombo, et al., 2013). A negative inhibition of CRF and POMC, as well as their receptors, by glucocorticoids in these regions has been studied, and it is still unclear whether the modulation by glucocorticoids occurs in extrahypothalamic regions (Zhou, Proudnikov, Yuferov, & Kreek, 2010; Zhou et al., 1996).

In early studies of the rodent HPA axis, an acute opiate challenge has a stimulatory effect on plasma ACTH and corticosterone release in opiate-naïve mice and rats (e.g., Gibson, Ginsburg, Hall, &

FIGURE 1 HPA axis. Stress increases the biosynthesis of hypothalamic corticotropin-releasing factor (CRF) and AVP and their release into the portal circulation, from which they act on CRF-R1 and V1b receptors, respectively, in the anterior pituitary. This drives the biosynthesis of proopiomelanocortin (POMC) mRNA and peptides in the corticotropes of the anterior pituitary and the release into the circulation of β-endorphin and ACTH, which are derived from the processing of POMC. ACTH acts on the ACTH receptor (MC2R) in the adrenal cortex to release the stress hormone cortisol (in humans and guinea pigs) or corticosterone (in rats and mice), which are primary mediators of the stress response. Cortisol or corticosterone exerts negative feedback regulation at both the hypothalamus and the pituitary via glucocorticoid receptors (GR). In addition to this classical negative feedback regulation by glucocorticoids, the endogenous opioid systems, especially the μ opioid receptors (MOP-r) and κ opioid receptors, inhibit and stimulate this axis, respectively.

(A)

CRF	SEEPPISLDLTFHLLREVLEMARAEQLAQQAHSNRKLMEII-NH$_2$
AVP	CYFQNCPRG-NH$_2$ (C$_1$-C$_6$ diS)

(B)

Orexin A	Pyr-PLPDCCRQKTCSCRLYELLHGAGNHAAGILTL-NH$_2$ (C$_6$-C$_{12}$ diS; C$_7$-C$_{14}$ diS)
Orexin B	RSGPPGLQGRLQRLLQASGNHAAGILTN-NH$_2$

(C)

Dynorphin A (1-17)	YGGFLRRIRPKLKWDNQ
Dynorphin B	YGGFLRRQFKVVT

FIGURE 2 The amino acid sequences of human AVP and CRF (A), human orexin A and B (B), and human dynorphin A and B (C).

Hart, 1979; Ignar & Kuhn, 1990; Martinez, Vargas, Fuente, Garcia, & Milanes, 1990; Zhou, Spangler, Maggos, Wang, Han, Ho, & Kreek, 1999). After chronic opiate administration for 1 to 2 weeks, however, the HPA axis is not activated by heroin or morphine but rather suppressed by them (Ignar & Kuhn, 1990; Martinez et al., 1990; Zhou et al., 2006; Zhou, Cui, et al., 2008; Zhou, Leri, et al., 2008; Zhou, Leri, Ho, & Kreek, 2013). In contrast, all short-acting opiates acutely or chronically administered result in suppression of the HPA axis hormones in humans, whereas withdrawal from these drugs causes a rapid rebound affect in the HPA axis and its associated hormones (Kreek, 1973a). Naloxone, acting as a short-acting opioid antagonist, causes HPA activation in humans and is an effective treatment during opioid overdose.

Heroin, morphine, and other short-acting opiates regulate the activity of endogenous opioid systems. In rats, chronic intermittent heroin administration by experimenters (Zhou, Leri, Ho, et al., 2013) or chronic heroin self-administration (Zhou, Leri, Cummins, & Kreek, 2015) resulted in decreased POMC mRNA levels in the hypothalamus, suggesting a hypothesis that long-term exposure to opiates leads to relative deficiency in the β-endorphin system. The presumed relative deficiency in endogenous β-endorphins (as reflected by decreased POMC mRNA levels in the hypothalamus (e.g., Zhou et al., 2015; Zhou, Leri, Ho, et al., 2013)) could lead to a hyperactive HPA axis and corresponding increases in hormone secretion during spontaneous opiate withdrawal. Indeed, increases in HPA hormonal levels have been observed both in rats (Ignar & Kuhn, 1990, Martinez et al., 1990; Zhou et al., 2006; Zhou, Cui, et al., 2008; Zhou, Leri, et al., 2008; Zhou, Leri, Ho, et al., 2013) and in humans (Culpepper-Morgan & Kreek, 1997; Kreek, 1973a; Kreek et al., 1984) during acute spontaneous opiate withdrawal.

Chronic exposure to short-acting opiates is a chronic stressor by virtue of the withdrawal that always occurs between exposures, and it may alter the responsivity of the HPA axis, as do many other stressors (e.g., Houshyar, Gomez, Manalo, Bhargava, & Dallman, 2003). Indeed, a low dose of heroin challenge decreased ACTH levels in rats during chronic withdrawal after HPA hormonal levels returned to baseline, which is in contrast to the stimulatory effect of acute heroin in opiate-naïve animals (Zhou, Leri, Ho, et al., 2013). The effects of opiates on HPA activity have been found to depend on the presence or absence of external stressors, although the mechanisms responsible for the interactions between heroin and stress are not well studied. In fact, the inhibitory effect of opiates on the HPA axis is consistently found in heroin addicts, and methadone, a long-acting opioid, has been shown to modulate the HPA axis during stress (Kreek, 1973a). It is also found that morphine suppresses cortisol release induced by surgical stress in humans (Cowen et al., 1982; George, Reier, Lanese, & Rower, 1974). In support of this concept, our animal studies found that while either acute morphine or water restriction stress alone increased ACTH levels as an independent stimulus, morphine decreases plasma ACTH levels elevated by water restriction stress (Zhou et al., 1999). The inhibitory effects of opiates on HPA activity in rats have been reported to occur at the PVN (Laorden, Milanes, Angel, Tankosic, & Burlet, 2003) or median eminence (Plotsky, 1986). It seems likely that opiates have no direct effect on the corticotropes (Buckingham, 1982), which is different from the potential direct inhibitory effect of alcohol or other drugs on the anterior pituitary (Zhou & Lapingo, 2014). Together, both human and animal studies demonstrate that morphine, heroin, or other short-acting opiates effectively reduce HPA activity caused by stress, indicating that opioids act in a counterregulatory role in modulating HPA stress responsivity under stress conditions.

ARGININE VASOPRESSIN AND V1b RECEPTOR SYSTEMS

The two central G-protein-coupled AVP receptor subtypes, V1a and V1b, are highly expressed in the rat extended amygdala. The V1b receptors are expressed prominently in the amygdala, PVN, hippocampus, and anterior pituitary (Lolait et al., 1995). Activation of V1b receptors in the amygdala is an important step involved in stress-related behaviors, including anxiogenic and depressive behaviors in rodent models (Griebel et al., 2002). To examine the role of AVP in heroin addiction, the expression of the AVP gene in the rat amygdala and hypothalamus was studied by our group after chronic intermittent escalating-dose heroin or during early and late spontaneous withdrawal. AVP mRNA levels expressed in the medial amygdala were increased during early heroin withdrawal (Zhou, Cui, et al., 2008; Zhou, Leri, et al., 2008). To further study the AVP system we designed experiments using foot-shock stress, a method that is well documented for the study of stress in rodents (Trentani et al., 2003). In a collaboration with the Leri laboratory at the University of Guelph in Canada, the results of these experiments demonstrated that foot-shock stress applied to rats increased the AVP mRNA levels in the medial amygdala after the rats were able to self-administer heroin, but not in heroin-naïve rats. These results demonstrate that AVP expression due to stress is increased in a population previously capable of heroin self-administration, implicating AVP in stress-related drug use. We then questioned if blockade of the central AVP receptors (V1a or V1b receptor) would attenuate the heroin-seeking behavior in experiments on the reinstatement of drug self-administration behavior induced by foot-shock stress and HPA hormonal responses to foot shock. The selective V1b receptor antagonist SSR149415 (but not a relatively selective V1a antagonist) dose-dependently attenuated foot-shock-induced reinstatement of drug behavior and lowered foot-shock-induced HPA activation (Zhou, Cui, et al., 2008; Zhou, Leri, et al., 2008). Additionally, we found increased AVP mRNA levels in the medial amygdala after acute withdrawal from chronic cocaine exposure (Zhou et al., 2005). These data suggest that stress-responsive AVP/V1b receptor systems (e.g., the medial amygdala) may be critical components of the neural circuitry underlying the averse emotional consequences of drug withdrawal and of negative emotional states on drug-seeking behavior (Zhou, Leri, Cummins, Hoeschele, & Kreek, 2008). To further investigate the involvement of AVP and V1b in alcohol-drinking behavior, we used genetically selected Sardinian alcohol-preferring (sP) rats, in collaboration with the Colombo laboratory at the CNR Institute of Neuroscience at the University of Cagliari in Italy (Colombo, Lobina, Carai, & Gessa, 2006). We found that pharmacological blockade of the V1b receptor reduced alcohol consumption in sP rats (Zhou, Colombo, et al., 2011). Consistently, V1b receptor antagonists blocked high alcohol drinking in alcohol-dependent rat models (Edwards, Guerrero, Ghoneim, Roberts, & Koob, 2012) (more details are given in a 2014 review on alcohol studies (Zhou & Kreek, 2014)).

Earlier studies showed that AVP in the parvocellular division of the PVN (pPVN) did not contribute to the acute stimulatory effects of cocaine on HPA activity. However, we found persistent elevations of both peripheral plasma ACTH levels and AVP mRNA levels in the pPVN of the rats during 14 days of protracted cocaine withdrawal, and V1b antagonists attenuated cocaine withdrawal-induced HPA activation (Zhou, Litvin, Piras, Pfaff, & Kreek, 2011). Interestingly, in AVP-enhanced green fluorescent protein (eGFP) transgenic mice, cocaine withdrawal increased the number of pPVN AVP neurons expressing GFP, further confirming that enhanced pPVN AVP gene expression is associated with persistent elevations in basal HPA activity. Our results indicate that AVP and its receptor system are involved in chronic stress and may be an attractive therapeutic target for treating anxiety and depressive symptoms associated with drug addiction. Therefore, it may be worthwhile to explore the value of both V1b and V1a receptor antagonists in the management of opiate, cocaine, or alcohol abuse (Zhou, Litvin, et al., 2011; Edwards et al., 2012).

ENDOGENOUS OPIOID SYSTEMS

Methadone is a selective MOP-r agonist (long-acting in humans) and widely used in the maintenance treatment of short-acting opiate (primarily heroin) addiction for 325,000 patients in the United States, around 365,000 patients in Europe, and approximately 700,000 patients throughout the rest of the world (personal communications, Mark W. Parrino, President of the American Association for the Treatment of Opioid Dependence, European Monitoring Center for Drugs and Drug Addiction, and Icro Maremmani, Professor of Addiction Medicine, University of Pisa, Italy). On an optimal daily oral dose of 80–150 mg/day, the range of plasma methadone levels is 74–732 ng/ml in humans (Borg, Ho, Peters, & Kreek, 1995; Kreek, 2000; Peles, Kreek, Kellogg, & Adelson, 2006). In contrast to in humans, methadone has a short half-life in rodents (about 60 min in mice and 90 min in rats) (Kreek, 1979). In our animal studies (e.g., Zhou et al., 1996), therefore, methadone is delivered through osmotic pumps to mimic steady-state methadone maintenance in humans (Kreek, 1973b). The steady-state administration of methadone in rats (10 mg/kg/day) by osmotic pumps achieves a mean plasma level of 123 ng/ml with a range of 100–150 ng/ml (Zhou et al., 1996), which is comparable to levels achieved and sustained at 24 h after the last administration during chronic methadone maintenance treatment at low doses in humans (60–80 mg/day). In this rat model, we did not find any effect of steady-state methadone treatment on the mRNA levels of hypothalamic CRF and POMC, anterior pituitary CRF-R1 and POMC, or circulating corticosterone levels (Zhou et al., 1996). Consistently, a very early study from our group found that chronic (36 days) methadone administration did not alter concentrations of immunoreactive β-endorphin in the rat amygdala and hypothalamus (Ragavan, Wardlaw, Kreek, & Frantz, 1983). These results demonstrate that steady-state occupancy of MOP-r with methadone does not have any significant effect on the rat CRF/CRF-R1 or POMC system in these studies. Our results support the hypothesis that there is no disruption of the HPA axis activity during steady-state administration of the exogenous opioid methadone (Kreek, 1973a, 1973b; Kreek et al., 1983, 1984).

β-Endorphin, primarily acting on MOP-r, is mainly expressed in the hypothalamic POMC neurons (major POMC/β-endorphin neurons in the brain). Since activation of MOP-r by β-endorphin is rewarding and modulates the NAc dopamine release (Spanagel, Herz, Bals-Kubik, & Shippenberg, 1991), β-endorphin is likely to be involved in the reinforcing effects and motivational behaviors of most drugs of abuse (Koob & Kreek, 2007). In cocaine self-administration or conditioned place preference (CPP) behavior, for example, MOP-r antagonists reduced the acquisition of self-administration and CPP, further supporting the possibility that β-endorphin acting on the MOP-r plays a functional role in the actions of cocaine. Indeed, cocaine CPP is blunted in β-endorphin-deficient mice (Marquez, Baliram, Dabaja, Gajawada, & Lutfy, 2008). We investigated whether cocaine CPP alters POMC gene expression in rat brains and found that cocaine CPP at 10 and 30 mg/kg doses increased the POMC mRNA levels in a dose-dependent manner in the hypothalamus, with no effect in the amygdala. Cocaine CPP had no effect on POMC mRNA levels in the anterior pituitary or on plasma ACTH or corticosterone levels. In rats that received cocaine at 30 mg/kg without conditioning, there was no such effect on hypothalamic POMC mRNA levels. Our results suggest that alterations in POMC gene expression in the hypothalamus are region-specific after cocaine CPP and dose-dependent, and such an increase may be involved in the reward/learning process of cocaine-induced conditioning (Zhou, Kruyer, Ho, & Kreek, 2012).

We hypothesized that dopamine D1 or D2-like receptors could play a role in the regulation of POMC gene expression in the hypothalamus. The selective antagonist sulpiride, when used to block the dopamine D2-like receptor, increased POMC mRNA levels in the hypothalamus, indicating that the dopamine D2-like receptor exerts a tonic inhibitory effect on hypothalamic POMC gene expression. Blockade by a selective antagonist, SCH23390, of the dopamine D1-like receptor, however, had no effect on hypothalamic POMC mRNA levels. These results suggest a specific role for the dopamine D2-like receptor in hypothalamic POMC expression (Zhou et al., 2004).

POMC-derived peptides, especially β-endorphin, are also distributed in the dopaminergic mesocorticolimbic regions, including the NAc, VTA, and frontal cortex. In addition to the arcuate nucleus, POMC mRNA has also been detected in the NAc and dorsal striatum at relatively low levels (Zhou, Colombo, et al., 2013). Therefore, the demonstration of POMC neuron distribution in the NAc region is an essential issue concerning the neural networks containing POMC mRNA and derived peptides in the NAc. Using POMC-eGFP transgenic mice in which POMC-expressing neurons were labeled with eGFP and enhanced by immunohistochemistry procedures, we found that POMC-eGFP-expressing neurons are present in modest amounts in the NAc core, NAc shell, and dorsal striatum of POMC-eGFP mice (Zhou, Colombo, et al., 2013). We also measured POMC mRNA levels in these two subdivisions of the NAc and the dorsal striatum of sP rats exposed to 17-day alcohol drinking and found that voluntary consumption of high amounts of alcohol by sP rats was associated with increases in POMC mRNA levels in the NAc shell, but not the NAc core or dorsal striatum. This result suggests that voluntarily consumed alcohol modulates POMC mRNA expression in the POMC neuron populations in the NAc shell. The POMC neurons in the shell are a region long considered to mediate processes of reward and reinforcement (e.g., Di Chiara, 2002), and our results support that they contribute to alcohol intake in this model (Zhou, Colombo, et al., 2013). Most relevant to this result, several studies have demonstrated that alcohol, cocaine, or cannabinoids are self-administered directly into the NAc shell, but not the core (e.g., Rodd-Henricks, McKinzie, Li, Murphy, & McBride, 2002). Therefore, the shell (not the core) may be the region in which alcohol and other drugs of abuse contribute to the reinforcing effects mediated through POMC neuronal activation. The involvement of NAc POMC in heroin- or cocaine-related behavior is under investigation.

In addition, polymorphisms in genes encoding the MOP-r have been found to be associated with drug addiction. A large number of association studies of the single-nucleotide polymorphism 118A → G have been reported (Bart et al., 2004; Bond et al., 1998) and reviewed (Kreek et al., 2012). For example, the most studied MOP-r gene (*OPRM1*) variant is 118A → G, which causes the replacement of an asparagine residue by aspartic acid. This change results in removal of an N-glycosylation site in the extracellular domain, which in turn results in altered receptor binding-site availability and reduced mRNA levels. It also leads to high-affinity binding of β-endorphin and altered signaling efficacy (e.g., Beyer, Koch, Schroder, Schulz, & Hollt, 2004; Bond et al., 1998). One mouse model of 118A → G has shown lower antinociceptive response and reward properties of morphine, as well as a reduction in the aversive effect of naloxone-precipitated morphine withdrawal, in a sex-dependent manner (Mague et al., 2009). In one study from our laboratory, the 118G variant was associated with heroin addiction in a sample of Swedish subjects with little genetic admixture (Bart et al., 2004). Of interest, the 118G allele has been associated with various phenotypes including atypical HPA axis response to objective measured stress: an association with a robust cortisol response to the MOP-r competitive antagonist naloxone, in a population-specific manner (e.g., Kreek, 2007; Wand et al., 2002), and a blunted ACTH response to metyrapone in healthy subjects (Ducat et al., 2013).

Laboratory studies in humans found that yohimbine increased HPA activity as well as subjective anxiety in normal subjects (e.g., Rosen et al., 1999; Vythilingam et al., 2000) and drug craving in abstinent opiate addicts (Stine et al., 2002). Yohimbine enhances central noradrenergic activity by acting as an antagonist at α2 adrenergic autoreceptors and reinstates heroin, methamphetamine, cocaine, alcohol, and food seeking and increases ACTH and corticosterone release in rats (e.g., Smythe, Bradshaw, & Vining, 1983; Zhou, Leri, Grella, Aldrich, & Kreek, 2013). Our work further found that the yohimbine-induced or food restriction stress-induced HPA activation was blunted by the selective KOP-r antagonist nor-BNI, providing further evidence that there is an involvement of the KOP-r system in modulation of HPA activity (Allen, Zhou, & Leri, 2013; Zhou, Leri, Grella, Aldrich, & Kreek, 2013). Of interest, pretreatment with the KOP-r antagonist nor-BNI blocked the heroin-taking and -seeking behavior in a rat self-administration model (Schlosburg et al., 2013; Zhou, Leri, Grella, Aldrich, & Kreek, 2013).

OREXIN AND DYNORPHIN IN THE LATERAL HYPOTHALAMUS

The orexins are expressed in the lateral hypothalamus, perifornical area, and dorsomedial hypothalamus, with extensive projections in the brain (de Lecea et al., 1998). Orexin A acts at orexin type 1 and 2 receptors (OX1R and OX2R) and orexin B acts on OX2R exclusively. Hypothalamic orexins are involved in the regulation of sleep–wakefulness, arousal, feeding, and stress. Orexins may also have a role in the modulation of drug reward and drug-seeking behaviors. Orexin receptor blockade in the VTA after acute morphine administration, for example, attenuated an increase in extracellular dopamine levels in the NAc (Narita et al., 2006). Orexins and orexin receptor interactions have been found to trigger morphine-motivated behaviors as well (Harris et al., 2005).

Because most of the lateral hypothalamic orexin neurons coexpress the dynorphin gene (Chou et al., 2001), we examined levels of both the orexin and the dynorphin mRNAs in the lateral hypothalamus. During the aversive state of acute withdrawal from chronic intermittent escalating-dose morphine, orexin mRNA levels were increased in rat lateral hypothalamus, indicating that the increased orexin neuronal activity in the lateral hypothalamus occurred during opiate withdrawal and could contribute to negative affective states in opiate withdrawal. Dynorphin mRNA levels remain unaltered in the lateral hypothalamus under the withdrawal-related stress condition (Zhou et al., 2006).

To investigate whether this observation held true with other drugs of abuse, we extended our research to cocaine. We were primarily interested in investigating whether orexin or dynorphin mRNA levels in rat lateral hypothalamus or medial hypothalamus (perifornical and dorsomedial areas) are altered following a cocaine rewarding process, using a cocaine-induced CPP model, and during chronic cocaine exposure in binge-pattern administration with steady-dose (45 mg/kg/day) and escalating-dose (45–90 mg/kg/day) regimens (Zhou, Cui, et al., 2008). The results of these experiments demonstrated that orexin mRNA levels in the lateral hypothalamus were decreased after cocaine CPP and chronic escalating-dose cocaine. The orexin mRNA levels did not increase in the CPP pattern regimen without a conditioning or steady-dose regimen. Of interest, acute withdrawal from the chronic escalating-dose cocaine administration resulted in increased orexin mRNA levels in the lateral hypothalamic, but not the medial hypothalamic region. In contrast to opiate withdrawal, acute cocaine withdrawal from chronic escalating-dose administration increased the dynorphin mRNA levels in the lateral hypothalamus (Zhou, Cui, et al., 2008). These results suggest that the orexin gene expression alterations in response to drug are lateral hypothalamus-specific after cocaine CPP and dose-dependent after chronic cocaine exposure. Gene expression of orexin and dynorphin in the lateral hypothalamus may also contribute to the enhanced negative affective states in cocaine withdrawal.

SUMMARY

As shown in this review, we have learned a great deal from selective animal models regarding pathways of action of psychoactive drugs, focusing on the neurobiology of heroin dependence and the modulation of stress using a bidirectional approach combining animal models and basic clinical research, particularly relative to the stress-responsive HPA axis. The endogenous opioid systems (including POMC/MOP-r and dynorphin/KOP-r systems) clearly play a major role in heroin addiction, and specific *OPRM1* gene variants contribute to stress responsivity and may affect vulnerability to developing heroin addiction. Other stress-response systems mentioned above (including vasopressin with its V1b receptors and orexin with its receptors), derived from animal studies of the brain stress-response systems, are also potentially involved in opiate addiction.

DEFINITION OF TERMS

Conditioned place preference CPP, a form of Pavlovian conditioning, is used to measure the motivational effects of objects or experiences in both mice and rats. With an identical procedure involving aversive stimuli, the CPP paradigm can be used to measure conditioned place aversion also. This procedure is applied to measuring extinction and reinstatement of the conditioned stimulus. Drugs of abuse (including heroin and morphine) are used in this CPP paradigm to measure their reinforcing properties. There are two different methods used to choose the chamber to be conditioned: biased and unbiased.

Self-administration In rodents, self-administration of drugs of abuse is a form of operant conditioning when the drug is rewarding. The drug (heroin or morphine) can be administered through an implanted intravenous tubing or intracerebroventricular injection. It is considered that self-administration of putatively addictive drugs is one of the most valid rodent models to study drug-seeking and -taking behavior. The more rewarding (and possibly addictive) the test substance is considered, the higher the frequency with which a test animal exhibits the operant behavior.

Extinction In the CPP paradigm, extinction is the process by which the association of the chamber with the paired rewarding or aversive stimulus is greatly reduced, thus lessening the place preference or aversion. Extinction occurs when the conditioned stimulus is presented on repeated trials without the presence of the rewarding or aversive stimulus.

Reinstatement This is a method used in rodents to test procedures including CPP and self-administration. Reinstatement is the reacquisition of an extinguished behavior, which is resultant from the presentation of the unconditioned stimulus, stress, or context cues. This shows that the process of extinction does not completely eliminate an association, since the association between the unconditioned stimuli and the conditioned stimuli can be rapidly reacquired. In the context of CPP, for instance, after a place preference has been extinguished, the behavior is reinstated when the animal quickly reacquires its place preference after repeated extinctions have caused the preference to be extinguished. It is used to model the behavior of drug craving and relapse in humans, although its validity is still a topic of debate.

Green fluorescent protein GFP is a protein that has 238 amino acid residues (26.9 kDa) and exhibits bright green fluorescence when exposed to light in the blue to ultraviolet range. In cell and molecular biology, and transgenic animal models, the GFP gene is frequently used as a reporter of gene expression. For instance, many transgenic mice have been created that express GFP as a proof of concept that a gene can be expressed throughout a given organism, for example, in the central nervous system.

Polymorphisms The term is used here by molecular biologists to describe certain point mutations in a genotype. Polymorphism in biology occurs when two or more clearly different genotypes or phenotypes exist in the same population of a species.

Sardinian alcohol-preferring and -nonpreferring rats These are rat lines selectively bred for high and low alcohol preference and consumption, respectively, under the home-cage, continuous two-bottle choice regimen. sP rats meet most of the fundamental criteria for an animal model of alcoholism, in that they voluntarily consume sufficient amounts of alcohol to achieve significant blood alcohol levels and produce psychopharmacological effects, including anxiolysis and motor stimulation. sP rats are also willing to "work" (such as lever-pressing) for alcohol. Chronic alcohol drinking in sP rats results in the development of tolerance to a given effect of alcohol and relapse-like drinking.

Vulnerability This refers to the ability to withstand the effects of addictive drugs.

KEY FACTS OF STRESS RESPONSES IN OPIATE ADDICTION

- Vulnerability to drug abuse is enhanced by stress, and the HPA response to stress is one of the critical factors influencing individual vulnerability to drug abuse.
- The involvement of the dysregulation of the stress-responsive HPA axis has been shown in opiate addiction.
- The stress-responsive AVP/V1b receptor system (including the amygdala) may be a critical component of the neural circuitry underlying the aversive emotional consequences of drug withdrawal and the effects of negative emotional states on drug-seeking and -taking behavior.
- Enhanced pPVN AVP gene expression is associated with persistent elevations of basal HPA activity after chronic stress.
- Enhanced lateral hypothalamic orexin neuronal activity, resulting from its increased gene expression, contributes to negative affective states in opiate withdrawal and cocaine withdrawal.
- Endogenous opioids are critical in the control of the HPA axis.
- Opiate effectively blunts the HPA activity caused by stress, indicating that opioids play a counterregulatory role in modulating HPA stress responsivity under stress conditions.
- Alterations in POMC gene expression in the hypothalamus are region-specific after cocaine CPP, and such an increase may be involved in the reward/learning process of cocaine-induced conditioning.
- Specific *OPRM1* gene variants contribute to stress responsivity and may affect vulnerability to developing heroin addiction.
- An involvement of the KOP-r system in the modulation of HPA activity has been shown.

SUMMARY POINTS

- HPA response to stress is one of the critical factors in opiate addiction.
- AVP/V1b receptor system underlies the aversive emotional consequences of drug withdrawal and negative emotional states on drug-seeking and -taking behaviors.
- Enhanced AVP is associated with persistent elevations in basal HPA activity after chronic stress.
- Opiate effectively blunts the HPA activity caused by stress, indicating a counterregulatory role under stress conditions.
- Enhanced lateral hypothalamic orexin neuronal activity contributes to negative affective states in opiate withdrawal and cocaine withdrawal.

ACKNOWLEDGMENT

This work was supported by HIH-NIDA R01DA032928 (Aldrich/subcontract Kreek), NIH-NCATS UL1 TR000043 (B. Coller), and the Dr. Miriam and Sheldon G. Adelson Medical Research Foundation (Kreek).

REFERENCES

Allen, C. P., Zhou, Y., & Leri, F. (2013). Effect of food restriction on cocaine locomotor sensitization in Sprague-Dawley rats: role of kappa opioid receptors. *Psychopharmacology, 226*, 571–578.

Bart, G., Heilig, M., LaForge, K. S., Pollak, L., Leal, S. M., Ott, J., & Kreek, M. J. (2004). Substantial attributable risk related to a functional mu-opioid receptor gene polymorphism in association with heroin addiction in central Sweden. *Molecular Psychiatry, 9*, 547–549.

Beyer, A., Koch, T., Schroder, H., Schulz, S., & Hollt, V. (2004). Effect of the A118G polymorphism on binding affinity, potency and agonist-mediated endocytosis, desensitization, and resensitization of the human mu-opioid receptor. *Journal of Neurochemistry, 89*, 553–560.

Bond, C., LaForge, K. S., Tian, M., Melia, D., Zhang, S., Borg, L., ... Yu, L. (1998). Single-nucleotide polymorphism in the human mu opioid receptor gene alters beta-endorphin binding and activity: possible implications for opiate addiction. *Proceedings of the National Academy of Sciences of the United States of America, 95*, 9608–9613.

Borg, L., Ho, A., Peters, J. E., & Kreek, M. J. (1995). Availability of reliable serum methadone determination for management of symptomatic patients. *Journal of Addictive Disease, 14*, 83–96.

Buckingham, J. C. (1982). Secretion of corticotrophin and its hypothalamic releasing factor in response to morphine and opioid peptides. *Neuroendocrinology, 35*, 111–116.

Chou, T. C., Lee, C. E., Lu, J., Elmquist, J. K., Hara, J., & Willie, J. T. (2001). Orexin (hypocretin) neurons contain dynorphin. *Journal of Neuroscience, 21*, RC168.

Colombo, G., Lobina, C., Carai, M. A., & Gessa, G. L. (2006). Phenotypic characterization of genetically selected Sardinian alcohol-preferring (sP) and -non-preferring (sNP) rats. *Addiction Biology, 11*, 324–338.

Cowen, M. J., Bullingham, R. E., Paterson, G. M., McQuay, H. J., Turner, M., Allen, M. C., & Moore, A. (1982). A controlled comparison of the effects of extradural diamorphine and bupivacaine on plasma glucose and plasma cortisol in postoperative patients. *Anesthesia Analgesia, 61*, 15–18.

Culpepper-Morgan, J. A., & Kreek, M. J. (1997). HPA axis hypersensitivity to naloxone in opioid dependence: a case of naloxone induced withdrawal. *Metabolism, 46*, 130–134.

Di Chiara, G. (2002). Nucleus accumbens shell and core dopamine: differential role in behavior and addiction. *Behavoiural Brain Research, 137*, 75–114.

Ducat, E., Ray, B., Bart, G., Umemura, Y., Varon, J., Ho, A., & Kreek, M. J. (2013). Mu-opioid receptor A118G polymorphism in healthy volunteers affects hypothalamic-pituitary-adrenal axis adrenocorticotropic hormone stress response to metyrapone. *Addiction Biology, 18*, 325–331.

Edwards, S., Guerrero, M., Ghoneim, O. M., Roberts, E., & Koob, G. F. (2012). Evidence that vasopressin V1b receptors mediate the transition to excessive drinking in ethanol-dependent rats. *Addiction Biology, 17*, 76–85.

Farren, C. K., O'Malley, S., Grebski, G., Maniar, S., Porter, M., & Kreek, M. J. (1999). Variable dose naltrexone-induced hypothalamic-pituitary-adrenal stimulation in abstinent alcoholics: a preliminary study. *Alcoholism, Clinical and Experimental Research, 23*, 502–508.

George, J. M., Reier, C. E., Lanese, R. R., & Rower, M. (1974). Morphine anesthesia blocks cortisol and growth hormone response to surgical stress in humans. *Journal of Clinical Endocrinology and Metabolism, 38*, 736–741.

Gibson, A., Ginsburg, M., Hall, M., & Hart, S. L. (1979). The effects of opiate receptor agonists and antagonists on the stress-induced secretion of corticosterone in mice. *British Journal of Pharmacology, 65*, 139–146.

Griebel, G., Simiand, J., Serradeil-Le Gal, C., Wagnon, J., Pascal, M., & Scatton, B. (2002). Anxiolytic- and antidepressant-like effects of the non-peptide vasopressin V1b receptor antagonist, SSR149415, suggest an innovative approach for the treatment of stress-related disorders. *Proceedings of the National Academy of Sciences of the United States of America, 99*, 6370–6375.

Harris, G. C., Wimmer, M., & Aston-Jones, G. (2005). A role for lateral hypothalamic orexin neurons in reward seeking. *Nature, 437*, 556–559.

Houshyar, H., Gomez, F., Manalo, S., Bhargava, A., & Dallman, M. (2003). Intermittent morphine administration induces dependence and is a chronic stressor in rats. *Neuropsychopharmacology, 28*, 1960–1971.

Ignar, D. M., & Kuhn, C. M. (1990). Effects of specific mu and kappa opiate tolerance and abstinence on hypothalamo-pituitary-adrenal axis secretion in the rat. *Journal of Pharmacology and Experimental Therapeutics, 255*, 1287–1295.

Koob, G., & Kreek, M. J. (2007). Stress, dysregulation of drug reward pathways, and the transition to drug dependence. *American Journal of Psychiatry, 164*, 1149–1159.

Kreek, M. J. (1973a). Medical safety and side effects of methadone in tolerant individuals. *JAMA, 223*, 665–668.

Kreek, M. J. (1973b). Plasma and urine levels of methadone. Comparison following four medication forms used in chronic maintenance treatment. *New York State Journal of Medicine, 73*, 2773–2777.

Kreek, M. J. (1979). Methadone disposition during the perinatal period in humans. *Pharmacology, Biochemistry and Behavior, 11* (Suppl.), 7–13.

Kreek, M. J. (2000). Methadone-related opioid agonist pharmacotherapy for heroin addiction. History, recent molecular and neurochemical research and future in mainstream medicine. *Annals of the New York Academy of Sciences, 909*, 186–216.

Kreek, M. J. (2007). Opioids, dopamine, stress, and the addictions. *Dialogues in Clinical Neuroscience, 9*, 363–378.

Kreek, M. J., & Koob, G. F. (1998). Drug dependence: stress and dysregulation of brain reward pathways. *Drug and Alcohol Dependence, 51*, 23–47.

Kreek, M. J., Levran, O., Reed, B., Schlussman, S. D., Zhou, Y., & Butelman, E. R. (2012). Opiate addiction and cocaine addiction: underlying molecular neurobiology and genetics. *Journal of Clinical Investigation, 122*, 3387–3393.

Kreek, M. J., Ragunath, J., Plevy, S., Hamer, D., Schneider, B., & Hartman, N. (1984). ACTH, cortisol and beta-endorphin response to metyrapone testing during chronic methadone maintenance treatment in humans. *Neuropeptides, 5*, 277–278.

Kreek, M. J., Wardlaw, S. L., Hartman, N., Raghunath, J., Friedman, J., & Schneider, B. (1983). Circadian rhythms and levels of beta-endorphin, ACTH, and cortisol during chronic methadone maintenance treatment in humans. *Life Sciences, 33*(Suppl. 1), 409–411.

Laorden, M. L., Milanes, M. V., Angel, E., Tankosic, K., & Burlet, A. (2003). Quantitative analysis of corticotrophin-releasing factor and arginine vasopressin mRNA in the hypothalamus during chronic morphine treatment in rats: an in situ hybridization study. *Journal of Neuroendocrinology, 15*, 586–591.

de Lecea, L., Kilduff, T. S., Peyron, C., Gao, X., Foye, P. E., & Danielson, P. E. (1998). The hypocretins: hypothalamus-specific peptides with neuroexcitatory activity. *Proceedings of the National Academy of Sciences of the United States of America, 95*, 322–327.

Lolait, S. J., O'Carroll, A. M., Mahan, L. C., Felder, C. C., Button, D. C., & Young, W. S., 3rd (1995). Extrapituitary expression of the rat V1b vasopressin receptor gene. *Proceedings of the National Academy of Sciences of the United States of America, 92*, 6783–6787.

Mague, S. D., Isiegas, C., Huang, P., Liu-Chen, L. Y., Lerman, C., & Blendy, J. A. (2009). Mouse model of OPRM1 (A118G) polymorphism has sex-specific effects on drug-mediated behavior. *Proceedings of the National Academy of Sciences of the United States of America, 106,* 10847–10852.

Marquez, P., Baliram, R., Dabaja, I., Gajawada, N., & Lutfy, K. (2008). The role of beta-endorphin in the acute motor stimulatory and rewarding actions of cocaine in mice. *Psychopharmacology (Berl), 197,* 443–448.

Martinez, J. A., Vargas, M. L., Fuente, T., Garcia, J. D. R., & Milanes, M. V. (1990). Plasma beta-endorphin and cortisol levels in morphine-tolerant rats and in naloxone-induced withdrawal. *European Journal of Pharmacology, 182,* 117–123.

Narita, M., Nagumo, Y., Hashimoto, S., Narita, M., Khotib, J., & Miyatake, M. (2006). Direct involvement of orexinergic systems in the activation of the mesolimbic dopamine pathway and related behaviors induced by morphine. *Journal of Neuroscience, 26,* 398–405.

Peles, E., Kreek, M. J., Kellogg, S., & Adelson, M. (2006). High methadone dose significantly reduces cocaine use in methadone maintenance treatment (MMT) patients. *Journal of Addictive Disease, 25,* 43–50.

Plotsky, P. M. (1986). Opioid inhibition of immunoreactive corticotropin-releasing factor secretion into the hypophysial-portal circulation of rats. *Regulatory Peptides, 16,* 235–242.

Ragavan, V. V., Wardlaw, S. L., Kreek, M. J., & Frantz, A. G. (1983). Effect of chronic naltrexone and methadone administration on brain immunoreactive beta-endorphin in the rat. *Neuroendocrinology, 37,* 266–268.

Rodd-Henricks, Z. A., McKinzie, D. L., Li, T. K., Murphy, J. M., & McBride, W. J. (2002). Cocaine is self-administered into the shell but not the core of the nucleus accumbens of Wistar rats. *Journal of Pharmacology and Experimental Therapeutics, 303,* 1216–1226.

Rosen, M. I., Kosten, T. R., & Kreek, M. J. (1999). The effects of naltrexone maintenance on the response to yohimbine in healthy volunteers. *Biological Psychiatry, 45,* 1636–1645.

Schlosburg, J. E., Whitfield, T. W., Jr., Park, P. E., Crawford, E. F., George, O., Vendruscolo, L. F., & Koob, G. F. (2013). Long-term antagonism of κ opioid receptors prevents escalation of and increased motivation for heroin intake. *Journal of Neuroscience, 33,* 19384–19392.

Schluger, J. H., Ho, A., Borg, L., Porter, M., Maniar, S., Gunduz, M., & Kreek, M. J. (1998). Nalmefene causes greater hypothalamic-pituitary-adrenal axis activation than naloxone in normal volunteers: implications for the treatment of alcoholism. *Alcoholism, Clinical and Experimental Research, 22,* 1430–1436.

Sinha, R., Garcia, M., Paliwal, P., Kreek, M. J., & Rounsaville, B. J. (2006). Stress-induced cocaine craving and hypothalamic-pituitary-adrenal responses are predictive of cocaine relapse outcomes. *Archives of General Psychiatry, 63,* 324–331.

Smythe, G. A., Bradshaw, J. E., & Vining, R. F. (1983). Hypothalamic monoamine control of stress-induced adrenocorticotropin release in the rat. *Endocrinology, 113,* 1062–1071.

Spanagel, R., Herz, A., Bals-Kubik, R., & Shippenberg, T. S. (1991). Beta-endorphin-induced locomotor stimulation and reinforcement are associated with an increase in dopamine release in the nucleus accumbens. *Psychopharmacology (Berl), 104,* 51–56.

Stine, S., Southwick, S., Petrakis, I., Kosten, T., Charney, D., & Krystal, J. (2002). Yohimbine-induced withdrawal and anxiety symptoms in opioid dependent patients. *Biological Psychiatry, 51,* 642–651.

Trentani, A., Kuipers, S. D., Meerman, G. J., Beekman, J., Horst, G. J., & den Boer, J. A. (2003). Immunohistochemical changes induced by repeated footshock stress: revelations of gender-based differences. *Neurobiology of Disease, 14,* 602–618.

Vale, W., Spiess, J., Rivier, C., & Rivier, J. (1981). Characterization of a 41-residue ovine hypothalamic peptide that stimulates secretion of corticotropin and beta-endorphin. *Science, 213,* 1394–1397.

Volavka, J., Bauman, J., Pevnick, J., Reker, D., James, B., & Cho, D. (1990). Short-term hormonal effects of naloxone in man. *Psychoneuroendocrinology, 5,* 225–234.

Vythilingam, M., Anderson, G. M., Owens, M. J., Halaszynski, T. M., Bremner, J. D., Carpenter, L. L., ... Charney, D. S. (2000). Cerebrospinal fluid corticotropin-releasing hormone in healthy humans: effects of yohimbine and naloxone. *Journal of Clinical Endocrinology and Metabolism, 85,* 4138–4145.

Wand, G. S., McCaul, M., Yang, X., Reynolds, J., Gotjen, D., Lee, S., & Ali, A. (2002). The mu-opioid receptor gene polymorphism (A118G) alters HPA axis activation induced by opioid receptor blockade. *Neuropsychopharmacology, 26,* 106–114.

Zhou, Y., Bendor, J., Hofmann, L., Randesi, M., Ho, A., & Kreek, M. J. (2006). Mu opioid receptor and orexin/hypocretin mRNA levels in the lateral hypothalamus and striatum are enhanced by morphine withdrawal. *Journal of Endocrinology, 191,* 137–145.

Zhou, Y., Bendor, J. T., Yuferov, V., Schlussman, S. D., Ho, A., & Kreek, M. J. (2005). Amygdalar vasopressin mRNA increases in acute cocaine withdrawal: evidence for opioid receptor modulation. *Neuroscience, 134,* 1391–1397.

Zhou, Y., Colombo, G., Carai, M. A., Ho, A., Gessa, G. L., & Kreek, M. J. (2011). Involvement of arginine vasopressin and V1b receptor in alcohol drinking in Sardinian alcohol-preferring rats. *Alcoholism, Clinical and Experimental Research, 35,* 1876–1883.

Zhou, Y., Colombo, G., Niikura, K., Carai, M. A. M., Femenía, T., García-Gutiérrez, M. S., ... Kreek, M. J. (2013). Voluntary alcohol drinking enhances proopiomelanocortin (POMC) gene expression in nucleus accumbens shell and hypothalamus of Sardinian alcohol-preferring rats. *Alcoholism, Clinical and Experimental Research, 37,* E131–E140.

Zhou, Y., Cui, C. L., Schlussman, S. D., Choi, J. C., Ho, A., Han, J. S., & Kreek, M. J. (2008). Effects of cocaine place conditioning, chronic escalating-dose "binge" pattern cocaine administration and acute withdrawal on orexin/hypocretin and preprodynorphin gene expressions in lateral hypothalamus of Fischer and Sprague-Dawley rats. *Neuroscience, 153,* 1225–1234.

Zhou, Y., & Kreek, M. J. (2014). Alcohol: a stimulant activating brain stress responsive systems with persistent neuroadaptation. *Neuropharmacology, 87,* 51–58.

Zhou, Y., Kruyer, A., Ho, A., & Kreek, M. J. (2012). Cocaine place conditioning increases pro-opiomelanocortin gene expression in rat hypothalamus. *Neuroscience Letters, 530,* 59–63.

Zhou, Y., & Lapingo, C. (2014). Modulation of pro-opiomelanocortin gene expression by ethanol in mouse anterior pituitary corticotrope tumor cell AtT20. *Regulatory Peptides, 192–193,* 6–14.

Zhou, Y., Leri, F., Cummins, E., Hoeschele, M., & Kreek, M. J. (2008). Involvement of arginine vasopressin and V1b receptor in heroin withdrawal and heroin seeking precipitated by stress and by heroin. *Neuropsychopharmacology, 33,* 226–236.

Zhou, Y., Leri, F., Cummins, E., & Kreek, M. J. (2015). Individual differences in gene expression of vasopressin, D2 receptor, POMC and orexin: vulnerability to relapse to heroin seeking in rats. *Physiology and Behavior, 139,* 127–135.

Zhou, Y., Leri, F., Grella, S., Aldrich, J., & Kreek, M. J. (2013). Involvement of dynorphin and kappa opioid receptor in yohimbine-induced reinstatement of heroin seeking in rats. *Synapse, 67,* 358–361.

Zhou, Y., Leri, F., Ho, A., & Kreek, M. J. (2013). Suppression of hypothalamic-pituitary-adrenal axis by acute heroin challenge in rats during acute and chronic withdrawal from chronic heroin administration. *Neurochemical Research, 38*, 1850–1860.

Zhou, Y., Litvin, Y., Piras, A. P., Pfaff, D. W., & Kreek, M. J. (2011). Persistent increase in hypothalamic arginine vasopressin gene expression during protracted withdrawal from chronic escalating-dose cocaine in rodents. *Neuropsychopharmacology, 36*, 2062–2075.

Zhou, Y., Proudnikov, D., Yuferov, V., & Kreek, M. J. (2010). Drug-induced and genetic alterations in stress-responsive systems: Implications for specific addictive diseases. *Brain Research, 1314*, 235–252.

Zhou, Y., Spangler, R., Maggos, C. E., LaForge, K. S., Ho, A., & Kreek, M. J. (1996). Steady-state methadone in rats does not change mRNA levels of corticotropin-releasing factor, its pituitary receptor or proopiomelanocortin. *European Journal of Pharmacology, 315*, 31–35.

Zhou, Y., Spangler, R., Maggos, C. E., Wang, X. M., Han, J. S., Ho, A., & Kreek, M. J. (1999). Hypothalamicpituitary-adrenal activity and pro-opiomelanocortin mRNA levels in the hypothalamus and pituitary of the rat are differentially modulated by acute intermittent morphine with or without water restriction stress. *Journal of Endocrinology, 163*, 261–267.

Zhou, Y., Spangler, R., Yuferov, V. P., Schlussmann, S. D., Ho, A., & Kreek, M. J. (2004). Effects of selective D1- or D2-like dopamine receptor antagonists with acute "binge" pattern cocaine on corticotropin-releasing hormone and proopiomelanocortin mRNA levels in the hypothalamus. *Molecular Brain Research, 130*, 61–67.

Chapter 95

Accelerated Aging in Heroin Abusers: Readdressing a Clinical Anecdote Using Telomerase and Neuroimaging

Gordon L.F. Cheng[1,2,3], Tatia M.C. Lee[1,2,3,4]
[1]*Laboratory of Neuropsychology, The University of Hong Kong, Pokfulam, Hong Kong;* [2]*Laboratory of Cognitive Affective Neuroscience, The University of Hong Kong, Pokfulam, Hong Kong;* [3]*Institute of Clinical Neuropsychology, The University of Hong Kong, Pokfulam, Hong Kong;* [4]*The State Key Laboratory of Brain and Cognitive Sciences, The University of Hong Kong, Pokfulam, Hong Kong*

Abbreviations

BOLD Blood oxygenated level dependent
DLPFC Dorsolateral prefrontal cortex
DTI Diffusion tensor imaging
MMT Methadone maintenance therapy
MRI Magnetic resonance imaging
MTL Medial temporal lobe
OFC Orbitofrontal cortex
PFC Prefrontal cortex
ROS Reactive oxygen species

INTRODUCTION

Half a century has passed since the notion of "premature aging" in substance abuse was first put forward to the scientific community (Fitzhugh, Fitzhugh, & Reitan, 1965). For a few decades thereafter, numerous studies that sought to address the validity of this notion were published, but inconsistent findings reduced empirical interest in this topic as quickly as it had emerged. However, there is a renewed interest in the hypothesis because of some evidence that substance abuse may accelerate the aging process. This chapter explicates the premature aging hypothesis as one that warrants scientific interest, provides a fertile ground for research, and deserves a thorough examination.

The Conceptual Need

Substance abuse has wide impacts on the biological system at biochemical, physiological, anatomical, and functional anatomical levels. Some current theories have been proposed about how substance abusers might be affected. For example, functional neuroanatomical accounts (Goldstein & Volkow, 2011) explain the cognitive impairment in substance abusers. However, there is no convincing, overarching theoretical framework that currently encapsulates the detrimental impact of substance abuse across various levels of understanding. The premature aging hypothesis may be one such framework because there is a relatively good understanding of the biological consequences of aging. This hypothesis may also explain the long-standing clinical anecdote that people with substance abuse appear physically older than their actual chronological age. There have been isolated quantitative reports that substance abusers show some overt signs of premature aging, such as greater degrees of hair graying (Reece, 2007a) and skin wrinkling (Yin, Morita, & Tsuji, 2001). However, the validity of the premature aging hypothesis remains elusive, perhaps because of a lack of adequate understanding and tools for thorough examination at the time of the hypothesis's inception.

The Initial Studies

Much of the early research on the addiction–premature aging phenomenon used neuropsychological testing to identify the cognitive impairment exhibited by alcoholic patients. Several of these studies reported supporting evidence for the hypothesis, which was characterized by greater similarities in profiles of cognitive impairment between alcoholic patients and older nonalcoholic controls than to age-matched nonalcoholic controls (Blusewicz, Dustman, et al., 1977; Blusewicz, Schenkenberg, et al., 1977; Hochla & Parsons, 1982). These findings suggested analogous effects of alcohol abuse and aging on neuropsychological functioning.

Conversely, later research presented data that were inconsistent with the premature aging hypothesis. These studies reported that the profile of impairment demonstrated by alcoholic patients was different from that of old-age nonalcoholic controls (Kramer, Blusewicz, & Preston, 1989; Shelton, Parsons, & Leber, 1984), which suggests that the behavioral impairments resulting from alcohol abuse may not be comparable to aging after all. Another study examined the effect of alcohol use as a function of age, but no alcohol-by-age interaction was evident (Grant, Adams, & Reed, 1984). This result was not consistent with the premature aging hypothesis.

The idea and interest in the theory failed to yield consistent supporting results. References to this theory in the empirical literature are rare after the initial wave of neuropsychological studies. This paucity of research may be due to a premature understanding of the biological impact of aging at the time and the lack of an adequate conceptual basis to proceed with further investigation. Another likely explanation was the lack of modern neuroimaging technologies to complement the neuropsychological methodology.

Opioids and Aging

Opioids are powerful analgesics that are used to satisfy both medical and abusive intentions. There is an emerging epidemic of the misuse of prescription opioids in the United States (Compton & Volkow, 2006; see Figure 1), and heroin (an abusive opioid) persists as the most prevalent illicit substance abused in China (Tang, Zhao, Zhao, & Cubells, 2006). Opioid addiction incurs enormous criminal, legal, and health costs (Wall et al., 2000), and therefore, this abuse presents severe and persistent socioeconomic burdens at the global level.

Heroin penetrates the blood–brain barrier to elicit rapid euphoria and transcendent relaxation (Rook, Huitema, van den Brink, van Ree, & Beijnen, 2006). These feelings, coupled with adverse withdrawal symptoms, make heroin one of the most habit-forming illicit drugs. Heroin addiction exhibits high comorbidity with other psychiatric conditions, such as depressive and anxiety disorders (Darke & Ross, 1997). Therefore, abusers succumb to problems coping with the demands of daily life and slip into the shadows of society. The life expectancy of heroin abusers is substantially lower than that of the general population (Smyth, Hoffman, Fan, & Hser, 2007) because heroin greatly limits their potential to live a fulfilling life.

Heroin is among the most addictive substances, but it is also one of the least well studied drugs. This lack of research may be due to the limited accessibility to the heroin-abusing population for research purposes. Our 2013 study attempted to overcome such practical challenges and examined whether premature aging was evident in a group of heroin abusers (Cheng et al., 2013).

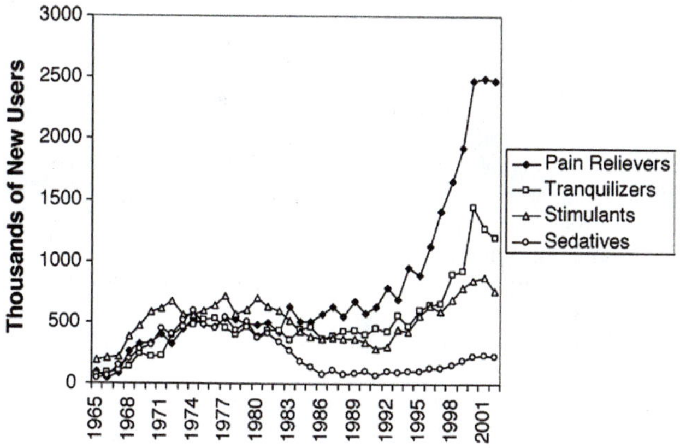

FIGURE 1 **The growing trend of opioid prescription use in the United States.** Among various psychotropic prescription medicines, the use of pain relievers (i.e., opioids) has been increasing at an exponential rate in the United States. *The figure was copied from Compton and Volkow (2006), with permission from the publishers.*

The concept for this study arose from two separate lines of studies: (i) the neuropathology and neuropsychology of heroin addiction and (ii) the biological indicators of aging.

NEUROPATHOLOGY AND NEUROPSYCHOLOGY OF HEROIN ADDICTION

Neuroimaging Studies of Heroin Addiction

Advances in technology since 1995 have significantly increased sophisticated and noninvasive methodologies to improve our understanding of the human brain. Investigators experienced in these approaches have studied the associations between psychopathology and neurobiological abnormalities in the brain. One of the most common methods used is magnetic resonance imaging (MRI), which is an in vivo neuroimaging technique that is capable of revealing relatively fine brain morphology and activation (see Table 1).

MRI is widely used to study the neural basis of generic psychiatric conditions, but it is rarely used to investigate the effects of heroin addiction on the integrity of brain structures. The minimal studies that used MRI established that heroin users had gray matter atrophy in extensive brain areas, including the prefrontal, insula, temporal, parietal, occipital, and cerebellar regions (Liu et al., 2009; Lyoo et al., 2006; Wang et al., 2012; Yuan et al., 2010, 2009). Atrophy was also discovered in the bilateral thalami of heroin abusers undergoing methadone maintenance therapy (MMT)

TABLE 1 Key Facts of Neuroimaging

- Neuroimaging includes a variety of noninvasive in vivo methodologies that measure many aspects of brain anatomy and functioning.
- Common neuroimaging techniques include MRI, positron emission tomography, single-photon-emission computed tomography, and electroencephalography.
- There are different MRI protocols that measure distinct aspects of the brain. These include T1 imaging, T2* imaging, and DTI.
- T1 imaging generates high-resolution structural images that may be used to analyze brain structure (typically gray or white matter) in terms of volume, cortical thickness, and surface area.
- T2* imaging is commonly referred as functional MRI.
- T2* imaging measures BOLD signals that can be computed to generate functional images reflecting brain activation.
- T2* images can be captured when a person is engaged in an experimental task or at rest to measure the brain's response to external demands or intrinsic functional organization, respectively.
- DTI measures the direction of water molecule flow, which is constrained in the brain owing to the anatomy of fiber tracts (i.e., bundles of white matter connections).
- DTI is generally the preferred imaging protocol for measuring the brain's structural connectivity.

This table lists the key facts of neuroimaging in terms of various imaging modalities. It includes information on T1 (structural) imaging, T2* (functional) imaging, and DTI (structural connectivity).

(Reid et al., 2008). These findings are consistent with the observation that opioid receptors, including μ, κ, and σ receptors, are spread across the entire brain (Pfeiffer, Pasi, Mehraein, & Herz, 1982). However, the cross-sectional nature of the relevant research prevented any determination of whether the observed neural deficits were an effect of heroin per se. There was a fair chance that the discrepancies had existed before the abuse occurred, i.e., they represented a predisposition factor that increased vulnerability to engage in addictive behaviors. The direct impact of heroin addiction on brain structural integrity was more confidently attributed in two studies that found negative correlations between the duration of heroin use and the gray matter volume in the brain, especially in the prefrontal cortex (Yuan et al., 2009, 2010).

The structural pathology in terms of volumes in localized brain areas is interesting, but it is equally important to investigate the effects of heroin abuse on structural connectivity. Diffusion tensor imaging (DTI) is sensitive to the direction of water molecule flow, which forms the basis of its ability to measure the integrity of fiber tracts (i.e., structural connectivity) in the brain (Sundgren et al., 2004). A study that utilized DTI on opiate users demonstrated that these individuals showed signs of structural connectivity deficits in prevalent brain areas (Bora et al., 2012). Another DTI study of abstinent heroin abusers found that fiber tract abnormalities were most prominent in anterior brain regions, i.e., prefrontal regions (Liu et al., 2008). The two investigations reported an association between the severity of abuse and the degree of fiber tract damage in frontal regions. These findings corroborate the structural volumetric studies and reinforce the observation that although widespread brain regions are implicated in heroin addiction, the prefrontal cortex (PFC) is particularly sensitive to the extent of heroin use.

The fundamental role of the PFC in the neuropathology of heroin addiction is also supported by MRI research that measures the brain's functional activation in response to various cognitive demands (Yang et al., 2008). Brain functional activation in these studies is characterized by blood oxygenated level-dependent BOLD signals based on the premise that higher levels of oxygenated blood are delivered to regions that are actively engaged. Dysregulation of the PFC underlies impulse control (Lee et al., 2005) and decision-making (Schultz, 2011) deficits in heroin abusers. Furthermore, the PFC was abnormally connected with other brain areas that subserve cognitive control and related processes (Ma et al., 2010). Neuroimaging studies of heroin addiction are consistent with theoretical models that posit prefrontal dysfunctions to be the neurobiological basis of compulsive drug taking (Goldstein & Volkow, 2011).

Neuropsychological Studies of Heroin Addiction

In addition to neuroimaging research, some studies examined the behavioral functioning of heroin abusers using neuropsychological tests that evaluate various aspects of cognition. Among the various cognitive domains, heroin abuse is perhaps most associated with impairments of executive functioning (Gruber, Silveri, & Yurgelun-Todd, 2007), which is the umbrella term for higher cognitive abilities that are subserved by the prefrontal system. Addicts demonstrated significantly worse performances than matched controls on tasks that measured impulse control (Lee & Pau, 2002; Pau, Lee, & Chan, 2002) and decision-making (Brand, Roth-Bauer, Driessen, & Markowitsch, 2008).

Studies that investigated a wider range of cognition in heroin abusers suggested that heroin is associated with a pervasive pattern of cognitive impairment. Darke, Sims, McDonald, and Wickes (2000) found that abstinent heroin users undergoing MMT performed poorly on several neuropsychological domains, including attention, memory, processing speed, and problem solving. These authors proposed that heroin and methadone use increases the risks of irreversible brain damage and permanent cognitive impairment, but other research has suggested that the recovery of cognitive functioning is possible through protracted abstinence from these drugs. One study compared methadone-maintained former heroin addicts with abstinent heroin users who were free from any drug, and the methadone group had a worse neuropsychological profile (Davis, Liddiard, & McMillan, 2002). The reversible nature of cognitive impairment in heroin abuse was further supported by another investigation that examined heroin abusers during the early stages of abstinence and found a positive correlation between the duration of abstinence and cognitive performance (Rapeli et al., 2006).

BIOLOGICAL INDICATORS OF AGING

Aging at the Biochemical Level

Aging is an immensely complicated process (see Table 2) that elicits a decline in various physical and mental functionalities. It is widely believed that these physical and mental declines are not best predicted by chronological age but by "biological age," the definition and measure of which constitute an active empirical

TABLE 2 Key Facts of Aging

- Aging is multidimensional. It may refer to chronological age or be understood in terms of physical/mental variables that show time-dependent changes.
- There are various biological processes that show time-dependent deterioration.
- Some of these biological processes have been coined "aging biomarkers," and they are typically biochemical variables that predict mortality.
- Biological aging is probably a better predictor of functionality than chronological age.
- Brain aging refers to the observation that the brain does not deteriorate in a uniform fashion. Different brain regions have different trajectories of aging decline.
- Similarly, cognitive aging refers to the observation that different aspects of cognition have different trajectories of aging decline.
- The PFC and medial temporal lobe show the most significant age-related decline.
- Executive functioning, memory, and processing speed undergo particularly pronounced deterioration with increasing age.
- The relationships among various aspects of aging are not well understood.

This table lists the key facts of aging in terms of its multidimensionality. It includes information on biochemical, brain, and cognitive aging.

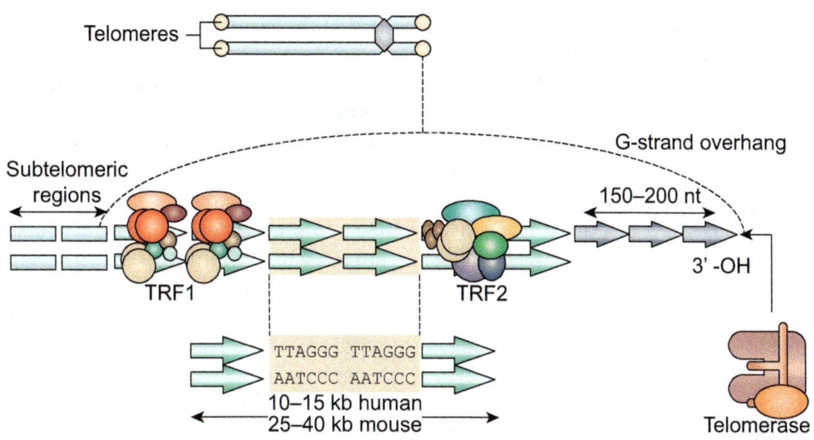

FIGURE 2 **Illustration of telomere and telomerase.** Telomeres contain a double-stranded DNA region of TTAGGG repeats (green arrows), which is typically 10–15 kb long in humans and 25–40 kb long in mice. Telomeres are characterized by a 150- to 200-nt-long single-stranded overhang of a G-rich strand (G-strand overhang; blue arrows). Note that the length of the telomere repeats is not drawn to scale. Telomerase recognizes the 3′-OH at the end of the G-strand overhang, which leads to telomere elongation. Two main protein complexes are bound to telomeres, the telomere repeat binding factor 1 and 2 complexes, TRF1 and TRF2. *Figure and legend were partially copied from Blasco (2005), with permission from the publishers.*

effort. However, significant strides in the identification of aging biomarkers have been made in the past several years. It is becoming increasingly apparent that aging is characterized by alterations in multiple, possibly interrelated, biochemical processes rather than a single chain of biological events. Many different factors have been proposed as aging biomarkers, but the relatively better understood markers include parameters sensitive to oxidative stress, inflammation, and cellular senescence.

Oxidative stress may be defined as an imbalance between the production of reactive oxygen species (ROS) and antioxidant defenses. ROS are a natural by-products of oxygen metabolism, and they are molecules (or "radicals") that may be generated both exogenously (i.e., at the extracellular level) and endogenously (i.e., at the intracellular level). Superoxide anions and hydrogen peroxide are examples of ROS (Simm et al., 2008). A homeostatic level of ROS is beneficial for cellular proliferation, but increased levels of ROS without the appropriate matching levels of antioxidant defense (in other words, oxidative stress) induce modifications to cellular proteins, lipids, and DNA (Finkel & Holbrook, 2000). Moreover, oxidative stress is intricately linked with compromised cellular metabolism because the mitochondria are the largest consumers of oxygen for the purpose of energy generation. Importantly, indices of oxidative stress, such as 8-hydroxy-2′-deoxyguanosine, correlate negatively with mammalian life span (Lopez-Torres, Gredilla, Sanz, & Barja, 2002).

Inflammation refers to the biological response that becomes activated at the invasion of pathogens that damage body tissues. This biological response induces biochemical processes with the primary function of tissue repair. However, inflammation also has the side effect of gradually damaging surrounding tissues, which may lead to long-term deterioration if the inflammatory response is prolonged. There is a fair amount of evidence that implicates the association of aging with chronic and low-grade inflammatory responses. This upregulation of inflammation is intricately linked with the time-dependent regression of the immune system and age-related diseases. The terms "inflamm-aging" and "immunosenescence" have appeared to describe the aging effects on inflammation and reduced immunological functioning (Licastro et al., 2005).

A range of biomarkers are also sensitive to these effects, which may predict human mortality (Simm et al., 2008).

Cellular senescence is a state wherein cells can no longer divide and replicate, i.e., a state of arrested cellular proliferation. Biomarkers of cellular senescence include telomere length and telomerase activity (Blasco, 2005; see Figure 2), which have been referred to as "psychobiomarkers" of aging because of their association with psychological stress (Epel, 2009). Telomeres are chromatin structures at the ends of chromosomes that shorten as cells divide because the method of replication is incomplete ("the end replication problem"). The enzyme telomerase is recruited to elongate telomeres and compensate for the telomere shortening. This strategy prevents telomeres from reducing to a critical length that leads to cell senescence and apoptosis. Therefore, telomerase activity is an important index of cellular age, which was confirmed by previous studies that demonstrated an association between telomerase activity and aging, aging-related diseases, and mortality (Blasco, 2005; Cawthon, Smith, O'Brien, Sivatchenko, & Kerber, 2003). Both oxidative stress (von Zglinicki, 2002) and immunological deterioration (Monteiro, Batliwalla, Ostrer, & Gregersen, 1996) are associated with increased cellular senescence, which suggests an interrelatedness of these different aspects of biochemical aging.

Aging at the Brain System Level

The effect of aging at the brain system level is examined using neuroimaging studies that measure brain integrity in people across their life span. These studies suggest that the PFC and medial temporal lobe (MTL) are the two broad brain areas that are most implicated in the aging process.

The developmental trajectory of distinct brain areas in the human brain seems to follow the phylogenetic order, such that the PFC is the last brain system to mature (Gogtay et al., 2004). The PFC is relatively immature during adolescent years, during which most of the other brain areas appear to have reached a structural plateau. However, the PFC is also one of the brain areas that show the earliest signs of age-related structural decline. Neuroimaging data suggest that gray matter in the PFC exhibits a linear reduction of 5% per decade after the age of 20 (Raz et al., 2005). The PFC subserves higher cognitive abilities, such as executive functioning. Therefore, it is not surprising that deficits in the PFC lead to impairments in executive functioning during old age. Raz, Gunning-Dixon, Head, Dupuis, and Acker (1998) were among the first researchers to observe this type of "cognitive aging" in relation to brain structural discrepancies. They found that PFC structural volume was negatively correlated with executive functioning within a group of elderly participants. Consistent with the structural volumetric literature, more recent studies using DTI revealed that the age-related degradation of structural connectivity is particularly pronounced in anterior (i.e., frontal) regions (Madden, Bennett, & Song, 2009).

Aging is also associated with PFC alterations at the functional level. It has been consistently reported that elderly participants show more decreased activation when engaged in tasks requiring higher cognitive abilities than younger participants (Prakash et al., 2009). Notably, high-functioning elderly participants (i.e., those without marked decline in behavioral performance) show extra activation in brain regions that are not commonly activated in younger participants (Cabeza, 2002). These brain regions are usually contralateral to commonly activated regions. This observation has been referred to as "hemispheric asymmetry" and indicates age-related neuroplasticity that allows the recruitment of additional brain regions to compensate for cognitive decline. Moreover, there are indications that the left PFC is more frequently found to counterbalance deficiencies in the right PFC, which supports the "right hemi-aging" hypothesis that the right PFC is particularly sensitive to the detrimental impact of aging (Rajah & D'Esposito, 2005).

The hippocampus is part of the MTL, which is also known to be implicated in the aging process. The hippocampus enjoys a relatively stable structural volume throughout much of the adult life span, but it undergoes rapid deterioration in people over the age of 60 (Raz et al., 2005). Hippocampal degeneration is the hallmark of age-related pathology, most notably in Alzheimer disease. Deterioration of the MTL is consistent with the clinical picture that elderly people often present a decline in memory functioning, which is dependent on hippocampal functioning. It is important to remember that the MTL is an inner brain area that encompasses several brain structures. These brain structures might play distinct roles in aging and age-related diseases. For example, atrophy in the entorhinal cortex (adjacent to the hippocampus) seems particularly implicated in age-related diseases. Specifically, the extent of entorhinal cortex atrophy is predictive of progression from a healthy to a diseased state in aging (Pennanen et al., 2004).

HEROIN, TELOMERASE, AND NEUROIMAGING: CROSSROADS BETWEEN MOLECULAR AND BRAIN SYSTEM PATHOLOGY

The empirical progress in biological aging coupled with the technological advances in neuroimaging have improved our position to critically evaluate the premature aging hypothesis in the context of heroin addiction. If heroin addiction accelerates the aging process, then abusers of heroin should present an altered level of the biochemical processes that suggest advanced aging. Furthermore, these altered biochemical processes should associate with brain system deficits related to the aging process. These hypotheses were addressed in a 2013 study (Cheng et al., 2013). This study was an ambitious but worthy endeavor because the findings could potentially push beyond the traditional study of brain–behavior relationships into a more basic insight into biochemical processes that enable the neuropathology of heroin addiction. If our predictions prove correct, then the demonstration of direct connections between multiple dimensions of biological aging at both the biochemical and the brain system levels would raise the possibility of clinical interventions that target an earlier point of addiction biological pathology, which might be more specific and malleable for change than interventions that target later stages of the pathology.

The following results are our main observations. Our study of 33 abstinent heroin abusers and 30 matched healthy controls demonstrated that telomerase activity (a molecular indicator reflective of cellular aging) measured from peripheral blood was significantly reduced in heroin abusers. This finding suggests, for the first time, that cellular aging is accelerated in past users of heroin. Our cellular aging observations echo a few previous reports that linked heroin addiction with other molecular indicators of aging. These studies include a postmortem study of nine former heroin abusers, which found that glutathione, which is a nonenzymatic antioxidant, was significantly reduced in various brain regions compared to those of matched controls (Gutowicz, Kazmierczak, & Baranczyk-Kuzma, 2011). Other research also discovered adverse effects of heroin on the body's inflammatory responses (McCarthy, Wetzel, Sliker, Eisenstein, & Rogers, 2001) and immune cell functioning (McCarthy et al., 2001; Reece, 2007b). Our observation that heroin abusers had reduced telomerase activity in combination with these findings provide suggestive evidence that heroin addiction may lead to premature biochemical aging due to oxidative stress, inflammation, and cellular senescence.

The next question we addressed was whether the cellular aging in heroin abusers leads to aging at the brain system level. We examined the interaction between heroin abuse and telomerase activity on brain structural volume using high-resolution MRI in the same group of participants. Voxel-based morphometry was used to study specific brain regions that might show disproportionate structural deterioration in relation to heroin abuse and telomerase activity. The brain region that emerged with a significant interaction was the right dorsolateral PFC (DLPFC; see Figure 3). This result was consistent with our prediction, which was based on the premise that the PFC is one of the brain areas most implicated in aging (Raz et al., 2005). This finding was also consistent with the "right hemi-aging" hypothesis that postulates particular age-related vulnerability in the right PFC (Rajah & D'Esposito, 2005). Therefore, the molecular and brain structural observations converged to suggest that heroin addiction is indeed associated with accelerated biological aging.

The observed brain structural effect in the right DLPFC strongly suggests an association between heroin addiction and accelerated aging at the brain system level. However, there is an increasing recognition that the brain should not be evaluated based solely on the integrity of isolated brain regions, but that the connectivity between distinct brain regions is another important index to examine. We built on the above finding that the right DLPFC structural integrity was implicated in heroin-associated accelerated aging and investigated the functional connectivity of the right

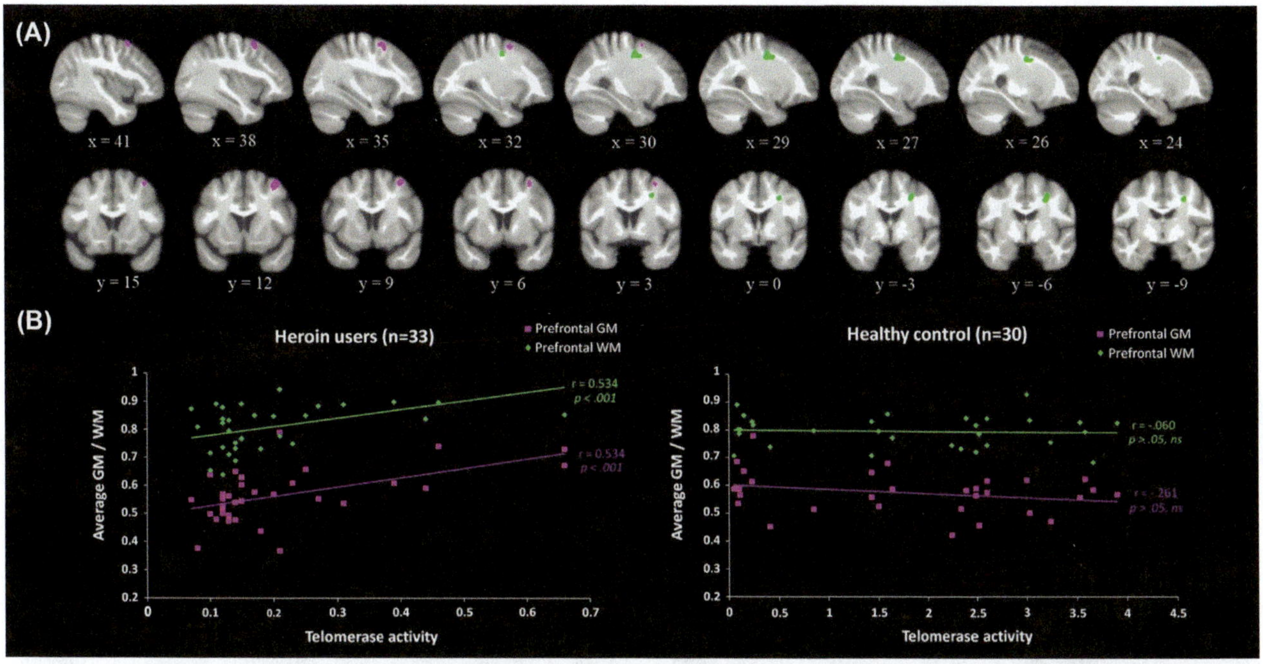

FIGURE 3 **Heroin–telomerase interaction on brain structural volume.** The right DLPFC showed a significant interaction between heroin abuse and telomerase activity. The pink patch depicts a significant gray matter cluster. The green patch depicts a significant white matter cluster. For the heroin abusers, DLPFC structural volume reduced as telomerase activity reduced. (A) The anatomical locations of each significant brain cluster. (B) The corresponding scatter plots from the interaction analysis. *Figure was copied from Cheng et al. (2013), with permission from the publishers.*

DLPFC using resting-state functional MRI (Fox & Raichle, 2007) within the same group of heroin abusers and healthy controls. The significant right DLPFC cluster from the structural result was extracted and used as the "seed" in the analysis to study its functional connectivity with every other brain region. The results demonstrated that several brain regions implicated in aging were abnormally connected with the right DLPFC in heroin abusers in relation to their telomerase activity level (see Figure 4). One such region was the orbitofrontal cortex (OFC), which was negatively coupled with the DLPFC, but this negative coupling diminished as telomerase activity decreased. Other brain regions that were abnormally connected with the right DLPFC in heroin abusers were the entorhinal cortex (part of the MTL), anterior cingulate cortex, superior occipitoparietal cortex, and superior temporal gyrus.

Based on the known functional roles of distinct brain regions, one could make informed speculations on the behavioral relevance of the brain clusters found to show a significant interaction between heroin abuse and telomerase activity. For example, the OFC, in connection with the DLPFC, is a core region for executive functioning and decision-making processes (Kringelbach & Rolls, 2004), and the entorhinal cortex is particularly relevant for enabling the sense of "familiarity" during memory recall (Diana, Yonelinas, & Ranganath, 2007). It could be speculated that the accelerated cellular aging in heroin abusers was associated with dysfunctional brain circuits that mediate executive functioning and memory processes. However, this hypothesis would have been only assumptive. Therefore, we directly assessed the behavioral relevance of the significant brain networks in our study.

We measured several different cognitive abilities using well-established neuropsychological tests in the same group of participants. The heroin abusers' behavioral performances on these tests were correlated with each of the functional brain networks that significantly interacted with heroin abuse and telomerase activity. The results indicated that in connection with the right DLPFC, the OFC correlated with executive functioning and the entorhinal cortex correlated with learning and memory, working memory, and executive functioning. The anterior cingulate cortex and superior occipitoparietal cortex correlated with attentional control. These findings are very important for understanding the behavioral relevance of the neuroimaging outcomes. The fact that both addiction and aging are implicated in executive functioning, memory, and attentional processes places these findings in striking consistency with the overall conclusion that heroin addiction accelerates the aging process.

The research outlined above provides converging evidence from different levels of analysis that heroin addiction is associated with an accelerated aging process. It was remarkable that direct connections between multiple dimensions of aging were identified. In fact, there are a few studies that found cocaine abusers to show seemingly accelerated brain system aging (Bartzokis et al., 2002, 2000; Ersche, Jones, Williams, Robbins, & Bullmore, 2012). Specifically, these reports consistently conveyed that cocaine abusers presented a steeper structural decline in frontal and temporal cortices as a function of chronological age. One of the differences between cocaine research and our study is that our group of heroin abusers was relatively homogeneous in terms of chronological age (mean age of 35 ± 4 years). However, the fact that the heroin abusers displayed clear signs of premature biological aging indicates that the effect of addiction was not merely a disproportionate vulnerability to deficits during old age, which was raised previously as an explanation of the premature aging

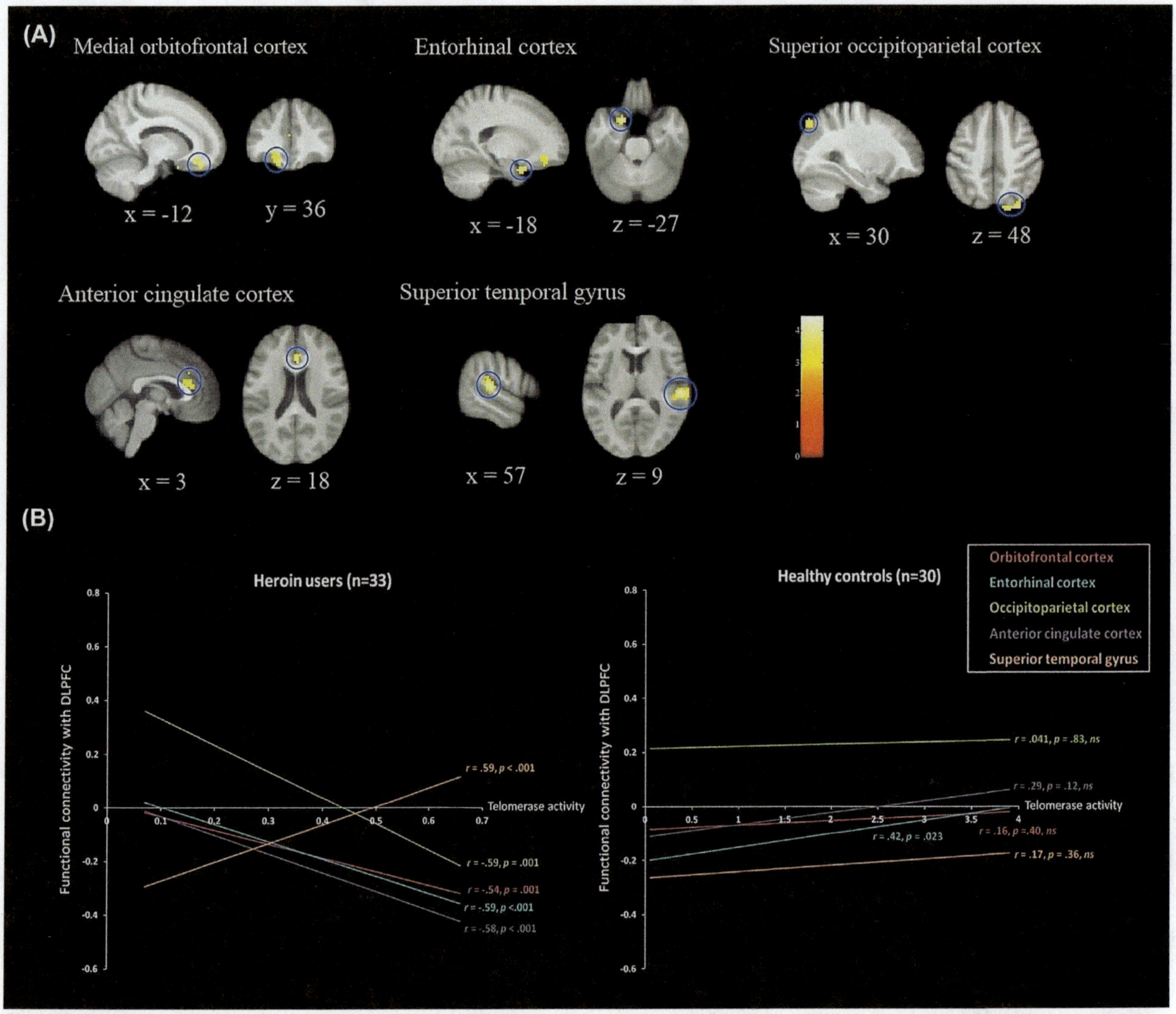

FIGURE 4 **Heroin–telomerase interaction on brain functional connectivity.** The right DLPFC was abnormally connected with several other brain regions in relation to the lower telomerase activity of the heroin abusers. These brain regions included the medial OFC, entorhinal cortex, superior occipitoparietal cortex, anterior cingulate cortex, and superior temporal gyrus. (A) The anatomical locations of each significant brain cluster. (B) The corresponding scatter plots from the interaction analysis. *Figure was copied from Cheng et al. (2013), with permission from the publishers.*

phenomenon (Noonberg, Goldstein, & Page, 1985). Our findings propose that substance abuse may trigger an early process of biological aging that puts abusers at risk of suffering age-related phenotypes ahead of their time.

A PARADIGM SHIFT IN THE UNDERSTANDING OF ADDICTION PATHOLOGY

The present observations, together with other research that suggests a link between substance abuse and premature aging (Ersche et al., 2012; Reece & Hulse, 2014), mark the emergence of a modern frontier in the study of addiction pathology. It is hoped that the increasingly solid knowledge of the addiction–premature aging relationship will foster a more holistic conceptual understanding of the biological processes that predispose and perpetuate addictive behaviors. However, our study provides only some proof-of-principle prompts that must be rigorously explored through larger and more concerted research efforts. Although an exhaustive discussion of all of the potential research directions is beyond the scope of this chapter, we list here several ideas that may be meaningfully pursued in future studies.

Cause or Effect, Multiaging, and Translational Value

One of the most important issues that should be addressed in future studies is the temporal nature of the effects observed in our study. The cross-sectional experimental design prevented any cause-and-effect inferences of the group differences in telomerase, neuroimaging, and behavioral findings. Animal models show a direct impact of substance abuse on immunological functioning (Coller & Hutchinson, 2012). Therefore, it is reasonable to assume that the observed effect on telomerase activity represented a causative effect of heroin addiction. Longitudinal designs are the gold standard to establish causal relationships in human studies. However, such studies are impractical in the context of substance abuse because no study could realistically track substance abusers from before they started the drug habit. Therefore, the question of cause or effect will probably need to be resolved by paralleling substance abusers with people who are at an increased, possibly familial, risk of substance abuse. A very different conceptualization of addiction pathology compared to the one depicted in this chapter would emerge if people at an increased risk of substance abuse also presented signs of premature aging.

The use of multiple aging biomarkers is necessary to fully characterize the acceleration of biological aging in heroin addiction. At the biochemical level, this would at least include biomarkers that measure the degree of oxidative stress and inflammation and the cellular senescence that our study investigated. Numerous biomarkers are reflective of aging, but an all-inclusive approach would be overly idealistic in any one study. Therefore, the appropriate selection of a few biochemical markers that are most sensitive to aging is essential. Further development in our knowledge of addiction–premature aging is partially reliant on advances in the biochemistry of aging. At the brain system level, it would be of interest to examine whether heroin addiction is associated with age-related alterations in other neuroimaging parameters indicative of the aging process, including the anterior-to-posterior gradient of white matter degradation and hemispheric asymmetry in task-driven functional activation (Madden et al., 2009; Rajah & D'Esposito, 2005).

Translational studies, in the broadest sense, will offer substantial expansion in the basic understanding and clinical implications of the premature aging hypothesis. At one end of the translational spectrum, genomic studies are also needed to identify the genetic determinants that contribute to the aging effects of heroin addiction. The identification of genetic moderators will open up new avenues in the understanding of addiction pathology, including the potential application of genetic screening and therapy. On the other end of the spectrum, animal models will be useful for testing the prospects of molecular interventions to induce adaptive changes that cease or possibly reverse the aging effects of addiction at the levels of brain and behavior.

Applications to Other Addictions and Substance Misuse

It should be of significant interest to scrutinize whether premature aging is evident in other substances of abuse and other forms of addiction. The fact that cocaine abusers show signs of brain system aging (Ersche et al., 2012) places this substance in a ready position for further testing of the potential aging effects at the biochemical level. Tobacco addiction may be another substance that deserves in-depth investigation because of the overt signs of aging displayed by chronic cigarette smokers (Yin et al., 2001). Reexamination of premature aging in alcohol abuse is also necessary because the theoretical and methodological advances since the initial theoretical conception are likely to shed new light on the detrimental impact of this substance. In fact the study of premature aging in all substances of abuse may be a fruitful endeavor because altered immunological functioning has been indicated in wide range of drug types (Reece, 2007b). Finally, there is a burgeoning interest in the scientific study of behavioral addictions, such as pathological gambling and Internet gaming. These behavioral problems may belong to the same type of syndrome as drug addiction (Shaffer et al., 2004). However, the nosology of behavioral addiction remains controversial. Perhaps investigations of premature aging will pave the way for better basic understanding of these behavioral problems and whether they truly represent different expressions of common etiological roots.

OVERALL SUMMARY

People with problems of substance abuse tend to appear older than their age-matched peers. This observation represents a long-standing clinical anecdote that substance abusers may suffer from premature aging. Some early neuropsychological studies submitted this premature aging phenomenon to testing, but they failed to yield convincing conclusions. Nevertheless, these neuropsychological studies were critical in documenting this theoretical possibility, which was, empirically speaking, ahead of its time. Scientific advances in biological aging and neuroimaging have generated new opportunities to reexamine the premature aging hypothesis. In a study that combined molecular, neuroimaging, and neuropsychological measures, we presented evidence that heroin addiction was associated with accelerated aging at the cellular and brain system levels. This finding, together with results reported by other research groups, supports some aspects of the validity of the premature aging hypothesis. New horizons have opened for future research efforts to expand our current knowledge of addiction pathology.

DEFINITION OF TERMS

Aging biomarker An aging biomarker is a biological measure that can be safely extracted from the human body and reflects longevity. There is no universal agreement on the best aging biomarker, but there exist several biomarkers that were shown to predict the human life span.

Blood oxygen level-dependent signal BOLD signals form the basis of measurement in functional MRI. Varying degrees of the BOLD signal are purported to reflect the extent of engagement in particular brain regions, based on the assumption that more oxygen is delivered to the brain regions being actively engaged.

Cellular senescence This refers to the state of a cell wherein it can no longer divide and replicate owing to the critical shortening of telomeres. Cellular senescence is now known to reflect the aging phenotype.

Medial temporal lobe This consists of a network of brain structures that are positioned deep within the brain. These brain structures include, but are not limited to, the hippocampus, entorhinal cortex, and perirhinal cortex. The MTL is critically related to learning and memory processes.

Methadone maintenance therapy This is a form of agonist replacement commonly used as pharmacological therapy for people with heroin addiction. Its use is controversial because there is evidence to suggest that methadone itself induces neurobiological deficits.

Neuropsychological testing Neuropsychological testing involves the use of standardized and psychometrically validated behavioral tests that measure different aspects of cognition. It is commonly used in the clinical setting to estimate and enumerate neurological damage. In the research context, some neuropsychological tests may be computerized to further improve standardization.

Prefrontal cortex This is the brain region that has probably evolved to the greatest extent in the human brain. The functional role of the PFC is heterogeneous, as it subserves various cognitive abilities such as executive functioning, attention, and memory.

Resting-state functional connectivity This refers to one specific modality of MRI. It is commonly used to measure the correlations of activation among various brain regions when the person being imaged is in a state of waking "rest," i.e., not being engaged in any task demand.

Telomerase This is an enzyme with the function of elongating telomeres, such that the cell remains viable. High telomerase activity may be reflective of a biochemical process that delays cellular senescence.

Voxel-based morphometry This refers to an analytic approach to MRI structural volumetric data. Generally speaking, it computes the amount of gray/white matter volume in various brain regions after controlling for overall brain size.

SUMMARY POINTS

- The premature aging hypothesis was first proposed 50 years ago.
- Initially, the hypothesis was tested by neuropsychological studies that failed to produce consistent support for the hypothesis.
- Advances in biological aging and neuroimaging technologies provided new tools that can be used to reexamine the hypothesis.
- Heroin is a form of opioid and one of the least well studied substances of abuse.
- Cellular senescence is a key biochemical process underlying biological aging.
- Telomerase activity is a molecular marker that reflects cellular aging.
- The PFC and MTL are the two brain areas most implicated in aging.
- Heroin addiction was found to associate with accelerated cellular and brain system aging.
- Premature aging might represent an overarching theoretical framework that encapsulates the various detrimental impact of substance abuse.

ACKNOWLEDGMENT

This work was supported by funding from The University of Hong Kong May Endowed Professorship and the KKHo International Charitable Foundation.

REFERENCES

Bartzokis, G., Beckson, M., Lu, P. H., Edwards, N., Bridge, P., & Mintz, J. (2002). Brain maturation may be arrested in chronic cocaine addicts. *Biological Psychiatry*, *51*(8), 605–611.

Bartzokis, G., Beckson, M., Lu, P. H., Edwards, N., Rapoport, R., Wiseman, E., & Bridge, P. (2000). Age-related brain volume reductions in amphetamine and cocaine addicts and normal controls: implications for addiction research. *Psychiatry Research*, *98*(2), 93–102.

Blasco, M. A. (2005). Telomeres and human disease: ageing, cancer and beyond. *Nature Reviews Genetics*, *6*(8), 611–622.

Blusewicz, M. J., Dustman, R. E., Schenkenberg, T., & Beck, E. C. (1977). Neuropsychological correlates of chronic alcoholism and aging. *The Journal of Nervous and Mental Disease*, *165*(5), 348–355.

Blusewicz, M. J., Schenkenberg, T., Dustman, R. E., & Beck, E. C. (1977). WAIS performance in young normal, young alcoholic, and elderly normal groups: an evaluation of organicity and mental aging indices. *Journal of Clinical Psychology*, *33*(4), 1149–1153.

Bora, E., Yucel, M., Fornito, A., Pantelis, C., Harrison, B. J., Cocchi, L., … Lubman, D. I. (2012). White matter microstructure in opiate addiction. *Addiction Biology*, *17*(1), 141–148.

Brand, M., Roth-Bauer, M., Driessen, M., & Markowitsch, H. J. (2008). Executive functions and risky decision-making in patients with opiate dependence. *Drug and Alcohol Dependence*, *97*(1–2), 64–72.

Cabeza, R. (2002). Hemispheric asymmetry reduction in older adults: the HAROLD model. *Psychology and Aging*, *17*(1), 85–100.

Cawthon, R. M., Smith, K. R., O'Brien, E., Sivatchenko, A., & Kerber, R. A. (2003). Association between telomere length in blood and mortality in people aged 60 years or older. *Lancet*, *361*(9355), 393–395.

Cheng, G. L. F., Zeng, H., Leung, M. K., Zhang, H. J., Lau, B. W., Liu, Y. P., … Lee, T. M. C. (2013). Heroin abuse accelerates biological aging: a novel insight from telomerase and brain imaging interaction. *Translational Psychiatry*, *3*, e260.

Coller, J. K., & Hutchinson, M. R. (2012). Implications of central immune signaling caused by drugs of abuse: mechanisms, mediators and new therapeutic approaches for prediction and treatment of drug dependence. *Pharmacology & Therapeutics*, *134*(2), 219–245.

Compton, W. M., & Volkow, N. D. (2006). Major increases in opioid analgesic abuse in the United States: concerns and strategies. *Drug and Alcohol Dependence*, *81*(2), 103–107.

Darke, S., & Ross, J. (1997). Polydrug dependence and psychiatric comorbidity among heroin injectors. *Drug and Alcohol Dependence*, *48*(2), 135–141.

Darke, S., Sims, J., McDonald, S., & Wickes, W. (2000). Cognitive impairment among methadone maintenance patients. *Addiction*, *95*(5), 687–695.

Davis, P. E., Liddiard, H., & McMillan, T. M. (2002). Neuropsychological deficits and opiate abuse. *Drug and Alcohol Dependence*, *67*(1), 105–108.

Diana, R. A., Yonelinas, A. P., & Ranganath, C. (2007). Imaging recollection and familiarity in the medial temporal lobe: a three-component model. *Trends in Cognitive Sciences*, *11*(9), 379–386.

Epel, E. S. (2009). Telomeres in a life-span Perspective: a new "Psychobiomarker"? *Current Directions in Psychological Science*, *18*(1), 6–10.

Ersche, K. D., Jones, P. S., Williams, G. B., Robbins, T. W., & Bullmore, E. T. (2012). Cocaine dependence: a fast-track for brain ageing? *Molecular Psychiatry*, *18*(2), 134–135.

Finkel, T., & Holbrook, N. J. (2000). Oxidants, oxidative stress and the biology of ageing. *Nature*, *408*(6809), 239–247.

Fitzhugh, L. C., Fitzhugh, K. B., & Reitan, R. M. (1965). Adaptive abilities and intellectual functioning of hospitalized alcoholics: further considerations. *Quarterly Journal of Studies on Alcohol, 26*(3), 402–411.

Fox, M. D., & Raichle, M. E. (2007). Spontaneous fluctuations in brain activity observed with functional magnetic resonance imaging. *Nature Reviews Neuroscience, 8*(9), 700–711.

Gogtay, N., Giedd, J. N., Lusk, L., Hayashi, K. M., Greenstein, D., Vaituzis, A. C., ... Thompson, P. M. (2004). Dynamic mapping of human cortical development during childhood through early adulthood. *Proceedings of the National Academy of Sciences of the United States of America, 101*(21), 8174–8179.

Goldstein, R. Z., & Volkow, N. D. (2011). Dysfunction of the prefrontal cortex in addiction: neuroimaging findings and clinical implications. *Nature Reviews Neuroscience, 12*(11), 652–669.

Grant, I., Adams, K. M., & Reed, R. (1984). Aging, abstinence, and medical risk factors in the prediction of neuropsychologic deficit among long-term alcoholics. *Archives of General Psychiatry, 41*(7), 710–718.

Gruber, S. A., Silveri, M. M., & Yurgelun-Todd, D. A. (2007). Neuropsychological consequences of opiate use. *Neuropsychology Review, 17*(3), 299–315.

Gutowicz, M., Kazmierczak, B., & Baranczyk-Kuzma, A. (2011). The influence of heroin abuse on glutathione-dependent enzymes in human brain. *Drug and Alcohol Dependence, 113*(1), 8–12.

Hochla, N. A., & Parsons, O. A. (1982). Premature aging in female alcoholics. A neuropsychological study. *The Journal of Nervous and Mental Disease, 170*(4), 241–245.

Kramer, J. H., Blusewicz, M. J., & Preston, K. A. (1989). The premature aging hypothesis: old before its time? *Journal of Consulting and Clinical Psychology, 57*(2), 257–262.

Kringelbach, M. L., & Rolls, E. T. (2004). The functional neuroanatomy of the human orbitofrontal cortex: evidence from neuroimaging and neuropsychology. *Progress in Neurobiology, 72*(5), 341–372.

Lee, T. M. C., & Pau, C. W. (2002). Impulse control differences between abstinent heroin users and matched controls. *Brain Injury, 16*(10), 885–889.

Lee, T. M. C., Zhou, W. H., Luo, X. J., Yuen, K. S., Ruan, X. Z., & Weng, X. C. (2005). Neural activity associated with cognitive regulation in heroin users: a fMRI study. *Neuroscience letters, 382*(3), 211–216.

Licastro, F., Candore, G., Lio, D., Porcellini, E., Colonna-Romano, G., Franceschi, C., & Caruso, C. (2005). Innate immunity and inflammation in ageing: a key for understanding age-related diseases. *Immunity & Ageing, 2*, 8.

Liu, H., Hao, Y., Kaneko, Y., Ouyang, X., Zhang, Y., Xu, L., ... Liu, Z. (2009). Frontal and cingulate gray matter volume reduction in heroin dependence: optimized voxel-based morphometry. *Psychiatry and Clinical Neurosciences, 63*(4), 563–568.

Liu, H., Li, L., Hao, Y., Cao, D., Xu, L., Rohrbaugh, R., ... Liu, Z. (2008). Disrupted white matter integrity in heroin dependence: a controlled study utilizing diffusion tensor imaging. *The American Journal of Drug and Alcohol Abuse, 34*(5), 562–575.

Lopez-Torres, M., Gredilla, R., Sanz, A., & Barja, G. (2002). Influence of aging and long-term caloric restriction on oxygen radical generation and oxidative DNA damage in rat liver mitochondria. *Free Radical Biology & Medicine, 32*(9), 882–889.

Lyoo, I. K., Pollack, M. H., Silveri, M. M., Ahn, K. H., Diaz, C. I., Hwang, J., ... Renshaw, P. F. (2006). Prefrontal and temporal gray matter density decreases in opiate dependence. *Psychopharmacology, 184*(2), 139–144.

Ma, N., Liu, Y., Li, N., Wang, C. X., Zhang, H., Jiang, X. F., ... Zhang, D. R. (2010). Addiction related alteration in resting-state brain connectivity. *NeuroImage, 49*(1), 738–744.

Madden, D. J., Bennett, I. J., & Song, A. W. (2009). Cerebral white matter integrity and cognitive aging: contributions from diffusion tensor imaging. *Neuropsychology Review, 19*(4), 415–435.

McCarthy, L., Wetzel, M., Sliker, J. K., Eisenstein, T. K., & Rogers, T. J. (2001). Opioids, opioid receptors, and the immune response. *Drug and Alcohol Dependence, 62*(2), 111–123.

Monteiro, J., Batliwalla, F., Ostrer, H., & Gregersen, P. K. (1996). Shortened telomeres in clonally expanded CD28-CD8+ T cells imply a replicative history that is distinct from their CD28+CD8+ counterparts. *Journal of Immunology, 156*(10), 3587–3590.

Noonberg, A., Goldstein, G., & Page, H. A. (1985). Premature aging in male alcoholics: "accelerated aging" or "increased vulnerability"? *Alcoholism, Clinical and Experimental Research, 9*(4), 334–338.

Pau, C. W., Lee, T. M. C., & Chan, S. F. (2002). The impact of heroin on frontal executive functions. *Archives of Clinical Neuropsychology, 17*(7), 663–670.

Pennanen, C., Kivipelto, M., Tuomainen, S., Hartikainen, P., Hanninen, T., Laakso, M. P., ... Soininen, H. (2004). Hippocampus and entorhinal cortex in mild cognitive impairment and early AD. *Neurobiology of Aging, 25*(3), 303–310.

Pfeiffer, A., Pasi, A., Mehraein, P., & Herz, A. (1982). Opiate receptor binding sites in human brain. *Brain Research, 248*(1), 87–96.

Prakash, R. S., Erickson, K. I., Colcombe, S. J., Kim, J. S., Voss, M. W., & Kramer, A. F. (2009). Age-related differences in the involvement of the prefrontal cortex in attentional control. *Brain and Cognition, 71*(3), 328–335.

Rajah, M. N., & D'Esposito, M. (2005). Region-specific changes in prefrontal function with age: a review of PET and fMRI studies on working and episodic memory. *Brain, 128*(9), 1964–1983.

Rapeli, P., Kivisaari, R., Autti, T., Kahkonen, S., Puuskari, V., Jokela, O., & Kalska, H. (2006). Cognitive function during early abstinence from opioid dependence: a comparison to age, gender, and verbal intelligence matched controls. *BMC Psychiatry, 6*, 9.

Raz, N., Gunning-Dixon, F. M., Head, D., Dupuis, J. H., & Acker, J. D. (1998). Neuroanatomical correlates of cognitive aging: evidence from structural magnetic resonance imaging. *Neuropsychology, 12*(1), 95–114.

Raz, N., Lindenberger, U., Rodrigue, K. M., Kennedy, K. M., Head, D., Williamson, A., ... Acker, J. D. (2005). Regional brain changes in aging healthy adults: general trends, individual differences and modifiers. *Cerebral Cortex, 15*(11), 1676–1689.

Reece, A. S. (2007a). Hair graying in substance addiction. *Archives of Dermatology, 143*(1), 116–118.

Reece, A. S. (2007b). Evidence of accelerated ageing in clinical drug addiction from immune, hepatic and metabolic biomarkers. *Immunity & Ageing, 4*, 6.

Reece, A. S., & Hulse, G. K. (2014). Impact of lifetime opioid exposure on arterial stiffness and vascular age: cross-sectional and longitudinal studies in men and women. *BMJ Open, 4*(6), e004521.

Reid, A. G., Daglish, M. R., Kempton, M. J., Williams, T. M., Watson, B., Nutt, D. J., & Lingford-Hughes, A. R. (2008). Reduced thalamic grey matter volume in opioid dependence is influenced by degree of alcohol use: a voxel-based morphometry study. *Journal of Psychopharmacology, 22*(1), 7–10.

Rook, E. J., Huitema, A. D. R., van den Brink, W., van Ree, J. M., & Beijnen, J. H. (2006). Pharmacokinetics and pharmacokinetic variability of heroin and its metabolites: review of the literature. *Current Clinical Pharmacology, 1*(1), 109–118.

Schultz, W. (2011). Potential vulnerabilities of neuronal reward, risk, and decision mechanisms to addictive drugs. *Neuron, 69*(4), 603–617.

Shaffer, H. J., LaPlante, D. A., LaBrie, R. A., Kidman, R. C., Donato, A. N., & Stanton, M. V. (2004). Toward a syndrome model of addiction: multiple expressions, common etiology. *Harvard Review of Psychiatry, 12*(6), 367–374.

Shelton, M. D., Parsons, O. A., & Leber, W. R. (1984). Verbal and visuospatial performance in male alcoholics: a test of the premature-aging hypothesis. *Journal of Consulting and Clinical Psychology, 52*(2), 200–206.

Simm, A., Nass, N., Bartling, B., Hofmann, B., Silber, R. E., & Navarrete Santos, A. (2008). Potential biomarkers of ageing. *Biological Chemistry, 389*(3), 257–265.

Smyth, B., Hoffman, V., Fan, J., & Hser, Y. I. (2007). Years of potential life lost among heroin addicts 33 years after treatment. *Preventive Medicine, 44*(4), 369–374.

Sundgren, P. C., Dong, Q., Gómez-Hassan, D., Mukherji, S. K., Maly, P., & Welsh, R. (2004). Diffusion tensor imaging of the brain: review of clinical applications. *Neuroradiology, 46*(5), 339–350.

Tang, Y. L., Zhao, D., Zhao, C., & Cubells, J. F. (2006). Opiate addiction in China: current situation and treatments. *Addiction, 101*(5), 657–665.

Wall, R., Rehm, J., Fischer, B., Brands, B., Gliksman, L., Stewart, J., ... Blake, J. (2000). Social costs of untreated opioid dependence. *Journal of Urban Health, 77*(4), 688–722.

Wang, X., Li, B., Zhou, X., Liao, Y., Tang, J., Liu, T., ... Hao, W. (2012). Changes in brain gray matter in abstinent heroin addicts. *Drug and Alcohol Dependence, 126*(3), 304–308.

Yang, Z., Xie, J., Shao, Y., Xie, C. M., Fu, L. P., Li, D., ... Li, S. J. (2008). Dynamic neural responses to cue-reactivity paradigms in heroin-dependent users: an fMRI study. *Human Brain Mapping, 30*(3), 766–775.

Yin, L., Morita, A., & Tsuji, T. (2001). Skin premature aging induced by tobacco smoking: the objective evidence of skin replica analysis. *Journal of Dermatological Science, 27*(1), S26–S31.

Yuan, K., Qin, W., Dong, M., Liu, J., Sun, J., Liu, P., ... Tian, J. (2010). Gray matter deficits and resting-state abnormalities in abstinent heroin-dependent individuals. *Neuroscience Letters, 482*(2), 101–105.

Yuan, Y., Zhu, Z., Shi, J., Zou, Z., Yuan, F., Liu, Y., ... Weng, X. (2009). Gray matter density negatively correlates with duration of heroin use in young lifetime heroin-dependent individuals. *Brain and Cognition, 71*(3), 223–228.

von Zglinicki, T. (2002). Oxidative stress shortens telomeres. *Trends in Biochemical Sciences, 27*(7), 339–344.

Chapter 96

Attention Deficit Hyperactivity Disorder, Substance Use Disorders, and Heroin Addiction

Saad Salman[1], Jawaria Idrees[2], Muhammad Anees[3], Fariha Idrees[4]

[1]*Department of Psychiatry and Drug Detoxification Centre, Lady Reading Hospital, Post Graduate Medical Institute, Peshawar, Khyber Pakhtunkhwa, Pakistan;* [2]*Department of Integrated Sciences, Post Graduate Nursing College, Peshawar, Khyber Pakhtunkhwa, Pakistan;* [3]*Department of Neurology and Neurosurgery, Lady Reading Hospital, Post Graduate Medical Institute, Peshawar, Khyber Pakhtunkhwa, Pakistan;* [4]*Department of Chemistry, Islamia College University, Peshawar, Khyber Pakhtunkhwa, Pakistan*

Abbreviations

ADD Attention deficit disorder
ADDS Attention deficit disorder screen
ADHD Attention deficit hyperactivity disorder
Adult ADHD SRS Adult ADHD self-report scale
CAARS Conners' adult ADHD rating scale
CBT Cognitive behavioral therapy
CD Conduct disorder
CIDI-SAM, MINI Mini-international neuropsychiatric interview
COMT Catechol-*O*-methyltransferase
DBH Dopamine β-hydroxylase
DICA Diagnostic interview for children and adolescents
DIS Diagnostic interview schedule for DSM-IV
DISC Diagnostic interview schedule for children
DRD4 Dopamine D4 receptor gene
DRD5 Dopamine D5 receptor gene
HD Heroin addiction disorder
HTR1B Serotonin 1b
ODD Oppositional defiant disorder
OPRM1 μ-Opioid receptor gene
SCID-I Structured clinical interview for *Diagnostic Statistical Manual*-IV for axis I disorders
SNP Single-nucleotide polymorphism
SSADDA Semi-structured assessment for drug dependence and alcoholism
SUD Substance use disorder
VNTR Variable number of tandem repeats polymorphism

INTRODUCTION

Attention deficit hyperactivity disorder (ADHD) is a neuropsychiatric disorder affecting 5% of children and 4% of adults of the general population (Kessler et al., 2006; Polanczyk & Rohde, 2007). Its prevalence in general population studies showed higher estimates in adolescents than in the adults (Biederman, Mick, & Faraone, 2000; Franke et al., 2012). Genetic studies started with the fact that ADHD symptoms and traits aggregate in families, and familial clustering was observed across generations (Wilens, 2004). Elevated rates of ADHD were seen among siblings and parents of ADHD progeny. The situation is similar in adolescent and adult twin studies. ADHD is associated with functional and cognitive deficits that run in first-degree relatives, indicating its highly heritable nature. Environmental factors do interact with genetic factors. Based on self-rated symptoms, a sample of adults demonstrated 30–40% heritability. Using a multiple-source evaluation battery, child and adult ADHD heritable estimates were very similar (Franke et al., 2012). Proband sibling samples and population twin models suggest that ADHD is a composite of one or more distributed traits. Twin and family studies demonstrated 70–80% heritability that is influenced by genetic and environmental factors. Genetic predisposition for the development of heroin addiction disorder (HD) is about 40–60%, influenced by complex inheritance, physiological, and environmental factors (Kendler, Jacobson, Prescott, & Neale, 2003; Tsuang et al., 1998). Like ADHD, the genetic variants of substance use disorder (SUD) and HD have been associated with family-based linkage studies (Kreek, Bart, Lilly, Laforge, & Nielsen, 2005; Zou et al., 2008).

The SUD lifetime prevalence is estimated to be 19.1%, constituting a very large proportion of hospital and mental health facility patients (Cherpitel and Ye, 2008; de Graaf et al., 2012). ADHD is a main risk factor for HD and other SUD development (Biederman et al., 2000; Charach, Yeung, Climans, & Lillie, 2011; Lee, Humphreys, Flory, Liu, & Glass, 2011), often mediated by psychiatric comorbidity with oppositional defiant disorder or conduct disorder (Fergusson, Horwood, & Ridder, 2007). Patients with comorbid SUD and ADHD develop addiction or dependence at a very tender age and are relatively more often hospitalized than patients with SUD but without ADHD (Arias et al., 2008). The relapse rate of ADHD is significantly associated with SUD dependence and addiction (Carroll & Rounsaville, 1993). Furthermore, except for one study (Schubiner et al., 2002), others consistently demonstrated that the pharmacotherapy of ADHD with atomoxetine and methylphenidate is not very effective in patients with comorbid SUD and ADHD (Carpentier, De Jong, Dijkstra, Verbrugge, & Krabbe, 2005; Castells et al., 2011; Konstenius, Jayaram-Lindström, Beck, & Franck, 2010; Wilens et al., 2008). An optimal therapeutic plan and pharmaceutical care for these patients with SUD and ADHD for operative identification and early adequate diagnosis are very critical, although various overlapping symptoms of drug withdrawal and intoxication make it even more difficult to diagnose (Levin, 2006).

ADHD is a risk factor for SUD and HD. Many patients suffering from HD may also have comorbid ADHD. In one study, a sample of 110 addicted patients was assessed for retrospective analysis of childhood ADHD. The data showed comorbidity among 41.5% of HD patients according to the Wender Utah Rating Scale (WURS) and 37.7% through the *Diagnostic and Statistical Manual of Mental Disorders*, fourth edition (DSM-IV), and opioid addicts demonstrated 46.6% (WURS) and 40% (DSM-IV) comorbidity. The Conners' adult ADHD rating scale (CAARS) identified a significant number of patients who fulfilled the diagnostic criteria of adult ADHD. The CAARS used to assess adult ADHD in the same group of patients demonstrated 35.8% with HD (Salman et al., 2012). In another study (Salman et al., 2014), comprising 137 HD patients only, the percentage of addicts comorbid with ADHD was 38.6% according to DSM-IV (text revision)and 41.6% and 37.9% according to WURS and CAARS, respectively (Table 1). It is presumed that the prevalence of ADHD in SUD patients is relatively higher than that of SUD in ADHD. However, these estimates of prevalence vary substantially, ranging from 2% in one study (Hannesdottir, Tyrfingsson, & Piha, 2001) to 83% in another (Matsumoto, Kamijo, Yamaguchi, Iseki, & Hirayasu, 2005). Apparently there is an uncertainty regarding differences of SUD in ADHD and vice versa. The co-occurrence and high rate of comorbidity of psychiatric disorders with SUDs are thoroughly studied and well proven. Various clinical trials and cohort studies have established the relationship time and again. The critical question that arises is why this comorbidity occurs so often. Are there any neurobiological or genetic mediator connections for such association? Another issue that is really of concern is that different studies report different psychiatric disorders with substance abuse. Among them, ADHD is a commonly reported high-rate disorder comorbid with SUD. The mechanistic connections of ADHD are particularly active with SUD. With the evolution of neuroscience and technology concerning emerging neurobiological findings in molecular biology, neurotransmitters and neural circuits intellectualize chronic distress as a mainstay of psychiatric association with SUD. Studies of the comorbidity of ADHD with HD are shown in Table 2, whereas that of ADHD with SUD is shown in Table 3. A meta-analysis of 29 studies and 6689 subjects estimated about 13% comorbidity of ADHD and HD and 23% comorbidity of ADHD and SUD, irrespective of gender, age, duration of abstinence, ethnicity, and setting. Meta-regression analyses demonstrated that the prevalence of ADHD in HD was relatively lower than in other SUDs but more than in cocaine abusers (Carpentier et al., 2012).

APPLICATIONS TO OTHER ADDICTIONS AND SUBSTANCE MISUSE

ADHD, Substance Abuse, and Stimulant Therapy

The persistent myth regarding pharmacotherapy of ADHD is that stimulant therapy can lead to and/or exacerbate symptoms of drug addiction. ADHD patients may use illegal drugs at one stage of life or another without succumbing to physical or emotional dependence, but the mounting ratios of substance dependence also depend on social background, poor stress responses, undivisional development, and genetic predisposition. The association of ADHD with SUD is not merely a "compulsion" but rather a constellation of biological and behavioral steps that periodically lead to comorbidity. ADHD and SUD cannot always be encapsulated together, but the chances of their comorbidity remaining clinical treatment. In one study (Biederman et al., 2008), the effect of stimulant pharmacotherapy was assessed in children with ADHD who were followed up to adulthood. The authors found no substantial evidence that the treatment can subsequently cause SUD. In fact, early treatment of children with ADHD reduced the risk of developing subsequent SUD (Wilens et al., 2008) (Table 4). The effects of ADHD treatment of adult patients who have already developed SUD are understudied, while the few medication trials so far conducted provide no clear effect of medication treatment on SUD (Levin, 2006; Thurstone, Riggs, Salomonsen-Sautel, & Mikulich-Gilbertson, 2010). One study (Levin, 2006) of cocaine-dependent patients with comorbid ADHD on methylphenidate treatment exceptionally demonstrated a decrease in symptoms of ADHD through reduction in the use of cocaine. In SUD patients, ADHD seems to be prevalent equally among both genders. However, ADHD and SUD are usually more frequently diagnosed in males in community samples. Symptoms of withdrawal and substance intoxication can easily be misinterpreted as symptoms of ADHD, so the timing of diagnosis is very important. Before diagnostic assessment, an adequate abstinence period is considered to be very useful in this regard.

Maternal and Paternal Smoking during Pregnancy: A Risk Factor for ADHD

SUD and ADHD present remarkable genetic and environmental associations with exacerbation of the symptoms due to maternal and paternal smoking during pregnancy (Linnet et al., 2003; Young, Stallings, Corley, Krauter, & Hewitt, 2000; Table 5). Despite heritability, numerous environmental factors and family influences are associated with the risk of developing ADHD. One such risk is maternal and paternal smoking during pregnancy (Button, Thapar, & Mcguffin, 2005).

Various studies have reported an association between prenatal smoking and maternal psychopathology such as addiction, anxiety, depression, and ADHD. This could be a possible compelling reason that maternal ADHD could lead to prenatal smoking because the mother could not give it up even while knowing the adverse effects and risk to the fetus (Kollins, McClernon, & Fuemmeler, 2005).

TABLE 1 Diagnostic and Statistical Manual, Wender Utah Rating Scale, and Conners' Adult ADHD Rating Scale for the Diagnosis of ADHD

Addiction, n (%) (Total n = 110)	DSM-IV, n (%)[a]	Inattentive Type, n (%)	Hyperactive–Impulsive Type, n (%)	Combined Type, n (%)	WURS, n (%)[b]	CAARS, n (%)[c]	With ADHD, n (%)	Without ADHD, n (%)
Heroin, 53(48.1)*	20(37.7)	2(10)	11(55)	7(35)	22(41.5)	19(35.8)	19(35.8)	34(64.1)
THC, 23(20.9)	7(30.4)	2(28.5)	1(14.2)	4(57.1)	8(34.7)	8(34.7)	8(34.7)	15(65.2)
Opium, 15(13.6)	6(40)	5(83.3)	–	1(16.6)	7(46.6)	7(46.6)	7(46.6)	8(53.3)
Alcohol, 8(7.2)	3(37.5)	–	2(66.6)	1(33.3)	4(50)	3(37.5)	3(37.5)	5(62.5)
Polydrug, 11(10)	6(54.5)	–	1(16.6)	5(83.3)	11(63.6)	5(33.3)	5(33.3)	6(54.5)
Heroin**, 137(100)	53(38.6)[d]	6(11.3)	27(50.9)	20(37.7)	57(41.6)	52(37.9)	53(38.6)	84(61.4)

THC, tetrahydrocannabinol; SD, standard deviation.
[a]DSM-IV, Diagnostic and Statistical Manual of Mental Disorders, fourth edition.
[b]WURS, The Wender Utah Rating Scale indicates ADHD with a score of more than 30.
[c]CAARS, Conner's adult ADHD rating scale.
[d]DSM-IV (text revision).
Adapted from *Salman et al. (2012), and **Salman et al. (2014).

TABLE 2 Methodological and Sample Characteristics of ADHD and Heroin Addicts

Study	n	Mean Age	Males (%)	Setting	Recruitment	Ethnicity (%)	ADHD Diagnosis	SUD Diagnosis	ADHD Prevalence (%)	Other Informer (e.g., Parent)
Eyre, Rounsaville, and Kleber (1982)	157	27.0	34	Treatment	Consecutive	–[b]	Cantwell criteria	–	22	No
King, Brooner, Kidorf, Stoller, and Mirsky (1999)[d]	125	37.0	46	Treatment	Consecutive	Caucasian (36)	DISC	SCID-I	19	Not all
Modestin, Matutat, and Würmle (2001)[d]	101	26.0	100	Treatment	–	–	DSM criteria[c]	DSM criteria[c]	11	No
Kolpe and Carlson (2007)	687	37	54.9	Treatment	Consecutive	Caucasian (62.5)	ADDS	–	19	No
Arias et al. (2008)[a]	1761[a]	38.4	51.9	Community	Random	Non-Hispanic Black (48.7)	SSADDA	SSADDA	5.22	Yes
Subramaniam and Stitzer (2009)	74	16.9	55	Treatment	–	Caucasian (89)	DICA-IV	CIDI-SAM	33	No
Behdani et al. (2011)	144	22.09	97.9	Treatment	Consecutive	Asian	WURS ADHD SRS DSM-IV-R	DSM-IV-R	30.6	Yes
Carpentier et al. (2012)	107	59.4	49.4	Treatment	Consecutive	Caucasian	DSM-IV-TR	CIDI-SAM	35.2	No
Salman et al. (2014)[d]	137	37.5	100	Treatment, community	Consecutive	Asian	WURS, CAARS	DSM-IV-TR	35.8	No

DISC, diagnostic interview schedule for children; CIDI-SAM, composite international diagnostic interview-substance abuse module; MINI, mini-international neuropsychiatric interview; DICA-IV, diagnostic interview for children and adolescents, fourth edition; ADDS, attention deficit disorder screen; ADHD SRS, adult ADHD self-report scale; CAARS, conners' adult ADHD rating scale; DIS, diagnostic interview schedule for DSM-IV; SSADDA, semi-structured assessment for drug dependence and alcoholism; SCID-I, structured clinical interview for DSM-IV for axis I disorders; DSM-IV, diagnostic and statistical manual of mental disorders, fourth edition; DSM-IV-TR, DSM-IV text revision.

[a]Non-treatment-seeking population with lifetime nicotine and/or opioid dependence diagnosis.
[b]'–' refers to data not reported.
[c]ADHD and SUD diagnosis was made according to criteria of DSM, but any specific tool was not reported.
[d]Studies that reported retrospective childhood diagnosis without mentioning of current symptoms or childhood-onset ADHD with persisting symptoms in adulthood.

TABLE 3 Methodological and Sample Characteristics of ADHD, SUD, and HD

Study	Design	Duration	Method	Results	Conclusions
van Emmerik-van Oortmerssen et al. (2011)[a]	Meta-analysis of 29 studies	Studies included from 1997 to 2009	ADHD mean 12.7 years, SUD mean 28.3 years	Overall ADHD prevalence in adolescents was 25.3% (95% CI=20.0–31.4%) and in adults, 21.0% (95% CI=15.9–27.2%)	HD was associated with higher ADHD prevalence than other addictions
Wilens et al. (2011)	Longitudinal, case-control	10-year follow-up; individuals with ADHD, $n=268$; controls, $n=229$	ADHD mean 10.9 years, controls mean age 11.9 years	ADHD was a significant predictor of SUD, HR=1.47 (95% CI=1.07–2.02)	ADHD is a significant risk factor for the development of SUDs
Salman et al. (2012[b], 2014)	Case series	110 individuals with SUD, 137 patients with HD	Adults with mean 37.1 years	Retrospective assessment of ADHD in childhood presented in 21% (DSM-IV) and 22.5% (WURS) ($p=0.029$), while 17.6% patients had persisting ADHD (CAARS)	ADHD is a risk factor associated with SUDs, expressed in the form of heroin, opium, alcohol, and polydrug dependence
Biederman, Petty, Hammerness, Batchelder, and Faraone (2012)	Two identically designed, longitudinal, case-control studies	10-year follow-up period	165 individuals with ADHD and 374 controls	Adolescents with ADHD who smoked cigarettes were more likely to develop SUD compared with youth with ADHD who did not smoke ($n=138$, $p<0.05$)	Cigarette smoking increases the risk for subsequent drug and alcohol use disorders among individuals with ADHD
Madsen and Dalsgaard (2014)	Case-control study	13–18 years, adolescents	117 with ADHD and 102 controls	Among alcohol users, 52% of ADHD probands vs 70% controls confirmed monthly alcohol intake ($p=0.014$); 4% vs 7% of controls used illicit drugs ($p=0.260$)	ADHD patients had a heavier use of cigarettes, but less consumption of alcohol and illicit drug use than controls

CAARS, Conners adult ADHD rating scale; WURS, Wender Utah Rating scale; HR, hazard ratio; CI, confidence interval.
[a]This meta-analysis did not include the findings of Biederman and Wilens.
[b]The study by Salman et al. (2014) comprised samples of heroin, opium, alcohol, and polydrug dependence.

TABLE 4 Studies Evaluating the Risk of SUD among Treated versus Untreated ADHD Individuals

Study	Duration/Target	Sample	Results	Conclusions
Biederman, Wilens, Mick, Spencer, and Faraone (1999)	4-year follow-up study/adolescents	Medicated, $n=56$; nonmedicated, $n=19$; controls, $n=137$	Reduction in risk of SUD in medicated subjects; adjusted OR=0.15 (95% CI=0.04–0.6)	Untreated ADHD was associated with elevated risk for SUD; ADHD treatment reduced up to 85% the risk for SUD
Wilens, Faraone, Biederman, and Gunawardene (2003)	Meta-analysis of six studies; follow-up of adolescents into adulthood	Medicated, $n=674$ (97% stimulants); nonmedicated, $n=360$	Reduction in risk of SUD in medicated subjects; OR=1.9 (95% CI=1.1–3.6)	Almost twofold reduction in SUD risk with stimulant therapy
Katusic et al. (2005)	17.2 years (mean) follow-up of adolescents into adulthood	Medicated with stimulants, $n=295$; nonmedicated, $n=84$	SUD 20% reduction in medicated and 27% in nonmedicated subjects; adjusted OR=0.6 (95% CI=0.3–1.0)	SUD risk was significantly reduced with stimulant therapy
Wilens et al. (2008)	5-year follow-up of adolescents into adulthood	Medicated, $n=140$; nonmedicated, $n=122$	Protective effects of stimulant therapy on SUD development, HR=0.27 (95% CI=0.125–0.60), cigarette smoking, HR=0.28 (95% CI=0.14–0.60)	Stimulant therapy does not increase, but rather reduces, the risk for SUD in adolescents with ADHD
Groenman et al. (2013)	Mean follow-up of 4.4 years, at a mean age of 16.4 years/adolescents into adulthood	Medicated, $n=505$ (probands and affected siblings), $n=223$ controls	Reduced risk for SUD (HR=1.12, 95% CI 0.45–2.96), even after controlling for CD and ODD, HR=1.9 (95% CI=1.10–3.36)	Stimulant therapy lowered the risk of SUD and does not develop nicotine dependence in adolescents with ADHD

ODD, oppositional defiant disorder; CD, conduct disorder; HR, hazard ratio; OR, odds ratio; CI, confidence interval.

Prenatal maternal smoking is considered to be the single largest risk factor for adversarial developmental disorders during childhood (Baler, Volkow, Fowler, & Benveniste, 2008). Several studies have shown that maternal smoking during pregnancy is due to psychiatric disorders, problematic relationships, being a single parent, or having low socioeconomic status and leads to intrauterine growth retardation, low birth weight, and premature birth. However, the association of childhood behavioral problems and prenatal exposure to smoking remains vague. A meta-analysis suggests that maternal smoking during pregnancy is associated with self-regulation, which can consequently cause aggression, impulsivity, and attention deficit (Linnet et al., 2003). One hypothesis by Tarter (2002) suggests that nicotine consumed by a mother can ultimately cause neurological damages that can affect the behavioral and cognitive competencies of the developing fetus. Therefore maternal or paternal smoking manifests itself clinically to precipitate various behavioral complications that can cause ADHD.

Risk of ADHD in Children of Parents with Substance Use Disorder

SUD is a major community health problem that is predominantly associated with commotion in life, delinquency, unadorned medical and psychiatric complications, and poor yield. The familial aggregation of SUDs has been established by a pool of studies since 1995 using SUD and HD probands from both community and treatment settings (Faraone et al., 2000; Vidal et al., 2012; Wilens et al., 2008). Early assessment of psychiatric manifestations in the offspring due to the impact of parental psychopathology of SUD is a potent strategy. Studies have reported the presentation of social malfunctioning and mental health problems among the offspring of SUD-dependent parents. Studies found higher rates of SUD among offspring with ADHD-affected parents and vice versa. Heroin-addicted parents demonstrate a risk of ADHD to their offspring that is three times higher and rates of SUD about 16 times higher than those of the offspring of controls after adjusting for the probands' comorbid disorders and demographic characteristics. The elevated rate of ADHD in the offspring of HD but not alcohol-dependent probands substantiates an analogous finding by Wilens et al. (2008). This increased risk could be attributed to genetic as well as environmental factors. The probands' SUD was also associated with exposure of the offspring to stress in terms of traumatic events in one study. However, by adjusting for co-parental variables, the ADHD association with SUD in the offspring was no longer significant (Vidal et al., 2012).

TABLE 5 Symptoms of ADHD Precipitated by Parental Smoking in Pregnancy

Study	Design	Duration	Sample	Results		Conclusions
				Fully Adjusted Risk[a]	Partially Adjusted[b]	
Kollins et al. (2009)	Population-based, case-control	First 7 years after birth, 5 and 12 years	244 offspring and 151 families	Postnatal maternal smoking significantly associated with inattentive symptoms (67.63 vs 63.45, $p=0.05$)	Postnatal paternal smoking was associated with higher levels of hyperactive–impulsive symptoms (log T-scores 4.18 vs 4.07, $p<0.05$)	Postnatal parental smoking increased risk for development of ADHD symptoms
Langley, Heron, Smith, and Thapar (2012)	Prospective cohort 1991–2000	Gestation to late adolescence, mean 7.5 years	8324 families and 5719 individuals	Maternal and paternal smoking were analyzed together (mothers, $\beta=0.18$, 95% CI 0.10, 0.26; fathers, $\beta=0.15$, 95% CI 0.09, 0.22)	Separate maternal and paternal smoking analyses during pregnancy (mothers, $\beta=0.25$, 95% CI 0.18, 0.32; fathers, $\beta=0.21$, 95% CI 0.15, 0.27)	ADHD symptoms were associated with both parents smoking during pregnancy
Silva, Colvin, Hagemann, and Carol Bower (2014)	Population-based, case-control-record linkage, 1981–2003	Children and adolescents aged 25 years	Stimulant therapy, $n=12,991$, controls, $n=30,071$	Smoking in pregnancy compared to controls, boys, OR=1.86 (95% CI 1.53–2.27); girls, OR=1.67, (95% CI 1.07–2.6)	Smoking in pregnancy compared to controls, boys, OR=2.06 (95% CI 1.74–2.44); girls, OR=1.73, (95% CI 1.21–2.48)	Smoking in pregnancy elevated the risk of ADHD symptoms

[a] Fully adjusted for all factors in the model including maternal age, year of birth, and data in model available.
[b] Partially adjusted by year of birth and socioeconomic indexes.

A study reported that ADHD probands were at increased risk for SUD. First-degree relatives of ADHD probands were at elevated risk for SUD, whereas relative risk in second-degree relatives was significantly lower compared to relatives of the controls. The study suggested that the co-occurrence of SUD and ADHD was due to genetic influences rather than detrimental effects of ADHD pharmacotherapy or general propensity for various psychiatric conditions (Burt, 2009; Franke et al., 2012). Owing to these common genetic influences, the symptoms continue from childhood to adolescence and later into adulthood. The comorbidity of ADHD and SUD is highly heritable and there is evidence of a genetic basis for this overlap (Groman, James, & Jentsch, 2009; Wilens, 2004). The prevalence of SUD is frequent among the family members of patients of ADHD (Faraone et al., 2000; Milberger, Faraone, Biederman, Chu, & Wilens, 1998), whereas the prevalence of ADHD is higher in proband families with SUD (Knopik, Jacob, Haber, Swenson, & Howell, 2009; Wilens et al., 2008). Owing to the unknown etiologies of the two disorders the genetic risk factors are still understudied, although the neurobiological factors of the comorbidity have been extensively considered, and numerous genes and neurotransmitters have been identified, including serotonin (Ross & Peselow, 2009) and dopamine, which play substantial roles in this comorbidity (Koob & Volkow, 2010). Direct comparisons of SUD and ADHD risk genotypes are rather scarce. Also the results of genetic determinants of the particular risk variants of this comorbidity are controversial because their individual impacts are small and can be attributed to the polygenic nature of these disorders (Faraone & Mick, 2010).

Genetic Susceptibility to Comorbid ADHD and Heroin Addiction

The *OPRM1* gene is a mainstay for morphine as well as endogenous opioids that codes for the μ-opioid receptor. Research studies showed that the receptor genome 118A→G (rs1799971) single-nucleotide polymorphism (SNP) is significantly associated with various addictive aspects of SUDs and opioid dependence (Drakenberg et al., 2006; van der Zwaluw et al., 2007), but is not well studied in ADHD (Olmstead, Martin, Brien, & Reynolds, 2009). However, a 2012 study reported that *OPRM1* 118A→G is associated with the combination of ADHD and SUD (Table 6) (Carpentier et al., 2012).

Serotonin is involved in alcohol and drug dependence, and the association of SNPs with SUD has already been reported (Yuferov, Levran, Proudnikov, Nielsen, & Kreek, 2010). It is hypothesized that the dysregulation of serotonin causes impulsive behaviors (Dalley, Mar, Economidou, & Robbins, 2008) that mediate ADHD symptoms. A significant association of the G4→C transition at nucleotide position 861 (861G→C; rs6296) in the serotonin 1B-receptor gene (*HTR1B*, G861C SNP; T1065G SNP), the serotonin transporter gene (*SLC6A4*, 44-bp insertion/deletion in the promoter region (5-HTTLPR)), and the synaptosomal-associated protein 25 gene (*SNAP25*) and novel latrophilin 3 (*LPHN3*) and cadherin 13 (*CDH13*) genes across the life span of ADHD has been identified (Franke et al., 2012; Hawi et al., 2003). Although the rs6296 SNP demonstrated no association with heroin addiction, a few other polymorphisms in the *HTR1B* genome have shown nominal associations with ADHD (Proudnikov et al., 2006).

The dopamine receptors D4 and D5 are present in the frontal-subcortical networks associated with the pathophysiology of ADHD and have been extensively studied via neuropsychological and neuroimaging cohorts (Curatolo, Paloscia, D'Agati, Moavero, & Pasini, 2009; Franke et al., 2012). Compared to the dopamine D5 receptor (DRD5), the DRD4 polymorphism has not been extensively studied in SUDs, but a major focus has been on the DRD4 variable number of tandem repeat polymorphism (VNTR) in ADHD. This VNTR codes an amino acid sequence that comprises repeated 48-bp units that are involved in G-protein coupling (DiMaio, Grizenko, & Joober, 2003). Hence, the association of the seven-repeat allele, DRD5 gene 148-bp, and dopamine β-hydroxylase gene (DBH; 50 taq1 A allele (Franke et al., 2012)) with ADHD is very common but not reported in SUDs (Gizer, Ficks, & Waldman, 2009; Maher, Marazita, Ferrell, & Vanyukov, 2002), whereas DBH-intron 5–C→T (rs2519152) showed significant association with both ADHD and SUD. These risk allele carriers in methadone maintenance patients demonstrated divergent patterns (Table 6) (Carpentier et al., 2012). On the other hand, a meta-analysis found that the catechol-*O*-methyltransferase gene (Val108Met-polymorphism) (Cheuk & Wong, 2006) has no shared genetic association with ADHD, while other studies demonstrate a rather weaker link of the dopamine transporter gene (*SLC6A3*, 480-bp VNTR in the 30-bp untranslated region) (Carpentier et al., 2012; Li, Sham, Owen, & He, 2006) with ADHD.

The implication of the endogenous opioid system in the neurobiology of ADHD is not thoroughly studied, so the risk genotype still remains indistinct. However, it is apparent that the opioid system is involved in the regulation of the reward center and the parent–child relationship (Carpentier et al., 2012; Copeland et al., 2011), which suggests that there might be a pathophysiological role of the opioid system that ultimately leads to behavioral disturbances expressed in the form of ADHD and SUD.

The majority of the studies have not tested associated aspects other than parental SUD and HD that could affect the psychopathological risk of ADHD in offspring. Many studies have reported data with large confidence intervals and odds ratios, which indicates that the relationship could not be established accurately owing to insufficient statistical power. It is difficult to establish that ADHD in the parents will lead to offspring with SUD and HD because parental concordance with such disorders is rare. However, SUD and HD in the offspring of parents with the same disorder are validated by many studies that recognized the psychopathology and association of affected parents with their children. But again, most of the studies established the association by merely going through the family reports rather than by direct involvement of the parent or offspring. Also, the sample size was too small, statistical analysis was too weak, and other genetic, environmental, and parental comorbidities were either ignored or not well studied. Very few studies separately recruited offspring with parental SUD and studied the precise effects of parental addiction on children.

Endophenotypes and the endogenous opioids system, based on neuroimaging, neurobiology, neuropsychology, genome analysis

TABLE 6 Frequency of Six Risk Allele Carriers (both Homozygous and Heterozygous Individuals) in Various Subgroups of Patients with ADHD and/or SUD

n % Male	ADHD_All	ADHD_Pure	ADHD and SUD_Pure	SUD_Pure	SUD_All	Controls
	212	176	36	71	107	500
	52.4	47.7	75.0	84.5	81.3	49.4
DRD4 VNTR48-bp VNTR risk group (7Rcarriers)	30.4	29.9	33.3	32.9	33.0	36.0
DRD5 VNTR2-bp VNTR risk group (148-bpcarriers)	66.5	66.9	64.7	79.7	74.8	73.7
HTR1B 861G→C (rs6296) risk group (C-carriers)	44.8	47.2	32.4	45.1	41.0	46.4
DBH-intron 5–C→T (rs2519152) risk group (C-carriers)	78.8	80.8	69.7	72.9	71.8	68.2
COMT-VALMET-G→A (rs4680) risk group (G→Val-carriers)	69.3	70.8	62.9	66.2	65.1	69.1
OPRM1 118A→G (rs1799971) risk group (G-carriers)	27.0	25.0	37.1	14.3	21.9	17.8

Odds Ratios of DBH and OPRM1 Risk Allele Carriers (Homozygous and Heterozygous Individuals) in Various Subgroups of Patients with ADHD and SUD Compared to the Control Group

	Genotype			Sex		
	Adj OR	95% CI	p	Adj OR	95% CI	p
DBH–Intron 5–C→T (rs2519152) Risk Group (C-Carriers)						
ADHD_all	1.73	1.15–2.59	0.008	1.13	0.80–1.60	0.48
ADHD-pure	1.97	1.25–3.10	0.003	0.94	0.65–1.37	0.75
ADHD and SUD_pure	1.05	0.49–2.28	0.89	2.76	1.26–6.07	0.011
SUD_pure	1.22	0.69–2.17	0.49	5.55	2.84–10.8	<0.001
SUD_all	1.15	0.71–1.86	0.57	4.29	2.55–7.21	<0.001
OPRM1 118A→G (rs1799971) Risk Group (G-Carriers)						
ADHD_all	1.71	1.17–2.50	0.006	1.12	0.81–1.54	0.50
ADHD-pure	1.54	1.02–2.32	0.040	0.94	0.67–1.33	0.73
ADHD and SUD_pure	2.75	1.32–5.72	0.007	2.98	1.36–6.51	0.006
SUD_pure	0.80	0.39–1.66	0.56	5.47	2.81–10.67	<0.001
SUD_all	1.35	0.79–2.30	0.28	4.38	2.61–7.35	<0.001

VNTR, variable number of tandem repeat polymorphisms; HTR1B, serotonin 1b; DBH, dopamine β-hydroxylase; COMT, catechol-O-methyltransferase; OPRM1, μ-opioid receptor gene; DRD4, dopamine D4 receptor gene; DRD5, dopamine D5 receptor gene.
With permission from Carpentier et al. (2012).

of next generation, and enhanced bioinformatics and statistical analysis, can identify additional variants and candidate genes involved in etiology and comorbidity of the disease.

ADHD and Substance Use Disorder: The Neurobiological Effects

Many prospective studies have shown increased risk of SUD onset among ADHD patients. A few studies have shown a retrospective analysis of SUD patients, in particular HD, who were assessed for ADHD in childhood. They were further analyzed as to whether the symptoms of ADHD persisted in adulthood. Comorbidity was higher among the peers who were involved in polydrug abuse.

Drug detoxification and withdrawal of SUD and HD often mimic the symptoms of ADHD. This often leads to differential diagnoses and this overlap consequently leads to higher rates of comorbidity in epidemiological studies. This caveat still cannot exclude many conservative and strict studies that demonstrate a high rate of ADHD comorbidity with SUD and HD.

The convincing data of an ADHD association with SUD and HD cannot fully explain the nature of this relationship. There are different subtypes of ADHD that may or may not have the same mechanism and nature with the various kinds of SUD and HD. ADHD may be a possible risk factor for modifying SUD/HD or for developing one. Theories of severity of drug addiction suggest that the specific psychotropic effects of SUD and HD are reasons to manage emotional distress. Chronic drug addiction is the cause of neurobiological alterations and neuroadaptations in brain stress circuits that induce various ADHD symptoms during protracted and acute withdrawal states.

Psychosocial, neurobiological, and genetic risk factors are involved in the comorbidity of SUD, HD, and ADHD. The pathophysiological development of SUD and HD to ADHD and vice versa is well studied. ADHD is characterized as an externalizing disorder (others are oppositional defiant disorder and conduct disorder). People with externalizing disorders are compelled by their personality traits and behavioral disinhibitions, such as high levels of impulsivity, poor self-control, and aggression.

Substantial evidence classifies ADHD as a cognitive dysfunction of the prefrontal and frontal cortex regions. Children with ADHD, SUD, and HD often demonstrate poor performance on various neuropsychological tests related to the prefrontal cortex region, such as attention, working memory, planning, self-monitoring, cognitive flexibility, and behavioral and motor control. Electrophysiological markers such as P3 amplitude demonstrate lower values for ADHD, SUD, and HD. Genetic factors are involved, too, in the development of comorbid ADHD and SUD. However, negative child–parent conflict, emotional distress, and high levels of negative affect may also lead to SUD, HD, and ADHD. The data of numerous studies show that ADHD and HD can develop from coping with family conflicts.

Preclinical studies suggest that stress severely affects prefrontal cortical function. Self-regulation processes and cognitive-conflict monitoring are controlled by the anterior cingulate cortex and prefrontal cortex. Prefrontal cortex functions are regulated by norepinephrine and dopamine. Lower self-regulation and response inhibition are symptoms of ADHD. Individuals with ADHD who were exasperated in the laboratory showed higher levels of behavioral aggression and anger and elevated heart rates.

They experienced lower cortisol levels, difficulty in self-control, and risk appetite, which may possibly be connected to SUD development in later life. ADHD and SUD have neurobiological, neuropsychopathological, and behavioral origins. Furthermore, preclinical studies suggest that prefrontal cortical functions are impaired by stress. Human clinical trials indicate that individuals with SUD and ADHD have poor self-regulation and poor stress-coping abilities.

The anterior cingulate cortex and prefrontal cortex play important roles in the regulation of "future reward" behaviors. Various primate studies demonstrate the assessment of future reward expectancy and motor responses through the anterior cingulate cortex and prefrontal cortex. This proves that the prefrontal cortex is involved in reinstatement, reinforcement, and self-administration of drugs (Brady and Sinha, 2005). ADHD individuals demonstrate disinhibited behavioral and physiological responses during cognitive tasks related to reward. This is due to increased activity in the sensory and posterior cortices and decreased striatal and prefrontal cortex activity. This problem can usually be solved through the prolonged use of methylphenidate. Symptoms of ADHD and prefrontal cortical activity can be improved by inhibiting catecholamine influx to the prefrontal cortex region. This can be done through the use of α_2-adrenergic agonists (e.g., guanfacine) that can centrally inhibit norepinephrine. Other α_2-adrenergic agonists (lofexidine and clonidine) severely decrease stress-induced relapse to drug-seeking behavior. The α_2-adrenergic agonists are beneficial in prefrontal cortex dysfunction in both SUD and ADHD.

CONCLUSION

The relationship between ADHD, SUD, and HD is somewhat complex and multifaceted.

Dysfunctional frontal and prefrontal cortex regions are related to behavioral disinhibition and deficits in self-monitoring that can be a possible cause of comorbidity. Neuroadaptations, mechanistic relationships, and overlapping neural circuitry pathways in this regard are yet to be reconnoitered. Offspring of HD parents demonstrate a higher risk of ADHD and higher rates of other SUDs than the offspring of controls. About one in every four SUD patients meets the diagnostic criteria of ADHD. Drug-seeking behaviors, ADHD, and prefrontal cortical activity can be improved using α_2-adrenergic agonists. It is critically important to devise adequate screening assessment tools and procedures to recognize patients with comorbid SUD and ADHD. It is apparent that the development of an effective therapeutic plan and cognitive behavioral therapy is very important for patients with ADHD comorbid with SUD and HD.

DEFINITION OF TERMS

- **Attention deficit disorder**
  This is an older name having the same meaning as ADHD. Attention deficit disorder is characterized by a problem with focus and inattention without hyperactivity.
- **Attention deficit hyperactivity disorder**
  The American Psychiatric Association's DSM-IV (text revision) officially named this condition. It is characterized by a problem with attention and hyperactivity.

There are three main types of ADHD, namely:

- Inattentive type;
- Hyperactive–impulsive type;
- Combined type.

The severity and symptoms of ADHD vary among children; usually boys have more severe and higher rates.

- **Inattentive type**
  Signs and symptoms of ADHD are as follows:
  - Often seems not to be listening or easily distracted by irrelevant sounds and sights;
  - Often leaves one task in the middle and jumps to another;
  - Seldom follows instructions, and frequently forgets or loses things like pencils, toys, bags, books required for a task;
  - Often makes careless mistakes and fails to pay close attention to minor details.
- **Hyperactive–impulsive type**
  Signs and symptoms of ADHD are as follows:
  - Climbing, running, or getting up in locations where quiet behavior or remaining seated are expected;
  - Having difficulty waiting to take turns or stay in line;
  - Often restless, fidgeting with feet or hands or while seated;
  - Bursts out with a response before even hearing the whole sentence or question.
- **Combined type**
  People with combined type show symptoms of both inattentive and hyperactive–impulsive types.
- **Adult ADHD**
  Children with a diagnosis of ADHD can continue to demonstrate persistent symptoms into adulthood. The symptoms of adult ADHD may include difficulty with concentration, relationship issues, difficulty with handling information, and problems with organizing various tasks. Similar to children, its treatment involves both medications and behavioral therapies.
- **Neurotransmitter**
  It is a chemical released by neurons that acts as a messenger to conduct nerve impulses (messages) from the brain to the body or vice versa.
- **Hyperactivity**
  This is activity (action/movement or behavioral response) that is higher than usual or above normal levels. Hyperactive individuals cannot sit still; they are always moving and talk continuously. They are intensely restless, always roam around with wiggling feet, and squirm in their chairs.
- **Impulsivity**
  This describes when an individual acts before thinking and is unable to curb his or her abrupt actions or behavioral reactions. Consequently, these individuals cannot wait for their turn, blurt out answers even before the question is even completed, and find it hard to wait for things.
- **Comorbid condition**
  This is also known as a coexisting or co-occurring condition. When two or more than two medical conditions are present in a patient, they are said to be comorbid conditions.
- **Medication holiday**
  For evaluation or medical purposes, the prescribed pharmacotherapy of ADHD is discontinued temporarily, for a brief period of time, under the strict supervision of a trained medical professional.
- **Multimodal treatment**
  This refers to pharmacotherapy of ADHD along with multiple interventions, social training, behavioral management, and counseling, tailored to the needs of an individual with ADHD.
  - **Social skills training:** This training helps a child to be socially acceptable, to behave normally, and to be less hyperactive and aggressive.
  - **Behavior management:** ADHD children can effectively be managed by a reward system to behave normally. Behavioral management skills for parents require an extensive counseling session to learn how to deal with their child.
  - **Family therapy:** All the members of a family are counseled, educated, and trained regarding the member with ADHD and how to deal with his or her behavior effectively.
- **Psychostimulants or stimulant medication**
  Stimulant medications excite the nervous system through the production of neurotransmitters. They ameliorate the symptoms of ADHD by helping an individual to ignore distractions and promote alertness, awareness, and focus.
- **Nonstimulant medication**
  This refers to those medications that are not classified by the US FDA as controlled substances and are approved to treat ADHD. Atomoxetine, clonidine, and guanfacine are medications approved by the FDA under the class of nonstimulants.
- **Amphetamine**
  This is a drug that has an effect on the central nervous system and can be physically and psychologically addictive when overused. Amphetamines have been much abused recreationally. The street term "speed" refers to stimulant drugs such as amphetamines.
- **Rebound effect**
  This refers to the propensity of a few medications for ADHD to show more severe symptoms of impulsivity, inattention, and hyperactivity when withdrawn from use.
- **Psychotherapy**
  This is the treatment of a behavioral disorder, mental illness, or any other medical condition by psychological means. Psychotherapy may utilize insight, persuasion, suggestion, reassurance, and instruction so that patients may see themselves and their problems more realistically and have the desire to cope effectively with them.

KEY FACTS OF ADHD AND OTHER PSYCHIATRIC DISORDERS

- In a classroom of 30 children, it is possible that at least one student is suffering from symptoms of ADHD.
- ADHD is not only a disorder of seeking attention. It is a disorder of paying attention and/or uncontrolled social behavior.
- ADHD can lead to learning disabilities, anxiety, depression, relationship problems, and drug addiction.
- Of about every four drug addicts, one meets the diagnostic criteria of ADHD.
- ADHD does not escalate an individual's risk for other disorders, but these individuals are likely to experience a variety of coexisting conditions that include learning disabilities, oppositional defiant disorder, conduct disorders, antisocial behavior, anxiety disorder, depression, bipolar disorder, Tourette syndrome, sleep disorders, and substance abuse.

- Boys usually display classical externalizing symptoms that may include:
  - inattentiveness or lack of focus;
  - impulsivity or "acting out";
  - hyperactivity, such as hitting and running;
  - physical aggression.
- In girls, symptoms of ADHD are different from in boys. They may include:
  - anxiety or low self-esteem;
  - inattentiveness or "daydreaming";
  - social withdrawal;
  - trouble with academics and intellectual impairment;
  - verbal aggression: taunting, teasing, and name-calling.

SUMMARY

- ADHD is a neuropsychiatric and behavioral disorder of impulsivity, inattention, and hyperactivity affecting 8–12% of children worldwide.
- Stimulant pharmacotherapy assessed in children with ADHD found no substantial evidence of subsequent SUD or HD but helped in its remittance.
- ADHD is a risk factor for HD and many patients suffering from HD may also have comorbid ADHD.
- Molecular imaging studies demonstrate anomalies in the dopamine transporter that lead to neurotransmission impairment in ADHD, SUD, and HD.
- Owing to repetitive intake of heroin, the brain chemistry adapts itself, which consequently leads to neurobiological changes that have elements in common with ADHD.
- The drug detoxification and withdrawal of SUD and HD often mimic the symptoms of ADHD.
- Early treatment of children with ADHD reduces the risk of developing subsequent SUD and HD.
- Prenatal maternal and paternal smoking is considered to be the largest risk factor for developmental disorders in children.
- Adoption, twin, and molecular genetic studies indicated that ADHD, SUD, and HD are highly heritable.
- Evidence from human and animal studies proposes pathophysiological dysregulation of frontal–subcortical–cerebellar–catecholamine circuits in patients with ADHD as well as SUD and HD.
- ADHD, SUD, and HD are symptomatic expressions of similar neurobiological disorders.

REFERENCES

Arias, A. J., Gelernter, J., Chan, G., Weiss, R. D., Brady, K. T., Farrer, L., & Kranzler, H. R. (2008). Correlates of co-occurring ADHD in drug-dependent subjects: prevalence and features of substance dependence and psychiatric disorders. *Addictive Behaviors, 33*(9), 1199–1207.

Baler, R. D., Volkow, N. D., Fowler, J. S., & Benveniste, H. (2008). Is fetal brain monoamine oxidase inhibition the missing link between maternal smoking and conduct disorders. *Journal of Psychiatry and Neuroscience, 33*(3), 187–195.

Behdani, M., Zeinali, S., Khanahmad, H., Karimipour, M., Asadzadeh, N., Azadmanesh, K., ... Muyldermans, S. (2011). Generation and characterization of a functional Nanobody against the vascular endothelial growth factor receptor-2; angiogenesis cell receptor. *Molecular Immunology, 50*(2), 35–41.

Biederman, J., Mick, E., & Faraone, S. V. (2000). Age-dependent decline of symptoms of attention deficit hyperactivity disorder: impact of remission definition and symptom type. *The American Journal of Psychiatry, 157*(5), 816–828.

Biederman, J., Petty, C. R., Fried, R., Kaiser, R., Dolan, C. R., Schoenfeld, S., ... Faraone, S. V. (2008). Educational and occupational underattainment in adults with attention-deficit/hyperactivity disorder: a controlled study. *Journal of Clinical Psychiatry, 69*(8), 1217–1222.

Biederman, J., Petty, C. R., Hammerness, P., Batchelder, H., & Faraone, S. V. (2012). Cigarette smoking as a risk factor for other substance misuse: 10-year study of individuals with and without attention-deficit hyperactivity disorder. *British Journal of Psychiatry, 201*(3), 207–214.

Biederman, J., Wilens, T., Mick, E., Spencer, T., & Faraone, S. V. (1999). Pharmacotherapy of attention-deficit/hyperactivity disorder reduces risk for substance use disorder. *Pediatrics, 104*(2), 1–5.

Brady, K. T., & Sinha, R. (2005). Co-occurring mental and substance use disorders: the neurobiological effects of chronic stress. *The American Journal of Psychiatry, 162*(3), 1483–1493.

Burt, S. A. (2009). Rethinking environmental contributions to child and adolescent psychopathology: a meta-analysis of shared environmental influences. *Psychological Bulletin, 135*(4), 608.

Button, T. M., Thapar, A., & Mcguffin, P. (2005). Relationship between antisocial behaviour, attention-deficit hyperactivity disorder and maternal prenatal smoking. *The British Journal of Psychiatry, 187*(2), 155–160.

Carpentier, P. J., De Jong, C. A., Dijkstra, B. A., Verbrugge, C. A., & Krabbe, P. F. (2005). A controlled trial of methylphenidate in adults with attention deficit/hyperactivity disorder and substance use disorders. *Addiction, 100*(12), 1868–1874.

Carpentier, P. J., Vasquez, A. A., Hoogman, M., Onnink, M., Kan, C. C., Kooij, J. J. S., & Buitelaar, J. K. (2012). Shared and unique genetic contributions to attention deficit/hyperactivity disorder and substance use disorders: a pilot study of six candidate genes. *European Neuropsychopharmacology, 23*(6), 448–457.

Carroll, K. M., & Rounsaville, B. J. (1993). History and significance of childhood attention deficit disorder in treatment-seeking cocaine abusers. *Comprehensive Psychiatry, 34*(2), 75–82.

Castells, X., Ramos-Quiroga, J. A., Rigau, D., Bosch, R., Nogueira, M., Vidal, X., & Casas, M. (2011). Efficacy of methylphenidate for adults with attention-deficit hyperactivity disorder. *CNS Drugs, 25*(2), 157–169.

Charach, A., Yeung, E., Climans, T., & Lillie, E. (2011). Childhood attention-deficit/hyperactivity disorder and future substance use disorders: comparative meta-analyses. *Journal of the American Academy of Child & Adolescent Psychiatry, 50*(1), 9–21.

Cheuk, D. K. L., & Wong, V. (2006). Meta-analysis of association between a catechol-O-methyltransferase gene polymorphism and attention deficit hyperactivity disorder. *Behavior Genetics, 36*(5), 651–659.

Copeland, W. E., Sun, H., Costello, E. J., Angold, A., Heilig, M. A., & Barr, C. S. (2011). Child μ-opioid receptor gene variant influences parent–child relations. *Neuropsychopharmacology, 36*(6), 1165–1170.

Curatolo, P., Paloscia, C., D'Agati, E., Moavero, R., & Pasini, A. (2009). The neurobiology of attention deficit/hyperactivity disorder. *European Journal of Paediatric Neurology, 13*(4), 299–304.

Dalley, J. W., Mar, A. C., Economidou, D., & Robbins, T. W. (2008). Neurobehavioral mechanisms of impulsivity: fronto-striatal systems and functional neurochemistry. *Pharmacology Biochemistry and Behavior, 90*(2), 250–260.

DiMaio, S., Grizenko, N., & Joober, R. (2003). Dopamine genes and attention-deficit hyperactivity disorder: a review. *Journal of Psychiatry and Neuroscience, 28*(1), 18–27.

Drakenberg, K., Nikoshkov, A., Horváth, M. C., Fagergren, P., Gharibyan, A., Saarelainen, K., & Hurd, Y. L. (2006). μ-Opioid receptor A118G polymorphism in association with striatal opioid neuropeptide gene expression in heroin abusers. *Proceedings of the National Academy of Sciences*, *103*(20), 7883–7888.

van Emmerik-van Oortmerssen, K., van de Glind, G., van den Brink, W., Smit, F., Crunelle, C. L., Swets, M., & Schoevers, R. A. (2012). Prevalence of attention-deficit hyperactivity disorder in substance use disorder patients: a meta-analysis and meta-regression analysis. *Drug and Alcohol Dependence*, *122*(1–2), 11–19.

Eyre, S. L., Rounsaville, B. J., & Kleber, H. D. (1982). History of childhood hyperactivity in a clinic population of opiate addicts. *The Journal of Nervous and Mental Disease*, *170*(9), 522–529.

Faraone, S. V., Biederman, J., Mick, E., Williamson, S., Wilens, T., Spencer, T., & Zallen, B. (2000). Family study of girls with attention deficit hyperactivity disorder. *American Journal of Psychiatry*, *157*(7), 1077–1083.

Faraone, S. V., & Mick, E. (2010). Molecular genetics of attention deficit hyperactivity disorder. *Psychiatric Clinics of North America*, *33*(1), 159–180.

Fergusson, D. M., Horwood, L. J., & Ridder, E. M. (2007). Conduct and attentional problems in childhood and adolescence and later substance use, abuse and dependence: results of a 25-year longitudinal study. *Drug and Alcohol Dependence*, *88*(2), 14–26.

Franke, B., Faraone, S. V., Asherson, P., Buitelaar, J., Bau, C. H. D., Ramos-Quiroga, J. A., & Reif, A. (2012). The genetics of attention deficit/hyperactivity disorder in adults, a review. *Molecular Psychiatry*, *17*(10), 960–987.

Gizer, I. R., Ficks, C., & Waldman, I. D. (2009). Candidate gene studies of ADHD: a meta-analytic review. *Human Genetics*, *126*(1), 51–90.

de Graaf, R., ten Have, M., van Gool, C., & van Dorsselaer, S. (2012). Prevalence of mental disorders and trends from 1996 to 2009. Results from the Netherlands mental health survey and incidence study-2. *Social Psychiatry and Psychiatric Epidemiology*, *47*(2), 203–213.

Groenman, A. P., Oosterlaan, J., Rommelse, N., Franke, B., Roeyers, H., Oades, R. D., ... Faraone, S. V. (2013). Substance use disorders in adolescents with attention deficit hyperactivity disorder: a 4-year follow-up study. *Addiction*, *108*(8), 1503–1511.

Groman, S. M., James, A. S., & Jentsch, J. D. (2009). Poor response inhibition: at the nexus between substance abuse and attention deficit/hyperactivity disorder. *Neuroscience & Biobehavioral Reviews*, *33*(5), 690–698.

Hannesdottir, H., Tyrfingsson, T., & Piha, J. (2001). Psychosocial functioning and psychiatric comorbidity among substance-abusing Icelandic adolescents. *Nordic Journal of Psychiatry*, *55*(1), 43–48.

Hawi, Z., Lowe, N., Kirley, A., Gruenhage, F., Nöthen, M., Greenwood, T., & Gill, M. (2003). Linkage disequilibrium mapping at DAT1, DRD5 and DBH narrows the search for ADHD susceptibility alleles at these loci. *Molecular Psychiatry*, *8*(3), 299–308.

Katusic, S. K., Barbaresi, W. J., Colligan, R. C., Weaver, A. L., Leibson, C. L., & Jacobsen, S. J. (2005). Psychostimulant treatment and risk for substance abuse among young adults with a history of attention-deficit/hyperactivity disorder: a population-based, birth cohort study. *Journal of Child and Adolescent Psychopharmacology*, *15*(5), 764–776.

Kendler, K. S., Jacobson, K. C., Prescott, C. A., & Neale, M. C. (2003). Specificity of genetic and environmental risk factors for use and abuse/dependence of cannabis, cocaine, hallucinogens, sedatives, stimulants, and opiates in male twins. *The American Journal of Psychiatry*, *160*(4), 687–695.

Kessler, R. C., Adler, L., Barkley, R., Biederman, J., Conners, C. K., Demler, O., & Zaslavsky, A. M. (2006). The prevalence and correlates of adult ADHD in the United States: results from the national comorbidity survey replication. *The American Journal of Psychiatry*, *163*(4), 716–723.

King, V. L., Brooner, R. K., Kidorf, M. S., Stoller, K. B., & Mirsky, A. F. (1999). Attention deficit hyperactivity disorder and treatment outcome in opioid abusers entering treatment. *Journal of Nervous & Mental Disease*, *187*(8), 487–495.

Knopik, V. S., Jacob, T., Haber, J. R., Swenson, L. P., & Howell, D. N. (2009). Paternal alcoholism and offspring ADHD problems: a children of twins design. *Twin Research and Human Genetics*, *12*(1), 53–62.

Kollins, S. H., Garrett, M. E., McClernon, F. J., Lachiewicz, A. M., Morrissey-Kane, E., FitzGerald, D., & Ashley-Koch, A. E. (2009). Effects of postnatal parental smoking on parent and teacher ratings of ADHD and oppositional symptoms. *The Journal of Nervous and Mental Disease*, *197*(6), 442–449.

Kollins, S. H., McClernon, F. J., & Fuemmeler, B. F. (2005). Association between smoking and attention-deficit/hyperactivity disorder symptoms in a population-based sample of young adults. *Archives of General Psychiatry*, *62*(10), 1142–1147.

Kolpe, M., & Carlson, G. A. (2007). Influence of attention-deficit/hyperactivity disorder symptoms on methadone treatment outcome. *American Journal of Addiction*, *16*(1), 46–48.

Konstenius, M., Jayaram-Lindström, N., Beck, O., & Franck, J. (2010). Sustained release methylphenidate for the treatment of ADHD in amphetamine abusers: a pilot study. *Drug and Alcohol Dependence*, *108*(1), 130–133.

Koob, G. F., & Volkow, N. D. (2010). Neurocircuitry of addiction. *Neuropsychopharmacology*, *35*(1), 217–238.

Kreek, M. J., Bart, G., Lilly, C., Laforge, K. S., & Nielsen, D. A. (2005). Pharmacogenetics and human molecular genetics of opiate and cocaine addictions and their treatments. *Pharmacological Reviews*, *57*(1), 1–26.

Langley, K., Heron, J., Smith, G. D., & Thapar, A. (2012). Maternal and paternal smoking during pregnancy and risk of ADHD symptoms in offspring: testing for intrauterine effects. *American Journal of Epidemiology*, *176*(3), 261–268.

Lee, S. S., Humphreys, K. L., Flory, K., Liu, R., & Glass, K. (2011). Prospective association of childhood attention-deficit/hyperactivity disorder (ADHD) and substance use and abuse/dependence: a meta-analytic review. *Clinical Psychology Review*, *31*(3), 328–341.

Levin, F. R. (2006). Diagnosing attention-deficit/hyperactivity disorder in patients with substance use disorders. *The Journal of Clinical Psychiatry*, *68*(4), 9–14.

Linnet, K. M., Dalsgaard, S., Obel, C., Wisborg, K., Henriksen, T. B., Rodriguez, A., & Jarvelin, M. R. (2003). Maternal lifestyle factors in pregnancy risk of attention deficit hyperactivity disorder and associated behaviors: review of the current evidence. *The American Journal of Psychiatry*, *160*(6), 1028–1040.

Li, D., Sham, P. C., Owen, M. J., & He, L. (2006). Meta-analysis shows significant association between dopamine system genes and attention deficit hyperactivity disorder (ADHD). *Human Molecular Genetics*, *15*(14), 2276–2284.

Madsen, A. G., Dalsgaard, S. (2014). Prevalence of smoking, alcohol and substance use among adolescents with attention-deficit/hyperactivity disorder in Denmark compared with the general population. *Nordic Journal of Psychiatry*, *68*(1), 53–59.

Maher, B. S., Marazita, M. L., Ferrell, R. E., & Vanyukov, M. M. (2002). Dopamine system genes and attention deficit hyperactivity disorder: a meta-analysis. *Psychiatric Genetics*, *12*(4), 207–215.

Matsumoto, T., Kamijo, A., Yamaguchi, A., Iseki, E., & Hirayasu, Y. (2005). Childhood histories of attention-deficit hyperactivity disorders in Japanese methamphetamine and inhalant abusers: preliminary report. *Psychiatry and Clinical Neurosciences, 59*(1), 102–105.

Milberger, S., Faraone, S. V., Biederman, J., Chu, M. P., & Wilens, T. (1998). Familial risk analysis of the association between attention-deficit/hyperactivity disorder and psychoactive substance use disorders. *Archives of Pediatrics & Adolescent Medicine, 152*(10), 945–951.

Modestin, J., Matutat, B., & Würmle, O. (2001). Antecedents of opioid dependence and personality disorder: attention-deficit/hyperactivity disorder and conduct disorder. *European Archives of Psychiatry and Clinical Neuroscience, 251*(1), 42–47.

Olmstead, M. C., Martin, A., Brien, J. F., & Reynolds, J. N. (2009). Chronic prenatal ethanol exposure increases disinhibition and perseverative responding in the adult guinea pig. *Behavioural Pharmacology, 20*(6), 554–557.

Polanczyk, G., & Rohde, L. A. (2007). Epidemiology of attention-deficit/hyperactivity disorder across the lifespan. *Current Opinion in Psychiatry, 20*(4), 386–392.

Proudnikov, D., LaForge, K. S., Hofflich, H., Levenstien, M., Gordon, D., Barral, S., & Kreek, M. J. (2006). Association analysis of polymorphisms in serotonin 1B receptor (HTR1B) gene with heroin addiction: a comparison of molecular and statistically estimated haplotypes. *Pharmacogenetics and Genomics, 16*(1), 25–36.

Ross, S., & Peselow, E. (2009). The neurobiology of addictive disorders. *Clinical Neuropharmacology, 32*(5), 269–276.

Salman, S., Idrees, M., Anees, M., Idrees, J., Idrees, F., & Badshah, S. (2014). Association of attention deficit hyperactivity disorder with heroin addiction. *Bangladesh Journal of Medical Science, 13*(2), 128–134.

Salman, S., Ismail, M., Anees, M., Badshah, S., Nazar, Z., & Awan, R. N. (2012). Substance abuse in patients with comorbid ADHD. *Journal of Pakistan Psychiatric Society, 9*(2), 91–96.

Schubiner, H., Downey, K. K., Arfken, C. L., Johanson, C. E., Schuster, C. R., Lockhart, N., & Pihlgren, E. (2002). Double-blind placebo-controlled trial of methylphenidate in the treatment of adult ADHD patients with comorbid cocaine dependence. *Experimental and Clinical Psychopharmacology, 10*(3), 286–292.

Silva, D., Colvin, L., Hagemann, E., & Carol Bower, C. (2014). Environmental risk factors by gender associated with attention-deficit/hyperactivity disorder. *Pediatrics, 133*(1), 1–9.

Subramaniam, G. A., & Stitzer, M. A. (2009). Clinical characteristics of treatment-seeking prescription opioid vs. heroin-using adolescents with opioid use disorder. *Drug and Alcohol Dependence, 101*(2), 13–19.

Tarter, R. E. (2002). Etiology of adolescent substance abuse: a developmental perspective. *The American Journal on Addictions, 11*(3), 171–191.

Thurstone, C., Riggs, P. D., Salomonsen-Sautel, S., & Mikulich-Gilbertson, S. K. (2010). Randomized, controlled trial of atomoxetine for attention-deficit/hyperactivity disorder in adolescents with substance use disorder. *Journal of the American Academy of Child & Adolescent Psychiatry, 49*(6), 573–582.

Tsuang, M. T., Lyons, M. J., Meyer, J. M., Doyle, T., Eisen, S. A., Goldberg, J., & Eaves, L. (1998). Co-occurrence of abuse of different drugs in men: the role of drug-specific and shared vulnerabilities. *Archives of General Psychiatry, 55*(11), 967–972.

Vidal, S. I., Vandeleur, C., Rothen, S., Gholam-Rezaee, M., Castelao, E., Halfon, O., … Preisig, M. (2012). Risk of mental disorders in children of parents with alcohol or heroin dependence: a controlled high-risk study. *European Addiction Research, 18*(5), 253–264.

Wilens, T. E. (2004). Attention-deficit/hyperactivity disorder and the substance use disorders: the nature of the relationship, subtypes at risk, and treatment issues. *Psychiatric Clinics of North America, 27*(2), 283–301.

Wilens, T. E., Adamson, J., Monuteaux, M. C., Faraone, S. V., Schillinger, M., Westerberg, D., & Biederman, J. (2008). Effect of prior stimulant treatment for attention-deficit/hyperactivity disorder on subsequent risk for cigarette smoking and alcohol and drug use disorders in adolescents. *Archives of Pediatrics & Adolescent Medicine, 162*(10), 916–921.

Wilens, T. E., Faraone, S. V., Biederman, J., & Gunawardene, S. (2003). Does stimulant therapy of attention-deficit/hyperactivity disorder beget later substance abuse? A meta-analytic review of the literature. *Pediatrics, 111*(1), 179–185.

Wilens, T. E., Martelon, M., Joshi, G., Bateman, C., Fried, R., Petty, C., & Biederman, J. (2011). Does ADHD predict substance-use disorders? A 10-year follow-up study of young adults with ADHD. *American Academy of Child and Adolescent Psychiatry, 50*(6), 543–553.

Young, S. E., Stallings, M. C., Corley, R. P., Krauter, K. S., & Hewitt, J. K. (2000). Genetic and environmental influences on behavioral disinhibition. *American Journal of Medical Genetics, 96*(5), 684–695.

Yuferov, V., Levran, O., Proudnikov, D., Nielsen, D. A., & Kreek, M. J. (2010). Search for genetic markers and functional variants involved in the development of opiate and cocaine addiction and treatment. *Annals of the New York Academy of Sciences, 1187*(1), 184–207.

Zou, Y., Liao, G., Liu, Y., Wang, Y., Yang, Z., Lin, Y., & Wang, Z. (2008). Association of the 54-nucleotide repeat polymorphism of HPER3 with heroin dependence in Han Chinese population. *Genes, Brain and Behavior, 7*(1), 26–30.

van der Zwaluw, C. S., van den Wildenberg, E., Wiers, R. W., Franke, B., Buitelaar, J., Scholte, R. H., & Engels, R. C. (2007). Polymorphisms in the mu-opioid receptor gene (OPRM1) and the implications for alcohol dependence in humans. *Pharmacogenomics, 8*(10), 1427–1436.

Chapter 97

Impaired Cognition Control and Inferior Frontal Cortex Modulation in Heroin Addiction

André Schmidt, Marc Walter, Stefan Borgwardt
Department of Psychiatry (UPK), University of Basel, Basel, Switzerland

Abbreviations

ACC Anterior cingulate cortex
DA Dopamine
DAM Diacetylmorphine
DTI Diffusion tensor imaging
FA Fractional anisotropy
IGT Iowa gambling task
MMT Methadone maintenance treatment
MRI Magnetic resonance imaging
OFC Orbitofrontal cortex
rIFG Inferior frontal gyrus
SSRT Stop signal reaction time

INTRODUCTION

Successful cognitive control is essential to process and integrate incoming perceptual information in a flexible fashion (Aron, Robbins, & Poldrack, 2014). Such a trial-by-trial adjustment of information processing is important for adapting ongoing behavior according to prevailing conditions in a highly dynamic environment. One of the key processes underlying goal-directed behavior is the ability to adequately suppress already initiated response tendencies. Inhibition is considered a high-order (executive) system responsible for the optimization and regulation of lower-order functions and is particularly required when actions are no longer relevant and have to be stopped (Aron et al. 2014). It thus enables the suppression of previously learned statistical regularities and the switch of attention allocation to unexpected, relevant events in the environment. The ability to inhibit and update initiated motor activities is essential for common daily activities such as crossing the road, waiting in line, or playing sports.

Impaired cognitive self-control is a major hallmark of drug addiction. Failures of cognitive self-regulation may contribute to an increased risk of exposure to the drug, increased vulnerability to the transition to addiction, and increased relapse (Everitt, Dickinson, & Robbins, 2001). The loss of inhibitory control functioning in these subjects is suggested to develop from strong stimulus–response habits, thereby increasing the motivation and desire for renewed drug intake (Robinson & Berridge, 2003). Motivational values need to be added to these learned associations as strong stimulus–response habits such as putting on clothes or mowing the lawn must not necessarily contribute to addictive behaviors (Robinson & Berridge, 2003). In this way drug-related cues acquire incentive-motivational value and evoke expectations of drug availability and memories of past drug-induced rewarding effects. The compulsive drug-seeking behavior induced by chronic self-administration is continuously strengthened by the acute drug-induced positive rewarding effect and occurs at the expense of other sources of reinforcement (Everitt et al., 2001). The result is an overvaluation of drug-related reinforcers and an undervaluation of alternative reinforcers. This imbalance is most pronounced during states of craving and acute withdrawal, in which the drug-driven behavior of addicted people claims a high amount of neural resources, while at the same time little capacity remains for the processing of non-drug-related events (Goldstein & Volkow, 2011).

Hence understanding the pathophysiological mechanisms underlying deficient inhibitory control in drug addiction implies the study of drug-induced sensitization that attributes incentive salience to reward-associated stimuli. The motivational value of drug consumption is mediated, at least in part, by the activation of midbrain dopamine (DA) areas (the ventral tegmental area and substantia nigra) and the ventral striatum to which they project, which are known to be involved in reward, conditioning, and habit formation. In particular, experimental evidence in animals has revealed that the subjective rewarding effect of drug intake is triggered at least in part by DA release into striatal brain regions. This has been supported in humans; the subjects who had the greatest striatal DA increases after acute drug administration were the ones who experienced the rewarding effects most intensely (for a review see Koob & Volkow, 2010). Furthermore, functional magnetic resonance imaging (MRI) studies have also demonstrated activation in the orbitofrontal cortex (OFC) when addicts, in this case cocaine-dependent subjects, are exposed to drug-related stimuli that elicit craving (Volkow et al., 2010). Accordingly, it has been shown that when heroin-dependent males were receiving heroin injections while viewing drug-related videos, cerebral blood flow in the OFC correlated with the urge to use the drug (Sell et al., 2000).

Such a relation between OFC activation and drug craving is evident across various substance abuse disorders (Jasinska, Stein, Kaiser, Naumer, & Yalachkov, 2014), indicating the general OFC involvement in drug wanting. Critically, when cocaine abusers were instructed to inhibit craving, metabolism in the right OFC decreased, whereas at the same time activation of the right inferior frontal gyrus (rIFG) increased (Volkow et al., 2010). The rIFG is key region for cognitive control, in particular for the inhibition of prepotent motor responses, as demonstrated by numerous functional MRI studies (Aron et al., 2014). It has been shown that lesions of the rIFG were predictive of inhibitory deficits during the stop-signal task in patients with brain damage and that transcranial magnetic stimulation to the rIFG transiently impaired the ability to stop an initiated action (Aron et al., 2014). Moreover, structural individual differences in white matter connection density of the rIFG covaries with individual differences in the proficiency of functional rIFG activation during response inhibition (Forstmann et al., 2008). These studies together indicate a distinction of prefrontal cortex regions that are more involved in the mediation of drug-driven emotional cognition, such as the OFC on the one side and regions that facilitate non-drug-related cognitive efforts such as the IFG on the other (Goldstein & Volkow, 2011). Impaired cognitive self-control in drug addiction can thus be understood as a failure of specific prefrontal brain regions, and in particular of the rIFG, to down-regulate hypersensitive striatal and specific prefrontal regions. Such abnormal frontostriatal mechanisms may provide promising targets to tailor new treatments for the regulation of inhibitory control in drug-dependent subjects as previously suggested (Feil et al., 2010).

The purpose of this chapter is to present a broad overview of how the cognitive control system is affected in heroin-dependent individuals, including chronic and abstinent users, as well as patients actively enrolled in a maintenance treatment. The chapter is structured in three sections: the first part indicates that heroin-dependent subjects, depending on their abstinence or maintenance duration, exhibit various degrees of behavioral deficits across a wide spectrum of cognitive control functions. In the second and main part of this chapter, we provide experimental evidence for a relation between impaired motor response inhibition in heroin addiction and structural and functional rIFG abnormalities. These findings are discussed in relation to other addictions and substance misuses by pointing out major consistencies and divergences among them. In the last section, we conclude by summarizing the reported findings and indicating current limitations. We also suggest potential developments for further studies addressing the neural mechanisms of impaired cognitive control functioning in heroin addiction. These kinds of research are intended not only to better understand pathophysiological mechanisms of impaired cognitive control but also to provide a platform for the development of novel behavioral and pharmacological therapies for the treatment of cognitive control deficits.

BEHAVIORAL IMPAIRMENTS OF COGNITIVE CONTROL IN HEROIN ADDICTION

Heroin-dependent individuals exhibit behavioral impairments across a wide range of cognitive control functions including deficits in attention, memory, and learning processes, as well as inhibitory control (Ersche & Sahakian, 2007). However, the considerable heterogeneity of the samples studied also provides controversial findings. In the following, we unveil in which specific cognitive control functions chronic heroin users reveal behavioral deficits and how they develop over the duration of abstinence and maintenance treatment.

Cognitive Control in Chronic Heroin Users

Using a huge battery of cognitive tests, it has been found that heroin-dependent individuals selected significantly more risky options during the Cambridge decision-making task and showed impairments in working memory processing as assessed with the delayed matching to sample task. The same study demonstrated that these subjects also revealed deficits in pattern recognition during the paired associates learning task, in problem-solving capacity during the stockings of Cambridge test, and in cognitive flexibility and conflict monitoring during the Stroop interference task compared with healthy controls (Fishbein et al., 2007). Furthermore, heroin-dependent persons exhibited impairments in learning the intradimensional shift component during the ID/ED task and have impairments in visual working memory, pattern recognition memory, and strategy learning (Ornstein et al., 2000). Heroin addicts also revealed higher discount rates for delayed rewards, an effect that positively correlated with the number of years of heroin using (Cheng, Lu, Han, González-Vallejo, & Sui, 2012).

Cognitive Control during Abstinence and Maintenance

Notably, most of these impairments in chronic heroin users persist even after several months of abstinence, while others may remit after protracted abstinence. In particular, polysubstance abusers of heroin with an abstinence of 25 weeks still showed deficits in shifting attention, spatial working memory, and pattern recognition memory. They also revealed increased task-switch costs during the go/no-go task as expressed by an increased number of omission errors (Verdejo-García, Perales, & Pérez-García, 2007). Further, heroin-addicted subjects with a dependency of at least 12 months and a current abstinence of 3–18 months demonstrated significantly lower impulse control than normal controls as measured with the Porteus Maze Test (Lee & Pau, 2002). However, although the impairments in impulse control during the Porteus Maze Test appear to persist even after an abstinence of 13 months, no deficits in attention or cognitive flexibility have been observed after 13 months abstinence (Pau, Lee, & Chan, 2002), suggesting that certain cognitive functions develop differentially with ongoing abstinence. Deficits in cognitive control functioning have also been observed in patients enrolled in methadone maintenance therapy (MMT). Specifically, MMT patients showed impairments relative to healthy controls in psychomotor speed, working memory, decision-making, and possibly also inhibitory mechanisms (Pirastu et al., 2006). Both methadone-maintained (>1 year) and abstinent subject groups (>6 months) performed worse than healthy controls on tasks that measured verbal function, visuospatial analysis and memory, and resistance to distractibility (Prosser et al., 2006).

Cognitive Control in Abstinent Compared to Maintained Heroin-Dependent Users

The behavioral impairments in cognitive control seem to vary depending on the duration of heroin consumption, abstinence, or maintenance treatment. By comparing currently abstinent former opioid abusers with opioid abusers enrolled in MMT, it was reported that the performance of the abstinent abusers fell between that of the maintained subjects and the controls on many cognitive measures, including psychomotor/cognitive speed, working memory, and decision-making (Mintzer, Copersino, & Stitzer, 2005). The authors suggested that methadone treatment may be associated with deficits in addition to those observed in long-term abuse and that recovery of functioning may occur during abstinence. This finding is consistent with evidence that MMT patients scored significantly less well than heroin users with an abstinent period of 6 weeks to 12 months on a word fluency task (Davis, Liddiard, & McMillan, 2002). The notion that some recovery of cognitive functioning may occur during abstinence is further supported by evidence showing that the performance of a maintenance group (both methadone and buprenorphine) was poorer than that of the abstinent group (with a period of 3–6 months) in planning, information processing speed, and verbal memory, while the performance of the abstinent group did not significantly differ from that of controls in any test (Darke, McDonald, Kaye, & Torok, 2012). In accordance with this, formerly opiate-dependent subjects in protracted abstinence for more than 6 months scored better than MMT subjects on sustained attention (Prosser, London, & Galynker, 2009). However, other investigations have reported controversial results. For example, subjects that were abstinent for circa 1 year performed worse than their MMT counterparts on tests measuring visual memory and construct formation (Prosser et al., 2006). Along this line, stop-signal reaction times (SSRTs) were significantly prolonged in abstinent opioid-dependent subjects compared to MMT subjects (Liao et al., 2014). Moreover, dependent participants with an abstinence of approximately 6 months showed more severe impairment in cognitive control, as indexed by longer SSRTs, than subjects with relatively short abstinence duration of up to 2 months (Liao et al., 2014). While this study revealed a negative association between duration of abstinence and cognitive processing, studies in MMT patients showed an opposite relation. In particular, although participants who had been enrolled in an MMT program for less than 2 months showed impaired response inhibition relative to healthy control subjects, after maintenance of therapy for more than 12 months, performance during response inhibition appeared to return toward control levels (Bracken et al., 2012). Another study has also demonstrated that 2 months of MMT significantly improved measures of verbal learning and memory, visuospatial memory, and psychomotor speed and reduced frequency of drug use relative to baseline (Gruber et al., 2006).

Taken together, these results show that heroin-dependent individuals reveal a wide range of cognitive control impairments, comprising deficits in cognitive flexibility, inhibitory control, sustained attention, decision-making, and memory and learning processes (Ersche & Sahakian, 2007). Although several lines of research point toward a recovery of these functions after abstinence by indicating intermediate performance values in these subjects compared with MMT patients and healthy control subjects, other evidence challenges this notion. This discrepancy can probably be explained either by different durations of heroin dependency, abstinence, or maintenance treatments across patients or by the fundamentally different cognitive paradigms applied in these studies. It is thus important for future research to systematically follow up how a specific cognitive control function (assessed with well-established and simple paradigms) develops across heroin-dependent subjects as a function of the duration of dependence, abstinence, and maintenance treatment.

INFERIOR FRONTAL GYRUS, RESPONSE INHIBITION, AND HEROIN ADDICTION

Inhibition and the Right Inferior Frontal Gyrus

Evidence from cognitive and clinical neuroscience emphasizes the pivotal role of the rIFG in response inhibition (Aron et al., 2014). In particular, using high-spatiotemporal-resolution electrocorticography to elucidate the functional role of the rIFG in subjects undergoing presurgical monitoring, rIFG responses were detected approximately 100–250 ms after the stop signal, a time frame consistent with a putative inhibitory control process (Swann et al., 2009). Numerous functional MRI studies have also demonstrated the essential contribution of the rIFG in stopping initiated responses (Aron et al., 2014) (see Figure 1(A)). Furthermore, temporary disruption of rIFG activity with transcranial magnetic stimulation selectively impaired the ability to stop an initiated action (Chambers et al., 2006). Dysfunctional IFG activity may arise from structural abnormalities in this region, given that lesions of the rIFG impaired inhibitory performance during the stop-signal task in patients (Aron, Fletcher, Bullmore, Sahakian, & Robbins, 2003) (see Figure 1(B)). In line with this, performance levels on inhibitory control tasks were correlated with gray matter volume in the IFG in healthy subjects (Tabibnia et al., 2011). Moreover, it has even been demonstrated that the function and structure of the rIFG predicted individual differences in response inhibition, reflecting a relation between structure, function, and behavior (Forstmann et al., 2008).

Structural Right Inferior Frontal Gyrus Imaging Findings in Heroin Addiction

Using structural MRI, reduced gray matter volume in the right prefrontal cortex was found in heroin-dependent subjects (Wang et al., 2012). Critically, the duration of heroin use correlated negatively with the density of gray matter volume in the rIFG (Brodmann area 47) of young lifetime heroin-dependent individuals (Yuan et al., 2009) (see Figure 2(A)). A multimodal imaging study in heroin-dependent patients showed that perfusion of the rIFG positively correlates with gray matter volume in the rIFG (Denier et al., 2013) (see Figure 2(B)).

Moreover, current heroin-dependent subjects also show reduced right frontal white matter integrity as assessed with diffusion tensor imaging (DTI) (Li et al., 2013), and an [^{18}F]fluorodeoxyglucose positron emission tomography study revealed that the severity of heroin consumption in substance-dependent individuals is negatively correlated with brain metabolism in the rIFG

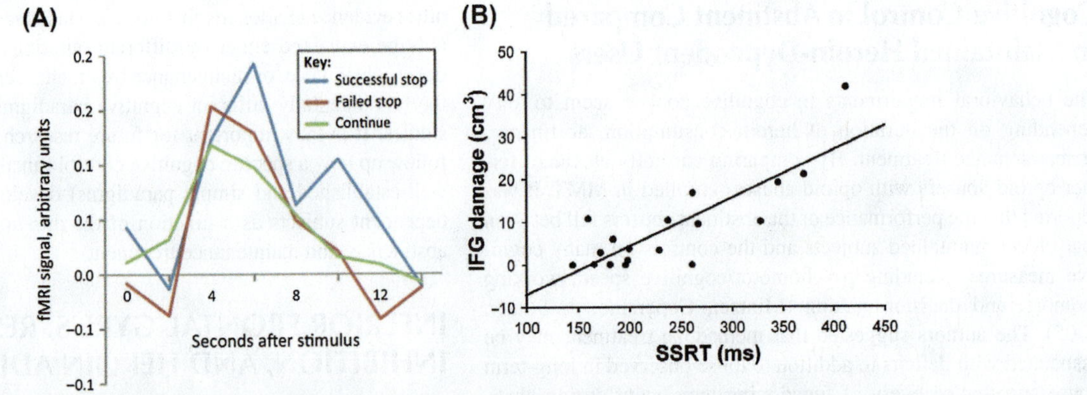

FIGURE 1 The contribution of the right inferior frontal gyrus to stopping initiated motor responses. (A) Neural activity in the rIFG as measured with functional MRI during successful and unsuccessful stop signals and also for continued trials that did not require stopping the signal. *(Reprinted from Aron et al. (2014), with permission from Elsevier.)* (B) Damage to the rIFG is associated with longer SSRTs. *(Reprinted by permission from Macmillan Publishers Ltd.: Nature Neuroscience (Aron et al. 2003), copyright (2003).)*

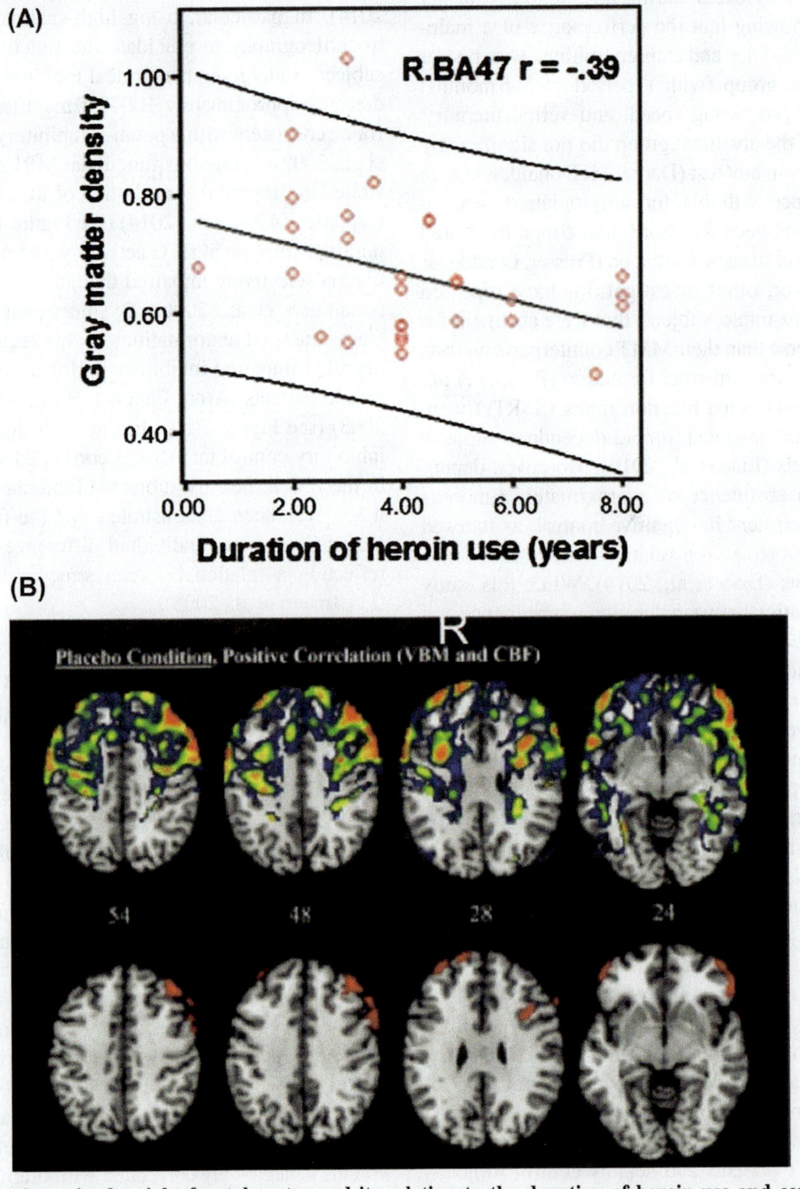

FIGURE 2 Gray matter volume in the right frontal cortex and its relation to the duration of heroin use and cerebral blood flow. (A) The correlation between gray matter density in the rIFG (BA47) and the duration of heroin use in dependent patients (r = −0.39, p < 0.05). *(Reprinted from Yuan et al. (2009), with permission from Elsevier.)* (B) Positive correlation between gray matter volume and cerebral perfusion in the rIFG in heroin-dependent patients. *(Reprinted from Denier et al. (2013).)*

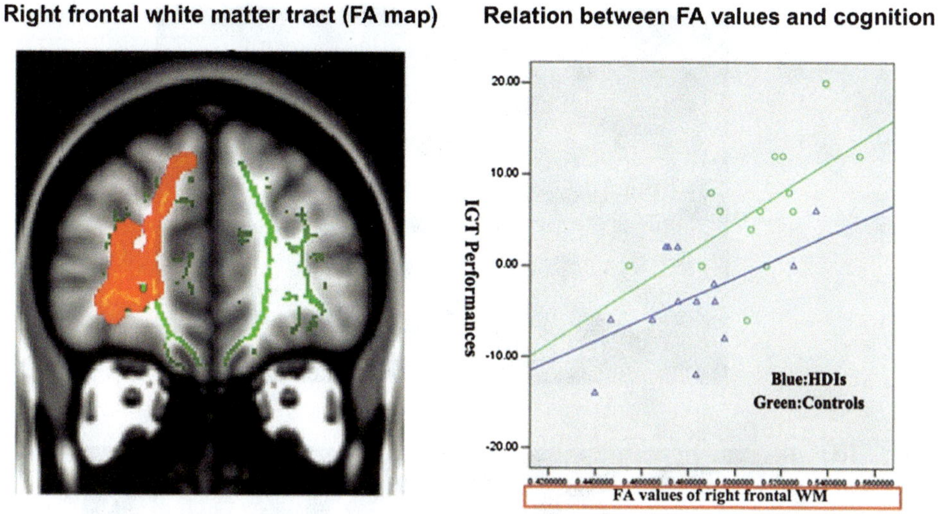

FIGURE 3 White matter volume in the right frontal cortex and its relation to cognitive performance. The positive relation between fractional anisotropy (FA) values in the right frontal white matter (depicted in the left image) and cognitive control performance (decision-making) during the Iowa gambling task in chronic heroin-dependent and healthy individuals is shown. *Reprinted from Qiu et al. (2013).*

(Moreno-López, Catena, et al., 2012; Moreno-López, Stamatakis, et al., 2012). Intriguingly, fractional anisotropy values (measured with DTI) of the right frontal cortex in chronic heroin-addicted patients correlate negatively with the duration of heroin dependency and with decision-making performance during the Iowa gambling task (Qiu et al., 2013) (Figure 3).

Significant reductions in gray matter volume of the rIFG, as well as reduced white matter integrity in the right frontal cortex, have also been reported in heroin-dependent individuals on MMT (Lyoo et al., 2006). The reductions in right frontal white matter thereby correlated positively with the duration of dependency (Bora et al., 2012). Furthermore, the accumulated consumption of methadone in MMT patients positively correlated with the degree of injury of white matter integrity in the right frontal cortex. This study further found no changes in white matter integrity after prolonged abstinence, suggesting that a normalization of white matter injury may occur during abstinence (Wang et al., 2011). These studies together demonstrate evidence for intra- and interregional structural abnormalities of the rIFG across various stages of heroin addiction and that these changes may be related to the duration of heroin use and methadone dosage. Such gray and white matter rIFG deficits might provide a pathophysiological substrate underlying functional and behavioral abnormalities during inhibitory control functioning in heroin dependence.

Functional Right Inferior Frontal Gyrus Imaging Findings in Heroin Addiction

Using resting-state MRI, a previous study found reduced functional connectivity of the right prefrontal cortex to the parietal cortex. Both functional and structural changes were negatively correlated with the duration of heroin use (Yuan et al., 2010) (see Figure 4(A)). These findings suggest that structural abnormalities in the right prefrontal may indeed mediate a functional deficit in heroin addiction, albeit the causal relationship between gray matter density and functional connectivity is unknown and needs to be disentangled in future investigations. Another resting-state MRI investigation found decreased functional connectivity between the bilateral IFG and the dorsal anterior cingulate cortex (ACC) in heroin-dependent subjects (Wang et al., 2013), perhaps reflecting a neurophysiological connectivity marker for deficits in cognitive control functioning. This is important evidence as the ACC exhibits increased connectivity to the rIFG during inhibitory processes (Bari & Robbins, 2013), and dysfunctional connectivity between the lateral prefrontal cortex and the ACC has been proposed to render patients with drug dependence susceptible to compulsive drug seeking (Li & Sinha, 2008). Consistent with this, previous studies showed that opiate addicts had an attenuated ACC activity in response to errors compared with a healthy control group (Forman et al., 2004).

Functional MRI studies have also begun to address how impaired response inhibition in heroin-dependent individuals is reflected in brain activity. It was shown that heroin-dependent subjects with an abstinence of at least 4 weeks had reduced activity in the IFG and ACC during the go/no-go task (Fu et al., 2008). Reduced ACC activity has also been found in newly admitted heroin-addicted patients while performing a modified version of the spatial-congruence task (Lee et al., 2005). Comparing functional brain activity in response to heroin-related cues in MMT and abstinent heroin-dependent patients, rIFG activity was significantly higher in the abstinent group than in the MMT group, whose activation was not significantly different compared with normal subjects (Tabatabaei-Jafari et al., 2014). Given that both of these therapeutic methods successfully reduced heroin craving, the authors concluded that the increased IFG activation in abstinent subjects might reflect a compensatory mechanism to control their undesirable craving behavior, while patients on MMT may not need to develop such an effortful mechanism, as this regulation may be replaced by the neurochemical effect of the powerful opioid agonist methadone (Tabatabaei-Jafari et al., 2014). This finding fits with a previous report on cocaine addiction, which showed that when cocaine abusers were instructed to inhibit craving, activity in the rIFG increased (Volkow et al., 2010). Furthermore, relative to a placebo treatment, it has been shown that acute administration of pharmaceutical heroin (diacetylmorphine; DAM) to

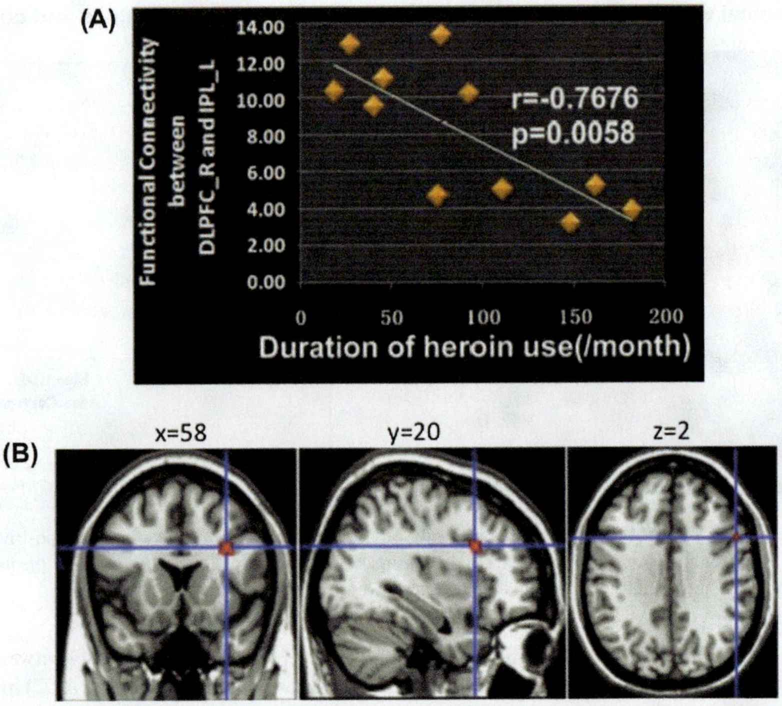

FIGURE 4 **Functional right frontal cortex abnormalities in heroin addiction during resting state and cognitive control.** (A) Negative relation between resting-state functional connectivity of the right dorsolateral prefrontal cortex (DLPFC) to the inferior parietal lobule (IPL) and the duration of heroin use in heroin-dependent patients. *(Reprinted from Yuan et al. (2010), with permission from Elsevier.)* (B) Significant acute heroin-induced reduction in rIFG activity during the go/no-go task (no-go versus go contrast; response inhibition) in heroin-dependent patients. *(Reprinted by permission from Macmillan Publishers Ltd.: Neuropsychopharmacology Schmidt et al. (2013), copyright (2013).)*

heroin-dependent patients reduced rIFG activity during response inhibition as operationalized by a go/no-go task (Schmidt et al., 2013) (see Figure 4(B)). Moreover, a modeling study extended this result by showing that a single DAM maintenance dose reduced not only ACC and rIFG activity during response inhibition during the go/no-go task, but also the inhibition-induced modulation of connectivity from the dorsal ACC to the rIFG (Schmidt et al., 2014). These findings indicate that unraveling the contribution of the rIFG to other brain regions, as for example, to the ACC, may provide valuable insights into pathophysiological mechanisms of impaired cognitive control in heroin addiction. This study further suggested that acute DAM administration reduced ACC activity in response to errors (Schmidt et al., 2014), consistent with other studies on heroin addiction (Forman et al., 2004). However, in addition to its connectivity to the ACC, the IFG is also functionally connected with the subthalamic nucleus during response inhibition (Aron et al., 2014). Indeed, it has been suggested that impaired sustained attention in MMT patients was mediated by a relative reduction in regional cerebral glucose metabolism in the thalamus, perhaps indicating that impaired cognitive control in heroin addiction might result not only from abnormal activity in the IFG and thalamus (Prosser et al., 2009), but also from disrupted thalamocortical connectivity. These findings further underpin the importance of considering the rIFG within a network of regions that underlie response inhibition and not as an isolated and independent mediator of response inhibition. It is thus crucial to study cognitive control functions and how they are affected in heroin addiction from a network connectivity perspective.

APPLICATIONS TO OTHER ADDICTIONS AND SUBSTANCE MISUSE

In this section, we relate these findings from heroin addiction to other drug dependencies. However, this is a very wide and complex field and it is beyond the scope of this chapter to adequately compare them in a comprehensive manner. Given that several reviews on inhibitory control functions already exist for cocaine (Spronk, van Wel, Ramaekers, & Verkes, 2013), alcohol (Bernardin, Maheut-Bosser, & Paille, 2014), amphetamine (Ersche & Sahakian, 2007), and cannabis dependence (Wrege et al., 2014), as well as across diverse substances (Ersche & Sahakian, 2007; Feil et al., 2010), here we provide a brief overview by highlighting major similarities and differences across different drugs (mainly between heroin and cocaine addiction) with respect to cognitive control impairments and how they are related to abnormalities of the rIFG.

In accordance with the deficits in heroin addiction (Ersche & Sahakian, 2007), impairments in sustained attention, response inhibition, memory and learning, decision-making, and psychomotor performance have also reported in long-term cocaine (Spronk et al., 2013), amphetamine (Ersche & Sahakian, 2007), cannabis (Wrege et al., 2014), and alcohol users (Bernardin et al., 2014), although there are also distinct differences in the long-term effects across these drugs in specific cognitive functions (Fernández-Serrano, Pérez-García, & Verdejo-García, 2011). While these negative effects on cognitive functions seem to be reversible after protracted abstinence in heavy alcohol drinkers (Bernardin et al., 2014) and cannabis smokers (Fernández-Serrano et al., 2011), they appear to be more persistent in chronic cocaine and heroin users even after years of abstinence (Feil et al., 2010).

Consistent with findings in heroin-dependent patients (Fu et al., 2008), reduced rIFG activity during response inhibition has been observed in chronic cocaine users (Hester & Garavan, 2004). As suggested in heroin addiction (Bora et al., 2012), reduced white matter integrity of the rIFG in stimulant-dependent individuals was significantly associated with poorer inhibitory control (Ersche et al., 2012). In patients with methamphetamine dependence, those with more drug craving had longer SSRTs and reduced gray matter intensity in the rIFG (Tabibnia et al., 2011). Cocaine addiction is also associated with reduced gray matter volume in the rIFG (Moreno-López, Catena, et al., 2012; Moreno-López, Stamatakis, et al., 2012), consistent with that observed in chronic heroin-dependent subjects (Yuan et al., 2009) and patients on MMT (Lyoo et al., 2006).

Although long-term consumption generally impairs cognitive control across different substances, acute effects diverge more between drugs. For example, while acute methadone infusion reduced cognitive control in chronic users (Bracken et al., 2012), acute cocaine intake improved task performance during motor response inhibition as expressed by the percentage of successful inhibitions for all no-go trials. Notably, this improvement was associated with concomitant increased activation in the rIFG (Garavan, Kaufman, & Hester, 2008). These findings contrast with evidence showing that an acute dose of DAM to heroin-dependent patients reduced rIFG activity during response inhibition (Schmidt et al., 2013). However, the relation between acute cocaine intake and inhibitory control seems to depend on the administered dose. While inhibitory performance on a go/no-go task decreased following oral cocaine administration in the 50- to 150-mg dose range, it improved following administration in the 100- to 300-mg dose range (Fillmore, Rush, & Hays, 2005). In accordance with the acute cocaine effect (Garavan et al., 2008), increased rIFG activity during response inhibition has also been observed in successful cocaine abstinence (>69 weeks). Given that their behavioral task performances did not differ from those of nonusers, the increased rIFG activation has been interpreted as a compensatory mechanism to adequately perform the task (Connolly, Foxe, Nierenberg, Shpaner, & Garavan, 2012). It would be very interesting to see whether this putative compensatory increase in rIFG activity may normalize with further abstinence. Nevertheless, these studies together indicate the crucial influence of treatment and abstinence durations on the rIFG activity during response inhibition.

Taken together, although the long-term consumption of various substances contributes to consistent impairments in cognitive control functions, which are often accompanied by reduced rIFG activation, acute effects diverge on the behavioral and neural levels across drug classes. The findings that chronic substance consumption consistently reduces rIFG activity during inhibitory control may point toward a common final pathway, while differences in their acute effects might reflect their distinct pharmacological modes of action. However, comparisons among drugs are severely hampered by major methodological differences among studies such as the application of fundamentally different cognitive paradigms for assessing the same function or the inclusion of totally different study samples with respect to the duration of substance consumption, abstinence or maintenance treatment, and the administered drug dose. Comparing subjects at different durations of addiction, abstinence, or maintenance treatment and even during different stages of addiction (intoxication, withdrawal, and craving) is delicate, as their neurochemistry is differentially affected (Koob & Volkow, 2010). These limitations result in considerable heterogeneity among the findings. Hence further reviews and meta-analyses are needed that compare studies with consistent methods in large samples and systematically address the influence of specific drug-related parameters (e.g., duration of dependence, drug doses) and therapeutic interventions (e.g., duration of abstinence and maintenance treatments).

CONCLUSIONS

To summarize, heroin addiction is characterized by behavioral impairments in a wide range of cognitive control functions, including inhibitory control, sustained attention, decision-making, and memory or learning processes. Neuroimaging studies indicate that these deficits may arise from abnormalities in the rIFG. Both behavioral and neural findings seem to depend on the duration of the dependence and the therapeutic scenario (abstinence versus maintenance treatments). While evidence on the long-term administration of heroin has consistently reported impaired cognitive functioning, how these impairments develop over the course of abstinence and maintenance treatment is matter of ongoing debate. Further research is needed to study how cognitive impairments evolve during abstinence and maintenance programs and how this development is related to the previous history of the dependence (i.e., duration, dose, consumption of other substances, etc.). Nevertheless, the reported findings suggest that chronic heroin consumption may contribute more to a general deficit in cognitive control rather than to a specific cognitive deficit. In the following, we discuss existing limitations of previous research and provide potential developments and challenges in this field.

SUGGESTIONS FOR FURTHER RESEARCH

It is proposed that the heroin-induced reduction in rIFG activation during the go/no-go task does not specifically reflect an impairment in response inhibition but rather a deficit in the allocation of attention to infrequently presented stimuli (Schmidt et al., 2013). This interpretation is supported by the fact that the rIFG is activated not only during response inhibition in the go/no-go task but also during stimulus-driven attention allocation depending on the activated subregion of the IFG. For a detailed discussion about the controversy concerning IFG and cognitive control see also Aron et al. (2014) and Bari and Robbins (2013). Future research on heroin addiction should therefore assess rIFG abnormalities with more anatomical precision and by using less complex and uniform paradigms to specify whether heroin addiction is associated with a general impairment in cognitive functioning or with a specific deficit in response inhibition or attention allocation. This distinction is of clinical relevance with respect to the application of specific cognitive training interventions.

Furthermore, the IFG is functionally connected with several cortical and subcortical structures during inhibitory motor control, including the ACC, the thalamus, and the striatum (Bari & Robbins, 2013), building a network comprising striatal and frontal regions. Abnormal frontostriatal mechanisms have been proposed to underlie impaired inhibitory control in drug addiction (Feil et al., 2010). For example, a 2013 study demonstrates that subjects with more severe alcohol dependence exhibit less frontal connectivity with the striatum during response inhibition, suggesting that aberrant frontostriatal connectivity mediates impaired response inhibition in alcoholism (Courtney, Ghahremani, & Ray, 2013). To further elucidate how IFG abnormalities affect cognitive control in heroin addiction, future studies should consider the IFG as a node within a network that navigates cognitive control and not as an isolated brain region that acts solely during inhibition.

Finally, to better tailor treatments for cognitive control impairments in heroin addiction, more studies are needed to address the underlying pharmacology of these impairments. Previous evidence has often proposed that the cognitive loss of inhibitory control in drug addiction derives from abnormal DA signaling within circuits involved in reward learning, conditioning, and habit formation (Everitt et al., 2001). Although the DA system critically influences response inhibition, more recent investigations have pointed out a complex interplay of dopaminergic, noradrenergic, and serotonergic neurotransmission in the inhibition of already initiated responses (Bari & Robbins, 2013). A challenge for the future will be to disentangle how these different neurotransmitter systems modulate frontostriatal connections in the healthy brain and how these relations are modified in heroin addiction.

In summary, a lot of challenges remain to better understand cognitive impairments in heroin addiction and to strengthen the clinical perception of cognitive impairments. First, longitudinal study designs are necessary to assess cognitive control impairments over the duration of dependence, abstinence, and maintenance treatment (including dosing). To clarify whether heroin induces a general or specific cognitive impairment, simple and unified experimental paradigms should be conducted that focus on a specific function and do not require the involvement of several different control functions simultaneously. This might facilitate the improvement of specific neurocognitive training and remediation strategies. Second, these behavioral effects should also be pursued with neuroimaging measurements using simple and established tasks to capture a priori defined cognitive function. Using high-resolution MRI techniques, the contribution of rIFG should be studied with more anatomical precision on one hand and from a cognitive networks perspective on the other. Finally, pharmacological challenges are needed to assess the relation between chemistry, frontostriatal connectivity, and behavioral performance. Neuroimaging of frontostriatal connectivity together with pharmacological challenges and neuropsychological assessments may serve to assess the efficacy of novel pharmacological treatments.

DEFINITION OF TERMS

Opioid maintenance treatment Opioid maintenance treatments are specific therapeutic programs in which drug-dependent patients are maintained with different opioids (in a clinical setting). MMT has been established as the gold standard of care. However, buprenorphine maintenance therapy (BMT) has also achieved promising results. Furthermore, for heroin addicts who have failed to benefit from MMT or BMT, the prescription of heroin has been found to be an effective treatment for severe heroin addiction.

Diacetylmorphine This is pharmaceutical heroin used for the maintenance of heroin-dependent patients.

Cognitive control functions This is an umbrella term for processes that are important to quickly and flexibly adapt behavior to changing situations. These higher-order processes exert top-down control over lower-order functions to regulate behavior. Cognitive control refers to numerous functions including attention and memory processes implicated, for example, in task switching or learning and reversing of stimulus–response associations. Inhibitory control is one these processes and can be separated in behavioral and cognitive inhibition.

Cognitive inhibition This is the psychological process that interrupts memories, thoughts, perceptions, or emotions. It is an important function to crack irrelevant associations and to newly allocate attention to relevant events. It is difficult to study owing to the absence of overt behavioral measures.

Behavioral/motor inhibition These processes refer to decision-making or delay discounting, but also to the inhibition of prepotent motor responses (i.e., motor response inhibition). How relevant behavioral inhibition is for cognitive inhibition is a matter of ongoing research. Experimental paradigms often assess behavioral inhibition. Two paradigms that are often used to study motor/behavioral response inhibition are the stop-signal task and the go/no-go task.

Go/no-go task The conventional "go" task is a choice reaction-time paradigm, in which arrows point either to the left or to the right side. During the frequently presented go trials, subjects are instructed to press a left or right response button according to the direction of the arrow. During the infrequently presented "no-go" trials, in which arrows point upward, participants have to inhibit their motor responses. Response inhibition is then calculated as the probability of executing a response on a no-go trial.

Stop-signal task During this task, subjects perform a choice reaction task on "go" trials (no-stop-signal), consistent with the go/no-go task. However, in the stop-signal task, the "stop" signal is randomly presented after a variable delay (stop-signal delay), which instructs subjects to withhold the response to the go stimulus on those trials. The first index of inhibitory control is the probability of responding on stop-signal trials, and the second index of inhibitory control is an estimate of the covert latency of the stop process (reaction time; SSRT).

Inferior frontal gyrus The IFG is a specific region of the prefrontal cortex and can be separated into the pars opercularis, pars triangularis, and pars orbitalis. The right IFG is critically involved in inhibitory control functions. In particular, evidence suggests that the subregion of the rIFG most commonly implicated is the pars opercularis, while the inferior frontal junction seems to be more related to attentional processes. However, how exactly IFG subregions are functionally dissociable is in the scope of current efforts.

Cognitive network connectivity Evidence has emphasized that the rIFG is functionally connected with other cortical and subcortical brain regions during the inhibition of motor responses. In particular, evidence revealed a network of the IFG, ACC, thalamus, and striatum. Future research in heroin addiction or drug addiction in general should address the impaired inhibitory control from a cognitive network perspective.

Magnetic resonance imaging MRI measurements can be used either to assess structural properties of the brain, such as gray matter volume or white matter tracts (using DTI), or to assess functional activity in response to specific stimuli (task-induced MRI, stop-signal task, or go/no-go task). These neural correlates can then be related to behavioral indices of inhibitory control.

KEY FACTS OF COGNITIVE CONTROL FUNCTIONS IN HEROIN ADDICTION AND THE RELATION TO THE INFERIOR FRONTAL GYRUS

- Cognitive control functions are important to process and integrate incoming perceptual information in a flexible fashion to adapt ongoing behavior according to prevailing conditions.
- One of the key processes underlying cognitive control is the ability to adequately suppress already initiated response tendencies (i.e., response inhibition).
- Heroin addiction is associated with deficits in a broad range of cognitive control functions, including deficits in response inhibition.
- These deficits contribute to an inability to exert control over heroin urges or to inhibit drug-seeking behavior.
- An important area in the brain for cognitive control is the rIFG.
- Heroin-dependent patients have structural and functional abnormalities in the rIFG. These abnormalities are often related to patients' behavioral deficits in cognitive control.
- More research is needed to study how abnormalities in the rIFG are related to specific cognitive control deficits and how this relation develops over the course of various therapeutic scenarios.

SUMMARY POINTS

- Heroin addiction is characterized by behavioral deficits in several cognitive control functions, including impairments in cognitive flexibility, inhibitory control, sustained attention, decision-making, and memory and learning processes.
- These behavioral deficits in cognitive control are probably mediated by structural and functional abnormalities of the rIFG.
- It is not clear so far whether the abnormalities of the rIFG contribute to specific cognitive control impairment such as response inhibition or attention allocation or to a more general cognitive control deficit.
- Both behavioral and rIFG findings of cognitive control depend on the duration of dependence, abstinence, or maintenance treatment.
- Longitudinal research is needed to understand how these impairments develop over the course of abstinence and maintenance treatment and whether they are reversible.
- The rIFG and its relation to cognitive control should be considered from a network perspective in future research.
- Pharmacological challenges are needed to understand the neurochemical base of response inhibition in the healthy brain and in addicted individuals.

REFERENCES

Aron, A. R., Fletcher, P. C., Bullmore, E. T., Sahakian, B. J., & Robbins, T. W. (2003). Stop-signal inhibition disrupted by damage to right inferior frontal gyrus in humans. *Nature Neuroscience, 6*, 115–116.

Aron, A. R., Robbins, T. W., & Poldrack, R. A. (2014). Inhibition and the right inferior frontal cortex: one decade on. *Trends in Cognitive Sciences, 18*, 177–185.

Bari, A., & Robbins, T. W. (2013). Inhibition and impulsivity: behavioral and neural basis of response control. *Progress in Neurobiology, 108*, 44–79.

Bernardin, F., Maheut-Bosser, A., & Paille, F. (2014). Cognitive impairments in alcohol-dependent subjects. *Frontiers in Psychiatry, 5*, 78.

Bora, E., Yücel, M., Fornito, A., Pantelis, C., Harrison, B. J., Cocchi, L., ... Lubman, D. I. (2012). White matter microstructure in opiate addiction. *Addiction Biology, 17*, 141–148.

Bracken, B. K., Trksak, G. H., Penetar, D. M., Tartarini, W. L., Maywalt, M. A., Dorsey, C. M., & Lukas, S. E. (2012). Response inhibition and psychomotor speed during methadone maintenance: impact of treatment duration, dose, and sleep deprivation. *Drug and Alcohol Dependence, 125*, 132–139.

Chambers, C. D., Bellgrove, M. A., Stokes, M. G., Henderson, T. R., Garavan, H., Robertson, I. H., ... Mattingley, J. B. (2006). Executive "brake failure" following deactivation of human frontal lobe. *Journal of Cognitive Neuroscience, 18*, 444–455.

Cheng, J., Lu, Y., Han, X., González-Vallejo, C., & Sui, N. (2012). Temporal discounting in heroin-dependent patients: no sign effect, weaker magnitude effect, and the relationship with inhibitory control. *Experimental and Clinical Psychopharmacology, 20*, 400–409.

Connolly, C. G., Foxe, J. J., Nierenberg, J., Shpaner, M., & Garavan, H. (2012). The neurobiology of cognitive control in successful cocaine abstinence. *Drug and Alcohol Dependence, 121*, 45–53.

Courtney, K. E., Ghahremani, D. G., & Ray, L. A. (2013). Fronto-striatal functional connectivity during response inhibition in alcohol dependence. *Addiction Biology, 18*, 593–604.

Darke, S., McDonald, S., Kaye, S., & Torok, M. (2012). Comparative patterns of cognitive performance amongst opioid maintenance patients, abstinent opioid users and non-opioid users. *Drug and Alcohol Dependence, 126*, 309–315.

Davis, P. E., Liddiard, H., & McMillan, T. M. (2002). Neuropsychological deficits and opiate abuse. *Drug and Alcohol Dependence, 67*, 105–108.

Denier, N., Schmidt, A., Gerber, H., Schmid, O., Riecher-Rössler, A., Wiesbeck, G. A., ... Borgwardt, S. (2013). Association of frontal gray matter volume and cerebral perfusion in heroin addiction: a multimodal neuroimaging study. *Frontiers in Psychiatry, 4*, 135.

Ersche, K. D., Jones, P. S., Williams, G. B., Turton, A. J., Robbins, T. W., & Bullmore, E. T. (2012). Abnormal brain structure implicated in stimulant drug addiction. *Science, 335*, 601–604.

Ersche, K. D., & Sahakian, B. J. (2007). The neuropsychology of amphetamine and opiate dependence: implications for treatment. *Neuropsychology Review, 17*, 317–336.

Everitt, B. J., Dickinson, A., & Robbins, T. W. (2001). The neuropsychological basis of addictive behaviour. *Brain Research. Brain Research Reviews, 36*, 129–138.

Feil, J., Sheppard, D., Fitzgerald, P. B., Yücel, M., Lubman, D. I., & Bradshaw, J. L. (2010). Addiction, compulsive drug seeking, and the role of frontostriatal mechanisms in regulating inhibitory control. *Neuroscience and Biobehavioral Reviews, 35*, 248–275.

Fernández-Serrano, M. J., Pérez-García, M., & Verdejo-García, A. (2011). What are the specific vs. generalized effects of drugs of abuse on neuropsychological performance? *Neuroscience Biobehavioral Reviews, 35*, 377–406.

Fillmore, M. T., Rush, C. R., & Hays, L. (2005). Cocaine improves inhibitory control in a human model of response conflict. *Experimental and Clinical Psychopharmacology, 13*, 327–335.

Fishbein, D. H., Krupitsky, E., Flannery, B. A., Langevin, D. J., Bobashev, G., Verbitskaya, E., ... Tsoy, M. (2007). Neurocognitive characterizations of Russian heroin addicts without a significant history of other drug use. *Drug and Alcohol Dependence, 90*, 25–38.

Forman, S. D., Dougherty, G. G., Casey, B. J., Siegle, G. J., Braver, T. S., Barch, D. M., ... Lorensen, E. (2004). Opiate addicts lack error-dependent activation of rostral anterior cingulate. *Biological Psychiatry*, 55, 531–537.

Forstmann, B. U., Jahfari, S., Scholte, H. S., Wolfensteller, U., van den Wildenberg, W. P., & Ridderinkhof, K. R. (2008). Function and structure of the right inferior frontal cortex predict individual differences in response inhibition: a model-based approach. *Journal of Neuroscience*, 28, 9790–9796.

Fu, L. P., Bi, G. H., Zou, Z. T., Wang, Y., Ye, E. M., Ma, L., & Yang, Z. (2008). Impaired response inhibition function in abstinent heroin dependents: an fMRI study. *Neuroscience Letters*, 438, 322–326.

Garavan, H., Kaufman, J. N., & Hester, R. (2008). Acute effects of cocaine on the neurobiology of cognitive control. *Philosophical Transactions of the Royal Society London. Series B, Biological Sciences*, 363, 3267–3276.

Goldstein, R. Z., & Volkow, N. D. (2011). Dysfunction of the prefrontal cortex in addiction: neuroimaging findings and clinical implications. *Nature Reviews. Neuroscience*, 12, 652–669.

Gruber, S. A., Tzilos, G. K., Silveri, M. M., Pollack, M., Renshaw, P. F., Kaufman, M. J., & Yurgelun-Todd, D. A. (2006). Methadone maintenance improves cognitive performance after two months of treatment. *Experimental and Clinical Psychopharmacology*, 14, 157–164.

Hester, R., & Garavan, H. (2004). Executive dysfunction in cocaine addiction: evidence for discordant frontal, cingulate, and cerebellar activity. *Journal of Neuroscience*, 24, 11017–11022.

Jasinska, A. J., Stein, E. A., Kaiser, J., Naumer, M. J., & Yalachkov, Y. (2014). Factors modulating neural reactivity to drug cues in addiction: a survey of human neuroimaging studies. *Neuroscience and Biobehavioral Reviews*, 38, 1–16.

Koob, G. F., & Volkow, N. D. (2010). Neurocircuitry of addiction. *Neuropsychopharmacology*, 35, 217–238.

Lee, T. M., & Pau, C. W. (2002). Impulse control differences between abstinent heroin users and matched controls. *Brain Injury*, 16, 885–889.

Lee, T. M., Zhou, W. H., Luo, X. J., Yuen, K. S., Ruan, X. Z., & Weng, X. C. (2005). Neural activity associated with cognitive regulation in heroin users: a fMRI study. *Neuroscience Letters*, 382, 211–216.

Li, C. S., & Sinha, R. (2008). Inhibitory control and emotional stress regulation: neuroimaging evidence for frontal-limbic dysfunction in psycho-stimulant addiction. *Neuroscience and Biobehavioral Reviews*, 32, 581–597.

Li, W., Li, Q., Zhu, J., Qin, Y., Zheng, Y., Chang, H., ... Wang, W. (2013). White matter impairment in chronic heroin dependence: a quantitative DTI study. *Brain Research*, 1531, 58–64.

Liao, D. L., Huang, C. Y., Hu, S., Fang, S. C., Wu, C. S., Chen, W. T., ... Li, C. S. (2014). Cognitive control in opioid dependence and methadone maintenance treatment. *PLoS One*, 9, e94589.

Lyoo, I. K., Pollack, M. H., Silveri, M. M., Ahn, K. H., Diaz, C. I., Hwang, J., ... Renshaw, P. F. (2006). Prefrontal and temporal gray matter density decreases in opiate dependence. *Psychopharmacology (Berl)*, 184, 139–144.

Mintzer, M. Z., Copersino, M. L., & Stitzer, M. L. (2005). Opioid abuse and cognitive performance. *Drug and Alcohol Dependence*, 78, 225–230.

Moreno-López, L., Catena, A., Fernández-Serrano, M. J., Delgado-Rico, E., Stamatakis, E. A., Pérez-García, M., & Verdejo-García, A. (2012). Trait impulsivity and prefrontal gray matter reductions in cocaine dependent individuals. *Drug and Alcohol Dependence*, 125, 208–214.

Moreno-López, L., Stamatakis, E. A., Fernández-Serrano, M. J., Gómez-Río, M., Rodríguez-Fernández, A., Pérez-García, M., & Verdejo-García, A. (2012). Neural correlates of the severity of cocaine, heroin, alcohol, MDMA and cannabis use in polysubstance abusers: a resting-PET brain metabolism study. *PLoS One*, 7, e39830.

Ornstein, T. J., Iddon, J. L., Baldacchino, A. M., Sahakian, B. J., London, M., Everitt, B. J., & Robbins, T. W. (2000). Profiles of cognitive dysfunction in chronic amphetamine and heroin abusers. *Neuropsychopharmacology*, 23, 113–126.

Pau, C. W., Lee, T. M., & Chan, S. F. (2002). The impact of heroin on frontal executive functions. *Archives of Clinical Neuropsychology*, 17, 663–670.

Pirastu, R., Fais, R., Messina, M., Bini, V., Spiga, S., Falconieri, D., & Diana, M. (2006). Impaired decision-making in opiate-dependent subjects: effect of pharmacological therapies. *Drug and Alcohol Dependence*, 83, 163–168.

Prosser, J., Cohen, L. J., Steinfeld, M., Eisenberg, D., London, E. D., & Galynker, I. I. (2006). Neuropsychological functioning in opiate-dependent subjects receiving and following methadone maintenance treatment. *Drug and Alcohol Dependence*, 84, 240–247.

Prosser, J., London, E. D., & Galynker, I. I. (2009). Sustained attention in patients receiving and abstinent following methadone maintenance treatment for opiate dependence: performance and neuroimaging results. *Drug and Alcohol Dependence*, 104, 228–240.

Qiu, Y., Jiang, G., Su, H., Lv, X., Zhang, X., Tian, J., & Zhuo, F. (2013). Progressive white matter microstructure damage in male chronic heroin dependent individuals: a DTI and TBSS study. *PLoS One*, 8, e63212.

Robinson, T. E., & Berridge, K. C. (2003). Addiction. *Annual Reviews of Psychology*, 54, 25–53.

Schmidt, A., Borgwardt, S., Gerber, H., Schmid, O., Wiesbeck, G. A., Riecher-Rössler, A., ... Walter, M. (2014). Altered prefrontal connectivity after acute heroin administration during cognitive control. *International Journal of Neuropsychopharmacology*, 1–11.

Schmidt, A., Walter, M., Gerber, H., Schmid, O., Smieskova, R., Bendfeldt, K., ... Borgwardt, S. (2013). Inferior frontal cortex modulation with an acute dose of heroin during cognitive control. *Neuropsychopharmacology*, 38, 2231–2239.

Sell, L. A., Morris, J. S., Bearn, J., Frackowiak, R. S., Friston, K. J., & Dolan, R. J. (2000). Neural responses associated with cue evoked emotional states and heroin in opiate addicts. *Drug and Alcohol Dependence*, 60, 207–216.

Spronk, D. B., van Wel, J. H., Ramaekers, J. G., & Verkes, R. J. (2013). Characterizing the cognitive effects of cocaine: a comprehensive review. *Neuroscience and Biobehavioral Reviews*, 37, 1838–1859.

Swann, N., Tandon, N., Canolty, R., Ellmore, T. M., McEvoy, L. K., Dreyer, S., ... Aron, A. R. (2009). Intracranial EEG reveals a time- and frequency-specific role for the right inferior frontal gyrus and primary motor cortex in stopping initiated responses. *Journal of Neuroscience*, 29, 12675–12685.

Tabatabaei-Jafari, H., Ekhtiari, H., Ganjgahi, H., Hassani-Abharian, P., Oghabian, M. A., Moradi, A., ... Zarei, M. (2014). Patterns of brain activation during craving in heroin dependents successfully treated by methadone maintenance and abstinence-based treatments. *Journal of Addiction Medicine*, 8, 123–129.

Tabibnia, G., Monterosso, J. R., Baicy, K., Aron, A. R., Poldrack, R. A., Chakrapani, S., ... London, E. D. (2011). Different forms of self-control share a neurocognitive substrate. *Journal of Neuroscience*, 31, 4805–4810.

Verdejo-García, A. J., Perales, J. C., & Pérez-García, M. (2007). Cognitive impulsivity in cocaine and heroin polysubstance abusers. *Addictive Behaviors, 32*, 950–966.

Volkow, N. D., Fowler, J. S., Wang, G. J., Telang, F., Logan, J., Jayne, M., ... Swanson, J. M. (2010). Cognitive control of drug craving inhibits brain reward regions in cocaine abusers. *Neuroimage, 49*, 2536–2543.

Wang, Y., Li, W., Li, Q., Yang, W., Zhu, J., & Wang, W. (2011). White matter impairment in heroin addicts undergoing methadone maintenance treatment and prolonged abstinence: a preliminary DTI study. *Neuroscience Letters, 494*, 49–53.

Wang, X., Li, B., Zhou, X., Liao, Y., Tang, J., Liu, T., ... Hao, W. (2012). Changes in brain gray matter in abstinent heroin addicts. *Drug and Alcohol Dependence, 126*, 304–308.

Wang, Y., Zhu, J., Li, Q., Li, W., Wu, N., Zheng, Y., ... Wang, Z. (2013). Altered fronto-striatal and fronto-cerebellar circuits in heroin-dependent individuals: a resting-state FMRI study. *PLoS One, 8*, e58098.

Wrege, J., Schmidt, A., Walter, A., Smieskova, R., Bendfeldt, K., Radue, E. W., ... Borgwardt, S. (2014). Effects of cannabis on impulsivity: a systematic review of neuroimaging findings. *Current Pharmaceutical Design, 20*, 2126–2137.

Yuan, K., Qin, W., Dong, M., Liu, J., Sun, J., Liu, P., ... Tian, J. (2010). Gray matter deficits and resting-state abnormalities in abstinent heroin-dependent individuals. *Neuroscience Letters, 482*, 101–105.

Yuan, Y., Zhu, Z., Shi, J., Zou, Z., Yuan, F., Liu, Y., ... Weng, X. (2009). Gray matter density negatively correlates with duration of heroin use in young lifetime heroin-dependent individuals. *Brain and Cognition, 71*, 223–228.

Section D

Methods

Chapter 98

Assay for Opium Alkaloids

Jelena Acevska[1], Gjoshe Stefkov[2], Svetlana Kulevanova[2], Aneta Dimitrovska[1]
[1]Faculty of Pharmacy, Institute of Applied Chemistry and Pharmaceutical Analysis, University Ss. Cyril and Methodius, Skopje, Republic of Macedonia;
[2]Faculty of Pharmacy, Institute of Pharmacognosy, University Ss. Cyril and Methodius, Skopje, Republic of Macedonia

Abbreviations

AC Acetylcodeine
ACN Acetonitrile
ADC Acetyldihydrocodeine
AmA Ammonium acetate
AmF Ammonium formate
APCI Atmospheric pressure chemical ionization
BSTFA N,O-bis(trimethylsilyl)trifluoroacetamide
CC Cocaine
CD Codeine
CE Capillary electrophoresis
DAD Diode array detector
EI Electron impact ionization
EMO Ethylmorphine
ESI Electron-spray ionization
FA Formic acid
FID Flame-ionization detector
FLD Fluorescence detection
FTIR Fourier transform infrared spectroscopy
GC Gas chromatography
HC Hydrocodone
HCC Hydroxyl cocaine
HFBA Heptafluorobutyric anhydride
HF-LPME Hollow-fiber liquid-phase microextraction
HM Hydromorphone
HPLC High-performance liquid chromatography
HR Heroin
INCB International Narcotics Control Board of the United Nations
IS Internal standard
LLE Liquid–liquid extraction
6-MAM 6-Monoacetylmorphine
M3G Morphine-3-glucuronide
M6G Morphine-6-glucuronide
MAE Microwave-assisted extraction
MEPS Microextraction by packed sorbent
ML Methadol
MN Methadone
MO Morphine
MOH Methanol
MPE Micropulverization extraction
MRM Multiple-reaction monitoring
MS Mass spectrometry
MSPD Matrix solid-phase dispersion
NCC Norcocaine
NCD Norcodeine
NIR Near-infrared spectroscopy
NMN Normethadone
NMO Normorphine
NS Noscapine
NX Naloxone
OC Oxycodone
OM Oxymorphone
PFPA Pentafluoropropionic anhydride
Ph.Eur. European pharmacopoeia
PLE Pressurized liquid extraction
PP Propoxyphene
PPT Protein-precipitation technique
PV Papaverine
RIA Radioimmunoassay
SAMHSA Substance Abuse and Mental Health Services Advisory
SBSE Stir-bar sorptive extraction
SDME Single-drop microextraction
SFE Supercritical fluid extraction
SIM Single-ion monitoring
SPDE Solid-phase dynamic extraction
SPE Solid-phase extraction
SPME Solid-phase microextraction
TB Thebaine
TC Thebacon
TEA Triethylamine
TFAA Trifluoroacetic anhydride
TMCS Chlorotrimethylsilane
UAE Ultrasonic-assisted extraction
USP US pharmacopoeia
UV Ultraviolet detection
WHO World Health Organization

INTRODUCTION

Along with various societal efforts to deal with the increasing incidence of drug abuse, there is an emerging need to develop analytical methods that are convenient for the determination of these substances in biological materials of drug addicts. Convenient and robust analytical methods for bioassays relating to drug addiction and substance misuse can help detect addiction and/or misuse; they also provide means of better understanding drug biometabolism pathways and consequently can help in the treatment of the addiction. Such analytical methods are important tools in the efforts to prevent and reduce the alarming issue of drug misuse.

Opium alkaloids represent several different classes of components isolated from *Papaver somniferum* L. Papaveraceae (Bernath & Nemeth, 2009). This plant contains more than 40 individual alkaloids in the form of salts of meconic acid or other simple plant acids. The major alkaloids are morphine, codeine, and thebaine (phenanthrenes or morphinans); papaverine (benzylisoquinolines); and noscapine (phthalide isoquinolines). Other classes of alkaloids are present in trace quantities. These alkaloids, including the analgesic and narcotic drug morphine, the cough suppressant codeine, the muscle relaxant papaverine, and the antitumor agent noscapine, are extracted from dried opium poppy capsules by "dry extraction" and are then used in medicine as pharmaceutical substances.

While the production and use of opium are strictly forbidden and well controlled by the International Narcotics Control Board of the United Nations (INCB), there continue to be seizures of this "pie" of opium latex containing high concentrations of opium alkaloids. Furthermore, most abuse incidents involve the consumption and distribution of opium. However, in terms of drug addiction, the predominantly mentioned opium alkaloid is morphine; this is an opioid analgesic used in the treatment of moderate to severe pain, as recommended by the World Health Organization (2008), and is misused because of its "side" effects: euphoria and sedation. Another opium alkaloid has now come under increased focus in forensic studies: codeine. Codeine can be purchased in the form of prescription and nonprescription cough medications; it is far cheaper than heroin, for example, and it can be quickly and easily chemically converted to desomorphine (Nelson, Bryant, & Aks, 2014).

The use of opioids (opium alkaloids, opium, heroin, synthetic prescription opioids) is an ever-persistent societal problem. Their controlled and limited availability in well-defined therapeutic fields has not eliminated the risk of their misuse.

Metabolism of Opioids

Morphine is predominantly metabolized to morphine-3-glucuronide (M3G) and morphine-6-glucuronide (M6G). M3G is the major metabolite and the minor portion is represented by M6G normorphine (produced by the demethylation of morphine). The individual rates of metabolism, distribution, and elimination influence the concentration of the drug in biological fluids. When a normal dose of morphine (about 10 mg per dose) is given to a person, only a few nanograms of morphine per milliliter of plasma can be detected. Therefore, detection of the two metabolites (M3G and M6G) is used as an indication of morphine (or heroin) intake (Goldberger, Caplan, Maguire, & Cone, 1991). Analyzing biological fluids for these two metabolites after heroin or morphine intake is important and useful. The reason for this is the pharmacological activity of M6G, which has a slightly different and probably increased respiratory depressant action over morphine. Additionally, in fatal cases, the time elapsed between morphine (or heroin) intake and death, as well as the severity of the intoxication, can be assessed by evaluating the ratio of morphine over its metabolites in biological fluids.

The occurrence of morphine in biofluids, tissues, and cell lines from mammals is not only due to the intake of morphine; it has been shown to also be related to the endogenous synthesis of the compound. The importance of endogenous morphine in the neuroendocrine system is still, however, under debate (Stefano & Scharrer, 1994).

After codeine ingestion, both morphine and codeine are generally found in biological fluids. Codeine is metabolized by conjugation and by its demethylation to morphine. The compound 6-monoacetylmorphine (6-MAM) has been shown to be a unique metabolite of heroin (Goldberger et al., 1991, 1993), but is difficult to detect in the hours after heroin intake because of its rapid hydrolysis to morphine (Bencharit, Morton, Xue, Potter, & Redinbo, 2003).

All of these issues indicate the need for an analytical method that is suitable for the assay of several opioids and their metabolites simultaneously. For this reason, and almost as an analytical rule, the methods of choice for assays of opium alkaloids and related opioid substances are the chromatographic methods combined with high-resolution detection systems. The guidelines issued by the Substance Abuse and Mental Health Services Advisory have so far addressed the requirements for collection and testing of urine specimens, and the development of additional guidelines for associated issues regarding alternative specimens (oral fluids, sweat patches, hair testing, etc.) are of interest to define a more comprehensive regulatory framework.

Bioassay of Opioids

The major reasons for performing bioanalysis of opioids in human samples are drug monitoring (clinical studies) and detection of drug abuse (toxicological and forensic studies). In any case, sensitive analytical procedures have to be used to detect and/or to quantify the drug and/or its metabolites (Saito et al., 2011). The samples include plasma and serum, urine, saliva, hair, etc.

The bioassay of opium alkaloids, as for any other drugs present in complex matrices, presents many analytical challenges. Owing to matrix effects, the high rate of drug–protein binding, and the presence of metabolites, analytical methods should have a high selectivity for the target analytes. Additionally, because of the very low drug concentration due to metabolism and excretion (Table 1), a high degree of method sensitivity is necessary. A convenient method of high selectivity and sensitivity may be achieved using an intelligent combination of selective sample preparation (including a concentration step if necessary) and a highly efficient separation system combined with a selective and sensitive detection system.

Comprehensive reviews on analytical methodologies for the bioanalysis of morphine and its related substances, with a special focus on sample preparation and separation techniques, were published by Espinosa Bosch, Ruiz Sanchez, Sanchez Rojas, and Bosch Ojeda (2007) and Honore Hansen (2009). In this chapter, developments in the bioanalysis of opioids in biosystems, covering the consequent years not included in the above reviews, are presented, additionally highlighting the work of Dams, Benijts, Lambert, and De Leenheer (2002), Dams, Murphy, Lambert, and Huestis (2003), Dams, Murphy, Choo, et al. (2003), Dams, Huestis, Lambert, and Murphy (2003), reflecting best practices in the field.

TABLE 1 Concentration Levels of Some Opioids in Various Biological Samples (Based on Case Study Reports)

Compound	Dosage	Plasma and Serum	Urine	Oral Fluid	Keratinized Matrices
MO	Daily dose range: 12–120 mg, separated in four daily doses	Therapeutic conc.: analgesia, 50–260 ng/ml; anesthesia, 2 μg/ml Fatal conc.: 0.2–2.3 μg/ml	50.2–3359.4 ng/ml (Dams, 2003)	0–200 ng/ml (Rohrig & Moore, 2003) 1.1–53.1 ng/ml (Dams, 2003)	Nail: 0.58–3.16 ng/mg (Shen, Chen, & Xiang, 2014) Hair: 0.17–1.36 ng/mg (Shen et al., 2014)
CD	Daily dose range: 60–240 mg, separated in four daily doses	Therapeutic conc.: 0.03–0.34 ng/ml Fatal conc.: 1.0–8.8 μg/ml	26.2–156.3 ng/ml (Dams, 2003)	1.3–6.4 ng/ml (Dams, 2003)	Nail: 0.11–0.27 ng/mg (Shen et al., 2014) Hair: 0.43–14.07 ng/mg (Shen et al., 2014) Hair: 0.71–8.7 ng/mg (Ropero-Miller, Goldberger, Cone, & Joseph, 2000)
HR intake (free MO, total MO, CD, 6-MAM)	/	Drug users: free MO, 0–128 ng/ml; total MO[a], 10–2110 ng/ml Fatal conc.: free MO, 0–2800 ng/ml; total MO, 33–5000 ng/ml (Meissner, Recker, Reiter, Friedrich, & Oehmichen, 2002) HR, 109 ng/ml; 6-MAM, 168 ng/ml; and MO, 1140 ng/ml (Rop, Fornaris, Salmon, Burle, & Bresson, 1997)	Free MO, 3650 ng/ml (Rop et al., 1997) Free CD, 0.028–39.4 μg/ml; total CD, 0.070–307 μg/mol; free MO, 1.74–218 μg/ml; total MO, 11.2–2870 μg/ml (Moriya, Chan, & Hashimoto, 1999)	1.0–295.2 ng/ml (Dams, 2003)	Hair: 0.14 ng/mg (Shen et al., 2014) Nails: 0.06–4.69 ng/mg (Lemos, Anderson, Valentini, Tagliaro, & Scott, 2000)

The table shows the approximate concentration ranges of some opioids in various types of samples (plasma/serum, urine, oral fluid, and keratinized matrices, like hair and nails), based on random forensic case reports; MO, morphine; CD, codeine; HR, heroin; 6-MAM, 6-monoacetylmorphine.
[a]Total MO, free MO plus MO conjugates.

TYPES OF BIOLOGICAL MATERIAL USED FOR DETERMINATION OF OPIOIDS

Plasma and Serum

The available bioactive fraction of a drug can be assessed by measuring the concentration of the analytes in plasma or serum, which reflects the amount available for the active sites in the biosystem at the time of sampling. In this way, these matrices are the most important for clinical studies. Measurement of the free fraction of the drug, which is considered to be the active part, may be difficult, and therefore the total amount of analyte is usually quantified (including the fraction of the drug bound to plasma proteins). Plasma may be injected directly into a high-performance liquid chromatography (HPLC) system, but owing to the lack of a concentrating step, the limit of quantification can be too high for the analysis of morphine in many samples. In this, as in the case of serum also, matrix complexity makes sample pretreatment a necessity (Fountain, Yin, & Diehl, 2009).

Urine

The principles of determining opioids and their metabolites in urine are similar to the analysis of plasma and serum. Advantages of using this type of sample include fewer matrix effects, noninvasive collection, and large sample sizes that enable suitable quantities of the analytes to be obtained and consequently direct injection (Dams, 2003; Edinboro, Backer, & Poklis, 2005). Urine analysis is routinely performed for verification of drug abuse and is mostly by liquid chromatography/mass spectrometry (LC/MS) (Low & Taylor, 1995; Nordgren & Beck, 2004). A comprehensive review regarding the determination of drugs of abuse and their metabolites in urine has been published by Pizzolato, Lopez de Alda, and Barceló (2007).

Alternative Biological Fluids

Oral fluid (saliva) and sweat have successfully been used as alternative fluid matrices for testing for drug abuse. Here, as with urine, the noninvasiveness and ease of sample collection are the main advantages of using these types of sample. However, while it may reflect the blood concentration of the drug, the use of these types of samples presents some built-in problems regarding sample collection, resulting in data variability (Langel et al., 2008).

Many of the methods for determining morphine and its related substances in urine can be used for the analysis of oral fluid and sweat samples too. Alen, Azad, Field, and Blake (2005) have replaced the immunoassay with LC/tandem MS (LC/MS/MS) for the routine measurement of drugs of abuse in oral fluids. To our best knowledge, sweat-patch specimens require further examination before being recommended as a sample for opioid bioassay.

Keratinized Matrices

It is considered that the analysis of hair to determine the contents of abused drugs started back in 1979, when Baumgartner, Jones, Baumgartner, and Black succeeded in detecting opiates in the hair of heroin abusers by radioimmunoassay. Opiate abuse histories could also be estimated by sectional hair analysis. Because of the deposition of the drugs in hair, each centimeter of hair that contains traces of the drug of interest during the time the hair was growing can be used to estimate the drug usage rate corresponding to the elapsed time scale, retrospectively.

As stated by Baciu, Borrull, Aguilar, and Calull (2015), the obvious advantages of this type of sample are the long detection window (months to years), noninvasive collection, robustness regarding sample adulteration, and ease of sample storage and transportation. However, considering the fact that the hair samples have to be collected some time after the person may have used the drug (given that the growth rate of hair is approximately 1 cm/month), this type of biomaterial cannot be used as a real-time sample. Additionally, the low concentration of drug in the hair (ng/mg sample) requires the use of the most sensitive analytical chemical techniques (Kronstrand, Nystrom, Strandberg, & Druid, 2004) and also appropriate preanalytical concentration amplification techniques.

Fingernails are another keratinized sample, alternative to hair. The use of this specimen is especially highlighted in cases of humans with alopecia, personal habits of tonsure, religious customs, etc. (Shen et al., 2014).

SAMPLE PREPARATION FOR BIOANALYSIS OF OPIOIDS

Owing to the similar polarities between morphine and its major metabolites, the selective extraction from biofluids is difficult. This holds true for other opioids also. The complex matrix of the real samples and the low concentration of opium alkaloids make the development of a simple and reliable method for their extraction and preconcentration the main challenge and a critical step for their analysis (Ahmadi-Jouibaria, Fattahib, Shamsipurc, & Pirsahebb, 2013). Many different approaches for sample preparation are available in the literature and comprehensive reviews on this subject have been published by Baumgartner and Hill (1993), Chen, Guo, Wang, and Qiu (2008), and Kole, Venkatesh, Kotecha, and Sheshala (2011). The aim of this stage of the analytical methodology is to obtain a sample that is as clean as possible for further analysis, but also to concentrate the analytes.

Sample preparation usually involves several steps: decontamination/washing, extraction/digestion, cleanup, and preconcentration. The washing step is especially important for hair and nail samples, because it helps avoid false-positive results by removing possible passive external environmental contamination (Baciu et al., 2015). Before solubilization of the sample, a miniaturization of the sample particles (by pulverization) is sometimes necessary (Baciu et al., 2015). To speed up this usually time-consuming step, ultrasonic-assisted extraction, microwave-assisted extraction, supercritical fluid extraction, and pressurized liquid extraction may be employed to accelerate and improve efficiency. Matrix solid-phase dispersion (Baciu, Borrull, Aguilar, & Calull, 2015) and micropulverized extraction (Favretto et al., 2011) have been reported for the same purpose. The advantages of frozen pulverization as a novel approach for sample pretreatment before extraction of opioids have been highlighted by Shen et al. (2014). This method is commonly used in chemistry, biology, and other fields, and in it samples are pulverized completely at cold temperatures generated by liquid nitrogen.

After pretreatment, samples need to be cleaned up and preconcentrated. Liquid–liquid extraction (LLE), protein precipitation technique (PPT), or solid-phase extraction (SPE) followed by a concentrating step is most often used to further improve method sensitivity because of the low concentration of analytes. The last is the preferred method for sample extraction, cleanup, and concentration of opioids from plasma and serum samples, as indicated in the reviews (Espinosa Bosch et al., 2007; Honore Hansen, 2009).

A trend in more recent publications is to employ simplified, miniaturized techniques and to minimize the amount of organic solvent used during sample preparation. Techniques such as solid-phase microextraction, solid-phase dynamic extraction, single-drop microextraction, stir-bar sorptive extraction, hollow-fiber liquid-phase microextraction, and microextraction by packed sorbent offer many advantages.

Considering the facts that LLE is time-consuming and tedious and uses large amounts of high-purity and hazardous organic solvents and that SPE has limitations relating to the pH of sample solutions and may require lengthy processing such as washing, conditioning, eluting, and solvent evaporation, the advantages of the miniaturized techniques over the conventional extraction methods are more than clear. Yet, with appropriate optimization, most of the disadvantages of LLE and SPE can be diminished, and economically viable methods that require very small amounts of samples and extraction solvents can be developed, requiring no additional specific equipment, and most importantly, they can cover a wide variety of applications.

Specific types of sample preparation have also been used by some research teams. Beike, Köhler, Brinkmann, and Blaschke, 1999 developed a method involving the use of immunoaffinity-based extraction to isolate M3G and M6G from human blood. Polyclonal antisera were coupled to an activated Trisacryl gel and used for extraction prior to analysis with HPLC–FLD. Similarly, a method using the multitarget immunoaffinity column prior to capillary electrophoresis separation has been developed for the determination of morphine and related compounds, codeine, acetylcodeine, 6-MAM, and M3G, in the urine of heroin abusers (Qi, Mi, Zhang, & Chang, 2005).

The choice for sample preparation techniques is closely connected to the respective sample matrices analyzed and the nature of the separation and detection system. All together the three parts of the analytical method will contribute to the selectivity and sensitivity and also to the reliability of the final data obtained.

DETECTION AND QUANTIFICATION OF OPIOIDS IN BIOLOGICAL SAMPLES

Opioids in biological samples can be determined with either a nonseparation technique or a separation technique. However, the applications of separation techniques (like HPLC and gas

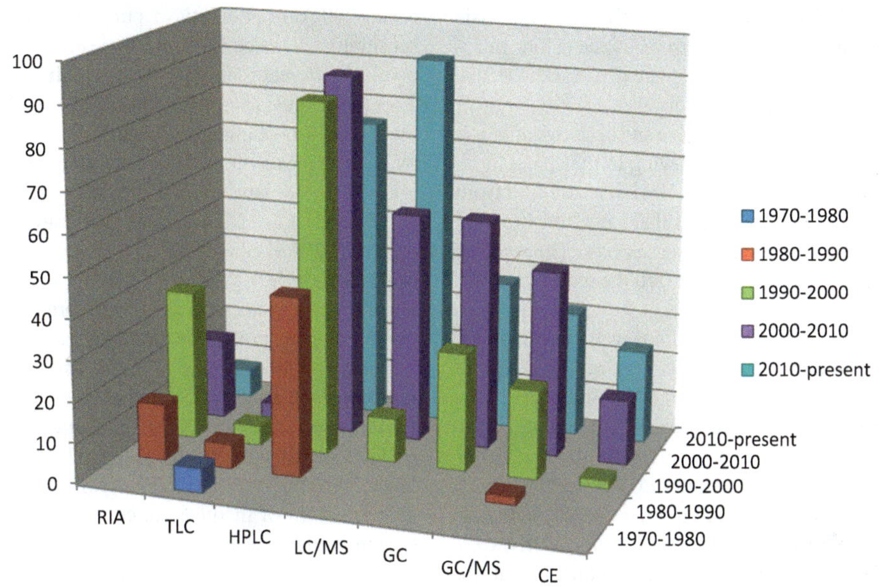

FIGURE 1 Trends of various application techniques for determination of opioids over the past decades (bar chart showing the number of publications containing "determination of opioids" and one of the given techniques as keywords). Each column of the bar chart indicates the sum of publications for a decade, revealing that HPLC was the dominant technique used for determination of opioids from the very introduction of this application. In more recent years, LC/MS has rapidly replaced other types of detection systems combined with HPLC. RIA, radioimmunoassay; TLC, thin-layer chromatography; HPLC, high-performance liquid chromatography; LC/MS, liquid chromatography/mass spectrometry; GC, gas chromatography; GC/MS, gas chromatography/mass spectrometry; CE, capillary electrophoresis.

chromatography (GC)) for bioassay of opium alkaloids and other related opioids, especially when combined with highly specific detection systems (like MS), are far more numerous than the applications of nonseparation techniques (Figure 1).

Nonseparation Techniques

Immunoassays may be sensitive, simple, and easy to use, and are often used to screen large numbers of samples, but often their selectivity is compromised as cross-reactivity with similar substances occurs (Ensinger & Doevendans, 1984). This is the main reason nonseparation techniques are rarely used for the detection of opiates in complex matrix systems such as whole blood, plasma, and serum.

Immunotechniques can, however, be used advantageously as a screening, first-phase analysis, and then the positive finding can be confirmed by a second independent method (that is as sensitive as the screening test) with built-in high selectivity (De Jong et al., 2005).

Nowadays, far more interesting applications of these techniques are found in sample preparation steps, as raised by Rashid, Aherne, Katmeh, Kwasowski, and Stevenson back in 1998.

Although Fourier transform infrared spectroscopy using a diamond-composite attenuated total reflectance crystal and near-infrared Fourier transform–Raman spectroscopy for the simultaneous identification and quantification of the most important alkaloids have proven to be efficient quality control methods and are useful in forensic analysis and in high-throughput evaluation of poppy breeding material (Schulz, Baranska, Quilitzsch, & Schutze, 2004), there are no published methods proving the benefits of these techniques for bioassay of opium alkaloids.

Novel approaches to improving the selectivity issues of nonseparation techniques have become available with the introduction of chemometrics for method development and result interpretation. A combination of electrochemistry with chemometrics was employed by Gholivand et al. (2015) to introduce an efficient analytical method for the simultaneous quantification of five opium alkaloids in complex matrices. A strong voltammetric overlapping, observed for the simultaneous analysis of these compounds, was successfully resolved using a chemometric approach for variable optimization and multivariate calibration.

Separation Techniques

Considering the high demands on the selectivity and sensitivity of methods for bioassay of opium alkaloids and related opioids, separation techniques and suitable detection systems are not only preferable, they are generally necessary to obtain reliable results.

When separation techniques of great selectivity (like GC and LC) are combined with highly sensitive and selective detection techniques (such as the various mass spectrometric techniques), the ideal methods become available, comprising all the advantages needed for suitable bioassay.

Gas Chromatography

GC is one of the most commonly used techniques in bioanalysis work. Its main advantages over other chromatographic techniques are a combination of speed, sensitivity, and high resolution power (Grob & Barry, 2004). The most commonly used detector for quantitative determinations in GC is the flame-ionization detector (FID). This is due to its sensitivity, fast response, wide dynamic range, and ease of application. However, the combination of a gas chromatograph with a mass spectrometer allows the identification and structural determination of new and/or known components in trace amounts (Kitson, Larson, & McEwan, 1996; Scott, 1997). Because of this, GC coupled with MS is particularly attractive for bioanalysis in drug addiction. In addition to mass spectrometric detectors, nitrogen–phosphorous detectors and electron capture detectors have also been used for determination of opiates in biofluids.

GC combined with MS is a very sensitive and selective analysis technique; it is often superior to LC/MS owing to the characteristic pattern of fragmentation of morphinan alkaloids resulting from electron impact ionization (EI), which facilitates the identification of unpredicted drugs of abuse present in the samples. The work of Rana, Garg, and Singla (2014) has emphasized the use of fast GC/MS for analysis of urinary opiates.

Although the separation of opium alkaloids using the GC/FID/MS method and avoiding any derivatization step was proved possible (Acevska, Stefkov, Petkovska, Kulevanova, & Dimitrovska, 2012) for the determination of the main alkaloids in opium poppy samples, its lack of sensitivity for poorly volatile trace components, such as morphine and related opioids, in biological matrices cannot be overcome unless a derivatization step is employed before the separation takes place. An evaluation of the various derivatization agents (pentafluoropropionic anhydride, heptafluorobutyric anhydride, trifluoroacetic anhydride, N,O-bis(trimethylsilyl)trifluoroacetamide with 1% chlorotrimethylsilane, a mixture of propionic acid anhydride with triethylamine (1:1, v/v), acetic anhydride, and propionyl chloride in pyridine) found that acetyl derivatives have the greatest stability. However, trimethylsilane derivatization is most commonly used for opioid derivatization (Espinosa Bosch et al., 2007).

What is worth stating is that the polar glucuronide metabolites can be determined only after their hydrolysis. It is therefore reasonable that LC/MS became the technique of choice here, as the polar metabolites could be analyzed simultaneously with the parent drug substance (Maralikova & Weinmann, 2004).

High-Performance Liquid Chromatography

HPLC is the method of choice for the simultaneous analysis of morphine, its metabolites, and related opioids in biological samples. Usually, a reversed-phase chromatography column with gradient elution, using ion pair reagents in an acidic buffer, is necessary, because of the large differences in polarity of the alkaloids of interest. This type of chromatographic system has also been recommended by the European and US pharmacopoeias, as an official method for the quality control of opium. Several modifications to the reversed-phase HPLC method were offered to improve the separation of these basic components (Acevska, Dimitrovska, et al., 2012; Ahmadi-Jouibari et al., 2013; Petkovska, Babunovska, & Stefova, 2011) and to make the commonly used chromatographic systems MS-compatible (Acevska et al., 2015).

Among the various detection systems used in bioanalysis work, ultraviolet (UV)–visible or diode array detection are rarely used because of their limited selectivity (Pravadali et al., 2013). Electrochemical detection can be used for morphine and M6G, but M3G is not electrochemically active (Rashid, Aherne, Katmeh, Kwasowski, & Stevenson, 1998). Another alternative may be fluorescence detection (Dams et al., 2002), but the sensitivity of MS detection (usually 0.1–0.5 ng morphine per milliliter of plasma) was not achieved.

Most importantly, if the glucuronides of morphine are to be covered by the bioanalysis, then LC/MS is the method of choice (Wood et al., 2006). In a variety of published methods, emphasizing the sensitivity and selectivity of MS detection of morphine and related opioids, different ionization types and different analyzers have been reported (Dams, Murphy, Choo, et al., 2003). The matrix effect in the bioanalysis of illicit drugs with LC/MS/MS was studied in detail by Dams, Huestis, et al. (2003), covering the influence of ionization (electron-spray ionization (ESI) and atmospheric pressure chemical ionization (APCI)), sample preparation type (direct injection, dilution, PPT, SPE), and biofluid (urine, oral fluid, and plasma). By evaluation of their synergistic effect on the presence of the matrix effect it was indicated that ESI was more susceptible than APCI and that acetonitrile protein precipitation provided both sample cleanup and concentration for oral fluid analysis, while SPE was necessary for extensive cleanup of plasma prior to LC–APCI–MS/MS.

An application of LC in hair analysis was demonstrated by Baciu, Borrull, Aguilar, and Calull (2015). De Castro, Concheiro, Shakleya, and Huestis (2009) applied LC/MS for simultaneous quantification of methadone, cocaine, opiates, and their metabolites in human placenta.

Capillary Electrophoretic/Electrokinetic Methods

Capillary electrophoresis (CE) is complementary to chromatography because of its different separation mechanism. The main advantage of CE over LC and GC is that all components of the sample elute quickly from the capillary, and the analysis ends immediately after measuring the analytes of interest, followed only by a simple washing procedure with no danger of analyte carryover occurring. Regarding the use of CE for opium analysis, by applying ordinary buffers, a poor resolution was achieved. Application of micellar electrokinetic capillary chromatography (MECC), nonaqueous CE, or the addition of complexing agents such as cyclodextrins in buffer improves selectivity. Further improvement is achieved by applying the microemulsion with MECC combined with injection of the sample into the short end of the capillary—this has resulted in analysis times shorter than 2 min (Sini Panicker, Wojno, & Ziska, 2007).

CE in combination with a mass spectrometer has become a popular technique suitable for the analysis of complex matrices. This combination has overcome the limitations of CE–UV in terms of sensitivity and selectivity, owing to the limited choice of wavelengths at which most of the compounds display sufficient molar absorptivity. CE could offer clear advantages over current chromatographic techniques for hair analysis, owing to the minimal need of sample mass for analysis, which in the case of hair can be a crucial point.

REVIEW OF ANALYTICAL METHODS FOR BIOASSAY OF OPIUM ALKALOIDS AND RELATED OPIOIDS

Applications of state-of the-art methods for bioassay of opium alkaloids and related opioids are summarized in Table 2. The data included here, along with the data included in the comprehensive reviews published by Saito et al. (2011) and Baciu et al. (2015), emphasize the superiority of LC/MS methods for determination of drug (ab)use in biological samples.

TABLE 2 Applications of State-of-the-Art Methods for Bioassay of Opium Alkaloids and Related Opioids

Sample	Analytes	Sample Preparation	Separation/Detection	References
Blood, urine	17 opium alkaloids, including MO, CD, NS, PV, AC, 6-MAM, NMN, NMO, NCD, HR, MN, NX, HC, EMO, ADC, TC, IS: Butorphanol	SPE	Fast **LC-DAD-FLD**, on phenyl column, 53 × 7.0 mm, 3 μm, mobile phase water/MOH containing TEA and FA, pH 4.5	Dams et al. (2002)
Urine, oral fluid, plasma	27 compounds including MO, CD, PV, NS, NMO, NCD, 6-MAM, AC, HR, CC and metabolites, NCC, HCC, MD, ML, PP	Comparison between direct injection, dilution, PPT, and SPE	**LC-(ESI)-MS/MS** and **LC-(APCI)-MS/MS**, using SRM on a polar RP column (150 × 2.0 mm, 4 μm, mobile phase AmF/water/ACN, pH 4.5	Dams et al. 2003, Dams, Murphy, Lambert, et al. (2003), Dams, Murphy, Choo, et al. (2003), and Dams, Huestis, et al. (2003)
Urine	MO, M3G, M6G, CD, NCD, 6-MAM, HC, HM, OM, OC	Direct injection, dilution following centrifugation	**LC-(ESI)-MS/MS**, on phenyl column, 150 × 2.1 mm, 5 μm, mobile phase AmF/FA/water/ACN	Edinboro, et al. (2005)
Human placenta	MN, CC, opiates, and their metabolites	Homogenization, centrifugation Extraction: SPE	**LC-(ESI+)-MS** on phenyl column, 75 × 2 mm, 4 μm, mobile phase FA/water/ACN	De Castro et al. (2009)
Plasma	MO, CD, PV, NS	DLLME-SFO	**LC-UV**, on a C-18 column (250 × 4.6 mm, 5 μm), mobile phase sodium phosphate/sodium dodecyl sulfate buffer and ACN	Ahmadi-Jouibari et al. (2013)
Fingernail	MO, CD, 6-MAM, AC, HR	Washing: Water/acetone Extraction/digestion: Frozen pulverization + LLE Preconcentration: Evaporation	**LC/MS/MS** coupled with an API 4000 QTRAP mass spectrometer; using ESI+ and MRM mode separation: PFP propyl column, 100 × 2.1 mm, 5 μm, mobile phase AmA/water/ACN	Shen et al. (2014)
Urine	MO, CD, HM, HC	Acid hydrolysis before SPE to release the free drugs from their conjugates Keto-opiates were converted to oximes with AmOH prior to extraction to eliminate interference with MO and CD Extraction: SPE Derivatization: BSTFA + 1% TMCS	Fast **GC/MS**, using SIM mode. Separation achieved on 10 m × 0.15 mm column, film thickness 0.12 μm, using hydrogen as a carrier gas	Rana et al. (2014)
Serum	MO, CD, NS, TB, PV	PPT	**Electrochemistry** in combination with chemometrics	Gholivand et al. (2015)

The table shows trends in separation techniques (e.g., liquid chromatography (LC), gas chromatography (GC)), coupled with highly specific detection systems (mass spectrometry (MS), tandem mass spectrometry (MS/MS), fluorescence detection (FLD)), used for opioid-like analytes in various body types. As sample preparation techniques, solid-phase extraction (SPE), liquid–liquid extraction (LLE), and dilution technique are dominantly used. APCI, atmospheric pressure chemical ionization; BSTFA, N,O-bis(trimethylsilyl)trifluoroacetamide; DAD, diode array detection; DLLME-SFO, dispersive liquid–liquid microextraction based on solidification of floating organic drop; ESI+, positive electron-spray ionization; MRM, multiple-reaction monitoring; PFP, pentafluorophenyl; RP, reversed phase; SIM, selected-ion monitoring; SRM, selected-reaction monitoring; TMCS, chlorotrimethylsilane. See the chapter abbreviations list for analytes.

APPLICATIONS TO OTHER ADDICTIONS AND SUBSTANCE MISUSE

The methods described above can also be used for assays of morphine, codeine, methadone, heroin, and related opiates.

DEFINITION OF TERMS

Bioanalysis The assay of compounds in all kinds of biological materials is defined as "bioanalysis." Although assay in a narrow sense means quantitative analysis, in a wider sense it is often considered enough to obtain only qualitative results.

Opium This is a mixture of substances that contains 8–17% morphine, 1–10% noscapine, 0.5–1.5% papaverine, and 0.7–5% codeine.

Opioids Opioids represent an important class of components used in medicine for their analgesic and antitussive activity and for other purposes. Their effect is obtained through interaction with opioid receptors located in the central nervous system, peripheral nervous system, and gastrointestinal tract. Various pharmacological actions and side effects of opioids are manifested depending on their selectivity for different receptors.

Opiate Although the term "opiate" is often used as a synonym for opioid, its use should be limited to the natural alkaloids found in the resin of the poppy, primarily morphine, codeine, and thebaine, but not papaverine and noscapine, which have different modes of action. Also, there are other natural opiates, such as salvinorin A isolated from *Salvia divinorum*. Possibly this term can be used for the semisynthetic opioids produced by them.

Semisynthetic opioid This is an opioid derived from natural opiates, produced by semisynthesis, such as hydromorphone, hydrocodone, oxycodone, oxymorphone, desomorphine, diacetylmorphine (heroin), nicomorphine, dipropanoylmorphine, benzylmorphine, ethylmorphine, and buprenorphine.

Synthetic opioids These are opioids produced by chemical synthesis, such as fentanyl, pethidine, methadone, tramadol, and dextropropoxyphene.

Illicit opioids These are opioids banned by international drug control treaties including, for instance, heroin and opium, as well as diverted pharmaceutical opioids (such as buprenorphine, methadone, and morphine).

Endogenous opioid peptides These are opioids produced in the human body, such as endorphins, enkephalin, dynorphins, and endomorphins.

Liquid chromatography This is a method of chromatographic separation based on a forced transport of a liquid (mobile phase) carrying an analyte mixture through a porous medium (stationary phase) and the differences in the interactions of the analytes with the surface of this porous medium resulting in different migration times for a mixture's components.

Gas chromatography This is a chromatographic method in which the stationary phase is a liquid distributed on an inert support or coated on a column wall and the mobile phase is a gas. The separation is based on the partition differences of the sample components by the two phases. It is considered that the revolution of GC started in the late 1950s with the introduction of the wall-coated open tubular capillary column.

Chemometrics This is a scientific discipline that uses mathematical and statistical principles to choose optimal procedures and experiments (design of experiments) and to provide chemical information by analysis (multivariate data analysis) and pattern recognition techniques.

KEY FACTS OF BIOASSAY OF OPIUM ALKALOIDS AND OPIOID-RELATED COMPOUNDS

- Cultivation of the opium poppy and opium production, as well as control and monitoring of the legal movement of narcotic drugs, are regulated by the INCB.
- This system, based on the Single Convention on Narcotic Drugs, 1961 (as amended by the 1972 protocol amending the Single Convention on Narcotic Drugs, 1961), issued by the INCB, limits the legal cultivation of narcotic plants for each country and the world as a whole, as well as the legal production and distribution of narcotic drugs in quantities needed for medical and scientific use.
- Most medicinal opium is used for the production of morphine, codeine, papaverine, and their derivatives, as starting materials for the production of pharmaceutical finished products, but more than 90% of the illegally produced opium is used by drug addicts.
- The Prescription Drug Monitoring Program is a tool that can be used to address prescription drug diversion and abuse.
- Although the validation of a bioanalytical method development has been regulated, there are still "hot" topics extensively discussed on an international level to reach a consensus on the extent of validation experiments and on acceptance criteria for validation parameters of bioanalytical methods in forensic (and clinical) toxicology, in order to obtain reliable data.
- In trend with the "green" chemistry, while developing a bioassay, methods should consider reducing the use of toxic organic solvents to minimize the environmental footprint.

SUMMARY POINTS

- Convenient and robust analytical methods for bioassays relating to drug addiction and substance misuse can help detect addiction and/or misuse; they also provide a means of better understanding drug biometabolism pathways and consequently can help in the treatment of addiction, thus presenting important tools in the efforts to prevent and reduce the alarming issue of drug misuse.
- The major reasons for performing bioanalysis of opium alkaloids and related opioids in human samples are drug monitoring (clinical studies) and detection of drug abuse (toxicological and forensic studies).
- To detect and/or to quantitate the drug and/or its metabolites in various biological samples simultaneously, there is a high demand for sensitivity and selectivity of the analytical procedures.
- A convenient method of high selectivity and sensitivity may be achieved using an intelligent combination of selective sample preparation (including a concentration step if necessary) and a highly efficient separation system combined with a selective and sensitive detection system.
- Plasma and serum are the most important matrices for clinical studies, as they reflect the amount available for the active sites in the biosystem at the time of sampling.

- Advantages of urine analysis include fewer matrix effects, noninvasive collection, and large sample sizes that enable suitable quantities of the analytes to be obtained and consequently direct injection into the separation system.
- The use of alternative matrices, like oral fluid (saliva), sweat, hair, fingernails, etc., may reflect the blood concentration of the drug and seems promising in terms of noninvasiveness and ease of sample collection, yet the use of these samples presents some built-in problems regarding sample collection, resulting in data variability.
- The nonseparation techniques are mainly used in sample preparation steps, but there are novel approaches introducing chemometrics for overcoming the insufficient selectivity of electrochemical methods.
- GC/MS is a very sensitive and selective analysis technique, often superior to LC/MS owing to a characteristic pattern of fragmentation of morphinan alkaloids resulting from EI, which facilitates the identification of unpredicted drugs of abuse present in the samples, but the polar glucuronide metabolites can be determined only after their hydrolysis and consequent derivatization.
- Application of state-of-the-art methods for bioassay of opium alkaloids and related opioids emphasizes the superiority of the LC/MS technique as a method of choice for determination of drug (ab)use in biological samples.

REFERENCES

Acevska, J., Dimitrovska, A., Stefkov, G., Brezovska, K., Karapandzova, M., & Kulevanova, S. (2012). Development and validation of RP-HPLC method for determination of alkaloids from *Papaver somniferum* L., Papaveraceae. *Journal of AOAC International, 95*, 399–405.

Acevska, J., Stefkov, G., Petkovska, R., Kulevanova, S., & Dimitrovska, A. (2012). Chemometric approach for development, optimization and validation of different chromatographic methods for separation of opium alkaloids. *Analytical and Bioanalytical Chemistry, 403*, 1117–1129.

Acevska, J., Stefkov, G., Cvetkovikj, I., Petkovska, R., Kulevanova, S., Cho, J. H., & Dimitrovska, A. (2015). Fingerprinting of morphine using chromatographic purity profiling and multivariate data analysis. *Journal of Pharmaceutical and Biomedical Analysis, 109C*, 18–27.

Ahmadi-Jouibaria, T., Fattahib, N., Shamsipurc, M., & Pirsahebb, M. (2013). Dispersive liquid–liquid microextraction followed by high-performance liquid chromatography–ultraviolet detection to determination of opium alkaloids in human plasma. *Journal of Pharmaceutical and Biomedical Analysis, 85*, 14–20.

Allen, K. R., Azad, R., Field, H. P., & Blake, D. K. (2005). Replacement of immunoassay by LC tandem mass spectrometry for the routine measurement of drugs of abuse in oral fluid. *Annals of Clinical Biochemistry, 42*(4), 277–284.

Baciu, T., Borrull, F., Aguilar, C., & Calull, M. (2015). Recent trends in analytical methods and separation techniques for drugs of abuse in hair. *Analytica Chimica Acta, 856*, 1–26.

Baumgartner, W. A., & Hill, V. A. (1993). Sample preparation techniques. *Forensic Science International, 63*(1–3), 121–135.

Baumgartner, A. M., Jones, P. F., Baumgartner, W. A., & Black, C. T. (1979). Radioimmunoassay of hair for determining opiate-abuse histories. *Journal of Nuclear Medicine, 20*, 748–752.

Beike, J., Köhler, H., Brinkmann, B., & Blaschke, G. (1999). Immunoaffinity extraction of morphine, morphine-3-glucuronide and morphine-6-glucuronide from blood of heroin victims for simultaneous high-performance liquid chromatographic determination. *Journal of Chromatography B, 726*, 111–119.

Bencharit, S., Morton, C. L., Xue, Y., Potter, P. M., & Redinbo, M. R. (2003). Structural basis of heroin and cocaine metabolism by a promiscuous human drug-processing enzyme. *Nature Structural and Molecular Biology, 10*(5), 349–356.

Bernath, J., & Nemeth, E. (2009). Poppy (1st ed.). *Handbook of plant breeding: Oil Crops* (Vol 4)New York: Springer. ISBN: 978-0-387-77594-4.

Chen, Y., Guo, Z., Wang, X., & Qiu, C. (2008). Sample preparation. *Journal of Chromatography A, 1184*, 191–219.

Dams, R., Benijts, T., Lambert, W. E., & De Leenheer, A. P. (2002). Simultaneous determination of in total 17 opium alkaloids and opioids in blood and urine by fast liquid chromatography–diode array detection–fluorescence detection, after solid-phase extraction. *Journal of Chromatography B, 773*, 53–61.

Dams, R., Murphy, C. M., Lambert, W. E., & Huestis, M. A. (2003). Urine drug testing for opioids, cocaine, and metabolites by direct injection liquid chromatography/tandem mass spectrometry. *Rapid Communications in Mass Spectrometry, 17*, 1665–1670.

Dams, R., Murphy, C. M., Choo, R. E., Lambert, W. E., De Leenheer, A. P., & Huestis, M. A. (2003). LC-atmospheric pressure chemical Ionization-MS/MS analysis of multiple illicit drugs, methadone and their metabolites in oral fluid following protein precipitation. *Analytical Chemistry, 75*, 798–804.

Dams, R., Huestis, M. A., Lambert, W. E., & Murphy, C. M. (2003). Matrix effect in bio-analysis of illicit drugs with LC-MS/MS: influence of ionization type, sample preparation, and biofluid. *Journal of the American Society for Mass Spectrometry, 14*(11), 1290–1294.

Dams, R. (2003). *Bio-analysis of opioids by liquid chromatography tandem mass spectrometry* (Ph.D. thesis). Ghent University.

De Castro, A., Concheiro, M., Shakleya, D. M., & Huestis, M. A. (2009). Simultaneous quantification of methadone, cocaine, opiates, and metabolites in human placenta by liquid chromatography–mass spectrometry. *Journal of Analytical Toxicology, 33*, 243–252.

De Jong, L. A. A., Krämer, K., Kroeze, M. P. H., Bischoff, R., Uges, D. R. A., & Franke, J. P. (2005). Development and validation of a radioreceptor assay for the determination of morphine and its active metabolites in serum. *Journal of Pharmaceutical and Biomedical Analysis, 39*(5), 964–971.

Edinboro, L. E., Backer, R. C., & Poklis, A. (2005). Direct analysis of opiates in urine by liquid chromatography-tandem mass spectrometry. *Journal of Analytical Toxicology, 29*(7), 704–710.

Ensinger, H. A., & Doevendans, J. E. (1984). Plasma levels of opioid analgesics determined by radioreceptor assay. *Drug Research, 34*, 609–613.

Espinosa Bosch, M., Ruiz Sanchez, A., Sanchez Rojas, F., & Bosch Ojeda, B. (2007). Morphine and its metabolites: analytical methodologies for its determination. *Journal of Pharmaceutical and Biomedical Analysis, 43*, 799–815.

Favretto, D., Vogliardi, S., Stocchero, G., Nalesso, A., Tucci, M., & Ferrara, S. D. (2011). High performance liquid chromatography–high resolution mass spectrometry and micropulverized extraction for the quantification of amphetamines, cocaine, opioids, benzodiazepines, antidepressants and hallucinogens in 2.5 mg hair samples. *Journal of Chromatography A, 1218*, 65–83.

Fountain, K. J., Yin, Z., & Diehl, D. M. (2009). Simultaneous analysis of morphine-related compounds in plasma using mixed-mode solid phase extraction and UltraPerformance liquid chromatography–mass spectrometry. *Journal of Separation Science, 32*(13), 2319–2326.

Gholivand, M. B., Jalalvand, A. R., Goicoechea, H. C., Gargallo, R., Skov, T., & Paimard, G. (2015). Combination of electrochemistry with chemometrics to introduce an efficient analytical method for simultaneous quantification of five opium alkaloids in complex matrices. *Talanta, 131*, 26–37.

Goldberger, B. A., Caplan, Y. H., Maguire, T., & Cone, E. J. (1991). Testing human hair for drugs of abuse. III. Identification of heroin and 6-acetylmorphine as indicators of heroin use. *Journal of Analytical Toxicology, 15*(5), 226–231.

Goldberger, B. A., Darwin, W. D., Grant, T. M., Allen, A. C., Caplan, Y. H., & Cone, E. J. (1993). Measurement of heroin and its metabolites by isotope-dilution electron-impact mass spectrometry. *Clinical Chemistry, 39*(4), 670–675.

Grob, R. L., & Barry, E. F. (2004). *Modern practice of gas chromatography* (4th ed.). New Jersey: John Wiley & Sons, Inc. ISBN: 0-471-22983-0.

Honore Hansen, S. (2009). Sample preparation and separation techniques for bioanalysis of morphine and related substances. *Journal of Separation Science, 32*, 825–834.

Kitson, F. G., Larson, B. S., & McEwan, C. N. (1996). *Gas chromatography and mass spectrometry – a practical guide*. New York: Academic Press. ISBN: 0-12-483385-3.

Kole, P. L., Venkatesh, G., Kotecha, J., & Sheshala, R. (2011). Recent advances in sample preparation techniques for effective bioanalytitical methods. *Biomedical Chromatography, 25*, 199–2017.

Kronstrand, R., Nystrom, I., Strandberg, J., & Druid, H. (2004). Screening for drugs of abuse in hair with ion spray LC-MS-MS. *Forensic Science International, 145*(2–3), 183–190.

Langel, K., Engblom, C., Pehrsson, A., Gunnar, T., Ariniemi, K., & Lillsunde, P. (2008). Drug testing in oral fluid–evaluation of sample collection devices. *Journal of Analytical Toxicology, 32*, 393–401.

Lemos, N. P., Anderson, R. A., Valentini, R., Tagliaro, F., & Scott, R. T. (2000). Analysis of morphine by RIA and HPLC in fingernail clippings obtained from heroin users. *Journal of Forensic Science, 45*(2), 407–412.

Low, A. S., & Taylor, R. B. (1995). Analysis of common opiates and heroin metabolites in urine by high-performance liquid chromatography. *Journal of Chromatography B, 663*, 225–233.

Maralikova, B., & Weinmann, W. (2004). Confirmatory analysis for drugs of abuse in plasma and urine by high-performance liquid chromatography-tandem mass spectrometry with respect to criteria for compound identification. *Journal of Chromatography B, 811*(1), 21–30.

Meissner, C., Recker, S., Reiter, A., Friedrich, H. J., & Oehmichen, M. (2002). Fatal versus non-fatal heroin "overdose": blood morphine concentrations with fatal outcome in comparison to those of intoxicated drivers. *Forensic Science International, 130*(1), 49–54.

Moriya, F., Chan, K. M., & Hashimoto, Y. (1999). Concentrations of morphine and codeine in urine of heroin abusers. *Legal Medicine (Tokyo), 1*(3), 140–144.

Nelson, M. E., Bryant, S. M., & Aks, S. E. (2014). Emerging drugs of abuse. *Disease-a-Month, 60*, 110–132.

Nordgren, H. K., & Beck, O. (2004). Multicomponent screening for drugs of abuse: direct analysis of urine by LC-MS-MS. *Therapeutic Drug Monitoring, 26*(1), 90–97.

Petkovska, A., Babunovska, H., & Stefova, M. (2011). Fast and selective HPLC-DAD method for determination of pholcodine and related substances. *Macedonian Journal of Chemistry and Chemical Engineering, 30*(2), 139–150.

Pizzolato, T. M., Lopez de Alda, M. J., & Barceló, D. (2007). LC-based analysis of drugs of abuse and their metabolites in urine. *Trends in Analytical Chemistry, 26*(6), 609–624.

Pravadali, S., Bassanese, D. N., Conlan, X. A., Francis, P. S., Smith, Z. M., Terry, J. M., & Shalliker, R. A. (2013). Comprehensive sample analysis using high performance liquid chromatography with multi-detection. *Analityca Chimica Acta, 803*, 188–193.

Qi, X. H., Mi, J. Q., Zhang, X. X., & Chang, W. B. (2005). Design and preparation of novel antibody system and application for the determination of heroin metabolites in urine by capillary electrophoresis. *Analytica Chimica Acta, 551*, 115–123.

Rana, S., Garg, R. K., & Singla, A. (2014). Rapid analysis of urinary opiates using fast gas chromatography–mass spectrometry and hydrogen as a carrier gas. *Egyptian Journal of Forensic Sciences, 4*, 100–107.

Rashid, B. A., Aherne, G. W., Katmeh, M. F., Kwasowski, P., & Stevenson, D. (1998). Determination of morphine in urine by solid-phase immunoextraction and high-performance liquid chromatography with electrochemical detection. *Journal of Chromatography A, 797*(1–2), 245–250.

Rohrig, T. P., & Moore, C. (2003). The determination of morphine in urine and oral fluid following ingestion of poppy seeds. *Journal of Analytical Toxicology, 27*(7), 449–452.

Rop, P. P., Fornaris, M., Salmon, T., Burle, J., & Bresson, M. (1997). Concentrations of heroin, 06-monoacetylmorphine, and morphine in a lethal case following an oral heroin overdose. *Journal of Analytical Toxicology, 21*.

Ropero-Miller, J. D., Goldberger, B. A., Cone, E. J., & Joseph, R. E., Jr. (2000). The disposition of cocaine and opiate analytes in hair and fingernails of humans following cocaine and codeine administration. *Journal of Analytical Toxicology, 24*(7), 496–508.

Saito, K., Saiko, R., Kikuchi, Y., Iwasaki, Y., Ito, R., & Nakazawa, H. (2011). Analysis of drugs of abuse in biological specimens. *Journal of Health Science, 57*(6), 472–487.

Schulz, H., Baranska, M., Quilitzsch, R., & Schutze, W. (2004). Determination of alkaloids in capsules, milk and ethanolic extracts of poppy (*Papaver somniferum* L.) by ATR-FT-IR and FT-Raman spectroscopy. *Analyst, 129*, 917–920.

Scott, R. P. W. (1997). *Tandem techniques (Separation science series)*. John Wiley & Sons Ltd. ISBN: 0-471-96760-2.

Shen, M., Chen, H., & Xiang, P. (2014). Determination of opiates in human fingernail —Comparison to hair. *Journal of Chromatography B, 967*, 84–89.

Sini Panicker, S., Wojno, H. L., & Ziska, L. H. (2007). Quantitation of the major alkaloids in opium from papaver setigerum DC. *Microgram Journal, 5*(1–4), 13–19.

Stefano, G. B., & Scharrer, B. (1994). Endogenous morphine and related opiates, a new class of chemical messengers. *Advances in Neuroimmunology, 4*(2), 57–67.

Wood, M., Laloup, M., Samyn, N., Ramirez Fernandez, M. M., de Bruijn, E. A., Maes, R. A. A., & De Boeck, G. (2006). Recent applications of liquid chromatography–mass spectrometry in forensic science. *Journal of Chromatography A, 1130*, 3–15.

World Health organization. (October 14, 2008). *WHO Treatment Guidelines on chronic non-malignant pain in adults* Adopted in WHO Steering Group on Pain Guidelines . www.who.org.

Chapter 99

Features of the Tridimensional Personality Questionnaire Associated with Heroin Users

Chen-Ying Wu[1], Wei-Lieh Huang[2,3], Yu-Hsuan Lin[3], Cheryl C.H. Yang[3,4]

[1]Department of Psychiatry, Maimonides Medical Center, Brooklyn, NY, USA; [2]Department of Psychiatry, National Taiwan University Hospital, Douliou City, Yunlin County, Taiwan; [3]Institute of Brain Science, National Yang-Ming University, Taipei, Taiwan; [4]Sleep Research Center, National Yang-Ming University, Taipei, Taiwan

Abbreviations

5-HTTVNTR Serotonin transporter intron 2 variable number tandem repeat
ALDH2 Aldehyde dehydrogenase 2
ANS Autonomic nervous system
ASPD Antisocial personality disorder
BPD Borderline personality disorder
COMT Catechol-*O*-methyltransferase
DAT1 Dopamine active transporter 1
DRD3 Dopamine D3 receptor
DSM-IV *Diagnostic and Statistical Manual of Mental Disorders,* fourth edition
HA Harm avoidance
HPA hypothalamic–pituitary–adrenal
HRV heart rate variability
HF high-frequency power
iv Intravenous
LF% normalized low-frequency power
LF/HF ratio of low-frequency power to high-frequency power
MAOA–LPR monoamine oxidase A gene promoter region
NET Norepinephrine transporter
NRXN3 Neurexin 3
NS Novelty seeking
OTI Opiate Treatment Index
PTSD posttraumatic stress disorder
RD reward dependence
SCL 90 Symptom Checklist-90
TCI Temperament and Character Inventory
TPQ Tridimensional Personality Questionnaire

INTRODUCTION

Heroin use, no matter whether it is viewed as a behavior or a disorder, is affected by multiple biological, psychological, and social factors. The relationship between heroin use and the short-term emotional status of the individual or the long-term traits of the individual is an interesting topic. Personality is an enduring set of qualities or characteristics that are related to an person's inner experience and behavior and these usually have developed before substance use behavior begins. Thus personality is considered to be a contributing factor when substance use behavior occurs. While this remains interesting to researchers, how personality traits are modulated by various neurotransmitters or are influenced by a range of environmental factors remains poorly understood. Researchers continue to explore how the above three types of factors interact with one another and would like to understand how personality traits might serve as a predictor in terms of future substance use behavior.

Personality can be explored from either a dimensional or a categorical perspective. The *Diagnostic and Statistical Manual of Mental Disorders,* fourth edition (DSM-IV), defined personality disorders from a categorical standpoint by grouping various sets of enduring psychological or behavioral patterns and linking them to a disorder. On the other hand, many other clinical instruments, such as the Five Factor Model, the Eysenck Personality Questionnaire, the Maudsley Personality Inventory, the Minnesota Multiphasic Personality Inventory, the Tridimensional Personality Questionnaire (TPQ), and the Temperament and Character Inventory (TCI), are based on the dimensional concept; these observe and quantify preselected (preidentified) common characteristics among individuals. The TPQ and the TCI were devised by the same research team, namely the Cloninger group. The TPQ is designed to assess hereditary and biological characteristics, which are then divided into three dimensions. The TCI, a revised and extended version of the TPQ, has seven dimensions, which include innate temperament (the basis of TPQ), but it also investigates characters that are acquired from the environment and through experience. Among all the above rating tools, the TPQ is the most applicable to heroin-use-related studies because it uses biological properties, and the results in relation to heroin users have been supportive. The following content introduces a range of TPQ studies that have been carried out as of this writing.

THE CONCEPTS OF THE TPQ AND THE TCI

The TPQ, a self-reported true/false questionnaire consisting of 100 questions, was designed by Claude Robert Cloninger, MD, in 1987 (Cloninger, Przybeck, & Svrakic, 1991). It has been widely used in studies of substance abuse over the past decades. The questionnaire consists of three underlying genetic dimensions of personality, namely, novelty seeking (NS), harm avoidance (HA), and reward dependence (RD). Each of these three fundamental dimensions is composed of subcategorical behavior patterns that are described below.

NS consists of NS1 (exploratory excitability), NS2 (impulsivity), NS3 (extravagance), and NS4 (disorderliness). NS refers to behavioral activation, which is a heritable tendency toward intense excitement in response to novel stimuli for potential rewards or relief of punishment; this leads to frequent exploratory activity, as well as the active avoidance of monotony (Hofmann & Loh, 2006). Evidence suggests that NS is associated with the neurotransmitter dopamine.

HA consists of HA1 (anticipatory worry), HA2 (fear of uncertainty), HA3 (shyness with strangers), and HA4 (fatigability and asthenia). HA refers to behavioral inhibition, which is a heritable tendency to respond intensely to aversive stimuli; it therefore results in learning to inhibit behavior to avoid punishment and frustrating nonrewarding activity. γ-Aminobutyric acid and serotonin activity, especially in the dorsal raphe nucleus, are believed to be related to HA (Hofmann & Loh, 2006).

RD is divided into RD1 (sentimentality), RD2 (persistence), RD3 (attachment), and RD4 (dependence). RD refers to behavioral maintenance, which is a heritable tendency to respond intensely to reward signals (especially verbal signals of social approval and sentiment) and to maintain behavior that has been associated with rewards in the past (Hofmann & Loh, 2006). RD has its strongest relationship with the neurotransmitter norepinephrine, but is also believed to be related to the release of serotonin in the median raphe nucleus. RD2 (persistence) was later proved to be an independent dimension and is usually excluded when calculating the total RD score. In the TCI, persistence reflects a behavior or a psychological status of "intermittent reinforcement" and "resistance to extinction" and is usually considered to be the fourth main dimension of temperament. The neurotransmitters that are linked to persistence are glutamate and serotonin in the dorsal raphe nucleus (Cloninger, 1994).

Research has demonstrated that various strong biological properties are associated with the three dimensions of the TPQ. Behavioral patterns similar to those described in the TPQ can be observed in animals. However, the TPQ is unable to explain all human personality qualities and behaviors, especially conscious-driven and acquired personality qualities. Therefore, Cloninger (1994) added the concept of "character" and introduced the TCI. Character represents the cognitive learning and environment-related components of the personality. It comprises self-directedness, cooperativeness, and self-transcendence. Self-directedness, or self-determination, refers to a tendency to achieve personal goals and values. Cooperativeness represents the tendency toward agreeableness and socially acceptable interaction. Self-transcendence stands for spiritual nature of the personality, such as considering oneself an integral part of universe (Cloninger, Svrakic, & Przybeck, 1993). Generally speaking, temperament corresponds to primitive brain functions, such as the limbic system and striatum, while character is more related to the functioning of the neocortex and hippocampus. One study has validated the above statement and revealed that there is 40–60% genetic heritability associated temperament but only 10–15% genetic heritability associated with character (Cloninger, 1994).

THE TPQ DIMENSIONS AND HEROIN USE

Research evidence supports the idea that the concept of the TPQ and TCI dimensions is able to help toward a better understanding of heroin-related diagnosis or behaviors. The link between the TPQ/TCI dimensions and heroin use behavior can be reviewed from either a psychosocial or a biological standpoint.

From a psychosocial perspective, research has shown that opiate-dependent individuals have a TPQ pattern that has a high NS, a low RD, and a high self-transcendence (Milivojevic et al., 2012). Furthermore, heroin-dependent individuals and eating disorder subjects were found to display similar TPQ combinations (high HA, low self-directedness, and low cooperativeness) when a male population was explored (Abbate-Daga, Amianto, Rogna, Fassino, 2007). Other studies have concluded that individuals with a high NS show an association with poor retention and adherence to appointments during methadone or buprenorphine treatment (Helmus, Downey, Arfken, Henderson, & Schuster, 2001; Roll, Saules, Chudzynski, & Sodano, 2004). In addition, patients with heroin dependence seem to exhibit a higher NS and self-directedness compared to those with alcohol dependence (Le Bon et al., 2004). Conway's study found that substance-dependent individuals who are involved in the use of more than two substances have a higher NS (Conway, Kane, Ball, Poling, & Rounsaville, 2003). Another study suggests that a high NS, a high RD, and a low self-directedness are related to an early age of onset of heroin abuse. In addition, cooperativeness may help with predicting community admission (Fassino, Daga, Delsedime, Rogna, & Boggio, 2004). Furthermore, heroin-dependent females with a high NS and a high HA show lower mental maturity (Gerdner, Nordlander, & Pedersen, 2002). Evren and colleagues concluded that, among heroin-dependent subjects, a high cooperativeness and a high self-transcendence are related to neurotic defense styles, while a low RD, a low self-directedness, and a high self-transcendence are linked to immature defense styles (Evren et al., 2012).

From the biological standpoint, it has been found that polymorphism affecting the catechol-O-methyltransferase (COMT) gene is related to both NS and the age of onset of heroin use (Demetrovics et al., 2010; Li et al., 2011). Lee et al. (2013) discovered that the aldehyde dehydrogenase 2 gene interacts with NS in heroin-dependent subjects. Polymorphism of the dopamine D3 receptor gene (a candidate gene associated with heroin dependence) was found to be unlinked to the features of the TPQ (Kuo et al., 2014). According to Yeh's research group (Yeh, Lu, Tao, Shih, & Huang, 2011; Yeh et al., 2010), there is no association between the norepinephrine transporter gene or the dopamine active transporter 1 gene and the results of the TPQ. Lin et al. (2013) suggested that the retention rate for methadone maintenance treatment is influenced by NS and the functioning of the hypothalamic–pituitary–adrenal axis. Furthermore, less craving was experienced by subjects with a low HA (De Los Cobos et al., 2011). In heroin-dependent individuals with comorbid depression, a study found that there was a better response to the antidepressant sertraline, a selective serotonin reuptake inhibitor, among individuals with low RD scores; furthermore, individuals with a high HA and persistence were less likely to achieve abstinence (Raby et al., 2006). Another study revealed that a profile consisting of high NS/high HA/low RD traits is common among heroin users and that this coincides with a higher rate of Taq1A polymorphism, which implies that the Taq1A polymorphism is related to vulnerability to addiction (Teh, Izuddin, Fazleen, Zakaria, & Salleh, 2012). Finally, the same personality pattern of high NS and high HA is also found among people who are addicted to both alcohol and heroin (Wang et al., 2013).

Huang et al. (2014) in a study that compared adult male heroin abusers and a healthy population concluded that heroin abusers have significantly lower total RD (RD2 was excluded from this calculation), NS1, and RD4 scores, but higher NS2 and HA4 scores, than non-heroin users (Table 1). High NS2 (impulsivity) and low NS1 (exploratory excitability) suggest that heroin users are impulsive and tend not to explore novel environmental stimuli; in other words, they have restricted interests. This specific NS behavior pattern coincides with the poor self-control, rigidity, and self-indulgent behavior patterns that can be observed in heroin users. A high HA4 (fatigability and asthenia) and heroin addiction are likely to mutually influence each other. People with high fatigability and low frustration tolerance are less likely to resist the temptation of heroin, while long-term intoxication and withdrawal contribute to exhaustion. A low total RD score implies heroin users have limited social interactions, while a low RD4 (dependence) reflects their social isolation, namely, the fact that they have insufficient or unstable interpersonal relationships with healthy individuals seems likely to be a factor impeding heroin abstinence (Huang et al., 2014).

The same study suggests that the TPQ can serve as a predictive factor for substance use behavior and social functioning and that this could be identified and confirmed by the Opiate Treatment Index (OTI) assessment (Huang et al., 2014). The OTI is a comprehensive evaluation instrument that examines opioid treatment from health, legal, and social perspectives. The OTI assesses heroin use behavior, methadone use behavior, other substance use behavior, social functioning, and criminality (Darke, Hall, Wodak, Heather, & Ward, 1992). In this context, RD2 (persistence) is positively correlated with the maximal methadone dose, while at the same time it is negatively correlated with the amount of recent heroin exposure. These findings suggests that high persistence is associated with superior compliance to methadone treatment because, hypothetically, a high dose of methadone would be prescribed only to patients who take it regularly. Methadone adherence therefore contributes to less recent heroin exposure. On integrating these findings across the various studies, it can be concluded that persistence is related to a lower rate of abstinence, but also a lower rate of recent heroin exposure due to methadone adherence.

Another important index to determine the severity of heroin dependence is the length of time since the last use of heroin. In this context a short period suggests frequent heroin exposure. NS, mainly NS2 (impulsivity) and NS4 (disorderliness), and HA, mainly HA1 (anticipatory worry) and HA4 (fatigability and asthenia), are negatively correlated with "the duration from the last use of heroin"; this finding indicates that an impulsive and disorderly individual with a negativistic viewpoint and higher fatigability has increased exposure to heroin.

Alcohol-drinking behavior among heroin abusers can be assessed by the "Q score of alcohol consumption," which is positively correlated with NS1, NS3, RD4, and the total scores for NS and RD. In summary, heroin abusers have a tendency to conform to peer pressure (RD4, dependence) and presumably they tend to drink with friends, resorting to alcohol for joyfulness (NS1, exploratory excitability), but then have difficulty saving or delaying gratification (NS3, extravagance). Other statistically significant findings of the TPQ when used to assess heroin abusers are a low RD1 (sentimentality), which refers to subjects who are insensitive to social cues related to approval or rejection and has also been shown to be related to greater use of tranquilizers. Furthermore, heroin abusers also have a high HA1 (anticipatory worry) and are less likely to show criminal behavior, as well as a high NS4 (disorderliness), which means that these subjects are unconventional rule-breakers who have better social functioning.

CATEGORICAL VIEWS OF PERSONALITY IN RELATION TO HEROIN USE: THE SWITCH FROM THE TPQ

Studies have shown that both personality traits and the environment can be contributing factors to the development of heroin dependence. One study has shown that there is a prevalence rate of 77% when examining personality disorder among methadone maintenance patients (Teplin, O'Connell, Daiter, & Varenbut, 2004). Common representative characteristics and behaviors are placed in different categories to define personality disorders using the DSM-IV criteria. Compared to dimensional personality disorders, categorical classification is better when carrying out the diagnosis of and evaluating the treatment response of a group of people who share similar characteristics. However, it is less diverse than the dimensional classification when elucidating the heterogeneity of individuals' personalities. Many studies have identified an association between heroin abuse and cluster B personality disorders based on the DSM-IV, especially borderline personality disorder (BPD) and antisocial personality disorder (ASPD).

The risk of substance use disorder, including heroin dependence, is increased among BPD and ASPD individuals (Feske, Tarter, Kirisci, & Pilkonis, 2006). In 2005, a study disclosed that up to 72% of study subjects met the criteria for ASPD and that, in addition, 47% of study subjects met the criteria for BPD (Ross et al., 2005). Another study showed a 75% prevalence for ASPD and a 51% prevalence for BPD among young Australian heroin-dependent subjects age 18 to 24 (Mills, Teesson, Darke, Ross, & Lynskey, 2004). Interestingly, a gender difference has also been found for the association between ASPD and heroin dependence (Yang, Mamy, Wang, et al., 2014; Yang, Mamy, Zhou, et al., 2014). Furthermore, the rate of BPD and ASPD comorbidity is high among intravenous (iv) drug users (Mackesy-Amiti, Donenberg, & Ouellet, 2012). However, another study that compared iv heroin users and non-iv heroin users found that the latter had a

TABLE 1 Comparison of Demographic and TPQ Data between Heroin Users and Controls

	Heroin Users n=42	Controls n=43	F/χ^2	p Value
Age (years)	36.14 (±6.26)	33.74 (±9.49)	6.202	0.172
Gender (male)	42 (100%)	43 (100%)	NA	NA
Educational level (>9 years)	42 (100%)	43 (100%)	NA	NA
Marital status (married)	8 (19.05%)	10 (23.26%)	0.179	0.672
Common substance use behaviors				
Cigarettes, daily amount	16.59 (±8.24)	3.19 (±6.66)	1.044	<0.001[b]
Duration of cigarette use (years)	10.88 (±7.55)	2.47 (±5.57)	2.614	<0.001[b]
Psychological state				
STAI state	45.24 (±9.38)	40.12 (±11.13)	2.768	0.024[a]
STAI trait	43.79 (±8.81)	40.98 (±10.28)	2.379	0.180
BDI	11.76 (±10.18)	8.63 (±10.14)	0.216	0.159
TPQ data				
NS1 (exploratory excitability)	3.56 (±1.40)	4.35 (±1.67)	1.610	0.021[a]
NS2 (impulsivity)	3.00 (±1.48)	2.26 (±1.76)	3.664	0.038[a]
NS3 (extravagance)	3.17 (±1.48)	3.16 (±1.60)	1.118	0.996
NS4 (disorderliness)	4.64 (±1.49)	4.60 (±1.60)	0.209	0.908
HA1 (anticipatory worry)	3.78 (±1.85)	4.07 (±2.35)	2.373	0.543
HA2 (fear of uncertainty)	3.45 (±1.38)	3.70 (±1.82)	2.765	0.488
HA3 (shyness with strangers)	2.81 (±1.67)	2.77 (±1.93)	1.780	0.915
HA4 (fatigability and asthenia)	5.00 (±1.93)	3.81 (±2.78)	9.559	0.025[a]
RD1 (sentimentality)	4.14 (±1.03)	4.10 (±1.13)	0.198	0.836
RD2 (persistence)	4.87 (±1.21)	5.40 (±1.53)	2.877	0.082
RD3 (attachment)	6.17 (±1.58)	6.70 (±2.09)	4.251	0.189
RD4 (dependence)	1.91 (±1.59)	2.70 (±1.42)	0.253	0.017[a]
NS total	14.37 (±3.58)	14.37 (±3.23)	0.267	0.998
HA total	15.04 (±4.86)	14.34 (±7.18)	5.041	0.603
RD total	12.21 (±2.57)	13.49 (±3.07)	0.766	0.041[a]

RD total does not include RD2.
STAI, State and Trait Anxiety Inventory; BDI, Beck Depression Inventory; TPQ, Tridimensional Personality Questionnaire; NS, novelty seeking; HA, harm avoidance: RD, reward dependence. Results are expressed as mean (±SD).
[a]$p<0.05$.
[b]$p<0.01$.
Adapted from Huang et al., (2014).

lower prevalence of comorbid posttraumatic stress disorder, while no differences were found in prevalence of BPD and ASPD between the two groups (Darke, Williamson, Ross, Teesson, & Lynskey, 2004).

Studies have shown that there is a higher percentage of antisocial personality characteristics, such as hostility and risk-taking behaviors, among heroin users compared to the general population (Chatham, Knight, Joe, & Simpson, 1995) and that these personality traits are associated with a continuation of substance use and poor psychosocial functioning (Joe, Simpson, & Hubbard, 1991). Furthermore, one study found that there was higher rate of suicide and attempted suicide among heroin-dependent subjects with BPD (Darke, Ross, Lynskey, & Teesson, 2004). Another study, which recruited heroin-dependent subjects with BPD, revealed that these

TABLE 2 Personality Patterns among Heroin Users and a Control Group

Personality Pattern (NS, HA, RD)	Heroin Users	Control Group	Total
Independent (schizoid, −−−)	4 (9.52%)	1 (2.33%)	5 (5.88%)
Reliable (staid, −−+)	4 (9.52%)	5 (11.63%)	9 (10.59%)
Methodical (obsessional, −+−)	8 (15.38%)	6 (13.95%)	14 (16.47%)
Adventurous (antisocial, +−−)	1 (2.38%)	7 (16.28%)	8 (9.41%)
Explosive (borderline, ++−)	17 (40.48%)	6 (13.95%)	23 (27.06%)
Passionate (histrionic, +−+)	3 (7.14%)	6 (13.95%)	9 (10.59%)
Cautious (avoidant, −++)	0 (0.00%)	3 (6.98%)	3 (3.53%)
Sensitive (narcissistic, +++)	5 (11.90%)	9 (20.93%)	14 (16.47%)
Total	42	43	85

Adapted from Huang et al., (2014).

subjects had a higher rate of committing crimes, a greater risk of undergoing heroin overdose, and poorer global mental health over a 3-year follow-up period (Darke et al., 2007). Furthermore, among comorbid BPD heroin-dependent individuals there are higher risks of suicide attempt, of needle sharing, and of psychopathology, while, on the other hand, these risks are lower in the ASPD group (Darke, Williamson, et al., 2004). Among female substance users, one study has a revealed a high correlation between BPD and Symptom Checklist-90 scores, which suggests that this specific population has prominent psychiatric symptoms (Jansson, Hesse, & Fridell, 2008). Another study demonstrated that there are consistent findings of higher criminality within the ASPD and BPD groups among heroin-dependent subjects (Marel et al., 2013). Overall, among heroin-dependent subjects, when there is comorbid BPD, there is an overall higher risk of the following: needle sharing, injection-related health problems, heroin overdose, major depressive disorder, attempted suicide, and suicide (Darke, Ross, & Teesson, 2005; Darke, Ross, Williamson, & Teesson, 2005).

When reviewed from the perspectives of therapeutic effectiveness and physiology, it was found that heroin-dependent subjects who receive buprenorphine treatment and those with comorbid major depressive disorders are at lower risk of illicit opioid use compared to those with BPD and ASPD (Gerra et al., 2006). Dialectical behavior therapy is known to be effective when treating heroin-dependent women with BPD (Linehan et al., 2002). The association between BPD and polymorphism of the neurexin 3 gene was established by the study of Panagopoulos et al. (2013). Furthermore, Yang, Mamy, Wang, et al. (2014) and Yang, Mamy, Zhou, et al. (2014), in a study investigating female heroin-dependent subjects, linked comorbid BPD with a serotonin transporter intron 2 variable number tandem repeat and with polymorphism of the monoamine oxidase A gene promoter region.

The above study results explore opiate addiction from a categorical perspective. However, it is intriguing to ask whether personality categories and personality dimensions are interchangeable or correspond with each other to some degree. Cloninger and Cloninger (2011) established eight possible personality characteristics that closely correspond to combinations of the three dimensions of the TPQ. Taking two of the cluster B personality traits as examples, the explosive (or borderline) pattern was found to display high NS, high HA, and low RD, while the adventurous (or antisocial) pattern was found to display high NS, low HA, and low RD. However, in this context it is important to note that TPQ personality patterns do not always match with DSM-IV criteria in terms of personality disorders.

To date, the most common and representative personality pattern among heroin users is the borderline trait (high NS, high HA, and low RD), which implies the presence of behaviors related to high impulsivity with limited and unstable social interactions (Table 2) (Huang et al., 2014; Teh et al., 2012). This TPQ combination leads to difficulty in sustaining stable relationships during stress and the display of maladaptive coping skills that lead to the abuse of substances. Assuming that the explosive personality pattern is a predisposing factor contributing to heroin dependence, an odds ratio of 4.19 (95% confidence interval is 1.45–12.10) was calculated in one study (Huang et al., 2014).

THE AUTONOMIC BASIS OF BORDERLINE PERSONALITY TRAIT AND HEROIN USE DISORDER

The phenomenological biological basis of TPQ and its clinical implications with respect to the borderline trait supports the influence of the borderline trait on heroin use from a physiological viewpoint. Numerous studies have been conducted to find the underlying physiological mechanisms that evolve into specific psychological responses; this has been done via various approaches including neurophysiology, neuroanatomy, and psychophysiology. Porges proposed the polyvagal theory in 1995 to explain the vagal pathways of the heart, specifically two sources of vagal efferents that terminate at the sinoatrial node (Porges, 1995). The polyvagal theory has evolved around how these two pathways are able to regulate heart rate in response to stressors. Specifically, the unmyelinated vagus originates from the dorsal motor nucleus and mediates reflexive cardiac activity, namely, the primitive threat response including fight or flight, orienting, and death feigning, all of which are common in reptiles. In contrast, the myelinated vagus originates from the nucleus ambiguus and is unique to mammals; it has a role in attention, vocalization, facial expression, and the coordination of motion, emotion, and communication.

In nonthreatening situations, the myelinated path from the nucleus ambiguus regulates heart rate by sending constant inhibitory signals; these allow individuals to engage in tasks with low metabolic demands (low adrenergic activation). When facing a threatening situation, a rapid drop in the inhibitory signal (a decrease in vagal tone) results in the sympathetically dominant flight-or-fight response. This vagal withdrawal phenomenon enables the mammal to detect novelty, communicate socially, and adapt to its environment (Porges, 1995). Any dysfunction of vagal withdrawal leads to inappropriate processing with respect to movement, emotion, and communication, which may explain the core symptoms of emotional dysregulation in BPD individuals. Weinberg and colleagues found that BPD subjects exhibit increased sympathetic activity and decreased parasympathetic activity compared to healthy controls; this agrees with the polyvagal theory (Weinberg, Klonsky, & Hajcak, 2009). An imbalance affecting sympathovagal activity could also contribute to a higher risk of stroke and of ischemic heart disease among the BPD population.

The autonomic nervous system (ANS) can be assessed by frequency domain analysis of heart rate variability (HRV); in this context cardiac parasympathetic modulation is measured by high-frequency power (HF) and cardiac sympathetic modulation is measured by normalized low-frequency power (LF%) and the LF/HF ratio. Studies have discovered that heroin abusers show decreased cardiac vagal activity compared to a healthy population (low HF). Immediate surges of HF after methadone treatment indicate that methadone facilitates vagal stimulation (Chang et al., 2012). These findings suggest that long-term opiate exposure reduces vagal modulation, while methadone treatment has the opposite effect; this may help to explain how methadone brings about relief of withdrawal symptoms, namely adrenergic hyperactivity.

A study of 44 male heroin users by Huang et al. was used to explore the polyvagal theory and TPQ; this found consistent evidence showing a relationship to the borderline pattern (Huang et al., 2012). After methadone treatment, borderline-pattern heroin abusers show a lower ΔHF but a higher ΔLF/HF compared to non-borderline-pattern heroin abusers (Figure 1); these findings suggest that, among individuals with opiate exposure, the borderline property further decreases parasympathetic activity, making opiate addiction more difficult to overcome when there is administration of methadone. Owing to inefficient vagal modulation in the BPD population, an overwhelming sympathetic tone is able to induce an intense flight-or-fight response when there is a stressful event. This intense but inappropriate engagement of the defense system results in affective instability, difficulty in anger management, and self-destructive behavior. The limited increase in HF brought about by methadone implies that BPD subjects receive partial relief from this intense sympathetic defense state. This partial relief serves as an incentive that reinforces borderline pattern patients to self-medicate. This may explain why borderline pattern heroin abusers have a low retention rate and poor prognosis when on methadone maintenance treatment. However, it should be noted that in this study, borderline personality pattern is defined according to the TPQ and this does not necessarily fulfill the criteria of the DSM-IV.

The same study pioneered the introduction of a quantitative index for severity of the borderline pattern as assessed by the TPQ (borderline index = NS total + HA total − RD total) and showed that there is a high correlation with ANS as measured using the HRV parameters. This index is negatively correlated with ΔHF, but positively correlated with ΔLF% and ΔLF/HF (Figure 2). At the same time, when examining the TPQ subdimensions individually, only RD, which is related to norepinephrine activity, shows a significant correlation with ΔLF% and ΔLF/HF (Huang et al., 2012).

APPLICATIONS TO OTHER ADDICTIONS AND OTHER SUBSTANCE MISUSE

The concept of TPQ can be applied to research related to other substances, no matter whether this involves the original dimensional approach or there is a switch to the categorical view. Specifically, there have been studies exploring the relationship between the TPQ and the use of alcohol and cannabis. Whether high persistence is also associated with a better prognosis in terms of other substance addiction awaits further clarification. Which TPQ pattern is the most common among other substance use disorders is another interesting issue.

Importantly, research has shown that the stabilizing effect of methadone, which is paralleled by a decrease in sympathetic tone and an increase in vagal activity, is less obvious among individuals with the borderline personality pattern. Methadone is a central nervous system (CNS) depressant. Based on this viewpoint, other CNS depressants (such as alcohol or benzodiazepines) ought to have similar effects. These aspects need to be explored using the HRV technique in the future.

CONCLUSIONS

The TPQ and the TCI have been applied in many studies related to personality trait research. The TPQ's main dimensions not only are able to be used to predict substance use behaviors, but also have proved to be associated with various genetic polymorphisms and with substance abuse treatment responses. Combinations of the various TPQ patterns have been found to correspond with various DSM-IV personality disorders. Among heroin-dependent subjects, the borderline (explosive) pattern of high NS/high HA/low RD is the most common TPQ combination. The psychopathology behind this combination can be explained by phenomena associated with the ANS. Further studies are warranted to elucidate the genetic, physiological, and psychopathological basis of substance use behavior using approaches that apply the TPQ and the TCI.

DEFINITION OF TERMS

Novelty seeking This is an innate temperament that represents a tendency toward behavioral activation and being curious about novel stimuli. It is related to impulsivity and exploration.

Harm avoidance This is an innate temperament that reflects a tendency toward behavioral inhibition and avoiding aversive stimuli. It is associated with anticipatory worry and fear of uncertainty.

Reward dependence This is an innate temperament that represents a tendency toward a response to reward signals, especially social messages. It is related to intimacy and dependence.

Persistence This is an innate temperament that represents a tendency toward maintaining a behavior and resistance to extinction.

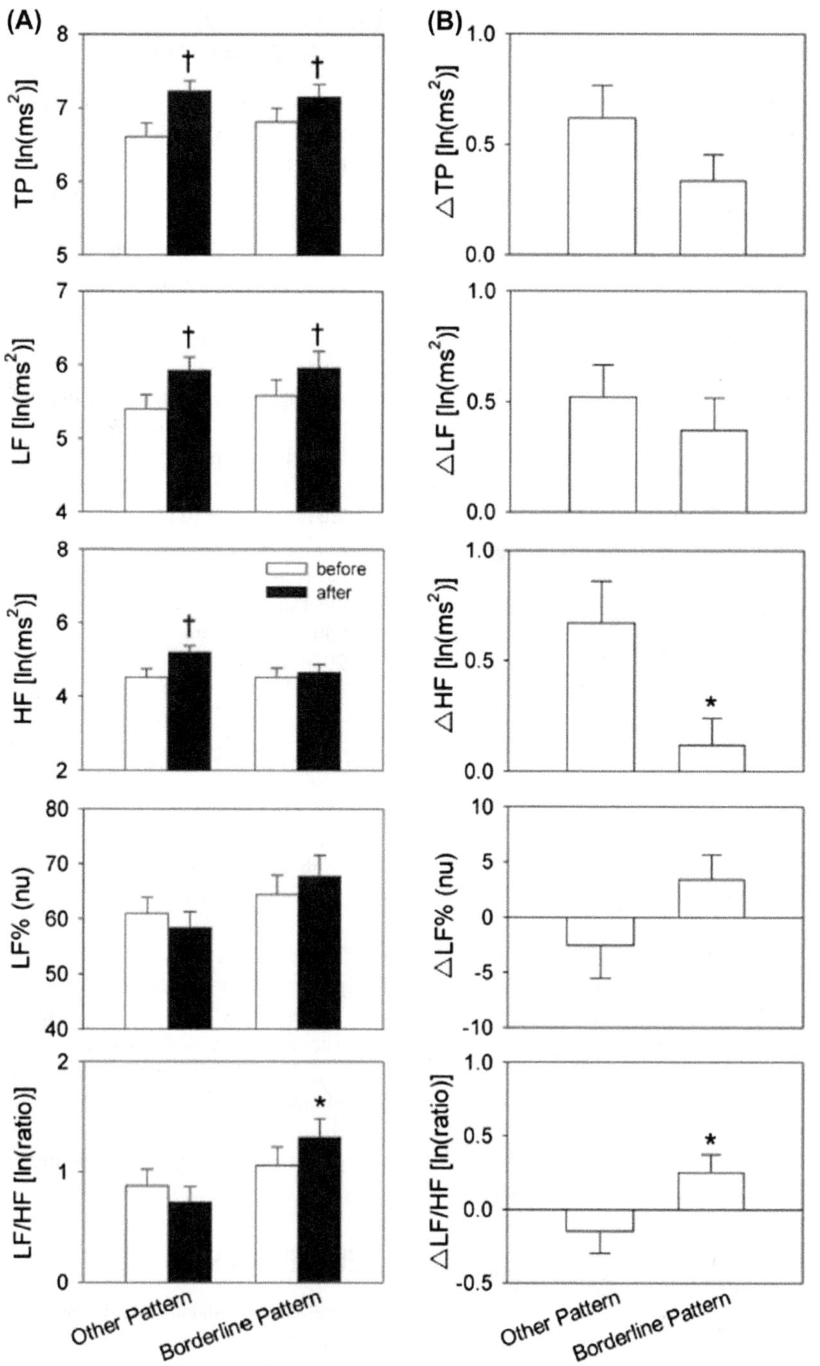

FIGURE 1 Comparison of autonomic activities (A) before and (B) after taking methadone. *Adapted from Huang et al. (2012).*

Borderline personality disorder This is a personality category defined by the DSM-IV. The core psychopathologies include fear of being abandoned, emptiness, emotional dysregulation, and impulsive and self-injurious behaviors.

Explosive (borderline) pattern This is a personality category based on switching from TPQ dimensions. The borderline pattern corresponds to high NS, high HA, and low RD. It shows a positive correlation with the BPD of DSM-IV.

Polyvagal theory This theory proposes that the vagal system of mammals has two different origins, the dorsal motor nucleus and the nucleus ambiguus, which regulate defensive and pro-social behaviors, respectively. Some research considers the emotional dysregulation of BPD to be associated with unstable vagal activity, which can be explained by the polyvagal theory.

Heart rate variability Heart rate is affected by both the sympathetic and the parasympathetic nervous systems. Therefore, autonomic functioning can be understood via an analysis of heart rate. Five minutes of frequency domain analysis of the HRV is able to separate sympathetic and vagal modulation.

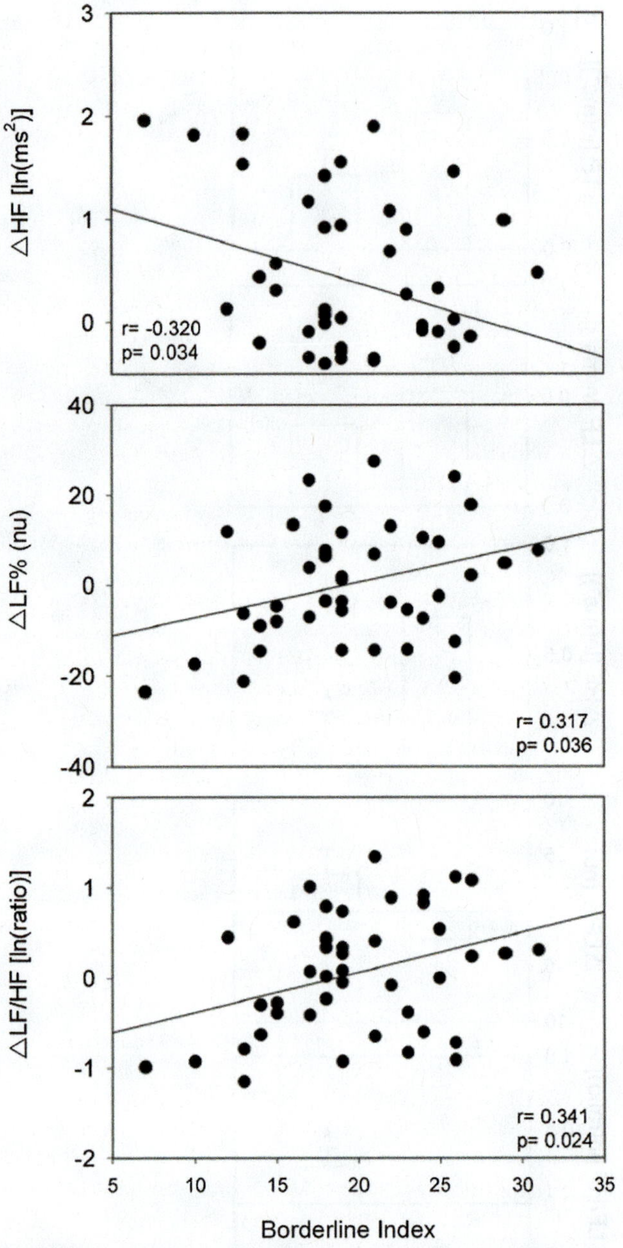

FIGURE 2 A simple linear regression model of the borderline index against ΔHF, ΔLF%, and ΔLF/HF comparing before and after taking methadone. *Adapted from Huang et al. (2012).*

High-frequency power This power is found between 0.15 and 0.40 Hz after frequency domain analysis. It represents parasympathetic activity.

Normalized low-frequency power LF is found between 0.04 and 0.15 Hz after frequency domain analysis and is influenced by both the sympathetic and the vagal systems. LF% shows a high correlation with sympathetic activity.

Ratio of low-frequency power to high-frequency power This is another index of sympathetic activity. Some researchers consider this to be an index of balance between the sympathetic and the vagal systems.

KEY FACTS OF PERSONALITY FEATURES IN HEROIN USERS

- Subjects with high NS show poor retention in methadone and buprenorphine maintenance therapy programs. They are also prone to be dependent on more than one substance (Conway et al., 2003; Helmus et al., 2001; Lin et al., 2013; Roll et al., 2004).
- NS in heroin users is associated with polymorphism of the COMT gene (Demetrovics et al., 2010; Li et al., 2011).
- Among subjects comorbid with heroin dependence and depression, low RD is related to a good response to sertraline, while individuals with high HA and high persistence are more likely to be abstinent (Raby et al., 2006).
- Among subjects with heroin dependence who have BPD, compared to those without BPD, individuals show more criminal behavior, are more prone to heroin overdose, are more likely to attempt suicide, show increased needle-sharing behavior, and are more prone to major depression (Darke, Ross, et al., 2004; Darke, Ross, & Teesson, 2005; Darke et al., 2007; Darke, Ross, Williamson, et al., 2005).
- Overall, 72–75% of heroin users are diagnosed with ASPD, while BPD is present in 47–51% of heroin users (Mills et al., 2004; Ross et al., 2005).

SUMMARY POINTS

- The TPQ has been widely applied in biological and psychosocial research related to heroin use.
- High persistence is related to low heroin consumption and good compliance among individuals undergoing methadone maintenance therapy.
- The TPQ can be switched to categorical concepts; for example, explosive (borderline) pattern, which is composed of high NS, high HA, and low RD, is similar to BPD.
- The explosive (borderline) personality pattern is the most common TPQ pattern among heroin abusers.
- Subjects with the explosive (borderline) personality pattern show a reduced sympathetic reduction and a lower parasympathetic elevation after taking methadone.

REFERENCES

Abbate-Daga, G., Amianto, F., Rogna, L., & Fassino, S. (2007). Do anorectic men share personality traits with opiate dependent men? A case-control study. *Addictive Behaviors*, 32(1), 170–174.

Chang, L. R., Lin, Y. H., Kuo, T. B., Ho, Y. C., Chen, S. H., Wu Chang, H. C., ... Yang, C. C. (2012). Cardiac autonomic modulation during methadone therapy among heroin users: a pilot study. *Progress in Neuro-Psychopharmacology and Biological Psychiatry*, 37(1), 188–193.

Chatham, L. R., Knight, K., Joe, G. W., & Simpson, D. D. (1995). Suicidality in a sample of methadone maintenance clients. *American Journal of Drug and Alcohol Abuse*, 21(3), 345–361.

Cloninger, C. R. (1994). *The temperament and character inventory (TCI): A guide to its development and use.* St Louis, Missouri: Washington University: Centre for Psychobiology of Personality.

Cloninger, C. R., & Cloninger, K. M. (2011). *Person-centered therapeutics*.

Cloninger, C. R., Przybeck, T. R., & Svrakic, D. M. (1991). The tridimensional personality questionnaire: U.S. normative data. *Psychological Reports*, 69(3 Pt 1), 1047–1057.

Cloninger, C. R., Svrakic, D. M., & Przybeck, T. R. (1993). A psychobiological model of temperament and character. *Archives of General Psychiatry*, 50(12), 975–990.

Conway, K. P., Kane, R. J., Ball, S. A., Poling, J. C., & Rounsaville, B. J. (2003). Personality, substance of choice, and polysubstance involvement among substance dependent patients. *Drug and Alcohol Dependence*, 71(1), 65–75.

Darke, S., Hall, W., Wodak, A., Heather, N., & Ward, J. (1992). Development and validation of a multi-dimensional instrument for assessing outcome of treatment among opiate users: the Opiate Treatment Index. *British Journal of Addiction*, 87(5), 733–742.

Darke, S., Ross, J., Lynskey, M., & Teesson, M. (2004). Attempted suicide among entrants to three treatment modalities for heroin dependence in the Australian Treatment Outcome Study (ATOS): prevalence and risk factors. *Drug and Alcohol Dependence*, 73(1), 1–10.

Darke, S., Ross, J., & Teesson, M. (2005). Twelve-month outcomes for heroin dependence treatments: Does route of administration matter? *Drug and Alcohol Review*, 24(2), 165–171.

Darke, S., Ross, J., Williamson, A., Mills, K. L., Havard, A., & Teesson, M. (2007). Borderline personality disorder and persistently elevated levels of risk in 36-month outcomes for the treatment of heroin dependence. *Addiction*, 102(7), 1140–1146.

Darke, S., Ross, J., Williamson, A., & Teesson, M. (2005). The impact of borderline personality disorder on 12-month outcomes for the treatment of heroin dependence. *Addiction*, 100(8), 1121–1130.

Darke, S., Williamson, A., Ross, J., Teesson, M., & Lynskey, M. (2004). Borderline personality disorder, antisocial personality disorder and risk-taking among heroin users: findings from the Australian Treatment Outcome Study (ATOS). *Drug and Alcohol Dependence*, 74(1), 77–83.

De Los Cobos, J. P., Siñol, N., Trujols, J., Bañuls, E., Batlle, F., & Tejero, A. (2011). Drug-dependent inpatients reporting continuous absence of spontaneous drug craving for the main substance throughout detoxification treatment. *Drug and Alcohol Review*, 30(4), 403–410.

Demetrovics, Z., Varga, G., Szekely, A., Vereczkei, A., Csorba, J., Balazs, H., ... Barta, C. (2010). Association between Novelty Seeking of opiate-dependent patients and the catechol-O-methyltransferase Val(158)Met polymorphism. *Comprehensive Psychiatry*, 51(5), 510–515.

Evren, C., Ozcetinkaya, S., Ulku, M., Cagil, D., Gokalp, P., Cetin, T., & Yigiter, S. (2012). Relationship of defense styles with history of childhood trauma and personality in heroin dependent inpatients. *Psychiatry Research*, 200(2–3), 728–733.

Fassino, S., Daga, G. A., Delsedime, N., Rogna, L., & Boggio, S. (2004). Quality of life and personality disorders in heroin abusers. *Drug and Alcohol Dependence*, 76(1), 73–80.

Feske, U., Tarter, R. E., Kirisci, L., & Pilkonis, P. A. (2006). Borderline personality and substance use in women. *American Journal on Addictions*, 15(2), 131–137.

Gerdner, A., Nordlander, T., & Pedersen, T. (2002). Personality factors and drug of choice in female addicts with psychiatric comorbidity. *Substance Use and Misuse*, 37(1), 1–18.

Gerra, G., Leonardi, C., D'Amore, A., Strepparola, G., Fagetti, R., Assi, C., ... Lucchini, A. (2006). Buprenorphine treatment outcome in dually diagnosed heroin dependent patients: a retrospective study. *Progress in Neuro-Psychopharmacology and Biological Psychiatry*, 30(2), 265–272.

Helmus, T. C., Downey, K. K., Arfken, C. L., Henderson, M. J., & Schuster, C. R. (2001). Novelty seeking as a predictor of treatment retention for heroin dependent cocaine users. *Drug and Alcohol Dependence*, 61(3), 287–295.

Hofmann, S. G., & Loh, R. (2006). The Tridimensional Personality Questionnaire: changes during psychological treatment of social phobia. *Journal of Psychiatric Research*, 40(3), 214–220.

Huang, W. L., Chang, L. R., Chen, Y. Z., Chang, H. C., Hsieh, M. H., Lin, C. H., & Lin, Y. H. (2014). The tridimensional personality of male heroin users treated with methadone in Taiwan. *Comprehensive Psychiatry*, 55(5), 1220–1226.

Huang, W. L., Lin, Y. H., Kuo, T. B., Chang, L. R., Chen, Y. Z., & Yang, C. C. (2012). Methadone-mediated autonomic functioning of male patients with heroin dependence: the influence of borderline personality pattern. *PLoS One*, 7(5), e37464.

Jansson, I., Hesse, M., & Fridell, M. (2008). Personality disorder features as predictors of symptoms five years posttreatment. *American Journal on Addictions*, 17(3), 172–175.

Joe, G. W., Simpson, D. D., & Hubbard, R. L. (1991). Unmet service needs in methadone maintenance. *International Journal of the Addictions*, 26(1), 1–22.

Kuo, S. C., Yeh, Y. W., Chen, C. Y., Huang, C. C., Chang, H. A., Yen, C. H., ... RB, H. S. (2014). DRD3 variation associates with early-onset heroin dependence, but not specific personality traits. *Progress in Neuro-Psychopharmacology and Biological Psychiatry*, 51, 1–8.

Le Bon, O., Basiaux, P., Streel, E., Tecco, J., Hanak, C., Hansenne, M., ... Dupont, S. (2004). Personality profile and drug of choice; a multivariate analysis using Cloninger's TCI on heroin addicts, alcoholics, and a random population group. *Drug and Alcohol Dependence*, 73(2), 175–182.

Lee, S. Y., Wang, T. Y., Chen, S. L., Huang, S. Y., Tzeng, N. S., Chang, Y. H., ... Lu, R. B. (2013). Interaction between novelty seeking and the aldehyde dehydrogenase 2 gene in heroin-dependent patients. *Journal of Clinical Psychopharmacology*, 33(3), 386–390.

Li, T., Yu, S., Du, J., Chen, H., Jiang, H., Xu, K., ... Zhao, M. (2011). Role of novelty seeking personality traits as mediator of the association between COMT and onset age of drug use in Chinese heroin dependent patients. *PLoS One*, 6(8), e22923.

Lin, S. H., Chen, W. T., Chen, K. C., Lee, S. Y., Lee, I. H., Chen, P. S., ... Yang, Y. K. (2013). Associations among hypothalamus-pituitary-adrenal axis function, novelty seeking, and retention in methadone maintenance therapy for heroin dependency. *Journal of Addiction Medicine*, 7(5), 335–341.

Linehan, M. M., Dimeff, L. A., Reynolds, S. K., Comtois, K. A., Welch, S. S., Heagerty, P., & Kivlahan, D. R. (2002). Dialectical behavior therapy versus comprehensive validation therapy plus 12-step for the treatment of opioid dependent women meeting criteria for borderline personality disorder. *Drug and Alcohol Dependence*, 67(1), 13–26.

Mackesy-Amiti, M. E., Donenberg, G. R., & Ouellet, L. J. (2012). Prevalence of psychiatric disorders among young injection drug users. *Drug and Alcohol Dependence*, 124(1–2), 70–78.

Marel, C., Mills, K. L., Darke, S., Ross, J., Slade, T., Burns, L., & Teesson, M. (2013). Static and dynamic predictors of criminal involvement among people with heroin dependence: findings from a 3-year longitudinal study. *Drug and Alcohol Dependence*, 133(2), 600–606.

Milivojevic, D., Milovanovic, S. D., Jovanovic, M., Svrakic, D. M., Svrakic, N. M., Svrakic, S. M., & Cloninger, C. R. (2012). Temperament and character modify risk of drug addiction and influence choice of drugs. *American Journal on Addictions*, 21(5), 462–467.

Mills, K. L., Teesson, M., Darke, S., Ross, J., & Lynskey, M. (2004). Young people with heroin dependence: findings from the Australian Treatment Outcome Study (ATOS). *Journal of Substance Abuse Treatment, 27*(1), 67–73.

Panagopoulos, V. N., Trull, T. J., Glowinski, A. L., Lynskey, M. T., Heath, A. C., Agrawal, A., ... Nelson, E. C. (2013). Examining the association of NRXN3 SNPs with borderline personality disorder phenotypes in heroin dependent cases and socio-economically disadvantaged controls. *Drug and Alcohol Dependence, 128*(3), 187–193.

Porges, S. W. (1995). Orienting in a defensive world: mammalian modifications of our evolutionary heritage. A Polyvagal Theory. *Psychophysiology, 32*(4), 301–318.

Raby, W. N., Carpenter, K. M., Aharonovich, E., Rubin, E., Bisaga, A., Levin, F., & Nunes, E. V. (2006). Temperament characteristics, as assessed by the tridimensional personality questionnaire, moderate the response to sertraline in depressed opiate-dependent methadone patients. *Drug and Alcohol Dependence, 81*(3), 283–292.

Roll, J. M., Saules, K. K., Chudzynski, J. E., & Sodano, R. (2004). Relationship between Tridimensional Personality Questionnaire scores and clinic attendance among cocaine abusing, buprenorphine maintained outpatients. *Substance Use and Misuse, 39*(6), 1025–1040.

Ross, J., Teesson, M., Darke, S., Lynskey, M., Ali, R., Ritter, A., & Cooke, R. (2005). The characteristics of heroin users entering treatment: findings from the Australian treatment outcome study (ATOS). *Drug and Alcohol Review, 24*(5), 411–418.

Teh, L. K., Izuddin, A. F., Fazleen, H. M. H., Zakaria, Z. A., & Salleh, M. Z. (2012). Tridimensional personalities and polymorphism of dopamine D2 receptor among heroin addicts. *Biological Research For Nursing, 14*(2), 188–196.

Teplin, D., O'Connell, T., Daiter, J., & Varenbut, M. (2004). A psychometric study of the prevalence of DSM-IV personality disorders among office-based methadone maintenance patients. *American Journal of Drug and Alcohol Abuse, 30*(3), 515–524.

Wang, T. Y., Lee, S. Y., Chen, S. L., Huang, S. Y., Chang, Y. H., Tzeng, N. S., ... Lu, R. B. (2013). Association between DRD2, 5-HTTLPR, and ALDH2 genes and specific personality traits in alcohol- and opiate-dependent patients. *Behavioural Brain Research, 250*, 285–292.

Weinberg, A., Klonsky, E. D., & Hajcak, G. (2009). Autonomic impairment in borderline personality disorder: a laboratory investigation. *Brain and Cognition, 71*(3), 279–286.

Yang, M., Mamy, J., Wang, Q., Liao, Y. H., Seewoobudul, V., Xiao, S. Y., & Hao, W. (2014). The association of 5-HTR2A-1438A/G, COMTVal158Met, MAOA-LPR, DATVNTR and 5-HTTVNTR gene polymorphisms and borderline personality disorder in female heroin-dependent Chinese subjects. *Progress in Neuro-Psychopharmacology and Biological Psychiatry, 50*, 74–82.

Yang, M., Mamy, J., Zhou, L., Liao, Y. H., Wang, Q., Seewoobudul, V., ... Hao, W. (2014). Gender differences in prevalence and correlates of antisocial personality disorder among heroin dependent users in compulsory isolation treatment in China. *Addictive Behaviors, 39*(3), 573–579.

Yeh, Y. W., Lu, R. B., Tao, P. L., Shih, M. C., & Huang, S. Y. (2011). A possible association of the norepinephrine transporter gene in the development of heroin dependence in Han Chinese. *Pharmacogenetics and Genomics, 21*(4), 197–205.

Yeh, Y. W., Lu, R. B., Tao, P. L., Shih, M. C., Lin, W. W., & Huang, S. Y. (2010). Neither single-marker nor haplotype analyses support an association between the dopamine transporter gene and heroin dependence in Han Chinese. *Genes, Brain and Behavior, 9*(6), 638–647.

Index

'*Note*: Page numbers followed by "f" indicate figures and "t" indicates tables.'

A

Abstinence-induced craving, fMRI studies on, 276
Abulia, 1000
ACC. *See* Anterior cingulate cortex (ACC)
ACD. *See* Alcoholic cerebellar degeneration (ACD)
Acetaldehyde, 119, 381–382, 516
  ability to inhibit cell growth, 555t
  to acetate, 382
  applications to other addictions and substance misuse, 559
  on brain cAMP-PKA signaling, 557, 558f
  chemical properties of, 553t
  chemical structures of, 552f
  classification, 552
  cytotoxicity of, 556–557
    cytotoxic effectiveness, 556, 557t
    mitochondrial toxicity, 557
    related to hydrophobicity and electrophilicity, 556
  damage to DNA by, 555–556
  definition of, 559
  endogenous, 552–553
    cytochrome P450, 552
    lipid peroxidation, 552–553
    myeloperoxidase, 552
  exogenous, 553–554
    dietary acetaldehyde, 553–554
    natural and manufactured aldehydes, 554
  and human health risks, 557–559
    Alzheimer disease, 557–559
    Parkinson disease, 557–559
    Wernicke encephalopathy, 559
  and neurotoxicity, 552
  producers of, 552
  production pathways, 552, 553f
  protein adducts, 554
    in cerebellum after ethanol treatment, 554f
    formation, 555, 558f
  reactivity, 554–555
    mutagenic potency, 554–555, 555t
    pathogenic effects, 554f
  at room temperature, 552
  vs. ethanol, 557
Acetaldehyde dehydrogenase 2 (*ALDH2*) gene, 548
  with *DRD2* gene and ANX/DEP ALC, 545–546
  with DRD2 gene and ASPD, 546

Acetate
  chemical structures of, 552f
  metabolism of, 382
  production from acetaldehyde metabolism, 553f
Acetylcholine (ACh), 94–95, 136, 163, 168, 175–176, 192
  nicotine effects, 150f
  release, measured by microdialysis, 496f
Acquired immunodeficiency syndrome, 53
ACTH. *See* Adrenocorticotropic hormone (ACTH)
Actin cytoskeleton, 126, 128–130
Action potentials (APs), 170
Addiction
  brain regions implicated in, 775f
  interaction model of, 26f
  neuroadaptations associated with, 651
  neurophysiological mechanisms, 25–27
    binge/intoxication stage, 25
    dopamine, role of, 26–27
    endogenous agonist opioids, role of, 27
    neurotransmitters and neuropeptides, role of, 25–26
    preoccupation/anticipation stage, 25
    transition to addiction, 25
    withdrawal/negative affect stage, 25
  pathophysiology of, 15–17
    brain areas involved, 15
    craving, withdrawal, and bingeing, 15–16
    effective connectivity changes in, 17f
    functional connectivity changes, 16–17, 16f
    from information theory perspective, 17
    key facts, 22
    as learning, 15
    Nacc lesions and, 21
    neuroplasticity and, 16
    stages, 15
  theories related to
    biological, 340–342
    comparison of, 341t
    hedonic dysregulation, 340–341
    implications, 344–345
    learning, 339–340
    sensitization-homeostasis theory, 341–342
  transition to, 28–29
    control theory, 29
    differential association theory, 28–29

emotional self-medication (ESM) and, 72–74
interface between biological and psychosocial theories, 28f
personality and drug habit of abuser, 29–33
secondary reinforcing effects of environment, 28
social environment of an addict and, 28
strain theory, 29
Addiction cycle model of SUD, 38, 39f
Addiction vulnerability
  definition of, 643
  hypothesis, 265–267
  sex differences in, 641–643
  social isolation model of, 638–640
Addiction vulnerability hypothesis, 265–267
Addiction-premature aging phenomenon
  initial studies, 1012–1013
  opioids and, 1013
Adducts, 559
Adenosine triphosphate-binding cassette (ABC), 910
ADH. *See* Alcohol dehydrogenase (ADH)
ADH alleles
  ADH1B, 511
    *ADH1B*1*, 511
    *ADH1B*2*, 511
    *ADH1B*3*, 511
    characteristics of, 512t
    frequencies by ethnicity, 512t
  and alcohol use disorders, 511
  application to other drugs of abuse, 516
  effects on neuropathology, 511
  and fetal alcohol spectrum disorders, 511–512, 512t
  neuropathology in, 513
  role in fetus, 513–516
ADH gene cluster, 521–528
  with risk of alcohol dependence, 527f
ADH1B, 516
*ADH1B*1* allele, 511
*ADH1B*2* allele, 511
*ADH1B*3* allele, 511, 515t
ADH1B alleles, 511
  *ADH1B*1*, 511
  *ADH1B*2*, 511
  *ADH1B*3*, 511
  characteristics of, 512t
  frequencies by ethnicity, 512t

1071

Adolescence, 597, 643
Adolescent AUD, 587
  brain regions involved in, 588f
  hippocampal volume reduction in, 588–592, 593f
  other brain regions with differential volumes in, 594–595
    insular cortex, 594–595
    striatum, 595
    temporal cortex, 594
  prefrontal cortex volume in, 593–594
  structural brain imaging in, 589t–591t
    findings with functional significance in, 592t
  VBM and related methods of, 587–588
Adrenalectomy, 614
Adrenocorticotropic hormone (ACTH), 610
  early life stress and, 640–641
Affect dysregulation theory, 265
A118G polymorphism, 406–407
Aggression
  alcohol associated with, 832
  animal, 828
  animal models of, 833
  cannabis withdrawal and, 832
  defensive, 833
  definition of, 827, 833
  displayed by mice lacking CB1 receptor, 832f
  human
    cannabis and, 828–831, 830f
    classification of, 827
    marijuana and, 828
  offensive, 833
  relation with CB1 receptor, in animal studies, 831–832
  social interaction test, 832f
Aging
  at biochemical level, 1014–1015
  at brain system level, 1015–1016
  key facts, 1013t
  opioids, impact of, 1013
  telomere length and telomerase activity, 1015, 1015f
Agonists, 709, 823
ALC + BP. See Alcoholism comorbid with bipolar disorder (ALC + BP)
Alcohol, Smoking, and Substance Involvement Screening Test (ASSIST), 649
  ASSIST-Lite protocol, 650–651
  development of, 649–650
    phase I, 650
    phase II, 650
    phase III, 650
Alcohol abuse, neurological consequences of, 445–446
  cortical spreading depression (CSD), 446–447
  on electrical activity, 446
Alcohol and other drug use disorder (AODUD) individuals, 104
  policy interventions to reduce tobacco-related mortality among, 110–111
  prevalence of tobacco use, 104–105, 105f
  smoking cessation treatment, 109
  tobacco-related mortality among, 107. See also Ethanol

Alcohol consumption
  adolescent and college-age, 457–458
  history of, 540
  impact on CNS pathology, 454
  and neuroimmune signaling, 459–461
    DNA methylation, 459, 461
    inflammasomes and, 459
    TLR4 signaling, 459, 622
  during pregnancy, 618
Alcohol dehydrogenase (ADH), 380, 381t, 516, 528
  alcohol metabolism and, 510–511, 511f
  alleles. See ADH alleles
  gastric activity of, 383
  genetics of metabolism, 382
Alcohol dependence (AD), 14, 520, 528
  ADH gene cluster with risk of, 527f
  ADH1B rs1229984 and ALDH2 rs671
    association analysis of, 522t
    genetic effect of, 521t
  comorbid with psychiatric disorders, 544
  definition of, 528
  GWAS and replication study of, 524t–526t
  OPRM1 gene and, 230
  pathophysiology of, 15–17
  premature aging and, 1019
  role of GR activity in transition to, 611–612
  tachykinins and, 196. See also Ethanol
Alcohol metabolism, 510–511, 511f
  ADH and, 510–511, 511f
  and genetic factors, 520–521
Alcohol misuse, 614
Alcohol use disorder (AUD), 389, 462, 575, 597, 634
  ADH alleles and, 511
  anti-inflammatory therapeutics for, 461
    minocycline, 461
    naltrexone, 461
    pioglitazone, 461
  and cerebellum, 577–579
    alcohol cues and alcohol-dependent individuals, 578
    cerebellar neuronal integrity in alcoholism, 578
    cerebellar white matter structure in alcoholism, 578
    functional connectivity in alcoholism, 579
    functional neuroimaging in alcoholics, 578
    MRI in alcohol-dependent individuals, 577–578
    resting state functional connectivity in alcoholics, 578–579
    structural and functional changes in AD, 579
  definition of, 454, 462
  endogenous opioids in, 406–408
  epidemiology by sex and, 638
  individual variations in vulnerability to, 406–407
  opioids in pharmacotherapy of, 407–408
  prevalence/incidence of, 454
  role of neuropeptides in, 401
Alcohol use in HIV population, 180
  adherence to HIV medication, impact on, 180

HIV pathogenesis and neuroAIDS, impact on, 180–181, 180f. See also Ethanol
Alcohol withdrawal syndrome (AWS), 466, 632
  ceftriaxone for, 472–473
  changes in activity of neurotransmitters, 632t
  GABA receptors in, 632
  glutamate signaling during, 469–471
    elevated extracellular glutamate levels, 471
    enhanced glutamate release, 469
    genetic variation, contribution of, 471–472
    impaired synaptic glutamate clearance, 471
    overactivated glutamate receptors, 470
    reduced glutamate synthesis, 469
  impact on astrocyte functionality, 470f
  pathophysiology of, 632
Alcoholic cerebellar degeneration (ACD), 633
Alcoholic hallucinosis, 633–634
Alcoholics, white matter and myelin alterations in, 423–425, 424t
  in adolescent drinkers, 425
  prenatal alcohol exposure and, 425
Alcohol-induced cognitive deficits, neuropathobiology of, 619–624
  activation of neuronal free radicals, 620
  astrocytes and microglia in alcohol-induced neurodegeneration, 621–622
  cell adhesion, 623
  disruption in neuronal glucose transport, 623
  future scope, 623
  growth-factor signaling, disruption in, 622–623
  inhibition of neurogenesis, 623
  neuroinflammatory signaling and cell death, 620–621
  NF-κB activation, 620–621
  N-methyl-D-aspartate-induced excitotoxicity, role of, 621
  pathways involved in, 621f
  TLR-4 activation, 622
Alcohol-induced dopamine release, 405f
Alcohol-induced myelin alterations, 425–427
Alcohol-induced neurodegeneration
  astrocytes and microglia in, 621–622
Alcoholism, 610
  antisocial. See Antisocial alcoholism (Antisocial ALC)
  anxiety/depression comorbid with, 544, 548
  cerebellar neuronal integrity in, 578
  cerebellar white matter structure in, 578
  dementia and, 624
  mixed, 545
  molecular genetics of, 543–544
  pure, 544, 548
  subtypes
    DRD2 gene associated with, 545–547
    identified in Han Chinese, 544–545
Alcoholism comorbid with bipolar disorder (ALC + BP), 548
  DRD2 gene associated with, 546–547
  genetic interaction in, 547
Alcohol-mediated oxidative stress
  CYP2E1, role of, 181f, 182
  HIV pathogenesis and neuroAIDS, 180f, 181–183. See also Ethanol
Alcohol-preferring rodents, 406

Alcohol-related behaviors, effects of mifepristone on, 612, 612t
Alcohol-related dementia, 632–633
Alcohol-related neurodevelopmental disorders (ARND), 565, 571
Aldehyde dehydrogenase (ALDH), 381, 381t, 516, 528
  genetics of metabolism, 381–382
ALDH. *See* Aldehyde dehydrognase (ALDH)
ALDH2 gene. *See* Acetaldehyde dehydrogenase 2 (*ALDH2*) gene
Aldoketoreductase 7A2 (AKR7A2), 557–559
Allostatic load, 63–64
Allosteric modulation, 756
Altered behavioral activation, model of, 37
Altered behavioral control, model of, 37
Altered emotion processing in smokers, fMRI studies on, 277
Alzheimer disease, 136
Alzheimer disease, acetaldehyde and, 557–559
AM251, 163
American Cancer Society's Cancer Prevention Study II (CPS-II), 105
American Society of Addiction Medicine (ASAM), 520
α-Amino-3-hydroxy-5-methyl-4-isoxazolepropionic acid (AMPA), 38, 201–202
α-Ketoglutarate dehydrogenase (α-KGDH), 627–628
α-KGDH. *See* α-ketoglutarate dehydrogenase (α-KGDH)
AMPA receptor, 467
Amphetamines, 273, 874
  chemical structures of, 683f
Amygdala, 63, 75, 202–203, 213–215, 273, 597, 774, 780, 783
  alcohol intake and, 405
  cue-induced cocaine craving and, 33
  internalizing pathway and, 37
  involvement in cognitive activities, 305
  location and physiological action of opiate receptors, 924–925
  nicotine effects on, 170
  reactivity in smokers, 274–275, 275t
Anandamide (*N*-arachidonoylethanolamine), 65–67, 160f, 709
  under conditions of chronic stress, 706
  metabolized by FAAH, 704–705
  synthesis, 704f
Anatomy of Melancholy, The, 815
Anhedonia, 895
Anilidopiperidines, 867
Animal research, addiction-related
  behavioral tests, 318–322
  cellular and molecular studies, 317–318
  inhalative/aqueous tobacco smoke exposure effects, 317–318
  locomotor sensitization, 318, 319t
  self-administration tests, 320–322
  tobacco particulate matter and extracts in, 317
Antagonist, 823
Anterior cingulate bundle (ACb), 338

Anterior cingulate cortex (ACC), 37–38, 44, 893, 895
  in heroin-dependent subjects, 1023–1024
  in regulation of "future reward" behaviors, 1032
  responds to smoking cues, 338
  response to cue-induced craving, 292
  response to smoking cues, 338–339, 343f, 344
  resting-state functional connectivity (rsFC), 342–343
Anterograde amnesia, 595–596
Antidepressant-like activity
  fatty acid amide hydrolase inhibition, 820–821
  WIN 55,212-2 treatment of, 818
Antidiuretic hormone (vasopressin), 601, 607
Antioxidants, 485
Antiretroviral therapy (ART), 179
Antisense oligomers, 905
Antismoking campaigns
  efficiency of graphic antismoking messages in curbing smoking behavior, 296–297
  persuasive impact of, 296–297
Antisocial alcoholism (Antisocial ALC), 548
  personality traits and genes in, interaction of, 548
Antisocial personality disorder (ASPD), 29t, 546, 978–979, 1063–1065
  comorbidity, 1063–1065
Antisocial personality disorders, 874
ANX/DEP ALC. *See* Anxiety/depression comorbid with alcoholism (ANX/DEP ALC)
Anxiety disorders, 36, 147
Anxiety/depression comorbid with alcoholism (ANX/DEP ALC), 544, 548
  DRD2 gene associated with, 545–546
  personality traits and genes in, interaction of, 547
Anxiolysis, 154, 539
Apoptosis, 485, 624
  definition of, 607
  oligodendrocyte, 602
Apparent motion, 745
Appetitive memory
  effects of THC and CBD on, 809–811
  reconsolidation, 808t
Appetitive smoking cues
  cerebral activations differences between aversive cues and, 300t
  clinical relevance of, 291–292
  functional imaging studies of, 293t–295t
  neural correlates of smokers to, 292–296
Arachidonoylethanolamine (AEA), 175, 204–205, 723
2-Arachidonoylglycerol (2-AG), 703, 709
Archicortex, 485
Arcuate nucleus (ARC), 134–135
Arginine vasopressin (AVP), 205, 1003
  and V1B receptor systems, 1005–1006
ARND. *See* Alcohol-related neurodevelopmental disorders (ARND)
ARRB2, 228
  in relation to smoking behavior, 230

ASAM. *See* American Society of Addiction Medicine (ASAM)
Ascending reticular activating system (ARAS), 282–283
Asparagine (Asn), 223
Aspartate (Asp), 223
Asp389Asn, 266
ASPD. *See* Antisocial personality disorder (ASPD)
ASSIST. *See* Alcohol, Smoking, and Substance Involvement Screening Test (ASSIST)
ASSIST-Lite, 650–651, 652f
  illicit drug user, brief intervention with, 656–657
  neuropathology for screening and intervention, 651–656
    alcohol consumption, 653–654
    cannabis, 654–655
    opioids, 656
    preoccupation/anticipation stage, 651
    psychostimulant use, 655
    reprogramming of neuronal circuits, 651
    sedatives, 655–656
    tobacco, 651–653
  stages of change model, 656f
Astrocytes, 181, 497, 624
  in alcohol-induced neurodegeneration, 621–622
  maintain balance between glutamatergic and GABAergic neurons, 467f
  and phospholipases D, 492, 492f
  PLD ablation and, 493f
  streptolysin O-permeabilized, 492–493
Astrocytic glutamate, 469
AT-1001, 322
Ataxia, 583
Atrophy, of nerve cells and brain shrinkage, 618, 624
Attention, effect of nicotine on, 282–285
  EEG measures, 283–284, 284f
  key facts, 285t
  P300 wave, 284, 284f
  P3b wave, 284
Attention deficit hyperactivity disorder (ADHD), 36, 45
  characteristics of, 1026t
  in children of parents with substance use disorder, 1028–1030
  DRD4 variable number of tandem repeat polymorphism (VNTR) in, 1030
  externalizing pathway and, 37
  frequency of six risk allele carriers, 1031t
  genetic susceptibility to comorbid, 1030–1032
  maternal and paternal smoking as risk factor for, 1024–1028, 1029t
  probands, 1030
  psychopathological risk of, 1030
  reward incentive and outcomes, 44
  as risk factor for SUD and HD, 1024, 1032
  stimulant therapy and, 1024
  symptoms, 1023
  Wender Utah Rating Scale, and Conners' Adult ADHD Rating Scale for, 1025t

Attentional impulsiveness, 975–976, 978
AUD. See Alcohol use disorder (AUD)
Aversive memory, 811
    ECS and reconsolidation of, 809
    effects of THC and CBD on, 809–811
Aversive smoking cues
    cerebral activations differences between appetitive cues and, 299f, 300t
    cerebral activations of chronic smokers, 298t
    clinical relevance, 296–297
    neural reactivity of smokers to, 297–301
Avoidant personality disorder (AvPD), 29t, 875
AVP. See Arginine vasopressin (AVP)
AvPD. See Avoidant personality disorder (AvPD)
Aztec gold, 55

B

Baclofen, 930
Bad trips, 53
Barrat Impulsiveness Scale-11 (BIS-11), 975–976, 978, 978t–979t
Basal ganglia, cue-induced cocaine craving and, 33
Bath salts, 55, 57
BBB. See Blood–brain barrier (BBB)
β-Caryophyllene (BCP), 675–676
    structure and key facts regarding, 675f
β2-Chimaerin gene (*CHN2*), 127–128, 129t
BCP. See β-caryophyllene (BCP)
BDNF. See Brain-derived neurotrophic factor (BDNF)
Bed nucleus of stria terminalis, 756
Behavioral sensitization, 934, 940
Benzodiazepine, 170
Benzodiazepine chlordiazepoxide, 77
Benzomorphans, 867
Beta-carbolines (harman/norharman), 324
β-Funaltrexamine, 222t
β-Pompilidotoxin, 170
Bicuculline, 169
Binge drinking (BD), 389
    alcohol dependence and, 392
    among adolescents, 390
    circadian typology in, 393
    defined, 389–390
    dysfunctional impulsivity scores of, 392–393, 393f
    endophenotypes of progression, 393–394
    executive functions, effects on, 394, 394f
    future research on prevention, 394
    key facts of neuropsychology in, 395
    learning and memory tasks, effects on, 393–394
    link between psychological distress and, 390–391
    mortality risk, 390
    negative consequences, 391t
    neurocognitive dysfunctions in, 391
    neuropsychological impact of, 393–394
    personality traits associated with, 392–393
    predictors of sustained, 392
    risk and protection factors associated with, 391–392, 391t–392t
    single episode of, effects of, 393

Biological theories of addiction, 340–342
Bipolar disorder (BD), 836, 838–839
    cannabis use impact on outcomes of, 836–837
    childhood trauma, cannabis and, 836
    smoking and cognitive function in, 311–312
    neuropsychological performance, 310t
Bipolar onset episode, longitudinal overview of, 838f
Black Mamba, 55
Blood ethanol concentration (BEC), 383–385
    beverage content *vs*, 383–384
    first-pass metabolism (FPM), 383
    food and, 383
Blood oxygen level derived (BOLD) analysis, 342
    of pain, 959
Blood oxygen level-dependent (BOLD) images, 983, 1014
    measurement of, 983
Blood–brain barrier (BBB), 349–351, 427, 601, 607
BOLD images. See Blood oxygen level-dependent (BOLD) images
Borderline personality disorder, 29t, 874
Borderline personality disorder (BPD), 978–979, 1063–1065
    comorbidity, 1063–1065
    emotional dysregulation in, 1066
Bottom up and top-down processing, 745
BP897, 203
BP_{ND}, 792
Brain, drug-induced gene expression changes in, 934
Brain alterations, 478
    in alcoholic patients, 479t
    brain atrophy, 478–480
    functional changes, 480
Brain areas, nicotine effects on, 169–174
    amygdala, 170
    cerebellum, 170
    cortex, 170–171
    hippocampus, 171
    hypoglossal nucleus, 174
    hypothalamus, 171
    laterodorsal tegmental nucleus, 173–174
    raphe nucleus, 172
    septum, 172
    spinal cord, 174
    striatum, 172
    ventral tegmental area (VTA), 173
Brain atrophy, 478–480
    gray matter atrophy, 478–479
    pathogenesis of neurotoxicity in alcoholics, 480
        activation of TLR, 480
        microglial activation of NFκB, 480
    white matter atrophy, 479–480
Brain cannabinoid system, 851
Brain cells, and PLD, 491–492
    astrocytes, 492, 492f
    neurons, 491
Brain damage
    ethanol-mediated
        in adolescents, 485

potential mechanisms leading to, 628f
    TLR-4 induced, 622
Brain growth spurt, 488–489, 497
Brain imaging, in adolescent AUD, 589t–591t
    findings with functional significance in, 592t
    for other addictions and substance misuse, 596–597
    VBM and related methods of, 587–588
Brain lesioning, 21–22
Brain plasticity, 934, 940
Brain reward system, 161–163, 165
Brain volumes, in adolescent AUD
    hippocampal volume, 588–592, 593f
    moderating and mediating factors, 595–596
        childhood trauma, 596
        gender, 595
        psychiatric comorbidities, 595–596
    other brain regions with differential volumes, 594–595
        insular cortex, 594–595
        striatum, 595
        temporal cortex, 594
    prefrontal cortex volume, 593–594
Brain-derived neurotrophic factor (BDNF), 391, 538
Brains of smokers and nonsmokers, 338–339
    hypothesis testing, 342–343
    resting-state functional connectivity (rsFC), 342–343
Buprenorphine, 903, 922, 949–951
Buprenorphine maintenance therapy (BMT), 1039
Bupropion, 256, 268
Butorphanol, 928

C

Caenorhabditis elegans, 536
Cajal-Retzius neurons, 478
Calcium channels, 958
cAMP. See Cyclic AMP (cAMP)
cAMP response element-binding protein (CREB), 620
Candida meningitis, 949–951
Cannabichromene (CBC), 675
Cannabidiol (CBD), 665, 668, 675, 685, 756
    effects of, 749
    on $5\text{-}HT_{1A}$-mediated neurotransmission, 753f
    effects on neurogenesis, 755
    endocannabinoid system interaction with, 755
    history of, 749
    and 5-hydroxytryptamine-1A receptor, 749–752
        controls anxiety-like behaviors, 751f
        decreases fear conditioning behavior, 752f
        and generalized anxiety and depression, 750–751
        motor control, 752
        nausea and vomiting, 752
        neuroprotection by, 751–752

panic and posttraumatic stress disorder, 751
impact on reconsolidation of drug, appetitive, and aversive memories, 809–811
increases intracellular calcium concentrations, 755
interact with $5HT_{1A}$ receptors, 753–754
mechanisms of action assessed in in vitro studies, 750t
for other addictions and substance misuse, 756
therapeutic potential, 749
TRPV1 channels and, 755
Cannabigerol, 672–674
Cannabigerol acid (CBGA), 672
Cannabinergic indoles, 723
Cannabinoid, 669, 728
Cannabinoid 1 receptor (CB1R), 161, 204–205, 665, 676, 685, 699, 718, 728, 783, 801
  activation of Na^+/H^+ exchanger in CHO cells, 764
  adenylate cyclase activity, 762
  aggression relation with, in animal studies, 831–833
  agonists and antagonists on nicotine reward, 163, 164f
  amino acids for, 717f
  in brain reward processing in nicotine administration, 161–163
  chronic administration for adolescence and adulthood, 821–822
  distribution in adult brain, 762
  distribution in brain, 797f
  expression pattern, 761–762
  in feeding behavior and mood change, 833
  induction of long-term potentiation (LTP) in, 169
  localization in lysosomes, 694f
  nicotine-induced behaviors, role of receptor antagonists on, 162–163, 162t
  receptor ligands, 163
  regulation of cellular growth, 763–764
  role during early brain development, 762
  signal transduction mechanism of, 762–764
  signaling
    complexes and transactivation of tyrosine kinase receptors in lipid rafts, 696–698
    to effector extracellular signal-regulated kinase, 689–690
    to ERK emanates from lipid rafts, 690–691
    filamentous-actin membrane cytoskeleton integrity, effects of, 694–696
    lipid raft cholesterol, effects of, 691–692
    microtubule integrity, effects of, 693–694, 696f
    neuronal development and function, 689
    out of lipid raft, 698f
    of synthetic cannabinoids found in Spice products, 766f
  signaling pathway, 763f
  and stress, 733
  subcellular distribution of, 693f
Cannabinoid 2 receptor (CB2R), 676, 685, 718
  amino acids for, 675f

reinforcing effects of drugs, impact on, 807
signaling, 764, 765f
and stress, 733
Cannabinoid agonists
  cannabinergic indoles, 723
  of CB1R, 728
  classical, 723
  classification, 722–723
  in clinic, 708
  conditioned place preference/avoidance assays, 727
  degradation, 704
  dependence and withdrawal studies in animals, 725
  drug discrimination, 725–726
  efficacy of, 727–728
  eicosanoids and derivatives, 723
  endocannabinoids, 702–707
  herbal incense, 723–724
  marijuana dependence studies in humans, 725
  nonclassical, 723
  for other addictions and substance misuse, 708–709, 728
  overview, 702
  preclinical studies of, 724–725
  receptor pool interactions with, 726f
  reinforcement of, 726–727
  structures of, 724f
  synthetic cannabinoids, 708
  Δ^9-THC, 707–708
  tolerance to, 725
Cannabinoid receptors, 732, 760–762, 767
  CB1 receptor, 718
  CB2 receptor, 718
  ion channels modulated by, 717
  ligands of
    binding experimental setup and work flow, 715f
    structure–activity relationship studies of, 713
    structures and definitions of, 714f
  modulation of intracellular calcium, 717–718
  mutational analysis of, 718
  for other addictions and substance misuse, 766
  and posttraumatic stress disorder, 65
  research on, 713
  signaling
    in cells, 717–718
    model of, 716f
  spice signaling, 764–766
Cannabinoid tetrad test, 669–670
Cannabinoids, 27, 676, 718, 783
  classification of, 682t
  electrophysiological effects on vision, 741–744, 742f
  and event-related potentials, 852–854
    mismatch negativity, 853
    P50, 853
    P300, 853
    P50 suppression deficits, 853
  functions of, 732

anti-inflammatory profile, 732
as antioxidant, 732
regulation of brain vasculature, 732
regulation of intracellular concentration of Ca^{2+}, 732
and neural oscillations, 854–859, 855f
  induced oscillations, 857
  quantified neural oscillations, types of, 855
  resting-state oscillations, 857–859
  steady-state evoked oscillations, 856–857
  transient evoked oscillations, 856
pharmacological properties, 27
psychophysical effects on vision, 739–741
in reward circuitry, 796–798
synthetic. *See* Synthetic cannabinoids
Cannabinol (CBN), 675
Cannabis, 665, 709, 731, 801, 823
  and bipolar disorder, 836
  and human aggression, 828–831, 830f
  incidence of schizophreniac, 84
  myelin alterations from abuse of, 427–428
  number of published articles on, 666f
  for other addictions and substance misuse, 669
  potential therapeutic value of, 668–669
  regulation, policy changes in, 669
Cannabis, 672, 685, 709
  and its trichomes, 673f
  structures of cannabinoids found in, 673f
  terpenophenolic compounds in, 676, 676t
Cannabis abstinence, 778, 780–781
Cannabis addiction, 801
Cannabis consumption, 833
Cannabis dependence, 798–799
  [^{11}C]PHNO BP_{ND} images in, 791f
  instruments employed to measure symptoms of, 779f
  PET neuroimaging study in, 787f
Cannabis misuse, and risk of addiction, 666–667
Cannabis sativa, 161, 707, 731, 738, 815
Cannabis use, 654–655
  acute and long-term effects of, 665–666
  age of starting, 665–666
  in animal models, translational research, 667
  effects on gamma-band auditory steady-state response, 856f
  effects on induced gamma activity during perception of coherent motion, 858f
  gateway drug theory, 667
  impact on bipolar disorder outcome, 836–837
  impact on mania, 836
  longitudinal overview of, 838f
  and manic symptoms, 838
  men and women percentage of, 666f
  and onset manic episode, 837
  prevalence of, 666t
  and psychosis, 837
  therapies for, 667–668
  white matter changes with, 844–847

Cannabis use disorder (CUD), 576, 665, 773, 783, 801
  assessment of, 775–778
  and cerebellum, 582–583
  cognitive deficits in, 774
  and comorbid psychopathologies, 781–782
  DSM-V criteria for, 773–774, 774t
  gray matter associated with, 782–783
  hippocampus and amygdala, 774
  neurocognitive aspects of, 774
  nosology of, 773–774
  regional alterations associated with, 781
  structural neuroimaging studies, 774–781
    and controls, 778–780
    gender effects, 781
    gray matter morphology, 776t–777t
    sample characteristics and diagnostic instruments, 775–778, 778f
    severity, 780
    treatment of, 780–781
    vs. controls, 778–780
      amygdala, 780
      cerebellum, 780
      hippocampus, 780
      prefrontal cortex, 780
      striatum, 778–780
Cannabis withdrawal, 778
  therapies for, 667–668
Carbonylation, 559
Catechol-O-methyltransferase (COMT) gene, 910, 1030, 1062–1063
Catha edulis, 55, 683, 685
Cathine, 685
Cathinone, 685
Cations, 158–159
Caudate, 168
CB1 agonist ACEA, effect of single dose of, 833f
CB1 receptor agonists, 833
CB1 receptor antagonists, 833
CB1 receptor-coupled G proteins, 718
CB1 receptor-mediated inhibition, of adenylyl cyclase, 717
CBC. See Cannabichromene (CBC)
CBD. See Cannabidiol (CBD)
CBGA. See Cannabigerol acid (CBGA)
CB1-GFP (green fluorescent protein)
  partially localizes along microtubules, 695f
  subcellular localization in COS-7 cells, 692, 694f
CBN. See Cannabinol (CBN)
CB1R mRNA, 761–762
CCL2. See Chemokine ligand 2 (CCL2)
CCL5, 181
CeA. See Central amygdala (CeA)
Cell adhesion, 623
Cell death, alcohol-induced, 620–621
Cell signaling pathways, 462
Cellular senescence, 1015–1016
Central amygdala (CeA), 614, 893, 895
Central cannabinoid receptor 1, 860
Central nervous system (CNS)
  CNS disorders in opiate addicts, 948t
  demyelination and remyelination processes in, 423
  HIV-associated malignancies, 949–951
  infections involving, 949–951
  myelin formation in, 422f
  neurokinin-mediated mechanisms in, 191–192
  neuropeptide distribution in, 190–191
  receptor distribution in, 191
Central neuronal sensitization, 957
Central nucleus of amygdala (CeA), 533
Central pontine myelinolysis (CPM), 600, 634
  chronic alcoholic male with, brain MRI in, 604f
  clinical presentation of, 605
  histopathology of pons in, 603f
Cerebellar abnormalities, in SUD, 579–582
  cerebellar functional differences associated with family history, 580–581
  cerebellar lobes and functional differences, 580
  cerebellum structure and familial risk, 580
  functional connectivity, 581
  premorbid risk factors in high-risk offspring, 579–580
Cerebellar lobes, 580
Cerebellar neuronal integrity, in alcoholism, 578
Cerebellar vermis, 583
Cerebellar volume
  in healthy subjects, 581f
  in high- and low-risk individuals, 581f
  and substance use outcomes, 581–582
Cerebellar white matter structure, in alcoholism, 578
Cerebellum, 168, 583, 780
  and addiction connection, 576
  anatomical divisions of, 577f
  AUD and, 577–579
    alcohol cues and alcohol-dependent individuals, 578
    cerebellar neuronal integrity in alcoholism, 578
    cerebellar white matter structure in alcoholism, 578
    functional connectivity in alcoholism, 579
    functional neuroimaging in alcoholics, 578
    MRI in alcohol-dependent individuals, 577–578
    resting state functional connectivity in alcoholics, 578–579
    structural and functional changes in AD, 579
  cannabis use disorder and, 582–583
  development of, 577
  functional abnormalities of, 582–583
  localization of, 576f
  nicotine effects on, 170
  region-of-interest tracings of, 580f
  structure and function of, 576–577
  tobacco use disorder and, 582
Cerebrovascular accidents, following heroin, 994–996
  damage to mitochondria, 996
  embolization, 995
  hypersensitivity reaction, 996
  platelet count and hemostasis, 996
  systemic blood pressure decrease, 995–996
c-Fos-positive neurons, 174–175
Chaperone proteins, 612–613
Chasing the dragon, 1000
Chemokine, 462
Chemokine ligand 2 (CCL2), 5
Children with familial substance use disorder, 38–40
  functional studies, 40–45, 41t–43t
  morphological studies, 40
  prefrontal activation changes in, 44
  reward task studies, 44–45
  studies using inhibitory control tasks, 40–44
  utilization of inhibitory control tasks, studies using, 40–44
  structure, 128f
Chimaerins
  α1- and α2-chimaerins, 126
  β1- and β2-chimaerins, 126–127
  association with smoking tobacco, 126
  biological role of β2-chimaerin gene (CHN2), 127–128, 129t
  nervous system and, 126–127
  receptors involved in regulation of, 127f
Chlordiazepoxide
  consummatory successive negative contrast on postsession consumption of, effects of, 78f
1-(3-Chlorophenyl)piperazine (mCPP), 56–57, 57f
Chronic opioid exposure
  CRF levels in, 893
  and hyperalgesia, 892
Cingulotomies, 21
Circadian genes, drug-induced alteration of, 940
Clinical endocannabinoid deficiency syndrome, 718
Cloud 9, 55
Clozapine, 86
Cobalamin deficiency, 482
Cocaine, 50–52, 273
  activity-dependent gene expression, 940–941
  expression of orexinergic system and, 140–141
  female users, impact on, 52
  gender differences in addiction, 930
  metabolic effects of, 27
  misusers, 33
  myelin alterations from abuse of, 427
  plasma concentration changes, 52
  premature aging and addiction of, 1019
  Rho GTPases and, 128
  as stimulant of central nervous system, 52
  tachykinins and, 196
  TM1 transcriptional activation for, 934–935
  toxic effects of, 52
  use among HIV-infected individuals, 185
  various names of, 52
Codeine, 1052
COGA. See Collaborative Studies on Genetics of Alcoholism (COGA)
Cognition, 618
Cognitive deficits, in CUD, 774
Cognitive dysfunction, 495–496

Cognitive functions
 long-term effect of nicotine on, 287
 major, 306t
 neurotransmitters and, 307t
Collaborative Studies on Genetics of Alcoholism (COGA), 528
Collapsin response mediator protein 2 (CRMP-2), 480–481
Comorbid, 905
Comorbid drug abuse, neurobiology of, 84–86
Comorbidity, 73–74, 78
 with schizophrenia, 83–84
Compartment syndrome, 952
Complex adaptive systems (CAS), 18
Complex trauma, 62
Comprehensive computerized battery (CANTAB), 312
Conditioned place avoidance, 728
Conditioned place preference (CPP), 669, 728, 807, 811, 905, 1008
 paradigms, 116, 162–163, 248
Conditioned stimulus, 811
Conduct disorder, 36
Consummatory extinction (cE), 75
Consummatory successive negative contrast, 78f
Continuous performance test (CPT), 283
Control theory, 29
Controlled substances
 act schedules, 868–869
 definition of, 868
Controlled Substances Act, 868–869
 Schedule I, 868
 Schedule II, 868
 Schedule III, 868–869
 Schedule IV, 869
 Schedule V, 869
Convention on Psychotropic Substances (of 1971), 679
Cooperativeness, 977–978
Cornu Ammonis, 485
Corpus callosum, 628–629
Correlational study, 833
CORT. *See* Cortisol (CORT)
Cortex, 168
 nicotine effects on, 170–171
Cortical cytoskeleton, 699
Cortical morphology, 566–568
 concept of, 566
 cortical morphometry, 567–568
  computational tools for, 567f
  object-based morphometry, 567–568
  surface-based morphometry, 567
 corticometry, 566–567
 definition of, 571
 in FASD, 568–571
  altered cortical morphometry, 569–570
  corticometry, 568–569
  relation to diagnosis, pathophysiology, and perspectives, 570–571
 high-resolution MRI of, 566f
 for other addictions and substance misuse, 571
Cortical morphometry, 567–568
 computational tools for, 567f
 object-based morphometry, 567–568
 surface-based morphometry, 567
Cortical spreading depression (CSD), 446–447
 antagonizing effect of astaxanthin against, 450f
 carotenoids and antagonizing effect of, 448
 ethanol and progression of, 447–449
 shrimp carotenoids and, 448, 449f
Cortical thickness
 analyses in FASD, 568f
 estimation, 566f
Corticometry, 566–567
Corticosterone, 643
 early life stress and, 640–641
Corticotropin-releasing factor (CRF), 202–203, 205, 610
 interactive nature of, 894f
 in opioid-induced hyperalgesia, 893–894
 signaling in chronic pain states, 894, 895f
Corticotropin-releasing hormone (CRH), 63, 535–536, 536f
Cortisol (CORT), 610–611
Cosmetic orthognathic surgery, 914f
Cotinine, 94f
 assays for detection of, 364–366, 366f, 366t
 chemical formula, 363
 ethnic differences and metabolism of, 364
 gas chromatography with mass spectrometry (GCMS) analysis, 364–365, 365f
 gender differences and metabolism of, 364
 high-performance liquid chromatography (HPLC) analysis, 364–365, 365f
 as marker of exposure to nicotine, 363
 metabolites of, 364f
 pharmacokinetics and pharmacodynamics, 363–364. *See also* Urine cotinine assay techniques
CpG islands, 507
CPM. *See* Central pontine myelinolysis (CPM)
CPP. *See* Conditioned place preference (CPP)
Crack, 52
Craving, 15, 25, 31, 273, 291–292, 337
 cocaine-induced, 33
 cue-induced, 33, 342, 344
 dorsal anterior cingulate cortex (dACC) and, 15
 from information theory perspective, 17
 nicotine, 273
 phasic desire, 291–292
 psychological experience of, 292
 tonic desire, 291–292
 withdrawal symptoms and, 31
 withdrawal-induced, 342–343
Craving generation system, 337, 341–342
CREB. *See* cAMP response element-binding protein (CREB)cAMP responsive element binding protein (CREB)
CRF. *See* Corticotropin-releasing factor (CRF)
CRF1R antagonism, 893–894
CRMP-2. *See* Collapsin response mediator protein 2 (CRMP-2)
Cryoglobulin-positive heroin addicts, 953
Cue-induced craving, 342, 344
 cocaine craving, 33
 nicotine craving, 273
Cue-induced nicotine-seeking behavior, 238–239, 241–242
 pharmacological substrates responsible for, 241–242
Cyclic adenosine monophosphate response element-binding protein (CREB), 205
cyclic AMP (cAMP), 762
2′:3′-cyclic nucleotide 3′-phosphodiesterase (CNP), 421–423
CYP1A1, 184–185
CYP2A6, 184–185, 364
CYP3A4 pathway, 185
CYP3A4-ART interactions, alcohol effects, 182–183
CYP2D6 gene, 910
Cytochrome P450 (CYP), 180, 180f, 552, 910
 in alcohol-mediated HIV pathogenesis and neuroAIDS, 180f, 181–183
 in smoking-mediated HIV infection, 184–185
Cytochrome P4502E1 (CYP2E1), 380–381
Cytokines, 485
 CX3CL1, 455
 CX3CR1, 455
 definition of, 462
Cytoskeletal proteins, 491
Cytotoxic edema, 607
Cytotoxicity, of acetaldehyde, 556–557
 cytotoxic effectiveness, 556, 557t
 mitochondrial toxicity, 557
 related to hydrophobicity and electrophilicity, 556

D

D-amphetamine, 247
DARPP-32, 169
DAT. *See* Dopamine transporter (DAT)
DAT-promoter methylation, 535
ΔBP_{ND}, 792
Death thought accessibility, 330
Decision-making, 583
Declarative learning, nicotine and, 285–286
Deep brain stimulation (DBS), 21–22
 efficacy and safety in addictions, 21
 improvement in addiction post, 21
 method, 21
 of Nacc for depression, 21
 as treatment for Parkinson disease, 21
Default mode network (DMN), 342–343, 343f
Default-mode network, 583
Defensive aggression, 833
Dehydroepiandrosterone (DHEA), 433–434
Delirium tremens, 531–532, 539
Dementia, 624
 alcohol-related, 632–633
Demyelination, 607
Dendritic spines, 126, 128
Dependence, 801
 development and expression of, 16
Dependent personality disorder, 29t, 875
Dephosphorylation, 507
Depolarization-induced suppression of inhibition/excitation, 801
Depression, 823–824
 CBD associated with, 750–751

Depressive disorders, 36
Desensitization, 905
Designer drugs, 54–57
　piperazine, 57
Desipramine, 27
Dexamethasone, 215
Dexamethasone suppression test, 643
Diacetylmorphine, 994
Diacetylmorphine (DAM), 1041–1042
Diacetylmorphinegam (diamorphine), 922
Diazepam, 170
Dickman Impulsivity Inventory (DII), 392–393
Dietary acetaldehyde, 553–554
Differential association theory, 28–29
Diffusion tensor imaging (DTI), 634, 848, 959, 1014
Dihydro-β-erythroidine (DHβE), 169, 241, 250, 414–415
2,3-Dihydroxy-6-nitro-7-sulfamoylbenzo[f]quinoxaline-2,3-dione (NBQX), 169
3,4-Dihydroxyphenyl-acetaldehyde (DOPAL), 544
3,4-Dihydroxyphenylacetic acid (DOPAC), 544
Diphenylpropylamine, 867
Distal defenses, 329f
Disulfiram, 930
dlPAG. See Dorsolateral periaqueductal gray matter (dlPAG)
DNA methylation, 507
DOPAC. See 3,4-Dihydroxyphenylacetic acid (DOPAC)
DOPAL. See 3,4-Dihydroxyphenyl-acetaldehyde (DOPAL)
Dopamine, 85, 136, 192, 533–535, 966–968
　acute cannabinoid administration on, effects of, 787–788
　chronic cannabis use on, effects of, 788–792
　　and comorbid psychosis, 790–792
　　imaging studies capacity for dopamine release, 790
　　imaging studies of dopamine transporter, 788–790
　　imaging studies of stress response, 788
　　PET studies of dopaminergic indices in, 789t
　D1 or D2 dopamine receptors, mechanism of, 15, 26–27, 962, 966–967
　dopamine D2 receptor, 533–535
　dopamine transporter, 535
　homeostatic adaptation, issue of, 27
　human imaging studies
　　positron emission tomography, 787
　　single-photon emission computed tomography, 787
　methadone treatment and, 967, 967f
　nicotine withdrawal syndrome, role in, 203
　opioid dependence and, 967
　for other addictions and substance misuse, 792
　overactivity issues, 27
　radiotracers used for imaging of, 791f
　release from VTA, 533
　release post acute THC administration, 788t
　role in addiction, 26–27
　role in mediating addictive behavior, 786–787
　transmission, 535, 535f
　uptake site in human beings, 27
Dopamine D2 receptor, 533–535
Dopamine D2 receptor (DRD2-TAQ-1A), 15
Dopamine D5 receptor (DRD5), 1030
Dopamine D2 receptor (DRD2) gene, 543, 548
　alcoholism and TaqIA polymorphism of, 544
　associated with ALC + BP, 546–547
　associated with alcoholism with conduct disorder, 545
　associated with ANX/DEP ALC, 545–546
　associated with ASPD, 546
　interaction with MAOA-uVNTR, 547
　temperament interaction in subtyped alcoholism, 547–548
　personality traits and genes with antisocial ALC, interaction of, 548
　personality traits and genes with ANX/DEP ALC, interaction of, 547
Dopamine β-hydroxylase gene (DBH), 1030
Dopamine receptor blockade, 933
Dopamine release capacity, 792
Dopamine (DA) reward pathway, 63
Dopamine transporter (DAT), 535, 967–968
δ-opioid receptor (DOR), 221, 899
　antagonists, 902
　function of, 903–904
　heterodimerization of, 904
　nontransgenic disruption of, 902t
Dopaminergic mesolimbic system, 795–796, 796f
　cellular elements mediating cannabinoid actions in, 798f
　long-term depression and, 795–796
　long-term potentiation and, 795–796
　for other addictions and substance misuse, 799–801
　ventral tegmental area and, 795–796
Dopaminergic nigrostriatal pathway, 116
Dorsal medial hypothalamus (DMH), 134–135
Dorsal medial nucleus (DMN), 134–135
Dorsal raphe, 823
Dorsal raphe nucleus (DRN), 146–147, 204, 756
　acetylcholine effects on glutamatergic excitatory postsynaptic currents (EPSCs) of serotonergic, 151f
　activity of 5-HT, 147
　electrical activity of 5-HT DRN neurons, 147
　electrophysiological and immunocytochemical characteristics of serotonergic and non-serotonergic, 148f
Dorsal root ganglion (DRG), 957
Dorsal striatum (dStr), 116
Dorsolateral frontal cortex, 38
Dorsolateral periaqueductal gray matter (dlPAG), 750–751, 756
Dorsolateral prefrontal cortex (DLPFC), 16, 20, 1016–1017
Dorsolateral prefrontal cortex (dlPFC), response to cue-induced craving, 292
Dorsomedial prefrontal cortex (dmPFC), response to cue-induced craving, 292
Downregulation, 63
　of CB$_1$ receptors, 65
Doxorubicin, 552–553
DRD2 gene. See Dopamine D2 receptor (DRD2) gene
Drive induction, 71
　instrumental reinforcement by, 72f
Drive reduction, 78
　instrumental reinforcement by, 72, 73f
Drug discrimination, 728
Drug discrimination chamber, 726f
Drug rewarding-related memories
　endocannabinoid system and reconsolidation of, 808
　postretrieval strategies to attenuate, 807f
Drug specificity, 72
Drug-induced gene expression changes
　activation of stress-related genes by drugs of abuse, 935–936
　alteration of circadian genes, 940
　aspects of cell functioning, 935
　in astrocytes, by glucocorticoids and GR, 936–937
　in brain, 934
　by chronic administration of psychostimulants or opioids, 937–938
　the first group of genes (TM1), 934–935
　IEG activation, 934–935
　in nucleus accumbens, by chronic methamphetamine and heroin treatment, 938
　regulation of activity-dependent genes, 934–935
　regulation of glucocorticoid receptor-dependent genes, 938–940
　in striatum, by chronic methamphetamine and heroin treatment, 938
Drug-induced regulation of activity-dependent genes, 934–935
Drug-induced rewarding effects, 1037
D4S1644, 915–916
DSM5, 86
Δ^9-Tetrahydrocannabinol (THC), 55, 65, 83, 161, 324, 665, 674–675, 709
　acute intraperitoneal administration of, 816–817
　acute intravenous administrations of variable doses of, 817f
　CB1/CB2 receptor agonist, 668
　definition of, 685, 783, 801, 823, 860
　effects of previous exposure on rewarding effects of psychostimulants/opioids, 668t
　effects on auditory P300b, 854f
　impact on reconsolidation of drug, appetitive, and aversive memories, 809–811
　repeated treatment, 817–818
　rewarding/reinforcing effects, 667t
　and serotonin firing activity, 816–818
Dualism, 17–18
Dynorphin, in lateral hypothalamus, 1007
Dysfunctional impulsivity, in binge drinking (BD), 392–393, 393f

Dysphoria, 895
Dysphoric mania, 838

E

EAATs. *See* Excitatory amino acid transporters (EAATs)
Early phase psychosis, 848
Eclipse, 57
EEG. *See* Electroencephalography (EEG)
Efficacy, 728
Efflux, 905
EGFR. *See* Epidermal growth factor receptor (EGFR)
Eicosanoids and derivatives, 723
Electroencephalogram (EEG), 44
    attention, effect of nicotine on, 283–284, 284f
    during cortical spreading depression, 446, 448f
    effects of Nicotine$_{PM}$, 352–354, 354f
    neural activity, nicotine-induced changes in, 348–349, 349f–350f, 354f
Electroencephalography (EEG), 851–852
    definition, 860
    event-related potentials, 851–852
    for other addictions and substance misuse, 859–860
Elevated-plus maze, 539
Elevated-plus maze test (EPM), 750–751
Elevated-T maze (ETM), 751
Embryogenesis, timeline of, 489f
Emergence, defined, 22
Emotional activation, 74
Emotional self-medication (ESM) hypothesis of addiction, 71
    addiction in humans and, 72–74
    applications, 79
    chronic stress among animals, 74–75
    comorbidity studies, 73–74
    consummatory extinction (cE), 75
    design of, 76f
    drug specificity postulate of, 72
    drug-taking behavior, relationship between, 75
    induced by reward loss, 75–77
    instrumental appetitive extinction (iE), 75
    instrumental successive negative contrast (iSNC), 75
    involvement in alcohol intake, 74
    issues of ecological validity, 75
    in nonhuman animals, 74–75
    partial reinforcement contrast effect (PRCE), 75
    partial reinforcement extinction effect (PREE), 75
    psychopathology postulate of, 72
    studies and outcome, 72–73
    SUDs and, 77–79
Endocannabinoid (eCB) signaling system, 65, 160–161
    key functions of, 161
    molecular structure of, 160f
    retrograde signaling, 161f

Endocannabinoid system (ECS), 204–205, 654, 676, 709, 719, 731–732, 756, 766, 828, 829f, 833
    and cannabidiol interactions, 755
    cannabinoid receptors, 732
    implicated in relapse and drug-seeking behavior, 808
    interaction with opioids, 674–675, 674t
    learning and memory processing, role in, 807–808
    main pharmacological targets of, 806f
    memory reconsolidation and extinction, role in, 807
    molecular biology, 714–717
    for other addictions and substance misuse, 745
    overview of, 805–807
    in primary rewarding effects, 807
    and reconsolidation of aversive memories, 809
    and reconsolidation of drug rewarding-related memories, 808
    role in memory reconsolidation and extinction, 807
    and stress, 732–733
Endocannabinoids, 669, 709
    in CNS, 703–707
    endocannabinoid system, 702–703
    metabolism, 703–705, 704f–705f
    modulate thalamic neuronal responses, 743f
    role in neuronal rhythmicity and synchrony, 744f
    and synaptic modulation, 705–707
    synthesis and retrograde signaling, 706f
Endogenous, 905
Endogenous acetaldehyde, 552–553
    cytochrome P450, 552
    lipid peroxidation, 552–553
    myeloperoxidase, 552
Endogenous agonist opioids, role in addiction, 27
Endogenous carbonylated proteins, 552
Endogenous opioid system, 63
    acute alcohol-induced effects on, 404
    amino acid sequences for peptides of, 403t
    in AUD, 406–408
    chronic alcohol intake, effects of, 405
    history, 402
    opioid precursor proteins, 402
    opioid receptors, 402–404
    peptide families, 403f
Endogenous opioid systems, 1006–1007
    β-endorphin, 1006
    cocaine CPP, 1006
    dopamine D1 or D2-like receptors, 1006
    methadone, 1006
    MOP-r, 1007
    OPRM1 gene, 1007
    POMC-derived peptides, 1006–1007
    yohimbine, 1007
Endokinin A (EKA), 190
Endorphins, 63
    β-endorphins, 27, 222t
Enkephalins, 27

    microdialysis studies of acute alcohol-induced effects on, 404
Environmental enrichment paradigm, on nicotine addiction
    behavioral effects of nicotine, 247–248
    dopamine transporter (DAT) function, 249–250
    neurochemical effects of nicotine, 248–250
Enzymatic reactions, of PLD, 491, 491f
    hydrolysis, 491
    transphosphatidylation, 491
Epidermal growth factor receptor (EGFR), 689–690
Epigenetics, 539
EPM. *See* Elevated-plus maze test (EPM); Extrapontinemyelinolysis (EPM)
ERK. *See* Extracellular signal-regulated kinase (ERK)
ERPs. *See* Event-related potentials (ERPs)
Estrous cycle, 643
    effects on ethanol self-administration, 642t
Ethanol, 74, 140–141, 206
    absorption and distribution of, 379
    additive effect in reinstating, 240
    associated burden of disease, 378–379
    blood concentrations, effects of, 383–385
    brain alterations induced by, 449
    brain electrophysiological effects of, 447t
    consummatory successive negative contrast on postsession consumption of, effects of, 78f
    content of an alcoholic beverage, 379
    CSD propagation, 447–449, 449f
    effects on brain, 384
    effects using IHC, 435
    elimination *in vivo*, 382
    first-pass metabolism (FPM), 383–384
    gabaergic neuroactive steroids interactions with, 434–436
    genetics of metabolism, 382
    impaired nutrition status and, 384–385
    influence on nAChRs, 174
    instrumental extinction on consumption, effects of, 76f
    international variations in consumption of alcohol, 378, 379t
    metabolism of, 380–382, 380f
    partial reinforcement (PR), effects of, 77f
    physical properties of, 377–378
    quantification of absorption, 384
    stomach, role in metabolism, 383
    stress and consumption of, 414
Ethanol (EtOH), 500, 516, 614
    actions at NMDAR, 500–501
    chemical structures of, 552f
    dose-dependent inhibition of NMDAR by, 501f
    effects of chronic administration of, 619f
    effects on astroglial development, 622
    NMDAR function, 500–501
    NMDAR subunits and
        NR1 subunit, 501–504, 502f
        NR2 subunit, 504–505
        NR3 subunit, 505–506
    vs. acetaldehyde, 557

Ethanol binding site
  NR1 subunit contain, 502
  residues comprising, 503f
Ethanol toxicity, of brain
  sites of, 494f
  timeline of, 489f
Ethanol-induced phosphorylation, of NR2B subunit, 505f
Ethanol-mediated brain damage, in adolescents, 485
ETM. See Elevated-T maze (ETM)
Etorphine, 922
Euphoric mania, 838
European Monitoring Centre for Drugs and Drug Addiction, 680
Event-related oscillations, 44
Event-related potentials (ERPs), 44, 851–854
  definition, 860
  mismatch negativity, 853
  P50, 853
  P300, 853
  P50 suppression deficits, 853
  prototypical, 852f
  and spectral analyses of neural oscillations, 852f, 860
Evoked excitatory postsynaptic potentials (eEPSPs), 170
Evoked inhibitory postsynaptic potentials (eIPSPs), 170
Excitatory amino acid transporter (EAAT), 467–468
  definition, 473
  subtypes in humans, 468t
  transportation of glutamate by, 469f
Excitatory amino acid transporters (EAATs), 734–735
Excitatory postsynaptic currents (EPSCs), 150–152, 151f
Excitotoxicity, 485, 735
  CB1R and, 734
  CB2R and, 734–735
  definition of, 624, 735
  glutamate, 634
  N-methyl-D-aspartate-induced, 621
  stress and cellular damage/death by, 733–734
  stress-induced, 734–735, 734f
Executive functioning, 583
Exogenous, 905
Exogenous acetaldehyde, 553–554
  dietary acetaldehyde, 553–554
  natural and manufactured aldehydes, 554
Externalizing pathway, 37
Extinction, 811, 1008
Extracellular signal-regulated kinase (ERK), 689–690
Extrahypothalamic stress systems, 612f
Extrapontinemyelinolysis (EPM), 600
  chronic alcoholic male with, brain MRI in, 604f
  clinical presentation of, 605
Extraversion, 393
Eysenck Personality Questionnaire, 1061

F

FAAH. See Fatty acid amide hydrolase (FAAH)
FAAH enzyme, 709
F-actin cytoskeleton, 699
  integrity of, 697f
Fagerström Test for Nicotine Dependence (FTND), 213, 273–276, 274t, 338–339
False marijuana, 55
FASD. See Fetal alcohol spectrum disorder (FASD)Fetal alcohol spectrum disorders (FASD)
Fatty acid amide hydrolase, 709
Fatty acid amide hydrolase (FAAH), 162–163, 704, 815–816
  formation of, 705
  isoforms, 705
Fatty acid ethyl esters (FAEE), 380, 382
Fentanyl, 909
Fetal alcohol spectrum (FAS), 488–489
Fetal alcohol spectrum disorder (FASD), 456–457, 497, 510–512, 516
  alteration of neuroimmune gene expression in, 457
  definition of, 462
  key facts related to, 512t
Fetal alcohol syndrome (FAS), 36, 497, 624
Fetal alcohol syndrome disorders (FASD), 571–572
  cortical biomarkers of, 571
  cortical morphology in, 568–571
    altered cortical morphometry, 569–570
    corticometry, 568–569
    relation to diagnosis, pathophysiology, and perspectives, 570–571
  cortical thickness analyses in, 568f
Fine mapping, 528
Five Factor Model, 1061
FK506-binding protein molecular weight (MW) 51 (FKBP51), 613
FK506-binding protein MW 52 (FKBP52), 613
FKBP51. See FK506-binding protein molecular weight (MW) 51 (FKBP51)
FKBP52. See FK506-binding protein MW 52 (FKBP52)
FKBP51 dysfunction, 613
Fluorodeoxyglucose (FDG), 960–961
1-(4-fluorophenyl)piperazine (pFPP), 56–57, 57f
fMRI. See Functional magnetic resonance imaging (fMRI)
Folic acid, 482–483
  in homocysteine metabolism, 484f
Food-addicted obese, 22
Forced swim test, 823
Fractional anisotropy, 848
Frontal cortex, 37, 40
Frontal lobe epilepsy, 136
Frontocerebellar circuitry, 583
Functional connectivity, 583
Functional impulsivity, 392–393
Functional magnetic resonance imaging (fMRI), 983
  on abstinence-induced craving, 276
  on altered emotion processing in smokers, 277
  of appetitive smoking cues, 293t–295t
  application to types of addictions, 983–987
  BOLD images, 983
  brain activity in nonsmokers and in smokers under conditions of nicotine satiety, 343f
  changes after motion, 987t
  children with familial substance use disorder, studies of, 40–44, 41t–43t
  cocaine-induced craving, 33
  on cue-induced nicotine craving, 273–276
  impaired response inhibition in heroin-dependent individuals, 1041–1042
  of pain, 959
  reactivity to smoking cues in, 339
  schizophrenia, smoking behavior in, 266–267
  of slice above corpus callosum
    in heroin addict of 5 years, in motion, 985f
    in heroin addict of over 10 years, in motion, 985f
    in normal individual at rest, 984f
    in normal individual in motion, 985f
  of slice at level of cerebellum
    of heroin addict in motion, 986f
    of normal individual in motion, 986f
  of slice at level of diencephalon
    of heroin addict in motion, 986f
    of normal individual in motion, 985f
  of superior parietal cortex
    of heroin addict in motion, 984f
    of normal individual at rest, 984f
    of normal individual in motion, 984f
  utilization of inhibitory control tasks during, studies using, 40–44
Functional polymorphisms, 516
Future reward expectancy, 1032

G

G proteins, CB1 receptor activation of, 27
$GABA_A$ receptor, 433, 533, 534f
Gabapentin, 930
GABRA2 gene, 439–440
γ-amino butyric acid (GABA), 82, 95, 116, 134, 201–202, 260, 956–957
  dopamine interaction with, 204. See also $GABA_A$ receptors
γ-amino butyric acid (GABA) receptor, 466
G861C SNP, 1030
Gender differences in addiction
  Cocaine, 930
  heroin, 928–930
  opiates, 927–928, 929t
Gene, 538
Gene expression, during alcohol withdrawal
  in cerebral cortices, 539f
  corticotropin-releasing hormone, 535–536, 536f
  dopamine, 533–535
    dopamine D2 receptor, 533–535
    dopamine transporter, 535

release from VTA, 533
transmission, 535, 535f
GABA$_A$ receptor, 533, 534f
glutamate, 531–533
   alcohol-related alterations, 532
   delirium tremens, 531–532
   GRIN1, 531–532
immune/proinflammatory factors, 538
myelin, 538
neuropeptide S, 537–538
nociceptin, 536–537, 537f
opioid receptor, 533
Gene microarray, 539
Gene promoter, 507
Generalized anxiety, CBD associated with, 750–751
Generalized anxiety disorder (GAD), 978
Genome-wide association studies (GWASs), 910
   applications, 917–918
   for heroin addiction, 917
   multistage, 915f
   opioid dependence, 915–917
   opioid sensitivity, 912–915, 913t, 916f
Genome-wide association study (GWAS), 523, 528
   ADH gene cluster, 521–527
   of German descent, 523
   of Korean population, 523–527
   and replication studies of African Americans, 523
   and replication study of AD, 524t–526t
GFP. *See* Green fluorescent protein (GFP)
G$_i$-coupled receptor, 756
Glia, definition of, 462
Gliosis, 634
Glucocorticoid receptor-dependent genes, drug-induced regulation of, 938–940
Glucocorticoid receptors (GR), 610–611
   gene transcriptional regulation by, 613f
   induction of transcriptional activation by, 613
   role in transition to alcohol dependence, 611–612
   signaling, 613
   structure, 612–613, 613f
Glucocorticoid receptors (GRs), 209–210
   coding gene (*NR3C1*), 211–212
   interdependent actions of MRs and, 215
   mechanisms of, 211f
   structure of, 212f
Glucocorticoid response elements (GREs), 612–613
Glucocorticoids (GCs)
   hypothalamic-pituitary-adrenal (HPA) axis, effects on, 209–211
Glucose, 623
Glucose transporter proteins, 623
Glutamate, 82, 136, 507, 531–533
   alcohol-related alterations, 532
   delirium tremens, 531–532
   GRIN1, 531–532
   nicotine withdrawal and, 203–204
   signaling. *See* Glutamate signaling
Glutamate excitotoxicity, 634

Glutamate signaling
   during alcohol withdrawal, 469–471
      elevated extracellular glutamate levels, 471
      enhanced glutamate release, 469
      genetic variation, contribution of, 471–472
      impaired synaptic glutamate clearance, 471
      overactivated glutamate receptors, 470
      reduced glutamate synthesis, 469
   genes encoding proteins in, 472t
   during normal physiological conditions, 467–469
      astrocytic glutamate, 469
      maintain synaptic glutamate concentration, 467–468
      synthesis, release, and receptors, 467
   in other addictive disorders, 473
Glutathione, 1016
Glycation, 559
GnRH. *See* Gonadotropin-releasing hormone (GnRH)
GNTI, 222t
Gonadotropin-releasing hormone (GnRH), 493
Go/no-go task, 1038, 1041–1043
Gonyautoxin III, 961–962
GPCR. *See* G-protein-coupled receptor (GPCR)
G-protein-coupled receptor (GPCR), 228, 689–690, 925
   structure and activation of, 926f
GR. *See* Glucocorticoid receptors (GR)
Gray matter, 485
   associated with CUD, 782–783
   definition of, 848
   dorsolateral periaqueductal, 756
Gray matter atrophy, 478–479
Gray matter morphology, 776t–777t
   of amygdala, 780
   cerebellar, 780
   hippocampal, 780
   PFC, 780
   striatal, 778–780
Green fluorescent protein (GFP), 1008
Group housing, 643
Growth-factor signaling, 622–623
GTPase activating proteins (GAPs), 125–126
Guanidinium toxins, 961–962
Guanine nucleotide exchange factors (GEFs), 125–126
GWAS. *See* Genome-wide association study (GWAS)

H

Hallucinogenic mushrooms, 52–53
   common physiological reactions, 53
   fatal intoxications due to, 53
Haloperidol, 84–85
Hamilton Depression Rating Scale, 968
Haplotype, 905
Hardy–Weinberg equilibrium, 913–914
Harm avoidance (HA), 978–979
Healthy subjects, smoking and cognitive function in, 305–307
Heaviness of Smoking Index (HSI), 230
Heavy smoking, defined, 260–261

Hedonic dysregulation theory of addiction, 340–341
Heightened reward sensitivity, 46, 46f
Herbal incense, 723–724
Heroin, 53–54, 909, 923, 970
   absorption of, 53
   activation of nucleus accumbens (NAc), 938
   alteration of circadian genes, 940
   altered perfusion, 995f
   appearance of, 869–870, 869f
   availability in world market, 878
   body packing of, 870, 870f
   case reports of leukoencephalopathy post exposure to, 991t–993t
   cerebral perfusion and, 994
   cerebrovascular accidents following, 994–996
      damage to mitochondria, 996
      embolization, 995
      hypersensitivity reaction, 996
      platelet count and hemostasis, 996
      systemic blood pressure decrease, 995–996
   CNS hypoperfusion-ischemic effect, 994–995
   complications from cervical intra-arterial heroin injection, 950f
   component analysis of, 870–873
   drug use trends in Georgia (from 2000 to 2014), 882–884
   elimination forms of, 53–54
   gene expression changes and, 937f
   historical review of, 868
   hypoxia and, 997–999
   ischemic stroke associated with, 989–990, 990t
   mode of administration, 53
   6-monoacetylmorphine, 869
   morphine, 869
   OPRM1 gene and dependence of, 230
   overdose deaths, 882f
   percentage of population using, 880f
   pharmacology of, 869
   production process, 868
   regulation of glucocorticoid receptor-dependent genes and, 938–940
   routes of administration for, 870
   spongiform degeneration, 996f
   spongiform leukoencephalopathy, 996–997
      hypoventilation and hyperemia, 997
      mitochondrial injury, 997
      toxin-related cerebral vasculitis, 996
   street
      adulterants and diluents in, 870–871
      adulterants used to cut, 872t
      heavy metals in, 871
      microbial contamination of, 871
      street names for drug combinations with, 871t
   street names and slang terms for, 870
   TM1 transcriptional activation for, 934–935
   tolerance and, 54
   transcriptional effects on striatum, 938
Heroin abuse
   harm reduction response to increasing, 884
   treatment for
      admissions for, in the United States, 883f
      of US military veterans, 883f

Heroin addiction, 14
　characteristics of addicts, 1026t
　deficits in, 1042–1043
　functional right inferior frontal gyrus imaging findings in, 1041–1042, 1042f
　gender differences in, 928–930
　gray matter volume, 1040f, 1041
　impaired cognitive controls and, 1038–1039
　neuroimaging studies of, 1013–1014, 1014t
　neuropsychological studies of, 1014
　pathophysiology of, 15–17
　premature aging and, 1013, 1016–1018
　premorbid neurological vulnerabilities fostering development in, 953
　relationships of personality dimensions with, 978–979
　rIFG impaired inhibitory performance in, 1039
　structural right inferior frontal gyrus imaging findings in, 1039–1041, 1040f
　telomerase interaction on brain structural volume, 1016–1018, 1017f
　white matter volume, 1041, 1041f
Heroin addiction disorder (HD), 1023–1024
　characteristics of, 1027t
Heroin-induced leukoencephalopathy, 990–994
　MRS for metabolites in brain, 998f
　MRSI findings in, 998f
Heterotrimeric G proteins, 926
Hexamethonium, 172, 349–351, 351f
HINT1, 228–230
　in relation to smoking behavior, 230
Hippocampal volume reduction, in adolescent AUD, 588–592, 593f
Hippocampus, 63, 75, 168, 485, 597, 774, 780, 783, 924–925
　aging process and, 1016
　binge drinking and, 393
　location and physiological action of opiate receptors, 924–925
　nicotine effects on, 171
Histrionic personality disorder, 874
HIV encephalitis (HIVE), 181
HIV-associated dementia (HAD), 181
HIV-infected macrophages, 181
HIV-infected microglia, 181
HKD motif (H, histidine; K, lysine; D, aspartic acid), 489
Homeostasis, 614, 895
Homer1a, 934
Homocysteine, 482–483
　metabolism, 484f
Hooked on Nicotine Checklist (HONC), 338
HPA axis. *See* Hypothalamic-pituitary-adrenal (HPA) axis
HPA axis dysfunction, 643
H-Ras, 507
5-HT DRN neuron firing rate and 5-HT release, effects of nicotine, 147–154
　microdialysis studies, 147–149
　stimulatory effects, 149f
　in vitro electrophysiological experiments, 149–153
　in vivo electrophysiological experiments, 153–154

5-HT firing activity
　administration of CB1R agonists in adolescence and adulthood, 821–822
　Effects of URB 597 on, 820f
　fatty acid amide hydrolase inhibition, 820–821
　mPFCv in, 818
　WIN 55,212-2 treatment of, 818
5-HT neurotransmission, alterations in, 822f
HTR1B, 1030
Human immunodeficiency virus (HIV), 949–951
　infected population, 179
　illicit drug use among, 179–180
　interaction of drugs of abuse and, 180
　prevalence of alcohol use among, 180
Hydrocodone (Vicodin), 922
Hydromorphone (Dilaudid), 922
Hydrosoluble vitamins, 482–484
　folic acid, 482–483
　homocysteine, 482–483
　thiamine deficiency, 484
　vitamin B12, 482–483
　vitamin B3 deficiency, 483–484
　vitamin C deficiency, 483
　Wernicke encephalopathy, 484
6-Hydroxydopamine (6-OHDA), 250, 404
4-Hydroxy-2-nonenal (HNE), 384
5-Hydroxytryptamine-1A (5-HT$_{1A}$) receptor, 749–752, 756
　cannabidiol effects mediated by, 754–755, 754t
　cannabidiol interaction with, 753–754
　expression, changes in, 753–754
　controls anxiety-like behaviors, 751f
　decreases fear conditioning behavior, 752f
　and generalized anxiety and depression, 750–751
　motor control, 752
　nausea and vomiting, 752
　neuroprotection by, 751–752
　panic and posttraumatic stress disorder, 751
Hyperalgesia
　chronic opioid exposure and, 892
　definition, 895
Hyperemia, 997
Hyperglutamatergic state, 473
Hyperhomocysteinemia, 483
Hyper-reward sensitivity, 37
Hypothalamic-pituitary-adrenal (HPA) axis, 63–64, 209, 435, 414, 643, 735, 935, 938–940, 1003
　adrenocorticotropic hormone release, 610
　alcohol-induced activation of, 610–611, 611f
　alcohol-induced dysregulation of, 612f
　endogenous opioid system to control, 1003–1004
　function of, chronic alcohol exposure impact on, 611
　functioning in humans, 210f
　glucocorticoids (GCs), effects of, 209–211
　in human heroin addiction and rodent models, 1004–1005
　internalizing pathway and, 37
　MR/GR interactions and, 215

opiate consumption and, 1003
　in other addictions and substance misuse, 613–614
　paraventricular nucleus in, 610
　stress-responsive, 1004f
Hypothalamus, 168
　internalizing pathway and, 37
　nicotine effects on, 171
　orexin receptors and, 134–135
Hypothesis-driven method, 910
Hypothesis-free method, 910
Hypoventilation, 997
Hypoxic-ischemic leukoencephalopathy, associated with heroin, 997–999
　cases in minors, 997–999
　mitochondrial injury, 997–999
　MRI finding in, 999, 999f

I

IGF. *See* Insulin-like growth factors (IGF)
IGF-1. *See* Insulin-like growth factor 1 (IGF-1)
Illicit drug abuse
　in high-school-age children, 761f
IL1RN gene, 912
Immediate early genes (IEGs), 934–935
Immunohistochemistry (IHC), 434
Impaired cognitive controls
　during abstinence and maintenance, 1038
　in abstinent compared to maintained heroin-dependent users, 1039
　in chronic heroin users, 1038
　in heroin-dependent individuals, 1038–1039
　self-control, 1037
Impaired response inhibition, 37
Impoverished condition (IC) and drug effects, 246–247
　behavioral effects, 247–248, 247f
　neural effects, 248–250
Impulsive behavior, 979
Impulsivity, 393, 974
　in alcohol dependents, 975
　"bottom-up" activity involving, 37
　choice, 37
　defined, 37
　heroin addicts and, 975–976
　heroin dependency and, 979
　motor impulsiveness (MI), 975–976
　nonplanning impulsiveness (NPI), 975–976
　predictors of severity of, 979t
　relationships of personality dimensions with, 978–979
　"top-down" behavioral control, 37
INCB. *See* International Narcotics Control Board (INCB)
Incentive sensitization model of SUD, 37–38
Incentive-salience hypothesis, 37
Incentive-sensitization theory of addiction, 29, 337
　incentive motivational properties of drugs, 32
　key points, 33t
　neural system and, 32
　problems with, 33, 33t
Index of Tobacco Treatment Quality (ITTQ), 110

Induced oscillations, 857
Inducible nitric oxide synthase (iNOS), 620
Inferior frontal cortex, 38
Inflammasomes, 459
Inflammation, 1015
iNOS. See Inducible nitric oxide synthase (iNOS)
Instrumental appetitive extinction (iE), 75
Instrumental successive negative contrast (iSNC), 75
Insular cortex, 594–595
Insulin-like growth factor 1 (IGF-1), 497, 622–623
Intergenerational trauma, 62
Internalizing disorders, 36
Internalizing pathway, 37
International Narcotics Control Board (INCB), 882
Internet, sale of psychoactive drugs, 50–58
 cocaine, 50–52
 common drugs sold, 52t
 designer drugs, 54–57
 hallucinogenic mushrooms, 52–53
 heroin, 53–54
 ketamine (DL-2-(O-chlorophenyl)-2-(methylamine)cyclohexane hydrochloride), 54
 lysergic acid diethylamine (LSD), 54
 mephedrone (4-methylmethcathinone), 56
 3,4-methylenedioxymethamphetamine (MDMA or ecstasy), 53
 methylone (3,4-methylenedioxi-N-methyl-cathinone), 55–56
 new phenylethylamine derivatives, 57
 phenylethylamine (phenylethylamine-N-benzyl derivatives), 57
 piperazine derivatives, 56–57
 Salvia divinorum, 57–58
 Salvinorin A, 58
 synthetic cannabinoids, 55
 synthetic cathinones, 55–56
 websites, 51t
Interpeduncular nucleus (IPN), 190–191, 195
Intra-basolateral amygdala (BLA), 163
Intracellular signaling cascades, 699
Intracranial self-stimulation (ICSS), 116, 727–728
2-(4-Iodo-2,5-dimethoxyphenyl)-N-[(2-methoxyphenyl)methyl, 57
Ischemic stroke, associated with heroin, 989–990, 990t
Ivory wave, 55

J
Jitai, 967
JWH-018, 685

K
K2, 55
κ-Opioid receptor (KOR), 204, 221, 899, 1003–1004
Ketamine (DL-2-(O-chlorophenyl)-2-(methylamine)cyclohexane hydrochloride), 54
 action of, 54
 forms available, 54
Ketanserin, 320
Knocking out beta-endorphin, effects of, 405
Knockout, 905
Korsakoff syndrome (KS), 629–630
 brain areas involved in episodic memory, 630
 deficit proposed in memory disturbances, 630
 imaging of, 630
 locations of lesions in, 629
 thiamine replacement in, 629
Korsakoff's psychosis, 384
Krebs Cycle. See Tricarboxylic acid (TCA) cycle

L
Lateral hypothalamic area (LHA), 134–135
Laterodorsal tegmental nucleus, nicotine effects on, 173–174
Latrophilin 3 (LPHN3), 1030
Learning and memory processing, 805
 acquisition and consolidation, 805
 emotional, phases of, 805f
 endocannabinoid system role in, 807–808
 extinction, 805, 805f
 reconsolidation, 805
 retrieval, 805
Learning theories of addiction, 339–340
Lempel–Ziv complexity, 860
Leukoencephalopathy
 delayed, 994
 heroin-induced, 990–994
 spongiform, 994, 1000
Lexical decision task, 286
Linkage disequilibrium (LD), 913–914
Linkage studies of human opioid dependence, 915–917
 European-American (EA) and African-American subjects, 915, 917
 heavy-opioid-use cluster and non-opioid-use cluster, 915
 nonparametric linkage (NPL) Z scores, 915–916
Lipid peroxidation (LPO), 552–553
Lipid raft asymmetry, 699
Lipid rafts, 699
 detection of CB1 in, 691t
 detergent-free, 692f
 outer and inner leaflets of, 690f
Liposoluble vitamins, 480–482
 vitamin A deficiency, 481
 vitamin D deficiency, 481–482
  in alcoholics, 482
  prevalence of, 482f
 vitamin E deficiency, 480–481
  chronic ethanol feeding and, 481
  CRMP-2 in, 480–481
  MAP-LC3 in, 480–481
Liquid chromatography-MS and immunoassay techniques, 94
Locked-in state, 607
Locomotor sensitization, 318
 effects of nicotine vs TPM on, 319t
 withdrawal effects, 318

Longitudinal study, 833
Long-term potentiation, 125
LY686017, 196
Lysergic acid diethylamine (LSD), 53–54
 effects, 54
 5-HT receptors and, 54

M
Mad Hatter, 55
MAGL. See Monoacylglycerol lipase (MAGL)
Magnetic resonance imaging (MRI), 597
 of heroin addiction, 1013–1014
Magnetic resonance spectroscopy (MRS), 959–960, 1000
Major depression, 816
Major depressive disorder (MDD), smoking and cognitive function in, 309–311
 neuropsychological performance, 310t
Malondialdehyde (MDA), 384
Manganese, 962
Mania, 838–839
 dysphoric, 838
 euphoric, 838
 impact of cannabis on, 836, 839t
 medications and substances induces, 837t
Manic episode, 838
MAOA. See Monoamine oxidase A (MAOA)
MAOA gene, 546–547
MAOA-uVNTR gene. See Monoamine oxidase A upstream variable number tandem repeat (MAOA-uVNTR) gene
MAP. See Mitogen-activated protein (MAP)
MAPK. See Mitogen-activated protein kinase (MAPK)
MAPK-8 signaling pathways, 958
MAP-LC3. See Microtubule-associated protein-light chain 3 (MAP-LC3)
Marchiafava–Bignami disease (MBD), 605, 630–632, 634
 chronic alcoholic female with, brain MRI in, 605f
 definition of, 607
 MRI findings in, 631f
 neuroimaging of, 631–632
 subtypes of, 632t
Marijuana, 760, 766
 dependence studies in humans, 725
 spice vs., 723–724
 synthetic, 728
 use among HIV-infected individuals, 185
Mary Herb, 57
Maudsley Personality Inventory, 1061
Maximum reaction rate (V_{max}), 516
Mazindol, 27
MBD. See Marchiafava–Bignami disease (MBD)
MDA- acetaldehyde-protein adduct (MAA), 384
MDMA. See 3,4-Methylenedioxymethamphetamine (MDMA)
MDPV. See 3,4-Methylenedioxypyrovalerone (MDPV)
Mean diffusivity, 848

Mecamylamine, 247–249, 320–322, 320f, 414–415
Medial habenula (MHb), 190–191, 195
Medial prefrontal cortex (mPFC), 249
Medial temporal lobe (MTL)
    aging process and, 1015–1016
Medium spiny neurons (MSNs), 796–798
    Golgi-Cox-stained, 800f
Medulla, 168
Membrane skeleton, 699
Memory, 811
    consolidation, 811
    effect of nicotine on, 285–286, 286t
Memory extinction, 805, 805f
    aversive, cannabinoid effects on, 810t
    rewarding drug and appetitive, cannabinoid effects on, 808t
    role of endocannabinoid system in, 807
Memory reconsolidation, 805
    aversive, cannabinoid effects on, 810t
    rewarding drug and appetitive, cannabinoid effects on, 808t
    role of endocannabinoid system in, 807
MEOS. See Microsomal ethanol-oxidizing system (MEOS)
Mephedrone (4-methylmethcathinone), 55–56
Mesocortical system, 204
Mesocorticolimbic brain reward pathway, 115–116, 119–120
Mesocorticolimbic circuit, 272–273
Mesocorticolimbic dopaminergic system
    nicotine abuse and, 159
Mesocorticolimbic pathway, 116
Mesocorticolimbic system, 933
Mesolimbic dopamine system, 38, 204
Messenger ribonucleic acid (mRNA), 539
Metabolic superimposition, 385
Methadone, 885, 903, 967
Methadone maintenance therapy (MMT), 1013–1014, 1038–1039
Methamphetamine, 935
    activation of nucleus accumbens (NAc), 938
    alteration of circadian genes, 940
    gene expression alterations in striatum, 936f
    gene expression changes and, 937f
    regulation of glucocorticoid receptor-dependent genes and, 938–940
    transcriptional effects on striatum, 938
Methamphetamine, use among HIV-infected individuals, 185
Methamphetamine dependence, 655
Methcathinone, 685
1-(4-Methoxyphenyl)piperazine (MeOPP), 56–57, 57f
Methyl-β-cyclodextrin (MCD), 691–692
    effects on distribution of CB1, 691t, 692, 693f
3,4-methylenedioxymethamphetamine (MDMA or ecstasy), 53, 163–165, 606–607
    commmercially available forms, 53
    common effects, 53
    critical effects, 53
    hallucinogenic effects, 53
3,4-Methylenedioxypyrovalerone (MDPV), 55–56, 684

chemical structures of, 683f
for other addictions and substance misuse, 684–685
Methyllycaconitine (MLA), 415
Methylone (3,4-methylenedioxi-N-methyl-cathinone), 55–56
Methylphenidate, 33, 37
Mexican mint magic, 57
mGluR2/3 receptors, 469
    alcohol impact on, 470f
Michaelis constant (K_m), 516, 629
Microcephaly, 494–495, 495f
    FAS-related, gyral pattern in, 570f
Microdialysis
    ACh release measured by, 496f
    definition of, 497
Microglia, 454–456, 485
    activated, 455, 455f
    alcohol effects on, 456f
        in adolescent and adult brain, 458f
        in fetal brain, 457f
    in alcohol-induced neurodegeneration, 621–622
    function of, 455f, 462
    innate immune responses, 459–461
    morphology of, 456f
    phenotypes of, 455–456, 455f
    ramified, 455f
    toll-like receptors on, 455–456
Microglial activation
    alcohol effects on, 460f
    definition of, 462
    of NFκB, 480
Microsomal ethanol-oxidizing system (MEOS), 380–381, 552
    inhibition of, 557
Microtubule organizing center (MTOC), 692
Microtubule-associated protein-light chain 3 (MAP-LC3), 480–481
Midazolam (MDZ), 170
Mifepristone, effects on alcohol-related behaviors, 612, 612t
Mineralocorticoid receptors (MRs), 209–211
    coding gene (NR3C2), 213–215, 214t
    interdependent actions of GRs and, 215
    mechanisms of, 211f
    structure of, 212f
Miniature excitatory postsynaptic currents (mEPSCs), 168–169
Minnesota Multiphasic Personality Inventory, 1061
Minor alkaloids, 119
Minor allele frequency, 905
miRNA, 461–462
Missense polymorphism, 126–127
Missing heritability, 36
Mitochondrial toxicity, of acetaldehyde, 557
Mitogen-activated protein (MAP), 717
    kinase pathways, 717
Mitogen-activated protein kinase (MAPK), 689–690
    signaling pathways, 958
Mitogen-activated protein kinase (MAPK)-1, 958
Monism, 17–18

6-Monoacetylmorphine (6-MAM), 53, 1052
Monoacylglycerol lipase (MAGL), 705
Monoamine oxidase A (MAOA), 548
Monoamine oxidase A upstream variable number tandem repeat (MAOA-uVNTR) gene, 546–548
Monoamine oxidases (MAOs), 116, 119, 205, 322
μ-Opioid receptor (MOR), 221, 222t, 228, 230, 899, 910, 1003–1004
    heterodimerization of, 904
MOR-DOR heterodimers, 904
    role in morphine use, 904f
Morphine, 869, 922–923, 1052
    analgesic actions of, 922–923
    dependence, OPRM1 gene and, 230
    exposure and expression in orexinergic system, 140–141
    TM1 transcriptional activation for, 934–935
Morphine equivalent, 885
Morphine-3-glucuronide (M3G), 1052, 1054
Morphine-6-glucuronide (M6G), 1052, 1054
Mortality
    awareness in human social behavior, 328
    conscious and nonconscious death-related cognitions, 328–329
    cultural worldview, 328f
    ICD-9 and ICD-10 codes and comparability ratios forconditions, 106t
    nonconscious, 331–333
    policy interventions to reduce tobacco-related, 110–111
    psychological defenses against conscious, 330–333
    rates among HIV/AIDS patients and, 183
    smoking-attributable fraction (SAF) of, for appendicitis and drug/alcohol cohort groups, 108f
    standardized mortality ratios (SMRs) from tobacco use, 106t, 107, 108f
    substance use and, 331–333
    tobacco-related, 107
Mortality salience, 328
Motor disinhibition, 37
Motor impulsiveness (MI), 975–976, 978t
MSNs. See Medium spiny neurons (MSNs)
Multiple sclerosis, 745
Multiplex family study, 583
Munich Gene Data Bank, 531–532
Mutational analysis, 719
Myelin, 538, 848
Myelin alterations
    alcohol-induced, 425–427
    amphetamine-induced, 427
    cannabis (marijuana)-induced, 427–428
    cocaine-induced, 427
    nicotine-induced, 428
    opiate-induced, 428
    polydrug abuse and, 428
    therapeutic approaches to, 428
Myelin basic protein (MBP), 421–423, 427
Myelin oligodendrocyte glycoprotein (MOG), 427
Myelin sheath, composition of, 421–423
Myelin staining, 843f

Myelin-associated glycoprotein (MAG), 421–423
Myelination process, 421
Myeloperoxidase, 552

N

NAC. *See* Nucleus accumbens (NAC)
N-acetylaspartate/choline (NAA/Cho), 494–495
N-acetylaspartate/creatine (NAA/Cr), 494–495
NADA. *See N*-arachidonoyldopamine (NADA)
Nalbuphine, 922, 928
Naloxonazine, 222t
Naloxone, 222t, 884, 922
Naloxone nasal administration unit, 885f
Naltrexone, 242, 922
Naltrindole, 222t
N-arachidonoyldopamine (NADA), 703
N-arachidonoylphosphatidylethanolamine (NArPE), 703–704
NArPE. *See N*-arachidonoylphosphatidylethanolamine (NArPE)
National Council on Alcoholism and Drug Dependence (NCADD), 520
National Epidemiology Survey on Alcohol and Related Conditions, 837
Natural and manufactured aldehydes, 554
Natural cathinone, chemical structures of, 683f
Natural opiates, 867
Nausea and vomiting, CBD associated with, 752
N-benzylpiperazine (BZP), 56–57, 57f
NBOMe, 57
NCADD. *See* National Council on Alcoholism and Drug Dependence (NCADD)
Near-infrared fluorescence spectroscopy (NIRS), 960
Needing, 337
Negative reinforcement, 614, 895
Negative reinforcement models of addiction, 29–30
 factor in maintaining nonaddictive behavior, 30
 key points, 30t
 problems with, 30–31
 withdrawal and tolerance, 30–31
 withdrawal distress and drug-seeking behavior, 30
Nervous anorexia, 147
N^2-ethyl-2′-deoxyguanosine (N^2-ethyl-dG), 555
N^2-ethyl-dG. *See* N^2-ethyl-2′-deoxyguanosine (N^2-ethyl-dG)
Neural activity, nicotine-induced changes in
 blood–brain barrier (BBB), blockade of, 349–351
 conditioned, by stimulation of peripheral nicotinic receptors, 355–356
 development of neural and behavioral sensitization, 354
 EEG studies, 348–349, 349f–350f, 354f
 locomotor effects, 353
 from peripheral and central actions, 349–353, 355
 physiological effects, 353–354
Neural oscillations, 854–859, 855f
 definition of, 860
 induced oscillations, 857
 quantified neural oscillations, types of, 855
 resting-state oscillations, 857–859
 steady-state evoked oscillations, 856–857
 transient evoked oscillations, 856
Neural oscillatory activity, 745
Neural progenitor cells (NPC), 492
Neural stem cells (NSCs), 493–494
Neurite outgrowth, 493–494, 497
Neuroactive steroids
 biosynthesis pathway, 434f
 ethanol drinking and
  in humans, 439
  in primates, 438–439
  in rats, 436–437
 ethanol drinking in humans and, 439
 ethanol drinking in primates and, 438–439
 ethanol drinking in rats and, 436–437
 gabaergic, and ethanol interactions, 437–438
 genetic polymorphisms and, 440
Neurobiology
 of drug use and addiction, 64
 of trauma, 63–64
Neurodegeneration, definition of, 624
Neurofeedback (NFB), 18–19
Neurogenesis, 624
 inhibition of, 623
Neuroimmune response, 462
Neuroimmune signaling, 459–461
 alcohol effects on, 460f
 DNA methylation, 459, 461
 inflammasomes and, 459
 TLR4 signaling, 459
Neuroinflammation, 462
 definition of, 462, 624, 735
 potential mechanisms of, in chronic alcoholics, 603t
 TLR-4 activation induced, 622
Neurokinin A (NKA), 190, 195
 immunoreactivity for, 190
Neurokinin B (NKB), 194–195
Neurokinin receptors (NKRs), 190
 neurokinin 1 receptor (NK1R), 194, 196
 neurokinin 2 receptor (NK2R), 195
 neurokinin 3 receptor (NK3R), 194–195
Neurokinins, interactions with nicotine
 cellular mechanisms, 193–194
 regulation of mood and affect, 194–195
 symptoms of withdrawal and, 195
Neuroleptic dysphoria, 84–86
Neurological withdrawal symptoms in opiate addicts, 952
Neuromodulation, 14, 21
 approaches to find targets for, 15
 defined, 14, 22
 key facts, 22
 link between brain and mind, 17–22
 nonsurgical technique, 20f
Neuronal glucose transport, alcohol-induced disruption in, 623
Neuronal nicotinic acetylcholine receptors (nAChRs), 146–147, 411–412
 distribution, 413f
Neuropeptide S (NPS), 537–538
Neuropeptide Y (NPY), 134, 205, 634
Neuropeptides
 distribution in CNS, 190–191
 role in alcohol use disorder (AUD), 401
 synthesis, release, and degradation, 402f
 as target systems, 401
Neuroplasticity, 14, 18, 506, 597
Neuropsychiatric disorder, 634
 alcohol withdrawal syndrome, 632
 alcoholic cerebellar degeneration, 633
 alcoholic hallucinosis, 633–634
 alcohol-related dementia, 632–633
 central pontine myelinolysis, 634
 Korsakoff syndrome, 629–630
 Marchiafava–Bignami disease, 630–632
 neuropathological spectrum of, 630t
 Wernicke encephalopathy, 627–629, 634
Neuropsychology, 597
Neuroreceptor, 538
Neurotoxicity
 in alcoholics, pathogenesis of, 480
 hypothesis, 634
Neurotransmitter system, 767
Nicotiana (tobacco), 93
 leaves, 93
 synthesis of, 93
Nicotinamide adenine dinucleotide phosphate oxidase (NOX), 480
Nicotine (3-(1-methyl-2-pyrrolidinyl) pyridine), 94, 133, 135–136, 651–653
 absorption, 97
 activation of NMDA receptors, 203
 acute administration of low doses, 154
 adaptive changes in glutamate transmission, 203–204
 alcohol codependence and, 416
 anxiogenic withdrawal effect of, 154
 anxiolytic effects, 154
 bioavailability, 94–95, 97
 biotransformation, 97–99
 cause of tobacco addiction, 272–273
 chemical structures, 94f
 chronic exposure, effect of, 272–273
 clinical conditions of exposure, 136–137
 concurrent effects of, 174–175
 DA supersensitivity in, 201–202
 dependence, 158–159
 desensitization and upregulation, 96–97
 distribution, 97–98
 dual effects of, 175–176, 175f
 effects on 5-HT DRN neuron firing rate and 5-HT release, 147–154
 effects on mesocorticolimbic pathway, 96f
 excretion, 98
 human studies of nicotine on orexin, 136
 indirect effects, 287
 inhibition of proteasomal function, 202
 inhibitory effect on eating, 154
 key facts, 165
 locomotor activity, effect on, 154
 maximal concentration (C_{max}) and time to maximal concentration (T_{max}), 94–95, 99t
 mEPSC frequency and amplitude, effect on, 170–171

Nicotine (3-(1-methyl-2-pyrrolidinyl) pyridine) (Continued)
  metabolism, 98, 100f
  myelin alterations from abuse of, 428
  neostriatum, effects in, 175–176
  orexinergic system, effects of, 136
  pharmacokinetics, 97–98, 98f
  pharmacology, 94–97
  regional differences and CNS effects, 95–96
  rewarding effects, 116–117, 158–159
  Rho GTPases and addiction of, 125–126
  sag and I_h modulation, 173f
  self-administration, effects of, 119–120, 162–163, 204–205, 238f, 320–322, 320f–322f, 323t
  serotonin release and, 154
  short-term potentiation (STP) of EPSPs and population spikes (PS), 171
  (S)-Nicotine and (R)-nicotine, 135, 136f
  tobacco particulate matter and extracts, 118–119
  tobacco-related diseases, 105–107
  vasoconstrictive effects of, 353. See also brain areas, nicotine's effects on; smoking tobacco. neurobiological effects of
Nicotine abstinence, assessment of, 366–367
Nicotine addiction, 254–255
  differences between addiction to other drugs and, 345
  neuropathology, 255
  personal and environmental factors, 255
  trajectory of, 345
Nicotine priming, 240–241, 320
Nicotine replacement therapy (NRT), 256, 267–268
Nicotine with alcohol, concurrent effects of, 174–175
Nicotine withdrawal syndrome
  brain parts involved in, 202f
  dopamine and, 203
  glutamate levels and, 203–204
  hypothalamic-pituitary-adrenal axis, role of, 202–203
  neural basis of, 202–203
  neurotransmitters and mediator systems involved in, 203–205, 203f
Nicotine-conditioned hyperlocomotor activity, 249
Nicotine-induced EEG desynchronization, 351–352
Nicotine-induced locomotor sensitization, 249
  nAChR density and, 250f
Nicotine-induced neuroadaptations, 201–202, 202f
Nicotine-induced neuroplasticity, 201–202
Nicotine$_{PM}$
  EEG and EMG effects, 353–354, 354f
  effects of, 349–353, 352f, 355f
  induced rapid EEG desynchronization and EMG activation, 352
Nicotine-seeking behavior
  cue-induced reinstatement, 238–239, 239f–240f

effects of stress exposure on, 240–241, 241f
  representative behavior profile, 238f
  response reinstatement paradigm, 237–238
Nicotinic acetylcholine receptors (nAChRs), 135–136, 146, 168, 174, 201–202, 241, 260, 272–273
  α3* nAChRs in VTA DA neurons, 175
  allostasis and allostatic load, role in maintaining, 416
  CB1 receptor ligands and, 163
  desensitization of, 172
  distribution, 412–413
  effects of enrichment on, 249
  gene knockout (KO) studies of subunits, 413–414
  homeostasis and, 97
  key facts, 120
  localization and function within neurons, 95
  mechanism of action, 117
  neuronal, 146–147, 411–412
  nicotine binding sites on, 95
  nicotine desensitized β2*, 170–171
  pharmacotherapeutic development based on, 414–415
  reward pathway, 413f, 414t
  role in alcohol addiction, 413
  structural diversity, 412
  structure of, 94–95, 94f, 115–116
  voltage-gated calcium channels (VGCCs) and, 152–153
Nigrostriatal dopamine system, 204
Nitric oxide (NO), 620
NMDA receptors, 957
NMDAR. See N-methyl-D-aspartate receptors (NMDAR)
N-methyl-D-aspartate (NMDA), 201–202
N-methyl-D-aspartate receptors (NMDAR), 241, 467, 500, 501f
  alcohol impact on, 470f
  amino-terminal domain, 501f
  carboxy-terminal domain, 501f
  phosphorylation, 502–504
  ethanol actions at, 501
  heterotetrameric structure of, 468f
  role in other drugs of, 506
  subunits, 500–506
    composition of, 501f
    NR1 subunit, 501–504, 502f
    NR2 subunit, 504–505
    NR3 subunit, 505–506
N-methyl-D-aspartate-induced excitotoxicity, 621
N-methyl-D-aspartic acid ionotropic receptor, 735
Nociceptin, 536–537, 537f
Nomifensine, 175–176
Nondeclarative learning, nicotine and, 285–286
Non-food-addicted obese, 22
Nonnucleoside reverse transcriptase inhibitors (NNRTIs), 181
Nonplanning impulsiveness (NPI), 976, 978t
Non-REM sleep, 134
Nonspecific simplified gyral pattern, in FAS-related microcephaly, 570f

Noradrenaline, 205
Nor-binaltorphimine, 222t
Norepinephrine, 192
Nornicotine, 94f
Note about challenge paradigms, 792
Novelty seeking, 392–393
Novelty seeking (NS), 977–978, 979t
Npas4, 934
NPC. See Neural progenitor cells (NPC)
$1,N^2$-PdG. See $1,N^2$-propano-2′-deoxyguanosine ($1,N^2$-PdG)
$1,N^2$-propano-2′-deoxyguanosine ($1,N^2$-PdG), 555–556
NPS. See Neuropeptide S (NPS)
NPS receptor (NPS-R), 538
NR3C1 gene, 211–212, 214t
  functional aspects of, 212–213
  polymorphisms effects on smoking behavior, 213
NR3C2 gene, 213–215, 214t
  functional aspects of, 215
  polymorphisms effects on smoking behavior, 215
Nucleoside reverse transcriptase inhibitors (NRTIs), 181
Nucleotide 118, 223
Nucleus accumbens (NAC), 37–38, 44–45, 71, 158, 204, 349, 532f, 926
  activation by chronic methamphetamine and heroin treatment, 938
  addictive spines in, 801
  associated ethanol-mediated dopamine release in, 414–415
  effects of enrichment on, 249
  Golgi-Cox-stained MSNs in, 800f
Nucleus hypoglossal, nicotine effects, 174
Nucleus tractus solitarius (NTS), 205

O

OARS (open-ended questions, affirmation, reflective listening, and summarizing), 657
Obesity, as addiction, 22
Object recognition test, 497
Object-based morphometry, 567–568
Obsessive-compulsive disorder (OCD), 147
  smoking and cognitive function in, 312
Obsessive-Compulsive Drinking Scale, 535
ODS. See Osmotic demyelination syndromes (ODS)
Offensive aggression, 833
Oligodendrocytes, 420–423, 425, 848
  apoptosis, 602
Onset mania, 838
Onset manic episode, cannabis use and, 837
Ophryocystis elektroscirrha, 74
Opiate abuse
  global trends in, 879
  trends in treatment for, 882–884
Opiate addiction, 1003
  stress responses in, 1008
Opiate addicts
  causes of neurological disease in, 947

cerebral bleeding in, 949
CNS disorders in, 948t
cognitive impairment in, 951
dependency of neurological abnormalities and opiate administration, 953
epilepsy in, 949
head trauma or traumatic brain injury in, 949
imaging studies of, 952
infections involving CNS or PNS in, 949–951
ischemic stroke in, 948
neurological abnormalities, 947
neurological disease in, 947
neurological withdrawal symptoms in, 952
neuropathology in, 952
parkinsonism in, 951–952
PNS disease in, 952
premorbid neurological vulnerabilities fostering development in, 952–953
secondary disease in, 948t
spongiform leukoencephalopathy in, 951–952
tertiary neurological abnormalities in, 953
transverse myelitis in, 951–952
Opiate receptors, 923
  binding studies, 923
  classification and localization of, 923–925, 924t
  opiate ligand specificity for, 925t
  signal transduction and, 925–926
  structure of, 925–926
  subtypes, 923–924
  tolerance, dependence, and addiction, 926–927
Opiate Treatment Index (OTI), 1063
Opiates, 878, 885, 895
  overdose deaths, 879–881, 881t
  peptides of, 923
  pharmacology of, 927–928
  semisynthetic, 867
  short-acting
    chronic exposure to, 1005
    heroin, morphine, 1005
  synthetic, 867
Opioid addiction, 576, 891, 895
  addicted phenotype, 891–892
  animal models of, 892
  definition, 895
  in relation to pain management, 892
  shared neural mechanisms/neurocircuitry underlying pain and, 892–893
Opioid dependence, transition to, 893–894
Opioid intake escalation, animal models of, 892
Opioid precursor proteins, 402
Opioid receptors, 27, 402–404, 533
Opioid sensitivity, 909
  candidate gene association studies, 910–912, 911t–912t
  European Pharmacogenetic Opioid Study (EPOS), 913–914
  genome-wide association studies (GWASs), 912–915, 913t
  pain relief and, 913–914
Opioid use disorder. See Opioid addiction
Opioid-induced hyperalgesia, 892, 895, 961f

role of corticotropin-releasing factor in, 893–894
Opioids, 885, 895, 909, 956
  abuse of prescription, 886
  acute alcohol-induced effects on, 404
  alcohol intake and, 405–406
  bioassay of, 1052
  biological material used for determination of, 1053–1054
  capillary electrophoretic/electrokinetic methods studies, 1056
  classification
    based on chemical structures, 867
    based on effects on opioid receptors, 868
    source based, 867
  concentration levels in various biological samples, 1053t
  detection and quantification of, 1054–1056, 1055f
  effects of variances in monozygotic and dizygotic twin pairs, 909–910
  Fourier transform infrared spectroscopy studies, 1055
  gas chromatography studies, 1055–1056
  genetic variations, effects of, 909–910
  high-performance liquid chromatography studies, 1056
  impact on aging, 1013
  individual differences in response to opioid antagonists, 407–408
  interactions between alcohol and, 404–406
  keratinized matrices for determination of, 1054
  mechanisms of dependence from, 195–196
  metabolism of, 1052
  myelin alterations from abuse of, 428
  nonseparation techniques studies, 1055
  OPRM1 gene and dependence of, 230
  oral fluid (saliva) and sweat analysis for determination of, 1053–1054
  for other addictions and substance misuse, 884–885
  overdose, 886
  overdose deaths, 879–881, 881t
  pharmacological interventions on, and nicotine-induced behavioral responses, 222t
  plasma or serum analysis for determination of, 1053
  prescription misuse, 956–957
  sample preparation for bioanalysis, 1054
  self-administration, effects of, 933
  semisynthetic, 878
  synthetic, 878, 886
  THC effects of previous exposure on rewarding effects of, 668t
  transcriptional changes, 934
  urine analysis for determination of, 1053
  use among HIV-infected individuals, 185
Opioids abuse
  global trends in, 879
  harm reduction response to increasing, 884
  percentage of population, 880f, 881t
  trends in treatment for, 882–884

  in Australia, 884
  in Canada, 884
  in the United States, 883f
  US military veterans, 883f
  in the United States, 881–882
Opioids use, 656
Opium, 867, 922
  alkaloids, 922, 1052
  review of analytical methods for bioassay of, 1056, 1057t
  seizures from, 1052
  percentage of population using, 880f
Opponent-process theory of addiction, 337
Oppositional defiant disorder, 36
OPRD1 gene, associated with addiction, 899–901
  locations of variants within, 900f
  rodent models, 901–903
    nontransgenic disruption of DOR in, 902t
  treatment efficacy, 900t
OPRD1-knockout mice, addiction in, 901–903, 901t
OPRM1 A118G SNP
  functional, 223–224
  in humans, 224
  nicotine dependence and withdrawal, polymorphism and, 227
  nicotine reward, polymorphism and, 227
  smoking heaviness, polymorphism and, 227
  smoking initiation, polymorphism and, 224–227, 225t–226t
OPRM1 gene, 223t, 910, 917, 1030
  alcohol dependence and, 230
  pharmacogenetic studies in smoking cessation, 227–228, 229t
  polymorphisms, 223–224
  smoking behaviors and, 224–227
Orbitofrontal cortex, 37–38
  internalizing pathway and, 37
Orbitofrontal cortex (OFC), 15, 63–64, 1037–1038
Orexin (hypocretin), 133
  addiction, role in, 139
  human studies of nicotine on, 136
  in lateral hypothalamus, 1007
  orexin A (OxA), 133
  orexin B (OxB), 133
  radioimmunoassays (RIA) studies, 136
Orexin receptor antagonists, 141t
Orexin receptors, 134–135
  receptors 1 and 2 (OX1R, OX2R), 133
Orexinergic system, 134
  cannabinoid exposure and expression of, 140–141
  cocaine exposure and expression of, 140–141
  expression of, nicotine exposure and, 140
  morphine exposure and expression of, 140–141
  nicotine, effects of, 136
  nicotine exposure and expression of, 137–139, 138t–139t
  production of, 134f

Osmotic demyelination syndromes (ODS), 600
  in alcoholics, 607
  clinical presentation, 605
  disease states associated with, 601t
  neuroimaging of, 603–605
    brain MRI, 604f–605f, 605
    CT scans, 603
    CT/MRI abnormalities, 605
    diffusion-weighted imaging, 604–605
    magnetic resonance imaging, 603–604
  outcomes in, 606
  pathology of, 602
  pathophysiology of, 600–602
    alcoholics, 602, 602f
    antidiuretic hormone secretion, 601
    blood-brain barrier damage, 601–602, 603f
    oligodendrocyte apoptosis, 602
  prognosis of, 606
  thyrotropin-releasing hormone for, 606
  treatment of, 606
Osmotic stress, 607
Oxidation, 559
Oxidative stress, 485, 559, 607, 624, 735, 1015
Oxidative stress pathways
  alcohol-mediated HIV pathogenesis and neuroAIDS, 181–183
  CYP2E1 in alcohol-mediated, 182
  in smoking-mediated HIV infection, 184–185
  via CYP pathways in monocytes/macrophages, astrocytes, and neurons, 184–185, 184f
Oxidative-nitrosative stress, 620
Oxycodone (Percodan), 884, 922
Oxymorphone (Opana), 922
Oxytocin, 970

P
Pain
  brain imaging methods, 958–960, 959f
  chronic, 957
  chronic noncancer, 956
  functional and molecular neuroimaging studies, 958–960
  imaging studies of, 957–958
  inflammatory and neuropathic, 961–962
  locating source of, 957
  matrix, 958, 958f
  molecular and cellular mechanisms of, 960f
  molecular and functional imaging studies, 960–962
  peripheral neuropathic, 961
Paired-pulse depression (PPD), 168–169
Paired-pulse facilitation (PPF), 170
Paleomammalian complex, 63
Papaver somniferum, 867, 922, 1052
Paranoid personality disorder, 29t, 874
Parasitism, 74
Paraventricular nucleus (PVN), 134–135, 205, 610
Parental SUDs, 36
  postnatal environmental effects of, 36
Parietal cortex, 597
Parkinson disease, acetaldehyde and, 557–559
Partial agonist, 823

Partial reinforcement contrast effect (PRCE), 75
Partial reinforcement extinction effect (PREE), 75
Pavlovian conditioning, 811
Pavlovian fear conditioning, 273, 337
Pentazocine, 928
Periaqueductal gray (PAG), 533
Perifornical area (PFA), 134–135
Peripheral neuropathic pain, 961
Peripheral vision, 745
Persistence, 978–979
Personality, temperament and character dimensions of, 976–977
Personality and drug habit of abuser, 29–33
  incentive-sensitization theory, 29
  negative reinforcement models, 29–30
  positive reinforcement models, 29
Personality disorders, 29t
PET. *See* Positron emission tomography (PET)
Pethidine, 909
Phenanthrenes, 867
Phenylpiperidines, 57, 867
Phosphatidic acid (PA), 497
Phosphatidylethanol (PEth), 491
Phospholipases D (PLDs), 488–490, 497
  ablation, 493f
  binding partners, 490t
  brain cells and, 491–492
    astrocytes, 492, 492f
    neurons, 491
  definition of, 497
  effects of ethanol in vitro, 493–494
    apoptosis, 494
    neurite outgrowth, 494
    proliferation, 493–494
    signaling, 494
  effects of ethanol in vivo, 494–496
    cognitive dysfunction, 495–496
    microcephaly, 494–495, 495f
    neurochemistry, 496
  enzymatic reactions of, 491, 491f
    hydrolysis, 491
    transphosphatidylation, 491
  function, 492–493
    apoptosis, 493
    neurite outgrowth, 493
    proliferation, 492–493
    signaling, 493
  isoforms, 489–490
  location, 490
  regulation of, 490–491
    cytoskeletal proteins, 491
    protein kinase C isoforms, 490–491
    protein kinases, 490
    small GTPases, 490
Phosphorylation, 507, 699
Physical dependence, 337–338
  defined, 337
  key facts, 345
  stages, 338t
Physical self-medication (PSM) hypothesis of addiction, 71
  conditions fulfilled for behavior, 74
  species-typical, 74

Physostigmine, 286
Phytocannabinoids, 65–67, 676, 709, 728
  β-caryophyllene, 675–676
  cannabichromene, 675
  cannabidiol, 675
  cannabigerol, 672–674
  cannabigerol acid, 672
  cannabinol, 675
  for other addictions and substance misuse, 676
  structures of, 673f
  Δ^9-tetrahydrocannabinol, 674–675
  tetrahydrocannabinol acid, 672
  tetrahydrocannabivarin, 675
  tetrahydrocannabivarinic acid, 672
Piperazine derivatives, 56–57
  commercial names, 56–57
Plasma membrane, 699
Plasmapheresis, 607
PNU-282987, 169
Polycyclic aromatic hydrocarbons (PAHs), 184
Polydrug abuse, 941
Polymorphisms, 1008
Polyvagal theory, 1065–1066
Ponto-geniculo-occipital (PGO) spikes, 146
Positive reinforcement, 614
Positive reinforcement models of addiction, 29, 31
  drugs as positive reinforcers, 31
  habitual self-administration, 31–32
  key points, 31t
  problems with, 31–32
Positron emission tomography (PET), 787, 792
  cocaine, metabolic effects of, 27
  of pain, 958–960
Posterior cingulate cortex (PCC), response to cue-induced craving, 292–296
Posttraumatic stress (PTS), 62
Posttraumatic stress disorder (PTSD), 61–62, 65, 72–73, 327
  CBD associated with, 751
Postural sway, 583
Potency, 728
Prefrontal cortex (PFC), 63, 75, 202–203, 273, 597, 780, 783, 926
  alterations due to aging, 1015–1016
  binge drinking (BD) and neurocognitive dysfunctions in, 391
  involvement in cognitive activities, 305
  in neuropathology of heroin addiction, 1014
  in regulation of "future reward" behaviors, 1032
Prefrontal cortex volume, in adolescent AUD, 593–594
Pregenual anterior cingulate cortex (pgACC), 16–17
Prelimbic medial prefrontal cortex, 756
Premature aging hypothesis, 1012
  effect on telomerase activity, 1018–1019
  link between substance abuse and, 1018–1019
  multiple aging biomarkers, 1019
Prenatal alcohol exposure (PAE), 495–497
Prenatal maternal smoking, as risk factor for ADHD, 1024–1028
Prenatal nicotine exposure, 137–139
  on orexin expression, 137

Prepro-orexin (PPO), 133
Prescription opioid abuse, 878
  percentage of population, 879f
  treatment, in military and veterans, 882
Prescription opioids, overdose deaths, 882f
Pro-opiomelanocortin (POMC), 209–210
Proponent-process theory, 32
Protease inhibitors (PIs), 181
Protein adducts, 554
  in cerebellum after ethanol treatment, 554f
  formation, 555, 558f
Protein kinase C isoforms, 490–491
Protein kinase C phosphorylates, 957
Protein kinases, 490
Proteolipid protein (PLP), 421–423
Proximal defenses, 329f
PSD-95 (postsynaptic density protein 95), 168–169
Psilocin (4-hydroxy-N,N-dimethyltryptamine), 52
Psilocybe cubensis (Earle) Strophariaceae, 52
Psilocybin (O-phosphoryl-4-hydroxy-N,N-dimethyltryptamine), 52
Psychological distress syndrome, 30
Psychopathology postulate, 72
Psychosis
  cannabis use and, 837
  early phase, 848
  longitudinal overview of, 838f
  PET imaging studies in comorbid cannabis use and, 790–792
  synthetic cannabinoids and, 681–683
Psychostimulant use, 655
Psychostimulants, THC effects of previous exposure on rewarding effects of, 668t
Pure alcoholism (Pure ALC), 544, 548
Purkinje cells, 583
Putative drug, modulates mood, 816
PVN. *See* Paraventricular nucleus (PVN)
Pyrolysate, 1000

Q

Quantified neural oscillations, 855

R

Raclopride, 175–176
RANTES, 181
Raphe nucleus, nicotine effects on, 172
Rapid eye movement (REM) sleep, 134, 146
Rapid visual information processing task (RVIP) in smokers, 283
Rat model of smoking relapse, 237–238
Rate-limiting step, 516
Reactive oxygen species (ROS), 1015
Receptor agonist, 614, 756
Receptor antagonist, 614, 756
Receptors, 614, 709
Reconsolidation, 811
  of drug, effects of THC and CBD, 809–811
Reduction, 559
Region-of-interest (ROI) analyses of amygdala reactivity to smoking, 274–275, 275t
Reinforcers, 801
Reinstatement, 1008
Relapse, 669

Response reinstatement of drug relapse, 237–238, 241
  interaction of nicotine cue with stress challenge in, 240, 240f
Resting-state functional connectivity (rsFC), 342, 1016–1017, 1041, 1042f
Resting-state network (RSN), 959
Resting-state oscillations, 857–859
Retinitis pigmentosa, 745
Retrieval, 811
Retrieval-induced forgetting, 286
Retrograde messenger, 801, 811
Retrograde signaling, 161f, 165
Reward, 669, 811
Reward circuit, 936–937
Reward circuitry, cannabinoids in, 796–798
Reward deficiency hypothesis, 37
Reward deficiency syndrome, 46, 46f
Reward dependence (RD), 977–979
Reward loss, negative emotional impact of, 75–77
  studies in inbred Roman high-avoidance (RHA) and Roman low-avoidance (RLA) rats, 75–76
Reward system, 71, 272–273, 801, 926
Rewarding effects of nicotine, 116–117
RHBDF2 gene, 913–914
Rho GTPases
  activation of, 125–126
  association of RhoA with smoking, 126
  cocaine and, 128
  nicotine addiction and, 125–126
  in regulation of dendritic spine dynamics and spine development, 126
  synaptic plasticity, role in, 125–126
Right hemi-aging hypothesis, 1016
Right inferior frontal gyrus (rIFG), 1037–1038
Rimonabant, 204–205
rs12112301 polymorphism, 127
rs186911567 polymorphism, 127–128, 130t
rs5522 SNP, 215
rs6198 SNP, 212–213
rs6296 SNP, 1030
rs950302 SNP, 917
rs1799971 SNP, 223
rs2070951 SNP, 215
rs2526303 SNP, 230
rs2952768 SNP, 916f
rs3864283 SNP, 230
rs10052957 SNP, 212–213
rs10494334 SNP, 917
rs12948783 SNP, 913–914
rs41423247 SNP, 212–213
Ryanodine receptor (RyR), 535f

S

S-adenosylmethionine, 482
Saint Salvia, 57
Salvia divinorum, 57–58, 680
Salvinorin A, 58
SAMHSA. *See* Substance Abuse and Mental Health Services Administration (SAMHSA)
SAP-102 (synapse-associated protein 102), 168–169

Sardinian alcohol-preferring and -nonpreferring rats, 1008
Saxitoxin (STX), 961–962
SB-277011A, 203
SCH 23390, 175–176, 242, 320
Schedule I US Controlled Substance Act, 728, 766–767
Schizoid personality disorder, 29t, 874
Schizophrenia, 82, 127, 136, 147, 254, 841
  correlates of comorbid drug abuse in, 83–84
  deficiency of DA, 204
  definition of, 848
  demographics, personality, and clinical characteristics, 83
  early cannabis use and development of, 83–84
  epidemiology, 82–83
  genetics and family studies, 84
  Khantzian model, 84
  neurobiology of comorbidity in drug abuse in, 84–86
  self-medication hypothesis, 84–86
  smoking and cognitive function in, 307–309
  smoking behavior in, 285
    addiction vulnerability hypothesis, 265–267
    affect dysregulation theory, 265
    clinical features, 262
    contingency management, 268
    epidemiology, 260–261
    etiology, 262–267
    frequencies of smoking-related prevalence, 263t
    genetic studies, 266
    neuroimaging studies, 266–267
    neurophysiological studies, 267
    nicotine replacement therapy (NRT), 267–268
    nonpharmacological treatments for, 268
    P50 deficits, 266–267
    pathophysiology, 266f
    pharmacological treatments for, 267–268
    prepulse inhibition (PPI), 266–267
    self-medication and affect dysregulation hypotheses in, 263–265, 265f
    smoking-related parameters of normal smokers and smokers with, 261t
    SPEM (smooth pursuit eye movements), 266–267
    theoretical models of, 264f
    treatment guidelines, 267
    white matter changes in, 844
Schizotypal personality disorder, 29t, 874
Secondhand smoke (SHS) exposure, 367
Sedative drug abuse, 655–656
Selective serotonin reuptake inhibitor (SSRI), 969–970
Self-administration, 1008
  effects of
    cocaine, 320
    fixed ratio (FR) schedule, 320
    nicotine, 119–120, 162–163, 204–205, 238f, 320–322, 320f–322f, 323t
    opioids and psychostimulants, 933
    progressive ratio (PR) schedule, 321

Self-directedness, 977–978
Self-esteem, 331–332, 334
Self-medication hypothesis, 72, 84–86
　affect dysregulation hypotheses and, 263–265, 265f
Self-transcendence, 978
Semisynthetic opiates, 867
Sensation seeking behavior, 37
Sensitization, 833
Sensitization-homeostasis theory of addiction, 341–342
Sensitization-related neuroadaptations, 32
Sensory gating, 267
Septum, 168
　nicotine effects on, 172
Sequela of responses, 63
　associated with traumatic stress, 62
Serotonin, 136, 192, 756, 968–970
　associations between serotonergic activity and other biological or psychological variables, 968–969
　opioid dependence and controls, 968
　in opioid use, 968
Serotonin (5-hydroxytryptamine), 204, 823
Serotonin firing activity, THC and, 816–818
Serotonin transporter (SERT), 968–969
Sex differences
　from acute response to alcohol to neurobiological consequences, 637–638
　in addiction vulnerability, 641–643
　in alcohol models, 641
　neuroendocrine responses, 638
Shadows (film), 330–331
Shepherd, The, 57
Signal transduction, 767
Single channel currents, 507
Single Convention on Narcotic Drugs (of 1961), 679
Single-nucleotide polymorphism (SNP), 126, 539, 910, 1030
　in ADH genes, 382
　COMT Val/Met, 910
　in *OPRM1*, 223
　OPRM1 A118G, 910
　pain relief phenotype and, 913–914
　resulting in amino acid substitutions in *OPRM1* gene, 223t
Single-photon emission computed tomography (SPECT), 787, 966, 969
　clinical implications, 969–970
　of dopamine function, 27
siRNA. See Small interfering ribonucleic acid (siRNA)
Site-directed mutagenesis, 507
SLC6A4, 1030
SLV330, 163
Small GTPases, 490
Small interfering ribonucleic acid (siRNA), 539
Smokers with mental illness
　neuropathology, 255
　nicotine users, 306f
　personal and environmental factors, 255–256
Smoking cessation treatment, 109
　access to, 110
　attitudes towards, 109–110
　barriers to implementation, 109

　benefits of rimonabant for, 164t
　collaborative initiative, 110
　counseling and psychosocial support, 256–257
　effectiveness, 110
　efficiency of graphic antismoking messages in curbing smoking behavior, 297
　health professionals, role of, 256
　impact of, 257
　nicotine replacement therapy (NRT), 256
　OPRM1 gene and, 227–228, 229t
　for people with mental illness, 256–257
　staff knowledge and attitudes about, 109
Smoking maintenance and relapse, 206f
Smoking relapse, rat model of, 237–238
Smoking tobacco
　addictive properties, 117–118
　among HIV-infected individuals, 183
　association of RhoA with, 126
　biological role of β2-chimaerin gene (*CHN2*), 127–128, 129t
　chimaerins and, 126
　conditioned place preference (CPP) paradigms and intracranial self-stimulation (ICSS) models, 116
　CYP3A4 pathway and HIV pathogenesis, 185
　cytochrome P450 and oxidative stress pathways in smoking-mediated HIV infection, 184–185
　HIV pathogenesis and NeuroAIDS, effect on, 183–184
　monoaminergic pathways relevant to smoking addiction, 116f
　mortality rates among HIV/AIDS patients and, 183
　neuroAIDS/HAND, effects on, 183–184
　neurobiological effects of, 115–116. See also Nicotine (3-(1-methyl-2-pyrrolidinyl) pyridine)
　neurologically active compounds found in, 117t
　nonadherence to ART, 183
　rewarding effects, 118t
Smoking-attributable fraction (SAF), 105–106
　of mortality for appendicitis and drug/alcohol cohort groups, 108f
Smoking-Attributable Mortality, Morbidity, and Economic Costs (SAMMEC), 105
Social condition (SC) and nicotine addiction, 246–247
　behavioral effects, 247
　hyperlocomotor activity, 249
Social instability, 643
Social instability stress, in female long Evans rats, 643f
Social isolation, 643
　in other addictions and substance misuse, 643
Social isolation model
　of addiction vulnerability in male rodents, 638–640
　of addiction-vulnerable phenotype in female rats, 641
　adolescent *vs.* group housing, 640t
　of chronic early life stress, 640t
　ethanol intake

　　in female long Evans SI/GH rats, 640f
　　in male long Evans SI/GH rats, 639f
　experimental procedural timeline in, 639f
　hormones in male and female rodents exposed to early life stress, 640–641
Social phobia, 147
Social recognition test, 497
Sodium channels, 957–958
Somatic symptoms of nicotine, 195
SPECT. See Single-photon emission computed tomography (SPECT)
Spectral analysis of gyrification, 571
Spice, 723–724, 728, 766
　products, 723, 767
　signaling, 764–766, 766f
　vs. marijuana, 723–724
Spinal cord, nicotine effects on, 174
Spongiform leukoencephalopathy, associated with heroin, 996–997, 1000
　heroin-induced, 996–997
　hypoventilation and hyperemia, 997
　mitochondrial injury, 997
　toxin-related cerebral vasculitis, 996
Spontaneous inhibitory postsynaptic current (sIPSC), 170
SR141716A, 163
SSVEPs. *See* Steady-state visual evoked potentials (SSVEPs)
Stage of Physical Dependence instrument, 337–338
Standard drinks chart, 654f
Standardized mortality ratios (SMRs) from tobacco use, 106t, 107, 108f
Steady-state evoked oscillations, 856–857
Steady-state visual evoked potentials (SSVEPs), 741
Stimulant use disorders, 576
Stop-signal task, 1037–1039
Strain theory, 29
Street heroin
　adulterants and diluents in, 870–871
　　neurological effects and neuropathology of, 873
　adulterants used to cut, 872t
　heavy metals in, 871
　microbial contamination of, 871
　street names for drug combinations with, 871t
Street heroin abuse, medical consequences of, 873–874
　bacterial infections, 873
　Brown–Séquard syndrome, 873
　cerebral edema, 873
　direct toxic effects, 873
　dopamine and, 874
　heart diseases, 873–874
　liver diseases, 873
　lung diseases, 873
　pathologic changes in kidney, 873
Stress, 614
　and cellular damage/death by excitotoxicity, 733–734
　and endocannabinoid system, 732–733
　　CB1R, 733
　　CB2R, 733
　and ethanol intake, 414

Stress, and ethanol intake, 414
Stress-induced CB1R upregulation, 734
Stress-induced excitotoxicity, 734–735, 734f
Stress-related genes by drugs of abuse, activation of, 935–936
Stress-related neuroendocrine response, 63–64
Stress-related neuropsychopathologies, therapeutic potential in, 735
Striatum, 83–86, 192, 273, 595, 778–780, 783, 792
　alcohol intake and, 405
　nicotine effects on, 172
　transcriptional effects on chronic methamphetamine and heroin treatment, 938
Structure–activity relationship, 728
Studies with Rimonabant and Tobacco Use (STRATUS), 163
Suboxone, 885
Substance Abuse and Mental Health Services Administration (SAMHSA), 722
Substance P (SP), 190–191, 194, 957
Substance use disorder (SUD), 36, 297, 783, 833, 836, 974–975, 1023–1024
　addiction cycle, 38, 39f
　alcohol use disorder, 575
　association of SNPs with, 1030
　behavioral activation and behavioral control, 37
　binge/intoxication cycle, 38
　cannabis use disorder, 576
　cerebellar abnormalities in, 579–582
　　cerebellar functional differences associated with family history, 580–581
　　cerebellar lobes and functional differences, 580
　　cerebellum structure and familial risk, 580
　　functional connectivity, 581
　　premorbid risk factors in high-risk offspring, 579–580
　cerebellar volume
　　in healthy subjects, 581f
　　in high- and low-risk individuals, 581f
　　and substance use outcomes, 581–582
　characteristics of, 1027t
　in *Diagnostic and Statistical Manual for Mental Disorders*, 71
　emotional self-medication (ESM) and, 77–79
　evaluation of, 1028t
　externalizing pathway, 37
　impulsive behaviors and, 37
　incentive sensitization model, 37–38
　internalizing pathway, 37
　model pathways of development, 46f
　models of vulnerability, 36–38
　neurobiological basis of, 36
　opioid use disorder, 576
　positive and negative reinforcement phases of, 38
　preoccupation/anticipation cycle, 38
　prevalence of, 575
　reward deficiency/reward sensitivity hypotheses, 37–40, 44–45, 46f
　role of neurobiological factors in, 37
　sensitivity to rewarding effects of drugs and, 37
　stimulant use disorders, 576
　tobacco use disorder, 575–576
　ventral striatum/dorsal striatum/thalamus circuit and, 38
　withdrawal/negative affect cycle, 38
Subutex, 885
　use trends in Georgia, 882–884
Sulcal-based morphometry, 571
Supraoptic nucleus (SON), 134–135
Surface-based morphometry, 567
Symptom Checklist-Revised, 978
Synaptic plasticity, 507
　role of Rho GTPases, 125–126
Synaptic/nonsynaptic cholinergic transmission, 305
Synaptosomal-associated protein 25 gene (*SNAP25*), 1030
Synaptosomes, 735
Synonymous polymorphisms, 127–128
Synthetic cannabinoids, 55, 680, 708–709, 760
　acute renal lesions and, 55
　CB1R activation by, 764
　chemical structures of, 56f, 680f–681f
　chemistry, 680–681
　chronic use, impact of, 55
　classification of, 682t, 761t
　clinical symptoms of exposure, 55
　common names, 55
　formulations, 55
　inhibition of N-type voltage-gated channels, 762–763
　known physiological effects of, 765t
　pharmacological properties of, 681
　and psychosis, 681–683
　sympathomimetic and hallucinogenic effects, 55
　toxicology, 681
Synthetic cathinones, 55–56, 683
　adverse effects, 55
　chemical structures of, 56f, 683f
　chemistry, 683–684
　general skeleton of, 683f
　pharmacological properties of, 684
　popularity, 55
　toxicology, 684
Synthetic marijuana, 55
Synthetic opiates, 867

T
Tachykinin genes, 190
Tachykinin peptides, 190f
Tachykinins, and addiction mechanism, 192
　alcohol, mechanisms of dependence from, 196
　cocaine, mechanisms of dependence from, 196
　opioids, mechanisms of dependence from, 195–196
Tagged SNP, 905
Temperament and Character Inventory (TCI), 976–977, 1061
　concepts of, 1062
　four traits in, 978
Temporal cortex, 594, 597
Teratogenicity, 516
Terror management health model (TMHM), 327
　conceptual foundation, 327–328
　graphic overview, 328f
　hypothesis, 328–329
　key facts, 334
　rationalist perspective, 333
Terror management theory (TMT), 327
　basic premise of, 327–328
　conscious and nonconscious death-related cognitions, 328–329
　cultural worldview, 328f
　graphic overview, 328f
　self-esteem and, 328f
Tetrahydrocannabinol acid (THCA), 672
Tetrahydrocannabivarin (THCV), 675
Tetrahydrocannabivarinic acid, 672
T1065G SNP, 1030
Thalamus, 168
THCA. *See* Tetrahydrocannabinol acid (THCA)
THCV. *See* Tetrahydrocannabivarin (THCV)
THDP. *See* Thiamine diphosphate (THDP)
Therapeutics, 462
Thiamine deficiency, 484
Thiamine diphosphate (THDP), 627–628, 634
Thiamine replacement, in Korsakoff syndrome, 629
Time × frequency transform, 860
Tobacco alkaloid myosmine, 93, 94f
Tobacco alkaloid nicotine, 115
Tobacco and Mental Illness Project in South Australia, 256
Tobacco effects
　on body weight, 136
　premature aging and addiction of, 1019
　on sleep, 136
Tobacco particulate matter and extracts, 118–119
　acetaldehyde, 119
　in animal research, 317
　locomotor sensitization with, 318
　minor alkaloids, 119
　monoamine oxidase inhibitors (MAO), 119
　production of, 118f
　self-administration, effects of, 320f–322f, 323t
Tobacco use, 651–653
Tobacco use disorder, 575–576
　and cerebellum, 582
Tobacco-related diseases, 105–107
　ICD-9 and ICD-10 codes and comparability ratios for mortality conditions, 106t
　policy interventions to reduce, 110–111
　in populations with psychiatric disorders, 111
Tolerance, 669, 709
Toll-like receptor 4 (TLR-4), 425–427, 426f, 538, 539f, 624
　induces neuroinflammation and brain damage, 622
　mechanism of action of ethanol-induced myelin disarrangement through, 426f
Toll-like receptors (TLR), 455–456, 458–459, 485
Torpedo nobiliana, 174
Toxin-related cerebral vasculitis, 996
TPQ. *See* Tridimensional Personality Questionnaire (TPQ)

Transcranial direct current stimulation (tDCS), 18, 20–21
Transcranial magnetic stimulation (TMS), 18, 448
Transcription factor, 485, 507
Transient evoked oscillations, 856
Transphosphatidylation, 491, 497
Trauma and neurological risks of addiction, 61
  adverse childhood experiences, 61
  brain regions involved, 64f–65f
  chronic toxic stress, 62–63
  complex trauma, 62
  intergenerational trauma, 62
  neurobiological and cognitive areas affected, 66t
  neurobiology of, 63–64
  posttraumatic stress (PTS), 62
  posttraumatic stress disorder (PTSD), 61–62, 65
  substance misuse and, 65–67
  war-related trauma, 61–62
Treatment Episode Data Set (TEDS), 104
Tricarboxylic acid (TCA) cycle, 473
  acetyl-CoA for, 469
  glucose-driven, 469
Tridimensional Personality Questionnaire (TPQ), 547, 549, 1061
  analysis of heart rate variability (HRV), 1066
  basis of borderline personality trait and heroin use disorder, 1065–1066
  concepts of, 1062
  dimensions and heroin use, 1062–1063, 1064t
  genetic dimensions of personality, 1062
  personality patterns, 1063–1065, 1065t
  Q score of alcohol consumption, 1063
Trier Inventory for the Assessment of Chronic Stress, 215
Trier Social Stress Test (TSST), 212–213, 215
1-(3-trifluoromethylphenyl)piperazine (TFMPP), 56–57, 57f
Tuberal mammillary nucleus (TMN), 134–135
Twin study, 528

U

Uncertainty
  defined, 17, 22
  link between brain and mind, 17–18, 18f
  rACC activity, 17
Unconditioned stimulus, 811
United Nations Office on Drugs and Crime (UNODC), 654, 656, 878
UNODC. See United Nations Office on Drugs and Crime (UNODC)
Upregulation of receptor/transporter, 473
URB597, 162–163
Urine cotinine assay techniques, 366
  accurate estimation of, 368
  in adolescent age group, 368
  advantages of, 368
  comparison with other biomarkers, 368
  for confirming tobacco abstinence, 366–367
  detecting presence of nicotine dependence, 368
  in pregnant women, 367–368
  as proxy marker of smoking cessation in treatment studies, 367
  research and clinical applications, 366–368, 367t
  secondhand smoke (SHS) exposure and, 367
  sensitivity and specificity issues, 368. See also Cotinine

V

Valine (Val) by methionine (Met) at codon 158 (Val158Met), 910
Vanilla sky, 55
Varenicline, 256, 268, 321–322, 415
  adverse side effects, 415
  effect on alcohol self-administration, 415
Vasoconstrictive effects of nicotine, 353
Vasogenic edema, 607
V1B receptor systems, arginine vasopressin and, 1005–1006
VBM. See Voxel-based morphometry (VBM)
Ventral medial hypothalamus (VMH), 134–135
Ventral medial nucleus (VMN), 134–135
Ventral striatum, 37–38, 44–45
  drug abuse, role in, 38
Ventral tegmental area (VTA), 159, 192, 204, 206, 272–273, 349, 414–415, 774, 926, 933
  alcohol intake and, 405
  DA neurons, 159
  effects of enrichment on, 249
  excitatory/inhibitory circuits among, 159f
  involvement in cognitive activities, 305
  nicotine effects on, 173
Vernier visual acuity, 745
Vision
  cannabinoids effects on, 740f
    electrophysiological effects, 741–744, 742f
    psychophysical effects, 739–741
  pathology of, 738
  peripheral, 745
Vitamin A deficiency, 481
  in cirrhotics and noncirrhotics, 482f
Vitamin B6, in homocysteine metabolism, 484f
Vitamin B12, 482–483
  according to Child's classification, 483f
  in homocysteine metabolism, 484f
Vitamin B3 deficiency, 483–484
Vitamin C deficiency, 483
Vitamin D deficiency, 481–482
  in alcoholics, 482
  prevalence of, 482f
Vitamin E deficiency, 480–481
  chronic ethanol feeding and, 481
  in cirrhotics and noncirrhotics, 481, 481f
  CRMP-2 in, 480–481
  MAP-LC3 in, 480–481
Voxel-based morphometry (VBM), 587–588, 597, 959, 1016
VTA. See Ventral tegmental area (VTA)
Vulnerability, 1008

W

Wallerian degeneration, 634
WaNting, 337
War-related trauma, 61–62
WAY-100635, 153–154
Wernicke encephalopathy (WE), 384, 478, 484, 627–629, 634
  acetaldehyde in, 559
  histopathological changes in, 628
  MRI brain findings in, 629f
  nutritional deficiency, 628
  permanent or chronic changes in, 628–629
  THDP in, 627–628
  topographical involvement in, 629
Wernicke-Korsakoff syndrome, 516
White ice, 55
White matter, 421, 423, 841
  age-dependent changes in, 421
  alcohol abuse and disturbances of, 423–425, 424t
  changes in schizophrenia, 844, 845f, 848–849
    diffusion tensor imaging studies, 847t
  changes with cannabis use, 844–847
    adolescent, 849
    diffusion tensor imaging studies, 846t–847t
    in people with schizophrenia, 847–848
  classification of, 841
  CNPase distribution, 843
  definition of, 848
  differences between males and females in, 421
  diffusion tensor FA map, 843f
  and endocannabinoid system, 421
  microstructure, 421
  structural connectivity, measurement of, 843
  studied in postmortem tissue, 843
  tracts in human brain, 842f
White matter atrophy, 479–480
White matter tracts, 848
WIN55,212-2 (synthetic cannabinoid), 163
  DR 5-HT neurons, impact on, 819f
  impact on 5-HT firing and antidepressant-like activity, 818
  incremental doses of, 818
  stimulation of DR 5-HT neurons, 818
  tPFC transection and, 818
Withdrawal syndrome, 669, 833
Withdrawal-induced craving, 342–343
Working memory, binge drinking and, 393–394

Y

Yale Food Addiction Scale, 22
Yohimbine, 240

Z

ZD7288, 171